ELECTRICAL ENGINEERS' HANDBOOK

Electric Power

WILEY ENGINEERING HANDBOOK SERIES

HANDBOOK OF ENGINEERING MATERIALS. Edited by DOUGLAS F. MINER and JOHN B. SEASTONE.

HANDBOOK OF ENGINEERING FUNDAMENTALS. Second Edition. Edited by OVID W. ESHBACH.

KENT'S MECHANICAL ENGINEERS' HANDBOOK. Twelfth Edition.

DESIGN AND PRODUCTION. Edited by COLIN CARMICHAEL.

POWER. Edited by J. K. SALISBURY.

ELECTRICAL ENGINEERS' HANDBOOK. Fourth Edition.

ELECTRIC POWER. Edited by HAROLD PENDER and WILLIAM A. DEL MAR.

ELECTRIC COMMUNICATION AND ELECTRONICS. Edited by HAROLD PENDER and KNOX McILWAIN.

MINING ENGINEERS' HANDBOOK. Third Edition. Edited by the late ROBERT PEELE.

HANDBOOK OF MINERAL DRESSING.

ORES AND INDUSTRIAL MINERALS. Edited by ARTHUR F. TAGGART.

ELEMENTS OF ORE DRESSING. ARTHUR F. TAGGART.

ELECTRICAL ENGINEERS' HANDBOOK

Electric Power

Prepared by a Staff of Specialists

HAROLD PENDER, Ph.D., Sc.D.
and
WILLIAM A. DEL MAR, F.C.G.I.
Editors

FOURTH EDITION

WILEY ENGINEERING
HANDBOOK SERIES

JOHN WILEY & SONS, INC.
NEW YORK · LONDON · SYDNEY

FOURTH EDITION

20 19 18 17 16 15 14 13 12 11 10

PRINTED IN THE UNITED STATES OF AMERICA

PREFACE

The first edition of Pender's *Handbook for Electrical Engineers,* compiled by a staff of specialists under the editorship of Harold Pender, appeared in 1914. The second edition, under the joint editorship of Pender and William A. Del Mar, was published in 1922. Both these editions covered all branches of electrical engineering as well as a large amount of material dealing with allied fields of interest to electrical engineers.

The third edition, published in 1936, was divided into two volumes: one on electric power under the editorship of Pender, Del Mar, and Knox McIlwain, the other on electrical communication and electronics under the editorship of Pender and McIlwain. Certain tables and fundamental theory were duplicated in the two volumes in order that each might be complete and independent of the other.

This plan met with such enthusiastic response that it has been continued in the fourth edition. The growth of knowledge and the greater degree of specialization in the various phases of electrical engineering have necessitated a considerable enlargement of both volumes. Careful selection and compression have been exercised in an effort to keep the books compact and readable. The treatment of subjects of decreased importance and those which are adequately treated by other handbooks of the Wiley Handbook Series has been either curtailed or left unchanged in length.

The bibliographies have been prepared with the idea of assisting the reader to further study of each subject, and they reflect each author's idea of this plan. The publications referred to are, in general, in the Engineering Societies Library, 29 West 39th Street, New York, N. Y. Most of them may be borrowed from the Library by members of its Founder Societies, the American Society of Civil Engineers, American Institute of Mining and Metallurgical Engineers, American Society of Mechanical Engineers, and American Institute of Electrical Engineers.

Seventy-one specialists in their respective fields have contributed to this fourth and entirely rewritten edition of the *Electrical Engineers' Handbook,* as compared with twenty-seven, forty-five, and fifty to successive previous editions. This reflects the increased importance which several subjects have assumed. Notable among these, on the theoretical side, are circuit stability and symmetrical components, and, among items of equipment, electronic rectifiers, aircraft equipment, heat pumps, servomechanisms, permanent magnets, plastic insulating materials, and induction and dielectric heating apparatus. The increased staff of authors also reflects a greater degree of specialization in the treatment of other subjects. Thus, whereas in the third edition, one author covered all types of rotating machinery, the fourth edition has five authors to discuss the various phases of that subject.

We take this occasion to thank the many well-known and busy men who have taken the time and trouble to contribute to this volume and make it a reliable guide to electrical power engineering practice.

HAROLD PENDER
WILLIAM A. DEL MAR

LIST OF CONTRIBUTORS

E. W. Allen, Manager of Engineering, Thomas A. Edison, Inc., Edison Storage Battery Division. *Alkaline-Type Storage Batteries.*

E. R. Ambrose, Air-Conditioning Engineer, American Gas & Electric Service Corp. *Electric Heating and Air Conditioning.*

F. S. Bennett, Mechanical Engineer, Construction Engineering Department, General Electric Co., Schenectady, N. Y. *Steam-Electric Power Stations.*

P. H. Brace, Consulting Metallurgist, Research Laboratories, Westinghouse Electric Corp. *Conductor Materials.*

J. G. Brainerd, Professor, Moore School of Electrical Engineering, University of Pennsylvania. *Units and Conversion Factors; Symbols and Abbreviations.*

A. S. Brookes, Senior Engineer, Public Service Electric & Gas Co., Newark, N. J. *Grounding of Electric Circuits.*

D. B. Brooks, Chief, Automotive Section, National Bureau of Standards. *Thermometry.*

H. B. Brooks, Consultant, former Chief Electrical Instrument Section, National Bureau of Standards. *Measurements and Measuring Apparatus.*

J. W. Butler, Formerly Application Engineer, Power Factor Capacitors, General Electric Co., Pittsfield, Mass. *Capacitors.*

Carl C. Chambers, Professor of Electrical Engineering, Moore School of Electrical Engineering, University of Pennsylvania. *Mathematics, Mathematical Tables.*

E. K. Clark, Manager of Engineering, Westinghouse Electric Corp., Mansfield, Ohio. *Household Heating and Cooking.*

J. J. Coleman, Chief Engineer, Burgess Battery Co., Battery Division. *Dry Batteries.*

A. G. Conrad, Chairman, Dept. of Electrical Engineering, Yale University. *Alternating-Current Commutator Motors.*

S. B. Crary, Assistant Manager, Central Station Engineering Divisions, General Electric Co. *Power-System Stability.*

C. M. Davis, Manager, Transportation Engineering Division, General Electric Co. *Electric Traction.*

W. A. Del Mar, Chief Engineer, Phelps Dodge Copper Products Corp., Habirshaw Cable and Wire Division. *Associate Editor-in-Chief; Parts of section on Power Transmission and Distribution; Weights and Thermal Properties of Materials.*

C. D. Dimity, Transmission and Distribution Engineer, American Copper Products Division, Phelps Dodge Copper Products Corp. *Rural Electrification Distribution Systems.*

R. B. Elsworth, Signal Engineer, New York Central Railroad, Albany, N. Y. *Railroad Signaling, Railroad Signaling Batteries.*

H. A. Enos, Distribution Engineer, American Gas & Electric Service Corp. *Pole and Tower Lines.*

L. J. Gorman, Assistant Engineer, Electrical Engineering Dept., Consolidated Edison Co. of New York. *Electrolysis of Underground Structures.*

J. H. Goss, Assistant Division Engineer, General Electric Co., Schenectady, N. Y. *Magnets and Magnetic Materials.*

F. K. Harris, Electrical Instruments Section, National Bureau of Standards. *Oscillographs.*

C. T. Hatcher, Division Engineer, Consolidated Edison Co. of New York. *Distribution.*

S. W. Herwald, Manager, Development Section, Special Products Engineering, Westinghouse Electric Corp. *Servomechanisms.*

W. F. Hess, Professor, Dept. of Metallurgical Engineering, Rensselaer Polytechnic Institute. *Electric Welding.*

M. H. Hobbs, Manager of Engineering, Switchgear and Control Division, Westinghouse Electric Corp. *Switchgear and Control Equipment.*

T. B. Holliday, Consulting Engineer, Chicago, Ill. *Airplanes.*

R. A. Hopkins, Head Electrical Engineer, Tennessee Valley Authority. *Hydroelectric Power Stations; Power-Station Circuits.*

P. L. Howard, Research Engineer, Graham, Crowley & Associates, Inc. *Reserve-Type Primary Batteries.*

J. P. Jordan, Electronics Laboratory, General Electric Co., Syracuse, N. Y. *Induction and Dielectric Heating.*

Martin Kilpatrick, Professor of Chemistry and Head of Department, Illinois Institute of Technology. *Principles of Electrochemistry.*

Mary Kilpatrick, Research Assistant Professor, Illinois Institute of Technology. *Principles of Electrochemistry.*

I. F. Kinnard, Manager of Engineering, Meter and Instruments Divisions, General Electric Co., West Lynn, Mass. *Magnets and Magnetic Materials.*

E. H. Kirkham, Assistant Engineer, Phelps Dodge Copper Products Corp., Habirshaw Cable and Wire Division. *Part of Section 14 on Carrying Capacity of Cables.*

W. A. Koehler, Head, Dept. of Chemical Engineering, West Virginia University. *Electrochemical Processes.*

T. C. Lloyd, Chief Engineer, Robbins & Myers, Inc. *Induction Motors.*

M. H. McGrath, Chief Engineer, General Cable Corp. *Electric Circuits.*

L. E. Markle, Manager Control Equipment Engineering, Switchgear and Control Division, Westinghouse Electric Corp. *Switchgear and Control Equipment.*

E. J. Merrell, Head of Electrical Research Laboratory, Phelps Dodge Copper Products Corp., Habirshaw Cable and Wire Division. *Resistors and Rheostats.*

A. M. Miley, Electrical Engineer, Electric Storage Battery Co. *Lead-Acid Storage Batteries.*

D. F. Miner, Carnegie Institute of Technology. *Weights and Thermal Properties of Materials.*

R. E. Morse, Cable Engineer, American Gas & Electric Service Corp. *Underground Conduits.*

Edmund B. Neil, Consulting Engineer, Columbus, Ohio. *Internal-Combustion Automobiles.*

J. H. Palmer, Electrical Engineer, Phelps Dodge Copper Products Corp., Habirshaw Cable and Wire Division. *Insulating Materials.*

Harold Pender, Dean of Moore School of Electrical Engineering, University of Pennsylvania. *Editor-in-Chief; Parts of Transmission and Distribution.*

H. O. Poland, Electrical Engineer, Transportation and Generator Engineering Dept., Westinghouse Electric Corp. *Alternating-Current Synchronous Machines.*

J. H. Reifenberg, Section Engineer, Appliance Engineering Dept., Westinghouse Electric Corp., Mansfield, Ohio. *Water Heaters.*

L. M. Roberts, Technical and Development Dept., Research Corp., Bound Brook, N. J. *Electric Precipitation.*

Willard Roth, Industrial Heating Engineer, Westinghouse Electric Corp., Meadville, Pa. *Industrial Furnaces and Ovens.*

R. L. Sanford, Chief, Magnetic Measurements Section, National Bureau of Standards. *General Characteristics of Magnetic Materials, Non-Retentive Magnetic Materials, Testing of Magnetic Materials.*

J. L. Schroyer, Consulting Engineer, Hotpoint, Inc., Chicago, Ill. *Commercial Cooking.*

Raymond C. R. Schulze, American Gas & Electric Service Corp. *Symmetrical Components.*

W. C. Sealey, Engineer-in-Charge, Transformer Division, Allis-Chalmers Mfg. Co. *Transformers.*

J. F. Sellers, Engineer-in-Charge, D-C. Machine Design, Allis-Chalmers Mfg. Co. *Direct-Current Machines and Rotary Energy Converters.*

F. B. Silsbee, Chief, Division of Electricity and Optics, National Bureau of Standards. *Measurements and Measuring Apparatus.*

J. W. Simpson, Engineering Section Manager, Switchgear and Control Division, Westinghouse Electric Corp., E. Pittsburgh, Pa. *Switchgear and Control Equipment.*

W. A. Sloan, Professor of Mechanical Engineering, University of Pennsylvania. *Internal-Combustion Power Stations.*

J. D. Stacy, Design Engineer, Capacitor Engineering Division, General Electric Co., Pittsfield, Mass. *Capacitors.*

R. P. Teele, Photometry and Calorimetry Section, National Bureau of Standards. *Photometry.*

H. A. Travers, Consulting Engineer, Switchgear and Control Division, Westinghouse Electric Corp., E. Pittsburgh, Pa. *Switchgear and Control Equipment.*

Irven Travis, Professor of Electrical Engineering, Moore School of Electrical Engineering, University of Pennsylvania. *Electric Circuits.*

M. S. Van Dusen, Temperature Measurement Section, National Bureau of Standards. *Pyrometers.*

F. J. Vogel, Professor of Electrical Engineering, Illinois Institute of Technology. *Reactors.*

C. E. Weitz, Nela Park Engineering Division, General Electric Co., Cleveland, Ohio. *Lighting.*

F. A. Westbrook, Consulting Engineer, Benson, Vt. *Industrial Applications of Motors.*

W. F. Wetmore, System and Stations Engineer, Detroit Edison Co. *Substations.*

E. C. Whitney, Electrical Engineer, Transportation and Generator Engineering Dept., Westinghouse Electric Corp. *Alternating-Current Synchronous Motors.*

H. Winograd, Engineer-in-Charge of Rectifier Design, Allis-Chalmers Mfg. Co. *Power Rectifiers and Inverters.*

J. L. Woodbridge, Formerly Chief Engineer, Electric Storage Battery Co. *Lead-Acid Storage Batteries.*

G. W. Zink, Electrical Engineer, Rubber-Insulated Wire Section, Engineering Dept., Phelps Dodge Copper Products Corp., Habirshaw Cable and Wire Division. *Wiring of Buildings.*

W. N. Zippler, Chief Electrical Engineer, Gibbs and Cox, Inc. *Ships.*

GENERAL TABLE OF CONTENTS

Detailed tables of contents are given at the beginning of each section. An alphabetical index appears after Section 19.

This book is divided into sections, each section carrying its independent sequence of page numbers. For example, 3–15 indicates Section 3, page 15.

SECTION 1

MATHEMATICS, UNITS, AND SYMBOLS

MATHEMATICS

By Carl C. Chambers

MATHEMATICAL TABLES AND CHARTS

UNITS AND CONVERSION FACTORS

By J. G. Brainerd and Carl C. Chambers

SYMBOLS AND ABBREVIATIONS

CONSTANTS

By Carl C. Chambers

MATHEMATICS, UNITS, AND SYMBOLS

MATHEMATICS

By Carl C. Chambers

1. ALGEBRAIC FORMULAS

MISCELLANEOUS FORMULAS

$$(a \pm b)^2 = a^2 \pm 2ab + b^2$$

$$(a \pm b)^3 = a^3 \pm 3a^2 b + 3ab^2 \pm b^3$$

$$(a \pm b)^n = \sum_{k=0}^{n} \frac{n!}{k!\,(n-k)!} a^k (\pm b)^{(n-k)}, \qquad M! = M(M-1)\ldots 3 \times 2 \times 1$$

$$a^2 - b^2 = (a+b)(a-b)$$

$$a^2 + b^2 = (a+jb)(a-jb), \quad j = \sqrt{-1}$$

$$a^x \times a^y = a^{(x+y)}, \quad a^\circ = 1 \text{ [for } a \neq 0], \quad (ab)^x = a^x b^x$$

$$\frac{a^x}{a^y} = a^{(x-y)}, \quad a^{-x} = \frac{1}{a^x}, \quad \left(\frac{a}{b}\right)^x = \frac{a^x}{b^x}$$

$$(a^x)^y = a^{xy}, \quad a^{1/x} = \sqrt[x]{a}, \quad \sqrt[x]{ab} = \sqrt[x]{a}\,\sqrt[x]{b}$$

$$\sqrt[x]{\sqrt[y]{a}} = \sqrt[xy]{a}, \quad a^{x/y} = \sqrt[y]{a^x}, \quad \sqrt[x]{\frac{a}{b}} = \frac{\sqrt[x]{a}}{\sqrt[x]{b}}$$

$$\log (a^x) = x \log a, \quad \log ab = \log a + \log b$$

$$\log \frac{a}{b} = \log a - \log b$$

$$\text{If } \frac{a}{b} = \frac{c}{d} \text{ then } \frac{a \pm b}{b} = \frac{c \pm d}{d} \text{ and } \frac{a-b}{a+b} = \frac{c-d}{c+d}$$

The sum of an arithmetical progression is given by

$$s = \frac{n}{2}(a + l) = \frac{n}{2}\{2a + (n-1)d\}$$

where $l = a + (n-1)d$ is the last term, a is the first term, d is the common difference, and s is the sum of the n terms.

The sum of a geometrical progression is given by

$$s = a\frac{(1 - r^n)}{1 - r} = \frac{lr - a}{r - 1}$$

where $l = ar^{n-1}$ is the last term, a is the first term, d is the common ratio, and s is the sum of the n terms. If n approaches infinity and r^2 is less than unity

$$s = \frac{a}{1 - r}$$

The multiple product represented by $n(n-1)(n-2)\ldots 3 \times 2 \times 1$ is designated by the symbol $n!$ or $\underline{|n}$ and is called " n factorial." The following list gives the value of $n!$ up to $n! = 10$

1! =	1	6! =	720
2! =	2	7! =	5,040
3! =	6	8! =	40,320
4! =	24	9! =	362,880
5! =	120	10! =	3,628,800

For large values of n a good approximation for $n!$ is, from Stirling's formula,

$$n! = (2\pi n)^{1/2}\left(\frac{n}{e}\right)^n, \quad e = 2.7182818$$

This formula is accurate to about $2^1/_2$ per cent at $n = 10$ and becomes more accurate very rapidly as n is increased.

The number of permutations or arrangements of n things taken p at a time is

$$P_p^n = \frac{n!}{(n-p)!}$$

The number of combinations of n things taken p at a time is then

$$C_p^n = \frac{1}{p!} P_p^n$$

QUADRATIC EQUATION. The solution of

$$ax^2 + bx + c = 0$$

is

$$x = \frac{-b \pm \sqrt{b^2 - 4ac}}{2a}$$

If a, b, and c are real, and the discriminant, $b^2 - 4ac$, is positive, the roots are real and unequal; if it is zero, the roots are real and equal; if it is negative, the roots are conjugate complex numbers.

CUBIC EQUATIONS. The solution of

$$ax^3 + 3bx^2 + 3cx + d = 0 \tag{1}$$

is obtained as follows: Put $x = \frac{1}{a}(y - b)$; then (1) becomes

$$y^3 - 3Hy + G = 0$$

where

$$H = b^2 - ac$$

$$G = a^2 d - 3abc + 2b^3$$

For a solution let

$$A = \sqrt[3]{\frac{-G}{2} + \sqrt{\frac{G^2}{4} - H^3}}, \quad B = \sqrt[3]{\frac{-G}{2} - \sqrt{\frac{G^2}{4} - H^3}}$$

then the values of y will be given by

$$y = A + B, \quad -\frac{1}{2}(A + B) + j\frac{\sqrt{3}}{2}(A - B), \quad -\frac{1}{2}(A + B) - j\frac{\sqrt{3}}{2}(A - B)$$

If a, b, c, d are real and if $G^2 - 4H^3$, the discriminant, is positive there are one real root and two conjugate complex roots; if $G^2 - 4H^3$ is zero there are three real roots, at least two of which are equal; if $G^2 - 4H^3$ is negative there are three real and unequal roots.

The solution may be written in three other forms.

(1) Put

$$\phi = \frac{1}{3}\sin^{-1}\left[\frac{G}{2H\sqrt{H}}\right]$$

then the roots are

$$y = 2\sqrt{H}\sin\phi, \quad 2\sqrt{H}\sin(\phi + 120°), \quad 2\sqrt{H}\sin(\phi - 120°)$$

Or (2) put

$$u = \frac{1}{3}\cosh^{-1}\left[\frac{G}{2H\sqrt{H}}\right]$$

then the roots are

$$y = -2\sqrt{H}\cosh u, \quad \sqrt{H}\cosh u + \sqrt{-3H}\sinh u, \quad \sqrt{H}\cosh u - \sqrt{-3H}\sinh u$$

Or (3) put

$$u = \frac{1}{3}\sinh^{-1}\left[\frac{G}{2H\sqrt{-H}}\right]$$

Then the roots are

$$y = 2\sqrt{-H}\sinh u, \quad -\sqrt{-H}\sinh u + \sqrt{3H}\cosh u$$

$$-\sqrt{-H}\sinh u - \sqrt{3H}\cosh u$$

SIMULTANEOUS EQUATIONS. Given n independent equations in n unknowns, these n equations usually fix one or more values for each of the n unknowns. To solve such a set of simultaneous equations in x, y, and z, say, solve each of the three equations for x in terms of y and z. Equating these three values for x gives two equations in y and z. Solving each of these two equations for y in terms of z and equating these two values of y gives a single equation in z. The solution of this last equation then gives the value of z. Then substitute this value of z in either of the equations in y and z, and solve for y. Then substitute these values of y and z in any one of the original equations and solve for x.

DETERMINANTS. In the case of linear simultaneous equations (i.e., when x, y, and z occur only to the first power), the equations may be solved by determinants. This method is a considerable time-saver when the number of unknowns is greater than three, but when the number of unknowns is three or less the straight substitution method is preferable.

The determinant of a set of simultaneous equations is formed by writing the equations one below the other with the same unknown in the same relative position in each. The block of numbers forming the coefficients of the unknowns is called the determinant. For example, the determinant of the equations

$$\begin{aligned} w + x + y + z &= 6 \\ w + y + 3z &= 4 \\ w + 2x + 3y &= 1 \\ w + 3x + z &= 3 \end{aligned}$$

is

$$D = \begin{vmatrix} 1 & 1 & 1 & 1 \\ 1 & 0 & 1 & 3 \\ 1 & 2 & 3 & 0 \\ 1 & 3 & 0 & 1 \end{vmatrix}$$

The values of any one of the unknowns, say y, is found by writing a second determinant, D_y, exactly like the determinant D, except that the constants forming the right-hand members of these equations are substituted for the coefficients of y in the determinant, that is

$$D_y = \begin{vmatrix} 1 & 1 & 6 & 1 \\ 1 & 0 & 4 & 3 \\ 1 & 2 & 1 & 0 \\ 1 & 3 & 3 & 1 \end{vmatrix}$$

Then

$$y = \frac{D_y}{D}$$

and similarly for the other unknowns.

The value of any determinant is found by making use of the following rules:

(1) If a determinant has two equal rows or columns, it is equal to zero.

(2) To any row or column one may add or substract any number of times any other row or column without altering the value of the determinant.

(3) To multiply any row or column by a number is the same as multiplying the determinant by that number.

(4) If all the terms in a row or column except one are zero, the determinant reduces to one of a lower order which may be obtained by striking out the row and column which intersect at the element of the row or column which is not zero, and multiplying the whole by that element, changing the sign of this element, however, if it is removed by an odd number of elements from the principal diagonal. The principal diagonal is the line of elements beginning at the upper left-hand corner and ending at the lower right-hand corner. Thus,

$$\begin{vmatrix} 1 & 2 & 8 & 5 \\ 3 & 4 & 6 & 9 \\ 0 & 3 & 0 & 0 \\ 6 & 7 & 4 & 3 \end{vmatrix} = -3 \begin{vmatrix} 1 & 8 & 5 \\ 3 & 6 & 9 \\ 6 & 4 & 3 \end{vmatrix}$$

the principal diagonal being that with the figures 1, 4, 0, and 3. It is immaterial whether the distance from the diagonal is counted along a row or a column.

(5) The value of a determinant of the second order is

$$\begin{vmatrix} a_1 & b_1 \\ a_2 & b_2 \end{vmatrix} = a_1 b_2 - a_2 b_1$$

The reduction of determinants is effected by altering the terms according to the above rules until a row or column is obtained in which all terms but one are zero. This enables a reduction of order to be effected in accordance with rule 4. Reductions are continued until one of the second order is obtained.

EQUATIONS OF COMMON CURVES. Straight Line.

$$\frac{x}{a} + \frac{y}{b} = 1$$

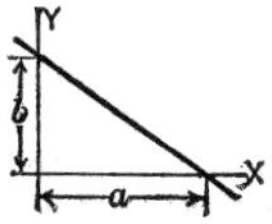

or

$$y = x \tan \theta + b.$$

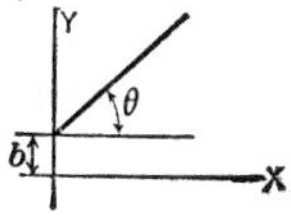

Circle.

$$x^2 + y^2 = R^2$$

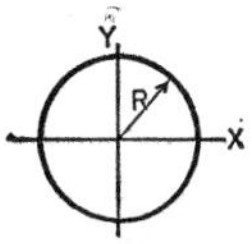

Ellipse

$$\frac{x^2}{a^2} + \frac{y^2}{b^2} = 1.$$

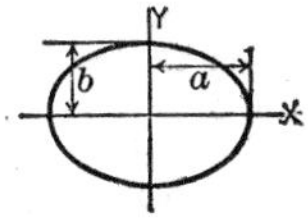

Parabola (Vertical).

$$y = kx^2$$

where k is a constant.

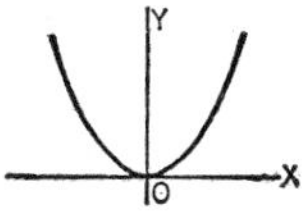

Parabola (Horizontal).

$$y = k\sqrt{x}$$

where k is a constant.

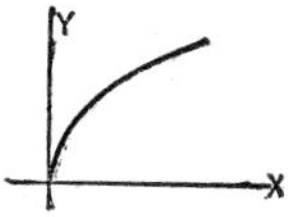

Hyperbola.

$$\frac{x^2}{a^2} - \frac{y^2}{b^2} = 1 \text{ (Horizontal)}$$

$$\frac{x^2}{a^2} - \frac{y^2}{b^2} = -1 \text{ (Vertical)}$$

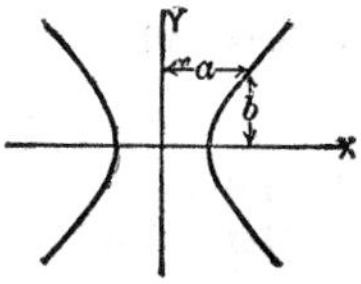

Rectangular or Equilateral Hyperbola.

$$y = \frac{k}{x}$$

where k is a constant.

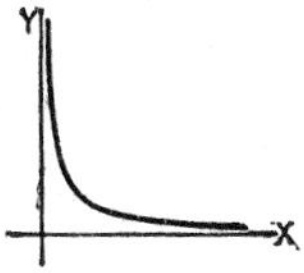

Catenary.

$$y = \frac{1}{k} \cosh kx - 1$$

where k is a constant. The length of arc from O to P is

$$= \frac{1}{k} \sinh (kx)$$

See tables of hyperbolic functions.

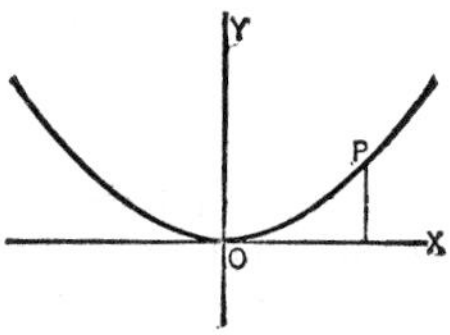

Sinusoid.

$$y = A \sin (ax + \theta).$$

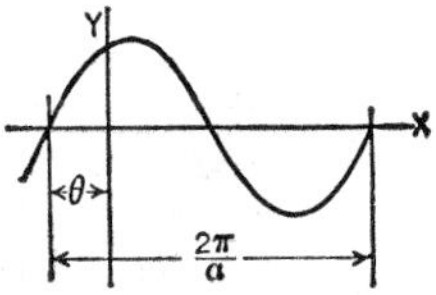

2. COMPLEX QUANTITIES

The square root of a negative quantity is called an " imaginary " quantity, or a pure imaginary. A quantity consisting of the sum or difference of a real quantity and an imaginary quantity is called a " complex " quantity. All the rules of ordinary algebra apply to pure imaginaries and complex quantities. The square root of minus one is called the imaginary unit and is usually represented by the symbol j (writers on pure mathematics use the symbol i), that is,

$$j = \sqrt{-1}$$

Any complex quantity may then be written

$$a + jb$$

where a and b are both real quantities.

GEOMETRICAL REPRESENTATION OF A COMPLEX QUANTITY. A positive real quantity may be represented by a line drawn on a plane in a given direction; a negative real quantity may be represented by a line drawn in the opposite direction. Multiplying a quantity by -1 then reverses its direction. Also, since multiplying a real quantity by $\sqrt{-1}$ twice is equivalent to multiplying it by -1, the operation of multiplying once by $\sqrt{-1}$ may be represented by turning the line representing the quantity through 90° in the positive direction of rotation. The positive direction of rotation is taken as the opposite direction to that in which the hands of a clock move. Hence, a complex quantity $a + jb$ may be represented by the line OP in the figure, where $OA = a$ and $AP = b$. The complex quantity $a + jb$ is then completely specified by a line of length $\sqrt{a^2 + b^2}$ making an angle θ, with the axis of reference OX where $\tan\theta = \frac{b}{a}$. The length $M = \sqrt{a^2 + b^2}$ is called the magnitude of the complex quantity, and the angle $\theta = \tan^{-1}\frac{b}{a}$ is called its angle. From the figure it is evident that the complex quantity $a + jb$ may also be written

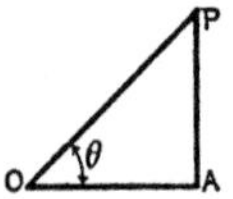

$$a + jb = M(\cos\theta + j\sin\theta)$$

Expanding $\cos\theta$ and $\sin\theta$ into series (see Series, Article 9) and adding, the resultant series obtained is the series for $e^{j\theta}$; hence

$$a + jb = Me^{j\theta} \tag{1}$$

From the above definitions and equation (1) it is evident that complex numbers possess the following properties:

ADDITION OF TWO COMPLEX QUANTITIES.

$$(a + jb) + (a_1 + jb_1) = (a + a_1) + j(b + b_1)$$

SUBTRACTION OF TWO COMPLEX QUANTITIES.

$$(a + jb) - (a_1 + jb_1) = (a - a_1) + j(b - b_1)$$

MULTIPLICATION OF A COMPLEX QUANTITY BY A COMPLEX NUMBER.

$$(a + jb)(a_1 + jb_1) = aa_1 - bb_1 + j(ab_1 + a_1 b)$$

or, putting

$$a + jb = Me^{j\theta} \quad \text{and} \quad a_1 + jb_1 = M_1 e^{j\theta_1}$$

where

$$M = \sqrt{a^2 + b^2}, \qquad M_1 = \sqrt{a_1^2 + b_1^2}$$

$$\tan\theta = \frac{b}{a}$$

and

$$\tan\theta_1 = \frac{b_1}{a_1}$$

we have

$$(a + jb)(a_1 + jb_1) = Me^{j\theta} M_1 e^{j\theta_1} = MM_1 e^{j(\theta + \theta_1)}$$

Hence the product of two complex quantities is in general a complex quantity which has a magnitude equal to the product of the magnitudes of the two quantities and an angle equal to the sum of the angles of the two quantities.

DIVISION OF A COMPLEX QUANTITY BY A COMPLEX NUMBER.

$$\frac{a + jb}{a_1 + jb_1} = \frac{(a + jb)(a_1 - jb_1)}{(a_1 + jb_1)(a_1 - jb_1)} = \frac{aa_1 + bb_1 - j(ab_1 - a_1 b)}{a_1^2 + b_1^2}$$

or

$$\frac{a + jb}{a_1 + jb_1} = \frac{Me^{j\theta}}{M_1 e^{j\theta_1}} = \frac{M}{M_1} e^{j(\theta - \theta_1)}$$

Hence the quotient of two complex quantities is in general a complex quantity which has a magnitude equal to the quotient of the magnitudes of the two quantities and an angle equal to the difference of the angles of the two quantities.

SQUARE ROOT OF A COMPLEX QUANTITY.

$$\sqrt{a+jb} = \pm\left[\sqrt{\frac{\sqrt{a^2+b^2}+a}{2}} + j\sqrt{\frac{\sqrt{a^2+b^2}-a}{2}}\right]$$

$$\sqrt{a-jb} = \pm\left[\sqrt{\frac{\sqrt{a^2+b^2}+a}{2}} - j\sqrt{\frac{\sqrt{a^2+b^2}-a}{2}}\right]$$

and in general

$$\sqrt[N]{a+jb} = M^{1/N}\,\epsilon^{j\theta/N}$$

Hence the nth root of a complex quantity is, in general, a complex quantity which has a magnitude equal to the nth root of the magnitude M of the quantity and an angle equal one-nth of the angle of the quantity.

EQUATIONS CONTAINING COMPLEX QUANTITIES. Since a real quantity cannot be equal to an imaginary quantity it follows that any equation of the form

$$A + jB = A_1 + jB_1$$

where A, B, A_1, and B_1 are all real quantities (which may, however, consist of any number of terms), is equivalent to the two equations

$$A = A_1$$

and

$$B = B_1$$

Also, if one member of an equation reduces to the form $A + jB$, then the other member of this equation must likewise contain an equal real and an equal imaginary part.

See K. S. Johnson, *Transmission Circuits for Telephonic Communication*, D. Van Nostrand.

3. TRIGONOMETRIC FORMULAS

The trigonometric functions of an angle are the ratios to one another of the various sides of a right triangle having the given angle as one of its angles. Referring to Fig. 1, let B, P, and H be the three sides of a triangle. Then the trigonometric functions of the angle x are

sine of x, abbreviated $\sin x = \dfrac{P}{H}$; cotangent of x, abbreviated $\cot x = \dfrac{B}{P}$

cosine of x, abbreviated $\cos x = \dfrac{B}{H}$; secant of x, abbreviated $\sec x = \dfrac{H}{B}$

tangent of x, abbreviated $\tan x = \dfrac{P}{B}$; cosecant of x, abbreviated $\csc x = \dfrac{H}{P}$

When B, P, and H are limited to the three sides of a right triangle, the above definitions are directly applicable only to angles lying between 0 and 90°. The definitions, however,

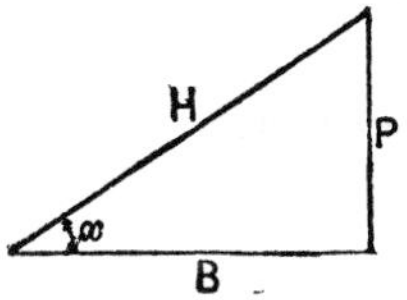

FIG. 1

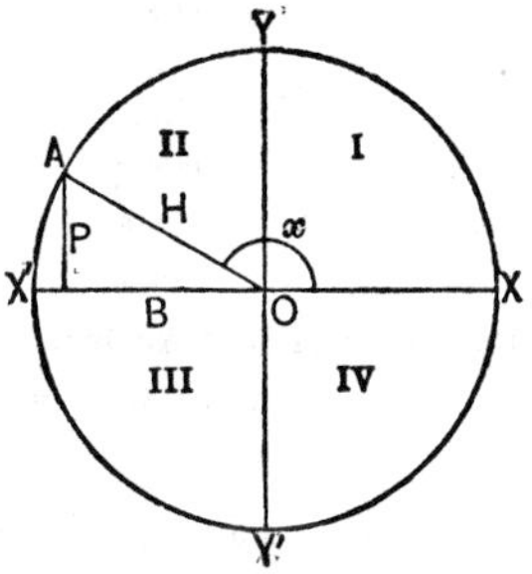

FIG. 2

may be extended by considering the point A (Fig. 2) as describing a circle of radius OA with the center at O. Let XX' be the horizontal diameter and YY' the vertical diameter of this circle, and call P the perpendicular distance from A to the line XX' and B the horizontal distance from A to YY'. P is to be considered positive when A lies above XX', negative when below. B is considered positive when A is to the right of YY' and negative when to the left. The four quarters of the circle are called quadrants, and are designated as the first, second, third, and fourth quadrants as indicated. The angle is said to lie in the quadrant in which the point A lies. In Fig. 2 the angle x is in the second quadrant.

Algebraic Signs of the Functions

	Sine	Cosine	Tangent
Angle in first quadrant........	+	+	+
Angle in second quadrant......	+	−	−
Angle in third quadrant.......	−	−	+
Angle in fourth quadrant......	−	+	−

Period. From the above definitions it is evident that adding 2π radians or 360° to an angle does not change the value of any of its functions, that is, these functions repeat themselves every time the angle increases by the 2π radians or 360°. They are therefore said to have a period equal to 2π radians or 360°.

Functions of Angles in Any Quadrant in Terms of Angles in First Quadrant.

$$\begin{aligned} \sin(-x) &= -\sin x & \sin(90+x) &= \cos x \\ \cos(-x) &= \cos x & \cos(90+x) &= -\sin x \\ \tan(-x) &= -\tan x & \tan(90+x) &= -\cot x \\ \sin(180-x) &= \sin x & \sin(180+x) &= -\sin x \\ \cos(180-x) &= -\cos x & \cos(180+x) &= -\cos x \\ \tan(180-x) &= -\tan x & \tan(180+x) &= \tan x \\ \sin(270-x) &= -\cos x & \sin(270+x) &= -\cos x \\ \cos(270-x) &= -\sin x & \cos(270+x) &= \sin x \\ \tan(270-x) &= \cot x & \tan(270+x) &= -\cot x \end{aligned}$$

Anti-functions. If $a = \sin x$, then x is the angle whose sine is a; this may be expressed symbolically $x = \sin^{-1} a$, which is read "x equals the angle whose sine is a." The angle x is also called the "anti-sine" or the "inverse sine" of a. Similar notation is used for the other functions; for example, $x = \cos^{-1} b$ is used to express the relation that x is the angle whose cosine is b. At least two "anti-functions" must be known to completely determine the quadrant in which an angle lies; for example, if $x = \sin^{-1} 0.5$ then x may be either 30° or 150°, but if we also have $x = \cos^{-1} 0.866$, then x must equal 30°, while if $x = \cos^{-1}(-0.866)$, then x must equal 150°.

Anti-functions may be taken from the Trigonometric Tables by finding the angle in the margin corresponding to the function in the table.

Example. $\sin^{-1} 0.319 = 18.6° \text{ or } 180° - 18.6° = 161.4°$.

Versine. The expression $(1 - \cos x)$ is called the "versine" of x.

Relations among Functions of the Same Angle.

$$\tan x = \frac{\sin x}{\cos x} = \frac{1}{\cot x} \qquad \sin^2 x + \cos^2 x = 1$$

$$\sec x = \frac{1}{\cos x} \qquad 1 + \tan^2 x = \frac{1}{\cos^2 x}$$

$$\csc x = \frac{1}{\sin x} \qquad 1 + \cot^2 x = \frac{1}{\sin^2 x}$$

$$\begin{aligned} \sin(90-x) &= \cos x & \sin(-x) &= -\sin x \\ \cos(90-x) &= \sin x & \cos(-x) &= \cos x \\ \tan(90-x) &= \cot x & \tan(-x) &= -\tan x \end{aligned}$$

Sum and Difference of Two Angles.

$$\sin(x+y) = \sin x \cos y + \cos x \sin y$$

$$\cos(x+y) = \cos x \cos y - \sin x \sin y$$

$$\tan(x+y) = \frac{\tan x + \tan y}{1 - \tan x \tan y}$$

$$\sin(x-y) = \sin x \cos y - \cos x \sin y$$

$$\cos(x-y) = \cos x \cos y + \sin x \sin y$$

$$\tan(x-y) = \frac{\tan x - \tan y}{1 + \tan x \tan y}$$

Product of the Functions of Two Angles.

$$\sin x \sin y = {}^1/_2 [\cos (x - y) - \cos (x + y)]$$
$$\sin x \cos y = {}^1/_2 [\sin (x + y) + \sin (x - y)]$$
$$\cos x \sin y = {}^1/_2 [\sin (x + y) - \sin (x - y)]$$
$$\cos x \cos y = {}^1/_2 [\cos (x + y) + \cos (x - y)]$$

Functions of Twice an Angle.

$$\sin 2x = 2 \sin x \cos x \qquad \cos 2x = \cos^2 x - \sin^2 x = 2 \cos^2 x - 1$$

$$\tan 2x = \frac{2 \tan x}{1 - \tan^2 x}$$

Functions of Half an Angle.

$$\sin \frac{x}{2} = \sqrt{\frac{1 - \cos x}{2}} \qquad \cos \frac{x}{2} = \sqrt{\frac{1 + \cos x}{2}} \qquad \tan \frac{x}{2} = \sqrt{\frac{1 - \cos x}{1 + \cos x}}$$

Functions of Three Times an Angle.

$$\sin 3x = 3 \sin x - 4 \sin^3 x \qquad \cos 3x = 4 \cos^3 x - 3 \cos x$$

$$\tan 3x = \frac{3 \tan x - \tan^3 x}{1 - 3 \tan^2 x}$$

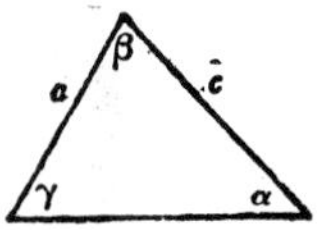

FIG. 3

TRIGONOMETRY. Any triangle is completely defined when (1) two sides and the included angle are known, (2) one side and two angles are known, (3) three sides are known. Let the sides and angles of a triangle be designated as in Fig. 3.

1. Given two sides a and b, and the included angle γ. Then

$$c = \sqrt{a^2 + b^2 - 2\,ab \cos \gamma}$$
$$\sin \alpha = \frac{a}{c} \sin \gamma$$
$$\beta = 180 - \alpha - \gamma$$

2. Given the side a and the two angles β and γ. Then

$$\alpha = 180 - \beta - \gamma$$
$$b = a \frac{\sin \beta}{\sin \alpha}$$
$$c = a \frac{\sin \gamma}{\sin \alpha}$$

3. Given the three sides a, b, and c. Put

$$s = {}^1/_2 (a + b + c)$$

Then

$$\sin \alpha = \frac{2}{bc} \sqrt{s(s - a)(s - b)(s - c)}$$
$$\sin \beta = \frac{b}{a} \sin \alpha$$
$$\gamma = 180 - \alpha - \beta$$

Relations between Sides and Angles. The following relations between the sides and angles of a triangle are sometimes useful:

$$\frac{a}{\sin \alpha} = \frac{b}{\sin \beta} = \frac{c}{\sin \gamma}$$
$$\cos \alpha = \frac{b^2 + c^2 - a^2}{2\,bc}$$
$$\sin \frac{\alpha}{2} = \sqrt{\frac{(s - b)(s - c)}{bc}}$$
$$\cos \frac{\alpha}{2} = \sqrt{\frac{s(s - a)}{bc}}$$

and similar relations for the other two angles.

4. EXPONENTIAL AND HYPERBOLIC FORMULAS

When the relation between any variable y and another variable x is such that x occurs as an exponent of one or more terms, y is said to be an exponential function of x. Of particular importance in connection with electric circuits are the exponential functions e^x and e^{-x}, where e is the base of the natural logarithms. Since x is the natural logarithm of e^x, the value of e^x can be obtained from the table of common logarithms as shown at the beginning of that table. In addition the values of e^x and e^{-x} are given in a separate table.

Hyperbolic functions are an extension of the trigonometric functions to those cases where the use of the latter gives rise to imaginary or complex angles. From the relations

$$\cos x = \frac{e^{jx} + e^{-jx}}{2}$$

$$\sin x = \frac{e^{jx} - e^{-jx}}{2j}$$

where $j = \sqrt{-1}$, it follows that, putting $x = jz$:

$$\cos jz = \frac{e^z + e^{-z}}{2} \tag{1}$$

$$-j \sin jz = \frac{e^z - e^{-z}}{2} \tag{2}$$

Expressions (1) and (2) are both real quantities when z is real, that is, when the angle jz is imaginary. The first expression is called the hyperbolic cosine of z, abbreviated and pronounced "cosh"; the second expression is called the hyperbolic sine of z, abbreviated sinh and pronounced "shin." Hence, using x for the variable,

$$\sinh x = \frac{e^x - e^{-x}}{2}$$

$$\cosh x = \frac{e^x + e^{-x}}{2}$$

The hyperbolic tangent, cotangent, secant, and cosecant are defined as follows:

$$\tanh x = \frac{\sinh x}{\cosh x}$$

$$\coth x = \frac{\cosh x}{\sinh x}$$

$$\operatorname{sech} x = \frac{1}{\cosh x}$$

$$\operatorname{csch} x = \frac{1}{\sinh x}$$

The hyperbolic angle x is a number analogous to radians in circular measure; it is never expressed in degrees.

Adding 2π to an angle does not change the value of the trigonometric functions; they are therefore said to have a period equal to 2π radians. Hyperbolic functions, however, have no true period, but adding $2\pi j$ to the hyperbolic angle does not change the values of the functions; hence these functions have an imaginary period, $2\pi j$.

For the value of the hyperbolic functions see tables of exponential and hyperbolic functions, Article 13.

Approximate Formulas. Note that, for x less than 0.1,

$$\text{Sinh } x = x \text{ with an error of less than 0.2 per cent}$$

$$\cosh x = 1 + \frac{x^2}{2} \text{ with an error of less than 0.09 per cent}$$

For x greater than 6,

$$\sinh x = \cosh x = \frac{e^x}{2} = \frac{1}{2} \log_{10}{}^{-1} (0.43429x)$$

with an error of less than 0.01 per cent.

Anti-functions. If $a = \sinh x$, then x is the angle whose hyperbolic sine is a; this may be expressed symbolically

$$x = \sinh^{-1} a$$

which is read "x equals the angle whose hyperbolic sine is a." The angle x is also called the "anti-hyperbolic sine" or the "inverse hyperbolic sine" of a. Similarly for the other hyperbolic functions. The following relations exist between the anti-hyperbolic functions and the natural logarithms:

$$\sinh^{-1} x = \log\,(x + \sqrt{x^2 + 1})$$
$$\cosh^{-1} x = \log\,(x + \sqrt{x^2 - 1})$$
$$\tanh^{-1} x = \frac{1}{2}\log\left(\frac{1 + x}{1 - x}\right)$$

Relations among Functions of the Same Angle.

$$\cosh^2 x - \sinh^2 x = 1$$
$$1 - \tanh^2 x = \frac{1}{\cosh^2 x}$$
$$\coth^2 x - 1 = \frac{1}{\sinh^2 x}$$
$$\sinh\,(-x) = -\sinh x$$
$$\cosh\,(-x) = \cosh x$$
$$\tanh\,(-x) = -\tanh x$$

See also the definitions given above.

Sum and Difference of Two Angles.

$$\sinh\,(x + y) = \sinh x \cosh y + \cosh x \sinh y$$
$$\cosh\,(x + y) = \cosh x \cosh y + \sinh x \sinh y$$
$$\tanh\,(x + y) = \frac{\tanh x + \tanh y}{1 + \tanh x \tanh y}$$
$$\sinh\,(x - y) = \sinh x \cosh y - \cosh x \sinh y$$
$$\cosh\,(x - y) = \cosh x \cosh y - \sinh x \sinh y$$
$$\tanh\,(x - y) = \frac{\tanh x - \tanh y}{1 - \tanh x \tanh y}$$

Product of the Functions of Two Angles.

$$\sinh x \sinh y = 1/2\,[\cosh\,(x + y) - \cosh\,(x - y)]$$
$$\sinh x \cosh y = 1/2\,[\sinh\,(x + y) + \sinh\,(x - y)]$$
$$\cosh x \sinh y = 1/2\,[\sinh\,(x + y) - \sinh\,(x - y)]$$
$$\cosh x \cosh y = 1/2\,[\cosh\,(x + y) + \cosh\,(x - y)]$$

Functions of Twice an Angle.

$$\sinh 2x = 2 \sinh x \cosh x$$
$$\cosh 2x = \sinh^2 x + \cosh^2 x = 2 \sinh^2 x + 1 = 2 \cosh^2 x - 1$$
$$\tanh 2x = \frac{2 \tanh x}{1 + \tanh^2 x}$$

Functions of Half an Angle.

$$\sinh \frac{x}{2} = \sqrt{\frac{\cosh x - 1}{2}}$$
$$\cosh \frac{x}{2} = \sqrt{\frac{\cosh x + 1}{2}}$$
$$\tanh \frac{x}{2} = \sqrt{\frac{\cosh x - 1}{\cosh x + 1}}$$

Functions of Three Times an Angle.

$$\sinh 3x = 3 \sinh x + 4 \sinh^3 x$$
$$\cosh 3x = 4 \cosh^3 x - 3 \cosh x$$
$$\tanh 3x = \frac{3 \tanh x + \tanh^3 x}{1 + 3 \tanh^3 x}$$

Relations between Hyperbolic and Trigonometric Functions.

$$\sinh\,(jx) = j \sin x \qquad \sin\,(jx) = j \sinh x$$
$$\cosh\,(jx) = \cos x \qquad \cos\,(jx) = \cosh x$$
$$\tanh\,(jx) = j \tan x \qquad \tan\,(jx) = j \tanh x$$
$$\sinh^{-1} jx = j \sin^{-1} x \qquad \sin^{-1} jx = j \sinh^{-1} x$$
$$\tanh^{-1} jx = j \tan^{-1} x \qquad \tan^{-1} jx = j \tanh^{-1} x$$

$$\cosh^{-1} jx = j \cos^{-1} jx = \log\,(x + \sqrt{1 + x^2}) - j\frac{\pi}{2}$$

Hyperbolic Functions of a Complex Angle.

$$\sinh (x + jy) = \sinh x \cos y + j \cosh x \sin y = Me^{j\theta}$$

where $$M = \sqrt{\frac{\cosh 2x - \cos 2y}{2}} \quad \text{and} \quad \tan \theta = \frac{\tan y}{\tanh x}$$

$$\cosh (x + jy) = \cosh x \cos y + j \sinh x \sin y = Ne^{j\phi}$$

where $$N = \sqrt{\frac{\cosh 2x + \cos 2y}{2}} \quad \text{and} \quad \tan \phi = \tanh x \cdot \tan y$$

$$\tanh (x + jy) = \frac{\sinh x \cos y + j \cosh x \sin y}{\cosh x \cos y + j \sinh x \sin y} = Pe^{j\psi}$$

where $$P = \sqrt{\frac{\cosh 2x - \cos 2y}{\cosh 2x + \cos 2y}} \quad \text{and} \quad \psi = \tan^{-1} \left[\frac{\sin 2y}{\sinh 2x}\right]$$

$$\tanh^{-1} (Ae^{j\alpha}) = B_1 + jB_2$$

where $$B_1 = \frac{1}{2} \tanh^{-1} \left[\frac{2A \cos \alpha}{1 + A^2}\right] \quad \text{and} \quad B_2 = \frac{1}{2} \tan^{-1} \left[\frac{2A \sin \alpha}{1 - A}\right]$$

5. CALCULUS FORMULAS

The formula for the integration by parts is:

$$\int_a^b u\, dv = [uv]_a^b - \int_a^b v\, du$$

The following table is used in the formulas

$$\frac{df(x)}{dx} = f'(x)$$

$$\int f'(x)\, dx = f(x) + C$$

where C is an arbitrary constant.

$f'(x)$	$f(x)$	$f'(x)$	$f(x)$
x^m	$\frac{1}{m+1} x^{m+1}$	$\frac{1}{\cos^2 ax}$	$\frac{1}{a} \tan ax$
$\frac{1}{ax}$	$\frac{1}{a} \log_e x$	$\frac{1}{\sin^2 ax}$	$-\frac{1}{a} \cot ax$
e^{ax}	$\frac{1}{a} e^{ax}$	$\frac{1}{\sqrt{a^2 + bx^2}}$	$\frac{1}{\sqrt{-b}} \sin^{-1} \sqrt{-b} \frac{x}{a}$
a^{bx}	$\frac{1}{b \log a} a^{bx}$	$\frac{1}{x\sqrt{x^2 + a}}$	$\frac{1}{\sqrt{-a}} \cos^{-1} \sqrt{-a} \frac{1}{x}$
$\cos ax$	$\frac{1}{a} \sin ax$	$\frac{x}{\sqrt{a^2 \pm x^2}}$	$\pm\sqrt{a^2 \pm x^2}$
$\sin ax$	$-\frac{1}{a} \cos ax$	$\frac{x}{\sqrt{x^2 - a^2}}$	$\sqrt{x^2 - a^2}$
$\cosh ax$	$\frac{1}{a} \sinh ax$	$\frac{u \frac{dv}{dx} - v \frac{du}{dx}}{u^2}$	$\frac{v}{u}$
$\sinh ax$	$\frac{1}{a} \cosh ax$	$\log x$	$x \log x - x$
$\tan ax$	$-\frac{1}{a} \log (\cos ax)$	$\sin^2 x$	$-\frac{1}{2}(\cos x \sin x - x)$
$\tanh ax$	$\frac{1}{a} \log (\cosh ax)$	$\cos^2 x$	$\frac{1}{2}(\sin x \cos x + x)$

MAXIMA AND MINIMA. Let y be any function of a variable x, then y will be a maximum or minimum for any value of x which satisfies

$$\frac{dy}{dx} = 0 \tag{1}$$

provided $\frac{d^2y}{dx^2}$ is not zero. If the second derivative $\frac{d^2y}{dx^2}$ is positive for this value of x, then the corresponding value of y is a minimum; if this second derivative is negative, the corresponding value of y is a maximum.

In case $\frac{d^2y}{dx^2}$ is also zero for the value of x which satisfies (1), the corresponding value of y is not a maximum or minimum unless $\frac{d^3y}{dx^3}$ is also zero and $\frac{d^4y}{dx^4}$ is not zero. When $\frac{d^3y}{dx^3} = 0$, y is a minimum if $\frac{d^4y}{dx^4}$ is positive and a maximum if $\frac{d^4y}{dx^4}$ is negative. In case $\frac{d^4y}{dx^4}$ is also zero, similar relations must hold for the fifth and sixth derivatives, etc.

6. DIFFERENTIAL EQUATIONS

Differential equations of the following forms are met with in the theory of alternating and transient currents.

The following notation is used: $e = 2.7183\cdots$ = base of natural system of logarithms; x, y, z are variables. A, ϕ, γ, and θ are constants of integration or arbitrary constants. Other letters represent known constants.

$$\frac{dy}{dx} = ay \tag{1}$$

Solution:

$$y = Ae^{ax}$$

$$\frac{dy}{dx} + ay = 0 \tag{2}$$

Solution:

$$y = Ae^{-ax}$$

$$\frac{dy}{dx} + ay = b \tag{3}$$

Solution:

$$y = \frac{b}{a}[1 - Ae^{-ax}]$$

$$\frac{d^2y}{dx^2} = -a^2y \tag{4}$$

Solution:

$$y = A \sin(ax + \phi)$$

$$\frac{d^2y}{dx^2} = a^2y \tag{5}$$

Solution:

$$y = A \sinh(ax + \phi)$$

$$\frac{d^2y}{dx^2} + 2u\frac{dy}{dx} + (u^2 + a^2)y = 0 \tag{6}$$

Solution:

Case I. a^2 positive: $y = Ae^{-ux} \sin(ax + \phi)$

Case II. a^2 negative: $y = Ae^{-ux} \sinh(ax + \phi)$

Case III. $a^2 = 0$: $y = A(x + \phi)e^{-ux}$

$$\frac{d^2y}{dx^2} + 2u\frac{dy}{dx} + (u^2 + a^2)y = B \sin(\omega x + \theta) \tag{7}$$

The complete solution of this equation consists of the solution of (6) plus the term

$$\left(\frac{B \sin \delta}{2u\omega}\right) \sin(\omega x + \theta - \delta) \tag{a}$$

where

$$\delta = \tan^{-1} \frac{2u\omega}{a^2 + u^2 - \omega^2}$$

For each additional sine term added to the right-hand member of the equation, there will be a corresponding term of the same form as (a) in the solution.

$$\frac{d^n y}{dx^n} + a_{n-1}\frac{d^{n-1}y}{dx^{n-1}} + \cdots a_1 \frac{dy}{dx} + a_0 y = B \sin(\omega x + \theta) \tag{8}$$

Solution:

$$y = A_1 e^{m_1 x} + A_2 e^{m_2 x} + \cdots A_n e^{m_n x} + KB \sin(\omega X + \theta + \delta)$$

where m_1, m_2, etc., are the n roots of the equation

$$m^n + a_{n-1}m^{n-1} + \ldots a_1 m + a_0 = 0$$

and K and δ are found by substituting the $KB \sin(\omega x + \theta + \delta)$ by itself in the given differential equation and equating the coefficients of $\sin(\omega x + \theta)$ and $\cos(\omega x + \theta)$ respectively on the two sides of the resulting equation. When the second member of the differential equation is a constant, B, the sine term in the solution becomes simply $\frac{B}{a_0}$.

Note that all the preceding equations are merely special cases of the general equation (8)

$$\frac{d^2y}{dx^2} + 2u\frac{dy}{dx} + (u^2 - q^2)y = \frac{1}{c^2}\frac{d^2y}{dz^2} \tag{9}$$

The complete solution of this equation contains an infinite number of terms of the form

$$y = e^{-(u-s)x}\,[A_1 e^{mz} \sin(\omega x + nz + \phi_1) + A_2 e^{-mz} \sin(\omega x - nz + \phi_2)] \tag{a}$$

where A_1, ϕ_1, A_2, ϕ_2 and two of the four constants ω, s, m, and n are integration constants (fixed by the terminal conditions). The values of m and n in terms of ω and s are

$$m = c\sqrt{ab}\cos\frac{\eta + e}{2}$$

$$n = c\sqrt{ab}\sin\frac{\eta + e}{2}$$

where $a = \sqrt{(s+q)^2 + \omega^2}$, $\qquad e = \tan^{-1}\left(\frac{\omega}{s+q}\right)$

$b = \sqrt{(s-q)^2 + \omega^2}$, $\qquad \eta = \tan^{-1}\left(\frac{\omega}{s-q}\right)$

The values of ω and s in terms of m and n are

$$\omega = \frac{\sqrt{FG}}{c}\cos\frac{\alpha + \beta}{2}$$

$$s = \frac{\sqrt{FG}}{c}\sin\frac{\alpha + \beta}{2}$$

where $F = \sqrt{(n + cq)^2 + m^2}$, $\qquad \alpha = \tan^{-1}\left(\frac{m}{n + cq}\right)$

$G = \sqrt{(n - cq)^2 + m^2}$, $\qquad \beta = \tan^{-1}\left(\frac{m}{n - cq}\right)$

The solution of eq. (9) may also be written as a series of terms of the form

$$y = Me^{-(u-s)x}\sin(\omega x + \phi + \mu) \tag{b}$$

where $M = \frac{A}{\sqrt{2}}\sqrt{\cosh 2(mz + \gamma) + \cos 2(nz + \theta)}$

$$\tan\mu = \tanh(mz + \gamma)\tan(nz + \theta)$$

where A, ϕ, γ, and θ are integration constants, and the relations between the other constants ω, s, m, and n are the same as above.

In the special case when $q = 0$, the solution of eq. (9) is

$$y = e^{-ux}\,[f_1(\omega x + nz) + f_2(\omega x - nz)] \tag{c}$$

where f_1 and f_2 are any two arbitrary functions and ω and n are connected by the relation

$$\frac{\omega}{n} = \frac{1}{c}$$

$$\frac{d^2y}{dx^2} + \frac{1}{x}\frac{dy}{dx} + \left(1 - \frac{n^2}{x^2}\right) y = 0 \tag{10}$$

This is known as Bessel's equation of order n. $J_n(x)$, Bessel's function of the first kind of order n, is a particular solution of this equation. It may be computed from the infinite series:

$$J_n(x) = \frac{x^n}{2^n\Gamma(n+1)}\left[1 - \frac{x^2}{2^2(n+1)} + \frac{x^4}{2^4 2!(n+1)(n+2)} - \frac{x^6}{2^6 3!(n+1)(n+2)(n+3)} + \cdots\right] \tag{a}$$

where $\Gamma(n+1)$ is the gamma function which reduces to unity for $n = 0$ and to $n!$ for n equal to any positive integer. In general, the function $J_n(x)$ is an oscillatory function of x having the value zero for $x = 0$, except for the case where $n = 0$. For values of n larger than 1, the slope of $J_n(x)$ is zero for $x = 0$ and the first maximum and the first zero occurs at successively higher values of x as n takes on larger values. For small values of n, the values of x for which $J_n(x)$ is a maximum or zero can be gotten from tables of Bessel functions. For large values of n, the first maximum, that is, the smallest x for which $J_n'(x) = 0$, is given by

$$n + 0.809\sqrt[3]{n} \tag{b}$$

with an error not larger than $1/\sqrt[3]{n}$, and the first zero, that is, the smallest x for which $J_n(x) = 0$, is given by

$$n + 1.856\sqrt[3]{n} \tag{c}$$

again with an error of the order of $1/\sqrt[3]{n}$.

For integral values of n greater than zero,

$$J_{n+1}(x) = \frac{2n}{x} J_n(x) - J_{n-1}(x) \tag{d}$$

which permits one to compute Bessel functions for successively higher order from tables of $J_0(x)$ and $J_1(x)$.

When n is an integer,

$$J_{-n}(x) = (-1)^n J_n(x) \tag{e}$$

BIBLIOGRAPHY

Watson, G. N., *A Treatise on the Theory of Bessel Functions*, 2nd Ed., New York, The Macmillan Company (1944).

7. ERRORS OF OBSERVATION

When a quantity is measured with all possible accuracy many times in succession, the numbers expressing the results are found to differ by amounts which, although generally small, are occasionally considerable in comparison with the quantity measured. Though these differences may be decreased by improved methods, better instruments, or greater skill, they can never be entirely removed. They are known as the errors of observation. The following formulas, which are derived from the theory of least squares, apply to such errors and not to errors which can be eliminated by correcting mistakes of the observer or defects of instruments or methods of observation. That is, they apply only to errors which may be either positive or negative, the chance of a positive error occurring being exactly the same as the chance of a negative error occurring.

WEIGHTED OBSERVATIONS. Sometimes, in spite of the care with which observations are taken, there are reasons for believing that some observations are better than others. In this case the observations are given different "weights" or numbers expressing their relative practical worth. A weighted observation is an observation multiplied by its weight.

PROBABLE VALUE OF SEVERAL OBSERVATIONS. The most probable value of a quantity which is observed directly several times with equal care is the arithmetical mean of the measurements.

The most probable value of a quantity which is observed directly several times, but the observations of which have different weights, is equal to the sum of the weighted observations divided by the sum of the weights.

PROBABLE ERROR OF ANY ONE OF SEVERAL OBSERVATIONS. The probable error or dispersion of a number of direct observations made with equal care is given by the following formula:

$$r = 0.6745\sqrt{\frac{\Sigma v^2}{n-1}}$$

where n = number of observations.
r = probable error of a single observation,
v = residual found by subtracting the arithmetical mean from each measurement.

The probable error of each of a number of direct observations, where the observations have different weight, is found by the following formula, in which p represents the per unit weight of an observation.

$$r_1 = 0.6745\sqrt{\frac{\Sigma pv^2}{n-1}}$$

PROBABLE ERROR OF THE ARITHMETICAL MEAN. If

r = probable error of a single observation,
n = number of observations,
r_0 = probable error of the arithmetical mean,

$$r_0 = \frac{r}{\sqrt{n}} \text{ for observations of equal weight}$$

or

$$r_0 = \frac{r_1}{\sqrt{\Sigma p}} \text{ for unequal weight}$$

It should be noted that the probable error of the mean decreases inversely as the square root of the number of observations.

PROBABLE ERROR IN A RESULT CALCULATED FROM THE MEANS OF SEVERAL OBSERVED QUANTITIES. Let Z = a sum or difference of several independent quantities.

Let r_1, r_2, r_3, etc., be the probable errors in these quantities. Then the probable error of Z is equal to

$$\sqrt{r_1^2 + r_2^2 + r_3^2 + \text{etc.}}$$

Let $Z = Az$, where z is an observed quantity, and A, a known number. Let r be the probable error in z. Then the probable error in Z is Ar.

Let Z be the product of two independently observed quantities z_1 and z_2 whose probable errors are r_1 and r_2 respectively. Then the error in Z is equal to

$$\sqrt{z_1^2 r_2^2 + z_2^2 r_1^2}$$

Let Z be any function of the independently obzerved quantities z_1, z_2, z_3, etc., whose probable errors are r_1, r_2, r_3, etc. Then the probable error in Z is equal to

$$\sqrt{\left(\frac{\partial Z}{\partial z_1}\right)^2 r_1^2 + \left(\frac{\partial Z}{\partial z_2}\right)^2 r_2^2 + \left(\frac{\partial Z}{\partial z_3}\right)^3 r_3^2 + \text{etc.}}$$

8. APPROXIMATIONS

If a is small

$$(1 \pm a)^m = 1 \pm ma$$

If m is nearly equal to n

$$\sqrt{mn} = \frac{m+n}{2}$$

If θ, expressed in radians, is small compared to a radian

$$\sin\theta = \tan\theta = \theta \text{ radians}$$

9. SERIES

Taylor's series is written

$$f(x+h) = f(x) + \frac{h}{1!}f'(x) + \frac{h^2}{2!}f''(x) + \cdots$$

$$= f(h) + \frac{x}{1!}f'(h) + \frac{x^2}{2!}f''(h) + \cdots$$

where the prime on the function means the derivative with respect to the argument.

The following series are frequently useful.

$$e^x = 1 + x + \frac{x^2}{2!} + \frac{x^3}{3!} + \cdots$$

$$a^x = 1 + x \log a + \frac{(x \log a)^2}{2!} + \frac{(x \log a)^3}{3!} + \cdots$$

$$\sin x = x - \frac{x^3}{3!} + \frac{x^5}{5!} - \frac{x^7}{7!} + \cdots$$

$$\cos x = 1 - \frac{x^2}{2!} + \frac{x^4}{4!} - \frac{x^6}{6!} + \cdots$$

$$\cos (x \sin \theta) = J_0(x) + 2\,\{J_2(x) \cos 2\theta + J_4(x) \cos 4\theta + \cdots$$

where $J_n(x)$ is Bessel's function of order n,

$$\sin (x \sin \theta) = 2\,\{J_1(x) \sin \theta + J_3(x) \sin 3\theta + \cdots$$

$$\cos (x \cos \theta) = J_0 - 2J_2(x) \cos 2\theta + 2J_4 \cos 4\theta + \cdots$$

$$\sin (x \cos \theta) = 2J_1(x) \cos \theta - 2J_3(x) \cos 3\theta + 2J_5(x) \cos 5\theta + \cdots$$

$$\begin{aligned} \sin (A + x \sin \theta) = {} & J_0(x) \sin A + J_1(x)[\sin (A + \theta) - \sin (A - \theta) \\ & + J_2(x)[\sin (A + 2\theta) + \sin (A - 2\theta)] \\ & + J_3(x)[\sin (A + 3\theta) - \sin (A - 3\theta)] \\ & + J_4(x)[\sin (A + 4\theta) + \sin (A - 4\theta)] + \cdots \end{aligned}$$

10. MENSURATION

The term mensuration is used in this article to include the relations between the areas and volumes of geometric figures and their linear dimensions.

Triangle.

$$\text{Area} = {}^1/_2\ (\text{Base}) \times (\text{Perpendicular height})$$

$$= \sqrt{s(s-a)(s-b)(s-c)}$$

where a, b, and c are the lengths of the three sides respectively, and $s = {}^1/_2\,(a + b + c)$

Trapezoid.

$$\text{Area} = \left(\frac{a+b}{2}\right) d$$

where a and b are the lengths of the parallel sides respectively, and d their distance apart.

Parallelogram.

$$\text{Area} = (\text{Base}) \times (\text{Perpendicular height})$$

Parabola.

$$\text{Area} = {}^2/_3\ (\text{Area of circumscribing rectangle})$$

Cycloid.

$$\text{Area} = {}^3/_4\ \pi x(\text{Altitude})^2$$

the altitude being the diameter of the rolling circle.

Circle.

$$\text{Circumference} = 2\pi r = \pi d$$

$$\text{Area} = \pi r^2 = \frac{\pi}{4} d^2$$

where r is the radius and d the diameter.

$$\text{Area of segment} = \frac{r^2}{2}(\theta - \sin\theta)$$

where θ is the angle in radians (see Angles) subtended by the arc of the segment. If n is the height of the segment, measured along the radius perpendicular to the chord,

$$\text{Area of segment} = \pi r^2 M - A(r - n)$$

where $$A = \sqrt{n(2r - n)} \quad \text{and} \quad M = \frac{1}{180} \sin^{-1}\left(\frac{A}{r}\right)$$

Ellipse.

$$\text{Area} = \pi ab$$

where a and b are the principal semi-axes.

Prism with Parallel Sides and Parallel Ends.

$$\text{Volume} = (\text{Area of end}) \times (\text{Perpendicular distance between ends})$$

Right Circular Cylinder.

$$\text{Volume} = \frac{\pi}{4} d^2 l$$

where d is the diameter and l the length.

$$\text{Total surface of right cylinder} = \pi d(l + {}^1/_2 d)$$

Right Circular Cone.

$$\text{Volume} = {}^1/_3\ (\text{Area of base}) \times (\text{Height})$$

$$= {}^1/_3\ (\text{Volume of circumscribing cylinder})$$

where r is the radius of base and h the height of the cone.

$$\text{Area of curved surface of a right circular cone} = \pi r \sqrt{h^2 + r^2}$$

Right Pyramid.

$$\text{Volume} = {}^1/_3\ (\text{Area ot base}) \times (\text{Height}).$$

$$\text{Volume of frustum of pyramid} = {}^1/_3\ (\text{Height})\ (A + \sqrt{aA}\,)$$

where A and a are the areas of the ends respectively.

Sphere.

$$r = \text{radius}$$

$$\text{Area of surface} = 4\pi r^2 = {}^2/_3\ (\text{total area of circumscribing cylinder})$$

Area of the surface of a zone of a sphere = area of zone of the same height as this zone projected on to a cylinder.

$$\text{Volume} = {}^4/_3\,\pi r^3 = {}^2/_3\ (\text{volume of circumscribing cylinder})$$

Volume of a frustum of a sphere $= \pi r^2(k \pm h) - \frac{\pi}{3}(k^3 \pm h^3)$, where k is the distance of its outer face from center and h the distance of its inner face from the center, the negative signs in the brackets to be used if both faces are on the same side of the center and the positive signs if on opposite sides of the center.

Ellipsoid.

$$\text{Volume} = {}^4/_3\,\pi abc$$

where a, b, and c are the three principal semi-axes, respectively.

Paraboloid. Volume of a paraboloid of revolution equals one-half that of the circumscribing cylinder.

MATHEMATICAL TABLES AND CHARTS

11. COMMON AND NATURAL LOGARITHMS OF NUMBERS

The common logarithm of a number is the index of the power to which the base 10 must be raised in order to equal the number.

The common logarithm of every positive number not an integral power of 10 consists of an *integral* and a *decimal part.* The integral part or whole number is called the *characteristic* and may be either *positive or negative.* The decimal or fractional part is a *positive* number called the *mantissa* and is the same for all numbers which have the same sequential digits.

The characteristic of the logarithm of any positive number greater than one is positive and is one less than the number of digits before the decimal point.

The characteristic of the logarithm of any positive number less than one is negative and is one more than the number of ciphers immediately after the decimal point.

A negative number or number less than zero has no real logarithm.

EXAMPLES: $\text{Log}_{10}\ 25400. = 4.404834$ $\text{Log}_{10}\ 0.0254 = \bar{2}.404834$ or $8.404834 - 10$

The two systems of logarithms in general use are the common or Briggsian logarithms, introduced in 1615 by Henry Briggs, a contemporary of John Napier, the inventor of logarithms, and the natural or less appropriately termed Napierian or hyperbolic logarithms, which developed somewhat accidentally from Napier's original work. The latter have a base denoted by e, an irrational number, which is:

$$e = \text{Lim}_{u=\infty}\left(1 + \frac{1}{u}\right)^u = 1 + 1 + \frac{1}{2!} + \frac{1}{3!} + \frac{1}{4!} + \ldots = 2.7182818$$

To obtain the natural logarithm, the common logarithm given below is multiplied by $\log_e 10$ which is 2.302585, or $\log_e N = 2.302585 \log_{10} N$.

N	0	1	2	3	4	5	6	7	8	9
0		**000000**	**301030**	**477121**	**602060**	**698970**	**778151**	**845098**	**903090**	**954243**
1	000000	041393	079181	113943	146128	176091	204120	230449	255273	278754
2	301030	322219	342423	361728	380211	397940	414973	431364	447158	462398
3	477121	491362	505150	518514	531479	544068	556303	568202	579784	591065
4	602060	612784	623249	633468	643453	653213	662758	672098	681241	690196
5	698970	707570	716003	724276	732394	740363	748188	755875	763428	770852
6	778151	785330	792392	799341	806180	812913	819544	826075	832509	838849
7	845098	851258	857332	863323	869232	875061	880814	886491	892095	897627
8	903090	908485	913814	919078	924279	929419	934498	939519	944483	949390
9	954243	959041	963788	968483	973128	977724	982271	986772	991226	995635
10	**000000**	**004321**	**008600**	**012837**	**017033**	**021189**	**025306**	**029384**	**033424**	**037426**
1	041393	045323	049218	053078	056905	060698	064458	068186	071882	075547
2	079181	082785	086360	089905	093422	096910	100371	103804	107210	110590
3	113943	117271	120574	123852	127105	130334	133539	136721	139879	143015
4	146128	149219	152288	155336	158362	161368	164353	167317	170262	173186
5	176091	178977	181844	184691	187521	190332	193125	195900	198657	201397
6	204120	206826	209515	212188	214844	217484	220108	222716	225309	227887
7	230449	232996	235528	238046	240549	243038	245513	247973	250420	252853
8	255273	257679	260071	262451	264818	267172	269513	271842	274158	276462
9	278754	281033	283301	285557	287802	290035	292256	294466	296665	298853
20	**301030**	**303196**	**305351**	**307496**	**309630**	**311754**	**313867**	**315970**	**318063**	**320146**
1	322219	324282	326336	328380	330414	332438	334454	336460	338456	340444
2	342423	344392	346353	348305	350248	352183	354108	356026	357935	359835
3	361728	363612	365488	367356	369216	371068	372912	374748	376577	378398
4	380211	382017	383815	385606	387390	389166	390935	392697	394452	396199
5	397940	399674	401401	403121	404834	406540	408240	409933	411620	413300
6	414973	416641	418301	419956	421604	423246	424882	426511	428135	429752
7	431364	432969	434569	436163	437751	439333	440909	442480	444045	445604
8	447158	448706	450249	451786	453318	454845	456366	457882	459392	460898
9	462398	463893	465383	466868	468347	469822	471292	472756	474216	475671
30	**477121**	**478566**	**480007**	**481443**	**482874**	**484300**	**485721**	**487138**	**488551**	**489958**
1	491362	492760	494155	495544	496930	498311	499687	501059	502427	503791
2	505150	506505	507856	509203	510545	511883	513218	514548	515874	517196
3	518514	519828	521138	522444	523746	525045	526339	527630	528917	530200
4	531479	532754	534026	535294	536558	537819	539076	540329	541579	542825
5	544068	545307	546543	547775	549003	550228	551450	552668	553883	555094

N	0	1	2	3	4	5	6	7	8	9
5	544068	545307	546543	547775	549003	550228	551450	552668	553883	555094
6	556303	557507	558709	559907	561101	562293	563481	564666	565848	567026
7	568202	569374	570543	571709	572872	574031	575188	576341	577492	578639
8	579784	580925	582063	583199	584331	585461	586587	587711	588832	589950
9	591065	592177	593286	594393	595496	596597	597695	598791	599883	600973
40	**602060**	**603144**	**604226**	**605305**	**606381**	**607455**	**608526**	**609594**	**610660**	**611723**
1	612784	613842	614897	615950	617000	618048	619093	620136	621176	622214
2	623249	624232	625312	626340	627366	628389	629410	630428	631444	632457
3	633468	634477	635484	636488	637490	638489	639486	640481	641474	642465
4	643453	644439	645422	646404	647383	648360	649335	650308	651278	652246
5	653213	654177	655138	656098	657056	658011	658965	659916	660865	661713
6	662758	663701	664642	665581	666518	667453	668386	669317	670246	671173
7	672098	673021	673942	674861	675778	676694	677607	678518	679428	680336
8	681241	682145	683047	683947	684845	685742	686636	687529	688420	689309
9	690196	691081	691965	692847	693727	694605	695482	696356	697229	698100
50	**698970**	**699838**	**700704**	**701568**	**702431**	**703291**	**704151**	**705008**	**705864**	**706718**
1	707570	708421	709270	710117	710963	711807	712650	713491	714330	715167
2	716003	716838	717671	718502	719331	720159	720986	721811	722634	723456
3	724276	725095	725912	726727	727541	728354	729165	729974	730782	731589
4	732394	733197	733999	734800	735599	736397	737193	737987	738781	739572
5	740363	741152	741939	742725	743510	744293	745075	745855	746634	747412
6	748188	748963	749736	750508	751279	752048	752816	753583	754348	755112
7	755875	756636	757396	758155	758912	759668	760422	761176	761928	762679
8	763428	764176	764923	765669	766413	767156	767898	768638	769377	770115
9	770852	771587	772322	773055	773786	774517	775246	775974	776701	777427
60	**778151**	**778874**	**779596**	**780317**	**781037**	**781755**	**782473**	**783189**	**783904**	**784617**
1	785330	786041	786751	787460	788168	788875	789581	790285	790988	791691
2	792392	793092	793790	794488	795185	795880	796574	797268	797960	798651
3	799341	800029	800717	801404	802089	802774	803457	804139	804821	805501
4	806180	806858	807535	808211	808886	809560	810233	810904	811575	812245
5	812913	813581	814248	814913	815578	816241	816904	817565	818226	818885
6	819544	820201	820858	821514	822168	822822	823474	824126	824776	825426
7	826075	826723	827369	828015	828660	829304	829947	830589	831230	831870
8	832509	833147	833784	834421	835056	835691	836324	836957	837588	838219
9	838849	839478	840106	840733	841359	841985	842609	843233	843855	844477
70	**845098**	**845718**	**846337**	**846955**	**847573**	**848189**	**848805**	**849419**	**850033**	**850646**
1	851258	851870	852480	853090	853698	854306	854913	855519	856124	856729
2	857332	857935	858537	859138	859739	860338	860937	861534	862131	862728
3	863323	863917	864511	865104	865696	866287	866878	867467	868056	868644
4	869232	869818	870404	870989	871573	872156	872739	873321	873902	874482
5	875061	875640	876218	876795	877371	877947	878522	879096	879669	880242
6	880814	881385	881955	882525	883093	883661	884229	884795	885361	885926
7	886491	887054	887617	888179	888741	889302	889862	890421	890980	891537
8	892095	892651	893207	893762	894316	894870	895423	895975	896526	897077
9	897627	898176	898725	899273	899821	900367	900913	901458	902003	902547
80	**903090**	**903633**	**904174**	**904716**	**905256**	**905796**	**906335**	**906874**	**907411**	**907949**
1	908485	909021	909556	910091	910624	911158	911690	912222	912753	913284
2	913814	914343	914872	915400	915927	916454	916980	917506	918030	918555
3	919078	919601	920123	920645	921166	921686	922206	922725	923244	923762
4	924279	924796	925312	925828	926342	926857	927370	927883	928396	928908
5	929419	929930	930440	930949	931458	931966	932474	932981	933487	933993
6	934498	935003	935507	936011	936514	937016	937518	938019	938520	939020
7	939519	940018	940516	941014	941511	942008	942504	943000	943495	943989
8	944483	944976	945469	945961	946452	946943	947434	947924	948413	948902
9	949390	949878	950365	950851	951338	951823	952308	952792	953276	953760
90	**954243**	**954725**	**955207**	**955688**	**956168**	**956649**	**957128**	**957607**	**958086**	**958564**
1	959041	959518	959995	960471	960946	961421	961895	962369	962843	963316
2	963788	964260	964731	965202	965672	966142	966611	967080	967548	968016
3	968483	968950	969416	969882	970347	970812	971276	971740	972203	972666
4	973128	973590	974051	974512	974972	975432	975891	976350	976808	977266
5	977724	978181	978637	979093	979548	980003	980458	980912	981366	981819
6	982271	982723	983175	983626	984077	984527	984977	985426	985875	986324
7	986772	987219	987666	988113	988559	989005	989450	989895	990339	990783
8	991226	991669	992111	992554	992995	993436	993877	994317	994757	995196
9	995635	996074	996512	996949	997386	997823	998259	998695	999131	999565
100	**000000**	**000434**	**000868**	**001301**	**001734**	**002166**	**002598**	**003029**	**003461**	**003891**

12. TRIGONOMETRIC TABLES

The following tables give the values of sin x, cos x, and tan x for values of x from 0 to 90° in intervals of 0.1 degree. By making use of the periodic character of these functions, the values can be determined from these tables for all values of x to an accuracy of 0.1 degree. (See Trigonometric Formulas.)

If the angle is given in radians multiply the number of radians by $\frac{180}{\pi}$ (57.295) to obtain the number of degrees.

Trigonometric Functions **0.0°–15.9°**

Angle in Degrees	Name of Function	Value of Function for Each Tenth of a Degree									
		0.0	0.1	0.2	0.3	0.4	0.5	0.6	0.7	0.8	0.9
0	sin	0.0000	0.0017	0.0035	0.0052	0.0070	0.0087	0.0105	0.0122	0.0140	0.0157
	cos	1.0000	1.0000	1.0000	1.0000	1.0000	1.0000	0.9999	0.9999	0.9999	0.9999
	tan	0.0000	0.0017	0.0035	0.0052	0.0070	0.0087	0.0105	0.0122	0.0140	0.0157
1	sin	0.0175	0.0192	0.0209	0.0227	0.0244	0.0262	0.0279	0.0297	0.0314	0.0332
	cos	0.9998	0.9998	0.9998	0.9997	0.9997	0.9997	0.9996	0.9996	0.9995	0.9995
	tan	0.0175	0.0192	0 0209	0.0227	0.0244	0.0262	0.0279	0.0297	0.0314	0.0332
2	sin	0.0349	0.0366	0.0384	0.0401	0.0419	0.0436	0.0454	0.0471	0.0488	0.0506
	cos	0.9994	0.9993	0.9993	0.9992	0.9991	0.9990	0.9990	0.9989	0.9988	0.9987
	tan	0.0349	0.0367	0.0384	0.0402	0.0419	0.0437	0.0454	0.0472	0.0489	0.0507
3	sin	0.0523	0.0541	0.0558	0.0576	0.0593	0.0610	0.0628	0.0645	0.0663	0.0680
	cos	0.9986	0.9985	0.9984	0.9983	0.9982	0.9981	0.9980	0.9979	0.9978	0.9977
	tan	0.0524	0.0542	0.0559	0.0577	0.0594	0.0612	0.0629	0.0647	0.0664	0.0682
4	sin	0.0698	0.0715	0.0732	0.0750	0.0767	0.0785	0.0802	0.0819	0.0837	0.0854
	cos	0.9976	0.9974	0.9973	0.9972	0.9971	0.9969	0.9968	0.9966	0.9965	0.9963
	tan	0.0699	0.0717	0.0734	0.0752	0.0769	0.0787	0.0805	0.0822	0.0840	0.0857
5	sin	0.0872	0.0889	0.0906	0.0924	0.0941	0.0958	0.0976	0.0993	0.1011	0.1028
	cos	0.9962	0.9960	0.9959	0.9957	0.9956	0.9954	0.9952	0.9951	0.9949	0.9947
	tan	0.0875	0.0892	0.0910	0.0928	0.0945	0.0963	0.0981	0.0998	0.1016	0.1033
6	sin	0.1045	0.1063	0.1080	0.1097	0.1115	0.1132	0.1149	0.1167	0.1184	0.1201
	cos	0.9945	0.9943	0.9942	0.9940	0.9938	0.9936	0.9934	0.9932	0.9930	0.9928
	tan	0.1051	0.1069	0.1086	0.1104	0.1122	0.1139	0.1157	0.1175	0.1192	0.1210
7	sin	0.1219	0.1236	0.1253	0.1271	0.1288	0.1305	0.1323	0.1340	0.1357	0.1374
	cos	0.9925	0.9923	0.9921	0.9919	0.9917	0.9914	0.9912	0.9910	0.9907	0.9905
	tan	0.1228	0.1246	0.1263	0.1281	9.1299	0.1317	0.1334	0.1352	0.1370	0.1388
8	sin	0.1392	0.1409	0.1426	0.1444	0.1461	0.1478	0.1495	0.1513	0.1530	0.1547
	cos	0.9903	0.9900	0.9898	0.9895	0.9893	0.9890	0.9888	0.9885	0.9882	0.9880
	tan	0.1405	0.1423	0.1441	0.1459	0.1477	0.1495	0.1512	0.1530	0.1548	0.1566
9	sin	0.1564	0.1582	0.1599	0.1616	0.1633	0.1650	0.1663	0.1685	0.1702	0.1719
	cos	0.9877	0.9874	0.9871	0.9869	0.9866	0.9863	0.9860	0.9857	0.9854	0.9851
	tan	0.1584	0.1602	0.1620	0.1638	0.1655	0.1673	0.1691	0.1709	0.1727	0.1745
10	sin	0.1736	0.1754	0.1771	0.1788	0.1805	0.1822	0.1840	0.1857	0.1874	0.1891
	cos	0.9848	0.9845	0.9842	0.9839	0.9836	0.9833	0.9829	0.9826	0.9823	0.9820
	tan	0.1763	0.1781	0.1799	0.1817	0.1835	0.1853	0.1871	0.1890	0.1908	0.1926
11	sin	0.1908	0.1925	0.1942	0.1959	0.1977	0.1994	0.2011	0.2028	0.2045	0.2062
	cos	0.9816	0.9813	0.9810	0.9806	0.9803	0.9799	0.9796	0.9792	0.9789	0.9785
	tan	0.1944	0.1962	0.1980	0.1998	0.2016	0.2035	0.2053	0.2071	0.2089	0.2107
12	sin	0.2079	0.2096	0.2113	0.2130	0.2147	0.2164	0.2181	0.2198	0.2215	0.2232
	cos	0.9781	0.9778	0.9774	0.9770	0.9767	0.9763	0.9759	0.9755	0.9751	0.9748
	tan	0.2126	0.2144	0.2162	0.2180	0.2199	0.2217	0.2235	0.2254	0.2272	0.2290
13	sin	0.2250	0.2267	0.2284	0.2300	0.2317	0.2334	0.2351	0.2368	0.2385	0.2402
	cos	0.9744	0.9740	0.9736	0.9732	0.9728	0.9724	0.9720	0.9715	0.9711	0.9707
	tan	0.2309	0.2327	0.2345	0.2364	0.2382	0.2401	0.2419	0.2438	0.2456	0.2475
14	sin	0.2419	0.2436	0.2453	0.2470	0.2487	0.2504	0.2521	0.2538	0.2554	0.2571
	cos	0.9703	0.9699	0.9694	0.9690	0.9686	0.9681	0.9677	0.9673	0.9668	0.9664
	tan	0.2493	0.2512	0.2530	0.2549	0.2568	0.2586	0.2605	0.2623	0.2642	0.2661
15	sin	0.2588	0.2605	0.2622	0.2639	0.2656	0.2672	0.2689	0.2706	0.2723	0.2740
	cos	0.9659	0.9655	0.9650	0.9646	0.9641	0.9636	0.9632	0.9627	0.9622	0.9617
	tan	0.2679	0.2698	0.2717	0.2736	0.2754	0.2773	0.2792	0.2811	0.2830	0.2849

Trigonometric Functions

16.0°–35.9°

Angle in Degrees	Name of Function	Value of Function for Each Tenth of a Degree									
		0.0	0.1	0.2	0.3	0.4	0.5	0.6	0.7	0.8	0.9
	sin	0.2756	0.2773	0.2790	0.2807	0.2823	0.2840	0.2857	0.2874	0.2890	0.2907
16	cos	0.9613	0.9608	0.9603	0.9598	0.9593	0.9588	0.9583	0.9578	0.9573	0.9568
	tan	0.2867	0.2886	0.2905	0.2924	0.2943	0.2962	0.2981	0.3000	0.3019	0.3038
	sin	0.2924	0.2940	0.2957	0.2974	0.2990	0.3007	0.3024	0.3040	0.3057	0.3074
17	cos	0.9563	0.9558	0.9553	0.9548	0.9542	0.9537	0.9532	0.9527	0.9521	0.9516
	tan	0.3057	0.3076	0.3096	0.3115	0.3134	0.3153	0.3172	0.3191	0.3211	0.3230
	sin	0.3090	0.3107	0.3123	0.3140	0.3156	0.3173	0.3190	0.3206	0.3223	0.3239
18	cos	0.9511	0.9505	0.9500	0.9494	0.9489	0.9483	0.9478	0.9472	0.9466	0.9461
	tan	0.3249	0.3269	0.3288	0.3307	0.3327	0.3346	0.3365	0.3385	0.3404	0.3424
	sin	0.3256	0.3272	0.3289	0.3305	0.3322	0.3338	0.3355	0.3371	0.3387	0.3404
19	cos	0.9455	0.9449	0.9444	0.9438	0.9432	0.9426	0.9421	0.9415	0.9409	0.9403
	tan	0.3443	0.3463	0.3482	0.3502	0.3522	0.3541	0.3561	0.3581	0.3600	0.3620
	sin	0.3420	0.3437	0.3453	0.3469	0.3486	0.3502	0.3518	0.3535	0.3551	0.3567
20	cos	0.9397	0.9391	0.9385	0.9379	0.9373	0.9367	0.9361	0.9354	0.9348	0.9342
	tan	0.3640	0.3659	0.3679	0.3699	0.3719	0.3739	0.3759	0.3779	0.3799	0.3819
	sin	0.3584	0.3600	0.3616	0.3633	0.3649	0.3665	0.3681	0.3697	0.3714	0.3730
21	cos	0.9336	0.9330	0.9323	0.9317	0.9311	0.9304	0.9298	0.9291	0.9285	0.9278
	tan	0.3839	0.3859	0.3879	0.3899	0.3919	0.3939	0.3959	0.3979	0.4000	0.4020
	sin	0.3746	0.3762	0.3778	0.3795	0.3811	0.3827	0.3843	0.3859	0.3875	0.3891
22	cos	0.9272	0.9265	0.9259	0.9252	0.9245	0.9239	0.9232	0.9225	0.9219	0.9212
	tan	0.4040	0.4061	0.4081	0.4101	0.4122	0.4142	0.4163	0.4183	0.4204	0.4224
	sin	0.3907	0.3923	0.3939	0.3955	0.3971	0.3987	0.4003	0.4019	0.4035	0.4051
23	cos	0.9205	0.9198	0.9191	0.9184	0.9178	0.9171	0.9164	0.9157	0.9150	0.9143
	tan	0.4245	0.4265	0.4286	0.4307	0.4327	0.4348	0.4369	0.4390	0.4411	0.4431
	sin	0.4067	0.4083	0.4099	0.4115	0.4131	0.4147	0.4163	0.4179	0.4195	0.4210
24	cos	0.9135	0.9128	0.9121	0.9114	0.9107	0.9100	0.9092	0.9085	0.9078	0.9070
	tan	0.4452	0.4473	0.4494	0.4515	0.4536	0.4557	0.4578	0.4599	0.4621	0.4642
	sin	0.4226	0.4242	0.4258	0.4274	0.4289	0.4305	0.4321	0.4337	0.4352	0.4368
25	cos	0.9063	0.9056	0.9048	0.9041	0.9033	0.9026	0.9018	0.9011	0.9003	0.8996
	tan	0.4663	0.4684	0.4706	0.4727	0.4748	0.4770	0.4791	0.4813	0.4834	0.4856
	sin	0.4384	0.4399	0.4415	0.4431	0.4446	0.4462	0.4478	0.4493	0.4509	0.4524
26	cos	0.8988	0.8980	0.8973	0.8965	0.8957	0.8949	0.8942	0.8934	0.8926	0.8918
	tan	0.4877	0.4899	0.4921	0.4942	0.4964	0.4986	0.5008	0.5029	0.5051	0.5073
	sin	0.4540	0.4555	0.4571	0.4586	0.4602	0.4617	0.4633	0.4648	0.4664	0.4679
27	cos	0.8910	0.8902	0.8894	0.8886	0.8878	0.8870	0.8862	0.8854	0.8846	0.8838
	tan	0.5095	0.5117	0.5139	0.5161	0.5184	0.5206	0.5228	0.5250	0.5272	0.5295
	sin	0.4695	0.4710	0.4726	0.4741	0.4756	0.4772	0.4787	0.4802	0.4818	0.4833
28	cos	0.8829	0.8821	0.8813	0.8805	0.8796	0.8788	0.8780	0.8771	0.8763	0.8755
	tan	0.5317	0.5340	0.5362	0.5384	0.5407	0.5430	0.5452	0.5475	0.5498	0.5520
	sin	0.4848	0.4863	0.4879	0.4894	0.4909	0.4924	0.4939	0.4955	0.4970	0.4985
29	cos	0.8746	0.8738	0.8729	0.8721	0.8712	0.8704	0.8695	0.8686	0.8678	0.8669
	tan	0.5543	0.5566	0.5589	0.5612	0.5635	0.5658	0.5681	0.5704	0.5727	0.5750
	sin	0.5000	0.5015	0.5030	0.5045	0.5060	0.5075	0.5090	0.5105	0.5120	0.5135
30	cos	0.8660	0.8652	0.8643	0.8634	0.8625	0.8616	0.8607	0.8599	0.8590	0.8581
	tan	0.5774	0.5797	0.5820	0.5844	0.5867	0.5890	0.5914	0.5938	0.5961	0.5985
	sin	0.5150	0.5165	0.5180	0.5195	0.5210	0.5225	0.5240	0.5255	0.5270	0.5284
31	cos	0.8572	0.8563	0.8554	0.8545	0.8536	0.8526	0.8517	0.8508	0.8499	0.8490
	tan	0.6009	0.6032	0.6056	0.6080	0.6104	0.6128	0.6152	0.6176	0.6200	0.6224
	sin	0.5299	0.5314	0.5329	0.5344	0.5358	0.5373	0.5388	0.5402	0.5417	0.5432
32	cos	0.8480	0.8471	0.8462	0.8453	0.8443	0.8434	0.8425	0.8415	0.8406	0.8396
	tan	0.6249	0.6273	0.6297	0.6322	0.6346	0.6371	0.6395	0.6420	0.6445	0.6469
	sin	0.5446	0.5461	0.5476	0.5490	0.5505	0.5519	0.5534	0.5548	0.5563	0.5577
33	cos	0.8387	0.8377	0.8368	0.8358	0.8348	0.8339	0.8329	0.8320	0.8310	0.8300
	tan	0.6494	0.6519	0.6544	0.6569	0.6594	0.6619	0.6644	0.6669	0.6694	0.6720
	sin	0.5592	0.5606	0.5621	0.5635	0.5650	0.5664	0.5678	0.5693	0.5707	0.5721
34	cos	0.8290	0.8281	0.8271	0.8261	0.8251	0.8241	0.8231	0.8221	0.8211	0.8202
	tan	0.6745	0.6771	0.6796	0.6822	0.6847	0.6873	0.6899	0.6924	0.6950	0.6976
	sin	0.5736	0.5750	0.5764	0.5779	0.5793	0.5807	0.5821	0.5835	0.5850	0.5864
35	cos	0.8192	0.8181	0.8171	0.8161	0.8151	0.8141	0.8131	0.8121	0.8111	0.8100
	tan	0.7002	0.7028	0.7054	0.7080	0.7107	0.7133	0.7159	0.7186	0.7212	0.7239

Trigonometric Functions

36.0°–55.9°

Angle in Degrees	Name of Function	0.0	0.1	0.2	0.3	0.4	0.5	0.6	0.7	0.8	0.9
		Value of Function for Each Tenth of a Degree									
36	sin	0.5878	0.5892	0.5906	0.5920	0.5934	0.5948	0.5962	0.5976	0.5990	0.6004
	cos	0.8090	0.8080	0.8070	0.8059	0.8049	0.8039	0.8028	0.8018	0.8007	0.7997
	tan	0.7265	0.7292	0.7319	0.7346	0.7373	0.7400	0.7427	0.7454	0.7481	0.7508
37	sin	0.6018	0.6032	0.6046	0.6060	0.6074	0.6088	0.6101	0.6115	0.6129	0.6143
	cos	0.7986	0.7976	0.7965	0.7955	0.7944	0.7934	0.7923	0.7912	0.7902	0.7891
	tan	0.7536	0.7563	0.7590	0.7618	0.7646	0.7673	0.7701	0.7729	0.7757	0.7785
38	sin	0.6157	0.6170	0.6184	0.6198	0.6211	0.6225	0.6239	0.6252	0.6266	0.6280
	cos	0.7880	0.7869	0.7859	0.7848	0.7837	0.7826	0.7815	0.7804	0.7793	0.7782
	tan	0.7813	0.7841	0.7869	0.7898	0.7926	0.7954	0.7983	0.8012	0.8040	0.8069
39	sin	0.6293	0.6307	0.6320	0.6334	0.6347	0.6361	0.6374	0.6388	0.6401	0.6414
	cos	0.7771	0.7760	0.7749	0.7738	0.7727	0.7716	0.7705	0.7694	0.7683	0.7672
	tan	0.8098	0.8127	0.8156	0.8185	0.8214	0.8243	0.8273	0.8302	0.8332	0.8361
40	sin	0.6428	0.6441	0.6455	0.6468	0.6481	0.6494	0.6508	0.6521	0.6534	0.6547
	cos	0.7660	0.7649	0.7638	0.7627	0.7615	0.7604	0.7593	0.7581	0.7570	0.7559
	tan	0.8391	0.8421	0.8451	0.8481	0.8511	0.8541	0.8571	0.8601	0.8632	0.8662
41	sin	0.6561	0.6574	0.6587	0.6600	0.6613	0.6626	0.6639	0.6653	0.6665	0.6678
	cos	0.7547	0.7536	0.7524	0.7513	0.7501	0.7490	0.7478	0.7466	0.7455	0.7443
	tan	0.8693	0.8724	0.8754	0.8785	0.8816	0.8847	0.8878	0.8910	0.8941	0.8972
42	sin	0.6691	0.6704	0.6717	0.6730	0.6743	0.6756	0.6769	0.6782	0.6794	0.6807
	cos	0.7431	0.7420	0.7408	0.7396	0.7385	0.7373	0.7361	0.7349	0.7337	0.7325
	tan	0.9004	0.9036	0.9067	0.9099	0.9131	0.9163	0.9195	0.9228	0.9260	0.9293
43	sin	0.6820	0.6833	0.6845	0.6858	0.6871	0.6884	0.6896	0.6909	0.6921	0.6934
	cos	0.7314	0.7302	0.7290	0.7278	0.7266	0.7254	0.7242	0.7230	0.7218	0.7206
	tan	0.9325	0.9358	0.9391	0.9424	0.9457	0.9490	0.9523	0.9556	0.9590	0.9623
44	sin	0.6947	0.6959	0.6972	0.6984	0.6997	0.7009	0.7022	0.7034	0.7046	0.7059
	cos	0.7193	0.7181	0.7169	0.7157	0.7145	0.7133	0.7120	0.7108	0.7096	0.7083
	tan	0.9657	0.9691	0.9725	0.9759	0.9793	0.9827	0.9861	0.9896	0.9930	0.9965
45	sin	0.7071	0.7083	0.7096	0.7108	0.7120	0.7133	0.7145	0.7157	0.7169	0.7181
	cos	0.7071	0.7059	0.7046	0.7034	0.7022	0.7009	0.6997	0.6984	0.6972	0.6959
	tan	1.0000	1.0035	1.0070	1.0105	1.0141	1.0176	1.0212	1.0247	1.0283	1.0319
46	sin	0.7193	0.7206	0.7218	0.7230	0.7242	0.7254	0.7266	0.7278	0.7290	0.7302
	cos	0.6947	0.6934	0.6921	0.6909	0.6896	0.6884	0.6871	0.6858	0.6845	0.6833
	tan	1.0355	1.0392	1.0428	1.0464	1.0501	1.0538	1.0575	1.0612	1.0649	1.0686
47	sin	0.7314	0.7325	0.7337	0.7349	0.7361	0.7373	0.7385	0.7396	0.7408	0.7420
	cos	0.6820	0.6807	0.6794	0.6782	0.6769	0.6756	0.6743	0.6730	0.6717	0.6704
	tan	1.0724	1.0761	1.0799	1.0837	1.0875	1.0913	1.0951	1.0990	1.1028	1.1067
48	sin	0.7431	0.7443	0.7455	0.7466	0.7478	0.7490	0.7501	0.7513	0.7524	0.7536
	cos	0.6691	0.6678	0.6665	0.6652	0.6639	0.6626	0.6613	0.6600	0.6587	0.6574
	tan	1.1106	1.1145	1.1184	1.1224	1.1263	1.1303	1.1343	1.1383	1.1423	1.1463
49	sin	0.7547	0.7559	0.7570	0.7581	0.7593	0.7604	0.7615	0.7627	0.7638	0.7649
	cos	0.6561	0.6547	0.6534	0.6521	0.6508	0.6494	0.6481	0.6468	0.6455	0.6441
	tan	1.1504	1.1544	1.1585	1.1626	1.1667	1.1708	1.1750	1.1792	1.1833	1.1875
50	sin	0.7660	0.7672	0.7683	0.7694	0.7705	0.7716	0.7727	0.7738	0.7749	0.7760
	cos	0.6428	0.6414	0.6401	0.6388	0.6374	0.6361	0.6347	0.6334	0.6320	0.6307
	tan	1.1918	1.1960	1.2002	1.2045	1.2088	1.2131	1.2174	1.2218	1.2261	1.2305
51	sin	0.7771	0.7782	0.7793	0.7804	0.7815	0.7826	0.7837	0.7848	0.7859	0.7869
	cos	0.6293	0.6280	0.6266	0.6252	0.6239	0.6225	0.6211	0.6198	0.6184	0.6170
	tan	1.2349	1.2393	1.2437	1.2482	1.2527	1.2572	1.2617	1.2662	1.2708	1.2753
52	sin	0.7880	0.7891	0.7902	0.7912	0.7923	0.7934	0.7944	0.7955	0.7965	0.7976
	cos	0.6157	0.6143	0.6129	0.6115	0.6101	0.6088	0.6074	0.6060	0.6046	0.6032
	tan	1.2799	1.2846	1.2892	1.2938	1.2985	1.3032	1.3079	1.3127	1.3175	1.3222
53	sin	0.7986	0.7997	0.8007	0.8018	0.8028	0.8039	0.8049	0.8059	0.8070	0.8080
	cos	0.6018	0.6004	0.5990	0.5976	0.5962	0.5948	0.5934	0.5920	0.5906	0.5892
	tan	1.3270	1.3319	1.3367	1.3416	1.3465	1.3514	1.3564	1.3613	1.3663	1.3713
54	sin	0.8090	0.8100	0.8111	0.8121	0.8131	0.8141	0.8151	0.8161	0.8171	0.8181
	cos	0.5878	0.5864	0.5850	0.5835	0.5821	0.5807	0.5793	0.5779	0.5764	0.5750
	tan	1.3764	1.3814	1.3865	1.3916	1.3968	1.4019	1.4071	1.4124	1.4176	1.4229
55	sin	0.8192	0.8202	0.8211	0.8221	0.8231	0.8241	0.8251	0.8261	0.8271	0.8281
	cos	0.5736	0.5721	0.5707	0.5693	0.5678	0.5664	0.5650	0.5635	0.5621	0.5606
	tan	1.4281	1.4335	1.4388	1.4442	1.4496	1.4550	1.4605	1.4659	1.4715	1.4770

Trigonometric Functions

56.0°–75.9°

Angle in Degrees	Name of Function	Value of Function for Each Tenth of a Degree									
		0.0	0.1	0.2	0.3	0.4	0.5	0.6	0.7	0.8	0.9
	sin	0.8290	0.8300	0.8310	0.8320	0.8329	0.8339	0.8348	0.8358	0.8368	0.8377
56	cos	0.5592	0.5577	0.5563	0.5548	0.5534	0.5519	0.5505	0.5490	0.5476	0.5461
	tan	1.4826	1.4882	1.4938	1.4994	1.5051	1.5108	1.5166	1.5224	1.5282	1.5340
	sin	0.8387	0.8396	0.8406	0.8415	0.8425	0.8434	0.8443	0.8453	0.8462	0.8471
57	cos	0.5446	0.5432	0.5417	0.5402	0.5388	0.5373	0.5358	0.5344	0.5329	0.5314
	tan	1.5399	1.5458	1.5517	1.5577	1.5637	1.5697	1.5757	1.5818	1.5880	1.5941
	sin	0.8480	0.8490	0.8499	0.8508	0.8517	0.8526	0.8536	0.8545	0.8554	0.8563
58	cos	0.5299	0.5284	0.5270	0.5255	0.5240	0.5225	0.5210	0.5195	0.5180	0.5165
	tan	1.6003	1.6066	1.6128	1.6191	1.6255	1.6319	1.6383	1.6447	1.6512	1.6577
	sin	0.8572	0.8581	0.8590	0.8599	0.8607	0.8616	0.8625	0.8634	0.8643	0.8652
59	cos	0.5150	0.5135	0.5120	0.5105	0.5090	0.5075	0.5060	0.5045	0.5030	0.5015
	tan	1.6643	1.6709	1.6775	1.6842	1.6909	1.6977	1.7045	1.7113	1.7182	1.7251
	sin	0.8660	0.8669	0.8678	0.8686	0.8695	0.8704	0.8712	0.8721	0.8729	0.8738
60	cos	0.5000	0.4985	0.4970	0.4955	0.4939	0.4924	0.4909	0.4894	0.4879	0.4863
	tan	1.7321	1.7391	1.7461	1.7532	1.7603	1.7675	1.7747	1.7820	1.7893	1.7966
	sin	0.8746	0.8755	0.8763	0.8771	0.8780	0.8788	0.8796	0.8805	0.8813	0.8821
61	cos	0.4848	0.4833	0.4818	0.4802	0.4787	0.4772	0.4756	0.4741	0.4726	0.4710
	tan	1.8040	1.8115	1.8190	1.8265	1.8341	1.8418	1.8495	1.8572	1.8650	1.8728
	sin	0.8829	0.8838	0.8846	0.8854	0.8862	0.8870	0.8878	0.8886	0.8894	0.8902
62	cos	0.4695	0.4679	0.4664	0.4648	0.4633	0.4617	0.4602	0.4586	0.4571	0.4555
	tan	1.8807	1.8887	1.8967	1.9047	1.9128	1.9210	1.9292	1.9375	1.9458	1.9542
	sin	0.8910	0.8918	0.8926	0.8934	0.8942	0.8949	0.8957	0.8965	0.8973	0.8980
63	cos	0.4540	0.4524	0.4509	0.4493	0.4478	0.4462	0.4446	0.4431	0.4415	0.4399
	tan	1.9626	1.9711	1.9797	1.9883	1.9970	2.0057	2.0145	2.0233	2.0323	2.0413
	sin	0.8988	0.8996	0.9003	0.9011	0.9018	0.9026	0.9033	0.9041	0.9048	0.9056
64	cos	0.4384	0.4368	0.4352	0.4337	0.4321	0.4305	0.4289	0.4274	0.4258	0.4242
	tan	2.0503	2.0594	2.0686	2.0778	2.0872	2.0965	2.1060	2.1155	2.1251	2.1348
	sin	0.9063	0.9070	0.9078	0.9085	0.9092	0.9100	0.9107	0.9114	0.9121	0.9128
65	cos	0.4226	0.4210	0.4195	0.4179	0.4163	0.4147	0.4131	0.4115	0.4099	0.4083
	tan	2.1445	2.1543	2.1642	2.1742	2.1842	2.1943	2.2045	2.2148	2.2251	2.2355
	sin	0.9135	0.9143	0.9150	0.9157	0.9164	0.9171	0.9178	0.9184	0.9191	0.9198
66	cos	0.4067	0.4051	0.4035	0.4019	0.4003	0.3987	0.3971	0.3955	0.3939	0.3923
	tan	2.2460	2.2566	2.2673	2.2781	2.2889	2.2998	2.3109	2.3220	2.3332	2.3445
	sin	0.9205	0.9212	0.9219	0.9225	0.9232	0.9239	0.9245	0.9252	0.9259	0.9265
67	cos	0.3907	0.3891	0.3875	0.3859	0.3843	0.3827	0.3811	0.3795	0.3778	0.3762
	tan	2.3559	2.3673	2.3789	2.3906	2.4023	2.4142	2.4262	2.4383	2.4504	2.4627
	sin	0.9272	0.9278	0.9285	0.9291	0.9298	0.9304	0.9311	0.9317	0.9323	0.9330
68	cos	0.3746	0.3730	0.3714	0.3697	0.3681	0.3665	0.3649	0.3633	0.3616	0.3600
	tan	2.4751	2.4876	2.5002	2.5129	2.5257	2.5386	2.5517	2.5649	2.5782	2.5916
	sin	0.9336	0.9342	0.9348	0.9354	0.9361	0.9367	0.9373	0.9379	0.9385	0.9391
69	cos	0.3584	0.3567	0.3551	0.3535	0.3518	0.3502	0.3486	0.3469	0.3453	0.3437
	tan	2.6051	2.6187	2.6325	2.6464	2.6605	2.6746	2.6889	2.7034	2.7179	2.7326
	sin	0.9397	0.9403	0.9409	0.9415	0.9421	0.9426	0.9432	0.9438	0.9444	0.9449
70	cos	0.3420	0.3404	0.3387	0.3371	0.3355	0.3338	0.3322	0.3305	0.3289	0.3272
	tan	2.7475	2.7625	2.7776	2.7929	2.8083	2.8239	2.8397	2.8556	2.8716	2.8878
	sin	0.9455	0.9461	0.9466	0.9472	0.9478	0.9483	0.9489	0.9494	0.9500	0.9505
71	cos	0.3256	0.3239	0.3223	0.3206	0.3190	0.3173	0.3156	0.3140	0.3123	0.3107
	tan	2.9042	2.9208	2.9375	2.9544	2.9714	2.9887	3.0061	3.0237	3.0415	3.0595
	sin	0.9511	0.9516	0.9521	0.9527	0.9532	0.9537	0.9542	0.9548	0.9553	0.9558
72	cos	0.3090	0.3074	0.3057	0.3040	0.3024	0.3007	0.2990	0.2974	0.2957	0.2940
	tan	3.0777	3.0961	3.1146	3.1334	3.1524	3.1716	3.1910	3.2106	3.2305	3.2506
	sin	0.9563	0.9568	0.9573	0.9578	0.9583	0.9588	0.9593	0.9598	0.9603	0.9608
73	cos	0.2924	0.2907	0.2890	0.2874	0.2857	0.2840	0.2823	0.2807	0.2790	0.2773
	tan	3.2709	3.2914	3.3122	3.3332	3.3544	3.3759	3.3977	3.4197	3.4420	3.4646
	sin	0.9613	0.9617	0.9622	0.9627	0.9632	0.9636	0.9641	0.9646	0.9650	0.9655
74	cos	0.2756	0.2740	0.2723	0.2706	0.2689	0.2672	0.2656	0.2639	0.2622	0.2605
	tan	3.4874	3.5105	3.5339	3.5576	3.5816	3.6059	3.6305	3.6554	3.6806	3.7062
	sin	0.9659	0.9664	0.9668	0.9673	0.9677	0.9681	0.9686	0.9690	0.9694	0.9699
75	cos	0.2588	0.2571	0.2554	0.2538	0.2521	0.2504	0.2487	0.2470	0.2453	0.2436
	tan	3.7321	3.7583	3.7848	3.8118	3.8391	3.8667	3.8947	3.9232	3.9520	3.9812

Trigonometric Functions

76.0°–89.9°

Angle in Degrees	Name of Function	Value of Function for Each Tenth of a Degree									
		0.0	**0.1**	**0.2**	**0.3**	**0.4**	**0.5**	**0.6**	**0.7**	**0.8**	**0.9**
76	sin	0.9703	0.9707	0.9711	0.9715	0.9720	0.9724	0.9728	0.9732	0.9736	0.9740
	cos	0.2419	0.2402	0.2385	0.2368	0.2351	0.2334	0.2317	0.2300	0.2284	0.2267
	tan	4.0108	4.0408	4.0713	4.1022	4.1335	4.1653	4.1976	4.2303	4.2635	4.2972
77	sin	0.9744	0.9748	0.9751	0.9755	0.9759	0.9763	0.9767	0.9770	0.9774	0.9778
	cos	0.2250	0.2232	0.2215	0.2198	0.2181	0.2164	0.2147	0.2130	0.2113	0.2096
	tan	4.3315	4.3662	4.4015	4.4374	4.4737	4.5107	4.5483	4.5864	4.6252	4.6646
78	sin	0.9781	0.9785	0.9789	0.9792	0.9796	0.9799	0.9803	0.9806	0.9810	0.9813
	cos	0.2079	0.2062	0.2045	0.2028	0.2011	0.1994	0.1977	0.1959	0.1942	0.1925
	tan	4.7046	4.7453	4.7867	4.8288	4.8716	4.9152	4.9594	5.0045	5.0504	5.0970
79	sin	0.9816	0.9820	0.9823	0.9826	0.9829	0.9833	0.9836	0.9839	0.9842	0.9845
	cos	0.1908	0.1891	0.1874	0.1857	0.1840	0.1822	0.1805	0.1788	0.1771	0.1754
	tan	5.1446	5.1929	5.2422	5.2924	5.3435	5.3955	5.4486	5.5026	5.5578	5.6140
80	sin	0.9848	0.9851	0.9854	0.9857	0.9860	0.9863	0.9866	0.9869	0.9871	0.9874
	cos	0.1736	0.1719	0.1702	0.1685	0.1668	0.1650	0.1633	0.1616	0.1599	0.1582
	tan	5.6713	5.7297	5.7894	5.8502	5.9124	5.9758	6.0405	6.1066	6.1742	6.2432
81	sin	0.9877	0.9880	0.9882	0.9885	0.9888	0.9890	0.9893	0.9895	0.9898	0.9900
	cos	0.1564	0.1547	0.1530	0.1513	0.1495	0.1478	0.1461	0.1444	0.1426	0.1409
	tan	6.3138	6.3859	6.4596	6.5350	6.6122	6.6912	6.7720	6.8548	6.9395	7.0264
82	sin	0.9903	0.9905	0.9907	0.9910	0.9912	0.9914	0.9917	0.9919	0.9921	0.9923
	cos	0.1392	0.1374	0.1357	0.1340	0.1323	0.1305	0.1288	0.1271	0.1253	0.1236
	tan	7.1154	7.2066	7.3002	7.3962	7.4947	7.5958	7.6996	7.8062	7.9158	8.0285
83	sin	0.9925	0.9928	0.9930	0.9932	0.9934	0.9936	0.9938	0.9940	0.9942	0.9943
	cos	0.1219	0.1201	0.1184	0.1167	0.1149	0.1132	0.1115	0.1097	0.1080	0.1063
	tan	8.1443	8.2636	8.3863	8.5126	8.6427	8.7769	8.9152	9.0579	9.2052	9.3572
84	sin	0.9945	0.9947	0.9949	0.9951	0.9952	0.9954	0.9956	0.9957	0.9959	0.9960
	cos	0.1045	0.1028	0.1011	0.0993	0.0976	0.0958	0.0941	0.0924	0.0906	0.0889
	tan	9.5144	9.6768	9.8448	10.02	10.20	10.39	10.58	10.78	10.99	11.20
85	sin	0.9962	0.9963	0.9965	0.9966	0.9968	0.9969	0.9971	0.9972	0.9973	0.9974
	cos	0.0872	0.0854	0.0837	0.0819	0.0802	0.0785	0.0767	0.0750	0.0732	0.0715
	tan	11.43	11.66	11.91	12.16	12.43	12.71	13.00	13.30	13.62	13.95
86	sin	0.9976	0.9977	0.9978	0.9979	0.9980	0.9981	0.9982	0.9983	0.9984	0.9985
	cos	0.0698	0.0680	0.0663	0.0645	0.0628	0.0610	0.0593	0.0576	0.0558	0.0541
	tan	14.30	14.67	15.06	15.89	15.46	16.35	16.83	17.34	17.89	18.46
87	sin	0.9986	0.9987	0.9988	0.9989	0.9990	0.9990	0.9991	0.9992	0.9993	0.9993
	cos	0.0523	0.0506	0.0488	0.0471	0.0454	0.0436	0.0419	0.0401	0.0384	0.0366
	tan	19.08	19.74	20.45	21.20	22.02	22.90	23.86	24.90	26.03	27.27
88	sin	0.9994	0.9995	0.9995	0.9996	0.9996	0.9997	0.9997	0.9997	0.9998	0.9998
	cos	0.0349	0.0332	0.0314	0.0297	0.0279	0.0262	0.0244	0.0227	0.0209	0.0192
	tan	28.64	30.14	31.82	33.69	35.80	38.19	40.92	44.07	47.74	52.08
89	sin	0.9998	0.9999	0.9999	0.9999	0.9999	1.000	1.000	1.000	1.000	1.000
	cos	0.0175	0.0157	0.0140	0.0122	0.0105	0.0087	0.0070	0.0052	0.0035	0.0017
	tan	57.29	63.66	71.62	81.85	95.49	114.6	143.2	191.0	286.5	573.0

13. EXPONENTIAL AND HYPERBOLIC TABLES

The following tables give values of e^x, e^{-x}, sinh x, cosh x and tanh x for values of x from 0.00 to 6.00 in intervals of 0.01.

To facilitate computations involving multiplication, the common logarithms of e^x, sinh x, cosh x, and tanh x are also given.

For values of x greater than 6, e^x may be computed from the relationship $e^x = \log^{-1}(x \log_{10} e) = \log^{-1} 0.43429x$; e^{-x} approaches zero; sinh x and cosh x are approximately equal and become 0.5 e^x; and tanh x and coth x have values approximately equal to unity.

Where more accurate values of the exponentials and functions are required they may be computed from the following relationships.

$$e = 2.71828\ 18285 \qquad \frac{1}{e} = 0.36787\ 94412$$

$$M = \log_{10} e = 0.43429\ 44819 \qquad \frac{1}{M} = \log_e 10 = 2.30258\ 50930$$

$$e^x = \log^{-1} Mx \qquad e^{-x} = \log^{-1} - Mx$$

$$\sinh x = \frac{e^x - e^{-x}}{2} \qquad \cosh x = \frac{e^x + e^{-x}}{2} \qquad \tanh x = \frac{e^x - e^{-x}}{e^x + e^{-x}}$$

$$\operatorname{csch} x = \frac{1}{\sinh x} \qquad \operatorname{sech} x = \frac{1}{\cosh x} \qquad \coth x = \frac{1}{\tanh x}$$

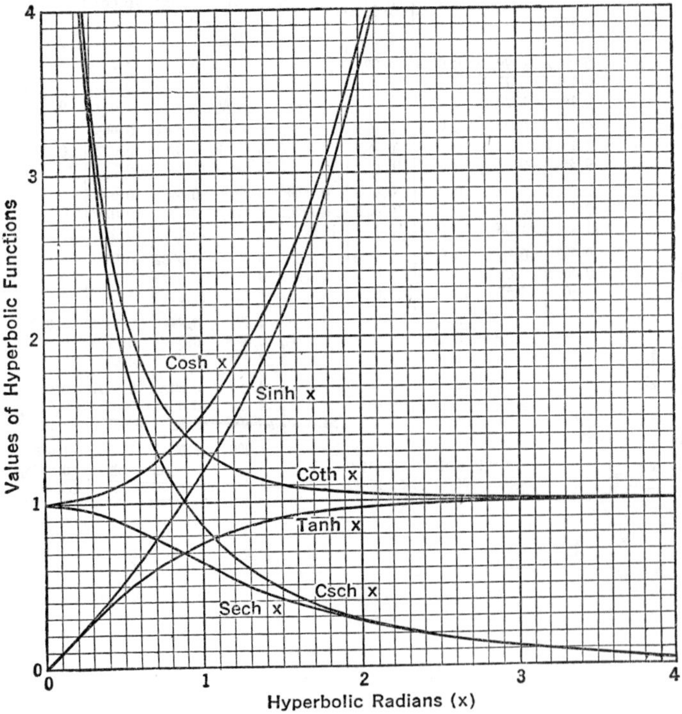

Chart of the Hyperbolic Functions.

x	Natural Values					Common Logarithms			
	e^x	e^{-x}	Sinh x	Cosh x	Tanh x	e^x	Sinh x	Cosh x	Tanh x
0.00	**1.0000**	**1.0000**	**0.0000**	**1.0000**	**.00000**	**0.00000**	$-\infty$	**0.00000**	$-\infty$
0.01	1.0101	.99005	0.0100	1.0001	.01000	.00434	$\bar{2}$.00001	.00002	$\bar{3}$.99999
0.02	1.0202	.98020	0.0200	1.0002	.02000	.00869	.30106	.00009	$\bar{2}$.30097
0.03	1.0305	.97045	0.0300	1.0005	.02999	.01303	.47719	.00020	.47699
0.04	1.0408	.96079	0.0400	1.0008	.03998	.01737	.60218	.00035	.60183
0.05	1.0513	.95123	0.0500	1.0013	.04996	.02171	.69915	.00054	.69861
0.06	1.0618	.94176	0.0600	1.0018	.05993	.02606	.77841	.00078	.77763
0.07	1.0725	.93239	0.0701	1.0025	.06989	.03040	.84545	.00106	.84439
0.08	1.0833	.92312	0.0801	1.0032	.07983	.03474	.90355	.00139	.90216
0.09	1.0942	.91393	0.0901	1.0041	.08976	.03909	.95483	.00176	.95307
0.10	**1.1052**	**.90484**	**0.1002**	**1.0050**	**.09967**	**0.04343**	**$\bar{1}$.00072**	**0.00217**	**$\bar{2}$.99856**
0.11	1.1163	.89583	0.1102	1.0061	.10956	.04777	.04227	.00262	$\bar{1}$.03965
0.12	1.1275	.88692	0.1203	1.0072	.11943	.05212	.08022	.00312	.07710
0.13	1.1388	.87810	0.1304	1.0085	.12927	.05646	.11517	.00366	.11151
0.14	1.1503	.86936	0.1405	1.0098	.13909	.06080	.14755	.00424	.14330
0.15	1.1618	.86071	0.1506	1.0113	.14889	.06514	.17772	.00487	.17285
0.16	1.1735	.85214	0.1607	1.0128	.15865	.06949	.20597	.00554	.20044
0.17	1.1853	.84366	0.1708	1.0145	.16838	.07383	.23254	.00625	.22629
0.18	1.1972	.83527	0.1810	1.0162	.17808	.07817	.25762	.00700	.25062
0.19	1.2092	.82696	0.1911	1.0181	.18775	.08252	.28136	.00779	.27357
0.20	**1.2214**	**.81873**	**0.2013**	**1.0201**	**.19738**	**0.08686**	**$\bar{1}$.30392**	**0.00863**	**1.29529**
0.21	1.2337	.81058	0.2115	1.0221	.20697	.09120	.32541	.00951	.31590
0.22	1.2461	.80252	0.2218	1.0243	.21652	.09554	.34592	.01043	.33549
0.23	1.2586	.79453	0.2320	1.0266	.22603	.09989	.36555	.01139	.35416
0.24	1.2712	.78663	0.2423	1.0289	.23550	.10423	.38437	.01239	.37198
0.25	1.2840	.77880	0.2526	1.0314	.24492	.10857	.40245	.01343	.38902
0.26	1.2969	.77105	0.2629	1.0340	.25430	.11292	.41986	.01452	.40534
0.27	1.3100	.76338	0.2733	1.0367	.26362	.11726	.43663	.01564	.42099
0.28	1.3231	.75578	0.2837	1.0395	.27291	.12160	.45282	.01681	.43601
0.29	1.3364	.74826	0.2941	1.0423	.28213	.12595	.46847	.01801	.45046
0.30	**1.3499**	**.74082**	**0.3045**	**1.0453**	**.29131**	**0.13029**	**1.48362**	**0.01926**	**$\bar{1}$.46436**
0.31	1.3634	.73345	0.3150	1.0484	.30044	.13463	.49830	.02054	.47775
0.32	1.3771	.72615	0.3255	1.0516	.30951	.13897	.51254	.02107	.49067
0.33	1.3910	.71892	0.3360	1.0549	.31852	.14332	.52637	.02323	.50314
0.34	1.4049	.71177	0.3466	1.0584	.32748	.14766	.53981	.02463	.51518
0.35	1.4191	.70469	0.3572	1.0619	.33638	.15200	.55290	.02607	.52682
0.36	1.4333	.69768	0.3678	1.0655	.34521	.15635	.56564	.02755	.53809
0.37	1.4477	.69073	0.3785	1.0692	.35399	.16069	.57807	.02907	.54899
0.38	1.4623	.68386	0.3892	1.0731	.36271	.16503	.59019	.03063	.55956
0.39	1.4770	.67706	0.4000	1.0770	.37136	.16937	.60202	.03222	.56980
0.40	**1.4918**	**.67032**	**0.4108**	**1.0811**	**.37995**	**0.17372**	**$\bar{1}$.61358**	**0.03385**	**$\bar{1}$.57973**
0.41	1.5068	.66365	0.4216	1.0852	.38847	.17806	.62488	.03552	.58936
0.42	1.5220	.65705	0.4325	1.0895	.39693	.18240	.63594	.03723	.59871
0.43	1.5373	.65051	0.4434	1.0939	.40532	.18675	.64677	.03897	.60780
0.44	1.5527	.64404	0.4543	1.0984	.41364	.19109	.65738	.04075	.61663
0.45	1.5683	.63763	0.4653	1.1030	.42190	.19543	.66777	.04256	.62521
0.46	1.5841	.63128	0.4764	1.1077	.43008	.19978	.67797	.04441	.63355
0.47	1.6000	.62500	0.4875	1.1125	.43820	.20412	.68797	.04630	.64167
0.48	1.6161	.61878	0.4986	1.1174	.44624	.20846	.69779	.04822	.64957
0.49	1.6323	.61263	0.5098	1.1225	.45422	.21280	.70744	.05018	.65726
0.50	**1.6487**	**.60653**	**0.5211**	**1.1276**	**.46212**	**0.21715**	**1.71692**	**0.05217**	**1.66475**
0.51	1.6653	.60050	0.5324	1.1329	.46995	.22149	.72624	.05419	.67205
0.52	1.6820	.59452	0.5438	1.1383	.47770	.22583	.73540	.05625	.67916
0.53	1.6989	.58860	0.5552	1.1438	.48538	.23018	.74442	.05834	.68608
0.54	1.7160	.58275	0.5666	1.1494	.49299	.23452	.75330	.06046	.69284
0.55	1.7333	.57695	0.5782	1.1551	.50052	.23886	.76204	.06262	.69942
0.56	1.7507	.57121	0.5897	1.1609	.50798	.24320	.77065	.06481	.70584
0.57	1.7683	.56553	0.6014	1.1669	.51536	.24755	.77914	.06703	.71211
0.58	1.7860	.55990	0.6131	1.1730	.52267	.25189	.78751	.06929	.71822
0.59	1.8040	.55433	0.6248	1.1792	.52990	.25623	.79576	.07157	.72419
0.60	**1.8221**	**.54881**	**0.6367**	**1.1855**	**.53705**	**0.26058**	**1.80390**	**0.07389**	**$\bar{1}$.73001**

x	Natural Values					Common Logarithms			
	e^x	e^{-x}	Sinh x	Cosh x	Tanh x	e^x	Sinh x	Cosh x	Tanh x
0.60	**1.8221**	**.54881**	**0.6367**	**1.1855**	**.53705**	**0.26058**	**$\bar{1}$.80390**	**0.07389**	**$\bar{1}$.73001**
0.61	1.8404	.54335	0.6485	1.1919	.54413	.26492	.81194	.07624	.73570
0.62	1.8589	.53794	0.6605	1.1984	.55113	.26926	.81987	.07861	.74125
0.63	1.8776	.53259	0.6725	1.2051	.55805	.27361	.82770	.08102	.74667
0.64	1.8965	.52729	0.6846	1.2119	.56490	.27795	.83543	.08346	.75197
0.65	1.9155	.52205	0.6967	1.2188	.57167	.28229	.84308	.08593	.75715
0.66	1.9348	.51685	0.7090	1.2258	.57836	.28663	.85063	.08843	.76220
0.67	1.9542	.51171	0.7213	1.2330	.58498	.29098	.85809	.09095	.76714
0.68	1.9739	.50662	0.7336	1.2402	.59152	.29532	.86548	.09351	.77197
0.69	1.9937	.50158	0.7461	1.2476	.59798	.29966	.87278	.09609	.77669
0.70	**2.0138**	**.49659**	**0.7586**	**1.2552**	**.60437**	**0.30401**	**$\bar{1}$.88000**	**0.09870**	**$\bar{1}$.78130**
0.71	2.0340	.49164	0.7712	1.2628	.61068	.30835	.88715	.10134	.78581
0.72	2.0544	.48675	0.7838	1.2706	.61691	.31269	.89423	.10401	.79022
0.73	2.0751	.48191	0.7966	1.2785	.62307	.31704	.90123	.10670	.79453
0.74	2.0959	.47711	0.8094	1.2865	.62915	.32138	.90817	.10942	.79875
0.75	2.1170	.47237	0.8223	1.2947	.63515	.32572	.91504	.11216	.80288
0.76	2.1383	.46767	0.8353	1.3030	.64108	.33006	.92185	.11493	.80691
0.77	2.1598	.46301	0.8484	1.3114	.64693	.33441	.92859	.11773	.81086
0.78	2.1815	.45841	0.8615	1.3199	.65271	.33875	.93527	.12055	.81472
0.79	2.2034	.45384	0.8748	1.3286	.65841	.34309	.94190	.12340	.81850
0.80	**2.2255**	**.44933**	**0.8881**	**1.3374**	**.66404**	**0.34744**	**$\bar{1}$.94846**	**0.12627**	**1.82219**
0.81	2.2479	.44486	0.9015	1.3464	.66959	.35178	.95498	.12917	.82581
0.82	2.2705	.44043	0.9150	1.3555	.67507	.35612	.96144	.13209	.82935
0.83	2.2933	.43605	0.9286	1.3647	.68048	.36046	.96784	.13503	.83281
0.84	2.3164	.43171	0.9423	1.3740	.68581	.36481	.97420	.13800	.83620
0.85	2.3396	.42741	0.9561	1.3835	.69107	.36915	.98051	.14099	.83952
0.86	2.3632	.42316	0.9700	1.3932	.69626	.37349	.98677	.14400	.84277
0.87	2.3869	.41895	0.9840	1.4029	.70137	.37784	.99299	.14704	.84595
0.88	2.4109	.41478	0.9981	1.4128	.70642	.38218	.99916	.15009	.84906
0.89	2.4351	.41066	1.0122	1.4229	.71139	.38652	0.00528	.15317	.85211
0.90	**2.4596**	**.40657**	**1.0265**	**1.4331**	**.71630**	**0.39087**	**0.01137**	**0.15627**	**$\bar{1}$.85509**
0.91	2.4843	.40252	1.0409	1.4434	.72113	.39521	.01741	.15939	.85801
0.92	2.5093	.39852	1.0554	1.4539	.72590	.39955	.02341	.16254	.86088
0.93	2.5345	.39455	1.0700	1.4645	.73059	.40389	.02937	.16570	.86368
0.94	2.5600	.39063	1.0847	1.4753	.73522	.40824	.03530	.16888	.86642
0.95	2.5857	.38674	1.0995	1.4862	.73978	.41258	.04119	.17208	.86910
0.96	2.6117	.38289	1.1144	1.4973	.74428	.41692	.04704	.17531	.87173
0.97	2.6379	.37908	1.1294	1.5085	.74870	.42127	.05286	.17855	.87431
0.98	2.6645	.37531	1.1446	1.5199	.75307	.42561	.05864	.18181	.87683
0.99	2.6912	.37158	1.1598	1.5314	.75736	.42995	.06439	.18509	.87930
1.00	**2.7183**	**.36788**	**1.1752**	**1.5431**	**.76159**	**0.43429**	**0.07011**	**0.18839**	**1.88172**
1.01	2.7456	.36422	1.1907	1.5549	.76576	.43864	.07580	.19171	.88409
1.02	2.7732	.36059	1.2063	1.5669	.76987	.44298	.08146	.19504	.88642
1.03	2.8011	.35701	1.2220	1.5790	.77391	.44732	.08708	.19839	.88869
1.04	2.8292	.35345	1.2379	1.5913	.77789	.45167	.09268	.20176	.89092
1.05	2.8577	.34994	1.2539	1.6038	.78181	.45601	.09825	.20515	.89310
1.06	2.8864	.34646	1.2700	1.6164	.78566	.46035	.10379	.20855	.89524
1.07	2.9154	.34301	1.2862	1.6292	.78946	.46470	.10930	.21197	.89733
1.08	2.9447	.33960	1.3025	1.6421	.79320	.46904	.11479	.21541	.89938
1.09	2.9743	.33622	1.3190	1.6552	.79688	.47338	.12025	.21886	.90139
1.10	**3.0042**	**.33287**	**1.3356**	**1.6685**	**.80050**	**0.47772**	**0.12569**	**0.22233**	**$\bar{1}$.90336**
1.11	3.0344	.32956	1.3524	1.6820	.80406	.48207	.13111	.22582	.90529
1.12	3.0649	.32628	1.3693	1.6956	.80757	.48641	.13649	.22931	.90718
1.13	3.0957	.32303	1.3863	1.7093	.81102	.49075	.14186	.23283	.90903
1.14	3.1268	.31982	1.4035	1.7233	.81441	.49510	.14720	.23636	.91085
1.15	3.1582	.31664	1.4208	1.7374	.81775	.49944	.15253	.23990	.91262
1.16	3.1899	.31349	1.4382	1.7517	.82104	.50378	.15783	.24346	.91436
1.17	3.2220	.31037	1.4558	1.7662	.82427	.50812	.16311	.24703	.91607
1.18	3.2544	.30728	1.4735	1.7808	.82745	.51247	.16836	.25062	.91774
1.19	3.2871	.30422	1.4914	1.7957	.83058	.51681	.17360	.25422	.91938
1.20	**3.3201**	**.30119**	**1.5095**	**1.8107**	**.83365**	**0.52115**	**0.17832**	**0.25784**	**$\bar{1}$.92099**

x	Natural Values					Common Logarithms			
	e^x	e^{-x}	Sinh x	Cosh x	Tanh x	e^x	Sinh x	Cosh x	Tanh x
1.20	**3.3201**	**.30119**	**1.5095**	**1.8107**	**.83365**	**0.52115**	**0.17882**	**0.25784**	**1.92099**
1.21	3.3535	.29820	1.5276	1.8258	.83668	.52550	.18402	.26146	.92256
1.22	3.3872	.29523	1.5460	1.8412	.83965	.52984	.18920	.26510	.92410
1.23	3.4212	.29229	1.5645	1.8568	.84258	.53418	.19437	.26876	.92561
1.24	3.4556	.28938	1.5831	1.8725	.84546	.53853	.19951	.27242	.92709
1.25	3.4903	.28650	1.6019	1.8884	.84828	.54287	.20464	.27610	.92854
1.26	3.5254	.28365	1.6209	1.9045	.85106	.54721	.20975	.27979	.92996
1.27	3.5609	.28083	1.6400	1.9208	.85380	.55155	.21485	.28349	.93135
1.28	3.5966	.27804	1.6593	1.9373	.85648	.55590	.21993	.28721	.93272
1.29	3.6328	.27527	1.6788	1.9540	.85913	.56024	.22499	.29093	.93406
1.30	**3.6693**	**.27253**	**1.6984**	**1.9709**	**.86172**	**0.56458**	**0.23004**	**0.29467**	**$\bar{1}$.93537**
1.31	3.7062	.26982	1.7182	1.9880	.86428	.56893	.23507	.29842	.93665
1.32	3.7434	.26714	1.7381	2.0053	.86678	.57327	.24009	.30217	.93791
1.33	3.7810	.26448	1.7583	2.0228	.86925	.57761	.24509	.30594	.93914
1.34	3.8190	.26185	1.7786	2.0404	.87167	.58195	.25008	.30972	.94035
1.35	3.8574	.25924	1.7991	2.0583	.87405	.58630	.25505	.31352	.94154
1.36	3.8962	.25666	1.8198	2.0764	.87639	.59064	.26002	.31732	.94270
1.37	3.9354	.25411	1.8406	2.0947	.87869	.59498	.26496	.32113	.94384
1.38	3.9749	.25158	1.8617	2.1132	.88095	.59933	.26990	.32495	.94495
1.39	4.0149	.24908	1.8829	2.1320	.88317	.60367	.27482	.32878	.94604
1.40	**4.0552**	**.24660**	**1.9043**	**2.1509**	**.88535**	**0.60801**	**0.27974**	**0.33262**	**$\bar{1}$.94712**
1.41	4.0960	.24414	1.9259	2.1700	.88749	.61236	.28464	.33647	.94817
1.42	4.1371	.24171	1.9477	2.1894	.88960	.61670	.28952	.34033	.94919
1.43	4.1787	.23931	1.9697	2.2090	.89167	.62104	.29440	.34420	.95020
1.44	4.2207	.23693	1.9919	2.2288	.89370	.62538	.29926	.34807	.95119
1.45	4.2631	.23457	2.0143	2.2488	.89569	.62973	.30412	.35196	.95216
1.46	4.3060	.23224	2.0369	2.2691	.89765	.63407	.30896	.35585	.95311
1.47	4.3492	.22993	2.0597	2.2896	.89958	.63841	.31379	.35976	.95404
1.48	4.3929	.22764	2.0827	2.3103	.90147	.64276	.31862	.36367	.95495
1.49	4.4371	.22537	2.1059	2.3312	.90332	.64710	.32343	.36759	.95584
1.50	**4.4817**	**.22313**	**2.1293**	**2.3524**	**.90515**	**0.65144**	**0.32823**	**0.37151**	**$\bar{1}$.95672**
1.51	4.5267	.22091	2.1529	2.3738	.90694	.65578	.33303	.37545	.95758
1.52	4.5722	.21871	2.1768	2.3955	.90870	.66013	.33781	.37939	.95842
1.53	4.6182	.21654	2.2008	2.4174	.91042	.66447	.34258	.38334	.95924
1.54	4.6646	.21438	2.2251	2.4395	.91212	.66881	.34735	.38730	.96005
1.55	4.7115	.21225	2.2496	2.4619	.91379	.67316	.35211	.39126	.96084
1.56	4.7588	.21014	2.2743	2.4845	.91542	.67750	.35686	.39524	.96162
1.57	4.8066	.20805	2.2993	2.5073	.91703	.68184	.36160	.39921	.96238
1.58	4.8550	.20598	2.3245	2.5305	.91860	.68619	.36633	.40320	.96313
1.59	4.9037	.20393	2.3499	2.5538	.92015	.69053	.37105	.40719	.96386
1.60	**4.9530**	**.20190**	**2.3756**	**2.5775**	**.92167**	**0.69487**	**0.37577**	**0.41119**	**$\bar{1}$.96457**
1.61	5.0028	.19989	2.4015	2.6013	.92316	.69921	.38048	.41520	.96528
1.62	5.0531	.19790	2.4276	2.6255	.92462	.70356	.38518	.41921	.96597
1.63	5.1039	.19593	2.4540	2.6499	.92606	.70790	.38987	.42323	.96664
1.64	5.1552	.19398	2.4806	2.6746	.92747	.71224	.39456	.42725	.96730
1.65	5.2070	.19205	2.5075	2.6995	.92886	.71659	.39923	.43129	.96795
1.66	5.2593	.19014	2.5346	2.7247	.93022	.72093	.40391	.43532	.96858
1.67	5.3122	.18825	2.5620	2.7502	.93155	.72527	.40857	.43937	.96921
1.68	5.3656	.18637	2.5896	2.7760	.93286	.72961	.41323	.44341	.96982
1.69	5.4195	.18452	2.6175	2.8020	.93415	.73396	.41788	.44747	.97042
1.70	**5.4739**	**.18268**	**2.6456**	**2.8283**	**.93541**	**0.73830**	**0.42253**	**0.45153**	**$\bar{1}$.97100**
1.71	5.5290	.18087	2.6740	2.8549	.93665	.74264	.42717	.45559	.97158
1.72	5.5845	.17907	2.7027	2.8818	.93786	.74699	.43180	.45966	.97214
1.73	5.6407	.17728	2.7317	2.9090	.93906	.75133	.43643	.46374	.97269
1.74	5.6973	.17552	2.7609	2.9364	.94023	.75567	.44105	.46782	.97323
1.75	5.7546	.17377	2.7904	2.9642	.94138	.76002	.44567	.47191	.97376
1.76	5.8124	.17204	2.8202	2.9922	.94250	.76436	.45028	.47600	.97428
1.77	5.8709	.17033	2.8503	3.0206	.94361	.76870	.45488	.48009	.97479
1.78	5.9299	.16864	2.8806	3.0492	.94470	.77304	.45948	.48419	.97529
1.79	5.9895	.16696	2.9112	3.0782	.94576	.77739	.46408	.48830	.97578
1.80	**6.0496**	**.16530**	**2.9422**	**3.1075**	**.94681**	**0.78173**	**0.46867**	**0.49241**	**$\bar{1}$.97626**

x	Natural Values					Common Logarithms			
	e^x	e^{-x}	Sinh x	Cosh x	Tanh x	e^x	Sinh x	Cosh x	Tanh x
1.80	**6.0496**	**.16530**	**2.9422**	**3.1075**	**.94681**	**0.78173**	**0.46867**	**0.49241**	**$\bar{1}$.97626**
1.81	6.1104	.16365	2.9734	3.1371	.94783	.78607	.47325	.49652	.97673
1.82	6.1719	.16203	3.0049	3.1669	.94884	.79042	.47783	.50064	.97719
1.83	6.2339	.16041	3.0367	3.1972	.94983	.79476	.48241	.50476	.97764
1.84	6.2965	.15882	3.0689	3.2277	.95080	.79910	.48698	.50889	.97809
1.85	6.3598	.15724	3.1013	3.2585	.95175	.80344	.49154	.51302	.97852
1.86	6.4237	.15567	3.1340	3.2897	.95268	.80779	.49610	.51716	.97895
1.87	6.4883	.15412	3.1671	3.3212	.95359	.81213	.50066	.52130	.97936
1.88	6.5535	.15259	3.2005	3.3530	.95449	.81647	.50521	.52544	.97977
1.89	6.6194	.15107	3.2341	3.3852	.95537	.82082	.50976	.52959	.98017
1.90	**6.6859**	**.14957**	**3.2682**	**3.4177**	**.95624**	**0.82516**	**0.51430**	**0.53374**	**$\bar{1}$.98057**
1.91	6.7531	.14808	3.3025	3.4506	.95709	.82950	.51884	.53789	.98095
1.92	6.8210	.14661	3.3372	3.4838	.95792	.83385	.52338	.54205	.98133
1.93	6.8895	.14515	3.3722	3.5173	.95873	.83819	.52791	.54621	.98170
1.94	6.9583	.14370	3.4075	3.5512	.95953	.84253	.53244	.55038	.98206
1.95	7.0287	.14227	3.4432	3.5855	.96032	.84687	.53696	.55455	.98242
1.96	7.0993	.14086	3.4792	3.6201	.96109	.85122	.54148	.55872	.98272
1.97	7.1707	.13946	3.5156	3.6551	.96185	.85556	.54600	.56290	.98311
1.98	7.2427	.13807	3.5523	3.6904	.96259	.85990	.55051	.56707	.98344
1.99	7.3155	.13670	3.5894	3.7261	.96331	.86425	.55502	.57126	.98377
2.00	**7.3891**	**.13534**	**3.6269**	**3.7622**	**.96403**	**0.86859**	**0.55953**	**0.57544**	**$\bar{1}$.98409**
2.01	7.4633	.13399	3.6647	3.7987	.96473	.87293	.56403	.57963	.98440
2.02	7.5383	.13266	3.7028	3.8355	.96541	.87727	.56853	.58382	.98471
2.03	7.6141	.13134	3.7414	3.8727	.96609	.88162	.57303	.58802	.98502
2.04	7.6906	.13003	3.7803	3.9103	.96675	.88596	.57753	.59221	.98531
2.05	7.7679	.12873	3.8196	3.9483	.96740	.89030	.58202	.59641	.98560
2.06	7.8460	.12745	3.8593	3.9867	.96803	.89465	.58650	.60061	.98589
2.07	7.9248	.12619	3.8993	4.0255	.96865	.89899	.59099	.60482	.98617
2.08	8.0045	.12493	3.9398	4.0647	.96926	.90333	.59547	.60903	.98644
2.09	8.0849	.12369	3.9806	4.1043	.96986	.90768	.59995	.61324	.98671
2.10	**8.1662**	**.12246**	**4.0219**	**4.1443**	**.97045**	**0.91202**	**0.60443**	**0.61745**	**$\bar{1}$.98697**
2.11	8.2482	.12124	4.0635	4.1847	.97103	.91636	.60890	.62167	.98723
2.12	8.3311	.12003	4.1056	4.2256	.97159	.92070	.61337	.62589	.98748
2.13	8.4149	.11884	4.1480	4.2669	.97215	.92505	.61784	.63011	.98773
2.14	8.4994	.11765	4.1909	4.3085	.97269	.92939	.62231	.63433	.98798
2.15	8.5849	.11648	4.2342	4.3507	.97323	.93373	.62677	.63856	.98821
2.16	8.6711	.11533	4.2779	4.3932	.97375	.93808	.63123	.64278	.98845
2.17	8.7583	.11418	4.3221	4.4362	.97426	.94242	.63569	.64701	.98868
2.18	8.8463	.11304	4.3666	4.4797	.97477	.94676	.64015	.65125	.98890
2.19	8.9352	.11192	4.4116	4.5236	.97526	.95110	.64460	.65548	.98912
2.20	**9.0250**	**.11080**	**4.4571**	**4.5679**	**.97574**	**0.95545**	**0.64905**	**0.65972**	**$\bar{1}$.98934**
2.21	9.1157	.10970	4.5030	4.6127	.97622	.95979	.65350	.66396	.98955
2.22	9.2073	.10861	4.5494	4.6580	.97668	.96413	.65795	.66820	.98975
2.23	9.2999	.10753	4.5962	4.7037	.97714	.96848	.66240	.67244	.98996
2.24	9.3933	.10646	4.6434	4.7499	.97759	.97282	.66684	.67668	.99016
2.25	9.4877	.10540	4.6912	4.7966	.97803	.97716	.67128	.68093	.99035
2.26	9.5831	.10435	4.7394	4.8437	.97846	.98151	.67572	.68518	.99054
2.27	9.6794	.10331	4.7880	4.8914	.97888	.98585	.68016	.68943	.99073
2.28	9.7767	.10228	4.8372	4.9395	.97929	.99019	.68459	.69368	.99091
2.29	9.8749	.10127	4.8868	4.9881	.97970	.99453	.68903	.69794	.99109
2.30	**9.9742**	**.10026**	**4.9370**	**5.0372**	**.98010**	**0.99888**	**0.69346**	**0.70219**	**$\bar{1}$.99127**
2.31	10.074	.09926	4.9876	5.0868	.98049	1.00322	.69789	.70645	.99144
2.32	10.176	.09827	5.0387	5.1370	.98087	.00756	.70232	.71071	.99161
2.33	10.278	.09730	5.0903	5.1876	.98124	.01191	.70675	.71497	.99178
2.34	10.381	.09633	5.1425	5.2388	.98161	.01625	.71117	.71923	.99194
2.35	10.486	.09537	5.1951	5.2905	.98197	.02059	.71559	.72349	.99210
2.36	10.591	.09442	5.2483	5.3427	.98233	.02493	.72002	.72776	.99226
2.37	10.697	.09348	5.3020	5.3954	.98267	.02928	.72444	.73203	.99241
2.38	10.805	.09255	5.3562	5.4487	.98301	.03362	.72885	.73630	.99256
2.39	10.913	.09163	5.4109	5.5026	.98335	.03796	.73327	.74056	.99271
2.40	**11.023**	**.09072**	**5.4662**	**5.5569**	**.98367**	**1.04231**	**0.73769**	**0.74484**	**$\bar{1}$.99285**

x	Natural Values					Common Logarithms			
	e^x	e^{-x}	Sinh x	Cosh x	Tanh x	e^x	Sinh x	Cosh x	Tanh x
2.40	**11.023**	**.09072**	**5.4662**	**5.5569**	**.98367**	**1.04231**	**0.73769**	**0.74484**	**$\bar{1}$.99285**
2.41	11.134	.08982	5.5221	5.6119	.98400	.04665	.74210	.74911	.99299
2.42	11.246	.08892	5.5785	5.6674	.98431	.05099	.74652	.75338	.99313
2.43	11.359	.08804	5.6354	5.7235	.98462	.05534	.75093	.75766	.99327
2.44	11.473	.08716	5.6929	5.7801	.98492	.05968	.75534	.76194	.99340
2.45	11.588	.08629	5.7510	5.8373	.98522	.06402	.75975	.76621	.99353
2.46	11.705	.08543	5.8097	5.8951	.98551	.06836	.76415	.77049	.99366
2.47	11.822	.08458	5.8689	5.9535	.98579	.07271	.76856	.77477	.99379
2.48	11.941	.08374	5.9288	6.0125	.98607	.07705	.77296	.77906	.99391
2.49	12.061	.08291	5.9892	6.0721	.98635	.08139	.77737	.78334	.99403
2.50	**12.182**	**.08208**	**6.0502**	**6.1323**	**.98661**	**1.08574**	**0.78177**	**0.78762**	**$\bar{1}$.99415**
2.51	12.305	.08127	6.1118	6.1931	.98688	.09008	.78617	.79191	.99426
2.52	12.429	.08046	6.1741	6.2545	.98714	.09442	.79057	.79619	.99438
2.53	12.554	.07966	6.2369	6.3166	.98739	.09877	.79497	.80048	.99449
2.54	12.680	.07887	6.3004	6.3793	.98764	.10311	.79937	.80477	.99460
2.55	12.807	.07808	6.3645	6.4426	.98788	.10745	.80377	.80906	.99470
2.56	12.936	.07730	6.4293	6.5066	.98812	.11179	.80816	.81335	.99481
2.57	13.066	.07654	6.4946	6.5712	.98835	.11614	.81256	.81764	.99491
2.58	13.197	.07577	6.5607	6.6365	.98858	.12048	.81695	.82194	.99501
2.59	13.330	.07502	6.6274	6.7024	.98881	.12482	.82134	.82623	.99511
2.60	**13.464**	**.07427**	**6.6947**	**6.7690**	**.98903**	**1.12917**	**0.82573**	**0.83052**	**$\bar{1}$.99521**
2.61	13.599	.07353	6.7628	6.8363	.98924	.13351	.83012	.83482	.99530
2.62	13.736	.07280	6.8315	6.9043	.98946	.13785	.83451	.83912	.99540
2.63	13.874	.07208	6.9008	6.9729	.98966	.14219	.83890	.84341	.99549
2.64	14.013	.07136	6.9709	7.0423	.98987	.14654	.84329	.84771	.99558
2.65	14.154	.07065	7.0417	7.1123	.99007	.15088	.84768	.85201	.99566
2.66	14.296	.06995	7.1132	7.1831	.99026	.15522	.85206	.85631	.99575
2.67	14.440	.06925	7.1854	7.2546	.99045	.15957	.85645	.86061	.99583
2.68	14.585	.06856	7.2583	7.3268	.99064	.16391	.86083	.86492	.99592
2.69	14.732	.06788	7.3319	7.3998	.99083	.16825	.86522	.86922	.99600
2.70	**14.880**	**.06721**	**7.4063**	**7.4735**	**.99101**	**1.17260**	**0.86960**	**0.87352**	**$\bar{1}$.99608**
2.71	15.029	.06654	7.4814	7.5479	.99118	.17694	.87398	.87783	.99615
2.72	15.180	.06587	7.5572	7.6231	.99136	.18128	.87836	.88213	.99623
2.73	15.333	.06522	7.6338	7.6991	.99153	.18562	.88274	.88644	.99631
2.74	15.487	.06457	7.7112	7.7758	.99170	.18997	.88712	.89074	.99638
2.75	15.643	.06393	7.7894	7.8533	.99186	.19431	.89150	.89505	.99645
2.76	15.800	.06329	7.8683	7.9316	.99202	.19865	.89588	.89936	.99652
2.77	15.959	.06266	7.9480	8.0106	.99218	.20300	.90026	.90367	.99659
2.78	16.119	.06204	8.0285	8.0905	.99233	.20734	.90463	.90798	.99666
2.79	16.281	.06142	8.1098	8.1712	.99248	.21168	.90901	.91229	.99672
2.80	**16.445**	**.06081**	**8.1919**	**8.2527**	**.99263**	**1.21602**	**0.91339**	**0.91660**	**$\bar{1}$.99679**
2.81	16.610	.06020	8.2749	8.3351	.99278	.22037	.91776	.92091	.99685
2.82	16.777	.05961	8.3586	8.4182	.99292	.22471	.92213	.92522	.99691
2.83	16.945	.05901	8.4432	8.5022	.99306	.22905	.92651	.92953	.99698
2.84	17.116	.05843	8.5287	8.5871	.99320	.23340	.93088	.93385	.99704
2.85	17.288	.05784	8.6150	8.6728	.99333	.23774	.93525	.93816	.99709
2.86	17.462	.05727	8.7021	8.7594	.99346	.24208	.93963	.94247	.99715
2.87	17.637	.05670	8.7902	8.8469	.99359	.24643	.94400	.94679	.99721
2.88	17.814	.05613	8.8791	8.9352	.99372	.25077	.94837	.95110	.99726
2.89	17.993	.05558	8.9689	9.0244	.99384	.25511	.95274	.95542	.99732
2.90	**18.174**	**.05502**	**9.0596**	**9.1146**	**.99396**	**1.25945**	**0.95711**	**0.95974**	**$\bar{1}$.99737**
2.91	18.357	.05448	9.1512	9.2056	.99408	.26380	.96148	.96405	.99742
2.92	18.541	.05393	9.2437	9.2976	.99420	.26814	.96584	.96837	.99747
2.93	18.728	.05340	9.3371	9.3905	.99431	.27248	.97021	.97269	.99752
2.94	18.916	.05287	9.4315	9.4844	.99443	.27683	.97458	.97701	.99757
2.95	19.106	.05234	9.5268	9.5791	.99454	.28117	.97895	.98133	.99762
2.96	19.298	.05182	9.6231	9.6749	.99464	.28551	.98331	.98565	.99767
2.97	19.492	.05130	9.7203	9.7716	.99475	.28985	.98768	.98997	.99771
2.98	19.688	.05079	9.8185	9.8693	.99485	.29420	.99205	.99429	.99776
2.99	19.886	.05029	9.9177	9.9680	.99496	.29854	.99641	.99861	.99780
3.00	**20.086**	**.04979**	**10.018**	**10.068**	**.99505**	**1.30288**	**1.00078**	**1.00293**	**$\bar{1}$.99785**

x	Natural Values					Common Logarithms			
	e^x	e^{-x}	Sinh x	Cosh x	Tanh x	e^x	Sinh x	Cosh x	Tanh x
3.00	**20.086**	**.04979**	**10.018**	**10.068**	**.99505**	**1.30288**	**1.00078**	**1.00293**	**$\bar{1}$.99785**
3.01	20.287	.04929	10.119	10.168	.99515	.30723	.00514	.00725	.99789
3.02	20.491	.04880	10.221	10.270	.99525	.31157	.00950	.01157	.99793
3.03	20.697	.04832	10.325	10.373	.99534	.31591	.01387	.01589	.99797
3.04	20.905	.04783	10.429	10.477	.99543	.32026	.01823	.02022	.99801
3.05	21.115	.04736	10.534	10.581	.99552	.32460	.02259	.02454	.99805
3.06	21.328	.04689	10.640	10.687	.99561	.32894	.02696	.02886	.99809
3.07	21.542	.04642	10.748	10.794	.99570	.33328	.03132	.03319	.99813
3.08	21.758	.04596	10.856	10.902	.99578	.33763	.03568	.03751	.99817
3.09	21.977	.04550	10.966	11.011	.99587	.34197	.04004	.04184	.99820
3.10	**22.198**	**.04505**	**11.077**	**11.122**	**.99595**	**1.34631**	**1.04440**	**1.04616**	**$\bar{1}$.99824**
3.11	22.421	.04460	11.188	11.233	.99603	.35066	.04876	.05049	.99827
3.12	22.646	.04416	11.301	11.345	.99611	.35500	.05312	.05481	.99831
3.13	22.874	.04372	11.415	11.459	.99618	.35934	.05748	.05914	.99834
3.14	23.104	.04328	11.530	11.574	.99626	.36368	.06184	.06347	.99837
3.15	23.336	.04285	11.647	11.689	.99633	.36803	.06620	.06779	.99841
3.16	23.571	.04243	11.764	11.807	.99641	.37237	.07056	.07212	.99844
3.17	23.807	.04200	11.883	11.925	.99648	.37671	.07492	.07645	.99847
3.18	24.047	.04159	12.003	12.044	.99655	.38106	.07927	.08078	.99850
3.19	24.288	.04117	12.124	12.165	.99662	.38540	.08363	.08510	.99853
3.20	**24.533**	**.04076**	**12.246**	**12.287**	**.99668**	**1.38974**	**1.08799**	**1.08943**	**$\bar{1}$.99856**
3.21	24.779	.04036	12.369	12.410	.99675	.39409	.09235	.09376	.99859
3.22	25.028	.03996	12.494	12.534	.99681	.39843	.09670	.09809	.99861
3.23	25.280	.03956	12.620	12.660	.99688	.40277	.10106	.10242	.99864
3.24	25.534	.03916	12.747	12.786	.99694	.40711	.10542	.10675	.99867
3.25	25.790	.03877	12.876	12.915	.99700	.41146	.10977	.11108	.99869
3.26	26.050	.03839	13.006	13.044	.99706	.41580	.11413	.11541	.99872
3.27	26.311	.03801	13.137	13.175	.99712	.42014	.11849	.11974	.99875
3.28	26.576	.03763	13.269	13.307	.99717	.42449	.12284	.12407	.99877
3.29	26.843	.03725	13.403	13.440	.99723	.42883	.12720	.12840	.99879
3.30	**27.113**	**.03688**	**13.538**	**13.575**	**.99728**	**1.43317**	**1.13155**	**1.13273**	**$\bar{1}$.99882**
3.31	27.385	.03652	13.674	13.711	.99734	43751	.13591	.13706	.99884
3.32	27.660	.03615	13.812	13.848	.99739	.44186	.14026	.14139	.99886
3.33	27.938	.03579	13.951	13.987	.99744	.44620	.14461	.14573	.99889
3.34	28.219	.03544	14.092	14.127	.99749	.45054	.14897	.15006	.99891
3.35	28.503	.03508	14.234	14.269	.99754	.45489	.15332	.15439	.99893
3.36	28.789	.03474	14.377	14.412	.99759	.45923	.15768	.15872	.99895
3.37	29.079	.03439	14.522	14.556	.99764	.46357	.16203	.16306	.99897
3.38	29.371	.03405	14.668	14.702	.99768	.46792	.16638	.16739	.99899
3.39	29.666	.03371	14.816	14.850	.99773	.47226	.17073	.17172	.99901
3.40	**29.964**	**.03337**	**14.965**	**14.999**	**.99777**	**1.47660**	**1.17509**	**1.17605**	**$\bar{1}$.99903**
3.41	30.265	.03304	15.116	15.149	.99782	.48094	.17944	.18039	.99905
3.42	30.569	.03271	15.268	15.301	.99786	.48529	.18379	.18472	.99907
3.43	30.877	.03239	15.422	15.455	.99790	.48963	.18814	.18906	.99909
3.44	31.187	.03206	15.577	15.610	.99795	.49397	.19250	.19339	.99911
3.45	31.500	.03175	15.734	15.766	.99799	.49832	.19685	.19772	.99912
3.46	31.817	.03143	15.893	15.924	.99803	.50266	.20120	.20206	.99914
3.47	32.137	.03112	16.053	16.084	.99807	.50700	.20555	.20639	.99916
3.48	32.460	.03081	16.215	16.245	.99810	.51134	.20990	.21073	.99918
3.49	32.786	.03050	16.378	16.408	.99814	.51569	.21425	.21506	.99919
3.50	**33.115**	**.03020**	**16.543**	**16.573**	**.99818**	**1.52003**	**1.21860**	**1.21940**	**$\bar{1}$.99921**
3.51	33.448	.02990	16.709	16.739	.99821	.52437	.22296	.22373	.99922
3.52	33.784	.02960	16.877	16.907	.99825	.52872	.22731	.22807	.99924
3.53	34.124	.02930	17.047	17.077	.99828	.53306	.23166	.23240	.99925
3.54	34.467	.02901	17.219	17.248	.99832	.53740	.23601	.23674	.99927
3.55	34.813	.02872	17.392	17.421	.99835	.54175	.24036	.24107	.99928
3.56	35.163	.02844	17.567	17.596	.99838	.54609	.24471	.24541	.99930
3.57	35.517	.02816	17.744	17.772	.99842	.55043	.24906	.24975	.99931
3.58	35.874	.02788	17.923	17.951	.99845	.55477	.25341	.25408	.99933
3.59	36.234	.02760	18.103	18.131	.99848	.55912	.25776	.25842	.99934
3.60	**36.598**	**.02732**	**18.285**	**18.313**	**.99851**	**1.56346**	**1.26211**	**1.26275**	**$\bar{1}$.99935**

x	Natural Values					Common Logarithms			
	e^x	e^{-x}	Sinh x	Cosh x	Tanh x	e^x	Sinh x	Cosh x	Tanh x
3.60	**36.598**	**.02732**	**18.285**	**18.313**	**.99851**	**1.56346**	**1.26211**	**1.26275**	**$\bar{1}$.99935**
3.61	36.966	.02705	18.470	18.497	.99854	.56780	.26646	.26709	.99936
3.62	37.338	.02678	18.655	18.682	.99857	.57215	.27080	.27143	.99938
3.63	37.713	.02652	18.843	18.870	.99859	.57649	.27515	.27576	.99939
3.64	38.092	.02625	19.033	19.059	.99862	.58083	.27950	.28010	.99940
3.65	38.475	.02599	19.224	19.250	.99865	.58517	.28385	.28444	.99941
3.66	38.861	.02573	19.418	19.444	.99868	.58952	.28820	.28878	.99942
3.67	39.252	.02548	19.613	19.639	.99870	.59386	.29255	.29311	.99944
3.68	39.646	.02522	19.811	19.836	.99873	.59820	.29690	.29745	.99945
3.69	40.045	.02497	20.010	20.035	.99875	.60255	.30125	.30179	.99946
3.70	**40.447**	**.02472**	**20.211**	**20.236**	**.99878**	**1.60689**	**1.30559**	**1.30612**	**$\bar{1}$.99947**
3.71	40.854	.02448	20.415	20.439	.99880	.61123	.30994	.31046	.99948
3.72	41.264	.02423	20.620	20.644	.99883	.61558	.31429	.31480	.99949
3.73	41.679	.02399	20.828	20.852	.99885	.61992	.31864	.31914	.99950
3.74	42.098	.02375	21.037	21.061	.99887	.62426	.32299	.32348	.99951
3.75	42.521	.02352	21.249	21.272	.99889	.62860	.32733	.32781	.99952
3.76	42.948	.02328	21.463	21.486	.99892	.63295	.33168	.33215	.99953
3.77	43.380	.02305	21.679	21.702	.99894	.63729	.33603	.33649	.99954
3.78	43.816	.02282	21.897	21.919	.99896	.64163	.34038	.34083	.99955
3.79	44.256	.02260	22.117	22.140	.99898	.64598	.34472	.34517	.99956
3.80	**44.701**	**.02237**	**22.339**	**22.362**	**.99900**	**1.65032**	**1.34907**	**1.34951**	**$\bar{1}$.99957**
3.81	45.150	.02215	22.564	22.586	.99902	.65466	.35342	.35384	.99957
3.82	45.604	.02193	22.791	22.813	.99904	.65900	.35777	.35818	.99958
3.83	46.063	.02171	23.020	23.042	.99906	.66335	.36211	.36252	.99959
3.84	46.525	.02149	23.252	23.274	.99908	.66769	.36646	.36686	.99960
3.85	46.993	.02128	23.486	23.507	.99909	.67203	.37081	.37120	.99961
3.86	47.465	.02107	23.722	23.743	.99911	.67638	.37515	.37554	.99961
3.87	47.942	.02086	23.961	23.982	.99913	.68072	.37950	.37988	.99962
3.88	48.424	.02065	24.202	24.222	.99915	.68506	.38385	.38422	.99963
3.89	48.911	.02045	24.445	24.466	.99916	.68941	.38819	.38856	.99964
3.90	**49.402**	**.02024**	**24.691**	**24.711**	**.99918**	**1.69375**	**1.39254**	**1.39290**	**$\bar{1}$.99964**
3.91	49.899	.02004	24.939	24.960	.99920	.69809	.39689	.39724	.99965
3.92	50.400	.01984	25.190	25.210	.99921	.70243	.40123	.40158	.99966
3.93	50.907	.01964	25.444	25.463	.99923	.70678	.40558	.40591	.99966
3.94	51.419	.01945	25.700	25.719	.99924	.71112	.40993	.41025	.99967
3.95	51.935	.01925	25.958	25.977	.99926	.71546	.41427	.41459	.99968
3.96	52.457	.01906	26.219	26.238	.99927	.71981	.41862	.41893	.99968
3.97	52.985	.01887	26.483	26.502	.99929	.72415	.42296	.42327	.99969
3.98	53.517	.01869	26.749	26.768	.99930	.72849	.42731	.42761	.99970
3.99	54.055	.01850	27.018	27.037	.99932	.73284	.43166	.43195	.99970
4.00	**54.598**	**.01832**	**27.290**	**27.308**	**.99933**	**1.73718**	**1.43600**	**1.43629**	**$\bar{1}$.99971**
4.01	55.147	.01813	27.564	27.583	.99934	.74152	.44035	.44063	.99971
4.02	55.701	.01795	27.842	27.860	.99936	.74586	.44469	.44497	.99972
4.03	56.261	.01777	28.122	28.139	.99937	.75021	.44904	.44931	.99973
4.04	56.826	.01760	28.404	28.422	.99938	.75455	.45339	.45365	.99973
4.05	57.397	.01742	28.690	28.707	.99939	.75889	.45773	.45799	.99974
4.06	57.974	.01725	28.979	28.996	.99941	.76324	.46208	.46233	.99974
4.07	58.557	.01708	29.270	29.287	.99942	.76758	.46642	.46668	.99975
4.08	59.145	.01691	29.564	29.581	.99943	.77192	.47077	.47102	.99975
4.09	59.740	.01674	29.862	29.878	.99944	.77626	.47511	.47536	.99976
4.10	**60.340**	**.01657**	**30.162**	**30.178**	**.99945**	**1.78061**	**1.47946**	**1.47970**	**$\bar{1}$.99976**
4.11	60.947	.01641	30.465	30.482	.99946	.78495	.48380	.48404	.99977
4.12	61.559	.01624	30.772	30.788	.99947	.78929	.48815	.48838	.99977
4.13	62.178	.01608	31.081	31.097	.99948	.79364	.49249	.49272	.99978
4.14	62.803	.01592	31.393	31.409	.99949	.79798	.49684	.49706	.99978
4.15	63.434	.01576	31.709	31.725	.99950	.80232	.50118	.50140	.99978
4.16	64.072	.01561	32.028	32.044	.99951	.80667	.50553	.50574	.99979
4.17	64.715	.01545	32.350	32.365	.99952	.81101	.50987	.51008	.99979
4.18	65.366	.01530	32.675	32.691	.99953	.81535	.51422	.51442	.99980
4.19	66.023	.01515	33.004	33.019	.99954	.81969	.51856	.51876	.99980
4.20	**66.686**	**.01500**	**33.336**	**33.351**	**.99955**	**1.82404**	**1.52291**	**1.52310**	**$\bar{1}$.99980**

x	Natural Values					Common Logarithms			
	e^x	e^{-x}	Sinh x	Cosh x	Tanh x	e^x	Sinh x	Cosh x	Tanh x
4.20	**66.686**	**.01500**	**33.336**	**33.351**	**.99955**	**1.82404**	**1.52291**	**1.52310**	**$\bar{1}$.99980**
4.21	67.357	.01485	33.671	33.686	.99956	.82838	.52725	.52745	.99981
4.22	68.033	.01470	34.009	34.024	.99957	.83272	.53160	.53179	.99981
4.23	68.717	.01455	34.351	34.366	.99958	.83707	.53594	.53613	.99982
4.24	69.408	.01441	34.697	34.711	.99958	.84141	.54029	.54047	.99982
4.25	70.105	.01426	35.046	35.060	.99959	.84575	.54463	.54481	.99982
4.26	70.810	.01412	35.398	35.412	.99960	.85009	.54898	.54915	.99983
4.27	71.522	.01398	35.754	35.768	.99961	.85444	.55332	.55349	.99983
4.28	72.240	.01384	36.113	36.127	.99962	.85878	.55767	.55783	.99983
4.29	72.966	.01370	36.476	36.490	.99962	.86312	.56201	.56217	.99984
4.30	**73.700**	**.01357**	**36.843**	**36.857**	**.99963**	**1.86747**	**1.56636**	**1.56652**	**$\bar{1}$.99984**
4.31	74.440	.01343	37.214	37.227	.99964	.87181	.57070	.57086	.99984
4.32	75.189	.01330	37.588	37.601	.99965	.87615	.57505	.57520	.99985
4.33	75.944	.01317	37.966	37.979	.99965	.88050	.57939	.57954	.99985
4.34	76.708	.01304	38.347	38.360	.99966	.88484	.58373	.58388	.99985
4.35	77.478	.01291	38.733	38.746	.99967	.88918	.58808	.58822	.99986
4.36	78.257	.01278	39.122	39.135	.99967	.89352	.59242	.59256	.99986
4.37	79.044	.01265	39.515	39.528	.99968	.89787	.59677	.59691	.99986
4.38	79.838	.01253	39.913	39.925	.99969	.90221	.60111	.60125	.99986
4.39	80.640	.01240	40.314	40.326	.99969	.90655	.60546	.60559	.99987
4.40	**81.451**	**.01228**	**40.719**	**40.732**	**.99970**	**1.91090**	**1.60980**	**1.60993**	**$\bar{1}$.99987**
4.41	82.269	.01216	41.129	41.141	.99970	.91524	.61414	.61427	.99987
4.42	83.096	.01203	41.542	41.554	.99971	.91958	.61849	.61861	.99987
4.43	83.931	.01191	41.960	41.972	.99972	.92392	.62283	.62296	.99988
4.44	84.775	.01180	42.382	42.393	.99972	.92827	.62718	.62730	.99988
4.45	85.627	.01168	42.808	42.819	.99973	.93261	.63152	.63164	.99988
4.46	86.488	.01156	43.238	43.250	.99973	.93695	.63587	.63598	.99988
4.47	87.357	.01145	43.673	43.684	.99974	.94130	.64021	.64032	.99989
4.48	88.235	.01133	44.112	44.123	.99974	.94564	.64455	.64467	.99989
4.49	89.121	.01122	44.555	44.566	.99975	.94998	.64890	.64901	.99989
4.50	**90.017**	**.01111**	**45.003**	**45.014**	**.99975**	**1.95433**	**1.65324**	**1.65335**	**$\bar{1}$.99989**
4.51	90.922	.01100	45.455	45.466	.99976	.95867	.65759	.65769	.99989
4.52	91.836	.01089	45.912	45.923	.99976	.96301	.66193	.66203	.99990
4.53	92.759	.01078	46.374	46.385	.99977	.96735	.66627	.66637	.99990
4.54	93.691	.01067	46.840	46.851	.99977	.97170	.67062	.67072	.99990
4.55	94.632	.01057	47.311	47.321	.99978	.97604	.67496	.67506	.99990
4.56	95.583	.01046	47.787	47.797	.99978	.98038	.67931	.67940	.99990
4.57	96.544	.01036	48.267	48.277	.99979	.98473	.68365	.68374	.99991
4.58	97.514	.01025	48.752	48.762	.99979	.98907	.68799	.68808	.99991
4.59	98.494	.01015	49.242	49.252	.99979	.99341	.69234	.69243	.99991
4.60	**99.484**	**.01005**	**49.737**	**49.747**	**.99980**	**1.99775**	**1.69668**	**1.69677**	**$\bar{1}$.99991**
4.61	100.48	.00995	50.237	50.247	.99980	2.00210	.70102	.70111	.99991
4.62	101.49	.00985	50.742	50.752	.99981	.00644	.70537	.70545	.99992
4.63	102.51	.00975	51.252	51.262	.99981	.01078	.70971	.70979	.99992
4.64	103.54	.00966	51.767	51.777	.99981	.01513	.71406	.71414	.99992
4.65	104.58	.00956	52.288	52.297	.99982	.01947	.71840	.71848	.99992
4.66	105.64	.00947	52.813	52.823	.99982	.02381	.72274	.72282	.99992
4.67	106.70	.00937	53.344	53.354	.99982	.02816	.72709	.72716	.99992
4.68	107.77	.00928	53.880	53.890	.99983	.03250	.73143	.73151	.99993
4.69	108.85	.00919	54.422	54.431	.99983	.03684	.73577	.73585	.99993
4.70	**109.95**	**.00910**	**54.969**	**54.978**	**.99983**	**2.04118**	**1.74012**	**1.74019**	**$\bar{1}$.99993**
4.71	111.05	.00900	55.522	55.531	.99984	.04553	.74446	.74453	.99993
4.72	112.17	.00892	56.080	56.089	.99984	.04987	.74881	.74887	.99993
4.73	113.30	.00883	56.643	56.652	.99984	.05421	.75315	.75322	.99993
4.74	114.43	.00874	57.213	57.222	.99985	.05856	.75749	.75756	.99993
4.75	115.58	.00865	57.788	57.796	.99985	.06290	.76184	.76190	.99993
4.76	116.75	.00857	58.369	58.377	.99985	.06724	.76618	.76624	.99994
4.77	117.92	.00848	58.955	58.964	.99986	.07158	.77052	.77059	.99994
4.78	119.10	.00840	59.548	59.556	.99986	.07593	.77487	.77493	.99994
4.79	120.30	.00831	60.147	60.155	.99986	.08027	.77921	.77927	.99994
4.80	**121.51**	**.00823**	**60.751**	**60.759**	**.99986**	**2.08461**	**1.78355**	**1.78361**	**$\bar{1}$.99994**

x	Natural Values					Common Logarithms			
	e^x	e^{-x}	Sinh x	Cosh x	Tanh x	e^x	Sinh x	Cosh x	Tanh x
4.80	**121.51**	**.00823**	**60.751**	**60.760**	**.99986**	**2.08461**	**1.78355**	**1.78361**	**$\bar{1}$.99994**
4.81	122.73	.00815	61.362	61.370	.99987	.08896	.78790	.78796	.99994
4.82	123.97	.00807	61.979	61.987	.99987	.09330	.79224	.79230	.99994
4.83	125.21	.00799	62.601	62.609	.99987	.09764	.79658	.79664	.99994
4.84	126.47	.00791	63.231	63.239	.99987	.10199	.80093	.80098	.99995
4.85	127.74	.00783	63.866	63.874	.99988	.10633	.80527	.80532	.99995
4.86	129.02	.00775	64.508	64.516	.99988	.11067	.80962	.80967	.99995
4.87	130.32	.00767	65.157	65.164	.99988	.11501	.81396	.81401	.99995
4.88	131.63	.00760	65.812	65.819	.99988	.11936	.81830	.81835	.99995
4.89	132.95	.00752	66.473	66.481	.99989	.12370	.82265	.82269	.99995
4.90	**134.29**	**.00745**	**67.141**	**67.149**	**.99989**	**2.12804**	**1.82699**	**1.82704**	**$\bar{1}$.99995**
4.91	135.64	.00737	67.816	67.823	.99989	.13239	.83133	.83138	.99995
4.92	137.00	.00730	68.498	68.505	.99989	.13673	.83568	.83572	.99995
4.93	138.38	.00723	69.186	69.193	.99990	.14107	.84002	.84006	.99995
4.94	139.77	.00715	69.882	69.889	.99990	.14541	.84436	.84441	.99996
4.95	141.17	.00708	70.584	70.591	.99990	.14976	.84871	.84875	.99996
4.96	142.59	.00701	71.293	71.300	.99990	.15410	.85305	.85309	.99996
4.97	144.03	.00694	72.010	72.017	.99990	.15844	.85739	.85743	.99996
4.98	145.47	.00687	72.734	72.741	.99991	.16279	.86174	.86178	.99996
4.99	146.94	.00681	73.465	73.472	.99991	.16713	.86608	.86612	.99996
5.00	**148.41**	**.00674**	**74.203**	**74.210**	**.99991**	**2.17147**	**1.87042**	**1.87046**	**$\bar{1}$.99996**
5.01	149.90	.00667	74.949	74.956	.99991	.17582	.87477	.87480	.99996
5.02	151.41	.00660	75.702	75.710	.99991	.18016	.87911	.87915	.99996
5.03	152.93	.00654	76.463	76.470	.99991	.18450	.88345	.88349	.99996
5.04	154.47	.00647	77.232	77.238	.99992	.18884	.88780	.88783	.99996
5.05	156.02	.00641	78.008	78.014	.99992	.19319	.89214	.89217	.99996
5.06	157.59	.00635	78.792	78.798	.99992	.19753	.89648	.89652	.99997
5.07	159.17	.00628	79.584	79.590	.99992	.20187	.90083	.90086	.99997
5.08	160.77	.00622	80.384	80.390	.99992	.20622	.90517	.90520	.99997
5.09	162.39	.00616	81.192	81.198	.99992	.21056	.90951	.90955	.99997
5.10	**164.02**	**.00610**	**82.008**	**82.014**	**.99993**	**2.21490**	**1.91386**	**1.91389**	**$\bar{1}$.99997**
5.11	165.67	.00604	82.832	82.838	.99993	.21924	.91820	.91823	.99997
5.12	167.34	.00598	83.665	83.671	.99993	.22359	.92254	.92257	.99997
5.13	169.02	.00592	84.506	84.512	.99993	.22793	.92689	.92692	.99997
5.14	170.72	.00586	85.355	85.361	.99993	.23227	.93123	.93126	.99997
5.15	172.43	.00580	86.213	86.219	.99993	.23662	.93557	.93560	.99997
5.16	174.16	.00574	87.079	87.085	.99993	.24096	.93992	.93994	.99997
5.17	175.91	.00568	87.955	87.960	.99994	.24530	.94426	.94429	.99997
5.18	177.68	.00563	88.839	88.844	.99994	.24965	.94860	.94863	.99997
5.19	179.47	.00557	89.732	89.737	.99994	.25399	.95294	.95297	.99997
5.20	**181.27**	**.00552**	**90.633**	**90.639**	**.99994**	**2.25833**	**1.95729**	**1.95731**	**$\bar{1}$.99997**
5.21	183.09	.00546	91.544	91.550	.99994	.26267	.96163	.96166	.99997
5.22	184.93	.00541	92.464	92.470	.99994	.26702	.96597	.96600	.99997
5.23	186.79	.00535	93.394	93.399	.99994	.27136	.97032	.97034	.99998
5.24	188.67	.00530	94.332	94.338	.99994	.27570	.97466	.97469	.99998
5.25	190.57	.00525	95.281	95.286	.99994	.28005	.97900	.97903	.99998
5.26	192.48	.00520	96.238	96.243	.99995	.28439	.98335	.98337	.99998
5.27	194.42	.00514	97.205	97.211	.99995	.28873	.98769	.98771	.99998
5.28	196.37	.00509	98.182	98.188	.99995	.29307	.99203	.99206	.99998
5.29	198.34	.00504	99.169	99.174	.99995	.29742	.99638	.99640	.99998
5.30	**200.34**	**.00499**	**100.17**	**100.17**	**.99995**	**2.30176**	**2.00072**	**2.00074**	**$\bar{1}$.99998**
5.31	202.35	.00494	101.17	101.18	.99995	.30610	.00506	.00508	.99998
5.32	204.38	.00489	102.19	102.19	.99995	.31045	.00941	.00943	.99998
5.33	206.44	.00484	103.22	103.22	.99995	.31479	.01375	.01377	.99998
5.34	208.51	.00480	104.25	104.26	.99995	.31913	.01809	.01811	.99998
5.35	210.61	.00475	105.30	105.31	.99995	.32348	.02244	.02246	.99998
5.36	212.72	.00470	106.36	106.36	.99996	.32782	.02678	.02680	.99998
5.37	214.86	.00465	107.43	107.43	.99996	.33216	.03112	.03114	.99998
5.38	217.02	.00461	108.51	108.51	.99996	.33650	.03547	.03548	.99998
5.39	219.20	.00456	109.60	109.60	.99996	.34085	.03981	.03983	.99998
5.40	**221.41**	**.00452**	**110.70**	**110.71**	**.99996**	**2.34519**	**2.04415**	**2.04417**	**$\bar{1}$.99998**

x	Natural Values					Common Logarithms			
	e^x	e^{-x}	Sinh x	Cosh x	Tanh x	e^x	Sinh x	Cosh x	Tanh x
5.40	**221.41**	**.00452**	**110.70**	**110.71**	**.99996**	**2.34519**	**2.04415**	**2.04417**	**$\bar{1}$.99998**
5.41	223.63	.00447	111.81	111.82	.99996	.34953	.04849	.04851	.99998
5.42	225.88	.00443	112.94	112.94	.99996	.35388	.05284	.05285	.99998
5.43	228.15	.00438	114.07	114.08	.99996	.35822	.05718	.05720	.99998
5.44	230.44	.00434	115.22	115.22	.99996	.36256	.06152	.06154	.99998
5.45	232.76	.00430	116.38	116.38	.99996	.36690	.06587	.06588	.99998
5.46	235.10	.00425	117.55	117.55	.99996	.37125	.07021	.07023	.99998
5.47	237.46	.00421	118.73	118.73	.99996	.37559	.07455	.07457	.99998
5.48	239.85	.00417	119.92	119.93	.99997	.37993	.07890	.07891	.99998
5.49	242.26	.00413	121.13	121.13	.99997	.38428	.08324	.08325	.99999
5.50	**244.69**	**.00409**	**122.34**	**122.35**	**.99997**	**2.38862**	**2.08758**	**2.08760**	**$\bar{1}$.99999**
5.51	247.15	.00405	123.57	123.58	.99997	.39296	.09193	.09194	.99999
5.52	249.64	.00401	124.82	124.82	.99997	.39731	.09627	.09628	.99999
5.53	252.14	.00397	126.07	126.07	.99997	.40165	.10061	.10063	.99999
5.54	254.68	.00393	127.34	127.34	.99997	.40599	.10495	.10497	.99999
5.55	257.24	.00389	128.62	128.62	.99997	.41033	.10930	.10931	.99999
5.56	259.82	.00385	129.91	129.91	.99997	.41468	.11364	.11365	.99999
5.57	262.43	.00381	131.22	131.22	.99997	.41902	.11798	.11800	.99999
5.58	265.07	.00377	132.53	132.54	.99997	.42336	.12233	.12234	.99999
5.59	267.74	.00374	133.87	133.87	.99997	.42771	.12667	.12668	.99999
5.60	**270.43**	**.00370**	**135.21**	**135.22**	**.99997**	**2.43205**	**2.13101**	**2.13103**	**$\bar{1}$.99999**
5.61	273.14	.00366	136.57	136.57	.99997	.43639	.13536	.13537	.99999
5.62	275.89	.00362	137.94	137.95	.99997	.44074	.13970	.13971	.99999
5.63	278.66	.00359	139.33	139.33	.99997	.44508	.14404	.14405	.99999
5.64	281.46	.00355	140.73	140.73	.99997	.44942	.14839	.14840	.99999
5.65	284.29	.00352	142.14	142.15	.99998	.45376	.15273	.15274	.99999
5.66	287.15	.00348	143.57	143.58	.99998	.45811	.15707	.15708	.99999
5.67	290.03	.00345	145.02	145.02	.99998	.46245	.16141	.16142	.99999
5.68	292.95	.00341	146.47	146.48	.99998	.46679	.16576	.16577	.99999
5.69	295.89	.00338	147.95	147.95	.99998	.47114	.17010	.17011	.99999
5.70	**298.87**	**.00335**	**149.43**	**149.44**	**.99998**	**2.47548**	**2.17444**	**2.17445**	**$\bar{1}$.99999**
5.71	301.87	.00331	150.93	150.94	.99998	.47982	.17879	.17880	.99999
5.72	304.90	.00328	152.45	152.45	.99998	.48416	.18313	.18314	.99999
5.73	307.97	.00325	153.98	153.99	.99998	.48851	.18747	.18748	.99999
5.74	311.06	.00321	155.53	155.53	.99998	.49285	.19182	.19182	.99999
5.75	314.19	.00318	157.09	157.10	.99998	.49719	.19616	.19617	.99999
5.76	317.35	.00315	158.67	158.68	.99998	.50154	.20050	.20051	.99999
5.77	320.54	.00312	160.27	160.27	.99998	.50588	.20484	.20485	.99999
5.78	323.76	.00309	161.88	161.88	.99998	.51022	.20919	.20920	.99999
5.79	327.01	.00306	163.51	163.51	.99998	.51457	.21353	.21354	.99999
5.80	**330.30**	**.00303**	**165.15**	**165.15**	**.99998**	**2.51891**	**2.21787**	**2.21788**	**$\bar{1}$.99999**
5.81	333.62	.00300	166.81	166.81	.99998	.52325	.22222	.22222	.99999
5.82	336.97	.00297	168.48	168.49	.99998	.52759	.22656	.22657	.99999
5.83	340.36	.00294	170.18	170.18	.99998	.53194	.23090	.23091	.99999
5.84	343.78	.00291	171.89	171.89	.99998	.53628	.23525	.23525	.99999
5.85	347.23	.00288	173.62	173.62	.99998	.54062	.23959	.23960	.99999
5.86	350.72	.00285	175.36	175.36	.99998	.54497	.24393	.24394	.99999
5.87	354.25	.00282	177.12	177.13	.99998	.54931	.24828	.24828	.99999
5.88	357.81	.00279	178.90	178.91	.99998	.55365	.25262	.25262	.99999
5.89	361.41	.00277	180.70	180.70	.99998	.55799	.25696	.25697	.99999
5.90	**365.04**	**.00274**	**182.52**	**182.52**	**.99998**	**2.56234**	**2.26130**	**2.26131**	**$\bar{1}$.99999**
5.91	368.71	.00271	184.35	184.35	.99999	.56668	.26565	.26565	.99999
5.92	372.41	.00269	186.20	186.21	.99999	.57102	.26999	.27000	.99999
5.93	376.15	.00266	188.08	188.08	.99999	.57537	.27433	.27434	.99999
5.94	379.93	.00263	189.97	189.97	.99999	.57971	.27868	.27868	.99999
5.95	383.75	.00261	191.88	191.88	.99999	.58405	.28302	.28303	.99999
5.96	387.61	.00258	193.80	193.81	.99999	.58840	.28736	.28737	.99999
5.97	391.51	.00255	195.75	195.75	.99999	.59274	.29171	.29171	.99999
5.98	395.44	.00253	197.72	197.72	.99999	.59708	.29605	.29605	.99999
5.99	399.41	.00250	199.71	199.71	.99999	.60142	.30039	.30040	.99999
6.00	**403.43**	**.00248**	**201.71**	**201.72**	**.99999**	**2.60577**	**2.30473**	**2.30474**	**.99999**

UNITS AND CONVERSION FACTORS

By J. G. Brainerd and Carl C. Chambers

14. SYSTEMS OF UNITS

The magnitude of a physical quantity has no tangible meaning except as the relative magnitude of that quantity as compared with some other quantity of the same nature. Thus, 50 ohms is a resistance having a magnitude 50 times the resistance of 1 ohm. Therefore, whenever it is necessary or desirable to talk about the magnitude of a physical quantity, it is necessary to have a basis for comparison. This basis for a quantity is called the *unit* of magnitude of that quantity. In order to communicate the idea of magnitude between different people, it is necessary that they at least know the relative magnitudes of their units. It is the purpose of this section to act as tool for the specification of the relative magnitudes of the more commonly used systems of units for physical quantities.

Because of the relations defining physical laws, there are relations between the magnitudes of physical quantities. It is desirable that these physical relations be expressed alike in the different systems of units. For instance, the relation mass × acceleration = force should be independent of the system of units. Therefore, unit mass times unit acceleration should equal unit force. This gives a relation among these three units.

Because of such physical relations, all the mechanical units can be derived from the units for three fundamental quantities. The three quantities ordinarily taken as fundamental are mass, length, and time. Thermal quantities are conveniently derived from these three quantities together with another fundamental quantity, temperature. Photometric quantities are derived from the three fundamental mechanical quantities together with luminous intensity as a fourth fundamental quantity.

Similarly, electrical and magnetic quantities are derived from the three fundamental mechanical quantities and one fundamental electrical or magnetic quantity.

Two systems of mechanical units are in use in English-speaking countries, the English and the metric systems. The metric system is used universally by physicists and to a great extent by engineers, although the English system is still very common in engineering. The English system uses the foot, the pound, and the second as the units for length, mass, and time, respectively. The metric system (as used in the current literature—see MKS system below) employs the meter, the kilogram, and the second as the units for length, mass, and time, respectively.

STANDARDS OF THE FUNDAMENTAL UNITS. The physical units upon which these fundamental units are based and the legalized standards of the United States and Great Britain are described below.

Standard of Length. The standard meter (100 cm) is the distance between two lines on a platinum-iridium bar carefully preserved at the Bureau of Weights and Measures, at Sèvres, France, when the bar is kept at a uniform temperature of 0 deg cent throughout. In the United States the yard (3 ft) was defined by Act of Congress, July 28, 1866, as

$$1 \text{ U. S. yard} = \frac{3600}{3937} \text{ meter}$$

and similarly the British imperial yard is defined by law as

$$1 \text{ British imperial yard} = \frac{3600}{3937.079} \text{ meter}$$

For engineering purposes the U. S. and British yards may be considered identical.

Standard of Mass and Force. The standard kilogram (1000 grams) is the mass of a cylinder of platinum preserved at the Bureau of Weights and Measures, at Sèvres, France. The U. S. pound avoirdupois is defined by law (Act of Congress, 1866) as $\frac{1}{2.2046}$ kg, but in 1893, the Superintendent of Weights and Measures, with the approval of the Secretary of the Treasury, declared the pound to be

$$1 \text{ U. S. lb} = \frac{1}{2.204622} \text{ kg}$$

The British imperial pound has the same value.

The same relations between the pound and kilogram hold whether these units be taken as units of mass or as units of force, the unit of force being defined in both cases as the pull of the earth on unit mass at 45 deg latitude and sea level.

Standard of Time. The standard second universally adopted is the $\frac{1}{86,400}$ part of a mean solar day. The solar day is the interval of time between two successive transits of the sun across a meridian of the earth at the point of observation; this interval varies in length at different times during the year, but the average length of the interval for one year is constant as far as can be determined by any known methods of observation.

Temperature Scales. Two units of temperature, or temperature scales, are commonly employed, viz., the centigrade and the fahrenheit units. The relation between these two units results solely from the manner in which they are defined. One degree centigrade = $^9/_5$ degree fahrenheit. Owing to the difference in the zeros of the two scales, a temperature of t_f degrees fahrenheit corresponds to a temperature of $t_c = {}^5/_9(t_f - 32)$ degrees centigrade, and vice versa, $t_f = {}^9/_5 t_c + 32$ degrees fahrenheit.

Standard of Luminous Intensity. Before Jan. 1, 1948, the standard of luminous intensity was the mean intensity in the horizontal plane from a group of incandescent lamps maintained by the National Bureau of Standards (U. S.), in cooperation with similar custodians in France, Great Britain, and Germany. The International candle was a point source of light having an intensity of a definite fraction of this standard intensity.

The National Bureau of Standards, in pursuance of decisions of the International Committee on Weights and Measures, decided that, beginning Jan. 1, 1948, it would take as the primary standard for the system of photometric units a black-body radiator operated at the temperature of freezing platinum. The "candle," unit of intensity, is defined as one-sixtieth of the intensity of one square centimeter of such a radiator. Other units are derived from the candle, with the provision that when differences of color are involved the evaluation shall be made by means of standard spectral luminosity factors which have been adopted by the International Commission on Illumination and the International Committee on Weights and Measures.

ELECTRIC UNITS. Three systems of electric and magnetic units are in general use, viz., (1) the cgs electrostatic system, (2) the cgs electromagnetic system, and (3) the practical system. In the cgs electrostatic system the dielectric coefficient, ϵ, of air * at 0 deg cent and 760 mm mercury pressure is arbitrarily chosen as unity. In the cgs electromagnetic system the magnetic permeability of air under the same standard conditions is arbitrarily chosen as unity. In the practical system a concrete standard of the unit of resistance (called the ohm) and of the unit of current (ampere) is arbitrarily chosen (it was stated above that only one electric or magnetic unit need be chosen; the choice of two leads to inconsistencies; see below); the unit of resistance is closely equal to 10^9 times the unit of resistance in the cgs electromagnetic system and the unit current is approximately 0.1 that in the latter system. Occasionally other (special) systems are used, most of which are designed to get rid of a factor 4π which frequently appears in the usual systems. The most popular of these others is the Heaviside–Lorentz system in which the unit of electric charge is $1/\sqrt{4\pi}$ of the unit in the electrostatic system. (See MKS system.)

Use of the Prefixes "Stat" and "Ab." To designate the electric and magnetic units in the electrostatic and electromagnetic systems of units respectively, the prefixes "stat" and "ab" may be used with the name of the corresponding practical unit. For example, the cgs electrostatic unit of electric charge may be called the statcoulomb and the cgs electromagnetic unit of electric charge may be called the abcoulomb, etc.†

Relations among the Three Systems of Electrical Units. The fundamental relations, experimentally determined, between the cgs electrostatic and the cgs electromagnetic system is that 1 abfarad = 8.9878×10^{20} statfarads, which may be approximated for engineering purposes to

$$1 \text{ abfarad} = 9 \times 10^{20} \text{ statfarads}$$

which, as a consequence of the definition of the various terms, is equivalent to

$$1 \text{ abcoulomb} = 3 \times 10^{10} \text{ statcoulombs}$$

the erg being the unit of energy in both systems. Rigorously,

$$1 \text{ abcoulomb} = 2.9979 \times 10^{10} \text{ statcoulombs}$$

(See the article by Birge, *Rev. of Mod. Phys.*, Vol. 1, 1 [July 1929].)

---

* Rigorously, ϵ_0 of free or empty space is chosen unity; for air at 0 deg cent and 76 cm mercury pressure $\epsilon_0 = 1.000585$; see *International Critical Tables*, Vol. 6, 77, for the value of ϵ for air under various conditions.

† This abcoulomb, the unit of quantity of electricity in the electromagnetic system, should not be confused with an "absolute coulomb," which is a unit closely equal to the coulomb and is what the latter would be if 1 international or practical ohm equaled *exactly* 10^9 abohms and 1 ampere equaled *exactly* 0.1 abamp. For engineering purposes, the difference between an absolute coulomb and a coulomb is negligible.

The fundamental relations between the cgs electromagnetic system and the practical system are

$$1 \text{ abcoulomb} = 10 \text{ coulombs}$$

$$1 \text{ erg} = 10^{-7} \text{ watt-second or joule}$$

the erg being the unit of energy in the cgs electromagnetic system and the joule (or watt-second) that in the practical system.

Practical Electrical Units. The former (see below) legal units of electrical measure in the United States are given in an Act of Congress, July 12, 1894. Unfortunately, the units there defined are not consistent with one another; for example, the unit of power (watt) there given is not equal to the unit of power derived from the units of current (ampere) and voltage (volt) as defined in the Act. The practical units (the so-called international units) in use before Jan. 1, 1948, are based on the following two definitions:

The unit of resistance is the (international) ohm and is equal to the resistance offered to an unvarying electric current by a column of mercury at the temperature of melting ice, 14.4521 grams in mass, of a constant cross-sectional area and 106.300 cm in length.

The unit of current is the (international) ampere and is equal to the unvarying electric current which, when passed through a solution of nitrate of silver in accordance with certain specifications, deposits silver at the rate of 0.00111800 gram per second.

The unit of electromotive force, the (international) volt, is derived from the above by Ohm's law. Other international units are derived from these.

The National Bureau of Standards, in agreement with decisions of the International Committee on Weights and Measures, decided to use as standard, beginning Jan. 1, 1948, the electrical units "derived from the fundamental mechanical units of length, mass, and time by use of accepted principles of electromagnetism, with the value of the permeability of space taken as unity in the centimeter-gram-second system or as 10^{-7} in the corresponding meter-kilogram-second system." The reference is to the unrationalized MKS system; in the rationalized MKS system, the permeability of space is $4\pi \times 10^{-7}$ henry per meter.

In explanation of the legal status of the new standard, the Bureau states, "When the electrical units were defined by law (Public Law No. 105, 53rd Congress) in 1894 it was supposed that the international units were practically identical with the corresponding multiples of the centimeter-gram-second electromagnetic system. Alternative definitions were given for most of the units, and those definitions which appear to be legally controlling were taken partly from one system and partly from the other. The joule and the watt, for example, are clearly defined as multiples of the cgs units. In brief, the absolute units have as good a legal basis under the terms of that act as do the present international units. New legislation is being proposed to remove the ambiguities of the old act, but there should be no objection on legal grounds to the general adoption of the absolute units even in advance of Congressional action."

Using "international" to refer to the previous standard, and "absolute" to refer to the new, the relations accepted by the International Committee on Weights and Measures at its meeting in Paris in October, 1946, are as follows:

$$1 \text{ mean international ohm} = 1.00049 \text{ absolute ohms}$$

$$1 \text{ mean international volt} = 1.00034 \text{ absolute volts}$$

The mean international units to which the above equations refer are the averages of units as maintained in the national laboratories of the six countries (France, Germany, Great Britain, Japan, U.S.S.R., and the United States) which took part in this work before the war. The units maintained by the National Bureau of Standards differ from these average units by a few parts in a million, so that the conversion factors for adjusting values of standards in this country will be as follows:

$$1 \text{ international ohm (U. S.)} = 1.000495 \text{ absolute ohms}$$

$$1 \text{ international volt (U. S.)} = 1.00033 \text{ absolute volts.}$$

Other electrical units will be changed by amounts shown in the following table:

$$1 \text{ international ampere} = 0.999835 \text{ absolute ampere}$$

$$1 \text{ international coulomb} = 0.999835 \text{ absolute coulomb}$$

$$1 \text{ international henry} = 1.000495 \text{ absolute henrys}$$

$$1 \text{ international farad} = 0.999505 \text{ absolute farad}$$

$$1 \text{ international watt} = 1.000165 \text{ absolute watts}$$

$$1 \text{ international joule} = 1.000165 \text{ absolute joules}$$

The Act of 1894 defined the international ohm as previously stated, but defined the ampere as 0.1 abampere. These units give rise to the so-called "semiabsolute" system, which is seldom used.

THE MKS SYSTEM OF UNITS. In 1904 Giorgi proposed a system of units in which the fundamental units were the meter, the kilogram, the second, and the ohm. Using this system of fundamental units, the permeability of free space is $\mu_0 = 4\pi \times 10^{-7}$ henry per meter, and the equations of electricity and magnetism, using the practical units, become equations without factors such as 10^8, etc. Such a system is similar to the so-called absolute systems such as the cgs electromagnetic and the cgs electrostatic systems. It follows from the theory of radiation of electromagnetic waves that the dielectric coefficient $\epsilon_0 = \dfrac{1}{\mu_0 c^2}$, where c is the ratio of electromagnetic to electrostatic units, which can be taken as the velocity of light in free space.

The International Committee of Weights and Measures, at its meeting in October 1946, decided that the actual substitution of this absolute system of electrical units for the international system should take place on January 1, 1948.

The units are then defined by a set of definitions such as follows:

(a) Ampere. The ampere is the constant current which, maintained in two parallel rectilinear conductors of infinite length separated by a distance of 1 meter, produces between these conductors a force equal to 2×10^{-7} mks (meter-kilogram-second) units of force per meter of length.

(b) Volt. The volt is the difference of electrical potential between two points of a conductor carrying a constant current of 1 ampere when the power dissipated between these points is equal to 1 mks unit of power (watt).

(c) Coulomb. The coulomb is the quantity of electricity transported each second by a current of 1 ampere.

(d) Ohm. The ohm is the electrical resistance between two points of a conductor when a constant difference of potential of 1 volt, applied between these points, produces in the conductor a current of 1 ampere, the conductor not being the seat of an electromotive force.

(e) Weber. The weber is the magnetic flux which, traversing a circuit of a single turn, would produce an electromotive force of 1 volt, if brought to zero in 1 second with uniform diminution.

(f) Henry. The henry is the inductance of a closed circuit in which an electromotive force of 1 volt is produced when the electric current traversing the circuit varies uniformly at the rate of 1 ampere per second.

(g) Farad. The farad is the electrical capacitance of a capacitor between the plates of which appears an electrical difference of potential of 1 volt, when charged with 1 coulomb of electric charge.

The original Giorgi MKS system chose the ohm as the fourth fundamental unit. This choice has not been confirmed. The electrical fundamental unit could be almost any of the electrical units. No particular unit has as yet been chosen as fundamental. The preferences seem to be divided between the ampere, the ohm, the permeability, and the coulomb.

The original Giorgi MKS system chose $\mu_0 = 4\pi 10^{-7}$ henry per meter, the 4π factor causing the electromagnetic formulas expressing rectilinear symmetry, such as the Maxwell equations, to be free of the factor 4π, and the electromagnetic formulas expressing circular symmetry, such as Coulomb's law, to contain the factor 4π. Such a system is called a rationalized system as contrasted with a non-rationalized system, examples of which are the electromagnetic and the electrostatic cgs systems. The non-rationalized MKS system corresponding to the original Giorgi system is defined by the choice of $\mu_0 = 10^{-7}$. This changes the values of some of the units as shown in the table below.

Rationalized MKS Units and Corresponding CGS Electromagnetic Units

Multiply mks units by F to obtain cgs units

Quantity	*Symbol*	MKS *Unit*	CGS *Unit*	F
Mechanical				
Length	L	m	cm	10^2
Mass	M	kg	g	10^3
Time	T	sec	sec	1
Area	S	sq m	sq cm	10^4
Volume	V	cu m (stere)	cu cm	10^6
Frequency	f	cycle per sec (hertz)	cycle per sec	1
Density	d	kg per cu m	g per cu cm	10^{-3}
Velocity	v	m per sec	cm per sec	10^2
Acceleration	a	m per sec per sec	cm per sec per sec	10^2
Force	F	newton (j per m)	dyne	10^5
Pressure	p	newton per sq m	dyne per sq cm	10
Angle	α, β	radian	radian	1
Angular velocity	ω	radian per sec	radian per sec	1
Torque	τ	j per radian	dyne cm	10^7
Moment of inertia	J	kg-sq m	g-sq cm	10^7
Energetics				
Work or energy	W	j	erg	10^7
Volume energy or energy density	w	j per cu m	erg per cu cm	10
Active power	P	w	erg per sec	10^7
Reactive power	Q	var	erg per sec	10^7
Thermal				
Quantity of heat	Q	kg cal	g cal	10^3
Temperature	θ	C or K	C or K	1
Luminous				
Intensity	I	candle	candle	1
Luminous flux	ψ	l	l	1
Illumination	E	lux	phot	10^{-4}
Brightness	b	candle per sq m	stilb	10^{-4}
Electrical				
Electromotive force	E	volt	abvolt	10^8
Potential gradient or electric field intensity	E	volt per m	abvolt per cm	10^6
Resistance	R	ohm	abohm	10^9
Resistivity	ρ	ohm-m	abohm-cm	10^{11}
Conductance	G	siemens, mho	abmho	10^{-9}
Conductivity	γ	mho per m	abmho per cm	10^{-11}
Quantity or displacement	Q	coulomb	abcoulomb	10^{-1}
Current	I	amp	abamp	10^{-1}
Electric flux	Ψ	coulomb	abcoulomb	10^{-1}
Flux density	D	coulomb per sq m	abcoulomb per sq cm	10^{-5}
Current density	i	ampere per sq m	abampere per sq cm	10^{-5}
Capacitance	C	farad	abfarad	10^{-9}
Specific inductive capacity	ϵ/ϵ_0	numeric	numeric	1
Dielectric coefficient for free space or space capacitivity	ϵ_0	$10^7/4\pi c^2 = 8.854 \times 10^{-12}$	$\frac{1}{c^2} = 1.113 \times 10^{-21}$	
Magnetic				
Magnetomotive force	$\mathfrak{F}$	amp-turn	gilbert	$4\pi 10^{-1}$
Magnetizing force or magnetic field intensity	H	amp-turn per m	oersted	$4\pi 10^{-3}$
Space permeability	μ_0	$4\pi 10^{-7} = 1.257 \times 10^{-6}$	1	
Relative permeability	μ/μ_0	numeric	numeric	1
Magnetic flux	Φ	weber	maxwell	10^8
Flux density	B	weber per sq m	gauss	10^4
Reluctance	$\mathcal{R}$	amp-turn per weber		$4\pi 10^{-9}$
Permeance	$\mathcal{P}$	weber per amp-turn		$10^9/4\pi$
Inductance	L	henry	abhenry	10^9
Pole strength	m	weber	maxwell/4π	$10^8/4\pi$
Magnetization	$\mathfrak{J}$	weber per sq m		$10^4/4\pi$
Magnetic moment	$\mathfrak{M}$	weber-m	maxwell-cm/4π	$10^{10}/4\pi$

Rationalized MKS Units and Corresponding Non-rationalized Units

Multiply non-rationalized mks units by F to obtain rationalized mks units

Quantity	*Symbol*	*Name of Rationalized* MKS *Units*	F
Electrical			
Electric flux	Ψ	coulomb	4π
Flux density	D	coulomb per sq m	4π
Space capacitivity	ϵ_0	farad per m	4π
Magnetic			
Magnetomotive force	M or $\mathfrak{F}$	amp-turn	$1/4\pi$
Magnetizing force	H	amp-turn per m	$1/4\pi$
Space permeability	μ_0	henry per m	4π
Permeance	P	weber per amp-turn	4π
Reluctance	R	amp-turn per weber	$1/4\pi$
Pole strength	m	weber	4π
Magnetic moment	$\mathfrak{M}$	weber-m	4π
Flux density	B	weber per sq m	4π

15. CONVERSION TABLES

Table 1. Length [L]

Multiply Number of → by to Obtain ↓	Centimeters	Feet	Inches	Kilometers	Nautical miles	Meters	Mils	Miles	Millimeters	Yards
Centimeters	1	30.48	2.540	10^5	1.853×10^5	100	2.540×10^{-3}	1.609×10^5	0.1	91.44
Feet	3.281×10^{-2}	1	8.333×10^{-2}	3281	6080.27	3.281	8.333×10^{-5}	5280	3.281×10^{-3}	3
Inches	0.3937	12	1	3.937×10^4	7.296×10^4	39.37	0.001	6.336×10^4	3.937×10^{-2}	36
Kilometers	10^{-5}	3.048×10^{-4}	2.540×10^{-5}	1	1.853	0.001	2.540×10^{-8}	1.609	10^{-6}	9.144×10^{-4}
Nautical miles		1.645×10^{-4}		0.5396	1	5.396×10^{-4}		0.8684		4.934×10^{-4}
Meters	0.01	0.3048	2.540×10^{-2}	1000	1853	1		1609	0.001	0.9144
Mils	393.7	1.2×10^4	1000	3.937×10^7		3.937×10^4	1		39.37	3.6×10^4
Miles	6.214×10^{-6}	1.894×10^{-4}	1.578×10^{-5}	0.6214	1.1516	6.214×10^{-4}		1	6.214×10^{-7}	5.682×10^{-4}
Millimeters	10	304.8	25.40	10^6		1000	2.540×10^{-2}		1	914.4
Yards	1.094×10^{-2}	0.3333	2.778×10^{-2}	1094	2027	1.094	2.778×10^{-5}	1760	1.094×10^{-3}	1

Metric Multiples

10^6 microns = 10^3 millimeters = 10^2 centimeters = 10 decimeters = 1 meter = 10^{-1} dekameter = 10^{-2} hectometer = 10^{-3} kilometer = 10^{-4} myriameter = 10^{-6} megameter = 10^{10} Angstrom Units.

Land Measure

7.92 inches = 1 link
25 links = 1 rod = 16.5 feet = 5.5 yards (1 rod = 1 pole = 1 perch)
4 rods = 1 chain (Gunther's) = 66 feet = 22 yards = 100 links
10 chains = 1 furlong = 660 feet = 220 yards = 1000 links = 40 rods
8 furlongs = 1 mile = 5280 feet = 1760 yards = 8000 links = 320 rods = 80 chains

Ropes and Cables

2 yards = 1 fathom
120 fathoms = 1 cable's length

Nautical Measure

6080.27 feet = 1 nautical mile = 1.15156 statute miles
3 nautical miles = 1 league (U. S.)
3 statute miles = 1 league (Gr. Britain)

(Note.—A nautical mile is the length of a minute of longitude of the earth at the equator at sea level. The British Admiralty uses the round figure of 6080 feet. The word " knot " is used to denote " nautical miles per hour.")

Miscellaneous

3 inches = 1 palm
4 inches = 1 hand
9 inches = 1 span
2 1/2 feet = 1 military pace

Table 2. Area $[L^2]$

Multiply Number of → by → to Obtain ↓	Acres	Circular mils	Square centimeters	Square feet	Square inches	Square kilometers	Square meters	Square miles	Square millimeters	Square yards
Acres	1			2.296×10^{-5}		247.1	2.471×10^{-4}	640		2.066×10^{-4}
Circular mils		1	1.973×10^{5}	1.833×10^{8}	1.273×10^{6}		1.973×10^{9}		1973	
Square centimeters		5.067×10^{-6}	1	929.0	6.452	10^{10}	10^{4}	2.590×10^{10}	0.01	8361
Square feet	4.356×10^{4}		1.076×10^{-3}	1	6.944×10^{-3}	1.076×10^{7}	10.76	2.788×10^{7}	1.076×10^{-5}	9
Square inches	6,272,640	7.854×10^{-7}	0.1550	144	1	1.550×10^{9}	1550	4.015×10^{9}	1.550×10^{-3}	1296
Square kilometers	4.047×10^{-3}		10^{-10}	9.290×10^{-8}	6.452×10^{-10}	1	10^{-6}	2.590	10^{-12}	8.361×10^{-7}
Square meters	4047		0.0001	9.290×10^{-2}	6.452×10^{-4}	10^{6}	1	2.590×10^{6}	10^{-6}	0.8361
Square miles	1.562×10^{-3}		3.861×10^{-11}	3.587×10^{-8}		0.3861	3.861×10^{-7}	1	3.861×10^{-13}	3.228×10^{-7}
Square millimeters		5.067×10^{-4}	100	9.290×10^{4}	645.2	10^{12}	10^{6}		1	8.361×10^{5}
Square yards	4840		1.196×10^{-4}	0.1111	7.716×10^{-4}	1.196×10^{6}	1.196	3.098×10^{6}	1.196×10^{-6}	1

Land Measure

30 1/4 square yards = 1 square rod = 272 1/4 square feet
16 square rods = 1 square chain = 484 square yards = 4356 square feet
2 1/2 square chains = 1 rood = 40 square rods = 1210 square yards
4 roods = 1 acre = 10 square chains = 160 square rods
640 acres = 1 square mile = 2560 roods = 102,400 square rods
1 section of land = 1 square mile; 1 quarter section = 160 acres

Architect's Measure

100 square feet = 1 square

Circular Inch and Circular Mil

A circular inch is the area of a circle 1 inch in diameter = 0.7854 square inch
1 square inch = 1.2732 circular inches
A circular mil is the area of a circle 1 mil (or 0.001 inch) in diameter = 0.7854 square mil
1 square mil = 1.2732 circular mils
1 circular inch = 10^6 circular mils = 0.7854×10^6 square mils
1 square inch = 1.2732×10^6 circular mils = 10^6 square mils

Metric Multiples

1 square meter = 1 centiare = 10^{-2} are = 10^{-4} hectare
= 10^{-6} square kilometer = 10^{-8} square myriameter

Table 3. Volume [L^3]

Multiply Number of → by to Obtain ↓	Bushels (dry)	Cubic centimeters	Cubic feet	Cubic inches	Cubic meters (steres)	Cubic yards	Gallons (liquid)	Liters	Pints (liquid)	Quarts (liquid)
Bushels (dry)	1		0.8036	4.651×10^{-4}	28.38			2.838×10^{-2}		
Cubic centimeters	3.524×10^{4}	1	2.832×10^{4}	16.39	10^{6}	7.646×10^{5}	3785	1000	473.2	946.4
Cubic feet	1.2445	3.531×10^{-5}	1	5.787×10^{-4}	35.31	27	0.1337	3.531×10^{-2}	1.671×10^{-2}	3.342×10^{-2}
Cubic inches	2150.4	6.102×10^{-2}	1728	1	6.102×10^{4}	46,656	231	61.02	28.87	57.75
Cubic meters (steres)	3.524×10^{-2}	10^{-6}	2.832×10^{-2}	1.639×10^{-5}	1	0.7646	3.785×10^{-3}	0.001	4.732×10^{-4}	9.464×10^{-4}
Cubic yards		1.308×10^{-6}	3.704×10^{-2}	2.143×10^{-5}	1.308	1	4.951×10^{-3}	1.308×10^{-3}	6.189×10^{-4}	1.238×10^{-3}
Gallons (liquid)		2.642×10^{-4}	7.481	4.329×10^{-3}	264.2	202.0	1	0.2642	0.125	0.25
Liters	35.24	0.001	28.32	1.639×10^{-2}	1000	764.6	3.785	1	0.4732	0.9464
Pints (liquid)		2.113×10^{-3}	59.84	3.463×10^{-2}	2113	1616	8	2.113	1	2
Quarts (liquid)......		1.057×10^{-3}	29.92	1.732×10^{-2}	1057	807.9	4	1.057	0.5	1

Acre-feet: multiply number of acre-feet by 4.356×10^4 to obtain number of cubic feet; multiply by 3.259×10^5 to obtain number of gallons.

Metric Multiples

10 milliliters = 1 centiliter = 0.338 fluid ounce
10 centiliters = 1 deciliter = 0.845 liquid gill
10 deciliters = 1 liter = 1.0567 liquid quarts
10 liters = 1 dekaliter = 2.6417 liquid gallons
10 dekaliters = 1 hectoliter = 2.8375 U. S. bushels
10 hectoliters = 1 kiloliter (or stere) = 28.375 U. S. bushels

Cubic Measure

1 cord of wood = a pile cut 4 feet long, piled 4 feet high and 8 feet on the ground = 128 cubic feet

1 perch of stone = a quantity 1 1/2 feet thick, 1 foot high and 16 1/2 feet long = 24 3/4 cubic feet

(NOTE.—A perch of stone is, however, often computed differently in different localities; thus, in most if not all of the States and Territories west of the Mississippi, stonemasons figure rubble by the perch of 16 1/2 cubic feet. In Philadelphia, 22 cubic feet are called a perch. In Chicago, stone is measured by the cord of 100 cubic feet. Check should be made against local practice.)

Board Measure

In board measure, boards are assumed to be one inch in thickness. Therefore, feet board measure of a stick of square timber = length in feet × breadth in feet × thickness in inches.

Shipping Measure

For register tonnage or measurement of the entire internal capacity of a vessel, it is arbitrarily assumed, to facilitate computation, that:

100 cubic feet = 1 register ton

For the measurement of cargo:

40 cubic feet = 1 U. S. shipping ton = 32.143 U. S. bushels
42 cubic feet = 1 British shipping ton = 32.703 Imperial bushels

Dry Measure

One U. S. Winchester bushel contains 1.2445 cubic feet or 2150.42 cubic inches. It holds 77.601 pounds distilled water at 62 deg fahr.

(NOTE.—The above is a *struck* bushel. A *heaped* bushel in general equals 1 $^1/_4$ struck bushels, although for apples and pears it contains 1.2731 struck bushels = 2737.72 cubic inches.)

One U. S. gallon (dry measure) = $^1/_8$ bushel and contains 268.8 cubic inches.

(NOTE.—This is not a legal U. S. *dry measure* and therefore is given for comparison only.)

One British Imperial bushel contains 1.2843 cubic feet or 2219.36 cubic inches. It holds 80 pounds distilled water at 62 deg fahr.

One British Imperial gallon = $^1/_8$ Imperial bushel and contains 277.42 cubic inches.

1 Winchester bushel = 0.9694 Imperial bushel
1 Imperial bushel = 1.032 Winchester bushels
Same relations as above maintain for gallons (dry measure)

[NOTE.—1 U. S. gallon (dry) = 1.164 U. S. gallons (liquid)].

U. S. Units

2 pints	= 1 quart	=	67.2	cubic inches
4 quarts	= 1 gallon * = 8 pints	=	268.8	cubic inches
2 gallons *	= 1 peck = 16 pints = 8 quarts	=	537.6	cubic inches
4 pecks	= 1 bushel = 64 pints = 32 quarts = 8 gallons *	=	2150.42	cubic inches

1 cubic foot contains 6.428 gallons (dry measure) *

Liquid Measure

One U. S. gallon (liquid measure) contains 231 cubic inches. It holds 8.336 pounds distilled water at 62 deg fahr.

One British Imperial gallon contains 277.42 cubic inches. It holds 10 pounds distilled water at 62 deg fahr.

1 U. S. gallon (liquid) = 0.8327 Imperial gallon
1 Imperial gallon = 1.201 U. S. gallons (liquid)

[NOTE.—1 U. S. gallon (liquid) = 0.8594 U. S. gallon (dry)].

U. S. Units

4 gills	= 1 pint	=	16 fluid ounces
2 pints	= 1 quart = 8 gills	=	32 fluid ounces
4 quarts	= 1 gallon = 32 gills = 8 pints	=	128 fluid ounces

1 cubic foot contains 7.4805 gallons (liquid measure)

Apothecaries' Fluid Measure

60 minims = 1 fluid drachm. 8 drachms = 1 fluid ounce

In the U. S. a fluid ounce is the 128th part of a U. S. gallon, or 1.805 cu in. or 29.58 cu cm. It contains 455.8 grains of water at 62 deg fahr. In Great Britain the fluid ounce is 1.732 cu in. and contains 1 ounce avoirdupois (or 437.5 grains) of water at 62 deg fahr.

* The *gallon* is not a U. S. legal dry measure.

Table 4. Plane Angle [*No Dimensions*]

Multiply Number of → by to Obtain ↓	Degrees	Minutes	Quadrants	Radians *	Revolutions * (Circumferences)	Seconds
Degrees	1	1.667 ×10⁻²	90	57.30	360	2.778 ×10⁻⁴
Minutes	60	1	5400	3438	2.16 ×10⁴	1.667 ×10⁻²
Quadrants	1.111 ×10⁻²	1.852 ×10⁻⁴	1	0.6366	4	3.087 ×10⁻⁶
Radians *	1.745 ×10⁻²	2.909 ×10⁻⁴	1.571	1	6.283	4.848 ×10⁻⁶
Revolutions * (Circumferences)	2.778 ×10⁻³	4.630 ×10⁻⁵	0.25	0.1591	1	7.716 ×10⁻⁷
Seconds	3600	60	3.24×10⁵	2.063×10⁵	1.296×10⁶	1

* 2π radians = 1 circumference = 360 degrees by definition.

Table 5. Solid Angle [*No Dimensions*]

Multiply Number of → by to Obtain ↓	Hemispheres	Spheres *	Spherical right angles	Steradians †
Hemispheres	1	2	0.25	0.1592
Spheres *	0.5	1	0.125	7.958 × 10⁻²
Spherical right angles	4	8	1	0.6366
Steradians †	6.283	12.57	1.571	1

* A sphere is the total solid angle about a point. † 4π steradians = 1 sphere by definition.

Table 6. Time [*T*]

Multiply Number of → by to Obtain ↓	Days	Hours	Minutes	Months (average)*	Seconds	Weeks
Days	1	4.167 ×10⁻²	6.944 ×10⁻⁴	30.42	1.157 ×10⁻⁵	7
Hours	24	1	1.667 ×10⁻²	730.0	2.778 ×10⁻⁴	168
Minutes	1440	60	1	4.380 ×10⁴	1.667 ×10⁻²	1.008 ×10⁴
Months (average) *	3.288 ×10⁻²	1.370 ×10⁻³	2.283 ×10⁻⁵	1	3.806 ×10⁻⁷	0.2302
Seconds	8.64 ×10⁴	3600	60	2.628 ×10⁶	1	6.048 ×10⁵
Weeks	0.1429	5.952 ×10⁻³	9.921 ×10⁻⁵	4.344	1.654 ×10⁻⁶	1

* One common year = 365 days; one leap year = 366 days; one average month = 1/12 of a common year.

Table 7. Linear Velocity $[LT^{-1}]$

Multiply Number of → by to Obtain ↓	Centimeters per second	Feet per minute	Feet per second	Kilometers per hour	Kilometers per minute	Knots *	Meters per minute	Meters per second	Miles per hour	Miles per minute
Centimeters per second	1	0.5080	30.48	27.78	1667	51.48	1.667	100	44.70	2682
Feet per minute	1.969	1	60	54.68	3281	101.3	3.281	196.8	88	5280
Feet per second	3.281×10^{-2}	1.667×10^{-2}	1	0.9113	54.68	1.689	5.468×10^{-2}	3.281	1.467	88
Kilometers per hour	0.036	1.829×10^{-2}	1.097	1	60	1.853	0.06	3.6	1.609	96.54
Kilometers per minute	0.0006	3.048×10^{-4}	1.829×10^{-2}	1.667×10^{-2}	1	3.088×10^{-2}	0.001	0.06	2.682×10^{-2}	1.609
Knots*	1.943×10^{-2}	9.868×10^{-3}	0.5921	0.5396	32.38	1	3.238×10^{-2}	1.943	0.8684	52.10
Meters per minute	0.6	0.3048	18.29	16.67	1000	30.88	1	60	26.82	1609
Meters per second	0.01	5.080×10^{-3}	0.3048	0.2778	16.67	0.5148	1.667×10^{-2}	1	0.4470	26.82
Miles per hour	2.237×10^{-2}	1.136×10^{-2}	0.6818	0.6214	37.28	1.152	3.728×10^{-2}	2.237	1	60
Miles per minute	3.728×10^{-4}	1.892×10^{-4}	1.136×10^{-2}	1.036×10^{-2}	0.6214	1.919×10^{-2}	6.214×10^{-4}	3.728×10^{-2}	1.667×10^{-2}	1

* Nautical miles per hour.

The Miner's Inch

(Used in Measuring Flow of Water)

An Act of the California legislature, May 23, 1901, makes the standard miner's inch 1.5 cu ft per minute, measured through any aperture or orifice.

The term miner's inch is more or less indefinite, for the reason that California water companies do not all use the same head above the center of the aperture, and the inch varies from 1.36 to 1.73 cu ft per minute, but the most common measurement is through an aperture 2 in. high and whatever length is required, and through a plank 1 1/4 in. thick. The lower edge of the aperture should be 2 in. above the bottom of the measuring-box, and the plank 5 in. high above the aperture, thus making a 6-in. head above the center of the stream. Each square inch of this opening represents a miner's inch, which is equal to a flow of 1.5 cu ft per minute.

Table 8. Angular Velocity $[T^{-1}]$

to Obtain ↓ / Multiply Number of → by	Degrees per second	Radians per second	Revolutions per minute	Revolutions per second
Degrees per second	1	57.30	6	360
Radians per second	1.745×10^{-2}	1	0.1047	6.283
Revolutions per minute	0.1667	9.549	1	60
Revolutions per second	2.778×10^{-3}	0.1592	1.667×10^{-2}	1

Table 9. Linear Acceleration * $[LT^{-2}]$

to Obtain ↓ / Multiply Number of → by	Centimeters per second per second	Feet per second per second	Kilometers per hour per second	Meters per second per second	Miles per hour per second
Centimeters per second per second	1	30.48	27.78	100	44.70
Feet per second per second	3.281×10^{-2}	1	0.9113	3.281	1.467
Kilometers per hour per second	0.036	1.097	1	3.6	1.609
Meters per second per second	0.01	0.3048	0.2778	1	0.4770
Miles per hour per second	2.237×10^{-2}	0.6818	0.6214	2.237	1

* The (standard) acceleration due to gravity (g_0) = 980.7 cm per sec per sec = 32.17 feet per sec per sec = 35.30 km per hour per sec = 9.807 meters per sec per sec = 21.94 miles per hour per sec.

Table 10. Angular Acceleration $[T^{-2}]$

to Obtain ↓ / Multiply Number of → by	Radians per second per second	Revolutions per minute per minute	Revolutions per minute per second	Revolutions per second per second
Radians per second per second	1	1.745×10^{-3}	0.1047	6.283
Revolutions per minute per minute	573.0	1	60	3600
Revolutions per minute per second	9.549	1.667×10^{-2}	1	60
Revolutions per second per second	0.1592	2.778×10^{-4}	1.667×10^{-2}	1

Table 11. Mass [*M*] and Weight *

Multiply Number of → by to Obtain ↓	Grains	Grams	Kilograms	Milligrams	Ounces †	Pounds †	Tons (long)	Tons (metric)	Tons (short)
Grains	1	15.43	1.543 $\times 10^{4}$	1.543 $\times 10^{-2}$	437.5	7000			
Grams	6.481 $\times 10^{-2}$	1	1000	0.001	28.35	453.6	1.016 $\times 10^{6}$	10^{6}	9.072 10^{5}
Kilograms	6.481 $\times 10^{-5}$	0.001	1	10^{-6}	2.835 $\times 10^{-2}$	0.4536	1016	1000	907.2
Milligrams	64.81	1000	10^{6}	1	2.835 $\times 10^{4}$	4.536 $\times 10^{5}$	1.016 $\times 10^{9}$	10^{9}	9.072 $\times 10^{8}$
Ounces †	2.286 $\times 10^{-3}$	3.527 $\times 10^{-2}$	35.27	3.527 $\times 10^{-5}$	1	16	3.584 $\times 10^{4}$	3.527 $\times 10^{4}$	3.2 $\times 10^{4}$
Pounds †	1.429 $\times 10^{-4}$	2.205 $\times 10^{-3}$	2.205	2.205 $\times 10^{-6}$	6.250 $\times 10^{-2}$	1	2240	2205	2000
Tons (long)		9.842 $\times 10^{-7}$	9.842 $\times 10^{-4}$	9.842 $\times 10^{-10}$	2.790 $\times 10^{-5}$	4.464 $\times 10^{-4}$	1	0.9842	0.8929
Tons (metric)		10^{-6}	0.001	10^{-9}	2.835 $\times 10^{-5}$	4.536 $\times 10^{-4}$	1.016	1	0.9072
Tons (short)		1.102 $\times 10^{-6}$	1.102 $\times 10^{-3}$	1.102 $\times 10^{-9}$	3.125 $\times 10^{-5}$	0.0005	1.120	1.102	1

* These same conversion factors apply to the *gravitational* units of force having the corresponding names. The dimensions of these units when used as gravitational units of force are MLT^{-2}; see table for *Force*.

† Avoirdupois pounds and ounces.

Metric Multiples

10^6 micrograms = 10^3 milligrams = 10^2 centigrams = 10 decigrams = 1 gram = 10^{-1} dekagram = 10^{-2} hectogram = 10^{-3} kilogram = 10^{-4} myriagram = 10^{-6} megagram

Avoirdupois Weight

(Used Commercially)

27.343 grains	= 1 drachm
16 drachms	= 1 ounce (oz) = 437.5 grains
16 ounces	= 1 pound (lb) = 7000 grains
28 pounds	= 1 quarter (qr)
4 quarters	= 1 hundredweight (cwt) = 112 pounds
20 hundredweight	= 1 gross or long ton *
2000 pounds	= 1 net or short ton

(* Note.—The long ton is used by the U. S. custom-houses in collecting duties upon foreign goods. It is also used in freighting coal and selling it wholesale.)

14 pounds = 1 stone; 100 pounds = 1 quintal

Troy Weight

(Used in weighing gold or silver)

24 grains	= 1 pennyweight (dwt)
20 pennyweights	= 1 ounce (oz) = 480 grains
12 ounces	= 1 pound (lb) = 5760 grains

The grain is the same in Avoirdupois, Troy and Apothecaries' weights. A carat, for weighing diamonds = 3.086 grains = 0.200 gram. (International Standard, 1913.)

1 pound troy = .8229 pound avoirdupois
1 pound avoirdupois = 1.2153 pounds troy

Apothecaries' Weight

(Used in compounding medicines)

20 grains = 1 scruple (℈)
3 scruples = 1 drachm (ʒ) = 60 grains
8 drachms = 1 ounce (℥) = 480 grains
12 ounces = 1 pound (lb) = 5760 grains

The grain is the same in Avoirdupois, Troy and Apothecaries' **weights.**

1 pound apothecaries = 0.82286 pound avoirdupois
1 pound avoirdupois = 1.2153 pounds apothecaries

Table 12. Density or Mass per Unit Volume $[ML^{-3}]$

Multiply Number of → by, to Obtain ↓	Grams per cubic centimeter	Kilograms per cubic meter	Pounds per cubic foot	Pounds per cubic inch
Grams per cubic centimeter	1	0.001	1.602×10^{-2}	27.68
Kilograms per cubic meter	1000	1	16.02	2.768×10^{4}
Pounds per cubic foot	62.43	6.243×10^{-2}	1	1728
Pounds per cubic inch	3.613×10^{-2}	3.613×10^{-5}	5.787×10^{-4}	1
Pounds per mil foot *	3.405×10^{-7}	3.405×10^{-10}	5.456×10^{-9}	$9.425 \times 10^{-}$[illegible]

* Unit of volume is a volume one foot long and one circular mil in cross-section area.

Table 13. Force * $[MLT^{-2}]$ or $[F]$

Multiply Number of → by, to Obtain ↓	Dynes	Grams	Joules per cm	Joules per meter (newtons)	Kilo-grams	Pounds	Poundals
Dynes	1	980.7	10^{7}	10^{5}	9.807×10^{5}	4.448×10^{5}	1.383×10^{4}
Grams	1.020×10^{-3}	1	1.020×10^{4}	102.0	1000	453.6	14.10
Joules per cm	10^{-7}	9.807×10^{-5}	1	.01	9.807×10^{-2}	4.448×10^{-2}	1.383×10^{-3}
Joules per meter (newtons)	10^{-5}	9.807×10^{-3}	100	1	9.807	4.448	0.1383
Kilograms	1.020×10^{-6}	0.001	10.20	0.1020	1	0.4536	1.410×10^{-2}
Pounds	2.248×10^{-6}	2.205×10^{-3}	22.48	0.2248	2.205	1	3.108×10^{-2}
Poundals	7.233×10^{-5}	7.093×10^{-2}	723.3	7.233	70.93	32.17	1

* Conversion factors between absolute and gravitational units apply only under standard acceleration due to gravity conditions.

Table 14. Torque or Moment of Force $[ML^2T^{-2}]$ or $[FL]$ *

Multiply Number of → by to Obtain ↓	Dyne-centimeters	Gram-centimeters	Kilogram-meters	Pound-feet	Newton-meter
Dyne-centimeters	1	980.7	9.807×10^{7}	1.356×10^{7}	10^{7}
Gram-centimeters	1.020×10^{-3}	1	10^{5}	1.383×10^{4}	1.020×10^{4}
Kilogram-meters	1.020×10^{-8}	10^{-5}	1	0.1383	0.1020
Pound-feet	7.376×10^{-8}	7.233×10^{-5}	7.233	1	0.7376
Newton-meter	10^{-7}	9.807×10^{-4}	9.807	1.305	1

* Same dimensions as energy.

Table 15. Pressure or Force per Unit Area $[ML^{-1}T^{-2}]$ or $[FL^{-2}]$

Multiply Number of → by to Obtain ↓	Atmospheres *	Baryes or dynes per square centimeter †	Centimeters of mercury at 0° C ‡	Inches of mercury at 0° C ‡	Inches of water at 4° C	Kilograms per square meter §	Pounds per square foot	Pounds per square inch	Tons (short) per square foot	Newtons per square meter
Atmospheres *	1	9.869×10^{-7}	1.316×10^{-2}	3.342×10^{-2}	2.458×10^{-3}	9.678×10^{-5}	4.725×10^{-4}	6.804×10^{-2}	0.9450	9.869×10^{-6}
Baryes or dynes per square centimeter †	1.013×10^{6}	1	1.333×10^{4}	3.386×10^{4}	2.491×10^{-3}	98.07	478.8	6.895×10^{4}	9.576×10^{5}	10
Centimeters of mercury at 0° C ‡	76.00	7.501×10^{-5}	1	2.540	0.1868	7.356×10^{-3}	3.591×10^{-2}	5.171	71.83	7.501×10^{-4}
Inches of mercury at 0° C ‡	29.92	2.953×10^{-5}	0.3937	1	7.355×10^{-2}	2.896×10^{-3}	1.414×10^{-2}	2.036	28.28	2.953×10^{-4}
Inches of water at 4° C	406.8	4.015×10^{-4}	5.354	13.60	1	3.937×10^{-2}	0.1922	27.68	384.5	4.015×10^{-3}
Kilograms per square meter §	1.033×10^{4}	1.020×10^{-2}	136.0	345.3	25.40	1	4.882	703.1	9765	0.1020
Pounds per square foot	2117	2.089×10^{-3}	27.85	70.73	5.204	0.2048	1	144	2000	2.089×10^{-2}
Pounds per square inch	14.70	1.450×10^{-5}	0.1934	0.4912	3.613×10^{-2}	1.422×10^{-3}	6.944×10^{-3}	1	13.89	1.450×10^{-4}
Tons (short) per square foot	1.058	1.044×10^{-6}	1.392×10^{-2}	3.536×10^{-2}	2.601×10^{-3}	1.024×10^{-4}	0.0005	0.072	1	1.044×10^{-5}
Newtons per square meter	1.013×10^{5}	10^{-1}	1.333×10^{3}	3.386×10^{3}	2.491×10^{-4}	9.807	47.88	6.895×10^{3}	9.576×10^{4}	1

* Definition: One atmosphere (standard) = 76 cm of mercury at 0 deg cent.

† Sometimes called a bar.

‡ To convert height h of a column of mercury at t degrees Centigrade to the equivalent height h_0 at 0 deg cent use $h_0 = h\left\{1 - \frac{(m-l)t}{1+mt}\right\}$ where $m = 0.0001818$ and $l = 18.4 \times 10^{-6}$ if the scale is engraved on brass; $l = 8.5 \times 10^{-6}$ if on glass. This assumes the scale is correct at 0 deg cent; for other cases (any liquid) see *International Critical Tables* Vol. 1, 68.

§ 1 gram per sq cm = 10 kilograms per sq m.

Table 16. Energy, Work and Heat * $[ML^2T^{-2}]$ or $[FL]$

Multiply Number of → by ↘ to Obtain ↓	British thermal units †	Centimeter-grams	Ergs or centimeter-dynes	Foot-pounds	Horsepower-hours	Joules ‡ or watt-seconds	Kilogram-calories †	Kilowatt-hours	Meter-kilograms	Watt-hours
British thermal units †	1	9.297×10^{-8}	9.480×10^{-11}	1.285×10^{-3}	2545	9.480×10^{-4}	3.969	3413	9.297×10^{-3}	3.413
Centimeter-grams	1.076×10^{7}	1	1.020×10^{-3}	1.383×10^{4}	2.737×10^{10}	1.020×10^{4}	4.269×10^{7}	3.671×10^{10}	10^5	3.671×10^{7}
Ergs or centimeter-dynes	1.055×10^{10}	980.7	1	1.356×10^{7}	2.684×10^{13}	10^7	4.186×10^{10}	3.6×10^{13}	9.807×10^{7}	3.6×10^{10}
Foot-pounds	778.0	7.233×10^{-5}	7.367×10^{-8}	1	1.98×10^{6}	0.7376	3087	2.655×10^{6}	7.233	2655
Horsepower-hours	3.929×10^{-4}	3.654×10^{-11}	3.722×10^{-14}	5.050×10^{-7}	1	3.722×10^{-7}	1.559×10^{-3}	1.341	3.653×10^{-6}	1.341×10^{-3}
Joules ‡ or watt-seconds	1054.8	9.807×10^{-5}	10^{-7}	1.356	2.684×10^{6}	1	4186	3.6×10^{6}	9.807	3600
Kilogram-calories †	0.2520	2.343×10^{-8}	2.389×10^{-11}	3.239×10^{-4}	641.3	2.389×10^{-4}	1	860.0	2.343×10^{-3}	0.8600
Kilowatt-hours	2.930×10^{-4}	2.724×10^{-11}	2.778×10^{-14}	3.766×10^{-7}	0.7457	2.778×10^{-7}	1.163×10^{-3}	1	2.724×10^{-6}	0.001
Meter-kilograms	107.6	10^{-5}	1.020×10^{-8}	0.1383	2.737×10^{5}	0.1020	426.9	3.671×10^{5}	1	367.1
Watt-hours	0.2930	2.724×10^{-8}	2.778×10^{-11}	3.766×10^{-4}	745.7	2.778×10^{-4}	1.163	1000	2.724×10^{-3}	1

* See note at the bottom of Table 17.

† Mean calorie and Btu used throughout. One gram-calorie = 0.001 kilogram-calorie; one Ostwald calorie = 0.01 kilogram-calorie.

The IT cal, 1000 international steam-table calories, has been defined as the 1/860th part of the international kilowatthour (see *Mechanical Engineering*, Nov., 1935, p. 710). Its value is very nearly equal to the mean kilogram-calorie, 1 IT cal = 1.00037 kilogram-calories (mean). 1 Btu = 251.996 IT cal.

‡ Absolute joule, defined as 10^7 ergs. The international joule, based on the international ohm and ampere, equals 1.0003 absolute joules.

Table 17. Power or Rate of Doing Work * $[ML^2T^{-3}]$ or $[FLT^{-1}]$

Multiply Number of → by, to Obtain ↓	British thermal units per minute	Ergs per second	Foot-pounds per minute	Foot-pounds per second	Horsepower *	Kilogram-calories per minute	Kilowatts	Metric horsepower	Watts
British thermal units per minute	1	5.689×10^{-9}	1.285×10^{-3}	7.712×10^{-2}	42.41	3.969	56.89	41.83	5.689×10^{-2}
Ergs per second	1.758×10^{8}	1	2.259×10^{5}	1.356×10^{7}	7.457×10^{9}	6.977×10^{8}	10^{10}	7.355×10^{9}	10^{7}
Foot-pounds per minute	778.0	4.426×10^{-6}	1	60	3.3×10^{4}	3087	4.426×10^{4}	3.255×10^{4}	44.26
Foot-pounds per second	12.97	7.376×10^{-8}	1.667×10^{-2}	1	550	51.44	737.6	542.5	0.7376
Horsepower *	2.357×10^{-2}	1.341×10^{-10}	3.030×10^{-5}	1.818×10^{-3}	1	9.355×10^{-2}	1.341	0.9863	1.341×10^{-3}
Kilogram-calories per minute	0.2520	1.433×10^{-9}	3.239×10^{-4}	1.943×10^{-2}	10.69	1	14.33	10.54	1.433×10^{-2}
Kilowatts	1.758×10^{-2}	10^{-10}	2.260×10^{-5}	1.356×10^{-3}	0.7457	6.977×10^{-2}	1	0.7355	10^{-3}
Metric horsepower	2.390×10^{-2}	1.360×10^{-10}	3.072×10^{-5}	1.843×10^{-3}	1.014	9.485×10^{-2}	1.360	1	1.360×10^{-3}
Watts	17.58	10^{-7}	2.260×10^{-2}	1.356	745.7	69.77	1000	735.5	1

1 Cheval-vapeur = 75 kilogram-meters per second
1 Poncelet = 100 kilogram-meters per second

* The "horsepower" used in these tables is equal to 550 foot-pounds per second by definition. Other definitions are one horsepower equals 746 watts (U. S. and Great Britain) and one horsepower equals 736 watts (continental Europe). Neither of these latter definitions is equivalent to the first; the "horsepowers" defined in these latter definitions are widely used in the rating of electrical machinery.

Table 18. Quantity of Electricity and Electric Flux [Q]

Multiply Number of → by, to Obtain ↓	Abcoulombs	Ampere-hours	Coulombs	Faradays	Stat-coulombs
Abcoulombs *	1	360	0.1	9649	3.335×10^{-11}
Ampere-hours	2.778×10^{-3}	1	2.778×10^{-4}	26.80	9.259×10^{-14}
Coulombs	10	3600	1	9.649×10^{4}	3.335×10^{-10}
Faradays	1.036×10^{-4}	3.731×10^{-2}	1.036×10^{-5}	1	3.457×10^{-15}
Statcoulombs *	2.998×10^{10}	1.080×10^{13}	2.998×10^{9}	2.893×10^{14}	1

* Conventionally, in the electrostatic and electromagnetic systems of units, the number of lines of electric flux emanating from a point charge is 4π times that charge (or quantity of electricity). The statcoulomb and the abcoulomb are units of charge, not flux.

Table 19. Charge per Unit Area and Electric Flux Density [QL^{-2}]

to Obtain ↓ / Multiply Number of → by	Abcoulombs per square centimeter *	Coulombs per square centimeter	Coulombs per square inch	Statcoulombs per square centimeter	Coulombs per square meter
Abcoulombs per square centi-centimeter *	1	0.1	1.550×10^{-2}	3.335×10^{-11}	10^{-5}
Coulombs per square centimeter	10	1	0.1550	3.335×10^{-10}	10^{-4}
Coulombs per square inch	64.52	6.452	1	2.151×10^{-9}	6.452×10^{-4}
Statcoulombs per square centimeter *	2.998×10^{10}	2.998×10^{9}	4.647×10^{8}	1	2.998×10^{5}
Coulombs per square meter	10^{5}	10^{4}	1550	3.335×10^{-6}	1

* See footnote to Table 18.

Table 20. Electric Current [QT^{-1}]

to Obtain ↓ / Multiply Number of → by	Abamperes	Amperes	Statamperes
Abamperes	1	0.1	3.335×10^{-11}
Amperes	10	1	3.335×10^{-10}
Statamperes	2.998×10^{10}	2.998×10^{9}	1

Table 21. Current Density [$QT^{-1}L^{-2}$]

to Obtain ↓ / Multiply Number of → by	Abamperes per square centimeter	Amperes per square centimeter	Amperes per square inch	Statamperes per square centimeter	Amperes per square meter
Abamperes per square centimeter	1	0.1	1.550×10^{-2}	3.335×10^{-11}	10^{-5}
Amperes per square centimeter	10	1	0.1550	3.335×10^{-10}	10^{-4}
Amperes per square inch	64.52	6.452	1	2.151×10^{-9}	6.452×10^{-4}
Statamperes per square centimeter	2.998×10^{10}	2.998×10^{9}	4.647×10^{8}	1	2.998×10^{5}
Amperes per square meter	10^{5}	10^{4}	1550	3.335×10^{-6}	1

Table 22. Electric Potential and Electromotive Force $[MQ^{-1}L^2T^{-2}]$ or $[FQ^{-1}L]$

Multiply Number of → by ↘ to Obtain ↓	Abvolts	Microvolts	Millivolts	Statvolts	Volts
Abvolts	1	100	10^5	2.998×10^{10}	10^8
Microvolts	0.01	1	1000	2.998×10^8	10^6
Millivolts	10^{-5}	0.001	1	2.998×10^5	1000
Statvolts	3.335×10^{-11}	3.335×10^{-9}	3.335×10^{-6}	1	3.335×10^{-3}
Volts	10^{-8}	10^{-6}	0.001	299.8	1

Table 23. Electric Field Intensity and Potential Gradient $[MQ^{-1}LT^{-2}]$ or $[FQ^{-1}]$

Multiply Number of → by ↘ to Obtain ↓	Abvolts per centimeter	Microvolts per meter	Millivolts per meter	Statvolts per centimeter	Volts per centimeter	Kilovolts per centimeter	Volts per inch	Volts per mil	Volts per meter
Abvolts per centimeter	1	1	1000	2.998×10^{10}	10^8	10^{11}	3.937×10^7	3.937×10^{10}	10^6
Microvolts per meter	1	1	1000	2.998×10^{10}	10^8	10^{11}	3.937×10^7	3.937×10^{10}	10^6
Millivolts per meter	0.001	0.001	1	2.998×10^7	10^5	10^8	3.937×10^4	3.937×10^7	1000
Statvolts per centimeter	3.335×10^{-11}	3.335×10^{-11}	3.335×10^{-8}	1	3.335×10^{-3}	3.335	1.313×10^{-3}	1.313	3.335×10^{-5}
Volts per centimeter	10^{-8}	10^{-8}	10^{-5}	299.8	1	1000	0.3937	393.7	10^{-2}
Kilovolts per centimeter	10^{-11}	10^{-11}	10^{-8}	0.2998	0.001	1	3.937×10^{-4}	0.3937	10^{-5}
Volts per inch	2.540×10^{-8}	2.540×10^{-8}	2.540×10^{-5}	761.6	2.540	2540	1	1000	2.540×10^{-2}
Volts per mil	2.540×10^{-11}	2.540×10^{-11}	2.540×10^{-8}	0.7616	2.540×10^{-3}	2.540	0.001	1	2.540×10^{-5}
Volts per meter	10^{-6}	10^{-6}	10^{-3}	2.998×10^4	100	10^5	39.37	3.937×10^4	1

Table 24. Electric Resistance $[MQ^{-2}L^2T^{-1}]$ or $[FQ^{-2}LT]$

Multiply Number of → by, to Obtain ↓	Abohms	Megohms	Microhms	Ohms	Statohms
Abohms	1	10^{15}	1000	10^9	8.988×10^{20}
Megohms	10^{-15}	1	10^{-12}	10^{-6}	8.988×10^5
Microhms	0.001	10^{12}	1	10^6	8.988×10^{17}
Ohms	10^{-9}	10^6	10^{-6}	1	8.988×10^{11}
Statohms	1.112×10^{-21}	1.112×10^{-6}	1.112×10^{-18}	1.112×10^{-12}	1

Electrical Conductance $[F^{-1}QL^{-1}T^{-1}]$

1 mho = 1 ohm^{-1} = 10^{-6} megmho = 10^6 micromho

Table 25. Electric Resistivity * $[MQ^{-2}L^3T^{-1}]$ or $[FQ^{-2}L^2T]$

Multiply Number of → by, to Obtain ↓	Abohm-centimeters	Microhm-centi-meters	Microhm-inches	Ohms (mil, foot)	Ohms (meter, gram) †	Ohm-meters
Abohm-centimeters	1	1000	2540	166.2	$\frac{10^5}{\delta}$	10^{11}
Microhm-centimeters	0.001	1	2.540	0.1662	$\frac{100}{\delta}$	10^8
Microhm-inches	3.937×10^{-4}	0.3937	1	6.545×10^{-2}	$\frac{39.37}{\delta}$	3.937×10^7
Ohms (mil, foot)	6.015×10^{-3}	6.015	15.28	1	$\frac{601.5}{\delta}$	6.015×10^8
Ohms (meter, gram) †	$10^{-5}\delta$	0.01δ	$2.540 \times 10^{-2}\delta$	$1.662 \times 10^{-3}\delta$	1	$10^{-6}\delta$
Ohm-meters	10^{-11}	10^{-8}	2.540×10^{-8}	1.662×10^{-9}	$\frac{10^{-6}}{\delta}$	1

* In this table δ is density in grams per cm.^{3} The following names, corresponding respectively to those at the tops of columns, are sometimes used: abohms per cm cube; microhms per cm cube; microhms per inch cube; ohms per mil-foot; ohms per meter-gram. The first four columns are headed by units of *volume* resistivity, the last by a unit of *mass* resistivity. The dimensions of the latter are $Q^{-2}L^6T^{-1}$; not these given in the heading of the table.

† One ohm (meter, gram) = 5710 ohms (mile, pound).

Table 26. Electric Conductivity * $[M^{-1}Q^2L^{-3}T]$ or $[F^{-1}Q^2L^{-2}T^{-1}]$

Multiply Number of → by to Obtain ↓	Abmhos per cm	Mhos (mil, foot)	Mhos (meter, gram)	Micromhos per cm	Micromhos per inch	Mhos per meter
Abmhos per cm	1	6.015×10^{-3}	$10^{-5}\delta$	0.001	3.937×10^{-4}	10^{-11}
Mhos (mil, foot)	166.2	1	$1.662 \times 10^{-3}\delta$	0.1662	6.524×10^{-2}	1.662×10^{-9}
Mhos (meter, gram)	$10^5/\delta$	$601.5/\delta$	1	$100/\delta$	$39.37/\delta$	$10^{-6}/\delta$
Micromhos per cm	1000	6.015	0.01δ	1	0.3937	10^{-8}
Micromhos per inch	2540	15.28	$2.540 \times 10^{-2}\delta$	2.540	1	2.54×10^{-8}
Mhos per meter	10^{11}	6.015×10^8	$10^6\delta$	10^8	3.937×10^7	1

* See footnote of Table 25, Electric Resistivity. Names sometimes used are abmho per cm cube, mho per mil-foot, etc. Dimensions of mass conductivity are $Q^2L^{-6}T$.

Table 27. Capacitance $[M^{-1}Q^2L^{-2}T^2]$ or $[F^{-1}Q^2L^{-1}]$

Multiply Number of → by to Obtain ↓	Abfarads	Farads	Microfarads	Statfarads
Abfarads	1	10^{-9}	10^{-15}	1.112×10^{-21}
Farads	10^9	1	10^{-6}	1.112×10^{-12}
Microfarads	10^{15}	10^6	1	1.112×10^{-6}
Statfarads	8.988×10^{20}	8.988×10^{11}	8.988×10^5	1

Table 28. Inductance $[MQ^{-2}L^2]$ or $[FQ^{-2}LT^2]$

Multiply Number of → by → to Obtain ↓	Abhenrys *	Henrys	Microhenrys	Millihenrys	Stathenrys
Abhenrys *	1	10^9	1000	10^6	8.988×10^{20}
Henrys	10^{-9}	1	10^{-6}	0.001	8.988×10^{11}
Microhenrys	0.001	10^6	1	1000	8.988×10^{17}
Millihenrys	10^{-6}	1000	0.001	1	8.988×10^{14}
Stathenrys	1.112×10^{-21}	1.112×10^{-12}	1.112×10^{-18}	1.112×10^{-15}	1

* An abhenry is sometimes called a "centimeter." See footnote to Table 30 on "Magnetic Flux Density."

Table 29. Magnetic Flux $[MQ^{-1}L^2T^{-1}]$ or $[FQ^{-1}LT]$

Multiply Number of → by → to Obtain ↓	Kilolines	Maxwells (or lines)	Webers
Kilolines	1	0.001	10^5
Maxwells (or lines)	1000	1	10^8
Webers	10^{-5}	10^{-8}	1

Table 30. Magnetic Flux Density $[MQ^{-1}T^{-1}]$ or $[FQ^{-1}L^{-1}T]$

Multiply Number of → by → to Obtain ↓	Gausses (or lines per square centimeter)	Lines per square inch	Webers per square centimeter	Webers per square inch	Webers per square meter
Gausses (or lines per square centimeter)	1	0.1550	10^8	1.550×10^7	10^4
Lines per square inch	6.452	1	6.452×10^8	10^8	6.452×10^4
Webers per square centimeter	10^{-8}	1.550×10^{-9}	1	0.1550	10^{-4}
Webers per square inch	6.452×10^{-8}	10^{-8}	6.452	1	6.452×10^{-4}
Webers per square meter	10^{-4}	1.550×10^{-5}	10^4	1550	1

Table 31. Magnetic Potential and Magnetomotive Force $[QT^{-1}]$

Multiply Number of → by, to Obtain ↓	Abampere-turns	Ampere-turns	Gilberts
Abampere-turns	1	0.1	7.958×10^{-2}
Ampere-turns	10	1	0.7958
Gilberts	12.57	1.257	1

Table 32. Magnetic Field Intensity, Potential Gradient, and Magnetizing Force $[QL^{-1}T^{-1}]$

Multiply Number of → by, to Obtain ↓	Abampere-turns per centimeter	Ampere-turns per centimeter	Ampere-turns per inch	Oersteds (gilberts per centimeter)	Ampere-turns per meter
Abampere-turns per centimeter	1	0.1	3.937×10^{-2}	7.958×10^{-2}	10^{-3}
Ampere-turns per centimeter	10	1	0.3937	0.7958	10^{-2}
Ampere-turns per inch	25.40	2.540	1	2.021	2.54×10^{-2}
Oersteds (gilberts per centimeter)	12.57	1.257	0.4950	1	1.257×10^{-2}
Ampere-turns per meter	10^3	10^2	39.37	79.58	1

Table 33. Specific Heat $[L^2T^{-2}t^{-1}]$

(t = temperature)

To convert specific heat in any unit given to any other unit multiply the number of original units by a factor obtained by dividing the factor in the last column for the final unit by the factor for the original unit.

© Unit of Heat or Energy	Unit of Mass	Temperature Scale*	Factor
Gram-calories	Gram	Centigrade	1
Kilogram-calories	Kilogram	Centigrade	1
British thermal units	Pound	Centigrade	1.800
British thermal units	Pound	Fahrenheit	1.000
Joules	Gram	Centigrade	4.186
Joules	Pound	Fahrenheit	1055.
Kilowatt-hours	Kilogram	Centigrade	1.163×10^{-3}
Kilowatt-hours	Pound	Fahrenheit	2.930×10^{-4}

* Temperature conversion formulas:

t_c = temperature in Centigrade degrees
t_f = temperature in Fahrenheit degrees.
1 deg fahr = (5/9) deg cent.
$t_c = \frac{5}{9}(t_f - 32)$
$t_f = \frac{9}{5} t_c + 32$

Table 34. Thermal Conductivity $[MLT^{-3}t^{-1}]$ and Thermal Resistivity $[M^{-1}L^{-1}T^{3}t]$

(t = temperature)

To convert thermal conductivity, in gram-calories transmitted per second from one face of a cube 1 cm on edge to the opposite face per degree centigrade temperature difference between these faces, to the units given in any line of the following table, multiply by the factor in the last column.

To convert thermal conductivity in any unit given to any other unit multiply the number of original units by a factor obtained by dividing the factor in the last column for the final unit by the factor for the original unit.

To convert thermal resistivity, in degrees centigrade between one face of a cube 1 cm on edge and the opposite face per gram-calories transmitted per second between these faces, to the units given in any line of the following table, divide by the factor in the last column.

To convert thermal resistivity in any given unit to any other unit multiply the number of the original units by a factor obtained by dividing the factor in the last column for the original unit by the factor for the final unit.

Surface emission resistance in thermal ohms per square centimeter is derived from degrees fahrenheit per Btu per hour per square foot by multiplying the number of the latter units by 1761.

Units of				Temperature Scale	Factor
Heat	Area	Thickness	Time		
Gram-calories	cm^2	cm	second	Centigrade	1
Kilogram-calories	m^2	cm	hour	Centigrade	3.6×10^4
British thermal units	ft^2	inch	hour	Fahrenheit	2903.
Joules *	cm^2	cm	second	Centigrade	4.186
Joules	ft^2	inch	second	Fahrenheit	850.6
Kilowatt-hours	m^2	cm	hour	Centigrade	41.86
Kilowatt-hours	ft^2	inch	hour	Fahrenheit	0.8506

* Thermal resistances in these units are known as *thermal ohms;* see Section 14, Article 69.

Table 35. Light

Multiply Number of → by to Obtain ↓	International candles	Hefners	10-cp pentanes	Carcels	Bougie decimales	English candles	German candles
International candles	1.00	0.90	10.0	9.61	1.00	1.04	1.055
Hefners	1.11	1.00	11.1	10.66	1.11	1.154	1.17
10-cp pentanes	0.10	0.09	1.00	0.96	0.10	0.104	0.105
Carcels	0.104	0.094	1.04	1.00	0.104	0.1	0.109
Bougie decimales	1.00	0.90	10.0	9.61	1.00	1.04	1.055
English candles	0.96	0.864	9.6	9.24	0.96	1.00	1.02
German candles	0.95	0.855	9.5	9.19	0.95	0.98	1.00

16. GAGES

SHEET METAL GAGES. The important sheet metal gages in use in the United States are: the United States Standard Gage for sheet and plate iron and steel, the Electrical Steel and Strip Gage, the American Wire Gage (also called the Brown and Sharpe W.G.) for copper, aluminum, and brass and other non-ferrous alloys, the Tin Plate Gage, the Galvanized Sheet Gage, the American Zinc Gage, and the Birmingham Wire (or Stubs' Iron Wire) Gage. In Canada and England the Birmingham Gage (different from the Birmingham Wire Gage) and the Imperial Standard Wire Gage (S.W.G.) are used. Still other gages are used elsewhere. In Japan standard thickness of sheet metal is denoted by the thickness in millimeters. A standard Decimal Gage, in which the standard thicknesses are denoted by decimal parts of an inch and not by gage numbers, has been used in the United States. Copper sheets may be obtained with thicknesses any integral multiple of $1/16$ of an inch up to 2 in. Heavy copper sheets may be obtained in definite weights per square foot. Each ounce of weight is equivalent to approximately 0.001352 in. thickness. Lead is usually ordered in this manner, each pound being equivalent to approximately 0.017 in. thickness.

The United States Standard Gage for sheet iron and steel (Act of Congress, March 3, 1893; formerly the legal standard for duties) is a *weight* gage based on a density for wrought iron of 480 pounds per cubic foot. Since 1893, steel (density of 489.6 lb per cu ft) has come into general use. A given gage number of this gage represents a fixed weight per unit area; hence a steel sheet will have a smaller thickness than a wrought iron sheet of the same gage number. Monel metal sheets are rolled to the thickness given for wrought iron without regard to its weight, which is about 552.2 lb per cu ft. Practice among steel manufacturers is irregular, some keeping the *thickness* constant for a given gage number irrespective of weight. If this practice is followed, the weight per square foot and per square meter given in the second and third columns of Table 37 will vary, whereas thickness will remain near that given for wrought iron.

For technical reasons it is undesirable to have the sheet thickness of electrical sheets vary with weight. The electrical steel and strip gage has been issued by the Association of American Steel Manufacturers' Technical Committee and is given in Table 36.

Table 36. Electrical Steel Sheet and Strip Gage Table

Gage Number	Gage Thickness, Inch	Gage Number	Gage Thickness, Inch
11	0.1250	22	0.0310
12	0.1090	23	0.0280
13	0.0940	24	0.0250
14	0.0780	25	0.0220
15	0.0700	26	0.0185
16	0.0625	27	0.0170
17	0.0560	28	0.0155
18	0.0500	29	0.0140
19	0.0435	30	0.0125
20	0.0375	31	0.0110
21	0.0340	32	0.0100

The American Wire Gage specifies thicknesses without regard to weight. For the basis of this gage see Wire Gages, p. 1-64, where are also given the Birmingham W.G. and the S.W.G.

Tables of Thickness and Weight corresponding to United States Standard gage and American Wire gage numbers are shown in Tables 37 and 38. These tables are taken from *Circular* 391 of the Bureau of Standards, in which are given all the gages mentioned above and the tolerances customary in commerce.

WIRE GAGES. The sizes of wires having a diameter less than $1/2$ in. are usually stated in terms of certain arbitrary scales called "gages." The size or gage number of a solid wire refers to the cross-section of the wire perpendicular to its length; the size or gage number of a stranded wire refers to the total cross-section of the constituent wires, irrespective of the pitch of the spiraling. Larger wires are usually described in terms of their area expressed in circular mils. A circular mil is the area of a circle 1 mil in diameter, and the area of any circle in circular mils is equal to the square of its diameter in mils.

Table 37. United States Standard Gage * for Sheet and Plate Iron and Steel, and Its Extension †

Gage No.	Weight per square foot		Weight per square meter	Approximate thickness			
				Wrought iron 480 lb/ft³		Steel and open-hearth iron 489.6 lb/ft³	
	Ounces	Pounds	kg	Inch	mm	Inch	mm
0000000	320	20.00	97.65	0.500	12.70	0.490	12.45
000000	300	18.75	91.55	.469	11.91	.460	11.67
00000	280	17.50	85.44	.438	11.11	.429	10.90
0000	260	16.25	79.34	.406	10.32	.398	10.12
000	240	15.00	73.24	.375	9.52	.368	9.34
00	220	13.75	67.13	.344	8.73	.337	8.56
0	200	12.50	61.03	.312	7.94	.306	7.78
1	180	11.25	54.93	.2812	7.14	.2757	7.00
2	170	10.62	51.88	.2656	6.75	.2604	6.62
3	160	10.00	48.82	.2500	6.35	.2451	6.23
4	150	9.375	45.77	.2344	5.95	.2298	5.84
5	140	8.750	42.72	.2188	5.56	.2145	5.45
6	130	8.125	39.67	.2031	5.16	.1991	5.06
7	120	7.500	36.62	.1875	4.76	.1838	4.67
8	110	6.875	33.57	.1719	4.37	.1685	4.28
9	100	6.250	30.52	.1562	3.97	.1532	3.89
10	90	5.625	27.46	.1406	3.57	.1379	3.50
11	80	5.000	24.41	.1250	3.18	.1225	3.11
12	70	4.375	21.36	.1094	2.778	.1072	2.724
13	60	3.750	18.31	.0938	2.381	.0919	2.335
14	50	3.125	15.26	.0781	1.984	.0766	1.946
15	45	2.812	13.73	.0703	1.786	.0689	1.751
16	40	2.500	12.21	.0625	1.588	.0613	1.557
17	36	2.250	10.99	.0562	1.429	.0551	1.400
18	32	2.000	9.765	.0500	1.270	.0490	1.245
19	28	1.750	8.544	.0438	1.111	.0429	1.090
20	24	1.500	7.324	.0375	.952	.0368	.934
21	22	1.375	6.713	.0344	.873	.0337	.856
22	20	1.250	6.103	.0312	.794	.0306	.778
23	18	1.125	5.493	.0281	.714	.0276	.700
24	16	1.000	4.882	.0250	.635	.0245	.623
25	14	.8750	4.272	.0219	.556	.0214	.545
26	12	.7500	3.662	.0188	.476	.0184	.467
27	11	.6875	3.357	.0172	.437	.0169	.428
28	10	.6250	3.052	.0156	.397	.0153	.389
29	9	.5625	2.746	.0141	.357	.0138	.350
30	8	.5000	2.441	.0125	.318	.0123	.311
31	7	.4375	2.136	.0109	.278	.0107	.272
32	6 1/2	.4062	1.983	.0102	.258	.0100	.253
33	6	.3750	1.831	.0094	.238	.0092	.233
34	5 1/2	.3438	1.678	.0086	.218	.0084	.214
35	5	.3125	1.526	.0078	.198	.0077	.195
36	4 1/2	.2812	1.373	.0070	.179	.0069	.175
37	4 1/4	.2656	1.297	.0066	.169	.0065	.165
38	4	.2500	1.221	.0062	.159	.0061	.156
39	3 3/4	.2344	1.144	.0059	.149	.0057	.146
40	3 1/2	.2188	1.068	.0055	.139	.0054	.136
41	3 3/8	.2109	1.030	.0053	.134	.0052	.131
42	3 1/4	.2031	.9917	.0051	.129	.0050	.126
43	3 1/8	.1953	.9536	.0049	.124	.0048	.122
44	3	.1875	.9155	.0047	.119	.0046	.117

* For the Galvanized Sheet Gage, add 2.5 ounces to the weight per square foot as given in the table. Gage numbers below 8 and above 34 are not used in the Galvanized Sheet Gage.

† Gage numbers greater than 38 were not in the standard as set up by law, but are in general use.

Table 38. American Wire Gage—Weights of Copper, Aluminum, and Brass Sheets and Plates

Gage No.	Thickness		Approximate weight * per sq ft in lb		
	Inch	mm	Copper	Aluminum	Commercial (high) brass
0000	0.4600	11.68	21.27	6.49	20.27
000	.4096	10.40	18.94	5.78	18.05
00	.3648	9.266	16.87	5.14	16.07
0	.3249	8.252	15.03	4.58	14.32
1	.2893	7.348	13.38	4.08	12.75
2	.2576	6.544	11.91	3.632	11.35
3	.2294	5.827	10.61	3.234	10.11
4	.2043	5.189	9.45	2.880	9.00
5	.1819	4.621	8.41	2.565	8.01
6	.1620	4.115	7.49	2.284	7.14
7	.1443	3.665	6.67	2.034	6.36
8	.1285	3.264	5.94	1.812	5.66
9	.1144	2.906	5.29	1.613	5.04
10	.1019	2.588	4.713	1.437	4.490
11	.0907	2.305	4.195	1.279	3.996
12	.0808	2.053	3.737	1.139	3.560
13	.0720	1.828	3.330	1.015	3.172
14	.0641	1.628	2.965	0.904	2.824
15	.0571	1.450	2.641	.805	2.516
16	.0508	1.291	2.349	.716	2.238
17	.0453	1.150	2.095	.639	1.996
18	.0403	1.024	1.864	.568	1.776
19	.0359	0.9116	1.660	.506	1.582
20	.0320	.8118	1.480	.451	1.410
21	.0285	.7230	1.318	.402	1.256
22	.0253	.6438	1.170	.3567	1.115
23	.0226	.5733	1.045	.3186	0.996
24	.0201	.5106	0.930	.2834	.886
25	.0179	.4547	.828	.2524	.789
26	.0159	.4049	.735	.2242	.701
27	.0142	.3606	.657	.2002	.626
28	.0126	.3211	.583	.1776	.555
29	.0113	.2859	.523	.1593	.498
30	.0100	.2546	.4625	.1410	.4406
31	.00893	.2268	.4130	.1259	.3935
32	.00795	.2019	.3677	.1121	.3503
33	.00708	.1798	.3274	.0998	.3119
34	.00630	.1601	.2914	.0888	.2776
35	.00561	.1426	.2595	.0791	.2472
36	.00500	.1270	.2312	.0705	.2203
37	.00445	.1131	.2058	.0627	.1961
38	.00397	.1007	.1836	.0560	.1749
39	.00353	.0897	.1633	.0498	.1555
40	.00314	.0799	.1452	.0443	.1383

* Assumed specific gravities or densities in grams per cubic centimeter; copper, 8.89; aluminum, 2.71; brass, 8.47.

There are a number of wire gages in use, the principal ones being the following:

American or Brown and Sharpe Wire Gage. This gage is the one commonly used in the United States for copper, aluminum, and resistance wires. The gage is designated by either of the abbreviations A.W.G. or B. & S.

Basis of the A.W.G. or B. & S. Gage. The diameters of wires having successive numbers on this gage are in the ratio of $\sqrt[39]{92}$(= 1.1229 approx.) to 1, and the No. 36 wire has a diameter of 5 mils. No. 35 A.W.G., therefore, has a diameter of $5 \times 1.1229 = 5.61$ mils and so on until No. 0000 is reached, having a diameter of 460 mils.

The ratio $\sqrt[39]{92}$ is approximately equal to $\sqrt[6]{2}$, which is 1.1225. This circumstance makes it possible to have a group of wires of regular gage size with an aggregate area approximately equal to that of another regular gage size. For example, a reduction of three gage numbers (as from gage No. 36 to No. 33) results in a new gage number representing a diameter approximately $\sqrt{2}$ times that represented by the original gage number—or an area approximately two times as great.

The following approximate relations are also useful:

An increase of 1 in the number increases the resistance	25 per cent.
An increase of 2 in the number increases the resistance	60 per cent.
An increase of 3 in the number increases the resistance	100 per cent.
An increase of 10 in the number increases the resistance	10 times.

A No. 10 A.W.G. copper wire has the following approximate characteristics:

Ohms per 1000 ft	1
Circular mils area	10,000
Weight, pounds per 1000 ft	32

A No. 10 A.W.G. aluminum wire has the following approximate characteristics:

Ohms per 1000 ft	1.6
Circular mils area	10,000
Weight, pounds per 1000 ft	9.5

Remembering these rules it is easy to find the approximate size, resistance, area, or weight of any size wire. For example, a No. 12 A.W.G. copper wire has a resistance of 1 plus 60 per cent = 1.6 ohms per 1000 ft approximately. Its area, being inversely as its resistance, is 10,000/1.6 = 6250 circular mils; its diameter is therefore $\sqrt{6250} = 79$ mils, and its weight 32/1.6 = 20 lb per 1000 ft.

U. S. Steel Wire Gage. This gage, known also as the "Washburn and Moen," "Roebling," "American Steel and Wire Co.'s gage," is the one usually employed in the United States for steel and iron wire. It is frequently abbreviated "S.W.G.," but to avoid confusion with the British Standard Wire Gage (*see below*) it should be abbreviated "Stl W.G." or "A. (steel) W.G."

Birmingham (or Stubs' Iron) Wire Gage. This gage is still used in the United States for some purposes, e.g., to designate the size of brass wire, and is also employed to a limited extent in Great Britain. It is usually abbreviated "B.W.G." It is sometimes referred to as the "Stubs' Iron Wire Gage," but it should not be confused with the Stubs' Steel Wire Gage.

British Standard Wire Gage. This gage, usually called simply the "Standard Wire Gage," and abbreviated "S.W.G.," is also known as the "New British Standard" (abbreviated "N.B.S."), the English Legal Standard, or the Imperial Wire Gage, and is the legal standard of Great Britain for all wires, as fixed by order in Council, August 23, 1883. It was constructed by modifying the Birmingham Wire Gage, so that the differences between successive diameters were the same for short ranges, i.e., so that a graph representing the diameters consists of a series of a few straight lines.

Edison Wire Gage. The size of a wire on this gage is equal to its cross-sectional area in circular mils divided by 1000. For example, a solid wire 0.2 in. in diameter has the number $(200)^2/1000 = 40$. This gage is now rarely used.

Metric Wire Gage. The gage number is ten times the diameter in millimeters.

Other Gages. In addition wire sizes are sometimes specified in terms of the "Old English Wire Gage," known also as the "London Gage," and the "Stubs' Steel Wire Gage." The Old English Wire Gage is the same as B.W.G. for all gage numbers under 20.

Comparison of Wire Gages. A comparison of the different gages, in terms of the diameters (in mils or thousandths of an inch) of solid wires corresponding to the various numbers, is given in Table 39. The cross-section in **circular mils** is the square of the diameter in mils.

Table 39. Comparison of Wire Gage Diameters in Mils

(Bureau of Standards, *Circulars* 31 and 67)

Gage No.	American wire gage (B. & S.)	Steel wire gage	Birmingham wire gage (Stubs')	Old English wire gage (London)	Stubs' steel wire gage	(British) Standard wire gage	Metric gage *	Gage No.
7–0	....	490.0	....	....	....	500	....	7–0
6–0	....	461.5	....	....	....	464	....	6–0
5–0	....	430.5	....	....	....	432	....	5–0
4–0	460	393.8	454	454	....	400	....	4–0
3–0	410	362.5	425	425	....	372	....	3–0
2–0	365	331.0	380	380	....	348	....	2–0
0	325	306.5	340	340	....	324	....	0
1	289	283.0	300	300	227	300	3.94	1
2	258	262.5	284	284	219	276	7.87	2
3	229	243.7	259	259	212	252	11.8	3
4	204	225.3	238	238	207	232	15.7	4
5	182	207.0	220	220	204	212	19.7	5
6	162	192.0	203	203	201	192	23.6	6
7	144	177.0	180	180	199	176	27.6	7
8	128	162.0	165	165	197	160	31.5	8
9	114	148.3	148	148	194	144	35.4	9
10	102	135.0	134	134	191	128	39.4	10
11	91	120.5	120	120	188	116	....	11
12	81	105.5	109	109	185	104	47.2	12
13	72	91.5	95	95	182	92	....	13
14	64	80.0	83	83	180	80	55.1	14
15	57	72.0	72	72	178	72	....	15
16	51	62.5	65	65	175	64	63.0	16
17	45	54.0	58	58	172	56	....	17
18	40	47.5	49	49	168	48	70.9	18
19	36	41.0	42	40	164	40	....	19
20	32	34.8	35	35	161	36	78.7	20
21	28.5	31.7	32	31.5	157	32	....	21
22	25.3	28.6	28	29.5	155	28	....	22
23	22.6	25.8	25	27.0	153	24	....	23
24	20.1	23.0	22	25.0	151	22	....	24
25	17.9	20.4	20	23.0	148	20	98.4	25
26	15.9	18.1	18	20.5	146	18	....	26
27	14.2	17.3	16	18.75	143	16.4	....	27
28	12.6	16.2	14	16.50	139	14.8	....	28
29	11.3	15.0	13	15.50	134	13.6	....	29
30	10.0	14.0	12	13.75	127	12.4	118	30
31	8.9	13.2	10	12.25	120	11.6	....	31
32	8.0	12.8	9	11.25	115	10.8	....	32
33	7.1	11.8	8	10.25	112	10.0	....	33
34	6.3	10.4	7	9.50	110	9.2	....	34
35	5.6	9.5	5	9.00	108	8.4	138	35
36	5.0	9.0	4	7.50	106	7.6	....	36
37	4.5	8.5	....	6.50	103	6.8	....	37
38	4.0	8.0	....	5.75	101	6.0	....	38
39	3.5	7.5	....	5.00	99	5.2	....	39
40	3.1	7.0	....	4.50	97	4.8	157	40
41	....	6.6	....	....	95	4.4	....	41
42	....	6.2	....	....	92	4.0	....	42
43	....	6.0	....	....	88	3.6	....	43
44	....	5.8	....	....	85	3.2	....	44
45	....	5.5	....	....	81	2.8	177	45
46	....	5.2	....	....	79	2.4	....	46
47	....	5.0	....	....	77	2.0	....	47
48	....	4.8	....	....	75	1.6	....	48
49	....	4.6	....	....	72	1.2	....	49
50	....	4.4	....	....	69	1.0	197	50

* For diameters corresponding to metric gage numbers, 1.2, 1.4, 1.6, 1.8, 2.5, 3.5, and 4.5, divide those of 12, 14, etc., by ten.

SYMBOLS AND ABBREVIATIONS

17. ABBREVIATIONS FOR ENGINEERING TERMS

NOTE: This list is a selection of American Tentative Standard abbreviations, for scientific and engineering terms, recommended by the American Standards Association. (See ASA, Z10.1—1941.)

Absolute	abs
Acre	acre
Alternating-current (as adjective)	ac
Ampere	amp
Ampere-hour	amp-hr
Angstrom unit	A
Atmosphere	atm
Atomic weight	at. wt.
Average	avg
Avoirdupois	avdp
Barometer	bar.
Barrel	bbl
Baumé	Bé
Boiler pressure	spell out
Boiling point	bp
Brake horsepower	bhp
Brake horsepower-hour	bhp-hr
Brinell hardness number	Bhn
British thermal unit	Btu or B
Calorie	cal
Candle	c
Candle-hour	c-hr
Candlepower	cp
Centigram	cg
Centiliter	cl
Centimeter	cm
Centimeter-gram-second (system)	cgs
Chemically pure	cp
Circular	cir
Circular mils	cir mils
Coefficient	coef
Cologarithm	colog
Concentrate	conc
Conductivity	cond
Constant	const
Cord	cd
Cosecant	csc
Cosine	cos
Cotangent	cot
Coulomb	spell out
Counter electromotive force	cemf
Cubic	cu
Cubic centimeter	cu cm, cm^3, cc
Cubic feet per minute	cfm
Cubic foot	cu ft
Cubic inch	cu in.
Cubic meter	cu m or m^3
Cubic yard	cu yd
Cycles per second	spell out or c
Decibel	db
Degree	deg or °
Degree Centigrade	C
Degree Fahrenheit	F
Degree Kelvin	K
Degree Réaumur	R
Diameter	diam
Direct-current (as adjective)	d-c
Dozen	doz
Dram	dr
Efficiency	eff
Electric	elec
Electromotive force	emf
Equation	eq
External	ext
Farad	spell out or f
Foot	ft
Foot-candle	ft-c
Foot-Lambert	ft-L
Foot-pound	ft-lb
Foot-pound-second (system)	fps
Freezing point	fp
Fusion point	fnp
Gallon	gal
Grain	spell out
Gram	g
Gram-calorie	g-cal
Henry	h
Horsepower	hp
Horsepower-hour	hp-hr
Hour	hr
Hundred	C
Hyperbolic sine	sinh
Hyperbolic cosine	cosh
Hyperbolic tangent	tanh
Inch	in.
Inch-pound	in-lb
Internal	int
Joule	j
Kilocycles per second	kc
Kilogram	kg
Kilogram-calorie	kg-cal
Kilogram-meter	kg-m
Kiloliter	kl
Kilometer	km
Kilovolt	kv
Kilovolt-ampere	kva
Kilowatt	kw
Kilowatthour	kwhr
Lambert	L
Latitude	lat or ϕ
Linear foot	lin ft
Liter	l
Liquid	liq
Logarithm (common)	log
Logarithm (natural)	$\log_e$ or ln
Longitude	long. or λ
Lumen	l
Lumen-hour	l-hr
Magnetomotive force	mmf
Maximum	max
Melting point	mp
Meter	m
Meter-kilogram	m-kg
Mho	spell out
Microampere	μa or mu a
Microfarad	μf
Micromicron	$\mu\mu$ or mu mu
Micron	μ or mu
Microwatt	μw or mu w
Mile	spell out
Milliampere	ma
Milligram	mg
Millihenry	mh

Abbreviations for Engineering Terms—*Continued*

Milliliter	ml
Millimeter	mm
Millimicron	$m\mu$ or m mu
Million	spell out
Millivolt	mv
Mean horizontal candlepower	mhcp
Miles per hour	mph
Minimum	min
Minute	min
Minute (angular measure)	′
Ohm	spell out Ω
Ounce	oz
Ounce-foot	oz-ft
Ounce-inch	oz-in.
Pint	pt
Potential	spell out
Pound	lb
Pound-foot	lb-ft
Pound-inch	lb-in.
Pounds per square foot	psf
Pounds per square inch	psi
Power factor	spell out or pf
Quart	qt
Radian	spell out
Reactive kilovolt-ampere	kvar
Reactive volt-ampere	var
Revolutions per minute	rpm
Revolutions per second	rps
Root mean square	rms
Secant	sec
Second	sec
Second (angular measure)	″
Sine	sin
Specific gravity	sp gr
Specific heat	sp ht
Spherical candlepower	scp
Square	sq
Square centimeter	sq cm or cm^2
Square foot	sq ft
Square inch	sq in.
Square kilometer	sq km or km^2
Square meter	sq m or m^2
Square micron	sq μ or sq mu or μ^2
Square root of mean square	rms
Standard	std
Tangent	tan
Temperature	temp
Thousand	M
Ton	spell out
Versed sine	vers
Volt	v
Volt-ampere	va
Watt	w
Watthour	whr
Weight	wt
Yard	yd
Year	yr

18. LETTER SYMBOLS FOR THE MAGNITUDES OF ELECTRICAL QUANTITIES

(Tentative American Standard Z10.8–1947) †

In the alphabetical order of the names of the quantities

Each quantity appears at only one place in this table (with a few exceptions), listed alphabetically under its preferred name. The non-preferred names appear in parentheses under the preferred names. Deprecated names are also in parentheses and in addition are asterisked thus: (electric force)*.

Names beginning with the qualifying adjectives, *electric, electrostatic, dielectric, magnetic, mutual, self,* and *relative,* are listed under the term that is so qualified.

Symbols for scalar quantities, whose values are expressed by real numbers, are printed in ordinary-face *italic letters.*

Symbols for vector quantities are printed in **bold-face Roman letters.**

Symbols for phasor quantities, whose values are expressed by complex numbers, are printed in ***bold-face italic letters.***

Item	Quantity	Symbol
1	admittance	$\boldsymbol{Y}$
2	attenuation constant	α
3	capacitance	C
	(capacity) *	
	(permittance) *	
4	capacitivity	ϵ
	dielectric constant	
	(permittivity) *	
	of evacuated space	ϵ_v
5	capacitivity, relative	ϵ_r
	relative dielectric constant	
	(specific inductive capacity)	
6	charge, electric or quantity of electricity	Q
	charge density	
7	line d. of charge	λ
8	surface d. of charge	σ
9	volume d. of charge	ρ
10	conductance	G
11	conductivity	γ
12	conductivity, equivalent	Λ
13	coupling coefficient	k
14	current	I
	(intensity of current) *	
15	current density	
16	sheet c.d. (linear c.d.)	A
17	damping constant or coefficient (decay constant)	δ

* Deprecated name.

† Reprinted by permission of the American Institute of Electrical Engineers.

Letter Symbols for the Magnitudes of Electrical Quantities—*Continued*

Item	Quantity	Symbol
18	dielectric constant	ϵ
	see capacitivity	
	dielectric, a qualifier	
	see term that it qualifies	
19	displacement, electric	**D**
20	efficiency	η
21	elastance	S
	mutual e. S_m, S_{rc}	
	self e. S, S_{cc}	
22	elastivity	σ
	electric, a qualifier	
	see term that it qualifies	
23	electronic charge	
	(absolute value of)	
24	electromotive force	E
	(electromotance)	
	(potential difference, electric)	
	(voltage) *	
25	energy	W
26	force	**F**
27	flux, displacement f.	Ψ
	(flux of e. displacement)	
28	flux, magnetic	Φ
	(flux of magnetic induction)	
29	flux-linkage	
30	frequency	f
31	frequency, angular	ω
	angular velocity	
32	frequency, rotational	n
33	impedance	Z
	mutual i. Z_m, Z_{rc}	
	self i. Z, Z_{cc}	
34	induction, magnetic	**B**
	(magnetic flux density)	
35	inductance	L
	mutual i. L_m, L_{rc}	
	self i. L, L_{cc}	
36	intensity, electric	**E**, **K**
	(electric field intensity)	
	(electric field strength)	
	(electric force) *	
	(electric field) *	
37	intensity, magnetic or magnetizing force	**H**
	(magnetic field strength)	
	(magnetic force) *	
	magnetic, a qualifier	
	see term that it qualifies	
38	magnetomotive force	M, $\mathfrak{F}$
	(m. potential difference)	
	magnetomotance	
39	moment, electric	**p**
40	moment, magnetic	**m**
41	number of conductors or turns	N
42	number of poles	p
43	number of phases	m
44	period	T
45	permeance	$\mathcal{P}$, Λ
46	permeability, magnetic	μ
	of evacuated space	μ_v
47	permeability, relative	μ_r
48	(permittivity) * (see capacitivity)	
49	phase angle	φ
50	phase constant	β
	wavelength constant	
	(wave number)	
51	polarization, electric	
52	polarization, magnetic	**B**i
	intrinsic induction	
	metallic induction	
53	pole strength	m
54	potential, electric	V
	(electromagnetic scalar p.)	
55	potential, retarded scalar	
56	potential, magnetic	M, $\mathfrak{F}$
	(magnetic scalar p.)	
	m. pot. difference	
57	potential, magnetic vector p.	**A**
58	potential, retarded vector p.	$\mathbf{A}_r$
59	power, active	P
60	power, reactive	Q
	volt-amperes, reactive	
61	power, apparent	S
	volt-amperes	
62	power factor	F_p
63	propagation constant	γ
64	Poynting vector	Π
65	quantity of electricity	Q
	charge, electric	
66	quality factor of a reactor	Q
	figure of merit of a reactor	
67	reactance	X
	capacitative r.	X_c
	inductive r.	X_L
	mutual r. X_m, X_{rc}	
	self r. X, X_{cc}	
68	reactive factor	F_q
69	reluctance	$\mathcal{R}$
70	reluctivity	ν
71	resistance	R
	mutual r. R_m, R_{rc}	
	self r. R, R_{cc}	
72	resistivity	ρ
73	resistance-temperature coefficient	α
	rotative operators	
74	90°, $\sqrt{-1}$	j
75	120°, $\sqrt[3]{1}$	a
	self, a qualifier	
	see term that it qualifies	
76	slip	s
77	susceptance	B
	susceptibility	
78	dielectric s.	η
	intrinsic capacitivity	
79	magnetic s.	κ
	intrinsic permeability	
80	symmetrical components	
81	temperature	t, (θ)
		T, (Θ)
82	time	t
83	time constant	τ
84	velocity of light	c
85	vibration constant	**p**
	(oscillation constant)	
86	wavelength	λ
87	wavelength constant	β
	phase constant	
88	work	W

* Deprecated name.

Note 1. Designation of maximum, instantaneous, rms, and average values.

Where distinctions between maximum, instantaneous, root-mean-square (effective), and average values are necessary, E_m, I_m, Q_m, and P_m are recommended for maximum values; e, i, q, and p for instantaneous values, E, I, and Q for root-mean-square values and E_a, I_a, Q_a, and P for average values.

Note 2. Quantities per unit volume, area, or length.

It is recommended that quantities per unit volume, area, length, etc., be represented as far as practicable by lower-case letters corresponding to the cap letters which represent the total quantities, or by the cap letters with the subscript 1, except for those quantities for which this table has symbols for the quantity per unit volume, area, etc.

Note 3. Distinction between the symbols V and E for potential and electromotive force.

The distinction between the use of V for potential and E for electromotive force is:

V is to be used for potentials or potential differences that are attributed solely to that distribution of electric field intensities which is computed (by the inverse square law of force) from the segregated charges of the field.

E is to be used for the emf along a path from a terminal A to a terminal B when in the region A to B one or more non-electrostatic types of electric intensities exist, or turbulent actions occur—as in voltaic cells, electrostatic generators, and electromagnetic sources of emf.

Note 4. The sequence of the double subscripts to multiplying operators.

The sequence of the double subscripts to the *multiplying* operators (mutual impedances, resistances, or elastances or transconductances, etc.) that occur in the fundamental equations of networks is to be determined by the following consideration:

Consideration. The set of fundamental equations (e.g., Kirchhoff's emf equations) should yield a determinant in which the subscript sequence conforms to the mathematician's convention for writing determinants; namely,

Convention. In the double subscripts of the elements of a determinant, the subnumber designating the "row" is to precede the subnumber designating the "column" to which the element belongs, or the order is e_{rc}. Thus

$$D(e) = \begin{vmatrix} e_{11} & e_{12} \\ e_{21} & e_{22} \end{vmatrix}$$

This consideration leads to the following rule:

Rule for writing double subscripts.

The first subnumber in the symbol for a *multiplying* operator designates the number of the circuit in which the *product* of the *multiplication* is measurable, while the second subnumber designates the number of the circuit in which the operand or *multiplicand* is measurable.

As an illustration, Kirchhoff's emf law for the emfs of the rth circuit due to the currents in all the circuits of a network is written:

$$E_{r,d} = Z_{r1}I_1 + Z_{r2}I_2 + \cdots\ Z_{rr}I_r + \cdots$$

or

$$E_{r,d} = \Sigma c\ Z_{rc}I_c$$

($E_{r,d}$ being the driving emf impressed in the rth circuit).

Note 5. Notation for symmetrical components.

The standard notation for designating the symmetrical components of the currents and potential differences in unbalanced polyphase systems is that subscript notation in which:

(a) double subscripts are added to the symbols for current and potential difference;

(b) the first and second subscripts designate, respectively, the phase and the sequence to which the component belongs;

(c) the first, or phase, subscript may be the phase number, or the phase letter, or a two-letter combination that designates (on a diagram) both the phase and the direction in the phase;

(d) the second, or sequence, subscript is *always* to be the number that designates the sequence to which the component belongs; the positive, negative, and zero sequence components in three-phase systems being designated by the numbers 1, 2, and 0, respectively.

Illustration of notation:

$$I_a = I_{a1} + I_{a2} + I_{a0}$$
$$I_b = I_{b1} + I_{b2} + I_{b0}$$
$$I_c = I_{c1} + I_{c2} + I_{c0}$$

Note 6. Policy relative to coordinate standard symbols for the same magnitude and alternates to the standard symbols of magnitude.

Coordinate Letter Standards (of the same order or rank).

The listing of two *coordinate* letter symbols (separated by a comma) for the same magnitude is only to be done when each symbol would be suitable for the standard symbol, and when the standardizing body finds itself (for the time being) unable to designate, as between the two, the *preferred*, or the *standard*, symbol. If it is in a position to do so, the body may tentatively recommend one letter as the preferred symbol.

Alternates to the Standard Symbols (letters designated for use in place of the standard in case of the unsuitability of the standard).

The listing of a second letter symbol (in parentheses) as an *alternate* to the standard symbol of magnitude is only to be done for those magnitudes for which it appears evident that the standard letter will be in frequent conflict with the use of the same letter as the standard for some other magnitude. The thought is that the alternate symbol shall be used only in equations or articles or fields in which the use of the standard letter would result in confusion because the same letter is therein used as the standard symbol for some other magnitude. The column of alternate symbols is not to become a harbor for the preservation of national or group preferences.

19. USE OF GREEK ALPHABET FOR SYMBOLS

Capital	Lower Case	Name	Commonly Used to Designate
Α	α	Alpha	Angles. Area. Coefficients. Attenuation constant.
Β	β	Beta	Angles. Flux density. Coefficients.
Γ	γ	Gamma	Conductivity. Specific gravity. Propagation constant.
Δ	δ	Delta	Variation. Density. Damping coefficient.
Ε	ε	Epsilon	Base of natural logarithms. Capacitivity.
Ζ	ζ	Zeta	Impedance. Coefficients. Coordinates.
Η	η	Eta	Hysteresis coefficient. Efficiency.
Θ	θ	Theta	Temperature. Phase angle.
Ι	ι	Iota	
Κ	κ	Kappa	Dielectric constant. Susceptibility.
Λ	λ	Lambda	Wavelength.
Μ	μ	Mu	Micro. Amplification factor. Permeability.
Ν	ν	Nu	Reluctivity.
Ξ	ξ	Xi	
Ο	ο	Omicron	
Π	π	Pi	Ratio of circumference to diameter = 3.1416.
Ρ	ρ	Rho	Resistivity.
Σ	σ	Sigma	Capital: sign of summation.
Τ	τ	Tau	Time constant. Time phase displacement.
Υ	υ	Upsilon	
Φ	φ or φ	Phi	Magnetic flux. Angles.
Χ	χ	Chi	
Ψ	ψ	Psi	Dielectric flux. Phase difference.
Ω	ω	Omega	Capital: ohms. Lower case: angular velocity, or $2\pi \times$ frequency.

20. GRAPHICAL SYMBOLS FOR ELECTRIC POWER AND CONTROL

(Taken from American Standard Graphical Symbols for Electric Power and Control, ASA Z 32.3—1946, approved by American Standards Association, March 1, 1946)

The basic symbols which follow are grouped under the following headings:

1. Alarms
2. Battery
3. Capacitor
4. Circuit Breakers
5. Coils
6. Connections
7. Contacts—Electrical
8. Contactors
9. Fuse
10. Indicating Lights
11. Instruments and Meters
12. Lightning Arresters
13. Machines (Rotating Motors and Generators)
14. Meters (*see* Instruments and Meters)
15. Plug Connections
16. Rectifiers
17. Relays
18. Relay Function Symbols
19. Resistors
20. Shielding
21. Switches
22. Thermal Elements
23. Thermocouples
24. Transformers
25. Windings
26. Wiring

BASIC SYMBOLS

Basic symbols are the subject of a separate ASA publication (Z32.12). Some of the basic symbols used in this standard are shown below.

The basic symbols given are those which can be combined into various adaptations and form the basis for symbols and adaptations which are given in the following numbered sections.

Capacitor

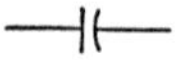

Note:. Where it is necessary to identify the capacitor electrodes, the curved element shall represent the outside electrode in fixed paper-dielectric and ceramic-dielectric capacitors, the negative electrode in electrolytic capacitors, and the movable element in variable and adjustable capacitors. When it is desired especially to distinguish trimmer capacitors, the letter T should appear adjacent to the symbol.

Coils

Operating

See Note

* This symbol must always be used with an identifying legend within or adjacent to the circle.

Note:. This symbol is customarily used when it is desired to show more detailed construction.

Blowout

Contact

Normally Open 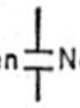 Normally Closed 

or

Normally Open 

Normally Closed 

Field

(Motors and Generators)

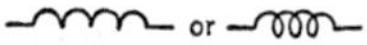

(Commutating, and/or compensating, series and shunt fields shall be indicated by two scallops or one loop (or), three scallops or two loops (or), and four scallops or three loops (or) respectively

Inductor

(Including Transformer Winding)

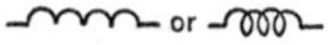

Note:. When it is desired especially to distinguish magnetic core inductors, lines parallel to the axis of the scallops or loops should be used.

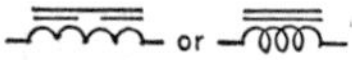

Resistor

(Including Rheostat and Voltage Divider)

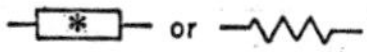

* This symbol must always be used with an identifying legend within or adjacent to the rectangle.

SYMBOLS AND ADAPTATIONS

1. Alarms

Annunciator

Bell

Buzzer

Horn

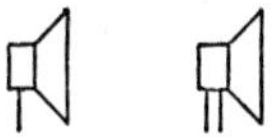

One Line Complete

2. Battery

General

(The long line is always positive.)

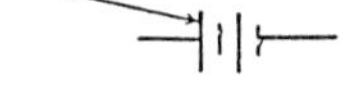

Adaptations

With Polarity Indicated

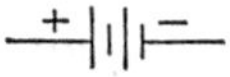

One Cell

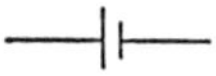

3. Capacitor

Fixed

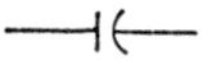

Adaptations

Adjustable by Fixed Taps

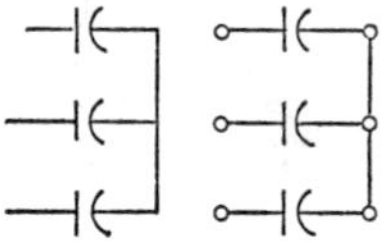

With Loads With Terminals

Shielded

Continuously Adjustable

Capacitance Bushing

Note:. Where it is necessary to identify the capacitor electrodes, the curved element shall represent the outside electrode in fixed paper-dielectric and ceramic-dielectric capacitors, the negative electrode in electrolytic capacitors and the moving element in variable and adjustable capacitors. When it is desired especially to distinguish trimmer capacitors, the letter T should appear adjacent to the symbol.

4. Circuit Breakers

Air

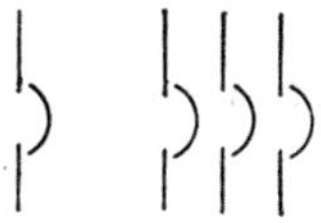

Oil (or Other Liquid)

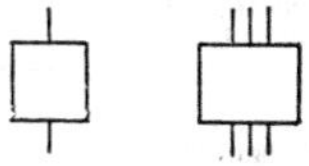

One Line Complete for Three Phase or Three Wire

Adaptations

Three Pole (Single Throw) (With Terminals)

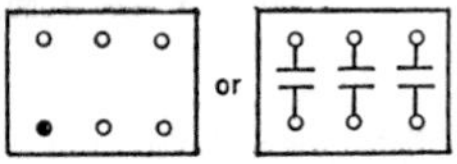

Three Pole (Double Throw) (With Terminals)

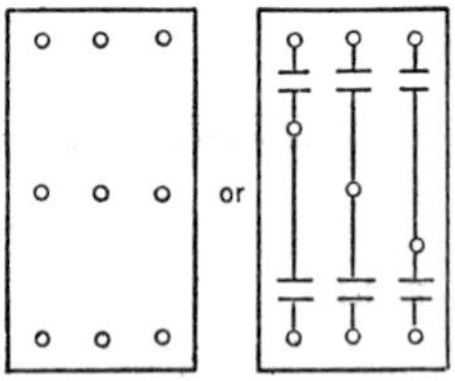

Auxiliary Switch (With Terminals)

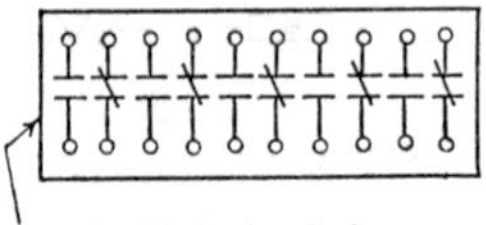

Operating Mechanism End

Thermal Trip Air Circuit Break (With Terminals)

5. Coils (Also See Windings)

Operating Coil

*This symbol must always be used with an identifying legend within or adjacent to the circle.

Blowout Coil

With Leads   With Terminals

6. Connections

Mechanical Connection

Adaptations

Mechanical Connection with Fulcrum

Mechanical Interlock

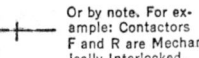

Or by note. For example: Contactors F and R are Mechanically Interlocked.

Direct Connected Units

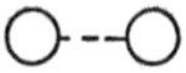

7. Contacts—Electrical

Note: NO—Normally Open, NC—Normally Closed, designates the position of the contacts when the main device is in the de-energized or non-operated position.

Normally Closed Contact (NC)

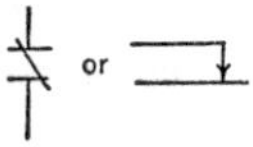

Normally Open Contact (NO)

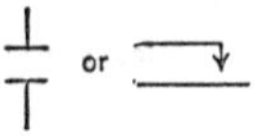

Adaptations (Also See Contactors, Relay, and Switches

NO Contact with Time Closing (TC) Feature

TC

NC Contact with Time Opening (TO) Feature

TO

8. Contactors

Fundamental Symbols for Contacts, Coils, Mechanical Connections, etc., Are the Basis of Contactor Symbols

Adaptations

Three-pole Manually Operated Contactor without Blowout Coils

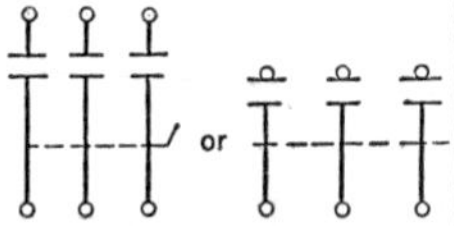

Three-pole Electrically, Operated Contactor with Blowout Coils and 2-NO and 1-NC Auxiliary Contacts (See Note under Contacts)

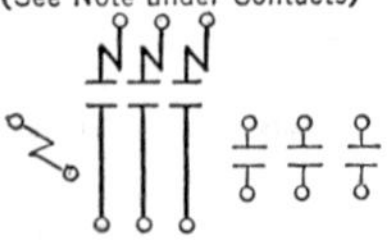

or

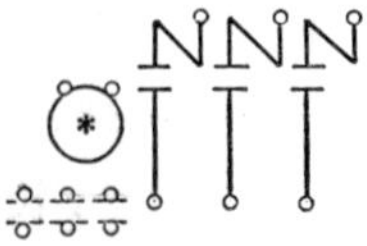

Three-pole Electrically Operated Contactor with Blowout Coils, 2-NO and 1 NC Main Contacts, and 1-NO Auxiliary Contact with Time Closing (See Note under Contacts)

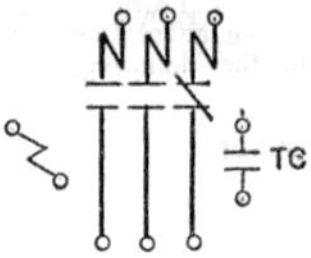

or

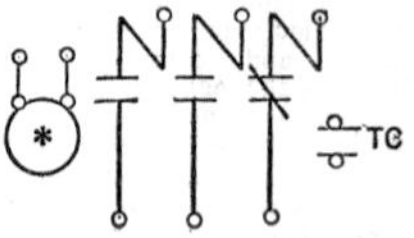

"Double-Throw" Electrically Operated Contactor with 1-NO and 1-NC Main Contacts with Common Connection 3-NO and 1-NC Auxiliary Contacts, Spring or Gravity Return (See Note under Contacts)

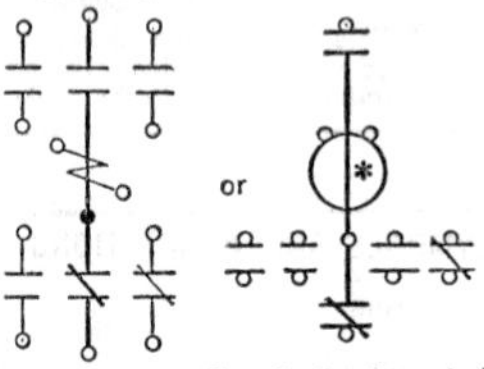

Three-pole Electrically Operated Contactor with Oil-immersed Contacts and Time-delay Undervoltage Release, Instantaneous Trip from Trip Coil, Time-delay Opening Auxiliary Contact, and 3-NO and 3-NC Auxiliary Contacts with Common Points (See Note under Contacts)

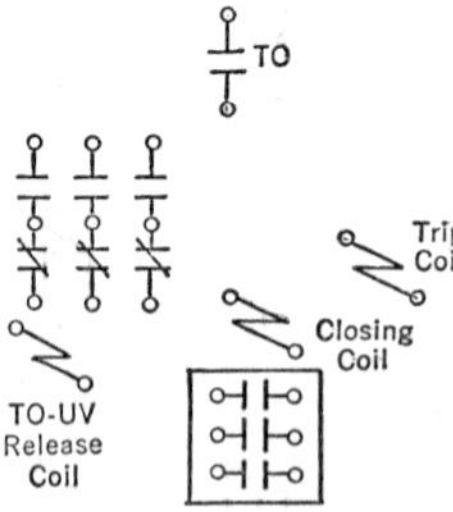

Single-pole Electrically Operated Contactor without Blowout Coil, with Operating Coil and Series Holding Coil 1-NC Auxiliary Contacts (See Note under Contacts)

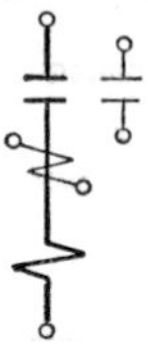

Single-pole Electrically Operated Contactor with Blowout Coil and 1 NO Auxiliary Contact (See Note under Contacts)

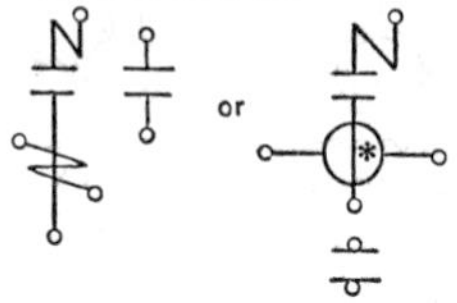

Two-pole Electrically Operated Contactor with Blowout Coils and 1-NO and 1-NC Auxiliary Contacts (See Note under Contacts)

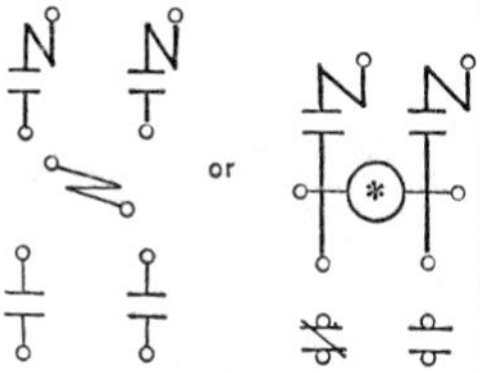

Two Single-pole Electrically Operated Contactors with Operating Coils and Separately Excited Blowout Coils, Shown Mechanically Interlocked

(Mechanical Interlocking may also be indicated by notes, for example: "Contactors F and R are mechanically interlocked")

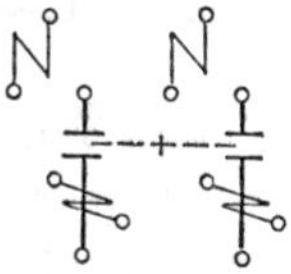

Multipole Electrically Operated Contactor with 5-NO and 2-NC Contacts (See Note under Contacts)

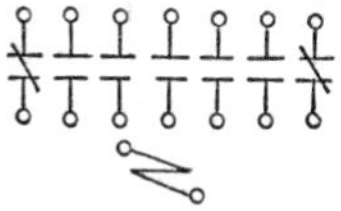

"Double-throw" Electrically Operated Contactor with 2-NO and 2-NC Main Contacts with Blowout Coils and 2-NO Auxiliary Contacts with a Common Connection (See Note under Contacts)

(The upper coil is the operating coil. The lower coil is the holding coil which, when excited, locks the contactor in the de-energized position.)

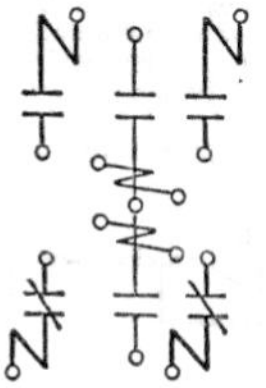

9. Fuse

General

Current Responsive Element (Fuse Element)

10. Indicating Lights

General

Adaptations

With Leads — With Terminals

Note: Letter or letters shall indicate color as follows:

A	Amber	P	Purple
B	Blue	R	Red
C	Clear	W	White
G	Green	Y	Yellow
O	Orange	FL	Fluorescent
OP	Opalescent		

In case of conflict with any other symbol, spell out.

11. Instruments and Meters

General

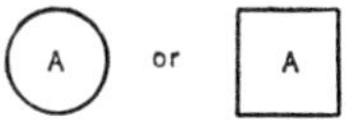

Note: Letter or letters shall be placed within circle or rectangle to indicate type of instrument.

A	Ammeter
AH	Ampere-hour Meter
CMC	Contact-making Clock
D	Demand Meter
F	Frequency Meter
G	Galvanometer
GD	Ground Detector
I	Indicating
M	Integrating
OHM	Ohmmeter
OSC	Oscillograph
PH	Phase Meter
PI	Position Indicator
PF	Power Factor Meter
RF	Reactive Factor Meter
REC	Recording
RD	Recording Demand Meter
S	Synchroscope
TLM	Telemeter
T	Temperature Meter
VRH	Var-hour Meter
VAR	Varmeter
V	Voltmeter
WH	Watt-hour Meter
W	Wattmeter

In case of conflict with any other symbol, spell out.

Adaptations

Instruments Showing Terminals: Ammeter

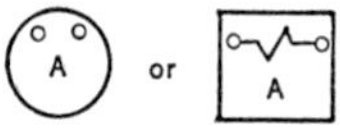

Voltmeter

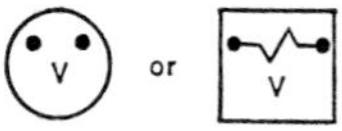

Wattmeter

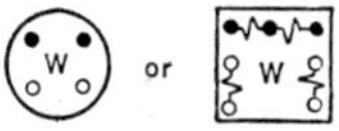

Direct-current Watthour Meter

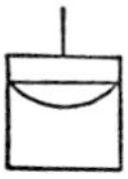

Note:. For complete symbol show view approximating outline of actual instrument, indicate terminals in relative locations, and show potential terminals with solid circle, current terminals with open circle. Scale range and manufacturers' type numbers may be marked adjacent to symbol if desired.

12. Lightning Arresters

General

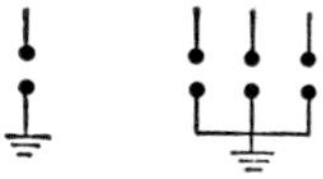

Adaptations

Auxiliary Series Sphere-gap Type

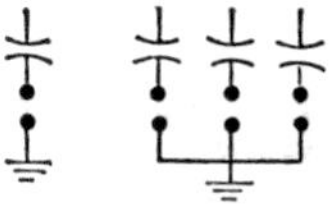

Electrolytic or Aluminum Cell Type

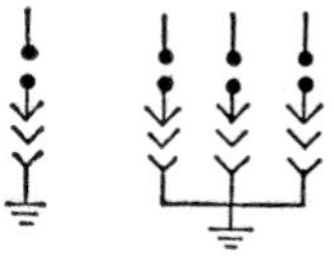

Valve or Film Type

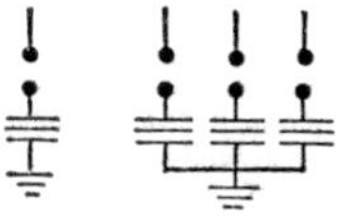

One Line — Complete for Three Phase or Three Wire

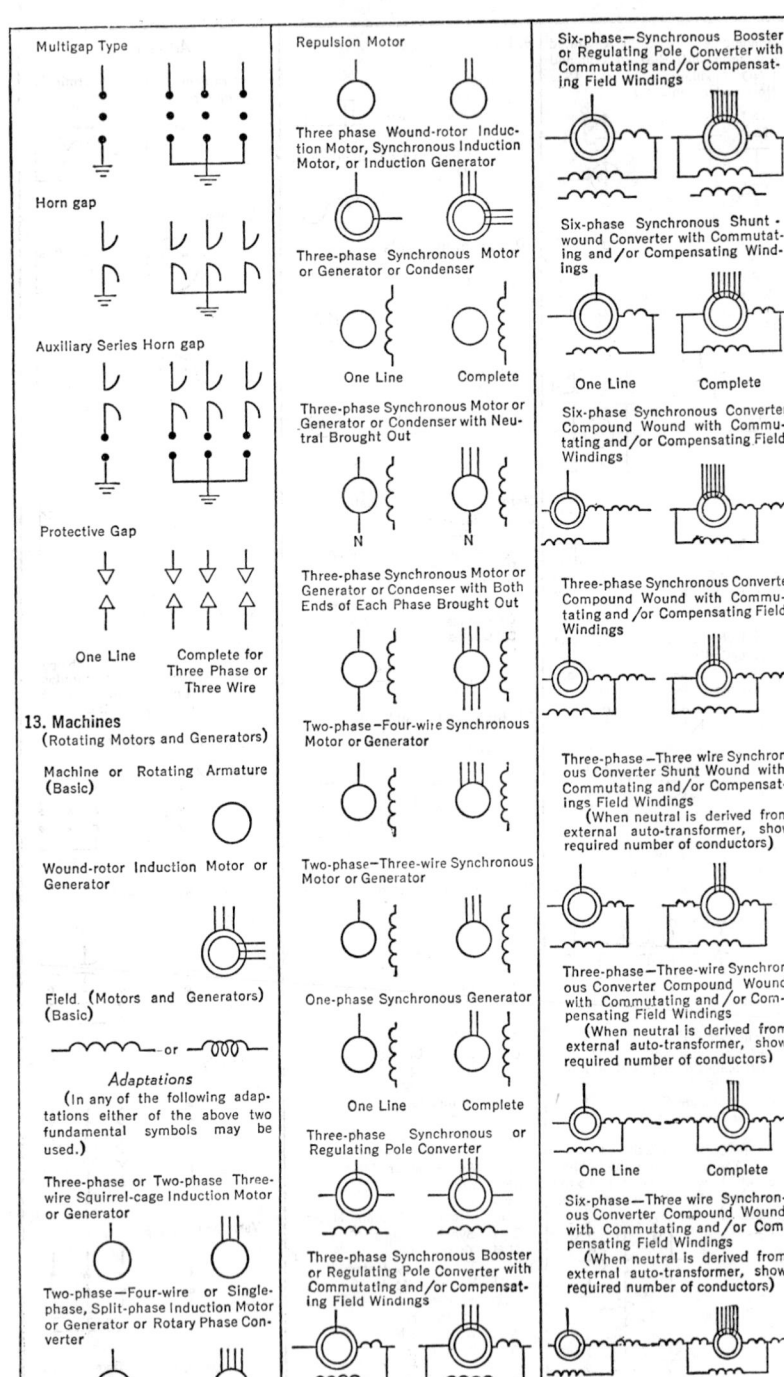

Multigap Type

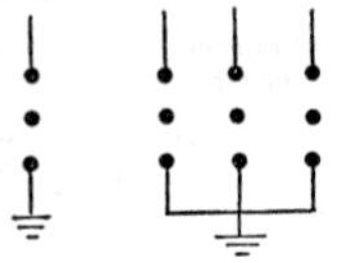

Horn gap

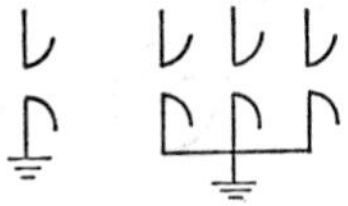

Auxiliary Series Horn gap

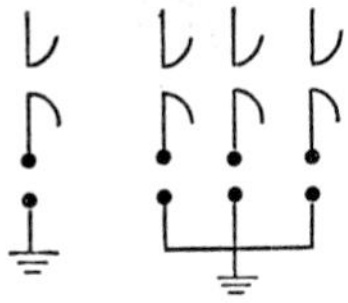

Protective Gap

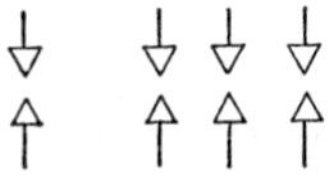

One Line — Complete for Three Phase or Three Wire

13. Machines

(Rotating Motors and Generators)

Machine or Rotating Armature (Basic)

Wound-rotor Induction Motor or Generator

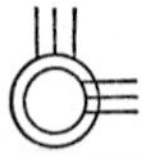

Field (Motors and Generators) (Basic)

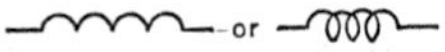

Adaptations

(In any of the following adaptations either of the above two fundamental symbols may be used.)

Three-phase or Two-phase Three-wire Squirrel-cage Induction Motor or Generator

Two-phase—Four-wire or Single-phase, Split-phase Induction Motor or Generator or Rotary Phase Converter

Repulsion Motor

Three phase Wound-rotor Induction Motor, Synchronous Induction Motor, or Induction Generator

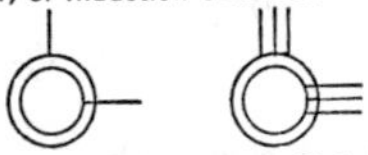

Three-phase Synchronous Motor or Generator or Condenser

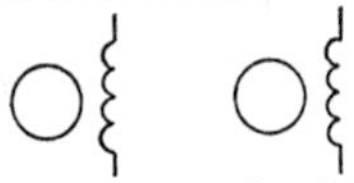

One Line — Complete

Three-phase Synchronous Motor or Generator or Condenser with Neutral Brought Out

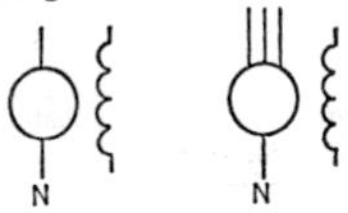

Three-phase Synchronous Motor or Generator or Condenser with Both Ends of Each Phase Brought Out

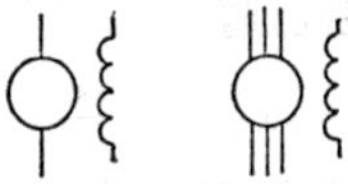

Two-phase—Four-wire Synchronous Motor or Generator

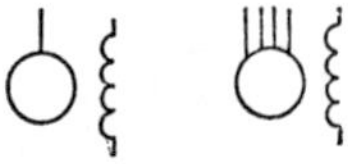

Two-phase—Three-wire Synchronous Motor or Generator

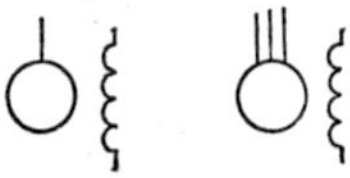

One-phase Synchronous Generator

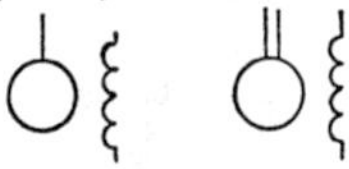

One Line — Complete

Three-phase Synchronous or Regulating Pole Converter

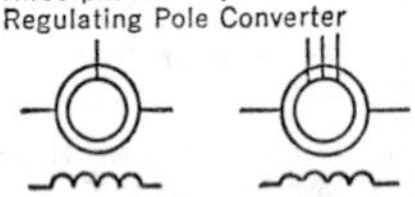

Three-phase Synchronous Booster or Regulating Pole Converter with Commutating and/or Compensating Field Windings

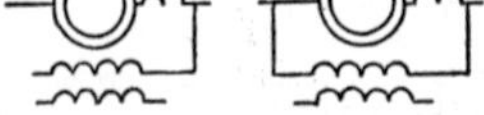

Six-phase—Synchronous Booster or Regulating Pole Converter with Commutating and/or Compensating Field Windings

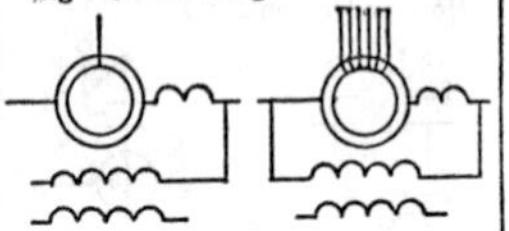

Six-phase Synchronous Shunt-wound Converter with Commutating and/or Compensating Windings

One Line — Complete

Six-phase Synchronous Converter Compound Wound with Commutating and/or Compensating Field Windings

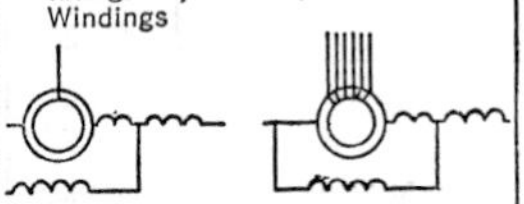

Three-phase Synchronous Converter Compound Wound with Commutating and/or Compensating Field Windings

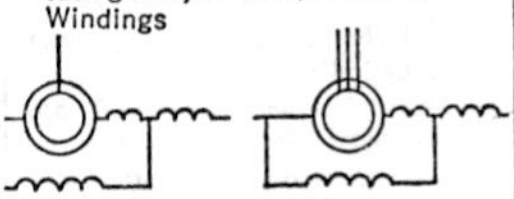

Three-phase—Three wire Synchronous Converter Shunt Wound with Commutating and/or Compensatings Field Windings

(When neutral is derived from external auto-transformer, show required number of conductors)

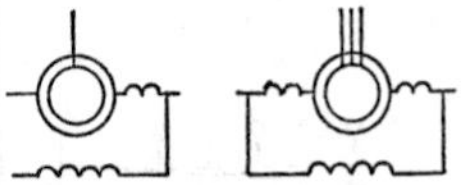

Three-phase—Three-wire Synchronous Converter Compound Wound with Commutating and/or Compensating Field Windings

(When neutral is derived from external auto-transformer, show required number of conductors)

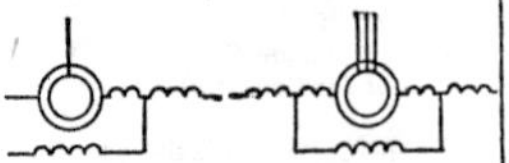

One Line — Complete

Six-phase—Three wire Synchronous Converter Compound Wound with Commutating and/or Compensating Field Windings

(When neutral is derived from external auto-transformer, show required number of conductors)

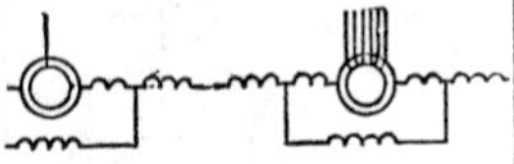

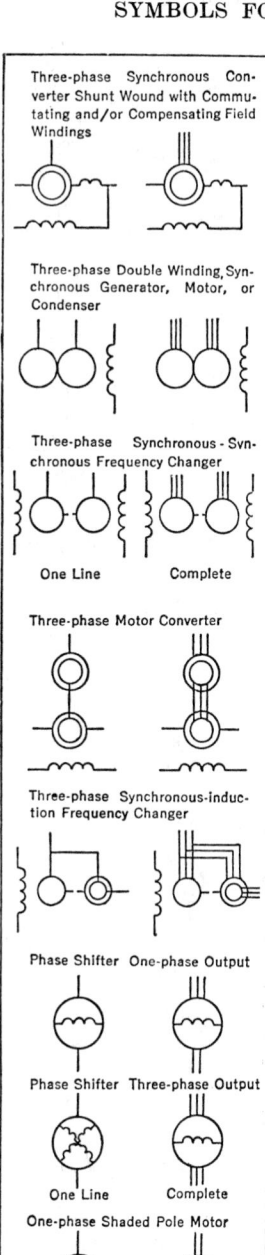

Three-phase Synchronous Converter Shunt Wound with Commutating and/or Compensating Field Windings

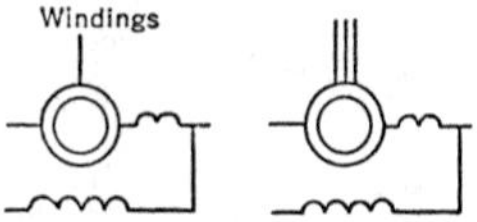

Three-phase Double Winding, Synchronous Generator, Motor, or Condenser

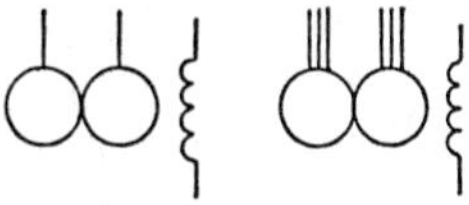

Three-phase Synchronous - Synchronous Frequency Changer

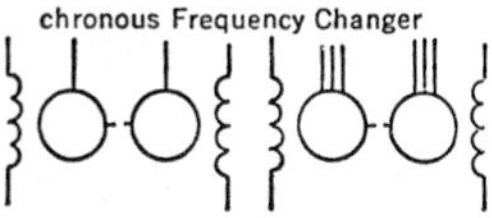

One Line Complete

Three-phase Motor Converter

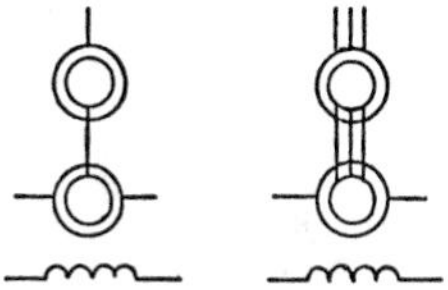

Three-phase Synchronous-induction Frequency Changer

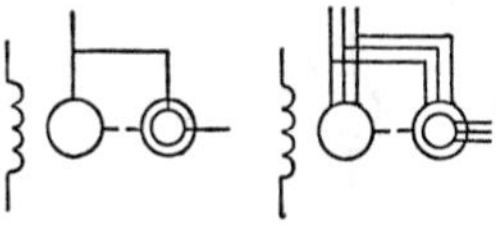

Phase Shifter One-phase Output

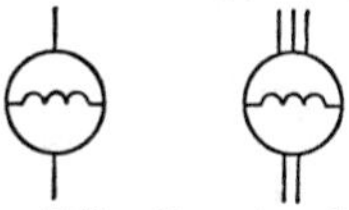

Phase Shifter Three-phase Output

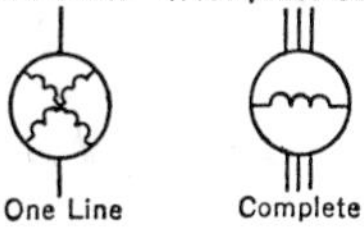

One Line Complete

One-phase Shaded Pole Motor

One-phase Repulsion Start Induction Motor

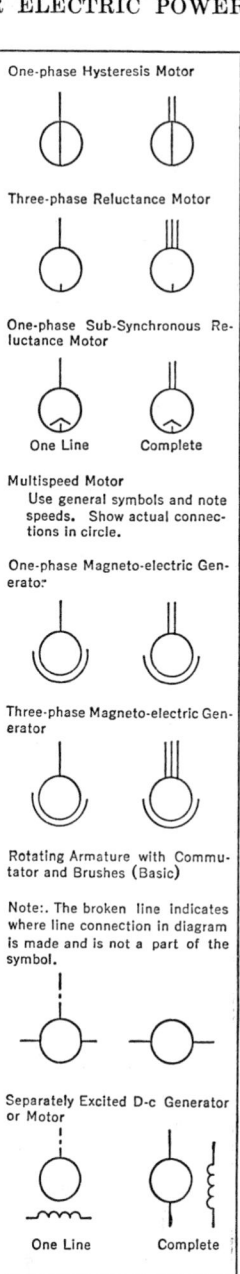

One-phase Hysteresis Motor

Three-phase Reluctance Motor

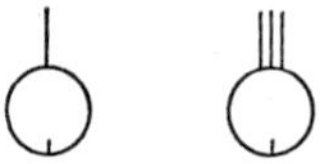

One-phase Sub-Synchronous Reluctance Motor

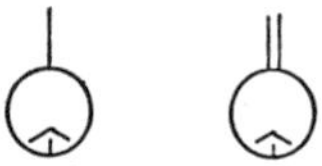

One Line Complete

Multispeed Motor

Use general symbols and note speeds. Show actual connections in circle.

One-phase Magneto-electric Generator

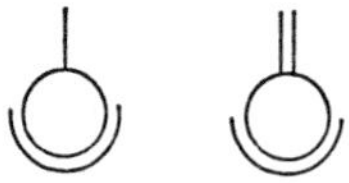

Three-phase Magneto-electric Generator

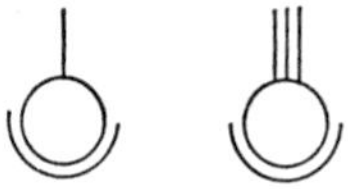

Rotating Armature with Commutator and Brushes (Basic)

Note: The broken line indicates where line connection in diagram is made and is not a part of the symbol.

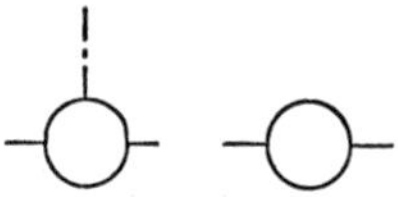

Separately Excited D-c Generator or Motor

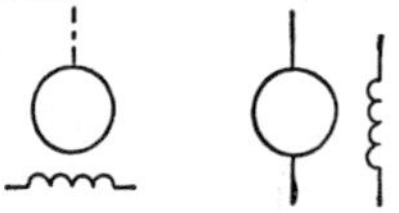

One Line Complete

Separately Excited D-c Generator or Motor with Commutating and/or Compensating Field Winding

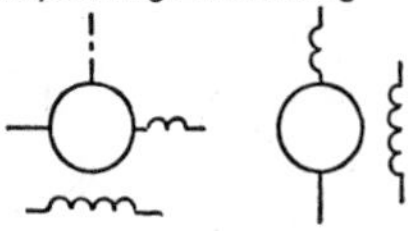

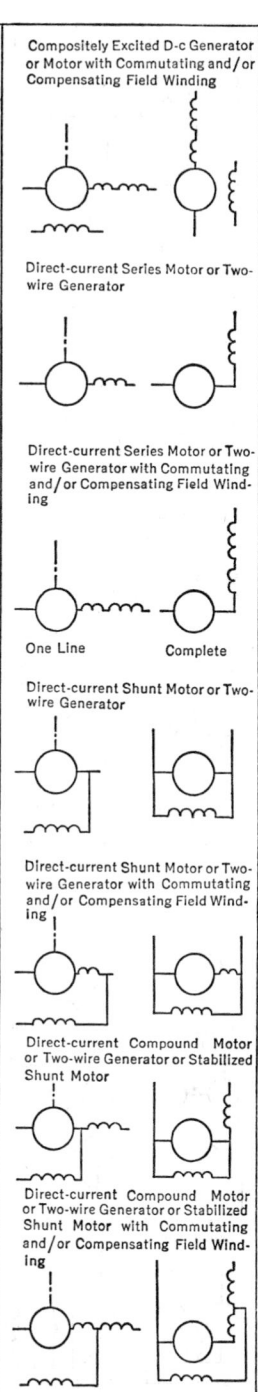

Compositely Excited D-c Generator or Motor with Commutating and/or Compensating Field Winding

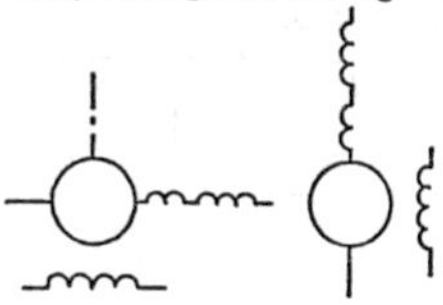

Direct-current Series Motor or Two-wire Generator

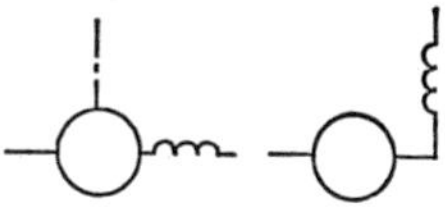

Direct-current Series Motor or Two-wire Generator with Commutating and/or Compensating Field Winding

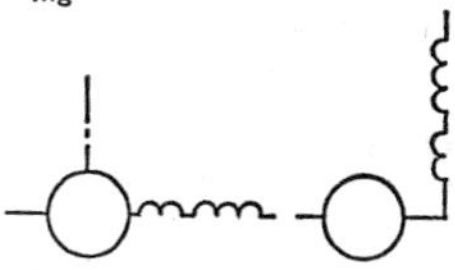

One Line Complete

Direct-current Shunt Motor or Two-wire Generator

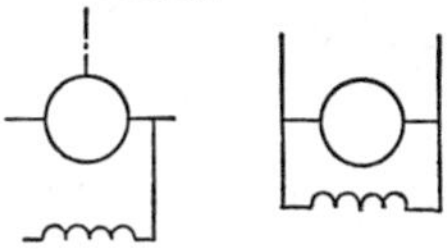

Direct-current Shunt Motor or Two-wire Generator with Commutating and/or Compensating Field Winding

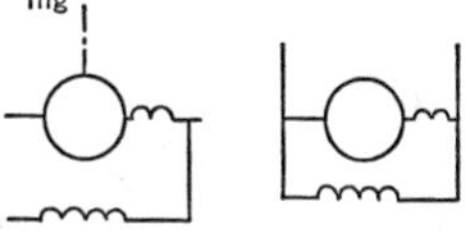

Direct-current Compound Motor or Two-wire Generator or Stabilized Shunt Motor

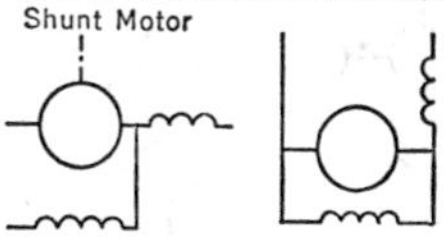

Direct-current Compound Motor or Two-wire Generator or Stabilized Shunt Motor with Commutating and/or Compensating Field Winding

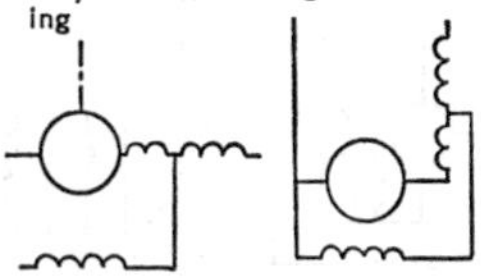

One Line Complete

Direct-current Three-wire Shunt Generator
(When neutral is derived from external auto-transformer show required number of conductors)

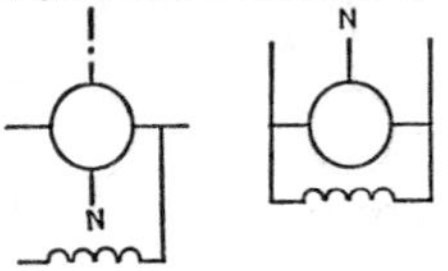

Direct-current Three-wire Shunt Generator with Commutating and/or Compensating Field Winding
(When neutral is derived from external auto-transformer show required number of conductors)

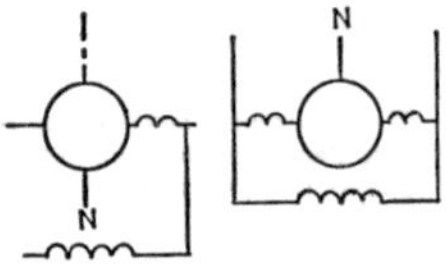

Direct-current Three-wire Compound Generator
(When neutral is derived from external auto-transformer show required number of conductors)

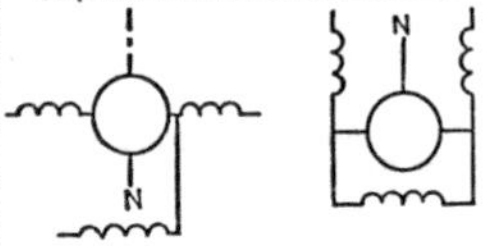

Direct-current Three-wire Compound Generator with Commutating and/or Compensating Field Winding
(When neutral is derived from external auto-transformer, show required number of conductors)

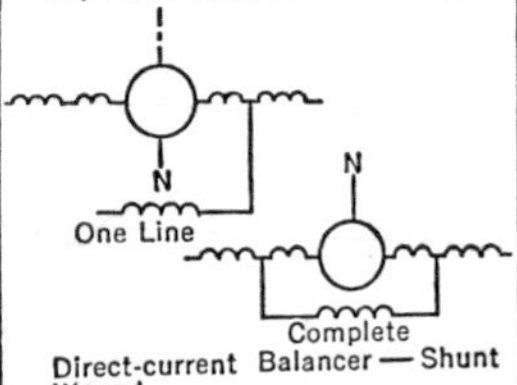

Direct-current Balancer — Shunt Wound

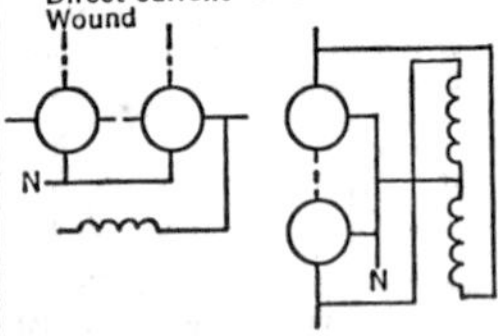

Direct-current Balancer — Compound Wound

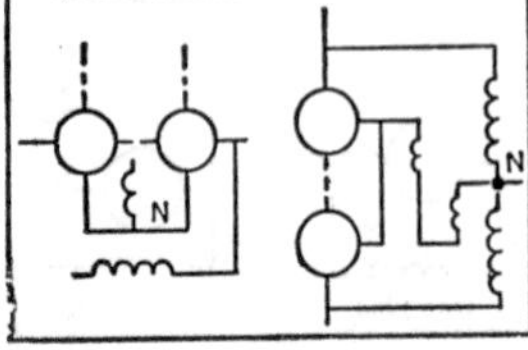

Dynamotor

One Line | Complete

Double-current Generator

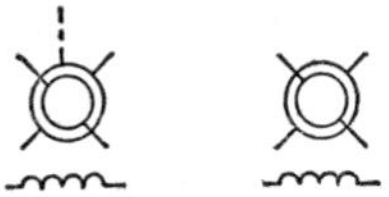

Acyclic Generator (Separately Excited)

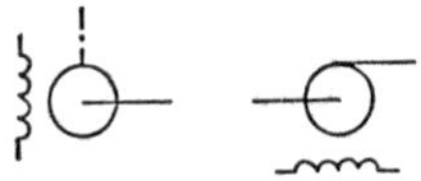

Regulating Exciter—Shunt Wound

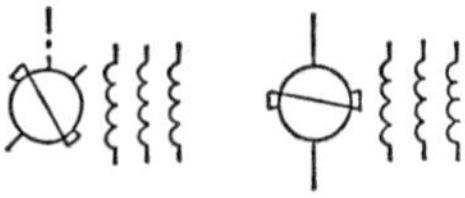

Regulating Exciter—Shunt Wound with Compensating Field Winding

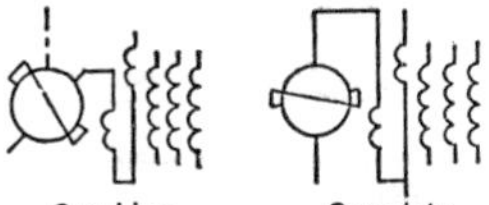

One Line | Complete

Three-phase or Two-phase—Three-wire Shunt Characteristic Brush-shifting Motor

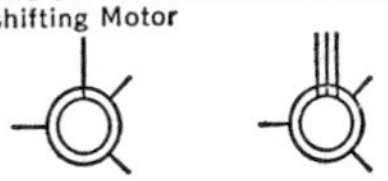

Two-phase—Four-wire Shunt Characteristic Brush-shifting Motor

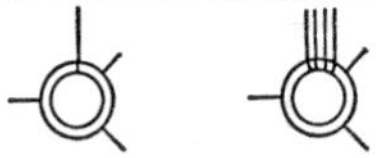

Three-phase or Two-phase—Three-wire Series Characteristic Brush-shifting Motor, with Three-phase Rotor

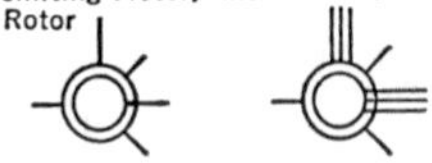

Two-phase—Four-wire Series Characteristic Brush-shifting Motor, with Three-phase Rotor

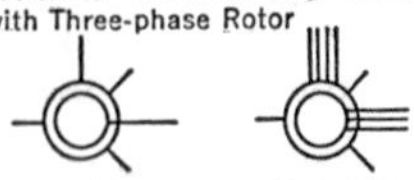

One Line | Complete

Three-phase or Two-phase—Three-wire Series Characteristic Brush-shifting Motor, with Six-phase Rotor

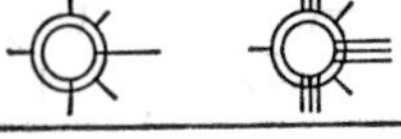

Two-phase—Four-wire Series Characteristic Brush-shifting Motor, with Six-phase Rotor

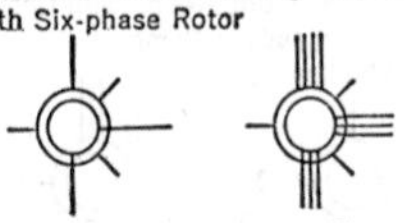

Three-phase or Two-phase—Three-wire Series Characteristic Brush-shifting Motor, with Eight-phase Rotor

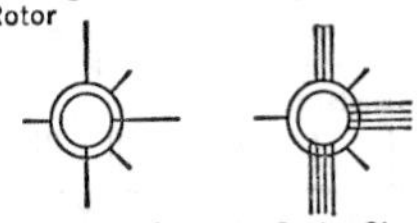

Two-phase—Four-wire Series Characteristic Brush-shifting Motor, with Eight-phase Rotor

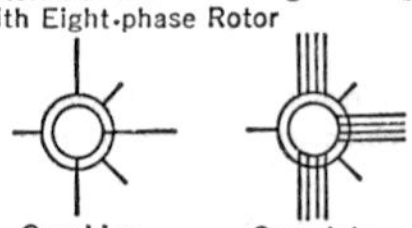

One Line | Complete

Ohmic Drop Exciter with Three-phase Input

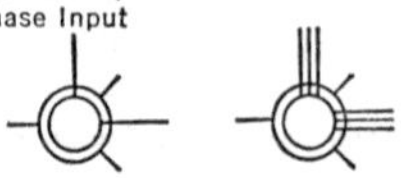

Ohmic Drop Exciter with Six-phase Input

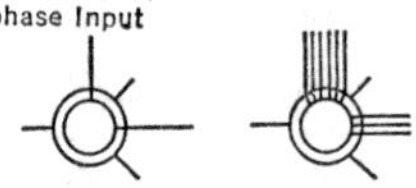

Ohmic Drop Exciter with Three-Input, and Six Output Leads Brought Out

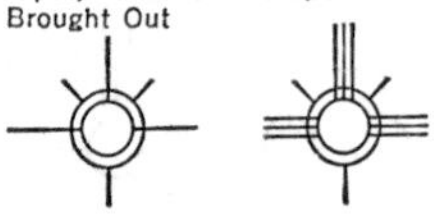

Ohmic Drop Exciter with Six-phase Input, and Six Output Leads Brought Out

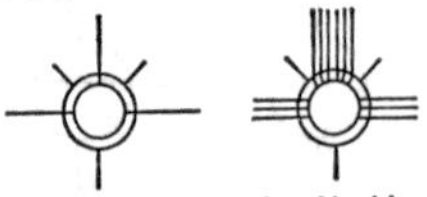

Three-phase Regulating Machine

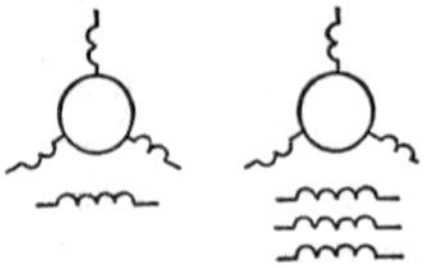

One Line | Complete

Winding Symbols (Winding symbols may be shown in circle, as follows, for all motor and generator symbols)

One-phase

Two-phase

Three-phase Wye (Ungrounded)

Three-phase Wye (Grounded)

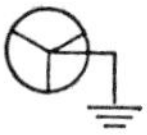

Three-phase Delta

Six-phase Diametrical

Six-phase Double Delta

14. Meters (See Instruments)

15. Plug Connections

Disconnecting Device (Coupling or plug-type contact)

Adaptations

Plug

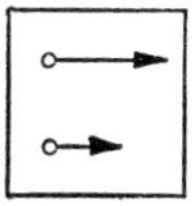

Receptacle

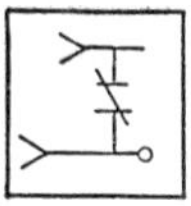

Receptacle

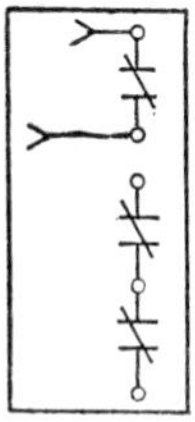

16. Rectifiers

Dry or Electrolytic Rectifier (For schematic or elementary diagram)

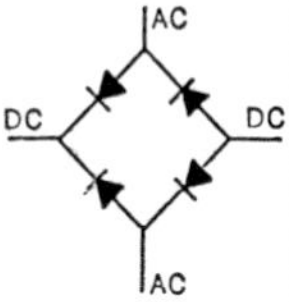

Half Wave Full Wave

Adaptations

Rectifier (On complete diagram rectifiers may be shown with terminals and polarity markings. Heavy line may be used to indicate nameplate or positive polarity end)

17. Relays

Fundamental symbols for contacts, coils, mechanical connections, etc., are the basis of relay symbols.

Adaptations

Overcurrent or Overvoltage Relay with One NO Contact

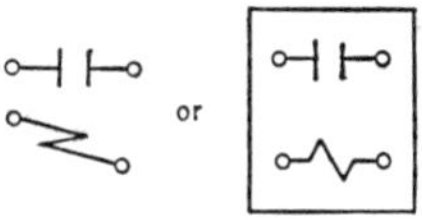

Overload Relay Having Two Current Coils and One—NC Contact with Time Opening

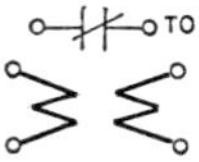

Thermal Overload Relay Having Two Series Heating Elements and One NC Contact

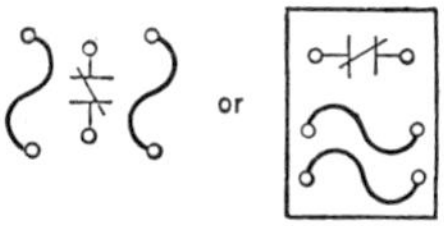

Open-phase or Phase-reversal Relay with Four Current Coils and One-NC Contact

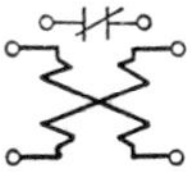

Relay Having Magnetic Time Delay Drop-out with Operating Coil and Bucking Coil; One-NO and One-NC Contact; Relative Position of Coils Indicated on Diagram

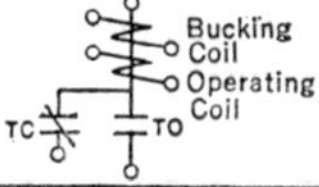

Magnetic Time Delay Drop-out Type Relay with Operating Coil; One-NC and Two-NO Contacts

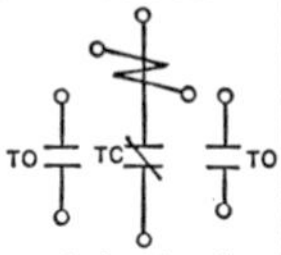

Field Forcing and Accelerating Relay; One-NO Contact, with Blow-out Coil; Shunt and Series Operating Coils Connected Cumulative

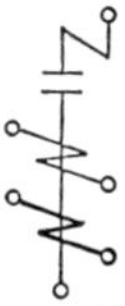

Multicircuit Relay with Two-NO and Two-NC Contacts, Operating Coil

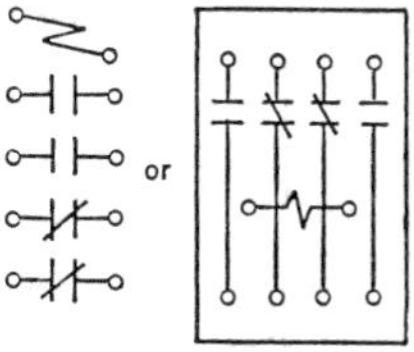

Motor-driven Timing Relay

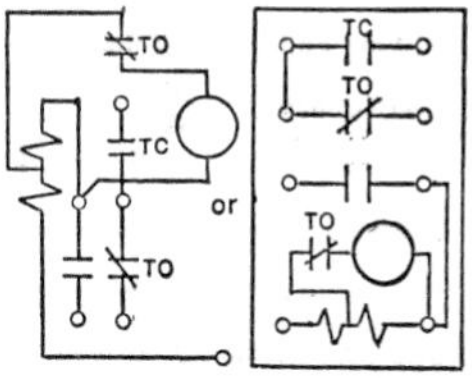

Multicircuit Time Relay with One-NO and One-NC Instantaneous Contacts; Four-NO and One-NC Timed Contacts; Timing on Each Contact is Individually Adjustable

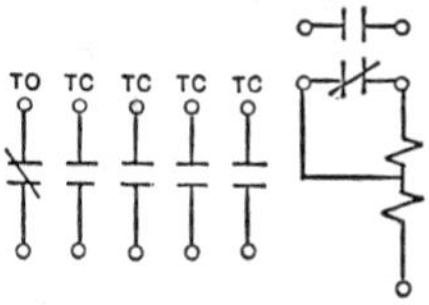

A Relay Designed to Close a Circuit When a Set Amount of Power Flows in a Given Direction through its Coils

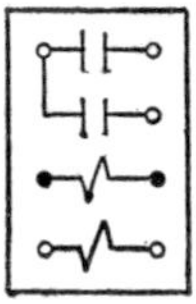

A Latched-type Lockout Relay

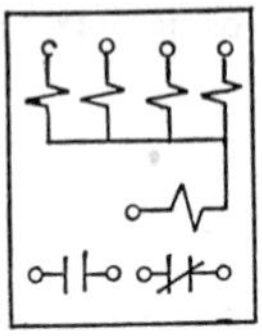

A Relay Which Furnishes Protection to a System from Overcurrent, Undercurrent and Reverse current

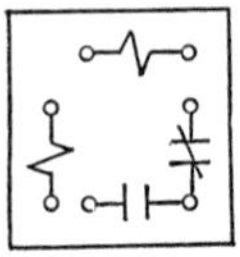

Relay with Operating and Release Coils; One-NO Main Contact with Blowout Coil and One-NO Interlock

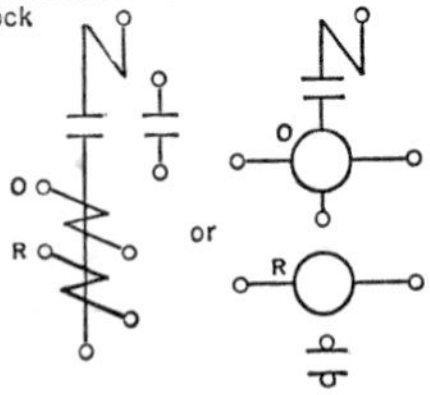

18. Relay Function Symbols

Over (General)

Under (General)

Direction (General) (Directional Over)

Balance (General)

Differential (General)

Pilot Wire (General)

Carrier Current (General)

Note:.

Operating Quantity. The operating quantity is indicated by the following letters or symbols placed either on or above center of the basic symbol given in the preceding column.

*Current	C
Distance	Z
Voltage	V
Power	W
Frequency	∼
Phase	ϕ
Temperature	T
Gas Pressure	GP
Synchronism	S

*The use of the letter may be omitted in the case of current and the absence of such letter presupposes that the relay operates on current.

Ground Relays Relays operative on residual current only are so designated by prefixing the ground symbol ⏚. (Note that the zero phase-sequence designation given below may be used instead when desirable.)

Phase Sequence Quantities Operation on phase sequence quantities may be indicated by the use of the conventional subscripts 1, 2, and 0 after the letter indicating the operating quantity.

Adaptations

Overcurrent

Directional Overcurrent

Directional Residual Overcurrent

Undervoltage

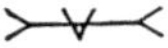

Power Directional

Balanced Current

Differential Current

Distance

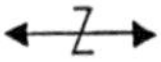

Directional Distance

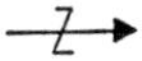

Overfrequency

Overtemperature

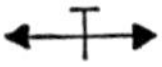

Phase Balance

Phase Rotation

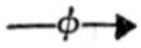

Pilot Wire (Differental Current)

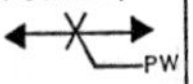

Pilot Wire (Directional Comparison)

Carrier Pilot

Positive-phase-sequence Undervoltage

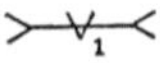

Negative-phase-sequence Overcurrent

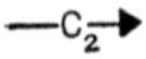

Gas-pressure Relay (Bucholz)

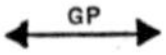

Out of Step

Note:.

Additional Information: Prefixes and/or suffixes to indicate relay types, inclusion of instantaneous trip attachments, time relay, etc., may be added at the discretion of the user.

19. Resistors

General

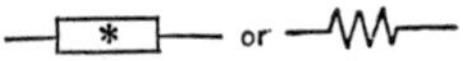

*This symbol must always be used with an identifying legend within or adjacent to the rectangle.

Adaptations

(In any of the following adaptations either of the above two fundamental symbols may be used.)

Resistor, Fixed

With Leads — With Terminals

Resistor, Adjustable by Fixed Taps

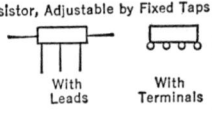

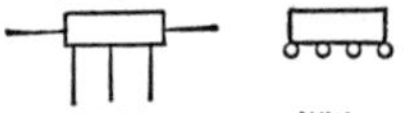

With Leads — With Terminals

Rheostat or Resistor, Adjustable Tap or Slide Wire

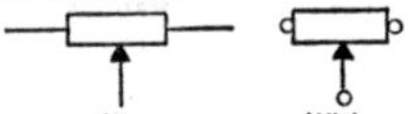

With Leads — With Terminals

Resistor, Continuously Adjustable

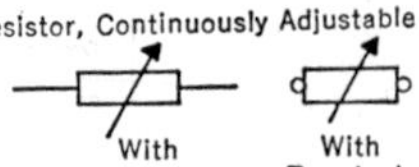

With Leads — With Terminals

Resistor, Voltage Divider

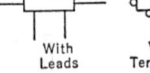

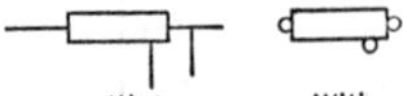

With Leads — With Terminals

Instrument or Relay Shunt

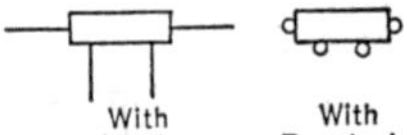

With Leads — With Terminals

Motor-driven Dial-type Rheostat (Motor Omitted) Detailed

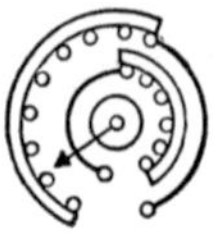

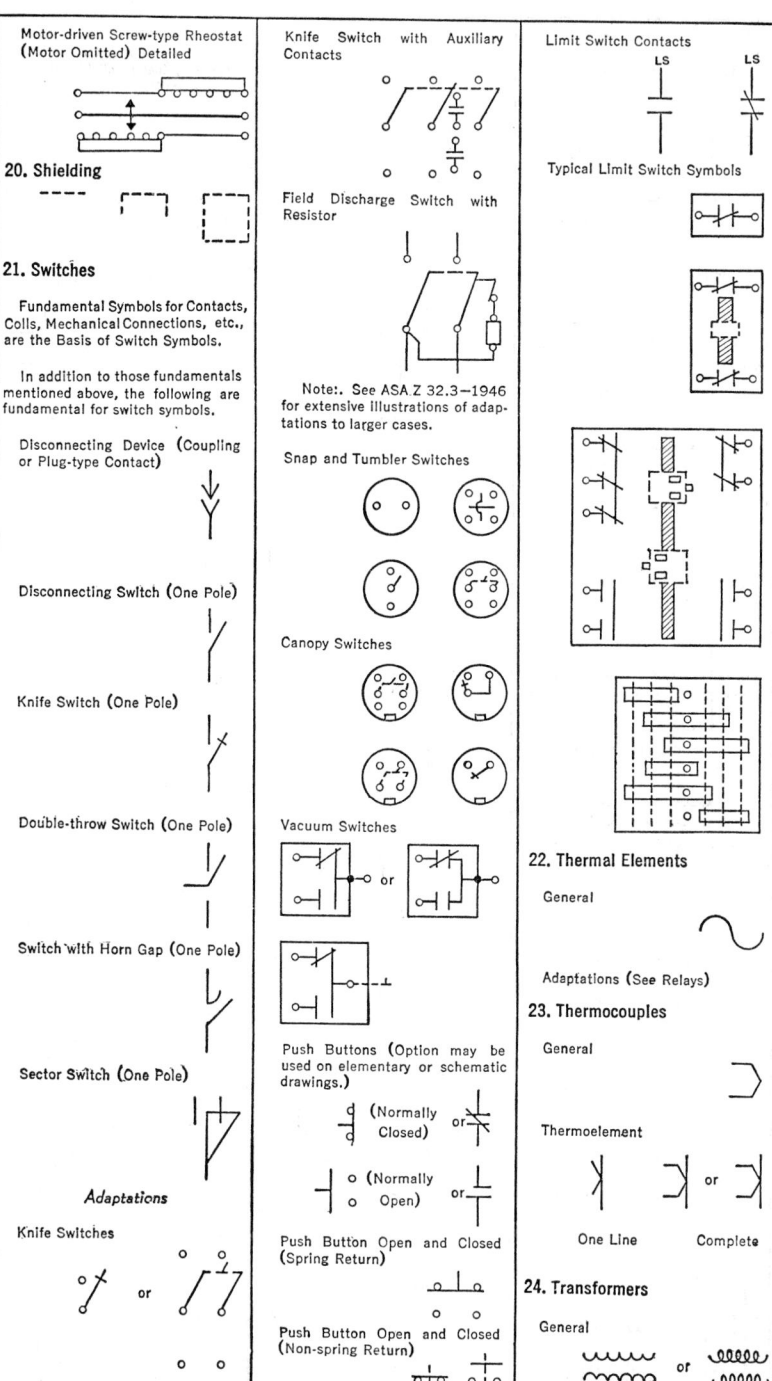

Motor-driven Screw-type Rheostat (Motor Omitted) Detailed

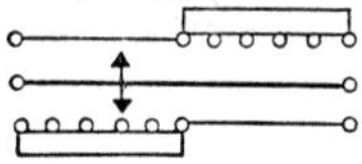

20. Shielding

21. Switches

Fundamental Symbols for Contacts, Coils, Mechanical Connections, etc., are the Basis of Switch Symbols.

In addition to those fundamentals mentioned above, the following are fundamental for switch symbols.

Disconnecting Device (Coupling or Plug-type Contact)

Disconnecting Switch (One Pole)

Knife Switch (One Pole)

Double-throw Switch (One Pole)

Switch with Horn Gap (One Pole)

Sector Switch (One Pole)

Adaptations

Knife Switches

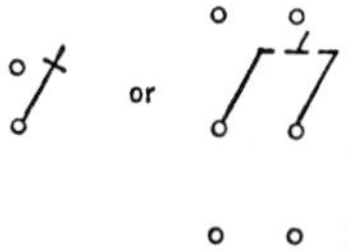

Knife Switch with Auxiliary Contacts

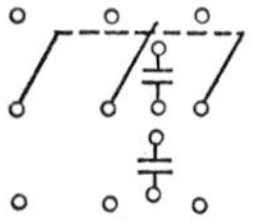

Field Discharge Switch with Resistor

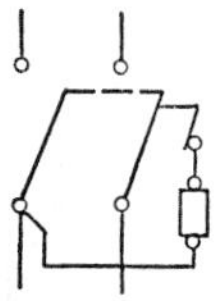

Note:. See ASA Z 32.3–1946 for extensive illustrations of adaptations to larger cases.

Snap and Tumbler Switches

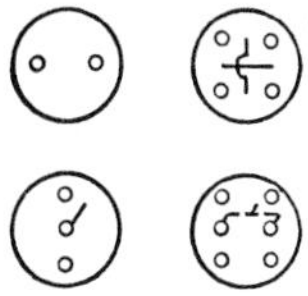

Canopy Switches

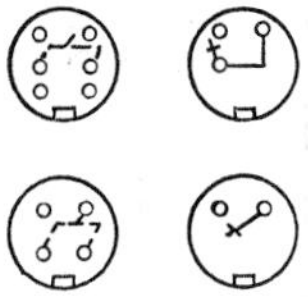

Vacuum Switches

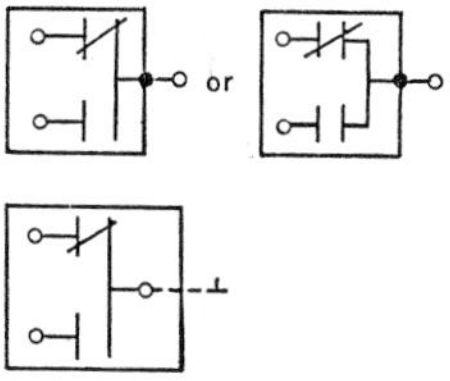

Push Buttons (Option may be used on elementary or schematic drawings.)

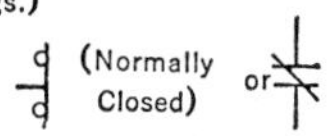

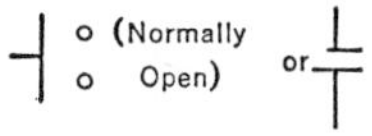

Push Button Open and Closed (Spring Return)

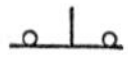

Push Button Open and Closed (Non-spring Return)

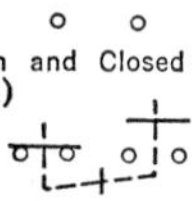

Limit Switch Contacts

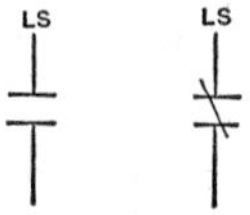

Typical Limit Switch Symbols

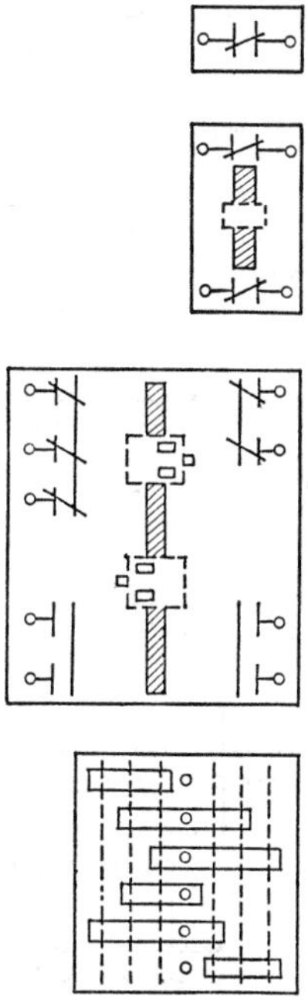

22. Thermal Elements

General

Adaptations (See Relays)

23. Thermocouples

General

Thermoelement

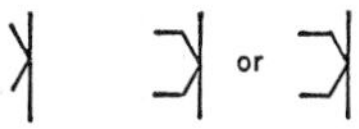

One Line Complete

24. Transformers

General

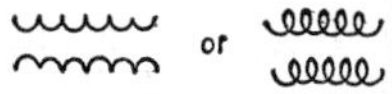

Adaptations

(In any of the following adaptations either of the two fundamental symbols may be used.)

One-phase Two-winding Transformer

Three-phase Bank of One-phase Two-winding Transformers

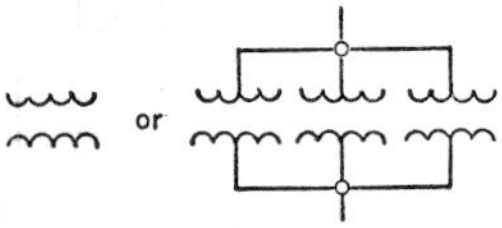

One Line

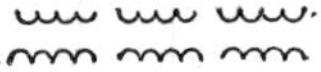

Complete

Polyphase Two-winding Transformer

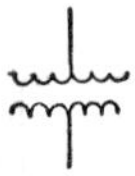

or

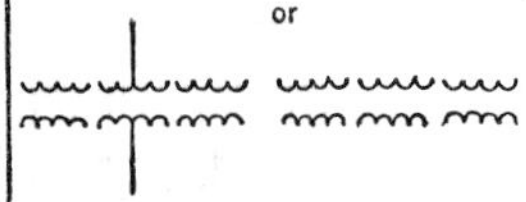

Three-winding Transformer, Single phase

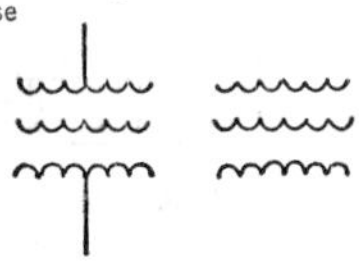

Transformer with Taps, Single Phase

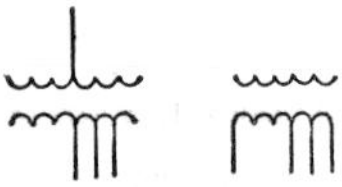

Constant-current Transformer, Single Phase

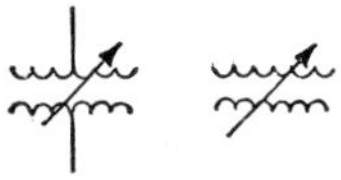

Auto-transformer, Single Phase

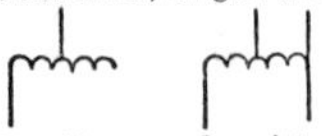

One Line Complete

Potential (Voltage) Transformer (s)

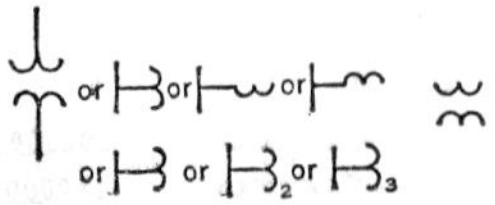

Current Transformer (s)

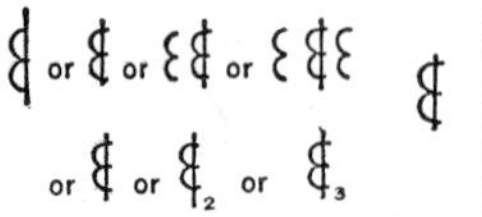

Bushing-type Current Transformer

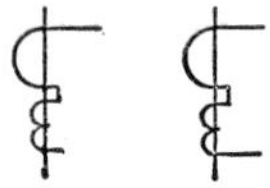

Outdoor Metering Outfit

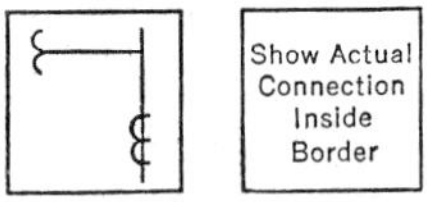

One Line Complete

Step-voltage Regulator or Load Ratio Control Auto-transformer

Load Ratio Control Transformer with Taps

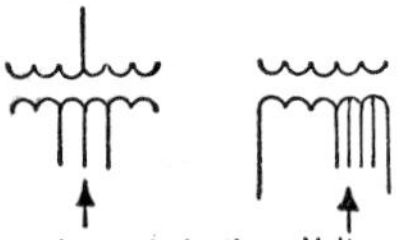

One-phase Induction Voltage Regulator

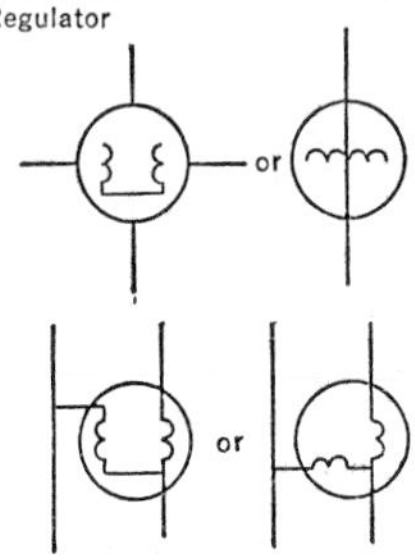

Three-phase Induction Voltage Regulator

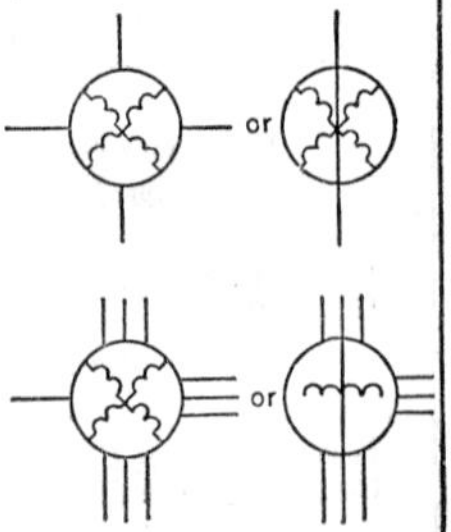

One Line Complete

Note:. When it is desired especially to distinguish magnetic core transformers, lines parallel to the axis of the scallops or loops be used.

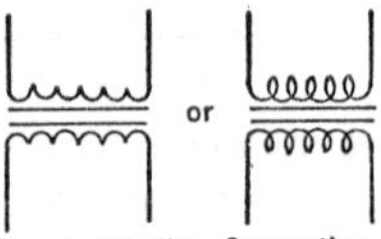

Transformer Winding Connection Symbols

Two-phase Three-wire Ungrounded

Two-phase Three-wire Grounded

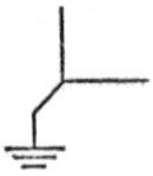

Two-phase Four-wire

Two-phase Five-wire Grounded

Three-phase Three-wire Delta (or Mesh)

Three-phase Four-wire Delta Ungrounded

Three-phase Four-wire Delta Grounded

Three-phase Wye Ungrouded (or Star)

Three-phase Wye Grounded Neutral

Three-phase Zig-zag Ungrounded

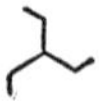

Three-phase Zig-zag Grounded

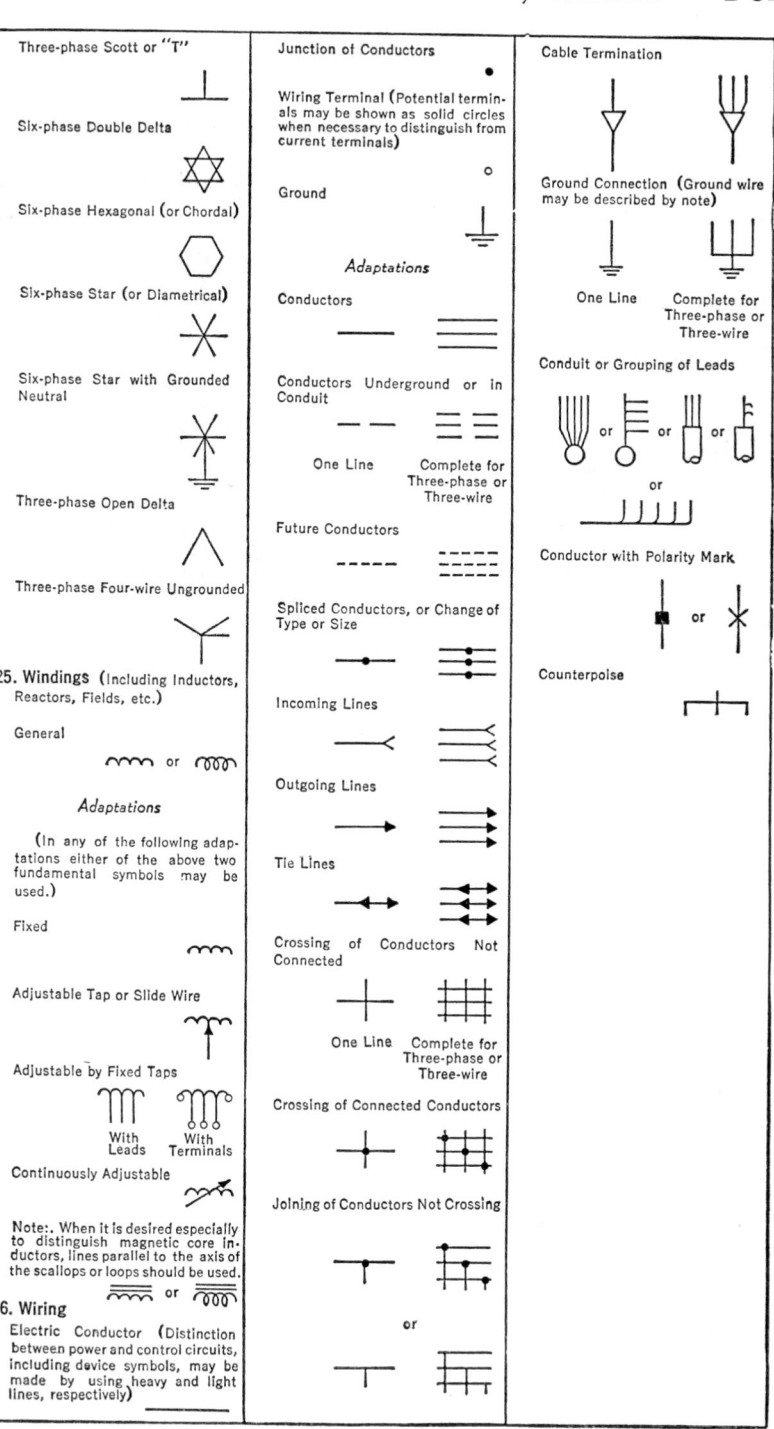

Three-phase Scott or "T"

Six-phase Double Delta

Six-phase Hexagonal (or Chordal)

Six-phase Star (or Diametrical)

Six-phase Star with Grounded Neutral

Three-phase Open Delta

Three-phase Four-wire Ungrounded

25. Windings (Including Inductors, Reactors, Fields, etc.)

General

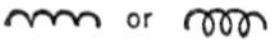

Adaptations

(In any of the following adaptations either of the above two fundamental symbols may be used.)

Fixed

Adjustable Tap or Slide Wire

Adjustable by Fixed Taps

With Leads

With Terminals

Continuously Adjustable

Note:. When it is desired especially to distinguish magnetic core inductors, lines parallel to the axis of the scallops or loops should be used.

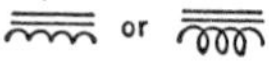

26. Wiring

Electric Conductor (Distinction between power and control circuits, including device symbols, may be made by using heavy and light lines, respectively)

Junction of Conductors

Wiring Terminal (Potential terminals may be shown as solid circles when necessary to distinguish from current terminals)

Ground

Adaptations

Conductors

Conductors Underground or in Conduit

One Line

Complete for Three-phase or Three-wire

Future Conductors

Spliced Conductors, or Change of Type or Size

Incoming Lines

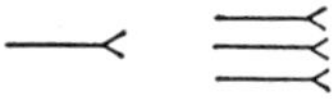

Outgoing Lines

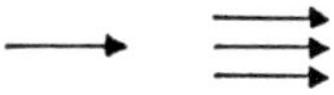

Tie Lines

Crossing of Conductors Not Connected

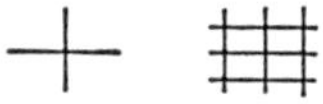

One Line

Complete for Three-phase or Three-wire

Crossing of Connected Conductors

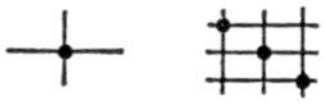

Joining of Conductors Not Crossing

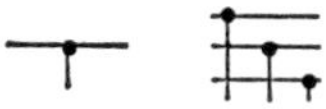

or

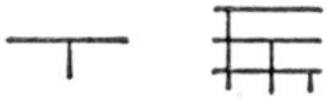

Cable Termination

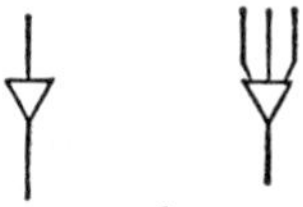

Ground Connection (Ground wire may be described by note)

One Line

Complete for Three-phase or Three-wire

Conduit or Grouping of Leads

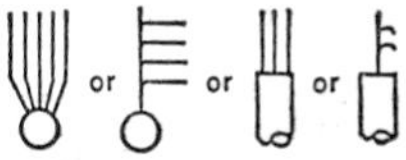

or

Conductor with Polarity Mark

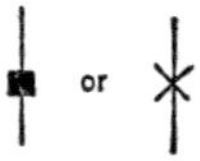

Counterpoise

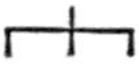

CONSTANTS

By Carl C. Chambers

21. PRINCIPAL PHYSICAL CONSTANTS AND RATIOS *

Velocity of light	$(2.99776 \pm 0.00004) \times 10^{10}$ cm sec^{-1}
Ratio of electrostatic to electromagnetic units	$(2.9971 \pm 0.0001) \times 10^{10}$ cm$^{1/2}$ sec$^{-1/2}$ (int ohms)$^{1/2}$ $(2.9978 \pm 0.0001) \times 10^{10}$ cm sec^{-1} (in absolute units)
Volume of a perfect gas (0 deg cent and normal atmospheric pressure)	$(22.4146 \pm 0.0006) \times 10^{3}$ cm^{3} mole^{-1}
Normal atmospheric pressure	$(1.013246 \pm 0.000004) \times 10^{6}$ dynes cm^{-2}
45 deg cent atmospheric pressure	$(1.013195 \pm 0.000004) \times 10^{6}$ dynes cm^{-2}
Ice point (absolute scale)	$273.18 \pm 0.01°$ K
Mechanical equivalent of heat (15 deg cent)	4.1855 ± 0.0004 abs joule cal^{-1}
Electrical equivalent of heat (15 deg cent)	4.1847 ± 0.0003 int joule cal^{-1}
Faraday constant	96494 ± 5 int coulombs g-equiv^{-1}
Electronic charge	$(4.8025 \pm 0.0010) \times 10^{-10}$ abs-es unit $(1.60203 \pm 0.00034) \times 10^{-20}$ abs-em unit
Planck constant	$(6.624 \pm 0.002) \times 10^{-27}$ erg sec
Acceleration of gravity	980.665 cm sec^{-2}
Electrochemical equivalent of silver	1.11800×10^{-3} g. int coulombs^{-1}
Wave length of red cadmium line (15 deg cent, normal atmospheric pressure)	6438.4696 I.A. †
Effective grating space of calcite (18 deg cent)	3.02904×10^{-8} cm
Avogadro's number	$(6.0228 \pm 0.0011) \times 10^{23}$ mole^{-1}
Boltzmann constant	$(1.3708 \pm 0.0014) \times 10^{-16}$ erg deg^{-1}
Stefan-Boltzmann constant	$(5.672 \pm 0.003) \times 10^{-5}$ erg cm^{-2} deg^{-4} sec^{-1}
Mass of the electron	$(9.1066 \pm 0.0032) \times 10^{-28}$ g
Ratio of mass of H to mass of electron (measured by deflection)	1837.5 ± 0.5

* Values taken from Birge, *Rev. of Mod. Phys.*, Vol. 13, No. 4 (October, 1941).

† This defines the international angstrom unit (I.A.). The unit is of the order of 1 part in several million different from 10^{-8} cm.

SECTION 2

PROPERTIES OF MATERIALS

PROPERTIES OF MATERIALS

WEIGHTS OF MATERIALS

By D. F. Miner
Revised by W. A. Del Mar

1. INTERNATIONAL ATOMIC WEIGHTS

1947

Published by the *Journal of the American Chemical Society*

	Sym- bol	Atomic Number	Atomic Weight		Sym- bol	Atomic Number	Atomic Weight
Aluminum	Al	13	26.97	Molybdenum	Mo	42	95.95
Antimony	Sb	51	121.76	Neodymium	Nd	60	144.27
Argon	A	18	39.944	Neon	Ne	10	20.183
Arsenic	As	33	74.91	Nickel	Ni	28	58.69
Barium	Ba	56	137.36	Nitrogen	N	7	14.008
Beryllium	Be	4	9.02	Osmium	Os	76	190.2
Bismuth	Bi	83	209.00	Oxygen	O	8	16.0000
Boron	B	5	10.82	Palladium	Pd	46	106.7
Bromine	Br	35	79.916	Phosphorus	P	15	30.98
Cadmium	Cd	48	112.41	Platinum	Pt	78	195.23
Calcium	Ca	20	40.08	Potassium	K	19	39.096
Carbon	C	6	12.010	Praseodymium	Pr	59	140.92
Cerium	Ce	58	140.13	Protactinium	Pa	91	231
Cesium	Cs	55	132.91	Radium	Ra	88	226.05
Chlorine	Cl	17	35.457	Radon	Rn	86	222
Chromium	Cr	24	52.01	Rhenium	Re	75	186.31
Cobalt	Co	27	58.94	Rhodium	Rh	45	102.91
Columbium	Cb	41	92.91	Rubidium	Rb	37	85.48
Copper	Cu	29	63.54	Ruthenium	Ru	44	101.7
Dysprosium	Dy	66	162.46	Samarium	Sm	62	150.43
Erbium	Er	68	167.2	Scandium	Sc	21	45.10
Europium	Eu	63	152.0	Selenium	Se	34	78.96
Fluorine	F	9	19.00	Silicon	Si	14	28.06
Gadolinium	Gd	64	156.9	Silver	Ag	47	107.880
Gallium	Ga	31	69.72	Sodium	Na	11	22.997
Germanium	Ge	32	72.60	Strontium	Sr	38	87.63
Gold	Au	79	197.2	Sulfur	S	16	32.066
Hafnium	Hf	72	178.6	Tantalum	Ta	73	180.88
Helium	He	2	4.003	Tellurium	Te	52	127.61
Holmium	Ho	67	164.94	Terbium	Tb	65	159.2
Hydrogen	H	1	1.0080	Thallium	Tl	81	204.39
Indium	In	49	114.76	Thorium	Th	90	232.12
Iodine	I	53	126.92	Thulium	Tm	69	169.4
Iridium	Ir	77	193.1	Tin	Sn	50	118.70
Iron	Fe	26	55.85	Titanium	Ti	22	47.90
Krypton	Kr	36	83.7	Tungsten	W	74	183.92
Lanthanum	La	57	138.92	Uranium	U	92	238.07
Lead	Pb	82	207.21	Vanadium	V	23	50.95
Lithium	Li	3	6.940	Xenon	Xe	54	131.3
Lutecium	Lu	71	174.99	Ytterbium	Yb	70	173.04
Magnesium	Mg	12	24.32	Yttrium	Y	39	88.92
Manganese	Mn	25	54.93	Zinc	Zn	30	65.38
Mercury	Hg	80	200.61	Zirconium	Zr	40	91.22

The following elements of engineering interest were not included in the table:

Neptunium	Np	93	239
Plutonium	Pu	94	239

2. WEIGHTS (MASSES) OF MATERIALS

The density of any substance is the mass of that substance per unit volume. Of, if weight is used in the ordinary sense as equivalent to mass, the density may also be defined as the *weight* per unit volume. The numerical value of the density of any substance depends upon the unit in which the mass or weight is expressed and also upon the unit of volume used; see Units and Conversion Factors, Section 1. It is quite common to state the density of a substance in grams per cubic centimeter, however, without naming the units, since when so expressed the density is numerically equal (practically) to the specific gravity.

The specific gravity of a substance is defined as the ratio of the weight (mass) per unit volume of that substance to the weight (mass), expressed in the same unit, of an equal volume of water. To make such a statement exact, the temperature of the water should be specified. There is no general agreement as to the temperature of reference, although water at 0 deg cent is commonly taken as the reference temperature. For gases, air at 0 deg cent and 760 mm mercury pressure is frequently taken as the reference substance instead of water.

VARIATION OF DENSITY OF WATER WITH TEMPERATURE. Table 1 gives the results of measurements by Thiesen, Scheel, and Diesselhorst (Landolt, Börnstein, and Roth, *Physikalisch-chemische Tabellen*, 1913).

Table 1. Density of Water; Grams per Cubic Centimeter

Degrees Cent	0	1	2	3	4	5	6	7	8	9
0	0.99987	0.99993	0.99997	0.99999	1.00000	0.99999	0.99997	0.99993	0.99988	0.99981
10	0.99973	0.99963	0.99953	0.99940	0.99927	0.99913	0.99897	0.99880	0.99862	0.99843
20	0.99823	0.99802	0.99780	0.99756	0.99732	0.99707	0.99681	0.99654	0.99626	0.99597
30	0.99567	0.99537	0.99505	0.99473	0.99440	0.99406	0.99371	0.99336	0.99299	0.99262
40	0.99224	0.99186	0.99147	0.99107	0.99066	0.99025	0.98982	0.98940	0.98896	0.98852
50	0.98807	0.98762	0.98715	0.98669	0.98621	0.98573	0.98525	0.98475	0.98425	0.98375
60	0.98324	0.98272	0.98220	0.98167	0.98113	0.98059	0.98005	0.97950	0.97894	0.97838
70	0.97781	0.97723	0.97666	0.97607	0.97548	0.97489	0.97429	0.97368	0.97307	0.97245
80	0.97183	0.97121	0.97057	0.96994	0.96930	0.96865	0.96800	0.96734	0.96668	0.96601
90	0.96534	0.96467	0.96399	0.96330	0.96261	0.96192	0.96122	0.96051	0.95981	0.95909
100	0.95838	0.95765	0.95693	....	....	....	....	....	....	....

Example: The density of water at 33 deg cent is 0.99473.

WEIGHTS PER CUBIC FOOT AND SPECIFIC GRAVITY. In Table 2 are given the values of the density in pounds per cubic foot of the more commonly used substances. The specific gravity, or density in grams per cubic centimeter, corresponding to any weight per cubic foot w is equal to $w/62.43$; for the conversion factors necessary to convert these figures into densities for other units of mass and volume, see Units and Conversion Factors, Section 1.

Table 2. Specific Gravity and Pounds per Cubic Foot of Various Materials at Room Temperatures

Specific gravities all referred to water at 0 deg cent. For special alloys see Article 15, and for insulating materials see Article 21. See Section 14, Article 76, for cable materials.

Material	Pounds per Cubic Foot		Average Specific Gravity	Material	Pounds per Cubic Foot		Average Specific Gravity
	From	To			From	To	
Air *	0.0807	....	0.00129	Lead	708	....	11.34
Acetylene gas *	0.0733	....	0.00117	Leather, dry	54	....	0.86
Aluminum, cast	160	161	2.57	Leather, greased	64	....	1.02
Aluminum wire	168	....	2.70	Lime	81	87	1.35
Antimony	414	....	6.64	Limestone	167	171	2.72
Asbestos	125	175	2.40	Loam	65	88	1.23
Asphaltum	69	94	1.30	Marble	160	177	2.72
Basalt	150	190	2.7	Masonry	100	165	2.12
Bismuth	604	618	9.78	Mercury at 0° C	849	....	13.6
Brass	511	542	8.45	Mercury at 20° C	846	....	13.55
Brick, red	111	128	1.92	Methyl methacrylate	69	78	1.18
Brick, fire	110	....	1.76	Mica	165	200	2.9
Bronze	545	555	8.80	Molybdenum	636	....	10.2
Carbon	125	144	2.25	Mortar, hard	103	111	1.75
Carbon dioxide *	0.124	....	0.00199	Muck	40	74	0.915
Carbon monoxide *	0.0781	....	0.00125	Mud	80	130	1.68
Caoutchouc (rubber)	57	62	0.955	Nickel	540	550	8.8
Cellulose acetate	79	82	1.30	Nitrogen *	0.0782	....	0.00125
Cellulose acetate butyrate	71	76	1.18	Nitrous oxide *	0.1234	....	0.00198
Cement, loose	72	105	1.42	Oil, cottonseed	57.8	....	0.926
Cement, set	170	190	2.85	Oil, gasoline	41	43	0.675
Charcoal	17	35	0.421	Oil, lard	57.4	....	0.920
Clay, hard	129	133	2.10	Oil, linseed	58.8	....	0.942
Clay, soft	118	....	1.89	Oil, mineral, lubricating	56.2	57.7	0.912
Coal, anthracite	81	106	1.50	Oil, petroleum	54.8	....	0.878
Coal, anthracite, piled loose	47	58	0.84	Oil, turpentine	54.2	....	0.873
Coal, bituminous	78	88	1.33	Osmium	1400	....	22.5
Coal, bituminous, piled loose	44	54	0.79	Oxygen *	0.0892	....	0.00142
Coal, lignite	52	....	0.83	Palladium	686	749	11.5
Cobalt	530	563	8.71	Paper	44	72	0.92
Coke	62	105	1.35	Paraffin	54	57	0.89
Coke piled loose	23	32	0.45	Pitch	67	....	1.07
Concrete, 1:2:4	146	....	2.34	Platinum	1320	1350	21.37
Concrete 1:1 1/2:3	139	....	2.23	Polystyrene (molded)	65	67	1.06
Concrete 1:3:6	156	....	2.50	Porcelain	143	156	2.4
Copper, cast	549	558	8.87	Pumice stone	23	56	0.63
Copper, wrought	552	558	8.90	Quartz	165	....	2.65
Copper wire	555	558	8.89 †	Rhodium	686	775	12.44
Cork	15.6	....	0.25	Salt	50	70	0.965
Ebonite	72	....	1.15	Sand	90	120	1.68
Ethyl cellulose	69	78	1.18	Sandstone	134	147	2.25
Flint	164	....	2.63	Selenium	300	....	4.82
German silver (52 Cu + 26 Zn + 22 Ni)	527	....	8.45	Silver	650	657	10.5
Glass, common	150	175	2.6	Slate	162	205	2.85
Glass, flint	180	370	4.4	Snow, fresh fallen	5	12	0.136
Gold, cast	1200	....	19.3	Snow, wet compact	15	50	0.520
Granite	165	172	2.7	Soapstone	162	175	2.7
Gravel	90	147	1.9	Steel (see Iron)	....	....	....
Gutta percha	61.1	....	0.980	Sulfur	120	130	2.05
Gypsum or plaster of Paris	142	145	2.26	Tantalum	1035	....	16.6
Hydrogen *	0.00561	....	0.0000900	Tar	62.4	....	1.00
Ice	57.2	....	0.917	Tile, hollow terra cotta, building block	26	38	0.51
Iridium	1400	....	22.42	Tile, flat and segmental arches	31	45	0.608
Iron, pure	490	492	7.86	Tile partitions ‡	12	26	....
Iron, gray cast	439	445	7.08	Tin	455	....	7.29
Iron, white cast	473	482	7.65	Traprock	187	190	3.02
Iron, wrought	487	492	7.85	Tungsten	1160	1190	18.8
Iron, steel	474	494	7.76	Turf	20	30	0.400
				Water, max. density	62.4	....	1.00

Table 2. Specific Gravity and Pounds per Cubic Foot of Various Materials at Room Temperatures—*Continued*

Material	Pounds per Cubic Foot		Average Specific Gravity	Material	Pounds per Cubic Foot		Average Specific Gravity
	From	To			From	To	
Water, sea	64.0	64.3	1.03	Wood, lignum vitae	73	83	1.25
Wax, bees	60.5	....	0.965	Wood, mahogany	41	53	0.75
Wood, ash	40	53	0.75	Wood, maple	39	47	0.68
Wood, butternut	24	....	0.38	Wood, oak	37	56	0.75
Wood, cedar	30	35	0.53	Wood, pine, white	22	31	0.42
Wood, chestnut	38	41	0.63	Wood, pine, yellow	23	37	0.50
Wood, cypress	32	37	0.55	Wood, poplar	22	31	0.42
Wood, elm	34	37	0.57	Wood, redwood	30	....	0.481
Wood, fir	34	35	0.55	Wood, spruce	25	32	0.457
Wood, hemlock	25	29	0.43	Wood, walnut	40	43	0.67
Wood, hickory	37	58	0.75	Zinc	428	448	7.10

* At a temperature of 0 deg cent and a pressure of 760 mm mercury.
† This value has been adopted internationally as representing the average density at 20 deg cent.
‡ Including air spaces.

3. BIBLIOGRAPHY

Eshbach, O. W., *Handbook of Engineering Fundamentals.* Wiley (1936).
Fowle, F. E., *Smithsonian Physical Tables.* Washington (1934).
Landolt, Börnstein, and Roth, *Physikalisch-chemische Tabellen.*
Lange, N. A., *Handbook of Chemistry.* Handbook Publishing Company, Sandusky, Ohio (1944).
Orr, S. W., and A. M. Mutersbaugh, *Investigation of Weights of Building Material,* Thesis, Mass. Inst of Tech. (1913).
Simonds, H. R., & C. Ellis, *Handbook of Plastics.* Van Nostrand, N. Y. (1943)
Copper Wire Tables, *Circ.* 31, Bur. Standards.
Publications of Forestry Division, U. S. Dept. of Agriculture: *Bull.* 10; *Circ.* 32; *Circ.* 115,
International Critical Tables.

THERMAL PROPERTIES OF MATERIALS

By D. F. Miner
Revised by W. A. Del Mar

4. THERMAL CAPACITY AND SPECIFIC HEAT

The thermal capacity of a body is defined as the heat absorbed by the body per unit increase in its temperature, there being during this change in temperature no change of state (e.g., no change from solid to liquid or from liquid to gaseous form or no chemical change) and no transfer of heat energy from the body in question to other bodies. The thermal capacity *per unit mass* of a substance is approximately constant, but increases slightly with increase in temperature; in iron the increase with temperature is quite marked. If C is called the thermal capacity per unit mass of a substance, the heat absorbed by a homogeneous mass M when its temperature increases from t_1 to t_2 is

$$H = CM(t_2 - t_1) \tag{1}$$

provided C is constant.

The mean thermal capacity per unit mass of water (between 0 and 100 deg cent), when expressed in mean gram-calories per gram per degree centigrade, is numerically equal to unity. The ratio of the thermal capacity per unit mass of any substance to the mean thermal capacity of water is called the specific heat of the substance. The specific heat of a substance does not depend upon the units in which the various quantities are measured; its thermal capacity per unit mass does. When heat is expressed in mean gram-calories, mass in grams and temperature in degrees centigrade, the thermal capacity per unit mass is equal to its specific heat (compare with density and specific gravity).

CALCULATION OF HEAT ABSORBED OR GIVEN OUT.

C = specific heat (gram-calories per gram per degree centigrade).
M = mass heated.
$t_2 - t_1$ = rise of temperature.

Then for any set of units the heat absorbed is

$$H = kCM(t_2 - t_1) \qquad (2)$$

where k has the values of the factors in Table 33, Section 1.

VALUES OF SPECIFIC HEAT. In Table 1 are given the specific heats of the more common substances used in engineering work. These number are also equal to the thermal capacity per unit mass, when mass is expressed in grams, temperature in degrees centigrade, and heat in gram-calories.

Table 1. Specific Heat of Some Common Substances

(From Landolt-Börnstein Tables). See Article 16 for special alloys, and Article 22 for insulating materials.

Substance	Temperature, °C	Mean Specific Heat, C	Substance	Temperature, °C	Mean Specific Heat, C
Air *	−102 to 440	0.237	Iridium	0 to 100	0.032
Aluminum	15 to 435	.236	Iron, cast	18 to 100	.113
Ammonia	23 to 216	.520	Lead	17 to 100	.031
Antimony	22 to 600	.052	Marble	0 to 100	.206
Asbestos	20 to 98	.195	Mercury	0	.0335
Bismuth	−79 to 100	.029	Mercury	100	.0326
Brass †	20 to 100	.092	Mica	20 to 98	.208
Bronze ‡	20 to 100	.104	Molybdenum	20 to 550	.072
Carbon, gas	20 to 1040	.315	Nickel	0 to 105	.108
Carbon, graphite	0 to 3000	.535	Nitrogen *	0 to 200	.244
Carbon dioxide *	−78 to 7	.184	Nitrous oxide *	13 to 172	.231
Carbon dioxide *	0 to 200	.215	Oil, transformer	20	.44
Carbon monoxide *	23 to 198	.243	Oxygen *	20 to 440	.224
Cement, Portland	28 to 30	.271	Osmium	19 to 98	.031
Chlorine	13 to 202	.124	Palladium	0 to 100	.059
Cobalt	15 to 350	.109	Palladium	0 to 1265	.071
Concrete, tamped	20 to 100	.156	Paraffin	25 to 30	.589
Concrete, tamped	800	.219	Petroleum	21 to 58	.511
Copper	−188 to 20	.080	Petroleum	18 to 99	.498
Copper	0 to 100	.094	Platinum	0 to 100	.032
Copper	300	.098	Rhodium	10 to 97	.058
Copper	900	.126	Silver	0 to 260	.057
Cork	....	.485	Steel	20 to 100	.118
Cotton	0 to 100	.362	Tantalum	−185 to 20	.033
Ebonite	....	.339	Tin	17 to 100	.056
German silver	0 to 100	.095	Tungsten	17 to 100	.056
Glass	0 to 19	.171	Wax, yellow	26 to 42	.820
Glass	56 to 78	.192	Wood's metal §	5 to 50	.035
Gold	0 to 100	.032	Wool	....	.393
Hydrogen *	−28 to 198	3.41	Zinc	20 to 100	.093
Ice	−78 to −18	0.463			

* At constant pressure of 1 atm.
† 60 Cu + 40 Zn.
‡ 88.7 Cu + 11.3 Al.
§ 25.85 Pb + 6.99 Cd + 52.43 Bi + 14.73 Sn.

5. MELTING OR FREEZING POINT AND HEAT OF FUSION

Certain chemically simple substances, when heated to a definite temperature, pass from the solid to the liquid state with no increase in temperature during this change in state, provided the solid and the liquid are kept thoroughly mixed, but the change is accompanied by a considerable absorption of heat. The temperature at which the change takes place is called the melting point or freezing point (the reverse change takes place at the same temperature), and the heat absorbed per unit mass is called the heat of fusion or heat of liquefaction; this same amount of heat is given out when the body solidifies. Many substances, however, have no definite melting point, the change from one state to the other being gradual; such substances begin to melt at a lower temperature than that at which solidification begins during cooling. The melting points and heats of fusion for some common substances are given in Table 2.

Table 2. Fusion and Vaporization

(At atmospheric pressure, i.e., 760 mm mercury)

From *International Critical Tables*

Substance	Melting Point, ° C *	Heat of Fusion, g-cal per g †	Boiling Point, ° C *	Heat of Vaporization g-cal per g †
Aluminum	657	87	1800	2000
Ammonia	−75	108	−33.04	327
Antimony	630	38.9	1380	373
Bismuth	270	10.2	1450	220
Brass	900±	....	....	....
Bronze	900±	....	....	....
Cadmium	321	11.1	767	206
Carbon	over 3600	....	over 4200	....
Carbon dioxide	−56.2	45.3	−79	....
Carbon monoxide	−206	8.0	−192	51.6
Cesium	28.5	3.6	670	132
Chlorine	−103.5	23	−34.6	69
Chromium	1610	31.6	2200	1470
Cobalt	1480	58.5	2900	1540
Copper	1084	41.6	2300	1750
German silver	1100±	....	....	....
Glass, flint	1300	....	....	....
Gold	1064	15.9	2600	446
Gutta-percha	100	....	....	....
Hydrogen	−260.6	14.1	−252.7	107
Iridium	2350(?)	....	4800	....
Iron	1535	23.0 to 50	3000	1625
Lead	327	5.6	1620	221
Manganese	1260	36.6	1900	1045
Marble	2500±	....	....	....
Mercury	−38.7	2.7	356.9	68.0
Molybdenum	2620	....	3700	1770
Nickel	1450	72.8	2900	1545
Nitrogen	−210	6.1	−195.8	47.65
Nitric oxide	−160.6	....	−153	....
Oxygen	−219	3.3	−183.0	50.97
Osmium	2700	....	5300	....
Palladium	1545	36.3	2200	....
Paraffin	52.4	35.1	....	....
Platinum	1755	26.9	4300	635
Rhodium	1955	....	2500	....
Rubber	100	....	....	....
Selenium	220	....	688	94
Silicon	1420	....	2600	....
Silver	961	25.9	1950	552
Steel	1300 to 1475	....	....	....
Sulfur	115 to 119	8.9 to 13.2	444.6	69.1
Tantalum	2850	....	4100	....
Tin	232	13.4	2260	655
Tungsten	3370	....	5900	1180
Vanadium	1710	....	3000	....
Water	0	80	100	539
Wax, bees	62	42.3	....	....
Wood's metal	75.5	8.40	....	....
Zinc	419	25.5	907	363

* Let t_c be the value in degrees centigrade; then the value in degrees fahrenheit is $t_f = 32 + 1.8\,t_c$.
† Let H be the value in gram-calories per gram; then the corresponding heat of fusion or of vaporization

In kg-cal per kg is	$1.000\,H$,
In watt-seconds per gram is	$4.183\,H$,
In kwhr per kg is	$1.162 \times 10^{-3}\,H$,
In kwhr per lb is	$5.271 \times 10^{-4}\,H$,
In kwhr per ton (2000 lb) is	$1.054\,H$.

References

U. S. Bur. Standards Circ. 35.
Smithsonian Physical Tables.
Landolt-Börnstein-Roth, *Physikalisch-chemische Tabellen.*
International Critical Tables.

6. TRANSFER OF HEAT

When a body is at a higher temperature than the surrounding bodies, energy is transferred from the hotter to the colder bodies, as is manifested by the changes in temperature or state which tend to, or actually do, take place, even though the intervening space is entirely void of matter. The energy thus transferred from one body to another through empty space is called radiant energy or radiant heat and is similar in nature to the energy radiated in the form of light waves and electromagnetic waves. The waves of radiant heat have a length greater than that of light waves and less than that of the ordinary electromagnetic waves used in wireless telegraphy. Radiant heat is absorbed by, transmitted through, and reflected by ordinary matter in much the same way that light waves are absorbed, transmitted, and reflected. Matter which is transparent to light waves, however, may be practically opaque to heat waves; e.g., water absorbs practically all the heat waves which fall upon it.

When a hot and a cold body are separated by a fluid which is free to circulate, heat is transferred from the hot to the cold body by currents of the fluid itself flowing from one to the other; similarly, all parts of a fluid which is being heated quickly come to approximately the same temperature. This transfer of heat by currents of the fluid itself is called convection.

When a hot and a cold body are separated by a solid, the transfer of heat, which may be very rapid, particularly when the separating medium is a metal, is probably due to an extremely rapid to-and-fro motion of the molecules which constitute the medium. In any event the process is essentially different from the transfer of heat either by radiation or by convection; it is described by the term *conduction* of heat. In a fluid heat is transferred both by conduction and convection.

RADIATION, ABSORPTION, AND REFLECTION OF HEAT. The rate at which a hotter body radiates heat and a colder body absorbs heat depends upon the state of the surfaces of the bodies as well as on their temperatures. The rates of radiation and of absorption are increased by darkness and roughness of the surfaces of the bodies and diminished by smoothness and polish. For this reason the covering of steam pipes and boilers should be smooth and of a light color; uncovered pipes and steam-cylinder covers should be polished.

The heat radiated by a body at a given temperature T to surrounding bodies at a lower temperature is equal to the heat which this body would absorb at this same temperature T from surrounding bodies at a higher temperature. When a given quantity of radiant heat strikes a body, only part of the heat is, as a rule, absorbed, the rest being reflected.

Let H_i be the incident heat, H_r the reflected heat, and H_a the absorbed heat, at temperature t, and let H_e be the heat which the body would emit at this same temperature to bodies at a lower temperature; then

$$H_i = H_r + H_a$$

$$H_a = H_e$$

DEFINITION OF "BLACK BODY"; STEFAN-BOLTZMANN LAW. A "black body" is defined as one that absorbs all radiations falling upon it, neither reflecting nor transmitting any. The radiation of such a body is a function of the temperature alone and is identical with the radiation inside an inclosure all parts of which have the same temperature. By heating the walls of an inclosure as uniformly as possible and observing the radiation through a very small opening, a practical realization of a black body is obtained.

RADIATING AND REFLECTING POWERS. The ratio of the heat radiated per unit area by any surface at a given temperature T to the heat radiated per unit area by an absolutely black surface at this same temperature T is called the radiating power, or relative emissivity, of the surface at that temperature. The difference between this ratio and unity is a measure of the heat which would be reflected by this surface at the same temperature, and is defined as the reflecting power of the surface. The radiating power of a surface depends upon the temperature of the surface. At very high temperatures the radiating power of every surface approaches the value of unity; i.e., at high temperatures the total energy radiated by any surface approaches in value the total energy radiated by an absolutely black body. Table 3, taken in part from Kent's *Mechanical Engineer's Handbook* and in part from Langmuir's paper [*Trans. AIEE* (1913)], gives the approximate value of the radiating power of some common surfaces at ordinary temperatures.

Oiling a polished surface may increase its radiating power from 2 to 3 times, but oiling does not seriously affect the radiating power of a rough surface.

CONDUCTION OF HEAT. Whenever a difference of temperature is maintained between any two parts of the same body, there is a transfer of heat from the hotter to the colder part by the process described by the term conduction, as distinguished from radiation and convection.

Table 3. Approximate Radiating and Absorbing Powers

Surface	Radiating or Absorbing Power	Surface	Radiating or Absorbing Power
Lampblack	1.00	Steel, polished	0.17
Water	1.00	Platinum, polished	0.24
Carbonate of lead	1.00	Platinum in sheet	0.17
Writing paper	0.98	Tin	0.15
Ivory, jet, marble	0.93–0.98	Brass, cast, dead polished	0.11
Ordinary glass	0.90	Brass, bright polished	0.07
Ice	0.85	Copper, varnished	0.14
Gum lac	0.72	Copper, hammered	0.07
Silver leaf on glass	0.27	Copper, oxidized	0.568
Cast iron, bright polished	0.25	Copper, calorized	0.27
Cast iron, oxidized	0.71	Monel metal, polished	0.40
Aluminum paint on cast iron	0.47	Monel metal, oxidized	0.44
Gold enamel on cast iron	0.39	Gold, plated	0.05
Mercury (approx.)	0.23	Gold on polished steel	0.03
Wrought iron, polished	0.23	Silver, polished bright	0.03
Zinc, polished	0.19		

Consider a flat layer *within* a substance, the two sides of the layer being parallel and its thickness small compared with its area. Let one side of this be maintained at a constant temperature T, the same at all points of this surface, and the other side of the layer be maintained at a constant temperature T_1; let A be the area of the layer (i.e., of one of its flat surfaces) and x the thickness of the layer. Then the amount of heat transferred through the layer in time t is

$$H = \frac{KA(T - T_1)t}{x} \tag{3}$$

where K, called the thermal conductivity, is approximately a constant for a given material and is independent of the temperature difference $T - T_1$, when this difference is small; K is not constant, however, for wide temperature variations, particularly in the case of gases (see Coefficient of Expansion of Gases, p. 2-12). Values of K are given in Tables 5 and 6. The reciprocal of the thermal conductivity, viz., $\rho = 1/K$, is called the thermal resistivity.

The values of K given in the tables below are the values of this factor when H is expressed in gram-calories, A in square centimeters, x in centimeters, $T - T_1$ in degrees centigrade, and t in seconds; i.e., K is the number of gram-calories transmitted per second through a cube 1 cm on each edge when a difference of temperature of 1 deg cent is maintained between opposite faces of this cube. For other units the value of K as given should be multiplied by the factor in Table 34, Section 1.

TEMPERATURE COEFFICIENT OF THERMAL CONDUCTIVITY. For metals the variation of the internal thermal conductivity with temperature may be expressed with a fair degree of approximation by the relation

$$K = K_0(1 + at) \tag{4}$$

where K_0 is the conductivity at 0 deg cent, say, K is the conductivity at any other temperature of t deg cent, and a is a constant. The coefficient a may be either positive or negative; its value for some of the common metals is given in Tables 5 and 6.

VALUES OF THERMAL CONDUCTIVITY (K) OF MATERIALS. In Tables 5 and 6 are given the thermal conductivities of certain common non-metallic and metallic substances, respectively. The data are from the following sources: Landolt-Börnstein, *Physikalisch-chemische Tabellen* (1923); C. P. Randolph, *Trans. Am. Electrochem. Soc.* (1912); *Smithsonian Physical Tables;* Langmuir, *Trans. AIEE*, Vol. 32, p. 301 (February, 1913); *International Critical Tables*, Vol. V. For building materials, see Eshbach, Fundamentals of Engineering, Section 12.

THERMAL CONDUCTIVITY OF LAMINATED STEEL. The thermal conductivity of a laminated iron core perpendicular to the laminations depends upon the conductivity of the iron and the oxide scale or varnish on the laminations and also upon the pressure on the laminations. G. E. Luke [*Elec. World*, Vol. 70, p. 562 (Sept. 22, 1917)] gives the following values for K in gram-calories per cubic centimeter per degree centigrade.

Material	Pressure, pounds per square inch			
	25	50	75	100
Silicon steel, 15.5 mils thick, 3 per cent oxide layer	0.00096	0.00115	0.00125	0.00132
Ordinary varnished steel, 18.35 mils thick, 5 per cent varnish layer	0.00094	0.00098	0.00109	0.00120

THERMAL CONDUCTIVITY OF GASES. The thermal conductivity of a gas depends not only upon the nature of the gas but also upon its specific heat at constant volume, C_v, and its absolute temperature, T. Langmuir [*Phys. Rev.*, Vol. 34, p. 408 (1912)] gives the following values for the *watts* conducted from a plane surface through an adhering film A square centimeters in area and B centimeters thick:

$$W = \frac{A}{B}(\phi_2 - \phi_1) \tag{5}$$

where ϕ_1 and ϕ_2 are functions of the absolute temperatures at the two surfaces of the film as given in Table 4.

Table 4. Values of ϕ

Abs. Temp. °C	Hydrogen	Air	Mercury Vapor	Abs. Temp. °C.	Hydrogen	Air	Mercury Vapor
0	0.0000	0.0000	......	1700	5.945	0.931	0.228
100	0.0329	0.0041	......	1900	7.255	1.138	0.284
200	0.1294	0.0168	......	2100	8.655	1.363	0.345
300	0.278	0.0387	......	2300	10.18	1.608	0.411
400	0.470	0.0669	......	2500	11.82	1.871	0.481
500	0.700	0.1017	0.0165	2700	13.56	......	0.556
700	1.261	0.189	0.0356	2900	15.54	......	0.636
900	1.961	0.297	0.0621	3100	17.42	......	0.719
1100	2.787	0.426	0.0941	3300	19.50	......	0.807
1300	3.726	0.576	0.1333	3500	21.79	......	0.898
1500	4.787	0.744	0.1783				

7. EXPANSION DUE TO HEATING

Most substances expand when heated, but water between 0 and 4 deg cent, quartz glass below −84 deg cent, and a few other substances contract with increase of temperature. When the temperature of a solid is changed by rapid cooling, slow changes in its dimensions continue long after it has attained the same uniform temperature throughout. This effect, which is particularly marked in glass, is known as thermal hysteresis. It can be largely eliminated by prolonged heating at a high temperature followed by a very gradual cooling, i.e., by annealing.

Table 5. Thermal Conductivity of Non-metallic Substances

For electrical insulating materials see Article 20.

Substance	Temp. Range, ° C		Therm. Conduct., K ‡		Temperature Coefficient per ° C, a
	From *	To	From †	To †	
Asbestos	100	500	0.00016	0.00050	
Asbestos paper			0.00043	0.00060	
Brick building	15	1100	0.00149		
Brick, fire	0	1300	0.00140	0.0054	
Carborundum, brick	150	1200	0.0032	0.027	
Cardboard	Below 0		0.000394		
Cement, Portland	35	90	0.000712	0.00217	
Chalk			0.00219		
Cloth, empire			0.00060		
Concrete	0		0.002		
Cork	20	200	0.000076	0.00717	
Cotton batting, loose			0.000096		
Cotton batting, tightly packed	−150	150	0.000091	0.00018	
Ebonite	6	90	0.00038		−0.0019
Felt	21	175	0.000087	0.000225	
Fullerboard			0.00034		
Glass			0.0014	0.002	
Glycerin			0.00068		
Granite	100	200	0.0045	0.0097	
Ice	−160	0	0.0053	0.0057	
Infusorial earth	20	350	0.00032	0.00040	
Lampblack	100	500	0.000074	0.000107	
Leather			0.00015	0.00042	
Limestone	40	350	0.0046	0.0035	
Linen			0.00021		
Liquids, hydrocarbons, oils, etc.			about 0.0003		
Magnesia, carb	20	188	0.00023		
Magnesia, calcined	20	155	0.000165	0.000173	
Magnesia, asbestos	100	400	0.000162	0.000178	
Marble	15	30	0.00770	0.00910	−0.0005
Mica, pure			0.00086	0.0012	
Mica, paper			0.00038		
Micanite		30	0.000050	0.00010	
Oil, paraffin			0.00033		
Oil, castor			0.000425		
Oil, petroleum	0	34	0.000355	0.000382	+0.0110
Oil, turpentine	13		0.000325		+0.0067
Paper			0.00011	0.00031	
Paper, treated			0.00014	0.00041	
Paraffin		30	0.000473	0.00062	
Pasteboard			0.000450		
Plaster			0.00130		
Plaster of paris	20	155	0.000425		
Plumbago	20	155	0.00100		
Porcelain	95		0.00249		
Quartz glass			0.0036		
Rubber, Para			0.00038		
Rubber, vulcanized			0.00034	0.00054	
Sand	20	155	0.000855	0.000867	
Sawdust		30	0.000143		
Shellac			0.00060		
Silk	50	100	0.000095	0.000141	
Slate	94		0.0048		
Snow		0	0.00038		
Tape, treated cloth			0.00036	0.00065	
Tape, rubber			0.00103		
Water	0	25	0.00150	0.00136	
Wood			0.000087	0.00086	
Wool, sheep's, loose		30	0.00010		
Wool, sheep's, tightly packed			0.000055		
Wool, mineral	0	175	0.0000930	0.000128	
Woolen	100		0.0000553	0.000119	

* When only one temperature is given, the measurement was made at that temperature.

† Range of determination by different experimenters except where only one value of K is given, in which case the range is that to which the given value of K applies.

‡ In gram-calories per centimeter-cube per degree centigrade per second; see Section 1 for multiplying factors when other units are employed.

Table 6. Thermal Conductivity of Metals and Various Forms of Carbon

See tables in Article 14 for electrical alloys.

Substance	Temp. Range, ° C		Therm. Conduct. K *		Temp. Coeff. per ° C
	From †	To	From ‡	To ‡	a
Aluminum	18	100	0.480	0.492	+0.0030
Aluminum	200	400	0.545	0.760	+0.0020
Aluminum	500	600	0.885	1.01	+0.0014
Brass, yellow	0	100	0.204	0.254	+0.0024
Brass, red	0	100	0.246	0.283	+0.0015
Calorite	20	250	0.038	....	....
Carbon, amorphous	100	360	0.089	....	....
Carbon, amorphous	100	842	0.129	....	....
Carbon, Ach. graph	100	390	0.338	....	....
Carbon, Ach. graph	100	914	0.291	....	....
Carbon, graphite brick	300	700	0.024	....	....
Charcoal	20	155	0.00019	....	....
Coal	....	....	0.00030	....	....
Constantan	18	100	0.054	0.064	+0.00227
Copper	−54	−14	0.921	1.059	+0.0053 to
Copper	74	167	0.914	1.024	+0.00047
Copper	100	197	0.908	....	....
Copper	100	837	0.858	....	....
Copper, commercial	18	....	0.835	....	....
German silver	0	100	0.070	0.089	+0.0027
Gold	100	....	0.703	....	−0.00007
Iron	0	....	0.167	0.207	−0.0008 to
Iron	100	....	0.142	0.163	−0.0001
Iron	200	....	0.136	....	....
Iron	100	727	0.202	....	....
Iron	100	1245	0.191	....	....
Iron, cast	100	....	0.096	....	....
Lead	0	100	0.0836	0.0764	{−0.00086 to −0.00016
Manganin	18	100	0.052	0.63	+0.0026
Mercury	0	50	0.0148	0.0189	+0.0055
Nickel	18	....	0.142	....	−0.00031
Nickel	300	....	0.126	....	−0.00095
Nickel	1200	....	0.058	....	−0.00047
Platinum	18	100	0.166	0.173	+0.00051
Platinoid	18	....	0.060	....	....
Silver	18	100	1.006	0.992	−0.00017
Steel	....	....	0.108	0.111	−0.0006
Steel, transformer §	20	250	0.077	....	....
Tin	0	100	0.155	0.145	−0.0007
Zinc	18	100	0.265	0.262	−0.00016

* In gram-calories per centimeter-cube per degree centigrade per second; see Section 1 for multiplying factors when other units are employed.

† When only one temperature is given, the measurement was made at that temperature.

‡ Range of determinations by different experimenters except where only one value of K is given, in which case the range is that to which the given value of K applies.

§ See also Thermal Conductivity of Laminated Steel, page 2-10.

COEFFICIENT OF EXPANSION OF GASES. Gases present a remarkable uniformity, all the so-called permanent gases expanding about $^1/_{273}$ part of their **initial volume at 0 deg cent** per degree centigrade increase of temperature, irrespective of the pressure, provided this remains constant throughout. That is, for any of the ordinary gases,

$$V = V_0\left(1 + \frac{t_c}{273}\right)$$

where V_0 is the volume at 0 deg cent, and V the volume at any other temperature t_c deg cent, the pressure being the same at both temperatures. If the temperature is expressed in fahrenheit degrees and V_1 is the volume at 0 deg fahr, then

$$V = V_0\left(1 + \frac{t_f}{460}\right)$$

Note that −273 and −460 are the absolute zeros on the centigrade and fahrenheit scales respectively.

Table 7. Coefficients of Linear Expansion

$l = l_0(1 + at + bt^2)$, temperature in deg cent *

a_{20} = "true" coefficient at 20 deg cent

See tables in Article 14 for electrical alloys, and Article 20 for electrical insulating materials.

(From Landolt-Börnstein's Tables)

Substance	Temp., °C		a	b	a_{20}
	From	To			
Aluminum	0	610	0.235×10^{-4}	0.707×10^{-8}	0.238×10^{-4}
Brass (73.7 Cu + 24.2 Zn + 1.5 Sn + 0.6 Pb)	0	80	0.179×10^{-4}	0.456×10^{-8}	0.181×10^{-4}
Bronze (81.2 Cu + 8.6 Zn + 9.9 Sn + 0.2 Pb)	0	80	0.176×10^{-4}	0.469×10^{-8}	0.177×10^{-4}
Carbon, gas-carbon	40	......	0.054×10^{-4}	......	......
Carbon, graphite	40	......	0.079×10^{-4}	......	......
Constantan (60 Cu + 40 Ni)	0	500	0.148×10^{-4}	0.402×10^{-8}	0.150×10^{-4}
Copper	0	625	0.167×10^{-4}	0.403×10^{-8}	0.169×10^{-4}
Glass, Jena	0	100	0.077×10^{-4}	0.350×10^{-8}	0.079×10^{-4}
Glass, French	2	100	0.072×10^{-4}	0.544×10^{-8}	0.075×10^{-4}
Gold	9	95	0.136×10^{-4}	1.12×10^{-8}	0.140×10^{-4}
German silver	0	100	0.184×10^{-4}	......	......
Ice	−27	−2	0.514×10^{-4}	......	......
Iron, cast	0	625	0.098×10^{-4}	0.566×10^{-8}	0.102×10^{-4}
Iron, wrought	0	500	0.117×10^{-4}	0.525×10^{-8}	0.119×10^{-4}
Lead	14	94	0.273×10^{-4}	0.74×10^{-8}	0.276×10^{-4}
Marble, white	15	100	0.117×10^{-4}	......	......
Mica, parallel to cleavage	5	80	0.077×10^{-4}	1.200×10^{-8}	0.082×10^{-4}
Mica, perpendicular to cleavage	4	82	0.076×10^{-4}	0.490×10^{-8}	0.079×10^{-4}
Nickel	0	1000	0.135×10^{-4}	0.332×10^{-8}	0.136×10^{-4}
Nickel steel (24% Ni)	0	38	0.175×10^{-4}	0.711×10^{-8}	0.178×10^{-4}
Phosphor bronze (97.6 Cu + 2.2 Sn + 0.2 P)	0	80	0.167×10^{-4}	0.462×10^{-8}	0.168×10^{-4}
Platinum	0	1000	0.0887×10^{-4}	0.1324×10^{-8}	0.0892×10^{-4}
Porcelain, Berlin	20	100	0.027×10^{-4}	0.306×10^{-8}	0.028×10^{-4}
Porcelain, Bayeux	0	600	0.034×10^{-4}	0.107×10^{-8}	0.035×10^{-4}
Rubber, hard	17	25	0.77×10^{-4}	......	......
Rubber, hard	25	35	0.84×10^{-4}	......	......
Silver	0	750	0.1827×10^{-4}	0.4793×10^{-8}	0.1846×10^{-4}
Steel	0	300	0.092×10^{-4}	0.336×10^{-8}	0.093×10^{-4}
Tin	8	95	0.203×10^{-4}	2.63×10^{-8}	0.214×10^{-4}
Vulcanite	0	18	0.636×10^{-4}	......	......
Zinc	9	96	0.274×10^{-4}	2.34×10^{-8}	0.284×10^{-4}

* When the temperature is expressed in fahrenheit degrees, the formulas become

$$l = l_{32}\left[1 + \frac{a\,(t_f - 32)}{1.8} + \frac{b\,(t_f - 32)^2}{3.24}\right]$$

$$a_{68} = \frac{a_{20}}{1.8}$$

where a, b, and a_{20} have the values given in Table 7.

Table 8. Coefficients of Cubical Expansion

$V = V_0(1 + \alpha t + \beta t^2)$, temperatures in degrees centigrade

α_{20} = "true" coefficient at 20 deg cent

(From Landolt-Börnstein's Tables)

Substance	Temp., °C		α	β	α_{20}
	From	To			
Caoutchouc, crude gray	0	75	6.62×10^{-4}	24.2×10^{-8}	6.80×10^{-4}
Gutta-percha, pure rolled	0	40	4.96×10^{-4}	496×10^{-8}	6.94×10^{-4}
Paraffin	0	33	5.84×10^{-4}	99.2×10^{-8}	5.88×10^{-4}
Petroleum, sp. gr. 0.8467	24	120	8.99×10^{-4}	140×10^{-8}	9.55×10^{-4}
Wax, white solid	10	57	10.7×10^{-4}	-5580×10^{-8}	3.06×10^{-4}

Mercury (− 10 to 300 deg cent.):

$$V = V_0\,(1 + 1.805553 \times 10^{-4}\,t + 1.2444 \times 10^{-8}\,t^2 + 2.539 \times 10^{-11}\,t^3)$$

8. FLASH POINT OF OILS

The flash point of an oil is the temperature to which the oil must be raised before the vapor immediately above it will take fire upon the application of a flame. This temperature depends to an appreciable extent upon the size of the flame, the method of applying it, and the shape and dimensions of the containing vessel. The following values of the flash point for various oils are taken from J. Lewkowitisch, *Chemical Technology and Analysis of Oils, Fats, and Waxes.*

Table 9. Flash Point of Oils

Oils	Spec. Grav. at 60° F	Flash Point, ° F
Mineral oils:		
Refined American	0.875 to 0.920	325 to 425
Refined Russian	0.895 to 0.915	300 to 425
Scotch	0.875 to 0.895	300 to 350
Natural (dark) American	0.880 to 0.895	325 to 425
Natural (dark) Russian	0.910 to 0.915	250 to 300
Natural filtered American	0.885 to 0.905	450 to 575
Animal oils:		
Sperm	0.8804 to 0.8807	446 to 457
Lard	0.9172	494
Tallow	0.951	265
Neat's foot	0.9178	470
White whale	0.9207	476
Vegetable oils:		
Castor	0.963	275
Linseed	0.930	285
Olive	0.914	305
Rape, crude	0.920	265
Rape, refined	0.911	305

9. BIBLIOGRAPHY

Fechheimer, C. J., Hydrogen a Successor to Air, *Elec. J.*, Vol. 26, p. 405.
Fowle, F. E., *Smithsonian Physical Tables.* Washington (1934).
King, W. J., The Basic Laws and Data of Heat Transmission, *Mech. Engr.*, Vol. 54, pp. 190, 275, 347, 410, 492, 560, 1932. (A series of articles.)
Lange, N. A., *Handbook of Chemistry.* Handbook Publishing Company, Sandusky, Ohio (1944).
Luke, G. E. Surface Heat Transfer in Electric Machines with Forced Air Flow, *AIEE*, p. 1036 (1926).
McAdams, W. H., *Heat Transmission.* McGraw-Hill (1933).

CONDUCTOR MATERIALS

By P. H. Brace

10. INTRODUCTORY MATERIAL

(For numerical references in parentheses, see Article 15.)

The materials that are of interest to the electrical engineer, considered with reference to their electrical resistivity, may be grouped as follows: (*a*) conductors, (*b*) semiconductors, and (*c*) dielectrics (insulators). In a few cases these groups overlap. Table 1 is illustrative of this classification.

Figures 1 and 2 show, respectively, the resistivity and the temperature coefficient of resistivity, against platinum, of the elements that can be considered conductors, plotted against atomic weight, as derived from standard sources, such as *International Critical Tables* and *International Annual Tables of Constants.*

The roughly periodic variation of electrical properties reflects the progressive complexity of the outer electronic configurations of the atoms with increasing atomic weight. It appears that those portions of the atoms are the ones that most directly determine the chemical and ordinary electrical properties of materials (1).

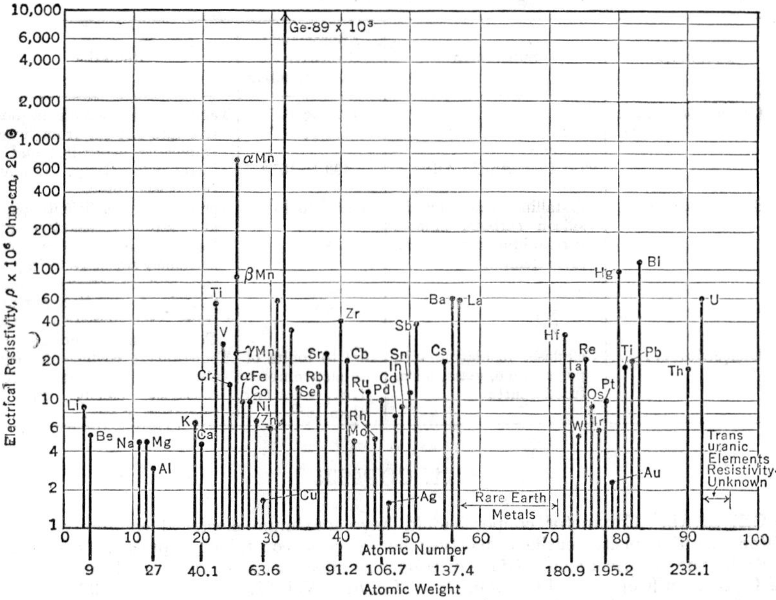

FIG. 1. Electrical Resistivity of Conductor Elements in Relation to Their Atomic Numbers and Atomic Weights

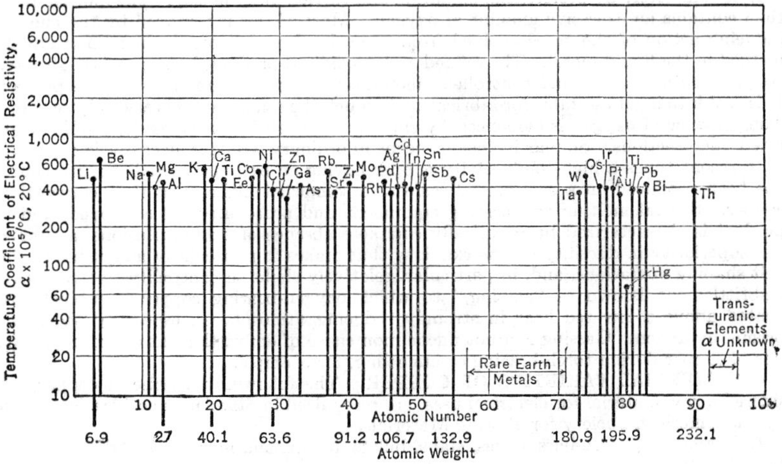

FIG. 2. Temperature Coefficient of Electrical Resistivity of Conductor Elements in Relation to Their Atomic Numbers and Atomic Weights

Table 1. Classification of Materials with Respect to Order of Magnitude of Electrical Resistivity at Ordinary Temperatures

Materials	Substances	Approximate Resistivity, ohm-centimeters	Remarks
a. Conductors.......	Silver 80–20 Nickel-chromium Alloy 99–1 Bismuth-tin Alloy	1.59×10^{-6} 112×10^{-6} 400×10^{-6}	Temperature coefficients positive and not far from linear. Impurities generally tend to increase resistivity. Conduction is electronic.
b. Semiconductors....	Crystalline lead sulfide, silicon carbide, silicon, germanium, cuprous oxide, selenium	10^{-3} to 10^{9}	Temperature coefficients generally strongly negative and far from linear. Resistivity extremely sensitive to impurities and in chemical compounds, to small departures from stoichiometric ratios. Conduction electronic.
c. Dielectrics........	Cellulose, insulating oils, polystyrene, glass, mica, fused quartz	10^{10} to limits of measurement, i.e., greater than 10^{20}	Temperature coefficients negative and non-linear. Conductivity chiefly due to migration of ions.

11. COPPER

GENERAL. Conductor copper should have maximum electrical conductivity consistent with requisite tensile properties; hence, close control of impurities is essential. Data on the electrical conductivity of copper as affected by alloying metals are given by L. L. Wyman [Copper and Oxygen, *Gen. Elec. Rev.*, Vol. 37, p. 123, Fig. 8 (1934)].

Copper can be electrodeposited as "cathode copper," to the practically complete exclusion of the deleterious impurities in the crude anodes. (See Section 18, Articles 4 and 5, for electrolytic refining procedures.) It is melted and cast before fabrication.

In conventional fire-refining, the crude copper is subjected to oxidizing treatments while molten in a reverberatory furnace. Objectionable impurities can be eliminated almost completely by preferential oxidation and carried away in slag. The molten copper then contains an excessive amount of oxygen. Most of this is removed by the chemical reducing action of the gases evolved from green hardwood "poles" when these are submerged in the liquid metal. The refined and deoxidized product, known as "tough pitch" copper, contains small and controlled amounts of oxygen, approximately 0.04 per cent. It is cast into ingots suitable for fabrication in much the same way as is electrolytic copper. The presence of oxygen is evidenced by small inclusions of cuprous oxide, visible on polished microsections. Similar inclusions occur in ordinary wrought electrolytic copper.

In recent times a high-conductivity copper having an extremely small oxygen content has been produced on a commercial scale at a somewhat higher cost than conventional copper. Special melting and casting equipment and procedures are employed, and the product is characterized by practically complete absence of oxide inclusions, a substantial superiority in ductility and workability that makes it particularly adaptable to difficult shaping operations, and, in particular, relatively slight susceptibility to that form of embrittlement known as "gassing," to which oxygen-bearing copper is subject when heated above a low red heat in strongly reducing atmospheres, especially those comprising hydrogen. Gassing is a not uncommon cause of embrittlement of ordinary varieties of copper during torch-brazing or annealing, for example.

COMMERCIAL VARIETIES OF COPPER. The following varieties of copper, considered a raw material for electrical uses, are recognized in Standard Specifications, issued by the American Society for Testing Materials.

1. Lake copper wire-bars, cakes, slabs, billets, ingots, and ingot bars (ASTM Designation B4-42).*
2. Electrolytic copper wire-bars, cakes, slabs, billets, ingots, and ingot bars (ASTM Designation B5-43).
3. Electrolytic cathode copper (ASTM Designation B115-43).

* "Lake" copper is copper originating in the northern peninsula of Michigan and may contain small but significant amounts of silver, arsenic, or both; from those alloy contents benefits may be obtained in the way of superior thermal endurance and creep resistance.

4. Oxygen-free electrolytic copper wire-bars, billets, and cakes (ASTM Designation B170-44T).

5. Fire-refined copper other than lake (ASTM Designation B72-33).*

In the specification for oxygen-free copper provision is made for the addition of controlled amounts of silver.

In all cases lower limits are set for resistivity.

The specifications designated as B4-42 and B115-43 are approved as American Standards by the American Standards Association.

CONDUCTIVITY AND RESISTIVITY. Standard Conductivity. The International Electrotechnical Commission specifies the standard value for the resistivity of annealed copper to be:

0.15328 ohm (meter, gram at 20 deg cent)

This is the International Annealed Copper Standard (IACS) and corresponds to the "100 per cent conductivity" of electrical engineering parlance.†

The IACS resistivity may be expressed in other units as follows:

1.7241	microhm-cm
0.017241	ohm (meter, mm^2)
0.6789	microhm, in.
10.371	ohms (mil, ft)
0.15328	ohm (meter, gram)
875.20	ohms (mil, pound)

The conductivity of highly purified annealed copper is reported by Heuer to be 102.1 per cent IACS at 20 deg cent.

Conductivity of Hard-drawn Copper. The conductivity of annealed copper may be decreased as much as 2.7 per cent by cold-working. Annealing restores the lost conductivity if the deformations have not been carried to the point of damage to the continuity of the metal. The magnitude of the cold-working effects depends on the drawing and annealing schedules employed. The uncertainties due to these factors and to impurities are recognized in commercial work by setting the conductivity (IACS) for acceptance of cold-drawn copper for electrical purposes at 97.16 per cent for nominal diameters 0.324 to 0.0403 (3). For medium-hard wire (4) the corresponding conductivity limits are 97.66 per cent and 96.66 per cent, and for soft or annealed copper wire, 100 per cent (5, 6).

Temperature Coefficient of Resistivity. The temperature coefficient of commercial copper is proportional to the conductivity.‡ The same is true of many other dilute alloys, i.e., that the change of resistivity due to change of temperature is characteristic of the basis metal and is superimposed on an added component of the resistivity due to alloying which is quite insensitive to temperature. This fact may be expressed by saying that the change of resistivity per degree centigrade of a sample of copper is 0.000597 ohm per meter, gram or 0.00681 microhm per centimeter cube. The 20 deg cent temperature coefficient of a sample of copper is found by multiplying the per cent

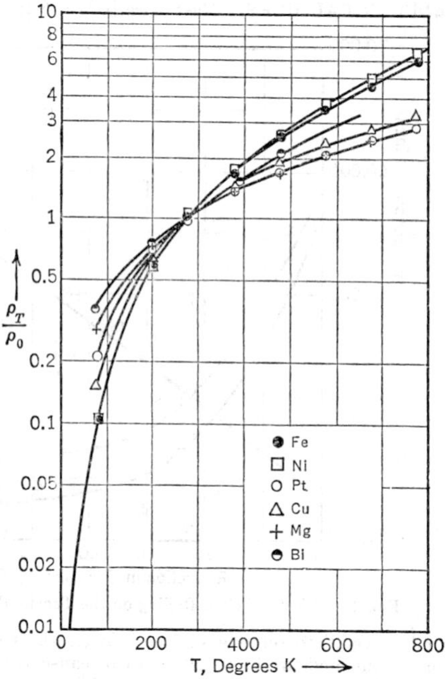

Fig. 3. Effect of Temperature on the Electrical Resistivity of Several Metals. Vertical ordinate shows ratio of resistivity at a given temperature (ρ_T) to the resistivity at 0 deg cent (ρ_0)

* 1946 Book of ASTM Standards, Part I-B, Non-ferrous Metals.

† Ratified at the Meeting of the International Electrotechnical Commission held in Berlin, Sept. 1 to 6, 1913; see *Trans. Amer. Inst. Elec. Eng.*, Vol. 32, p. 2156 (1913).

‡ Bureau of Standards, *Bull.* 147, Vol. 7, No. 1 (1910).

conductivity by 0.00393 and dividing by 100. These rules apply only to copper furnished for electrical use and to the temperature range of 10 to 100 deg cent, over which range the temperature coefficient was found to be linear. Table 2 gives values of the temperature coefficient α_T in the formula:

$$R_t = R_T[1 + \alpha_T(t - T)]$$

for various conductivities and temperatures. Figure 3 shows the effect of temperature on the resistivity of copper and a few other metals over the range from 90 to 773 deg K (−183 to 500 deg cent).

Table 2. Temperature Coefficient, α_T, in Formula $R_t = R_T[1 + \alpha_T(t - T)]$

Ohms per Meter-gram at 20 deg cent	Per Cent Conductivity	Values for α at Temperatures Noted				
		0° C	15° C	20° C	25° C	30° C
0.16134	95	0.00403	0.00380	0.00373	0.00367	0.00360
0.15966	96	0.00408	0.00385	0.00377	0.00370	0.00364
0.15802	97	0.00413	0.00389	0.00381	0.00374	0.00367
0.15753	97.3	0.00414	0.00390	0.00382	0.00375	0.00368
0.15640	98	0.00417	0.00393	0.00385	0.00378	0.00371
0.15482	99	0.00422	0.00397	0.00389	0.00382	0.00374
0.15328	*100*	0.00427	0.00401	*0.00393*	0.00385	0.00378
0.15176	101	0.00431	0.00405	0.00397	0.00389	0.00382

The italicized values in the table have been adopted as standard by the American Institute of Electrical Engineers.

The resistivity of copper at high temperatures is given by E. F. Northrup [The Resistivity of Copper from 20° to 1450° C, *J. Franklin Inst.*, Vol. 177, p. 1 (1914)].

MECHANICAL AND THERMAL PROPERTIES OF COPPER AS PREPARED FOR ELECTRICAL USES. Cast wire-bars are frequently machined all over, i.e. "scalped"

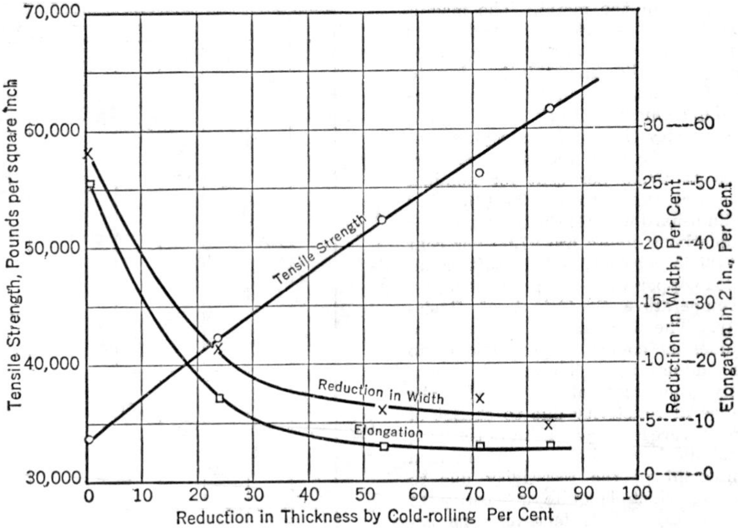

Fig. 4. Effects of Cold Rolling on the Tensile Properties of Commercial Electrical Copper

or "shaved," to remove surface imperfections that could cause laps, seams, slivers, etc., in the finished product. Slivers can cause very expensive difficulties by perforating the insulation between turns in coils. The cast bars are usually shaped roughly by hot-rolling, then cleaned by pickling in hot 10 per cent sulfuric acid, rinsed, dried, and finally finished by a series of cold-drawing or rolling operations, interspersed with cycles of annealing in the range between 450 and 600 deg cent. Controlled-atmosphere furnaces are frequently used to minimize oxidation and thereby decrease pickling costs.

The elastic strength of the copper in its soft state is very low but may be increased by a factor of 20 or more when cold-worked, as by wire-drawing or rolling. This strengthening and hardening effect is the result of the distortion and fragmentation, without rupture, of the original crystalline structure of the metal and is accompanied by considerable loss of ductility. Annealing, properly done, causes "recrystallization," greatly decreases the hardness, and restores the ductility. By recrystallization is meant the coalescence and reorientation of the crystalline fragments comprising the cold-wrought metal, the development of a new set of crystal boundaries bearing no relation to the old, and the formation of a crystal pattern in which, on the average, the dimensions of the individual grains are the same along all directions, i.e., an "equiaxial" structure takes the place of

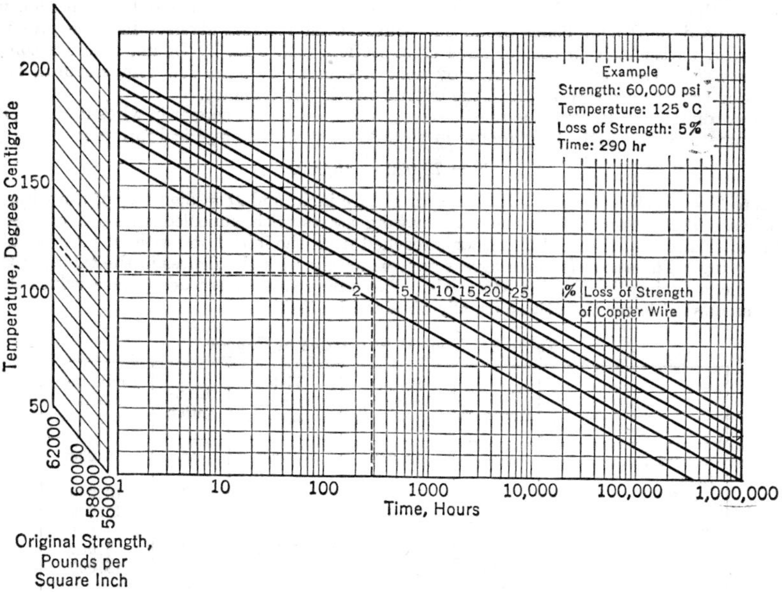

FIG. 5. Relations between Initial Tensile Strength of Cold-drawn Copper Wire and Loss of Strength with Time at Elevated Temperatures

the directional one resulting from plastic deformation while cold. Annealing temperatures in excess of approximately 600 deg cent may cause undesirably coarse crystals with no gain in ductility.

The mechanical properties of copper (and of the ductile metals and alloys generally) may be adjusted over a considerable range by controlling the amount of cold reduction and by "process annealing" or "tempering" at temperatures below those required for complete annealing. See Table 3. The more severe the cold-working, the lower is the temperature at which softening occurs, and the more rapid the loss of hardness at a given temperature, i.e., the less the "thermal endurance" (7), (8). Figures 4 (7) and 5 (9) illustrate the relations between cold-working, annealing temperatures, and strength of copper as ordinarily found in conductors.

MECHANICAL PROPERTIES OF COPPER. The elastic strength of a single crystal of highly purified copper is so slight as to be scarcely measurable, and even that of fully annealed commercial copper wire is hardly more than some tens of pounds per square inch. When a specimen of annealed copper, e.g. wire, is subjected to tension test, however, the stress at maximum load generally approximates 30,000 psi, calculated from the maximum test load and the original cross-section of the test piece.* For standard wire specimens 10 in. long plastic elongations of 25 per cent or more will occur, and the reduction of area at the "neck" of the fracture may exceed 70 per cent. Thus, although the breaking load may be substantially less than the maximum load, the actual maximum stress at rupture may be somewhat more than 70,000 psi. (15). If soft copper is subjected to extensive cold-reduction, as by cold-drawing or cold-rolling, the elastic strength is increased manyfold, while the ductility is very greatly reduced; failure in the tension test

* Tensile strength is defined as the result, expressed as stress, obtained by dividing the number representing the maximum load by that representing the original cross-sectional area of the test specimen.

will occur with relatively little plastic deformation, and the yield strength and the "tensile strength" will much more closely approach the actual stress at rupture. This stress at rupture has practically the same value for annealed as for hard-drawn copper.

Tensile Strength and Elongation of Soft Annealed Copper. The tensile strength of soft annealed copper wire is about 30,000 to 33,000 psi, and the overall elongation is approximately 25 per cent (in 10 in.) at fracture. It has no true elastic limit, permanent elongation being produced by very small stresses.

Tensile Strength and Elongation of Hard-drawn Copper. In modern hard-drawn copper the effects of cold-drawing are substantially uniform throughout the section, no pronounced hard skin being produced. Substantial residual stresses may be present, however, particularly in bars of large cross-section, e.g., 3 in. by 1/2 in., and it is sometimes worth while to relieve these by heat treatment at quite low temperatures, e.g., 100 deg cent, in order to improve the overall elastic performance.

According to D. R. Pye, the tensile strength of hard-drawn copper in wires up to 1/2 in. in diameter varies with the diameter in accordance with the empirical formula:

$$T = 70{,}000 - 45{,}000D$$

where T = tensile strength, in pounds per square inch; and D = diameter of the wire, in inches. This assumes that the cold-drawing starts with annealed wire-rod of 1/2-in. diameter and that the final diameter is reached from the initial one without intermediate annealing. The above formula agrees approximately with the figures in the tables of the American Society for Testing Materials. These figures represent acceptance values that are somewhat lower than those usually obtained by test.

Table 3. Typical Properties of Copper

	Brinell Hardness	Tensile Strength, lb per sq in.	Yield Strength, lb per sq in.	Elongation	Endurance Limit, lb per sq in.
As cast	36	30,000	17,000	45	6,000
Bars					
Soft	40	32,000	12,000	38	10,000
Hard	70	50,000	46,000	18	15,000
Sheet					
Soft	40	32,000	12,000	37	10,000
Hard	100	51,000	48,000	4	15,000
Wire					
Soft	....	38,000	12,000	36	10,000
Hard	....	60,000	39,000	3	15,000

The elongation at fracture is approximately represented by the empirical formula

$$E = \sqrt{4D}$$

where E = per cent elongation at fracture.

Tests by G. C. Batson on 50-ft lengths of hard-drawn copper showed a tensile strength only 1/2 per cent less than that of 10-ft lengths, a fact which indicates that the material can be quite uniform.

Acceptance values for physical properties of soft, medium-hard and hard copper wire will be found in Section 14, Article 1.

Modulus of Elasticity of Hard-drawn Copper. The modulus of elasticity of hard-drawn copper varies somewhat with diameter, higher values being shown by small wires. The usual figure taken is 16×10^6 as an average for all copper forms. Because of the poor elastic properties of copper, however, deflections under load may be somewhat greater than would be expected from calculations based on the above value for modulus.

Compression Test. Copper of good quality does not fracture under compression; it yields and flattens. According to Thurston, the resistance to compression may be calculated, within the limits $e < 1/2$, from the formula

$$C = 145{,}000\sqrt[3]{e}$$

where C = resistance in pounds per square inch of original area, and e = fractional compression.

CREEP CHARACTERISTICS OF COPPER. The deformation of a ductile metal under constant load and constant temperature may be regarded as taking place in four stages, namely:

a. *Elastic.* Strain is directly proportional to stress during the brief interval required for application of load, up to the "elastic limit."

b. *First-stage Creep.* Initial plastic strain rate is relatively rapid and decreases markedly with time.

c. *Second-stage Creep.* Plastic strain rate remains constant for a limited period.

d. *Third-stage creep.* Plastic strain increases with time at an accelerated pace until rupture occurs under the increasing stress due to contraction of cross-section.

Figure 6 shows an idealized strain curve representing the "stage" conception. In any real material, however, the stages merge, making it impossible to determine exactly the positions of the transition points.

The creep characteristics of copper are very markedly affected by its thermal and mechanical history and by differences in composition within the limits encountered in the high conductivity varieties. Very substantial advantages in creep resistance may be gained at the expense of relatively small sacrifices of electrical conductivity by judicious

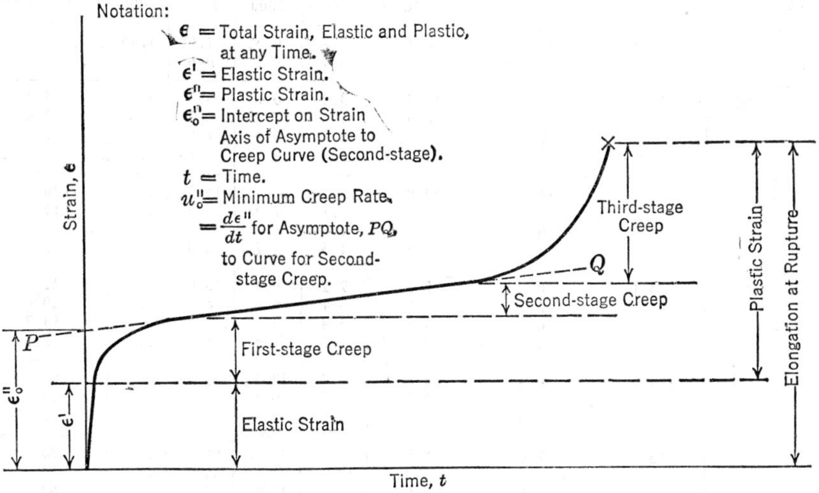

Fig. 6. Idealized Strain-time Curve for Creep of a Metal in Tension

alloying, cold-working by wire-drawing or rolling, and prestressing sufficiently to cause plastic deformations of several per cent along the directions in which the working strains will occur.

However, thermal endurance decreases as cold-working is increased, and there is a tendency toward corresponding decreases in creep resistance. In the copper ordinarily supplied for electrical conductors, it is usually not advantageous to apply cold-reductions exceeding 15 per cent when the maximum creep resistance is desired at temperatures above approximately 100 deg cent. Greater amounts of cold-working may be useful with alloyed copper.

Creep Data for Certain Varieties of Copper. Table 4, based on the work of Davis (12) and Burghoff and Blank (14), as well as unpublished data from Davis, show how the creep characteristics of a few types of copper are related to stress and to temperature. Davis' data on oxygen-free copper may be expressed by the equation

$$U_0 = 2Ae^{-1/BT} \sinh \frac{\sigma}{11{,}020 - 15.8T}$$

where U_0 = minimum creep rate, in inches per inch per hour.
A = material constant = 0.1075 for oxygen-free copper.
e = base of natural logarithms = 2.7183.
B = material constant = 1.513×10^{-4}.
T = absolute temperature, in centigrade degrees.
σ = stress, in pounds per square inch.

The equation given above, like others proposed for the same purpose (16), is an empirical one which, with the constants as given, is valid only for oxygen-free copper and for temperatures between approximately 100 and 235 deg cent. The general form of the expression, however, has some foundation in that the hyperbolic sine relation between stress and creep has been found applicable to a considerable variety of creep data (17, 18,

19), while the factor $e^{-1/BT}$ appears in chemical rate theory concerned with the effect of temperature on speed of reaction. The estimation of long-time results from short-time tests is, however, attended with considerable risk. Only by extended tests can the numerical values for the constants A and B be accurately determined for any particular substance, but valuable guidance can be obtained from the literature (20, 21) and by intelligent use of the results of short-time creep tests.

Table 4. Creep Data on Oxygen-free Copper [After Davis (12) and Burghoff and Blank (14)]

Test Temp., °C	Modulus, E, psi × 10^{-6}	Stress, α, psi × 10^{-3}	Nominal Initial Strain, ϵ in. per in. × 10^4 *	Total Creep, ϵ in. per in. × 10^4, Hours				Minimum Creep Rate, U_0, in. per in. per hr. × 10^8
				500	1000	2000	3000	
30	15.4	16	10.4	0.7	0.85	1.05	1.15	....
		12	7.8	0.45	0.5	0.6	0.65	....
80	15.0	16	10.7	3.0	3.6	4.3	4.9	5.9
		12	8.0	1.6	1.9	2.3	2.5	2.6
		7.7	5.1	0.8	0.9	1.0	1.15	1.3
		4	2.7	0.5	0.52	0.55	0.55	....
130	14.8	16	10.8	....	....	15.5	.....	36
		13	8.8	5.7	7.2	8.6	.....	15
		10	6.8	3.6	4.4	5.3	.....	8
		5	3.4	1.3	1.7	2.0	.....	3
165	14.1	10	7.1	7.2	9.4	13.0	.....	32
		8	5.7	5.1	6.8	8.7	.....	20
		6	4.3	3.4	4.5	5.8	.....	13.7
		4	2.8	2.0	2.6	3.4	.....	7.2
200	14.0	6.4	4.6	7.4	10.5	....	.....	66
		4.5	3.2	4.2	6.5	....	.....	42
		2.5	1.8	2.0	3.2	....	.....	22
235	13.9	7.0	5.0	31	46	68	.....	230
		5.5	4.0	18	26	40	.....	150
		4.0	2.9	11	16	23	.....	82
		2.5	1.8	5	7	9	.....	32

* Calculated from E and test stress.

Relaxation. Complementary to creep is the relaxation of stress in a loaded member held to constant length at constant temperature between fixed supports. Like creep, the relaxation process is a complex one that converts initial elastic strains to equivalent plastic ones, while concurrently microstructural readjustments and sometimes metallurgical reactions are in progress. Materials showing high resistance to first-stage creep are likely to have low relaxation rates.

Directional Effects of Prestressing. The elastic and creep characteristics of copper and of plastic metals in general under tensile stresses may be considerably improved over those of ordinarily annealed or cold-drawn material by prestretching sufficiently to produce some plastic elongation. The tensile advantage thus gained, however, is, to a considerable extent, purchased at the expense of corresponding sacrifices of compressive strength. This result, sometimes called the Bauschinger effect, is of considerable importance when structural parts are subject to reversal of stress. Similar relations hold for material conditioned by plastic precompression; i.e., the tensional properties will be reduced as the compressional ones are increased.

12. COPPER ALLOYS

EFFECTS OF IMPURITIES AND ALLOYING ELEMENTS ON THE ELECTRICAL RESISTIVITY OF COPPER. For the present purpose four general classes of alloys may be recognized, namely:

Solid-solution Alloys. The atoms of alloying metal substitute for those of the basis metal without altering the character of the crystal lattice, e.g., nickel in copper.

Eutectiferous Alloys. These consist of two distinct constituents physically distinguishable and intimately mingled, e.g., silver in copper.

Precipitation-hardening Alloys. In these alloys the amount of added metal that can be held in solid solution is dependent on temperature, e.g., chromium in copper.

Intermetallic Compounds. These are produced by the reaction of two metals in definite proportions, e.g., Cu_3Si, containing approximately 13 per cent Si.

In alloys, particularly those based on copper, the resistivity usually is substantially greater than that of the base metal.

HIGH-STRENGTH, HIGH-CONDUCTIVITY COPPER-BASE ALLOYS. The elastic strength of copper is relatively low, and much of the increment of strength that may have been conferred upon it by cold-working may be lost with time at temperatures no higher than those frequently encountered in electrical equipment in normal service. The thermal endurance of cold-worked pure copper is so poor that noticeable changes in strength and hardness occur even at temperatures as low as 100 deg cent. Alloying provides considerable improvements of thermal endurance, and a number of commercial materials have elastic properties and thermal endurance superior to those of copper without prohibitive sacrifice of electrical conductivity.

Silver-bearing Copper ("Lake Copper"). The electrical conductivity of copper is not significantly impaired by the presence of silver in proportions up to approximately 0.15 per cent. Even considerably smaller amounts of silver, however, produce very substantial increases in thermal endurance over that of pure copper. The ductility is not impaired, and the tensile properties attainable by cold-working are but slightly superior to those available from silver-free copper.

Copper-chromium Alloys. When chromium in amounts between approximately 0.4 and 0.8 per cent is dissolved in molten copper or copper-silver alloy (approximately 0.1 per cent silver), alloys are produced which, when cast or wrought and then quenched from temperatures between 1000 and 1050 deg cent, consist of a solid solution of chromium in the basis alloy of silver and copper. If the quenched alloy is held at temperatures near 450 deg cent for approximately 24 hr, the chromium "precipitates" in the form of submicroscopic crystallites scattered throughout the mass of the alloy, with the result that *both* the electrical conductivity and the hardness are greatly *increased* while the ductility remains high. Furthermore, the thermal endurance of the precipitation-hardened alloy is much greater than that of other high-conductivity alloys.

The properties of a typical chromium-hardened copper-base alloy are summarized in Table 5 and are described in more detail by Brace (27).

Table 5. Typical Properties of Copper-chromium Alloys *

	Wrought and Quenched	Precipitation-hardened	Precipitation-hardened and Cold-rolled 50 Per Cent
Electrical conductivity, per cent (IACS)	30–35	80–90	80–90
Ultimate tensile strength, pounds per square inch	Approximately 35,000	50,000	65,000
Elongation in 2 in., per cent	30–35	30	15
Fatigue strength, pounds per square inch		Approximately 20,000 †	Approximately 25,000
Thermal endurance–hardness loss, per cent per day at 200 deg cent	Tends to harden	Nil ‡	Approximately 0.5 percent

* As manufactured by Westinghouse Electric Corporation under the trade name Cupaloy.

† Compared with 12,000 psi for copper cold-drawn 15 per cent.

‡ Compared with approximately 50 per cent for hard-drawn copper containing 0.0012 per cent silver (0.35 oz per ton) and 20 per cent for similarly prepared copper containing 0.036 per cent silver (10.5 oz per ton).

Beryllium-copper Alloys. While, strictly speaking, not "high-conductivity," certain copper-base alloys containing beryllium in amounts between approximately 0.4 per cent and 2 per cent, together with relatively small amounts of other metals, notably silver, nickel, or cobalt, do provide especially high elastic strength, as compared with previously known alloys of like conductivity. These alloys are of the precipitation-hardening type, and their favorable characteristics become evident after heating to 790 to 800 deg cent for 1/2 to 3 hr to bring the beryllium into the solid-solution state, quenching in water or oil from the solution temperature, and then reheating and holding at temperatures in the neighborhood of 315 deg cent for approximately 3 hr.

After quenching, these alloys are in their softest condition, and, because the beryllium is in solid solution, the conductivity is at its lowest, being approximately 17 per cent for beryllium contents near 2 per cent. The figures given in Table 6 show, in a general way, what may be expected of commercially available beryllium-copper alloys. In view of the variety not only of compositions available but also of heat and mechanical treatments possible, the foregoing information must be regarded as indicative rather than definitive, and when precise data are required reference should be made to the literature or aid sought from suppliers.

Table 6. Approximate Values for the Physical Properties of Beryllium-copper Alloys of the 21 Per Cent Beryllium Class

Condition	Solution-treated "Annealed"	Solution-treated and Cold-worked	Solution-treated and Precipitation-hardened	Solution-treated, Cold-worked, and Precipitation-hardened
Electrical conductivity, per cent (IACS), at 20 deg cent.	17	17	20–25 * 32–38 †	20–25 * 32–38 †
Tensile strength, pounds per square inch	70,000	90,000	160,000 * 130,000 †	180,000 * 140,000 †
Yield strength, pounds per square inch, at 0.5 per cent elongation under load			150,000 *	175,000 *
Elongation in 2 in., per cent	35		3.0 *	2.0 *
Modulus of elasticity, pounds per square inch	16×10^6		$18.4–19.4 \times 10^6$	Approximately 18×10^6
Endurance limit, pounds per square inch, at 10^8 reversals of stress		23,000	28,000	28,000

* Heat-treated for maximum hardness.
† Heat-treated for maximum conductivity.

Solid-solution Alloys. There have come into commercial use a number of copper-base alloys in which conductivity has been deliberately sacrificed in the interests of superior tensile properties. The most used of these alloys are covered by Standard Specifications for Hard-drawn Copper-alloy Wires for Electrical Conductors, Designation B105-39, issued by the American Society for Testing Materials (28). The limits of composition of the different alloys that may be supplied under the above specifications are shown in Table 7. Table 8 (29) shows the grading system and acceptance limits for the electrical and mechanical properties of this group of alloys in the form of hard-drawn wire.

Table 7. Limits of Composition of Copper Alloys Covered by ASTM Specification B105-39

Element	Percentage	Element	Percentage
Iron, max	0.75	Aluminum, max	3.50
Manganese, max	0.75	Tin, max	5.00
Cadmium, max	1.50	Copper, max	94.00
Silicon, max	3.00	Sum of the above elements, min	99.5

Table 8. Electrical and Mechanical Properties of Hard-drawn Copper-alloy Wire from ASTM Specifications (29) for Hard-drawn Copper-alloy Wires for Electrical Conductors

Alloy Type	Grade No.	Electrical Resistivity at 20° C, maximum		Wire Diameter, In.	Tensile Requirements, Minimum			Density Nominal, grams per cu cm
		Microhm-cm	Ohms per mil foot		Tensile Strength, Psi	Elongation, %		
						10 in.	60 in.	
Copper, silicon, iron	8.5	20.276	122.00	0.2893	97,500	2.17		8.78
Copper, silicon, manganese				0.1443	115,300		1.09	
Copper, silicon, zinc				0.07196	121,200		0.92	
Copper, silicon, tin, iron				0.03196	123,500		0.82	
Copper, silicon, tin, zinc								
Copper, aluminum, tin	13	13.259	79.78	0.2893	102,500	2.17		8.78
Copper, aluminum, silicon, tin				0.1443	120,300		1.09	
Copper, silicon, tin				0.07196	126,200		0.92	
				0,03196	128,500		0.82	
Copper, aluminum, silicon	15	11.491	69.14	0.2893	109,500	2.17		8.54
Copper, aluminum, tin				0.1443	126,500		1.09	
Copper, aluminum, silicon, tin				0.07196	132,600		0.92	
Copper, silicon, tin				0.03196	135,000		0.82	
Copper, tin	20	8.6191	51.86	0.2893	109,500	2.17		8.89
				0.1443	126,500		1.09	
				0.07196	132,600		0.92	
				0.03196	135,000		0.82	
Copper, tin	30	5.7455	34.57	0.2893	74,000	2.17		8.89
				0.1443	96,500		1.09	
				0.07196	103,000		0.92	
				0.03196	108,000		0.82	
Copper, tin	40	4.3059	25.93	0.2893	73,400	2.17		8.89
Copper, tin, cadmium				0.1443	81,800		1.09	
				0.7196	89,600		0.92	
				0.03196	95,000		0.82	
Copper, tin, cadmium	55		18.86	0.2893	76,000	2.17		8.89
				0.1443	85,500		1.09	
				0.07196	93,800		0.92	
				0.03196	99,500		0.82	
Copper, tin	65	2.6525	15.96	0.2893	67,500	2.17		8.89
Copper, tin, cadmium				0.1443	72,900		1.09	
				0.07196	76,800		0.92	
				0.03196	79,600		0.82	
Copper, cadmium	80	2.1539	12.96	0.2893	72,000	2.17		8.89
				0.1443	79,800		1.09	
				0.07196	86,800		0.92	
				0.03196	92,200		0.82	
Copper, cadmium	85	2.0276	12.20	0.2893	68,500	2.17		8.89
				0.1443	75,800		1.09	
				0.07196	82,500		0.92	
				0.03196	87,600		0.82	

COPPER ALLOYS AS RESISTORS IN ROTATING MACHINES. Damper bars in the fields of a-c machines and rotor bars in some designs of a-c motors make use of resistance material composed chiefly of copper. Some of the alloys commonly used for these purposes are given in Table 9.

Table 9. Copper Alloys for Resistors in Rotating Machines

Name	Chemical Composition (parts per 100)								Resistivity, ohms per cir mil-ft	Conductivity, Per Cent IACS
	Cu	Zn	Pb	Sn	Si	Mn	Ni	Fe		
Copper	100								10.371	100
Brass	62	35	3						41.46	25.0
Gun bronze	92			8					79.8	13.0
Silicon bronze	96				3	1			155.0	6.7
Telephone bronze	98.25			1.75					29.62	35.0
Monel metal	24					3	60	3	268.0	3.9
PMG 3	98.2				1.2			0.6	69.1	15.0
PMG 10	95.6				3.2			1.2	159.5	6.5
PMG 94	93.0	4.0			2.5			0.5	129.6	8.0

13. ALUMINUM

The response of aluminum to cold-working and annealing is governed by the same general laws that apply to copper. The resistivity is increased slightly by cold-working and markedly by the presence of alloying elements.

CONDUCTIVITY AND RESISTIVITY. The Aluminum Company of America gives, as the average of many thousands of separate determinations, 2.828 microhms per centimeter cube, or 17.002 ohms per circular mil-foot, at 20 deg cent, as the resistivity of commercial aluminum wire. This corresponds to a conductivity of 61 per cent IACS. The conductivity of annealed superpurity (99.997 per cent) aluminum is reported to be 65.45

Table 10. Effect of Alloying on Electrical Resistivity of High-purity Aluminum

Purity, %	Fe	Si	Cu	Electrical Conductivity, %, IACS, 20° C	Resistivity, Ohm-cm × 10^6
99.997	0.0012	0.0015	0.0003	65.45	2.63
99.991	0.0013	0.0037	0.0035	65.35	2.64
99.97	0.012	0.008	0.008	64.90	2.66

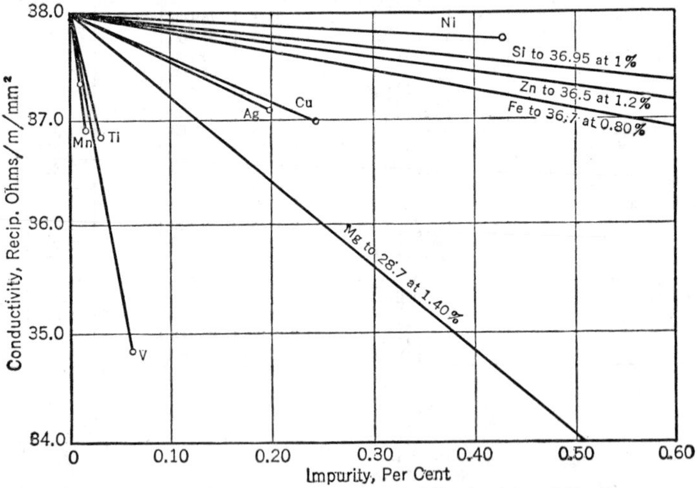

FIG. 7. Effects of Impurities on the Electrical Conductivity of Aluminum

per cent (2.63×10^{-6} ohm-cm) by Gauthier (32), who reported also on the effects of various impurities on the conductivity. Table 10 and Fig. 7 (33) summarize Gauthier's findings relating to the effects of impurities.

EFFECT OF COLD-WORKING ON RESISTIVITY. The electrical resistivity of aluminum is affected by cold-working. For example, the resistivity of cold-drawn aluminum wire having a tensile strength of about 20 kg per sq mm (28,500 psi) is approximately 2 per cent greater than that of the same wire when thoroughly annealed and consequently soft.

TEMPERATURE COEFFICIENT OF RESISTIVITY. The 20 deg cent temperature coefficient of annealed aluminum wire of 61 per cent conductivity is 0.00403 per degree centigrade, and the 0 deg cent coefficient is 0.0044. The temperature coefficient of aluminum is proportional to its conductivity, as expressed in per cent of that of the pure metal.

DISADVANTAGES OF LOW-TENSILE STRENGTH. The low-tensile strength of aluminum, as compared with copper, for equal length and conductance, and also the high coefficient of thermal expansion affect the cost of an aerial line in two ways: (1) by making it necessary to erect the spans with a greater sag or less length in order to reduce the stresses, thereby increasing either the height or the number of poles; and (2) by making it necessary to increase the distance between wires on account of the increased sag. The increase in the height of poles for the same spacing amounts to about 10 per cent. Aluminum cables with steel core (ACSR type) are now used and overcome these difficulties.

EFFECT OF LARGE ELONGATION OF ALUMINUM. The extraordinarily great ductility of aluminum enables even cold-drawn wire to yield to severe mechanical overloads by stretching and thus increasing the sag. In dealing with a single solid wire, however, this cannot be relied on, because even one small imperfection, such as a transverse scratch or nick, will often cause the wire to break without significant elongation. This is one reason why cables are to be preferred to single-wire conductors.

ACTION OF ARCS ON ALUMINUM CABLES. Because of the lower melting point of aluminum (660 deg cent, compared to 1085 deg cent for copper), it was once believed that aluminum transmission cables would be subject to more damage from flashover arcs on transmission lines. Tests have shown that aluminum and copper cables are equally good in resisting arcs. In some tests it was noted that the arc wandered about more on aluminum than on copper, and the effect was therefore less concentrated. Furthermore, the steel core of the ACSR type provides ample strength even though the aluminum is considerably burned.

ALUMINUM CONDUCTORS IN ELECTRICAL MACHINES. The increased bulk of aluminum required for a given current rating entails larger and heavier iron parts than where copper is used, and the net result is an increase in weight per unit of output. For this reason aluminum has found but little application in the windings of electrical machines. Table 11, which gives a comparison of copper and aluminum wire for equal resistance per unit length, is of interest in this connection. Figure 8 gives a general idea

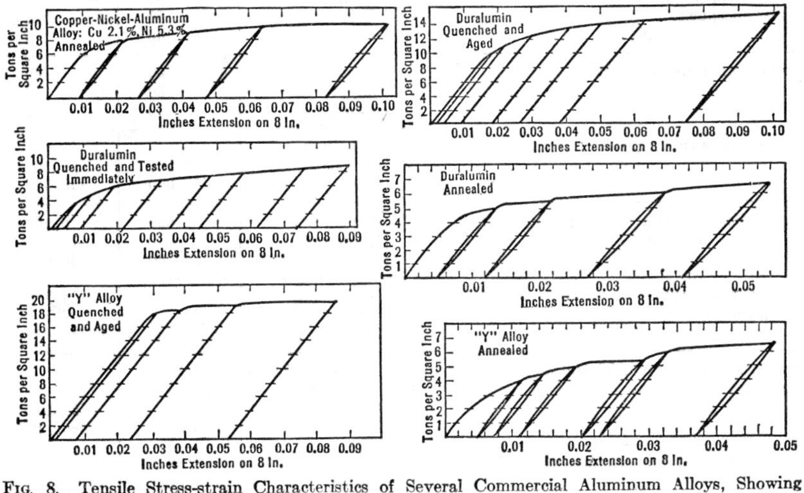

FIG. 8. Tensile Stress-strain Characteristics of Several Commercial Aluminum Alloys, Showing Effects of Plastic Tensile Strain on Tensile Elastic Properties

of the stress-strain characteristics of a few aluminum alloys that have found commercial use. Advantages may sometimes be gained by the substitution of strong alloys of aluminum for steel in structural parts of electrical apparatus, e.g., in the frames of motors.

Table 11. Comparison of Copper and Aluminum Wires for Equal Resistances per Unit Length

Item	Copper	Aluminum
Cost	1	0.488 × p/P
Cross-section	1	1.61
Diameter	1	1.27
Volume	1	2.04
Weight	1	0.488
Breaking strength	1	0.64 annealed

p = unit price by weight of aluminum wire.
P = unit price by weight of copper wire.

MECHANICAL PROPERTIES OF ALUMINUM. Aluminum is a softer metal than copper and has a lower elastic modulus. Table 12 gives a comparison of the nominal physical properties of copper, aluminum, and steel. For more detailed information reference may be made to the publications listed in Article 15.

Table 12. Nominal Physical Properties of Copper, Aluminum, and Steel

Property	Material	Conditions: Annealed	Conditions: Cold-drawn
Density, grams per cubic centimeter at 20 deg cent	Aluminum	2.7048	2.7019
	Copper	8.94	8.92
	Steel	7.85	7.83
Modulus of elasticity, pounds per square inch	Aluminum	....	10 × 10⁶
	Copper	....	16 × 10⁶
	Steel	30 × 10⁶	
Yield strength, pounds per square inch, 0.2% elongation	Aluminum	3,000	12,000
	Copper	7,000	48,000
	Steel	30,000	65,000
Tensile strength, pounds per square inch	Aluminum	13,000	25,000
	Copper	30,000	52,000
	Steel	50,000	80,000
Elongation, per cent on 2 in. of 0.505 St. ASTM test specimen	Aluminum	45	7
	Copper	50	5
	Steel	40	5
Reduction of area at fracture	Aluminum	80	40
	Copper	80	70
	Steel	50	30
Approximate true stress at rupture	Aluminum		
	Copper	70,000	70,000
	Steel	150,000	250,000
Endurance strength, pounds per square inch	Aluminum	5,000	8,500
	Copper	10,000	14,000
	Steel	25,000	40,000
Coefficient of expansion, mean, 20–200 deg cent, per deg cent × 10^6	Aluminum	24.7	24.7
	Copper	17.7	17.7
	Steel	Approx. 12.5	Approx. 12.5
Thermal conductivity, 20–200 deg cent, watts per cm^2 per deg per cm	Aluminum (at 50 deg cent)	2.21	2.21
	Copper	3.86	3.86
	Steel	0.55	Approx. 0.55
Electrical resistivity, microhm-cm at 20 deg cent	Aluminum	2.68	2.78
	Copper	1.682	1.716
	Steel	Approx. 16	Approx. 16

Effect of Cold-working on Tensile Properties. Cold-working increases the tensile strength and decreases the ductility of aluminum. Table 13, based on data from the Aluminum Company of America, reported in *U.S. Bur. Standards Circ. No.* 346 (34) gives a general idea of the magnitude of the effect.

Table 13. The Effect of Cold-working on the Tensile Properties of Aluminum Sheet

Ratio of: Thickness of Sheet Tested / Thickness of Sheet from Which Rolled	Tensile Strength, psi $\times 10^{-3}$	Elongation, Per Cent in 2 in.
1.0	13.6	35
0.8	16.7	17
0.6	20.1	8
0.4	23.8	5
0.2	27.8	6.7
0.1	31.0	6.3
0.57	35.0	5.7

Effects of Temperature on Mechanical Properties. The tensile properties of metals generally increase as temperature is decreased. Even mercury displays quite respectable physical properties when frozen. If the properties at room temperature are taken as a basis for comparison, however, it is usually true that the rate of change of properties with temperature is greater the lower the melting point of the metal. Table 14 shows the effect of temperature on the torsional and (calculated) tensile modulus of aluminum (35) and gives a general idea of the relatively great sensitivity of aluminum to elevated temperatures.

Table 14. Modulus of Torsion of Aluminum at Elevated Temperatures

Temperature, deg cent	Torsional Modulus of Elasticity, F, psi $\times 10^{-6}$	Tensile Modulus of Elasticity, E *, calc. psi
20	3.87	10.5
100	3.73	10.1
200	3.45	9.4
300	3.10	8.43
400	2.63	7.16
450	2.03	5.53
500	0.68	1.85

* $E = 2.72F$ (approximately).

Table 15 (*a*) shows the short-time tensile properties of commercially pure aluminum, 2S (57–59 per cent conductivity, IACS), as reported by the Aluminum Company of America, and Table 15 (*b*) shows those of the alloy 24-S, whose nominal composition is: copper, 4.5 per cent; manganese, 0.6 per cent; magnesium, 1.5 per cent; aluminum, balance. The alloy develops relatively high strength when processed by water-quenching from 910 deg fahr (488 deg cent) to 930 deg fahr (499 deg cent) and cold-working and subsequently precipitation-hardening by reheating to 370 deg fahr (187 deg cent) to 380 deg fahr (193 deg cent) for 11 to 13 hr. In this condition the material is designated as 24-S-T. The electrical conductivity in the annealed (soft, "O") condition is given as 50 per cent IACS, whereas that of the heat-treated material is but 30 per cent IACS. This reflects the fact that the high-tensile properties have been obtained by bringing the alloying elements into solid solution and subsequently enhancing the hardness of the solid solution by precipitating a portion of the alloy content by reheating, as noted above. The corrosion characteristics are affected by the condition of the precipitate formed from the quenched solid solution. The best results are obtained by fast quenching (water) and prolonged tempering (11 to 13 hr).

In Table 15 (*b*) it will be noted that the "hard" material, i.e., that which received the greatest amount of cold-working, showed lower yield and tensile strengths than did the half-hard material for test temperatures above 400 deg fahr (204.5 deg cent), although the room-temperature strength of the hard material was the higher by 50 per cent or more.

Table 15. Typical Short-time Tensile Properties at Elevated Temperatures

Condition	Test Temperature		Strength, psi		Elongation in 2 in., %
	deg fahr	deg cent	0.2% Yield	Tensile	
a. Commercial Aluminum, 2S (Purity Nominally 99.6 Per Cent)					
Soft	75	23.9	5,000	13,000	45
	300	149.0	3,500	7,500	65
	400	204.5	3,000	6,000	70
	500	260.0	2,000	3,500	85
	600	313.5	1,500	2,500	90
	700	389.0	1,000	1,500	95
Half-hard	75	23.9	14,000	17,000	20
	300	149.0	10,000	13,000	22
	400	204.5	6,500	9,500	25
	500	260.0	2,000	3,500	85
	600	313.5	1,500	2,500	90
	700	389.0	1,000	1,500	95
Hard	75	23.9	21,000	24,000	15
	300	149.0	14,000	17,500	16
	400	204.5	3,000	6,000	70
	500	260.0	2,000	3,500	85
	600	313.5	1,500	2,500	90
	700	289.0	1,000	1,500	95
b. Heat-treatable Alloy "24 S-T" (Nominal Composition; Copper, 4.5%; Manganese, 0.6%; Magnesium 1.5%)					
Heat-treated, i.e., quenched and precipitation-hardened	75	23.9	46,000	68,000	22
	300	149.0	37,000	42,000	20
	400	204.5	16,000	20,000	26
	500	260.0	10,000	12,000	40
	600	313.5	5,500	7,000	70
	700	289.0	3,500	5,000	100

CREEP CHARACTERISTICS OF ALUMINUM. The creep resistance of aluminum is even lower than that of copper; there is much less published information. Figure 9 (36) presents data on the creep of aluminum wire of the conductor grade at room temperature, as reported by Sturm, Durmont, and Howell (37).

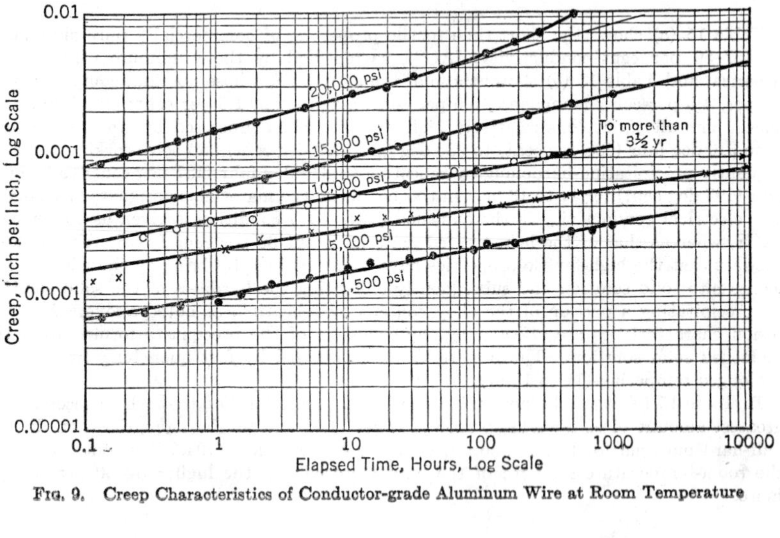

FIG. 9. Creep Characteristics of Conductor-grade Aluminum Wire at Room Temperature

Von Zeerleder and Irmann present data on an extensive investigation on the effects of time at various temperatures on the short-time tensile properties of aluminum of commercial purity and of several aluminum alloys. Curves are given that show the effects of continuous heating for periods of 1 year upon the yield strength (stress at 0.2 per cent yield) for test temperatures and heating temperatures from 20 to 250 deg cent. The general indication of these tests is that the tensile properties of aluminum and most of its alloys tend to decline with increasing rapidity as the temperatures of treatment and test increase from approximately 130 to 160 deg cent, and that after 250 deg cent yield strengths less than 25 per cent of the room-temperature values are to be expected.

Further information on the properties of aluminum and its alloys may be found in *Metals Handbook*, published by the American Society for Metals.

14. ELECTRICAL RESISTANCE ALLOYS AND OTHER SPECIAL-PURPOSE CONDUCTOR MATERIALS

GENERAL. MATERIAL. A considerable number of alloys, which may be grouped under the term electrical alloys, owe their importance to their exceptionally low conductivity or other unusual physical, electrical, or magnetic characteristics.

Among the metals that have been particularly important as basic materials for making "electrical alloys" the following may be mentioned: nickel, chromium, copper, manganese, platinum, palladium, gold, silver, and iron.

There is a second, smaller group of conductor materials, including certain elementary metals, which, because of their relatively high melting points, special chemical properties, or both are used in substantially pure state and have become indispensable in some uses, particularly incandescent lamps, thermionic devices, and contacts: tungsten, tantalum, molybdenum, and platinum.

Thermostatic bimetals, which act as current-sensitive elements in many electrical control devices, are also based on metals in the groups noted above.

There is also a third group of metals that, although used in relatively small proportions, play exceedingly important roles by virtue of their effectiveness in modifying or stabilizing the electrical, mechanical, or other significant properties of the basic alloys. Among such special-purpose metals the following should be mentioned:

Metal	*Purpose*
Aluminum	Controls resistivity and increases resistance to oxidation at high temperature.
Titanium	Controls effects of carbon and confers precipitation-hardenability.
Beryllium	Precipitation-hardening element in beryllium-copper alloys.
Zirconium	Improvement of strength and oxidation resistance in nickel-chromium base alloys.
Cerium	Same as zirconium.
Silicon	Metallurgical reagent, hardener, to increase resistivity, and to increase durability at high temperatures.
Vanadium	Metallurgical reagent, hardener. Increases resistivity.
Manganese	Metallurgical reagent, to increase resistivity and to improve resistance to attack by certain atmospheres sometimes found in industrial electric resistance furnaces.
Rhodium	Controls thermoelectric power, is a hardener, and improves performance of precious-metal contact alloys in certain cases.
Iridium	Same as rhodium.
Osmium	Hardener in platinum-metal contact alloys.

PROPERTIES. Table 16 shows the nominal compositions and physical properties of a number of the "electrical alloys" that have found significant engineering applications. Table 17 gives the properties of a few metals of the second group noted above—those used in substantially pure state for certain purposes. Figures 10, 11, and 12 show some of the electrical and mechanical properties of a number of platinum-base alloys that are interesting in connection with electrical applications (45).

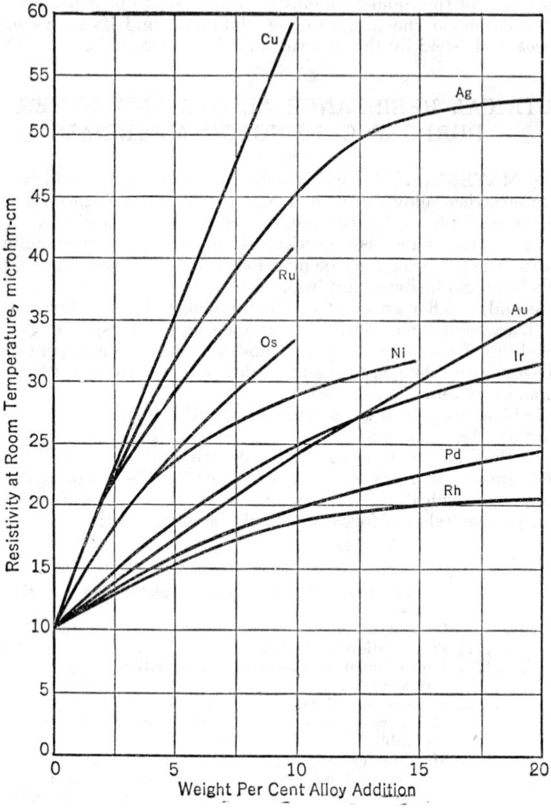

Fig. 10. Relations between Alloy Additions and Electrical Resistivity for Several Binary Platinum-base Alloys

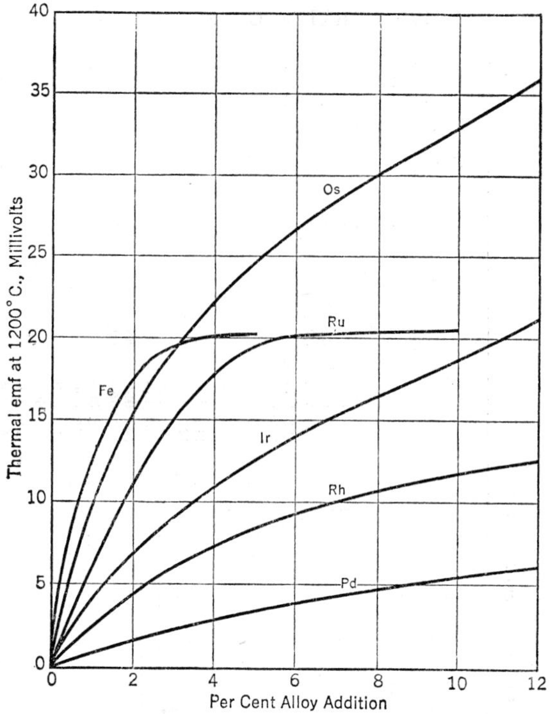

FIG. 11. Thermoelectric Properties of Several Binary Platinum-base Alloys

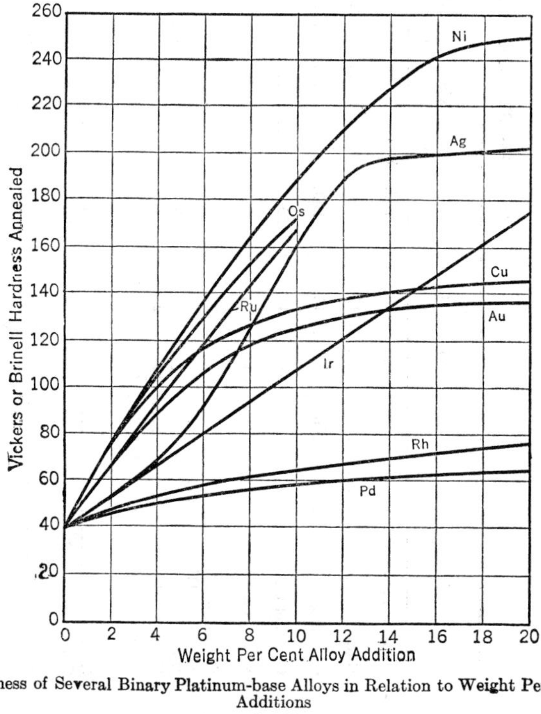

FIG. 12. Hardness of Several Binary Platinum-base Alloys in Relation to Weight Percentage of Alloy Additions

Table 16. Physical Properties of Electrical Alloys

Material	C	Ni	Cr	Fe	Mn	Cu	Zn	Al	Si	Sn	Tensile Strength, Psi	Yield Point	% Elongation	% R.A.	Brinell No.
Ohmax	......	......	20	Bal.	......	......	......	8.5	......	......	130,000	......	......	......	......
											100,000				
Radiohm	......	......	12–13	Bal.	......	......	......	4–5	......	......	128,000	......	5	......	......
											85,000		33		
Ohmalloy	0.6	......	13	Bal.	......	......	......	4	......	......	128,000	103,800	5	......	......
											85,700	67,640	33		
Chromel C	......	61	16	23	......	......	......	......	......	......	88,000	......	30	......	......
											107,000		25		
Nichrome I	......	65	11.18	22.36	0.70	......	......	......	0.26	......	100,000	......	......	......	157
															187
Nichrome II	......	66	22	Bal.	......	......	......	......	......	......	100,000	......	......	......	......
Nichrome III	......	86	12	......	......	......	......	......	......	......	110,000	......	......	......	......
Nichrome IV	......	82.36	15.9	0.84	0.48	......	......	......	......	......	120,000	60,000	25	35	183
Nichrome V	......	80	20	......	......	......	......	......	......	......	95,000	50,000	35	......	......
Tophet C	1.53	59.8	15.44	23.17	......	......	......	......	0.06	......	100,000	......	100	16	......
											140,000	85,000	3	45	......
											100,000	60,000	25	......	165
Comet (a)	......	30	5	65	......	......	......	......	......	......	85,000	47,000	37	69	......
Comet (b)	......	30.4	2.2	66.9	0.8	0.4	......	......	......	......	75,000	......	......	......	......
											160,000				
Nilvar	......	36	......	Bal.	......	......	......	......	......	......	150,000	......	......	......	......
											70,000				
Advance	......	44–46	......	......	54–55	......	......	......	......	......	62,000	40,000	40	75	......
Copel	......	45	......	......	55	......	......	......	......	......	60,000	......	40	......	......
Cupron	......	45	......	......	......	55	......	......	......	......	62,000	......	......	......	......
Constantan (a)	0.1	40–55	......	Bal.	0–1.4	60–54	......	......	......	......	60,000	20,000	40	50	100
											69,500	30,000	60	70	120
Constantan (b)	......	40	......	......	......	60	......	......	......	......	62,000	......	51	67	......
Constantan (c)	0.10	55	......	Bal.	1.0	43.9	......	......	......	......	141,000	125,000	1	5	300
Ideal	......	40–45	......	1.0–0.66	0.45–1.0	53.4–58	......	0–0.66	......	......	......	......	......	......	......

Manganin (a)	...	4	...	...	12	84	...	...	...	...	70,000	...	...	...	...
Manganin (b)	...	3–4	...	...	11–13	Bal.	...	...	...	...	...	...	...	...	...
Manganin (c)	...	2.5	...	...	1.7	53	39	0.2	...	2.7	...	...	...	...	...
Manganin (d)	...	5	...	...	25	70	...	...	...	...	...	...	...	...	...
Manganin (e)	...	2–12	...	...	4–15	86–82	...	...	...	...	...	...	...	...	...
Manganin (f)	...	3–4	...	...	9.5–10.5	Bal.	...	...	...	...	...	...	...	...	...
Lucero (a)	...	70	...	...	...	30	...	...	...	...	140,000	...	...	...	...
Lucero (b)	...	65	...	...	5	30	...	...	...	...	50,000	...	35	...	...
Monel	Max. 0.30	Min. 60	...	Max. 3.5	Max. 3.5	Min. 23	...	Max. 0.5	Max. 0.5	...	180,000 65,000 95,000 125,000	30,000 65,000 95,000	35 35 15	35 50 50	105 175 260
German Silver	...	20–31	...	...	...	46–50	29–34	...	...	...	...	...	...	...	...
Midohm	...	22–23	...	...	...	Bal.	...	...	...	...	100,000	...	...	...	...
Hytemco	...	70	...	Bal.	...	...	...	...	...	...	150,000	...	...	...	...
Chace Manganese Alloy No. 720	...	20	...	...	20	60	...	...	...	...	70,000	...	...	...	...

Material	Specific Resistance at 20° C (68° C)		Temperature Coeff. of Resistance		Coefficient of Linear Expansion	
	Microhm-cm	Ohms per Cir Mil-foot	Temp. Coeff.	Diff. in Temp., ° C	Coeff. of Linear Exp.	Diff. in Temp., ° C
Ohmax	167	1000	−0.000066	20–500	0.0000158	20–1000
Radiohm	133	800	0.0007	20–500	0.0000155	20–1000
Chromel C	...	675	2.60×10^{-4} (20°–100° C)	...	13.2×10^{-r} (20°–100° C)	...
Nichrome I	112	675	0.00017	20–100	0.000017	20–1000
Nichrome V	108	650	0.00013	20–100	0.000017	20–1000
Comet (a)	95	570	0.00088	20–500	0.000015	20– 500
Nilvar	80.5	484	0.00135	20–100	0.000001	20– 100
Advance	49	294	±0.00002	20–100	0.0000149	20– 100
Copel	...	294	$\pm 0.2 \times 10^{-4}$ (20°–100° C)	...	14.9×10^{-6} (20°–100° C)	...
Manganin (a)	48.2	290	±0.000015	15–35	0.0000187	15– 35
Lucero (a)	48.2	290	0.0010	20–250	0.0000125	20– 100
Midohm	30	180	0.00018	20–100	0.0000175	20– 500
Hytemco	20	120	0.0045	20–100	0.000015	20–1000
Chace Manganese Alloy No. 720	...	(Soft) 500–525 (at 80° F)	−0.000125	25–150	...	...
	...	(Hard) 300–450 (at 80° F)	±0.00024	25–150	...	...

Table 16. Physical Properties of Electrical Alloys—*Continued*

Material	Specific Heat, Gram-cal per deg. cent*	Thermal Conductivity, Watts per Cm C°	Approx. Melting Point, deg cent	Specific Gravity	Pounds per Cubic Inch	Uses	Remarks
Ohmax	....	....	1500	6.80	0.248	Electrical resistances	Heat resistant to 1200° C, discontinued
Radiohm	....	....	1480	7.30	0.263	Radio-tube parts, resistors	Heat resistant to 1000° C
Ohmalloy	....	....	....	....	....	Resistance wire, resistors, rheostats	Brittle above 1600° F, superseded by Fecraloy
Chromel C	0.107	0.134	1400	8.24	0.2977	Flat irons, toasters, heating elements	Corrosion and heat resistant
Nichrome I	0.107	0.132	1350	8.247	2.979	Electrical resistance units, annealing pots, heating elements	Heat and corrosion resisting, discontinued
Nichrome II	....	....	....	....	....	Same as Nichrome I	Discontinued
Nichrome III	....	....	....	....	....	Same as Nichrome I	Heat and corrosion resisting
Nichrome IV	....	....	....	....	....	Same as Nichrome I	Heat and corrosion resisting, discontinued
Nichrome V	0.104	0.136	1400	8.412	0.3039	Heating elements in furnaces, electric ranges	Max. operating temperature: 1100° C
Tophet C	....	....	....	....	....	Resistance wire, heating elements, percolators, toasters	Resists some common acids, max. operating temperature: 1920° F
Comet (a)	0.114	0.135	1480	8.15	0.294	Dipping baskets, electric resistance wire, rheostats	Max. operating temperature: 700° C, heat resisting
Comet (b)	....	....	....	....	....	Heavy-duty rheostats and electric control machinery	Heat resisting
Nilvar	0.123	0.110	1425	8.08	0.292	Bimetals, measuring tapes, length standards, thermostats	Low expansion alloy, max. operating temperature: 200° C
Advance	0.094	0.218	1210	8.9	0.321	Rheostats, electrical instrument work, resistance metal	Similar to Constantan and Ideal

Copel	0.094	0.218	1275	8.86	0.3200	Electrical instruments, rheostats, hot resistances	Resistance alloy up to 800° F
Cupron	...	...	...	...	...	Rheostats, voltmeters, shunt resistances	Similar to Advance, Iaeal, and Constantan
Constantan (a)	...	...	...	...	...	Electric resistance and thermocouples	High temperature and heat resistance
Constantan (b)	...	...	...	...	...	Electric resistance	...
Manganin (a)	...	...	1020	8.192	0.296	Electrical instruments	High resistivity and low temp. coeff. of resistance
Manganin (b)	...	...	...	...	...	Same as above	Same as above
Manganin (d)	...	...	...	...	...	Electrical instruments	...
Manganin (e)	...	...	...	...	...	Same as above	Max. operating temperature: 100° C
Manganin (f)	...	...	...	...	...	Instrument shunts	Max. operating temperature: 100° C
Lucero (a)	0.127	0.250	1350	8.90	0.3215	Springs, electrical resistance wire, bolts, valves	Resists heat to 500° C
Lucero (b)	...	...	...	...	...	Szrings, electrical resistance wire, pump rods	Corrosion and heat resistant
Monel	...	...	...	...	...	Turbine blading, airplane parts, pump rods, valve stems, valve seats, fittings	Resists corrosion, high impact resistance
German Silver	...	...	...	...	...	Springs and contact points in electrical work	...
Midohm	0.092	...	1100	8.9	0.321	Resistances, rheostats	Heat resistant, max. operating temperature: 200° C
Hytemco	0.125	0.289	1425	8.46	0.305	Immersion heaters, heater pads, electrical resistances	Max. operating temperature: 500° C
Chace Manganese Alloy No. 720	0.12 0.11 (15 to 35 C°)	0.047 (cal/sec/cm²/°C/cm)	1910° F	8.23 8.29	...	...	...

* At room temperature.

Table 17. Some Special-purpose Conductor Metals and Alloys

Material	Approximate Composition, %	Specific Gravity	Specific Heat, cal / gm × °C	Melting Point, deg cent	Mean Linear Coefficient of Expansion, 10^{-6} cm / deg cent × cm	Thermal Conductivity, cal × cm / cm^2 × sec × °C	Electrical Resistivity, microhms × cm / cm	Electrical Conductivity, % of Copper	Tensile Modulus of Elasticity, 10^6 psi	Magnetic Properties
Molybdenum	Min. 99.9	10.3	0.065	2620	5.4	0.346	5.7	29	42	Paramag.
Tungsten	Min. 99.9	19.3	0.034	3370 *	4.4	0.476	5.5	30	53	Paramag.
Tantalum	Min. 99.9	16.6	0.036	2850	6.5	0.130	15.5	11	27	Paramag.
Platinum	99.9	21.4 *	0.032 †	1775	8.7	0.165	10.6	16	24	Paramag.
Silver	99.99	10.5	0.056	960	19.0	0.974 *	1.5 †	112 *	10	Paramag.
Invar	36 Ni 64 Fe	8.1	0.12	1425	1.07 †	0.02 †	80	2	21	Mag. below 200° C, paramag. above 260° C
Westinghouse Kovar A	29 Ni 17 Co 0.3 Mn Bal. Fe	8.2		1450	4.5–5.1 (30–400° C)	0.046	49	3	25	Ferromag. below 400° C

Material	Typical Physical Properties at 20 deg cent: Form and Condition	Tensile Strength, 10^3 psi	Yield Strength, 10^3 psi	Elongation, %	Brinell Hardness No.	High-temp. Properties, Short-time Tests: Temperature, deg cent	Tensile Strength, 10^3 psi	Elongation, %	Brinell Hardness No.
Molybdenum	Unworked	60	70–90% of T.S.	Up to 25	150	20	150	....	....
	Sheet	70–140		Up to 25	175–250	600	110	....	....
	Wire	Up to 350		Up to 25		700	107	....	....
						900	94	....	....
Tungsten	Unworked	....	Approx. 90% of T.S.	....	285	20	400		
	Sheet	Up to 400			Up to 450	500	260 Drawn wire		
	Wire	Up to 500				900	85		
						1300	70		
Tantalum	Sheet and wire	50–170		25–45	60–120	....	....	....	....
Platinum	Annealed	24	10	24	40	....	....	....	....
	Cold-rolled	26	27	2	100				

Silver	Annealed Cold-worked	23 54	8 40	50 5	30 90	....	....	....	....
Invar	Annealed Hot-rolled	70 75	42 50	41 37	130 140	....	....	....	....
Westinghouse Kovar A	Annealed sheet	75	55	35	140–160	....	....	....	....

Material	Corrosion Resistance	Oxidation Resistance	Typical Applications
Molybdenum	Resists atmosphere and some acids, attacked by dilute HNO_3 and conc. H_2SO_4	Oxidizes slowly at 400° C, forms dense white smoke at 700° C	General for ultra-high-temperature parts (with protection); corrosion resistance
Tungsten	Resists atmosphere and most acids, attacked by mixed HF and HNO_3, aqua regia, conc. hot KOH	Oxidizes at 600° C	Lamp filaments, miscellaneous electronic devices
Tantalum	Excellent in all media—noble metal	....	Filament wire and plates for incandescent lamps, laboratory ware, cathodes for electrochemical analysis, alloys
Platinum	Same as tantalum	....	Corrosion resistance, electrodes, thermocouples, furnace windings, laboratory ware, electrical contacts, alloys
Silver	Good in atmosphere, resists Cl_2, HCl, alkalies, soluble in HNO_3, hot H_2SO_4, alkali cyanides, attacked by sulfur compounds	....	Electrical contacts, galvanometer coils, aircraft bearings, electrical conductors, lining material in chemical industry
Invar	....	....	Bimetal thermostats, instruments requiring fixed distances between points to be independent of temperature
Westinghouse Kovar A	Better than carbon steel but inferior to stainless steel	....	Metal-to-glass seals

* Indicates highest value of each constant for materials listed.
† Indicates lowest value of each constant for materials listed.

15. BIBLIOGRAPHY

General

1. Seitz, Frederic, *Modern Theory of Solids*, 1st Ed., pp. 9–12, 40–42, 43, 55–56, 60–72, 139–194, 420–469, 516–575. McGraw-Hill (1940).

Copper and Copper Alloys

2. Gregg, J. L., *Arsenical and Argentiferous Copper. American Chemical Society Monograph Series*, The Chemical Catalog Co. (1934).
3. ASTM Standard Specifications for Hard-Drawn Copper Wire, ASTM Designation B1-40, ASTM Standards, Part 1-B, Non-ferrous Metals, American Society for Testing Materials (1946).
4. *Ibid.*, Designation B2-40.
5. *Ibid.*, Designation B3-45.
6. Smart, J. S., A. A. Smith, and A. J. Philips, Preparation and Some Properties of High-purity Copper, *Trans. Amer. Inst. Mining Metal. Eng.*, Inst. Metals Div., Vol. 143, pp. 272–286 (1941).
7. Pilling, N. B., and G. P. Halliwell, *Proc. ASTM*, 25-II, 97, Fig. 1, p. 102 (1925).
8. Burghoff, H. L., and A. I. Blank, Creep Characteristics of a Phosphorized Copper, *Trans. Amer. Inst. Mining Metal. Eng.*, Vol. 161, pp. 420–438 (1945).
9. Seely, H., unpublished information (reproduced by permission) presented to the Transmission and Distribution Committee of Edison Electrical Institute.
10. Wilkins, R. A., and E. S. Bunn, *Copper and Copper-base Alloys*, 1st Ed., McGraw-Hill (1943). Brace, P. H., Physical Metallurgy of Copper-base Alloys, *Elec. Eng.*, Vol. 63, p. 16, Fig. 18 (1944).
11. Irwin, P. L., Unpublished data, Mechanics Department, Westinghouse Research Laboratories.
12. Davis, E. A., Creep and Relaxation of Oxygen-free Copper, *J. Applied Mechanics*, Vol. 65, pp. A-101–A-105 (June, 1943).
13. Gregg, J. L., *Arsenical and Argentiferous Copper*. The Chemical Catalog Co., pp. 120 and 122 (1934).
14. Burghoff, H. L., and A. I. Blank, Creep Characteristics of a Phosphorized Copper, *Trans. Amer. Inst. Mining Metal. Eng.* Vol. 161, pp. 420–438 (1945).
15. Peterson, R. E., Discussion, *Proc. ASTM*, Vol. 40, p. 585 (1940).
16. Marin, J., *Mechanical Properties of Materials and Design*, 1st Ed. McGraw-Hill (1942).
17. Nadai, A., and P. G. McVetty, Hyperbolic Sine Chart for Estimating Working Stresses at Elevated Temperatures, *Proc. ASTM*, Vol. 43, pp. 735–745 (1943).
18. Nadai, A., *The Influence of Time upon Creep; The Hyperbolic Sine Creep Law.* Stephen Timoshenko Anniversary Volume. Macmillan (1938).
19. McVetty, P. G., Creep of Metals at Elevated Temperatures, the Hyperbolic Sine Relation between Stress and Creep Rate, *Trans. Amer. Soc. Mech. Eng.*, Vol. 65, pp. 761–769 (1943).
20. Tapsell, H. J., *Creep of Metals.* Oxford University Press, London (1931).
21. *Symposium on Effect of Temperature on the Properties of Metals*, Joint Publication ASTM and Amer. Soc. Mech. Eng. (1932).
22. Boyd, John, Relaxation of Copper at Normal and Elevated Temperatures, *Proc. ASTM*, Vol. 37, pp. 218–232 (1937).
23. Seitz, Frederic, *Modern Theory of Solids*, 1st Ed., p. 41, Fig. 49. McGraw-Hill (1940).
24. Wyman, L. L., Copper and Oxygen, *Gen. Elec. Rev.*, Vol. 37, p. 123, Fig. 8 (1934).
25. Brace, P. H., Physical Metallurgy of Copper and Copper-base Alloys, *Elec. Eng.*, Vol. 62, p. 14, Fig. 12 (1944).
26. Brace, P. H., Physical Metallurgy of Copper and Copper-base Alloys, *Elec. Eng.*, Vol. 63, p. 13, Fig. 6 (1944).
27. Brace, P. H., Physical Metallurgy of Copper and Copper-base Alloys, *Elec. Eng.*, Vol. 63, pp. 11–17, Figs. 8, 9, and 17 (1944).
28. ASTM Standards, Part 1-B, Non-ferrous Metals, pp. 32–36 (1946).
29. Standard Specifications for Hard-Drawn Copper-alloy Wires for Electrical Conductors. ASTM Standards, Part 1-B, Non-ferous Metals, Designation B105–39, pp. 32–36 (1946).
30. Seitz, Frederic, *Modern Theory of Solids*, 1st Ed., p. 177, Fig. 11. McGraw-Hill (1944).
31. Wilkins, R. A., and E. S. Bunn, *Copper and Copper-base Alloys*, 1st Ed. McGraw-Hill (1943). Brace, P. H., Physical Metallurgy of Copper and Copper-base Alloys, *Elec. Eng.*, Vol. 63, p. 17, Fig. 23 (1944).

Aluminum

32. Gauthier, G. G., The Conductivity of Super-purity Aluminum: the Influence of Small Metallic Additions, *J. Inst. Metals* (London), Vol. 59, 2, pp. 129–146 (1936).
33. Gauthier, *loc. cit.*
34. Light Metals and Alloys, *U. S. Bur. Standards Circ. No.* 346, Fig. 16, p. 65 (1927).
35. Koch, K. R., and C. Dannecker, Elasticity at High Temperatures, *Ann. d. Physik*, Vol. 47 (IV), p. 197; *U. S. Bur. Standards Circ. No.* 346, p. 71 (1927).
36. McVetty, P. G., Interpretation of Creep Data, *Proc. ASTM*, Vol. 43, pp. 707–734 (1943).
37. Sturm, R. G., C. Durmont, and F. M. Howell, A Method of Analyzing Creep Data, *Trans. Amer. Soc. Mech. Eng.*, *J. Applied Mechanics*, Vol. 58, pp. A62–A66 (1936).
38. Light Metals and Alloys—Aluminum and Magnesium, *U. S. Bur. Standards Circ.* (1927).
39. "*Alcoa*" *Aluminum and Its Alloys.* Trade publication of Aluminum Company of America, Pittsburgh (1946).
40. "*Alcoa*" *Aluminum Bus Conductors.* Trade publication of Aluminum Company of America, Pittsburgh (1940).
41. *Electrical Characteristics of A.C.S.R.*, Revised. Trade publication of Aluminum Company of America, Pittsburgh (1936).

Resistance Alloys and Special-purpose Alloys

42. Hoyt, S. L., *Metals and Alloys Data Book;* Mechanical Properties of Metals and Alloys, *U. S. Bur. Standards Circ.* C-447.
43. Silver: Its Properties and Industrial Uses, *U. S. Bur. Standards Circ.* C–412.
44. Woldman and Dornblatt, *Engineering Alloys.* American Society for Metals (1936).
45. Vines and Wise, Ed., *The Platinum Metals and Their Alloys.* The International Nickel Co.
46. Smithells, C. J., *Tungsten—Metallurgy, Properties and Application.* Van Nostrand (1936).
47. Smithells, C. J., *Impurities in Metals.* John Wiley (1931).

INSULATING MATERIALS

By John H. Palmer

16. SOLID MATERIALS

The following material on Classification of Insulating Materials, Limiting Insulation Temperatures ("Hottest-spot" Temperatures), Ambient Temperature, Limiting Values of Insulation Temperature Rise, Conventional "Hot-spot Allowances," and Limiting Observable Temperature Rises, is abstracted from the AIEE pamphlet No. 1, "Introduction to AIEE Standards, General Principles upon Which Temperature Limits Are Based in the Rating of Electrical Machinery and Apparatus" (June, 1940).

TEMPERATURES. For the purpose of establishing temperature limits, insulating materials are classified as shown in Table 1.

Table 1. Classification of Insulating Materials

Class	*Description of Material*
O	Class O insulation consists of cotton, silk, paper, and similar organic materials when neither impregnated * nor immersed in a liquid dielectric.
A	Class A insulation consists of: (1) cotton, silk, paper, and similar organic materials when either impregnated * or immersed in a liquid dielectric; (2) molded and laminated materials with cellulose filler, phenolic resins, and other resins of similar properties; (3) films and sheets of cellulose acetate and other cellulose derivatives of similar properties; and (4) varnishes (enamel) as applied to conductors.
B	Class B insulation consists of mica, asbestos, fiber glass, and similar inorganic materials in built-up form with organic binding substances. A small proportion of Class A materials may be used for structural purposes only.†
C	Class C insulation consists entirely of mica, porcelain, glass, quartz, and similar inorganic materials.

* An insulation is considered to be "impregnated" when a suitable substance replaces the air between its fibers, even if this substance does not completely fill the spaces between the insulated conductors. The impregnating substance, in order to be considered suitable, must have good insulating properties; must entirely cover the fibers and render them adherent to each other and to the conductor; must not produce interstices within itself as a consequence of evaporation of the solvent or through any other cause; must not flow during the operation of the machine at full working load or at the temperature limit specified; and must be not unduly deteriorate under prolonged action of heat.

† The electrical and mechanical properties of the insulated winding must not be impaired by application of the temperature permitted for Class B material. (The word "impair" is here used in the sense of causing any change which could disqualify the insulating material for continuous service.) The temperature endurance of different Class B insulation assemblies varies over a considerable range, in accordance with the percentage of Class A materials employed and the degree of dependence placed on the organic binder for maintaining the structural integrity of the insulation.

LIMITING INSULATION TEMPERATURES ("HOTTEST-SPOT" TEMPERATURES). From the results of experience with apparatus in service and of laboratory tests on various insulating materials, limiting insulation temperatures (called "hottest-spot" temperatures) have been assigned **for purposes of standardization.** The "hottest-spot" temperature is, therefore, the primary point of reference, or the "bench-mark" temperature. It is not employed in commercial transactions because it is not directly measurable in the ordinary course of testing or operation of electrical machines. The values of "hottest-spot" temperatures established for this purpose are as follows:

	Degrees Centigrade
For Class O material	90
For Class A material	105
For Class B material	130
For Class C material	No limit selected

The life at the limiting temperature for any one class of insulation may vary widely according to the quality of the material used, the thoroughness with which it is constructed, and its application in service. Materially higher "hottest-spot" temperatures may be satisfactory for particular forms of Class B insulation, e.g., thin sections of built-up mica flakes, as recognized in the standards for specific types of apparatus.

Ambient Temperature. Ambient temperature, as herein used, is the temperature of the medium used for cooling, either directly or indirectly, and is to be subtracted from the measured temperature of the machine to determine the temperature rise under specified test conditions.

LIMITING VALUES OF INSULATION TEMPERATURE RISE. The limiting values of "hottest-spot" temperature rise of the insulation are obtained by subtracting the 40 deg cent value of limiting ambient temperature from the limiting "hottest-spot" temperature.

The values of "hottest-spot" temperature rise so obtained are:

	Degrees Centigrade
For Class O material...............	50
For Class A material...............	65
For Class B material...............	90
For Class C material...............	No limit selected

Conventional "Hot-spot Allowances." Limiting "observable" temperature rises of insulated windings and parts immediately adjacent thereto are obtained from the limiting "hottest-spot" temperature rise values by subtracting therefrom a specified number of degrees which, **for purposes of standardization,** is the difference established between the limiting "hottest-spot" and the limiting "observable" temperature. Since the "hot-spot" allowance increases roughly in proportion to the losses generated in the insulated winding, it should be considered as a percentage of the "hot-spot" temperature rise, from which the allowance in degrees may be derived.

In practice, the difference between the "hottest-spot" and the "observable" temperature under steady state conditions will vary considerably, depending on such factors as the following:

1. Accessibility of the hottest spot.
2. Uniformity of cooling.
3. Thickness and kind of insulation.
4. Form of winding.
5. Rate of heat flow.
6. Relative locations of heat generation and dissipation.

The time lag of the thermometer or embedded-detector temperature behind the actual copper temperature is also an important factor under varying load conditions.

In view of these variable factors, no single value of the "hot-spot" allowance, or difference by which the "observable" temperature is lower than the "hottest-spot" temperature, will apply exactly to different types or sizes of apparatus. Experience indicates, however, that for any given machine the thermometer method gives the lowest temperature rise, and the embedded-detector method the highest, while the resistance method usually gives an intermediate value, which is normally closer to the embedded-detector than to the thermometer value.

For purposes of standardization, the normal difference between the "hottest-spot" temperature rise and the temperature rise measured by thermometer will be taken as approximately 25 per cent of the "hottest-spot" temperature rise, and the corresponding figure for measurements by the resistance or embedded-detector method will be taken as 10 per cent. Applying these percentages to the limiting values of temperature rise and rounding off the resulting figures to the nearest multiple of 5 deg, Table 2 is derived:

Table 2. Conventional "Hot-spot" Allowances (Degrees Centigrade)

Method	Class O Material	Class A Material	Class B Material
Thermometer..................	15	15	20
Resistance or embedded-detector..	5	5	10

Different values of "hot-spot" allowances may be established for particular machines and methods of measurement in the standards for specific types of apparatus.

Limiting Observable Temperature Rises. Limiting observable temperature rises, **for the purpose of assigning a rating,** are obtained by subtracting the proper "hot-spot" allowances from the limiting values of "hottest-spot" temperature rise. The values obtained in this way that are applicable to the great majority of types of continuous rated apparatus are given in Table 3.

Table 3. Limiting Observable Temperature Rise (Degrees Centigrade) for Insulated Windings of Continuous Rated Apparatus

Method	Class O Material	Class A Material	Class B Material
Air or gas-immersed apparatus			
Thermometer method...........	35	50	70
Resistance or embedded-detector method *.....................	45	60	80
Mineral-oil-immersed apparatus			
Resistance method.............	....	55 †	....

* In a given machine, the embedded-detector method normally gives a higher temperature than the resistance method. However, the practice here indicated of allowing the same rise for both methods is correct if AIEE-recommended measurement methods are followed. For the larger machines to which the embedded-detector method is normally applied, the difference between the "hottest-spot" and the embedded-detector values is usually of the same order as the difference between the "hottest-spot" and the resistance values for machines to which the resistance method is applied.

† This lower value is selected in recognition of the life temperature characteristics of mineral oil. Consideration has also been given to the fact that this value is chiefly applicable to transformers which are usually installed outdoors, or in other locations where the average ambient temperature is materially lower than the 40 deg cent value selected as the limiting ambient temperature.

17. SIGNIFICANCE OF PROPERTIES OF SOLID INSULATING MATERIALS

The following material describes the significance of the properties which are given in Tables 4 to 12 and is in part abstracted, by permission, from ASTM Significance Statements. These tables have been chosen as those most commonly needed in the design of electrical apparatus, and the units used are those which the author believes to be most convenient for this purpose.

ELECTRICAL PROPERTIES. Power Factor. Power factor of insulation is the cosine of the angle between vectors representing the voltage across the insulation and the total current through it when a sinusoidal alternating voltage is applied. Power factor may also be defined as the ratio of the leakage or inphase current through the insulation, to the total current through it when a sinusoidal alternating voltage is applied.

Power factor is a property of the material and is used in calculating the dielectric energy loss in an insulation, this loss being directly proportional to power factor. It is, therefore, considered an important criterion of quality. Power factor is also often used as a control test to determine the uniformity and/or purity of materials, since it is very sensitive to changes in purity or composition.

Dielectric Constant (Specific Inductive Capacity or Permittivity). Dielectric constant may be defined as the ratio between the capacitance of a capacitor using the insulating material under consideration as dielectric, to the capacitance of the same capacitor using vacuum (air gives a close approximation) as dielectric, measured at a specified frequency.

Dielectric constant is used in the calculation of capacitance, charging current, and dielectric loss in dielectrics, all of which are directly proportional to it. It is usually of advantage to use insulating materials of high dielectric constant for capacitors and of low dielectric constant for other types of electrical equipment.

Resistivity (Volume). (Partially abstracted from ASTM Significance Statements.) Volume resistivity of a material is the ratio of unidirectional potential gradient in volts per unit length parallel to the current flow in the material, to current density in amperes per unit area. It is numerically equal to the volume resistance when measured between two electrodes which cover opposite faces of a unit cube of the material. Centimeter units are usually used.

Resistivity has long been employed as a criterion of quality for electrical insulating materials for several reasons. Insulation resistance of electrical equipment depends largely upon the resistivity of the material used. The insulation-resistance test was one of the first developed to test the suitability of electrical insulation and is a quick, simple test to make, requiring only inexpensive and compact equipment.

This property does not give a complete picture of materials which are to be used in connection with a-c circuits. Resistivity is, therefore, being largely replaced by power-factor tests made at or near the voltage stresses and frequencies to be used in service.

Resistivity is, however, still used extensively as a quality-comparison test. For many materials it is sensitive to changes in composition or purity and will usually detect very small variations in moisture content. It is, therefore, essential that these conditions be taken into account when interpreting results of the test.

Dielectric Strength. The dielectric strength of an insulating material may be defined as the minimum voltage gradient at which electrical failure or breakdown occurs under

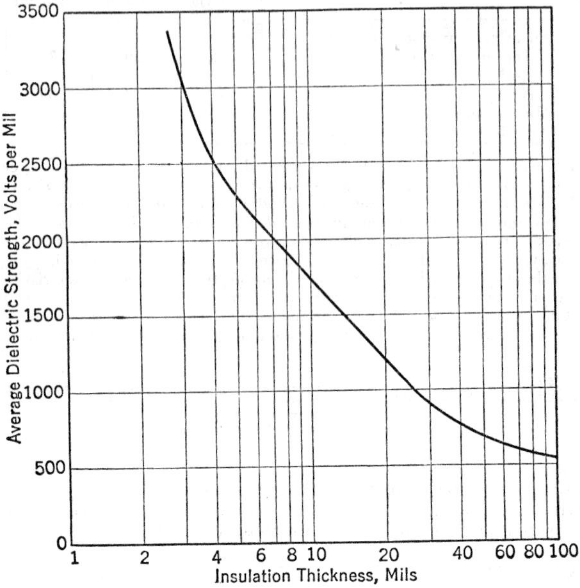

FIG. 1. Dielectric Strength as a Function of Thickness

prescribed conditions. For practical purposes, however, it is obtained by dividing the breakdown voltage by the thickness of the dielectric between the test electrodes, regardless of whether this value corresponds to the actual maximum voltage gradient that may exist, such as at the edges of electrodes or at points of maximum gradient in a non-uniform field.

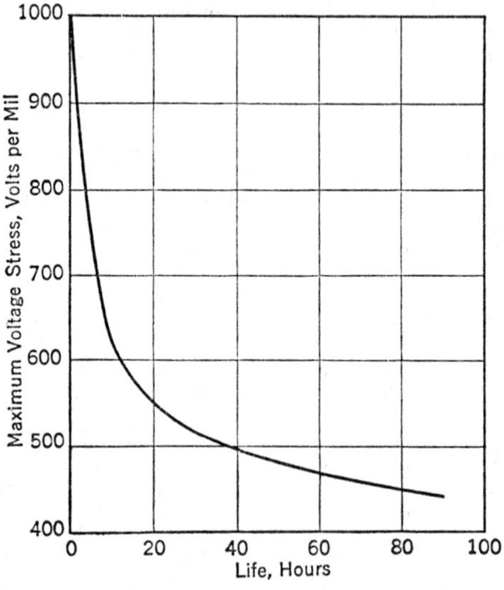

FIG. 2. Dielectric Strength as a Function of Time

The test values for dielectric strength of an insulating material vary to an extent not generally appreciated, since dielectric strength is a function of the thickness of the dielectric, being greater the less the thickness, as illustrated for polyethylene in Fig. 1. It is also a function of the time of application of voltage, being less the longer the time, as illustrated by Fig. 2 for impregnated-paper insulation. The form and size of electrode are also factors, the dielectric strength being greater with small than with large electrodes. The frequency and wave shape of the applied voltage are other factors. The SIC of the material surrounding the electrodes is also an important factor, the dielectric strength being higher the greater the SIC. Temperature of the test specimen has its effect too, most materials having lesser dielectric strength the higher the temperature.

Hence it is not practicable to determine, by means of laboratory tests of materials, a value for the dielectric strength which will apply accurately under service conditions. Definite numerical values for dielectric strength are needed, however, for specification purposes in the purchase of materials, for comparison of different materials, and for use in design and must, therefore, be associated with definite conditions of the dielectric.

Impulse Dielectric Strength. This property is similar to dielectric strength, except that the test voltage is applied in the form of an impulse, a unidirectional voltage surge, during which the voltage rises to its peak value in a very short time and then decreases to zero during a longer period, both of these times depending on the constants of the circuit, which includes the sample under test. The impulse wave form is usually designated by the time required for the voltage to rise to its peak value and by the time required for the voltage to decrease from the peak to half of this value. For instance, a 1 $^1/_2$/40-μsec impulse is one in which the voltage rises to its peak value in 1 $^1/_2$ μsec, then decreases to half this value in 40 μsec. This 1 $^1/_2$/40-μsec wave has been standardized for most impulse testing. Specimens are usually tested by applying repeated impulses to the sample, starting at a peak voltage well below the expected impulse breakdown strength of the sample and increasing the voltage in small increments on each successive impulse until failure of the specimen occurs. Experience indicates that these repeated impulses below the breakdown strength do not damage the specimen and so do not affect the final result. The final breakdown strength is usually expressed in volts per mil of thickness of the insulation under test, at the point of failure.

This test is of considerable significance, since nearly all electrical circuits are subject to surges or impulses produced by switching, discharges, lightning, sudden changes in load, etc. In many instances necessary insulation thickness is determined by the impulse strength of the insulation rather than by its normal dielectric strength.

Arc Resistance (Resistance to Electrical Discharge). Arc resistance is a measure of the ability of the exposed surface of electrical insulating materials to withstand electrical discharge produced over the surface between two pointed electrodes by means of a high-voltage transformer, usually operated from a 60-cycle power source. Length of arc, arc current, and time of application are carefully controlled, current and time being increased until failure occurs, usually by formation of a charred conducting path on the surface of the insulation between the pointed electrodes, or in some cases by melting of the insulation in the path of the arc.

Insulation used in high-voltage equipment is often subjected to arc conditions. Such conditions may be caused by poor design, by dirt and moisture collected on insulation surfaces, or by ionization of voids, gas pockets, or films within the body of the insulation or between the insulation and metallic conductors which it insulates.

Knowledge of the arc resistance of the material is, therefore, important in order that the proper material may be chosen to resist arcing or in order that proper design steps may be taken to eliminate the possibility of arcing.

Frequency Range over Which the Material May Be Used Efficiently. This property gives the frequency range of the alternating voltage applied across the insulation, through which the material may be used without excessive dielectric loss. The electrical properties of many insulating materials, particularly the specific inductive capacity and the power factor, change with frequency.

CHEMICAL PROPERTIES. Use in Water. If water is absorbed to any extent, it usually seriously impairs the electrical properties of the insulation, causing a decrease in its resistance, an increase in dielectric power factor, and a decrease in dielectric strength. Insulation used around negatively charged conductors of a d-c system is more susceptible to water damage than that around positively charged ones because of endosmose, a phenomenon which tends to draw water through the insulation to the negative conductor.

Use in High Humidity. Insulation exposed to atmosphere, even indoors, is at times exposed to high humidity. Although the effect is not usually as great as that of entire immersion in water, it will in general be the same if water is absorbed. High humidity also may have a catalytic effect on oxidation. Also, if acid fumes or other impurities are present in the air, they may form solutions with condensed water vapor, attacking the surfaces of insulators. Conducting particles from the air may also be deposited with condensed moisture on insulation surfaces to form semiconducting films, which will lower the surface resistance and sparkover potentials of insulators. This often contributes to failures of bushings, terminals, line insulators, and other similar equipment.

Use in Petroleum Oil. Many types of electrical equipment depend in part upon solid insulation and in part upon liquid insulation, such as petroleum oil, which serves a double purpose; as insulation and as a cooling medium. In such equipment the solid and liquid insulations must be compatible, i.e., neither should be deleteriously affected by the other. Typical examples of such equipment are transformers, switchgear, and static condensers.

Use in Acids or Alkalies. Water or oil in contact with solid insulations may become either weakly or strongly acid or alkaline because of surroundings or, in the case of oils, by-products of deterioration. Many solid insulations attacked by these agents suffer severe losses in mechanical as well as electrical properties.

Weathering Characteristics. Many insulating materials are deleteriously affected by sunlight, humidity, rain, extremes of temperature, or combinations of these, such as occur during exposure to the elements. In choosing materials for outdoor use it is, therefore, necessary to know whether they will withstand such exposure or, possibly, to protect them against the particular types of exposure which they will not withstand. One common effect of weather is to lower the surface resistance of insulators, causing excessive leakage, discharge or possibly sparkover of the insulators.

Corrosion of Metals. Electrical insulation nearly always surrounds or is surrounded by metallic conductors or coverings. The most common conductors are aluminum, copper, and copper coated with lead or tin or an alloy of the two to protect it from corrosion. Lead is also a common sheath material for a protective conducting covering on electric cables. Corrosion of these metals reduces their conductivity, as well as makes them brittle and subject to breakage or splitting. It also produces products of deterioration which may in turn have a deleterious effect upon the surrounding electrical insulation.

Flammability. Electrical failure of equipment usually produces flame and very high local temperature. The insulation involved may consequently burn. If it supports combustion, the fire may be carried to other parts of the equipment and cause considerable damage. If, on the other hand, the insulation will not support combustion, damage may be confined to the failure area.

It is not always possible to choose incombustible materials with other suitable properties but this characteristic should be considered and given proper weight, depending upon its relative importance in the use of the equipment.

Ozone Resistance. Many insulating materials are adversely affected by exposure to ozone, even of very low concentration. Rubber compounds in particular are severely affected, since ozone causes cracking of the insulation, often entirely through its thickness, causing electrical failure. Such compounds are particularly affected if they are under strain due to binding or stretching or to inherent strains caused by the method of application.

Ozone is often formed around electrical apparatus because of electrical discharge or ionization of surrounding air. It is, therefore, important to consider the ozone resistance of any exposed insulation which is to be used in or near high-voltage equipment.

Solvents. Solvents are often used as vehicles in the application of insulating materials, such as saturating varnishes, or waxes and films, such as varnish or lacquer. Solvents are also likely to be present in surrounding materials and to affect the insulation adversely, causing softening or complete disintegration of the material.

THERMAL PROPERTIES. **Maximum Operating Temperature.** Electrical apparatus is always heated by the electrical energy losses occurring within it. Most of these losses are a function of the amount of current supplied by, or passing through the conductors of, the apparatus, whereas other losses are produced by leakage currents passing through or over the surfaces of the electrical insulation of the apparatus. These latter losses are a function of the temperatures to which the insulating materials are subjected.

The maximum operating temperature of the insulating material consequently is often the determining factor in calculating the maximum amount of electrical energy or load which the apparatus can deliver or transmit.

Brittle Point. Electrical apparatus is often installed or operated at extremely low temperatures. If the insulation is to be flexed during either installation or operation, it is important to know at what temperature it will become brittle.

Softening Point. Insulation is often applied to conductors in a molten or plastic form, at high temperature. This softening temperature must, therefore, be known in order that application temperatures may be properly chosen. Such softening temperatures must also be safely above the maximum permissible operating temperatures, so that there will be no danger of deformation of the insulation during either normal operation or short-time emergency overload conditions which may occur.

Specific Heat. This property is the ratio of the amount of heat energy necessary to increase the temperature of a unit mass of material 1 deg, to the amount of energy necessary to raise a unit mass of water 1 deg in its liquid state. Being a ratio, it is independent of units used. Specific heat is of value in calculating thermal capacity.

Thermal Capacity. This is a measure of the amount of heat necessary to raise the temperature of a given mass of material 1 deg. Unlike specific heat, it depends on the amount of material being considered and the units used.

Thermal capacity is used in calculating the rate of heating of electrical equipment and is of particular value in calculating short-time emergency overloads of such equipment.

Thermal Cubic Expansion Coefficient. This coefficient gives the proportional volume increase, per degree change in temperature, based on the volume at the original temperature.

It is useful in calculating the volume of space required for expansion of insulation, as well as of all other parts of electrical equipment, over the operating temperature range. Also, solid parts which must remain in intimate contact should have identical expansion coefficients.

Thermal Resistivity. Thermal resistivity is a measure of the temperature difference required across opposite faces of a unit cube to force one unit of energy through the cube. This property is useful in determining temperature differences across insulating materials for given loads carried by electrical equipment. It is also used to determine the maximum permissible loads which electrical equipment can carry to produce maximum permissible operating temperatures. (See Section 14, Article 69.)

MECHANICAL PROPERTIES. Tensile Strength. Tensile strength is a measure of the force per unit area of original cross-section (applied at a definite rate) necessary to produce rupture of the material under test.

Many rigid insulating materials, such as porcelain, phenolic materials, and glass, are used as supports for electrical conductors, and as such are often under tensile stress. Tensile strength of the material gives the designer of such insulators knowledge of the amount of the tensile stresses which may be applied in service, using, of course, suitable factors of safety.

Tensile strength is also often used as a measure of quality of materials, both when new and after various types of aging tests. Loss of tensile strength after such aging tests or after natural aging is considered an important criterion of quality of flexible insulating materials.

This property is also used for flexible insulating materials to give an idea of the amount they may be safely stressed in service.

Elongation. This property is a measure of the elongation or stretch of flexible insulating materials during tensile-strength tests, usually expressed as a percentage of the original length. It gives an indication of the amount of stretch or strain which may be allowed in service, using a suitable factor of safety. Change of elongation with various types of aging is also considered an important criterion of quality. Small change of elongation is, of course, desirable.

Compression Strength. Compression strength is a measure of the force per unit of cross-sectional area, measured in a plane perpendicular to the force, necessary to deform permanently or to crush the material.

Rigid insulating materials are often used as supports and are thus placed under compressive stress. Compression strength is, therefore, useful as an indication of the compressive load which may be applied to a material, using a suitable factor of safety.

Hardness. There are two standard hardness tests in common use for insulating materials, the Durometer method and the Rockwell test. The former is generally used for more flexible materials, such as rubber and rubberlike materials. The Rockwell tester is designed for more rigid materials, such as harder plastics, laminated phenolic materials, and hard rubber. Both tests measure depth of indentation of an indentor or presser having a definite area of contact and applied with a definite pressure.

Impact Strength (Extracted from ASTM Significance Statements). Impact strength is an expression of the relative toughness of materials as indicated by the energy used in breaking a standard test specimen in one blow. Two methods of test are used, the cantilever beam or Izod test and the Charpy test. The Izod test requires a notched specimen, whereas the Charpy specimen may or may not be notched.

The purpose of the notched bar impact test is to show the resistance of the material to shock loads. The basic assumption of the test is that resistance to shock depends upon the ability of materials rapidly to equalize dangerously concentrated stresses by flow or deformation without failure. The notch and the speed of test combine to produce a concentration of stress as a test condition.

There is no known correlation between the various static tests, such as compression, tension, and flexure, and the impact test. Several materials may show the same values on static tests, and from the results it may be difficult to differentiate between them, whereas the impact test may give decidedly different values.

Specific Gravity. Specific gravity may be defined as the ratio of the weight of a given volume of material to the weight of an equal volume of water at a given reference temperature (usually taken as 0 deg cent).

This property is used in calculating total weights of materials and costs based on weight, and in checking purity of materials of which the correct specific gravity is known.

18. SIGNIFICANCE OF PROPERTIES OF INSULATING LIQUIDS

CHEMICAL PROPERTIES. Neutralization Number (Abstracted from ASTM Significance Statements with minor changes). Neutralization number is defined as the weight in milligrams of potassium hydroxide required to neutralize 1 gram of the product.

Practically all liquid dielectrics contain natural organic substances which are either weak acids or weak alkalies or which, upon oxidation, form compounds acidic or alkaline in nature or weak alkalies. The presence of these substances is commonly referred to as organic acidity and alkalinity. A properly refined product is free from mineral acids or alkalies.

Neutralization number is important in the examination of unused liquid dielectrics only as a check against the presence of mineral acidity or appreciable organic acidity.

In used liquid dielectrics, neutralization number may be pertinent, if compared to values obtained on the new products, to detect contamination by substances with which the liquid has been in contact or to reveal tendency toward increase in acidity due to inherent chemical change. An increase in neutralization number may be employed as a general guide for determining when a particularly well-known dielectric should be replaced by fresh material to prevent further decomposition and consequent sludging, provided suitable rejection limits have been established and other tests confirm the need for replacement.

Saponification Number (Partially extracted from ASTM Test Methods). Saponification number of an insulating liquid is an indication of the saponifiable organic material present in the liquid. Literally, it is the number of milligrams of potassium hydroxide consumed by 1 gram of the liquid under the specific conditions of test.

This property is of significance in the testing of insulating liquids, since the soaps formed by saponification have poor electrical properties and form deposits in the oil, which would tend to impede flow of the liquid if used in a circulating system.

Sludge (Partially extracted from ASTM Test Methods). The sludge-accumulation test is for mineral transformer oils only and evaluates the propensity of an oil to deposit an insoluble sludge under definitely prescribed conditions of oxidation in the presence of a copper catalyst for various specified times.

This test is of particular value for transformer oils, since sludge formation in a transformer coats the windings and core, impeding circulation of the oil and adding considerably to the thermal resistance.

Inorganic Ions (as Chlorides and Sulfates). This is a qualitative test to determine the presence of inorganic ions (Cl^- or $SO_3^=$) in mineral oils, the presence of which causes poor electrical properties, such as low dielectric strength, high power factor, and low resistivity, especially after long exposure to electric stress.

Free and Corrosive Sulfur (Extracted from ASTM Significance Statements). This test gives a measure of the amount of free corrosive sulfur present in mineral oil.

If a liquid dielectric containing free and corrosive sulfur is maintained in contact with certain metals, the sulfur may attack the metal surface and form a film of sulfide. In extreme cases this action may continue to a degree of corrosion whereby the strength and character of the metal are seriously impaired. It is, therefore, important that the amount of this impurity in the oil be known.

Steam Emulsion. The steam-emulsion test is for the purpose of determining the resistance of an insulating liquid to emulsification or mixing of the liquid with water, which is introduced by bubbling steam through it under controlled conditions. The result is expressed in the number of seconds required for a specified volume of emulsion to be formed.

Many types of electrical equipment, such as transformers and switchgear, contain liquid insulation which is exposed to atmosphere. In many cases the tanks containing this equipment and liquid insulation are allowed to "breathe," or to suck in and expel air, as the liquid expands and contracts due to temperature changes, thus permitting condensation of considerable quantities of water in the tanks. If the insulating liquid and water easily emulsify, the insulating properties quickly become poorer, resulting in low breakdown strength, high power factor, and low resistivity. The liquid would also act as a vehicle to carry the water to the solid insulation, deteriorating this in turn. On the other hand, if the insulating liquid and water do not easily emulsify, the water will cause no serious harm and may be drained off periodically. It is, therefore, of advantage to use in such equipment insulating liquids which have high emulsification numbers or which are slow to emulsify.

PHYSICAL PROPERTIES. Specific Gravity (Extracted from ASTM Significance Statements). The specific gravity of a liquid is, according to the general custom in this country, the ratio of the weights of equal volumes of the liquid and of water, both weights being at 60 deg fahr and corrected for the buoyant effect of air. The temperature limitations are indicated conveniently by the expression, "Specific gravity 60/60 deg fahr."

Liquid dielectrics are usually sold on the basis of volumes delivered at 60 deg fahr. For mineral oils, if the specific gravity is determined at a temperature other than 60 deg fahr, volume corrections are made by the use of approved tables, which show the correction as a function of the specific gravity of the product, regardless of its source or character. As for askarels, either the specific gravity must be determined at 60 deg fahr, or the proper correction for the particular material must be applied.

The specific gravity of the liquid dielectric must, therefore, be known whenever volume corrections are to be made. In some cases, e.g., the delivery of liquid dielectric in drums, it is customary to determine weights and calculate volumes by dividing the weight of the contents of the drum by the weight per gallon. The latter figure is, of course, a function of the specific gravity of the product.

Color (Extracted from ASTM Significance Statements). The color of a liquid dielectric is determined by means of transmitted light. The chief significance of color, as applied to a liquid dielectric, lies in the fact that it is a generally accepted index of the degree of refinement for unused liquids and an approximate measure of deterioration of products in service.

Although, in general, there may be some connection between the color of a mineral insulating oil and its degree of refinement, there has never been established a relationship between this characteristic and physical or electrical stability. Therefore, color, by itself, can be considered of little inherent value. Increase in color number of an askarel may be indicative of the presence of impurities.

Viscosity (Extracted from ASTM Significance Statements). The viscosity of a fluid is the measure of its resistance to flow. For liquid dielectrics an expression involving the time required for a measured volume of liquid to flow, under specified conditions, through a carefully standardized tube is known as its viscosity. As viscosity changes rapidly with temperature, a numerical value of viscosity has no significance unless both the temperature and the type of instrument are specified.

Supplemented by specific gravity data, calculation of absolute viscosities permits a more accurate comparison of materials possessing widely different densities.

Since the rate of change of viscosity with temperature varies with different liquids, viscosity tests should, in general, be made at a temperature which approximates that of normal use.

In connection with a liquid dielectric for impregnating a fibrous material, viscosity at a temperature approximating that of impregnation is a function of the rate of impregnation. When determined at operating temperature, it is a partial guide to the mobility of the liquid phase of the dielectric under performance conditions.

For transformers, switchgear, and similar apparatus viscosity of the liquid dielectric reflects the ease or rate of its circulation in the casing and consequently its relative heat-carrying characteristics.

THERMAL PROPERTIES. Flash Point (Extracted from ASTM Significance Statements). The flash point of a liquid which emits combustible vapor may be defined as the temperature to which the liquid must be heated in order to give off sufficient vapor to form a flammable mixture with air. The flash point of an insulating liquid reveals the limit to which it may be heated before the vapors emitted become a fire hazard. An unusually low flash point for a given product usually indicates contamination. From these standpoints the test has some significance. As an indication of other insulating qualities, however, it has no significance.

For askarels, the pseudo-flash, which is typical of this class of products, bears no relationship to the ultimate *fire* point or to the explosibility of the vapors or gases evolved from the liquids when heated to the boiling point. Detection of a *fire* point signifies that the dielectric is either not an askarel or that it is contaminated by a flammable impurity.

Pour Point (Extracted from ASTM Significance Statements). Liquid dielectrics become more or less plastic solids when sufficiently cooled, because of either the separation of wax or the congealing of the hydrocarbons or other derivatives composing the product. The temperature at which the liquid just flows under the prescribed conditions of test is known as the pour point.

The pour point of a liquid dielectric gives an indication of the temperature below which it may not be possible to pour or remove it from its container. It reveals the temperature at which fluidity has decreased to a point where the liquid will no longer flow freely through a restricted area.

Pour point is important, for liquid dielectrics used in transformers and other similar equipment, as an index of the lowest temperature to which the equipment may be cooled without seriously limiting the degree of circulation of the liquid.

Specific Heat (see Article 17).
Thermal Capacity (see Article 17).
Thermal Cubic Expansion Coefficient (see Article 17).

ELECTRICAL PROPERTIES. Power Factor (see Article 17). A note should be added concerning liquids used as saturants for fibrous materials, such as in impregnated paper used in capacitors and cables. Power factor of an impregnated fibrous product may be of a much lower magnitude than that of the liquid alone. For instance, impregnating oil for paper-insulated cable may have a power factor of the order of 10 at 80 deg cent without unduly affecting either the power factor or the electrical stability of the finished product, which may have a power factor of the order of 2 per cent at 80 deg cent. The effect, of course, depends upon the cause for the high power factor of the oil. Oxidation of the oil seems to have very little influence on final power factor of the impregnated paper, whereas moisture in the oil will go into the paper fibers and have considerable effect, especially at elevated temperature. A combination of moisture and oxidation has more effect than moisture alone.

Resistivity (Megohm-cm) (see Article 17). The above remarks concerning power factor apply also to resistivity or insulation resistance; i.e., low-resistivity liquid does not necessarily indicate that the insulation which it saturates will also have low resistivity.

Dielectric Strength. In liquids the dielectric strength is usually taken as the 60-cycle voltage at which breakdown occurs when applied at a specified rate between standardized electrodes at a specified temperature.

This property is of importance as an indication of ability to withstand high voltage stress without failure. Its greatest value, however, is to indicate the presence of contaminating agents, such as water, dirt, conducting particles, or products of oxidation. A high breakdown value, however, is not a certain indication of the absence of all contaminants.

Breakdown stress varies greatly with electrode spacing and temperature, so that these factors should be considered when applying test results to service conditions.

19. SIGNIFICANCE OF PROPERTIES OF INSULATING GASES

GENERAL. Gases have many uses as electrical insulation in connection with electrical apparatus, sometimes acting as primary insulation and sometimes acting incidentally as a part of, or in solution with, the primary insulation. Air, of course, is the most common insulating gas, since it surrounds all bare wires, commutators, and slip rings, terminals, bushings, and pin-type insulators.

Because of its high thermal capacity and lightness, hydrogen is circulated in closed sealed systems for cooling electric generators, even though it is highly inflammable when mixed with oxygen. It consequently becomes a part of the insulation of such machines.

Carbon dioxide and nitrogen are often used in contact with insulating oils which are to be degassed, so that any residual gas left in the oil will be one of these materials. Nitrogen is also used in conjunction with dry paper in telephone cables. These gases are chosen because they are chemically inert, cheap, and easily available.

Helium is used to a limited extent as a means of applying pressure to oil insulation where the gas is in direct contact with the oil, because of the low solubility of helium in oil. The most common applications of carbon dioxide, nitrogen, and helium are in connection with impregnated-paper insulation of high-voltage cables of the solid type, oil-filled type, and oil-compression type.

The various "Freon" gases have very high dielectric strengths compared to other gases but have not as yet been used to any great extent as electrical insulation, probably because of the fact that they are slightly corrosive to copper and iron when in water solution.

CHEMICAL PROPERTIES. Chemically Inert or Reactive? Inert gases may be safely used in contact with either solid or liquid insulations without fear of changing their chemical and consequently their electrical properties and also without fear of explosion when mixed with air or oxygen. On the other hand, a chemically reactive gas may cause serious deterioration of certain types of either solid or liquid insulations and may produce violent explosions or fires if allowed to mix with oxygen in proper proportions.

Corrosion of Metals When in Water Solution. It is usually difficult and expensive to eliminate all traces of moisture from the insulation of electrical equipment. Moreover, any small leaks or porosity in sealed containers will permit infiltration of moisture, even though high positive liquid or gas pressures are maintained within the containers, forming mixtures with the gas which would damage the metal parts if corrosive.

PHYSICAL PROPERTIES. Molecular Weight. The molecular weight of an element or chemical compound is equal to the sum of the weights of the atoms contained in the molecule and may be obtained from the atomic weights when the molecular formula of the element or compound is known. These atomic weights are values based upon the relations of the weights of the atoms of elements under consideration, to the weight of an atom of oxygen, which is taken as 16.

The molecular weight of a gaseous element or compound is proportional to its density if not near its critical temperature or pressure and may be converted to pounds per cubic foot at 0 deg cent and atmospheric pressure, by multiplying by the constant 0.00279. Thus the density of any gas may be calculated if the chemical formula and atomic weights of the elements of which it is composed are known.

Solubility in Water, Petroleum, and Pyranol. Gases, as stated above, are often used in electrical apparatus in contact with water, petroleum, or Pyranol. Chemically inert gases in solution in these liquids have little or no effect upon the electrical properties. However, if for any reason the gas comes out of solution in the form of bubbles at a location where the liquid is under high electrical stress, electrical breakdown of the gas in the bubbles may occur, bombarding the surrounding liquid and causing either serious deterioration of it or complete electrical failure. The solubility of gases in mineral oil increases with rising temperature.

Some bubble formation is bound to occur with changes in pressure or temperature of liquids in contact with gases, and the use of gases of low solubility keeps such bubble formation at a minimum.

Thermal Capacity. Thermal capacity is the amount of heat energy necessary to raise the temperature of a given mass 1 deg of temperature.

The thermal capacity per unit mass also is a function of the thermal conductivity of a gas. It is also used in calculating the rate of heating of equipment and the amount of gas flow necessary to carry off a given amount of power loss in the form of heat.

Thermal Resistivity. This may be defined as the temperature difference necessary to maintain unit power flow between opposite boundaries of a unit cube of material, under constant temperature conditions.

This property is used in calculating power ratings or temperature rises for electrical equipment.

Critical Temperature and Pressure. The critical temperature of a gas is the highest temperature at which it is possible to liquefy the gas by application of pressure alone.

The critical pressure is the pressure at which liquefaction of a gas occurs at its critical temperature.

Insulating gases must be used in temperature and pressure ranges which do not even approach the liquefaction point, since all properties change when near the critical region.

ELECTRICAL PROPERTIES. Dielectric Strength. Dielectric strength of gases is usually taken as the voltage stress required to produce sparkover of a specified sphere gap under specified conditions. It is important from two standpoints, as follows:

1. When a high potential surface is surrounded by gas, the dielectric strength is one important factor in the determination of the voltage at which corona will form around this surface, producing electrical bombardment of the surface and generating ozone if oxygen is present. Such bombardment, besides causing dielectric loss, may, if intense enough, erode many of the solid insulations and may cause chemical changes to occur in the material. Ozone formed may also produce chemical changes in the material, as well as causing serious cracking of rubber insulating compounds (see Article 17).

2. The dielectric strength of gases is also an important factor in determining the voltage at which breakdown or ionization of gas-filled voids or films in solid dielectrics occurs. Such ionization produces electrical bombardment of the surrounding surfaces of insulation, causing chemical changes to occur (charring and melting in the cases of some plastics) and sometimes causing ultimate breakdown of the dielectric wall. If oxygen is present, ozone is also produced, with its usual results as described above.

Ionization of voids also produces an increase in power factor of the overall dielectric, causing increase in dielectric loss. Determination of this power-factor increase is often used to detect ionization. The dielectric loss, of course, produces in its turn heating of the dielectric, which, if severe, may decrease the power rating of the equipment involved.

NOTE. The dielectric strength of gases varies greatly with the thickness of gas film involved, increasing as the thickness decreases. It is, therefore, necessary that this film thickness be known and taken into consideration when using the dielectric strength of a gas (see Tables 15 and 16). This phenomenon is governed in accordance with a law known as Paschen's law.

Dielectric Constant (see Section 2, Article 17). When gases are used in combination with other insulating materials under alternating voltage stress, the potential distribution varies inversely as their respective capacitances, neglecting leakage, which is usually negligible. This distribution is, therefore, also a function of the dielectric constants of the materials involved.

20. TABLES OF PROPERTIES OF ELECTRICAL INSULATING MATERIALS

The properties given in Tables 4 to 16 were obtained as far as possible from the suppliers of the various materials. It will be noted that in many cases these tables are incomplete. Spaces are left, however, so that individuals may fill them in for their own reference if these data become available.

Table 4. Properties of Insulating Materials—Solid

	Ceramics and Similar Materials					Fibrous-base Flexible Sheet and Tape Materials
Name	G.E. Mycalex	Porcelain—Dry Process	Porcelain—Wet Process	Steatite (Unglazed)	Zircon Porcelain	Impregnated Paper (Oil Impregnant)
General Properties						
Chemical name or components	Mica and glass				$ZrO_2 \cdot SiO_2$	Paper (usually wood pulp) + light mineral oil
General uses as insulation	Molded parts for electronic and power equipment	Low-voltage insulation, such as wiring devices	High-voltage bushings, line insulators, post insulators, etc.	High-frequency insulation (rigid)	High-frequency insulators	Capacitors and cables—oil-filled type
Electrical Properties						
Power factor (% at 60 cycles) at 20 deg cent	0.007		2.0 to 3.0	1.5	0.04	0.3 to 0.6
Power factor at max. operating temperature						0.5 to 1.0
Dielectric constant (SIC) at 60 cycles, 20 deg cent	7.6	6 to 7	6 to 7	6.1 to 6.5	10	3.5
Resistivity (megohm — cm) 1-min elect. at 15.5 deg cent	4.8×10^{13}		1×10^{14}	$>10^{14}$		1.32×10^{9}
Dielectric strength (volts/mil), min. thickness 20 deg cent	440, 1/8 in. thickness	40, 0.5 in. thickness	135, 0.5 in. thickness; 175, 0.375 in. thickness	250, 0.2 in. thickness	290, 0.125 in. thickness	Approx. 1,500, 0.006 in. thickness
Dielectric strength (volts/mil), min. thickness, max. operating temperature						Approx. 1,200
Dielectric strength (volts/mil), 100 mils, 20 deg cent			350			Approx. 800
Dielectric strength (volts/mil), 100 mils, max. operating temperature						Approx. 600
Impulse dielectric strength (volts/mil), room temperature						1,700 to 2,000 (200 mils)
Resistance to electrical discharge			Excellent		Good	Good
Frequency range over which used efficiently	25 to 30 megacycles	25 to 60 cycles	25 to 100,000 cycles	25 cycles to 10 megacycles	Entire range	0 to 100 kc

Table 4. Properties of Insulating Materials—Solid—*Continued*

	Ceramics and Similar Materials					Fibrous-base Flexible Sheet and Tape Materials
Name	G.E. Mycalex	Porcelain—Dry Process	Porcelain—Wet Process	Steatite (Unglazed)	Zircon Porcelain	Impregnated Paper (Oil Impregnant)
Chemical Properties						
Can it be used in water?	Yes	Yes, with some adverse effect	Yes	Yes	Yes	No
Can it be used in high humidity?	Yes	Yes, with some adverse effect	Yes	Yes	Yes	No
Can it be used in petroleum oil?	Yes	Yes	Yes	Yes	Yes	Yes
Resistance to weak inorganic acids	Poor	Good	Excellent except for hydrofluoric acid	Good	Good	Poor
Resistance to strong inorganic acids	Poor	Good	Excellent except for hydrofluoric acid	Good	Poor	Poor
Resistance to weak organic acids	Poor	Good	Unaffected	Good	Good	Poor
Resistance to strong organic acids	Poor	Good	Unaffected	Good	Good	Poor
Resistance to weak alkalies	Poor	Good	Good	Good	Poor	Poor
Resistance to strong alkalies	Poor	Good	Fair, affected slowly by caustic alkalies	Good	Poor	Poor
Is it usable exposed to weather indefinitely?	Yes	Yes	Yes	Yes	Yes	No
Does it corrode copper?	No	No	No	No	No	No
Does it corrode lead?	No	No	No	No	No	No
Does it corrode aluminum?	No	No	No	No	No	No
Does it corrode tin?	No	No	No	No	No	No
Flammability	Does not support combustion	Does not support combustion	Does not support combustion	Does not support combustion	Does not support combustion	Readily inflammable
Ozone resistance (IPCEA)	Excellent	Excellent	Excellent		Good	Excellent
Solvents		None	None	None	Strong alkalies	

Table 4. Properties of Insulating Materials—Solid—*Continued*

	Ceramics and Similar Materials					Fibrous-base Flexible Sheet and Tape Materials
Name	G.E. Mycalex	Porcelain—Dry Process	Porcelain—Wet Process	Steatite (Unglazed)	Zircon Porcelain	Impregnated Paper (Oil Impregnant)
Thermal Properties						
Maximum operating temperature (deg cent)	325 to 375	1,000	150	150	Approx. 1,000	70 to 85 continuous, 90 to 110 short time
Softening-point temperature (deg cent)	400	1,225 to 1,285	1,000 to 1,285	1,450	>1,800	
Specific heat	0.16		0.20, 0 to 200° C; 0.22, 200 to 400° C		0.13	0.45
Thermal capacity (watt-hr per °C per lb)	0.084		0.11, 0 to 200° C; 0.12, 200 to 400°C		0.07	0.24
Thermal cubic expansion coefficient (per °C)	3×10^{-5}		1.59×10^{-5} to 1.71×10^{-5}	2.07×10^{-5}	3.7×10^{-6} to 4.9×10^{-6}	43×10^{-5}
Thermal resistivity (°C per watt per cm cubed) at 20° C	207		20.8 to 96	39.7		400 to 750
Thermal resistivity at max. operating temperature						500 to 750
Mechanical Properties						
Tensile strength (lb per square inch) at 20° C	13,000		5,000 to 8,500	8,500	12,700	
Elongation (% at break)			0.1			
Compression strength (lb per square inch) at 20° C	39,000 to 52,000	30,000	50,000 to 55,000	75,000	90,000	
Hardness, 20°C	22-L Rockwell		6 to 7 Mohs Scale	7.5 Mohs Scale	7.5 Mohs Scale	
Hardness at max. operating temperature						
Impact strength (ft-lb) (ASTM 256–43T)	1.86	1.2	1.5 Charpy	4.5 Charpy	17,800 Modified Charpy	
Specific gravity	2.64 to 3.44	Approx. 2.4	Approx. 2.4			1.17

Table 5. Properties of Insulating Materials—Solid—*Continued*

	Fibrous-base Flexible Sheet and Tape Materials					Glass	
Name	Impregnated Paper (Oil or Oil-base Impregnant)	Varnished Cambric (Cable Cloth)	Varnished Paper	Varnished SIC and Rayon Cloths	Varnished Silk Cloth	Glass CGW No. 001 Clear	Silicone-varnished Fiberglas
General Properties							
Chemical name or components	Paper (usually wood pulp) + heavy mineral oil or oil + rosin compound	Straight-cut cotton cloth + oleo resinous varnish	Condenser Bond Kraft or Red Rope Paper + oleo resinous varnish	Thin cotton or rayon cloth + oleo resinous varnish	Silk cloth + oleo resinous varnish	Potash, soda, lead	Glass cloth coated with silicon resin varnish
General uses as insulation	Cables—solid, gas pressure, and compression types	Power-cable insulation	Layer insulation for coils, sheet insulation—high mechanical strength	Thin insulation, high dielectric strength—cable joints, etc.	Thin insulation with high dielectric strength—cable joints, etc.	Lamp parts, neon tubes, pressed parts, insulators	Where heat resistance is required
Electrical Properties							
Power factor (% at 60 cycles) at 20 deg cent	0.4 to 0.9	6.0 to 8.0		Approx. 1.0	Approx. 1.0	0.33	
Power factor at max. operating temperature	1.0 to 2.5	11.0 to 18.0		Approx. 2.0	Approx. 2.0		1.0 to 1.4 at 100° C
Dielectric constant (SIC) at 60 cycles, 20 deg cent	4.0	5.0		5.0		6.7	3.4 to 3.5
Resistivity (megohm-cm) 1-min elect. at 15.5 deg cent	2.18×10^8	Approx. 2.0×10^6		Approx. 3.0×10^6	Approx. 3.0×10^6		
Dielectric strength (volts/mil), min. thickness 20 deg cent	Approx. 1,500 (6 mil)	Approx. 1,050	500 to 2,000, 20 to 3/4-mil thickness	1,200, 3-mil thickness	1,800, 3-mil thickness	Approx. 8,000 volts/mil	
Dielectric strength (volts/mil), min. thick., max. operating temperature	Approx. 1,000 (6 mil)	750 at 75° C					
Dielectric strength (volts/mil), 100 mils, 20 deg cent	Approx. 600					3,000 to 8,000	
Dielectric strength (volts/mil), 100 mils, max. operating temperature	Approx. 500						
Impulse dielectric strength (volts/mil), room temperature	1,200 to 1,600 (200 mils)	1,400					
Resistance to electrical discharge	Good	Good	Good	Good	Good	Excellent	
Frequency range over which used efficiently	0 to 100 kc	Power frequencies	Power frequencies	Power frequencies	Power frequencies	High	Power frequencies

Table 5. Properties of Insulating Materials—Solid—*Continued*

	Fibrous-base Flexible Sheet and Tape Materials					Glass	
Name	Impregnated Paper (Oil or Oil-base Impregnant)	Varnished Cambric (Cable Cloth)	Varnished Paper	Varnished SIC and Rayon Cloths	Varnished Silk Cloth	Glass CGW No. 001 Clear	Silicone-varnished Fiberglas
Chemical Properties							
Can it be used in water?	No	No	No	No	No	Yes	
Can it be used in high humidity?	No	Yes, although electrical properties affected	No	Yes, although electrical properties affected	Yes, although electrical properties affected	Yes	Yes
Can it be used in petroleum oil?	Yes	Yes	Yes	Yes	Yes	Yes	Not in hot oil
Resistance to weak inorganic acids	Poor	Good	Good	Good	Good	Excellent	Good
Resistance to strong inorganic acids	Poor	Good	Good	Good	Good	Unaffected	
Resistance to weak organic acids	Poor	Good	Good	Good	Good	Unaffected	Good
Resistance to strong organic acids	Poor	Good	Good	Good	Good	Unaffected	
Resistance to weak alkalies	Poor					Unaffected	
Resistance to strong alkalies	Poor					Good	Good
Is it usable exposed to weather indefinitely?	No	No	No	No	No	Yes	Yes
Does it corrode copper?	No	No	No	No	No	No	No
Does it corrode lead?	No	No	No	No	No	No	No
Does it corrode aluminum?	No	No	No	No	No	No	No
Does it corrode tin?	No	No	No	No	No	No	No
Flammability	Readily inflammable	Slow burning	Slow burning	Slow burning	Slow burning	Does not support combustion	Does not support combustion
Ozone resistance (IPCEA)	Excellent	Good	Good	Good	Good	Excellent	Excellent
Solvents		None				None	

Table 5. Properties of Insulating Materials—Solid—*Continued*

	Fibrous-base Flexible Sheet and Tape Materials					Glass	
Name	Impregnated Paper (Oil or Oil-base Impregnant)	Varnished Cambric (Cable Cloth)	Varnished Paper	Varnished SIC and Rayon Cloths	Varnished Silk Cloth	Glass CGW No. 001 Clear	Silicone-varnished Fiberglas
Thermal Properties							
Maximum operating temperature (deg cent)	60 to 85 continuous, 90 to 110 short time	70 to 85	90	90	90	397 (strain point)	175
Softening-point temperature (deg cent)						626	
Specific heat	0.45					Approx. 0.20	
Thermal capacity (watt-hr per °C per lb)	0.24					Approx. 0.105	
Thermal cubic expansion coefficient (per °C)	43×10^{-5}					2.73×10^{-5}	
Thermal resistivity (°C per watt per cm cubed) at 20° C	Approx. 450 to 700	600					
Thermal resistivity at maximum operating temperature	Approx. 450 to 700						
Mechanical Properties							
Tensile strength (lb per square inch) at 20° C		2,400 to 3,000 depending on thickness		4,000 to 10,500 depending on thickness	2,250 to 9,000 depending on thickness		100 to 175
Elongation (% at break)						0	
Compression strength (lb per square inch) at 20° C						>150,000	
Hardness, 20° C						5.5 Mohs Scale	
Specific gravity		1.11	Approx. 1.15	Approx. 1.15	Approx. 1.2	2.85	1.41 to 1.42

Table 6. Properties of Insulating Materials—Solid
Glass—*Continued*

Name	Glass CGW No. 008 Clear	Glass CGW No. 774 Clear	Glass CGW No. 707 Clear	Glass CGW No. 790 Clear and Multiform	Glass CGW No. 7052	Glass CGW No. 7761 Multiform
			General Properties			
Chemical name or components	Soda, lime	Borosilicate	Borosilicate	High silica (96%)		
General uses as insulation	Lamp bulbs, electronic tubes, pole insulators, etc.	Outdoor, high-temperature insulators	Low-loss insulation for special purposes	High temperature, high thermal shock, low loss	Sealing to low-melting metal	Insulators—low loss
			Electrical Properties			
Power factor (% at 60 cycles) at 20° C	6.2	1.2	0.065			
Dielectric constant (SIC) at 60 cycles, 20° C	8.4	4.8	4.0	Approx. 3.85		
Resistivity (megohm-cm), 1-min Elect. at 15.5° C						$1 \times 10^{11}+$
Dielectric strength (volts/mil), min. thickness, 20° C	Approx. 8,000	Approx. 8,000	Approx. 8,000	Approx. 8,000	Approx. 8,000	Approx. 8,000
Dielectric strength (volts/mil), 100 mils, 20° C	3,000 to 8,000	3,000 to 8,000	3,000 to 8,000	3,000 to 8,000	3,000 to 8,000	3,000 to 8,000
Resistance to electrical discharge	Excellent	Excellent	Excellent	Excellent	Excellent	Excellent
Frequency range over which used efficiently	Power and audio ranges	Through broadcast frequencies	Entire range	Entire range	Entire range	Entire range

Table 6. Properties of Insulating Materials—Solid Glass—*Continued*

Name	Glass CGW No. 008 Clear	Glass CGW No. 774 Clear	Glass CGW No. 707 Clear	Glass CGW No. 790 Clear and Multiform	Glass CGW No. 7052	Glass CGW No. 7761 Multiform
			Chemical Properties			
Can it be used in water?	Yes	Yes	Yes	Yes	Yes	Yes
Can it be used in high humidity?	Yes	Yes	Yes	Yes	Yes	Yes
Can it be used in petroleum oil?	Yes	Yes	Yes	Yes	Yes	Yes
Resistance to weak inorganic acids	Excellent	Excellent	Excellent	Excellent	Excellent	Excellent
Resistance to strong inorganic acids	Excellent	Excellent	Excellent	Excellent	Excellent	Excellent
Resistance to weak organic acids	Excellent	Excellent	Excellent	Excellent	Excellent	Excellent
Resistance to strong organic acids	Excellent	Excellent	Excellent	Excellent	Excellent	Excellent
Resistance to weak alkalies	Excellent	Excellent	Good	Excellent	Excellent	Excellent
Resistance to strong alkalies	Good	Good	Good	Good	Good	Good
Is it usable exposed to weather indefinitely?	Yes	Yes	Yes	Yes	Yes	Yes
Does it corrode copper?	No	No	No	No	No	No
Does it corrode lead?	No	No	No	No	No	No
Does it corrode aluminum?	No	No	No	No	No	No
Does it corrode tin?	No	No	No	No	No	No
Flammability	Does not support combustion	Does not support combustion	Does not support combustion	Does not support combustion	Does not support combustion	Does not support combustion
Ozone resistance (IPCEA)	Excellent	Excellent	Excellent	Excellenr	Excellent	Excellent
Solvents	None	None	None	None	None	None

Table 6. Properties of Insulating Materials—Solid Glass—*Continued*

Name	Glass CGW No. 008 Clear	Glass CGW No. 774 Clear	Glass CGW No. 707 Clear	Glass CGW No. 790 Clear and Multiform	Glass CGW No. 7052	Glass CGW No. 7761 Multiform
Thermal Properties						
Maximum operating temperature (deg cent)	475 (strain point)	510 (strain point)	455 (strain point)	820 (strain point)	442 (strain point)	463 (strain point)
Softening-point temperature (deg cent)	696	820	746	1,500	708	820
Specific heat	Approx. 0.20	Approx. 0.20	Approx. 0.20	Approx. 0.20	Approx. 0.20	Approx. 0.20
Thermal capacity (watt-hr per °C per lb)	Approx. 0.105	Approx. 0.105	Approx. 0.105	Approx. 0.105	Approx. 0.105	Approx. 0.105
Thermal cubic expansion coefficient (per °C)	2.76×10^{-5}	0.99×10^{-5}	0.96×10^{-5}	0.24×10^{-5}	1.41×10^{-5}	0.84×10^{-5}
Thermal resistivity (°C per watt per cm cubed) at 20° C		Approx. 85				
Thermal resistivity at max. operating temperature		65 at 250° C				
Mechanical Properties						
Elongation (% at break)	0	0	0	0	0	0
Compression strength (lb per square inch) at 20° C	>150,000	>150,000	>150,000	>150,000	>150,000	>150,000
Hardness, 20° C	5.5 Mohs Scale	5.5 Mohs Scale	5.5 Mohs Scale	5.5 Mohs Scale	5.5 Mohs Scale	5.5 Mohs Scale
Specific gravity	2.45	2.23	2.13	2.18	2.29	2.16

Table 7. Properties of Insulating Materials—Solid—*Continued*

	Glass	Rigid-molded and Laminated-molded Materials				Rubber and Rubberlike Compounds
Name	Glass CGW No. 8870	Mica—New England Mica Co. Y-26 High-heat Mica Plate	Phenolic	Phenolite	Vulcanized Fiber	Code Rubber Compound
General Properties						
Chemical name or components		Built-up mica bonded with fused inorganic salts	Laminated phenolic—paper base	Laminated phenolic—fabric base	Regenerated cellulose	New or reclaimed GRS or natural rubber + fillers + vulcanizing agent (variable)
General uses as insulation	Capacitors, X-ray shields, thin sheets	Toasters, flatirons, and other heating appliances	Insulating sheets, tubes, rods, and parts machined from these	Insulating sheets, tubes, rods, and parts machined from these	Molded and machined insulating parts	Wire and cable insulation
Electrical Properties						
Power factor (% at 60 cycles) at 20° C		0.2 at 1,000 kc 0.3 at 95% relative humidity at 1,000 kc	3 to 5	5 to 10	3 to 8 at 550 kc	4 to 8
Power factor at max. operating temperature						4 to 8
Dielectric constant (SIC) at 60 Cycles, 20° C		5.11			4 to 7	5 to 7
Resistivity (megohm-cm), 1-min elect. at 15.5° C		2×10^6				Approx. 6.2×10^7
Dielectric strength (volts/mil), min. thickness, 20° C	Approx. 8,000	800 to 1,000, 0.015-in. thickness				
Dielectric strength (volts/mil), min. thickness, max. operating temperature		750 to 900 at 260° C				
Dielectric strength (volts/mil), 100 mils, 20° C	3,000 to 8,000	500, 0.062-in. thickness	350 to 500	350 to 500	150 to 250	300
Dielectric strength (volts/mil), 100 mils, max. operating temperature		500, 0.062-in. thickness				
Resistance to electrical discharge	Excellent	Excellent				Poor
Frequency range over which used efficiently	Entire range	25 to 1,000 cycles				Power frequencies

Table 7. Properties of Insulating Materials—Solid—*Continued*

	Glass	Rigid-molded and Laminated-molded Materials				Rubber and Rubberlike Compounds
Name	Glass CGW No. 8870	Mica—New England Mica Co. Y-26 High-heat Mica Plate	Phenolic	Phenolite	Vulcanized Fiber	Code Rubber Compound
Chemical Properties						
Can it be used in water?	Yes	No			No	No
Can it be used in high humidity?	Yes	No	Yes	Yes	No	No
Can it be used in petroleum oil?	Yes	Yes	Yes	Yes	Yes	No
Resistance to weak inorganic acids	Poor	Poor	Fair	Fair	Poor	Fair
Resistance to strong inorganic acids	Poor	Poor	Poor	Poor	Poor	Poor
Resistance to weak organic acids		Poor	Fair	Fair	Poor	Poor
Resistance to strong organic acids		Poor	Poor	Poor	Poor	Poor
Resistance to weak alkalies	Excellent	Poor	Fair	Fair	Poor	Poor
Resistance to strong alkalies	Good	Poor	Poor	Poor	Poor	Poor
Is it usable exposed to weather indefinitely?	Yes	Yes			No	Yes, if protected from sunlight
Does it corrode copper?	No	No	No	No	No	Yes
Does it corrode lead?	No	No	No	No	No	No
Does it corrode aluminum?	No	No	No	No	No	No
Does it corrode tin?	No	No	No	No	No	No
Flammability	Does not support combustion	Does not support combustion				Slow burning
Ozone resistance (IPCEA)	Excellent		Excellent	Excellent	Excellent	Poor
Solvents	None	Water (slow)	None	None	Cupra-monium-hydroxide (Triton B)	Most organic solvents extract fractions

Table 7. Properties of Insulating Materials—Solid—*Continued*

	Glass	Rigid-molded and Laminated-molded Materials				Rubber and Rubberlike Compounds
Name	Glass CGW No. 8870	Mica—New England Mica Co. Y-26 High-heat Mica Plate	Phenolic	Phenolite	Vulcanized Fiber	Code Rubber Compound
Thermal Properties						
Maximum operating temperature (deg cent)	398 (strain point)	650	127	107	80	60
Brittle-point temperature (deg cent)		None				Approx. −30
Softening-point temperature (deg cent)	580	625	No	No	None	
Specific heat	Approx. 0.20		Approx. 0.4			Approx. 0.5
Thermal capacity (watt-hr per ° C per lb)	Approx. 0.105					0.26
Thermal cubic expansion coefficient (per °C)	2.73×10^{-5}					
Thermal resistivity (°C per watt per cm cubed) at 20° C			Approx. 600			Approx. 500
Thermal resistivity at max. operating temperature			Approx. 600			Approx. 500
Mechanical Properties						
Tensile strength (lb per square inch) at 20° C			8,000 to 12,000	9,000 to 11,000	6,000 to 12,000	600
Elongation (% at break)	0				Depends on moisture content	200
Compression strength (lb per square inch) at 20° C	>150,000		34,000	38,000	20,000 to 30,000	
Hardness, 20° C	5.5 Mohs Scale				Rockwell R-60 to R-100	
Hardness at max. operating temperature						
Impact strength (ft-lb) (ASTM 256-43T)					1 to 8 Izod	
Specific gravity	4.23	2.78	1.35	1.35	0.9 to 1.5	1.35 to 1.50

Table 8. Properties of Insulating Materials—Solid Rubber and Rubberlike Compounds—*Continued*

Name	Heat-resistant Grade Natural-rubber Compound	Heat-resistant Grade Synthetic-rubber Compound	Neoprene Compound	Ozone-resisting Grade Rubber Compound	Performance Grade Natural-rubber Compound	Performance-Grade Synthetic-rubber Compound
General Properties						
Chemical name or components	Natural rubber + fillers + vulcanizing agent	GRS + fillers + vulcanizing agent	Polymerized chloroprene + plasticizers + fillers, etc.	Vegetable-oil base + GRS or natural-rubber compound + vulcanizing agent	Natural rubber + fillers + vulcanizing agent	GRS + fillers + vulcanizing agent
General uses as insulation	Wire and cable insulation	Wire and cable insulation	Jackets for wire and cable, building-wire insulation	Wire and cable insulation for high voltage or for exposure to ozone	Wire and cable insulation	Wire and cable insulation
Electrical Properties						
Power factor (% at 60 cycles) at 20° C	3 to 5	3 to 5	0.9 to 6.0	Approx. 5	4 to 6	4 to 6
Power factor at max. operating temperature	3 to 5	3 to 5		Approx. 5	4 to 6	4 to 6
Dielectric constant (SIC) at 60 cycles, 20° C	5	5	6.5 to 9	Approx. 5	5	5
Resistivity (megohm-cm), 1-min elect. at 15.5° C	Approx. 1.8×10^9	Approx. 1.7×10^8	3×10^{11} to 6×10^{12}	Approx. 4.1×10^8	Approx. 1.8×10^9	Approx. 1.7×10^8
Dielectric strength (volts/mil), min. thickness, 20° C			500 to 800, 25-mil thickness			
Dielectric strength (volts/mil), 100 mils, 20° C	300	300		400	300	300
Dielectric strength (volts/mil), 100 mils, maximum operating temperature	300	300			300	300
Impulse dielectric strength (volts/mil), room temperature	Approx. 1,000	Approx. 1,000		Approx. 1,000	Approx. 1,000	Approx. 1,000
Resistance to electrical discharge	Poor	Poor	Excellent	Poor	Poor	Poor
Frequency range over which used efficiently	Power frequencies	Power frequencies	Power frequencies	Power frequencies	Power frequencies	Power frequencies

Table 8. Properties of Insulating Materials—Solid Rubber and Rubberlike Compounds—*Continued*

Name	Heat-resistant Grade Natural-rubber Compound	Heat-resistant Grade Synthetic-rubber Compound	Neoprene Compound	Ozone-resisting Grade Rubber Compound	Performance Grade Natural-rubber Compound	Performance-Grade Synthetic-rubber Compound
Chemical Properties						
Can it be used in water?	No	No	Yes	Yes	No	No
Can it be used in high humidity?	No	No	Yes	Yes	No	No
Can it be used in petroleum oil?	No	No	Yes	No	No	No
Resistance to weak inorganic acids	Fair	Fair	Excellent	Fair	Fair	Fair
Resistance to strong inorganic acids	Fair	Fair	Fair	Fair	Poor	Poor
Resistance to weak organic acids	Poor	Poor	Good	Poor	Poor	Poor
Resistance to strong organic acids	Poor	Poor		Poor	Poor	Poor
Resistance to weak alkalies	Fair	Fair	Good	Poor	Fair	Fair
Resistance to strong alkalies	Poor	Poor	Fair	Poor	Poor	Poor
Is it usable exposed to weather indefinitely?	Yes, if protected from sunlight	Yes, if protected from sunlight	Yes	Yes	Yes, if protected from sunlight	Yes, if protected from sunlight
Does it corrode copper?	Yes	Yes	No	Yes	Yes	Yes
Does it corrode lead?	No	No	No	No	No	No
Does it corrode aluminum?	No	No	No	No	No	No
Does it corrode tin?	No	No	No	No	No	No
Flammability	Slow burning	Slow burning	Does not support combustion	Slow burning	Slow burning	Slow burning
Ozone resistance (IPCEA)	Poor	Poor	Excellent	Good	Poor	Poor
Solvents	Most organic solvents extract fractions	Most organic solvents extract fractions	None	Most organic solvents extract fractions	Most organic solvents extract fractions	Most organic solvents extract fractions

Table 8. Properties of Insulating Materials—Solid Rubber and Rubberlike Compounds—*Continued*

Name	Heat-resistant Grade Natural-rubber Compound	Heat-resistant Grade Synthetic-rubber Compound	Neoprene Compound	Ozone-resistant Grade Rubber Compound	Performance Grade Natural-rubber Compound	Performance-Grade Synthetic-rubber Compound
Thermal Properties						
Maximum operating temperature (deg cent)	75	75	60	70	60	60
Brittle-point temperature (deg cent)	−40	−40	−40 to −60	−20	−40	−40
Specific heat	0.5	0.5		0.5	0.5	0.5
Thermal capacity (watt-hr per °C per lb)	0.26	0.26		0.26	0.26	0.26
Thermal resistivity (°C per watt per cm cubed) at 20° C	500	500			500	500
Thermal resistivity at max. operating temperature					500	500
Mechanical Properties						
Tensile strength (lb per square inch) at 20° C	1,500	750	2,500 to 4,500	450	1,200	700
Elongation (% at break)	400	300	500 to 1,000	250	400	300
Hardness, 20° C			20 to 90			
Specific gravity	1.60 to 1.85	1.60 to 1.85		1.35 to 1.45	1.60 to 1.85	1.60 to 1.85

Table 9. Properties of Insulating Materials—Solid—*Continued*

	Rubber and Rubberlike Compounds		Thermoplastic Materials			
Name	Silicone Rubber	Submarine Grade Rubber Compound	Ethocel	Lucite	Nylon FM-1	Plastacele
General Properties						
Chemical name or components		Natural rubber + fillers + vulcanizing agent	Ethyl cellulose	Polymethyl methacrylate + minor modifying ingredients	Long-chain synthetic polymeric amide	Cellulose acetate, coloring ingredients, and plasticizer
General uses as insulation	Gaskets, bushings, etc.	Wire and cable insulation, for use under water	Wire insulation	Molded parts and sheets	Extruded low-voltage insulation and cable jackets	Molded insulators, tapes, etc.
Electrical Properties						
Power factor (% at 60 cycles) at 20° C	0.045 to 0.37, depending on formulation	Approx. 1.0	0.5 to 2.0, depending on formulation	5	4.5	3 to 6, depending on formulation
Power factor at max. operating temperature		Approx. 1.0				
Dielectric constant (SIC) at 60 cycles, 20° C	3.25 to 7.40, depending on formulation	3	2.5 to 4.0, depending on formulation	4.5	4.5	3.5 to 6.5, depending on formulation
Resistivity (megohm-cm) 1-min elect. at 15.5° C	1.5×10^5 at 25° C	Approx. 2.0×10^9	10^4 to 10^6, depending on formulation	$>10^9$	4×10^8	10^4
Dielectric strength (volts/mil), min. thickness, 20° C	260 to 450, depending on formulation, 1/8-in. thickness		1,400 to 1,800, 10-mil thickness			
Dielectric strength (volts/mil), 100 mils, 20° C		350		500	500	400
Dielectric strength (volts/mil), 100 mils, max. operating temperature		350				
Impulse dielectric strength (volts/mil), room temperature		Approx. 1,000				
Resistance to electrical discharge	Excellent	Poor	Good			
Frequency range over which used efficiently	Up to 1 megacycle	Up to 10,000 cycles	Entire range			

Table 9. Properties of Insulating Materials—Solid—*Continued*

	Rubber and Rubberlike Compounds		Thermoplastic Materials			
Name	Silicone Rubber	Submarine Grade Rubber Compound	Ethocel	Lucite	Nylon FM-1	Plastacele
			Chemical Properties			
Can it be used in water?	Yes	Yes	Limited use	Yes	Yes	Yes
Can it be used in high humidity?	Yes	Yes	Yes	Yes	Yes	Yes
Can it be used in petroleum oil?	Yes, but not if oil contains aromatics	No	No	Yes	Yes	Yes
Resistance to weak inorganic acids	Good	Fair	Excellent	Good	Good	Good
Resistance to strong inorganic acids	Good, except to conc. H_2SO_4	Fair	Poor	Poor	Poor	Poor
Resistance to weak organic acids		Poor	Poor to good	Good	Good	Good
Resistance to strong organic acids		Poor	Poor	Poor	Good	Poor
Resistance to weak alkalies	Good	Fair	Excellent	Good	Good	Good
Resistance to strong alkalies	Good to poor, depending on formulation	Poor	Good	Poor	Good	Poor
Is it usable exposed to weather indefinitely?	Yes	Yes, if protected from sunlight	Yes	Yes	Yes	
Does it corrode copper?	No	Yes	No			
Does it corrode lead?	No	No	No			
Does it corrode aluminum?	No	No	No			
Does it corrode tin?	No	No	No			
Flammability	Does not support combustion	Slow burning	Slow burning	Slow burning	Does not support combustion	Does not support combustion
Ozone resistance (IPCEA)	Excellent	Poor		Excellent		
Solvents	Alcohol, acetone, aromatics and CCl_4	Most organic solvents extract fractions	Most organic solvents	Esters, ketones, aromatics, chlor. hydrocarbons, conc. acids	Phenol, formic acid, and conc. mineral acids	Strong acids and alkalies, ketones, and esters

Table 9. Properties of Insulating Materials—Solid—*Continued*

	Rubber and Rubberlike Compounds		Thermoplastic Materials			
Name	Silicone Rubber	Submarine Grade Rubber Compound	Ethocel	Lucite	Nylon FM-1	Plastacele
Thermal Properties						
Maximum operating temperature (deg cent)	300	60	50 to 85, depending on formulation	70	>150	70
Brittle-point temperature (deg cent)	−55	−40	None			
Softening-point temperature (deg cent)			None	71		
Specific heat		0.5	0.3 to 0.5	0.35	0.43	0.35
Thermal capacity (watt-hr per °C per lb)		0.26		0.18	0.23	0.18
Thermal cubic expansion coefficient (per °C)				2.6×10^{-4}	3.1×10^{-4}	2.6×10^{-4}
Thermal resistivity (°C per watt per cm cubed) at 20° C		500		345	400	400
Mechanical Properties						
Tensile strength (lb per square inch) at 20° C	200 to 650, depending on formulation	2,000	3,000 to 8,000 depending on formulation	8,400	10,500	4,700
Elongation (% at break)		400	5 to 50, depending on formulation	4.8	54	36
Compression strength (lb per square inch) at 20° C			3,000 to 12,000 depending on formulation	11,000	18,000	
Hardness, 20° C	40 to 80, depending on formulation		R-50 to R-110 (Rockwell), depending on formulation	M-98 Rockwell	M-90 Rockwell	R-112 Rockwell
Hardness at max. operating temperature	40 to 80, depending on formulation					
Impact strength (ft-lb) (ASTM 256–43T)			1 to 7 per inch of notch, depending on formulation	0.5	0.9	1.2
Specific gravity		1.15 to 1.30			1.14	1.27 to 1.37

Table 10. Properties of Insulating Materials—Solid—*Continued*

	Thermoplastic Materials					Varnishes
Name	Polyethylene (DE 3401) or Polythene	Saran	Teflon	Vinylite (VE 5901 Series)	Vinylite (VE 5507 Series) Jacket Stock	Varnish (Insulating) GE #1678—Clear
General Properties						
Chemical name or components	Polyethylene resin + antioxidant	Polyvinylidene chloride	Polytetrafluoroethylene	Copolymer of vinyl chloride and vinyl acetate with plasticizers	Copolymer of vinyl chloride and vinyl acetate + plasticizer	Phenol, oil
General uses as insulation	Wire and cable insulation		Primary insulation for temperatures up to 200° C	Flame-resistant wire and cable insulation	Flame-resistant wire and cable insulation and jackets	Electrical insulation
Electrical Properties						
Power factor (% at 60 cycles) at 20° C	0.02	0.030 to 0.045	<0.03	10.5	6.6	
Power factor at max. operating temperature	0.03			5.8 at 70° C	5.5 at 70° C	
Dielectric constant (SIC) at 60 cycles, 20° C	2.3	4.5 to 6.0	2.1	6.8	8.3	
Resistivity (megohm-cm), 1-min elect. at 15.5° C	$>10^9$	Approx. 10^9	$>10^9$	$>10^9$	$>10^7$	
Dielectric strength (volts/mil), min. thickness 20° C	Approx. 1,000, 25-mil thickness	1,000, 1.5- to 10-mil thickness	>1,000, 5-mil thickness	750, 25-mil thickness	750, 25-mil thickness	1,900 to 2,100
Dielectric strength (volts/mil), min. thickness, max. operating temperature	Approx. 1,000, 25-mil thickness			525 at 60° C, 25-mil thickness	525 at 60° C, 25-mil thickness	
Dielectric strength (volts/mil), 100 mils, 20° C	550	300 to 350	500	300	300	
Dielectric strength (volts/mil), 100 mils, max. operating temperature	550			150 at 60° C	150 at 60° C	
Resistance to electrical discharge	Good	Poor				Excellent
Frequency range over which used efficiently	Unlimited	Entire range	Unlimited	Power frequencies	Power frequencies	

Table 10. Properties of Insulating Materials—Solid—*Continued*

	Thermoplastic Materials					Varnishes
Name	Polyethylene (DE 3401) or Polythene	Saran	Teflon	Vinylite (VE 5901 Series)	Vinylite (VE 5507 Series) Jacket Stock	Varnish (Insulating) GE #1678—Clear
			Chemical Properties			
Can it be used in water?	Yes	Yes	Yes	Yes	Yes	Yes
Can it be used in high humidity?	Yes	Yes	Yes	Yes	Yes	Yes
Can it be used in petroleum oil?	Only under special conditions	Yes	Yes	Yes, at temperatures below 60° C	Yes, at temperatures below 60° C	Yes
Resistance to weak inorganic acids	Excellent	Excellent	Excellent	Excellent at 25° C	Excellent at 25° C	Excellent
Resistance to strong inorganic acids	Excellent	Good	Excellent	Good at 25° C	Good at 25° C	Excellent
Resistance to weak organic acids	Excellent	Good	Excellent	Good at 25° C	Good at 25° C	Excellent
Resistance to strong organic acids	Excellent	Good	Excellent			Excellent
Resistance to weak alkalies	Excellent	Good	Excellent	Excellent at 25° C	Excellent at 25° C	Excellent
Resistance to strong alkalies	Excellent	Poor	Excellent			Excellent
Is it usable exposed to weather indefinitely?	Yes, if protected from sun and oxidation by proper compounding	Good, darkens color		Yes	Yes	
Does it corrode copper?	No	No		No	No	No
Does it corrode lead?	No	No		No	No	No
Does it corrode aluminum?	No	No		No	No	No
Does it corrode tin?	No	No		No	No	No
Flammability	Slow burning	Does not support combustion	Does not support combustion	Does not support combustion	Does not support combustion	
Ozone resistance (IPCEA)	Excellent		Excellent	Excellent	Excellent	Good
Solvents	Aromatic aliphatic or halogenated hydrocarbons above 50 to 60° C	None	Molten alkalies	Ketones, esters, and aromatic hydrocarbons	Ketones, esters, and aromatic hydrocarbons	

Table 10. Properties of Insulating Materials—Solid—*Continued*

	Thermoplastic Materials					Varnishes
Name	Polyethylene (DE 3401) or Polythene	Saran	Teflon	Vinylite (VE 5901 Series)	Vinylite (VE 5507 Series) Jacket Stock	Varnish (Insulating) GE # 1678—Clear
Thermal Properties						
Maximum operating temperature (deg cent)	85	80	200	60	80	
Brittle-point temperature (deg cent)	<−70		<−70	−17	−28	
Softening-point temperature (deg cent)	105 to 115	115 to 138	327, loses strength			
Specific heat	0.55 at 25° C, 1.5 at 102° C	0.32 at 25° C	0.25	0.33		
Thermal capacity (watt-hr per °C per lb)	0.28	0.168	0.13	0.18		
Thermal cubic expansion coefficient (per °C)	9×10^{-4} at 25° C, 12.1 $\times 10^{-4}$ at 80° C	5.7×10^{-4}	3×10^{-4}			
Thermal resistivity (°C per watt per cm cubed) at 20° C	295 at 0 to 15° C, 385 at 25 to 40° C	1,090	410	650	650	
Thermal resistivity at max. operating temperature	580					
Mechanical Properties						
Tensile strength (lb per square inch) at 20° C	1,900	3,000 to 6,000 unoriented, to 40,000 oriented	3,200	3,000	2,200	Very high toughness
Elongation (% at break)	600	20 to 300	350	200	225	Medium flexibility
Compression strength (lb per square inch) at 20° C	4,200 to 5,100	7,500 to 8,500	1,700—1% deformation			
Hardness, 20° C	98 Durometer A, 70 Durometer C, 45 Durometer D	M-50–65 Rockwell	55 Durometer D	85 Durometer A at 25° C	80 Durometer A at 25° C	
Hardness at max. operating temperature	84 Durometer A					High penetration strength
Impact strength (ft-lb) (ASTM 256–43T)	>4 Izod	0.3 to 1.0 Izod Notched	4 Izod			
Specific gravity		1.65 to 1.72		1.19 to 1.24	1.17 to 1.22	0.932 after baking at 150° C

Table 11. Properties of Insulating Materials—Solid—*Continued*

	Varnishes			Miscellaneous		
Name	Varnish (Insulating) GE #2480-Clear	Varnish (Insulating) GE #9435-Black	Varnish (Insulating) GE #9535-Clear	Styraloy	Styron	Titania
General Properties						
Chemical name or components	Glyptal	Synthetic resin, asphalt	Phenol, glyptal, oil	Styrene, butadiene, copolymer	Polystyrene	Titanium dioxide
General uses as insulation	Electrical insulation	Electrical insulation	Electrical insulation	Wire and cable insulation	Connectors, insulators (stand-off), coil forms, switch parts, etc.	Condenser insulation
Electrical Properties						
Power factor (% at 60 cycles) at 20° C				0.05 to 0.10	0.01 to 0.03	0.005 to 0.007
Dielectric constant (SIC) at 60 cycles, 20° C				2.5 to 2.6	2.5 to 2.6	100 to 110
Resistivity (megohm-cm), 1-min elect. at 15.5° C				10^8 to 10^{10}	10^{11} to 10^{13} at 25° C	10^7 to 10^8
Dielectric strength (volts/mil), min. thickness, 20° C	1,200 to 1,300	1,150 to 1,860	1,790 to 2,100	3,000, 5-mil thickness	500 to 700, 0.125-in. thickness	150
Dielectric strength (volts/mil), 100 mils, 20° C				Approx. 700		
Dielectric strength (volts/mil), 100 mils, max. operating temperature				Approx. 600 at 90° C		
Impulse dielectric strength (volts/mil), room temperature						
Resistance to electrical discharge	Excellent		Excellent	Excellent	Good	Good
Frequency range over which used efficiently				Entire range	Entire range	Entire range

Table 11. Properties of Insulating Materials—Solid—*Continued*

	Varnishes			Miscellaneous		
Name	Varnish (Insulating) GE #2480-Clear	Varnish (Insulating) GE #9435-Black	Varnish (Insulating) GE #9535-Clear	Styraloy	Styron	Titania
			Chemical Properties			
Can it be used in water?	Yes	Yes	Yes	Yes	Yes	Yes
Can it be used in high humidity?	Yes	Yes	Yes	Yes	Yes	Yes
Can it be used in petroleum oil?	Yes	Yes	Yes	No	No	Yes
Resistance to weak inorganic acids	Excellent	Excellent	Excellent	Excellent	Excellent	Good
Resistance to strong inorganic acids	Excellent	Excellent	Excellent	Good	Good	Good
Resistance to weak organic acids	Excellent	Excellent	Excellent	Excellent	Good	Good
Resistance to strong organic acids	Excellent	Excellent	Excellent	Good		Good
Resistance to weak alkalies	Excellent	Excellent	Excellent	Excellent	Excellent	Poor
Resistance to strong alkalies	Excellent	Excellent	Excellent	Excellent	Good	Poor
Is it usable exposed to weather indefinitely?				Yes	No	Yes
Does it corrode copper?	No	No	No	No	No	No
Does it corrode lead?	No	No	No	No	No	No
Does it corrode aluminum?	No	No	No	No	No	No
Does it corrode tin?	No	No	No	No	No	No
Flammability				Slow burning	Slow burning	Does not support combustion
Ozone resistance (IPCEA)	Good	Good	Good	Excellent		Good
Solvents				Swells in most organic solvents	Most solvents	Strong alkalies

Table 11. Properties of Insulating Materials—Solid—*Continued*

	Varnishes			Miscellaneous		
Name	Varnish (Insulating) GE 2480-Clear	Varnish (Insulating) GE 9435-Black	Varnish (Insulating) GE 9535-Clear	Styraloy	Styron	Titania
Thermal Properties						
Maximum operating temperature (deg cent)				60 to 95	70 to 85	Approx. 1,000
Brittle-point temperature (deg cent)				<−80 C		
Softening-point temperture (deg cent)				Indefinite	240 to 280	Approx. 1,600
Specific heat				0.4 to 0.56	0.32 at 25° C	0.15
Thermal capacity (watt-hr per °C per lb)					0.168	0.08
Thermal cubic expansion coefficient (per °C)					2.1×10^{-4}	
Thermal resistivity (°C per watt per cm cubed) at 20° C					1,260	
Mechanical Properties						
Tensile strength (lb per square inch) at 20° C	High toughness	High toughness	High toughness	700 to 1,000	5,000 to 9,000	
Elongation (% at break)	Very high flexibility	Very high flexibility	High flexibility	50 to 300	1.5 to 3.5	
Compression strength (lb per square inch) at 20° C					11,500 to 15,000	
Hardness, 20° C				95 to 100 Durometer A	M-80–90 Rockwell	
Hardness at max. operating temperature	High penetration strength	High penetration strength	Very high penetration strength			
Impact strength (ft-lb) (ASTM 256-43T)				1.5 to 2.0 Izod notched	0.5 to 0.9 Izod notched	
Specific gravity	0.990 after baking at 150° C	0.880 after baking at 130° C	0.965 after baking at 150° C		1.05 to 1.07	

Table 12. Properties of Insulating Materials—Solid Waxes (Microcrystalline)—*Continued*

Name	S/V Cerese Wax Yellow	S/V Cerese Wax AA	S/V Cerese Wax Brown	S/V Petrosene A	S/V Petrosene B	S/V Petrosene C	S/V Product 2310	S/V Product 2305	S/V Product 2300
Chemical Properties									
Can it be used in water?	Yes	Yes	Yes	Yes	Yes	Yes	Yes	Yes	Yes
Can it be used in high humidity?	Yes	Yes	Yes	Yes	Yes	Yes	Yes	Yes	Yes
Can it be used in petroleum oil?	No	No	No	No	No	No	No	No	No
Resistance to weak inorganic acids	Excellent	Excellent	Excellent	Excellent	Excellent	Excellent	Excellent	Excellent	Excellent
Resistance to strong inorganic acids	Excellent	Excellent	Excellent	Excellent	Excellent	Excellent	Excellent	Excellent	Excellent
Resistance to weak organic acids	Poor	Poor	Poor	Poor	Poor	Poor	Poor	Poor	Poor
Resistance to strong organic acids	Poor	Poor	Poor	Poor	Poor	Poor	Poor	Poor	Poor
Resistance to weak alkalies	Good	Good	Good	Good	Good	Good	Good	Good	Good
Resistance to strong alkalies	Good	Good	Good	Good	Good	Good	Good	Good	Good
Does it corrode copper?	No	No	No	No	No	No	No	No	No
Does it corrode lead?	No	No	No	No	No	No	No	No	No
Does it corrode aluminum?	No	No	No	No	No	No	No	No	No
Does it corrode tin?	No	No	No	No	No	No	No	No	No
Solvents	Petroleum and organic solvents	Petroleum and organic solvents	Petroleum and organic solvents	Petroleum and organic solvents	Petroleum and organic solvents	Petroleum and organic solvents	Petroleum and organic solvents	Petroleum and organic solvents	Petroleum and organic solvents
Physical Properties									
Specific gravity (at 60° F) (liquid phase)	0.840	0.840	0.840	0.845	0.845	0.845	0.845	0.845	0.845
Flash (closed) (deg cent)	232	232	232	232	232	232	232	232	232
Flash (open) (deg cent)	238	238	238	238	238	238	238	238	238
Fire (deg cent)	272	272	272	272	272	272	272	272	272
Viscosity at 210° F (centistokes)	12.9	12.9	12.9	13.0	13.0	13.0	13.0	13.0	13.0
Melting point (° C minimum)	77	77	77	75	75	75	68	68	68
Color (Lovibond 1/4-in. cell)	5	40	200	5	40	200	5	40	200
Penetration (needle at 77° F)	10–15	10–15	10–15	15–25	15–25	15–25	25–35	25–35	25–35
Maximum operating temperature (deg cent)	63	63	63	61	61	61	55	55	55

Table 13. Properties of Insulating Materials—Liquids

	Askarels		Petroleum Oils					
Name	Pyranol 1476	Pyranol 1467	Transformer Oil	Sunoco Transformer Oil	Sun Capacitor and Light-cable Oil	Sun XX Heavy-cable Impregnating Oil	Sun XXX Heavy-cable Impregnating Oil	Switch Oil
General uses	Dielectric liquid for capacitor impregnation	Dielectric and cooling medium for transformers, bushings, regulators, and the like	Insulation and coolant for transformers, terminals, etc.	Insulation and coolant for transformers, terminals, etc.	Liquid dielectric and oil-filled cable impregnant	Liquid dielectric and solid-type cable impregnant	Liquid dielectric and solid-type cable impregnant	Insulation and coolant for switchgear
Chemical Properties								
Neutralization number, strong acid	0.00	0.00	0.00	0.00	0.00	0.00	0.00	0.00
Neutralization number, strong base	0.00	0.00	0.03	0.00	0.00	0.00	0.00	0.03
Saponification number	0.00	0.00	Trace	0.00	0.00	0.00	0.00	Trace
Sludge, ASTM D-670-42T, bomb %	0.00	0.00	0.00	0.05				0.00
Chlorides and sulfates	0.00	0.00	0.00	0.00	0.00	0.00	0.00	0.00
Free and corrosive sulfur	0.00	0.00	0.00	0.00	0.00	0.00	0.00	
Steam emulsion number				15–25				
Physical Properties								
Specific gravity, 60/60° F	1.500	1.565	0.892	0.904	0.907	0.934	0.946	0.885
Color, transmitted light	Yellow tinge	Light yellow	Light	Pale yellow	Pale yellow	Med. yellow	Light red	
Viscosity/100° F (centistokes)				11.1	21.0	58.8	1,090	
Viscosity/210° F (centistokes)	6.7	8.8		2.22	3.31	19.3	29.6	
Thermal Properties								
Flash point, open cup (deg cent)	None	None	135	135	155	240	265	152
Pour point (deg cent)	10	Lower than −35	−40	−50	−50	−10	−5	−40
Specific heat/15.5° C (60° F)	0.42	0.42	0.43	0.43	0.43	0.42	0.41	0.43
Thermal capacity (watt-hr/° C/lb)	0.22	0.22	0.23	0.23	0.23	0.22	0.22	0.23
Thermal coefficient of cubic expansion/° C (0 to 100° C)	0.0007	0.0007	0.0007	0.0007	0.0007	0.0007	0.0007	0.0007
Electrical Properties								
Power factor at 100° C (%)	1.0 or less	1 or less	0.1–1.0	0.1–1.0	0.1–1.0	0.1–1.0	0.1–1.0	0.1–1.0
Resistivity (megohm-cm) at 85° C	Greater than $1,000 \times 10^9$	Greater than $1,000 \times 10^9$	1×10^{12} to 30×10^{12}	1×10^{12} to 30×10^{12}	1×10^{12} to 30×10^{12}	1×10^{12} to 30×10^{12}	1×10^{12} to 30×10^{12}	1×10^{12} to 30×10^{12}
Dielectric strength (volts/mil), 0.1-in. gap, at 20° C	350 or higher	350 or higher	300–400	300–400	300–400	300–400	300–400	300–400
Dielectric strength (volts/mil), 0.1-in. gap, at 85° C	350 or higher	350 or higher	300–400	300–400	300–400	300–400	300–400	300–400

NOTE. Data are based on ASTM procedures and are typical of the oils listed.

Table 14. Properties of Insulating Materials—Gases

Name	Air	Argon	Carbon Dioxide	"Freon 11"	"Freon 12"	"Freon 21"
			General Properties			
Components (using chemical symbols)	(N_2, 78.03%) + (O_2, 20.99%) + (A, 0.933%) + traces CO_2, H_2, Ne, He, Kr, Xe	A	CO_2	CCl_3F	CCl_2F_2	$CHCl_2F$
Uses as insulation	All conductors surrounded by air, traces or bubbles in other insulations		Residual gas in liquid dielectrics		Dielectric in compressed gas condensers, sometimes mixed with nitrogen	
			Electrical Properties			
Dielectric strength,* 60 cycles, RMS volts/mil, spheres, atmospheric pressure, 0.001-in. gap	330	200	400	1,140	910	
0.005-in. gap	146	60	175	510	410	
0.010-in. gap	121		145	465	340	
0.100-in. gap	66	723	79	230	185	
			Chemical Properties			
Is it inert or active?	Active	Inert	Inert	Inert	Inert	Inert
Is it explosive with oxygen?	No	No	No	No	No	No
Does it accelerate corrosion of iron when in water solution?	Yes	No	Yes, at temperatures above 150° F	Slightly	Slightly	Slightly
Does it accelerate corrosion of copper when in water solution?	No	No	No	Slightly	Slightly	Slightly
Does it accelerate corrosion of lead when in water solution?	No	No	Yes			
Does it accelerate corrosion of tin when in water solution?	No	No	Yes			

* The dielectric-strength values for air are probably quite accurate. Values for other gases, however, are questionable, since in most cases ratios were used and may not be constant for the full range of gaps given. It is probable, however, that the values are accurate enough for most engineering purposes.

NOTE. Dielectric constants of gases are substantially 1.00 at pressures and temperatures well removed from the critical values, which is accurate enough for most engineering purposes. There are some differences in the third and fourth decimal places.

Table 14. Properties of Insulating Materials—Gases—*Continued*

Name	Air	Argon	Carbon Dioxide	"Freon 11"	"Freon 12"	"Freon 21"
Physical Properties						
Molecular weight	Approx. 29	39.944	44.010	137.4	120.9	102.9
Solubility, % by volume at atmospheric pressure, 20° C, in water		4.00	93.6		4.8	146
In petroleum oil			Approx. 1.5			
Specific heat, C_p	0.2407 at 19.5° C	0.125 at 17.8° C	0.1998 at 20° C	0.134	0.145	0.139
Specific heat, C_v	0.1708 at 16° C	0.075 at 15° C	0.1543 at 18° C			
Thermal resistivity (C°/watt/cm^3)	4,140	6,240	6,880	11,900 at 30° C	10,400 at 30° C	10,200 at 30° C
Viscosity (centipoises)	0.018 at 20° C	0.0221 at 23° C	0.014 at 20° C	0.0111 at 30° C	0.0127 at 30° C	0.0116 at 30° C
Critical temperature (deg cent)	−140.7	−122	31.34	198	111.5	178.5
Critical pressure (lb/sq in.)	5,470	706	1,070	635	582	750
Specific gravity at 20° C atmospheric pressure compared to water	Approx. 0.0012	0.00165	Approx. 0.00181	0.0057	0.0050	0.0042

Table 15. Properties of Insulating Materials—Gases—*Continued*

Name	"Freon 22"	"Freon 113"	"Freon 114"	Helium	Hydrogen	Neon
General Properties						
Components (using chemical symbols)	$CHClF_2$	$C_2Cl_3F_3$	$C_2Cl_2F_4$	He	H_2	Ne
Uses as insulation					Hydrogen-cooled generators	Gas-tube lights, signs, etc.
Electrical Properties						
Dielectric strength,* 60 cycles, RMS volts/mil, spheres, atmospheric pressure, 0.001-in. gap	500	990	1,060		287	252
0.005-in. gap	225	440	475		127	
0.010-in. gap	185	370	400		105	
0.100-in. gap	101	200	215		57	
Chemical Properties						
Is it inert or active?	Inert	Inert	Inert	Inert	Active	Inert
Is it explosive with oxygen?	No	No	No	No	Yes	No
Does it accelerate corrosion of iron when in water solution?	Slightly	Slightly	Slightly	No	No	No
Does it accelerate corrosion of copper when in water solution?	Slightly	Slightly	Slightly	No	No	No
Does it accelerate corrosion of lead when in water solution?				No	No	No
Does it accelerate corrosion of tin when in water solution?				No	No	No

* The dielectric-strength values for air are probably quite accurate. Values for other gases, however, are questionable, since in most cases ratios were used and may not be constant for the full range of gaps given. It is probable, however, that the values are accurate enough for most engineering purposes.

NOTE. Dielectric constants of gases are substantially 1.00 at pressures and temperatures well removed from the critical values, which is accurate enough for most engineering purposes. There are some differences in the third and fourth decimal places.

Table 15. Properties of Insulating Materials—Gases—*Continued*

Name	"Freon 22"	"Freon 113"	"Freon 114"	Helium	Hydrogen	Neon
Physical Properties						
Molecular weight	86.5	187.4	170.9	4.003	2.0160	20.183
Solubility, % by volume at atmospheric pressure, 20° C, in water	31		1.8	1.47	1.94	
In petroleum oil				Approx. 5.0	Approx. 5.5	
Specific heat, C_p	0.151	0.154	0.159	1.251 at 18° C	3.415 at 15° C	
Specific heat, C_v				0.752 at 18° C	2.430 at 21.2° C	$C_p/C_v = 1.642$
Thermal resistivity (C°/watt/cm^3)	8,500 at 30° C	12,800 at 30° C	8,900 at 30° C	6,930	5,770	2,150
Viscosity (centipoises)	0.0131 at 30° C	0.0104 at 30° C	0.0116 at 30° C	0.0197 at 23° C	0.008872 at 20.8° C	0.0297
Critical temperature (deg cent)	96.0	214.1	145.7	−268	−240	−228.75
Critical pressure (lb/sq. in.)	716	495	474	33.2	188	3,960
Specific gravity at 20° C (Atmospheric pressure compared to water)	0.0036	0.0077	0.0070	0.000165	0.0000833	0.000833

Table 16. Properties of Insulating Materials—Gases—*Continued*

Name	Nitrogen	Sulfur Hexafluoride
General Properties		
Components (using chemical symbols)	N_2	SF_6
Uses as insulation	Residual gas in paper-insulated power cables and other electrical equipment using liquid dielectrics	Cooling of electrical apparatus or machinery
Electrical Properties		
Dielectric strength,* 60 cycles, RMS volts/mil, spheres, atmospheric pressure, 0.001-in. gap	380	640
0.005-in. gap	170	290
0.010-in. gap	142	240
0.100-in. gap	77	130
Chemical Properties		
Is it inert or active?	Inert	Inert
Is it explosive with oxygen?	No	No
Does it accelerate corrosion of iron when in water solution?	No	No
Does it accelerate corrosion of copper when in water solution?	No	No
Does it accelerate corrosion of lead when in water solution?	No	No
Does it accelerate corrosion of tin when in in water solution?	No	No

Name	Nitrogen	Sulfur Hexafluoride
Physical Properties		
Molecular weight	28.016	146
Solubility, % by volume at atmospheric pressure, 20° C, in water	1.63	Very low
In petroleum oil	Approx. 10.0	
Specific heat, Cp	0.2481 at 20° C	0.148 at 0° C
Specific heat, Cv	0.1774 at 20° C	
Thermal resistivity (C°/watt/cm³)	4,110	
Viscosity (centipoises)	0.01765 at 23° C	
Critical temperature (deg cent)	−147	54
Critical pressure (lb/sq in.)	493	640
Specific gravity at 20° C, atmospheric pressure compared to water	0.00116	0.00603

* The dielectric-strength values for air are probably quite accurate. Values for other gases, however, are questionable, since in most cases ratios were used and may not be constant for the full range of gaps given. It is probable, however, that the values are accurate enough for most engineering purposes.

NOTE: Dielectric constants of gases are substantially 1.00 at pressures and temperatures well removed from the critical values, which is accurate enough for most engineering purposes. There are some differences in the third and fourth decimal places.

21. BIBLIOGRAPHY

AIEE Standard #1.
ASTM Standards on Electrical Insulating Materials.
Curtis, H. L., Electrical Resistivity of Insulating Materials, *Trans. AIEE*, Vol. 46, p. 1039 (1927). (Bibliography of 74 items.)
Dawson, T. R., and B. D. Porritt, *Rubber: Physical and Chemical Properties*. Research Association of British Rubber Manufacturers, Croyden, England (1935).
DelMar, W. A., W. F. Davidson, and R. H. Marvin, Electric Strength of Solid and Liquid Dielectrics (with bibliography), *Trans. AIEE*, Vol. 46, p. 1061 (1927).
Fowle, F. E., *Smithsonian Physical Tables*. Smithsonian Institution, Washington, D. C.
Gemant, A., *Liquid Dielectrics*. John Wiley (1933).
Hoover, P. L., Mechanism of Breakdown of Dielectrics, *Trans. AIEE*, Vol. 45, p. 983 (1926).
Kauppi, T. A., and G. L. Moses, Organo-silicon Compounds for Insulating Electrical Machines, *Trans. AIEE*, Vol. 64, p. 90 (1945).
Kline, D. M., and Staff, *Plastics Catalog*. Plastics Catalog Corporation, N. Y.
Littleton, J. T., and G. W. Morey, *The Electrical Properties of Glass*. John Wiley (1933).
Loeb, L. B., *The Nature of a Gas*. John Wiley (1933).
Memmler, K., *The Science of Rubber*. Reinhold (1934).
National Research Council, *International Critical Tables*. McGraw-Hill
Simonds, H. R., *Industrial Plastics*. Pitman (1939).
Simonds, H. R., and Ellis, *Handbook of Plastics*. VanNostrand (1943).
Wagner, K. W., Physical Nature of Electrical Breakdown of Solid Dielectrics, *Trans. AIEE*, Vol. 41, p. 288 (1922).
Whitehead, J. B., *Impregnated-paper Insulation*. John Wiley (1935).

GENERAL CHARACTERISTICS OF MAGNETIC MATERIALS

By R. L. Sanford

22. CLASSIFICATION

Materials may be assigned to one of three classes on the basis of magnetic properties, namely:

a. **Diamagnetic,** having a permeability less than that of a vacuum. Bismuth is a material in this class.

b. **Paramagnetic,** having a permeability slightly greater than that of a vacuum, and approximately independent of the magnetizing force.

c. **Ferromagnetic,** having a permeability considerably greater than that of a vacuum, which will in general be a function of the previous magnetic history of the material.

Only the ferromagnetic materials find general commercial application on the basis of their magnetic properties.

DEFINITIONS. The following general concepts and definitions, which have been adopted by the ASTM and are quoted by permission, are useful in considering the magnetic properties of materials.

Aging, magnetic.* The normal or accelerated change in magnetic properties of a magnetic material resulting from the effects of time and temperature.

Aging coefficient, core loss.† The percentage change in the specific core loss resulting from continued heating at 100 deg cent for 600 hr.

Coercive force, H_c. The magnetizing force required to bring the induction to zero in a magnetic material which is in a symmetrically cyclically magnetized condition.

Coercive force, intrinsic, H_{ci}. The value of H_c at the point on the hysteresis loop where B_i equals zero.

Coercivity. That property of a material measured by the maximum value of the coercive force.

Core loss (iron loss), P_c. The power expended in a magnetic material subjected to a varying magnetizing force.

Core loss, apparent, P_a. The product of the rms induced voltage and rms exciting current for a circuit containing a ferromagnetic core. The induced voltage must be approximately sinusoidal. Specific apparent core loss may be expressed in volt-amperes (apparent watts) per unit weight.

Core loss, incremental, P_Δ. The core loss in a magnetic material when subjected simultaneously to a biasing and an incremental magnetizing force.

* When used with reference to core loss, this term, unless otherwise modified, implies an increase in loss. When used with reference to permeability or remanence, the term, in a positive sense, indicates a decrease in these quantities.

† The term, in a positive sense, indicates an increase in loss.

Core loss, specific, $P_{B;f}$. The total power in watts expended per unit weight of magnetic material in which there is a symmetrical harmonically varying induction of a specified maximum value B in kilogausses at a specified frequency f.

Cyclically magnetized condition. A magnetic material is in a cyclically magnetized condition when, under the influence of a magnetizing force which varies cyclically between two specific limits, its successive hysteresis loops are identical.

Demagnetization curve. That portion of the hysteresis loop which lies between the residual induction point, B_r, and the coercive force point, H_c. Points on this curve are designated by the coordinates B_d and H_d.

Demagnetizing force, H_d. A magnetizing force applied in such a direction as to reduce the remanent induction in a magnetized body. (*See* Demagnetization Curve.)

Diamagnetic material. A material having a permeability less than that of a vacuum.

Eddy current loss, P_e. That portion of the core loss due to currents circulating in the magnetic material as a result of electromotive forces induced by varying induction.

Electrical sheet or strip. A thin magnetic material of such composition and so processed as to be suitable for the construction of laminated magnetic cores.

Energy-product curve, magnetic. The curve obtained by plotting the product of the coordinates of the demagnetization curve (B_dH_d) as abscissas against the induction B_d.*, †

Ferromagnetic material. A material which, in general, exhibits hysteresis phenomena, and whose permeability is dependent on the magnetizing force.

Form Factor, *ff*. The ratio of the effective value of a symmetrically alternating quantity to its half-period average value.

Gauss (plural gausses). The cgs unit of magnetic induction. (*See* Induction, magnetic.)

Gilbert. The cgs unit of magnetomotive force. (*See* Magnetomotive force.)

Hysteresis, magnetic. The property of a magnetic material by virtue of which the magnetic induction for a given magnetizing force depends upon the previous conditions of magnetization.

Hysteresis loop. A curve (usually with rectangular coordinates) which shows, for a magnetic material in a cyclically magnetized condition, for each value of the magnetizing force, two values of the magnetic induction, one when the magnetizing force is increasing, the other when it is decreasing.

Hysteresis loss, P_h. The power expended in a magnetic material, as a result of magnetic hysteresis, when the magnetic induction is cyclic.

Hysteresis loss, incremental. The hysteresis loss occurring in a magnetic material when subjected simultaneously to a biasing and an incremental magnetizing force.

Induction curve, normal. A curve depicting the relation between normal induction and magnetizing force.

Induction, incremental, B_Δ. One half the algebraic difference of the maximum and minimum values of the magnetic induction during a cycle in a magnetic material which is subjected to a biasing magnetizing force. Twice the incremental induction is indicated by ΔB, so that:

$$B_\Delta = \frac{\Delta B}{2}$$

Induction, intrinsic (ferric induction), B_i. The excess of the induction in a magnetic material over the induction in vacuum, for a given value of the magnetizing force.

The equation for intrinsic induction is:

$$B_i = B - \mu_v H$$

Induction, magnetic (magnetic flux density), B. Flux per unit area through an element of area at right angles to the direction of the flux.

The cgs unit of induction is called the gauss (plural gausses) and is defined by the equation:

$$B = \frac{d\phi}{dA}$$

Under sinusoidal conditions $B_{max.}$ may be calculated as follows:

$$B_{max.} = \frac{E \times 10^8}{4\pi NAf}$$

where E is in rms volts, and N is the number of turns.

Induction, normal, B. The limiting induction, either positive or negative, in a magnetic material which is in a symmetrically cyclically magnetized condition.

* $(B_dH_d)_{max.}$ corresponds to the maximum value of the external energy.

† The demagnetization curve is usually plotted to the left of the vertical axis (negative values of H_d), and the energy-product curve to the right.

Induction, remanent, B_d. *See* Remanence.

Induction, residual, B_r. The magnetic induction corresponding to zero magnetizing force in a magnetic material which is in a symmetrically cyclically magnetized condition.

Induction, saturation, B_s. The maximum intrinsic induction possible in a material.

Magnetic field strength, H. *See* Magnetizing force.

Magnetic flux, ϕ. A condition in a medium produced by a magnetomotive force, such that when altered in magnitude a voltage is induced in an electric circuit linked with the flux.

The cgs unit of magnetic flux is called the maxwell and is defined by the equation:

$$e = -N\frac{d\phi}{dt} \times 10^{-8}$$

where e = induced emf in volts.

$d\phi/dt$ = time rate of change of flux in maxwells per second.

Flux density, magnetic, B. *See* Induction, magnetic.

Magnetizing force, H. Magnetomotive force per unit length. The cgs unit is called the oersted and is defined by the equation:

$$H = \frac{d\mathfrak{F}}{dl}$$

where $\mathfrak{F}$ is in gilberts and l in centimeters. For a toroid, or at the center of a long solenoid, the magnetizing force in oersteds may be calculated as follows:

$$H = \frac{0.4\pi NI}{l}$$

where I is in amperes and l is in centimeters.

Magnetizing force, incremental, H_Δ. One-half the algebraic difference of the maximum and minimum values during a cycle of a periodic magnetizing force. The whole difference or twice the incremental magnetizing force is indicated by the symbol ΔH. Thus:

$$H_\Delta = \frac{\Delta H}{2}$$

Magnetomotive force, $\mathfrak{F}$. That which tends to produce a magnetic field. In magnetic testing it is most commonly produced by a current flowing through a coil of wire, and its magnitude is proportional to the current and to the number of turns, N.

The cgs unit of magnetomotive force is called the gilbert and is defined by the equation:

$$\mathfrak{F} = 0.4\pi NI$$

where I is in amperes.

Magnetomotive force may also result from a magnetized body.

Maxwell. The cgs unit of magnetic flux.

$$1 \text{ maxwell} = 10^{-8} \text{ webers}$$

Oersted, H. The cgs unit of magnetizing force.

Paramagnetic material. A material having a permeability which is slightly greater than that of a vacuum, and which is approximately independent of the magnetizing force.

Permeability, a-c, μ_{ac}. Alternating-current permeability is variously defined, and the values obtained for a given material depend on the methods and conditions of measurement. As measured by the Standard Methods of Test for Magnetic Properties of Iron and Steel (ASTM designation: A 34), it is the ratio of the maximum value of induction to the maximum value of the magnetizing force for a material in a symmetrically cyclically magnetized condition.

It is sometimes defined as the ratio of the rms flux density to the rms magnetizing force. Some of the factors which affect a-c permeability are thickness of laminations, frequency, and resistivity.

Permeability, differential, μ_d. The slope of the normal induction curve.

Permeability, incremental, μ_Δ. The ratio of the cyclic change in magnetic induction to the corresponding cyclic change in magnetizing force when the mean induction differs from zero.

Permeability, initial, μ_o. The slope of the normal induction curve at zero magnetizing force.

Permeability, intrinsic, μ_i. The ratio of intrinsic normal induction to the corresponding magnetizing force.

Permeability, maximum, μ_m. The maximum value of the normal permeability for a given material.

Permeability, normal, μ.* The ratio of the normal induction to the corresponding magnetizing force.

In the cgs system the flux density in a vacuum is numerically equal to the magnetizing force, and consequently the magnetic permeability is numerically equal to the ratio of the flux density to the magnetizing force. Thus:

$$\mu = \frac{B}{H}$$

Permeability, relative, μ. Permeability of a body relative to that of a vacuum. In the cgs system the relative permeability is the same as the normal permeability.

Permeability, reversible, μ_r. The incremental permeability when the cyclic change in induction is vanishingly small.

Permeability, space, μ_v. The factor that expresses the ratio of magnetic induction to magnetizing force in a vacuum. In the cgs electromagnetic system of units the permeability of a vacuum is arbitrarily taken as unity.

Permeance, $\mathcal{P}$. The ratio of the flux through any cross-section of a tubular portion of a magnetic circuit bounded by lines of force and by two equipotential surfaces to the magnetic potential difference between the surfaces, taken within the portion under consideration.

Reluctance, $\mathcal{R}$. The reciprocal of permeance

$$\mathcal{R} = \frac{\mathcal{F}}{\phi}$$

For uniform μ and A

$$\mathcal{R} = \frac{l}{\mu A}$$

where A is area in square centimeters, and l is length in centimeters.

Reluctivity, ν. The reciprocal of the permeability of a medium.

Remanence, remanent induction, B_d. The magnetic induction which remains in a magnetic circuit after the removal of an applied magnetomotive force.†

Residual induction. *See* Induction, residual.

Retentivity. The property of a magnetic material measured by the maximum value of the residual induction.

Stabilization. A treatment of a magnetic material designed to increase the permanency of its magnetic properties or condition.

Symmetrically Cyclically Magnetized Condition. A magnetized material is in a symmetrically cyclically magnetized condition when it is cyclically magnetized and the limits of the applied magnetizing forces are equal and of opposite sign, so that the limits of induction are equal and of opposite sign.

Weber. The practical unit of magnetic flux. It is the amount of magnetic flux which, when linked at a uniform rate with a single-turn electric circuit during an interval of 1 sec, will induce in this circuit an electromotive force of 1 volt.

$$1 \text{ weber} = 10^8 \text{ maxwells}$$

23. CRITERIA OF MAGNETIC QUALITY

In judging the suitability of magnetic materials for various applications it is customary to make use of various standard forms of data.

Normal induction curves or tables constitute the locus of the extremes of a series of hysteresis loops (O–a–a'–a'', etc., in Fig. 1). A typical normal induction curve for 1 per cent silicon steel is shown in Fig. 2. In using such data the effect of air gaps or joints in the magnetic circuit should be considered, since these joints may have appreciable reluctance.

Normal permeability curves or tables are obtained from the normal induction curves by dividing the flux density in gausses by the magnetizing force in oersteds.

$$\mu = \frac{B}{H}$$

A typical normal permeability curve is shown in Fig. 2.

* In a nonisotropic medium the permeability is a function of the orientation of the medium, since, in general, the magnetizing force and the magnetic flux are not parallel.

† If there is an air gap in the magnetic circuit, the remanence will be less than the residual induction.

Hysteresis loops represent the relation, in an initially demagnetized material, between induction and magnetizing force, as the induction is varied from a positive maximum to a negative maximum and back to the initial point. Figure 1 shows a family of such hysteresis loops taken at various values of peak induction. The areas of such loops represent the loss due to magnetic hysteresis and are sometimes used in comparing magnetic materials. The values of the residual induction and coercive force, criteria particularly useful in selecting permanent magnet materials, may also be found from such loops.

Core loss curves or tables are usually prepared to show the relation between the summation of the hysteresis and eddy current components of the core loss, and maximum induction, at a definite frequency, and when the induction is varying harmonically. In using core loss data obtained from tests on small samples, it is important to remember that extra losses, not present in the test sample, will be found in most apparatus because of eddy currents in various parts of the magnetic path or in the structural members, stray flux in structural parts, and possibly harmonics in the flux wave.

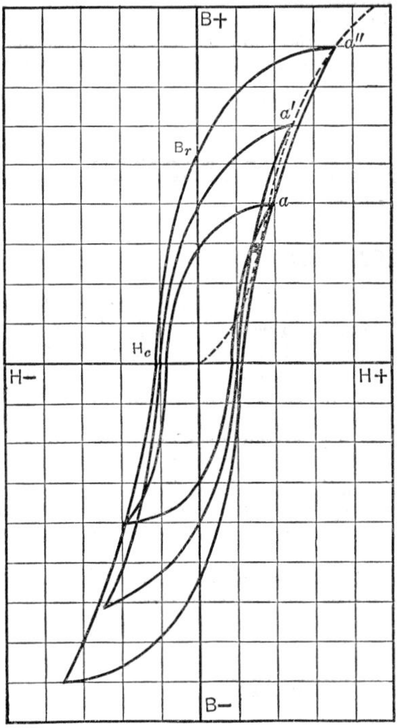

FIG. 1. Typical Hysteresis Loops

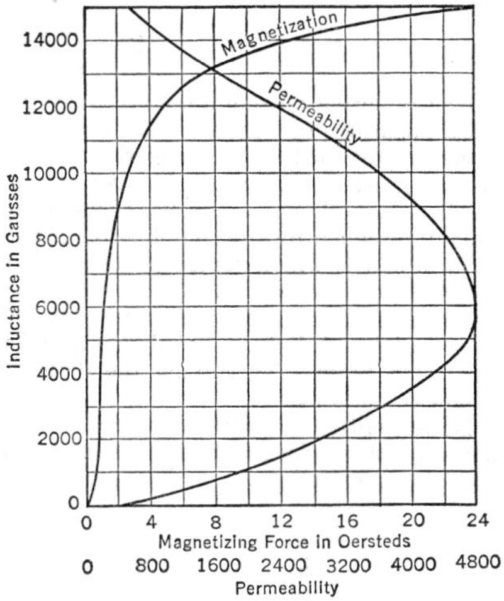

FIG. 2. Normal Induction and Permeability Curves for Low-silicon Steel

Incremental permeability data are frequently useful in the design of equipment where an alternating flux is superposed on a steady flux. The length and number of the joints or air gaps in the magnetic circuit must be carefully considered in using such data, which are frequently of much value in the design of radio equipment, such as audio transformers and filter reactors.

The value of permanent magnet materials is frequently judged on the basis of the value of the product of the induction and the magnetizing force at that point on the demagnetization curve where this product is a maximum. This figure represents the maximum amount of useful energy that a magnet material can deliver; it is sometimes known as the **maximum energy product** and written $(B_d H_d)$ max.

24. CLASSES OF COMMERCIAL MAGNETIC MATERIAL

Commercial magnetic materials may be divided into three groups for convenience of description, namely: (*a*) the **non-retentive** or magnetically "soft" materials, (*b*) the **retentive** or magnetically "hard" materials, and (*c*) those **special** magnetic materials having properties suited only to certain special applications.

a. The non-retentive or magnetically "soft" materials are by far the most important, on the basis of the amount used. Materials falling in this class range from cast iron, which is one of the poorest, to Hipernik, which is one of the best, and include also cast steel, ordinary low-carbon sheet or plate, ingot iron, and the electrical (silicon) sheet or strip materials. These materials are in general characterized by low coercive force, comparatively low hysteresis loss, relatively high permeability, and, usually, a fairly high saturation induction; they are used in the magnetic circuits of electrical apparatus.

b. The retentive or magnetically "hard" materials find application in the permanent magnets of meters and relays, sound-reproducing units, magnetos, and other equipment where a steady magnetic field is required and it is not convenient to obtain this field electromagnetically. These materials are characterized by high coercive force, comparatively high hysteresis loss, and, usually, low permeability.

c. Special alloys include those materials having special temperature-permeability relations, used in compensating for changes in flux in a magnetic circuit due to temperature changes, materials having unusually good properties only at very low inductions. materials especially suited to operation at very high flux densities, and materials which can be used at very high frequencies.

NON-RETENTIVE MAGNETIC MATERIALS

By R. L. Sanford *

25. CLASSIFICATION

Non-retentive magnetic materials are used in the cores and magnetic circuits of electrical apparatus and equipment. The choice of material to be used for a particular application is governed by the properties required and cost. The basic ferromagnetic metals are iron, nickel, and cobalt, but it rarely happens that any one of these metals by itself has the magnetic characteristics required for a given purpose. The desired properties are usually obtained by the combination of one or more of these metals with other elements to form alloys, which generally have to be subjected to certain heat treatments to bring about the desired results.

Commercial manufacturers issue technical information to assist in the proper selection of material and design of equipment, and it is always advisable to obtain the latest information before undertaking the design or development of new apparatus involving magnetic materials.

Non-retentive materials employed in commercial practice may be considered under the following three classifications: (*a*) solid core materials, (*b*) electrical sheet and strip, (*c*) special-purpose alloys.

26. SOLID CORE MATERIALS

These materials are used for the cores of d-c electromagnets, relays, field frames of d-c machines, etc. The principal requirement is high permeability, particularly at relatively high inductions. For the majority of uses it is also desirable that the hysteresis be low. The principal materials employed are soft iron, relay steel, cast steel, cast iron, and an alloy of approximately 35 per cent cobalt and 65 per cent iron, known as ferrocobalt. Ferrocobalt is characterized by very high permeability in the upper part of the normal induction range and a saturation induction approximately 10 per cent greater than that of pure iron. Its cost is relatively high, however, and its use is limited in general to pole pieces in which a very high induction is required. Cast iron has a relatively low permeability and is used principally in field frames when cost is of primary importance and extra weight is not objectionable. Several varieties of soft iron are available, such as Norway iron, Armco iron, and Swedish charcoal iron. Relay steels contain from 1 to 3.25 per cent of silicon to reduce aging. Electrolytic iron may also be used. All these materials are relatively pure iron, low in carbon and other impurities. Cast steel should be low in carbon, not over 0.1 to 0.2 per cent and should contain only the usual small amounts of the ordinary impurities. Cast iron is high in carbon, about 3 per cent, and also contains about 3 per cent of silicon and varying percentages of phosphorus, manganese, and sulfur.

The best magnetic properties are obtained by a suitable annealing treatment. The properties of cast iron can be greatly improved by malleablizing, a process that converts a large part of the carbon to the amorphous form.

Typical normal induction curves for solid core materials are given in Fig. 1. Other properties are given in Tables 1 to 5.

* With some material from previous edition by D. F. Miner.

Table 1. General Data on Magnetic Properties of Commercial Materials

Figures represent approximate average properties; individual samples may differ somewhat

Property	Perm-alloy Western Electric	Hipernik Westing-house Electric	Si Steel 0.60 Loss	Si Steel 0.66 Loss	Si Steel 0.72 Loss	Si Steel 0.82 Loss	Si Steel 1.01 Loss	Si Steel 1.17 Loss	Si Steel 1.30 Loss	Pure Iron Armco or Norway	Low-carbon Sheet or Shapes	Cast Steel An-nealed	Cast Iron An-nealed
Specific gravity, grams per cc	8.6	8.3	7.5*	7.5*	7.5*	7.5*	7.5*	7.7*	7.7*	7.85	7.85	7.8	7.0
Sheet weight, lb per sq ft in 29 gage†	0.63	0.60	0.54	0.54	0.54	0.54	0.54	0.55	0.55	0.56	0.56	....	....
Ultimate strength, lb per sq in.	....	65,000	75,000	75,000	72,000	70,000	60,000	45,000	40,000	40,000	45,000	60,000	25,000
Yield point, lb per sq in.	....	20,000	55,000	55,000	52,000	50,000	35,000	22,000	20,000	20,000	25,000	30,000	....
Per cent elongation (2 in.)	....	50	3	3	3.5	5	14	18	20	25	18	20	....
Erichsen draw, mm (ductility index)	....	8	2.5	2.5	2.5	4.5	5	7	7	10	9	....	....
Resistivity, microhms per cc	22	45	60	58	56	48	41	26	18	10.7	13	15	100
Steinmetz, hysteresis coefficient	0.0001	0.00015	0.00046	0.00051	0.00056	0.00065	0.00081	0.00088	0.001	0.002	0.003	0.005	0.012
Typical coercive force, oersteds B (max) = 10,000 gausses	0.04	0.06	0.32	0.36	0.39	0.46	0.56	0.70	0.85	1.0	2.0	5	11
W 10/60–29 gage iron loss, watts per lb‖	....	0.25	0.60	0.66	0.72	0.82	1.01	1.17	1.30	....	....	....	....
Maximum aging, core loss	....	Nil	Nil	Nil	Nil	2%	3%	5%	....	....	....	....	....
Typical maximum permeability	100,000	90,000	8,000	8,000	6,500	6,000	5,500	5,200	5,000	4,500	2,500	1,500	500
Saturation (ferric induction)	11,000	15,600	19,500	19,500	19,800	20,200	20,500	21,000	21,200	21,600	21,200	21,000	14,000
Typical initial permeability	9,000	6,000	750	700	600	500	400	350	325	275	250	175	125
Approximate relative cost, in percentage	....	2,500	100	93	80	73	64	50	43	36	32	20	17
Approximate percentage silicon	‡	§	4.5	4.5	4.4	3.5	2.5	1.0	0.5	low	low	0.4	2.0
Typical application	Telephone equipment	High-quality audio and instrument transformers, relays	Distribution and power transformers		High-efficiency rotating machines and small transformers		Small motors, a-c magnets, starting transformers			Pole pieces, relays	Fields and frames of d-c and synchronous machines	Frames and solid poles	Frames

* A.S.T.M. Standard, not quite actual.

† For magnetic sheet of low silicon content (2% and less) the table of sheet iron gages and weights should be used. For material of medium and high silicon grades (over 2%) the gage dimensions in the sheet iron table are satisfactory but the weights should be multiplied by 0.975.

‡ Seventy-eight per cent nickel.

§ Fifty per cent nickel.

‖ The guaranteed maximum loss at 10,000 gausses, 60 cycles, in 29 gage, is used to designate the grade of electrical sheet steel.

Table 2 (Part 1). Approximate Frequency and Induction Core Loss Factors for 29 Gage (0.014 in.) Material

Induction, Gausses	15 Cycles	25 Cycles	40 Cycles	50 Cycles	60 Cycles	80 Cycles	100 Cycles	160 Cycles	250 Cycles	Induction Lines per Sq In.
10	0.35×10^{-6}	0.60×10^{-6}	0.10×10^{-5}	0.13×10^{-5}	0.16×10^{-5}	0.22×10^{-5}	0.28×10^{-5}	0.47×10^{-5}	0.80×10^{-5}	64
16	$.10\times10^{-5}$	$.17\times10^{-5}$	$.30\times10^{-5}$	$.36\times10^{-5}$	$.46\times10^{-5}$	$.64\times10^{-5}$	$.80\times10^{-5}$	$.14\times10^{-4}$	$.24\times10^{-4}$	103
25	$.29\times10^{-5}$	$.50\times10^{-5}$	$.80\times10^{-5}$	$.10\times10^{-4}$	$.12\times10^{-4}$	$.18\times10^{-4}$	$.23\times10^{-4}$	$.40\times10^{-4}$	$.72\times10^{-4}$	161
40	$.80\times10^{-5}$	$.14\times10^{-4}$	$.23\times10^{-4}$	$.28\times10^{-4}$	$.35\times10^{-4}$	$.51\times10^{-4}$	$.64\times10^{-4}$	$.12\times10^{-3}$	$.21\times10^{-3}$	258
64	$.22\times10^{-4}$	$.41\times10^{-4}$	$.65\times10^{-4}$	$.82\times10^{-4}$	$.10\times10^{-3}$	$.15\times10^{-3}$	$.18\times10^{-3}$	$.32\times10^{-3}$	$.58\times10^{-3}$	423
100	$.58\times10^{-4}$	$.10\times10^{-3}$	$.18\times10^{-3}$	$.22\times10^{-3}$	$.26\times10^{-3}$	$.38\times10^{-3}$	$.47\times10^{-3}$	$.80\times10^{-3}$	$.15\times10^{-2}$	645
160	$.15\times10^{-3}$	$.28\times10^{-3}$	$.44\times10^{-3}$	$.56\times10^{-3}$	$.66\times10^{-3}$	$.92\times10^{-3}$	$.12\times10^{-2}$	$.21\times10^{-2}$	$.38\times10^{-2}$	1,030
250	$.36\times10^{-3}$	$.64\times10^{-3}$	$.10\times10^{-2}$	$.13\times10^{-2}$	$.16\times10^{-2}$	$.22\times10^{-2}$	$.30\times10^{-2}$	$.50\times10^{-2}$	$.90\times10^{-2}$	1,610
400	$.90\times10^{-3}$	$.15\times10^{-2}$	$.26\times10^{-2}$	$.30\times10^{-2}$	$.38\times10^{-2}$	$.56\times10^{-2}$	$.72\times10^{-2}$	$.12\times10^{-1}$	$.22\times10^{-1}$	2,580
640	$.20\times10^{-2}$	$.35\times10^{-2}$	$.62\times10^{-2}$	$.75\times10^{-2}$	$93.\times10^{-2}$	$.13\times10^{-1}$	$.17\times10^{-1}$	$.30\times10^{-1}$	$.53\times10^{-1}$	4,230
1,000	$.43\times10^{-2}$	$.74\times10^{-2}$	$.13\times10^{-1}$	$.16\times10^{-1}$	$.22\times10^{-1}$	$.28\times10^{-1}$	$.36\times10^{-1}$	$.61\times10^{-1}$	.11	6,450
1,600	$.10\times10^{-1}$	$.16\times10^{-1}$	$.28\times10^{-1}$	$.36\times10^{-1}$	$.43\times10^{-1}$	$.60\times10^{-1}$	$.78\times10^{-1}$	.14	.26	10,300
2,000	$.14\times10^{-1}$	$.24\times10^{-1}$	$.40\times10^{-1}$	$.52\times10^{-1}$	$.62\times10^{-1}$	$.88\times10^{-1}$	.11	.20	.38	12,900
2,500	$.20\times10^{-1}$	$.34\times10^{-1}$	$.60\times10^{-1}$	$.75\times10^{-1}$	$.92\times10^{-1}$	.13	.16	.29	.55	16,100
3,000	$.29\times10^{-1}$	$.46\times10^{-1}$	$.81\times10^{-1}$	.10	.12	.18	.22	.37	.74	19,400
4,000	$.45\times10^{-1}$	$.73\times10^{-1}$	.13	.17	.20	.28	.36	.64	1.25	25,800
5,000	$.66\times10^{-1}$	.11	.19	.25	.29	.41	.54	.96	1.80	32,300
6,000	$.90\times10^{-1}$	.15	.26	.34	.40	.56	.72	1.30	2.50	38,700
7,000	.12	.20	.35	.44	.52	.72	.96	1.70	3.30	45,200
8,000	.15	.25	.44	.55	.66	.92	1.20	2.10	4.1	51,600
9,000	.19	.31	.54	.68	.82	1.14	1.48	2.60	5.1	58.100
10,000	.23	.38	.64	.82	1.00	1.40	1.80	3.3	6.2	64,500
11,000	.28	.46	.76	.98	1.20	1.70	2.25	4.0	7.7	71,000
12,000	.33	.55	.91	1.15	1.45	2.05	2.8	4.9	9.5	77,400
13,000	.39	.64	1.07	1.37	1.70	2.5	3.3	5.9	11.3	83,900
14,000	.45	.74	1.25	1.60	2.00	2.90	3.9	7.1	13.0	90,300
15,000	.52	.86	1.47	1.90	2.35	3.40	4.6	8.5	15.0	96,800

NOTE: To find the approximate iron loss at any induction or frequency, multiply the loss, at 60 cycles, 10,000 gausses, by the factor shown in the table. Values are most accurate for about 4 per cent silicon steel, but are reasonably exact for lower silicon steels and may be used to give reasonably approximate values for Hipernik.

Table 2 (Part 2). Approximate Frequency and Induction Core Loss Factors for 29 Gage (0.014 in.) Material

Induction, Gausses	400 Cycles	640 Cycles	1000 Cycles	1600 Cycles	2500 Cycles	4000 Cycles	6400 Cycles	10,000 Cycles	15,000 Cycles	Induction Lines per Sq In.
10	0.14×10^{-4}	0.30×10^{-4}	0.56×10^{-4}	0.13×10^{-3}	0.39×10^{-3}	0.79×10^{-3}	0.16×10^{-2}	0.38×10^{-2}	0.80×10^{-2}	64
16	$.47\times10^{-4}$	$.95\times10^{-4}$	$.20\times10^{-3}$	$.43\times10^{-3}$	$.10\times10^{-2}$	$.24\times10^{-2}$	$.56\times10^{-2}$	$.11\times10^{-1}$	$.30\times10^{-1}$	103
25	$.14\times10^{-3}$	$.28\times10^{-3}$	$.58\times10^{-3}$	$.14\times10^{-2}$	$.32\times10^{-2}$	$.80\times10^{-2}$	$.19\times10^{-1}$	$.44\times10^{-1}$	.10	161
40	$.40\times10^{-3}$	$.85\times10^{-3}$	$.18\times10^{-2}$	$.44\times10^{-2}$	$.95\times10^{-2}$	$.24\times10^{-1}$	$.58\times10^{-1}$	.14	.31	258
64	$.11\times10^{-2}$	$.24\times10^{-2}$	$.51\times10^{-2}$	$.11\times10^{-1}$	$.27\times10^{-1}$	$.64\times10^{-1}$	.15	.38	.84	423
100	$.28\times10^{-2}$	$.59\times10^{-2}$	$.13\times10^{-1}$	$.29\times10^{-1}$	$.67\times10^{-1}$	.15	.38	.96	2.0	645
160	$.70\times10^{-2}$	$.15\times10^{-1}$	$.32\times10^{-1}$	$.70\times10^{-1}$	.16	.38	.95	2.3	5.0	1,030
250	$.17\times10^{-1}$	$.36\times10^{-1}$	$.74\times10^{-1}$	.17	.38	.93	2.2	5.3	11.0	1,610
400	$.41\times10^{-1}$	$.86\times10^{-1}$	.18	.39	.88	2.2	5.2	13.0	27.0	2,580
640	$.95\times10^{-1}$	.20	.42	.90	2.2	5.1	13.0	30.0	62.0	4,230
1,000	.22	.45	.94	2.2	5.0	12.0	29.0	72.0	....	6,450
1,600	.48	1.0	2.15	4.9	11.5	28.0	65.0	....	....	10,300
2,000	.71	1.5	3.2	7.4	17.0	42.0	....	....	....	12,900
2,500	1.10	2.2	4.6	11.0	25.0	63.0	....	....	....	16,100
3,000	1.40	3.1	6.4	15.0	35.0	....	....	....	....	19,400
4,000	2.30	5.0	11.0	26.0	58.0	....	....	....	....	25,800
5,000	3.50	7.7	16.0	38.0	82.0	....	....	....	....	32,300
6,000	5.0	10.0	22.0	50.0	....	....	....	....	....	38,700
7,000	6.6	14.0	29.0	65.0	....	....	....	....	....	45,200
8,000	8.5	18.0	37.0	....	....	....	....	....	....	51,600
9,000	10.9	22.0	46.0	....	....	....	....	....	....	58,100
10,000	13.4	27.0	58.0	....	....	....	....	....	....	64,500
11,000	16.0	32.0	....	....	....	....	....	....	....	71,000
12,000	19.0	38.0	....	....	....	....	....	....	....	77,400
13,000	22.0	....	....	....	....	....	....	....	....	83,900
14,000	25.	....	....	....	....	....	....	....	....	90,300
15,000	29.	....	....	....	....	....	....	....	....	96,800

NOTE: To find the approximate iron loss at any induction or frequency, multiply the loss, at 60 cycles, 10,000 gausses, by the factor shown in the table. Values are most accurate for about 4 per cent silicon steel, but are reasonably exact for lower silicon steels and may be used to give reasonably approximate values for Hipernik.

Table 3. Loss in Watts per Pound of Electrical Sheets as a Function of Gage Commercial Grades in Common Use

The grade is designated by the guaranteed maximum iron loss when tested at 10,000 gausses, 60 cycles, ASTM method

Induction in Gausses

Grade	Gage	3000	4000	5000	6000	7000	8000	9000	10,000	11,000	12,000	13,000	14,000	15,000
0.72	30	0.096	0.160	0.237	0.317	0.402	0.495	0.60	0.71	0.84	1.00	1.18	1.38	1.63
.72	29	.096	.160	.237	.317	.402	.495	.60	.71	.84	1.00	1.18	1.38	1.63
.72	28	.104	.172	.250	.329	.417	.52	.63	.75	.89	1.05	1.24	1.45	1.71
.72	27	.112	.184	.255	.344	.438	.54	.66	.79	.94	1.11	1.31	1.53	1.80
.72	26	.124	.196	.271	.360	.464	.57	.70	.83	.99	1.17	1.38	1.62	1.91
.72	25	.137	.214	.302	.392	.492	.61	.74	.89	1.06	1.25	1.48	1.74	2.05
.72	24	.152	.231	.320	.410	.53	.66	.80	.97	1.17	1.36	1.59	1.86	2.19
.82	30	.098	.158	.237	.318	.415	.53	.64	.78	.94	1.10	1.27	1.47	1 68
.82	29	.098	.162	.242	.330	.430	.55	.66	.81	.97	1.13	1.32	1.53	1.76
.82	28	.104	.173	.253	.347	.450	.57	.69	.85	1.01	1.18	1.38	1.61	1.85
.82	27	.112	.183	.264	.368	.477	.60	.73	.90	1.06	1.24	1.45	1.69	1.95
.82	26	.123	.198	.285	.390	.51	.64	.77	.96	1.11	1.30	1.52	1.78	2.07
.82	25	.137	.218	.313	.423	.55	.68	.83	1.03	1.17	1.38	1.61	1.89	2.21
.82	24	.154	.240	.350	.468	.60	.74	.89	1.10	1.26	1.48	1.72	2.02	2.38
1.01	30	.129	.209	.313	.423	.54	.66	.81	.98	1.16	1.36	1.58	1.86	2.15
1.01	29	.132	.214	.319	.435	.56	.69	.84	1.01	1.19	1.40	1.62	1.91	2.20
1.01	28	.139	.224	.331	.452	.58	.72	.87	1.06	1.24	1.47	1.70	1.99	2.30
1.01	27	.150	.237	.346	.475	.61	.75	.91	1.12	1.30	1.54	1.79	2.08	2.42
1.01	26	.164	.253	.363	.498	.64	.79	.96	1.18	1.37	1.62	1.89	2.20	2.56
1.01	25	.180	.271	.384	.53	.68	.84	1.01	1.24	1.45	1.71	2.02	2.34	2.72
1.01	24	.198	.293	.408	.56	.73	.90	1.09	1.30	1.55	1.82	2.16	2.51	2.91
1.17	30	.144	.247	.350	.450	.58	.74	.91	1.10	1.31	1.55	1.82	2.11	2.50
1.17	29	.155	.258	.360	.460	.61	.76	.93	1.15	1.34	1.58	1.86	2.16	2.55
1.17	28	.165	.268	.370	.480	.63	.78	.96	1.17	1.37	1.62	1.90	2.22	2.61
1.17	27	.178	.285	.388	.50	.65	.80	1.00	1.21	1.43	1.69	2.00	2.35	2.74
1.17	26	.197	.308	.425	.53	.70	.87	1.07	1.28	1.52	1.82	2.18	2.55	2.98
1.17	25	.224	.342	.47	.60	.78	.97	1.19	1.43	1.70	2.04	2.43	2.84	3.31
1.17	24	.260	.385	.53	.72	.91	1.15	1.39	1.65	1.96	2.33	2.78	3.32	3.96
1.30	30	.172	.280	.395	.53	.68	.85	1.03	1.24	1.49	1.77	2.10	2.47	2.91
1.30	29	.178	.290	.410	.55	.71	.88	1.07	1.28	1.53	1.81	2.15	2.55	3.00
1.30	28	.186	.305	.435	.58	.74	.92	1.12	1.33	1.58	1.88	2.22	2.67	3.14
1.30	27	.198	.325	.465	.62	.79	.98	1.19	1.41	1.67	2.00	2.37	2.86	3.38
1.30	26	.216	.350	.50	.67	.85	1.06	1.28	1.53	1.82	2.18	2.60	3.11	3.67
1.30	25	.235	.375	.55	.75	.95	1.17	1.41	1.69	2.02	2.42	2.90	3.47	4.10
1.30	24	.260	.420	.61	.85	1.09	1.37	1.68	2.03	2.44	2.94	3.54	4.25	5.10

Table 4. Apparent Incremental Permeability, Standard E and I Laminations

Length of magnetic circuit = 5.6 in.
Two lap joints
29 gage material, 0.66 grade

A-c Induction	D-c Magnetizing Force, in oersteds							
	0	.5	1.0	1.5	2	3	4	5
10 gausses....	650	600	480	390	330	260	220	210
100 gausses....	1230	1000	750	580	475	360	300	280
1000 gausses....	2400	1450	1070	850	700	560	500	480

Data by American Rolling Mill Company, Middletown, Ohio.
Data shown are for 60 cycles; 1000 cycles shows almost identical result.

Table 5. Incremental Permeability D-c Tests

29 gage material, 0.60 grade

	$B=10$	$B=30$	$B=100$	$B=300$	$B=1000$	$B=3000$
Steady magnetizing force = 0.0 oersted. .	1000	1440	1970	2770	4460	7320
Steady magnetizing force = 0.1 oersted. .	1000	1350	1910	2550	4030	6650
Steady magnetizing force = 0.3 oersted. .	840	1090	1470	1985	3120	5200
Steady magnetizing force = 1.0 oersted. .	578	740	934	1130	1570	2750
Steady magnetizing force = 3.0 oersteds.	200	204	214	250	450	1000
Steady magnetizing force = 10.0 oersteds.	62	63	65	70	100	310

Data from Westinghouse Electric tests.
Ring samples, no air gaps.
60 cycle a-c tests with no air gaps have checked these figures closely.

27. ELECTRICAL SHEET AND STRIP

The terms **electrical sheet** and **electrical strip** are commonly used to designate silicon-iron alloys produced in sheet or strip form and used as core materials in a-c apparatus, such as transformers, generators, motors, electromagnets, or relays. The principal requirements are high permeability, low hysteresis, and high resistivity. The several grades differ mainly with respect to their silicon content, which ranges from about 0.5 per cent to approximately 4.5 per cent. Alloys containing the higher percentages of silicon are practically non-aging; i.e., the permeability and losses do not change with time. The required magnetic properties are produced by annealing.

By a suitable combination of cold-rolling and heat treatment, materials are produced in which the crystal axes are given a definite orientation. Such material has considerably better properties, when magnetized in the preferred direction, than the ordinary grades.

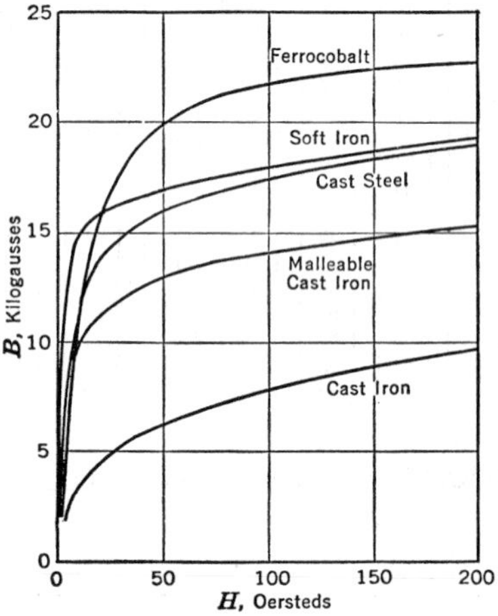

FIG. 1. Typical Normal Induction Curves for Solid Core Materials

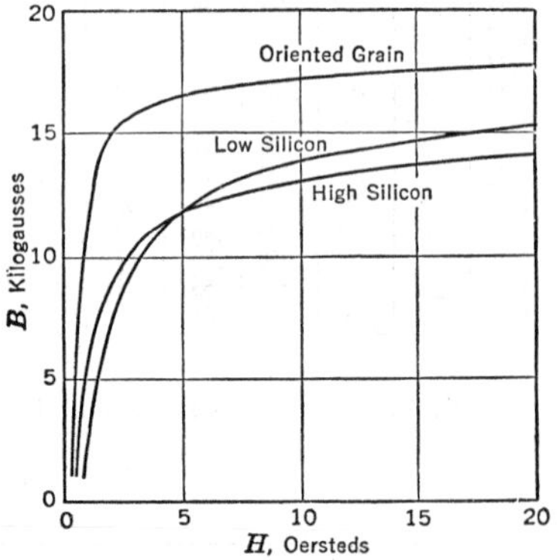

FIG. 2. Typical Normal Induction Curves for Three Types of Electrical Sheet

Figure 2 shows typical normal induction curves for two grades of electrical sheet and orientated-grain material. The improvement in the orientated-grain material is particularly conspicuous in the upper part of the normal induction curve.

The various grades of electrical sheet and strip are usually sold on the basis of guaranteed maximum values of total core loss, as determined in accordance with the specifications of the American Society for Testing Materials. The common designations of the various grades are armature, electrical, motor, dynamo, and transformer. The transformer grades are further subdivided into classes denoted by numerals corresponding to the core loss under standard conditions. Armature, electrical, and motor grades are used principally in small motors, a-c magnets, and starting transformers. The dynamo grade is used in high-efficiency rotating machines and small transformers. The transformer grades are used in power and radio transformers.

The thickness of electrical sheets is ordinarily given in terms of the U. S. Standard sheet gage (see Section 1). For magnetic sheet of low silicon content (2 per cent and less) the table of sheet-iron gages and weights should be used. For material of medium- and high-silicon grades (above 2 per cent) the gage dimensions in the sheet-iron table are satisfactory, but the weights will be somewhat less. The approximate weight can be obtained by using the sheet-iron figures multiplied by a correction factor of 0.975.

28. SPECIAL-PURPOSE ALLOYS

For certain applications special alloys have been developed which, after proper heat treatment, have superior properties in certain ranges of magnetization. For instance, alloys of nickel and iron with possible small percentages of molybdenum or chromium have very high values of initial and maximum permeability. Alloys of this class, which may have from 70 to 80 per cent of nickel, are called Permalloys. An alloy of 50 per cent nickel and 50 per cent iron is called Hipernik. Another alloy having a small percentage of copper in its composition is called Mumetal. The characteristics of these alloys differ in detail, but in general they have high initial and maximum permeability, low hysteresis, and low saturation values. They are particularly applicable for use at low inductions Typical permeability curves are given in Fig. 3.

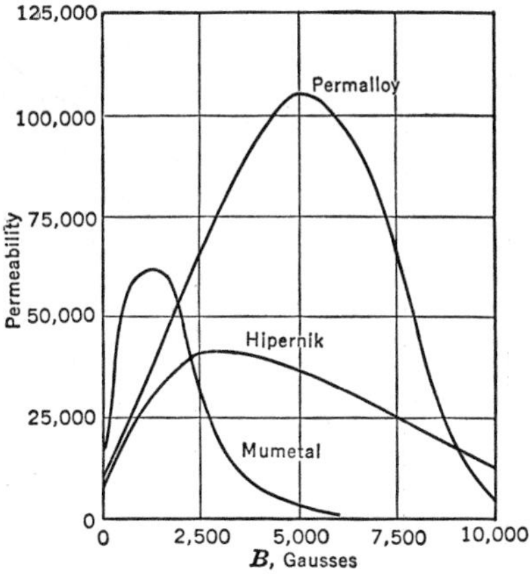

FIG. 3. Typical Permeability Curves for Three Special-purpose Alloys

A certain alloy of nickel, cobalt, and iron, after suitable heat treatment, has very nearly constant permeability for inductions below 1,000 gausses and is called Perminvar. The 50–50 nickel-iron alloy can also be heat-treated so as to have similar characteristics.

An alloy of equal proportions of iron and cobalt, called Permendur, has high permeability, which persists at higher values of induction than that of the nickel-iron alloys.

Another series of magnetic alloys of copper, nickel, and iron is temperature sensitive, having an approximately linear relation between permeability and temperature. These alloys are called Thermalloys. Their principal use is in the compensation of watt-hour meters for temperature variations. They are also used in certain types of thermal relays.

29. BIBLIOGRAPHY

The best information on the properties of commercial magnetic materials is to be obtained from latest technical bulletins issued by the various producers of magnetic materials, such as Allegheny Ludlum Steel Corp., American Rolling Mill Co., Carnegie-Illinois Steel Corp., Carpenter Steel Co., Follansbee Steel Corp., Republic Steel Corp., and Wheeling Steel Corp.

Magnetic properties of sheet steel at frequencies of several hundred cycles per second are given by Robert Pohl, *J. IEE* (London), Vol. 94–2, p. 118 (1947).

PERMANENT MAGNET MATERIALS *

By I. F. Kinnard and J. H. Goss

30. TYPES AVAILABLE

Before 1930 the magnetic properties of the commerically available permanent magnet materials depended, as do ordinary steels, on the hardening of carbides. Since then,

* See Section 6. Articles 23 to 26, for permanent magnet construction and design.

alloys with superior magnetic properties hardened by other metallurgical effects have been developed. The maximum external energy has been increased twenty times over the available energy in the original carbon steel. High-energy-value ductile alloys have been developed, as well as an electrically non-conducting material composed of metallic oxides, making a total of over twenty different materials, each with a number of possible variations for special purposes. These are summarized in Table 1.

TYPES OF ALLOYS. Quench-hardening Steels. The alloys shown in Fig. 1 may be classified as high-carbon, quench-hardening steels. They are usually produced in the form of hot-rolled bar or sheet stock but may be cast to final shape. Magnets of these steels may be forged or formed hot from bar stock. They may also be blanked or machined to a limited extent in the annealed conditions. Magnetic properties are developed by

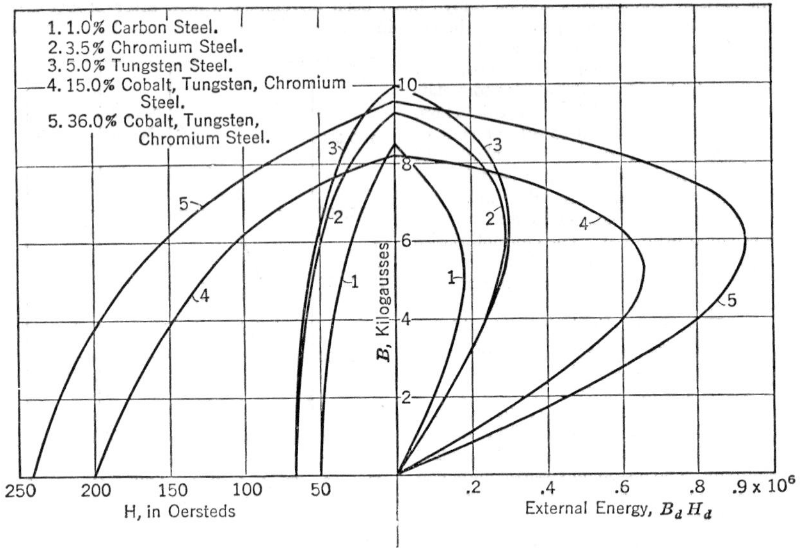

Fig. 1. Demagnetization and External Energy Curves for Quench-hardening Steels

quenching, which is usually followed by tempering at a low temperature to stabilize the material. Temperatures must be closely controlled in heat treatments for consistent results, and care must be taken to avoid overheating, decarburization, and cracking.

Al-Ni-Fe Carbon-free Alloys. The aluminum-nickel-iron alloys, commonly called Alnicos, (Fig. 2), offer external magnetic energy values up to approximately five times those of the best quench-hardening alloys. These are carbon-free alloys whose magnetic hardness is realized by a controlled cooling from a high temperature. This yields a critically stressed internal structure, which is developed as intermetallic compounds start to separate out from a previously solid-solution structure. In addition, the controlled cooling is supplemented by drawing, as indicated in Table I, to develop the properties fully. In Alnico V and its modifications the outstanding properties are developed by a directional heat treatment consisting of cooling the material in a magnetic field through a region just below its Curie point. These alloys are comparatively weak and brittle and in the cast state usually exhibit a very coarse crystalline structure. Because of their physical properties they are generally cast or sintered as close to their final shape as possible and critical dimensions ground. Holes in such magnets are usually cast by means of cores, or inserts of steel are used for mounting or other purposes.

Newer Machinable Alloys. The alloys shown in Fig. 3 differ from both the above groups in that they are machinable and all except Vectolite and Remalloy are ductile in their optimum magnetic state. Desired magnetic properties are obtained in these alloys by a heat treatment at a high temperature followed by quenching and subsequent drawing at a lower temperature. The Cunifes and Vicalloy require cold-working between quenching and drawing to develop full magnetic properties. These alloys are therefore made into rod, wire, or strip stock, from which magnets are fabricated. Cunico is usually produced in these forms but may also be cast to shape and machined to final dimensions.

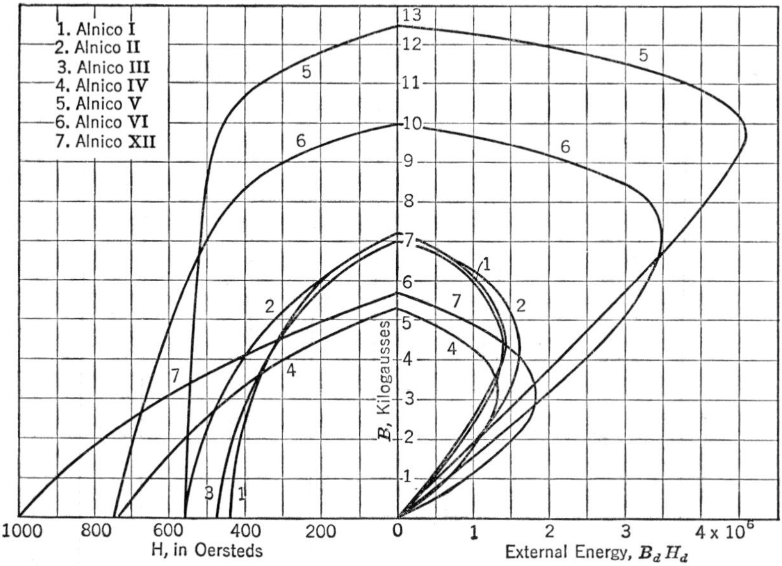

Fig. 2. Demagnetization and External Energy Curves for Al-Ni-Fe Carbon-free Alloys

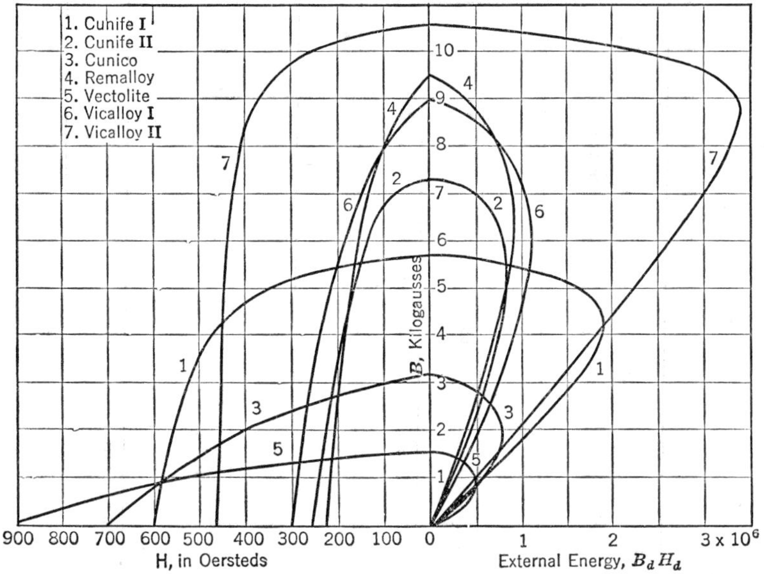

Fig. 3. Demagnetization and External Energy Curves for Newer Machinable Alloys

Table 1. Properties of Retentive Magnetic Materials

Material	Typical Composition, Per Cent	Coercive Force, Oersteds	Residual Induction, Gausses	Maximum Energy Product ($B \times H$)	Relative Magnetic Energy per Unit Weight	Forging Temperature, deg cent	Annealing Temperature, deg cent	Hardening Temperature, deg cent	Quenching Medium	Magnetizing Force, Oersteds	Mechanical Prop. in Magnetic State	Commercial Methods of Fabrication
1.00% carbon...	1.00 C, bal. Fe	50	8,500	180,000	1.0	850	875, cool in air	830	Oil	300	Hard, strong	Hot-forge, punch, machine
3 1/2% chromium	3.5 Cr, 1 C, bal. Fe	65	9,000	290,000	1.6	1,000	900, cool in air	830	Oil	300	Hard, strong	Hot-forge, punch, cast, machine
5% Tungsten...	5 W, 1 C, bal. Fe	65	10,000	300,000	1.6	1,000	850, cool in air	840	Oil	400	Hard, strong	Hot-forge, punch, machine
15% Cobalt chromium	0.9 C, 10 Cr, 15 Co, bal. Fe	200	8,200	650,000	3.4	1,000	750, cool in air	*	Oil	1,000	Hard, strong	Hot-forge, punch, cast, machine
36% Cobalt chromium tungsten	0.85 C, 36 Co, 3.5 Cr, 3 W, bal. Fe	210	9,000	930,000	4.9	1,000	950, cool in air	950	Oil	1,000	Hard, strong	Hot-forge, punch, cast, machine
Alnico I........	12 Al, 20 Ni, 5 Co, bal. Fe	400	7,100	1,400,000	8.7	Cannot be forged or machined, is cast to shape and ground	Heat to 1,100, cool in air			2,000	Hard, brittle	Cast, sinter, grind
Alnico II.......	10 Al, 17 Ni, 12.5 Co, 6 Cu, bal. Fe	540	7,200	1,600,000	9.7	See Alnico I	Heat to 1,100, cool in air			2,000	Hard, brittle	Cast, sinter, grind
Alnico III......	12 Al, 25 Ni, bal. Fe	480	7,100	1,400,000	8.7	See Alnico I	Heat to 1,200, quench, age at 700			2,000	Hard, brittle	Cast, sinter, grind
Alnico IV.......	12 Al, 28 Ni, 5 Co, bal. Fe	700	5,200	1,300,000	5.1	See Alnico I	Heat to 1,200, cool in air			2,000	Hard, brittle	Cast, sinter, grind

Alnico V........	8 Al, 14 Ni, 3 Cu, 24 Co, bal. Fe	575	12,000	4,500,000	26.4	See Alnico I	Heat to 1,300, cool in 1,000 H magnetic field, age at 600			2,000	Hard, brittle	Cast, grind
Alnico VI.......	8 Al, 14 Ni, 3 Cu, 24 Co, 1 Ti, bal. Fe	750	10,000	3,500,000	20.6	See Alnico V				2,000	Hard, brittle	Cast, grind
Alnico XII......	8 Ti, 18 Ni, 35 Co, 6 Al, bal. Fe	1,000	5,800	1,800,000	10.1	See Alnico I				3,000	Hard, brittle	Cast, grind
Cunife I........	20 Ni, 60 Cu, 20 Fe	600	5,800	1,960,000	9.8	Requires cold-working to develop properties	1,100	Age at 600	Oil	2,400	Ductile, malleable	Cold-roll, machine, punch
Cunife II	20 Ni, 50 Cu, 27.5 Fe, 2.5 Co	250	7,300	780,000	3.9	See Cunife I	1,100	600	Oil	2,400	Ductile, malleable	Cold-roll, machine, punch
Cunico..........	29 Co, 21 Ni, 50 Cu	710	3,200	850,000	4.4		1,100	650	Oil	3,200	Ductile, malleable	Cold-roll, machine, punch, cast
Vectolite.......	44 Fe_3O_4, 30 Fe_2O_3, 26 Co_3O_4	900	1,600	500,000	7.8	Heat-treat in magnetic field at 315				4,800	Soft, brittle	Sinter, grind, machine
Vicalloy I, II....	52 Co, 10 V, bal. Fe	300 470	9,000 10,600	1,000,000 3,330,000	5.5	See Cunife for Vicalloy II	Quench from 800 to 1,300, age from 600 to 800			1,500	Ductile, malleable	Cold-roll, machine, punch
Remalloy.......	12 Co, 17 Mo, bal. Fe	240	10,300	1,100,000	5.8	1,100	Quench from 1,300, age at 650			1,000	Hard, brittle	Forge, cast, grind, machine
Silmanal........	86.7 Ag, 8.8 Mn, 4.5 Al	H_{ci} 6,000	B_r 550	85,000	0.4	See Cunife I	Quench and age			20,000	Ductile, malleable	Cold-roll, machine, punch
Platinum iron...	77.8 Pt, 22.2 Fe	H_{ci} 1,570	5,830	2,280,000	5.4					20,000	Hard, brittle	Cast, grind
Platinum cobalt.	76.7 Pt, 23.3 Co	H_{ci} 2,650	4,530	2,400,000	5.6					20,000	Hard, brittle	Cast, grind

* Heat rapidly to 1,150, cool in air, heat slowly to 725, cool in air, heat to 1,000, cool in moving air to 300, and quench in oil (according to Darwins, Ltd., Sheffield, England).

Remalloy is hard and brittle in its final state but is machinable when quenched or annealed. It may be hot-rolled, formed, forged, or cast and machined to final shape. Vectolite differs from all other magnetic materials in that it is made from electrically non-conducting metallic oxides. The powdered material is pressed into shape and sintered at a high temperature. This is followed by heat treatment in a magnetic field, which gives it directional properties. Vectolite is relatively brittle at all stages but may be ground or machined.

Precious-metal Alloys. A few permanent magnet materials have been made with precious metals (Fig. 4). These obtain their magnetic properties through the formation of a superlattice structure and are essentially non-directional.

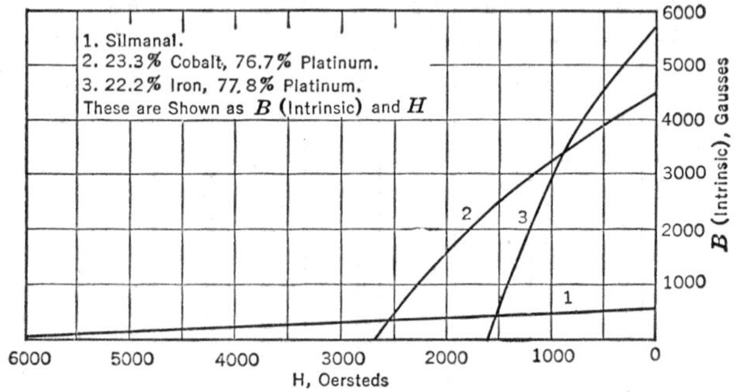

FIG. 4. Demagnetization and External Energy Curves for Precious-metal Alloys

Silmanal is a ductile, relatively soft, machinable alloy which has found application as the rotor of permanent magnet magnetometers and as polarizing magnets in relay applications, because of its nearly straight-line demagnetization characteristic and the very high magnetizing forces required to magnetize or demagnetize the material.

31. PROPERTIES OF PERMANENT MAGNET MATERIALS

MAGNETIC PROPERTIES. Most of the information required in permanent magnet design is obtained from the upper left-hand quadrant of the hysteresis loop of a material (Fig. 5). The properties usually of the most interest are the residual induction (B_r), the

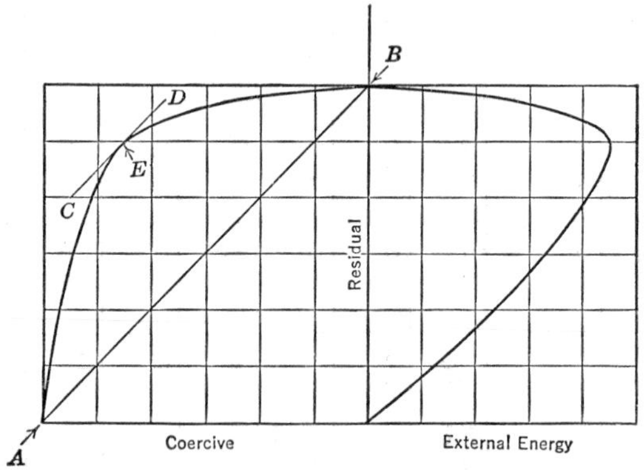

FIG. 5. Graphical Location of Maximum External Energy Point on Demagnetization Curve

coercive force (H_c), and the maximum product of B and H. This is the point of maximum energy value and will be located approximatelv as shown graphically in Fig. 5. These values are listed in Table 1. H_{ci} is defined in Article 22.

The flux output of magnets varies inversely with temperature. In the usual working range this variation is approximately 0.025 per cent flux per degree centigrade; in Alnico V, however, this figure varies with temperature and may be as high as 0.05 per cent per degree centigrade.

PHYSICAL PROPERTIES. The physical properties have been discussed under the main heading, Types of Alloys. The range from hard and brittle materials to relatively soft and ductile materials (Table 1) affects the method of fabrication and consequently is an important economic consideration.

32. BIBLIOGRAPHY

Jellinghaus, New Alloys with High Coercive Forces, *Z. f. tech. Physik*, Vol. 17, No. 2, p. 33 (1935).
Kato and Takei, Permanent Oxide Magnet and Its Characteristics, *J. IEE (Japan)*, Vol. 53, p. 408 (1933) and U. S. Patent No. 1,976,230 (1934).
Koster, Permanent Magnet Materials on the Basis of Precipitation Hardening, *Stahl und Eisen*, Vol. 53, p. 849 (1933).
Nesbitt, E. A., Vicalloy—A Workable Alloy for Permanent Magnets, *Metals Tech.* (Feb. 1946).
Rabbitt and Fujwara, Report on New Magnetic Steel Discovered by Dr. T. Mishima, *Far Eastern Rev.*, Vol. 28, p. 451 (1932), and Mishima, U. S. Patents Nos. 2,027,994–5–6–7–8–9 and 2,028,000 (1936).
Webb, C. E., Recent Developments in Magnetic Materials, *J. IEE*, Vol. 82, p. 303 (1938).
Zumbusch, W., Permanent Magnets Made from Iron-nickel-cobalt-(titanium) Alloys with Preferred Magnetic Direction, *Archiv f. Eisenhüttenwesen* (Sept., 1942).
Nickel Alloys Steels, International Nickel Co., New York.
Dahl, Cunife, Patent 2,196,824.
H. T. Faus, Silmanal, Patent 2,274,804.
Howe, Alnico VI, Patent 2,264,038.
Jonas, Alnico V, Patent 2,295,082.

SECTION 3

ELECTRIC CIRCUITS AND ELECTRIC LINES

ALTERNATING CURRENTS

By Irven Travis

LINEAR CIRCUITS

By Irven Travis

NON-LINEAR CIRCUITS

By Irven Travis

POLYPHASE SYSTEMS

By Irven Travis

SYMMETRICAL COMPONENTS

By Raymond C. R. Schulze

CIRCUITS OF ELECTRICAL MACHINES

By Irven Travis

GENERAL EQUATIONS OF TRANSMISSION LINES

By Irven Travis

POWER-SYSTEM STABILITY

By S. B. Crary

(Mr. M. H. McGrath has assisted in checking this section.)

ELECTRIC CIRCUITS AND ELECTRIC LINES

ALTERNATING CURRENTS

By Irven Travis

1. GENERAL DEFINITIONS, PERIODIC QUANTITIES

The definitions given below are applicable to currents, electromotive forces, potential differences, or any other function of time. These definitions are those recommended by the Sectional Committee on Electrical Definitions of the AIEE.

a. An **oscillating quantity** is a quantity which as a function of some independent variable (such as time) alternately increases and decreases in value, always remaining within finite limits.

b. A **periodic quantity** is an oscillating quantity the values of which recur for equal increments of the independent variable.

c. The **period** of a periodic quantity is the smallest value of the increment of the independent variable which separates recurring values of the quantity.

d. A **cycle** is the complete series of values of a periodic quantity which occur during a period.

e. The **frequency** of a periodic quantity in which time is the independent variable is the reciprocal of the period.

f. The **angular velocity** of a periodic quantity is the frequency multiplied by 2π.

g. An **alternating quantity** is a periodic quantity which has alternately positive and negative values.

In order to avoid repetition the following statements will be given in terms of electric currents; they apply equally well to electromotive forces, drops of potential, etc.

If a given current is represented by the equation

$$i = \phi(t) \tag{1}$$

and if the function $\phi(t)$ has the property that

$$\phi(t) = \phi(t + T) \tag{2}$$

in which T is a constant, then the current is said to be periodic in time and T is the period. If time is measured in seconds, then $1/T$ represents the number of periods per second and is usually denoted by f. The frequency f may also be said to be the number of cycles per second. If, as often happens, the current is expressible more simply as

$$i = \phi(\omega t) \tag{3}$$

in which ω is defined by

$$\phi(\omega t) = \phi(\omega t + 2\pi) \tag{4}$$

then the constant ω, being equal to 2π divided by the period or 2π times the frequency, is the angular velocity as defined in (*f*); that is

$$\omega = 2\pi f = \frac{2\pi}{T} \tag{5}$$

2. MAXIMUM, AVERAGE, AND RMS OR EFFECTIVE VALUES

The **instantaneous value** of an alternating current is the value of the current at any instant. Throughout this section instantaneous values of current, potential difference, and electromotive force will be designated by lower-case letters i, v, and e, respectively.

The **maximum value** of an alternating current is the numerical value of its maximum instantaneous value. Maximum values will be designated by capital letters with the subscript m.

The **half-period average value** of a symmetrical alternating current * is the absolute value of the algebraic average of the values of the current taken throughout a half period, beginning with a zero value. If the current has more than two zeros during a cycle, that zero shall be taken which gives the largest half-period average value. The expression for the half-period average of a symmetrical alternating current $i(t)$ having a period T is

$$I_{av} = \frac{2}{T}\int_{t_0}^{t_0+\frac{1}{2}T} i(t)\, dt \tag{6}$$

where t_0 is chosen such that $i(t) = 0$ at t_0, and I_{av} is the largest value obtainable.

The square root of the means of the squares of the instantaneous values of an alternating current over a complete period is called the rms, or the *effective* value of the alternating current. In specifying the value of an alternating current as so many amperes, this rms value is always meant unless specifically stated otherwise. In the same manner the square root of the mean of the squares of the instantaneous values of an alternating potential difference over a complete period is called the rms value of the alternating potential difference. When the value of an alternating potential difference is specified as so many volts, this rms value is always meant unless specifically stated otherwise.

The reason for selecting this particular function of the instantaneous values of an alternating current or a potential difference as a measure of the current or the potential difference is that the deflection of all instruments used in a-c measurements is a function of this rms value. (See Measurements, Section 5.) Moreover, the average power dissipated as heat in a resistance r, when an alternating current of rms value I flows through it, is rI^2.

Throughout this section rms values will be designated by capital letters without subscripts.

The general expression for the rms value of an alternating current is

$$I = \sqrt{\frac{1}{T}\int_0^T i^2\, dt} \tag{7}$$

and similarly for an alternating potential difference.

3. FORM FACTOR, CREST OR PEAK FACTOR, DEFORMATION FACTOR

The **form factor** of a symmetrical alternating current is the ratio of the effective value of the current to its half-period average value.

The **peak** or **crest factor** of an alternating current is the ratio of the maximum value of the current to its effective value; this is also called the **amplitude factor.**

The **equivalent sinusoidal current** of a given alternating current is a sinusoid having the same period and the same effective value of the given alternating current.

The **deformation factor** of an alternating current is the ratio *to* the maximum value of the equivalent sinusoidal current *of* the maximum difference between the corresponding values of the current considered and the equivalent sinusoid, when the two are superimposed in such a way as to make this difference a minimum.

4. POWER, POWER FACTOR, VOLT-AMPERES, REACTIVE POWER

Let v be the value at any instant of the potential drop from any point 1 to any other point 2, and let i be the instantaneous value of the current from 1 to 2 at this same instant; then the **power input** at this instant is

$$p = vi \tag{8}$$

When v and i are both positive (i.e., in the direction from 1 to 2, say) or when they are both negative, the power input is positive; but when v is positive and i negative, or vice versa, the power input is negative, i.e., there is an actual power output.

The average value of the product vi over a complete period for both v and i (or over any whole number of periods) is the **average power** input or output, usually called simply the power input or output (input when the average of vi is positive, output when the average

* A symmetrical alternating quantity is one of which all values separated by a half-period have the same magnitude but opposite sign. The term half-period average has no meaning for alternating currents which are not symmetrical.

of *vi* is negative), the word average being understood. That is, the average power input is

$$P = \frac{1}{T}\int_0^T p\,dt = \frac{1}{T}\int_0^T vi\,dt \tag{9}$$

T being a complete period. For the actual measurement of a-c power see Measurements, Section 5.

Only in certain special cases (see below) is the average power input P equal to the product of the rms value V of the potential difference by the rms value I of the current; it can never be greater and as a rule is less. The ratio of the average power P to the product of the rms value V of the potential difference by the rms value I of the current is called the **power factor** of the circuit between the terminals considered, i.e.,

$$\text{Power factor} = \frac{P}{VI} \tag{10}$$

When V is expressed in volts and I in amperes, then P must be in watts; when V is expressed in kilovolts and I in amperes, P must be in kilowatts.

The product of the rms volts across the terminals of a circuit and the rms amperes through it is called the **volt-amperes** or **apparent power** taken by the circuit; this product divided by 1000 is called the kilovolt-ampere input. Or, when V is in volts and I in amperes

$$\text{Volt-amperes} = VI \tag{11}$$

$$\text{Kilovolt-amperes} = \frac{VI}{1000} \tag{11a}$$

Kilovolt-amperes is usually abbreviated kva, kv-a, or KVA, the first form being that recommended by the American Institute of Electrical Engineers and used in this book.

Reactive power has no accepted definition when either the current or the emf is non-sinusoidal. (See below.)

5. SINUSOIDAL CURRENTS AND VOLTAGES, PHASE ANGLE

A **simple sinusoidal current** (simple harmonic current) is an alternating current the instantaneous values of which are equal to the product of a constant and the sine of an angle having values varying linearly with time. Thus

$$i = I_m \sin(\omega t + \theta) \tag{12}$$

where t represents time in seconds, measured from any arbitrarily chosen instant; I_m the maximum value of the current; $\omega = 2\pi f = 2\pi/T$, where f is the frequency in cycles per second and T the period as a fraction of a second; and θ a constant which depends upon the instant chosen as the zero of time.

The quantity I_m is often called the **amplitude** of the sinusoidal current.

The ratio of I_m to the rms value I (the amplitude factor) for a sinusoidal current is $\sqrt{2}$.

The **phase** of a periodic current, for a particular value of the independent variable, is the fractional part of a period through which the independent variable has advanced measured from an arbitrary origin. In the case of a simple sinusoidal current, the origin is usually taken as the last previous passage through zero from the negative to positive direction. The **phase angle** is the angle obtained by multiplying the phase by 2π if the angle is to be expressed in radians, or by 360 deg if the angle is to be expressed in degrees.

The phase angle is thus equal to $\omega t + \theta$. In electric-circuit theory the phase differences between various quantities having the same frequency are of importance. Since these differences in phase are independent of the instant at which they are evaluated, it is customary to refer to the angle θ (which is the phase angle when $t = 0$) as the phase angle of the current given by eq. (12).

In general, when a sine-wave electromotive force is impressed on a circuit, the resulting current is likewise a sine function of time after a very brief interval (see *Handbook of Engineering Fundamentals*, by Eshbach, John Wiley, "Transients"), having the same frequency, but the emf and current do not reach their maximum values simultaneously. Let the current be represented by eq. (12) and let the voltage be given by

$$v = V_m \sin \omega t \tag{12a}$$

where t is the time measured from the instant when $v = 0$ and is increasing in the positive direction. The voltage reaches its maximum value when $t = \pi/2\omega$, while the current reaches its maximum value when $t = (\pi/2\omega) - (\theta/\omega)$. Hence, when θ is positive, the voltage reaches its maximum value θ/ω sec after the current reaches its maximum, or the current reaches its maximum value θ/ω sec before the voltage reaches its maximum; when θ is negative, the current reaches its maximum value θ/ω sec after the voltage reaches its maximum. In the first case the current is said to "lead" the voltage, and in the second case the current is said to "lag" the voltage. The angle θ is called the angular phase difference between the current and voltage.

When the phase difference is zero, the current and voltage are said to be "in phase"; when the phase difference is $\pi/2$ radians or 90 deg, the current and voltage are said to be "in quadrature," when the phase difference is π radians or 180 deg, the current and voltage are said to be "in opposition."

6. POWER AND POWER FACTOR FOR SINUSOIDAL CURRENT AND VOLTAGE

Let the voltage drop from terminal 1 to terminal 2 through any piece of apparatus be $v = \sqrt{2}V \sin (\omega t + \theta_v)$, and the current from terminal 1 to terminal 2 be $i = \sqrt{2}I \sin (\omega t + \theta_i)$, where V and I are the rms values and therefore $\sqrt{2}V$ and $\sqrt{2}I$ are the maximum values. Then the instantaneous power input is

$$p = vi = VI[\cos (\theta_v - \theta_i) - \cos (2\omega t + \theta_v + \theta_i)] \tag{13}$$

A study of Fig. 1 will show the physical meaning of this expression. The average power input is

$$P = VI \cos (\theta_v - \theta_i) \tag{13a}$$

where $\theta_v - \theta_i$ is the angular difference in phase between the current and voltage. If we put θ for this difference in phase, viz., $\theta = \theta_v - \theta_i$, eq. (13a) may be written

$$P = VI \cos \theta \tag{13b}$$

Whence the power factor of the load supplied to the apparatus is, from eq. (10),

$$\cos \theta = \frac{P}{VI} \tag{14}$$

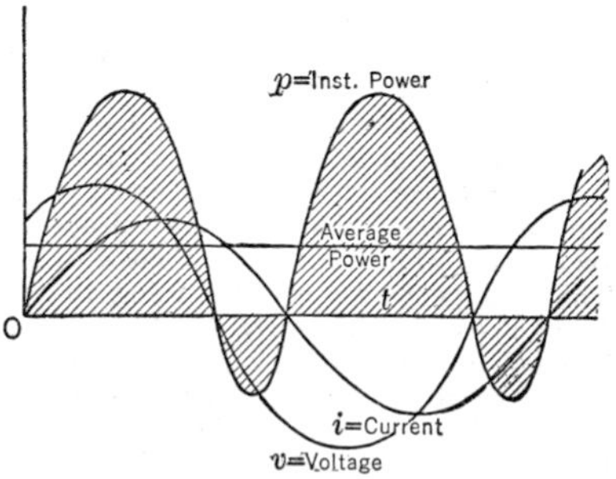

FIG. 1. Power versus Time in the Case of a Sinusoidal Current and Voltage

Since in sine-wave currents and voltages the power factor is equal to the cosine of the angle which expresses the difference in phase between them, this difference in phase is frequently called the power-factor angle. When the wave shape is not a pure sine curve, the power factor cannot be interpreted as the cosine of the phase difference, for phase difference has no definite meaning except in reference to sine waves (see definitions above). A non-sinusoidal voltage and current may both reach their zero values at the same instant, and in a sense may be said to be "in phase," but the power factor as defined by eq. (10) may be far from unity.

The **reactive power** in a circuit in which a sinusoidal current is flowing is equal to the effective emf times the effective current times the sine of the phase difference between them. When the emf and current are in volts and amperes respectively, the reactive power is in *vars*.

7. VECTOR REPRESENTATION OF SINUSOIDS

Consider any sine function

$$i = I_m \sin \omega t$$

The value of i at any instant may be represented graphically (see Fig. 2) by the vertical projection, i.e., the vertical distance from P_1 to OX, of a point P_1 at the end of a radius $OP_1 = I_m$ which revolves * at a constant angular velocity ω about a fixed point O, the angle ωt being measured from the horizontal fixed line OX. Similarly, any other sine function

$$v = V_m \sin (\omega t + \theta)$$

* Counterclockwise rotation has been adopted (1911) as standard by the International Electrotechnical Commission.

may be represented by the vertical projection of the point P_2 at the end of a radius $OP_2 = V_m$, also revolving about O with a constant angular velocity ω, the angle between OP_1 and OP_2, when the frequency of both i and v is the same, remaining fixed in value and equal to the difference in phase θ between v and i. That is, v and i may be represented by rotating vectors. When v and i are of the same frequency, the relative position of the two vectors remains fixed. Similarly any number of currents and voltages of the same frequency may be represented by rotating vectors which remain fixed with respect to one another.

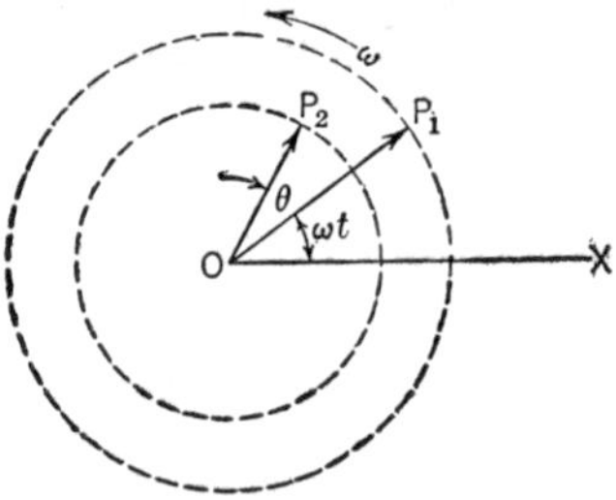

FIG. 2. Vector Representation of Sinusoidal Quantities

Instead of referring the various rotating vectors to a fixed line OX, this line of reference may also be considered as rotating with the same speed as the various vectors, or any one of the vectors may be chosen as the line of reference, e.g., the vector OP_1 in Fig. 2. The rotating vectors referred to this rotating line of reference are then fixed with respect to this line of reference, and the entire diagram may be considered fixed, as in Fig. 3.

Instead of making the vectors equal in length to the maximum values of the sine functions, they may be chosen equal in length to their rms values. This merely introduces a factor $\sqrt{2}$, so that the instantaneous value of the quantity which any rotating vector is considered to represent is equal to $\sqrt{2}$ times the perpendicular distance from the end of the vector to the fixed line of reference.

Since the rms values and phase relations of sine-wave currents and voltages may be represented by vectors, sine-wave currents may be added in exactly the same manner as vectors, or forces, are added, and similarly for sine-wave voltages. To add any two sine-wave currents or voltages not only their effective values but also their phase relation must be known; the resultant of two alternating voltages of rms values V_1 and V_2 is never the arithmetical sum of V_1 and V_2, except when the two voltages are exactly in phase, and similarly for alternating currents.

In Fig. 3, considering OI as equal to the rms value I of the current and OV as representing the rms value V of the voltage, the voltage V may be considered as made up of two components, viz.: $V_1 = V \cos\theta$ in phase with I, $V_2 = V \sin\theta$ in quadrature with I. The average power corresponding to the component $V_1 = V \cos\theta$ is, from eq. (13*b*), $V_1 I = VI \cos\theta$, and is equal to the total power corresponding to V and I. The average power corresponding to the component $V_2 = V \sin\theta$, since the angle between the current and this component of the voltage is 90 deg, is equal to zero. The voltage component $V_1 = V \cos\theta$ is therefore frequently called the "power" component of the voltage, and the component $V_2 = V \sin\theta$ is frequently called the "wattless" component of the voltage. These terms, however, are not recommended. It is preferable to refer to these two components as the in-phase and quadrature components, respectively. The terms active and reactive components are also used.

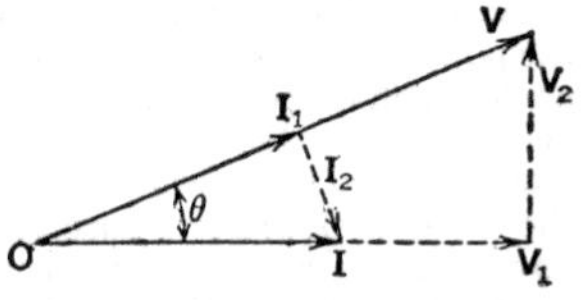

FIG. 3. Components of Vectors

Similarly, the current I may be considered as made up of two components, viz.:

$$I_1 = I \cos\theta \text{ in phase with } V$$

$$I_2 = I \sin\theta \text{ in quadrature with } V$$

The first component is called the in-phase component of the current, and the second the quadrature component of the current.

8. FOURIER SERIES, HARMONICS

It can be shown that any single-valued, periodic function (which fulfils certain mathematical conditions always fulfilled by electric currents) may be represented by a sum of sinusoids of which the frequencies form an arithmetical progression. Hence any current $i(t)$ can be written

$$i(t) = A_1 \sin 2\pi \frac{t}{T} + A_2 \sin 2\pi \frac{2t}{T} + \cdots + A_n \sin 2\pi \frac{nt}{T} + \cdots$$
$$+ B_0 + B_1 \cos 2\pi \frac{t}{T} + B_2 \cos 2\pi \frac{2t}{T} + \cdots + B_n \cos 2\pi \frac{nt}{T} + \cdots \quad (15)$$

where the values of the coefficients A and B are given by the integrals:

$$A_n = \frac{2}{T}\int_0^T i(t) \sin 2\pi \frac{nt}{T}\,dt \tag{15a}$$

$$B_n = \frac{2}{T}\int_0^T i(t) \cos 2\pi \frac{nt}{T}\,dt \tag{15b}$$

The series of eq. (15) is called a *Fourier series.*

If the terms involving the same frequencies are combined, and if $2\pi/T = \omega$, eq. (15) may be written

$$i(t) = I_0 + I_1 \sin(\omega t - \theta_1) + I_2 \sin(2\omega t - \theta_2) + \cdots + I_n \sin(n\omega t - \theta_n) + \cdots \tag{16}$$

The term $I_1 \sin(\omega t - \theta_1)$ is called the **fundamental,** and the terms $I_n \sin(n\omega t - \theta_n)$ are called **harmonics.** The **order** of a harmonic is the ratio of the harmonic frequency to the frequency of the fundamental. Each harmonic is characterized by two constants, its amplitude I_n and the angle θ_n. It should be noted that this angle may be given in either of two ways:

$$I_n \sin(n\omega t - \theta_n) \tag{16a}$$

$$I_n \sin n(\omega t - \theta'_n) \tag{16b}$$

In the first notation the angle θ_n is taken relative to the period of the harmonic, whereas in the second notation the angle θ'_n is taken relative to the period of the fundamental. It follows that

$$\theta_n = n\theta'_n \tag{16c}$$

The mathematical statement of the definition given above for a symmetrical alternating current is

$$i(\omega t + \pi) = -i(\omega t) \tag{17}$$

from which it follows by substitution of eq. (16) that

$$a_2 = a_4 = a_6 = \cdots = a_{2k} = \cdots = 0 \tag{18}$$

i.e., all the even harmonics of a symmetrical alternating current are zero.

If the two half periods are alike and if, further, each is symmetrical about an axis erected at a quarter of a period from a null point, then not only are the even harmonics zero but all the θ's in eq. (16) are either zero or π. In this case all the odd harmonics are in phase or in direct opposition. This type of symmetry often exists in commercial alternating currents. If this more restricted type of symmetry holds, the current may be written

$$i(t) = I_1 \sin \omega t + I_3 \sin 3\omega t + I_5 \sin 5\omega t + \cdots \tag{19}$$

or in terms of cosine functions,

$$i(t) = I_1 \cos \omega t + I_3 \cos 3\omega t + I_5 \cos 5\omega t + \cdots \tag{19a}$$

The odd harmonics may be divided into two groups, those of order $4k - 1$ and those of order $4k + 1$. Harmonics of the first class have the same sign in (19*a*) as in (19); those of the second class have opposite signs in (19) and (19*a*). Thus harmonics of order $4k - 1$ are in phase at their maximum values if they are in phase at their zero values; harmonics of order $4k + 1$ are in opposition at their maximum values if they are in phase at their zero values, and conversely. These facts are of importance in the consideration of harmonics in synchronous machines and induction motors.

If an analytic expression for $i(t)$ is available and if the integrations (15*a*) and (15*b*) can be performed, the terms of the Fourier series (15) can be written. In practice the function $i(t)$ will usually be obtainable only in the form of a plotted curve, and the evaluation of the A's and B's must be carried out by graphical means. This process is called **harmonic analysis.** Various means of harmonic analysis have been devised, the Fisher-Hinnen method being perhaps the most convenient in ordinary cases.

This method is quite simple when the wave contains only the fundamental and the third, fifth, and seventh harmonics. When even harmonics are present, or when higher odd harmonics than the seventh exist, certain corrections must be applied. Since voltage and current waves usually contain only odd harmonics, and seldom contain higher harmonics than the seventh, the simple method without corrections is usually sufficiently accurate.

WAVES CONTAINING ONLY THE THIRD, FIFTH, AND SEVENTH HARMONICS. To determine the nth harmonic (n equals 3, 5, or 7), divide the base of a half wave into $2n$ equal parts and measure the ordinates of the wave at the beginning of each of these sections of the base. Call these ordinates $y_1, y_2, y_3 \ldots y_{2n}$, taking the y's positive if above the base line, negative if below. y_1 will be zero, since the first section begins where the resultant wave crosses the base line.

Then the ordinates of this harmonic at the points 1 and 2 are, respectively,

$$A_n = \frac{1}{n}[(y_5 + y_9 + \cdots y_{2n-1}) - (y_3 + y_7 + \cdots y_{2n-3})]$$

$$B_n = \frac{1}{n}[(y_2 + y_6 + \cdots y_{2n}) - (y_4 + y_8 + \cdots y_{2n-2})]$$

The maximum value of this harmonic is

$$Y_n = \sqrt{A_n^2 + B_n^2}$$

and the phase angle, calling the wavelength of the fundamental 360 deg, is

$$\phi_n = \frac{1}{n}\tan^{-1}\frac{A_n}{B_n}$$

These formulas give the third, fifth, and seventh harmonics ($n = 3$, 5, and 7, respectively). The fundamental is found by calculating

$$A_1 = -(A_3 + A_5 + A_7)$$

$$B_1 = y_0 + B_3 - B_5 + B_7$$

where y_0 is the mid-ordinate of the half wave. Then

$$Y_1 = \sqrt{A_1^2 + B_1^2}$$

$$\phi_1 = \tan^{-1}\frac{A_1}{B_1}$$

The equation of the given wave is then

$$y = Y_1 \sin(x + \phi_1) + Y_3 \sin 3(x + \phi_3) + Y_5 \sin 5(x + \phi_5) + Y_7 \sin 7(x + \phi_7)$$

The **effective** value of the given wave is

$$Y = \sqrt{\frac{Y_1^2 + Y_3^2 + Y_5^2 + Y_7^2}{2}}$$

and the average value is

$$Y_{av} = \frac{2}{\pi}[Y_1 \cos\phi_1 + {}^1\!/_3 Y_3 \cos 3\phi_3 + {}^1\!/_5 Y_5 \cos 5\phi_5 + {}^1\!/_7 Y_7 \cos 7\phi_7]$$

In using the above formulas strict attention must be paid to algebraic signs.

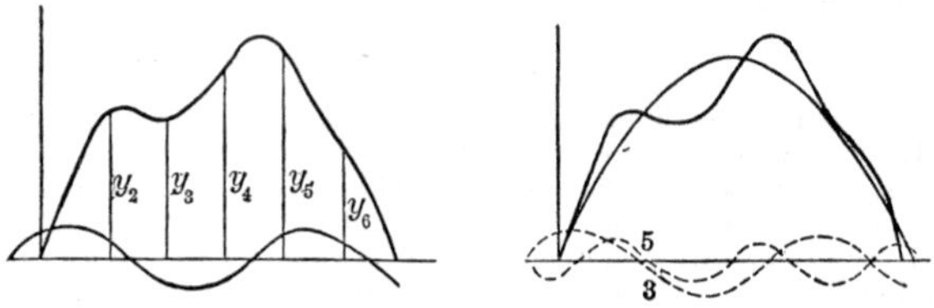

FIG. 4. Fisher-Hinnen Method of Wave Analysis

Example. Find the third harmonic in the wave shown in Fig. 4. The values of the six ordinates are found by measurement to be 0, 676, 660, 940, 1004, and 554, respectively. Then

$$A_3 = {}^1\!/_3(1004 - 660) = 114.7$$

$$B_3 = {}^1\!/_3(676 + 554 - 940) = 96.7$$

$$Y_3 = \sqrt{(114.7)^2 + (96.7)^2} = 150$$

$$\phi_3 = {}^1\!/_3 \tan^{-1}\frac{114.7}{96.7} = 16.6°$$

Similarly, for the fifth harmonic,

$$A_5 = -92.8 \text{ and } B_5 = 37.4$$

$$Y_5 = 100 \quad \text{and} \quad \phi_5 = -13.6°$$

For the fundamental

$$A_1 = -114.7 + 92.8 = -21.9$$

$$B_1 = 940 + 96.7 - 37.4 = 999.3$$

$$Y_1 = \sqrt{(21.9)^2 + (999.3)^2} = 1000$$

$$\phi_1 = \tan^{-1} \frac{-21.9}{999.3} = -1.25°$$

Hence the complete expression for the given wave, taking as the origin the point at which the resultant wave crosses the base line in the rising direction, is

$$y = 1000 \sin (x - 1.25°) + 150 \sin 3(x + 16.6°) + 100 \sin 5(x - 13.6°)$$

The **effective** value is then 718, and **average** value 673.

WAVES CONTAINING ANY NUMBER OF HARMONICS, ODD OR EVEN. See *Elec. J.*, Vol. 5, p. 386 (1908); *Electrotechn. Zeit.*, Vol. 34, p. 936 (1913).

9. RMS VALUE AND POWER OF NON-SINUSOIDAL CURRENTS

The rms value of a non-sinusoidal current can be obtained directly from eq. (3), or if the rms values I_1, I_2, I_3, etc., of the fundamental and harmonics are known, the rms value of the resultant current is

$$I = \sqrt{I_1^2 + I_2^2 + I_3^2 + \cdots} \tag{20}$$

A like relation holds for the rms value of a non-sinusoidal voltage. Similarly, if, for example, a 25-cycle emf, say E_{25}, and a direct emf, say E_d, are acting in series in the same circuit, the resultant emf of the combination is

$$E = \sqrt{E_{25}^2 + E_d^2} \tag{21}$$

The power corresponding to non-sinusoidal currents and voltages may be computed as follows:

Let I_1 be the rms value of the fundamental of the current, V_1 the rms value of the fundamental of the voltage, and θ_1 the difference in phase between these two fundamentals, both being of the same frequency; let I_2, V_2, and θ_2 be the corresponding quantities for the second harmonic; let I_3, V_3, and θ_3 be the corresponding values for the third harmonic, etc. Then the average power is

$$P = V_1 I_1 \cos \theta_1 + V_2 I_2 \cos \theta_2 + V_3 I_3 \cos \theta_3 + \cdots \tag{22}$$

That is, each harmonic contributes to the total power an amount equal to the power it would develop were the other harmonics not present. If, for example, the third harmonic is not present in the current wave, then this harmonic contributes nothing to the average power, even though there may be a large third harmonic in the voltage wave. Again, when a 25-cycle alternating emf E_{25} and a direct emf E_d are acting in series on the same circuit, the power developed is the sum of the powers which each would develop if they acted separately, but the resultant emf of the combination is not $E_{25} + E_d$, but, as noted above,

$$\sqrt{E_{25}^2 + E_d^2}$$

LINEAR CIRCUITS

By Irven Travis

10. CIRCUIT EQUATIONS AND VECTOR IMPEDANCES

Practically all steady-state calculations of electric-circuit problems are carried out by means of the **complex number** or **vector** method based on the differential equations for the electric charge in terms of the physical parameters of the electric circuit and the impressed voltages. This method is given in detail in *Handbook of Engineering Fundamentals*, "Electricity and Magnetism," by Eshbach, John Wiley, and is abstracted here. It is valid only when the resistances, inductances, and capacitances in the network under consideration are constants. Various means are available for extending the method in order to obtain good approximations when these parameters are not constants. Some of these are given in "Non-linear Circuits" and "Circuits of Electrical Machines."

The following are the rules for finding the steady-state currents produced in a network by a group of sinusoidal electromotive forces of a given frequency:

a. Write the differential equations for the instantaneous values of the currents in the network. These will be of the form

$$r_{11}i_1 + L_{11}\frac{di_1}{dt} + \frac{1}{C_{11}}\int i_1\,dt - r_{12}i_2 - L_{12}\frac{di_2}{dt} - \frac{1}{C_{12}}\int i_2\,dt \pm M_{12}\frac{di_2}{dt}$$

$$+ \text{ other terms involving the other currents in the network}$$

$$= \text{impressed voltage in mesh 1}$$

b. Replace each electromotive force in the network by a complex number having a modulus equal to the rms value of the given electromotive force, and a phase angle equal to the phase angle of the given electromotive force (when expressed in cosine form).

c. Replace d/dt by $j\omega$ and $\int dt$ by $1/j\omega$, where ω is the angular velocity of the given sinusoidal electromotive forces and j is the square root of -1.

d. Replace each current in the network by an unknown complex number, and solve for these complex numbers.

e. The rms values of the various currents are the moduli of the corresponding complex numbers, and the phase angles of the currents (when expressed in cosine form) are the phase angles of these complex numbers.

The coefficients of each of the complex number currents in the equations constructed as indicated above will also be complex numbers; these are called *vector impedances* or often simply *impedances*.

Example. The differential equation for the current in a single isolated mesh having resistance, inductance, and capacitance when a 60-cycle voltage having a modulus value E is impressed is

$$ri + L\frac{di}{dt} + \frac{1}{C}\int i\,dt = \sqrt{2}E\cos 120\pi t \tag{1}$$

The complex number * equation is

$$r\mathbf{I} + j120\pi L\mathbf{I} + \frac{1}{j120\pi C}\mathbf{I} = \mathbf{E} \tag{1a}$$

hence the vector impedance is

$$\mathbf{Z} = r + j\left(120\pi L - \frac{1}{120\pi C}\right) \tag{2}$$

The current in the circuit is

$$i = \sqrt{2}I\cos(120\pi t - \theta) \tag{3}$$

in which

$$I = \frac{E}{\sqrt{r^2 + \left(120\pi L - \dfrac{1}{120\pi C}\right)^2}} \tag{3a}$$

and

$$\theta = \tan^{-1}\frac{120\pi L - \dfrac{1}{120\pi C}}{r} \tag{3b}$$

11. IMPEDANCE, ADMITTANCE, REACTANCE, SUSCEPTANCE

The **impedance** of a portion of an electric circuit to a completely specified periodic current and potential difference is the ratio of the effective value of the potential difference between the terminals to the effective value of the current, there being no source of emf in the portion under consideration. Impedance defined in this way is a real number, which in the case of a sinusoidal current and voltage reduces to the modulus value of the vector impedance defined above.

Admittance is defined as the reciprocal of the impedance. The **vector admittance** is the reciprocal of the complex number representing the vector impedance.

Both impedance and admittance, as thus defined, depend upon the frequency and wave shape of the current. Impedance is expressed in the same units as resistance (e.g., ohms), and admittance in the same units as conductance (e.g., mhos).

* Vectors are printed in boldface roman letters, the modulus value of a vector **A** being printed in lightface italics, *A*.

The **reactance** of a portion of a circuit for a sinusoidal current, and hence for any one of the frequencies of a periodic current, is the ratio of the quadrature component of the potential difference for a particular frequency to the value of the current for that frequency, there being no source of emf in the portion of the circuit under consideration.

The **susceptance** of a portion of a circuit for a sinusoidal potential difference, and hence for any one of the frequencies of a periodic potential difference, is the ratio of the quadrature component of the current for a particular frequency to the value of the potential difference for that frequency, there being no source of emf in the portion of the circuit under consideration.

From the definitions given above it may be shown that for sine-wave currents and voltages of a given frequency the following relations hold for any portion of a circuit:

$$\left.\begin{aligned} \mathbf{Z} &= r + jx & \mathbf{Y} &= g - jb \\ Z &= \sqrt{r^2 + x^2} & Y &= \sqrt{g^2 + b^2} \\ \mathbf{Z} &= \frac{1}{\mathbf{Y}} & \mathbf{Y} &= \frac{1}{\mathbf{Z}} \\ r &= \frac{g}{Y^2} & g &= \frac{r}{Z^2} \\ x &= \frac{b}{Y^2} & b &= \frac{x}{Z^2} \end{aligned}\right\} \qquad (4)$$

where r = effective resistance, x = reactance (taken positive when inductive and negative when capacitive), g = effective conductance, b = susceptance (taken positive when inductive and negative when capacitive), Z = impedance, Y = admittance, $\mathbf{Z}$ = vector impedance, and $\mathbf{Y}$ = vector admittance, all for the given portion of circuit.

12. IMPEDANCES IN SERIES AND PARALLEL

In general any vector impedance can be written in the form

$$\mathbf{Z} = r + jx \qquad (5)$$

where r is the sum of the resistances, and x is the algebraic sum of the reactances of the several portions of the network which make up this impedance.

The vector impedance may also be expressed in the form

$$\mathbf{Z} = Z\angle\theta \qquad (6)$$

where

$$Z = \sqrt{r^2 + x^2} \qquad (7)$$

is the modulus of the complex number $(r + jx)$ and

$$\theta = \tan^{-1}\left(\frac{x}{r}\right) \qquad (8)$$

is the phase angle of this complex number.

In eq. (6) the symbol $\angle$ is read "at an angle" (or "phase") and has the same mathematical significance as e^j. Thus $Z\angle\theta$ is *identical* with $Ze^{j\theta}$.

When two or more vector impedances $\mathbf{Z}_1$, $\mathbf{Z}_2$, etc., are connected in **series**, so that the *same current* **I** *flows through each*, the total impedance drop through the group is

$$\begin{aligned} \mathbf{V} &= \mathbf{Z}_1\mathbf{I} + \mathbf{Z}_2\mathbf{I} + \cdots \\ &= (\mathbf{Z}_1 + \mathbf{Z}_2 + \mathbf{Z}_3 + \cdots)\mathbf{I} \end{aligned} \qquad (9)$$

hence two or more vector impedances in series are equivalent to a single vector impedance equal to the sum of several impedances.

When two or more vector impedances $\mathbf{Z}_1$, $\mathbf{Z}_2$, etc., are connected in **parallel,** so that the *voltage drop is the same for each,* the total current through the group is

$$\begin{aligned} \mathbf{I} &= \frac{\mathbf{V}}{\mathbf{Z}_1} + \frac{\mathbf{V}}{\mathbf{Z}_2} + \cdots \\ &= \left(\frac{1}{\mathbf{Z}_1} + \frac{1}{\mathbf{Z}_2} + \cdots\right)\mathbf{V} \end{aligned} \qquad (10)$$

It is customary to express this law in terms of the vector admittances, $\mathbf{Y}_1$, $\mathbf{Y}_2$, $\cdots$. Two or more vector admittances in parallel are equivalent to a single vector admittance equal to the sum of the several admittances.

13. VECTOR POWER

In Article 4 average power is defined, and in Article 6 the average power due to a sinusoidal current and voltage of the same frequency was shown to be

$$P = VI \cos (\theta_v - \theta_i) \tag{11}$$

The vector voltage and vector current in this case are

$$\mathbf{V} = Ve^{j\theta_v}$$

$$\mathbf{I} = Ie^{j\theta_i}$$

of which the product **is not** the average power. The average power is given by the **real part** of the product of the vector voltage by the **conjugate** of the vector current.* The **complex power** is given by

$$\begin{aligned}\mathbf{P} &= \mathbf{V} \times \text{Conjugate of } \mathbf{I} \\ &= VIe^{j(\theta_v - \theta_i)} \\ &= VI \cos (\theta_v - \theta_i) + jVI \sin (\theta_v - \theta_i)\end{aligned} \tag{12}$$

As stated above, the real part of the complex power is the average power. The imaginary part of the complex power is the reactive power. The average power is the actual average rate of transformation by the receiver of electrical energy into some other form of energy. The reactive power measures the rate of interchange of unused electrical energy between the source and receiver and is usually expressed in **kilovars** (the word is a contraction of kilovolt-amperes).

This power surges back and forth between source and receiver but is not transformed into any other type of energy; it does not therefore represent a rate of dissipation of energy in the receiver.

When the current and voltage are written in the form

$$\mathbf{V} = V_1 + jV_2$$

$$\mathbf{I} = I_1 + jI_2$$

the average power is given by

$$P = V_1I_1 + V_2I_2 \tag{13}$$

14. EQUIVALENT IMPEDANCES OF A RECEIVER

The receiver at the end of a transmission line is often a motor or other device which contains a back-electromotive force. The vector impedance equations for a network containing such a device therefore involve this back-emf as one of the emf forces in the network. It is often convenient, however, to express the emf in terms of an equivalent impedance drop.

Let the back-emf be denoted by **E**, and let the current through the device be **I**; then the equivalent impedance of the back-emf is simply $\mathbf{E}/\mathbf{I} = \mathbf{Z}_e$. If the true impedance of the remainder of the circuit is **Z**, then the impedance equation becomes

$$(\mathbf{Z}_e + \mathbf{Z})\mathbf{I} = \mathbf{V} \tag{14}$$

where **V** is the voltage impressed across the device.

If the emf leads the current by an angle θ, then an amount of electric power

$$P = EI \cos \theta \tag{15}$$

is converted into some other form of power. A resistance which will absorb this same amount of power when the current I flows through it is

$$r_e = \frac{P}{I^2} = \frac{E}{I} \cos \theta \tag{16}$$

---

* To conform with the present (1948) AIEE standards, vector power must be defined as the product of the current and the conjugate of the voltage. General practice, however, is to consider vector power the product of the voltage and the conjugate of the current. This practice conforms with a recent recommendation of a subcommittee of the Standards Committee to change the sign of reactive power from minus to plus. This proposal not only validates practice but also brings the definition of reactive power into conformity with the AIEE standard form of equation for admittance, as given in Article 11, eq. group (4).

This resistance is called the equivalent resistance of the back-emf; it is the real part of the vector impedance $\mathbf{Z}_e$. The equivalent reactance of the back-emf is

$$x_e = r_e \tan \theta \tag{17}$$

It should be noted that the equivalent impedance of a back-emf, as given above, is a function of the current through the receiver. The quantity $\mathbf{Z}_e + \mathbf{Z}$ is the total equivalent impedance of the receiver.

15. COMPLICATED CIRCUITS

When an electrical network is extremely complicated, the process of combining series and parallel impedances in order to solve for the currents in the various branches is quite difficult. In such cases it is usually easier to set up the problem in terms of **mesh currents** instead of branch currents. If currents are assumed to flow around each of the meshes of a network, so that the current in any branch is the difference of the mesh currents in the two meshes on either side of that branch, then the impedance equations resulting from the application of Kirchhoff's second law will contain the Kirchhoff point equations. See *Handbook of Engineering Fundamentals*, by Eshbach, John Wiley, "Direct-current Circuits" and "Alternating-current Circuits."

The mesh equations for a network having n meshes, each mesh containing impressed voltages, are

$$\left.\begin{array}{l} \mathbf{Z}_{11}\mathbf{I}_1 + \mathbf{Z}_{12}\mathbf{I}_2 + \cdots + \mathbf{Z}_{1n}\mathbf{I}_n = \mathbf{V}_1 \\ \mathbf{Z}_{21}\mathbf{I}_1 + Z_{22}\mathbf{I}_2 + \cdots + \mathbf{Z}_{2n}\mathbf{I}_n = \mathbf{V}_2 \\ \cdot \quad \cdot \quad \cdot \quad \cdot \quad \cdot \quad \cdot \quad \cdot \quad \cdot \\ \mathbf{Z}_{n1}\mathbf{I}_1 + \mathbf{Z}_{n2}\mathbf{I}_2 + \cdots + \mathbf{Z}_{nn}\mathbf{I}_n = \mathbf{V}_n \end{array}\right\} \tag{18}$$

in which $\mathbf{Z}_{jj}$ is the self-impedance of the jth mesh, and $\mathbf{Z}_{jk}$ is the mutual impedance of the jth mesh with respect to the kth mesh. $\mathbf{Z}_{jk}$ is always equal to $\mathbf{Z}_{kj}$ in passive networks and in active networks in which there is no valve action.

A representative network of this type is shown in Fig. 1; in this network the number of meshes is easy to count. In general, for example, when the network cannot be drawn on the surface of a sphere, the number of mesh currents necessary to formulate the problem completely is

$$B - V + S$$

where B = the number of branches, V = the number of intersections or vertices, and S = the number of separate parts. A circuit which closes on itself (single mesh) is assumed to have one intersection.

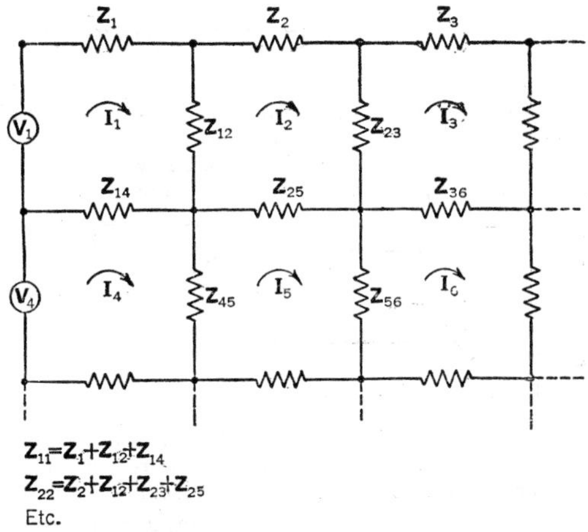

FIG. 1. Complicated Network. The network shown is *not* perfectly general but is given here for illustration; the network corresponding to eqs. (18) cannot be drawn on a plane.

These impedance equations can be solved by the substitution method or by the method of determinants. For numerical calculations it is usually simpler to solve by substitution and elimination of currents. In theoretical work, however, particularly in the development of the general properties of electrical networks, the method of determinants is much more powerful.

The values of the currents in eqs. (18) can be calculated by means of Cramer's rule, which is developed in the theory of linear algebraic equations and determinants.

Let $\mathbf{D}$ denote the determinant of order n, whose elements are the $\mathbf{Z}$'s in eqs. (18), i.e.,

$$\mathbf{D} = \begin{vmatrix} \mathbf{Z}_{11} & \mathbf{Z}_{12} & \cdots & \mathbf{Z}_{1n} \\ \mathbf{Z}_{21} & \mathbf{Z}_{22} & \cdots & \mathbf{Z}_{2n} \\ \cdot & \cdot & \cdot & \cdot \\ \mathbf{Z}_{n1} & \mathbf{Z}_{n2} & \cdots & \mathbf{Z}_{nn} \end{vmatrix} \tag{19}$$

and let $\mathbf{M}_{jk}$ denote the determinant of order $n - 1$, obtained from **D** by deleting the jth row and the kth column of that determinant. $\mathbf{M}_{jk}$ is called a minor of **D**. Further let

$$\mathbf{A}_{jk} = (-1)^{j+k}\mathbf{M}_{jk} \tag{20}$$

$\mathbf{A}_{jk}$ is called a cofactor of **D**.

Then the current $\mathbf{I}_k$ is given by

$$\mathbf{I}_k = \sum_{k=1}^{n} \frac{\mathbf{A}_{jk}\mathbf{V}_j}{\mathbf{D}} \tag{21}$$

16. RECIPROCAL THEOREM

Let a single impressed voltage **V** be applied to, say, the jth mesh of a network, all other impressed voltages being zero. The current in the kth mesh is

$$\mathbf{I}_k = \frac{\mathbf{A}_{jk}\mathbf{V}}{\mathbf{D}}$$

Now let the voltage be removed from the jth mesh and inserted in the kth mesh. The current in the jth mesh in this case is

$$\mathbf{I}_j = \frac{\mathbf{A}_{kj}\mathbf{V}}{\mathbf{D}}$$

Since $\mathbf{Z}_{jk} = \mathbf{Z}_{kj}$, it follows also that $\mathbf{A}_{jk} = \mathbf{A}_{kj}$, hence the theorem commonly called the **reciprocal theorem:**

The current in a given branch of a network due to an electromotive force in another branch of the network is the same as the current which would be produced in the second branch of the network if the electromotive force were transferred to the first branch.

The quantity $\mathbf{A}_{kj}/\mathbf{D}$ is called the **transfer admittance** from the kth to the jth mesh. The reciprocal of this quantity is the **transfer impedance.**

17. THÉVENIN'S THEOREM

A theorem due to Thévenin, which can be deduced from eqs. (18), is extremely valuable in network problems, inasmuch as it allows the calculation of the performance of a device from its terminal properties only, without inquiring into its internal make-up.

Consider any complicated network having internal emf's and having two available points of entry, such as that shown in Fig. 2. Such a network is capable of acting as a source of electrical energy. Let **E** be the open-circuit terminal voltage of this source, and let the impedance *looking into* the source terminals be $\mathbf{Z}_s$. (By impedance "looking into" a pair of terminals is meant the input impedance measured at these terminals with the rest of the network disconnected, and all emf's made equal to zero.) Let this source be connected to a receiver having two points of entry and containing no internal emf's. The impedance looking into the receiver terminals will be taken as $\mathbf{Z}_r$. Then Thévenin's theorem states that the current which flows from source to receiver is given by

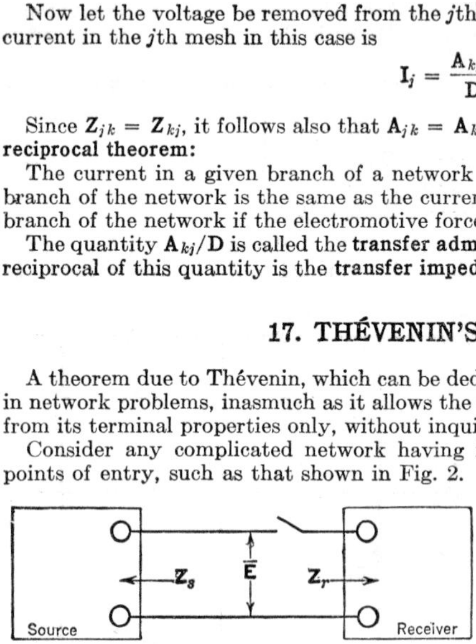

FIG. 2. Junction of a Source and Receiver

$$(\mathbf{Z}_r + \mathbf{Z}_s)\mathbf{I} = \mathbf{E} \tag{22}$$

A generalization of this equation may be stated thus:

If two networks are connected together by means of a two-conductor line, the line current is equal to the algebraic sum of the open-circuit voltages which would appear if the line were cut, divided by the sum of the impedances looking toward the source and toward the receiver from the cut.

A simple use of this theorem is the calculation of the current delivered to a motor by a generator. In this case the generator emf minus the motor back-emf divided by the sum of the two internal impedances plus the line impedance yields the line current.

18. EQUIVALENT TRANSDUCERS (T AND π)

It is often desirable to study the performance of a network at two pairs of its terminals when various voltages or various impedances are connected to them. In such cases it is convenient to call one pair of terminals the input terminals and the other pair the output terminals. The input terminals are those usually connected to a source of energy, and the output terminals are those usually connected to a receiver. Both terminals may, of course, be connected to sources. A device of this type having one pair of input terminals and one pair of output terminals is called a **transducer.** A transformer, an underground cable, a transmission line together with its terminal transformers, etc., are transducers.

Any **passive** transducer, i.e., one which does not contain a source of electromotive force, may, for a given frequency, be replaced by an equivalent circuit containing only three independent impedances. These impedances may be arranged as in Fig. 3(*a*) or as in Fig. 3(*b*). The former arrangement is called an equivalent T; the latter arrangement, an equivalent π of the given transducer.

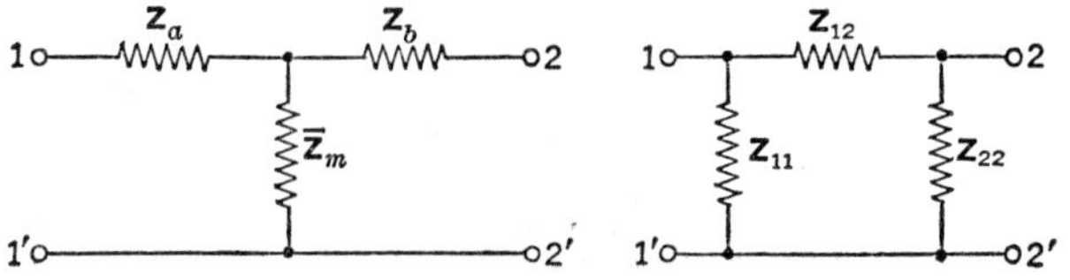

FIG. 3. Equivalent T and π Sections

The constants of the equivalent T may be computed from measurements made at the terminals of the given network. Let:

Z_{10} = impedance measured at the input terminals when the output terminals are open.
Z_{20} = impedance measured at the output terminals when the input terminals are open.
Z_{1s} = impedance measured at the input terminals when the output terminals are short-circuited.
Z_{2s} = impedance measured at the output terminals when the input terminals are short-circuited.

These quantities are not all independent, the relation

$$Z_{10}Z_{2s} = Z_{20}Z_{1s} \tag{23}$$

providing a means for calculating any one when the other three are known.

In terms of these open- and short-circuit impedances the parameters of the equivalent T are:

$$Z_m = \sqrt{Z_{20}(Z_{10} - Z_{1s})} \tag{24}$$

$$Z_a = Z_{10} - Z_m \tag{25}$$

$$Z_b = Z_{20} - Z_m \tag{26}$$

Other forms for these equations may be obtained by eq. (23). It is sometimes convenient to make use of the transfer impedance of the transducer in these calculations. The transfer impedance Z_{ts} can of course be measured directly; it is related to the open- and short-circuit impedances by:

$$Z_{ts} = Z_{1s}\sqrt{\frac{Z_{20}}{Z_{10} - Z_{1s}}} \tag{27}$$

The shunt arm Z_m of the equivalent T may also be expressed as:

$$Z_m = \frac{Z_{10}Z_{2s}}{Z_{ts}} \tag{28}$$

The impedances in the equivalent π may be calculated from those of the equivalent T by the formulas:

$$Z_{11} = Z_aZ_m\left(\frac{1}{Z_a} + \frac{1}{Z_b} + \frac{1}{Z_m}\right) \tag{29}$$

$$Z_{22} = Z_bZ_m\left(\frac{1}{Z_a} + \frac{1}{Z_b} + \frac{1}{Z_m}\right) \tag{30}$$

$$Z_{12} = Z_aZ_b\left(\frac{1}{Z_a} + \frac{1}{Z_b} + \frac{1}{Z_m}\right) \tag{31}$$

19. GENERAL CIRCUIT CONSTANTS

Another method of specifying a transducer is by means of the four constants **A**, **B**, **C**, and **D**, defined by the equations:

$$\mathbf{E}_s = \mathbf{A}\mathbf{E}_r + \mathbf{B}\mathbf{I}_r \tag{32}$$

$$\mathbf{I}_s = \mathbf{C}\mathbf{E}_r + \mathbf{D}\mathbf{I}_r \tag{33}$$

in which the subscript *s* stands for source, and the subscript *r* stands for receiver.* These constants are no more "general" than those described in Article 18; they have the advantage, however, of expressing the source conditions directly in terms of the receiver conditions. The constants **A**, **B**, **C**, and **D** are not independent. Any one can be calculated, when the other three are known, by the equation:

$$\mathbf{AD} - \mathbf{BC} = 1 \tag{34}$$

This relation may be used to deduce from eqs. (32) and (33) the alternative forms:

$$\mathbf{E}_r = \mathbf{D}\mathbf{E}_s - \mathbf{B}\mathbf{I}_s \tag{35}$$

$$\mathbf{I}_r = -\mathbf{C}\mathbf{E}_s + \mathbf{A}\mathbf{I}_s \tag{36}$$

These equations have the advantage of expressing the receiver conditions directly in terms of the conditions which obtain at the source.

General circuit constants are of great value in the calculation of the performance of power systems and in studies of stability (see Articles 63 to 68 of this section).

The general circuit constants may be calculated from the constants of the equivalent T by the formulas:

$$\left.\begin{aligned}\mathbf{A} &= 1 + \frac{\mathbf{Z}_a}{\mathbf{Z}_m}\\ \mathbf{B} &= \mathbf{Z}_a + \mathbf{Z}_b + \frac{\mathbf{Z}_a\mathbf{Z}_b}{\mathbf{Z}_m}\\ \mathbf{C} &= \frac{1}{\mathbf{Z}_m}\\ \mathbf{D} &= 1 + \frac{\mathbf{Z}_b}{\mathbf{Z}_m}\end{aligned}\right\} \tag{37}$$

The general circuit constants may also be calculated from the constants of the equivalent π, thus:

$$\left.\begin{aligned}\mathbf{A} &= 1 + \frac{\mathbf{Z}_{12}}{\mathbf{Z}_{22}}\\ \mathbf{B} &= \mathbf{Z}_{12}\\ \mathbf{C} &= \frac{1}{\mathbf{Z}_{11}} + \frac{1}{\mathbf{Z}_{22}} + \frac{\mathbf{Z}_{12}}{\mathbf{Z}_{11}\mathbf{Z}_{22}}\\ \mathbf{D} &= 1 + \frac{\mathbf{Z}_{12}}{\mathbf{Z}_{11}}\end{aligned}\right\} \tag{38}$$

It is interesting to note that **A** and **D** are pure numerics; from eqs. (32) and (33) it is apparent that **A** is the open-circuit voltage ratio and that **D** is the short-circuit current ratio of the transducer.

The constant **B** has the dimensions of impedance, whereas the constant **C** has the dimensions of admittance. From eq. (32) the constant **B** may be recognized as the transfer impedance of the transducer. The constant **C**, from eq. (33), may be seen to be the ratio of source current to receiver voltage when the receiver terminals are open. These physical meanings which attach themselves to the general circuit constants provide means for their direct measurement.

If two transducers are connected in series, the general circuit constants of the resulting transducer are easily expressible in terms of the general circuit constants of the separate transducers. Let $\mathbf{A}_1$, $\mathbf{B}_1$, $\mathbf{C}_1$, and $\mathbf{D}_1$ be the general circuit constants of the first, and $\mathbf{A}_2$, $\mathbf{B}_2$, $\mathbf{C}_2$, and $\mathbf{D}_2$ the general circuit constants of the second transducer; then, if $\mathbf{A}_s$, $\mathbf{B}_s$, $\mathbf{C}_s$, $\mathbf{D}_s$ are the general circuit constants of the transducer formed by the two in series, we have:

$$\left.\begin{aligned}\mathbf{A}_s &= \mathbf{A}_1\mathbf{A}_2 + \mathbf{B}_1\mathbf{C}_2\\ \mathbf{B}_s &= \mathbf{A}_1\mathbf{B}_2 + \mathbf{B}_1\mathbf{D}_2\\ \mathbf{C}_s &= \mathbf{C}_1\mathbf{A}_2 + \mathbf{D}_1\mathbf{C}_2\\ \mathbf{D}_s &= \mathbf{C}_1\mathbf{B}_2 + \mathbf{D}_1\mathbf{D}_2\end{aligned}\right\} \tag{39}$$

* $\mathbf{E}_s$ and $\mathbf{E}_r$ are voltage **rises** in source and receiver, respectively; $\mathbf{I}_s$ is the current flowing **into** the network from the source; and $\mathbf{I}_r$ is the current flowing **out** of the network tc the receiver.

If $\mathbf{A}_p$, $\mathbf{B}_p$, $\mathbf{C}_p$, $\mathbf{D}_p$ are the general circuit constants of the transducer formed by the two transducers in parallel, we have:

$$\left.\begin{aligned} \mathbf{A}_p &= \frac{\mathbf{A}_1\mathbf{B}_2 + \mathbf{B}_1\mathbf{A}_2}{\mathbf{B}_1 + \mathbf{B}_2} \\ \mathbf{B}_p &= \frac{\mathbf{B}_1\mathbf{B}_2}{\mathbf{B}_1 + \mathbf{B}_2} \\ \mathbf{C}_p &= \mathbf{C}_1 + \mathbf{C}_2 + \frac{(\mathbf{A}_1 - \mathbf{A}_2)(\mathbf{D}_2 - \mathbf{D}_1)}{\mathbf{B}_1 + \mathbf{B}_2} \\ \mathbf{D}_p &= \frac{\mathbf{B}_1\mathbf{D}_2 + \mathbf{D}_1\mathbf{B}_2}{\mathbf{B}_1 + \mathbf{B}_2} \end{aligned}\right\} \quad (40)$$

Since **A**, **B**, **C**, and **D** for a given transducer are not independent, slightly more work is done in calculation of these constants than if three independent constants were used; this, however, affords an extremely valuable check of the numerical computation. Equation (34) may be checked at various stages throughout a lengthy calculation, it being extremely unlikely that this equation will be satisfied if a numerical error has been made.

20. GENERAL n-TERMINAL NETWORK

It is possible, given an electrical network with any number of points of entry, to obtain a circuit having the same number of points of entry which cannot be distinguished from the given circuit by any single-frequency electrical tests made at the terminals. This equivalence is independent of any arbitrary pairing of terminals and hence is more general than the transducer equivalence discussed in Articles 18 and 19.

The performance of a network having $(n + 1)$ points of entry is determined by $n(n + 1)/2$ independent impedance links. This becomes apparent if the mesh equations are written down. Referring to the network of Fig. 4, assume that equations have been written for all the currents in the network and that the resulting system of equations has been reduced by elimination of currents until there remain n equations in n of the $(n + 1)$ branch currents indicated. The remaining branch current is then determined, since the sum of all $(n + 1)$ currents must be zero. The impedance equations are of the form given by eqs. (18). Since $\mathbf{Z}_{jk} = \mathbf{Z}_{kj}$, the number of independent impedances in eqs. (18) is $n(n + 1)/2$.

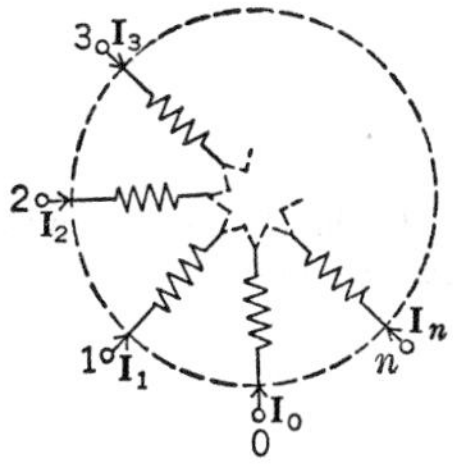

Fig. 4. General Network Having n Points of Entry

If a new network is set up, having the same number of points of entry as the given network, in which each point of entry is connected with every other point of entry through an impedance, the total number of impedances in the new network is the number of combinations of $(n + 1)$ things taken 2 at a time, which is $n(n + 1)/2$.

It is possible to solve for the $n(n + 1)/2$ new impedances in terms of the $n(n + 1)/2$ impedances of the given network so as to preserve equality of branch currents in any set of single-frequency impressed voltages. The method is illustrated in Article 21 for a four-terminal network; it may readily be generalized to any number of terminals.

An equivalent network of this type is sufficiently general for any steady-state calculations and is valuable in the representation of complex networks on calculating boards.

21. GENERAL FOUR-TERMINAL NETWORK

Let the general four-terminal network be represented by Fig. 5(*a*), and let the network which we propose to make equivalent to it be represented by Fig. 5(*b*). Terminal 0 will be considered a reference or "ground" terminal in both cases. In other words the impressed voltages will be thought of as being $\mathbf{V}_{10}$, $\mathbf{V}_{20}$, and $\mathbf{V}_{30}$, all other voltages being differences of these. All quantities in (*a*) will be differentiated from corresponding quantities in (*b*) by a prime. It is required that the currents $\mathbf{I}'_1$, $\mathbf{I}'_2$, and $\mathbf{I}'_3$ be equal to $\mathbf{I}_1$, $\mathbf{I}_2$, and $\mathbf{I}_3$, respectively, when $\mathbf{V}'_{10}$, $\mathbf{V}'_{20}$, and $\mathbf{V}'_{30}$ are equal to $\mathbf{V}_{10}$, $\mathbf{V}_{20}$, and $\mathbf{V}_{30}$, respectively.

The equations for the currents in the given network are of the form

$$\left.\begin{aligned} \mathbf{A}_{11}\mathbf{I}'_1 + \mathbf{A}_{12}\mathbf{I}'_2 + \mathbf{A}_{13}\mathbf{I}'_3 &= \mathbf{V}'_{10} \\ \mathbf{A}_{12}\mathbf{I}'_1 + \mathbf{A}_{22}\mathbf{I}'_2 + \mathbf{A}_{23}\mathbf{I}'_3 &= \mathbf{V}'_{20} \\ \mathbf{A}_{13}\mathbf{I}'_1 + \mathbf{A}_{23}\mathbf{I}'_2 + \mathbf{A}_{33}\mathbf{I}'_3 &= \mathbf{V}'_{30} \end{aligned}\right\} \quad (41)$$

The solution of these equations may be found by a simple transformation to be

$$\left.\begin{aligned} I'_1 &= B_{11}V'_{10} + B_{12}V'_{20} + B_{13}V'_{30} \\ I'_2 &= B_{12}V'_{10} + B_{22}V'_{20} + B_{23}V'_{30} \\ I'_3 &= B_{13}V'_{10} + B_{23}V'_{20} + B_{33}V'_{30} \end{aligned}\right\} \quad (42)$$

in which the **B**'s are equivalent admittances calculable from the actual network impedances by ordinary algebra of complex numbers.

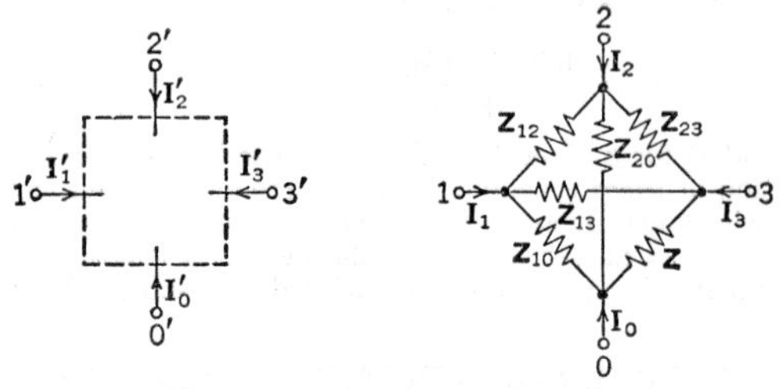

FIG. 5. General Four-terminal Networks

If all terminals except the first are shorted to the reference terminal as shown in Fig. 6, it follows that in the (*a*) network the currents are given by

$$\left.\begin{aligned} I'_1 &= B_{11}V'_{10} \\ I'_2 &= B_{12}V'_{20} \\ I'_3 &= B_{13}V'_{30} \end{aligned}\right\} \quad (43)$$

In the (*b*) network a current entering the common bus from any terminal is simply the current in that branch of the network which joins the given terminal with terminal 1.

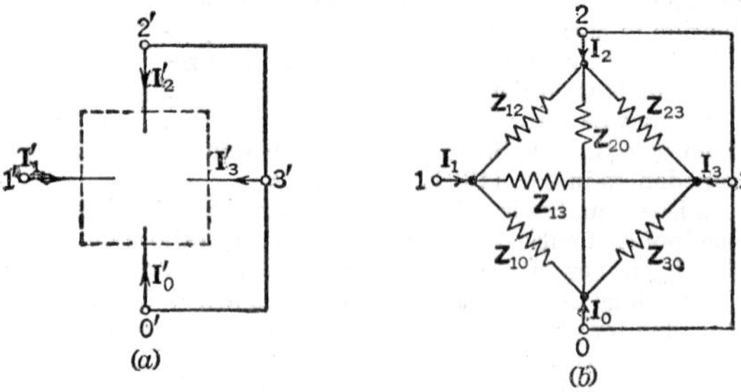

FIG. 6. General Four-terminal Networks

This is due to the fact that no current flows in any of the branches which are short-circuited by the common bus. The currents are

$$\left.\begin{aligned} I_2 &= -\frac{V_{10}}{Z_{12}} \\ I_3 &= -\frac{V_{10}}{Z_{13}} \\ I_0 &= -\frac{V_{10}}{Z_{10}} \end{aligned}\right\} \quad (44)$$

When eqs. (43) and (44) are compared, the two simple relations

$$Z_{12} = -\frac{1}{B_{12}}$$

and

$$Z_{13} = -\frac{1}{B_{13}} \quad (45)$$

are observed. By shorting the terminals in other groupings the following general equations may be verified:

$$\mathbf{Z}_{jk} = -\frac{1}{\mathbf{B}_{jk}} \tag{46}$$

It may be shown that the equivalent circuit of a network having *any* number of points of entry follows the same scheme, and that eq. (46) also holds.*

Example. The general equivalent circuit of a two-winding transformer is easily found by the method outlined. Let the primary terminals be 0 and 1, and let the secondary terminals be 2 and 3. The mesh equations, when all terminals except terminal 1 are strapped together, are [refer to Fig. 7(*a*)]

$$\begin{aligned}(r_a + j\omega L_a)\mathbf{I}_a - j\omega M\mathbf{I}_b &= \mathbf{V}_a \\ -j\omega M\mathbf{I}_a + (r_b + j\omega L_b)\mathbf{I}_b &= 0\end{aligned} \tag{47}$$

in which $\mathbf{I}_a$ and $\mathbf{I}_b$ are the primary and secondary mesh currents. The following relations are apparent, comparing the transformer with Fig. 6:

$$\left.\begin{aligned}\mathbf{I}_a &= \mathbf{I}'_1 = -\mathbf{I}'_0 \\ \mathbf{I}_b &= \mathbf{I}'_3 = \mathbf{I}'_2 \\ \mathbf{V}_a &= \mathbf{V}'_{10}\end{aligned}\right\} \tag{48}$$

Hence we have

$$\mathbf{B}_{12} = -\frac{j\omega M}{\mathbf{D}} \tag{49}$$

and

$$\mathbf{B}_{13} = \frac{j\omega M}{\mathbf{D}} \tag{50}$$

in which

$$\mathbf{D} = (r_a + j\omega L_a)(r_b + j\omega L_b) + \omega^2 M^2 \tag{51}$$

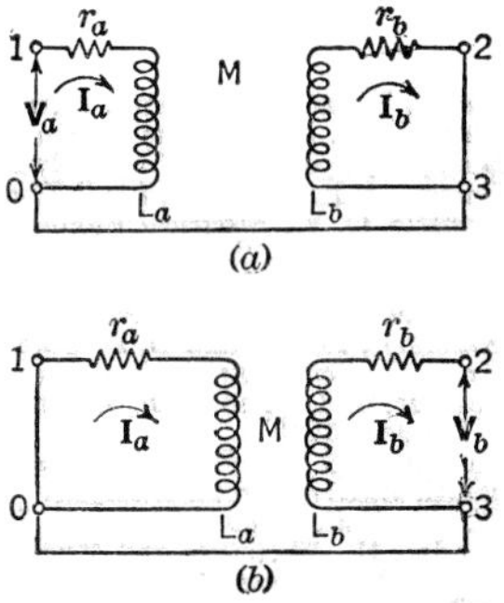

FIG. 7. Air-core Transformer

Similarly, when all terminals except terminal 2 are strapped together, the mesh equations are [see Fig. 7(*b*)]:

$$\begin{aligned}(r_a + j\omega L_a)\mathbf{I}_a - j\omega M\mathbf{I}_b &= 0 \\ -j\omega M\mathbf{I}_a + (r_b + j\omega L_b)\mathbf{I}_b &= \mathbf{V}_b\end{aligned} \tag{52}$$

In this case the relations between $\mathbf{I}_a$ and $\mathbf{I}_b$ and the currents of Fig. 6 remain unchanged, but $\mathbf{V}_b$ now corresponds to $\mathbf{V}_{02}$. Hence we have

$$\mathbf{B}_{12} = -\frac{j\omega M}{\mathbf{D}} \tag{53}$$

$$\mathbf{B}_{23} = -\frac{r_a + j\omega L_a}{\mathbf{D}} \tag{54}$$

From the relation $\mathbf{B}_{jk} = -\dfrac{1}{\mathbf{Z}_{jk}}$, and from symmetry the $\mathbf{Z}$'s of the equivalent network are

$$\mathbf{Z}_{10} = \frac{(r_a + j\omega L_a)(r_b + j\omega L_b) + \omega^2 M^2}{r_b + j\omega L_b}$$

$$\mathbf{Z}_{23} = \frac{(r_a + j\omega L_a)(r_b + j\omega L_b) + \omega^2 M^2}{r_a + j\omega L_a}$$

$$\mathbf{Z}_{12} = \mathbf{Z}_{30} = \frac{(r_a + j\omega L_a)(r_b + j\omega L_b) + \omega^2 M^2}{j\omega M}$$

$$\mathbf{Z}_{13} = \mathbf{Z}_{20} = \frac{(r_a + j\omega L_a)(r_b + j\omega L_b) + \omega^2 M^2}{-j\omega M}$$

The circuit is shown in Fig. 8. This circuit cannot be realized in practice because negative resistances would be required. In order to eliminate this difficulty and to obtain a circuit

* F. M. Starr, Equivalent Circuits. I, *Trans. AIEE* (Feb., 1932).

which can be used on a calculating board it is desirable to transform this network, if possible, into one having positive resistances only. This has been done by Starr; * the equivalent circuit is shown in Fig. 9.

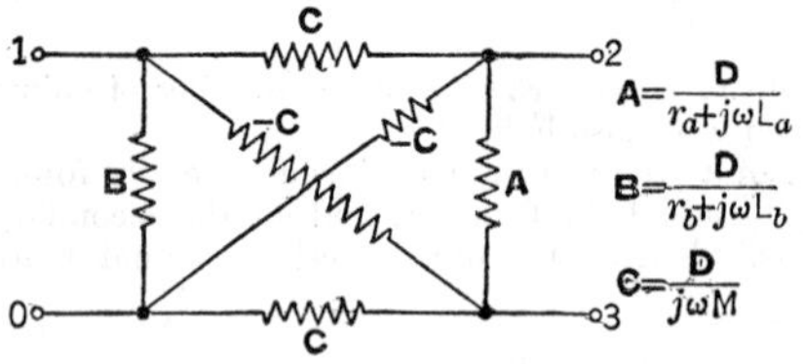

FIG. 8. Equivalent Circuit of a Two-winding Transformer

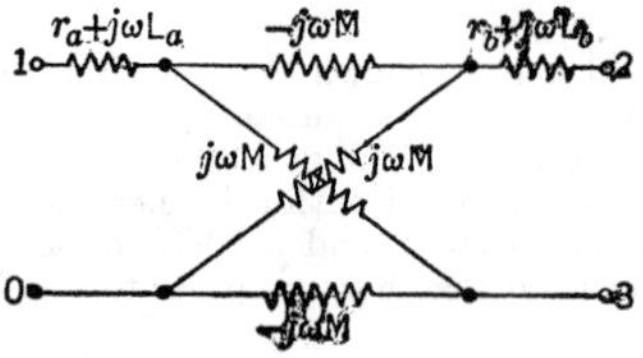

FIG. 9. Equivalent Circuit of a Two-winding Transformer (having positive resistance only)

22. BIBLIOGRAPHY

Bedell, F., and A. Crehore, *Alternating Currents.* McGraw-Hill (1909).
Blondel, A., *Les courants alternatifs.* J. B. Ballière et Fils, Paris (1933).
Bush, V., *Operational Circuit Analysis.* John Wiley (1937).
Carson, J. R., *Electric Circuit Theory and the Operational Calculus.* McGraw-Hill (1927).
Christie, C. V., *Electrical Engineering.* McGraw-Hill (1925).
Drysdale, C. V., *The Foundations of Alternating-current Theory.* Edward Arnold, London (1910).
Franklin, W. S., and W. Esty, *The Elements of Electrical Engineering.* Macmillan (1914).
Karapetoff, V., *The Electric Circuit.* McGraw-Hill (1912).
Kerchner, R. M., and G. F. Corcoran, *Alternating Current Circuits.* John Wiley (1943).
Kron, G., *Tensor Analysis of Networks.* John Wiley (1939).
LaCour, J. L., and O. S. Bragstad, *Theory and Calculation of Electric Currents.* Longmans, Green, London (1913).
Lawrence, R. R., *Principles of Alternating Currents.* McGraw-Hill (1922).
Malti, *Electric Circuit Analysis.* John Wiley (1930).
M.I.T. Staff, *Electric Circuits.* John Wiley (1940).
Pender, H., *Electricity and Magnetism for Engineers.* McGraw-Hill (1918). 2 vols.
Rosen, A., A New Network Theorem. *J.I.E.E.*, Vol. 62, p. 916.
Russell A., *The Theory of Alternating Currents.* Cambridge University Press (1914).
Starr, F. M., Equivalent circuits—I. *AIEE Trans.* (February, 1932).
Steinmetz, C. P., *Theory and Calculation of Alternating-Current Phenomena.* 5th ed. McGraw-Hill (1916).
Steinmetz, C. P., *Transient Electric Phenomena and Oscillations.* McGraw-Hill (1911).
Thévenin, M. L., Sur un nouveau théorème d'électricité dynamique. *Comptes rendus*, 97, p. 159 (1883).

NON-LINEAR CIRCUITS

By Irven Travis

A non-linear circuit is one in which one or more of the circuit elements, resistance, inductance, or capacitance, depends upon one or more of the currents flowing in the circuit. Variable electric circuits in which the resistances, inductances, or capacitances are functions of time or position in space, but not dependent upon the currents, are not non-linear but require special methods for their solutions. Some of these are given in circuits of Electrical Machines, Articles 50 to 53.

The outstanding feature of a non-linear circuit is the production of component frequencies of current other than those corresponding to impressed voltages or natural periods.

In the present discussion the most important types of non-linearity arising in electric power circuits are described; the various methods for solving networks involving them are outlined.

23. FERROMAGNETISM

A ferromagnetic material is one having a permeability greater than that of a vacuum. The permeability of such materials is ordinarily a function of the magnetizing force and dependent upon the previous history of the sample under consideration.

When any completely demagnetized ferromagnetic substance is subjected to an increasing magnetizing force H, a curve such as the dotted curve in Fig. 1(a) results. Such a curve is variously termed a **rising characteristic,** a **virgin curve,** or a **neutral curve;** all these names are reserved for the special case of a magnetization curve of a **demagnetized** body.

* F. M. Starr, Equivalent Circuits, I, *Trans. AIEE* (Feb., 1932).

If, from any positive value of H, the magnetizing force is decreased, the flux density will also decrease, but not along the same curve as the original rising characteristic. For the decreasing values of H, B is greater than for the corresponding values on the rising characteristic. The magnetized body retains a part of its magnetization. It will be noticed that, when the magnetizing force is reduced to zero, there is still a positive value of B, which is dependent upon the maximum positive value of H on the rising characteristic, before H is decreased. The flux density which remains when H has been reduced to zero is called the **residual magnetism** and is in the nature of permanent or semi-permanent magnetism. If now the magnetizing force is reversed (e.g., by reversing the direction of the current through the magnetizing coil), a value of H will be found which will reduce B to zero. This value of H is called the **coercive force.** By proceeding through a cycle of values for the magnetizing force—from zero to a positive maximum,

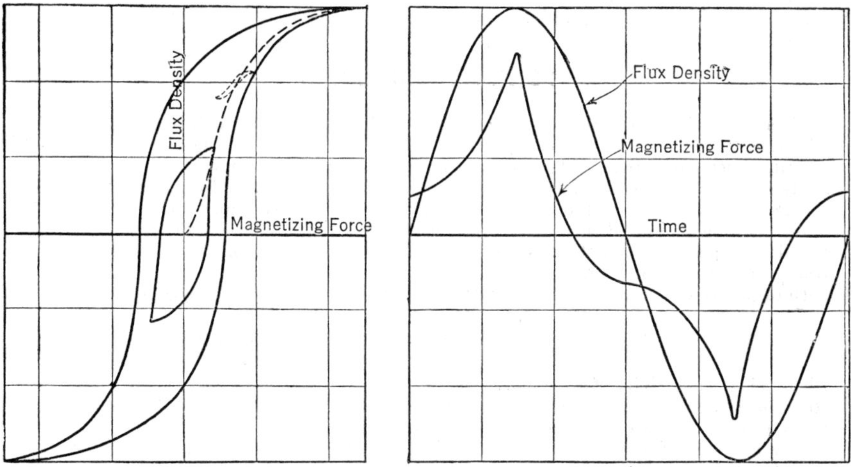

Fig. 1. **Magnetization Curves and Wave Form of Magnetizing Force**

to zero, to a negative maximum, and back to zero—a loop of flux densities will be obtained. Such a loop is called a **hysteresis loop;** each substance is capable of having an infinity of different hysteresis loops depending on the conditions of magnetization. During part of the magnetizing cycle the electric circuit is transferring energy to the magnetic circuit, and during the remainder of the cycle energy is retransferred from the magnetic to the electric circuit. When any ferromagnetic substance is present in the magnetic circuit, more energy is transferred to the magnetic circuit than is retransferred to the electric circuit. The difference between these two amounts of energy is a loss and is dissipated as heat. Part of this loss is called the **hysteresis loss** and is presumably due to molecular friction accompanying the change in orientation of the elementary molecular magnets within the ferromagnetic substance. It may be shown that the hysteresis loss per cycle per cubic centimeter in any magnetic substance due to a complete cycle of changes of the flux density and the magnetizing force is equal to $1/4\pi$ times the area of the corresponding hysteresis loop, provided this loop is determined under such conditions that there is no mechanical motion and no change in the relative distribution of the lines of force. This relation is based on the further assumption that unit distance on one scale represents 1 gauss and that unit distance on the other scale represents 1 oersted. The following empirical formula is useful in determining hysteresis loss:

$$w = \eta B_m^{1.6} \text{ ergs per cubic centimeter per cycle} \tag{1}$$

where η is a coefficient depending upon the substance, and B_m is the maximum flux density, in gausses, reached during the cycle.

The power loss per unit weight of magnetic material is given by

$$P_h = kfB_m^{1.6} \text{ watts per pound} \tag{2}$$

where f is the frequency in cycles per second, and k is constant.

If the value B varies sinusoidally with time, as is approximately true in many practical cases, the value of H at any instant of time can be found graphically from the hysteresis loop as shown in Fig. 1(*b*). This is of value in calculating exciting currents of transformers and machines.

If, as sometimes happens, the variation of magnetizing force with time is such that there is more than one maximum and one minimum value per cycle, the above formulas for power loss and the above graphical construction are not valid. In such cases the hysteresis loop has interior loops, as indicated by the dotted loop in Fig. 1(*a*). The shape of such interior loops and the power loss corresponding to them will depend upon the value of H at which the loop occurs and the width of the loop. No simple method has been devised to calculate circuits in which this occurs, even when all the data are known, since ordinarily the character of the B versus H curve cannot be found except by direct measurement under the specified conditions.

For detailed information, including tables and curves, refer to Section 2, Articles 22–29.

24. EDDY CURRENTS

Since a varying magnetic field induces an emf in any closed circuit which it links, it will in general cause a flow of current in the magnetic materials composing the magnetic circuit. Such currents are called **eddy currents** or **Foucault currents** and cause rI^2 losses. Eddy currents can be distinguished from ordinary induced currents by the fact that they are due solely to the lines of force which pass **through** the space occupied by the conducting mass. Ordinary induced currents are produced by lines of force which **link** the conductor but do not pass through the space occupied by the conducting mass. It is customary wherever possible to laminate magnetic cores in order to increase the resistance of the path of induced eddy currents. The eddy-current loss in metal sheets is given by

$$P_e = \frac{\pi^2}{6 \times 10^{16}} \frac{1}{\rho} (afB_m)^2 \text{ watts per cubic centimeter} \tag{3}$$

where ρ is the resistivity of the material in ohm-centimeters ($\rho = 0.25 \times 10^{-4}$ for ordinary steel), a is the thickness of each sheet in centimeters, f is the frequency of the eddy currents, in cycles per second, and B_m is the maximum value of the flux density, in gausses, during a cycle. Equation (3) holds, provided that (*a*) the lines of force are parallel to the planes of the laminations, (*b*) the laminations are thoroughly insulated from each other, and (*c*) the thickness of each lamination is small in comparison with its other dimensions. The combination of hysteresis loss and eddy current is called the **total core loss.** A method of calculating circuits in which eddy currents are of importance is given in Article 30.

25. SKIN EFFECT

A conductor of finite cross-section may be looked upon as made up of separate filaments, each carrying its portion of the total current. When the same potential gradient is established through all the filaments of the conductor, the exterior filaments are linked by fewer flux lines than the interior filaments. If the emf producing the potential gradient through the wire is an alternating one, the induced back-emf in the interior filaments will be greater than in those nearer the surface. Since the potential drop is the same across all the filaments, the resistance drops in the internal filaments are less than in the external filaments. This can be brought about only by the current distributing itself over the cross-section of the conductor in such a manner that the current density in the interior of the wire will be less than at the surface; i.e., the current is forced toward the surface filaments or "skin" of the wire. Hence the term skin effect is applied to this phenomenon.

The self-induced emf depends not only upon the amount of flux set up but also upon the rapidity of its variations; hence the skin effect becomes more pronounced the greater the frequency of the impressed emf. It is also greater the larger the cross-section of the conductor, the greater the conductivity of the conductor, and the greater its magnetic permeability. It also depends slightly upon the temperature, since the conductivity changes with temperature.

As a consequence of the skin effect the effective resistance of a conductor to alternating currents is greater than to direct currents, but the **internal** inductance **decreases** with the frequency; the external inductance is not altered. Whereas, however, the internal inductance approaches a limiting value with increasing frequency, the resistance increases indefinitely as the frequency approaches an infinite value. The change of resistance is always relatively much larger than the change in the total inductance.

The effects just described are, for the most part, negligible at low frequencies, except in heavy conductors and in coils wound with stout wire in several layers. In the latter case, however, the diminution of the inductance, due to the irregular distribution of the current, is masked, to a greater or less degree, by the effect of the capacitance between the windings of the coil, which gives rise to an **increase** of the inductance with the frequency. For the same reason the resistance is increased more than it would be by the eddy currents alone.

Unfortunately, the rigorous or approximate solution of the problem at high frequencies for the various cases which arise in practice is in many instances very difficult, if not impossible.

In the above discussion the redistribution of the current stream lines due to the current in the conductor itself has been considered. If there are other current-carrying conductors in the vicinity, a further redistribution of the current stream lines will occur, producing an additional change in resistance and inductance. This is called the **proximity effect.**

For a mathematical analysis of skin effect refer to A. Russell, *Phil. Mag.*, Vol. 17, p. 524 (1909); J. R. Carson, *Phil. Mag.*, Vol. 41, p. 607 (April, 1921); Sallie Pero Mead, *B.S.T.J.*, Vol. 4, No. 2 (April, 1925).

For tables, curves, and empirical formulas refer to H. B. Dwight, Skin Effect in Tubular and Flat Conductors, *Trans. AIEE*, Vol. 27, pp. 1379–1403 (1918); Skin Effect and Proximity Effect in Tubular Conductors, *Trans. AIEE*, Vol. 41, pp. 189–198 (1922); Proximity Effect in Wires and Thin Tubes, *Trans. AIEE*, Vol. 42, pp. 850–859 (1923); Chester Snow, Alternating Current Distribution in Cylindrical Conductors, *Sci. Paper Bur. Standards* (1925); E. Grünwald, Current Displacement in Tubular Cylindrical Conductors, *Arch. f. Electrot.*, Vol. 26, pp. 513–517 (July 5, 1932); H. C. Forbes, and L. J. Gorman, Skin Effect in Rectangular Conductors, *Trans. AIEE*, Vol. 52, pp. 516–519; Disc., pp. 519–520 (June, 1933); *Elect. Eng.*, Vol. 52, pp. 636–639 (Sept., 1933).

SKIN EFFECT IN STRAIGHT ROUND WIRES. An accurate solution for the case of straight **solid** * wires of circular cross-section has been given in a number of different forms by various scientists. A summary of the formulas is given in a paper by Rosa and Grover, *Bull. Bur. Standards*, Vol. 8, p. 172 (1912). The calculations are most conveniently made by the use of the tables in Rosa and Grover's paper, which are given in a condensed form in Table 1. Let

f = frequency in cycles per second.
μ = permeability of the wire, **assumed constant.**
R = d-c resistance, in ohms, of 1000 ft of the wire; see tables in Bare Wires and Cables, Section 14, Article 54.
L = d-c inductance, in millihenrys per 1000 ft, of a **non-magnetic** wire of the same **cross-section** as that of the given wire, and at the given spacing between wires; see tables in Section 14, Article 21.

Calculate the quantity

$$x = 0.02768 \sqrt{\frac{\mu f}{R}} \tag{4}$$

and take from Table 1 the corresponding values of K_1 and K_2. Then, in the absence of proximity effect, the a-c resistance at the frequency f is

$$R' = K_1 R \text{ ohms per 1000 ft} \tag{5}$$

and the a-c inductance of the given wire at the frequency f is

$$L' = L + 0.01524(\mu K_2 - 1) \text{ millihenrys per 1000 ft} \tag{6}$$

For x greater than 7 the following relations hold to within less than 1 per cent, the error being less the greater the value of x:

$$R' = \left(\frac{x}{2.828} + 0.25\right) R \tag{7}$$

$$L' = L + 0.01524 \left(\frac{2.828\mu}{x} - 1\right) \tag{8}$$

* Tests at the Massachusetts Institute of Technology [*Proc. AIEE* (Sept., 1915)], indicate that the formulas for solid wires also apply to round **stranded** wires of the **same cross-section of metal,** not the same overall diameter, for frequencies up to 1200, but above this frequency there is an increase in the skin effect, due to the spiraling of the component wires of the cable.

Table 1. Skin-effect Factors for Solid Round Wires

x	K_1	K_2	x	K_1	K_2	x	K_1	K_2	x	K_1	K_2
0.0	1.000	1.000	2.9	1.286	0.860	6.6	2.603	0.424	17.0	6.268	0.166
0.1	1.000	1.000	3.0	1.318	0.845	6.8	2.673	0.412	18.0	6.621	0.157
0.2	1.000	1.000	3.1	1.351	0.830	7.0	2.743	0.400	19.0	6.974	0.149
0.3	1.000	1.000	3.2	1.385	0.814	7.2	2.813	0.389	20.0	7.328	0.141
0.4	1.000	1.000	3.3	1.420	0.798	7.4	2.884	0.379	21.0	7.681	0.135
0.5	1.000	1.000	3.4	1.456	0.782	7.6	2.954	0.369	22.0	8.034	0.128
0.6	1.001	1.000	3.5	1.492	0.766	7.8	3.024	0.360	23.0	8.387	0.123
0.7	1.001	0.999	3.6	1.529	0.749	8.0	3.094	0.351	24.0	8.741	0.118
0.8	1.002	0.999	3.7	1.566	0.733	8.2	3.165	0.343	25.0	9.094	0.113
0.9	1.003	0.998	3.8	1.603	0.717	8.4	3.235	0.335	26.0	9.447	0.109
1.0	1.005	0.997	3.9	1.641	0.702	8.6	3.306	0.327	28.0	10.154	0.101
1.1	1.008	0.996	4.0	1.678	0.686	8.8	3.376	0.320	30.0	10.861	0.094
1.2	1.011	0.995	4.1	1.715	0.671	9.0	3.446	0.313	32.0	11.568	0.088
1.3	1.015	0.993	4.2	1.752	0.657	9.2	3.517	0.306	34.0	12.275	0.083
1.4	1.020	0.990	4.3	1.789	0.643	9.4	3.587	0.299	36.0	12.982	0.079
1.5	1.026	0.987	4.4	1.826	0.629	9.6	3.658	0.293	38.0	13.689	0.074
1.6	1.033	0.983	4.5	1.863	0.616	9.8	3.728	0.287	40.0	14.395	0.071
1.7	1.042	0.979	4.6	1.899	0.603	10.0	3.799	0.282	42.0	15.102	0.067
1.8	1.052	0.974	4.7	1.935	0.590	10.5	3.975	0.268	44.0	15.809	0.064
1.9	1.064	0.968	4.8	1.971	0.579	11.0	4.151	0.256	46.0	16.516	0.061
2.0	1.078	0.961	4.9	2.007	0.567	11.5	4.327	0.245	48.0	17.223	0.059
2.1	1.094	0.953	5.0	2.043	0.556	12.0	4.504	0.235	50.0	17.930	0.057
2.2	1.111	0.945	5.2	2.114	0.535	12.5	4.680	0.226	60.0	21.465	0.047
2.3	1.133	0.935	5.4	2.184	0.516	13.0	4.856	0.217	70.0	25.001	0.040
2.4	1.152	0.925	5.6	2.254	0.498	13.5	5.033	0.209	80.0	28.536	0.035
2.5	1.175	0.913	5.8	2.324	0.481	14.0	5.209	0.202	90.0	32.071	0.031
2.6	1.201	0.901	6.0	2.394	0.465	14.5	5.386	0.195	100.0	35.607	0.028
2.7	1.228	0.888	6.2	2.463	0.451	15.0	5.562	0.188			
2.8	1.256	0.875	6.4	2.533	0.437	16.0	5.915	0.176			

SKIN EFFECT IN THIN STRIPS AND TUBES. The following formulas are exact for a flat strip of infinite width and at an infinite distance from any other conductor carrying a current; they also apply with a close degree of approximation to a tube which has a circumference 10 or more times its thickness, provided no other conductor carrying a current is closer than a distance of 10 times the thickness of the strip or tube.

Let t = thickness of strip, or **twice** the thickness of the wall of a tube, in centimeters.
w = width of strip or **half** the mean circumference of tube, in centimeters.
l = length of strip or tube in centimeters.
ρ = specific resistance of conductor, in microhms per centimeter cube, at the given temperature.
μ = magnetic permeability of conductor in absolute units.
f = frequency in cycles per second.

$$x = 0.1987t\sqrt{\frac{\mu f}{\rho}}.$$

Then d-c resistance is

$$R = \frac{10^{-6}\rho l}{wt} \text{ ohms}$$

The d-c internal inductance is

$$L_i = 1.047 \times 10^{-6} l \frac{\mu t}{w} \text{ millihenrys}$$

The ratio of the a-c to the d-c resistance is

$$\frac{R'}{R} = \frac{x}{2}\left(\frac{\sinh x + \sin x}{\cosh x - \cos x}\right)$$

and the ratio of the a-c to the d-c internal inductance is

$$\frac{L'_i}{L_i} = \frac{3}{x}\left(\frac{\sinh x - \sin x}{\cosh x - \cos x}\right)$$

For x less than unity these ratios are unity to within 0.6 and 0.2 per cent, respectively; i.e., the a-c resistance is practically equal to the d-c resistance, and the a-c internal inductance is practically equal to the d-c internal inductance. For x greater than 6 the following formulas are accurate to within 0.5 per cent:

$$\frac{R'}{R} = \frac{x}{2} \quad \text{and} \quad \frac{L'_i}{L_i} = \frac{3}{x}$$

To a very rough degree of approximation the skin effect in a conductor of any shaped cross-section may be approximated by using the above formulas for a strip, taking for the effective width w one-half the perimeter of the section, in centimeters, and for its effective thickness twice the area of the section, in square centimeters, divided by its perimeter. In general, then, for the same area the skin effect will be less the greater the perimeter of the section. If the section approaches more nearly that of a solid circle than that of an elongated rectangle, the formulas for a solid round wire will give more accurate results.

SKIN EFFECT IN BUS-BARS (STRIPS OF FINITE WIDTH). The formulas given above are for strips of infinite width; the skin effect in narrow strips, such as those used for bus-bars, is much greater than the theoretical value for infinite strips, because of the crowding of the current to the edge of the strip. This latter effect has been called the "edge effect." Experimental results obtained by Kennelly, Laws, and Pierce [*Proc. AIEE* (September, 1915)] for strip conductors, outgoing and return conductor 60 cm apart, are given in Fig. 2. For other spacings the original paper should be consulted.

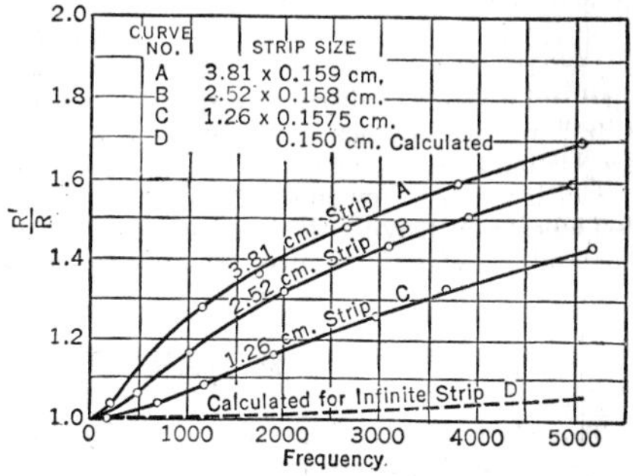

FIG. 2. Skin Effect in Finite Strips. Return conductor 60 cm away.

In the *Journal of the Franklin Institute* for March, 1918, are reported tests on copper bus-bars 4 in. by 1/4 in., showing a ratio of a-c to d-c resistance of 1.3 at 60 cycles and 1.1 at 24 cycles. It is stated that this increase in resistance is due chiefly to the crowding of the current toward the edge of the strip, and not to the sides. Three such bars in parallel with an air space of 0.25 in. between them gave a ratio of a-c to d-c resistance of 2.2 at 60 cycles.

SKIN EFFECT IN TRANSMISSION LINES. For a comprehensive treatment of effective resistance and inductance of bimetallic (copper-clad steel) and also of iron wires, see J. M. Miller, *Bull. Bur. Standards*, 12, p. 207 (1915). For tables of skin-effect ratios for concentric-lay and annular conductors see Section 14, Article 59.

26. ELECTRIC ARCS

An electric arc is an incandescent vapor bridge consisting of material electrically impelled from a negative to a positive electrode. A spark is also an incandescent vapor bridge, but differs from an arc in not depending upon the electrodes for its material medium. The establishment and maintenance of an arc require the expenditure of energy for the latent heat of evaporation of the electrode and for the motion of the vapor stream. As no energy can be expended for these purposes until a current flows, an arc cannot start spontaneously. The following expedients are, therefore, adopted to start arcs:

1. Bringing the electrodes into contact with each other and separating them after the current has commenced to flow.
2. Stressing the dielectric between the conductors by overvoltage until it breaks down electrically and becomes conducting.
3. Using a subsidiary arc to furnish the initial vapor bridge.

The first of these methods is used in carbon-arc lamps, and the third in mercury-arc rectifiers.

A peculiarity of the carbon arc is that, with any particular length of arc, if the current is increased the difference of potential across the carbons will decrease. This occurs continuously up to a certain point, when in the open arc the voltage drops quite suddenly. If the current is still increased, the voltage will again become steady at a much lower value. Between the values before and after the drop the arc is unstable and emits a hissing sound. In inclosed arcs the hissing point is absent and the curve is continuous.

The relation between the drop of potential across the arc and the current flowing through it depends upon whether the arc is in the steady state corresponding to the current flowing or whether the current has been changed without giving time to the vapor column and electrodes to accommodate themselves to the new strength of current. Let v be the drop of potential across an arc, and i be the current flowing. If the current is increased at a rate rapid enough so that a steady state is not set up, then dv/di is the resistance of the arc to the rapidly changing current. Throughout ordinary ranges of frequency (less than 100 kc) the ratio dv/di is negative. Thus the a-c resistance of an arc is negative and is a function of the current.

27. CORONA

When the voltage between two conductors becomes sufficiently great, the electric field, which is greatest at the surface, ionizes the air at the surfaces of the conductor. The increase in the conductivity of the air or other gas surrounding the conductor is equivalent to an increase in the effective diameter of the conductor to the value at which the decreasing electric field is balanced by the dielectric strength of the air. The phenomenon is known as **corona.** When the corona is formed, a faint violet light is visible near the conductor. If the voltage is raised, a **brush discharge** occurs in which bluish streaks like the bristles of a brush are visible near the surface of the conductor. If the voltage is still further increased, a disruptive spark discharge takes place between conductors.

In transmission lines (q.v.), corona is accompanied by power losses, frequently of serious proportions. It can be controlled by the use of larger-diameter conductors and by operating with voltages low enough to prevent its formation.

The current leaving the line as corona is a function of the instantaneous voltage on the line and therefore introduces non-linearity into the network. See Section 14, Article 18.

28. VARIABLE INDUCTANCE

The electromotive force produced by electromagnetic induction in a closed circuit having no sliding contacts is

$$e_1 = \frac{d\lambda_1}{dt} \tag{9}$$

in which λ_1 is the total flux linkages in the circuit. The flux linkages λ_1 will in general be produced by the current in the given circuit, by the currents in neighboring circuits, and by permanent magnets in the vicinity of the given circuit. The flux linkages may change, owing to changes in the currents or to motion. If there is motion, the space coordinates giving the relative positions of the circuits and magnets in the group are functions of time and can be eliminated in the expression for λ. Hence λ_1 may be written

$$\lambda_1 = \lambda_1(i_1, i_2, \cdots, i_n, t) \tag{10}$$

where i_1 is the current in the circuit under consideration, $i_2, \cdots, i_n$ are the currents in neighboring circuits, and t is the time.

It follows that

$$e = \frac{\partial\lambda_1}{\partial i_1}\frac{di_1}{dt} + \frac{\partial\lambda_1}{\partial i_2}\frac{di_2}{dt} + \cdots + \frac{\partial\lambda_1}{\partial t} \tag{11}$$

The coefficient of di_1/dt in this expression for the emf is called the **coefficient** of **self-induction** or simply the **inductance** of the circuit.

The coefficient of di_n/dt in this expression for the emf is called the **coefficient** of **mutual induction** or simply the **mutual inductance** of the nth mesh with respect to the first.

There result the following definitions:

The **self-inductance** of a given circuit is the factor L by which the rate of increase of current in that circuit must be multiplied to give the induced emf in that circuit when at rest, the currents in all neighboring circuits being zero.

The **mutual inductance** of a given circuit with respect to a second circuit is the factor M by which the rate of increase of current in the first circuit must be multiplied to give the induced emf in the second circuit when these circuits are at rest, the currents in the second and all neighboring circuits being zero.

In terms of these definitions the general expression for the induced emf is

$$e_1 = L_1\frac{di_1}{dt} + M_{12}\frac{di_2}{dt} + M_{13}\frac{di_3}{dt} + \cdots + M_{1n}\frac{di_n}{dt} + \frac{\partial\lambda_1}{\partial t} \tag{12}$$

There are several special cases which arise in practical problems, each of which will be considered separately.

a. If λ_1 *is a function of* i_1 *only* (as in an isolated iron-cored choke coil at rest),

$$\frac{\partial \lambda_1}{\partial t} = \frac{di_2}{dt} = \frac{di_3}{dt} = \cdots = \frac{di_n}{dt} = 0$$

from which it follows that

$$e_1 = L_1 \frac{di_1}{dt} \tag{13}$$

If the permeance of the core is constant, L is a constant; in general, however, L will be a function of the current i_1.

b. If λ_1 *is a function of* i_n *only* (as in an open winding of a multiwinding transformer in which another winding carries current),

$$\frac{\partial \lambda_1}{\partial t} = \frac{di_1}{dt} = \frac{di_2}{dt} = \cdots = \frac{di_{n-1}}{dt} = 0$$

from which it follows that

$$e_1 = M_{1n} \frac{di_n}{dt} \tag{14}$$

If the permeance of the core is constant, M_{1n} is a constant; in general, however, M will be a function of the current i_n.

c. If λ_1 *is composed of a sum of terms each of which is directly proportional to one of the currents,* and is also a function of t (as when several circuits move with respect to each other but have constant permeances, e.g., an air-core alternator)

$$\lambda_1 = i_1 A_1(t) + i_2 A_{12}(t) + \cdots + i_n A_{1n}(t) \tag{15}$$

The electromotive force is, from eq. (11),

$$e_1 = A_1(t) \frac{di_1}{dt} + A_{12}(t) \frac{di_2}{dt} + \cdots + A_{1n}(t) \frac{di_n}{dt}$$
$$+ \frac{\partial}{\partial t} [i_1 A_1(t) + i_2 A_{12}(t) + \cdots + i_n A_{1n}(t)] \tag{16}$$

and comparing coefficients with eq. (12) it is seen that

$$A_1(t) = L_1$$
$$A_{12}(t) = M_{12}$$

and

$$A_{1n}(t) = M_{1n} \tag{17}$$

hence

$$e_1 = L_1 \frac{di_1}{dt} + M_{12} \frac{di_2}{dt} + \cdots + M_{1n} \frac{di_n}{dt} + i_1 \frac{dL_1}{dt} + i_2 \frac{dM_{12}}{dt} + \cdots + i_n \frac{dM_{1n}}{dt} \tag{18}$$

Equation (18) may be written

$$e_1 = \frac{d}{dt} (L_1 i_1 + M_{12} i_2 + M_{13} i_3 + \cdots + M_{1n} i_n) \tag{19}$$

Equation (19) holds when the self and mutual inductances are functions of t only, but does not hold when they are functions of the current. In other words, eq. (19) may be used for generator and motor calculations if the magnetization curve can be assumed to be linear; whereas, for calculations when the magnetization curve cannot be assumed to be straight (even in a transformer), eq. (12) must be used.

29. SPECIAL MATHEMATICAL METHODS

There is no general method of solution applicable to differential equations which are not linear. Therefore the electric-circuit equations arising from non-linear circuits are not capable of solution by any standard scheme. Various special methods, satisfactory for limited classes of problems, have been devised, some of which are outlined in this article. The methods may be classified as follows:

REDUCTION OF THE NETWORK EQUATIONS TO APPROXIMATE LINEAR EQUATIONS. In many circuits the non-linearity is in the nature of a second order effect. In such circuits it is desirable to solve for those currents which would flow if the network were linear, and to superimpose upon them correction currents to take account of the non-linearity. An excellent example is the iron-core transformer which is given in some detail in Article 30. The method of solution outlined in that article may be applied to any circuit falling in this classification.

SUBSTITUTION OF AN EMF FOR THE NON-LINEAR ELEMENT. If the non-linearity consists in the sudden opening or closing of a circuit, the currents may be found by making use of the two principles:

a. The effect at any succeeding instant of time of a short circuit at time $t = 0$ is precisely the same as the insertion at $t = 0$ of a voltage $-v(t)$, equal and opposite to the voltage $v(t)$ which exists at that instant across the terminals to be short-circuited.

The resultant currents are therefore composed of two components: the currents which would exist in the invariable network in the absence of the short circuit and the currents due to the emf $v(t)$ inserted at $t = 0$. Both these components are calculable by usual methods.

b. The effect at any succeeding instant of time of opening a branch of a network at time $t = 0$ is the same as the insertion at $t = 0$ of a voltage $v(t)$ which produces in the branch to be opened a current $-i(t)$ equal and opposite to the current $i(t)$ which would exist in the branch in the absence of the open circuit.

The resultant currents are therefore composed of two components: the currents which would flow in the network in the absence of the open circuit and the currents which flow because of the emf $v(t)$ inserted at $t = 0$. Both these components are calculable by usual methods.

In a network having a non-linear circuit element, the mesh equation for the mesh involving this element is

$$Z(p)i + \phi(i) = e(t) \tag{20}$$

in which $Z(p)$ is the operational impedance of the linear part of the mesh, $\phi(i)$ is the voltage drop across the non-linear circuit element, and $e(t)$ is the impressed emf in the mesh.

If this equation is rewritten as

$$Z(p)i = e(t) - \phi[i(t)] \tag{20a}$$

it is possible to solve formally for the current by treating the right-hand member as the impressed voltage. Hence

$$i(t) = \frac{d}{dt}\int_0^t h(t - \lambda)\,\{e(\lambda) - \phi[i(\lambda)]\}\,d\lambda \tag{21}$$

which may be written

$$i(t) = i_0(t) - \frac{d}{dt}\int_0^t h(t - \lambda)\phi[i(\lambda)]\,d\lambda \tag{21a}$$

in which $i_0(t)$ = the current which would flow in the invariable part of the circuit due to $e(t)$; $h(t)$ = the indicial admittance of the invariable part.

This is an integral equation, which can be solved in general only in the form of an infinite series. For methods of solution see Lovitt, *Linear Integral Equations*, McGraw-Hill.

SUCCESSIVE APPROXIMATIONS, NUMERICAL INTEGRATION. When the non-linear differential equations for the currents in a non-linear network are written down, their exact solution becomes a purely mathematical problem. The process of obtaining an infinite series solution by successive approximations is given in any standard text on differential equations. For the numerical integration of differential equations refer to Runge and König, *Numerisches Rechnen*, Julius Springer (1924). These methods are usually very laborious and should be avoided if possible.

A method of solution of non-linear differential equations of the particular type found in vacuum-tube problems was developed by Carson. (See Bibliography and Pender-McIlwain, *Electrical Engineers' Handbook*, Section on Electric Circuits, Lines, and Fields.) This method is based on a Taylor's series expansion and is of great practical value to the communication engineer. It is not easily extended to other problems.

Another special scheme which has been applied to non-linear communication problems is the *isoclyne* method. This method involves the plotting of families of curves, using the derivative as a parameter; it has been used by Van der Pol in the analysis of relaxation oscillators.

THE DIFFERENTIAL ANALYZER. Perhaps the most powerful tool available for the numerical solution of non-linear circuit problems is a machine called the differential analyzer. This device is capable of solving mechanically differential equations in which the coefficients are functions of the dependent variables. The solution may be obtained as a plotted curve or in tabular form, the latter giving the greater accuracy. The precision is adequate for all engineering purposes. Several such machines are at present available in this country, including one at the Massachusetts Institute of Technology and one at the Moore School of Electrical Engineering, University of Pennsylvania. For a complete description of its principle of operation see V. Bush, The Differential Analyzer, *J. Franklin Inst.*, Vol. 212, No. 4 (Oct., 1931).

30. IRON-CORED TRANSFORMERS

The mesh equations of a two-winding transformer upon the primary of which is impressed a voltage v_1, and which supplies a voltage v_2 to a receiver connected to its secondary, are:

$$r_1 i_1 + L_1 \frac{di_1}{dt} - M \frac{di_2}{dt} = v_1 \tag{22}$$

$$r_2 i_2 + L_2 \frac{di_2}{dt} - M \frac{di_1}{dt} = -v_2 \tag{22a}$$

in which i_1 and i_2 are the primary and secondary mesh currents. Because of the variable permeability of the iron core the values of the inductances depend upon the currents. The equations are therefore non-linear and cannot be solved by ordinary means. In the design of power transformers for usual purposes, an attempt is made to keep the leakage as small as possible, whereas the mutual inductance is made relatively large. This fact makes possible an approximate method of solution which yields results sufficiently accurate for most practical purposes.

The self-inductances of the windings may be expressed in terms of the leakage and mutual inductances:

$$L_1 = L_1' + aM \tag{23}$$

$$L_2 = L_2' + \frac{M}{a} \tag{23a}$$

in which L_1' and L_2' are the primary and secondary leakage inductances, and a is the turn ratio. L_1' and L_2' are constants and are quite small in comparison to M. Substitution of eqs. (23) and (23*a*) in eqs. (22) and (22*a*) gives the following mesh equations:

$$aM \frac{d}{dt}\left(i_1 - \frac{i_2}{a}\right) = v_1 - r_1 i_1 - L_1' \frac{di_1}{dt} \tag{24}$$

$$M \frac{d}{dt}\left(i_1 - \frac{i_2}{a}\right) = v_2 + r_2 i_2 + L_2' \frac{di_2}{dt} \tag{24a}$$

It is apparent that, if r_1 and L_1' are small, the right-hand side of eq. (24) does not differ greatly in wave form from v_1. Hence the quantity $M \frac{d}{dt}\left(i_1 - \frac{i_2}{a}\right)$ does not differ greatly in wave form from v_1. It should be noted that, since M is variable, the quantity $i_1 - (i_2/a)$ may be considerably different in wave form from v_1. The secondary current i_2 will be essentially the same in wave form as v_1, since the term $M \frac{d}{dt}\left(i - \frac{i_2}{a}\right)$ may be thought of as an emf impressed upon a linear mesh in which the impedance drop is

$$v_2 + r_2 i_2 + L_2' \frac{di_2}{dt}$$

The primary current can thus be treated as the sum of two components, i_2/a and $i_1 - (i_2/a)$, the first component having essentially the same wave shape as the impressed voltage, and therefore representable as a current flowing in an equivalent linear network, and the second component being distorted in wave shape and representing the non-linear effect of the transformer. The first component, i_2/a, is often called the load component of the primary current; the second component, $i_1 - (i_2/a)$, is called the magnetizing or exciting current of the transformer.

The scheme used in eqs. (24) and (24*a*) of splitting mesh equations into two parts, one containing all the non-linearity and the other being entirely linear, is extremely useful in problems involving iron cores, especially electrical machines.

If the core of a transformer is conducting, as it always is in practical cases, there are an infinite number of conducting paths, each having mutual inductances with respect to the primary and secondary. The currents flowing in these paths are called eddy currents.

Let M_{1i} be the mutual inductance between the ith eddy-current path and the primary winding, and let M_{2i} be the mutual inductance between the ith eddy-current path and the secondary winding. If, as in eqs. (22) and (22*a*), we take M_{12} with a negative sign, then M_{1i} and M_{2i} must have opposite signs.

If we adopt the convention that all the M_{1i}'s have negative signs, then all the M_{2i}'s have positive signs. The mesh equations for the primary and secondary currents (neglecting non-linearity of the core) are

$$(r_1 + j\omega L_1)\mathbf{I}_1 - j\omega M_{12}\mathbf{I}_2 - j\omega \sum_{i=3}^{\infty} M_{1i}\mathbf{I}_i = \mathbf{V}_1 \quad (25)$$

$$(r_2 + j\omega L_2)\mathbf{I}_2 - j\omega M_{12}\mathbf{I}_1 + j\omega \sum_{i=3}^{\infty} M_{2i}\mathbf{I}_i = -\mathbf{V}_2 \quad (26)$$

Corresponding to the eddy-current paths there will exist an infinite number of equations. If in these the terms involving $\mathbf{I}_1$ and $\mathbf{I}_2$ are transposed to the right-hand side, it is possible, at least formally, to solve for each of the currents $\mathbf{I}_i$ in terms of $\mathbf{I}_1$ and $\mathbf{I}_2$. If these values are then substituted back in eqs. (25) and (26), two equations will result which involve only parameters of the network and the currents $\mathbf{I}_1$ and $\mathbf{I}_2$. These are the equations of an equivalent T network, in which the quantity multiplying $\mathbf{I}_2$ in the first equation will be the same as the quantity multiplying $\mathbf{I}_1$ in the second equation. This quantity will be a complex number, both real and imaginary parts of which are functions of the frequency ω.

If we make the assumptions that (*a*) the resistance of a given eddy-current path is large in comparison with its inductance, (*b*) the mutual inductance between an eddy-current path and either the primary or secondary is large in comparison with the mutual inductance between two eddy-current paths, and (*c*) the leakage is small, the mesh equations for the primary and secondary currents become

$$(r_1 + j\omega L_1')\mathbf{I}_1 + a(r_e + j\omega M_{12})\left(\mathbf{I}_1 - \frac{\mathbf{I}_2}{a}\right) = \mathbf{V}_1 \quad (27)$$

$$(r_2 + j\omega L_2')\mathbf{I}_2 - (r_e + j\omega M_{12})\left(\mathbf{I}_1 - \frac{\mathbf{I}_2}{a}\right) = -\mathbf{V}_2 \quad (28)$$

in which L_1' and L_2' are the primary and secondary leakage inductances, a is the turn ratio, and r_e is a resistance representing the effect of the eddy currents. The resistance r_e is proportional to the square of the frequency of the impressed voltage.

31. BIBLIOGRAPHY

Bozorth, R. M., Present Status of Ferromagnetic Theory. *Elec. Eng.*, Vol. 54, p. 1251 (Nov., 1935).
Bush, V., *Operational Circuit Analysis*. John Wiley (1929).
Campbell, G. A., and R. M. Foster, Fourier Integrals for Practical Applications. *Bell System Tech. Pub.* B-584 (Sept., 1931).
Carson, J. R., A Theoretical Study of the Three Element Vacuum Tube. *Proc. I.R.E.*, Vol. 7, p 187 (April, 1919).
Carson, J. R., *Electric Circuit Theory and the Operational Calculus*. McGraw-Hill (1927).
Lovitt, W. V., *Linear Integral Equations*. McGraw-Hill (1924).
Van der Pol, Balth., Relaxation Oscillations. *Phil. Mag.* (Nov., 1926) and *Zeits. f. Hochfrequenztechnik*, Vol. 24, pp. 114–118 (April, 1927).
Spooner, Thomas, *Properties and Testing of Magnetic Materials*. McGraw-Hill (1927).
Suits, Chauncey Guy, Studies in Non-linear Circuits. *Trans. AIEE* (1931).

POLYPHASE SYSTEMS

By Irven Travis

32. POLYPHASE CURRENTS AND VOLTAGES

The term polyphase alternating voltages is applied to the voltages of a group in which the modulus values are approximately the same and in which the difference in phase between successive voltages in the group is approximately equal to 2π divided by the number of voltages in the group. If these conditions are exactly instead of only approximately true, the voltages are said to be **balanced** polyphase voltages. If these conditions are not even approximately true, the voltages are still termed polyphase voltages (but badly unbalanced) if they arise in a system in which the normal or ideal voltages are polyphase.

Polyphase currents are defined in the same way as polyphase voltages.

A polyphase circuit is a circuit in which the normal or ideal currents are polyphase currents. Thus a circuit having three meshes may be a three-phase circuit or not, depending upon its use.

33. MESH AND STAR CONNECTIONS

Consider n separate coils or windings, which may be mounted on a common armature or be entirely distinct, for example, n groups of lamps. When these n windings are connected end to end so that they are all in series, forming a closed chain, as in Fig. 1, and terminals are brought out from the n junctions, they are said to be connected in **mesh.** When one terminal of each of these windings is connected to a common junction point, as in Fig. 2, and terminals are brought out from the free ends, the windings are said to be connected in **star,** and the common point is called the **neutral point.**

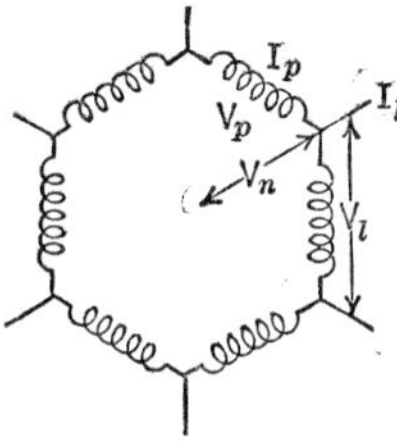

FIG. 1. Mesh-connected Network

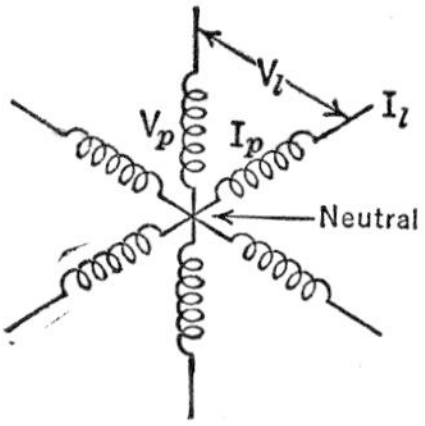

FIG. 2. Star-connected Network

When such a group of n windings, as shown in Fig. 1 or 2, is connected to a generator or other source of emf having n separate windings and therefore developing n different emf's which differ in phase from one another, the system is called an n-phase system, each winding being called a **phase.** For example, when there are three separate windings on the generator, three line wires and three windings constituting the load, the system is a three-phase system.

A system is said to be balanced if the phase impedances are equal and the impressed voltages are balanced.

34. PHASE AND LINE CURRENTS AND VOLTAGES

The current I_p in any winding or phase of an n-phase system (see Figs. 1 and 2) is called the **phase current,** and the drop (or rise) of potential V_p through this winding is called the **phase voltage.** The current I_l in the line leaving any terminal of an n-phase system is called the **line current,** and the voltage V_l between adjacent line wires or terminals is called the **line voltage,** except in the special case of a two-phase connection (see below) when the voltage between diametrically opposite terminals is called the line voltage. In a star connection the voltage between any terminal and the neutral point is called the **voltage to neutral;** in a balanced mesh connection, by voltage to neutral is meant the voltage which would exist between any terminal and the neutral of a star connection connected to the terminals of the actual device, the impedance of all the legs of the star connection forming this **artificial neutral** being equal and sufficiently large not to take an appreciable current.

The relations between these various currents and voltages for a balanced n-phase system are as follows:

	Mesh	Star
Number of phases	n	n
Line current	I_l	I_l
Line voltage	V_l	V_l
Phase current	$I_p = \dfrac{1}{2\sin\frac{\pi}{n}} I_l$	$I_p = I_l$
Phase voltage	$V_p = V_l$	$V_p = \dfrac{1}{2\sin\frac{\pi}{n}} V_l$
Voltage to neutral	$V_n = \dfrac{1}{2\sin\frac{\pi}{n}} V_p$	$V_n = V_p$
Total volt-amperes	$nV_p I_p = \dfrac{n}{2\sin\frac{\pi}{n}} V_l I_l$	$nV_p I_p = \dfrac{n}{2\sin\frac{\pi}{n}} V_l I_l$

35. POWER IN POLYPHASE SYSTEMS

The total power in any polyphase system is the sum of the powers in the separate phases of the system. Thus for an unbalanced n-phase system the total average power is

$$P = \sum_{k=1}^{n} V_k I_k \cos \theta_k \tag{1}$$

in which $\cos \theta_k$ is the phase power factor of the kth phase, and V_k and I_k are the phase voltage and current respectively of that phase. This is true for either a star or a mesh connection.

In the special case of a balanced n-phase system the power may be expressed as

$$P = nV_pI_p \cos \theta_p = \frac{nV_lI_l \cos \theta_p}{2 \sin (\pi/n)} \tag{2}$$

Note that, even when the power is expressed in terms of line voltages and currents, the power factor is the phase power factor.

36. TWO-PHASE (OR QUARTER-PHASE) SYSTEM

Strictly, the so-called single-phase system is a star-connected two-phase system, since the currents from the two terminals are in opposite directions at any instant, the current leaving by one and entering by the other. However, in practice the name two-phase system is used for a system supplied from a generator or other source of emf having two windings in which are developed two emf's differing in phase by 90 deg; i.e., a two-phase system is in reality two distinct single-phase systems, each with two terminals. Two of the four terminals may be connected to each other, in which case but three line wires are required. Or the two single-phase systems may be connected at their middle points; in this case the two-phase system may be considered a four-phase, or, as it is usually called, a "quarter-phase," system.

37. THREE-PHASE SYSTEM, DELTA (Δ) and WYE (Y) CONNECTIONS

For a three-phase system the generators and motors are designed with three windings or phases, which may be connected either in mesh, usually called a **delta connection,** since the diagram of the three windings forms a Greek delta, or in star, usually called a **wye connection,** since the diagram of the three windings forms a Y. The relations between line and phase currents and voltages for a balanced three-phase system are as follows:

	Delta	Y
Line current	I_l	I_l
Line voltage	V_l	V_l
Phase current	$I_p = \frac{I_l}{\sqrt{3}}$	$I_p = I_l$
Phase voltage	$V_p = V_l$	$V_p = \frac{V_l}{\sqrt{3}}$
Voltage to neutral	$V_n = \frac{V_l}{\sqrt{3}}$	$V_n = V_p$
Total volt-amperes	$3V_p I_p = \sqrt{3} V_l I_l$	$3V_p I_p = \sqrt{3} V_l I_l$
Phase power factor	$\cos \theta_p$	$\cos \theta_p$
Total power	$\sqrt{3} V_l I_l \cos \theta_p$	$\sqrt{3} V_l I_l \cos \theta_p$

As was pointed out in Article 20, any network having n points of entry can be completely represented by a network having n points of entry in which each point is connected by an impedance to every other point. If $n = 3$, the equivalent network is a delta. Hence any wye, balanced or unbalanced, may be transformed into an equivalent delta. The inverse transformation also exists. The transformation formulas are given by

$$\mathbf{Z}_1 = \frac{\mathbf{Z}_{31}\mathbf{Z}_{12}}{\mathbf{Z}_{12} + \mathbf{Z}_{23} + \mathbf{Z}_{31}}$$

$$\mathbf{Z}_2 = \frac{\mathbf{Z}_{12}\mathbf{Z}_{23}}{\mathbf{Z}_{12} + \mathbf{Z}_{23} + \mathbf{Z}_{31}}$$

$$\mathbf{Z}_3 = \frac{\mathbf{Z}_{23}\mathbf{Z}_{31}}{\mathbf{Z}_{12} + \mathbf{Z}_{23} + \mathbf{Z}_{31}}$$

$$\mathbf{Z}_{12} = \frac{\mathbf{Z}_1\mathbf{Z}_2 + \mathbf{Z}_2\mathbf{Z}_3 + \mathbf{Z}_3\mathbf{Z}_1}{\mathbf{Z}_3}$$

$$\mathbf{Z}_{23} = \frac{\mathbf{Z}_1\mathbf{Z}_2 + \mathbf{Z}_2\mathbf{Z}_3 + \mathbf{Z}_3\mathbf{Z}_1}{\mathbf{Z}_1}$$

$$\mathbf{Z}_{31} = \frac{\mathbf{Z}_1\mathbf{Z}_2 + \mathbf{Z}_2\mathbf{Z}_3 + \mathbf{Z}_3\mathbf{Z}_1}{\mathbf{Z}_2}$$

In which $\mathbf{Z}_1$, $\mathbf{Z}_2$, and $\mathbf{Z}_3$ are the Y impedances from lines 1, 2, and 3 to neutral, and $\mathbf{Z}_{12}$, $\mathbf{Z}_{23}$, $\mathbf{Z}_{31}$ are the Δ impedances between lines 1 and 2, 2 and 3, and 3 and 1.

38. BALANCED THREE-PHASE CIRCUITS

Any problem in regard to a **balanced** three-phase circuit may be solved by reducing all parts of the circuit to an equivalent Y connection, provided the currents and emf's are sine waves. The transformations are made as follows:

Any Δ-connected motor or generator is considered as equivalent to a Y-connected generator or motor in which

$$E_y = \frac{E_\Delta}{\sqrt{3}}, \qquad r_y = \frac{r_\Delta}{3}, \qquad x_y = \frac{x_\Delta}{3}$$

where the quantities E_y, r_y, and x_y are the emf, resistance, and reactance per phase of the Y-connected machine equivalent to the emf, resistance, and reactance per phase of the actual Δ-connected machine.

Each of the line wires is in series with a corresponding phase of the equivalent Y-connected machine.

When all parts of the circuit have thus been reduced to equivalent Y's, each of the three phases may be treated as a single-phase circuit, each circuit considered completed by a wire having zero impedance and connecting all the neutrals together, since all the neutrals are at the same potential.

The voltages thus calculated are the voltages to neutral, and the currents are line currents. To find the line voltage multiply the calculated voltage by $\sqrt{3}$; similarly, to find the actual phase current in the Δ-connected generator or load divide the calculated current by $\sqrt{3}$.

SYMMETRICAL COMPONENTS

By Raymond C. R. Schulze

39. APPLICATIONS AND BASIC ASSUMPTIONS OF THE METHOD

The method of symmetrical components is a mathematical tool for the calculation of unbalanced conditions (faults or loads) on a polyphase power system. Theoretically, the method can be applied to systems of any number of phases, but actually, since the three-phase system is used almost universally, the application of the method to three-phase systems only will be considered here.

Unbalanced three-phase circuit problems are hard to solve by direct methods, whereas balanced three-phase circuits are susceptible of comparatively easy treatment. The method of symmetrical components takes advantage of this fact by substituting for each set of unbalanced vectors (current, voltage, or reactance) three sets of balanced vectors which are their vectorial equivalents. The several solutions of these balanced networks are then combined in the manner required for the particular type of unbalance. The method is based on the following assumptions:

1. For any type of fault, all three phases are connected together and to ground at the point of fault; fault resistance may be included, if wanted, as explained in Article 45.
2. The impedances of the three phases are balanced.
3. All vector rotation is in the counterclockwise (positive) direction.
4. Rms values are used throughout.
5. Unless specifically mentioned otherwise, all zero-sequence current returns through the ground. This, of course, does not apply either to overhead lines with ground wires or to underground cables with sheaths or in metallic pipes.

As mentioned above, one fundamental assumption is that the impedances of the three phases of the system are equal to each other. In actual practice, however, the neglect of the usual unbalances of the impedances of the three phases will give answers close enough for most engineering problems. Where extreme accuracy is needed, however, the unbalance of the impedances of the three phases can be taken into account by following the procedure outlined in C. F. Wagner and R. D. Evans, *Symmetrical Components*, Chapter 18. In the present discussion, however, it will be assumed that the impedances of the three phases are equal to each other in both resistance and reactance.

40. POSITIVE-, NEGATIVE-, AND ZERO-SEQUENCE COMPONENTS

DEFINITIONS. Any system of unbalanced three-phase vectors—current or voltage—can be resolved into three balanced sets of vectors known as the positive-, negative-, and zero-phase-sequence components.* All the vectors rotate in the positive direction, i.e., counterclockwise. The three vectors of the positive phase-sequence system are equal to each other in magnitude and are 120 deg apart in phase, and their maximum values occur in the positive sequence of a, b, c. The three negative-sequence vectors are also equal to each other in magnitude and 120 deg apart in phase, and they rotate positively, just like the positive-sequence vectors, but their maximum values occur in the *negative* sequence of a, c, b. The zero-phase-sequence vectors are likewise equal to each other in magnitude, but the three vectors are in phase with each other: there is a 0-deg angle or zero sequence between the three vectors.

The three vectors of the positive-phase-sequence group are equal to each other; likewise the negative- and the zero-phase-sequence groups. However, the vectors of the positive-, negative, and zero-phase-sequence groups may or may not be equal to each other, depending on the type of fault. This point will be clearer after a review of the material on the calculation of each type of fault in Article 44.

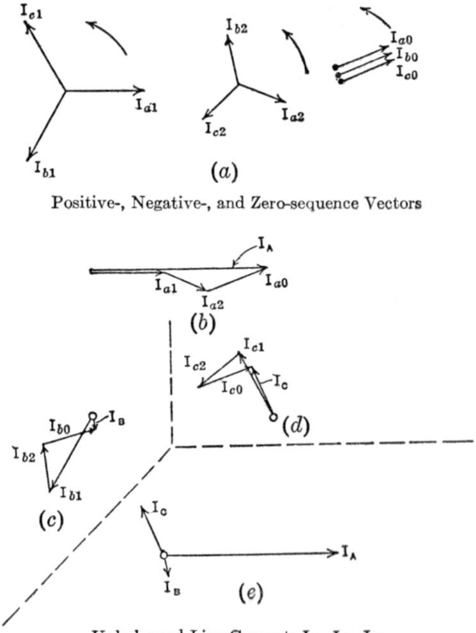

Positive-, Negative-, and Zero-sequence Vectors

Unbalanced Line Currents I_A, I_B, I_C

FIG. 1. Combination of Phase-sequence Components into Unbalanced Line Currents

PHYSICAL MEANING. An idea of the physical meaning of each of these phase-sequence components may help in understanding them. The positive-phase-sequence components are the currents (and voltages) which occur in any balanced system; they are the values used in the calculation of any balanced load or voltage problem—the only difference is that here they are given a longer name. When there is

* The subscripts 1, 2, and 0, applied to voltages, currents, reactances or impedances, indicate positive-, negative-, and zero-sequence values, respectively.

an unbalance, as with a single-phase load connected between two phase wires, but there is *no* connection to ground, then both positive- and negative-sequence currents will flow. In any case where the ground or a neutral conductor is involved, however, the current through the ground (or neutral) is made up of the three zero-sequence components, and the currents in the phase wires are therefore made up of positive-, negative-, and zero-sequence components.

RELATION BETWEEN UNBALANCED VECTORS AND SYMMETRICAL COMPONENTS. Perhaps the clearest way to show the relation between the phase-wire currents and the several sequence components is to start with a set of phase-sequence components, combine them into the line currents, and then resolve those line currents back into the phase-sequence components. This is done in Figs. 1 and 2, where Fig. 1(*a*) shows a set of positive-, negative-, and zero-phase-sequence current vectors (currents will be assumed, for convenience; voltage vectors would be treated in exactly the same way). The current in phase *A* is obtained readily in Fig. 1(*b*) by adding vectorially the three phase-sequence components of that phase: the positive component I_{a1}, the negative I_{a2}, and the zero-phase-sequence component I_{a0}. Similarly, the line currents in *B* and *C* phases are obtained in Figs. 1(*c*) and 1(*d*), the resulting unbalanced line currents being shown in one diagram in Fig. 1(*e*).

This procedure is reversed in Fig. 2, which shows how the unbalanced line currents of Fig. 1(*e*) can be resolved into the phase-sequence components. The easiest to obtain is the zero-phase-sequence component. Since the positive-sequence components of I_A, I_B, and I_C are equal to each other in magnitude and 120 deg apart in phase [see Fig. 1(*a*)], their sum is zero; likewise, the summation of the negative-sequence components is zero; the three zero-sequence components are in phase with each other, however, so their sum is not zero. Therefore, when I_A, I_B, and I_C are added vectorially, as in Fig. 2(*a*), since the positive- and the negative-sequence components add up to zero, the resultant must be the sum of the three zero-sequence components, and one-third of this resultant must be I_{a0}, likewise I_{b0} and I_{c0}. The resultant, incidentally, is equal to the ground current, or

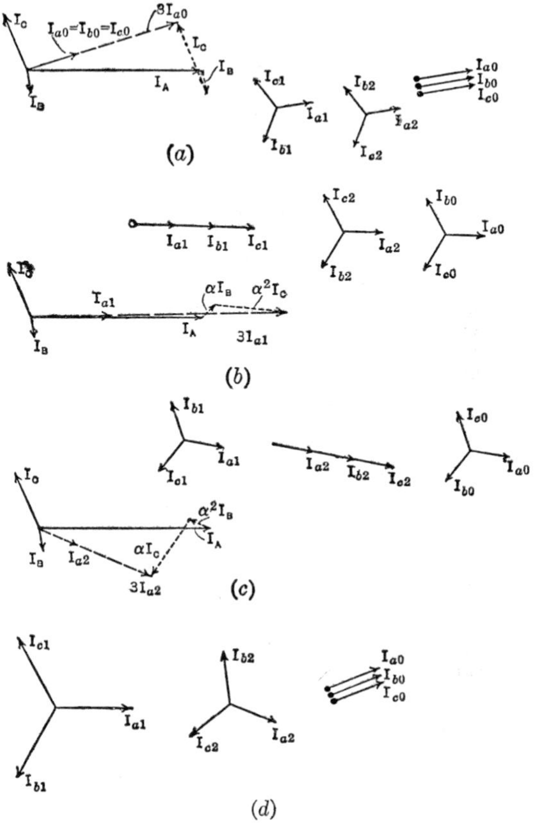

FIG. 2. Resolution of Unbalanced Line Currents of Fig. 1(*e*) into Phase-sequence Components of Fig. 1(*a*)

$$I_g = 3I_{a0} = I_{a0} + I_{b0} + I_{c0} \tag{1}$$

Mathematically, this determination of I_{a0} can be written as the vectorial sum:

$$I_{a0} = \tfrac{1}{3}(\mathbf{I}_A + \mathbf{I}_B + \mathbf{I}_C) \tag{2}$$

This description of the graphical (and also mathematical) determination of the zero-sequence components of I_A, I_B, and I_C should give a clue to the means for determining the positive- and the negative-sequence components. The method involves the rotation of the three line currents to positions such that the three desired components (say, the positive) are in phase with each other, whereas the three negative components are 120 deg apart, and the three zero-sequence components are also 120 deg apart. This can be accomplished by rotating current I_B counterclockwise (positive direction of rotation) through 120 deg, and current I_C through 240 deg, thus placing the several components in the positions indicated in Fig. 2(*b*). This is accomplished by the use of the vector operator α, which is explained below, and the resultant summation of I_A, αI_B, and $\alpha^2 I_C$ is equal to $3I_{a1}$ [Fig. 2(*c*)]. When I_{a1} is thus located, the other two positive-sequence components are readily determined.

In a similar manner, by rotating I_B through 240 deg and I_C through 120 deg, it is possible to determine the magnitude and direction of the negative-sequence component I_{a2}, and then I_{b2}, and I_{c2} [Fig. 2(*d*)].

Expressed mathematically, the positive- and the negative-phase sequence components are

$$\mathbf{I}_{a1} = \tfrac{1}{3}(\mathbf{I}_A + \alpha\mathbf{I}_B + \alpha^2\mathbf{I}_C) \tag{3}$$

$$\mathbf{I}_{a2} = \tfrac{1}{3}(\mathbf{I}_A + \alpha^2\mathbf{I}_B + \alpha\mathbf{I}_C) \tag{4}$$

The operator α used above may require some explanation. Where the operator j rotates a vector through 90 deg, the operator α rotates it through 120 deg. Numerically it is equal to $-0.50 + j0.866$, and $\alpha^2 = -0.50 - j0.866$. Also, α is one of the cube roots of 1.0

41. POSITIVE-, NEGATIVE-, AND ZERO-SEQUENCE REACTANCES

Another fundamental assumption might have been stated above: that positive-sequence currents and voltages occur only in positive-sequence reactance networks; similarly, negative-sequence values occur only in negative-sequence networks, and zero-sequence values only in zero-sequence networks. A knowledge of the physical meaning of each phase-sequence reactance will help toward understanding the symmetrical-component method.

POSITIVE-PHASE-SEQUENCE REACTANCE. This is the ordinary value of reactance normally encountered in balanced power flow or voltage drop calculations, only here it is given a more descriptive name. It should be noted particularly that the positive-sequence reactance of an overhead transmission line is due, at any instant, to current going out over one conductor and returning over the other two, each "return" conductor being separated by 3 to 30 ft from the "outgoing" conductor.

NEGATIVE-PHASE-SEQUENCE REACTANCE. Since the negative-phase-sequence currents are equal to each other in magnitude and 120 deg apart, the negative-sequence reactance of a transmission line, and of any stationary equipment, will be the same as the positive value, the sequence of the phases having no effect. In rotating equipment, however, the sequence of the phases does affect the reactance, so that the negative-phase-sequence reactance of a generator will generally be between 0.7 and 1.3 times the positive-sequence reactance. In the great majority of cases, however, it is sufficiently accurate to assume that the positive- and the negative-sequence reactances of any rotating machine are equal to each other.

ZERO-PHASE-SEQUENCE REACTANCE. The zero-phase-sequence reactance is somewhat different, but it can be readily understood. In the transmission line, the three zero-sequence components of current go "out" over the three conductors and return through the ground, the equivalent depth of the earth return path being 500 to 20,000 ft, so that the separation between the "outgoing" and the "return" conductors of an overhead line is perhaps 3000 ft (an average value for this country), instead of 3 to 30 ft, as in the positive-sequence reactance. Consequently, the zero-sequence reactance of a transmission line (without ground wires) will be four to five times the positive-sequence reactance.

In a rotating machine, such as a turbogenerator, the zero-sequence reactance will be 0.2 to 0.5 of the positive-sequence reactance. In transformers the zero-sequence reactance of a single-phase unit will be the same as the positive, since the magnetic field of the unit is affected by the current in that one phase only; in a three-phase transformer, however, because of the presence of magnetic fields due to the currents in all phases in the same core at the same instant, the zero-sequence reactance may or may not be the same as the positive value; generally, in modern transformers, it will be the same.

Typical values of the phase-sequence reactances, taken from several sources, are given in Table 1. These are per cent reactance on the name-plate kva.

Table 1. Typical Values of X_1, X_2, X_0

	X_1	X_2	X_0
Rotating Machines			
Turbine generators			
Two-pole	10	10	3
Four-pole	14	14	4
Salient-pole machines			
With dampers	22	22	6
Without dampers	30	50	7
Synchronous condensers	25	25	8
Transformers (on self-cooled basis)			
Distribution		3	
Network		5	
Power			
Up to 66 kv		5–7	
88–110 kv		6–9	
132–154 kv		8–10	
220 kv		10–14	

	1 3-phase Circuit	2 3-phase Circuits
X_0 of Transmission Lines		
Non-magnetic ground wire (if ACSR, 2 or more layers of aluminum)	$X_{01} = 2X_{11}$	$X_{02} = 1.5X_{11}$
Magnetic ground wire	$X_{01} = 3.5X_{11}$	$X_{02} = 2.5X_{11}$
No ground wire	$X_{01} = 3.5X_{11}$	$X_{02} = 2.5X_{11}$

Where X_{01} = zero-sequence reactance of one circuit.
X_{02} = zero-sequence reactance of two circuits in parallel.
X_{11} = positive-sequence reactance of one circuit.

ZERO-SEQUENCE REACTANCE OF A TRANSMISSION LINE, WITHOUT GROUND WIRES. The calculation of the zero-sequence reactance of a transmission line without ground wires is relatively simple. The equation is:

$$X_0 = 0.01397f \log \frac{H}{R_0} \text{ ohms per phase per mile} \tag{5}$$

where f = frequency in cycles per second.
H = equivalent depth of earth return in feet = $2160\sqrt{\rho/f}$.
ρ = earth resistivity in meter-ohms (see Table 2).
R_0 = equivalent mean radius of the conductor group.
All logs are to the base 10; all quantities must be expressed in the same units.

To understand the meaning of this equation, refer to Fig. 3. The three-phase conductors are replaced by an equivalent tubular conductor of radius R_0 (being a hollow tube, there is no internal flux), and the calculated reactance is that of this equivalent conductor with the current returning through a fictitious conductor at the equivalent depth of the earth return circuit.

Table 2. Typical Values of Earth Resistivity

	Meter-ohms
Average	100
Sea water	0.01–1.0
Swampy ground	10–100
Pure water	1000
Dry earth	1000
Pure slate	10^7
Sandstone	10^9

It was stated that eq. (5) gives the zero-sequence reactance of the transmission line without ground wires. The impedance of this line must also include a resistance component. Actually, the theory gives the total impedance as the resistance of the conductors plus a reactance term plus a rather complex $(R + jX)$ term, but with the aid of certain simplifying assumptions which introduce a negligible error, this exact expression reduces to

$$Z_0 = R_{\text{condr.}} + 0.00477f + j0.01397f \log_{10} \frac{H}{R_0} \text{ ohms per phase per mile} \tag{6}$$

Equation (6) is entirely satisfactory for practically all power work. Generally, the exact method need be used only for inductive coordination problems, and then only when a high degree of accuracy is needed.

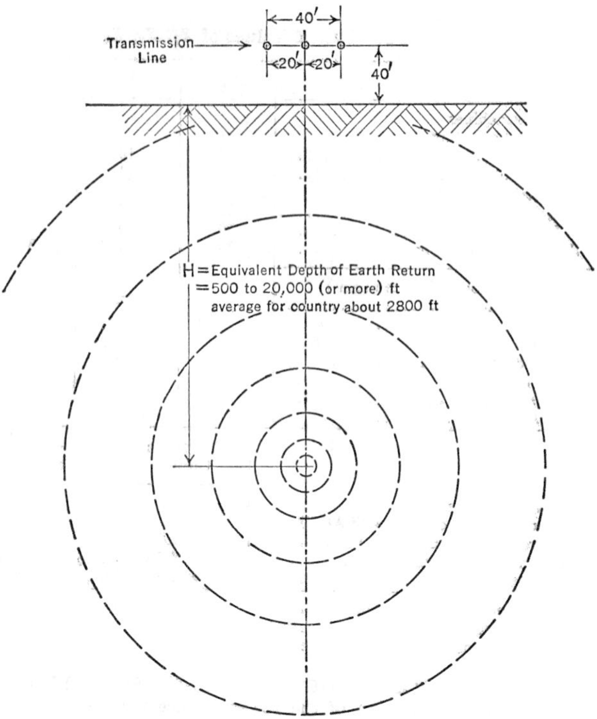

FIG. 3. Relative Positions of Transmission Line and Earth Return

EQUIVALENT RADIUS. The usual equation for the reactance of a transmission line considers the magnetic flux both inside and outside the conductor. The computation can be simplified if only the flux outside the conductor need be considered, as can be done if the actual conductor is replaced by an equivalent one in the form of a tube of negligible

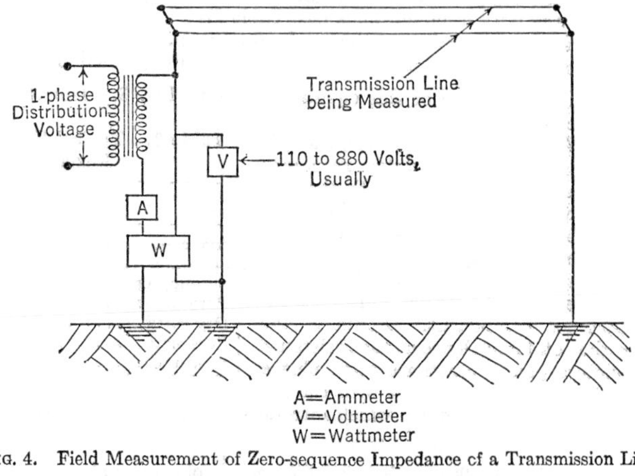

FIG. 4. Field Measurement of Zero-sequence Impedance of a Transmission Line

thickness and having a radius of about 0.779 of the real copper conductor, or roughly 0.81 of the real ACSR conductor. The value of 0.779 strictly applies only to a copper conductor with a smooth surface (it is only 0.726 for one with seven strands) but is very generally used for all conductors. (See Section 14, Article 22.)

The equivalent radius of the equivalent hollow conductor which will represent the actual conductor group of N conductors is equal to the N^2 root of the product of N^2 terms, these N^2 terms consisting of N subgroups of N terms each, the N terms in each subgroup being the equivalent radius of one conductor and the N-1 possible distances from that conductor to each other conductor.

As a simple example, consider the transmission line of Fig. 3. Assume 500,000-cm copper conductors, with a diameter of 0.814 in. and therefore a radius of

$$\frac{0.814}{2 \times 12} = 0.0339 \text{ ft}$$

The equivalent radius of this circuit is

$$\begin{aligned} R_{01} &= \sqrt[9]{0.779 \times 0.0339 \times 20 \times 40 \times 0.779 \times 0.0339 \times 20 \times 20 \times 0.779 \times 0.0339 \times 40 \times 20} \\ &= \sqrt[9]{(0.779 \times 0.0339)^3 \times 20^4 \times 40^2} \\ &= \sqrt[9]{4700} \\ &= 2.55 \text{ ft.} \end{aligned}$$

Now, if $f = 60$ cycles, and assuming $H = 2800$ ft, by eq. (5),

$$\begin{aligned} X_0 &= 0.01397 \times 60 \times \log_{10} \frac{2800}{2.55} \\ &= 0.839 \times \log_{10} 1099 = 0.839 \times 3.040 \\ &= 2.55 \text{ ohms per phase per mile} \end{aligned}$$

Note that this answer is in ohms **per phase** per mile; if the zero-sequence reactance were measured in the usual manner, as indicated in Fig. 4, the measured impedance of the loop would be for the three phases in parallel, or one-third of the calculated **per phase** value.

The computation of the zero-sequence **impedance** of a transmission line and of an underground cable will be discussed in Articles 47 and 48, respectively.

42. ZERO-SEQUENCE REACTANCES OF TRANSFORMERS

Briefly, the zero-sequence reactance of a transformer is generally either equal to its positive-sequence reactance or to infinity, the choice depending on whether the following two conditions are met: (1) the winding on the "faulted" side must be Y-connected (or zig-zag) and grounded; and (2) on the unfaulted side there must be a path for "compensating" amperes equal to the current in the faulted side, to permit current flow in the faulted-side windings. If these two conditions are met, the zero-sequence reactance will be equal to the positive value; if either condition is not met, the zero-sequence reactance will be infinity. The second point can be understood more easily by considering the magnetizing and the load currents in the primary and secondary of a power transformer.

Figure 5 shows several typical transformer connections and the zero-sequence reactance of each. Case A, for instance, shows a Y-grounded winding on the "faulted" side, with a delta on the "unfaulted" side. Since the three zero-sequence current components are in phase, it can be seen that these components on the Y side, all flowing from the neutral, will cause a compensating current to circulate in the delta; there will be no zero-sequence current in the leads connected to the delta. Since there are equal currents in both windings of each phase, these currents encounter the normal reactance of the transformer, and the zero-sequence reactance of the transformer will be equal to its positive-sequence reactance.

In case B condition 1 of the first paragraph is met, but since the neutral on the unfaulted side is not grounded, there can be no compensating current flow, and therefore the zero-sequence impedance of this Y-grounded–Y-*un*grounded bank is infinity.

In case C again condition 1 is met, and the unfaulted side has its neutral grounded, but there is no complete circuit for the compensating current on the unfaulted side, with the "unfaulted" winding connected to either a delta or an ungrounded Y of another transformer. Consequently this Y-grounded–Y-ungrounded bank has infinite impedance to zero-sequence current.

In case D, however, both conditions 1 and 2 are met, condition 2 by the fact that the Y-grounded–Δ bank completes the path for the compensating current needed to permit the flow of zero-sequence current in the winding on the faulted side. The zero-sequence impedance, however, is the sum of three values: the two transformers and the circuit between them.

The transformers in cases E and F obviously have infinite zero-sequence impedance, since there is no connection to ground on the faulted side with either connection, for the flow of zero-sequence current to the fault.

Case G shows a special condition, with a zig-zag winding with the neutral grounded. Since the currents in the two windings on the same leg of the transformer core are equal and flowing in opposite directions, these currents "compensate" each other, and X_0 is therefore equal to X_1.

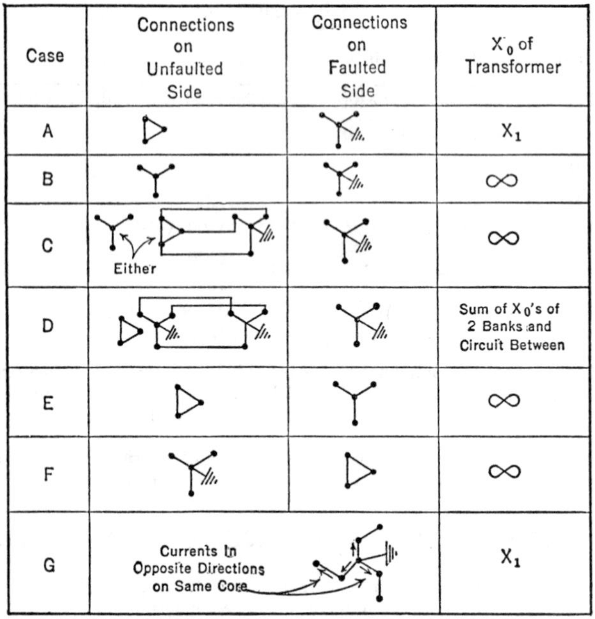

Case
Connections on Unfaulted Side
Connections on Faulted Side
X'0 of Transformer
A
X1
B
∞
C
Either
∞
D
Sum of X0's of 2 Banks and Circuit Between
E
∞
F
∞
G
Currents in Opposite Directions on Same Core
X1

FIG. 5. Zero-sequence Reactances of Transformers

Other transformer connections are possible, but the examples given here illustrate the principles involved in the determination of the zero-sequence reactance of a given transformer under specified conditions.

43. REPRESENTATION OF NEUTRAL IMPEDANCE

All the current and impedance figures discussed so far have been **per phase**, and these are entirely satisfactory in all conditions except when representing a neutral impedance. Consider a grounding resistor, for example, of resistance R_n; since the zero-sequence currents of all three phases pass through the grounding resistor, the voltage drop across it is

$$E_R = (3I_{a0}) \times R_n = I_{a0}(3R_n) \tag{7}$$

and therefore any neutral impedance should be represented by three times its actual value.

44. CALCULATION OF CURRENTS IN UNBALANCED FAULTS

As mentioned before, this method of symmetrical components is a tool for calculating the currents in unbalanced faults, and the resulting components must be put together in a definite way to obtain the solution for a given type of fault. This problem of arrangement of the phase-sequence networks will be discussed here.

Figure 6 consists of sketches illustrating these network arrangements. For comparison and completeness a balanced three-phase fault is included also.

The balanced three-phase fault will be considered first. The desired result is three currents, one in each phase, equal in magnitude and 120 deg apart. These can be obtained if there are no zero- or negative-sequence components, as will occur if only the positive-sequence reactance network is used, as shown in Fig. 6(*a*), and

$$I_{a1} = \frac{E_n}{X_1} \tag{8}$$

where E_n is the voltage from line to neutral.

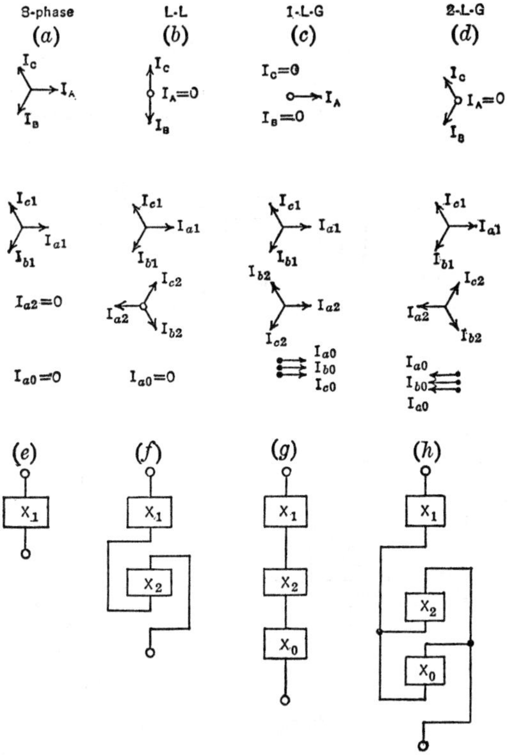

Fig. 6. Relations between Phase-sequence Components and Connections of Reactance Networks to Represent Different Types of Faults

If the fault involves only two phase wires, then there must be no current in one phase (assumed to be phase A) and none in the ground, and the result is as shown in Fig. 6(*b*). Since there is no ground current, the zero-sequence components must be equal to zero, and the zero-sequence reactance network does not enter the calculation. The result is obtained if I_{a1} and I_{a2} are equal and opposite, and this will result from the connections of Fig. 6(*f*).

In a one-line-to-ground fault, the result in the faulted circuit will be current in phase A (assumed to be faulted to ground), but none in phases B and C. This condition is obtained if $I_{a1} = I_{a2} = I_{a0}$, as shown by Fig. 6(*c*); these values will result from the connections of Fig. 6(*g*).

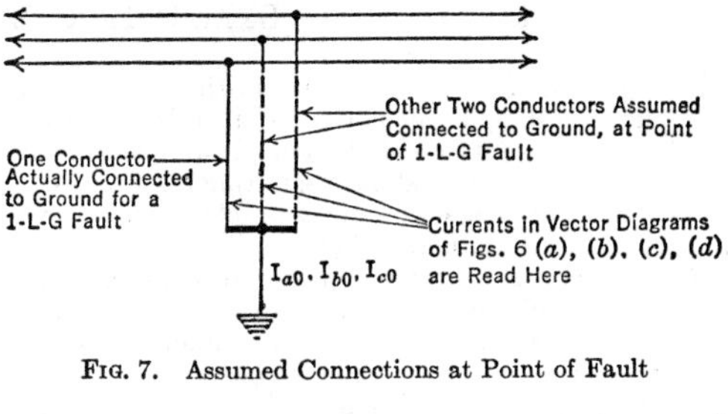

Fig. 7. Assumed Connections at Point of Fault

Finally, with a two-line-to-ground fault, the current in the unfaulted phase must be zero, and there must be currents in the two faulted phases and through the ground. This condition is obtained with the phase-sequence components of Fig. 6(*d*), which are obtained with the connections indicated in Fig. 6(*h*).

The currents mentioned above are the values in the assumed connections from the point of fault to ground, as shown in Fig. 7. In any actual case, where the fault current divides between two or more circuits, and these circuits have different ratios of X_1/X_0, there will be phase currents in the unfaulted phase or phases in these "branch" circuits, as well as in the faulted phase or phases. This point will be clearer after a review of the numerical examples on the following pages.

45. FAULT IMPEDANCE

So far, it has been assumed that the fault impedance is zero. There are some cases, however, where some impedance value in addition to the system impedance must be considered, e.g., a neutral resistor or fault impedance or an unbalanced load represented as a high impedance fault.

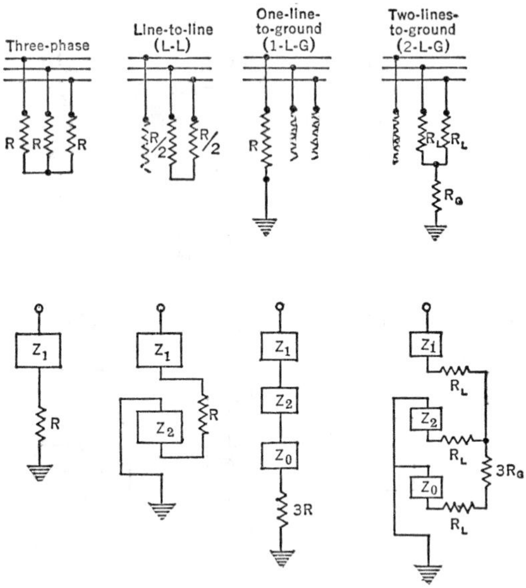

FIG. 8. Representation of Fault Impedance

The proper means for connecting this fault impedance into the network is shown in Fig. 8.

46. NUMERICAL EXAMPLES OF THE APPLICATION OF THE METHOD OF SYMMETRICAL COMPONENTS

The various points brought out in the preceding discussion are put to practical use in the following sample calculations, which develop the phase-sequence components and the total phase currents for each type of fault (three-phase, L-L, 1-L-G, and 2-L-G) at one location on a sample system, neglecting resistances.

The positive-, negative-, and zero-phase-sequence reactances of each part of the system are given in Fig. 9, and in Fig. 10 these values are combined to give the total positive-, negative-, and zero-phase-sequence reactances to the point of fault. Figures 11 to 14 show the calculation of the current values for the three-phase, single-phase line-to-line, one-line-to-ground, and two-line-to-ground faults, respectively.

There is no particular problem concerned with the calculation of the three-phase fault. This is a balanced condition, and the positive-sequence values only are needed for its solution. Since the system is balanced with respect to the three phases, there will be no current in the ground even if the fault is grounded. Consequently, the method of solution is the same whether the three-phase fault is grounded or ungrounded.

The single-phase line-to-line fault requires the use of both the positive- and the negative-sequence values. I_B and likewise I_C are equal to the vector sum of the positive- and the

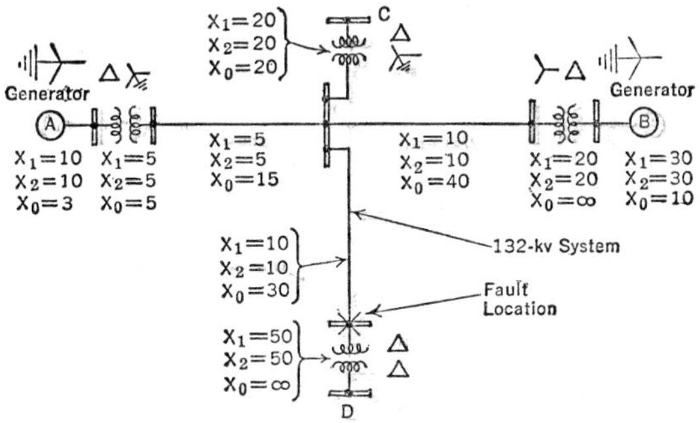

FIG. 9. Phase-sequence Reactance Values for Sample Calculations

negative-sequence components, which are 60 deg apart. Although Fig. 12 shows the rigorous solution, a shorter method is available which gives the same answer, when $X_2 = X_1$ for each part of the system:

$$I_{b1} = I_{b2} = \frac{E_n}{X_1 + X_2} = \frac{E_n}{2X_1} \tag{9}$$

Since $I_B = 1.732 I_{b1}$, then

$$I_B = I_C = \frac{1.732 E_n}{2X_1} = \frac{E_{L-L}}{2X_1} \tag{10}$$

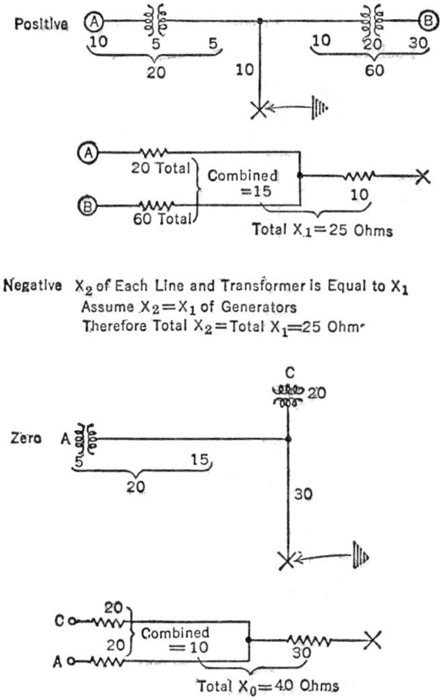

FIG. 10. Phase-sequence Reactance Networks for Indicated Fault Location

The solution of the one-line-to-ground fault illustrates quite clearly the use of all three phase-sequence components and the computation of the phase currents from the sequence components. It should be noticed, first, that the zero-sequence network can be quite different from the positive- or the negative-sequence networks, the "sources" for the zero-sequence network being properly connected transformer banks with grounded neutrals on the faulted side, rather than the generators or synchronous condensers. Also, where

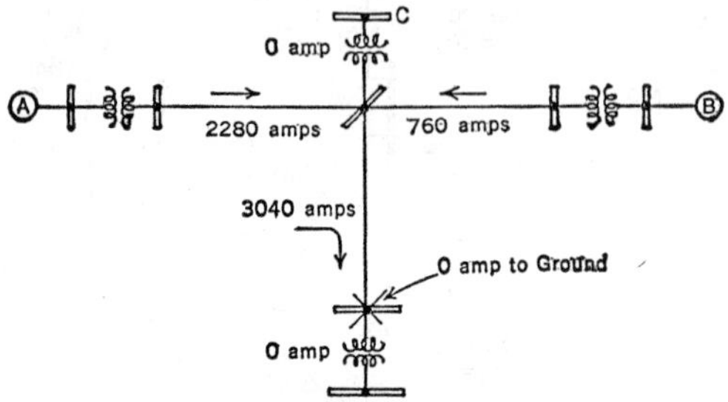

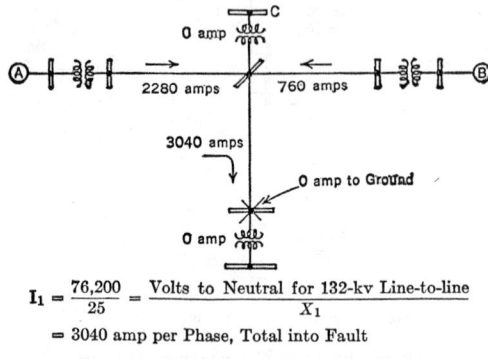

$$I_1 = \frac{76{,}200}{25} = \frac{\text{Volts to Neutral for 132-kv Line-to-line}}{X_1}$$

= 3040 amp per Phase, Total into Fault

FIG. 11. Calculation of Three-phase Fault

the fault current divides between two or more circuits having different ratios of X_0/X_1 (1.6 for circuit from A, ∞ from B, ∞ from C), there will be some current in the two unfaulted phases as well as in the faulted phase. The steps in the graphical determination of the

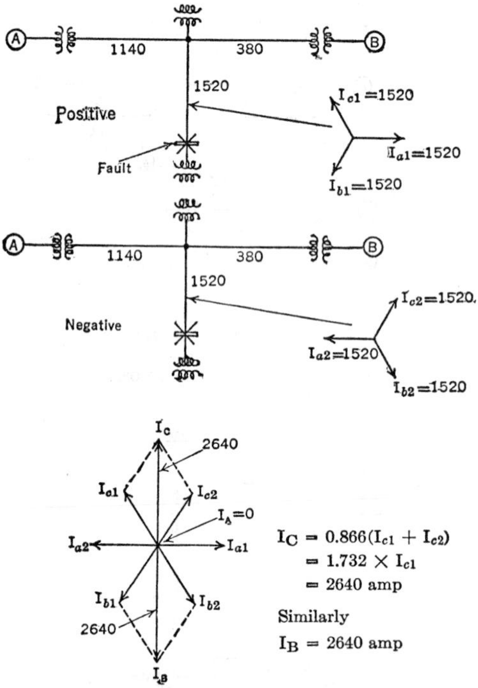

$$I_C = 0.866(I_{c1} + I_{c2})$$
$$= 1.732 \times I_{c1}$$
$$= 2640 \text{ amp}$$

Similarly

$$I_B = 2640 \text{ amp}$$

$$I_{a1} = I_{a2} = \frac{E_n}{X_1 + X_2} = \frac{76{,}200}{25 + 25} = 1520 \text{ amp}$$

FIG. 12. Calculation of Line-to-line Fault

phase currents from the phase-sequence components of current are shown clearly in Fig. 13(b), while Fig. 13(c) shows these phase currents on a three-line diagram. Since the three-phase load currents have been neglected, the one-line-to-ground fault currents are those of a single-phase system—they are either in phase or 180 deg out of phase.

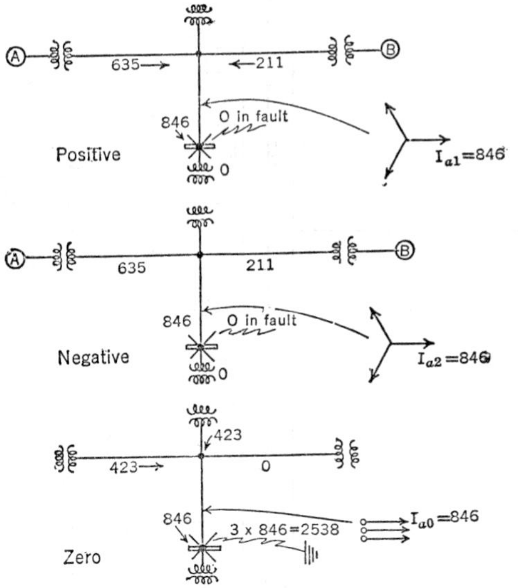

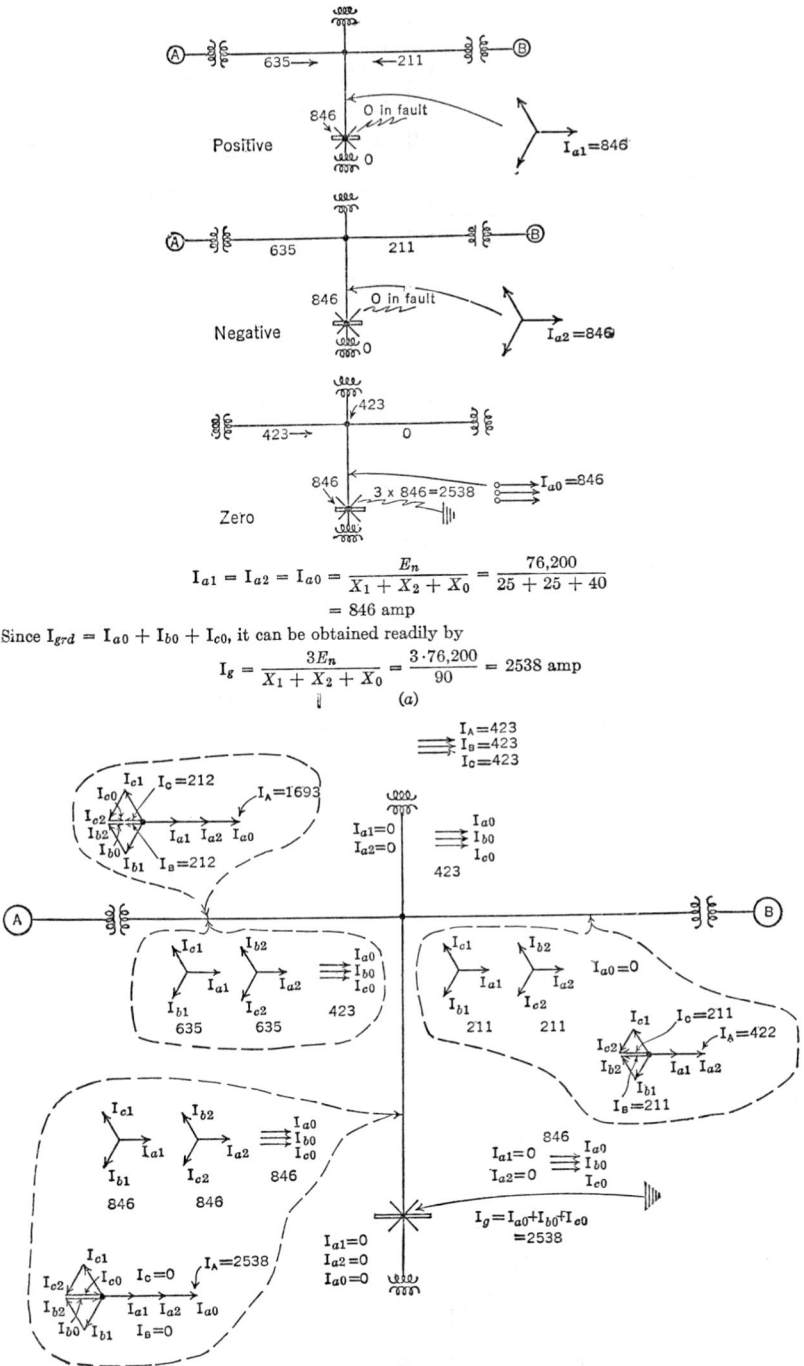

$$I_{a1} = I_{a2} = I_{a0} = \frac{E_n}{X_1 + X_2 + X_0} = \frac{76{,}200}{25 + 25 + 40}$$

$$= 846 \text{ amp}$$

Since $I_{grd} = I_{a0} + I_{b0} + I_{c0}$, it can be obtained readily by

$$I_g = \frac{3E_n}{X_1 + X_2 + X_0} = \frac{3 \cdot 76{,}200}{90} = 2538 \text{ amp}$$

(*a*)

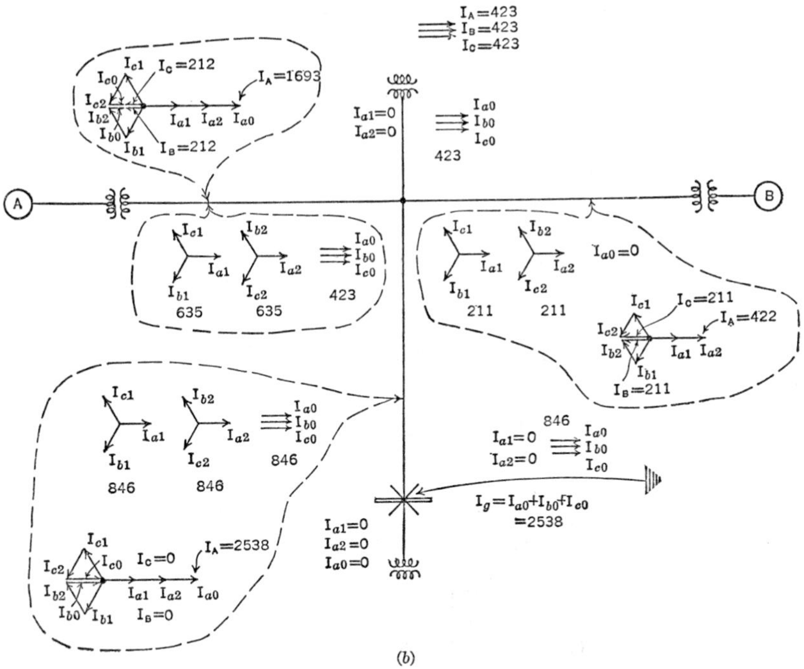

(*b*)

FIG. 13. Calculation of One-line-to-ground Fault

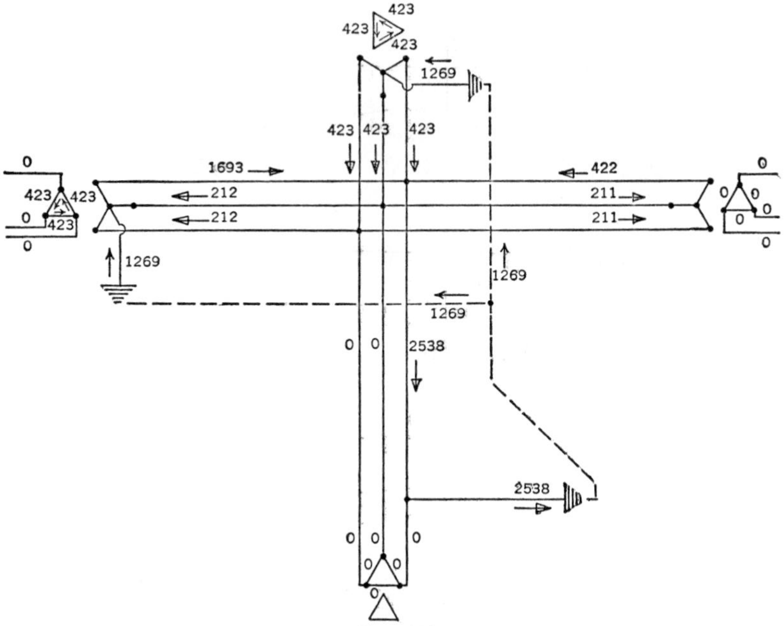

Fig. 13(c)

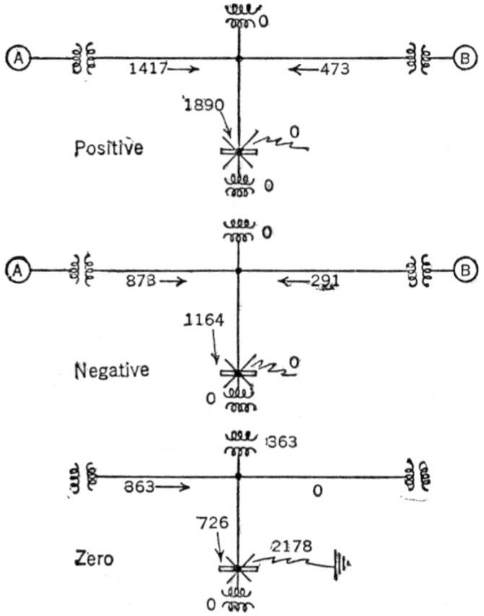

$$\mathbf{I}_{a1} = \frac{E_n}{X_1 + \dfrac{X_2 X_0}{X_2 + X_0}} = \frac{76{,}200}{25 + \dfrac{25 \cdot 40}{25 + 40}} = \frac{76{,}200}{40.4}$$

$$= 1890 \text{ amp}$$

$$\mathbf{I}_{a2} = \mathbf{I}_{a1} \times \frac{15.4}{25} = 1164 \text{ amp}$$

$$\mathbf{I}_{a0} = \mathbf{I}_{a1} \times \frac{15.4}{40} = 726 \text{ amp}$$

Fig. 14 (a)

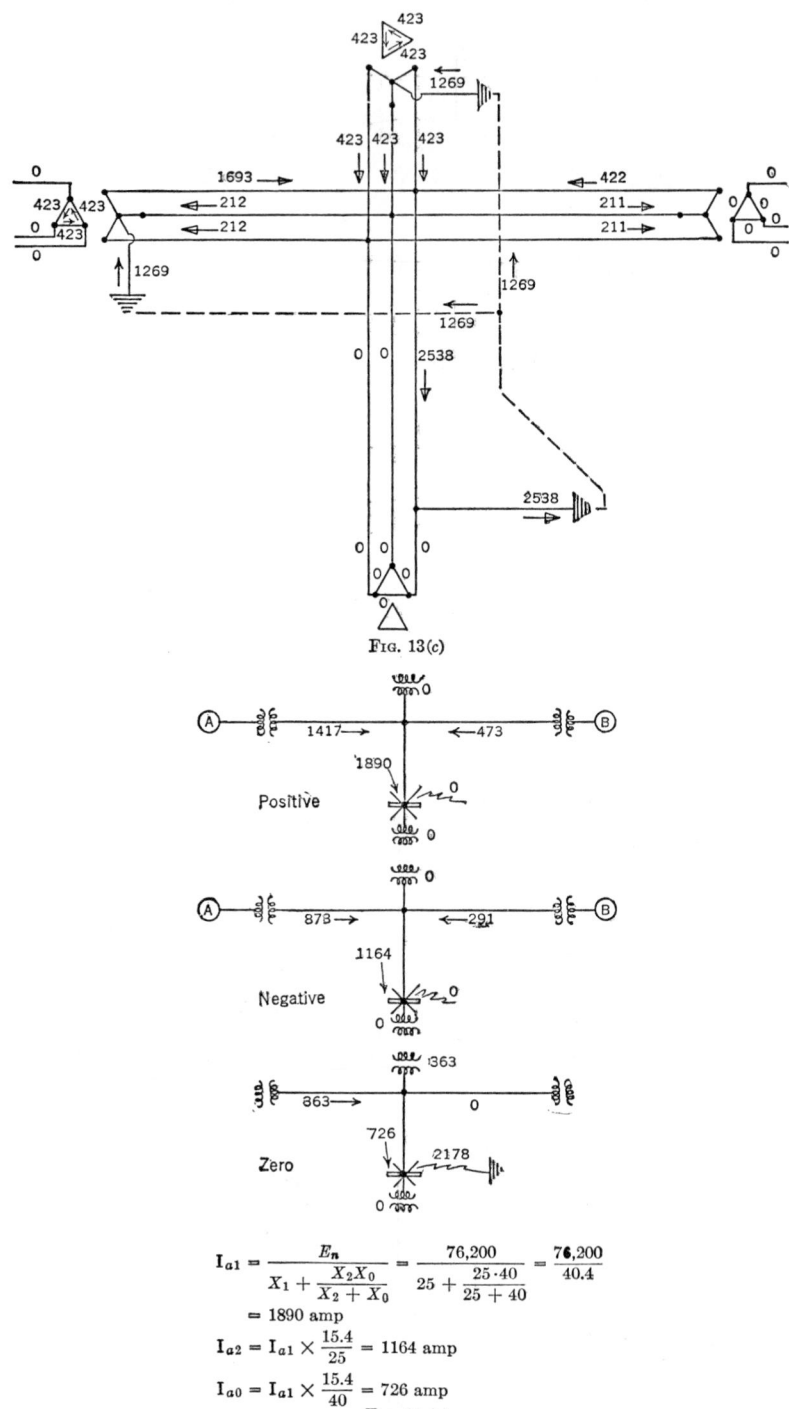

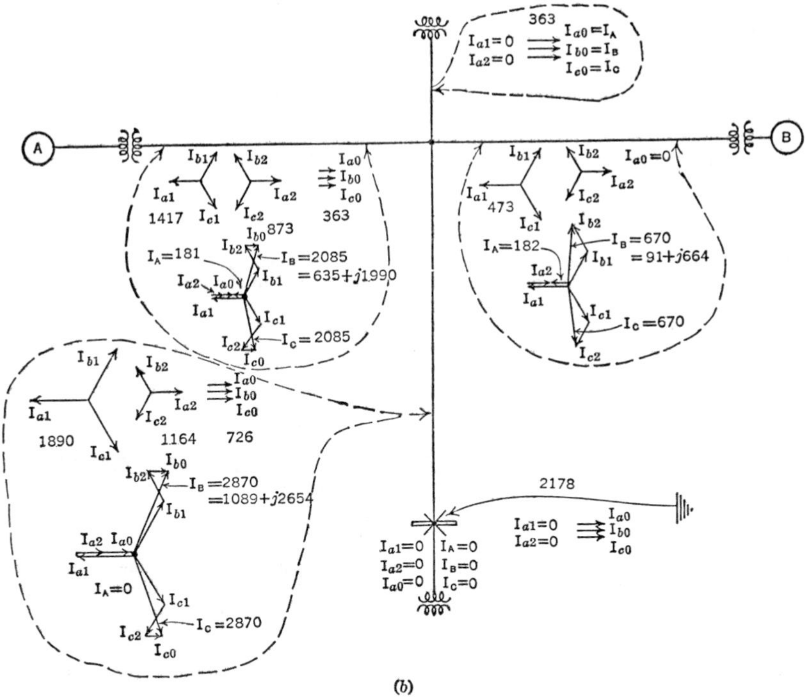

(b)

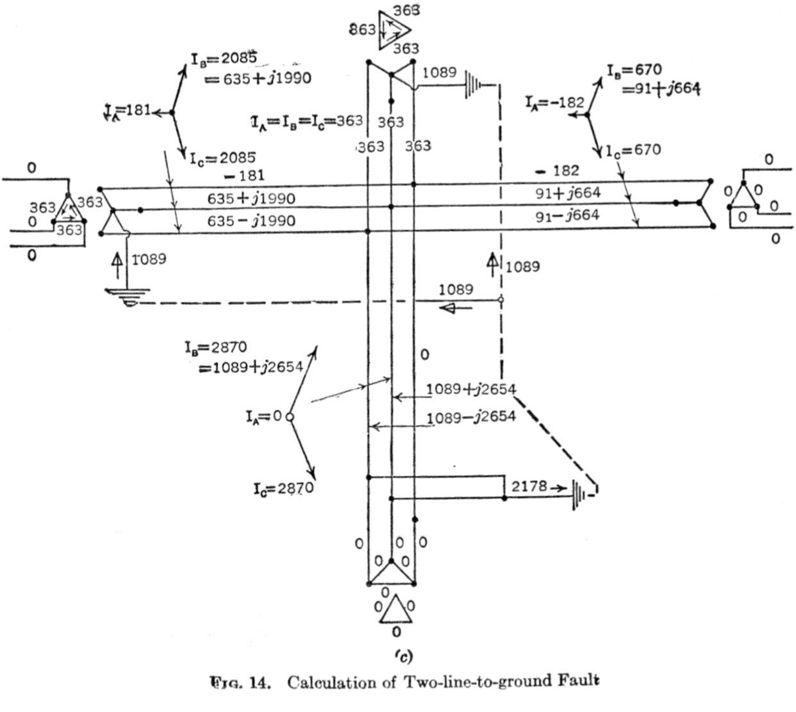

(c)

FIG. 14. Calculation of Two-line-to-ground Fault

The two-line-to-ground fault is worked out in the same detail as the one-line-to-ground. The phase-sequence components of the fault current are first calculated, remembering the connections of the phase-sequence reactance networks shown in Fig. 6. The several phase-sequence components in each circuit and the combination of these components to give the phase currents are shown in Fig. 14(*b*), and these phase currents are then presented in clearer fashion on Fig. 14(*c*).

Determination of the phase currents is not always necessary. In fact, in many cases where ground relays are used, it is necessary only to determine the values of the ground currents,

$$I_g = 3I_{a0} = \frac{3E_n}{X_1 + X_2 + X_0} \text{ (1-L-G fault)} \tag{11}$$

thus considerably reducing the necessary amount of calculation.

Also, this explanation has dealt only with currents, and in the great majority of cases this is sufficient. However, in those few problems where a knowledge of voltage values is necessary, they can be determined readily with the knowledge of symmetrical components outlined here, including the fact that phase-sequence currents and voltages of one kind (positive, negative, or zero) occur only in the phase-sequence reactance network of the same kind, plus perhaps some help from such works as Wagner and Evans' *Symmetrical Components* and Monseth and Robinson's *Relay Systems.*

47. ZERO-SEQUENCE IMPEDANCE OF A TRANSMISSION LINE WITH GROUND WIRES

(See also Section 14, Article 22)

The determination of the zero-sequence impedance of a transmission line with ground wires or of an underground cable, and the solution of an inductive coordination problem, are similar in that in each case the phase conductors and ground can be compared to the primary of a transformer, whereas the ground wires, the cable sheath, or the communication conductors, and the ground represent the transformer secondary.

Consider the condition shown in Fig. 15. The grounded transformer at the left supplies zero-sequence current which flows over the phase conductors to the one-line-to-ground

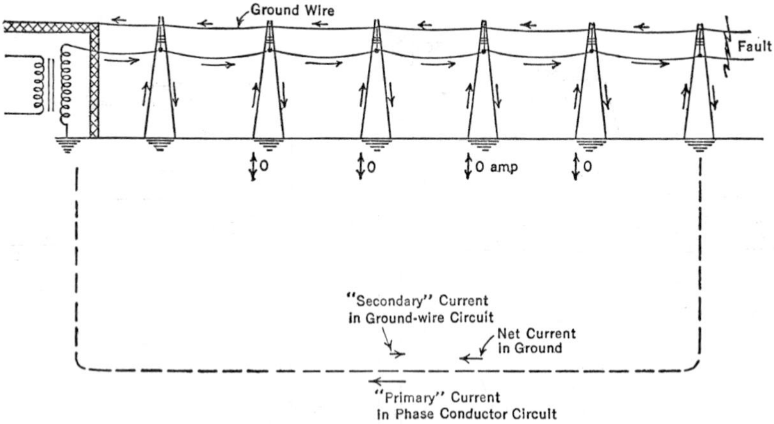

FIG. 15. Example of One-line-to-ground Fault

fault at the right, where, neglecting the ground wire for the moment, the fault current flows to the tower, down the tower to ground, and then returns through the ground to the transformer neutral.

The ground wire and the towers form a number of grounded loops, but, since in any one tower the current of one loop flowing up is practically neutralized by the downward-flowing current of the next loop, the ground wire can be considered a through conductor from the fault back to the transformer neutral. The voltage induced in this ground wire is equal to $I_g \times Z_{0m}$, where I_g = ground current in phase conductors, and Z_{0m} = zero-sequence mutual impedance between the phase-conductor group and the ground-wire group.

The current in the ground wire is, therefore, the voltage induced in it, divided by the impedance of the ground-wire circuit, or

$$I_{gw} = \frac{I_g \times Z_{0m}}{R_{gw} + jX_{gw}} \tag{12}$$

The magnetic field set up by this ground-wire current is in opposition to the field due to the ground current in the phase conductors. The resistance R_{gw} of the ground-wire circuit is of importance, however, because a high value of this resistance throws the ground-wire current considerably out of phase from the phase-conductor current, thus reducing the effectiveness of the ground-wire current. This is indicated by the vector diagram of Fig. 16.

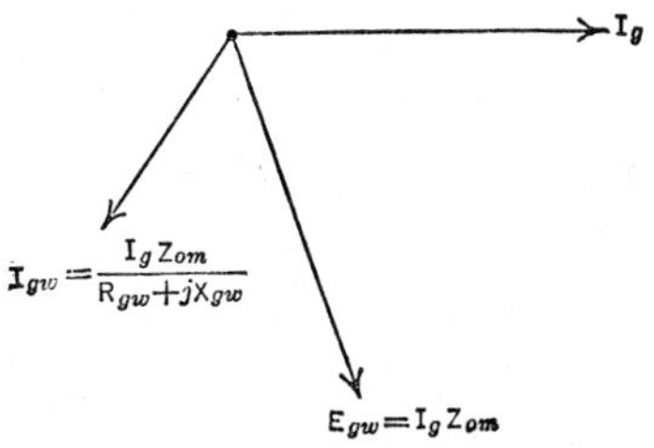

FIG. 16. Vector Diagram for Case of Fig. 15

The pictorial diagram of Fig. 15 is considerably simplified in Fig. 17(*a*) which shows simply the phase conductors, the ground wire, and the mutual impedance Z_{0m} between them. Since both the phase-conductor group and the ground-wire group are connected to ground at the fault, the impedances of the several branches can be arranged as in Fig. 17(*b*). The two grounded points are then connected together, giving the arrangement of Fig. 17(*c*).

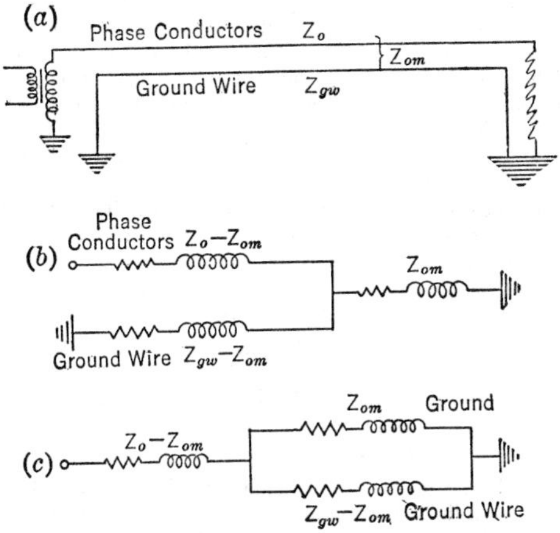

FIG. 17. Simplified Diagrams for Case of Fig. 15

The diagram of Fig. 17(*c*) is expanded in Fig. 18 to show the complete equations for each branch, and then, by neglecting the resistances, which are usually small, the relatively simple equation in Fig. 18 is obtained. For most power-system calculations, es-

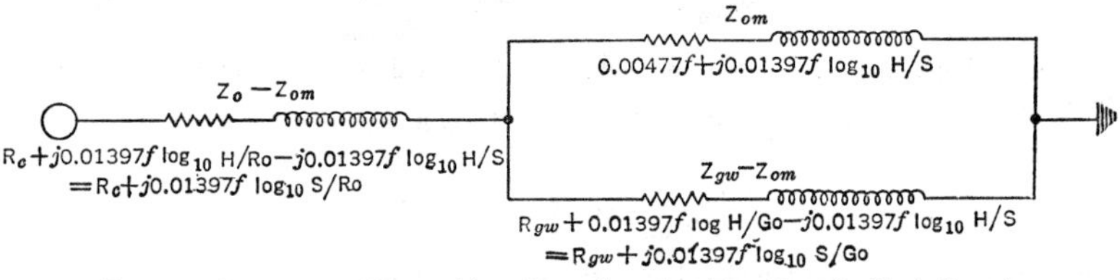

FIG. 18. Expansion of Fig. 17(*c*) to Show Complete Equations for Each Branch

Neglecting resistances of all three branches, this diagram reduces to

$$X_0 = 0.01397f \left[\log_{10} \frac{S}{R_0} + \frac{\log_{10} \frac{H}{S} \times \log_{10} \frac{S}{G_0}}{\log_{10} \frac{H}{G_0}} \right]$$

The various quantities are defined in the text on page 3-50.

pecially where high-conductivity ground wires are used, this equation is entirely satisfactory. The various quantities are defined below:

f = frequency.

S = (gmd) geometric mean distance between the conductors of the phase group and those of the ground-wire group, equal to the $n \cdot m$ root of the $n \cdot m$ distances between the n phase conductors and the m ground wires.

R_0 = equivalent radius of the phase-conductor group, defined in Article 41.

H = equivalent depth of earth return, defined in Article 41.

G_0 = equivalent radius of the ground-wire group, calculated in the same way as R_0.

48. ZERO-SEQUENCE IMPEDANCE OF AN UNDERGROUND CABLE

The zero-sequence impedance of an underground cable is obtained by following the same principles as outlined above. Because of the relatively high resistivity of the lead of the sheath, compared to copper, the resistance becomes the major element of the zero-sequence impedance. The exact value of zero-sequence impedance is difficult to calculate, because of the complexity of the ground structure. In any given case there are two limits to the value of Z_0: with no connection from the sheath to ground, or with the sheath solidly grounded; the two scalar values of Z_0 are about the same, but the values of R_0 and X_0 of any cable with and without the connection from the sheath to ground will vary widely. Very approximately, R_0 will be 5 to 10 times R_1, and X_0 will be 1 to 4 times X_1.

Although the value of the zero-sequence reactance is affected by the presence of other cable sheaths, pipes, and other ground conductors, the zero-sequence resistance will be affected to a greater degree by such other conductors. Consequently, although the zero-sequence impedance of an underground cable may be calculated, the only way to determine this value with certainty is to measure it, using the procedure outlined in Fig. 4.

The procedure for calculating the X_0 of an underground cable is outlined in Figs. 19(*a*) to 19(*e*). The three conductors and the sheath are shown in Fig. 19(*a*), and the make-up of the impedance values in Figs. 19(*b*) and 19(*c*); the similarity between these and Figs. 15, 17(*a*), 17(*b*), and 17(*c*) should be noted.

The development of the equation is shown in Fig. 19(*d*). The value of D_{ab}, the geometric mean distance between the phase-conductor group and the sheath, is given as the average radius of the sheath—a very close approximation. The value of R_s, the equivalent radius of the sheath, is likewise represented by the average radius, closely enough for all practical purposes. Consequently, the value of the resistance per mile of the sheath branch is $\dfrac{0.20}{r_y^2 - r_x^2}$ for a typical value of sheath resistivity (r_y and r_x being the outside and inside radii, in inches), or, multiplying by 3 to obtain ohms per phase,

$$\frac{0.60}{r_y^2 - r_x^2} = \frac{0.60}{(r_y + r_x)(r_y - r_x)}$$

The development in Fig. 19(*d*) is summarized in Fig. 19(*e*). Since the ground and the sheath branches are in parallel, and the ground branch has a relatively high value, it may be neglected (thus considerably simplifying the calculation) where a reasonably approximate answer is sufficient.

Where the knowledge of the ground conditions and the need for accuracy require the inclusion of the ground path, the paralleled value of the sheath and the ground path can be found most readily by using the form

$$Z_{\text{combined}} = \frac{(Z_{\text{sheath}})(Z_{\text{ground}})}{(Z_{\text{sheath}}) + (Z_{\text{ground}})}$$

using a log-log-vector slide rule to convert impedance values to the polar form, remembering that

$$(Z_a\underline{/A \text{ deg}})(Z_b\underline{/B \text{ deg}}) = Z_aZ_b\underline{/A \text{ deg} + B \text{ deg}}$$

and

$$\frac{Z_a\underline{/A \text{ deg}}}{Z_b\underline{/B \text{ deg}}} = \frac{Z_a}{Z_b}\underline{/A \text{ deg} - B \text{ deg}}$$

and then finding the combined value of the two impedances in this way:

$$A + jB = C\underline{/\theta_1 \text{ deg}}$$
$$D + jE = F\underline{/\theta_2 \text{ deg}}$$
$$\overline{(A + D) + j(B + E)} = G\underline{/\theta_3 \text{ deg}}$$

$$\text{Combined value} = \frac{CF\underline{/\theta_1 \text{ deg} + \theta_2 \text{ deg}}}{G\underline{/\theta_3 \text{ deg}}} = \frac{CF}{G}\underline{/\theta_1 \text{ deg} + \theta_2 \text{ deg} - \theta_3 \text{ deg}}$$

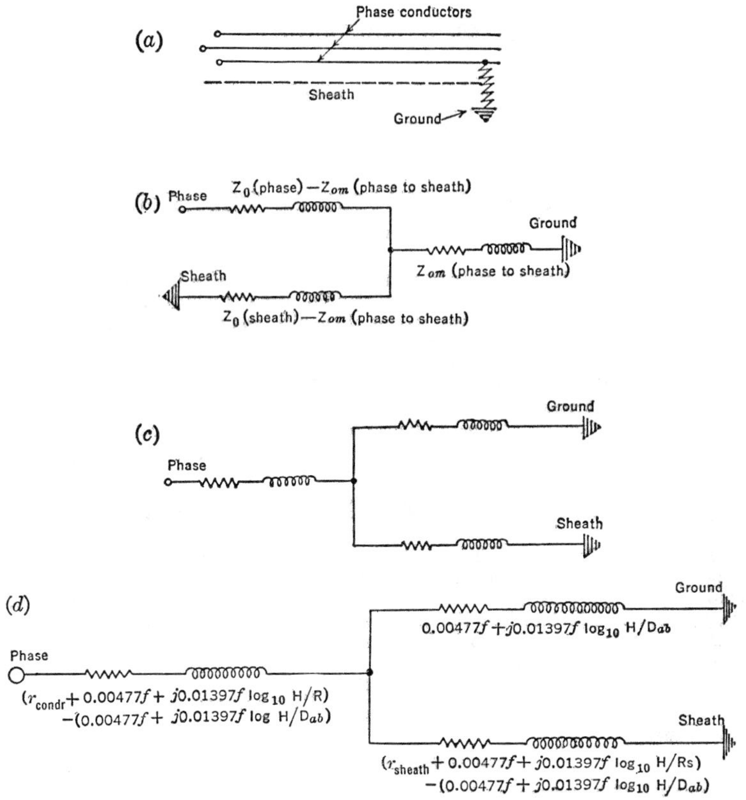

where

r_{condr} = resistance of conductor (ohms per mile per phase).

R = geometric mean radius of phase-conductor group.

$\sqrt[3]{0.779 \times \text{Radius} \times (\text{Axial spacing})^2}$

D_{ab} = geometric mean distance between phase-conductor group and sheath

$= \dfrac{r_y + r_x}{2}$.

r_y = outside radius of sheath.

r_x = inside radius of sheath.

H is defined in Article 41

Impedance of phase-conductor group, therefore, is

$$r_{condr} + j0.01397 f \log \frac{r_y + r_x}{2R}$$

where

r_{sheath} = resistance of sheath (ohms per mile per phase).

$$R_S = \text{equiv. radius of sheath} = \frac{R_y + R_x}{2}.$$

D_{ab} also equals $\dfrac{R_y + R_x}{2}$.

Therefore impedance of sheath branch reduces to

$$r_{sheath} = \frac{0.60}{r_y^2 - r_x^2} = \frac{0.60}{(r_y + r_x)(r_y - r_x)}$$

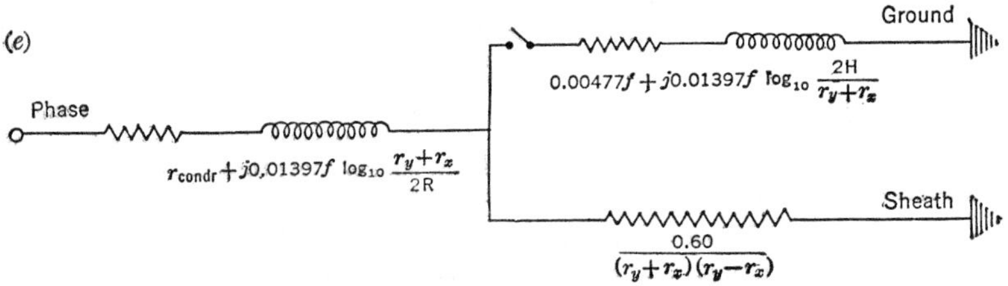

FIG. 19. Zero-sequence Impedance of an Underground Cable

49. BIBLIOGRAPHY

Edison Electrical Institute, *Relay Engineer's Handbook*, Supplement, Chapter 8.
Monseth, I. T., and P. H. Robinson, *Relay Systems*. McGraw-Hill.
Wagner, C. F., and R. D. Evans, *Symmetrical Components*. McGraw-Hill.
See also Bibliography, Section 14, Article 23.

CIRCUITS OF ELECTRICAL MACHINES

By Irven Travis

The calculation of the performance of any electrical machine involves the solution of the electric-circuit equations for a group of circuits, some of which move with respect to the others. There are in general two groups of circuits, one group fixed in space (usually called the stator winding) and one group in motion (usually called the rotor winding). Since the self and mutual inductances involved are in general functions of the various currents and of time, many simplifying assumptions are ordinarily made, and certain special methods of calculation must be used. Only fundamental considerations are treated in this chapter; for design and performance of machines refer to the particular type of machine of interest.

50. MAGNETOMOTIVE FORCES OF WINDINGS

When a current flows through a constant self- or mutual inductance, the voltage produced can easily be calculated by multiplying di/dt by the inductance in question. When the inductance is not constant, it is often more convenient to make this calculation in two parts, namely, to calculate first the magnetomotive forces produced by the currents in the various windings and then to find the flux linkages which these mmf produce. Thus instead of using the expression

$$e_1 = L_1 \frac{di_1}{dt} + M_{12} \frac{di_2}{dt} + M_{13} \frac{di_3}{dt} + \cdots \tag{1}$$

mmf of the type

$$\text{Mmf} = 0.4\pi(N_1 i_1 + N_2 i_2 + \cdots) \tag{2}$$

are calculated. The saturation in any particular flux path can then be determined and the flux in that path evaluated by the relation

$$\phi = \text{Mmf} \times \text{Permeance} \tag{3}$$

The voltage in the winding designated as No. 1, due to this flux, is then

$$e = 10^{-8} \frac{d}{dt} (N_1 \phi) \tag{4}$$

This method of calculating emf, coupled with the methods used normally in electric-circuit theory, offers a valuable tool for determining the performance of electrical machines.

In most machines the windings are distributed along the air gap in such a way that the mmf may be assumed to be proportional to the cosine of the angle from the center of a given coil. For example, the mmf produced at the air gap of a three-phase induction motor or synchronous machine by a direct current I flowing in the stator (armature) windings may be assumed to be given by

$$\left.\begin{aligned} A_a &= NI \cos x \\ A_b &= NI \cos (x - 120°) \\ A_c &= NI \cos (x - 240°) \end{aligned}\right\} \tag{5}$$

in which x is the distance in electrical degrees measured from the center of one of the coil groups of phase a, and N is a factor depending upon the number and arrangement of turns. The resultant mmf at any point on the stator is the sum of these three mmf's. Thus with direct current in the windings the resultant mmf A is

$$A = NI[\cos x + \cos (x - 120°) + \cos (x - 240°)] = 0 \tag{6}$$

If the armature windings are excited by means of three alternating currents which form a balanced positive-sequence group, the mmf will be a function of the time as well as of the angle x and may be written as $A_1(x, t)$. The three currents may be written as

$$\left.\begin{aligned} i_{a1} &= \sqrt{2}\, I_{a1} \sin(\omega t - \beta_1) \\ i_{b1} &= \sqrt{2}\, I_{a1} \sin(\omega t - \beta_1 - 120°) \\ i_{c1} &= \sqrt{2}\, I_{a1} \sin(\omega t - \beta_1 - 240°) \end{aligned}\right\} \qquad (7)$$

The armature mmf becomes

$$\left.\begin{aligned} A_1(x, t) &= \sqrt{2}\, NI_{a1}[\sin(\omega t - \beta_1)\cos x + \sin(\omega t - \beta_1 - 120°)\cos(x - 120°) \\ &\quad + \sin(\omega t - \beta_1 - 240°)\cos(x - 240°)] \\ A_1(x, t) &= \frac{3NI_{a1}}{\sqrt{2}} \sin(\omega t - x - \beta_1) \end{aligned}\right\} \qquad (8)$$

51. ROTATING FIELDS

In eq. (8), $A_1(x, t)$ is the resultant mmf at a point on the stator at a distance x from the center of one of the coil groups of phase a, expressed as a function of time. If it is desired to find the resultant mmf at some point on the rotor in terms of the distance from a datum on the rotor structure, say the center of a pole, x may be put equal to the distance from the center of a pole to the point in question plus the distance between the centers of the pole and the coil group. Let the origin of time be taken as the instant at which the center of a coil group of phase a coincides with the center of a pole, so that the distance between the center of the coil group and the center of the pole is simply $\omega_r t$. Let the distance between the center of the pole and the point on the rotor at which the mmf is desired be y. That is, put

$$x = \omega_r t + y \qquad (9)$$

then we have

$$A_1(y,t) = -\frac{3NI_{a1}}{\sqrt{2}} \sin[y + \beta_1 - (\omega - \omega_r)t] \qquad (10)$$

in which $A_1(y,t)$ is the mmf at a distance $(\omega_r t + y)$ from the center of coil a, or at a distance y from the center of a pole. This quantity is not a function of time when ω and ω_r are equal. Physically this means that the armature currents produce an mmf which is sinusoidal in space and which rotates at the angular velocity ω. When ω_r is equal to ω, there is no relative motion between the armature flux and the field conductors. The result is not limited to the case of three phases; it is readily shown that *any* polyphase winding through which balanced positive-sequence currents flow produces a similar rotating mmf. The more general theorem is known as the theorem of Ferraris.

This principle makes possible certain graphical solutions of synchronous machine problems which are of use when non-linearity of the saturation curve is not neglected. The resultant mmf of the field and armature can be found; and since this resultant is independent of time so far as the field iron is concerned, the flux which this mmf produces can be found from a d-c saturation curve. In this way, the problem of **armature reaction** (see Section 9, Article 8) in the synchronous machine is rendered relatively simple. It is important to stress the fact that **armature reaction** is simply a special name applied to a component mmf involved in the simultaneous solution of two or more mesh equations. Armature reaction is not used in transformer theory because the mmf of the secondary is not constant with time and is not a convenient quantity to use in transformer calculations. Ordinary circuit theory methods are more suitable for that problem.

52. FLUX LINKAGES IN SYNCHRONOUS MACHINES

The space wave form of flux around the periphery of the armature of a synchronous machine is the same as that of the mmf only if the air gap is uniform and if saturation is negligible. Since flux linkages, and not mmf, determine the emf's in the machines, flux linkage is the quantity which must be evaluated. An approximate method applicable to machines having salient poles was developed by Blondel, on the empirical theory of alternators (*L'Industrie Électrique*, Nov. 10 and 25, 1899). He resolved the resultant mmf acting across the air gap into two sinusoidal components, one having its maximum value at the center of a pole and the other having its maximum value at a point midway between the poles.

It is well known that in d-c machines the armature reaction is transverse when the brushes are in the neutral position, and direct when they are shifted 90 elec deg from the neutral. It is customary to treat the two reactions separately and to split any armature reaction occurring with the brushes in some intermediate position into two parts. The reaction in a salient pole alternator is fundamentally the same phenomenon. Blondel assumed that the resultant flux could be found by multiplying the mmf due to the first component by the permeance p_d of the path along the polar axis, multiplying the second component by the permeance p_q of the path along an axis midway between the poles, and adding the results. There is no necessity to restrict this process to balanced positive-sequence armature currents, or indeed to restrict it to the armature at all. All mmf's acting in the air gap can be split into direct and quadrature components, and the flux linkages which these mmf's produce can then be found. The subsequent analysis follows closely, except for his use of per-unit quantities, the work of Park.*

The components of mmf in the direct and quadrature axes due to the field will be denoted by F_d and F_q. Then the field flux in maxwells per electrical degree is given by

$$F = p_dF_d \cos y + p_qF_q \sin y \tag{11}$$

The mmf in ampere-turns due to the armature was shown in Article 50 to be

$$A(x,t) = N[i_a \cos x + i_b \cos (x - 120°) + i_c \cos (x - 240°)] \tag{12}$$

Putting $x = \omega t + y$ and expanding, there results

$$\begin{aligned} A(x,t) = {} & N \cos y[i_a \cos \omega t + i_b \cos (\omega t - 120°) + i_c \cos (\omega t - 240°)] \\ & - N \sin y[i_a \sin \omega t + i_b \sin (\omega t - 120°) + i_c \sin (\omega t - 240°)] \end{aligned}$$

which is seen to be separated into components, one in the direct polar axis and the other in the quadrature polar axis. The resultant flux density in maxwells per electrical degree in the air gap is

$$\begin{aligned} b_r = {} & p_dF_d \cos y + p_qF_q \sin y \\ & + 0.4\pi p_dN \cos y[i_a \cos \omega t + i_b \cos (\omega t - 120°) + i_c \cos (\omega t - 240°)] \\ & - 0.4\pi p_qN \sin y[i_a \sin \omega t + i_b \sin (\omega t - 120°) + i_c \sin (\omega t - 240°)] \end{aligned} \tag{13}$$

which may be rewritten in the form

$$\begin{aligned} b_r = {} & p_dF_d \cos (\omega t - x) - p_qF_q \sin (\omega t - x) \\ & + 0.4\pi N \frac{p_q + p_d}{2} [i_a \cos x + i_b \cos (x - 120°) + i_c \cos (x - 240°)] \\ & + 0.4\pi N \frac{p_d - p_q}{2} [i_a \cos (2\omega t - x) + i_b \cos (2\omega t - x - 120°) \\ & + i_c \cos (2\omega t - x - 240°)] \end{aligned} \tag{14}$$

The flux linkages in phase a due to the flux b_r are proportionate to the flux density at the axis of coil a. This is the value of eq. (14) when $x = 0$. If N' is the constant of proportionality between flux density and flux linkages, we have

$$\begin{aligned} \psi_a = {} & N'p_dF_d \cos \omega t - N'p_qF_q \sin \omega t + 0.4\pi NN' \frac{p_d + p_q}{2} \left(i_a - \frac{i_b + i_c}{2}\right) \\ & + 0.4\pi NN' \frac{p_d - p_q}{2} [i_a \cos 2\omega t + i_b \cos (2\omega t - 120°) + i_c \cos (2\omega t - 240°)] \end{aligned} \tag{15}$$

The value of ψ_a given by eq. (14) represents the total flux linkages in coil a due to flux which crosses the air gap. There will be additional flux linkages due to flux which links coil a only, and due to flux which links coils b and c but which does not cross the air gap. This flux is mostly slot leakage and flux encircling the end connections, and hence has its paths mostly in the air.

The total flux linkages in coil a may be written

$$\begin{aligned} \psi_a = {} & N'p_dF_d \cos \omega t - N'p_qF_q \sin \omega t + \frac{x_{l\sigma}}{\omega 10^{-8}} i_a - \frac{x_m}{\omega 10^{-8}} (i_b + i_c) \\ & + 0.4\pi NN' \frac{p_d + p_q}{4} \left(i_a - \frac{i_b + i_c}{2}\right) \\ & + 0.4\pi NN' \frac{p_d - p_q}{2} [i_a \cos 2\omega t + i_b \cos (2\omega t - 120°) + i_c \cos (2\omega t - 240°)] \end{aligned} \tag{16}$$

where $x_{l\sigma}$ = self-leakage reactance of one phase of the armature, and x_m = mutual leakage reactance of any two phases.

* R. H. Park, Definition of an Ideal Synchronous Machine, *Gen. Elec. Rev.* (June, 1928).

It is convenient to write the armature currents in terms of their symmetrical components and to split the armature flux linkages into four parts, the flux linkages due to the field flux and those due to each of the symmetrical components of the armature currents, thus

$$\psi_a = \psi_{af} + \psi_{a0} + \psi_{a1} + \psi_{a2} \tag{17}$$

The flux linkages due to the field are given by

$$\psi_{af} = N'p_dF_d \cos \omega t - N'p_qF_q \sin \omega t \tag{18}$$

The flux linkages due to positive-sequence armature currents are

$$\psi_{a1} = \left(\frac{x_{l\sigma}}{\omega 10^{-8}} + \frac{x_m}{\omega 10^{-8}} + 1.2\pi NN' \frac{p_d + p_q}{4}\right) \sqrt{2} I_{a1} \sin (\omega t - \beta_1)$$
$$- 1.2\pi NN' \frac{p_d - p_q}{4} \sqrt{2} I_{a1} \sin (\omega t + \beta_1) \tag{19}$$

The flux linkages due to negative-sequence armature current are

$$\psi_{a2} = \left(\frac{x_{l\sigma}}{\omega 10^{-8}} + \frac{x_m}{\omega 10^{-8}} + 1.2\pi NN' \frac{p_d + p_q}{4}\right) \sqrt{2} I_{a2} \sin (\omega t - \beta_2)$$
$$+ 1.2\pi NN' \frac{p_d - p_q}{4} \sqrt{2} I_{a1} \sin (3\omega t - \beta_2) \tag{20}$$

The flux linkages due to zero-sequence armature currents are

$$\psi_{a0} = \left(\frac{x_{l\sigma}}{\omega 10^{-8}} - \frac{2x_m}{\omega 10^{-8}}\right) \sqrt{2} I_{a0} \sin (\omega t - \beta_0) \tag{21}$$

53. NOMINAL EMF AND ARMATURE REACTANCES

The emf's due to the various flux linkages are easily found. The open-circuit or nominal emf is

$$e_a = -10^{-8} \frac{d}{dt} \psi_{af} \tag{22}$$

The net induced emf in phase a of the armature is equal to the derivative with respect to time of ψ_a; hence the total armature inductive drop is equal to $\frac{d}{dt}(\psi_a - \psi_{af})$. The total inductive drop per unit armature current is the armature reactance. It is apparent that the character of the armature reactance will depend upon whether there are salient poles and whether the armature currents are balanced. It is customary to ascribe several different "reactances" to a synchronous machine armature, each reactance suitable for a particular type of computation.

Consider a machine having non-salient poles and having equal receiver impedances connected to the three phases. If there are no salient poles, the permeance is the same along any radius and hence

$$p_d = p_q$$

If the receiver impedances are equal, the currents form a positive-sequence group; hence

$$I_{a0} = I_{a2} = 0$$

It follows, therefore, that the total armature inductive drop is

$$-[x_{l\sigma} + x_m + {}^3/_2(4\pi\omega NN'p_d 10^{-9})]\sqrt{2} I_{a1} \cos (\omega t - \beta_1) \tag{23}$$

The quantity in the bracket in eq. (23) is called the **synchronous reactance** of the machine. The synchronous reactance is the effective leakage reactance of the armature plus the combined reactive effect, on one phase of the armature, of all the armature flux which traverses the polar path. The first part is usually called **armature reactance,** and the second part is usually called **armature reaction.** Denote the synchronous reactance by X_s, then

$$X_s = x_{l\sigma} + x_m + {}^3/_2\omega(4\pi NN'p_d 10^{-9}) \tag{24}$$

The **synchronous impedance** of the machine is an impedance having a reactive part X_s and a resistive part equal to the armature effective resistance.

$$\mathbf{Z}_s = r_a + jX_s \tag{25}$$

The vector equation for the terminal voltage is

$$\mathbf{E}_a - \mathbf{Z}_s\mathbf{I}_a = \mathbf{V}_a \tag{26}$$

The synchronous reactance given by eq. (24) is the value of reactance for a salient pole machine when the power factor is such that the armature current is in quadrature with the nominal emf. It does not depend on the permeance of the quadrature path and hence is called the **direct axis synchronous reactance** and is usually denoted by X_d.

The **quadrature axis synchronous reactance,** denoted by X_q, is defined by the equation

$$X_q = x_{l\sigma} + x_m + {}^3/_2\omega(4\pi N N' \, p_q 10^{-9}) \tag{27}$$

and may readily be seen to be the value of the synchronous reactance for a salient pole machine when the armature current is in phase with the nominal emf.

The reactance to armature currents having zero sequence is often called the **zero-sequence reactance** and is denoted by X_0. From eq. (21) this reactance is

$$X_0 = x_{l\sigma} - 2x_m \tag{28}$$

54. BIBLIOGRAPHY

Arnold, E., *Die Wechselstromtecnik.* Berlin, Julius Springer (1909–13). (5 vols.)
Doherty, A Simplified Method of Analyzing Short-circuit Problems. *AIEE Trans.*, Vol. 42, pp. 841–849 (June, 1923).
Hague, B., *Electromagnetic Problems in Electrical Engineering.* London, Oxford University Press (1929).
Hahn and Wagner, Standard Decrement Curves. Presented at the Winter Convention of the AIEE, New York (Jan. 25–29, 1932).
Jackson, D. C. and J. P., *Alternating Currents.* Macmillan (1913).
Kron, Gabriel, The Application of Tensors to the Analysis of Rotating Electrical Machinery. *Gen. Elec. Rev.*, Vol. 38, pp. 181, 230, 282, 339, 386, 434, 473, 527, 582 (April to Dec., 1935).
Ku, Y. H., Transient Analysis of A-c Machinery. *Pub. M.I.T.*, Vol. 64, No. 94 (April, 1929).
Mauduit, A., *Machines électriques.* Paris, Dunod (1931). (2 vols.)
Park, R. H., Two-reaction Theory of Synchronous Machines—I. Presented at the Winter Convention of the AIEE, New York (1929).
Park, R. H., Two-reaction Theory of Synchronous Machines—II. *AIEE Trans.*, Vol. 52, pp. 352–354 (June, 1933).
Park, R. H., and B. L. Robertson, Reactances of Synchronous Machines. *AIEE J.*, Vol. 47, pp. 345–348 (May, 1928).
Waring and Crary, Operational Impedance of Synchronous Machines. *Gen. Elec. Rev.*, pp. 578–582 (Nov., 1932).

GENERAL EQUATIONS OF TRANSMISSION LINES

By Irven Travis

Transmission lines and cables, in contrast to electric circuits such as transformers and machines, have parameters which are **distributed.** The **lumped-parameter** electric-circuit theory outlined above cannot be used for transmission-line calculations except in special cases in which the lines are electrically short. (The **electrical length** of a line is defined in Article 60.) A **smooth** line has series resistance, series inductance, shunt conductance or leakance, and shunt capacitance, uniformly distributed along the line. The series constants are usually termed the conductor impedance; the shunt constants are usually called the dielectric admittance.

55. TRANSMISSION-LINE EQUATIONS

Consider a transmission line consisting of two parallel wires and having constants **per unit length** given by

Resistance	= r ohms
Inductance	= L henrys
Leakance	= g mhos
Capacitance	= C farads

Let x be the distance measured from the receiving end of the line. Consider the change in voltage and current in a differential element dx (Fig. 1). The impedance drop across dx is $\left(r\,dx\,i + L\,dx\,\frac{\partial i}{\partial t}\right)$, and the current through the admittance is $\left(g\,dx\,e + C\,dx\,\frac{\partial e}{\partial t}\right)$; hence the two equations

$$ri + L\frac{\partial i}{\partial t} = \frac{\partial e}{\partial x} \tag{1}$$

$$ge + C\frac{\partial e}{\partial t} = \frac{\partial i}{\partial x} \tag{2}$$

give the current and voltage at any point on the line at any instant. These partial differential equations are perfectly general and therefore are applicable to both steady-state and transient calculations. Although these equations were established for two parallel wires, they may be used for polyphase as well as single-phase calculations. In the polyphase case the series constants are taken per unit length of **wire** instead of per unit length of **line,** and the shunt constants are taken per unit length of wire to neutral. The current is then **per wire** and the voltage is **wire to neutral.**

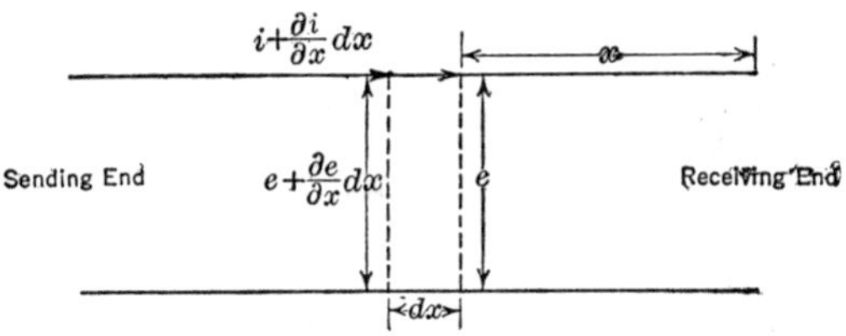

FIG. 1. Section of a Transmission Line

56. STEADY-STATE CURRENT AND VOLTAGE DISTRIBUTION

If the voltages impressed at the ends of the line have the angular velocity ω, then in the steady state eqs. (1) and (2) become

$$(r + j\omega L)\mathbf{I} = \frac{d\mathbf{E}}{dx} \tag{3}$$

$$(g + j\omega C)\mathbf{E} = \frac{d\mathbf{I}}{dx} \tag{4}$$

in which **E** and **I** are the vector voltage and vector current at a distance x from the receiver. A more symmetrical form may be obtained by differentiating eqs. (3) and (4) with respect to x and solving for **E** and for **I** separately:

$$\frac{d^2\mathbf{E}}{dx^2} = (r + j\omega L)(g + j\omega C)\mathbf{E} \tag{5}$$

$$\frac{d^2\mathbf{I}}{dx^2} = (r + j\omega L)(g + j\omega C)\mathbf{I} \tag{6}$$

The solutions of these equations may be written in either of two alternative forms: *

$$\mathbf{E} = \mathbf{M}e^{\gamma x} + \mathbf{N}e^{-\gamma x} \tag{7}$$

$$\mathbf{I} = \mathbf{M}'e^{\gamma x} + \mathbf{N}'e^{-\gamma x} \tag{7a}$$

or

$$\mathbf{E} = \mathbf{M}_1 \cosh \boldsymbol{\gamma} x + \mathbf{N}_1 \sinh \boldsymbol{\gamma} x \tag{8}$$

$$\mathbf{I} = \mathbf{M}'_1 \cosh \boldsymbol{\gamma} x + \mathbf{N}'_1 \sinh \boldsymbol{\gamma} x \tag{8a}$$

where

$$\boldsymbol{\gamma} = \sqrt{(r + j\omega L)(g + j\omega C)}$$

The constants **M** and **N** in eqs. (7) and (8) are determined by the terminal currents and voltages. Let $\mathbf{E}_r$ and $\mathbf{I}_r$ be the voltage and current at the receiver, and let $\mathbf{E}_s$ and $\mathbf{I}_s$ be the voltage and current at the source. The **E**'s are both voltage rises; the **I**'s are currents *into* the line from the source and *out* of the line to the receiver. Only the form eqs. (8) and (8*a*) will be given; these are usually most convenient for calculation.

$$\mathbf{E} = \mathbf{E}_r \cosh \boldsymbol{\gamma} x + \mathbf{Z}_c\mathbf{I}_r \sinh \boldsymbol{\gamma} x \tag{9}$$

$$\mathbf{I} = \mathbf{I}_r \cosh \boldsymbol{\gamma} x + \frac{\mathbf{E}_r}{\mathbf{Z}_c} \sinh \boldsymbol{\gamma} x \tag{9a}$$

$$\mathbf{E} = \mathbf{E}_s \cosh \boldsymbol{\gamma} y - \mathbf{Z}_c\mathbf{I}_s \sinh \boldsymbol{\gamma} y \tag{10}$$

$$\mathbf{I} = \mathbf{I}_s \cosh \boldsymbol{\gamma} y - \frac{\mathbf{E}_s}{\mathbf{Z}_c} \sinh \boldsymbol{\gamma} y \tag{10a}$$

in which $\mathbf{Z}_c$ is defined by

$$\mathbf{Z}_c = \sqrt{\frac{r + j\omega L}{g + j\omega C}} \tag{11}$$

and in which y is defined by

$$y = \text{Length of line} - x \tag{12}$$

* In the following γ is a complex number; because of printing limitations it appears in lightface type when in an exponent.

Thus the current and voltage at any point in the line may be expressed in terms of the conditions at and distance from the sending end or receiving end, whichever is the more convenient.

57. GENERAL CIRCUIT CONSTANTS

If in eqs. (9) and (9a) x is put equal to the length of the line l, the current **I** and voltage **E** become those of the source; hence

$$\mathbf{E}_s = \mathbf{E}_r \cosh \boldsymbol{\gamma} l + \mathbf{I}_r \mathbf{Z}_c \sinh \boldsymbol{\gamma} l \tag{13}$$

$$\mathbf{I}_s = \mathbf{E}_r \frac{1}{\mathbf{Z}_c} \sinh \boldsymbol{\gamma} l + \mathbf{I}_r \cosh \boldsymbol{\gamma} l \tag{14}$$

Comparing these with eqs. (32) and (33), Article 19, for linear circuits, the general circuit constants are

$$\left.\begin{aligned} \mathbf{A} &= \cosh \boldsymbol{\gamma} l \\ \mathbf{B} &= \mathbf{Z}_c \sinh \boldsymbol{\gamma} l \\ \mathbf{C} &= \frac{1}{\mathbf{Z}_c} \sinh \boldsymbol{\gamma} l \\ \mathbf{D} &= \cosh \boldsymbol{\gamma} l \end{aligned}\right\} \tag{15}$$

By means of these constants the transmission line can be treated in the same way as any other impedance link in a given transmission system. The usual laws for combination in series and in parallel can, of course, be used. (See Article 19.) When lines of different characteristics are connected together, e.g., an aerial line and an underground cable, the general circuit constants may be calculated separately and then combined.

58. PROPAGATION CONSTANT, CHARACTERISTIC IMPEDANCE

Although less useful in practical calculations, the exponential form of solution of the transmission-line equations is more valuable in the development of theory. Consider eq. (7).

$$\mathbf{E} = \mathbf{M} e^{\boldsymbol{\gamma} x} + \mathbf{N} e^{-\boldsymbol{\gamma} x} \tag{7}$$

Both **M** and **N** must have the dimensions of voltage; **M** is often called the **incident** voltage and **N** the **reflected** voltage. The incident voltage increases in magnitude as we go farther and farther away from the receiver, and the reflected voltage decreases as we go farther and farther away from the receiver. Viewed from the source, the incident voltage is **attenuated** as it progresses along the line toward the receiver, and the reflected wave builds up, i.e., is attenuated from the receiver. Let $\boldsymbol{\gamma}$ be written in the form

$$\boldsymbol{\gamma} = \alpha + j\beta \tag{16}$$

then the factor $e^{\boldsymbol{\gamma} x}$ may be written

$$e^{\boldsymbol{\gamma} x} = e^{\alpha x} \angle \beta x \tag{17}$$

Hence α is the quantity which causes a change in magnitude of the incident or reflected voltage, and β is the quantity which causes a phase shift in the voltage.

For this reason α is called the **attenuation** constant, β the **phase-shift** constant, and $\boldsymbol{\gamma}$ the **propagation** constant of the line.

In the special case in which the receiver impedance is equal to the impedance $\mathbf{Z}_c$ defined by eq. (11),

$$\mathbf{I}_r \mathbf{Z}_c = \mathbf{E}_r$$

and eqs. (9) and (9a) become

$$\mathbf{E} = \mathbf{E}_r(\cosh \boldsymbol{\gamma} x + \sinh \boldsymbol{\gamma} x) = \mathbf{E}_r e^{\boldsymbol{\gamma} x} \tag{18}$$

$$\mathbf{I} = \mathbf{I}_r(\cosh \boldsymbol{\gamma} x + \sinh \boldsymbol{\gamma} x) = \mathbf{I}_r e^{\boldsymbol{\gamma} x} \tag{19}$$

If such is the case, the reflected voltage and current disappear, and the line is said to be **smoothly terminated.**

The impedance at any point in the line, from eqs. (9) and (9a), is

$$\mathbf{Z} = \frac{\mathbf{E}}{\mathbf{I}} = \frac{(\mathbf{E}_r + \mathbf{I}_r\mathbf{Z}_c)e^{\boldsymbol{\gamma} x} + (\mathbf{E}_r - \mathbf{I}_r\mathbf{Z}_c)e^{-\boldsymbol{\gamma} x}}{(\mathbf{E}_r + \mathbf{I}_r\mathbf{Z}_c)e^{\boldsymbol{\gamma} x} - (\mathbf{E}_r - \mathbf{I}_r\mathbf{Z}_c)e^{-\boldsymbol{\gamma} x}} \mathbf{Z}_c \tag{20}$$

For an infinite line the impedance is

$$\mathbf{Z}_\infty = \lim_{x \to \infty} \mathbf{Z} = \mathbf{Z}_c \tag{21}$$

The impedance $\mathbf{Z}_c$ is therefore the impedance at any point of an infinite line; it is also the impedance at any point of a smoothly terminated line. This impedance is called the **characteristic** or **surge impedance** of the line.

59. SHORT- AND OPEN-CIRCUIT IMPEDANCE

If the receiving end of the line is short-circuited, $\mathbf{E}_r$ is zero, and the input impedance becomes

$$\mathbf{Z}_{ss} = \mathbf{Z}_c \tanh \boldsymbol{\gamma} l \tag{22}$$

If the receiving end is open, $\mathbf{I}_r$ is zero, and the input impedance becomes

$$\mathbf{Z}_{s0} = \mathbf{Z}_c \coth \boldsymbol{\gamma} l \tag{23}$$

From these relations the following simple expression for the characteristic impedance is derived:

$$\mathbf{Z}_c = \sqrt{\mathbf{Z}_{ss}\mathbf{Z}_{s0}} \tag{24}$$

We have also

$$\boldsymbol{\gamma} = \frac{1}{l} \tanh^{-1} \frac{\mathbf{Z}_{ss}}{\mathbf{Z}_c} = \frac{1}{l} \coth^{-1} \frac{\mathbf{Z}_{s0}}{\mathbf{Z}_c} \tag{24a}$$

60. ELECTRICAL LENGTH

One **wavelength** is a length of line such that, when smoothly terminated, the phase shift in current or voltage is 360 deg. The same phase shift will, of course, occur in the same length of an infinite line having the same constants. Thus in general the wavelength λ of a line is

$$\lambda = \frac{2\pi}{\beta} \tag{25}$$

A line whose physical length is short in comparison with one-quarter wavelength is said to be electrically short. If the length of line is an appreciable fraction of one-quarter wave, it is said to be electrically long. Most power-transmission lines are electrically short.

61. TRANSIENTS IN TRANSMISSION LINES

The calculation of transmission-line transients is extremely specialized and will be considered in outline only.

The steady-state transfer impedance from the sending end of the line to any other point at a distance y from the sending end may be shown to be

$$\mathbf{Z}_y = \mathbf{Z}_c \left[\frac{(\mathbf{Z}_r + \mathbf{Z}_c)e^{\gamma l} + (\mathbf{Z}_r - \mathbf{Z}_c)e^{-\gamma l}}{(\mathbf{Z}_r + \mathbf{Z}_c)e^{\gamma(l-y)} - (\mathbf{Z}_r - \mathbf{Z}_c)e^{-\gamma(l-y)}} \right] \tag{26}$$

In "Electric Circuits, Transient State" (*Handbook of Engineering Fundamentals*, by Eshbach, John Wiley) the expression

$$\frac{1}{\beta Z(\beta)} = \int_0^\infty e^{-\beta t}\, h(t)\, dt \tag{27}$$

is given for the **indicial admittance** $h(t)$. In this expression the quantity $Z(\beta)$ is the transfer impedance as a function of the parameter β.

Hence for a transmission line the indicial admittance is given by the Laplace integral

$$\int_0^\infty e^{-\beta t}\, h_y(t)\, dt = \frac{1}{\beta Z_c(\beta)} \left[\frac{1 + U(\beta)}{e^{-y\gamma(\beta)} - U(\beta)e^{y\gamma(\beta)}} \right] \tag{28}$$

where $h_y(t)$ = transfer indicial admittance from the sending end to a point y from the sending end.

$Z_c(\beta) = \sqrt{\dfrac{r + \beta L}{g + \beta C}}$.

$\gamma(\beta) = \sqrt{(r + \beta L)(g + \beta C)}$.

$U(\beta) = \left[\dfrac{Z_r(\beta) - Z_c(\beta)}{Z_r(\beta) + Z_c(\beta)} \right] e^{-2l\gamma(\beta)}$.

$Z_r(\beta)$ = impedance of the receiver impedance when the angular frequency $j\omega$ is replaced by β.

For methods of solving eq. (28) refer to J. R. Carson, *Electric Circuit Theory and Operational Calculus*, McGraw-Hill; E. Clarke, *Circuit Analysis of A-C Power Systems*, John Wiley (1943); and Campbell and Foster, Fourier Integrals for Practical Applications, *B.S.T.J.*, (September, 1931).

When the indicial admittance has been determined, the current due to any applied emf is

$$i_y(t) = \frac{d}{dt}\int_0^t h_y(t - \lambda)E(\lambda)\,d\lambda \tag{29}$$

where $i_y(t)$ = current at a distance y from the sending end.
$E(t)$ = sending-end voltage. λ = a variable of integration.

62. BIBLIOGRAPHY

Bewley, *Traveling Waves on Transmission Systems*. John Wiley (1932).
Dwight, H. B., *Transmission Line Formulas*. Van Nostrand (1925).
Franklin, W. S., and F. E. Terman. *Transmission Line Theory*. Franklin and Charles, Lancaster, Pa. (1926).
Kennelly, A. E., *Electric Lines and Nets*. McGraw-Hill (1927).
Kennelly, A. E., *The Application of Hyperbolic Functions to Electrical Engineering Problems*. London, University Press (1912).
Lewis, W. W., *Transmission Line Engineering*. McGraw-Hill (1928).
Woodruff, L. F., *Principles of Electric Power Transmission and Distribution*. John Wiley (1938).

POWER-SYSTEM STABILITY

By S. B. Crary

63. INTRODUCTION

Power-system stability is that property of a power system which insures that it will remain in operating equilibrium through normal and abnormal conditions. This operating equilibrium is the balance which must be maintained between the various mechanical driving forces as well as the driven load forces through the electrical transmitting system. One of the essential requirements for maintaining this balance is the maintenance of synchronism between the synchronous machines of the system.

Stability studies usually are made to determine the ability of the major elements to remain in synchronism during and after transient faults and circuit changes, as well as during sustained gradual load changes. The most common and probably the most important measure of stability is the maximum power which can be carried for a given set of system and operating conditions. This critical value of power, below which the system is stable and above which it is unstable, is termed the **stability power limit** or the **stability limit.**

The classification of conditions under which the system is operating when tested or analyzed for stability has provided a basis for the development of the theory as well as for its application. The two most important divisions of the theory are as follows:

1. **Steady State.** The operating state of a power system which is characterized by small or gradual changes.

2. **Transient State.** The operating state of a power system which is characterized by large or sudden changes in load or circuit conditions.

A further classification of steady state distinguishes between operation with and without automatic control devices, although practically all power systems now use voltage regulators, speed governors, and various supplementary controls. A subgrouping of the phenomena under transient conditions which describes the extent of the analysis can also be made, such as **first-swing** and **multiswing,** although stability is desired before, during, and after both slow and sudden changes.

Table 1. Classification of Power-system Stability

Overall	Steady	Without automatic regulators With automatic regulators
	Transient	First-swing Multiswing
	Hunting	Natural Forced

Table 1 shows diagrammatically three different classifications, with the term **overall** used to indicate the ability of a system to be stable in the different states. In general, a hunting or sustained oscillatory state may be classified as either **natural** or **forced,** depending on whether it is the result of the inherent system properties or of a forced stimulus such as a periodic load application. Some of these classifications of stability will be found in Article 69, Reference 1, as defined by the American Standards Association.

The methods of analysis of **steady state without automatic regulators** and **first-swing transient stability** constitute the most important tools for judging system performance. The reasons for their relative importance are:

a. Simplicity of application compared with a treatment involving automatic regulating devices and subsequent swings.

b. Steady-state limits are, in general, conservative if automatic regulation is neglected.

c. Transient stability on the first swing and steady-state stability for the subsequent steady-state condition will for most cases be indicative of stability during the intermediate period.

d. Objectionable hunting represents a special problem which has become relatively rare with the use of modern apparatus and system design practices.

The published material on the subject of power-system stability is extensive. (See Article 69, References 3 and 54, for a selected English language bibliography.) The methods of analysis range from simple to involved calculations, depending on the nature of the investigation.

Article 64 gives an outline of methods of analysis, with references from which detailed information may be obtained for making a stability study.

Articles 65, 66, 67, and 68 correlate the results of some of the extensive theoretical, experimental, and operational analyses which apply to system design.

References to numbered items in the bibliography, Article 69, will assist those who wish to go into greater detail.

64. ANALYSIS

An analysis of the stability characteristics of a power system may employ many of the modern methods of analysis. Since a power system is a dynamical system with many degrees of freedom regulated by diverse control mechanisms, rigorous solution is impractical. Simplifying assumptions must necessarily be made. The extent of such simplifications depends on the accuracy requirements and the nature of the particular stability analysis, as well as the experience and judgment of the analyst. Quite accurate estimates may be made of the power limit of a particular link of a system from very simple calculations, whereas another problem may require a very comprehensive and detailed study. Alternating-current network analyzers (References 4, 5, 6) and differential analyzers (References 7, 8, 9) may be used to assist in obtaining the necessary information.

The following is an outline of the essential features and steps for a stability analysis. Additional refinements may be necessary for special investigations.

I. *Steady-state stability* (References 10, 2).
   1. Reduction of system to a simplified equivalent circuit representing the synchronous machines by equivalent synchronous reactances (Reference 11).
   2. Determination of initial quantities for system loading and operating conditions at which analytical test for stability is to be made.
   3. Calculation of the power angle characteristics between the major machine groups, including, if necessary, the response of composite loads of differing characteristics. If an a-c network analyzer is used, the power angle characteristics can be obtained directly rather than by equations (Reference 12).
   4. Testing the system for stability by one or more of the various criterions for stability. These may be applied entirely by numerical methods or with the assistance of an a-c network analyzer, which greatly reduces the numerical work.
   5. Repetition of the above steps for different loadings and system conditions, in order to obtain an understanding of the essential steady-state stability characteristics of the system (Reference 13).

II. *Transient stability* (References 10, 3).
   1. Reduction of system to a simplified equivalent circuit representing the synchronous machines by their transient reactances and grouping machines together into major groups which are known to swing together in angular displacement for the particular disturbance to be studied.
   2. Determination of initial quantities for system loading and operating conditions for which analytical test for stability is to be made.
   3. Calculation of power angle characteristics between equivalent machine groups for the various conditions of the disturbance, such as fault on, first breaker cleared, second breaker cleared. Unbalanced faults may be represented by symmetrical phase sequence equivalent circuits. An a-c network analyzer gives the required power angle characteristics directly.
   4. Calculation by a step-by-step process of the angular swing of the machine groups during the transient disturbance to determine whether loss of synchronism will result. See Fig. 1 for a typical swing curve.
   5. Repetition of these steps for other loadings and system conditions to obtain a quantitative understanding of the transient-stability characteristics of the system (Reference 14).

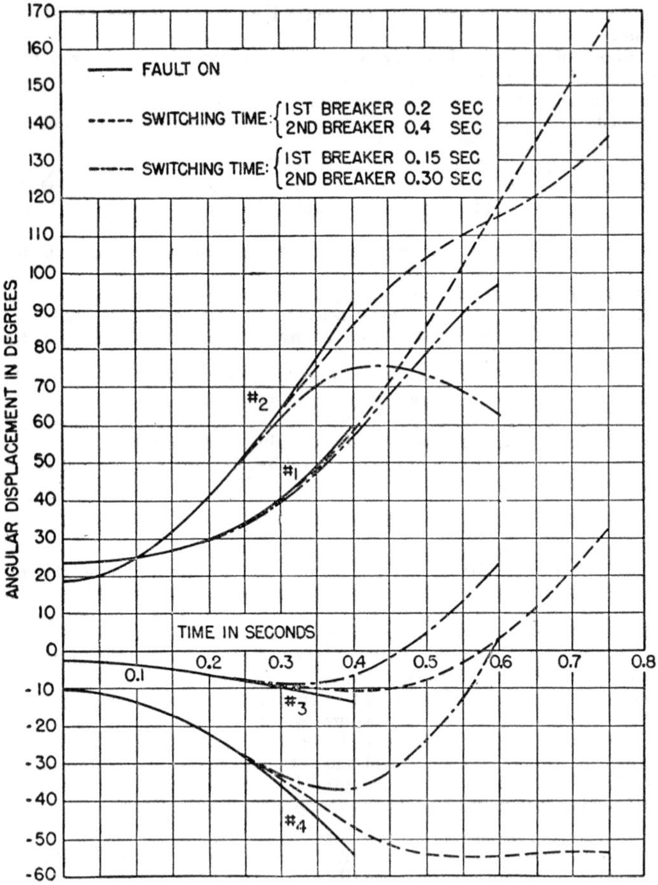

FIG. 1. Calculated Four-machine Swing Curve

III. *Hunting* (Reference 15).
1. Reduction of problem to essential elements involved.
2. Expression of the dynamical relations of these essential elements by differential equations.
3. Determination of the characteristics and constants of these elements required for the solution of the differential equations.
4. Solution of the dynamical relations by numerical methods or by the use of differential (References 16, 17) or digital analyzers (Reference 18) to determine the boundaries between the stable and unstable regions or the oscillation magnitudes as a function of the important parameters. (For an example see Reference 19.)

65. SYSTEM DESIGN FOR STABILITY

The steady-state stability characteristics fix an upper limit of power transfer, which is determined by the fundamental system design, such as size and short-circuit ratio of generators, circuit voltage, number and arrangement of circuits. The transient limits, on the other hand, are a function of clearing time, fault location, and severity, which need not always be limiting in temporary emergency system power requirements or which can be improved by modernization of protective devices. The steady-state limit with the faulted section cleared from the system is determined by the conditions which exist after the fault has been cleared and the subsequent oscillations have been damped out. This is an upper limit of power transfer which is approached by using quick relaying and switching to clear transient faults.

TRANSMISSION. Transmission lines have a strength, or electrical stiffness, which is directly proportional to the square of the circuit voltage and inversely proportional to the length.

VOLTAGE. As the voltage required for a transmission-line link of moderate length tends to vary directly with its length, the power delivered by a circuit when economically used tends to vary directly with the voltage; i.e., when 220 kv instead of 110 kv is used for moderate distances, it will generally deliver about twice as much power twice as far. In general, the system high-voltage short-circuit requirements (kva) tend to be determined chiefly by the amount of connected generation at the sending end and therefore for the same number of outgoing circuits tend to vary directly with the transmission-line voltage level. This requires of the circuit breaker that its kva interrupting capacity increase directly with voltage. Since the maximum current which can be interrupted is approximately constant with respect to voltage, for a given stage of circuit-breaker development, it may be said in general that the circuit-breaker interrupting capacity (kva) also tends to increase directly with the circuit voltage. Therefore, the inherent system requirements tend to be met by the inherent circuit-breaker capabilities.

Table 2. Approximate Surge Impedance Transmission Loadings per Circuit of Conventional Overhead A-c Lines

Receiver Voltage, kilovolts	(Receiver) Surge Impedance Loading,* kw = 2.5 $(kv)^2$
345	300,000
275	190,000
220	120,000
154	60,000
132	44,000
110	30,000
66	11,000

* Based on surge impedance per phase to neutral of 400 ohms.

The foregoing reasoning is based on the assumption that the number of circuits per station and circuit loadings in kilowatt-miles per circuit voltage squared or kw-miles per $(kv)^2$ are kept constant. However, there are other factors which increase the required interrupting capacity of circuit breakers for a given voltage level.

1. Increased circuit loadings per $(kv)^2$ to obtain more economical transmission.
2. Increased number of paralleled circuits.
3. Tendency to remain at outgrown lower voltage levels before superimposing higher voltage transmission.

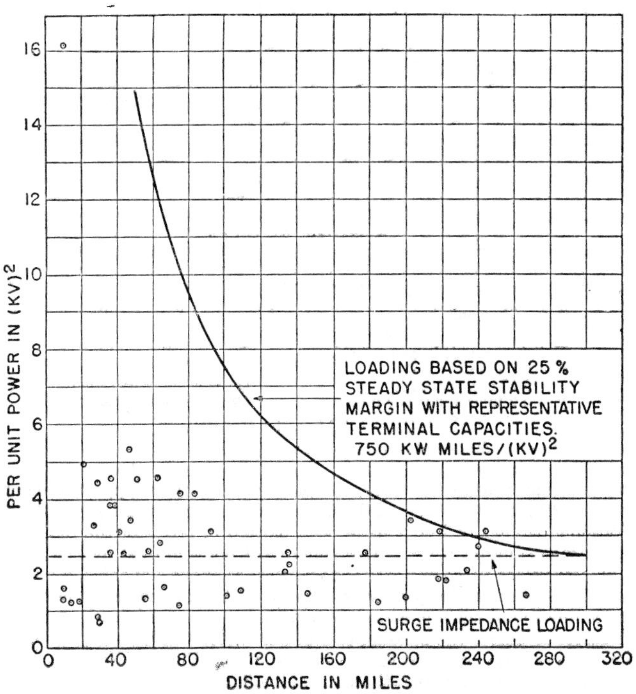

FIG. 2. Major Transmission-line Loadings in the United States before World War II

At present, circuit breakers having interrupting ratings of 5,000,000 kva (References 20, 21) at 230 kv can be built. Such circuit breakers allow for a greater concentration of closely coupled generating capacity, using the highest voltage circuits as the backbone for the interconnection of the major system generation. Emphasis, as regards stability, is accordingly placed primarily on the high-voltage transmission line, with consequent decrease in the need of emphasis on the stability of the lower voltage and of the individual generating elements or stations. This allows for a more ready adoption of the unit type of station and a simplification of the generating plant, with a corresponding reduction in generating-station costs.

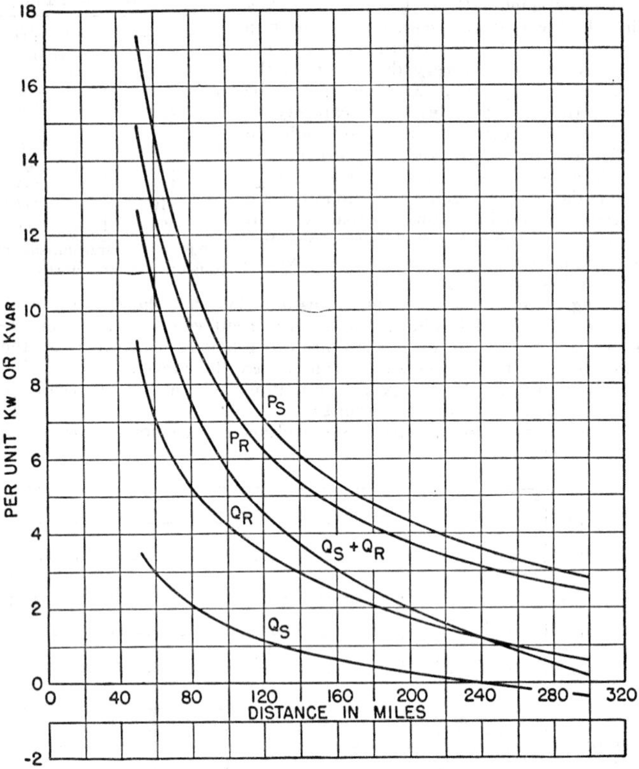

Fig. 3. Representative Transmission-line Loadings and Required Kvar for Transmission for Loading of 750 kw-miles per $(kv)^2$

Unit kw = unit kvar = $(kv)^2$	P_S = sending-end kw
x = 0.8 ohm per mile	P_R = receiving-end kw
r = 0.16 ohm per mile	Q_S = sending-end kvar
y = 5.2 micromhos per mile	Q_R = receiving-end kvar

$Q_S + Q_R$ = total kvar for transmission

Equal sending- and receiving-end voltages $E_S = E_R$

Most conventional transmission lines, except those of 60 cycles approaching or exceeding 300 miles in length, have ample steady-state margins for normal power transfer. The transmission-line loadings of Fig. 2 have been plotted in terms of the rated transmission line-to-line voltage squared $(kv)^2$. Also since most conventional overhead transmission lines have an equivalent surge impedance to neutral of about 400 ohms, a kilowatt loading of 2.5 $(kv)^2$ independent of frequency will be found to approximate the condition when the I^2X of the line equals the line-charging volt-amperes. At this loading, if the receiving-end power factor is unity, the sending-end power factor will be unity also, and the line will have a voltage drop equal to the resistance drop. This loading has been variously called the surge impedance loading, the natural loading, or the unity power factor loading. Table 2 gives the surge impedance loadings for different circuit voltages. The **kilovar** (kvar) required for the transmission of power goes up rapidly when the loadings are increased appreciably above the surge impedance loading of about 2.5 $(kv)^2$, as shown in Fig. 3. Steady-state stability limits in terms of receiver power for typical single-circuit transmission lines are given in Fig. 4. (See Section 9, Article 6, for per unit system.)

Lines 50 to 100 miles in length will be limited primarily in their loading because of I^2X and I^2R losses and thermal limitations, and not necessarily because of steady-state stability. During emergency overloads, such lines, particularly tie lines, may have low voltage at their terminals for lack of sufficient system kvar capacity.

VOLTAGE	SPACING
345 KV	29' FLAT
287.5	29' FLAT
230	23' FLAT
161	20' eq Δ
138	16.5' eq Δ
115	14' eq Δ
69	8' eq Δ

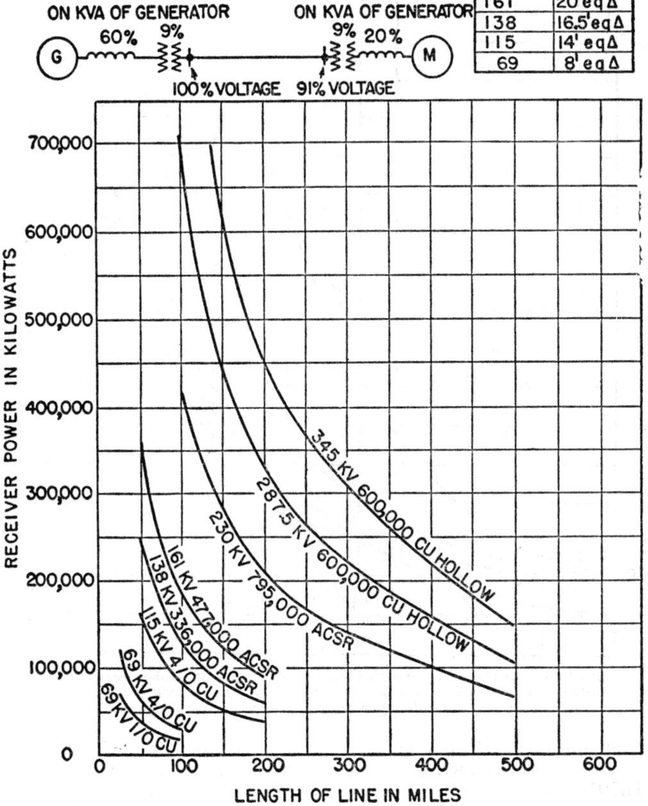

Fig. 4. Typical Steady-state Stability Characteristics (Single-circuit transmission line, 60 cycles)

Proper system design should provide for sufficient controlled kvar capacity to withstand successfully the emergency transmission-line loadings which the line will be required to carry. Hydroelectric generating plants which are planned as reliable power supplies have their transmission lines arranged so as to result in a minimum loss of line capacity due to a fault at any location and have sufficient margin in synchronous condenser capacity to maintain normal system voltages when the faulted section is switched out.

The voltage of a line determines to a considerable extent its kilowatt loading. The design kilowatt loading of a conventional overhead 60-cycle line less than 300 miles in length will usually lie between the surge impedance loading of 2.5 $(kv)^2$, with no kvar required for transmission, and 750 kw-miles per $(kv)^2$. This latter figure, which includes the distance, is indicative of the relative stability. For a 300-mile line, these limits are equal $[2.5\ (kv)^2 \times 300 = 750\ (kv)^2]$, and a 220-kv line of this length could accordingly be expected to carry about 120,000 kw ($2.5 \times \overline{220^2}$). For a 100-mile, 220-kv line, the loading could be expected to be between 120,000 kw, corresponding to 2.5 $(kv)^2$, and 360,000 kw, corresponding to 750 kw miles per $(kv)^2$. Whereas no kvar is required for transmission at a loading of 120,000 kw at 360,000 kw, 200,000 to 300,000 kvar would be required for transmission. Figure 5 shows the kvar requirement of 100-mile lines for different loadings in terms of a $(kv)^2$ base. In general, it becomes uneconomical to design a line of this length for operation at the higher loadings because of the cost of the controlled kvar. A

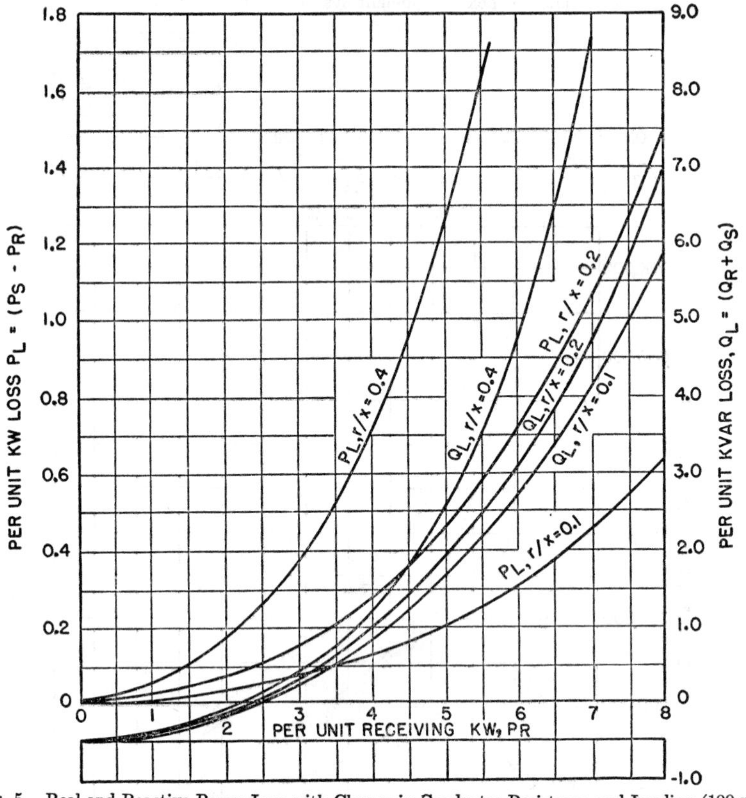

Fig. 5. Real and Reactive Power Loss with Change in Conductor Resistance and Loading (100-mile line, equal sending- and receiving-end voltages. Unit kw = unit kvar = $(kv)^2$. x = 0.8 ohm per mile, y = 5.2 mhos per mile. No load losses neglected.)

moderate amount of controlled kvar, however, may be economical, since thereby the line loading can be appreciably increased above its surge impedance loading.

REACTANCE. Reactance is essential for transmission of power in a constant potential system, but it decreases the stability limit and increases the kvar required. Many proposals have been made to reduce the amount of reactance in a line to approach an optimum amount. One of these methods is to use multiple conductors in each phase (References 22, 23), and thereby increase the effective radius of the phase conductor and decrease the reactance. Such a method, however, may be expensive when compared with the advantages so obtained (Reference 24). The use of cable has been proposed (Reference 25) because of its low reactance, but, for long distances, the cable charging kva becomes large and requires shunt inductive compensation, the cost of which becomes comparable to the cost of series capacitive compensation for an overhead line. Therefore, in order for cable to compete economically with overhead lines for long-distance a-c transmission, the cable cost per mile should be comparable with the cost per mile of overhead line.

RESISTANCE. Line resistance produces line loss, reduces the delivered kilowatt, and the stability limit, and increases the reactive kva requirements. A high-resistance line will have a lower stability limit in terms of delivered power and will have a higher generated power at the stability limit. Lines which are operated with heavy loadings require low resistance to keep the losses within economic limits. For main transmission lines operating with high load factors, it becomes economical to use low-resistance conductors so as to obtain resistance-reactance (r/x) ratios of 0.05 to 0.15. Such low-resistance conductors reduce the total kvar required for transmission of a given amount of power and also reduce the proportion of the total kvar required at the receiving end relative to that required at

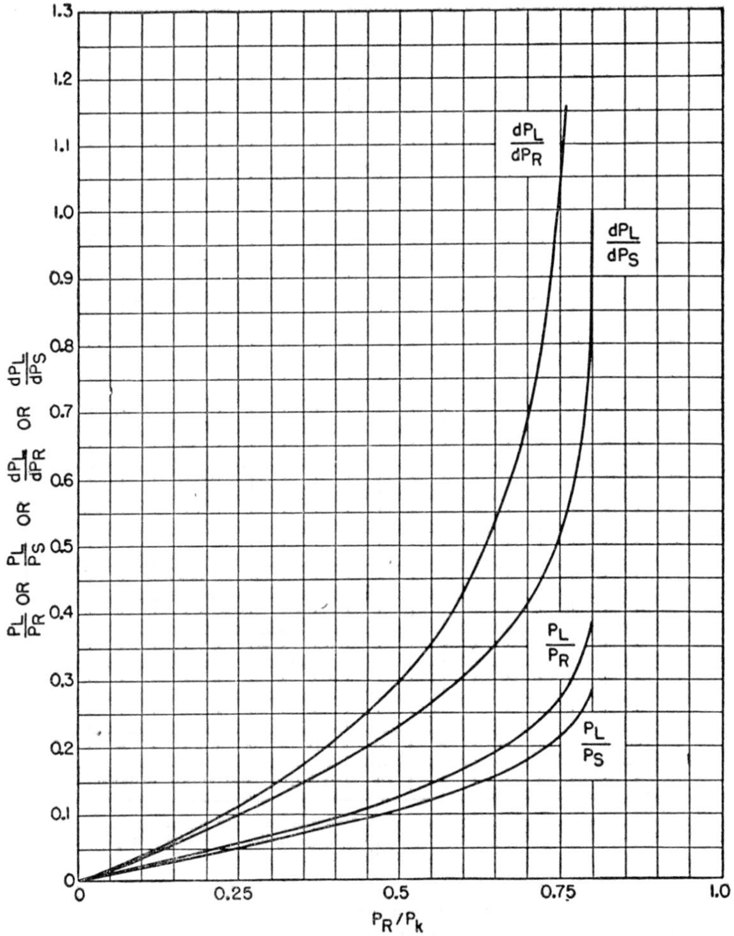

FIG. 6. Total and Incremental Loss Ratios for $r/x = 0.20$ (No-load loss neglected)

P_L = power losses $\quad$ P_S = sending-end power
P_R = receiving-end power

the sending end, for given sending- and receiving-end voltages. This latter advantage is of importance, since the receiving end is usually more restricted in its kvar capacity, as it must also provide the kvar requirement of the load. The sending end may obtain some of the kvar from generator capacity at a relatively lower cost per kvar. Table 3 shows the relative sending- and receiving-end overexcited kvar required for a typical 100-mile conventional transmission line loaded at 500 kw-miles per $(kv)^2$ as a function of the r/x ratio, with $E_1 = E_2 = 1.0$.

Many interconnected tie lines are built as emergency tie lines and the real losses are not of prime concern. However, the stability limit is low and the required kvar for transmission is large for a given load transfer if a high r/x ratio is used. Lines originally designed for emergency tie-line capacity may be required at some future time to carry substantial base loads, and a high-resistance line is then quite undesirable. Also, with the growth of interconnected systems, future loadings cannot always be predicted. Sometimes a high-resistance line is paralleled with additional line capacity, and the high-resistance line limits the system power flow because of its losses and heating restrictions, since the parallel current division is determined primarily by the relative reactances rather than the resistances.

In an interconnected system, if a large block of power is to be transferred a considerable distance, the incremental losses may become so great as to make such a block power-

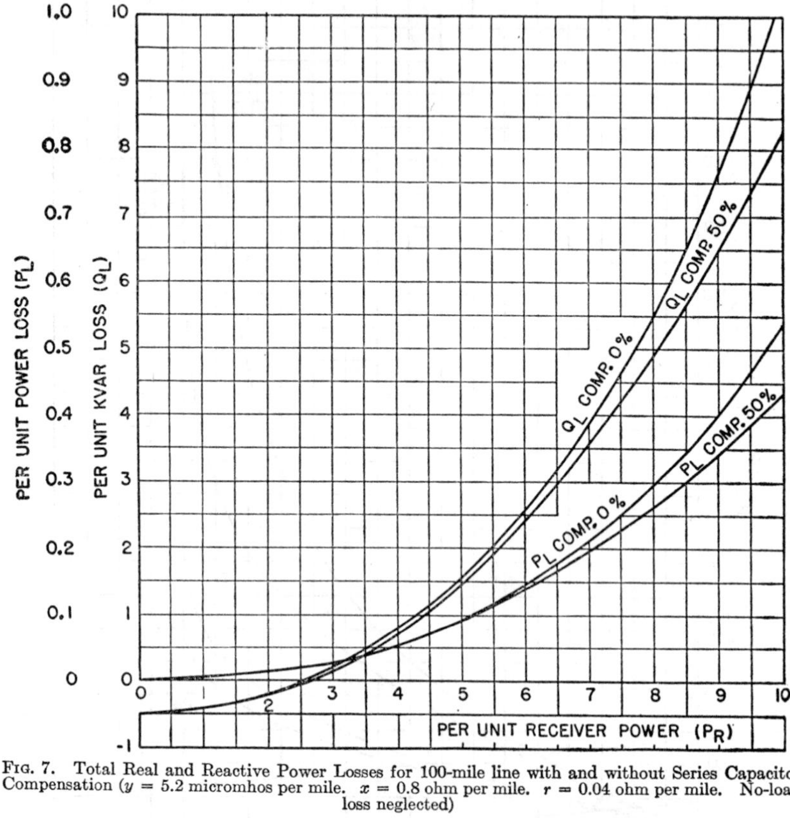

FIG. 7. Total Real and Reactive Power Losses for 100-mile line with and without Series Capacitor Compensation (y = 5.2 micromhos per mile. x = 0.8 ohm per mile. r = 0.04 ohm per mile. No-load loss neglected)

transfer uneconomical, although the system is not overloaded from a stability point of view. Figure 6 shows the relation between the total real losses and the incremental losses for a typical 100-mile conventional transmission line.

Table 3. Relative Sending-end and Receiving-end Overexcited Kvar as Function of Resistance-reactance Ratio for Typical Line

	r/x	0.1	0.2	0.4
Receiver kvar........	Q_r	+1.3	+1.9	+3.4
Sending kvar.........	Q_s	+0.5	+0.1	−0.8
Kvar losses..........	$Q_l = Q_r + Q_s$	+1.8	+2.0	+2.6

SERIES CAPACITOR LINE COMPENSATION. Series capacitors may be used to compensate for part of the line reactance. It is usually found more economical, however, to use the corrective kvar in shunt relation (with controlled shunt capacitors or synchronous condensers), except possibly for cases in which stability or voltage dip due to suddenly applied loads is limiting. The advantage of corrective kvar in shunt relation is that it is more flexible and is able to make use of the diversity of kvar requirements at a switching station; the corrective capacity available is then independent of the number of lines in operation and their individual loadings. Kvar control can be obtained with a synchronous condenser by using a voltage regulator, and methods have been developed for automatic control of shunt capacitor kvar. The steady-state stability limit of conventional transmission lines above 300 miles may be less than the surge impedance loading of the line, and, for these longer lines, series compensation becomes an effective means of overcoming the stability limitations (References 26, 27, 28). Figures 7 and 8 show the total real and reactive power losses for a 100- and a 300-mile line with and without series

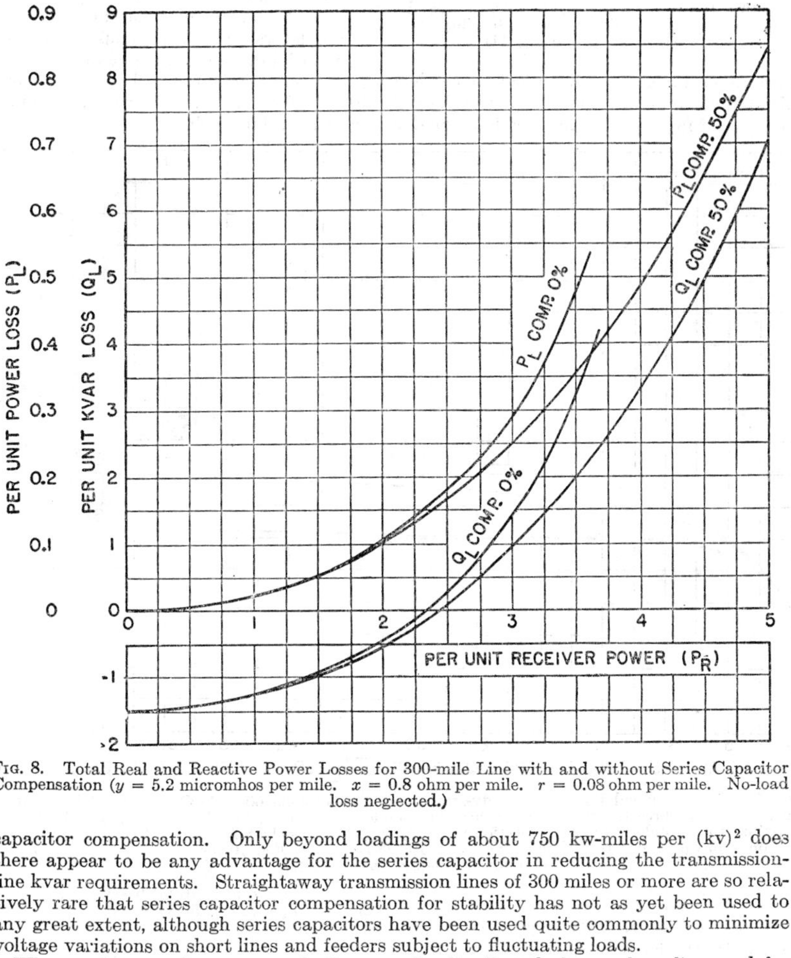

FIG. 8. Total Real and Reactive Power Losses for 300-mile Line with and without Series Capacitor Compensation (y = 5.2 micromhos per mile. x = 0.8 ohm per mile. r = 0.08 ohm per mile. No-load loss neglected.)

capacitor compensation. Only beyond loadings of about 750 kw-miles per $(kv)^2$ does there appear to be any advantage for the series capacitor in reducing the transmission-line kvar requirements. Straightaway transmission lines of 300 miles or more are so relatively rare that series capacitor compensation for stability has not as yet been used to any great extent, although series capacitors have been used quite commonly to minimize voltage variations on short lines and feeders subject to fluctuating loads.

When series capacitors are used, they can be distributed along a long line, and for parallel lines can be located at the intermediate switching stations in the manner shown

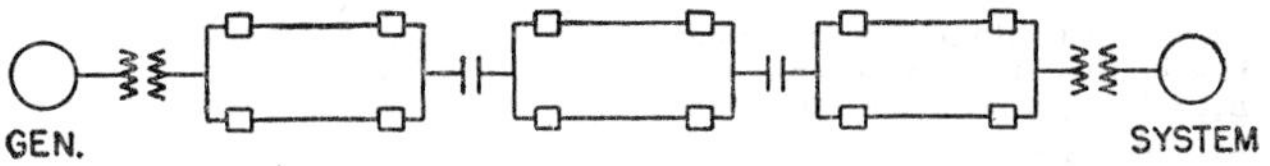

FIG. 9. Series Line Compensation with Capacitors

in Fig. 9. This arrangement will reduce the magnitude of voltage change at any one location and will not necessitate the disconnection of capacitor capacity when a line section is opened.

INTERMEDIATE SYNCHRONOUS CONDENSERS. Intermediate synchronous condensers may be used to increase the stability limit of long-distance power-transmission systems by supporting the voltage along the line (References 29, 30), as illustrated in Figs. 10 and 11. The increase in power limit is a function of the effective reactance of the intermediate synchronous condenser and its step-up transformer to changes in high-tension terminal voltage. With special continuously acting voltage regulators and synchronous condensers of low transient reactance, the effective reactance of the intermediate condenser equipment can be reduced so as to make this method comparable in cost with series capacitors (Reference 30). Otherwise, without such special features, the intermediate condenser cost is prohibitively large.

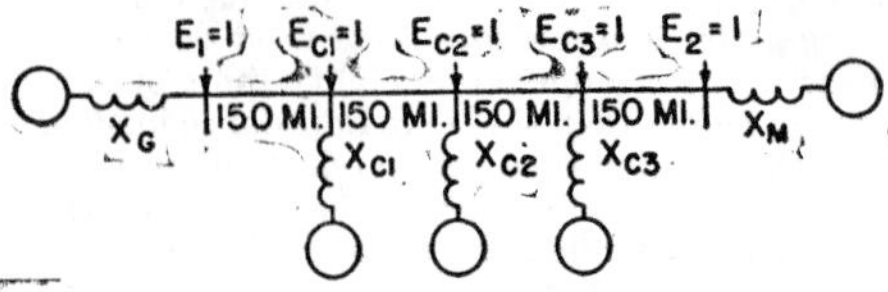

FIG. 10. System Arrangement with Three Intermediate Synchronous Condensers (Stability characteristics shown on Fig. 11)

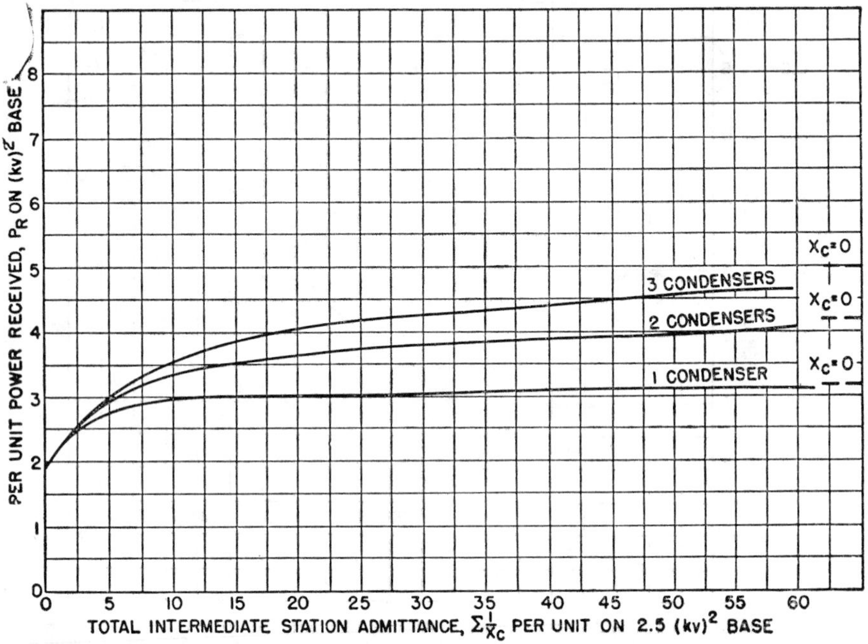

FIG. 11. Receiving-end Power at Steady-state Limit for 600-mile 60-cycle Conventional Line with One, Two, and Three Intermediate Synchronous Condenser Stations, Equally Spaced along the Line, as Affected by Total Intermediate Station Capacity Expressed as the Summation of the Intermediate Station Equivalent per Unit Admittance on 2.5(kv)^2 Base (See Fig. 10 for three intermediate synchronous condenser station system)

x = 0.8 ohm per mile
r = 0.08 ohm per mile
$b = 5.2 \times 10^{-6}$ mho per mile

X_G = 40 per cent [2.5(kv)^2 base]
X_M = 25 per cent [2.5(kv)^2 base]

Curve 1: One intermediate condenser station.
Line in two 300-mile sections.
$E_1 = E_2 = E_{c1} = 1.0$
$X_c = X_{c1}$

Curve 2: Two intermediate condenser stations.
Line in three 200-mile sections.
$E_1 = E_2 = E_{c1} = E_{c2} = 1.0$
$X_c = X_{c1} = X_{c2}$

Curve 3: Three intermediate condenser stations.
Line in four 150-mile sections.
$E_1 = E_2 = E_{c1} = E_{c2} = E_{c3} = 1.0$
$X_c = X_{c1} = X_{c2} = X_{c3}$

CIRCUIT BREAKERS. Relays are now used with less than 1-cycle operating time. Circuit breakers with total clearing times as low as 3 to 5 cycles, capable also of quick reclosure, are available. Their use makes possible considerable improvement in reliability and the increase in reliable kilowatt loading per circuit of transmission systems. A circuit-breaker and relay modernization program can be most effectively carried out by making use of a comprehensive stability study showing the most important or critical locations which require quick fault clearing, thus obtaining greatly improved reliability at relatively small cost. It will be found that most systems usually require, from a stability standpoint, quick-opening circuit breakers only for those circuits electrically close to generators or large synchronous motor loads. Severe faults which temporarily interrupt the electrical supply to synchronous motors may ordinarily require clearing times as low as 6 cycles, from hydrogenerators 6 to 9 cycles, and from steam generators 9 to 12 cycles. These are only general or average values, but they indicate the relative order of magnitude. The logical method of determining the required fault-clearing times is by a stability study.

The use of breakers of higher interrupting capacity can be as important as quick fault clearing in further raising the stability limits, since they permit a more solidly interconnected system with a resulting decrease in severity of the disturbance caused by clearing the faulted circuit.

RECLOSING. System requirements led to the development of high-speed reclosing circuit breakers. Quick-opening and quick-reclosing breakers of high interrupting capacities allow for an increase in the electrical loadings of a system, a simplification of the generating stations, and corresponding reductions in system costs.

With the more general use of high-speed reclosing and the ability to operate systems at higher loadings and with smaller stability margins, it becomes desirable to prevent tripping of circuit elements during heavy power swings. Experience has indicated that many major system disturbances which were at first thought to be due to poor system stability were primarily due to tripping the lines on large swings, which would not have caused instability if the unnecessary tripping had not occurred. Unnecessary tripping of a circuit element during a system swing or emergency condition may start a cascading of cumulative switch operations and result in a major disturbance with load interruptions. Quick reclosing assists in preventing such cumulative operations and also in general simplifies the system operations during system disturbances. However, it is also desirable that the important circuit elements be provided with relaying which will not trip on large swings.

OTHER MEANS TO INCREASE STABILITY LIMITS. Various methods have been proposed to remove the stability limitations of long lines, such as:

1. Quadrature boost transformers.
2. Series synchronous condensers.
3. Tuned transmission lines operated at higher frequency.
4. Low-frequency transmission.
5. Wave transmission, variable voltage with surge impedance loading.
6. Direct-current transmission.

The first two methods are impractical technically. The third, fourth, and fifth schemes are possible technically but impractical economically (compared with series capacitors or special intermediate synchronous condensers) because of the cost of the conversion equipment. Direct-current transmission is an effective way to remove the synchronous stability limitation and possibly to decrease the transmission-line costs, but its future use depends on further development and trial operating experience.

GENERATORS. The generators of a system are usually the most important synchronous machines and as such constitute essential elements in a stability analysis. With the development of quick-operating relays and circuit breakers the stability requirements make fewer demands on the generator design. Generators can accordingly be designed and developed in keeping with the normal electrical, mechanical, and thermal requirements of a well-balanced design and need be only slightly modified to meet the stability requirements. This is an important factor in permitting the standardization of characteristics, particularly of large 3600-rpm steam-driven alternators (Reference 31).

Reactance. Steady-state-stability performance of synchronous generators is usually specified by their short-circuit ratio,* which provides means for comparing the relative electrical stiffness of different designs. Higher short-circuit ratio generators are more expensive. The cost of greater short-circuit ratio for waterwheel generators increases, in general, as shown in Fig. 12.

Waterwheel generators are usually located at a distance from their loads and are there-

* The short-circuit ratio is the ratio of field current for rated open-circuit armature voltage and rated frequency to the field current for rated armature current on sustained symmetrical short circuit at rated frequency.

fore likely to require relatively high short-circuit ratios. Small changes, such as 5 or 10 per cent, have a relatively negligible effect on the stability limit. Generators which are to transmit power 200 or 300 miles over high-voltage transmission lines can, in general, economically justify a generator short-circuit ratio of at least 1.5 to 2.0.

Steam turbine generators, which will be referred to as turbine generators or turbogenerators, however, are usually required to transmit power relatively short distances, or into an interconnected system of loads and generation, and the equivalent transmission distance is usually much less than for hydrogenerators. Therefore, turbine generators may be required to operate at lower power factors than waterwheel generators. Furthermore, turbine-generator sets of considerable capacity are now built to run at 3600 rpm, which necessarily places severe restrictions on the design because of the high stresses resulting from large rotor dimensions. All these factors have combined to result in a short-circuit ratio of 0.8 to 1.0 for the usual design of turbine generator in the United States. Further reduction appears to be entirely feasible with the more general use of voltage regulators. Generators of high short-circuit ratio can be obtained as special machines when system conditions warrant their use. Figures 13 and 14 show the steady-state power limit of turbogenerators with various short-circuit ratios operating at 0.8 and 1.0 power factor, respectively, delivering power to an equivalent system which is represented by a reactance x_e.

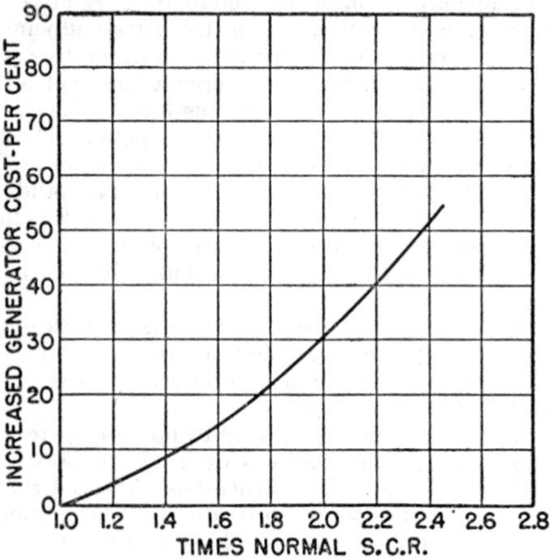

FIG. 12. Approximate Cost of Increasing Short-circuit Ratio (S.C.R.) of Typical Large Waterwheel Generators

Power Factor. The generator power factor should, in general, be selected to meet the system requirements. A generator with a low power-factor rating used at high power factor is more unstable than a generator of the same size or cost rated more nearly to the actual power factor required.

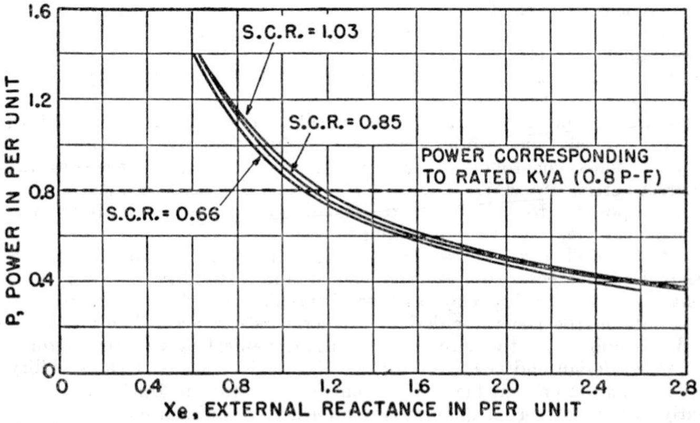

FIG. 13. Steady-state Power Limits of Turbogenerators with Various Short-circuit Ratios Operating at Normal Terminal Voltage and 0.8 Power Factor

Hydroelectric generators, which are located at a distance from their load, will in general require a power factor higher than turbine generators, and are normally rated between 0.9 and 1.0 power factor. Turbine generators have usually been built for 0.8 power factor, with the possibility of carrying a kilowatt load up to their full kva rating at unity power factor. With hydrogen-cooled machines it is possible to increase the kva loading and provide additional kvar capacity at maximum kilowatt output by an increase in the hydrogen pressure (Reference 31). Some kvar capacity at maximum turbine flow and maximum generator output is desirable, since the generator is a relatively inexpensive source of kvar when operated at the higher power factors. Also there usually exists a system requirement of kvar for transmission of power at the generator end under the condition of heavy transmission-line loadings as well as for supplying local loads. These factors, combined with a more general use of power-factor corrective equipment, have resulted in turbogenerators operating under full load at power factors between 0.8 and 0.9. In some cases, for plant economy, efficient units of a station may be operated up to unity power factor at close to maximum turbine flow, with the less efficient units operating at lower power factors.

Methods for Increasing Generator Steady-state Stability. On rare occasions generators have been installed where the system has developed in such an unpredictable manner as to result in too-small steady-state-stability margins. If the rated power factor of the generator is lower than the operating power factor, it is possible to use shunt reactors at the generator terminals to increase the field loading and the steady-state-stability limits.

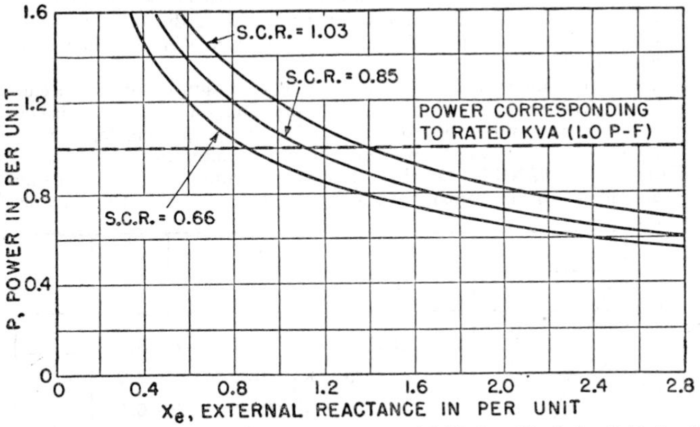

FIG. 14. Steady-state Power Limits of Turbogenerators with Various Short-circuit Ratios Operating at Normal Terminal Voltage and 1.0 Power Factor

It may also be possible to increase the terminal voltage of the generator if there is sufficient margin in the field requirements and to improve the stability limit thereby.

VOLTAGE REGULATORS. Voltage regulators have been used almost as a necessity for hydroelectric generators because of the large inherent regulation of the transmission lines to which waterwheel generators are connected. Until recently, voltage regulators have not been used so generally for turbine generators, but there is a growing recognition of the benefits to be derived by their use even for turbine generators. The advantages of voltage regulators are:

1. Better system voltage regulation.
2. Improved stability.
3. Less overvoltage on loss of load.

Improved stability of turbine generators allows the use of generators having lower short-circuit ratio with greater capacity per unit at 3600 rpm and less short-circuit current.

Voltage regulators for generators connected to an interconnected system of large capacity may make it desirable to use upper and lower limits on the excitation current. The upper limit need only be an alarm to indicate excitation above field currents that are safe from a heating standpoint. The lower limit should prevent reduction of field current to values which may result in loss of synchronism. This lower limit may be set arbitrarily or varied with change in load, as indicated by steam conditions or electrical power output (References 32, 33).

The regulator may be provided with compensation to regulate the voltage at some point in the system beyond the generator terminals, and at the same time to equalize the reactive load properly between generators of the same station. It is also possible to divide the reactive load by means of reactive droop compensation. With increase in reactive load the bus voltage is allowed to droop slightly, thus securing an equalization of kvar between units. The bus voltage can be brought back to a desired value by manual adjustment. The relative simplicity of the reactive droop compensation, requiring no control circuit interconnection between generator units, makes it highly desirable for some types of station where line-drop compensation or completely automatic regulation is not necessary.

Voltage regulators tend to increase the stability limit above the ordinary steady-state limit by allowing for operation in a region of "dynamic stability" which is dependent on the regulator. Power systems are not usually designed to operate normally in this region, as absolute dependence on voltage-regulator performance has not been considered necessary. During disturbances and emergencies, however, the additional margin provided by the voltage regulator in allowing for operation in this region may be valuable. Theoretically, it is possible in effect to reduce the reactance determining the generator stability from a value of about 70 per cent of the synchronous reactance to a value approaching the transient reactance. This then corresponds to a doubling or tripling of the effective short-circuit ratio of the generator. Accordingly, voltage regulators incorporating in their design features allowing for operation in the region of dynamic stability provide additional stability to a system.

Continuously acting voltage regulators capable of operating in the dynamic region and intermediate synchronous condensers with low transient reactance provide an alternative method to that of series capacitors for compensating long-distance transmission lines (Reference 30). This method provides a means for regulating the line voltage automatically for any load condition.

The principle of dynamic stability has been very successfully applied to electric ship-propulsion drives (Reference 34), where there is a premium on weight and size of electrical equipment. In this case the benefits to be derived are very appreciable, because the total impedance determining the power limit is entirely in the motor and generator of the drive.

TRANSIENT REACTANCE. Quick fault clearing, combined with the inherent low transient reactance of turbine generators, has practically eliminated the requirement for

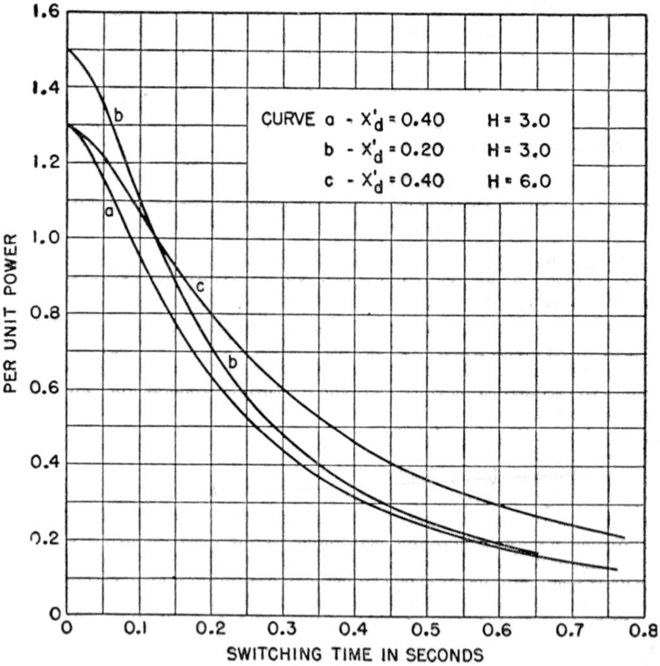

Fig. 15. Power versus Required Fault-clearing Time

X'_d = per unit transient reactance
H = per unit inertia constant

lower-than-normal transient reactance for this type of generator. Quick fault clearing has also greatly reduced the necessity for lower-than-normal transient reactance for water-wheel generators. In most cases there is little need for requiring lower-than-normal transient reactance as obtained with the required short-circuit ratio. For systems with low stability margins, however, a lower-than-normal transient reactance may be a practical means for obtaining a further increase in the stability limit after the clearing of a

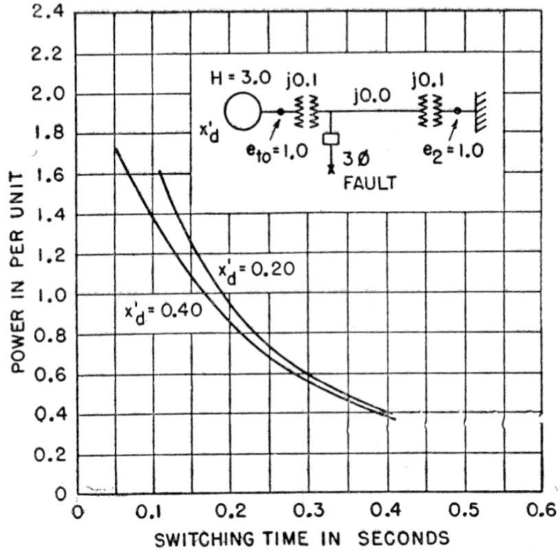

FIG. 16. Switching-time Curves Showing Effect of Change in x'_d with Stub Feeder Fault (Low impedance tie line)

faulted line section, realizable even with quick fault clearing. Figures 15, 16, and 17 show the gains to be realized in transient stability by a reduction in the transient reactance from 0.40 to 0.20 for three typical system conditions.

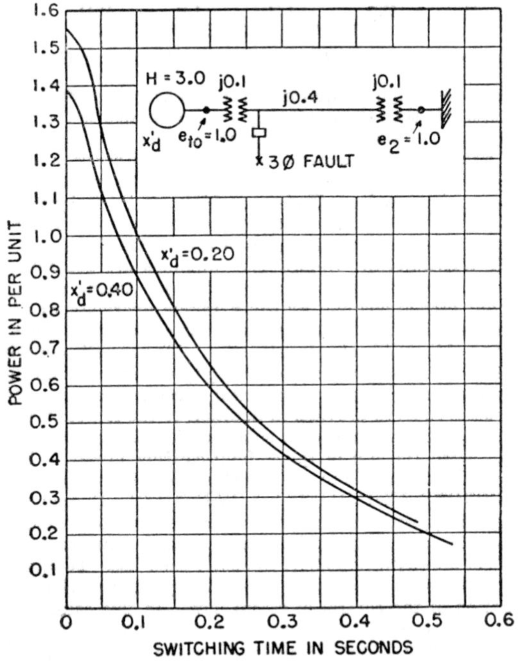

FIG. 17. Switching-time Curves Showing Effect of Change in x'_d with Stub Feeder Fault (High impedance tie line)

INERTIA. Quick clearing of faults has considerably reduced the necessity, for stability purposes, of additional WR^2 above that normally required by the governor and hydraulic systems. Compare curves (*a*) and (*c*) of Fig. 15. Large inertia is effective in improving the stability when the switching time is long or in allowing a longer transmission-line fault deionization time with high-speed reclosing. An impractically large inertia, however, would be required to make effective the use of high-speed reclosing for severe faults of a single-circuit line carrying the full output of hydroelectric generators (Reference 35).

EXCITATION. The per-unit excitation system response need only be in the normal range of 0.5 to 1.0 in order to realize the most important benefits from automatic voltage regulation. Figure 18 shows that a reduction in switching time is much more effective than an increase in excitation response for a typical hydroelectric system. For generators it is generally desirable to use a variable lower limit as a function of prime mover torque in order to realize a maximum range of regulator control. For waterwheel generators,

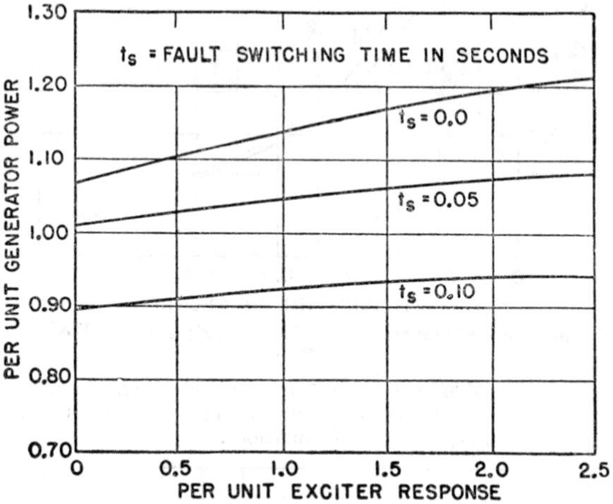

FIG. 18. Effect of Excitation System Response and Switching Time on Generator Power Limit

which usually have a line-charging requirement as well as a relatively high overspeed upon loss of load, it is desirable that the regulating system be able to force the voltage down rapidly and to control the voltage at or near zero slip ring voltage.

Exciters may be either direct- or separately driven and either separately or self-excited. In general, the larger units require separate excitation of the exciter field, usually from a pilot exciter. Motor-driven exciters (Reference 36) may be designed with high inertia and with a high torque induction motor to make the exciter-set reliability comparable with that of the other motor-driven auxiliaries of the station.

AMORTISSEUR WINDINGS. Amortisseur windings are generally beneficial additions to all salient-pole generators (References 37, 38). Among other advantages they assist in damping out oscillations and reducing system overvoltages due to unbalanced faults. None is normally required for the usual turbine generator, which has an effective amortisseur by virtue of its solid-rotor construction. The amortisseur winding may be of the incomplete, rather than the complete, end ring type and maintain the chief electrical advantages with the simpler type of mechanical construction. Double-winding amortisseurs have been used, so as to provide, in addition to the low-resistance amortisseur, a high-resistance winding which produces a substantial braking torque during periods of unbalanced faults. With the use of quick fault-clearing breakers, however, the fault time has been so greatly reduced as to make such an additional winding of little value.

BRAKING RESISTORS AND SPECIAL GOVERNORS. Special means for increasing the generator stability, such as braking resistors and quick-operating watt-responsive governors, have not, in general, given important enough improvements, in comparison with other methods of improving stability, to justify their use. Topping or superimposed steam units introduce some special types of control problems which are most directly solved by coordination of the control of steam flow. In this case, as for generators in general, automatic voltage regulators insure that any abnormal increase in load on any one unit, due to a change in either the electrical or the steam flow, will result in a corresponding increase in the field strength of that unit.

66. LOADS

The effect on the system of its composite loads is in general stabilizing, in that a sudden increase in load or the loss of a unit of generation does not require that the total amount of the applied load change be picked up immediately by the other generators (References 39, 40). If the emergency is so severe that no generation is immediately available to pick up the load, the system may suffer only a temporary drop in frequency and voltage until additional generation may be added. This inherent resiliency of a loaded system increases the stability margin for suddenly applied loads. During transient shocks it has been noted that the generator power oscillations are damped out in many cases more quickly than would be expected from the inherent damping available in the generator rotor. This indicates that the load or system is providing additional damping torque.

Induction and synchronous motor loads usually tend to slow down or pull out of synchronism if the sustained voltages are reduced below about 70 per cent of normal. At such a point the reactive current increases rapidly, producing a cumulative effect in the collapsing of the system voltage. At somewhat lower voltages, however, many of the contactors held in by line voltage will drop out and relieve the system temporarily of some of the load. Such a low range of sustained voltage will, however, usually result in the pulling out of synchronism of important heavily loaded ties. It is advisable to avoid sustained voltage drops likely to reduce the system voltages to 80 per cent or less. Some systems have a definite established procedure of relieving their system by opening previously specified feeders when the voltage or frequency has dropped a given amount. This allows a more rapid recovery to normal conditions, since it reduces the possibility of reaching the point where the difficulty becomes cumulative.

Automatic voltage regulators are helpful in such an emergency by supplying the system with as much reactive current as possible. This helps maintain the voltage and prevents its cumulative reduction, with the resulting loss in synchronism of important transmission-line links.

SYNCHRONOUS MOTORS. Synchronous motors are usually provided with out-of-step relays to remove the field excitation and in some cases to disconnect them from the system quickly upon loss of synchronism. This is desirable in order to prevent mechanical damage produced by the severe pulsations in shaft power caused by the field excitation torque (Reference 41), as well as to provide the first step in the quick resynchronization of the motor, by either manual or automatic means.

Motors which can be started unloaded with small shaft load usually are not capable of resynchronizing loaded, so that in addition to means for removing the field excitation, the load must also be removed before the motor can be pulled back into synchronism. Automatic and prompt removal of both field and load upon loss of synchronism is required for such motors in order that they can be quickly pulled back into synchronism without substantial loss in speed. Some processes do not allow for prompt removal of load, and for these applications it is necessary, in order to avoid complete disconnection of the motor from the system and a possible serious shut-down, either to have a motor capable of pulling back into step when loaded or to design a supply system so that such forced shut-downs are very infrequent. Motors which are started and resynchronized fully loaded will ordinarily be capable of resynchronizing by prompt removal of field excitation and its subsequent reapplication after the disturbance without disconnection.

There may exist, therefore, motor installations where it becomes not only desirable, but necessary, to design the supply system so that loss of synchronism will not occur for the usual system faults. A synchronous motor is particularly liable to loss of synchronism for the following reasons:

1. When the fault is on, the synchronous motor tends to decelerate because of its shaft load, plus the losses due to the flow of fault current. This is not the case for a generator, as the torque produced by the losses due to the flow of fault current oppose the prime mover torque, thereby decrease the acceleration, and are accordingly beneficial.
2. The size of the individual synchronous motors is usually relatively small, so that the field flux decays more rapidly than in a generator, because of the armature reaction created by the flow of fault current and the swing or oscillation after the fault is cleared.
3. The per-unit effective inertia of the rotating element is usually much less than that of generators. Exceptions exist, of course, for high-inertia loads. This means that the angular displacement, in general, increases more rapidly.
4. Synchronous motors are not usually provided with automatic voltage regulators. This lack, combined with the low field time constant, reduces their synchronous strength appreciably during the period the fault is on and immediately after it is cleared.

5. The problem of stability and the effect of the system are not always fully recognized at the time synchronous motors are purchased. Quite often this leads to the installation of motors with inadequate margin between the pull-out torque and the load torque. A large group of motors of only ordinary characteristics, connected through a long line to a system, may have very little stability margin even under steady-state conditions.

All these factors combine to make the synchronous motor load particularly susceptible to pulling out of step. A most effective manner of preventing loss of synchronism is by quick clearing of system faults. Because of the nature of synchronous motor loads, the fault-clearing times for severe faults and locations relative to the synchronous motor load must be briefer than those required in general for corresponding severe faults relative to generating stations. Clearing times of 6 cycles or less for severe faults relative to the synchronous motor load may become highly desirable for important loads which require a high degree of reliability (Reference 42).

Synchronous motors may be provided with automatic voltage regulators which will improve the stability characteristics, as well as contribute to better system performance by providing controlled kvar. Large motors for driving varying loads, such as the motor-generator sets of steel strip mills, will be found to require a very large swing in kvar if they are operated without automatic control of their field excitation. For these cases, it becomes highly desirable to use some automatic means for controlling the excitation, so as to reduce the swing in system current and voltage. Also, such **excitation control** may allow the use of a smaller motor, since it will not be required continuously to carry the full field excitation required for the maximum load condition (Reference 43).

Some of the larger motors are used for driving **irrigation pumps.** Installations have been designed which require a unit motor capacity of 65,000 hp. These larger installations require special considerations in order to assure proper coordination of the characteristics of the system with those of the pump and motor. When the system is limited in capacity with respect to the size of the pump, it may be necessary to start the pump unwatered by using compressed air to depress the water in the pump casing. With the pump unloaded in this manner, reduced voltage starting and synchronizing by means of a starting auto-transformer may be used. Reduced frequency or synchronous starting (Reference 44) may be used when it is possible to assign separate generating capacity for the motors to be started.

If the system capacity is not too limited, it is possible to use full voltage starting and synchronizing. The primary consideration for the normal type of system is usually the maximum allowable voltage drop on starting and synchronizing. A discharge valve may be used and operated in the closed position during starting and synchronizing to reduce the starting torque requirements. The synchronizing characteristics of large motors should be sufficiently ample so that they can be synchronized with the system when it is operated with its normal minimum of generating capacity.

The stability characteristics, both transient and steady state, should also be given proper consideration to allow for reliable operation. Particularly careful coordination is required for the larger motors, where the supply system impedances external to the motor have an appreciable effect on the motor torques that can be developed.

Some types of variable speed drive may be of large capacity and essentially synchronous in their performance from a system standpoint. One of these types uses a synchronous motor connected to the rotor winding of a wound-rotor induction motor (Reference 45). The synchronous motor of this drive, which absorbs the slip energy, drives a d-c generator and must be of sufficiently low impedance so that it will not pull out of synchronism when it is operated at low speed. The external impedance of the synchronous motor in this case is represented by the induction-motor primary and secondary impedances, as well as the supply-system impedances to the drive.

INDUCTION-MOTOR LOADS. Induction motors are used to drive many varying loads. Their effect in general is beneficial on system performance because of the reduction in kilowatt requirement with small decreases in voltage and frequency. This is particularly true for the lightly loaded motors and for motors driving fans and pumps which do not have a constant torque speed characteristic.

LIGHTING LOADS. Lighting loads have, in general, a desirable characteristic from a stability standpoint, since a reduction in voltage results in a decrease in current.

RECTIFIER LOADS. For reductions of sustained voltage in the range of values considered for steady-state system operation, the rectifier characteristics are such as to reduce the current required in almost the same way as a constant impedance load. This is a desirable characteristic from the standpoint of system stability. Large changes in voltage, as occasioned by short circuits or severe emergencies, may result in such misfiring or opening of the rectifier breakers as to cause a more rapid reduction in rectifier loads, with a drop in voltage. This also represents a desirable characteristic from a system-stability point of view, unless some complication results from the sudden reapplication of the load in too large a block after it has been disconnected.

CONVERTER LOADS. For small changes in applied voltage synchronous converters have characteristics somewhat corresponding to rectifiers. These depend on the type of d-c load, but in general the current drawn by the converter, as in the case of the rectifier, drops off with decrease in system voltage.

SYNCHRONOUS CONDENSERS. Synchronous condensers provide a source of controlled kvar to meet both the varying demands of the load and the requirements for transmission of power. The size of a synchronous condenser for a given overexcited kva rating is determined primarily by the sum of the overexcited and underexcited capacity. The normal amount of lagging capacity for an air-cooled synchronous condenser is 50 per cent of the overexcited rated kva, whereas with hydrogen cooling at atmospheric pressures the lagging capacity is about 40 per cent of the overexcited rated kva. In some cases, however, 100 per cent lagging condensers are used. Increased lagging capacity may be necessary for charging long high-voltage lines, or for those cases in which the number of non-switched capacitors is more than is required for the compensation of the lagging reactive under minimum load conditions.

Synchronous condensers increase the steady-state and transient-stability limits of systems in that they tend to support the voltage. Special synchronous condensers with properly coordinated regulators may be used at intermediate switching stations of long-distance transmission lines as a means of overcoming the stability limitations.

The theoretical idea of intermediate synchronous condensers for supporting the transmission of power between areas is also accomplished in practice by connection to intermediate systems of appreciable capacity. This is an effective and practicable realization of the notion of using synchronous capacity along the line to improve system stability.

Synchronous condensers are also useful in reducing the voltage dip due to suddenly applied loads, e.g., those due to arc furnaces and steel strip mill loads. For the major interconnected systems only the largest loads of this type are likely to give trouble from a light-flicker standpoint, and it has been found that synchronous condensers with automatically controlled excitation systems tend to relieve this condition.

Synchronous condensers have a certain inherent overload ability in case of an emergency, since the excitation system is capable of applying considerably more than rated full-load field current. This capacity can be used only during short momentary overloads because of the thermal limitations of the field and armature, but it does provide a margin in the steady-state and transient-stability limits and gives the system an additional stiffness above that required for the normal load transfer.

Conversely, the synchronous condenser is capable of holding voltages down on the system in case of loss of load or overspeeding of the prime movers. A network analyzer study provides a useful means for the determination of the optimum location and capacity of the synchronous condenser as well as of shunt capacitors, so as to meet present and future load conditions.

Synchronous condensers, which are usually located at the receiving end near the load areas of transmission systems, may offer some special stability considerations. Because of their proximity to the load, a severe disturbance may result in loss of synchronism of the condenser. Since they are usually provided with voltage regulators, their out-of-step operation will be normally accompanied by an increased field excitation. Experience has demonstrated that under these conditions the condenser may fall off in speed and operate at some subsynchronous speed until it is disconnected from the system or until the field is removed. This phenomenon can be explained (Reference 46) by the proximity of the load and the rise in field strength, which increases the braking torques relative to the induction-motor torques. Subsynchronous operation may occur under these conditions at a speed where the braking and induction-motor torques balance each other. A further requirement for continued subsynchronous operation is that this point of equilibrium be sufficiently below system speed so that the slip frequency speed pulsation due to the variation in synchronizing torque will not be sufficient at any time to reach system speed. If this were to occur, resynchronization would result (Reference 47). For a synchronous condenser such operation can be avoided by removal of the field excitation and its subsequent prompt reapplication after the system disturbance. Automatic means may be provided to accomplish such resynchronization.

SHUNT CAPACITORS. Shunt capacitors, either unswitched or controlled, are finding increased application. Studies (Reference 48) have indicated that for some types of systems the use of shunt capacitors instead of synchronous condensers may result in only small difference in the stability performance. A synchronous condenser, particularly during a period of low voltage, is capable of contributing more corrective kvar than a shunt capacitor of equal capacity. Accordingly, the system voltage will, in general, be better maintained by synchronous condensers than by shunt capacitors, which may or may not be an important effect, depending upon the type of system and amounts of corrective kva to be supplied. The relative effects of these two types of kvar supply can be evaluated by a stability study. Because of their low cost per kvar capacitors have found general and broad application in the load areas for improving the power factor at the point at which the load kvar is required. This has relieved the transmission system of the need of carrying at least part of the reactive requirements of the load. Because of the variations in the load demand, however, it has not been found advisable to use unswitched shunt capacitors for the compensation of all the load kvar requirements. It is more economical of kvar capacity to apply part of the correction at load substations by using controlled kvar in the form of either controlled shunt capacitors or synchronous condensers. At the high-capacity substations, where controlled kvar is required for both load and transmission, synchronous condensers have found general application because they are easily controlled by voltage regulators and also because in the larger sizes they cost less than capacitors. For a given load area or system an optimum ratio exists between unswitched capacitors and controlled kvar. Also there will probably exist an optimum ratio between the amount of synchronous condenser and shunt-capacitor capacity of the controlled type. Because of the inherent lagging capacity of synchronous condensers, their use will in many cases allow for increased amounts of unswitched shunt capacitors.

FREQUENCY-CHANGER SETS. *a.* **Synchronous-synchronous.** The synchronous-synchronous frequency-changer set is the most common type of frequency changer (Reference 49). It may be used to tie together systems of relatively large capacity, compared with the frequency-changer-set capacity, if care is taken to control the load swings through the frequency-changer set by proper supplementary control, similar to that used for tie lines. Many systems do not have large, rapidly varying load changes, and it is not necessary to provide the system governing with additional supplementary control other than the careful attention of the station operators. In any case, governor supplementary control on either of the two systems which are tied together by the frequency-changer sets may be used to reduce the danger of overloading and the possibility of pulling out of step. Frequency-changer sets are usually provided with voltage regulators and sufficient pull-out torque to ride through the system disturbances and load swings to which they may be subjected.

One of the important types of low-frequency system is that required for single-phase railroad electrification. Synchronous-synchronous frequency-changer sets are able to handle successfully this type of load, although it may be quite variable, because of the relatively small amount of low-frequency generation which is ordinarily used, in comparison with the total frequency-changer capacity; i.e., the ratio of frequency-changer capacity to total load is high.

System stability may be reduced when synchronous-synchronous frequency-changer sets of limited capacity are used to tie two relatively large systems together. Where the system steady-state stability margins are more than ample, this may not be an important limitation. However, if one of the systems has a long transmission line which is loaded at the time the frequency-changer set at the receiving end is also delivering power to another system, the steady-state stability limit of the system having the long transmission lines may be appreciably reduced. An increase in the steady-state stability limit of the system with the long line may be obtained by operating the frequency-changer set connected to only a part of the other systems, thereby reducing the inertia of the system to which the frequency changer is delivering power. The amount of the increase in stability obtained can be determined by analysis.

Because the load on frequency changers depends on the balance of the generation and loading between two systems of differing frequency, they represent a special problem upon loss of synchronism. Synchronous-synchronous frequency changers are usually provided with out-of-step relays, which will quickly disconnect the set upon loss of synchronism from either end. This is desirable because the natural mechanical frequency of the shaft is usually much lower than that of all other types of generators. During out-of-step operation with ceiling excitation applied, the set may be subjected to large torque pulsations at slip frequency. If the slip approaches the natural frequency of the shaft, mechanical damage may result. Since the load on the set is indeterminable during such an emergency operation, it usually is necessary to disconnect the set from the system upon loss of synchronism rather than just to remove the field excitation without disconnection, as is possible with the synchronous condenser.

b. **Variable Frequency Ratio.** The variable-ratio or **Scherbius** type of frequency changer does not provide so rigid a link between the two systems as does the synchronous-synchronous type; i.e., it provides a non-synchronous tie. The variable-ratio set has very desirable stability characteristics when used to tie two systems together; therefore it may be used when the frequency-changer capacity is very small in relation to the total capacity of the smaller of the two interconnected systems. One end, which may be at either frequency, has an induction-motor slip torque characteristic, and the variable-ratio set can thereby continue in successful operation even though one system frequency changes in relation to the other system frequency. If regulating means are provided to maintain a given load and power factor, the performance then depends on the inherent induction-motor characteristics, the regulating scheme used, and its response.

The use of variable-ratio sets will generally mean that added frequency-changer capacity will be of the same type or asynchronous. The low-frequency systems, however, except those for railroad electrification and some steel mills, are generally reducing in size rather than increasing, and the requirement for an increase in the frequency-changer capacity does not ordinarily exist.

c. **Electronic.** Electronic frequency changers have been built and are in use in both the United States (Reference 50) and Europe. They provide a non-synchronous tie between the systems and allow for a flexible control of power flow. In this respect they have desirable steady-state stability and control characteristics. Their operation under transient voltage and system disturbances differs from that of the rotating types, since proper commutation depends upon the system voltages and frequency. For some systems this may not be a particularly great disadvantage, but for others it may be very undesirable.

67. SYSTEM INTERCONNECTION

There are three general types or classifications of system arrangements: (*a*) synchronized at the load; (*b*) having low-tension busing; (*c*) having high-tension busing.

System designs of the metropolitan type (Reference 51) are particularly affected by the requirement to keep to reasonable values the short-circuit interrupting duty of circuit breakers. In order to meet this problem, systems have been designed so that the ties between generating units are synchronized at or through the load areas, which is known as "synchronizing at the load." This type of system arrangement accomplishes the reduction in short-circuit kva, but to do so requires greater line and transformer capacity. Also, as the system develops and incorporates generation from outlying areas, it becomes more and more difficult to keep the original design plan without a considerable increase in transmission-line and transformer capacity. As a result, most systems which started to adopt this plan have found it necessary to compromise and incorporate features of the other types or to abandon the principle altogether. It is then necessary to interconnect the major stations of the system by tie lines of capacity sufficient for interchange and to provide each generating station or substation with its own load, with no interconnection between the load areas except at the tie-line transmission voltage.

For transmission systems which do not involve great distances, such as may be found in developed industrial areas, the high-tension short-circuit kva can be kept down by busing the transmission lines only on the low-tension side. This increases the amount of transformer kva required, but for relatively short distances this type of system has appreciably less short-circuit duty at the tie-line voltages. Transmission between generation and load areas is provided by the lines with their individual transformers. The generating station of this type of system may use ring buses with reactors between bus sections.

High-tension busing of the transmission lines becomes desirable at distances at which stability is a problem, and where the switching out of any line section results in an appreciable increase in the demand for kvar capacity. Most long-distance transmission systems are of this type. High-tension busing is more applicable to the long-distance high-tension system, whereas the low-tension busing arrangement is used for shorter lines interconnecting large local load and generating areas. In a large system, however, both these types will be found interconnected.

Although high-tension busing has advantages for long distances, this type of system is also suited for much shorter distances. A comparison of low-tension versus high-tension busing is shown in Fig. 19, which indicates the relative advantages of each type.

INTERMEDIATE SWITCHING STATION. With higher voltage and more heavily loaded interconnections, it becomes advisable also to bus the transmission-line circuits together at intermediate switching stations in order to minimize the shock or disturbance due to the switching out of line capacity. If parallel lines are used, the number of intermediate stations can be increased as the loadings increase and need be installed only when required in a later stage of the development. The advantage of the sectionalized parallel line is the ability to switch off a faulted line section with only a small loss of line capacity and therefore with much less disturbance in the system. This permits operation up toward the steady-state limit with the faulted section switched out, which for most systems is sufficiently high for the transfer of not only normal but also emergency power requirements.

As contrasted with this type of transmission system, it is desirable to avoid tapped lines or looping, which usually require the switching out of relatively larger amounts of line capacity. Because of the poor stability characteristics of looped or tapped lines, they are used in general only for relatively lightly loaded or "shoe string" lines.

As is well recognized, when one of two parallel lines between interconnecting systems of equal impedance is switched out, the resulting peak power swing on the remaining line will be from two to three times its original load. If the lines are unequally loaded, possibly due to unequal length, the switching out of the more heavily loaded line will result in an even greater proportionate swing on the remaining line. These swings, if sufficiently severe, may cause loss of synchronism of the unfaulted line. The location of parallel lines

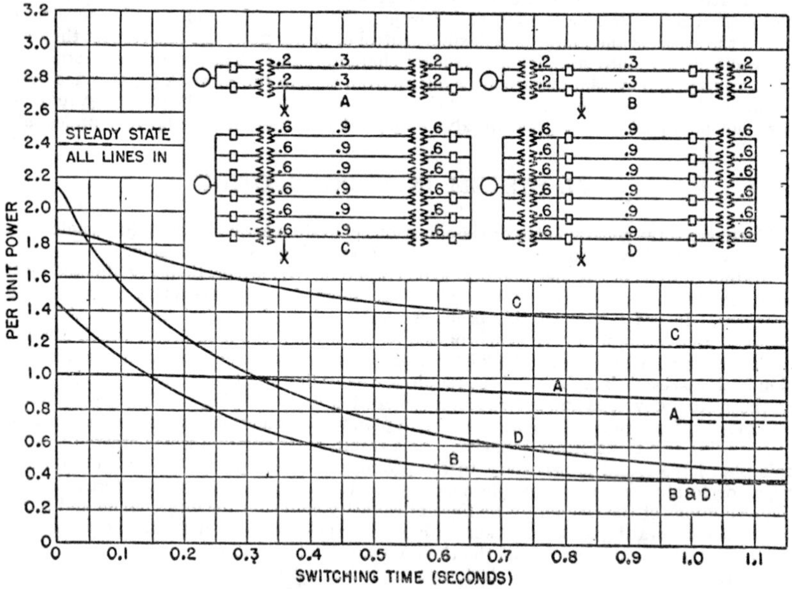

Fig. 19. Power versus Required Switching Time for Various Busing Arrangements

and the possible means of providing switching stations at a future time should be considered in the early stages of a power-system development, when the need for bus sectionalization may not be so evident. The loads to be transmitted may be relatively small when the new extension or development is considered, compared to what the future loadings may be.

BUS-TIE REACTORS. High-voltage bus-tie reactors have been used at large generating stations in order to retain the essential advantages of high-tension busing and to keep the circuit-breaker interrupting requirements down when they would otherwise exceed the capacity of the circuit breakers. This affords a practical compromise, as it is possible to retain the unit type of generating-station arrangement, each generator with its own step-up transformer, and to avoid a low-tension bus. In selecting reactors for such a high-tension bus, the value of reactance is relatively easily determined in order to meet a given circuit-breaker interrupting capacity. The continuous rating of the reactor is more difficult to determine, however, as it will depend more upon the emergency than the normal system requirements. Once selected and installed, it becomes difficult to increase the current rating for a given effective reactance between buses without, for all practical purposes, replacing the reactor. For this reason the reactor may become a limitation for the flow of power between sections, particularly under emergency conditions.

SUPPLEMENTARY CONTROL. Supplementary control of the governors of the prime movers provides means for loading more reliably the interconnecting transmission lines of a power system (References 52, 53). In general, this control makes slow adjustments in the prime mover power inputs in order to hold frequency and loads. This makes it possible to keep the tie line loading within specified limits under conditions of normal operation and therefore allows for realization of larger overload capacity of these lines during system emergencies. Since most interconnected systems are of relatively large capacity in comparison with the magnitude of the load changes, it is possible by this means to control properly both frequency and transmission-line load. Governor control gives better stability margins and a reduction in the controlled kvar capacity required to support the system voltage under emergency transmission-line overloadings.

An extreme example of those cases in which the loads are of large magnitude and suddenly applied is the steel strip mill load requiring 20,000 or 40,000 kw periodically every few seconds. It has been found necessary to allow the frequency of the interconnected system to sag slightly. If the frequency is held constant, large and rapid changes in prime mover torque, corresponding to the load change, are required. The difficulty of accomplishing this, even if the capacity were available, is so great that holding the frequency is less advantageous than allowing a slight deviation in frequency.

68. GROUNDING

In the early days of slow fault-clearing, line-to-ground faults were taken as the basis for design for stability. Later, but before quick-opening circuit breakers and relays were generally applied, systems designed for reliable power supply were built to withstand double-line-to-ground faults near the important buses rather than three-phase faults. When double-line-to-ground or line-to-ground faults of long or moderate duration were used as a basis for system design, stability advantages could be found for neutral impedance grounding, either reactor or resistance, in order to reduce the severity of faults during the fault period. In some cases a resistor was used at the generating-station transformer location, with a neutral reactor at the receiving end. In this way faults involving ground were reduced in severity and provided some kilowatt loading on the portion of the system tending to advance in angular displacement in relation to the receiving end, which tended to be retarded. Accordingly, an additional benefit was obtained in stability.

With the use of quick-opening circuit breakers of high interrupting capacity, however, the necessity for grounding through impedance for stability improvement has practically disappeared, and three-phase faults may be taken as a basis of design. Systems may be effectively grounded to take full advantage of the operating and economic advantages afforded by high-voltage transformers with reduced neutral insulations. Low neutral grounding impedances may, however, be used to limit the line-to-ground short circuits, so that they will not exceed those obtained for three-phase short circuits. In this way the short-circuit interrupting capacity requirements are kept down to the three-phase short-circuit values.

69. BIBLIOGRAPHY

1. "American Standard Definitions of Electrical Terms," ASA C42-1941, published by the American Institute of Electrical Engineers, Def. 35.20.200 to 35.20.208.
2. Crary, S. B., *Power System Stability*, Vol. I: *Steady State Stability*. John Wiley (1945).
3. Crary, S. B., *Power System Stability*, Vol. II: *Transient Stability*. John Wiley (1947).
4. Hazen, H. L., O. R. Schurig, and M. F. Gardner, Network Analyzer of the Massachusetts Institute of Technology. Design and Application to Power System Problems, *AIEE J.*, Vol. 49, pp. 872–875 (Oct., 1930).
5. Travers, H. A., and W. W. Parker, An Alternating-current Calculation Board, *Elec. J.*, p. 260 (May, 1930).
6. Kuehni, H. P., and R. G. Lorraine, A New A-c Network Analyzer, *AIEE Trans.*, Vol. 57, pp. 67–73 (1938).
7. Bush, V., The Differential Analyzer, a New Machine for Solving Differential Equations, *J. Franklin Inst.*, Vol. 212, No. 4, pp. 447–488 (Oct., 1931).
8. Travis, Irven, Differential Analyzer Eliminates Brain Fag, *Machine Design*, pp. 15–18 (July, 1935).
9. Kuehni, H. P., and H. A. Peterson, A New Differential Analyzer, *AIEE Trans.*, Vol. 63, pp. 221–228 (1944).
10. Dahl, O. G. C., *Electric Power Circuits—Theory and Application*, Vol. II: *Power System Stability*. McGraw-Hill (1938).
11. Chapter 5, pp. 114–150, of Reference 2.
12. Chapters 3 and 4, pp. 74–113, of Reference 2.
13. Chapter 6, pp. 151–172, of Reference 2 for examples.
14. Chapter 5, pp. 88–128, of Reference 3.
15. Section 61, pp. 263–273, of Reference 3.
16. Maginniss, F. J., Differential Analyzer Applications, *Gen. Elec. Rev.*, Vol. 48, No. 5, pp. 54–59 (May, 1945).
17. Peterson, H. A., and C. Concordia, Analyzers for Use in Engineering and Scientific Problems, *Gen. Elec. Rev.*, Vol. 48, No. 9, pp. 29–37 (Sept., 1945).
18. Aiken, H. H., and Grace M. Hopper, The Automatic Sequence Controlled Calculator—I, *Elec. Eng.*, pp. 384–391 (Aug.–Sept., 1946).
19. Concordia, C., S. B. Crary, and E. E. Parker, Effect of Prime-mover Speed Governor Characteristics on Power-system Frequency Variations and Tie-line Power Swings, *AIEE Trans.*, Vol. 60, pp. 559–567 (1941).
20. Byrd, H. L., and E. B. Rietz, A High-capacity High-voltage Oil Circuit Breaker, *AIEE Trans.*, Vol. 64, pp. 160–163 (1945).

21. Hill, A. W., and W. M. Leeds, The Next Step in Interrupting Capacity—5,000,000 kva, *AIEE Trans.*, Vol. 64, pp. 317–323 (1945).
22. Clarke, Edith, Three-phase Multiple-conductor Circuits, *AIEE Trans.*, Vol. 51, pp. 809–821 (1932).
23. Dwight, H. B., and E. B. Farmer, Double-conductors for Transmission Lines, *AIEE Trans.*, Vol. 51, pp. 803–808 and 823 (1932).
24. Discussion by W. W. Lewis of Edith Clarke, Three-phase Multiple-conductor Circuits, *AIEE Trans.*, Vol. 51, pp. 822 and 823 (1932).
25. Hobart, H. M., Transmission of Power in Compressed Gas Atmospheres, *J. Franklin Inst.*, Vol. 234, pp. 251–273 and 331–354 (Sept.–Oct., 1942).
26. Clarke, Edith, and S. B. Crary, Stability Limitations of Long-distance A-c Power-transmission Systems, *AIEE Trans.*, Vol. 60, pp. 1051–1059 and 1299–1303 (1941).
27. Starr, E. C., and R. D. Evans, Series Capacitors for Transmission Circuits, *AIEE Trans.*, Vol. 61, pp. 963–973 (1942).
28. Butler, J. W., J. E. Paul, and T. W. Schroeder, Steady-state and Transient-stability Analysis of Series Capacitors in Long Transmission Lines, *AIEE Trans.*, Vol. 62, pp. 58–65 (1943).
29. Wagner, C. F., and R. D. Evans, Static Stability Limits and the Intermediate Condenser Station, *AIEE Trans.*, Vol. 47, pp. 94–121 (1928).
30. Concordia, C., S. B. Crary, and F. J. Maginniss, Long-distance Power Transmission as Influenced by Excitation Systems, presented at the AIEE Pacific Coast Convention, Seattle, Washington (Aug., 1946).
31. Preferred Standards for Large 3600-rpm, Three-phase, 60-cycle, Condensing Steam Turbine-generators (Larger than 10,000 kw Rated Capacity), prepared by Joint Committee on Steam Turbine-generators of the American Institute of Electrical Engineers and American Society of Mechanical Engineers.
32. Chapter 12, pages 271–277, of Reference 2.
33. McClure, J. B., S. I. Whittlesey, and M. E. Hartman, Modern Excitation Systems for Large Synchronous Machines, presented at the AIEE Summer Convention, Detroit, Michigan (June, 1946).
34. Concordia, C., Steady-state Stability of Synchronous Machines as Affected by Voltage Regulator Characteristics, *AIEE Trans.*, Vol. 63, pp. 489–490 (1944).
35. Crary, S. B., L. F. Kennedy, and C. A. Woodrow, Analysis of the Application of High-speed Reclosing Breakers to Transmission Systems, *AIEE Trans.*, Vol. 61, pp. 339–348 (1942).
36. Bodine, R. B., S. B. Crary, and A. W. Rankin, Motor-driven Exciters, presented at the AIEE Midwinter Convention, New York City (Jan., 1946).
37. Wagner, C. F., Damper Windings for Water-wheel Generators, *AIEE Trans.*, Vol. 50, pp. 140–151 (1931).
38. Crary, S. B., and W. E. Dungan, Amortisseur Windings for Hydrogenerators, *Elec. World* (June 28, 1941).
39. Bauman, H. A., O. W. Manz, Jr., J. E. McCormack, and H. B. Seeley, System Load Swings, *AIEE Trans.*, Vol. 60, p. 541 (1941).
40. *Discussion of* Supplementary Control of Prime-mover Speed Governors, by P. B. Juhnke, *AIEE Trans.*, Vol. 61, pp. 395–396 (1942).
41. Lauder, A. H., Salient-pole Motors out of Synchronism, *AIEE Trans.*, Vol. 55, pp. 636–649 (1936).
42. Sporn, Philip, and H. P. St. Clair, Tests on and Performance of a High-speed Multibreak 138-kv Oil Circuit Breaker, *AIEE Trans.*, Vol. 57, pp. 696–703.
43. Boice, W. K., B. H. Caldwell, and M. N. Halberg, Synchronous Motors with Controlled Excitation for Suddenly Applied Loads, *AIEE Trans.*, Vol. 62, pp. 113–118 (1943).
44. Concordia, C., S. B. Crary, C. E. Kilbourne, and C. W. N. Weygandt, Jr., Synchronous Starting of Generator and Motor, *AIEE Trans.*, Vol. 64, pp. 629–634 (1945).
45. Clymer, C., Large Adjustable-speed Wind-tunnel Drive, *AIEE Trans.*, Vol. 61, pp. 156–158 (1942).
46. Kroneberg, A. A., Out-of-step Condition on a Synchronous System, *Elec. Eng.*, Vol. 51, pp. 769–772 (Nov., 1932).
47. Crary, S. B., A. H. Lauder, and D. R. Shoults, Pull-in Characteristics of Synchronous Motors, *AIEE Trans.*, Vol. 54, pp. 1385–1395 (1935).
48. Butler, J. W., W. Ridgway, and T. W. Schroeder, Capacitors, Condensers, and System Stability, *AIEE Trans.*, Vol. 63, pp. 1130–1138 (1944).
49. Easley, R. M., and S. B. Crary, Frequency Changers—Characteristics, Applications, and Economics, *AIEE Trans.*, Vol. 64, pp. 351–358 (1945).
50. Cramer, F. W., F. W. Morton, and A. G. Darling, The Electronic Converter for Exchange of Power, *AIEE Trans.*, Vol. 63, pp. 1059–1069 (1944).
51. Ellis, W. R., S. B. Griscom, and W. A. Lewis, Generalized Stability Solution for Metropolitan-type Systems, *AIEE Trans.*, Vol. 51, pp. 363–372 (1943).
52. Wild, Earle, Methods of System Control in a Large Interconnection, *AIEE Trans.*, Vol. 60, pp. 232–236 (1941).
53. Crary, S. B., and J. B. McClure, Supplementary Control of Prime-mover Speed Governors, *AIEE Trans.*, Vol. 61, pp. 209–214 (1942).
54. Report on Stability, *Trans. AIEE*, Vol. 56, p. 280 (1937). Contains a bibliography of 180 items.

SECTION 4

PRINCIPLES OF ELECTROCHEMISTRY

BY

MARTIN KILPATRICK AND MARY L. KILPATRICK *

* Several pages of the article by H. M. Goodwin, from an earlier edition, have been retained.

PRINCIPLES OF ELECTROCHEMISTRY

By Martin Kilpatrick and Mary L. Kilpatrick

Electrochemistry is the science which deals with the phenomena resulting from the direct transformation of electrical into chemical energy or the converse transformation of chemical into electrical energy. By general usage the term, especially as applied to industrial processes, has come to include also those thermochemical phenomena which occur at temperatures produced in electric furnaces, although in such processes electrical energy frequently plays no other role than that of generating heat.

1. DEFINITIONS

The following terms are commonly used:

Element. An element is a substance which cannot be decomposed by the ordinary process of chemical analysis. (See Table 1, Electrochemical Equivalents, p. 4-06, and International Atomic Weights, Section 2.) At present 96 elements are known. In connection with the definition given, it should be added that the radioactive elements undergo spontaneous decomposition into other elements, and that recently certain of the elements have been decomposed by bombardment with alpha-rays, hydrogen nuclei, and neutrons.

Atoms and Atomic Weights. An atom is the smallest mass of an element that can enter into chemical combination with another element or with itself. As the chemical standard, the atomic weight of oxygen as it occurs in nature has been arbitrarily chosen as 16.0000. As the physical standard, the atomic weight of the most abundant isotope of oxygen has been chosen as 16.0000; since the isotope of mass 17 is present in ordinary oxygen to the extent of 0.04 per cent, and that of mass 18 to the extent of 0.20 per cent, the mean atomic weight of oxygen on the physical scale becomes 16.0044, and all atomic weights on the physical scale are thus slightly higher. Unless otherwise stated, the atomic weights used in this chapter are on the chemical scale, and all quantities that involve the mol or the gram-equivalent are on the chemical scale. The combining or equivalent weight of an element is the weight which combines with 8 parts by weight of oxygen. The atomic weight is identical with, or a simple multiple of, the combining weight, depending on the number of atoms of the element which combine with 1 atom of oxygen. The number of equivalent weights of an element which constitute its atomic weight is determined from a study of the composition of the gaseous compounds of the element or from a study of its specific heat. On the atomic theory the atomic weight is proportional to the mass of the atom.

Molecules and Molecular Weight. A molecule is the smallest mass of a substance which can exist and preserve its chemical properties. The molecular weight of an element or of a chemical compound is equal to the sum of the atomic weights of the atoms contained in the molecule and may readily be calculated from the atomic weights when the molecular symbol of the compound is known. Thus the molecular weight of oxygen, O_2, is $2 \times 16.00 = 32.00$, as there are two atoms in the molecule. The molecular weight of silver nitrate, $AgNO_3$, is $107.88 + 14.01 + 3 \times 16.00 = 169.89$.

Radical. A radical is a combination of two or more elements which persists as a group in chemical reactions; e.g., the SO_4 in H_2SO_4 is a radical, as exemplified by the reaction $H_2SO_4 + Zn = ZnSO_4 + H_2$.

Formula Weight. The formula weight of an atom, radical, or molecule is the sum of the atomic weights of the elements of which it is formed. For example, the formula weight of an oxygen atom, O, is 16; the formula weight of an oxygen molecule, O_2, is 32; the formula weight of the radical NO_3 is $14.01 + 3 \times 16.00 = 62.01$.

Valence. An adequate consideration of valence would involve a discussion beyond the limits of this article. As a simple definition, the valence of an element may be taken as the number of hydrogen atoms, or of atoms of an element such as chlorine which combines with hydrogen atom for atom, combining with an atom of the element in question. Thus from the formulas H_2O, HCl, KCl, $BaCl_2$ the valences of oxygen, chlorine, potassium, and barium are seen to be 2, 1, 1, and 2, respectively. The valence of a radical is similarly defined; e.g., the formulas KNO_3 and H_2SO_4 show that the nitrate radical has a valence of 1, the sulfate radical a valence of 2. Acids are classified as mono-, di-, or tribasic according to the number of replaceable hydrogen atoms which they contain (cp. HCl, H_2SO_4, H_3PO_4). Bases are classified as mono-, di-, or triacidic (KOH, $Ba(OH)_2$, $La(OH)_3$). Salts are classified as uni-univalent (KCl), bi-univalent ($BaCl_2$), uni-bivalent (K_2SO_4), bi-bivalent ($BaSO_4$), etc.

The valence of an element may vary, cp. the cupric compounds (CuO, $CuCl_2$, etc.), in which the valence of copper is 2, and the cuprous compounds (Cu_2O, CuCl, etc.), in which the valence of copper is 1.

Equivalent Weight—Chemical Equivalent. The equivalent weight of an atom, ion, or radical is defined as its formula weight divided by its valence. The equivalent weight of an element or compound is also referred to as its chemical equivalent. The equivalent weight of an acid, base, or salt is its molecular formula weight divided by the valence of the ion of higher valence. As hydrogen exhibits no other valence than unity, its equivalent weight and atomic weight are identical. Similarly, the equivalent weight and molecular weight of hydrochloric acid, HCl, are each 36.47. Copper in copper sulfate, $Cu(SO_4)$, and in cupric chloride, $CuCl_2$, has a valence of 2; hence equivalent weights of these salts are one-half their molecular weights respectively.

In computations of electrochemical reactions it is frequently convenient to take, as the unit of mass (or weight) of a substance, a mass in grams equal to the number expressing the atomic weight, ionic weight, molecular weight, or equivalent weight of the substance. The following names have been given these units.

Gram-atom or Gram-atomic Weight. The number of grams of an element equal to its atomic weight; e.g., 1 gram-atom of oxygen is 16 grams.

Gram-molecule or Mol. The number of grams of a substance equal to its molecular weight; e.g., 1 mol of silver nitrate, $AgNO_3$, is 169.89 grams.

Gram-equivalent. The number of grams of an atom, radical, or molecule equal to its equivalent weight. For example, a gram-equivalent of sulfuric acid is $^1/_2\ H_2SO_4 = {}^1/_2\ 98.09 = 49.04$ grams.

Electrolyte, Electrolysis. When an electric current is passed through certain substances known as electrolytic conductors, a chemical reaction occurs at the places where the current enters and leaves, and there is a transfer of matter through the substance. Fused salts and aqueous solutions (or solutions in other ionizing solvents) of acids, bases, and salts belong in the class of electrolytic conductors. The acid, base, or salt in question is called an electrolyte, and the process of the passage of a current through an electrolytic conductor is known as electrolysis. It is important to note that an electrolyte is ionized before the current is passed.

Electrodes. Electrodes are the conductors by which the current enters and leaves the electrolyte or its solution. The electrode at which the current enters is called the anode, and that by which the current leaves is called the cathode.

Ions. Anions and Cations. The constituents of an electrolytic conductor which conduct the current through the solution or the fused salt are called ions. They exist in the conductor before the passage of the current; during the passage of the current the positively charged ions, or cations, travel toward the cathode, and the negatively charged ions, or anions, travel toward the anode.

The charge of an ion is proportional to its valence. Each individual univalent anion bears a charge of 1 electron; each bivalent anion, a charge of 2 electrons; etc. Each univalent cation has a deficit of 1 electron or a positive charge of magnitude equal to that of an electron; each bivalent cation has a deficit of 2 electrons; etc. The value of the electron is $(4.8025 \pm 0.0010) \times 10^{-10}$ abs esu; see "A New Table of Values of the General Physical Constants" by Raymond T. Birge, *Rev. of Mod. Phys.*, Vol. 13, 233–239 (1941).

To distinguish ions from neutral particles, small plus or minus signs are written after the symbol, the number of such signs being equal to the valence of the ion. Thus Cl^- represents a gram-atom of chlorine bearing a charge of an equivalent of negative electricity or N electrons, N being Avogadro's number. An equivalent of negative electricity is designated as e. Similarly, Ca^{++} represents a gram-atom of calcium having a deficit of $2N$ electrons, or $2e$.

Ionic Weight. The formula weight of an ion is called its ionic weight.

Gram-ion. The number of grams of an ion equal to its ionic weight; e.g., 1 gram-ion of hydrogen is 1.008 grams, and 1 gram-ion of SO_4 is 96.07 grams.

Electrochemical Equivalent of an Ion. The mass in grams of the ion liberated or deposited by 1 coulomb of electricity.

Electrochemical Constant or Faraday. See Article 4.

Solvent, Solute, and Solution. A solution is formed if, upon mixing substance A with substance B. a homogeneous or one-phase system results. For dilute solutions. it is customary to refer to the substance present in great excess as the solvent, and to the other as the solute.

Osmotic Pressure. The osmotic pressure of a solution is the pressure which must be exerted upon it in order that the vapor pressure of the solvent from the solution will be the same as the vapor pressure of the pure solvent at the temperature in question. See Article 13, Theory of Solutions.

Specific Conductance (κ). The specific conductance of a solution is the reciprocal of its specific resistance; i.e., it is the conductance of a column of liquid 1 cm long and 1 sq cm in cross-section. It is expressed in reciprocal ohms, or mhos, and is denoted by κ.

Concentration. The composition of a solution may be expressed in a number of ways, the more common of which are listed below.

a. For a solution composed of n_1 gram-molecules of X_1, n_2 of X_2, . . . n_r of X_r, the mol fraction of the *i*th component is given as $N_i = \frac{n_i}{\sum_{j=1}^{r} n_j}$.

b. The number of gram-molecules of X per liter of solution is called the molarity of X.

c. The number of gram-molecules of X dissolved in 1000 grams of solvent is called the **molality** of X.

d. The concentration of X may be expressed as the number of grams of X contained in unit volume of the solution.

e. The number of gram-equivalents of X in unit volume of solution is called the equivalent concentration. The symbol η is used here to represent the number of equivalents per cubic centimeter of solution.

Dilution (ϕ). The dilution of a solution is the number of cubic centimeters of solution in which 1 gram-equivalent of solute is dissolved. It is denoted by ϕ. Hence $\phi = 1/\eta$ and $\eta = 1/\phi$.

Normal Solution. A normal solution is a solution containing 1 gram-equivalent of solute per liter. For such a solution $\eta = 10^{-3}$ or $\phi = 1000$.

Equivalent Conductance (Λ). The equivalent conductance of a solution at the dilution ϕ is the conductance which a volume (in cubic centimeters) of the solution containing 1 gram-equivalent of the solute would have, if placed between parallel plate electrodes 1 cm apart. It is denoted by Λ_ϕ or Λ_η, according to whether the dilution or concentration of the solution is given. Hence $\Lambda_\phi = \phi\kappa$, or $\Lambda_\eta = \kappa/\eta$.

The dilution or concentration of a solution should always be stated in connection with its equivalent conductance; otherwise the expression is indefinite.

Heat of Reaction. The heat of reaction, by analogy with the heat of fusion and the heat of evaporation, is best defined as the heat absorbed in a chemical reaction. It is usually expressed in calories or kilogram-calories (1 kg-cal = 1000 cal).

The two calories in common use, the 15 deg calorie (the amount of energy required to raise 1 gram of water from 15 to 16 deg cent) and the mean calorie (the amount of energy necessary to raise 0.01 gram of water from 0 to 100 deg cent), differ only slightly, the mean calorie being 0.017 per cent larger than the 15 deg calorie. The mean calorie is represented by cal in this article.

Exothermic Reactions. Reactions in which heat is given out to the surrounding bodies.

Endothermic Reactions. Reactions in which heat is absorbed from the surrounding bodies.

2. NOTATION

For convenience of reference the symbols most frequently used are tabulated below.

κ = specific conductance in reciprocal ohms.
η = concentration, expressed as gram-equivalents of solute per cubic centimeter of solution.
ϕ = dilution, expressed as cubic centimeters of solution per gram-equivalent of solute.
Λ = equivalent conductance.
Λ_0 = equivalent conductance at infinite dilution.
α = degree of ionization.
E = electrical potential difference.
I = current strength.
r = resistance.
R = gas constant per mol.
F = the faraday = 96,500 coulombs per gram-equivalent.
T = absolute temperature in degrees centigrade.
t = centigrade temperature.

3. CHEMICAL EQUATIONS

A chemical equation, such, for example, as that representing the solution of zinc in a dilute hydrochloric acid solution (Aq = water),

$$Zn + 2HClAq = ZnCl_2Aq + H_2$$

is to be interpreted as follows: 1 gram-atom (65.37 grams) of zinc reacting with 2 mols (2 × 36.47 = 72.94 grams) of hydrochloric acid in aqueous solution forms 1 mol (65.37 + 2 × 35.46 = 136.3 grams) of zinc chloride in solution and 1 mol (2 × 1.008 = 2.016 grams) of hydrogen gas. Since atomic weights are relative numbers, any unit of weight, as the pound or kilogram, may be substituted for gram in this statement.

If the reaction is intended to express not only the chemical change which takes place but also the thermal effect involved, one must write

$$Zn + 2HClAq = ZnCl_2Aq + H_2 + 34.2 \text{ kg-cal}$$

This thermochemical equation signifies that, when the reaction occurs, 34.2 kg-cal of heat are evolved. If the pressure on the reacting system is kept constant, one has, by the first law of thermodynamics,

$$\Delta u = Q - P\,\Delta V$$

where Δu is the change in the internal energy of the system, Q the heat absorbed, P the pressure, and ΔV the change in volume accompanying the reaction. The heat content, or enthalpy, of the system, H, is by definition

$$H = u + PV$$

from which it follows that for a reaction at constant pressure

$$Q = \Delta H$$

In the case cited, $\Delta H = -34.2$ kg-cal if the reaction occurs at constant pressure.

4. FARADAY'S LAWS

The quantitative relations between the magnitude of an electric current and its chemical effect were discovered by Faraday in 1834 and are known as Faraday's laws. They may be stated as follows:

FIRST LAW. The amount of chemical change produced by an electric current is directly proportional to the total quantity of electricity which passes through the electrolytic cell. The amount of chemical change is independent of the voltage and intensity of the current, size of electrodes, and concentration of electrolyte, so long as the total *quantity* of electricity flowing through the circuit remains constant. (These factors, however, affect the *ultimate products* of an electrolysis.)

SECOND LAW. A given quantity of electricity always decomposes *equivalent* weights of different electrolytes irrespective of their nature.

Faraday's laws are the expression of the results of direct experiment and have been tested to the limit of precision with which physical and chemical measurements can be carried out at the present time. All evidence indicates that they hold rigidly, i.e., that they are exact laws of nature. Where exceptions have seemed to exist, they have been shown to arise from secondary causes. See Kraus, *Trans. Am. Electrochem. Soc.*, Vol. 21, 119 (1912) for special cases of electronic and ionic conduction.

VALUE OF THE ELECTROCHEMICAL CONSTANT OR FARADAY. The determination of the number of coulombs required to deposit 1 gram-equivalent of an ion, i.e., the constant connecting the quantity of electricity and the mass of a substance liberated, involves two distinct investigations: first, the measurement of the number of coulombs which pass through a given electrolytic cell; and second, the determination of the amount of the chemical decomposition. For a résumé of work on this subject, see *Bull. Bur. Standards*, Vol. 13, 497 (1916). The value at present accepted is 96,501 ± 10 int. coulombs per gram-equivalent; see Birge, *Rev. of Mod. Phys.*, Vol. 13, 238 (1941). The best round value is taken as 96,500 coulombs.

5. ELECTROCHEMICAL EQUIVALENTS OF THE ELEMENTS

From the definitions of the electrochemical equivalent and the faraday, it follows that the electrochemical equivalent of any ion, expressed in grams per coulomb, is equal to

$$\frac{\text{Gram-equivalent of the ion}}{96{,}500}$$

Table 1 contains the values of the electrochemical equivalents of the more common elements, which may be found useful in certain calculations. The electrochemical equivalent of any radical may be readily calculated from its formula; for example, the electrochemical equivalent of SO_4 is

$$\frac{32.06 + 4(16.00)}{2(96{,}500)} = 0.0004978 \text{ gram}$$

the factor 2 in the denominator being the valence of the ion.

Table 1. Electrochemical Equivalents of the More Important Elements

Based on the atomic weights of 1941 [*J. Am. Chem. Soc.*, Vol. 63, 850 (1941)] and on F = 96,500 coulombs

Element	Symbol	Atomic Weight	Valence	Milligrams Deposited by 1 amp in 1 sec	Grams Deposited by 1 amp in 1 hr
Aluminum	Al	26.97	3	0.09316	0.3354
Antimony	Sb	121.76	3	0.4206	1.514
Arsenic	As	74.91	3	0.2588	0.9315
Barium	Ba	137.36	2	0.7117	2.562
Bismuth	Bi	209.00	3	0.7219	2.599
Bromine	Br	79.916	1	0.8282	2.981
Cadmium	Cd	112.41	2	0.5824	2.097
Calcium	Ca	40.08	2	0.2077	0.7476
Carbon	C	12.010	4	....	....
Cerium	Ce	140.13	3	0.4840	1.745
Chlorine	Cl	35.457	1	0.3674	1.323
Chromium	Cr	52.01	2	0.2695	0.9701
Chromium	Cr	52.01	3	0.1797	0.6467
Cobalt	Co	58.94	2	0.3054	1.099
Cobalt	Co	58.94	3	0.2036	0.7329
Copper	Cu	63.57	1	0.6588	2.372
Copper	Cu	63.57	2	0.3294	1.186
Fluorine	F	19.00	1	0.1969	0.7088
Gold	Au	197.2	1	2.044	7.357
Gold	Au	197.2	3	0.6812	2.452
Hydrogen	H	1.0080	1	0.01045	0.03761
Iodine	I	126.92	1	1.315	4.735
Iron	Fe	55.85	2	0.2894	1.042
Iron	Fe	55.85	3	0.1929	0.6945
Lead	Pb	207.21	2	1.074	3.865
Lithium	Li	6.940	1	0.07192	0.2589
Magnesium	Mg	24.32	2	0.1260	0.4536
Manganese	Mn	54.93	2	0.2846	1.025
Manganese	Mn	54.93	3	0.1897	0.6831
Mercury	Hg	200.61	1	2.079	7.484
Mercury	Hg	200.61	2	1.039	3.742
Nickel	Ni	58.69	2	0.3041	1.095
Nickel	Ni	58.69	3	0.2027	0.7298
Oxygen	O	16.0000	2	0.08290	0.2984
Platinum	Pt	195.23	2	1.012	3.642
Platinum	Pt	195.23	4	0.5058	1.821
Potassium	K	39.096	1	0.4052	1.459
Silver	Ag	107.880	1	1.118	4.025
Sodium	Na	22.997	1	0.2383	0.8579
Strontium	Sr	87.63	2	0.4540	1.635
Sulfur	S	32.06	....	....	....
Thallium	Tl	204.39	1	2.118	7.625
Thallium	Tl	204.39	2	1.059	3.812
Tin	Sn	118.70	4	0.3075	1.107
Titanium	Ti	47.90	4	0.1241	0.4467
Tungsten	W	183.92	2	0.9530	3.431
Zinc	Zn	65.38	2	0.3388	1.220

6. CONDUCTIVITY OF ELECTROLYTES

SOLUTIONS. (See also articles on properties of materials in Section 2.) The specific conductance of an electrolytic solution varies between wide limits. It depends upon the nature of the solute and of the solvent, on the temperature, and on the concentration of the solution. The effect of these factors is discussed below. For numerical values see Kohlrausch and Holborn, *Leitvermögen der Elektrolyte;* Landolt-Börnstein, *Tabellen;* Tables annuelles internationales de constantes; and the International Critical Tables.

FUSED ELECTROLYTES. Most inorganic salts conduct electrolytically when heated to a sufficiently high temperature to cause them to pass into the liquid state. For such electrolytes Faraday's, Ohm's, and Joule's laws hold as they do for solutions. The conductance increases in general from 1 to 1 1/2 per cent per degree centigrade. An exhaustive résumé of all matters relating to fused electrolytes may be found in Lorenz's *Elektrolyse der Geschmolzene Salze.* On account of their enormous concentration (100 per cent) the specific conductance of fused salts is generally very high. This factor, together with the difficulty of obtaining a non-conducting chemically inert vessel of suitable shape to contain the salt and the difficulty of regulating high temperatures to a fraction of a degree, makes the measurement of the conductivity of fused electrolytes far more difficult than that of solutions. Fused silica and natural quartz crystals have been successfully used for conductivity cells. See Jaeger and Kapma, *Z. anorg. Chem.*, Vol. 113, 27 (1920); Arndt and Ploetz, *Z. physik, Chem.*, Vol. 110, 237 (1924); Biltz and Klemm, *ibid.*, Vol. 110, 318 (1924); Lorenz, *Z. Elektrochem.*, Vol. 30, 371 (1924); and for further references Taylor, *Treatise on Physical Chemistry*, Vol. 1, pp. 624–628.

7. VOLTAIC AND ELECTROLYTIC CELLS

When a cell formed by two electrodes and one or more electrolytes gives out electric energy, the cell is called a voltaic cell. All ordinary chemical batteries are voltaic cells. If across the terminals of a cell an external electromotive force, greater than the electromotive force developed within the cell, is impressed, a current will flow through the cell in the opposite direction, and the cell will absorb electric energy. A cell which absorbs electric energy is usually referred to as an electrolytic cell. The cells used in electrolytic refining and similar processes are electrolytic cells. An electrolytic cell may or may not have an electromotive force on open circuit, but it may develop an electromotive force by virtue of the chemical changes or changes in concentration which take place because of the current forced through it from some external source. A voltaic cell becomes an electrolytic cell when the electromotive force impressed across its terminals exceeds (and opposes) its own electromotive force.

8. POLARIZATION

When an electric current passes through either a voltaic or an electrolytic cell, in general, there is developed at the electrodes of the cell, as a result of the chemical actions and changes in concentration which take place, a *back* or *counter electromotive* force, in addition to its open-circuit electromotive force (if any), and an *increase in the resistance* of the cell, in addition to the change in resistance due to the heating effect of the current. Both these effects cause a decrease in the strength of the current through the cell. The counter electromotive force is said to be due to "polarization," and the increase in resistance to "transition resistance."

Even where the electrolyte, as a whole, suffers no change in concentration, as in the electrolysis of copper sulfate between copper electrodes, a counter emf of polarization is produced in consequence of the difference in concentration of the copper ions in the neighborhood of the anode and cathode. Vigorous stirring of the electrolyte tends to reduce this concentration difference and the resulting polarization, but, except for very low current densities, the difference cannot be completely eliminated. It is for this reason, in refining processes, that energy is required to transfer the metal from anode to cathode in addition to that necessary to overcome the internal resistance of the cell.

A transition resistance results from the formation of a film of poorly conducting material over the electrode and may or may not be present, according to the character of the electrolysis.

MEASUREMENT OF EMF OF POLARIZATION AND TRANSITION RESISTANCE. As both transition resistance and polarization tend to cut down the current flowing through the cell, they cannot be distinguished by this effect alone. The transition resistance may be determined by measuring the ohmic resistance of the cell before and after the passage of the current by the usual alternating-current method. The existence of an emf of polarization in a cell which normally has no open-circuit emf may be qualitatively demonstrated by short-circuiting the cell through a galvanometer immediately after breaking the primary circuit; if the cell is polarized, a current which diminishes to zero will flow through it in the reverse direction. A voltmeter or, better, an electrometer connected across the electrolytic cell will indicate the polarization voltage the instant after the current is broken. This voltage will diminish as the polarization disappears, the rate depending upon the current through the voltmeter and the rate of diffusion of the electrolytic products causing the polarization. Because of the rapidity with which the emf of polarization falls off after the exciting cause is removed, special precautions must be observed in its measurement. One of the best methods is to connect the circuit containing the applied emf and electrolytic cell with a tuning-fork interrupter and electrometer so that at the instant the battery current is broken at each vibration of the fork, the circuit containing the cell and electrometer or voltmeter is closed, and vice versa. By this arrangement the emf of polarization is measured during the fraction of a second that the battery circuit is open, and before it has had time to diminish sensibly. The polarizing current may also be regarded as practically constant.

It is frequently of importance to know, not the polarization of the cell as a whole, but the polarization at each electrode. This value is obtained by measuring the drop in potential at each electrode against a normal hydrogen or calomel electrode while the impressed emf is acting on the electrolytic cell, or immediately after the current through the cell is interrupted.

9. REVERSIBLE ELECTRODES—DEPOLARIZERS

If the chemical and thermal actions which take place at a given electrode when a given quantity of electricity passes in one direction through a cell are exactly the reverse of the actions which take place when the same quantity of electricity passes through it in the opposite direction, the electrode and the solution with which it is in contact are said to form a reversible electrode. Such an electrode does not polarize provided that the concentration of the solution is kept constant.

There are two types of reversible metal-liquid electrodes, namely:

Electrodes of the First Type, consisting of a metal in contact with a solution of one of its own salts, e.g., copper in copper sulfate, zinc in zinc sulfate. The two electrodes of a Daniell cell are of this type.

Electrodes of the Second Type, consisting of a metal in contact with a solution containing one of its difficultly soluble salts and a second soluble salt of some other metal, having the same anion. The difficultly soluble salt, called the depolarizer, must be present in excess as solid. Such an electrode is mercury in contact with a solution of zinc sulfate containing mercurous sulfate in excess as solid; this is one of the electrodes of the Clark cell.

Reversible Gas Electrode. An electrode consisting of "platinum black" saturated with an atmosphere of hydrogen gas and dipping partially into a solution containing hydrogen ions is also a reversible electrode. This is a special form of electrode of the first type in which a gas, by being occluded in platinum, is made to play the role of a metal.

Other "gas electrodes," e.g., chlorine, may be similarly prepared.

10. CONTACT POTENTIALS; ELECTROMOTIVE FORCE

The electromotive force of a voltaic cell, the back-electromotive force of an electrolytic cell, and the electromotive force of a thermocouple are all due to differences of potential which always exist at the junction of dissimilar substances. For a general discussion of this subject see Langmuir, *Trans. Am. Electrochem. Soc.*, Vol. 29, 125 (1916); Bridgman, *Phys. Rev.*, Vol. 14, 306 (1919); and Sommerfeld and Bethe, *Handbuch der Physik*, Vol. 24, pp. 333–622 (Springer, Berlin, 1933).

THE VOLTA EFFECT. When pieces of two different metals are brought into metallic contact, it is found that between points in the surrounding medium (gas or vacuum) immediately outside the two metals a difference of potentials exists. When the surrounding medium is a perfect vacuum, the potential difference is known as the true Volta contact potential difference between the metals in question. Investigations on the thermo- and photoemission of electrons have demonstrated that an amount of work W_A is required to expel an electron from the metal A; from measurements on the thermo- and photoemission of electrons from A and B the Volta contact potential difference between A and B, K_{AB}, may be obtained. It is assumed that the potential difference at the junction of the metals is relatively small.

K_{AB} may also be measured directly by the condenser method. Considerable difficulty has been experienced in the experimental determination of K_{AB}, chiefly because of the fact that the adsorption of minute traces of chemically active gases upon the surface of the metals greatly affects the results. Table 2, taken from the International Critical Tables, Vol. 6, p. 57, gives values of K_{AB} for a number of pairs of conductors.

Table 2. Contact Potential of Miscellaneous Pairs of Conductors, Condenser Method, Room Temperature

$K_{AB} = V_A - V_B$, unit = 1 volt

A	B	K_{AB}	A	B	K_{AB}
Al *	Fe	+0.87	C + NH_3 ¶	Cu	+0.079
Al *	Zn	+0.29	C + H_2 ¶	Cu	+0.096
Fe *	Zn	−0.60	C + N_2 ¶	Cu	+0.129
CuO *	Li	−1.52 †	C + CO_2 ¶	Cu	+0.130
		−1.11 ‡	C + NO ¶	Cu	+0.136
CuO *	Na	−2.52	C + O_2 ¶	Cu	+0.142
Cd §	Hg	−0.22	C + O_3 ¶	Cu	+0.155
Hg §	Sb	−0.26			

* In vacuum.
† Fresh surface.
‡ Old surface.
§ Initial value in dry air, pressure 0.05 mm Hg. K_{AB} varies with the time.
¶ Coconut charcoal saturated with the gas indicated.

THE PELTIER EFFECT. Heat is absorbed or evolved when a current passes through the junction of two metals, and reversal of the current causes the reversal of the heat effect. Conversely, if the two junctions of a closed circuit composed of metals A and B in series are kept at different temperatures, current flows around the circuit. Let the metals be so designated that current flows from A to B at the hot junction, and let P_{AB} be the heat absorbed when unit quantity of positive electricity flows from A to B at the temperature T. It may be shown thermodynamically that the temperature coefficient of the emf of the circuit is related to the Peltier heat P_{AB} by the equation

$$T\frac{dE_{AB}}{dT} = P_{AB}$$

where the subscripts indicate the direction of flow of the current at the hot junction.

P_{AB} is also known as the **Peltier coefficient** and as the Peltier electromotive force. The use of the term Peltier emf, with the understanding that the heat absorbed is equal to the emf located at the junction between the metals, is open to criticism; see Bridgman, *Phys. Rev.*, Vol. 14, 318 (1919).

Numerous data for the computation of the Peltier heat and the thermoelectric emf of a bimetallic circuit are to be found in the International Critical Tables, Vol. 6, pp. 213–229. For most pairs of metals P_{AB} is of the order of magnitude of a few millivolts; for antimony and bismuth, however, it is approximately 0.03 volt. Fair agreement exists between the values of P_{AB} directly observed and the values computed from the equation

$$P_{AB} = \mathrm{T}\frac{dE_{AB}}{dT}$$

LIQUID-LIQUID POTENTIALS. When two dissimilar electrolytic solutions are brought into contact, the phenomenon of diffusion takes place between them, and a difference of potentials is produced between the solutions, the magnitude of which depends upon the velocity of migration and the relative concentration of the ions taking part in the diffusion, the charge which they carry, and the temperature. The exact determination of liquid-liquid (also called liquid-junction) potentials is not possible, since the measured emf of a cell, e.g.,

$$\mathrm{Ag} \mid \mathrm{AgCl}(s) \quad \mathrm{HCl}\ (0.1\ N) \mid \mathrm{KCl}\ (0.1\ N) \quad \mathrm{AgCl}(s) \mid \mathrm{Ag}$$

is distributed between the electrode potentials and the liquid-junction potential. For a discussion of the problem see P. B. Taylor, *J. Phys. Chem.*, Vol. 31, 1478 (1927) and MacInnes, *The Principles of Electrochemistry*, Chapter 13.

If the liquid junction is formed in accordance with certain specifications, and if certain simplifying assumptions are made, it is possible to derive expressions for the liquid-junction potential. The Henderson equation applies to a junction of the "continuous mixture layer" type [Henderson, *Z. physik. Chem.*, Vol. 59, 118 (1907); Vol. 63, 325 (1908)], and the Planck equation to a "constrained diffusion" junction, i.e., a junction in which a steady distribution of concentrations is maintained by artificial means [Planck, *Ann. Physik*, Vol. 40, 561 (1890)].

For cells of the type

$$\text{Ag} \mid \text{AgCl}(s) \quad \text{M}'\text{Cl}(C) \mid \text{M}''\text{Cl}(C) \quad \text{AgCl}(s) \mid \text{Ag}$$

where M′Cl and M″Cl represent chlorides with univalent cations M′ and M″, present at concentration C, both equations give for the liquid-junction potential

$$E_L = \frac{RT}{F} \ln \frac{\Lambda'}{\Lambda''} = E$$

where Λ' and Λ'' are the equivalent conductances of the solutions. Some values of the emf (E_{obs}) found by MacInnes and Yeh [*J. Am. Chem. Soc.*, Vol. 43, 2563 (1921)] for such cells are listed in Table 3, together with the calculated values of the liquid-junction potential (E_L). For the junction between two solutions of the same electrolyte, of concentrations C_1 and C_2, e.g.,

$$\text{NaCl}(C_1) \mid \text{NaCl}(C_2)$$

both the Henderson and the Planck equation give

$$E_L = (2n_c - 1)\frac{RT}{F} \ln \frac{C_1}{C_2}$$

where n_c is the transference number of the cation.

Table 3. Liquid-junction Potentials

C = 0.1 mol per liter
T = 25 deg cent

Junction	E_{obs}, mv	E_L, mv
HCl:KCl.	+26.78	+28.52
HCl:NaCl. . . .	33.09	33.38
HCl:LiCl.	34.86	36.14
HCl:NH_4Cl. . .	24.80	28.57
KCl:LiCl.	8.78	7.62
KCl:NaCl. . . .	6.42	4.86
KCl:NH_4Cl. . .	2.16	0.046
NaCl:LiCl. . . .	2.62	2.76
NaCl:NH_4Cl. .	− 4.21	− 4.81
LiCl:NH_4Cl. .	− 6.93	− 7.57

Guggenheim [*J. Am. Chem. Soc.*, Vol. 52, 1315 (1930)] has discussed the reproducibility of potential of the "continuous mixture layer" junction and the "constrained diffusion" junction, as well as that of a third type, the "free diffusion" junction (here the transition layer is initially short in comparison to the distance between the electrodes, and unrestrained diffusion takes place). He found the potential of the constrained diffusion junction less reproducible than those of the other two types. He also found that, if the junction is made at the end of a thin tube dipping into a wider one, irregular fluctuations of several millivolts are obtained, whereas, if the junction is formed in a tube of uniform bore, giving cylindrical symmetry, the potentials are much more stable and reproducible.

In order to minimize the liquid-junction potential, it is customary to introduce a strong solution of some electrolyte, known as a salt bridge, between the solutions otherwise in direct contact, e.g., to replace the junction

$$\text{HCl}(C_1) \mid \text{NaCl}(C_2)$$

by

$$\text{HCl}(C_1) \mid \text{KCl (saturated)} \mid \text{NaCl}(C_2)$$

Saturated potassium chloride solution is selected for the salt bridge because of the fact that, with the potassium chloride solution much more concentrated than either solution in contact with it, the Henderson equation gives small values of E_L. It should be stated, however, that the assumptions on which the equation is based are not valid over a wide range of concentration; see MacInnes, *The Principles of Electrochemistry*, Chapter 13. By application of the "swamping salt" principle of Brönsted [Brönsted, *Z. physik. Chem.* Vol. 98, 239 (1921); Brönsted and Petersen, *ibid.*, Vol. 103, 307 (1922)] it is sometimes possible to cause two liquid-junction potentials to cancel. An illustration is furnished by the following cell [see Chase and Kilpatrick, *J. Am. Chem. Soc.*, Vol. 53, 1732 (1931)]:

Pt	quinhydrone HCl(c) KCl($m - c$)	KCl (saturated)	quinhydrone KCl($m - b$) NaBz(b) HBz (saturated)	Pt

Here Bz represents the benzoate ion, and HBz, benzoic acid; the quantities in parentheses are the concentrations, in mols per liter, of the various solutes. It will be seen that in each half-cell the ionic strength is (m); moreover, in each half-cell potassium chloride constitutes at least 9/10 of the total electrolyte. Chase and Kilpatrick assumed the liquid-junction potentials to cancel and computed the hydrogen-ion concentration in the buffer solution from the measured emf of the cell, and the concentration of hydrochloric acid (c).

A pair of vertical lines ‖ is commonly used to indicate that the liquid-junction potential is not included in the emf of the cell as stated, i.e., that a correction has already been made for the liquid-junction potential.

METAL-LIQUID POTENTIALS. From what has been said regarding the magnitude of metal-metal and liquid-liquid potentials, it follows by the process of elimination that the main seat of the electromotive force of a voltaic cell must reside at the junctions between metals and liquids. The determination of the *absolute* value of the potential difference between a metal and the solution with which it is in contact is problematical. In practice electrode potentials are stated relative to the potential of the normal hydrogen electrode, which is by definition zero.

NORMAL HYDROGEN ELECTRODE. The normal hydrogen electrode consists of a strip of platinum coated with a thin deposit of platinum black and saturated with hydrogen gas at atmospheric pressure. The electrode is mounted partly surrounded by hydrogen gas and partly dipping into an acid solution in which the mean ion activity (see below) is unity. By convention the potential of the normal hydrogen electrode is taken as zero at all temperatures.

CALOMEL ELECTRODES. Three calomel electrodes are in common use as reference electrodes, viz., the decinormal, the normal, and the saturated. The decinormal calomel electrode consists of mercury in contact with a 0.1 N solution of potassium chloride saturated with mercurous chloride, the mercurous chloride being present in excess as a solid. In the normal and the saturated electrodes the potassium chloride solution is 1 N and saturated, respectively. The potentials of these electrodes, referred to the normal hydrogen electrode, are as follows:

Calomel Electrodes

0.1 N KCl	$E = -0.3335 + 0.00007(t - 25)$
1.0 N KCl	$E = -0.2810 + 0.00024(t - 25)$
Saturated KCl	$E = -0.2420 + 0.00076(t - 25)$

where t is the temperature in degrees centigrade. The emf, E_c, of the cell

Electrode | Electrolytes | Calomel electrode

is converted to the scale of the normal hydrogen electrode by means of the equation

$$E_h = E_c + E_{\text{ref}}$$

where E_{ref} is the potential of the proper calomel electrode, as given above.

SIGN OF ELECTROMOTIVE FORCE. Two conventions in regard to the sign of emf are unfortunately in common use. The one used here is employed in most textbooks on thermodynamics and physical chemistry published in the United States.

If upon shortcircuiting the cell

Electrode 1 | Electrolytes | Electrode 2

the current (i.e., the positive ions) passes from left to right through the cell, so that the right-hand electrode is positive, the emf is positive. This means that the cell reaction, written as taking place from left to right,

$$a\text{A} + b\text{B} \rightarrow l\text{L} + m\text{M}$$

(i.e., a mols of A react with b mols of B to give l mols of L and m mols of M) is accompanied by a decrease in free energy. If the spontaneous reaction is the reverse reaction, the emf is negative.

SPECIFIC OR STANDARD ELECTRODE POTENTIALS. The emf of a cell (without liquid junction, or corrected for the liquid-junction potential) may be expressed as $E_1 - E_2$, where E_1 and E_2 are the potentials of the electrodes in the cell above, and the cell reaction may be written as the difference between the two electrode reactions. It is important to note that the reaction at each electrode is written as an oxidation, i.e., with the electron as a product. To illustrate, consider the cell

$$H_2\ (\text{Pt}) \mid \text{HCl} \mid Cl_2\ (\text{Pt})$$

$$1/2\ H_2 \rightarrow H^+ + e \qquad (1)$$

$$E_1 = E_1° - \frac{RT}{F} \ln \frac{a_{H^+}}{a_{H_2}^{1/2}}$$

$$Cl^- \rightarrow 1/2\ Cl_2 + e \qquad (2)$$

$$E_2 = E_2° - \frac{RT}{F} \ln \frac{a_{Cl_2}^{1/2}}{a_{Cl^-}}$$

the cell reaction being

$$1/2\ H_2 + 1/2\ Cl_2 \rightarrow H^+ + Cl^-$$

and the emf

$$E = E_1 - E_2 = E_1° - E_2° - \frac{RT}{F} \ln \frac{a_{H^+}a_{Cl^-}}{(a_{H_2}a_{Cl_2})^{1/2}}$$

When the ions are present at unit activity (mean ion activity, see below) and the gases at a pressure of 1 atm, the second term on the right drops out, and $E_1 = E_1°$, $E_2 = E_2°$, where $E_1°$ and $E_2°$ are the "standard" or "specific" electrode potentials. The great advantage in their use lies in the fact that by combination of standard electrode potentials the reversible emf and the decrease in free energy (nFE) may be set down for a new cell, in the same manner that thermal data for known reactions are combined to give the change in heat content for a new reaction.

Table 4, in which most of the values are taken from the International Critical Tables, Vol. 6, p. 332, gives the potential at an electrode dipping into a solution in which the molal activity (mean ion activity) of the ions in equilibrium with the electrode is unity. All gases are present at a pressure of 1 atm.

Table 4. Electrodes in Equilibrium with Aqueous Solutions in Which the Molal Activity of the Ions Indicated Is Unity

$T = 25$ deg cent

Electrochemical Reaction	Volts	Electrochemical Reaction	Volts
$Li \rightarrow Li^+ + e$	+2.96	$1/2\,H_2 \rightarrow H^+ + e$	0.0
$K \rightarrow K^+ + e$	+2.92	$Br^- + Ag \rightarrow AgBr + e$	−0.07
$Na \rightarrow Na^+ + e$	+2.72	$Cl^- + Ag \rightarrow AgCl + e$	−0.22
$Mg \rightarrow Mg^{++} + 2e$	+1.55	$Bi \rightarrow Bi^{+++} + 3e$	−0.23
$Al \rightarrow Al^{+++} + 3e$	+1.33	$Cu \rightarrow Cu^{++} + 2e$	−0.34
$Mn \rightarrow Mn^{++} + 2e$	+1.1	$2OH^- \rightarrow 1/2\,O_2 + H_2O + 2e$	−0.40
$Zn \rightarrow Zn^{++} + 2e$	+0.76	$I^- \rightarrow 1/2\,I_2(s) + e$	−0.54
$Fe \rightarrow Fe^{++} + 2e$	+0.44	$Fe^{++} \rightarrow Fe^{+++} + e$	−0.75
$Cd \rightarrow Cd^{++} + 2e$	+0.40	$Ag \rightarrow Ag^+ + e$	−0.80
$Co \rightarrow Co^{++} + 2e$	+0.28	$Hg \rightarrow 1/2\,Hg^{++} + e$	−0.80
$Ni \rightarrow Ni^{++} + 2e$	+0.23	$Br^- \rightarrow 1/2\,Br_2(l) + e$	−1.07
$Sn \rightarrow Sn^{++} + 2e$	+0.14	$Cl^- \rightarrow 1/2\,Cl_2 + e$	−1.36
$Pb \rightarrow Pb^{++} + 2e$	+0.12	$Au \rightarrow Au^{+++} + 3e$	−1.36
$1/2\,H_2 \rightarrow H^+ + e$	±0.0	$F^- \rightarrow 1/2\,F_2 + e$	−2.85

VARIATION OF ELECTRODE POTENTIALS WITH CONCENTRATION. The change in the value of an electrode potential with a change in the concentration of the ion with which the electrode is in equilibrium follows from the formulas given in the preceding section. The change may be generalized as

$$E' - E = \mp \frac{RT}{nF} \ln \frac{a'}{a} = \mp \frac{0.000198T}{n} \log_{10} \frac{m'\gamma'\pm}{m\gamma\pm}$$

Where the minus sign is to be used for cations, and the plus sign for anions; n is the valence of the ion; a' and a are the values of the activity of the ion; $\gamma'\pm$ and $\gamma\pm$ are the values of the activity coefficient (see below) at concentrations m' and m. An approximate value of $E' - E$ may be obtained by replacing the activity by the concentration of the ion.

ELECTROCHEMICAL SERIES—NOBILITY OF THE ELEMENTS. The elements, as arranged in Table 4, constitute the "electrochemical" series. Those elements with a negative standard electrode potential are said to be more "noble" than hydrogen; those with a positive standard electrode potential, to be less noble than hydrogen. The alkali metals, having the greatest tendency to form ions in water, are at one end, and the noble metals, such as gold, platinum, and palladium, having but a very slight tendency to form ions, are at the other end. The halogens, which go into solution as negative instead of positive ions, stand at the lower end of the series. Other series have been given which differ from the above in that the elements are not compared under similar conditions in regard to the concentration of the solution with which they are in contact. Changing the electrolyte, e.g., to potassium cyanide, not only alters the numerical values of the potentials but also may completely change the order in which certain elements occur in the series.

Any metal, if placed in a solution of a salt of a metal standing below it in the series, will tend to replace it; thus zinc precipitates iron, copper, silver, etc., from their solutions but does not displace the alkali metals from solutions of their salts. Any metal standing above hydrogen will tend to displace it from an acid solution with evolution of hydrogen. Metals below hydrogen in the series will not dissolve in acid by a simple replacement of hydrogen. From a chemical standpoint the adoption of the hydrogen electrode as a standard has the advantage over the calomel standard that it divides the metals into two groups according to their behavior toward acids.

The approximate emf of a battery consisting of two metals dipping into molal solutions of their salts may be computed at once from Table 4. Thus for the Daniell cell $E_{zn} - E_{cu} = 0.76 - (-0.34) = 1.10$ volts, and the positive sign indicates that zinc is oxidized and copper reduced:

$$Zn + Cu^{++} \rightarrow Zn^{++} + Cu$$

11. ELECTROMOTIVE FORCE AND HEAT OF REACTION

The electromotive force of a reversible cell maintained at constant temperature may be readily calculated from the first and second laws of thermodynamics. By a reversible cell is meant a cell which does not polarize and which operates under such conditions that changes which take place within the cell constitute a thermodynamically reversible process. The discussion which follows also applies, with a rough degree of approximation, to most commercial cells, which, as a rule, are not strictly reversible.

GIBBS-HELMHOLTZ EQUATION. In a reversible cell, such, for example, as the **Daniell cell,*** shown in Fig. 1, the relation between the heat of reaction and the electromotive force of the cell is calculated as follows.

From Faraday's laws the quantity of electricity which must flow through the cell in order to deposit or liberate 1 gram-ion at either electrode is nF coulombs, where n is the valence of the ion of highest valence involved in the chemical reaction, and F is the electrochemical constant (see above). Hence the external work done by the cell (when there is no other external work than electrical work) is nFE joules, where E is the emf of the cell. If the cell is kept at the same temperature (T degrees, absolute scale) as the surrounding bodies, then the external work nFE done by the cell is the maximum external work it can do at this temperature T. Hence

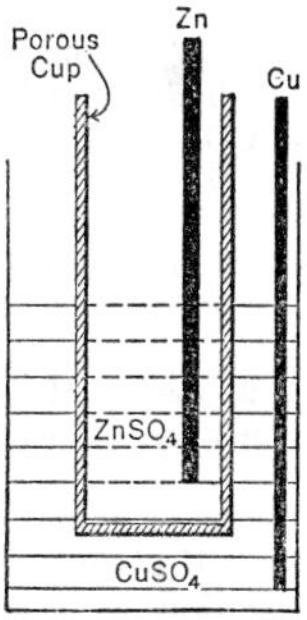

FIG. 1. Daniell Cell

$$nFE = nFT\left(\frac{\delta E}{\delta T}\right)_V - \Delta u$$

or

$$E = -\frac{\Delta u}{nF} + T\left(\frac{\delta E}{\delta T}\right)_V$$

where Δu is the heat of reaction (heat absorbed) at constant volume, expressed in joules per gram-ion, corresponding to the chemical change which takes place within the cell. Δu represents the gain in internal energy accompanying the reaction; it can be determined by an independent calorimetric measurement at constant volume.

If the cell operates at constant pressure rather than at constant volume,

$$E = -\frac{\Delta H}{nF} + T\left(\frac{\delta E}{\delta T}\right)_P$$

where ΔH, the heat of reaction at constant pressure, can be determined by an independent calorimetric measurement at constant pressure. For a condensed system like the above, Δu and ΔH are practically equal, but for reactions involving gases this is not the case.

TEMPERATURE COEFFICIENT OF ELECTROMOTIVE FORCE. From the foregoing it follows that the energy developed by the cell is equal to the heat energy of the reaction taking place within it, plus another term which may be positive or negative as the emf of the cell increases or decreases with a rise in temperature. If the emf on open circuit increases with a rise in temperature, then on closed circuit the cell will tend to cool; and, as its temperature falls, its emf will decrease. If the emf on open circuit decreases with a rise in temperature, then on closed circuit the cell will tend to become heated; and, as its temperature rises, its emf will decrease. A cell with temperature coefficient other than zero can work at constant temperature only by absorbing heat from the surroundings or by giving out heat to them.

Only for the unique conditions that the temperature coefficient of electromotive force is zero, or the absolute temperature is zero, is the electrical work equal to the heat energy of the reaction. It happens that the first of these conditions practically obtains in the Daniell cell, the emf being nearly independent of temperature. The observed emf is 1.10 volts, and the value computed from the heat of reaction, setting the temperature coefficient of the emf equal to zero, is 1.09 volts.

THOMSON'S RULE. In computing the emf of a battery from heats of reaction it is usually necessary to be content with a first approximation and to neglect the second

* The special construction of this cell renders it reversible.

term in the foregoing equations, since few data are available from which the temperature coefficient can be evaluated. The approximate equation

$$E = -\frac{\Delta H}{nF}$$

is known as Thomson's rule. If ΔH is expressed in kilogram-calories, it follows that, since $F = 96{,}500$ coulombs and 1 calorie = 4.185 joules,

$$E = -\frac{\Delta H \text{ (in kilogram-calories)}}{23.06n}$$

This rule also applies approximately to most commercial cells, although they are not strictly reversible. Of course, the heat of reaction used must correspond to the actual chemical changes which take place in the cell.

NERNST'S THERMODYNAMIC THEOREM. By assuming that the values of the intrinsic energy and free energy of a system not only become equal at the absolute zero but also approach equality asymptotically at this point, Nernst has been able to express the electromotive force (E) of a cell explicitly in terms of the thermal properties of the substances taking part in the reaction. So far as the formula has been tested, a satisfactory agreement has been found between theory and experiment. The principles involved are of fundamental importance and are fully discussed in Nernst's Silliman Lectures, "Applications of Thermodynamics to Chemistry." See also Nernst's *Theoretical Chemistry* and Lewis and Randall's *Thermodynamics*, Chapter 31.

CONCENTRATION CELLS. Concentration cells may be divided into two classes: (*a*) those without liquid junction (without transference), and (*b*) those with liquid junction (with transference).

As an illustration of the first class the cell

$$\text{Ag} \mid \text{AgCl}(s) \quad \text{HCl}(m'') \mid \text{H}_2 \mid \text{HCl}(m') \quad \text{AgCl}(s) \mid \text{Ag}$$

will serve. Here m'' represents the concentration of the hydrochloric acid in the more concentrated solution. The cell reaction is

$$\text{HCl}(m'') \rightarrow \text{HCl}(m')$$

and, if μ' and μ'' are the values of the chemical potential of hydrochloric acid in the two solutions,

$$\mu' - \mu'' = -nFE = RT \ln \frac{a'}{a''}$$

Since $a_{\text{HCl}} = a_{\text{H}^+}a_{\text{Cl}^-} = m_{\text{H}^+}m_{\text{Cl}^-}\gamma_\pm^2$, where $\gamma_\pm$ is the mean activity coefficient (see below), and since $n = 1$,

$$E = \frac{2RT}{F} \ln \frac{m''\gamma_\pm''}{m'\gamma_\pm'} \approx \frac{2RT}{F} \ln \frac{m''}{m'}$$

The same result may be obtained by setting down the emf of each half cell, with the cell reaction written with the right-hand electrode positive, and adding the two emfs.

As an illustration of the second class, the cell

$$\text{Ag} \mid \text{AgCl}(s) \quad \text{HCl}(m'') \mid \text{HCl}(m') \quad \text{AgCl}(s) \mid \text{Ag}$$

will serve. Let us suppose that to the left of the point A in the solution the concentration is constant and equal to m'', and to the right of the point B the concentration is constant and equal to m'. Between the two points the concentration varies continuously. If we let n_{H^+} be the transference number of the hydrogen ion, it can be shown that during the passage of 1 faraday of electricity n_{H^+} mols of hydrochloric acid are transferred from the region to the left of A through the boundary and into the region to the right of B, so that

$$-nFE = \int_A^B n_{\text{H}^+} d\mu$$

If we put in $n = 1$ and replace $d\mu$ by $RT\, d \ln a_{\text{HCl}} = 2RT\, d \ln m\gamma_\pm$ (see below),

$$E = -\frac{2RT}{F}\int_A^B n_{\text{H}^+} d \ln(m\gamma_\pm)$$

and, if n_{H^+} and $\gamma_\pm$ are taken as constant over the range of concentration involved,

$$E \approx \frac{2RT}{F} n_{\text{H}^+} \ln \frac{m''}{m'}$$

where the approximation is the better, the smaller the difference in concentration. It should be noted that the transference number appearing in the equation is that of the ion to which the electrode is not reversible.

For a discussion of the general case of cells with transference see Taylor, *J. Phys. Chem.*, Vol. 31, 1478 (1927). For the determination of transference numbers from cells with transference, see MacInnes and Beattie, *J. Am. Chem. Soc.*, Vol. 42, 1117 (1920).

CARBON GENERATOR. The problem of converting the energy liberated in the combustion of carbon to carbon dioxide directly into electric energy has not yet been practically solved. The reaction, $C + O_2 = CO_2$, liberates a very large quantity of heat, namely, 961 large calories per gram-atom of carbon consumed. Of this amount only about 10 per cent is converted into electrical energy through the agency of a steam engine and dynamo. The difficulties encountered in devising a commercial carbon generator are the following: (1) the velocity of the foregoing reaction is small, except at high temperatures; (2) it is difficult to utilize a gas as an electrode, except through the agency of a conducting medium which occludes it, such as platinum black; (3) a satisfactory high-temperature electrolyte which will not deteriorate in the presence of CO_2 is difficult to obtain; and (4) carbon has a slight tendency to form ions.

12. ELECTROLYTIC DECOMPOSITION

Faraday's laws enable one to determine the amount of the substances primarily liberated or deposited at the electrodes of any electrolytic cell, but they do not enable one to predict the nature of the secondary chemical actions which may take place. The factors which determine what products will be formed in any given cell are numerous. The chemical process occurring at the cathode is always of the nature of a reduction; at the anode, of an oxidation, these terms being used in their most general sense. It may be that the electrolyte, taken as a whole, undergoes no change, as, for example, when a copper sulfate solution is electrolyzed between copper electrodes. Here the chemical reaction at the cathode is the exact reverse of that at the anode. Generally, however, this is not the case; different chemical products are usually formed at the two electrodes and appear either in the form of gas or as a solid precipitate, or remain dissolved in the electrolyte.

DECOMPOSITION VOLTAGE OF ELECTROLYTES. When a gradually increasing difference of potential is applied to the electrodes of an electrolytic cell, electrolysis does not in general begin at once but only after a certain minimum electromotive force is reached, even though the cell has no open-circuit emf. It should be noted that this condition does not apply when the electrodes are such that the electrolyte as a whole undergoes no change as a result of electrolysis. The minimum voltage necessary to decompose a compound electrolytically is called its "decomposition voltage." It is influenced by a variety of factors and conditions.

Dilute solutions of most acids and bases have approximately the same decomposition voltage; this is due to the fact that it is the ions of water which are discharged at the electrodes. Salt solutions show considerable variation in decomposition voltage.

MEASUREMENT OF DECOMPOSITION VOLTAGE. Decomposition voltages have been experimentally determined in several ways. Certain investigators have taken the value necessary to produce visible electrolysis as the criterion; others have followed the polarization at the electrodes until it became constant and assumed this value as that at which electrolysis begins. Results obtained by the so-called ammeter-voltmeter method must be accepted with caution. In this method the value of the electromotive force applied to the cell is gradually increased from zero and plotted as abscissa, and the corresponding current through the cell is plotted as ordinate. The resulting curves have the general form shown in Fig. 2.

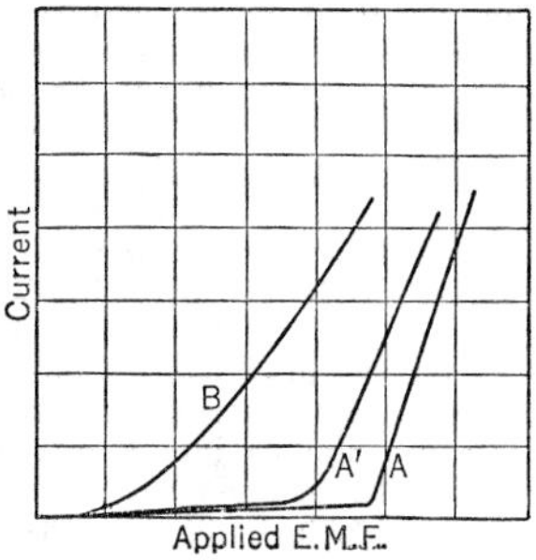

FIG. 2. Conductivity of Electrolytic Cells

The slope of these current-voltage curves depends upon the internal resistance of the cell and the size and the distance apart of electrodes. The critical voltage at which the current suddenly increases in value is, however, quite sharply defined with some electrolytes, e.g., silver nitrate between platinum electrodes, and other salts from which a metal is deposited. Below the critical voltage very little or no current passes through the cell, and no visible decomposition at the electrodes is apparent. When the point A (Fig. 2) is reached, visible electrolysis begins.

The sharpness of the bend in the curve and its course below A depend in large measure upon the sensitiveness of the galvanometer used for detecting the current, the size of electrodes, and the tendency of the products of the electrolysis to go back into solution.

RESIDUAL CURRENTS. With a sensitive reflecting galvanometer a steady deflection may be observed for days without any apparent electrolysis when an emf of a few tenths of a volt (far below the "decomposition voltage") is applied to platinum electrodes dipping into acidulated water. It might seem from this fact that Faraday's law is here violated. These slight currents are, however, to be explained as follows. The initial liberation of very small amounts of the products of reaction at anode and cathode sets up a back emf; but, as the products diffuse away from the electrodes, more ions discharge, yielding the small currents observed. At the decomposition voltage the back emf has reached its maximum value, and the current thereafter increases rapidly. Currents produced in this manner are called "residual" or "diffusion" currents; they play an important role in the electrolysis of fused salts and probably also in conduction in solid electrolytes. The larger the residual current the less sharply marked is the decomposition point (see A', Fig. 2). For many fused salts the current-voltage curve is of the form shown in B. It would be difficult to say from this curve at what voltage decomposition actually began.

OVERVOLTAGE. When a gas is evolved at either electrode, the relations which have been described are usually more complicated because of the phenomenon known as overvoltage. Thus in the electrolysis of an acid between polished platinum or other metal electrodes, hydrogen gas is evolved, to liberate which the cathode potential must be raised to a value in excess of that required to liberate hydrogen at a reversible (platinum black) electrode.

This excess of voltage which is necessary to liberate a gas at any electrode over that required to liberate the same gas against a *reversible* gas electrode is called overvoltage. Its value depends on the character of the material of the electrode against which the gas is set free, upon the nature of the gas, and upon the current density employed. Moreover, the value obtained depends to a considerable extent upon the method of measurement used. In the earlier work on overvoltage a continuous current was employed; see Caspari, *Z. physik. Chem.*, Vol. 30, 89 (1899); Müller, *ibid.*, Vol. 50, 641 (1905); Tafel, *ibid.*, Vol. 65, 226 (1909). Most of the recent work has been done using an intermittent current; see Newbery, *J. Chem. Soc.*, Vol. 109, 1051, 1066 (1916); Glasstone, *ibid.*, Vol. 123, 1745 (1923); Knobel, *J. Am. Chem. Soc.*, Vol. 46, 2613, 2751 (1924). The overvoltages obtained with a continuous current are higher than those obtained with an intermittent current and are probably more nearly correct; see Tartar and Keyes, *J. Am. Chem. Soc.*, Vol. 44, 557 (1922).

Varying conditions make the usual reproducibility of overvoltage determinations not better than 0.05 volt. For tables showing the overvoltages of hydrogen, oxygen, and the halogens against various metals see the International Critical Tables, Vol. 6, p. 339.

A study of the change of overvoltage with time has been made by MacInnes and Adler, *J. Am. Chem. Soc.*, Vol. 41, 194 (1919), and by Bowden and Rideal; see Bowden and Rideal, *Proc. Roy. Soc.*, Vol. A120, 59 (1928); Bowden, *ibid.*, Vol. A125, 446 (1929), Vol. A126, 107 (1929); Bowden and O'Connor, *ibid.*, Vol. A128, 317 (1930). The results of MacInnes and Adler indicate that small values of hydrogen overvoltage can be explained by concentration polarization; at high current densities the slow rate of formation of molecular hydrogen from atomic hydrogen is generally regarded as the major cause of the overvoltage. For theories of overvoltage see Lewis and Jackson, *Z. physik. Chem.*, Vol. 56, 193 (1906); Butler, *Trans. Farad. Soc.*, Vol. 19, 734 (1924), Vol. 28, 379 (1932); Erdey-Grúz and Volmer, *Z. physik. Chem.*, Vol. A150, 203 (1930); Hammett, *Trans. Farad. Soc.*, Vol. 29, 770 (1933); Gurney, *Proc. Roy. Soc.*, Vol. A134, 137 (1931), Vol. A136, 378 (1932); Fowler, *Trans. Farad. Soc.*, Vol. 28, 368 (1932); Kimball, Glasstone, and Glassner, *J. Chem. Phys.*, Vol. 9, 91 (1941).

ELECTROLYTIC SEPARATION OF METALS. The reason that two or more metals can be electrolytically separated from each other by a regulation of voltage follows from the foregoing discussion. Consider a solution containing two salts, say silver and copper nitrates, each at normal concentration. If a gradually increasing emf is applied to inert electrodes dipping into this solution, electrolysis will begin when a voltage is reached sufficient to set free simultaneously any anions and cations present in the solution. Reference to Table 4 will show that copper stands 0.46 volt above silver in the electrochemical series, and hence silver can be deposited from the solution with emf nearly half a volt less than copper.

No copper ions can separate until the applied emf has been increased 0.46 volt above that necessary to first separate the silver. The decomposition emf necessary will, of course, depend on the nature and the concentration of the anions present. As electrolysis proceeds, the solution becomes weaker and weaker with respect to silver ions, and the voltage necessary to separate them from the solution increases. If the electrolysis is continued until the silver remaining in solution has a concentration only 1/1,000,000 that at the start (the limit of analytical determinations), the change in voltage will be approximately 0.35 volt (see material on Variation of Electrode Potentials with Concentration, p. 4-12). This is still insufficient to raise the applied emf to a value sufficient to permit the copper to deposit, and hence, for all practical purposes, silver and copper may be completely separated from each other. This process may be greatly accelerated by violently stirring the solution, e.g., by a rapidly rotating electrode.

ELECTRODE REACTIONS. Below are listed a number of reactions of various types by which the current is conducted from the solution to the electrode, or vice versa. The ultimate products are often quite different from the ions which conduct the current through the solution.

At the cathode:

$$2H^+ + 2e \rightarrow H_2$$

$$Cu^{++} + 2e \rightarrow Cu$$

$$8H^+ + 2SO_4^= + 2e \rightarrow SO_3^= + S + 5H_2O$$

$$Ag(CN)_2^- + e \rightarrow Ag + 2CN^-$$

$$4H^+ + PbO_2 + 4e \rightarrow Pb + 2H_2O$$

$$\text{Quinone} + 2H^+ + 2e \rightarrow \text{Hydroquinone}$$

At the anode:

$$2Cl^- \rightarrow Cl_2 + 2e$$

$$4OH^- \rightarrow 2H_2O + O_2 + 4e$$

$$2S_2O_3^= \rightarrow S_4O_6^= + 2e$$

$$\begin{array}{l} COO^- \\ | \\ COO^- \end{array} \rightarrow 2CO_2 + 2e$$

$$2CH_3COO^- \rightarrow 2CO_2 + C_2H_6 + 2e$$

At the anode an oxidation (i.e., a withdrawal of electrons) takes place; at the cathode, a reduction (i.e., an addition of electrons).

Some of the foregoing electrode reactions are reversible, e.g., the reaction $2H^+ + 2e \rightarrow H_2$ at a platinized platinum electrode in an atmosphere of hydrogen, and the reaction Quinone $+ 2H^+ + 2e \rightarrow$ Hydroquinone at a quinhydrone electrode (quinhydrone is a compound of one molecule of quinone and one of hydroquinone); some, e.g., the oxidation of oxalate ion, are not. The principle of the reversible reactions is the same as that involved in the electrolytic dissolution and separation of metals. At high current densities, however, the reaction in question is often not sufficiently fast to carry the entire current, and evolution of gases occurs at the electrodes. What happens in a given case depends on the concentration of the solution, the current density, and the temperature, as well as on the electrode potential. Sometimes it is possible, by changing the material of the electrode and consequently the overvoltage, to avoid secondary reactions. Thus, when a platinized platinum electrode dips into titanous-titanic ion solution, the reversible emf is approximately −0.05 volt, which is not far below the standard electrode potential of hydrogen; and, if it is attempted to reduce the titanic ion cathodically to titanous, considerable hydrogen is evolved. By substituting an electrode with a large hydrogen overvoltage it is possible to avoid discharge of hydrogen and secure complete reduction of the titanic ion.

Most oxidations and reductions of organic compounds are irreversible. Those organic compounds which undergo reversible oxidation and reduction have as a rule a quinoid structure in the oxidized form and a benzenoid structure in the reduced form; *cf.* the foregoing quinone-hydroquinone reaction. An interesting fact was brought to light about 1930 by Michaelis [*J. Biol. Chem.*, Vol. 92, 211 (1931); Vol. 96, 703 (1932); *Chem. Rev.*, Vol. 16, 243 (1935)] and by Elema [*Rec. trav. chim.*, Vol. 50, 807 (1931)], namely, that the oxidation of an organic compound sometimes proceeds by loss of one electron rather than by the loss of two (*cf.* hydroquinone). For instance, the oxidation of tetramethyl-*p*-phenylenediamine yields the substance known as Wurster's blue, which is paramagnetic, indicating that one electron has been removed from the unshared pair of a nitrogen atom leaving a free radical known as a semiquinone [Michaelis, *loc. cit.*, and Michaelis and Granick, *J. Am. Chem. Soc.*, Vol. 65, 1747 (1943)]. It is possible that all oxidations and reductions proceed stepwise, one electron being acquired or given up in each step.

The electrolytic oxidation and reduction of certain inorganic ions and of many organic compounds are irreversible, and these processes are usually characterized by a marked change in efficiency with any variation in temperature, electrode material, and addition of catalysts and poisons, as well as with a variation in electrode potential and current density. It is believed that in these cases anodic oxidation is often brought about by the intermediate formation of hydrogen peroxide at the anode through the discharge of hydroxyl ions which are always present in aqueous solution,

$$OH^- \rightarrow OH + e$$

(where OH is the hydroxyl radical)

$$2OH \rightarrow H_2O_2$$

and that the hydrogen peroxide oxidizes the substance in question, e.g.,

$$H_2O_2 + 2S_2O_3^{=} \rightarrow S_4O_6^{=} + 2OH^-$$

(see Glasstone and Hickling, *Chem. Rev.*, Vol. 25, 407 (1939)]. The extent of oxidation or reduction usually increases with the electrode potential. Thus at a current density of 0.001 amp per sq cm and an electrode potential of -0.6 volt, nitrobenzene is reduced to azoxybenzene at a platinum cathode in an alkaline alcoholic solution, whereas at a current density of 0.0035 the electrode potential is -1.0 volt and hydrazobenzene is formed. Cathodic reductions can often be facilitated by the presence in solution of titanic, vanadic, or ferric ions, which are themselves reduced at the cathode and in turn reduce the substance in question. The action of these ions, whose oxidation and reduction involve one electron, is considered to be an argument for the theory that all oxidations and reductions proceed by one-electron steps.

PASSIVITY. A metal is said to assume the "passive" state when it comports itself toward acids like a noble metal, i.e., becomes insoluble. Iron affords a very striking example of this phenomenon, although it is exhibited by other metals as well. Thus, if dipped in strong nitric acid or if anodically polarized, iron becomes passive. Several theories have been advanced to account for the phenomenon, but it is at present generally accepted that passivity is caused by the presence of a thin insoluble film on the surface of the metal. It has been possible to isolate the film in some cases. Much of the experimental evidence supporting the theory of the protective film has been furnished by U. R. Evans and coworkers; see *J. Chem. Soc.*, Vol. 1929, 2651; Vol. 1930, 1361, 1773; *Nature*, Vol. 128, 1062 (1931); see also *Metallic Corrosion Passivity and Protection* by U. R. Evans, Arnold and Company, London. It is of interest that amalgamation causes iron, chromium, and nickel to become passive toward oxidizing agents; see A. S. Russell, D. C. Evans, and S. W. Rowell, *J. Chem. Soc.*, Vol. 1926, 1872. For a review of the field see Desch, *The Chemistry of Solids*, Cornell University Press, Chapter 7. For a presentation of the theories of passivity see *Trans. Am. Electrochem. Soc.*, Vol. 29, 217 (1916), and Förster's *Elektrochemie wässeriger Lösungen*, Barth, Leipzig, 1921.

ALTERNATING-CURRENT ELECTROLYSIS. The decomposition of an electrolyte and the solution of metal electrodes by electrolysis can be effected under certain conditions by an alternating current as well as by a direct current. The chief determining factors are the velocity with which the chemical reaction takes place and the periodicity of the alternating current.

13. THEORY OF SOLUTIONS

SOLUTIONS OF NON-ELECTROLYTES. From the work of Raoult on the depression of the freezing point, the elevation of the boiling point, and the lowering of the vapor pressure of solutions, the following general law was deduced for a binary system:

$$p_1 = p_1^\circ N_1$$

where p_1° represents the vapor pressure of pure solvent, and p_1 its vapor pressure when present in a solution in which its mol fraction is N_1.

The law of Henry, as originally stated in 1803, is

$$\frac{c_2}{p_2} = k$$

where c_2 is the concentration of the volatile solute, p_2 its partial pressure above the solution, and k a constant. In terms of mol fractions this becomes

$$p_2 = k'N_2$$

The third law found applicable to solutions of non-electrolytes is van't Hoff's law of osmotic pressure.

OSMOTIC PRESSURE. The earliest recorded observations of the process of osmosis seem to have been those of Abbé Nollet in 1748; he found that, if a glass vessel was filled with alcohol, the opening covered with a bladder, and the vessel immersed in water, the bladder expanded and burst. In terms of a semipermeable membrane the osmotic pressure may be defined as the equivalent of hydrostatic pressure produced when the solution and solvent, separated by a membrane permeable to the solvent, are in equilibrium, or as the equivalent of the excess pressure which must be imposed on a solution to prevent the passage into it of solvent through a semipermeable membrane.

SEMIPERMEABLE MEMBRANES. Membranes possessing the property of semipermeability are not only theoretically conceivable but may actually be prepared. Such a membrane is formed by the precipitate of copper ferrocyanide deposited within the walls of an unglazed porous cell when copper sulfate and potassium ferrocyanide solutions are allowed to diffuse into the cell from within and without respectively. An improved membrane is produced by previously depositing some inert material within the pores of the cell; see Grollman and Fraser, *J. Am. Chem. Soc.*, Vol. 45, 1710 (1923). The copper ferrocyanide membrane is permeable to water but impermeable to sugar and many other organic and inorganic substances. Most membranes, in fact, possess the property of semipermeability to a limited degree.

MEASUREMENT OF OSMOTIC PRESSURE. Figure 3 shows a simple apparatus for the measurement of osmotic pressure. The solution is contained in the closed vessel A, the walls of which are made semipermeable by depositing a membrane of copper ferrocyanide within the pores of the cell, and this vessel is immersed in the solvent B; an open manometer (M) of small bore is inserted in the top of A. Solvent passes from B into A, causing the liquid to rise in M. When equilibrium is established, the hydrostatic pressure gives a measure of the osmotic pressure of the solution then existing in the cell. As the entrance of an appreciable amount of solvent into the cell reduces the concentration of the solution, the pressure measured is less than that of the original solution; hence for quantitative work the open manometer is replaced by a closed mercury manometer. Another method of measuring osmotic pressure consists of applying to the solution gradually increasing pressure until the entry of the solvent is just prevented.

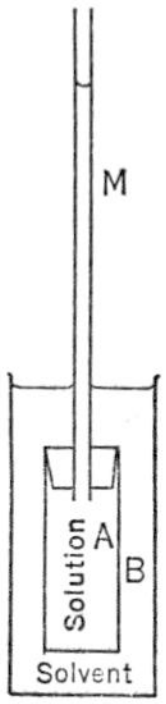

FIG. 3. Measurement of Osmotic Pressure

DEFINITION OF OSMOTIC PRESSURE WITHOUT REFERENCE TO A MEMBRANE. Osmotic pressure can also be defined in the following way. Let us consider a solution made up of the substance A and any number of other components. In a second vessel we have pure A. The partial pressure of A from the solution is less than from pure A at the same temperature, so that A will tend to distil from the pure substance into the solution. Distillation can be prevented by increasing the external pressure on the solution until the partial pressure of A from the solution is equal to that of pure A, or by decreasing the external pressure on pure A until the vapor pressures are equal. The difference in total pressure on the two vessels is called the osmotic pressure, and it can be shown that

$$\pi = \frac{RT}{\overline{V}_A} \ln \frac{P_A{}^\circ}{P_A}$$

where π is the osmotic pressure; $\overline{V}_A$ is the partial molar volume of A in the solution, which for dilute solutions can be taken as equal to the molar volume of pure liquid A; and P_A is the partial pressure of A above the solution, $P_A{}^\circ$ that above pure liquid A.

See Townend, *J. Am. Chem. Soc.*, Vol. 50, 2958 (1928), for a method of measuring osmotic pressure which is based upon these principles.

LAWS OF OSMOTIC PRESSURE IN NON-ELECTROLYTIC SOLUTIONS. The early experiments on the osmotic pressure of dilute aqueous solutions of many organic substances and of many solutions of non-electrolytes in organic solvents showed that the osmotic pressure (*a*) is independent of the *nature* of the solute, (*b*) is proportional, at a given temperature, to the concentration, and (*c*) is proportional, at a given concentration, to the absolute temperature. These facts suggest the gas laws, and van't Hoff demonstrated that an equal number of mols of an ideal gas, confined to a volume equal to that of the solution, will exert the same pressure as the osmotic pressure, i.e.,

$$\pi V = RT$$

where V is the volume of solution containing 1 mol of solute. This equation can be derived from the one given in the preceding section provided that the solvent obeys Raoult's law and the solution is quite dilute.

THE LAWS OF DILUTE SOLUTION; DEVIATIONS THEREFROM FOR SOLUTIONS OF NON-ELECTROLYTES. Activity. On the assumption that the vapors of solvent and solute obey the gas laws, any two of the following laws—Raoult's, Henry's, van't Hoff's—may be derived from the third. This may be done by means of Carnot cycles or by the general principles and formulas of Gibbs; see Guggenheim, *J. Phys. Chem.*, Vol. 34, 1751 (1930).

Most actual solutions, except at quite low concentrations, show deviations from these ideal laws, and in order to express the deviations concisely G. N. Lewis introduced the idea of activity. If the solvent obeys Raoult's Law, one may write

$$\mu_1 = \mu_1^\circ + RT \ln \frac{p_1}{p_1^\circ} = \mu_1^\circ + RT \ln N_1$$

where μ_1 is the chemical potential, or partial molar free energy of the solvent in the solution, and μ_1° that of pure liquid solvent at the same temperature and pressure. For a non-ideal solution the concentration of solvent is replaced by its activity

$$a_1 = N_1 f_1$$

where f_1 is the activity coefficient of the solvent, concentrations being expressed on the mol fraction scale, and $f_1 \rightarrow 1$ as $N_1 \rightarrow 1$. Similar equations may be set up using the other scales of concentration.

Although the activity and activity coefficient of a volatile solute may be expressed in the same way, it is customary to employ a different convention. Many cases are known where the solvent obeys Raoult's law when its mol fraction is large, say 0.9 to 1.0. It can be shown that in these solutions the solute obeys Henry's law, and that the chemical potential of the solute is given by the expression

$$\mu_2 = (\mu_2^\circ)' + RT \ln N_2$$

where $(\mu_2^\circ)'$ is a function of temperature, pressure, and the solvent and solute species but is independent of concentration over the range of concentration in question. At higher concentrations the solvent departs from Raoult's law, and for the solute

$$\mu_2 = (\mu_2^\circ)' + RT \ln N_2 f_2 = (\mu_2^\circ)' + RT \ln a_2$$

where $f_2 \rightarrow 1$ as $N_2 \rightarrow 0$.

The activity coefficient of the solvent in the solution may be obtained from any of the colligative properties of the solution, e.g., the vapor pressure, the osmotic pressure. The activity coefficient of the solute may be obtained from that of the solvent by means of the Gibbs-Duhem equation, as well as by measurement of the solubility of the solute, its distribution between two immiscible solvents, or its partial pressure. See Lewis and Randall, *Thermodynamics*.

SOLUTIONS OF ELECTROLYTES. The Theory of Electrolytic Dissociation. Arrhenius studied the conductivities of a large number of acids, bases, and salts in aqueous solution and noted that the specific conductivity did not vary directly with the concentration. When the data of Raoult on the lowering of the freezing point were published, Arrhenius put forward the theory of electrolytic dissociation. See Arrhenius, *Z. physik. Chem.*, Vol. 1, 631 (1887). He showed that α, the degree of dissociation of the electrolyte, calculated from the ratio of conductivities

$$\alpha = \Lambda/\Lambda_0$$

where Λ is the equivalent conductivity of the solution in question, and Λ_0 the equivalent conductivity of an infinitely dilute solution, could be explained on the basis that salts, acids, and bases dissociated into ions. Thus, for an electrolyte of the RX type,

$$\mathrm{RX} \rightleftharpoons \mathrm{R}^+ + \mathrm{X}^-$$

To express the fact that the osmotic pressure, freezing-point lowering, etc., for a solution of an electrolyte are greater than for a solution of a non-electrolyte, van't Hoff had introduced the coefficient i:

$$\pi V = iRT, \qquad i = \frac{\Delta T_f}{1.86}, \qquad \text{etc.}$$

and Arrhenius showed that i could be related to α by the equation

$$i = 1 + (k - 1)\alpha$$

where k is the number of ions formed by dissociation of a molecule. The agreement between α calculated from this equation and α taken from the conductance ratio was originally the chief support of the theory of electrolytic dissociation.

Ostwald showed that the dissociation of weak monobasic acids in aqueous solution followed the mass law,

$$\frac{\alpha^2}{(1-\alpha)v} = k$$

where α is the degree of dissociation and v the reciprocal of the total concentration.

It is to be noted in what follows that the experimental basis and subsequent support of the idea of electrolytic dissociation were largely confined to work in aqueous solutions. The arguments in favor of electrolytic dissociation may be summarized as follows:

1. Solutions of electrolytes show two independent sets of chemical properties, as if there were two independent constituents present. This fact is of course exceedingly important in analytical chemistry, particularly in qualitative analysis. Related to this is the fact that the heat of neutralization of a strong acid by a strong base, in dilute solution, is constant.

2. Certain physical properties of solutions (e.g., specific volume, density, color, index of refraction) are additive with respect to the ionic components. For instance, in a colorless solvent, solutions of salts with one colorless ion and a common colored ion show, at high dilution, the same absorption spectrum.

3. Solutions of electrolytes conduct electricity, and the equivalent conductance increases with dilution but approaches a limiting value.

4. The abnormally high freezing-point depressions, boiling-point elevations, and osmotic pressures of solutions of electrolytes are accounted for by the theory.

5. Those acids which give the most highly conducting solutions are also the most active chemically, i.e., they react upon metals the most rapidly, and they show the greatest catalytic effect in such reactions as the inversion of sucrose.

Lewis gives as the principal weaknesses of the theory [see Lewis, *Z. physik. Chem.*, Vol. 70, 212 (1910)]:

1. Lack of agreement in some cases between values for the degree of dissociation calculated from freezing points on the one hand and from conductivities on the other. Thus for $N/2$ LiCl, $MgCl_2$, and $Ca_2Fe(CN)_6$ the values of α from freezing-point data are 94, 99, and 2 per cent, respectively, whereas from conductivities they are 71, 62, and 20 per cent.

2. Additivity of properties of ions is observed even when considerable quantities of undissociated molecules are supposed to be present.

3. The most vulnerable point in the theory is the fact that there is not the slightest tendency to follow the classical dilution law in dilute solutions of strong electrolytes. In non-aqueous solutions the law is still less applicable.

Realizing that strong electrolytes in dilute aqueous solution do not conform to the classical mass action law, G. N. Lewis proposed that the concentration of the ions be replaced by their activity; and, since neither the concentration nor the activity of the undissociated molecules in a solution of a strong electrolyte can be determined, he proposed that by convention the activity of the undissociated molecules be set equal to the product of the activities of the ions, i.e., for an electrolyte of the RX type

$$\mu = \mu^+ + \mu^- = \mu^\circ + RT \ln a_+a_- = \mu^\circ + RT \ln m_+m_-\gamma_\pm^2$$

$$a = a_+a_- = m_+m_-\gamma_+\gamma_- = m_+m_-\gamma_\pm^2$$

$$\gamma_\pm = \sqrt{\frac{a_+a_-}{m_+m_-}}; \qquad \gamma_\pm \to 1 \quad \text{as} \quad m \to 0$$

where a is the activity of the undissociated molecule, m_+ and m_- are the stoichiometric molalities, and $\gamma_\pm$ is known as the mean activity coefficient. To illustrate, the activity of the hydrochloric acid in a solution 0.1 M in hydrochloric acid, 0.9 M in potassium chloride, is

$$a_{\mathrm{HCl}} = a_{\mathrm{H^+}}a_{\mathrm{Cl^-}} = (0.1)(1.0)\gamma_\pm^2$$

Either the molality or the molarity scale of concentration is generally employed for solutions of electrolytes; for the molality scale the activity coefficient is represented by $\gamma_\pm$, for the molarity scale, by $f_\pm$.

The methods employed in determining the activity coefficient of a strong electrolyte are the electrometric method, the solubility method (if the electrolyte is sparingly soluble), and the application of the Gibbs-Duhem equation as already mentioned. See Lewis and Randall, *Thermodynamics*, and Harned and Owen, *Physical Chemistry of Electrolytic Solutions*.

It is realized that the physical properties of solutions of electrolytes can frequently be explained on the basis that there are practically no molecules of the electrolyte present in the solution. This topic is discussed on p. 4-28 after the presentation of the section on electrolytic conduction.

14. THEORY OF ELECTROLYTIC CONDUCTION

The external effects of a current flowing through an electrolyte cannot be distinguished from those produced by a current of the same strength conducted by a metal. Thus the magnetic effect of a current flowing through a helical glass tube filled with electrolyte is the same as that produced by the same current flowing through an equivalent circuit of an equal number of turns of copper wire. A current may be induced in a closed ring of electrolyte just as in a ring of metal. Ohm's law and Joule's law hold for conduction in electrolytes as well as in metals. The mechanism of the conduction in the two media, however, is very different, as is shown by the phenomena produced at the junction of two conductors, in one of which the conduction is metallic or "electronic," and in the other electrolytic or "ionic."

EQUIVALENT CONDUCTIVITY. When a solution containing one equivalent weight of electrolyte is placed between parallel electrodes 1 cm apart and is diluted with solvent, the conductivity increases as the volume of the solution increases. Table 5 shows the increase in equivalent conductivity with dilution for a number of electrolytes.

For electrolytes like KCl, NaCl, and $AgNO_3$, i.e., for strong electrolytes, the value of the equivalent conductivity in infinitely dilute solution, Λ_0, is determined by extrapolation. Extrapolation is best made by means of the square-root law of Kohlrausch:

$$\Lambda = \Lambda_0 - a\sqrt{c}$$

i.e., the equivalent conductivity, plotted against the square root of the concentration, yields a straight line in dilute solution. This was established as an empirical rule by Kohlrausch; Onsager and Debye and Hückel have given it a theoretical basis. The extrapolation may be made from data obtained over a greater range of concentration if Shedlovsky's equation is employed [Shedlovsky, *J. Am. Chem. Soc.*, Vol. 54, 1405 (1932); Shedlovsky and Brown, *ibid.*, Vol. 56, 1066 (1934)].

That increase in the concentration should cause a decrease in conductivity appears reasonable enough; the increase may be thought of as due to incomplete ionization, to increase in viscosity of the solution and consequent decrease in the speed of the ions, or to retardation of the ions because of electrical effects. Of course, a combination of causes may be operating.

Table 5. Equivalent Conductance Λ of Typical Electrolytes Dissolved in Different Quantities of Water, at 18 Deg Cent

Concentration in Gm-equivalents per Liter = m = 1000η	Dilution in Liters per Gm-equivalent = ν = $\phi/1000$	KCl	NaCl	KNO_3	$AgNO_3$	1/2 $CuSO_4$	1/2 H_2SO_4	HCl	CH_3COOH	KOH	NH_3
0.0001	10,000	129.07	108.10	125.50	115.01	109.95	......	(378)	107	......	(66)
0.0002	5,000	128.77	107.82	125.18	114.56	107.90	......	(378)	80	......	53
0.0005	2,000	128.11	107.18	124.44	113.88	103.56	(368)	377	57	......	38.0
0.001	1,000	127.34	106.49	123.65	113.14	98.56	361	376	41	(234)	28.0
0.002	500	126.31	105.55	122.60	112.07	91.94	351	375	30.2	(233)	20.6
0.005	200	124.41	103.78	120.47	110.03	80.98	330	373	20.0	230	13.2
0.01	100	122.43	101.95	118.19	107.80	71.74	308	369	14.3	228	9.6
0.02	50	119.96	99.62	115.21	......	62.40	286	366	10.4	225	7.1
0.05	20	115.75	95.71	109.86	99.50	51.16	253	358	6.48	219	4.6
0.1	10	112.03	92.02	104.79	94.33	43.85	225	351	4.60	213	3.3
0.2	5	107.96	87.73	98.74	......	37.66	214	342	3.24	206	2.30
0.5	2	102.41	80.94	89.24	77.5	......	205	327	2.01	197	1.35
1	1	98.27	74.35	80.46	67.6	25.77	198	301	1.32	184	0.89
2	0.5	92.6	64.8	69.4	......	......	183.0	254	0.80	160.8	0.532
3	0.33	88.3	56.5	(61.3)	......	......	166.8	215.0	0.54	140.6	0.364
5	0.2	......	42.7	......	......	......	135.0	152.2	0.285	105.8	0.202

Note.—The 1/2 before $CuSO_4$ and H_2SO_4 indicates 1 gram-equivalent = 1/2 mol.

For complete tables consult Landolt and Börnstein, Kohlrausch and Holborn, and Tables annuelles internationales de constantes.

Because of the great increase of equivalent conductivity for electrolytes like NH_4OH and CH_3COOH when the solution is more dilute than 0.001 N, it is clear that, in order to obtain values of Λ from which to get Λ_0 by extrapolation, one would have to work with impossibly dilute solutions. The limiting value of the equivalent conductivity of a weak electrolyte is accordingly obtained in another way.

THE LAW OF INDEPENDENT MIGRATION OF IONS. Table 6 shows the limiting values of the equivalent conductivity for a series of potassium salts and for a series of calcium salts at 18 deg cent. The third and sixth columns of the table give the difference between the Λ_0 values of two salts with a common cation.

Table 6. Limiting Equivalent Conductivities of Some Salts of Potassium and Calcium

Salt	Λ_0	Difference	Salt	Λ_0	Difference
KCl	130.1		$CaCl_2$	116.5	
		− 1.5			− 1.5
KBr	131.6		$CaBr_2$	118.0	
		5.3			5.3
KNO_3	126.3		$Ca(NO_3)_2$	112.7	
		26.7			26.7
$KC_2H_3O_2$	99.6		$Ca(C_2H_3O_2)_2$	86.0	
		1.1			1.1
KIO_3	98.5		$Ca(IO_3)_2$	84.9	
		−34.1			−34.1
K_2SO_4	132.6		$CaSO_4$	119.0	

Since the differences between the values of Λ_0 for KCl and KBr, and $CaCl_2$ and $CaBr_2$, are the same, although the individual values are quite different, Λ_0 must in each case be the sum of two constant quantities. That is to say, in infinitely dilute solution and at a given temperature, the current-carrying capacity of one equivalent of potassium ions is a fixed and constant quantity independent of the anion present. The limiting equivalent ionic conductance of the potassium ion is represented as l_{K^+}. The equivalent conductivity of an electrolyte at infinite dilution is equal to the sum of two constants, the equivalent conductances of anion and cation at infinite dilution,

$$\Lambda_0 = l_a + l_c$$

a law established by Kohlrausch and known as the law of the independent migration of ions.

This law immediately suggests a method of evaluating Λ_0 for electrolytes of the NH_4OH—CH_3COOH type, i.e., for weak electrolytes.

$$\Lambda_0 \text{ for } CH_3COOH = l_{H^+} + l_{CH_3COO^-}$$

$$\Lambda_0 \text{ for } CH_3COOH = \Lambda_0 \text{ for } HCl + \Lambda_0 \text{ for } CH_3COONa - \Lambda_0 \text{ for } NaCl$$

From such a table of limiting equivalent conductivities as Table 6 for the potassium and the calcium salts, it is evident that, if the absolute value of the limiting ionic conductance were known for one ion, all the other limiting ionic conductances might at once be calculated.

TRANSFERENCE NUMBERS OR TRANSPORT RATIOS. The current is carried through the solution by the simultaneous movement of the anions toward the anode, the cations toward the cathode. If the anions are faster moving than the cations, they carry the greater part of the current. For q coulombs of electricity passing, $q/96{,}500$ = number of equivalents of electricity passing = $N_t = N_a + N_c$, the number of equivalents carried by the anion, plus the number carried by the cation. In the solution in question the ratio N_a/N_t is called the transference number or the transport ratio of the anion, and the ratio N_c/N_t is called the transference number or the transport ratio of the cation. The sum of the transference numbers is unity.

The transference numbers are found from the changes in concentration which take place about the electrodes when a direct current is passed through the solution of an electrolyte. Consider the electrolysis of a solution of $AgNO_3$ between silver electrodes. When N_t equivalents of electricity have passed, N_t equivalents of silver have gone from the anode into the solution surrounding the anode, and the same number of equivalents of silver have plated out onto the cathode from the solution surrounding the cathode. N_c equivalents of silver have during this time moved out from the anode portion of the solution, and the same number have entered the cathode portion. The anode portion has gained $N_t - N_c$ equivalents of silver ion, and the cathode portion has lost the same number. The concentration of nitrate ion in the two portions has changed in the same way. Analysis of either electrode portion, together with the coulometer reading, yields directly N_c/N_t, the transference number of the silver ion in a silver nitrate solution of the concentration given. For uni-univalent salts in solutions less concentrated than 0.01 N there is very little change in the transference numbers of the ions with change in concentration. Table 7 gives the transference numbers of the anions of a number of electrolytes in aqueous solution at 18 deg cent.

Table 7. Transport Ratios n_a of Anions in Aqueous Solutions at about 18 Deg Cent

Values in parentheses are somewhat uncertain. (From Le Blanc's *Electrochemistry*)

Gram-equivalents per liter Liters per Gram-equivalent	0.01 100	0.05 20	0.2 5	1 1	2 0.5	5 0.2
K{Cl, Br, I}, NH_4Cl	0.506	0.506	......	......	......	......
$NaCl$	0.604	0.604	......	......	......	......
$LiCl$	0.670	0.680	0.697	......	......	......
KNO_3	......	......	0.496	0.487	0.479	......
$NaNO_3$	......	......	0.614	(0.611)	(0.608)	0.585
$AgNO_3$	0.528	0.528	0.527	0.501	0.476	......
$KC_2H_3O_2$	......	......	0.33	(0.331)	(0.332)	0.335
$NaC_2H_3O_2$	......	......	(0.43)	(0.425)	0.421	......
KOH	......	......	0.736	(0.740)	......	......
$NaOH$	......	(0.81)	(0.82)	0.825	......	......
$LiOH$	......	......	0.85	(0.873)	......	......
HCl	0.167	0.165	......	......	......	......
HNO_3	0.170	0.170	0.170	......	......	......
1/2 $BaCl_2$	0.553	......	......	......	......	......
1/2 $CaCl_2$	(0.58)	(0.61)	(0.66)	0.686	(0.700)	0.737
1/2 $MgCl_2$	......	(0.63)	0.68	0.709	(0.729)	0.776
1 2 $CdCl_2$	0.570	0.570	(0.65)	(0.72)	0.745	0.865
1/2 CdI_2	0.558	0.606	0.86	......	......	......
1/2 $Ba(NO_3)_2$ at 25°	......	0.544	0.545	......	......	......
1/2 K_2CO_3	......	(0.39)	(0.41)	0.434	0.413	(0.380)
1/2 Na_2CO_3	......	(0.52)	(0.53)	0.548	(0.542)	......
1/2 K_2SO_4	0.505	0.512	......	......	......	......
1/2 Na_2SO_4	0.610	0.624	......	......	......	......
1/2 $CdSO_4$	0.616	0.635	0.672	0.746	......	......
1/2 $MgSO_4$	0.620	0.633	(0.66)	0.74	(0.76)	......
1/2 $CuSO_4$	......	0.625	0.657	......	......	......
1/2 H_2SO_4	0.176	0.175	......	......	......	......

Note.—The 1/2 before the various bivalent electrolytes indicates that 1 gram-equivalent = 1/2 mol.

Transference numbers may also be evaluated by measurement of the electromotive force of concentration cells with transference [see, for instance, MacInnes and Parker, *J. Am. Chem. Soc.*, Vol. 37, 1445 (1915); and MacInnes and Beattie, *ibid.*, Vol. 42, 1117 (1920)], and by the moving boundary method [see MacInnes, Cowperthwaite, and Blanchard, *J. Am. Chem. Soc.*, Vol. 48, 1909 (1926)]; a review of the problem of the determination of transference numbers is given by MacInnes and Longsworth, *Chem. Reviews*, Vol. 11, 172 (1932) and by MacInnes, *Principles of Electrochemistry*, Chapter 4. For a more complete table of transference numbers and a comparison of the results obtained by the different methods, see the International Critical Tables, Vol. 6, pp. 309–311.

LIMITING IONIC CONDUCTIVITIES. Since N_c/N_t represents the fraction of the current carried by the cation, $\Lambda N_c/N_t$ is the equivalent conductance of the cation. The limiting value of $\Lambda N_c/N_t$, i.e., the conductance of one equivalent of the cation in an infinitely dilute solution, is l_c. From conductivity data and transference data, the tables of limiting equivalent ionic conductivities are built up. Table 8 gives values of the equivalent conductivities of some typical ions at infinite dilution and 18 deg cent.

THE VELOCITY OF MIGRATION OF IONS. A charged body in an electric field is acted upon by a force equal to the product of the charge and the intensity of the field. If the field changes in one direction only, the force acting is

$$f = q\,\frac{dE}{dx}$$

where q is the charge of the particle, and dE/dx the gradient of the potential. The speed with which a small body moves through a medium of great frictional resistance is proportional to the force acting upon it. Therefore the speed u_c with which the cation moves through the solution is

$$u_c = kq\frac{dE}{dx}$$

The proportionality factor kq is written U_c and is the velocity of migration of the ion under unit potential gradient, or the mobility.

Table 8. Conductivities of Typical Ions at Infinite Dilution and 18 Deg Cent

Values at $t°$ may be computed by the formula $l_t = l_{18}[1 + \alpha(t - 18) + \beta(t - 18)^2]$. (Kohlrausch)

Anions	l_{18}	α	β	Cations	l_{18}	α	β
F	46.6	0.0232	0.000094	Li	33.4	0.0261	0.000142
Cl	65.5	0.0215	0.000067	Na	43.5	0.0245	0.000116
Br	67.0	....	....	K	64.6	0.0220	0.000075
I	66.5	0.0206	0.000052	Rb	67.5	0.0217	0.000069
SCN	56.6	....	....	Cs	68.0	....	....
ClO_3	55.0	0.0207	0.000054	NH_4	64.0	0.0223	0.000079
BrO_3	46.0	....	....	Tl	66.0	....	....
IO_3	33.9	0.0233	0.000096	Ag	54.3	0.0231	0.000093
ClO_4	64.0	....	....	H	315.0	0.0154	−0.000033
IO_4	48.0	....	....	1/2/Ba	55.0	0.0239	0.000106
NO_3	61.7	0.0203	0.000047	1/2 Sr	51.0	....	....
MnO_4	53.4	....	....	1/2 Ca	51.0	....	....
OH	174.0	0.0179	0.00008	1/2 Mg	45.0	0.0255	0.000132
CHO_2	47.0	....	....	1 2 Zn	46.0	0.0256	0.000133
$C_2H_3O_2$	35.0	0.0236	0.000101	1/2 Cd	46.0	....	....
1/2 SO_4	68.0	0.0226	0.000084	1/2/Cu	46.0	0.0240	0.000107
1/2 Cr_2O_7	72.0	....	....	1/2 Pb	61.0	0.0244	0.000114
1/2 CO_3	60.0	0.0269	0.000155				
1/2 $(COO)_2$	63.0	....	....				

In the solution of an incompletely dissociated electrolyte, one may say that only part of the ions are free, part being combined into molecules, or one may say that each ion is free only part of the time, and, if α is the degree of dissociation, the *measured* velocity of migration of the cation is

$$\alpha u_c = \alpha U_c\frac{dE}{dx}$$

In this case αU_c is called the mobility of the cation constituent.

Let a constant current I pass for t seconds through a solution of a uni-univalent electrolyte of concentration C equivalents per liter. Let the solution be contained in a cylindrical tube of cross-section A, and let the distance between the electrodes be l. Then the number of equivalents of the ion constituents passing through any cross-section is

$$N_c = \alpha U_c At(0.001C)\frac{dE}{dx}$$

$$N_a = \alpha U_a At(0.001C)\frac{dE}{dx}$$

and the total number of equivalents of electricity passing is $N_t = N_c + N_a$, which is N_tF coulombs, where F is the faraday. Since for constant current $dE/dx = E/l$, and since

$$It = FN_t = \frac{E}{r}t = Et\kappa\frac{A}{l}$$

where κ is the specific conductivity of the solution, it follows that

$$\kappa = F(\alpha U_c + \alpha U_a)(0.001C)$$

or

$$\Lambda = 1000\frac{\kappa}{C} = F(\alpha U_c + \alpha U_a)$$

The transference numbers are

$$n_c = \frac{N_c}{N_t} = \frac{\alpha U_c}{\alpha U_c + \alpha U_a}; \qquad n_a = \frac{N_a}{N_t} = \frac{\alpha U_a}{\alpha U_c + \alpha U_a}$$

and since, by the equation above,

$$(\alpha U_c + \alpha U_a) = \frac{\Lambda}{F}$$

the mobilities of the ion constituents are

$$\alpha U_c = n_c \frac{\Lambda}{F}; \qquad \alpha U_a = n_a \frac{\Lambda}{F}$$

The mobilities thus calculated are in fair agreement with those obtained by the moving boundary method; for example, in a 0.1 N silver nitrate solution

$$(\alpha U_c)_{\text{calc}} = 4.6, \qquad (\alpha U_c)_{\text{obs}} = 4.9 \times 10^{-4} \text{ cm/sec per volt per cm}$$

and in 0.07 N acetic acid solution

$$(\alpha U_c)_{\text{calc}} = 0.48, \qquad (\alpha U_c)_{\text{obs}} = 0.65 \times 10^{-4} \text{ cm/sec per volt per cm}$$

[Whetham, *Phil. Trans.*, *A*, Vol. 184, 340 (1893)].

It is frequently desired to estimate the specific conductance of a solution containing several electrolytes. Proceeding as for the uni-univalent solute above, we can show that, if z_A, z_B, $\cdots$ are the valences of the various ionic species present in the solution at concentrations m_A, m_B, $\cdots$ mols per liter (where for incompletely dissociated electrolytes the extent of dissociation is taken into account in computing the ionic concentrations), and U_A, U_B, $\cdots$ are their speeds under unit potential gradient,

$$1000\kappa = F\Sigma z_A m_A U_A + z_B m_B U_B + \cdots$$

The speed of any ion is affected by the presence of the other ions, but for many purposes it is sufficiently accurate to set $FU_J = l_J$ (the equivalent conductance of an ion of species J in an infinitely dilute solution), and as a first approximation

$$1000\kappa = \Sigma z_A m_A l_A + z_B m_B l_B + \cdots$$

THE CONDUCTANCE RATIO FOR STRONG AND FOR WEAK ELECTROLYTES. If the conductance ratio, Λ/Λ_0, is examined for a large number of strong and weak electrolytes, it is at once evident that the strong electrolytes possess large values of the ratio Λ/Λ_0 as compared with the weak. For most uni-univalent salts and for strong uni-univalent acids and bases in aqueous solution, $\Lambda_{0.1\,N}/\Lambda_0$ is about 0.8. For bi-bivalent salts $\Lambda_{0.1\,N}/\Lambda_0$ is 0.4–0.5. Among the weak electrolytes, for NH_4OH and CH_3COOH in aqueous solution, $\Lambda_{0.1\,N}/\Lambda_0 = 0.01$; for HClO and HCN, $\Lambda_{0.1\,N}/\Lambda_0 < 0.002$; and for the mercuric halides, $\Lambda_{0.1\,N}/\Lambda_0 < 0.001$.

In the older textbooks the conductance ratios are regarded as measures of the degree of ionization, uni-univalent electrolytes of the KCl–$AgNO_3$ type (strong electrolytes) being said to be about 80 per cent ionized in 0.1 N solution. The weak electrolytes of the same valence type possess very much lower $\Lambda_{0.1\,N}/\Lambda_0$ values and are therefore considered much less ionized. There is far wider variation in the $\Lambda_{0.1\,N}/\Lambda_0$ ratio for weak electrolytes than for strong electrolytes of the same valence type. Examination of the ratio for a large number of uni-univalent salts shows that the deviation from the mean seldom exceeds 1 or 2 per cent, whereas changes in the conductance ratio for weak organic acids amount to hundreds of per cent.

It is obvious that the ratio $\Lambda_{0.1\,N}/\Lambda_0$ represents the degree of ionization of the electrolyte in 0.1 N solution only if in 0.1 N solution both ions move with the same speed as they do at infinite dilution.

Many of the Λ/Λ_0 ratios recorded in the literature are subjected to a viscosity correction. By Stokes's law the speed with which a small particle moves through a medium of great frictional resistance is inversely proportional to the viscosity of the medium. When one corrects for the effect of viscosity on the speed of the ions, the so-called degree of ionization becomes $\Lambda\eta/\Lambda_0\eta_0$, where η is the viscosity, rather than Λ/Λ_0.

APPLICATION OF THE MASS LAW TO STRONG AND TO WEAK ELECTROLYTES. There is another important difference in behavior between strong and weak electrolytes. For a uni-univalent electrolyte the dilution law of Ostwald (the law of mass action in terms of concentrations) is

$$K_c = \frac{c_{A^+}c_{B^-}}{c_{AB}} = \frac{\alpha^2 C}{1 - \alpha}$$

where α is the degree of ionization and C the stoichiometric concentration. This differs from the expression of the mass law by lack of the activity coefficient ratio, $f_{A^+}f_{B^-}/f_{AB}$.

Now when α is represented by Λ/Λ_0 or, if there is a considerable change in the viscosity, by $\Lambda\eta/\Lambda_0\,\eta_0$, for weak electrolytes in aqueous solution fairly constant values are obtained for K_c. Perhaps the classical example is acetic acid. As the concentration of acetic acid increases from 0.0001 to 1 M, K_c decreases from 1.87×10^{-5} to 1.74×10^{-5} (13 parts in 200, 7 per cent).

A strong electrolyte of the same valence type over the same concentration range makes no pretense of obeying the dilution law. Nor does the inclusion of the ratio of activity coefficients

$$K_a = \frac{c_{A^+}c_{B^-}f_{A^+}f_{B^-}}{c_{AB}f_{AB}} = \frac{K_c f_{A^+}f_{B^-}}{f_{AB}}$$

yield a constant. Since the law of mass action is a thermodynamic law, it can only be concluded that the ratio Λ/Λ_0 does not represent with any accuracy the degree of ionization of strong electrolytes.

MODERN THEORIES OF SOLUTION. Van Laar [*Arch. Teyler* (2) 7, I, 1 (1900)], Sutherland [*Phil. Mag.*, Vol. 14, 1 (1907)], and Bjerrum [*Z. Elektrochem.*, Vol. 24, 231 (1918)], from different points of view, suggested that the dissociation of strong electrolytes was much greater than appeared from the usual calculation on the basis of conductivity data. Milner [*Phil. Mag.*, Vol. 23, 551 (1912); Vol. 25, 743 (1913)] attempted to calculate quantitatively the effect of interionic forces, but because of mathematical difficulties did not arrive at a strict solution of the problem.

In 1923 Debye and Hückel [*Physik. Z.*, Vol. 24, 185 (1923); Vol. 25, 97 (1924)] presented what appears to be a satisfactory solution of the problem. Briefly, their reasoning is as follows. Because of the Coulomb forces the "ionic atmosphere" surrounding any selected positive ion contains an excess of negative electricity. Consequently an average electrical potential and an average electrical density different from zero exist at any distance from the selected ion, the total potential and density being interrelated by the Boltzmann distribution principle and the Poisson equation. From this it is possible to calculate the increase in energy attending the removal of the n molecules constituting a gram-molecule of any kind of ion of valence z from the "ionic atmosphere" (all other ions being simultaneously removed, their removal being attended by no other energy effects) by infinitely diluting the solution. Without following the mathematical derivation, it may simply be stated that Debye and Hückel find that, for water as a solvent, and at 25 deg cent

$$-\log_{10} f = 0.5z^2\sqrt{\mu}$$

where f represents the activity coefficient, z the valence of the ion, and μ the ionic strength defined by the equation $\mu = {}^1/_2\Sigma c_i z_i^2$. This limiting equation has been verified experimentally for aqueous solutions.

Debye and Hückel extended the theory to the problem of conductivity also [*Physik. Z.*, Vol. 24, 305 (1923)]. They showed that two factors cause a decrease in the mobility of the ions. The first is an "ion effect" due to the fact that the thermal equilibrium of the ions in the different configurations is disturbed by the migration of the ions, equilibrium being restored with finite velocity. The unsymmetrical arrangement of the ions creates an electric field which opposes the applied potential difference. The second effect arises from the tendency of the ions to drag along solvent molecules, according to the hydrodynamic laws for viscous media. About a given ion is found an excess of ions of opposite sign, so that the given ion has to move relative to a current of medium flowing in the opposite direction.

Taking into consideration the diminution in the mobility of an ion arising from these causes, Onsager [*Physik. Z.*, Vol. 27, 388 (1926); Vol. 28, 277 (1927)] derived the square-root law of Kohlrausch

$$\Lambda = \Lambda_0 - a\sqrt{c}$$

and showed that a, the slope of the line, can be evaluated theoretically, the calculated and observed values of a being in good agreement.

For weak electrolytes the proper correction of the mobility for the effect of the "ionic atmosphere" has made possible a more accurate evaluation of the concentration of the ions. See, for example, Davies, *Phil. Mag.* (7), Vol. 4, 244 (1927); Shedlovsky, Brown, and MacInnes, *Trans. Am. Electrochem. Soc.*, Vol. 66, 165 (1934), and MacInnes, *Principles of Electrochemistry*, pp. 342–348.

Further support for the interionic attraction theory is furnished by studies of conductance in electric fields of very high intensity and of very high frequency. Wien [*Ann. Physik.* (5), Vol. 1, 400 (1929)] found that in fields of the order of 10^6 volts per cm Ohm's law fails, and the conductance of strong electrolytes increases with field strength; this phenomenon is explained as due to the fact that the ions are moving so rapidly that there is not time for the ionic atmosphere to form. As had been predicted by Debye and Falkenhagen [*Physik. Z.*, Vol. 29, 121, 401 (1928)], it was demonstrated by Sack [*ibid.*, Vol. 29, 627 (1928)] that at frequencies of the order of 10^8 sec.$^{-1}$ the conductance increases with the frequency of the field; this increase is explained as due to the fact that with the rapid alternation of motion of the ions the ionic atmosphere about a given ion has not time to acquire asymmetry, and thus one of the above-mentioned causes of retardation is removed.

To summarize: Because of interionic forces the older method of determining the degree of dissociation of strong electrolytes is incorrect. All the evidence seems to indicate that in aqueous solutions strong uni-univalent electrolytes are practically completely dissociated. At the present time the corresponding dissociation constants are not known, because there is no physical property which is unquestionably due to undissociated molecules. In regard to weak electrolytes, correction of the mobility for the retardation caused by interionic forces leads to a more exact evaluation of the ionic concentrations.

EFFECT OF TEMPERATURE ON IONIZATION. The percentage to which a substance is ionized in solution may increase or decrease with the temperature. The sign and magnitude of the effect depend upon whether the ionization reaction is accompanied by an absorption or evolution of heat. Substances which dissociate with evolution of heat become less ionized with increasing temperature; substances which dissociate with absorption of heat become more ionized with rising temperature. Water belongs to the latter class, and phosphoric acid (first dissociation) to the former; for acetic acid and many other carboxylic acids the dissociation constant exhibits a maximum in the range of temperature investigated. The figures in Table 9 are taken from the work of Harned and his associates [for acetic acid, see Harned and Ehlers, *J. Am. Chem. Soc.*, Vol. 54, 1350 (1932); Vol. 55, 652 (1933); for phosphoric acid, Nims, *ibid.*, Vol. 56, 1110 (1934); and for water, Harned and Hamer, *ibid.*, Vol. 55, 2194 (1933)].

Table 9. Effect of Temperature on Ionization

Temperature, Degrees Centigrade							
0	10	20	25	30	40	50	60
		Acetic Acid, $K \times 10^5$					
1.657	1.729	1.753	1.754	1.750	1.703	1.633	1.542
		Phosphoric Acid, $K_1 \times 10^3$					
8.983	8.519	7.896	7.537	7.152	6.330	5.475	......
		Water, $K \times 10^{14}$					
0.115	0.293	0.681	1.008	1.471	2.916	5.476	9.614

The effect of temperature on α may be computed through its effect on K by means of van't Hoff's thermodynamic relation

$$\frac{d \log K}{dt} = \frac{Q}{RT^2}$$

where K is the equilibrium constant of the ionization reaction, Q the heat of the reaction, R the gas constant, and T the absolute temperature.

NEGATIVE TEMPERATURE COEFFICIENTS OF ELECTRICAL CONDUCTIVITY. The fact that certain electrolytes, e.g., a phosphoric acid solution, may have a negative temperature coefficient is readily explained in terms of the foregoing relations. If the increase in the velocity of migration of the ions with rising temperature is more than offset by a diminution in the average number of free ions, resulting from a decrease in ionization, the conductivity of the solution will diminish. By combining solutions having positive and negative temperature coefficients in suitable proportions, electrolytes have been prepared which have nearly a zero temperature coefficient over quite a range of temperature. The following mixture, proposed by Manganini, has this property: 121 grams mannite, 41 grams boracic acid, 0.06 gram potassium chloride dissolved in sufficient water to make 1 liter. Its specific conductance at 18 deg. cent is $\kappa = 0.00097$. Such a solution is well adapted for a liquid resistance, just as manganine wire is adapted for resistance coils.

THE DEFINITION OF *p*H. Originally the term pH was introduced by Sörensen because of the form of the equation relating the free energy to the hydrogen-ion concentration, and $\log_{10} 1/c_{H^+}$ was called the hydrogen-ion exponent or pH. However, since the electrometric method was used in the determinations, and since the pH of the standard was based upon calculations of the hydrogen-ion concentration from conductivity data, the quantity measured was not that defined by the equation $p\text{H} = \log 1/c_{H^+}$. Accordingly a new definition, designated as p_aH, was later brought forward, p_aH being defined as $\log 1/a_{H^+}$, where a_{H^+} represents the hydrogen-ion activity. It is now realized that ionic free energies and activities cannot be evaluated without certain non-thermodynamic assumptions. The result is that pH's are really arbitrary values with the scale often varying from worker to worker, depending on: (1) the standard solution chosen and the value assigned to it, and (2) the value of E_0 chosen for the cell used in referring to the standard. The cell employed is of the type

$$\text{Pt, H}_2 \mid \text{HCl, KCl} \mid \text{KCl} \mid \text{KCl, HgCl} \mid \text{Hg}$$

In order to determine E_0 it is necessary to evaluate the liquid-junction potential.

The definitions of pH may be classified under three headings:

1. $pH = -\log c_{H^+}$, the definition in terms of concentration.
2. $pH = -\log 1.1a_{H^+}$, the "Sörensen" definition.
3. $p_aH = -\log a_{H^+}$, the definition in terms of activity.

The second definition refers to pH's determined by use of Sörensen's value of the 0.1 N calomel electrode or from the pH values of his standard buffer solutions, obtained therefrom. Although there are certain advantages in the consistent use of the first definition, most of the data in the literature are reported in terms of the second. See Kilpatrick and Kilpatrick, *J. Chem. Educ.*, Vol. 9, 1010 (1932); MacInnes, *Principles of Electrochemistry*, Chapter 15; and Harned and Owen, *The Physical Chemistry of Electrolytic Solutions*, pp. 316–325.

THE DEFINITION OF AN ACID. The usual definitions of an acid as a substance containing hydrogen which in solution forms hydrogen ions, and of a base as a hydroxyl compound which forms hydroxyl ions in solution, were a natural result of the development of the ideas of Arrhenius and Ostwald and the emphasis on aqueous solutions. A much more general and fruitful definition is that put forward by Brönsted [*Rec. trav. chim.*, Vol. 42, 718 (1923)] and Lowry [*Chem. & Ind.*, Vol. 42, 43 (1923)]. Brönsted defines an acid by the formal equation

$$A \rightleftharpoons B + H^+$$

where H^+ represents the proton, A an acid of any charge whatsoever, and B the conjugate base of charge less by one than that of A. According to this definition, the most common carrier of acid properties in aqueous solution, the hydrogen ion (H_3O^+), is only one of a number of acids.

$$H_3O^+ \rightleftharpoons H_2O + H^+$$
$$CH_3COOH \rightleftharpoons CH_3COO^- + H^+$$
$$NH_4^+ \rightleftharpoons NH_3 + H^+$$
$$H_2O \rightleftharpoons OH^- + H^+$$

Attention is also called to the fact that, according to this definition, the hydroxyl ion is only one of many bases.

The definition given applies to any solvent, emphasizes the acid, basic, amphoteric or inert nature of the solvent, and brings simplicity and order into the field. The definition has led to a number of important catalytic studies [see Brönsted, *Chem. Reviews*, Vol. 5, 284 (1928)].

A still broader definition of acids and bases has been proposed by G. N. Lewis [*J. Franklin Inst.*, Vol. 226, 293 (1938)]. According to Lewis, a base is a substance capable of donating a pair of unshared electrons, and an acid a substance capable of accepting the electron pair, e.g.,

$$\underset{\text{Acid}}{(CH_3)_3B} + \underset{\text{Base}}{:NH_3} \rightleftharpoons (CH_3)_3B:NH_3$$

Here the dots represent the pair of lone electrons in the valence shell of the nitrogen atom of ammonia which are shared with the boron atom of trimethyl boron to form a co-ordinate covalent bond in the addition compound. It will be seen that Lewis's definition includes that of Brönsted and of Lowry.

15. ELECTROKINETIC PHENOMENA

The junctions previously considered have been formed between two conductors of electricity. Now the phenomena occurring at the junction of a conductor and a non-conductor or at the junction of two non-conductors, viz., electro-osmosis or endosmose, electrophoresis, and streaming potential, will be briefly discussed. These phenomena indicate that a difference of potential exists at the junction of the two phases, one phase having one charge and the other the opposite charge, the "Helmholtz double layer" thus formed constituting a condenser. See MacInnes, *Principles of Electrochemistry*, Chapter 23; also H. Müller, *Cold Spring Harbor Symposia on Quantitative Biology*, Vol. 1, 1 (1933), and Abramson, *Electrokinetic Phenomena and Their Application to Biology and Medicine*.

ELECTRO-OSMOSIS OR ENDOSMOSE. If a current is passed along a film of water on a solid insulating surface, the water will be attracted to the negative electrode. Similarly an electric current flowing through a porous partition or a capillary tube filled with conducting liquid causes a flow of the liquid through the partition or tube. The phenomenon is known as electro-osmosis or endosmose. For a review of the experimental and theoretical investigation of endosmose see MacInnes, *loc. cit.*, and Schönfeldt, *Z. Elektrochem.*, Vol. 39, 103 (1933).

The quantity of a given liquid carried through a given porous diaphragm in a definite time varies directly with the current strength and is independent of the area and thickness of the diaphragm. The quantity varies with the nature of the solution, being greater with liquids of high specific resistance and high dielectric constant. The direction of flow is generally from positive to negative electrode, but under certain conditions the flow may be in the opposite direction.

APPLICATIONS OF ENDOSMOSE. Endosmose may be utilized to remove water from wet substances. Thus, if a wooden box is equipped with perforated metallic plates at opposite ends and filled with wet turf, and if current is circulated through the turf from one end to the other, water will ooze out of the perforations of the cathode plate. Endosmose is utilized in tanning processes to accelerate the passage of tanning fluids through the hides. For a review of the industrial applications of endosmose see Schönfeldt, *Z. Elektrochem.*, Vol. 38, 744 (1932).

ENDOSMOSE OF NEGATIVE FEEDERS. If electric conductors covered with a saturated braid or a number of such braids are made the cathode of a water bath and, say, 100 volts applied between anode and cathode, the braid will blister in a few minutes and the blisters will finally burst, grounding the conductor to the water. The same action goes on at lower voltages at a proportionately lower rate, but with equal certainty. No action is observable if the wire is made the anode instead of the cathode. Because of endosmose, insulation of low specific resistance should never be used on negative feeders. For the same reason it is not practicable to maintain a negative contact rail on electric railways below earth potential, as the insulators soon become saturated or covered with moisture and thereby become conductors [see Fortenbaugh, *Trans. AIEE*, Vol. 28, 1215 (1908)].

When testing wires having insulation of low specific resistance, endosmose may further lower the resistance if the wire is negative to the water bath.

STREAMING POTENTIAL. When a liquid is forced by pressure through a porous diaphragm or capillary tube, and an electrode is inserted in the liquid on each side of the diaphragm, the cell formed exhibits an emf known as the streaming potential. Streaming potential is thus the reverse of electro-osmosis.

For the streaming potential there has been derived the expression

$$E = \frac{PD\zeta}{4\pi\kappa\eta}$$

where P is the pressure applied, D the dielectric constant of the liquid, η the viscosity of the liquid, κ the specific conductance of the liquid, and ζ is the potential drop at the Helmholtz double layer, the so-called "zeta potential." Kruyt and van der Willigen, using this formula, found zeta potentials of the order of 10^2 mv for potassium chloride solution in glass capillary tubes [Kruyt and van der Willigen, *Kolloid-Z.*, Vol. 45, 307 (1928)]. The zeta potential between egg albumin and an aqueous buffer or acid solution was determined from measurements of the streaming potential at a quartz diaphragm saturated with egg albumin [Briggs, *J. Am. Chem. Soc.*, Vol. 50, 2358 (1928)] and from measurements of the electrophoretic mobility (see the following section) of quartz particles coated with egg albumin [Abramson, *ibid.*, Vol. 50, 390 (1928)]; the two methods gave values in good agreement, ranging from +30 mv at pH = 3 to *ca.* −20 at pH = 7.

ELECTROPHORESIS OR CATAPHORESIS. Electrophoresis, or cataphoresis, is the migration of colloidal particles in an electric field. It can be demonstrated by forming a boundary between a clear solution and a colloidal suspension in an electrolytic cell and observing the movement of the boundary when an emf is applied. If the colloidal suspension is colorless, the movement of the boundary may be followed by a photographic method; for the Toepler "schlieren" method, see Toepler, *Handwörterbuch der Naturwissenschaften*, Vol. 8, 924 (1913); for the Lamm "scale" method, see Lamm, *Z. physik. Chem.*,

Vol. 138, 313 (1928). Many of the older data on electrophoretic mobility are of little value because of the disturbance of the boundary by convection currents. In the cell devised by Tiselius (see MacInnes, *Principles of Electrochemistry*, Chapter 23) convection currents are largely avoided, and by its use electrophoresis has become an important tool in the study of proteins; see, for instance, Tiselius, "Electrophoretic Analysis and the Constitution of Native Fluids," in *The Harvey Lectures*, the Science Press Printing Company, Lancaster, Pa., 1940.

Another method of measuring electrophoretic mobility is available when the colloidal particles are visible in a microscope, so that the progress of an individual particle can be followed.

The charge carried by the colloidal particle is in some cases attributed to its acidic or basic nature; in others, to its preferential adsorption of ions. The molecule of a protein is considered to exist chiefly as zwitterion; thus, letting R represent the portion of the chain lying between the amino and carboxyl groups,

$$NH_2—R—COOH \rightleftharpoons \overset{+}{N}H_3—R—COO^-$$

whereas in acid solution the protein is present to some extent as cation

$$\overset{+}{N}H_3—R—COOH$$

and in alkaline solution, as anion

$$NH_2—R—COO^-$$

The electrophoretic mobility of a protein changes with the pH of the solution in conformity with the scheme presented, the direction of electrophoretic migration reversing upon passing through the isoelectric point.

Helmholtz derived for the mobility of a cylindrical colloidal particle, situated in a motionless liquid in which there is unit potential gradient, the expression

$$U = \frac{\zeta D}{4\pi\eta}$$

where the terms on the right have the same significance as above. As already mentioned, Briggs, using a quartz diaphragm saturated with egg albumin, obtained, by measurement of the streaming potential, values of ζ in agreement with those obtained by Abramson by measurement of the electrophoretic mobility of quartz particles coated with egg albumin, computed by the foregoing formula.

16. BIBLIOGRAPHY

Electrochemical Data

Landolt, Börnstein and Roth, *Tabellen.*
Kohlrausch and Holborn, *Leitvermögen der Elektrolyte.*
Tables annuelles internationales de constantes.
The International Critical Tables.

Books

Allmand, A. J., *Applied Electrochemistry*. Longmans, Green (1924).
Creighton, H. J. M., and W. A. Koehler, *The Principles and Applications of Electrochemistry*. John Wiley (1943–44).
Davies, C. W., *The Conductivity of Solutions*, 2nd ed. John Wiley (1933).
Dole, M., *The Principles of Experimental and Theoretical Electrochemistry*. McGraw-Hill (1935).
Förster, F., *Elektrochemie wässeriger Lösungen*, 2nd ed. Barth (1915).
Glasstone, S., *The Electrochemistry of Solutions*, 2nd ed. Methuen, London (1937).
Glasstone, S., *An Introduction to Electrochemistry*. Van Nostrand (1942).
Harned, H. S., and B. B. Owen, *The Physical Chemistry of Electrolytic Solutions*. Reinhold (1943).
LeBlanc, M. J. L., *Electrochemistry*. Macmillan (1907).
Lehfeldt, R. A., *Electrochemistry*. Longmans, Green (1904).
Lorenz, R., and F. Kaufler, *Elektrochemie geschmolzener Salze*. J. A. Barth (1909).
MacInnes, D. A., *The Principles of Electrochemistry*. Reinhold (1939).

Journals

Consult also the publications of the Bunsen Society (*Zeitschrift für Elektrochemie*), of the Faraday Society (*Transactions of the Faraday Society*) and of the American Electrochemical Society (*Transactions of the American Electrochemical Society*).

SECTION 5

MEASUREMENTS AND MEASURING APPARATUS

BY

H. B. BROOKS AND F. B. SILSBEE

This section was written for the previous edition by Dr. H. B. Brooks and revised for this edition by Dr. F. B. Silsbee in collaboration with the following members of the scientific staff of the National Bureau of Standards:

Mr. Donald B. Brooks	Thermometry
Dr. F. K. Harris	Oscillographs
Mr. R. L. Sanford	Testing of Magnetic Materials
Mr. R. P. Teele	Photometry
Dr. M. S. Van Dusen	Pyrometers

MEASUREMENTS AND MEASURING APPARATUS

By H. B. Brooks and F. B. Silsbee

1. UNITS AND STANDARDS

In the measurement of electrical, magnetic, thermal, and optical quantities, reference must ultimately be made to established **units** and verified **standards.** Experience has indicated the most suitable units and standards and also the most practicable procedures for carrying out the precise laboratory measurements upon which rest the less accurate but indispensable practical measurements.

DISTINCTION BETWEEN UNITS AND STANDARDS. The distinction between unit and standard should be carefully noted. A **unit** is a quantity in terms of which similar quantities are to be measured; a **standard** is a physical structure by which the unit is realized and made available. For example, the **unit** of electromotive force, the volt, is an invisible, intangible quantity, the very nature of which is unknown; but the **standard** of electromotive force is a particular kind of physical structure known as a standard cell.

UNITS. Since January 1, 1948, the units used for the measurement of electric quantities are those of the "absolute practical," i.e., mks, system. This system makes the electric units consistent with the mechanical units, and was approved in principle in 1933 by the General Conference of Weights and Measures. In October, 1946, the International Committee on Weights and Measures set the above date and, on the basis of the available experimental data, assigned values to the ratios of the new absolute units to the so-called "international" units which were in use from 1911 to 1947 inclusive.

These earlier "international" units had been defined by the London Conference of 1908 as those derived from a unit of resistance equal to the resistance at 0 deg cent of a column of pure mercury 106.300 cm long and weighing 14.4521 grams, and from a unit of current such that it deposits 0.118 mg of silver per second under standard conditions. In 1910 a meeting was held in Washington at which delegates from Great Britain, France, Germany, and the United States agreed on the values in international units of the cells and resistors which they had first intercompared. Thereafter the electrical units of the various countries were derived from the international volt and the international ohm, as maintained at national standardizing laboratories by means of groups of standard cells and of standard resistors.

The magnitude of the change in any unit can be derived from the two relations:

1 international ohm as formerly certified by the National Bureau of Standards = 1.000495 absolute ohms.

1 international volt as formerly certified by the National Bureau of Standards = 1.000330 absolute volts.

BASIC STANDARDS FOR ELECTRICAL MEASUREMENTS. All measurements of electric current, voltage, power, and energy, in either d-c or a-c circuits, are ultimately referred to resistance standards, standard cells, and time standards. Standards of mass (standard "weights") require some form of balance for their comparison. Similarly, the Wheatstone bridge and other bridges are used in the comparison of resistance standards, and potentiometers serve to measure electromotive forces in terms of the known electromotive force of a standard cell. In the use of bridges and potentiometers, adjustments are made in one or more arms of a network of conductors until a particular condition of balance is established, as shown by a suitable detector, usually a galvanometer.

2. GALVANOMETERS

A galvanometer is an instrument for detecting or measuring a small electric current. A d-c galvanometer consists essentially of a permanent magnet and a coil or wire through which the current may flow. In the **moving-magnet galvanometer** the coil is fixed and relatively large; the magnet is very small and light and is suspended within the magnetic field of the coil so that it may readily turn as a result of the interaction of the field of the magnet and that of the coil. In the **moving-coil galvanometer** (often called d'Arsonval galvanometer) the coil is very light and is arranged to rotate within the field of the relatively large fixed permanent magnet. The **string galvanometer** is a variant of the moving-coil galvanometer, and has a conducting "string" (wire or metal-coated fiber) stretched within the field of a strong magnet. The string is deflected laterally when a current flows through it. (See Article 18, Oscillographs.) In all forms of galvanometer the motion is opposed and brought to rest by a counter torque (or counter force, in the string galvanometer), by electromagnetic damping, and by friction. Friction is made up of air friction, molecular friction due to imperfect elasticity of the suspensions or control springs, and (in pivoted galvanometers) bearing friction.

MOVING-MAGNET GALVANOMETERS have certain advantages over moving-coil galvanometers which cause them to be used for some purposes in spite of their inherent disadvantages. Their advantages are (*a*) the moving element is light; (*b*) there is an absence of current in the moving element, permitting the use of very delicate suspensions which (like silk and quartz fibers) may be non-conducting; (*c*) they have very high sensitivity as a result of (*a*) and (*b*); (*d*) the sensitivity may be changed by regroupingsections of the fixed coil, as well as by varying the degree of astaticism of astatic galvanometers; (*e*) the damping is virtually unaffected by changes in resistance of the external circuit. Their disadvantages are (*a*) they are highly susceptible to disturbance by variations of the local magnetic field; (*b*) the damping is not as readily controlled as in the moving-coil galvanometer. Disturbances resulting from variation of the local magnetic field are reduced either by using an astatic construction or by surrounding the galvanometer with an iron shield or a number of concentric shields, or by both precautions. In the **astatic galvanometer** the moving system is double, with magnetic polarities arranged in such a way that the moving system would be unaffected by a change in the strength of the local magnetic field, provided this change were such as to affect both elements of the astatic system equally. In practice such an ideal condition is seldom realized, and the user of astatic galvanometers (or other nominally astatic instruments) should take all reasonable care to avoid exposing them to stray magnetic fields.

MOVING-COIL GALVANOMETERS. Originally much inferior to moving-magnet galvanometers in sensitivity, moving-coil galvanometers have been refined and improved until they now serve for all but extreme requirements. They have the great advantages of being almost completely unaffected by even large variations in the local magnetic field and of being readily damped by proper choice and adjustment of external circuit conditions. In some forms the damping may be regulated by a magnetic shunt.

MOTION OF GALVANOMETER MOVING ELEMENT. If the moving coil (or moving magnet) of a galvanometer is deflected from its zero position and then released, it may come to rest at the zero position only after one or more swings through this position. For this case the motion is said to be **periodic.** Under other conditions the coil or magnet will return to its equilibrium position without passing through it. This constitutes **aperiodic** motion. When the damping is just sufficient to make the motion aperiodic, the galvanometer is said to be **critically damped.**

When critically damped, the moving system will arrive within 0.1 per cent of its final deflection in about 1.5 times the free period. Although theoretical calculations usually assume exact critical damping, in practical work a slight underdamping is usually preferable because it allows the coil to overshoot the rest position slightly and then return to it. This gives assurance that sticking is not present.

The **damping ratio** of a galvanometer is the ratio, expressed as a positive number, of a given deflection to the next deflection in the opposite direction. The greater this ratio, the greater the degree of damping. The natural logarithm of this ratio is called the **logarithmic decrement.**

Much time will be lost, in using moving-coil galvanometers, if the total resistance to the galvanometer circuit is widely different from the value which gives aperiodic motion. If one is constrained to choose between using a galvanometer in a circuit having, for example, only one-half the resistance for critical damping, or in a circuit of twice the critical value, the latter condition, giving underdamping, is preferable, because with it the moving system comes to apparent rest in a shorter time.

SENSITIVITY OF GALVANOMETERS. The sensitivity of a galvanometer may be defined either as its response (in specified deflection units) to unit stimulus or as the magnitude of the stimulus required to produce a specified unit deflection. The second form of definition gives a numerical value of sensitivity (for a given galvanometer) which is the reciprocal of the value given by the first definition. For example, the sensitivity of a given galvanometer may be stated either as 10 mm per μv or as 0.1 μv per mm. Both systems of definition are in successful use, and the ease with which the sensitivity may be translated from one form to the other evidently explains the lack of confusion which would ordinarily attend the use of radically different definitions for the same thing.

The second form of definition has been found more convenient in American manufacturing practice. In galvanometers having attached scales the specified unit deflection is assumed to be one scale division. In reflecting galvanometers for use with separate reading devices the standard unit deflection is assumed to be 1 mm, with a scale distance of 1 meter. The **current sensitivity** is usually expressed as the number of microamperes to produce the specified unit deflection. The **voltage sensitivity** is expressed as the number of microvolts that must be impressed on the circuit consisting of the galvanometer coil and the external resistance for critical damping to produce the specified unit deflection.

The **ballistic sensitivity** is expressed as the number of microcoulombs which must be discharged through the galvanometer to produce the specified unit deflection. The **megohm sensitivity** is expressed as the number of megohms in the galvanometer circuit for which an impressed emf of 1 volt gives the specified unit deflection.

Particular care is necessary in making and in interpreting statements concerning the microvolt sensitivity of galvanometers, because some makers state the microvolt sensitivity as the microvolts per millimeter (or reciprocally, as the millimeters per microvolt) at the galvanometer terminals. Most modern moving-coil galvanometers are so designed that they cannot be effectively used with the zero value of external resistance implied in this definition, because this condition causes excessive overdamping. All statements of microvolt sensitivity should contain definite information as to the total resistance to which they refer. When such a statement is not given in the maker's catalog, his basis for the definition of microvolt sensitivity can often be discovered from other data given, namely, the microampere sensitivity and the resistance of the galvanometer. The external resistance for critical damping should always be included in makers statements concerning their moving-coil galvanometers.

The use of the above-defined second method of stating sensitivity gives numerical values (for galvanometers of relatively high sensitivity) in the form of very small decimal fractions. To avoid this, some foreign makers who follow this principle in defining sensitivity use units of current and voltage smaller than the microampere and microvolt respectively. For example, in a single circular of one maker the current sensitivity of various galvanometers is given in terms of 10^{-7}, 10^{-8}, 10^{-9} and 10^{-10} amp, and the voltage sensitivity is given in millimeters for 10^{-8} volt. This lack of uniformity in the units used in stating galvanometer sensitivity would be a source of inconvenience if it were not so easy to convert a given sensitivity from one form to another. The need for a named unit of current smaller than the microampere (and, continuing the sequence, ampere, milliampere, microampere) has induced some foreign authors to use the word **nanoampere** to signify 10^{-9} amp.

The attempt is sometimes made to increase the effective sensitivity of a galvanometer by placing the scale at a comparatively great distance from it. In doing this, some users overlook the fact that the instability of the zero is magnified in the same ratio as the deflections. An increased scale distance is justifiable only when the stability, including the effect of vibrations, is in excess of the precision of reading.

GALVANOMETER SHUNTS. For many purposes it is essential to reduce the sensitivity of a galvanometer to a definite fraction of its value. This is conveniently done by connecting a resistance across the galvanometer terminals to bypass the greater part of the current. If the resistance of this bypass ("galvanometer shunt") is $^1/_9$, $^1/_{99}$, or $^1/_{999}$ of the galvanometer resistance, the current sensitivity will thereby be reduced to $^1/_{10}$, $^1/_{100}$, or $^1/_{1000}$ of its original value, respectively. A simple shunt of this sort has two drawbacks, namely, it must be adjusted for a particular galvanometer, and unless it is wound with copper wire it will not have the correct shunting effect except at some one temperature. To overcome these defects, the Ayrton-Mather shunt was designed (see Fig. 1). It may be shown that, if the combination of galvanometer and shunt has a given value of current sensitivity with the slider on the stud b, the current sensitivity with the slider on any other stud x will be lower in the ratio of r to R, regardless of the resistance of the galvanometer. In choosing an Ayrton-Mather shunt for a moving-coil galvanometer, the value of R should be selected with reference to the external critical damping resistance and the probable value of resistance of the circuit to be connected to the terminals A and X so that the damping will be satisfactory.

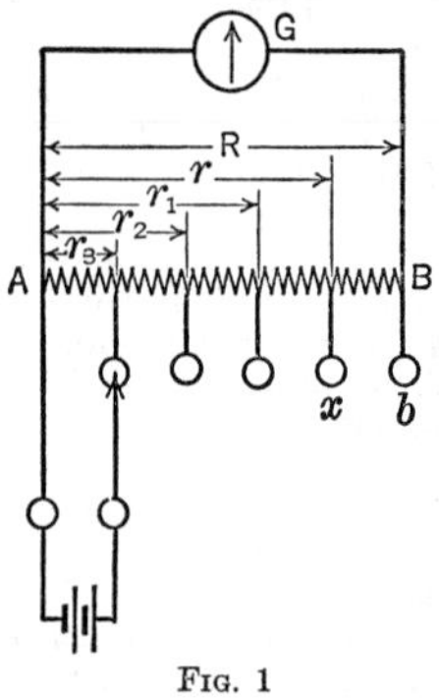

FIG. 1

The Ayrton-Mather shunt was originally devised for circuits of high resistance, such as were encountered in measurements of insulation resistance. When this shunt is used in such a circuit, the variations of resistance produced by the operation of its range-changing switch have an inappreciable effect on the magnitude of the current in the external circuit. However, when the Ayrton-Mather shunt is used with low-resistance external circuits, large variations in the current in the external circuit may be caused by changing the range of the shunt. If necessary, compensating resistances may be provided, in series with each input tap, so as to maintain a constant input resistance.

CHOICE OF A TYPE OF GALVANOMETER. A moving-coil galvanometer should always be preferred to a moving-magnet galvanometer except for some very special purposes. For any application the most sturdy, simple, and inexpensive galvanometer which has sufficient sensitivity with a suitable period should be chosen. In general, superfluous sensitivity increases the initial cost of the galvanometer and tends to increase delicacy and hence fragility, as well as to necessitate a longer period and consequently greater time required for observations. For the most precise work of the standardizing laboratory it is necessary to use delicate, expensive galvanometers of relatively long period and to employ a reading arrangement consisting of a telescope and scale or a lamp and scale. Such moving-coil galvanometers are made in this country in two grades of sensitivity and at corresponding prices. For work requiring only moderate sensitivity, with shorter periods, portable lamp-and-scale (box-type) galvanometers are supplied as compact, self-contained units. For work requiring still lower sensitivity, sturdy portable suspended-coil pointer galvanometers are available, which have the advantages of low price and ease of exchange or replacement of the coil system. For some purposes for which moderate sensitivity is sufficient and the deflection should be proportional to the current over a relatively large angle, pivoted pointer galvanometers are made which embody the standard features of construction of d-c indicating instruments.

Certain old theorems state that for maximum sensitivity the coil of the galvanometer should be wound to have a resistance equal to the "internal resistance" of the source of the emf which is to produce the deflection. Although this can be proved to be true as a matter of mathematics, the statement is largely of academic interest, as one can depart very widely from the theoretically correct value of coil resistance with but little loss of sensitivity. Furthermore, most moving-coil galvanometers as now made are designed to be used with an external resistance of, say, 3 to 15 times the coil resistance in order to get suitably damped motion. (This statement does not apply to galvanometers having damping frames or damping windings to increase still further the value of external resistance for critical damping.) A more practical rule for choosing a moving-coil galvanometer is as follows: If a given small emf in an external circuit of resistance R is to be detected or measured, select the galvanometer which has a value of external critical damping resistance most nearly equal to R; if no available galvanometer has just this value, select the galvanometer for which the external resistance for critical damping falls short of R by the smallest amount. The resulting motion will be underdamped, which is much to be preferred to overdamping; if desired, the damping may be improved by suitably shunting the galvanometer.

PARASITIC THERMAL EMF IN GALVANOMETERS. Moving-coil galvanometers, as made without special precautions against parasitic emf, contain a number of junctions of dissimilar metals. The resultant thermal emf will depend upon thermal conditions and may amount to 10 or 20 μv under very bad conditions, for example, in winter weather when a descending current of cold air strikes the top of a galvanometer but does not sensibly affect the temperature of the bottom. Older forms of galvanometers with a long tube at the top containing the suspension are particularly liable to this defect, which is of consequence chiefly when relatively low voltages are to be detected or measured by the galvanometer. The thermal emf may be minimized by enclosing the galvanometer in a case made (*a*) of metal, such as copper, aluminum, or brass, of high thermal conductivity; or (*b*) of material of high thermal resistivity (cork, balsa wood, etc.). In (*a*) the metal enclosure will have nearly the same temperature in all its parts; in (*b*) the galvanometer itself, protected from radiation and convection, will have a nearly uniform temperature in all its parts. Some makers supply galvanometers in which not only the coil but also the suspensions, binding posts, and other parts are made of copper to minimize thermal emf. It is important that ballistic galvanometers should be so constructed.

BALLISTIC GALVANOMETERS. Ballistic galvanometers are used to measure a quantity of electricity, such as is given up by a capacitor during discharge. They are usually constructed to have a relatively long period for one or both of the following reasons: first, because a longer period makes it easier to read the momentary maximum deflection; second, because in some applications the time required for the passage of the electrical quantity is relatively great enough to introduce an error if a galvanometer of short period should be used.

Ballistic galvanometers may be used either undamped or damped. When critically damped, they have only about one-third the sensitivity when undamped, but for most applications the sensitivity is so ample that the reduction which accompanies critical damping can well be tolerated for the much greater convenience in operation.

The **ballistic constant** of a galvanometer is conveniently expressed as the number of microcoulombs per millimeter deflection with a scale distance of 1 meter. It is usually determined by discharging through the moving coil a known quantity of electricity from a charged capacitor or from the secondary winding of a mutual inductor. If a charged capacitor is used, the quantity in microcoulombs is the product of the capacitance of the capacitor in microfarads and the difference of potential in volts to which it is charged. Ballistic galvanometers have important applications in magnetic testing.

When used to measure the change of flux linkages in a coil connected to the galvanometer terminals, the sensitivity in maxwell-turns per millimeter deflection is equal to $100RQ$, where Q is the ballistic constant in microcoulombs per millimeter, and R is the resistance in ohms of the total circuit, i.e., coil and galvanometer in series. For a discussion of the use of ballistic galvanometers in this field and the method of calibrating them by means of a mutual inductor, see Article 19, Testing of Magnetic Materials.

A-C GALVANOMETERS. The two forms of a-c galvanometer most used in this country are the vibration galvanometer and the moving-coil separately excited galvanometer. In its usual form the moving-coil vibration galvanometer differs from the d-c moving-coil galvanometer chiefly in having a very narrow coil and stiff suspensions. The stiffness of the suspensions is adjustable to bring the natural frequency of the moving system into resonance with the frequency of the a-c supply circuit. For any other frequency the sensitivity is relatively very small, and harmonics present in the wave form are substantially without effect on the deflection. This valuable property of the vibration galvanometer often makes possible the use of the simple mathematical relations based on the assumption of a sine-wave form, even though the available a-c supply voltage contains harmonics.

The moving-coil separately excited a-c galvanometer differs from the ordinary d-c moving-coil galvanometer mainly in having a laminated electromagnet and a laminated core in place of a permanent magnet and solid core. The theory of this class of galvanometer has been thoroughly treated by Weibel [*Bull. Bur. Standards*, Vol. 14, p. 23 (1918); reprinted as *Sci. Paper* 297]. Unlike the vibration galvanometer, the separately excited a-c galvanometer is not restricted to a particular frequency (or frequency range). It must be excited by a current of the same frequency as that used in the measurement circuit (bridge or potentiometer). It responds only to that component of the current in its moving coil which is in time phase with the magnetic flux in which the coil is located, and consequently for maximum sensitivity the current in the coil must be in phase with the flux. It is also possible to overload and damage the coil without producing a large deflection, if the current is nearly in quadrature with the flux. In some cases the in-phase condition of current and flux is readily and simply attained; for example, in using the galvanometer with a resistance bridge, the exciting coil of the galvanometer is connected to the supply circuit through a suitable non-inductive resistance, and the current for the bridge is taken from potential taps on this resistance. To obtain proper damping in reflecting galvanometers of this type, the circuit external to the galvanometer must have suitable values of resistance, inductance, and capacitance. When these conditions are fulfilled, the separately excited a-c galvanometer has a possible sensitivity but little inferior to that of d-c moving-coil galvanometers of corresponding construction. In the use of the sturdy pointer-type a-c galvanometers no particular attention has to be paid to the external inductance and capacitance in order to obtain good damping in the usual circuits.

Electronic detectors consisting of one or more stages of amplification and an indicator formed by a plate milliammeter, an "electric eye" tube, or an oscilloscope are often used.

SUPPORTS FOR GALVANOMETERS. Sensitive reflecting galvanometers should preferably be mounted on stable masonry walls or piers free from vibration. Vibration galvanometers are particularly susceptible to mechanical vibrations set up in buildings by the operation of motors or alternators at the frequency for which the galvanometers are tuned. When trouble is experienced with either d-c or a-c galvanometers mounted on the best available firm supports, recourse must be had to expedients, of which the Julius suspension is among the oldest. See Bibliography, paper by W. P. White in *Phys. Rev.*, vol. 19, p. 323 (1904), and book by Werner [*Empfindliche Galvanometer für Gleich- und Wechselstrom*, de Gruyter (1928)]. Numerous useful antivibration mountings containing rubber "buttons" in shear are on the market. Hartsough [*Phys. Rev.*, 2nd series, Vol. 29, p. 910 (1927)] has stated that the shielding of a sensitive apparatus from vibration is very nearly perfect when it is supported on thin inflated rubber bags. In his work three interconnected bags were used, containing air at a pressure of about 50 cm (20 in.) of water, and a mass of about 4 lb was placed on each bag.

3. STANDARD CELLS

Standard cells are primary cells made of pure materials in accordance with exact specifications. They have the special characteristic of maintaining a very constant emf when suitably cared for and used. They are of great and increasing importance as standards of emf.

The only kinds of standard cell of technical importance at present are the **saturated cadmium cell,** called by international agreement the Weston normal cell, which serves as the primary standard of emf in national and other important standardizing laboratories, and the **unsaturated cadmium cell,** used generally (except in England) in engineering laboratories and industrial plants as a secondary or working standard of emf.

WESTON NORMAL CELL. Figure 2 shows the construction of this cell. The H-form glass vessel, with a platinum lead-in wire at the bottom of each limb, contains mercury covered with mercurous sulfate paste as the depolarizer in the positive limb, cadmium amalgam in the negative limb, and cadmium sulfate solution with an excess of cadmium sulfate crystals above the paste and amalgam as shown. For the exceedingly exact work for which these cells are used, they must be kept at a very constant temperature. In the standard-cell laboratory of the National Bureau of Standards the normal cells are kept in baths of oil maintained very accurately at 28 deg cent by a sensitive thermostat. This temperature was chosen because of climatic conditions in Washington, in order to avoid the condensation of moisture from the air upon parts of the oil bath in humid summer weather. Aside from such considerations, other temperatures would be equally suitable. The normal cells of England, France, and Germany are kept at about 20 deg cent.

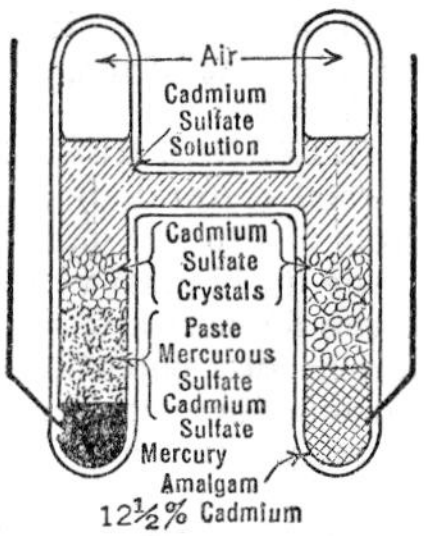

FIG. 2

For very exact work the saturated cells should be maintained at as nearly a constant temperature as practicable for at least several days before electrical comparisons are made, because, in addition to having an appreciable temperature coefficient, these cells do not immediately assume their true emf after a change in temperature.

Saturated standard cells made with "heavy water" (i.e., deuterium oxide) have less voltage by 388 μv than those using ordinary water, but are equally constant and reproducible.

UNSATURATED CADMIUM CELL. This cell differs from the normal cell in having no excess crystals of cadmium sulfate. The concentration of the cadmium sulfate solution is intended to be such that it will reach saturation when the temperature is reduced to approximately 4 deg cent, a feature which insures a minimum effective temperature coefficient of the cell within the temperature range permissible in normal use. The unsaturated cell also differs from the normal cell shown in Fig. 2 in being made portable by the use of some form of porous plug or septum above the active material in each limb to keep these materials in place even when the cell is inverted.

CHARACTERISTICS OF THE TWO KINDS OF CELL. The Weston normal (saturated) cell, made to specifications from carefully purified materials, has a highly reproducible value of emf which usually remains very constant for years if the cell receives proper care. It has, however, an appreciable temperature coefficient of emf. The unsaturated cell is not as reproducible in emf, and there is usually a slow decrease of emf with time. Only very general statements may be made in regard to the rate of decrease of emf. Some exceptional unsaturated cells have decreased over a long period of years by only 10 to 15 μv per year. A decrease of 25 to 50 μv per year may be taken as average performance. Some cells have shown a decrease of 100 μv per year. Any rate of decrease much in excess of this figure may be taken as an indication of either abuse of the cell or defect in materials or workmanship.

The advantages of the unsaturated cadmium cell are its portability and the very low temperature coefficient of cells which are within certain limits of emf. To be acceptable, new cells should have values of emf within the range 1.0188 to 1.0198 absolute volts; such cells have extremely small temperature coefficients, provided that all parts of the cell are at the same temperature and no abrupt changes of temperature have recently occurred. As the emf of the cell decreases with age and use, however, an appreciable temperature coefficient of emf develops. The National Bureau of Standards does not issue certificates for cells having an emf lower than 1.0183 absolute volts. For very accurate work it is advisable to use cells for which the emf is not lower than 1.0186 absolute volts.

It is important to realize that the low temperature coefficient of the unsaturated cadmium cell is actually the small difference between two relatively large temperature coefficients of opposite signs (those of the two limbs), and consequently depends upon the existence of a high degree of temperature equality between the two limbs.

MEASUREMENT OF EMF OF STANDARD CELLS. The emf of a standard cell is measured by comparing it with that of another cell, the "reference cell," which is taken as a basis. In national standardizing laboratories the reference cell is compared at suitable intervals with the cells composing the primary standards of the institution. In industrial and engineering laboratories which do not maintain a group of Weston normal (saturated) cells, the reference cell is one for which the emf has been certified by a standardizing laboratory.

Standard cells must be measured by the null (potentiometer) method to avoid taking any current from them. Although it is possible to measure the entire emf of the cell under test with a potentiometer, the opposition method is much more accurate and can be carried out with a very much simpler and inherently less accurate potentiometer. In the opposition method the emf of the cell under test is opposed to that of the reference cell, and their small difference is measured with a potentiometer. Attention must be paid to the algebraic sign of this difference, so that its numerical value may be added to, or subtracted from, the value of the emf of the reference cell to obtain that of the cell under test.

INTERNAL RESISTANCE OF STANDARD CELLS. The internal resistance of the normal (saturated) cell, of usual construction and in good condition, may be taken as from 600 to 800 ohms. The internal resistance of the portable (unsaturated) cell, when made up in glass vessels of the size used by the principal national standardizing laboratories for normal cells, is about 500 ohms. For some purposes, particularly for use with a deflection potentiometer, this high value of resistance is a drawback. One American manufacturer regularly supplies portable cells with an internal resistance of about 100 ohms, whereas another maker supplies two forms having nominal internal resistances of 500 ohms and 100 ohms, respectively. Cells with 100-ohm resistances are particularly desirable for use with the deflection potentiometer.

To **measure the internal resistance** of a standard cell, its emf is first measured in terms of that of any other cell used as a reference cell in the manner described, then the cell is shunted under test with a resistance of 1 megohm, and the resulting apparent decrease in emf is noted. The numerical value of this decrease, expressed in microvolts, is (with sufficient accuracy) the numerical value of the internal resistance in ohms. The current through the 1-megohm resistor should be allowed to flow only long enough to obtain the reading. This resistor should be of the wire-wound type.

HIGH RESISTANCE RESULTING FROM GAS BUBBLE. In time, a gas bubble may form between the amalgam and the septum of a portable cell. Although harmless at first, the bubble may grow until it has forced out the solution between the amalgam and the septum and thereby open-circuited the cell. Before this extreme condition is reached, the internal resistance increases sufficiently to impair the usefulness of the cell by reducing the working sensitivity of the galvanometer through which the emf of the cell is opposed to some other emf or potential difference.

If periodic measurements of the internal resistance of a cell show increasing values which are variable through rather wide limits, depending upon whether the cell is in the normal upright position or slightly inclined, the existence of a gas bubble at the amalgam limb is indicated. This test may be applied without the necessity of breaking the maker's seals.

PRECAUTIONS IN USING STANDARD CELLS. No appreciable current can be taken from a standard cell without some alteration of its emf. If the current is small and flows for only a short time, the change is not permanent, and the cell gradually recovers its original emf; of course, the cell is unreliable until recovery is complete. Consequently, standard cells should be used only in methods in which their emf is opposed to an equal emf, as in connection with potentiometers. They should always be protected by a key or set of keys, which are closed only momentarily, and for preliminary adjustments a resistance of at least several thousand ohms should be in series with the cell. When an approximate balance has been obtained, this resistance may be cut out and the final balance then made.

In using portable standard cells, care should be taken to protect them from drafts of air, from sunlight, and from heat radiation from lamps, rheostats, radiators, and steam pipes to prevent unequal heating of the two limbs of the cell. Even the heat of the hand, applied for several minutes, may so unbalance the temperature as to produce a noticeable change in the emf. The magnitude of this change, for various kinds of temperature disturbance which may occur in practice, has been determined by J. H. Park [*Bur. Standards J. Research*, Vol. 10, p. 89 (1933); reprinted as *Res. Paper* 518]. Cells may be protected by a thick wrapping of felt or other heat-insulating material. In the electrical instrument laboratory of the National Bureau of Standards each cell is kept in a thick-walled copper pot jacketed with heat-insulating material. Such precautions are justified when one wishes to avoid errors of the order of 0.01 per cent, as is necessary when acceptance tests are made on instruments with a guaranteed accuracy of 0.1 per cent.

At least three portable standard cells should be provided for a laboratory having a single potentiometer. The cells should be intercompared at least once a month, oftener if they are frequently used. A change of emf of one of the cells, from improper handling or other cause, will be revealed by its failure to check with the other two.

It is preferable not to ship standard cells in extremely cold weather, although, when the cells are properly packed, damage from exposure to low temperatures during shipment seems to be infrequent. It is much more objectionable to *carry* the cells from place to place in cold weather, either without packing or so insufficiently packed as to afford little protection from the cold. After exposure to temperatures outside the permissible range (4–40 deg cent, i.e., 39–104 deg fahr) the cell may recover its normal emf after a rest of, say, several weeks at a reasonably constant temperature within the permissible range.

A standard cell should never be connected to a voltmeter in the attempt to measure its emf or to check the reading of the voltmeter. The voltmeter reading will be lower than the emf of the cell by an indeterminate amount; and, unless the voltmeter is of relatively high resistance, the emf of the cell will be altered, at least for a time.

4. TIME STANDARDS

Time enters into electrical measurements in passing from the measurement of power (rate of flow of energy) to the measurement of energy itself. Time also enters in a-c measurements; for example, the frequency of an alternating voltage or current is the reciprocal of the time required for a complete cycle. Time measurements in the laboratory cover an extremely wide range as to the magnitude of the time, the accuracy required, and the time-measuring devices available for various purposes. Among these devices may be mentioned clocks, stop watches, synchronous timers, tuning forks, and piezoelectric oscillators, the last-named being capable of an accuracy which exceeds that of the most reliable astronomical clocks.

For many purposes, especially for relatively long time intervals, the frequency of large electric power systems is an adequate time standard and seldom is more than 0.1 per cent in error. For greater accuracy the radio signals emitted by station WWV of the National Bureau of Standards can be used. These emissions have carrier frequencies of 2.5, 5, 10, and 15 megacycles per second. Two standard audio frequencies, 440 cps and 4000 cps, are broadcast on the radio carrier frequencies. In addition there is a pulse of 0.005-sec duration which occurs at intervals of precisely 1 sec. The pulse consists of five cycles each of 0.001-sec duration. On the fifty-ninth second of each minute the pulse is omitted.

5. RESISTANCE STANDARDS

FORMS OF CONSTRUCTION OF RESISTANCE STANDARDS. These standards, for resistances of 0.1 ohm and above, are usually wound with manganin wire and, for the most accurate work, are designed for immersion in an oil bath or contain oil, surrounding the coil, in a sealed case. They are generally provided with copper terminals for immersion in mercury cups formed in copper connection blocks. In the older Reichsanstalt type (Fig. 3) the distance between centers of the copper terminals is 160 mm (6.3 in.), and the case is perforated to permit the oil of the bath to circulate around the coil. In the type developed at the National Bureau of Standards by E. B. Rosa (Fig. 4), the case contains purified oil and is hermetically sealed to prevent the entrance of moisture. The distance between the centers of the terminals of the Rosa-type coil is 75 mm (3.0 in.). In the more recent resistance standard devised by J. L. Thomas at the National Bureau of Standards (see Fig. 5), the general dimensions and the distance between centers of terminals are the same as in the Reichsanstalt type, but the coil is hermetically sealed in the space between two coaxial brass tubes; it is wound on (and hence in good thermal contact with, though electrically insulated from) the inner tube, which is in contact with the oil of the bath. The manganin wire is thus protected from chemical attack which might occur if the

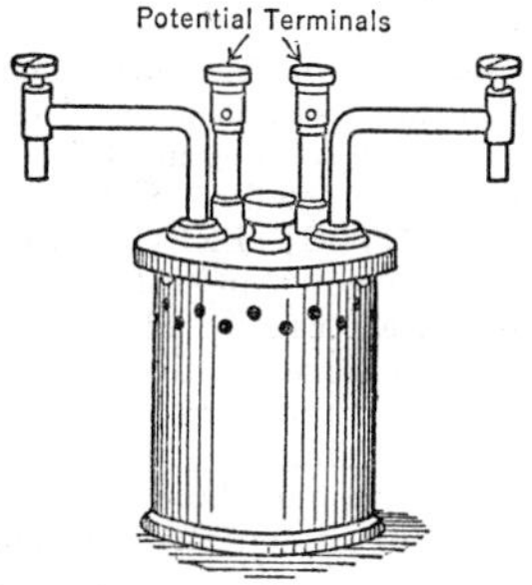

Fig. 3

oil should deteriorate. The wire of the Thomas standard is annealed at a red heat after winding and is not bent or otherwise deformed after this annealing. Coils of the Rosa type and the Thomas type are exempt from change of resistance, which occurs in unsealed coils as a result of variation of atmospheric humidity. Gold-chromium alloys have also been developed which are well suited for use in standard resistors.

Resistance standards of values lower than 0.1 ohm are usually made of manganin sheets hard-soldered to copper terminals and are of the four-terminal type to avoid contact-resistance errors at the terminals; that is, the resistance is adjusted to a stated value between two potential terminals which are connected to points on the resistor intermediate between those at which the current enters and leaves the standard. These

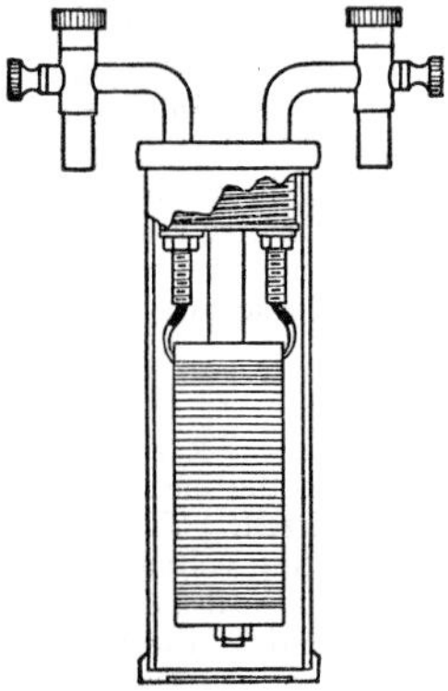

Fig. 4

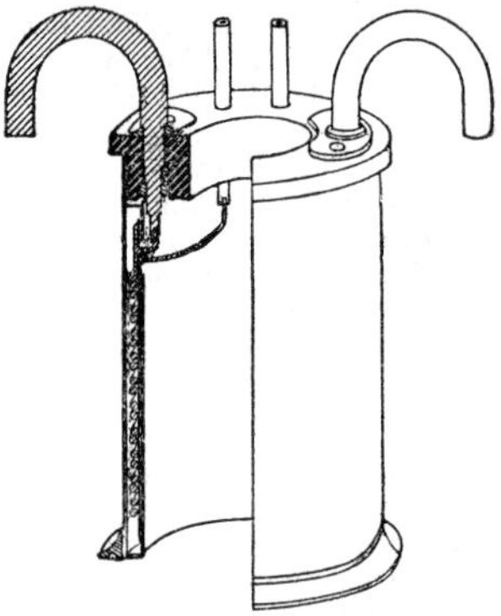

Fig. 5

standards, when used with a potentiometer for the measurement of current, are often colloquially called shunts because of their similarity to ammeter shunts.

AIR-COOLED VS. OIL-COOLED RESISTANCE STANDARDS. For a given difference of potential at rated current the power loss (and hence the rate of liberation of heat) increases with the current. Some forms of resistance standard for large currents are made for oil immersion, the oil being cooled by circulating water. Other forms are designed primarily for use in still air. To avoid excessive weight and cost, standards of the latter type are made with progressively lower values of potential difference as the rating in amperes increases. Oil-immersed heavy-current standards may have a larger value of this potential difference, with which certain sources of error (such as thermal emf) have correspondingly less effect on the accuracy of the measurements. The use of oil makes it possible to determine the temperature of the manganin more accurately. These advantages are accompanied by the usual disadvantages of oil immersion, among which are the necessity for providing for cooling water, the possibility of leakage or creeping of oil, the tendency for oil to become rancid in time and to attack the manganin. Both oil-cooled and air-cooled standards for large currents are used, according to the properties upon which the most emphasis is laid. Air-cooled standards may on occasion be used in oil or in a forced current of air. The Bureau of Standards uses non-inductive resistance standards, cooled by a forced air current, in testing current transformers up to 500 amp. Beyond this value it is necessary to resort to oil cooling to obtain an adequate voltage drop with a sufficiently low temperature rise and suitable dimensions.

LIMIT OF ERROR AND RATED CURRENT OF RESISTANCE STANDARDS. Single-value standards of 1 ohm or decimal multiples of this value, not exceeding 10,000 ohms, are usually adjusted to a stated limit of error of 0.01 per cent when used with a current corresponding to 0.1 watt loss. For 0.1 ohm and lower decimal submultiples of the ohm, the stated limit is generally 0.02 per cent, but this figure refers to a current corresponding to 1 watt loss. Standards of 1 ohm and above may be used with a load of 1 watt, and those of 0.1 ohm and below with 10 watts, with reduced accuracy during use but without any permanent change in resistance as a result of the load. The statements in this paragraph apply to standards having terminals for mercury-cup connections, and used in an oil bath.

Resistance standards ("shunts") of 0.1 to 0.00001 ohm for measuring large currents have various values of permissible power loss. The user should obtain information on this point from the maker or from experiments to show the temperature rise of the manganin and the consequent change in the resistance of the standard.

6. RESISTANCE BOXES

CONSTRUCTION OF RESISTORS. A resistance box or precision rheostat is a group of resistors assembled in a case and provided with means for varying the resistance in circuit between two terminals. The resistors are usually of manganin wire and differ from those used in the best grade of resistance standards chiefly in being of smaller dimensions and lower accuracy of adjustment. It is standard practice to wind the resistors non-inductively. For use with direct current and with alternating current of usual power frequencies, the ordinary bifilar winding is usually satisfactory. Its use results in a coil having a time-constant nearly equal to zero, for coils of 100 ohms. For lower values the inductance predominates; for higher values the coil behaves like a pure resistance shunted by a capacitance. The bifilar winding is objectionable for coils of high resistance (1000 ohms and above) when used with frequencies above those for power and lighting, and various special types of winding have been devised to minimize both inductance and capacitance. The greatly increased use of frequencies in the audio range and higher has stimulated the development of resistors and boxes of very low time-constant; for example, a recent line of resistance boxes by Leeds & Northrup Company is stated to have a time-constant of less than 0.1 μsec for all settings above 5 ohms.

The coils of resistance boxes are usually wound on spools of wood, brass, porcelain, lava, or molded insulating material. In some cases the wire is wound on cards of mica or other sheet insulating material or woven, with threads of silk or other insulating materials, to form a fabric. These constructions give a very low time-constant and good facility for the escape of heat from the winding.

ARRANGEMENT OF RESISTORS. The resistance between the terminals of a resistance box is varied by changing the manner of connection of its resistors, either by shifting plugs or turning rotary switches. The 1, 2, 3, 4 and the 1, 2, 2, 5 plug arrangements are the most usual. With the former, a resistance box of 1110 ohms total contains resistors of the following resistances: 1, 2, 3, 4, 10, 20, 30, 40, 100, 200, 300, 400 ohms. The resistors are in series between the terminals, and the junctions of adjacent coils are connected to brass blocks which are reamed to receive taper plugs between them (see Fig. 6). The insertion of a plug short-circuits a resistor. The disadvantages of these arrangements are the time and mental effort required to add up the values of the resistors unplugged, the likelihood of errors, and the risk of loss of plugs. A further disadvantage of this construction is that there are a maximum number of plug contacts in circuit when the resistance setting is a minimum. Thus the lower the resistance setting, the greater is the likelihood of error or variability of resistance arising from imperfect contacts at the plugs. **The decade form** of plug resistance box avoids all these disadvantages except the risk of loss of plugs; the amount of resistance in circuit can be read off directly from the position of a single plug in each decade; the decade plan is usually preferred. Although the original decade arrangements had 9 (or 10) resistors per decade, special arrangements have been devised to obtain 9 (or 10) steps with a smaller number of resistors. The object is to reduce the cost of the box for a given total resistance. Among these arrangements may be mentioned Northrup's, using coils of 1, 2, 3, and 3 units, and Smith's, using coils of 1, 2, 2, 2, and 2 units; these and other arrangements are discussed by Northrup [*Methods of Measuring Electrical Resistance*, McGraw-Hill (1912)]. When some of these plans are used with the rotary-switch construction, the resistance of the decade goes through some undesired value as the brush passes from one stud to the adjacent one. Behr's new arrangement for a 10-step decade, free from this disadvantage, consists of six resistors of two units each, two being connected in parallel when the resistance of the decade is to have an odd value.

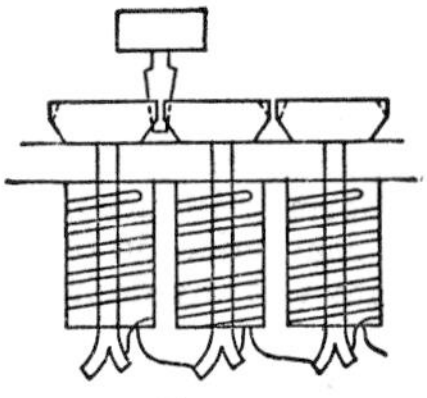

Fig. 6

PLUGS VS. SWITCH CONTACTS. Plug contacts, when carefully made, cared for, and used, give (in general) a lower and more definite contact resistance than switch contacts. However, well-made switch contacts, when properly cared for, have a satisfactory definiteness of contact resistance and have the advantages of more rapid manipulation and reading, with no loose parts to be mislaid or lost. They require more care than plug contacts. Switch contacts are now generally preferred for most purposes. They permit placing all the live parts inside the box and inside the metal shield in the case of shielded resistance boxes for a-c use. The location of the live parts where the operator cannot come in contact with them is an advantage when the box is used in a high-voltage circuit. For resistance boxes for use on high frequencies the switch construction makes it possible to keep the geometry of the internal circuit fixed. In the most recent boxes for this purpose the number of resistors in circuit is constant for all settings, even the connectors in the zero position being in effect resistors which are balanced as to resistance and inductance.

The **angle of taper** of the plug is an important detail. A taper of 1 in 12 in diameter is very satisfactory. Taper reamers having this angle ("1 in. per ft") are supplied as standard articles by United States tool manufacturers. A taper of 1 in 10 is much used. If the taper of the plug is too pronounced, the plug is likely to loosen spontaneously; if the taper is much less pronounced than the foregoing values, the lateral forces tending to spread the contact blocks apart will be excessive, and the plug may seat so tightly as to require undue force for its removal.

MAINTENANCE OF CONTACTS. In both the plug and the switch constructions the contact surfaces must be kept clean and properly lubricated if the contact resistance is to be kept low and definite. It has been shown that the use of a suitable petroleum oil on these surfaces not only avoids cutting and scoring but also reduces the contact resistance and makes it more definite. The purified petroleum oils used internally in medicine (such as "Nujol") are very suitable. Keinath states that to give a satisfactory contact the force between the brush and the contact segment should be of the order of magnitude of 1 to 2 kg (2 to 4 lb), and that, to avoid the resulting danger of scoring, the contact surfaces should be kept greased with vaseline. Vaseline is satisfactory when the contacts are effectively protected from dust, but for open-switch contacts vaseline becomes thickened by the dust and eventually makes the contact resistance high and irregular. When exposed contacts are lubricated with a suitable oil, this difficulty does not arise.

It is an advantage to have the switches enclosed to hinder the access of dirt, but the construction should allow ready access for cleaning and oiling.

Further information concerning plug and switch contacts as used in resistance boxes is given in Article 7, Bridges, under the heading Accuracy of Measurements with Wheatstone Bridge.

7. BRIDGES

GENERAL. Instruments embodying forms of networks known as **bridge circuits** are used for the measurement of resistance, inductance, capacitance, and frequency. The operation of a bridge consists in varying the constants of its component circuits or "arms" until the response of a suitable detector connected between two points on the network is reduced to zero or to an amount measurable by the detector for the purpose of interpolation.

SLIDE-WIRE BRIDGE (Fig. 7). This very simple form of bridge, of low cost and moderate accuracy, is identical in principle with the Wheatstone bridge (see below), the difference being that a simple wire AB forms the two ratio arms. It may be a wire of uniform resistance per unit length, stretched along a meter scale divided into millimeters. All connections in the X and R circuits are of low resistance. A slider is arranged to make contact at any desired point along the wire. When the bridge is balanced,

$$X = \frac{AR}{1000 - A} \tag{1}$$

where A is the distance in millimeters between the slider and the left-hand end of the slide wire. The accuracy will be greater when the slider is nearer the center of the wire. For this case the error may be minimized by taking the mean of two values of X, between which the relative positions of X and R are interchanged. The accuracy may be increased by adding end coils to the slide wire, but at the expense of reduction of the range of measurement.

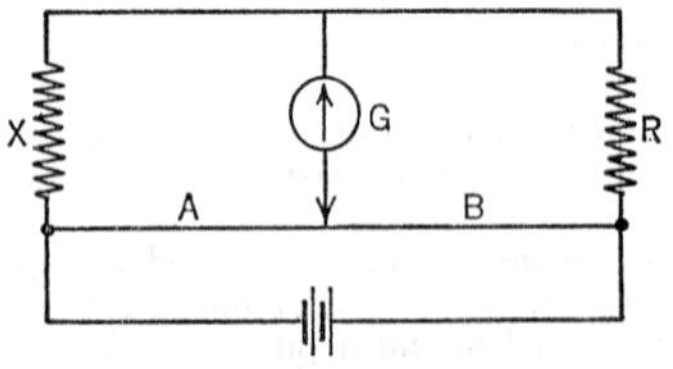

FIG. 7

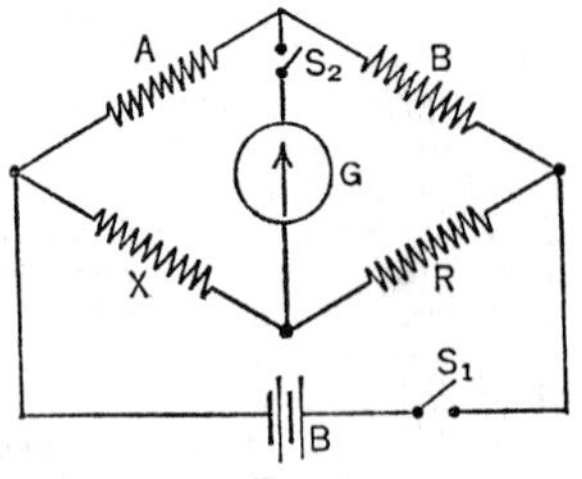

FIG. 8

WHEATSTONE BRIDGE. Figure 8 shows diagrammatically the arrangement of resistances in an ordinary Wheatstone bridge. A and B represent two coils or sets of coils, usually called the ratio coils or ratio arms. Their relative resistances must be known, but their absolute accuracy of adjustment is immaterial. R is usually a set of coils which can be connected to give any value of resistance, by small equal steps, from zero to the maximum; X is the unknown resistance, G a galvanometer connected in one diagonal of the bridge, and B the battery in the other diagonal. The operation of the bridge consists in successively adjusting R until the galvanometer shows no deflection, or until two values of R are found which differ by the smallest step and between which lies the value of R for no deflection, obtained by interpolation of the galvanometer deflections. The value of the unknown resistance is then

$$X = \frac{AR}{B} \tag{2}$$

where the letters represent the resistances of the four arms.

Forms of Wheatstone Bridge. Wheatstone bridges are made in a variety of forms, sizes, and degrees of accuracy to meet various requirements. For miscellaneous testing where the highest accuracy is not required, low cost is essential, and the bridge may be or must be taken to the object of the measurement, small portable bridges ("testing sets") are usually provided with self-contained battery and galvanometer. Their usual error limits are 0.1 per cent and 0.05 per cent for rheostat coils and ratio arms respectively. Larger bridges for laboratory purposes usually require an outside battery and galvanometer, although all the coils and the keys are usually mounted in one box. The coils are generally adjusted to a smaller limit of error than those in the portable bridges. Characteristic differences in laboratory bridges relates to (a) the use of plug or dial contacts; (b) the arrangement of the ratio coils; (c) the number and arrangement of the coils in the rheostat arms.

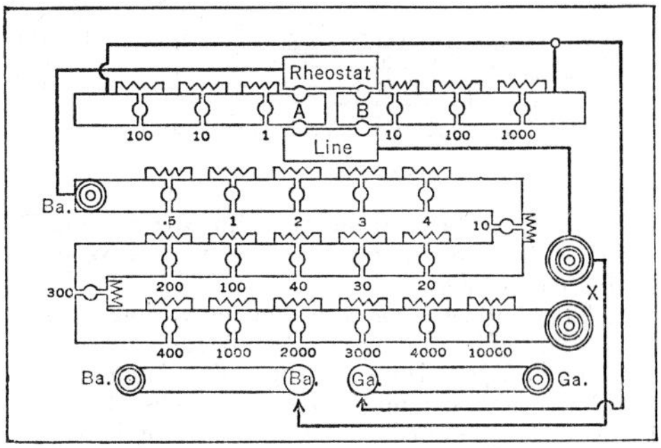

FIG. 9

The differences of arrangement of ratio coils and rheostat coils may best be brought out by brief descriptions of several classical types of bridge. Figure 9 shows the so-called post office type, of English origin, with the rheostat coils arranged on the 1, 2, 3, 4 plan. The advantage of this design is the small number of coils; the disadvantage is the necessity for adding the values of the coils unplugged in the rheostat arm. Figure 10 shows a decade bridge of the type originally devised by Professor Anthony. The resistance in the rheostat arm may be read off by

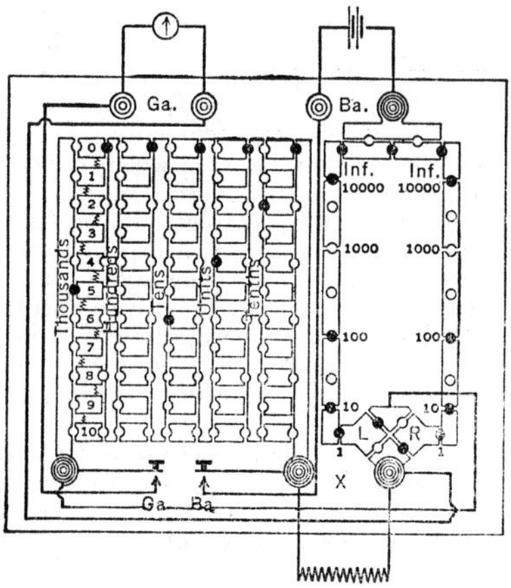

FIG. 10

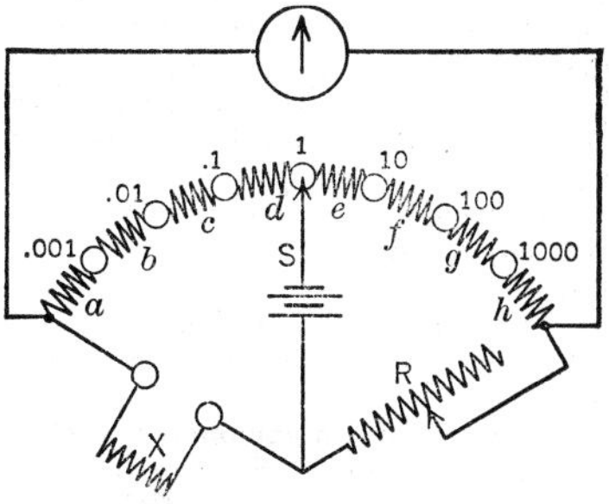

FIG. 11

inspection—an advantage which is obtained at the expense of a relatively large number of coils. An important advantage of the Anthony form of bridge is that the relative values of the coils may be checked by the user. To obtain the direct-reading feature with a relatively small number of coils, special methods of connection are used in other forms of bridge by which a nine-coil "decade" is obtained with as few as four coils.

Special Arrangements of Ratio Coils. In the Wheatstone bridge shown diagrammatically in Fig. 11 the ratio is changed by a single sliding contact in the battery circuit where variations of its resistance have no effect on the accuracy of measurement. A comparison of the improved arrangement of ratio coils indicated in Fig. 12 with the old post office arrangement of Fig. 13 shows that the former has only 2 plugs to manipulate, whereas the latter has 6.

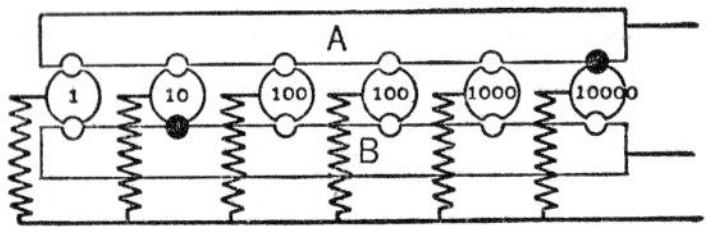

FIG. 12

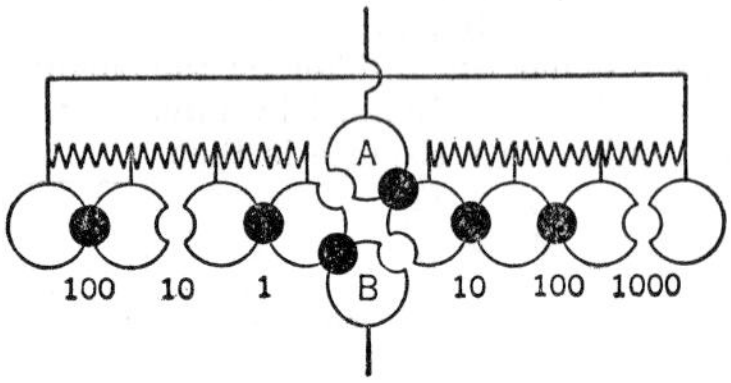

FIG. 13

Accuracy of Measurements with Wheatstone Bridge. The degree of accuracy with which resistance can be measured with a Wheatstone bridge is determined by the following conditions:

1. The limit of error to which the various coils are adjusted to their stated resistances. Limits stated by the maker usually range from 0.2 to 0.02 per cent.

2. The relative values of the coil resistances and the contact resistances at plugs, contact brushes, and binding posts. A well-fitting clean plug has a contact resistance which is definite to the order of 0.0001 ohm or better; a poorly fitting or dirty plug may fail to repeat by 0.01 ohm or more. The contact resistance between a binding post and a well-clamped wire may range from 0.0001 to 0.001 ohm. The magnitude and definiteness of the contact resistance of rotary switches depend greatly on their design. Each leaf of the laminated brush should press independently against the contact block. Rotary switches in which the thrust of the brushes is balanced about the axis of rotation are preferable, other things being equal, to those in which a considerable unbalanced force exists.

3. The relative values of the coil resistances and the insulation resistances between the contact blocks. Dust and moisture on the hard-rubber top may cause an appreciable error in high-resistance coils. Hard rubber (ebonite), which is largely used for the tops of bridges, resistance boxes, potentiometers, and volt boxes, undergoes surface **deterioration when exposed to light.** Such apparatus should never be exposed to direct sunlight and should be covered when not actually in use. The deterioration goes on, though more slowly, even in diffused daylight. Dust and other impurities present in the air of cities and manufacturing centers aggravate the deterioration, which may reach the stage of a surface film of dilute acid formed from excess sulfur present in the hard rubber.

4. The effects of changes of temperature in modifying the resistances of the coils and in setting up parasitic thermal emf's. Small errors from the latter source may be eliminated by taking the mean of the results of two measurements, between which the direction of flow of the battery current through the bridge is reversed.

5. Changes of resistance resulting from heating by the battery current. For any given type and size of resistor and given environment the temperature rise depends on the watts loss. Only very general figures can be given as a guide. Northrup states that the safe watt rating of the coils of a Wheatstone bridge ranges from 0.25 to 4 watts per coil, according to its construction, and that, as a rule, 1 watt should be the limit. If many coils are in use, lower values of watts per coil must be employed. It is better to obtain sensitivity by using a sensitive galvanometer than by using large currents through the coils.

6. The sensitivity of the galvanometer and its relative suitability (as regards its value of external resistance for critical damping) for use with given values of resistance in the bridge arms.

Best Value of Galvanometer Resistance and Preferable Location of Galvanometer. When a choice may be made between two or more moving-coil galvanometers which differ only in the resistance of their coils, the galvanometer should be selected for which the external resistance for critical damping most nearly equals the resultant resistance R_r of the arms $A + X$ in parallel with $B + R$ (Fig. 8); namely,

$$R_r = \frac{(A + X)(B + R)}{A + B + X + R} \tag{3}$$

This is the resistance of the bridge network between the points to which the galvanometer is connected, because for the condition of balance the resistance in the battery diagonal has no effect on the value of R_r. Tradition states that the galvanometer should be selected for which the coil resistance most nearly equals the value of R as given by eq. (3); that is, for example, if A, B, R, and X are each 100 ohms, a galvanometer with a 100-ohm coil is most suitable. Such a rule could be followed when moving-magnet galvanometers were used, but to apply it to moving-coil galvanometers as now made would either result in excessive overdamping or require the insertion of additional resistance with consequent loss of sensitivity. Present-day moving-coil galvanometers which are damped only through the external circuit usually require that the external resistance for critical damping must be from, say, 3 to 15 times the coil resistance. If no galvanometer is available which complies sufficiently well with this requirement, the one which most nearly meets it may be used, resistance being placed in series with it to correct any overdamping or in parallel with it to correct underdamping. Whenever one must choose between a moderate degree of overdamping or an equal degree of underdamping, underdamping saves time and has other advantages.

In practice, the user is limited to a relatively small number of values of galvanometer resistance, and galvanometer sensitivities are usually more than ample for all ordinary purposes.

The resistance of modern batteries is so low that the old question of "best location of the battery" no longer arises; in fact, it is often advisable to insert some resistance in series with the battery to prevent damage to low-resistance bridge arms. For example, particular care in this respect is necessary when the resistance of a millivoltmeter is being measured, lest the pointer be bent and the moving coil and springs be damaged by too great a current.

Precautions in Using a Wheatstone Bridge. The following rules will conduce to accuracy of measurement and maintenance of the bridge in good condition.

1. See that all binding posts are screwed down tight and that the plugs and switch contacts are clean and properly lubricated. Insert and remove plugs with a twisting motion. Use only moderate force in inserting them. After withdrawing a plug, retighten adjacent plugs, if necessary. The contact surfaces of the battery key and galvanometer key may need cleaning at infrequent intervals. This is conveniently done with a fine flat file.

2. For measuring all but very high resistances, use a battery of only a few volts, for example, one to three dry cells in series. Use the minimum voltage required to obtain the necessary sensitivity. To obtain sufficient sensitivity when measuring high resistances (say, 100,000 ohms up to several megohms) it is generally necessary to use a greater number of cells, together with very unequal ratio arms; for example, 1 ohm and 1000 ohms. In such a case care should be taken not to use a battery voltage which will pass too great a current through any coil of the bridge, and the battery should be connected in that diagonal of the bridge which gives minimum current in the bridge coils.

3. For a preliminary balance use a 1 to 1 ratio and protect the galvanometer by shunting it or preferably by inserting a resistance of several hundred ohms (or more, according to the circumstances) in series with the battery. This resistance may later be reduced or cut out entirely.

4. Avoid touching the metal parts of keys, plugs, blocks, etc., forming part of the bridge circuit. When the circuit under measurement contains appreciable capacitance or inductance, always close the battery key first, then tap the galvanometer key momentarily to note the direction of the swing; when balance is more closely approached, the galvanometer key may be held down. Always open the galvanometer key before opening the battery key when measuring coils of high time-constant, such as motor, transformer, and electromagnet windings. For these, allow time for the current to build up before closing the galvanometer key. When circuits which are substantially non-inductive are under measurement, particularly where resistances are low, it is sometimes advantageous to close the galvanometer key first, to allow the galvanometer coil to assume a "false zero" position under the action of thermal emf, and then to close the battery key and adjust the rheostat arm to produce the condition of balance with respect to this zero position. This procedure gives the same general result as the conventional one of reversing the direction of the battery current through the bridge but has the advantages of saving time and of eliminating the effect of all thermal emf, including that in the galvanometer.

5. If it is necessary to use connecting leads between the X posts of the bridge and the object of measurement, clamp the outer ends of the leads together and measure the lead resistance before making measurements on the object. This precaution is especially necessary when the leads are of flexible stranded conductor with attached terminal lugs. Defective soldered joints or the breaking of strands of wire may make the lead resistance variable, and this fact should be discovered at the beginning. If the leads are moved while the keys are depressed, small deflections of the galvanometer may occur because

the leads cut the earth's magnetic field. This effect should be carefully distinguished from deflections caused by an actual variation of the lead resistance.

6. When measuring resistances greater than 100,000 ohms, take care to avoid leakage between the leads which connect the bridge to the resistor being measured.

Care of Bridge. The hard-rubber top should be kept free from moisture and dirt, especially between the blocks; it should **never be exposed to direct sunlight** and should be **covered** when not in use. The plugs should be free from tarnish and dirt. When necessary, they should be cleaned with whiting (never with sandpaper or emery paper) and then lightly oiled; see Article 6, Resistance Boxes, under the heading Plug vs. Switch Contacts.

THOMSON (KELVIN) DOUBLE BRIDGE. When the resistance of a conductor is 0.1 ohm or less, an accurate measurement of its resistance cannot be made with a Wheatstone bridge, used in the ordinary way, because of the relatively large uncertainty introduced by contact resistances at the binding posts. Furthermore, in applications where the resistance of a conductor between certain points must be accurately measurable and remain definite in use, it is necessary to attach so-called potential leads to these points, thus converting the simple two-terminal conductor (or resistor) into a **four-terminal resistor.** Its "four-terminal resistance" is defined as the ratio of e to i, where e is the difference of potential set up between the two potential leads by a current i which enters and leaves the conductor at the "current terminals."

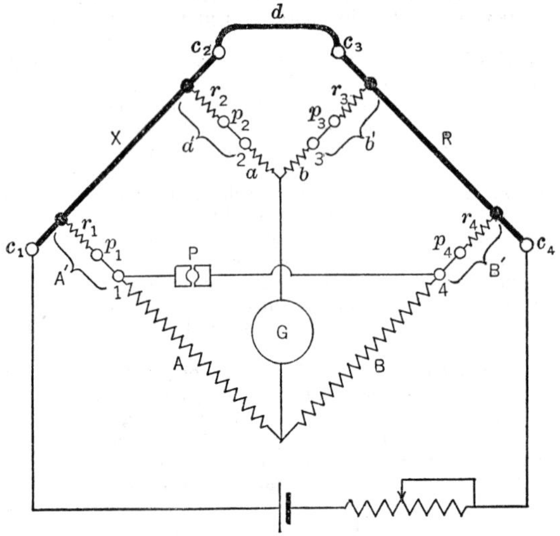

FIG. 14

A bridge network for the accurate comparison of two four-terminal resistances was devised by William Thomson (Lord Kelvin) and is known as the Thomson (or Kelvin) **double bridge.** It is shown diagrammatically in Fig. 14, in which the four-terminal resistors X and R to be compared are connected in series by the link d, which joins their current terminals c_2 and c_3. Current from a battery enters and leaves at the current terminals c_1 and c_4, respectively. The Kelvin bridge consists of the main ratio arms A and B and the auxiliary ratio arms a and b. In several forms of Kelvin bridge made by Leeds & Northrup Company these ratio arms may be selected from a comparatively few values, and the four-terminal resistance of the standard R is continuously adjustable to balance the bridge by moving the potential taps. In other forms the ratio arms B and b may be selected from a few fixed values (for example, 50, 100, and 200 ohms and values made up of two or more of these), and A and a are adjustable from 0 to 1000 ohms by steps of 0.1 ohm. The arms A and a, in some constructions, are mechanically connected so that both must have the same nominal value at all times. The binding posts 1, 2, 3, 4 in Fig. 14 are the terminals of the double bridge itself, and laboratory checks of its coils give values of resistance of A, B, a, and b up to these binding posts.

The four-terminal resistance X is actually the resistance of that part of the conductor c_1c_2 lying between two points within its structure where the leads to the potential terminals p_1p_2 effectively branch off. The resistances of the conductors from these branch points to p_1 and p_2 are denoted in Fig. 14 by r_1 and r_2. These resistances are usually very small, but in some special cases may be relatively large, for example, in some external shunts for d-c ammeters in which the four-terminal resistance is intentionally made too high and calibrating resistance (sometimes as much as several ohms) is added at r_1 or r_2. In any event it is necessary to determine the effect on the measurement, not only of r_1, r_2, r_3, and r_4, but also of the resistances of the leads joining p_1 to 1, p_2 to 2, and so on. Let the resistances of r_1 plus the lead to terminal 1, r_2 plus the lead to terminal 2, and so on, be denoted by A', a', b', and B', respectively.

It may be shown that the necessary and sufficient **condition for zero current** through the galvanometer is

$$X = \frac{A + A'}{B + B'} R + \frac{(b + b')d}{a + a' + b + b' + d} \left(\frac{A + A'}{B + B'} - \frac{a + a'}{b + b'} \right) \tag{4}$$

This expression for X contains five unknown quantities, A', B', a', b', and d, and in its present form requires a knowledge of a and b also. Most double bridges of the type under discussion are made with the intention that a/b shall equal A/B, but the lack of perfect adjustment of the coils makes these ratios slightly unequal. Certain **auxiliary procedures** have been devised which make it unnecessary to determine the values of A', B', a', b', d, and the exact relation between the ratios A/B and a/b. Each of these procedures makes it possible to adjust the lead resistances to give the condition that

$$\frac{A'}{B'} = \frac{A}{B} = \frac{a + a'}{b + b'} \tag{5}$$

whereupon the expression for X in eq. (4) assumes the convenient form

$$X = \frac{A}{B} R \tag{6}$$

One of these auxiliary procedures requires the provision of a low-resistance connection (shown in Fig. 14) which will join the terminals 1 and 4 when a plug is inserted at P. The procedure is as follows:

a. With link d in place, joining c_2 and c_3, and plug out at P, balance the bridge by varying A and a, or B and b, so as to keep the ratios A/B and a/b nominally equal. If the extra resistances A' and B' are small compared with A and B respectively, this balance will give nearly the correct value of A/B.

b. With link d and plug P in place, and A and B left unchanged, adjust the lead resistance forming part of A' (or of B') until balance again exists; this makes A'/B' equal to A/B, to a very close approximation.

c. With link d out and plug out, and previous adjustments unchanged, adjust the lead resistance forming part of a' (or of b') until balance exists; this makes $(a + a')/(b + b')$ very closely equal to $(A + A')/(B + B')$ and therefore to A/B.

d. With link replaced and plug removed, readjust the settings of A and a to get a new and more accurate balance. If the amount of readjustment required is relatively small, this balance may be regarded as final, and the value of X may be found directly from the simple relation (6). If a considerable amount of readjustment is required, as may occur when a resistance of several ohms is present in one of the potential leads of X, steps (*b*), (*c*), and (*d*) may be repeated until the balance obtained in any step (*d*) agrees closely with that obtained in the previous step (*d*).

Although the procedure outlined eliminates the effect of the resistance of the link, it is important to keep this resistance as low and as definite as possible. It is customary to use a heavy copper link with its ends arranged for mercury contacts.

Other methods of making the auxiliary adjustments in measurements with the Kelvin bridge are given in papers by Wenner and Weibel [*Bull. Bur. Standards*, Vol. 11, p. 65. (1915); reprinted as *Sci. Paper* 225].

As may be seen from a consideration of step (*c*) in the foregoing procedure, the coils composing the arms a and b need to be adjusted to only a very moderate degree of accuracy, and there is no object in knowing their corrections. It is also obvious that the unit in terms of which A and B are adjusted is immaterial because only the ratio of A to B is significant. Nothing would be gained, for example, by readjusting the coils of such a double bridge in terms of the "absolute" ohm to be adopted in the near future.

In addition to its use for very accurate measurements of low resistances in the laboratory, the Kelvin bridge, in simplified form, has been adapted by several American makers for shop and field measurements of generator and transformer windings, samples of wire and cables, etc. Such instruments usually have a range of measurement of 0.0001 to about 25 ohms. The Kelvin bridge has been used by Leeds & Northrup Company in its apparatus for continually measuring the resistance of the copper field windings of generators while in operation. The indicating (or indicating and recording) part of the apparatus is calibrated to show the mean temperature of the field winding.

HOOPE'S CONDUCTIVITY BRIDGE. This bridge is a modification of the Kelvin bridge, designed for the rapid determination of the relative conductivity of samples of wire. It is extensively used in wire factories. A diagram of the connections is shown in Fig. 15. The standard A–B and the unknown C–D are of the same metal; consequently, if care is taken that they are at the same temperature, all corrections for temperature are avoided. The arms p, r, p_1, r_1 are in the same case and are made of material of low temperature coefficient so that their relative values will not change. They are adjusted so that $p = r$ and $p_1 = r_1$; consequently at balance the resistance of c–d equals the resistance of a–b.

The sample C–D is placed alongside of scale I divided into 100 parts so that the graduations represent percentages of the total length of the scale. Accompanying the standard wire A–B is a scale H, on which are laid off a number of points corresponding to the weights of the standard length (38 in.) of a range of sizes of sample wires.

To **make a conductivity reading,** the weight of the standard length of the sample C–D is found to an accuracy of $^1/_{20}$ per cent. The contact b is set at the point on scale H corresponding to this weight, the contacts a and c being at the zero points of their respective scales. After the case has been closed a sufficient length of time to allow the standard and sample to assume the same temperature, the contact d is moved until the galvanometer shows no deflection; this will occur when the resistance between c and d is equal to that between a and b. The scale reading corresponding to the position of d

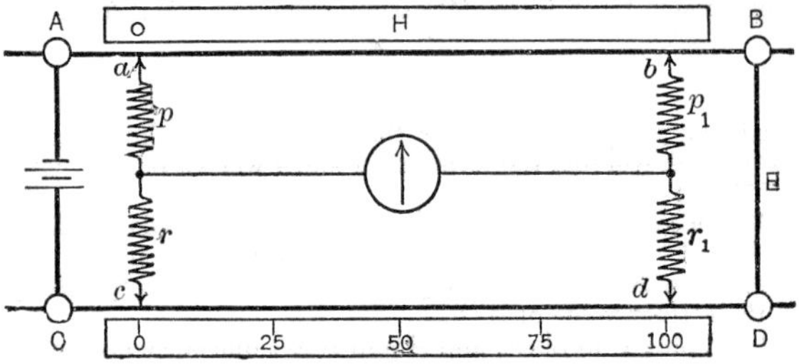

FIG. 15

for a balance is equal to the percentage conductivity. The bridge is designed for use with standard wire samples, each of which covers a range of sizes equal to 3 numbers of the American Wire Gage. Any number of standards can be supplied with a bridge, so that it can cover an extensive range of sizes and can also be used for wires of different materials. In order to keep the standard wire and the test wire at the same temperature the bridge is mounted in a metal-lined case, and the scale is read through a glass window in the case, the window being closed by a metal screen when readings are not being taken.

BRIDGES FOR MEASURING THE RESISTIVITY OF ELECTROLYTES. The resistance of an electrolyte contained in a suitable cell can be measured by using a Wheatstone bridge with an a-c detector and an a-c source substituted for the battery. Apparatus is on the market in various grades ranging from that suitable for research of the highest precision (0.001 per cent) to industrial control equipment (1 per cent). Alternating current must be used on account of the polarizing action of a direct current. Suitable **a-c sources** are: ordinary 25-cycle or 60-cycle circuits; a small 1000-cycle generator; a "microphone hummer" which is a combination of electrically operated 1000-cycle fork, a microphone, and a transformer, the secondary of which supplies 1000-cycle current; or an audio-frequency oscillator.

A **telephone receiver** is a suitable detector when a 1000-cycle source is used, but is not suitable for low frequencies such as 60 cycles. For low frequencies, an a-c galvanometer of the separately excited moving-coil type or of the vibration type is necessary. With the moving-coil galvanometer separately excited by a current in phase with the bridge currents it is sufficient if the three arms of the bridge (other than the one formed by the cell containing the electrolyte) are non-reactive. With the vibration galvanometer or the telephone receiver there must be a condition not only of resistance balance but also of phase balance, and the coils must have good a-c and good d-c characteristics. One way to obtain the phase balance is to use an adjustable air condenser connected across one arm of the bridge.

Cells for containing the electrolyte are usually made of glass and have fixed electrodes of platinum or gold. Each cell is calibrated by one or more measurements with the cell filled with liquid of known resistivity, in order that the measured *resistance* of any sample of liquid may be reduced to its *resistivity*. A cell of the immersion type for approximate measurements may be extemporized from a glass tube of known internal diameter, in the ends of which circular electrodes are secured at a measured distance apart. There should be several small holes in the wall of the tube. In use, the cell is immersed in a vessel containing the electrolyte to be tested. The resistivity is the product of the cross-sectional area of the tube and the observed resistance, divided by the distance between the electrodes.

A-C BRIDGES. Bridges for the measurement of inductance, capacitance, and other quantities of importance in a-c circuits have been developed in a great variety of forms and for use over widely varying ranges of frequency. With few exceptions they have hitherto usually been assembled in the laboratory from component pieces of apparatus. Recently, however, a-c bridges as self-contained units have been developed and are being manufactured. Space limitations prevent an extended treatment of this subject, for which the reader is referred to Hague's book, *Alternating-current Bridge Methods*, Pitman (1943) and numerous articles in *The Radio Experimenter*, a publication of the General Radio Company. See also Schering Bridge in Article 20, Testing of Insulating Materials.

8. POTENTIOMETERS

DEFINITIONS. A potentiometer is an instrument for measuring an unknown emf or potential difference by balancing it, wholly or in part, by a known potential difference produced by the flow of known currents in a network of circuits of known electrical constants. In d-c potentiometers the only network constants involved are the resistances, but in a-c potentiometers the capacitances and the self- and mutual inductances of the network also enter into the result, as well as the frequency.

In potentiometers of the **null type** the unknown emf or potential difference is completely balanced, and the galvanometer (or other detector) merely indicates the absence of a current. In the **deflection potentiometer** a part of the unknown voltage produces a current through the galvanometer, which is calibrated to read directly the corresponding part of the measured value.

USES OF THE D-C POTENTIOMETER. For the checking of d-c ammeters, voltmeters, and wattmeters the d-c potentiometer is recognized as the most accurate instrument available. In most forms it has the unique advantage that only the *relative* values of its coils are of importance, and these values, in other than complicated designs, can often be adequately checked by the user even though the absolute values of his standards are not known. With suitable resistance standards of appropriate current ratings, the potentiometer may be used to measure direct currents over a wide range; and with suitable "volt boxes," it will measure voltages up to several thousand volts.

By careful attention to the elimination of spurious sources of emf a potentiometer for measuring voltages up to 10 μv with an accuracy of 0.01 μv has been developed by R. P. Teele [*J. Research Nat. Bur. Standards*, Vol. 27, p. 217 (1941)].

Auxiliary devices for the potentiometer proper for d-c measurements are a cell or battery to supply the auxiliary current, a rheostat to adjust this current, a standard cell, and a galvanometer. In some forms of potentiometer, some or all of these auxiliary devices are built into the instrument.

PRINCIPLE OF THE D-C NULL-TYPE POTENTIOMETER. The d-c potentiometer in its simplest form (Fig. 16) consists of a uniform wire stretched over a scale divided into a number of equal divisions, for example, 1500. A storage cell *B*, having an emf of about 2 volts, and a rheostat *R* are connected in series with the wire, and two contact points are provided, *A* being fixed at the zero point and *S* being movable. *G* is a galvanometer, and *K* is a key for the galvanometer circuit. Between the terminals at *P* there may be connected at will a standard cell or any unknown voltage. At *P* is first connected a standard cell, for example, an unsaturated cadmium cell having an emf of 1.0190 volts. The terminal of *B* and that of the standard cell which are connected to the point *A* must have like polarities. The contact *S* is set at 1019 divisions on the scale, and the rheostat *R* is adjusted by trial until the closing of the key *K* produces no deflection of the galvanometer.

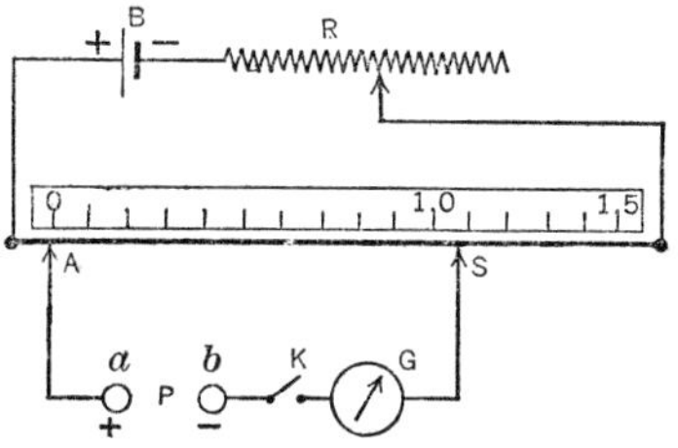

Fig. 16

To measure any other potential difference the standard cell is disconnected from the terminals *a* and *b*, and the unknown potential difference, for example, that of an ammeter shunt, is connected to *a* and *b*. The contact *S* is then moved until a position is found such that the closing of *K* produces no deflection. The corresponding position of *S* on the wire, as read from the scale, gives the value of the unknown potential difference.

Although the simple slide-wire potentiometer is sufficiently accurate for some technical purposes, its range and accuracy are inadequate for precision laboratory measurements, such as the checking of high-grade indicating instruments. Many forms of precision potentiometer have been devised, of which only a few typical ones can be described.

POTENTIOMETERS WITH ONE DIAL AND A SLIDE WIRE. An obvious modification of the above simple potentiometer is the insertion, in series with the slide wire, of a set of coils, each having a resistance equal to that of the slide wire between the zero and the point 1.0. Figure 17 shows such a potentiometer (**Leeds & Northrup type K**). The slide wire *DB* is eleven turns of manganin wound in an external screw thread cut on a Bakelite cylinder. It is shunted to a resistance of 5.5 ohms and is in series with fifteen 5-ohm coils connected to studs, over which runs the brush *M*. The rheostats *P* are for adjustment of the auxiliary current. The other slide wire, over which runs the slider *T*,

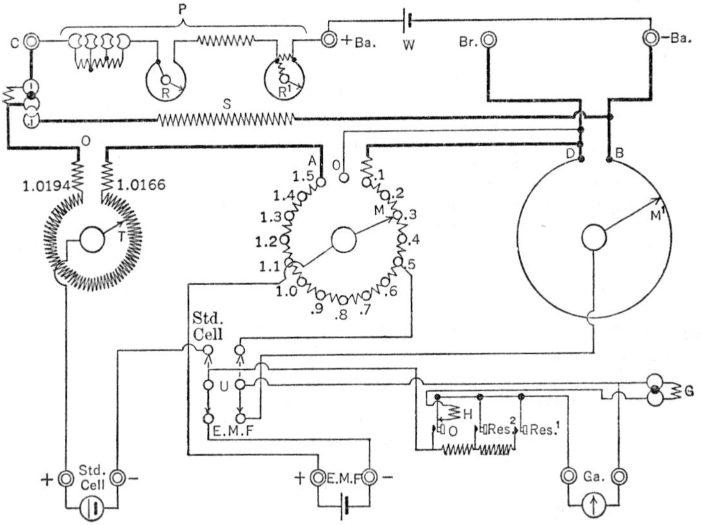

FIG. 17

adapts the potentiometer for direct reading with any standard cell having an emf within the slide-wire range. Although this range permits the use of standard cells of emf as low as 1.0166 volts, cells having an emf below 1.0180 should not be used for precise measurements because unsaturated cadmium cells of lower value are apt to be unreliable under ordinary conditions of varying room temperature.

A double-throw switch *U* inserts the galvanometer, with its keys and protective resistors, either in the standard-cell circuit or in the unknown-emf circuit. Keys Res. 1, Res. 2, and 0 are closed in the order stated to protect the galvanometer from violent deflections and the standard cell from currents which might (at least temporarily) alter its emf.

The function of coil *K*, the plug switch near it, and the coil *S* is to permit, after adjusting the battery current properly, with the plug set at 1, the shunting of nine-tenths of the current through coil *S* while keeping the total battery current unchanged. The potentiometer current is thus reduced to one-tenth of its normal value, and the normal range of 0 to 1.61 volts is thereby reduced to 0 to 0.161 volt.

The procedure of operation just outlined is typical for d-c potentiometers. For any particular potentiometer the maker's instructions should be consulted for details.

The **Type K-2 potentiometer** (Leeds & Northrup Company) retains the slide wire and set of coils of the Type K potentiometer but includes a new low range of 0 to 0.016 volt. The constancy of the auxiliary current can be checked, by reference to a standard cell, regardless of which range factor (1, 0.1, or 0.01) is in use.

The Type K-2 potentiometer has three extra taps from the main-dial coils to external binding posts. Two of these taps include exactly 1 volt and are used with a volt box in **wattmeter testing** where it is required to hold the test voltage constant at 100 volts. This feature makes it possible to check the voltage without regard to the positions of the dial switch and the slider and in effect makes the potentiometer read directly in watts.

One of the preceding taps and a third tap serve in the checking of the eleven turns of the slide wire against the first eleven coils of the main dial.

The Type K-2 potentiometer differs from the Type K in the following structural details: All contacts are enclosed; the battery rheostats are all dial-controlled; the change from emf to standard-cell check and the change of range are effected by dial switches instead of the rocker switch and plug switch, respectively, of the Type K.

The Rubicon Type B potentiometer is very similar in purpose and performance to the Type K but uses a slide wire of only one turn with a third dial of intermediate value to cover the range of the long Type K slide wire.

WHITE COMBINATION POTENTIOMETER. This is a low-resistance instrument designed to minimize parasitic thermal emf and is intended particularly for precise measurements of temperature with thermocouples. A simplified diagram of the White "single potentiometer" is shown in Fig. 18. In the White "double potentiometer" an additional set of dials is provided, to permit measurements with two separate thermocouples at widely different temperatures without extensive and repeated changes of the dial settings.

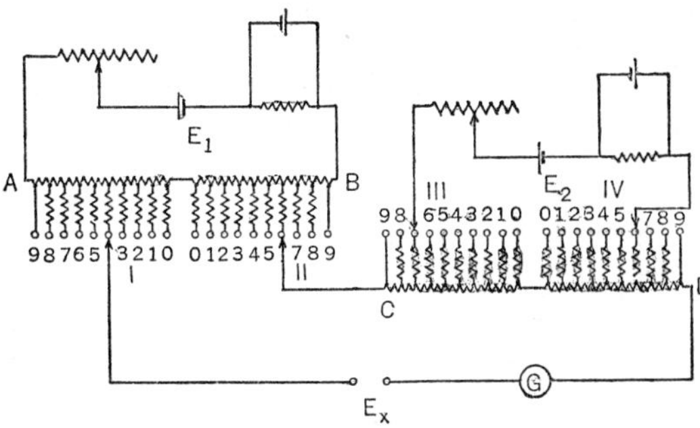

Fig. 18

WENNER POTENTIOMETER. This is a recent low-resistance five-dial potentiometer, in which the galvanometer circuit resistance is maintained substantially constant at either 40 or 174 ohms. The instrument has two ranges, 0 to 1.9111 and 0 to 0.19111 volt. It is described by Behr in *Rev. Sci. Insts.*, Vol. 3, p. 109 (1932).

WOLFF FIVE-DIAL POTENTIOMETER, FEUSSNER TYPE. In this potentiometer, shown diagrammatically in Fig. 19, the auxiliary current entering at $B+$ flows through the nine 100-ohm coils of the upper right-hand dial, then through the coils of the lower halves of the 0.1-, 1-, and 10-ohm double dials, through the 1000-ohm coils, then through the upper halves of the 10-, 1-, and 0.1-ohm double dials. These three double dials vary the resistance between the zero studs of two upper dials from 0 to 99.9 ohms by steps of 0.1 ohm while keeping the total resistance between the terminals *B* constant at 14,999.9 ohms. With the auxiliary current commonly used (0.0001 amp)

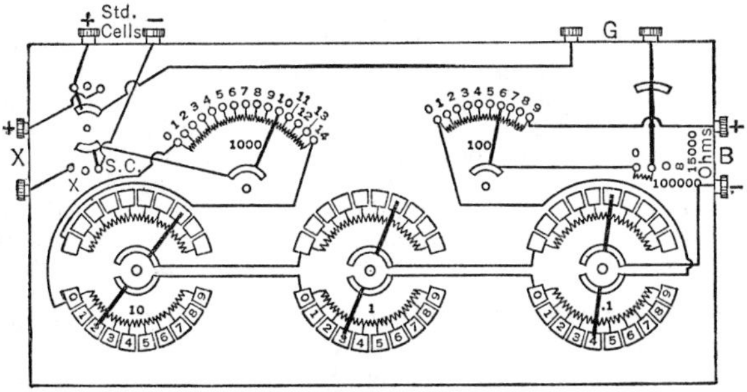

Fig. 19

the range of measurement is 0 to 1.5 volts, approximately. An external rheostat is required for adjusting the current, which may be supplied by a single storage cell or by two dry cells in series. In either case it is not necessary to open the battery circuit when the potentiometer is out of use. In one form of Wolff potentiometer, in order to check the value of the auxiliary current, it is necessary to set the dials to the value of the emf of the standard cell and to set the double-pole double-throw switch in the upper left-hand corner to S.C. In a later form a separate standard-cell dial is provided, having eleven steps for 1.0180 to 1.0190 volts. This feature makes it possible to check the auxiliary current quickly without the necessity of setting the regular five dials to the value of the standard cell.

EPPLEY FIVE-DIAL POTENTIOMETER, FEUSSNER TYPE. This potentiometer is similar to the Wolff potentiometer in plan of circuits and values of coils but has a number of refinements, including: the elimination of six of the twelve contact resistances in the measuring circuit, an additional standard-cell dial by which the standard-cell setting can be made to one more decimal place (namely, to 0.00001 volt), copper binding posts to minimize parasitic thermal emf, built-in rheostats for the adjustment of the auxiliary current, and provision for oil immersion and control of oil temperature. The additional standard-cell dial is obtained by bringing out taps from one of the 1000-ohm coils.

INCREASING THE RANGE OF VOLTAGE MEASUREMENT. To measure voltages greater than the maximum directly measurable with the potentiometer, an accessory variously known as a "multiplier," "potential divider," or "volt box" is used. (The first two names imply respectively that the accessory *multiplies* the range of the potentiometer or *divides* the line voltage to get a fraction which can be measured.) It consists essentially of a high resistance to be connected to the points between which the voltage is to be measured, with the potentiometer connected across a fractional part of the resistance. The reading of the potentiometer, multiplied by the reciprocal of this fraction, gives the value of the voltage to be measured. Volt boxes are usually multirange. Two general types of volt-box circuit are shown in Fig. 20. In (*a*) the total resistance is varied to change the range, and in (*b*) the resistance across which the potentiometer is connected is changed. In (*a*) there is the danger of burning out the coils by applying too high a voltage to a part of the total resistance meant for measurements of relatively low voltage. In (*b*) this danger does not exist but, when the lower ranges are in use, the volt box may introduce an excessive amount of resistance into the measurement circuit, thereby reducing the sensitivity. There is the further disadvantage in arrangement (*b*) that, if a high voltage is applied to the line posts while the switch is set for a low ratio, an abnormally high voltage will be impressed on the potentiometer, and the galvanometer is very easily damaged. This trouble is possible even with arrangement (*a*), but experience shows that it is very rare. Experience shows also that separate "line" binding posts for the various ranges are safer than a rotary switch.

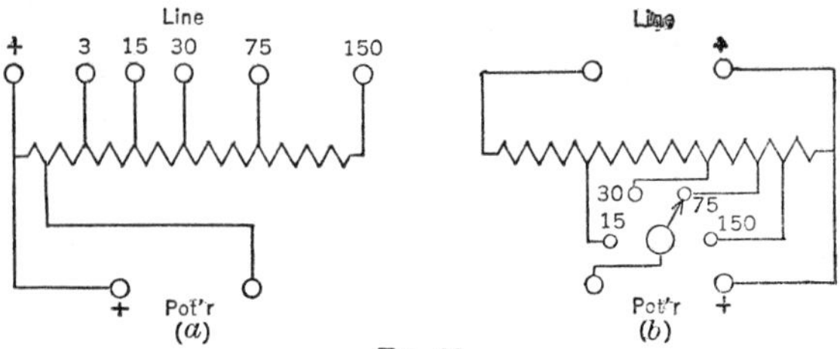

FIG. 20

The combination of a potentiometer with a volt box is **not a null instrument,** and for accurate work a correction must be made for the resistance of the leads connecting the line terminals of the volt box to the points between which the voltage is to be measured. For a given resistance in the leads, this correction is smaller as the resistance of the volt box is greater. Furthermore, a high value of "ohms per volt" is desirable because it reduces the self-heating of the volt box. The considerations which act to set an upper limit to the ohms-per-volt are the greater cost and lower stability of high-resistance coils, the greater effect of leakage over the insulating surfaces, and the reduction in sensitivity caused by the large resistance which the volt box contributes to the measurement circuit [see Silsbee and Gross, *J. Res. Nat. Bur. Standards*, Vol. 27, p. 269 (1941); reprinted as *Res. Paper* 1419].

MEASUREMENT OF DIRECT CURRENTS. The current to be measured is passed through a four-terminal resistance standard having its potential terminals joined to the emf terminals of the potentiometer. The measured potential difference, divided by the four-terminal resistance of the standard, gives the value of the current in amperes. For convenience, the values of the four-terminal resistance are often decimal submultiples of an ohm (0.1, 0.01, etc.), in which case the numerical result of the measurement in volts, multiplied by 10, 100, etc., gives the value of the current in amperes.

DEFLECTION POTENTIOMETERS. In the potentiometers which have been described the galvanometer is commonly used merely as a detector, although it is feasible to interpolate between adjacent steps on the lowest dial by noting the corresponding galvanometer deflections. In the deflection potentiometer the greater part of the measured result is read from the dial, and the galvanometer is calibrated to indicate directly the last two figures of the result. The only difference between the circuits of an ordinary potentiometer and those of a deflection potentiometer is that the deflection potentiometer has supplementary coils which maintain a constant resistance in the galvanometer circuit for all positions of the switch of the potentiometer dial and those of the battery rheostats, as well as for all values of resistance in the volt box and in the shunts used for current measurements.

The accuracy and precision of the deflection potentiometer are intermediate between those of an ordinary ammeter or voltmeter and those of a high-precision null-type potentiometer. It is used for the checking of ammeters, voltmeters, and wattmeters and for current and voltage measurements in photometry.

PRECAUTIONS IN THE USE OF D-C POTENTIOMETERS. The emf of a storage cell is unsteady immediately after charging and decreases rapidly for a time. The cell is unfit for use with a potentiometer during this period. The emf of storage cells and of dry cells changes with temperature, and for the most accurate results the cells should be protected from abrupt changes of temperature. The temperature coefficient of emf of dry cells is about twice that of storage cells, the values being about 0.02 per cent and 0.01 per cent per degree centigrade, respectively.

Potentiometers should not be unnecessarily exposed to dust, moisture, or chemical fumes and should be *kept covered* when not in use. If the top is of hard rubber, it should *not be exposed to direct sunlight,* which will cause deterioration of its surface and greatly impair its surface resistivity.

The galvanometer and all the other accessories of the potentiometer should be carefully insulated to avoid **leakage currents** from other d-c circuits. It is often necessary or desirable to apply the Price "guard-wire" method of protection, in which the potentiometer and each of its auxiliaries, including the galvanometer, are supported by, but insulated from, metal plates which in turn are preferably insulated from the table or other support. These guard plates are connected together and to that pole of the circuit, supplying the current or the voltage under measurement, which is more nearly at earth potential. The line terminal of the volt box (the + terminal in Fig. 20) which in use is connected directly to the potentiometer is to be connected to this pole of the supply circuit, and the shunt used for current measurement should be connected in this side of the line. If the potential of this side of the line unavoidably differs appreciably from earth potential, it is advisable to apply the guard plates to the observer's chair also.

The wires connecting the guard plates together should be protected against breakage. W. P. White has proposed a series connection of the guard plates, so that a single test of the resulting loop checks all the wires for continuity.

When the atmosphere is very dry, as in heated buildings during the winter, disturbances of the galvanometer may result from static charges on the clothing of the observer. Particular difficulty has been found experienced with fur-trimmed garments.

The following practice should be observed in order to avoid accidental short-circuiting of the standard cell. In connecting a standard cell to a potentiometer, first attach *both* wires to the potentiometer, then connect the outer ends of the wires to the cell. In disconnecting the cell, first remove the wires from it. It is well to disconnect one wire from the cell at the end of the day even though the apparatus will be used the next morning. The cell should never be removed with the two wires still attached to it.

A-C POTENTIOMETERS. In the simple potentiometer of Fig. 16 let a milliammeter be inserted in circuit to measure the current in the slide wire. It should be of such type as to give equally accurate readings on direct current and on alternating current of the frequency to be used. After the direct current through the slide wire has been adjusted to the standard value, the reading of the milliammeter should be noted; if this reading depends somewhat on the direction of current flow through the milliammeter, the mean of two readings, between which the direction of flow through the milliammeter is reversed, should be taken. The battery B is then replaced by an a-c source, and the rheostat R is adjusted until the reading of the milliammeter is the same as on direct current. The unknown alternating voltage is then connected at P, and the d-c galvanometer replaced by a suitable a-c galvanometer or other detector. The unknown voltage must have the same frequency as the current through the potentiometer. This condition is ordinarily fulfilled by taking both currents from the same source or from generators which are mechanically coupled. To obtain a balance, the potential difference taken from the slide wire not only must have the same magnitude as the unknown potential difference but also must be in phase with it. Means must therefore be provided for shifting the phase of the current through the slide wire. In the **Drysdale potentiometer** this is done by means of a phase-shifting transformer. This potentiometer is of the polar-coordinate type; i.e., it gives the magnitude of the unknown emf and its phase with respect to a reference vector.

The first a-c potentiometer of the cartesian-coordinate type was devised by **Larsen,** and its principle of operation is shown in Fig. 21. An alternating current i of a standard value flows through the slide wire AB and the primary winding of the adjustable mutual inductor M. The potential difference which balances the unknown potential difference E_x is made up of a component taken from the slide wire, in phase with i, and a quadrature component $2\pi fiM$ induced in the secondary winding of N. Since the potentiometer must usually be available for a range of frequencies, the scale of M must be marked in terms of mutual inductance, and its readings must be multiplied by $2\pi fi$ to reduce them to volts.

Erlang [*J. Inst. Elec. Engrs.* (London), Vol. 51, p. 794 (1913)] **and Pedersen** [*Electrician* (London), Vol. 83, p. 525 (1919)] proposed other forms of a-c potentiometer, especially suitable for investigations of telephone apparatus, but not so well adapted for power frequencies. The **Gall a-c potentiometer,** made by H. Tinsley & Company, is an assembly of two potentiometers to form a single piece of apparatus. In use, the two potentiometers are traversed by currents mutually in quadrature. The **Geyger a-c potentiometer,** made by Hartmann & Braun, consists of two center-zero slide wires, one fed from the secondary of an insulating transformer through a non-inductive resistance, the other from the secondary of an air-core transformer having its primary in series with the first slide wire. The **Campbell-Larsen a-c potentiometer,** made by Cambridge Instrument Company, is shown diagrammatically in Fig. 22. The operation is like that of the Larsen potentiometer, but the troublesome multiplication by $2\pi fi$ is avoided. To

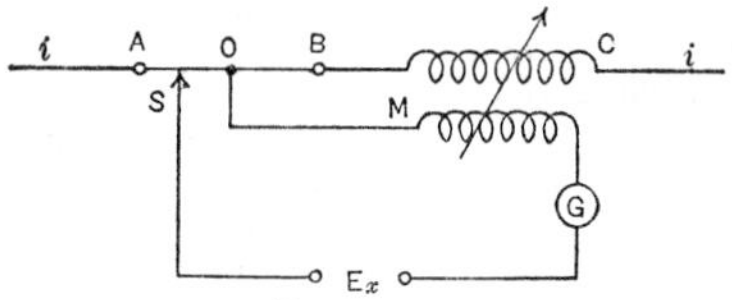

FIG. 21

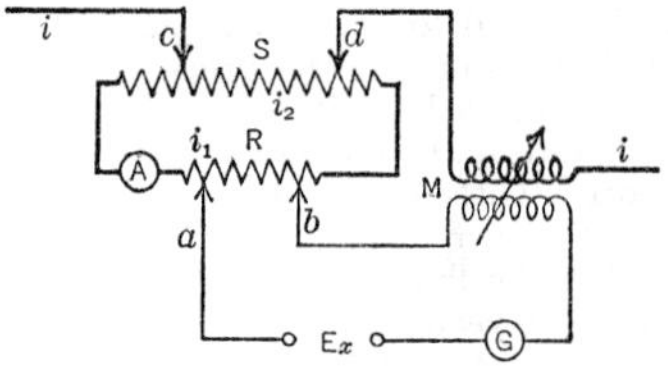

FIG. 22

balance the unknown voltage E_x, two components are required; one is taken from the non-inductive resistance R by adjusting the sliders a and b, and the other by adjusting the value of the mutual inductance M. The current i through the primary of M is composed of the current i_1 through R plus the current i_2 through S. The current i_1 is set at a standard value by means of the device A, which consists of a heating device, a thermocouple, a resistor, and a galvanometer to indicate when the effective value of i_1 is equal to an auxiliary direct current of standard value. By adjusting the position of the contacts c and d the total current i corresponding to the standard value of i_1 may be made such that the scale of M indicates directly the emf induced in the secondary of M, in millivolts, at the given frequency. The range of frequency for which this can be done is 25 to 2000 cps. The scales for the sliders a and b are also graduated in millivolts.

Uses of the A-C Potentiometer. Although the a-c potentiometer was originally advocated as a standard instrument for checking ammeters, voltmeters, and wattmeters, the inherent limitations on its accuracy and the other improved apparatus now available for a-c instrument checking make it inadvisable to purchase an a-c potentiometer for this purpose. The accuracy of measurement of the effective value of a given alternating emf or current with the average a-c potentiometer may be assumed to be from 0.5 to 1 per cent for the usual power and lighting frequencies, when the wave form of the quantity to be measured and that of the current through the potentiometer are closely alike and sinusoidal. The real **usefulness of the a-c potentiometer** lies in engineering research, testing, and investigation, where an accuracy of about 1 per cent is adequate and where the measurements either could not be made at all by other means or would be very difficult to carry out. The a-c potentiometer has been used with good results for studying the actions which take place in induction meters and other a-c measuring apparatus; in magnetic analysis; and for measuring the magnetic properties of laminated steel, the ratio and phase angle of current transformers, and the amplification factors of amplifiers. The success of work of this kind depends very much on the manner of setting up and manipulating the potentiometer and the related apparatus, and those who wish to apply the a-c potentiometer method to their problems may consult with advantage papers by Spooner, Geyger, and von Krukowski (see Bibliography, p. 5-110).

9. LABORATORY MEASUREMENTS OF D-C POWER AND ENERGY

For the precise measurement of d-c power in the laboratory, where the current and voltage are supplied by batteries and can be kept very constant, the accepted technique is to measure the current and the voltage by means of a potentiometer, volt box, resistance standard, and standard cell, as indicated in Article 8 on Potentiometers. Measurements of d-c power with such equipment are made in checking the correctness of wattmeters. The same procedure, with the addition of the measurement of time intervals, is the accepted method for the checking of portable standard d-c energy meters (watthour meters).

10. LABORATORY MEASUREMENTS OF ALTERNATING CURRENT, VOLTAGE, AND POWER

GENERAL. The measurement of the **instantaneous values** of alternating current, voltage, and power requires instruments such as are described in Article 18 on Oscillographs. The majority of a-c measurements, however, refer to conventionally defined and universally recognized **effective** (or **root-mean-square**) values. An alternating current of 1 amp, effective value, flowing through a circuit having a resistance of 1 ohm, will liberate heat at the same rate as if 1 amp of direct current were flowing through this circuit. It follows that electrothermic instruments are suitable in principle for the measurement of effective values of alternating currents. Although such instruments are necessary for measurements at high frequencies, they have limitations which have caused two other types to be preferred for precision laboratory measurements at the frequencies used for power and lighting. These are the electrostatic and the electrodynamic types.

METHODS USED BY NATIONAL LABORATORIES. Methods employing electrostatic instruments are preferred for precision a-c measurements by the British National Physical Laboratory. The National Bureau of Standards at Washington prefers instruments of the electrodynamic type. These instruments are of the reflecting type, with the moving-coil system carried by upper and lower suspension strips. The deflection is observed by means of a lamp-and-scale arrangement. The moving-coil system has two coils of unlike instantaneous polarity in order to make the instrument astatic as to uniform stray fields. The arrangement and connection of the windings in order to serve for current, voltage, and power measurements are the same (in the older instruments long used at the Bureau) as the corresponding features in the portable and laboratory-standard electrodynamic instruments so widely used, and the reader is referred to the subsequent treatment of such instruments. In Bureau practice, these instruments are checked on direct current, using a potentiometer, standard cells, and resistance standards, immediately before and after a pair of a-c observations; i.e., they serve to carry over or **transfer** the a-c measurements to the units embodied in these basic d-c standards. They are accordingly spoken of as "transfer instruments." This method avoids most of the sources of error which attend the use of standard instruments which are checked at only occasional intervals (of weeks or months).

In recent years **other transfer instruments** have been devised at the Bureau to overcome these limitations. One is the Harris suppressed-zero electrodynamic voltmeter, the other the Silsbee composite-coil electrodynamic ammeter. In the Harris voltmeter the entire deflecting torque is balanced by the torque of the suspensions, as in the older transfer instruments, but the coil system is restrained by two stops so that it cannot return to zero when the current ceases to flow. The continual, nearly constant stress in the suspensions eventually results in a steady state such that "zero drift" is eliminated to a very high degree. In the Silsbee composite-coil ammeter each of the fixed coils and each of the moving coils is wound with a pair of wires, and in operation the deflecting torque resulting from the alternating current is nearly balanced by the counter torque set up by a precisely adjusted direct current in the other circuit through the coil system. Small variations in the magnitude of the alternating current are measured by observing the deflection of the coil system to the right or to the left of its central zero position.

TRANSFER INSTRUMENTS FOR GENERAL USE. Reflecting electrodynamometers are used for the measurement of very small amounts of power at low power factor, as in the testing of samples of cable for dielectric loss. For many purposes, however, in commercial laboratories electrodynamic voltmeters, wattmeters, and ammeters of the pivoted type, with pointer and scale, are of satisfactory accuracy and are much more convenient. The inductance of their circuits can be kept small by suitable design, and corrections are usually slight and not difficult to apply. For the measurement of very small amounts of power at high voltage, however, or for measurements at frequencies much above the usual power and lighting frequencies, the effects of the unavoidable inductances of the electrodynamic instrument become sufficiently pronounced to warrant the use of an electrometer in spite of its more delicate construction and much longer period.

11. ELECTRICAL INDICATING INSTRUMENTS

GENERAL. In an **electrical indicating instrument** the present value of the quantity under observation is indicated by the position of a pointer relative to a scale. The term instrument may include the instrument proper, as defined below, or it may include the instrument proper together with any necessary auxiliary devices, such as shunts, shunt leads, resistors, reactors, capacitors, or instrument transformers. The **instrument proper** consists of the actuating mechanism together with those auxiliary devices (scale, resistors, shunts, etc.) which are built into the instrument case or otherwise made an integral part of its structure.

A Standard covering many features of instrument performance and terminology was issued by the American Standards Association as C39.1–1938. This has been supplemented by ASA War Standard C39.2–1944, covering in great detail $2^1/_2$-in. and $3\ ^1/_2$-in. panel instruments; C39.4–1943, giving dimensions of thermocouple converters; C39.5–1943, covering external ammeter shunts; and C39.3–1943, describing the shock-testing mechanism for electrical indicating instruments. These Standards are in substantial agreement with Joint Army-Navy Spec. JAN–I–6–1944.

DISTINCTION BETWEEN ERROR AND CORRECTION; ACCURACY OF AN INSTRUMENT. The **error of indication** of an instrument is found by subtracting from the value which it indicates the true value of the quantity measured by it, as determined by reference to some standard of a higher order than the instrument itself. It is the quantity which must be **algebraically subtracted from** the indication to obtain the true value of the quantity. A positive error thus denotes that the instrument "reads high," i.e., that its indication is greater than the true value. The **correction to the indication** is found by subtracting the indicated value from the true value; i.e., the correction has the same numerical value as the error of indication, but the opposite sign. **The correction, algebraically added** to the indication, gives the true value.

The **accuracy of an instrument** may be expressed in various ways, some of which are ambiguous and likely to cause disagreement between maker and purchaser. It is generally defined as the ratio of the error of indication to some like quantity taken as a basis. If the indication were taken as the basis, it would be necessary to have a sliding scale of accuracy with the permissible percentage error increasing in some manner as the deflection decreased. The majority of American manufacturers consequently favor the simple definition of accuracy as the percentage ratio of the limit of error at any point on the scale to the full-scale value. In instruments having uniform scales this is equivalent to saying that no division mark is out of its proper position by more than a specified distance. This definition of accuracy applies to the usual case in which the zero point is at the lower end of the scale. For instruments in which the zero point is in an intermediate position on the scale, the sum of the full-scale readings to the right and to the left of the zero point should be used as the basis for reckoning the accuracy.

The accuracy of an electrical instrument depends on the operating principle; the quality of the design, materials, and workmanship; and the conditions of use. In general, the accuracies stated by American instrument makers are about as follows: switchboard instruments, 4 in. diameter and larger, 1 per cent; 3 in. diameter and smaller, 2 per cent; laboratory-standard instruments, 0.1 per cent; portable instruments, highest-grade, 0.2 to 0.25 per cent, intermediate-grade, 0.5 to 0.75 per cent; miniature instruments, 2 per cent. Exceptions: high-grade portable polyphase wattmeters, 0.5 per cent; rectifier instruments, for which relatively large errors must be tolerated; frequency meters, power-factor meters, phase-angle meters, and capacitance meters, the accuracy of which must be stated in special ways for which the makers' catalogs should be consulted.

Although a high degree of accuracy is important in much laboratory work and in such field work as acceptance tests of generating units, too much stress should not be laid on accuracy requirements for switchboard instruments used for operation and control of equipment. Accuracy in such instruments is usually subservient to reliability and ruggedness.

CLASSIFICATION OF INDICATING INSTRUMENTS. Electrical indicating instruments are classified **as to use** into portable instruments, switchboard (back-connected) instruments, wall-type (front-connected) instruments, and laboratory standard instruments. These terms are all self-defining. As to the **kind of protection** afforded to the instrument mechanism by the case, instruments are classified as dustproof, moisture-proof, weather-proof, gas-proof, drip-proof, splash-proof, and hermetically sealed.

As to the **principle of operation,** indicating instruments are divided into the following classes: (1) electrodynamic; (2) permanent-magnet moving-coil; (3) moving-iron; (4) induction; (5) electrothermic; (6) electrostatic; (7) electronic (thermionic); (8) rectifier. These terms will be defined separately.

Electrodynamic instruments depend for their operation on the reaction between the current in one or more moving coils and the current in one or more fixed coils. Electrodynamic instruments may be used for measuring current, voltage, power, and other quantities, in circuits carrying direct, alternating, or rectified currents, provided that for alternating and rectified currents the **effective** (rms) values are desired. Variations of electrodynamic instruments include "iron-clad" types, having magnetic shielding which does not form a part of the torque-producing system, and "iron-cored" types, in which magnetic material is used to increase the operating torque or to provide greater angular scale deflection. For the range of frequencies used in power and lighting, electrodynamic instruments are either independent of the frequency or may readily be compensated to avoid the small corrections which would otherwise be necessary. Air-core instruments have no hysteresis ("magnetic lag") and are the most suitable pointer-type transfer instruments for relating a-c measurements at power frequencies to the basic d-c standards. Electrodynamic voltmeters are affected by variation in frequency to a greater extent than moving-iron voltmeters of corresponding grade. In general, they are suitable for frequencies of not more than 150 cycles, although they can be specially calibrated for measurements at a higher frequency, up to 3000 cycles. They are made for ranges of 1 volt to 750 volts.

Permanent-magnet moving-coil instruments depend for their operation on the reaction between the current in a coil and the field of a permanent magnet within which the coil is arranged to move. These instruments are widely used for the measurement of direct current and voltage and are the only instruments suitable for the measurement of the **average value** of rectified current and voltage. They have the advantages of a uniform (linear) scale, high ratio of torque to weight of moving element, and small amount of power required to sustain a given deflection. The intensity of the magnetic field in which the coil moves and the nearly closed magnetic circuit make these instruments, when unshielded, much less sensitive to disturbance from stray magnetic field than corresponding instruments of the electrodynamic and the moving-iron types. Their only disadvantage appears to be that their smaller clearances make them more expensive to manufacture and to repair.

The development of modern magnet materials such as Alnico (an alloy of aluminum, nickel, cobalt, and iron) has permitted many variations in the magnetic circuit, with increased sensitivity and immunity to stray fields as well as reduction in size and weight. Notable among these newer designs are the concentric style, using segments of alnico set inside a steel ring, and the concentric scale type, adapted for long angular scales.

For the measurement of direct currents of the order of 0.1 amp and below, the entire current to be measured flows through the moving coil. The upper limit is set by the maximum current which can be carried by the two spiral springs which serve the dual purpose of providing the counter torque and of carrying the current to and from the coil. For higher ranges a part of the current to be measured must be by-passed around the moving coil through a "shunt" which may be within the instrument for moderate ranges (25 to 60 amp, or higher when special facilities are provided for the escape of heat from the shunt). For higher ranges an external shunt is used, and the fact that this shunt may be located in the line wherever convenient and may be connected to the instrument proper by very small wires constitutes one of the great advantages of the permanent-magnet moving-coil ammeter.

To maintain an acceptably low temperature coefficient in shunted ammeters requires care in the design, choice of materials, and construction to keep substantially constant the ratio of the current in the shunt to the current in the moving coil.

In the permanent-magnet moving-coil voltmeter a resistor is connected in series with the moving coil to bring the total resistance up to a value such that the desired upper limit of voltage (for example, 150 volts) produces full-scale deflection. The current required for full-scale deflection is small; consequently the power loss in watts is relatively low, and the series resistor may be conveniently mounted inside the instrument proper up to 300 to 750 volts, depending on the design and physical dimensions of the instrument. For higher values the voltmeter is used in series with an additional resistor, which is called an **external series resistor** if the voltmeter is not calibrated for use without it. If this additional resistor extends the voltage range beyond some particular value for which the voltmeter is already complete, it is properly called a **multiplier.**

The widespread use of d-c ammeters and voltmeters for radio purposes has stimulated the development of small inexpensive instruments of good operating characteristics. In particular, the current required for full-scale deflection has been greatly reduced, so that d-c voltmeters may be had as stock articles which require only 1 ma or even 0.2 ma for full-scale deflection. It has become usual practice, however, to state the reciprocal of this current, which (for the voltmeters just mentioned) is 1000 or even 5000 "ohms per volt." This form of statement has two advantages, namely, the numerical value of the ohms per volt is always greater than unity, and the more sensitive the voltmeter the larger is the number.

Direct-current ammeters and voltmeters are conveniently checked by means of a potentiometer and suitable accessories, as described in Article 8 on Potentiometers.

Moving-iron instruments depend for their operation upon the reactions resulting from the current in one or more fixed coils acting upon one or more pieces of soft iron (or magnetically similar material) located within the field of the coils. Various forms of moving-iron instrument are distinguished chiefly by mechanical features of their construction and operation. For example, in moving-iron instruments of the **plunger type** a core or plunger of magnetic material is drawn into a solenoid when current flows through it. In the **magnetic-vane** instrument one or more thin vanes of magnetic material are mounted within one or more coils in such a way that, when current flows, the vanes are drawn closer to the coils, where the magnetic field is more intense. In the **repulsion type** of moving-iron instrument the flow of current through the coil magnetizes one or more fixed pieces of magnetic material and also one or more pieces which are free to move, and the force of repulsion between like magnetic poles causes relative motion of the pieces to take place. Some types of moving-iron instrument utilize both the force of repulsion and that of attraction to produce a usable scale over a wide angle.

Moving-iron instruments have the advantages of simplicity, ruggedness, low cost of construction, and usefulness for some d-c measurements and the measurement of **effective values** of alternating and of rectified currents and voltages. They were relatively crude and inaccurate in the early days of instrument construction, but have been greatly improved by skilful design and the use of modern nickel-iron alloys of very high permeability and low hysteresis. Inherently, their scales are approximately quadratic, but a great deal of variation from the quadratic law is possible by suitable shaping of the coils and the moving-iron parts and by varying the initial (zero) position of the iron parts with respect to the coils. Moving-iron ammeters not only are inherently very free from heating errors but also require only a small amount of power for full-scale deflection, namely, from 1 to 2 watts for the standard forms in ranges of, say, 25 amp and less. They have small frequency errors and may be used with frequencies ranging from 15 to 500 cycles.

The **moving-iron voltmeter** is limited in frequency variation because of its relatively high time-constant. This limitation is not felt over the range of present power and lighting frequencies (15 to 60 cycles) but becomes important at higher frequencies. The frequency error varies as the square of the product of the time-constant and the frequency and can be greatly reduced by using a capacitor in shunt with part of the series resistor. Moving-iron voltmeters may be used on 500-cycle circuits, for example, when specially calibrated for this frequency.

Induction instruments depend for their operation on the reaction between the magnetic field of one or more fixed coils and the currents induced in a moving conducting part such as a disk or a drum. Induction instruments have some pronounced advantages but have the manufacturing drawback of needing to be made and adjusted for one value (or at most two values) of frequency. Induction ammeters and voltmeters have been displaced in this country by moving-iron instruments which can be used interchangeably on any commercial power or lighting frequency. Similarly, the induction wattmeter, limited to a single frequency, has been superseded by the electrodynamic wattmeter for the same reason.

Electrothermic instruments depend for their operation on the heating effect of a current. Two markedly distinct types are (1) the **expansion type,** commonly known as the hot-wire type, and (2) the **thermocouple type.** In the hot-wire instrument the indication is effected by mechanically multiplying the expansion of a wire or strip which is heated by the current. Hot-wire instruments have been virtually superseded by instruments of the thermocouple type, in which one or more thermojunctions are heated (directly or indirectly) by the current and supply a direct current which actuates a suitable d-c instrument mechanism. Both the hot-wire and the thermocouple types may be made as ammeters, voltmeters, and wattmeters. Among the **advantages of the thermocouple instrument** over the hot-wire instrument are: much lower power loss, greater sensitivity, and ability to measure currents down to very much smaller values. Like the hot-wire instruments, the thermocouple instruments are **easily damaged by overload.** The burn-out currents of vacuum-type thermocouples are normally considered a function of the maximum current rating. Typical values are 25 ma for couples rated 4 ma and 800 ma for couples rated 400 ma maximum. The burn-out currents of typical air-type thermocouples are usually in the order of 1.6 times the maximum current rating and about 2.5 times the normal rating. The "thermal inertia" which brings most instruments through momentary inadvertent overloads without thermal damage is so small in the thermocouple instrument that even careful users must count on burning out a heating element occasionally.

Ammeters of the thermocouple type are suitable for use on alternating current of frequencies ranging from the lowest up to those employed in radio circuits. They are not ordinarily used at power frequencies but have advantages in special cases where minimum resistance and negligible inductance are necessary. They are indispensable for measurements at audio and higher frequencies. Self-contained thermocouple ammeters are made in ranges from 2 ma up to 50 amp. They are supplied with external heating elements up to 1000 amp. Multiple-range milliammeters, because of the necessity for passing part of the current through a shunt around the heating element, are limited to a maximum frequency of 8000 to 25,000 cps, depending upon the shunt design. This limitation does not apply to multiple-range instruments in which a separate heating element is used for each range.

Voltmeters of the thermocouple type are made in sensitivities at least as high as 500 ohms per volt and may be used with frequencies up to 7000 cycles without objectionable frequency errors.

Electrostatic instruments depend for their operation on the forces of attraction or repulsion between electrically charged fixed metal parts and moving metal parts. The relatively weak forces developed in this way tend to cause an undesirably long period and other operating drawbacks and limit the application of electrostatic instruments to particular purposes for which their special advantages make them preferable or even indispensable. They are useful when voltage measurements are to be made without drawing any appreciable current from the source, or when the cost of series resistors or of voltage transformers for high voltages would be prohibitive. They are designed either for direct connection to the points between which the voltage is to be measured, or through an electrostatic voltage divider analogous to the volt box used with a potentiometer. The use of such voltage dividers is feasible for a-c voltage measurements only. Special forms of electrostatic instruments are used as ground detectors on high-voltage lines.

Electronic instruments (vacuum-tube voltmeters) utilize the amplifying and rectifying properties of electronic tubes and make possible measurements of adequate precision with little or no drain of energy from the circuit being measured. Many types are also applicable at radio frequencies and are therefore widely used in the communication field. The book by Rider [*Vacuum Tube Voltmeters*, John F. Rider Publications, New York (1941)] contains an excellent summary of the status as of 1941. Development in this field is still very rapid. By suitable circuit modifications electronic instruments can function as microammeters (down to 10^{-14} amp) or as megohmmeters.

The indications of an electronic voltmeter may be a measure of either (*a*) the rms value, (*b*) the average value, or (*c*) the crest value of one-half wave of the applied a-c voltage, depending on the part of the tube characteristic on which the instrument operates. Those of type (*a*) necessarily have square-law scales compressed near zero and are seldom used, although they read correctly the effective value for any wave form. Those of type (*c*) are commonly calibrated to read the rms value of a sine-wave which has the same crest value as that of the applied voltage.

RECTIFIER INSTRUMENTS. A rectifier instrument is the combination of an instrument responding only to direct current and a device for rectifying the alternating current to be measured. Instruments of this type are usually of the permanent-magnet moving-coil type in combination with **copper oxide rectifiers** in the Wheatstone-bridge full-wave arrangement. They have been developed to meet the need for low-range a-c ammeters and voltmeters of very small power consumption and are used for those a-c measurements for which the relatively large power consumption of electrodynamic, thermocouple, and moving-iron instruments makes them ill-adapted or even inapplicable, also for a-c measurements where only moderate accuracy (say 5 per cent) is sufficient and ruggedness and ability to sustain overloads are important. The deflection of a rectifier instrument is proportional to the **average value** of the rectified wave, but, since the user is generally concerned with the effective value, it is customary to mark the scales in terms of the effective (rms) values for an assumed (usually sine) wave form. For a sine-wave form the ratio of the average to the rms value is 0.9 when using a perfect full-wave rectifier and 0.45 when using a perfect half-wave rectifier. These ratios are reduced when using practical commercial rectifiers, which invariably have, in the conducting direction, a resistance characteristic which is a function of current density, and, in the non-conducting direction, leakage which is a function of the inverse voltage across the rectifier. This means that a full-scale sensitivity of, for example, 1 ma a-c requires the use of a d-c instrument rated approximately 0 to 0.88 ma with a full-wave rectifier or approximately 0 to 0.44 ma with a half-wave rectifier. Low-range voltmeters, say 10 volts or below, are characterized by an a-c scale which is crowded toward zero because the variation in effective resistance of the rectifier is not masked by the series resistance; and, accordingly, the current applied to the moving coil is a function of both the rectifier-current ratio and the rectifier resistance. The resistance of a single 0.23-sq-cm typical instrument-rectifier disk in terms of the applied voltage is shown in Fig. 23.

Milliammeters and high-range voltmeters are usually characterized by essentially linear scales. For certain applications it is possible to make a-c voltmeters of the rectifier type with as high as 20,000 ohms per volt, and microammeters giving full-scale deflection for as little as 50 μa. Rectifier instruments not only are subject to relatively large **wave-form errors and frequency errors** but also are noticeably affected by changes in room temperature.

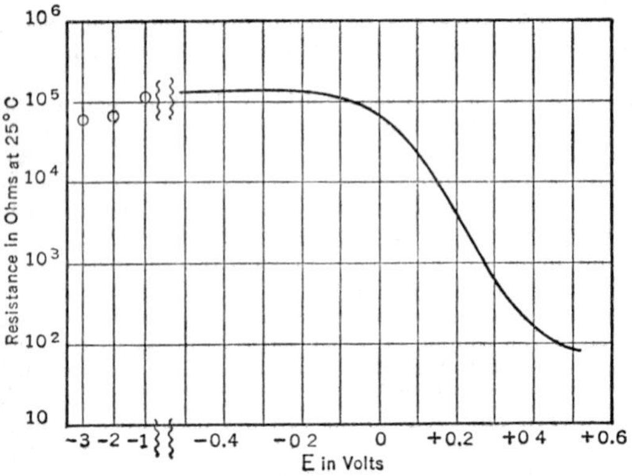

Fig. 23

Rectifier instruments, particularly those using copper oxide rectifiers, tend to read low with increasing frequency, principally because of the capacitive shunting effects of the rectifier disks which parallel various parts of the circuit and because of the increase in effective resistance of the multiplier resistors. For details concerning the magnitude of these effects, methods of compensation for temperature, and other factors, reference should be made to the maker's instructions and to the various items in the Bibliography, p. 5-111. Rectifier ammeters are available in ranges of from 50 μa to 750 μa. Higher current ratings can be made by shunting, but this practice is not recommended. Rectifier voltmeters are made in ranges of from 0.5 to 300 volts. Higher ranges usually employ external multiplier resistors. Ratings of 300 volts and higher are frequently built in volt-ohm-milliammeters and special test kits.

Rectifier instruments are of relatively recent origin and have not reached the stage of standardization which characterizes most other types. The makers should be consulted for the latest information about rectifier instruments for specific requirements, particularly since the accuracy of these instruments is so dependent on temperature and wave form.

WATTMETERS. General. A wattmeter measures the average value of the product ei of the instantaneous voltage by the instantaneous current, i.e., the average power. Types of instrument which may be adapted for use as wattmeters include the electrodynamic, induction, electrostatic, hot-wire, and thermocouple. Even the moving-iron instrument may be so used, but, although patents for such wattmeters have been issued, no practical use appears to have been made of them. For the use of electrostatic wattmeters, see **Electrometers,** Article 14. For most purposes the electrodynamic wattmeter is preferred, and the discussion in this section is limited to this type.

The **electrodynamic wattmeter** consists essentially of two coils, one of relatively coarse wire carrying the load current, and the other, of fine wire, in series with a non-inductive resistor, connected in parallel with the load. The current in the fine-wire coil being very nearly in phase with the voltage e across the load, the instantaneous torque will be closely proportional to the instantaneous power. Because the natural period of vibration of the moving coil is so much greater than the period of an alternating current, the coil does not follow the rapid cyclic changes of the instantaneous power but assumes a deflected position which indicates the average power. The torque produced by the currents in the coils is opposed and balanced by a counter torque, usually that of a spring. In nearly all wattmeters the current coil is made the fixed coil because the current in it is ordinarily too great to be carried by the springs. Usually the fixed coils surround the moving coil.

Current and Voltage Ranges of Wattmeters. It was formerly customary to construct portable wattmeters for currents up to 200 to 400 amp, but the present tendency is to avoid such high current ranges. It is very difficult to construct a 200-amp current coil so that the distribution of an alternating current over the cross-section of the coil will be accurately identical (as to production of torque) with the distribution of the direct current with which the wattmeter must be checked. Also, as the current range is made greater, it becomes increasingly difficult to guard against errors occasioned by stray field from the current leads to the wattmeter. Recent great improvements in the ratio and phase-angle accuracy of current transformers make it possible to obtain better accuracy from a 5-amp wattmeter and a current transformer than from a wattmeter of large current rating connected directly in the line. However, when a wattmeter is to be used to measure the total power in an a-c circuit in which flows a d-c component, the current transformer cannot be used. Although single-phase switchboard wattmeters are made for currents up to 400 amp, there is a recent tendency to restrict the current range to 20 amp or less.

Wattmeters are relatively expensive instruments; and, when they are used to check portable standard watthour meters, the accuracy of their indications is of great commercial importance. It is therefore usual to build portable wattmeters with two current ranges and with two (sometimes three) voltage ranges, thus facilitating the obtaining of the largest possible deflection for the amount of power under measurement. Abroad, wattmeters are frequently made with three current ranges in the ratio 1 : 2 : 4 and, as an extreme case, with six current ranges, which, however, are obtained at a sacrifice of 50 per cent in torque.

A **polyphase wattmeter** contains two single-phase wattmeter mechanisms with the moving coils attached to a common shaft. Its use makes it possible to measure power in two-phase or three-phase three-wire circuits with a single instrument.

Methods of Connecting Wattmeter in Circuit; Correction for Power Loss in Wattmeter. A wattmeter may be connected to the load which it is to measure in either of the two ways illustrated in Figs. 24 and 25. When connected as in Fig. 24, the current

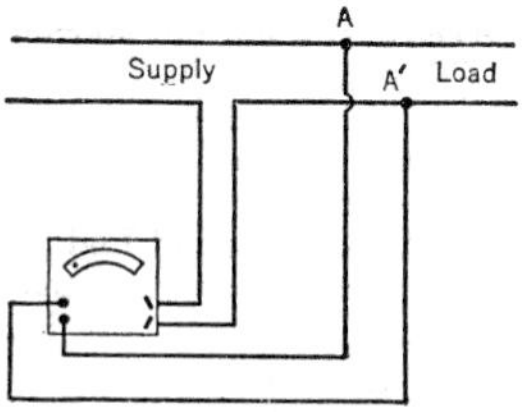

FIG. 24

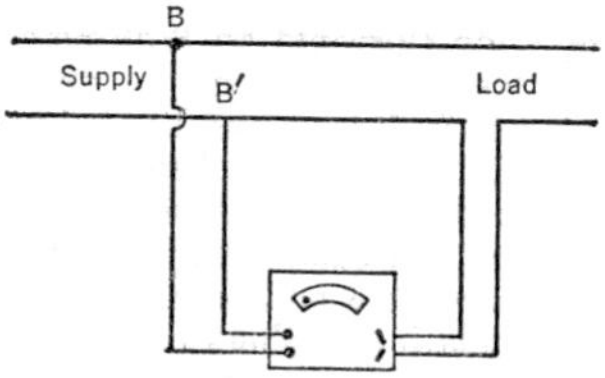

FIG. 25

through the current circuit of the wattmeter is equal to the vector sum of the current in the load and that in the voltage circuit of the wattmeter; hence the wattmeter will read the sum of the watts in the load and the watts loss in the voltage circuit, which is equal to E^2/R_p, where R_p is the resistance of the voltage circuit and E the voltage across the load. For precise work this correction should always be made, unless the wattmeter is of relatively high current range or is "compensated"; see below.

In the second scheme of connection (Fig. 25) the current through the current circuit is the same as that taken by the load, but the voltage across the voltage circuit of the wattmeter is higher than the voltage across the load by the drop through the wattmeter current coil, and the wattmeter reads too high by an amount equal to the loss in this coil. This loss is equal to R_cI^2, where R_c is the resistance of the current circuit and I the load current; it is usually less than the loss in the voltage circuit. Hence, if no correction is made for the wattmeter loss, the scheme of connections shown in Fig. 25 should be used. If the highest accuracy is required, especially with a wattmeter of low current rating, the connections shown in Fig. 24 should be used, and a correction should be applied.

When a voltmeter is connected to the load across AA' in Fig. 24, while the wattmeter reading is being taken, a correction may need to be made for the power taken by the voltmeter. If we call R_v the resistance of the voltmeter and multiplier, if any, and E the voltage across the load, the loss in the voltmeter is E^2/R_v. In most cases the voltage-circuit losses in the wattmeter and voltmeter are best determined by a direct measurement with the wattmeter, using the test voltage and leaving the load circuit open. If the voltmeter is connected across BB' in Fig. 25, the power loss in it is not read by the wattmeter, and the voltmeter reading equals the load voltage plus the impedance drop in the wattmeter current coil.

Compensation for Loss in Voltage Circuit of Wattmeter. In the so-called compensated wattmeters a stationary compensating coil, in series with the moving coil, is placed so that the current through it produces on the moving coil a torque equal and opposite to that produced by this same current flowing through the current coil. To make this compensation accurate for all positions of the moving coil, the compensating coil must be intermingled with the current coil. A compensated wattmeter should always be connected to the circuit, as shown in Fig. 24; when it is so connected, no correction for the loss in the wattmeter is necessary. As a check on the correctness of connections, the load circuit may be opened, whereupon the wattmeter reading should be zero.

Phase Angle of Voltage Circuit of a Wattmeter. The inductance of the moving coil of a simple (uncompensated) wattmeter tends to cause the alternating current in the voltage circuit to lag slightly behind the impressed voltage. Distributed capacitance in the series resistor tends to counteract this tendency, but only slightly in the flat-card resistors now generally used. The resultant lag of the current is known as the phase angle α of the wattmeter. It is usually about 2 to 4 min (at 60 cycles) in high-grade wattmeters of the deflection type having a voltage circuit rated at 125 to 150 volts; in the Westinghouse precision wattmeter, 150-volt range, it is about 12 min; it varies directly as the frequency and inversely as the voltage rating. The error which it introduces into the measurement of a-c power becomes relatively greater as the power factor is lowered. The angle α is taken as positive when the positive reactance of the wattmeter moving coil predominates over any negative (capacitive) reactance in the voltage circuit. Such a wattmeter will give a larger reading on an inductive load than on a noninductive load, for the same actual power.

Effect of Mutual Inductance on Wattmeter Reading. In most wattmeters of the deflection type the mutual inductance between the fixed coils and the moving coil changes sign at approximately half-scale deflection. For any other relative position of the coils there is an error caused by mutual inductance which is negligible in good wattmeters of current ratings of 5 amp or more and voltage ratings of 75 volts or over, used with ordinary power frequencies. For higher frequencies, lower current ranges, or lower voltage ranges it may be advisable to check the magnitude of the mutual-inductance error; formulas for this purpose are given in *Trans. AIEE*, Vol. 39, pp. 563–567 (1920).

Effect of Eddy Currents on Wattmeter Reading. Eddy currents in fixed conducting masses near the fixed coils will set up magnetic fluxes which combine with the flux of the current coil to form a resultant flux which differs from that of the current coil in both magnitude and phase. There is no simple way to correct for the resulting error, and it is up to the maker to use as little metal as possible, and that of high resistivity, and to arrange it in such a way as to minimize the presence of closed paths for eddy currents. Appreciable eddy-current errors resulted in one case from the inadvertent omission of insulating washers in a metal assembly holding the field coils. In another, short-circuited turns in the compensating winding caused a similar result. One or more short-circuited turns in a current coil would produce the same kind of result but to a greater extent; such a condition, however, should be revealed by its effect on the d-c performance of the wattmeter.

Compensation for Inductance of Voltage Circuit of Wattmeter. The phase angle of a wattmeter may be reduced to zero (i.e., the inductance of its voltage circuit may be compensated) by shunting a suitable capacitor around a fraction, n, of the resistance of the voltage circuit. If the voltage circuit has a total resistance of R ohms and an inductance of L henrys, the required capacitance in farads is given by

$$C = \frac{Ln^2}{R^2} \tag{7}$$

The practical application of this formula requires precautions which are summarized in *Trans. AIEE*, Vol. 39, pp. 562–563 (1920).

Correction Factor for Self-inductance of the Voltage Circuit of a Wattmeter. If a wattmeter is used to measure the power of a sinusoidal current I which lags by an angle θ behind the emf E, and if the inductance and resistance of the wattmeter voltage circuit are L and R, so that $\tan \alpha = \omega L/R$ is the tangent of the angle of lag in the voltage circuit, the correction factor C by which the reading of the wattmeter must be multiplied to correct it for the effect of inductance in the voltage circuit is

$$C = \frac{1 + \tan^2 \alpha}{1 + \tan \alpha \tan \theta} \tag{8}$$

In a wattmeter which is fit to use, $\tan \alpha$ will be small and its square negligible. Consequently the term $\tan^2 \alpha$ may be omitted from the numerator. This formula becomes useless for power factors near zero, at which point tan θ becomes infinite. To avoid this difficulty, **Drysdale** proposed an **additive correction** for the self-inductance of the voltage circuit. It is outlined, with numerical examples, in *Trans. AIEE*, Vol. 39, pp. 560–562 (1920).

The tangent formula in eq. (8) has been much used in America in a form which is trigonometrically identical with it but in which the wattmeter phase angle α is extended to include also the phase displacements introduced by a current transformer and a voltage transformer.

Correction of Wattmeter Reading for Phase Angles of Wattmeter and of Instrument Transformers. Let α denote the phase angle of the wattmeter as above defined, and β

Table 1. Phase-Angle Correction Factors: for Lagging Current When $(\alpha + \beta - \gamma)$ Is Positive; for Leading Current When $(\alpha + \beta - \gamma)$ Is Negative

Phase Angle $(\alpha+\beta-\gamma)$	Apparent Power Factor ($\cos\theta_2$)													
	0.10	0.15	0.20	0.25	0.30	0.40	0.50	0.60	0.70	0.80	0.90	0.95	0.99	1.00
5′	0.9855	0.9904	0.9929	0.9944	0.9954	0.9967	0.9975	0.9981	0.9985	0.9989	0.9993	0.9995	0.9998	1.0000
10′	0.9711	0.9808	0.9857	0.9887	0.9907	0.9933	0.9950	0.9961	0.9970	0.9978	0.9986	0.9990	0.9996	1.0000
15′	0.9566	0.9712	0.9786	0.9831	0.9861	0.9900	0.9924	0.9942	0.9955	0.9967	0.9979	0.9986	0.9994	1.0000
20′	0.9421	0.9616	0.9715	0.9775	0.9815	0.9867	0.9899	0.9922	0.9940	0.9956	0.9972	0.9981	0.9992	1.0000
25′	0.9276	0.9520	0.9643	0.9718	0.9768	0.9833	0.9874	0.9903	0.9926	0.9945	0.9965	0.9976	0.9989	1.0000
30′	0.9131	0.9424	0.9572	0.9662	0.9722	0.9800	0.9848	0.9883	0.9911	0.9934	0.9957	0.9971	0.9987	1.0000
40′	0.8842	0.9232	0.9429	0.9549	0.9629	0.9733	0.9798	0.9844	0.9881	0.9912	0.9943	0.9961	0.9983	0.9999
50′	0.8552	0.9040	0.9286	0.9436	0.9536	0.9666	0.9747	0.9805	0.9851	0.9890	0.9929	0.9951	0.9978	0.9999
1° 0′	0.8262	0.8848	0.9143	0.9323	0.9444	0.9599	0.9696	0.9766	0.9820	0.9868	0.9914	0.9941	0.9974	0.9998
10′	0.7972	0.8656	0.9000	0.9209	0.9350	0.9531	0.9645	0.9726	0.9790	0.9845	0.9899	0.9931	0.9969	0.9998
20′	0.7682	0.8464	0.8857	0.9096	0.9257	0.9464	0.9594	0.9687	0.9760	0.9823	0.9885	0.9921	0.9964	0.9997
30′	0.7392	0.8271	0.8714	0.8983	0.9164	0.9397	0.9543	0.9648	0.9730	0.9800	0.9870	0.9911	0.9959	0.9997
40′	0.7102	0.8079	0.8571	0.8869	0.9071	0.9329	0.9492	0.9608	0.9699	0.9778	0.9855	0.9900	0.9954	0.9996
50′	0.6812	0.7886	0.8428	0.8756	0.8978	0.9262	0.9441	0.9568	0.9668	0.9755	0.9840	0.9890	0.9949	0.9995
2° 0′	0.6521	0.7694	0.8284	0.8642	0.8884	0.9194	0.9389	0.9529	0.9638	0.9732	0.9825	0.9879	0.9944	0.9994
10′	0.6231	0.7501	0.8141	0.8529	0.8791	0.9127	0.9338	0.9489	0.9607	0.9709	0.9810	0.9869	0.9939	0.9993
20′	0.5941	0.7308	0.7997	0.8415	0.8697	0.9059	0.9287	0.9449	0.9576	0.9686	0.9795	0.9858	0.9934	0.9992
30′	0.5650	0.7115	0.7854	0.8301	0.8603	0.8991	0.9235	0.9409	0.9545	0.9663	0.9779	0.9847	0.9928	0.9990
40′	0.5360	0.6923	0.7710	0.8187	0.8510	0.8923	0.9183	0.9369	0.9515	0.9640	0.9764	0.9836	0.9923	0.9989
50′	0.5069	0.6730	0.7566	0.8073	0.8416	0.8855	0.9132	0.9329	0.9483	0.9617	0.9748	0.9825	0.9917	0.9988
3° 0′	0.4779	0.6537	0.7422	0.7959	0.8322	0.8787	0.9080	0.9288	0.9452	0.9594	0.9733	0.9814	0.9912	0.9986
10′	0.4488	0.6344	0.7279	0.7845	0.8228	0.8719	0.9028	0.9248	0.9421	0.9570	0.9717	0.9803	0.9906	0.9985
20′	0.4198	0.6151	0.7135	0.7731	0.8134	0.8651	0.8976	0.9208	0.9390	0.9547	0.9701	0.9792	0.9900	0.9983
30′	0.3907	0.5957	0.6991	0.7617	0.8040	0.8583	0.8924	0.9167	0.9359	0.9523	0.9686	0.9781	0.9894	0.9981
40′	0.3616	0.5764	0.6847	0.7503	0.7946	0.8514	0.8872	0.9127	0.9327	0.9500	0.9670	0.9769	0.9888	0.9980
50′	0.3326	0.5571	0.6702	0.7388	0.7852	0.8446	0.8820	0.9086	0.9296	0.9476	0.9654	0.9758	0.9882	0.9978
4° 0′	0.3035	0.5378	0.6558	0.7274	0.7758	0.8377	0.8767	0.9046	0.9264	0.9452	0.9638	0.9746	0.9876	0.9976
10′	0.2744	0.5185	0.6414	0.7160	0.7663	0.8309	0.8715	0.9005	0.9232	0.9429	0.9622	0.9735	0.9870	0.9974
20′	0.2453	0.4991	0.6270	0.7045	0.7569	0.8240	0.8663	0.8964	0.9201	0.9405	0.9605	0.9723	0.9864	0.9971
30′	0.2163	0.4798	0.6125	0.6930	0.7474	0.8171	0.8610	0.8923	0.9169	0.9381	0.9589	0.9711	0.9857	0.9969
40′	0.1872	0.4604	0.5981	0.6816	0.7380	0.8103	0.8558	0.8882	0.9137	0.9357	0.9573	0.9699	0.9851	0.9967
50′	0.1581	0.4411	0.5837	0.6701	0.7285	0.8034	0.8505	0.8841	0.9105	0.9333	0.9556	0.9687	0.9844	0.9964
5° 0′	0.1290	0.4217	0.5692	0.6586	0.7191	0.7965	0.8452	0.8800	0.9073	0.9308	0.9540	0.9675	0.9838	0.9962
10′	0.0999	0.4024	0.5548	0.6472	0.7096	0.7896	0.8400	0.8759	0.9041	0.9284	0.9523	0.9663	0.9831	0.9959
20′	0.0708	0.3830	0.5403	0.6357	0.7001	0.7827	0.8347	0.8717	0.9008	0.9260	0.9507	0.9651	0.9824	0.9957

Interpolation for correction factors corresponding to values of $(\alpha + \beta - \gamma)$ lying between those given in the table may be made without error. Interpolation for correction factors corresponding to values of $\cos\theta_2$ lying between those given in the table may be made without exceeding an error of 0.0010 in the sections of the table lying between the heavy black lines; outside of these sections, and in all cases where the adjacent values of $\cos\theta_2$ are separated by the heavy black lines, the maximum error in interpolation will exceed 0.0010.

and γ the phase angles of the current transformer and the voltage transformer, respectively. Let P_2 denote the wattmeter reading corrected for scale error and multiplied by the product of the corrected ratios of the current and voltage transformers, E the voltmeter reading corrected for scale error and multiplied by the ratio of the voltage transformer, and I the ammeter reading corrected for scale error and multiplied by the corrected ratio of the current transformer. Then the **apparent power factor** is

$$\cos \theta_2 = \frac{P_2}{EI} \tag{9}$$

and the **true power** is

$$P = P_2 \frac{\cos (\theta_2 + \alpha + \beta - \gamma)}{\cos \theta_2} \tag{10}$$

The angle θ_2 is to be taken as positive when the current lags behind the voltage, and as negative when the current leads. In the case of β and γ a positive angle denotes that the reversed secondary quantity leads the primary quantity. When this definition is used, formula (10) shows that a leading phase angle γ tends to offset the effect of a leading phase angle β. Values of the correction factor $\cos (\theta_2 + \alpha + \beta - \gamma)/\cos \theta_2$ are given in Tables 1 and 2. Note carefully the conditions, stated at the head of each table, to which each table applies. It should be noted that formula (10), like the tangent formula (8), becomes useless as the power factor of the load approaches zero.

Table 2. Phase-Angle Correction Factors: for Lagging Current When $(\alpha + \beta - \gamma)$ Is Negative; for Leading Current When $(\alpha + \beta - \gamma)$ Is Positive

Phase Angle $(\alpha+\beta-\gamma)$	Apparent Power Factor ($\cos \theta_2$)													
	0.10	0.15	0.20	0.25	0.30	0.40	0.50	0.60	0.70	0.80	0.90	0.95	0.99	1.00
5′	1.0145	1.0096	1.0071	1.0056	1.0046	1.0033	1.0025	1.0019	1.0015	1.0011	1.0007	1.0005	1.0002	1.0000
10′	1.0289	1.0192	1.0142	1.0113	1.0092	1.0067	1.0050	1.0039	1.0030	1.0022	1.0014	1.0010	1.0004	1.0000
15′	1.0434	1.0288	1.0214	1.0169	1.0139	1.0100	1.0075	1.0058	1.0044	1.0033	1.0021	1.0014	1.0006	1.0000
20′	1.0579	1.0383	1.0285	1.0225	1.0185	1.0133	1.0101	1.0077	1.0059	1.0043	1.0028	1.0019	1.0008	1.0000
25′	1.0723	1.0479	1.0356	1.0281	1.0231	1.0166	1.0126	1.0097	1.0074	1.0054	1.0035	1.0024	1.0010	1.0000
30′	1.0868	1.0575	1.0427	1.0338	1.0277	1.0200	1.0151	1.0116	1.0089	1.0065	1.0042	1.0028	1.0012	1.0000
40′	1.1157	1.0766	1.0569	1.0450	1.0369	1.0266	1.0201	1.0154	1.0118	1.0087	1.0056	1.0038	1.0016	0.9999
50′	1.1446	1.0958	1.0711	1.0562	1.0461	1.0332	1.0251	1.0193	1.0147	1.0108	1.0069	1.0047	1.0020	0.9999
1° 0′	1.1735	1.1149	1.0853	1.0674	1.0553	1.0398	1.0301	1.0231	1.0177	1.0129	1.0083	1.0056	1.0023	0.9998
10′	1.2024	1.1340	1.0995	1.0787	1.0645	1.0464	1.0351	1.0269	1.0206	1.0151	1.0097	1.0065	1.0027	0.9998
20′	1.2313	1.1531	1.1137	1.0898	1.0737	1.0530	1.0400	1.0308	1.0235	1.0172	1.0110	1.0074	1.0030	0.9997
30′	1.2601	1.1722	1.1279	1.1010	1.0829	1.0596	1.0450	1.0346	1.0264	1.0193	1.0123	1.0083	1.0034	0.9997
40′	1.2890	1.1913	1.1421	1.1122	1.0921	1.0662	1.0500	1.0384	1.0292	1.0214	1.0137	1.0091	1.0037	0.9996
50′	1.3178	1.2104	1.1562	1.1234	1.1012	1.0728	1.0549	1.0421	1.0321	1.0235	1.0150	1.0100	1.0040	0.9995
2° 0′	1.3466	1.2294	1.1704	1.1346	1.1104	1.0794	1.0598	1.0459	1.0350	1.0256	1.0163	1.0109	1.0044	0.9994
10′	1.3755	1.2485	1.1845	1.1457	1.1195	1.0859	1.0648	1.0497	1.0379	1.0276	1.0176	1.0117	1.0047	0.9993
20′	1.4043	1.2675	1.1986	1.1569	1.1286	1.0925	1.0697	1.0535	1.0407	1.0297	1.0189	1.0126	1.0050	0.9992
30′	1.4331	1.2866	1.2127	1.1680	1.1377	1.0990	1.0746	1.0572	1.0435	1.0318	1.0202	1.0134	1.0053	0.9990
40′	1.4618	1.3056	1.2268	1.1791	1.1469	1.1055	1.0795	1.0610	1.0464	1.0338	1.0215	1.0142	1.0055	0.9989
50′	1.4906	1.3246	1.2409	1.1902	1.1560	1.1120	1.0844	1.0647	1.0492	1.0359	1.0227	1.0150	1.0058	0.9988
3° 0′	1.5194	1.3436	1.2550	1.2013	1.1650	1.1185	1.0893	1.0684	1.0520	1.0379	1.0240	1.0158	1.0061	0.9986
10′	1.5481	1.3626	1.2691	1.2124	1.1741	1.1250	1.0942	1.0721	1.0548	1.0399	1.0252	1.0166	1.0063	0.9985
20′	1.5768	1.3816	1.2832	1.2235	1.1832	1.1315	1.0990	1.0758	1.0576	1.0419	1.0265	1.0174	1.0066	0.9983
30′	1.6056	1.4005	1.2972	1.2346	1.1923	1.1380	1.1039	1.0795	1.0604	1.0439	1.0277	1.0182	1.0068	0.9981
40′	1.6343	1.4195	1.3113	1.2456	1.2013	1.1445	1.1087	1.0832	1.0632	1.0459	1.0289	1.0190	1.0071	0.9980
50′	1.6630	1.4384	1.3253	1.2567	1.2103	1.1509	1.1136	1.0869	1.0660	1.0479	1.0301	1.0197	1.0073	0.9978
4° 0′	1.6916	1.4573	1.3393	1.2677	1.2194	1.1574	1.1184	1.0906	1.0687	1.0499	1.0313	1.0205	1.0075	0.9976
10′	1.7203	1.4763	1.3533	1.2788	1.2284	1.1638	1.1232	1.0942	1.0715	1.0519	1.0325	1.0212	1.0077	0.9974
20′	1.7489	1.4952	1.3673	1.2898	1.2374	1.1703	1.1280	1.0979	1.0742	1.0538	1.0337	1.0220	1.0079	0.9971
30′	1.7776	1.5141	1.3813	1.3008	1.2464	1.1767	1.1328	1.1015	1.0770	1.0558	1.0349	1.0227	1.0081	0.9969
40′	1.8062	1.5329	1.3953	1.3118	1.2554	1.1831	1.1376	1.1052	1.0797	1.0577	1.0361	1.0234	1.0083	0.9967
50′	1.8348	1.5518	1.4092	1.3228	1.2644	1.1895	1.1424	1.1088	1.0824	1.0596	1.0373	1.0241	1.0085	0.9964
5° 0′	1.8634	1.5707	1.4232	1.3337	1.2733	1.1959	1.1472	1.1124	1.0851	1.0616	1.0384	1.0248	1.0086	0.9962
10′	1.8920	1.5895	1.4371	1.3447	1.2823	1.2023	1.1519	1.1160	1.0878	1.0635	1.0396	1.0255	1.0088	0.9959
20′	1.9205	1.6083	1.4510	1.3557	1.2912	1.2086	1.1567	1.1196	1.0905	1.0654	1.0407	1.0262	1.0089	0.9957

Interpolation for correction factors corresponding to values of $(\alpha + \beta - \gamma)$ lying between those given in the table may be made without error. Interpolation for correction factors corresponding to values of $\cos \theta_2$ lying between those given in the table may be made without exceeding an error of 0.0010 in the sections of the table lying between the heavy black lines; outside of these sections, and in all cases where the adjacent values of $\cos \theta_2$ are separated by the heavy black lines, the maximum error in interpolation will exceed 0.0010.

The above convention in regard to the algebraic sign of the phase angle γ of a voltage transformer (positive sign for secondary voltage *leading* the reversed primary voltage) is now universally adopted (see ASA Standard C57). It has always been used by the National Bureau of Standards and since 1930 by the International Electrotechnical Commission. From 1925 to 1940 the AIEE officially used the opposite convention. With this opposite convention the correction factor becomes $\cos(\theta_2 + \alpha + \beta + \gamma)/\cos\theta_2$. In both these forms of the correction factor the + and − signs in the numerator are addition (or subtraction) *operators*, and the angles α, β, and γ may be intrinsically either + or −. The angles α and β, however, are nearly always intrinsically positive. The following examples illustrate the use of the correction tables.

Example 1. Given a single-phase circuit with lagging current in which the wattmeter reading corrected for scale error and multiplied by the corrected ratios of current and voltage transformers equals 24,520 watts, and the product of the voltmeter and ammeter readings, similarly corrected, equals 35,600 va. Then $\cos\theta_2 = 24{,}520/35{,}600 = 0.689$. If the equivalent phase angle α of the wattmeter is $+4'$ and if from examination of characteristic curves the current-transformer phase angle is found to be $+48'$ and the voltage-transformer phase angle is $+10'$, then $\alpha + \beta - \gamma = +42'$, and from Table 1 the correction factor is 0.9871. Whence the true power equals $24{,}520 \times 0.9871 = 24{,}204$ watts.

Example 2. Given a single-phase circuit with leading current, in which the wattmeter reading, corrected as in example 1, equals 12,266 watts and the product of the voltmeter and ammeter readings, similarly corrected, equals 24,532 va. Then $\cos\theta_2 = 12{,}266/24{,}532 = 0.5$. If the equivalent phase angle α of the wattmeter is $+5'$, the phase angle β of the current transformer is $+2°\,33'$, and the phase angle γ of the voltage transformer is $-38'$, then $\alpha + \beta - \gamma = +3°\,16'$, and therefore from Table 2 the correction factor to be used is 1.0971. Whence the true power equals $12{,}266 \times 1.0971 = 13{,}457$ watts.

In measuring badly unbalanced three-phase loads with two wattmeters, corrections should be applied separately, as above, to each wattmeter. When the circuit is balanced, a single correction based on $\cos\theta_2$ for the whole circuit may be used.

Accuracy of Wattmeters. High-grade portable wattmeters have a stated accuracy of 0.2 to 0.25 per cent of full-scale value for single-phase or 0.5 per cent for polyphase wattmeters. It is not practicable to obtain as high accuracy with a polyphase wattmeter on account of the impossibility of making both elements follow exactly the same scale law. This theoretical disadvantage of the polyphase wattmeter is offset by its greater convenience and the reduction in number of observations necessary. High-grade switchboard wattmeters have a stated accuracy of 1 per cent of full-scale value. Laboratory-standard wattmeters have a stated accuracy of 0.1 per cent of full-scale value.

Wattmeters are usually tested for accuracy on direct current, using a potentiometer and suitable accessories. For convenience in computing the values of watts, a voltage of 100 volts may be used for wattmeters for 110-volt circuits. The current required to produce a given deflection having been recorded, a second observation is made after reversing the currents in both windings of the wattmeter. The mean of two such readings which do not differ more than several per cent is free from error due to stray magnetic field in unshielded wattmeters, or from accidental magnetic polarity in the iron shield of shielded wattmeters, such as may arise from an accidental large overload in the current coil. This test on "reversed direct current" having been made at a sufficient number of points on the scale, it is necessary, for a new wattmeter, and advisable, for one that has been in service for some time, to determine the difference between the a-c power and the d-c power required to produce a given deflection. For this test another wattmeter must be used, of known or calculable a-c performance. This test may be made only at power factor 0.5, for wattmeters of proved construction, but for a new design the test should be made at zero power factor also.

Measurement of Three-phase Power. The power supplied to a three-phase load may be measured in one or more of the following ways. The connections may be made either directly or through proper instrument transformers.

Single-element Wattmeter on Three-wire Three-phase Load. The current circuit of the wattmeter is connected in series with one of the mains supplying the load, and the voltage circuit of the wattmeter is connected between the corresponding terminal of the load and the neutral. If the neutral point of the load, or of the transformers supplying the load, is not available, a "Y-box" (see below) can be used to establish an artificial neutral. If the load is perfectly balanced, the power input is then three times the wattmeter reading. However, a three-phase load, even a three-phase motor load, is seldom sufficiently well balanced to render this method of measurement accurate.

Y-box. The simplest form of Y-box consists of two equal non-inductive resistors connected in series, each free end and the junction point being connected to a binding post.

Each resistor has a resistance equal to that of the voltage circuit of the wattmeter. One terminal of the voltage circuit of the wattmeter is connected to the junction terminal of the Y-box, and the other terminal of the voltage circuit to the line wire in which the current circuit of the wattmeter is connected. The other two terminals of the Y-box are connected to the other two line wires respectively.

In wattmeters designed especially for use with a Y-box, part of the resistance of the potential circuit of the wattmeter is placed in the Y-box, being connected permanently to the junction point between the other two resistors. A similar arrangement may be used as a multiplier. The connections of such a Y-box, wattmeter, and instrument transformers are shown diagrammatically in Fig. 26.

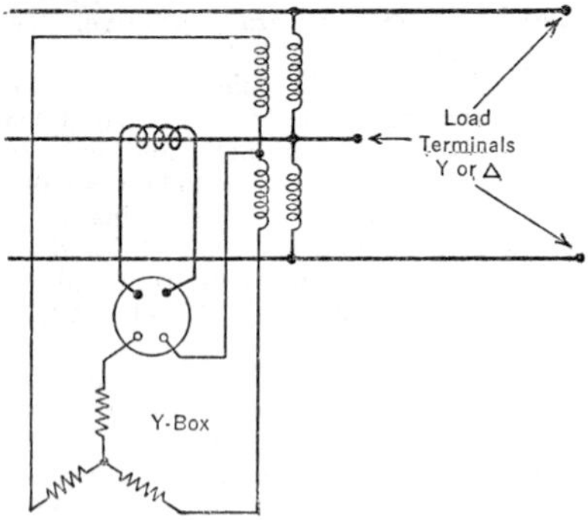

Fig. 26

Two-wattmeter Method for a Three-wire Three-phase Load. The simple arrangement of two wattmeters shown in Fig. 27 will give exactly the total power in any three-wire circuit, provided each wattmeter by itself gives accurate indications. Aside from the sources of error, as noted above, which may affect the accuracy of a wattmeter on a single-phase circuit, the arrangement shown in Fig. 27 will indicate the true power for any condition of unbalancing, wave form, frequency, etc. It is also immaterial whether the load be Y or Δ connected. The connections may be made directly as shown or through two current transformers and two voltage transformers, the connections in the latter arrangement being the same as in Fig. 28 except that two separate wattmeters instead of a polyphase wattmeter are used.

Rule for Adding or Subtracting Readings. With the arrangement shown in Fig. 27 the total power is always the *algebraic* sum of the readings of the two wattmeters. Since a wattmeter reads in only one direction, usually to the right, the two instruments must be so connected to the line that the needle of each instrument is deflected over the scale. For a balanced three-phase load having a power factor greater than 50 per cent, the sum of the two wattmeter readings, when the connections are thus made, gives the total power; for a power factor less than 50 per cent the difference of the two readings must be taken. When the power factor is not known, one can determine whether the sum or the difference of the readings should be taken by interchanging the two wattmeters, leaving unaltered the potential connections to the third wire (C in Fig. 27); if the pointers of the two wattmeters deflect in the same direction as before, add the two readings; if one of the pointers deflects in the opposite direction (i.e., against the stop), take the difference. In making this test, care must be exercised to connect the source side of each of the two lines (in which the current coil of each wattmeter is connected in succession) to the same binding post of the given wattmeter in each of the two positions.

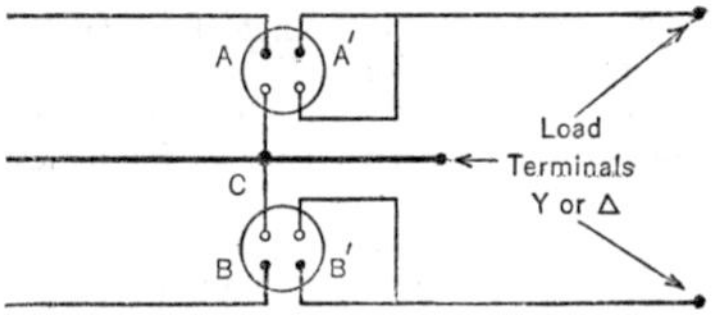

Fig. 27

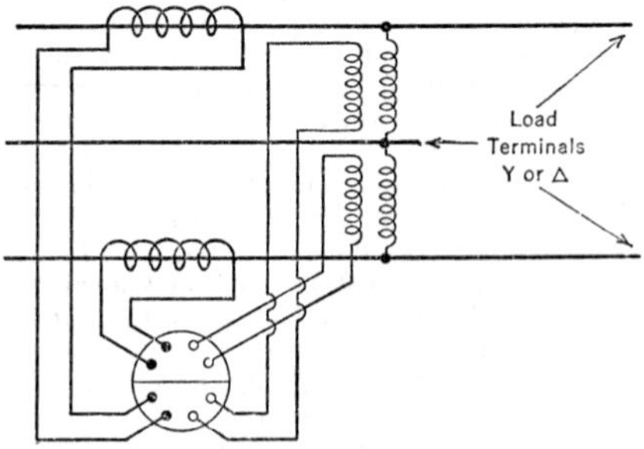

Fig. 28

When the load is balanced, and the two wattmeters are connected, as in Fig. 27, so that each gives a positive reading, the question of whether the power factor is above or below 50 per cent can be determined by changing the potential lead of the lower-reading wattmeter from the common connection C to the line in which the current coil of the other wattmeter is connected. If the reading thus obtained is positive, the power factor is more than 50 per cent, and the readings should be added; if the reading is negative, the power factor is less than 50 per cent, and the readings should be subtracted.

When three wattmeters are used for the measurement of a **three-phase four-wire load,** they should be connected in a manner similar to that used for two wattmeters on a three-wire system; i.e., each of the three voltage circuits should be connected between a main wire and the neutral wire, and each of the three current circuits should be connected in series with the corresponding main wire.

Special Precautions in the Use of Wattmeters. In addition to the general precautions given under the heading Precautions in Placing and Using Instruments, p. 5-44, the following special points must be noted. A wattmeter has **three distinct limits,** namely, current, voltage, and power, no one of which should be exceeded. The potential of the moving coil must not be greatly different from that of the fixed coil, and the connections must always be such as to ensure this condition, both in the use and in the checking of a wattmeter. This precaution is especially necessary when an external multiplier is used for high voltages. Unshielded wattmeters, not of astatic construction, are very susceptible to the effect of **stray magnetic field** because of the relatively weak operating field of the current in the current coil. In using such wattmeters, special care should be taken to locate them where this effect will be negligible. To check this point, current should be allowed to flow in the voltage circuit only. Any deflection will then be due to stray field. In applying this test to a wattmeter with more than one current range which is compensated for the loss in its voltage circuit, the switch for changing the compensating-coil connections (as the current range is varied) must be correctly set, or an error will result. To minimize stray-field error in wattmeters of the deflection type, it is the practice of American makers to surround the wattmeter coils with a laminated **magnetic shield.**

FREQUENCY METERS. Indicating frequency meters have been made in a great variety of types, of which relatively few are now used in American practice. The following may be mentioned.

In the **vibrating-reed frequency meter** a number of thin steel reeds, tuned to respond to a closely graded series of frequencies, are mounted so that their free ends are aligned with a scale marked in cycles per second. An electromagnet is arranged to act upon the reeds, either directly or by vibrating a common frame supporting all the reeds. The electromagnet is so designed that the normal line voltage will produce such moderate forces on the reeds that only those reeds with frequencies very close to the line frequency will respond. The ends of the reeds are provided with white flags which appear to widen as the amplitude of vibration increases. Frequency meters of this type cannot be read as closely as those of the pointer type, but they have some advantages which make them valuable, especially for use in the laboratory. These advantages are as follows: The indications being dependent on the fundamental frequency, the effect of distortion of the voltage wave is very small. The indications are not influenced by moderate changes of voltage; nevertheless extreme changes must be avoided. The indications are only very slightly dependent on the temperature of the reeds, the change in frequency being 0.01 per cent per degree centigrade. An increase of temperature reduces the normal frequency of the reeds, i.e., causes the indicated frequency to be too high. Vibrating-reed frequency meters must be protected from intense mechanical vibrations.

In American practice the vibrating-reed frequency meter has been restricted to laboratory uses. For switchboards some form of indicating frequency meter has always been preferred. Frequency meters of the induction type (Westinghouse), the resistance-reactance type (General Electric), and the bridge type (Weston) have been superseded in recent years by modern types capable of the greater sensitivity and accuracy demanded by present-day practices, including system interconnection and the operation of synchronous-motor clocks.

The **tuned-circuit frequency meter** resembles the earlier pointer-type frequency meters but may be made much more sensitive and much less susceptible to error from variation of wave form because the tuned circuits act as filters to pass only the fundamental. The extreme sensitiveness obtainable is shown by the fact that switchboard frequency meters are made by the General Electric Company in which the entire length of the scale corresponds to the range 59.5–60.5 cycles. Even narrower ranges would be possible under this principle of construction, the practical limitations arising from the temperature coefficient and other minor changes in the available capacitors. The **General Electric frequency meter** has a pair of coaxial fixed coils which are composed of two equal interwound windings of opposite polarity. There is a single moving winding which carries the sum of the currents in the two fixed windings. The torque depends upon the product of the vector sum of the two currents by their vector difference, and changes greatly for a small change in the relative value of the two currents. To make the instrument indicate frequency, a control is necessary; and, to make the indications independent of the **line voltage,** this control must vary as the square of the line voltage. Such a control is realized by the addition to the moving element of a nickel-iron-alloy vane which tends to align itself with the field of the fixed coils. Another variation of the tuned-circuit frequency meter is the General Electric long-scale instrument shown in Fig. 29. Tuning is accomplished in the manner which has been described, and the iron-cored field

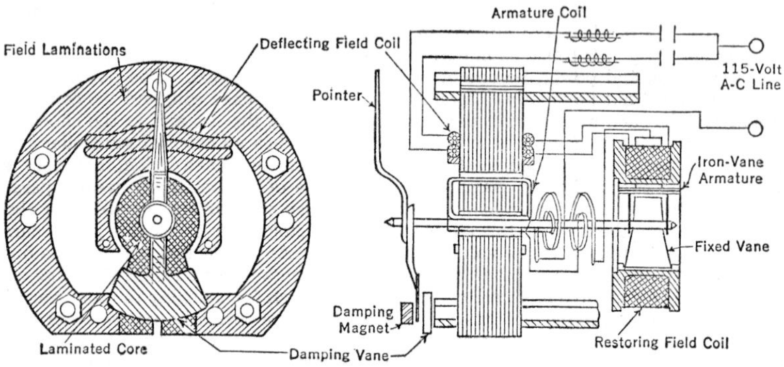

Fig. 29

acts upon the moving coil with a force proportional to the difference in the currents in the two branches. Control is established by an iron-vane repulsion type of restoring system acting upon the same shaft.

The **Westinghouse resonance-type frequency meter** (Fig. 30) consists of a pair of stationary field coils within which rotate two moving coils fixed on a common shaft at right angles with each other. A reactor *8* is connected in series with the fixed coils, and a capacitor *10* is shunted across the fixed coils and the moving coil *3*. The relation between capacitance and inductance is such that parallel-circuit resonance occurs at the normal frequency. The moving coil *2* is connected in series with the parallel-resonance circuit. There is no mechanical control of the moving system, which assumes a position depending on the relation between the forces set up by the currents in the coils. The sensitivity to frequency changes is such that a range of 58 to 62 cycles is obtainable for use on 60-cycle systems. The resonant circuit employed makes the instrument very insensitive to temperature changes and keeps the burden down to a low figure, namely, 3.2 va at 110 volts, 60 cycles.

The **Weston frequency meter** is of the moving-iron type. Two fixed coils mounted at right angles to each other act upon a single iron needle carried by a shaft, which also carries the pointer and a damping vane. There is no spring or other mechanical control. The connections are shown in Fig. 31. Current from the upper line wire passes through the reactor X_1, resistor R, and field coils *1*; at the junction point P_1 the current divides into two components mutually displaced in phase, namely, a lagging component through field coils *2* and reactor X, and a leading component through capacitor C. At the junction point P_2 these components unite and flow through reactor X_2 to the lower line wire.

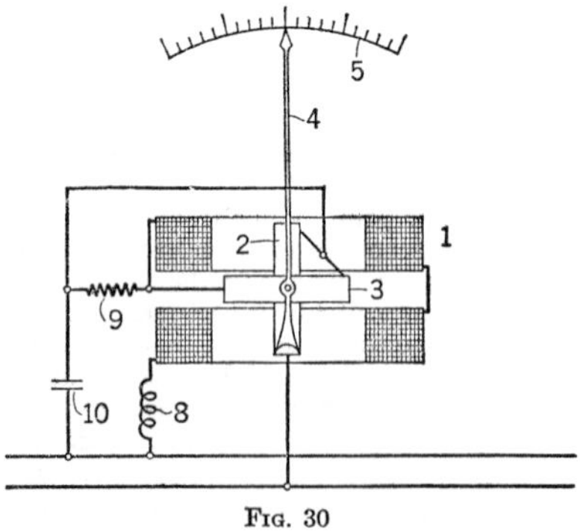

Fig. 30

The undivided current in coils *1* produces a magnetic field tending to displace the needle from the central position in which it is shown and which corresponds to the normal operating frequency. The lagging component of current flowing through coils *2* tends to hold the needle in its central position. At normal frequency the effect of the fundamental of this current is very great because the circuit *2*, X, C becomes resonant. For lower frequencies the lagging component of the current in coils *1* preponderates because of the decrease in the reactance of X and the increase in the reactance of C, and the needle is deflected to the left. The converse effect takes place when the frequency exceeds the normal value. The reactors

X_1, X_2, are used only when the wave form of the line voltage is very much distorted. For ordinary commercial wave forms the action of the resonant circuit in responding selectively to the fundamental makes the accuracy of indication independent of the wave form.

The **Leeds & Northrup frequency recorder** records and indicates the frequency. It contains a bridge network connected to the line voltage. Two arms are resistors, a third is a capacitor shunted by a resistor, and the fourth is a capacitor in series with a resistor. Sliding contacts, actuated by the auxiliary motor of the recorder in accordance with the position of the galvanometer pointer, move with a change of frequency until a balance of the bridge is re-established. This instrument is very sensitive to changes of frequency and very insensitive to changes of temperature and line voltage. In one 60-cycle recorder the pen traverses the 10-in. chart for the change from 58 to 62 cycles, and in another for the change from 59 to 61 cycles.

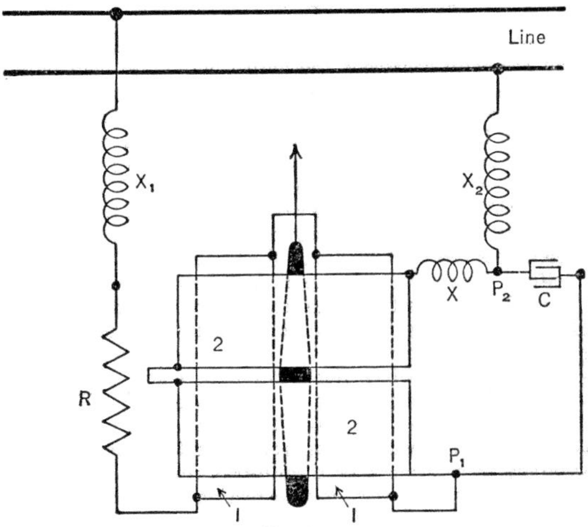

FIG. 31

POWER-FACTOR METERS. The phase relation between the current and the voltage in a circuit or a polyphase system can be measured by indicating instruments of a type in which the mechanical angle between the instrument pointer and a reference mark is equal to the electrical phase angle between current and voltage. If the scale is graduated in equal angular intervals, the instrument is called a **phase meter.** If the scale is graduated according to the cosine of the phase angle, the instrument is called a **power-factor meter.** If the graduations are marked according to the sine of the phase angle, it is called a **reactive-factor meter.**

Strictly speaking, the power factor as defined for single-phase systems is the quotient obtained by dividing the power by the product of the effective values of the voltage and the current. The power factor may therefore be computed directly from the readings of a wattmeter, a voltmeter, and an ammeter; but, if the wave forms are not sinusoidal, the power factor will not be measured exactly by an instrument whose indication depends only on phase displacement.

For a polyphase system the power factor has been defined by arbitrary convention as the ratio of the total power to the total vector volt-amperes, the total vector volt-amperes, in turn, being defined as the square root of the sum of the squares of the total active power and the total reactive power. For sine waves this definition is consistent with that for the single-phase case, but for non-sinusoidal wave forms the polyphase definition is theoretically inconsistent with any logical generalization of the single-phase definition.

Power-factor meters of the indicating type depend for their operation on the interaction of a pulsating single-phase magnetic field produced by a single coil and a rotating magnetic field produced by two or more other coils, which are so placed that the mechanical (space) angle between their axes is equal to the electrical (time-phase) angle between the currents in them. The actual moving element may be either the single coil or the two or more other coils mentioned, or an iron vane magnetized by the single coil. In any event, the moving element will take up a position such that at the instant the pulsating field has its maximum value the rotating field has the same direction as that of the pulsating field.

In all types of these instruments the moving element is intended to be free from all mechanical constraint, and hence on open circuit it will point indifferently in any direction. Because of the relatively weak electrical restoring torque, the indication becomes definite only when the product of the applied current and voltage is at least one-fifth of normal value, and the best accuracy is obtained only when this product is at least two-fifths (40 per cent) of normal value.

In instruments for use on single-phase circuits the rotating magnetic field is obtained (usually from the voltage circuit) by supplying one coil through a non-inductive resistance while the other is in series with a highly inductive iron-cored reactance coil. When used on polyphase circuits the rotating field can be obtained more easily (usually from the voltage circuits) by using a non-inductive resistor in series with each of the two or three coils spaced 120 deg apart which produce the rotating field. If a single current coil connected in series with one line is used, the instrument reading will depend only on the phase displacement between that current and the corresponding star voltage and will correctly represent the entire polyphase load only if this is balanced. An instrument having three such elements will correctly indicate the polyphase power factor even on an unbalanced load if the wave form is sinusoidal.

Some instruments, especially those of the magnetic-vane type, are provided with a scale covering the complete 360 deg and serve to indicate the direction of flow of power as well as the phase relation of current to voltage.

VARMETERS. These instruments, also called reactive-power meters or reactive component meters, are essentially wattmeters so connected in an a-c line as to make the voltage across the voltage circuit (or circuits) of the instrument 90 deg out of phase with the voltage which would be applied in power measurement. If a polyphase reactive-power meter is connected in circuit so that the shift of phase is obtained by suitable connections to the secondary terminals of ordinary voltage transformers, the indication in watts which the instrument would give if connected as wattmeter must be multiplied by a suitable factor (which depends on the connection used) to obtain the reactive power in "vars." (At the meeting of the International Electrotechnical Commission in Oslo in 1930 it was voted to adopt "the name or term **var** for designating the unit of reactive power"; i.e., 1 var results from 1 volt acting in quadrature with 1 amp). If special voltage transformers, or if auto-transformers connected to the secondary of the voltage transformers, are used, the correction factor can be automatically allowed for, and the meter will read directly in vars.

Any reactive-power measuring device which depends on cross-connection of the voltage circuits will read incorrectly if both the voltages and the currents are unbalanced. Those in which the number of current coils is less by two than the number of wires in the circuit will be incorrect on unbalanced currents, even if the voltages are symmetrical.

SYNCHROSCOPES. A synchroscope is essentially a robust form of phase meter in which the pointer is free to rotate. It is provided with a single scale mark which denotes synchronism, that is, equality of frequency and coincidence of phase of two a-c sources. In use, one of the two windings is connected to the bus, the other to a machine which is to be connected to the bus. Synchroscopes have superseded, in great measure, the various forms of synchronizing devices employing lamps.

In the **General Electric synchroscope** two split-phase windings are located on an iron core arranged to rotate. The windings are connected through slip rings to the incoming machine, and the phase displacement is obtained by a reactor in one winding and a resistor in the other. A stator surrounding the rotor has its windings connected to the bus. When exact synchronism is reached, a pointer attached to the rotor shaft remains at rest in a vertical position. The synchroscope is usually connected through voltage transformers; on polyphase circuits only one phase is used. Iron-vane type synchroscopes made by General Electric Company have much lighter moving systems than the iron-cored type and require only a fraction of the power to operate. The construction is essentially the same as that of the iron-vane type of power-factor meter but has the central magnetizing coil wound with many turns of fine wire and excited from the bus. The incoming machine is connected to the stator through a resistance-capacitance network that produces a rotating field.

In the **Westinghouse synchroscope** an iron vane, free to rotate, is magnetized by a fixed coil connected to the incoming machine. The vane is located within a rotating field produced by current from the bus passing through two coils which are fixed at approximately right angles with each other. A reactor in series with one of the coils and a resistor in series with the other produce a phase displacement between the two currents. The pointer attached to the rotating vane indicates at any moment the phase displacement between the voltages of the bus and the incoming machine, and the speed and direction of its rotation depend on the difference between the two frequencies.

The **Weston synchroscope** consists of an instrument mechanism similar to that of an electrodynamic voltmeter. The fixed winding is connected to the bus through a resistor, and the moving coil is connected to the incoming machine through a capacitor. When the voltage of the incoming machine is either in phase with or in opposition to the bus voltage, no torque is exerted on the moving coil, and the pointer stands in its central (vertical) position. For any other phase relation the pointer will be deflected to the right or to the left when the incoming machine is ahead or behind, respectively. The pointer is behind a translucent screen, and only its shadow, cast by a synchronizing lamp, can be seen. This lamp is operated by a double-primary transformer connected for synchronizing "light." When the shadow of the pointer is in the vertical position, the incoming machine is in phase. The pointer will be vertical for phase opposition also but cannot be seen then because for this condition the synchronizing lamp is dark. When the incoming machine is too fast or too slow, a succession of shadows of the pointer will be seen, and the shadow will appear to rotate in the clockwise or counter-clockwise direction, respectively.

INSTRUMENTS FOR MEASURING RECTIFIED CURRENTS. Ammeters of different types will not give identical indications when used to measure the same rectified current. A rectified current may be regarded as equivalent to the superposition of a direct current upon an alternating current. The proper **type of ammeter to use** depends entirely upon the function to be performed by the rectified current. For charging storage batteries or for electroplating, the desired result is proportional to the d-c component of the rectified current, and a permanent-magnet moving-coil instrument should be used because it responds only to this component. For rough measurements the less expensive polarized-vane instrument is adequate if it has been calibrated for use on the particular wave form. If the rectified current is used for incandescent lamps or for electric heating devices, its effective (rms) value is significant, and ammeters of either the electrodynamic type or the electrothermic type will theoretically give the correct value for the given purpose. Electrodynamic and electrothermic ammeters, however, are not generally available, so that it is desirable to measure the effective value of rectified currents with a moving-iron ammeter of good quality if an accuracy of, say, 1 or 2 per cent is sufficient. If the wave form of the rectified current is very much distorted in comparison with a half-cycle of a sinusoidal current, or if doubt exists as to the quality of the design of the moving-iron ammeter and the magnetic material used in it, the ammeter should be checked on the given current against an electrodynamic or electrothermic ammeter.

When the rectified current is used to operate an **arc lamp** and is to be adjusted to produce the same light output as would be given by a stated direct current, the permanent-magnet moving-coil ammeter gives approximately correct results and is the only kind to be used. However, its indications will usually be very erroneous as regards the **heating effect** of the current in the carbons, the rheostats, cables, etc.; and, if overheating is to be avoided, a moving-iron ammeter should also be in circuit. This precaution is needed also in using rectified currents for battery charging and electroplating.

For a full description of the measurement of rectified currents for various purposes, see paper by W. N. Goodwin, Jr. [*Trans. Soc. Motion Picture Operators*, Vol. 12, p. 364 (1928); reprinted in *Instruments*, (Pittsburgh), Vol. 2, p. 115 (1929)].

The choice of a voltmeter for voltage measurements with rectified currents is based on the preceding considerations for the ammeter. For battery charging or electroplating, a permanent-magnet moving-coil voltmeter should be used; for regulating the voltage applied to incandescent lamps, an electrodynamic, electrothermic, or (with reservations as for the ammeter) a moving-iron voltmeter; the product of the effective voltage, so measured, by the effective current will give the watts only if the circuit is non-reactive. If reactance is present, the power must be measured with a wattmeter.

The **a-c component of a rectified current** may be measured by a "selective ammeter" if the d-c component is 5 times the a-c component, or less. This instrument is of the iron-core electrodynamic type and has a self-contained current transformer with an air gap in its magnetic circuit.

OHMMETERS. The name ohmmeter is applied to two radically different types of portable instrument for the rapid and simple measurement of resistance with moderate accuracy. One is a simplified Wheatstone bridge, and the other is essentially a voltmeter with an ohm scale. The **Roller-Smith ohmmeters** are slide-wire bridges. They are built in three forms. Self-contained batteries are the source of current in each form. In two forms a built-in galvanometer is the only detector provided. In the third form either a galvanometer or a telephone receiver may serve as the detector, and the battery may be used either as a d-c source or to operate an induction coil, the primary of which may be used as an a-c source. This latter ohmmeter serves for the usual resistance measurements, for the measurement of inductance, capacitance, and resistivity of electrolytes, and also for fault location.

In the **Leeds & Northrup d-c ohmmeter** the slide wire is mounted within the instrument on the edge of a circular disk which may be rotated by means of a handle above the panel. A d-c galvanometer and a battery are mounted within the case. The **Leeds & Northrup a-c ohmmeter** closely resembles the d-c ohmmeter but has an a-c galvanometer and an insulating transformer. In the ordinary applications of these ohmmeters any one of five built-in standard resistors may be used, according to the magnitude of the resistance under measurement. Provision is also made for an external standard, such as a copper coil, against which nominally like coils are to be checked. In this case the scale reads the ratio of the resistance of the unknown coil to that of the known coil.

Permanent-magnet moving-coil ohmmeters are made in various types which differ somewhat radically in principle of operation, including: volt-ohmmeters, ohmmeters with differentially wound moving coil, and quotient-type ohmmeters.

Volt-ohmmeters differ from ordinary voltmeters mainly in having an extra scale figured in ohms. It is usual to have the full-scale value of voltage coincide with the zero of the ohm scale. In use, a battery of such voltage as to give full-scale deflection is connected to the instrument; if the unknown resistance is now introduced into the circuit, the pointer will drop back to a position corresponding to the reduced value of the current, when the resistance may be read off directly from the ohm scale. As it is obviously difficult to ensure that an available battery will produce exactly full-scale deflection, the more accurate instruments of this type, intended only for use as ohmmeters, have an adjustment, usually a magnetic shunt, by which the pointer may be brought to the full-scale mark in spite of normal variation of the battery voltage.

Any **d-c voltmeter may be used as an ohmmeter** in connection with a constant-voltage source. If the deflection is d_1 with the voltmeter connected directly to the source, and d_2 with the unknown resistance x in series with the voltmeter, the unknown resistance is given by the relation

$$x = \left(\frac{d_1}{d_2} - 1\right) r \tag{11}$$

where r is the resistance of the voltmeter.

Ohmmeters with differentially wound moving coil are exemplified by the Weston Model 1 direct-reading ohmmeter. This permanent-magnet moving-coil instrument has two windings in the moving coil, the effect of the current in one winding being to deflect the pointer up scale and that of the other to deflect it down scale. Suitable resistances in the instrument, controlled by a plug switch, provide for three ranges, for example, 0–200, 0–1000, and 1000–2000 ohms. To take care of variations in battery voltage, a magnetic shunt is adjusted to bring the pointer to the full-scale mark with no connection between the X posts and with the plug in the checking position.

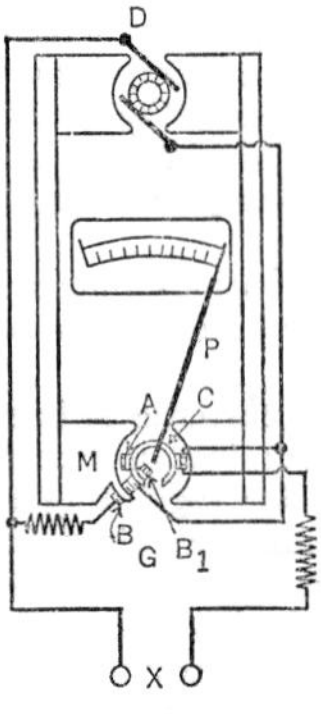

Fig. 32

The **quotient-type ohmmeter,** in principle, is independent of the value of the battery voltage because its indication depends upon the ratio of two oppositely directed turning moments produced by currents flowing in two coils. One of these coils is in a circuit of fixed resistance; the other contains the resistance under measurement. A change of the voltage changes both the currents but does not alter their ratio. The **Evershed megger** is an ohmmeter of the quotient type, intended principally for measuring relatively high resistances in megohms, whence the name. Its construction is indicated diagrammatically in Fig. 32, in which D is a small hand-driven d-c magneto and G a special form of moving-coil instrument. One pair of bar magnets supplies the flux for both. The galvanometer contains three coils rigidly attached at a fixed angular distance apart to the shaft which carries the pointer P. The unknown resistance is connected at X.

The coils BB_1, called the pressure coils, are permanently connected through a resistance across the terminals of D. The current coil A is made to move through an annular gap in such a manner that the field in which it moves is uniform, whereas the pressure coils BB_1 move from a position midway between the poles, where the field is at a minimum, into a stronger field, the connections being such that the torque due to the current in the current coil is opposed by the torque due to the current in the pressure coil. When there is no current in the current coil, that is to say, when the resistance to be measured is infinite, current through the pressure coils will cause them to come to rest with their plane at right angles to the magnetic field. When the current through the current coil is increased by putting in lower resistances, the current coil drags the moving moving system round in a clockwise direction; since the pressure coils come into a stronger and stronger field, the resistance to this motion becomes greater and greater. Hence a definite position is assumed by the system for the particular resistance at X. An increase in voltage would increase the current in both current and pressure coils in the same proportion; consequently the instrument is independent of the voltage of the generator.

In the less-expensive meggers the generator gives a variable voltage depending upon the speed of rotation of the handle. In higher-priced meggers the generator is driven through a slipping clutch to give a constant voltage for a considerable range in the speed of driving. The voltage at which an insulation test is made is often of some importance. Constant-voltage meggers are especially useful for testing insulation resistance associated with a large capacitance, where a varying test voltage would cause charging and discharging currents to flow, with resulting fluctuations in the readings. In a similar device made by Holtzer-Cabot the voltage is regulated by neon discharge tubes. In any such device the terminal voltage falls off materially as current is drawn from the generator because of its internal resistance.

Meggers are made with generators giving a maximum of 2500 volts and with a maximum measurement range of 4 to 10,000 megohms. The successful operation of these instruments to measure such high values of resistance is possible because the Price guard-wire scheme prevents leakage of current across and through insulating supports. Meggers may be checked for accuracy by using known resistances having only a small fraction of the full-scale value; see paper by H. B. Brooks, *Elec. World*, Vol. 85, p. 973 (1925).

The most recent form of megger is of particular interest to instrument designers and users because its small size and light weight (3 lb) are obtained by the use of modern magnetic alloy for the permanent magnets of the 500-volt generator and of the ohmmeter.

An alternative type of instrument for checking insulation resistance is the "vibrometer," which uses a vibrating reed to "chop" the current from a dry battery. The pulsating voltage is stepped up by a transformer and then rectified by contacts on the reed, which operate in synchronism with the interrupter. The rectified voltage is filtered and applied to the unknown resistance with a microammeter in series.

For very high resistances bridge circuits with electronic detectors, such as the General Radio megohmmeters and megohm bridges, can be used.

MICROFARADMETERS. Direct-reading pointer instruments for capacitance measurements are made by a number of manufacturers under the names of condenser meters, capacitance meters, and microfaradmeters. They are usually made to be operated by alternating current from ordinary lighting circuits. For details as to their operating principles and internal circuits, see references under Microfaradmeters in Bibliography, p. 5-112.

FLUXMETERS. The Grassot fluxmeter is essentially a moving-coil galvanometer, the suspension fiber of which exerts a negligible torque on the moving system. The motion of the coil is very heavily overdamped. Figure 33 is a diagram showing its construction. The coil C swings in a uniform field in the air gap between the pole pieces NS of a permanent magnet and the soft-iron core A. The system is supported at the top by a single cocoon fiber, the upper end of this fiber being attached to a flat spiral spring R to minimize the effect of shocks. The torsional force exerted by such a fiber is extremely small. S and S_1 are very thin silver strips, serving to lead the current to and from the coil C. These strips are in the form of springs, which, however, are extremely weak and exert the minimum practicable torque on the coil. An index attached to the coil swings over a calibrated scale. The fluxmeter is also provided with a mirror in addition to the pointer so that it may be used in conjunction with a lamp and scale or with a telescope and scale.

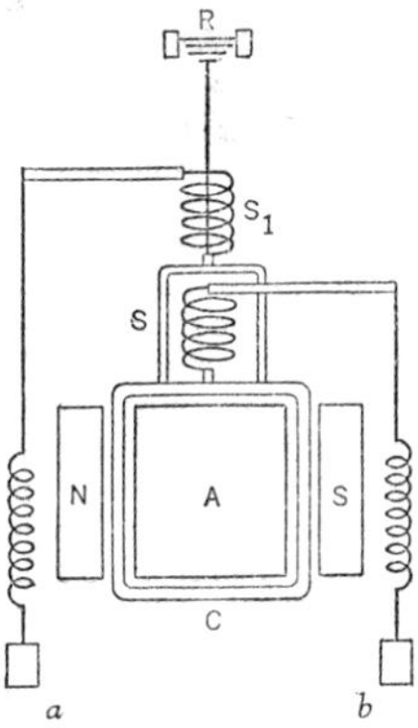

Fig. 33

When a given quantity of electricity is discharged through the moving coil, for example, by changing the flux threading an exploring coil connected to the terminals a and b, a force is exerted upon the coil, tending to deflect it. The only opposing forces (neglecting the small torsional forces) are the mechanical friction to motion and the back emf induced in the coil by its motion through the field of the permanent magnet. The back emf is proportional at any instant to the velocity of the coil, and the frictional force is also approximately proportional to this velocity. If both forces are directly proportional to the velocity, when a given quantity of electricity is discharged through the circuit, the coil comes to rest at a definite point, depending only upon its initial or "zero" position and the total quantity discharged through it. As the quantity of electricity discharged through an exploring coil, when the magnetic flux threading it is changed, is proportional to the change in flux, the instrument can be calibrated to read directly the change in the flux produced in the region occupied by the coil. In practice the friction is not exactly proportional to the velocity, and the fiber and leading-in strips usually exert an appreciable force on the coil. The instrument therefore has not proved altogether satisfactory.

In motor or in dynamo work an exploring coil, consisting of one or more turns of wire, may be fixed or wound in position, and the change in flux of induction observed on exciting the field magnet. Even in the largest work, where some minutes may be taken to reach the limit of magnetization, no serious error is thus introduced. The fluxmeter can also be used for the measurement of magnetic field strength, pole strength, and the distribution of magnetism in bar magnets. It is also adapted to the measurement of permeability and hysteresis.

PROTECTION OF INSTRUMENT MECHANISM FROM STRAY MAGNETIC FIELD. Because of the low reluctance of iron as compared with air, a soft-iron shield placed around the instrument mechanism will greatly reduce the disturbing effect of an external magnetic field on the reading of an instrument. Many types of switchboard and portable instruments are thus shielded. Permanent-magnet moving-coil instruments are not seriously affected except by relatively strong external fields and ordinarily do not need shielding other than that afforded by the iron case used with switchboard instruments.

The modern high-permeability nickel-iron alloys, such as permalloy C and mumetal, are very suitable for magnetic shields. A given degree of shielding ratio can be obtained with very much thinner and lighter shields of such alloys than when silicon steel is used.

PRECAUTIONS IN PLACING AND USING INSTRUMENTS. The weight of the moving element of an electrical instrument or meter is carried by a steel pivot resting on a jewel bearing, and the area of the surfaces in contact is so minute that the pressure is very great. The operating forces are very small. Accidental shocks which would not damage a steam gage or a gas meter may blunt the pivot and destroy the usefulness of an electrical instrument. If portable instruments are to be used where vibration is unavoidably present, they should be placed on pads of folded cloth, felt, or similar material. **Do not hammer on a table or bench on which instruments are placed.** Switchboard instruments should preferably not be put in place until all work on or near the boards has been finished.

Instruments to be transported by automobile should preferably be packed in excelsior in suitable boxes. If this is not done, they should be placed on the seat cushions, upside down or on edge, in order to minimize the liability of damage to the lower pivot and its jewel. Instruments to be shipped by rail should be upside down when the packing case is right side up.

Tapping an instrument while it is being read is usually unnecessary if it is properly constructed and in good order, but, when slight friction is present, its effect can be eliminated by gently tapping the instrument. Hard tapping defeats its object, and may cause damage to pivots and jewels.

Portable instruments are nearly always calibrated with the scale in a horizontal plane. For the most accurate results they should be used in this position. The moving system should be mechanically well balanced so that no appreciable error will be introduced if the instrument is not exactly horizontal. New instruments should be examined for balance by tipping them in various directions about 10 deg from the horizontal. If the zero reading varies more than a few tenths of a division during this test, the moving system is not well balanced. Tipping an instrument by too great an angle (say 45 deg or more) may give misleading results by causing the point of contact of pivot and jewel to shift, thereby changing the axis of rotation.

Avoid wiping the glass cover over the pointer of an instrument before taking readings because the resulting electrostatic charge on the glass will attract the pointer. Very large errors from this cause are possible with some types of instrument. If for any reason the cover glass has recently been rubbed, breathe on it to dispel the charge.

Wires attached to instruments should preferably extend back over the table and away from the observers. When such wires are draped over the front of the table, they are sometimes accidentally caught by the observer, and the instrument is thrown to the floor. If it is impossible to avoid having the wires placed in this way, they should be clamped to the table near the instrument.

Instruments should be kept at a distance from motors, generators, transformers, or cables carrying large currents, in order to avoid error caused by **stray magnetic field.** Unshielded instruments should not be placed close together; for example, a particular type of unshielded d-c instrument, widely used, will be affected by about 1 per cent if a similar instrument is placed alongside it, and about 2 per cent if the instruments are in contact with their magnetic poles facing each other. This disturbance occurs regardless of whether the second instrument is in circuit.

Instruments should not be used in **extreme temperatures,** if this can be avoided. Even though no permanent injury may result, the indications of all instruments change to some extent with temperature. This effect is very small in d-c voltmeters and in moving-iron ammeters, but may be large in d-c ammeters and millivoltmeters.

In any current-carrying part of an instrument the heating increases at least as the square of the current. Great care should be used to **avoid passing excessive currents** through instruments. All set-ups should be checked carefully before the current is turned on, to see that no instrument will be overloaded. In using **multiple-range instruments** with unknown or doubtful values of current or voltage, always begin with the highest range. Special care is necessary in the use of **millivoltmeters,** which will usually be damaged by connection to a single dry cell. Wattmeters have three operating limits, namely, current, voltage, and power, only the last of which is shown by the position of the pointer. The fact that the current or the voltage is excessive, where the power is relatively small, can reveal itself only by smoke or other sign of thermal damage. In using a **millivoltmeter with shunts,** always **attach both its leads to the shunt** before connecting the other ends to the millivoltmeter, and in disconnecting **take the leads off the millivoltmeter first.** When the reverse procedure is followed, and only one end of one of the leads remains to be connected to the shunt, an accidental contact between this end and a switch blade or other live part, or a grounded metal object, may burn out the millivoltmeter instantly.

In connecting instruments into circuit, see that all **contact surfaces** of wires, lugs, binding posts, etc., are clean and that connections are firmly made. If the test is to last for a number of hours, it is advisable to provide **switches for by-passing** the current around ammeters and the current coils of wattmeters. (Modern watthour meters are made to carry three times the rated current.) The switches are to be opened when readings are to be taken. Switches may be provided also to disconnect a-c voltmeters and the voltage circuits of wattmeters, if these instruments do not have keys. The power required by d-c voltmeters is so small that this precaution is unnecessary for them.

If an ammeter must be connected directly in a line, it is advisable to put it in the **grounded wire,** if available, to protect the operator and the instrument.

If an instrument has been dropped, overloaded, or exposed to abnormal temperatures, it should be checked, perhaps by comparing it with a similar instrument not so exposed, before continuing to rely on its indications.

A heavy short circuit in nearby conductors, even though of extremely short duration, is likely to partially demagnetize the magnet of a d-c instrument. After such an exposure the instrument should be rechecked, even though it has sustained no visible injury.

Repairs to instruments, if at all extensive, are usually best made at the factory. Electrical instruments are so delicate that repairs elsewhere should not be attempted unless equipment and some experience are available. Instruments should be opened only in a clean dry place free from lint, dust, and metal filings. Direct-current instruments have very strong permanent magnets and very small clearances for the moving coil and should never be opened on a workbench where iron filings may be present.

Small displacements (say a few tenths of a division) of the pointer from the zero mark, when no current is flowing through the instrument, may be corrected by turning the zero adjuster. Large displacements (several divisions) may indicate bending of the pointer or other derangement and consequent need for repairs. The use of the zero corrector to bring a bent pointer back to zero usually introduces appreciable errors.

The heads of binding posts should be screwed down before instruments are put away. Instruments are sometimes picked up by the binding posts, and this always unsafe procedure reaches a maximum of insecurity when the binding-post heads happen to be loose.

DETAILS OF DESIGN AND STRUCTURE OF ELECTRICAL INSTRUMENTS. Space limitations permit only references to sources of information on the numerical design of the windings, springs, magnets, damping mechanisms, and other parts of electrical instruments. The more important ones are Drysdale and Jolley's volumes, *Electrical Measuring Instruments*, Benn (1924); Edgcumbe and Ockenden's book, *Industrial Electrical Measuring Instruments*, Pitman (1933); Keinath's book, *Die Technik der elektrischen Messgeräte*; and the technical journal *Archiv für technisches Messen*. The *Review of Scientific Instruments* (New York) and the *Journal of Scientific Instruments* (London) publish from time to time numerical data on the construction and functioning of electrical measuring instruments.

12. ELECTRICAL INTEGRATING METERS

AMPEREHOUR METERS. Amperehour or quantity (coulomb) meters may be classified as electrolytic meters and motor meters. They are used in this country chiefly for storage-battery work. Amperehour meters of small current rating are used abroad as substitutes for watthour meters and are calibrated to read directly in kilowatthours at some assumed voltage.

Electrolytic Amperehour Meters. This type of meter is used extensively in Europe but rarely (if ever) in this country. The meter generally consists of a glass vessel or U-tube containing an electrolyte, such as water, or an aqueous solution of a salt of mercury or of caustic soda. The solution is decomposed by the current; and, since the rate of decomposition is proportional to the current, the deposit of metal or the quantity of gas evolved can be used to measure the electrical quantity in amperehours. In some types the entire current to be measured is passed through the electrolyte, and in others the bulk of the main current passes through a shunt and only a very small part through the electrolyte.

Motor Amperehour Meters. The essential parts of the motor amperehour meter are the motor element, the brake or speed-regulating device, and the register. The motor consists of an armature rotating in the field of a permanent magnet. Usually only a shunted portion of the current flows through the armature. Motor amperehour meters are intended to be cheaper and, if possible, more rugged than the ordinary watthour meter. The absence of a constantly excited voltage circuit in the meter avoids the constant loss of energy in such a circuit.

Commutator Amperehour Meter. The commutator amperehour meter in its simplest form usually consists of one or two permanent magnets which provide a driving field within which rotates an armature with three coils and a three-part commutator. In some designs the armature coils are mounted flat on an aluminum disk which, in addition to supporting the coils, produces the retarding action in connection with the field magnet. In another design there is a stationary iron core between the poles of the field magnet and a drum armature rotating in the gap between this iron core and the pole faces of the field magnet. No braking device is used in this latter type, and the motor runs at a speed proportional to the impressed emf, taking only sufficient power to overcome the friction in the bearings and in the gear train.

Amperehour meters of the commutator type are gradually passing out of use because of their inferior accuracy at light loads and the difficulty of adequately insulating the windings from the framework of the meter.

Mercury-motor Amperehour Meter. In the mercury-motor amperehour meter, current is carried to the armature by mercury, which takes the place of the brushes and commutator of the ordinary d-c motor. One form consists of a copper disk armature mounted in jeweled bearings and immersed in mercury. The current is led into and out of the mercury chamber through suitable terminals; and, since mercury has about 55 times the resistivity of copper, the greater part of the current goes diametrically across the disk. A device to compensate for the increase of fluid friction with increase of speed takes some such form as an auxiliary coil, in series with the motor element, which produces a flux to augment the flux cutting the driving disk.

Special Features of Amperehour Meters. Amperehour meters used for showing the state of charge of storage batteries are sometimes provided with special features. For example, a "compensation for overcharge" causes the meter to run more slowly on a given value of charging current than on the same value of current during discharge. The percentage difference in the two cases is chosen to provide for the excess that must be put into the battery because it is not 100 per cent efficient. This is accomplished by a "variable resistor element" in parallel with the main mercury chamber. A pivoted copper bar in this element is immersed in mercury and is cut by some of the flux from the driving magnets. It will thus change its position with the reversal of current direction and will vary the resistance of the element. The "compensation for light-load accuracy and internal battery losses" consists of a thermocouple heated by a potential winding and delivering a small current continually to the motor element in the direction corresponding to battery discharge. This action may be adjusted to cause the meter to record the loss of charge when the battery is not in use. The "thermal shunt method" of compensation for high discharge rates consists in making the shunt to the motor element of iron or other material of high temperature coefficient, and so designing it that it will heat up at the heavier loads. This causes the amperehour meter to have a higher percentage registration as the load increases.

Amperehour meters are sometimes provided with contacts which actuate auxiliary devices, such as circuit breakers, bells, or other signaling devices. This feature is used in connection with battery charging and electroplating.

WATTHOUR METERS. General. A watthour meter consists essentially of (1) a small electric motor, which may be either of the commutator type, mercury and disk type, or induction type; (2) a brake system composed of a disk of non-magnetic conducting material (usually aluminum) mounted on a spindle and arranged to rotate between the poles of one or more permanent magnets; and (3) a register indicating a number proportional to the number of revolutions of the disk. One winding, called the voltage coil, is connected in shunt with the load; and the other winding, called the current coil, is connected in series with the load, the connections being the same as for an indicating wattmeter. These connections may be made either directly or through suitable instrument transformers.

Principle of Operation of Watthour Meters. Watthour meters are so constructed that the average torque exerted by the motor is proportional to the average power in the load. The brake system is so designed that the opposing torque, due to the eddy currents induced in the disk, is proportional to the speed. When the speed is steady, the driving torque must be equal to the opposing torque and therefore proportional to the speed. Hence the speed of the disk is proportional to the average power in the load, and the total number of revolutions which the disk makes during any interval must be proportional to the total energy which has passed through the meter during this interval, whether the power remains constant or varies. To determine the energy input to the load in watthours or kilowatthours during a given interval, it is therefore only necessary to take the difference between the dial readings at the beginning and end of the interval and to multiply by the proper constant if the meter is not direct reading.

Sources of Error in Watthour Meters. In a d-c watthour meter the chief sources of error are the somewhat variable friction of the brushes, friction of the bearings and register, and air friction. In a-c watthour meters, the lack of exact 90-deg phase relation between the impressed voltage and the magnetic flux due to the current in the voltage coil may cause additional errors which vary with the frequency and power factor and also with the distortion of the wave form; in well-designed modern meters these errors are practically negligible. Instrument transformers introduce additional sources of error which may be undesirably large if the transformers are not properly designed or if the burden is excessive.

Commutator Type of Watthour Meter. The commutator type of meter, used only on d-c circuits, consists of a set of stationary field coils which carry the load current and of an armature wound with fine wire which is connected in shunt with the supply circuit. A resistor is usually connected in series with the armature, and the *ir* drop in it and in the armature itself is almost exactly equal to the line voltage at any practicable speed of revolution. In other words, the counter emf of the armature is too small to have any appreciable effect on the speed, which explains why an increase in the current in the fixed (field) windings increases the speed instead of reducing it as in ordinary shunt motors. The connection to the armature is by a commutator and brushes, as in an ordinary d-c commutator motor. Special designs are sometimes adopted for very large capacities, such as double armatures astatically arranged, and damping magnets enclosed in a laminated iron shield in order to reduce stray-field errors where heavy currents are used and heavy short circuits may occur. A four-pole construction with a single armature is also used for the same purpose.

Friction in a well-designed d-c watthour meter should have a noticeable effect only on light loads. To compensate for it, a "light-load adjusting coil" is added, which is an auxiliary or compounding coil connected in series with the armature. It is placed adjacent to the field coils so that its field strengthens the main field and produces a slight torque independent of the load and just sufficient to compensate for friction.

The design of the modern d-c commutator watthour meter includes close electromagnetic coupling between the armature and field coils, a light-weight armature, a commutator of very small diameter, and brushes having very small friction but able to stand the wear and carry the current without undue sparking and pitting of the contact parts. The lower bearing, which carries all the weight, usually consists of a steel pivot supported by a cupped and polished jewel, generally of sapphire. In some meters, particularly high-capacity ones with astatic armatures and a correspondingly heavy weight, diamond jewels are used.

Commutator-type d-c meters are built with current ratings ranging from 5 to 10,000 amp, 100 to 600 volts. Meters of this type are furnished with double current circuits for three-wire service, the maximum ampere capacity for three-wire meters being about 6000 amp. Special meters have been supplied for 1200- and 2400-volt railway circuits.

Tests on d-c commutator meters show that a meter of good design should start on 2 per cent of full load and should give accuracies about as follows: to within 3 $^1/_2$ per cent from 5 per cent to $^1/_4$ load, 2 per cent from $^1/_4$ to full load. The Code for Electricity Meters (see Bibliography, p. 5-112) requires d-c watthour meters to run continuously on normal voltage and 2 per cent of rated current, and higher loads to be accurate within the following limits:

Percentage rated current..........	5	10	20	50	100	150
Maximum deviation, percentage...	6.0	3.0	2.0	2.0	2.0	2.0

Mercury-motor Watthour Meter. The mercury-motor d-c watthour meter has a copper disk armature submerged in mercury. The line current enters and leaves the mercury through two electrodes diametrically apart. Since mercury has about fifty-five times the resistivity of copper, the major part of the current traverses the disk. A laminated iron electromagnet having many turns of fine wire is connected across the line and sends through the copper disk a magnetic flux proportional to the line voltage. A shaft attached to the disk passes through tubes constructed to prevent the mercury from escaping when the meter is inverted. Externally, this shaft carries a worm which engages with the register train and an aluminum damping disk moving in the field of a permanent magnet. The submerged copper disk carries a float so proportioned that there is a slight upward thrust. In consequence, the lower bearing is a ring-stone jewel, and the upper end of the shaft is carried in a ring-stone end-stone bearing. The driving torque is proportional to the power supplied to the load. Nearly all the damping at small and moderate loads is caused by the drag disk, but at full load and overloads the fluid friction of the mercury increases at a rate which would cause a drop in the accuracy curve unless compensated. Compensation is effected by passing the line current through a few turns of heavy wire around the voltage electromagnet. The copper disk is slotted radially to reduce the damping effect which would be caused by the flux from the voltage electromagnet. This expedient reduces the influence of variation of voltage on the accuracy of the meter.

The compensation for friction is effected by a thermocouple shunted around the mercury chamber and heated from a resistance coil connected in series with the voltage circuit. The small emf generated in the thermocouple sends a current through the armature which is sufficient to overcome the friction of the moving parts.

On account of the low resistance of the current circuit, mercury-motor watthour meters are particularly well adapted for use with external shunts and are used for switchboard work where large currents must be metered. They withstand vibration and shocks better than commutator meters and are therefore invariably used on railroads and other types of transportation equipment.

Direct-current mercury-motor watthour meters have a full-load current rating of 10 amp. For all higher ranges external shunts are used. Shunts with ratings up to 60,000 amp have been furnished. Meters designed for voltages up to 700 volts are regularly furnished, and special meters have been supplied for 1200- to 3000-volt d-c railway circuits.

The load-accuracy curve at normal voltage is said to deviate from a straight line by less than 0.5 per cent from 5 per cent to 200 per cent of rated load.

Induction Watthour Meter. The essential elements of a typical induction watthour meter are shown in Fig. 34. A laminated electromagnet has three cores carrying a voltage coil and two current coils. The voltage coil, of many turns of fine wire, is connected across the line; the current coils, of a few turns of heavy wire, carry the load current. The magnetic flux set up by the voltage coil would be in quadrature with the applied voltage but for the iron loss and copper loss in the voltage electromagnet. Exact quadrature is obtained by expedients of which that shown in Fig. 34 is typical, namely, the use of a so-called lag coil closed through an external resistance (not shown). This resistance is adjusted by trial to give the quadrature relation at rated frequency. In some designs a metal plate forming a closed loop is used instead of a wound lag coil. The flux of the current electromagnet is practically in phase with the load current.

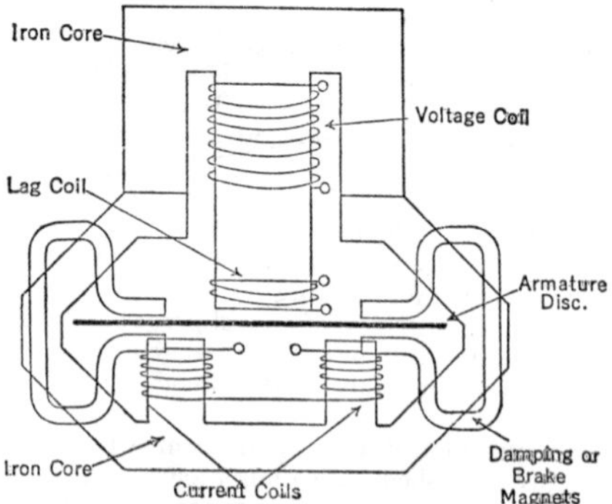

Fig. 34

Currents induced in the disk flow about each pole in the general form of concentric circles. The driving torque results from the interaction of each induced current with the flux which does not induce it. When the load power factor is unity, each induced current will be in phase with the flux with which it reacts to produce torque; and, as the power factor of the load becomes less than unity, this torque will decrease in proportion to the cosine of the angle of lag (or lead) in the load. Hence the driving torque is proportional to the power in watts taken by the load. The retarding torque of the damping magnets is proportional to the speed of rotation, and the speed will therefore be proportional to the power taken by the load. The total number of revolutions over any given period will be proportional to the energy in watthours.

To maintain accuracy down to very light loads, the **friction** of the bearings and of the register is usually compensated by placing a metal plate, forming a closed loop, near the pole of the voltage electromagnet with means for adjusting it in a plane at right angles to the axis of the voltage coil. The current induced in this loop produces a flux out of phase with the flux of the voltage coil; and the interaction between each of these two fluxes and the current induced in the disk by the other produces a slight turning moment, independent of the load current, which is adjusted by shifting the light-load plate until the frictional moment is just balanced.

Changes in temperature, if not compensated for, may introduce a number of errors which can be grouped into two main classes. Class 1 temperature errors include all effects which are equal at all loads and power factors and may be considered equivalent to a change in the permanent magnet strength. At usual operating temperatures, the braking magnet strength varies inversely with the temperature, causing over-registration at higher temperatures. This is overcome by placing a piece of temperature-sensitive magnetic alloy near the braking magnet air gap, thus by-passing a portion of the total flux. As the temperature rises, the braking magnet strength decreases; however, the reluctance or magnetic resistance of the shunt increases and by-passes less flux, thus allowing more to cross the air gap and making the braking effect remain constant. An alternative method is to mount the magnet so that the thermal expansion of the support increases its effective radius of action.

Class 2 temperature errors are those which vary in magnitude with the power factor of the load and may be considered equivalent to a change in the potential coil resistance. As temperature increases, the resistance of the potential coil increases, which means that the voltage flux is not in exact quadrature with the voltage. Considerable error at low power factor results from this change. By placing a piece of temperature-sensitive alloy (thermalloy) on the lag plate, the reactance of the lag-plate circuit will change with temperature in such a way that it will compensate for Class 2 errors. Alternatives are to mount the lag plate on a bimetallic strip or to use a lag coil of high temperature coefficient on the current poles.

Multi-element Induction Watthour Meters. Polyphase induction meters are usually made by combining two (or three) single-phase meter elements with their disks mounted on a common shaft. Two-element meters are used for 3-wire 3-phase circuits, and also for metering 3-wire single-phase loads with an accuracy which is independent of current and voltage unbalance. Three-element meters are used to meter 3-phase 4-wire loads. Although it is possible to meter 3-phase 4-wire loads with a special 2-element meter equipped with a third current circuit divided between the two current electromagnets, such meters are subject to errors from unbalance of the voltages. The use of shields of modern high-permeability alloys and the development of a slotted and laminated disk have made it possible to build polyphase induction meters with a single disk and hence with less weight on the bearing and less total volume.

Recent Improvements in Induction Watthour Meters. Among the more important improvements are (1) extension of range of accurate registration up to 400 per cent rated load; (2) use of high-strength alloys in damping magnets to replace the lower coercive chrome steel formerly used; and (3) socket-type meters which are easier to install and maintain than the older base-connected types.

Accuracy of Induction Watthour Meters. For the induction meters now in American production the makers' stated accuracy is 0.5 per cent or better from "light load" of, say, 5 per cent of rated watts up to 400 per cent of rating. Such a performance is needed for residence meters because electrically operated household appliances are widely used. Meters of the older designs of larger current rating, say 15 amp, would not correctly register on very light loads.

Current transformers do not have a large continuous overload capacity; for example, a secondary current of 7.5 amp would probably be considered an upper limit for a secondary rating of 5 amp. A modern 5-amp watthour meter with a current transformer is inefficient, and meters rated at 2.5 amp are coming into use for this purpose.

Performance Data of Induction Watthour Meters. For the single-phase watthour meters in production the following representative data are given by the manufacturers: weight of moving element, 13 to 16 grams; torque at rated load in watts, 50 to 55 mm-grams; loss in voltage coil at rated voltage, 0.9 to 1.4 watts; and loss in current coil at rated current for 5 amp meters, 0.15 to 0.30 watts.

Standardization of Induction Watthour Meters. As a result of cooperative planning between the meter manufacturers and the meter committees of the electric power industry, great strides have been made toward standardization of metering equipment.

Some of the more important features which are the same for the meters of all manufacturers are: dimensions of base and sockets of single-phase and some polyphase meters, counterclockwise direction of disk rotation as seen from above, terminal board connections for single-phase meters, and standardization of some mounting accessories. Two of the four makers have adopted a disk constant of $^1/_3$ watthours per disk revolution for the 115 volt, 5-amp service meter, whereas the other two manufacturers use a disk constant of 0.6. The nominal ratings are standardized for the four makers in the single-phase line, namely, 5, 15, and 50 amp, with an overload capacity of 400 per cent of rated current. With the standardization of meter mountings and sockets, the outdoor socket-type installation is becoming increasingly popular.

MAINTENANCE AND TESTING OF WATTHOUR METERS. The use of a portable standard watthour meter ("portable test meter" or "rotating standard") is the accepted method not only for checking customers' meters in position but also for testing repaired or exchanged meters in the meter laboratory. These standard meters should be checked against an indicating wattmeter and a time standard at intervals depending upon the conditions of use. To facilitate the testing of customers' meters, a large number of artificial-load and other devices are available.

A piece of equipment that has been widely adopted by utility-company meter shops is the photoelectric testing unit. This device facilitates meter testing by automatically counting the disk revolutions of the meter under test; through a system of relays the rotating standard is either started or stopped. This method eliminates most of the human errors of testing.

For details, consult the *Handbook for Electrical Metermen* (see Bibliography, p. 5-112). Additional valuable information on devices and methods for watthour-meter testing and on demand meters, indicating instruments, potentiometers, etc., is contained in the subsequent Reports of the N.E.L.A. Meter Committee under the heading, "New Developments."

DEMAND METERS. The charge to any customer should logically be made up of a part fixed by the maximum power in kilowatts taken at any time during a certain definite period, and another part fixed by the total energy used during the same period. The demand system of charging tends to improve the load factor of the station and to distribute its fixed charges more equitably among its customers.

A large number of variables are involved in the problem of demand metering, including the rating of generators, lines, and transformers, and the diversity factor, load factor, and power factor of the system. No practical meter has yet been devised to take into account all these variables; consequently charges based on the use of demand meters in addition to the usual energy meters must necessarily be somewhat arbitrary.

Requirements of a Demand Meter. The energy supplied to the customer is measured by a watthour meter; the maximum power (in kilowatts) taken by him is measured by some form of demand meter. For d-c supply at practically constant voltage a device which measures the maximum current is as satisfactory as one which measures maximum power, but when the voltage or power factor (in case of an a-c system) varies, the maximum-current indicator is not suitable. In any case, the demand meter should not measure the instantaneous peak but rather the average of the power demanded by the customer over a specified time interval, for the maximum demand recorded should not be unduly influenced by short circuits, excessive starting current of motors, or any other abnormal demand over a time too short to have any real effect on the generator and line capacity which must be provided. This time interval should theoretically differ with the character of the installation and the relation of its maximum power demand to the maximum station capacity. In relatively large customers' installations the time should be carefully chosen with reference to the time that the central station can endure an overload successfully. Hydroelectric plants have a very definite power limit which is reached with maximum gate opening, while steam-power plants have considerable thermal inertia and their boilers and generating apparatus can usually carry relatively large overloads for a considerable time. These considerations require much shorter time intervals for large users of hydroelectric power than would be necessary in steam plants of corresponding magnitude.

Demand intervals of 15 min, 30 min, and 1 hr are those in general use at the present time. For very special installations, intervals as short as 1 min and as long as 3 hr have been used. The general tendency is toward the use of a single interval for a given system, and toward the use of intervals larger than those formerly employed.

To improve the load factor, the off-peak system of rate schedules has been developed, which gives a preferential rate to customers utilizing current during the times of light load. Demand meters giving time of day as well as demand are used for this class of metering.

Classification of Demand Meters. Demand meters may be classified into (1) **curve-drawing instruments,** giving the load-time curve of the installation; (2) **integrated-demand meters,** consisting of an integrating meter combined with a device which registers the energy consumption from time to time in such a way as to indicate or record the maximum demand; (3) **lagged-demand meters,** which require a certain time interval for the indication to reach the value corresponding to the load. Under each of these types there are a number of further distinctions, such as how long the demand interval is, whether the demand intervals begin at specified times of the day or may be so chosen as to include the maximum average load during any time interval of the given length, and whether the maximum demand is simply indicated without reference to time or is recorded so that the time of its occurrence may be determined.

Curve-drawing Instruments as Demand Meters. The most complete knowledge of the conditions attending the customer's use of energy may theoretically be obtained by using a curve-drawing wattmeter, with which the average demand for any time interval may be found and also the time at which it occurred. In practice, however, unless the paper speed is fairly high, the lines will run together so as to make it difficult or impossible to determine the demand with any approach to accuracy. Curve-drawing wattmeters are expensive, and their indications are difficult of comprehension for the ordinary customer. They are used largely by central stations to accumulate demand data. Curve-drawing ammeters may be used on the assumption of constant voltage and power factor, but they do not give as accurate information as wattmeters.

Integrated-demand Meters Indicating Maximum Demand Only. The Type M-16 demand meter of the General Electric Company and the Type G demand meter of the Sangamo Electric Company consist essentially of a demand-registering and a timing mechanism mechanically connected and mounted within the same case. The demand-registering element is electrically operated by means of a contact actuated by the watthour meter in conjunction with which the demand meter is used. The registering element consists of a train of gears which drive a demand pointer forward over a dial. This train is actuated by a ratchet-and-pawl mechanism operated by an electromagnet which receives an impulse every time a certain amount of energy is registered by the watthour meter. The speed of advance of this pointer is therefore proportional to the power. The pointer-driving mechanism is reset to zero position at the end of each time interval, leaving the pointer itself in the highest position reached. This resetting is accomplished by a mechanism controlled either by a clock or by a constant-speed motor. These demand meters are arranged for intervals of 15, 30, or 60 min. The Type M-16 demand meter, for a-c service, is timed by a Warren synchronous motor; the Sangamo Type G, for d-c service, has a motor-wound spring-driven clock. The Types M-20 and M-21 demand registers, when used in appropriate General Electric a-c watthour meters instead of the regular register, serve the same purpose as the M-16 demand meter and also have the usual four dials showing the energy registration in kilowatthours.

The Westinghouse Types RW and RW-2, the Sangamo Types HG and HGC, and the Duncan Types T3 and T3C demand registers, when installed in the corresponding watthour meters in place of the usual registers, operate in the same general way and serve the same general purpose as the General Electric Types M-20 and M-21 demand registers. In the case of each manufacturer one of these two registers is an accumulative type, i.e., the demand is read and recorded on a set of dials rather than from the indication of a pointer on a scale. Each period, when the meter reader resets the pointer to zero, the accumulative dials automatically advance the amount of the period's maximum demand as indicated by the pointer.

The value of demand is obtained by subtracting the previous dial reading from the present dial indication. Dials can be read more accurately than the pointer and scale and have the added advantage of retaining their reading through the rate period.

Integrated-demand Meters with Arbitrary Time Interval, Recording All Demands and Time of Occurrence. Printing demand meters are used in combination with a watthour meter or with a totalizing device to provide a printed record of demand and time of occurrence. They are of the contact-operated block interval type and print the demands for each time interval.

The earliest instrument of this class was the printometer, later made by the General Electric Company as the Type P demand meter. This instrument printed on a paper tape at regular time intervals the time of record and the energy consumption registered up to that time by a watthour meter. Thus the difference between consecutive recorded values on the tape was proportional to the energy consumption in kilowatthours in the circuit during a definite time interval. The average demand in kilowatts over a period of time, corresponding to the interval between successive operations of the printing solenoid, was obtained by dividing the kilowatthours of energy consumption during a time interval by the length of the interval in hours.

The General Electric Type PD demand meter is based on the same general principles as the type P but embodies new features which eliminate the labor of computing the maximum demand and shows it directly along with a printed record of all the demands and the time at which each demand occurred since the meter was last manually reset. Another dial shows the present interval demand, that is, the number beneath the printing platen at that particular instant.

A third dial (six-digit cyclometer) integrates the total number of impulses delivered to the demand meter, thus providing a means of verifying the demand record by comparing it with the register reading of the watthour meter to which it is electrically coupled. Printing-type demand meters are inherently more accurate than graphic or recording devices and are used where the importance of the demand indications dictates the use of an especially reliable and accurate meter.

The operation of the Type PD meter is essentially as follows: A three-wire contact device geared to the watthour-meter shaft provides electrical impulses to two operating coils in the demand meter. These coils actuate a pair of advancing pawls which, by means of a ratchet arrangement, cause the printing-type wheel to revolve. At the end of the time interval a timing mechanism closes a circuit which energizes both a printing and a reset coil. The printing coil actuates a printing platen which records the position of the printing wheel before it is reset to zero. The reset coil operates a system of levers which disengages the advancing pawls, thereby allowing the printing wheel to return to zero and advances the record and carbon tapes.

The timing mechanism may be either an external contact-making clock, as in the Types PD-5 and PD-6, or a self-contained, synchronous timing motor, as in the Types PD-7 and PD-8, depending upon the particular application.

The Type RA recording-demand watthour meter of the Westinghouse Company consists of a polyphase watthour meter combined with a mechanism for obtaining a permanent graphic record of the integrated demand in kilowatts over successive predetermined time intervals. The total energy consumption is shown in the usual way by the register of the watthour meter. The time interval is controlled either by a hand-wound 35-day clock or by a synchronous-motor clock. A separate spring mechanism advances the record paper. The pen is moved across the paper by the gear train of the watthour meter, and at the end of each time interval a release mechanism frees it from the gear train and allows a weight to move it back to zero, where it is again meshed with the gear train to repeat its advance during the next time interval. Just before the release occurs, the record paper is advanced slightly so that the pen makes a distinct record of the maximum travel, which is a measure of the integrated demand. The time of occurrence is shown by the time figures printed on the paper. This meter is made with time intervals of 5, 15, 30, or 60 min. It is a self-contained structure, with the metering and demand-recording mechanisms mechanically connected.

The Westinghouse Type RB recording-demand watthour meter is a development from the Type RA and functions in the same general manner. It employs OB watthour-meter elements and has a synchronous motor which drives the timing device and the chart-driving mechanism.

The General Electric Type G-9 demand meter is used in connection with a watthour meter and gives a graphic record on a circular chart of the demands integrated over definite time intervals, with the time of day and day of week of their occurrence. It consists essentially of a registering element and a timing element. The registering element is electrically connected to the watthour meter, which has a contact device either on one of the spindles of the register or in a separate mechanism operated by the rotating element. This contact device alternately makes and breaks the circuit of a solenoid in the registering element each time a definite amount of energy is registered by the watthour meter. An armature operates a ratchet-and-pawl mechanism to advance the stylus by a small step for each operation of the contacts. At the end of each time interval a cam mechanism operated by the timing device allows the stylus to return to the zero position, where it is re-engaged ready to integrate the energy used in the following interval. The timing mechanism rotates the chart continuously, giving a record in the form of a saw-tooth polar curve. The chart gives a record for 1 week or 1 month. The time interval is 15, 30, or 60 min.

The General Electric Type GM-10 demand meter is a development from the G-9 demand meter and serves the purpose of assisting the plant operator to keep the maximum demand at the lowest practicable level. Alarm contacts used for this purpose in previous demand meters had the serious limitation that they closed after the predetermined maximum demand had been reached. The GM-10 demand meter avoids this limitation by having the alarm contacts so arranged that they will close if for any part of the time interval the integrated load (number of impulses transmitted by the watthour meter to the demand meter) has exceeded a predetermined and adjustable value for the elapsed part of the time interval. To accomplish this, a so-called "ideal-demand" mechanism has been added to the G-9 demand meter. Type GM-10 demand meters are made for use on a-c circuits only and employ the Warren clock motor as the timing element.

The Type DG-1 watthour demand meter of the General Electric Company is a combination of a two-element polyphase watthour meter and a block-interval demand meter which records the demand for each definite time interval on a strip chart. The watthour-meter disk shaft has the usual worm at the top which drives the register and a second worm at the bottom which drives the recording pen mechanism through a gear train and clutch. The timing is accomplished by a Warren motor which also drives and re-rolls the chart, winds a resetting spring, and actuates the resetting mechanism. Standard time intervals are 15, 30, and 60 min. This same design of polyphase watthour demand meter is furnished also in three-element construction as General Electric Type DG-2.

Thermal-storage Lagged-demand Meters. These meters all show maximum demand only, and the indication depends on the duration of the load as related to the demand interval of the meter, as well as on the previous history of the load. The demand interval of this type of meter is usually taken as the time required for the indication to reach 90 per cent of the full value of a steady load which is thrown suddenly on it.

The General Electric Type HI-1 demand meter is a watt demand instrument activated by a pair of bimetal spiral strips. When these strips are heated, their relative deflection causes a pointer to move across the scale indicating watt demand. The strips are heated by the secondary currents of two instrument transformers, one connected across the line and the other in series with the load current. The meter indicates a value of watts, with a time lag introduced to give the desired demand. On constant load the meter will reach 90 per cent of its indication in 15 min, and the ultimate indication will be obtained in 45 min.

The Lincoln Type A.D. demand meter indicates maximum ampere demand on either d-c or a-c circuits. In the two-wire Type A.D. meter a shaft carries the demand pointer and two bimetallic spiral coils which tend to rotate the shaft in opposite directions when their temperature is increased. A heater carrying the line current (or a definite part of it, when shunts or current transformers are used for the larger capacities) is arranged so as to heat only one of the spirals. The angular deflection of the pointer, when a steady state is reached, is proportional to the square of the current, but the scale is marked either in terms of the first power of the current or in kilowatts for an assumed value of line voltage. The three-wire Type A.D. demand meter operates in a similar manner but has three heating coils, two of equal resistance connected in the outside wires and acting on one bimetallic spiral, and the third, having one-half the resistance of each of the others, connected in the neutral wire and acting on the second spiral. This arrangement ensures that the maximum ampere demand of the three-wire circuit will be correctly measured, regardless of the degree of unbalance of the load.

The Lincoln Type W.D. demand meter is constructed on the principle of the hotwire wattmeter and indicates demand in kilowatts. In construction it resembles the two-wire Lincoln Type A.D. meter just described but differs in that two heaters are used, one acting on each of the two bimetallic spirals. The meter contains a small transformer supplying a low-voltage current to the two heating coils connected in series. The line current is brought in at the central point on the secondary winding of the transformer and flows through the two heating coils in parallel and on to the load. One heating coil therefore carries one-half the line current plus a current proportional to the line voltage, and the other heating coil carries one-half the line current minus the current which is proportional to the line voltage. It may be shown that the resulting difference in rate of heat liberation in the heaters, and therefore the angular deflection of the pointer, will be proportional to the load in kilowatts, for any power factor. The Type W.D. meter is internally shunted in the larger capacities.

Demand Meters Taking Account of Power Factor. The increasing appreciation of the effect of low power factor on the cost of energy supply has resulted in the development of demand devices taking power factor into account. The Westinghouse Type RI kilovolt-ampere recording-demand meter resembles the Type RA recording-demand meter which has been described but has also a second polyphase meter which is connected by means of phasing transformers to register the kilovarhours (reactive kilovolt-ampere hours). This meter and the energy meter drive two small aluminum disks on which rests an aluminum ball. A third disk, in contact with the ball, is caused to rotate at a speed which is equal to the vector sum of the speeds of the other two disks. The total kilowatthours and total kilovarhours are shown by the usual registers on the energy meter and the reactive-component meter. The graphic record shows the kilowatt demand and the kilovolt-ampere demand during the predetermined time interval, from which the average power factor during any time interval may be found.

The Sangamo Type S kilovolt-ampere demand meter is controlled by two watthour meters which measure the energy in kilowatthours and the reactive component (kilovarhours), respectively. It gives a record on a single chart from which the following information about the load can be obtained: (1) the kilowatt demand; (2) the kilovolt-ampere demand; (3) the maximum kilowatt demand; (4) the power factor at time of demand; (5) the reactive-component demand; (6) the time of occurrence of the demand; (7) the energy component in kilowatthours; (8) the reactive component in kilovarhours (reactive kilovolt-ampere hours). The power triangle for any given demand interval is drawn on a strip chart. The energy is measured by a watthour meter, and the registration is transferred by electrical means to move a stylus horizontally across a vertical paper chart which is driven (by electrical means) by a second watthour meter connected so that it integrates the reactive component. Both the kilowatthours and the kilovarhours are shown by registers in the kilovolt-ampere meter. A pusher arm mechanically connected to the energy stylus advances a friction pointer which shows the maximum kilowatt demand. A small synchronous motor drives an auxiliary stylus across thirteen equally spaced time (hour) divisions at the right of the chart and also trips the energy stylus and returns it to zero at the end of each predetermined time interval.

13. INSTRUMENT TRANSFORMERS

GENERAL. An instrument transformer is a transformer in which the conditions of current or voltage and of phase position in the primary circuit are represented with acceptable accuracy in the secondary circuit. An instrument transformer may be either an instrument current transformer or an instrument potential (voltage) transformer. These terms are usually abbreviated to current transformer and potential (voltage) transformer, respectively. Both kinds of transformer are applicable for measurement and control purposes. The current transformer is intended to have its primary winding connected in series with the circuit carrying the current to be measured or controlled, and the voltage transformer is intended to be connected in parallel with the circuit.

The use of instrument transformers separates the secondary measurement and control devices from the dangerous line voltage and also permits all these devices to have a single rated current and a single rated voltage. For measurement purposes high accuracy is important; for the operation of control and protective devices a very moderate accuracy is sufficient, but reliability and ruggedness are of the highest importance. It is therefore frequently advisable to use separate transformers for the two functions, even at an increased cost.

DEFINITIONS. The **true ratio** of a current transformer or a voltage transformer is the ratio of the rms primary current or voltage, as the case may be, to the rms secondary current or voltage, under specified conditions. The quotient obtained by dividing the true ratio by the nominal ratio (as marked on the name plate) is often called the ratio factor or ratio correction factor.

The **phase angle** of a current transformer or a voltage transformer is the angle between the primary current or voltage vector and the secondary current or voltage vector reversed. This angle, by international agreement, is considered as positive when the reversed secondary current or voltage vector leads the primary current or voltage vector. The phase angle of a current transformer is often denoted by the letter β, that of a voltage transformer by γ. The phase angles β and γ may be inherently either positive or negative, depending on the characteristics of the particular transformer and its burden.

If a wattmeter is connected to the secondary circuits of a voltage transformer of ratio R_v and phase angle γ, and of a current transformer of ratio R_c and phase angle β, the true power P in the primary circuit is given by

$$P = P_s R_v R_c \frac{\cos (\theta_2 + \alpha + \beta - \gamma)}{\cos \theta_2} \tag{12}$$

where P_s is the power indicated by the wattmeter; θ_2 is the apparent phase displacement by which the voltage leads the current as deduced from the indications of the instruments in the secondary circuits; and α is the phase angle of the voltage circuit of the wattmeter. Tables of values of the correction factor $\cos (\theta_2 + \alpha + \beta - \gamma)/\cos \theta_2$ are given in the section on Wattmeters.

The effect of a single current transformer on the measurement of power, or energy, is characterized by a "transformer correction factor," abbreviated TCF and defined as

$\frac{R_c \cos(\theta_2 + \beta)}{R_{nom} \cos \theta_2}$. The correction factor for energy measurements, when a watthour meter is used with instrument transformers, is the same as that given above for power measurements.

The **burden** of an instrument transformer is that property of the circuit connected to its secondary winding which determines the active power and reactive power required at the secondary terminals of the transformer. It is expressed either as total ohms impedance, together with the effective resistance and reactance components of the impedance, or as the total volt-amperes and power factor of the secondary devices and connecting leads. The values expressing the burden apply conventionally to the condition of rated secondary current or voltage of the instrument transformer and a stated frequency, both of which must be included with the burden expression. The impedance expression is the more convenient for current transformers, the volt-ampere power-factor expression for voltage transformers.

POLARITY MARKINGS. The primary terminals and secondary terminals of instrument transformers should be so marked that, when a lead is connected to a marked secondary terminal, the polarity will be the same as if the primary conductor running to the marked primary terminal were detached from the transformer and connected directly to the lead; that is, the relation of the marked terminals is such that, if the instantaneous direction of the current in the marked primary lead is toward the transformer, the instantaneous direction of current in the marked secondary lead is away from the transformer, or vice versa.

RATIO AND PHASE ANGLE OF CURRENT TRANSFORMER. Refer to Fig. 35 and let

n = ratio of the number of secondary to the number of primary turns.
I_p = primary current (reversed).
I_s = secondary current.
k = ratio of exciting current (referred to secondary) to the secondary current.
η = phase angle between the exciting current and primary induced emf.
θ = phase angle between the secondary current and secondary induced emf (positive for lagging current).
β = the "phase angle" of the transformer, i.e., the angle between the primary current and the secondary current reversed.

Then the ratio of transformation is

$$\frac{I_p}{I_s} = n[(\cos\theta + k\cos\eta)^2 + (\sin\theta + k\sin\eta)^2]^{1/2}$$
$$= n[1 + k\cos(\eta - \theta) \cdots] \text{ approx.} \tag{13}$$

and the phase angle is

$$\beta = \tan^{-1}\frac{k\sin(\eta - \theta)}{1 + k\cos(\eta - \theta)} \tag{14}$$

The value of the ratio k for a given primary current depends upon the impedance of the burden connected to the secondary; the higher this impedance, the greater the value of k. Therefore increasing the secondary burden without changing its phase angle tends to increase the difference between the ratio of turns and the true ratio of the primary and secondary currents; increasing the secondary burden also tends to increase the phase angle β of the transformer. The true ratio I_p/I_s and the phase angle β also depend upon the power factor of the burden and on the internal impedance of the transformer. The core flux required for a given induced voltage is inversely proportional to the frequency; consequently the exciting current and hence the phase angle and the departure of the true ratio from the ratio of turns will also vary approximately in inverse proportion to the frequency.

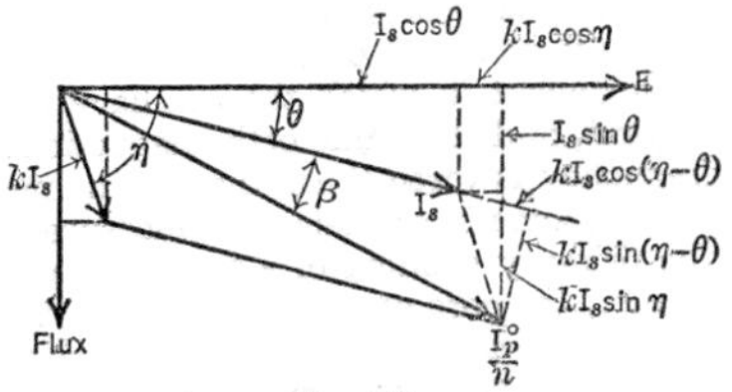

FIG. 35

In order to use these formulas for numerical results it is necessary to know the amount and power factor of the exciting current under the particular conditions of burden. Since this is difficult to predetermine, it is customary to use curves similar to those given in Fig. 38, obtained by plotting actual test results.

SPECIAL DESIGNS OF CURRENT TRANSFORMERS. Although the design of current transformers is necessarily based on the same fundamental principles as that of power transformers, the conditions under which they operate (nearly short-circuited secondary and varying terminal voltage) are so special that the resulting design is very different. The flux density under normal conditions is very low (a few hundred gausses). The ampere-turns at rated primary current should be at least 1000 for good accuracy under normal conditions if silicon steel is used. In high-range current transformers this condition is met by the use of a single primary conductor running through the center of the core opening. A transformer of this type has the great advantage that, when subjected to severe overloads, there is no mechanical force tending to distort the primary. Moreover, in high-voltage circuits a single-turn transformer can be made by slipping a core and secondary winding over the bushing of a power transformer or oil switch, thus saving the cost of additional insulation. In ratings below 600 amp the ratio error and phase angle of single-turn transformers become large unless some special means is provided for reducing them.

The ratio error at any chosen burden may be made small for one value of current by choosing a number of secondary turns slightly less than that corresponding to the nominal ratio of the transformer. The phase angle for a given burden can be reduced somewhat by providing a closed tertiary circuit of suitable impedance.

Much greater improvement applicable over a considerable range of current, burden, and frequency can be obtained by certain other procedures. One is the use of a **two-stage current transformer** [see paper by Brooks and Holtz, *Trans. AIEE*, Vol. 41, p. 382 (1922)], which is provided with an auxiliary magnetic circuit encircled by primary and secondary windings having a turn ratio equal to the nominal transformer ratio. Any departure from perfect performance of the main (or first-stage) unit induces a corrective current in a third winding on the auxiliary core. This corrective current, passed through an auxiliary current coil in a secondary instrument, will add its effect vectorially to that of the main current in the regular instrument coil, with the result that the effective ratio of primary current to the vector sum of the main and auxiliary secondary currents is very close to the nominal value. By the use of a suitably adjusted impedance network, two-stage transformers can be used with ordinary instruments and meters having single current windings.

Another method for improving the accuracy of current transformers is the use of **special nickel-iron alloys** [such as Hipernik; see paper by Spooner, *Trans. AIEE*, Vol. 45, p. 701 (1926)] which have very low losses and high permeability. The errors are reduced roughly in proportion to the losses, and high accuracy with low rated ampere-turns or moderate accuracy with a very small and light transformer can be obtained.

A third scheme is to provide a hole through the core splitting the magnetic circuit into two parallel paths over part of its length [see paper by Wilson, *Trans. AIEE*, Vol. 48, p. 783 (1929)]. A few turns of the secondary circuit are wound through this hole and cause one path to operate at a flux density different from that in the other. A change in the current changes the relation between the reluctances of the paths, and the effective secondary linkages can be made to vary in such a way as to maintain a nearly constant ratio. A short-circuited turn through the hole tends to compensate the phase angle in a similar fashion. A saturated-core reactor of suitable design in parallel with the secondary burden can be used with similar results.

A fourth and basically similar scheme involves the use of auxiliary magnetization which causes the core material to work in a range where it has a higher effective permeability. By using two cores any direct coupling between the auxiliary circuit and the secondary circuit is balanced out. The name biased core has recently been suggested for such an arrangement in which the auxiliary magnetization has the same frequency as the main current, and the name orthomagnetic has been proposed for those in which the auxiliary frequency is three times that of the main current.

It is sometimes desired to measure the current in a cable without cutting it to insert the primary winding of a transformer. This can be done by the use of a current transformer, the core of which is made in two parts hinged together, and provided with a secondary winding. Such a transformer can be clamped around the cable, which thus forms a single-turn primary winding. Although not capable of very high accuracy, such transformers are very useful in approximate measurements, for example, of the current taken by motors or the distribution of load among transformers banked in parallel.

Most portable current transformers and certain types of switchboard transformers are provided with two or more sections of primary winding which can be connected in series or in parallel to give various primary ranges. The use of such switchboard transformers makes it possible to take care of growth or redistribution of the station loads without replacing the transformers.

Certain types of current transformers, especially oil-filled high-voltage transformers, are made with duplicate cores and duplicate secondary windings, placed in a common oil tank and linked with the same primary winding. Such "double-secondary" transformers constitute in effect two transformers, one of which can be used with a light burden for accurate metering, while the other supplies relays or other control equipment of higher burden. This construction avoids the duplication of the expensive high-voltage bushing and the oil tank.

OPERATION AT HIGH OVERCURRENTS. An important application of current transformers is to operate the relays which serve to clear the circuits during severe overloads or short circuits. For this reason they must be so constructed that momentary overcurrents of 20 or even 100 times the normal will not injure the transformer. Also in some cases its ratio should remain sufficiently close to its nominal value to provide for the proper differential action of the relay circuits in isolating the faulted portion of the system. At other times it is better if the saturation of the transformer core keeps its secondary current from rising to a value which would injure the relays.

For these applications current transformers are classified as having either (1) high internal reactance (H), in which case the rise of terminal voltage or overcurrent is somewhat limited by the internal drop, or (2) low internal reactance (L), in which case this effect is absent. [See report of AIEE Committee on Protective Devices, "Current and Potential-transformer Standardization," *Trans. AIEE*, Vol. 61, p. 698 (1942), and also ASA Standard for Transformers C-57]. In general transformers of the bushing, bar, or window types are in class (L).

Because of the experimental difficulties involved it is customary to calculate the ratio (and, when desired, the phase angle) of current transformers at large overcurrents from data supplied by the maker on the magnetic characteristics of the core (see ASA Standard C-57).

TESTING CURRENT TRANSFORMERS FOR RATIO AND PHASE ANGLE.

General. For accurate measurements the ratio error and the phase angle may not be negligible. It is also possible for damage to the transformer windings to be caused by surges, overloads, or other disturbance, and it has therefore become the practice of many power companies and the rule of a number of public utility commissions to require the periodic testing, at installation and at intervals of 5 or 10 years thereafter, of all instrument transformers used in important metering installations.

If the core of a current transformer becomes permanently magnetized, as by the passage of direct current through its winding or by the opening of its secondary circuit while primary current is flowing, the ratio and phase angle may be seriously changed. The original values may be restored by **demagnetizing the core.** This is conveniently done by sending an alternating current of 5 amp through the secondary winding, while the primary is open, and gradually reducing this current to zero. An alternative method is to send rated current through the primary winding while the secondary is connected to a resistance of several hundred ohms and gradually to reduce this resistance to zero.

Methods of test fall into two classes, direct and relative. In the direct method the ratio and phase angle of a single transformer are determined directly; in the relative the transformer under test is compared with a standard transformer of the same nominal range, the ratio and phase angle of which are already known as the result of some previous test. In general, relative methods are simpler, quicker, and more convenient. Direct methods can be made more accurate but are usually less convenient and require the use of more sensitive and hence more delicate equipment. They are, however, a necessary link in the chain of measurement. Equipment for making direct measurements on instrument transformers is maintained in the laboratories of the National Bureau of Standards, of certain state public utility commissions, of the larger manufacturing companies, and of a few of the larger power companies. Relative testing methods are regularly used by other state commissions and by a large number of power companies, their standard transformers being checked at regular intervals by reference to one of the direct equipments.

Direct Methods for Ratio and Phase Angle. In these methods the IR drops in two known non-inductive resistors, carrying the primary current and the secondary current respectively, are adjusted to be equal and opposite. The ratio of currents is then the reciprocal of the ratio of resistances. The value of the phase angle is obtained from the value of mutual inductance or of capacitance required to bring the two IR drops into exact opposition.

These direct null methods can be forced to a precision of setting of 0.01 per cent, which is enough to reveal unavoidable changes in the performance of most current transformers caused by self-heating, magnetization, etc. The principal sources of error are uncertainties as to the residual inductance and the resistance at rated current of the primary resistor and the effects of the stray magnetic field produced by the large primary current of the transformer under test. The design of suitable non-inductive primary resistors becomes very difficult for currents above 1000 amp. Current transformers of high ranges are probably best tested by a relative method, using as a standard a special transformer provided with a number of equal sections of primary windings. Such a transformer can be standardized by a direct method on a low range with the sections of primary winding in series and then used as a standard of higher range with its sections in parallel. The ratio can be considered as strictly inversely proportional to the number of series primary turns, and the phase angle as strictly constant, provided only that each section of the primary winding carries its equal share of the total current when the sections are in parallel, or that all sections have equally close magnetic coupling to the secondary winding. This latter condition can be very closely approximated by the use of a ring core with a uniformly distributed secondary winding. This same principle is utilized in the Baker test-ring method for the direct testing of current transformers.

Relative Methods for Ratio and Phase Angle. Of the relative methods the simplest is that in which the primary windings of the test and standard transformers are connected in series while each secondary winding is connected to one of a pair of similar measuring instruments, the difference in the readings of which gives directly the difference in the two transformers. If only the ratio is desired, the two instruments may be 5-amp ammeters. If a phase-angle comparison also is desired, two wattmeters or two watthour meters are used. In any case the effect of errors in calibration of the instruments can be almost entirely eliminated by interchanging them and using the mean of the two sets of readings thus obtained. The wattmeter or watthour-meter readings made with the voltage circuits excited in phase with the primary current give directly a comparison of the ratios of the two current transformers. The readings made with the voltage circuits excited in quadrature with the current provide a measure of the difference in phase angle of the two transformers.

The accuracy of methods of this type is limited by the precision of reading the ammeters or wattmeters and becomes very poor at low currents. By using two watthour meters and allowing the registration to accumulate over a longer time at small currents the precision can be considerably increased.

The advantages of the relative method of test are much greater when balance methods are used. As indicated in Fig. 36, if the primary windings of two transformers T and S are connected in series, the secondary windings may also be connected in series so that their induced emf's aid in sending current through the main circuit $ABCD$. If, for instance, the transformer under test T has a smaller ratio (and hence a larger secondary current I_T) than the standard transformer S, the excess secondary current $I_T - I_S$ will flow through the bridge circuit R_1 between points A and C. This vector difference in current may be measured directly by inserting in R_1 one coil of a separately excited electrodynamic instrument (a 1-amp wattmeter fixed coil or a 30-volt wattmeter moving coil is convenient; a 10-amp ammeter should be first tried as a precaution against damage to the wattmeter if the transformers are connected with the wrong polarity). The difference in ratio and the phase angle (in radians) of the two transformers are given directly with ample accuracy by the ratio of the in-phase and quadrature components, respectively, of the difference in current, to the total secondary current of either transformer.

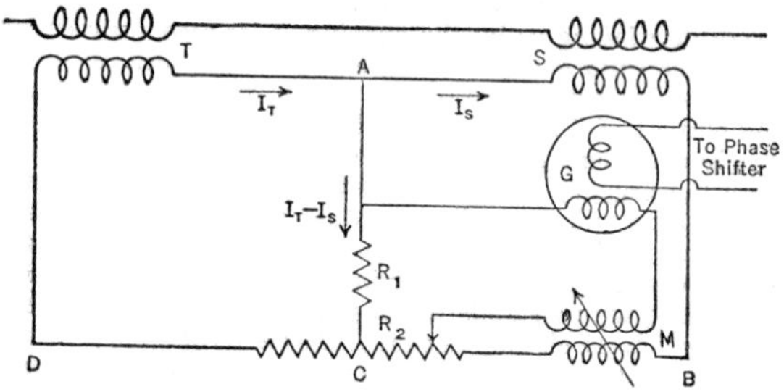

FIG. 36

A more convenient modification of this method is to measure the vector difference $(I_T - I_S)$ in current by a null method. This is done by inserting a fixed resistance R_1 (of, say, 0.2 ohm) in the bridge circuit AC and balancing the IR drop of the difference current in it against the sum of the drop of the secondary current flowing in a slide wire of resistance R_2 and the emf induced in the secondary of the mutual inductor M. The detector at G is thus brought to a zero deflection, and its calibration need not be known. The difference in ratio of the two transformers is then R_2/R_1, and the phase angle (in radians) is $\omega M/R_1$. Portable test sets (known as the Silsbee Current Transformer Test Set) embodying the principles of Fig. 36 are widely used for routine current transformer testing.

RATIO AND PHASE ANGLE OF POTENTIAL (VOLTAGE) TRANSFORMERS. The relations of the quantities involved are shown in the vector diagram, Fig. 37.

$$\text{Ratio} = \frac{E_p}{E_s} = n\left[1 + \frac{IR\cos\theta + IX\sin\theta}{nE_s} + \frac{(IR\sin\theta - IX\cos\theta)^2}{2n^2E_s^2} + \frac{I_0R_p\cos\eta + I_0X_p\sin\eta}{nE_s}\right] \quad (15)$$

$$\text{Phase angle } \gamma \text{ in minutes} = \frac{3438}{E_p}[IR\sin\theta - IX\cos\theta + I_0R_p\sin\eta - I_0X_p\cos\eta] \quad (16)$$

where n = ratio of the primary turns to the secondary turns.
R and X = equivalent resistance and reactance of the transformer, referred to the primary side.
R_p and X_p = resistance and reactance of the primary winding.
E_p = primary impressed voltage reversed.
E_s = secondary terminal voltage.
E = primary induced voltage.
$I = I_s/n$ = load current in the primary winding, I_s denoting the secondary current.
I_0 = exciting current.
η = angle by which secondary induced voltage leads exciting current.

It is assumed that

$$\frac{R_p}{R} = \frac{X_p}{X} \quad (17)$$

because the reactance of the primary and secondary windings cannot be directly determined. The third term within the brackets in eq. (15) is very small and can be omitted with very little error. Hence it is evident that at constant power factor and constant secondary voltage the ratio and phase angle will vary linearly with the secondary current. Also, if the impedance of the burden is kept constant as the voltage is changed, the ratio and phase angle of high-grade voltage transformers are found to change only to the very slight extent which results from lack of exact proportionality between I_0 and E_s.

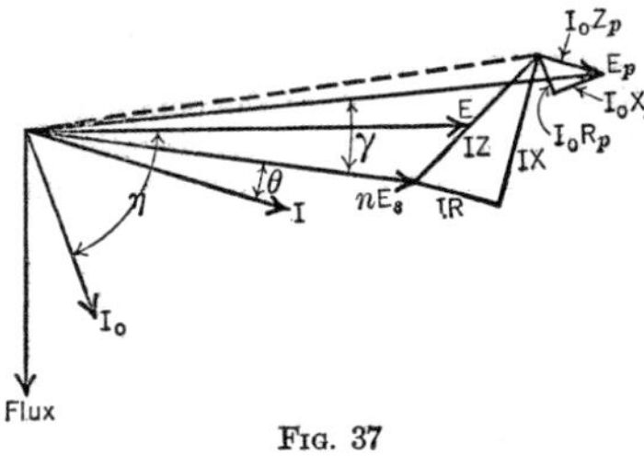

Fig. 37

The phase angle γ of a voltage transformer, i.e., the angle between the primary terminal voltage and the reversed secondary terminal voltage, may range from a negative angle (reversed secondary voltage lagging behind primary voltage) to a considerable positive angle (reversed secondary voltage leading primary voltage), depending on the exciting current, power factor, and impedance of the secondary burden and the impedance of the transformer windings. Under the no-load condition this angle is nearly always positive, and at high core densities, where the exciting current is large and of very low power factor, it may be very large. The general tendency of non-inductive secondary burdens is to cause γ to vary in the negative direction (secondary voltage to lag) and of inductive (lagging current) burdens to cause it to vary in the positive direction (secondary voltage to lead). It is frequently possible to bring γ practically to zero for a single voltage and burden by adding a non-inductive burden in suitable amount.

SPECIAL VOLTAGE-TRANSFORMING DEVICES. For high-voltage circuits the size and hence the cost of voltage transformers of the conventional type increase very rapidly (roughly in proportion to the square of the rated primary voltage) because of the large amount of insulation required and the need for using in the primary winding a size of wire of reasonable mechanical strength, much larger than would be required to carry the load current. To avoid this expense European manufacturers have developed several alternative types of construction. See Bibliography, papers by Imhof [*Bull. Schweiz. Electrot. Ver.*, Vol. 23, p. 742 (1928)], Keinath [*Siemens Zeitschr.*, Vol. 8, p. 629 (1928)], and Goldstein [*Electrot. Zeitschr.*, Vol. 52, p. 378 (1931)]; for a summary of these developments in English see chapter 24 of Edgcumbe and Ockenden's book, *Industrial Electrical Measuring Instruments*, and Appendix IV of Hague's book, *Instrument Transformers*.

TESTING VOLTAGE TRANSFORMERS FOR RATIO AND PHASE ANGLE. The phase angle and the departure from nominal ratio are usually less in voltage transformers than in current transformers but are not always negligible. Methods for testing voltage transformers also can be classified as direct or relative, the direct method being more accurate but slower and requiring more expensive and delicate equipment. The relative methods have the further advantage that the circuits which the operator has to handle are completely insulated from the high voltage, which is not true of direct methods.

Resistance Method. This method requires the use of a high-resistance non-inductive resistor which can be connected across the line supplying the full rated primary voltage of the transformer under test. The current flowing in the resistor is usually 0.1 or 0.05 amp. The secondary voltage of the transformer under test is balanced against the *IR* drop in a known portion of the high resistance at its grounded end. The ratio of primary voltage to secondary voltage is then given by the ratio of the total resistance to the resistance of the portion across which the secondary is balanced. The effect of the transformer phase angle can be balanced by inserting an adjustable self-inductor in one portion of the high-resistance circuit, by shunting part of the other portion with a capacitor, or by connecting a mutual inductor with its primary in series with the resistor and its secondary in series with the detector. An alternative procedure for determining phase angle is first to make the ratio balance with an electrodynamic instrument excited in phase with the applied voltage and then to measure the phase angle by noting the deflection of a second calibrated electrodynamic instrument excited in quadrature.

Because of the high applied voltage this method is capable of very great precision even when used with insensitive and rugged detectors. It also has the advantage that the transformer can be tested with zero burden. The principal sources of error arise from the self-heating of the main resistor and from the effects of stray capacitance between sections of the main resistor and ground. For voltages above 5000 volts it is essential to minimize these effects by subdividing the resistor into a number of sections, each enclosed in a metal shield. An auxiliary guard circuit (usually having the same resistance as the main resistor) is connected in parallel across the line. Each shield of the main resistor is connected to a tap on the guard circuit so located that the potential of the shield is midway between the potentials of the ends of that section of the main resistor which it encloses. Resistors of this type can be made for use up to 50 kv, and by using a set of autotransformers in place of the guard resistance an equipment has been built for 132 kv.

Capacitance Methods. By using a pair of capacitors in place of the subdivided high-voltage resistor much of the complexity of the shielding is avoided. One capacitor, however, must be of special construction, capable of withstanding the full primary voltage without the occurrence of corona discharge, and provided with a shield on its low-voltage plate.

In one such method the two windings of the voltage transformer under test are connected in series opposition, so that each winding supplies current to one of the capacitors. The currents in the capacitors are compared by balancing the voltage drops produced in two adjustable resistors, one of which is in series with each capacitor. The capacitors are then interchanged with respect to the resistors, and the resulting circuits connected in parallel to form a Schering bridge. The balance of this bridge as an auxiliary measurement serves to fix the ratio of the capacitances and thereby of the transformer. Small adjustable capacitors shunting the resistors serve to measure the phase angle of the transformer and to compensate for residual phase displacements in the equipment. By this auxiliary balance the determination of the ratio of the transformer is made to depend on the ratio of the two resistors, while the phase angle is based on that of the high-voltage capacitor.

In an earlier arrangement the two windings of the voltage transformer under test are connected in series aiding. The two capacitors are connected in series across the outside terminals of the transformer circuit, and a detector is bridged from the junction of the two capacitors to the junction of the two transformer windings. When a balance is obtained by varying the magnitude of the secondary capacitor, the ratio of the transformer is equal to the inverse ratio of the capacitances. The phase angle can be obtained in a variety of ways by inserting resistors or mutual inductors in suitable locations in the circuit.

Relative Methods. The use of two voltmeters, two wattmeters, or two watthour meters connected respectively to the standard and the test voltage transformers offers, of course, a possible method of test analogous to those used with current transformers. These methods are little used, however, because it is simpler to compare the two transformers by an opposition method. In this method [see paper by Brooks, *Bull. Bur. Standards*, Vol. 10, p. 419 (1914)] the primary windings of the two transformers are supplied in parallel at rated voltage. The secondary windings are connected in series opposition, and the small difference in voltage is measured by a suitable separately excited electrodynamic instrument. This instrument may be a low-range wattmeter or one having fine-wire windings on both fixed and moving coils as in the Weston "comparator voltmeter." The difference in voltage as read with the instrument excited in phase with the supply voltage, when divided by the secondary voltage of either transformer, gives the difference in ratio factor of the two transformers; the corresponding quotient obtained when the excitation is in quadrature gives (in radians) the difference in the two phase angles.

This procedure can be made into a null method of high precision. A resistance of about 2000 ohms is connected across the secondary of the transformer which has the higher secondary voltage. The secondary voltage of the other transformer is then balanced against the IR drop in this resistance between one end and an adjustable sliding tap near the other end. The phase angle can be balanced by using a mutual inductor, the primary of which, in series with about 4000 ohms, is connected across the secondary of the standard transformer while the secondary of the inductor is in series with the detector. If the standard transformer is chosen to have a ratio, say, 2 per cent lower than nominal, the two measuring circuits will always be connected to it and will not constitute a burden on the transformer under test. This method avoids the error in reading any indicating instruments; and, if the transformers have nearly the same ratio, even a large error in the 2000-ohm resistance has a negligible effect on the result. The main limitation on the accuracy, therefore, lies in the calibration of the standard transformer. Another and somewhat similar relative method is used in the Leeds & Northrup voltage transformer testing set.

SERIES-PARALLEL CONNECTION OF INSTRUMENT TRANSFORMER WINDINGS. *a.* If the **primary** winding of an instrument transformer consists of two or more sections, all having the same number of turns, then at the same frequency, secondary burden, and secondary current (or voltage) the ratio correction factor and the phase angle will be the same, to a high degree of precision, whether the primary sections are connected in series or in parallel. This relation often furnishes a very useful basis for determining the ratio and phase angle of multiple-range transformers on ranges which have not been tested. It permits the extension of precise measurements to larger currents (and to higher voltages) than can be tested experimentally and affords a valuable check on experimental data on multiple-range transformers.

Deviations from this principle will arise, for a current transformer, only in proportion to the product of two factors, each of which is unlikely to be large: (1) the relative difference in the distribution of current among the several primary sections when they are in parallel; and (2) the relative difference in the coupling (mutual inductance) between each of the several primary sections and the secondary winding. If these relative differences do not exceed 1 per cent each, the principle will be valid to 0.01 per cent. In voltage transformers the presence of capacitance currents between the several sections of the winding may cause a failure of the principle in proportion to the product of the capacitance currents and the leakage reactance of the primary winding regarded as an auto transformer. No such failures have been reported.

b. If the **secondary** winding of an instrument transformer consists of two or more sections, all having the same number of turns, then at the same frequency and the same primary current (or voltage) the ratio correction factor and the phase angle will be the same, to a high degree of precision, whether the secondary sections are connected in series or in parallel, provided that the vector impedance of the secondary burden is kept proportional to the square of the number of effective secondary turns of the transformer.

SECONDARY VOLT-AMPERE RATING OF INSTRUMENT TRANSFORMERS. Current transformers intended for use with a single measuring instrument usually have a secondary rating of 25 volt-amperes, and for use with more than one measuring instrument a rating of 50 volt-amperes, at rated secondary current.

Voltage transformers usually have a rating of 50 or 200 volt-amperes at rated secondary voltage.

COMPENSATION FOR RATIO ERROR. By properly proportioning the number of turns in the windings of a current transformer, it is possible to raise the secondary current to overcome the ratio error, with a given condition of burden. It is usual to compensate a current transformer for one-half its rated secondary burden, with secondary power factor of 80 per cent and frequency of 60 cycles.

The actual ratio of turns in a voltage transformer differs from the marked ratio by an amount sufficient to make up the voltage drop in the transformer at a specified burden. Voltage transformers are usually compensated for one-fifth their volt-ampere rating, with a secondary power factor of 80 per cent, and at rated frequency. The effect of phase displacement, however, cannot be compensated for, as it depends not only on the constants of the transformer itself, but on the power factor of the burden on the voltage transformer, and on the power factor of the load which is being measured with the help of the transformers.

ACCURACY OF INSTRUMENT TRANSFORMERS. Figure 38 gives curves typical of the performance of modern current transformers under operating conditions as indicated in the following schedule:

Curve	Grade of Transformer	Frequency	Burden	
			Ohms	Microhenrys
A	Small and cheap	60	0.68	920
B	Average	60	0.68	920
C	"	60	0.97	4020
D	"	25	0.68	920
E	Excellent	60	0.68	920

The burden for curve *C* is typical of a relay circuit; that for the other curves includes a Silsbee current transformer testing set and indicating instruments. For a watthour meter only, or for indicating instruments only, the burden would be materially less.

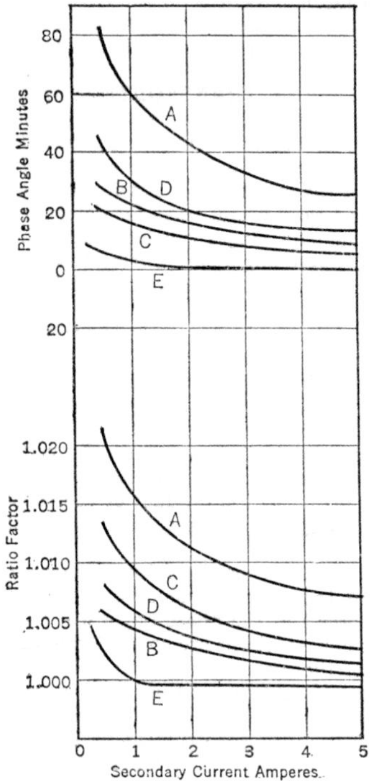

FIG. 38

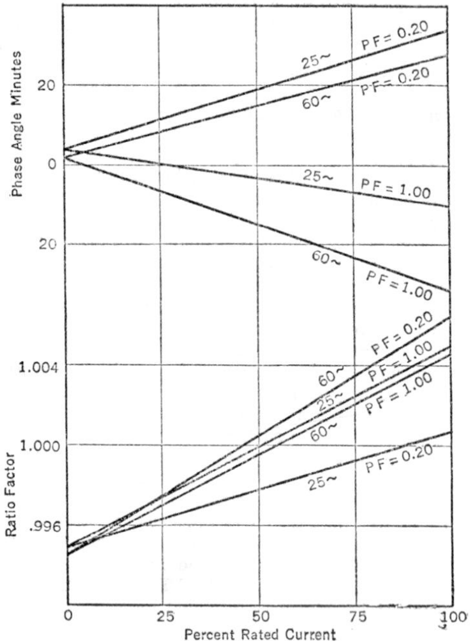

FIG. 39

Figure 39 gives performance curves of a typical voltage transformer, ratio and phase angle being plotted as ordinate against per cent of rated current as abscissa, for the operating conditions indicated on the curves.

For rating purposes, ASA Standard C-57 sets up certain standard accuracy classes in which instrument transformers can be grouped according to their ratio and phase-angle errors when supplying one or another of certain standard burdens. The accuracy classes are designated by numbers (viz., 1.2, 0.6, 0.3) which in current transformers indicate the maximum percentage error at 5 amp. The error at 0.5 amp may be double this value. An additional accuracy class, 0.5, for current transformers has been set up which permits no error exceeding 0.5 per cent from 5 amp to 0.5 amp. These stipulated errors are those in the measurement of power or energy at any load power factor exceeding 0.6, due to the combined effect of the ratio and the phase-angle errors of the instrument transformer. (For the standard burdens, originally designated "W," "X," . . . and later "B0.1" . . . "B-1" . . . , see the latest edition of ASA Standard C-57.)

14. ELECTROMETERS

GENERAL. An electrometer is primarily an instrument for measuring potential differences but under certain conditions may also be used as a wattmeter; in the latter case it is usually called an electrostatic wattmeter. The deflection of the instrument is due to the attraction or repulsion of electrostatic charges. An electrometer may be used for either d-c or a-c measurements.

There are many forms of electrometers. On account of the care required in its use, the ordinary quadrant electrometer is seldom employed except for laboratory purposes, but various modifications of the electrometer, provided with pointer and scale so as to be direct reading, are used commercially for high-voltage measurements. Such instruments are known as electrostatic voltmeters.

The advantage of the electrometer over the galvanometer or electrodynamometer is that it takes no current when used for d-c voltage measurements. Because of its capacitance, however, it takes a certain amount of charge, which should always be allowed for if the capacitance of the electrometer is appreciable compared with any other capacitance which affects its reading. Also, when used for a-c measurements, the charging current taken by the electrometer should be allowed for, if appreciable.

APPLICATIONS. Electrometers are used in the measurement of the very small currents, 10^{-8} amp or less, dealt with in measuring insulation resistance and gaseous ionization, especially in studying x-rays and atmospheric electricity; in chemical work when electrode potentials are used to indicate the end-points of certain titrations; and in investigations of static electricity. Electrostatic wattmeters are used in measuring small amounts of power at low power factor and high voltage, as in cable testing, and by the (British) National Physical Laboratory as a basic transfer instrument for the measurement of a-c power.

QUADRANT ELECTROMETER FOR VOLTAGE MEASUREMENTS. In this common type (Fig. 40) the two opposite quadrants a and a' are permanently connected together, as are b and b'. Each pair is insulated from the other pair and from the case. The "needle" n (a light aluminum vane shaped like the figure 8) is suspended symmetrically between the upper and lower faces of the quadrants by a fiber of silvered quartz and is insulated from the case and quadrants. The complete expression for the electrostatic torque τ acting on the moving system is

$$\tau = \alpha V_n^2 + \beta_a V_a^2 + \beta_b V_b^2 + \gamma_a V_n V_a + \gamma_b V_n V_b + \delta V_a V_b + \eta_n V_n + \eta_a V_a + \eta_b V_b + \eta_0 \quad (18)$$

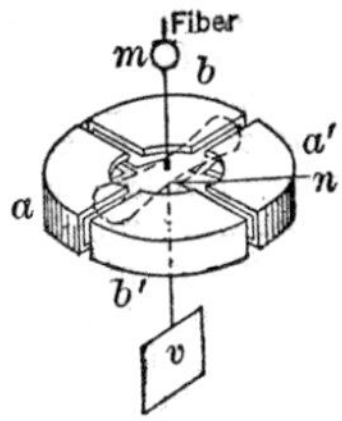

FIG. 40

where V_n, V_a, and V_b are the potentials, relative to the case, of the needle and of the quadrant pairs a and b respectively. The coefficients η result from the contact potentials of the metals used and are generally small. The effect of η_0 is merely to stress the suspension slightly when the zero reading is made. The other η terms, if appreciable, can be eliminated by suitable reversals of polarity. The coefficient δ is usually extremely small. If the instrument is perfectly symmetrical, α is also zero and

$$\beta_b = -\beta_a = \frac{\gamma_a}{2} = -\frac{\gamma_b}{2} = \kappa \quad (19)$$

and for these conditions the torque becomes

$$\tau_s = \kappa(V_a - V_b)\left(V_n - \frac{V_a + V_b}{2}\right) \quad (20)$$

In general the coefficients α, β, etc., will depend on the angular position of the needle as well as upon its size, shape, etc. If, for instance, the coefficient $\alpha = b\theta$ and the stiffness of the suspending fiber is U, the other coefficients having the values given in (19), the deflection is given by the relation

$$\theta = \frac{\kappa}{U - bV_n^2}(V_a - V_b)\left(V_n - \frac{V_a + V_b}{2}\right) \quad (21)$$

When a small difference of potential is to be measured, it is usually applied between the quadrants, while a suitable larger potential V_n from an auxiliary battery is applied to the needle. This is known as the heterostatic connection. The term $(V_a + V_b)/2$ in eq. (21) is then negligible, and it will be seen that the deflection will be proportional to the first power of the measured difference $V_a - V_b$ and will increase rapidly with V_n

both because V_n is a factor in the numerator and because (if b is positive) an increase in V_n decreases the denominator. For too large values of V_n the instrument becomes unstable. In the Compton electrometer one quadrant is made adjustable so that b can be given any desired value.

When relatively large voltages are to be measured, V_a and V_n are made zero, and the voltage to be measured is applied between the b quadrants and the case. This is known as the idiostatic connection. The deflection is then proportional to the square of the applied voltage.

EXTENSION OF RANGE OF ELECTROMETER BY USE OF AUXILIARY CAPACITORS. By connecting the potential difference to be measured across two capacitors in series and measuring the voltage across only one of them, the instrument may be used for the measurement of high voltages of practically any magnitude. The connections are as shown in Fig. 41. If the condensers have no leakage, then

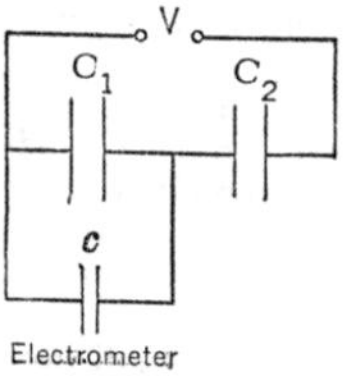

FIG. 41

$$V = V_1 \frac{C_1 + C_2 + c}{C_2} \tag{22}$$

where V_1 is the voltage read by the electrometer, C_1 and C_2 the capacitances of the two capacitors, and c the capacitance of the capacitor formed by the electrometer quadrants and needle. If C_1 and C_2 are large compared with c, then c may be neglected; but the larger C_1 and C_2 are, the greater will be the charge (and therefore the charging current, in an a-c measurement) taken by the measuring circuit.

QUADRANT ELECTROMETER AS ELECTROSTATIC WATTMETER (Fig. 42). The quadrant electrometer may be used to measure, with a fair degree of precision, very small amounts of power, of the order of 1 watt, when the voltage giving this power is 5000 volts or more. It therefore serves as a very convenient means of measuring the power loss in small samples of insulating materials at high voltages. See Bibliography, p. 5-113.

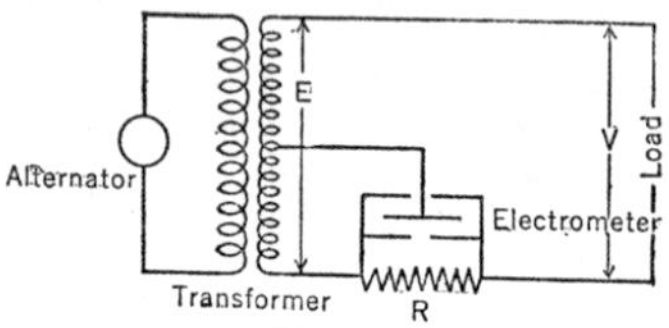

FIG. 42

STRING (FIBER) ELECTROMETERS. In this class of instruments the moving element consists of one or two fine fibers (usually of silvered quartz) held taut by a thicker quartz filament. Each fiber moves sideways under the influence of the electrostatic forces and is viewed through a microscope provided with a scale in the eyepiece. Instruments of this class have the advantage of being quickly responsive and well damped and of having relatively low capacitance. The scale length and voltage sensitivity, however, are rather less than can be obtained with the quadrant type.

15. ROTARY VOLTMETER FOR HIGH D-C VOLTAGES

The **rotary voltmeter** ("generating voltmeter") for measuring high d-c voltages consists of an armature formed of two semicylindrical metal plates rotating in the electrostatic field between two electrodes which are connected to the points between which the voltage is to be measured. The two plates are connected through a two-part commutator to a galvanometer. The armature is rotated by a motor at a known constant speed. The current through the galvanometer varies directly as the voltage between the two inducing electrodes. The results of a calibration made at low and moderate voltages may therefore be extrapolated to high voltages, provided the electric field in the vicinity of the rotor is not distorted by space charges such as might result from corona at high voltages. The range is limited only by the clearances between live parts. One terminal of the galvanometer is grounded, and the galvanometer may be located at any desired distance from the high-voltage circuit. The sensitivity may be varied by altering a shunt around the galvanometer.

16. SPARK GAPS FOR MEASURING HIGH VOLTAGES

GENERAL. The voltage (crest value) required to break down a given gap depends principally upon: (1) the shape, size, and arrangement of the electrodes; (2) the presence of other conductors or insulators in the vicinity of the gap; (3) the time of application of the voltage; and (4) the temperature, pressure, and humidity of the air. Other

Table 3. Sphere-gap Spark-over Crest Voltages*

(At 760 mm barometric pressure; one sphere grounded)

(For corrections at other air densities see Table 4. Corrections to be applied to 25 C values only)

Sphere-gap Spacing, cm	Sphere Diameter, cm: 6.25 — 60-Cycle and Negative Impulse, kv Crest, 25 C	6.25 — 60-Cycle and Negative Impulse, kv Crest, 20 C	6.25 — Positive Impulse, kv Crest, 25 C	6.25 — Positive Impulse, kv Crest, 20 C	12.5 — 60-Cycle and Negative Impulse, kv Crest, 25 C	12.5 — 60-Cycle and Negative Impulse, kv Crest, 20 C	12.5 — Positive Impulse, kv Crest, 25 C	12.5 — Positive Impulse, kv Crest, 20 C
0.5	17.0	17.3	17.0	17.3				
1.0	31.3	31.8	31.3	31.8	31.7	32.2	31.7	32.2
1.5	44.5	45.3	44.8	45.6	44.9	45.7	44.9	45.7
2.0	57.0	58.0	57.4	58.4	58.0	59.0	58.0	59.0
2.5	68.8	70.0	69.3	70.5	70.8	72.0	70.8	72.0
3.0	78.8	80.1	79.4	80.7	83.5	84.9	83.5	84.9
3.5	86.6	88.1	88.0	89.5	95.0	96.6	95.3	96.9
4.0	93.6	95.2			106.0	107.8	108.0	109.8
4.5	99.8	101.5			117.0	119.0	120.0	122.0
5.0	105.5	107.3			127.0	129.2	132.3	134.6
5.5	110.0†	112.0†			135.3	137.6	142.5	144.9
6.0	114.0†	116.0†			143.5	145.9	153.8	156.4
6.25	117.0†	119.0†			147.5	150.0	158.0	160.7
7.0					157.7	160.4	171.0	173.9
8.0					170.5	173.4		
9.0					182.0	185.1		
10.0					192.0†	195.0†		
11.0					200.0†	203.0†		
12.0					208.0†	212.0†		
12.5					211.0†	215.0†		

Sphere-gap Spacing, cm	Sphere Diameter, cm: 25 — 60-Cycle and Negative Impulse, kv Crest, 25 C	25 — 60-Cycle and Negative Impulse, kv Crest, 20 C	25 — Positive Impulse, kv Crest, 25 C	25 — Positive Impulse, kv Crest, 20 C	50 — 60-Cycle and Negative Impulse, kv Crest, 25 C	50 — 60-Cycle and Negative Impulse, kv Crest, 20 C	50 — Positive Impulse, kv Crest, 25 C	50 — Positive Impulse, kv Crest, 20 C	75 — 60-Cycle and Negative Impulse, kv Crest, 25 C	75 — 60-Cycle and Negative Impulse, kv Crest, 20 C	75 — Positive Impulse, kv Crest, 25 C	75 — Positive Impulse, kv Crest, 20 C
2.5	72	73	72	73								
5.0	136	138	136	138	136	138	136	138	136	138	136	138
7.5	192	195	196	199	197	200	197	200	200	203	200	203
10.0	241	245	252	256	260	264	260	264	261	265	261	265
12.5	278	283	296	301	317	322	319	324	324	329	324	329
15.0	309	314	334	340	367	373	374	380	380	386	380	386
17.5	338	344	364	370	411	418	426	433	433	440	443	451
20.0	362	368	390	397	451	459	474	482	484	492	499	507
22.5	379	385	409	416	486	494	511	520	528	537	548	557
25.0	393	400	426	433	519	528	547	556	573	583	597	607
30.0					573	583	605	615	653	664	687	699
35.0					615	625	655	666	721	733	755	768
40.0					651	662	698	710	777	790	816	830
45.0					681	693	732	744	827	841	870	885
50.0					707	719	758	771	870	885	917	933
55.0									910	925	960	976
60.0									945	961	999	1016
65.0									977	994	1031	1049
70.0									1003	1020	1058	1076
75.0									1025	1042	1081	1099

Note: These values were adopted in 1936 and were based on standardization work by a number of laboratories in the United States (see bibliography). The experimental data were analyzed and correlated in accordance with accepted theory. The International Electrotechnical Commission is now attempting to prepare universal standards because of differences in several national standards. By the latest agreement it is expected that the final IEC values will not differ from the various national standards by an amount exceeding 1 1/2 per cent. For most of the gap spacings in these tables there will be no significant changes.

* From AIEE Standard No. 4 (ASA Standard C68.1-1942, "Measurement of Test Voltage in Dielectric Tests").

† These values are to be used with power frequencies only.

Table 3. Sphere-gap Spark-over Crest Voltages—*Continued*

(At 760 mm barometric pressure; one sphere grounded)

(For corrections at other air densities see Table 4. Corrections to be applied to 25 C values only)

Sphere-gap Spacing, cm	Sphere Diameter, cm: 100 — 60-Cycle and Negative Impulse, kv Crest, 25 C	100 — 60-Cycle and Negative Impulse, kv Crest, 20 C	100 — Positive Impulse, kv Crest, 25 C	100 — Positive Impulse, kv Crest, 20 C	150 — 60-Cycle and Negative Impulse, kv Crest, 25 C	150 — 60-Cycle and Negative Impulse, kv Crest, 20 C	150 — Positive Impulse, kv Crest, 25 C	150 — Positive Impulse, kv Crest, 20 C	200 — 60-Cycle and Negative Impulse, kv Crest, 25 C	200 — 60-Cycle and Negative Impulse, kv Crest, 20 C	200 — Positive Impulse, kv Crest, 25 C	200 — Positive Impulse, kv Crest, 20 C
10	261	265	261	265	261	265	261	265	261	265	261	265
20	504	513	504	513	505	514	505	514	506	515	506	515
30	700	712	715	727	736	749	736	749	746	759	746	759
40	862	877	888	903	947	963	955	971	973	990	973	990
50	985	992	1024	1041	1120	1139	1140	1159	1172	1192	1178	1198
60	1084	1102	1124	1143	1254	1275	1293	1315	1346	1369	1364	1387
70	1163	1183	1209	1230	1360	1383	1400	1424	1505	1531	1533	1559
80	1234	1255	1284	1306	1458	1483	1502	1528	1635	1663	1671	1700
90	1295	1317	1344	1367	1552	1579	1597	1624	1752	1782	1788	1818
100	1338	1361	1390	1414	1628	1656	1678	1707	1857	1889	1896	1928
110					1695	1724	1755	1785				
120					1760	1790	1824	1855				
130					1815	1846	1880	1912				
140					1865	1897	1920	1953				
150					1900	1932	1944	1977				

factors of lesser significance are: (1) the composition and surface condition of the electrodes; (2) the circuit arrangements and constants; (3) the wave form and frequency of the voltage; (4) the extent to which the air in the gap is ionized; (5) the degree of irradiation of the spheres. Used under specified conditions, the spark gap is a commercially accurate device for measuring crest voltage in the testing of insulating materials and insulating structures.

In American practice the **sphere gap** and, for less accurate work, the rod gap have been used. The former AIEE sphere-gap tables have been extended downward to include spheres 6.25 cm in diameter to cover the range for which AIEE standards formerly recommended the needle gap.

For higher voltages the sphere gap has the advantages of occupying less space, of being free from time lag, and of breaking down with a sharply defined spark discharge without previous formation of corona, provided the gap length does not greatly exceed the sphere diameter. The sphere gap has the further advantage that the electrodes do not have to be renewed after each discharge, as they do with the needle gap.

CONSTRUCTION OF SPHERE SPARK GAP. Figure 43 shows the arrangement recommended by Farnsworth and Fortescue. The top sphere is stationary but slightly adjustable in height so as just to make contact with the lower sphere when it is set for zero separation. The lower sphere is mounted on a piece of brass tubing which carries a threaded bushing on its lower end. This bushing works on a carefully threaded rod having a pitch of two threads per centimeter. The bushing being graduated to fiftieths on its circumference, separation may be measured to the nearest 0.01 cm directly.

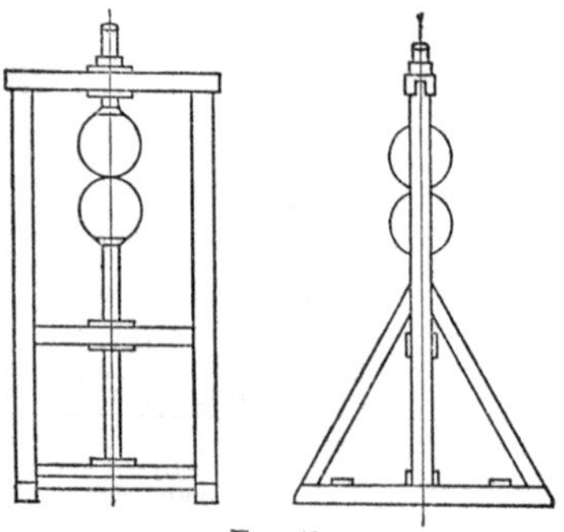

Fig. 43

AIEE STANDARD SPHERE SPARK GAP. The standard sphere spark gap should be between two suitably mounted spheres. No extraneous conducting body, other than the supporting shanks, should be nearer the gap than twice the sphere diameter. Insulating bodies, such as supporting structures, should not be nearer than one sphere diameter to the gap. The shanks should not be greater in diameter than 1/5 the sphere diameter. Metal collars, etc., through which the shanks extend, should be as small as practicable and should not, during any measurement, come closer to the spheres than the maximum gap length used in the measurement.

The sphere curvature measurement by a spherometer should not vary more than 1 per cent from that of a true sphere of the required diameter.

SPHERE-GAP SPARKING DISTANCES. For various distances between spheres the crest voltages shall be assumed to be as shown in Table 3 when the air density corresponds to a barometric pressure of 760 mm mercury and a temperature of 25 deg cent. For any other air density the voltage at which the gap breaks down, for any given spacing, is equal to the breakdown voltage for this spacing at standard pressure and standard temperature multiplied by the correction factor given in Table 4. The relative air density is found from the formula:

$$\text{Relative air density} = \frac{0.392b}{273 + t} \qquad (23)$$

where b = barometric pressure in millimeters and t = temperature in degrees centigrade. To determine the gap spacing for a required spark-over voltage, divide the required

Table 4. Air-density Correction Factors for Sphere Gaps

Relative Air Density	Diameter of Standard Spheres in Millimeters					
	20	62.5	125	250	500	750
0.50	0.573	0.547	0.535	0.527	0.519	0.517
0.55	0.617	0.594	0.583	0.575	0.567	0.565
0.60	0.661	0.640	0.630	0.623	0.615	0.613
0.65	0.705	0.686	0.677	0.670	0.663	0.661
0.70	0.748	0.732	0.724	0.718	0.711	0.709
0.75	0.791	0.777	0.771	0.766	0.759	0.757
0.80	0.833	0.821	0.816	0.812	0.807	0.805
0.85	0.875	0.866	0.862	0.859	0.855	0.854
0.90	0.917	0.910	0.908	0.906	0.904	0.903
0.95	0.959	0.956	0.955	0.954	0.952	0.951
1.00	1.000	1.000	1.000	1.000	1.000	1.000
1.05	1.041	1.044	1.045	1.046	1.048	1.049
1.10	1.082	1.090	1.092	1.094	1.096	1.097

voltage by the correction factor obtained from Table 4 and use the new voltage thus obtained to find the corresponding spacing from Table 3, using a graph of this table if more convenient.

PRECAUTIONS IN SPHERE-GAP MEASUREMENTS. With the exception of its own support or parts, all objects at ground potential should be separated from all live parts of the apparatus under test by a distance in inches not less than the quotient obtained from dividing by 5000 the maximum test voltage which will be applied.

As a precaution against overvoltage oscillations, and in order to limit the resulting current at the time of breakdown, a non-inductive resistance of about 1 ohm per volt of test voltage to be used should be inserted in series with the sphere gap. If the test is made with one terminal grounded, the entire resistance should be in series with the non-grounded electrode; if neither terminal is grounded, one-half of the resistance should be in series with each electrode. In either case this resistance shall be electrically as near the spark gap as possible and not in series with the apparatus under test. Water-tube resistors are used for this purpose. Carbon resistors of high resistivity should not be used because their resistance may become very low at high voltages.

The sphere gap should be set up for use in a space comparatively free from external dielectric fields. Conducting bodies forming part of the circuit, or at circuit potential, should not be so located with reference to the gap that their dielectric fields are superimposed on that of the gap. For example, the protective resistor should not be arranged to present large surfaces toward the gap, even at a distance of twice the sphere diameter. If one sphere is grounded, the spark point of this sphere should be not less than 5 and not more than 10 times its diameter from any grounded surface.

It is important that the surfaces of the spheres be kept free from dust and moisture, and that the spheres do not become heated appreciably by repeated discharges. When sphere gaps are used as specified, with corrections properly applied, the accuracy of voltage measurement should be approximately 2 per cent.

Irradiation of the gap with ultraviolet light considerably reduces the scattering of the values of breakdown voltage and is particularly useful with high frequencies and essen-

Table 5. Gap Spacings of 1/2 in. Rods for Alternating Spark-over Voltages *

(At 25 deg cent, 760 mm barometric pressure, and 1.55 cm vapor pressure)

Gap Spacing		Gap Breakdown, Kilovolts Crest
Centimeters	Inches	
2	0.8	26
4	1.6	47
6	2.4	62
8	3.1	72
10	3.9	81
15	5.9	102
20	7.9	124
25	9.8	147
30	11.8	172
35	13.8	198
40	15.7	225
50	19.7	278
60	23.6	332
70	27.6	382
80	31.5	435
90	35.4	488
100	39.4	537
120	47.2	642
140	55.1	744
160	63.0	847
180	70.9	950
200	78.7	1054
220	86.6	1160

Rod-gap spark-over voltage increases with increasing relative air density and also increases with increasing humidity.

Relative air density shall be calculated in accordance with the formula applying to sphere gaps.

If the humidity is other than standard 0.6085 in. (1.55 cm) vapor pressure, corrections should be made in accordance with Table 7.

* From AIEE Standard No. 4 (ASA Standard C68.1-1942, "Measurement of Test Voltage in Dielectric Tests").

Table 6. Gap Spacings of 1/2 in. Rods for Impulse Spark-over Voltages *

(At 25 deg cent, 760 mm barometric pressure, and 1.55 cm vapor pressure)

Gap Spacing		Critical Spark-over, Kilovolts			
		(1 × 5) Wave		(1.5 × 40) Wave	
Centimeters	Inches	Positive	Negative	Positive	Negative
2	0.8	32	32	32	32
3	1.2	42	42	42	42
4	1.6	51	51	51	51
5	2.0	60	62	60	62
6	2.4	65	70	65	70
8	3.1	80	86	77–78 †	86
10	3.9	94	102	89–93 †	101
12	4.7	113	119	102–109 †	118
14	5.5	132	136	117–128 †	135
16	6.3	150	152	132–145 †	150
18	7.1	167	168	142–156 †	164
20	7.9	185	185	157–164 †	180
25	9.8	230	228	188	222
30	11.8	272	269	222	255–266 †
35	13.8	315	311	255	290–313 †
40	15.7	356	352	287	320–355 †
45	17.7	396	396	316	355–383 †
50	19.7	436	440	346	390–400 †
60	23.6	515	525	400	465
70	27.6	595	610	460	535
80	31.5	675	695	520	600
90	35.4	750	775	580	665
100	39.4	830	865	640	730
120	47.2	975	1025	750	855
140	55.1	1125	1195	870	985
160	63.0	1285	1365	985	1115
180	70.9	1460	1555	1125	1265
200	78.7	1585	1695	1220	1370
220	86.6	1740	1865	1340	1500
240	94.5	1900	2045	1460	1640

Rod-gap spark-over voltage increases with increasing relative air density and also increases with increasing humidity.

Relative air density shall be calculated in accordance with the formula applying to sphere gaps.

If the humidity is other than standard 0.6085 in. (1.55 cm) vapor pressure, corrections should be made in accordance with Table 7.

* From AIEE Standard No. 4 (ASA Standard C68.1-1942, "Measurement of Test Voltage in Dielectric Test").

† Dual values are due to unstable conditions, the cause not being known.

tial with impulse voltages. For this purpose an open arc or a quartz mercury-vapor lamp may be used.

The sphere gap is sensitive to momentary rises of voltage, and the voltage required to spark over the gap should be obtained by slowly closing the gap under constant voltage or by slowly raising the voltage with a fixed setting of the gap.

THE ROD GAP. This gap consists of two one-half in. square rod electrodes cut off squarely and mounted on supports so that a length of rod equal to or greater than one-half the gap spacings overhangs the inner edge of the support. Spark-over values for rod gaps are given in Tables 5 and 6 and humidity corrections for them in Table 7.

Table 7. Humidity Corrections for Rod Gap Spark-over Voltages *

Vapor Pressure in		Humidity Correction, H, percentage				
		60 Cycles	(1 × 5) Wave		(1.5 × 40) Wave	
Cm Hg	In. Hg		Positive	Negative	Positive	Negative
0.254	0.10	−14.2	−6.4	−6.4	−11.3	−9.8
0.5	0.20	−11.6	−5.2	−5.2	− 9.2	−8.0
1.0	0.39	− 6.2	−2.6	−2.6	− 4.7	−4.0
1.5	0.59	− 0.5	−0.2	−0.2	− 0.4	−0.3
2.0	0.79	+ 4.6	+2.2	+2.2	+ 3.8	+3.6
2.5	0.98	+ 5.6	+4.2	+4.2	+ 7.4	+6.9
3.0	1.18	+11.2	+5.8	+5.6	+ 9.8	+8.8

To obtain the spark-over voltage value for a specified gap spacing at a vapor pressure other than standard, first correct for relative air density, then add or subtract in accordance with the plus or minus signs a percentage H of the corrected tabular value.

To obtain the correct gap spacing for a specified voltage at a vapor pressure other than standard, first derive a reference voltage by dividing the specified voltage by a factor.

$$\frac{100 \text{ plus or minus } H}{100}$$

Correct this reference voltage for relative air density and refer to Tables 5 and 6 for gap spacing.

These values apply to spark-over voltages above 141 kv crest. For lower voltages, reduce H in proportion

$$H_1 = \frac{H \times \text{kv crest}}{141}$$

For vapor pressures 0.79 in. (2 cm) and above, correction values are approximate.

* From AIEE Standard No. 4 (ASA Standard C68.1-1942, "Measurement of Test Voltage in Dielectric Tests").

For additional detailed information on the characteristics of the testing equipment, its arrangement, and methods for voltage control, consult the latest edition of AIEE Standards for the Measurement of Test Voltage in Dielectric Tests (ASA No. C68.1, AIEE No. 4).

17. CORONA VOLTMETER

The corona voltmeter, originated by Whitehead, has certain advantages over the sphere gap for the measurement of crest voltages. For a discussion of this instrument and a bibliography, see Brooks and Defandorf, *Bur. Standards J. Research,* Vol. 1, p. 589, 1928; reprinted as *Bur. Standards Research Paper* 21.

18. OSCILLOGRAPHS

GENERAL. An oscillograph is essentially an instrument for observing rapid variations in voltage or current. These variations are automatically plotted (usually against time) on a suitable screen or, if a permanent record is desired, on a photographic film.

The moving system of the oscillograph may be a conducting loop or an iron vane (the galvanometer oscillograph); a single conducting filament (Einthoven's string galvanometer); two insulated conducting filaments (the electrostatic oscillograph); or an electron beam (the cathode-ray oscillograph). The moving system is situated in a magnetic or electric field and is actuated either by changes in its own current or potential or by changes in the field in which it lies. The inertia of the moving system is kept small in order that it may follow very rapid variations in the phenomenon to be observed. An oscillograph will record without appreciable distortion only those events which have frequencies much less than the natural frequency of its moving element. Distortion of high-frequency harmonics in a wave varies with the type of moving system and with its damping but is generally negligible for frequencies up to $^1/_5$ of the natural frequency of the vibrator.

The **moving-coil oscillograph** is ordinarily used for frequencies up to 2000 cps and, where high sensitivity is required, may be used with a suitable amplifier. The **cathode-ray oscillograph** may be used at any frequency, but its chief application is at frequencies beyond the range of the moving-coil oscillograph. The **string galvanometer** is useful as an oscillograph at low frequencies where extremely high sensitivity is required.

MOVING-COIL OSCILLOGRAPH. The moving-coil oscillograph is essentially a galvanometer of very short period. It is used for voltage observations with a suitable multiplier, or for current observations with a shunt. Single-phase and three-phase wattmeter oscillograph elements give directly a record of instantaneous power. Current and watt-meter elements are generally interchangeable in the same mountings.

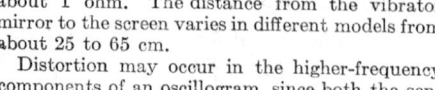

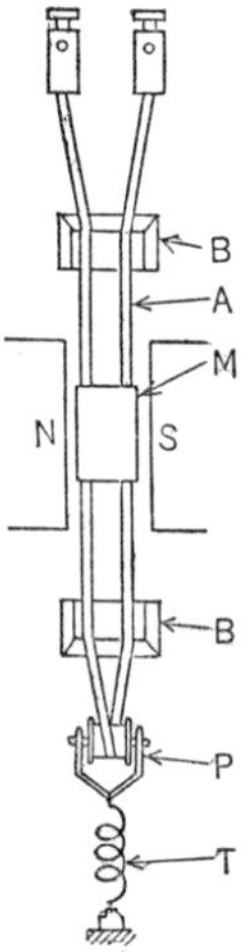

Fig. 44

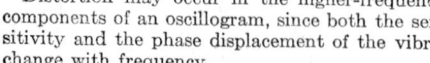

The vibrator element (see Fig. 44) is usually a loop A of very light ribbon of silver alloy or aluminum stretched across ivory bridges BB and passed around a pulley P. Tension is applied to the loop and may be varied by a spring T and screw arrangement attached to the pulley. A small mirror M is cemented to the loop between the bridges. The vibrator element with its mounting is immersed in a damping liquid and may be removed from the oscillograph as a unit for repair or replacement. It is placed in the narrow gap between the pole pieces NS of a magnet. Older designs use an electromagnet, but recently these have been replaced by small permanent magnets which provide an intense field with a large reduction in weight and with the advantage of requiring no d-c source.

Instead of the carbon arc formerly used, most recent designs use a low-voltage filament lamp operated at a considerable overvoltage only at the instant of making a photographic record. Light from this source is reflected by the vibrator mirror to the slit of the recording camera or to an oscillating mirror, driven by a synchronous motor, which projects the oscillogram as a standing wave, with linear time scale, on a screen or tracing table. Some models are equipped for simultaneously viewing and photographing a wave. A zero line is obtained on the oscillogram by interposing a small stationary mirror in the light beam just above and in front of the vibrator mirror. A photographic record may be made either with a film wound around a rotating drum or with film or sensitized paper fed at constant speed from a magazine past the camera slit.

Oscillographs are made with from 1 to 24 vibrator elements. Completely self-contained portable instruments are available with from 1 to 6 elements and weighing up to 60 lb. Oscillographs which are completely automatic in operation are made for use in recording switching and other disturbances in power systems. The disturbance may be made to start the record in less than a single cycle. The duration of the record may be predetermined. Such oscillographs are used with magazines taking 200 ft of sensitized paper and will record without attention as many as 100 disturbances.

SENSITIVITY OF MOVING-COIL OSCILLOGRAPH. The sensitivity of the oscillograph vibrator varies greatly with its free period and damping, increasing with greater free period and with diminished damping. Table 8 gives the sensitivity of some available elements together with their natural (undamped) frequencies.

Table 8. Sensitivity of Oscillograph Elements

Natural Frequency, cycles per second	Sensitivity, radians per ampere
8000	0.2
5000	0.4
3000	1.2
1200	12.0

The resistance of a vibrator element is ordinarily about 1 ohm. The distance from the vibrator mirror to the screen varies in different models from about 25 to 65 cm.

Distortion may occur in the higher-frequency components of an oscillogram, since both the sensitivity and the phase displacement of the vibrator with respect to the applied voltage change with frequency.

Assuming unit deflection for a given steady direct voltage, the maximum deflection θ for a sine-wave voltage of any frequency and having the same crest value is given by the formula

$$\theta = \frac{1}{[p^4 + p^2(4N^2 - 2) + 1]^{1/2}} \qquad (24)$$

where $p = f/f_r$ is the ratio of the applied frequency to the resonance frequency of the vibrator, and $N = n/n_c$ is the relative damping.

The phase displacement of the vibrator with respect to the applied voltage is given by the formula

$$\tan \Phi = \frac{2Np}{1 - p^2} \tag{25}$$

The phase displacement of a harmonic of frequency f_h with respect to its fundamental of frequency f_0 is a measure of phase distortion in an oscillogram. This distortion δ, measured in degrees of the fundamental, is given by the formula

$$\delta = \frac{f_0}{f_h} \Phi_h - \Phi_0 \tag{26}$$

The resonance frequency to be used in these formulas is not that which would be found in air (without appreciable damping) but about half that value. The damping liquid is to some extent moved with the vibrator, adding to its effective inertia and thus increasing its natural period. This increase in period depends on the material of the vibrator element, being about 40 per cent for the usual silver-alloy loop and nearly 60 per cent for aluminum, and should be added to the air-damped period, which is usually stated by the manufacturers.

Table 9 gives values computed from the above formulas for a damping (70 per cent of critical damping) which gives nearly the minimum distortion in oscillograms. The symbols are explained above. The last column gives typical values for the nth harmonic ($n = 50p$) of a 60-cycle oscillogram for the element commonly used, having a resonance frequency in air of about 5000 cps.

Table 9. Performance of Oscillograph Vibrator for Damping Equal to 70 Per Cent of Critical Damping

Ratio of Applied Frequency to Resonant Frequency p	Ratio of Deflection at Applied Frequency to Deflection with Direct Current θ	Phase Displacement of Vibrator in Degrees Φ	Phase Distortion $\frac{0.02}{p}\Phi - 1.60$ in Degrees δ
0.01	1.000	0.80	0.00
0.02	1.000	1.60	0.00
0.1	1.000	8.05	0.01
0.2	1.000	16.2	0.02
0.3	0.998	24.8	0.05
0.4	0.991	33.7	0.08
0.5	0.975	43	0.12
0.6	0.947	53	0.16
0.7	0.905	63	0.19
0.8	0.850	72	0.20
0.9	0.785	81	0.21
1.0	0.714	90	0.20
1.5	0.394	121	0.01
2.0	0.243	137	−0.23

The actual damping of the vibrator element depends on the viscosity of the damping liquid and therefore on its temperature. It can be determined for any particular operating condition with a single observation, as follows: A constant direct voltage is impressed on the element and then suddenly removed. If the element is underdamped, it will deflect past its equilibrium zero position and will oscillate about zero. The amount of the first overshoot past zero is measured, and its ratio to the initial d-c deflection is computed. If this ratio is x, the relative damping is given by the formula

$$N = \frac{-\log_e x}{[\pi^2 + (\log_e x)^2]^{1/2}} = \frac{-\log_{10} x}{[1.86 + (\log_{10} x)^2]^{1/2}} \tag{27}$$

CATHODE-RAY OSCILLOGRAPH. Description. The moving system of the cathode-ray oscillograph is a beam of electrons. It is therefore practically without inertia and will respond equally to any frequency.

In an evacuated vessel (see Fig. 45) electrons emitted from a cathode A are accelerated by a suitable voltage toward an anode B which has a small hole in its center. Those electrons passing through the hole form an "electron beam" of sensibly uniform velocity coinciding with the axis of the tube. The beam passes between parallel metal deflection plates C and strikes a fluorescent screen D, where it gives rise to a luminous spot. If a photographic film is substituted for the screen, a latent image (subject to development by ordinary means) appears where the beam strikes the active emulsion of the film. In the presence of an electric or magnetic field at C, at right angles to the axis of the tube, the electrons are accelerated perpendicularly to their direction of motion in the beam, resulting in the deflection of the spot at D. Phenomena to be examined are made to produce such a deflecting field at C. A time axis is supplied to the oscillogram by a second field, perpendicular to the first, and varying in a known manner with time.

There are two general types of cathode-ray oscillograph. The **hot-cathode** tube utilizes the electron emission of an incandescent filament or of an indirectly heated Wehnelt cathode. The beam is formed by a relatively low voltage, usually less than 3000 volts, impressed between cathode and anode. The **cold-cathode** tube utilizes the electrons freed from the surface of the cathode under positive-ion bombardment in a gas discharge tube. The voltage required to generate the beam is high, frequently 50 kv or more.

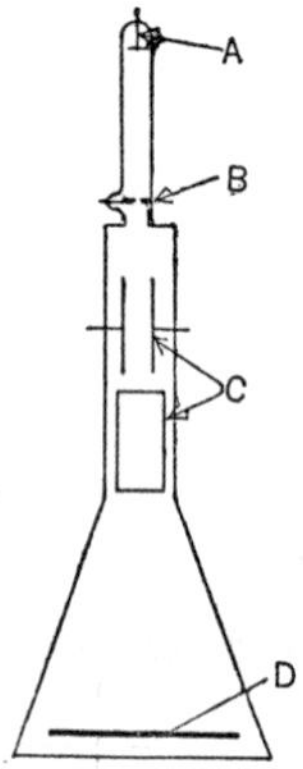

FIG. 45

The diameter of the beam, and consequently that of the recording spot, is determined by the focusing action of axially symmetric electric or magnetic fields and by diaphragms in the beam path. Electric focusing is usually accomplished by voltages impressed on a series of cylindrical electrodes arranged along the axis of the tube. Combinations of these electrodes supplied at suitable potentials act to produce axial as well as radial fields and thus accelerate the electrons in the beam in addition to focusing them. In the hot-cathode tube a control electrode may be interposed between the cathode and the first accelerating electrode to regulate the electron current in the beam, and hence the brightness of the fluorescent spot. Beam intensity is controlled by changing the potential of this electrode. Electric focusing is almost universally used in modern hot-cathode, high-vacuum tubes.

Magnetic focusing, produced by the fields of coils located with their axis of symmetry coincident with the beam axis, is commonly used in the cold-cathode tube.

Formerly, focusing was accomplished in hot-cathode tubes by the action of the field of positive ions formed in the residual gas left in the tube for this purpose. Gas focusing is now obsolete, and the modern hot-cathode tube operates in high vacuum.

Tubes are available in which two or more beams are formed by independent electrode assemblies. Each beam makes use of separate sets of deflection plates, so that two or more phenomena can be simultaneously viewed or photographed on the screen. In other oscillographs the deflecting plates are switched rapidly between two voltage sources so that, through persistence of vision, two phenomena can be observed simultaneously.

Sensitivity of Cathode-ray Oscillograph. The electrostatic sensitivity of the cathode-ray tube is inversely proportional to the beam-accelerating voltage and is given approximately by the formula

$$S = \frac{LA}{2VD} \tag{28}$$

in which S = deflection per unit difference of potential between the deflection plates.
L = length of deflection plates.
D = separation of plates.
A = distance from plates to screen.
V = potential drop along the beam from the cathode to the plates.

The sensitivity of available hot-cathode tubes ranges from 0.1 mm per volt to 1 mm per volt; and for high-voltage cold-cathode tubes is about 0.04 mm per volt.

Oscillographs are available which incorporate not only a cathode-ray tube and appropriate circuits for controlling beam intensity and focus, but also amplifiers having controllable gain ranging upward to 2000, with a response which is flat to within 10 per cent between 2 cps and 100 kcps. With such an amplifier an input of 0.01 volt (rms) will produce a deflection of 1 in. at the oscillograph screen. Also tubes are available in which the accelerating voltage is divided into two stages. The first and smaller voltage is impressed on the beam before its passage through the deflecting field. A second, higher accelerating voltage is applied to the beam between the deflecting plates and the screen. In this way the deflection sensitivity of the low voltage tube is retained while the brightness of the trace on the screen is increased to correspond to the higher voltage. In one type of sealed tube produced by the Allan B. DuMont Laboratories a total beam-accelerating voltage of 25 kv can be used, of which less than 4 kv precedes the deflection.

Electromagnetic sensitivity is inversely proportional to the square root of the beam voltage. It depends on the shape, size, and position of the deflecting coils and may amount to 1 mm per ampere-turn for a low-voltage tube and 0.1 mm per ampere-turn for a 50,000-volt tube.

Recording Speed. Photographic recording is possible with the lowest voltage tubes only where a recurrent pattern makes time exposures possible. Photographic speed is a function of beam voltage, screen material, and speed at which the fluorescent spot moves across the screen, as well as of the speed of the camera lens and the type of film used. A number of fluorescent materials are used in cathode-ray tubes, differing in color of excitation, photographic intensity, and persistence of after-luminescence. All these factors are of importance in photographic applications. The Radio Manufacturer's Association has assigned code numbers to certain screen materials. The characteristics of these materials as found in DuMont tubes are given in Table 10.

Table 10. Characteristics of Certain Screen Materials as Found in DuMont Tubes

RMA Screen Type	P1	P2	P4	P5	P7	P11
Excitation color	Green	Blue fluorescence; green phosphorescence	White	Blue	Blue fluorescence; yellow phosphorescence	Blue
Spectral range A.U.	4850–5740	4280–6080	3980–6880	3470–6100	4140–6210	3770–5690
Spectral peak A.U.	5220	4550; 5300	4600; 5550	4280	4500; 5700	4400
Persistence for decay to 10%	30μ sec	Long	Medium	15μ sec	Long	60μ sec
Visual observation	Recurrent phenomena	Recurrent phenomena; high-speed transients	Television	Not recommended in general	Recurrent; transient at low frequency	Recurrent; high-speed transients
Photography	Still; moving film to 40 cps	Still; moving film with blue filter to 10 kcps	....	High-speed moving film to 200 kcps	Still; moving film with blue filter	Very high speed; moving film to 200 kcps
Photographic efficiency factor	0.3	0.6	....	0.1	....	1.0
Photographic film recommended	Orthochromatic: Eastman 5211; Eastman Super XX; Ansco Triple S Ortho					

With the RMA Type P11 screen, which is highly actinic, and with a beam voltage of 25 kv, transients have been recorded in which the fluorescent spot moved on the screen at a rate of 400 in. per μ sec.

In transient recording the beam must be kept off the screen until the instant of recording to prevent injury to the screen from continuous exposure of one spot to the high-energy electron beam and to prevent fogging the film. This is conveniently done by applying to the intensity-control electrode a biasing voltage sufficient to reduce the beam intensity to zero. This biasing voltage is removed at the time of recording.

High-voltage cold-cathode tubes, in which the photographic film is placed in the evacuated chamber and comes into direct contact with the electron beam, are capable of somewhat higher recording speed than is attainable at the present time with the sealed, hot-cathode tube. In the cold-cathode type, voltage is applied continuously between cathode and anode. Electrons are focused on a hole in the anode by a coil situated between cathode and anode, providing a magnetic field parallel to the tube axis. The vacuum in the discharge tube is controlled by an adjustable leak, and its value is indicated by reading the discharge current. An intense beam is formed in this way and is kept off the photographic film by an electric field or by some suitable beam-blocking device (e.g., the Norinder relay) until the instant the record is to be made. For this type of tube, operating under the best conditions, no practical limit to the recording speed has been found. Records have been made in which the recording spot moved across the film at a velocity of 10^{10} cm per sec ($^1/_3$ the velocity of light).

A theoretical limiting condition for recording a cathode-ray oscillogram without distortion is that no appreciable change shall take place in the field deflecting the electron beam during the time taken by an electron to pass through the field. This time varies with the voltage used to generate the beam and with the length of the deflection plates. If plates 1 cm long are used in a tube in which the beam is generated at 50,000 volts, this time of passage through the deflecting field is about 10^{-10} sec. Hence this theoretical limit need not be considered in any ordinary engineering applications.

Life of Cathode-ray Tube. The life of a hot-cathode, sealed tube is determined by the cathode life which, in modern tubes, may amount to 2000 hr or more. The beam intensity of the high-voltage cold-cathode tube falls off as the cathode becomes pitted by the discharge. It is found in practice that there is an initial rise in intensity and in steadiness of the discharge as a newly polished cathode "burns in." This is followed by a period of some 10–20 hr of satisfactory operation, after which the beam intensity begins to decrease. Finally, after perhaps 50 hr, the cathode must be removed from the tube. If the pit formed by the discharge is removed by grinding the face of the cathode, which is then repolished, the tube may be restored to its initial condition.

Circuits Used with Cathode-ray Oscillograph. When using low-voltage, high-sensitivity oscillographs, the maximum voltage which may be impressed on the deflection plates is about 50 to 100 volts. The maximum deflection-plate voltage for the high-voltage oscillograph is roughly 1000 volts. Above these limits, imposed by the size of the screen or film, a potential divider, made of resistors or capacitors or some combination of the two, must be used. The choice of a divider requires detailed examination to ensure that two conditions are satisfied; namely, (1) the divider must make no appreciable change in the character of the circuit; that is, the impedance of the divider should be very great compared with that of the circuit; (2) the form of the wave impressed on the deflection plates must be the same as that impressed on the potential divider. Account must also be taken of distributed capacitance, capacitance to ground, and capacitance of the deflection plates; or the parts of the divider must be so arranged that the effect of these capacitances is negligible. The divider must be arranged so that the deflection plates do not differ greatly in potential from their surroundings in order to avoid spurious induced effects in the oscillogram.

In general, capacitance dividers are used for high-frequency and impulse phenomena; resistance dividers are used for low and intermediate frequencies, being limited in their application by two considerations: (1) at high frequencies, unavoidable stray capacitance will change the ratio of voltage division; (2) for impulse work, especially with steep wave fronts, the rate at which potential can change on the deflection plates is limited by the resistance in series with them. The oscillogram recorded may show only the exponential charge and discharge, through a series resistor, of the capacitor formed by the deflection plates, regardless of the form of the voltage wave impressed on the divider. This effect can be reduced by decreasing the resistance of the divider; but, as the resistance decreases, the effect of the divider on the circuit becomes greater.

At high frequencies and for surge measurements, leads to the deflection plates must be shielded to eliminate coupling. Coaxial cables may be used for this purpose, but the characteristic impedance of the cables must be well matched at their terminations in order to eliminate reflections at these points.

A number of auxiliary circuits are needed in operating a cathode-ray oscillograph. A low-current, high-voltage d-c source of potential is used for generating the beam. If the tube operates with electrostatic focusing, this d-c source must be provided with suitable taps for controlling beam intensity and focus. The time axis of the oscillogram may be furnished by the voltage generated by an oscillator. A saw-tooth wave is used as a linear time base. This may be developed by a relaxation oscillator in which a capacitance is allowed to charge through a resistance. Only a small portion of the charging curve of the RC combination is used, and the voltage therefore rises almost linearly with time. The capacitor is quickly discharged through a gas-discharge tube, when its potential reaches the breakdown voltage of the tube. When the capacitor has discharged to a value equal to the extinction voltage on the tube, the cycle repeats. Sweep time can be continuously controlled over wide limits by varying the product RC in the charging circuit; and, if a recurrent signal is impressed on the grid of the discharge tube, the sweep can be synchronized with this signal so that a standing pattern can be obtained on the oscillograph screen. Such an oscillator may have a useful range from 2 to 50,000 cps. The higher frequency limit is set by the de-ionization time of the gas-triode tube, and the lower limit is set by the leakage resistance of the charging capacitor, which will reduce the charging rate as voltage rises across the capacitor and thus distort the time base. The time base need not be restricted to a linear function but can be a sinusoidal, circular, or spiral function as needed in special applications by appropriate circuit changes.

In hot-cathode tubes which have an electrode in front of the cathode for intensity control it is possible to put time markers on the trace by periodically impressing a small voltage on the intensity-control electrode. Thus the trace may be blanked out at the desired time intervals.

In photographic recording it is often convenient to provide a linear time base by means of a continuously moving film camera. In this case no electrical sweep circuit is used.

In recording transients it is often convenient to use a single linear or exponential sweep which may be obtained from the charge or discharge of an *RC* circuit. If the time of occurrence of the observed phenomenon cannot be predetermined, the timing sweep must be synchronized with the event under observation either by a mechanical switch or by an auxiliary electrical circuit. Another relay circuit may be required to operate the beam-blocking device at the proper instant. Such timing and relay circuits are actuated by the breakdown of spark gaps or of gas-discharge tubes. Timing circuits may be made to operate by these means within 10^{-8} sec.

For details of timing and relay circuits see the Bibliography, p. 5-114.

STRING GALVANOMETER USED AS OSCILLOGRAPH. The string galvanometer can be used for low-frequency phenomena where extremely high sensitivity is required. It consists essentially of a fine conducting filament, stretched between the pole pieces of a powerful electromagnet. The filament is illuminated through a microscope mounted in a hole bored in one pole piece. Its motion is observed through a second microscope mounted in the opposite pole piece, and it appears as a shadow against a luminous background. Magnifications up to 1000 diameters are used in observing the motion.

For a photographic record the image of the string is projected on a slit placed parallel to the direction of motion of the image. A sensitized film behind the slit is moved at constant speed to provide a time axis. The film may be mounted on a rotating drum for short records, or fed across the slit from a reel for continuous records. An incandescent filament lamp may be used for film speeds up to 400 mm per sec at a magnification of 500 diameters. At higher speeds an arc lamp must be used.

As a rule the string is a gold-coated glass or quartz fiber about 0.001 mm in diameter, with a resistance greater than 1000 ohms. It is stretched over bridges, and its tension is varied by a suitable slow-motion screw and lever arrangement. The string and mounting may be removed from the galvanometer as a unit. Models are available in which two or more strings can be used simultaneously with the same optical system, and in one model six strings are mounted in one optical system for simultaneous use.

Sensitivity of String Galvanometer. By changing the tension of the string, the sensitivity and damping may be varied over wide limits, the sensitivity being increased by decreasing the tension with a corresponding increase in the free period. The sensitivity of a typical instrument with a 14-cm string, using a magnification of 600 diameters, is 3000 mm per μa when the string is adjusted to a period of 0.5 sec and 100 mm per μa at a period of 0.008 sec. Instruments are available with strings as short as 0.5 cm with a free period of less than 5×10^{-5} sec. Some instruments are so constructed that a number of strings of varying length can be used interchangeably, covering a wide range of sensitivity and free period.

19. TESTING OF MAGNETIC MATERIALS

MAGNETIC CHARACTERISTICS. Routine magnetic testing is usually done for the purpose of determining one or more of the following characteristics of magnetic materials: (1) normal induction, (2) normal hysteresis, (3) core loss, (4) a-c permeability.

Normal Induction. The induction produced by a given magnetizing force depends upon the previous magnetic condition of the specimen and upon the mode of approach to the given magnetizing force. In order to obtain consistent and reproducible results, it is therefore necessary to follow a definite test procedure. The effect of previous magnetic history can be removed by demagnetization, that is, by subjecting the specimen to a succession of reversals of magnetizing force gradually decreasing from a certain maximum value to one somewhat lower than the lowest at which a determination is to be made. Demagnetization should be started from an initial value of magnetizing force well above that corresponding to maximum permeability, but not necessarily higher than any previously experienced by the specimen. After demagnetization, points on the induction curve are obtained by observing the induction caused by a given magnetizing force after a sufficient number of reversals to bring the material to a cyclic condition, thus closing the hysteresis loop. After a determination has been made, further measurements can be made at higher values of magnetizing force without demagnetizing again; but, if a lower point is desired, the demagnetization process must be repeated. Values of induction so obtained are called "normal induction" and are reproducible for a given specimen within the limits of experimental error. The normal-induction curve is the locus of the tips of a succession of hysteresis loops, as illustrated in Fig. 46.

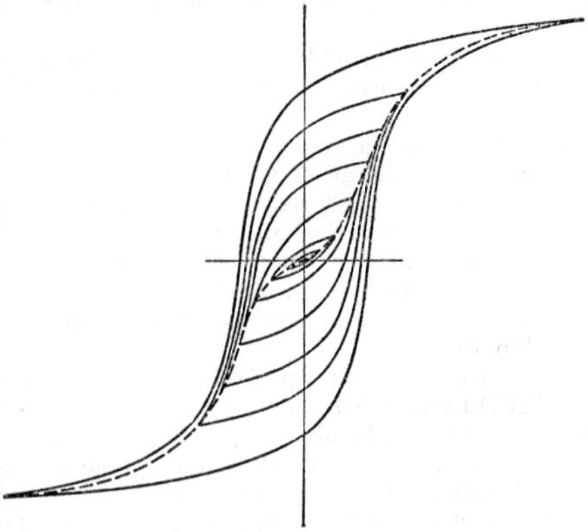

FIG. 46

Normal Hysteresis. When points on a normal hysteresis loop are to be determined, cyclic condition is first obtained by a number of reversals of the magnetizing force corresponding to the desired tip, and the induction is noted when the magnetizing force is reduced to some lower value either in the same or the opposite direction. Before each determination the material is brought back to a cyclic condition by reversals of the maximum magnetizing force. This procedure differs somewhat from the "step-by-step" method but gives more consistent and reliable results.

CORE LOSS. When magnetic materials are subjected to varying magnetizing force, as in the cores of a-c apparatus, power is expended in overcoming magnetic hysteresis and in maintaining induced eddy currents in the core material. The total amount of power thus expended is called core loss. If the excitation has a d-c biasing component, so that the average induction in the core material differs from zero, the hysteresis loop is not symmetrical. The corresponding core loss is called incremental core loss.

A-C PERMEABILITY. Values of permeability determined with alternating current depend upon the methods and conditions of measurement. The differences lie in the values taken for induction and magnetizing force. In some methods the a-c permeability is taken as the ratio of the maximum value of induction to the maximum value of magnetizing force in a symmetrical cycle. In other methods the rms value of the magnetizing force is used. A-c permeability is also sometimes defined as the ratio of the rms value of induction to the rms value of magnetizing force. If the magnetizing current has a d-c biasing component, the ratio of the cyclic change in induction to the cyclic change in magnetizing force is called the incremental a-c permeability.

MEASUREMENT OF NORMAL INDUCTION AND HYSTERESIS. Classification of Methods and Apparatus. Normal induction and hysteresis are ordinarily determined by methods which may be classified according to the way the magnetic induction is determined as (*a*) magnetometric, (*b*) traction, (*c*) air-gap, and (*d*) ballistic.

In the magnetometric method the induction is measured in terms of the deflection of a suspended magnetic needle placed in a definite position with respect to the test sample. The method is especially suitable for observing slow changes in induction, for instance, such as occur when a specimen is heated or cooled. The magnetometer, being very delicate and sensitive to variations in the earth's field, is not recommended for general magnetic testing.

In the traction method the induction is estimated in terms of the mechanical force of attraction between the surfaces of two magnetized bodies, one of which is the test specimen. Careful machining of the specimens is required, and the results are not of high accuracy. Traction methods are not in general use.

In the air-gap method the magnetic circuit is closed, except for a transverse air gap in which is located some measuring device, such as a deflecting coil, a rotating armature, or a bismuth spiral. The method is subject to rather large and uncertain errors.

The ballistic method employs a ballistic galvanometer (or a fluxmeter) for the measurement of magnetic induction. Ballistic methods are most generally used for both laboratory and shop testing.

In commercial practice the specifications of the American Society for Testing Materials are generally followed. As these specifications are revised periodically, the latest edition of the ASTM Book of Standards should be consulted for details.

Ring Method. The ring method is simple in principle and the easiest to apply, but it is not often used for routine testing because of the cost of preparing and winding the specimens. A diagram of connections is shown in Fig. 47. The magnetizing winding is uniformly distributed around the ring, and the magnetizing force in oersteds is calculated from the formula

$$H = \frac{0.4\pi NI}{L} \tag{29}$$

in which N = total number of turns.
I = current in amperes.
L = mean circumference of ring in centimeters.

The difference in concentration of the winding over the inner and the outer part of the ring causes the magnetizing force to vary over the cross-section, with an error in the result which depends upon the ratio of the mean diameter to the radial width of the ring. If this ratio is greater than 10, however, the error can usually be neglected.

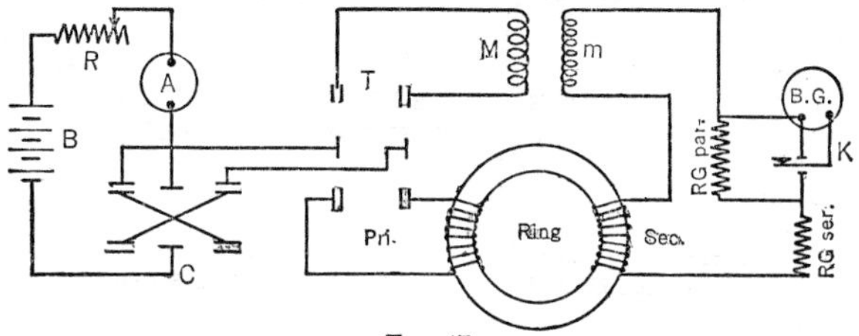

FIG. 47

The magnetic induction is measured by the deflection of a ballistic galvanometer connected to a suitable test coil wound on the specimen, preferably under the magnetizing winding. When the flux linked with this test coil is changed, an electromotive force is induced, and a certain quantity of electricity, proportional to the change in flux, flows through the galvanometer. If the flux is reversed in direction by reversing the magnetizing current, the galvanometer deflection will be proportional to twice the flux.

The galvanometer is calibrated by means of a standard mutual inductor, the secondary of which is permanently connected in series with the test coil. It is usually convenient to adjust the sensitivity of the galvanometer by means of suitable series and parallel resistances so that the scale is direct reading in terms of magnetic induction. It is customary to make 1 cm deflection correspond to the reversal of an induction of 1000 gausses. The calibrating current to be reversed in the primary of the mutual inductor depends upon the value of the mutual inductance, the number of turns in the test coil, and the cross-section of the specimen, and is calculated from the following formula:

$$I_c = \frac{BAN}{M \times 10^8} \tag{30}$$

in which I_c = calibrating current in amperes.
B = induction in gausses.
A = cross-sectional area in square centimeters.
N = number of turns in the test coil.
M = mutual inductance in henrys.

It is often convenient to use 100 turns in the test coil. When points on the hysteresis loop are to be determined, an auxiliary series resistance is suddenly inserted in the magnetizing circuit. In this case, since the galvanometer is calibrated for reversal of the induction, the readings must be multiplied by 2.

Burrows Permeameter. It is generally most convenient to prepare samples in the form of straight bars of moderate length and uniform cross-section. Of the various testing methods adapted to such samples the most accurate one, for samples having uniform permeability along their length and a maximum permeability not exceeding 5000, is the Burrows compensated double-yoke method. In this method auxiliary compensating coils provide the extra magnetomotive force required for the yoke and joints in the magnetic circuit, the current in them being adjusted so as to produce a uniform induction in the test bar. Figure 48 shows the magnetic circuit of the Burrows permeameter and the relative positions of the magnetizing and the test coils. The test rod and its auxiliary, which should be of the same size and material, are joined at the ends by soft-iron yokes which make good magnetic joints. The magnetizing coils T and A are located over the test rod and auxiliary, respectively. Coil J is in four sections, connected in series, and located over the ends of the rods as near the joints as possible. In operation, the currents in these three windings are so adjusted before each reading that there is equal flux in the two rods and no leakage from the greater part of the test rod; for this condition the value of the applied magnetizing force can be calculated from the current and number of turns per centimeter in the solenoid surrounding the test rod. For testing the compensation and determining the value of the induction when the compensation is properly adjusted, there are three test coils designated t, a, and j, respectively. These coils are each of the same number of turns and are distributed as shown in Fig. 48; t is wound over the middle of the test bar, a over the middle of the auxiliary bar, and j half over one end and half over the other end of the test bar far enough away from the yokes and joints to avoid disturbances from these causes. When the reversal of the magnetizing currents causes no deflection of the galvanometer, whether connected to t and a or to t and j in series-opposition, the magnetizing currents are properly adjusted, the induction may be measured in terms of the deflection of the galvanometer connected to t alone, and the

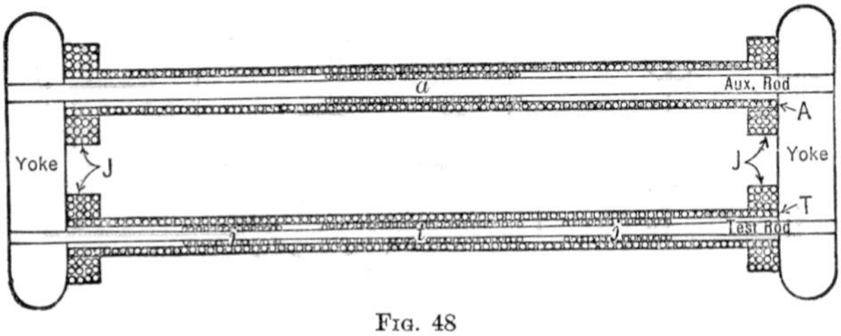

FIG. 48

magnetizing force is proportional to the current in the winding T. As usually constructed, the magnetizing coil is so wound that the magnetizing force is 100 times the current in amperes.

In all ballistic methods a correction must be made to the observed value of induction because of the flux in the space between the specimen and the test coil. The amount to be subtracted from the observed value of induction is proportional to the magnetizing force H and depends upon the relative areas of the test coil and specimen. It is equal to kH where $k = (a - A)/A$, a = area of test coil, and A = area of specimen. This "air correction" may be eliminated by means of an adjustable mutual inductor having its primary in series with the main magnetizing winding and its secondary in series with the main test coil. If desired, the two main magnetizing coils may be connected in series, and the average values for the two samples determined by connecting the two middle test coils in series and using them as one. In this arrangement four auxiliary test coils near the ends of the specimens are connected in series and used as one coil for adjusting the current in the compensating magnetizing coils.

The Burrows permeater is an absolute instrument in that its constants can be derived from measurements of its dimensions, and it is therefore suitable for standardization work under proper conditions. It is extremely sensitive to variations in permeability along the length of the specimen, and the results obtained on inhomogeneous material should be interpreted with this fact in mind. It is limited to magnetizing forces not exceeding 300 oersteds.

Fahy Simplex Permeameter. This permeameter requires but a single specimen. The magnetizing force is applied by means of an electromagnet, across the poles of which the specimen is clamped. A uniformly wound test coil extends over the whole length of the specimen. A ballistic galvanometer connected to this test coil indicates the induction in the specimen when the magnetizing current is reversed. The magnetizing force is measured by means of a test coil uniformly wound on a non-magnetic form and extending between two iron blocks which are clamped to the ends of the specimen.

When the magnetizing force is changed in value, the deflection of a ballistic galvanometer connected to this coil is proportional to the change in magnetic potential between the ends of the coil. The function of the iron blocks is in effect to transfer the ends of the test coil to the ends of the specimen. When the magnetic circuit is properly constructed of suitable materials, this method is capable of giving very satisfactory results. It does not require an auxiliary specimen, and no compensation is necessary. It is also less sensitive than the Burrows permeameter to the effect of magnetic inhomogeneity along the length of the specimen. The galvanometer is calibrated by means of a standard mutual inductor as in the methods described above, and the correction for air-flux linked with the test coil but not in the specimen is applied. The Fahy permeameter is limited to magnetizing forces not exceeding 3000 oersteds.

Full-range Permeameter. This instrument is a compensated single yoke. The specimen is clamped between the poles of a laminated U-shaped yoke, and the magnetizing coil surrounds the specimen. An auxiliary coil on the yoke serves to provide the magnetomotive force required for the joints and yoke. An *H*-coil mounted within the main magnetizing coil, adjacent to the specimen but not surrounding it, is connected in series and opposing the secondary of a mutual inductor whose primary is in series with the magnetizing coil. The current in the yoke winding is adjusted so that a ballistic galvanometer connected to the test coils shows no residual deflections upon reversal of the currents. The magnetizing force is then determined either from the magnetizing current or from the deflections of the ballistic galvanometer connected to the *H*-coil alone. The intrinsic induction $(B - H)$, is determined by means of a test coil surrounding the middle part of the specimen in series with another *H*-coil having the same value of area-turns. The total induction is obtained by adding to the value of intrinsic induction thus obtained the value of the corresponding magnetizing force. The range of this instrument is from 0.01 to 1000 oersteds.

Babbitt (or J) Permeameter. This permeameter is designed especially for testing in the range 40 to 1000 oersteds. The specimen is clamped between the poles of a laminated yoke made up of different kinds of material, such as permalloy, silicon steel, and ordinary iron, in such proportions that the resultant permeability is practically constant throughout the range of flux density carried by the yoke. Compensation for the yoke reluctance is effected by a winding surrounding the yoke and connected in series with the main magnetizing winding around the test specimen. The value of induction is determined ballistically by means of a test coil surrounding the specimen. The air correction is eliminated by means of an auxiliary test coil placed within the magnetizing coil but not surrounding the specimen. The area-turns of this auxiliary coil are adjusted to equal the area-turns of the main test coil to which it is connected in series-opposition. The readings thus give values of intrinsic induction $(B - H)$. The value of magnetizing force is determined either in terms of the magnetizing current (using an experimentally determined constant) or by means of an air coil mounted within the magnetizing solenoid. Care must be exercised in the use of this instrument to avoid excessive heating of the specimen.

Saturation Permeameter. This permeameter was originally developed for the determination of saturation values. Its magnetic circuit is essentially the same as that of the Full-range or Babbitt permeameter, but there is no compensating winding on the yoke, since the compensation is less important at the higher magnetizing forces. The test coil arrangement is similar to that in the Full-range or the Babbitt permeameter. The magnetizing winding is so designed as to permit the application of magnetizing forces up to 4000 oersteds, and there is provision for cooling the apparatus to prevent overheating.

The N.B.S. High-*H* Permeameter. In this permeameter the specimen is inserted between pole pieces which are spanned by two similar U-shaped laminated yokes. The purpose of using two yokes is to provide a symmetrical magnetic circuit and thus produce a more uniform distribution of induction across the section of the test specimen. The main magnetizing coils surround the pole pieces, and there are auxiliary magnetizing windings on the yokes connected in series with the main windings. The pole pieces are adjustable so that the gap length can be chosen with reference to the size and properties of the specimen in order to obtain the maximum degree of uniformity in the magnetic field immediately adjacent to the specimen. The degree of uniformity of this field and its value, which is equal to the magnetizing force acting on the specimen, are determined by means of a small double *H*-coil which is rotated through 180 deg by a motor. The induction in the specimen is measured by means of a test coil wound on a thin brass form, which closely surrounds the specimen. With this apparatus measurements can be made at magnetizing forces up to 5000 oersteds.

Fahy Super-*H* Adapter. This is an attachment to be used with the Fahy Simplex permeameter. It consists of specially shaped pole pieces between which the specimen is clamped, thus confining the test to a short length of the specimen. The pole pieces are clamped edge-on to the permeameter in order to equalize the magnetic potential on the two sides of the specimen. The usual *B*-coil and *H*-coil are mounted in the gap, the faces of which are cut at such an angle as to make the field between them as nearly uniform as possible. A magnetizing coil surrounds the pole pieces and gap and is used in conjunction with the main winding of the permeameter. With this arrangement magnetizing forces up to about 2500 oersteds can be obtained.

General Precautions. In magnetic testing the following precautions should be observed: (1) in order to obtain consistent and reproducible results, the specimen (and yoke) must be thoroughly demagnetized before the measurements are made; (2) stray fields, especially from parts of the electrical circuit in which current changes during an observation, should be avoided (this can generally be accomplished by twisting together the current-carrying conductors and also the secondary circuits); (3) specimens should be clamped so as to be free from mechanical strain; (4) heating of the specimen should be avoided; (5) specimens should be protected from mechanical vibration.

Accuracy of Magnetic Tests. The most common sources of error in magnetic testing are (1) lack of uniformity in permeability along the length of the specimen; (2) mechanical strain; and (3) temperature effects. Errors due to these causes may be large and cannot be estimated with any degree of certainty. The methods mentioned above may be expected to give the following accuracies for specimens having a satisfactory degree of uniformity, clamped so as to be free from strain and kept at a constant temperature within 5 deg cent: For normal induction up to the point of maximum permeability, values of magnetizing force corresponding to given values of induction should be accurate within 3 per cent provided that the permeability does not exceed 10,000. This limitation as to permeability does not apply to the ring method. Values of induction corresponding to given values of magnetizing forces above the point of maximum permeability should be accurate within 1 per cent. For residual induction, values should be accurate within 2 per cent, and for coercive force within 5 per cent. For magnetically inhomogeneous material the accuracies specified above cannot be obtained with the Burrows permeameter. The Fahy Simplex permeameter and the Babbitt permeameter, each used within its proper range, give values more nearly representative of the average properties of non-uniform material. Unless a specimen has been tested for uniformity, it is usually safe to assume that it is non-uniform.

TESTING MATERIALS OF LOW PERMEABILITY. General Laboratory Method. The methods described above are not adapted to the testing of materials of very low permeability. The commercial importance of such tests lies in the relationship that has been found between the magnetic permeability and the resistance to corrosion of certain types of corrosion-resistant steel. The permeability of such materials ranges from about 1.01 to about 5.0. A ballistic method is generally employed. A straight solenoid is used which has a test coil of several hundred turns mounted within it, and in which the specimen can be inserted. Because of the very low value of magnetization in the specimen, the self-demagnetizing effect is so small that it can be neglected. A variable mutual inductor is used, with its primary connected in series with the magnetizing solenoid and its secondary in series-opposition with the test coil. This mutual inductor is so adjusted as to balance the mutual inductance between the solenoid and the test coil. This adjustment is made with no specimen in the test coil, so that there is no deflection of the ballistic galvanometer when the primary current is reversed. The galvanometer can then be used at its maximum sensitivity. When a specimen is inserted within the test coil and the magnetizing current is reversed, the deflection of the galvanometer is proportional to the intrinsic induction of the specimen. The magnetizing force is calculated from the current and the constants of the solenoid in the usual way.

The Fahy-Low Mu Permeameter. This instrument is a modification of the one which has been described. The variable mutual inductor is replaced by an air coil mounted within the magnetizing solenoid and adjacent to the test coil. The value of area-turns of this air coil is slightly greater than that of the test coil. The value of effective area-turns of the air coil is adjusted by means of a variable resistance connected in parallel with it. The adjustment of the shunt resistance is correct when the galvanometer connected to the two secondary coils is not deflected upon reversal of the magnetizing current with no specimen in place. The deflection obtained upon reversal with a specimen inserted in the test coil is proportional to the intrinsic induction ($B - H$). The magnetizing force is measured by means of the auxiliary coil with the shunt resistance disconnected. Calibration is carried out with a standard mutual inductor in the usual way as described under the ring method.

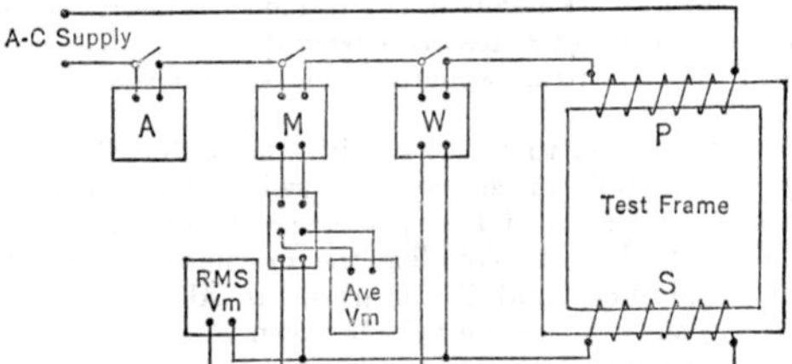

Fig. 49

TESTS WITH ALTERNATING CURRENTS. Alternating-current tests are ordinarily made on commercial materials in accordance with the specifications of the American Society for Testing Materials. The latest edition of the Book of Standards should be consulted for details. The material to be tested is cut into strips of specified dimensions and assembled in the form of a square in a set of coils usually referred to as a test frame. The dimensions of the strips and test frame and the nature of the joints at the corners of the magnetic circuit depend upon the type of test to be made. Three different types of test are included in the standard specifications: (1) the 50-cm Epstein test for core loss, (2) the 25-cm Epstein test for core loss and a-c permeability, and (3) the 25-cm Epstein test for normal or incremental permeability or core loss at low inductions. Diagrams of connections for these tests are given in Figs. 49, 50, and 51. The tests are ordinarily made at a frequency of 60 cps, but tests can be made with the 25-cm test frame at frequencies up to 4000 cps, or even higher, by using a suitable power supply and appropriate instruments, provided that suitable precautions are taken.

The 50-cm Epstein Test for Core Loss. The test specimen for this test consists of strips 50 cm (19 11/16 in.) long and 3 cm (1 3/16 in.) wide. The standard sample weighs 10 kg (22 lb), but 5 kg (11 lb) may be used. For ordinary hot-rolled sheet half the strips are cut parallel and half at right angles to the directions of rolling. Materials produced in strip form or those having pronounced directional properties, however, are tested with strips all cut in the same direction. The strips are assembled in four equal bundles and inserted in the test frame so as to form a square with butt joints. The bundles are ordinarily arranged so that opposite sides of the square consist of material cut in the same direction. The strips are held in place at the corners by clamps. The coil forms of the test frame are uniformly wound with two sets of coils connected in series to form primary and secondary windings of 600 turns each. For the core-loss test the mutual inductor is not used and need not be connected. (See Fig. 49.)

An electromotive force having a wave form approximating a sine curve is applied to the primary winding and adjusted until the voltage of the secondary circuit is that determined by the equation

$$E = \frac{4\mathit{ff}NfBM}{4lD10^5} \tag{31}$$

where ff = form factor of primary emf = 1.11 for sine wave.
N = number of secondary turns = 600.
f = number of cycles per second = 60.
B = maximum induction in kilogausses.
M = total mass in grams.
l = length of strips in centimeters = 50 cm.
D = density in grams per cubic centimeter.

Standard tests are made at specified values of maximum induction. For tests at 15 kilogausses or higher, or whenever the form factor of the applied electromotive force departs from the value 1.11 by more than 1 per cent, a voltmeter reading average volts is used in parallel with the rms voltmeter. The scale of such an instrument is conveniently marked in terms of the average volts times 1.11, in which case the voltage, as calculated for a sine wave, is held on the average voltmeter.

The density of silicon steel is usually assumed, the value depending upon the silicon content and ranging from 7.55 to 7.85.

For nickel-iron alloys the density is assumed from the nickel content as given by the straight lines joining the points defined as follows: *

Nickel, percentage	Density, grams per cm^3
0	7.85
30	8.00
100	8.90

For other alloys a density corresponding to the actual measured or estimated value is used.

When the voltage and frequency have been adjusted to the proper values, the wattmeter indicates the total loss, including the loss in the secondary circuit. The loss in the secondary circuit can be calculated in terms of the rms voltage and the resistance. Subtracting this correction from the total loss gives the net loss in the steel, and dividing this value by the mass in kilograms gives the core loss in watts per kilogram.

The 25-cm Epstein Test for Core Loss and A-c Permeability. In view of the fact that in recent years manufacturers have been able to improve the uniformity of the magnetic properties of the various commercial materials to such an extent that it is now possible to obtain a sufficiently representative sample of a given heat or "lift" of steel weighing not more than 2 kg, a smaller specimen and test frame can be used for testing. The 25-cm Epstein test is designed for specimens weighing from 0.5 kg to 2 kg, but for general commercial testing the 2-kg sample is standard. On account of the shorter length of the magnetic circuit, a better joint at the corners is required than the butt joint employed in the 50-cm test. A double lap joint has been found to give best results, and consequently a minimum length of 28 cm is needed. However, longer strips up to a maximum of 50 cm can be used. The width of the strips is 3 cm.

The connections and testing procedure for measuring core loss are essentially the same as for the 50-cm test (Fig. 49); but, on account of the smaller cross-sectional area, the number of turns in the test-frame windings is increased to 700, and somewhat more sensitive instruments are required. Also, in calculating the core loss it is necessary to take into account the additional material at the corners. This is done by using the "active weight," calculated on the assumption that the effective length of the magnetic circuit is 94 cm. This value was experimentally determined.

When testing specimens of very small cross-sectional area relative to the area of the secondary winding, or when the magnetizing force is high enough to make the air-flux

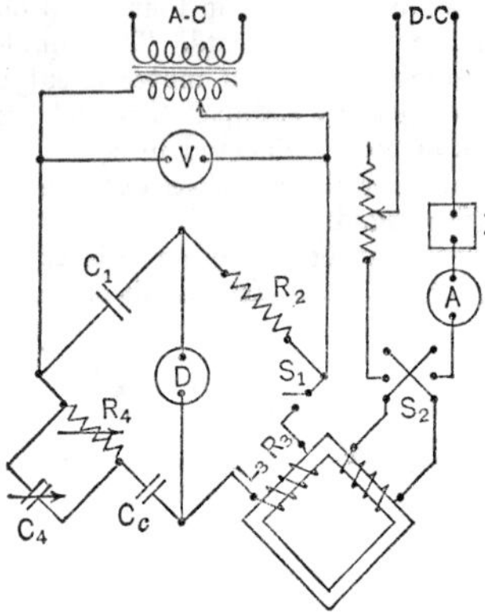

FIG. 50

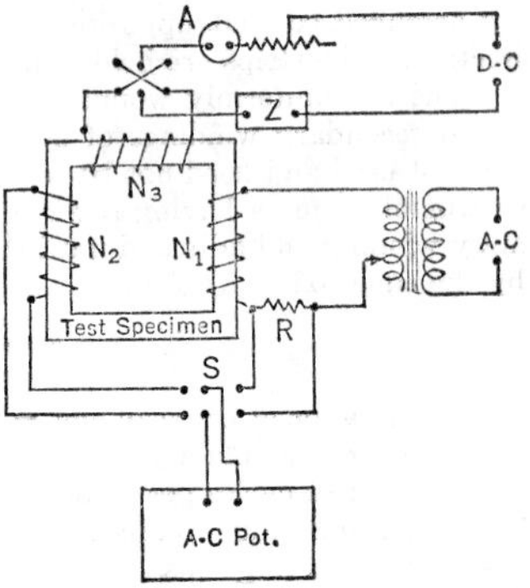

FIG. 51

voltage more than 0.33 per cent of the total voltage, it is necessary to make a correction for the air-flux voltage. This can be done by calculation or, preferably, by means of a compensating mutual inductor. The compensating mutual inductor is adjusted to have the same value of mutual inductance as that between the primary and secondary windings of the test frame with no specimen inserted. Its primary is connected in series with the primary winding of the test frame, and its secondary is connected in series-opposition with the secondary test-frame winding. The measured voltage is thus proportional to the intrinsic induction, B_i. If the form factor, as indicated by the ratio of the rms voltage to the average voltage, departs from 1.11 (the value for a sine wave) by more than 1 per cent, it is necessary to make a correction to the observed value of core loss to account for the fact that, whereas the hysteresis component of the total loss is a function of the average voltage, the eddy current component is a function of the rms voltage.

For the measurement of magnetizing force when determining a-c permeability the mutual inductor M (Fig. 49) is used; the secondary voltage measured with the average voltmeter is proportional to the crest value of the magnetizing current. The magnetizing force is calculated from the equation

$$H = \frac{0.4\pi N I_c}{l_2} = 10 I_c$$

where H = magnetizing force in oersteds.
N = number of turns = 700.
I_c = crest current in amperes.
l_2 = assumed length of specimen (88 cm in this case).

The value of B in gausses is determined from the secondary voltage as for the core-loss test, and the permeability, μ, is $\mu = B/H$.

The 25-cm Epstein Test for Normal or Incremental Permeability or Core Loss at Low Inductions. The methods which have been described are not sufficiently sensitive for testing laminated core materials at the low inductions usually employed in apparatus

* For 50 per cent nickel content the density is 8.26 grams per cubic centimeter. For 80 per cent nickel content the density is 8.64 grams per cubic centimeter.

used in many types of communications equipment. Furthermore, these methods do not provide for testing under the conditions in which a unidirectional magnetizing force is applied in addition to the alternating magnetizing force. The American Society for Testing Materials specifies two different methods for testing at low alternating inductions either with or without a biasing direct current; (*a*) an a-c bridge method, or (*b*) an a-c potentiometer method. The tests are made at frequencies of either 60 or 1000 cps. The standard induction with no biasing magnetizing force is either 10 or 1000 gausses at 60 cycles or 10 gausses at 1000 cycles. With the biasing magnetizing force the standard induction is 10 gausses at either frequency. Values of biasing magnetizing force up to 2 oersteds are employed. The test frame is similar to that for the ordinary core-loss test except for the windings. Three windings are provided, an inner coil of 100 turns, an intermediate coil of 1000 turns, and an outer winding of 100 turns, which is used for the application of the biasing direct current. The specimens are cut and assembled in the same way as for the ordinary 25-cm Epstein test which has been described. The test frame and specimen are the same for either the bridge or potentiometer method.

Figure 50 is a diagram of connections for the bridge method. The a-c supply should be of approximately a sine-wave form having not more than 10 per cent of total harmonics and should be effectively insulated from the bridge circuit by a suitable coupling transformer as shown. A storage battery provides a steady source of direct current. As shown in the diagram, one winding of the test frame constitutes one arm of the bridge. The opposite arm is a fixed capacitor, C_1. The other two arms consist of a fixed resistor, R_2, and a network in which a fixed capacitor, C_c, is in series with an adjustable capacitor, C_4, in parallel with which is the adjustable resistor, R_4. The a-c voltage applied to the bridge is measured by the voltmeter, V. The detector, D, is a null-point indicator. For the 60-cycle test a vibration galvanometer is recommended; for the 1000-cycle test a telephone receiver may be used. The detector may be preceded by a suitable amplifier.

Before making the test, the specimen is thoroughly demagnetized. In view of the fact that the permeability of most materials at low inductions drifts to an appreciable extent with time after demagnetization, this should be done at least 24 hr before making the test. During the interval between demagnetization and testing, the specimen should be protected against stray magnetic fields or mechanical vibration.

The constants of the bridge circuit depend upon the kind of material to be tested. For the 60-cycle test on materials which have exceptionally high permeability and low loss, such as cold-reduced silicon steel or nickel-iron alloys, the inner 100-turn winding of the test frame is used, and R_2 is 10 ohms. For the 60-cycle test on ordinary silicon steel or materials having similar magnetic properties, the 1000-turn coil is used, and R_2 is 100 ohms. In either case C_1 is 1 μf. Capacitor C_c balances the ohmic resistance of the test-frame coil and has a value calculated from the equation

$$C_c = \frac{R_2 C_1}{R_3}$$

where R_3 is the ohmic resistance of the test frame winding. R_4 is nonreactive and has a range of 10,000 ohms in steps of 1 ohm. C_4 is an adjustable capacitor having a range up to 2 μf in steps of 0.001 μf.

Measurements of incremental permeability or core loss are always made in the order of increasing values of biasing magnetizing force. Direct current is supplied to the outer 100-turn winding through the reactor Z, which should have an inductance of 1 henry or more when carrying current equivalent to 2 oersteds. This reactor is for the purpose of limiting the a-c current in the d-c circuit to a negligible value. The d-c current is adjusted to the required value with switch S_1 open and reversed by means of the reversing switch S_2 several times to establish a cyclic condition in the specimen. Switch S_1 is then closed, and the test made. Values of normal permeability or core loss are made with the d-c circuit open.

The d-c magnetizing force is calculated from the equation

$$H_{dc} = \frac{0.4\pi N I_{dc}}{l} \tag{32}$$

in which H_{dc} = biasing magnetizing force in oersteds.
N = number of turns.
I_{dc} = direct current in amperes.
l = assumed length of the magnetic circuit.

When N = 100 and l = 94, this reduces to

$$H_{dc} = 1.34\ I_{dc} \tag{33}$$

or

$$I_{dc} = 0.748 H_{dc} \tag{34}$$

In making a test, either with or without the biasing current, the voltage across the bridge is set to equal

$$E = 0.0707AB\ * \tag{35}$$

where A = the cross-sectional area of the specimen in square centimeters
B = the maximum induction in gausses.

The resistance R_4 and capacitance C_4 are adjusted to balance the bridge. When the bridge is balanced, the permeability is calculated from the equation

$$\mu \text{ (or } \mu_\Delta) = \frac{0.748R_4}{A} \times \frac{R_2}{N_2} \times 10^4 \tag{36}$$

in which μ = normal permeability.
μ_Δ = incremental permeability.
R_4 = resistance in ohms.
A = cross-section in square centimeters
R_2 = resistance in ohms.
N = number of turns.

The loss is calculated from the equation

$$P_c \text{ (or } P_\Delta) = E^2C_4\omega^2C_1R_2 \tag{37}$$

in which P_c = total loss in watts without biasing current.
P_Δ = total loss in watts with biasing current.
E = electromotive force in volts.
C_4 = capacitance in farads.
ω = 2π times the frequency in cycles per second.
C_1 = capacitance in farads.
R_2 = resistance in ohms.

The total loss in watts, divided by 84 per cent of the mass of the specimen in kilograms, is the value of core loss in watts per kilogram.

Figure 51 is a diagram of connections for the a-c potentiometer method. The a-c potentiometer is of the coordinate type, which measures voltage in terms of two components having a quadrature phase relation. A phase-shifting device (not shown) is used to adjust the phase of either the potentiometer current or the magnetizing current. The d-c circuit and procedure for applying a biasing magnetizing force are the same as for the bridge method.

In making a test, the 100-turn coil is used as the primary winding, and the 1000-turn coil is the secondary unless the secondary voltage exceeds the range of the potentiometer, in which case the two windings must be interchanged. The potentiometer is connected to the secondary coil, and the magnetizing current and phase relations are adjusted so that the total voltage is read on the in-phase dial of the potentiometer, the other dial being set at zero. The voltage corresponding to a given induction is calculated from the equation

$$E = \sqrt{2}\pi f N_2 AB \times 10^{-8} \tag{38}$$

where E = electromotive force in volts.
f = frequency in cycles per second.
N_2 = number of turns in the secondary winding.
A = cross-sectional area in square centimeters.
B = induction in gausses.

With the secondary voltage set at the proper value, the two components of the magnetizing current are measured by observing the drop across the non-inductive resistor R, which is connected in series with the magnetizing winding. The power component is in phase with the voltage, and the magnetizing component is in quadrature. Permeability and core loss are calculated in terms of secondary voltage, and the quadrature and in-phase components of the current respectively by using the equations

$$\mu \text{ (or } \mu_\Delta) = 52.9 \frac{B}{N_1 I_q} \tag{39}$$

$$P_c \text{ (or } P_\Delta) = \frac{N_1}{N_2} EI_p \tag{40}$$

* Under certain conditions it may be necessary to make a correction to the voltage thus calculated. The ASTM specifications should be consulted for details.

in which μ = normal permeability.
μ_Δ = incremental permeability.
B = induction in gausses.
N_1 = number of primary turns.
I_q = quadrature component of the current in amperes.
P_c = total loss in watts without biasing magnetizing force.
P_Δ = total loss in watts with biasing magnetizing force.
N_2 = number of secondary turns.
E = secondary electromotive force in volts.
I_p = in-phase component of the current in amperes.

The loss in watts per kilogram is found by dividing the total loss by 84 per cent of the mass of the specimen in kilograms.

20. TESTING OF INSULATING MATERIALS

The most important electrical properties of an insulating material are its dielectric strength, resistivity, dielectric constant, and dielectric loss. Some of the common methods of measuring these quantities are described below.

MEASUREMENT OF DIELECTRIC STRENGTH. The **dielectric strength** of an insulating material is defined as the puncturing voltage per unit thickness, the thickness being measured either in centimeters or in mils (1 mil = 0.001 in.). The voltage required is in general higher than can be conveniently obtained from d-c sources of emf, and alternating emf's are therefore used in such tests. The materials to be tested are placed between electrodes connected to the high-voltage terminals of a transformer which receives power from a low-voltage alternator.

Transformers and Alternators for Dielectric Testing. For making routine tests of dielectric strength, any well-designed high-voltage transformer connected to a suitable a-c supply may be used. For the purpose of comparison and computation of the maximum emf from the effective value, it is essential that the high-voltage transformer deliver as nearly a pure sine-wave voltage as possible. Both transformer and alternator should be large enough to operate with good voltage regulation at all testing loads, so that no serious distortion of the wave form will be produced by the charging current. The transformer and source of supply should have a rating of at least 2 kva for voltages under 50,000 volts, and at least 5 kva for voltages above 50,000 volts. For ordinary tests the frequency should not exceed 100 cps.

Special tests, or tests calling for extremely high voltages or frequencies, will require special equipment.

Control of Voltage. The method of voltage control should be such that the high voltage from the secondary of the testing transformer can be raised gradually from any point. In no case should the control of voltage be by discontinuous steps of more than 500 volts each. The control may be effected by means of generator field regulation, an induction regulator, a variable-ratio auto-transformer, or a combination of these methods. By using a variable-ratio auto-transformer and varying the field excitation of the alternator, the variation of field saturation in the alternator may be reduced to a minimum, and the shape of the voltage wave kept almost constant over a wide range of secondary voltage. Any method of regulating the voltage may be considered satisfactory for ordinary work which does not distort the wave more than 10 per cent from a sinusoidal shape.

Measurement of Voltage. The voltage impressed upon the material under test may be determined by an adjustable spark gap shunted across the electrodes, a voltmeter and multiplier connected across the electrodes, an electrostatic voltmeter connected across the electrodes, a voltmeter connected across the low-voltage terminals of the transformer, the readings of which are to be multiplied by the step-up ratio of transformation, a voltage transformer with a voltmeter connected across its low-voltage terminals, or a special voltmeter winding placed in the middle of the high-voltage windings of the step-up transformer. The needle-spark-gap method, though convenient because of its indication of the maximum rather than the effective value of the voltage, is tedious in use, its readings are dependent upon the circulation of air, the condition of the needles, the proximity of neighboring objects, and other factors. The sphere spark gap is preferable; see Article 16, Spark Gaps for Measuring High Voltages. The use of the voltmeter and multiplier, though flexible and convenient, is not recommended because of the load which is placed upon the transformer and the possibility of leakage in the multiplier. The use of an electrostatic voltmeter is desirable, except that at high voltages the moving element must be immersed in oil, making the instrument sluggish in action. The instrument is also liable to breakdown. In any method involving the ratio of transformation of the transformer, results which are based on the assumption that this ratio is constant are questionable.

The most satisfactory and accurate measurements of voltage are obtained, as a rule, with the separate step-down voltage transformer and voltmeter, or with a voltmeter connected to a special voltmeter coil in the testing transformer.

Protection of Apparatus. Some protection of the measuring apparatus is necessary when using voltages such as are commonly employed in making breakdown tests. A circuit breaker should be provided in the low-voltage side of the test transformer to prevent damage to the transformer or alternator when the specimen breaks down. In addition, some protection is needed in the high-voltage circuit of the test transformer, when the voltage is 25,000 volts or over, to prevent dangerous surges of current in this circuit when the specimen punctures. It is desirable, however, to have as much energy available as possible when the puncture occurs. If impedance in the form of choke coils is used in series with the high-voltage terminals, it should not be greater than that which will limit the high-voltage current to double the normal rated current of the transformer. See Precautions in Sphere-gap Measurements, p. 5-67.

Forms of Electrodes. The breakdown voltage of a specimen depends to a marked extent upon the size and shape of the electrodes used and the nature of the surrounding medium. Inge and Walther (see Bibliography, p. 5–115) have made extensive experimental investigations of the effects of these and other variables upon the breakdown voltage of various substances. If the edges of the electrodes used are not rounded, the increased flux density at the edges produces an excessive stress on the dielectric, and failure of the specimen almost invariably occurs at these points. At higher voltages, say over a few thousand volts, some means must always be provided to prevent corona and flashover at the edges of the electrodes; otherwise the resultant heating of the specimen would entirely vitiate the results.

The American Society for Testing Materials recommends that the electrodes for testing sheet materials be 2-in. cylinders, 1 in. in length, with edges rounded to a radius of 1/4 in. For oils the electrodes recommended are circular disks of polished copper or brass 1 in. in diameter and having square edges. The oil electrodes are mounted in a test cup with their axes horizontal and coincident, with a gap of 0.1 in. between their adjacent faces. By far the most complete single summary of current practice with respect to testing electrical insulating materials will be found in the specifications of this Society (see Bibliography, p. 5-115).

Method of Applying Voltage. Materials may be tested by impressing the puncturing voltage, (1) instantaneously; (2) as a continuously increasing voltage, at a specified rate, beginning with an initial low voltage; (3) in steps of a specified value held for a given duration of time. The method of application of voltage is a significant factor in the comparison of breakdown voltages. It is customary to use method (2), with a rate of increase of voltage of approximately 1000 volts per second, or method (3), with steps of 500 volts applied for 1 min each. Procedure with respect to these last two points is not as yet completely standardized.

Causes of Variations in Results. The puncturing voltage of an insulating material is affected by its previous history, its precise condition when tested, its size and thickness, its temperature, the nature of the electric field to which it is subjected, the frequency of the applied voltage, and the method by which the puncturing voltage is applied. For crystalline materials such as mica the puncturing voltage (expressed as volts per centimeter or volts per mil) is independent of temperature and increases with decreasing thickness of the specimen. For fibrous and synthetic materials the puncturing voltage probably is affected as much by temperature as by any other factor. In any case it is necessary to prevent corona discharge and flashover at the edges of the electrodes. This is commonly accomplished by immersing the specimen in oil. The nature of the electric field to which the specimen is subjected is largely determined by the size of the specimen, the size and shape of the electrodes used, and the nature of the surrounding medium. Results of dielectric-strength tests are not immediately comparable unless obtained under strictly identical experimental conditions.

MEASUREMENT OF INSULATION RESISTANCE. The method commonly used is one of substitution; that is, the deflection of a galvanometer is noted when a standard resistance R_s (0.1 or 1 megohm as the case may be) is inserted in series with the galvanometer and a source of emf and again when the unknown resistance R_x is inserted in the same circuit. A galvanometer should be used of such sensitivity as to give deflections of the order of 1 mm at a scale distance of 1 meter, for a current of 10^{-9} amp. The deflections of the galvanometer should be proportional to the current. The galvanometer should be provided with an Ayrton-Mather universal shunt (see Galvanometer Shunts, p. 5-04), such that the current under measurement will be a known multiple, K, of the galvanometer current. The shunt factor K may be 10, 100, etc. Voltages from 100 to 500 volts may be used in determining the unknown insulation resistance. Because of the difficulty of producing and maintaining resistance standards of extremely high values it will generally be necessary not only to decrease the sensitivity of the galvanometer, but also to decrease the voltage used when calibrating the galvanometer with the standard resistance, in order to keep the galvanometer deflection within the scale limits.

The test is made by observing the shunt factor, the voltage, and the galvanometer deflection when the standard resistance is inserted in the circuit and again when the unknown is inserted. The unknown resistance is then calculated from the equation

$$R_x = R_s \frac{K_s E_x D_s}{K_x E_s D_x} \qquad (41)$$

where, when measuring the specimen, K_x is the shunt factor, E_x the applied voltage, and D_x the galvanometer deflection; K_s, E_s, and D_s represent the corresponding quantities when employing the standard resistance; and R_s represents the value of the standard resistance. With such an arrangement resistances of 10^4 to 10^{11} ohms can be measured, the accuracy being of the order of 10 per cent.

Where the resistance is too high to make the deflection method feasible, the leakage current through the specimen is allowed to charge a condenser for a known interval. The condenser is then discharged through the galvanometer. Then

$$R_x = \frac{Et}{KD} \qquad (42)$$

where E is the applied voltage, t the time of charge, K the ballistic constant, and D the deflection. By this method resistances from 10^{11} to 10^{15} ohms can be measured.

Resistances from 10^{15} to 10^{17} ohms can be measured if a quadrant electrometer is used to measure the current through the unknown resistance. The equation for R_x is identical in form with that given for the ballistic-galvanometer method; E is the applied voltage, t is the time required to produce a deflection V; and K is the quantity of electricity necessary to produce unit deflection. By this method R_x can be determined at any instant.

The **volume resistivity** of a material is defined as the resistance to the current flowing through the material between two opposite faces of a centimeter cube. The **surface resistivity** of a material is defined as the resistance to the current flowing between opposite edges of a surface film which is 1 cm square. When results are to be expressed in terms of volume resistivity or of surface resistivity, it is necessary to use electrodes of known dimensions, in good contact with the specimen. Metallic electrodes, wrapped in tin foil and carefully pressed against the specimen, may be used. Where the nature of the specimen permits, the specimen may be floated on a pool of mercury with an upper electrode of mercury contained within a suitable dam. Great care must be taken to provide suitable **guard rings** in making such measurements; for a discussion of this topic see paper by H. L. Curtis, Insulating properties of solid dielectrics, *Bull. Bur. Standards*, Vol. 11, p. 359 (1915), reprinted as *Bur. Standards Sci. Paper* 234. In any case it is essential to note the temperature of the material, the relative humidity, the voltage used, and the time of application of the voltage. Since these resistance values in general vary with time after the application of voltage, it has become customary to take the resistance 1 min after the application. Since surface resistivity is largely a surface-film phenomenon, the past history of the sample, such as exposure to sunlight, moisture, and corrosive vapors, will materially affect its value.

Self-contained instruments are available for measuring insulation resistance (see Ohmmeters, p. 5-41).

MEASUREMENT OF DIELECTRIC CONSTANT, DIELECTRIC LOSS, AND POWER FACTOR. The **dielectric constant** of any substance is defined as the ratio of the capacitance of a capacitor in which the substance in question fills the space between the plates to the capacitance of the same capacitor in vacuum, that is, to the geometric capacitance of the capacitor. This ratio, designated by K, may vary with the temperature of the substance, the time after the application of voltage to the dielectric, and therefore with the frequency when tested with alternating voltages. The dielectric constant of most substances decreases as the temperature increases. The dielectric constant in general increases with time after the application of voltage and hence decreases as the frequency increases.

The dielectric losses in a substance are caused by direct conduction through the substance, by true dielectric absorption in the substance, and in some cases by electrolytic conduction through the substance. For most engineering purposes, however, it is unnecessary to differentiate between these various sources of dielectric loss. The total dielectric loss of a substance is expressed in terms of the power factor of that substance. The power factor of a capacitor made up with a good dielectric between its plates will be small; that is, the current and voltage will be almost in quadrature. For this reason this property of a dielectric is often defined by the phase-defect angle or dielectric-loss angle, which is the small angle by which the current through and the voltage across the test capacitor depart from exact quadrature.

The determination of dielectric constant and power factor requires the measurement of the capacitance and power factor of a test capacitor in which the space between the plates is first filled with air and then with the substance under investigation. The dielectric constant will be the ratio of the latter capacitance to the former, and the power factor will be the power factor of the capacitor when the dielectric is the substance under investigation. The power factor of a properly constructed air capacitor will be zero to a high degree of precision. For solids it is preferable to use a test capacitor whose capacitance can be computed. It should be remembered that the computed capacitance will be in electrostatic units and the measured capacitance will be in electromagnetic units. For a parallel-plate capacitor the dielectric constant is given by the formula

$$K = \frac{11.3Ct}{S} \tag{43}$$

where K is the dielectric constant, C is the measured capacitance (in micromicrofarads) of the test capacitor made from the substance under test, t is the thickness of the specimen in centimeters, and S is the area of the capacitor plates in square centimeters.

MEASUREMENT OF DIELECTRIC CONSTANT AND POWER FACTOR AT RADIO FREQUENCIES. The measurement of dielectric constant and power factor at radio frequencies calls for special technique. In Fig. 52, which shows a circuit for this purpose, L represents a suitable radio-frequency inductor, C_s a standard adjustable capacitor of known or negligible losses, C_x a capacitor made from the material under test, T and G a shielded thermoelement galvanometer. The galvanometer G should be a sensitive low-resistance instrument whose deflections within reasonable limits are proportional to the square of the current in the heating element. The points 1, 2, 3, 4, 5, 6 are points of a special mercury-cup switch into which are inserted short-circuiting links or the terminals of special non-inductive resistors.

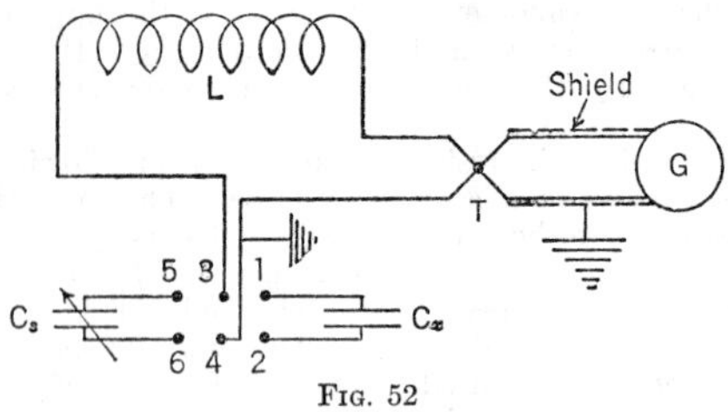

FIG. 52

Short-circuiting links are inserted between the points 1–3 and 2–4; a suitable radio-frequency generator of known frequency is coupled inductively to the coil L; and the circuits are tuned to resonance. The deflection of the galvanometer is then noted when the short-circuiting links are in place and again when a series of suitable non-inductive resistors are inserted between the points 1–3 and 2–4. The equivalent resistance of the circuit is then computed from the equation

$$R = \frac{R_1}{[d_0/d_1 - 1]^{1/2}} \tag{44}$$

where R is the equivalent resistance of the circuit, R_1 is the resistance inserted, d_0 is the galvanometer deflection with the short-circuiting links inserted, and d_1 the galvanometer deflection with R_1 inserted.

Short-circuiting links are now inserted between 3–5 and 4–6 and the circuit tuned to resonance with C_s, the generating circuit being left unchanged. The capacitance of C_s now gives the capacitance of C_x. The equivalent resistance of the circuit containing C_s is now determined as before. The difference between the two equivalent resistances just determined will give the equivalent resistance of the capacitor C_x, since C_s is assumed to have negligible resistance.

The power factor in percentage is then computed from the equation

$$\text{Power factor (in percentage)} = 6.283 \times RCf \times 10^{-7} \tag{45}$$

where R is the equivalent resistance of the sample capacitor in ohms, C is the capacitance of the sample capacitor in micromicrofarads, and f is the test frequency in kilocycles per second.

The dielectric constant K is computed from the equation

$$K = \frac{11.3Ct}{S} \tag{46}$$

where C is the capacitance of the sample capacitor in micromicrofarads, t is the thickness of the specimen in centimeters, and S is the area of the sample capacitor (parallel-plate capacitor) in square centimeters. For further details of the measurements see the Bibliography, p. 5-115.

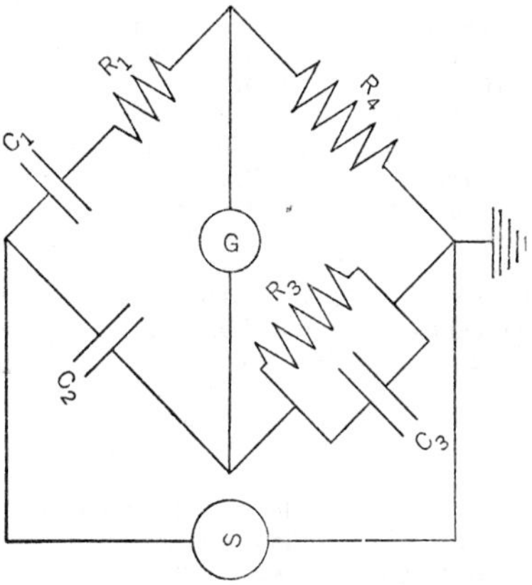

FIG. 53

MEASUREMENT OF DIELECTRIC LOSS AT HIGH VOLTAGES: THE SCHERING BRIDGE. One of the most important measurements in modern high-voltage technology is the determination of the dielectric losses in samples of insulating materials under high dielectric stress. For such measurements the Schering-bridge method has great technical advantages and may now be regarded as the standard method. In Fig. 53, C_1 is the effective capacitance and R_1 the equivalent series resistance of the imperfect capacitor to be tested; C_2 is a standard no-loss capacitor which must be capable of withstanding the full test voltage; C_3 is an adjustable capacitor; and R_3 and R_4 are non-reactive resistors. When the bridge is balanced so that no current flows through the detector circuit G, it can be shown that

$$R_1 = \frac{C_3}{C_2} R_4 \tag{47}$$

and

$$\tan \theta_1 = \omega R_3 C_3 \tag{48}$$

Here θ_1 is the loss angle of the tested capacitor, by which its phase angle differs from 90 deg, and ω is 2π times the frequency. A knowledge of the capacitance of C_2 is not necessary for a determination of the power factor of C_1, but it is essential that the losses of C_2 be negligible or known. In actual practice C_1 and C_2 are small, and consequently these arms are of high impedance in comparison with the two remaining arms. If E is the voltage applied to the bridge, the power loss in the branch containing the unknown capacitor (C_1, R_1) is given to a high degree of approximation by the expression

$$\text{Power in watts} = E^2\omega^2 R_3^2 \frac{C_2 C_3}{R_4} \tag{49}$$

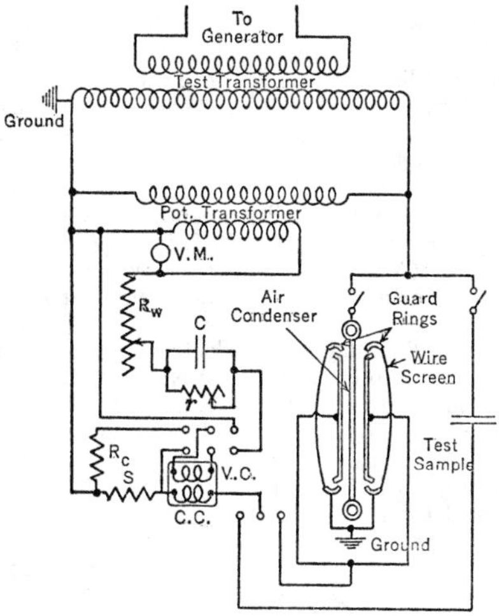

FIG. 54

MEASUREMENT OF DIELECTRIC LOSS WITH COMPENSATED ELECTRO-DYNAMIC WATTMETER. It is sometimes convenient to measure the dielectric loss of a specimen with a compensated electrodynamic wattmeter. The circuit is shown in Fig. 54. The test sample is subjected to a high voltage from a test transformer, and the power absorbed by it is measured by means of a sensitive reflecting wattmeter. The phase angles of the voltage transformer and the wattmeter voltage coil VC are compensated by means of a capacitor C. The effective capacitance of this capacitor may be varied by changing the value of the shunting resistance r. Before making a test a zero-loss air capacitor is substituted for the test sample, and the resistance r varied until the watt-meter shows zero deflection. The test sample is then substituted for the air capacitor and the test repeated, the deflection indicating the power lost in the sample. To measure the current, the voltage coil of the wattmeter is connected across the resistance S through the variable resistance R_c. This puts the current in the two coils in phase, and the instrument will read as an ammeter. The wattmeter is calibrated by direct current, a potentiometer and standard cell being used.

SHIELDING OF BRIDGE CIRCUITS. In using such a-c bridges as have been described, both at radio frequencies and at power frequencies, the proper shielding of the bridge circuit is of as great importance as the choice of a suitable bridge circuit. See references on Shielding and Guarding Instruments and Circuits in the Bibliography, p. 5-112.

USE OF CATHODE-RAY OSCILLOGRAPH IN DIELECTRIC RESEARCH. Although not capable of as great precision as the Schering bridge, the cathode-ray oscillograph is a powerful instrument for the investigation of short-time phenomena in dielectric researches. In using it, a voltage in phase with the voltage across the specimen is applied to one of the pairs of deflection plates, and a voltage in quadrature with the current through the specimen is applied to the other pair of plates. The figure traced by the electron beam is in general an ellipse, and for sinusoidal currents its area is proportional to the dielectric loss. The trace also gives data from which the dielectric constant can be deduced. The oscillograph can also reveal ionization and show the effect of transient voltages, as well as any resulting momentary changes in properties of the dielectric.

21. PHOTOMETRY

If light is considered radiant energy evaluated according to its capacity to produce visual sensation, the chief aspects of photometry are the determination of (*a*) luminous intensity (candlepower) of light sources in a single direction; (*b*) candlepower distribution of light sources, usually in the horizontal and vertical planes; (*c*) luminous flux from light sources; (*d*) illumination of surfaces; (*e*) brightness of surfaces; (*f*) reflecting properties of surfaces; and (*g*) transmitting properties of substances. Of these (*f*) and (*g*), together with (*b*), can be made without reference to any standards, since they can be expressed as ratios and not in terms of any units. All the other determinations, however, must be made directly or indirectly against primary or fundamental standards of light.

PRIMARY STANDARD OF LIGHT. A primary standard of light is one by which the unit of light is established and from which the values of other standards are derived. A satisfactory primary standard must be reproducible from specifications.

Waidner-Burgess Standard of Light. Electric incandescent lamps maintain their values over long periods of time and are convenient to use, but their use as primary or fundamental standards is not justifiable in theory. The National Bureau of Standards proposed, as a suitable primary standard, the Waidner-Burgess standard of light [see *Bur. Standards J. Research*, Vol. 6, pp. 1103-1117 (June, 1931), reprinted as *Bur. Standards Research Paper* 325]. It consists of a black-body radiator, immersed in platinum, heated to a temperature slightly above its melting point and allowed to cool. When the freezing point is reached, the temperature remains constant thereafter until all the platinum is frozen; the temperature of the black body and its brightness also remain constant. Intercomparisons of results obtained at national standardizing laboratories throughout the world have been carried out. To provide a universal system, it has been internationally agreed to assign a value of 60 candles per square centimeter (60 stilb) for the brightness of a black body at the freezing point of platinum. (A bill legalizing this new unit in the United States has been drafted but, at the time of printing of this edition, has not been considered by Congress.)

Electric Lamps. Precise photometry dates from the adoption of electric incandescent lamps as standards in 1909. An agreement, entered into in 1909 by the National Bureau of Standards, the National Physical Laboratory, and the Laboratoire Central d'Électricité, brought the photometric units of the United States, England, and France to a single value within the limits of experimental error. These national standardizing laboratories agreed to maintain the new unit constant and to call it the international candle. This name was formally adopted by the International Commission on Illumination in 1921. It was not specified in the agreement of 1909 how the new unit was to be maintained, but the agreement had been reached by the consideration of results of international comparisons of carbon-filament lamps, and the unit has since been maintained by means of carbon lamps. It is assumed that the average luminous intensity of the lamps maintained by each laboratory has not changed since individual values were assigned to the lamps in 1909. Numerous intercomparisons have shown that this assumption is not ill-founded.

Flame Standards. Standard candles are no longer of practical importance. The name of the unit of luminous intensity was derived from these earliest standards. Candles were superseded by the Hefner lamp, used in Germany and other continental European countries, and the Harcourt 10-cp pentane lamp, used in England. The Hefner standard lamp is a wick lamp burning amyl acetate of very high purity. The flame must be adjusted very accurately to the specified height of 40 mm. The Harcourt 10-cp pentane standard lamp is essentially an argand burner supplied with pentane-air gas and preheated air. For flame standards it is necessary to make corrections for barometric pressure, humidity, and the amount of carbon dioxide in the atmosphere. (For details of these corrections earlier editions of the Handbook may be consulted.)

SECONDARY STANDARDS OF LIGHT. A secondary standard is one calibrated by comparison with a primary standard. For want of a satisfactory reproducible primary standard the unit of light, the international candle, has been maintained at the National Bureau of Standards by means of 45 carbon-filament lamps, which are loosely designated as the "primary" standards. Photometric measurements, as generally made, are all in terms of secondary standards which derive their assigned values from the "primary" lamps, whose values were agreed upon in 1909. Certified lamps can be obtained from the National Bureau of Standards and from testing laboratories.

WORKING STANDARDS OF LIGHT. A working standard of light is any standardized luminous source for daily use in photometry. Electric incandescent lamps as working standards are superior to all other types of lamps. After being properly aged, lamps are standardized for candlepower and current at a definite voltage. The standardization can then usually be relied upon as long as the current taken by the lamps has not changed when the lamps are operated at the originally assigned voltages.

VISUAL PHOTOMETRY. In visual photometry the eye is used to compare the brightnesses of two surfaces. Because of its power of adaptation to different brightnesses, the eye cannot *measure* light with any degree of accuracy. Measurements must therefore depend on a *judgment of equality* of brightness of two surfaces. These surfaces are presented, usually side by side, to the eye for comparison by means of a photometer. Visual photometers now in general use are of three general types, according to the method of comparison employed: (1) by equality of brightness of two surfaces visible simultaneously, (2) by equality of contrast between two pairs of surfaces visible simultaneously, and (3) by disappearance of flicker when two surfaces are viewed in rapid succession. Besides serving the purpose of presenting two surfaces to the eye for comparison, photometers must supply a means for varying the brightness of one or both of these surfaces according to some known law, so that a photometric balance may be obtained. The most common methods of doing this are: (1) varying the distance of one or both light sources from the surface compared, according to the inverse-square law of illumination; (2) interposing a rotating sector disk or other absorbing media between one of the light sources and the surface it illuminates; and (3) polarizing the light from the two surfaces in planes perpendicular to each other and using a Nicol prism, capable of rotation, to obtain a photometric balance. Combinations of these methods are commonly used.

PHOTOMETER HEADS OR SIGHT BOXES. An optical device, variously called a photometer head, sight box, or simply "the photometer," is used to make a precise judgment of the equality of brightness or contrast of two surfaces illuminated by different sources. It consists of two diffusing surfaces, each illuminated by one of the sources of light, with accessories to facilitate the comparison and to enable the eye to view the two surfaces simultaneously. The surfaces are often called photometer screens. Many photometer heads or sight boxes have been devised.

INVERSE-SQUARE METHOD OF PHOTOMETRY. If a source of candlepower I_1 is at a distance d_1 from a surface S_1, the illumination on S_1 is I_1/d_1^2 (if the light is incident perpendicularly). If the reflection factor of S_1 is ρ_1, the brightness of S_1 is $I_1\rho_1/d_1^2$. Similarly, if a source of candlepower I_2 is at a distance d_2 from a surface S_2 of reflection factor ρ_2, the brightness of S_2 is $I_2\rho_2/d_2^2$. If these two surfaces are of equal brightness,

$$\frac{I_1\rho_1}{d_1^2} = \frac{I_2\rho_2}{d_2^2} \tag{50}$$

Hence, if ρ_1 and ρ_2 are equal,

$$\frac{I_1}{I_2} = \frac{d_1^2}{d_2^2} \tag{51}$$

If I_2 is known, and d_1 and d_2 are measured, I_1 can be calculated. This method of comparing directly I_1 against I_2 assumes that the sources are point sources, that the light is incident normally on the surfaces, and that the reflection factors of the surfaces are equal. These assumptions sometimes would lead to appreciable errors. This difficulty may be obviated by the use of the substitution method. In this a third source whose candlepower must be constant, but need not be known, is used as a comparison lamp on one side of the photometer head, while the two sources to be compared are placed in turn on the other side of the photometer. A photometric balance with the comparison lamp is made in each case. If the candlepower of the comparison lamp is assumed to be I_c, then the required ratio

$$\frac{I_1}{I_2} = \frac{I_1}{I_c} \cdot \frac{I_c}{I_2} \tag{52}$$

I_c cancels out, and the required ratio is obtained independent of any lack of symmetry in the photometer. Usually, also, the distance between the photometer head and comparison lamp is fixed, so that the brightness of the comparison surface is constant. If, then, photometric balance is obtained with the photometer head at distances d_1, d_2, and d_3, from sources of candlepower I_1, I_2, and I_3, respectively, $I_1/d_1^2 = I_2/d_2^2 = I_3/d_3^2$. Much calculation is thereby avoided. Such photometric measurements are made on a photometer bench.

ROTATING SECTOR DISK: TALBOT'S LAW. By Talbot's law, if a disk with an angular aperture in it is rotated between a lamp and a photometer head so that the light from the lamp reaches the photometer screen for only a certain fraction of the time, and if the rotation is so fast that the eye perceives no flicker, the effective candlepower of the lamp is reduced in the ratio of the time of exposure to the total time. If the aperture of the disk is a sector of angle θ degrees, the effective candlepower is reduced by the factor $\theta/360$. The sector disk has advantages over other absorbing media in that it is not affected by time and is independent of color.

SQUARED-TANGENT LAW OF POLARIZATION. The use of this law in photometry involves the production of images of the two surfaces to be compared, by means of an optical train which includes a device for plane polarizing the light from the two surfaces in planes perpendicular to each other. If then an anaylzing Nicol prism is interposed in the path of these polarized planes, the intensity of one image will be reduced by the factor $\cos^2 \theta$ and that of the other image by the factor $\sin^2 \theta$, where θ is the angle between the polarizing plane of the Nicol prism and the plane of polarization of the light forming the first image. If, therefore, the analyzing Nicol is rotated until a photometric balance is obtained at the angle θ, it follows that the ratio of the brightnesses of the two images, with no Nicol interposed, would be $\tan^2 \theta$.

BUNSEN PHOTOMETER. In the Bunsen photometer the screen is a disk of white diffusing paper, a well-defined region of which is made translucent by impregnation with oil or paraffin. The disk is set transversely in a sight box, as shown on the plan in Fig. 55. The interior of the sight box is blackened. Light from the lamps to be compared enters the apertures A–A, and falls normally on the disk surfaces. Dihedral mirrors M_1 and M_2 enable both sides of the disk to be viewed at the sight tube T. The opaque portion of the disk reflects diffusely, while the translucent region partially reflects and partially transmits the light received. A photometric balance exists when the two sides of the disk appear alike. If both lamps are alike in color and both regions of the disk have the same absorption, the boundary disappears and both sides appear uniformly bright. With unequal absorption, balance exists when equal contrast exists between the opaque and translucent regions on both sides of the disk. The contrast principle is of distinct advantage with slight color differences. The sensitiveness of the screen depends largely on the definition of the boundary of the impregnated portion.

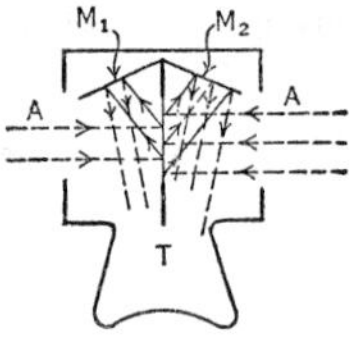

FIG. 55

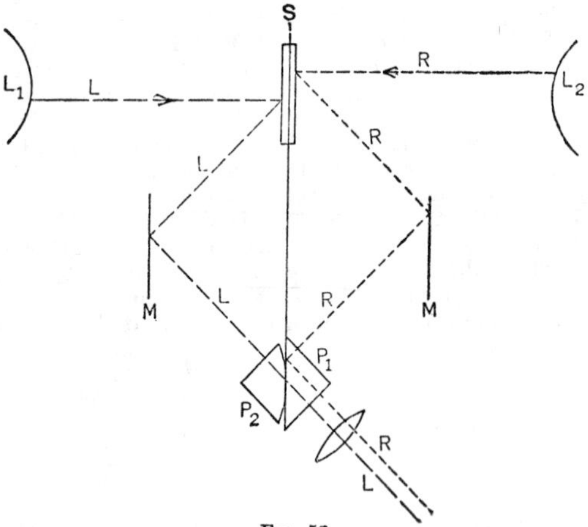

FIG. 56

The Bunsen photometer was the first really accurate photometer devised. It is not as sensitive, however, as later types of photometers, chiefly because each surface receives light from both sources being compared. This difficulty is avoided in the photometers described below. The Leeson disk is a modification of the Bunsen screen.

LUMMER-BRODHUN HEAD OR SIGHT BOX. The plan of this box is shown in Fig. 56. An opaque, diffusely reflecting screen S receives light from the sources to be compared and reflects it along the paths indicated by aid of the mirrors M–M and the prisms P_1–P_2, often referred to as a Lummer-Brodhun cube. The prisms present to the eye a composite field in which the brightnesses of the two sides of S can be conveniently compared. Figure 56 also shows the arrangement for equality-of-brightness working. The prisms are in optical contact over an elliptical portion of their hypotenusal faces, and the remainder of one is cut away. The central portion of the field is illuminated by direct transmission through the contact area, the outer portion by total reflection from the face of the uncut prism.

For equality-of-contrast working, the arrangement of the prisms is as shown in Fig. 57. The hypotenusal face of P_2 is recessed over an area that provides a pattern as shown in (B). The face of P_1 is plane. Two thin glass absorbing strips C_1 and C_2 are set before the faces of P_1 and P_2 as shown in the plan. The two trapezoids L and R are darkened by the absorption of light in C_1 and C_2. In a state of balance L' and R' appear equally bright and L and R equally dark in contrast. The degree of contrast created by C_1 and C_2 is ordinarily about 8 per cent. Accuracy of adjustment and cleanliness of all parts are essential in photometers of the Lummer-Brodhun type.

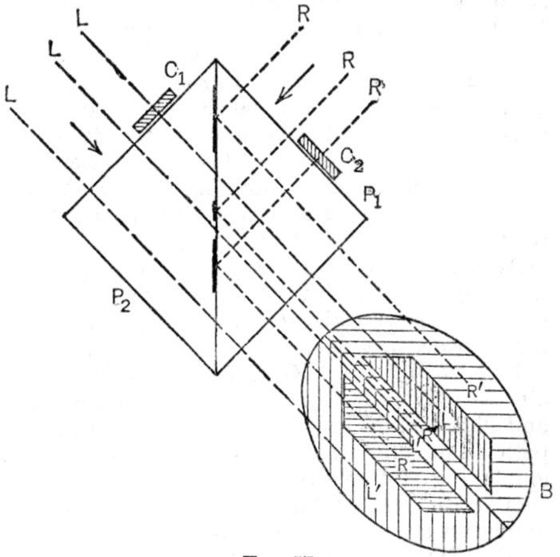

FIG. 57

PHOTOMETER BAR FOR MEASUREMENT OF LUMINOUS INTENSITY (CANDLEPOWER). The devices which have been described are best suited for use in connection with a fixed photometer bar in a laboratory. With the substitution method of photometry the most convenient arrangement of the equipment is as shown in Fig. 58. The photometer head P and the comparison lamp C are clamped to a movable carriage so that the distance between them is constant. The carriage can be moved with respect to the lamp socket T, which is fixed on the bar. Lamp C is maintained at constant voltage throughout any series of measurements. Working standard lamps and

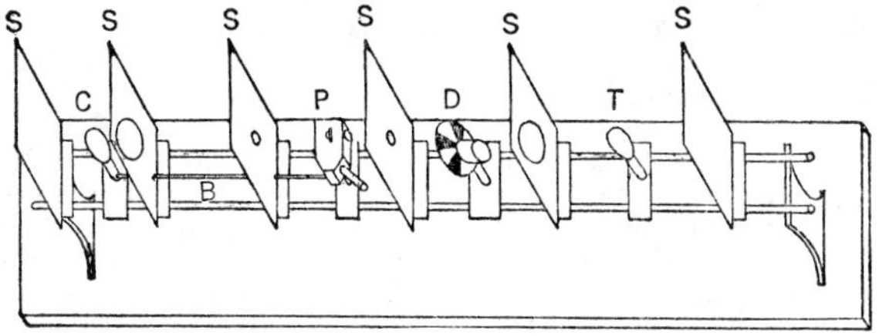

FIG. 58

test lamps whose candlepower is to be determined are in turn placed in socket T and maintained at any desired voltage while photometric balance is obtained by setting the movable carriage. In precision photometry more than one standard is used. A common procedure is to use a set of six working standards, reading three at the start of a run and the other three after the test lamps have been read.

For measurements of candlepower in a fixed direction the lamps must be mounted so that they may be oriented on the bar in that direction. For measurements of mean horizontal candlepower, socket T is rotated at about 120 rpm. The bar should be provided also with a series of screens S of dead black material (preferably velvet) having graded apertures along the photometric axis and with solid screens at the ends. These screens should completely exclude all extraneous light from the photometer head and should protect the eye of the observer from the direct light of the lamps. If these conditions are met, the photometric room need not be blackened, although it is advisable to keep it as dark as is convenient.

ACCURACY AND CRITICAL FACTORS IN VISUAL PHOTOMETRY. The accuracy of visual photometry depends on the brightness of the field, the color of the field, the visual angle of the field, and the departure from normal of the observer's eyesight. The accuracy is maximum for a field brightness of about 3 ft-L. At brightness levels below 1 ft-L, the maximum sensitivity of the eye to radiant energy shifts toward shorter wavelengths (Purkinje effect). The visual angle of the photometric field has been standardized at 2 deg diameter, compared to a value of 8 to 10 deg in the usual Lummer-Brodhun sight box. This smaller photometric field surrounded by a field of approximately equal brightness has been found more accurate for measurements in heterochromatic photometry.

Some practice is required before an observer can match brightnesses within 1 per cent by use of the Lummer-Brodhun sight box. Under favorable conditions an experienced observer can attain accuracy within 0.02 per cent in the mean of a large number of readings.

Accurate voltage control of light sources is most important because the luminous intensity varies with the square of the voltage. Secondary standard electrical indicating instruments are calibrated within 0.2 per cent, but many workers prefer potentiometers for most accurate measurement of lamp voltage.

PORTABLE PHOTOMETERS. Many forms of portable photometers and illuminometers are available for use both in the laboratory and for measurement of candlepower and illumination in the field. Such instruments are the Sharp-Millar photometer, Macbeth illuminometer, Holophane light meter, and footcandle meter. Illuminometers employing photoelectric cells or photoelectric tubes have become very common, since they are more portable than the earlier types. The Weston and General Electric illuminometers are examples of this newer type of instrument.

PHOTOMETERS AND ILLUMINOMETERS BASED ON INVERSE-SQUARE LAW. The Sharp-Millar photometer, the Macbeth illuminometer, and the Weber photometer are portable inverse-square-law photometers, consisting essentially of a tube a foot or more in length, a comparison lamp of low voltage, designed to operate on a storage battery or dry cells, a Lummer-Brodhun cube of special design, and a diffusing glass photometer screen. In the Sharp-Millar and Macbeth photometers the comparison lamp is movable; in the Weber photometer the screen is movable. The instruments are provided with scales (graduated according to the inverse-square-law), the range of which may be extended by means of neutral filters that can be placed on the test side or comparison side of the cube. An external test plate or standard surface is placed at the point at which illumination is to be measured. The illuminated test plate then serves as the second comparison screen. For details of these photometers see the papers listed in the Bibliography, p. 5-116.

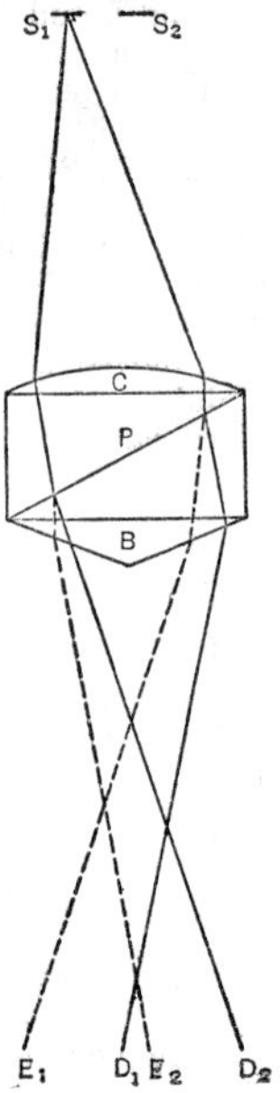

FIG. 59

MARTENS POLARIZATION PHOTOMETER. In this type of photometer a Fresnel biprism is used to secure contiguous images of the surfaces to be compared. In Fig. 59 such a biprism, together with a Wollaston prism, is shown diagrammatically. Here S_1 and S_2 are the surfaces to be compared. The light from S_1 passes through a plano-convex lens C into a Wollaston prism P. Here it is split up into two beams: one, shown by the full line, is polarized in the plane of the paper; the other, shown by the broken line, is polarized in a plane perpendicular to the paper. The light from S_2 is treated similarly, and the angle of the biprism is designed so that images of S_1 and S_2 oppositely polarized coincide at the eye, while the remaining six images are stopped by a diaphragm. It follows that one-half the biprism B appears bright because of the light from S_1 which is polarized in the plane of the paper, whereas the other half appears bright because of light from S_2, which is polarized in a perpendicular plane. A Nicol prism and graduated circle capable of rotation about the axis of the instrument complete the photometer.

FOOTCANDLE METER. A footcandle meter is a convenient, inexpensive, and self-contained device for making approximate measurements of illumination. It consists of a small metal box, part of whose top is a paper screen, opaque except for a row of Bunsen spots on it. The under surface of this screen is illuminated by a small lamp fed by three tubular dry cells. The lamp is at one end of the box, so that the Bunsen spots decrease in brightness from the end where the lamp is placed. Measurement is made by observing which spot disappears when the scale is illuminated from above.

HETEROCHROMATIC PHOTOMETRY. The illuminants in use today, i.e., vacuum and gas-filled tungsten-filament lamps, arc lamps, mantle gas lamps, and gaseous-discharge lamps, give light different in color from that of the fundamental reference-standard lamps. The measurement of the light emitted by modern lamps therefore involves, somewhere in the process, heterochromatic photometry, i.e., photometry with a color difference.

If the eye is required to decide when equality of brightness or equality of contrast exists between the comparison surfaces, with a color difference existing, it encounters a difficulty which increases as the color difference becomes greater.

Cascade Method. One method of bridging the color difference is the cascade method. It involves dividing up the color difference by making the photometric settings in a number of steps. When lamps of efficiencies different from the efficiency of carbon lamps are to be measured against carbon-filament standards, it is possible to divide the color difference into a number of steps by using lamps operating at intermediate efficiencies. Tungsten lamps are blue in color in comparison with the carbon-filament standards, but by reducing the voltage on tungsten lamps a series of color steps can be interposed between the carbon standard and the tungsten test lamp.

Compensation Method. In the compensation or mixture method the color difference is reduced by illuminating one or both of the comparison sources with light from both the sources being compared.

Flicker-photometer Method. A reliable means of comparing two lights of different colors is the flicker photometer, in which the comparison field is alternately illuminated by the two sources of light to be compared. This alternate illumination causes a flicker which is due to difference of color and difference in brightness. Above a certain moderate rate of alternation the color sensations blend, and the disappearance of flicker is a true indication of equal brightness. The following conditions are required for high precision: (*a*) a field illumination of about 25 meter-candles; (*b*) a photometric field of 2 deg diameter; and (*c*) a background field about 25 deg in diameter surrounding the photometric field and about equal to it in brightness.

The **Ives-Kingsbury flicker photometer** consists of an ingenious attachment to a Lummer-Brodhun contrast-type photometer, described above, with the contrast glasses removed. It consists essentially of an optical train including (*a*) a lens for magnifying the field, (*b*) a rotating prism, (*c*) a disk with a hole limiting the field to 2 deg, (*d*) an eye lens, and (*e*) a lamp for illuminating the background field. For a complete description of this photometer see *J. Franklin Inst.*, Vol. 180, p. 215 (1915).

The Ives-Kingsbury photometer and most of the other flicker photometers have so many optical parts that the brightnesses of the comparison surfaces are very materially reduced. Where only low illuminations of these surfaces are available, this is a distinct disadvantage. The **Guild flicker photometer** obviates this difficulty. It has no optical parts between the eye and the comparison surfaces. These surfaces are (1) a stationary surface of magnesium oxide illuminated by one source and (2) a rotating sector disk of magnesium oxide illuminated by the second source. The stationary surface is viewed through the rotating disk, whose speed of rotation can be adjusted. Thus the eye sees alternately the stationary surface and the sector disk, the two comparison surfaces. The sight tube is arranged so as to supply the background illumination. For a complete description of the Guild flicker photometer, see *J. Sci. Instruments*, Vol. 1, p. 182 (1924).

Color-equalizing Filters. Where a large color difference exists between two sources to be compared directly, flicker photometry gives results which are more consistent and dependable than those obtained by equality-of-contrast photometers, which are, however, much more convenient and easy to use. If two sources of light to be compared are of different color, the color difference can be removed by placing a suitably colored filter between the photometer and one of the sources and so causing the light from this source to match that from the other source. The problem of heterochromatic photometry is thus reduced to one of homochromatic photometry and can be handled very conveniently with a Lummer-Brodhun photometer. It then becomes necessary to determine the transmission factor of the filter for light of the color given by the source with which it is used. The transmission factor of the filter can be determined by means of a flicker photometer or by a spectrophotometer.

SPECTROPHOTOMETRY. In ordinary photometry the light emitted by a source is measured as a whole, regardless of its spectral distribution. In spectrophotometry the light emitted by the source under examination is compared, wavelength for wavelength, with that given by a standard source. If the spectral distribution of the light given by the standard source is known, the spectral distribution of the light from the test source can be obtained. The acetylene flame has frequently been used as a standard, but a tungsten-filament lamp is a more convenient standard of reference. A tungsten lamp is slightly selective as a radiator; and, when vacuum lamps are used as reference standards, they are operated at a definite "color temperature," as, for example, at a color temperature of 2360 deg K, and gas-filled lamps at 2680 or 2848 deg K. The "color temperature" of a lamp is the temperature of a black body that gives light of the same color as is given by the lamp.

To make a measurement, it is necessary to disperse the light by some device, such as a prism, and then measure the intensity of each wavelength, or in practice, a small group of wavelengths, as compared with the reference standard. A spectrophotometer consists essentially of (1) a device for analyzing the light of each source and making a photometric comparison of the two sources for each small group of wavelengths, and (2) auxiliary equipment for changing the brightness of the components of the field to obtain photometric balance.

Spectrophotometric measurements are particularly applicable to the determination of the spectral transmission of colored filters. From such measurements the overall transmission of a filter for light of known distribution of energy can be calculated if the spectral characteristics of the average eye are known, for the eye responds very differently for different wavelengths, its limits of sensitivity being about 400 and 750 mμ, with the maximum sensitivity at 550 mμ. The most generally accepted values of the relative sensitivity of the "average eye" are those proposed by Gibson and Tyndall, which have been provisionally accepted for international use by the International Commission on Illumination. [See Illuminating Engineering Nomenclature and Photometric Standards, ASA, Z7-1932, p. 8, for Table of Values of Relative Visibility ($V\lambda$).]

MEASUREMENT OF CANDLEPOWER DISTRIBUTION AND LIGHT FLUX. No actual light source is a point source; therefore no light source used in practice has the same luminous intensity (candlepower) in all directions. Consequently the measurement of the candlepower of a source of light in any one direction gives no information as to the light output of the source or the distribution of the light. The determination of the luminous intensity of an incandescent lamp in, say, 18 directions (10-deg intervals) in the horizontal plane and in the vertical plane is sufficient for plotting polar curves of light distribution. To make such determinations, various arrangements of mirrors and means for rotating the lamp and mirrors have been devised. A typical three-mirror device is shown diagrammatically in Fig. 60. The lamp may remain stationary or be rotated about its vertical axis. The mirrors may be turned about the horizontal photometric axis in steps.

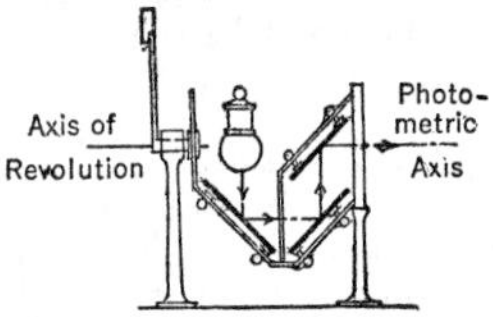

Fig. 60

The mean spherical candlepower or lumen output of an incandescent lamp may be determined: (1) by measuring the candlepower of the lamp in many directions in space and computing the mean; (2) by measurement of mean horizontal candlepower, and computation by using the reduction factor; and (3) by measurement of total flux or mean spherical candlepower in an integrating-sphere photometer.

Total flux of luminaires may be measured with sufficient accuracy for general purposes by making candlepower measurements on a distribution photometer at the middle of each 10-deg zone, from 0 to 180 deg, angles being measured counterclockwise from the "six o'clock" position. If then the mean candlepower for each 10-deg zone is multiplied by zone factors, which are the zonal areas on the unit sphere, total lumens for any zone or the complete sphere can be computed. Zone factors may be calculated from the following formula:

$$\text{Zone factor} = 2\pi(\cos\theta_1 - \cos\theta_2) \qquad (53)$$

where θ_1 and θ_2 are the angles bounding the zone.

INTEGRATING-SPHERE PHOTOMETER. An integrating-sphere photometer (Ulbricht sphere) provides the means of making measurements of mean spherical candlepower or total flux of light in lumens from a lamp or luminaire by a single observation, if a standard of luminous flux is available for comparison. Sphere photometers range in size from 15 in. diameter to 100 in. or even more. The inside of the sphere is painted with a specially prepared white paint, as non-selective as possible and of highest diffusing power obtainable.

When a lighted lamp is put into the sphere, the illumination on any element of area A of the sphere wall is made up of two parts: (1) the direct light from the lamp, E_d; and (2) the light diffusely reflected, E_r, from all parts of the sphere wall. The component E_r of the total illumination on A is proportional to the total flux of the source and independent of the location of the source within the sphere. It is sufficient therefore to measure the component E_r. This is done by inserting a small translucent glass window flush with the inner surface of the sphere wall and screening the window from the direct light of the source by the smallest screen D that will completely cut off all direct light from the window. The sphere window then serves as a transmitting test plate, and its brightness may be measured by a suitable arrangement of a bar photometer, with Lummer-Brodhun cube, or by a portable photometer.

Figure 61 is a diagrammatic sketch showing a commonly used method for visual readings. The mirror M in the Lummer-Brodhun sight box makes the external sphere-window surface W one of the comparison-screen surfaces. The photometric balance is obtained by moving the comparison lamp, O, keeping the photometer head fixed. Other arrangements are also used.

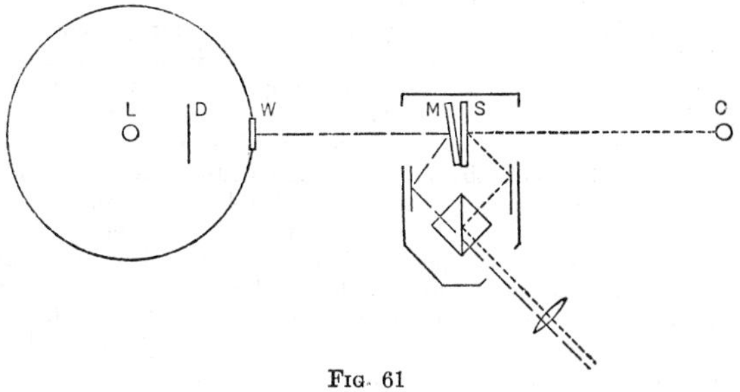

FIG. 61

The substitution method is used in making measurements, standards of luminous flux and the test lamps or luminaires to be measured being placed in the sphere in turn, and photometer settings made, either visually or by physical photometry (photoelectric-cell methods). Auxiliaries such as sector disks and colored glass filters are easily applied.

Attention to details is essential. The sphere window should be non-selective in transmission, have good diffusing qualities, and be no larger than about one-tenth the sphere diameter and preferably smaller. The shield in the sphere between the source and window must be so located and of such size as to reduce to a minimum the area of the sphere wall from which first reflections cannot reach the area, but fully screen all direct light of the source. If the source is at the center of the sphere, the best position of the screen is at a point 0.387 times the radius of the sphere from the window. All objects in the sphere introduce errors, including the lamps to be measured, particularly if such lamps are blackened from use, or if the test lamps differ much in size from the standards. The paint should be non-selective and renewed periodically.

Sphere-photometry errors are discussed at length by Weigel and Knoll in a paper in *Licht und Lampe*, Vol. 17, p. 753 (1928).

MEASUREMENT OF BRIGHTNESS, REFLECTION FACTOR, AND TRANSMISSION FACTOR. Measurement of Brightness. Brightness is defined as the quotient of the luminous intensity of a surface, measured in a given direction, by the area of this surface projected on the plane perpendicular to the direction considered. The brightness of an illuminated surface or transmitting medium may therefore be obtained by measuring the candlepower of the surface in a given direction and dividing this value by the projected area. However, this direct method is applicable practically only to very bright sources. For illuminating engineering purposes the brightness of surfaces may be measured satisfactorily with a Macbeth or Sharp-Millar portable photometer. The scale of the photometer is ordinarily calibrated in footcandles to read illumination on a test plate. If the reflection factor of the test plate is known, the photometer may be calibrated to read brightness directly, if scale readings are multiplied by the quantity ρ/π, where ρ is the reflection factor of the test plate. Test plates of magnesium oxide or magnesium carbonate (reflection factor 98 per cent), freshly prepared, are useful in such calibrations.

Reflecting Properties of Surfaces. Ability to reflect light not only varies from surface to surface, but for a particular surface it varies with the direction of the incident beam. Thus a description of the geometry of this incident beam must be a part of any description of a measurement of reflectance. The reflectance of a surface, R, for a specified incident beam, is the fraction of the incident light reflected. To measure reflectance, an instrument such as the Taylor reflectometer, which has means for gathering the light reflected in all directions, must be used. This reflected light, however, may be specularly reflected by the surface in the direction of mirror reflection, it may be diffusely distributed in all possible directions of view, or it may be a mixture of both these types. Reflectometers which measure light reflected in only certain directions of view are widely used. Instruments which measure the fraction of an incident beam reflected in the direction of mirror reflection are used for evaluating the reflectance of mirrors and the specular gloss of surfaces. Instruments which measure light reflected in selected directions not involving mirror reflection (such as 45 deg illumination, normal viewing) are widely used for approximate evaluations of diffuse reflectance.

Transmitting Properties of Substances. The transmission of light by substances varies in the same manner as reflection; and, as in that case, it is not possible, in general, to give a transmittance for a particular substance without stating the complete conditions. For a given set of conditions the transmittance of a substance is the ratio of the light transmitted by the substance to the light incident upon it. The transmittance may be regular, diffuse, or a superposition of the two.

PHOTOMETRY OF PROJECTION APPARATUS. The photometric measurement of light projectors, such as automobile headlamps, airway beacons, airport and hangar floodlights, locomotive headlamps, searchlights, and general floodlighting equipment, constitutes a special problem. The approval by state authorities of automobile headlamp devices has brought about the development by testing laboratories of special equipment for the measurement of candlepower distribution in headlamp beams. A photometric range of 100 ft is desirable. The Sharp-Millar, Macbeth, or Bechstein portable photometers with accessory equipment are suitable for making candlepower-distribution measurements to determine compliance of types of headlamps with special state requirements or with the standard specification of the American Standards Association covering the laboratory testing of automobile-headlamp devices. [See Specifications of Laboratory Tests for Approval of Electric Headlighting Devices for Motor Vehicles, Illuminating Engineering Society, New York. See also Joint IES-SAE Standard, *Trans. Illum. Engng. Soc.*, Vol. 25, p. 835 (1930). Current testing specifications appear in the latest SAE Handbook.]

In making candlepower measurements on projectors, usually the whole aperture of the projection device is filled with light and appears equally bright all over. The area of the projector opening is therefore the source, and the values determined are in terms of apparent candlepower, measured at a specified distance. The apparent candlepower of an extended source of light, measured at a specified distance, is the candlepower of a point source of light which would produce the same illumination at that distance.

PHOTOELECTRIC PHOTOMETRY. The very rapid development of photoelectric cells and tubes has made photoelectric photometry practical for use in testing laboratories and in lamp factories, and, by the application of certain types of cells to portable instruments (illuminometers or light meters), for use in making illumination measurements in the field.

The photoelectric cell, known as blocking layer or barrier layer, and the photoelectric (electron emission) tube, together with a sensitive meter (and batteries for the photoelectric-tube type), make a readily portable meter for measuring illumination. A number of these are manufactured. Some of them, such as the one made by General Electric, have the cell and microammeter mounted together in a compact case, whereas others, such as the Weston Electrical Meter Corporation's product, have a target and a flexible lead connected to the microammeter. The manufacturing tolerances result in an overall uncertainty of reading of about ± 7.5 per cent. The need for a cover glass and the rim of the holder for the cell result in rather large errors when light is incident at large angles from the normal to the cell or target surface. If light at such angles is important, the target should be oriented perpendicular to the light, and the reading multiplied by the cosine of the angle between the normal to the surface where the illumination is desired and the direction of the incident light. The spectral response curve of the photoelectric cells and tubes differs markedly from that of the human eye, and hence the errors in measuring illumination from sources differing in color from that used in calibrating may be quite large. Filters can be supplied, however, which make the spectral response approach quite closely to that of the average human eye. When thus corrected, the meters evaluate sources that are not too selective in their spectral emission sufficiently closely for purposes of illuminating engineering practice.

A voltage indicator consisting of a cylindrical hood, a properly selected incandescent lamp to operate on 115-volt circuits, and a cord and attachment plug can be used with a footcandle meter. The device is adjusted to read 20 footcandles when the voltage is exactly 115. A conversion scale on the hood shows the relationship between volts and footcandles and enables the user to ascertain voltages between 100 and 130.

In order to measure the irradiation from germicidal lamps an attachment consisting of a frame holding together plates of fused quartz and of glass, with a layer of fluorescent phosphor between them, is placed over the target or cell. The quartz plate is uppermost.

A photoelectric-cell photometer, in its simplest form, consists of a light-sensitive cell, one or more electric batteries, a suitable protective resistor (10 to 100 megohms), and a galvanometer. Both gas-filled and vacuum photoelectric cells are used in photometry. The electric current produced is very small, and for photometers for routine use amplification is desirable. An ideal photoelectric cell for photometric use should have, among others, the following characteristics: (*a*) The electric current produced by the cell when illuminated should be directly proportional to the illumination of the cell. This property is commonly called "linearity of response." (*b*) The response of the cell from day to day should be constant. (*c*) The color sensitivity should be the same as the color sensitivity of the average eye. By careful selection of cells and by the use of filters all these conditions can be met satisfactorily for practical use. Manufacturers of photometric apparatus now supply photometers equipped with photoelectric cells and auxiliary devices complete for laboratory installation as well as several types of portable photometers.

Reproducibility of results by photoelectric-cell photometers is much better than by visual photometers. Great care is necessary, however, for the elimination of disturbing influences. The cell and electron tubes must be carefully and completely shielded, and leakage caused by moisture must be guarded against. Dark rooms are not necessary for photoelectric-cell photometry. Candlepower distribution, transmission, and color temperature methods are being developed.

ACCURACY AND CRITICAL FACTORS IN PHYSICAL PHOTOMETRY. Some makes of photovoltaic cells are very stable and maintain calibration for a period of years. Pocket-size lightmeters and exposure meters usually employ predetermined scales and are accurate within 10 per cent, which is adequate for the intended purpose. Cells and instruments calibrated together may be accurate within 1 per cent.

It has been general practice in the industry to calibrate lightmeters from a point source, a tungsten lamp operating at about 2700 deg K being used. A totally diffused source will give almost 20 per cent less scale deflection for the same illumination. Uncorrected lightmeters will also read about 20 per cent low in daylight illumination. To improve the usefulness and accuracy of lightmeters, color-correcting filters are inserted over the cell. The filter changes the response to equal that of the eye and eliminates errors due to the color of the illuminant. Because of the thickness of the filter glass, the filter factor for oblique light is greater than for rays of normal incidence. To obtain compliance with the cosine function for oblique light, the filtered cells are mounted in a cluster behind a diffusing plate. Precision lightmeters of this type should be accurate within 1 per cent under most conditions.

When non-selective primary detectors (thermopile, bolometer, etc.) are used to measure radiant energy, the visible light may be computed by multiplying the energy in each wavelength by the corresponding relative luminosity factor. (For table see ASA Z7.1, Illuminating Engineering Society.)

22. ELECTRICAL RECORDING INSTRUMENTS

A recording instrument is defined as an instrument which makes a graphic record of the value of an electrical quantity as a function of time. The term recording instrument may be replaced by the shorter term recorder. For other definitions and some of the more general requirements to be met, see AIEE Standard for Electric Recording Instruments, No. 40, July, 1947.

An electrical recording instrument produces or controls the motion of a marking device which leaves a record on a chart advanced at a controlled speed by clockwork, a synchronous motor, or equivalent. Recorders are classified as (1) direct-acting, in which the measuring mechanism itself moves the pen or other marking device over the chart; and (2) relay-type, in which the measuring mechanism controls a secondary mechanism, which supplies the power necessary to move the marking device across the chart. Recorders are also classified as (*a*) continuously marking, having a moving pen or other marking unit continually in contact with the chart; and (*b*) intermittently marking, in which the marking unit is periodically pressed against the chart, producing a record.

The operating forces set up by nearly all indicating-instrument mechanisms are so small that the attempt to make them move a pen over a chart, without serious errors from pen friction, is beset with difficulties. In direct-acting recorders of the continuously marking type, the instrument-mechanism design is modified to secure a relatively high operating torque (or force), even at the cost of increased power required for operation. This difficulty is overcome in intermittently marking recorders, in which a low operating torque may be used by leaving the pointer free to move except during periodic short-time intervals when a marking medium, chart, and pointer are pressed together by an auxiliary striker mechanism, producing a trace on the chart. Relay-type recorders may use very sensitive measuring circuits or mechanisms, making it possible to measure electrical quantities too small to operate a direct-acting recorder.

Recording instruments are made in many forms and for an ever-increasing list of applications in industrial processes for which accurate control of conditions is essential to the maintenance of quality, minimum cost of production, or both. For detailed descriptions and operating instructions the makers' publications should be consulted. Detailed treatments of the construction of the various types of recording instruments are given in the books and articles listed in the Bibliography, p. 5-117.

23. TELEMETERING

The AIEE Committee on Instruments and Measurements defines telemeter as "measuring, transmitting and receiving apparatus for indicating, recording or integrating at a distance by electrical means the value of a quantity." The art of telemetering ("remote metering" or, in British usage, "distant indication") came into existence as a matter of necessity as the interconnection of systems became more widespread and the control of scattered power plants became more centralized. In 1926 the importance of the subject prompted the setting up by the AIEE Committee on Instruments and Measurements of a subcommittee to survey the field and collect information. The reports of this subcommittee should be consulted (see Bibliography, p. 5-117) for recommended definitions of terms and the classification of remote-metering systems. A subcommittee of the AIEE Automatic Stations Committee cooperated with the subcommittee on telemetering in the preparation of a report on telemetering and supervisory control systems and related communication systems (see Bibliography, p. 5-117).

Telemetering for distances less than 50 miles is accomplished by the use of d-c selsyns, a-c selsyns, torque balance telemeters (producing d-c currents for transmission), impulse systems, time-duration impulse systems, and variable-frequency systems operating directly over private wire lines or leased-wire lines. For distances over 50 miles carrier currents over transmission lines are generally used as a vehicle for the measurement signal. Telemetering systems used with carrier currents include time-duration, impulse, and variable-frequency types.

24. MEASUREMENT OF NON-ELECTRICAL QUANTITIES BY ELECTRICAL AND MAGNETIC MEANS

Electrical and magnetic means and measuring instruments have been applied for a vast number of kinds of measurement of non-electrical quantities. These have been classified as follows:

1. Measurement of temperature.
2. Measurement of stress, strain, or changes in physical dimensions.
3. Measurement of flow of liquids and gases.
4. Measurement of velocity, angular or linear acceleration, and torque.
5. Measurement of work, mechanical power, and force.
6. Measurement of radiant energy, light, and color.
7. Measurement of time.
8. Measurement of vibration.
9. Measurement of pressure and vacuum.
10. Measurement of liquid level.
11. Measurement of humidity and moisture content.
12. Measurement of dimension.
13. Measurement of weight.
14. Measurement of tool wear.
15. Measurement of deflection.
16. Measurement of sound.
17. Determination of metal identity.
18. Determination of chemical and molecular condition.
19. Determination of engine synchronism.
20. Navigational measurement and detection of hidden conditions.
21. Physiological and allied measurements.

The number of measuring devices coming under these twenty-one headings is so great that reference must be made to the Bibliography, p. 5-117, particularly the book by Smith and the paper by Borden, which contains a valuable bibliography of 330 entries.

25. MEASUREMENT OF TEMPERATURE

THE PRIMARY STANDARD SCALE OF TEMPERATURE. The International Temperature Scale, adopted by thirty-one nations in 1927, is based upon a number of fixed and reproducible equilibrium temperatures to which numerical values are assigned, and upon the indications of specified interpolation instruments calibrated according to a specified procedure at the fixed temperatures. This scale conforms, as closely as is possible with present knowledge, with the Thermodynamic **Centigrade Scale,** on which the temperature of melting ice and the temperature of condensing water vapor, both under a pressure of 1 atm, are numbered 0 and 100 deg, respectively. Temperatures on the international scale are ordinarily designated by the abbreviations deg cent, °C, or C. By a recently approved American Tentative Standard (ASA Z10.4-1943) the abbreviation C denotes a temperature scale or points thereon, but numbers expressing *intervals* on this scale are followed by the abbreviation °C.

THE FAHRENHEIT SCALE. This scale is now derived from the international scale by means of the relations

$$t_f = \frac{9}{5} t_c + 32 \tag{54}$$

$$t_c = \frac{5}{9} (t_f - 32) \tag{55}$$

where t_f = temperature in degrees fahrenheit.
t_c = temperature in degrees centigrade.

A degree fahrenheit is abbreviated deg fahr, °F, or F.

OTHER TEMPERATURE SCALES. A number of other temperature scales have been used or proposed, of which only the Réaumur survives to any extent in the United States. In this scale the interval between the ice and steam points is taken as 80 deg, the ice point being 0 deg R.

ABSOLUTE TEMPERATURE SCALES. Thermodynamic reasoning introduces the concept of absolute temperature, on which basis a temperature of −273.16 deg on the International Temperature Scale would be called zero temperature absolute. The two absolute temperature scales are: the centigrade absolute or Kelvin scale, abbreviated deg K, °K, or K, and the fahrenheit absolute or Rankine scale, abbreviated °R′. A temperature in degrees centigrade absolute is numerically equal to the same temperature in degrees centigrade plus 273.16, and a temperature in degrees fahrenheit absolute is numerically equal to the same temperature in degrees fahrenheit plus 459.69.

THERMOMETERS. Choice of Thermometers. The selection of a thermometer for a particular purpose is governed by the temperature range to be covered, the precision required, the nature of the substance whose temperature is to be measured, the space available, and the type of reading (indication or record) desired.

Liquid-in-glass Thermometers. For the measurement of temperatures below about 500 deg cent., in applications not requiring the high accuracy obtainable with precision thermocouples and resistance thermometers, liquid-in-glass thermometers are extensively used. The fluids generally used are **alcohol, pentane, toluene,** and **mercury,** the first three being suitable for temperatures below the freezing point of mercury. For all temperatures above its freezing point, mercury is much superior to the other liquids.

Mercury-in-glass thermometers are made in various ranges and are of all grades of glass and of workmanship. For engineering and laboratory measurements, thermometers of the laboratory type (sometimes called chemical thermometers), having the scale marked directly on the tube, are preferable to those in which the scale is not integral with the tube. The latter type, if of good workmanship, is suitable for industrial use where only moderate accuracy is required.

Mercury-in-glass thermometers are made to cover the range −35 to 550 deg cent. If fused silica is substituted for hard glass, the upper limit is extended to 800 deg cent, but such thermometers are very little used. They are fragile and expensive, their emergent-stem correction is large and uncertain, and some other form of pyrometer is much to be preferred.

When using thermometers **calibrated for total immersion** (i.e., calibrated with the entire liquid column of the thermometer at one temperature), a **correction** must be applied to the reading if part of the column is in the stem of the thermometer at a temperature materially different from that of the bulb. For mercury-in-glass centigrade thermometers the approximate "emergent-stem correction" is given by the formula

$$\text{Correction} = +0.00016n(t_b - t_s) \tag{56}$$

where the temperatures are in degrees centigrade, and n = number of degrees of the column of mercury exposed to the emergent-stem temperature t_s, and t_b = temperature of the bulb. For more precise corrections see Smithsonian Physical Tables.

Resistance Thermometers. For precise laboratory work the resistance thermometer consists of a coil of platinum wire wound on a mica frame and enclosed in a protecting tube, through which leads are brought out for connection to a thermometer bridge for measuring the resistance of the coil. Such thermometers are particularly useful in calorimetry and in laboratory work where high accuracy is essential. For engineering purposes a nickel resistance thermometer with a direct-reading bridge, or some form of thermocouple, is usually preferable. In the form of **embedded temperature detectors,** resistance thermometers are built into the slots of generators to measure the temperature of the windings during operation.

Pressure-gage Thermometers. These instruments are widely used where distant reading, recording, and control units are desired and are manufactured in a great variety of designs for special applications. They consist essentially of a metal bulb coupled to a pressure gage through a capillary tube. Where very distant readings are required, electrical transmitters are employed. In the vapor-pressure type the bulb contains a liquid which has a measurable vapor pressure in the temperature range for which the thermometer was designed. In the fluid-expansion type the bulb, capillary tube, and gage are filled with a liquid or gas. The vapor-pressure thermometer, which is unaffected by slight leakage of the vapor, has a non-uniform scale, more open at the higher end. The fluid-expansion thermometer, which is affected by any leakage, has a uniform scale.

Bimetallic Thermometers. Material improvement in the accuracy of bimetallic thermometers has been achieved recently by improvement in design and the development of materials used in these instruments. The bimetal strip, formed into a spiral or helix, is fixed at one end, and the pointer is attached directly to the other end, thus obviating mechanical friction.

Checking Thermometers. The graduation errors of a thermometer can best be determined by comparing it with a secondary standard mercury-in-glass thermometer which has been tested and certified by a competent laboratory, such as that of the National Bureau of Standards in Washington. Only the zero point need be checked subsequently on mercury-in-glass thermometers. This can be done by noting the reading of the thermometer after total immersion for several minutes in a mixture of crushed ice and distilled water. Pressure-gage thermometers, however, may need to be recalibrated frequently over the working range.

Precision and Range of Measurement of Types of Thermometers. Typical figures, obtainable with good technique and intended as a guide in the selection of suitable types, are given in Table 11.

Table 11. Precision and Range of Measurement of Types of Thermometers

Type of Thermometer	Range, deg cent	Precision, deg cent	
		Best Grade	Ordinary
Mercury			
Soft glass	0 to 100	0.02	0.5
	− 35 to 300	0.2	2
Hard glass	0 to 550	0.5	5
Alcohol	−100 to 40	2	5
Electrical resistance	−100 to 100	0.002	0.05
	−200 to 500	0.02	0.5
Thermoelectric	−200 to 300	0.1	0.5
	300 to 1200	0.3	2 to 5
Bimetallic	− 50 to 500	0.5 to 1% of range	
Pressure gage	−100 to 500	1 to 2% of range	

Location of Thermometers. In locating a thermometer, the fact must be realized that it indicates its own temperature, which may or may not be that of its surroundings. To indicate the temperature of its surroundings, a thermometer must be in good thermal contact with them; it must not be exposed to radiation from hotter bodies nor able to radiate to cooler ones. In short, it must not "see" anything which is at a different temperature. Equal precaution should be taken to prevent loss of heat by conduction or convection.

PYROMETERS. General. A pyrometer is a device intended primarily for measuring temperatures beyond the range of a mercury-in-glass thermometer, say above 500 deg cent. Certain types of pyrometers are frequently found useful in measuring lower temperatures also.

Numerous types of pyrometers, now of historical interest only, are treated in detail in Burgess and Le Chatelier's *The Measurement of High Temperatures,* Wiley (1912). This section is restricted to current pyrometric data and devices.

Thermoelectric Pyrometers. The most widely used pyrometric device consists of a thermocouple and an instrument capable of indicating or recording electromotive force. Although the standard thermocouple, one wire platinum and the other wire an alloy consisting of 10 per cent rhodium and 90 per cent platinum, is used in defining the temperature scale, the majority of couples in actual use are of less expensive materials ("base metals"). The commonly used thermocouples and their ranges of usefulness are given in Table 12. Constantan is also sold under a number of trade names, such as Ia Ia, Ideal, Advance, Copel, Eureka, Cupron, Copnic.

Table 12. Range of Usefulness of Thermocouples

Type of Thermocouple	Useful Range, deg cent	
	For Long-time Service	* For Short-time Service
Platinum vs. 90% platinum—10% rhodium..	0 to 1450	1700
" " 87% " —13% "	0 to 1450	1700
Chromel vs. Alumel......................	−200 to 1100	1350
Iron vs. constantan......................	−200 to 800	1000
Chromel vs. constantan..................	−200 to 800	1000
Copper vs. constantan....................	−200 to 350	600

* Thermocouples which have been subjected to these temperatures cannot be depended upon to retain their calibrations and should not be used for accurate work subsequently without being recalibrated.

Temperature-emf Relations for Various Thermocouples. In Table 13 are given representative calibration data for couples of the six types commonly used. These data may be used to obtain approximate temperatures, good to the equivalent of about 0.5 per cent in emf, except for couples having constantan as one element. For more accurate work each individual couple must be calibrated. Values of thermal emf for couples having

Table 13. Representative Data on Thermocouples with Reference Junction at 0 deg cent

Temperature, deg cent	Emf, Millivolts					
	Platinum to Platinum—10% Rhodium *	Platinum to Platinum—13% Rhodium *	Copper to Constantan †	Iron to Constantan †	Chromel P to Alumel ‡	Chromel P to Constantan §
0	0.mv	0.mv	0.mv	0.mv	0.mv	0.mv
25	0.144	0.143	0.990	1.32	1.00	1.50
50	0.299	0.297	2.034	2.66	2.02	3.04
100	0.643	0.646	4.276	5.40	4.10	6.32
200	1.436	1.464	9.285	10.99	8.13	13.42
300	2.316	2.394	14.859	16.56	12.21	21.03
400	3.251	3.398	20.865	22.07	16.39	28.94
500	4.219	4.454		27.58	20.64	36.99
600	5.222	5.561		33.27	24.90	45.09
700	6.260	6.720		39.30	29.14	53.14
800	7.330	7.927		45.72	33.31	61.09
900	8.434	9.177		52.29	37.36	68.86
1000	9.569	10.470		58.22	41.31	76.44
1100	10.736	11.811			45.14	
1200	11.924	13.181			48.85	
1300	13.120	14.563			52.41	
1400	14.312	15.940			55.81	
1500	15.498	17.316				
1600	16.674	18.680				
1700	17.841	20.032				

* *Nat. Bur. Stand. Res. Paper* 530.
† *Nat. Bur. Stand. Res. Paper* 1080.
‡ *Nat. Bur. Stand. Res. Paper* 767.
§ General Electric Company.

constantan as one element may differ by as much as 2 per cent from those in the table. However, most thermocouple manufacturers take pains to supply a uniform quality of constantan, and the calibrations of individual couples by a reputable maker can be depended to agree with the curve which he issues within about 0.5 per cent in emf.

Cold-junction Corrections. The data in Table 13 give the relation between the temperature of the measuring junction and the emf of the couple, when the reference junction is at 0 deg cent. In many measurements the reference junction is not at 0 deg cent; for example, when the thermocouple wires are connected directly to the measuring instrument, the temperature of the reference junction is that of the instrument. Although the table is based on a reference-junction temperature of 0 deg cent, it may be used to correct **any** observed emf to that which would have been observed if the reference junction had been at 0 deg cent. For example, if the observed emf of a Chromel P-constantan couple is 27.44 mv with the reference junction at 25 deg cent, the table shows that the emf would have been 1.50 mv greater, i.e., 28.94 mv, if the reference junction had been at 0 deg cent. The temperature of the measuring junction was accordingly that corresponding to 28.94 mv in the table, i.e., 400 deg cent. It can be seen that an incorrect value is obtained when one adds the cold-junction temperature, 25 deg cent, to 381 deg cent, the value interpolated from the table, corresponding to the measured emf, 27.44 mv. If the indicator is graduated in temperature, so that the emf is not observed, correction for the reference-junction temperature may be made by manually setting the pointer, on open circuit, to indicate the reference-junction temperature.

Cold-junction Compensators. Various devices are used to avoid the necessity for making the calculations described in the preceding paragraphs. Simple indicators may be set on open circuit to read the emf corresponding to the cold-junction temperature, or the cold-junction temperature itself, depending upon the way the scale is graduated. In many deflection-type instruments this setting is done automatically by means of a bimetallic spiral which moves the outer end of one of the control springs of the moving coil as the temperature of the instrument changes. Potentiometric instruments may be equipped with either manually or automatically operated cold-junction compensators. Evidently, the automatic compensators are effective only if the temperature of the cold junction is the same as that of the instrument. The cold junction may be brought to the instrument by the use of extension leads, which for base-metal couples are of the same materials as the couple. For rare-metal couples the extension leads consist of one wire of copper and one of a suitable alloy of copper and nickel. The extension leads may be carried directly to the indicator, or they may be brought to some point at constant temperature, for example, the bottom of a pipe driven about 10 ft into the ground. When the cold junction is not at the instrument, the compensator setting must correspond to the temperature of the cold junction.

Measurement of Emf. The instruments used to indicate or record the emf of thermocouples may be classed as potentiometers (null instruments) and millivoltmeters (galvanometers). Potentiometers are of two general types: (1) those using a galvanometer to indicate balance, and (2) those employing electron tubes without a galvanometer. The second type is self-balancing. Instruments using a combination of the galvanometric and potentiometric principles have been devised but are seldom used.

Since the emf's are small, the **galvanometers** must be very sensitive to voltage. This could be accomplished by making the galvanometer coil of low resistance, but on account of the variable line resistance of a thermoelectric circuit it is necessary to make the resistance of the galvanometer as high as possible, consistent with substantial construction. A large part of the resistance should be that of a manganin series resistor in order that the indicated emf may be nearly independent of the temperature of the instrument. If the indicator is graduated to read emf at its terminals, the relation between scale reading e_0 and true emf of the couple e becomes

$$e_0 = \frac{R_g e}{(R_g + R_c + R_l)} \tag{57}$$

where R_g, R_c, and R_l denote the resistances of the galvanometer, couple, and lead wires, respectively. If R_g is large compared with $R_c + R_l$, equation (57) reduces to $e_0 = e$ approximately, and the indicator reads correctly. When R_g is low, it is possible to graduate the scale to read true emf for a definite line resistance; but, if this resistance changes on account of deterioration of the couple or for any other reason, serious error may result. Thus a 300-ohm indicator designed for use with a 2-ohm line resistance will be in error by only 7 deg at 1000 deg cent if the line resistance changes to 4 ohms, whereas a 10-ohm indicator under similar conditions will read 140 deg cent too low. This emphasizes the importance of using galvanometers of high resistance. Most pyrometer manufacturers can supply galvanometers (millivoltmeters) with a resistance of several hundred ohms.

The advantages of the galvanometer method are quick reading, simplicity, and moderate initial cost. The disadvantages lie largely in the effects due to thermocouple and lead resistance which have been mentioned.

The most accurate instrument for measuring thermoelectric emf's is the **potentiometer.** Portable potentiometers are available for either rare-metal or base-metal couples, or for both with a double scale. There are several important advantages in the potentiometer method. The emf or temperature scale is easily made very open, thus permitting accurate readings. The calibration of the scale is in no way dependent upon the constancy of magnets, springs, or jewel bearings or upon the level of the instrument. From the pyrometric standpoint, however, the greatest advantage is the complete elimination of any error due to ordinary changes in the resistance of the couple or of the lead wires. Thus, if a potentiometer is correctly balanced and the resistance of the thermocouple is then greatly increased, the balance is unchanged. However, excessive resistance in the couple circuit reduces the sensitivity of a setting. The objections to the potentiometer are the greater initial cost and the fact that usually a manual adjustment must be made to obtain a setting.

Thermocouples for Special Uses. Many special types of thermocouple adapted for special uses are available. One is the pipe-type couple, in which one element is a pipe or tube of some metal such as iron or chromel, and the other element is a wire passing through the pipe and insulated from it, except at the closed end, where it is welded to the pipe to form the hot junction. There are a number of special types of couple designed for measuring the temperature of molten metals and alloys.

Pyrometer Protection Tubes. In the measurement of high temperatures the choice of a protection tube is nearly as important as the selection of the thermocouple. All thermocouples should be protected from reducing gases, sulfur, and metallic vapors. Only porcelain, quartz, and Pyrex-glass protection tubes should be used for platinum-rhodium thermocouples. Pyrex glass is usable up to 600 deg cent, quartz (fused silica) up to 1000 deg cent, and American-made porcelain (mullite) up to 1550 deg cent. In many cases it is advisable to protect these tubes by an outside or secondary protection tube made of materials such as chromel, nichrome, carborundum, graphite, alundum, nickel, iron, or calorized iron. All these tubes have their special uses. Base-metal thermocouples are ordinarily protected by metal protection tubes, but under severe conditions porcelain tubes are used alone or in combination with a metal tube. The thermocouple wires are usually insulated by two-hole porcelain insulating tubes.

Automatic Recorders. The types of pyrometers ordinarily used in the automatic recording of temperatures are (1) resistance thermometers, (2) radiation pyrometers, and (3) thermoelectric pyrometers. Of these the last has the greatest applicability. The type of curve ordinarily required is temperature versus time, in which case the instrument is equipped with a mechanism for periodically recording its indications upon a chart which moves with a uniform speed. Single- or multiple-point recorders operating on either the galvanometric or potentiometric principle are available. Multiple-point recorders are equipped with a commutator switch, which automatically connects various thermocouples into the circuit successively.

High-speed electronic self-balancing potentiometers and recorders are now available. These instruments contain no galvanometers and are not affected by vibration. Recorders and indicators can be made to maintain automatically a prescribed temperature for industrial process control or for laboratory research purposes and can be equipped to operate lights, high and low alarms, or remote indicators.

Standard Samples for Checking Thermocouple Pyrometers. In order to enable the user of thermocouple pyrometers to check their indications, the National Bureau of Standards supplies standard samples of metals of certified melting points. Information concerning the samples which could be purchased from the Bureau in May, 1944, is given in Table 14.

Table 14. Standard Samples of Metals for Checking Thermocouple Pyrometers

Sample Number	Metal	Melting Point		Weight of Sample in Grams	Price per Sample
		deg cent	deg fahr		
42d	Tin	231.90	449.42	350	$2.00
43e	Zinc	419.50	787.10	350	2.00
44c	Aluminum	660.15	1220.3	200	2.00
45b	Copper	1083	1981.4	450	2.00
49b	Lead	327.40	621.32	1650	2.00

Radiation Pyrometry and Optical Pyrometry. Radiation pyrometry and optical pyrometry are based on the intensity of total radiation and of visible radiation from a hot body, respectively. Since different substances at the same temperature do not necessarily emit radiation of the same intensity or spectral distribution, this fact must be taken into account in the use of either radiation pyrometers or optical pyrometers. There is one form of radiator, the so-called black body, the radiation from which depends only on its temperature. A uniformly heated enclosure with a small opening, or even a furnace with not too large an opening, constitutes a good approximation to a black body. If the object the temperature of which is to be measured is inside such a black body, or if the radiation comes from a deep cavity in the object, the radiation will depend only on the temperature.

Leeds & Northrup Rayotube. Two types of Rayotubes are in use. In both the sensitive element is a multiple-junction thermopile with a small receiving disk. In one the radiation is focused on the disk by a stainless steel mirror, which images a front diaphragm opening on the disk, not completely covering the disk. It is adaptable for measurement of temperatures as low as 300 deg fahr. In the other a fused silica lens images the source on the disk, which must be completely covered. The output of the thermopile is measured potentiometrically and is either indicated or recorded on a direct-reading temperature scale.

Brown Radiamatic. This instrument uses a fused silica or a Pyrex lens to focus the radiation on the blackened hot junctions of a thermopile. No receiving disk is used. It is compensated to reduce the errors due to changes in ambient temperature.

Thwing Pyrometer. In the Thwing pyrometer the reflecting mirror is replaced by an aluminum cone which by multiple reflections concentrates the radiation at its apex on one or more small thermocouples in series with a portable galvanometer. The instrument requires no focusing, the front diaphragm acting as a source. The object sighted upon must be large enough to cover the projection of the cone through this diaphragm.

Pyro Radiation Pyrometer. In the Pyro radiation pyrometer the radiation is focused by means of a quartz objective lens upon a thermocouple mounted in an evacuated bulb. A "receiving disk" fixed over the thermojunction can be seen in the field of view and must be fully covered by the image of the radiating source. The instrument is very compact and direct reading, the millivoltmeter being mounted in the telescope itself and graduated in temperature directly.

The chamber containing the sensitive element is usually sealed against dust, fumes, and moisture. Exposed lens or window surfaces must be kept clean and free from dust. Sufficient time must be allowed for the indication to reach a maximum after exposure to the radiation. This time may be as short as 1.5 sec or as long as 1 min, depending upon the type of sensitive element used, but for the instruments in most frequent use the time will fall between 3 and 10 sec. Care must be taken that the source is large enough to "fill" the aperture of the pyrometer completely and that no obstructions cut into the cone of radiation from the source which enters the pyrometer. If sighted into a closed-end tube, care must be taken that the image of the closed end covers the thermopile disk and is focused upon it. The atmosphere sighted through, either in a furnace or in a closed-end tube, should be free of smoke or fumes. Considerable errors may be introduced by sudden changes in ambient temperature. The effect of slow changes in ambient temperature of moderate amplitude is usually negligible. If sudden changes in ambient temperature must be guarded against, the effect can be made vanishingly small by suitable compensation.

Large corrections must be applied to temperature readings made on non-black bodies. These corrections are not well determined and are dependent on the surface condition and the surroundings of the hot body. A usual practice is to measure the temperature of the body with an optical pyrometer, for which the corrections are smaller and better known, and to adjust the pyrometer or its circuit to make its reading agree with the true temperature thus determined. Its subsequent readings will then be close to the true temperature.

Optical Pyrometry. An optical pyrometer is essentially a photometer for comparing the brightness of the red light (λ = approximately 0.65μ) radiated by a body with that radiated by a comparison source, such as an electric lamp. Temperatures on the international scale above 1063 deg cent are defined in terms of Wien's law of radiation in the following form:

$$\log \frac{J_2}{J_1} = \frac{c_2 \log_e}{\lambda} \left(\frac{1}{1336} - \frac{1}{T} \right)$$

where J_2/J_1 is the ratio of brightness at wavelength λ between a black body at the temperature T deg K and a black body at 1336 deg K, the melting point of gold.

Leeds & Northrup Optical Pyrometer. The filament F of a small electric lamp (see Fig. 62) is placed at the focus of the objective lens L, which superimposes upon the lamp filament an image of the source to be measured. This image and filament are viewed through the eyepiece R, which is equipped with a red glass screen. In making a setting the current through the lamp is adjusted until the filament and source appear equal in brightness.

The calibration of the lamp will remain remarkably constant if it is not operated above 1500 deg cent. For higher temperatures an absorption screen is provided between the lamp and objective, which may be thrown in or out of the field of view by a lever. This screen reduces the brightness of the source, so that any temperature up to 1800 deg cent can be matched by the lamp at 1400 deg cent or less. For higher temperatures a second absorption screen is used in combination with the first. Without the absorption screen the calibration gives a relation between lamp current and temperature. With the absorption screen calibration gives a second relation between lamp current and temperature.

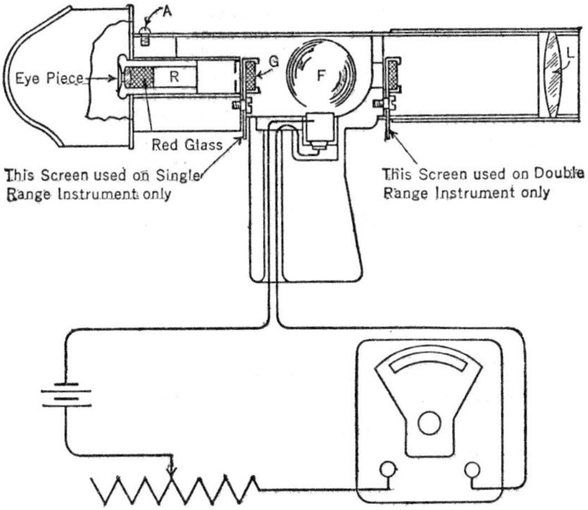

Fig. 62

The Leeds & Northrup Company has replaced the optical pyrometer, as diagrammed in Fig. 62, by another which is optically analogous to it, but in which the filament current is measured by means of a potentiometer having a scale graduated in degrees fahrenheit or centigrade. This affords a much more open scale, and temperatures can be read more accurately. The pyrometer lamp has also been improved by providing a bulb with flat windows and using a fine tungsten ribbon for a filament.

The **Pyro optical pyrometer** employs an absorption wedge to vary the image of the source and match it against the pyrometer lamp, the current in which is kept constant at some "normal" value. The wedge carries a pointer moving over a scale graduated directly in temperature. This instrument is made with two ranges, the same wedge being used alone or in combination with a fixed absorption glass.

Emissivity Corrections for Optical Pyrometers. Optical pyrometers indicate the true temperature of black bodies only. A peephole in a furnace or a porcelain tube with the closed end immersed in molten metal or otherwise surrounded by the temperature to be measured may be considered a black body. When an optical pyrometer is sighted upon objects in the open, corrections must be made in accordance with Table 15.

Table 15. Correction to Observed Temperature, Using Red Light

Emissivity	Add Corrections Below for the Following Observed Temperatures, deg cent						
	800	1000	1200	1400	1600	1800	2000
0.30	67	95	129	168	213	264	322
0.40	50	71	96	125	158	195	237
0.50	37	53	71	93	117	144	175
0.60	27	39	52	67	85	104	126
0.70	19	27	36	47	59	72	87
0.80	12	17	22	29	36	44	54
0.90	6	8	10	14	17	21	25
1.00	0	0	0	0	0	0	0

Values of the emissivity of a few of the more common materials are given in Table 16.

Table 16. Monochromatic Emissivity, E_λ, for Red Light

($\lambda = 0.65\mu$)

Material	Monochromatic Emissivity E_λ	Material	Monochromatic Emissivity E_λ
Silver	0.07	Nichrome, 900° C	0.90
Gold, solid	0.14	Nichrome, 1200° C	0.80
Gold, liquid	0.22	Cuprous oxide	0.70
Platinum, solid	0.33	Iron oxide, 800° C	0.98
Platinum, liquid	0.38	Iron oxide, 1000° C	0.95
Palladium, solid	0.33	Iron oxide, 1200° C	0.92
Palladium, liquid	0.37	Nickel oxide, 800° C	0.96
Copper, solid	0.10	Nickel oxide, 1300° C	0.85
Copper, liquid	0.15	Nickel, solid and liquid	0.36
Tantalum, 1100° C	0.60	Iridium	0.30
Tantalum, 2600° C	0.48	Rhodium	0.24
Tungsten, 1000° C	0.46	Graphite powder (estimated)	0.95
Tungsten, 2000° C	0.43	Carbon	0.85
Tungsten, 3000° C	0.41	Porcelain	0.25 to 0.50
Nichrome, 600° C	0.95		

26. BIBLIOGRAPHY

The following general sources may be consulted in addition to those specifically listed: *Science Abstracts* (Section A, Physics; Section B, Electrical Engineering); *Engineering Index; Industrial Arts Index;* and the List of Publications of the National Bureau of Standards.

Textbooks in the English Language

NOTE: The abbreviations Benn, Chapman, Griffin, Macmillan, McGraw, Pitman, Van Nostrand, and Wiley denote the following publishers: Ernest Benn, Ltd., London; Chapman & Hall, Ltd., London; C. Griffin & Co., Ltd., London; Macmillan & Co., Ltd., London; McGraw-Hill Book Co., Inc., New York; Sir Isaac Pitman & Sons, Ltd., London; D. Van Nostrand Co., New York; and John Wiley & Sons, Inc., New York.

Archer, R. M., *Commercial Electrical Measuring Instruments.* Pitman (1928).

Curtis, H. L., *Electrical Measurements*, McGraw (1937).

Drysdale, C. V., and A. C. Jolley, *Electrical Measuring Instruments*, 2 vols. Part I, *Commercial and Indicating Instruments;* Part II, *Induction Instruments, Supply Meters and Auxiliary Apparatus.* Benn (1924).

Edgcumbe, K., and F. E. J. Ockenden, *Industrial Electrical Measuring Instruments* (based on Edgecumbe's book of same name but entirely rewritten and enlarged). Pitman (1933).

Glazebrook, *Dictionary of Applied Physics*, Vol. II, Electricity. Macmillan (1922).

Golding, E. W., *Electrical Measurements and Measuring Instruments.* Pitman (1936). (Revised in 1942 as third edition.)

Hund, August, *High-frequency Measurements.* McGraw (1933).

Hague, B., *Alternating-current Bridge Methods*, 5th ed. Pitman (1943).

Karapetoff, V., *Experimental Electrical Engineering and Manual of Electrical Testing*, Vol. I, 4th ed. (1933); Vol. II, 3rd ed. (1922). Wiley.

Kerchner, R. M., and G. F. Corcoran, *Alternating Current Circuits*, Wiley, New York, and Chapman & Hall, Ltd., London (1938).

Laws, F. A., *Electrical Measurements*, 2nd ed. McGraw.

Michels, W. C., *Advanced Electrical Measurements*, 2nd ed. Van Nostrand (1941).

Moullin, E. B., *Radio-frequency Measurements*, 2nd ed. Griffin (1931).

Pyrometer Symposium, Chicago (a series of papers in book form). American Institute of Mining and Metallurgical Engineers, New York, 1920.

Smith, Arthur W., *Electrical Measurements in Theory and Application*, 2nd ed. McGraw (1924).

Stubbings, G. W., *Commercial A-C Measurements.* Chapman (1930).

Textbooks in Other Than the English Language

NOTE: The abbreviations de Gruyter, Leiner, Oldenbourg, and Springer denote the following publishers: Walter de Gruyter & Co., Berlin and Leipzig; Oskar Leiner, Leipzig; R. Oldenbourg, Munich and Berlin; and J. Springer, Berlin.

Brion, G., and V. Vieweg, *Starkstrommesstechnik.* Springer (1933).

Brückmann, H. W. L., *Elektrizitätszähler und Wandler*, 2nd ed. Leiner (1926).

Fabry, C., Introduction générale à la photométrie, *Revue d'Optique*, Paris (1927).

Goldstein, I., *Die Messwandler, ihre Theorie und Praxis.* Springer (1928).

Hund, A., *Hochfrequenzmesstechnik.* Springer (1928).

Keinath, G., *Elektrische Temperaturmessgeräte.* Oldenbourg (1933).

Keinath, G., *Die Technik der elektrischen Messgeräte*, 3rd ed., 2 vols. Bd. I, *Messgeräte und Zubehör;* Bd. II, *Messverfahren.* Oldenbourg (1928).

Krukowski, W. von, *Grundzüge der Zählertechnik.* Springer (1930).

Ribaud, G., Traité de pyrometrie optique, *Revue d'Optique Théorique et Instrumentale*, Paris (1931).

Ulbricht, R., *Das Kugelphotometer.* Oldenbourg (1920).

Werner, O., *Empfindliche Galvanometer für Gleich- und Wechselstrom.* de Gruyter (1928).

Galvanometers

Brooks, H. B., Sensitivity of a galvanometer as a function of its resistance. *Bur. Standards J. Research*, Vol. 7, p. 289 (1931), reprinted as *Research Paper* 342.
Coley, W. R., An all-copper d'Arsonval galvanometer with small thermal effect. *J. Optical Soc. Am.*, Vol. 11, p. 419 (1925).
Hartsough, R. C., Shielding from vibrations, *Phys. Rev.*, 2nd series, Vol. 29, p. 910 (1927).
Ising, G., A natural limit for the sensibility of galvanometers, *Phil. Mag.*, Vol. 1, p. 827 (1926). (Brownian movements.)
Lamson, H. W., An electronic null detector for impedance bridges, *Rev. Sci. Instruments*, Vol. 9, No. 9 (September, 1938).
Leeds & Northrup Company, Notes on moving-coil galvanometers, Note Book 2 (1925).
Pestarini, Galvanometer for alternating currents, *Electrician* (London), Vol. 80, p. 154 (1917).
Smith, F. E., article on galvanometers in Glazebrook's *Dictionary of Applied Physics*, Vol. II, p. 366.
Stewart, O. N., The damped ballistic galvanometer, *Phys. Rev.*, Vol. 16, p. 158 (1903).
Weibel, E., A study of electromagnet moving-coil galvanometers for use in a-c measurements, *Bull. Bur. Standards*, Vol. 14, p. 23 (1918), reprinted as *Scientific Paper* 297.
Wenner, F., Theoretical and experimental study of the vibration galvanometer, *Bull. Bur. Standards*, Vol. 6, p. 347 (1910), reprinted as *Scientific Paper* 134. Characteristics and applications of vibration galvanometers, *Trans. AIEE*, Vol. 31, p. 1243 (1912). General design of critically damped galvanometers, *Bull. Bur. Standards*, Vol. 13, p. 211 (1916), reprinted as *Scientific Paper* 273. Methods of varying the sensitivity of galvanometers, *J. Optical Soc. and Rev. Sci. Instruments*, Vol. 11, p. 495 (1925).
White, W. P., Sensitive moving-coil galvanometers, *Phys. Rev.*, Vol. 19, p. 305 (1904). Some properties of the moving-coil galvanometer, *Phys. Rev.*, Vol. 22, p. 371 (1906). Everyday problems of the moving-coil galvanometer, *Phys. Rev.*, Vol. 23, p. 382 (1906).
Zeleny, A., Precision measurements with the moving-coil ballistic galvanometer, *Phys. Rev.*, Vol. 23, p. 399 (1906).

Standard Cells

Eppley, Marion, Standard electrical cells, *J. Franklin Inst.*, Vol. 201, p. 17 (1926).
Eppley, Marion, International standard of electromotive force and its low-temperature-coefficient form, *Trans. AIEE*, Vol. 50, p. 1293 (1931).
Park, J. H., Effect of service temperature conditions on the electromotive force of unsaturated portable standard cells. *Bur. Standards J. Research*, Vol. 10, p. 89 (1933), reprinted as *Research Paper* 518.
Shaw, Reilley, and Clark, The aging of standard cells, *Phil. Trans.* A229, p. 125 (1929–30).
Vinal, G. W., Maintenance of the volt, *Trans. Am. Electrochem. Soc.*, Vol. 54, p. 255 (1928).
Vinal, G. W., D. N. Craig, and L. H. Brickwedde, Standards of electromotive force, *Trans. Am. Electrochem. Soc.*, Vol. 68, p. 139 (1935).
Vosburgh, Warren C., and Marion Eppley, The temperature coefficients of unsaturated Weston cells *J. Am. Chem. Soc.*, Vol. 45, p. 2268 (1923). Portable unsaturated Weston cells, *J. Optical Soc. Am.*, Vol. 9, p. 65 (1924).
Vosburgh, Warren C., Conditions affecting the reproducibility and constancy of Weston standard cells. *J. Am. Chem. Soc.*, Vol. 47, p. 1255 (1925). The decrease in the electromotive force of unsaturated cells, *J. Optical Soc. Am.*, Vol. 11, p. 59 (1925). Eppley unsaturated standard cells at high temperature, *J. Optical Soc. Am.*, Vol. 12, p. 393 (1926).

Time Standards and Measurements

Brown, Ernest W., Time and its measurement, *Trans. AIEE*, Vol. 51, p. 523 (1932).
Ellis, A. L., The limitations of the stopwatch as a precision instrument, *Trans. AIEE*, Vol. 40, p. 479 (1921).
Loomis, Alfred L., and W. A. Marrison, Modern developments in precision clocks, *Trans. AIEE*, Vol. 51, p. 527 (1932). Contains bibliography of 24 entries.
Warren, H. E., Utilizing the time characteristics of alternating current, *Trans. AIEE*, Vol. 38, p. 767 (1919).
Warren, H. E., A new time standard, *Trans. AIEE*, Vol. 59, pp. 137–141 (March, 1940).
Wood, A. B., and J. M. Ford, The phonic chronometer, *J. Sci. Instruments* (London), Vol. 1, p. 161 (1924).

Resistance Standards and Other Resistance Apparatus

Cabras, A., Mathematical theory of the Wheatstone bridge, *Rev. Gén. de l'Elec.*, Vol. 26, p. 959 (1929). Deals with the most general case, in which the bridge arms carry currents of variable intensity and of any form, the arms being any complex impedances.
Callendar, H. L., Conditions of sensitiveness (in Wheatstone-bridge measurements), *Proc. Phys. Soc. London*, Vol. 22, p. 221 (1909–10).
Dutch Testing Service for Electrotechnical Materials, Application of the Thomson double bridge for alternating currents, *Bull. Schweiz. Elek. Verein*, Vol. 19, p. 784 (1928).
Eppley, Marion, Modifications of the resistance thermometer bridge (Mueller bridge) and of the commutator selector for use with it, *Rev. Sci. Instruments*, Vol. 3, p. 687 (1932). See also under Measurement of Large Direct Currents.
Jaeger, W., Comparative discussion of the sensitivity of various methods of measuring resistance, *Zeitschr. für Instrumentenkunde*, Vol. 26, pp. 69, 360 (1906).
Leeds & Northrup Company, Notes on the Kelvin bridge, Note Book 4 (1929).
Manley, J. J., Lubrication of Resistance-Box Plugs, *Phil. Mag.*, 6th series, Vol. 33, p. 211 (1917).
Mueller, E. F., and Frank Wenner, The Waidner-Wolf and other adjustable electrical resistance elements, *J. Res. Nat. Bur. Standards*, Vol. 15, p. 477, (1935), reprinted as *Research Paper* 842.
Northrup, *Methods of Measuring Electrical Resistance*. McGraw (1912).
Rayner, E. H., Precision resistance measurements with simple apparatus, *Proc. Phys. Soc. London*, Vol. 27, p. 384 (1914–15), abstracted in *Electrician* (London), Vol. 75, p. 286 (1915).
Rose, G. M., Jr., Method for measuring very high values of resistance, *Rev. Sci. Instruments*, Vol. 2, p. 810 (1931). Uses a pliotron; measures up to about 10^{17} ohms.
Schuster, A., On the measurement of resistance, *Phil. Mag.*, 5th series, Vol. 39, p. 175 (1895).
Searle, G. F. C., On resistances with current and potential terminals, *Electrician* (London), Vol. 66, pp. 999, 1029 (1911); Vol. 67, pp. 12, 54 (1911).
Silsbee, F. B., Notes on the design of 4-terminal resistance standards for alternating currents, *Bur. Standards J. Research*, Vol. 4, p. 73 (1930), reprinted as *Research Paper* 133.

Smith, F. E., On methods of high precision for the comparison of resistances. *Electrician* (London), Vol. 57, pp. 976, 1009 (1906).
Thomas, J. L., A new design of precision resistance standard, *Bur. Standards J. Research*, Vol. 5, p. 295 (1930), reprinted as *Research Paper* 201.
Thomas, J. L., Gold-chromium resistance alloys, *J. Res. Nat. Bur. Standards*, Vol. 13, p. 681 (1934), reprinted as *Research Paper* 737.
Thomas, J. L., Gold-cobalt resistance alloys, *J. Res. Nat. Bur. Standards*, Vol. 14, p. 589 (1935), reprinted as *Research Paper* 789.
Thomas, J. L., Electrical resistance alloys of copper, manganin, and aluminum, *J. Res. Nat. Bur. Standards*, Vol. 16, p. 149 (1936), reprinted as *Research Paper* 863.
Wenner, F., The four-terminal conductor and the Thomson bridge, *Bull. Bur. Standards*, Vol. 8, p. 559 (1912), reprinted as *Scientific Paper* 181.
Wenner, F., and E. Weibel, Adjustments of the Thomson bridge in the measurement of very low resistances, *Bull. Bur. Standards*, Vol. 11, p. 65 (1915), reprinted as *Scientific Paper* 225.
Wenner, F., and A. Smith, Measurement of low resistance by means of the Wheatstone bridge, *Scientific Papers Bur. Standards*, Vol. 19, p. 297 (1923–24), reprinted as *Scientific Paper* 481.
Wenner, F., Methods, apparatus, and procedures for the comparison of precision standard resistors, *J. Res. Nat. Bur. Standards*, Vol. 25, p. 229 (1940), reprinted as *Research Paper* 1323.
Whitehead, J. B., Measurement of high values of insulation resistance, *Elec. World*, Vol. 82, p. 1007 (1923).

D-c Potentiometers

Behr, L., Multiple-range potentiometers, *J. Optical Soc. Am.*, Vol. 7, p. 665 (1923); The Wenner potentiometer, *Rev. Sci. Instruments*, Vol. 3 (new series), p. 109 (1932).
Brooks, H. B., papers on deflection potentiometers, *Bull. Bur. Standards*, Vol. 2, p. 225 (1906); Vol. 4, p. 275 (1908); Vol. 8, pp. 395, 419 (1912); reprinted as *Bur. Standards Scientific Papers* 33, 79, 172, and 173, respectively.
Brooks, H. B., Les potentiomètres, *Comptes Rendus du Congrès International d' Électricité*, Paris, Vol. 3, p. 275 (1932). Contains concise history of the potentiometer from its inception by Poggendorff, with selected bibliography covering the entire period.
Brooks, H. B., and A. W. Spinks, Multirange potentiometer and its application to the measurement of small temperature differences, *Bur. Standards J. Research*, Vol. 9, p. 781 (1932), reprinted as *Research Paper* 506.
Brooks, H. B., The standard-cell comparator, a specialized potentiometer, *Bur. Standards J. Research*, Vol. 11, p. 211 (1933), reprinted as *Research Paper* 586.
Diesselhorst, H., Five-dial potentiometer of constant low resistance, free from thermal emf, *Zeitschr. für Instrumentenkunde*, Vol. 28, p. 1 (1908).
Eppley, Marion, and W. R. Gray, A simple direct-reading potentiometer for standard-cell comparisons, *J. Optical Soc. and Rev. Sci. Instruments*, Vol. 6, p. 859 (1922); An improved Feussner-type potentiometer, *Rev. Sci. Instruments*, Vol. 2 (new series), p. 242 (1931).
Silsbee, F. B., and F. J. Gross, Testing and performance of volt boxes, *J. Res. Nat. Bur. Standards*, Vol. 27, p. 269 (1941), reprinted as *Research Paper* 1419.
Stein, I. Melville, Design of potentiometers, *Trans. AIEE*, Vol. 50, p. 1302 (1931).
Teele, R. P., and Shuford Schulmann, Potentiometer for measuring voltage of 10 microvolts to accuracy of 0.01 microvolts; *J. Res. Nat. Bur. Standards* Vol. 22, p. 431 (1939), reprinted as *Research Paper* 1195.
Wenner, F., and E. Weibel, The testing of potentiometers, *Bull. Bur. Standards*, Vol. 11, p. 1 (1914), reprinted as *Scientific Paper* 223.
White, W. P., papers on potentiometers, *Zeitschr. für Instrumentenkunde*, Vol. 27, p. 210 (1907); Vol. 34, pp. 71, 107, 142 (1914); Papers on thermoelements, potentiometers for use with them, and thermoelement installations, *Phys. Rev.*, Vol. 25, p. 334 (1907); *J. Am. Chem. Soc.*, Vol. 36, pp. 1856, 1868. 2011, 2292 (1914); *Am. Inst. Mining and Metallur. Engrs.*, *Bull.* 153, p. 1763 (1919).
Wolff, Otto, New form of Feussner potentiometer, *Zeitschr. für Instrumentenkunde*, Vol. 21, p. 227 (1901).

A-c Potentiometers

Brooks, H. B., Les potentiomètres, *Comptes Rendus du Congrès International d' Électricité*, Paris, Vol. 3, p. 275 (1932).
Campbell, A., New a-c potentiometers of Larsen type, *Proc. Phys. Soc. London*, Vol. 41, p. 94 (1928–29).
Drysdale, C. V., Use of the potentiometer on a-c circuits, *Proc. Phys. Soc. London*, Vol. 21, p. 561 (1909); *Electrician*, Vol. 62, p. 723 (1909); *Phil. Mag.*, Vol. 17, p. 402 (1909).
Erlang, A. K., New a-c compensation apparatus for telephonic frequencies, *J. Inst. Elec. Engrs.* (London), Vol. 51, p. 794 (1913).
Gall, D. C., A new a-c potentiometer, *Electrician* (London), Vol. 90, p. 360 (1923).
Gall, D. C., A new a-c potential comparator, *J. Sci. Instruments* (London), Vol. 9, p. 132 (1932). Testing nickel-iron alloy by means of the a-c potentiometer, *ibid.*, p. 219.
Geyger, W., Various papers on a-c potentiometers, *Archiv für Elektrotechnik*, Vol. 13, p. 80 (1924); Vol. 14, p. 560 (1925); Vol. 15, p. 174 (1925); Vol. 17, p. 213 (1926); also *Elektrot. Zeitschr.*, Vol. 45, p. 1348 (1924).
Geyger, Wilhelm, Measurement of the higher harmonics by the potentiometer method, *Archiv für Elektrot.*, Vol. 18, p. 629 (1927). Measurement of the a-c component of rectified currents by the potentiometer method, *Archiv für Elektrot.*, Vol. 18, p. 629 (1927).
Larsen, A., Der komplexe Kompensator, *Elektrot. Zeitschr.*, Vol. 31, p. 1039 (1910), abstracted in *Electrician* (London), Vol. 66, p. 736 (1911).
Pedersen, P. O., New a-c potentiometer for measurements on telephone circuits, *Electrician* (London), Vol. 83, p. 525 (1919).
Spooner, T., Some applications of the a-c potentiometer, *J. Optical Soc. and Rev. Sci. Instruments*, Vol. 12, p. 217 (1926).

Electrical Measuring Instruments, General Discussion

Brooks, H. B., Accuracy of commercial electrical measurements, *Trans. AIEE*, Vol. 39, p. 495 (1920).
Drysdale, C. V., Progress in the design and construction of electrical measuring instruments, *J. Sci. Instruments* (London), Vol. 4, pp. 177, 209, 241, 288 (1927).
Drysdale, C. V., Electrical measuring instruments other than integrating meters, *J. Inst. Elec. Engrs.*, Vol. 66, p. 596 (1928); Vol. 69, p. 170 (1931); Indicating electrical instruments, *J. Sci. Instruments* (London), Vol. 9, p. 209 (1932).

Edgcumbe, K., and F. E. J. Ockenden, Some recent advances in a-c measuring instruments, *J. Inst. Elec. Engrs.*, Vol. 65, p. 553 (1927).
Hoare, S. C., Accuracy of a-c test instruments, *Trans. AIEE*, Vol. 44, p. 618 (1925).
Keinath, G., New guiding lines for the construction of electrical measuring apparatus, *Zeitschr. Ver. deut. Ingen.*, Vol. 72, p. 1784 (1928); Progress of the technique of measurements in 1917, *Elektrot. Zeitschr.*, Vol. 49, p. 1 (1928).
Lenehan, B. E., and Paul MacGahan, Indicating instrument quality, *Elec. J.*, Vol. 26, p. 520 (1929).

Measurement of Large Direct Currents

Besag, Long-distance measurement of large direct currents, *Elektrot. Zeitschr.*, Vol. 40, p. 436 (1919).
Dahlgren, F. A., Meter shunts for heavy direct currents, *Electrician* (London), Vol. 97, p. 499 (1926).
Field, M. B., Multiple-unit shunts for the measurement of very heavy currents, *J. Inst. Elec. Engrs.*, Vol. 58, p. 661 (1920).
Gonnard, P., Measurement of large direct currents, *Rev. Gên. de l' Élec.*, Vol. 27, p. 95 (1930).
Melsom, S. W., and H. C. Booth, Efficiency of end connections and short-period ratings of large-current shunts, *J. Inst. Elec. Engrs.*, Vol. 63, p. 299 (1925).

Rectifier Instruments

Boekels, H., and A. Brosh, Measuring instruments using dry rectifiers, *E. T. Z.*, Vol. 59, No. 11 (1938).
Geffoken, H., Rectifier ammeters and voltmeters, *A.T.M.*, Vol. 64, No. 11, pp. 133–134 (1936).
Gieringer, C. K., A new alternating-current meter, *Rev. Sci. Instruments*, Vol. 7, p. 414 (1936).
Goodwin, W. N., Jr., Rectifier-type instruments, *Instruments* (Pittsburgh), Vol. 3, p. 706 (1930).
Grondahl, L. O., and P. H. Geiger, A new electronic rectifier, *Trans. AIEE*, Vol. 46, p. 357 (1927).
Grondahl, L. O., The copper-cuprous-oxide rectifier and photoelectric cell, *Rev. Modern Phys.*, Vol. 5, p. 141 (1933).
Hamann, C. E., and E. A. Harty, Fundamental characteristics and applications of the copper oxide rectifier, *Gen. Elec. Rev.*, Vol. 36, p. 342 (1933).
Hartshorn, L., The frequency errors of rectifier instruments of the copper-oxide type for a-c measurements, *Proc. Phys. Soc. London*, Vol. 42, p. 521 (1930).
Harty, E. A., Aging in copper oxide rectifiers, *Gen. Elec. Rev.*, Vol. 39, No. 5 (1936).
Oatley, C. W., Copper-oxide rectifiers (Some notes on their use in the construction of a-c instruments), *Elec. Rev.* (London), Vol. 107, p. 90 (1930).
Sahagen, Joseph, Use of the copper-oxide rectifier for instrument purposes, *Proc. Inst. Radio Engrs.*, Vol. 19, p. 233 (1931).
Singh, Kartar, Copper-oxide rectifiers, *Elec. Rev.* (London), Vol. 109, p. 247 (1931).
Williams, A. L., and L. E. Thompson, Metal rectifiers, *J. Inst. Elec. Engrs.*, Vol. 88-I, p. 353 (1941), Vol. 89, No. 20.

Electronic (Thermionic) Instruments

Henney, Keith, *Electron Tubes in Industry*, 2nd ed., pp. 84–124, McGraw (1937).
Hoare, S. C., A new thermionic instrument, *Trans. AIEE*, Vol. 46, p. 541 (1927).
Moullin, E. B., A direct-reading thermionic voltmeter and its applications, *J. Inst. Elec. Engrs.* (London), Vol. 61, p. 295 (1923).
Moullin, E. B., Some developments of the thermionic voltmeter, *J. Inst. Elec. Engrs.*, Vol. 68, p. 1039 (1930). Includes bibliography of 16 entries.
Rider, John F., *Vacuum Tube Voltmeters.* John F. Rider Publications, Inc. (1941).

Electrothermic Instruments

Gainsborough, G. F., Experiments with thermocouple milliammeters at very high radio frequencies, *J. Inst. Elec. Engrs.*, Vol. 91, part 3, pp. 156–161 (1944).
Goodwin, W. N., Jr., The compensated thermocouple ammeter, *Trans. AIEE*, Vol. 55, p. 23 (1936).
Meahl, H. R., P. C. Michel, N. W. Sheldorf, and T. M. Dickinson, Measurements at radio frequencies, *Trans. AIEE*, Vol. 59, p. 654 (1940).

Wattmeters

Auchincloss, John, Various methods of connecting indicating wattmeters, *Gen. Elec. Rev.*, Vol. 31, pp. 257, 376 (1928).
Doyle, E. D., Correcting wattmeter readings for phase-angle errors, *Elec. World*, Vol. 77, p. 314 (1921).
Edgerly, H. S., Connections for a three-current-coil wattmeter, *Elec. World*, Vol. 90, p. 70 (1927).
Ellis, A. L., Compensating wattmeters, *Trans. AIEE*, Vol. 31, p. 1579 (1912).
Geyger, W., New ironless electrodynamic precision wattmeter, *Archiv für Elektrot.*, Vol. 19, p. 132 (1927–28).
H & B Promille, (0.1 per cent) wattmeter, *Archiv für technisches Messen*, No. 29, Sheet F29 (1933).
Lee, E. S., Wattmeter connections, *Gen. Elec. Rev.*, Vol. 19, p. 212 (1916).
Werres, C. O., A simplified method of applying instrument transformer correction factors to wattmeter readings, *Gen. Elec. Rev.*, Vol. 36, p. 462 (1933).

Frequency Meters

MacGahan, Paul, A new indicating frequency meter, *Instruments* (Pittsburgh), Vol. 3, p. 197 (1930).
Pratt, W. H., and D. R. Price, Resonant-circuit frequency meters, *Trans. AIEE*, Vol. 31, p. 1595 (1912).
St. Clair, B. W., Frequency meters, *Instruments* (Pittsburgh), Vol. 2, p. 374 (1930).

Power-factor Meters

Edgcumbe and Ockenden, *Industrial Electrical Measuring Instruments*, 3d ed. Pitman (1933). Chapter entitled Power-factor or phase meters.
Rowell, R. M., Commercial power-factor measurements, *Gen. Elec. Rev.*, Vol. 36, p. 493 (1933).

Synchroscopes

Anderson, C., and L. J. Lunas, A low-energy synchroscope, *Elec. J.*, Vol. 30, p. 167 (1933).
Edgcumbe and Ockenden, *Industrial Electrical Measuring Instruments.* Pitman (1933). Chapter entitled Paralleling and synchronizing apparatus.
Haraden, J. A., The operation and installation of switchboard synchronism indicators, *Gen. Elec. Rev.*, Vol. 16, p. 262 (1913).

Measurement of Rectified Currents

Geyger, Wilhelm, Measurement of the a-c component of rectified currents by the potentiometer method. *Archiv für Elektrot.*, Vol. 18, p. 629 (1927).
Goodwin, W. N., Jr., Measurement of pulsating currents, *Trans. Soc. Motion Picture Operators*, Vol. 12, p. 364 (1928), reprinted in *Instruments* (Pittsburgh), Vol. 2, p. 115 (1929).
MacGahan, Paul, Measurement of pulsating currents, *Elec. J.*, Vol. 27, p. 596 (1930).
Todd, Victor H., Selecting ammeters for various current measurements, *Elec. Rev.* (Chicago), Vol. 77, p. 683 (1920).

Ohmmeters

Brooks, H. B., Accuracy tests for Meggers, *Elec. World*, Vol. 85, p. 973 (1925).
Geyger, W., Moving-iron quotient meters with circular moving iron, *Archiv für technisches Messen*, No. 5, Sheet T74 (1931). See same journal, No. 8, Sheet F7 (1932) for additional details.

Microfaradmeters

Bercovitz, D., The Weston microfaradmeter, *Elektrot. Zeitschr.*, Vol. 46, p. 312 (1925). See also *Archiv für technisches Messen*, Sheet F2 (July, 1931). In this latter article Figs. 1a, 1b, 2a, 2b should be corrected to agree with Figs. 9 and 11 of the article in *Elektrot. Zeitschr.*
Blamberg, E., New iron-cored a-c electrodynamometer without mechanical control and its various possible applications, Part 3, Direct-reading capacitance meter independent of voltage and frequency, *Archiv für Elektrot.*, Vol. 17, p. 308 (1926–27); New H & B capacitance meter, *Helios*, Vol. 36, pp. 241, 249 (1930). This instrument is further described in *Archiv für technisches Messen*, Sheet F5 (Feb., 1932).

Miscellaneous Measuring Instruments

Brooks, H. B., Temperature compensation of millivoltmeters, *J. Res. Nat. Bur. Standards*, Vol. 17, p. 497 (1936), reprinted as *Research Paper* 926.
Camilli, G., A flux voltmeter for magnetic tests (giving indications directly proportional to the maximum flux density, regardless of the wave form of the voltage), *Trans. AIEE*, Vol. 45, p. 721 (1926).
Camilli, G., Reduction of transformer exciting current to sine-wave basis, *Trans. AIEE*, Vol. 46, p. 692 (1927). Describes a "crest ammeter" for indicating the instantaneous maximum value of current.
Dannott, C., and N. Holt, A precision average voltmeter for power frequencies, *World Power*, Vol. 7, p. 241 (1927).
Doggett, L. A., J. W. Heim, and M. W. White, A new wave-shape factor and meter, *Trans. AIEE*, Vol. 45, p. 435 (1926).
Doyle, E. D., A differential electrodynamometer, *J. Optical Soc. Am.*, Vol. 6, p. 281 (1922).
Hueter, E., Direct-reading instrument for measuring voltage harmonics, *Elektrot. Zeitschr.*, Vol. 52, p. 471 (1931).
Ormondroyd, J., The use of vibration instruments on electrical machinery, *Trans. AIEE*, Vo. 45, p. 443 (1926).
Wilson, C. T. R., A micro-voltmeter, *Proc. Cambridge Phil. Soc.*, Vol. 19, p. 345 (1919).
Wilson, L. T., A novel alternating-current voltmeter, *Trans. AIEE*, Vol. 43, p. 220 (1924).

Shielding and Guarding Instruments and Circuits

Ayrton, W. E., and T. Mather, Insulation of coils and terminals of galvanometers, and the use of Price's "guard wire," *Phil. Mag.*, 5th series, Vol. 46, p. 367 (1898); Some developments in the use of Price's guard wire in insulation tests, *Phil. Mag.*, 5th series, Vol. 49, p. 343 (1900).
Curtis, Harvey L., Shielding and guarding electrical apparatus used in measurements; general principles, *Trans. AIEE*, Vol. 48, p. 1263 (1929). Contains bibliography of 18 entries.
Dawes, C. L., P. L. Hoover, and H. H. Richard, Some problems in dielectric-loss measurements, *Trans. AIEE*, Vol. 48, p. 1271 (1929).
Dye, D. W., Note on the use of stalloy ring stampings for magnetic shielding purposes, *J. Sci. Instruments*, Vol. 3, p. 65 (1925–26).
Ferguson, John G., Shielding in high-frequency measurements, *Trans. AIEE*, Vol. 48, p. 1286 (1929).
Gokhale, S. L., Magnetic shielding (shielding of magnetic instruments from steady stray fields), *Trans. AIEE*, Vol. 48, p. 1307 (1929).
Salter, E. H., Guarding and shielding for dielectric-loss measurements on short lengths of high-tension power cable, *Trans. AIEE*, Vol. 48, p. 1295 (1929).
Silsbee, Francis B., Precautions against stray magnetic fields in measurements with large alternating currents, *Trans. AIEE*, Vol. 48, p. 1301 (1929).

Electrical Integrating Meters

AIEE Committee on Instruments and Measurements, Progress in the art of metering electric energy, *Elec. Engineering* (Sept.–Dec., 1941).
American Standards Association, *Code for Electricity Meters*, 4th ed. Edison Electric Inst. (1941).
Boland, E. J., Metering kilovars and kilovolt amperes, *Gen. Elec. Rev.* (May–August, 1941).
Canfield, D. T., *Measurement of Alternating-Current Energy*. McGraw (1940).
Carr, J. L., Recent developments in electricity meters, with particular reference to those for special purposes, *J. Inst. Elec. Engrs.*, Vol. 67, p. 859 (1929).
Coleman, O. K., Timing method for meter laboratories, *Elec. World*, Vol. 90, p. 20 (1927).
Corum, E. A., and F. L. Cristenberry, Four-wire metering, *Elec. World*, Vol. 91, p. 95 (1928).
Edison Electric Institute, *Electrical Metermen's Handbook*, 5th ed. (1940).
Fawsett, E., Integrating electricity meters, *J. Inst. Elec. Engrs.*, Vol. 66, p. 443 (1928).
Frampton, W. R., Testing 100,000 meters yearly, *Elec. World*, Vol. 85, p. 1067 (1925); Electrical standardizing laboratory, *Elec. World*, Vol. 86, p. 411 (1925).
Johnson, H. M., Meter maintenance on scattered systems, *Elec. World*, Vol. 80, p. 923 (1922).
King, Harold W., Metering combined power and lighting loads, *Elec. World*, Vol. 96, p. 774 (1930).
Kinnard, I. F., and H. E. Trekell, Development of a modern watthour meter, *Trans. AIEE*, Vol. 56, p. 172 (1937).
Kinnard, I. F., and J. H. Goss, Watthour meter bearings, *Trans. AIEE*, Vol. 56, p. 129 (1937).
Knowlton, A. E., Calibration of test meters, *Elec. World*, Vol. 82, p. 645 (1923); Sustained accuracy of watthour meters, *Elec. World*, Vol. 92, p. 1147 (1928).
Knowlton, A. E., *Electric Power Metering*. McGraw (1934).
Newman, A. F., and M. Minikes, Polyphase meter connections checked by timing rotations, *Elec. World*, Vol. 102, p. 204 (1933).
Richardson, H. R., Automatic timing for rotating standards, *Elec. World*, Vol. 94, p. 1033 (1929).

Instrument Transformers

Arnold, A. H. M., Precision testing of current transformers, *J. Inst. Elec. Engrs.*, Vol. 68, p. 898 (1930).
Baker, H. S., Current-transformer ratio and phase error. *Proc. AIEE*, Vol. 37, p. 1173 (September, 1918). See also *Elec. Rev.* (Chicago), Vol. 73, p. 766 (1918).
Bousman, H. W., and R. L. Ten Broeck, A capacitance bridge for determining the ratio and phase angle of potential transformers, *Trans. AIEE*, Vol. 62, p. 541 (1943).
Boyajian, A., and W. F. Skeats, Bushing-type current transformers for metering, *Trans. AIEE*, Vol. 48, p. 949 (1929).
Boyajian, A., and G. Camilli, Orthomagnetic bushing current transformer for metering, *Trans. AIEE*, Vol. 64, p. 137 (1945).
Brooks, H. B., Testing potential transformers, *Bull. Bur. Standards*, Vol. 10, p. 419 (1914), reprinted as *Scientific Paper* 217.
Bruges, W. E., Methods of testing current transformers, *J. Inst. Elec. Engrs.*, Vol. 68, p. 305 (1930).
Camilli, G., New developments in current transformer design, *Trans. AIEE*, Vol. 59, p. 835 (1940).
Camilli, G., New developments in potential transformer design, *Trans. AIEE*, Vol. 62, p. 483 (1943).
Churcher, B. G., Measurement of the ratio and phase displacement of high alternating voltages, *J. Inst. Elec. Engrs.*, Vol. 65, p. 430 (1927).
Crothers, H. M., Field testing of instrument transformers, *Elec. World*, Vol. 73, p. 516 (1919); Vol. 74, p. 119 (1919); Vol. 75, pp. 319, 732 (1920). Concluding article of the series by F. A. Kartak, *Elec. World*, Vol. 75, p. 1369 (1920).
Gibbs, J. B., Connecting meters and relays to transformers, *Power*, Vol. 63, p. 326 (1926).
Gibbs, J. B., Principles, construction and operation of ring-type current transformers, *Power Plant Engng.*, Vol. 33, p. 520 (1929).
Gibbs, J. B., Accuracy of current transformers, *Elec. J.*, Vol. 27, p. 204 (1930).
Goldstein, J., A new voltage transformer for very high voltages, *Elektrot. Zeitschr.*, Vol. 52, p. 378 (1931).
Hague, B., *Instrument Transformers*. Pitman (1936).
Hammond, Calculation of instrument-transformer burdens, *Gen. Elec. Rev.*, Vol. 34, p. 115 (1931).
Imhof, A., A new voltage transformer, *Bull. Schweiz. Elektrot. Ver.*, Vol. 23, p. 742 (1928).
Keinath, G., Voltage transformers in cascade connection for very high voltage, *Siemens Zeitschr.*, Vol. 8, p. 629 (1928).
Laib, D. R., Instrument transformer corrections made simple, *Elec. World*, Vol. 103, p. 332 (1933).
Leeds & Northrup Company, Potential transformer testing set, Bulletin 716 (1930).
Martin, James S., Design of instrument transformers, *Instruments* (Pittsburgh), Vol. 1, p. 359 (1928).
Park, J. H., Accuracy of high-range current transformers, *J. Res. Nat. Bur. Standards*, Vol. 14, p. 367 (1935), reprinted as *Research Paper* 775.
Park, J. H., Effect of wave form upon the performance of current transformers, *J. Res. Nat. Bur. Standards*, Vol. 19, p. 517 (1937), reprinted as *Research Paper* 1041.
Schotter, G. F., New null method of testing instrument transformers, *J. Inst. Elec. Engrs.*, Vol. 68, p. 873 (1930).
Schwaiger, A. C., and V. A. Treat, Shaping of magnetizing curves and the zero-error current transformer, *Trans. AIEE*, Vol. 52, p. 45 (1933).
Silsbee, F. B., A method for testing current transformers, *Bull. Bur. Standards*, Vol. 14, p. 317 (1917), reprinted as *Scientific Paper* 309. See also Leeds & Northrup Company Bulletin 715 (1930).
Silsbee, F. B., Methods for testing current transformers, *Trans. AIEE*, Vol. 43, p. 282 (1924). Contains bibliography of 35 entries.
Silsbee, F. B., A shielded resistor for voltage-transformer testing, *Bull. Bur. Standards*, Vol. 20, p. 489 (1926), reprinted as *Scientific Paper* 516.
Silsbee, F. B., Ray L. Smith, Nyna L. Forman, and John H. Park, Equipment for testing current transformers, *Bur. Standards J. Research*, Vol. 11, p. 93 (1933), reprinted as *Research Paper* 580.
Silsbee, F. B., and F. M. Defandorf, A transformer method for measuring high alternating voltages and its comparison with an absolute electrometer, *J. Res. Nat. Bur. Standards*, Vol. 20, p. 317 (1938), reprinted as *Research Paper* 1079.
Specht, T., Biased core current transformer design method, *Trans. AIEE*, Vol. 64, p. 635 (1945).
Spooner, T., Current transformers with nickel-iron cores, *Trans. AIEE*, Vol. 45, p. 701 (1926).
Stubbings, G. W., Testing for the permanence of the ratio and phase errors of series transformers, *J. Sci. Instruments* (London), Vol. 4, p. 207 (1927).
Weller, C. T., Revised tables of phase-angle correction factors, *Gen. Elec. Rev.*, Vol. 28, p. 202 (1925).
Weller, C. T., 132-kv shielded potentiometer for determining the accuracy of potential transformers, *Trans. AIEE*, Vol. 48, p. 790 (1929).
Werres, Current-transformer design and application from the overcurrent standpoint, *Gen. Elec. Rev.*, Vol. 35, p. 544 (1932).
Wiggins, A. M., Parallel operation of current transformers for totalizing two or more circuits, *Elec. J.*, Vol. 26, p. 376 (1929).
Wiggins, A. M., Improved current transformers with nickel-iron cores, *Elec. J.*, Vol. 26, p. 152 (1929).
Wilson, M. S., A new high-accuracy current transformer, *Trans. AIEE*, Vol. 48, p. 783 (1929).

Electrometers

Berndt, G., Electrometers (with special reference to constructions for radioactive and atmospheric-electricity measurements), *Helios*, Vol. 26, pp. 429, 437, 449, 464 (1920); Accessory apparatus for use with electrometers, *Helios*, Vol. 27, pp. 253, 271, 281 (1921). These valuable series of articles contain many illustrations and many references.
Compton, A. H., and K. T. Compton, A sensitive modification of the quadrant electrometer: its theory and use, *Phys. Rev.*, 2nd series, Vol. 14, p. 85 (1919).
Kouwenhoven, W. B., and Paul L. Betz, Zero method of measuring power with the quadrant electrometer, *Trans. AIEE*, Vol. 45, p. 649 (1926).
Lindemann, F. A. and A. F., and T. C. Keeley, A new form of electrometer, *Phil. Mag.*, 6th series, Vol. 47, p. 577 (1924).
Paterson, C. C., E. H. Rayner, and A. Kinnes, The use of the electrostatic method for the measurement of power, *J. Inst. Elec. Engrs.*, Vol. 51, p. 294 (1913).
Simons, D. M., and W. S. Brown, The quadrant electrometer for the measurement of dielectric loss, *Trans. AIEE*, Vol. 43, p. 311 (1924).
Starke, H., and R. Schroeder, An electrometer for the measurement of very high direct and alternating voltages, *Archiv für Elektrot.*, Vol. 20, p. 115 (1928).
Whitehead, S., and D. Barham, A null method for the measurement of rms voltages using a quadrant electrometer, *J. Sci. Instruments* (London), Vol. 7, p. 337 (1930).

Spark Gaps

Bechdolt, H., Calibration of sphere spark gaps, *Elektrot. Zeitschr.*, Vol. 50, p. 1394 (1929).
Binder, L., Spark retardation in measurement and protective spark gaps, *Elektrot. Zeitschr.*, Vol. 47, p. 1511 (1926).
Campbell, Norman, Time-lag in the spark discharge, *Phil. Mag.*, Vol. 38, p. 214 (1919).
Clark, J. Cameron, and Harris J. Ryan, Sphere-gap discharge voltages at high frequencies, *Trans. AIEE*, Vol. 33, p. 973 (1914).
Duffendack, O. S., R. A. Wolfe, and D. W. Randolph, The development of an electron-emitting alloy, *Trans. Am. Electrochem. Soc.*, Vol. 59, p. 181 (1931).
Farnsworth, S. W., and C. L. Fortescue, The sphere spark gap, *Trans. AIEE*, Vol. 32, p. 733 (1913).
Fisher, H. W., Spark distances corresponding to different voltages, *Trans. Int. Elec. Congress*, St. Louis, Vol. II, p. 294 (1904).
Franck, Siegfried, Calibration voltages for sphere spark gaps, *Elektrot. Zeitschr.*, Vol. 51, p. 778 (1930).
Goodlet, B. L., F. S. Edwards, and F. R. Perry, Dielectric phenomena at high voltages, *J. Inst. Elec. Engrs.* (London), Vol. 69, p. 695 (1931).
Kampschulte, Josef, Breakdown of air and flashover with alternating voltages at 50 cycles and 100,000 cycles, *Archiv für Elektrot.*, Vol. 24, p. 525 (1930).
Kastenbein, H., and W. Kellermeyer, On guarded "sphere-grounded plate" spark gaps, *Elektrot. Zeitschr.*, Vol. 52, p. 969 (1931).
Marvin, R. H., A source of error when using the sphere gap, *Elec. World*, Vol. 67, p. 649 (1916).
Pedersen, P. O., Concerning the electric spark. Part I, Spark retardations; Part II, Experimental investigations on spark retardation and spark formation, *Annalen der Physik*, 4th series, Vol. 71, p. 317 (1923); Vol. 75, p. 827 (1924).
Reukema, L. E., The relation between frequency and spark-over voltage in a sphere-gap voltmeter, *Trans. AIEE*, Vol. 47, p. 38 (1928).
Viehmann, H., The disruptive breakdown of air, from investigations with the cathode-ray oscillograph, *Archiv für Elektrot.*, Vol. 25, p. 253 (1931).
Weicker, Contribution to the knowledge of sparking voltage for industrial alternating current, *Elektrot. Zeitschr.*, Vol. 32, p. 436 (1911).

Miscellaneous High-voltage Measuring Apparatus

Brooks, H. B., F. M. Defandorf, and F. B. Silsbee, An absolute electrometer for the measurement of high alternating voltages, *J. Res. Nat. Bur. Standards*, Vol. 20, p. 253 (1938), reprinted as *Research Paper* 1078.
Churcher, B. G., and C. Dannatt, Use of air condensers as high-voltage standards, *J. Inst. Elec. Engrs.*, Vol. 69, p. 1019 (1931).
Clark, Harry, A double-range electrostatic voltmeter for 200 kilovolts, *Rev. Sci. Instruments*, Vol. 1, p. 615 (1930).
Davis, R., G. W. Bowdler, and W. G. Standring, The measurement of high voltages, with special reference to the measurement of peak voltages, *J. Inst. Elec. Engrs.* (London), Vol. 68, p. 1222 (1930).
Kirkpatrick, P., and O. Miyaki, A generating voltmeter for the measurement of high potentials, *Rev. Sci. Instruments*, Vol. 3, pp. 1, 430 (1932); further described in *Elec. Engng.* (*J. AIEE*), Vol. 51, p. 863 (1932).
Palm, A., An absolute voltmeter for 250,000 volts effective value, *Zeitschr. für tech. Phys.*, Vol. 1, p. 137 (1920).
Palm, A., New high-voltage measuring devices, *Elektrot. Zeitschr.*, Vol. 47, pp. 873, 904 (1926).
Rayner, E. H., Design and use of an air condenser for high voltages, *J. Sci. Instruments* (London), Vol. 3, pp. 33, 70, 104 (1925–26).
Ryall, L. E., Peak voltage measurement by means of neon glow lamp, *World Power*, Vol. 1, p. 288 (1924).
Taylor, L. S., Apparatus for the measurement of high constant or rippled voltages, *Bur. Standards J. Research*, Vol. 5, p. 609 (1930), reprinted as *Research Paper* 217.
Thornton, W. M., M. Waters, and W. G. Thompson, The Ionic-wind voltmeter and thermoelectrostatic relay, *J. Inst. Elec. Engrs.*, Vol. 69, p. 533 (1931).

Moving Coil Oscillograph

Churcher, B. F., Use of the oscillograph in the testing of heavy electrical plant, *Metropol. Vickers Gaz.*, Vol. 10, p. 280 (1927).
Dovjikov, A., Application of automatic oscillographic equipment for power systems, *Elec. J.*, Vol. 26, p. 175 (1929).
Geiser, K. R., and J. E. Hancock, Magnetic oscillographs, *Gen. Elec. Rev.*, Vol. 46, p. 289 (1943).
Hathaway, C. M., Strain analyzing and recording instruments, *Electronics Industries*, Vol. 4, p. 74 (1945).
Howe, G. W. O., The amplitude and phase of higher harmonics in oscillograms, *J. Inst. Elec. Engrs.* (London), Vol. 54, p. 19 (1916).
Irwin, J. T., *Oscillographs*. Pitman (1925).
Kennelly, A. E., R. N. Hunter, and A. A. Prior, Oscillographs and their tests, *Trans. AIEE*, Vol. 39, p. 443 (1920).
Legg, J. W., A six-element portable oscillograph, *Elec. J.*, Vol. 22, p. 109 (1925).
Powell, G. A., and R. E. Walsh, Special uses for the automatic oscillograph, *Elec. Engng.*, Vol. 56, pp. 438, 476 (1937).
Rusher, M. A., New permanent-magnet oscillographs, *Gen. Elec. Rev.*, Vol. 33, p. 49 (1930).
Thomander, V. S., Characteristics of the oscillograph-galvanometer, *J. Franklin Inst.*, Vol. 213, p. 41 (1932).
Waldorf, Sigmund F., An amplifier to adapt the oscillograph to low-current investigations, *Trans. AIEE*, Vol. 47, p. 1418 (1928); Amplifiers for precise oscillographic measurements, *J. Franklin Inst.*, Vol. 213, p. 605 (1932).

Cathode-ray Oscillograph

Ackermann, O., A cathode-ray oscillograph with Norinder relay, *Trans. AIEE*, Vol. 49, p. 467 (1930).
Bedell, Fredrick, and Jackson G. Kuhn, A stabilized oscilloscope with amplified stabilization, *Rev. Sci. Instruments*, Vol. 1, p. 227 (1930).
Dudley, B., Applications of cathode-ray tubes, *Electronics*, Vol. 15, p. 49 (October, 1942).
Johnson, J. B., The cathode-ray oscillograph, *J. Franklin Inst.*, Vol. 212, p. 687 (1931).
Krug, W., Cathode-ray oscillograph circuits, *Elektrot. und Masch.*, Vol. 49, p. 233 (1931).

Kuehni, H. P., and S. Ramo, A new high-speed cathode-ray oscillograph, *Elec. Engng.*, Vol. 56, p. 721 (1937).
Lee, Everett S., Cathode-ray oscillographs and their uses, *Gen. Elec. Rev.*, Vol. 31, p. 404 (1928).
Macgregor-Morris, J. T., and R. Mines, Measurements in engineering by means of cathode rays, *J. Inst. Elec. Engrs.* (London), Vol. 63, p. 1056 (1925).
Maloff, I. G., and D. W. Epstein, *An Introduction to Electron Optics.* McGraw-Hill (1938).
Parr, Goeffrey, *The Low-Voltage Cathode-Ray Tube*, Chapman and Hall (1942).
Rogowski, W., E. Flegler, and R. Tamm, A new cathode-ray oscillograph, *Archiv für Elektrot.*, Vol. 18, p. 513 (1927).
Rogowski, W., O. Wolff, and H. Klemperer, Voltage dividers for cathode-ray oscillographs, *Archiv für Elektrot.*, Vol. 23, p. 579 (1930).
Rogowski, W., E. Flegler, and K. Buss, Useful range of the cathode-ray oscillograph, *Archiv für Elektrot.*, Vol. 24, p. 563 (1930).
Rohats, N., High-speed oscillograph, *Electronics*, Vol. 19, p. 135 (April, 1946).
Ruhlemann, E., Investigation of the electron current of the cold-cathode oscillograph, *Archiv für Elektrot.*, Vol. 25, p. 505 (1931).
Symposium on Lightning, *Trans. AIEE*, Vol. 49, p. 857 (1930).

String Galvanometer as Oscillograph

Trowbridge, Augustus, String-galvanometer oscillograph with automatic photography, *J. Optical Soc. Am.*, Vol. 9, p. 557 (1924).
Williams, Horatio B., The Einthoven string galvanometer, a theoretical and experimental study, Parts I and II. *J. Optical Soc. Am.*, Vol. 9, p. 129 (1924); Vol. 13, p. 313 (1926).

Electrostatic and Hot-wire Oscillographs

Ho, H., and Koto, S., An electrostatic oscillograph, *Electrician* (London), Vol. 72, p. 290 (1913).
Irwin, J. T., Hot-wire wattmeters and oscillographs, *Electrician* (London), Vol. 59, pp. 266, 306 (1907).
Irwin, J. T., *Oscillographs.* Pitman (1925).

Testing of Magnetic Materials

American Society for Testing Materials, Standard methods of test for magnetic properties of iron and steel, ASTM designation A-34. (See latest edition of book of standards.)
Babbitt, B. J., An improved permeameter for testing magnet steel, *J. Optical Soc. Am. and Rev. Sci. Instruments*, Vol. 17, p. 47 (1928).
Burgwin, S. L., Measurement of core loss and a-c permeability with the 25-cm Epstein frame, *Proc. A.S.T.M.*, Vol. 41, p. 779 (1941).
Burrows, C. W., The determination of the magnetic induction in straight bars, *Bull. Bur. Standards*, Vol. 6, p. 31 (1909), reprinted as *Scientific Paper* 117.
Camilli, G., flux voltmeter for magnetic tests, *Trans. AIEE*, Vol. 45, p. 721 (1926).
Cheney, W. L., Magnetic testing of straight rods in intense magnetic fields, *Sci. Papers Bur. Standards*, Vol. 15, p. 625 (1919–20), reprinted as *Scientific Paper* 361.
Chubb, L. W., Method of testing transformer core losses, giving sine-wave results on commercial circuits, *Trans. AIEE*, Vol. 28, p. 417 (1909).
Fahy, F. P., A permeameter for general magnetic analysis, *Chem. and Met. Engng.*, Vol. 19, p. 339 (1918).
Fahy, F. P., The determination of low permeability, *Instruments* (Pittsburgh), Vol. 4, p. 475 (1931).
Gokhale, S. L., The saturation permeameter, *J. Am. Inst. Elec. Engrs.*, Vol. 47, p. 196 (1928).
Sanford, R. L., Standards for testing magnetic permeameters, *Bur. Standards J. Research*, Vol. 4, p. 177 (1930), reprinted as *Research Paper* 140.
Sanford, R. L., and E. G. Bennett, An apparatus for magnetic testing at high magnetizing forces, *Bur. Standards J. Research*, Vol. 10, p. 567 (1933), reprinted as *Research Paper* 548.
Sanford, R. L., Drift of magnetic permeability at low inductions after demagnetization, *Bur. Standards J. Research*, Vol. 13, p. 371 (1934), reprinted as *Research Paper* 714.
Sanford, R. L., and E. G. Bennett, Determination of magnetic hysteresis with the Fahy Simplex permeameter, *Bur. Standards J. Research*, Vol. 15, p. 517 (1935), reprinted as *Research Paper* 845.
Sanford, R. L. and E. G. Bennett, An apparatus for magnetic testing at magnetizing forces up to 5000 oersteds, *Bur. Standards J. Research*, Vol. 23, p. 415 (1939), reprinted as *Research Paper* 1242.
Sanford, R. L., Magnetic testing, *Circular No.* 456, *Nat. Bur. Standards* (1946).
Smith, B. M., and C. Concordia, Measuring core loss at high densities, *Trans. AIEE*, Vol. 51, p. 36 (1932).
Smith, B. M., A new full-range permeameter, *Gen. Elec. Rev.*, Vol. 38, p. 520 (1935).
Spooner, T., *Properties and Testing of Magnetic Materials.* McGraw (1927).

Testing of Insulating Materials

A.S.T.M. Standards on Electrical Insulating Materials (with relative information); issued annually by American Society for Testing Materials, 260 S. Broad St., Philadelphia 2, Pa.
Conference on Electrical Insulation, National Research Council, Washington, D. C. See annual reports and bibliographies of its committees.
Curtis, H. L., Insulating properties of solid dielectrics, *Bull. Bur. Standards*, Vol. 11, p. 359 (1915), reprinted as *Scientific Paper* 234.
Dellinger, J. H., and J. L. Preston, Methods of measurement of properties of electrical insulating materials, *Sci. Papers Bur. Standards*, Vol. 19, p. 39 (1923–24), reprinted as *Scientific Paper* 471.
Inge, Lydia, and Alexander Walther, papers in *Archiv für Elektrot.* as follows: Breakdown of glass in homogeneous and non-homogeneous electric fields, Vol. 19, p. 257 (1927–28); Breakdown of solid insulators at high frequency, Vol. 21, p. 209 (1928–29); Breakdown of solid insulators in homogeneous and non-homogeneous electric fields in tests of long and of short duration, Vol. 22, p. 410 (1929); Breakdown of liquid insulators, Vol. 23, p. 279 (1929–30); Field distribution and breakdown voltage of solid insulators, I, Vol. 24, p. 88 (1930); same, instalment II, Vol. 25, p. 21 (1931); Breakdown of solid insulators in air, Vol. 26, p. 409 (1932); Is there an intermediate region between heat breakdown and purely electrical breakdown?, Vol. 26, p. 716 (1932); Breakdown of solid insulators in transformer oil, Vol. 27, p. 99 (1933).
Minton, J. P., An investigation of dielectric loss with the cathode-ray tube, *Trans. AIEE*, Vol. 34, p. 1627 (1915).
Rogowski, W., and W. Grösser, A bright-spot hot-cathode oscillograph for externally recording rapidly varying phenomena, *Archiv für Elektrot.*, Vol. 15, p. 377 (1925).

Ryan, H. J., A power-diagram indicator for high-tension circuits, *Trans. AIEE*, Vol. 30, p. 1089 (1911)
Wood, A. B., The cathode-ray oscillograph, *Proc. Phys. Soc. London*, Vol. 35, p. 109 (1922–23)

Photometry

Bechstein, W., Photometer with proportional graduation and decimally extended measuring range, *Zeitschr. für Instrumentenkunde*, Vol. 27, p. 178 (1907).
Benford, F., The parabolic mirror, *Trans. Illum. Engng. Soc.*, Vol. 10, p. 905 (1915).
Benford, Frank, A coordinated system of optical filters for color temperature determination, *J. Optical Soc. America*, Vol. 25, No. 5 (May, 1935).
Benford, Frank, The blackbody, Parts 1 and 2, *Gen. Elec. Rev.*, Vol 46, No. 7, 8 (July, August, 1943).
Benford and Ruggles, Some characteristics of metal mirrors and a new gonioreflectometer, *J. Optical Soc. America*, Vol. 32, No. 3 (March, 1942).
Cady, F. W., and H. E. Dates, *Illuminating Engineering*. Wiley (1928).
Committee on Motor-Vehicle Lighting of the Illuminating Engineering Society, Interim Report, *Trans. Illum. Engng. Soc.*, Vol. 17, p. 103 (1922).
Committee on Illumination of the Association of Railway Electrical Engineers, General procedure for photometry of incandescent floodlights, *Rwy. Elec. Engr.*, Vol. 18, p. 315 (1927).
Davis and Gibson, Filters for the reproduction of sunlight and daylight and the determination of color temperature, *Misc. Pub. Bur. Standards* 114 (1931).
Dows, C. L., and C. J. Allen, The lightmeter and its uses, *Trans. IES*, Vol. 31, No. 7 (July, 1936).
Dows, C. L., Committee on Lighting Service, Illumination measurements on the job with light-sensitive cells, *Illum. Engr. Soc.* (1941).
Forsythe, W. E., Editor, *Measurement of Radiant Energy*. McGraw-Hill.
Forsythe, W. E., and E. Q. Adams, Establishing and maintaining a color temperature scale, *J. Sci. Lab., Dennison University Bull.*, Vol. 38, No. 1 (April, 1943).
General Electric Co., Mazda Lamps, *Bull.* LD-1, Lamp Dept., Nela Park, Cleveland, Ohio.
Goodwin, W. N., Jr., The (Weston) photronic illumination meter, *Trans. Illum. Engng. Soc.*, Vol. 27, p. 828 (1932).
Guild, J., A new flicker photometer, *J. Sci. Instruments* (London), Vol. 1, p. 182 (1924).
Hardy and Perrin, *The Principles of Optics*. McGraw-Hill.
Illuminating Engineering Society and Society of Automotive Engineers (joint authors), Automotive lighting specifications, *Trans. Illum. Engng. Soc.*, Vol. 25, p. 835 (1930).
Illuminating Engineering Society, Illuminating engineering nomenclature and photometric standards, ASA Z7.1 (1942).
Kingsbury, E. F., A flicker-photometer attachment for the Lummer-Brodhun contrast photometer. *J. Franklin Inst.*, Vol. 180, p. 215 (1915).
Little, W. F., and C. E. Horn, Practical experiences in photoelectric photometry, *Trans. Illum. Engng. Soc.*, Vol. 23, p. 419 (1928).
Lummer, O., and Brodhun, E., A new photometer, *Zeitschr. für Instrumentenkunde*, Vol. 9, p. 41 (1889).
Martens, F. F., New polarization photometer for white light, *Verhand. der deutschen physik. Gesell.*, Vol. 1, p. 204 (1899); A new portable photometer for white light, *Verhand. der deutschen physik. Gesell.*, Vol. 5, p. 149 (1903).
McNicholas, H. J., Absolute methods in reflectometry, *Bur. Standards J. Research*, Vol. 1, p. 29, (1928), reprinted as *Research Paper* 3.
Raumgartner, G. R., A light-sensitive cell reflectometer, *Gen. Elec. Rev.* (Nov., 1937), reprinted as *Gen. Elec. Pub.* GEA 2855.
Redding, C. S., A new illumination photometer (the Macbeth illuminometer), *Lighting J.*, Vol. 3, p. 8 (1915).
Rosa, E. B., and A. H. Taylor, Theory, construction, and use of the photometric integrating sphere, *Bur. Standards Sci. Papers*, Vol. 18, p. 281 (1922–23); reprinted as *Scientific Paper* 447.
Sharp, C. H., and P. S. Millar, A new universal photometer, *Elec. World*, Vol. 51, p. 181 (1908).
Sharp, C. H., Construction of a simple illumination tester (the G. E. foot-candle meter), *Elec. World*, Vol. 68, p. 569 (1916).
Sharp, C. H., and W. F. Little, Measurement of reflection factors, *Trans. Illum. Engng. Soc.*, Vol. 15, p. 802 (1920).
Sharp, C. H., and H. A. Smith, Further developments in photoelectric photometers, *Trans. Illum. Engng. Soc.*, Vol. 23, p. 428 (1928).
Sharp, C. H., Notes on photoelectric photometry, in booklet entitled *Photoelectric cells and their applications*, p. 110 (1930). Published by Physical and Optical Societies, London. Same publication, T. H. Harrison, The photoelectric cell as a precision instrument in photometry, p. 118.
St. John, G. H., Discussion of photoelectric photometers, *Trans. Illum. Engng. Soc.*, Vol. 23, p. 439 (1928).
Taylor, A. H., A simple portable instrument for measuring reflection and transmission factors in absolute units, *Trans. Illum. Engng. Soc.*, Vol. 15, p. 811 (1920); The measurement of diffuse reflection factors, and a new absolute reflectometer, *Bur. Standards Sci. Papers*, Vol. 16, p. 421 (1920), reprinted as *Scientific Paper* 391; A simple portable instrument for the absolute measurement of reflection and transmission factors, *Bur. Standards Sci. Papers*, Vol. 17, p. 1 (1922), reprinted as *Scientific Paper* 405.
Taylor, A. H., Brightness and brightness meters, *Illum. Engr.*, Vol. 37, No. 1 (Jan., 1942).
Taylor, A. H., Measuring germicidal energy, *Gen. Elec. Rev.*, Vol. 47, No. 10 (Oct., 1944).
Teele, Ray P., A physical photometer, *J. Res. Nat. Bur. Standards*, Vol. 27, p. 217 (1941), reprinted as *Research Paper* 1415.
Tuck, Davis, A new form of portable light meter (the holophane light meter), *Trans. Illum. Engng. Soc.*, Vol. 15, p. 539 (1920).
Waldram, J. M., Precise measurements of optical transmission, reflection, and absorption factors, *Proc. Internat. Cong. Illum.*, p. 1020 (1928).
Walsh, J. W. T., *Photometry*. Van Nostrand (1926).
Walsh, J. W. T., Everyday photometry with photoelectric cells, *Illum. Engr.*, Vol. 26, p. 64 (1933).
Walsh, John W. T., *Photometry*. Constable, London.
Weber, L., A photometric apparatus, *Annalen der Physik*, Vol. 20, p. 326 (1883); The photometric comparison of light sources of different colors, *Elektrot. Zeitschr.*, Vol. 5, p. 166 (1884).
Wensel, H. T., Wm. F. Roeser, L. E. Barbrow, and F. R. Caldwell, The Waidner-Burgess standard of light, *Bur. Standards J. Research*, Vol. 6, p. 1103 (1931), reprinted as *Research Paper* 325.

See also articles on photometry and illumination in the following reference books: Glazebrook's *Dictionary of Applied Physics*, Vol. 4, Macmillan (1922); Gehrcke's *Handbuch der Physik. Optik*, Vol. 1,

Barth, Leipzig (1927); Geiger and Scheel's *Handbuch der Physik*, Vol. 19, Springer (1928); and Müller-Pouillet's *Lehrbuch der Physik*, Vol. 2, Vieweg und Sohn, Braunschweig (1928).

Electrical Recording Instruments

See books by Drysdale and Jolley (Part II), Edgcumbe and Ockenden; also Keinath's *Die Technik der elektrischen Messgeräte*, Vol. 1; *AIEE Standard for Electrical Recording Instruments*, No. 40 (1947); Chute's *Electronics in Industry*, McGraw-Hill (1946); and the following articles in *Archiv für technisches Messen:*

Hartmann & Braun Recorders, No. 9, Sheet F8 (1932).

Keinath, G., Recorder for disturbances in power networks, No. 6, Sheets T90, 91 (1931). Contains bibliography of 12 entries, of which 8 relate to American publications.

Keinath, G., Methods of making the record in recording instruments, No. 20, Sheets T23–25 inclusive (1933); Charts for recording instruments, No. 21, Sheets T40, 41 (1933); Straight-line guide for recording instruments, No. 22, Sheets T54, 55 (1933); Chart-driving devices for recording instruments, No. 24, Sheet T81 (1933).

La Pierre, C. W. (edited by Keinath), Photoelectric recording apparatus, No. 10, Sheet T56 (1932).

Siemens and Halske, A. G., Siemens compensograph (relay-type recording potentiometer), No. 19, Sheet F1 (1933).

Nielsen, D. M., Electric measuring instruments, *Elec. Engng.*, Vol. 65, pp. 66–74 (February, 1946).

Williams, A. J., Jr.; W. R. Clark, and R. E. Tarpley, Electronically balanced recorder for flight testing and spectroscopy, *Trans. AIEE*, Vol. 65, p. 205 (1946).

Telemetering

American Institute of Electrical Engineers, Bibliography on automatic stations, 1930–1941, *Spec. Pub.* 5 (December, 1942).

Bristol, F. B., and G. S. Lunge, Metameter system of telemetering, Gen. Elec. Rev., Vol. 42, No. 11, pp. 466–472 (November, 1939).

Joint Committee of AIEE Committees on Automatic Stations and on Instruments and Measurements, Telemetering, supervisory control, and associated circuits, *Trans. AIEE*, Vol. 60, p. 1411 (1941).

Renfro, L. E., and A. P. Peterson, Modulated-frequency telemetering, *Trans. AIEE*, Vol. 64, pp. 45–47 (1945).

Measurement of Non-electrical Quantities by Electrical and Magnetic Means

An experimental machine for measuring (the diameter of) fine wire, *J. Sci. Instruments* (London), Vol. 9, p. 256 (1932).

Bailey, Benj. F., Measurement of noise in electrical machinery, *Trans. AIEE*, Vol. 50, p. 1039 (1931).

Ballard, R. G., and C. P. Hall, Recent advances in aircraft tachometer design, *Trans. AIEE*, Vol. 63, pp. 646–8 (Sept., 1944).

Borden, Perry A., Electrical measurement of physical values, *Trans. AIEE*, Vol. 44, p. 238 (1925). Contains bibliograhy of 330 entries. Supplementary bibliographies by Borden were published in the annual reports of AIEE Committee on Instruments and Measurements as follows: *Trans. AIEE*, Vol. 46, p. 709 (1927), Vol. 47, p. 1168 (1928).

Castner, T. G., E. Dietz, G. T. Stanton, and R. S. Tucker, Indicating meter for measurement and analysis of noise, *Trans. AIEE*, Vol. 50, p. 1041 (1931).

Curtis, Harvey L., Use of an oscillograph in mechanical measurements, *Trans. AIEE*, Vol. 45, p. 264 (1925).

Geyger, W., Remote integration of flow with motor meters, *Archiv für technisches Messen*, No. 23, Sheets 62, 63 (1933).

Goodwin, W. N., Jr., Heat flow meter, Model 553, *Instruments* (Pittsburgh), Vol. 3, p. 59 (1930).

Janovsky, W., Magnetoelastic measurement of compressive, tensile and torsional forces, *Archiv für technisches Messen*, No. 22, Sheets 46, 47 (1933).

Keinath, G., Electrical measurement of force by the change of an inductance, *Archiv für technisches Messen*, No. 16, Sheets 145, 146 (1932).

Krönert, Josef, Measurement of gas flow by thermal methods, *Archiv für technisches Messen*, No. 25, Sheets 87, 88 (1933).

Marriott, Robert H., Electrical aids to navigation, *Trans. AIEE*, Vol. 48, p. 753 (1929).

Marvin, H. B., The measurement of machinery noise, *Trans. AIEE*, Vol. 50, p. 1048 (1931).

Mershon, A. V., Precision measurements of mechanical dimensions by electrical measuring devices, *Gen. Elec. Rev.*, Vol. 35, p. 139 (1932).

Miner, D. F., and W. B. Batten, Use of the oscillograph for measuring non-electrical quantities, *Trans. AIEE*, Vol. 48, p. 742 (1929).

Moll, F., Measurement of moisture in wood, *Archiv für technisches Messen*, No. 17, Sheet 159 (1932).

Oplinger, K. A., An electrical ear, *Elec. J.*, Vol. 28, p. 474 (1931).

Pratt, W. H., Electrical instruments used in the measurement of flow, *Trans. AIEE*, Vol. 48, p. 737 (1929).

Sanford, Raymond L., Magnetic analysis, *Trans. AIEE*, Vol. 48, p. 731 (1929).

Smith, E. S., *Automatic Control Engineering*. McGraw-Hill (1944).

Spooner, T., and J. P. Foltz, Study of noises in electrical apparatus, *Trans. AIEE*, Vol. 48, p. 747 (1929).

Measurement of Temperature

British Standard Code, *Temperature Measurement*, B.S. 1041:1943, British Standards Institution (1943).

Burgess, G. K., and H. LeChatelier, *The Measurement of High Temperatures*. Wiley (1912).

Burgess, George K., and Paul D. Foote, Characteristics of radiation pyrometers, *Bull. Bur. Standards*, Vol. 12, p. 91 (1915–16), reprinted as *Scientific Paper* 250.

Burgess, George K., The international temperature scale, *Bur. Standards J. Research*, Vol. 1, p. 635 (1928), reprinted as *Research Paper* 22.

Emmons, H., Theory and application of extended surface thermocouples, *J. Franklin Inst.*, 229, 29–52 (1940).

Foote, Paul D., C. O. Fairchild, and T. R. Harrison, Pyrometric practice, *Bur. Standards Technologic Paper* 170 (1921).

Forsythe, W. F., Ed., *Measurement of Radiant Energy*. McGraw-Hill, p. 452 (1937).

The Journal of Applied Physics, Vol. 11, No. 6 (1940). (Special issue on measurement and control of temperature.)

Rhodes, T. J., *Industrial Instruments for Measurement and Control*. McGraw-Hill (1941).

Roeser, Wm. F., Thermoelectric temperature scales, *Bur. Standards J. Research*, Vol. 3, p. 343 (1929), reprinted as *Research Paper* 99.
Roeser, Wm. F., The passage of gas through the walls of pyrometer protection tubes at high temperature, *Bur. Standards J. Research*, Vol. 7, p. 485 (1931), reprinted as *Research Paper* 354.
Roeser, Wm. F., and H. T. Wensel, Reference tables for platinum to platinum-rhodium thermocouples, *Bur. Standards J. Research*, Vol. 10, p. 275 (1933), reprinted as *Research Paper* 530.
Roeser, W. F., A. I. Dahl, and G. J. Gowens, Reference tables for chromel to alumel thermocouples, *J. Res. Nat. Bur. Standards*, Vol. 14, p. 239 (1935), reprinted as *Research Paper* 767.
Roeser, W. F., and A. I. Dahl, Reference tables for copper to constantan and iron to constantan thermocouples, *J. Res. Nat. Bur. Standards*, Vol. 20, p. 337 (1938), reprinted as *Research Paper* 1080.
Schwab, F. W., and E. Wickers, Freezing temperature of benzoic acid as a fixed point in thermometry, *J. Res. Nat. Bur. Standards*, Vol. 34, p. 333 (1945), reprinted as *Research Paper* 1647.
Swanger, Wm. H., and Frank R. Caldwell, Special refractories for use at high temperatures, *Bur. Standards J. Research*, Vol. 6, p. 1131 (1931), reprinted as *Research Paper* 327.
Symposium on temperature control and determination in the non-ferrous foundry, *Trans. Am. Foundrymen's Assn.*, Vol. 34, pp. 611, 640, 649, 658, 663, 670, 675 (1926).
Temperature, Its Measurement and Control in Science and Industry. Reinhold Publishing Corporation, New York (1941).
Weber, Robert L., *Temperature Measurement and Control.* The Blakiston Company, Philadelphia (1941).
Wensel, H. T., and Wm. F. Roeser, Temperature measurements in molten cast iron, *Trans. Am. Foundrymen's Assn.*, Vol. 36, p. 191 (1928).
Wensel, H. T., and Wm. F. Roeser, The freezing point of nickel as a fixed point on the international temperature scale, *Bur. Standards J. Research*, Vol. 5, p. 1309 (1930), reprinted as *Research Paper* 258.

SECTION 6

RESISTORS, RHEOSTATS, CAPACITORS, REACTORS, ELECTROMAGNETS, AND PERMANENT MAGNETS

* Revision of article by A. Dexter Hinckley.

RESISTORS, RHEOSTATS, CAPACITORS, REACTORS, ELECTROMAGNETS, AND PERMANENT MAGNETS

RESISTORS AND RHEOSTATS

By Edwin J. Merrell

1. DEFINITIONS AND GENERAL PRINCIPLES

RESISTOR. A resistor is a device to introduce resistance into an electric circuit. It consists essentially of an electric conductor, formed so that a relatively high resistance is contained in a relatively small volume. It may be a single unit or several units connected in series or in parallel. Tables 1 and 2 list pertinent characteristics of several types according to the different materials used as the resistance element. Table 3 gives the properties of the metals which may be used in wire-wound, metal ribbon, and cast metal grid resistors. The resistance element of the conducting-film type may be metallic, applied by plating, spraying, etc., or a semiconducting material based upon carbon, silicon carbide, metallic oxides, and the like may be employed. Ceramic resistors usually are formed from silicon carbide compositions. Water, or electrolytes made by adding an acid, salt or alkali to water, are suitable heavy-duty resistors.

Table 1. Electrical Characteristics of Resistors Available Commercially

Resistor Type (1)	Resistance Range, Ohms (2)	Current Range, Amperes (3)	Voltage Range, Volts (4)	Power * Range, Watts (5)	Maximum † Operating Temperature, Degrees Centigrade (6)	Inductive or Non-inductive (7)
Wire..........	0.06 to 7×10^6	0.0005 to 37	0.3 to 75,000	0.5 to 1000	35 to 400	Either
Metal ribbon..	0.01 to 124	1 to 75	....	90 to 1500	400	Inductive
Cast grid......	0.01 to 0.32	20 to 180	115 to 550		400	Inductive
Conducting film	0.24 to 10^{12}	0.001 to 0.01	250 to 75,000	0.33 to 90	100	Either
Molded carbon.	50 to 20×10^6	Low	500	0.5 to 5	100	Non-inductive
Molded plastic.	1000 to 10^{12}	Low	7000	1	77	Non-inductive
Ceramic.......	0.1 to 15×10^6		8000	3 to 118	170	Non-inductive
Electrolyte.....	See Note 1	1 per sq in.	Up to 1000	High	90	Non-inductive
Water.........	See Note 1	1 per sq in.	Above 1000 See Note 2	High	90	Non-inductive

* Power rating may be increased one-hundred-fold by auxiliary cooling, such as by water-submerged resistors, for which bare wire can be used up to 500 volts.

† Nichrome-wire heater units operate at 1150 deg cent; Globar-ceramic heaters, at 1500 deg cent.

NOTES: 1. Resistivity of tap water ranges from 1200 to 12,000 ohm cm; that of electrolytes, from 1 to 60.

2. Water resistors may be operated at a voltage gradient of 100 volts per inch approximately.

Table 2. Physical Characteristics of Resistors Available Commercially

Resistor Type (1)	Composition of Support (2)	Composition * of Insulation and Coverings (3)	Terminal (4)	Mounting (5)	Approximate Dimensions, Inches (6)
Wire	Ceramic, asbestos, bakelite, mica, self	Air, oxide, varnish, enamel, glass, wax, cement	Stud, post, bracket, lug, wire, strap, ferrule, Edison, prongs	Terminal, bolt, clip, eyelet	1 watt: 1.25 × 0.5 10 watts: 1.75 × 0.5 100 watts: 7 × 1.1 1000 watts: 20 × 2.5
Metal ribbon	Ceramic, metal, mica	Air, enamel	Stud, lug, clamp	Terminal, bolt	100 watts: 3.5 × 1.5 1000 watts: 15 × 2
Cast grid	Self	Air	Clamp	Bolt	100 watts: 11 × 3
Conducting film	Ceramic	Air, enamel, varnish	Lug, wire, ferrule	Terminal	1 watt: 1.75 × 0.3 10 watts: 4.5 × 0.75 100 watts: 18.5 × 2.0
Molded carbon	Self, glass	Varnish, glass	Lug, wire	Terminal	1 watt: 0.9 × 0.3
Molded plastic	Self	Air	Lug, wire	Terminal	1 watt: 1.5 × 0.3
Ceramic	Self	Air	Metallized lug, bracket, wire, ferrule	Terminal, clip	10 watts: 5 × 0.75 100 watts: 14 × 1.0
Electrolyte	Self	None, container	Carbon, lead, copper, iron	Bolt	
Water	Self	None, container	Metal	Bolt	

* Fungicide may be applied in varnish.

RHEOSTAT. A rheostat is an adjustable resistor, so constructed that its resistance may be changed without opening the circuit in which it is connected. Adjustment is made by a movable conducting arm which contacts the rheostat element, either directly, as when the winding is exposed, or through auxiliary contacts wired to the element at several points. Rheostats are available with uniform winding throughout or with tapered windings, i.e., two or more sizes of wire. The tapered winding provides more nearly linear control, decreases size, permits a higher resistance within given dimensions, and makes possible special resistance versus rotation curves. The several standard types of rheostats are listed in Table 6, together with a summary of their characteristics. A tubular rheostat consists of wire or metal ribbon wound on a ceramic tube. The wire is oxidized, forming ample insulation between turns. The resistive element of the ring-type rheostat is wound upon a toroidal form or sometimes upon a flat sheet of insulating material, which is bent into a toroidal form after winding, and then covered with a vitreous enamel, except for a small area along one side where contact is made. The resistive element is assembled in a refractory or a metal base to which it is intimately bonded. Circular-plate rheostats are so named because the resistive element is molded or cemented into a circular plate, the element being connected to auxiliary metal contacts, which likewise are integral with the plate. The unit and cast-grid types are assembled with standard resistors mounted on panels and connected to auxiliary contacts. In the compression carbon rheostats, carbon disks or blocks are supported in insulating tubes or other forms, and resistance is varied by turning a compression screw.

CONTACTS. Contactors consist of metal, graphite-metal combinations, and metal lubricated by an auxiliary graphite contactor. Proper lubrication of contacts is of prime importance in the maintenance of rheostat contacts, since it inhibits oxidation of the metal surfaces and prevents scoring and resultant arcing.

Table 3. Properties of Various Metals and Alloys

Material	Ohms per Circular-mil Foot at 20° C (68° F)	Relative Resistance With Copper = 1	Temperature Coefficient 20° C	Approximate Melting Point, Degrees Centigrade	Maximum Working Temperature, Degrees Centigrade	Specific Gravity	Weight, Pounds per Cubic Inch
Silver	9.796	0.95	0.0038	960		10.5	0.3793
Copper	10.37	1.00	0.00393	1085		8.89	0.3212
Aluminum	17.0	1.64	0.00446	660		2.70	0.0975
No. 30 alloy	30.00	2.89	0.00118	1100	350	8.92	0.322
Brass (Spring)	36.30	3.50	0.0020	965		8.55	0.309
Beryllium copper (heat-treated)	41.5 to 57.6	4.0 to 5.55		955		8.21	0.297
Phosphor bronze, 5% (Grade A)	56.5						
Nickel	58.0	5.45	0.0018	1050		8.88	0.320
Pure iron	61.1	5.60	0.0048	1445	500	8.90	0.321
No. 60 alloy	60.0	5.90	0.0062	1575		7.7	0.278
Platinum	63.8	5.78	0.00046	1100	350	8.9	0.321
No. 90 alloy	90.0	6.15	0.0030	1755		21.45	0.775
Lead	132.0	8.68	0.00038	1100	400	8.96	0.324
Everdur No. 1010	155.0	12.7	0.0039	327		11.4	0.412
No. 180 alloy	180.0	15.0	0.00034	1019		8.52	0.308
18% nickel silver	190.0	17.3	0.00016	1130	400	8.95	0.323
Monel	256.0	18.3	0.00019	1110	260	8.50	0.307
Manganin (copper, manganese,		24.7	0.00145	1360	500	8.9	0.321
nickel) *	290.0	28.0	±0.00002	1020	100	8.39	0.303
Cupron, Constantan, Advance (copper, nickel) *	294.0	28.4	±0.00002	1290	500	8.9	0.321
Stainless steel, No. 302	430	41.5		1430		8.026	0.290
Nichrome V, Tophet A (nickel, chromium) *	650.0	62.7	0.00013	1400	1150	8.412	0.304
Nichrome, Tophet C (nickel, chromium, iron) *	675.0	65.0	0.00017	1350	1000	8.247	0.298

* These alloys have limited usable temperature ranges within which their low temperature coefficient is retained: the Nichromes or Tophets and Cupron, Constantan, or Advance, −55° C to +110° C, Manganin, +20° C to +35° C. Manganin wire should be selected only where exceptional accuracy and stability are required in the above temperature ranges and where low thermal emf against copper is a requisite. Where great accuracy and stability are required and thermal emf is not a factor, copper-nickel alloys should be selected.

2. RESISTOR DESIGN

In general, three factors determine the design of a resistor: permissible temperature rise above ambient, thermal resistance to ambient, and energy to be dissipated. Exceptions occur in some instances: when the voltage drop across the resistor is so great that the length is dictated by the flashover voltage, or when the resistance is so great that the length of the resistor element results in dimensions far in excess of those required thermally.

CALCULATION OF TEMPERATURE RISE. The temperature rise above ambient may be obtained from the following formula:

$$T = WR_{ths}$$

where T is the rise in degrees centigrade, W is the watts dissipated, and R_{ths} is the thermal surface resistance of the resistor in thermal ohms. (See Section 14, Article 69.)

Calculation of the exact thermal resistance to ambient may be a very complicated procedure if all variables are taken into account. Usually, however, the thermal surface resistance of the resistor is the essential factor in the thermal resistance to ambient, particularly for steady-state current. This may be calculated from the following formula:

$$R_{ths} = \frac{\rho_{ths}}{A}$$

where R_{ths} is the thermal surface resistance, ρ_{ths} is the effective thermal surface resistivity in thermal ohms per square inch, and A is the surface area of the resistor in square inches. For resistors operated in "free air," ρ_{ths} is approximately 50. Where circulation of air is restricted, as when a resistor is enclosed or when proximity heating effects are caused by the installation of several resistor units close together, the effective ρ_{ths} may be taken as approximately 100.

It should be noted also that the manner of mounting a resistor may reduce the effective thermal surface resistance; thus, if a resistor element of the flat type contained in a metal cover is mounted flush on an enclosing metal case, the effective thermal area A is much increased over the surface area of the resistor.

The wattage to be dissipated in a resistor may be calculated by means of the following equations:

$$W = I^2R = \frac{E^2}{R} = EI$$

where W is the steady-state power dissipated in watts, I is the current flowing in amperes, R is the resistance in ohms, and E is the voltage across the resistor in volts.

When a resistor is loaded intermittently, with current I amperes, the steady-state watts, W, may be ascertained as follows:

$$W = \frac{NI^2Rt}{60}$$

where N is the number of duty cycles per minute, t is the time in seconds that the current flows through the resistor per duty cycle, and R is the resistance in ohms.

TEMPERATURE-RISE LIMITS. The permissible temperature rise is governed by the resistor materials and by the effect on surrounding materials. Table 1, Column 6, outlines maximum temperature limits imposed by resistor construction, and the National Electrical Manufacturers Association Standards for Industrial Control provide a guide for design in Section IC4-22, "Temperature of Resistors," as follows:

1. For bare resistive conductors the temperature rise shall not exceed 375 deg cent as measured by a thermocouple in contact with the resistive conductor.
2. For resistor units, rheostats, and wall-mounted rheostatic dimmers, having the resistive conductor embedded, the temperature rise shall not exceed 300 deg cent as measured by a thermocouple in contact with the surface of the embedding material.
3. For rheostatic dimmers having embedded resistive conductors and arranged for mounting on switchboards or in noncombustible frames, the temperature rise shall not exceed 350 deg cent as measured by a thermocouple in contact with the surface of the embedding material.
4. The temperature rise of the issuing air shall not exceed 175 deg cent as measured by a mercury thermometer at a distance of 1 in. from the enclosure.

3. RESISTOR SELECTION

The selection of a resistor may require consideration of a number of factors, as outlined in Article 2 and Tables 1, 2, and 3, depending upon any specific standard that must be met. For example, Underwriters' Laboratories Industrial Control Standards specify the minimum leakage distance from resistor lug to nearest ground as follows: 51- to 150-volt circuit, $1/4$ in.; 151- to 300-volt circuit, $3/8$ in.; 301- to 600-volt circuit, $1/2$ in. For many applications, however, all other factors are subordinate to electrical requirements, and an approximate selection may be made on the electrical basis alone. Thus, if the resistance and current required are known, the wattage to be dissipated may be calculated. This figure may be multiplied by 2, 3, 4, or more, as a factor of safety to allow for overload or for reduction in heat dissipation due to enclosure, proximity, accumulation of dust, etc. With this multiplied wattage as the free air rating, a suitable resistor may be selected from manufacturers' standard designs.

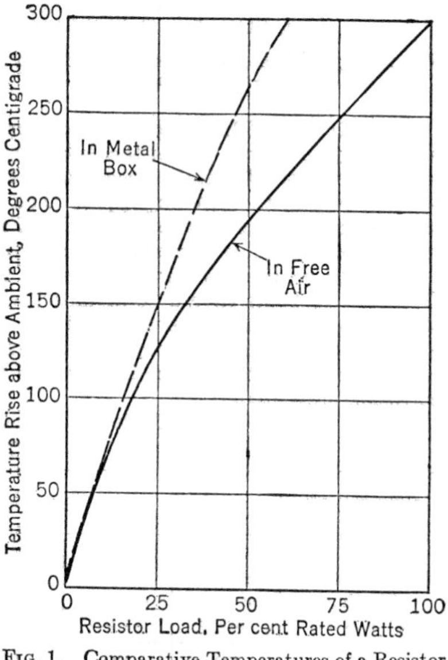

FIG. 1. Comparative Temperatures of a Resistor in Free Air and Enclosed

EFFECT OF ENCLOSURE. Figures 1 and 2 show thermal data obtained on tubular-type vitreous enameled resistors and illustrate the loss in rating caused by enclosure and by lowering the permissible temperature rise. The "In Free Air" curve of Fig. 1 illustrates how the temperature rise of a resistor above the ambient temperature varies with the percentage of rated watts. The relation is not a direct one; i.e., reducing the wattage by 50 per cent lowers the temperature by only 35 per cent, etc. The curve is an average of many sizes and is only approximate for a particular size of resistor. The second curve in Fig. 1 illustrates how a typical resistor enclosed in a metal box with restricted circulation of air attains its specified maximum temperature at a wattage much below its free air rating. The curve "In Free Air," Fig. 2, simplifies the calculation of the resistor rating for any desired temperature rise. To determine the nominal or "free air rating" of a resistor to operate approximately at any given temperature:

1. Start at the desired temperature, and follow a horizontal line across to where it meets the proper curve ("In Free Air" or "In Enclosure," as the case may be).
2. Follow a vertical line from the intersection down to the axis.
3. Obtain the required nominal or free air watt rating by multiplying the actual watts by the number located in Step 2.

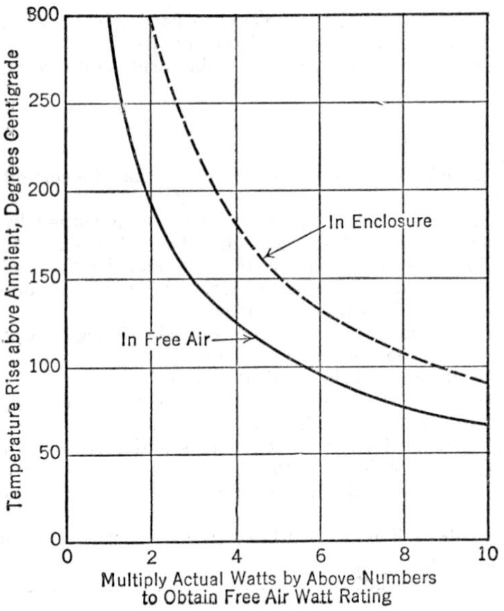

FIG. 2. Approximate Temperature Rise of a Resistor at Various Load Ratios

Several standards for particular types of resistors have been approved by the American Standards Association, and the National Electrical Manufacturers Association Standards for Industrial Control contains a number of sections useful in selecting resistors. Table 4 shows the NEMA classification of resistors with respect to duty cycles and current loading, and Table 5 shows the resistor classifications which in general are correct for various installations.

CAST-IRON-GRID RESISTORS. For heavy-duty service cast-iron-grid resistor units are usually employed, except that where severe vibration is a factor, non-breakable ribbon or edgewise-wound units are recommended. There are two styles of grid resistor units, the narrow and the wide lug types. The narrow lug type is for intermittent duty, since the construction permits the assembly of a large mass of metal in a small space, thereby making possible maximum utilization of the thermal capacity of the grids. The wide lug type is for continuous service, because the wider spacing provides better ventilation. Grid resistor boxes should be mounted with grids in a vertical plane, stacked not more than six boxes high, each box spaced 6 in. apart, and each stack 12 in. apart. It is advisable to place 10-mil soft-copper washers between all current-carrying surfaces before bolting them together, since in this way moisture can be excluded from the joints and corrosion prevented.

Table 4. Classification of Resistors
(NEMA)

Approximate Per Cent of Full-load Current on First Point Starting from Rest	Class Numbers Applying to Duty Cycles						
	5 sec on out of each 80 sec	10 sec on out of each 80 sec	15 sec on out of each 90 sec	15 sec on out of each 60 sec	15 sec on out of each 45 sec	15 sec on out of each 30 sec	Continuous duty
25	111	131	141	151	161	171	91
50	112	132	142	152	162	172	92
70	113	133	143	153	163	173	93
100	114	134	144	154	164	174	94
150	115	135	145	155	165	175	95
200 or over	116	136	146	156	166	176	96

Table 5. Resistor Application Standards
(NEMA)

Installation	Resistance Class Number
Blowers	
Centrifugal	133-93
Constant-pressure	135-95
Brick plants	
Augers	135
Conveyors	135
Dry pans	135
Pug mills	135
By-products coke plants	
Door machine	153
Leveler ram	153
Pusher bar	153
Rev. machines	153
Cement mills	
Conveyors	135
Crushers	145
Elevators	135
Rotary dryers	145-95
Grinders—pulverizers	135
Kilns	135-95
Coal and ore bridges	
Bridge	153
Closing	162
Holding	162
Trolley	163
Coal mines	
Car hauls	162
Conveyors	135-155
Cutters	135
Crushers	145
Fans	134-93
Hoists—slope	172
Vertical	162
Jigs	135
Picking tables	135
Rotary car dumpers	153
Shaker screens	135
Compressors	
Constant speed	135
Varying speed—centrifugal	93
Plunger type	95
Concrete mixers	135
Flour mills	
Line shafting	135
Cranes—general-purpose	
Hoist	153
Bridge—sleeve bearing	153
Bridge or trolley—roller bearings	152
Trolley—sleeve bearing	153
Food plants	
Butter churns	135
Dough mixers	135
Hoists	
Winch	153
Mine slope	172
Mine vertical	162
Contractor's	152
Larry cars	153
Lift bridges	152
Machine tools	
Bending rolls—Non-rev.	163-164
Rev.	163-164
Boring mills	135
Bulldozers	135
Drills	115
Gear cutters	115
Machine tools—*Continued*	
Grinders	135
Hobbing machines	115
Lathes	115
Milling machines	115
Presses	135
Punches	135
Saws	115
Shapers	115
Metal mining	
Ball—rod—tube mills	135
Car dumpers—rotary	153
Converters—copper	154
Conveyors	135
Crushers	145
Tilting furnace	153
Paper mills	
Beaters	135
Calenders	154-92
Pipeworking	
Cutting and threading	135
Expanding and flanging	135-95
Power plants	
Clinker grinders	135
Coal crushers	135
Conveyors—belt	135
Screw	135
Pulverized-fuel feeders	135
Pulverizers—ball type	135
Centrifugal	134
Stokers	135-93
Pumps	
Centrifugal	134-93
Plunger	135-95
Rubber mills	
Calenders	155
Crackers	135
Mixing mills	135
Washers	135
Steel mills	
Accumulators	153
Casting machines—pig	153
Charging machines—bridge	153-163
Peel	153-163
Trolley	153-163
Coiling machines	135
Converters—metal	154
Conveyors	135-155
Cranes—ladle-bridge-trolleys—	
Hoists	153-163
Roller bearings	152-162
Sleeve bearings	153-163
Crushers	145
Furnace doors	155
Gas valves	155
Gas washers	155
Hot-metal mixers	163
Ingot buggy	153
Kickoff	153
Levelers	153
Manipulator fingers	153-163
Pickling machine	153
Pilers—slab	153
Racks	153
Reelers	135
Saws—hot or cold	155
Screw downs	153-163

Table 5. Resistor Application Standards—*Continued*
(NEMA)

Installation	Resistance Class Number	Installation	Resistance Class Number
Steel mills—*Continued*		Steel mills—*Continued*	
Shears	155	Tilting furnace	153
Shuffle bars	155	Wire-stranding machine	153
Side guards	153-163	Woodworking plants	
Sizing rolls	155	Boring machines	115
Slab buggy	155	Lathe	115
Soaking-pit covers	155	Mortiser	115
Straighteners	153	Molder	115
Tables—approach	153	Planers	115
Lift	153-163	Power trimmer and mitre	115
Roll	153	Sanders	115
Shear approach	153-163	Saws	115
Transfer	153	Shapers	115
Main roll	153-163	Shingle machine	115

ENAMELED WIRE RESISTORS. Vitreous enameled wire wound resistors are available with an adjustable lug for use where final or periodic adjustments in resistance are necessary. A narrow strip of winding along one side is left bare so as to provide a conducting surface for the contact button on the adjustable lug.

NON-INDUCTIVE RESISTORS. All resistor types, except cast and ribbon, are available as non-inductive resistors. The wired units must be specially wound. The hairpin or bifilar winding, although non-inductive, is characterized by high capacitance and high voltage between turns, either of which may be a disadvantage. Pie or multiple-layer windings may reduce the power rating for a given size. The Aryton-Perry winding, which consists of two windings in parallel wound in opposite directions, overcomes the above-mentioned disadvantages. This design, however, requires more space for a given resistance.

Where the ultimate in precision and stability is required, hermetically sealed resistors are employed.

4. RHEOSTAT DESIGN AND SELECTION

Rheostat design is governed for the most part by the factors discussed in Article 2. Thus the usual methods of assembling and installing rheostats place them in the class of resistors with restricted air circulation, resulting in a thermal surface resistivity, ρ_{ths}, of 100 thermal ohms per sq in. of surface.

Table 6. Characteristics of Rheostats Available Commercially

Rheostat * Type	Resistance Range, Ohms	Current Range, Amperes	Power Range, Watts	Inductive or Non-inductive	Number of Steps	Control	Approximate Dimensions, Inches
Tubular	0.3 to 32,000	0.1 to 25	65 to 1200	Either	20 to 6000	Slider, vernier	200 watts: 8 × 2 diam. 1000 watts: 20 × 3 diam.
Ring	0.5 to 10,000	0.07 to 31	25 to 1000	Inductive	17 to 1375	Knob, screw, motor	100 watts: 1.75 × 3 diam. 1000 watts: 3 × 12 diam.
Circular plate	2 to 2000	0.2 to 44		Inductive	28 to 124	Knob, motor	
Unit	1.6 to 4400	0.1 to 10		Either	20 to 100	Knob, motor	
Cast grid	0.25 to 40	5 to 640		Inductive	60	Knob, motor	
Compression carbon	0.001 to 3000	6 to 800	20 to 5400	Non-inductive	Infinite	Knob	1000 watts: 15 × 11 × 7 5000 watts: 37 × 19 × 10

* See Tables 1 and 2 for properties of water and electrolytes which may be used as conducting mediums for liquid rheostats.

For a given set of rheostat conditions (resistance, maximum and minimum currents) any rheostat with a wattage rating greater than the required summation watts ($W = I_{max.} \times I_{min.} \times R_{rheo.}$) can be used. The choice among models will depend upon space requirements, mounting conditions, and quantities ordered. Since tapered windings involve extra manufacturing operations, they generally cost more than uniformly wound rheostats unless made in large quantities, when the material savings dominate. Characteristics of commercially available types are given in Table 6.

5. BIBLIOGRAPHY

Manufacturers' catalogs.
National Electrical Manufacturers Association Standards for Industrial Control.
American Standards Association Standards.

CAPACITORS

By J. W. Butler and J. D. Stacy

6. DESIGN

GENERAL. A capacitor, formerly known as a condenser or static condenser, is a device which has the capability of storing energy by the rearrangement of electrons in the dielectric material which forms its active element when a potential difference is applied to its electrodes, and of releasing this energy under the control of changes in the applied potential difference over a very wide time range. This concept of energy storage is useful in the majority of capacitor applications. Another concept, that the capacitor is a device which develops, by the rearrangement of electrons in the dielectric material, a counter electromotive force equal to the electromotive force applied to its electrodes, is useful in such applications as the blocking of direct current from a circuit through which a-c flow is desired.

The amount of energy stored in a capacitor is given by the equation:

$$W \text{ (watt-seconds or joules)} = \frac{C \text{ (microfarads)} \times E^2 \text{ (kilovolts)}}{2}$$

CAPACITOR TYPES. Capacitors are broadly classified by the major dielectric material used in their construction: (1) paper, (2) mica, (3) electrolyte, (4) gas, (5) liquid, (6) vacuum, (7) resin film, (8) ceramic. Some of these may be further subclassified on the basis of a minor dielectric material used in commercial manufacture. For instance, paper capacitors are commercially available with the following dielectric materials combined with paper in the dielectric: mineral waxes, chlorinated naphthalene,* mineral oils, vegetable oils, chlorinated diphenyl,† hydrogenated castor oil,‡ and others of present minor importance. See Fig. 1 for a sectionalized view of a typical capacitor, and Table 1 for physical data on a few popular dielectrics.

Table 1. Electrical Characteristics of Some Dielectrics

(General Electric Works Laboratory, Pittsfield, Mass.)

	Physical Condition at 25° C	Dielectric Constant		Approximate Per Cent Power Factor at Low Frequency	
		25° C	75° C	25° C	75° C
Chlorinated naphthalene	Solid	4.30	4.15	0.4	10
Chlorinated diphenyl	Viscous liquid	5.10	4.50	0.005	0.40
Mineral oil	Liquid	2.17	2.10	0.0014	0.01
Hydrogenated castor oil	Solid	10.2	4.28	5.3	11.7
Mica (ruby) *	Solid	7.06–7.22		0.06–0.08	0.15–0.20

* C. Dannat, *J. Inst. Elec. Eng.* (April, 1931).

Mica dielectric capacitors are suited to many high-frequency applications and are available in a variety of styles according to their application. Major field of application is in communications and induction heating equipment.

* Trade name: 1001 Halowax.
† Trade names: Pyranol, Inerteen, Dykanol A, etc.
‡ Trade name: Opal Wax.

Electrolytic capacitors employ as dielectric an anodic film built up by electrolytic action on the surface of aluminum electrode plates. This dielectric is unidirectional, making it necessary to have the film built upon both electrodes of a capacitor for a-c operation, whereas a d-c capacitor may use one filmed electrode and one without film. Alternating-current electrolytic capacitors are used mainly for starting low-voltage (110–220 volts) single-phase motors where the capacitor is connected momentarily in series with a starting winding which provides a flux characteristic similar to that obtained in a two-phase motor and, thereby, a relatively high starting torque. To avoid overheating and damage to the capacitor the capacitor circuit is opened as the motor comes up to speed.

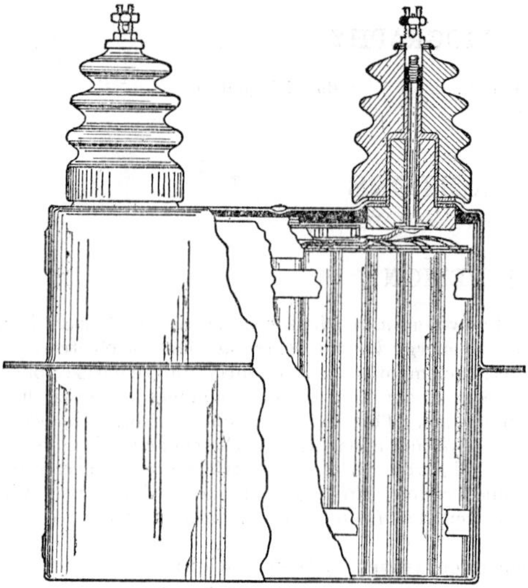

Fig. 1. Fifteen-kva, 2400-volt Capacitor

Gas dielectric is employed in the metal-plate types of variable tuning capacitors. In the majority of cases the gas is air at atmospheric pressure, but for high-voltage and high-power radio-frequency circuits the capacitors may be built into sealed containers, operate at high pressure, and use gases other than air.

Liquids may be employed as the dielectric in the metal-plate type of capacitors for applications to high-frequency circuits where considerable variation of capacitance with temperature is permissible.

Vacuum capacitors find application in very-high-frequency equipment for industrial heating.

Resin films are used to some extent as alternates for mica and paper in small capacitors. Some resins have excellent dielectric properties and, when combined with paper by coating or impregnation, produce materials which have suitable characteristics for commercial capacitors.

Ceramic capacitors are usually made from compounds of titanium, magnesium, barium, or strontium, although other materials, such as glass and porcelain, have been used in the past in limited quantities. The major application of ceramic capacitors is for electronic equipment, where they are employed with a mica capacitor to secure low capacitance variation with temperature.

Table 2 shows the approximate ranges of capacitance and voltage which may be obtained commercially in the most frequently used types of capacitors. It should be understood that the combinations of capacitance and voltage rating are limited by established standards, which should be consulted in the selection of suitable ratings for any application.

Table 2. Approximate Ranges of Capacitors

Type	Capacitance Range, Microfarads	Voltage Range, Volts
Paper	0.0001 to 375	100 to 100,000 d-c 110 to 13,800 a-c
Mica	0.000005 to 0.1	500 to 35,000 d-c
Electrolytic (a-c)	10 to 500	25 to 250 a-c
Electrolytic (d-c)	5 to 15,000	5 to 450 d-c
Gas	0.000001 to 0.1	200 to 60,000 d-c
Liquid	0.001 to 0.05	1000 to 15,000 a-c
Vacuum	0.000001 to 0.001	5000 to 50,000 d-c or peak a-c
Resin film	0.0005 to 0.1	500 to 10,000 d-c
Ceramic	0.0000005 to 0.002	500 to 15,000 d-c

CAPACITANCE. Capacitance is a function of the dielectric characteristics and dimensions of the materials used in the capacitor dielectric. Capacitance determines the amount of energy which can be stored with unit charging voltage and is therefore obviously independent of voltage applied. The primary variables determining capacitance are the area of dielectric between electrode plates (a), the thickness of the dielectric material (d), and the dielectric constant of the material (k), which may be combined into the following:

$$C \propto \frac{ak}{d}$$

The commonly used units of measurement of capacitance are the microfarad (mf or μf) and the micromicrofarad (mmf) or $\mu\mu$f). Introducing the appropriate constants gives the following practical equation for a plane capacitor:

$$C \text{ (microfarads)} = \frac{2248 \times 10^{-10} a \text{ (square inches) } K}{d \text{ (inches)}} \qquad (1)$$

Note from this equation that capacitance has the dimensional formula of length (L) and that capacitance is sometimes expressed in centimeters, where 1 cm = 1.1263 $\mu\mu$f.

The dielectric constant K used in many types is a composite constant having components which are characteristic of the spacer medium and impregnant. For this reason tables of the dielectric constant of individual materials are of little value to the designer until he knows the amount of each material entering the composition.

Dielectric constants of materials are usually **constant** for only one set of conditions. They may have a variation with temperature and frequency of applied voltage which impart to capacitors made from them a characteristic variation with temperature and an interrelated variation with frequency. Moisture absorption changes the dielectric constant of many materials, and therefore it is always necessary to dry materials thoroughly to obtain their true characteristics.

From the microfarad and voltage rating of a capacitor, other ratings can be determined by the following formulas:

$$\text{Ohms} = \frac{10^6}{2\pi f C}$$

where f = frequency in cycles per second, and C = microfarads.

$$\text{Kilovars} = 2\pi f C E^2 10^{-3}$$

where E = kilovolts.

In the manufacture of capacitors the control of capacitance depends upon the control of dimensions of materials. The difficulty of such control causes wide differences in the readily available tolerances in the different classes of capacitors. Air capacitors with large electrode spacing d and large electrode area a can be held to tolerances of a fraction of 1 per cent, whereas paper capacitors involving dielectric thickness of about 0.001 in. and small electrode area may require a tolerance of ± 50 per cent to secure a satisfactory yield of product from manufacture. In larger paper capacitors tolerances of ± 10 per cent are usually acceptable to the manufacturer. NEMA Standards for power-factor capacitors allow a variation of 0 to +15 per cent in capacitance, which is equivalent to $\pm 7\,^1/_2$ per cent.

POWER LOSS IN CAPACITORS. In practical construction of capacitors of all types, several factors may be introduced which will cause variation from the ideal pure capacitance. These factors, which cause loss of power in the energized capacitor, may be due to one or more of the following causes:

1. Resistance of electrodes, leads, and terminals through which the charging current flows.

2. Frictional losses which occur within the dielectric material with each rearrangement of the electrons in the molecular structure. These losses are manifested by an absorption of charge for a considerable period after application of a steady charging voltage and likewise the failure to deliver all the energy stored in the capacitor upon instantaneous discharge. This characteristic is known as dielectric absorption. The losses occurring from this cause are frequently called dielectric hysteresis losses.

3. Leakage current or ionic conduction occurring in the dielectric material. This current is usually expressed in terms of the resistance of the capacitor or insulation resistance.

4. Ionization of a portion of the dielectric and bombardment of adjacent dielectric by this ionized portion.

For a-c applications it is customary to consider all the factors of power loss as a group, since all of them cause a displacement of the current vector so that it is less than 90 deg from the voltage vector. The power lost is given by the equation

$$P = EI \cos \theta$$

where $\cos \theta$ is known as the power factor of the capacitor. It is sometimes convenient to express this loss as a function of the angle ϕ by which the current vector has been displaced from the ideal of 90 deg. Power loss is then given by

$$P = EI \sin \phi$$

For small angles $\sin \phi = \phi = \tan \phi$. Some measuring equipment measures $\tan \phi$, which is known as dissipation factor. Power factor and dissipation factor may be considered equal for many types of capacitors where the power loss in watts is a small percentage of the volt-amperes measured at the capacitor terminals. It is sometimes convenient to express the power loss in terms of the loss which would occur in a combination of pure capacitance and an equivalent series resistance as shown in Fig. 2. This relation is shown by the equation

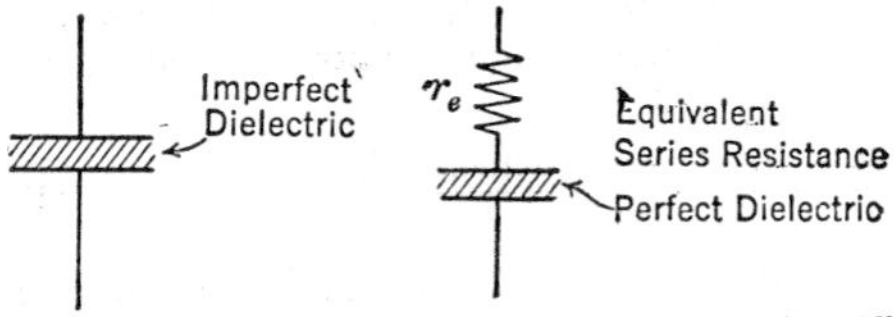

FIG. 2. Equivalent Circuit of Capacitor, Showing All Loss Effects Lumped into One Equivalent Series Resistance

$$P = EI \cos \theta = EI \sin \phi = wCE^2 \sin \phi = I^2 r$$

Power factors of the finished capacitor vary greatly with the type and may range from 0.00001 for a well-designed gas or vacuum capacitor to perhaps 0.70 for certain electrolytic and ceramic types.

Since power factor is a measure of the power lost as heat in a capacitor, it constitutes a major limitation on the volt-ampere rating which can be placed on a capacitor design for a-c application. The power factor of a capacitor may be a function of applied voltage, temperature, and applied frequency. It is, therefore, essential to know these characteristic variations in order to be certain that a capacitor is properly applied in a circuit.

CHARACTERISTICS OF CAPACITORS. As examples of the variations with temperature which occur in power factor and insulation resistance, Figs. 3 and 4 are shown for typical paper-dielectric capacitors impregnated with different liquids. Both these characteristics are used as indicators of quality. In so doing, however, care must be exercised, since they may be misleading. The capacitor with the lowest power factor or the highest insulation resistance may not have the longest life in service. These characteristics are useful tools of the manufacturer for process control but should have a relatively minor place in a capacitor specification. Allowance must be made for inherent differences caused by the characteristics of the impregnating material or the spacer material.

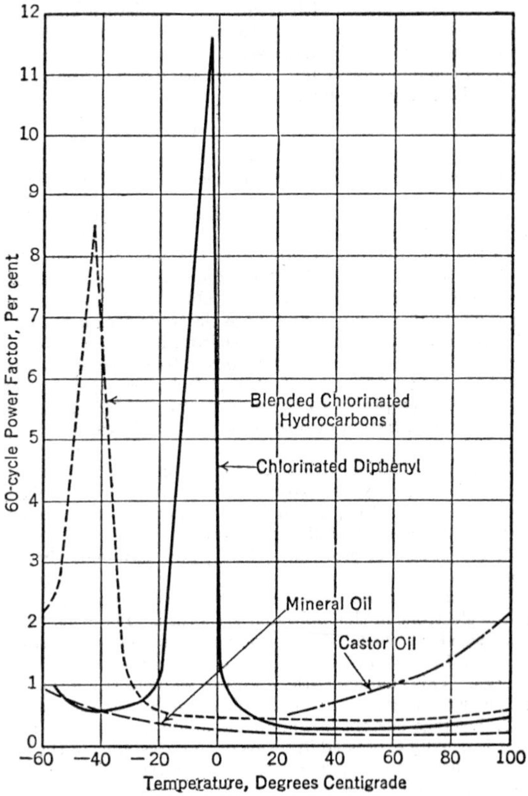

FIG. 3. Variations of Power Factor with Temperature for Four Popular Dielectrics

Life tests are the practical means by which capacitors of the many commercial types can be adequately compared. Even in making life tests care must be taken not to select accelerated conditions which will adversely affect one type with respect to another or one size with respect to another.

For a-c tests moderate ambient temperatures (30 to 85 deg cent) and moderate overvoltage (20 to 50 per cent above rating) are desirable, with the duration of the test extended as much as possible—certainly not less than 1000 hr. Measurements of power factor made initially and at regular intervals along the course of the test will indicate the relative stability of the dielectric in the types tested. Heat generated within the capacitor must be removed to avoid destructive interior temperatures, and the conditions for such removal in the capacitor design will have an important bearing on behavior in the life test. For this reason large capacitors should be tested under very moderate conditions of ambient temperature and overvoltage for a long time, perhaps 1 yr, whereas small capacitors may be subjected to a greater degree of temperature and voltage acceleration for a shorter time.

For d-c life tests an ambient temperature for many types of capacitors was established in the temporary standards during World War II. It is expected that these temperatures will also be adopted as RMA and NEMA standards. The amount of overvoltage to be applied may vary with the energy-storage capacity and the voltage rating of the design, with the higher-capacity and higher-voltage designs receiving the lesser voltage. For liquid-impregnated paper capacitors, the ambient testing temperature is 85 deg cent, the test voltage may vary from 80 to 190 per cent of 40 deg cent rated voltage as stated in the test standard, and the usual duration is 250 hr. It is usual to allow one failure from twelve samples tested in judging results from such tests.

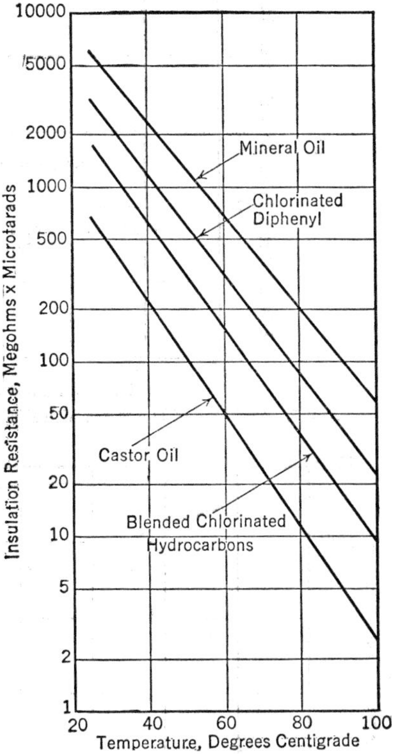

FIG. 4. Variation of Insulation Resistance with Temperature for Four Popular Dielectrics

VOLTAGE RATINGS AND LIFE. Voltage ratings are selected from a consideration of the service conditions under which the capacitor is to operate and the desired life expectancy. For a gas or vacuum capacitor, establishment of voltage rating involves determination of flashover and corona voltages and the current-carrying ability of the parts at the maxima of frequency, ambient temperature, and altitude to be encountered, with adequate factors of safety based on life-test experience applied. For a liquid-filled paper capacitor, consideration should be given to allowable factory-test voltage, corona starting voltage, heating characteristics, and life-test experience at the maxima of frequency, ambient temperature, and altitude to be encountered. Life expectancy is largely dependent upon the nature of the dielectric, the voltage stress on the dielectric, and the hot-spot temperature of the dielectric. The effect of stress and temperature on life is illustrated by Figs. 5 and 6.

In a large capacitor such as that used for power-system power-factor correction, the hot-spot temperature has a major effect on the allowable voltage stress and consequent physical size of an adequate design. Capacitor power factor and the heat-dissipation characteristics of the design determine the hot-spot temperature. A large capacitor unit usually has a higher temperature gradient between hot spot and ambient than a small one. Power factor depends upon the selection, processing, and control of materials and adequate drying and impregnation. It follows, therefore, that the manufacturer who exercises the greatest care in the selection of materials and the control of processes can reap the benefit of using a relatively high voltage stress in his design and thereby secure smaller physical size with better heat-dissipation characteristics and equal or greater life expectancy.

MISCELLANEOUS VARIABLES IN CAPACITOR DESIGN. In addition to the major variables, which are largely a function of the dielectric, the proper design of a capacitor for any application must involve consideration of several other requirements.

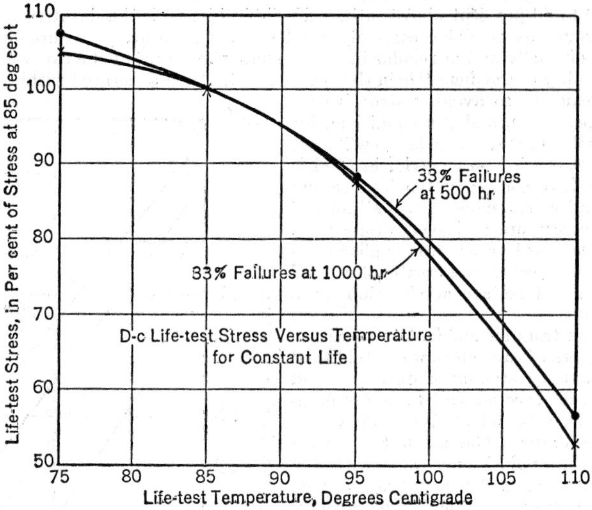

FIG. 5. Variation of D-c Stress with Dielectric Temperature for Constant Capacitor Life

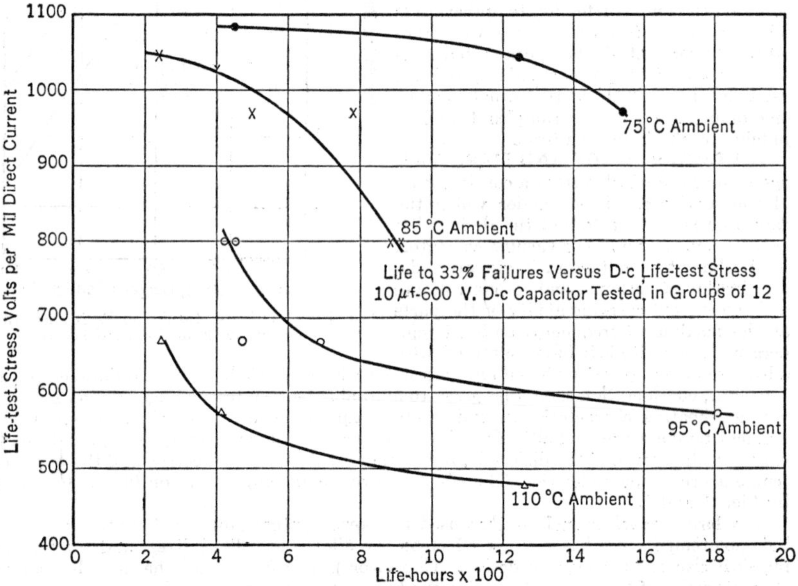

FIG. 6. Variation of D-c Stress with Capacitor Life for Constant Ambient Temperatures

Current-carrying Capacity of Parts. In many applications high surge currents or high-frequency currents are involved, and the design of conductors must be such that damaging temperatures will not be produced or mechanical damage will not result from large magnetic forces set up in the parts.

Inductance of Capacitor Structure and Current-carrying Leads. Every capacitor may be considered a pure capacitance in series with a resistance and an inductance. The magnitude of the inductance is subject to a wide range of variation by differences in mechanical design, and great attention must be given to these factors in applications where unwanted high-frequency currents must be drained from critical circuits.

Series Resistance of Current-carrying Parts. Some applications in resonant circuits require low resistance in order that minimum impedance (series resonance) or maximum impedance (parallel resonance) of the circuit may be obtained. Proper consideration must be given to the distribution of current in conductors, as well as their length and cross-section.

Effects of Atmospheric Conditions. Humidity, temperature variations, corrosive materials, and pressure changes may have serious effects on the service life of capacitors. The sealing of capacitors to avoid deterioration due to the high humidity and temperature ranges encountered in the military operations of World War II was a major problem. Adequate protection must be given to metallic surfaces when capacitors are to encounter salt water or corrosive gases. Terminal bushings resistant to fungus growth may be required. Bushings and internal connections must be designed for the low pressures of high altitude when that service is expected.

Case Insulation. Normally, capacitors in metallic cases have, between the active electrodes and the case, insulation which is commensurate with the terminal-to-terminal voltage rating of the capacitor. In some applications the terminals may be operating at a much higher voltage with respect to the metal on which the capacitor would be mounted than such voltage rating. Under these circumstances it is necessary either to raise the insulation level of the capacitor terminals by using larger bushings and more case insulation or to mount the capacitor on adequate insulation. In the second of these alternatives it is good practice to connect the case to one terminal of the capacitor to avoid trouble from faulty voltage distribution between the case insulation and the mounting insulation.

Shape. Capacitors of all types can be made in a large variety of shapes, so that a tremendous number of commercial designs have resulted. Both capacitor users and capacitor manufacturers are making a serious effort to standardize sizes and reduce the number of available items for the economic benefit of all.

7. APPLICATIONS

KILOVAR SUPPLY. The price trend of capacitors for kilovar supply (power-factor improvement) since 1930 is shown by Fig. 7. This decline in price has obviously greatly

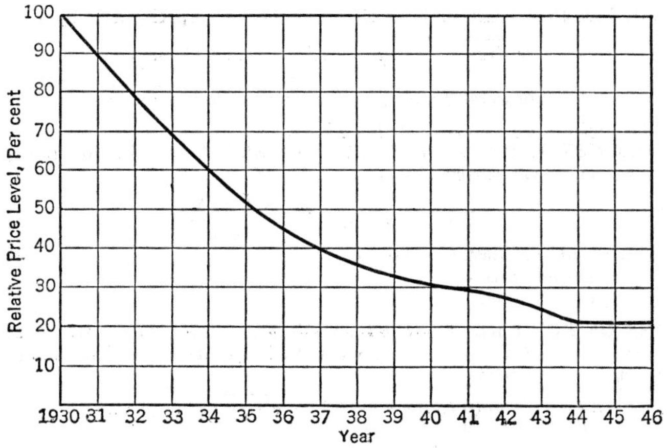

Fig. 7. Relative Price Level of Power-factor Capacitors Since 1930

stimulated their use. It is estimated that over 9,000,000 kvar are now supplied in this country by capacitors. The use of capacitors for kilovar supply started with the application of small equipments in industrial plants, later followed by individual units mounted on poles by power companies. As their application developed and prices were reduced, capacitors were built into housed equipments and used at substations like synchronous condensers. Banks up to 25,000 kvar switched in 2500-kvar increments have been installed. Some of the economies and benefits obtained by applying capacitors may be determined by the following selected formulas.

VOLTAGE IMPROVEMENT. Voltage rise on a three-phase circuit due to capacitors may be determined by the following relation; for single-phase circuits multiply by 2 the value given by eq. (2).

$$\text{Per unit voltage rise} = \frac{KvarXd}{1000Kv^2} \tag{2}$$

where $Kvar$ = three-phase capacitor rating.
X = reactance in ohms per mile to neutral.
d = length of circuit in miles.
Kv = line-to-line kilovolts.

ENERGY-LOSS REDUCTION. Assuming the kilovars leaving the generators take substantially the same paths through the system as the kilowatts that leave the same generator, the energy loss due to kilovar transmission may be readily determined by the relation:

$$Wq = \frac{1}{(Lp/Lq)\tan\theta + 1} = \text{Energy loss due to kilovars in per unit of total system } I^2R \text{ losses} \tag{3}$$

where Lp = annual loss factor of kilowatts.
Lq = annual loss factor of kilovars.
θ = peak load power-factor angle.

Savings through power-factor improvement by supplying kilovars near the load with capacitors can readily be determined by this formula, which is shown plotted in Fig. 8.

ECONOMIC OPERATING POWER FACTOR. When the system investment used by kilovars at the time of peak load equals the price of capacitors, the system can be said to be operating at its economical power factor. Two different bases for allocating system investment against kilovars are generally used in such an analysis.

a. Allocate investment used on basis of kilowatt capacity used by kilovars, e.g.,

$$Kva - Kw = \text{Capacity used by kilovars}$$

Then

$$\text{Economic operating power factor} = \frac{1 - (c/s)^2}{1 + (c/s)^2} \tag{4}$$

where c = cost of capacitors — \$/*kvar*.
s = value of system — \$/*kw*.

b. Allocate investment used on basis of kva or thermal capacity used by kilovars, e.g.,

$$\sqrt{Kva^2 - Kw^2} = \text{Capacity used by kilovars}$$

Then

$$\text{Economic operating power factor} = \sqrt{1 - (c/s)^2} \tag{5}$$

Relations (4) and (5) are shown plotted in Fig. 9.

VOLTAGE RISE THROUGH A TRANSFORMER. A capacitor on the secondary side of a transformer will cause a rise of approximately

$$\text{Per cent voltage rise} = \frac{Kvar\ X}{Kva} \tag{6}$$

where X = per cent transformer reactance.
Kva = transformer rating.
$Kvar$ = capacitor rating.

RELEASE OF EQUIPMENT THERMAL OR KVA CAPACITY. By improving the power factor from $\cos\theta_1$ to $\cos\theta_2$, the per unit kva capacity released for additional load at the original power factor is

$$\Delta T_c = \sin\theta_1 (\Delta \tan) \cos\theta_1 - 1 + \sqrt{1 - \cos^4\theta_1 (\Delta \tan^2 \theta_1)}$$

where

$$\Delta \tan = \tan\theta_1 - \tan\theta_2$$

For ready reference this relation is plotted in Fig. 10.

OTHER USEFUL RELATIONS.

1. Kilovar output of a capacitor varies as the square of the voltage.
2. Kilovar output varies directly with the frequency.
3. Current output varies directly with the voltage.

INDUSTRIAL APPLICATION. Industrial plants install capacitors primarily to reduce their power bill if the rate schedule has a power factor or kva demand clause. There are, however, applications where released capacity, reduced losses, or improved voltage conditions justify the application of capacitors. In these cases the formulas given apply.

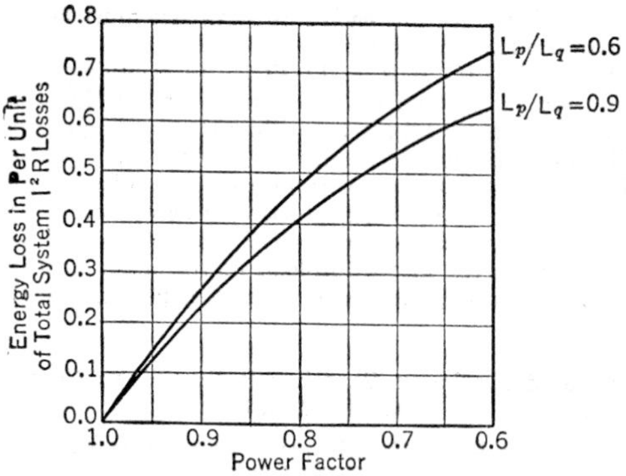

FIG. 8. Energy-loss Reduction by Power-factor Improvement in Terms of Total System I^2R Losses

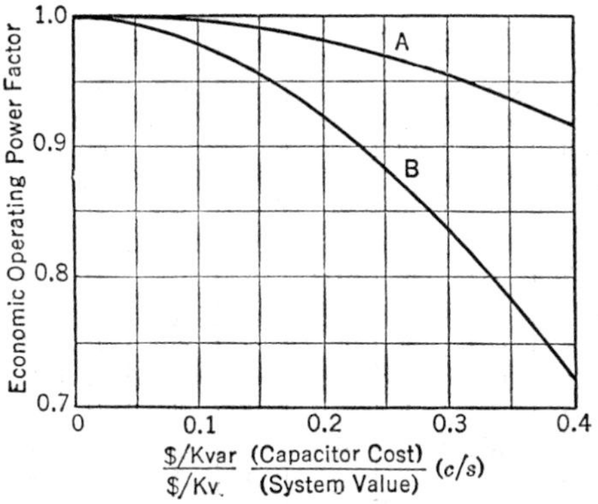

FIG. 9. Operating Power Factor at Which System Investment Usage by Kilovars Equals the Cost of Capacitors

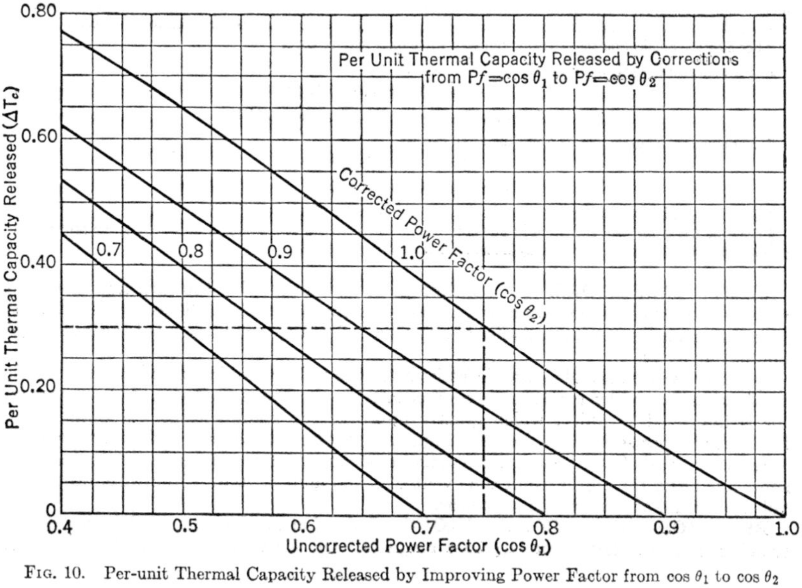

FIG. 10. Per-unit Thermal Capacity Released by Improving Power Factor from $\cos\theta_1$ to $\cos\theta_2$

Where power-bill reduction is sought, each rate structure must be considered on its own merits as it applies to the load in question.

SWITCHING WITH INDUCTION MOTORS. In applying capacitors to induction motor terminals and switching them with the motors, a certain precaution is necessary. If a capacitor is applied that will supply more than the exciting current of the motor, the motor voltage will build up, when the motor is disconnected, because of what is called self-excitation. This can harm the motor from excessive voltage, or in transferring from starting to running tap (if partial-voltage starting is used) destructive transient torques may obtain as a result of the phase differences of the system- and motor-generated voltages. In general, improving the motor full-load power factor to 95 per cent or less prevents this difficulty. Table 3 gives recommended capacitor sizes for different motor ratings.

Table 3. Recommended Maximum Capacitor Rating When Capacitor and Motor Are Switched as a Unit *

Representative Data for Three-phase 60-cycle General-purpose Open-type Induction Motors of 220-, 440-, 550-, and 2300-volt Rating †

Induction Motor, Horsepower Rating	Motor Speed, Revolutions per Minute											
	3600		1800		1200		900		720		600	
	Ckva ‡	Per Cent AR §	Ckva	Per Cent AR	Ckva	Per Cent AR	Ckva	Per Cent AR	Ckva	Per Cent AR	Ckva	Per Cent AR
10	2.5	9	4	11	4	12	5	17	5	23	7.5	28
15	2.5	9	5	11	5	11	7.5	16	7.5	21	10	26
20	5	9	5	10	5	11	7.5	15	10	20	12.5	24
25	5	9	7.5	10	7.5	10	10	14	10	19	15	22
30	7.5	9	10	9	10	10	10	13	12.5	18	15	21
40	10	9	10	9	10	10	12.5	12	15	16	17	19
50	12.5	9	12.5	9	12.5	9	15	12	20	15	22.5	17
60	15	9	15	8	15	9	17.5	11	22.5	14	25	16
75	17.5	9	17.5	8	17.5	8	20	11	27.5	13	30	15
100	22.5	9	22.5	8	22.5	8	25	10	35	12	37.5	14
125	25	9	27.5	8	27.5	8	30	9	40	11	47.5	13
150	32.5	9	35	8	35	8	37.5	9	47.5	11	55	13
200	42.5	9	42.5	8	42.5	8	45	9	60	10	67.5	12

* When a manual reduced-voltage auto-transformer type of starter, such as is shown in Fig. 2, is used, refer to the method described in the text for selecting the capacitor rating if load data are appreciably higher than those of the motor. The motor full-load power factor, with capacitor ratings as listed in this table, will range from 95 to 98 per cent.

† For 50-cycle application the following representative data may be used:

1. For standard 60-cycle motors operating at 50 cycles:

$$\text{Ckva} = \text{Ckva from table} \times 1.3$$

$$\text{Per cent AR} = \text{Per cent AR from table} \times 1.1$$

2. For standard 50-cycle motors operating at 50 cycles:

$$\text{Ckva} = \text{Ckva from table} \times 1.1$$

$$\text{Per cent AR} = \text{same as table}$$

‡ Ckva is rated kilovolt-amperes of capacitors connected at motor terminals.

§ Per cent AR is per cent reduction in line current due to capacitors and is helpful for selecting the proper motor-overload setting when the overload relay is located ahead of the capacitor.

SERIES CAPACITORS. A capacitor placed in series with the line can be used to compensate for the line reactance to improve voltage regulation. These applications are generally made by power companies to correct a bad voltage condition resulting from highly fluctuating loads.

Series capacitors are also used in resistance-type welders by industrial plants. In this application the welder reactance is compensated for by the capacitor, and the power factor is improved to unity, which greatly improves the plant voltage. Resistance-type furnaces also use series capacitors for improving the power factor in a similar manner.

In either type of application a protective device in the form of a gap and contactor for short-circuiting the capacitor in the event of a line fault is used. Otherwise, destructive voltages would appear on the capacitor.

BETATRONS AND SYNCHROTRONS. Induction accelerators used in nuclear research require excitation for large a-c magnets. On the order of 50,000 kvars may be required by these machines. Capacitors have been used exclusively to supply the bulk of these requirements, with a motor-generator set to supply the kilowatts and provide a small margin of control of the kilovars for fine tuning.

INDUCTION HEATING. Capacitors are widely used to supply exciting current to induction heating equipment. The frequency of this application may be from 25 cycles to several hundred kilocycles. A variety of different capacitor designs is required to cover this range, such as the water-cooled paper and parallel-plate designs described in Article 6.

MISCELLANEOUS APPLICATIONS.

Rotating machinery: for surge protection of windings.

Motors. 1. *Starting Duty.* Almost without exception electrolytic capacitors are used for starting single-phase motors. Single-phase motors up to 3 hp are built using split-phase starting with capacitors. The capacitors are connected in the motor winding to "split the phase" and give in effect a two-phase rotating field, which starts the motor. After the motor is started, a centrifugal, current, or thermal switch takes the capacitor out of the circuit, and the motor runs single-phase, the capacitors having an intermittent rating.

2. *Running Duty.* Where a high starting torque is not required, capacitors of the paper-foil type may be used for running as well as starting duty. In this case the motor operates as a permanently split-phase motor. This application also permits of a smaller and lighter-weight motor, which may be advantageous in such applications as hermetically sealed refrigerators or fans. Another advantage is to provide a better air-gap flux distribution, resulting in a larger "breakdown" torque for a given motor rating.

Fluorescent ballasts: for providing power-factor improvement and ballast service in fluorescent-light fixtures.

Carrier-current couplers: for providing coupling between power-circuit and carrier-current-terminal equipment.

Tank circuits: for power packs, radio transmitters, etc.

By-pass and blocking: in all types of electronic circuits.

Energy storage: for discharge duty on welders, flash photography, etc.

8. BIBLIOGRAPHY

Bloomquist, W. C., and W. K. Boice, Application of Capacitors for Power-factor Improvement of Induction Motors, *AIEE Tech. Paper No.* 45-72.

Brotherton, M., *Capacitors—Their Use in Electronic Circuits.* Van Nostrand.

Butler, J. W., The Economies of Using Capacitors in Amounts to Require Automatic Switching, *AIEE Tech. Paper No.* 47-55.

Butler, J. W., and E. B. Pope, Why Not Switch Capacitors with the Load? *Gen. Elec. Review* (August, 1938).

REACTORS

By F. J. Vogel *

9. DEFINITION AND CLASSIFICATION

A reactor is defined by the ASA Standard, C-57-1 as "a device used for introducing inductance in a circuit for purposes such as motor starting, paralleling transformers, and control of current." Similarly, it is stated: "A current-limiting reactor is a form of reactor intended for limiting the current that can flow in a circuit under short-circuit conditions." These are only part of the possible applications, some others being surge and lightning protection; shunts for field circuits, lamps, and other devices; ripple suppression in filters and welding equipment; current control for lamps and motor starters; and for grounding circuits, such as Petersen coils.

Reactors are generally classified as iron-core or air-core reactors. Iron-core reactors usually have air gaps within the core. The term inductor is used commonly where a device with inductance is introduced into a d-c circuit rather than an a-c circuit, or where a more general term than reactor is desired.

10. AIR-CORE REACTORS OR INDUCTORS

INDUCTANCE FORMULAS. Air-core reactors and inductors are commonly made of copper cable or strap wound into circular coils, either in a single layer or in several layers.

* Some material used from original article by A. Dexter Hinckley.

Collections of formulas for the reactance of such coils are found in several publications of the Bureau of Standards, one of the most general being "Radio Instruments and Measurements," *Bureau of Standards Circular No.* 74.

In the design of reactors it is usually desired to obtain the desired inductance with the least length of conductor. If a single-layer coil is to be used, the ratio of the diameter to the length will be approximately 2.46. For a multiple-layer square section coil it has been also found that the mean diameter of the coil should be 2.95 times the side of the square cross-section. This factor, however, is affected very slowly by a change in the ratio of the mean diameter to the square section.

Where great accuracy is not required, graphical solutions are feasible, and Fig. 1 can be used very readily for the design and the determination of the inductance of coils. Corrections for the spacing of conductors and other refinements necessitate using the formulas

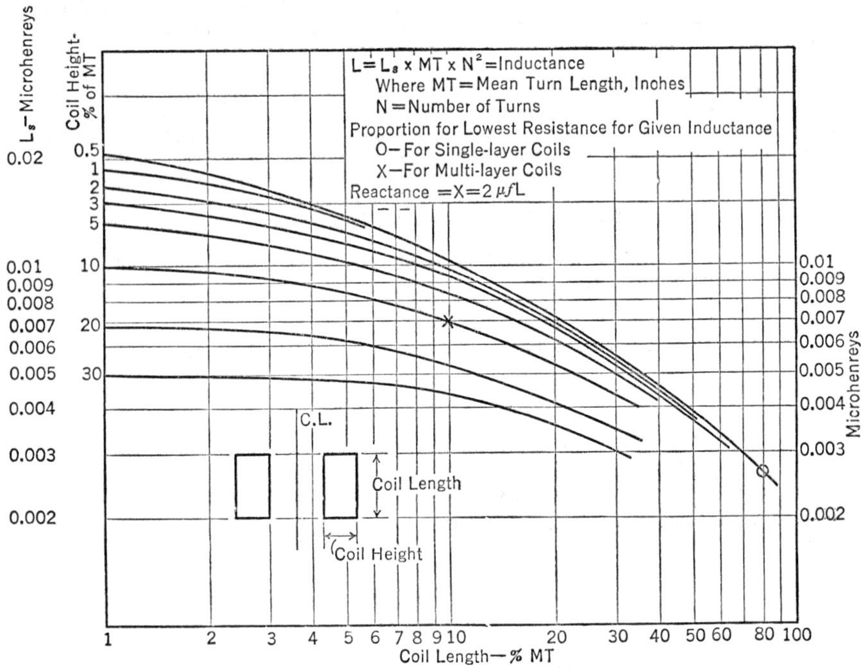

Fig. 1. Inductance Curve for Air-core Solenoids of Various Proportions

and corrections given in the various references. Each curve is based on a constant percentage of the mean turn and refers to the thickness or height of the coil. The abscissas are also in percentage of the mean turn and refer to the length of the coil. Values plotted are inductance in microhenrys. For the inductance of a given coil, the curve values are multiplied by the mean turn in inches and the number of turns squared. For example, the inductance of a coil with 100-in. mean turn and 100 turns, with 10-in. length and thickness, is $0.007 \times 100 \times 100^2 = 7000$ microhenrys.

CONSTRUCTION. The size of the conductor and the ventilation required are determined by the current, the guaranteed losses if any, and the temperature rise. The materials used will be determined by the insulation, whether Class A or Class B. If the currents are heavy, care will be required in the mechanical bracing. Also, the insulation used be required to stand the voltages arising in service as well as meeting the usual requirements stated in ASA Standard C-57-1 for transformers, regulators, and reactors.

Where the device is used for the elimination of steep wave fronts or high-frequency voltages, particular care is required in choosing the insulation clearances. Further extremely high voltages are often encountered when opening circuits with high inductance, and protective means are required at the circuit interrupter in such cases.

11. IRON-CORE REACTORS OR INDUCTORS

INDUCTANCE FORMULAS. The use of iron cores increases the flux within a coil and hence increases the reactance of coils many times. The general formula for the flux in the magnetic circuit is

$$\phi = \frac{0.4\pi NI}{\Sigma(l/\mu A)}$$

In this formula ϕ is expressed in maxwells, or total lines of flux, N is the number of turns in the coil, I is the current in the coil in amperes, l is the length of each part of the circuit in centimeters, μ is its permeability, and A is its area in square centimeters. Therefore, since

$$V = L\frac{di}{dt} = N\frac{d\phi}{dt} \times 10^{-8} = \frac{0.4\pi N^2(di/dt) \times 10^{-8}}{\Sigma(l/\mu A)}$$

$$L = \frac{0.4\pi N^2 \times 10^{-8}}{\Sigma(l/\mu A)} \text{ henrys}$$

This last formula holds accurately as a constant value only over regions where μ is constant.

In the usual design, air gaps are used, and if the major part of the reluctance is in the air gaps,

$$L = \frac{0.4\pi N^2 \times A}{l} \times 10^{-8}$$

where A and l are in square centimeters and centimeters, or

$$L = \frac{3.2N^2A}{l} \times 10^{-8}$$

where A and l are in square inches and inches, l in either case being the entire length of the air gaps. If an appreciable part of the reluctance is in the iron, the complete formula should be used.

Where air gaps are used, they should not, in general, exceed 1/4 in. in length and should be spaced with reasonable uniformity along the winding. There should be some allowance for "fringing" of the flux in estimating the cross-section of the gap, particularly if the core area is small.

Where no gaps are used, the reactor is similar to the primary of a transformer. The current is similar to that shown in Fig. 2. Where the major part of the "exciting current" is required for gaps, the current may be nearly sinusoidal and in proportion to the applied voltage. Where this is the situation, the reactor is said to have straight-line characteristics, and this characteristic or limits for deviation from it are often specified where over-

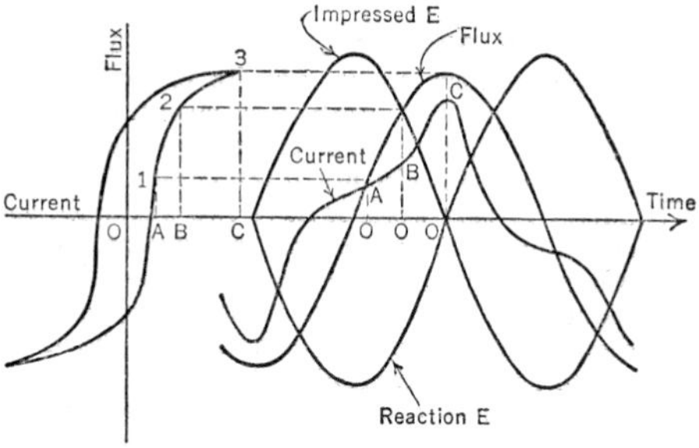

FIG. 2. Prediction of Current Wave Form through Hypothetical Inductor on Which Sinusoidal Emf is Applied

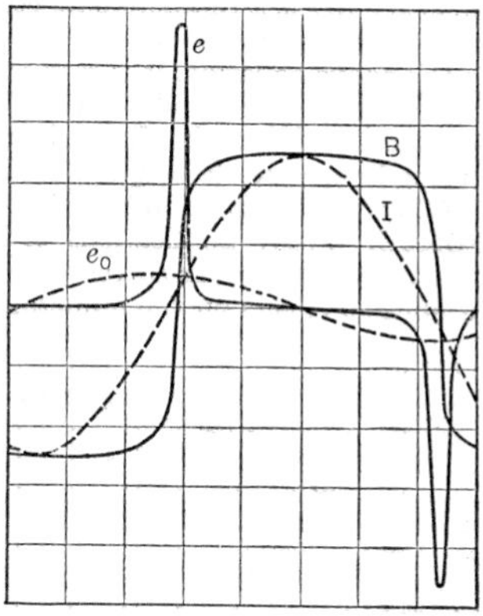

FIG. 3. Cyclical Variation of Flux and Induced Emf in a Saturating Reactor through Which a Sinusoidal Current Is Flowing

currents of appreciable magnitude are required. Where no gaps are used, and a sine wave of current passes through the reactor, very high peak voltages may occur across the reactor, as shown in Fig. 3. In the curves of Fig. 3 are shown the wave of magnetic flux of the induced voltage e, the sine wave of current, and the voltage e_0 which would be induced if the flux density B were a sine wave of the same maximum value. The variation of the induced voltage e approximately represents the cyclical variation of the instantaneous inductance. It is apparent from the curves of Figs. 2 and 3 and from preceding paragraphs that the voltages, currents, inductance, and resistance which may describe the actions of an iron-core inductor are far from sinusoidal or constant quantities. Therefore measurements on devices of this type are generally interpreted in terms of equivalent or effective values. Even where gaps are used, and the reactor is in series with the line, short-circuit conditions could result in conditions similar to those of Fig. 3.

TIME CONSTANT. In many circuits involving the use of an inductor there are periodic changes in the flow of direct current. It is sometimes required that the inductor control the rate of change of current in such a circuit. The growth of current in an inductor of constant inductance L is expressed as

$$i = \frac{E}{r}\,(1 - \epsilon^{-rt/L})$$

where i = current at any time t seconds after the circuit is closed; E = constant impressed voltage; r = resistance of the winding; ϵ = Napierian logarithmic base 2.7183. The curve of Fig. 4 shows the current rise in an air-core inductor and follows very closely the equation above. With the iron-core inductor, however, the inductance is not constant

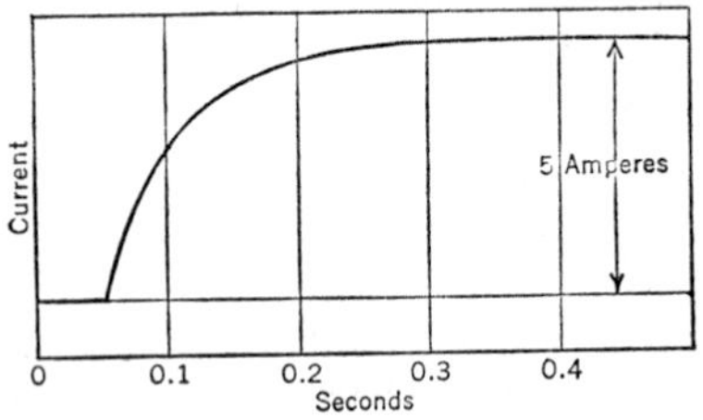

FIG. 4. Current Rise in an Air-core Inductor, $L = 0.62$ henry, $r = 13.2$ ohms (After Morecroft)

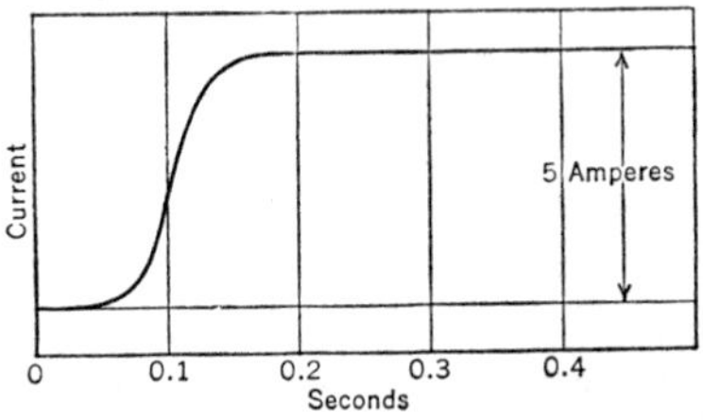

FIG. 5. Current Rise in Saturating Reactor, L variable, $r = 18.7$ ohms (After Morecroft)

and varies with increase of current. This effect is seen in the curve of Fig. 5, the form being very unlike that of Fig. 4. These effects must be considered when it is required that the inductor control the current growth. The magnitude of the exponent in the expression above determines the rate of current rise in the inductor. For this reason the quantity L/r is a characteristic constant (when L and r are constant) of the inductor and is given the name time constant. If the numerical value of L/r, in seconds, is substituted in the above equation, it is seen that this value is the time taken for the current to reach 63.2 per cent of its final value. The time constant is generally used in the comparison of inductors and from its dimensions indicates the effectiveness of a given winding in producing inductance for a given resistance.

ACTION OF D-C MAGNETOMOTIVE FORCE WITH A-C MAGNETOMOTIVE FORCE. Many applications of inductors require their use in circuits carrying both direct and alternating currents of more than one frequency. Both these situations result in unusual effects on the inductance of the device. Since some of the effects are used in certain control circuits, it is common for an inductor in this work to have two windings, one carrying the circuit alternating current and the other carrying a direct current. The effective inductance offered to the alternating current depends upon the magnitude of both the direct current and the alternating current. The state of magnetic saturation due to d-c magnetomotive force, depends upon the magnitude and direction of the direct current. For any particular direct current the iron core is brought to some operating point on the hysteresis loop. When an alternating current superimposes its magnetomotive force upon this as an origin, a small hysteresis loop is executed. The average slope and the area of this small displaced hysteresis loop give a measure of the permeability and

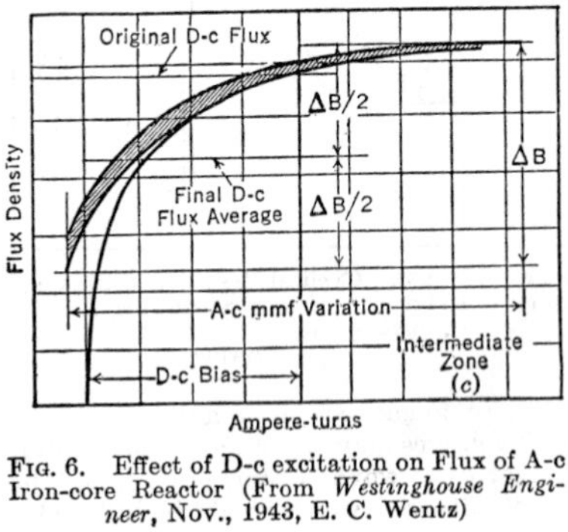

FIG. 6. Effect of D-c excitation on Flux of A-c Iron-core Reactor (From *Westinghouse Engineer*, Nov., 1943, E. C. Wentz)

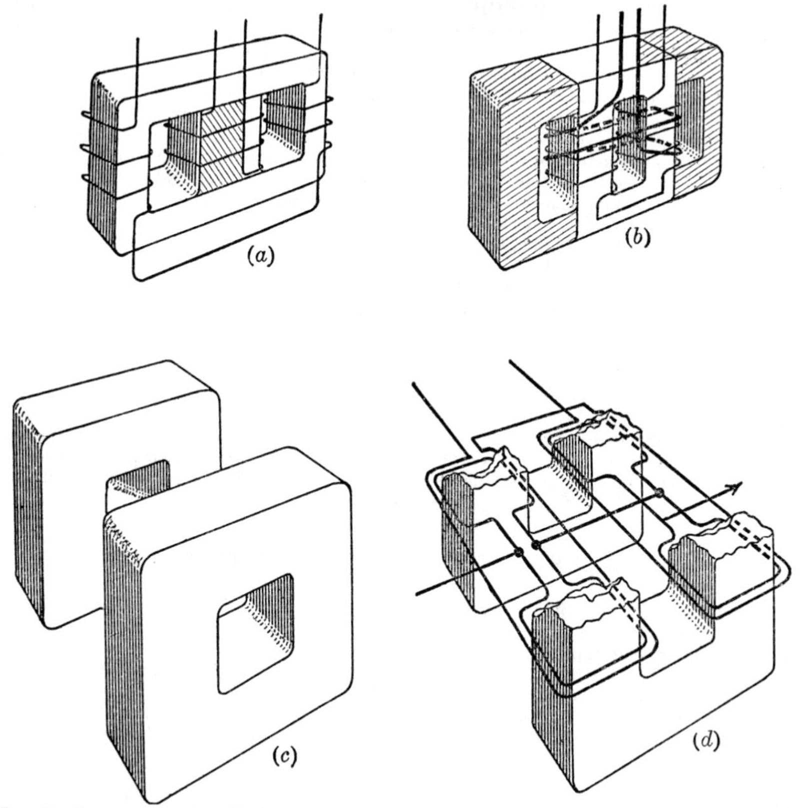

FIG. 7. Arrangements for Combining D-c Excitation with A-c Excitation for Iron-core Variable Reactances (From *Westinghouse Engineer*, Nov., 1943, E. C. Wentz)

In Fig. 7(*a*), alternating current is applied to the coils on the outer legs, and direct current to the middle leg, so that no net a-c flux links d-c coils. In Fig. 7(*b*), the coil shown heavy carries alternating current, while the light coils carry direct current. The same result is accomplished with two cores in Fig. 7(*c*), with coils mounted as in Fig. 7(*d*). The two leads at the top left carry direct current, and the alternating current is carried by the leads from left to right.

losses of the iron under these conditions, and these determine the effective inductance and resistance of the inductor for the particular alternating current. Figure 6 shows the effect obtained. The instantaneous inductance of the reactor at any point on the curves can be found from the slope of the curve times the area of the core times the number of turns times 10^{-8}. Figure 7 shows ways in which cores and coils may be assembled so that there is freedom from interaction between the d-c and the a-c circuits. Figure 8 shows how the apparent reactance can be changed, by changing the d-c excitation, and used for the control of power. This control can be obtained with as little as 1 per cent of the power rating of the reactor used in the d-c excitation.

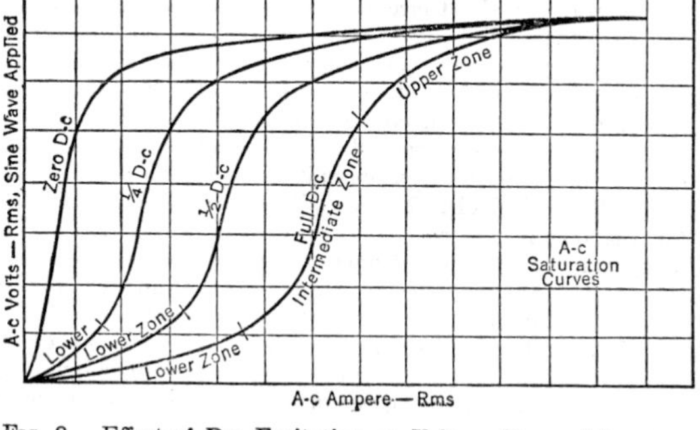

FIG. 8. Effect of D-c Excitation on Voltage Drop of Iron-core Reactor (From *Westinghouse Engineer*, Nov., 1943, E. C. Wentz)

PRACTICAL DESIGN CONSIDERATIONS—SHUNT REACTORS—SPECIAL REACTORS. **Shunt reactors** are connected across the lines or from the lines to neutral. They are used to compensate for the leading current taken on long, lightly loaded lines. The usual commercial shunt reactor is built with a laminated silicon-steel core with air gaps, usually set by preliminary test before varnishing or treating the entire assembly. Spacers in the gaps and treatment are necessary to prevent undue noise. The design is usually very similar to that of a transformer, the core being about the same in area as is a transformer of half the reactor rating. The unit may be of either shell or core type, dry or oil immersed, and use Class A or Class B insulation. The windings should conform to transformer practice in so far as insulating assembly, standards, and tests are concerned. Since shunt reactors do not experience heavy short-time currents due to short circuits or heavy loads, higher current densities in the copper can be used for reactors than for transformers. Loss guarantees usually limit the amount of copper used.

Petersen coils and **neutral reactors** are used to ground the neutrals of some systems. The purpose is either to limit the currents under certain conditions or to prevent outage in the event of single-phase-to-ground faults. In the case of Petersen coils, iron cores with gaps, and windings with taps for adjustment are used; the insulation is based on line to neutral voltage of the system and may require protection against lightning surges transmitted through the transformers. Since such coils normally carry practically no current, the copper size is determined by arbitrary duty cycles. Since the cycle durations are short, higher temperatures and less ventilation are required than in usual transformer practice. Neutral reactors usually are used only on the lower-voltage circuits and are dry type; except for duty cycles, the requirements are very similar to those for current-limiting reactors.

12. ADJUSTABLE INDUCTORS, VARIOMETERS

Inductance may be adjusted by several means. One way is to vary the number of effective turns in the winding, and this is frequently done by arranging a sliding contact on the winding or on tap points to the winding.

A very convenient continuously variable inductor is formed of two windings, generally in series and mounted so that one can rotate within the other. Although the self-inductance of the two coils is constant, the total inductance of the combination changes with the change of the mutual induction between the two coils. The inductance of this device is

$$L_{\text{total}} = L_1 + L_2 \pm 2M$$

where L_1 and L_2 are the individual self-inductances, and M is the mutual induction.

Another variometer effects the inductance adjustment by means of a variation of the overlap between adjacent flat coils. In variometers of these two types it is possible to adjust the inductance in a range of values of about 10 to 1. Adjustable inductors of the several forms have wide use in laboratories, and the principle of this device is applied in the development of the induction-type feeder voltage regulator.

13. CURRENT-LIMITING (CL) REACTORS FOR POWER SYSTEMS

(See also Section 12, Article 18)

Reactors for short-circuit limitation in power systems are generally installed at one or more of the following locations in a system: (1) generator reactors on the leads of generators feeding bus-bars; (2) feeder reactors on feeder circuits fed from bus-bars to protect switching equipment. The prevention of heavy currents on short circuits avoids severe mechanical stresses in connected apparatus. The placing of reactors in feeder circuits prevents a serious voltage drop in feeders other than the one in which the short circuit occurs and reduces the probability of apparatus falling out of step.

SPECIAL APPLICATION AND DESIGN REQUIREMENTS. The requirements for CL reactors in general are given in the Standards for CL Reactors, ASA C57-1, Part 7. They may be either dry type or oil immersed and use either Class A or Class B insulation. Where they are dry type, care must be taken in choosing the clearances to other apparatus or metallic structures to prevent trouble due to stray fields. Laminated silicon-steel shields are often used with oil-insulated reactors to prevent flux leakage to the tank walls and consequent eddy current losses. Figure 9 shows a typical oil-insulated reactor. A short-circuited turn of copper inside and adjacent to the tank wall is used for the same purpose.

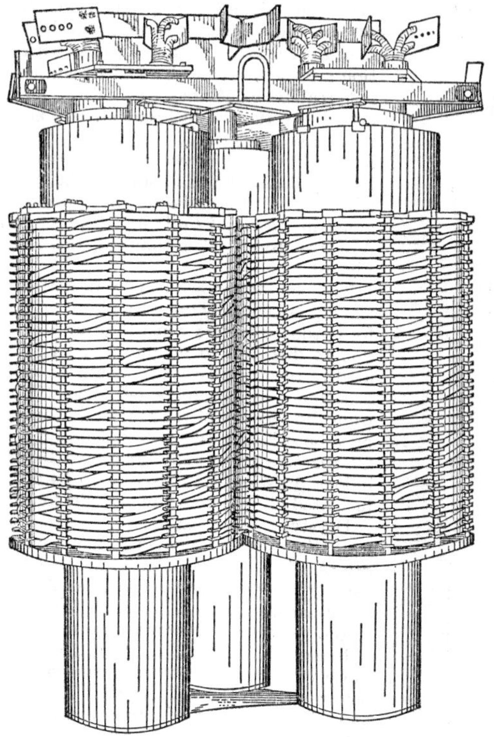

FIG. 9. Three-phase Oil-insulated CL Reactor Out of Case (Courtesy of Westinghouse Electric Corporation)

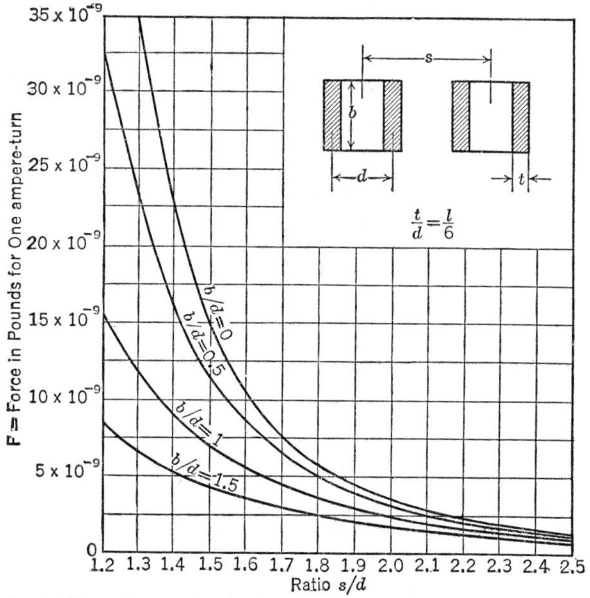

FIG. 10. Mechanical Forces between Parallel Reactors (From *Electric Coils and Conductors*, by H. B. Dwight, McGraw-Hill, 1945)

The requirements for short circuit are given in ASA Standard C57-1. There are severe compressive stresses on the end conductors, which should be calculated on the basis of off-set current waves. Because of the fact that stranded cables with considerable insulation are used, the mechanical strength of such assemblies is not readily calculable. Empirical formulas, obtained by actually testing typical reactors, are necessary. The entire reactor is also subject to severe mechanical forces, and the feet or legs must be designed to withstand them. Likewise, the foundation for reactors must be strong enough to stand these forces. Formulas for the stresses between reactors are given in *Electrical Coils and Conductors* by H. B. Dwight. A curve useful for estimating these stresses is shown in Fig. 10.

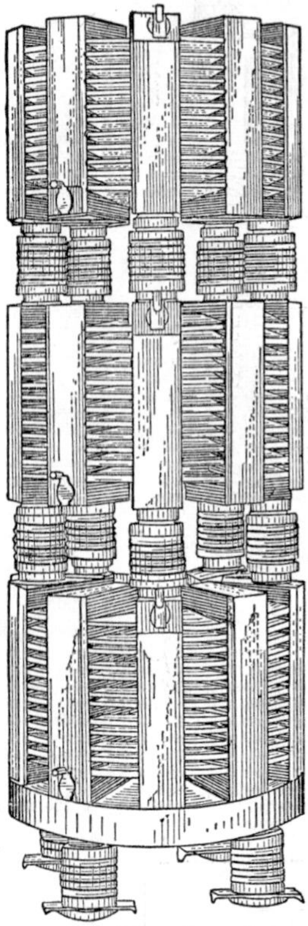

FIG. 11. Three-phase Indoor CL Reactor, Cast in Concrete (Courtesy of the General Electric Company)

The conductors are usually insulated. This is not essential but is generally done to obtain insulation between turns for outdoor reactors when they are wet or to prevent failure due to dirt and to magnetic materials drawn into the reactor. Incidentally, care should be taken to prevent this occurrence by keeping small magnetic machine parts away from the vicinity of the reactors. Current-limiting reactors may be either outdoor or indoor type. Outdoor dry-type CL reactors are usually limited to 15,000 volt circuits, and indoor dry-type CL reactors to 34,500-volt circuits. Oil-insulated CL reactors can be built for any circuit voltage.

Dry-type CL reactors, particularly those with Class B insulation, use porcelain or ceramic spacers between turns or are cast in concrete. The principal insulation to ground is usually porcelain insulators. The cable is stranded, and each individual strand is often enameled to reduce eddy current losses. A typical design is shown in Fig. 11.

Inasmuch as a reactor acts as a reflection point for impulse voltages or lightning, care should be taken to provide larger air-jump clearances than for most other equipment and also to provide by-pass resistance units or lightning arresters adjacent to the reactor, depending upon the circumstances. This caution is particularly applicable to dry-type reactors and is possibly the reason why the dielectric tests, as specified by ASA Standard C-57-1, are higher for CL reactors than for other equipment.

STANDARD RATINGS. The rating of a CL reactor is expressed as the kva and voltage drop at rated current and frequency on a circuit of specified voltage rating, and in per cent reactance. Preferred ratings are given in ASA Standard C-57 as follows.

Preferred Ratings of Voltage Drop. For CL reactors these are 3, 5, 7.5, or 10 per cent of the line-to-neutral voltages corresponding to the following circuit voltages:

2,400	23,000	138,000
4,160	34,500	161,000
4,800	46,000	196,000
7,200	69,000	230,000
12,000	92,000	287,000
13,800	115,000	345,000

Preferred Current Ratings. For CL reactors these are as follows:

Rating in Amperes

100	250	500	1200
150	300	600	1600
200	400	800	2000

Per Cent Reactance. This is the ratio of the kilovolt drop to the line-to-neutral kilovoltage of a three-phase system, and the line-to-line kilovoltage of a single-phase system, times 100:

$$\text{Per cent } X = \frac{\text{Kilovolt drop} \times 100}{\text{Kilovoltage}}$$

14. BIBLIOGRAPHY

Aggers, C. V., and W. E. Pakala, Direct-current Controlled Reactors, *Elec. J.* (Feb., 1937).
Boyajian, A., Theory of D-c-excited Iron-core Reactors and Regulators, *Trans. AIEE*, Vol. 43, p. 958 (1924).
Dwight, H. B., *Electric Coils and Conductors.* McGraw-Hill (1945).
Fahnoe and A. J. Maslin, Peak Voltages across Saturating Reactances, *Elec. J.*, Vol. 29, p. 120 (1932).
Hanna, C. R., Design of Reactances and Transformers Which Carry Direct Current, *Trans. AIEE*, Vol. 46, p. 155 (1927).
Harder, E. L., Peak Voltages on Saturating Reactances in Three-phase Circuits, *Trans. AIEE*, Vol. 29 (1932).
Holubow, H., D-c Saturable Reactors, *Electronics Industries* (March, 1945).
Hunter, E. M., Some Engineering Features of Petersen Coils and Their Application, *Trans. AIEE*, Vol. 57, p. 11 (1938).
Montsinger, V. M., Temperature Limits for Short-time Overloads for Oil-insulated Neutral Grounding Reactors and Transformers, *Trans. AIEE*, Vol. 57, p. 39 (1938).
Morecroft, J. H., *Principles of Radio Communication.* John Wiley (1933).
Rader, L. T., and E. C. Litscher, Some Aspects of Inductance When Iron Is Present, *Trans. AIEE*, Vol. 63, p. 133 (1944).
Specht, T., and E. C. Wentz, Peak Voltages Induced by Accelerated Flux Reversal in Cores Operating above Saturation Density, *Trans. AIEE*, Vol. 65, p. 254 (1946).
Spooner, T., Effect of Superposed Alternating Field on Apparent Magnetic Permeability and Hysteresis Loss, *Phys. Review*, Vol. 25, p. 527 (1925).
Thomson, W. T., Similitude of Critical Conditions in Ferroresonant Circuits, *Trans. AIEE*, Vol. 58, p. 127 (1939).
Thomson, W. T., The Generalized Solution for the Critical Conditions of the Ferroresonant Parallel Circuit, *Trans. AIEE*, Vol. 58, p. 743 (1939).
Wellings, J. G., and R. V. Wheeler, *J. IEE*, Vol. 89, Part II, p. 473 (1942).
Wentz, E. C., *Westinghouse Engineer* (Nov., 1943).
"Standards for Transformers, Regulators, and Reactors," ASA C57-1, C57-2, and C57-3 (1942).
"Standard Methods for Test of Magnetic Properties," ASTM Standards A33-44 (1944).
"Radio Instruments and Measurements," *Bureau of Standards Circ.* 74, and other bulletins and publications of the National Bureau of Standards.

ELECTROMAGNETS

Revised by I. F. Kinnard and J. H. Goss *

15. ELECTROMAGNET WINDINGS

The electromagnet is usually composed of a winding and an iron-air magnetic circuit more or less permanently related to the winding. The winding, when carrying an electric current, sets up a magnetic field, and to intensify it a core, a shell, or both, of magnetic material are added to the winding. When the core is fixed within a solenoidal winding, a bar electromagnet results. If the core is movable, the electromagnet is of the plunger variety.

The function of the electromagnet is to provide a magnetic field in space. This magnetic field may be used in the operation of electrical rotating machinery or may serve to magnetize materials for other purposes. In addition to these fundamental uses there is the action utilized in almost an infinite number of devices—that of the attraction of magnetic objects so as to perform useful work.

In switches, relays, and kindred magnetic devices the iron attracted in the magnetic field is usually in the form of an armature or clapper linked or hinged to the shell or core of the structure. The attracted magnetic material may also be in the form of a free body, such as the material lifted by an ore separator or lifting magnet.

16. CORE MATERIAL IN ELECTROMAGNETS

An electromagnet is generally intended for intermittent use, i.e., magnetizing current is applied when it is desired that the magnet act. When it is desired that the electromagnet cease to act, the magnetizing current is disconnected. Ideally, then, when the current is reduced to zero, it is desired that the attractive force of the electromagnet immediately cease and become zero. This would require the material being acted on by the magnetizing force to have zero residual induction and zero coercive force. An attempt to secure this ideal condition is made by the manufacture and use of magnetically soft materials, such as low-carbon steel, silicon steel, and certain alloys of nickel and cobalt with iron. (See Section 2, Articles 25 to 29.)

In addition to the magnetic requirements of low coercive force and residual induction

* Original article by A. Dexter Hinckley.

it is generally desired that the core and shell material have the possibility of a high maximum induction. The low-carbon and silicon steels have adequately high maximum induction for most applications. Special nickel-cobalt alloys are used when the ultimate is required, as in relays built to fit into the smallest practicable space.

17. FUNDAMENTAL ACTION IN ELECTROMAGNETS

Magnetic bodies within the influence of an electromagnet tend to move when current is passed through the winding. The moving body may be external to the electromagnet, such as a weight being lifted, or it may be a plunger or some other part of the electromagnet. The magnitude of the force acting on the magnetic body depends upon a large number of factors and usually can be determined satisfactorily for a new design only by a model study. Particular equations for simple theoretical cases are helpful in understanding the general effects of various factors. For example, in the theoretical case shown in Fig. 1, it is assumed that the flux density is uniform throughout the gap, the core is unsaturated, and there is no leakage flux. The force P tending to move the core may be expressed as follows:

$$P = \frac{B^2A}{8\pi} \quad (1a), \qquad P = \frac{F^2\mu A}{8\pi x^2} \quad (1b), \qquad F = 0.4\pi NI \quad (1c) \qquad (1)$$

From eq. (1a) it may be seen that the force (P) in dynes is proportional to the square of the flux density (B) in gausses, and to the area (A) in square centimeters that is perpendicular to the flux lines. The force (P) is also proportional to the square of the magnetomotive force (F) in gilberts and the permeability of the air gap (μ) and inversely proportional to the square of the length of the air gap (x) in centimeters [see eq. (1b)]. The relationship between the magnetomotive force (F) in gilberts and the ampere-turns (NI) is shown in eq. (1c). See Fundamental Theory of Electromagnetic Forces, in Article 20, for a more comprehensive description of the force.

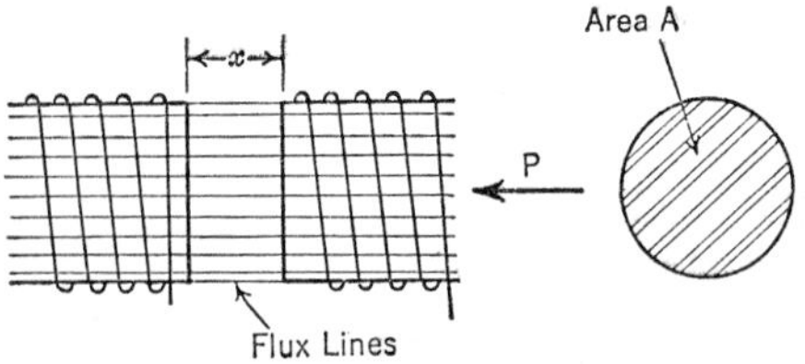

FIG. 1. Schematic Diagram of Theoretical Simple D-c Electromagnet

The general effect of other design factors may be learned from a comparative study of previous designs with established characteristics. A few of these will be introduced later in this chapter; others may be found in such books as *Electromagnetic Devices* by H. C. Roters, John Wiley (1941).

18. CLASSIFICATION OF ELECTROMAGNETS

Electromagnets may be grouped in several ways, but a general classification may be made on the following basis:

1. No moving parts (operate on direct current):

   Electrical machine field structures, electromagnets for physical and chemical research.

2. Portative (for attracting and holding magnetic materials; operate on direct current or with alternating current and shading coil):

   Lifting magnets, magnetic chucks, clutches, couplings and brakes, ore and trash separators, low-voltage release on starting rheostats.

3. Tractive (with associated moving parts; operate on direct current and alternating current; various combinations of size of armature, length of stroke, speed of action):

   Bells and buzzers, telegraph and telephone relays, current-, time-, and voltage-limit switches and relays, mechanical switch and valve control, and many forms of indicators and electromechanical controls.

19. COIL DESIGN AND CONSTRUCTION

From eq. (1a), (1b), and (1c) it may be seen that the principal problem encountered in the design of the coil is to provide the necessary flux density to make the electromagnet work. The factors in the coil design that influence the flux density are the number of turns, the size of the coil, and the current. The space available usually limits the number of turns and the size of the coil. The ability of the insulation to resist breakdown from temperature rise limits the magnitude of the current that may be used.

DESIGN OF BOBBIN. Most electromagnetic coils are wound on a spool, consisting of a tube with flanges at the end (see Fig. 2). Many different materials are used, including metals, plastics, and fiber.

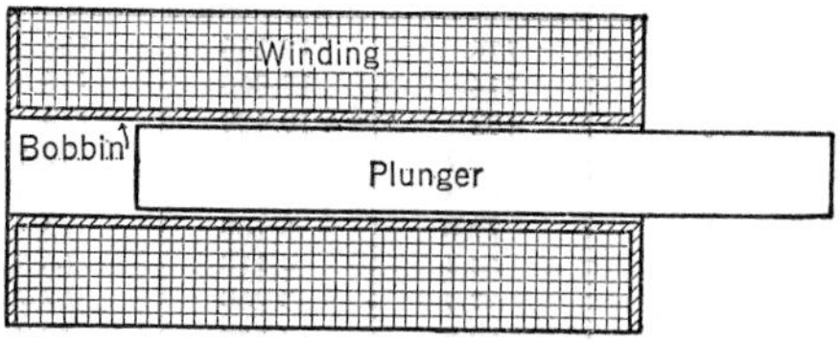

FIG. 2. The Principal Parts of a Simple D-c Electromagnet

WIRE INSULATION. Wire for winding coils may be insulated with enamel, cotton, glass, asbestos, or any of a number of other materials, such as the new silicone products. Present-day enamels are very durable and can be relied upon in many applications that formerly required other types of insulation. The wire may be carefully wound in layers, or it may be random wound, the latter method being the only one economical when using fine wire without paper interlayers. When layer winding is used, extra protection may be obtained by placing treated paper or fabric between layers. After winding, the coil is usually impregnated with varnish and then baked. This further protects the coil from short circuits, improves the mechanical strength, and increases the resistance to ageing. Solventless varnishes developed in recent years have greatly improved the effectiveness of the varnish treatment.

SPACE FACTOR. The **space factor** may be defined in several ways. Fundamentally it is the ratio of the space occupied by the metal in the wire to the total space provided for the winding. A second term, **winding space factor,** is frequently used in industry to indicate the efficiency with which the winding uses the space available. In Fig. 3 the space factor S may be expressed:

$$S = \frac{\pi d^2 N}{4L'_c T'} \tag{2}$$

In contrast to this the winding space factor S_w may be expressed:

$$S_w = \frac{\pi d^2 N}{4(L'_c T' - a)} \tag{3}$$

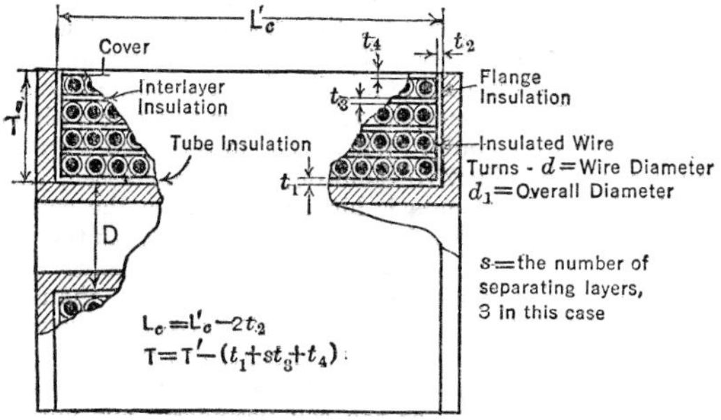

FIG. 3. Electromagnetic Coil, Showing Dimensions Used for Calculating Winding Space Factor

where a is the cross-section of space occupied by materials other than the insulated wire. The space a may be readily calculated from the dimensions of the materials. For the construction shown in Fig. 3 the value of S_w is:

$$S_w = \frac{\pi d^2 N}{4L_c T} \tag{4}$$

Equations (2), (3), and (4) are most useful for understanding the meaning of the terms space factor and winding space factor. A number of different ways of calculating these values may be readily visualized. In design work, when the space factor is used to determine the size of the coil required, a reasonable approximation of the winding space factor may be obtained from the following equation:

$$S_w = \frac{0.85\pi d^2}{4d_1^2} \tag{5}$$

where d_1 = overall diameter of the insulated wire. Tables showing the diameters of standard insulated wires may be found in the NEMA and other standards. In using this type of design data, the latest edition of the standards should be obtained. Table 1 is taken from a wire standards publication.

Table 1. Data on Bare and Insulated Round Copper Magnet Wire

A.W.G. Size	Bare Wire Diameter, Mils	Area, Circular Mils	Overall Diameter, Mils			
			Enamel	Cotton	Glass	Asbestos
8	128	16,380	130	136	135	140
9	114	13,000	116	121	121	126
10	102	10,400	104	108	108	113
11	90.7	8,230	92.5	95.7	96.7	101.7
12	80.8	6,530	82.5	85.8	86.8	90.8
13	72.0	5,180	73.7	77.0	78.0	82.0
14	64.1	4,110	65.8	69.1	70.1	73.6
15	57.1	3,260	58.7	62.1	63.1	66.1
16	50.8	2,580	52.3	55.8	56.8	59.3
17	45.3	2,050	46.8	50.3	51.3	53.8
18	40.3	1,620	41.7	45.3	46.3	48.3
19	35.9	1,290	37.3	40.9	41.9	43.9
20	32.0	1,020	33.3	37.0	38.0	40.0
21	28.5	812	29.8	33.5	34.5	36.5
22	25.3	640	26.5	29.8	31.3	33.3
23	22.6	511	23.7	27.1	28.6	30.6
24	20.1	404	21.2	24.6	26.1	28.1
25	17.9	320	19.0	22.4	21.9	25.9
26	15.9	253	16.9	20.4	19.9	23.9
27	14.2	202	15.2	18.7	18.2	22.2
28	12.6	159	13.5	17.1	16.6	20.6
29	11.3	128	12.2	15.8	15.3	19.3
30	10.0	100	10.8	14.5	14.0	18.0

HEAT DISSIPATION. The heat developed in the winding must be dissipated from the exposed surfaces at a rate sufficient to keep the temperature below the point where the insulation is impaired. The heat flow from the ends of conventional coils and from the inner surfaces is usually blocked by insulating material that has low thermal conductivity, and for that reason it is customary to base the temperature-rise calculations on the heat-radiating ability of the peripheral area only. When a coil makes good thermal contact with an iron shell that encases it, the effective radiating area is increased, and the heat dissipation per square inch of peripheral area is higher than when the coil is exposed to the air. An encased coil that is separated from its case by a substantial air space, however, may have less heat-dissipating ability than is present when the case is omitted. In some designs the ends of coils are as capable of dissipating heat as the coil periphery, but not in the more common designs. Any unusual conditions provided by the design for heat to get out of the coil will, of course, improve the heat transfer.

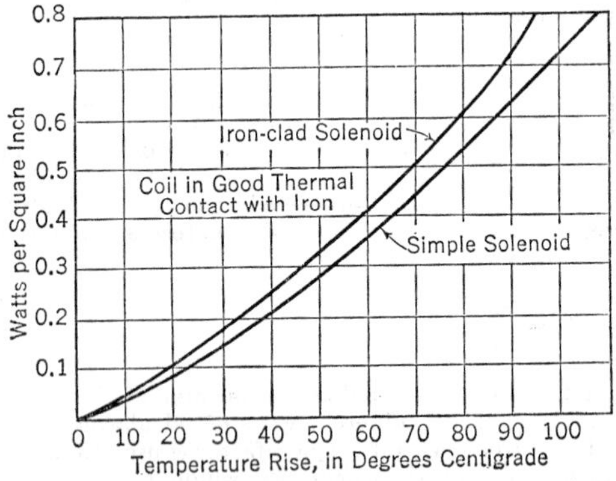

FIG. 4. Ratio of Power Imput to Peripheral Area for a Common Type Enamel-wire Coil

The amount of heat developed in a winding is a function of the power input, and the amount dissipated is a function of the radiating area, thus the **temperature rise** for a particular coil is a function of the ratio of these two quantities. Figure 4 shows the relationship between the temperature rise and the ratio of power input in watts to the square inches of peripheral area for a common enamel-wire type of coil. Table 2 lists the allowable temperature rise for a number of insulating materials.

From Table 2 and Fig. 4 it may be seen that the watts per square inch for this particular type of enamel coil should not exceed 0.58, or 0.67 if encased in an iron shell.

For short-time service, i.e., uses where the solenoid is energized only for short intervals with long periods between the applications of power, the thermal capacity of the solenoid will permit a greater dissipation of energy in the winding without overheating it.

Table 2. Allowable Temperature Rise for Insulating Materials

Class	Typical Materials in the Class	Allowable Temperature Rise, Degrees Centigrade
O	Cotton, silk, paper	70
A	Treated cotton, silk, or paper, enamel	85
B	Mica, asbestos, fiber glass, with organic binders	105
C	Entirely mica, porcelain, glass, or quartz	No set limit

Temperature rise above the temperature of cooling air. Temperature determined by resistance method. (NEMA Standards for Industrial Control Pub. 45–97, IC2-70, September, 1945.)

WINDING CALCULATIONS FOR D-C SOLENOIDS. Round solid wires are assumed throughout. Let

S_w = winding space factor, from eq. (5).

k = ratio of specific resistance of conductor used to that of standard annealed copper at 20 deg cent. For copper of 100 per cent conductivity at 20 deg cent, $k = 1$. For copper of any other per cent conductivity, say C per cent, at any other temperature, say t deg cent, $k = \frac{100}{C} + 0.004(t - 20)$. See also Section 14, Article 54.

A = cross-section of wire in circular mils (= square of diameter in thousandths of an inch).

$n = \frac{1,270,000S_w}{A}$ = number of conductors per square inch, the square inch being taken perpendicular to the direction in which the wire is wound.

$\rho = \frac{1,100,000S_wk}{A^2}$ = resistance of the winding per cubic inch of the winding space, excluding the space, if any, occupied by extra insulation between layers; ρ is in ohms.

$w = 0.271S_w + 0.040$ = weight of the winding (copper and cotton insulation) per cubic inch of the winding space, exclusive of the space, if any, occupied by extra insulation between layers; w is in pounds.*

E = impressed volts.

NI = ampere-turns, where N is the total number of turns and I the current in amperes.

p = allowable watts per square inch of radiating surface.

l = mean length of turn in inches.

T = depth of winding space in inches (see Fig. 3), excluding the space occupied by extra insulation, if any, between layers.

L_c = length of winding space in inches (see Fig. 3), excluding the space occupied by extra insulation at flanges.

S_r = radiating surface, in square inches, equal to $\pi L_c(2T + D)$. See Fig. 3.

$V = L_clT$ = volume of winding space in cubic inches, excluding space occupied by the extra insulation, if any, between layers.

The problem is usually to find the size of wire and necessary winding space for a coil which will give, without overheating, a required number of ampere-turns (NI) at a given impressed voltage E. The diameter of the core or spool is also usually known, or at least must not exceed certain fairly well-defined limits, depending upon the service for which it is to be used. The procedure is then to assume a reasonable value for the mean length of turn l. The size of wire is then immediately fixed by the relation

$$A = \frac{kl(NI)}{1.16E} \tag{6}$$

From eq. (5) winding-space factor S_w may then be found.

---

* This formula for w also holds approximately for other kinds of insulation and for most alloy resistance wires. If we call g_c the specific gravity of the conductor and g_i the specific gravity of the insulation, the exact formula is

$$w = 0.036E(g_c - g_i)S_w + 0.0284g_i$$

Underhill gives g_i as 1.6 for asbestos, 1.4 for cotton, and 1.0 for silk.

The dimensions L_c and T of the coil must then be chosen so as to satisfy the relation

$$\frac{L_c T S_r}{l} = \frac{k(NI)^2}{1{,}470{,}000 p S_w} \tag{7}$$

As a rule this can be done only by cut and try. Note that changing the depth T of the winding space will also change the mean length of turn l, unless the diameter of the core or spool is so altered as to keep l constant. The cross-section of the wire varies directly as l as shown by eq. (6), and therefore the value of the winding space factor S_w to be used in eq. (7) will depend upon l, but only to a slight extent, except in very small wires.

When the cross-section of the wire A, the mean length of turn l, and the dimensions L_c and T of the winding space have been determined so that both eq. (6) and eq. (7) are satisfied, the total number of turns in the winding will be

$$N = nL_cT = \frac{1{,}270{,}000 S_w L_c T}{A} \tag{8}$$

and the current is then equal to the given number of ampere-turns divided by N, and the total length of wire is equal to Nl.

The total resistance r and total weight W (including insulation) of the wire may then be found directly from a wire table or may be calculated from the formulas

$$r = \rho V = \frac{klN}{1.16A} = \frac{1{,}100{,}000 S_w k L_c T l}{A^2} \tag{9}$$

$$W = wV = (0.271 S_w + 0.040) L_c T l \tag{10}$$

See note on preceding page regarding w.

CALCULATION OF OPEN SOLENOID OF CIRCULAR CROSS-SECTION. For a coil wound on a spool of diameter D (see Fig. 3) the mean length of turn is $l = \pi(D + T)$. If we assume that the outside cylindrical surface of the winding is the only radiating surface, which is only approximately true, as pointed out above,

$$S_r = \pi(D + 2T)L_c$$

Whence, if we put

$$Q = \frac{K(NI)^2}{1{,}470{,}000 p S_w} \tag{11}$$

the required length of coil for a given diameter of core D and depth of winding T is

$$L_c = \sqrt{\frac{(D + T)Q}{(D + 2T)T}} \tag{12}$$

When L_c is given instead of T, then the required depth of winding is

$$T = \frac{1}{4}\left[\left(\frac{Q}{L_c^2} - D\right) + \sqrt{\left(\frac{Q}{L_c^2} - D\right)^2 + \frac{8QD}{L_c^2}}\right] \tag{13}$$

In either case, the number of turns, total resistance, weight, etc., are found as described above for the general case.

Example. It is required to design an open solenoid 10 in. long and having an internal diameter of 1.5 in. to give approximately 10,000 ampere-turns at 110 volts, the heat developed not to exceed 0.45 watt per sq in. of radiating surface (taken as the outside cylindrical surface of the coil). Assume a mean length of turn equal to 12 in.; then from eq. (6) the required cross-section of the wire, assuming copper at 70 deg cent, is $A = 1130$ cir mils. The insulation adds 5 mils to the diameter of the wire, assuming single cotton-covered wire (see Table 1). From eq. (5), then, the winding space factor is $S_w = 0.51$. From eq. (11) the value of Q is then 359, whence from eq. (13) the required depth of winding T is 2.25 in. The actual mean length l of turn is then 11.8 in., which, substituted in eq. (6), gives, for the cross-section of wire required, $A = 1110$ cir mils. The nearest commercial size of wire is No. 20 A.W.G., having a cross-section of 1020 cir mils. If this size of wire is used, the winding space factor from eq. (5) is 0.50, which agrees practically with the value $S_w = 0.51$ used above, and the actual ampere-turns will be $NI = 9200$. From eq. (8) the total number of turns will be $N = 14{,}000$, and from eq. (9) the total resistance will be $r = 168$ ohms; the current is then $I = 0.66$ amp. The total weight of the winding is, then, from eq. (10), $W = 46.5$ lb.

20. ELECTROMAGNETS, LIFTING AND PLUNGER

FUNDAMENTAL THEORY OF ELECTROMAGNETIC FORCES. In a tractive magnet, work is done as the configuration of the magnetic circuit is altered by the motion of a plunger or armature. At the same time the stored magnetic energy of the system increases. The energy appearing in both these forms comes from the electrical circuit.

If a plot is made of flux linkages (line-turns) versus current for each of a number of positions of the armature (Fig. 5), the external work done between any two successive positions, such as 3 and 4, is represented by the area between the corresponding curves (shown shaded in Fig. 5). The average force during this travel is equal to the work divided by the distance moved between positions 3 and 4. Thus, if we define

$$W = \int_0^I (N\phi)\, di \tag{14}$$

for any general armature position, $x = A$, and coil current I, the force on the armature at this position is

$$P \propto \frac{dW}{dx} \tag{15}$$

This relation holds despite saturation of the iron, but does require some slight modification if the hysteresis effects in the iron are to be considered. For soft iron, however, these effects are rather small.

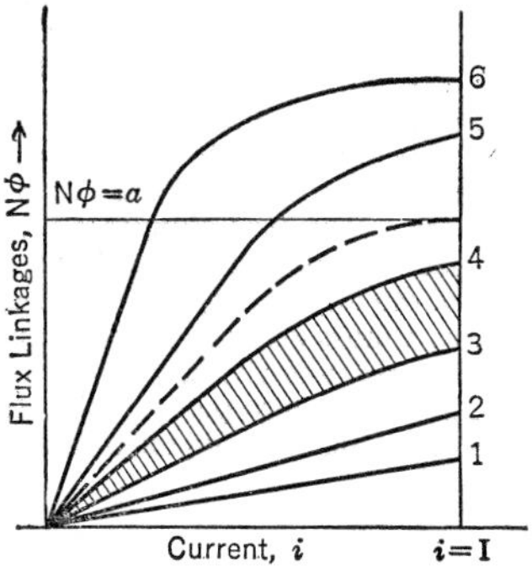

FIG. 5. Flux Linkages vs. Current Curves for Six Different Positions of the Plunger in a Simple Electromagnet

When $N\phi$ is always proportional to i (no iron saturation), eq. (15) may be reduced to

$$P = \frac{I^2}{2}\frac{dL}{dx} \tag{16}$$

where L is the inductance of the electrical circuit for the armature position in question.

Equations 14, 15, and 16 are not often used directly for calculating forces, since it is difficult to predict the flux linkages or the inductance except under certain simplifying assumptions, and then it is just as easy to use particular equations for the force without introducing the energy concept. From a study of the foregoing equations, however, it is possible to predict, at least qualitatively, the forces in more complex configurations where the simple formulas do not apply. Model studies or studies of previous designs with established characteristics are the only reliable methods for obtaining accurate evaluations of the forces developed by electromagnets. The following formulas for particular cases may be used by the designer for rough approximation of the forces.

SIMPLE SOLENOID AND PLUNGER. The simplest type of electromagnet consists of a solenoid and a plunger, as shown in Fig. 2. Its use is confined to applications where the stroke is long, and in these cases the plunger becomes saturated after it has travelled a short distance into the coil; from then until it approaches the other end the pull is approximately constant. This force-stroke relationship could be predicted from eq. (16) by reasoning that the change in the inductance of the electric circuit with respect to displacement dL/dx increases as the plunger moves into the coil until the plunger is saturated; from then on dL/dx is constant. Figure 6 is a typical force-stroke diagram for this type of electromagnet. As may be seen from eqs. (14) and (15), the pull will be zero just as the plunger enters the coil (just before it is under the influence of the coil flux) and again when the respective magnetic centers of the solenoid and the plunger coincide.

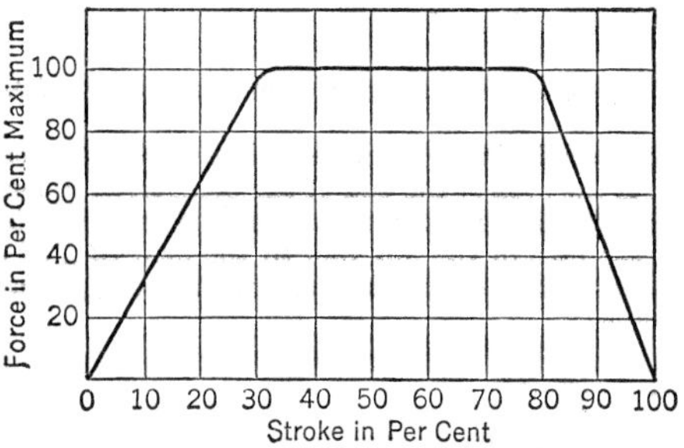

FIG. 6. Force-stroke Diagram for a Typical Simple Solenoid and Plunger Electromagnet

Calculation of Pull of Simple Solenoid. The calculation of the exact pull of such a magnet is very complicated, being dependent not only upon the ampere-turns, but also upon the shape of the coil, the size and length of the plunger, and the induction in the plunger. For practical purposes the maximum pull P_1 when the plunger is at least as long as the coil and saturated can be represented by the formula

$$P_1 = \frac{cA(NI)}{L_c} \tag{17}$$

where c = the pull in pounds per square inch per ampere-turn per inch of coil length.
A = the area of the cross-section of plunger in square inches.
I = the current in the coil in amperes.
N = the number of turns in the coil.
L_c = the length of the coil in inches.

It has been found that c varies between 9×10^{-3} and 10.5×10^{-3} when the plunger is magnetically saturated. For practical purposes sufficiently close results for a saturated plunger are obtained if c is taken equal to 10^{-2}. The formula shows that the maximum pull for a saturated plunger is directly proportional to the current, which makes this type of magnet especially suitable for relays and instruments which should be very sensitive. When the flux density in the plunger is well below the knee of the B-H curve, the factor c varies almost directly as the ampere-turns, and the pull, therefore, varies approximately as the square of the ampere-turns.

The pull at any point in the stroke for a saturated plunger, provided the plunger is in the coil a distance equal to at least three times its diameter, may be expressed by the formula

$$P_1 = \frac{10^{-2}kANI}{ls} \tag{18}$$

where k is a factor which gives the ratio of the pull at any point of the stroke to the maximum pull; Figure 6 gives the approximate value of k at various points of the stroke.

IRON-CLAD ELECTROMAGNET WITH FLAT-END PLUNGER. The simple solenoid is a very inefficient type of magnet except for very long strokes, because a large percentage of the reluctance of the magnetic path is found in the long air path outside the coil. Therefore the plunger magnet is modified by putting an iron return circuit around the outside of the coil, thus reducing considerably the reluctance of the path and increasing the work which can be obtained with a certain expenditure of energy in the coil. Figure 7 shows this type of iron-clad magnet.

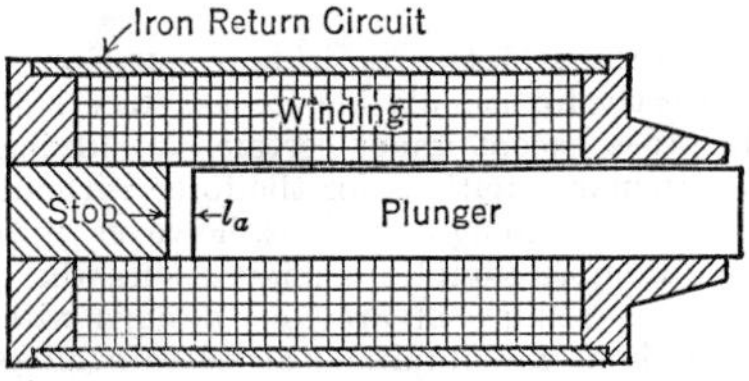

FIG. 7. Iron-clad Electromagnet with Flat-end Plunger

The stop may be of any length or may be omitted entirely. Although for short strokes the axial position of the gap is not critical, a somewhat higher pull can be obtained if the gap is located at the center of the coil.

Pull on Stop of Plunger: "Air-gap Pull." If we assume that the flux between the pulling surfaces (end of plunger and its seat) goes straight across the gap with no fringing, the pull per square inch of plunger area may be expressed as

$$\frac{P}{A} = \frac{B^2}{72 \times 10^6} \text{ lb [the same as eq. (1}a\text{) except for the units of measurement]} \tag{19}$$

where B is the flux density in the air gap perpendicular to the end surface of the plunger, in lines (maxwells) per square inch, and A is the area of the end of the plunger, in square inches. The value of B may be calculated from the ampere-turns of the coil and the dimensions of the magnetic circuit, as in the case of the flux due to the field coils of a generator. By neglecting leakage and the reluctance of the iron part of the path, B may be calculated approximately from the formula

$$B = \frac{3.19(NI)}{l_a} \tag{20}$$

where NI is the total ampere-turns of the coil, and l_a is the length of the air gap in inches.

Thus, when the magnetic circuit is below the point of saturation, the pull may be calculated from the expression

$$P = 1.4 \times 10^{-7} A\left(\frac{NI}{l_a}\right)^2 \text{ lb} \tag{21}$$

Total Pull of Iron-clad Electromagnet with Stop. In addition to the pull accounted for by the flux passing directly between plunger and stop, there is a component of pull due to the leakage flux from the plunger to the shell. This is referred to as the "solenoid pull," since it arises from the same action that produces force in the simple solenoid (Fig. 22). Unless the stroke is comparatively long, however, its contribution is small compared to the direct pull between the plunger and stop.

IRON-CLAD ELECTROMAGNET WITH CONED PLUNGER. It may be seen from eqs. (19) and (20) that P/A changes as $1/(l_a)^2$ for a flat-end plunger. This fact makes the flat-end plunger efficient for short strokes only, and, as discussed before, the simple solenoid is used for long-stroke applications. For intermediate strokes a plunger with a cone end (Fig. 8) of suitable angle and a corresponding seat in the stop finds application. The conical shape reduces the value of l_a for a given stroke and thus increases the flux density between the mating surfaces. Although only the axial component of the resulting pull is useful for doing external work, the pull perpendicular to the surfaces is so greatly increased that the net result is a gain in pull over what would exist if the flat-end plunger were used for the same stroke. The following paragraph will make this fact clearer.

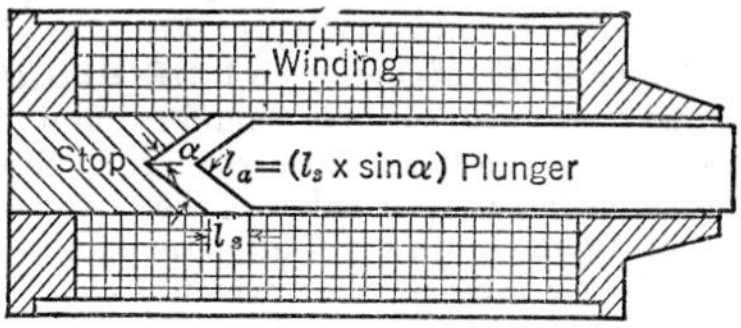

FIG. 8. Iron-clad Electromagnet with Coned Plunger

"Air-gap Pull" on Coned Plunger. In the following discussion it is assumed for the sake of simplicity that a uniform flux at right angles to and distributed over the entire surfaces of the cones passes between plunger and plug. This is not strictly correct, especially for long strokes, and the pull calculation is therefore only approximately correct.

Let l_s = length of stroke, in inches.
A = total cross-section of plunger, in square inches, = πr^2, where r is the radius of the plunger (see Fig. 8).
α = angle of the cone, in degrees (see Fig. 8).
NI = total ampere-turns of coil.

Then the flux density in the plunger is

$$B_i = \frac{3.19NI}{l_s \sin^2 \alpha} \text{ lines per square inch} \tag{22}$$

and the air-gap pull in the direction of the stroke is

$$P_2 = 1.4 \times 10^{-7} A \left(\frac{NI}{l_s \sin \alpha}\right)^2 \text{ lb} \tag{23}$$

Note that $l_s \sin \alpha = l_a$ is the "effective" length of the air gap, i.e., the length perpendicular to the surface of the cone. Hence, it is seen that the pull on a coned plunger for the same effective air gap is the same as on a flat-end plunger, but the length of the stroke is increased in the ratio of $1/\sin \alpha$. If we compare eq. (22) with eq. (20), it is seen that this advantage is gained by increasing the flux density in the plunger by the square of $1/\sin \alpha$.

Total Pull of Iron-clad Electromagnet with Coned Plunger. In addition to the air-gap pull there is a "solenoid pull" that is practically the same as in the flat-end plunger. This pull is difficult to calculate accurately, however, and is usually small compared with the pull at the air gap if the angle is properly chosen. If the angle is made smaller than 20 to 25 deg, saturation in the plunger becomes an important factor in limiting pull.

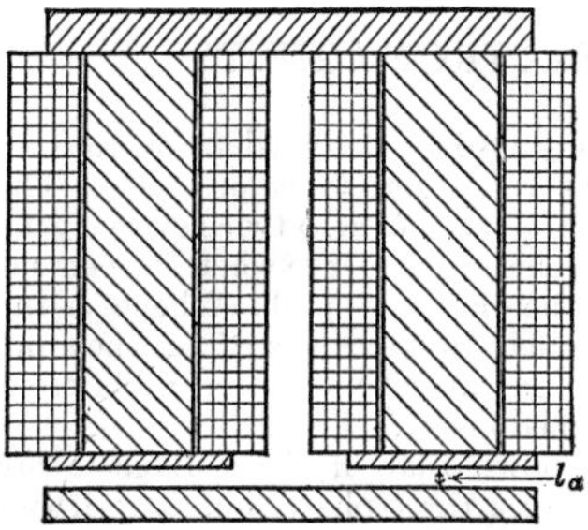

FIG. 9. Horseshoe-type Electromagnet

PLUNGERS OF OTHER SHAPES. A variety of other shapes of plungers is used in iron-clad electromagnets. Among these are the truncated cone, tapered, cylindrical-faced, and stepped-cylindrical-faced. The force-distance diagrams vary for the different types and are the basis for the selection of the type suitable for a given application.

HORSESHOE TYPE OF ELECTROMAGNETS. The horseshoe type of bipolar magnet shown in Fig. 9 is a common type of electromagnet. The equations for the pull of an iron-clad solenoid with a flat-end plunger are applicable to the horseshoe magnet if due allowance is made for the reluctance of the iron path, which may be treated as an additional air gap l'. Equation 21, as applied to the horseshoe magnet, then becomes

$$\frac{P}{A} = 2.8 \times 10^{-7} \left(\frac{NI}{2l_a + l'}\right)^2 \text{ lb} \tag{24}$$

where A = cross-section of each pole in square inches NI = total ampere-turns, l_a = length of each air gap in inches, and l' = length of air gap equivalent to the reluctance of the iron. This last depends on the permeability and dimensions of the iron part of the circuit; if we call l_i the mean length of this iron circuit and μ its permeability, and assume the mean cross-section of this path to be the same as the cross-section A of each pole, then $l' = l_i/\mu$. A more exact calculation of the pull may be made by determining, as for a generator field, the ampere-turns required to establish a given flux density of B lines per square inch in the gap and then applying eq. (19) to determine the pull.

SPEED OF MOVEMENT OF PLUNGER. In order to obtain quick action, the flux, upon closing the circuit, should reach its full value in as short a time as possible. The flux being a function of the current, the speed depends upon the rapidity with which the current reaches its full value. The time required for the current to reach its full value depends upon the quotient inductance L divided by the "effective" resistance r of the circuit. The larger this ratio L/r, which is called the "time constant" of the circuit, the longer the time required for the current to reach its full value. The inductance L is proportional to the square of the number of turns in the solenoid winding and inversely proportional to the total reluctance of the circuit. The effective resistance r depends not only upon the d-c or ohmic resistance of the winding, but also upon the eddy currents and hysteresis loss set up when the current is changing.

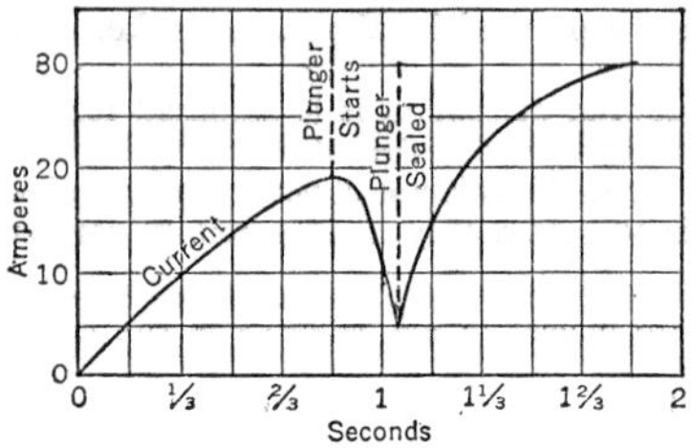

FIG. 10. Current-time Characteristic of D-c Electromagnet

In Fig. 10 is given a typical current-time curve, showing the change of the current during the switching-in period of a d-c magnet. It will be seen that, after the closure of the circuit, the current first rises to a certain value, which corresponds to a flux just sufficient to cause a movement of the armature or the plunger. As the plunger moves, the flux increases, thereby causing a counter electromotive force, which tends to reduce the current. This counter emf depends upon the speed of the plunger. In the case shown the current drops off continuously until the plunger strikes against the stop, at which moment its value is approximately one-third of that which started the motion of the plunger. After the plunger has come to rest, the current again increases and gradually reaches the value which is dependent upon the terminal voltage and the resistance of the coil.

Methods of Obtaining Quick Action. To reduce the eddy-current effect, the cross-section of the magnet frame should be as small as is consistent with other considerations, and a high-permeability steel should be chosen. Silicon steel is usually used for this purpose. Also the steel may be laminated as for a-c magnets.

Another method of obtaining quick action is to impress at the start a high voltage on the coil and insert resistance into the circuit of the coil as the plunger rises, in order to protect the magnet from overheating. This reduces the ampere-turns at the end of the stroke, which is permissible in most cases, because usually the pull of the magnet increases very rapidly toward the end of the stroke, as is indicated by the formulas which have been given.

21. ALTERNATING-CURRENT ELECTROMAGNETS

Electromagnets for producing a mechanical pull may also be designed to operate on alternating current. The flux in an a-c magnet passes through zero twice per cycle. The pull, which varies with the square of the current, therefore becomes zero twice every cycle, and it can be shown that it also varies according to a displaced sine curve when the current is sinusoidal. The average effective pull is one-half of the maximum pull. Whenever the pull is less than the load, there is a tendency for the plunger to move away from the stop, causing rattling or humming of the magnet. This humming may be overcome in different ways, one method being to use a "shading coil," described below.

POLYPHASE ELECTROMAGNETS. It can be shown that, if three magnets are used and each is supplied with current from one phase of a three-phase source, or if two magnets are used and each is supplied with current from one phase of a two-phase source, then the resultant pull will be constant at any moment, and, if the plungers are rigidly connected, there will then be no chattering. The point of application of the resultant force, however, will shift from midway between poles 1 and 2 to midway between 2 and 3 at twice the supply frequency. The most common form of three-phase magnet is shown in Fig. 11. It consists of a core having three poles and a plunger of similar construction.

Over each pole is wound a coil which is supplied from one of the three phases of the circuit. There are various modifications of the polyphase magnet, but their general principle is the same. In calculating the total pull, the pull of each pole is figured separately, and the several pulls, which are equal to one another, are combined vectorially, since they differ in time phase.

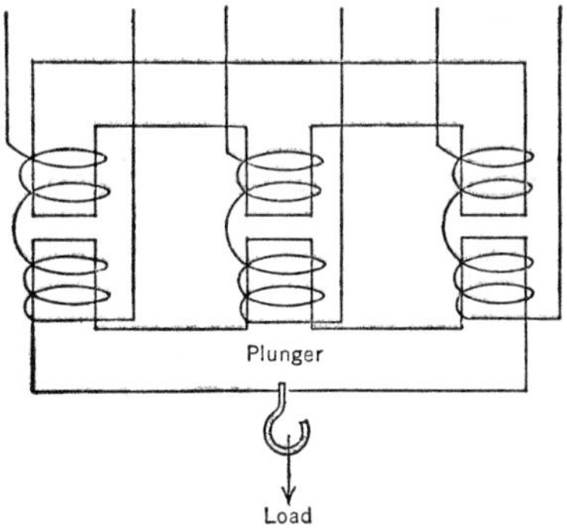

FIG. 11. Polyphase Electromagnet

CALCULATION OF PULL OF SINGLE-PHASE ELECTROMAGNETS OR OF ONE PHASE OF POLYPHASE ELECTROMAGNETS. The formulas for pull given for d-c electromagnets hold also for the average pull of a-c magnets, provided I is taken as the effective (rms) value of the current, B as the effective (rms) value of the flux density, and P as the average or effective value of the pull.

In contrast to the d-c electromagnet, the flux in an a-c electromagnet for a given impressed emf is approximately constant irrespective of the length of the air gap. The reason is that the opposition to the flow of current through the winding is due almost entirely to the back emf resulting from the alternation of the flux, and only to a very small extent to the resistance of the winding, the action in this respect being similar to that of a transformer (see Section 10) or induction motor. The back emf, being practically equal to the impressed emf, is constant.

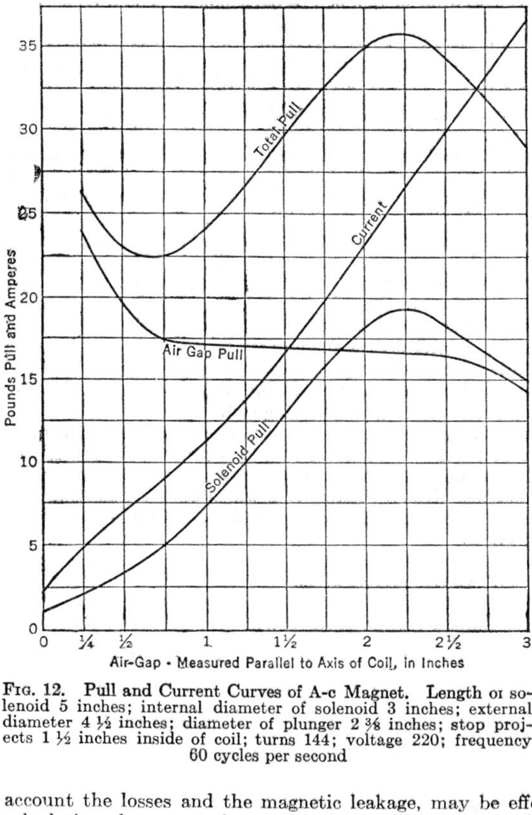

FIG. 12. Pull and Current Curves of A-c Magnet. Length of solenoid 5 inches; internal diameter of solenoid 3 inches; external diameter 4 ½ inches; diameter of plunger 2 ⅜ inches; stop projects 1 ½ inches inside of coil; turns 144; voltage 220; frequency 60 cycles per second

Since the flux remains practically constant, the current I varies approximately proportionally to the length of the air gap l_a; since not all the flux which generates a back emf in the coil crosses the gap, however, the current does not have to increase quite in proportion to the gap. The actual variation of the current with the length of the air gap for a particular a-c electromagnet is shown in Fig. 12.

If we use the same notation as above for d-c electromagnets, and in addition put E = effective value of impressed voltage per phase, and f = frequency of impressed voltage in cycles per second, the current taken by each phase of the magnet, neglecting the iron losses, leakage, and reluctance of the iron part of the magnetic circuit, will be

$$I = \frac{10^7 E l_a}{2fNA^2} \text{ amperes} \qquad (25)$$

A more accurate calculation of the current, taking into account the losses and the magnetic leakage, may be effected by the method used for calculating the current in a transformer. This calculated value of current may be substituted in eq. (19), (20), (21), and (22) to determine values of pull for a-c electromagnets.

Figure 12 shows the pull curve of an a-c magnet with flat-end stop and plunger. It will be noted that the current is roughly proportional to the air gap. The solenoid pull reaches a maximum at an air gap of about 2 1/4 in. and then falls again approximately proportional to the air gap; the air-gap pull is constant over a wide range of the travel. The result is a total pull curve which has a maximum at approximately 2 1/4 in., drops until an air gap of approximately 3/4 in. is reached, and from there on increases again, because of the increase of the air-gap pull, which is caused by the diminution of the leakage flux as the air gap decreases.

LIMITING FLUX DENSITY AND LOSSES IN A-C ELECTROMAGNETS. Attention must be paid to the fact that the iron losses, like those of the transformer, increase with the flux density, and therefore the design must be such that the flux density is not too high. The rms flux density in the iron in a flat-end plunger, neglecting the magnetic leakage and the resistance of the electrical circuit, is

$$B = \frac{1.6 \times 10^7 E}{fNA} \text{ maxwells per square inch} \tag{26}$$

and in a coned plunger, under the same assumptions, is

$$B = \frac{1.6 \times 10^7 E}{fN\,A \sin \alpha} \text{ maxwells per square inch} \tag{27}$$

where α is the cone angle (see Fig. 8).

These relations are only approximate. The magnetizing component of the exciting current, and from it the flux density, can be more accurately calculated by the methods employed in the design of transformers or induction motors.

The iron and copper losses can also be calculated by similar methods, and from these losses the energy component of the exciting current can be determined. The total current and power factor of the electromagnet can then be deduced.

SHADING COIL FOR SINGLE-PHASE ELECTROMAGNETS. The humming of single-phase magnets may be greatly reduced by introducing a so-called "shading-coil" in the pole face. This shading coil is nothing more than a short-circuited secondary winding, consisting of one or more turns, which encloses only part of the total flux passing through the plunger. Because of the leakage reactance of this turn, the current induced in it is out of phase with the inducing flux, so that at the moment when the inducing flux due to the main winding is zero, a flux due to the current in the shading coil still remains and produces a pull. The result is that the combined pull from the main flux and the shading-coil flux never becomes zero. Figure 13 shows the arrangement of the shading coil. This coil is mounted in the plunger or in the plug close to the pole face in order to reduce the length of the path for the magnetic lines, which are interlinked with the shading coil. Naturally the shading coil has no effect with long air gaps, and it is therefore imperative that a good magnetic contact be obtained when the plunger is in the sealed position to get the greatest possible effect of the shading coil.

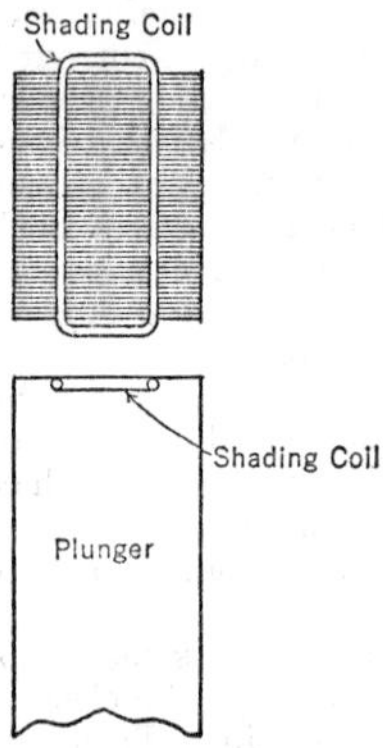

FIG. 13. Shading Coil

Incidentally, the shading coil also increases considerably the maximum pull for a given impressed emf, as the maximum pull is also due to the vectorially combined main end local flux. For example, a plunger magnet without shading coil, which gave a pulsating pull between zero and 28 lb, had, after the introduction of the optimum shading coil, a pull which pulsated between 18 and 143 lb.

22. BIBLIOGRAPHY

Allegheny Steel Company, *Selection of Magnetic Core Materials* (1937).
Armstrong, G. C., General Magnet Design, *Elec. J.*, Vol. 35, p. 414 (1938).
Erikson, B. W., Magnet Coils and Related Data, *Power Plant Eng.*, Vol. 45, pp. 62, 71, and 74 (1941)
Horstmann, H. C., *Practical Armature and Magnet Winding*, F. J. Drake & Co. (1941).
Moses, G. L., Magnet Coil Life as Affected by Higher Operating Temperatures, *Product Eng.*, Vol. 14, p. 22 (1943).
Moses, G. L., Magnet Coil Proportioning and Testing, *Product Eng.*, Vol. 11, p. 47 (1940).
Nachod, N. P., Winding Magnet Coils, *Elec. J.*, Vol. 31, p. 457 (1934).
Nachod, N. P., Nomograms for the Design of D-c Magnet Coils, *G. E. Rev.*, Vol. 37, p. 563 (1934).
Rader, L. T., Factors Affecting the Design of D-c Magnets, *Trans. AIEE*, Vol. 62, p. 307 (1943).
Roters, H. C., *Electromagnetic Devices*. John Wiley (1941).
Underhill, C. R., *Solenoids and Electromagnets*. Van Nostrand (1914).
Windred, G., *Electromagnets and Windings*. Geo. Newnes, London (1943).
Magnet Coils, Examples of Design Calculations for Various Types, *Electrician*, Vol. 115, p. 442 (1935)
Magnet Coils, Three Limb Magnets, *Electrician*, Vol. 115, p. 578 (1935).

PERMANENT MAGNETS

By I. F. Kinnard and J. H. Goss

(See Section 2 for Permanent-magnet Materials)

23. MAGNETIZATION

To deliver its full strength a magnet must be magnetized in a field of sufficient intensity to saturate the material. The field, especially in the case of high coercive materials, must have the correct shape. Handling of the magnet must not demagnetize it more than is desired in the final assembly.

FIELD STRENGTH. A field of four to five times the coercive force of the material is usually sufficient. Workable saturating values are listed in Table 1, Section 2, Article 31.

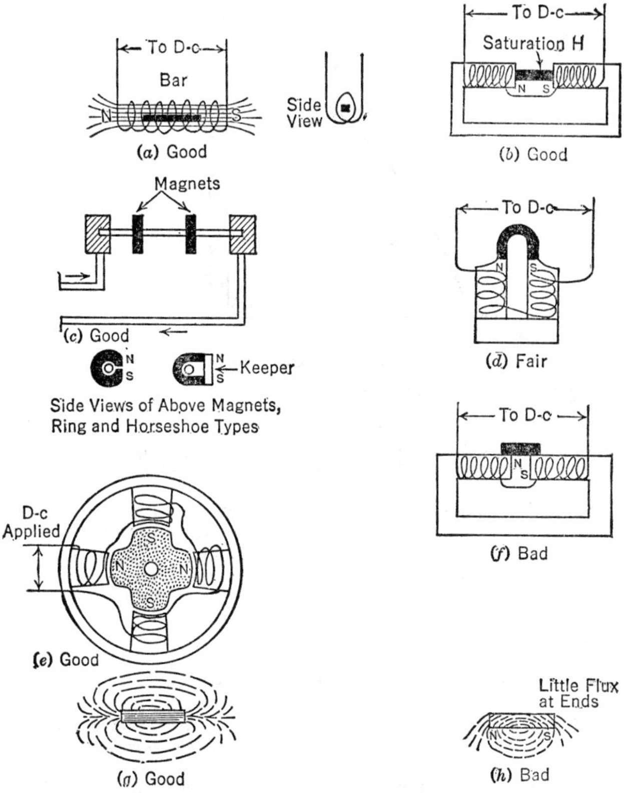

Fig. 1. Magnetizing Set-ups for Permanent Magnets

(a) Solenoid method for straight magnet
(b) Electromagnet method for straight magnet
(c) Single-bar method for ring magnets
(d) Electromagnet method for horseshoe magnet
(e) Magnetizing of four-pole rotor
(f) Improper magnetizing of straight bar magnet
(g) Bar magnetized as in (a) or (b)
(h) Bar magnetized as in (f)

High currents for magnetizing fields may be obtained by a steady direct current, a half wave impulse from an a-c source, or a condenser discharge. If an impulse of short duration is used, the peak current required for saturation is two to three times (depending on material and size of magnet) that required with direct current because of eddy currents set up in the magnetic material. Where facilities are not available for producing a field of sufficient strength with either alternating or direct current alone, the two fields may be used in combination by superimposing an a-c field while the direct current is being applied. In this case half or more of the total field should be supplied by direct current, and the remainder made up with alternating current. The alternating current must never exceed the direct current and must be applied after and removed before the direct current.

FIELD SHAPE. The magnetizing force may be applied in different ways, but, however it is applied, the shape of the field should coincide as closely as possible with that of the magnet to avoid cross-magnetization of parts of the magnet. Figure 1 shows schematically several methods of magnetizing that have been found useful, particularly with the high coercive materials. A simple shape like a straight bar magnet may be magnetized between the poles of an electromagnet or inside a solenoid with direct current applied. C-shaped and similar magnets are usually placed on a single bar through which a high current is passed.

HANDLING. Where possible, magnets should be magnetized in their final assembled condition. If this is not possible, a shunt or "keeper" of soft iron should be placed in the working air gap of the magnet while magnetizing. As the magnet is placed in the final

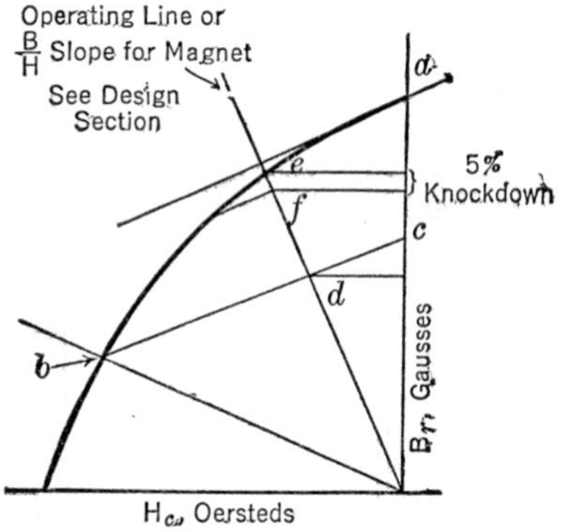

FIG. 2. Behavior of Magnet after Magnetizing

1. At removal of magnetizing force, flux in magnet is at point *a*.
2. Removal of keeper drops flux to point *b*.
3. Assembling of magnet in device returns flux along minor loop *b-c* to intersection of operating line at *d*.
4. If assembly were accomplished without introduction of an air gap, flux in magnet would drop only to point *e*, after which a 5 per cent stabilizing knockdown would leave the magnet operating at point *f*.

assembly, this keeper is removed without the introduction of a gap larger than the working air gap. Introduction of a larger air gap will reduce the output of the assembled magnet, as shown in Fig. 2. In special cases where removal of the keeper before assembly is provided for in the design, this limitation, of course, does not apply, as in electric-motor design, where allowance is made for removal of the permanent magnet rotor for repair work without using a keeper or requiring remagnetizing.

24. STABILIZATION OF MAGNETS

The term stability, as referred to permanent magnets, is used to designate the constancy of output of the magnet, either with the passage of time or under various demagnetizing influences.

METALLURGICAL STABILITY. The quench hardening alloys of the first group (see Fig. 1, Section 2, Article 1) are metallurgically unstable as quenched, since small structural changes take place with the passage of time. These are shown as permanent changes in the magnetic properties of the material. By low-temperature tempering, changes may be made to take place in a short time to yield a stable structure which does not change further.

The alloys of all the other groups are generally so stable metallurgically after their final heat treatment that changes of this nature are not evident. In Alnico III some metallurgical instability may be observed if the material is not properly drawn, probably because of the rapid cooling necessary to develop its properties.

Figure 3 shows, in terms of flux output, the ageing of tempered and untempered chromium-steel C-shaped magnets.

MAGNETIC STABILITY. If the conditions for metallurgical stability mentioned in the foregoing paragraphs are satisfied, the magnet will remain stable, after it is magnetized, until it is subjected to some demagnetizing effects. External fields and temperature change are the usual demagnetizing influences. If no specific external field strength must be withstood, it is good practice to subject a magnet to a field sufficient to produce approx-

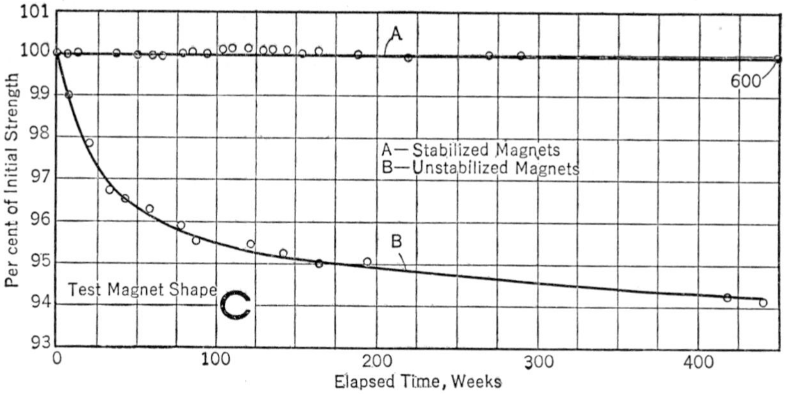

FIG. 3. Ageing of Tempered and Untempered Chromium-steel Magnets

imately 5 per cent reduction of flux output. If resistance to a given field is required, sufficient stability may be obtained by subjecting the magnet to a field strength slightly in excess of the required value. The effect of knockdown is graphically illustrated in Fig. 4.

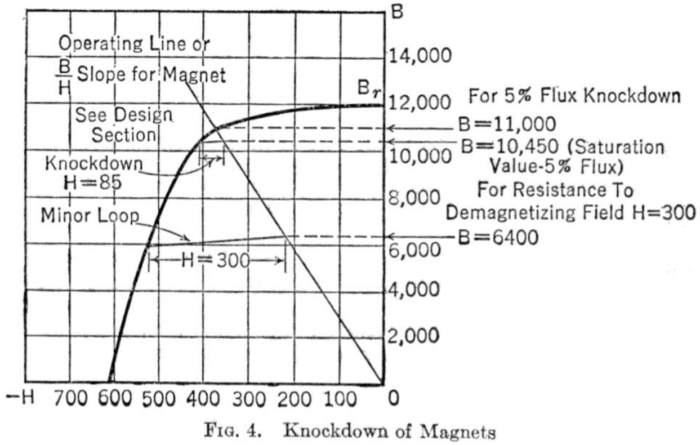

FIG. 4. Knockdown of Magnets

After the knockdown, in order to have a magnet stable over a temperature range, it is desirable to cycle it one or more times over a range in excess of that expected to be encountered in service. The temperature effect is relatively small, usually less than 0.5

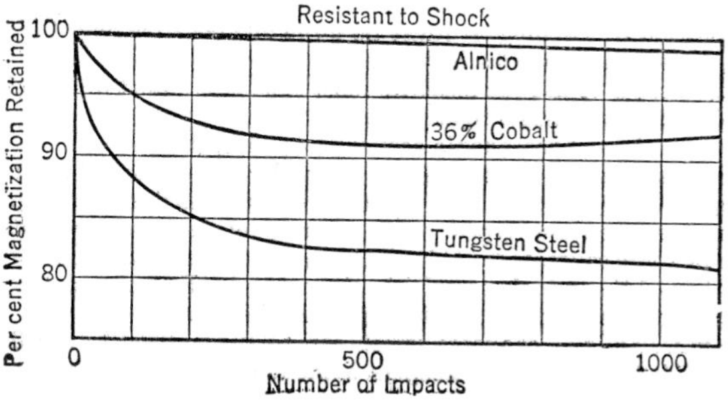

FIG. 5. Effect of Shock on Magnets

per cent permanent decrease in flux after repeated exposure to −50 deg cent and +100 deg cent. The major part of the change occurs during the initial cycle.

The effect of vibration and shock is very difficult to report in general terms, because of widely varying test conditions. In general, the carbon-hardened alloys appear to be noticeably affected, whereas the other alloys are essentially unchanged. See Fig. 5.

APPARENT INSTABILITY. It sometimes appears that magnets treated as above are still unstable. This lack of stability, however, is due to unwitting further demagnetization, which may be locally caused by touching the magnet with magnetic materials or by changes in its operating state as assembled with greater or less air gaps (open or closed circuits). See Fig. 6.

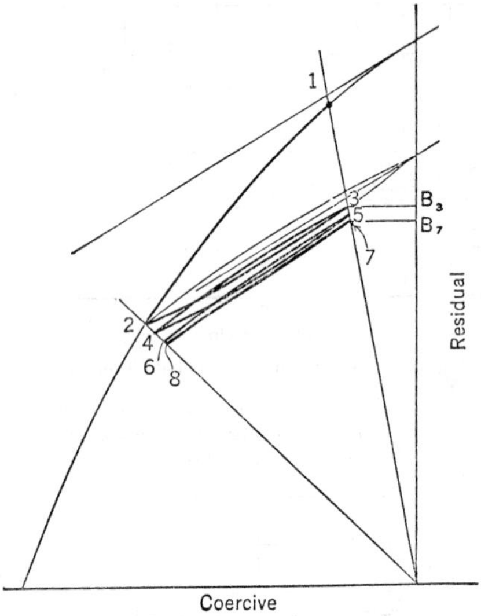

FIG. 6. Knockdown of Magnet as Assembled with Varying Air Gaps (Numbers indicate sequence of operating points as magnet is moved from closed- to open-circuit condition repeatedly. Points 7 and 8 show the end condition attained as the magnet is following an essentially closed loop and shows no further change of B.)

Permanent magnets are not subject to erratic changes, and their performance is readily predictable if the conditions to which they are subjected are known.

25. DESIGN OF PERMANENT MAGNETS

GENERAL PRINCIPLES. In selecting the optimum material for a given application the following factors should be considered:

1. The magnet must have sufficient developed length to supply the required magnetic potential to the air gap.
2. The cross-section of the magnet must be large enough to supply the necessary flux to the air gap, plus an allowance for leakage.
3. The design, in order to use the material most efficiently, must operate it at the maximum energy point on the demagnetization curve.
4. The magnet must meet reasonable economic considerations.

For purposes of calculation most magnets fall under one of three types: (1) fixed gap, (2) varying gap, (3) straight bars. The following paragraphs describe simplified methods of calculating magnets of these types. The accuracy of the results will depend on how well the leakage factor is estimated and how closely the demagnetization curve used represents the properties of the actual material in the magnet. For more detailed and precise methods of calculation, refer to the Bibliography, Article 26.

FIXED-GAP MAGNETS. Magnets which can be magnetized in the final assembled condition or assembled by the use of keepers without the introduction of a larger air gap are covered in this group.

a. Determination of magnet size.

It is desired to have a flux density B_g in an air gap of a specific cross-sectional area A_g and length L_g. To determine the dimensions of the magnet required, the following formulas are used:

$$\text{Area of magnet } A_m = \frac{f\, B_g A_g}{B_m}$$

$$\text{Length of magnet } L_m = \frac{B_g L_g}{H_m}$$

where B_g = flux density in air gap, in gausses.
A_g = area of gap, in square centimeters.
Φ_g = total flux in gap = $B_g \times A_g$, in maxwells.
L_g = length of gap, in centimeters.
H_g, in oersteds = magnetizing force in gap = B_g, in gausses.
B_m = flux density in neutral section of magnet, in gausses.
H_m = magnetizing force in neutral section of magnet, in oersteds.
L_m = length of magnet, in centimeters.
A_m = area of magnet, in square centimeters.
Φ_t = total flux in magnet = $B_m \times A_m$, in maxwells.
f = leakage factor = $\dfrac{\Phi_t}{\Phi_g}$.

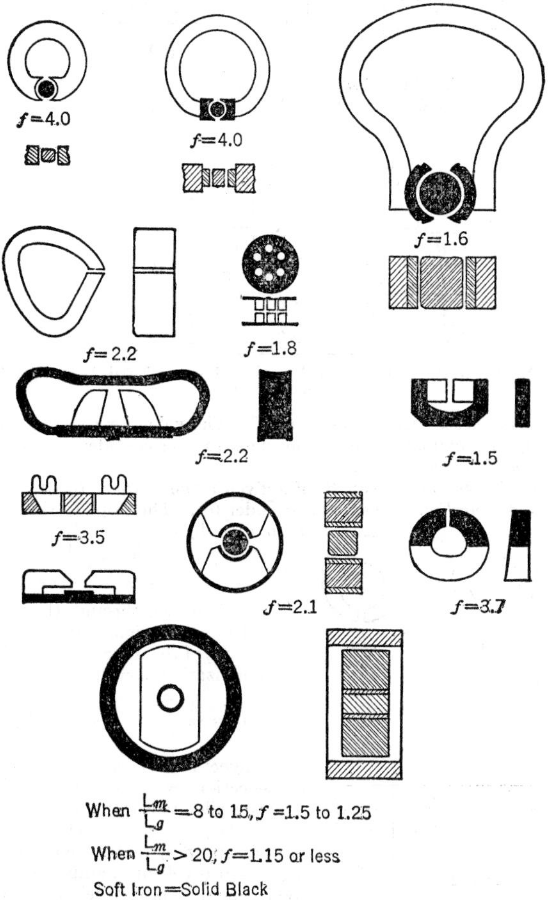

FIG. 7. Typical Magnet Assemblies and Leakage Factors

B_m and H_m are the coordinates of a point on the demagnetization curve of a possible material. For the most efficient use of material this point is at the point of maximum energy value but for the greatest flux output is chosen as high on the curve as feasible. For convenience Table 1 shows reasonable design values for 22 available materials.

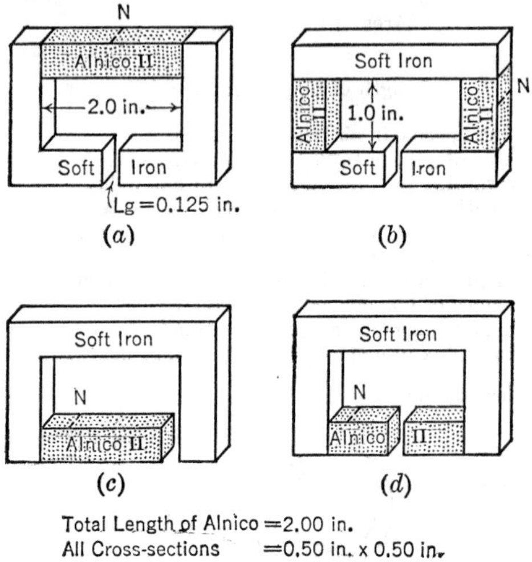

Measured Values in Saturated Condition

	B_g	f	ϕ_t
a	1,400	4.8	10,730
b	1,770	3.1	8,900
c	2,180	3.2	11,100
d	2,720	2.2	9,550

FIG. 8. Various Arrangements of a Magnet Assembly, Using a Fixed Amount of Permanent-magnet Material

The value for f is selected from Figs. 7 and 8. The value for the magnet arrangement which most nearly approximates the magnet design under consideration is chosen.

b. Determination of magnet output.

To find the flux density in the gap B_g of a given magnet where L_g, A_g, L_m, and A_m are known, f is selected from Fig. 7, as explained under (*a*). The slope of the operating line is then given by

$$\frac{B_m}{H_m} = f\frac{A_g L_m}{A_m L_g}$$

A line is drawn through the demagnetization curve from the origin having this slope. The coordinates of the intersection of this line and the demagnetization curve are B_m and H_m. Then

$$B_g = \frac{A_m B_m}{f A_g}, \qquad \text{or} \qquad B_g = \frac{L_m H_m}{L_g}$$

A magnet is not necessarily made with uniform cross-section from end to end. The value A_m given applies to the neutral section. Portions nearer the poles carry less flux because of loss by leakage and may be tapered to obtain a smaller area. An excellent example of the advantages to be gained by proper design can be seen from a study of Fig. 8.

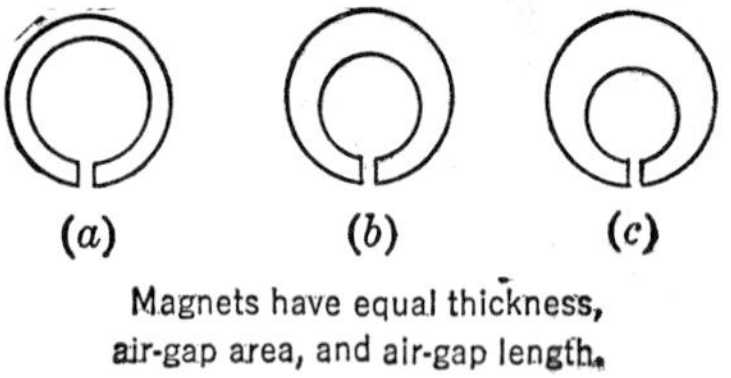

	A_m/A_g	L_m	B_g	ϕ_m	f
a	1.0	14.9	3000	12,300	3.2
b	1.5	14.0	3600	16,350	3.5
c	2.0	12.8	3400	15,600	3.5

FIG. 9. Effect of Cross-section Increases on Magnet Output

Figures 9 and 10 have been included to show the effect of making the magnet cross-section or length greater than is required.

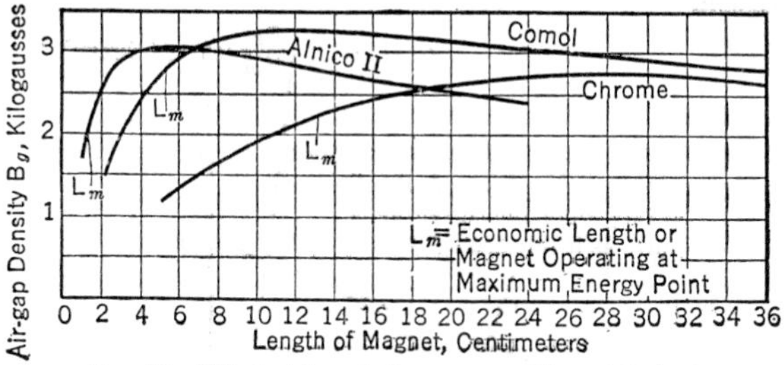

FIG. 10. Effect of Length Increases on Magnet Output

Table 1. Design Values of Permanent-magnet Materials

	B_m	H_m	$B \times H \times 10^6$	B_m/H_m
Silmanal	290	290	0.083	1.0
Vectolite *	1000	500	0.50	2.0
Cobalt platinum	2400	1000	2.400	2.4
Vicalloy II *	9000	330	3.07	2.7
Iron platinum	3000	760	2.28	3.9
Cunico	2000	400	0.80	5.0
Alnico XII	3100	580	1.80	5.3
Alnico IV	3200	405	1.30	7.9
Cunife I *	4200	400	1.68	10.5
Alnico II	4300	380	1.63	11.3
Alnico III	4500	315	1.42	14.3
Alnico I	4500	315	1.42	14.3
Alnico VI *	7200	485	3.50	14.8
Alnico V *	9900	445	4.40	22.2
Vicalloy I *	6200	180	1.12	34.4
36% cobalt	5700	162	0.92	35.6
Cunife II *	5150	148	0.78	35.8
15% cobalt	5100	120	0.61	42.5
Remalloy	6500	142	0.92	45.8
5% tungsten	6200	48	0.30	129.0
1% carbon	5000	35	0.18	143.0
3.5% chrome	6400	44	0.28	145.0

* These alloys have directional properties.

VARYING-GAP MAGNETS. In some cases it may be desirable to subject permanent magnets to widely varying air gaps during assembly or service. A typical example is a permanent-magnet rotor for a generator which may be removed from the assembled generator. In the generator it operates in a closed circuit with a small air gap, but when removed from the generator must operate in an open circuit with a much longer air gap. The effect of demagnetizing influences, such as operating in the generator itself, may be evaluated by reference to Article 24 on stabilization of magnets and Fig. 8. Other examples of this type include holding magnets with large air gaps when not in use and small gaps when in use.

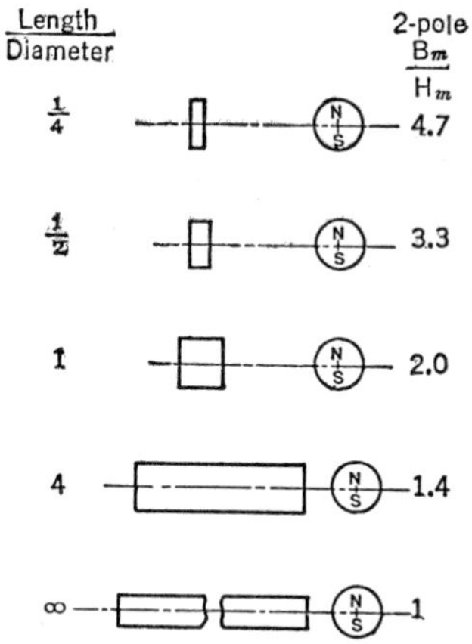

FIG. 11. Demagnetization Factors for Diametrically Magnetized Cylinders

To estimate the performance of these magnets, B_m/H_m is calculated for the condition of the small air gap, as set forth in the foregoing paragraph, and B_m/H_m is calculated for the large air gap by the same method or with the aid of Table 1 and Fig. 11. The resulting B_m/H_m slope lines are plotted on the demagnetization curve of a possible permanent-magnet material. As shown in Fig. 12, B_m of the magnet in the closed-circuit condition will be the intercept of the minor loop terminated by the intersection of

B_m/H_m open circuit with the demagnetization curve and the B_m/H_m line for the closed circuit. The gap density B_g of the magnet may be determined from B_m as shown in the foregoing paragraph.

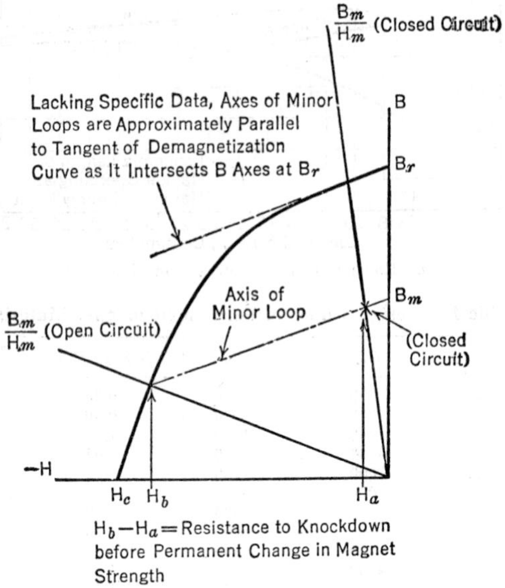

FIG. 12. Behavior of Magnet in Open- and Closed-circuit Condition

STRAIGHT-BAR MAGNETS. The flux density B_m in the neutral section of a round rod magnet in air may be estimated from the ratio of length to diameter. The slope corresponding to this ratio, as shown in the table below, may be used to determine B_m by the method illustrated in the paragraphs on fixed-gap magnets.

$\frac{L}{D}$ = 0.5	1	2	3	4	5	6	7	8	10	14	22	38	70
$\frac{B_m}{H_m}$ = 1.4	2.8	6.0	9.6	14	20	28	40	55	96	200	450	1200	3700

For rectangular sections approaching squares the equivalent diameter may be estimated as $2\sqrt{A_m/\pi}$.

26. BIBLIOGRAPHY

Edgar, R. F., Permanent Magnets, *Gen. Elec. Rev.*, p. 466 (Oct., 1935).

Evershed, S., Permanent Magnets in Theory and Practice, *J. Inst. Elec. Eng.*, Vol. 58, p. 780 (Sept., 1920).

Hornfeck, A. J., and R. F. Edgar, The Output and Optimum Design of Permanent Magnets Subjected to Demagnetizing Forces, *Trans. AIEE*, Vol. 59, p. 1017 (1940).

Kinnard, I. F., and J. H. Goss, Braking Magnets for Watthour Meters, *Trans. AIEE*, Vol. 60, p. 431 (1941).

SECTION 7

BATTERIES

DRY BATTERIES

BY J. J. COLEMAN

RESERVE-TYPE PRIMARY BATTERIES

BY PAUL L. HOWARD

RAILWAY SIGNALING BATTERIES

BY R. B. ELSWORTH

LEAD-ACID STORAGE BATTERIES

BY J. LESTER WOODBRIDGE AND A. MILTON MILEY

ALKALINE-TYPE STORAGE BATTERIES

BY E. W. ALLEN

BATTERIES

DRY BATTERIES

By J. J. Coleman

1. GENERAL DESCRIPTION

A dry cell consists of a central carbon rod embedded in a cylinder of "mix" (a mass of manganese dioxide and carbon black), which is contained in a cylindrical zinc can and separated from the can by paper or by a layer of starch gel.

The carbon rod is made by extruding a thermoplastic mixture of oil, coke, and carbon black into rods and then calcining these rods until sufficient carbonization and graphitization have taken place to render them strong and electrically conductive.

MIX. The composition of the "mix" varies widely among manufacturers. The manganese dioxide may be a natural ore or may be artificial. Only a few of the many different kinds of crystal structures prove active in a dry cell. Active natural ores are found in Montana and Africa. The artificial manganese dioxides used are more active than the best natural ores. One type is prepared electrolytically. The black is either graphite or acetylene black. Acetylene black is better for most dry cell uses because it can hold and render immobile a larger mass of electrolyte.

ELECTROLYTE. The electrolyte is basically a mixture of zinc chloride and ammonium chloride in water and is contained in the mix and in the starch gel or paper separating the cylinder of mix from the zinc can.

ZINC. The zinc can is usually formed by drawing or extruding. A very pure zinc containing a small amount of cadmium and lead is usually used.

CHEMICAL REACTIONS. The electrical potential developed is the result of the tendency of zinc to dissolve in the electrolyte as zinc ions and of the tendency of the manganese dioxide to be reduced to a lower oxide. When zinc ions leave the zinc, an excess of electrons remains, and a potential is developed. These electrons are slowly neutralized by hydrogen ions in the electrolyte. This process is the chief cause of the deterioration of batteries "on shelf," i.e., when not in use. It also results in a slow evolution of hydrogen gas, as a result of which the cell cannot be sealed completely gas tight. Because the cell must be vented, a slow evaporation of water results, which also decreases "shelf life." At higher temperatures both these processes are accelerated, and the pressure of hydrogen gas may lift the wax seal usually employed to close off the top of the cell. In use, the excess of electrons in the zinc flows through the load to the central carbon rod and through the carbon black to the manganese dioxide particles. The manganese dioxide particles absorb the electrons and either emit negative ions or absorb positive ions; the exact action is not known. Whichever process takes place acts to make the electrolyte more basic. This lessens the contribution of the manganese dioxide to the potential developed by the cell and is one of the chief causes of the fall in potential as the cell is discharged. The zinc chloride acts as a buffer with the ammonium chloride and thus to some extent counterbalances this action. The addition of zinc to the electrolyte and its growing basicity finally result in the precipitation of basic zinc salts. The ammonium chloride employed in the electrolyte retards this action. These salts block the flow of current through the cell and further reduce the potential.

Cells are incorporated into batteries by connecting in series to get higher potentials and in parallel to meet heavier drain requirements.

2. FACTORS AFFECTING SERVICE LIFE

The service life of a battery depends on the following factors: quality of manufacture, storage time required before the battery is used, temperature during storage, length and frequency of rest periods (if any) during discharge, size of the battery, power (or drain) required, and percentage drop in potential which is permissible.

STORAGE. Most batteries of good quality can be stored for periods of 6 months or more at temperatures of 90 deg fahr or less and 12 months or more at temperatures of

70 deg fahr or less. Intervals of a few days or even weeks at 130 deg fahr are not usually harmful. Lower storage temperatures greatly increase the life of batteries; when they are stored at temperatures below 0 deg fahr, they can be kept indefinitely. Storage temperatures down to −40 deg fahr are not harmful. Data are lacking on lower temperatures.

OPERATING TEMPERATURES. The open-circuit potential of a dry cell is decreased about 0.02 volts when the temperature is decreased from 75 deg fahr to −5 deg fahr. The "flash" (short-circuit current) is approximately a linear function of temperatures from 0 deg fahr to 130 deg fahr. If the flash at 70 deg is represented as 100, then the slope of the flash-temperature curve is about 0.7 per degree fahrenheit. The capacity of dry cells is also approximately a linear function of temperatures between 0 deg fahr and 130 deg fahr. The capacity-temperature slope varies, however, for different batteries and different loads. If the capacity at 70 deg is represented as 100, the slope will be from 1 to 1.5 per degree fahr. All these statements should be qualified with the observation that dry batteries are likely to fail suddenly at temperatures in the neighborhood of 130 deg fahr and above.

REST PERIODS. Rest periods are most effective when the drain is heavy and give little, if any, additional service life when the drain is very light. For a rough judgment as to the effect of rest periods, any drain which results in a service life of 10 hours or less can be considered a "heavy" drain, and any which results in a service life of 100 hours or more, a "light" drain.

The size of the battery, power (or drain) required, and percentage drop in potential permissible are dealt with in an approximate fashion by Table 1.

Table 1. Relation of Size, Power, and Service Life in Fresh Batteries of Good Quality

Power Required, Watts per Cubic Inch Volume of Battery	Watthours Obtainable with a 33% Potential Drop
	Watthours per Cubic Inch Volume of Battery
0.10	0 to 0.3
0.08	0.1 to 0.5
0.06	0.2 to 0.6
0.04	0.3 to 0.8
0.03	0.4 to 1.1
0.02	0.5 to 1.5
0.01	0.7 to 1.8
0.005	1.0 to 2.0

This table may be applied to all kinds of batteries and all types of services. Many tests which correspond rather closely to the services for which the batteries are employed, however, are made on dry batteries. Where applicable, they give information much more precise than that obtainable from Table 1. Such tests are described in Article 5.

3. DIMENSIONS OF CYLINDRICAL CELLS

Most manufacturers make their batteries from cylindrical cells which have the dimensions given in Table 2. A description of some common batteries, along with other useful

Table 2. Sizes of Standard Cylindrical Cells

Cell Designation	Nominal Diameter, Inches	Nominal Height over Can, Inches	Cell Designation	Nominal Diameter, Inches	Nominal Height over Can, Inches
No. 6	2 1/2	6	BB	3/4	1 5/16 *
J	1 1/4	5 7/8	A	5/8	1 7/8
G	1 1/4	4	AA	17/32	1 7/8
F	1 1/4	3 7/16	R	17/32	1 5/16
E	1 1/4	2 7/8	P	17/32	1
D	1 1/4	2 1/4	N	7/16	1 1/16
CD	1	3 3/16	M	17/32	3/4
C	15/16	1 13/16	K	17/32	1/2
B	3/4	2 1/8			

* This dimension applies to flashlight cells primarily for export. In B batteries, cells approximately 3/16 inch higher are commonly used.

data, can be found in *American Standard Specifications for Dry Cells and Batteries*, Circular of the National Bureau of Standards, C435, obtainable from the Superintendent of Documents, Washington, D. C.

4. SERVICE LIFE OF BATTERIES

The capacity (life) requirements listed in the *American Standard Specifications for Dry Cells and Batteries* are shown in Table 3.

Table 3. Capacities of Dry Cells

Cell Size *	Test †	Life	
		Initial	6 Months Delayed
No. 6 Dry Cells			
No. 6 general purpose	Light intermittent: 6 2/3 ohms	200 days	60 hr
	Heavy intermittent	70 hr	
No. 6 industrial	Light intermittent: 6 2/3 ohms	275 days	90 hr
	Light intermittent: 16 2/3	500 days	
	Heavy intermittent	100 hr	
General-purpose Flashlight Cells			
D	Light industrial	400 min	550 min
	Household intermittent	600 min	
C	Household intermittent	300 min	250 min
AA	Household intermittent	65 min	50 min
Industrial Flashlight Cells and Batteries			
			3 Months Delayed
D heavy industrial	Heavy industrial	750 min	650 min
D light industrial	Light industrial	850 min	750 min
4F railroad lantern	Railroad lantern	45 hr	40 hr (6 months)

* See Table 2. † See Table 4.

5. STANDARD DRY-BATTERY TESTS

STANDARD TEST REQUIREMENTS. These are given in Table 4.

INTERNAL RESISTANCE AND FLASH. The d-c characteristics of a dry cell can be represented approximately by the equation

$$E = RI + V \tag{1}$$

In this equation I is the current and V is the potential developed across the cell. E is a quantity having the characteristics of an emf. It cannot be defined as *the* emf of the dry cell because the manganese dioxide particles in the mix are reduced at different rates, and hence the contributions of different particles to the potential V will, in general, be

Table 4. Standard Dry-battery Flashlight Tests

	Discharge Time	Resistance	Average Current Drain	End-point	Practical Use
Flashlight intermittent	5 min per day, every day	4 ohms per cell	0.28 amp	0.75 volt per cell	Average city home use
Heavy industrial	4 min every 15 min, 8 hr every day	4 ohms per cell	0.29 amp	0.9 volt per cell	Meter readers, watchmen, etc.
Light industrial	4 min per hr, 8 hr every day	4 ohms per cell	0.30 amp	0.9 volt per cell	Farmers, campers, sportsmen, etc.
Railroad lantern	1/2 hr per hr, 8 hr every day	8 ohms per cell	0.15 amp	0.9 volt per cell	Railroad-type lantern

different. E may be thought of as the emf of an equivalent cell (i.e., one developing the same V) in which all the dioxide particles are reduced at the same rate. Likewise, R has the characteristic of a d-c resistance. The direct current flows through the mix partially as an electronic current in the web of black and partially as an ionic current through the web of electrolyte. The web of electrolyte and the web of black are connected, so far as the flow of direct current is concerned, only through the particles of manganese dioxide. The mix is thus a complicated network of resistors, and no single resistance can represent accurately the action of these resistances on the flow of direct current. Equation (1) is a good approximation because E and R are practically independent of the instantaneous values of I. They are affected by the duration and rate of discharge of the cell and any rest period to which the cell may have been subjected. If an a-c component of the current is present, an a-c potential is developed across the cell. The ratio of this potential to this current component is not necessarily equal to R in eq. (1). This is partially due to the fact that alternating current probably does not follow the same path in the mix as does direct current and to the fact that the electrolyte-zinc interface does not present the same resistance to alternating current as it does to direct current.

Flash. The value of I in eq. (1) when $V = 0$ is called the "flash" of a cell or battery. The flash is measured by putting an ammeter across the cell.

The flash is often used to detect defective cells in a lot identical except for chance variations in manufacture. This is possible because many such chance variations which lower the service capacity of a cell also lower its flash.

The flash **cannot** be used to judge the quality of one manufacturer's product as against that of another. Many changes in dry-cell manufacture which improved the service capacity of dry cells lowered their flash. An example is the substitution of acetylene black for graphite.

The flash **cannot** be used to judge the amount of service capacity remaining in a partially discharged cell; in fact, no known physical measurement will do this job. The remaining service capacity depends too much on the manner of the previous discharge.

6. BIBLIOGRAPHY

Letter Circular LC677, U.S. Dept. of Commerce, National Bureau of Standards, Washington, D.C.
Circular of the National Bureau of Standards C435.

RESERVE-TYPE PRIMARY BATTERIES

By Paul L. Howard

7. DESCRIPTION

TYPES. The reserve-type primary battery is a unit which is maintained in the dry state until the time of use, when the electrolyte is added to the battery elements and the battery discharged within a reasonable period thereafter. These units are not reversible. This type of battery has application in special equipment where indefinite shelf life, high rates of discharge, minimum size, and weight are of paramount importance. These batteries are classified in the following manner:

Type 1. Metallic halide: Metal couple activated by plain water or sea water.
Type 2. Metallic oxide: Metal couple using an alkaline electrolyte.
Type 3. Metallic peroxide: Metal couple using perchloric acid, hydrofluoboric acid, or hydrofluosilicic acid as electrolyte.
Type 4. Metallic oxide: Metal couple using an acid electrolyte.

These batteries are made up of a positive electrode, with a separator of either absorbent paper or an inert material, and the metal negative electrode in such a way that minimum spacing is allowed between plates. The cells either are assembled so that they are combined with the electrolyte chamber with a means of admitting the electrolyte at the time of using or are constructed in such a way that the electrolyte is admitted from a separate container.

METALLIC HALIDE TYPE. This type of battery is now in the form of the silver chloride magnesium couple, which may be activated by ordinary water or sea water. No electrolyte reservoir is necessary with this type of battery. Units may be designed to deliver various discharge rates for high-discharge-rate applications where extremely light weight per unit volume is required. In Table 1 the types of batteries are summarized, their sizes, weights (assembled), and discharge characteristics being given.

The discharge curve of this type of battery is practically a straight line for more than 80 per cent of its life. These units may be used as a basis for larger units by assembling them in series parallel construction.

Table 1. Characteristics of Type 1 Batteries

Type	Voltage	Current	Time
4mc96	6.3	1.6 amp	1 hr to 5.5 volts
4mc192	6.3	1.6 amp	2 hr to 5.5 volts
4cc42	6.2	0.7 amp	65 min to 5.8 volts
4cc25	6.3	0.5 amp	60 min to 5.8 volts
4cc12	6.3	0.2 amp	30 min to 5.8 volts
3K1	3.0	7.0 amp	Flash current for 20 sec duration
1cc18A	1.55	0.3 amp	66 min to 1.4 volts
2cc15	3.1	0.38 amp	55 min to 2.9 volts
2cc15	3.1	1.00 amp	18 min to 2.9 volts
2cc10	3.1	0.32 amp	36 min to 2.9 volts
2cc7.5	3.1	0.32 amp	28 min to 2.9 volts
2cc7.5	2.8	1.00 amp	5 min to 2.8 volts (flat)
1cc7.5	1.35	1.00 amp	5 min to 1.34 volts
30B4.5	45	34 ma	150 min to 36 volts
30B3	45	105 ma	38 min to 36 volts
30B1.5	45	8 ma	190 min to 39 volts
2mc30–60B3, A	3.2	210 ma	2.5 hr to 2.7 volts
B	92	30 ma	2.5 hr to 75 volts
18mc510	27	4.25 amp	2 hr to 24 volts
4cc167	6	10 amp	15 min
4cc143	5.6	25 amp	4 min
4cc200	6.0	1.8 amp	2 hr
5cc133	7.5	8 amp	12 min
4cc195	6.0	1.6 amp	2 hr
4–2cc304	12.0	40 amp	6 min to 8.8 volts

Type	Volts	General Characteristics Size (approx.), Inches	Dry Weight (approx.)	Wet Weight (approx.)
4mc96	6	2 3/16 × 2 × 3	190 gm	260 gm
4mc192	6	3 1/4 × 3 1/4 × 3	280 gm	400 gm
4cc42	6	1 3/4 diam × 3 1/4 (less socket)	74 gm	110 gm
4cc25	6	1 1/4 diam × 3 1/4 (less socket)	60 gm	75 gm
4cc12	6	11/8 diam × 1 5/8	27 gm	40 gm
3K1	3	Semicircular, 1 1/8 radius × 2 1/4 × 1/4 thick	15 gm	20 gm
1cc18A	1.5	7/8 diam × 1 11/16 long	20 gm	25 gm
2cc15	3.1	15/16 diam × 2 3/4 (over socket)		
2cc15 fuse	3.1	29/32 diam × 1 11/16 long	23 gm	28 gm
2cc10	3.1	7/8 diam × 2 5/8 long (socket)	22 gm	26 gm
2cc7.5	3.1	9/16 diam × 2 5/8 long (socket)	8 gm	12 gm
1cc7.5 plain	1.5	3/8 × 1 3/4 long	3 gm	5 gm
30B4.5	45	1 × 3 1/2 × 1 3/4	74 gm	100 gm
30B3	45	1 3/4 × 1 7/8 × 1 3/4	70 gm	90 gm
30B1.5	45	1 × 1 1/8 × 1 3/4	45 gm	60 gm
2mc3v–60B3, A	3	2 5/8 × 3 7/16 × 1 3/4 (high)	175 gm	225 gm
B	90			
18mc510	27	2 sections * 2 13/16 × 5 5/16 × 7 3/16	Each section: 3 lb	Each section: 4 lb
4cc167	6	2 7/8 diam × 3 5/8 long	350 gm	420 gm
4cc143	6	2 7/8 diam × 3 1/2 long	350 gm	420 gm
4cc200	6	6-diam shell, 4 11/16 center hole × 4 3/8; 12 lamps on top	1 lb 9 oz	
5cc133	7.5	2 7/8 diam × 3 5/8	300 gm	360 gm
4cc195	6	7.65 diam with 6.5 center hole × 4	2 lb	
4–2cc304	12	2 7/8 diam × 11	1200 gm	

* 2–9mc510 sections used and placed back to back.

For activation the units are either soaked in water for a short period and discharged in air or remain in water throughout the discharge. As long as the battery can be wet in water, it may then be operated in air where temperatures are as low as −40 deg fahr without serious loss in capacity. These units, stored in hermetically sealed containers with sufficient dehydrant, will last indefinitely in this state and give full capacity at the time of use.

METALLIC OXIDE TYPE WITH ALKALINE ELECTROLYTE. Metal couple using an alkaline electrolyte is at present found in the silver peroxide-zinc-alkaline battery. Batteries may be assembled in series or parallel from the units shown in Table 2 to give various desired voltages.

Table 2. Characteristics of Type 2 Batteries

Voltage	Current, amperes	Discharge Time	Approximate Size, inches	Weight, pounds
4.5	5.0	8.5 hr	2 1/4 × 6 × 6 1/8 (no electrolyte chamber)	5.7
1.5	166.0	8 min	3/4 × 4 × 5 (no electrolyte chamber)	2
1.35	2000.0	6.5 min	1 7/8 × 6 1/4 × 13 1/2 (includes electrolyte chamber)	15
5.6	50.0	6 min	2 7/8 diam × 4 1/4 (includes electrolyte chamber)	1.5

At the present time this type of battery is constructed with the reserve chamber as an integral part of the unit and a suitable means of breaking the diaphragm between the electrolyte chamber and the cell-element chamber. Designs can be made in which the electrolyte is added to the battery at the time of use, thus eliminating this additional space.

In general, this type of battery will give approximately 20 amp-hr per pound over a wide range of discharge rates. When hermetically sealed from external air conditions, this battery has an indefinite shelf life. With proper electrolyte concentrations fairly good low-temperature-discharge capacities may be obtained. Discharge characteristics of this unit are practically a flat voltage over the major portion of the discharge.

METALLIC PEROXIDE TYPE. This essentially is the lead peroxide-lead perchloric acid battery, which is designed with the reserve electrolyte chamber incorporated in high-rate designs and removable in low-rate designs. The types described in Table 3 have been made:

Table 3. Characteristics of Type 3 Batteries

Voltage	Current, amperes	Discharge Time	Approximate Size, inches	Weight
6.3	1.0 5.0 40.0	4.6 hr 40 min 3 min	1 7/8 × 3 5/8 × 2 1/4 (exclusive of electrolyte chambers)	1.6 lb
3.6 A 88.8 B	0.25 0.015	2.5 hr 2.5 hr	2.8 × 2 15/32 × 1 13/16 (electrolyte chamber excluded)	350 g
1.8	700.0	8 min	2 × 6 1/4 × 13 1/4 (electrolyte chamber included)	13 lb
1.8	2.0	20 min	1 5/8 diam × 2 (electrolyte chamber included)	3 oz
3.6	0.35	50 min	1/2 × 1 × 7 1/2 (electrolyte chamber included)	50 g

This battery is primarily known for its good characteristics at temperatures as low as −50 deg fahr. Shelf life of the unit is indefinite as long as the elements are kept dry. This battery may be operated not only with perchloric acid but also with hydrofluoboric or hydrofluosilicic acid. With hydrofluoboric acid the capacity is reduced by 25 per cent and the voltage by about 10 per cent, and with hydrofluosilicic acid the capacity and voltage are further reduced. These two acids, however, are less corrosive. This type of battery has a slightly drooping voltage curve during discharge regardless of the types of electrolyte used.

METALLIC OXIDE TYPE WITH ACID ELECTROLYTE. This type is a lead peroxide-zinc-sulfuric acid battery primarily developed for higher-rate applications. A cell of this type, weighing approximately 18 lb with reserve electrolyte chamber and connections, delivers 2.1 volts at 750 amp for 9 min. This battery may be adapted for moderately low temperature operation. The discharge curve rises during the early portion of the discharge but flattens out and finally drops rapidly at the end of the discharge.

8. OPERATION

The closed-circuit voltages of these types are as follows:

Type 1: 1.5 volts per cell.
Type 2: 1.5 volts per cell.
Type 3: 1.8 to 1.6 volts per cell.
Type 4: 2.1 volts per cell.

In general, these batteries must be used immediately after activation. Modifications have been made in Types 1 and 2, however, whereby reasonable standing periods may be obtained before discharge. The general characteristic of all these types is that, once the battery is started on discharge, the discharge should be completed in order to obtain maximum efficiency of the couples. It is not recommended that these units be discharged intermittently.

9. USES

At the present time these batteries are limited in use to applications where indefinite shelf life, light weight, and high capacity per unit volume are essential. Thus the end uses are for highly specialized equipment or equipment in which the present batteries are not suitable because of short shelf life. Since these batteries are in general entirely new developments during the past few years, maximum utilization or even a full coverage of types of batteries has not been attained.

At the present time batteries are being designed as the application arises. Thus a completely standardized line of batteries is not available at this writing.

RAILWAY SIGNALING BATTERIES

By R. B. Elsworth

10. DESCRIPTION

A copper oxide-zinc-caustic soda cell, also known as an Edison-Lalande cell, consists of copper oxide and zinc electrodes suspended from a porcelain cover in a solution of caustic soda (sodium hydroxide) and water. The soda in cube or flake form is supplied in cans and mixed with clean, soft water when the cells are originally set up or renewed. A layer of special mineral oil is placed on top of the solution to prevent evaporation and keep the cells free from salt formations. The positive and negative terminals are connected to the copper oxide and zinc electrodes, respectively. The copper oxide electrodes also serve as depolarizing agents.

Complete cells include jars of heat-resisting glass or enameled steel. Glass jars are preferable when cells are installed in permanent, protected locations because of the convenience in observing the electrodes, the electrolyte, and the visual indications of exhaustion. The use of steel jars is confined to installations where service conditions require greater mechanical strength, as well as protection against the spillage of electrolyte, provided by the splash-proof covers and gaskets supplied with these jars.

11. USES AND ADVANTAGES

USES. Primary batteries of the copper oxide-zinc caustic soda type are widely used either alone or as an emergency power supply for operating and lighting railway semaphore and light signals, track circuits, highway-crossing signals and gates, low-voltage switch machines, line-control circuits, low-voltage mains in interlocking plants, electrically lighted switch lamps, and telephone and telegraph circuits.

ADVANTAGES. Their chief advantages for these services are: large ampere-hour capacity, ability to deliver consistently full-rated capacity, uniform voltage under continuous discharge, low internal resistance, no loss of capacity on open circuit, ability to function entirely independently of, and without aid from, a power line, adaptability to open- or closed-circuit service, ease of installation and maintenance, visual indication of approaching exhaustion. This last feature makes possible the operation of a circuit indefinitely without battery failures.

APPLICATIONS. On light- and moderate-traffic railway lines, signals and track circuits may be operated directly from primary batteries in order to secure the simplest installation and keep down initial costs and operating expenses. Direct primary-battery operation may also be used for heavy-duty service at isolated locations, where a-c power cannot be obtained at the site without an excessive investment in pole lines and related apparatus. Normal a-c energy, with primary-battery emergency standby, is used when service conditions warrant the additional cost of providing commercial power for normal operation of the circuits, either directly or through rectifiers. If the a-c voltage becomes too low for satisfactory operation or fails completely, the entire load is automatically transferred to the standby battery of primary cells until normal a-c service is restored. This system provides two entirely separate sources of power and eliminates the necessity of continuously charging a standby battery, which is seldom used, to handle the normal load. Other features of this system are its large ampere-hour reserve for emergencies, an accurate visual means for determining available battery capacity, simplicity of maintenance, and efficient use of a-c power, made possible by elimination of battery-charging losses.

12. DIMENSIONS AND RATINGS

RATINGS. Primary batteries for signaling services are standardized in capacities of 500 and 1000 amp-hr for light, medium, heavy, and extra-heavy duty. Table 1 shows the maximum current rates at which the cells can be discharged without decreasing their

Table 1. Discharge Rates with Different Types of Battery Housings (See Article 13)

Ampere-hour Capacity	Type of Duty	Maximum Discharge, Amperes		
		Battery Box	Deep Well	Heated Buildings
500	Light	1.0	1.4	2.0
500	Medium	1.3	1.7	2.2
500 *	Heavy	3.6	5.0	6.7
1000	Medium	2.5	3.1	4.0
1000 *	Heavy	4.5	7.0	9.5
1000	Extra-heavy	8.0	12.0	19.2

* Some heavy-duty cells have progressive indications of exhaustion which show when 50, 75, and 100 per cent of capacity have been delivered. These cells may be used for higher current requirements than the rates given in the table by renewing the cells at either 50 or 75 per cent exhaustion.

voltage below the recommended minimum value. The average effective voltage of all cells on continuous discharge at the maximum recommended rates is between 0.6 and 0.65. If cells are discharged at lower rates than those in the table, the minimum effective voltage will be slightly higher.

DIMENSIONS. Primary cells of 500-amp-hr capacity are furnished in cylindrical, rectangular, and barrel-shaped heat-resisting glass jars and cylindrical steel jars; 1000-amp cells are furnished in cylindrical and rectangular heat-resisting glass jars. Light-, medium-, and heavy-duty cells take the same jars but different plate assemblies. Dimensions are as given in Table 2.

Table 2. Dimensions of Cells

Ampere-hour Capacity	Jar		Type of Duty	Overall Dimensions of Complete Cells, Inches
	Kind	Shape		
500	Glass	Round	Light, medium, heavy	6 3/4 diam × 12 3/4 high
500	Glass	Rectangular	Light, medium, heavy	6 3/4 long × 5 3/4 wide × 12 1/4 high
500	Glass	Barrel	Light, medium, heavy	7 diam × 11 5/8 high
500	Steel	Round	Light, medium, heavy	6 7/8 diam × 12 1/4 high
1000	Glass	Round	Light, medium, heavy	8 1/8 diam × 14 5/8 high
1000	Glass	Rectangular	Medium, heavy	8 1/4 long × 6 1/2 wide × 14 3/4 high
1000	Glass	Rectangular	Extra-heavy	8 1/2 long × 6 3/4 wide × 17 3/4 high

Present types of caustic soda primary cells range in capacity from 75 to 1000 amp-hr and are capable of delivering continuous currents up to 19 amp. Their low internal resistance fits them for heavy currents, and effective depolarization makes them desirable for continuous work. Their initial or open-circuit voltage is 0.95 and, as in all types of batteries, is higher than the effective working voltage when cells are on discharge. The average effective voltage at recommended continuous discharge rates is 0.60 and 0.65.

The electrochemical equivalents of zinc, copper oxide, and caustic soda are approximately 1.2, 1.5 and 1.5 grams, respectively, per ampere-hour. The actual weights of zinc, copper oxide, and caustic soda used in practical batteries vary among different manufacturers and for different types.

Special consideration may be given to cells for intermittent service of short duration and abnormally high current requirements.

The number of cells needed is found by dividing the required operating voltage by 0.5, the minimum recommended voltage of cells. The size of cells required is determined by the maximum current requirements and the number of months or years of service desired between renewals.

Figure 1 shows a typical caustic soda-zinc-copper oxide cell having a rectangular jar of heat-resisting glass.

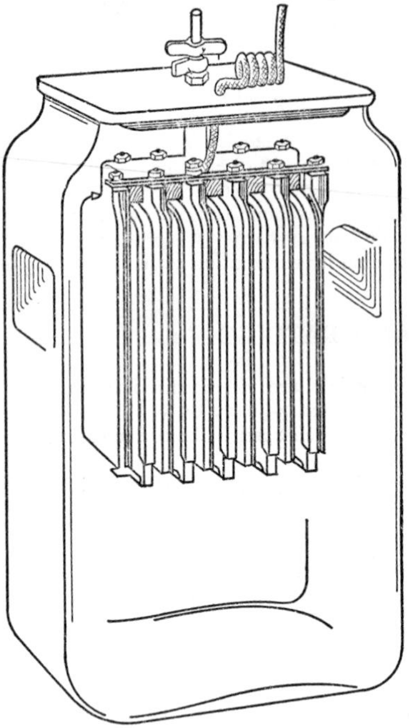

Fig. 1. Caustic Soda-zinc-copper Oxide Cell

13. HOUSING

Batteries for wayside signals and track circuits are generally housed in boxes or wells set in the ground to provide protection against low temperatures. Cold increases the internal resistance of all types of batteries, causing a decrease in voltage at all rates of discharge, particularly at high rates of long duration. Only in warm climates, or when cells are normally discharged at relatively low rates, is it considered good practice to house signal batteries where they will be directly subject to outside temperatures.

14. EXHAUSTION INDICATORS

In most classes of service where primary cells are used, it is extremely important to know when they are approaching exhaustion. In one type of cell, this is accomplished by means of small panels, molded in the lower portion of the outside zinc electrodes, which are designed to perforate gradually as the capacity is delivered. This arrangement affords an accurate means for determining the actual capacity remaining in the cell and therefore shows when renewal should be made, before the cell exhausts and before the voltage falls below normal.

Cells for light and medium duty have two panels of the same thickness in each outer zinc electrode. First indications appear in these panels when about 85 per cent of the rated capacity has been taken from a cell. Complete perforation of both panels indicates that a cell has reached the end of its rated life and should be renewed.

In 500- and 1000-amp-hr heavy-duty cells three separate and distinct panels in each outer zinc electrode provide progressive indications of exhaustion. Each of these three panels is of different thickness and designed to perforate gradually until it is completely eaten out when a definite percentage of the cell capacity has been used. Progressive indications start at about 40 per cent exhaustion, and complete perforation of a panel indicates that 50, 75, or 100 per cent of rated capacity has been delivered.

The extra heavy-duty cells have two panels that provide progressive indications of exhaustion. When perforations start in the left-hand panel, about 50 per cent of the capacity has been delivered; complete perforation indicates 75 per cent delivery. When the right-hand panel begins to perforate, the cell has delivered about 90 per cent; when this panel is entirely eaten out, the full rated capacity has been taken from the cell. As a factor of reliability, primary cells are designed to deliver 10 per cent more than rated capacity.

In some types of cells a state of exhaustion may be indicated by reduction of the entire lower portion of the zinc electrode or by the breaking of a reinforcing band on the zinc electrode.

RENEWAL. When exhausted, copper oxide-zinc-caustic soda primary cells can be readily restored to their original capacity by renewing their electrodes, electrolyte, and oil. This is the only maintenance that the cells normally require, except for occasional visual inspection. Jars and covers are permanent equipment and may be used indefinitely.

15. BIBLIOGRAPHY

American Association of Railroads, *Principles and Practices; Batteries.*
American Association of Railroads, *Signal Section Manual.*
Lincoln, E. S., *Primary and Storage Batteries.*

LEAD-ACID STORAGE BATTERIES

By J. Lester Woodbridge and A. Milton Miley

16. GENERAL AND DESCRIPTIVE MATERIAL

The action of a lead-acid cell when discharging is similar to that of a primary cell, except that the materials of the electrodes which undergo the chemical reactions are not dissolved but remain in much the same mechanical condition as before discharge. When this requirement is fulfilled, the materials of the electrodes can be restored to their original condition by "recharging" or passing a current through the cell in the reverse direction to that of the discharge current. In this way the electrical energy of the charging current is transformed into chemical energy and stored as such in the electrodes.

As there is no loss of material during the discharge of such a battery to expose fresh material to chemical action, it is necessary that the active material of the electrodes either be disposed in a very thin layer over a large surface or be porous to permit the use of the interior part of thick material.

A lead-acid cell consists essentially of positive plates containing lead peroxide and negative plates containing pure lead immersed in an electrolyte of dilute sulfuric acid.

(In accordance with common engineering practice the terms positive and negative plates are employed in this article to denote the plates which are connected to the positive and the negative terminals, respectively, of the external circuit on both charge and discharge. The term battery is employed to denote a group of two or more cells which constitutes an operating unit.)

CHEMICAL REACTIONS. The active elements of the lead-acid type of battery consist of lead peroxide (PbO_2) on the positive plate, sponge lead (Pb) on the negative plate, and dilute sulfuric acid (H_2SO_4) for the electrolyte.

Whatever the secondary reactions may be, it is agreed that the final result on discharge is the formation of lead sulfate ($PbSO_4$) on both the positive and negative plates, the SO_4 radical of the sulfuric acid combining with the lead of both plates to form this compound. As a result some water (H_2O) is formed, with a consequent decrease in the specific gravity of the electrolyte. On charge, the electric current splits up the lead sulfate ($PbSO_4$), returning the SO_4 radical to the electrolyte, oxidizes the positive plate to its original condition of lead peroxide (PbO_2), and reduces the negative plate to its original condition of sponge lead (Pb). This action may be represented as follows:

$$\underset{+\text{Plate}}{PbO_2} + 2H_2SO_4 + \underset{-\text{Plate}}{Pb} \rightleftarrows \underset{+\text{Plate}}{PbSO_4} + 2H_2O + \underset{-\text{Plate}}{PbSO_4}$$

This equation, read from left to right, is the equation of discharge; read from right to left, it is the equation of charge. In practice, toward the end of charge some of the water (H_2O) is split up by the current into its component parts, hydrogen (H) and oxygen (O), the hydrogen being liberated at the negative plate and the oxygen at the positive plate. This occurs whenever the density of charging current is greater than can be utilized in decomposing the lead sulfate remaining in the plates.

TYPES OF LEAD-ACID CELLS. Two general types of lead-acid batteries are in use, the Planté and the pasted plate, each being subject to some modifications. The distinction between these types is based on the plate structure.

PLANTÉ TYPE. The Planté plate is essentially a pure lead plate having a highly developed surface presenting a large superficial area which is oxidized electrochemically until the entire surface is covered with a thin porous layer of lead peroxide (PbO_2). This constitutes a positive plate. It may, however, be converted into a negative plate by reversal of the current, which reduces the lead peroxide to pure lead in a porous form. Planté negative plates are little used at this time.

Planté Plates. Of the various types of Planté plates, among those most commonly in use may be mentioned the central-web type and the cast-lead having no central web. The main purpose underlying both these methods of manufacture is to produce a large surface on which to form the active material.

Central-web Type. In the central-web type there is a solid sheet or "web" of pure lead on which the ribs are formed. This web prevents the circulation of the electrolyte through the plate. The ribs are formed from the original lead plate by rolling, spinning, or cutting. In the rolling and spinning processes the plate is formed from a lead blank by means of a number of steel disks placed side by side and separated by small spacers on a shaft. The lead blank is passed between two sets of disks by forward and backward movements. The disks gradually work deeper into the plate and squeeze lead up into the spaces between the adjacent steel disks. In the cut type of plate the ribs are formed from the lead sheet by a tool, which at each stroke turns up one complete rib. The cutting edge works at an angle so that the finished ribs stand out from the surface. The ribs may incline upward from the central web and thus form pockets to hold the active material and prevent its falling away.

The disadvantage of the web type is the web itself. There is invariably a tendency toward unequal work on the two sides of the positive plate, this tendency being caused by difference in the plate spacing, unequal capacity of the negative plate on either side, inequality in the shape of the ribs, or some other factor. Where the active material and the active surface of a plate are disposed in planes perpendicular to the face of the plate and extend through the plate, excessive action on one side simply works the plate a little further through from that side, the effect on the active material, however, being uniform throughout. With a plate provided with a central web, which prevents such action, any inequality of work will charge or discharge one side more than the other, producing a tendency to buckle.

Cast-lead Type. The type of plate which has no central web is made by casting pure soft lead in a mold, casting having the advantage of allowing for the distribution of metal in the plate without limitations in the manufacturing process. As the plate comes from the mold, it consists of a great number of vertical ribs running entirely through the plate and bound together by transverse ribs to give the plate strength. In this manner a large surface can be obtained, and in a plate having no central web the electrolyte can circulate through the plate. The active material will be uniformly worked throughout even if the amount of work on the two sides of the plate is unequal.

Modified Planté or Manchester Positive Plate. The so-called Manchester positive plate is made by rolling corrugated strips of pure lead ribbon into rosettes or "buttons" which are forced into circular openings in a rigid grid of cast lead-antimony alloy. This grid is strong, stiff, and subject to very little electrolytic corrosion; it therefore provides a plate structure immune from the growth and buckling which frequently occurs in Planté plates composed entirely of pure lead. The pure lead buttons constitute the active part of the plate and function in every way similarly to the original Planté positive plate.

"Forming" of Planté Plates. In making Planté positive plates, after the ribs or corrugations have been formed on the lead blank or the pellets have been placed in the hard grid, the plates are assembled in a sulfuric acid bath containing some corrosive chemical, called a forming agent, together with dummy lead plates. The forming agents used by various manufacturers are usually kept as trade secrets; the nature and method of using such agents largely determine the quality of the battery. To form the positive plates the dummies are connected as the cathodes, and a current is passed through the couple thus formed. The electrolytic action of this current causes lead peroxide (PbO_2) to be formed from the lead of the ribs. The strength and duration of the current produce the desired thickness of lead peroxide on the ribs or pellets.

In recent years the Planté plate has declined in popularity in this country, whereas the Manchester positive plate has continued in extensive use. The Planté and modified Planté types are used chiefly where space, weight, and initial cost are of secondary importance and where a battery more durable than the pasted plate type is desired.

PASTED-PLATE TYPE. The pasted plate is made by applying to a rigid grid of cast lead-antimony alloy the active material, in the form of a paste composed largely of lead oxides, usually litharge (PbO), red lead (Pb_3O_4), or both, and a solution containing sulfuric acid or a sulfate which acts as a binder. The grid usually has the form of a double gridiron composed of a series of vertical ribs which extend through the plate, giving it strength and providing conductivity to carry the current to plate lugs. These vertical ribs are tied together by short cross-bars which are flush with the surface but extend only partly through the plate and are staggered on opposite sides. These short bars are usually horizontal, but one manufacturer places them diagonally, making the so-called diamond grid. In either case the active material, when applied to the grid, is disposed in the form of corrugated strips or ribbons extending from the top to the bottom of the plate between the vertical ribs and held in place by the cross-bars.

As the active material has a depth equal to the thickness of the plate, composition of the paste must be such as to result in the necessary porosity but still maintain sufficient strength and cohesion of the material to assure the desired life of the plates in service. The composition of the paste is varied for different services, and the variations are trade secrets.

Plates of this type are very extensively used as both positive and negative plates, but in the composition of the paste, red lead usually predominates for positive plates and litharge for negative plates.

After the grid is filled with the paste, the plate is dried to allow the paste to harden and then is "formed" electrochemically. Positive plates are formed by immersing them in a forming bath and passing current through them while they are connected as anodes, thus further oxidizing the lead oxides in the dried paste and forming lead peroxide (PbO_2). Negative plates are formed by passing the current in the opposite direction, so that the lead oxide is reduced to sponge lead. After the forming charge the plates are washed and dried and are ready for assembly.

Processes are now used which permit the dried pasted plates to be assembled into cells, the plates being formed while in the assembled cells. This method differs from the formation procedure described in the previous paragraph in that the plates are not washed and dried, since the cells remain charged and wet suitable for immediate use.

Modified Pasted Plate. In one type of positive plate the active material is prepared from lead oxides of composition similar to that used for pasted positive plates, although the plate structure differs entirely from the usual flat plates of that type. The frame of this plate consists of a series of parallel vertical core rods joined at each end to horizontal top and bottom bars, all of lead-antimony alloy. The active material completely surrounds each core rod and is held in place by finely slotted rubber tubes. The core rod is thus in the center of a cylindrical pencil of active material, where it is well placed to conduct the current to the top bars and the cell terminals. The surface of the active material exposed to the electrolyte is greater than in a flat-plate type of equal superficial dimensions. The slotted rubber tube replaces the retainer of plastic, matted glass or perforated rubber used with flat plates and acts more effectively to retain the active material in place, since it softens with use and thus makes a plate having a longer service life. The first cost of this plate is greater than that of the usual flat plate, but it is claimed by its manufacturer to give twice the life in severe service.

This type of plate has come into very extensive use in motive-power work, railway-train lighting, and marine service.

ELECTROLYTE. The electrolyte used with the lead type of battery is always a dilute solution of sulfuric acid. The specific gravity of the electrolyte, when the battery is fully charged, varies from about 1.210 for stationary batteries to 1.300 for automobile-ignition batteries. These values have been adopted as standard by all leading manufacturers.

The proper specific gravity varies with the conditions. The variation of the resistance of 1 cc of electrolyte with specific gravity shows that the resistance of dilute sulfuric acid is least at a specific gravity of 1.190 to 1.235, depending on the temperature, this resistance increasing if the specific gravity is either increased or decreased. A 30 per cent solution of sulfuric acid by weight increases in resistance with decreasing temperature so that, when 77 deg fahr is used as a reference point, the resistance increases 58 per cent at 32 deg fahr, 166 per cent at 0 deg fahr, and 420 per cent at −30 deg fahr. Numerous other conditions influence the selection of the proper specific gravity.

The curves in Fig. 1 show the specific gravity of various mixtures, both by weight and by volume, of 1 part of 1.840-specific-gravity acid with from $^1/_{10}$ to 7 parts of water. There is also a curve showing the percentage, by weight, of 1.840-specific-gravity acid in mixtures of various specific gravities. These curves are approximately correct at 60 deg fahr. Unless a compensating hydrometer is used in determining the specific gravity, allowance must be made for temperature variation, on the basis of an increase of approximately 1 point (i.e., 0.001) in gravity for each decrease of 3 deg fahr in temperature, and vice versa; for instance, an electrolyte that has a specific gravity of 1.210 at 77 deg fahr will have a specific gravity of 1.213 at 68 deg fahr and 1.207 at 86 deg fahr.

Impurities in Electrolyte. The electrolyte should be of known purity, harmful ingredients such as iron, aluminum, sodium, potassium, manganese, calcium, sulfates, bicarbonates, chlorine, and nitrogen oxides being strictly limited. The various battery manufacturers issue exact instructions to the acid manufacturers, specifying the maximum amount of different impurities which the acid may contain. An electrolyte that is not approved by the company furnishing the battery should never be used.

Preparation of Electrolyte. In preparing the electrolyte, sulfuric acid approved by the battery manufacturer should be diluted with sufficient pure distilled water to bring the mixture to the required specific gravity. The acid should be poured into the water; *the water should never be poured into the acid.* If the water is poured into the acid, the heat formed by the mixture is sufficient to cause sputtering, and damage may ensue.

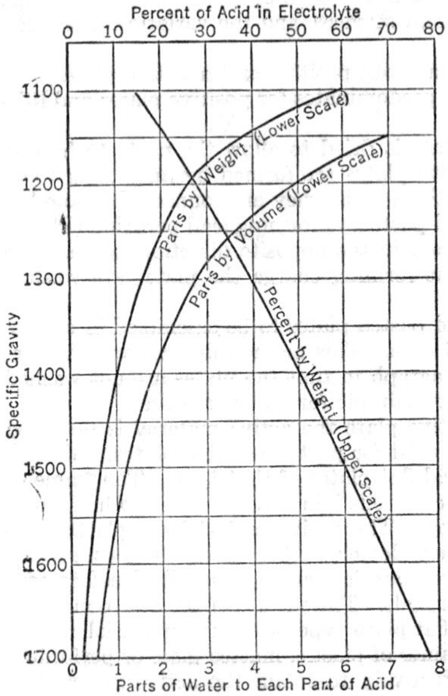

FIG. 1. Variation of Specific Gravity with Concentration

The sulfuric-acid manufacturing companies furnish electrolyte for battery work in such large quantities that they carry a stock of various standard mixtures. It will usually be found cheaper and more convenient to purchase the electrolyte ready-mixed than to purchase the concentrated sulfuric acid and prepare the mixture at the installation. The latter course, however, is sometimes adopted where the amount of acid used is considerable and where the item of freight saving is appreciable.

CONTAINERS. The containers for holding the battery plates and electrolyte may be made of hard rubber, asphaltic compound, glass, plastic, porcelain, or lead-lined wood, depending on the size of the cells and on the nature of the service.

Rubber Containers. Hard-rubber containers are used when the cells must be portable or are subject to mechanical shocks while in service. Typical examples of such cells are those used for automobile and aircraft starting and lighting, electric street and industrial trucks, railway-train lighting, and submarine boat propulsion.

In these batteries the plates are usually supported on ribs in the bottom of the jars, the ribs being of sufficient height to allow ample space for the accumulation of sediment, deposited as a result of the gradual loss of material from the plates, without danger of short-circuiting the plates. The chemical composition of the rubber compound is subject to the specifications of the battery manufacturer in order to prevent contamination of the electrolyte.

Asphaltic Compound Containers. Composition containers, as they are commonly termed, are used in portable service, almost exclusively in automobiles.

Their construction is similar to that of hard-rubber containers. Because of the porosity of the composition it is necessary to coat the inside of the container with an acid-proof paint to prevent the electrolyte from entering the composition and subsequently destroying its mechanical strength.

Glass Jars. Glass-jar containers are used in preference to rubber containers whenever the type of service and the size of the cells permit, on account of the greater ease of inspection and maintenance. The present tendency is toward the almost exclusive use of the cells assembled, sealed, and charged at the manufacturer's plant and shipped filled with electrolyte and ready for service as soon as they are installed and connected. This assembly is available in cells up to approximately 1100-amp-hr capacity at the 8-hr rate for pasted-plate types and up to about 600 amp-hr for the formed-plate types. Thoroughly annealed, molded glass containers for from 1 to 3 cells are used for the smaller sizes and to a limited extent by some manufacturers for the larger sizes of single cells. Internal strains may remain even after the most careful annealing. The higher cost of the molded jars has resulted in the much more extensive use of machine-blown jars, also thoroughly annealed, for the larger cells.

In brown-jar assemblies the plates are usually supported from the cover of the cell, but separate supporting ribs are sometimes placed on the bottom of the jars. In molded-jar assemblies ribs may be molded on the jar bottoms to support the plates.

Plastic Jars and Containers. During the past decade plastics have been employed because of their light weight, but they are generally brittle. Because of the susceptibility of plastics to acid attack and their high cost, research and development will be required before they can be used extensively.

Porcelain Jars. These jars are used for intermediate-sized cells, since porcelain is more rugged and economical in such sizes. One of the greatest disadvantages is the dimensional instability of these jars during manufacture.

Lead-lined Tanks. In the larger sizes of batteries lead-lined tanks have been used. These tanks are generally made from specially selected yellow pine, dovetailed together without nails or other metallic fastenings. The wood must be protected with several coats of acid-resisting paint. The tanks should be lined with lead of 3 to 4 lb per sq ft, depending on their size. All seams should be burned with a suitable flame with pure lead, without the use of any flux. Each tank should be self-supporting without any braces or reinforcements, so that any tank in the battery can be removed and replaced without affecting the remaining tanks in any way. A poorly constructed wood tank is bound to cause trouble. Special attention should be paid to this detail in preparing specifications.

In cells of this type it is of course necessary to insulate the plates from the lead lining. The plates are therefore supported on vertical sheets of glass resting on the bottoms of the tanks. The tank linings under the glass sheets are reinforced. The glass sheets are ground smooth on the top and bottom edges, and the bottom corners are beveled at 45 deg to avoid injury to the lead lining during installation.

Cell Covers. Batteries in lead-lined tanks and in the old forms of open glass-jar assemblies are usually provided with covers of double-thick window glass, which rest on the top of the cells. While the battery is charging, these covers serve the purpose of intercepting much of the fine spray thrown off from the surface of the electrolyte during "gassing" and draining it back into the cell. They also keep dirt out of the cells, this function being quite essential when the air is likely to contain particles of metal or other matter injurious to the battery.

Both these purposes are served very much more completely with the sealed type of cell in glass, plastic, porcelain, or rubber jars. In the blown-glass-jar assembly in which the plates are supported from the cover, the cover must be of sufficient strength to carry this weight even during shipment. The cover must also be impervious to the action of the electrolyte and not affected by any temperatures which are likely to be encountered either during shipment or in subsequent service. Molded glass and hard rubber or similar materials are most frequently used.

For sealed-glass-jar cells the covers are provided with a groove on the underside which fits over the edge of the jar, this groove being partially filled with a sticky plastic sealing compound which, under the weight of the plates, forms a tight seal between the cover and the jar. The joints between the cover and the cell terminal posts must be sealed to prevent creepage of electrolyte over the top of the cell. For this purpose sealing compound or some form of gasket clamped between the cover and the terminal post by means of a threaded seal nut is effective.

For rubber- and porcelain-jar cells the covers are so shaped as to provide between the cover and the jar sides a V-shaped groove, which is filled with sealing compound. The seal between the terminal posts and the cover may be as described for glass-jar cells, but in some types of automotive batteries a lead thimble is molded into the cover through which the terminal post passes. This thimble is then "burned" to the terminal post to complete the seal.

Separators and Separation. Since the positive and negative plates of the cells are assembled alternately, it is necessary that "separators" be provided to prevent their coming into contact with each other and so short-circuiting the cell. These separators must maintain their mechanical strength throughout the life of the battery and must be porous to allow diffusion of the electrolyte through them and the passage of the current through the electrolyte in their pores. Grooved wooden diaphragms and grooved microporous rubber are commonly used. The woods most popular for separators include cedar, cypress, redwood, and Douglas fir, cut as veneers, quarter sawed, or sliced. They are usually planed to provide the exact thickness required and to furnish grooves for the circulation of the electrolyte and free passage of loosened material to the sediment space at the bottom of cell. Wood separators are treated chemically to remove ingredients which, when in contact with the sulfuric acid in the electrolyte, will form acetic acid, which is injurious to the plates.

Glass-wool retainers are being used quite extensively by many manufacturers in starting, lighting, ignition, and standby batteries. This form of separator is composed of fine glass fibers laid at different angles and cemented on the surface with a soluble or insoluble cement to permit handling before assembly in the cell. It is placed against the positive plate where, because of its compressibility, it comes into intimate contact with the plate and prevents or retards the softened material on the surface of the plate from falling to the bottom of the cell. It is used in conjunction with wood or with wood and perforated rubber or plastic separators.

Perforated rubber retainers are used adjacent to the positive plates in conjunction with wood. In some cases triple insulation is used with glass wool adjacent to the positive plate, and the perforated rubber and wood against the negative plate.

Perforated plastic retainers are often used as a substitute for perforated rubber retainers.

A type of separator known as Mipor is manufactured from latex and can be classified as a microporous rubber separator. The pores are of microscopic dimensions (in the order of 0.00004 cm or 0.000016 in.) which prevent the penetration of even the smallest particles of active material, but the pores are so numerous as to permit ample diffusion of the electrolyte and ready passage of the electric current. The body of the separator, being of hard rubber, is unaffected by the electrolyte or by some high temperatures which are injurious to wood. This separator has a conductivity somewhat better than that of the corresponding wood separator.

17. CELL CHARACTERISTICS

VOLTAGE ON OPEN CIRCUIT. A lead-acid cell has an open-circuit voltage of approximately 2 volts regardless of the size of the cell. This voltage is, however, somewhat dependent on the specific gravity of the electrolyte used, being approximately 2.06 volts per cell with 1.210 specific gravity and 2.10 volts with 1.280 specific gravity. A sufficient number of cells are connected in series to get the desired voltage of a battery.

CAPACITY. The capacity of a cell or battery is usually expressed in amperes and time, as 20 amp for 8 hr, or 160 amp-hr at the 8-hr rate, which is the same thing. The capacity desired and the type of service (expected life) determine the size of cells used in a battery. To define the capacity of a cell or battery completely, it is also necessary to specify the initial minimum temperature of the battery and the minimum voltage permissible at the end of the discharge period, as 20 amp for 8 hr at 80 deg fahr to a final voltage of 1.75 volts per cell. Figure 2 shows the usual form of capacity data available from manufacturers. They are given in terms of amperes and ampere-hours per positive plate, and the voltages are for a single cell. These data are applicable to all sizes of cells using that particular size and type of plate.

In this country batteries are generally designated by a series of numbers and letters, as 60-EM-9, meaning 60 cells each having 9 plates, the letters being a code designating the details of the cell assembly. As there is usually one more negative than positive plate in a cell, and as the capacity is usually limited by that of the positive plate, a graph of the type given in Fig. 2 may be used for determining the capacity in ampere-hours and the initial, average, and final voltage of any size of battery of the designated type at any rate of discharge.

The capacity of the battery as given in ampere-hours is to be taken as the capacity which the manufacturer undertakes to deliver after the battery has been installed and is ready for an acceptance test. Practically all types of lead-acid batteries increase their capacity during the first few months of service and maintain this increased capacity during the greater part of their useful life. If capacities are interpreted from a curve showing cell performance, allowance must be made, when establishing the capacity of a battery, for connector voltage drop between cells.

The final voltage values as given define the minimum voltage at the end of a continuous discharge at any particular rate. It is true that some additional capacity can be obtained to lower final voltages, but the increase will not be great, as the voltage will fall quite rapidly if the discharge is continued. Although the final voltage values may be considered somewhat arbitrary, definite values are necessary for test purposes, and those assigned meet all practical requirements.

The average voltages given are those which can be computed from frequent periodic readings during a continuous discharge at the specified current to the final voltage given. These values are used principally for determining the size of battery required to carry loads expressed in kilowatts and kilowatt-hours.

The initial voltages given are again somewhat arbitrary, as the instantaneous voltage that can be observed on closing a circuit may change quite rapidly during the first few seconds, depending on the work the battery has been doing for some time before closing the circuit. The duration of this rapid change also depends on the rate of discharge. These transient voltages are seldom of engineering importance, and the values as given represent substantially the stabilized values which will exist after a short interval.

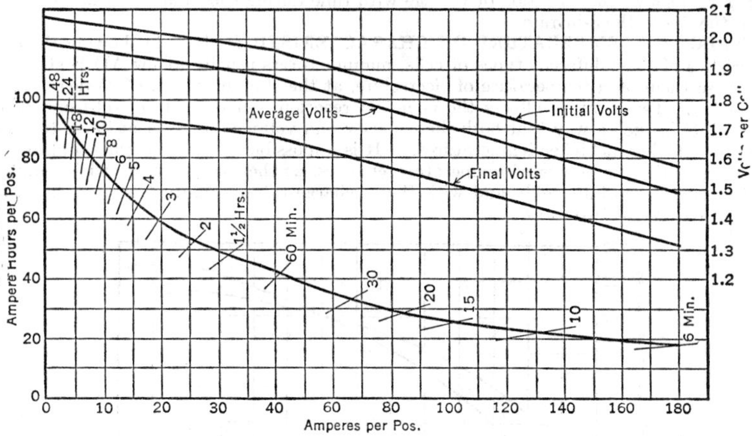

FIG. 2. Rated Discharge Characteristics of Lead-acid Cells, 1.210 Specific Gravity, 77 deg fahr

CAPACITY CHARACTERISTICS. The relation between capacity and rates of continuous discharge is given in graphs of the kind shown in Fig. 2. The reduced capacity at higher rates of continuous discharge is due to the depletion of acid in the electrolyte in the pores of the plates. During discharge the acid in the electrolyte combines with the active material of the plates, and water is formed, thus reducing the strength of the electrolyte inside the plate. This effect is counteracted to a varying extent by diffusion between this weakened electrolyte in the plate and the stronger electrolyte outside the plate. At higher rates of discharge the electrolyte inside the plate is weakened faster than it can be replaced by diffusion. It is this depletion of acid in the pores of the plates that limits the capacity at high discharge rates, not any limitation in the plates themselves. If the discharge is interrupted, diffusion will of course continue, bringing about the well-known recuperation of a battery during idle periods. The available capacity of a battery is a function of the total elapsed time of discharge or the rate of discharge. The capacities are the same for continuous discharge, but for intermittent discharge the capacity varies with the total elapsed time. Inasmuch as it is a lack of available acid that reduces the capacity at high rates of continuous discharge, obviously the plates cannot be damaged by excessive discharge at such rates. For high rates of discharge ample conductivity must be provided in the conducting parts of a battery.

VOLTAGE CHARACTERISTICS. The initial, average, and final voltage lines as shown in Fig. 2 are nearly straight lines, showing a drop in voltage which is proportional

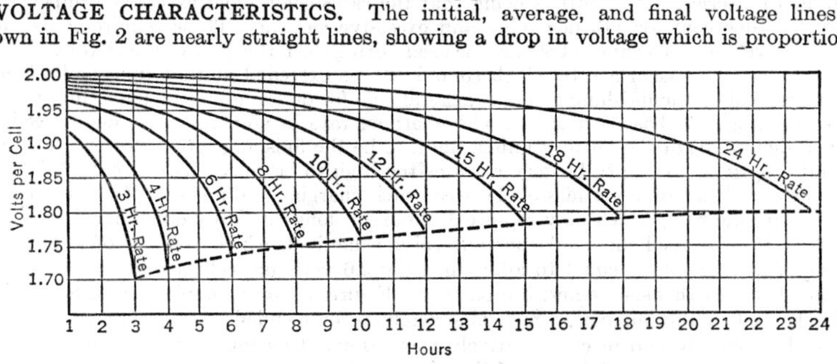

FIG. 3. Approximate Volt-time Curves for Batteries in 1.200–1.220 Specific Gravity, 70 to 80 deg fahr

to the current. The internal resistance of a cell is of little engineering importance in its application, and in fact, when measured by different methods in a laboratory, widely varying results may be obtained. It is generally accepted that during continuous discharge the internal resistance increases slowly during the first two-thirds of the discharge and then at an increasing rate to a value at the end of the complete discharge approximately twice as great as at the start. The internal resistance increases rapidly as the cell temperature falls below zero.

Figure 3 shows the variation of voltage with time during continuous discharge at rates from the 24- to the 3-hr rate.

EFFECT OF TEMPERATURE ON CHARACTERISTICS. In data showing the characteristics of their different types of cells, manufacturers usually follow AIEE standards and base them on a temperature of electrolyte, at the start of a test, of 25 deg cent (77 deg fahr). Temperature affects the chemical reaction and the rate of diffusion of the electrolyte in such a way that both the capacity and voltage are lower at lower temperatures and higher at higher temperatures. It is impossible to make a general statement of the quantitative effects of temperature changes, as they vary greatly with details of plate structure and assembly and with the discharge rates. Figure 4 gives data on one type of battery at two rates.

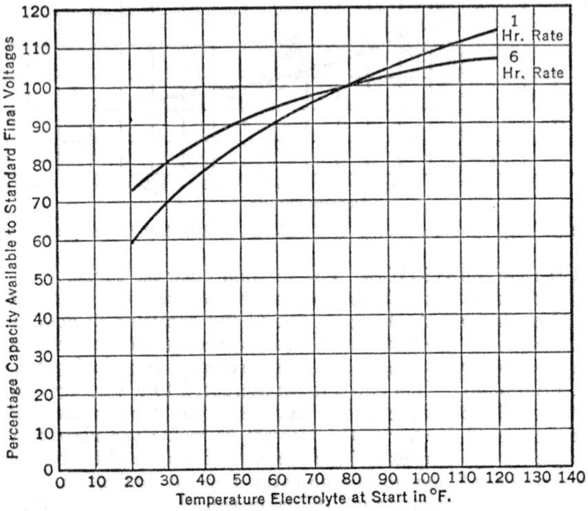

Fig. 4. Variation of Capacity with Temperature. Ambient temperatures are constant at values shown. Capacities shown are to standard final voltages. Additional useful capacity will be obtained at the lower temperatures if discharges are continued to the knee in the voltage curve.

CHARGING CHARACTERISTICS. The electrical characteristics of a battery on charge are usually of little interest except in choosing charging equipment that will be both economical and convenient. A wide latitude in charging rates is permissible with lead-acid batteries, the essentials being that the rates shall avoid excessive temperatures (about 45 deg cent, or 113 deg fahr) and excessive gassing. These requirements mean that relatively low charging rates must be used during the latter part of the charge. Manufacturers usually assign a rate which should not be exceeded at that time and which is commonly called the finishing rate. A general rule for determining the maximum permissible charging rate has been given: The charging rate in amperes must not exceed the ampere-hours out of the battery unless the rate be the finishing rate or lower. This rule permits charging an empty battery at rates too high to be convenient or economical but shows the wide latitude permissible between such charging and a constant-current charge at the finishing rate, which may require 16 hours or longer. Automatic charging may be done by the so-called modified constant-voltage method, which consists of a constant-voltage charging source equal to approximately 2.6 volts per cell of battery and a fixed resistor between this source and the battery. Figure 5 shows complete records of a battery charged in this manner, which is strongly recommended when batteries are to be charged frequently and in comparatively short time. Constant-current charging at the finishing rate or less is satisfactory if time is available.

When batteries are to be charged infrequently, manual control of the charging current is often employed, using a constant current at the finishing rate or a current about 2.5 times the finishing rate during the early part of the charge, then reducing the rate to the finishing rate. During the past 10 years automatic charging equipment has been placed on the market to reduce the starting charge rate when the battery is approximately

85 per cent charged, as determined by a voltage temperature-compensated relay that is responsive to battery voltage. When the charge rate is reduced, a timing device automatically stops the charge after a predetermined time interval. Figure 6 shows the variation of voltage during such a two-rate charge.

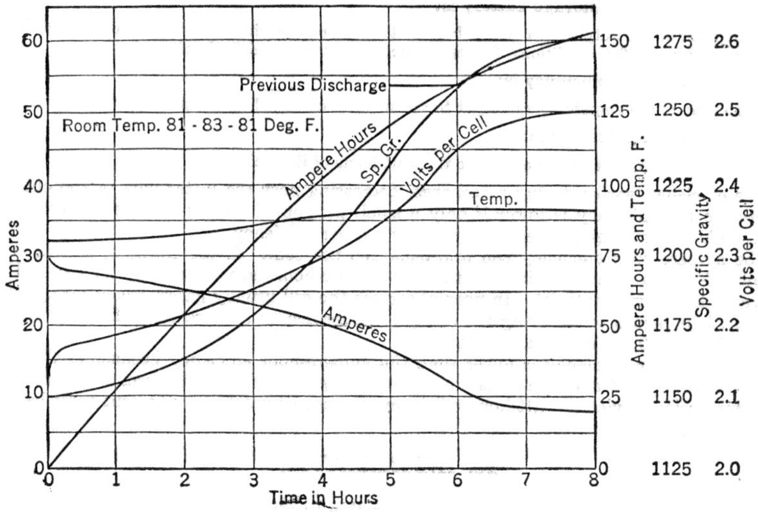

FIG. 5. Modified Constant-voltage Charge. Four cells, 136 amp-hr capacity in 6 hr. Charged from 10.52-volt bus, 2.63 volts per cell, through fixed resistance of 0.0665 ohm. Ohm per cell = 0.00166.

Trickle charging is a term applied to maintaining a very low charge rate of the proper value to compensate for the internal losses which occur in an idle battery. It is employed to keep a battery in a fully charged condition over long intervals (*see also* Standby Batteries, p. 7-24). The proper value of the trickle-charge current depends on the size and number of plates per cell and their spacing, the specific gravity of the electrolyte, the temperature, and the age and condition of the battery. It is usually in the order of 0.1 to 0.8 per cent of the 8-hr rate in amperes for 1.210 specific gravity and temperature near 25 deg cent (77 deg fahr), but a more accurate value should be obtained from the manufacturer.

Floating. A more satisfactory way to keep a battery in a fully charged condition is to maintain a constant voltage of the proper value across the battery terminals; this is known as floating. A low charging rate, which will vary with varying conditions to

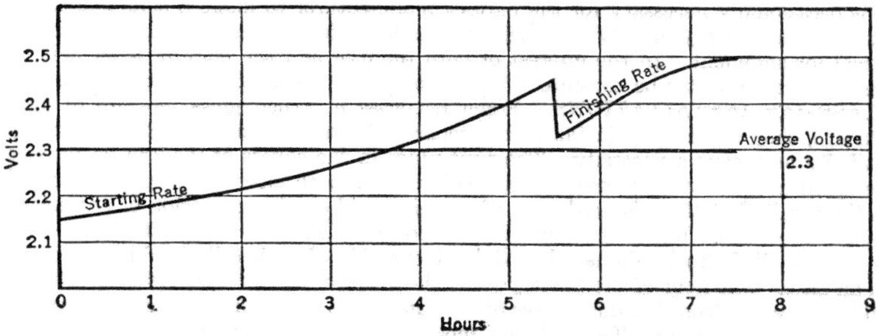

FIG. 6. Constant-current Charge, Characteristic Voltage Curve. This curve is only approximate, since the temperature condition of the cell and other factors will cause variations for the values shown.

accomplish the desired results, will then automatically flow to the battery. The proper value of the floating voltage varies with the specific gravity of the electrolyte. For 1.210 specific gravity the voltage should average 2.15 volts per cell and should not be allowed to vary outside the limits of 2.10 and 2.20 for sustained periods. For 1.280 specific gravity the average should be 2.22, and the corresponding limits 2.17 and 2.27 volts per cell. If batteries are to be floated continuously in service, cells having low-gravity electrolyte should be used, as they will have a longer service life.

EFFICIENCY. The following definitions of efficiency are taken from the AIEE Standards No. 36 on storage batteries.

The **efficiency of a storage battery** is:

The ratio of the output of a cell or battery to the input required to restore the initial state of charge under specified conditions of temperature, current rate, and final voltage.

The **ampere-hour efficiency** is:

The ratio of the ampere-hours output to the ampere-hours of the recharge.

The **watt-hour efficiency** is:

The ratio of the watt-hours output to the watt-hours of the recharge.

The efficiency varies during a recharge, decreasing near the latter part of the charge. Therefore the efficiency is lower if only a small portion of the capacity is removed. It is possible by prolonging a recharge to show a lower efficiency.

The ampere-hour efficiency can approach 100 per cent. If the charging rate is kept below the point where gassing occurs, the ampere-hours are almost totally utilized in charging the active material. In normal service the cell usually gasses freely for approximately 1 hr before terminating the charge. In service an ampere-hour efficiency in the order of 85 to 90 per cent may be obtained.

The watt-hour efficiency is lower than the ampere-hour efficiency because of the fact that the voltage of a cell falls off during discharge (output) and increases on charge. In commercial service a watt-hour efficiency of approximately 75 per cent may be obtained.

18. INSTALLATION

The essential features that require consideration in planning the details of all battery installations are: first, ample ventilation and moderate temperature; second, accessibility of each cell for inspection and maintenance. There are other features, depending on the type of battery and the nature of the service, which will be discussed in later paragraphs. Ventilation is required to dissipate the oxygen and hydrogen which are given off from the battery in considerable quantities during the latter part of charge and to some extent at all times, in order to prevent the possibility of an accumulation of these gases in such proportions as to form an explosive mixture, and also to remove any acid spray which may be thrown into the air if open-type cells are used. Control of the battery temperature is also essential, especially if there is any possibility that the battery may be subjected to high rates of charge or discharge. If the battery is to receive such attention as is outlined under Care and Operation, p. 7-27, it must be accessible to inspection and maintenance. Special attention should be given to the ease of keeping electrolyte at the proper level (by adding water to the cells to replace evaporation and that lost by electrolysis during overcharge), taking hydrometer readings, and keeping the battery clean and dry.

It is occasionally necessary for the battery manufacturer to make a complete detailed drawing of the battery exactly as it will be installed before he can prepare the necessary material for shipment. In the interest of both the user and the manufacturer it is very desirable that the latter be consulted in time to permit adequate provisions to be made.

GLASS-JAR INSTALLATIONS. When batteries in sealed glass jars are of appreciable size, they are usually installed in a separate battery room, although this arrangement is by no means a necessity. The vent plugs in the covers of modern batteries are so designed as to allow the escape of gases but to trap all electrolyte spray effectively, with the result that there is no danger of escaping acid causing corrosion of any near-by metal surfaces. The batteries should be located in a clean, dry, ventilated place where they are not likely to be damaged. At least one side of each cell should be accessible from an aisle. The floor of the battery room should be of smooth concrete. The battery should be shielded from the direct rays of the sun by the use of diffusing glass in windows or skylights if necessary. The cells are mounted on wooden or steel racks for one or more tiers of cells; for the larger sizes of cells, single-tier racks are recommended. When the cells are arranged in more than one tier, no cell should be located behind any of the uprights unless space is rigidly limited. To permit hydrometer readings, ample space (about 15 in.) should be allowed between the top of a cell cover and the bottom of the rack above.

RUBBER-JAR INSTALLATIONS. Batteries assembled in rubber jars are generally used only when portability is necessary or where other types of jars will not withstand the mechanical shocks to which the battery is likely to be submitted. Portable batteries should be so placed that battery temperatures in excess of 113 deg fahr are not encountered. In those installations where the battery is in motion it is necessary to provide against a shift in position of the battery and against the shocks due to sudden changes in motion. Springs have been tried and then abandoned. Where the battery is mounted in a moving vehicle, such as an automobile or an airplane, the battery should be relatively fixed to the supporting structure. Excessive clamping pressures should not be placed on batteries, especially containers, as the seal between the container and cover may be disturbed, and the hard rubber distorted. Where vibration or shock is encountered, as in tractors, a hold-down on top of the intercell connectors is recommended because they are connected mechanically to the element which contains the greatest mass of weight in a battery. Insulation must be used on the intercell connectors to prevent a short circuit; soft rubber, which should be backed with a rigid member, is recommended. The soft rubber adjusts for small variations in height of the intercell connectors. On electric street and industrial trucks, where the body springs are relatively stiff, the batteries are usually protected against bouncing by some form of cleat or by holding-down bolts and against lateral or longitudinal shifting by positive mechanical stops. In locomotive and marine work no hold-downs are necessary. In no installation should the battery be wedged tightly in the compartment, as any pressure applied in this way is added to other stresses that may be applied to the rubber jars because of weaving of the compartment or mechanical shocks during service.

Compartments. Practically all rubber-jar installations are made under conditions where the space required for the battery is also desirable for other purposes. This fact should not result in the installation of the battery in a manner which will prevent successful operation or cause excessive battery-maintenance cost. Ventilation and accessibility are essential. Ample air inlets at the bottom of the compartments and outlets at the top should be provided to permit free circulation of air. The trays should be separated from each other by at least $^1/_2$ in. to allow passage of air between them and to permit drainage when the battery is washed down. Rubber or porcelain spacing insulators are used for this purpose. Drainage should be provided for the wash water.

The size of the compartment should be such that the battery is a snug fit, but in no case should it move more than $^1/_2$ in. in any direction. If blocking is required, it should be placed next to the compartment walls, for if it is placed between trays the intertray connectors may be subjected to strain.

In street and industrial truck service, where the number of cells in series rarely exceeds 44 and where the vehicle is mounted on rubber tires, no special insulation of the battery from the compartment is required, but in other services where the number of cells in series exceeds 16 and where the compartment is grounded, rubber or porcelain insulators are used between the trays and the bottom and sides of the compartment. For marine installations lead pans or water-tight lead-covered floors are sometimes provided beneath the insulated cells.

BATTERIES IN LEAD-LINED TANKS should always be installed in a specially prepared battery room, the details of which must be worked out jointly by the user and the manufacturer before preparation of the battery for shipment. The special features which may be required include a forced ventilation system, which should be of the exhaust type with inlets near the floor at one end of the room and the exhaust outlet near the ceiling at the other end, through a series of baffle plates to intercept any fine particles of electrolyte thrown into the air when the cells are gassing.

It is impossible in the limits of this section to give complete specifications for the floor construction, but such specifications, based on many years of experience, may be obtained from battery manufacturers. In general, a smooth, level, cement foundation surface on which one or more layers of acid-proof fabric are laid and protected by a bituminous compound mastic floor is required.

Provision must be made for supporting copper conductors to the battery terminals and to the end cells if these cells are used. All exposed conductors must be lead-plated or painted with acid-resisting paint. Lead-covered cable in the battery room is not recommended. If a supporting structure for end-cell switches is used, it must be accurately installed, and all exposed metal protected against acid spray.

19. BASIC APPLICATIONS OF STORAGE BATTERIES

The advantages of the storage battery as a device for storing energy, especially electrical energy, may be enumerated as follows:

1. It is the lightest and most efficient device for storing energy in portable form, and

it is therefore employed wherever energy is to be stored at one time and place for use at another.

2. Its energy is instantly available, no time being required to prepare for delivering the energy thus stored.

3. Its energy may be stored at any convenient rate and delivered at any other rate required.

4. It may be employed as a voltage transformer, storing electrical energy at one voltage and delivering it at another, by series-paralleling groups of cells.

5. The constancy of voltage of a storage battery is another characteristic which is often of pronounced importance.

6. The storage battery is a very reliable piece of equipment within its limitation of capacity and life.

Advantage is taken of one or more of these characteristics in the various commercial applications of the storage battery. The variety of applications covers such a wide range that only typical examples can be presented in this section. These examples are classified below under applications based primarily on (*a*) portability, (*b*) reliability of supply of electrical energy, and (*c*) economies effected, although in some cases these classifications overlap.

20. APPLICATIONS BASED ON PORTABILITY

AUTOMOTIVE BATTERIES. Undoubtedly the largest single application of storage batteries is for the starting, lighting, and ignition of automobiles and buses. These batteries are charged automatically from a generator driven by the engine while it is running, and they supply power for the lights while the engine is shut down and for ignition and cranking when the engine is started. (For details of system as a whole see Section 17.)

The essentials for a battery for this service are ability to discharge at high rates with well-sustained voltage, minimum effect from low temperature, light weight, small space, and as long a life as is compatible with the other requirements. Pasted-plate batteries are universally used. Three-cell (6-volt) batteries are the recognized standard for automobiles; 6 to 12 cells (12 to 24 volts) are used for buses. In order to reduce the weight and minimize the likelihood of freezing in cold weather, electrolyte of high specific gravity (nominal 1.280) is used in temperate climates; lower specific gravity (nominal 1.240) is recommended for tropical climates. The cells are generally assembled in three-compartment rubber or asphaltic composition cases. The size of battery is usually determined by the cranking requirements, which depend on the size and number of cylinders of the engine, the minimum firing speed, the engine-to-starter gear ratio, and the design of the cranking motor. The arrangement of the compartments in the container and the location of the battery terminals are subject to the requirements of each model of car. As the life of these batteries is considerably shorter than the life of the vehicle, leading manufacturers have provided the required variety of batteries to make a replacement on any current model of automobile.

AIRCRAFT BATTERIES. With an increase in commercial and private planes and the expansion of the air forces of the military services during World War II the need for aircraft batteries has increased tremendously. These batteries are charged automatically from a generator, usually driven by the engine, whose output voltage within the generator capacity is held constant by a voltage regulator, which changes the generator field excitation. The battery supplies power for the radio, lights, instruments, feathering the propeller, retractable landing gears, turrets, ignition, and cranking when the engine is started. The battery supplies peak loads beyond the generator capacity, thereby reducing the generator size.

The essentials of a battery for this service are ability to discharge at high rates with well-sustained voltage, minimum weight and size, ability to be inverted without spilling, resistance to the effects of low temperature, dependability, and reasonable life commensurate with the other requirements. Pasted-plate batteries are used universally in 6-cell (12 volt) and 12-cell (24 volt) sizes, electrolyte of high specific gravity (nominal 1.280) being used. Additional data are given in Section 17.

MOTIVE-POWER BATTERIES are used for the propulsion of electric street trucks, industrial trucks and tractors, and mine and industrial locomotives. These vehicles are used in short-haul or frequent-stop service. Long-haul, high-speed service is left to the gasoline truck in road work and to the trolley locomotive in mine work; the economies which accrue with the use of battery-driven vehicles in frequent-stop and therefore slow-speed service seem to justify their use.

Batteries in this service usually have sufficient capacity to perform the work of a shift and are charged when the vehicle is out of service, usually at night. To a limited extent, exhausted batteries are replaced by charged ones at the end of a working shift, and the vehicle is kept in service for two or even three shifts a day. The essentials of a battery for this service are a minimum of space and weight, high electrical efficiency, mechanical strength to withstand road shocks, and long life in relation to the cost of purchase and maintenance. Pasted or modified pasted (such as Exide-Ironclad) plates are used to the total exclusion of Planté plates in this country, although Planté plates are used to some extent in Germany.

The cells are assembled in individual high-grade hard-rubber jars or monobloc units, and the jars are in turn assembled in a number of hardwood trays arranged to suit the compartment in the vehicle. If the battery consists of more than 44 cells, the trays are equipped with porcelain or rubber insulators to insulate the battery from the frame of the vehicle. The size of the battery is determined by the total amount of work to be performed and is usually specified by the truck manufacturer to conform to this work and to the characteristics of the motor he selects to drive the vehicle.

A user of motive-power batteries should exercise the same care in selecting charging equipment as in selecting a vehicle, and in properly charging and maintaining the battery as in inspecting and maintaining the running gear. The battery is likely to be subjected to greater hazards in the charging room than in service unless the charging equipment is properly selected and operated. Fully automatic charging equipment, as described under Charging Characteristics (Modified Constant Potential, p. 7-19), is recommended by battery manufacturers.

RAILWAY-TRAIN LIGHTING AND AIR CONDITIONING. Storage batteries are used to supply current for lighting railway cars when the train is standing or moving at slow speeds. When the train is moving above a certain speed (about 20 miles an hour), the lighting current is furnished by a generator mechanically driven from the car axle; this generator also recharges the battery automatically. In order to provide the constant voltage necessary for satisfactory lighting at variable train speeds, the generator voltage is controlled by an automatic voltage regulator.

Axle systems for lighting only are generally designed for 32 volts and include 16 cells of lead-acid battery, ranging in capacity from about 200 to 450 amp-hr. The much heavier loads due to air-conditioning equipment call for greater capacities up to 1250 amp-hr for 32-volt systems; in some cases 64-volt or 110-volt air-conditioning equipments have been adopted to reduce the size of conductors and facilitate the design of generators and control apparatus.

For the smaller sizes of batteries used for car lighting only, weight and space requirements have been considered to be of secondary importance, and many Planté and modified Planté (Manchester) plates have been used and are still being maintained by some railroads. There is, however, a trend toward the use of flat-plate and modified pasted-plate batteries to facilitate handling during service inspection and maintenance. For the larger batteries required for air conditioning, only flat plate and modified pasted plates are being used, as space is not available on the car for the heavier types.

The cells are assembled in high-grade rubber monobloc units or jars, the jars being assembled in hardwood trays of dimensions suitable for installing in battery boxes of dimensions standardized by the American Railway Association.

RAILWAY CONTROL BATTERIES. These batteries are used to supply a continuous low-voltage source of energy for a variety of purposes on electric, oil, diesel, or gasoline-electric locomotives and rail cars and on multiple-unit electric trains. On electric locomotives the batteries usually are operated in parallel with constant-voltage motor-generator sets to supply an unfailing source of energy for control circuits and marker and cab lights. They may also be used in connection with cab signals or automatic train control or with regenerative braking equipment when these features are embodied in the locomotive. On oil, diesel, or gasoline-electric locomotives and rail cars they are also used for engine starting, compressor operation, generator excitation, and train lighting. On multiple-unit electric trains the present tendency is to equip each motor car with a battery and constant-voltage motor-generator for the operation of control circuits, marker and emergency car lighting, and, if employed, cab signals or automatic train control. These circuits are paralleled throughout the train by means of "train lines."

The battery essentials for this service are similar to those for railway-train lighting except where engine starting is required; in this case provision must be made for the high-rate momentary discharges required for that purpose, and this feature will probably determine the size of the battery. Weight and space are of more importance than in train lighting because of the larger amount of equipment required on each unit. For this reason the modified pasted-plate (Exide-Ironclad) battery has found most general application. In all these applications the battery is kept practically fully charged at all times, receiving sufficient current from the generator to replace any discharge, usually of short duration, which may occur and also to replace any internal losses.

21. APPLICATIONS BASED ON RELIABILITY

STANDBY BATTERIES are installed to provide a reserve source of energy for use only in an emergency. They must therefore be kept in a state of full charge at all times, either by floating or by trickle charging (see Charging Characteristics, p. 7-18).

STANDBY BATTERIES IN DIRECT-CURRENT DISTRIBUTION SYSTEMS are connected to the station bus at all times without the interposition of circuit breakers or fuses and thus are always available for instant use in any emergency. The batteries are so floated as to keep them fully charged at all times.

The middle point of the battery is connected directly to the neutral of the three-wire system; the positive and negative ends are connected to their respective buses through end-cell switches, by means of which the number of cells in use may be varied at will to meet the operating requirements. These switches permit the use at all times of the right number of cells to maintain the proper floating voltage across the cells in service regardless of variation in bus voltage. They also permit additional cells to be placed in circuit as may be required to control the voltage during emergency discharges. These switches are the only movable piece of apparatus used in connection with the battery at such times; their dependability is therefore of prime importance. They must offer no possibility of interrupting the discharge; they must be able to carry any currents that the battery can deliver for as long a time as the battery can deliver them; and they must be capable of being moved from point to point while carrying these currents without opening the circuit and without destructive arcing. Such switches have been made having a continuous carrying capacity up to 10,000 amp and a 6-min carrying capacity of 40,000 amp. Two or more of these switches may be operated in multiple when more carrying capacity is required.

Charging. To charge these batteries after an emergency discharge, they are usually disconnected from the bus and charged from a motor-generator or synchronous converter whose voltage is raised as may be required until its maximum is reached, after which a booster generator is connected into the circuit to raise the voltage further to the value required to complete the charge. This booster is usually a three-unit motor-generator set having one motor which takes its power from the station bus and two generators, one for each side of the battery, thus permitting independent control of the charge of the two halves to compensate for any inequality in discharge which may have occurred. The charging is done at a time when emergency protection is of minor importance.

Type of Battery. Direct-current distribution systems are largely confined to the more densely populated sections of the larger cities where real estate is very valuable. The space occupied by these batteries is therefore of great importance, and they are required to do very little work during their lifetime. Both these reasons indicate that the flat-plate pasted type of battery should be used, and this type has been installed exclusively in recent years, although some of the older modified Planté (Manchester) batteries are still in service.

Size of Batteries. As the purpose of these batteries is to prevent a failure of power on the distribution system, they must be of sufficient capacity to carry the maximum load for which the station is designed for a long enough period to permit re-establishment of the normal source of power. The length of this interval is largely a matter of judgment and varies with different stations, but 20-min protection for the maximum load is usually considered sufficient. When the rate and duration of the discharge are known, the size of cells required is easily determined from data similar to those shown in Fig. 2, p. 7-17.

The number of cells is determined from the final voltage per cell at the maximum discharge rate and the minimum allowable voltage. The number of end cells must be sufficient to permit enough cells to be cut out of circuit to allow proper floating when the station bus voltage is at its minimum value. In order to keep the end-cell switches as small as possible and to reduce the amount of copper required for connections between these switches and the end cells, it is customary to use groups of two end cells each between switch contacts within the floating range and groups of three or four cells between contacts for those end cells, which are employed only to maintain voltage during emergency discharges.

STANDBY BATTERIES FOR EMERGENCY LIGHTING AND POWER are installed to accomplish the same general purposes as the larger batteries used with central-station d-c distribution systems, but the installations are so much smaller and the local conditions so different as to require very different treatment.

Emergency Lighting Batteries, in some places required by law or ordinance, are employed in theaters, schools, hospitals, stores, banks, and similar public institutions to provide a second source of power, which is automatically applied to selected circuits in case of failure of the normal power, and thus maintain illumination in corridors and exits and in special locations, such as operating rooms in hospitals and banking rooms and vaults in banks. These batteries must receive sufficient charge from the normal source of power to maintain them in a fully charged state, but usually they are not operatively connected to the emergency circuits except in a power failure, when an automatic switch disconnects the emergency circuits from the normal supply and connects them to the battery. In some cases automatic return to normal connections on return of power, and automatic recharging of the battery, are provided. Where the normal power is alternating current, the battery is trickle-charged by means of a small rectifier and recharged by a rectifier or motor-generator set. Where the normal power is direct current, the battery is usually trickle-charged from the main power supply in a manner similar to that described below under Emergency Power Batteries.

It is impossible within the limits of this section to discuss the details of the electrical connections used in this service, as they must differ widely with the nature of the normal supply, i.e., whether it is single-phase 2 or 3 wire, double-phase 4 or 5 wire, or triple-phase 3 or 4 wire. These details have been worked out, however, and suitable apparatus is available for any of these conditions. Ordinarily no provision is made for the control of voltage during emergency discharges. A suitable number of cells is used, generally 60 for a nominal 115-volt circuit, and the voltage is allowed to fall as the discharge progresses. If a certain minimum voltage must be maintained, a size of cell large enough to meet the requirements is selected.

Type of Battery. As these batteries are not subject to daily work, the pasted-plate type has been most generally used, particularly in installations required by law or ordinance, when low initial cost has been the deciding factor.

Emergency Power Batteries are employed in industrial plants to insure continuity of power for such a period as is necessary to prevent loss of material in process in the event of a power failure, as in glass mills to clear the mill before the glass cools or in steel mills to right a ladle which is pouring molten metal. The motors to be protected must be of d-c types, although recent development in "inverters" may make this protection available for a-c motors. In d-c service the batteries are trickle-charged in two multiple circuits through resistors from the normal supply circuit and recharged in the same manner through other resistors. In the event of a power failure a low-voltage relay drops, making contacts which close a contactor connecting the two halves of the battery in series and trips a contactor which disconnects the protected circuit from the incoming power circuit. Two trickle-charge resistors are allowed to remain in circuit, as the energy lost in them during the emergency is negligible; if, however, the battery is receiving a high-rate charge, the circuit through the charging resistors is automatically opened. If the low-voltage relay will pick up on return of normal power, normal operation will be resumed automatically when the power returns. Connections for recharging the battery are usually made manually.

Because of local conditions it is sometimes considered preferable to float and recharge the battery from a separate motor-generator set in order that the battery may be entirely isolated from the shop circuit under normal conditions. In such a case a simple throw-over switch to disconnect the protected circuit from the shop supply and connect it to the battery is all that is required for automatic protection.

Type of Battery. Pasted-plate batteries are ordinarily used for strictly emergency protection, but in some cases a battery installed primarily for this purpose may also be employed to perform some work during normal operation. Such applications are discussed in the following paragraph.

FLOATED WORKING BATTERIES. There are many electrical installations which justify the use of a standby battery for emergency protection but which also present opportunities for effecting economies if the battery is allowed to do some work during normal operation. These batteries are permanently connected to the circuits which they protect. During normal operation an external source of power, usually a motor-generator set or rectifier, is also connected to the circuit and in parallel with the battery. The voltage across the circuit is maintained by the generator at an average value suitable for "floating" the battery. The generator delivers to the circuit the average power requirements, and the battery discharges automatically at times of high load and recharges automatically when the load is below the average. The total amount of the work done by the battery per day during normal operation is usually relatively small, and the battery is kept in a state of nearly full charge and therefore always available for emergency use. In some applications the battery may become appreciably discharged at times during normal operation; when this is contemplated, the battery must have enough capacity to perform both services.

Type of Battery. In this service the life of the battery depends largely on the accuracy with which the floating voltage is maintained but sometimes also on the total amount of work which the battery must perform. If local conditions are such that the floating voltage may be erratic or if a considerable amount of work is required while floating, Planté, modified Planté, or modified pasted batteries are selected. If the floating voltage can be well maintained and the amount of work is small, flat pasted-plate batteries are used.

CONTROL BUS BATTERIES are employed in practically all generating and substations to insure in any emergency a continuous source of power for the operation of circuit breakers and other remotely controlled apparatus and also for emergency lights. The control bus is usually completely isolated from station buses so as not to be affected by any disturbance which may occur on them. The battery is permanently connected to this bus at all times, and a small motor-generator set (or rectifier) is provided to carry the continuous load, composed principally of indicating lights, and to keep the battery fully charged. The emergency lights ordinarily receive their energy from the "house" bus, but, if that bus fails, the emergency light circuit is automatically disconnected from the house bus and connected to the control bus by a throw-over switch. The normal load on the control bus is thus very small, as the indicating lamps consume only about 0.035 amp for each oil circuit breaker; therefore a small motor-generator set can be used if the battery is allowed to supply the high momentary loads required for operation of the circuit breakers. This is readily accomplished by employing a generator whose load-voltage characteristics are such as to present at overload a droop in voltage sufficient to cause the battery to take practically all the excess demand.

The design of circuit-breaker mechanisms has been standardized by NEMA for the operation of nominal 125-volt mechanisms within the limits of 130 to 90 volts for closing and 140 to 70 volts for tripping (for nominal 250-volt mechanisms these values are doubled). For 125-volt mechanisms 60 cells has been standardized, as well as an allowance of 15 volts for copper drop between the battery terminals and the circuit-breaker mechanisms. The battery must therefore be of such a size as to be able to deliver the required loads with the voltage at its terminals not less than 105 volts, or an average of 1.75 volts per cell for 60 cells. Battery manufacturers publish in their trade catalogs the ratings of their cells to this voltage. The correct floating voltage for 60 cells, using electrolyte of 1.210 specific gravity, is 129 volts, which is therefore the normal operating voltage of the control bus. Consequently the generator must deliver this voltage when carrying the normal control bus load.

In order to charge the battery after an emergency discharge, the generator voltage is raised from time to time as necessary, but not above 140 volts, which is the maximum voltage permissible on the circuit-breaker mechanisms and indicating lamps. In order to prevent excessive overload on the generator when heavy demands are thrown on the bus for closing oil circuit breakers, the generator voltage must fall to 105 volts in order to permit the battery to take a large part of the increased load. In manually operated stations where there are no prolonged changes in load on the control bus specially designed shunt-wound generators are extensively used for this purpose. In unattended stations where holding coils and relays may introduce prolonged changes in load, these changes introduce changes in the voltage of the control bus due to the drooping characteristics of the shunt generator, which interfere with the proper floating of the battery unless artificial loads are introduced when these variable loads are disconnected. In order to avoid this complication, generators have been designed and are now available which have a very flat voltage characteristic at all loads up to a predetermined point and a sharply drooping characteristic at higher loads. During normal operation such a generator maintains a favorable floating voltage even under moderately variable loads but throws all excess load due to circuit-breaker operation onto the battery.

Size of Battery. In an emergency the battery must be able to carry the indicating lamp load throughout the duration of any loss of normal power. It is now common practice to consider possible a 12-hr interruption in manually operated stations and a 24-hr interruption in non-attended stations. Emergency lighting protection is, however, usually provided for a period of 1 to 3 hr only. In addition to carrying these loads the battery must be able to deliver the current required for circuit-breaker operation at any time during the emergency. Trade catalogs give methods of approximating the size of battery required to meet any such desired conditions, but it is usually good practice to submit the conditions to the manufacturers for their recommendations.

RAILWAY SIGNALING BATTERIES. Storage batteries are used extensively in the so-called a-c floating battery system for railway automatic signaling. The functions of the battery are to supply a continuous source of energy for operation of the signaling devices in case of failure of either the power supply or the rectifier and also to maintain approximately constant voltage on the signal circuits under varying current demands. At each signal installation a battery is operated in parallel with a rectifier, which is adjusted to maintain the proper voltage across the battery and to supply the average load required for operation of the necessary relays and signal lights. The general method of operation is similar to that which has been described. The drooping characteristic of the rectifiers throws heavy demands for current on the battery, which is automatically restored during times of lighter load. The batteries have sufficient capacity to carry the entire load between periods of inspection, if the normal supply of current from the rectifier fails.

YACHT AND MOTORSHIP BATTERIES. Floated batteries on diesel-engine-driven yachts and motorships are used in much the same way as has already been described. Their most important function is to assure a reliable supply of energy at all times for lights and for the operation of important electrically operated auxiliaries, such as steering gears, winches, pumps, and radio. On a larger yacht they also furnish all necessary auxiliary power and light when the yacht is at anchor and the noise and vibration of a running engine are objectionable. They also relieve the auxiliary engine-generator sets of fluctuations in load caused by the simultaneous operation of the auxiliaries and so permit the use of smaller engine-generator sets, thus effecting saving in cost, weight, and operating expenses as well as adding to safety and convenience. Fifty-six-cell batteries are ordinarily used for nominal 115-volt installations; the floating voltage is 125 volts. When it is necessary to recharge the battery, the generator voltage is raised as may be required to not more than 135 volts, which is permissible for the auxiliary motor equipment. To protect the lights and some of the more sensitive auxiliaries, such as gyroscopic compass and radio equipment, automatic voltage regulators are employed in those circuits. The generators must be able to deliver full load at 112 to 135 volts, and, to assure stability of operation of the generators and battery in multiple, the generator voltage characteristic should under no operating condition show a rise in voltage with increasing load. With this condition met, the voltage characteristic will necessarily droop at higher loads. This fact, together with the drop in speed of the diesel engine with overloads, causes the battery to relieve the engine of excessive loads and permits the use of smaller auxiliary generating sets which operate at better load factors. Pasted- or modified pasted-plate batteries in 1.280 specific gravity* are used for engine-starting and combination standby batteries; batteries in 1.210 specific gravity are used only for standby service.

DRAWBRIDGE BATTERIES may be taken as typical of applications where protection is required against failure of the normal power supply and where the average load is low but short-time demands are heavy. The normal source of power may be an internal-combustion engine-generator set or a public utility whose power bills include a maximum-demand charge. In either case the battery supplies energy for the operation of the draw, thus reducing the demand on the normal source of supply and improving the economy of operation, as well as providing emergency protection. When the heavy-load demands are fairly frequent, the battery is floated across the generator in the same manner as described under Yacht and Motorship Batteries, except that the generator may have a more sharply drooping characteristic in order to throw a greater percentage of the load on the battery and thus maintain a more nearly constant load on the generator. If the heavy loads are sufficiently infrequent, greater economy may be effected by cycle operation of the battery, i.e., by using the battery alone for normal operation and running the generator only when it is necessary to recharge the battery.

TELEPHONE BATTERIES. The telephone companies rank among the largest users of storage batteries, as they employ one or more batteries in practically every main and branch exchange. These batteries are operated to provide sufficient reserve capacity for continuity of service if the normal source of power fails. Frequently, however, they serve as the sole source of power during intervals of light demand, this use depending on local operating conditions and the type of equipment employed. The batteries are floated at all times when not carrying the load or being charged.

CYCLED BATTERIES are used in some services where continuity is of such importance that entire isolation of the circuit is desirable in order that it may not be affected by any external disturbances.

22. CARE AND OPERATION

The essential features of the general care of storage batteries are simple and few, but they are essential.

1. Keep the battery clean and dry.
2. Add water (which has been approved by the manufacturer) from time to time to keep the level of the electrolyte within the limits specified by the manufacturer.

* Batteries in 1.280 full charge specific gravity electrolyte.

3. Never add electrolyte or acid except to replace a loss known to be due to spillage. The added acid should then be of the same specific gravity as that in adjacent cells.

4. Never allow any "special solutions," powders, or jellies or any metals or other foreign matter to get into the cells.

5. Stop discharge, except in emergency, before the voltage falls too low for satisfactory operation.

6. Charge at rates low enough to keep the cell temperature below 45 deg cent (113 deg fahr), and while the cells are gassing never charge at rates higher than those allowed by the manufacturer.

7. Keep sparks and flames away from the vent of a cell so that the hydrogen and oxygen gas produced within a cell is not ignited.

All manufacturers furnish instructions for the care and operation of their different types of batteries, the systematic following of which will pay for itself by prolonging the life of the battery and by assuring reliable performance.

Excessive overcharge in ampere-hours should be avoided, even if the rate of charge is low enough to avoid excessive temperature and gassing.

23. BIBLIOGRAPHY

Arendt, M., *Storage Batteries: Theory, Manufacture, Care and Operation.* D. Van Nostrand (1928).
Bermbach, W., *Die Akkumulatoren.* J. Springer, Berlin (1929).
Dolezalek-Von Ende, *Theory of the Lead Accumulator.* John Wiley (1904).
Jumau, L., *Les accumulateurs électriques.* Dunod, Paris (1904).
Jumau, L., *Étude résumée des accumulateurs électriques.* Dunod, Paris (1919).
Jumau, L., *Piles et accumulateurs électriques.* Colin, Paris (1928).
Vinal, G. W., *Storage Batteries.* John Wiley (1940).
Witte, O. A., *The Automobile Storage Battery: Its Care and Repair.* American Bureau of Engineers, Chicago (1926).

ALKALINE-TYPE STORAGE BATTERIES

By E. W. Allen

24. DEFINITION AND APPLICATIONS

Alkaline storage batteries are so termed because of the fact that their electrolyte is an alkaline solution, either potassium or sodium hydroxide. Alkaline batteries of present commercial manufacture employ active materials of either the nickel-iron or the nickel-iron-cadmium combination. Such batteries are the Alconum, Alklum, Britannia, Deac, Edison, Jungner, Nife, and Saft, all originally of foreign manufacture except the Edison. An alkaline battery (Drumm) using the nickel-zinc principle has been in experimental use in Ireland for some years. The nickel-iron-alkaline storage battery was invented by Edison in 1901, and is the best-known form of the alkaline cell in this country.

The more important uses to which Edison alkaline storage batteries are applied include: electric industrial trucks and tractors, storage-battery mine and industrial locomotives, railway-car lighting and air conditioning, railway signal systems, multiple-unit control, marine services, electric street trucks, isolated lighting plants, emergency lighting, time clocks, police and fire-alarm systems, oil circuit-breaker control, miners' safety-cap lamps, airway beacons, and telephone service.

25. THEORY AND ELECTROCHEMICAL REACTIONS

The fundamental principle of the Edison storage battery is the oxidation and reduction of metals in an electrolyte which neither combines with nor dissolves either the metals or their oxides, and which, notwithstanding its decomposition by the action of the battery, is immediately reformed in equal quantity and is therefore a practically constant element without change of density or conductivity over long periods of time. The active materials being insoluble in the electrolyte, no chemical deterioration can take place. The chemical reactions in charging the Edison storage battery are: (1) oxidation from a lower to a higher oxide of nickel in the positive plate, and (2) reduction from ferrous oxide to metallic iron in the negative plate. The oxidation and reduction are performed by the oxygen and hydrogen set free at the respective poles by the electrolytic decomposition of water during charge. The charging of the positive plate is therefore simply a process of increasing the proportion of oxygen to nickel. The discharge of the cell is merely the reverse of this process, the hydrogen reducing the higher oxides of nickel to lower oxides and the oxygen oxidizing the iron to ferrous oxide. The electrolyte is an aqueous solution of potassium and lithium hydroxide and does not at any stage cause a disintegration or solution of the active materials. Its specific gravity does not change appreciably on charge or on discharge. The changes that occur when an Edison nickel-iron-alkaline cell is charged and subsequently discharged are represented by the following equations:

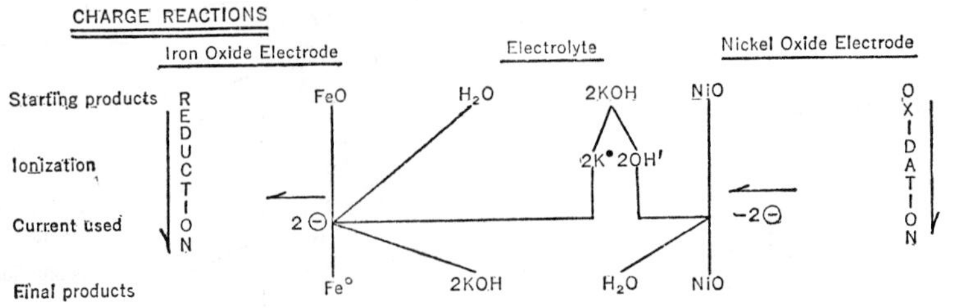

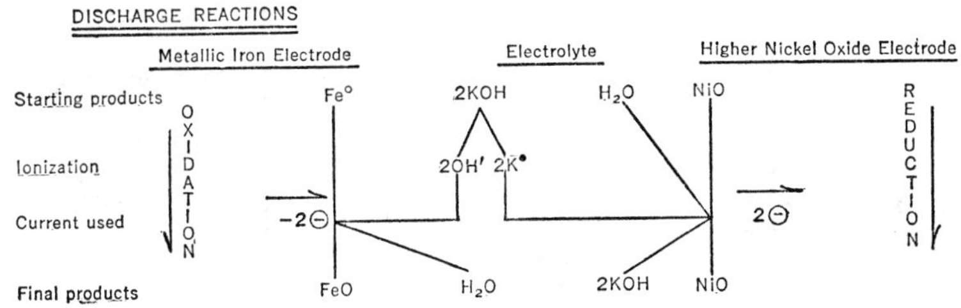

26. DESIGN AND CONSTRUCTION

POSITIVE PLATE. The positive plate consists of a nickel-plated steel grid holding nickel-plated steel tubes which contain the positive active material. When inserted in the tubes, the active material is in the form of nickel hydrate, but this changes to an oxide of nickel after the formation treatment to be described. In order to give the electrolyte free access to the active material the tubes have numerous perforations. To obtain improved conductivity the active material is alternated with layers of pure nickel flake at the time it is tamped into the tubes. The tubes are reinforced by eight seamless steel rings, spaced at equal distances. The tubes are mounted on the positive grid and forced into permanent position under heavy pressure.

NEGATIVE PLATE. The negative plate is of similar construction to the positive except that a finely divided oxide of iron is employed as active material and is contained in rectangular perforated steel pockets instead of tubes. The pockets are forced into place under hydraulic pressure. The grid and pockets, like the positive grid and tubes, are of nickel-plated steel. The Jungner battery has cadmium oxides.

Plate Assembly. The positive and negative plates are assembled into positive- and negative-plate groups (see Fig. 1) by passing a steel connecting rod through a hole in the top of each plate. Steel washers are employed to obtain proper spacing, the middle spacer being the base of the pole piece. A lock washer and nut are drawn up tightly at each end of the connecting rod, binding the plate groups firmly together. All washers, nuts, connecting rods, and terminal posts, like the plates, are of nickel-plated steel.

The plate groups are assembled into complete elements by intermeshing the plates so that they are alternately negative and positive. The negative group always contains one more plate than the positive, so that both outside plates are negative. Adjacent positive and negative plates are insulated from each other by vertical hard-rubber pins running their entire length. Thin rubber sheets insulate the outside negatives from the sides of the container. The edges of the plates are insulated from the bottom and from the other sides of the container by hard-rubber frames. These frames which also serve the purpose of separating the plates and holding them in correct alignment, are so designed as to permit free circulation of the electrolyte.

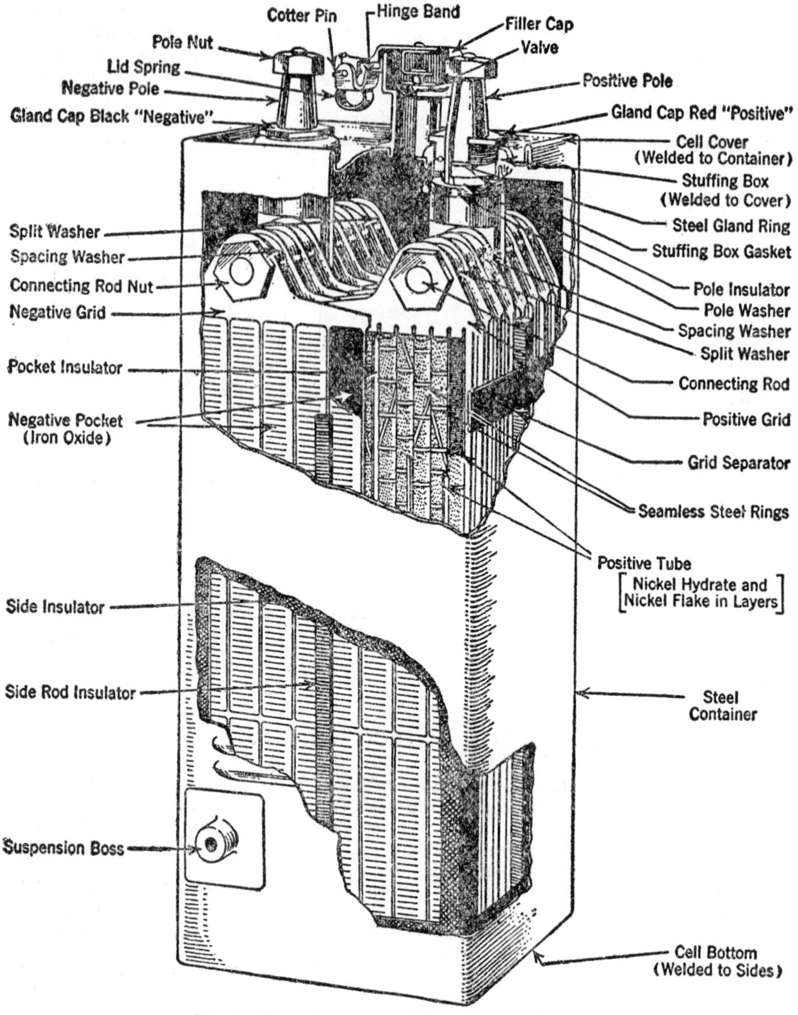

FIG. 1. Plate Assembly of Edison Storage Battery

CONTAINER. Caustic potash does not attack steel, and this fact is utilized in the Edison nickel-iron-alkaline cell in order to obtain the strength, durability, and light weight of steel in the container as well as in the elements. The container, like the elements, is made of nickel-plated steel with all seams welded. After the elements are inserted, the cell cover, also of nickel-plated steel, is welded in place. Soft-rubber bushings, expanded by steel rings and hard-rubber gland caps, insulate the pole pieces where they project through the cover of the cell and provide an air- and liquid-tight packing. The gland caps are colored red for the positive and black for the negative.

Projecting from the cell cover is the filler body, on which is mounted a hinged filler cap. The filler cap is held positively in either a completely open or completely closed position by means of a flat steel spring. Suspended from the underside of the filler cap is a hard-rubber hemispherical valve, which seats by gravity when the cap is closed. This valve, although permitting gas to escape, excludes the external air and reduces evaporation.

ELECTROCHEMICAL FORMATION. The electrolyte is then added and brought to a level of 1/2 in. to 3 in. above plate tops, depending upon the type of cell. The electrolyte consists of a solution of potassium hydroxide in water. A small quantity of lithium hydroxide is added to increase the capacity and life of the cell. The electrolyte has a normal specific gravity of 1.200 to 1.210 at 60 deg fahr. Throughout the useful life of the cell, as a result of spillage, loss through gassing, and the introduction of impurities through flushing (i.e., addition of water to make up for losses due to evaporation and electrolysis), the electrolyte gradually weakens and may need renewal one or more times. When the specific gravity reaches a low limit of 1.160 at 60 deg fahr, the electrolyte should be renewed. Standard renewal solution obtained from the manufacturer must be used.

When the assembly is completed and the cell filled with electrolyte, it is given several cycles of charge and discharge. This is known as the formation treatment, and its purpose is to stabilize the electrochemical characteristics of the cell. It is continued until tests show normal operation and an excess over rated capacity. All cells are coated with a heavy insulating paint (Esbalite), after which a protective film of rosin petroleum jelly (Esbaline) is applied to the cell tops.

ANNEALING PROCESS. Before assembly all nickel-plated steel parts are subjected to high temperature in an atmosphere of hydrogen. During this operation the nickel plating is actually bonded to the steel, becoming an integral part of its surface.

BATTERY ASSEMBLY. Nickel-iron-alkaline cells are assembled into batteries in painted hardwood trays to facilitate handling (see Fig. 2). Steel bosses, previously spot-welded to the sides of the cell containers, fit into hard-rubber buttons recessed in the tray

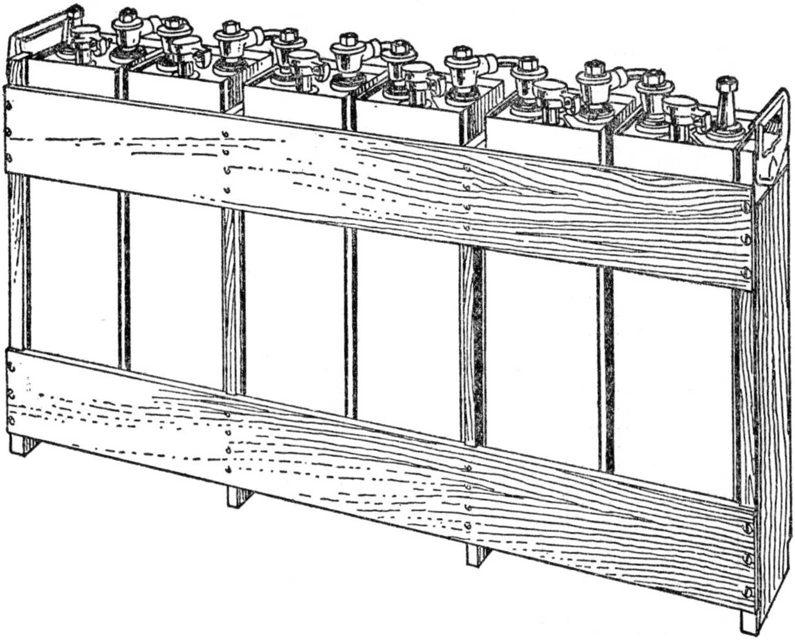

Fig. 2. Battery Assembly

slats. This type of assembly supports each cell in its proper place, insulates it from the tray and from adjacent cells, and provides proper ventilation. Tray assemblies are available with from 1 to 20 cells per tray, depending upon the size of the cells and the requirements of the installation. The usual application uses trays having somewhere between 3 and 8 cells.

The tops of the pole pieces are threaded. Immediately below they are tapered to fit the lugs of the intercell and intertray connectors. The connector lugs are steel forgings bored to fit the taper on the pole pieces and are swedged upon heavy copper connecting links. Connectors are secured by tightening nuts and are removed by means of a small jack after the nut is loosened. All lugs, links, and nuts are nickel-plated.

TYPES AND SIZES. The type of cell is determined by the size of the positive tubes and by the number of tubes per plate. The type is indicated by a letter, present manufacture including A, B, C, D, F, and N types. A, B, C, D, and F positive plates have tubes 1/4 in. in diameter and a little over 4 in. long. Tubes of this size have a normal discharge rate of 1/4 amp per tube for 5 hr. The A plate contains 2 rows of 15 tubes each, the B plate 1 row of 15 tubes, the C plate 3 rows of 15 tubes each, the D plate 4 rows of 15 tubes each, and the F plate 1 row of 4 tubes. The N plate contains 1 row of 7 tubes 3 in. long and 1/4 in. in diameter.

The sizes of the various types of cells are determined by the number of positive plates, which is indicated by a number following the type letter. For example, a cell containing 4 A-type positive plates is designated A4; a cell containing 6 B-type positive plates is designated B6. Type and size designation is stamped on the cover of each cell. In duty such as railway-car lighting and air conditioning, where regular flushing is not always practicable, containers of extra height or both extra height and width are often used to increase the relative quantity of electrolyte. Such cells are designated as high and high-wide types, the letters H and HW being added to the designation stamped on the cell covers, as A4H, A4HW.

Table 1. Ratings and Weights

Cell Type	Normal Ratings *			Weights †	
	Amp-hr	Amp	Watt-hr	Standard Type	High Type
N2	11.25	2.25	13.5	1.94	(Cell only)
B1, B1H	18.75	3.75	22.5	5.3	6.7
B2, B2H	37.50	7.5	45	6.0	7.2
B4, B4H	75.0	15	90	9.5	10.9
B6, B6H	112.5	22.5	135	13.0	15.2
B12H	225	45	270	....	29.0
A3, A3H	112.5	22.5	135	13.5	16.0
A4, A4H	150	30	180	16.5	19.3
A5, A5H	187.5	37.5	225	19.6	22.5
A6, A6H	225	45	270	22.4	25.5
A7, A7H	262.5	52.5	315	25.8	28.6
A8, A8H	300	60	360	31.2	35.9
A10, A10H	375	75	450	38.1	43.8
A12, A12H	450	90	540	47.3	53.5
A14, A14H	525	105	630	56.6	60.8
A16, A16H	600	120	720	62.9	68.0
A20H	750	150	900	....	86.5
A24H	900	180	1080	....	104.5
C4	225	45	270	24.3	....
C5	281.25	56.25	337	29.3	....
C6	337.5	67.5	405	34.4	....
C7	393.75	78.75	473	39.8	....
C8	450	90	540	45.5	....
C10	562.5	112.5	675	61.3	....
C12	675	135	810	71.0	....
D4	300	60	360	33.4	....
D5	375	75	450	40.0	....
D6	450	90	540	45.6	....
D7	525	105	630	53.4	....
D8	600	120	720	60.0	....
D10	750	150	900	77.3	....
D12	900	180	1080	91.6	....

* Ratings based on 5-hr discharge for A, B, C, D, and N types; average voltage 1.20.
† Pounds per cell, completely assembled, including trays, connectors, electrolyte, etc.

27. CHARGING CHARACTERISTICS

When only alternating current is available, a suitable motor-generator set or other current-rectifying device must be used to provide direct current. Specific-gravity readings are no indication of the state of charge and are therefore not necessary except to indicate when solution should be renewed. Although Edison batteries may be successfully charged at various rates, it is first necessary to establish a standard or normal rate for reference purposes. The normal rate of charge has been so chosen by the manufacturer that for an output of approximately the rated number of ampere-hours, when the battery is discharging at normal rate, the efficiency is higher than for a charge rate which is considerably higher or lower than the normal. The normal charge rate for the different types and sizes of Edison batteries is the same as the amperes shown in Table 1.

CONSTANT-CURRENT CHARGING. In constant-current charging, the charge rate, which should approximate the normal rate, is held fairly constant throughout the charge. The battery voltage rises from approximately 1.50 to 1.85 volts per cell during the charge, and adjustments are needed from time to time to maintain the current rate. A voltage of at least 1.85 volts per cell must be available at battery terminals in order to complete the charge at normal rate. Voltage regulation can be obtained by use of a variable resistance in series with the battery or by variation of the field rheostat of the charging generator if an individual motor-generator set is used. The length of time required to charge depends upon the extent to which the battery was previously discharged, the battery being considered fully charged when the terminal voltage ceases to rise for a period of 30 min during charge with constant current flowing.

MODIFIED CONSTANT POTENTIAL CHARGING. Modified constant potential is taper-current charging, the current during the first part of charge being higher than the normal rate; the rate then decreases until at the end of charge it is less than normal. With this method a fixed and permanent resistance is inserted in series with the battery, and no adjustments are made during charge. A charging line voltage of at least 1.84 volts per cell must be available and maintained throughout the charge, but higher voltages may be used. The higher the voltage, the more closely will the taper of the charging rate approach the straight constant-current method as a limit. When using the proper fixed resistance, the following charge rates will be obtained for various charging voltages:

Line volts per cell..........	1.84	1.90	1.95	2.00	2.10	2.20
Initial rate (% normal).....	165	155	147	140	128	124
Final rate (% normal)......	65	70	74	78	84	86

In each case the result will be average normal rate for a complete charge. The length of charge required will depend upon the state of charge of the battery when it was placed on charge. The object is to put back into the battery 25 per cent more ampere-hours than have been taken out. The state of charge can be determined by the use of an ampere-hour meter or by a special device made by the battery manufacturer and known as an Edison charge-test fork.

The following formula shows how to determine the proper fixed series resistance mentioned:

$$\frac{V - (1.69 \times N)}{A} = R$$

where V is the voltage of the charging line, N is the number of cells in series in the battery, A is the normal ampere rate of the battery, and R is the ohms of fixed resistance required. EXAMPLE: with a 45-volt charging line, what resistance is required to charge a 24-cell A8 Edison battery? ANSWER: 45 volts − (1.69 × 24 cells)/60 amp = 0.074 ohm. It is to be noted that the series resistance thus calculated is the total resistance required between the line and the battery; the charging leads, switches, etc., form part of the resistance, the balance to be added. The current-carrying capacity of the resistance should be sufficient to take care of the initial current rate which will flow into the battery at the beginning of charge.

BOOST CHARGING. Although not recommended as a regular practice, occasional emergencies will require that the batteries be charged faster than ordinarily, thus necessitating higher rates. Permissible ampere rates are: 1 hr at two times normal rate, or $^1/_2$ hr at three times normal, or $^1/_4$ hr at four times normal, or 5 min at five times normal. Frothing at the filler cap or an electrolyte temperature in excess of 115 deg fahr is an indication that boosting has been carried too far and should be discontinued at once.

LOW-RATE CHARGING. Where discharge requirements necessitate a low constant rate, or where intermittent or variable rates are such as to average a low value over a given period of time, then charge rates may be less than normal. It must be remembered, however, that 25 per cent more ampere-hours than have been used on discharge must still be put back into the battery. The length of time for a complete charge will thus be longer because the current is put in more slowly.

TRICKLE CHARGING. Edison batteries may be trickle-charged when used in applications where trickle charging is ordinarily employed. The voltage required will normally be between 1.50 and 1.55 per cell, depending upon the temperature, age of the battery, the rate used, and other factors. The proper trickling rate for any given installation can be approximated by the following formula:

$$\frac{(C \times 0.16) + (D \times 1.10)}{24 - H} = A$$

where C is the rated ampere-hour capacity of the battery, D is the ampere-hours used per day, H is the total hours of discharge per day, and A is the trickle-charge rate in amperes.

RETENTION OF CHARGE. The curve of Fig. 3 shows the loss of charge when standing idle after a complete charge.

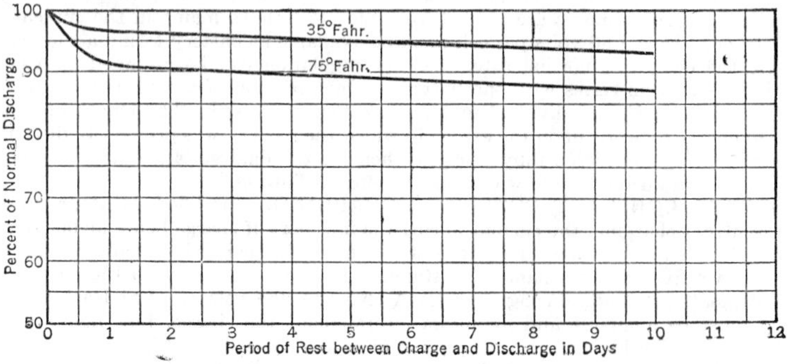

FIG. 3. Loss of Charge When Idle

28. DISCHARGE CHARACTERISTICS

At the normal rating, Edison cells have an average voltage of 1.20 volts. Like that of all other storage batteries, however, its voltage will be less when discharged at high ampere rates; when discharged at low rates the voltage characteristics are better than normal. The curves of Fig. 4 show the discharge-voltage characteristics of an Edison cell

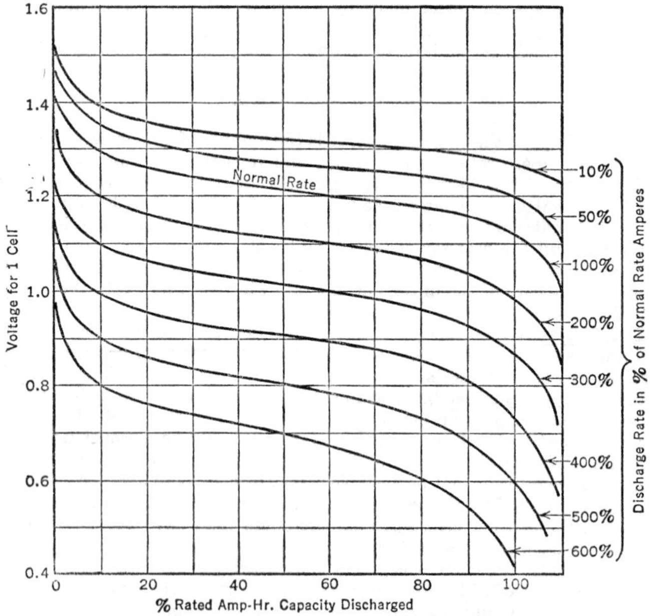

FIG. 4. Edison Alkaline Storage-battery Discharge Voltage at Battery Terminals. Allowance has been made for losses in connectors and jumpers.

in good condition when discharged continuously at various rates. For instance, suppose that it is desired to know what voltage to expect from a 30-cell A8 Edison battery after discharging for 1 1/2 hr at 120 amp. The ampere-hours taken out would be (120 × 1 1/2)

= 180; and, since the rated capacity (see Table 1) of an A8-type cell is 300, the cell would be in a condition of 60 per cent discharged. The 120-amp discharge rate is twice the normal (60 amp) rating of an A8. By referring to the 200 per cent normal-discharge curve, at the point of being 60 per cent discharged the voltage is found to be 1.10 per cell; therefore a 30-cell A8 Edison battery in good condition would deliver 33 volts at the end of $1\,^1/_2$ hr of discharge at a 120-amp rate.

CAPACITY. The capacity of a nickel-iron-alkaline cell depends upon the number and type of its positive plates. For ready reference the ampere-hour capacities are shown in Table 1. Normal watthour capacity is the normal ampere-hour capacity multiplied by the average voltage. The rated capacity is based on a 5-hr discharge to 1 volt per cell for A, B, C, D, F, and N types. These values are on the basis of discharge at normal rate. It will be noted from the discharge curves that there is no appreciable difference in ampere-hour capacity at several times the normal discharge rate because the characteristics of the cell do not necessitate a limit to the final voltage. The cell may be discharged to zero voltage without injury.

During the first 125 duty cycles the capacity increases gradually above its initial capacity. This increase may be as great as 15 per cent above the rated capacity. After this point the capacity gradually decreases. When the specific gravity of the electrolyte drops to 1.160 at 60 deg fahr, the capacity is renewed, and immediately after, another increase in capacity occurs. This is usually the point of maximum capacity during the life of the cell. In most services the useful life is considered at an end when the ampere-hour capacity drops to less than 80 per cent of the rated capacity. The Edison alkaline cell maintains its rated capacity for the greater part of its normal life.

EFFICIENCY. Although under laboratory conditions ampere-hour efficiencies ranging up to 95 per cent can be obtained, the practical operating ampere-hour efficiency of an Edison alkaline battery is approximately 80 per cent, thus requiring an input of about 25 per cent more than has been taken out on discharge. The laboratory voltage efficiency ranges up to 80 per cent, but this has no meaning as far as practical operation is concerned because the voltage efficiency of a battery installation will depend upon the methods of charging used, the charging line losses, the efficiency of the charging-current source, and similar factors. For any given installation the efficiency of battery operation can be computed by determining the watthour input required at the service meter as against the watthours obtained from the battery on discharge. In those services where alkaline batteries are applied, however, the cost of charging batteries is usually a relatively small item in the total cost of battery operation, and a comparatively wide range in the efficiency of batteries makes but a slight difference in total operating cost.

TEMPERATURES. High temperatures are to be avoided, as all batteries are damaged by high electrolyte temperatures. It is recommended that Edison alkaline batteries be operated at temperatures under 115 deg fahr. To assist in holding down electrolyte temperatures of Edison batteries used in heavy-duty service, such as motive power, the battery compartments should have provisions for ventilation in warm weather. Evenly spaced slots or holes should be made at the top and in the bottom of the battery box; top ventilation is usually placed on the sides of the box as close to the top as possible. A certain range of ventilation, depending on the severity of the work and on the climate, has been found by experience to be satisfactory and may be determined by the following formula:

$$F \times N \times C = A$$

where F is a factor of 0.01 to 0.02, depending on severity of condition; N is the number of cells in the battery; C is the rated ampere-hour capacity of the battery; and A is the area in square inches of ventilation required. This amount of ventilation should be applied to the top, and a like amount to the bottom, of the battery compartment. Adequate ventilation will result in both better operating performance and longer battery life.

Low temperatures will not harm an Edison alkaline cell. At an electrolyte temperature of approximately −20 deg fahr the electrolyte will become slushily crystallized, but it will not freeze hard and cause injury. The capacity of the cell decreases as the electrolyte temperature drops. The extent of the decrease varies considerably with the rate of discharge, being relatively small at low rates and large at high rates. At normal rate a reduction of electrolyte temperature from 115 to 50 deg fahr will reduce the capacity 10 per cent. Below 50 deg fahr the reduction in capacity is important. The electrolyte temperature of a working cell is always higher than the atmospheric temperature because of the heat generated by the passage of current. The amount of heat varies with the rate of charge and discharge, being relatively small at low rates and larger at high rates. In practice, electrolyte temperatures in a cell discharging at normal rate do not generally fall below 60 deg fahr. Under cold-weather conditions battery compartments should be closed, and under extreme conditions it is sometimes good practice to insulate the compartment. Even though the air temperature is very low, it takes a long time for the electrolyte temperature to drop to that of the outside air if the battery compartment is closed, as the curves of Fig. 5 show.

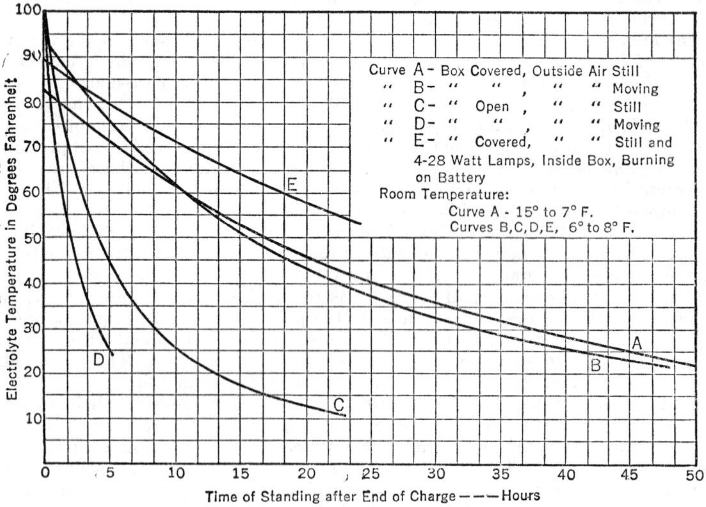

FIG. 5. Rate of Cooling of Electrolyte of Standard Battery of 40 Type A–6 Cells, Housed in a Regular Battery Box. Battery was charged at normal rate (45 amp) with the air temperature at about 65 deg fahr and then allowed to stand idle to cool under various conditions as noted above. Test made by Electrical Testing Laboratories.

29. LIFE

The life of the cell is from 7 to 20 years, depending upon the service. The years of life depend upon the average percentage of a complete cycle occurring between charging and the number occurring per year, as well as upon other factors, including the kind of duty in which the cell is operated and the care it receives.

ACCIDENTAL CONDITIONS. Accidental conditions arising from use or from the care given the battery are generally without injury to the nickel-iron-alkaline cell. Short circuiting, accidental charging in the reverse direction, accidental overcharging, or standing idle for an indefinite period in a discharged condition does not cause permanent injury. Vibration and jolting are not generally causes of injury. Permanent injury results from prolonged overheating above 115 deg fahr and prolonged operation with electrolyte level below plate tops.

MAINTENANCE. The principal points in caring for Edison alkaline batteries are as follows:

1. Maintain the solution at the proper level with distilled water in each cell, as recommended in the manufacturer's instruction booklet 850X (obtainable from Thomas A. Edison, Inc., West Orange, N. J.).
2. Avoid electrolyte temperatures exceeding 115 deg fahr.
3. Keep clean and dry the cells, trays, and the surface on which trays are placed.

30. BIBLIOGRAPHY

Arendt, Morton, *Storage Batteries*. D. Van Nostrand, New York (1928).
Baker & Company (Newark, N. J.). Publications on cadmium-nickel (Jungner) battery.
Crennell and Lea, *Alkaline Accumulators*. Longmans, Green, New York (1928).
Lyndon, Lamar, *Storage Battery Engineering*. McGraw-Hill, New York (1911).
Thomas A. Edison, Inc., West Orange, N. J., various publications.
Vinal, George, *Storage Batteries*. John Wiley, New York (1940).

SECTION 8

DIRECT-CURRENT MACHINES AND ROTARY ENERGY CONVERTERS

DIRECT-CURRENT GENERATORS

By J. F. Sellers

DIRECT-CURRENT MOTORS

Revision by J. F. Sellers of Article by W. I. Slichter

ROTARY ENERGY CONVERTERS

Revision by J. F. Sellers of Article by W. I. Slichter

DIRECT-CURRENT MACHINES AND ROTARY ENERGY CONVERTERS

DIRECT-CURRENT GENERATORS

By J. F. Sellers

1. BASIC THEORY

FLEMING'S RULE. A direct-current generator converts mechanical energy to electrical energy by the movement of a system of conductors relative to a magnetic field. A definite relation exists between the conductor movement, the direction of lines of magnetic flux, and the direction of induced voltage, illustrated by Fleming's **right**-hand rule for generator action. With the thumb, forefinger, and middle finger set at right angles to each other (Fig. 1) the thumb will point in the direction of conductor movement, the forefinger in the direction (north to south) of the lines of magnetic flux, and the middle finger in the direction of the induced voltage. Any generator will operate as a motor,

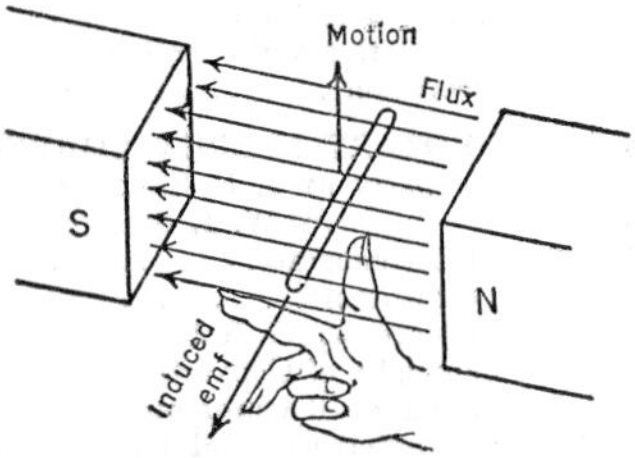

FIG. 1. Fleming's Right-hand Rule for Generators

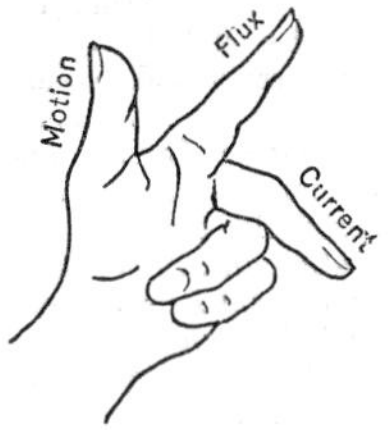

FIG. 2. Fleming's Left-hand Rule for Motors

and to study voltage and magnetic relations in a motor Fleming's **left**-hand rule may be used (Fig. 2). In this rule the thumb points in the direction of motion, the index finger in the direction of the magnetic lines, and the middle finger in the direction of current in the conductor, or the applied voltage. A conductor in a motor will have a voltage generated in it in accordance with the right-hand rule, so that the steady-state current in the conductor will be proportional to the difference between the applied voltage and the counter voltage, divided by the total armature-circuit resistance.

RELATION OF FLUX AND CURRENT IN A CONDUCTOR OR COIL. Two further relationships exist between current in a conductor and magnetic flux, whether in an isolated coil or in a rotating machine. The first relation is that of direction of flux surrounding a current-carrying conductor. As shown in Fig. 3, with the right hand grasping a

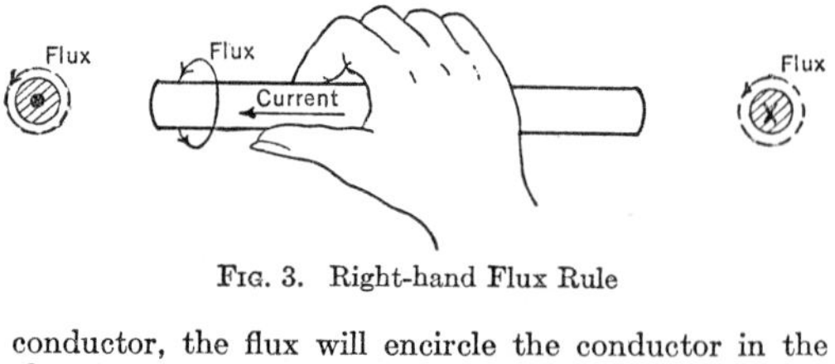

FIG. 3. Right-hand Flux Rule

current-carrying conductor, the flux will encircle the conductor in the direction of the fingers when the thumb points in the direction of the current flow. The conductor end views further illustrate this principle, that with the current flowing toward the observer (illustrated conventionally by the dot in the center of the conductor) the flux will flow in a counterclockwise direction. The second relation between flux and current in a coil of wire is shown in Fig. 4. With the right hand grasping the coil and the fingers pointing in the direction of current flow, the thumb will point in the direction of the flux inside the coil.

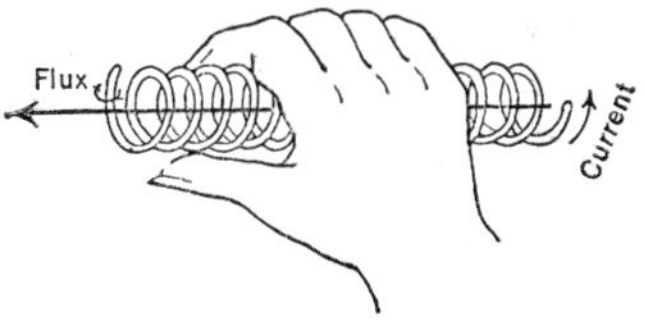

FIG. 4. Right-hand Magnet Rule

INDUCED VOLTAGE. With these relations fixed in mind, the principle of the direct-current generator can be established. When the amount of magnetic flux linking a coil undergoes a change, an electromotive force having an average value of $(n \times \phi)/(t \times 10^8)$ volts, where n is the number of turns in the coil and ϕ/t is the number of flux lines changed in 1 sec. The instantaneous voltage in the coil is $n \times (d\phi/dt) \times 10^{-8}$, or is 10^{-8} times the time rate of change of flux linkages through the coil. Expressed in another way, the instantaneous voltage in a coil moving in a magnetic field is $e = Blv \times 10^{-8}$, where B is flux density in the gap of the magnetic field in lines per square inch, l is total length in inches of active conductor in the field, and v = conductor velocity in inches per second at right angles to the flux. Figure 5(*a*) shows a single-turn coil rotating counterclockwise in a magnetic field of uniform density. If just one conductor *c* of this loop is considered, the voltage generated in it in the various positions shown from 1 to 8 while traveling at uniform velocity will be as shown in Fig. 5(*b*). At position 1 the number of lines through the coil is maximum, but the rate of change is zero, so the voltage generated is zero. If it is considered from the *Blv* formula, B is zero, so e is zero. At position 3, B and v are maximum, so the generated emf is maximum. At position 5 the voltage is again zero, and to the right of 5 it occurs in the opposite direction, since the motion of the conductor *c* is reversed with respect to the lines of magnetic force.

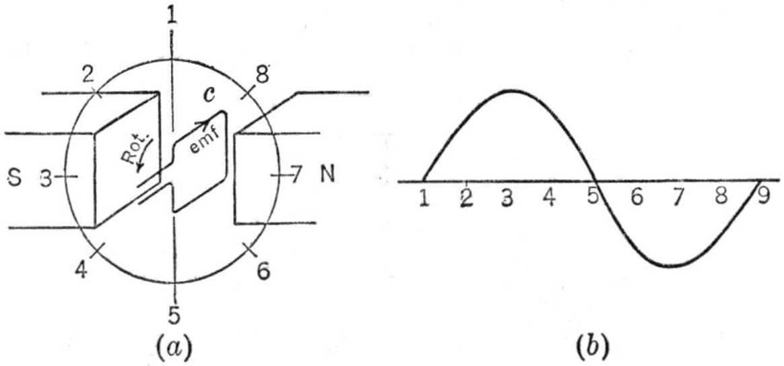

FIG. 5. Wave-form of Voltage Generated in Single Conductor

UNIDIRECTIONAL TERMINAL VOLTAGE. Figure 6 illustrates how the two ends of the coil *c* shown in Fig. 5(*a*) can be connected to two collector rings, on which metal or carbon-brush contacts, connected to an external circuit, bear. A voltage of alternating direction, as shown in Fig. 5(*b*), will then be impressed on the external circuit. To obtain a voltage on the external circuit which does not alternate but remains in one direction, it is necessary to reverse the connection of the armature coil to the external circuit twice each revolution. This is accomplished as shown in Fig. 7(*a*), in which each of the two coil ends are connected to half a collector ring, the two halves being insulated from each other. The resulting terminal voltage across the external circuit will be as shown in Fig. 7(*b*), although the coil voltage is still alternating in direction. The current in the coil reverses in the zone midway between the pole tips, and in this zone, called the neutral or commutating zone, the brushes make contact with the coil leads. To make a practical machine,

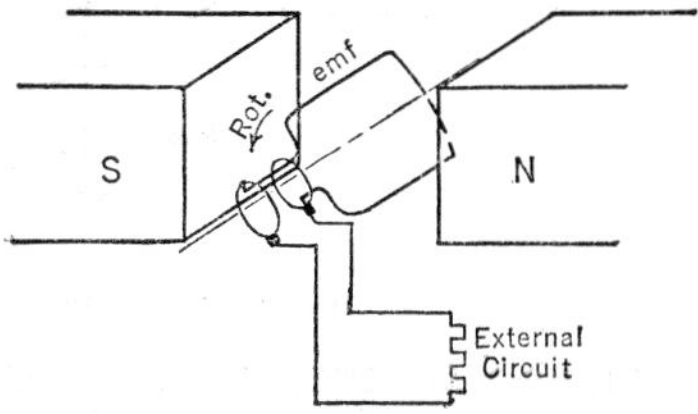

FIG. 6. Alternating Emf Applied to External Circuit

more than one armature coil is used (all being connected together in a closed circuit), and means for effecting commutation of the coil current without sparking at the brushes is provided.

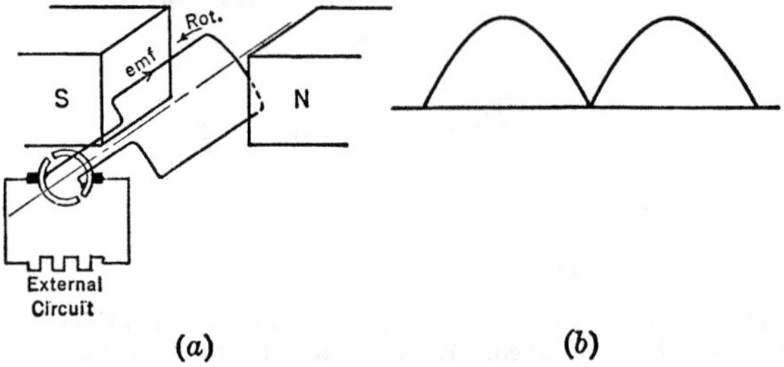

Fig. 7. Unidirectional Emf Applied to External Circuit

2. DEFINITIONS

SELF-EXCITED SHUNT MACHINE. A shunt machine, either generator or motor, is one in which the entire field excitation is derived from a circuit of many turns and high resistance connected in **shunt** or multiple with the armature circuit. The characteristic of a shunt machine is poor regulation; i.e., the voltage of a shunt generator decreases as the load increases. This is so marked that some shunt generators may be short-circuited, their terminal voltage dropping to zero, without resultant harm (Fig. 8).

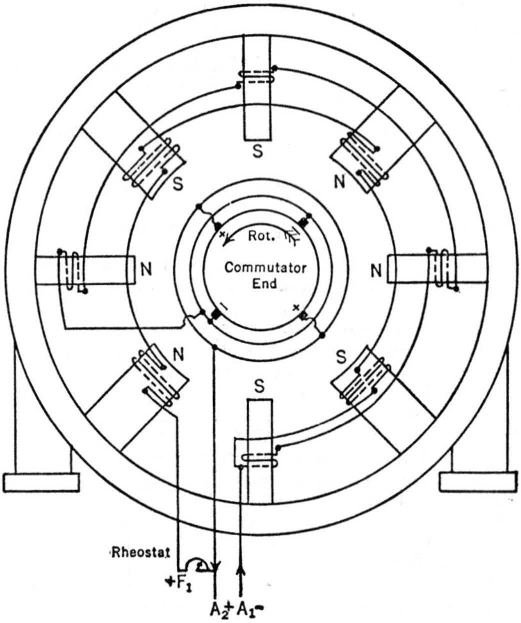

Fig. 8. Self-excited, Shunt-wound Interpole Generator

SERIES MACHINE. A series machine, generator or motor, is one in which the entire armature current flows through the field winding. Series generators are usually built to supply a constant current to an external circuit irrespective of the effective resistance of the circuit.

SEPARATELY EXCITED MACHINE. This type of machine is sometimes used in large stations where the exciting current is readily obtained from separately excited bus-bars and special characteristics are required.

COMPOUND-WOUND MACHINE. This type of machine has on each field pole, in addition to its shunt winding, a few turns of thick wire which carry the load current and are known as the series winding. This winding causes the excitation to increase as the load increases and tends to keep the terminal voltage of a generator constant or even to increase it. If the field windings are proportioned to cause the voltage at full load to be higher than the voltage at no load, the machine is said to be **over-compound.** A **flat-compound** machine has the same voltage at full load and at no load; an **under-compound** machine has a lower voltage at full load than at no load. An equalizer lead is always provided between series fields and armatures to obtain good parallel operation with other compound generators. A compound motor has poor speed characteristics (Fig. 9).

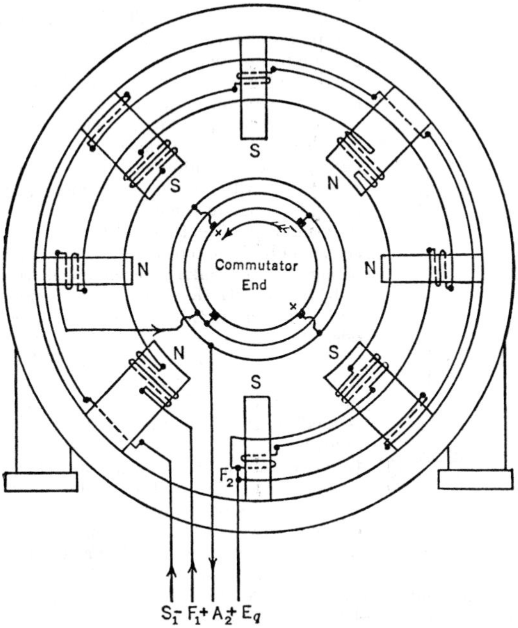

FIG. 9. Self-excited, Compound-wound Interpole Generator (Short Shunt)

SHORT- AND LONG-SHUNT CONNECTIONS. A compound-wound machine may be connected short shunt, as in Fig. 9, or long shunt, so that F_2 is connected to S_1 instead of the equalizer. The choice is merely a matter of convenience of station wiring.

COMMUTATING POLE OR INTERPOLE MACHINES. These machines have small auxiliary poles alternately placed with respect to the main poles and excited by a few turns in series with the load. The effect of these poles is to improve the operation of the machine in the matter of commutation, and they are used with all types of main-pole windings (Figs. 8 and 9).

3. ARMATURE WINDINGS, INSULATION, AND COMMUTATORS

RING WINDING. The first d-c machines were made with two main poles and a laminated smooth armature core around which was wound the armature winding, as shown in Fig. 10. To the inside of each loop was placed a tap which was connected to one segment of a switching device called the commutator. Two brushes were placed approximately midway between the main-pole tips. This type of winding, called a Gramme ring winding, is no longer used because it is uneconomical, only about 1/3 of the armature winding being in the voltage-generating zone.

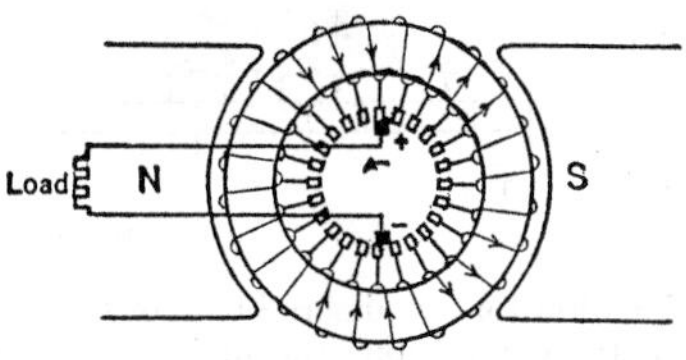

FIG. 10. Gramme Ring Armature Winding

FORMED COILS. This winding was superseded by the form-wound type of armature coil, in which the coils are made up of a suitable number of turns and formed into a two-layer diamond shape [Fig. 11(*a*)] before being wound in a slotted core armature. One side of the coil must be under one polarity of main pole and the other side under the opposite polarity, so that the two voltages will be additive. For this reason the two coil sides are always very nearly a pole pitch apart. The greatest deviation from full pitch (one-pole pitch) should be no more than one slot, long or short, because of the detrimental effect from main-pole flux that occurs during commutation with more than this amount of short pitch.

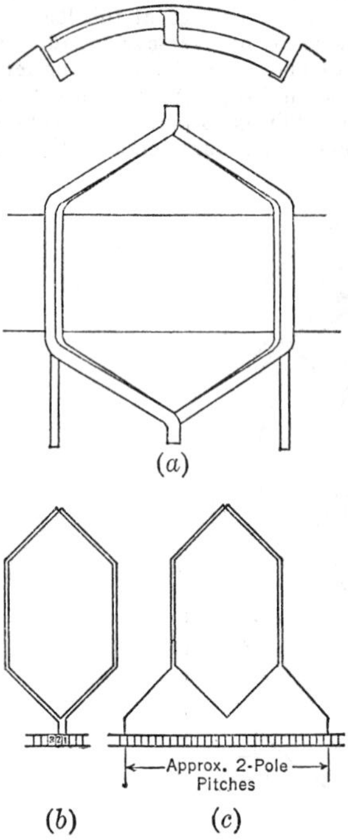

FIG. 11. Armature Coil Types

COILS PER SLOT. The formed-coil bundle may have one or more turns per coil, and there may be from 1 to 5 commutator segments for each armature slot. The main purpose of a multiple number of commutator segments per slot is to reduce the number of turns between segments and thereby keep below a certain safe limit the voltage that occurs between adjacent segments. Another reason is to reduce the inductance voltage in the armature coil connected to two adjacent segments, which opposes the current change in the coil during commutation. A multiple number of coils per slot is also required in some designs to obtain a tooth at the root mechanically strong enough and to make an economical winding.

LAP WINDING. The coil leads may be connected to each other and to the commutator by two different methods. (For two-pole machines the windings become the same.) In multipolar machines, if the coil ends are connected to adjacent commutator bars 1 to 2 [Fig. 11(*b*)], a multiple or lap winding is obtained. This type of winding is used on all large or low-voltage machines and usually gives better commutation than the next winding (wave) to be described. Simplex windings have as many circuits as there are poles on the machine and require as many brush arms as poles. In order to secure good division of current between the parallel circuits of this winding, it is necessary to connect together a number of points in the winding which should be at the same electrical potential. It is usually sufficient to have one such cross-connector or equalizer per slot with a cross-section approximately one-half that of the armature conductor. Figure 12 shows the development of a simplex lap winding on a four-pole machine, with equalizers at the commutator end of the machine. From a functional viewpoint this is the most desirable

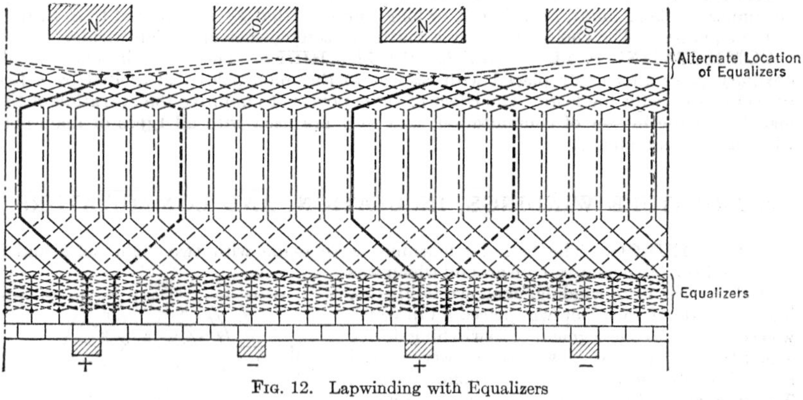

FIG. 12. Lapwinding with Equalizers

location for equalizers but has the disadvantage that in case of short circuits or grounds in the connections the main winding must be removed or disconnected to obtain access to the equalizers. An alternative location for the equalizers at the coil U-bends is shown at the coil ends opposite the commutator. This arrangement places the connections in a location easily accessible for inspection and repair.

The heavy lines of Fig. 12 indicate the fundamental principle of equalizers. Inequalities in current division between the parallel armature circuits are caused usually by variation in the amount of flux per pole. The equalizers thus connect together in parallel two or more coils with highly inductive characteristics as compared to their resistance, and the difference in alternating emf's between the coils will set up a magnetizing current around the flux from the weaker poles and a demagnetizing current around the stronger poles, producing a practical equality of flux per pole. One pole may even be left unexcited, and the distribution of flux will still be about 95 per cent symmetrical if at least one equalizer for each slot is used.

The foregoing description of equalizer action demonstrates the fact that in a lap winding every pair of poles must be symmetrical in regard to the number of slots and the commutator segments, so that the requirement for a lap winding is:

$$\frac{Z}{2 \times t \times c} \times \frac{2}{p} = \text{a whole number which is the number of armature slots per pair of poles}$$

where Z = total number of armature conductors, t = turns per coil between adjacent commutator segments, p = number of poles, and c = number of commutator segments per armature slot.

If the number of slots per pair of poles is odd, the armature coil will be corded 1/2 slot pitch; i.e., will be short of full pitch by that amount. If the number is even, the coil may be either full pitch or one slot long or short.

MULTIPLEX LAP WINDINGS. These are sometimes used in heavy-current, low-voltage machines or may be utilized to change the terminal voltage of an existing machine to one-half and the current rating to twice that of the original design. In this winding the coil leads are connected 1 to 3, and the winding has twice as many circuits as poles. This winding is not recommended for general practice, as it is difficult to keep in good commutating condition and it is necessary to keep the equalizers correctly located.

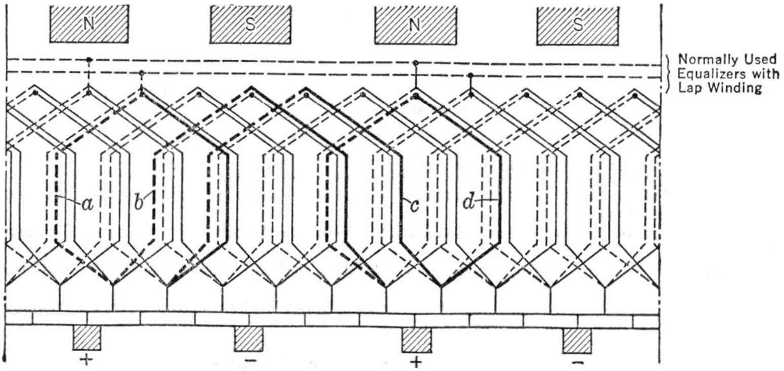

FIG. 13. Frog-leg Armature Winding

WAVE WINDINGS. If the two coil leads shown in Fig. 11(*a*) are shaped as in Fig. 11(*c*), it is possible to choose a number of slots and commutator segments so that the winding passes successively once under every north and south pole before returning to a commutator segment adjacent to the start. The winding then continues in this sequence until all the conductors are included before returning to the starting point. This winding is called a **wave** or **series** winding and has, for a simplex winding, but two parallel circuits between brushes of opposite polarity. The simplex-wave winding is used on all small multipolar machines and for higher-voltage machines when the maximum voltage between commutator segments does not exceed 35 at normal load. Wave windings require that the number of commutator segments $\pm 1 = k \times (p/2)$ where k is any whole number, and p is the number of poles. When $+1$ is used, the winding is termed progressive; when -1 is used, the winding is retrogressive. If a suitable design cannot be obtained otherwise, it is permissible to use a dummy coil in the winding to obtain a combination that satisfies the above relation. This dummy coil is a standard coil left in to balance the armature mechanically, but it is not connected.

Experience has shown that dead coils should not be used in motors of 100 hp and larger that are subject to repeated heavy overloads or must operate over more than 2-to-1 speed range by shunt field control. The commutator pitch of a simplex-wave winding is k of the foregoing equation.

Multiplex-wave windings are possible with 4, 6, or more circuits, regardless of the number of poles. They are called Arnold windings and allow, for instance, a 4-circuit armature winding on a 6-pole machine, but they are not practicable commercially, as their operation is very critical.

FROG-LEG WINDING. A combination of a simplex-lap winding and a multiplex-wave winding, each having the same number of circuits as main poles, is being used and has the name **frog-leg** armature winding. This winding requires no equalizers, yet all commutator bars of equal potential are connected together. The wave winding occupies the first and fourth layers in the slot, and the lap winding the second and third layers. Figure 13 shows the equalizing circuit of the winding in heavy lines. Dotted lines indicate equalizer connections if only a lap winding were used. This winding may be substituted for any cross-connected lap winding and gives 100 per cent cross-connection. The active equalizing coils for the section of winding shown are *a*, *b*, *c*, and *d*.

ARMATURE-COIL INSULATION. The insulation required on armature coils is quite thin, since more than 2500 volts dielectric test is seldom required, and most insulation can be used up to 50 volts per mil thickness with a good safety factor. Breakdown occurs

Table 1. Class A Insulation Allowance

	Class A Insulation to 300 Volts	
	Width, inches	Depth, inches
Coils per slot	1	1
1. Half-lapped 0.003-in. glass tape	0.014	0.014
2. Half-lapped 0.005-in. cotton tape	0.024	0.024
3. Butt-joint 0.005-in. cotton tape	0.012	0.012
4. Varnish	0.010	0.010
	0.060	0.060 ×2
		0.120
5. 0.010-in. cell lapped under stick	0.024	0.036
Clearance	0.020	0.020
Total insulation allowance	0.104	0.176

NOTE 1. Add above width and depth insulation allowances to copper-size and slot-stick dimensions to obtain slot dimensions.

NOTE 2. *a*. For 300- to 700-volt insulation add 0.024 to the width and 0.048 to the depth, for 2 layers of item 2.

b. For 2, 3, and 4 conductors per bundle add 0.014, 0.028, and 0.042 respectively to the width only, for item 1.

c. For coils of 2 or more turns omit item 1 and substitute the insulation covering on the conductor, taking into account the number of turns per coil.

Table 2. Class B Insulation Allowance

	Class B Insulation to 300 Volts	
	Width, inches	Depth, inches
Coils per slot	1	1
1. Half-lapped 0.003-in. glass tape	0.014	0.014
2. Half-lapped 0.005-in. mica tape	0.024	0.024
3. Butt-joint 0.010-in. glass tape	0.020	0.020
4. Varnish	0.010	0.010
	0.068	0.068 ×2
		0.136
Clearance	0.020	0.020
Total insulation allowance	0.088	0.156

NOTE 1. Add above width and depth insulation allowances to copper-size and slot-stick dimensions to obtain slot dimensions. NOTE 2. Same as NOTE 2, Table 1.

at 250 to 500 volts per mil thickness for new, unwrinkled tape. The applied insulation will be either Class A, for a 40-deg cent temperature rise, or Class B, for a 60- or 70-deg cent temperature rise. Class A insulation consists of cotton tape, fishpaper, or other similar organic materials, and Class B insulation of mica, woven glass, or asbestos material. The standard puncture test voltage for d-c machines is twice the rated voltage plus 1000.

INSULATION SPACE. A coil with two or more turns between commutator bars is usually insulated with double cotton or glass covering. Coils with one turn between commutator bars are taped with one layer of 1/2-lapped 0.003-in. cotton or glass tape. All the separate coils of a bundle are then taped together, using one layer of 1/2-lapped 0.005-in. varnished cotton or mica tape for voltages up to 300 volts, and two layers for voltages of 300 to 700. A protective covering of butt-joint 0.005-in. cotton tape or 0.010-in. glass tape is applied around the bundle. Insulating varnish is applied after each layer of tape, and the complete coil is dipped in varnish and baked twice. A coil with mica taping must be hot-pressed to eliminate looseness. A slot cell of 0.005-in. varnished cloth and 0.005-in. fishpaper is usually used when winding Class A coils. Tables 1 and 2 give the required slot insulation and clearance required for Class A and Class B insulated coils.

ARMATURE LAMINATIONS. The number of armature slots and the slot proportions are determined by several factors. Economically, the fewer slots the better, but a minimum number of slots per pole for a practical electrical design is 6 for small machines and up to 12 for machines of 25-in. armature diameter and larger. The tooth is usually equal to the slot at the armature surface or as much as 50 per cent wider. Slot depth varies from 3/4 in. in small machines to 1 3/4 in. in machines of 25-in. armature diameter and larger. To avoid breakage in handling and stacking, a tooth should not be less than 0.20 in. wide at its minimum. Armature laminations are usually made of silicon-steel sheet from 0.014 to 0.0185 in. thick, having from 0.75 to 0.95 watts per pound loss at 10,000 gausses at 60-cycle frequency. Each lamination is varnished on both sides, and the core is divided into 2 1/2- to 3-in. sections by 3/8-in.-wide ventilating ducts, which are made by spot-welding small 3/8-in.-wide I-beam sections to a 0.025-in.-thick lamination. The armature core is usually from 1/8 to 1/2 in. longer than the main poles, to allow for variations in assembly caused by shop tolerances.

COMMUTATORS. Commutator diameter usually varies from 60 to 80 per cent of the armature diameter. The larger sizes are used where many segments are required to keep the maximum value of volts per coil below the critical value of 30 to 35. If this value is exceeded on machines which have the usual maintenance, a continuous ring of fire will form around the commutator and, if not interrupted, will become a heavy power arc between several brush arms, so that considerable metal will be melted. Commutators of smaller diameter are used when the peripheral speeds approach 5000 ft per min for continuous speed of rotation. It is usual, however, to keep the commutator to such a diameter that the brushes will not cover more than 15 per cent of the periphery. High-speed turbine-driven exciters, with shrink-rings to hold the segment assembly against centrifugal force, operate at a peripheral speed of 8500 ft per min, and commutators for traction motors may attain speeds of 10,000 ft per min.

The amount of current collected per brush arm should not exceed 1000 amp at full load, and the number of brushes should be chosen so that the current density at full load will be from 55 to 65 amp per sq in. of contact surface for voltages above 50.

Commutator risers extend from commutator segment to coil ends. They may be either an extension of the segment (solid type) or attached. The attached type should have fiber spacers every 6 to 9 in. of radial length to prevent fatigue failure caused by windage or other vibration. Attached risers are usually made of hard-drawn strip copper and are riveted and soldered to the commutator segment for machines of 40-deg cent rise and brazed on with silver solder of 1100- to 1300-deg fahr melting point for silicone-insulated machines or those of 60-deg cent rise.

Commutator segments are made of commercially pure hard-drawn copper with a hardness of Rockwell 70 to 77 (F scale). Copper for use in high-temperature or silicone-insulated machines is usually specified with 15 to 40 oz of silver per ton, which gives better physical characteristics for brazing on the riser.

Commutator segments range from a minimum of 1/8 in. thick in small sizes to 3/16 in. in 12-in.-diameter and larger sizes. They are insulated from each other by mica, ranging in thickness from 0.020 in. in small machines to 0.060 in. in large machines where the maximum volts between segments may exceed 35. Mica should always be cut down about 1/32 in. below the surface of the segments, since the electrographitic carbon material usually used for current collection has no abrasiveness to wear down the mica if it becomes flush. Molding mica 1/16 in. thick is used for ground insulation in the V.

Wherever possible, armature laminations are stacked on a sleeve, and the commutator is pressed on an extension of the sleeve. This results in an integral construction which permits pressing in a shaft, allowing the shaft to be replaced in case of breakage without disturbing the commutator or armature winding.

4. FIELD STRUCTURE AND BEARINGS

FRAME. The stationary part of a d-c machine consists of a magnet frame to which are bolted main poles and, when required, commutating poles. Commutating poles are used in all high-speed machines of 1-kw rating and larger and in low-speed machines where the reactance voltage between adjacent commutator segments exceeds 2 volts. It is possible to move the brushes in machines which do not have commutating poles, so that current reversal occurs in the armature coils in a field from the main pole, which neutralizes the reactance volts at a certain load. This field intensity, however, will not be right for other loads and usually causes severe sparking at no load. Motors with speed control by shunt field adjustment also require commutating poles. Exciting coils for furnishing the necessary magnetomotive forces are mounted on the poles.

The field frame is usually a single welded cylinder of 0.10 to 0.20 per cent carbon steel plate or a steel casting with similar permeability. To facilitate assembly it is usual to have a split yoke when the armature diameter is 40 in. or larger. The yoke may, however, be split in machines of as little as 100-kw rating when installed in restricted locations. Supporting feet are welded to the frame, and the feet doweled into position after the coupling is aligned.

MAIN POLES. Main poles are made of high-permeability sheet-steel laminations, usually $^1/_{16}$ in. thick, riveted together. Tap bolts hold the pole to the field frame. It is customary to use about $^1/_{32}$ in. of steel shims between pole and frame to permit an adjustment of air gap, so that there is no variation from the specified or average gap greater than ± 0.007 in. The variation in spacing between main-pole tips should not exceed $^1/_{16}$ in. The radial length of main poles is from $^1/_2$ to $^2/_3$ of the main-pole pitch.

COMMUTATING POLES. The number of commutating poles in small machines is usually one half the number of main poles. Commutating poles are usually made of $^1/_{16}$-in. sheet-steel laminations, although solid poles of 0.10 to 0.20 per cent carbon steel are used when rapid or wide load fluctuations do not occur. Air-gap and spacing tolerances are held the same as for main poles, although it is customary to use $^1/_8$ to $^1/_4$ in. of shims between pole and magnet frame to permit adjustment of air-gap density during test. The width of a commutating pole is such as to give about 40,000 to 45,000 lines per square inch maximum density at full load. If the density is higher than this figure, the air-gap flux for commutation at overloads will not be proportional to armature current, as saturation in the pole will disturb the required linear relation.

BEARINGS. Bearings are usually of the antifriction (AF) type for low- and moderate-speed machines having as much as 100 hp or kw. Above this rating many AF bearings are used also, but the load and speed requirements of the particular application determine the choice between AF or sleeve-type bearings. Sleeve bearings require shoulders on the shaft journal to keep the armature centered, and deflecting flanges to prevent oil escape. For generators of more than 25-kw rating, most sleeve bearings are the split babbitted type with soft-brass oil rings of approximately twice the shaft diameter dipping into a reservoir of oil. Machines with armatures having diameters up to 25 in. usually have the bearings integral with the end housings. With self-lubrication a journal speed of 2500 ft per min is seldom exceeded with sleeve bearings, nor a loading of 150 lb per sq in. of projected bearing area. With forced lubrication journal speeds may be 16,000 ft per min.

SHAFT CURRENTS. When the full-load output of a machine, whether motor or generator, exceeds 0.50 hp per revolution, one bearing of pedestal-bearing machines is insulated to prevent circulating currents. This requires all attachments, such as oil piping and tachometer lead conduits, to have an insulating section.

BRUSH RIGGING. The brush rigging is conveniently an integral assembly either mounted on the bearing housing or supported otherwise from the field frame. It must be possible to rotate all brush arms at once and to lock them in position so that no vibration occurs. Brush holders should be between $^3/_{32}$ to $^1/_8$ in. from the commutator to obtain maximum stability of brush operation. Springs should give 2 to 3 lb per sq in. of pressure on brush-contact surface. For good general performance an electrographitic carbon brush with 0.0015- to 0.0025-ohm specific resistance should be used on machines above 50 volts, a metal-graphite mixture being used at less than 50 volts. A contact current density of 55 to 65 amp per sq in. at full-load rating is used for electrographitic grades, and 65 to 90 amp per sq in. for metal-graphite. If too low a density is used, a non-uniform threaded and grooved surface develops on the commutator, and frequently a high friction coefficient occurs which sets the brushes into a noisy chatter. Brush arms should be spaced around the commutator with a maximum variation from the correct location of one-half of a mica thickness. It is assumed that the commutator likewise maintains this tolerance.

BRUSH POLARITY. When an armature has symmetrically formed coils, the relation shown in Table 3 exists between brush polarity and rotation. In addition, commutating-coil polarity must be arranged so that, in a generator, a conductor on the armature will

Table 3. Armature Polarity

Rotation	Type of Winding	+ Brush in line with Center Line of:
Clockwise	Lap	South main pole
Counter-clockwise	Lap	North main pole
Clockwise	Retrogressive wave	South main pole
Counter-clockwise	Retrogressive wave	North main pole
Clockwise	Progressive wave	North main pole
Counter-clockwise	Progressive wave	South main pole

pass under a commutating pole of the opposite polarity to that of the immediately preceding main pole. In a motor, the commutating pole is of the same polarity as the preceding main pole.

SHAFT DEMAGNETIZATION. When a machine has both a series main-pole field and a commutating field, the connections from pole to pole on each winding are arranged to neutralize each other around the shaft. If both windings are in two parallel circuits, each pair of circuits can be individually neutral with respect to shaft magnetization. No more than 500 net ampere-turns of shaft magnetization is allowed before compensation is made by bringing a return connection back around the shaft.

5. ARMATURE REACTION

When current flows in the main-pole coils but none flows in the armature, a flux is set up on a path directly across the armature from pole to pole, as in Fig. 14(*a*). When current

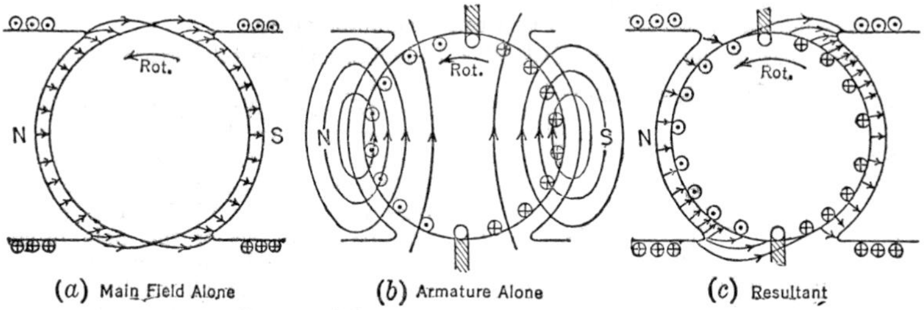

Fig. 14. Generator Air-gap Flux Distribution

flows in the armature but none flows in the field, a flux is set up across the axis of the poles, as in Fig. 14(*b*). When both armature and field have current, the resultant, Fig. 14(*c*), is a crowding of flux to the trailing-pole tip in a generator and the leading-pole tips in a motor. If the brushes are on the geometric axis between main poles, this is no longer a position of zero flux density. When no interpoles are used, as in fractional-horsepower motors and slow-speed machines up to 10 kw, the brushes are moved forward in the direction of rotation in a generator (against rotation in a motor) to the actual magnetic neutral, or slightly beyond to obtain an induced voltage in the coils under commutation that will neutralize the counter-voltage of self- and mutual induction.

DEMAGNETIZING EFFECT. Figure 15 illustrates how a band of armature conductors, 1 to 2 and 3 to 4, exerts a direct demagnetizing effect on the main field flux when the brushes are moved forward (in the direction of rotation in a generator) by the angle α electrical degrees from geometrical neutral. All the remaining conductors, 1 to 4 and 2 to 3,

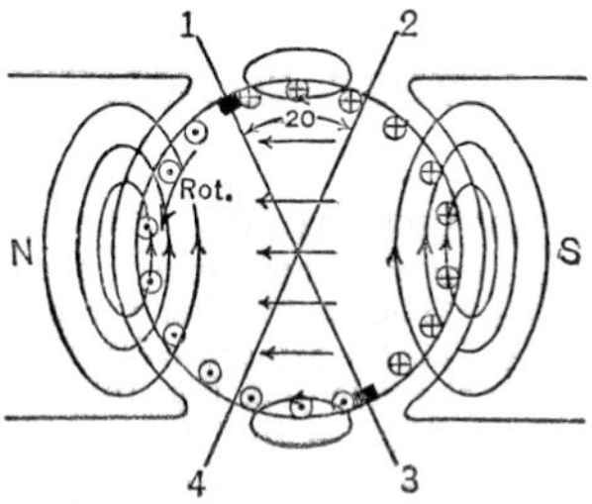

Fig. 15. Armature Demagnetizing with Brush Shift

act to produce a field at right angles to the main pole flux. The demagnetizing ampere-turns per pole will be:

$$\frac{2\alpha}{180} \times \frac{Zi_c}{2 \times p}$$

where Z = total number of armature conductors, i_c = current per conductor, and p = number of poles.

RESULTANT REACTION. When there are no interpoles and the normal ratio of 2 is used between armature ampere-turns per pole at rated load, and gap plus armature teeth ampere-turns, then the combined effects of cross-magnetizing and demagnetizing from the armature-winding ampere-turns result in a net combined demagnetizing effect of 30 to 35 per cent of the armature ampere-turns per pole.

When commutating poles are used and the brushes are on the electrical neutral, the cross-magnetizing effect results in a flux reduction equivalent to 10 to 15 per cent of the armature ampere turns per pole. A graphical method of actually calculating the resultant flux distribution is shown in most textbooks. Although commutating poles have no appreciable effect on armature reaction, they prevent sparking when the brushes are on neutral. The ratio of armature ampere-turns per pole at rated load, to gap plus teeth ampere-turns, should be from 1.1 to 1.3 to maintain good voltage regulation in an interpole generator and to give good stability in a motor. This ratio should be from 1.5 to 2 for non-interpole machines.

6. ELECTROMOTIVE FORCE INDUCED IN ARMATURE

In Article 1 the average voltage generated in a coil was found to be $(n \times \phi)/(t \times 10^8)$, where n is the number of turns in the coil, and ϕ the number of lines of flux change in the time t.

Let Z = total number of armature conductors.
p = number of field poles.
m = parallel paths between positive and negative brushes.
ϕ = useful air-gap flux per pole.
N = revolutions per minute.
n = number of turns per coil between commutator bars.

The volts per conductor $E_c = Np\phi/60 \times 10^8$ volts, and the total average voltage between positive and negative brushes is $E = (Z/m) \times (Np\phi/60 \times 10^8)$ volts. The average voltage between adjacent commutator bars is $(E \times 2 \times n \times p)/Z$, since $Z/(2 \times n \times p)$ = number of commutator bars in one pole pitch. The pole face will be about $^2/_3$ of the pole pitch, so that at no load the maximum value of volts between bars will be $^3/_2$ of the average volts per bar. This is evident in Fig. 16, where the no-load flux distribution in the air gap is shown as a, b, and c. The area under line d is the same as that under a, b, and c and corresponds to the average voltage per bar, which is $^2/_3$ of the maximum value b.

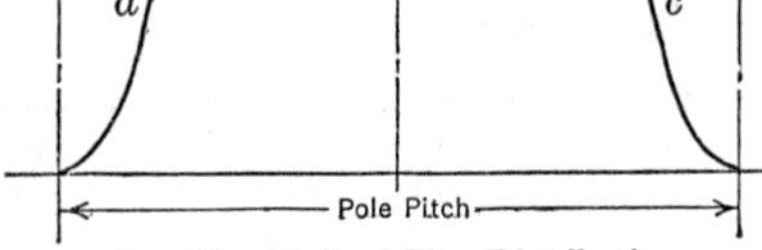

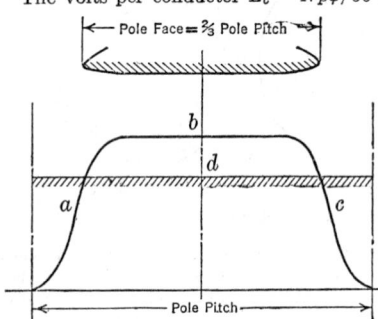

FIG. 16. No-load Flux Distribution

7. OUTPUT COEFFICIENT, ARMATURE HEATING, AND NUMBER OF POLES

OUTPUT FACTOR. The output of a d-c generator is ZE_cI_c, where E_c is the average volts per conductor, and I_c the current per conductor. It was found in Art. 6 that E_c is proportional to the flux per pole as one factor. The flux per pole is $^2/_3 \times (\pi DLB_g/p)$, where D is the armature diameter, L is the armature-core length, and B_g is the flux density under the pole (b in Fig. 16). Thus $E_c = Np/(60 \times 10^8) \times (DLB_g/p) \times (2\pi/3)$, and the output $ZE_cI_c = Z \times I_c \times N \times D \times L \times B_g/(60 \times 10^8) \times (2\pi/3)$. The quantity ZI_c is the total ampere-conductors on the armature surface; and if this is divided by the armature periphery, a quantity σ is obtained, which is the ampere-conductors per inch of periphery of the armature and is the armature loading. The armature heating and commutation are proportional to this factor. The loading factor is therefore $ZI_c/\pi D$. Substituting this value in the foregoing watts-output equation, we have:

$$ZE_cI_c = \sigma\pi D \times \frac{NDLB_g}{60 \times 10^8} \times \frac{2\pi}{3} = D^2L \times \sigma \times NB_g \times \frac{0.11}{10^8}$$

or

$$\frac{\text{Watts output}}{D^2LN} = \frac{0.11}{10^8} \times \sigma \times B_g$$

which is commonly termed the output coefficient of a machine. Expressed in another way, the watts output per cubic inch of armature-core volume per revolution of a machine,

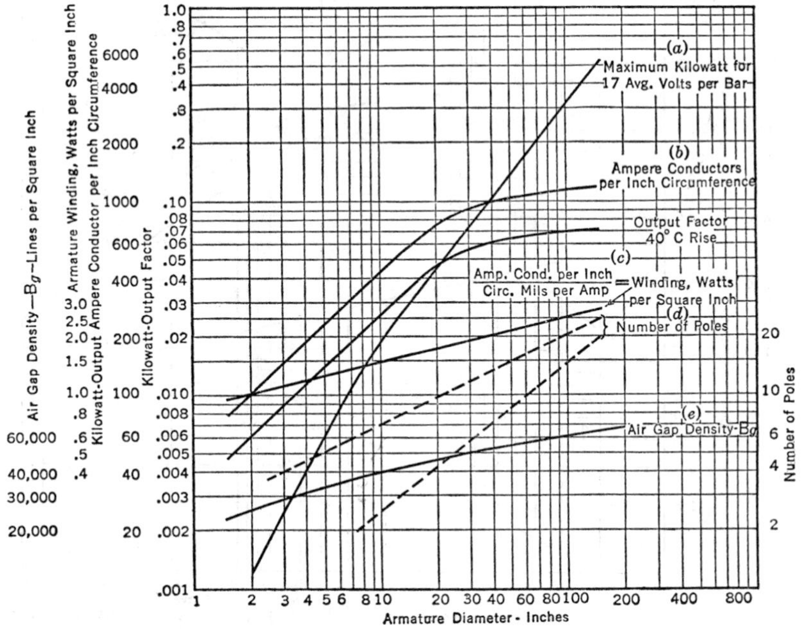

FIG. 17. Factors Determining Armature Output

varies only with σ and B_g. The armature loading, σ, the kilowatt-output factor, and the average densities B_g are shown in Fig. 17 for 40-deg cent machines. Machines with a 55-deg cent rise for 1 hr have an approximately 35 per cent greater σ factor than continuous 40-deg cent machines. Totally enclosed machines having armature diameters up to 21 in. without external ventilation have about 40 per cent of the σ factor for 40-deg cent open machines.

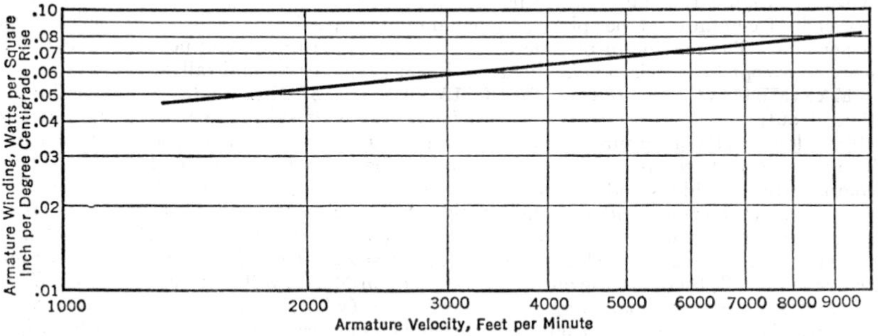

FIG. 18. Variation of Armature Heating with Armature Velocity

A factor of watts per square inch of armature surface caused by the heat loss in the armature coils is obtained by dividing the ampere-conductors per inch by the circular mils per ampere in the armature-conductor. The resultant heating with variable armature velocity is shown in Fig. 18. Curves *A* and *B* of Fig. 19 show variation in the weight of generators (including bearings, but no base) with speed, and curves *C*, *D*, *E*, and *F* of Fig. 19 show variation in generator output with speed.

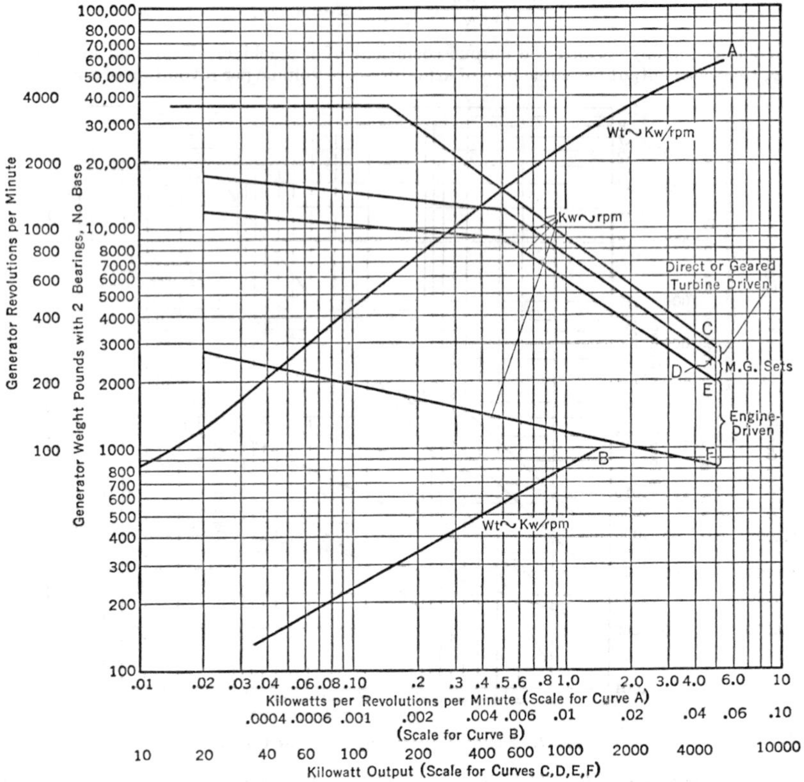

FIG. 19. Generator Speed and Velocity

NUMBER OF POLES. It is usual to keep the frequency of a d-c machine at less than 60 cps, the frequency being $pN/120$; therefore curve *d* of Fig. 17 shows the usual variation in number of poles with armature diameter. The smaller number of poles is used for high-voltage machines, and the larger number of poles for high-current machines. Not more than a 1000-amp output per brush arm should be used, nor more than 75 volts per inch of commutator periphery, the commutator speed usually being limited to 5000 ft per min. The maximum peripheral speed of the armature should not exceed 10,000 ft per min except for special applications, such as turbine drives and railway motors.

MAXIMUM VOLTS BETWEEN COMMUTATOR SEGMENTS. A further limitation of output from any armature diameter must be considered when it is required to obtain low-inertia armatures for applications to rapidly changing speed requirements. The average volts between adjacent commutator bars should be kept to 12 to 15 for non-compensated machines and not more than 19 for machines with interpoles and pole face windings. (Interpoles and pole face windings eliminate the effect of armature cross-magnetization.) Thus average volts per bar = $(E \times p/\text{Bars})$, or $E = (\text{Bars}/p) \times$ avg. volts per bar. The armature loading, $\sigma = Zi_c/\pi D = ZI/\pi Dm$, where I = the line current, or $I = \sigma Dm\pi/Z$; therefore the output $EI = (\text{Bars}/p) \times$ avg. volts per bar $\times\ (\sigma Dm\pi/Z) = (Z/2np) \times (\pi Dm/Z) \times \sigma \times$ avg. volts per bar $= (\pi/2) \times (Dm/np) \times \sigma \times$ avg. volts per bar. To obtain the lowest possible average volts per bar, a one-turn simplex lap winding is used, so that the m/p and n are 1. Then $EI = (\pi D/2) \times \sigma \times$ avg. volts per bar, showing that the maximum output of any armature diameter is independent of speed and depends entirely on the maximum loading factor and maximum average volts per bar (see curve *a*, Fig. 16). For 17 volts per bar and 900 ampere-conductors per inch loading, a 25-in.-diameter armature would have a maximum permissible output of $(\pi/2) \times 25 \times 900 \times 17 = 600$ kw. The armature length would be made as required for the air-gap flux density and speed.

8. MAGNETIC CIRCUIT AND NO-LOAD SATURATION CURVE

PATH. The flux flows from a north pole across the gap to the armature and divides so that one-half goes to each of the adjacent south poles through the armature core and then across the air gap, through a south pole, and from there through the yoke to the original north pole. Thus each pole piece carries two magnetic paths in parallel, and the armature core and field yoke carry one flux path.

LEAKAGE FACTOR. In addition there is, from each pole piece, a leakage of flux across the interpolar air space to the adjacent pole pieces. This flux, called the leakage flux, never gets into the armature nor cuts the armature conductors. The ratio (Useful flux + Leakage flux)/Useful flux is called the leakage coefficient. It may be determined by the method given under A-c Generators, Section 9, Article 8, or it may be assumed to be 1.20, which is a very close average value for all armature diameters up to 96 in.

DENSITIES. From experience it has been found advisable to operate the various parts of the magnetic circuit at certain conventional densities in order to keep the core loss and excitation within reasonable limits. Usual densities are given in Table 4. The upper limits apply to a large or a slow-speed machine.

Table 4. Conventional Densities

Section	Material	Density, lines per Square Inch
Magnet frame......	Low-carbon Steel	65,000 to 85,000
Main poles.........	1/16 laminations	85,000 to 95,000
Air gap............	Air	40,000 to 65,000
Armature teeth.....	0.014 laminations	90,000 to 150,000
Armature core......	0.014 laminations	60,000 to 90,000

CALCULATION. It is more convenient to analyze only one-half of the complete series circuit because the two halves are identical and the mmf for each half is supplied by each separate field coil. Thus, starting at a plane placed half-way between adjacent field poles, the main flux passes through the following parts in series: one-half the length

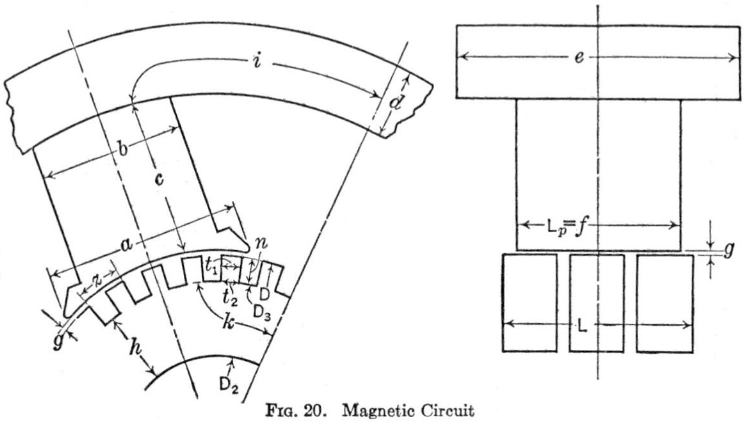

FIG. 20. Magnetic Circuit

in the yoke (*i*, Fig. 20), the pole piece (*c*), the air gap (*g*), the armature teeth (*n*), and one-half the length of the path in the armature core (*k*).

The number of ampere-turns required for excitation for any particular voltage or flux is calculated by means of a table, such as Table 5.

Table 5. Ampere-turns for Excitation

1	2	3	4	5	6	7	8
Part	Flux	Cross-section	Material	Flux Density	Ampere-turns per Inch	Length of Path	No-load Ampere-turns per Pole
Field yoke......	$\nu\phi/2$	de	....	....	....	i	....
Pole core........	$\nu\phi$	fb	....	....	....	c	....
Air gap..........	ϕ	aLp	....	....	$0.313B$	Cg	....
Arm. teeth......	ϕ	Tt_ol_i	....	....	....	n	....
Arm. core.......	$\phi/2$	hl_i	....	....	....	k	....

Net ampere-turns for flux: ......

Explanation of Table 5:

ϕ = useful flux per pole in maxwells.
ν = leakage coefficient.
a = pole arc in inches.
l_i = effective length of steel in armature core.
t_o = effective width of tooth.
T = number of teeth carrying flux.
C = Carter coefficient. (See below.)

In the armature teeth and air gap, the total flux ϕ is equal to the useful flux per pole. In the armature core this flux is divided into two halves. The flux in the pole cores is greater than that in the air gap by the amount which leaks from pole tip to pole tip and from pole to pole. By far the largest part of this leakage is between neighboring pole tips; thus the pole core carries the useful flux and the leakage flux, or the useful flux multiplied by the leakage coefficient.

In Table 5, Column 2 gives the total flux passing through each particular part.

Column 3 gives the cross-section of each part in square inches; the symbols refer to dimensions taken from a drawing similar to Fig. 20.

Column 4 calls attention to the fact that different materials have different permeabilities and magnetization curves.

Column 5 gives the density in lines or maxwells per square inch.

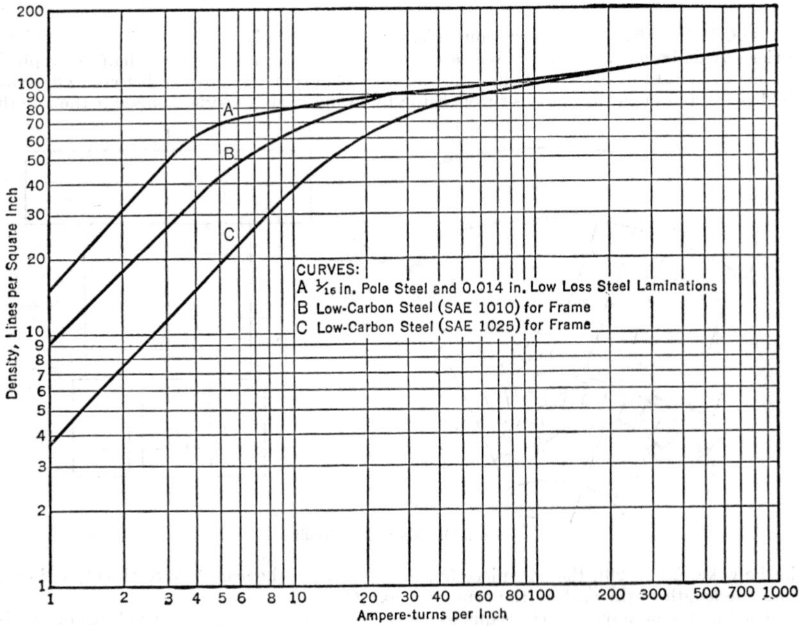

FIG. 21. Steel Magnetization Curves

Column 6 contains the specific value of ampere-turns per inch for the particular material and density, as taken from appropriate magnetization curves (Fig. 21).

Column 7 gives the length of the respective paths in inches, as taken from Fig. 20.

Column 8 gives the ampere-turns required for each part of the path and is obtained by multiplying the values in Column 6 by the corresponding values in Column 7. The sum of all values in Column 8 is the excitation required in each field coil to establish the assumed value of useful flux. At no load this excitation would correspond to a definite voltage as given by the formula in Article 6.

MAIN POLES. Field poles are usually of laminated sheet steel 1/16 in. thick, and the effective section of metal in them is usually 95 per cent of the actual cross-section. The magnetic density in the pole face is usually so much lower than that in the pole core that it may be neglected. The purpose of the pole tips is to make the area of the path of the flux across the air gaps as large as possible and larger than the path in the pole core.

GAP SECTION. The cross-section of the face of the pole face determines the cross-section of the path across the air gap. The pole face spans an arc of 2/3 of the pole pitch; the length parallel to shaft is about 0.5 in. less than the armature core, since the flux fans out at the ends.

CARTER COEFFICIENT. The effect of the constriction of the flux in the air gap caused by the slots in the armature was analyzed mathematically by F. W. Carter in a paper in *Trans. (Br.) I.E.E.* of 1901. The result is expressed in what is commonly known as the Carter coefficient, which is the ratio of the true reluctance across the gap with irregular distribution of flux to the simple geometrical reluctance (used in the text above). The Carter coefficient is a factor, greater than unity, by which the real length of the gap, g, may be multiplied to obtain an effective length which may be used with the simple cross-section to obtain the correct value of the mmf required.

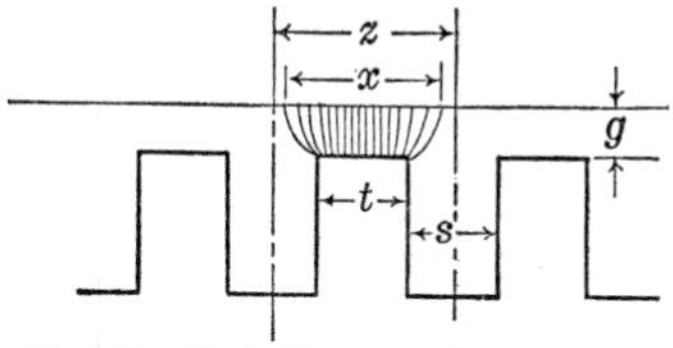

FIG. 22. Equivalent Length of Air-gap

On Fig. 22 let g = actual or physical length of air gap.
s = width of slot at opening.
t = width of tooth face.
z = pitch of tooth and slot $= t + s$.
f = a numerical factor as in table.

In addition, let g' = equivalent length of air gap, and l = length of pole face parallel to shaft. The band of flux from any tooth spreads to a width greater than t but less than z. Let x represent this indefinite width; then let $x = t + fs$.

The apparent or simple reluctance of each band is g/lz, and the real reluctance is $g/lx = g/l(t + fs)$. The ratio of the real to the apparent is $z/t - fs = g'/g = C$, the Carter coefficient. The empirical constant f has been determined by Carter by mathematical analyses checked by tests and found to depend upon the ratio s/g, slot width divided by gap length. The values are as follows:

s/g	0	0.5	1	1.5	2	3	4	5	6	7	8	9	10
f	1	0.95	0.84	0.78	0.71	0.62	0.55	0.49	0.44	0.40	0.38	0.35	0.32

If the apparent gap density is obtained by means of the gross pole face, and the mmf is found by means of the true gap g, the correct mmf is found by multiplying this value by the Carter coefficient, or by multiplying the ampere-turns per inch for the apparent density by $g' = Cg$.

APPARENT TOOTH DENSITY. At any given instant only a portion of the teeth carry flux. These are known as the teeth under a pole (T), and the number is slightly greater than those actually in the section subtended by the pole face, because of the spreading at the pole tips. This spreading is such that the number of teeth under a pole is $T = (a - 2g)/z$. T may be a fractional number, as it is a time average. The pole tips, however, are flared about 1/4 of the air-gap length, as shown in Fig. 20. The increase in air-gap reluctance caused by the armature slots almost completely offsets the spreading of the flux, so that actually the effective tooth area is that directly subtended by the pole arc; therefore $T = a/Z$.

The magnetic density in the teeth varies uniformly from a maximum at the root to a minimum at the face, and the mmf per unit length varies by a greater amount. Since the mmf required must be found, it is usual to consider the "effective" section which gives average mmf. This is taken at a point one-third of the distance from the narrowest end; thus the equivalent tooth section is

$$t_0 = \frac{t_1 + 2t_2}{3}$$

The effective length of iron (l_i) is the net length of magnetic material, excluding the space occupied by the ducts and by the insulation between laminations, and thus equals $0.9 \times (L - \text{space of ducts})$. The stacking factor is 0.9 and applies to sheet having a thickness of 14 mils.

ACTUAL TOOTH DENSITY. The tooth density thus calculated is still an apparent density, since an appreciable amount of useful flux which passes through the air of the vent ducts and between laminations will be cut by the armature conductors. If the amount of air flux per square inch obtained with a given mmf through the steel laminations is calculated, and the actual steel area is assumed to be 80 per cent of the air-gap area, the true tooth density of $^2/_3$ of the tooth depth will be as shown in Fig. 23. This

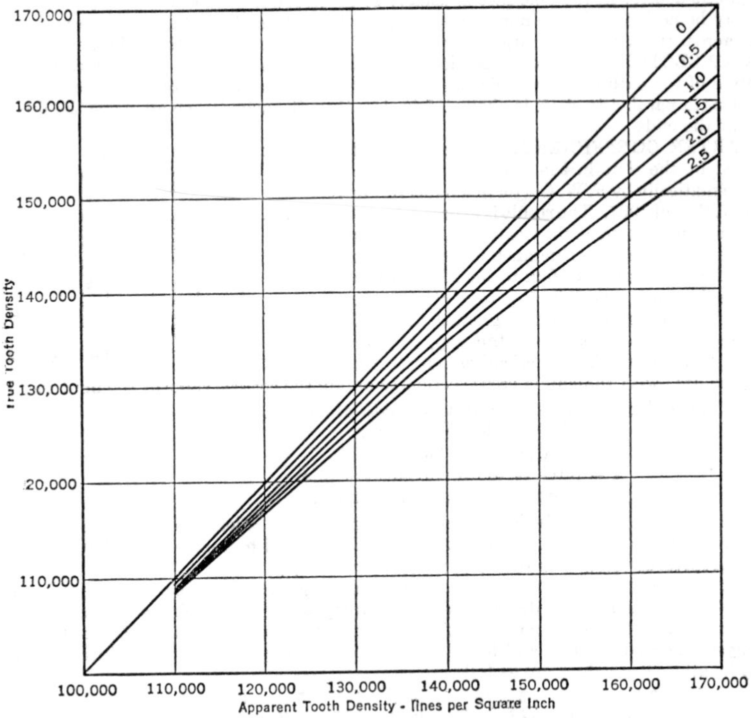

Fig. 23. True Tooth Density. (Tooth correction factor $= \dfrac{\text{Gap area}}{\text{Tooth area}} - 1$.)

density is the value to use for armature teeth in Column 5 of Table 5. The tooth correction factor = (gap area ÷ tooth area) − 1.

CORE DENSITY. The length of the armature core between heads (L) is usually greater than the length of the pole piece by about $^1/_2$ in. The core is made up of 14-mil steel, as are the teeth, and the stacking factor is 0.9. The space for air ducts is to be subtracted from L before calculating the area.

NO-LOAD SATURATION CURVE. The relation between the actual voltage and the field mmf in ampere-turns or amperes for various voltages is plotted to give the no-load saturation curve and may be predetermined by repeating the foregoing calculation for different assumed voltages, so that a curve like A in Fig. 25 is obtained.

9. LOAD SATURATION CURVES AND VOLTAGE REGULATION

COIL LENGTH. The full-load saturation curve is derived in three steps from the no-load saturation curve. The first step is to calculate the voltage drop, with full-load current, in the armature, interpole and series field windings, and brushes. The armature resistance is first calculated by assuming the end connection e to make a 35 deg angle with the center line of the slot (Fig. 24), or $e = 5\pi D/8p$. The length of mean turn of the coil will then be $LMT = 2L + 4e + 6$ in., where 6 in. is the total of the straight part of the coil extensions beyond the armature core and behind the commutator and the U-bend of the coil. The armature resistance at 60 deg cent will be $(Z \times LmT)/(2 \times m^2 \times a)$, where Z = total number of armature conductors, m = number of parallel paths in armature, and a = area of conductor in circular mils.

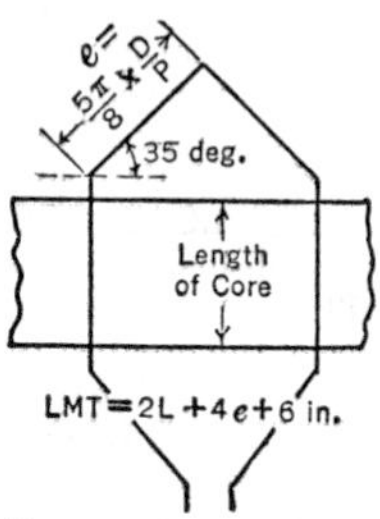

Fig. 24. Length of Armature Coil Ends

***IR* DROPS.** The armature IR drop will be the line current times the calculated resistance times 1.05 for an operating temperature of 75 deg cent. The IR drop in the interpole winding will be from 25 to 40 per cent of the armature IR drop, depending on whether half or the full number of interpoles are used. When both interpole and pole face windings are used, their total IR drop will be approximately 90 per cent of that of the armature. An ordinary series field will have about 15 per cent of the armature IR drop, but in a full series field motor the field has approximately 50 per cent of the armature IR drop. Positive and negative brushes have a total drop of 2 volts. The total IR drop of the windings carrying line current plus 2 volts for the brushes will be the length of the line ab, Fig. 25.

ARMATURE REACTION. Line bc (Fig. 25) is drawn at right angles to ab, starting at its termination and extending to the right a number of ampere-turns equal to the

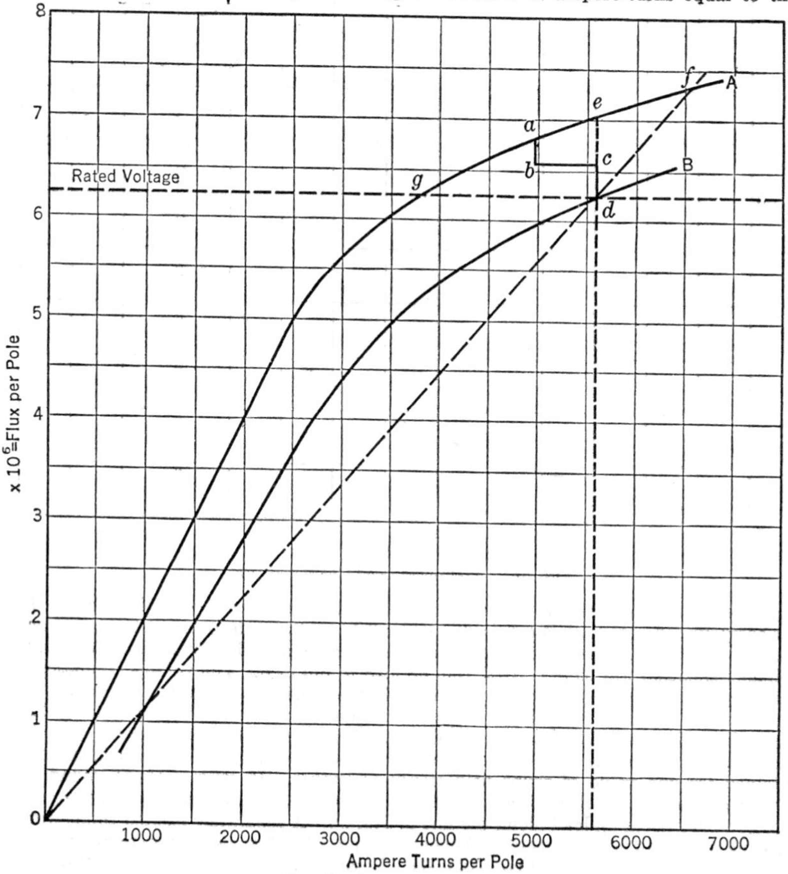

Fig. 25. Saturation Curves

armature reaction from cross-magnetizing and demagnetizing effects. As stated before, this will be 10 to 15 per cent of armature ampere-turns per pole for interpole machines and 30 to 35 per cent for non-interpole machines. A machine with pole face winding will have no appreciable length for *bc*.

SPEED CHANGE. Line *cd* represents the loss in volts due, in a generator, to the speed regulation of the prime mover. For instance, if the speed drop from no load to full load is 5 per cent, then line *cd* is 5 per cent of the number of no-load volts at the ampere-turns corresponding to the right end of line *bc*. When the prime mover has zero per cent speed regulation, such as in a synchronous motor, this factor is zero.

FULL-LOAD CURVE. By constructing several such diagrams as Fig. 25, three or four full-load points can be found, and a curve *B* drawn through them. Later discussion will show that the length of the line *bc* varies with the operating point on the saturation curve, the greatest percentage of armature reaction occuring at the knee of the saturation curve (the point of greatest rate of change of slope).

VOLTAGE REGULATION. A line drawn from the origin through the full-load curve at the rated voltage ϕ and extending to the no-load curve, Fig. 25, at *f* will give the voltage

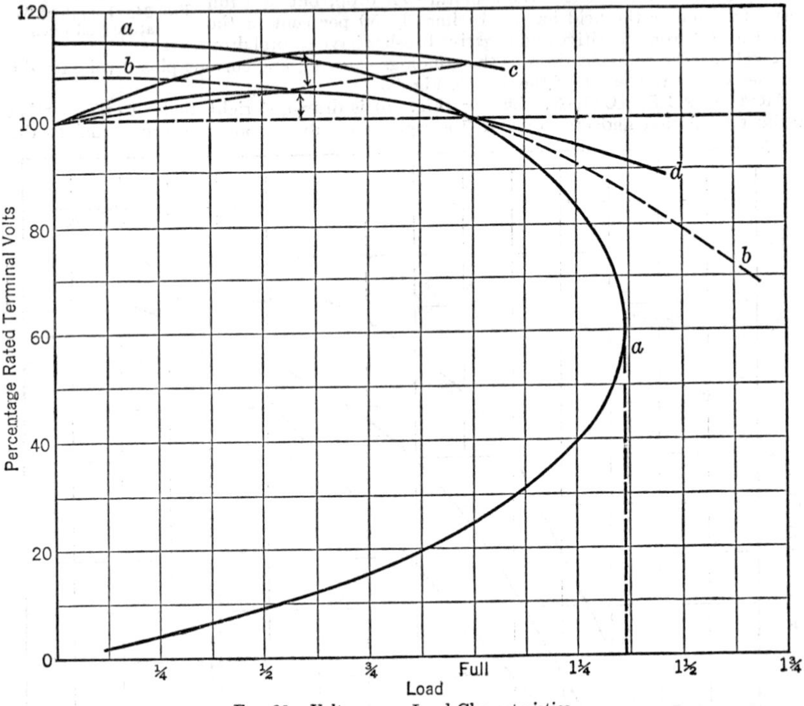

FIG. 26. Volt-ampere Load Characteristics

regulation of the generator when self-excited. In this case it is $(f - d)/d$ per cent. If a flat-compounded generator is required, a series field on the main poles with ampere-turns per pole of *gd* will be used. If the generator is shunt-wound with separate excitation, the voltage regulation will be $(e - d)/d$ from no load to full load. The voltage regulation for flat-compound or over-compound generators is the maximum variation, expressed as a percentage of the full-load voltage, of the volt-ampere characteristic from a straight line drawn between no-load and full-load voltages. If intermediate and overload saturation curves are constructed, the complete volt-ampere characteristic of the machine can be obtained for one value of shunt field circuit resistance. If, with a shunt-wound self-excited generator, the field circuit resistance is not changed, the terminal voltage will change with increasing load along the line *a*, Fig. 26. A separately excited shunt-wound machine will follow *b*, and a compound machine *c* or *d*. A self-excited shunt-wound machine will not carry any more load than the value *e* because of the combined effects of decreasing exciting ampere-turns, armature reaction, and winding *IR* drop.

VOLTAGE STABILITY. A self-excited shunt-wound generator becomes unstable when operated on the straight part of the saturation curve of the machine. This is shown in Fig. 27, where *a* is the no-load saturation curve and *b* is a line drawn from the origin through the curve at the desired operating voltage d. This line represents a value of E/I_f which is equal to the total resistance in the field circuit at voltage d. This same field resistance, however, would be required to operate at point e, so the generator voltage would be unstable in operation. It is therefore necessary to saturate a short length of the magnetic circuit at a flux value corresponding to voltage e, and this is accomplished by milling out a slot about $^1/_8$ in. deep in the top of the main pole and leaving about 10 per cent of

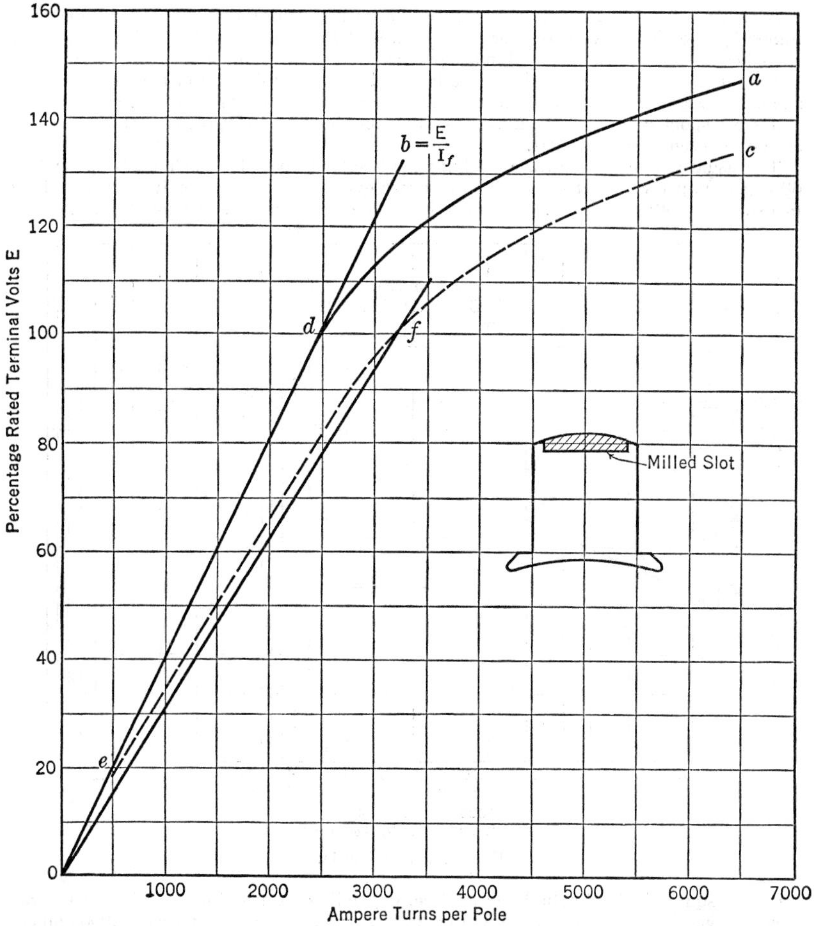

FIG. 27. Stabilized Generator

the area of the pole body left. This procedure will give a saturation curve such as c, requiring more total ampere-turns for rated voltage if the same armature air gap is used. The field-resistance line for the voltage used previously will now go from the origin through curve c at f, and the generator will be stable, since only one value of generated voltage is obtained with this value of shunt field resistance.

10. SHUNT AND SERIES FIELDS

SIZE OF WIRE. The proper size of the wire in the shunt field coil is definitely set by three arbitrary quantities: the ampere-turns desired, the average length of a turn around the spool, and the voltage drop on the spool, thus:

$$q = \frac{plF}{e}$$

where q = cross-section of proper size of conductor in circular mils.
p = 12.6 for copper at 75 deg cent (the resistance of 1 ft of copper wire of 1-cir mil area).
l = the average length of the turns on a shunt field in feet.
F = the ampere-turns per spool which are desired.
e = the voltage drop to be allowed on each spool for normal no-load conditions.

NUMBER OF TURNS. The proper number of turns t in the shunt field spool is determined by the allowable current or the allowable I^2R loss in the coil. This factor affects the efficiency of the machine and, more particularly, the heating of the coils. The greater the number of turns, the smaller is the loss and the less is the rise in temperature of the spools. Assigning a definite allowable value to this loss in watts or in field current, since watts per spool = ei, we have a fixed relation for the number of turns of the size of wire already chosen: $t = eF/\text{Watts}$.

The usual rise in temperature of shunt field coils of open d-c machines is about 50 to 60 deg cent per watt per square inch of outer cylindrical surface (omitting the end surfaces). For a 40-deg rise on the spools, it is customary practice to allow from 0.6 to 0.9

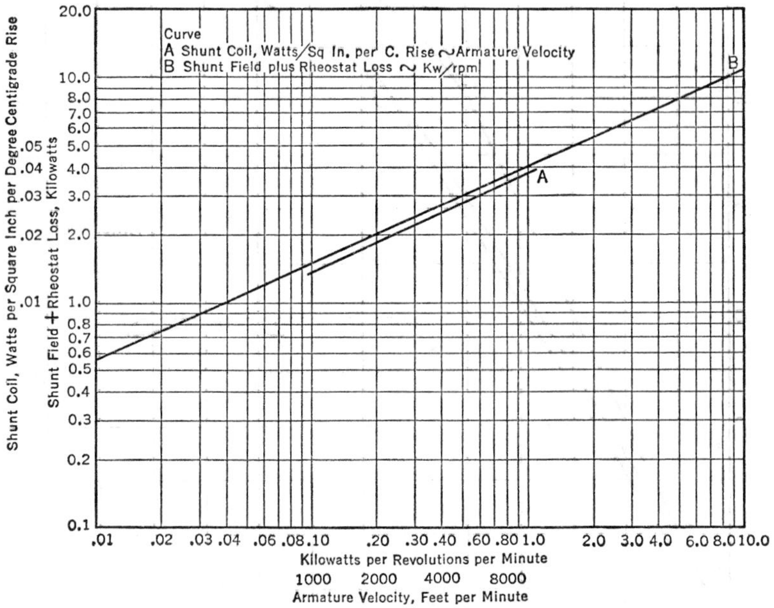

FIG. 28. Shunt Coil Data

watt per square inch of cylindrical surface, depending on armature velocity or separate means of ventilation. (See Curve *A*, Fig. 28.) A coil space of 1 in. for each 1000 ampere-turns is required on the pole for a 40-deg cent rise. In generators the allowable voltage per spool is taken as 10 or 20 per cent below the rated voltage per spool to allow a margin in emergencies. Curve *B*, Fig. 28, shows a normal value of shunt field plus rheostat loss for machines of varying size.

SERIES FIELD DESIGN. The ampere-turns required on a series field is determined for a generator as described under Voltage Regulation in Article 9. Since the required number of turns very seldom comes out to an even number, it is necessary to use the next larger whole number of turns and to shunt out current from the series winding by means of nickel-chrome resistance strips mounted between the equalizer and the negative machine terminals, the series field in modern practice being on the negative side of the armature. The series field is usually wound around the outside of the shunt coil, 1/4-in. spacing blocks being used to provide support and ventilation space. A current density of 1000 to 1200 amp per sq in. is used in the series field conductor.

The series field shunt is usually a strip of No. 15-gage nickel-chrome which dissipates 0.40 watts per square inch, using the total surface of the strip. The specific resistance of a 1-in. wide, No. 15-gage (0.057-in. thick) nickel-chrome ribbon is 0.0093 ohm at 20 deg cent per foot. The resistance of the series field winding without shunt, in ohms at 60° C, is:

$$\frac{LMT \times \text{No. turns per pole} \times \text{No. poles}}{m^2 \times \text{Cir mils of conductor area}}$$

where m is the number of parallel paths in the series field winding, and LMT is the length of mean turn in inches. Two or more circuits are used where current is heavy or where only 1/2 turn or less per pole of effective strength is required.

11. COMMUTATION AND INTERPOLES

CURRENT REVERSAL. Each coil of the armature carries a current flowing in one direction as it travels from a positive brush to a negative brush and in the opposite direction as the coil travels from a negative to a positive brush. Thus the direction of the current in each coil reverses during the time the commutator bars to which the coil is connected pass under the brush. To reverse the current in any circuit it is necessary to take out the energy stored in the form of magnetic flux linked with the circuit and to put back an equal amount represented by a flux in the opposite direction. In doing this, there is induced in the circuit a voltage which is proportional to the time rate of change of the flux and which opposes any change in the value of the current. This voltage is the emf of self-inductance of the coil.

To minimize the voltage of self-inductance it is necessary to keep as small as possible the number of ampere-turns in each coil and also the slot-leakage flux. It is possible to practically neutralize the voltage of self-inductance by introducing an opposing voltage, which is accomplished in practical machines by giving the brushes a forward lead or by using interpoles.

The brushes are moved in the direction of rotation in a generator (opposite direction in a motor) away from the true neutral until the coil undergoing reversal is moving in a flux, coming from an adjacent pole tip, of such a density that it induces by rotation in the coil a voltage that opposes and neutralizes the voltage. (See also discussion under Armature Reaction, in Article 9.)

RESISTANCE COMMUTATION. Increasing resistance in the path of the short-circuit current also aids commutation. This changes the time constant of the circuit; i.e., some of the stored energy of the magnetism is dissipated in I^2R loss in this resistance instead of in sparking at the brush. The usual method of increasing resistance is to use carbon brushes, which have a higher resistance of contact than metal brushes. Another method is to introduce a high resistance in the connection between the winding and the commutator segment.

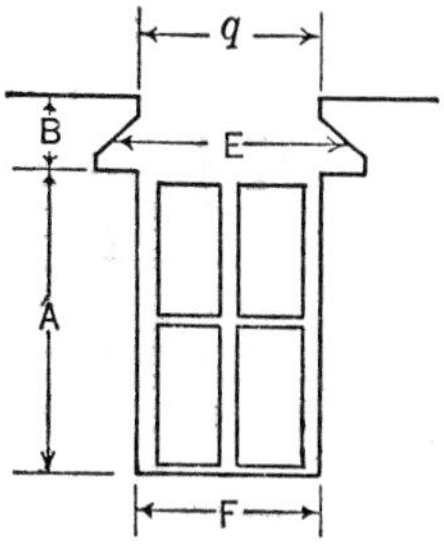

FIG. 29. Slot Diagram

LEAKAGE FLUX. For convenience in analysis the armature-leakage flux per ampere-conductor may be resolved into three parts. These three fluxes are: P_s flux inside the slot itself, P_t flux from tooth tip to tooth tip, and P_c flux about the end connections. With reference to the dimensions in Fig. 29, these fluxes are given by the following expressions:

$$P_s = 3.2 \times 1 \times \left(\frac{A}{3F} + \frac{B}{E}\right)$$

$$P_t = 2.35 \times 1 \times \log_{10}\left(\frac{\pi m}{2q}\right) \quad \text{(for non-interpole machines)}$$

$$P_t = \frac{0.798W}{\text{Effective gap}} \quad \text{(for interpole machines)}$$

$$P_c = 0.58l_c \times \log_{10} \times \left(\frac{21_c}{U_c}\right)$$

where l = length of iron core.
m = interpolar distance between pole tips.
q = width of slot opening at face.
l_c = length of end connection at one end.
U_c = perimeter of one coil of end connections.
W = width of interpole face in the direction of rotation.

All dimensions are in inches.

Also let t = turns in series between adjacent commutator segments.
= turns per coil in a multiple winding.
= $p/2$ times turns per coil in a series winding.

$$C = kt\left(\frac{\text{Brush thickness}}{\text{Segment pitch}}\right).$$

k = 2 for a full pitch winding and approaches unity as the pitch decreases.

$L = 2Ct(P_s + P_t + P_c)\ 10^{-8}$ henry.

$$f_c = \text{frequency of commutation} = \frac{12V_c}{120(\text{Brush thickness})}.$$

V_c = peripheral speed of commutator, feet per minute.

$$I_c = \text{amperes per circuit in armature} = \frac{(I\ \text{line})}{\text{No. of mult. cir.}}.$$

REACTANCE VOLTS. If we assume that the current in the short-circuited coil varies according to the sine law, the maximum value of the reactance voltage is: $e_x = 2\pi f_c L I_c$ volts.

Commutating poles are used when the air gap and teeth ampere-turns are less than 1.2 times the armature ampere-turns per pole at full load, or when the reactance volts are more than 2. Machines with armatures more than 3 in. in diameter should always have interpoles.

With a one-turn lap winding, an approximate ratio of 2.5 to 3.5 occurs between average volts per bar and reactance volts per bar, the lower ratio being for large machines and the larger ratio for small machines.

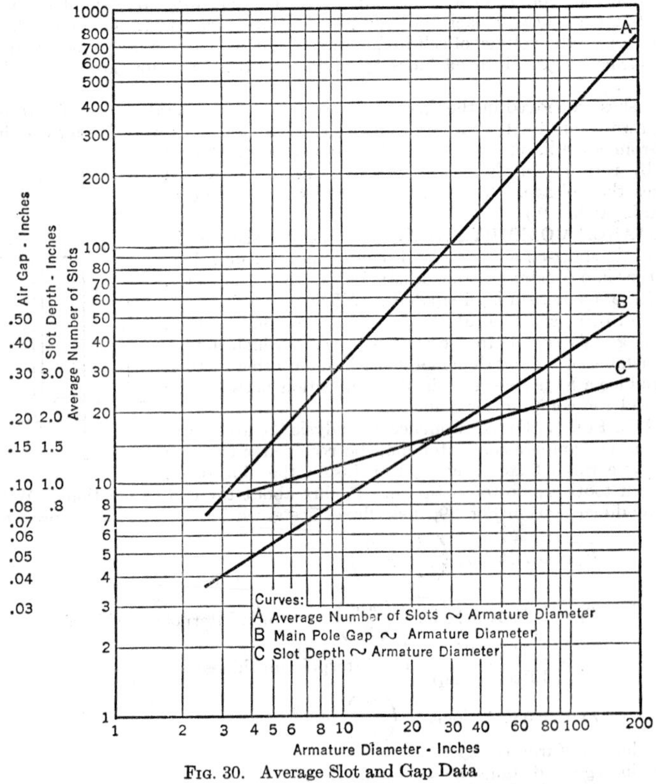

FIG. 30. Average Slot and Gap Data

SLOT PROPORTIONS. It is desirable to keep the reactance voltage as small a value as possible, since the flux distribution from the interpole face cannot be made to suit the exact requirement of each conductor when there is more than one coil per slot. This fact leads to the relation, shown in Fig. 30, between, on the one hand, number of slots, depth of slot, and main pole air gap, and, on the other, armature diameter.

The relation of the main field mmf ampere-turns A, for the gap and teeth, and the armature mmf ampere-turns B in a generator is shown in Fig. 31. The maximum value of B occurs at a brush and is equal to $Znc/2p$. The brushes are always kept very near neutral on an interpole machine, so that the interpole mmf is symmetrically distributed with the armature mmf. When the full number of interpoles is used, the value of the interpole

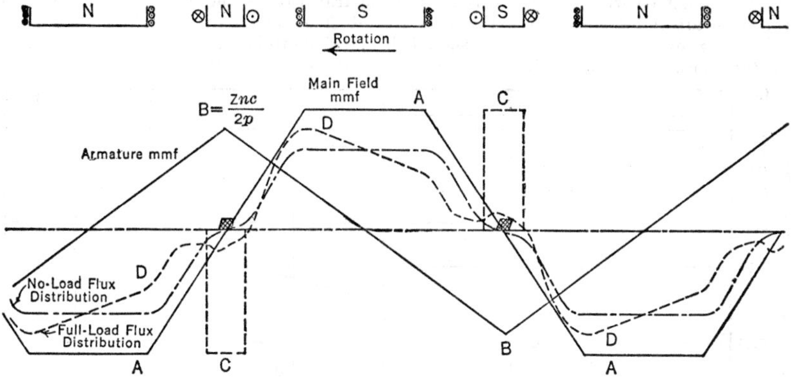

FIG. 31. Magnetomotive-force (mmf) and Flux Distribution in a Generator

mmf is made from 120 to 140 per cent of the armature mmf per pole. When only half the number of interpoles is used, the interpole has 140 to 160 per cent of the armature mmf. Thus the value of C is approximately 1.2 times B. The resultant flux distribution, shown as line D, is found by making saturation curves for several points along the contour of the pole and the interpolar space and reading off the flux density from these curves, the algebraic sum of the mmf's acting at the particular location being known.

POLE FACE WINDING. When the maximum value of D (Fig. 31) approaches 30 volts, it is necessary to neutralize the armature mmf, thus keeping the flux distribution at any load practically the same as at no load. Neutralization is accomplished by taking approximately $^1/_2$ the turns off the interpole and spreading them out into slots in the face of the main poles. Thus, for a six-pole machine having an armature with a one-turn lap winding of 108 slots and 216 commutator segments, the equivalent number of armature turns per pole with line current is $216/6^2 = 6$ turns. The interpole will have 8 turns, which is $1\ ^1/_3$ times the armature strength. If this is a 600-volt machine, the average voltage per bar is $(6 \times 600)/216 = 16\ ^2/_3$. This voltage per bar will be a maximum at no load of $16\ ^2/_3/^2/_3 = 25$, with a pole face covering $^2/_3$ of the pole pitch. At full load the

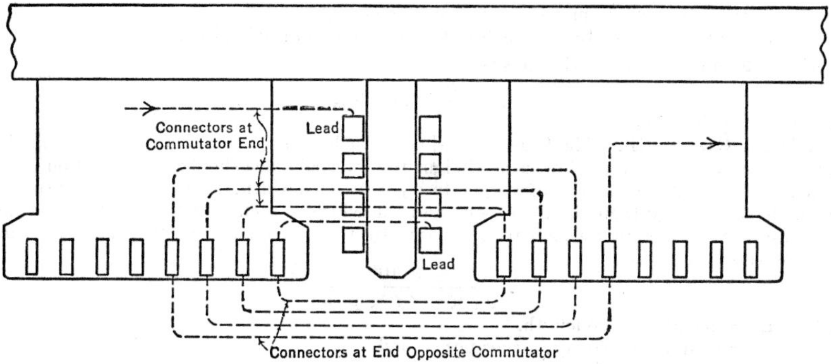

FIG. 32. Compensating and Interpole Connections

armature field will cause a maximum voltage per bar of approximately $2 \times 16\ ^2/_3 = 33\ ^1/_3$ volts. This is dangerously near a voltage which will sustain a continuous arc around the commutator, once the air has become ionized, and the armature field should be neutralized across the pole face by a compensating winding of $^2/_3 \times 6 = 4$ turns. This winding would be placed in 8 slots in the pole face, one conductor per slot (Fig. 32), and the interpole coil would have 8 − 4 = 4 turns. For special motor application, where extreme stability is necessary, compensating windings have been made with as many as 16 conductors per slot, with wide speed range obtained by variation of shunt field strength.

SLOT TRAVEL. The contour of the face of the interpole must be designed to give a flux distribution suited to neutralizing the total slot reactance voltage at each position. The face must be wide enough to cover the full travel of all the conductors in one slot while they are under one brush, so that they are not affected by main pole fringing flux while undergoing reversal. Consideration should also be given to keeping the air-gap reluctance constant for all armature positions so that no pulsation in total flux will occur through the

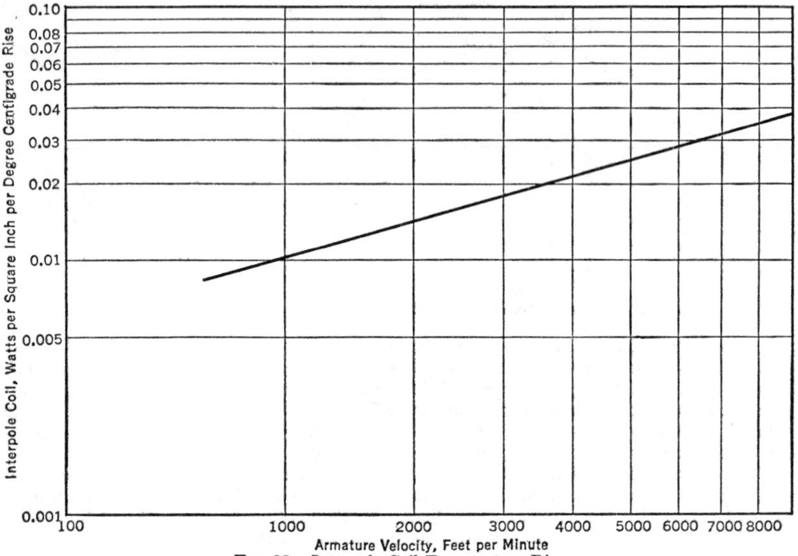

FIG. 33. Interpole Coil Temperature Rise

interpole magnetic circuit. For lap windings the distance traveled by one slot of armature conductors during commutation is

$$B - b + T(1 + S)$$

where B = brush thickness reduced to armature diameter.
b = bar pitch reduced to armature diameter.
T = tooth pitch reduced to armature diameter.
S = number of teeth by which coil pitch is off from pole pitch.

and the distance for wave windings is

$$B - \frac{b}{n} + T(1 + S)$$

where n = pairs of poles.

INTERPOLE FACE. The brush should cover at least as many commutator segments as there are coils per slot. For usual designs the interpole face should be beveled on each side at a 45-deg angle on a chamfer equal to 1/4 of the face width. The interpole is usually 1 in. shorter axially than the main pole, and the width is such that the pole will not be saturated at the maximum load to be carried by the machine. The flux density B_i in the interpole air gap is

$$\frac{e_x \times 10^8}{L_i \times v \times t \times p/m}$$

where L_i = armature-core length.
v = armature velocity, inches per second.
t = turns per coil.
p/m = 1 for lap windings, and *poles*/2 for wave windings.
e_x = maximum reactance voltage.

INTERPOLE DENSITY. The total interpole air-gap flux, then, is B_i times l_i times effective interpole face. The pole flux will be this amount plus the leakage, which is quite high in comparison to the percentage of leakage from the main poles. A non-compensated machine has a leakage factor of 2.5 to 3.5, and a compensated machine a leakage of 1.8 to

2.5. The leakage around the compensating bars has no effect on interpole leakage, which is caused only by turns on the interpole. This leakage is calculated just as for main poles. The usual limit of reactance voltage per coil is 8.5 volts for compensated machines and 5.0 for non-compensated machines. The interpole air-gap density at full load varies from 10,000 to 18,000 lines per square inch, and the interpole should not have a higher density at the frame than 45,000 lines per square inch at full load. This limit can be obtained, if necessary, by using a 2 to 1 taper in the width of the interpole.

Interpoles are usually laminated but may be solid if rapid changes in load are not liable to occur.

INTERPOLE SHUNTS. When shunts are placed around interpoles to adjust the airgap density, the ratio of inductance to resistance should be kept the same in the shunt as in the interpole circuit. The inductance of the interpole can easily be found by passing a 60-cycle current through the armature and interpole in series and measuring the voltage drop across the interpole. From the current, voltage, and resistance values the inductance is calculated.

COIL HEATING. Interpole coils are usually of bare copper except the end turns, which must be insulated to obtain sufficient creepage to ground. The current density is from 1000 to 1200 amp per sq in., and from 1000 to 1200 ampere-turns per inch of coil space is a practical limit for a 40-deg cent rise. Figure 33 shows the variation in heating of interpole coils with armature velocity. Compensating bars are usually taped with mica tape to obtain single-insulation wall thickness of 0.040 in. A density of 1800 to 2000 amp per sq in. is used for a 40-deg cent rise. Connections between compensating rods are usually of bare copper with a density of 2000 to 2200 amp per sq in. Soldered connections have a contact density of 200 to 250 amp per sq in.

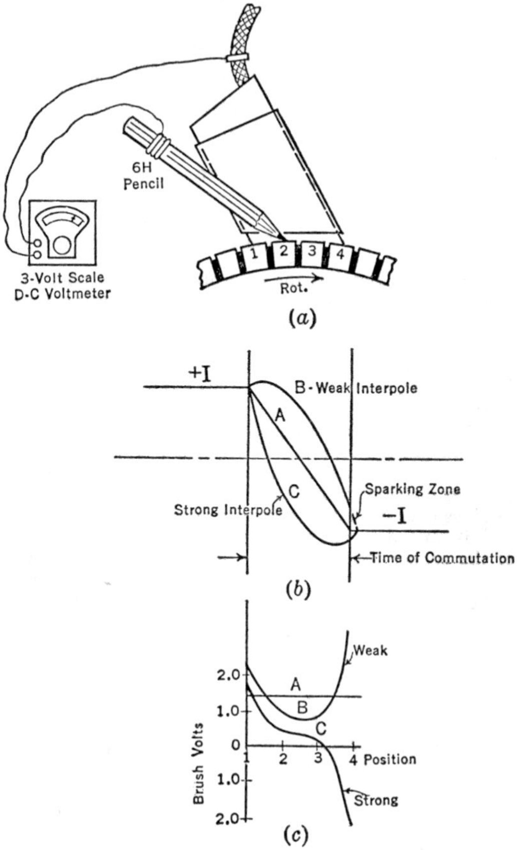

Fig. 34. Brush Potential Curve Test

EFFECT OF INTERPOLE FLUX ON VOLTAGE REGULATION. When the brushes on a generator are ahead in the direction of rotation of electrical neutral, not only has the armature a direct demagnetizing effect on the main pole flux, but also the interpole flux subtracts from the total voltage generated between brushes, thus dropping the voltage with increasing load. The opposite effect (increase of voltage with load) occurs when the brushes are moved behind electrical neutral.

INTERPOLE STRENGTH ADJUSTMENT. Since it is not possible in practice to compensate completely for all instantaneous reactance voltages in all conductors, the interpole-gap adjustment becomes very important if the commutation is to be satisfactory over the whole load range in which the machine must operate. Three methods of commutation investigation are used, all of which are complementary in a difficult case. In order of usual use, these methods are:

Brush Potential Curve. Figure 34(*a*) shows the arrangement of brush, pencil, and meter to obtain the potential distribution, under the brush-contact face, in the direction of rotation. Figure 34(*b*) shows the change of conductor current with time when the interpole strength is weak (Curve *B*), strong (Curve *C*), or correct (Curve *A*). Figure 34(*c*) shows the resulting voltage distribution at the brush contact with the three conditions of interpole field.

Black Band Method. Figure 35(*a*) shows a low-voltage generator, driven by any available means, connected in parallel with the interpole winding. By raising or lowering the field of this generator, the current in the interpole winding can be made greater or less than the armature current until all sparking (or a satisfactory minimum) is obtained. When a machine has properly adjusted interpoles and brush position, the interpoles can be boosted or bucked at open circuit by the same amount of current before visible sparking appears. The amount of buck or boost amperes will remain about the same, as shown in Fig. 35(*b*), before sparking occurs at all loads up to the limit of interpole density or other close design factors affecting commutation. The effect of a weak interpole is shown in Fig. 35(*c*), and of a strong interpole in Fig. 35(*d*).

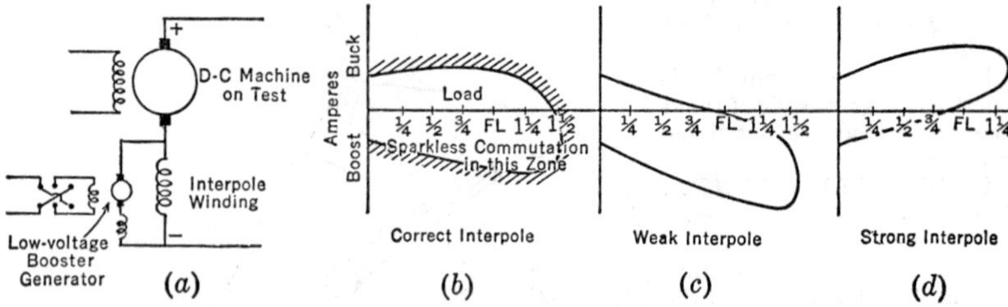

FIG. 35. Black Band Commutation Test

Oscillograph Method. This method is used to obtain a record of the instantaneous voltage in an armature conductor and thereby to judge the effectiveness of compensation and flux distribution in the commutating zone. Two temporary copper strips are fastened as rings over insulation around a convenient place on the shaft of the machine. Then a lead is tapped on one end of a single armature conductor ($^1/_2$ a coil) and brought to one of the copper strips. Another lead connects from the second strip to the opposite end of the armature conductor. By fashioning bronze strips as sliding contacts on the rings, the conductor voltage is applied to an oscillograph element, and a record such as Fig. 36 is obtained. By this means it is possible to discover transient or oscillating disturbances in the commutating zone which would not otherwise be apparent. In some such cases the cure is not too simple, a change in armature or field design being required.

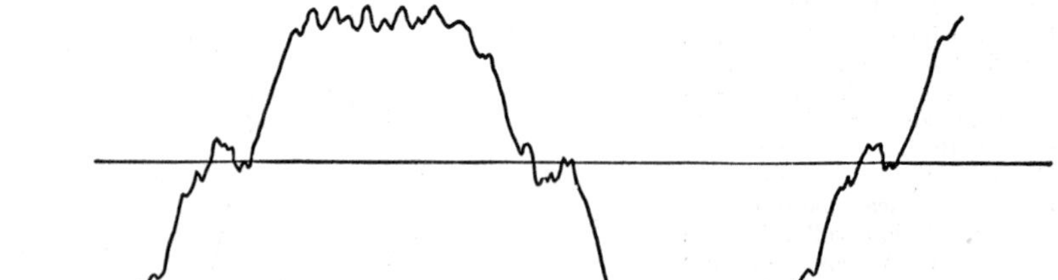

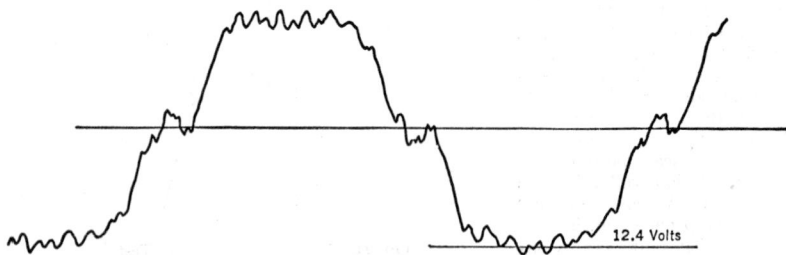

FIG. 36. Volts across One Armature Conductor of a 2500-kw Compensated Generator at 125 Per Cent Load

12. COMMUTATOR AND BRUSH OPERATION

The diameter of a commutator is so chosen that a velocity not greater than 5000 ft per min is used for continuous operation, except in turbine-driven machines, where 8000 to 9000 ft per min may be necessary. The lower the peripheral speed, the less brush vibration and chatter are obtained. The commutator length must be sufficient to mount all the brush holders and to give:

1. $^1/_{64}$-in. clearance between brush holders.
2. $^1/_8$-in. overlap of consecutive pairs of brush tracks.

3. 1/2- to 1-in. clearance between side of brush holder and commutator riser.
4. 1/4- to 1/2-in. space from edge of end brush to end of commutator for variation in manufacture and for float.

Brushes are normally electrographitic, with a total + and − contact drop of 2 volts. The value of 2 × armature current is taken as the loss at the contact of the brush on the commutator. When metal graphite brushes are used on machines with less than 40 volts, the total contact drop is taken as 0.50 volt.

Brush friction is a variable factor depending on pressure, humidity, and atmosphere, so that the friction coefficient, for electrographitic brushes, is arbitrarily taken to give 8

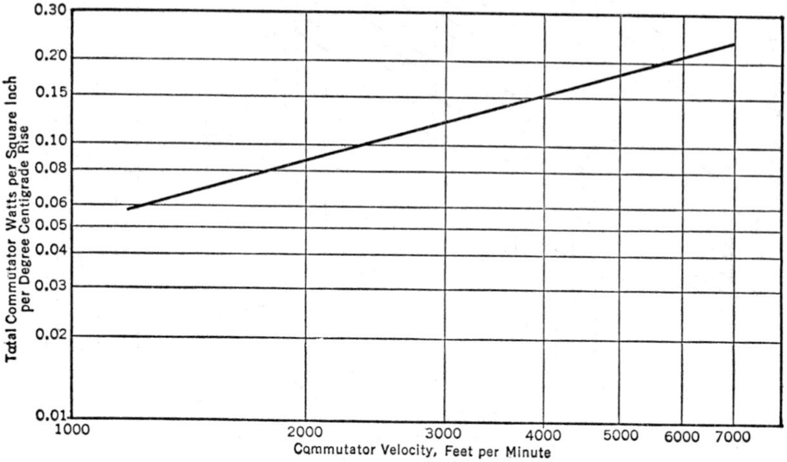

FIG. 37. Commutator Velocity, Feet per Minute

watts per square inch of brush contact per hundred feet per minute of commutator velocity, and, for metal graphite brushes, is taken as a factor of 5. Test results show in Fig. 37 the relation between commutator temperature rise and peripheral speed.

13. LOSSES AND EFFICIENCY

The efficiency of d-c machines is determined either by the input-output method, used in fractional horsepower sizes, or by the conventional method of measuring the separate losses. The component losses are as follows:

a. Core loss. This factor varies considerably, even from one duplicate armature to another. The variation is caused by filing of the slots or burring of the edges of the laminations when punching. Figure 38 shows the watts per pound of hysteresis and eddy-current loss in the laminations at various frequencies. A multiplier which is proportional to armature diameter (line *B* on this chart) must be used.

b. Brush friction. This is arbitrarily calculated as shown in Article 12.

c. Windage and bearing friction. Figure 39 shows typical windage and bearing friction losses for 2-sleeve-bearing machines without attached fans. Attached fans usually have about a 3-kw loss per inch of width at a tip velocity of 10,000 ft per min, and the loss varies as the cube of the tip velocity.

d. Shunt field and rheostat. This is equal to the applied excitation voltage multiplied by the field current.

e. Brush-contact loss. For electrographitic brush material this is 2 times the line current.

f. Winding I^2R loss. This is calculated at 75 deg cent and is equal to the line current squared times the sum of the various winding resistances at 75 deg cent.

g. Stray load losses. The mmf of the armature sets up a component of flux of its own, having a magnitude closely proportional to the armature current (because, purposely, its path is largely in air), which distorts the main flux and causes additional losses under load. These losses are due to eddy currents in the armature conductors and in any masses of metal near the conductors, such as the pole faces, and increased hysteresis in the teeth and pole faces. With no armature current hysteresis loss is zero; it increases rapidly with

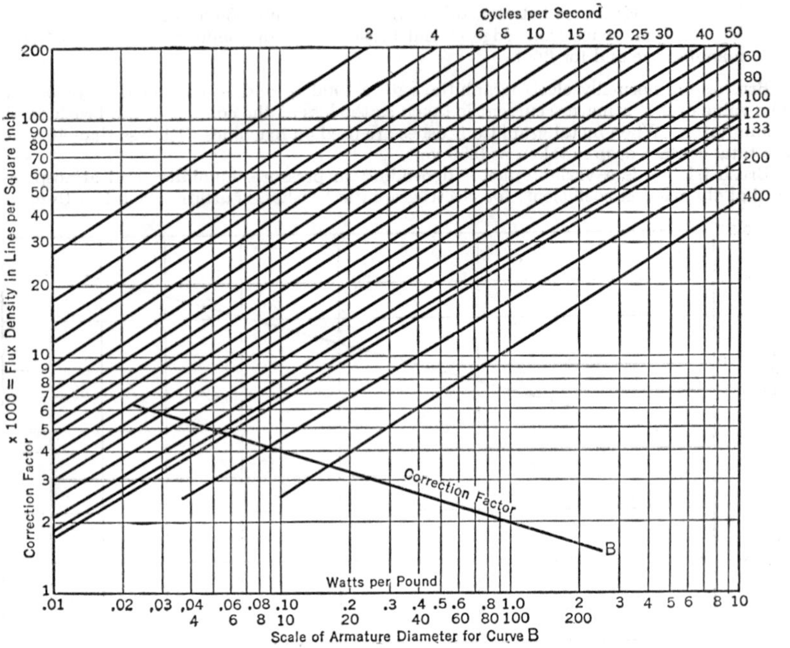

FIG. 38. Watts Loss in Low Loss Steel Material. (For actual loss, multiply calculated total watts by correction factor from Curve B.)

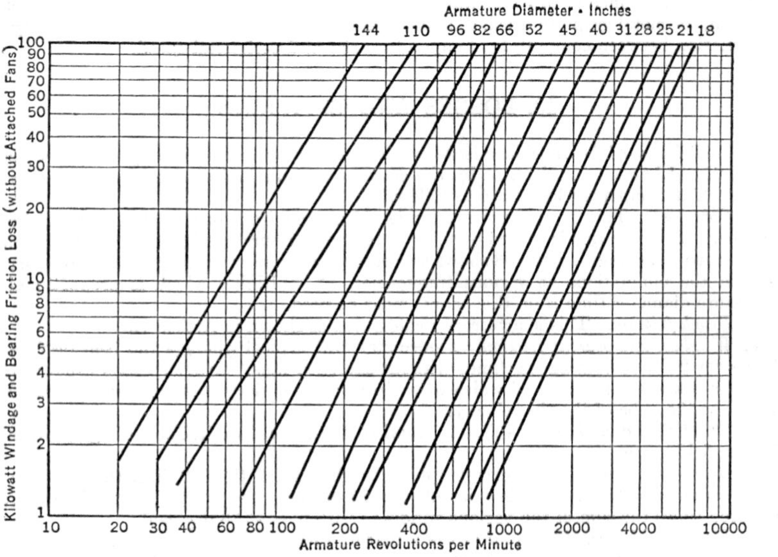

FIG. 39. Windage and Bearing Friction Loss, without Attached Fans. (These curves are for armature core length not more than 125 per cent of pole pitch.)

the armature current. The rate of increase depends upon how much of the loss is hysteresis and how much is eddy loss.

The AIEE Standards state that, for calculating the efficiencies of d-c generators or motors, the load losses at any load shall be taken as 1 per cent of that load, except in motors of 200 hp at 575 rpm and smaller, in which these losses shall be ignored. Some European standards say that the load losses shall be taken as proportional to the square of the load. There is no recognized method of calculating or testing for them.

THE CONVENTIONAL EFFICIENCY. This is the ratio of the output to the sum of the output and all losses. It is predetermined by estimating for the particular load or output under consideration the value of each of the seven losses which have been described.

If P = the output in watts, and the letters A, B, C, D, and E represent, respectively, five of the various losses which have been described, then the efficiency for any load P is:

$$\text{Per cent efficiency} = \frac{100P}{P + A + B + C + D + E}$$

The efficiency is a maximum for that load at which the variable losses $(D + E)$ are equal to the constant losses $(A + B + C)$; hence it is desirable to keep the constant losses small in a machine which is to be operated most of the time at partial load.

USUAL EFFICIENCIES AND LOSSES. An idea of a reasonable value for the efficiency of d-c machines of various sizes and the distribution of the losses is given in Table 6. It must be remembered that the efficiency of a machine of any given size may vary over a wide range, depending upon the speed, weight, and cost.

Table 6. Efficiency and Distribution of Losses

Rating, kw	Efficiency, per cent	Friction (Total), per cent	Excitation, per cent	Core Loss, per cent	Arm. RI^2, per cent
1	80	6	6	4	4
5	84	5	4.2	3.2	3.6
10	86	4	3.6	3	3.4
20	88	3	3	2.8	3.2
50	90	2.6	2.2	2.2	3
100	91.4	2.3	2	1.7	2.6
200	92	2.2	1.8	1.6	2
500	93	2	1.6	1.4	2

14. TESTING OF DIRECT-CURRENT MACHINES

Direct-current machines are judged by four characteristics: efficiency, regulation, commutation, and heating. It is necessary to subject a machine to actual full-load conditions to determine its heating and commutation, and it is desirable to do so to determine the regulation. The efficiency is determined more accurately, however, by the "separate loss method," which does not involve loading the machine.

The customary tests on d-c machines (see also AIEE Test Code) are: (1) resistance measurements, cold and hot; (2) saturation curve; (3) core loss and friction test; (4) load run for commutation and regulation; (5) heat runs; (6) compounding test; (7) insulation test.

RESISTANCE MEASUREMENTS. The first measurements are made when the machine is at the same temperature as the room, i.e., after it has been idle from 12 to 24 hr, depending upon its size. This is done in order that the relation between the resistance and the temperature may be accurately known. The resistance of the shunt field, series field, interpole field when present, and armature winding proper is measured.

Resistance of Lap Winding. To find the resistance of the armature winding, it is preferable to measure between two diametically opposite points of the commutator of a multiple-wound armature. The effective resistance of the armature is then calculated by the formula $r_a = 4r'/p^2$, where r' is the resistance measured as above, and p is the number of poles.

Resistance of Wave Winding. It is necessary to measure two-circuit series-drum windings between two commutator segments separated by a distance equal to the periphery of the commutator divided by the number of poles. This procedure will give the effective resistance of the armature.

Brush-contact Drop. This is arbitrarily taken as 2 volts total.

SATURATION CURVE. The saturation curve is not of immediate interest in determining the quality of a machine but is more particularly of interest to the designing engineer. It is made by driving at the proper speed and supplying current to the shunt fields from a separate source of potential. As the current in the shunt field increases, the potential generated by the armature varies and is measured by a voltmeter.

Because of the existence of the hysteresis loop it is necessary to increase the field current gradually and never to reduce the value while changing from point to point until the maximum value has been reached. Then the field current is reduced, step by step, to zero. This procedure gives two curves, one with "rising field current" and one with "falling field current." The curves plotted with volts as ordinates and field current or ampere-turns as abscissas show the typical "knee" of all saturation curves, as seen in Fig. 27.

CORE LOSS AND FRICTION, OR STRAY POWER. These tests, which are made with the same arrangement as is used in determining the saturation curve, may be made at the same time. The machine to be tested is driven by a small motor having a capacity about 10 per cent of the ratin, of the machine under test. All the losses and constants of this motor must be known.

First the machine under test is driven at the desired speed with no current in the fields, and the power taken by the driving motor is noted. The mechanical output of the driving motor is calculated to obtain the friction loss of the large machine.

The large machine is then excited, and the power to drive it is determined. This power represents the core loss plus the friction loss. By deducting the friction loss already determined, the values of the core loss for various values of excitation are found. (See also Magnetic Materials, Section 2.)

The combined core loss and friction loss constitute the stray power loss.

LOAD RUNS. The machine is run and the excitation adjusted to give the proper voltage at no load. The load is then added, and the terminal voltage noted. If E_0 = voltage at no load and E = voltage at full load, the regulation is $(E_0 - E)/E$.

If the machine does not have commutating poles, it is necessary to make a preliminary test in order to set the brushes at the position which provides the best compromise in sparking at no load and full load. Commutation is judged by observing the action of the brushes at full load and 150 per cent load. The conclusions are a matter of judgment and experience, although degrees of sparking have been arbitrarily agreed upon and are represented in a chart or series of pictures.

HEAT RUNS. Heat runs are made by operating the machine at full load until the temperatures of the various parts that can be noted during operation have become constant. The greater the capacity of the machine, the longer will be the time necessary.

Dead-load and Pumping-back Methods. Heat runs may be made either by (1) the "dead-load" method, in which a resistance load, such as a water rheostat, is connected to

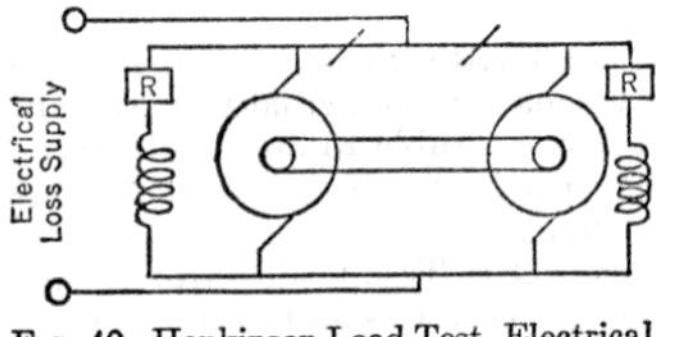

FIG. 40. Hopkinson Load Test, Electrical Supply of Losses

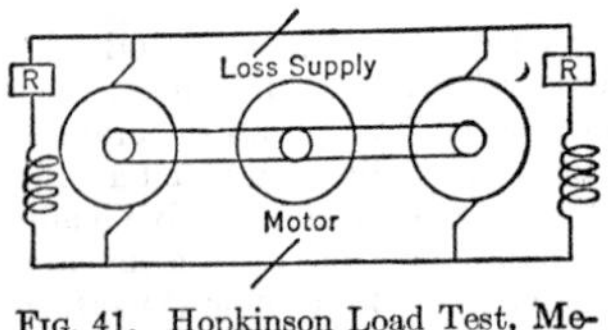

FIG. 41. Hopkinson Load Test, Mechanical Supply of Losses

the terminals and full rated-load power is required to drive the machine, or (2) the Hopkinson or "pumping-back" method, in which two machines having similar characteristics are run together, one acting as a generator to supply electrical power to the motor, which in turn drives the other by means of a belt or similar mechanical connection. For this test a "loss supply," which may consist of a source of either electrical power or mechanical power is required. The amount of power required is from 10 to 20 per cent of the rating of one of the machines. The connections are shown for electrical loss supply in Fig. 40, and for mechanical loss supply in Fig. 41.

Thermometer Readings. During the heat run there are taken, at stated periods, readings of thermometers which show the temperatures of the frame, field coils, bearings, and surrounding air. After the heat run the thermometers are placed on various parts of the machine, the bulbs being protected from radiation by small pads of cotton, and the temperatures of the following parts are noted:

Armature-core surface	Commutator surface	Bearings
Armature-core ventilating ducts	Pole tips	Frame
Armature conductors or winding	Field coils	Room

The resistance of all electrical circuits should be measured, and the average temperature of the copper calculated from the formula $t_1 = (r_1/r_0)(234.5 + t_0) - 234.5$, where r_1 and r_0 are the hot and cold resistances, respectively, and t_1 and t_0 the hot and cold temperatures, respectively, in degrees centigrade, copper of 98 per cent conductivity being assumed.

COMPOUNDING TEST. In order to adjust the current in the series field so that a compound-wound generator will give specified voltages at no load and full load, the machine is first operated at no load, and the current in the shunt field is adjusted to give the desired no-load voltage. The load is then put on; usually it will be found that at full load the terminal voltage is too great.

Strips of German silver or other resistance metal are then connected in multiple with the series field until, by shunting current from the series field, the voltage is reduced to the value which is desired. This shunt resistance is then made up in permanent form and connected in circuit. Before making the no-load adjustment, it is desirable to overexcite the shunt field for a moment in order to overcome the hysteresis. (See also Parallel Operation, in Article 15.)

INSULATION TESTS. Before any voltage is applied to a machine, it is customary to measure the insulation resistance by means of a source of potential of 500 or 600 volts and a voltmeter reading of 600 volts. With the voltmeter in series the potential is applied between the conducting windings and to the frame of the machine.

$$\text{Insulation resistance in ohms} = \frac{E_0 - E_v}{E_v} \times r_v$$

where E_0 is the potential of the source, E_v is the reading of the voltmeter in series, and r_v is the resistance of the voltmeter.

The AIEE Standards specify the proper values for various machines, but in general the insulation resistance should be more than 1 megohm. A lower value indicates moisture or a damaged winding.

After the heat runs it is customary to apply a high potential to test the insulation of the machine. This procedure consists of applying a 60-cycle alternating potential between each electrical circuit and the frame of the machine for 1 min. The magnitude of the potential depends upon the capacity and rated voltage of the machine under test and is specified in the Standards of the AIEE as twice the rated voltage plus 1000.

15. INSTALLATION

In installing a d-c machine the following precautions must be observed:

1. On large machines with foundations the foundation bolts must be provided in accordance with drawings.
2. Bearings must be lined up and well cleaned before being filled with oil. Insulation should be checked before setting in the armature.
3. The armature must be properly centered so that the air gap is correct at all points. Taper wedges are used to measure the gap. The magnet frame should be bolted to the base.
4. The field coils must be properly connected. A test for polarity should be made with a compass in order to make sure that no field coils are reversed. For a self-exciting generator there is one particular connection of the field to the armature for each direction of rotation.
5. The commutator must be smooth and polished with sandpaper, never an emery cloth.
6. The brushes must be properly and accurately spaced around the commutator. They must be sandpapered and fitted to the curvature of the commutator. The pressure on the brushes must be adjusted to the correct value, which is usually 2.5 to 2 lb per sq in. of contact surface.
7. The machine must be thoroughly dried out by heating, and the insulation measured as a check. It should be at least 1 megohm.

OPERATION. In starting a single generator, it is sometimes necessary to "charge" the field by separately exciting the shunt fields for a moment to set up residual magnetism.

To cause the machine to "pick up" or generate voltage by self-excitation, it is necessary to cut out or short-circuit most of the resistance of the regulating rheostat connected in series with the shunt field. If the total resistance of the shunt-field circuit exceeds a certain critical value, the machine will not "pick up," however much time is allowed.

The terminals of the shunt field must be connected to the terminals of the armature with a definite polarity, depending upon the direction of rotation; otherwise the tendency will be to demagnetize the field rather than to build it up. This polarity is usually found by trial and error.

PARALLEL OPERATION. In order to operate a power station under economical conditions it is necessary to have a number of machines, whose aggregate capacity is equal to the maximum demand on the station. As the demand varies, the number of machines in operation is adjusted so that the machines running are operating at a load near their rating and, therefore, at a good efficiency.

SHUNT GENERATORS. In order to operate shunt generators in parallel, i.e., feeding the same bus-bars, it is only necessary to adjust them all to the same polarity and voltage, connect them to the bus-bars, and adjust the division of load by strengthening the field of the underloaded machine if the voltage of the bus-bars is low. If the voltage of the bus-bars is high, the field of the overloaded machine should be weakened.

COMPOUND GENERATORS, EQUALIZER CONNECTION. In order to operate compound-wound machines in parallel, it is necessary to provide an equalizer connection which makes a common connection on all the machines at the point between the armature and the series field, as shown in Fig. 42.

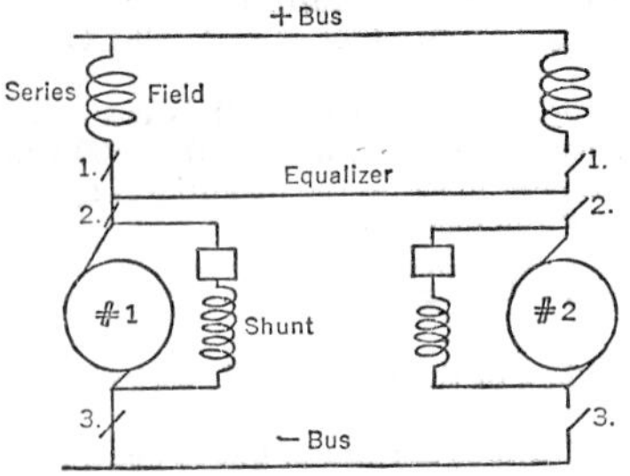

FIG. 42. Parallel Operation of Compound-wound Generators

The function of the equalizer is to divide the load current at all times in the proper proportion between the series fields of the different generators. This procedure prevents the machines from acting as series generators or differential motors, which would cause short circuits.

For compound-wound machines to operate successfully in multiple, all machines connected to one set of bus-bars must have the same amount of compounding, as well as the same voltage at no load. Because of saturation in the magnetic circuit it is not always possible to make the compounding curve a straight line, i.e., the increase in voltage directly proportional to the load. It is, therefore, necessary to investigate the compounding curves of machines before they are operated in parallel. If they have unlike compounding curves, one machine may become overloaded while another is underloaded, unless the field current of one machine is adjusted.

In connecting a machine in parallel with those in operation it is necessary to see that it has the proper polarity and voltage, that the equalizer circuit is made, and that the switches are closed in the order 1, 2, 3, as shown in Fig. 42. If any other order is used, the effect is the same as having no equalizer. In shutting down one machine, switch 3 must be opened first, then 2 and 1.

16. SPECIAL MACHINES

HOMO-POLAR GENERATORS. This type of machine is also sometimes called acyclic and was formerly incorrectly called unipolar. Its method of operation is based on the principle of the Faraday disk, which consists of a copper disk revolving about an axis and projecting between the poles of a magnet. By this rotation in a magnetic field an emf is set up between the axis and the periphery of the disk; and, if brushes bearing on these two parts are connected to an external circuit, a current flows. The peculiar characteristic of a homo-polar machine is that each conductor always cuts the flux in the same direction; consequently the emf induced in it is always in the same direction and is not alternating, as in the usual d-c machine. Thus no commutator is required.

This absence of a commutator is the feature which makes the homo-polar machine attractive. The commutator presents many difficulties in high-speed machines to be driven by steam turbines. It is on this application that recent attempts to develop a successful homo-polar machine have been centered. Instead of a commutator, collector rings with brushes are used to collect the current from the moving conductors. These collector rings, however, present difficulties in construction and operation on account of the high peripheral speed at which they must run. The rings are subject to a considerable centrifugal force, and there is a tendency for the current to arc between the brush and the collector on account of the high rubbing speed. Spiral grooves in the collector rings have reduced this difficulty.

VOLTAGE. The voltage of such a machine is not only unidirectional, as in all d-c machines, but is really constant. Since there can be only a few conductors in series, however, the voltage generated is very low. The voltage generated per disk or inductor is $E = Blv\ 10^{-8}$, where B = magnetic lines per square centimeter, l = length of conductor in centimeters, and v = velocity of conductor in centimeters per second. This is more conveniently expressed as $E = NZ\phi 10^{-8}/60$ volts, where N = revolutions per minute, Z = conductors in series, ϕ = total flux traversing gap.

RADIAL AND AXIAL TYPES. There are two types of homo-polar machines, the radial and the axial. The radial type (Fig. 43) is like the Faraday disk and consists of a disk revolving between the two poles of a cylindrical magnet. Brushes bear on the outer rim and the shaft to collect the current. The disk may be made of steel to reduce the reluctance of the magnetic path. The voltage of such a machine is limited to 10 or 15 volts, but the current may be very large. A variation of this type has two disks on the same shaft and the magnetic path so arranged that the voltage of the two disks may be added in series by brushes bearing the peripheries of the two disks. The axial type (Fig. 44) consists of a cylindrical steel armature with copper bars in the surface, the whole revolving in a cylindrical field so arranged that the magnetic flux flows outward from the armature in a radial direction at all points. The several conductors on the armature are connected to slip rings at both ends, and, by means of brushes and stationary conductors, these conductors are connected in series. The voltage of such a machine may be from 40 to 50 volts per conductor.

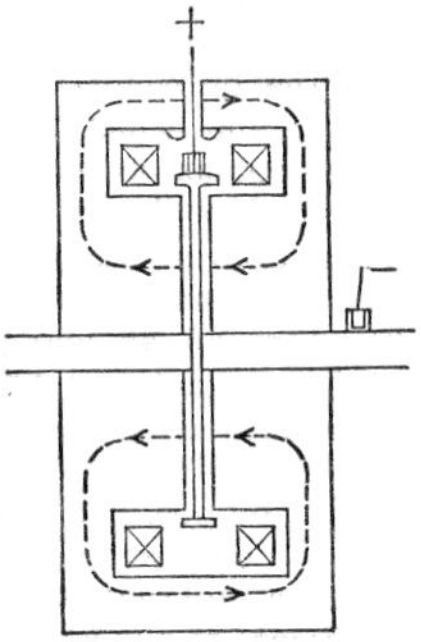

FIG. 43. Radial Type of Homo-polar Generator

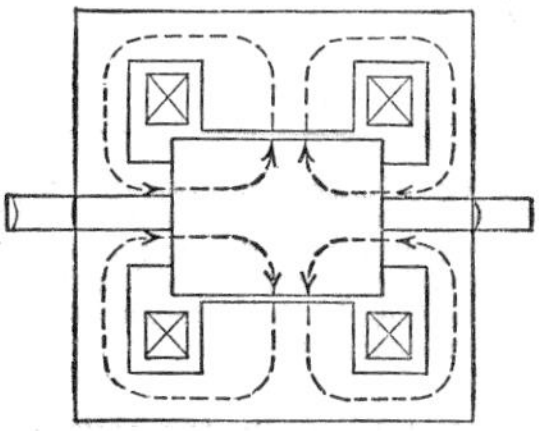

FIG. 44. Axial Type of Homo-polar Generator

DATA ON LARGE AXIAL-TYPE MACHINE. In the *Trans. AIEE*, Vol. 24, Noeggerath describes a machine of the axial type rated at 300 kw for 500 volts at 3000 rpm. The armature has 12 conductors connected in series for 500 volts. The diameter of the armature is 19 in., and the length 12 in. The peripheral velocity is 15,000 ft per min. The armature is of cast steel and has 24 cast-steel collector rings on it. The stationary conductors connecting together the collector rings are placed in the face of the pole; thus their mmf may be used to balance the armature reaction.

EXCITATION. The armature reaction of such machines is very high and has only a distorting effect. It weakens the field, however, since the cross-magnetization weakens one part more than it strengthens another, because of saturation. By a proper arrangement of the movable connections an mmf may be set up which will strengthen the field, and thus the machine may be compounded. Although these machines may be made self-exciting, the resistance of the shunt fields must be very low in order that the machine may pick up on starting, as the resistance in the brush contacts is high. The drop in voltage in each brush contact is about 0.8 volt at full load but is higher before the current flows. From 10 to 20 times the normal voltage of brush contact is sometimes required to start the current.

LOSSES. Such a machine as has been described has an efficiency of approximately 90 per cent at full load. The losses are made up principally of friction and I^2r_a in the brushes. The field I^2r_a is low, as the air gap is small for mechanical consideration. The armature I^2r_a is almost negligible because of the few turns. The eddy losses are low if the flux density is constant in one zone around the armature, but the density may vary along an element of the cylinder. The total weight of the machine is about 2 to 3 times that of the usual d-c machine, but the proportion of copper is much lower.

THREE-WIRE D-C GENERATORS. The 3-wire generators, used for supplying a 3-wire d-c system and taking care of a reasonable unbalance of the loads on the 2 legs, is really a double-current generator with collector rings as well as a commutator. It may have 2, 3, or 4 rings. The more phases, the less is the internal heating caused by the unbalanced current. The neutral of the distribution system is connected to the neutral or middle point of each of the "balance coils" or single-circuit transformers, and the terminals of these coils are connected to the collector rings.

For a 2- or 4-ring connection the coils must have a tap at their middle point, but for a 3-ring machine a Y connection may be used.

The unbalanced or neutral current divides among the coils inversely as the resistances, therefore equally. When the current enters the armature winding from the ring, it divides

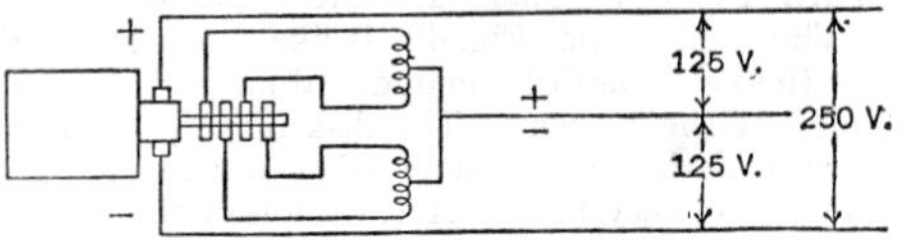

FIG. 45. Three-wire Direct-current Generator with Four Collector Rings and Two Coils

again inversely proportionally to the resistance of the path to the brush, whence the unbalanced component came originally. Thus the unbalanced current in any one coil of the armature is continually changing and reverses as the coil passes under a brush. The variation of current is a straight-line function, and the unbalanced component is superimposed on the armature current due to the balanced portion of the load. A coil halfway between taps has a lower peak of varying current than a coil next to a tap, as in a synchronous converter, and hence less heating. (See Figs. 45, 46, and 47.)

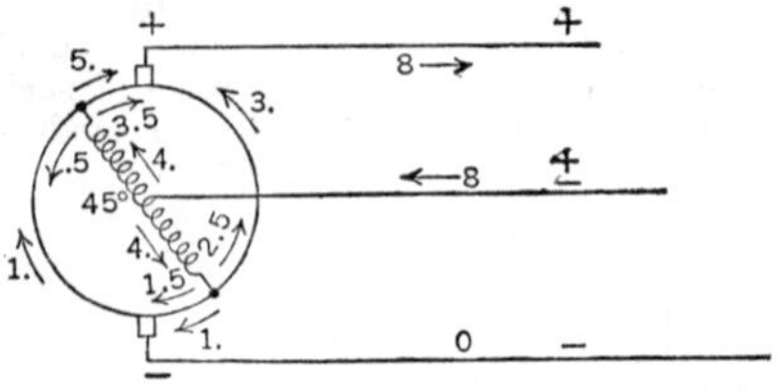

FIG. 46. Three-wire Generator: Path of Unbalanced Currents in a Two-ring Machine. Inside figures show the mathematical components of each 4 amp from coil terminal. Outside figures show the actual currents in the armature coils at that instant.

An alternating magnetizing current of small magnitude is continually flowing from the collector rings through the coils, whether the load is balanced or not. In some commercial machines the coils are placed inside the rotating armature, and then only one collector ring is needed; but generally the coils are separate and external to the generator and encased in a structure like a transformer.

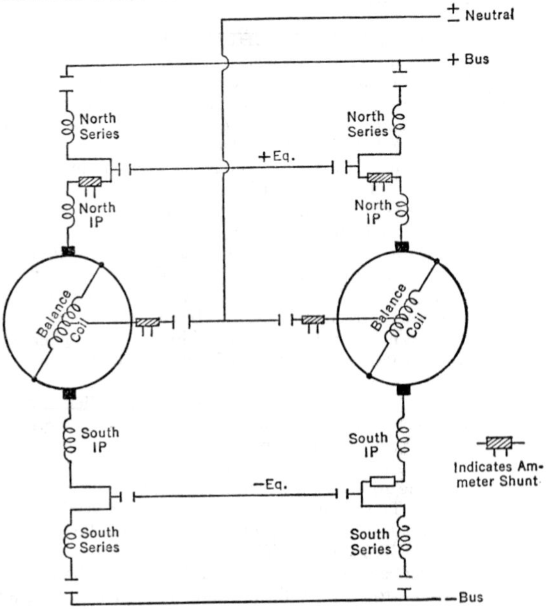

FIG. 47. Parallel Connection of Three-wire Compound-wound Generator (IP = interpole)

VARIABLE-SPEED GENERATORS. The most common form of this type of generator is the third brush generator used for battery charging in automobiles. Figure 48 shows the general arrangement, where the shunt field is excited between the positive brush and a point x about halfway around the commutator. When load tends to increase on the armature because of increased speed, the armature distortion reduces the flux in the trailing pole tip and lowers the excitation voltage between the + brush and the point x. To obtain a higher output current from the generator, the brushes should be shifted forward in the direction of rotation, but poorer regulation will result.

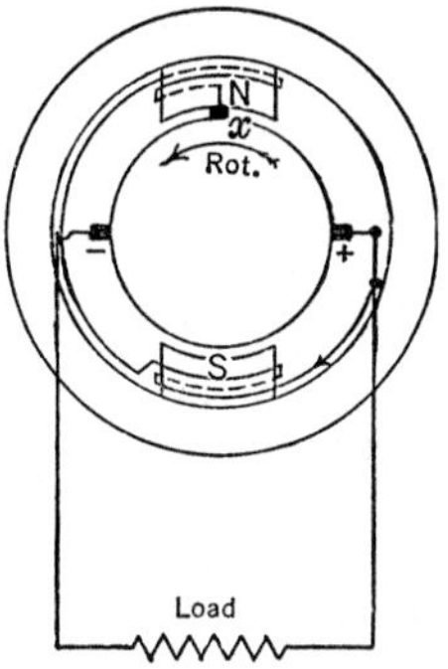

FIG. 48. Third-brush Generator

Another type of variable-speed generator is the Rosenberg car-lighting generator (Fig. 49), which has a constant-current-output characteristic. The fundamental regulating action occurs from the difference between an initial in-phase flux and an opposing flux produced by the armature cutting a quadrature flux set up by the quadrature field of the armature and produced by the auxiliary set of short-circuited brushes located 90 deg away from the main brushes. Based upon this principle, a rotating amplifier has been developed and widely applied as a regulator for voltage, current, or speed. Pole face windings are used on the output axis to eliminate reaction between the 2 axes. For

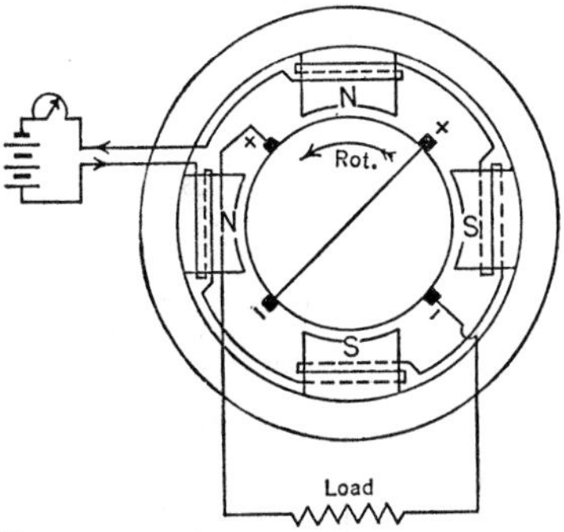

FIG. 49. Rosenberg Car-lighting Generator

improved performance a 2-pole armature winding is used in a 4-pole field frame, which is completely laminated.

EXCITERS. Small d-c machines acting as exciters require the following special considerations when self-excited:

1. They must be stable down to the no-load excitation requirement of the a-c machine in which no field rheostat is used. This is accomplished as described in Article 9.

2. They must have a nominal response of at least 0.50 (see AIEE rules). This requires the shunt field voltage to be about 50 per cent of the terminal voltage at rated exciter load. The rate of change of exciter voltage is calculated by:

$$\frac{dE}{dt} = -\frac{K(e - ir) \times 10^8}{np}$$

where E = induced volts.
K = ratio of induced volts to air-gap flux.
e = voltage across shunt field (constant if separately excited).
ir = instantaneous excitation current times field resistance.
k = leakage factor (approximately 1.2).
n = turns per pole.
p = number of poles.

A more general expression for rate of change of exciter field current during a transient when *separately* excited is:

$$i = I\epsilon^{-rt/L} \quad \text{(for decreasing current, field short-circuited)}$$

$$i = I(1 - \epsilon^{-rt/L}) \quad \text{(for increasing current)}$$

where i = the instantaneous current.
I = the maximum value of field current.
r = the total field circuit resistance.
L = the total circuit inductance.

The inductance of the shunt field circuit is $nd\phi/10^8di$, where n is the total number of turns per coil × the number of coils in series × the leakage factor, and $d\phi/di$ is the average value of flux per ampere of field current in a coil.

3. They should have a high value of amperes per square inch of brush cross-section at rated exciter load, since an exciter operates a great portion of the time at much less than rated load. A low current density in the brushes is liable to cause threading and grooving of the commutator surface.

4. Brushes should be replaceable during normal operation, since some a-c generators are not shut down for periods of a year or more.

17. EXAMPLES OF DESIGN AND PERFORMANCE

In Tables 7 and 8 will be found complete design and test data of four representative d-c machines.

Table 7. Mechanical Data on Typical D-C Machines

(Dimensions in inches, weights in pounds)

Type	1 Motor	2 Motor	3 Generator	4 Generator
Poles	2	4	6	12
Rating	1 hp	40 hp	500 kw	1,750 kw
Revolutions per minute	1,750	1,750	1,200	514
Volts	230	230	240	600
Current	4.4	145	2,080	2,917
Armature, diam out	3.75	12	25	66
Armature, diam in	1	4.25	14.5	48
Armature, total length	3.25	4.5	13.5	13
Air ducts	0	1 × 0.375	4 × 0.375	5 × 0.375
Slots, number	18	37	99	222
Slots, dimensions	5/8 × 0.385/0.165	1.370 × 0.35	1.81 × 0.30	1.90 × 0.46
Conductors per slot	84	8	2	6
Conductor, size	#22 dcc	0.05 × 0.50	0.20 × 0.35	0.70 × 0.16
Conductors in par	2	2	6	12
Type winding	14 TW	1 TW	1 TL	1 TL
Pitch of coils, slots	9	9	16	18
Air-gap, length	0.03	0.125	0.31	0.312
Pole arc, inches	3.66	6.31	9	12.25
Pole length, axial	3.0	4.5	13	13
Magnet core, length	3.0	4.5	13	13
Magnet core, width	2.375	4.0	6.5	9
Magnet, radial length	1.188	4.875	10.5	12
Yoke, length	3.00	11	23	20
Yoke, radial depth	1.0	1.25	2.25	4
Shunt spool, turns	2,679	1,400	642	392
Shunt conductor, size	#29 scc	#18 scc	0.07 × 0.15	#6 sq dcc
Series spool, turns	0	0	1	0
Series cond., size	0	0	1.5 × 0.625	0
Commutator, diameter	2.75	8.5	16.5	38
Commutator, length	1.625	5.25	15.5	22.5
Commutator, segments	54	147	99	444
Studs × brushes	2 × 2	4 × 4	6 × 10	8 × 16
Dimensions, each brush	5/16 × 5/8	0.5 × 1.0	1.125 × 1.25	0.75 × 1.00
Interpole, arc	5/8	1.25	2	1.75
Interpole, core	0.625 × 3.5	1.25 × 4.5	2 × 12	1.75 × 12
Interpole, turns	316	30	3.5	4
Interpole, cond.	#18 scc	0.10 × 1.0	2.9 × 0.65	1.2 × 1.5
Interpole, gap	0.050	0.189	0.56	0.50
Weight	120	1,150	10,000	45,000

NOTE. scc = single cotton covered, dcc = double cotton covered.

Table 8. Electrical and Magnetic Data on Typical D-C Machines

(Percentages are all in terms of rated or full-load values)

	1 Motor	2 Motor	3 Generator	4 Generator
Rating	1 hp	40 hp	500 kw	1,750 kw
Revolutions per minute	1,750	1,750	1,200	514
Rated voltage, volts	230	230	240	600
Rated current, amp	4.4	145	2,080	2,917
Flux per pole, 10 maxw	0.425	1.25	6.2	8.3
Leakage coefficient	1.12	1.15	1.18	1.18
Excitation, no load, amp-turns	1,025	3,600	5,790	8,620
Excitation, full load, amp-turns	1,025	3,600	6,620	8,870
Armature reaction, amp-turns	760	1,330	5,720	9,000
Stability factor	1.35	2.25	1.03	0.99
Excitation to balance arm reaction, amp-turns	91	107	545	0
Excitation at 110% voltage, amp-turns	1,190	4,220	7,450	11,800
Volts per bar, volts	8.6	6.3	14.6	16.2
Reactance voltage, volts	6.0	3.5	3.4	4.8
Brush friction, watts	8	30	3,480	5,970
Friction, other, watts	18	1,390	4,820	19,000
Core loss, watts	10	80	2,900	12,700
Armature resistance, 25 deg cent, ohms	4.62	0.0345	0.00084	0.00160
Series field resistance, ohms	....	....	0.00025	....
Interpole field resistance, ohms	1.95	0.007	0.00042	0.00132
Shunt field resistance, ohms	535	65.2	123	6.70
Friction loss, total, per cent	3.49	4.93	1.66	1.81
Core loss, per cent	1.34	0.80	0.58	0.73
Shunt field loss, per cent	11.8	2.55	0.46	0.32
Armature, *IP*, and Series, I^2R, per cent	18.2	3.36	1.56	1.71
Brush ($2 \times I_{\text{Arm}}$) loss, per cent	1.07	0.95	0.83	0.33
1% load loss, per cent	1.0	1.0	1.0	1.0
Efficiency at rating, per cent	73.0	88.0	94.2	94.5
Rise in temperature by thermometer	....	....	....	....
Armature, deg cent	37	32	31	30
Field, deg cent	30	34	32	22
Field rise by resistance, deg cent	40	45	41	26

DIRECT-CURRENT MOTORS

Revision by J. F. Sellers of Article by W. I. Slichter

A motor is a dynamo-electric machine for converting electrical power into mechanical power; i.e., it performs the converse function of a generator. Direct-current generators and motors are always interchangeable in function, although a machine which is designed specifically for a motor would probably not make a first-class generator, and vice versa. Motors of less than 5 hp are usually bipolar; larger machines are multipolar.

The applications of motors are treated in detail in Section 16, Industrial Applications of Motors and Servomechanisms (q.v.). The chief applications of d-c motors are the following:

Shunt Motor. Driving shafting, machine tools, blowers, reciprocating pumps; motor generators.

Series Motors. Railway and all other transportation work; hoists; cranes.

Compound Motor. Elevators, hoists, and machinery that must be started often.

Differential Motor. Very special applications of small units for peculiar speed conditions.

18. CLASSIFICATION

There are four types of d-c motors, differentiated by their characteristics and the connection of the exciting windings or circuits.

SHUNT MOTOR (Fig. 1). This motor has only one exciting winding, which is connected across the armature terminals and is thus in parallel or in shunt with the armature. The field winding consists of a large number of turns of fine wire on each pole, and usually the windings on all the poles are connected in series in one circuit. The current in the field depends upon the line voltage and upon the resistance of the field winding. The resistance of the field winding is purposely made high so that the field current will be between 1 and 5 per cent of the full-load current of the motor. The characteristic of the shunt motor is a fairly constant speed for all reasonable values of load.

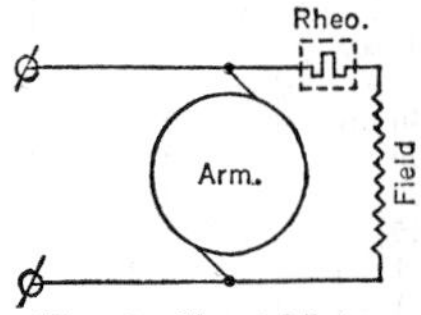

FIG. 1. Shunt Motor

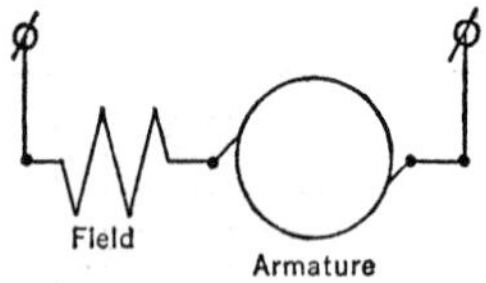

FIG. 2. Series Motor

SERIES MOTOR (Fig. 2). This motor has only one exciting winding, which is connected in series with the armature, so that all the current flows through the field as well as the armature. The field winding consists of a few turns of thick wire on each pole, and the windings on all poles are connected in series. The current in the field depends upon the load and is thus large with heavy load and small with light load. The resistance of the field winding is purposely made low so that the loss of voltage and power in that circuit will be small. The characteristics of a series motor are a speed varying with every change in load, high speed at light load and low speed at heavy load. The efficiency is high throughout a wide range of speed. The speed is dangerously high at no load; thus a series motor must always be connected rigidly to its load. Since the torque is high at low speeds, this motor is particularly adapted to work requiring frequent starting.

COMPOUND (OR CUMULATIVE) MOTOR. This motor has both a series winding and a shunt winding on each pole, wound and connected so that the two windings assist each other in the production of magnetism. It is a combination of a shunt and a series motor, designed to give the good starting qualities of the series motor and to avoid the danger of excessive speed at light loads. See also Section 16, Industrial Applications of Motors and Servomechanisms.

DIFFERENTIAL MOTOR. This motor has a shunt and a series winding connected so that they oppose each other in the production of magnetism. The motor therefore has poor starting qualities; it increases in speed with increase in load but has no tendency to run at a dangerously high speed. The applications of this motor are very limited.

ENCLOSED VS. OPEN TYPE. These terms refer to the mechanical housing of the motor. The open type has all its parts freely exposed to the air and is therefore well ventilated. It is intended to be used indoors and in protected places. The enclosed type is for use in exposed locations where there may be dampness or dirt. Special means must be provided to circulate the air inside the machine, but even then an enclosed motor is larger and more expensive than an open motor of the same capacity.

The relative capacities in output of open, semienclosed, and totally enclosed motors are shown by the data in Table 1 on one of a line of typical commercial motors. In general, an enclosed motor weighs about 50 per cent more than an open motor of the same capacity in spite of the fact that it is allowed to operate at 15 deg cent higher temperature by commercial convention.

Table 1. Data on Typical Commercial Motors

Type	Output, hp	Temp. Rise, deg cent	Weight for 700 rpm and Given Temp. Rise, lb
Open	10	40	970
Semienclosed	8	40	1000
Totally enclosed	5.75	55	1100

19. RATINGS

See Standards of the AIEE for limiting rise in temperature and for classification of types, Constant Speed, Multispeed, Adjustable Speed, and Variable Speed.

VOLTAGE AND CURRENT. Usual values of voltage for d-c motors are:

110–125 for small motors on lighting circuits.

220–250 for motors in factories, shops, etc.; on power mains or on the outside mains of a three-wire system.
500–600 for general railway work.
1200 for special railway installations.

The current required for any motor is found by the relation

$$\text{Current} = \frac{\text{Output in hp} \times 746}{\text{Efficiency} \times \text{Voltage}}$$

20. PRINCIPLES

The principles upon which a d-c motor operates are the same as those upon which a d-c generator operates (see Direct-current Generators). These principles are briefly as follows:

FORCE ACTING ON CONDUCTOR. A conductor of length l carrying a current I and placed in a magnetic field having a flux density B is acted upon by a force which is proportional to BlI and is in a direction at right angles to the direction of the magnetic flux and at right angles to the length of the conductor.

This in practical form gives the relation

$$T = \frac{p\phi ZI}{852m \times 10^6}$$

where T = torque of an armature in pounds at 1-ft radius.
p = number of poles.
m = number of armature paths between brushes.
ϕ = flux per pole in armature (lines).
Z = number of active conductors or inductors on armature.
I = current taken by the armature from line.

This torque is exerted whenever a current flows and is independent of the speed. The core loss and friction absorb some of the torque, so that the torque at the pulley is slightly less than the value given by the formula.

COUNTER EMF IN CONDUCTOR. A conductor of length l moving with a velocity V in a magnetic field of density B has induced in it an electromotive force E proportional to BlV. In practice, as soon as the armature starts to move, a counter emf is induced in its conductors, which has the value

$$E = \frac{p\phi Zn}{m \times 10^8}$$

when n = revolutions per second.

Thus, as soon as the armature moves, this counter emf tends to stop the flow of current, and the impressed emf must be increased to maintain the flow of current.

The relation between current and counter emf is given by the equation

$$E_i = E + IR$$

where E_i = impressed emf, E = counter or generated emf, and R = resistance of armature circuit.

In practice R is made as small as possible so that E and E_i are as nearly equal as possible.

REVERSING ROTATION. From a consideration of the equation for the torque it is evident that torque is proportional to the product ϕI, i.e., to the product of the flux in the armature by the current. If the direction of the current through the armature is reversed, i.e., if I becomes negative, the product becomes negative and the torque is in the opposite direction. If the direction of the flux is changed (the armature current being unchanged), the direction of torque is reversed. If both ϕ and I are reversed, however, the torque is not reversed. From this follows the rule for reversing the direction of rotation of any d-c motor: Change the direction of flow of current in either the field or armature winding, but not in both.

SPEED CONTROL. From the equation for counter emf it follows that the speed is proportional to E/ϕ; i.e., the speed varies directly as the counter emf and inversely as the flux. Thus, to reduce the speed, decrease the counter emf by decreasing the emf impressed on the armature or increase the flux by increasing the field current. To increase the speed, perform the converse. To decrease the emf impressed on the armature, insert a resistance between the source of potential and the armature terminals. This is the customary manner of controlling the speed during the starting of motors. (See also Starting of Direct-current Motors, Article 23.)

21. SHUNT MOTOR

Since the flux in the armature of a shunt motor is practically independent of load, the characteristics of the motor are: approximately constant speed for all reasonable variations of load, torque directly proportional to the armature current irrespective of speed, efficiency high throughout, a wide range of load but for only a small range of speed (see Fig. 3).

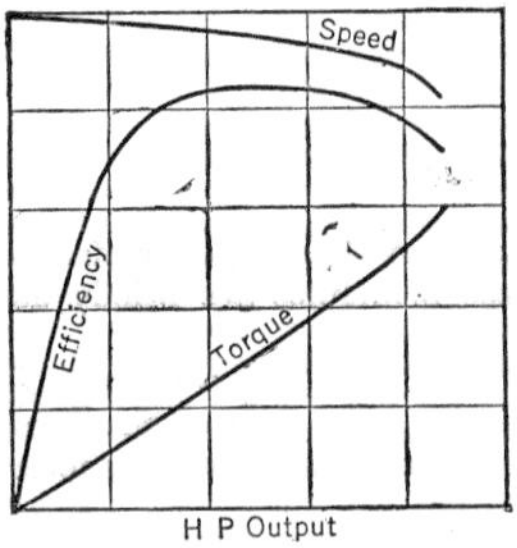

Fig. 3. Speed, Torque, and Efficiency Characteristics of a Shunt Motor

DESIGN OF SHUNT MOTOR. The method of design and calculation of shunt motors is the same as for d-c generators (see Direct-current Generators) except for the minor details noted below.

The armature reaction of a motor is in the opposite direction to that of a generator running in the same direction, and thus the field is distorted in the opposite direction. Hence, if the brushes are to be moved to assist commutation, they must be moved in a direction opposite to the direction of rotation of the armature.

The effect of armature reaction is to weaken the field. This causes a tendency to increase the speed.

The stability factor of a motor must be greater than that of a generator because, when the motor drops in speed as the load comes on, the field must be weakened to increase the speed. Hence the field is likely to be operated at an excitation less than normal.

TESTING OF SHUNT MOTORS. (See also Standards of the AIEE.) The tests on shunt motors may be divided into two classes: (*a*) commercial, to determine the qualities and serviceability of particular motors; and (*b*) special, to determine the general characteristics and actions of a type of motor. Commercial tests are the following:

Resistance measurements.
Stray-power test, including core loss and friction.
Input-output test for heating, efficiency, commutation, and speed regulation.
Insulation test.

RESISTANCE MEASUREMENTS are made first with the machine cold and later after the heat run. The resistances of the armature winding and field winding are measured, and the brush-contact resistance may be measured but is usually calculated (see Direct-current Generators). For any value of current the resistance losses (RI^2) are calculated from the measured hot resistances. (See Article 14.)

STRAY-POWER TEST. The term stray power is applied to the lumped sum of the core loss and the loss due to friction, bearings, and windage. The stray power of a d-c machine can be determined approximately by impressing normal voltage on the field and letting the machine run as a motor without load, varying the voltage impressed on the armature from about 10 per cent above normal to about 10 per cent below normal. The speed, armature voltage, and armature current for each adjustment are observed. Then the stray power for any induced voltage is equal to the armature input less the corresponding RI^2, where R is the armature resistance and I the armature current. The value of stray power for any given load on the motor is then equal to the measured value corresponding to the same counter emf E, where E is calculated from the impressed voltage E_i by the relation $E = E_i - RI$.

If it is desired to determine the stray power more accurately by taking into account the effect of armature reaction, the field current may be adjusted so that the speed on the above run is the same as the load speed. This gives a flux of the same average value as when the machine is under load.

CALCULATION OF EFFICIENCY FROM LOSSES. Let P = total output in kilowatts; R_a = resistance of armature, including brushes; I_a = armature current; I_f = field current; E_i = impressed voltage; S = stray power. Then the per cent efficiency is

$$\epsilon = \frac{100P}{E_i(I_a + I_f)} = \frac{100P}{P + E_iI_f + I_a^2R_a + S + 1\%\text{ load loss}}$$

INPUT-OUTPUT TEST WITH PRONY BRAKE. The input-output test may be made either by means of a prony or band brake on a pulley, or by using a d-c generator as a load. If the brake test is made, the output is

$$\text{Watts} = \frac{PLN}{7.04}$$

where P = net pull in pounds, L = lever arm or radius in feet, N = revolutions per minute

INPUT-OUTPUT TEST WITH GENERATOR AS LOAD. With large motors it is desirable to use a generator as a load in making an input-output test. In this case it is necessary to know the resistance of the generator armature circuit. It is also desirable to have the generator separately excited and to maintain a constant excitation throughout the entire test.

The input of the motor and the output of the generator, together with the speeds of both machines, are observed. A "counter-torque" test must also be made to determine the belt friction loss and the core loss and friction of the generator. This is performed by making two tests as follows:

a. The motor input is observed when driving the unloaded generator at normal speed, first through the regular leather belt and second through a light cotton belt. The difference in input to the motor in the first and second cases gives the belt-friction loss. As this loss is comparatively small, it may frequently be neglected.

b. A regular stray-power test (see above) is made on the generator when entirely disconnected from the motor. This gives the core loss and friction of the load machine (generator).

Then, for any load during the load run, the output of the motor under test is

$$P_1 = P_2 + R_2I^2 + S + F + 1\%\ \text{load loss,}$$

where P_1 = output in watts of motor, machine 1.
P_2 = output in watts of generator, machine 2.
R_2I^2 = loss in armature winding of generator.
S = stray power of generator for speed and induced voltage at observed load.
F = belt-friction or gear loss.

The ratio of this motor output to the electrical input as observed gives the efficiency of the motor.

HEAT RUN. From the input-output test it is also possible to determine the speed regulation, commutation features, and heating. The heat run may also be made by "bucking" two machines as described in Article 14. Small motors will reach a constant temperature in a short time, and the heat run need only last 5 or 6 hr for a 100-hp motor. A thermometer is usually placed on the machine in a safe and accessible place and read every half hour until it indicates no further rise in temperature.

INSULATION TEST. If the motor has been exposed to dampness, it may be desirable to make the test after the motor has been thoroughly dried out. The method is indicated in the Standards of the AIEE.

SPECIAL TESTS. As a special test there may be obtained a saturation curve of the machine and possibly the distribution of potential around the commutator. These are of particular interest in an adjustable-speed motor. In some of these motors with commutating poles some very high voltages may exist between bars which are not evident except in the bar-to-bar potential test.

22. SERIES MOTOR

Since the flux in the series motor is produced by the load current, the flux increases with the current. The torque is proportional to the product ϕI and therefore increases more rapidly than the current. Thus four times full-load torque can be obtained with from two to three times full-load current.

The characteristics of the series motor are: increase of torque faster than increase of current, variation in speed inversely as the load, and high efficiency throughout a wide range of speed as well as load (see Fig. 4).

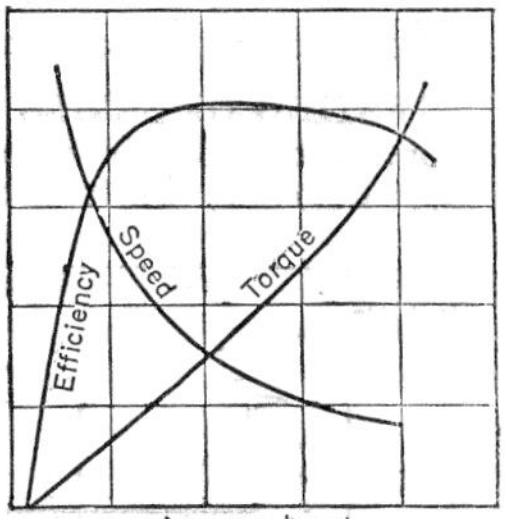

Fig. 4. Speed, Torque, and Efficiency Characteristics of a Series Motor

DESIGN OF SERIES MOTORS. In general the method of calculation is the same as for a d-c generator. The special considerations are:

A series motor is usually designed to have a large output and low speed at the 1-hr rating. At any lesser output the speed will be higher, so the peripheral velocity must be quite moderate at the rated load and speed.

Since the speed is very nearly inversely proportional to the flux, the speed curve depends on the shape of the saturation curve, to which very careful attention is paid in designing. By exactly fixing the flux for two extreme values of current, the speed for these two values of current is fixed.

The relations between the speed and current of a series motor are shown by the formulas

$$E_i = E + IR$$

$$E = \frac{p\phi Zn}{m\,10^8}$$

$$n = \frac{m(E_i - IR) \times 10^8}{p\phi Z}$$

where E_i = impressed voltage.
E = counter emf induced in armature.
R = total resistance of armature and field.
I = current taken by motor.
p = number of poles.
m = number of parallel paths between positive and negative brush sets.
Z = total number of conductors on armature.
n = speed of armature in revolutions per second.
ϕ = total flux per pole in maxwells.

Since the current in the field of a series motor is the same as that in the armature, the ratio of the turns in each is the same as the ratio of ampere-turns or mmf's. Thus, if the mmf per pole of the field is to be 1.5 times that of the armature, the number of turns will be 1.5 times the number of turns in series in the armature.

Since a series motor is usually an enclosed motor with a 1-hr rating, its rise in temperature and rating are a direct function of the watts lost and the ability of the mass of the motor to store up this heat energy. In a 1-hr run the amount of energy radiated is only about 10 per cent of the amount stored in the mass. For a rise in temperature of 75 deg cent in 1 hr there should be about 0.4 lb of material for each watt of loss. This assumes reasonable provision in the construction of the motor for the transfer of the heat from the armature to the field and frame.

Much attention has been directed recently to the ventilation of these motors by drawing air from outside the motor by means of fan blades on the armature and by circulation of the air inside the motor through definite paths. This has considerably increased the weight efficiency of these motors.

In railway motors, which are the most general application of the series motor, commutating poles are very generally used, as this construction makes it possible to obtain a much greater momentary output from a motor of a given size (see Table 2).

Table 2. Design Constants of a Typical Series Motor for Railway Service

Rated horsepower, 1 hr	50	Commutating pole turns	58
" voltage	600	Armature reaction, amp-turns per pole	3750
" current	80	Magnetic density in gap	34,000
" rpm	660	Amp × conds per inch, sigma	720
Number of main poles	4	Friction inc. gearing, watts	2400
" " commutating poles	2	Core loss	4000
Armature diameter × length	13.25 × 14	Copper loss	4400
" slots	25	Efficiency inc. gears, at rating	77.5
" conds over slot	30	Resistance of armature	0.322
" winding	S.D.	" " main field	0.190
Useful flux per pole	3.3×10^6	" " commutating field	0.139
Field turns per pole	60	Weight of motor, lb	2600

TESTING OF SERIES MOTORS. (See also Standards of the AIEE.) To determine properly the speed and torque characteristics of a series motor an "input-output" test must be made, which involves subjecting the motor to actual load and overload conditions. This may be accomplished by running the motor with a prony brake as a load or with a direct-connected generator as a load (see Testing of Shunt Motors, in Article 21).

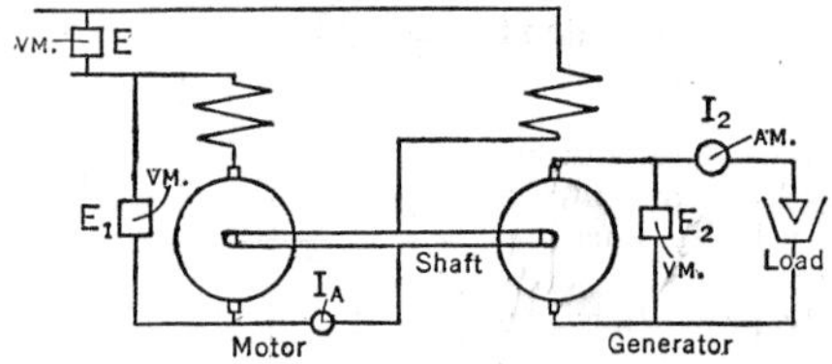

FIG. 5. Railway Motor Load Test

RAILWAY MOTOR TEST. When two similar motors are available, the method used by the manufacturers of railway motors is most desirable. The two motors are direct connected, or geared to each other, and the electrical connections made as in Fig. 5. The test is run by keeping constant rated voltage on the motor and regulating the load on the motor by changing the load on the generator.

As the two machines are operating under almost exactly the same conditions, their efficiencies are very nearly the same. Thus

$$\text{Efficiency of set} = \frac{E_2I_2}{EI}$$

$$\text{Efficiency of each motor} = \sqrt{\frac{E_2I_2}{EI}}$$

The speed and torque curves should be made for both directions of rotation of the armature, as an incorrect brush setting will give results differing with the direction of rotation. The direction of rotation is changed by reversing the connections of either the field or the armature of the motor.

Commutation is observed during the speed and torque test.

The heat run is made with the same arrangement as the speed-torque test. In making the heat run, the motor must start cold or at room temperature. The covers of the inspection openings of railway motors are customarily left open during the heat run.

LOSSES AND EFFICIENCY OF SERIES MOTORS. For a more accurate determination of the efficiency and losses the following special tests are made:

1. Resistance of armature, brushes, and field. These tests are similar to those for a shunt motor (see Article 21).

2. Core-loss test. On account of the variable speed and variable field of a series motor this test consists in repeating, at several different speeds, the usual core-loss test as described for a generator. The field strength is varied step by step throughout the maximum range for each speed. Figure 6 shows the curves for these different runs, and the dotted line connects the points on the different curves that apply to the normal speed curve of the motor.

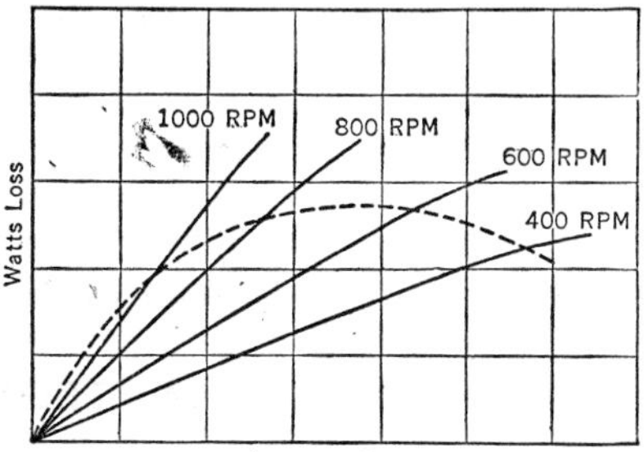

FIG. 6. Core-loss Curves of Series Motor

INSULATION TESTS. See Direct-current Generators and Standards of the AIEE.

23. STARTING OF DIRECT-CURRENT MOTORS

A starting box (Fig. 7) or rheostat is always employed in starting d-c motors in order to reduce the voltage impressed on the motor when it is not running at a high enough speed to generate the proper counter emf.

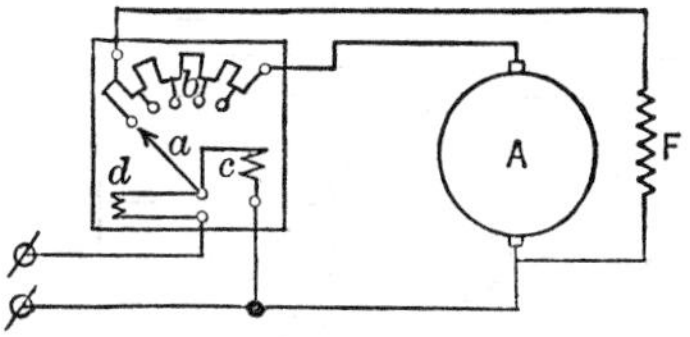

FIG. 7. Starting-box Connections for Shunt Motor

Let E_i = line voltage.
I^2 = armature current at full load.
N_0 = speed at full load (shunt motor).
E_0 = counter emf at full load.
I = $1.5I_0$, usually accepted peak starting current.
r = resistance of the armature circuit.
R = resistance of starting box or rheostat.
E = counter emf at any other speed, N.

Then, in general:

$$\text{Current} \frac{E_i - E}{r + R} \quad \text{and} \quad N = \frac{N_0E}{E_0}$$

The total resistance in the rheostat on the first step should be:

$$R_1 = \frac{E_i}{I} - r$$

the counter emf at the end of the first step:

$$E_1 = E_i - I_0(R_1 + r)$$

and the constant speed:

$$N_1 = \frac{N_0E_1}{E_0}$$

For the second step the total resistance in the rheostat should be:

$$R_2 = \frac{E_i - E_1}{I} - r$$

At the end of the second step:

$$E_2 = E_i - I_0(R_2 + r), \quad \text{and} \quad N_2 = \frac{N_0 E_2}{E_0}$$

Succeeding steps are determined until the value of R_2 decreases to zero. In general, for $I = 1.5I_0$, we have $1.5(R_2 + r) = (R_1 + r)$, and $R_n = {}^2/_3 R_{n-1} - {}^1/_3 r$.

Figure 8 shows the sudden rise in current when the resistance is changed and the gradual decrease in current as the speed increases. The number of steps necessary depends upon the ratio of the maximum allowable instantaneous value of the current to the final constant value, upon the value of the armature resistance, and upon the inertia of the load.

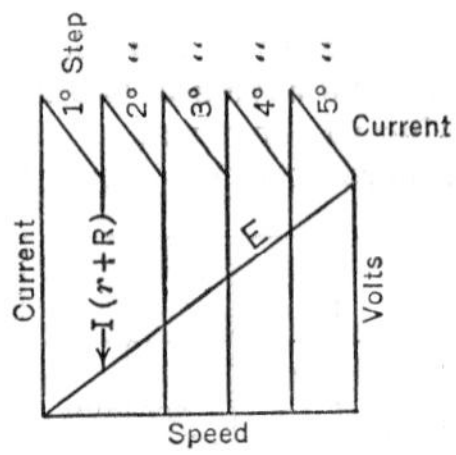

FIG. 8. Motor Current during Starting

STARTING BOX. (See article on Rheostats, Section 6.) A starting box usually contains the following features, as indicated in Fig. 7: (*a*) a means of opening and closing the circuit supplying all the current to the motor, including the field current; (*b*) a set of resistance steps in series with the motor armature and a means of short-circuiting this resistance step by step; (*c*) a magnet coil connected across the motor terminals to open the circuit if the impressed voltage fails or falls below a specified value (low-voltage release); (*d*) a magnet coil carrying the main current to ctuate a spring and open the circuit if the current exceeds a specified value (overload release). The usual connections of a starting box to the line and motor are shown in Fig. 7.

24. SPEED CONTROL

There are three methods of varying and controlling the speed of d-c motors, namely, potential, rheostatic, and field systems.

POTENTIAL CONTROL OR MULTIVOLTAGE SYSTEM. By means of several generators and several wires various definite voltages are made available, such as 240, 180, and 120. By connecting the motor to the 240-volt circuit, full speed is obtained; by connecting to the 180-volt circuit, $^3/_4$ speed is obtained, etc. The shunt field circuit is left connected at all times to a circuit of the proper voltage. A shunt motor with normal field excitation will be stable, that is, it will operate constantly, at the fractional speed.

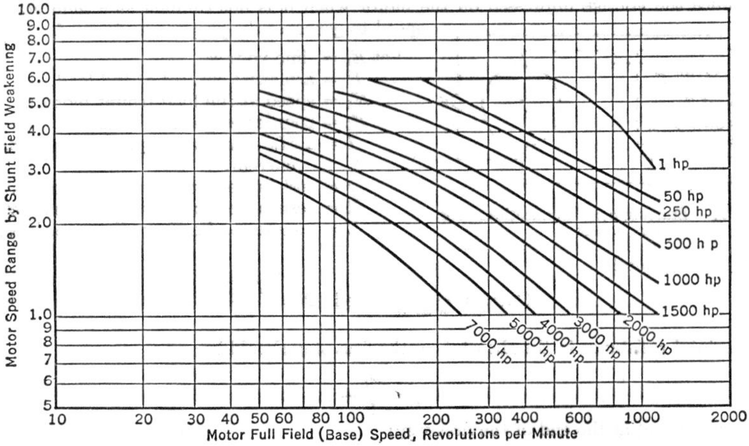

FIG. 9. Maximum Speed Range by Shunt Field Weakening for Standard High-voltage Motors

The efficiency will be good at the fractional speeds. A series motor controlled in this manner will be unstable, but for a given torque the speed will be roughly proportional to the voltage.

RHEOSTATIC CONTROL. A rheostat in series with the armature will reduce the voltage impressed on the armature by an amount proportional to the current and thereby reduce the speed. The speed is unstable with this arrangement, changing with every change of load, and the efficiency is poor.

FIELD CONTROL. By increasing the resistance in series with the field of a shunt motor, the speed is increased because of the weakening of the field. If the motor has commutating poles to assure good commutation, the speed may be varied in a ratio of 1 to 4, and even 1 to 6 in small sizes (see Fig. 9). The shunt motor is stable with this method of control, and the efficiency is good. In a series motor the field may be shunted by a resistance to increase the speed.

25. LOW-INERTIA MOTORS

INERTIA OF VARYING-SPEED MOTORS. Many d-c motors are now being used on rolling mill, calendar, and continually varying-speed applications in which the stored

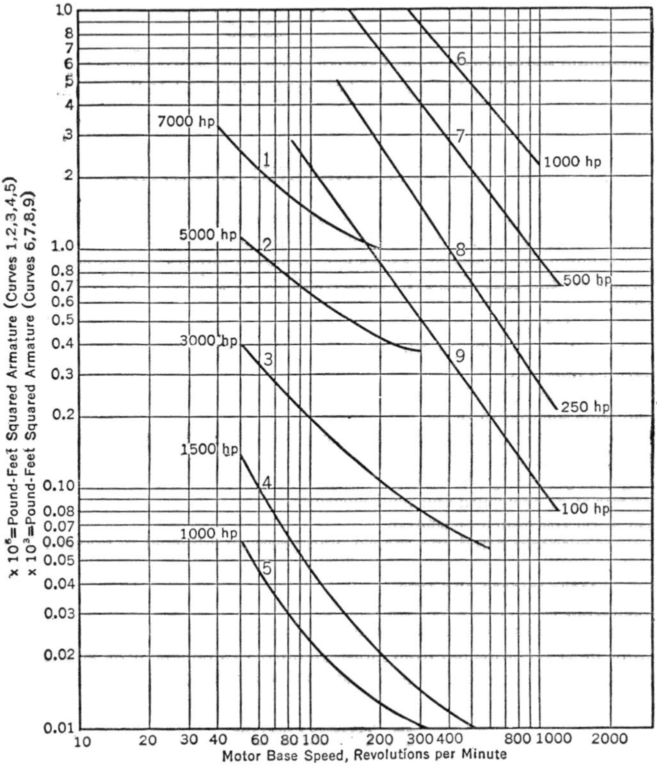

FIG. 10. Pound-feet-squared of Armatures of Typical Low-inertia Motors

energy in the motor armature is an important factor in limiting production. In normal d-c designs the armature core length is not more than 150 per cent of the pole pitch, but it is possible to extend the core length to as much as 400 per cent of the pole pitch in large machines (with armature diameters larger than 50 in.). Figure 10 gives an idea of the pound-feet-squared inertia of representative low-inertia motors.

To obtain the percentage of rated armature current to accelerate a motor uniformly by applying a constantly increasing armature voltage (the motor field current being constant from a separate source), the following formula is used:

$$\%\ \text{rated armature amperes} = \frac{100}{\text{Rated hp}} \times \frac{WR^2 \times \overline{RPM}^2}{t_{sec} \times 1{,}615 \times 10^6}$$

where WR^2 = motor armature inertia in pounds feet^2.

RPM = speed attained in t seconds with the armature current held constant.

To accelerate a motor with armature voltage held constant and the shunt field gradually weakened to maintain constant armature current, the time of acceleration in seconds, $t_{sec} = (WR^2 \times RPM^2)/(\text{hp applied} \times 3.23 \times 10^6)$.

The WR^2 of an armature, in pound-feet-squared, is

$$\text{Approximately its weight in pounds} \times \left(\frac{0.65 \text{ to } 0.75 \times \text{Inches of arm. radius}}{12}\right)^2$$

The stored energy in a rotating mass is

$$\frac{WR^2 \times RPM^2}{3.23 \times 10^6} \text{ in horsepower-seconds}$$

Also,

$$\text{Horsepower} = \frac{\text{Foot-pounds of work}}{t_{sec} \times 550} = \frac{2\pi \times \text{Torque} \times RPM}{33{,}000}$$

and the maximum horsepower, when accelerating a mass uniformly, is

$$\frac{\text{Stored energy (in horsepower-seconds)} \times 2}{t_{sec}}$$

26. INSTALLATION AND ERECTION

In the installation and erection of a d-c motor certain features must receive careful attention in order that the machine shall operate properly and not deteriorate with undue rapidity. Although this procedure varies with different motors according to their mechanical construction, the following brief memorandum of points to be looked after will be found useful:

1. Base bolted down.
2. Bearings clean and filled with oil.
3. Bearings lined up.
4. Magnet frame bolted to base.
5. Field coils secured in place.
6. Field coils tested for open circuit, wrong connection, and polarity.
7. Armature in place.
8. Air gap adjusted by shimming.
9. Resistance of armature and field measured.
10. Insulation resistance measured.
11. Brushes properly fitted and spaced, and pressure adjusted to about 1.5 to 2 lb per brush.
12. Commutator smooth and true.
13. Substantial connections of field circuit.
14. Field adjusted for correct direction of rotation.

The motor must be protected from moisture during shipment; if by accident it becomes damp, it must be dried out before it is subjected to a voltage.

27. OPERATION

In the operation of a d-c motor several factors should be considered.

CARE. All motors should be frequently inspected and the following points noted:

1. Bearings filled with proper amount of oil.
2. Brushes securely held in proper position.
3. Brushes fitted properly.
4. Commutator smooth: danger of "high mica" or the insulation between commutator bars projecting above the bars.
5. Air gap true.
6. Commutator not worn in grooves.

TROUBLES. In the following paragraphs is a concise list of the troubles that may be experienced in operating d-c motors and their causes, as given by Crocker and Wheeler in *Management of Electrical Machinery.*

1. *Sparking at the Commutator.* Causes: Armature carrying overload. Brushes improperly spaced. Brushes not at proper position. Rough commutator. Poor brush contact. Internal short or open circuit. Field too weak. Unequal strength of poles. Vibration.
2. *Heating of Commutator and Brushes.* Sparking. Bearing trouble. Bad connections. Brush friction too great.
3. *Heating of Armature.* Overload. Internal short circuit, moisture or ground. Reversed coil. Excessive eddy currents
4. *Heating of Field.* Internal short circuit.
5. *Heating of Bearings.* Bearings dry or dirty. Shaft out of true. Bearings out of line. Thrust due to belt. Unbalanced magnetic pull.
6. *Noise.* Armature not balanced. Brushes dry or not set at proper angle. Armature strikes.
7. *Speed Too Low.* Wrong voltage. Overload. Armature strikes. Bearing too tight.
8. *Speed Too High.* Wrong voltage. Field too weak.
9. *Motor Stoppage or Failure to Start.* Overload. Open circuit. Wrong connection.

ROTARY ENERGY CONVERTERS

Revision by J. F. Sellers of Article by W. I. Slichter

28. MOTOR-GENERATORS

A motor-generator set is a combination of a motor and a generator having separate fields and armatures but mounted on the same shaft with common base and bearings. Combinations of various types of motors and generators are used; some of the more important combinations and their applications are described below.

DIRECT-CURRENT MOTOR DRIVING DIRECT-CURRENT GENERATOR. These sets are used when it is desired to convert low-voltage direct current into high-voltage direct current, or vice versa; they are used in preference to a dynamotor (q.v.) when good regulation is desired in the secondary circuit.

DIRECT-CURRENT MOTOR DRIVING ALTERNATING-CURRENT GENERATOR. These sets are used for converting direct into alternating current. See also Art. 30 on Synchronous Converters.

INDUCTION MOTOR DRIVING DIRECT-CURRENT GENERATOR. The induction motor may be wound for potentials as high as 13,000 volts, and at this voltage the transformation from alternating current to direct current may be made without the use of transformers. Since the induction motor has a decreasing speed with increasing load, the d-c generator must be compounded to give good regulation; with proper compounding excellent regulation may be obtained. The induction-motor-generator set is sometimes used in preference to a synchronous converter when the service requires especially good regulation. However, the efficiency of such a set (about 85 per cent at rated load) is less than that of a synchronous converter, even if no transformers are required by the motor-generator set. The motor-generator set also occupies from 50 to 80 per cent more floor space, weighs from 30 to 50 per cent more, and costs from 25 to 50 per cent more than a synchronous converter, in spite of the fact that the set is designed to operate at the highest practicable speeds.

SYNCHRONOUS MOTOR DRIVING DIRECT-CURRENT GENERATOR. This combination is preferable in many instances to the preceding one, because it operates at constant speed and because the field of the synchronous motor may be adjusted to make use of line compounding or to compensate for low power factor in other apparatus on the circuit. Provision must be made for the d-c excitation for the synchronous motor. If the d-c generator is wound for too high a potential, a special exciter must be provided. Since there is no load on the set at starting, the synchronous motor may be started in the usual manner.

29. DYNAMOTORS

A dynamotor is a d-c device combining both motor and generator action in one magnetic field. It has an armature having two separate windings and two separate commutators, one at each end of the armature. Either winding may be used as the motor winding, and the other as the generator winding. Such a machine performs the same function in a d-c circuit that a power transformer does in an a-c circuit, i.e., serves as a means of transforming high-voltage direct current into low-voltage direct current, or vice versa.

PERFORMANCE CHARACTERISTICS. The device corresponds to two machines in which there is only one core loss, one friction loss, and one excitation loss, but two losses due to rI^2 in the two armature circuits. It is therefore more efficient than a motor-generator set, but less efficient than one machine. Since the currents in the two windings flow in opposite directions, their resultant magnetic effect is zero. The machine has therefore no armature reaction (except for the slight amount due to the current to overcome the losses in the machine). It is not subject to the troubles of field distortion and bad commutation that occur in either motors or generators. It is impossible to compound a dynamotor, since any increase in field strength intended to increase the voltage of the generator would decrease the speed of the motor by the same amount and no change would result. The ratio of the two voltages is therefore fixed by the number of turns and varies from this only by the loss due to rI drop in both windings. These two drops are additive, and therefore the regulation of such a machine is not very good.

APPLICATIONS. Dynamotors are used largely to provide large currents to start other motors or to supply low voltages or a fractional voltage in a multivoltage system for speed control. The motor of the combination may be wound for the line voltage and the generator for any fraction of the line voltage. Thus the combination supplies a large current at a low voltage, which will give a good starting torque in motors connected to it with a reasonable consumption of power. The combination is used as equalizer or "balancer" in three- or multiwire circuits but is not as desirable for this work as motor-generator sets with compound-wound machines, on account of poor regulation. Dynamotors are also used to supply a low voltage for such purposes as telephone and telegraph systems and the low voltage and large currents for electrolytic work.

30. SYNCHRONOUS OR ROTARY CONVERTERS

Since in general it is more economical to transmit electrical energy in the form of alternating currents and is frequently more convenient to utilize it in the form of direct currents, some means of converting from one form of electrical energy to the other is desirable. For this purpose synchronous converters and motor-generator sets (q.v.) are available. Synchronous converters are also called rotary converters.

A synchronous converter is a machine very similar to a d-c generator, in which certain commutator segments, or the conductors connected to them, are connected to 2, 3, 4 or 6 collector rings, as the case may be. When the movable member is caused to rotate, the voltage between any two collector rings is alternating. Such a machine, when driven by an engine or a motor, may be operated as an a-c generator or as a "double-current" generator giving alternating current from its collector rings and direct current from its commutator. If the collector rings are connected to a source of alternating currents, the machine will run as a synchronous motor, and direct current may be obtained from the brushes on the commutator; i.e., the machine, with but one set of windings, acts simultaneously as an a-c motor and a d-c generator. It has therefore the friction, core loss, and excitation loss of one machine instead of two; and, since the motor and generator currents flow in the same winding and during at least the major part of each cycle are in opposite directions, they more or less balance each other, and the armature rI^2 loss is much less than in either a motor or generator alone.

SYNCHRONOUS CONVERTER VS. MOTOR GENERATOR. A converter is much more efficient and weighs and costs less than a motor-generator set of the same capacity. It also occupies less space. Since only one winding is used, however, there is a definite relation between the emf's of the a-c and d-c terminals. The maximum value of the alternating wave bears a definite relation to the direct emf (see below). It is therefore necessary to supply the converter with a voltage of the same order as the direct voltage, and this involves the use of transformers if a high-voltage transmission line supplies the converter. Motors operating at voltages as high as 13,000 volts can be used in motor-generator sets.

RELATIVE EFFICIENCIES. The efficiency of a converter is in the neighborhood of 93 per cent and of the transformers 97 per cent; thus the efficiency of the combination is about 90 per cent. The efficiency of a synchronous motor is in the neighborhood of 93 per cent and of a d-c generator 92 per cent; thus the combination motor-generator set has an efficiency of 85.5 per cent. If the supply voltage is greater than 13,000 volts, transformers will also be needed for the motor-generator set, and the net efficiency would then be 83 per cent.

SYNCHRONOUS CONVERTER VS. RECTIFIERS. A converter differs from a rectifier (q.v.), since the converter gives a direct emf of constant and uniform value, and the rectifier a pulsating unidirectional voltage and current. In the converter the energy is stored in the form of magnetism for an instant, whereas in the rectifier there is no magnetic field and no storage of energy. A rectifier will not work satisfactorily on an inductive d-c circuit, but a converter will.

APPLICATION OF CONVERTERS. The most common application of synchronous converters is in electric railway work. The great majority of motors for electric traction are d-c series type, operating at 500 to 600 volts. The energy for these motors must be transmitted over long distances, which requires a high-voltage a-c transmission line and converters to link the d-c distribution with the a-c transmission. Converters have also been developed for high-voltage d-c traction systems, these converters giving 1500 volts of direct current. Two such converters may be connected in series for 3000-volt d-c distribution [*Elect. J.*, Vol. 12, p. 154 (Apr. 1915)]. Synchronous converters are also used for lighting and power service and for electrolytic work.

TERMINOLOGY. The following terms are used to describe certain characteristic features of the various kinds of synchronous converters.

Phases and Rings. A single-phase converter has two collector rings, and each ring is connected to the windings by as many equally spaced taps as there are pairs of poles. The taps for the two rings alternate at equal spaces. A single-phase converter is therefore a two-ring converter.

A three-phase converter has three rings and three equally spaced taps (one for each ring) for every pair of poles. A four-phase or quarter-phase converter has four rings and four taps for every pair of poles. A six-phase converter has six rings and six taps per pair of poles.

Shunt and Compound-wound Converters. A converter may be shunt or compound wound, depending upon the service for which it is intended. The series winding is intended to make the converter take leading current when the load increases and thus automatically increase the voltage at the a-c terminals, but the ratio of the a-c *terminal* voltage to the d-c voltage remains unaltered.

Converter with Series A-c Booster. A synchronous converter with series booster is frequently used when exact hand regulation of the d-c voltage is required, as for lighting circuits. It is merely a converter with a separately excited a-c generator on the same shaft, and each phase of the armature of this booster is in series with one phase of the converter and the line. This generator acts as a booster and raises the voltage impressed upon the armature of the converter.

Inverted Converter. Sometimes a converter is operated to convert from direct to alternating current. It is then called an "inverted converter." The machine will operate satisfactorily in this manner, but its speed depends upon the nature of the a-c load. An inductive load in the a-c circuit causes the armature to demagnetize the fields, with a resultant increase in speed. It is therefore dangerous to operate an inverted converter on an inductive load unless it is provided with a speed-limit device. This effect does not occur when the machine is operating as an a-c motor, since its speed is fixed by the frequency of the supply circuit.

Cascade Converter or Motor Converter. This is a combination of an induction motor and converter connected in series or concatenation (see Motor-converters, p. 8-49. The converter receives half the power in mechanical form from the shaft and half the power inductively in the form of alternating current at half frequency from the secondary of the induction motor. By this means the steadiness of a 30-cycle converter is obtained in a 60-cycle unit.

Split-pole Converter. The "split-pole" or "regulating-pole" converter is designed to give a variable ratio of alternating emf to direct emf for operation in parallel with storage batteries and similar purposes. The field poles are divided into sections, the excitation of which may be controlled independently. The effect of the split pole is primarily to produce a third harmonic in the flux distribution curve, which increases or decreases the *total* flux per pole and therefore the d-c voltage, but does not change the a-c counter emf between slip rings, since these rings are connected to the armature winding at points 120 electrical degrees apart. Split pole converters give a voltage variation of about 20 per cent.

Commutating-pole Type. In most large converters, 100 kw and above, it is customary to provide commutating poles to improve commutation and reduce the cost by making it possible to run with a higher value of "volts per bar." See Direct-current Generators, p. 8-23.

RATING AND PERFORMANCE. The standards of the AIEE specify a maximum rise of 50 deg cent on any part of the windings and 60 deg on the commutator for continuous operation. Twenty-five-cycle converters give momentarily an output of three times their normal rating without damage or injurious sparking at the commutator, but 60-cycle converters are more sensitive to overloads.

31. PRINCIPLES OF CONVERTER ACTION

In this section are briefly treated those features in which a synchronous converter differs in action from an a-c or d-c generator; see Table 1, p. 8-54, for a summary of the voltage, current, and capacity relations.

CONNECTIONS AND VOLTAGE RATIOS. The ratio of voltage on the a-c side to that on the d-c side depends upon the number of rings and the type of connection employed.

Two-ring Converter. The two collector rings are connected by taps to the same winding as the commutator; hence the alternating emf has the same values as the direct emf at the instant that the taps pass the brushes. As this is also the maximum value of the alternating emf, the effective value (i.e., the value to be indicated by a voltmeter) will be 0.707 times the maximum, or 0.707 times the direct emf. Figure 1 shows in a simple manner the connection of a single-phase converter. The circle represents a two-pole armature winding, and inside is shown the supply transformer connected to two taps diametrically opposite each other. The voltage across this transformer would be $0.707E$, where E is the voltage between the positive and negative brushes on the d-c side.

Four-ring Converter. If two additional collector rings are connected to conductors spaced halfway between the former taps, there results the quarter-phase converter shown

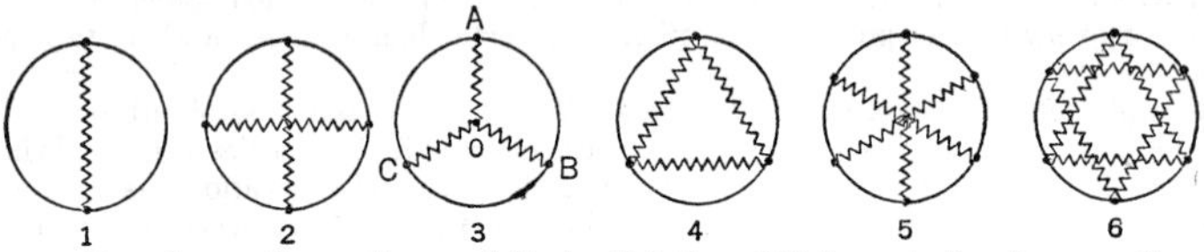

FIGS. 1–6. Transformer Connections and Vector Relation of Voltages in Synchronous Converter

in Fig. 2. The voltage across each supply circuit or transformer is the same as before, but the two voltages will differ in phase by 90 deg.

Three-ring Converter. If three collector rings are connected to taps spaced 120 deg apart, the emf between adjacent collector rings will be the vector sum of AO and OB in Fig. 3. Since AO and OB each equal $0.5 \times 0.707E$, the voltage AB will be $\sqrt{3} \times 0.5 \times 0.707E = 0.612E$. Thus in a three-ring or three-phase converter the voltage between adjacent taps is 0.612 times the direct voltage, and the connections of the transformer are as in Figs. 3 and 4.

Six-ring Converter, Diametrical and Double Delta Connections. A six-ring converter may be connected diametrically, as in Fig. 5, in which case the voltage of each transformer will be $0.707 \times E$ and the result will be like the combination of three single-phase groups. A six-ring or six-phase converter may also be connected "double delta" as shown in Fig. 6, which is similar to the combination of two groups of three-phase delta transformers. In both cases the voltage between adjacent taps of a six-phase converter is $0.355E$.

***n*-Ring Converter.** In general the voltage between taps of any converter having n equally spaced taps per pair of poles is

$$E_{ac} = \frac{E \sin \frac{\pi}{n}}{\sqrt{2}}$$

CURRENT RATIOS. To determine the ratio of the continuous current I to the alternating current I_3 per collector ring in a three-phase converter, for example, assume the d-c output equal to the a-c input with unity power factor; then

$$\sqrt{3}\, E_3 I_3 = EI$$

and, from the preceding paragraph,

$$E_3 = 0.612E$$

where E_3 = a-c voltage between lines; I_3 = alternating current per line; E = direct voltage; I = direct current. From these two relations

$$I_3 = 0.94I$$

The ratios of currents for other converters are obtained similarly and are given in Table 1 on p. 8-54. In actual practice the current on the input side must be greater than that given by these relations in order to supply the losses in the converter, and the alternating current will also vary inversely as the power factor, which is taken as unity in Table 1.

RESULTANT COIL-CURRENT. In any machine acting simultaneously as a motor and a generator, the two currents must flow in opposite directions, and the current in any particular conductor will be the difference between the two. In any particular coil the direct current is constant in amount and direction from the instant the commutator segment connected to this coil passes the positive brush to the instant it passes the negative brush, and conversely from negative to positive brush. In any coil midway between the a-c taps the alternating current is a maximum when this coil is halfway between brushes (for unity power factor), and the current falls to zero as the coil reaches the interpolar position. Therefore, for the period of time that the direct current in a coil remains constant in amount and direction, there is also in it a variable current changing from zero to a maximum and back to zero again. The net or resultant current will therefore have a wave shape and frequency somewhat as shown at *R* in Fig. 7. The heating in this particular coil will

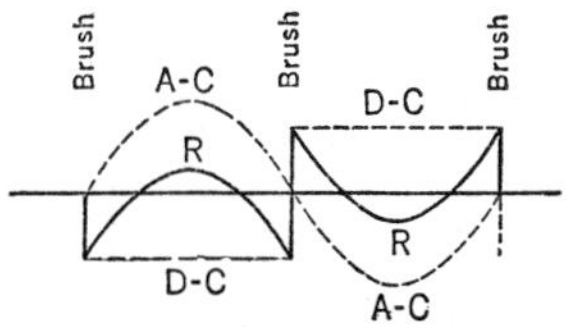

FIG. 7. Current in Coil Midway between Taps

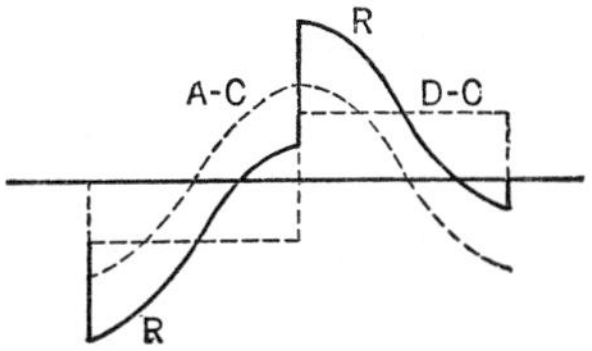

FIG. 8. Current in Coil at Tap in Two-ring Convertor

therefore be proportional to the product of the resistance of the coil and the square of this current, and it is readily seen that the power lost is less than that due to either the direct or the alternating current alone.

A coil situated very near one of the a-c taps will carry a direct current subject to the law before mentioned, but the alternating current in this coil is the same as that in the middle coil and has its maximum value when the middle coil is at the middle of the poles and not when the particular end coil is at the middle of the pole. The alternating current in this coil may therefore reach its maximum value soon after the coil has passed a brush, as shown in Fig. 8, and the resultant of the two currents is greater than that in the middle coil. The farther a coil is situated from the middle coil of a group, the greater is the phase displacement of its alternating current, the greater is the value of the resultant current, and the greater is the heating effect. Thus, although the heating effect in a rotary-converter winding is less than that of a d-c machine having the same output, it is different in each and every coil, is minimum in the central coil (when the power factor is unity), and is maximum at the coil nearest the tap, and the greater the angle between taps the greater is the total heating effect.

DISTRIBUTION OF HEAT LOSSES IN ARMATURE WINDING. In Figs. 9 to 12 inclusive, representing a two-ring converter, the numbers outside the circles represent

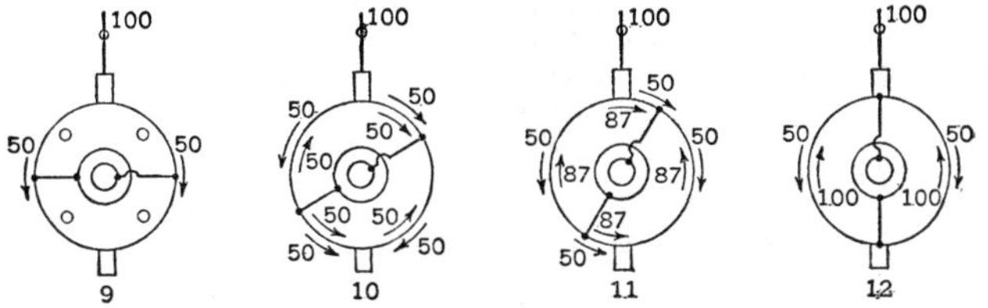

FIGS. 9–12. Component Currents in a Two-ring Converter

the direct current in the winding (corresponding to 100 amp in the external d-c circuit and unit power factor on a-c side), and the numbers inside the circles the instantaneous alternating current in the winding. It will be seen that the resultant current in a coil midway between taps (Figs. 9 and 12) never exceeds 50 amp. A coil 30 deg from the middle (Fig. 10) has a maximum current of 100 amp. A coil 60 deg from the middle (Fig. 11) has a maximum current of $50 + 87 = 137$ amp, and a coil 90 deg from the middle (tap coil) may have a current of $100 + 50 = 150$ amp. The heating of the armature winding is therefore not uniformly distributed, although the conduction of heat from one part of the winding to another tends to equalize the temperature rise.

If the converter operates at a power factor different from unity, the position of minimum resultant current is no longer at the middle coil but is moved one way with leading current and the opposite way with lagging current. Thus one end coil has improved heating conditions, but the middle coil and the other end coil have much worse heating conditions, the result being that the heating as a whole has increased and is more non-uniformly distributed than with unity power factor.

The power lost in each individual coil of a converter and the effect of the power factor on this loss is very well shown in Figs. 13 and 14, taken from a paper by J. E. Woodbridge (*AIEE*, 1908). In Fig. 13 the curved line shows the relative rI^2 loss in a coil having any position throughout 120 deg of one phase of a three-phase converter when the power factor is unity. The curve shows the ratio of the loss compared to the loss in the same machine acting as a d-c generator of the same capacity. It will be noted that the middle coil has a loss of 22 per cent of that of the generator and the end coils 120 per cent, and

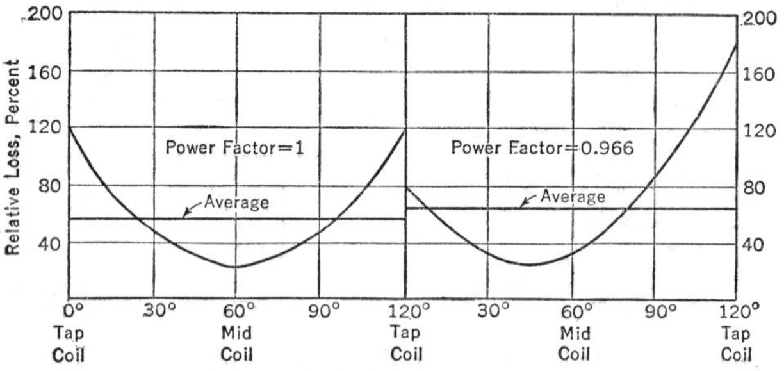

FIG. 13. Relative rI^2 in Individual Coils of Three-ring Converter

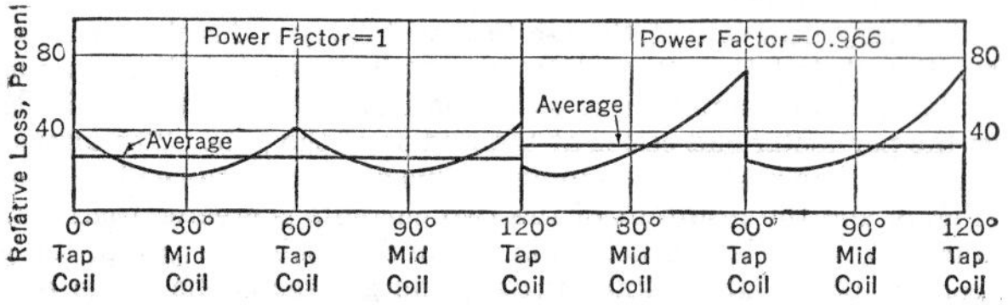

FIG. 14. Relative rI^2 in Individual Coils of Six-ring Converter

that the average loss is 57 per cent. For a power factor of 0.966, representing a phase displacement of current of 15 deg, the loss in individual coils ranges from a maximum value of 180 per cent at one tap to a minimum value of 22 per cent in the coil shifted 15 deg to one side of the middle coil. The other end coil has a loss of a little over 80 per cent of the generator loss. The average value has been increased to 65 per cent.

In Fig. 14 the same ratios are shown for a six-phase converter, in which the winding of one phase is distributed over only 60 deg. Consequently the conditions are better, and the maximum loss due to low power factor is less. (See Fig. 15.)

DEPENDENCE OF OUTPUT UPON NUMBER OF PHASES. As a result of these conditions we have the following relations of the capacity of a given armature with various numbers and connections of taps, the capacity or load-ability being based on an equal total amount of rI^2 loss.

It should be remembered that there are other losses besides coil losses in the converter, and that therefore practical figures are slightly different from those given in Table 1.

Table 1. Voltage, Current, and Output Ratios

	D-c Generator	Converters					
		2-ring	3-ring	4-ring	6-ring diametrical	6-ring double delta	12-ring
D-c volts	100	100	100	100	100	100	100
A-c volts between lines	...	71	61.2	71	71	61.2	71
A-c volts between rings	...	71	61.2	50	35	35	18
D-c amperes	100	100	100	100	100	100	100
A-c amperes in line	...	141	94	71	47	47	24
A-c amperes in winding	...	71	55	50	47	47	45
Relative RI^2 loss	100	137	55	37	26	26	20
Relative output:							
Unity power factor	100	85	134	165	197	197	224
87 per cent power factor	...	...	99	115	129	129	135

FIELD EXCITATION. The variation of the field excitation of a synchronous converter has much the same effect as in a synchronous motor (see Synchronous Motors, Section 9); i.e., if its field is underexcited, the armature will draw a lagging current which assists the field and sets up the necessary flux; whereas, if the field is overexcited, the armature will draw a leading current which opposes the field and reduces the flux. By this means the converter may be made to take either a leading or lagging current. A leading current flowing over a line having inductance tends to raise the voltage at the receiving end of the line.

Line Compounding. If there is sufficient leading current and sufficient inductance in the line, the voltage at the receiving end may be greater than at the sending end in spite of the resistance of the line. This is sometimes called "line compounding." The relation between the voltages at the two ends of the line may be expressed as follows:

Let E = voltage to neutral at sending end.
V = voltage to neutral at receiving end.
I_1 = component of line current in phase with V, i.e., the power component of the line current.
I_2 = component of line current at 90 deg to V, i.e., the *leading* reactive component of the line current.
r = resistance of one line.*
x = inductive reactance of one line.*

Then, noting that I_2 is to be taken positive when leading,

$$E^2 = (V + rI_1 - xI_2)^2 + (xI_1 + rI_2)^2$$

Use of Series Field. In practice a series field is added to the converter, and the shunt excitation is adjusted so that the armature current is lagging at no load. As the load increases, the series field increases, the adjustment being such that at about 3/4 load the proper excitation for unity power factor is given. Hence at all loads over 3/4 the field will be overexcited, the current will be leading, and the voltage will be raised or compounded.

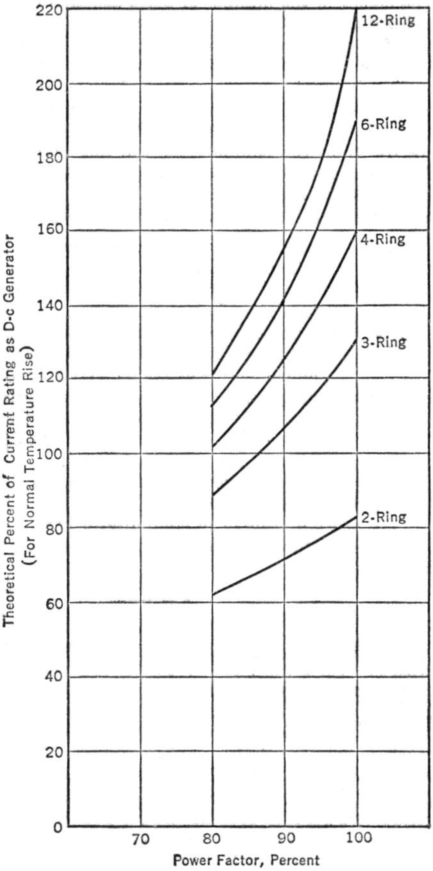

FIG. 15. Comparative Current Ratings of Convertors for Normal Temperature Rise

32. DESIGN OF SYNCHRONOUS CONVERTERS

DESIGN. The design of a synchronous converter is very similar to that of a d-c generator (see p. 8-38), except for certain special conditions due to the fact that the frequency is fixed by the frequency of the supply system, and that greater latitude is allowed in the choice of the nominal value of the d-c armature reaction and copper density, because the real value of these quantities is the difference between their nominal d-c values and their a-c values.

* r and x should include respectively the resistance and reactance not only of the line wire, but also of the transformers and reactance coils through which the line current passes.

SPEED AND NUMBER OF POLES. The revolutions per minute and the number of poles are definitely related to the frequency of the supply circuit, in cycles per second, in accordance with the formula:

$$120 \times \text{Frequency} = \text{Number of poles} \times \text{rpm}$$

The choice of the number of poles usually depends upon the commutator.

COMMUTATOR. The design of the commutator is usually the limiting feature and therefore the first to be considered, particularly in either high-frequency (60-cycle) machines or those in which the d-c voltage is 600 or over. Three factors must be considered in the design to secure successful commutation and life of the commutator, namely, peripheral speed, voltage between bars, and thickness of bars. If the peripheral speed or the voltage between bars is too high, commutation will be bad. If the commutator bars are too thin, the commutator will not retain its shape, and commutation will be bad. These three limiting factors are very closely related and in a high-voltage machine give very little choice. The diameter of the commutator depends upon the voltage and frequency, and its minimum value is given by the three following relations.

Let s = pitch of commutator bars in inches. This ranges from 0.15 in small machines to 0.20 in high-voltage or high-frequency machines and to 0.40 in liberally designed machines. These values include the width of bar and about 0.03-in. insulation between bars.

V = peripheral speed of commutator in feet per minute. This ranges from 3500 in liberally designed low-voltage, 25-cycle machines to 5000 in 60-cycle machines and is extended to 6000 on occasions.

e = average volts per bar = machine voltage divided by the number of bars between brush studs. Normal values are from 8 to 16. The maximum voltage between bars is about 1.57 times this value.

f = frequency in cycles per second.

E = voltage between d-c terminals.

p = number of poles.

n = total number of commutator bars.

d = diameter of commutator in inches.

Then

$$s = \frac{Ve}{10Ef}, \qquad n = \frac{pE}{e}, \qquad d = \frac{sn}{\pi}$$

ARMATURE. Most converters are now made 6-phase diametrical, with interpoles, although the largest are 12-phase. In all cases it is necessary to choose a number of armature slots and commutator segments per pair of poles divisible by the number of phases. Since the a-c and d-c armature currents are in opposite directions, the resultant heating is less than that caused by the direct current alone, and the effective armature reaction is extremely low, pulsating between 8 and 20 per cent of the d-c armature winding alone. This results in very little distortion of the main field flux distribution. It is possible, for these reasons, to use from 60 to 80 per cent more electrical loading (ampere-conductors per inch of armature periphery) on the armature than would be used on a standard d-c machine of the same physical size. The apparent copper watts per square inch can be approximately 3 times that of the d-c armature alone, since the resultant coil heating is actually 27 per cent that of a similar d-c winding alone. The number and proportions of the slots are the same as in a standard d-c machine.

The armature winding is nearly always a fully cross-connected lap winding, with the winding turns under each pair of poles tapped at equidistant points, the number of which is the same as the number of phases. One tap from each pair of poles connects to a collector ring, these taps being in the same relative locations from the armature.

Flux densities in the air gap and teeth follow the same values with respect to armature diameter that exist in standard d-c machines.

MAIN FIELDS. The determination of shunt field ampere-turns is the same as for a d-c machine, but the series field ampere-turns, when used for flat compounding, depend on the following factors:

1. Reactance of transformer and any other coils in the a-c circuit. The series field acts to give a compounding effect by producing a leading power factor in the same way as overexciting a synchronous motor, which in turn increases the a-c voltage applied to the converter by reason of the voltage rise from the leading current through the reactance of the transformer. Thus the actual voltage rise is dependent to a great extent on the transformer reactance, which is made as large as economically possible—about 15 per cent.

2. Ratio between shunt field ampere-turns and armature ampere-turns per pole. This is kept to a value of 1.0 to 1.1.

3. Ratio of series field ampere-turns to shunt field ampere-turns. With the transformer reactance of 15 per cent the series field ampere-turns required will be 25 to 30 per cent of the shunt field ampere-turns.

4. The resistance and reactance voltage drop in the converter itself reduces the d-c terminal voltage, this drop being usually in the order of 4 to 5 per cent of the terminal voltage.

From the foregoing considerations it can be seen that the range of voltage adjustment and compounding is quite limited, since excessive heating develops when the power factor drops much below 96 per cent. Also hunting is liable to develop from excessive leading power factor, so that no overcompounding is practical on a converter.

DAMPER WINDING. A damper winding is required in the pole face, as in any synchronous motor, to provide starting torque and to prevent hunting. These bars are uninsulated and have a cross-section of about 25 per cent of the armature winding. It is desirable to have heavier damping bars through the pole tips than through the central part of the pole face, since the bars at the tip enclose practically all the main pole flux and therefore have the greatest magnetic effect. The connections between rods and end ring should be silver-soldered, and the connections between poles made with very ample area.

COMMUTATING POLES. Converters of 100 kw and larger are provided with commutating poles for improved operation. Because of the pulsation in armature reaction, the pole is usually provided with ampere-turns equal to 75 to 100 per cent of the armature ampere-turns per pole. Thus with a large air gap the effect of the pulsation is minimized, and the flashing associated with quick load changes is reduced.

SHAFT, BEARINGS, ETC. Since the transfer of energy is in the conductors themselves, there is no mechanical torque other than that to overcome friction and core loss. Therefore the shaft, bearings, and mechanical housing of a rotary converter are quite light and present no particular mechanical difficulties in mechanical design. For the same reason converters do not require very elaborate foundations. Machines of less than 1000-kw capacity are usually supplied with a base and are complete in one piece; larger sizes are furnished with foundation plates.

Table 2. Efficiency and Losses of Converters

(All values are in percentage of input at full load)

Kilowatt rating	500	300
Frequency	25	60
Core loss	1.00	1.75
Armature I^2	0.55	0.60
Shunt field I^2	0.70	0.60
Brush I^2	0.40	0.40
Bearing friction	0.55	1.50
Brush friction	0.30	0.65
Efficiency	96.50	94.50

EFFICIENCY AND LOSSES. The efficiency and distribution of losses of a typical 25-cycle and 60-cycle converter are shown in Table 2.

EXAMPLES OF DESIGN. In Table 3 are given design data on four representative converters of different capacities.

Table 3. Design Data of Synchronous Converters

Item	Unit	1	2	3
Rating	Kilowatts	200	500	1,000
Frequency	Cycles per second	60	25	60
Speed	Revolutions per minute	1,200	750	900
D-c volts	Volts	250	250	600
Poles		6	4	8
Number of phases		6	6	6
Armature diam	Inches	18	28	31
Armature-core length	Inches	9	10 1/2	15
Comm. (diam. × length)	Inches	14 × 8	21 × 18 1/4	21 × 12 1/4
Comm. segments		252	120	384
Avg. volts per bar	Volts	6.0	8.3	12.5
Comm. speed	Feet per minute	4,400	4,130	4,950
Amp. cond. per inch loading on armature		1,200	1,365	1,650
Flux density in gap	Lines per square inch	46,300	57,000	55,200
Armature slots		84	120	96
D-c brushes		5(3/4 × 1 1/4)	12(1 1/8 × 1 1/4)	8(3/4 × 1 1/4)
Brush density	Amperes per square inch	57.0	59.2	55.5

33. CHARACTERISTICS AND OPERATION OF CONVERTERS

THREE-WIRE SYSTEMS. Converters can be made for 3-wire systems, with the same features of construction as are found in a d-c generator. All north series coils and interpole coils are on the positive side of the armature, and all south coils on the negative side. The neutral for the system comes from the center taps of the transformer secondaries, which act as the center-tapped balance coil used with a standard d-c generator. (See Fig. 16.)

STARTING. The customary method of starting a converter is to apply 50 per cent a-c voltage to the collector rings from the center taps of the transformer secondary. On converters larger than 300 kw all the brushes must be lifted from the commutator while starting to prevent serious deterioration from high circulating currents. Also a break-up switch is used in the shunt field of larger converters, which is open when starting. A voltage approximately 5 times normal is induced in each shunt coil when the converter is

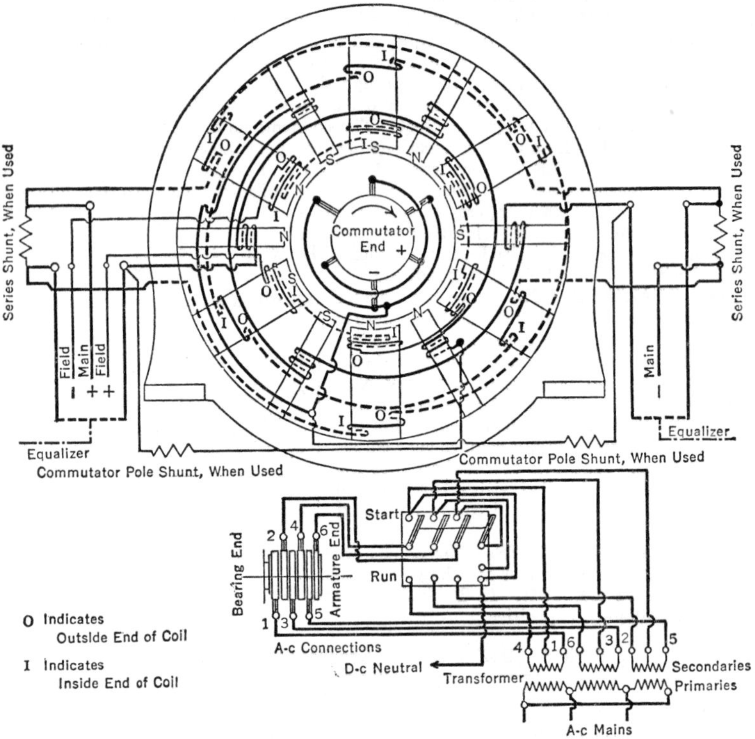

FIG. 16. Three-wire Synchronous Converter Connection, Showing Starting and Running Arrangements

connected to the a-c line, and to reduce the insulation stress the usual series connection of shunt fields is changed when the break-up switch is opened so that not more than 2 coils are left connected in series. One narrow pilot brush is used in one brush arm of each polarity, to check the polarity of d-c voltage when the converter falls into step. If the polarity is not correct for connecting shunt fields to the armature, the converter will produce violent spitting if the field switch is closed. To bring the converter up in the proper polarity, it must be made to slip a pole by opening and reclosing the starting switch or reversing the field switch quickly. When the starting switch is closed, the shunt field is sometimes connected across a low-voltage battery, with the correct direction of field current flowing, and this always brings up the armature with the correct polarity. An alternate method of starting from the a-c end is to connect the primary of the transformer to the line first through a Y connection; then, when the armature is up to speed, the primary is taken off the line and put back through a delta connection.

Another method of starting is to use a variable d-c potential applied to the d-c end of the converter, with the a-c side disconnected from the transformer. This requires the usual procedure of synchronizing two a-c machines, when the converter is connected to the a-c line after attaining full speed.

A small starting motor of either the a-c or d-c type may be used, and the machine placed on the line by the same synchronizing process as above. The d-c motor is a convenient maintenance tool to have available for turning the armature when the commutator has to be stoned or resurfaced with a lathe tool.

A speed-limiting device is provided on most converters to prevent overspeeding if the a-c line breaker should open and leave the machine connected to the d-c line. This would reverse the current in the series field and cause some speed increase.

Voltage adjustment is always provided on the a-c side by taps on the transformer primary. Usually four taps are provided on each side of the theoretical operating ratio,

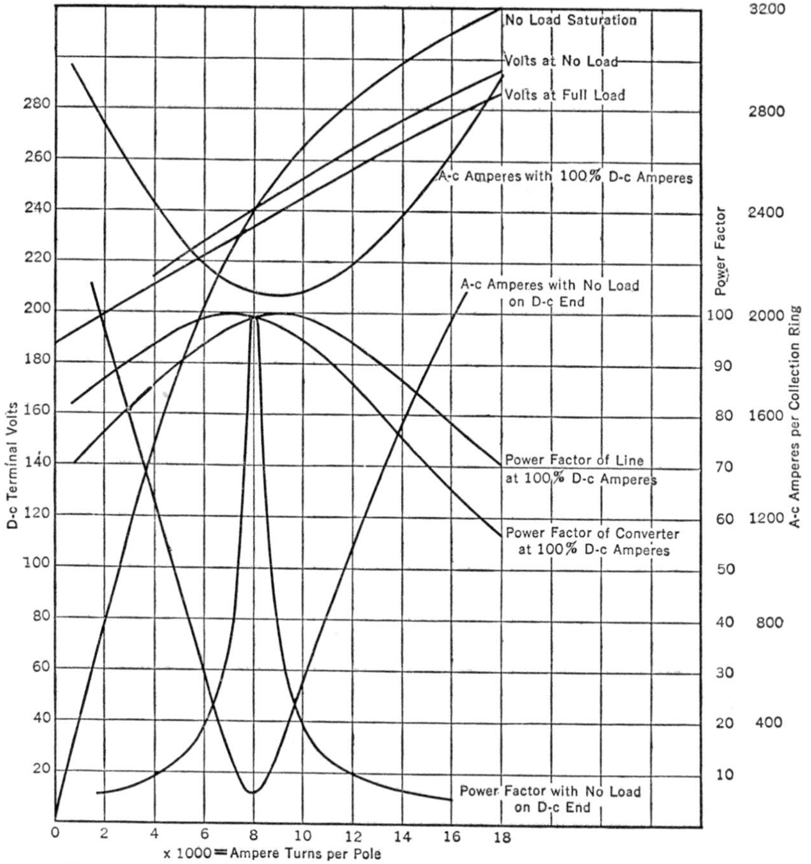

Fig. 17. Test Curves of 500-kw 240/240-Volt, 1200-rpm, 6-Phase, 60-Cycle Converter

with $2\,^1/_2$ per cent of the winding turns between each pair of taps. Tap changing requires opening the circuit or short-circuiting a portion of the transformer at each change.

A synchronous generator booster, which is a small a-c generator having a winding in series with each phase of the converter, may be used for voltage adjustment. By varying the booster excitation, the a-c input voltage to the converter, and thereby the d-c voltage, is changed.

In small converters of 25 kw and less the d-c voltage can be changed over a range of approximately 20 per cent by dividing the main poles into two magnetic paths, one of which normally carries $^2/_3$ and the other $^1/_3$ of the air-gap flux. By manipulation of the excitation of the small pole only, just the d-c potential will be affected, since the third harmonic voltage cancels any resultant effect on the a-c voltage.

The tests and performance curves of converters are conducted just as for d-c machines, except that readings on the a-c side are also recorded. Efficiency is calculated in exactly the same way, except that the armature winding i^2R loss is 27 per cent of the apparent value for 6-phase machines. (See Fig. 17.)

Improved performance of metal graphite brushes on the collector rings is obtained by cutting a spiral groove across the rings. Many rings are machined with a 1/8-in.-wide × 1/8-in.-deep groove with a pitch of 5/8 in. This grooving reduces selective action of the brushes and results in cooler operation. The collector-ring construction must be carefully considered from a maintenance standpoint, as the metal graphite dust from the brushes is liable to collect in inaccessible places and may seriously lower the insulation creepage resistance to ground. Easily wiped creepage surfaces must be provided.

High-voltage converters of large capacity are always protected by high-speed d-c circuit breakers which will open on overload in 3 to 6 cycles on a 60-cycle system. Also, flash barriers are used between the brush arms to break up the flow of ionized air that might start under severe flashing conditions.

Parallel-operation problems for converters are handled in much the same way as for d-c generators. Compounding from no load to full load must be the same, and the voltage drops across the series fields must be within 15 per cent of each other at rated load. In order to prevent hunting, each converter must have its own transformer to provide sufficient reactance between machines in parallel.

The speeds, weights, and efficiencies of 250 and 600 converters of 25- and 60-cycle frequencies are shown in Table 4.

Table 4. Typical Synchronous Converter Data

Phases	Kilowatts	Frequency	Volts	Poles	Revolutions per Minute	Weight	Full-load Efficiency
6	150	60	250	6	1,200	3,900	92.6
6	150	60	600	6	1,200	3,800	93.8
6	300	60	250	6	1,200	7,050	93 7
6	300	60	600	6	1,200	6,100	94.3
6	300	25	250	4	750	11,000	94.2
6	300	25	600	4	750	10,250	94.8
6	500	60	250	6	1,200	9,200	94.4
6	500	60	600	6	1,200	8,600	94.8
6	500	25	250	4	750	14,600	94.5
6	500	25	600	4	750	15,600	95.4
6	1,000	60	250	8	900	22,000	94.1
6	1,000	60	600	8	900	16,350	95.2
6	1,000	25	250	6	500	30,500	94.8
6	1,000	25	600	4	750	23,300	95.8
6	2,000	60	250	16	450	44,300	94.1
6	2,000	60	600	14	514	36,400	95.2
6	2,000	25	250	10	300	57,200	94.8
6	2,000	25	600	8	375	49,700	95.9

34. BIBLIOGRAPHY

Gray, A., *Electrical Machine Design*. McGraw-Hill (1926).
Hellmund, R., Commutation. *AIEE*, Vol. 51, p. 465.
Karapetoff, V., *Experimental Electrical Engineering*. Wiley.
Kuhlman, J. A., *Design of Electric Apparatus*. Wiley (1930).
Langsdorf, A. S., *Principles of Direct-current Machines*. McGraw-Hill (1931).
Luke, G. E., Cooling of Electrical Machinery. *AIEE*, Vol. 42, p. 636.
National Carbon Co., Inc., *Modern Pyramids*.
Slichter, W. I., *Principles of Design of Electrical Machinery*. Wiley (1926).
Still, A., *Elements of Electrical Design*. McGraw-Hill (1932).

SECTION 9

ALTERNATING-CURRENT GENERATORS AND MOTORS

ALTERNATING-CURRENT SYNCHRONOUS MACHINES

By E. C. Whitney and H. O. Poland

POLYPHASE INDUCTION MOTORS

By T. C. Lloyd

SINGLE-PHASE INDUCTION MOTORS

By T. C. Lloyd

ALTERNATING-CURRENT COMMUTATOR MOTORS

By A. G. Conrad

ALTERNATING-CURRENT GENERATORS AND MOTORS

ALTERNATING-CURRENT SYNCHRONOUS MACHINES

By E. C. Whitney and H. O. Poland

In the generation and use of alternating current, the synchronous machine is the basic means of conversion of mechanical energy of rotation into electric energy, and vice versa. The basic principle of operation is the relative motion of a constant magnetic field with respect to coils which are placed in the magnetic field. The constant magnetic field is produced by coils of alternating polarity in which a direct or continuous current flows.

Table 1. Relation of Electrical Characteristics in A-c Systems

Phase	Kva	Voltage Relation	Line Amperes	Diagram
3	$= \frac{\sqrt{3}E_{L-L}I}{1000}$	$E_{L-L} = E_{L-N}\sqrt{3}$	$= \frac{kva \cdot 1000}{\sqrt{3}E_{L-L}}$ $= \frac{kva \cdot 1000}{3E_{L-N}}$	Y or Star 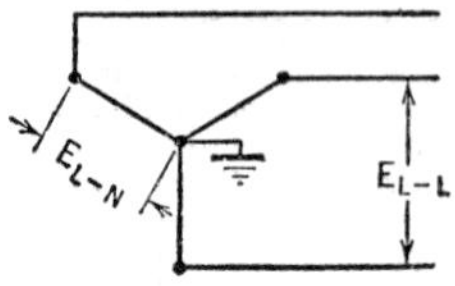
3	$= \frac{\sqrt{3}E_{L-L}I}{1000}$	$E_{L-L} = E_{Leg}$	$= \frac{kva \cdot 1000}{\sqrt{3}E_{L-L}}$, $I_{Line} = \sqrt{3}I_{Leg}$	Delta 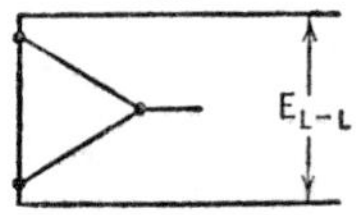
2	$= \frac{2E_{L-L}I}{1000}$	....	$= \frac{kva \cdot 1000}{2E_{L-L}}$	Two-phase or Quarter-phase 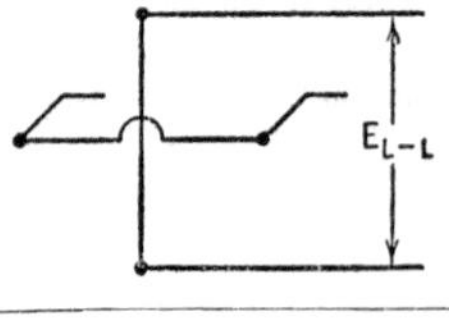
2	$= \frac{2E_{L-L}I}{1000}$	$E_{L-L} = 2E_{L-N}$	$= \frac{kva \cdot 1000}{2E_{L-L}}$	Two-phase Five-wire 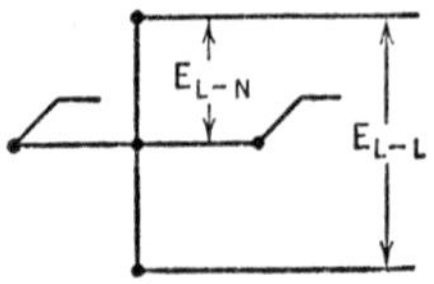
1	$= \frac{EI}{1000}$	....	$= \frac{kva \cdot 1000}{E}$	Single-phase Two-wire 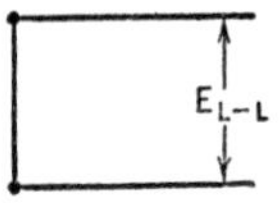

An alternating voltage is thus produced in the coils which move through the magnetic field. The alternating voltage repeats with relative movement of one pair of poles. For various synchronous machines to work together in parallel on an a-c system, it is necessary for each machine to rotate exactly at two pole pitches each cycle, hence the term synchronous. In the following discussion, items common to both synchronous generators and motors will be outlined, and details which apply specifically to each will be covered later.

1. RATING AND CONSTRUCTIONS

All a-c generators are rated in kilovolt-amperes (kva), and a-c motors are usually rated in horsepower but may also have a kva rating stamped on the nameplate. Table 1 indicates the relationship between kva, voltage, and current for different common systems. The kilowatt (kw) output or input is equal to kva times power factor.

The common frequency in the United States is 60 cycles, with 50 cycles on a part of the West Coast. Twenty-five cycles is used on railway installations and some other early installations. Fifty cycles is the standard frequency throughout most of the rest of the world. Some other frequencies used on small systems are 40 cycles, 30 cycles, and 16 $^2/_3$ cycles. For aircraft, 400 cycles is becoming a standard.

Three-phase systems are the commonest method of distribution of power throughout the world, since this system provides the maximum economy in both distribution and generation equipment. Two-phase or quarter-phase and single-phase power are also used in a few installations. For three-phase the angle spacing between phases of the three currents and voltages is 120 deg when referred to 1 cycle equalling 360 deg. For two-phase the angle is 90 deg between phases. The power factor is the cosine of the angle between voltage and current.

2. TYPES OF SYNCHRONOUS MACHINES

SALIENT POLE (Fig. 1). This type of machine normally consists of a stationary armature and a rotating field. The field has individual salient poles of alternate polarity. The stationary armature is composed of laminations with slots wound with insulated coils.

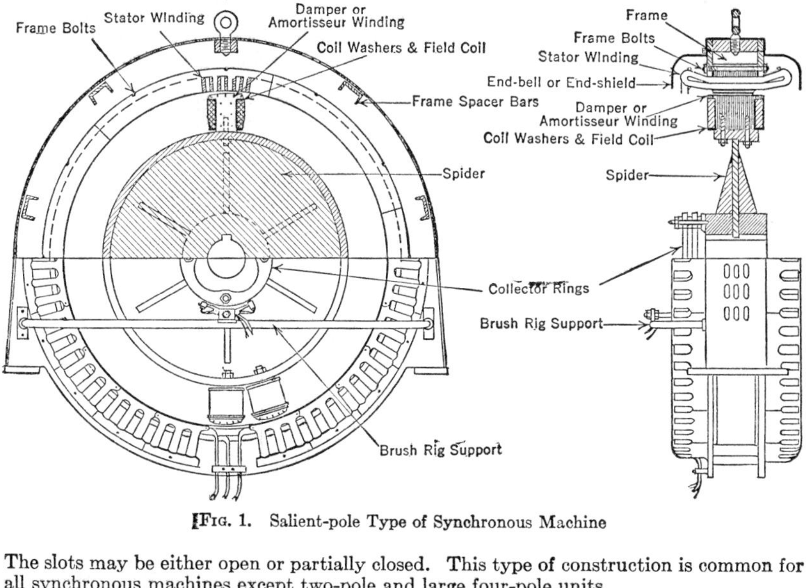

FIG. 1. Salient-pole Type of Synchronous Machine

The slots may be either open or partially closed. This type of construction is common for all synchronous machines except two-pole and large four-pole units.

TURBINE GENERATORS (Fig. 2). The same general physical arrangement of parts exists in the turbine generator as in the salient pole, with the exception that the rotor is constructed of a solid steel forging or from a series of circular plates. The surface of the rotor is round and free from projections. The unit tends to be very long and small in diameter in order to limit mechanical stresses and is characteristic of two-pole, 3600-rpm machines and very large four-pole machines.

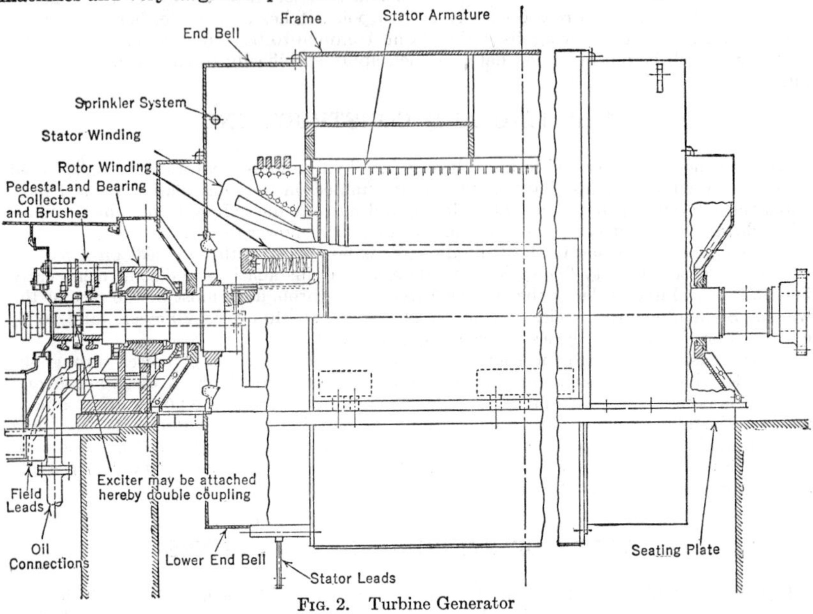

Fig. 2. Turbine Generator

INDUCTOR TYPE (Fig. 3). The inductor-type machine is now used for high-frequency generation applied chiefly to inductive heating. This type of machine operates on a variation of air gap or magnetic reluctance to cause a change in flux linking the armature coils. The standard frequencies in this country for this type of unit are 960, 3000, and 9600 cycles per second. The wave shape of the voltage generated is roughly a

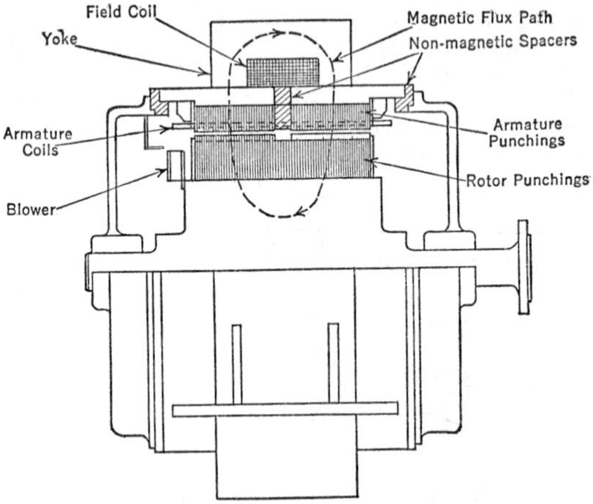

Fig. 3. Inductor-type Generator

sine wave and is satisfactory for inductive heating. A characteristic of these machines is a small air gap and consequently the necessity to maintain close alignment. Figure 3 shows one arrangement of the parts in an inductor machine, but a number of other arrangements are in common use today.

3. STRUCTURAL TYPES

Alternating-current synchronous machines may be broadly classified as to structure into two main divisions: (1) bearing arrangement, and (2) protective enclosure or frame structure.

BEARING ARRANGEMENT. Several different arrangements of bearings are employed, each being suited to a particular kind of prime mover or driven machine.

Engine-type generators or motors are furnished without bearings or shaft. These parts are provided by the manufacturer of the coupled machine, which is the prime mover in the case of a generator. If the machine is a motor, the shaft and bearings are furnished by the builder of the compressor, vacuum pump, or d-c generator. To facilitate mounting on the shaft, the rotors of engine-type motors are usually split diametrically with provision for clamping rigidly to the shaft by means of tightening heavy bolts in the hub. In certain cases, special fastenings are required in the rim of the spider where the two rotor halves are joined together. Rotors of engine-type generators usually are not split.

In the usual application of an engine-type machine the generator and prime mover, or the motor and its driven machine as the case may be, are sold together as a complete unit. For this reason the designs are closely coordinated from the standpoint of both mechanical arrangement and operating performance. Engine-type machines are usually of relatively low speed and therefore are generally seen to have many poles and comparatively large ratios of diameter to length. Besides compressors, vacuum pumps, and d-c generators, engine-type motors are used for driving certain centrifugal pumps and wood chippers.

One-bearing machines are built with their bearing on the outboard end, which may be either of bracket or pedestal type, depending on the size of the machine. The larger machines have the pedestal bearing. This type of construction is used largely for small diesel-driven generators.

Two-bearing bracket-type machines comprise the largest percentage of machines in service. In this type of construction a bracket is attached to each end of the frame, this bracket being of cast or fabricated construction and sometimes incorporating certain protective features, as explained later. The bracket includes the bearing housing, which can be designed to accommodate either sleeve bearings and oil reservoir or antifriction bearings. Oil fillers or greasing devices, oil-level indicators, inspection covers, and bearing-temperature thermostats may be added according to requirements. Sleeve bearings are in general oil lubricated, the oil being pumped up over the journals by large oil rings which rotate with the shaft and whose lower portion is immersed in the reservoir of oil.

The sleeve bearing proper is lined with babbitt metal, consisting of various alloys of tin, antimony, copper, and lead, the exact composition varying with different manufacturers and with the severity of service. As the babbitt wears, the bearings can be rebabbitted, and at this time the journals are usually reconditioned and the new babbitting dimensions made to suit. Properly designed sleeve bearings exclude dust and eliminate oil leakage. Special thrust surfaces can be provided in bearings when high axial thrust loads are encountered. Machines designed for belted service often have a larger-than-normal bearing on the pulley end to carry the additional load due to belt pull. Such bearings can be specially designed to suit the particular requirements of a given application if the facts regarding direction of belt pull, sheave dimensions, and speeds are at hand. Oil-lubricated bearings must be operated within fixed limits of inclination, and in general the brackets cannot be rotated about the axis of the machine. Specific data on these points for any given case may be obtained from the manufacturer. Antifriction bearings are usually grease lubricated, some types being designed for lubrication for life. Oil lubrication is not recommended for antifriction bearings because these bearings are more subject to the effects of corrosion during extended periods of idleness. The two-bearing bracket-type machines are of higher speeds than the engine type. Motors of this type are used for such applications as driving fans and blowers, pumps, high-speed compressors, fans-line shafts, grinding and crushing machinery, and pulp beaters. In addition to direct, coupled and belted classes of service, these machines are also sometimes geared.

Pedestal-bearing machines may be of high or low speed. In this construction the frame of the machine is usually arranged so that the frame feet rest on a steel bedplate, with foot height such that the periphery of the frame extends below the top of the bedplate or floor level. In this way a relatively large frame diameter may be used in applications requiring a low foot height. Obviously, some clearance is necessary beneath the frame, so a pit or opening in the foundation is required. The bearings, either sleeve or antifriction type, are housed in pedestals of suitable height, which are mounted on cross-members, or end-members, of the bedplate. Servicing a pedestal-type machine is facilitated in certain cases by making the bedplate and shaft long enough so that the frame can be shifted axially until the rotor is accessible. The pedestal-type machine finds wide application in motor-generator sets, in driving steel and metal rolling mills, rubber mills, plasticators, and banbury mixers, tube and ball mills, chippers and wood hogs, as well as the drives commonly employing bracket-type motors.

Three-bearing construction is used for belted applications where the sheave must have a wide face or the belt pull is particularly severe. The third or outboard bearing is of

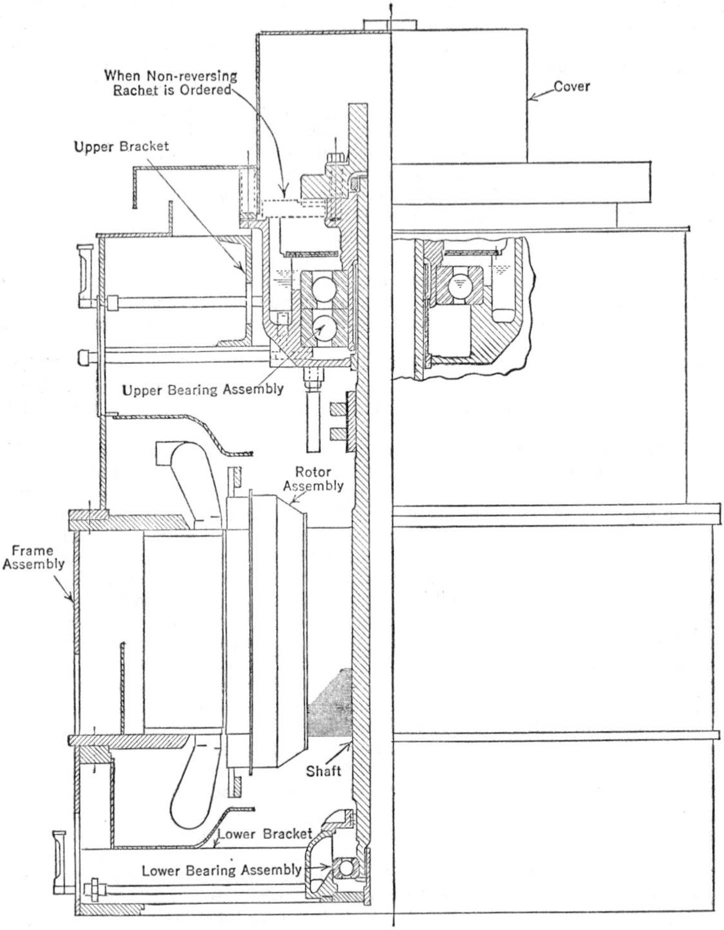

FIG. 4. Typical Vertical Synchronous Motor

pedestal type, and a common bedplate is used for mounting the machine and its outboard bearing. The other two bearings may be either bracket or pedestal type, depending upon the specific requirements of the application. Examples of three-bearing motors may be found in drives for fans and compressors, grinding and crushing machinery, pulp beaters, and line shafts.

Vertical construction of motors is used to a very great extent for centrifugal-pump drives, where the motor is mounted directly on top of the pump or in line with it at a higher level. A typical vertical synchronous motor is shown in section in Fig. 4. Obviously in this construction the bearing requirements differ from those in horizontal machines. Here the rotor weight is carried by one bearing, usually the top one, and to the weight of the rotor is added the downward pull or thrust due to the static weight of the pump impeller and the water thrust acting on the blades. This top, or thrust, bearing is designed specifically for thrust loading and may be a special ball bearing, either single or tandem depending on load conditions, or it may be a shoe-type thrust bearing in which segmental pivoted shoes carry the thrust load exerted by the bearing runner. The shoe-

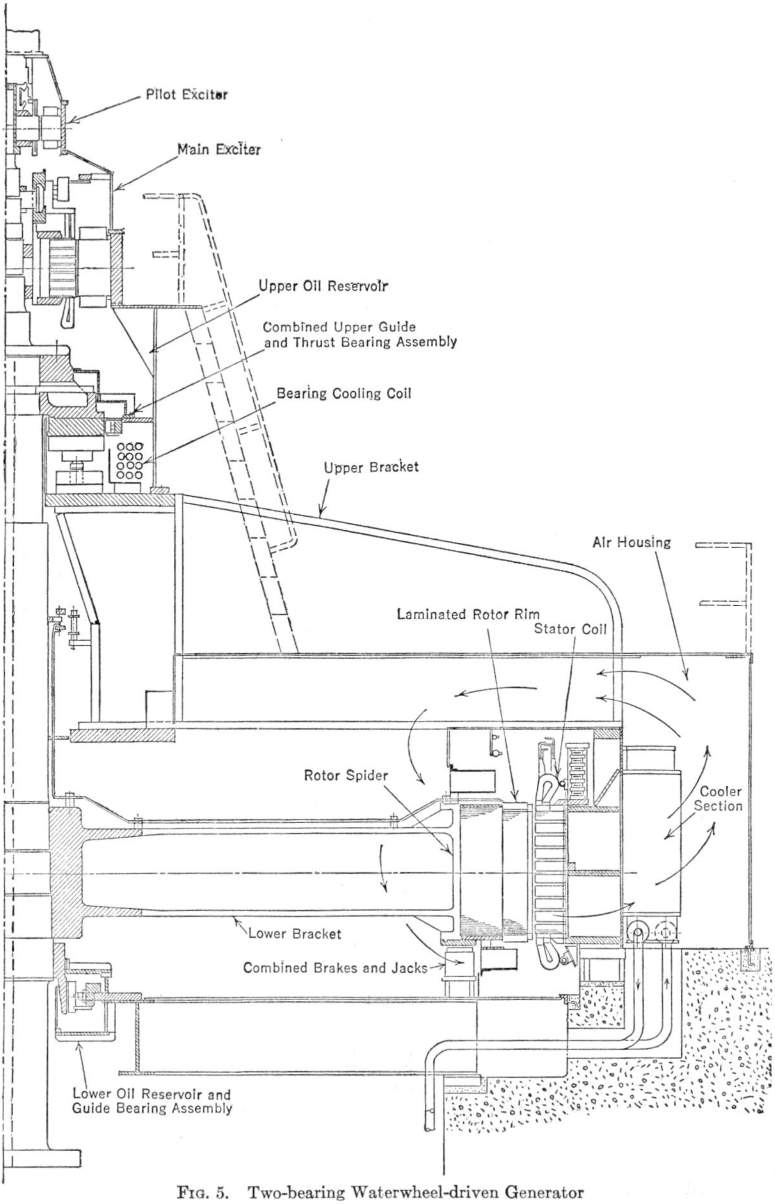

Fig. 5. Two-bearing Waterwheel-driven Generator

type bearing is generally used when the thrust loading is too high for a ball bearing. The lower bearing usually acts simply as a guide bearing to hold the shaft in its central position and carries no thrust load. This bearing usually is a ball bearing when the thrust bearing is of the ball type and a sleeve bearing when the thrust bearing is of the shoe type.

The shaft of a vertical motor is often made hollow so that the pump shaft can be brought up through its center and the coupling located at the top. This construction also permits the pump impeller to be adjusted up or down as required, simply by moving the

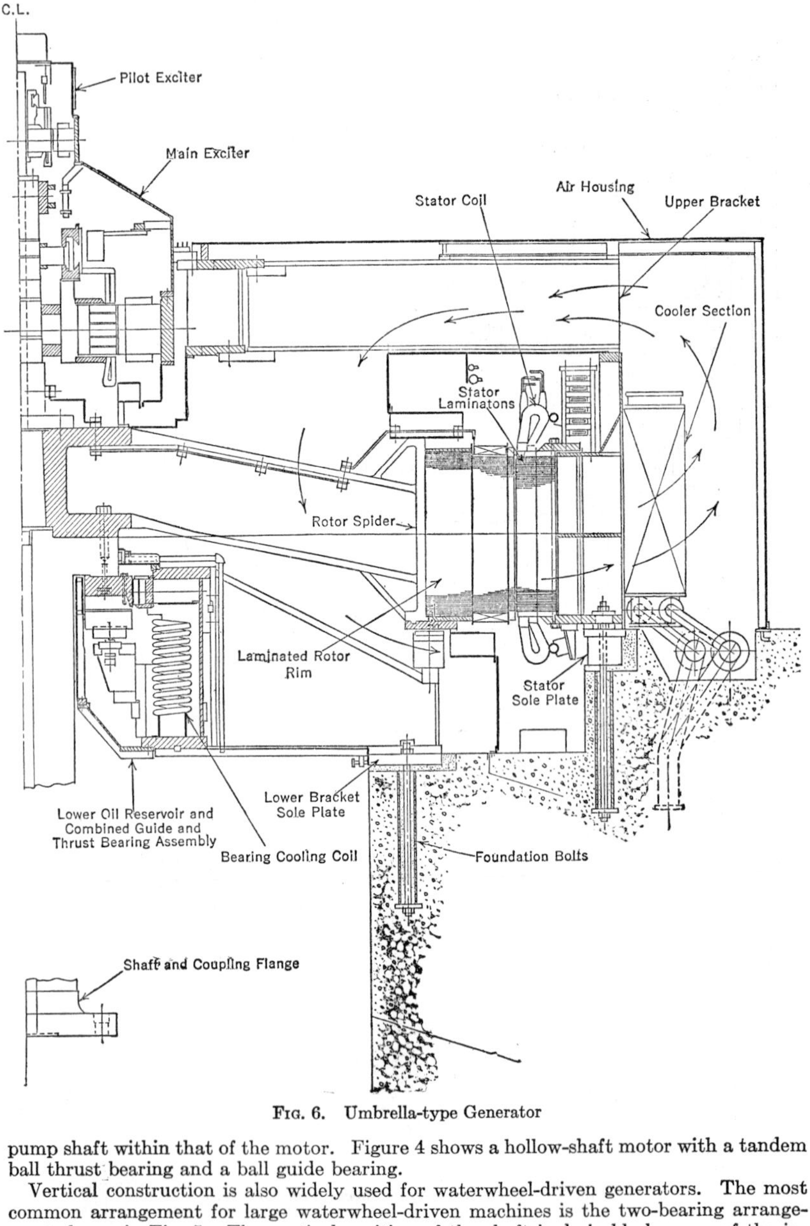

FIG. 6. Umbrella-type Generator

pump shaft within that of the motor. Figure 4 shows a hollow-shaft motor with a tandem ball thrust bearing and a ball guide bearing.

Vertical construction is also widely used for waterwheel-driven generators. The most common arrangement for large waterwheel-driven machines is the two-bearing arrangement shown in Fig. 5. The vertical position of the shaft is desirable because of the inherent characteristics of low-speed waterwheels. The thrust bearing and one guide bearing are located above the generator rotor, and another guide bearing is below the rotor.

The second common arrangement for large vertical units is the umbrella generator, in which the combined thrust and guide bearing is located below the generator rotor, as shown in Fig. 6. This type of unit is limited to large-diameter, low-speed, and short-core-length machines, where the weight of the rotor and low center of gravity compared to the bearing diameter are sufficient to overcome the upsetting force of magnetic pull.

PROTECTIVE ENCLOSURE. The second broad structural classification of synchronous machines is concerned with the frame structure or protective enclosure. Generally, the type of frame structure to be used depends upon the conditions of the location, the object being to protect the vital parts of the machine so as to ensure longest possible life with best operating efficiency and without excessive maintenance. There is some overlapping of enclosure types on a given application, due to individual preferences and interpretation of requirements.

The **open-type** machine is of course the oldest and probably the most widely used, although the tendency in recent years is to favor some type of enclosure. Factors involved in this tendency are the growing consciousness of good appearance and minimum noise, two considerations which favor enclosing the machine. In open construction it can generally be said that the parts are purely functional mechanically. That is, the frame and brackets are designed as supporting structures and the machine is self-ventilating, with no restriction to ventilation other than that necessitated by mechanical construction. This type of machine is used in inside locations which are free from splashing or falling water and where there is no danger of foreign material falling into the machine.

The **drip-proof** construction involves the addition of covers so as to exclude falling liquid or objects from the machine. These covers may be added over a standard open machine or, as is more often the case in recent years, may be built integral with the fabricated frame. The enclosing covers must be so arranged that liquid or solid matter falling on the machine at an angle 15 deg or less from the vertical cannot enter, either directly or by running along any surface. This type of construction finds wide application in power stations or other locations subject to dripping water. Also the integral fabricated design is sometimes used where improvement of appearance over an open motor is the only factor.

In **splashproof** construction the enclosure is still more complete, so that liquid or solid particles falling onto or approaching the machine at an angle 100 deg or less from the vertical will be excluded, whether by direct entrance or by running along a surface. Splashproof construction is used where there is a great deal of splashing water, e.g., in a paper mill. Many outdoor applications also use this construction for protection from the weather.

Both drip-proof and splashproof machines frequently have screens over the ventilation openings to guard further against the entrance of foreign material. Obviously, in either of these types of construction there is some restriction to ventilation, as compared with open design. Therefore the amount of active material in the machine is slightly increased, or the allowable temperature rise is increased to 50 deg cent instead of 40 deg in case of motors.

Structural arrangement may also be modified as required by the system of ventilation. For instance, an enclosed machine may be arranged for self-ventilation or separate ventilation, according to whether the ventilating air is circulated by means integral with the machine or by means of an external fan or blower. In either case the enclosure may have provision for the attachment of ducts so that the ventilating air may be taken from a source some distance away. This is often desirable when the immediately surrounding air is too hot or is contaminated with dust or other undesirable matter.

Enclosure may be made even more complete to the point of **total enclosure.** In this construction there is no exchange of air between the inside and outside of the enclosure, although actual airtightness is unnecessary. In the **totally enclosed fan-cooled** machine the cooling air is circulated externally over the enclosure and heat-transfer surfaces by means of fans, which are integral with the machine but external to the enclosure. Higher temperature rise is allowed in this construction, 55 deg for Class A insulation and 75 deg for Class B insulation, as measured by thermometer. (Very specialized construction is used in totally enclosed fan-cooled motors which are used in explosive atmospheres. This application is seldom encountered for synchronous motors but is rather frequent in induction-motor practice.)

In many cases the ventilating system for a machine forms a closed circuit. That is, the machine is totally enclosed, so that no external air is supplied and the heated exhaust air is passed through a water cooler and constantly recirculated through the machine. In this way the heat is transferred to the cooling water, which is piped to the cooler, and the condition of the surrounding air is of minor importance.

4. CONSTRUCTION OF WINDINGS

TYPES OF WINDINGS. Practically, only one type of winding is in use in modern machines today. This is the lap type, where two coil sides per slot are normally used and the number of coils is equal to the number of slots, as in Fig. 7. Usually the coils are of a diamond shape and may be inserted directly into open slots. Some small synchronous machines today employ the mush type of coil with partially closed slots, but the lap type of winding is usually employed. In some old machines and a few modern units of specialized ratings, the chain type of winding is employed, as in Fig. 8. This type usually involves several different types of coils of varying pitch, depending on the number of phases and the slots per phase per pole. There is usually one coil side per slot in this type of winding.

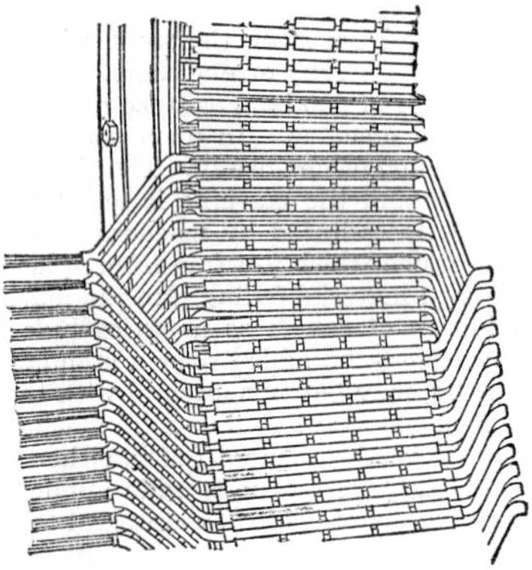

FIG. 7. Lap Winding

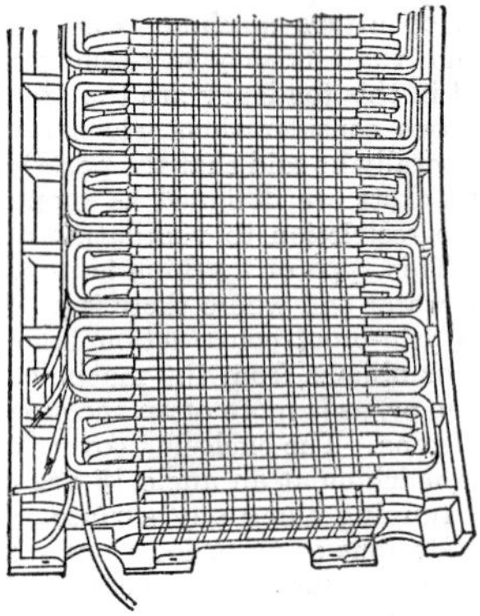

FIG. 8. Chain Winding

INSULATION. **Class A.** Class A insulation consists of (1) cotton, silk, paper, and similar organic materials when either impregnated or immersed in a liquid dielectric; (2) molded and laminated materials with cellulose filler, phenolic resins, and other resins of similar properties; (3) films and sheets of cellulose acetate and other cellulose derivatives of similar properties; and (4) varnishes (enamel) as applied to conductors.

Class B. Class B insulation consists of mica, asbestos, fiber glass, and similar inorganic materials in built-up form with organic binding substances. A small proportion of Class A materials may be used for structural purposes only.

Class C. Class C insulation consists entirely of mica, porcelain, glass, quartz, and similar inorganic materials. Although a Class C winding is theoretically possible, practically it is almost impossible to build.

Class H. Class H insulation consists of inorganic materials such as mica, asbestos, and fiber glass in built-up form, with organo-silicon compounds as the impregnating and

Table 2. Allowable Temperatures and Temperature Rises

	Class A	B	H *
Continuous allowable hot-spot temperature	105	130	180
Hot-spot allowance above observable temperature			
Method 1: thermometer........................	15	20	40
2: resistance †........................	5	10	20
3: detector ‡........................	5	10	20
Continuous permissible rise above 40 deg cent ambient			
Armature method 1............................	50	70	100
2............................	60	80	120
3............................	60	80	120
For 50 deg cent ambient, all rises are 10 deg cent less			
Field method 1................................	50	70	100
2................................	60	80	120

* Tentative AIEE recommendation.

† Used almost universally for determination of rotating field coil temperatures but not used practically for determination of armature temperatures except on small machines, usually 100 kva or less.

‡ Detector 10 ohm at 25 deg cent coil, or thermocouple placed at hottest available spot, usually between top and bottom coil side in the armature slot.

binding substance. Silicone compounds are thermally stable for temperatures far above impregnating organic varnishes. Tentative AIEE standards are now set up as indicated in Table 2. At present, these materials are still in the developmental stage and are quite limited in application. At the present time practically all applications are limited to the 600-volt class of stationary armature coils. The application to rotating field coils is still in the developmental stage. These limits will be increased as development progresses.

Temperature Limits. Table 2 lists the permissible hot-spot temperatures and permissible temperature rises of three classes of insulation, as given by AIEE standards.

COIL STRUCTURE. Stator coils are made up of individual copper strands, each insulated from the other. A number of strands are grouped together and insulated to form a conductor. The ground wall is then applied over the entire coil. In small coils the strand insulation and conductor insulation are usually one and the same thing. The wall thickness or ground insulation varies with the voltage and class of insulation. See Table 3 for average wall thicknesses for various voltage levels in use today. Rotor field coil insulation generally consists of a washer at the inner and outer surfaces of the field coil and a pole cell which may be either molded directly to the inner surfaces of the field coil, or in many wire field coils, may be wound around the pole itself before winding a field coil directly on the pole. Adequate creepage must be maintained at all points to enable the coil to stand service conditions and the high potential tests required by AIEE standards.

Table 3. Wall Thickness of Armature Coil Insulation

Voltage Class	Thickness, Inches
600	0.06
2,500	0.10
4,160	0.12
6,600	0.14
11,000	0.17
13,800	0.20

INSULATION RESISTANCE. Insulation resistance is the resistance of the winding to ground and is usually measured by an ohmmeter or megger. The insulation resistance of a machine will vary with the condition of the winding with respect to moisture absorption, cleanliness, and temperature. Any sharp or gradual reduction of the insulation resistance measured during periodic checks on a machine will aid materially in determining when and what preventive maintenance is required. In order to be useful, insulation-resistance readings must be taken at the same temperature, preferably promptly after shutdown. Insulation resistance in general approximately doubles for each 10 deg cent or 18 deg fahr for Class A windings and 14 deg cent or 25 deg fahr for Class B windings. A given machine may vary appreciably from this figure, and it is well to calibrate the variation of any unit for correcting readings to a given temperature for comparison purposes. Insulation resistance also varies with the time of application of voltage. The slope of the resistance-versus-time curve is indicative to a limited extent of the winding condition. If a unit has been standing idle for a long time with moist atmospheric conditions, it may be necessary to dry out the winding before use to prevent possible breakdown upon application of voltage. The usually recognized minimum acceptable value for machine operation is $\dfrac{\text{Rated voltage of the machine}}{(\text{Rating in kva}/100) + 1000}$. For certain high-voltage machines treated with surface conducting paints for corona protection, the minimum acceptable value may be appreciably less than that given by this formula.

5. TERMINAL MARKINGS AND ROTATION

The terminals of synchronous generators and motors are usually marked, and if this marking is in accord with an established standard considerable help is provided in making connections to the rest of the system so as to avoid unsatisfactory operation or possible damage. An established standard may not invariably be observed, however, for in some cases it may be found that terminals are marked without system or according to some non-standard system. Moreover, in certain cases connections may have been changed within the machine or errors may have passed unnoticed. Therefore it is recommended, when connecting synchronous machines to power systems, that the usual tests be made for phase rotation and relation as well as polarity and equality of voltages.

The terminal leads are usually marked with numerals, but the order of the numerals does not necessarily indicate the phase sequence. The phase sequence may be defined as the order in which the voltages successively reach their maximum positive values between the terminals. The phase sequence is determined by the direction of shaft rotation relative to the connection end of the coil winding. The numerals 1, 2, 3, etc., indicate the order in which the voltage at the terminals reach their maximum positive values *for clockwise rotation facing the connection end.* If the shaft rotation then is counterclockwise facing the connection end, the phase sequence is 1, 3, 2. The American Standards Association has specified that, when facing the end of the machine opposite the coupling end the standard direction of shaft rotation shall be:

For synchronous motors: counterclockwise.
For synchronous generators: clockwise.

From the above it is seen that, when synchronous generators are driven counterclockwise facing the connection end of the coil windings, the voltages generated give rise to terminal phase sequence 1, 3, 2. Furthermore, synchronous motors are operated with counterclockwise shaft rotation facing the connection end by connecting to leads in which the phase sequence is 1, 2, 3, as follows:

Power leads..............	1, 2 3
Machine terminals........	1, 3, 2

6. PER UNIT SYSTEM

In examining the overall performance of an electric power system or of any of its component elements, such as generators, motors, or transformers, the various constants may be expressed directly as ohms, volts, amperes, etc., as the case may be. This practice, however, does not truly represent the comparative importance of the factors involved. For instance, a stator resistance of 1 ohm in a 100-kva machine would have exactly half the effect of a 1-ohm resistance in a 200-kva machine. One way to overcome this disadvantage in handling the various constants is to express them in percentages, referred to a selected base. Thus, a loss would be expressed as a percentage of the base kilowatts, regulation would be expressed as a certain percentage of the base voltage, and so on. The base value is usually taken as the rating of the system or machine under consideration and is equal to 100 per cent. In this way a better concept is obtained of the relative effect of the various factors. In the case cited above, the 1-ohm resistance might be expressed as 10 per cent resistance for the 100-kva machine whereas for the 200-kva machine 1 ohm would be equal to 20 per cent resistance.

This method, too, however, can give rise to erroneous conclusions through multiplying together percentage quantities. For example, a 20 per cent current flowing through a 10 per cent resistance would, by simple multiplication, give 200, which could be erroneously considered 200 per cent voltage drop, whereas the correct result is a voltage drop of 2 per cent. To overcome this disadvantage the base or reference value can be taken as unity instead of 100 per cent, and all other quantities are then taken as decimal fractions or "per unit" quantities. Thus the basis of comparison is maintained without the danger of misleading results. In the above case a 0.20 per unit (pu) current flowing through a 0.10 pu resistance gives 0.20 × 0.10 or 0.02 voltage drop. The per unit system is now widely used in the analysis of machine and system performance and will be employed throughout this chapter except in cases noted.

EXAMPLE. A given synchronous motor is rated 500 hp, 2300 volts, 100 amp and is star-connected. The voltage rating per phase is then 1328, and this is considered unit or base volts. Unit or base ohms would then be that value of ohms which, when multiplied by unit current, 100 amp, gives unit volts or 1328. Unit ohms is therefore 1328/100 or 13.28. The armature leakage reactance is 1.328 ohms, and this is seen to be 0.10 in per unit notation.

Another way of finding the pu value simply is to take

$$\text{Pu reactance (or resistance)} = \text{Ohms} \times \frac{\text{Unit current}}{\text{Unit voltage}} = 1.328 \times \frac{100}{1328} = 0.10$$

To convert pu values to ohms it is simply necessary to use the same reasoning in reverse sense, thus:

$$\text{Ohms} = \text{Pu reactance} \times \frac{\text{Unit voltage}}{\text{Unit current}}$$

The above expressions use per phase values for current and voltage. Another expression which uses total kva and line-to-line voltage is

$$\text{Pu impedance} = \text{Ohms} \times \frac{\text{Kva}}{1000 \times (\text{kv})^2} = 1.328 \times \frac{398.5}{1000 \times (2.3)^2} = 0.10$$

In this example unit kilowatts is obviously 500 × 0.746 or 373, and a 14.92-kilowatt loss would then be regarded as an 0.04-pu loss.

In the per unit system troublesome coefficients are eliminated, but there is some disadvantage in the loss of a dimensional check. This is more than offset, however, by the advantages realized.

7. OPERATING PRINCIPLES

EXCITATION SYSTEMS. There are in use today a large variety of methods of excitation for synchronous machines. The most common is the direct connected or belted exciter for small machines and a direct-connected main exciter with a pilot exciter for large machines. For some synchronous motors, excitation may be received from an exciter bus, which is supplied either by motor-generator sets or rectifier equipment. The bus is usually kept at constant voltage, and the excitation to the machines varied by a main field rheostat. The direct-connected exciter or the belted exciter is usually shunt wound but may or may not have a series winding, depending on the manufacturer's wishes or the need to match previously determined characteristics. These exciters may be operated at constant voltage, but when used with synchronous generators are normally operated with variable voltage under the control of a voltage regulator. Where difficult stability or voltage-regulation problems exist, very high speed of exciter response may be required. The rotary amplifier type of exciter exemplified by the General Electric Amplidyne or the Westinghouse Rototrol provides high rates of field forcing and quick response to any voltage disturbance on the main machine. The mercury-arc rectifier has been used as an exciter in a few applications with large generators.

In the early days of the electrical industry, it was common to have so-called self-compounding a-c generators, which were compounded for nearly flat voltage regulation for a specific power factor. The present-day use of the modern voltage regulator makes these self-compounding schemes unnecessary. In general, these schemes are limited to satisfactory operation at a single power factor. A few small machines of this type are being made today.

The use of a main generator field rheostat to vary the field current of a synchronous generator is limited because of the additional losses, the bulky size of the rheostat, and its necessarily sluggish performance. Its chief use today occurs when the voltage regulator is inoperative, and continuous operation must be maintained. With high-power-factor loads on the generator and consequently low voltage drop across the field, a particular exciter may be in the range where it is unstable, so that the rheostat is used to increase the exciter voltage required to a point where the exciter is stable. Modern exciters are generally sufficiently stable for synchronizing the machine under hand control without a main field rheostat, but the exciter voltage is quite likely to drift with thermal heating and time.

ROUND ROTOR THEORY; LAGGING AND LEADING OPERATION. The performance of synchronous machines can be most simply analyzed by making a few reasonable assumptions. In the first place, it can be assumed that the armature winding is so distributed and the coil pitch so selected that its space distribution of mmf is sinusoidal, and its wave form essentially free from harmonics. Moreover it will be assumed that the field form due to excitation is also sinusoidal and devoid of harmonics. This condition can be approximated by suitably shaping the contour of the pole faces or, if the machine is round rotor type, by properly distributing the exciting coils on the rotor. If a round or turbo type rotor is assumed, further simplification is possible, for in this case the reluctances of the magnetic circuits acted upon by field and armature mmf's are practically equal. Therefore the armature and field mmf's may be added vectorially to obtain resultant mmf's and fluxes. This is done in Figs. 9, 10, 11, and 12, which are the vector diagrams for a generator and motor respectively, each shown with leading and with lagging operation. The following nomenclature is used for these figures:

V = terminal voltage.
I_a = armature current.
r_e = effective resistance.
x_l = leakage reactance.
E = induced voltage under load.
E_f = excitation or internal voltage.
Φ = resultant flux.
F_r = resultant mmf.
F_a = armature reaction mmf.
F_f = field excitation mmf.
X_a = armature reaction.
ϕ = power-factor angle.
δ = internal or displacement angle.

The reader will recall that vectorial representation of electrical quantities carries the concept that the system of vectors rotates counterclockwise, but that for a given condition of operation each vector is fixed in its relation to the others. The system of vectors can therefore be drawn on paper, with leading components represented as advanced in a counterclockwise direction. Each voltage or current vector is considered per phase.

In Fig. 9, which represents a generator supplying leading current, the angle of lead between current and voltage is obviously ϕ. The impedance drop in the machine can then be added vectorially to the terminal voltage V to obtain the internal or generated voltage E. This voltage must obviously be induced by action of a flux Φ, which is established by an mmf F_r. Since an induced voltage lags its inducing flux by 90 deg, the flux is shown as 90 deg in advance of E. Because a rotating machine has a relatively large air gap, the mmf F_r and flux Φ can be considered as directly in phase. The mmf which establishes the flux must be considered the resultant of two other mmf's, one, F_a, being due to the current I_a flowing in the armature winding, and the other, F_f, to the field excitation. This field excitation can be considered as contributing a definite component to the resultant mmf and, were it not for the armature current, would constitute the only mmf in the machine. A voltage E_f would then be induced, lagging F_f by 90 deg. The vector difference between E_f and E then represents a reactive drop, I_aX_a, which is due to armature reaction. The sum of X_a and the leakage reactance X_e is the synchronous reactance of the machine, and this, when added in quadrature to effective resistance, r_e, gives the synchronous impedance. The same explanation applies to Fig. 10, except that lagging operation is represented, since the current I_a lags the terminal voltage V. Synchronous generators normally operate with lagging current, whereas synchronous motors usually operate with either leading or in-phase current.

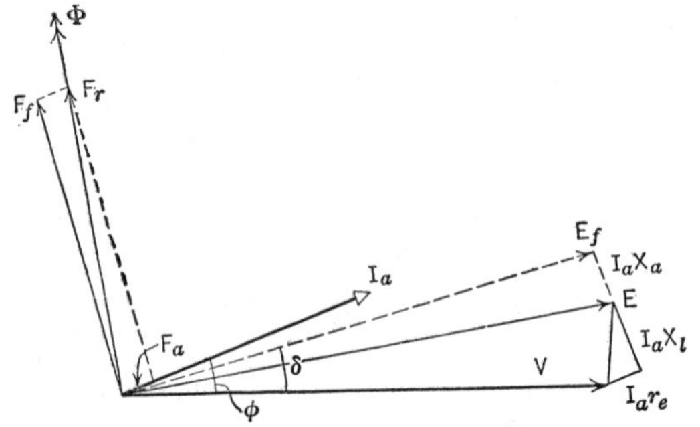

Fig. 9. Generator Supplying Leading Current

In considering motor action, it must be understood that in reference to its supply circuit the induced voltage must be considered in phase opposition and is therefore visualized as a counter voltage. The terminal voltage must be considered a component of the equal and opposite counterpart of the induced voltage and is equal to the difference after subtracting vectorially the impedance drop in the machine. These facts are represented in Figs. 11 and 12. Since the induced voltage lags its inducing flux by 90 deg, it is clear that the reverse of this voltage must lead the flux by 90 deg, and hence for motor action the flux vector takes the position shown. In other respects the explanation given for the generator diagrams applies equally well to the motor diagrams, where $-E_f$ is the reversed counter voltage and $-E$ is the reversed counterpart of the voltage actually generated in the armature under load.

In considering these diagrams it is to be noted that for both conditions of operation the current I_a and terminal voltage V, as well as the resistance and reactance components and phase angle, are considered to have the same magnitude. It can be seen from Figs. 9 and 10, applying to generator action, that, when the armature current is leading, the resultant mmf, F_r, is greater than the field excitation F, whereas with lagging current the resultant mmf is less than the field excitation. The correct conclusion, therefore, is that, in generators, a leading current magnetizes the field poles, whereas a lagging current demagnetizes them. In a motor, as shown by Figs. 11 and 12, the reverse is true. When we note that the angle ϕ and vectors V and I are the same in both cases, it is apparent that

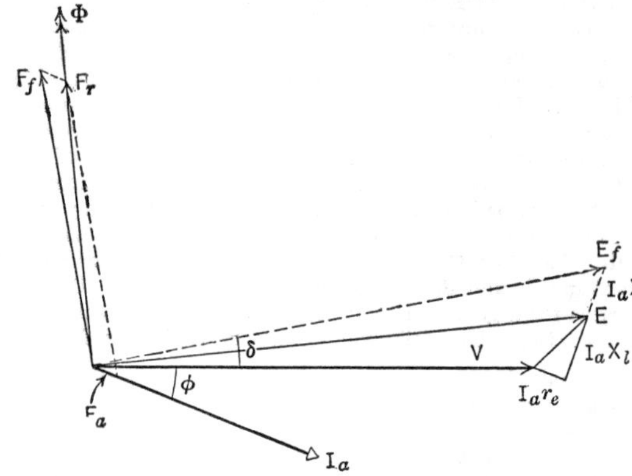

Fig. 10. Generator Supplying Lagging Current

the same power is absorbed. Hence a sufficient increase in the field excitation of a motor will cause it to change from lagging to leading operation. In other words, an overexcited motor draws a leading current, and this property is very often utilized in so operating

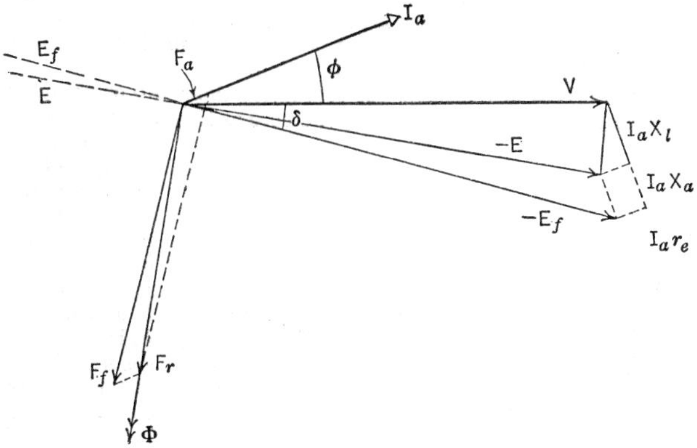

FIG. 11. Motor Taking Leading Current

motors on a system as to help compensate for the lagging current drawn by induction apparatus, and this improves the system power factor.

An angle δ is indicated on the diagrams as the angle between the terminal voltage V and the field excitation voltage E_f. This angle is observable physically and is the angle by which, in a given machine under load, the rotor poles are displaced from the consequent poles of armature mmf. It is variously called displacement angle, torque angle, power angle, coupling angle, or internal angle and is discussed in detail in other paragraphs of this section.

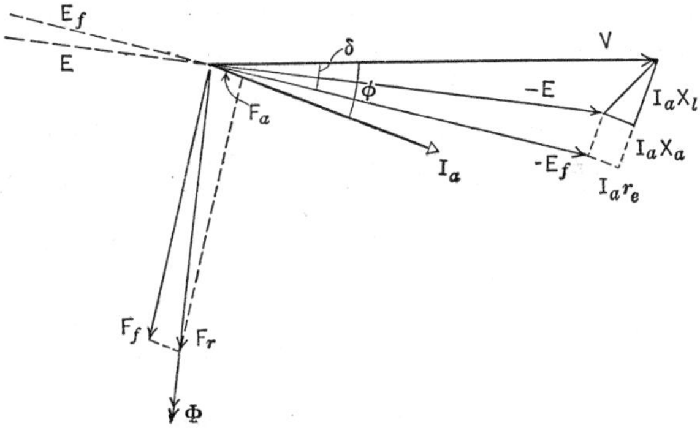

FIG. 12. Motor Taking Lagging Current

TWO-REACTION THEORY. In a salient-pole machine the reluctances acted upon by the field and armature mmf's are not equal, and therefore some means must be available for taking into account the dissymmetry. This means, known as the two-reaction method, was suggested by Blondel.

Armature reaction in a salient-pole machine not only modifies the field strength, as in a round rotor machine, but also, if the armature current has an in-phase component with the excitation voltage, distorts the field by crowding the flux toward one pole tip and away from the other. If the armature current is exactly in phase, the effect is purely distortion without modifying the strength of the field. If the current is 90 deg out of phase, no distortion results, but the field strength is modified. Between these two extremes, as in the usual case of operation, it is apparent that both effects are present. Blondel found that armature reaction could therefore be resolved into two components, and since armature current and the resulting mmf are in phase, these components can be represented as

$$I_a \sin \theta \quad \text{and} \quad I_a \cos \theta$$

where θ is the phase angle between armature current and excitation voltage. The first component, $I_a \sin \theta$, acts on the direct axis, or the same magnetic circuit as the field coils, and therefore has the effect of adding or taking away ampere-turns from the field winding, according to whether, for generators, the current leads or lags, respectively, the excitation voltage. For motors the effect is respectively opposite. The component $I_a \cos \theta$ acts on the quadrature axis midway between poles, the magnetic circuit being one of high reluctance because of the large air gap in this region.

Expressions are derived in textbooks for evaluating these two components of armature reaction. The final equations only will be given here as they apply to the vector diagram (Fig. 13). The generated voltage E is of course found by vectorially adding the resistance and leakage-reactance drops to the terminal voltage V. Blondel resolves this generated emf E into its two components. E_c is the component voltage generated by the cross or distorting flux set up by $I_a \cos \theta$. This representation is possible, since for any given armature current this flux is constant and fixed with respect to the poles, revolving with them to induce a voltage E_c in the armature winding. The second component of E is E_D, the voltage generated by the flux from the main poles.

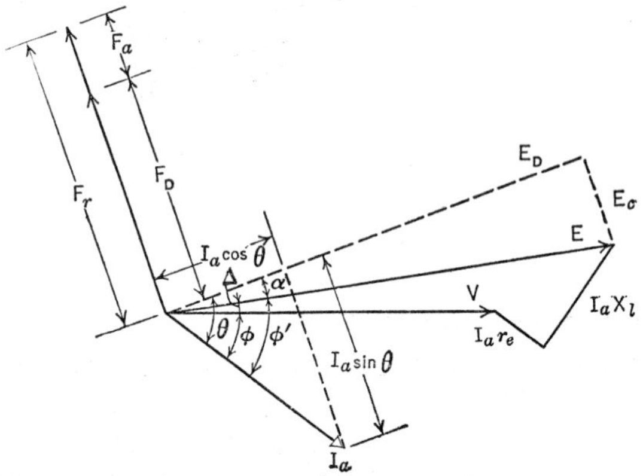

FIG. 13. Vector Diagram for Two-reaction Theory

The two components E_c and E_D can be found by first determining the angle α. Angle ϕ' is obviously determined by the operating power factor angle ϕ and the resistance and reactance drops. Then

$$\tan \alpha = \frac{V_l N_e K_c I_a \cos \phi'}{E + V_l N_e K_c I_a \sin \phi'}$$

where V_l = volts per field ampere-turn from lower part of open-circuit magnetization curve.
N_e = effective turns per phase in armature winding.
K_c = a factor depending upon form factor and ratio of pole arc to pole pitch; taken as 0.252 for the average machine.

With E_D known, the excitation required to produce it, F_D, can be found from the open-circuit magnetization curve. To this excitation must be *algebraically* added a component F_A, which balances the direct, or strength-modifying, component A_D (due to $I_a \sin \theta$) of armature reaction. The sign of F_A is plus when θ is an angle of lag and minus when θ is an angle of lead in a generator, and opposite to this in a motor. A_D is evaluated as

$$A_D = N_e K_D I_a \sin \theta$$

where K_D = a factor depending upon the ratio of pole arc to pole pitch and equal to 0.72 for the average machine.

The value of F_A corresponding to A_D is found from the open-circuit characteristic. For the impressed field F_r the voltage E' can likewise be found. The regulation then is

$$\frac{E' - V}{V}.$$

HUNTING AND NATURAL FREQUENCY. Hunting caused by electrical phenomena is rare in modern units. In a generator so-called hunting is more often caused by instability in the governor of the prime mover. When hunting occurs, it is generally caused by pulsations in the load in a motor and torque pulsations in the prime mover in a generator. If these pulsations coincide with the natural frequency of the machine with respect to the system to which it is connected, oscillations of appreciable magnitude may occur. The natural frequency of a machine with respect to an infinite bus or large system is

$$F = \frac{35{,}200}{\text{Rpm}} \sqrt{\frac{P_r \times f}{WK^2}}$$

where F is in cycles per minute, P_r equals the kilowatts per electrical radian of the machine, f is the rated frequency in cycles per second (i.e., 60, 50, etc.), and WK^2 is the rotor weight in pounds times the rotor radius of gyration in feet squared. The usual range of frequencies encountered by machines is between 100 and 250 cycles per minute, depending on the inertia and P_r value. P_r is normally given for generators in applications where the prime-mover torque pulsations are appreciable; it varies in a machine from no load to full load, the value at no load being approximately 60 per cent of the value at full load for an 80 per cent power-factor generator. Normally, P_r is given only at full load. Dampers or amortisseur windings are generally provided on generators driven by reciprocating engines to reduce the magnitude of the hunting pulsations in case they should occur. A discussion of synchronous motor applications with compressors will be found in Article 15.

SHORT CIRCUITS. A short circuit on a synchronous machine causes a large current to flow for a short time. The symmetrical a-c component of the sudden large current may be from 4 to 10 times normal rms a-c amperes. Figure 14 shows a typical current-time or current-decrement curve. The peak current, in case of the maximum asymmetry,

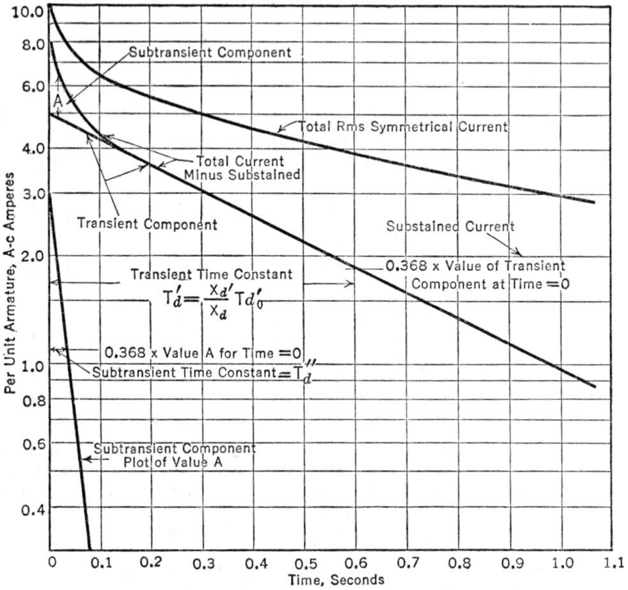

FIG. 14. Typical Current-decrement Curve for Short Circuit from Full Load

will be twice the square root of 2 or 2.82 times the rms symmetrical current. This peak current makes it necessary to brace machine windings and the powerhouse bus work to withstand the magnetic forces imposed by these peak currents. All modern machines are designed to stand a sudden short circuit at the terminals in accordance with ASA standards, provided the duration of the fault is not sufficient to cause injurious heating. The initial rms symmetrical current on short circuit for a three-phase fault is equal to the initial internal voltage of the machine divided by the subtransient reactance X_d''. Average values of reactance are listed in Table 4. Article 10 contains further definitions of reactances.

Standard synchronous generators are designed for operation at either plus or minus 5 per cent normal voltage with safe temperatures at full load kva. For synchronous motors, plus or minus 10 per cent normal voltage is allowable. Standard machines are guaranteed to operate within stated temperature limits only at full-load rated voltage and power factor. Operation at other than rated voltage causes increased heating, changes the reactances of the machine, and may impose hardships in efficiency and regulation. In general, there are two limiting factors in the output of a machine: (1) the voltage and current output from the stator or armature, and (2) the field heating limit. In normal machines it is satisfactory to operate up to the kva rating of the stator for all power factors above rated power factor without exceeding either the field-heating or the armature-heating

Table 4. Typical Constants of Three-phase Synchronous Machines

(Reactances are per unit; time constants are in seconds. Values below the lines give the normal range of values; those above give an average value.)

	Two-pole Turbine Generators	Four-pole Turbine Generators	Salient-pole Generators and Motors (with Dampers)	Salient-pole Generators (without Dampers)	Condensers
X_d unsat.	1.10 / 0.95–1.45	1.10 / 1.00–1.45	1.15 / 0.60–1.45	1.15 / 0.60–1.45	1.80 / 1.50–2.20
X_q rated current	1.07 / 0.92–1.42	1.08 / 0.97–1.42	0.75 / 0.40–1.00	0.75 / 0.40–1.00	1.15 / 0.95–1.40
X_d' rated voltage	0.155 / 0.12–0.21	0.23 / 0.20–0.28	0.37 / 0.20–0.50	0.35 / 0.20–0.45	0.40 / 0.30–0.60
X_d'' rated voltage	0.090 / 0.07–0.14	0.15 / 0.12–0.17	0.24 / 0.13–0.35	0.32 / 0.18–0.41	0.25 / 0.18–0.38
X_2 rated current	$= X_d''$	$= X_d''$	0.24 / 0.13–0.35	0.55 / 0.30–0.70	0.24 / 0.17–0.37
X_0* rated current	0.01 – 0.08	0.015–0.14	0.02–0.20	0.04–0.25	0.02–0.15
X_p	0.09 / 0.07–0.14	0.17 / 0.12–0.24	0.32 / 0.17–0.40	0.31 / 0.17–0.38	0.34 / 0.23–0.45
Td'_o	4.4 / 2.8–6.2	6.2 / 4.0–9.2	5.6 / 1.5–9.5	5.6 / 1.5–9.5	8.0 / 5.0–11.0
Td'	0.6 / 0.35–0.90	1.3 / 0.9–1.8	1.8 / 0.5–3.3	1.8 / 0.5–3.3	2.0 / 1.0–2.8
Td''	0.035 / 0.02–0.05	0.035 / 0.02–0.05	0.035 / 0.01–0.05		0.035 / 0.02–0.05
Ta	0.09 / 0.04–0.15	0.2 / 0.15–0.35	0.15 / 0.03–0.25	0.30 / 0.1–0.5	0.17 / 0.1–0.3

* X_0 varies so critically with armature winding pitch that an average value can hardly be given. Variation is from 0.1 to 0.7 of X_d''.

limits. Operation below rated power factor imposes a limit on the kva output of the machine to prevent exceeding the field-heating limit. Figure 15 shows a curve of allowable kva plotted against power factor for normal machines of 1.0, 0.9, 0.8, and 0.7 power

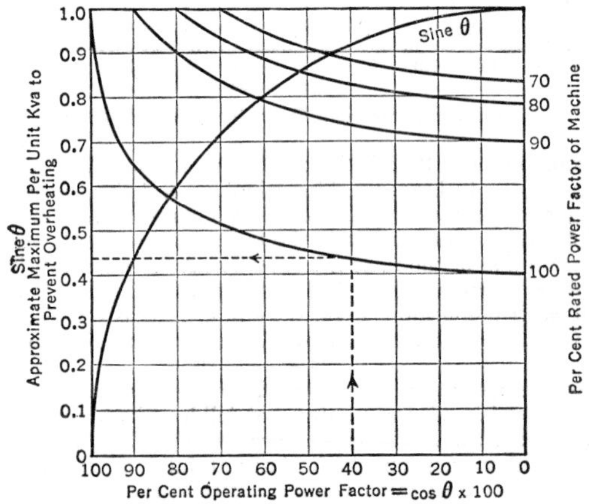

FIG. 15. Allowable Kva for Various Power Factors

factor and 0.9 short-circuit ratio. Machines of other short-circuit ratios will vary slightly in the heating limit plotted with respect to power factor. The maximum reactive kva (rkva) for any operating power factor equals maximum kva × sine θ.

8. BASIC PRINCIPLES IN DESIGN

D^2L AND DENSITIES. The size of a machine depends on the electrical characteristics built into the unit as well as the mechanical characteristics and speed. As a rough

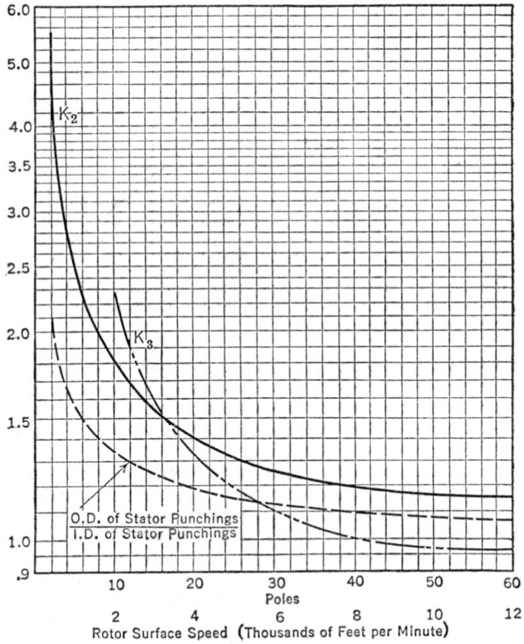

FIG. 16. Curves for Assisting in Determining the Approximate Size of the Active Parts of a Synchronous Machine

approximation the size of the active parts of machines with normal characteristics may be estimated by the following formula for two- or three-phase machines.

$$D^2L \text{ rpm} = K_1 \text{ kva } 10^4 K_2 K_3$$

where D = the outside diameter of the stator punching in inches.
L = the stacked length of the stator in inches.
K_2 = a function of the number of poles (see curve, Fig. 16).
K_3 = is a function of peripheral speed (see curve, Fig. 16).

For large standard 60 deg cent rise waterwheel generators, K_1 = 1.2 –1.4.
For large hydrogen-cooled two- and four-pole turbine generators, K_1 = 1.00–1.1.
For industrial air-cooled two-pole turbine generators, K_1 = 1.25–1.6.
For large air-cooled synchronous condensers, K_1 = 1.1 –1.4.
For hydrogen-cooled synchronous condensers, K_1 = 0.92–1.10.
For large synchronous motors, K_1 = 1.3 –1.5.

Special characteristics, such as unusual voltages, short-circuit ratio, reactance, WR^2, generally result in an increase in size. Small machines have a K_1 from 20 to 80 per cent greater than large machines. Small units are usually made with greater flexibility to take care of a wide range of characteristics and hence cannot be redesigned and reproportioned for each specified job to take advantage of minimum size.

Two main constants of design are the gap density and the ampere conductors per inch of periphery at the gap. The normal peak-gap density and ampere conductors per inch of periphery for modern machines are given in Table 5. These factors must be selected by the designer to suit the proportions selected in order to prevent overheating.

Table 5. Normal Densities and Ampere Conductors

Machine	Gap Density, Kilolines per Square Inch	Ampere Conductors per Inch of Periphery	Average Tooth Density *	Core Density *	Pole Density Full Load
Large waterwheels: 50 and 60 cycles					
Periph. vel., 3,000–5,000 ft per min.	42–48	800–1,000	90–100	60–75	95–110
Periph. vel., 5,000–7,500 ft per min.	45–51	900–1,200	95–105	65–85	100–115
Periph. vel., 7,500–14,000 ft per min.	48–54	1,200–1,500	100–110	75–90	105–115
Condensers					
Air-cooled	42–48	1,500–1,800	90–105	65–85	100–115
Hydrogen-cooled	43–49	1,700–2,200	95–110	70–89	100–115
Air-cooled turbine generators	2 & 4 pole	900–1,200†	85–98	80–95	....

* Tooth and core densities are based on 3 to 4 per cent silicon iron. With punchings with less than 1/2 per cent silicon, the densities are usually 85 to 90 per cent of the above.

† For hydrogen-cooled units increase 15 per cent.

SATURATION CURVES. Among the most used design data applying to synchronous machines are the saturation curves, which show terminal voltage plotted against field excitation over the range from zero volts to a voltage somewhat above the rated value or until considerable saturation of the magnetic circuits is shown. In Fig. 17 the heavy lines show a set of saturation curves.

The no-load saturation curve shows open-circuit terminal volts and, were it not for saturation, would be a straight line as shown by the dotted or air-gap line of Fig. 17. As the flux densities in the materials comprising the magnetic circuit increase because of higher field currents, however, the materials tend to become saturated, which means that the flux is no longer in proportion to the magnetizing force but increases less and less for the same increase in field current. Therefore the saturation curve shows a continuous curvature away from the air-gap line. The air-gap line is so called because at very low flux densities there is practically no saturation in the iron parts, and the magnetizing force required to force the flux through them is negligible in comparison to that required to establish the field across the air gap. The entire magnetizing force can thus be considered as expended on the air gap, and since air has a constant permeability the flux, and therefore the generated voltage, is always in proportion to the field current, giving rise to the so-called air-gap line. Predetermination of the no-load saturation curve is a design procedure that is beyond the scope of this article. Suffice it to say here that for any given voltage the necessary flux is calculated, and then the summation taken of the components of magnetizing force necessary to force the flux through the series circuit comprising effective gap, stator teeth, stator core, field pole, and spider. The effective gap is greater than the actual gap by a factor known as Carter's coefficient, which allows for the tendency of the flux to spread out or "fringe" at corners such as the tooth corners and ventilating-duct corners. The magnetizing force for the air gap is calculated as

$$\text{Ampere-turns} = \frac{\text{Flux density in gap}}{3.19} \times \text{Effective length of gap}$$

where the flux density is in lines per square inch, and the gap in inches. The magnetizing forces for the rest of the circuit are obtained from the magnetization curves applying to the material used. Points on the no-load saturation curve for a machine already built are easily obtained by test.

The zero-power-factor saturation curve is of considerable importance, since it is useful in obtaining the load-saturation curve, and is obtained relatively easily, either by test or by calculation. The excitation characteristic of a synchronous machine operating at very low power factor, in the order of 20 per cent, is essentially no different from that for zero power factor. Hence a generator may be loaded on an underexcited synchronous motor and the excitation so adjusted as to circulate rated load current, or a fraction thereof, at some specified voltage and very low power factor. Several such points can be taken to make up a curve.

At zero power factor the terminal voltage, including saturation effects, is essentially equal to the arithmetical difference between the actual generated voltage and the reactance voltage, and the net excitation is equal to the arithmetical difference between the magnetizing force due to field ampere-turns and that due to armature reaction for the current used. If (referring to Fig. 17) a value of field excitation *OF* is assumed, the net excitation is represented by *OD*, where *DF* is the calculated field amperes corresponding to armature reaction. The generated voltage corresponding to the net excitation is from the no-load curve *OE*, and the calculated voltage drop I_aX_p due to Potier reactance is *EV*. Subtracting *EV* from *OE* gives the terminal voltage *OV*, and a point *P* is located on the zero-power-factor curve for excitation *OF*. For a constant armature current the reactance drop and the field amperes equivalent to armature reaction are essentially constant. Thus, if the no-load curve is dropped a distance *BA* = *EV* and moved to the right a distance *AP* = *DF*, the zero-power-factor curve is obtained. It is implied that any different value of armature current results in a new zero-power-factor curve. The armature current is thus always proportional to the length of side *BP* of triangle *ABP*. This method is more accurate for high-speed machines than for those of low speed.

The Potier reactance referred to varies considerably with flux density and is most accurate as the density increases. At low densities it becomes much greater than the true leakage reactance. However, since it is useful, when intelligently used, in obtaining load-saturation characteristics, the manner of obtaining it will be outlined. As the above paragraph implies, it is obtained from the no-load saturation and zero-power-factor curves, which must be available. A horizontal line is drawn through the point *P* (Fig. 17), which is located well up on the zero-power-factor curve, and a distance *CP* laid off equal to *OG*. Then through *C* a straight line is drawn parallel to the air-gap line. The intersection of this line with the no-load curve at *B* determines the hypotenuse of the Potier triangle *ABP*. The reactance voltage drop *BA* divided by the current corresponding to the zero-power-factor curve gives the Potier reactance of the machine.

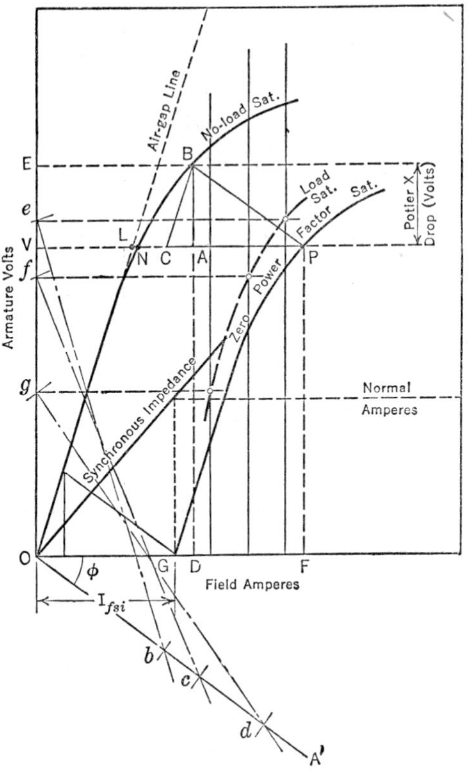

FIG. 17. Saturation Curves for Synchronous Machine

The most useful of the saturation curves in actual practice is the load-saturation curve, which shows the field excitation required for various voltages at a given armature current and power factor. Among the methods for obtaining field excitation at a given load and power factor are the following two, which are prescribed by the AIEE.

In the **Potier reactance method** (see Article 10), considering the case of an overexcited generator, it is assumed that the no-load saturation curve and air-gap line are known. The excitation for voltage *E* (Fig. 18) will be determined. The line *OB* is laid off equal in length to *OE* and at the power factor angle from the horizontal. The resistance drop E_R for the load current is added to *OB* horizontally to the right, and to this is added in quadrature the Potier reactance drop E_x, where the Potier reactance has been determined as described previously. Now radius *OC* can be considered equal to the internal voltage E_i for the load current considered. At this voltage I_{fs} is the difference between the air-gap excitation and the no-load excitation. In Fig. 18(*b*), which is reproduced from the AIEE Test Code, I_{fg} is the field current from the air-gap line at voltage *E*. From the extremity of this line, I_{fsi}, the field current required to circulate the load current through the short-circuited armature (from the synchronous impedance test) is laid off at the power factor angle ϕ, with the vertical, to the right as shown for overexcitation or to the left for underexcitation. Then the vector sum of I_{fg} and I_{fsi}, plus the excitation I_{fs}, is the required load excitation I_{fl} for voltage E. In Fig. 18(*b*), considering a motor, the drop E_R would extend to the left, and, considering underexcitation, drop E_x would be directed downward. Obviously a constant scale must be used when laying off vectors of the same

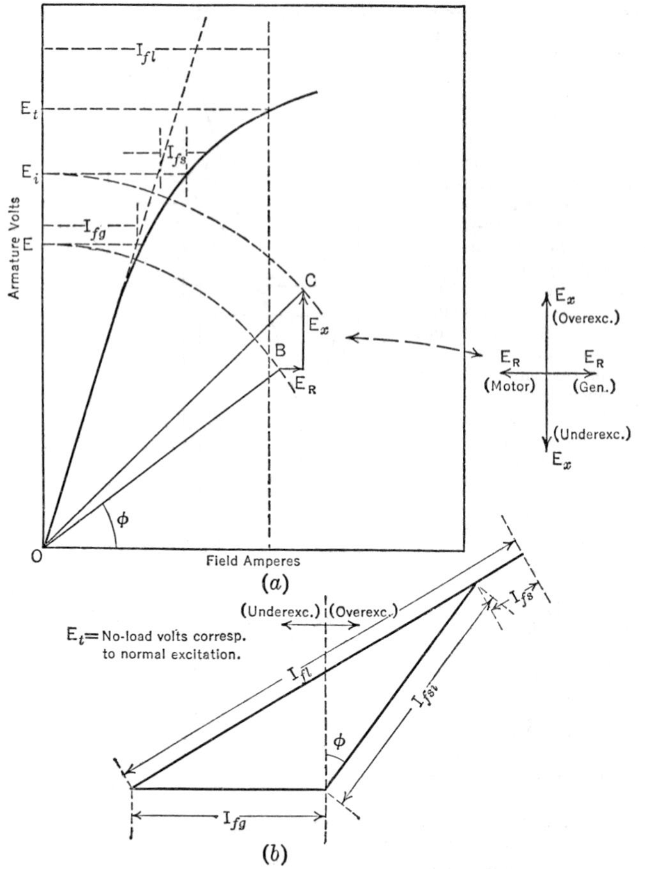

FIG. 18. Field Current and Regulation from Potier Reactance

kinds. If the voltages are plotted to per unit (pu) scale with voltage E = 1.0, the pu values may be used for resistances and reactances.

In the **adjusted-synchronous-reactance method** it is assumed that the no-load saturation and zero-power-factor saturation curves for desired armature current are known. A line OA' (see Fig. 17) is laid off at the power factor angle with the horizontal above for underexcitation and below for overexcitation. Three or more field currents are selected, such that the desired load excitation curve will include the desired voltage. Along OA' the distance O_b, O_c, O_d, etc., is laid off equal to the difference in voltage between the no-load and zero-power-factor curves at the respective field currents. The corresponding no-load voltages are used as radii, with b, c, d as centers, to swing arcs which intersect the voltage axis at e, f, g. The distances Oe, Of, and Og are then load voltages and, when plotted against their respective field currents, give the load excitation curve.

Direct calculation of the excitation required for any load or power factor can be accomplished, knowing the armature resistance and leakage reactance and having available the no-load saturation curve. The procedure consists of the use of complex numbers to solve for F in a vector diagram such as that shown in Fig. 9. The voltage V and power factor angle ϕ are known quantities. Examination of the diagram will identify the following steps:

Consider current vector I_a as an axis of reference.

Find $V(\cos\phi + j\sin\phi)$.

Add $I_a(r_e + jX_2)$.

The sum is $E(\cos\alpha + j\sin\alpha)$. Referring to the no-load saturation curve, find the excitation F_r for the voltage E.

Express $F_r(-\sin\alpha + j\cos\alpha)$.

Find the calculated value of F_a for the current I_a, and add it algebraically to $-F_r\sin\alpha$ to obtain one component of F, the load excitation sought. The other component of F, in quadrature, is obviously $F_r\cos\alpha$.

Similar procedure can be used for the conditions shown in Figs. 10, 11, and 12, with proper observance of complex algebra.

V CURVES. If all the possible values of field current corresponding to a given electrical power input or output at constant voltage are plotted against armature current, the resulting curve is seen to have a characteristic shape similar to the letter V, with the apex denoting minimum armature current for the given power. The apex therefore represents unity-power-factor operation. It will be noted that for a given armature current two field excitation values are possible, one corresponding to lagging power factor operation and the other to leading power factor. A family of such curves may be determined for various values of power and is often found useful in checking various conditions of performance, particularly with synchronous motors, where benefits of leading kvar are desired. By reference to the circle diagram (Fig. 29) it can be seen that for a given excitation and power value a second value of armature current is obtained corresponding to the second intersection of the two circles, which falls in the unstable region. If these are plotted on the same sheet with the V curves for the same power values, closed curves or O curves will be obtained. A family of such curves is shown in Fig. 19, with the V curves made in heavy lines. Since the second values of armature current shown by the lightly drawn portion of the O curves represent unstable conditions, they are of academic interest only.

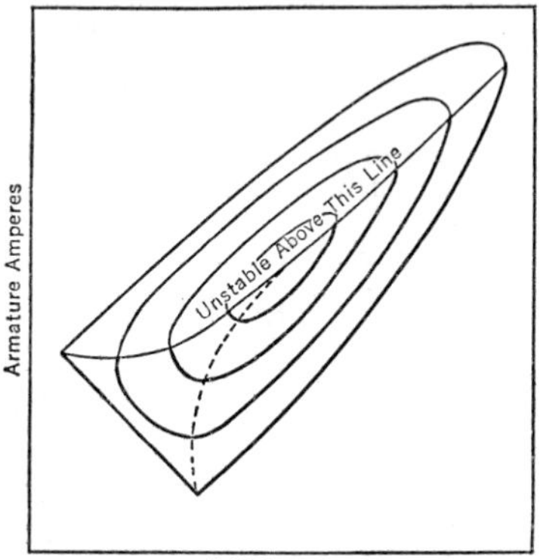

FIG. 19. Armature Currents for Various Field Currents (V and O Curves)

9. DESIGN OF WINDINGS

STATOR WINDINGS. As stated previously, the common type of winding is the two-coil-side-per-slot diamond-type lap winding (Fig. 7).

This type of winding usually consists of a group of one or more adjacent coils per phase per pole in series. Each phase group normally occupies a portion of the pole pitch equal to the pole pitch divided by the number of phases.

DISTRIBUTION FACTOR. The separation of the slots in a phase belt causes a reduction in the total voltage generated, since the voltages in adjacent coils are not in phase. The reduction factor called distribution or breadth factor k_2 is given in Table 6.

PITCH FACTOR. The two sides of one coil may lie in slots not one pole pitch apart. This also causes a reduction in voltage generated equal to K_p = sine (Y/mq) 90 deg, where Y = throw of coil (if coil sides lie in slots 1 and 8, the throw is 7), m = number of phases, q = slots per phase per pole. It is usually desirable to "short pitch" a winding a small

Table 6. Winding Distribution Factor (k_2)

Slots per Pole	Value of k_2		
	1 phase	2 phase	3 phase
1	1.000	.....	.....
2	0.707	1.000	.....
3	0.666	.....	1.000
4	0.65	0.925	.....
6	0.64	0.912	0.966
8	0.64	0.905	.....
9	0.64	.....	.....
12	0.64	0.90	0.960
18	0.64	0.90	0.960
24	0.635	0.90	0.958
∞	0.632	0.90	0.958

amount in order to reduce end turn length or space requirements and losses, as well as to improve wave form while adding only slightly to the flux burden in the iron. The throw of the coil is varied in new machines and in rewinding old machines to vary the regulation, reactances, short-circuit ratio, pull-out torque, inrush, and other characteristics. In effect the pitch factor acts as a fine adjustment in the effective turns per coil. It is preferable to use coils where $Y/mq = 0.66$ or higher in all but two-pole machines, where the range is usually from 0.5 to 0.66 because of difficulties in winding.

FRACTIONAL SLOT WINDINGS. For numerous reasons, including manufacturing, noise, wave form, and locking torques, it is usually desirable on machines of eight poles and more to use fractional slots per phase per pole. If for a machine with Q/mP slots per phase per pole, where Q is the total number of machine slots, m the number of phases, and P, the number of poles, both numerator and denominator are divided by the largest common factor, the resulting fraction is N/D. If trick windings, which are sometimes used to accomplish certain purposes, are eliminated, the number of slots per phase per pole should have a denominator D not greater than one-half the number of poles P and should not be divisible by the number of phases m. In order to have a balanced winding (one in which all the phase voltages are equal and are spaced exactly 120 or 90 elec deg apart for a three-phase or two-phase winding, respectively) the winding must be carried through a number of poles equal to the denominator D of the slots per phase per pole. The maximum number of parallels available in the machine is equal to the poles divided by the denominator D.

In order to connect the coils of such a machine, a simple layout chart may be used to decide on the number of coils in any phase group in the machine. Figure 20 is a layout chart for the case of $N/D = {}^7/_4$. The grouping of the coils around the machine is then 2221, 2221, 2221, or in other words 2, 2, 2, 1 is repeated three times in the four poles when reading the number of coils per phase group in the order of phases A, C, B, A, C, B. This repeats again each four poles around the machine. The connections to the phase C coils must be for opposite polarity to those in phases A and B.

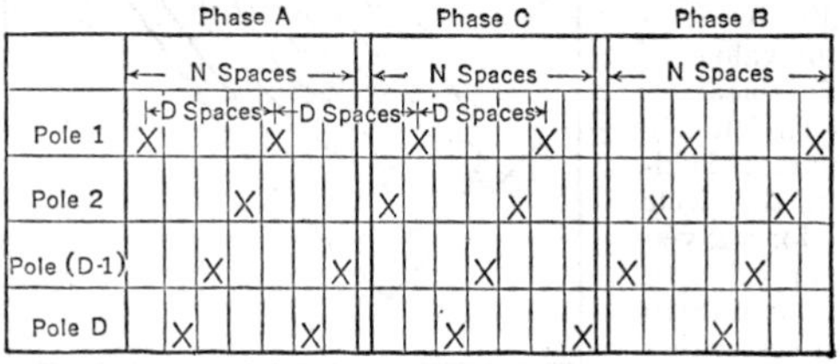

Fig. 20. Typical Layout Chart for Windings

Naturally all connections for phase belts opposite even-numbered poles are opposite in polarity to those for odd-numbered poles. For a study of a more complete type of winding diagram, see Reference 9 of the Bibliography, Article 23.

In reconnecting machines for lower voltages than the original, the maximum number of parallels C available will be P/D, as discussed in the previous paragraph, and the unit may be reconnected with any number of parallels which is an integral factor of C. In reconnecting machines for voltages higher than the original, the same rules concerning the number of parallels must be observed. Unless care is exercised, insulation failure will result in due time, if not immediately. Normal voltage classes for standard synchronous machines are 600, 2500, 4300, 6900, 11,500, 13,800 volts. For special applications intermediate insulation may occasionally be used. The delta connection may be employed under certain conditions to meet particular voltage combinations. The delta winding, however, must be used with caution since it is possible in some machines to burn up the winding at no load because of circulating currents. The delta connection should be used only with a rather complete knowledge of the machine, and checks should be made to determine the magnitude of the circulating current and machine heating before the machine is put into regular service. In general, if the circulating current at no load is 15 per cent or less of the rated line current, no serious extra coil heating will result. Any machine with a stator coil pitch of $^2/_3 = Y/mq$ may be connected in delta without appreciable circulating currents.

The resistance of a winding circuit may be calculated by the following formula:

$$\text{Resistance in ohms} = \frac{(\text{Mean turn length})(\text{Turns per coil})(\text{Coils in series}) \times K}{C \times \left(\begin{array}{c}\text{Total area of conductor strands in parallel in one}\\ \text{of the parallel circuits}\end{array}\right)}$$

where C is the number of parallels, and K is the constant of the material (for copper at 75 deg cent $= 0.825 \times 10^{-6}$) when all dimensions are given in inches.

TEMPERATURE-RISE ESTIMATES. The criterion for estimating temperature rise in synchronous machine windings is the surface per watt, or heat-dissipating area, computed as square inches per watt. The figure so obtained must be tempered according to several factors, such as type of ventilation, length of coil extension beyond core (longer for high-voltage machines), ratio of core length to diameter, and speed. Because of the many influencing factors, the surface per watt is usually significant only in comparison with temperature tests on similar machines. The usual range is from 1 to 2.5 sq in. per watt. In computation, for stator windings, the watts loss and square inches of area per inch length of winding are taken, where the area is the outside surface of the group of conductors filling a slot. Thus:

$$A = \frac{2(b + n_s d) \times a}{0.825 \left(\frac{I}{1000c}\right)^2 \times n_s} \text{ sq in. per watt}$$

where a = cross-section per conductor in square inches.
b = width of bare conductor in inches.
d = depth of bare conductor in inches.
n_s = number of conductors per slot.
I = amperes current per phase.
c = number of parallel circuits in winding.

The surface per watt for the field winding is computed similarly, taking the total uninsulated outside surface of coil cross-section and multiplying by the product of the average length of turn times the number of poles. The loss to be dissipated is of course the total loss in the field winding, not including outside losses, such as in rheostats.

Temperature rise of core can be estimated by reference to tests on similar machines, taking into account the watts loss and the surface area available for dissipating the loss, including the surface of both sides of all vent ducts and the two ends, as well as the back of the core.

NOISE CONSIDERATIONS. The reduction of noise in electrical apparatus is becoming of increasing importance to manufacturer and customer alike. Although the theoretical background of this problem is beyond the scope of this article, some general considerations will be given.

Noise produced by synchronous machines arises from irregular air movements within the machine (windage) and from varying magnetic forces which cause slight vibrations of mechanical parts, such as teeth, end bells, covers. In high-speed machines the windage noise usually predominates, whereas in low-speed machines most of the noise is of magnetic origin. The unpleasant effect of noise is due more to its frequency than to its intensity, high-pitched noises of relatively low intensity being more unpleasant than those of low frequency and high intensity.

Installation conditions are of the utmost importance in controlling noise. Hard surfaces, such as walls, reflect noises readily, resulting in standing waves which augment the effect of a machine which would otherwise appear quiet. Noise components of minor magnitude in a given machine may be transmitted through the foundation or building structure and become objectionable at a remote point. The effect becomes very pronounced if parts of the mounting or structure are resonant near an exciting frequency in the motor. Usually the machine mounting should have a natural frequency of one-fourth to one-third the exciting frequency. In observations of noisy installations proper allowance must be made for noise originating in the prime mover or, for motor installations, in the driven machine.

It is seen that readings taken under test floor conditions are no criterion for estimating the noise of a proposed installation. Moreover, it is seen that there is no adequate basis on which noise guarantees for an installed machine can be made.

Windage noises are reduced by controlling the air movement through the machine and by installing mufflers or baffles at the proper locations. Magnetic noises, usually ranging in frequency from 500 to 3000 cycles per second, are more difficult to reduce while maintaining other necessary operating characteristics, but manufacturers are making some progress along this line.

In any installation noise effects can be minimized by correct design of foundation and mounting and by provision, where possible, of sound-absorbing surfaces rather than hard, reflecting surfaces. Lining air ducts to or from the machine with absorbent material will be of appreciable benefit.

The following tabulation gives a comparison of common noise levels in decibels, the unit of measurement of sound intensity:

a. Country roadside	28
b. Average dwelling	30–40
c. Average office	50–60
d. Conversation	62
e. Typewriter	72
f. Average large motor or generator	85–90
g. Heavy street traffic	90
h. Average power station	95
i. Elevated train	100
j. Punch press	110

The decibel is proportional to the logarithm of the actual intensity, thus approximating the response of the human ear. Therefore, in the above tabulation, the actual intensity of *d* is about 2500 times that of *a*.

AIR GAP-POLE PITCH RATIOS. It can be observed that in all synchronous machines meeting usual conditions of performance, certain proportions fall within definite ranges. Variations within the established range are due to functions of the design, such as power factor, short-circuit ratio, and torque. One of the most significant of these proportions is the ratio of air gap to pole pitch, the range covered by most synchronous machines of average design being from 0.01 to 0.03.

10. REACTANCES, DEFINITIONS, AND SIGNIFICANCE

In a synchronous machine there are two axes, normally described as the direct axis and the quadrature axis. The direct axis is defined as that coinciding with the centerline of the poles on the rotor. The quadrature axis corresponds to the centerline between poles. Many of the following reactances vary with the position of the rotor and consequently will have two limits, normally defined as the direct-axis reactance and the quadrature-axis reactance.

LEAKAGE REACTANCE. The first reactance to be described is the leakage reactance. This reactance is normally considered the reactance of the stationary member or armature of the unit to alternating current; i.e., it includes only that reactance resulting from leakage flux, which is confined to the stator leakage paths, including slot and end turn leakage paths. This leakage reactance is a rather theoretical and arbitrary value but has practical significance when calculating other reactances. It is impossible to measure this leakage reactance directly from test, as any other reactance except zero sequence includes the leakage reactance as an integral part of the total. This reactance normally has a value of about 6 to 20 per cent, depending on the speed of the unit. The higher the speed, the lower is the reactance.

SYNCHRONOUS REACTANCE. Direct-axis synchronous reactance is the sum of the armature leakage reactance and the reactance to armature reaction. In other words, it is the sum of the leakage flux produced by rated current in the armature and in the stator leakage paths and the flux which is produced and crosses the air gap through the rotor poles, when the armature is excited, in such a position as to be lined up with the direct axis of the rotor, and the total flux linkage compared with rated flux in the machine at no-load rated voltage. This may be also determined from the normal saturation curves (Fig. 17) by using the field current required to circulate rated armature current at short circuit, divided by the field current required to produce rated voltage on the air-gap line. This is done by taking *OG* divided by *VL* in Fig. 17. The quadrature-axis synchronous reactance is practically the same as the direct-axis synchronous reactance on wound-rotor or turbine-generator type machines. However, on the salient-pole type, where the poles are definite projections in the rotor and the interpolar spaces are not magnetic, the flux produced by rated armature current in the stator which crosses the air gap is much less than in the direct axis. The quadrature synchronous reactance normally varies between approximately 0.5 and 0.7 times the direct-axis synchronous reactance. The quadrature synchronous reactance is the sum of the armature leakage reactance previously described and the flux due to rated armature current flowing in the quadrature axis in the stator which will cross the air gap in the interpolar space. This value is rather difficult to obtain in the normal test, but it may be calculated with a sufficient degree of accuracy for almost all normal requirements.

Both the direct-axis synchronous reactance and the quadrature-axis synchronous reactance are used in calculating steady-state loading limits, as described in Section 3, on Power-system Stability. The quadrature-axis synchronous reactance is also used in the calculation of the power angle curve of the machine and P_r, described later in this section.

TRANSIENT REACTANCE. Transient reactance is the reactance to sudden three-phase short circuit when the machine has no damper windings. This reactance is approached only in salient-pole machines without damper windings and is the sum of the leakage flux in the armature or stator and the flux which is produced by rated armature current applied on the direct axis but which does not link the rotor windings, or, in other words, the flux which crosses the air gap and is not permitted to pass through the field windings because of the induced current in the field. Thus the flux must pass from the armature across the air gap to the pole tip and across the interpolar spaces and into the next pole and return. In salient-pole machines the direct-axis transient reactance is much lower than the direct-axis synchronous reactance, whereas the quadrature transient reactance is effectively equal to the quadrature synchronous reactance.

The normal transient reactance varies considerably, depending upon the speed of the machine, but for very high-speed machines may be as low as 12 to 13 per cent and for low-speed machines may be as high as 40 to 50 per cent. Two values of the transient reactance are normally recognized today. The first is the rated current value of transient reactance $X'du$. This is obtained from tests by short-circuiting the machine from such a voltage that rated current will result in the transient component, as defined in the AIEE test code. The rated voltage value of direct-axis transient reactance $X'd$ is that reactance obtained from a short circuit made at rated voltage and at no load, neglecting the high decrement of the first few cycles. The rated voltage value varies from approximately 0.75 to 0.9 times the rated current value, but in most normal machines is 0.88 times the rated current value. The factor controlling the ratio of rated voltage value to rated current value of direct-axis transient reactance is chiefly the saturation in the stator leakage paths and in the pole tip on the rotor. Naturally, if these paths are saturated, less flux will result from a given unit of armature current and therefore the reactance is less.

It was the practice several years ago to use the rated voltage value as the "saturated" value of transient reactance, and the rated current value as the "unsaturated" value of transient reactance. However, this nomenclature was not sufficiently accurate, and practice varied so much between individual manufacturers that it was necessary to standardize on the new terminology given above. Unless otherwise specified, it is usual practice to give the rated current value of transient reactance in data concerning generators and motors.

SUBTRANSIENT REACTANCE ($X''d$). The subtransient reactance is that reactance which applies in the machine at the instant of short circuit and is equal to the sum of the leakage reactance and the reactance due to the flux which flows in the paths above the damper winding. There is a direct and quadrature axis, a subtransient reactance, and a rated voltage and a rated current value of each reactance. However, the rated voltage and the rated current values are, in general, not too far different on salient-pole machines. It is common practice to give the rated voltage value of subtransient reactance, since this is the lower reactance and determines the initial current flowing in the circuit after a fault, thus determining the maximum current and the interrupting capacity of circuit breakers. The rated current value of subtransient reactance has little practical application. The value of rated voltage, subtransient reactance, varies from approximately 6 per cent in old turbine generators or 8 per cent to 10 per cent in modern turbine generators up to approximately 30 per cent in some low-speed salient-pole machines with dampers.

POSITIVE-SEQUENCE REACTANCE. The positive-sequence reactances in the synchronous machine are the subtransient reactance, the transient reactance, and the synchronous reactance (see Section 3 on symmetrical components). These individual reactances apply at different times after a fault or under different conditions, such as in studies of stability and voltage dips.

NEGATIVE-SEQUENCE REACTANCE (X_2). The negative-sequence reactance is that reactance to a wave which rotates at rated speed in a direction opposite to that of the rotor. Consequently, this rotating wave meets the reactances on the direct axis and the quadrature axis alternately at relatively high speed. Thus the negative-sequence reactance is usually taken as the average between the direct and quadrature subtransient reactance.

ZERO-SEQUENCE REACTANCE (X_0). The zero-sequence reactance is the reactance to currents which flow in all three phases simultaneously and flow out the neutral. The vector position of the zero-sequence currents flowing in the three phases is identical. This reactance is used in the study of ground faults. Its value normally varies from about 2 per cent for a two-thirds pitch winding up to approximately the value of subtransient reactance for a full-pitch winding.

POTIER REACTANCE (X_p). Potier reactance is a reactance determined from the no-load saturation curve and the zero-power-factor overexcited saturation curve. It is useful primarily in estimating the excitation for power factors other than zero (see p. 9-21) and in estimating the zero-power-factor saturation curves at loads other than the rated load. For the method of determination of Potier reactance, refer to Fig. 17 under Saturation Curves in Article 8.

11. TIME CONSTANTS

In machine performance, any change in the stored magnetic energy of the unit follows an exponential law whose general equation is $Y = e^{(-t/T)}$ for decreasing stored energy and $Y = 1 - e^{(-t/T)}$ for increasing stored energy, to a specified level. The time constant T is that value of time in which the current or voltage, when starting from any initial value, will change by $(1 - 1/e)$ or 0.632 times the increment between the initial value and its final value. Another characteristic of the exponential curve is that the initial rate of change in the circuit is that corresponding to a 100 per cent change in a time equal to one time constant (see Fig. 14). In the practical case of a machine at zero voltage, if an excitation voltage corresponding to that necessary to supply rated no-load field current continuously were suddenly applied, the voltage would rise to 1 − 0.368 or 63.2 per cent of the full voltage value in a time corresponding to the open-circuit time constant of the machine. Also, if the machine were running at rated voltage and no load and the excitation voltage were suddenly short circuited, the voltage of the machine would decay to 36.8 per cent of its original value in a time equal to one time constant. The time constant applying to this condition is the open-circuit time constant Td'_0. The value of the open-circuit time constant depends chiefly on the kva capacity and the speed of the machine. This value varies from less than 1 sec in very small machines to as much as 12 sec in very large machines.

In addition to the open-circuit time constant, there is the transient time constant, which is equal to the product of the open-circuit time constant and the transient reactance over the synchronous reactance, or $= (Xd'/Xd)\ Td'_0 = Td'$. This time constant applies to transient phenomena, such as the decay of the transient component of current under short circuit as it approaches the sustained value. It is also used in the calculation of voltage dips and voltage recovery on a system in conjunction with the excitation system (see Reference 4 of Bibliography, Article 23). In addition, there is the subtransient time constant which defines the decay or the change in the subtransient component of current at the instant of fault. This time constant is usually very short and is normally in the range of 0.02 to 0.04 sec, the average being from 0.03 to 0.035. For a more complete description of the use of time constants and the method of securing them from tests, see Reference 4.

12. TESTS OF SYNCHRONOUS MACHINES

INSULATION HIGH POTENTIAL TEST. The standard high potential test voltage for the armature or a-c windings on synchronous machines is twice rated rms voltage of the circuit plus 1000 volts. This voltage must be an rms a-c voltage whose crest value is equal to the $\sqrt{2}$ times the test voltage specified. This is characteristic of commercial sine wave voltages. The frequency, unless otherwise specified, shall be between 25 and 60 cycles. This test voltage shall be applied continuously for a period of 1 min after completion of the manufacturer's other shop tests. Unless otherwise specified or unless the windings have been disturbed after shipment from the factory, the test voltage after installation and unless otherwise agreed upon shall be 75 per cent of the value specified above. The test voltage for field windings of synchronous generators shall be 10 times the exciter voltage but in no case less than 1500 volts. The test voltage for field windings of other synchronous machines which are started as a synchronous motor with the field short circuited or through a starting resistor shall be not less than 2500 volts or more than 5000 volts and not less than 10 times the rated excitation voltage.

INSULATION-RESISTANCE TEST. The insulation resistance may be measured by a self-contained instrument, such as a direct-reading ohmmeter of a generator, battery, or electronic type; a resistance bridge; or a milliammeter with voltmeter and d-c supply. The resistance measurements should be taken for a period of at least 1 min and sometimes several minutes, to eliminate the electrification effect in the insulation. The insulation resistance is read by applying a voltage from the winding to ground, either with a self-contained instrument or by the voltmeter method shown in Fig. 21.

EFFICIENCY AND LOSS TESTS. The efficiency of certain small synchronous machines may be determined by the input minus output method, whereby the subject machine is driven by a calibrated prime mover, and the input kilowatts are measured and compared with the output kilowatts. This method of testing is usually limited to machines of approximately 100 kva and smaller. For machines larger than this rating it is common to determine the efficiency by the segregated loss method. This loss method involves the determination of the individual component losses of the machine, which include friction and windage, core loss, armature I^2R loss at 75 deg cent, load loss, and field I^2R loss. The friction and windage loss is usually determined by driving the machine at rated speed with no excitation or voltage applied and measuring the input power to the shaft. The core loss is measured by driving the machine at rated speed and at a number of different voltages while exciting the unit from a separate source of power. By measuring the input power under this test and subtracting the friction and windage loss, the no-load core loss may be determined. The armature I^2R loss is determined by measuring the armature-winding resistance at room temperature and correcting for 75 deg cent. To determine the load loss, the machine should be driven in the same manner as for the friction and windage and core-loss tests, but with the armature terminals short circuited. A series of readings may be taken at various armature currents with various excitation values, and the load loss is then determined by subtracting from the power input, the friction and windage, and the armature I^2R with the resistance corrected to the temperature of the winding during the test. The field I^2R loss is calculated for the rated field current for rated voltage, power factor, and kva with the field resistance at 75 deg cent. During all these tests it is essential that the speed be held absolutely constant to prevent large errors in the readings of input power.

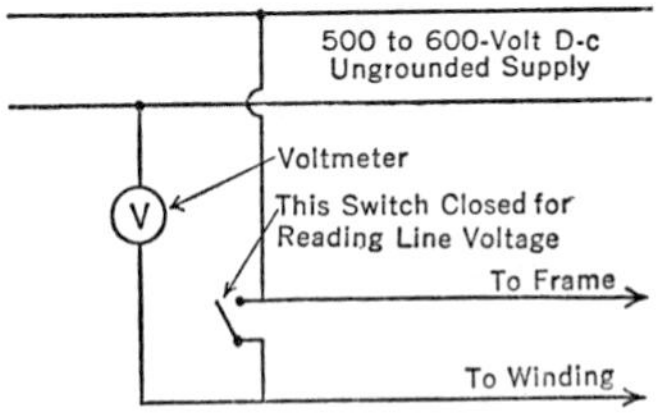

V = Line Voltage
V′ = Voltage Reading with the Insulation in Series with Voltmeter
R = Resistance of Voltmeter (Preferably 50,000 Ohms or More)
R′ = Resistance of the Insulation in Megohms

$$R' = \frac{R(V - V')}{V' \, 10^6}$$

FIG. 21. Insulation Resistance Test Circuit

Segregated loss tests can also be taken by various other means, including the synchronous motor or electrical input method. To obtain accurate results, these other methods are generally more complicated and are outside the scope of this article. For more complete methods of testing see the AIEE Test Code.

SATURATION CURVES. During the measurement of losses, as described in the previous paragraph, a plot of the voltage versus field current and armature current may be made, which determines the no-load saturation curve and the short-circuit saturation curve or synchronous impedance curve. The zero-power-factor saturation curve mentioned previously may be taken by connecting the unit to another synchronous machine and adjusting the excitation on the two units until full-load armature current is circulated between the two units at almost zero power factor. The zero-power-factor saturation curve is normally plotted for circulation of rated armature current at a series of different voltages, as shown in Fig. 17.

TEMPERATURE TESTS. The full-load temperature rise of synchronous machines in sizes of approximately 50 to 100 kva or less may often be taken by loading the machine at rated voltage and power factor and kva or horsepower output against a dynamometer, and the temperature rise measured by thermometer after a period of approximately 5 to 6 hr, when the temperature has become stabilized. On larger ratings it is usual practice to run temperature tests by loading the units at rated kva, zero per cent power factor, overexcited. For machines of 80 per cent power factor and below, this test can normally be taken without seriously overheating the field for the period of the test. For machines of power factors near unity, however, it is often necessary to take the temperature test at full-load armature current and reduced voltage in order to prevent seriously overheating the fields and causing damage to the machine. The temperature rise of the field winding may be corrected from the test value to the rated load condition by adjusting the temperature in accordance with the kilowatt input to the field at rated condition, as compared to the kilowatt input to the field under the test condition. The temperature rise of the windings and core on the stator or armature should be adjusted in proportion to the kilowatt input or loss in the stationary portions at the test and rated load conditions.

There are various methods of determining the temperature rise on machines. For machines under approximately 2000 kva or horsepower, the most commonly used method is by thermometer. Temperature rise is determined by thermometers placed in the machine at strategic points on the stator winding and on the back of the core punching, with the bulb suitably protected under a pad of felt or similar material. Readings on these thermometers are taken during normal operation of the machine and also immediately after shutting down the machine. The highest reading thermometer observed on a given part during running and shutdown is taken as the temperature of that part. The temperature rise is then determined by subtracting from this temperature the ambient or room air temperature determined by a series of thermometers placed at machine level not less than 3 or more than 6 ft from the machine under test and in a position to be free from drafts or abnormal heat radiation. Temperature tests on the rotating member may be made by applying a thermometer immediately after shutdown to the various parts of the rotating element. However, the temperature rise of the rotor winding is normally obtained by the voltmeter-ammeter method, by means of which the resistance of the field winding is determined while in operation or immediately after shutdown. To get a correct value of voltage applied across the field winding of the machine, the voltage must be read on the collector rings and not at the brushes. The winding temperature is then calculated by means of the following formula:

$$t_2 = \frac{R_2}{R_1}(K + t_1) - K$$

where R_1 = resistance at t_1° C.
R_2 = resistance at t_2° C.
K = 234.5 for 100 per cent volume conductivity copper.
$= \left(\frac{25{,}450}{n} - 20\right)$ for n per cent volume conductivity copper.

This is known as the resistance method of temperature determination. The resistance of the armature winding immediately after shutdown may be measured on all machines where the leads may be disconnected very rapidly after the machine is shut down. However, this method is very impractical on large machines, where the leads may not easily be handled. It is usual to supply such machines with temperature detectors embedded between the coils of the armature, the temperature of the winding then being determined by means of the detector. Detectors commonly used are either 10-ohm resistance coils whose resistance varies with temperature or thermocouples. For the allowable temperature rise by each of these methods, see Table 2.

13. SYNCHRONOUS GENERATORS

STANDARD DATA. Regulation. Regulation in a synchronous generator is defined as the rise in voltage, expressed in per cent of rated voltage, when rated load at rated power factor is reduced to zero with excitation constant. The standard value of voltage regulation for a 50 deg cent 80 per cent power factor synchronous generator is 40 per cent, and for a 40 deg cent 80 per cent power factor synchronous generator is 34 per cent. This inherent regulation of a synchronous generator has little importance when the machine is operated in conjunction with a voltage regulator. Regulation is not to be confused with the voltage dip encountered when a given load is added to the machine. The voltage dip is calculated by a rather involved process (see Reference 4), including the transient reactance and speed of response of the exciter and regulator. The regulation of a synchronous generator may be calculated by $100(E_1 - E)/E$, where E_1 is the no-load voltage corresponding to rated load and power factor field current, and E is rated voltage.

Overspeed. All standard generators are expected to withstand the stresses corresponding to 25 per cent above normal speed with the exception of waterwheel generators, which are constructed to withstand the stresses incident to operation at runaway speed. The runaway speed must be calculated for each hydraulic condition, and the maximum overspeed specified for the waterwheel generator.

Efficiency. The efficiency of the machine is the ratio of useful power output to the total power input. Because of the different forms of construction of a-c generators, however, the calculations of efficiency vary for different types of machines. Various losses which may be chargeable against the a-c generator are as follows:

a. Friction and windage loss.
b. Core loss.
c. Armature I^2R loss at 75 deg cent.
d. Stray load loss.
e. Field I^2R loss at 75 deg cent.
f. Rheostat loss.
g. Exciter loss.

For a more complete definition of each of these losses, see discussion of Tests, Article 12. For a-c generators not supplied with bearings by the generator manufacturer, it is usual to guarantee an engine-type efficiency, which includes losses *b*, *c*, *d*, and *e* above. For machines normally supplied with bearings, it is usual to include losses *a*, *b*, *c*, *d*, and *e*. For those cases where the complete loss estimate is required from the generator manufacturer, it is common to guarantee overall efficiency, which includes all the above losses. In specifying the efficiency for a new machine, the type of efficiency required should always be given.

Short-circuit Ratio. Short-circuit ratio (see Fig. 17) is defined as rated voltage no-load field current divided by the field current required to circulate rated armature current at short circuit, i.e., (VN/OG). Short-circuit ratio is an indication of the ability of the machine to maintain stability and withstand load variations. For most machines other than waterwheel generators, it is not common practice to specify any particular short-circuit ratio, since it is unusual to require any definite value of this ratio in the operation of the machine. Most ordinary machines, however, have short-circuit ratios between 0.7 and about 0.9, except waterwheel generators, which have short-circuit ratios in the region of unity for standard machines.

Telephone Influence Factor. Telephone influence factor, commonly called TIF, is a measure of the effect of various harmonics in the voltage or current wave form on near-by telephone circuits if exposure is permitted. The standard values recommended by NEMA, as compared to the capacity of machines, are given in Table 7. The telephone influence value is normally read only at no-load and rated voltage. There are two kinds of telephone influence factors: (1) the "balanced TIF," which is read from line to line, and (2) the "residual TIF," which is read in the open delta of the machine when the windings are rearranged so as to form an open delta. The residual component has no particular significance except where machines are operated with a grounded neutral. The residual TIF is the result of harmonics of the triple series, which flow only from line to neutral or in the closed delta of a machine. Special TIF consideration may be necessary where trouble exists or may be anticipated from difficult exposure conditions. In such cases a special low telephone influence factor may be purchased at increased price.

Table 7

Kva	Balanced TIF	Residual TIF
62.5– 299	300	....
300 – 699	200	....
700 – 999	150	....
1,000 –2,499	125	....
2,500 –9,999	60	30
above 10,000	50	30

The TIF harmonic waiting curve is now up for revision, and after republication of the new waiting curves the above values may be re-examined and revised as necessary.

Wave-form Deviation. Standard generators are all guaranteed for a wave-form deviation not to exceed 10 per cent of the equivalent sine wave voltage when the equivalent sine wave is adjusted so that the deviation is at a minimum.

VENTILATION SYSTEMS. General. Most synchronous machines today are built with radial ventilation, which employs ducts arranged in the armature core so that air or gas flows radially outward from the rotor through the stationary portion of the core and into the room or coolers or outside the powerhouse. Axial ventilation, in which the gas flows from one end of the machine to the other through ducts in the armature punching and in the air-gap zone, was formerly used and is still employed to some extent.

Turbine Generators. Most turbine generators below 20,000 to 25,000 kva are air cooled, and the internal ventilation system of the machine is determined by the manufacturer. The most common practice is to have the air flow from the pit into the outer part of a double set of end bells on each end of the machine, through the gap and core, and either vertically downward or upward from the frame to suitable exhaust ducts or cooler. In the larger units, however, it is very common practice to supply air-to-water heat exchangers to cool the air as it leaves the machine, thereby reducing maintenance cost by preventing large amounts of dirt from entering the machine. Most turbine generators above 20,000 to 25,000 kva, 3600 rpm, are hydrogen cooled and come with gas-to-water heat exchangers included in the frame of the machine.

Salient-pole Generators. Salient-pole generators have quite a variety of ventilation schemes, depending upon their construction. Most small machines are room ventilated, taking air from the room around the shaft and discharging it through the frame to the room. Larger horizontal machines such as condensers and larger horizontal waterwheel generators often are equipped with coolers or are arranged for connection to discharge ducts, which are taken outside the powerhouse in a manner corresponding to that described for the turbine generators. To reduce both noise and dirt it is desirable on larger, high-speed units to have a closed system requiring coolers. If cooling water is limited, some type of filter is generally employed in the intake air duct. For vertical generators it is common practice to use open construction on small units below approximately 3000 kva and coolers and recirculating systems on units larger than this rating, depending upon station conditions. In vertical generators the coolers are normally located around the periphery of the machine, a housing being furnished to return the cooled air to the top and bottom of the machine. Some units may be arranged for all the air to enter either the top or the bottom of the machine.

Hydrogen Cooling. The use of hydrogen cooling in machines theoretically reduces the windage loss to one-fourteenth that in air. Practically, however, because of moisture content in the gas and a slight component of air mixed with the hydrogen, the usual windage loss is approximately one-tenth that of air. Hydrogen also has greater thermal-conductivity and heat-transfer rates than air; consequently, more rating may be secured from a machine of given size when using hydrogen. The disadvantages of hydrogen are the possibility of explosion and the necessity for supplying the gas to take care of small leaks. All machines with hydrogen cooling are constructed to withstand the explosion pressure based on ½ psi operation, without serious permanent deformation of the housing. Accidents have been very rare on all hydrogen machines.

DAMPER WINDING. Damper windings, or amortisseur windings, are usually supplied with Diesel engine generators and other generators which will be used with a pulsating load or, in a few cases, which will be required to start as a synchronous motor. It has also become common in recent years to supply damper windings on waterwheel generators where they may be feeding long transmission lines or operating in parallel with other machines driven by prime movers which have bad pulsating torque characteristics. It is also common to supply the non-connected type, i.e., dampers which are not connected between poles, on Diesel engine generators and on many small waterwheel generators. In large waterwheel generators it sometimes becomes necessary for the quadrature subtransient reactance to be nearly equal to the direct-axis subtransient reactance in order to avoid resonance with the transmission line at harmonic frequency in cases of line-to-line or line-to-ground faults. This condition usually does not exist except where the machine feeds a rather long transmission line. On large waterwheel generators, where it may become necessary to keep the ratio of quadrature-axis subtransient reactance to direct-axis subtransient reactance low, it is possible to purchase either non-connected or connected dampers, depending on the ratio required. In most applications a ratio of quadrature-axis subtransient reactance to direct-axis subtransient reactance of approximately 1.35 is satisfactory to prevent undue resonance voltages from building up to dangerous values. The connected type of damper will almost always have a ratio of quadrature-axis subtransient reactance to direct-axis subtransient reactance more nearly approaching 1 than will the non-connected damper. As the ratio of reactance approaches 1, the harmonic content of the unfaulted phase is reduced to a minimum under resonance conditions with a transmission line.

MAXIMUM RATING OF MACHINES. In the past 10 or 15 years the maximum rating of turbine generators in a single unit at 3600 rpm has been growing rapidly, and present-day limits will probably be exceeded in the future. It is now possible to build a 100,000-kw turbine generator in one single unit at 3600 rpm, whereas at 1800 rpm it is possible to build 150,000 to 170,000 kw. It is possible to build waterwheel generators up to about 150,000 kva at speeds up to 200 rpm. With salient-pole machines the maximum capacity which it is desirable to build in one unit decreases as the speed goes up. The normal economical limit for waterwheel generators is about 40,000 kva with a ten-pole 720-rpm machine. When units do not require high overspeed, it is possible to build 50 to 75 per cent more kva at a given speed than when a high overspeed is needed, as in waterwheel generators.

VOLTAGE DIP ON SUDDENLY APPLIED LOADS. On any a-c machine, when the load is suddenly applied to the armature, the voltage will drop. The amount of the instantaneous drop is a function of the kva load thrown on and the power factor and reactance of the machine. After the initial drop the voltage will continue to decrease slightly until the voltage regulator and its associated exciter can increase the excitation sufficiently to bring the voltage back to normal. Conversely, when a load is suddenly dropped from an a-c generator, the voltage will rise until the excitation system can reduce the excitation to such a point that the voltage is again restored to normal. The ratio of the maximum voltage dip to the initial voltage dip is a function of the regulating system and the time constants of the regulating machine alone. The voltage dip encountered on a machine is not definitely related to the so-called inherent regulation of the machine and must be considered separately. Where large motors must be started from a single synchronous generator or a pair, the voltage dip at the time of start must be limited to a value such that contactors on the rest of the system do not drop out, thus permitting a shutdown of all other loads on the system. In general, this requires that the voltage must not dip below approximately 70 to 75 per cent of the rated voltage. For this reason it is usual to limit the horsepower rating of the largest motor to be started on a given synchronous generator to approximately one-fifth of the kva rating of the generator, assuming that the generator has a transient reactance of approximately 25 per cent and is also provided with a good degree of response in the regulating system. For a complete discussion of voltage dip and recovery systems, see Reference 4, Article 23.

TESTS. On large generators, it is frequently impossible to test the units at the factory because of the size of the unit, both in kva and in physical dimensions. Large waterwheel generators, because of their size, must necessarily be shipped as disassembled parts. Therefore, factory testing of the complete unit is impracticable. These generators are usually tested at the customer's site after complete erection. Certain other large generators might be set up and tested at the factory, but because of large capacity they exceed the testing facilities of the manufacturer. In these cases it is frequently necessary to resort to compromise tests in order to obtain an estimate of the full-load temperature rise of the unit under actual load.

INDUCTOR GENERATORS AND APPLICATION. The reader is referred to Article 2 on types of synchronous machines.

SINGLE-PHASE GENERATORS. Single-phase generators inherently are larger and more costly than three-phase generators of the same kva rating. Single-phase generators are also subject to vibration problems not encountered in three-phase generators and may even need to be spring mounted to prevent transmission of undue vibration to the foundation. Also, all single-phase generators are normally supplied with damper windings to carry the negative-sequence component of current resulting from the single-phase load. The damper winding must be proportioned with special regard to taking care of the additional heating in the bars caused by the negative-sequence current components. Single-phase generators are today mostly used in very small isolated applications or for welding or railway service.

LOAD CURRENT BALANCE. Most three-phase generators can carry some unbalanced current. If the three currents or the currents in the various phases are not balanced, it becomes a special case of a single-phase generator with heating in the damper or amortisseur winding in addition to extra heating in the phases carrying the largest current. Turbine generators are especially sensitive to current unbalance, since with the round rotor construction and solid forging the additional surface losses on the rotor are extremely high. On turbine generators it is usual to limit the load current unbalance to 20 per cent of the three-phase rated current at no load, with this unbalance tapering off to zero at full-load and rated power factor. In general, the salient-pole machines should be limited to approximately the same unbalance as the turbine generators, unless they are specifically designed for use with higher single-phase current and unless they are provided with a connected type of low-resistance damper winding.

SYNCHRONOUS CONDENSERS. A synchronous condenser is a special case of a synchronous motor which drives no mechanical load and is used primarily to supply reactive kva to or from a transmission line, either to correct power factor or to regulate the voltage of the transmission line. Figure 22 shows a typical saturation curve for a synchronous condenser. In the past small synchronous condensers were used quite frequently to correct power factor. With the introduction of reliable static capacitors, however, the small installations have become more uneconomical. As a result the static capacitor has largely replaced the small synchronous condenser in new applications for the purposes of correcting power factor alone, due to its lower loss and maintenance. In larger installations, however, where some voltage regulation is needed with large load variations in an infinite number of steps, and where lagging capacity is required, it is common to use the synchronous condenser. Today, condensers ranging from 5000 to 70,000 kva are quite common. These condensers are generally used to maintain voltage regulation on a transmission line and increased system stability in addition to supplying reactive kva for power-factor correction. Synchronous condensers can be made for either indoor or outdoor service. For ratings above approximately 25,000 kva and where water is available for cooling purposes, it is common to use hydrogen-cooled condensers to reduce the loss. Hydrogen-cooled condensers are inherently suitable for outdoor operation, thus eliminating the necessity for a powerhouse or substation building. Synchronous condensers are usually started as a synchronous motor by applying low voltage obtained from an auto-transformer with taps in the range of 20 to 35 per cent voltage. The starting kva under these circumstances is usually approximately 20 to 25 per cent of the rated kva of the machine. For condensers above approximately 10,000 kva the units are usually provided with oil lift to reduce the voltage necessary for start and consequently the inrush kva drawn from the line. Units are usually brought up to speed and the field is energized while still running on the auto-transformer tap. The following method will give the optimum value of field current to reduce the line current at transfer to a minimum:

Curve *A*, Fig. 22, is the V curve for a synchronous condenser having no-load and full-load saturation curves as shown.

The standard air-cooled synchronous condenser is designed for 50 per cent lagging or underexcited capacity at rated voltage, with slightly more than zero field current.

Since the V curve is a plot of the variation in armature current with field current for a given voltage, the curve is fixed by the following points: (1) field current for rated stator current and rated voltage 0 per cent power factor leading (overexcited), (2) field current for no-load rated voltage and zero stator current, (3) field current for one-half (50 per cent) rated stator current and zero field current. These points determine the V curve, *A*. Since we are interested only in the leading or overexcited part of the curve, we will ignore the lagging portion or that portion having field-current values below no load.

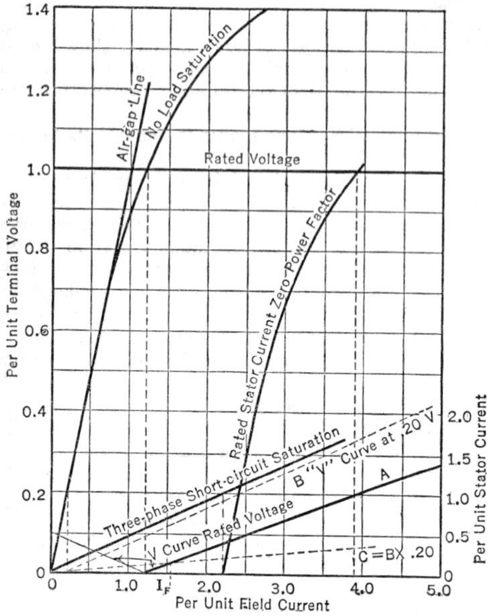

FIG. 22. Characteristic Curves for Air-cooled Synchronous Condenser

Another no-load V curve based on the starting tap (in this case 20 per cent) can be plotted, such as curve *B*. If the ordinates of this curve are referred to line voltage by multiplying by the per cent starting voltage (in this case 20 per cent), curve *C* will be obtained.

The intersection of curves *A* and *C* will give the optimum field current I_F to be used before transfer to line voltage.

When the switching transfer of the condenser from starting voltage to full voltage requires that the machine be disconnected from the line for a period of more than about 1/2 sec, it is desirable to prolong application of the field and synchronizing with the line until after the transfer has been effected. Otherwise the machine may drop far enough out of step to cause large transient torques and currents to be drawn from the line at the instant of transfer. Where it is necessary to get the starting kva down to an absolute minimum, a special motor may be provided to start the condenser. The inrush to starting motors is generally in the neighborbood of 50 per cent of the kva required for the condenser with auto-transformer starting, and the power factor of the current, instead of being approximately 30 to 40 per cent, as in auto-transformer starting, will be in the neighborhood of 75 to 85 per cent. This combination of increased power factor and decreased kva tends to reduce the line disturbance appreciably with a starting motor.

14. SYNCHRONOUS MOTORS

COMPARISON WITH SYNCHRONOUS GENERATOR. A synchronous generator will operate also as a motor, neglecting starting performance. If two equal generators are connected to the same bus-bars and the driving power is removed from one of them, the current or power flow in that one will reverse, so that it draws power from the other generator and delivers mechanical power from its shaft. Its direction of rotation and speed will be unchanged, provided the bus-bar voltage is maintained. This speed is

$$N = \frac{120f}{p} \text{ rpm}$$

where f is the line frequency in cycles per second, and p is the number of poles. The voltage generated by the second machine is now a counter-voltage, or back-emf, and is considered negative with respect to the circuit between the two synchronous machines. This counter-voltage actually limits the current drawn by the second machine, or synchronous motor, in that it combines vectorially with the impressed or line voltage to produce a resultant voltage, which equals the synchronous impedance drop of the machine. If a change in the current is required by a change in the load, then this counter-voltage must change in vector position (or phase) or in magnitude or both. If the excitation is unchanged, the counter-voltage changes in phase position only, and there are definite limits for this change. This subject will be explained at greater length in Article 16.

Although a synchronous generator will function also as a motor, there are certain differences in the machines because they are designed for specific purposes. Since, as will be explained later, a synchronous motor must develop torque from its pole-face (damper or amortisseur) winding, this winding is connected between poles by heavy end rings. The synchronous generator usually does not require a connected damper winding, and in some cases no damper winding at all is used. Moreover, the damper windings in generators are made of copper, and this, in connection with the above considerations, results in poor starting performance if the machine is operated as a motor. Also, the inherent electrical constants are somewhat different for machines designed as generators.

Although synchronous motors may be built single-phase, they are not self-starting, and the efficiency is comparatively low. In two- or three-phase construction, which is by far the most common, a synchronously revolving field established by the polyphase stator winding induces currents in the damper winding to produce positive torque, and self-starting is obtained.

A source of direct current is required for excitation, as in synchronous generators.

Since the armature, or stator, construction is stationary with solid connections, it may be wound for high voltages, sometimes going as high as 13.8 kv in the larger machines.

Compared with induction motors, synchronous motors are advantageous in operating with a controllable power factor, usually higher and leading to help compensate for lagging currents in the power system. Their constant synchronous speed is very often desirable, and they usually operate at higher efficiency; in many cases higher voltage ratings are possible. They are less sensitive to voltage fluctuations than are induction motors. The principal disadvantages are the requirement for d-c excitation and the possibility of hunting under certain conditions, as discussed in Article 16.

Voltage and current relationships for synchronous motors are as given in the articles applying to synchronous machines in general.

15. APPLICATIONS OF SYNCHRONOUS MOTORS

TYPES OF SERVICE. Synchronous motors are applicable to practically any kind of drive where the excitation requirement does not prohibit their use, since they are obtainable in a wide range of torque characteristics and in speeds from 80 rpm or lower to 3600 rpm. They are obtainable for coupled, belted, or geared service and in most of the various enclosure types. Where constant-speed and/or power-factor correction is desirable, the use of synchronous motors is dictated. Table 8 shows the range of typical applications with speeds and ratings.

Synchronous motors may be operated without shaft load purely for the benefit of leading reactive kva and in such cases are called synchronous condensers (see Article 13 on this topic).

TORQUES. A balanced polyphase stator winding, when energized with balanced polyphase currents, sets up a magnetic field which revolves at synchronous speed relative to the structure of the winding. The number of consequent poles comprised by this field is determined by the arrangement of the winding and corresponds to the number of poles in the movable field structure or rotor. If the field poles on the rotor are magnetized by excitation of the field winding, it is clear from fundamental principles of magnetism that, as the revolving stator field sweeps past the rotor poles, there is a tendency for the two fields to lock together, a north pole of the stator field locking with a south pole of the rotor field, and vice versa. When the rotor is at standstill, however, it is apparent that, since it has appreciable inertia, the rapidly revolving stator field must attempt first to drag the rotor in one direction and then to force it in the opposite direction as the north and south stator field poles sweep past the rotor field structure. The result is that the average torque is practically zero, and some other means must be provided for making the motor self-starting. This is accomplished by embedding in the pole faces heavy bars of relatively low-resistance material and connecting them at the ends by continuous end rings, so as to form a cage winding, variously called a damper winding or amortisseur winding. Then the revolving field induces in this damper winding currents which react with the armature field to produce a positive torque. In this way, therefore, the syn-

Table 8. Typical Applications of Synchronous Motors

Construction	Speed (Revolutions per Minute) and Horsepower Ratings	Typical Applications	
Cast frame Brackets with sleeve bearings Laminated or fabricated rotor Poles dovetailed or bolted to spider	1800 and 1200; up to 800 900 and 720; up to 600 600 and 514; up to 500 450 and 360; up to 400	Coupled	Fans and blowers Compressors Pumps
		Belted	Fans Line shafts Grinding and crushing machinery Beaters (pulp)
		Geared	Line shafts Grinding and crushing machinery
Fabricated frame Pedestal bearings Laminated rotor Poles dovetailed to spider With or without fabricated bedplate	1800 and 1200; 400 and above 900; 800 and above 720 and 600; 1500 and above 514; 1000 and above 450; above 2500 400; above 4000	Coupled	Fans, blowers, and pumps Grinding and crushing machinery Jordans (pulp)
		Belted	Fans and compressors Grinding and crushing machinery Beaters (pulp) Line shafts
		Geared	Grinding and crushing machinery Tube and ball mills Rubber mill line Plasticators and banbury mixers Steel and metal rolling mills Line shafts
Fabricated frame Fabricated rotor Pedestal bearing Poles bolted to spider With or without fabricated bedplate	720; 500–1500 600–514; 400– 900 450; up to 2500 400–360; up to 4000 327 and below; all ratings	Coupled	Jordans (pulp) Chippers and wood hogs Fans and compressors Pumps Plasticators and banbury mixers Tube and ball mills Steel and metal rolling mills Line shafts
Fabricated frame Fabricated rotor Split rotor hub or completely split spider Variable flywheel effect	450 and below; 40 hp and up	Direct-connected	Compressors (air, gas, and refrigeration) Vacuum pumps Centrifugal pumps Chippers Direct-current generators
Fabricated frame Fabricated rotor Pedestal bearings Special damper winding With bedplate	Polyphase damper winding 450 to 150; 300 and above Single-phase damper winding 720 to 150; 200 and above	Coupled	Line shafts (flour mills) Tube and ball mills (cement and mining) Rock and ore crushers Cold roll steel mills

chronous motor starts as an induction motor, and its starting and pull-in torques are really induction-motor torques which are proportional to the square of the armature voltage.

Various application requirements demand different characteristics of torques and starting current inrush. Since these are a function of the damper winding, many various designs are used, involving different sizes of bars in the pole faces and made of different materials, usually copper and various grades of bronze. In ordinary starting procedure the rotor field winding is not energized until the motor is nearly up to speed. While the motor is accelerating from standstill to full speed, it is obvious that the revolving stator field will induce a voltage in the rotor winding. Since this consists of many turns of wire around each field pole, the induced voltage can, for the whole rotor circuit, become of dangerously high magnitude. To minimize this danger the field circuit is sometimes broken up into sections during the acceleration period and connected in series just before full speed is reached. The more common practice is to connect a starting and discharge resistor across the field circuit. Since the currents induced in the field windings contribute a share of the total torque during acceleration, this starting resistor can be designed so as to obtain maximum benefit from the field at the pull-in point, where it is most desirable. The manner in which the field winding contributes to torque is shown by Fig. 23.

As the rotor reaches nearly synchronous speed, excitation is applied to the field winding, the revolving field poles established by the stator winding lock in with those of opposite polarity in the rotor, and the motor then operates at synchronous speed. Under synchronous operation the only function of the pole-face winding is to minimize hunting or damp-out oscillations; hence the term "damper" winding. Although the motor now operates synchronously, the interlock between the armature and rotor fields is not rigid. As in the pull between any magnetic poles of opposite polarity, the lines of magnetic force tend to act like elastic bands, which stretch as the pull increases. Therefore, as the shaft load, or required torque, on a synchronous motor increases, the lines of force holding the armature and rotor fields together are stretched, and the rotor poles now lag slightly behind the stator field. They assume a definite position, depending upon the magnitude of the pull or torque, and from then on revolve in synchronism with the stator field until the

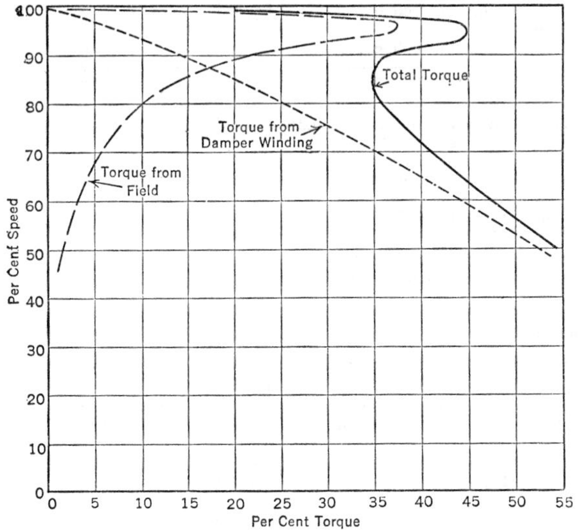

FIG. 23. Effect of Field on Speed-torque Curve

torque load again changes. It is apparent that fluctuation in torque load can cause oscillations of the rotor position, since for a given value of torque there must be a corresponding amount of "stretch" in the lines of force or a definite position of the rotor poles in relation to the stator field. Accompanying this change in rotor position is a change in the vector relation between the impressed voltage and counter-voltage or excitation voltage. These changes must stay within definite limits, and the torque corresponding thereto is known as the pull-out torque of the motor (see Article 16 on Stability). This is the limiting synchronous-motor torque and is directly proportional to the applied voltage, which explains why a synchronous motor is less sensitive to voltage variations than an induction motor, induction-motor torques being proportional to the square of the applied voltage.

The **starting torque,** or locked-rotor torque, is the torque developed at standstill and is a measure of the ability of the motor to start its load.

The **pull-in torque** may be defined as either nominal or actual. Nominal pull-in torque is the torque developed without excitation at 95 per cent of synchronous speed. For loads with normal inertia a motor will pull into synchronism from this speed. To consider both the inertia and required torque of the load, actual pull-in torque is specified as the maximum constant torque under which the motor will pull its connected inertia load into step.

The pull-out torque is the maximum sustained torque the motor will develop for 1 min with the excitation corresponding to that for rated power factor and load.

Torques for which synchronous motors are designed are expressed in per cent of rated torque, where rated torque in pounds at 1-ft radius is given as

$$\text{Pound-feet} = \frac{5250 \times \text{Hp}}{\text{Rpm}}$$

Table 9 gives a list of torque requirements and starting conditions for various types of loads.

Table 9. Application Data for Synchronous Motors

Application	Method of Connecting Motor to Load	Starting Conditions	Starting Torque, Per Cent	Pull-in Torque, Per Cent	Pull-out Torque, Per Cent
		Cement, rock products and metal mining			
Ball mills—mining	Coupled to mill pinion	Mills loaded at start	175–200	110–120	175
Crushers Gyratory	Coupled or belted	Loaded	150–200	100–125	200
Jaw	Coupled or belted	Loaded	150–225	100–125	250
Roll	Coupled or belted	Loaded	225	100	225
Bradley-Hercules	Coupled	Unloaded	80	100	175
Hercules mills Cement	Coupled to mill pinion shaft by resilient flexible coupling good for severe misalignment and vibration	Loaded	150–200	100–110	175
Rod Mills—mining	Coupled or belted (Lenix drive) to gear	Loaded	200	120	175
Tube mills	Coupled by flexible coupling to reducing gear	Loaded	190	120	175
Crushers—hammer mill	Coupled	Loaded	150	115	250
		Compressors and Blowers			
Blowers Cycloidal positive	Coupled or engine type	Unloaded	40– 60	40– 60	150
Blowing engines—reciprocating	Engine type	Unloaded	40	40– 60	150
Blowers High-speed	Direct-connected or step-up gear	Unloaded discharge closed	40	60	150
Compressors Air	Engine type	Unloaded	40	40	150
Ammonia and ammonia booster	High speed belted; low speed usually engine type, occasionally coupled	Unloaded by by-pass	40	40	150
Freon	High-speed belted; low speed usually engine type, occasionally coupled	Unloaded by by-pass	45	60	150
Gas-reciprocating	High speed belted; low speed engine type	Unloaded by by-pass	40	40	150
		Fans			
Exhaust and Ventilating	Coupled or belted	Usually loaded	50	60–125	150
		Flour Mills			
Line shaft	Coupled or belted	Loaded	175–225	100–125	175
		Lumber			
Bandsaws and resaws	Coupled or belted	Unloaded; torques depend on WR^2 of saw	80–125	75–115	200–250

Table 9. Application Data for Synchronous Motors—*Continued*

Application	Method of Connecting Motor to Load	Starting Conditions	Starting Torque, Per Cent	Pull-in Torque, Per Cent	Pull-out Torque, Per Cent
		Pulp and Paper			
Jordans..........	Coupled with flexible coupling	Unloaded	50	50	150
Beaters..........	Belted	Loaded	125	100	175
Pulp grinders......	Coupled by flexible coupling	Unloaded	50	50	150
Wood hogs and chippers	Coupled or engine type	Unloaded; torques depend on WR^2 of machine	80–125	80–110	225
		Pumps			
Centrifugal, water, oil, sewage	Coupled by flexible coupling	Valves either open or closed, usually closed	Valves open or closed 35–20	Valves closed 35–50 Valves open 100–120	150
Reciprocating......	Coupled by flexible coupling to pump pinion shaft	Unloaded or by-passed	50– 75	25– 50	150
Reciprocating—vacuum	Belted or direct-connected engine type	Unloaded	30– 50	60– 85	150
Rotating—vacuum.	Belted or direct-connected engine type	Loaded	40	55–100	150
		Rubber			
Banbury mixers—and plasticators	Direct drive or through reduction gear	Usually unloaded, may reverse under load in emergency	125–150	100–110	250
Mill lines—rubber..	Low speed—direct-coupled; High speed—coupled to reducing gear	Usually loaded, may reverse under load in emergency	110–150	110–125	175–250
		Steel			
Piercing mills......	Direct-connected by flexible coupling; sometimes coupled to reducing gears	Unloaded; may be reversed under load in emergency	80–100	50– 60	300–350
Continuous mills—steel rolling sheet, bar, billet, skelp, strip mills	Direct-connected by flexible coupling usual practice; sometimes coupled to reducing gear using higher-speed motors	Unloaded; may be reversed under load in emergency	50–100	50–100	250–350
Rod mills.........	Direct-connected by flexible coupling	Unloaded; may reverse under load in emergency	60–100	50– 60	225–275
Sizing and reducing mills	Direct-connected by flexible coupling; sometimes coupled to reducing gears	Unloaded; may reverse under load in emergency	80–100	50– 60	300–350

STARTING INRUSH. Starting current inrush or locked kva is a characteristic of great importance for motors installed on lines or systems of limited capacity, since a sudden relatively high current demand may so lower the line voltage as to affect adversely the operation of the connected load. The per unit (pu) starting inrush is the ratio of the

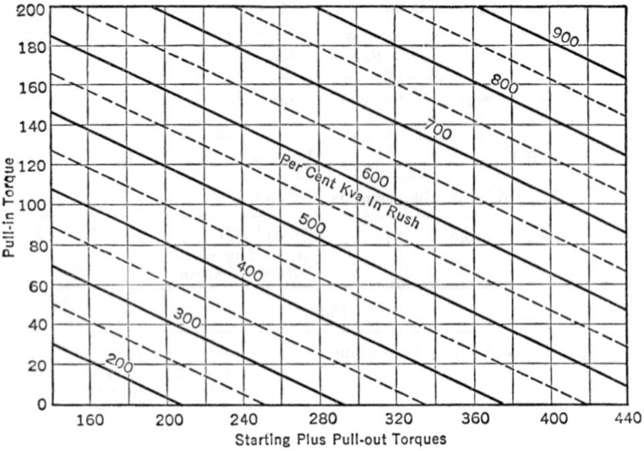

FIG. 24. Starting Inrush Kva; 100 Per Cent Power Factor Motors. (All values in percentage.)

kva drawn by the motor at full voltage and with the rotor locked to the rated full-load kva. The inrush current is directly proportional to the applied voltage, and therefore the inrush kva is proportional to the square of the applied voltage. For this reason, torque conditions permitting, large motors are often started on reduced voltage through an auto-transformer or reactor.

Since at standstill and without excitation there is no counter-voltage, the current drawn is limited only by the impedance of the motor. This current ranges from two and a half to seven times the rated current of the motor, depending on the torque characteristics and to some extent on speed, and occurs at power factors in the order of 25 to 35 per cent. Since high-speed motors have lower impedances than low-speed motors, a high-speed motor will in general have higher inrush than a low-speed motor and also higher pu

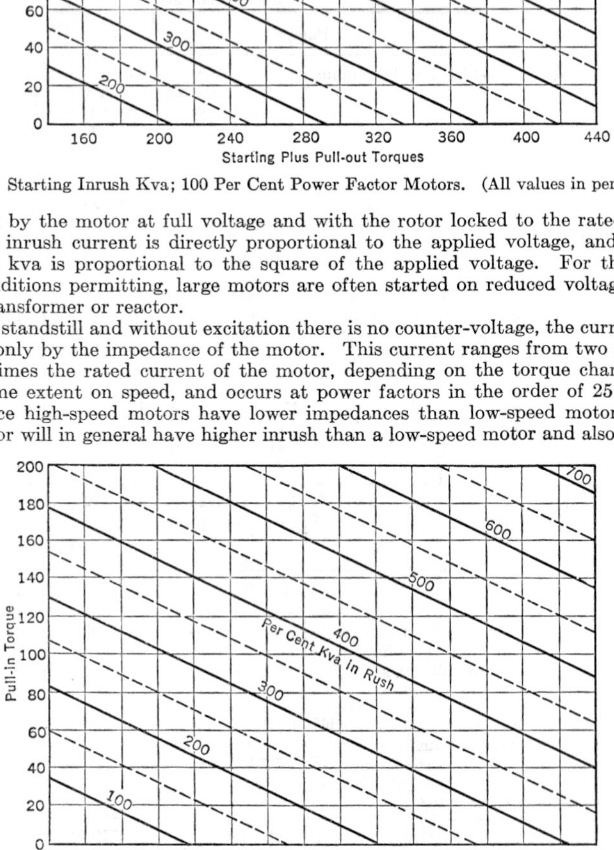

FIG. 25. Starting Inrush Kva; 80 Per Cent Power Factor Motors. (All values in percentage.)

torques. Figures 24 and 25 show the inrush values that may be expected from motors rated unity power factor and 80 per cent power factor and with various torques. In either motor, at unity power factor or 80 per cent power factor, the actual kva drawn from the line may be nearly the same, but, since the rated kva of an 80 per cent power factor motor is higher than that of a unity power factor motor of the same horsepower, the per cent current inrush will be lower for the 80 per cent power factor motor. In referring to Figs. 24 and 25, it must be remembered that low torques cannot always be obtained with certain high-speed motors, and therefore the expected obtainable torques must be used.

SPECIAL DESIGNS OF MOTORS. When particularly high torques or low starting inrush or both are required, it is often necessary to resort to some special construction that is more complicated and more expensive than the standard. One method of obtaining high torques is to build an oversized motor with low reactance, but this results in high inrush current. The increase in torque over normal, however, will not generally result in the inrush increasing by the same ratio over normal. In some cases the starting inrush with high-torque motors is reduced by opening the field circuit during starting so as to eliminate the low-power-factor current drawn by it. The voltage drop caused by this current flowing in the stator winding is thus eliminated, and starting torque is increased, while the inrush is reduced. In such cases, to minimize the dangers of high induced voltages in the field circuit, the field is broken up into sections.

Phase-wound Damper Windings. By bringing connections to the pole-face windings out to slip rings, so that external resistance can be connected in series, instead of short-circuiting them directly through end rings, some of the high-torque, low-inrush characteristics of wound-rotor induction motors can be obtained. In this polyphase type the damper winding is connected to three slip rings, and the external resistances are decreased in steps as the motor gains speed, so that desirable torque characteristics are obtained. The field circuit is left open during starting. Obviously this method is very expensive in both motor and control equipment. A somewhat simpler and less expensive construction using a single-phase winding which occupies one or two slots near the pole center and is connected to two slip rings gives better starting performance than a standard motor. The remaining slots in the pole face are occupied by a conventional short-circuited damper winding. This type of motor starts with closed field circuit.

Synchronous-induction Motors. This type of construction is somewhat similar to the motor with a polyphase damper winding but more closely resembles a wound-rotor induction motor in that no salient field poles are provided. The air gap is longer than that of a normal induction motor. The motor is brought up to speed in the same manner as an induction motor with wound rotor, and as synchronous speed is approached the rotor winding is excited with direct current applied to two of its terminals. This motor has a low inrush but has an inherent disadvantage in the limited space for copper to carry d-c excitation, which results in low efficiency, low excitation voltage, and lack of adaptability to leading power factor and high pull-out torque operation.

Clutch-type Motors. In some applications it is possible to start the motor unloaded, and after synchronizing to apply and accelerate the load through a clutch. Such motors have normal starting performance but require high pull-out torque.

"Supersynchronous" Motors. In this construction both stator and rotor are supported on bearings, so that, as the motor starts, the rotor is held stationary by the load and the stator rotates and accelerates to full speed. Then a brake is applied to the stator, and as it decelerates, a strong torque is applied to the load, which accelerates it to full speed as the stator comes to rest in normal operating position.

Double-deck Damper-winding Motors. When high pull-in torques with normal or low starting torques are required, as in centrifugal-pump or blower drives, some form of double-deck effect is incorporated in the damper winding. One method of accomplishing this is to provide two damper windings. One, which is of low reactance and high resistance, carries most of the current during starting, when the induced currents are of nearly line frequency, and provides most of the required starting torque. The other winding is of high reactance and low resistance and offers relatively high impedance to the currents of line frequency induced at standstill. As the rotor accelerates, this impedance decreases with reduction in frequency, so that current flows in the winding to give high pull-in torque.

Table 10 gives a comparison of the various special designs discussed.

Two-speed Synchronous Motors. Synchronous motors may be designed for two-speed operation where the speeds are in the ratio of 1 to 2. In such case the stator winding is arranged to give consequent poles on the second speed, and the field circuit is divided with four slip rings in such a manner that two adjacent poles may be connected for the same polarity, thereby in effect halving the number of poles.

Synchro-tie or Selsyn Systems. Synchro-tie, or Selsyn, machines are not strictly synchronous machines, but by virtue of their synchronous operation will be mentioned here. They are bipolar machines with rotor windings excited from single-phase alternating current. The stators have normal three-phase windings. In operation the stator windings of the transmitter and receiver machines are connected together, and any movement of one rotor (whether several revolutions or only a few degrees) is immediately followed by an equal movement of the other rotor. Such systems are used for various indicating and control functions, the machines usually being located at a remote distance apart, the only coupling required being the electrical connection between the stator windings.

Table 10. Special Types of Synchronous Motors *

Type	Starting Torque		Pull-in Torque	Pull-out Torque		Inrush, Per Cent of Full-load Kva Unless Otherwise Stated
	First Point	Second Point		100% PF	80% PF	
1. Standard low speed	50	....	40	150	200	Approx. 275
2. High torque, standard construction	200		110	175	200	Approx. 675 for 1.0 PF Approx. 550 for .8 PF
3. Open-field starting	200		110	175	200	80 to 90 of No. 2
4. Polyphase starting winding and open-field starting	150	200	125	175	200	Approx. 30% more than a wound-rotor induction motor
5. Polyphase starting winding and closed-field starting	130	195	120	200	200	Between No. 2 and No. 4, approx. 350 on 1st point and 400 on 2nd point for 100 P.F. motor
6. Synchronous-induction	150	200	150	140	175	Approx. 15 more than a wound-rotor induction motor
7. Single-phase starting winding and closed-field starting	150	200	115	175	200	Between No. 2 and No. 4
8. Double-deck starting winding for centrifugal-pump drive	60	....	100	150	200	Lower than for standard construction, approx. 425 for 100 PF motor

Values are given in per cent of rated values, and are typical only.

* From "High-torque Synchronous Motors," by M. R. Lory, *Elec. J.* (May, 1937).

LOAD INERTIA. If the load has high inertia, this must be taken into account in the application of the motor. During the relatively short period of acceleration the combined kinetic energies of the load and motor in the form of heat must be stored in the damper winding. Ample thermal capacity must therefore be provided in the damper winding so that excessive temperatures will not be reached, especially if the motor must start its load at frequent intervals. Most materials have poorer mechanical properties at high temperatures, and therefore the problem becomes more acute in the double-deck type of damper windings or in windings comprised of dissimilar materials, where differential expansion results. It is thus seen that formidable design problems are encountered in those applications which require high starting torque and have inherently high inertia. The energy which must be stored in the damper winding is given by the expression:

$$\text{Kw-sec} = 2.306 \times WK^2 \times N^2 \times 10^{-7}$$

where WK^2 is the sum of the inertias of the load and the motor rotor at speed N, which is the synchronous rpm of the motor. The expression WK^2 is the product of the weight of the rotating member in pounds times the square of its radius of gyration in feet, and is often written WR^2.

Not only are thermal problems involved with high inertia loads but also torque problems are encountered, in that this type of load often cannot be pulled into synchronism from 95 per cent speed by action of the field. That is, after the damper winding has brought the motor up to its maximum speed as an induction motor, the field is applied, and acceleration to full speed has to be accomplished by action of the field winding during a very short interval of time. If the load inertia is high, this synchronizing force may be insufficient, and the remedy is to accelerate the load to a higher speed before applying the field, sometimes to as high as 98 per cent speed. At this higher speed the induction-motor torque is less, and therefore for high inertia loads the actual pull-in torque developed is less than the nominal pull-in torque. It is, however, a better measure of the motor performance, since it is the torque against which the motor will actually accelerate its inertia and torque load and pull it into step.

RECIPROCATING LOADS. One of the most frequently used applications of low-speed synchronous motors is for driving loads, such as reciprocating pumps and compressors. The pulsating-torque requirement of such a load is likely to start hunting in the synchronous motor and pulsation of the line current, as the rotor tends to swing back and forth in its angular position relative to the stator field in order to follow the torque demand. Obviously this would be an undesirable condition, and to avoid or minimize it inertia of the proper magnitude should be incorporated in the motor or its load. This inertia, or WK^2, for a given condition of operation, can be neither too little nor too great, and the determination of its optimum value forms the basis for considerable theoretical development which is beyond the scope of this article.

For a given compressor a torque curve must be available for each condition of operation which is to be investigated. The irregular torque required by the compressor is due to the gas forces and the inertia forces of the cylinders and is a periodic function of time, repeating for every revolution of the crankshaft. By the method of Fourier the four significant sinusoidal components, consisting of fundamental, second, third, and fourth harmonics, are found, with values for the A and B coefficients, and the phase angles. These can be substituted in the following expression for Y:

$$Y = \sum \frac{\sqrt{A_n^2 + B_n^2}}{K_n} \sin\left(n\omega t + \tan^{-1}\frac{B_n}{A_n} - \tan^{-1}\frac{n\lambda}{1 - n^2 Z}\right)$$

in which $K_n = \sqrt{(1 - n^2Z)^2 + (n\lambda)^2}$.

n = order of harmonic.

λ = a constant depending on the damping torque and speed of the motor, usually taken as an average value of 0.1.

ω = angular velocity $= 2\pi N/60$.

Z = a function of the motor design

$$= \frac{1}{9.25} \times \frac{0.746N^4}{P_r \times f \times 10^8} \times WK^2.$$

where N = motor rpm, f = line frequency, and P_r = the motor synchronizing coefficient. The term $(0.746N^4/P_r \times f \times 10^8 \times WK^2$ is denoted by the American Standards association by the constant C and is often denoted also by the constant X, so that $Z = X/9.25$. Also, the reciprocal of the expression $0.746N^4/P_r \times f \times 10^8$ has been given the name flywheel constant, F_c, and is a characteristic of the motor design.

$$F_c = \frac{P_r \times f \times 10^8}{0.746N^4}$$

so that $X = WK^2/F_c$.

The synchronizing power, P_r, is the power at synchronous speed which corresponds to the torque tending to restore the rotor to its no-load position relative to the line voltage. It is determined by dividing the shaft power by the corresponding angular displacement of the rotor and is thus expressed in kilowatts per electrical radian. This displacement angle may be expressed either in electrical degrees or electrical radians and is defined as

$$\delta = \tan^{-1}\frac{IX_q \cos\theta}{e_t + IX_q \sin\theta}$$

and

$$P_r = 57.3\,\frac{\text{shaft } KW}{\delta}$$

where δ is expressed in electrical degrees.

I = per unit armature current.

e_t = per unit armature terminal voltage.

X_q = per unit quadrature-axis synchronous reactance.

θ = power factor angle.

The displacement angle can be observed directly on a given machine by stroboscopic means.

The value of Y determined above represents the torque pulsation in percentage, based on the theoretical average torque of the compressor, and, since the speed of the motor is essentially constant, it also represents the power pulsation in percentage, based on the indicated horsepower determined from theoretical indicator cards applying to the given compressor. With constant power factor, this is also equal to the current pulsation.

Expressed in percentage of the nameplate rated current, the value of Y for a given limit in current pulsation is found as

$$Y = \frac{\text{Motor rated hp}}{\text{Theoretical ihp} \times \text{PF}} \times \text{Pulsation limit in per cent of nameplate rating}$$

In the usual case a limit of 66 per cent pulsation is allowed.

It is therefore seen that the pulsation Y is a function of the inertia WK^2 and the flywheel constant F_c in a rather involved mathematical relationship. Since the inertia value is usually sought in a given analysis, values of Z are assumed, and the corresponding values of Y determined so that X–Y curves can be plotted for each desired condition of operation for which compressor-torque curves are available. It is then not difficult to determine the required WK^2 to stay within the desired limit of current pulsation, since $WK^2 = X \times F_c$. The practical problem is simplified by having available families of curves for the constant K for the four harmonics plotted as functions of Z. Families of curves are also used for the phase angle $\tan^{-1} n\lambda/1 - n^2Z$ plotted as functions of Z. Tabulations can then be made for the various constants, and angles for the four harmonics and the harmonics plotted to scale in their correct phase relationships for one crankshaft revolution. When these are added together, a representation of the current pulsation of the motor is obtained. The distance between the maximum points on the two sides of the zero axis then gives the value of Y for the assumed value of Z. Figure 26 shows a typical compressor-torque curve, and Fig. 27 the X–Y curve obtained from it.

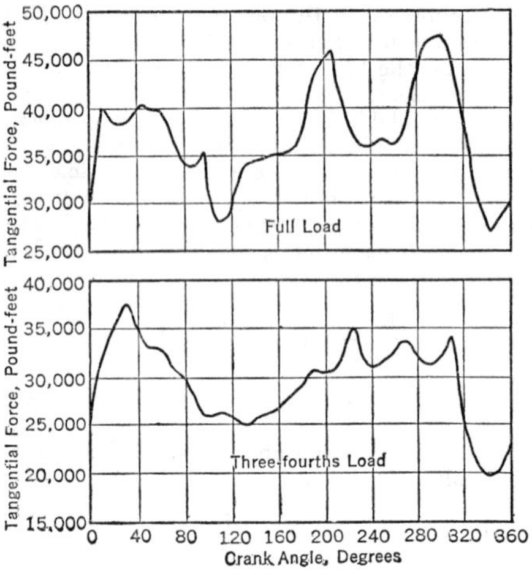

FIG. 26. Example of Compressor Torque Curves

Compressors are operated at partial loads by various arrangements of valves, and the compressors themselves are built in a wide variety of crank arrangements, cylinder numbers, and compression stages, depending on the pressure and kind of gas, so that in modern practice a complexity of operating conditions exists for the application of synchronous-motor drive. However, the majority of these conditions have been analyzed, and for each condition a standard application number has been adopted with the corresponding required values of X. Then, the flywheel constant F_c of the motor being known, the required WK^2 can be determined at once. This inertia may be incorporated directly in the motor rotor, or a separate flywheel may be added. Such application numbers have been assigned to compressors for ammonia, air, and carbon dioxide in many different arrangements. When a new compressor design is brought out, however, it is necessary to make a detailed analysis of its torque characteristic in order that the compressor-motor combination may be designed to avoid undesirable current pulsation or resonance effects.

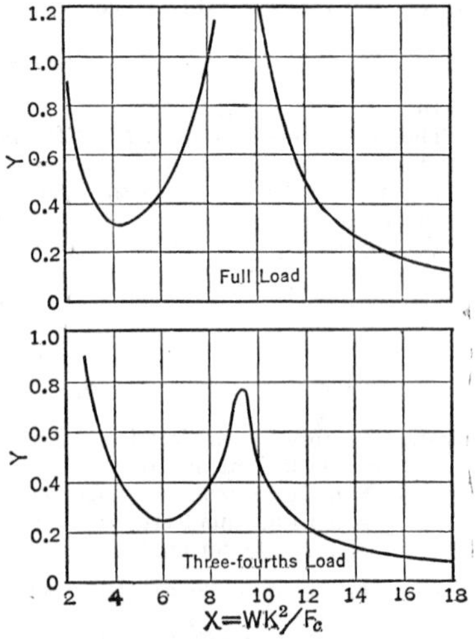

FIG. 27. X–Y Curve Derived from Fig. 26

DYNAMIC BRAKING. In certain synchronous-motor applications, such as steel- and rubber-mill drives, it becomes desirable at times to stop rotation as quickly as possible in order to prevent injury to personnel, product, or equipment. To accomplish this, either plugging or dynamic braking may be employed. In plugging, the phase rotation of the supply voltage is reversed. In this method, when the speed reaches zero, the supply voltage must be removed in order to prevent reversal of rotation. The control circuits are thereby complicated, and the method is also disadvantageous in that the switching may introduce troublesome system disturbances and the braking torque is relatively low, thus limiting the rate at which rotation can be stopped.

In dynamic braking, the stator winding is short-circuited through an external resistor while field excitation is maintained. The rotational energy is thus dissipated as heat in the resistor, the rate of dissipation and the reduction of speed depending upon the resistance used.

The value of external resistance, in ohms per leg, for stopping in the minimum possible number of revolutions is given by the formula:

$$R = 1.2\frac{X_d'}{\sqrt{3}} \times \frac{E}{I}$$

where X_d' = transient reactance, per unit.
E = rated volts per phase.
I = rated amperes per phase.

The energy rating of this resistance is:

$$2.3\ WK^2 \times (\text{Rpm})^2 \times 10^{-4} \text{ watt-sec}$$

and the number of revolutions required for stopping is:

$$R_s = \frac{WK^2 \times (\text{Rpm})^3 \times X_d'}{83.9 \times \text{Hp} \times E_0^2} \times 10^{-6}$$

where WK^2 = sum of inertias in both motor and driven machine, in pound-feet-squared, and E_0 is calculated from:

$$\sqrt{(1 + X_d' \sin \phi)^2 + (X_d' \cos \phi)^2}$$

ϕ being the power factor angle of the motor.

16. HUNTING AND STABILITY; CIRCLE DIAGRAM

HUNTING. In the discussions of torque and reciprocating loads it has been mentioned that hunting can be started by torque pulsations. Hunting can also be initiated by disturbances in the power system. In either case the rotor slows down or speeds up momentarily as it delivers some of its kinetic energy to supply the increased load or as the power drawn by the load becomes less than that being developed by the motor. The phenomenon can be illustrated by a mass attached to the end of a rod and supported by a spring, as shown in Fig. 28. It can be seen from ordinary physical considerations that, if an additional mass M is suddenly applied, the movable end of the rod will oscillate up and down a few times until the deflection of the spring attains its new steady value corresponding to the increased mass. Oscillation will likewise be caused by a sudden removal of mass. Obviously, for any magnitude of mass there is a definite deflection in the spring and a definite displacement angle δ. Moreover, there must be a limit in the mass a given spring can support. It can be seen also that, if the rod is oscillating and an intermittent force is applied in time with the oscillation and in the same direction, the displacement can be made to increase very greatly, to the point of causing the spring to deflect beyond its elastic limit. These same physical facts apply to the operation of a synchronous motor, wherein the torque load is represented by the mass, and the motor torque by the spring. The motor, likewise, has a characteristic displacement angle δ for any given torque load, and this is the angular displacement between the rotor and armature fields. Any change in load calls for a new angle, and as the rotor speeds up or slows down momentarily to attain the new position, it over-reaches the required angle, because of its appreciable inertia, and oscillation of the rotor about this angle, or so-called hunting, results. If the oscillation is in resonance with the pulsating-torque or system frequency or a strong harmonic, it is likely to become very severe. Such oscillation may be great enough to cause the displacement angle to overswing its maximum allowable magnitude, and the motor becomes unstable, pulls out of synchronism, and comes to rest. This maximum allowable angle approaches 90 elec deg, which is one-half of the physical angle between poles. A slight amount of hunting always accompanies a change in load, but the design of modern motors is such that the hunting is of slight

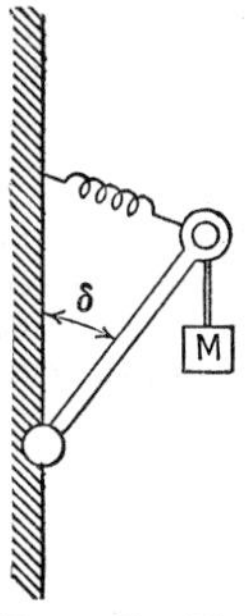

FIG. 28. Mechanical Model to Illustrate Hunting

magnitude and of little consequence unless resonance conditions are present. Steps are generally taken to avoid resonance effects in the application of any synchronous motor.

The free period of hunting, or natural frequency, neglecting damping, of a given motor is defined by the formula in Article 7, under Hunting and Natural Frequency. Should this frequency come very close to the frequency of torque pulsation, the electrical frequency of the supply, or any strong harmonic thereof, resonance occurs, and the hunting is likely to become very severe. As the formula for natural frequency shows, addition of WK^2 lowers the natural frequency and thus generally decreases the tendency to hunt. However, the more common method in use with motors, except in reciprocating loads, is to take advantage of the damper winding and thus avoid adding extra weight. As long as the motor operates under steady conditions, the rotor revolves synchronously, and the damper winding is inactive. If a momentary change in speed occurs, however, currents are induced in the damper bars with a period equal to the period of oscillation. The direction of these currents is such as to oppose the change of angular velocity in the rotor, and energy is dissipated in the damper winding because of the attendant copper loss in the bars, thereby damping out the oscillation. Some damping is also provided by eddy currents induced in the pole faces.

STIFFNESS OF COUPLING. The performance of a synchronous motor is considerably affected by its so-called stiffness of coupling, which is a measure of its tendency to follow closely any irregularities in the speed of the generators from which it is supplied or to maintain its displacement angle against load fluctuations. The combination of a large air gap and low leakage reactance gives stiff coupling; conversely, a small air gap and high reactance will result in soft coupling. It is undesirable to have too stiff coupling, since irregularities in the load or power supply will then impose undue mechanical strains on the moving parts and windings. Too soft coupling is also undesirable, since it promotes instability and danger of dropping out of synchronism with relatively small disturbances, as well as large power-factor variation with load under constant excitation. In practical designs a compromise is made. The stiffness of coupling is a function of the displacement angle δ and is zero when δ is equal to $\tan^{-1}(X_s/r_e)$, where X_s is the synchronous reactance and r_e the effective resistance, including the effect of load current. This is the angle corresponding to the stability limit, which means that the displacement angle can never exceed 90 elec deg, since the angle $\tan^{-1} X_s/r_e$ is seldom much less than 90 deg.

CIRCLE DIAGRAM. The circle-diagram method of representing synchronous motor performance, due to Blondel, assumes constancy of synchronous reactance and effective resistance. Therefore, it cannot be used for accurate predetermination of performance, but it does assist in visualizing some of the operating characteristics. Such a diagram is shown in Fig. 29, in which the proportions have been distorted from actual values for the sake of clarity.

The equal sides of the isosceles triangle OCA are equivalent in length to $V/2r_e$, which is the radius of a circle representing zero power output, P_0. The third side or chord, CA, is equivalent in length to V, the terminal voltage, and the angles θ are determined as $\tan^{-1}(X_s/r_e)$. Point A is the center for circles representing constant excitation, and point O is the center for circles representing constant power. Change in operating conditions of the motor is represented by change in the position of point B, which is always located at the intersection of a power circle with an excitation circle. The voltage E_z is then the vector difference between excitation voltage E and terminal voltage V. The displacement angle δ is the angle between E and V. Since X_s and r_e are assumed constant, the current I lags resultant voltage E_z by a constant angle

$$\theta = \tan^{-1}\frac{X_s}{r_e}$$

If E_z coincides with CO, then it is apparent that current I is in phase with voltage V, and unity-power-factor operation results. If E_z falls to the left of OC (as shown), then I leads V by angle ϕ for leading-power-factor operation; if E_z falls to the right of OC, lagging-power-factor operation is represented.

For increasing power, the radii of the constant power circles P_1, P_2, P_3, etc., decrease.

If constant excitation, as represented by circle (a), is assumed, an increase in load will be represented by movement of the point B to a power circle P_2 of smaller radius. It is seen that, as this happens, the displacement angle δ becomes larger, and the resultant voltage E_z changes in magnitude and phase, also causing a change in magnitude and phase of current I, so that lagging power factor results. Should the load increase sufficiently to cause B to move onto a power circle P_3, the excitation voltage will then coincide with OA. At this point it is seen that displacement angle $\delta = \theta = \tan^{-1}(X_s/r_e)$. Any further increase in power demand will cause the angle δ to increase still more, which will result in point *B* swinging to the right of *OA* onto a circle representing lower power. The cross-hatched area to the right of *OA* therefore represents the region of instability, and it is obvious that $\delta = \theta$ represents a condition of maximum output, or the stability limit. Consideration of the diagram shows that theoretically for any change in load the excitation can be adjusted to maintain any desired power factor angle or to maintain a constant displacement angle if desired. Practical limits occur because of temperature rise in the windings. The theoretical limits occur when *B* falls on *O* corresponding to maximum power (equal $V^2/4r_e$) for any excitation, and when *B* falls on *C* corresponding to zero power. It is obvious that in the practical case these limits are unattainable, since excitation equal to *AO* would result in internal voltages many times the impressed voltage *V* and currents of such magnitude as to burn up the motor windings. Zero power is likewise unattainable, since some power is always necessary to overcome friction and windage and other losses. In the practical case the displacement angle does not closely approach θ, since then the motor would operate so close to the unstable region as to break down with a very slight change in load or change of system conditions. The diagram shows that overexcitation increases stability by permitting greater power changes in the stable region.

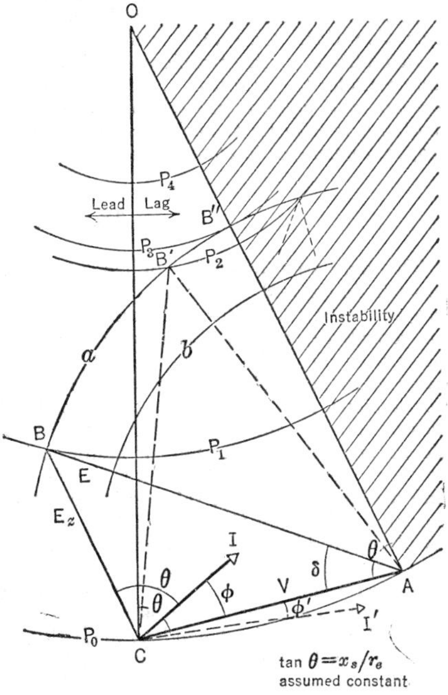

FIG. 29. Circle Diagram for Synchronous Motor

Automatic excitation control used on large motors carrying varying loads improves their stability and helps control the kvar. The stability characteristics of large motors are of importance in the operation of the power system and are discussed in Section 3.

17. STARTING AND CONTROL OF SYNCHRONOUS MOTORS

Synchronous motors are started either at full voltage or at reduced voltage. Reduced voltage is used to decrease the starting inrush current drawn from the line, and, when this method is used, there must be assurance that the motor will develop adequate torque to start and accelerate the load at the voltage used. The reduced voltage is obtained by connecting the motor across the secondary terminals of an auto-transformer during the starting period or by connecting a reactor in series with the motor. The torque developed by the motor at start and during acceleration is proportional to the square of the voltage at the motor terminals. The current drawn from the line in auto-transformer starting is proportional to the square of the voltage at the motor, whereas in reactor starting the line current is directly proportional to the voltage at the motor. Obviously, in reactor starting the decrease in current as the motor accelerates tends to increase the motor voltage, which results in further increasing the current and torque. The exact starting characteristic depends upon the design constants of the motor and reactor. Open or closed transition may be used in switching to full voltage or to a higher voltage when more than two voltage steps are required.

Series resistors can be used to lower the voltage at the motor terminals. The action is similar to that of the series reactor, except that the resistor entails considerable I^2R loss and is therefore seldom used.

As the motor reaches nearly synchronous speed, the field excitation is applied. Since application of excitation is somewhat critical, especially with high inertia loads, it is normally done automatically by a timing relay or by other types of relays, depending upon the frequency of the induced field current or upon the magnitude of the stator current in conjunction with a timing sequence. A feature of control known as angle switching recognizes the most favorable angular position of the field pole with respect to the stator mmf wave and takes advantage of it in applying the excitation, resulting in minimum line disturbance.

Synchronous-motor control may be fully magnetic, being actuated from push-button, master switch, float switch, or similar devices, or it may be semimagnetic, requiring manual operation of the control line switch.

In addition to the switching mechanism synchronous-motor controls usually include ammeters for line and field currents and protective relays for thermal overload, a-c undervoltage, and pull-out protection. Also included are the field rheostat and starting and discharge resistor for the field circuit. Some controls may include provision for automatic resynchronizing or in some cases stopping of the motor after pull-out.

18. VENTILATION

The ventilation or cooling of synchronous motors is one of the most important factors in their design and application, since it helps to determine the load they can carry with proper consideration for dependability and life of insulation. Most motors in use are air cooled, the air being drawn in and circulated through the motor by the action of blowers within the machine. Not all machines have specific blower vanes, since in many cases other parts of the rotor structure provide the required blower action. In certain applications, as mentioned under Protective Enclosures in Article 3, motors may be enclosed, and the ventilating air supplied by means of an external blower. In other cases the air is recirculated within the motor enclosure by successively passing through a water cooler mounted within the enclosure. The air usually enters the motor itself at the ends and discharges from the sides or top or bottom. Within the machine the air is directed across and between the end windings and radially through ventilating ducts, which are usually spaced in the core structure at approximately 2-in. intervals and vary from $^1/_4$ to $^1/_2$ in. in width. Some air also passes between salient rotor poles and across the outside of the core laminations. The function of the air is therefore to take up heat from the sources of its generation and carry it to the outside for dissipation. The amount of air required is obviously dependent upon the kilowatts of loss to be dissipated; a common allowance is 125 cu ft per min per kw loss. This figure is based on allowing the air temperature to rise 14 deg cent in passing through the machine.

In estimating ventilation for force-ventilated motors, allowance must be made for the pressure drops through the complete ventilating circuit, since pressure must be available for forcing air through the motor. A positive pressure of at least $^1/_2$ in. of water is usually required at the motor intake. This consideration is all the more critical when planning ventilation for self-pipe-ventilated motors, and in such cases the manufacturer should be consulted for his recommendation.

19. TESTS OF SYNCHRONOUS MOTORS

Tests which may be made on synchronous motors, either in the factory or in the field, are the same as those discussed in Article 12 for synchronous machines in general. With a synchronous motor, however, tests are usually made also for starting current and torques.

LOCKED ROTOR TEST. The locked rotor test is usually made with a prony brake or beam clamped rigidly to the shaft of the motor, its free end resting on a platform scale. Obviously, the scale reading in pounds times the length of the beam in feet gives the pound-feet torque developed with locked rotor, or the starting torque. Because both stator and damper windings heat very rapidly with locked rotor, the motor is not energized any longer than necessary for taking readings. Readings of volts, amperes, kilowatts, and torque are taken at several successive values of voltage, the highest voltage used being that which gives an armature current of about two and a half times rated value. The "locked current" varies as the voltage and the "locked kilowatts" as the square of the voltage, so it is a simple matter to extrapolate curves through rated voltage, having several test readings at lower voltages. Since the starting torque may in some cases vary appreciably over a pole pitch, it is often desirable to take readings for several different positions of the rotor throughout a pole pitch. If no scale is available, the rotor can be locked against turning, and the same electrical readings taken. The locked torque in synchronous kilowatts at any voltage is then calculated as the difference between the kilowatts input and the sum of the stator copper loss and stray load loss at that voltage. Torques at other voltages vary as the square of the voltage.

In making the locked rotor test, the field is either short circuited through its discharge resistance or left open, according to the method of starting for which the motor is designed. Factory tests usually include readings of voltage and current induced in the field winding.

PULL-IN TORQUE. Where new damper-winding designs are used, or in certain critical applications, tests are usually made to obtain the nominal pull-in torque, which is the torque at 95 per cent speed. In some cases torque readings are taken at lower speeds, so that a speed-torque curve can be plotted. The most common method of accomplishing this is to couple the motor to a d-c generator and operate it without excitation at as high a voltage as heating will permit, usually not less than 50 per cent of rated voltage. For any given voltage the speed is set by adjusting the load on the d-c generator. At each speed setting readings are taken of volts, amperes, kilowatts, and speed for the motor and field current, and line volts and amperes for the d-c generator. The field of the synchronous motor is short circuited through its normal starting resistor. The losses of the d-c generator should be known. The per unit torque of the motor at any speed point is then given as

$$T = \frac{\text{Kw output} + \text{Kw losses (d-c gen.)} \times 1.34}{\text{Rated hp of motor}} \times \frac{\text{Rated speed of motor}}{\text{Speed at test point}}$$

This value should, of course, be corrected to rated voltage, the torque varying as the square of the voltage.

For other methods, which are less often used for determining speed-torque curves, reference should be made to the AIEE Test Code for Synchronous Machines.

20. REACTIVE KVA (KVAR)

As shown in the paragraphs under Round Rotor Theory in Article 7, a synchronous motor when overexcited draws a leading current; this is a very useful characteristic. By

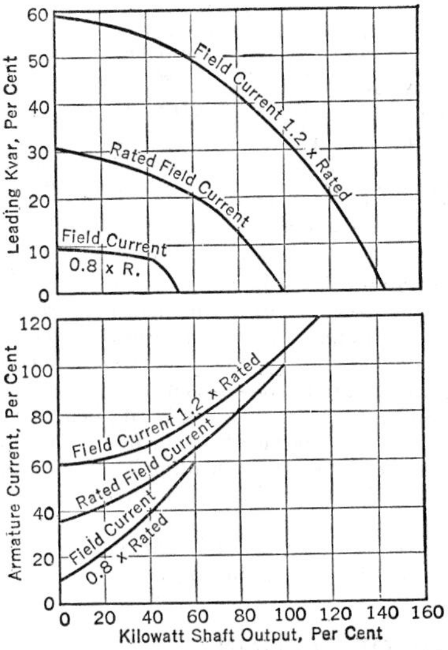

FIG. 30. Reactive Kva, Typical of 100 Per Cent Power Factor Motors

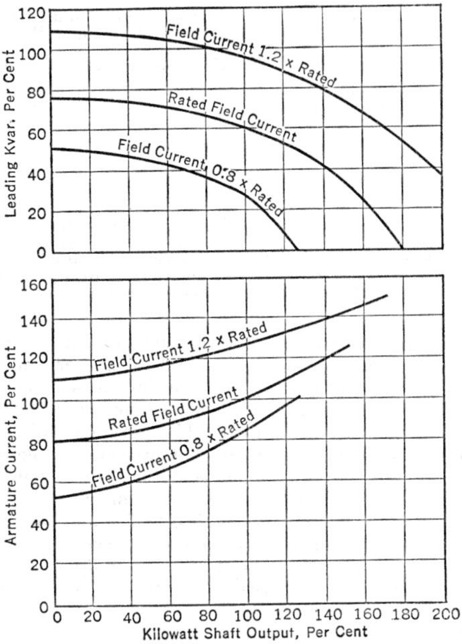

FIG. 31. Reactive Kva; Typical of 80 Per Cent Power Factor Motors

this means the motor not only supplies a mechanical load but also benefits the system to which it is connected by neutralizing in whole or in part the lagging current which lowers the power factor of the system. The result of eliminating wattless current is lower copper loss in the system and greater system efficiency, as well as improved voltage regulation for the generators supplying the load. Another advantage gained from the elimination of wattless current consists in making more of the system capacity available for handling useful load.

A synchronous machine operating as a motor (neglecting synchronous condensers, which are treated in Section 13) delivers mechanical power from its shaft at the same time that it supplies corrective effect to the system. Its total electrical load, therefore, is limited because of temperature rise, as determined by the vector sum of the kilowatt shaft load and the kvar load, this sum being the total kva drawn from the line.

At the same time a limit is established by the allowable temperature rise in the field winding, so that maximum leading kvar must also be dependent upon the capacity of the field to furnish overexcitation. Motors sold under standard ratings are designed for operation at either 80 per cent or unity power factor. It is obvious that, neglecting any overload margins there may be in the machine, a unity-power-factor motor operating at rated conditions furnishes only mechanical output from its shaft and does not provide any corrective kvar to the system. It does, however, provide some benefit to the system

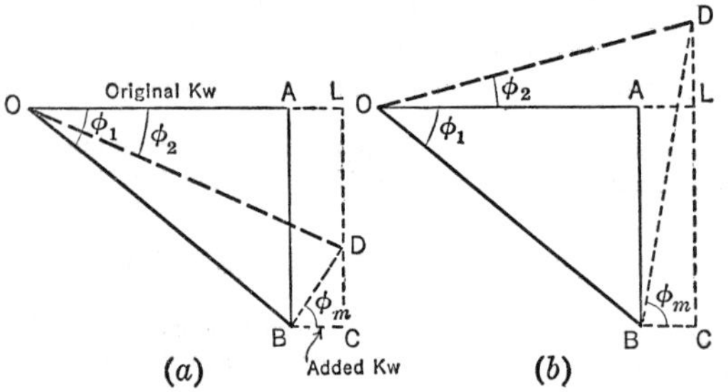

Fig. 32. Vector Diagram to Determine Required Motor Rating to Accomplish a Desired Change in System Power Factor

Synchronous Motor in Power Factor Improvement *

Original power factor = $OA/OB = \cos\phi_1$
Desired power factor = $\cos\phi_2$
Original kva = $OB = OA/\cos\phi_1$
New kva = $OD = OL/\cos\phi_2$

Original kvar = $AB = OB\sin\phi_1$
New kvar † = $LD = OD\sin\phi_2$
Motor kvar † = $AB - LD = CD$
Rated kva of motor (neglecting efficiency) = BD
$= \sqrt{\overline{BC}^2 + \overline{CD}^2}$
Motor power factor = $BC/BD = \cos\phi_m$

* This diagram is based on the present (1948) standard vector direction for vector power. (See Section 3, Article 13.) If the ASA should reverse this, as is now being recommended, the inductive kvar *AB* would be directed vertically upward, instead of downward, while the capacitive kvar *CD* would be directed vertically downward. Real power *OA* and *BC* remain horizontal.

† Consider *LD* to be negative in sign if the new system power factor $\cos\phi_2$ is leading as in (*b*).

power factor, since kilowatt loading is added with no additional lagging current. On the other hand, a motor designed for 80 per cent leading power factor will furnish its rated shaft output and at the same time draw 60 per cent of its nameplate kva from the system as leading kvar. If either machine is relieved of some or all of its mechanical load, it can draw more leading kvar subject to the condition that its rated armature current and rated normal field current are not exceeded. Figures 30 and 31 are essentially applicable to any standard synchronous motor of unity power factor or 80 per cent power factor rating, respectively, and show how the armature current and capacity in leading kvar change with variation in mechanical load. For specific cases differences in design constants may cause slight deviations from the curves.

The determination of required motor rating to accomplish a desired change in system power factor is illustrated graphically in Fig. 32, which is self-explanatory.

21. INSTALLATION AND OPERATION

FOUNDATION AND ALIGNMENT. To minimize vibration and misalignment during operation, it is necessary to have a rigid foundation, preferably of solid concrete carried down far enough to rest on a solid sub-base. When steelwork is used, the beams or girders should be adequately braced and supported by columns. When setting a steel bedplate, the top of the concrete foundation should be roughened and washed in order to provide a good bonding surface for the grout. The grout is made of sand and Portland cement, 1 to 1, made thin enough so that it can be tamped thoroughly under the base of the bedplate and up against the web. Rails are grouted to within a half inch of the top. Wooden forms must, of course, be used to hold the grout inside and outside the bedplate until it sets.

A template made to the dimensions of the outline drawing furnished with the machine will be helpful in locating foundation bolts. Some variation should be allowed for by locating the bolts in a steel pipe embedded in the foundation.

Accurate alignment with the prime mover or driven machine is most important in order to minimize shaft stresses, vibration, and coupling wear. The bedplate or soleplates of the assembled machine must be level, and adjustment is made by means of steel plates and shims before grouting to the foundation. Rigid couplings should be checked for alignment and trueness by loosening the bolts and measuring with thickness gages between the coupling faces, taking four equally spaced angular positions of the shafts. The maximum variation between the faces should not be over 0.002 in. per 12 in. of face diameter. The pilot fit in the coupling must not be so tight as to affect alignment measurements. In aligning flexible couplings, the recommendations of the manufacturer of the coupling should be followed. All alignment and air-gap measurements must be made with the foundation bolts pulled tight.

CONNECTIONS AND GROUNDING. The rotation, or phase sequence, of a three-phase three-wire machine can be reversed by interchanging two of the leads. In a two-phase four-wire machine one phase is connected to terminals T-1 and T-3, and the other phase to T-2 and T-4. Reversal is accomplished by interchanging the leads of one phase. In two-phase three-wire systems, the common lead is connected to T-3 and T-4 joined together, or to T-3 if there are only three terminals. For reversal, the outside leads to T-1 and T-2 are interchanged. Frames of machines should always be grounded for protection to personnel.

LUBRICATION. Bearings must be adequately lubricated at all times with good clean oil of the viscosity recommended by the manufacturer. An occasional inspection of oil rings should be made to see that they are functioning properly in supplying oil to the bearings. Oil reservoirs should be drained, flushed out, and refilled at intervals of 6 months to a year, depending on severity of service. In making necessary oil-pipe connections, care must be taken to see that any insulation used on the bearing is not short circuited. Front bearings are sometimes insulated against the flow of current between shaft and bearings which arises from certain electrical combinations in the machine. Continued flow of such currents would cause pitting and damage to the bearing.

If oil pumps are used, their operation should be periodically checked, and the entire lubricating system inspected for proper operation.

Antifriction bearings are usually grease lubricated and are packed with grease at the factory. The addition every 3 months of about 1 oz of grease to each bearing is recommended for large industrial machines. The bearing should be thoroughly cleaned and repacked about once every 2 years.

STARTING AND PARALLELING. Before starting any machine for the first time, a few fundamental rules must be observed:

1. Check all connections to see that they are "right and tight."
2. See that bearings are filled with oil to correct level.
3. Be sure that brush shunts are not touching and other electrical parts have adequate clearance.
4. Examine the area in and near the machine to see that there are no loose parts or tools that could be drawn into the machine and cause damage.
5. Check mechanical clearances between moving and stationary parts.
6. As soon as rotation starts, observe the oil rings to see that they are turning and feeding oil to the bearings.

The field excitation of a synchronous motor is not applied until the machine is well up to speed, and with full magnetic control this is done automatically. A synchronous generator is excited before connecting to the power system, so that the voltage and frequency will be correct for synchronizing. The generator must be connected to the system only when the voltage and frequency match the system voltage and frequency and at a point when the voltage is in phase with that of the system. This point is indicated either by a synchroscope, the pointer of which also indicates "slow" or "fast," or by lamps.

After synchronization the excitation and mechanical input are adjusted to make the generator carry its part of the electrical load. The excitation of a motor should be adjusted to rated nameplate value for full load, unless a change in operating conditions is desired, as discussed in Article 20 on reactive kva.

DIVISION OF LOAD. The division of load between synchronous generators operating in parallel is dependent upon the speed characteristics of the prime movers or upon the governors. For stable operating conditions, the prime movers must have a drooping speed characteristic or a drop in speed from no load to full load in the order of 3 per cent. If the prime movers driving the generators on a given system have dissimilar speed characteristics, the division of load between the generators as the system load changes is dissimilar. As the system load increases, the generators driven by prime movers having flatter speed characteristics take on the load faster than those driven by machines with more drooping characteristics. Also, as the system load decreases, the machines with drooping characteristics are the slowest in dropping their load. For equal division of load the prime movers must have equal speed-load characteristics, where outputs are plotted in percentage of full load. Since synchronous generators operating in parallel on a system must all operate at the same output frequency, changes in speed cannot be allowed. The change in load is therefore taken care of by changing the driving power or torque through adjustment of the governors, so that more or less steam, water, or fuel, as the case may be, is admitted at the same speed. At the same time the generator excitation is changed to suit the new conditions.

This change in excitation has no effect on the division of load as such but does change the division of reactive current between the machines; i.e., as the load conditions change, a circulatory current tends to flow to equalize the terminal voltages. This current transfers the excitation from the machine which would be overexcited to the one which would be underexcited. Correct adjustment of the excitations to suit the loads carried thus minimizes this flow of wattless current between the generators. The common terminal voltage is therefore dependent on the total excitation of the machines and on the amount of load and its power factor.

ROUTINE MAINTENANCE. In order to obtain long and satisfactory service from rotating machines it is important they they be given periodic routine checks, which should include the following items:

Bearings. Proper attention must be paid to lubrication, and an occasional inspection made to see that no pitting, wiping, or undue wear is taking place because of undesirable conditions of operation.

Air Gap. Air gap should be checked occasionally to see that it is even around the machine. Bearing wear shows up by a decrease in gap on the lower side.

Temperature Rise. An occasional observation of temperature rise, not necessarily in a quantitative sense, will show whether the ventilation is functioning satisfactorily and whether vent ducts and coil vents are becoming obstructed with dirt or foreign material.

Insulation Resistance. Measurement of insulation resistance is important in determining the condition of the winding and should always be made when drying out a winding after exposure to moisture or long periods of idleness. At operating temperature the insulation resistance of the stator winding in megohms should be equal to not less than

$$R = \frac{\text{Terminal voltage}}{(\text{Rating in kva}/100) + 1000}$$

The field winding should show an insulation resistance of $1/2$ to 1 megohm.

Cleaning. Periodic cleaning of end windings, ventilation ducts, and the machine in general promotes trouble-free service by preservation of insulation, adequate creepage distances, and ventilation performance. This item is especially important for open machines installed in dirty locations. The cleaning should include flushing out the oil reservoirs and renewing the oil supply.

Vibration. An occasional observation to detect any undue vibration of frame or brackets will give indication of possible misalignment, inadequate foundation, or electrical troubles within the machine.

22. SPECIFICATIONS

TECHNICAL ELEMENTS OF A SPECIFICATION. The best specification for a machine is brief but complete. A specification should not be so exacting as to limit the manufacturer to one style of construction; it should, however, include a list of accessories to be supplied as required and all special electrical and mechanical limitations. Where the manufacturer's standard is adequate, nothing need be specified. A statement of the manufacturer's standard may be requested if desired. Table 11 gives a list of data usually required by the manufacturer before quoting a machine.

Table 11. List of Data Required in Inquiry for Motor or Generator

	Horizontal Generator Other Than W.W.*	Horizontal W.W.* Generator	Vertical W.W.* Generator	Condensers	Horizontal Synchronous Motors	Vertical Synchronous Motors
Kva or hp	+	+	+	+	+	+
Voltage	+	+	+	+	+	+
Phases	+	+	+	+	+	+
Rpm	+	+	+	+	+	+
Frequency	+	+	+	+	+	+
Power factor	+	+	+	....	+	+
Type of cooling or enclosure	o	o	x	x	x	x
Inrush	o	o	o	x	x	x
Number of bearings	+	+	o	....	+	o
Extra load to be taken by electrical machine bearings	o	x	+	....	x	x
Type of drive	x	x	....	....	x	x
WR^2 of driven member and type of load	....	....	....	....	x	x
Pull-in torque	....	....	....	....	x	x
Starting torque	....	....	....	....	x	x
Pull-out torque	o	o	o	o	x	x
Efficiency if special	o	o	o	o	o	o
Type and voltage of excitation, i.e., direct drive, belt drive, bus, etc.	+	+	o	o	+	x
Starting of unit—whether at full voltage or by a specific alternate method	....	....	....	x	+	+
Overspeed	+	+	+	....	+	+
WR^2 of machine	o	x	x	....	o	o
Pit diameter	....	....	x	....	....	x
Short-circuit ratio	o	o	x	o	o	o
Transient reactance	o	o	o	o	o	o
Underexcited capacity	o	o	o	x	o	o

+ Definitely required.
x Usually necessary.
o Unnecessary unless manufacturing standard is not satisfactory because of special conditions.

* W.W. signifies waterwheel.

23. BIBLIOGRAPHY

1. AIEE Test Code for Synchronous Machines (June, 1945).
2. American Standards Association, Std. C50, Rotating Electrical Machinery.
3. American Standards Association, Std. C6, Rotation, Connections, and Terminal Markings for Electric Power Apparatus.
4. Wagner, C. F., Machine Characteristics; and Evans, R. D., Coordination of Power and Communication Systems, in *Electric Transmission and Distribution Reference Book*, Westinghouse Electric Corporation, East Pittsburgh, Pa. (1943).
5. Goss, H. R., and H. V. Putnam, Calculation of Flywheels for Air Compressors, Trans. ASME, Vol. 51, pp. APM-51-12-117 (1929).
6. Kilbourne, C. E., and I. A. Terry, Dynamic Braking of Synchronous Machines, *Trans. AIEE*, Vol. 51, p. 1007 (1932).
7. Lory, M. R., High-torque Synchronous Motors, *Elec. J.*, (May, 1937).
8. Poland, H. O., Power-factor Correction with Synchronous Motors, *Power Magazine* (Oct. 1940).
9. Taylor, T. S., New Armature Winding Diagrams, *Power Plant Engineering Magazine* (April, June, July, Aug., and Nov., 1943, and June, 1945).

POLYPHASE INDUCTION MOTORS

By T. C. Lloyd

24. GENERAL CHARACTERISTICS

BRIEF STATEMENT OF OPERATION. A polyphase induction motor depends for its action upon the rotating magnetic field built up by the stator or primary winding. This field crosses the air gap and is set up through the magnetic circuit of the rotor.

With the field windings separated in **space** around the periphery of the air gap, and connected suitably to a polyphase source of voltages which are displaced in **time** (by definition of polyphase), the resultant of these magnetic fields is a rotating mmf which gives the motor its action.

The rotor of the polyphase motor has a squirrel cage, usually of copper bars, all connected at the ends by end rings. As the flux cuts by these bars, voltages are induced in them, which cause circulating currents. Reaction of the bar currents and the rotating field produce the motor torque.

The speed of rotation of the magnetic field is fixed by the frequency, f, and the number of poles, P:

$$\text{Rpm} = \frac{120f}{P} \tag{1}$$

This is called the synchronous speed. If the rotor were turning at this speed, its conductors would not be cut by the air-gap flux, and hence no rotor voltage and current would flow. The developed torque would be zero. The rotor must turn at slightly less than synchronous speed, so that there is relative motion between magnetic field and rotor bars. This results in cutting of flux and induced rotor currents. An increased mechanical load on the rotor shaft would bring about further reduction in speed and a greater relative speed between synchronously rotating field and the rotor conductors. A higher voltage is thereby induced in these conductors, causing increased rotor current and torque. The reduction in speed continues until the developed torque is equal to the torque demand of the load.

The difference between synchronous speed and actual speed of the rotor is known as the **slip.** It may be expressed in rpm or as a **decimal fraction** of the synchronous speed.

$$\text{Slip } (s) = \frac{(\text{Syn.})\text{ rpm} - \text{rpm}}{(\text{Syn.})\text{ rpm}} \tag{2}$$

WINDING ARRANGEMENTS. In Fig. 1 is shown a schematic arrangement of a partially wound stator or primary winding embedded in slots of the laminated field core. Twenty-four slots are shown, giving 6 slots per pole for this 4-pole winding. A single coil spans 6 teeth, or covers a complete pole pitch. This is a **full pitch** winding. The coil in slots 2 and 8 is connected in series with that in 1 and 7, and the two form a **phase group.** This group is connected in series with a similar group under each of the other poles. The entire winding is then a **phase winding** or **leg.**

In slots 3 and 4 are the initial coil sides of another group of full pitch coils, their opposite coil sides being in slots 9 and 10. This phase group connects in series with all other similarly located coils, under the other poles, forming a second phase winding.

A similar set of coils lies in slots 5 and 6, extending to slots 11 and 12. The three phase windings can be connected in either delta or wye for external connections to the three-phase supply.

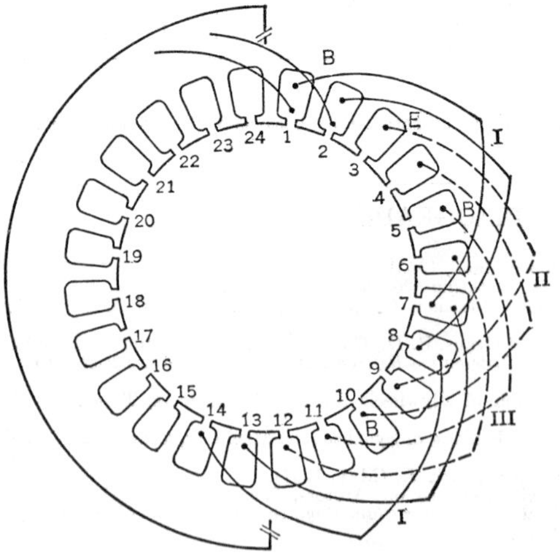

Fig. 1. Schematic Arrangement of a Partially Wound Stator

A point-by-point summation of the instantaneous values of currents in these sets of coils reveals a resultant mmf in the gap which is constant in magnitude but moving forward at synchronous speed.

In a two-phase motor each phase group under a pole covers 90 elec deg, as compared with the 60-deg belts considered above.

EFFECT OF PITCH AND DISTRIBUTION; HARMONIC FIELDS. When all the windings connected to one phase under any one pole lie in one slot, this becomes a concentrated winding. The mmf of such a coil acting alone may vary sinusoidally in time, but its plot will be a flat-topped "block" in space. For NI ampere-turns the wave can be expressed as:

$$y = \frac{4NI}{\pi} (\sin \chi + {}^1/_3 \sin 3\chi + {}^1/_5 \sin 5\chi \cdots) \tag{3}$$

If the winding is distributed into several slots per pole, with θ electrical degrees between them, the fundamental waves as denoted by $\sin \chi$ will add at the angle θ. The harmonic components will add vectorially at the angles $n\theta$, with n being the order of the harmonic. Distributing the windings results in more nearly sinusoidal space waves in the gap. All third, and multiples of third, harmonics are zero in Y-connected three-phase motors, as the mmf's of the three phases are 120 deg apart.

Table 1. Harmonic Poles, Three-phase Motors

Harmonic Field	Rotation	Relative Speed
3 (none)	....	....
5	Backward	$^1/_5$
7	Forward	$^1/_7$
9 (none)	....	....
11	Backward	$^1/_{11}$
13	Forward	$^1/_{13}$
15 (none)	....	....
17	Backward	$^1/_{17}$
.		
.		
.		

Space harmonics in the air gap have the effect of producing multiple poles, with mmf's rotating with or against the direction of the fundamental. See Table 1.

The fundamental mmf produces a speed-torque curve such as is shown in Fig. 2. The effect of the other forward revolving fields is to produce, in miniature, similar curves superposed, which may cause troublesome dips or "cusps" in the available accelerating torque of the motor.

Distributing a winding or shortening its pitch produces less effective mmf per ampere-turn (see Tables 7 and 8), but has the beneficial effect of reducing harmonics which might cause motor noise, unusual heating, or unfavorable accelerating torque.

CONSTRUCTION. The essential magnetic circuits of an induction motor consist of a field of laminated silicon steel, with slots spaced uniformly around the inner periphery, and a rotor to be described later. On motors below 50- to 75-hp ratings these slots are usually semiclosed, and form-wound coils are inserted through the comparatively narrow openings, resulting in a "random lay." Paper or combined paper and cloth form a "slot-cell" of insulation between the coil and the lamination edges. In larger motors the slots are wide open with parallel sides. Preformed coils of rectangular wire, layer insulated, varnished dipped, and baked, are then inserted in the slots to form the stator winding.

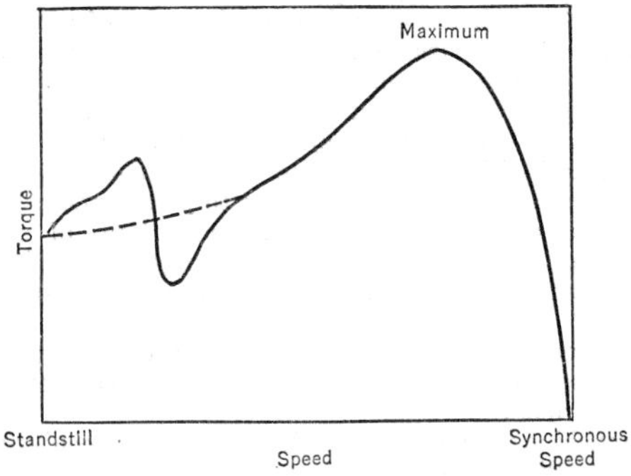

FIG. 2. Speed-torque Curve

Fields for smaller motors are dipped in varnish after all coils are inserted and connections made between coils and phase groups. The complete assembly is then baked.

For Class B motors (see Article 26) glass insulation is used on the wires, and glass cloth or sheet mica and glass are used as the slot cells. End connections are held in place with glass tape, and leads are of asbestos- or glass-insulated wires.

For large Class B motors glass fabric and mica sheets are used between the layers of rectangular conductors in building up the coils and in insulating the complete coils.

The laminated field structure, before or after winding, is pressed into a body of cast iron or rolled steel. In some cases the rolled steel shell is wrapped about the field in a complete welded assembly before the windings are inserted.

End-bells, or the motor heads, serve to hold the bearings and provide ventilating openings. Ventilation, except in very large, artificially ventilated machines, is provided by a fan or fans on the rotor shaft which may: (*a*) draw air in one end of the motor, circulating it over the end coils of the field and between the laminating field core and body and hence out the openings in the opposite head; or (*b*) draw in air from openings in both end-bells, circulating it over the end coils, between the body and field core, and hence out through openings provided in approximately the center of the body.

THE ROTOR. The magnetic requirements of the rotor are met by building it of laminated silicon steel, in which closed or semiclosed slots are punched around the outer periphery. These slots are filled with copper bars (most commonly), brazed or welded at the ends to comparatively heavy sections of end rings. The rotor is given a twist or *skew*, which reduces magnetic disturbance. Space harmonics in the air-gap flux give the effect of many poles around the gap. The number is a function of the order of the harmonic. If the skewed bar spans such a harmonic pole pair, no corresponding harmonic can appear in the rotor current, and the harmful effect of that harmonic on the torque is eliminated. The correct skew may result in quieter operation.

The completed rotor assembly is keyed to a rigid shaft. Bars are placed in these slots and brazed or welded to rings on each end of the rotor. Rotor bars and end rings are usually copper, but other conducting materials are frequently used to meet certain motor-performance requirements.

Most induction motors are built as described above, with some exceptions arising through the use of a squirrel cage consisting of integrally cast bars and end rings of copper or, more commonly, aluminum. The conductivity of the aluminum so employed is from 45 to 50 per cent that of copper.

For high-slip motors or some designs requiring frequent starting, brass or bronze bars, end rings, or both are employed to give comparatively higher resistance or equivalent resistance with greater mass. This latter is desirable from the standpoint of better heat dissipation.

Wound rotors are used where rheostatic speed control is desired. In such cases the rotor slots are shaped similarly to those used for the field windings. Wound or preformed coils (depending upon size) are inserted in the slots, and the winding leads connected to slip rings on the shaft for external connection through brushes.

STANDARDIZED FRAME SIZES. The National Electrical Manufacturers Association has standardized important induction-motor frame dimensions on a series of frames having mounting heights of 5 to 12.5 in. These correspond to horsepower ratings of 1 to 125 at 1725 rpm. Mounting height (center of shaft above base), shaft diameter and length, and relation of shaft extension to base mounting holes are the important dimensions which are standardized. They provide for interchangeability of motors built by various manufacturers.

Single-phase and d-c motors are also standardized in NEMA frames but not over as great a range of horsepower.

ENCLOSURE AND DEGREE OF PROTECTION. The greatest industrial demand is for open, ventilated motors. The present trend, however, is toward a type of motor protection with openings only in the lower half of the frame and end bells and so shielded as to protect the motor interior from falling objects and mechanical injury in general.

Other motor types, as classified by the degree of protection and ventilation, are:

Protected Machine "(formerly called semi-enclosed machine). A protected machine is one in which all ventilating openings are limited in size and shape. Such openings shall not exceed 1/2 sq in. (323 sq mm) in area and are of such shape as not to permit the passage of a rod larger than 1/2 in. (12.7 mm) in diameter, except where the distance of exposed live parts from the guard is more than 4 in. (101.7 mm) the openings may be 3/4 sq in. (484 sq mm) in area and must be of such shape as not to permit the passage of a rod larger than 3/4 in. (19 mm) in diameter." *

Semi-protected Machine. "A semi-protected machine is one in which part of the ventilating openings in the frame, usually in the top half, are protected, as in the case of a 'protected machine,' but the others are left open." *

Drip-proof Machine. "A drip-proof machine is one in which the ventilating openings are so constructed that drops of liquid or solid particles falling on the machine at any angle not greater than 15 deg from the vertical cannot enter the machine, either directly or by striking and running along a horizontal or inwardly inclined surface." *

Splash-proof Machine. "A splash-proof machine is one in which the ventilating openings are so constructed that drops of liquid or solid particles falling on the machine or coming toward it in a straight line at any angle not greater than 100 deg from the vertical cannot enter the machine, either directly or by striking and running along a surface."

Explosion-proof Machine. "An explosion-proof machine is one in an enclosing case which is designed and constructed to withstand an explosion of a specified gas or vapor which may occur within it, and to prevent the ignition of the specified gas or vapor surrounding the motor by sparks, flashes, or explosions of the specified gas or vapor, which may occur within the machine casing." *

Submersible Machine. "A submersible machine is one so constructed that it will operate successfully when submerged in water under specified conditions of pressure and time." *

* Reprinted from NEMA Publication No. 45-102.

25. THEORY AND ANALYSES OF OPERATION

THE VECTOR DIAGRAM. The vector diagram of the polyphase induction motor, laid out per phase, is very similar to that for the transformer. Diagrams are shown in Figs. 3 and 4.

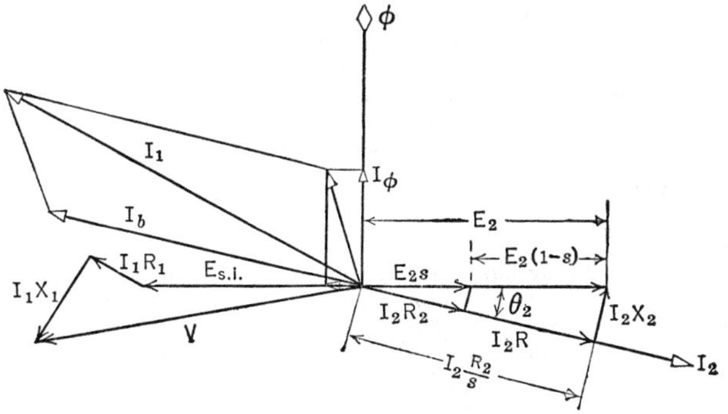

Fig. 3. Vector Diagram of a Polyphase Induction Motor

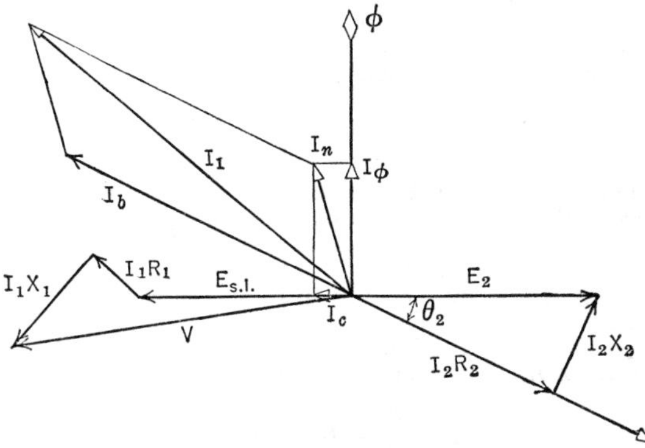

Fig. 4. Vector Diagram of Polyphase Induction Motor with Rotor Locked

Rotor Blocked. With the rotor at standstill and the field excited, the rotating air-gap flux cuts the rotor conductors, inducing in them emf's practically opposite in time phase to the emf's of the adjacent stator conductors. The rotor current is limited by the rotor impedance.

$$I_2 = \frac{E_2}{R_2 + jX_2} \tag{4}$$

where E_2 is the voltage induced in the rotor.

The currents about the rotor periphery build up as many poles as are present in the stator. The frequency of the rotor emf will be the same as that of the stator.

The applied voltage is used up in overcoming the stator winding impedance drop and the counter emf ($E_{s.i.}$) caused by the mutual flux linking of the stator turns. Fundamentally the mutual flux must satisfy the relationship

$$E_{s.i.} = 4.44\, fN\phi_m k_p k_d\, 10^{-8} \tag{5}$$

and as vectors (Fig. 4)

$$V \text{ per phase} = E_{s.i.} + IR_1 + IX_1 \tag{6}$$

where N = the number of series turns per phase.
k_pk_d = the product of pitch and distribution factors.
R_1 = the stator winding resistance per phase.
X_1 = the stator winding leakage reactance per phase.

To maintain this flux per pole requires a magnetizing current I_ϕ. This, together with the core-loss component I_c, makes up the total exciting current I_n.

The ampere-turns of the rotor tend to demagnetize the stator, but to maintain the mutual flux the stator draws more current I_b from the line, so that (see Fig. 3)

$$I_b N' = I_2 N_2' \tag{7}$$

N' and N_2' are **effective** turns of the stator and rotor, respectively. The load component of the stator current, combined with the exciting component, as vectors, forms the total primary current I_1.

Rotor Free. When the rotor is free to turn, the relative rpm between the rotating flux and the rotor is:

$$\text{Syn. rpm} - \text{Rotor rpm}$$

The frequency of the rotor currents [see Eqs. (1) and (2)] becomes

$$f_2 = fs \tag{8}$$

The speed of any rotating flux resulting from a properly connected polyphase winding is $120f/P$. Then the speed of the rotor mmf relative to the rotor is $120fs/P$. The rotor speed with respect to the stator is $120f(1 - s)/P$. [See Eq. (1).]

To obtain the speed of the rotor mmf with respect to the stator

$$\frac{120fs}{P} + \frac{120f(1 - s)}{P} = \frac{120f}{P} \tag{9}$$

Hence the rotor frequency, when referred to the stator, is able to react through its mmf at the same frequency as that of the stator.

The voltage induced in the rotor at standstill is E_2; when running at any slip, this becomes $E_2 s$. The rotor current, when running, is:

$$I_2 = \frac{E_2 s}{R_2 + jsX_2} = \frac{E_2}{R_2/s + jX_2} \tag{10}$$

This leads to the important concept that, at any speed, the induction motor has an apparent rotor resistance of R_2/s with other terms unaffected.

With the rotor voltage varying from E_2, at standstill, to $E_2 s$, when running, the concept of a counter-emf $E_2(1 - s)$, bringing about the reduction, is useful on the vector diagram. For each load the slip determines the effective E_2 and hence the rotor current and the balancing component of stator current.

The relationships are shown in Fig. 4.

POWER DEVELOPED. The input to the rotor is:

$$I_2^2 \frac{R_2}{s} = E_2 I_2 \cos \theta_2 \tag{11}$$

Rotor input minus the rotor copper loss equals the rotor power converted into mechanical form; that is:

$$\begin{aligned} \text{Rotor input} - \text{Copper loss} &= I_2^2 \left(\frac{R_2}{s} - R_2\right) \\ &= I_2^2 R_2 \left(\frac{1 - s}{s}\right) \end{aligned} \tag{12}$$

The term $(1 - s)/s$ is a *numeric*, and hence the power converted to mechanical form can be represented as a resistance load in the rotor circuit.

It will be noted that the rotor copper loss bears the same relation to the total power transferred across the air gap as the slip bears to the synchronous speed. Then

$$\text{Power across gap} \times \text{Slip (as decimal fraction)} = I_2^2 R_2 \text{ loss} \tag{13}$$

This fact is useful in determining the copper loss in squirrel-cage rotors.

All the above values are per phase and in stator terms.

EQUIVALENT CIRCUITS. Various equivalent networks can be laid out for the polyphase induction motor, differing in simplifying assumptions and accuracy. Two common forms are combined in Fig. 5, using the constants for one phase of the winding and all rotor values in stator terms. The theoretically exact circuit adds the magnetizing branch between stator and rotor constants, as shown in solid lines. The resistance component of this branch is so chosen as to yield the core loss per phase. The fact that the mechanical load on one phase can be replaced by an equivalent resistance, $R_2[(1 - s)/s]$, accounts for this component of the circuit.

For greatest accuracy, solution of this circuit forms the best means of predicting performance for various loads. Although the approximate equivalent circuit and the circle diagram which will be shown offer various advantages in ease of solution, accuracy suffers in varying degrees. Small motors or motors with comparatively large no-load currents are usually more affected by the approximate analyses. This is especially true of motors with many poles.

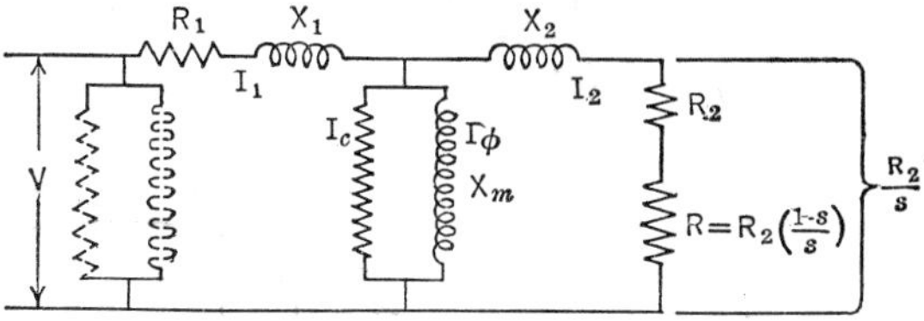

FIG. 5. Equivalent Network of Polyphase Induction Motor

The approximate equivalent network assumes that the magnetizing branch is connected at the phase terminals, as shown dotted in Fig. 5, instead of between stator and rotor constants. Stator impedance drop, caused by this component, is hence neglected, and power input to the rotor is exaggerated. Greater simplicity of solution is the only justification for this viewpoint.

THE CIRCLE DIAGRAM. The approximate equivalent circuit, with various constants lumped, and neglecting saturation, can be reduced to that shown in Fig. 6(*a*). In this circuit:

$$I' = \frac{V}{Z'} = \frac{V}{X'} \sin \theta'$$

This is the polar equation of a circle having a diameter of V/X', and the trace of I' for various values of R' is a circle. Such a condition is shown in the vector diagram of Fig. 6(*b*).

To make use of these facts in predicting induction-motor performance it is necessary to know: (*a*) the usual machine constants from a design sheet, or (*b*) the no-load current and its phase position, the rotor blocked current and its phase position, and the resistance of the stator winding. Per phase values are used.

Table 2. Calculation Procedure, Exact * Equivalent Circuit

1. Assume a slip, s.
2. $Z_2 = R_2/s + jX_2$
3. $G_2 = R_2/s \div Z_2^2$ $\quad B_2 = X_2 \div Z_2^2$
4. $Y_2 = G_2 - jB_2$
5. $B_m = 1 \div X_m$
6. Add (4) $- jB_m = G_2 - jB = Y$.
7. $R = G_2 \div Y^2$ $\quad X = B \div Y^2$
8. $Z = R + jX$
9. Add (8) and $R_1 + jX_1 = Z_2$.
10. $Z_e = R_e + jX_e = \sqrt{R_e^2 + X_e^2}$
11. $I_1 = V$ per phase $\div Z_e$
12. $I_2 = V$ per phase $\times Z \div Z_2$
13. Power factor $= R_e \div Z_e$
14. Gross output $= I_2^2(R_2/s)(1 - s) \times$ Number of phases
15. Net output $=$ (14) $-$ Core loss $-$ Friction Subtract stray load loss from above, if known.
16. Input $= I_1^2 R_e \times$ Number of phases
17. Efficiency $=$ (15) $\div$ (16)

Symbols without subscript refer to the combined values included in, and to the right of, the magnetizing branch.

Those with subscript *e* refer to total for the entire circuit.

* Agrees with the exact circuit as shown, except that the core-loss component of the exciting current is omitted. $I_c = 0$.

Constructing the Diagram. The voltage per phase is laid out vertically to a suitable voltage scale, as in Fig. 6(*c*). The no-load current *MO* is laid out to a suitable current scale at the proper angle behind the applied voltage. The rotor blocked current *Ma* is drawn to the same scale, in its proper phase position.

The vertical distance of the point *O* above the base line in Fig. 6(*c*) represents core loss plus friction and windage. The vertical *Ca* represents total copper loss, of which *Cd* is that in the stator, under blocked conditions.

Od is drawn, and various input currents such as *MP* are assumed. The line perpendicular to the base passing through *S*, *R*, *Q*, and *P* is drawn.

Identifying Values. The rotor current is OP; power factor is θ_1; watts input per phase are *TP*; stator copper loss per phase is *SR*; rotor copper loss per phase is *RQ*; useful output watts per phase are *QP*.

The slip is the ratio *RQ*/*RP*, and the efficiency is *QP*/*TP*. The torque is proportional to *RP* in amperes, but when multiplied by volts per phase and number of phases is the useful torque in synchronous watts.

Maximum torque is proportional to the line $h''e''$, where e'' is located by the tangent parallel with *Od*. Starting torque is proportional to *da*.

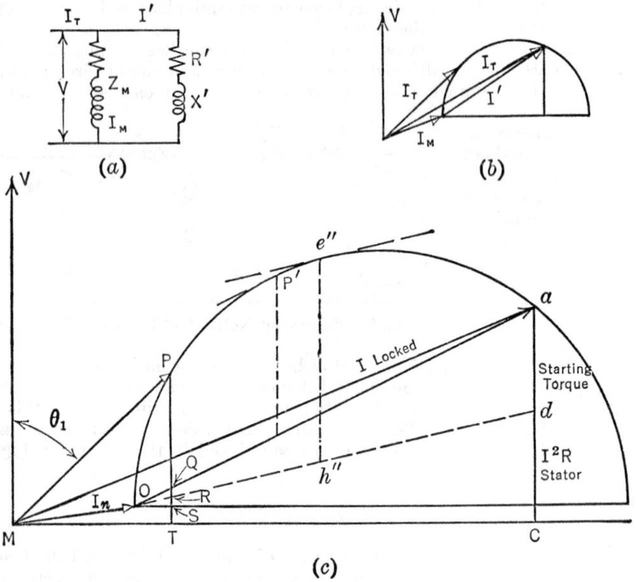

FIG. 6. Polar Diagram of Induction Motor

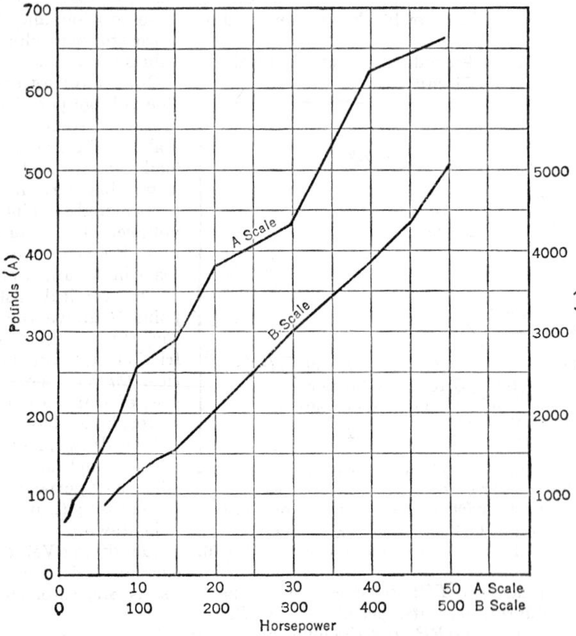

FIG. 7. Typical Induction Motor Weights

The maximum power output is obtained when the length QP is a maximum. This will occur when P' is on the point of tangency of the circle and line drawn parallel with Oa.

The maximum power factor is represented by that current MP which is drawn tangent to the circle.

This diagram is only one of many which have been used. It suffers all the inaccuracies of the approximate equivalent circuit, less serious on large motors without excessive no-load current. When the diagram is drawn from test data, some difficulty is introduced by the skin effect in the rotor bars and rapid heating of the rotor. More important, in many cases, is the effect of saturation on the reactances. The combined influence of these factors is to introduce inaccuracies in the value and phase of the rotor-blocked current.

TYPICAL VALUES OF MOTOR CONSTANTS. Table 3 shows typical motor constants and items for small and medium ratings. Note that r_1 and r_2 have their usual significance, X is the total leakage reactance, and X_0 is the magnetizing and primary leakage reactance combined. Thus the no-load current can be calculated on these tables as $127 \div X_0$. All winding constants are per phase. Typical motor weights are shown in Fig. 7.

FORMULAS AND DISCUSSION OF FACTORS. Formulas indicating various performance items vary in their accuracy, depending upon the type of equivalent circuit from which they are derived or the other simplifying assumptions made. Those which follow are identified as to their accuracy.

In these formulas the following symbols are used:

V = the applied volts per phase.
s = the slip as a decimal fraction.
R_1 = the primary resistance per phase.
R_2 = the rotor or secondary resistance in primary terms.
X = the total leakage reactance per phase.
n = the number of phases.
K_r = a leakage factor.
R_e = the equivalent primary and secondary resistance.
Z_e = the equivalent primary and secondary impedance.

Developed torque at any slip (approximately):

$$T = \frac{V^2 s R_2 n}{(R_1 s + R_2)^2 + s^2 X^2} \text{ synchronous watts} \tag{14}$$

$$\text{Pound-feet} = 7.04 \frac{\text{Syn. watts}}{\text{Syn. rpm}} \tag{14a}$$

Slip at point of maximum torque (approximately):

$$s = \frac{R_2}{\sqrt{R_1^2 + X^2}} \tag{15}$$

Maximum torque (approximately):

$$T_m = \frac{V^2 n}{2[R_1 + \sqrt{R_1^2 + X^2}]} \text{ synchronous watts} \tag{16}$$

$$= \frac{3.52 n V^2}{\text{Syn. rpm } [R_1 + \sqrt{R_1^2 + X^2}]} \text{ pound-feet} \tag{17}$$

More accurately:

$$T_m = \frac{3.52 V^2 n}{\text{Syn. rpm } [R_1 + \sqrt{R_1^2 + X^2}]} K_r \text{ pound-feet} \tag{18}$$

$$K_r = \frac{X_0 - X}{X_0} \tag{19}$$

Maximum power (approximately):

$$P_m = \frac{n V^2}{2(R_e + Z_e)} \text{ watts} \tag{20}$$

This occurs at a slip (approximately):

$$s = \frac{R_2}{R_2 + Z_e} \tag{21}$$

Table 3. Typical Values

Hp	r_1	r_2	X	X_0	Core Loss	Friction and Windage	Eff.	PF	Gap Length, Inches
2 Poles, 25 Cycles, 220 Volts, 3 Phases									
1/4	12.9	7.97	8.77	193	22	15	70	82	0.014
1/3	8.5	5.4	7.50	162	29	15	73	82	0.015
1/2	7.8	3.8	5.78	143	34	15	75.6	84	0.015
3/4	5.2	2.6	3.85	95	58	15	79.5	86	0.020
1	4.0	2.1	3.70	91	50	40	78.0	89	0.022
1 1/2	2.8	1.3	3.2	82	59	40	80.0	89	0.022
2	1.9	1.1	2.2	59	76	50	81.0	89	0.025
3	1.02	0.84	1.76	44.2	84	50	83.8	90	0.025
5	0.58	0.35	0.97	20.5	150	100	83.9	85	0.027
7 1/2	0.37	0.24	0.78	19.9	178	150	86.1	93	0.027
10	0.25	0.18	0.64	18.2	195	200	87.0	93	0.030
15	0.13	0.15	0.51	16.5	259	200	88.5	94	0.030
2 Poles, 60 Cycles, 220 Volts, 3 Phases									
1/3	7.93	6.85	13	316	28	20	74.5	92	0.014
1/2	5.75	3.84	10	200	36	50	73.0	90	0.020
3/4	3.52	2.19	6.1	122	52	50	76.1	87	0.020
1	2.45	1.66	4.7	102	60	50	79.5	88	0.020
1 1/2	2.16	1.23	3.2	66	59	60	79.5	88	0.020
2	1.41	1.13	2.7	66	77	75	81.2	90	0.020
3	0.80	0.58	1.8	44	102	100	84.2	91	0.022
5	0.43	0.36	1.2	34	140	100	87.0	92	0.022
7 1/2	0.21	0.24	0.9	25	194	100	89.1	92	0.025
10	0.18	0.18	0.7	21	233	320	89.3	94	0.027
15	0.11	0.13	0.5	15	325	500	85.0	94	0.027
20	0.08	0.10	0.4	11	384	600	87.0	92	0.027
4 Poles, 60 Cycles, 220 Volts, 3 Phases									
1/4	12.5	7.8	14.2	265	27	10	72	84	0.015
1/3	8.4	5.4	10.5	212	35	10	75	86	0.015
1/2	6.2	4.0	8.2	154	41	20	75	84	0.015
3/4	4.3	2.4	6.2	90	49	30	76	81	0.015
1	3.0	2.2	5.6	84	46	50	78	86	0.020
1 1/2	1.9	1.3	3.6	60	58	50	81	85	0.020
2	1.2	1.1	3.4	52	84	50	82	85	0.022
3	1.0	0.6	2.2	31	120	50	83	86	0.022
5	0.6	0.3	1.4	21	144	80	85	87	0.025
7 1/2	0.3	0.3	0.9	18	182	120	85	89	0.027
10	0.25	0.2	0.74	18	230	250	85	91	0.027
15	0.16	0.14	0.50	12	315	250	87	91	0.027
20	0.09	0.10	0.37	9	440	250	89	92	0.030
25	0.07	0.09	0.35	9	445	350	89	93	0.030
30	0.05	0.07	0.26	7	563	400	89	91	0.030
6 Poles, 60 Cycles, 220 Volts, 3 Phases									
1/4	12.8	9.1	19.2	212.0	29	10	71	77	0.012
1/3	9.4	7.6	17.2	154.0	30	10	70	76	0.012
1/2	6.3	4.5	10.9	92.0	38	12	73	72	0.015
3/4	3.86	2.5	7.0	60.5	45	15	76	70	0.020
1	3.52	2.1	5.7	55.8	52	20	77	76	0.020
1 1/2	1.92	1.4	4.5	38.3	74	20	79	74	0.022
2	1.53	1.1	3.5	30.2	74	25	78	76	0.022
3	0.80	0.56	2.3	22.8	112	25	84	77	0.025
5	0.48	0.41	1.6	19.4	148	40	86	84	0.027
7 1/2	0.25	0.23	1.1	15.9	206	75	88	86	0.027
10	0.20	0.16	0.82	12.1	263	75	88	86	0.027
15	0.13	0.18	0.60	8.7	338	100	87	87	0.030
20	0.09	0.13	0.43	6.5	428	170	87	86	0.030

More accurately, the maximum power output is:

$$\frac{nV^2}{1500(R_e + Z_e)} K_r \text{ horsepower} \tag{22}$$

The slip for this output is slightly higher than given above. The slip at any torque (approximately) equals:

$$s = \frac{R_2}{y + R_2} \tag{23}$$

$$y = \frac{V^2 n \times 5200}{746 \times T \times \text{rpm}} - 2(R_1 + R_2) \tag{24}$$

In terms of ratios, the slip at any torque is approximately:

$$s = \frac{\frac{R_2}{X}}{2\frac{T_m}{T}\left(\frac{R_1}{X} + \sqrt{\left(\frac{R_1}{X}\right)^2 + 1}\right) - 2\frac{R_1}{X} - \frac{R_2}{X}} \tag{25}$$

This type of equation is significant. For example, a polyphase motor with windings chosen so that R_1/X and R_2/X are each 0.3, and a ratio of maximum torque to full-load torque of 250 per cent, will always operate with a full-load slip of approximately 5.12 per cent, as calculated from this equation.

The starting torque (approximately) using the rotor locked current:

$$T_s = nI_l^2R_2 = \frac{nV^2R_2}{Z_e^2} \text{ synchronous watts} \tag{26}$$

More accurately:

$$T_s = \frac{7.04nV^2R_2}{\text{Syn. rpm } [Z_e^2]}\left(\frac{X_0 - X_1}{X_0}\right) \text{ pound-feet} \tag{27}$$

The ratio of starting torque to maximum torque (approximately):

$$\frac{T_s}{T_m} = \frac{2R_2[R_1 + \sqrt{R_1^2 + X^2}]}{Z_e^2} \tag{28}$$

In terms of ratios:

$$T_s = aT_m \tag{29}$$

where

$$a = \frac{\frac{2R_2}{X}\left[\frac{R_1}{X} + \sqrt{\left(\frac{R_1}{X}\right)^2 + 1}\right]}{\left(\frac{R_1}{X} + \frac{R_2}{X}\right)^2 + 1} \tag{30}$$

For the motor considered above, with R_1/X and R_2/X each equal to 0.3, this ratio is 0.595. Hence such a motor must have a starting torque 59.5 per cent of the maximum or 149 per cent of the full-load value.

In considering starting torque and starting currents, the skin effect in the rotor bars increases the apparent resistance. On deep-bar rotors, built especially for this effect to obtain low-current motors, the value must be calculated by designers. On motors over about 50 hp the value may be great enough to require special calculation on the basis of bar dimensions. On motors below this size a rough factor of 1.3 for 60-cycle motors is applicable. Values obtained from test results will already contain such a multiplier. Values obtained from design data should be multiplied by 1.3 to obtain the approximate resistance under locked conditions.

The starting current (approximately):

$$I = \frac{V}{Z_e} \tag{31}$$

More accurately:

$$Z_e = \sqrt{X^2 + (R_1 + K_rR_2)^2} \tag{32}$$

The volt-amperes required at starting and the volt-amperes per horsepower are significant in determining the selection of the proper wiring and control capacity. Locked kva per horsepower are coded by letter and used on the nameplates of motors, as per NEMA and Underwriters' Standards.

$$\text{Starting volt-amperes} = \frac{V^2n}{\sqrt{(R_1 + R_2)^2 + X^2}} \tag{33}$$

When the maximum torque and the rotor resistance are fixed, so also is the starting volt-amperes. Thus:

$$\frac{V_a}{T_m} = \frac{\text{Syn. rpm}}{3.52} \times \frac{R_1 + \sqrt{R_1^2 + X^2}}{Z_e} \text{ (approx.)} \tag{34}$$

$$= \text{Syn. rpm} \times (b) \tag{35}$$

Values of b are plotted in Fig. 8, or

$$\text{Starting volt-amperes} = (b) \times T_m \times \text{Syn. rpm} \tag{36}$$

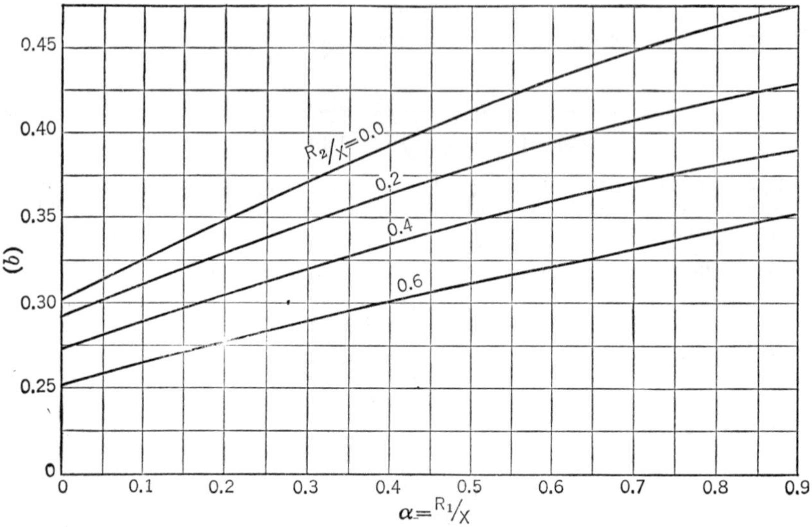

FIG. 8. Values of b Used in Eq. (35)

These formulas, most of which are derived from the approximate equivalent circuit, indicate the interdependence of motor characteristics. Performance of industrial motors represents a compromise among various trends.

26. MOTOR TYPES AND STANDARDS

Polyphase induction motors are classified into various groups according to a variety of factors and constructions. Some of the more common classifications are given below.

GENERAL-PURPOSE MOTOR STANDARDS. "A general-purpose motor is any motor of 200 hp or less and 450 rpm or more, having a continuous rating, and designed . . . in standard ratings for use without restriction to a particular application." *

Although motors built for special purposes may display certain unusual operating characteristics, general-purpose motors are expected to operate with the following characteristics to meet the standards of the National Electrical Manufacturers Association.

Maximum torque should be at least 200 per cent of the full-load torque for Class A and B motors. For Class E and F, this ratio should be not less than 150 per cent. (See below for definition of classes.)

Locked rotor torque (60 or 25 cycles) should be not less than:

150 per cent of full-load torque for	2 and 4 poles
135 " " " " " " "	6 poles
125 " " " " " " "	8 "
120 " " " " " " "	10 "
115 " " " " " " "	12 "
110 " " " " " " "	14 "
105 " " " " " " "	16 "

* From NEMA MG50-110.

Locked rotor currents should not exceed values given in Table 4.

Table 4. Locked Rotor Currents *

Horse-power	Cur-rent	Horse-power	Cur-rent	Horse-power	Cur-rent
1	27	15	220	75	1085
1 1/2	37	20	290	100	1450
2	47	25	365	125	1815
3	60	30	435	150	2170
5	90	40	580	200	2900
7 1/2	120	50	725	....	....
10	150	60	870	....	....

* For 220 volts, 3 phase. For other voltages use inverse proportion. These currents are for motor Classes B, C, and D.

MOTOR CLASSES. Squirrel-cage induction motors are classified by NEMA as:

Class A: Normal starting torque, normal starting current.

Class B: Normal starting torque, low starting current.

Class C: High starting torque, low starting current.

Class D: High slip.

Class E: Low starting torque, normal starting current.

Class F: Low starting torque, low starting current.

See Fig. 9 for typical speed-torque curves.

The usual squirrel-cage industrial motor is the Class A type. Its higher starting current may require a reduced voltage or resistance starter for motors above 7 1/2 hp, although this depends largely upon local power-source limitations.

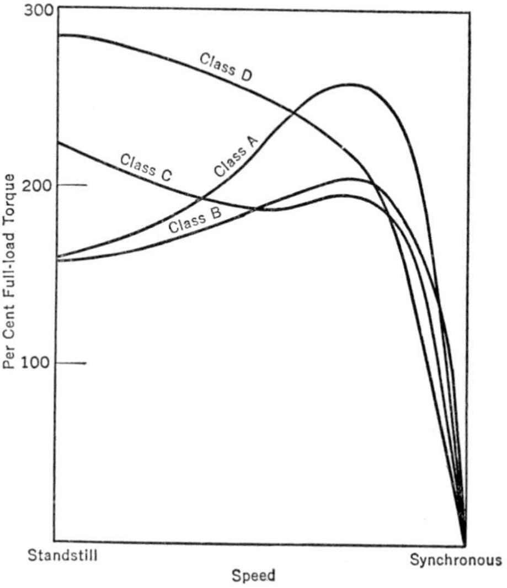

Fig. 9. Typical Speed-torque Curves of Polyphase Induction Motors

Deep-bar Rotors. To obtain the desired characteristics of Class B motors, it is customary to make use of comparatively deep, narrow rotor bars, giving a somewhat higher reactance but chiefly an increased apparent resistance at starting. With primary frequency on the rotor bars at the instant of starting, the skin effect of such bars increases the a-c resistance, which reduces at normal rotor frequencies. Thus high starting torque or lower starting current can be achieved without excessive full-load slip.

In some cases, starting torques high enough to satisfy the requirements for Class C service can be obtained by deep-bar rotors, but in general such motors use double-cage rotors.

Double-cage Rotors. With one cage of comparatively low resistance and high reactance buried beneath an outer cage of high resistance, high starting torque with low current can be obtained. Primary frequency induced in the rotor at starting gives a skin effect and high enough reactance to the buried cage so that the outer cage carries most of the starting current. With reduction of slip (and frequency) both cages become operative. The division of this current is shown in Fig. 10.

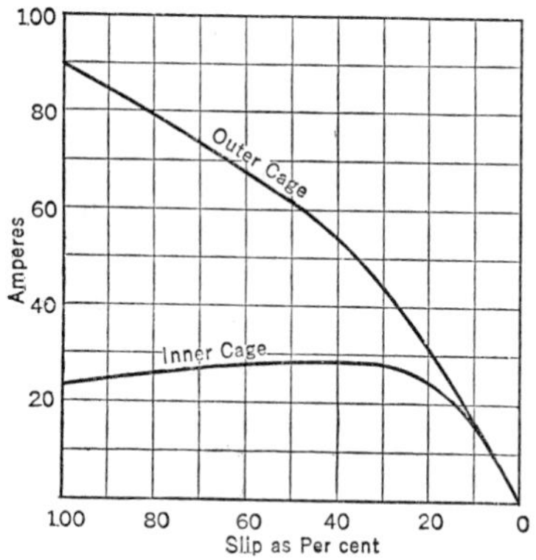

Fig. 10. Division of Current in Double-cage Rotors

Analyses and design methods for such motors are given in Hobart's translation of Punga and Raydt.

VENTILATION METHODS. (For classification of degrees of protection or enclosure, see Article 24.) An open machine is defined by NEMA as "a self-ventilated machine having no restriction to ventilation other than that necessitated by mechanical construction."

Self-ventilated Machine. "A self-ventilated machine is one which has its ventilating air circulated by means integral with the machine." *

Totally Enclosed Machine. "A totally enclosed machine is one so enclosed as to prevent exchange of air between the inside and the outside of the case but not sufficiently enclosed to be termed airtight." *

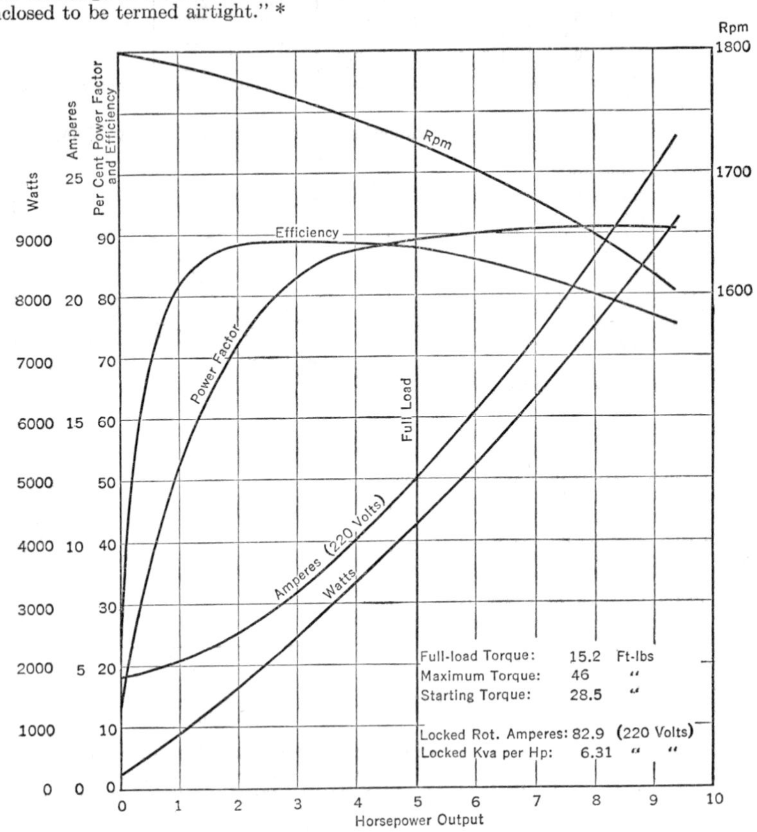

FIG. 11. Characteristic Curves of a 5-hp 1725-rpm Motor

Totally Enclosed Fan-cooled Machine. "A totally enclosed fan-cooled machine is a totally enclosed machine equipped for exterior cooling by means of a fan or fans, integral with the machine but external to the enclosing parts." *

Separately Ventilated Machine. "A separately ventilated machine is one which has its ventilating air supplied by an independent fan or blower external to the machine." *

TEMPERATURE RISE AND INSULATION CLASSES. The temperature rise of the windings of a motor is determined by the watts loss, the volume of cooling air, the frame area exposed for heat dissipation, and other minor factors of construction.

The air required for cooling in cubic feet per minute is

$$Q = \frac{K \times \text{Watts loss}}{\text{Temperature rise } (^\circ\text{ C})} \tag{36a}$$

where K = 1.8 for air at 25 deg cent.
= 1.95 for air at 50 deg cent.

* NEMA definitions.

Combined radiation and convection make up the heat-dissipation effects from the frame. Color, surface condition, and air density affect these factors. The effect of air density is recognized in recommending allowable temperature rises for machines intended to operate at high altitudes (see American Standard C-50). With an ambient of 25 deg cent, and at sea-level conditions, approximately 0.14 watt can be dissipated per square inch of surface having a temperature rise of 20 deg cent. This goes to 0.229 watt at 30 deg cent and approximately 0.5 watt at 55 deg cent.

As an example of heat elimination, consider a 5-hp 1725-rpm motor with total losses of 750 watts (Fig. 11). The frame area totals 490 sq in. and operates at an average rise of 25 deg. The solid portion of the end-bells totals 190 sq in. and averages 20 deg rise. Hence 115 watts are dissipated from the body, leaving 635 to be carried off by the cooling air. From eq. (36*a*) it follows that 68 cu ft of air per minute are required to cool the motor if the temperature of the incoming air increases 17 deg cent.

In most motors of small or medium size, the incoming air picks up only from 10 to 20 deg of temperature in passing through the motor. Smooth paths and high velocity are not conducive to effective pick-up.

From an operating standpoint it can readily be seen that any interference with ventilation by dust-clogged ducts or external devices which choke off free flow can vitally affect temperature rise and hence insulation life.

The majority of motors are insulated with **Class A** materials, defined as cotton, silk, paper, and similar organic materials impregnated (or immersed in a liquid dielectric), or molded and laminated materials with cellulose filler; or films and sheets of cellulose acetate; or varnishes as applied to conductors. (See Table 5.)

Class B materials consist of mica, asbestos, fiber glass, and similar inorganic materials.

Developments in the silicone field are opening the way to a new standard of insulation, which will doubtless permit operation at still higher temperatures than are feasible for the above methods.

Experience indicates the harmful effects of heat on the life of insulating materials. The life of organic compounds is cut in half for each increase of 8 to 12 deg cent in operating temperature. As a protection against reduced life through excessive temperature rise,

Table 5. Limiting Observable Temperature Rises *

Item	Machine Part	Method of Temperature Determination to be Employed	General-purpose Motors		Totally Enclosed, Totally Enclosed-Fan-cooled, Explosionproof, Waterproof Dust-tight, and Submersible Motors		Motors Other Than Those in Columns 1, 2, 3, and 4	
			Class A Insul.	Class B Insul.	Class A Insul.	Class B Insul.	Class A Insul.	Class B Insul.
			Col. 1	Col. 2	Col. 3	Col. 4	Col. 5	Col. 6
1	Windings	Therm.† Resis.‡	40 50	....	55 60	75 80	50 60	70 80
2	Cores and mechanical parts in contact with or adjacent to insulation	Therm.	40	....	55	75	50	70
3	Collector rings and commutators §	Therm.	55	75	65	75	65	85
4	Squirrel-cage windings and miscellaneous parts, such as brush holders, brushes, pole tips, may attain such temperatures as will not injure the machine in any respect.							

* The temperature rises in degrees centigrade in this table apply to the several classes of machines, as given at the heads of the respective columns.

† The method of temperature determination to be used shall be optional with the manufacturer unless otherwise agreed upon. Only one method of temperature determination shall be required in any particular case.

‡ Measurement of the core temperature shall not be required when the resistance method is used for the windings.

§ The class insulation refers to insulation affected by the heat from the commutator or collector rings, which insulation is employed in the construction of the commutator or collector rings or is adjacent thereto.

all industrial standards limit the maximum permissible operating temperatures for various equipment types and classes of insulation. Observable temperature rises based on 40 deg cent ambient are shown in Table 5.

SHELL-TYPE MOTORS. The machine-tool industry frequently uses shell-type motors, which are really polyphase, squirrel-cage motor parts consisting of a wound field and a rotor without shaft. The field may or may not be contained in a thin steel shell. The manufacturer of the machine on which such parts are used, supplies a shaft, bearings (for direct drive), ventilating equipment, and housing.

Shell-type motors are standardized by NEMA into three diameters:

1 1/2–10 hp: 8 in.
7 1/2–20 hp: 10 in.
15 –25 hp: 12.375 in.

Maximum lengths over end coils are also standardized.

The advantages to the user lie in the reduced size per horsepower, better integrated design of motor and driven device, and better control of balance and vibration.

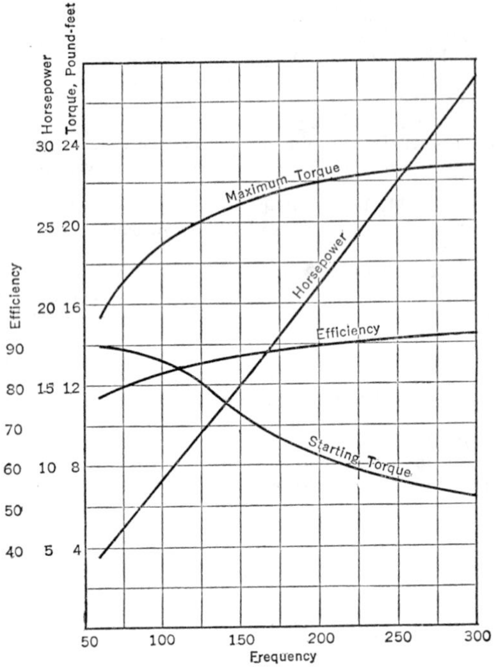

FIG. 12. Effect of Frequency Change

HIGH-FREQUENCY PARTS AND MOTORS. Shell-type motors, as described above, have been used for many years as built-in drives for wood-working machinery, operating at high speeds through higher frequencies. Frequencies of 180, 360, and 540 cycles are common, the last resulting in a motor speed of approximately 30,000 rpm in two-pole designs.

High frequencies are used chiefly for two reasons: (*a*) the higher speed is necessary for the operation, as in precision grinding equipment, or (*b*) the great reduction in bulk and weight are necessary for portable tool service. In the latter case a gear reduction to the necessary operating speed may be justified. Speeds of 60,000 to 120,000 rpm have been used successfully on grinder applications with motors of 2 to 3 in. in diameter.

Table 6. Effect of Frequency on Ratings and Performance

Freq.	Syn. Speed	Hp	T_m/F.L.T., Per Cent	T_{st}/F.L.T., Per Cent	Effic.	Volts
Same parts redesigned for each frequency. OD = 3.68 in.; core length = 3.0 in.						
60	3,600	1.0	250	220	0.81	220
180	10,800	3.0	233	106	0.83	220
500	30,000	5.0	258	143	0.82	220
Same parts redesigned for each frequency. OD = 2.875 in.; core length = 2.75 in.						
60	3,600	0.20	200	155	0.63	220
180	10,800	1.00	225	140	0.76	220
360	21,600	3.00	210	100	0.73	220
420	25,200	3.50	215	140	0.74	220
540	32,400	3.00	215	120	0.79	220

As the frequency increases, horsepower usually increases faster than the speed, and losses shift from the stator copper to the magnetic core. In spite of high efficiencies, the losses are concentrated in such little volume that unusual cooling means are required. Some motors of this type are water-jacketed.

The effect of frequency change on one motor is shown in Fig. 12. In this case a given motor was connected to various frequency sources, with the applied voltage increasing directly with the frequency. Nominal horsepower ratings at each frequency were selected on the basis of maximum torque being 225 per cent of full-load torque, and the efficiency was determined at that load. It will be noted that starting torque always reduces under such circumstances.

In contrast to using the identical motor on various frequencies, the results of designing various motors in the same cores, with individual windings for each frequency, are shown in Table 6.

27. TESTING

Tests on polyphase induction motors are usually made to check guaranteed performance items or design values. Complete performance items can be determined directly or through conventional efficiency calculations by the processes described here. Data are presented chiefly for investigating the standard type of squirrel-cage, polyphase induction motor. Many other special machines, such as slip-ring or wound-rotor motors, double cage machines, and high-slip motors, require modifications in the methods shown. Useful reference standards in this field are:

Rotating Electrical Machinery, American Standard C50.
Standards of the AIEE.
NEMA Motor and Generator Standards 45-102.

BRAKE TESTS (applicable on motors of fractional sizes and sometimes up to about 5 hp). A balanced, steady source of correct voltage and frequency should be used, and input watts, current, voltage, and frequency measured. About six load points from $1/4$ to $1\,1/2$ times full load are recommended. Torque, output, and slip should be read as well as the above values. The ambient and stator winding temperatures should be recorded.

DYNAMOMETER TESTS (recommended for all motors up to the limit of dynamometer capacity available). Values read are the same as those listed under Brake Tests. The AIEE Test Code for polyphase induction machines suggests that bearing friction in the dynamometer be investigated by noting any difference in scale readings when the load is decreasing from those obtained on increasing load. For greater accuracy, data should be obtained for torque measured versus watts input, under both conditions, and the average used.

Output watts of the motor are:

$$W_0 = 0.1420T \times \text{Rpm} \tag{37}$$

where T is the torque of the dynamometer in pound-feet.

INPUT MEASUREMENT. The motor to be tested should be belted or coupled to a variable load. Watts input and current should be measured in six steps from $1/4$ to $1\,1/2$ times full load. Slip readings must be read carefully at each load point, as ordinary-speed reading methods are not accurate enough to use with these data.

In addition to the above test values it is necessary to obtain the following data:

Primary resistance: see under Resistance Measurements.

No-load current and losses: see under No-load Test.

Stray-load losses (if motor is over 10 hp): see under Stray Power Loss.

Calculation of the efficiency at the temperature t_s involves the following steps:

For each load point the stator I^2R loss is calculated for the observed currents and the resistance at a temperature of t_s.

(Watts input $-$ Stator I^2R loss) $\times$ Corrected slip as decimal fraction $=$ Rotor I^2R loss.

$$\text{Corrected slip} = \text{Observed slip}\,\frac{234.5 + t_s}{234.5 + t} \tag{38}$$

t_s = 25 deg cent + Stator winding temperature rise (t) while operating load test.

Total losses are:

Stator and rotor I^2R losses as obtained above.

Friction and windage and core losses.

Stray-load loss if obtained.

Output $=$ Observed input $-$ Total losses.

$$\text{Power factor} = \frac{\text{Watts input}}{\text{Volt-amperes}}.$$

Performance curves of watts and amperes input, current, power factor, efficiency, and speed are usually plotted against horsepower output.

EQUIVALENT CIRCUIT. If the constants of the machine are obtained from test data, using sufficient care and suitable corrections, the performance can be calculated by the exact equivalent circuit with acceptable accuracy.

Tests required are:

Primary resistance: see under Resistance Measurements.

No-load or running-light: see under No-load Test.

Locked rotor: see under Locked Rotor Test.

Data from these tests can be used with the circuit and calculation procedure shown above under Equivalent Circuits.

RESISTANCE MEASUREMENTS. The voltmeter-ammeter method of resistance measurement, with direct current, is recommended for the stator windings. Care should be taken in determining the exact temperature of the windings, and to prevent increase it is suggested that the direct current used should not exceed 25 or 50 per cent of the rated value.

A suitable bridge can be used in place of the voltmeter-ammeter method suggested above.

This value, as read, is used on the equivalent circuit.

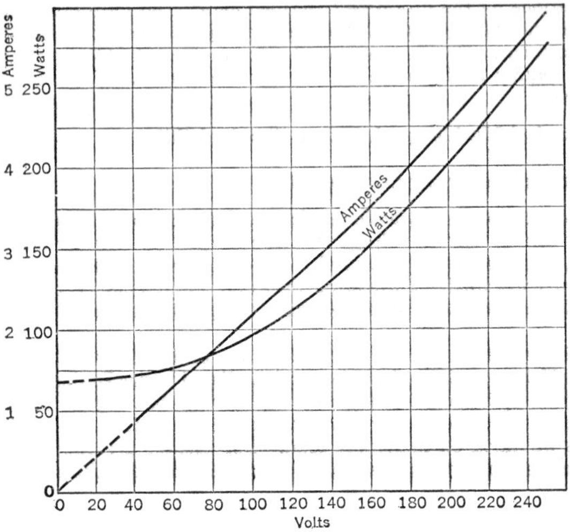

FIG. 13. No-load Test of Induction Motor

NO-LOAD TEST. Normal frequency and voltage are applied, and the motor should be run until the reduction in input due to warming up of the bearings has ceased. On motors up to 25 hp this may involve from 10 to 20 min for ball-bearing motors to double that time for sleeve-bearings.

Applied volts, input watts, and amperes are read (Fig. 13).

The input watts represents the sum of the stator I^2R losses, friction, windage, and core losses. Calculating the stator I^2R losses leaves the so-called constant losses, which are usually not separated further. If, however, it is desired that the core loss be separated from the friction components, then a further "saturation" run is made as described below.

Volts, amperes, and watts are read from about 125 per cent rated voltage down to the point where further voltage reduction increases the current. For rough approximation the watts input at this point represents the friction and windage loss. For greater accuracy the AIEE Test Code recommends that the watts input be plotted against volts squared as abscissa. The lower part of this curve can be extended as a straight line; the correct value of friction and windage loss is then read from the intercept with the zero-voltage axis.

A source of well-balanced voltages with good wave form is important on these tests.

LOCKED ROTOR TEST. The rotor is blocked so that it cannot rotate. Volts, amperes, and watts input are read. If rated voltage is applied, rapid heating occurs. Furthermore the flux is forced through saturated leakage paths. Both these effects interfere with the accuracy of the readings. It is recommended, therefore, that about half voltage be used on such tests, with the readings taken quickly and efforts made to read winding temperatures before and after the test. If a variable voltage source is available, the voltage should be quickly adjusted until rated current is obtained, using this point instead of the half-voltage point for the readings.

Normal starting current at rated voltage can be predicted by voltage proportion.

The above readings are used to obtain values of rotor resistance and total leakage reactance. If these are to be used in the equivalent circuit, however, special tests, as described below, are required for greater accuracy.

A reduced-frequency test is recommended with rotor blocked and a frequency of 15 to 25 cycles applied. The voltage should be such as to give full-load current. See eq. (44).

STRAY POWER LOSS. When all measured loss components (copper, core, friction, and windage) are added, their sum is frequently less than the total loss observed through other tests. The excess represents stray losses caused by modified magnetic paths and the resulting core losses, flux pulsations causing rotor I^2R losses, eddy losses in the conductors, and eddy losses in the motor frame.

Although the statement is made (under Input Measurement) that motors under about 10 hp may show no appreciable stray power loss, this may depend considerably on design. Stray load losses of 50 to 100 watts have been measured in motors below that rating.

Stray load losses can be measured indirectly by loading and measuring input and output. The measurement of losses by methods already described, checked against input minus output, indicates their value.

A direct method of measurement (as recommended by the AIEE) involves excitation of the stator windings by direct current. (Two legs used for Y connection.) The motor is run at synchronous speed by a calibrated driving motor, so that the driving power in watts, W, can be determined. $I_{d\text{-}c}$ and W are read over a range equivalent to no-load to about 150 per cent of full-load current, in seven steps.

To obtain the excitation current (d-c) equivalent to the a-c no-load current the following ratio is used:

$$I_{d\text{-}c} = 1.225\, I_{a\text{-}c} \tag{39}$$

The driving watts corresponding to the correct no-load current are designated W_o.

The induced I^2R loss in the rotor (P) is obtained from the rotor-blocked test by the calculation:

$$P = 0.142 \times \text{Syn. rpm} \times \text{Pound-feet torque} \tag{40}$$

The stray-load loss at any load is given by the calculation:

$$(W - W_o) - (P - P_o) \tag{41}$$

The subscript o refers to no-load values.

CIRCUIT CONSTANTS FROM TEST DATA. Circuit parameters, which may be checked against design data or used on the equivalent circuit to predict performance, are calculated from the previously described tests.

Total primary reactance from no-load test:

$$X_o = \frac{V \text{ per phase}}{I_{\text{no-load}}} \tag{42}$$

Magnetizing reactance:

$$X_m = X_o - X_1 \tag{43}$$

Total leakage reactance from rotor-locked test:

$$X = \frac{f}{f_t}\sqrt{\frac{V^2}{3I^2} - \left(\frac{W}{3I^2}\right)^2} \tag{44}$$

where f = rated frequency, and f_t = frequency at which test was made.
V = line voltage.
I = rotor-locked current.
W = rotor-locked input, three phases.

To divide total leakage reactance use:

$x_1 = 0.5X$, $x_2 = 0.5X$ for single squirrel-cage motors.
$x_1 = 0.4X$, $x_2 = 0.6X$ for low-starting current motors.

To determine the apparent rotor resistance in terms of stator windings (per Y-leg):

$$R\text{ (effective)} = \frac{W}{3I^2}\text{ (locked test)} \tag{45}$$

$$R_2 = R\text{ (effective)} - R_1 \tag{46}$$

The rotor resistance so obtained is the standstill value. For increased accuracy on the equivalent circuit calculations it will be noted that, not R_2, but R_2/s is needed for the circuit (Fig. 5). If data from the load test (see Input Measurement) are available, then for any value of R_2/s, as assumed for the circuit, values of input current are finally calculated. The slip readings of the load test indicate slip at various input currents. At these corresponding currents

$$R_2 = \left(\frac{R_2}{s}\text{ assumed for circuit}\right) \times (s \text{ read on test})$$

TORQUES. In addition to the tests already listed, the motor manufacturer usually tests production motors for starting torques and maximum torques.

Starting-torque tests usually accompany the locked rotor test, with rated voltage and frequency applied to the motor terminals. A brake arm and scale are utilized for reading torque, with the rotor turned to different positions. Because of "cogging," values of torque vary for various positions of the rotor, but the minimum is the correct one, by definition.

Maximum torques are obtained by loading to the breakdown point, with rated voltage (or not less than 75 per cent of rated voltage) applied. Maximum torque varies as the square of the voltage.

Maximum torque can be calculated by the equivalent circuit, using

$$s = \frac{R_2}{\sqrt{R_1^2 + X^2}} \tag{47}$$

as the slip at which maximum torque occurs.

TEMPERATURE TESTS. Tests are made under full load at rated voltage and frequency to determine the final, continuous rise in temperature of the windings. Any practicable loading method is permissible. The test is continued until final temperature is reached. Several methods are recommended, with the choice left to the motor manufacturer.

Thermometer. Thermometers or thermocouples are placed on the end coils of the windings and held by putty. Readings are taken every half hour until the observed rise is essentially constant. Because many motors are built with end-bell construction, which does not permit easy access to the winding surfaces (totally enclosed motors may prevent any such access), it is permissible to substitute a procedure about as follows:

Apply thermometers to the motor frame adjacent to the core. Read the rise, and when it has become constant, remove the end-bells or cover plates and apply thermometers to the exposed winding surfaces, which in some cases may be rotating parts. American Standard C50 recommends that the time elapsed between shutdown and reading be limited to 1 min for machines up to 50 kw; 2 min for 60 to 200 kw; and 3 min for ratings above 200 kw. Under such conditions no correction for temperature change is necessary.

For the usual squirrel-cage rotor some of the difficulties of the above procedure can be eliminated by the use of thermocouples instead of mercury bulb thermometers. Thermocouples can more readily be applied, even on enclosed motors. It must be kept in mind that when, so used, they are surface-temperature indicators and are not embedded. The entire question of relative readings by thermocouples and bulb thermometers is still under consideration by various standardizing groups, one point at issue being the corrections necessary as a function of motor size.

Resistance by Temperature Rise. For totally enclosed machines, or those for which thermometers cannot readily be applied, the rise in resistance of the windings with temperature increase forms a convenient method. Winding resistance is measured at the beginning of the temperature run to make sure that it is uniform with room temperature. The resistance and the room temperature are measured again at the conclusion of the run, and the rise is figured from:

$$T = \left(\frac{R_f}{R} - 1\right)(234.5 + t) \tag{48}$$

where R_f = final resistance in ohms.
R = cold resistance in ohms.
t = cold temperature in degrees centigrade.

Such a test is a valuable check on the usual thermometer method. It indicates the **average** temperature rise; end coil readings by thermometer give still a different value; and the two are both different from the actual **hot-spot** rise on which standards are based.

The limiting observable temperature rises, as set up by the American Standards for Class A and Class B insulation, are given in Table 5.

DIELECTRIC TESTS. During the process of manufacture, after final assembly, or after final installation, the windings of a motor are connected to one terminal of a high-voltage source, with the other terminal connected to the motor frame. In the equivalent shop test (NEMA) this voltage is applied for 1 min or, if used at 120 per cent of its recommended value, for only 1 sec. In general the recommended effective value of voltage is 1000 volts plus twice the rated voltage of the circuit.

VIBRATION TEST. As the degree of balance is important to many motor applications requiring smooth operation, a standard method of measuring vibration has been set up (NEMA). The motor is run idle on a resiliently mounted platform constructed for certain limits of deflection and period of vibration. A vibrometer measures deflections at three different locations on the motor hubs. Standard dynamic balance should provide for deflections of not more than:

0.001 in. for motor frames 203–254.
0.0015 in. " " " 284–364.
0.002 in. " " " 365–505.

28. DESIGN

METHODS. No single method can be recommended for the design of a polyphase induction motor, as the approach can be modified by many factors and circumstances.

1. In most cases the industrial designer may start with a given frame and laminations, and his problem is then to determine (*a*) the core length, (*b*) the primary winding, and (*c*) the cage material and the end-ring section.

These are the usual variables under his control to achieve a desired motor rating.

2. In other cases the actual frame in which the motor is to be built may not be provided, but the diameter may nevertheless already be fixed by competitive designs, customers' requirements, or industrial standardization. The available factors under the designer's control are again those mentioned above.

3. The desired design and rating may be such that previous designs of both larger and smaller ratings are already available to the designer, who may be aware of both the lamination diameter and the core length required. His problem is then to determine (*a*) the primary winding, (*b*) the cage material and the end-ring section.

4. In some cases the designer may be called upon to design a rating outside the range of any previous calculations, with no dimensions fixed. The designer must then determine (*a*) the principal lamination dimensions, including the complete magnetic circuit contour, (*b*) stator and rotor windings as above.

It can readily be seen from the above outline of a designer's functions that design methods must vary with the nature of the problem.

DESIGN OF LAMINATIONS. It is assumed that the horsepower and speed of the polyphase motor are known and that the outside diameter of the laminations is fixed or can be temporarily chosen for a trial design. In such case the following relations will give good average laminations for the stator and rotor, inch units being used throughout:

The **inside diameter of the stator:**

$$D = \frac{\mathrm{OD} - 0.647}{1.175 + (1.03/P)} \tag{49}$$

where P = the number of poles.

The **diameter to the bottom of the slots** (Fig. 14):

$$D_1 = 1.175D + 0.647 \tag{50}$$

The **tooth width** (see Fig. 17):

$$t_w = \frac{1.35D}{S_1} \tag{51}$$

where S_1 = number of stator slots.

Up to about 50 or 75 hp partially closed slots are commonly used, and the tooth sides are parallel. Above these sizes the slot sides are parallel, and the tooth width can be figured for about one-third of the distance from its base.

The number of stator and rotor slots affects noise, cogging, and the possibility of "cusps" in the speed-torque curve.

For two-pole designs, combinations of 10/12 and 24/34 have proven useful. For higher numbers of poles 24/34, 36/48, 48/60, and 72/79 are commonly satisfactory. The first number represents stator slots in all cases. See Figs. 14 and 15.

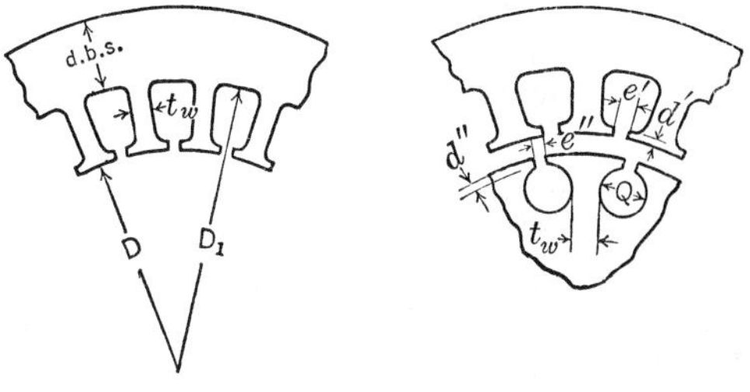

FIG. 14. Lamination Dimensions

For **partially closed slots:**

$$e' = 0.0143D + 0.0643 \tag{52}$$

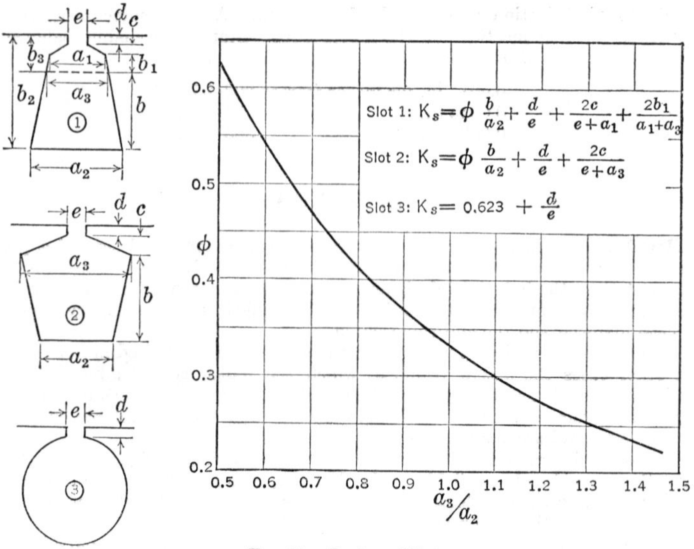

FIG. 15. Design of Slots

The **depth** ***d*** can be chosen so that

$$\frac{d'}{e'} = 0.3 \text{ (approximately)} \tag{53}$$

For the **radial length of the air gap** (all values in inches):

$$\Delta = 0.0016D + 0.001L + 0.0072 \tag{54}$$

L = the axial length of core in inches.

For the **rotor slot opening** in inches:

$$e'' = 0.01D + 0.045 \tag{55}$$

and

$$d'' = 0.00677D + 0.0304 \tag{56}$$

The **diameter of the rotor slot** in inches:

$$Q = \frac{1.95D - 0.236}{S_2 + \pi} \tag{57}$$

where S_2 is the number of rotor slots. This results in a **tooth width** (at the neck) of

$$t_w'' = 1.15 \frac{D}{S_2} \tag{58}$$

The above rules are really suggestions as to likely dimensions for a trial design. They have been investigated for rotor diameters of 3 to 15 in. Values can be modified for special applications. If followed, they result in maximum densities in the stator teeth of 120 per cent of the average stator yoke densities, and rotor tooth densities 118 per cent of the stator value.

DESIGN PROCEDURE. 1. *a.* Select a set of laminations or design trial laminations, as shown in the preceding paragraphs.

b. Select a trial stacking length of core, L.

c. Calculate the series conductors per phase, C, for a desired maximum flux density in the teeth and for a winding factor calculated from the distribution and assumed pitch.

$$C = \frac{1}{L} \times \frac{VP}{B_t f S_1 t_w k_w} 70.9 \times 10^6 \tag{59}$$

where L = axial length of stack in inches.
V = voltage per leg.

P = number of poles.
B_t = maximum stator tooth density in lines per square inch. (Use from 80,000 to 100,000 for 60-cycle motors.)
S_1 = number of stator slots.
t_w = tooth width. (Use the value at about one-third of the distance from the root for tapered teeth.)
k_w = winding factor or product of $k_p k_d$. (See Tables 7 and 8.)

It is assumed throughout that the winding is star connected, with possibility of later conversion to delta if desired.

d. Calculate, in the order given below, the following significant parameters and factors.

2. Permeances of Leakage Flux Paths.

a. Primary and secondary slots.

$$P_s = 3.60L\left(\frac{K_s'}{S_1} + \frac{K_s''}{S_2}\right)(0.625p + 0.375) \qquad (60)$$

where S_1 = number of stator slots.
S_2 = number of rotor slots.
p = pitch as a decimal fraction.
K_s' and K_s'' = stator and rotor slot constants, respectively, calculated by the formulas of Fig. 15 for the slot shapes used. It is assumed that a stator slot is full of winding to the point indicated by the length, b.

Table 7. Distribution Factor, k_d

Total Slots per pole	k_d (Two-phase)	k_d (Three-phase)
2	1.000	....
3	....	1.000
4	0.924	....
6	0.911	0.966
8	0.906	....
9	....	0.960
12	0.903	0.958

Table 8. Pitch Factor, k_p

Pitch	k_p	Pitch	k_p	Pitch	k_p
12/12	1.000	......	......	......	......
11/12	0.991	9/10	0.988	7/8	0.981
10/12	0.966	8/10	0.951	6/8	0.923
9/12	0.923	7/10	0.891	5/8	0.831
8/12	0.866	6/10	0.809		
7/12	0.793				
6/12	0.707				

b. Air-gap length, Δ, and gap constant, K_1. The radial length of air gap in inches can be calculated from eq. (54).

The effective gap length (allowing for fringing at the sides of the teeth) can be obtained through the use of the correction factor K_1, thus:

$$K_1 = \frac{\lambda' \times \lambda''}{T_c' \times T_c''} \qquad (61)$$

where λ' = the stator tooth pitch in inches.
λ'' = the rotor tooth pitch in inches.
T_c' = the stator tooth face corrected for fringing
= actual face + F. See Table 9.
T_c'' = a similar value for the rotor.

Table 9. Effective Tooth Face from Fringing

(Effective tooth face T_c = Actual face $T + F$)
Values of F

Slot Opening e' or e''	Radial Gap Length, Inches					
	0.010	0.015	0.020	0.025	0.030	0.040
0.04	0.021	0.024	0.026	0.027	0.028	0.029
0.05	0.023	0.028	0.031	0.032	0.033	0.035
0.06	0.026	0.031	0.035	0.037	0.039	0.041
0.07	0.028	0.034	0.038	0.041	0.043	0.046
0.08	0.030	0.036	0.041	0.045	0.048	0.051
0.09	0.031	0.039	0.044	0.049	0.053	0.057
0.10	0.032	0.041	0.047	0.052	0.056	0.062
0.11	0.034	0.043	0.049	0.055	0.059	0.066
0.12	0.035	0.044	0.052	0.058	0.062	0.070
0.13	0.036	0.046	0.054	0.060	0.065	0.073

c. The values in Table 9 are necessary to calculate the permeance of the zig-zag leakage flux paths.

$$P_{zz} = \frac{8.5D}{S_1S_2\Delta}\left(\frac{T_c' + T_c''}{\lambda' + \lambda''} - 0.5\right)^2 L(0.8p + 0.2) \tag{62}$$

d. The permeance of the leakage flux paths around the end coils of the stator and the end rings of the rotor is:

$$P_e = \frac{\lambda}{P} K_e(1.25p - 0.25) \tag{63}$$

where K_e = 0.45 for a ring comparatively great in axial length and extending considerably beyond the rotor core.
K_e = 0.65 for an end ring of comparatively great radial depth.
λ = the pole pitch in inches.

e. The total leakage permeance is then

$$P_L = P_s + P_{zz} + P_e \tag{64}$$

The total leakage reactance (both stator and rotor) is

$$X = 2\pi fC^2nP_L10^{-8} \tag{65}$$

3. Magnetizing Reactance.

The permeance of the mutual flux path is

$$P_m = 0.322\,\frac{LK_w^2}{P\Delta K_1(\text{SF})} \tag{66}$$

The magnetizing reactance is

$$X_m = 2\pi fC^2nP_m10^{-8} \tag{67}$$

The saturation factor (SF) is defined as the ratio of total ampere-turns to ampere-turns for the air gap only. Usual values for average designs are 1.1 to 1.2. Other symbols used are:

L = length of core in inches.
K_w = the winding factor, product of pitch and distribution factors.
P = the number of poles.
Δ = the radial air-gap length in inches.
K_1 = the gap constant.
C = the total series conductors per phase.

4. Resistances.

a. The average length conductor (one-half of the mean length of turn) on the stator, in inches:

$$l_c = 1.4p\left[\frac{(D + \text{Tooth length})\,\pi}{P}\right] + L + 0.15(\text{OD}) \tag{68}$$

b. Stator winding resistance per Y-leg:

$$r_1 = (1 + 0.004t_c^\circ)\;9.7\,\frac{l_c}{12}\,\frac{C}{Am} \tag{69}$$

where t_c° = operating temperature in degrees centigrade.
C = series conductors per phase.
A = area of one conductor in circular mils.
m = number of parallel paths.

or

$$r_1 = \frac{C^2n}{A_skS_1} \text{ (at 60 deg cent.)} \tag{70}$$

where n = number of phases.
A_s = area of one stator slot in circular mils.
k = a space factor of 0.26 to 0.28.
S_1 = number of stator slots.

c. The resistance of a squirrel-cage rotor, in stator terms:

$$r_2 = C^2k_w^2n\,[\text{Bar} + \text{Ring}] \text{ (at 60 deg cent)} \tag{71}$$

$$\text{Bar} = \frac{l_2k_b}{A_bS_2} \tag{72}$$

$$\text{Ring} = \frac{2D_rk_rK_R}{\pi P^2A_r} \tag{73}$$

where l_2 = rotor bar length in inches.
k_b = resistivity of bar material compared to that of copper.
A_b = area of each bar in circular mils. The bar diameter selected should be slightly smaller than the slot diameter.
D_r = mean diameter at which bars fasten to end rings.
k_r = resistivity of end-ring material compared to that of copper.
A_r = area of end-ring section in circular mils.
K_R = correction factor (see Fig. 16).
S_2 = number of rotor bars.

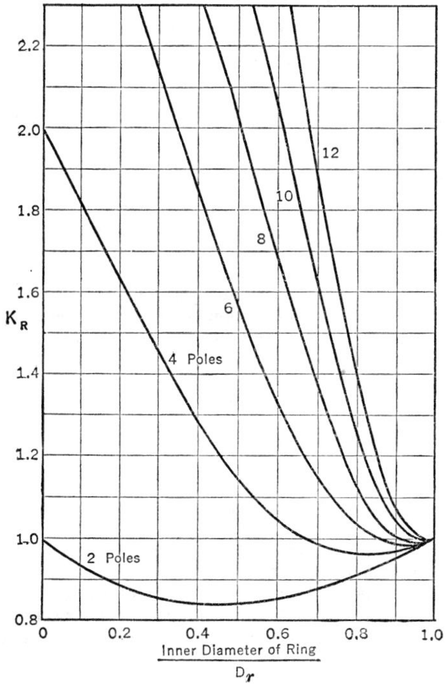

FIG. 16. Correction Factor for Eq. (73)

5. Flux Factors and Ratios.

a. The total reactance from which the quadrature component of the no-load current can be obtained is

$$X_0 = X_m + X_1 \tag{74}$$

$$X_1 = X_2 = {}^1/_2 X \tag{75}$$

or

$$P_0 = P_m + \frac{P_L}{2}$$

Then (approximately)

$$I_0 = \frac{V}{X_0} \tag{76}$$

A useful flux factor is

$$K_r = \frac{X_0 - X}{X_0} \tag{77}$$

and

$$K_p = \frac{X_m}{X_0'} \tag{78}$$

$$K_s = \frac{X_m}{X_0''} \tag{79}$$

With X_1 assumed equal to X_2

$$X_0' = X_0'' = X_0$$

and

$$K_p = K_s = \sqrt{K_r}$$

b. Factors which can be calculated independently of the number of conductors are

$$K_r = \frac{P_0 - P_L}{P_0} \tag{80}$$

$$\frac{r_1}{X} = \frac{10^8}{A_s k S_1 (2\pi f P_L)} \tag{81}$$

$$\frac{r_2}{X} = \frac{(\text{Bar} + \text{Ring})\, k_w^2 10^8}{2\pi f P_L} \tag{82}$$

Such ratios are useful in one approach to design discussed under Design for a Specified Maximum Torque.

6. Performance Items.

a. By means of some of the above calculations, plus the iron loss (calculated as shown under Magnetic Saturation and Core Loss), it is possible to use the equivalent circuit for determining complete motor performance. However, at this point, as the design may have undesirable features or otherwise fail to meet standards, it would be well to determine certain characteristics, such as maximum torque, starting torque, and starting amperes. The first of these items is independent of rotor resistance and should be determined and corrected if required. As a modification of eq. (17)

$$T_m = \frac{3.52 V^2 n K_r}{X(\text{Syn. rpm})} K \text{ (pounds-feet)} \tag{83}$$

$$K = \frac{1}{\beta} + 0.1(1 - K_r^2) \tag{84}$$

$$\beta = \frac{R_1}{X} + \sqrt{\left(\frac{R_1}{X}\right)^2 + 1} \tag{85}$$

Equation (83) is of semiempirical nature but is more accurate than that which is derived from the approximate equivalent circuit.

Maximum torque should be at least 200 per cent of the full-load torque (see under General-purpose Motors), but a ratio of approximately 250 per cent is often more suitable for general-purpose motors. If the design shows an undesired ratio, procedure should be as follows:

$$\frac{\text{Desired } T_m}{\text{Actual } T_m} = R_T$$

where T_m = the maximum torque in pound-feet.

Multiply the original number of conductors by $1/\sqrt{R_T}$. The new value of torque is then R_T times the original value. The flux densities will equal $\sqrt{R_T}$ times the original value. Leakage reactance will be $1/R_T$ times the original value.

Stator resistance will vary as X, if the slots contain the same cross-sectional area of copper in both cases. Hence R_1/x remains unchanged. R_2/x will also remain unchanged.

Except for the effect of saturation, K_r will remain unchanged.

Magnetizing reactances X_0 and X_m will be $1/R_T$ times the original value.

If the above change shows too low a flux density, so that the assumed core length can be reduced, or too high a flux density, so that additional core length is required, proportions can again be used for obtaining a rough check on maximum torque. Increase in L increases T_m at a rate higher than the linear relationship assumed. For small changes, however, it can be assumed that

$$\frac{\text{Desired } T_m}{\text{Actual } T_m} = \frac{L(\text{new})}{L(\text{original})} = R_w$$

where L = the core length in inches.

Multiplying the original core length by R_w gives a desired torque at the same flux density, if the series conductors are multiplied by $1/R_w$. Conductor cross-section should be changed in the proportion of R_w, if possible. As such an exact proportion is rarely possible, the machine parameters should usually be recalculated after the change in core length for maximum torque adjustment.

b. Minimum standards for starting torque should be investigated at this point, with the idea of modifying either the bar or ring material or the section of the end ring.

Let (a) = ratio of starting torque to maximum torque.

Then for the correct value of R_1/X read the required value of R_2/X and calculate R_2 (see Fig. 17).

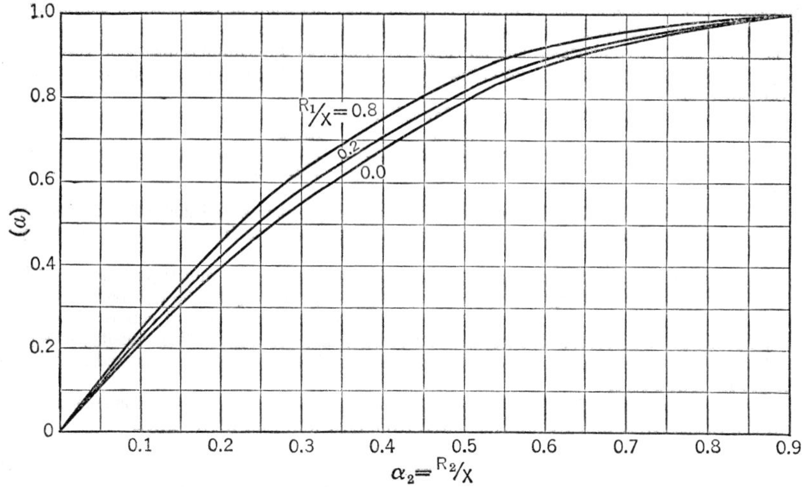

FIG. 17. Ratio of Starting Torque to Maximum Torque

c. Before attempting to design for the above value of R_2, it is well to determine the "locked kva per hp" (see Formulas and Discussion of Factors, in Article 25), from which locked amperes can also be determined.

$$\text{Locked volt-amperes} = (b) \times T_m \times \text{Syn. rpm} \tag{86}$$

where (b) is read from Fig. 8. (Data are based on the approximate equivalent circuit.) Also

$$\frac{\text{Locked kva}}{\text{Horsepower}} = 5.46 \times (b) \times \frac{T_m}{\text{Full-load torque}} \tag{87}$$

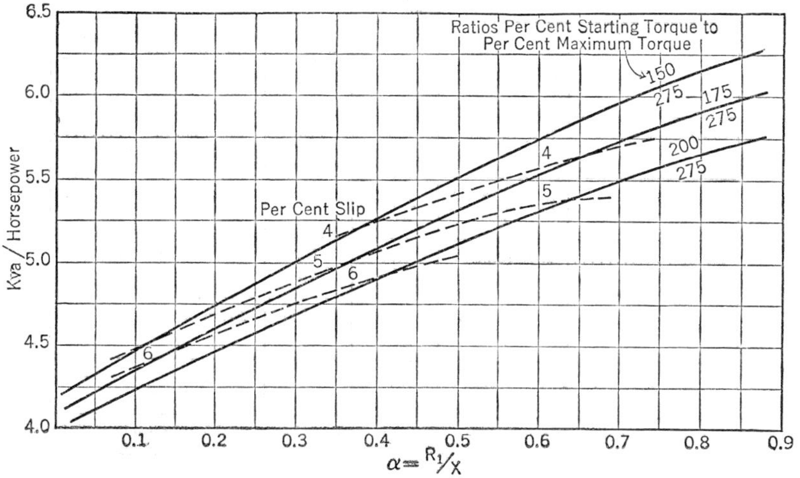

FIG. 18. Locked Kva per Horsepower in terms of Ratio of Primary Resistance per Phase to Total Leakage Reactance per Phase

If the locked kva per horsepower or the starting current is undesirable, read required values of R_2/X from these curves, and recheck the starting torque that would result.

d. As a further check on useful rotor resistance, the resulting full-load slip can be predicted approximately at this point by eq. 25.

This relationship of machine parameters can be combined with those represented by the curves of Figs. 8 and 17 to yield the type of curves shown in Fig. 18, which are only those for a maximum torque ratio of 275 per cent of full-load torque. Similar curves, plotted for other useful ratios, enable a quick determination of the suitable combination of constants to yield desired torques, slips, and starting kva.

e. With the above checks made on rotor resistance to determine a desirable value, it is then possible to determine the end-ring section necessary to yield R_2. "Bar" is calculated from eq. (72).

Cross section of the end ring in circular mils:

$$A_r = 0.636 \frac{D_r k_r K_R}{P^2} \times \frac{C^2 k_w{}^2 n}{(R_2 - \text{Bar } C^2 kw^2 n)} \tag{88}$$

See eq. (71) for symbols.

Too small a section of end ring results in high temperatures and distortion on frequent motor starting. High-resistance alloys are often desirable to increase mass and dissipation area.

Design for a Specified Maximum Torque. In calculating the motor parameters it has been shown that K_r and R_1/X can be calculated independently of the number of conductors. These ratios are, in fact, functions of the magnetic circuit and can be influenced only by the pitch selected. Hence, instead of designing for a given flux density, it is possible to select the winding on the basis of that required to produce a desired maximum torque in pound-feet.

With the laminations, length of core, and winding pitch assumed, the various permeances and the ratios R_1/X and K_r are calculated from eqs. (80) and (81). Then

$$X = \frac{3.52 n V^2 K_r K}{\text{Syn. rpm } T_m} \tag{89}$$

and the series conductors to yield this desired maximum torque are

$$C = \sqrt{\frac{X \times 10^8}{2\pi f n P_L}}$$

Conductor area is determined as

$$\frac{A_s k C n}{S_1} \tag{90}$$

Table 10. Calculating Ampere-turns

$$\phi = \frac{V \times 10^8}{2.22 C k_w f} = \text{Flux per pole}$$

Part	Area of Path (*A*)	Density (*B*)	Length (*C*)	Ampere-turns per inch (*D*)	Total Ampere-turns ($C \times D$)
Stator yoke.......	$L \times$ (d.b.s)′ *	$\frac{0.5\phi}{\text{Area}}$	$\frac{\pi D_y'}{2P}$	From curve	....
Stator teeth.......	$\frac{S_1 \times L \times t_w'}{P}$	$\frac{1.57\phi}{\text{Area}}$	Actual	From curve	....
Rotor yoke........	$L \times$ (d.b.s)″	$\frac{0.5\phi}{\text{Area}}$	$\frac{\pi D_y''}{2P}$	From curve	....
Rotor teeth.......	$\frac{S_2 \times L \times t_w''}{P}$	$\frac{1.57\phi}{\text{Area}}$	Actual	From curve	....
Saturation factor $= \frac{\text{Ampere-turns total}}{\text{Ampere-turns gap}}$				Total	——

* See Fig. 14 for depth below slot (d.b.s.) of stator. For the rotor this represents radial distance from bottom of the slot to the shaft or to the inside diameter of the rotor lamination.

V = volts per leg.

D_y' represents the diameter to the center of the yoke, or D_1 + d.b.s. A corresponding mean yoke diameter is selected for the rotor (D_y'').

This is adjusted to the nearest wire size and for the number of parallel paths desired. Thus the stator winding is completely determined, and by checking required values of rotor resistance to yield desired starting torques and currents, the design is fixed, ready for checks on flux density. As a quick check, the maximum stator tooth density can be determined as

$$B_t = \frac{70.9 \times V \times P \times 10^6}{Cfk_w t_w L S_1} \tag{91}$$

See eq. (59) for symbols.

Excessive values may require either a modification in the maximum torque (through adjustment of the number of conductors) or an increase in core length, using proportions as shown previously.

Magnetic Saturation and Core Loss. Once the core length and number of stator conductors have been determined, a check on the magnetic densities and ampere-turns is desirable. The method to be followed is outlined in Table 10. It will be noted that average densities in the yoke and maximum densities in the teeth are calculated, and that

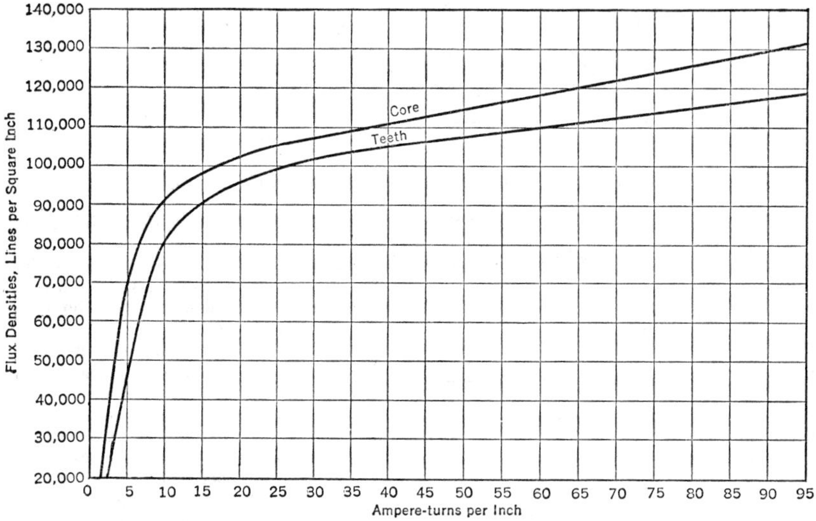

Fig. 19. Relation of Flux Densities to Ampere-turns per Inch

several convenient approximations are utilized in determining areas and lengths of paths. Reasonable accuracy is obtained by making use of the curves shown in Fig. 19 for reading ampere-turns. These curves are adjusted for a stacking factor of 0.95.

Core-loss calculations involve losses in stator yoke and teeth and surface losses due to slot openings. The procedure is shown in Table 11 with losses per cubic inch read from Fig. 20. These curves are sufficiently accurate for either armature-grade core-plated

Table 11. Calculating Core and Surface Losses

Part	Volume, Cubic Inches (1)	Density (2)	Core Loss per Cubic Inch (3)	Core Loss (4)
Stator yoke...	$\pi(\text{dbs})'LD_y'$	From Table 10	Fig. 23	(1) × (3)
Stator teeth...	Tooth length × t_w' × L × S_1	From Table 10	Fig. 23	(1) × (3)
Slot openings.	Surface, square inches	Factor q	Watts per square inch	
	$L \times T'' \times S_2$	See Table 12	From Table 12	(1) × (3)
Total core loss in watts				——

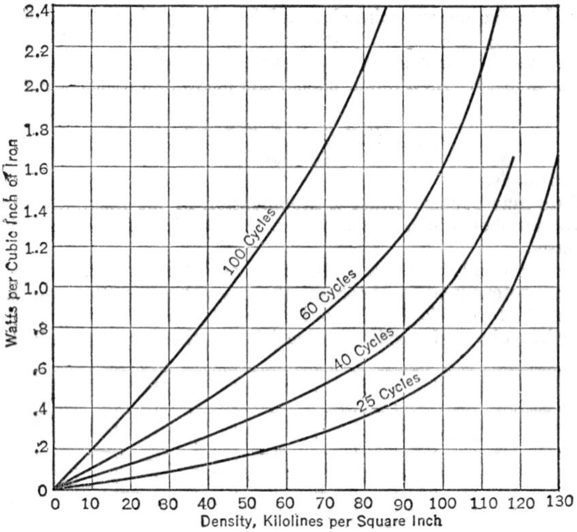

FIG. 20. Core Losses per Cubic Inch in Terms of Flux Density

lamination stock or electrical grade. Surface losses are calculated by determining the factor q and reading losses per square inch from Table 12.

Summary of Design Method. Regardless of whether trial laminations must be designed or laminations are already available, the above methods involve assumption of a core length and calculation of the constants. A check on such items as starting and maximum torques, starting current, and full-load slip should usually be made before completing the design. If such items are found to be unsatisfactory, a modification of stacking or winding, or both, can be made by the proportions indicated. A recheck of the motor constants is then followed by the calculation of the magnetic circuit and the core losses.

With the above data, the circle diagram can be laid out, or the equivalent circuit can be evaluated for calculation of the performance at various load points.

Table 12. Surface Losses

Frequency =	25 Cycles	50 Cycles	60 Cycles
Values of q	Watts per Square Inch		
10	0.20	0.35	0.45
20	0.25	0.45	0.50
30	0.30	0.50	0.60
40	0.35	0.60	0.70
50	0.40	0.70	0.85
60	0.50	0.90	1.05
70	0.55	1.10	1.30

$$q \doteq \frac{\text{Flux per pole} \times e'}{L \times \Delta \times T'} \times 1.57 \times 10^{-5}$$

T' = Actual stator tooth face in inches

29. STARTING AND SPEED CONTROL

(See also Section 12)

GENERAL. The determination of the starting equipment for polyphase induction motors depends to some extent on the starting torque required, the starting-current limitations, and the motor type. Although it is ordinarily recommended that a reduced voltage starter be used above 7 1/2 hp for Class A motors, actually local power-supply conditions may be such as to permit an increase in capacities before reduced voltage starters are used, or the use of a Class B or C motor may result in starting currents low enough to be acceptable even up to 30 hp.

MAGNETIC CONTACTORS. The simplest motor-starting device which offers any degree of protection is the magnetic contactor, consisting fundamentally of a set of stationary contacts and a set of matching moving contacts of the clapper type, which are brought together by magnetic means. A separate circuit actuating the electromagnet, which closes the contacts, gives remote control through a push-button station if desired.

Such a switch offers protection against low voltage if the clapper-magnet circuit is properly wired to open the main circuit if the supply voltage goes too low. The main circuit may or may not be restored on re-establishment of full voltage, depending upon the method used.

Protection against overload is offered by an auxiliary device, usually consisting of a heater coil actuating a thermally sensitive contact, which operates on the circuit of the holding magnet. If the heater coil is properly selected with reference to the motor load, excessive current causes the holding magnet circuit to open, releasing the main contacts. Resetting is necessary for subsequent operation.

AUTO-TRANSFORMER STARTERS OR COMPENSATORS. Where starting current limitations require that a motor be started on reduced voltage, auto-transformers form the chief means of reduction.

For three-phase circuits, auto-transformers are usually connected V and provided with standardized 80, 65, and 50 per cent taps.

Starting torque varies as the square of the applied voltage, and hence starting on the 80 per cent tap gives a starting torque 64 per cent of normal. If the motor design provides for a starting torque of at least 157 per cent of full-load torque, starting on 80 per cent voltage provides full-load torque.

Starting current reduces directly with the voltage. Because of the transformer ratio, however, the line current resulting from reduced motor voltage will be only 64 per cent of the full-voltage motor current when the 80 per cent tap is used. Similarly the 65 per cent tap gives a line current of 42 per cent, and the 50 per cent tap a line current of 25 per cent.

Auto-transformer starters are either manual or magnetic.

WOUND-ROTOR POLYPHASE MOTORS. Examination of the fundamental equations for the polyphase induction motor shows that the speed will reduce as the rotor resistance is increased. Furthermore, an increase in rotor resistance results in an increase in starting torque, up to the limit of breakdown torque. These two facts are utilized in applying the wound-rotor motor to two classes of duty, involving (1) unusual starting conditions and (2) speed control.

Construction. A distributed and insulated winding is provided in the rotor slots, with ends brought out to slip rings. Usually three-phase windings are provided, although two-phase windings will work equally well with a three-phase stator. Stationary brushes bear on the slip rings; with brushes short circuited the rotor represents a completely closed circuit operating very similarly to a squirrel cage.

Starting Duty. The simplest form of starting controller involves two rheostats connected in V across the slip rings of the motor. These may be tied in with a magnetic starter in the supply leads, so that only with full resistance in the rotor circuit can the push-button control circuit be actuated to close the line leads. As the motor gains speed, the starting rheostats are gradually cut out.

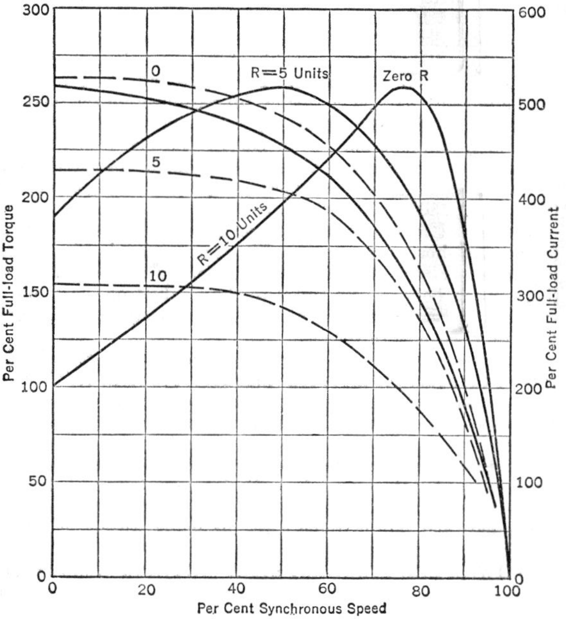

FIG. 21. Effect of Change of Rotor Resistance upon Starting Torque, Starting Current, and Full-load Speed

This provides good starting and accelerating torque with a minimum of inrush current. Furthermore, the excess secondary loss required for high torque is dissipated largely in the external resistors, simplifying the problem of motor cooling on frequent-starting duty.

Speed Control. If the external resistances of the secondary are left in the circuit during the operation of the wound-rotor motor, operation occurs at a reduced speed.

Such resistors and auxiliaries must be built for continuous duty, rather than for the intermittent-starting classification.

Figure 21 indicates that, for any setting of resistance, the speed decreases with load increase. Hence such a motor is of the adjustable varying-speed type. (See below for definitions.)

Frame sizes for wound-rotor motors are standardized by NEMA, providing essentially the same frame size for a given horsepower and speed as is recommended for squirrel-cage motors.

Motors are usually operated from near synchronous to no greater than 50 per cent synchronous speed by this method. A reduction in speed is always accompanied by an equivalent reduction in efficiency, so that at half speed the overall efficiency could be only 50 per cent if there were no other losses. Nevertheless, the need for adjustable-speed drive on many applications makes this a useful system.

SPEED CLASSIFICATION. Constant-speed Motor. A constant-speed motor is one having constant or practically constant normal-operation speed, e.g., an induction motor with small slip.

Varying-speed Motor. A varying-speed motor is one the speed of which varies with the load, ordinarily decreasing when the load increases, such as a series motor, or an induction motor with large slip.

Adjustable-speed Motor. An adjustable-speed motor is one the speed of which can be varied gradually over a considerable range but, when once adjusted, remains practically unaffected by the load.

Adjustable Varying-speed Motor. An adjustable varying-speed motor is one the speed of which can be adjusted gradually but, when once adjusted for a given load, varies in considerable degree with change in load, such as a compound-wound d-c motor adjusted by field control or a slip-ring induction motor with rheostatic speed control.

Multispeed Motor. A multispeed motor is one which can be operated at any one of two or more definite speeds, each being practically independent of the load, e.g., an induction motor with windings capable of various pole groupings.

MULTISPEED MOTORS. Polyphase induction motors of the squirrel-cage type can be built for several speeds by (*a*) reconnecting the field windings for double the number of poles (1800/900 rpm); (*b*) providing the stator with two independent windings, each for a different number of poles (1800/300 rpm); or (*c*) a combination of the two methods, so that four speeds can be obtained. In the third case, one speed must always be half of the other, as reconnecting any one winding can only result in doubling the poles (thus, 3600/1800/1200/600 rpm).

Multispeed motors can be classified as (*a*) constant-torque, (*b*) variable-torque, and (*c*) constant-horsepower types. The regrouped pole winding type is inherently a constant-torque motor and at half speed therefore is capable of delivering only half horsepower (horsepower being proportional to the product of speed and torque). Compressors and stokers are typical examples of equipment requiring approximately constant torque to operate, regardless of speed.

Variable-torque motors are standardized on a system having the available horsepower vary as the square of the speed (thus 1800/900 rpm; 40/10 hp). Although fans, centrifugal pumps, and similar equipment may not reduce in horsepower demand exactly as the square of the speed, selection of the motor size based on top-speed horsepower provides adequate power.

Constant-horsepower motors are capable of developing the same power at all speeds. As slow-speed motors are inherently large, motor size is largely dictated by the low-speed rating.

Standardized motor-frame sizes for many ratings of all types mentioned above are established by NEMA.

SLOW-SPEED MOTORS; GEAR MOTORS. Industrial processes frequently demand extremely slow speeds without adjustment. Sixty-cycle motors wound for many poles (such as 16 poles with a synchronous speed of 450 rpm) provide one means of obtaining useful speeds. Such motors, however, are always large and expensive; for instance, a 2-hp 16-pole motor requires the same frame as a 30-hp 2-pole motor.

Table 13. Comparative Sizes: Slow-speed and Gear Motors (Continuous Duty)

Horsepower	Speed, Rpm	Weight, Pounds		Overall Height, Inches	
		Motor	Gear Motor	Motor	Gear Motor
1	420	250	95	16	10
5	420	660	150	19 1/2	12 1/2
10	420	1000	375	22	16
30	420	1500	855	29	19 1/2

A second satisfactory means of obtaining slow-speed drive involves gear motors, with one end-head built as an integral part of a gear reducer unit. These drives are built around worm-and-wheel, helical gears with main and countershafts parallel—and internal planetary systems.

In comparison with standard electric motors, gears have a somewhat limited life. High stresses, poor alignment, and lubrication difficulties may all shorten the period of usefulness of a gear drive. Many applications may require such intermittent operation that "standard" designs are entirely satisfactory. Others, for continuous operation or with occasional overload or shock load, may require gear designs with much lower working stresses for long life. "Average load" is not a safe measure for determining the rating required of a gear drive, as peak torque demands determine tooth stresses. Hence the selection of a suitable gear motor involves more than ordinary considerations. Complete lines of gear motors offer various service factors, calculated as the ratio of gear rating to motor rating, so that proper selection will result in maximum overall economy.

30. BIBLIOGRAPHY

General (Books)

Dudley, A. M., *Connecting Induction Motors.* McGraw-Hill.
Kuhlmann, J. H., *Design of Electrical Apparatus.* John Wiley (1940).
Langsdorf, A. S., *Theory of Alternating Current Machinery.* McGraw-Hill.
Lawrence, R. R., *Principles of Alternating-Current Machinery.* McGraw-Hill.
Puchstein, A. F., and T. C. Lloyd, *Alternating Current Machines.* John Wiley (1942).
Punga, F. and O. Raydt, *Modern Polyphase Induction Motors.* (Translation by H. M. Hobart). Pitman.
Shoults, D. R., C. J. Rife, and T. C. Johnson, *Electric Motors in Industry.* John Wiley (1942).
Veinott, C. G., *Fractional Horsepower Electric Motors.* McGraw-Hill.

Theory, Tests, and Operation (Articles)

Calvert, J. F., Amplitudes of Magnetomotive Force for Fractional Slot Windings, *Trans. AIEE*, Vol. 57, p. 777 (1938).
Gault, J. S., Rotor-bar Currents in Squirrel-cage Induction Motors, *Trans. AIEE*, Vol. 60, p. 784 (1941).
Graham, Q., Dead Points in Squirrel-cage Motors, *Trans. AIEE*, Vol. 59, p. 637 (1940).
Hellmund, R. E., and C. G. Veinott, Irregular Windings in Wound Rotor Induction Motors, *Trans. AIEE*, Vol. 53, p. 342 (1934).
Kron, G., Slot Combinations of Induction Motors, *Trans. AIEE*, Vol. 50, p. 737 (1931).
Lloyd, T. C., Polyphase Induction Motor Design to Meet Fixed Specifications, *Trans. AIEE*, Vol. 63, p. 14 (1944).
Lloyd, T. C., and H. B. Stone, The Design and Properties of the Magnetic Circuit, *Trans. AIEE*, Vol. 65, p. 812 (1946).
Merrill, W. J., Harmonic Theory of Noise in Induction Motors, *Trans. AIEE*, Vol. 59, p. 474 (1940).
Norman, H. M., Induction Motor Locked Saturation Curves, *Trans. AIEE*, Vol. 53, p. 536 (1934).
Trickey, P. H., Induction Motor Resistance Ring Width, *Trans. AIEE*, Vol. 55, p. 144.
Ware, D. H., Measurement of Stray-load Loss in Induction Motors, *Trans. AIEE*, Vol. 64, p. 194 (1945).

SINGLE-PHASE INDUCTION MOTORS

By T. C. Lloyd

31. TYPES AND THEORY

GENERAL. The outstanding characteristic of the single-phase motor is the complete absence of starting torque. Once started by one of several methods, it is capable of running without auxiliary devices. The starting method is used to distinguish one type of single-phase induction motor from another. These types are:

Shaded-pole.
Resistance split-phase.
Reactance split-phase.
Capacitor-start induction-run.
Permanent split capacitor.
Repulsion-start induction-run.

In general, the starting circuit of the motor is automatically opened at about two-thirds of final speed by means of a centrifugal switch or, in larger capacitor motors, by means of relays. Final acceleration and operation under load are accomplished by plain single-phase motor action. Exceptions are noted in the shaded-pole motor, in which the shading coil continues in action during normal operation, and in some reactance motors, as well as all permanent split capacitor motors, in which the auxiliary windings used for starting are also effective in producing running torque.

SHADED-POLE MOTORS. These are usually the smallest and least efficient of the single-phase induction motors. Most manufacturers list ratings of $^1/_{200}$ to $^1/_{20}$ hp, with a few up to $^1/_8$ or $^1/_6$ hp capacity. Efficiencies vary from 0.10 to 0.30 in the larger sizes.

Starting torque is low, usually about 25 to 50 per cent of full-load torque. Ratings are usually chosen so that maximum torque may be only 125 to 175 per cent of full-load torque. This ratio is lower on motors applied to fans and blowers because of the speed torque characteristics of the load, resulting in little chance of overloading and pull-out.

Construction of the shaded-pole motor is comparatively simple. Salient poles are used with concentrated stator windings. A copper band covering from one-third to one-half of the pole arc forms the shading coil. Rotors are of the usual squirrel-cage type with copper bars and end rings or with complete aluminum die-cast cages.

Speeds and Speed Control. Shaded-pole motors are usually available in speeds resulting from 2, 4, and 6 poles, but many are built for slow-speed fans, for about 24- or 28-pole speeds.

When used for fan and blower applications, fairly stable operation at reduced speeds is possible by motor "breakdown"; i.e., a series resistance or reactance is used with the stator winding to reduce the applied voltage, resulting in operation at speeds as low as half synchronous. Taps on the series impedance result in multispeed operation.

Theory of Operation. As the flux of the poles of the motor goes through its cyclic change, the shading coil gives a choking action, causing the flux of the shaded portion of the pole to delay in time. This results in a difference in phase between the fluxes in the shaded and the unshaded portions of the pole, and an effective flux sweep across the pole face. Small as this is, it is capable of providing a small starting torque.

No suitable method of design or motor analysis has ever been made for the concentrated winding type of shaded-pole motor. All designs are made experimentally by adjusting the degree of shading, resistance of shading coil, rotor resistance, length of air gap, and number of stator winding turns.

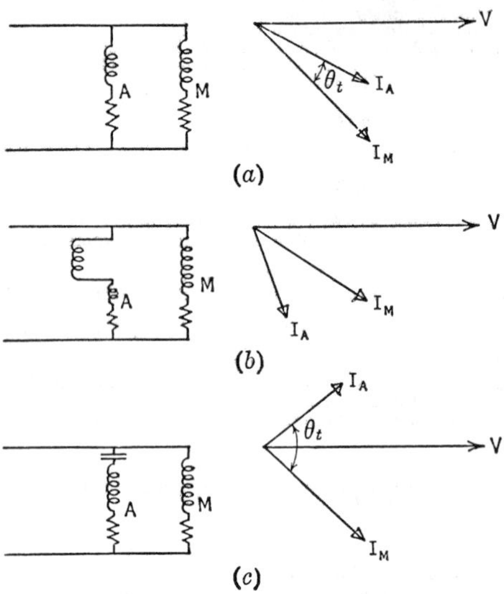

Fig. 1. Diagrams of Split-phase Motors

SPLIT-PHASE MOTORS. The principle of "splitting the phase" to produce single-phase motor-starting torque can be applied in any one of several ways.

Figure 1 (*a*) represents two field windings on the single-phase motor stator. The main winding, distributed (in concentric coils) around the periphery, gives the effect of the desired number of poles. The auxiliary winding (*A*) is distributed so that its poles are midway between the main poles.

The main winding is of comparatively high reactance and low resistance.

Resistance Split-phase. In this case the auxiliary winding has high resistance and comparatively low reactance. The currents in the two windings, arising from the same source, are then out of phase, as shown in the vector diagram.

With the windings displaced and the currents displaced in time, a rotating but pulsating magnetic field results in the gap. The motor gives a poor imitation of a two-phase action. Such a motor is usually built with a centrifugal switch cutting out the auxiliary winding, so that normal operation is on the main winding only.

Reactance-type Motor. By means of an external reactance connected in series with the auxiliary winding, reactance phase-splitting is accomplished as shown in Fig. 1 (*b*). Such motors commonly use the auxiliary phase for running as well as starting and, when built in small sizes for fan loads, achieve speed control by tapping the reactance coil. Because the so-called constants are greatly affected by saturation and ratings are usually small, calculation analyses for such motors are unsatisfactory.

Capacitor Motors. As shown in Fig. 1(*c*), the phase displacement in time, brought about through the use of a capacitor in the auxiliary phase, can give close to perfect two-phase action. More detailed analyses of these motors are given in later paragraphs. In every form, however, they represent a special case of splitting the supply phase.

STARTING-TORQUE CALCULATIONS; SPLIT-PHASE. The starting torque of the resistance split-phase motor in ounce-feet, using the blocked rotor currents of both the main, I_M and starting I_S windings, is:

$$T_s = \frac{225.4}{\text{Syn. rpm}} I_M I_S \frac{Z_s'}{Z_M'} a \sin \theta_t (R_2 K_r) \tag{1}$$

where $Z_s' = \left[\left(\frac{R_2K_r}{a^2}\right)^2 + X_s^2\right]^{1/2}$ (2) and $Z_M' = [(R_2K_r)^2 + X_M^2]^{1/2}$ (3)

R_2 = rotor resistance in main winding terms.

K_r = flux factor, $\frac{X_o - X}{X_o}$

X_s = leakage reactance of the start winding and rotor.

X_M = leakage reactance of the main winding and rotor.

a = ratio of effective main winding conductors to effective start winding conductors.

$a = \sqrt{X_M/X_s}$

θ_t = angle between main and start winding currents. See Fig. 1.

Expressed entirely in circuit constants:

$$T_s = \frac{225.4V^2R_2K_r}{\text{Syn. rpm}} \times \frac{Z_s'}{Z_M'} \times a \times \frac{X_MR_s - R_MX_s}{Z_M^2Z_s^2} \tag{4}$$

where $R_M = R_1 + R_2K_r$ (5)

$$R_s = R_{1s} + \frac{R_2K_r}{a^2} \tag{6}$$

$$Z_M = [R_M^2 + X_M^2]^{1/2} \tag{7}$$

$$Z_s = [R_s^2 + X_s^2]^{1/2} \tag{8}$$

As an approximation, with error increasing slightly with differences in winding distributions between main and start windings, $(Z_s'/Z_M')a$ in eq. 4 can be replaced by $1/a$.

Table 1. Typical Motor Constants (115 Volts, 60 Cycles)

Split-phase

Hp	Rpm	R_1	X	X_0	R_{1s}	X_s	K_r
1/20	1725	9.60	20.0	166	24.0	7.90	0.88
1/8	1725	4.10	7.75	108	10.4	4.28	0.93
1/3	1725	1.50	3.98	67	4.66	1.83	0.94
1/20	1140	14.1	18.0	134	20.2	5.02	0.86
1/4	1140	1.96	5.4	40	8.81	4.25	0.86

Capacitor-start

Hp	Rpm	R_1	X	X_0	R_{1s}	X_s	K_r	X_c
1/2	1725	0.78	2.87	44.6	2.72	2.96	0.93	7.4
1/4	1140	0.188	6.22	51.6	12.5	10.5	0.88	16.0

CAPACITOR-MOTOR TYPES. There are four types of capacitor motors, of which the first two listed below are more common.

Capacitor-start, induction-run motors use a high-capacity electrolytic-type capacitor for starting service only. A centrifugal switch or relay is employed in the start winding circuit. Starting torques of 200 to 350 per cent of full-load torque are feasible with acceptable starting currents. Ratings from 1/8 to 5 or 10 hp are used industrially.

Permanent-split capacitor motors use oil-paper-foil capacitors, permanently connected in the auxiliary winding circuit for both starting and running duty. Comparatively low capacitances are necessary for successful running performance, and they result in low starting torques. Values of 35 to 50 per cent of full-load torque are common. Greatest volume production is for fans, unit heaters, and similar applications.

Two-value capacitor motors use both types of capacitors described, with the electrolytic type cut out near final speed. This provides good starting torque with quiet, pulsation-free running operation.

Capacitor-transformer motors have the primary of a transformer connected in series with the auxiliary winding. The secondary leads connect to an oil-paper-foil capacitor, with high voltage impressed for starting duty and reduced voltage for running.

ELECTROLYTIC CAPACITORS. Effective industrial application of the capacitor-start motor depended upon the development of the electrolytic capacitor, resulting in large capacity with very little bulk. Favorable phase positions for high torque, and high torque per starting ampere, require capacities too bulky and expensive when provided by the usual capacitor construction. The development of the electrolytic type (see Section 6) reduced bulk from 25 to 50 times for a given capacity.

Electrolytic capacitors for motor-starting duty are usually limited to about 20 starts per hour of 3-sec duration and are given a double rating covering a manufacturing tolerance. See Table 2 for typical values.

Table 2. Typical Capacitor Mfd [Electrolytic Type for Starting Duty (115 Volts)]

Horsepower	3450 Rpm	1725 Rpm	1140 Rpm
1/6	124–149	124–149	124–149
1/4	124–149	124–149	233–281
1/3	158–191	158–191	233–281
1/2	233–281	233–281	233–281
3/4	341–412	460–550	341–412
1	460–550	460–550	682–824

STARTING TORQUE OF THE CAPACITOR MOTOR. With the circuit constants known, the starting torque in ounce-feet can be calculated:

$$T_s = \frac{225.4V^2R_2K_r}{\text{Syn. Rpm}} \times \frac{Z_s'}{Z_M'} \times a \times \frac{X_M R_s - R_M(X_s - X_c)}{Z_M^2 Z_{sc}^2} \tag{9}$$

where Z_M, Z_s', Z_M', a, X_s, X_M, R_M, and R_2 have been defined with the split-phase motor equation.

X_c = the capacity reactance of the capacitor.

r_c = the resistance of the capacitor.

$$R_s = R_{1s} + r_c + \frac{R_2K_r}{a^2} \tag{10}$$

where R_{1s} = resistance of starting winding.

$$Z_{sc} = [R_s^2 + (X_s - X_c)^2]^{1/2} \tag{11}$$

It is possible to modify the value of the capacitor until maximum starting torque is obtained for a given set of windings. This value, as well as other relationships, can best be shown by a circle diagram, illustrated in Fig. 2. On this diagram:

OM is the main winding current at standstill.

MC is the circle diameter and represents the resonance current in the auxiliary winding circuit.

MS is the current taken by the auxiliary winding if this were a split-phase motor, i.e., $X_c = 0$.

As the capacity reactance varies, the auxiliary winding current swings around a circular locus. With the tangent through P parallel with OM, MP represents the auxiliary winding current for the point of maximum starting torque. The total current is OP, and the reactive volt-amperes of the capacitor is the vertical S_1P. Starting torque is proportional to the perpendicular T_1P.

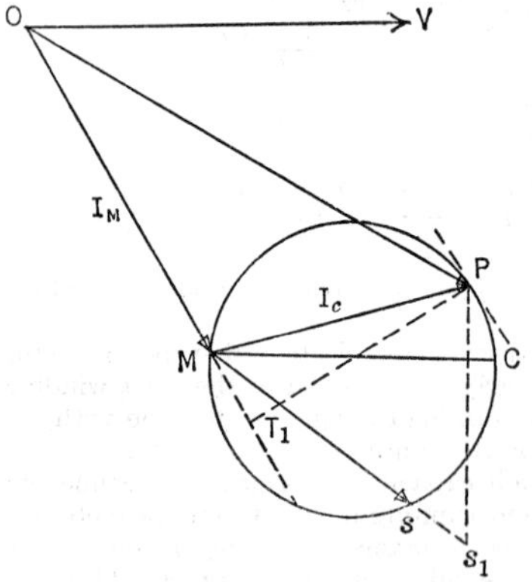

FIG. 2. Circle Diagram of Capacitor-start Motor

Table 3. Typical Performance

Shaded-pole Motors

Hp	Approx. Input, Watts	Efficiency, Per Cent	Starting Torque, Ounce-feet	Max. Torque, Ounce-feet	Speed, Rpm
1/250	28	10.7	0.156	0.203	3200
1/150	35	14.3	0.187	0.382	3200
1/100	45	16.7	0.203	0.437	3200
1/60	58	21.6	0.250	0.593	3200
1/25	120	25.0	0.468	1.35	3200
1/250	27	11.1	0.250	0.500	1600
1/150	31	16.2	0.310	0.720	1600
1/100	37	20.3	0.375	1.000	1600
1/60	48	26.0	0.500	1.470	1600
1/25	101	29.7	0.688	3.060	1600
1/150	36	13.9	0.350	0.950	1050
1/100	44	17.1	0.640	1.200	1050
1/60	58	21.5	0.900	1.900	1050
1/25	113	26.5	1.900	4.250	1050

Split-phase Motors

Hp	Locked Amperes	Starting Torque / Full-load Torque	Max. Torque / Full-load Torque	Efficiency, Per Cent	Speed, Rpm
1/12	13.9	2.15	3.45	48	3450
1/12	17.1	2.15	3.10	52	1725
1/8	16.9	1.75	2.35	55	3450
1/8	17.2	1.53	7.55	63	1725
1/4	25.0	1.30	2.60	66	3450
1/4	30	1.65	2.50	66	1725

Capacitor-start Motors (1725 Rpm)

Hp	Locked Amperes	Starting Torque / Full-load Torque	Max. Torque / Full-load Torque	Efficiency, Per Cent	Capacitor, Approx. Microfarads
1/8	12.1	3.80	2.40	57	140
1/4	22.0	3.50	3.00	63	140
1/2	35.4	3.75	2.50	67	260
1	60.0	2.50	2.75	75	390
3	113.0	2.60	2.60	75	594
5	202.0	2.50	2.25	77.5	956

REPULSION-START INDUCTION-RUN MOTORS. The performance of the repulsion motor (see Article 36) makes this construction highly desirable as a starting method for single-phase motors. Higher starting torques per starting ampere are possible on this type of motor.

The armature is provided with a winding and commutator of the d-c type, with brushes short circuited. The field winding is distributed, usually with concentric-type coils, and is connected to the supply line. With the brushes shifted with reference to the field, one portion of the field winding is conveniently considered as inducing electromotive forces and resulting currents in the armature winding which react with the flux of the other component of the field to produce torque. The torque and current are affected by the brush-shift angle, which varies from about 15 to 20 deg for various designs.

As the motor reaches about two-thirds of synchronous speed, a centrifugal mechanism short-circuits the bars of the commutator so that the armature coils are equivalent to a squirrel-cage winding. Reactions between the field and the rotor are then similar to those of other single-phase motors and can be determined by the analyses which follow.

To avoid wear on the brushes and the resulting noise, some repulsion-start induction-run motors are constructed so that the brushes are lifted from the commutator surface by the same mechanism which short-circuits the commutator. These are the so-called "brush-lifting" type.

These motors are essentially constant speed and are usually built in ratings of 1/8 to **5 hp.** High starting torque is the chief basis for using them on such applications as compressor duty. The inherent high starting torque results in quick acceleration, especially useful on medium-sized grinders, floor polishers, sanders, and other portable tools.

METHODS OF ANALYSIS. Two theories are used to explain the action of the single-phase induction motor: the doubly revolving field theory and the cross-field theory. Both may involve various simplifying assumptions, which have some effect on their accuracy. They can be proven to be exactly equivalent in their final expressions for torque, output, efficiency, etc.

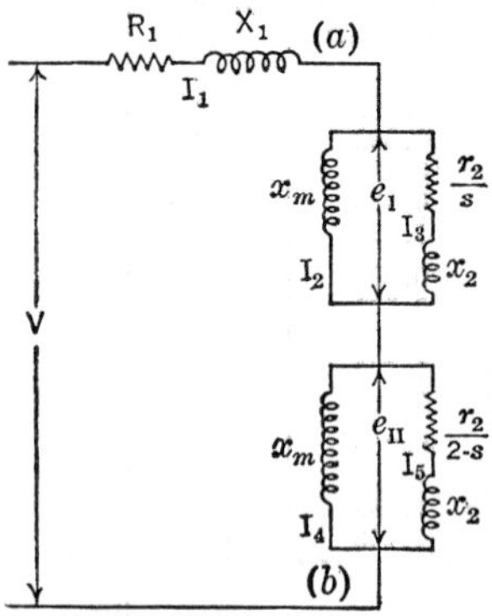

FIG. 3. Equivalent Circuit for Doubly Revolving Field Analysis

THE DOUBLY REVOLVING FIELD THEORY. This theory assumes that the "reciprocating" field built up between unlike poles of the stator winding degenerates into two oppositely rotating fields. The rotor follows one or the other, depending upon in which direction it is started, and finally operates with a slip of s with respect to the **forward** field, and a slip of $(2 - s)$ with respect to the **backward** field. The relative magnitude of the field varies with the slip.

To apply this theory it is assumed that the actual rotor is replaced by two imaginary rotors, each with half of the actual rotor resistance and leakage reactance. The magnetizing reactance of the motor is halved also when applied to an equivalent circuit, as shown in Fig. 3.

This approach to motor analysis is very closely allied to that developed by the symmetrical components method.

Solution of the Circuit. The equivalent circuit, as shown, neglects the core and friction losses. The friction loss is usually subtracted from the output, treating it as a motor load. The core loss can be handled in a variety of ways, as will be shown later. (See Fig. 3 for symbols.)

Forward torque in synchronous watts:

$$T_f = I_3^2 \frac{r_2}{s} \tag{12}$$

s = slip as decimal fraction, used throughout this method

Backward torque in synchronous watts:

$$T_b = I_5^2 \frac{r_2}{2 - s} \tag{13}$$

Resulting useful torque:

$$T = T_f - T_b$$

Impedance of the "forward" rotor:

$$Z_I = \frac{jX_m \left(\frac{r_2}{s} - jx_2\right)}{\frac{r_2}{s} + j(X_2 + X_m)} \tag{14}$$

Impedance of the "backward" rotor:

$$Z_{II} = \frac{jX_m \left(\frac{r_2}{2 - s} + jx_2\right)}{\frac{r_2}{2 - s} + j(x_2 + X_m)} \tag{15}$$

Total impedance:

$$Z_T = Z_1 + Z_I + Z_{II} \tag{16}$$

where Z_1 = stator impedance.

For the assumed slip, the above values of impedance enable one to calculate primary current and power factor.

$$e_I = I_1 Z_I \quad \text{and} \quad I_3 = \frac{e_I}{Z_3}, \qquad e_{II} = I_2 Z_{II} \quad \text{and} \quad I_5 = \frac{e_{II}}{Z_5}$$

From these values component currents and torques can be calculated.

Output watts = Syn. watts $(1 - s)$ (from eqs. 12 and 13) (17)

Net output = Output watts − Friction and windage (18)

Efficiency (see Fig. 3) = Net output ÷ (VI_1 × Power factor) (19)

Treatment of Core Loss. The above method completely neglects the core loss, which is relatively large in small, fractional-hp motors. The least exact method of considering this loss would be to add it directly to the input watts as a constant. A second approximation would involve an equivalent resistance added between points *a* and *b* on Fig. 3. A more exact method involves the determination of an equivalent core-loss resistance based on about 90 per cent of the applied voltage and the use of such a resistance in both halves of the circuit.

THE CROSS-FIELD THEORY. Although the initial concepts underlying this analysis may be more difficult to grasp than those of the previous theory, actual calculation is probably no more lengthy.

The following steps give an outline of the factors involved:

When the squirrel-cage rotor turns through the magnetic field of the air gap, it sets up a generated emf, short circuited in the rotor cage. This component of current flowing in the rotor is magnetizing, with its resulting poles midway between the main stator poles.

As a result, we must deal with two flux systems around the air gap: (1) that of the main poles, and (2) that originating in the rotor and more or less in quadrature with the first. These are the main (or y-axis) and the cross (or x-axis) fluxes, respectively.

Speed action, resulting from the rotor bars cutting the y-axis flux, generates a rotor component of voltage, E_{SM}, in phase with the flux causing it.

Rotation of the rotor bars through the x-axis flux generates a component voltage, E_{sc}, in phase or in phase opposition to the flux cut.

Pulsation of the y-axis flux induces, by transformer action, a counter-emf in the main field winding, E_{TM}. This is 90 deg behind the flux causing it.

Pulsation of the x-axis flux induces, by transformer action, a voltage in the rotor, E_{TC}. This is also 90 deg behind the flux causing it.

Any speed voltage equals S times the voltage set up by pulsation of its originating flux. These component voltages result in current components which give local impedance drops. Then, by Kirchhoff's law, it can be shown for the y-axis components:

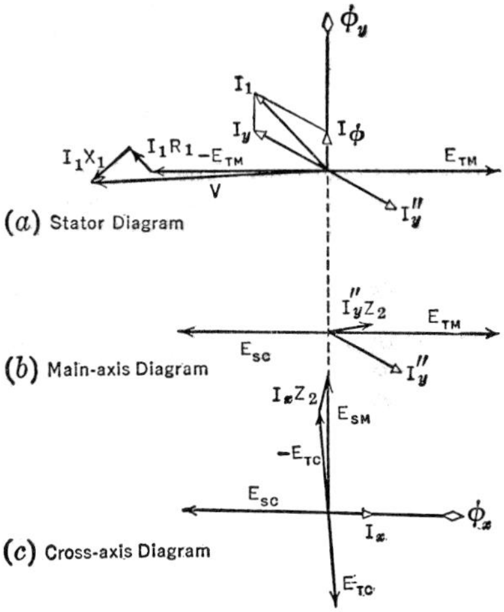

FIG. 4. Vector Diagrams for Cross-field Theory

$$I_yZ_2 = jX_m(I_1 - I_y) + SI_x(X_m + X_2) \tag{20}$$

where I_y = the rotor load component of current in stator terms.
I_1 = the total stator current.
$I_\phi = I_1 - I_y$ (as vectors).
= the magnetizing current in the stator winding.
I_x = the x-axis rotor current in stator terms.

Similarly, for the x axis:

$$I_xZ_2 = SX_m(I_1 - I_y) - jX_mI_x - SI_yX_2 \tag{21}$$

For the stator circuit:

$$V = I_1Z_1 + jX_m(I_1 - I_y) \tag{22}$$

In these equations $Z_1 = R_1 + jX_1$
= the stator winding impedance.
$Z_2 = R_2 + jX_2$
= the rotor winding impedance in stator terms (and twice the values used for the doubly revolving field theory).
X_m = the magnetizing reactance for the main winding, assumed to be the same in the x axis. (This is twice the value used in the doubly revolving field theory.)
S = the ratio of actual speed to synchronous speed, as a decimal fraction.

The above equations can be solved simultaneously to obtain working formulas for currents at various speeds. See Fig. 4.

$$I_1 = -V \frac{-R_2^2 + (1 - S^2)X_0^2 - j2R_2X_0}{U + jW} \quad (23)$$

$$I_y = -VX_m \frac{(1 - S^2)X_0 - jR_2}{U + jW} \quad (24)$$

$$I_x = +V \frac{SX_mR_2}{U + jW} \quad (25)$$

$$X_0 = X_m + X_1$$

$$U = -R_1R_2^2 + 2R_2X_1X_o + R_2X_m(X_o + X_2) + (1 - S^2)R_1X_o^2 \quad (26)$$

$$W = -R_2^2X_1 - 2R_1R_2X_0 - R_2^2X_m + (1 - S^2)(X_1X_0^2 + X_2X_mX_0) \quad (27)$$

Torque. The interaction of the y-axis current and the x-axis flux produces pulsating but useful torque. The interaction of the x-axis current and the y-axis flux produces a retarding torque, which is a measure of the x-axis losses. This also pulsates. The net torque in synchronous watts is:

$$T = \frac{V^2X_m^2R_2S[(1 - S^2)X_0^2 - R_2^2]}{U^2 + W^2} \quad (28)$$

No-load Conditions. The ideal no-load speed of a single-phase induction motor is not synchronous speed. Use of the theoretically correct value results in cumbersome expressions for no-load current, and so as a close approximation:

$$I_o = \frac{2V}{X_0(2 - K_r)} \quad (29)$$

Also the x-axis magnetizing current at no-load is approximately:

$$I_{x0} = \frac{V}{X_0} \times \frac{K_r}{2 - K_r} \quad (30)$$

where

$$K_r = \frac{X_0 - X}{X_0} \quad (31)$$

and

$$X = X_1 + X_2 \quad (32)$$

$$X_0 = X_m + X_1 \quad (33)$$

EQUIVALENT CIRCUIT. Although complete performance can be obtained through solution of the above equations, it is also possible to set up an equivalent network, which can be solved for various values of speed. Such a circuit, with core loss considered, is

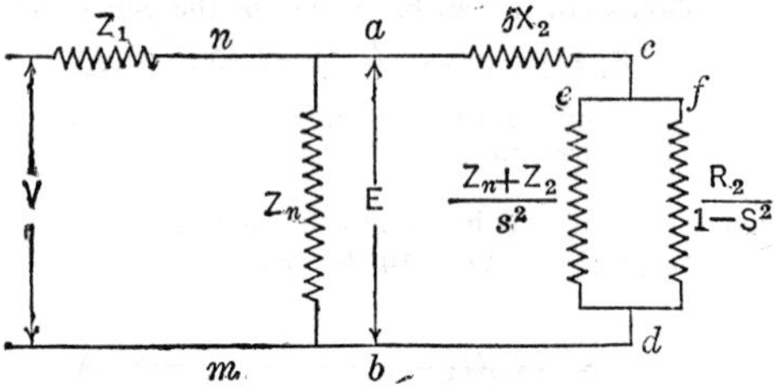

FIG. 5. Equivalent Circuit for the Cross-field Analysis

shown in Fig. 5. It is assumed that the core loss will be equally divided between y- and x-axes. Let the equivalent resistance, in which one-half of the core loss would appear, be R_F. Then:

$$Z_n = \frac{1}{(1/R_F) + (1/jX_m)} = R_n + jX_n \quad (34)$$

Output is obtained from input minus losses.
Main axis rotor loss = $I_{fd}^2R_2$.
x-axis core loss = $I_{ed}^2R_n$.
x-axis copper loss = $I_{ed}^2R_2$.
Stator copper loss = $I_1^2R_1$.
Stator core loss = $I_\phi^2R_n$.
Friction and windage losses are subtracted from the apparent output.

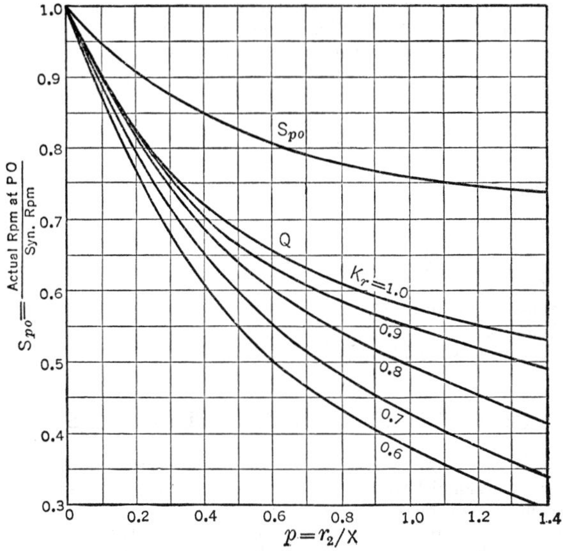

FIG. 6. Diagram for Determining Speed Ratio at Which Maximum Torque Occurs

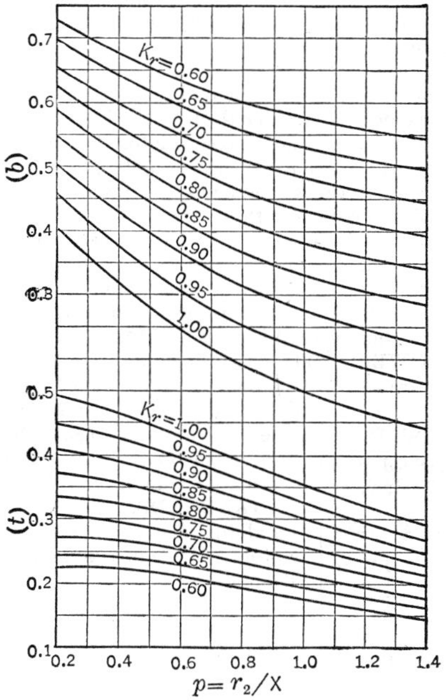

FIG. 7. Diagram for Calculation of Maximum Torque

MAXIMUM TORQUE. No simple expression exists for the maximum torque of the single-phase induction motor. It can be calculated by reading the speed ratio (S_{p0}) at which maximum torque occurs (from Fig. 6) and substituting in the previous analyses. A more direct, but still cumbersome, method is given below.

For the correct value of r_2/X and K_r read Q, t, and b from Figs. 6 and 7. Calculate Q'.

$$Q' = \frac{1}{(1 + p_1 t)^2 + (p_1 b)^2} \tag{35}$$

Then the maximum torque of the single-phase induction motor in ounce-feet is:

$$T_m = \frac{112.6}{\text{Syn. rpm} \times S_{p0}} \left[\frac{V^2}{X} t \frac{Q}{Q'} - F - F_e(c) \right] \tag{36}$$

where $p_1 = R_1/X$.
X = total leakage reactance.
F = friction loss.
$F_e(c)$ = x-axis core loss, usually one half of the total.

32. CONSTRUCTION AND STANDARDS

FRAMES. Single-phase motors can be divided into two groups by size. The majority of such motors are fractional horsepower, with integral-horsepower sizes limited largely to capacitor-start and repulsion-start types.

Fractional-horsepower motors of approximately 4-in. diameter or less are constructed with two steel "cups" fitting over the field to form a combination body and head. In such construction porous bronze bearings of the swivel or self-aligning type are commonly used, with felt washers forming the oil reservoir.

Frames for motors rated from about $^1/_8$ to 1 hp are usually constructed of rolled steel (wrapped around the field core) with a base or foot construction welded or screwed to the body. Turned fits on the ends of the body match the fits on the cast-iron heads to assure bearing and air-gap alignment.

Bearings are usually of bronze or steel-backed babbitt, with wool yarn packing as an oil reservoir. For extreme conditions of belt pull or for vertical operation, ball bearings are commonly used because of their greater thrust capacity.

The present tendency in fractional-horsepower motors is toward a protected construction, with openings in only the bottom half of the motor, and so shielded as to be drip-proof. Totally enclosed construction is commonly required also.

STANDARD FLANGES. There are three standardized frames in fractional-horsepower sizes. Motors, as manufactured by various concerns, have the same height of shaft above base, base mounting dimensions, and other features which affect interchangeability. Many non-standard frames, however, are also in production.

A flange and a face-type mounting, which are widely used in the oil-burner, pump, and other industries, have been standardized by the NEMA.

INTEGRAL HORSEPOWER MOTORS. Single-phase motors of 1 hp or above are usually built in the heavy frame constructions common to polyphase machines, as already described. NEMA-standardized frames are available in these ratings. Drip-proof, splashproof, totally enclosed, and totally enclosed, fan-cooled types are common.

STANDARDS. The ASA and the NEMA have standardized many items pertaining to single-phase motors. Some of the more important are given in Table 4.

Because of the variation commonly found in the maximum torques of similarly rated motors of different manufacture, definition of horsepower by torques has recently been completed by NEMA. If the maximum torque of a motor falls within a certain range, this automatically fixes the horsepower rating. Along with each rating is a service factor, as indicated in Table 5.

Table 4. Locked Rotor Currents *

2-, 4-, 6-, and 8-pole, 60-cycle Single-phase Motors, Speeds 900–3600 Rpm, Inclusive

Rated Horsepower	115 Volts	230 Volts
1/6 and smaller	20	10
1/4	23	11 1/2
1/3	31	15 1/2
1/2	45	22 1/2
3/4	61	30 1/2

* Exceptions are made for split-phase motors used on washing machines and ironing machines.

Table 5. Basis of Rating—Single-phase Motors *

Small-power single-phase induction motors shall be rated primarily on the basis of breakdown torque. The value of breakdown torque for the purpose of defining horsepower rating shall fall within the following ranges:

Synchronous rpm.....	3600	3000	1800	1500	1200	900
Approx. full-load rpm.	3450	2850	1725	1425	1140	850
Brake-hp Rating	Breakdown Torque, Ounce-feet					
1/20	2.0–3.7	2.4–4.4	4.0–7.1	4.8–8.5	6.0–10.4	8.0–13.5
1/12	3.7–6.0	4.4–7.2	7.1–11.5	8.5–13.8	10.4–16.5	13.5–21.5
1/8	6.0–8.7	7.2–10.5	11.5–16.5	13.8–19.8	16.5–24.1	21.5–31.5
1/6	8.7–11.5	10.5–13.8	16.5–21.5	19.8–25.8	24.1–31.5	31.5–40.5
1/4	11.5–16.5	13.8–19.8	21.5–31.5	25.8–37.8	31.5–44.0	40.5–58.0
1/3	16.5–21.5	19.8–25.8	31.5–40.5	37.8–48.5	44.0–58.0	58.0–77.0
1/2	21.5–31.5	25.8–37.8	40.5–58.0	48.5–69.5	58.0–82.5	....
3/4	31.5–44.0	37.8–53.0	58.0–82.5	69.5–99.0	....	....
1	44.0–58.0	53.0–69.5	....	....	....	....

* From NEMA publication 47-121.

"When authorized by the manufacturer, . . . a motor may be operated (at rated voltage and frequency and in an ambient temperature not exceeding 40 deg cent) at a continuous load greater than rated load determined by the service factor given in Table 6 with possible differences in efficiency and power factor from those at rated load." *

Table 6. Service Factors

Horsepower	Service Factor
1/20	1.4
1/12	1.4
1/8	1.4
1/6	1.35
1/4	1.35
1/3	1.35
1/2	1.25
3/4	1.25
1 at 3600 rpm only	1.25

Table 7. Torques as a Percentage of Full-load Torque (Split-Phase)

Horsepower Rating	Locked Rotor	Pull-up	Breakdown
60 Cycles, 1725 Rpm			
1/8	150	150	200
1/6	150	150	200
1/4	90	90	185
60 Cycles, 1140 Rpm			
1/8	125	125	175
1/6	125	125	175
1/4	75	75	175

Old standards, common to the fractional-horsepower industry and still useful for showing minimum values of performance to be expected from various motor types, are given in Tables 7 and 8.

Table 8. Capacitor-start and Capacitor Motor

Horsepower Rating	Locked Rotor	Pull-up	Breakdown	Horsepower Rating	Locked Rotor	Pull-up	Breakdown
High Torque 60 Cycles, 1725 Rpm				Repulsion-start Induction 60 Cycles, 1725 Rpm			
1/8	350	200	200	1/8	350	200	200
1/6	350	200	200	1/6	350	200	200
1/4	350	200	200	1/4	350	200	200
1/3	325	200	200	1/3	350	200	200
1/2	300	200	200	1/2	350	200	200
3/4	275	200	200	3/4	350	200	200
60 Cycles, 1140 Rpm				60 Cycles, 1140 Rpm			
1/8	300	185	185	1/8	300	150	185
1/6	300	185	185	1/6	300	150	185
1/4	300	185	185	1/4	300	150	185
1/3	300	185	185	1/3	300	150	185
1/2	300	185	185	1/2	300	150	185

* From NEMA Publication 47-121.

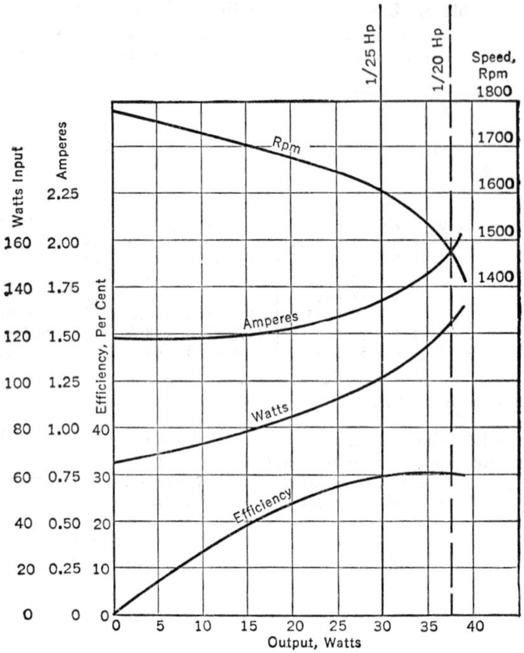

FIG. 8. General Characteristics of a Shaded-pole Motor

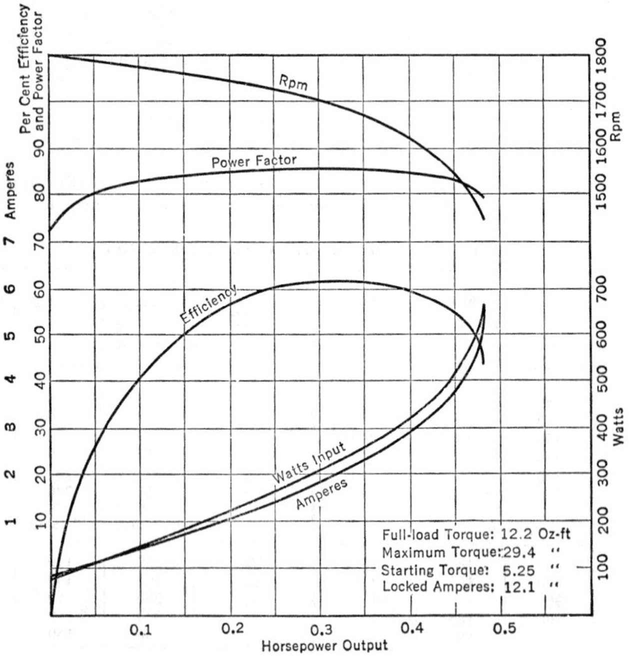

FIG. 9. Characteristics of a Permanent Split Capacitor Motor

SINGLE-PHASE MOTOR CHARACTERISTICS. The general characteristics of six types of single-phase motors are listed in Table 9 and Figs. 8, 9, and 10.

Table 9. Single-phase Motor Characteristics

Type	Usual Range of Horsepower	Speed Control	Approx. Efficiency, Per Cent	Reverse	Characteristics
Shaded pole	1/200–1/20	Series impedance	10–30	No	1
Reactance	1/200–1/8	Reactance *	10–40	By leads	2
Split-phase	1/20 –1/3	Constant †	40–70	" "	3
Permanent capacitor	1/100–1	Reactance *	30–75	" "	4
Capacitor-start	1/8 –5	Constant †	55–85	" "	5
Repulsion-start	1/8 –5	Constant †	55–85	Brush-shift	6

* The external reactance used for speed control is tapped so that coils may be switched from auxiliary to main winding circuits. Reconnection of two halves of auxiliary winding from parallel to series is also used for speed reduction.

† For fan and blower duty, split-phase and capacitor start motors are sometimes operated at reduced voltage for lower speed. Special design is often required for good "breakdown" characteristics. The first two types of these motors are sometimes provided with two sets of windings for different speeds by pole changing.

1. Inexpensive. Poor starting torque. Used on fans.
2. Inexpensive. Poor starting torque. Quiet. Used on fans.
3. Least expensive of higher starting-torque types. General purpose.
4. Poor starting torque. Quiet, vibrationless. Speed control. Used on fans.
5. Good starting torque. General-purpose above split-phase range.
6. Best starting torque. Low starting current. General purpose.

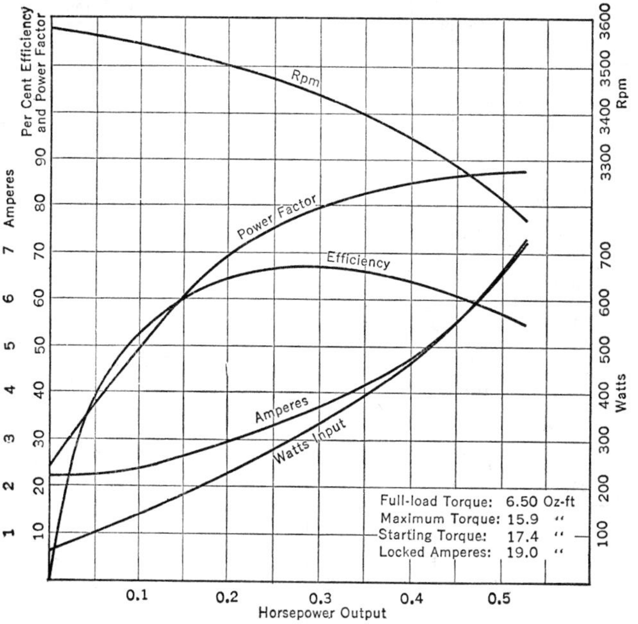

FIG. 10.

33. DESIGN

GENERAL. Certain motor types, such as the shaded-pole and reactance split-phase, are "designed" largely by the use of incomplete empirical relationships, plus trial and error. The complexities of the permanent split-capacitor analyses complicate design procedures to such an extent that final designs frequently depend upon trial and error.

A recent equivalent circuit development by S. S. L. Chang has greatly simplified this process. (See Article 34.)

Split-phase, capacitor-start, and repulsion-start motors are subject to comparatively straightforward design methods too lengthy for this presentation. An indirect method of design is to assume the laminations, core length, and windings and calculate the motor constants. Analyses already presented then enable the performance to be calculated. If these requirements are not met, adjustments by proportion can be made until a satisfactory motor is obtained. Such a method depends inherently on the ability to calculate winding constants as presented below.

STATOR WINDING DISTRIBUTION. The percentage of total turns per pole used in each of the concentric coils which make up the field winding can vary over a wide range. To obtain the minimum disturbing effects of space harmonics in the mmf wave, however, the relative distribution of turns in the concentric coils should follow the schedule shown in Table 10.

For instance, with 9 slots per pole, using 3 coils, the fullest slot is listed as the 100 per cent coil and spans 8 teeth; the inner coil should have 88.4 per cent as many turns; and

Table 10. Concentric-type Windings *

(Per Cent Distribution for Sinusoidal Wave Form)

Tooth Span of Coil	4 Coils per Pole	3 Coils per Pole	2 Coils per Pole
9 Slots per Pole			
2	35.0%		
4	65.5%	65.5%	
6	88.4%	88.4%	88.4%
8	100.0%	100.0%	100.0%
K_w	0.793	0.855	0.929
F	5.778	5.078	3.768
K_0	1.34	1.56	2.03
3	53.3%		
5	81.2%	81.2%	
7	100.0%	100.0%	100.0%
9	53.0%	53.0%	53.0%
K_w	0.821	0.893	0.961
F	5.760	4.690	3.066
K_0	1.43	1.72	2.59
8 Slots per Pole			
2	41.6%		
4	76.1%	76.1%	
6	100.0%	100.0%	100.0%
8	54.1%	54.1%	54.1%
K_w	0.795	0.870	0.950
F	5.436	4.604	3.082
K_0	1.34	1.64	2.35
3		70.4%	
5		84.7%	84.7%
7		100.0%	100.0%
K_w		0.815	0.913
F		5.102	3.694
K_0		1.42	1.89

Tooth Span of Coil	4 Coils per Pole	3 Coils per Pole	2 Coils per Pole
6 Slots per Pole			
2		57.7%	
4		100.0%	100.0%
6		57.7%	57.7%
K_w		0.804	0.914
F		4.308	3.154
K_0		1.38	1.90
3			73.4%
5			100.0%
K_w			0.856
F			3.468
K_0			1.56
4 Slots per Pole			
2			100.0%
4			64.5%
K_w			0.822
F			3.29
K_0			1.43

Effective conductors $= CK_w$
Equivalent full slots $= F \times P$

* After Appleman (see Bibliography, Article 34).

the center coil should have 65.5 per cent as many turns. When this is followed, the winding factor K_w (defined as the ratio of effectiveness of any winding to a full-pitch, concentrated winding) is then 0.855.

WINDING CONSTANTS. With a trial distribution and total number of series conductors, C, assumed as described, the winding constants can be calculated as follows. All symbols are the same as those given in the polyphase design procedure (Article 28).

$$\text{Flux per pole} = \frac{V \times 10^8}{2.22CK_wf} \tag{37}$$

Slot permeance:

$$P_s = 3.60L\left(\frac{K_s'}{S_1} + \frac{K_s''}{S_2}\right) \tag{38}$$

Zig-zag permeance:

$$P_{zz} = \frac{8.5D}{S_1S_2\Delta}\left(\frac{T_c' + T_c''}{\lambda' + \lambda''} - 0.5\right)^2 L \tag{39}$$

End-coil permeance:

$$P_e = \frac{\lambda}{P}\cdot K_e \tag{40}$$

$K_e = 0.8$ for average end-ring construction, or 0.7 for complete disk end rings.

Total leakage permeance:

$$P_L = P_s + P_{zz} + P_e \tag{41}$$

Total leakage reactance:

$$X = 2\pi fC^2P_LK_0 10^{-8} \tag{42}$$

K_o is the distribution correction for all components of leakage reactance as read from Table 10.

Magnetizing Reactance. Permeance of mutual flux path:

$$P_m = 0.644\frac{\lambda LK_w^3}{P\Delta K_1(\text{S.F.})} \tag{43}$$

$$X_m = 2\pi fC^2P_m10^{-8} \tag{44}$$

$$X_0 = X_m + {}^1\!/_2X \tag{45}$$

Flux Factors.

$$K_r = \frac{X_0 - X}{X_0} \tag{46}$$

$$K_p = K_s = \sqrt{K_r} \tag{47}$$

No-load Current. The above terms enable the no-load current, I_0, to be calculated by eq. (29).

Winding Resistances. It is necessary to estimate the length of turn on each coil of the field winding and obtain a weighted average for the entire winding. The length of one conductor (one-half turn) is l_c inches. Then:

$$R_1 = \frac{l_cC^2}{A_skFP}\ \text{(at 60 deg cent)} \tag{48}$$

where F = the number of equivalent full slots per pole. See Table 10.

In selecting values for k it should be kept in mind that the product A_sk represents useful slot area in circular mils allowable for the main winding. Some of the same slots must also hold start-winding conductors. See eq. (70), Article 28, for definition of other terms.

As an alternate method, if the number of conductors has been chosen for trial, and the available slot area is then calculated, the allowable wire size can be determined from the fullest slot. When the wire size, the length of conductor (l_c), and the number of conductors are given, resistance can be calculated in the usual manner.

Rotor resistance in stator terms:

$$R_2 = 2C^2k_w^2(\text{bar and ring})\ \text{(at 60 deg cent)} \tag{49}$$

See eqs. (72) and (73) for terms.

NO-LOAD LOSSES. These consist of copper losses in both stator and rotor, core loss, friction and windage losses.

$$\text{Stator copper loss} = I_0^2R_1$$

$$\text{Approximate rotor copper loss} = I_0^2(0.5K_r)R_2$$

Core loss, including the surface losses, can be calculated in the manner shown for polyphase motors. K_r is the leakage factor.

SUMMARY. Calculation of the motor-winding constants by the formulas given above makes it possible to evaluate an equivalent circuit for a check on performance. Maximum torque can be calculated directly as shown.

The addition of trial windings for starting methods (split-phase or capacitor) can be followed by a calculation of these winding constants and the determination of starting torque and current.

34. BIBLIOGRAPHY

(For books, see Bibliography for Polyphase Motors, Article 30.)

Ager, R. W., An Application of Deceleration Test Methods to Induction Motors, *Trans. AIEE*, Vol. 58, p. 72 (1939).
Alger, P. L., and T. C. Johnson, Rating of General-purpose Induction Motors, *Trans. AIEE*, Vol. 58, p. 445 (1939).
Appleman, W. R., The Cause and Elimination of Noise in Small Motors, *Trans. AIEE*, Vol. 56, p. 1359 (1937).
Branson, W. J., Single-phase Induction Motors, *Trans. AIEE*, Vol. 31, p. 1749 (1912).
Button, C. T., Single-phase Motor Theory—Correlation of the Cross-field and Revolving-field Concepts, *Trans. AIEE*, Vol. 60, p. 507 (1941).
Chang, Sheldon S. L., The Equivalent Circuit of the Capacitor Motor, *AIEE Tech. Paper* 47-101.
Hildebrand, L. E., Duty Cycles and Motor Rating, *Trans. AIEE*, Vol. 58, p. 478 (1939).
Himebrook, F. S., Single-phase Induction-motor Performance Calculation, *Trans. AIEE*, Vol. 60, p. 55 (1941).
Koch, C. J., Measurement of Stray Load Loss in Polyphase Induction Motors, *Trans. AIEE*, Vol. 51, p. 756 (1932).
Kuhn, C. W., A-c Motor Protection, *Trans. AIEE*, Vol. 56, p. 589 (1937).
Leader, C. C., and F. D. Phillips, Efficiency Tests of Induction Machines, *Trans. AIEE*, Vol. 53, p. 1628 (1934).
Lloyd, T. C., and Sheldon S. L. Chang, A Design Method for Capacitor Motors, *AIEE Tech. Paper* 47-102.
Lloyd, T. C., and J. H. Karr, Design of Starting Windings for Split-phase Motors, *Trans. AIEE*, Vol. 63, p. 9 (1944).
Lyon, W. V., and Charles Kingsley, Jr., Analysis of Unsymmetrical Machines, *Trans. AIEE*, Vol. 55, p. 471 (1936).
Macmillan, C., and G. K. Carter, Overvoltages in Polyphase Induction Motors during Single-phase Operation, *Trans. AIEE*, Vol. 60, p. 819 (1941).
Morgan, T. H., W. E. Brown, and A. J. Schumer, Test for Stray Load Loss in Induction Machines, *Trans. AIEE*, Vol. 58, p. 319 (1939).
Morgan, T. H., and P. M. Narbutovskih, Stray Load Loss Test on Induction Machines, *Trans. AIEE*, Vol. 53, p. 286 (1934).
Morrill, W. J., Calculating Single-phase-motor Performance, *Trans. AIEE*, Vol. 60, p. 1037 (1941).
Morrill, W. J., Characteristic Constants of Single-phase Induction Motors, Air-gap Reactances, *Trans. AIEE*, Vol. 56, p. 333 (1937).
Morrill, W. J., Revolving Field Theory of the Capacitor Motor, *Trans. AIEE*, Vol. 48, p. 614 (1929).
Potter, C. P., Measurement of Temperature in Induction Motors, *Trans. AIEE*, Vol. 58, p. 468 (1939).
Puchstein, A. F., and T. C. Lloyd, Capacitor Motors with Windings Not in Quadrature, *Trans. AIEE*, Vol. 54, p. 1235 (1935).
Puchstein, A. F., and T. C. Lloyd, Cross-field Theory of the Capacitor Motor, *Trans. AIEE*, Vol. 60, p. 58 (1941).
Puchstein, A. F., and T. C. Lloyd, Single-phase Induction-motor Performance, *Trans. AIEE*, Vol. 56 (1937).
Reed, H. R., and R. J. W. Koopman, Induction Motors on Unbalanced Voltages, *Trans. AIEE*, Vol. 55, p. 1206 (1936).
Trickey, P. H., An Analysis of the Shaded-pole Motor, *Trans. AIEE*, Vol. 55, p. 1007 (1936).
Trickey, P. H., Equal Volt-ampere Method of Designing Capacitor Motors, *Trans. AIEE*, Vol. 60, p. 990 (1941).
Trickey, P. H., Performance Calculations on Capacitor Motors, *Trans. AIEE*, Vol. 60, p. 73 (1941).
Trickey, P. H., Performance Calculations on Repulsion Motors, *Trans. AIEE*, Vol. 60, p. 67 (1941).
Veinott, C. G., Segregation of Losses in Single-phase Induction Motors, *Trans. AIEE*, Vol. 54, p. 1302 (1935).
Committee, Hot-spot Temperatures in Fractional-horsepower Motors, *Trans. AIEE*, Vol. 64, p. 128 (1945).
Committee, Hot-spot Temperatures in Integral-horsepower Motors, *Trans. AIEE*, Vol. 64, p. 124 (1945).

ALTERNATING-CURRENT COMMUTATOR MOTORS

By A. G. Conrad

(References to numbered items in the bibliography, Article 49, will assist those who wish to go into greater detail. These references are in the form of numbers enclosed in parentheses.)

35. THE SERIES MOTOR

The series motor used on single-phase alternating current possesses excellent starting torques. Its speed-torque characteristics can be made comparable with those of the d-c series motor. An ordinary d-c series motor with laminated field structure will run when supplied with single-phase alternating current, but its characteristics are most unsatisfactory. The circuit for such a motor is shown in Fig. 1 and its corresponding vector diagram in Fig. 2. The current I is common to both the armature and field winding, and

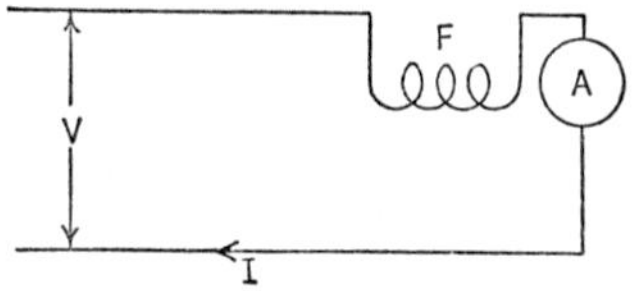

FIG. 1. Circuit of Uncompensated Series A-c Motor. (F) main field, (A) armature.

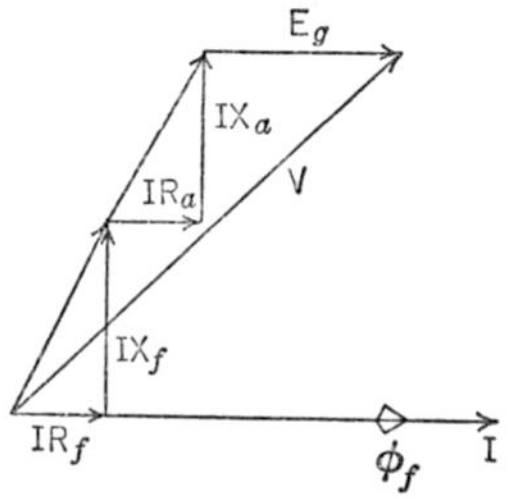

FIG. 2. Characteristic Vector Diagram of Uncompensated Series A-c Motor

it produces the corresponding voltage drops IR_a, IX_a, IR_f, and IX_f (R_a and R_f are the resistances of the armature and the field windings respectively; X_a and X_f are their reactances The voltage E_g is the speed voltage due to rotation of the armature in the main field flux. It is evident from this diagram that the speed voltage E_g is small in comparison to the

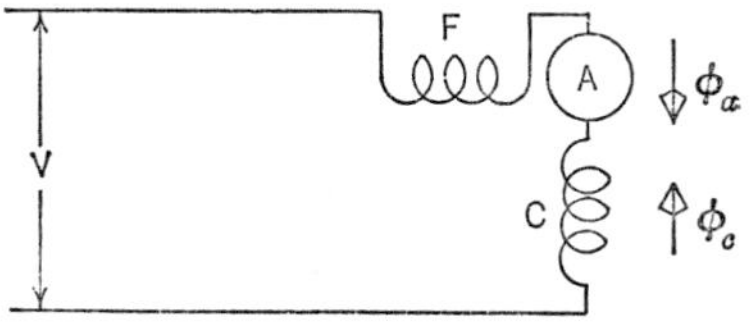

FIG. 3. Circuit of Series Compensated Motor. (F) main field, (A) armature, (C) compensating winding.

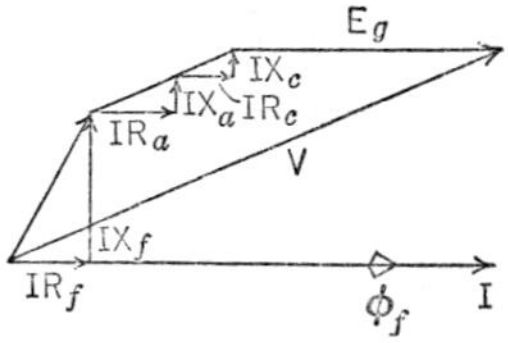

FIG. 4. Characteristic Vector Diagram of Series Compensated A-c Motor

applied voltage V, because of the large reactance drops in the motor windings. Consequently this machine has poor speed regulation and low power factor.

THE COMPENSATED SERIES MOTOR. The characteristics of the straight series motor can be improved by minimizing the reactances of its winding. This is usually done by (a) using fewer turns on the field than would be used on the d-c motor, (b) increasing the turns on the armature to keep the speed from increasing because of the changes indicated in (a), and (c) the insertion of a compensating winding on the stator in quadrature with the main field winding to neutralize the flux produced by the armature. The circuit for this motor is shown in Fig. 3 and its corresponding vector diagram in Fig. 4. Since the flux ϕ_c produced by the compensating winding opposes the flux ϕ_a produced by the armature, the resultant reactance of the armature and compensating winding X_{a+c} will be much less than the reactance of the armature acting alone. Thus the compensating winding reduces the machine reactance, and a greater voltage E_g will result for the same applied voltage and the same line current than when uncompensated. This larger voltage E_g is associated with a greater speed. This machine has speed regulation that is comparable to the d-c series motor. Its low reactance permits large currents upon starting and corresponding large torques. The effect of the compensating winding on the characteristics of the motor is shown in

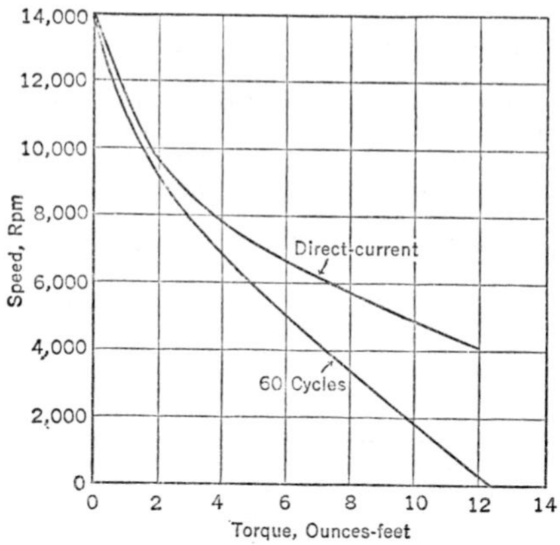

FIG. 5. Characteristics of Uncompensated ¼-hp Series Motor

the curves of Figs. 5 and 6, which are for the kind of motors used in vacuum cleaners, hand drills, etc.

For best results on the compensated series motor represented in Fig. 3 the number of ampere-turns on the compensating winding should be equal to the effective ampere-turns of the armature. Excessive turns on the compensating winding will not make the machine take a leading current. On the contrary, it will make the total impedance of the armature and compensating winding more inductive.

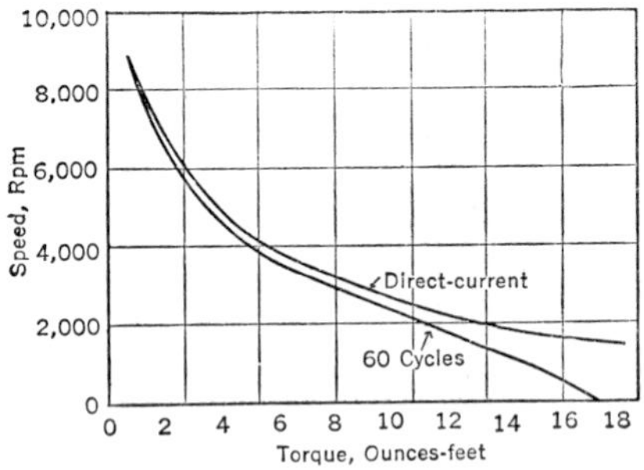

Fig. 6. Characteristics of Compensated ¼-hp Series Motor

INDUCTIVE COMPENSATION. Inductive compensation of the single-phase series motor is illustrated in Fig. 7. Here the short-circuited compensating winding neutralizes the armature flux, as in Fig. 3. The amount of compensation provided by the compensating field is slightly less than in Fig. 3, but the overall characteristics of the motor for these two connections are essentially the same.

The vector diagram of the motor for this connection is shown in Fig. 8. With inductive compensation the number of compensating-winding turns need not be the same as the number of armature turns, since the compensating-winding current will adjust itself to give essentially the compensation needed by the armature.

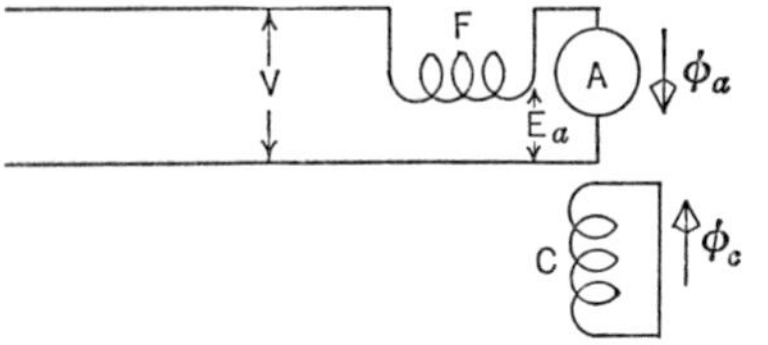

Fig. 7. Circuit of Inductively Compensated Series A-c Motor. (*F*) main field, (*A*) armature, (*C*) compensating winding.

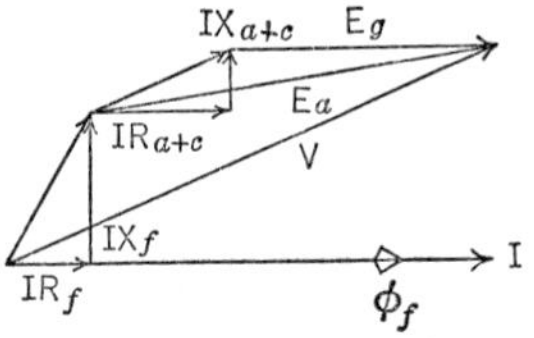

Fig. 8. Characteristic Vector Diagram of Inductively Compensated A-c Motor

36. THE REPULSION MOTOR

Another circuit common to single-phase commutator motors is shown in Fig. 9. Here the current is induced in the armature circuit by transformer action from the compensating winding. The characteristics of this motor are essentially the same as those illustrated in Fig. 3 and Fig. 7. In this machine the voltage across the compensating winding is made up of the impedance drops of the compensating winding and of the armature and the speed voltage of the armature, which is now reflected to the compensating winding.

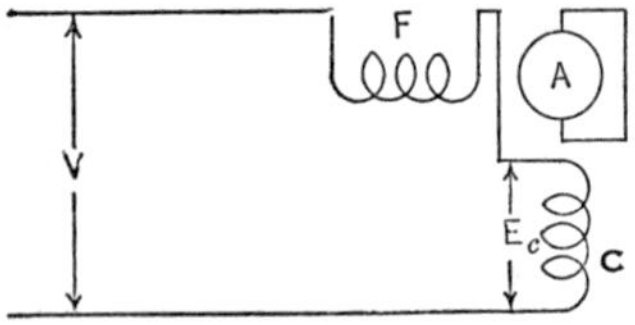

Fig. 9. Circuit of the Repulsion Motor. (*F*) main field, (*A*) armature, (*C*) compensating winding.

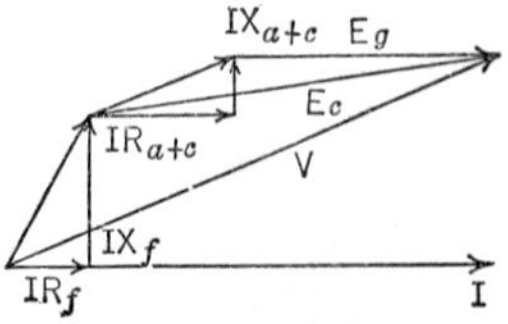

Fig. 10. Characteristic Vector Diagram of Repulsion Motor Illustrated in Fig. 9

The number of turns on the armature of this machine need not match the number of turns on the compensating winding, since the armature current will adjust itself to neutralize the ampere-turns of the compensating windings. This circuit permits the use of high-voltage stator windings with low-voltage armatures. The low-voltage armature is desirable, since it provides better commutation than that obtainable on high-voltage armatures.

The repulsion motor possesses characteristics similar to the machines illustrated in Figs. 3 and 7. Since the fluxes produced by the field and the compensating winding of the motor illustrated in Fig. 9 are in time phase and proportional to the line current (Fig. 10), these two windings can be replaced by one stator winding (*FC* in Fig. 11). This winding is spaced on the stator so as to provide a space angle between the total stator flux and that of the armature of approximately 20 elec deg, as shown in Fig. 11. The field winding of this motor provides a main field flux (a component of the total flux), which is not coupled magnetically to the armature.

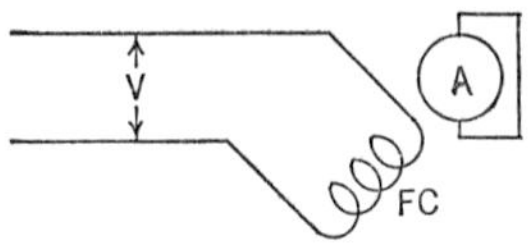

FIG. 11. Repulsion Motor Connection equivalent to that of Fig. 9. (*A*) armature, (*FC*) winding producing main field and compensating field.

CHARACTERISTICS OF SINGLE-PHASE COMMUTATOR MOTORS. The series compensated and repulsion motors described above possess the variable-speed characteristics of the d-c series motor. Adjustments of speed can be made by changing applied voltage. These characteristics are illustrated in Fig. 12.

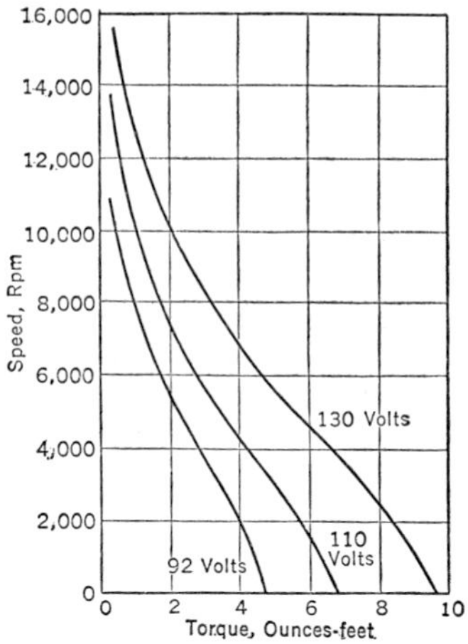

FIG. 12. Characteristics of Universal Motor (Appliance Type) with Different Operating Voltages

Reversal of the motors referred to as series machines can be made by reversing the field connections with respect to those of the armature circuit. The repulsion motor can be reversed by changing the position of the brushes on the commutator to the opposite side of the magnetic neutral. In most motors this is accomplished by changing the position of the stator winding with respect to the brush supports.

37. UNIVERSAL MOTORS

Universal motors are of the series type. The machines illustrated in Figs. 3 and 7 can be so designed that they will operate as series motors on either direct or alternating current with approximately the same characteristics. Their operation on direct current is usually more satisfactory than on alternating current; i.e., the speed for a given voltage and torque is slightly higher, the horsepower is a little greater, and the commutation is much better.

GOVERNORS FOR SMALL MOTORS. Governors are used on small universal motors to limit no-load speed and to provide speed adjustment. This feature is characteristic of many appliance motors. The governor consists essentially of a centrifugal switch, connected in the supply line, that opens and thereby inserts a resistance in the line when certain speeds are attained. The speed at which this switch opens is determined by the governor setting. While the governor is operating, the current to the motor is continually established and partially interrupted by the centrifugal switch. The torque developed by the motor is then a function of the ratio of the duration of the current pulse to the duration of the current interruption. The characteristics of this universal motor thus governed are shown in Fig. 13.

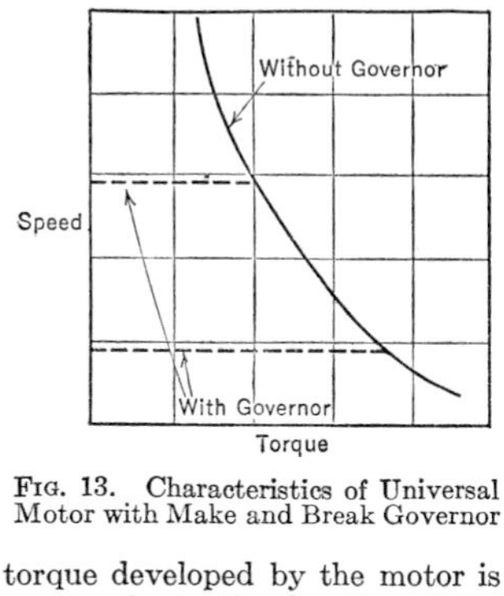

FIG. 13. Characteristics of Universal Motor with Make and Break Governor

38. SHUNT AND BRUSH-SHIFTING MOTORS

THE A-C SHUNT MOTOR. The shunt motor, when operated on single-phase alternating current, is most unsatisfactory because the inductance of the field circuit causes the field flux to lag the applied voltage by approximately 90 elec deg. This flux is therefore 90 deg out of phase with the power component of the armature current. The torque resulting from the action of the field flux on the armature current pulsates, and its average value is very low. For this reason the single-phase a-c shunt motor has not been considered practical.

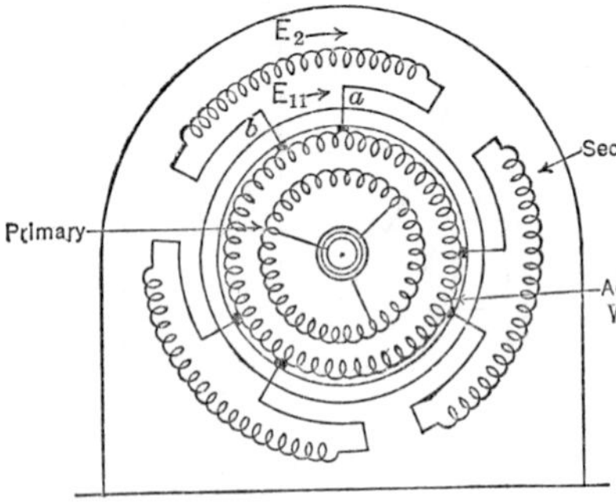

FIG. 14. Winding Connections of the BTA Motor

THE BRUSH-SHIFTING MOTOR. The brush-shifting a-c motor (BTA) (1) is a polyphase motor that provides adjustable-speed characteristics similar to those of the d-c shunt motor. For this reason the machine is sometimes called the a-c shunt motor. These motors are provided with three windings. The primary winding is built into the rotor and supplied with power through slip rings. The stator winding is phase-wound and is the secondary (Fig. 14). The third winding, which is magnetically coupled to the primary, is located on the rotor and is provided with a commutator. Voltages collected from the commutator are inserted into the secondary winding. The brushes are mechanically coupled so that they are spaced at the same distance for each secondary phase winding, thereby ensuring equal voltages conducted to each secondary phase. When the brushes of each secondary phase are adjusted so that they are in contact with the same commutator segment, the secondary is short-circuited, and no voltage is supplied from the adjusting winding. Under these conditions the motor behaves like an ordinary induction motor, and the adjusting winding is in no way operative. Voltage collected by the brushes for a given brush spacing is essentially constant in magnitude. The frequency of the adjusting winding voltage E_{11} is always equal to the frequency of the voltage E_2 induced in the secondary. Speed adjustment is obtained by changing the brush settings.

For speed adjustments below synchronous speed the brushes are spaced to make the voltage E_{11} opposite to E_2 at standstill. For these low-speed adjustments no secondary current will exist (no load on the motor), and no torque will be produced when E_2 and E_{11} are equal and opposite. Loading the motor will cause an increase in its slip, increasing the voltage E_2 (E_{11} remaining constant). The unbalance of these voltages results in a secondary current and an increase in torque. For these low-speed adjustments power flows from the primary to the secondary through transformer action, thereby producing motor torque. Some of the power delivered to the secondary is converted to the mechanical output, and the remainder is fed back to the adjusting winding, which returns it to the primary through transformer action.

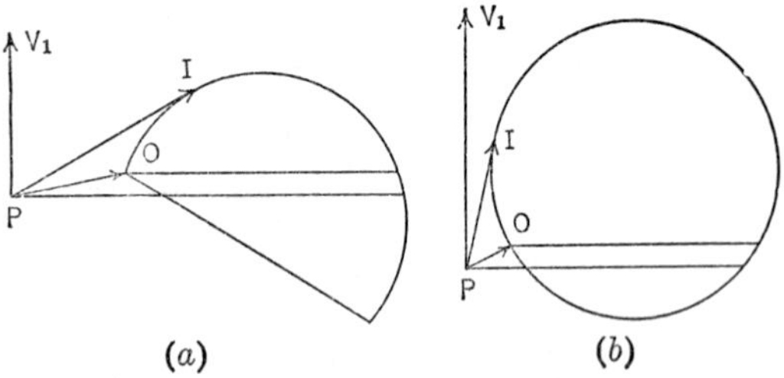

FIG. 15. Vector Diagrams of the BTA Motor: (*a*) for speeds below synchronism, (*b*) for speeds above synchronism. (*PO*) no-load current, (*PI*) input current under load, (V_1) primary voltage.

The machine can be operated above synchronism by shifting the brushes, i.e., interchanging their positions (*a* to *b*, and *b* to *a*, Fig. 14), so that the voltage collected by these brushes is in such a direction as to force a current through the secondary which will produce positive torque.

For speed adjustments above synchronism the adjusting-winding voltage E_{11} is larger than the secondary voltage E_2 caused by induction when the machine is loaded. At no load the voltage E_2 and E_{11} are equal. Increases in torque cause E_2 to reduce in magnitude, whereas E_{11} is essentially constant for a given brush setting. This results in an increase in the secondary current which produces the necessary torque.

The vector diagrams for the motor, showing the loci of the primary current for conditions of low-speed adjustment and for high-speed adjustment, are shown in Fig. 15. Methods of determining the vector diagrams (2) and prediction of machine constants from them have been worked out. Some correction of power factor by brush shift can be accomplished on these machines when they are used for unidirectional operation (3–4).

Commercial machines of this type are constructed so that the maximum commutator voltage E_{11} is approximately one-half the standstill voltage per phase on the stator E_2. This provides speed changes of 50 per cent of synchronous speed either above or below the synchronous speed, i.e., a speed range of 3 to 1. Characteristic speed-torque curves are shown in Fig. 16 for the type of motor used to control automatically the speeds of draft fans, paper machines, conveyors, etc.

The horsepower available from this machine is directly proportional to its operating speed. The advantages of this motor are its excellent speed control and high power factor for high-speed settings (2). Its disadvantages (2) are commutation limitations, low power factor at low-speed settings, relatively low efficiencies at low-speed settings, and cost.

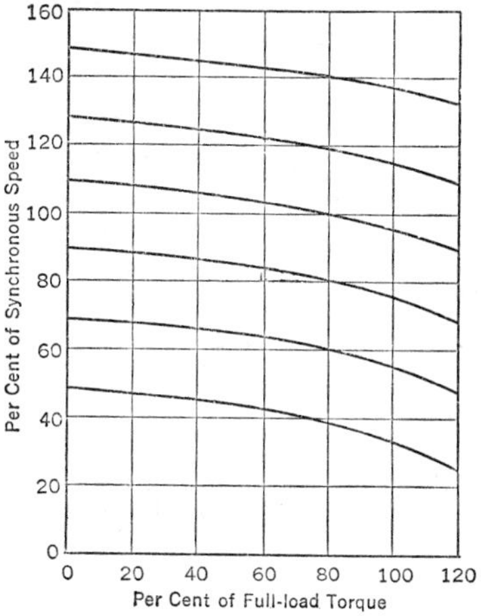

Fig. 16. Characteristics of BTA Motor for Different Speed Adjustments

39. SPEED ADJUSTMENT WITH POLYPHASE REGULATOR

A commutator motor with polyphase regulator has been used successfully for speed adjustment. The principle of operation is similar to that of the BTA motor, the primary difference being that the brushes on the adjusting winding are fixed in position. The output of these brushes is fed into a polyphase regulator. The output of the regulator is then supplied to the stator winding. In this machine the adjustable voltage is provided by the adjustment of the regulator rather than the adjustment of the brush positions, as on the BTA motor.

40. THE FLYNN-WEICHSEL MOTOR

The Flynn-Weichsel polyphase induction-synchronous motor starts as an induction motor but, when once brought up to speed, runs as a self-excited synchronous motor. Its principal features are its good power factor and its ability to correct power factor. The rotor carries two windings, the true polyphase primary connected to the supply by slip rings, and a d-c winding connected to a commutator on which bear two brushes per pair of poles. The stator carries the secondary, which is a two-phase winding with the terminals brought out for connection to a starting resistance, as in all induction motors with phase-wound secondaries. On one phase of the stator is impressed the voltage from the brushes on the commutator. As in the BTA motor, this voltage is of slip frequency, hence of zero frequency at synchronism. This slip-frequency voltage provides a flux which generates a voltage in the primary winding of fundamental frequency. Under conditions of no load the motor runs at synchronous speed. For light loads synchronous speed is maintained from the electrical torque produced by the stator phase, which is excited from the commutator. For longer loads the machine runs at speeds below synchronism, and both stator windings produce torques. The efficiency in a 15-hp motor ranges from 80 to 90 per cent, and the power factor may be made capacitive in nature. The starting torque and current are like those of any polyphase inductive motor started with resistance in the secondary.

41. SINGLE-PHASE MOTOR ON POLYPHASE ALTERNATING CURRENT

A single-phase motor operated on a polyphase system has been developed and successfully applied where wide speed changes are required (5–6). The circuit for such a motor is shown in Fig. 17, and the vector diagram for this motor in Fig. 18. In this machine it is necessary to provide the proper phase relation of the field voltage with that of the armature voltage (proper angle α) for the particular motor. The constants of the motor R_a (armature circuit resistance) and X_a (armature circuit reactance) determine the proper value of the angle α. With proper adjustments this motor takes a leading current at no load. At full load the current lags slightly the armature voltage (power factor approximately 95 per cent).

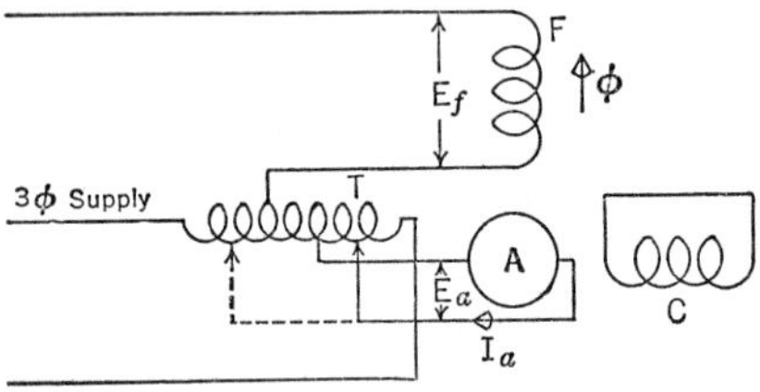

FIG. 17. Circuit for Operation of Single-phase Motor on Polyphase Supply

The motor is started by first adjusting the armature voltage to zero. With three-phase power supplied to the field and to the transformer, it is then possible to raise the speed of the machine by increasing the armature voltage. The direction of the armature voltage with respect to the field voltage determines the direction of rotation of the machine. Line disturbances created by starting are exceedingly small, since the starting transformer or speed-adjusting transformer will need little primary current to supply the large armature current. The advantages of the drive are its good speed adjustment, low initial cost, low starting current, high power factor at full load, leading currents at light loads, ease of reversing, and regenerative braking throughout its speed range. Its disadvantages are that it draws unbalanced currents from the polyphase supply, and for this reason it is limited in its application to small sizes. The speed-torque characteristics of such a motor are shown in Fig. 19.

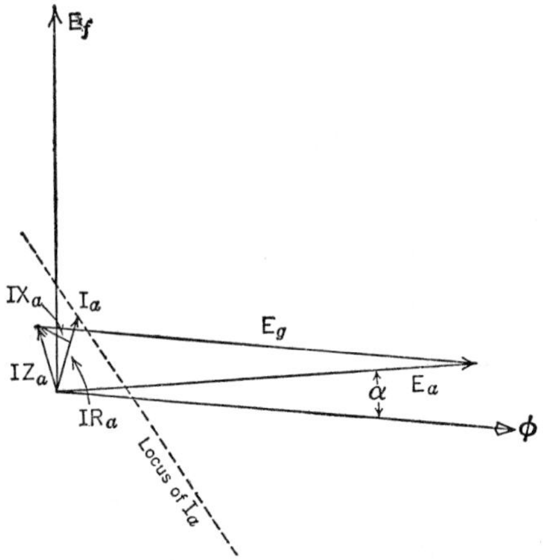

FIG. 18. Vector Diagram of Motor Illustrated in Fig. 17

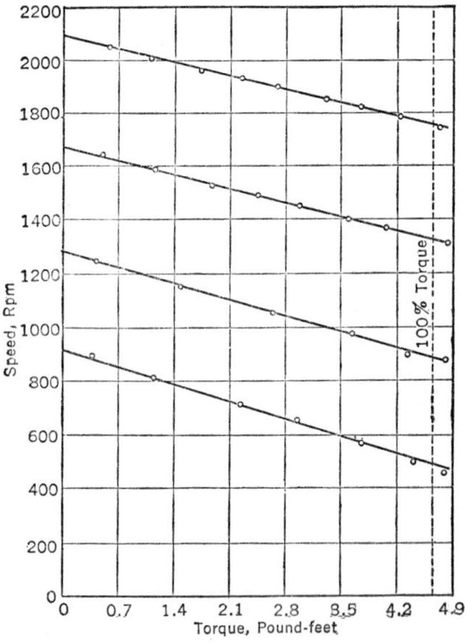

FIG. 19. Characteristics of Adjustable-speed Motor Illustrated in Fig. 17

42. FREQUENCY CONVERTERS FOR SPEED ADJUSTMENT

Frequency converters can be used to control the speed of wound-rotor induction motors. Two types of converters are commonly used for such purposes. The polyphase-commutator type of converter can be used to provide either a constant-horsepower drive or a constant-torque drive. The drives using the polyphase-commutator type of regulator are commonly called **Scherbius systems.** The same types of drives can be obtained by the use of a rotary converter as a frequency changer. These systems using the rotary converter are commonly called **Krämer drives.**

The polyphase-commutator type of frequency converter in its simplest form is similar to a rotary converter, except that there is no winding on the stator. The rotor is built with both slip rings and a commutator. Three brush arms per pair of poles are used for the collection of polyphase voltages from the commutator. One method of using this machine to control speed is shown in Fig. 20. Here the slip rings of the converter are connected to the same source of supply as the induction motor. The voltage applied to these rings produces a flux which rotates at synchronous speed with respect to the rotor. The rotor, however, is turned by the induction motor in a direction opposite to that of the rotating flux. The frequency of the voltage E_{11} collected on the commutator is therefore the same as the frequency of the voltage E_2 induced in the rotor of the induction motor. If the voltage applied to the rings of the converter is held constant, then the commutator voltage E_{11} is constant regardless of the speed of the system. By proper spacing of the brushes the phase position of the commutator voltage E_{11} can be made opposite to the voltage E_2 collected from the rings of the motor. Thus the motor secondary is provided with a back-electromotive force which is constant in magnitude and in a given phase position with respect to the voltage induced in the rotor of the motor. When these two voltages E_2 and E_{11} are equal, no rotor current flows, and the drive operates at no load. Loading the machine will cause the speed to decrease, thereby increasing the voltage E_2, which forces a current against the voltage E_{11}. Thus power is fed from the motor secondary to the regulator; it then goes through the transformer and is fed back to the supply. The regulator used in this manner does not produce torque. This system is known as a **constant-torque drive.** This means that the rated torque for any speed adjustment is constant.

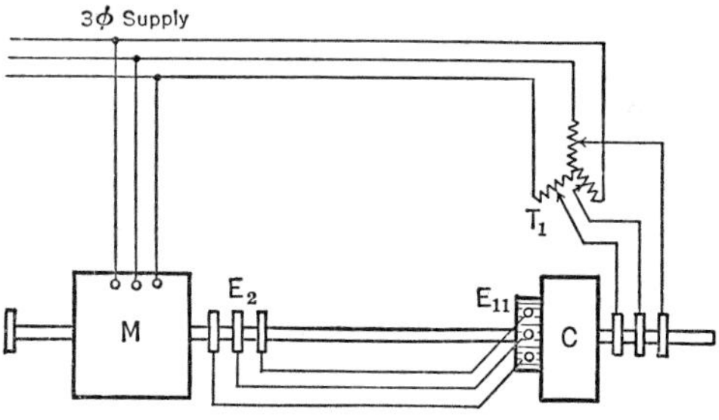

FIG. 20. Frequency Converter Used with Induction Motor for Speed Control. (*M*) main motor, (*C*) frequency converter, (T_1) control transformer

This type of frequency converter has not been considered practical when wide speed ranges are required. The limitation of its application is the characteristic poor commutation of this machine when conducting large currents at high voltage, which is required at speeds well above or well below synchronism. This machine has been used successfully on double-range Scherbius drives as an exciter for polyphase commutator regulators. See Figs. 23 and 24.

43. CONSTANT-HORSEPOWER SCHERBIUS SYSTEM

A Scherbius system providing a constant horsepower drive can be obtained by the use of a polyphase-commutator regulator having the standard polyphase rotor which operates in a polyphase stator. The connections for such a drive are shown in Fig. 21.

The electrical circuit of the rotor of this regulating machine is similar to that of a d-c armature. It usually is designed for low voltage in order to obtain good commutation of the alternating currents supplied to it from the brushes. The spacing of the brushes on the commutator is determined by the number of phases on the rotor of the motor, that is, on a three-phase machine the angle between brushes is 120 elec deg. The stator or field winding of the regulator may be a normal distributed polyphase winding such as that found on a polyphase induction motor. However, for good commutation the main field windings are usually placed on salient poles between which interpoles can be provided. A compensating winding is usually placed on the stator and connected in series with the rotor or armature. The field may be designed for operation either in series or in shunt with the armature. The drive using the shunt connection is the more common and is the one which is discussed here. This connection is shown diagrammatically in Fig. 21. A transformer bank is normally provided to control the voltage applied to the field winding and thereby govern the speed. Excellent speed regulation can be obtained when shunt excitation of the regulator is employed.

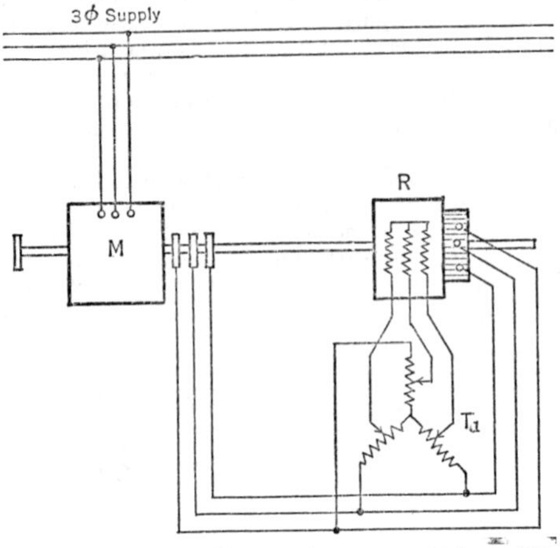

FIG. 21. Scherbius Constant-horsepower Drive. (M) main induction motor, (R) regulator, (T_1) speed-control transformer.

In this system torque is developed by the regulating machine as well as the induction motor. High starting torques can be obtained from this system. Its speed range is limited to speeds below synchronism. The size of the regulating machine is proportional to the speed reduction below synchronism provided in the design. The power factor of the system is inherently low (8).

44. CONSTANT-TORQUE SCHERBIUS SYSTEM

A Scherbius system (9–13) for providing a constant-torque drive is illustrated in Fig. 22. This system provides adjustable speed for the main induction motor. The performance of the regulating machine is similar to that described above. Since it possesses both rotor and stator windings, it can deliver torque to the induction machine coupled to it. Under conditions of load the induction machine will feed power back on the system. For light loads the induction machine may act as a motor to supply rotational losses of the motor-generator set. It is impossible to accelerate this machine to speeds above synchronism, since the excitation current of the regulator is zero at synchronous speed. The size of the regulator and the induction machine to which it is coupled is determined by the amount of speed change desired from the main motor. In general, if the speed must be adjusted from zero to 100 per cent of synchronous speed, both the regulator and the induction machine coupled to it must have kva ratings corresponding to that of the main motor. If a speed range of 50 per cent of synchronous speed is desired, the capacity of the regulator and the capacity of its motor must be 50 per cent of that of the main motor. These drives usually have good speed adjustment but poor power-factor characteristics.

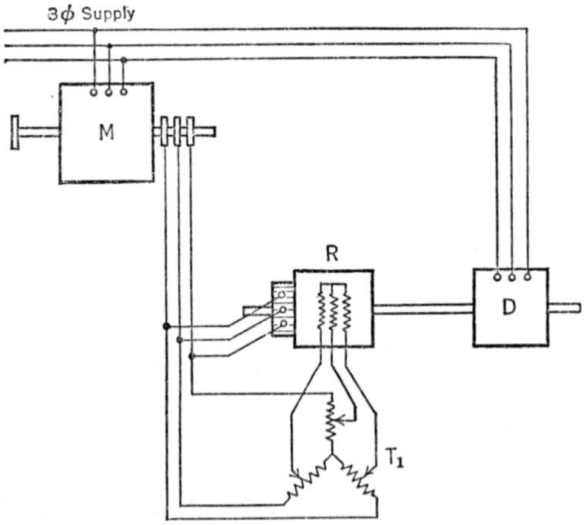

FIG. 22. Scherbius Constant-torque Drive. (M) main induction motor, (R) regulator, (T) control transformer, (D) induction motor or induction generator.

45. DOUBLE-RANGE SCHERBIUS DRIVE

Speeds both above and below synchronous (13–14) speed can be obtained on an induction motor if provision is made to supply power to the secondary of the motor at slip frequency for speeds that are less than, equal to, or greater than synchronous speeds. In the Scherbius drives heretofore described the power in the secondary of the motor becomes zero at synchronous speed. Consequently, these systems cannot be accelerated through synchronous speed. If, however, an auxiliary device is used to force a torque-producing current through the induction motor secondary at any slip, speeds above synchronous speed can be produced. This auxiliary device used with the Scherbius system is commonly called an exciter or, more frequently, an **ohmic-drop exciter.** This exciter has an armature similar to that of a rotary converter, except that brushes are provided to collect polyphase voltages from the commutator. The stator of the exciter is an iron ring surrounding the armature. It has no winding on it and is used to conduct flux. This exciter produces no torque during operation (Fig. 23).

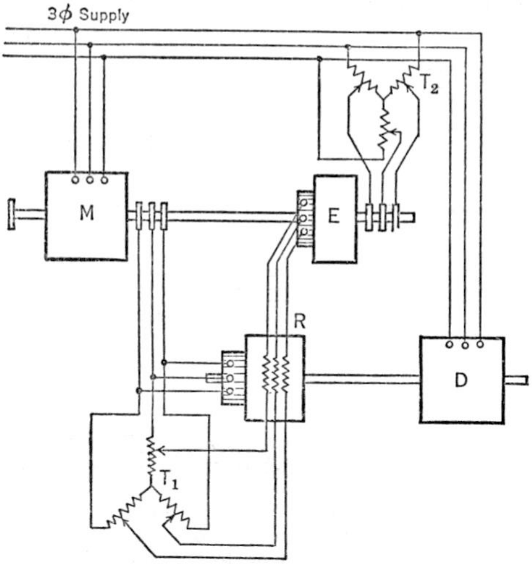

FIG. 23. Double-range Scherbius Adjustable-speed Drive. (*M*) main induction motor, (*R*) speed regulator, (T_1) speed-control transformer, (*E*) ohmic drop exciter, (T_2) exciter transformer, (*D*) motor or generator

One method of using the ohmic-drop exciter to raise the speed of the induction motor above synchronism is shown in Fig. 24. This method of speed control provides a constant-torque drive. In normal operation at speeds below synchronism, the electrical power leaving the secondary of the main motor *M* is fed back to the power supply through the auxiliary motor-generator set. For speeds above synchronism the motor-generator set feeds power into the main motor secondary, and the output power at the shaft of the main motor is the sum of power that it receives from the stator and that received from the regulator connected to the secondary of the main motor. The method by which this system permits the motor to pass through synchronous speed is as follows:

Suppose that the induction motor speed approaches synchronism when the exciter transformer T_2 is adjusted to provide voltage at the rings of the exciter. This will cause a low voltage of slip frequency at the brushes of the exciter. This voltage will then produce a magnetic field in the regulating machine which is rotating, causing slip-frequency voltage to appear at the brushes of the regulator. This voltage forces a current through the secondary of the induction motor, thus developing torque and acceleration of the main motor. These exciters are usually provided with brush-shifting devices, so that the phase position of the voltages at the commutator can be altered. This feature provides the possibility of collecting a current that will produce torque in the main motor but also provide power-factor correction of the main motor.

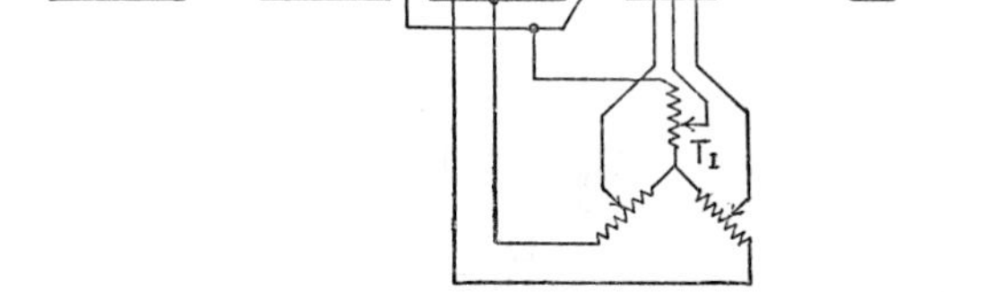

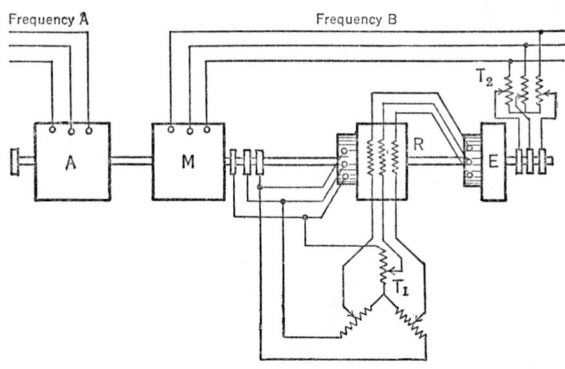

FIG. 24. Variable-ratio Frequency-changer. (*M*) main induction motor, (*R*) regulator, (*E*) ohmic drop exciter, (T_1) speed-control transformer, (T_2) exciter transformer.

46. VARIABLE-RATIO FREQUENCY-CHANGERS

Variable-ratio frequency-changers (10, 11, 12) can be built which operate on the Scherbius principle. These consist of a standard alternator (see Fig. 24), connected to one of the standard frequency supplies. This alternator is driven mechanically by the induction motor M. Speed and power-factor control are provided through the transformers T_1 and T_2. This drive operates in essentially the same manner as that shown in Fig. 21, except that provision is made for excitation of the regulator R by the exciter E when running at synchronous speed. In this case the exciter produces no torque. It does, however, transmit sufficient electrical energy to change the speed of the main induction motor slightly, thereby changing the frequency of the alternator A with respect to the frequency of the system B.

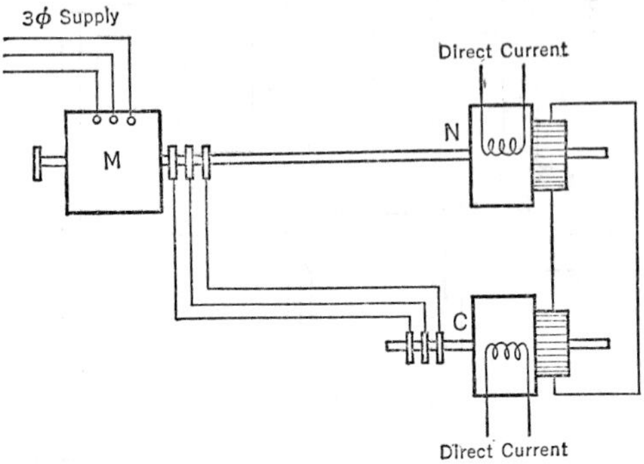

FIG. 25. Krämer Constant-horsepower Drive. (M) main induction motor, (C) rotary converter, (N) d-c motor.

47. CONSTANT-HORSEPOWER KRÄMER DRIVES

The Krämer systems of speed control provide either constant-horsepower or constant-torque drives. Figure 25 shows the connection for constant horsepower. This drive is useful for speeds below synchronism. It develops no torque at synchronous speed. Power from the secondary of the main motor M is supplied to the slip rings of the rotary converter C. The output of the converter is supplied to the d-c motor N. The developed torque is therefore the sum of the torques produced by M and N. Reduction of speed is accomplished by increasing the field current of the motor N. Separate excitations are used on N and C. Changes in the excitation of C modify the power factors of the main motor M.

48. CONSTANT-TORQUE KRÄMER DRIVES

The Krämer system of speed control providing constant torque is illustrated in Fig. 26. In this case an extra induction motor O has been added to those used in the constant-horsepower drive illustrated in Fig. 25. The functions of the d-c fields are the same as described above. For speeds below synchronism power is fed from the secondary of M to the converter C, to the d-c motor N, and back to the supply through O, which acts as a generator. For speeds above synchronism the machine O becomes a motor supplying power to the secondary of M through the converter C and the generator N.

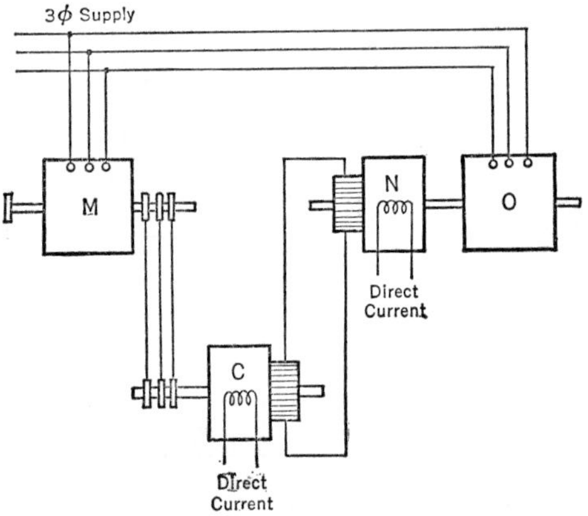

FIG. 26. Double-range Krämer System for Constant-torque Drive. (M) main induction motor, (C) rotary converter, (N) d-c machine (motor or generator), (O) a-c machine (motor or generator).

In both Krämer drives the converter changes its speed with speed adjustments of the main motor M. It always operates at a speed corresponding to the slip frequency of the main motor. This feature adds inertia to the moving system, which makes it sluggish in its response. However, it has advantages over the Scherbius drives where large changes in speed are desired. The constant-horsepower Krämer drive, because of its simplicity, is usually less costly than the corresponding Scherbius drive. The Scherbius system is superior to the Krämer drive when fine speed control is needed near synchronous speed.

49. BIBLIOGRAPHY

1. H. K. Schrage, Ein neuer Drehstrom-Kommutator Motor mit Nebenschlussregulierung durch Bürstenverschiebung, *E. T. Z.*, Vol. 35, p. 89 (Jan. 22, 1914).
2. A. G. Conrad, J. G. Clarke, F. Zweig, Theory of the Brush-shifting A-c Motor, Part I, *Elec. Eng.* (August 1941).
3. A. G. Conrad, J. G. Clarke, F. Zweig, Theory of the Brush-shifting A-c Motor, Part II, *Elec. Eng.*, (August 1941).
4. A. G. Conrad, J. G. Clarke, F. Zweig, Theory of the Brush-shifting A-c Motor, Part III, *Elec. Eng.* (July 1942).
5. A. G. Conrad, S. T. Smith, P. F. Ordung, A New Type of Adjustable-speed Alternating-current Drive, Part I, *Trans. AIEE*, Vol. 62 (Jan. 1943).
6. A. G. Conrad, S. T. Smith, P. F. Ordung, A New Type of Adjustable-speed Alternating-current Drive, Part II, *Trans. AIEE*, Vol. 62 (Aug., 1943).
7. A. F. Puchstein and T. C. Lloyd, *Alternating-current Machines*, 2nd Ed., pp. 306–307. John Wiley (1942).
8. A. G. Conrad, F. E. Brooks, R. G. Fellers, Polyphase Commutator for Speed Control, *Trans. AIEE*, Vol. 65 (June 1946).
9. Liwschitz, Basic Problems in Electrical Machinery, *Elec. Eng.*, Vol. 63 (Feb., 1944).
10. E. S. Bundy, A. Van Niekirk, W. H. Rogers, Variable-ratio Frequency-converter Installations, *Trans. AIEE*, Vol. 49, p. 245 (1930).
11. L. Encke, Interconnection of Power and Railroad Systems by Means of Frequency Changers, *Trans. AIEE*, Vol. 47 (Oct., 1928).
12. C. W. Kincaid, Variable-ratio Frequency-changer Sets, *Elec. J.*, Vol. 26 (June 1928).
13. L. A. Umanski, Adjustable-speed Drives for Rolling Mills, *Iron and Steel Eng.* (Sept. 1924).
14. J. D. Wright, Speed Control of Induction Motors for Steel Mill Drive, *Gen. Elec. Review*, Vol. 20, p. 104 (1917).

SECTION 10

TRANSFORMERS

BY

WILLIAM C. SEALEY

TRANSFORMERS

By William C. Sealey

PRINCIPLES OF CONSTRUCTION AND OPERATION

1. DEFINITION AND ESSENTIAL PARTS

DEFINITION. A transformer is a static electrical device which electromagnetically transforms a-c energy from one circuit to another. It usually changes the original voltage to a higher or lower voltage, at the same frequency.

ESSENTIAL PARTS. A transformer usually consists of two or more coils of insulated copper wire wound around a laminated iron core. Any two sets of turns of insulated conductor which are magnetically coupled may be made to function as a transformer. The magnetic coupling is performed by a common magnetic circuit, which may be in air or may consist of an iron core, through which alternating magnetic flux flows when alternating current flows in the insulated conductor.

In operation, an a-c supply is connected to one set of turns which thereby becomes the primary winding; the other set of turns to which the load is connected is called the secondary winding. Since for most transformers either winding may be used as a primary winding or as a secondary winding, common practice is to designate the windings, not by the terms **primary** and **secondary,** but by the terms **high-voltage** and **low-voltage,** based on a comparison of the rated voltage of the windings.

In order to perform the functions described, the essential parts of a transformer are: a high-voltage winding to carry the high-voltage current, a low-voltage winding to carry the low-voltage current, insulation between turns and usually to the other windings and ground so that the current will flow in the conductors of the windings, and a core (usually of laminated iron) to carry the alternating magnetic flux.

In addition to these essential parts, commercial transformers usually require the following in order to secure the desired operating characteristics: mechanical bracing for core and coils, a cooling medium such as air or oil, a tank to protect the transformer and contain the oil, and bushings for bringing the leads out of the tank for connection to the transmission lines.

2. OPERATION WITH AND WITHOUT LOAD

NO-LOAD OPERATION. When one winding of a transformer is connected to a suitable a-c supply, with the other winding open, it behaves like a reactance coil and draws exciting current from the line to magnetize the core. When an a-c voltage is applied, sufficient magnetic flux flows in the core to induce a voltage in the winding to equal (and oppose) the applied voltage (less the small voltage drop due to the exciting current). If the applied voltage has a sine-wave shape, the flux will have a sine-wave shape. In other words, the transformer assumes the voltage which is impressed on it and adjusts its flux so as to produce the impressed voltage. Sufficient exciting current flows from the line at every instant to produce the required value of magnetic flux at that instant. Since the permeability of the iron core is not constant, a non-sinusoidal current is required in order to produce a sinusoidal flux wave, which in turn produces a sinusoidal voltage.

When the transformer has an iron core, power is lost in the core because of the continuously changing flux. The loss in the iron may be separated into hysteresis loss and eddy-current loss. The hysteresis loss is the power required to change the position of the individual molecules as they rotate because of the influence of the magnetizing force. This loss is proportional to the frequency and is a function of the flux density in the iron, increasing in a complex manner with the flux density. The eddy-current loss is the loss due to circulating currents in the core iron, caused by the magnetic flux in the iron cutting the iron, which is a conductor. The eddy-current loss is proportional to the square of the frequency and the square of the flux density. In order to reduce the eddy-current loss the core is built up of thin laminations (usually 0.014 in. thick) insulated from each other by an insulating coating on the iron. Special grades of steel alloyed with silicon and processed to produce low eddy current and hysteresis loss are generally used in transformers. The loss in the core produces heat, which causes the temperature of the core to rise above its surroundings. The total no-load loss of a transformer is composed of the core loss, plus the (usually very small) I^2R loss in the primary winding due to the exciting current flowing in it, plus a negligible dielectric loss.

Most commercial transformers are designed so that the exciting current is a small percentage of the full-load current of the transformer. The applied voltage distributes itself on the turns of the primary winding practically uniformly, so that each turn on the transformer has very nearly the same voltage across it. The open secondary winding will have the same voltage across each turn as the primary winding has. Consequently, the voltage of a winding is equal to the voltage per turn of the transformer multiplied by the number of turns in the winding; it is proportional to the number of turns in the winding. The voltage ratio of most transformers, including distribution and power transformers, is based on their turns ratio, which is practically the same as the no-load voltage ratio of the transformer. The rated voltages which appear on the transformer nameplate are proportional to the number of turns in the windings.

OPERATION WITH LOAD. If the secondary winding of a transformer which has its primary winding connected to a suitable a-c supply is connected to an impedance, load current will flow in the secondary winding. Because the primary winding is magnetically coupled to the secondary winding, a current will flow in the primary winding in the opposite direction, such that at any instant the sum of the load ampere-turns of all windings of the transformer is equal to zero. This means that the primary-load ampere-turns are equal to the secondary-load ampere-turns. If the exciting volt-amperes and the impedance volt-amperes are neglected, the primary volt-amperes are equal to the secondary volt-amperes.

Because the primary and secondary windings of the transformer do not occupy exactly the same space, some of the flux which links one winding does not link the other winding, so that the voltages per turn of the two windings are not exactly the same. The amount of this leakage flux is proportional to the current in the windings. For purposes of calculation, it may be assumed that the voltages per turn of both windings are the same, and the effect of the leakage flux can be taken care of by reactances considered as placed in series with the windings. The winding current flowing through this reactance produces the **reactance voltage drop** of the transformer. There is a similar resistance voltage drop due to the resistance of the windings through which the current flows.

The combined effect of resistance and reactance may be represented by a series impedance, which is called the **impedance** of the transformer. It is measured by short-circuiting one winding and applying to the other winding a-c voltage at rated frequency of such value as to circulate full-load current in it. The impedance of the transformer in ohms is equal to the applied test voltage divided by the current. The **percentage impedance** is equal to 100 × Applied test voltage/Rated voltage of the winding to which voltage was applied. The impedance of a transformer is represented by a constant value, since practically all the reluctance of the leakage flux path is in air, or equivalent. Since a transformer is really a combination of two coils with self- and mutual inductances, the impedance of a transformer is a fictitious quantity, without real physical existence, for use in equivalent circuits of transformers. This also applies to any separation of the transformer impedance into components in series with each winding. Nevertheless, the impedance of a transformer is of such importance that its value is usually stamped on the transformer nameplate.

3. EQUIVALENT CIRCUITS

REASON FOR EQUIVALENT CIRCUITS. Several equivalent circuits are used for transformers, general practice being to choose the simplest one which will give the required degree of accuracy for the application. Since a transformer is made up of two coils with mutual inductance, the most general circuit consists of a primary coil with self-inductance and mutual inductance to the secondary winding, and a secondary coil with self-inductance and mutual inductance to the primary winding. Because of the complexity of the voltage and current relations of such circuits, and because the inductances of coils with iron cores are not constant but variable, the general circuit is seldom used except occasionally for transient conditions of current and voltage.

One may conceive of an idealized transformer, without exciting current or voltage drop, in which the ratio of primary to secondary voltage is the same as the turns ratio, the winding currents are inversely proportional to the turns ratio, and the secondary kva is equal to the primary kva. In constructing the equivalent circuits of transformers, one assumes an idealized transformer with these characteristics and uses series impedances to take care of the drop in voltage from primary to secondary (the fact that the voltage ratios are different from the turns ratio) and shunt impedances to represent the exciting current (the fact that the secondary ampere-turns are not exactly equal to the primary ampere-turns).

GENERAL EQUIVALENT CIRCUIT. The most complex equivalent circuit in general use and the circuit from which the others may be derived is shown in Fig. 1. In this figure r_1 and x_1 represent a resistance and a reactance in series with the primary winding. The coils with turns N_1 and N_2 represent an idealized transformer with zero impedance to the flow of load current and zero exciting current when voltage is applied to a winding.

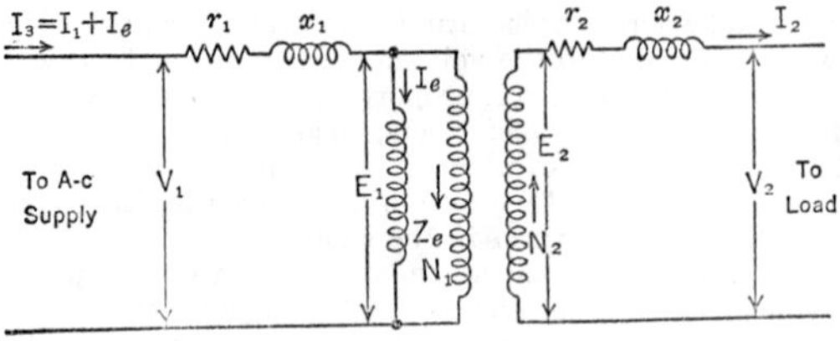

FIG. 1. Basic Circuit of a Transformer

The current I_e which flows through the impedance Z_e is the exciting current, which is the same as the no-load current of the transformer. Since I_e is not proportional to the winding voltage, Z_e is not a constant but varies greatly with the voltage. The impedances r_1, r_2, x_1, x_2, and Z_e are fictitious impedances which have no physical existence but, when used in the equivalent circuit, correctly represent the transformer and can be used to determine current and voltage relations in the transformer and the circuits to which it is connected. The circuit of Fig. 1 has been checked both experimentally and analytically and has been found to represent closely the actual transformer. The arrows on the figure represent the relative instantaneous directions of current flow.

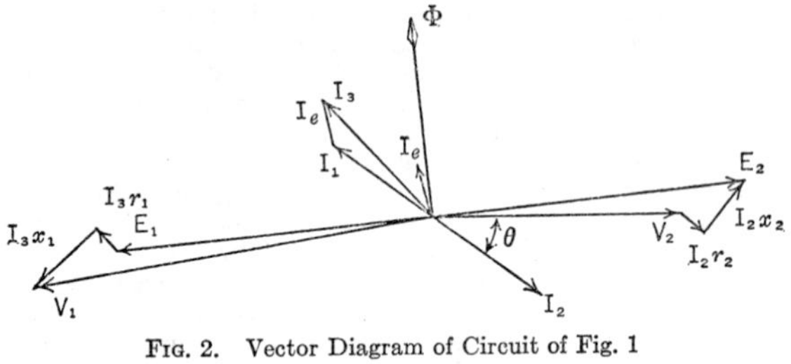

FIG. 2. Vector Diagram of Circuit of Fig. 1

The vector diagram for the equivalent circuit of Fig. 1 is shown in Fig. 2. It may be drawn in the following sequence:

The secondary terminal voltage V_2 is drawn in any convenient position.

The secondary-load current I_2 is drawn lagging V_2 by the power factor angle Θ.

The resistance drop I_2r_2 is drawn in phase with I_2.

The reactance drop I_2x_2 is drawn leading I_2 by 90 deg.

The sum of $V_2 + I_2r_2 + I_2x_2$ is E_2.

E_1 is drawn 180 deg from E_2; $E_1 = E_2N_1/N_2$.

Main core flux ϕ is drawn lagging E_1 by 90 deg.

I_1 is drawn 180 deg from I_2; $I_1 = I_2N_2/N_1$.

The exciting current I_e is drawn lagging E_1 by the power factor angle of the exciting current.

The total primary current I_3 is drawn as the sum of $I_1 + I_e$.

The resistance drop I_3r_1 is drawn in phase with I_3.

The reactance drop I_3x_1 is drawn leading I_1 by 90 deg.

The sum of $E_1 + I_3r_1 + I_3x_1$ is V_1, which is the voltage applied to the primary terminals.

The vector diagram for a load of any power factor on any transformer may be drawn in similar fashion.

SIMPLIFIED EQUIVALENT CIRCUIT. Figure 3 is a simpler equivalent circuit for a transformer; it has the same accuracy as Fig. 1 and may be derived from it by assuming that $N_1 = N_2$, so that $E_1 = -E_2$. If the impedances, currents, and voltages are expressed on a percentage or per unit basis, no change in their values is necessary when transforming from Fig. 1 to Fig. 3. If these quantities are expressed in ohms, amperes, and volts, the conversion factor for converting from secondary terms to primary terms for resistance and reactance is $(N_1/N_2)^2$. For current the factor is N_2/N_1; for voltage, N_1/N_2.

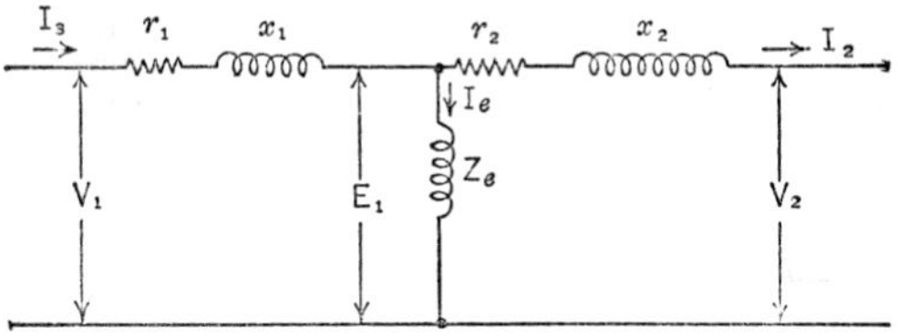

FIG. 3. Equivalent Circuit of a Transformer

The vector diagram for Fig. 3 is shown in Fig. 4 and may be derived in the same manner as Fig. 2 was derived. The equivalent circuits of Figs. 1 and 3 are used mainly where small errors are significant, such as in determining the accuracy of metering transformers. For power and distribution transformers, when the exciting current is to be considered, Figs. 5 and 6 are used; when the exciting current is to be neglected, as it generally may be, Fig. 7 is used.

The value of Z_e is determined from a measurement of the wattage loss and exciting current with no load on the secondary winding. The values of x and r are determined from a measurement of the voltage and wattage loss for circulating full-load current in one winding with the other winding short-circuited. The values of r_1, x_1, r_2, and x_2 for Figs. 1 and 3 may be determined by using the principles outlined for three-winding transformers, in combination with information on the design and construction of the transformer.

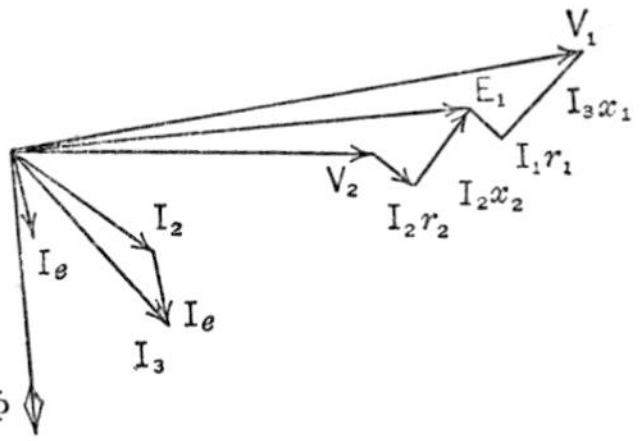

FIG. 4. Vector Diagram of Circuit of Fig. 3

EQUIVALENT CIRCUIT FOR THREE-WINDING TRANSFORMER. The equivalent circuit of Fig. 8(*a*) is commonly used for three-winding transformers. A three-winding transformer has three windings on the same core. This circuit assumes the excit-

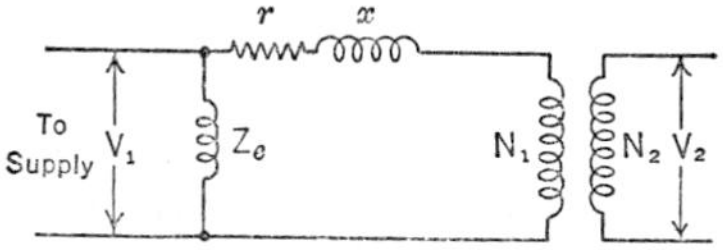

FIG. 5. Simplified Equivalent Circuit of a Transformer

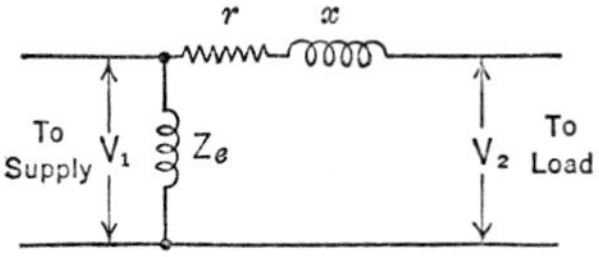

FIG. 6. Simplified Equivalent Circuit of Power and Distribution Transformer

ing current to be zero, with all quantities reduced to a 1 to 1 ratio of transformation, just as in the circuit of Fig. 7(*b*). The terminals of the three windings are:

For winding 1, 1 and N; for winding 2, 2 and N; for winding 3, 3 and N.

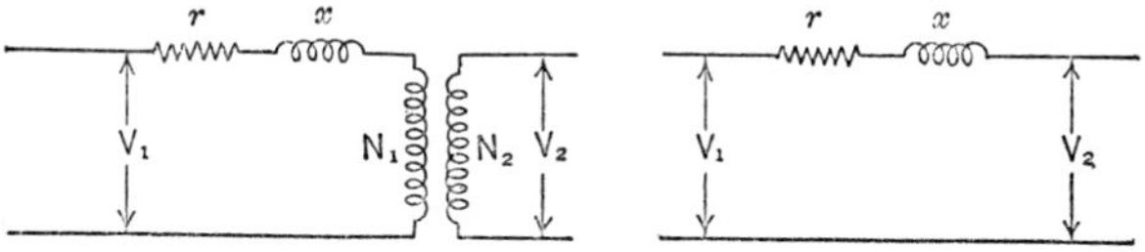

(*a*) (*b*)

FIG. 7. Simplified Equivalent Circuit of a Transformer When Exciting Current May Be Neglected

As in other equivalent circuits, the impedances Z_1, Z_2, and Z_3 of the equivalent circuit of Fig. 8(*a*) have no physical existence but are fictitious impedances which may be used to determine current and voltage relations in the transformer and the circuits to which the transformer is connected. In Fig. 8(*a*) winding 1 is shown connected to a source of power, and windings 2 and 3 are shown connected to loads. Actually any of the windings can be connected to either sources or loads.

The impedances of the equivalent circuit are determined from measurements made on the transformers as follows: The impedance measured from winding 1 to winding 2, with winding 3 open, is called $Z12$; that from winding 2 to winding 3, with winding 1 open, is called $Z23$; that from winding 1 to winding 3, with winding 2 open, is called $Z13$. All impedances should be expressed as complex quantities. If these measurements are applied to the circuit of Fig. 8(a), it is evident that

$$Z_1 + Z_2 = Z12$$
$$Z_2 + Z_3 = Z23$$
$$Z_1 + Z_3 = Z13$$

Solving these three simultaneous equations for Z_1, Z_2, and Z_3, we obtain

$$Z_1 = \tfrac{1}{2}(Z12 + Z13 - Z23)$$
$$Z_2 = \tfrac{1}{2}(Z12 + Z23 - Z13)$$
$$Z_3 = \tfrac{1}{2}(Z13 + Z23 - Z12)$$

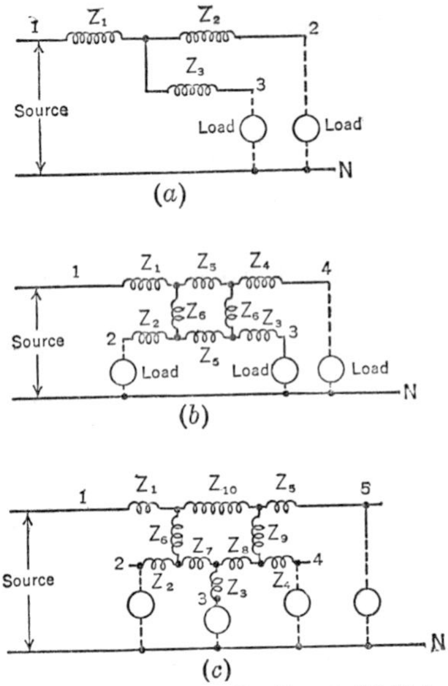

FIG. 8. Equivalent Circuits of Multiple-winding Transformers

The last three equations are used to determine the impedance values for the equivalent circuit of Fig. 8(a). This circuit may be used to determine currents, voltages, losses, or other circuit relations in circuits involving three-winding transformers. Other equivalent circuits may be drawn for three-winding transformers, but the circuit shown is the simplest and most generally useful.

EQUIVALENT CIRCUIT FOR FOUR-WINDING TRANSFORMER. The simplest equivalent circuit for a four-winding transformer is shown in Fig. 8(b), in which any of the windings shown can be connected to sources or loads. The simplest equivalent circuit for a five-winding transformer is shown in Fig. 8(c). Equivalent circuits for transformers with six or more windings become extremely complicated and are seldom used.

4. POLARITY

POLARITY MARKING. The relative directions around the core of the windings between transformer terminals must be known when two or more transformers are to be connected together in a bank or when transformers are connected to wattmeters, power-factor meters, etc. Polarity is a designation of such relative directions between terminals. There are two common methods of designating polarity: (1) by a marker on a primary lead and a similar marker on a secondary lead, indicating that they are of the same polarity, (2) by the use of the terms **additive polarity** and **subtractive polarity.** A primary and a secondary lead have the same polarity when, at every instant, the load current enters the primary lead and leaves the secondary lead in the same direction, as if the two leads formed a continuous circuit. In terms of winding direction, a primary and a secondary lead have the same polarity when the turns attached to those leads proceed in the same direction around the mutual core in their progress to the other terminals of the windings. For example, in Figs. 9(a) and 9(b) leads H_1 and X_1 have the same polarity. Instrument transformers commonly have a similar mark, such as a white dot, on leads of like polarity.

If a transformer has its leads of like polarity adjacent to each other, as in Fig. 9(a), the polarity is subtractive; if the leads of unlike polarity are adjacent, as in Fig. 9(b), the polarity is additive. In terms of voltage, if the instantaneous direction of voltage between the leads (bushings) of each winding as they leave the case is in the same physical direction for both windings, the polarity is subtractive; if the instantaneous direction of voltage between the leads of each winding is in opposite physical directions for the two windings, the polarity is additive. In both Fig. 9(a) and Fig. 9(b) the instantaneous direction of voltage is the same from H_1 to H_2 as from X_1 to X_2.

TEST. One common test for polarity is to connect adjacent leads, one from each winding together, and, with suitable a-c voltage applied to one winding, to measure the voltage between the other two leads. The polarity is subtractive if the measured voltage is less than that of the high-voltage winding, and additive if the measured voltage is greater than that of the high-voltage winding.

Polarity refers to the leads as they are brought out of the transformer case. In both Figs. 9(*a*) and 9(*b*), the transformer windings are identical, the only difference being the arrangement of the individual leads on the cover. According to the ASA standards, leads are to be so marked that the instantaneous direction of voltage from a lower numbered lead to a higher numbered lead of the same winding is the same for all windings. The H_1 lead should be located on the extreme right when facing the high-voltage side of the transformer.

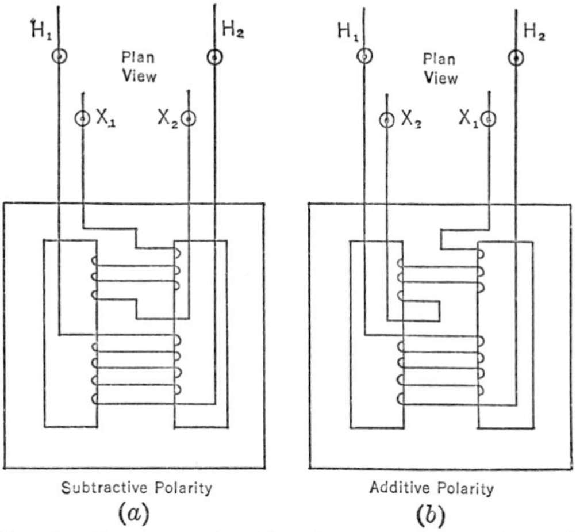

FIG. 9. Connections for Transformers of Subtractive and Additive Polarities

VOLTAGE VECTOR DIAGRAMS. The polarity relations described provide sufficient information to permit connections to be made satisfactorily on single-phase lines. For three-phase connections and three-phase transformers additional information, such as that furnished by a voltage vector diagram, is required.

Voltage vector diagrams for transformer connections are commonly drawn with arrows omitted. The following principles are employed:

1. The line of the vector represents the physical winding.
2. The length of the line is proportional to the magnitude of the voltage.
3. The direction of the line in space represents the time the voltage reaches its maximum value, in comparison to other vectors of voltage or current. (Counterclockwise rotation of vectors is assumed as standard.)

The vector line is so marked that the line 1–2 represents the voltage from terminal 1 to terminal 2. The vector 2–1 represents the voltage from terminal 2 to terminal 1. A voltage vector may be moved to any position in space so long as it is kept parallel to its original position, with reference to other vectors of the system of which it is a part.

For example, Fig. 10(*a*) shows windings 1–2 and 3–4 drawn to represent physical windings of different transformers. They are drawn so that their directions in space represent their relative phase relations. Voltage vectors representing these windings are shown in Fig. 10(*b*), the outlined principles being followed. If terminal 4 of one transformer is physically connected, as in Fig. 10(*c*), to terminal 1 of the other transformer, the magnitude and phase position of the voltage from 2 to 3 may be found by moving the vector 3–4, keeping it parallel to its original position, until the point 4 falls on 1. The dotted line 2–3 is a voltage vector representing the voltage from 2 to 3. By use of these principles voltage vector diagrams of any desired degree of complexity can be drawn. Any

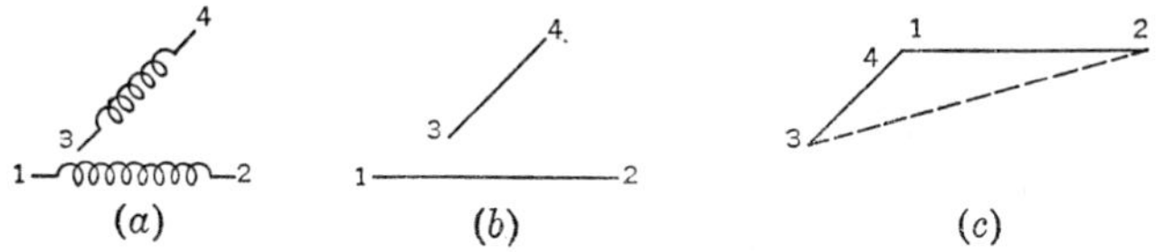

FIG. 10. Winding Vectors

figure drawn with straight lines, no matter how complex, represents a possible voltage vector diagram. Voltage vector diagrams of transformer connections are drawn for no-load conditions; consequently the vectors for two windings on the same core are parallel lines. Three-phase transformers have three core legs on which windings are placed. The fluxes in these cores are 120 deg out of phase with each other; consequently the winding voltages of the three phases are 120 deg out of phase with each other, but the windings on the same core leg are in phase with each other.

5. THREE-PHASE CONNECTIONS

BALANCED THREE-PHASE CONNECTIONS FOR TRANSFORMERS. Most power and distribution transformers are connected to three-phase lines, since three-phase systems are commonly used for power distribution. Although a wide variety of connections are possible and are used where they are advantageous, the most common three-phase connections are delta-star and delta-delta. A balanced three-phase system has three transmission lines with the voltages between the lines equal in magnitude and 120 elec deg apart in phase relation.

The various three-phase connections for transformers are represented by voltage vector diagrams, in which each straight line represents a voltage vector and the physical winding

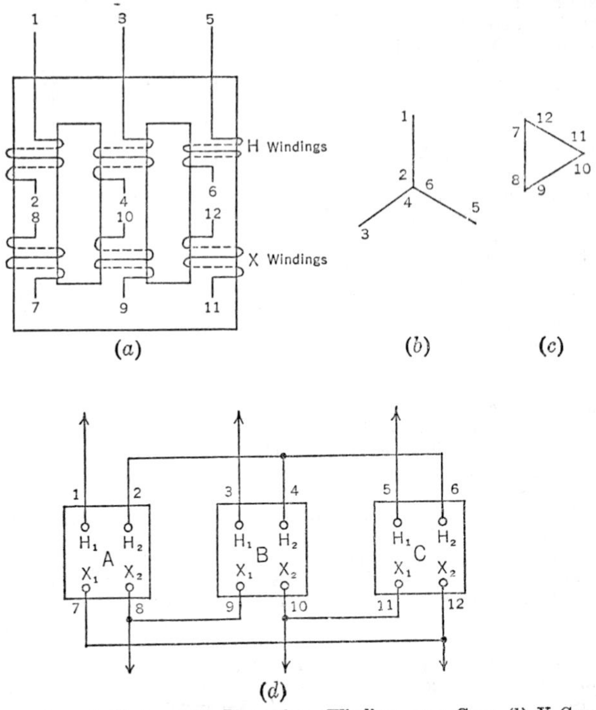

FIG. 11. Connection of Windings: (*a*) Three-phase Windings on a Core; (*b*) Y Connection of H Windings; (*c*) Delta Connection of X Windings; (*d*) Typical Y-delta Connection of Single-phase Transformers

producing that voltage, as outlined under voltage vector diagrams. Either three single-phase transformers or a three-phase transformer may be used for any of the connections. In the following figures the terminals which are to be connected to the three-phase lines are indicated by solid dots for convenience; actually they will have numbers corresponding to the numbers on the physical transformers, which will vary depending on the physical connections.

The method of constructing the vector diagrams for polyphase connections will be demonstrated, starting from the physical transformer windings in one instance and from the marked winding terminals on the outside of the cases of single-phase transformers in another. The directions of the windings on a three-phase core are shown in Fig. 11(*a*). The windings are arbitrarily given numbers so that the winding direction on the core is always the same in going from a lower to a higher number on each leg. It is desired to connect the windings 1–2, 3–4, and 5–6 in Y, with the windings 7–8, 9–10, and 11–12 in delta. If lead 8 is connected to 9 to one three-phase line conductor, lead 10 is connected to 11 to another of the three-phase line conductors, and lead 7 is connected to 12 to the third three-phase line conductor, the voltage vector diagram of Fig. 11(*c*) will represent the phase relations of those windings. The delta can be placed in any convenient position which one wishes to assume.

Figure 11(*c*) represents not only the vector relations but also the physical windings and the connections between them. To connect the other winding in star and draw its diagram, line 1–2 is drawn in the same direction as 7–8, since the voltage 1–2 is in phase with the voltage 7–8. If lead 4 is connected to lead 2, the line 4–3 is drawn in the same direction as line 10–9, starting at point 2, since the voltage 4–3 is in phase with the voltage 10–9. If lead 6 is connected to lead 2, the line 6–5 is drawn in the same direction as line 12–11, since the voltage 6–5 is in phase with the voltage 12–11. Figures 11(*b*) and 11(*c*) represent the voltage vector diagram and connection diagram of the transformer of Fig. 11(*a*).

Voltage vector diagrams for single-phase transformers may be drawn in the same manner, starting with the winding voltage and polarity. For example, the three single-phase transformers of Fig. 11(*d*) have been given arbitrary numbers in the same manner as the windings of Fig. 11(*a*). If these arbitrary numbers and the connections shown between transformers are used, and exactly the same procedure is followed as for Fig. 11(*a*), it will be found that the voltage vector diagram of Figs. 11(*b*) and 11(*c*) apply to Fig. 11(*d*) also. There are several other ways to connect the bank of three transformers of Fig. 11(*d*) in Y-delta, but only one way is shown to illustrate the principles which apply.

The voltage vector diagrams of Fig. 12 and the following figures show transformer connections which can be applied to any suitable single-phase or three-phase transformers, following the method outlined. Usually only one of several possible angular phase displacements between primary and secondary windings is shown, if the other characteristics are the same.

The angular phase shift is the angle between the voltage to neutral of each high-voltage winding lead and the corresponding low-voltage winding lead. The obtainable phase shifts listed may be either positive or negative (leading or lagging); consequently, to avoid repetition, angles greater than 180 deg are omitted. The outstanding characteristics of each connection are described briefly. Either winding may be used for the primary.

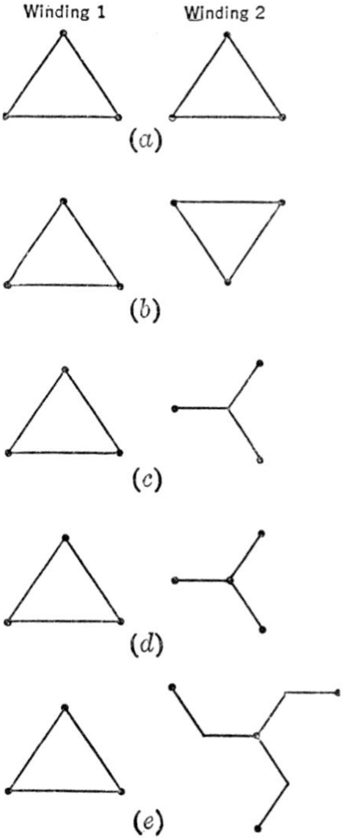

FIG. 12. Three-phase Transformer Connections

Delta-delta Connection. [Figures 12(*a*) and 12(*b*): 0-, 60-, 120-, or 180-deg phase shift.] If one single-phase transformer in a three-phase bank fails, the remaining two may be operated open-delta [Fig. 14(*b*)] at 58 per cent of three-transformer bank capacity. If one phase of a three-phase shell type delta-delta transformer fails, it may be disconnected and shorted, and the other two phases operated open-delta. The delta connection is especially suitable for heavy currents, since the winding current is 58 per cent of the line current. A delta winding on a symmetrical system must be fully insulated from ground at every point. A delta winding does not provide a neutral for grounding the neutral of the line, or a neutral for three-phase four-wire service. Triplen * harmonic exciting current can circulate in the delta winding.

Delta-Y Connection. [Figure 12(*c*): 30-, 90-, or 150-deg phase shift. Y connection is also called star connection.] The Y winding is especially suitable for small currents because of fewer turns (58 per cent) and greater current (173 per cent) than a delta winding, resulting in better space factor. Triplen harmonic exciting current can circulate in the delta winding. If one single-phase transformer in a bank fails, the remaining two cannot be used to provide three-phase power until a spare transformer is installed or the transformer is repaired, except in special cases where the connection of Fig. 14(*c*) may be used.

Delta-Y Connection with Neutral Out. [Figure 12(*d*): 30-, 90-, or 150-deg phase shift.] The characteristics are the same as for Fig. 12(*c*) with the following additions: The neutral may be used for grounding the neutral of the line, either solidly or through an impedance, or for providing a neutral wire for three-phase four-wire service. Grounding the transformer neutral permits the use of transformers with graded insulation (reduced

* The term "triplen" denotes frequencies which are multiples of three.

insulation to ground for the portion of the winding near the grounded end), considerably reducing the cost of transformers above the 69-kv class.

Delta-zigzag Connection. [Figure 12(*e*): 0-, 60-, 120-, or 180-deg phase shift.] The zigzag connection is sometimes called the interconnected star connection. Triplen harmonic currents can circulate in the delta winding. The neutral may be used for grounding the neutral of the line or for providing a neutral for three-phase four-wire service. A zigzag winding has 15 per cent more turns than a Y-connected winding and carries the same current; consequently it has 15 per cent greater kva parts. It is more expensive than a Y winding, particularly for higher voltages.

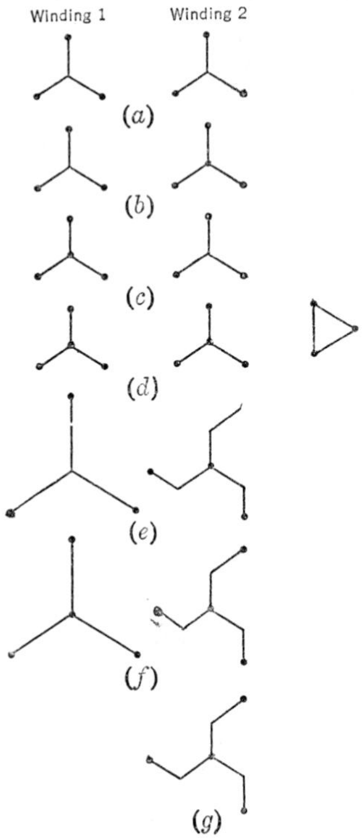

FIG. 13. Three-phase Transformer Connections

Y-Y Connection. [Figure 13(*a*): 0-, 60-, 120-, or 180-deg phase shift.] This connection is suitable only for low-voltage, three-phase core type transformers, because triplen harmonic exciting current cannot flow in the windings. A triplen harmonic voltage to neutral is present in each phase.

Y-Y Connection with One or Two Neutrals Out. [Figures 13(*b*) and 13(*c*): 0-, 60-, 120-, or 180-deg phase shift.] This connection is used principally where a Y neutral can be connected to a neutral established by a generator neutral or grounding transformer neutral, thereby providing a path for the triplen harmonic current to flow and reducing to a negligible value the triplen harmonic voltage to neutral. When one neutral is connected to an established neutral, the neutral of the other winding may be used to provide a neutral for grounding the line or for three-phase four-wire service. When used on a three-phase core type transformer without its neutral connected to an established neutral, a triplen harmonic voltage will be present between each line and neutral; only small loads in percentage of rated kva may be taken from line to neutral because of the high impedance to such loads.

Y-Y Connection with Delta (Tertiary) Winding. [Figure 13(*d*): 0-, 60-, 120-, or 180-deg phase shift between Y lines.] The delta winding is sometimes called a stabilizing winding. Either neutral or both of them may be used for grounding or four-wire service, within the limits of capacity of the delta winding. The impedance for line to neutral current may be reduced to any desired value, since opposing current can circulate in the delta winding. The usual practice is to make the minimum rated capacity of the delta winding 35 per cent of the capacity of the other windings. The third harmonic current can circulate in the delta winding, which may or may not be connected to a line. If not connected to a line, one terminal of the delta winding is usually grounded in order to fix its potential definitely. This connection has most of the characteristics of a delta-Y connection.

Y-zigzag Connection. [Figure 13(*e*): 30-, 90-, or 150-deg phase shift.] This connection is seldom used, being suitable only for low-voltage, three-phase core type transformers, because triplen harmonic exciting current cannot flow in the windings. A triplen harmonic voltage is present from line to neutral of the Y winding, but not from line to neutral of the zigzag winding. The neutral of the zigzag winding can be used for grounding or four-wire service.

Y-zigzag Connection with Y Neutral Out. [Figure 13(*f*): 30-, 90-, or 150-deg phase shift.] This connection is used principally where the Y neutral can be connected to a neutral established by a generator neutral or grounding transformer neutral, thereby providing a path for the triplen harmonic current to flow and reducing to a negligible value the triplen harmonic voltage to neutral. The zigzag neutral can be used for grounding or four-wire service.

Zigzag Connection, One Winding. [Figure 13(*g*).] This connection is principally used to provide a neutral for grounding where a suitable neutral is not otherwise available. This is the smallest and least expensive type of grounding transformer; its kva parts are only 58 per cent of those of a Y-delta grounding transformer of the same kva output.

An impedance may be connected in series with the neutral to limit the grounding current. In a short circuit from one line to ground, equal in-phase currents flow to ground through each winding of the transformer.

THREE-PHASE CONNECTIONS WITH TWO TRANSFORMERS. Two transformers are sometimes used for providing three-phase power from a three-phase line, as an emergency connection when one single-phase transformer fails in service, as a permanent connection to use available transformers, or as a temporary connection to be changed to a balanced three-phase connection when load growth requires. The disadvantages are: the secondary voltages of the three phases are unequal because of the impedance drop through the transformers, resulting in an unbalanced three-phase line on the secondary; the kva parts required for the transformers are greater than the load kva.

T-T Connection. [Figure 14(*a*): 0-, 60-, 120-, or 180-deg phase shift.] If both transformers are wound for full line voltage, as they usually are, the connection may be changed to delta-delta by the addition of a third transformer, to provide an increased capacity of 173 per cent of the capacity of the T–T bank of two transformers. The transformer windings should be so arranged on the transformer cores that the impedance to each other of the primary and secondary half windings carrying opposing currents will not be too high. (This means that half windings carrying line currents which are in phase or 180 deg out of phase with each other should have normal impedance to each other.) Both high-voltage and low-voltage windings of at least one transformer must have 50 per cent taps. The other transformer may have 86.6 per cent taps on both windings or may be operated at reduced voltage (and increased impedance) on the full windings.

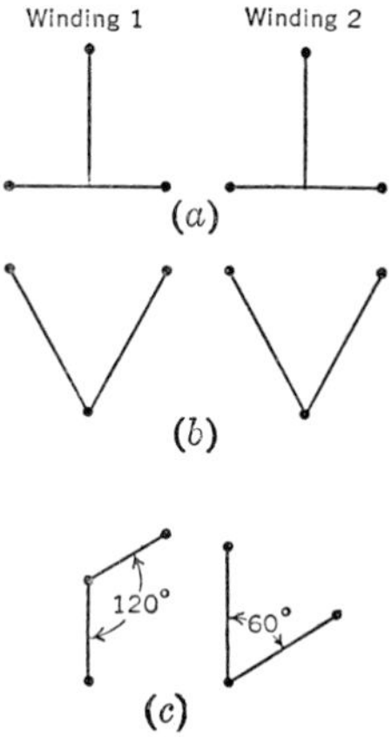

FIG. 14. Three-phase Connections with Two Transformers

Open Delta Connection on High Voltage and Low Voltage. [Figure 14(*b*): 0-, 60-, 120-, or 180-deg phase shift.] Standard transformers without special taps may be used for this connection. Two transformers connected in open delta on a balanced three-phase load can deliver 86.6 per cent of the sum of their kva ratings, or 57.7 per cent of the combined kva rating of three transformers, without exceeding the rating of the transformers. By the addition of a third transformer, the bank may be connected delta-delta to increase the bank rating to 173 per cent of its rating with two transformers.

Open Y-open Delta Connection. [Figure 14(*c*): 30-, 90-, or 150-deg phase shift.] This connection requires a three-phase four-wire supply, to which the open Y connection is made. It permits the use of standard distribution transformers designed to be connected between line and neutral of a four-wire system. Two transformers connected open Y-open delta on a balanced three-phase load can deliver 86.6 per cent of the sum of their kva ratings, or 57.7 per cent of the combined kva rating of three transformers. By the addition of a third transformer the bank may be connected Y-delta to increase the bank rating to 173 per cent of its rating with two transformers.

6. SINGLE-PHASE CONNECTIONS

SINGLE-PHASE CONNECTIONS TO THREE-PHASE SYSTEMS. Single-phase loads are frequently connected to one phase of a three-phase transformer bank or sometimes from one line to neutral. For supplying only single-phase loads, the most common connection is a single-phase transformer connected on a single-phase line, or between lines or from line to neutral of a three-phase system. Although various other connections are used to supply single-phase power from three-phase circuits, such connections usually require larger and more expensive transformers and have at best only a negligible effect in balancing the load among the three phases. A single-phase load is inherently unbalanced and cannot be converted into a balanced three-phase load by transformer connections. Special connections to more than one phase are sometimes used as a means of obtaining a desired voltage ratio with available transformers.

TYPES OF SINGLE-PHASE CONNECTIONS. Simple Single-phase Connection. [Figure 15(*a*).] This is the most common connection for single-phase power supplied at a single voltage and requires the least expensive transformers.

Single-phase–Single-phase 3-wire Connection. [Figure 15(*b*).] This is the common connection used to supply single-phase three-wire service for domestic use, with, for example, 240 volts across the two line wires and 120 volts from each line wire to the ground wire. By balancing the loads on the 120-volt lines, the economy of 240-volt transmission can be secured. This connection provides 120 volts for lamps and 240 volts for heavier loads, such as motors. Transformers for three-wire service must be designed and constructed so that the impedance from each individual 120-volt winding to the total high-voltage winding will be normal.

Special Single-phase Connection. [Figure 15(*c*).] This connection utilizes two transformers, with one set of windings, which are suitably insulated, connected in series, and the other windings connected in parallel. The advantage of this connection lies in obtaining a voltage ratio different from that obtainable with one transformer. When transformer windings are used in series to obtain special connections, it is desirable to parallel the other windings in order that the series voltage may be distributed equally on all the transformers used. In Fig. 15 any difference in the exciting current of the two transformers will circulate in the paralelled windings, so that both transformers operate at the same voltage.

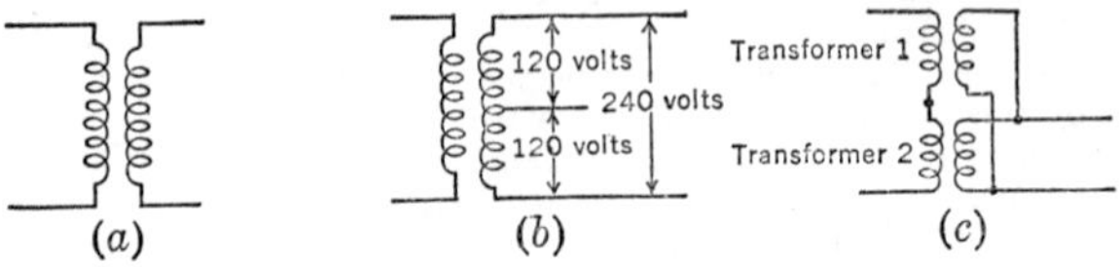

FIG. 15. Types of Single-Phase Connections

7. TWO-PHASE CONNECTIONS

INDEPENDENT AND CONNECTED PHASES. A two-phase system consisting of two equal voltages 90 deg apart in phase may have the two phases independent and insulated from each other, or the two systems may be connected together conductively. Transformers connected to a two-phase system may have the high-voltage windings connected in any one of the various ways and the low-voltage windings also connected in any one of these ways, regardless of the connection used for the high-voltage winding except for phase relations. Figure 16(*a*) shows the transformer connections for two independent phases, and Fig. 16(*b*) the transformer connections to a two-phase three-wire system, in which one line is used as a conductor for both phases. With a balanced two-phase load, the common line has 1.41 times the current in the other lines, and the voltage drops will be different on the two phases for most loads. This condition produces unequal voltages on the two phases, producing undesirable unbalance in the line if the line has considerable voltage drop. Figure 16(*c*) shows the connections for four-wire interconnected transformers in which a neutral point is established by connecting the midpoints of the windings together. This connection makes provision for grounding the neutral of the system with minimum operating voltage from line to ground. Figure 16(*d*) shows two windings without center taps connected to a four-wire interconnected system.

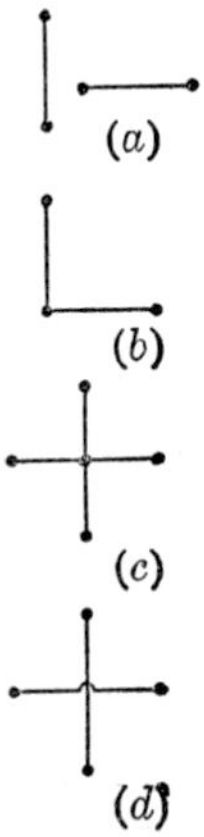

FIG. 16. Types of Two-phase connections

8. THREE-PHASE TO TWO-PHASE CONNECTIONS

PRINCIPLE. A wide variety of connections may be used to transform from three-phase to two-phase and vice versa, because there are many ways of securing two voltage vectors at right angles to each other, starting with a Y or delta voltage vector diagram, or alternatively of securing balanced three-phase voltages, starting with 90-deg vectors. Only some of the possible connections are shown. In calculating the winding currents for the various connections with balanced loads, the three-phase currents and the two-phase currents can each be calculated on the basis of the ine kva and line voltage. For some of the more complicated connections a calculation of the currents with a single-phase load on either the two-phase line or the three-phase line may be made. If this calculation is repeated for load on the other phases, the total current in the windings may be found by superposition.

CONNECTIONS. T Three-phase–Two-phase Connection. [Figure 17(*a*).] The T or **Scott connection,** which employs two transformers, is the most frequently used connection for three-phase to two-phase transformation. Any of the connections of Fig. 16 may be used for the connections of the two-phase side of the transformers. One transformer, which is called the main transformer, has a 50 per cent tap on the three-phase side; the other transformer, which is called the teaser transformer, has an 86.6 per cent tap on the three-phase side. (Usual construction practice is to make both transformers alike, with both a 50 per cent and an 86.6 per cent tap, so that they may be used interchangeably as main and teaser transformers.) The windings for the three-phase side are designed to carry full three-phase line current; those for the two-phase side are designed to carry the two-phase line current. This results in the kva size of the main winding for the three-phase side (as the product of voltage and current) being 115 per cent of the kva rating of the winding for the two-phase side, that is, 57.7 per cent of the kva transformed. The two halves of the main winding must be interlaced with each other, so that the reactance between them is a minimum to avoid excessive voltage drop and consequent voltage unbalance between phases.

T Three-phase–Modified T Two-phase Connection. [Figure 17(*b*).] This connection is similar to that in Fig. 17(*a*), except that the windings on the two-phase side are provided with taps and interconnected so that three-phase power can be obtained between leads 1–2–3 at the same time that two-phase power is available at terminals 1–4 and 2–3. By substituting any of the connections of Fig. 16 for the left-hand diagram of Fig. 17(*b*), a connection is secured which can be used to transform from two-phase to both two-phase and three-phase simultaneously.

Delta Three-phase–Tapped Delta Two-phase Connection. [Figure 17(*c*).] This connection employs three single-phase transformers or one three-phase transformer. The windings for the three-phase side may be connected either star or delta, with current ratings corresponding to the output kva. For the two-phase side, delta-connected windings having a voltage rating of 111.5 per cent of the required two-phase voltage, each provided with a tap to divide the winding in the proportion of 81.6 to 29.9 as shown, and having a current rating of 71.5 per cent and 81.6 per cent respectively of the two-phase line current, are required. The winding currents for a balanced load are shown on the figure.

When the three-phase side is connected in delta, this connection may be used with the untapped transformer omitted. In this case the current capacity of the winding for the three-phase side is three-phase line current; for the two-phase side the winding current is 141 per cent of the two-phase line current for the shorter part of the winding and two-phase line current for the longer part.

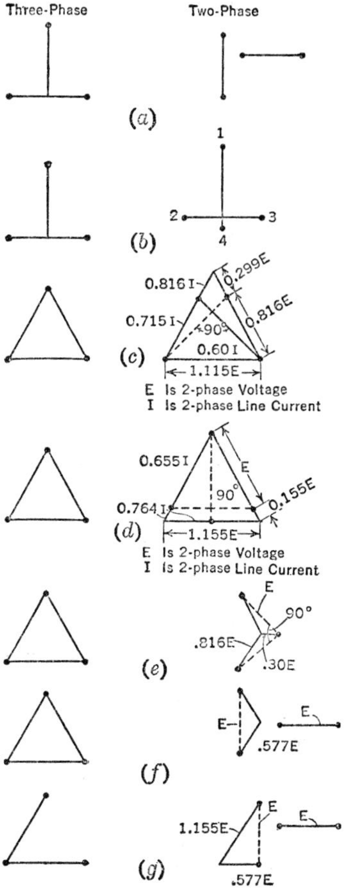

FIG. 17. Three-phase to Two-phase Connections

Delta Three-phase–Tapped Delta Two-phase Connection. [Figure 17(*d*).] This connection employs three single-phase transformers or one three-phase transformer. The windings for the three-phase side may be connected either star or delta. For the two-phase side, delta-connected windings having a voltage rating of 115.5 per cent of the desired two-phase voltage are required. Two of the windings require a tap to divide the winding in the proportion of 100 to 15.5; the other winding requires a center tap. The winding current ratings for the two-phase side are 76.4 and 65.5 per cent of the two-phase line current. The winding currents for a balanced polyphase load are shown on the figure. The winding current rating for the three-phase side when Y connected is the same as the three-phase line current corresponding to the load kva. When the winding is delta connected, the winding current is 57.7 per cent of the three-phase line current. This connection requires more taps than the connection of Fig. 17(*c*).

Other Connections for Three-phase to Two-phase Operation. [Figures 17(*e*), 17(*f*), 17(*g*).] There are many other possible connections for three-phase to two-phase transformation, such as are shown in Figs. 17(*e*), 17(*f*), and 17(*g*), where the windings have the voltages indicated.

9. TRANSFORMER CONNECTIONS FOR SIX-PHASE SYNCHRONOUS CONVERTERS

Common secondary connections used for transformers supplying power to synchronous converters are shown in Figs. 18 and 19. The primary three-phase connections for the six-phase secondary connections of Fig. 18 may be either Y or delta, although the delta connection is preferable in Figs. 18(*a*), 18(*b*), and 18(*c*) to eliminate the production of harmonic circulating currents in the synchronous converter windings.

Figure 19 shows the common secondary connection used with two transformers for two-phase to six-phase operation. The two-phase connections may be any of those shown in Fig. 16.

The neutral of any of these connections, where it is available, may be used as the neutral wire for three-wire d-c service. The dotted-line diagrams indicate the relative location of the voltage vectors when operating connected to the slip rings of a six-phase rotary converter.

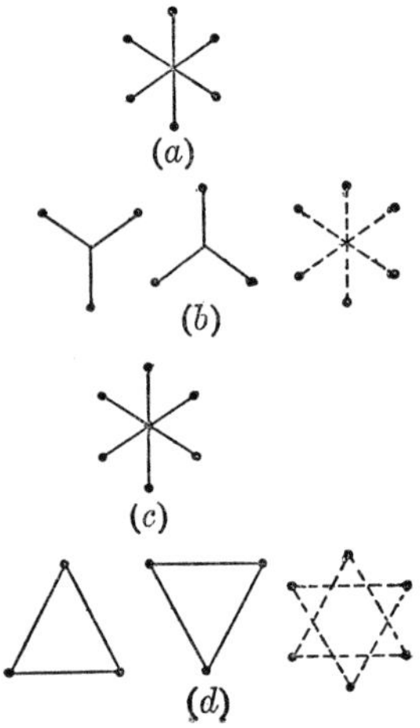

FIG. 18. Three-phase to Six-phase Connections

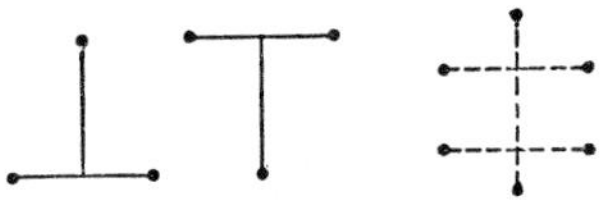

FIG. 19. Two-phase to Six-phase Connections

10. TRANSFORMER CONNECTIONS FOR POLYPHASE MERCURY ARC RECTIFIERS

Figure 20 shows some of the common connections used for the d-c windings of transformers used with mercury arc rectifiers. The d-c winding of a transformer supplying a rectifier is the winding conductively connected to the rectifier. Usually each d-c winding is connected to an anode. The a-c windings may be either delta connected or Y connected; with a Y connection a delta-connected tertiary winding is generally desirable to eliminate triplen harmonic voltages. Figures 20(*a*), 20(*b*), and 20(*c*) show six-phase connections. In the connections of Figs. 20(*a*) and 20(*c*) direct current flows through each d-c winding and the anode to which it is connected for only one-sixth of each cycle. In the connection of Fig. 20(*b*) direct current flows through each d-c winding and the anode to which it is connected for one-third of each cycle. The two Y-connected groups which are connected to each other through the interphase transformer marked *f* operate in parallel, their

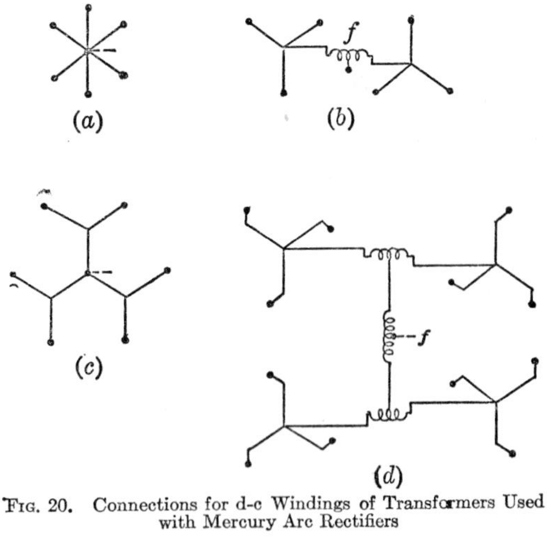

FIG. 20. Connections for d-c Windings of Transformers Used with Mercury Arc Rectifiers

difference in voltage being the voltage across the interphase transformer, so that the maximum direct current in the anodes is only 50 per cent of the maximum current for the connections shown in Figs. 20(*a*) and 20(*c*). The voltage across the interphase transformer is principally at three times the primary frequency of the transformer. The connection of Fig. 20(*b*) is the most commonly used polyphase rectifier transformer connection.

The winding of Fig. 20(*a*) is simpler than that of 20(*c*), but in operation the reactance between adjacent d-c terminals is less for the winding of Fig. 20(*c*) and is used more than that of Fig. 20(*a*). Figure 20(*d*) is a twelve-phase connection employing a modified zigzag connection and three-interphase transformers. Another method of securing twelve-phase operation is by using two main transformers, each having a six-phase connection of the d-c winding, one with a delta-connected a-c winding, the other with a Y-connected a-c winding.

Higher numbers of phases may be secured by using phase shifters, which are two winding transformers or autotransformers, in the a-c (or the d-c) lines of the separate rectifier transformers. Because the d-c windings of rectifier transformers carry current for a smaller portion of a cycle than the a-c windings, the total rated kva of the d-c windings is greater than the rated kva of the a-c windings. Both are larger than the kilowatt output of the rectifier.

11. HARMONIC EXCITING CURRENTS AND VOLTAGES

A sine-wave of voltage applied to a single-phase transformer produces a sine-wave of flux in its core. If the permeability of the iron core were constant, the magnetizing current required at every instant would be proportional to the flux density and would be a sine-wave also. Higher values of flux near the crest of the flux wave, however, require much greater than proportional instantaneous values of magnetizing current, and the curve of the exciting current against time will be peaked in comparison to a sine-wave. In this manner a sine-wave of voltage applied to a transformer requires a peaked wave of exciting current to produce it. For convenience in analysis, we may separate this current wave into a fundamental sine-wave and higher-frequency sine-waves, with frequencies which are usually odd multiples of the fundamental, so that the sum of all the waves at any instant is equal to the actual exciting current. The fundamental sine-wave of current has the highest value, with successive higher harmonics of lesser magnitude, the relative magnitudes varying greatly for different transformers.

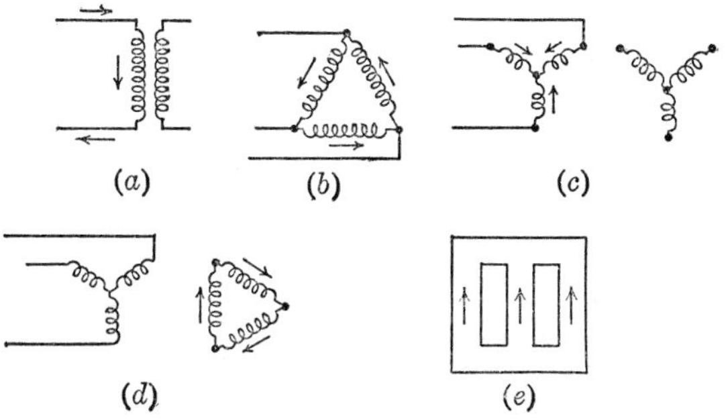

FIG. 21. Path of Exciting Current and Its Harmonics and Path of Triplen Harmonic Fluxes

In Fig. 21(*a*), which shows a single-phase transformer connected to a single-phase line, the exciting current path for the fundamental and its harmonics is shown by the arrows.

THREE-PHASE DELTA-CONNECTED PRIMARY. In a three-phase delta-connected primary, as in Fig. 21(*b*), all components of the exciting current which are not multiples of 3 flow in the three-phase lines. The exciting currents which are multiples of 3 would flow in the lines, as in Fig. 21(*a*), if only one transformer were connected to the lines at one time. When three transformers are connected in delta to the lines, as in Fig. 21(*b*), the triplen (multiples of 3) harmonic currents flow in opposite directions in each line. (If fundamental waves are 120 deg apart, as they are on a balanced three-phase system, triplen harmonic waves are 3 × 120, or 360, deg apart, that is, in phase with each other.) If the triplen harmonic exciting currents of the three transformers are equal, these currents cancel out in each line, and no triplen harmonic exciting current will flow in the lines. The path of the triplen current is shown by the arrows of Fig. 21(*b*), where it is said to circulate in the delta. If the triplen exciting currents of the three transformers are unequal, the difference in the currents will flow in the lines.

THREE SINGLE-PHASE TRANSFORMERS IN Y. If three single-phase transformers are connected in Y to a three-phase line, as in Fig. 21(*c*), all components of the exciting current which are not multiples of 3 flow in the three-phase lines. The triplen harmonic currents, however, are in phase and tend to flow as shown by the arrows of Fig. 21(*c*), which represent the instantaneous direction (not vector direction) of current flow. If the transformer neutral is connected to a supply neutral, these currents can flow in the neutral. If the transformer neutral is not connected to an established neutral, the triplen currents cannot flow, and the exciting current will lack some of the harmonic components necessary to produce a sine-wave of flux, which in turn is necessary to produce a sine-wave of voltage. Under these conditions the flux wave will be flat on top, producing a peaked wave of voltage, since the instantaneous voltage is proportional to the rate of change of flux. The voltage peak at no load may be as high as 160 per cent of the peak of the fundamental, which is the voltage when the triplen currents are provided with a suitable path. With a resonant capacity load, connected between each line and neutral, it may reach several times this value.

TRIPLEN HARMONIC CURRENTS. A suitable path may be provided by connecting the neutral to an established neutral or by providing the transformers with delta-connected windings, as in Fig. 21(*d*). The triplen currents will flow around the delta as shown by the arrows in this figure (whether the windings are connected to a line or not), the result being similar to that of Fig. 21(*b*). The triplen currents flowing in the delta provide the necessary harmonic components of magnetizing current to produce a sine-wave of flux. There will be some triplen harmonic voltage between each Y line and neutral, the magnitude depending on the impedance between the Y and delta windings and the magnitude of the triplen harmonic currents. In the event that more than one path is provided for the flow of triplen harmonic currents, they will divide between the paths in inverse proportion to their impedances.

TRIPLEN HARMONIC VOLTAGES. The core of a three-phase core-type transformer has an action in suppressing triplen harmonic voltages similar to that of a high-reactance delta-connected winding, because the triplen harmonic fluxes which flow in the iron core must return through an air path. (Like the triplen harmonic waves of current, if fundamental waves of flux are 120 deg apart on a balanced three-phase system, the triplen harmonic waves are 3×120, or 360, deg apart, that is, in phase with each other.) The instantaneous directions of the triplen harmonic fluxes in the iron of a three-phase core-type transformer are shown in Fig. 21(*e*), where it is obvious that the only return path is through the air. The suppression obtained by a three-phase core may be satisfactory in a low-voltage transformer but generally is not satisfactory in high-voltage transformers.

Triplen harmonic voltages are undesirable because, if they appear on transmission lines, they may electrostatically induce voltages on adjacent exposed telephone lines, or the charging current they cause to flow in the lines, because of the capacity between lines and ground, may induce by electromagnetic induction voltages in adjacent exposed telephone lines. In either case **telephone interference** may result. If large, the triplen harmonic voltages may cause the line and the transformers to operate with higher voltage stresses than those for which they were designed. Because triplen harmonic voltages appear between lines and ground, and triplen harmonic currents circulate in phase in the lines and return through the ground, their effect cannot be neutralized by transposing the conductors of the transmission line, as can the effect of the fundamental frequency currents and voltages. For these reasons the usual practice is to provide a path in which the triplen harmonic currents can flow in order to eliminate undesirable triplen harmonic voltages. Triplen harmonic currents of large magnitude flowing in the transmission line are undesirable, since, as has been mentioned, they may cause telephone interference. Accordingly, when they are allowed to flow in the neutral (and lines), it is usually in a short unexposed section, for example, within the same station when the established neutral and the transformer Y neutral are in the same station. Because of limitations with neutral connections, the most common path provided for the flow of triplen harmonic currents is a delta-connected winding on the transformer.

12. PARALLEL OPERATION

SINGLE-PHASE TRANSFORMERS. Transformers are often connected in parallel (on both the high-voltage and low-voltage sides) in order to furnish power in excess of the capacity of one transformer alone. For satisfactory parallel operation the circulating current between units should be small, and the load division proportional to their kva ratings. Parallel operation is generally considered to be satisfactory when the kva output of the bank is within 5 to 10 per cent of the combined kva ratings of the transformers in the bank, without the load on any transformer exceeding its kva rating.

In order for satisfactory parallel operation to be possible without auxiliary devices, transformers should have the following characteristics:

Approximately the same voltage ratings.

The same ratio of high-tension volts to low-tension volts.

The same percentage impedance (based on the kva rating of each transformer and the same voltage base).

Theoretical requirements are that both transformers have the same percentage resistance and the same percentage reactance, but practically, if the transformers are somewhere near the same size and have the same percentage impedance, the percentage resistance and percentage reactance will be sufficiently close for satisfactory parallel operation.

Division of Load. The load currents of paralleled transformers divide inversely as the impedances of the two transformers, when the impedances are expressed in ohms on the same voltage base or in percentage on the same kva base. If Z_1 and Z_2 are the impedances of paralleled transformers 1 and 2, respectively, and I is the total load current, the load current in transformer 1 $= IZ_2/(Z_1 + Z_2)$, and the load current in transformer 2 $= IZ_1/(Z_1 + Z_2)$. For most applications the load division will be obtained with sufficient accuracy if the impedances Z_1 and Z_2 are expressed as real numbers, but if close accuracy is desired, the impedances may be expressed as complex quantities, e.g., $(r + jx)$, where r is the resistance and x the reactance.

Three-winding Transformers. For best parallel operation, three-winding transformers must have the same impedance between windings of the same voltage for the two transformers; in other words, the percentage impedances in the equivalent circuits must be the same for the two transformers. The kva ratings of windings of the same voltage must be the same for the two transformers or be the same percentage of each other for all windings of the transformer. Otherwise one or more windings must be derated (or carry less than its full load in operation). An example of two transformers which would parallel successfully is:

	Transformer No. 1	*Transformer No. 2*
H 66,000-volt winding	10,000 kva	5,000 kva
X 13,800-volt winding	6,000 kva	3,000 kva
Y 2,400-volt winding	7,000 kva	3,500 kva
Impedance H to X winding	7% on 6,000-kva base	7% on 3,000-kva base
Impedance H to Y winding	8% on 7,000-kva base	8% on 3,500-kva base
Impedance X to Y winding	12% on 6,000-kva base	12% on 3,000-kva base

It naturally follows that uniform loading will not be secured for all conditions of operation when a two-winding transformer is operated in parallel with a three-winding transformer with loads on all three windings. For any combination of three-winding transformers and three- or two-winding transformers connected in parallel, the load division for any combination of loading may be determined by using the equivalent circuits for the transformers. Although the division of load may not be perfect, it still may be satisfactory if the total load is less than the sum of the kva ratings of the transformers. The test is whether any winding is overloaded when the desired loads are carried.

Transformers of Different Voltages. When transformers of different voltage ratios are operated in parallel, with their primary and secondary winding terminals connected together, the voltage impressed on the primary windings will be the same for both transformers. If their secondary windings were not connected together in parallel, the secondary winding voltages would be different. When the windings are connected together in parallel, this difference in voltage causes a current to circulate in the primary and secondary windings, the value of the current being limited by the sum of the impedances of the two transformers. Accordingly, the circulating current at no load for transformers of different voltage ratios operating in parallel may be determined by the formula: Circulating current $= (E_1 - E_2)/(Z_1 + Z_2)$, where E_1 and E_2 are the secondary-open-circuit voltages of the two transformers not connected in parallel, when the primary-circuit voltage is applied to their respective primary windings. If E_1 and E_2 are in volts, and Z_1 and Z_2 are in ohms referred to the secondary windings, the circulating current will be in amperes in the secondary windings. If E_1 and E_2 are in percentage of the no-load secondary voltage of one transformer arbitrarily selected as the reference transformer, and Z_1 and Z_2 are the impedances expressed in percentage, both based on the reference-transformer kva, secondary voltage, and current, the circulating current will be in per unit terms of the base current. The kva supplied by the line because of the circulating current is equal to $(E_1 - E_2) \times$ Circulating current in amperes/1000. Only a small difference in voltage ratio is required to produce a large circulating current, so it is important that the voltage ratios of paralleled transformers be nearly the same. The current flowing in the primary line is much smaller than the current in the windings, since it is the difference between the currents in the primary windings.

To determine the load on individual transformers being operated in parallel, for any given case, the division of load current and the circulating current may be determined independently as outlined. For practical purposes the current in each transformer will be the vector sum of its load current and the no-load circulating current, each calculated independently.

When two transformers have different impedances, satisfactory parallel operation is sometimes secured by connecting a reactor in series with one winding of the transformer having low impedance, so that the sum of its percentage impedance and the transformer percentage impedance is equal to the percentage impedance of the other transformer. For reasons of convenience and cost, a paralleling reactor is usually an iron core reactor (with air gaps in its core to produce constant impedance) in series with the low-voltage winding of the transformer.

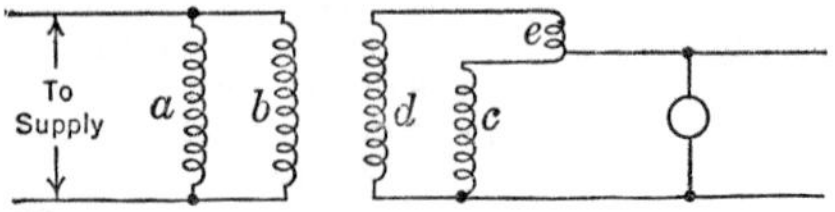

FIG. 22. Parallel Operation of Transformers with Different Voltages

When the difference in voltage ratios between transformers is too great for satisfactory parallel operation, an auto-transformer, sometimes called a balance coil or balancing transformer, may be used to secure satisfactory paralleling. A balance coil will also correct for differences in impedance between transformers. The auto-transformer fixes the ratio between the loads carried by the two transformers, so that the load ampere-turns in the two windings of the auto-transformer are equal at all times; that is, the current in each winding of the auto-transformer is inversely proportional to the number of turns in the winding.

In Fig. 22, *a* and *b* are the primary windings of two transformers connected in parallel with secondary windings *c* and *d* respectively, the voltage of winding *d* being higher than that of winding *c*. The auto-transformer *e* must be made for a total voltage across its terminals equal to the maximum difference in voltage which will occur in operation because of any combination of the effects of differences in transformer voltage ratios and impedances. The tap is placed on the auto-transformer in such location as to force the desired division of current between the transformers, so that the turns on each side are inversely proportional to the desired load current. For example, if one transformer winding should carry 40 per cent of the total load current, 40 per cent of the total turns of the auto-transformer would be placed in series with the winding of the other transformer. As indicated in Fig. 22, the voltage applied to the load will be intermediate in value between the secondary voltages of the two transformers. The auto-transformer is constructed like any conventional transformer with a closed iron core to provide low exciting current.

THREE-PHASE TRANSFORMERS. Three-phase transformers must meet the same requirements for parallel operation as single-phase transformers, and in addition their voltage vector diagrams must be such that they can be made to coincide on both high- and low-voltage sides, and the transformers must be connected so that their vector diagrams do coincide. In moving vector diagrams from one location to another, the vectors of a set (one high-voltage winding group and its corresponding low-voltage winding group on the same transformers) must not be twisted in relation to each other, but vectors representing windings in phase with each other must be kept parallel and in the same direction.

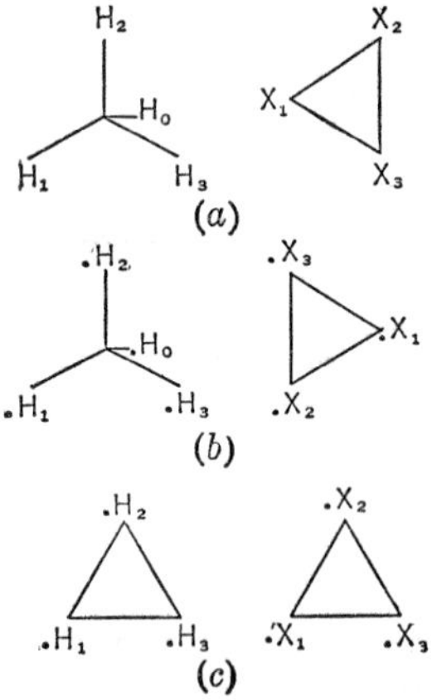

FIG. 23. Parallel Operation of Three-phase Transformers

In the examples of Fig. 23, dots are placed before some of the terminal designations on the transformer nameplate as a convenient means of identifying leads. Any other device for identifying leads may be arbitrarily used. Figure 23(*a*) represents a three-phase transformer having the same percentage impedance and voltage ratio as the three-phase transformers of Figs. 23(*b*) and 23(*c*) but different phase relations, as indicated by the voltage vector diagrams, which represent the physical windings in addition to the vector relations. The transformer of Fig. 23(*a*) cannot be connected in parallel with the transformer of Fig. 23(*b*) by connecting H_1 to $.H_1$, H_2 to $.H_2$, H_3 to $.H_3$, because with these high-voltage connections the low-voltage diagrams cannot be made to coincide when they are kept in their relative directions with respect to their high-voltage windings, in accordance with the general principles of voltage vectors. By connecting H_1 to $.H_3$, H_3 to $.H_1$,

H_2 to $.H_2$, however, the low-voltage winding voltage vectors may be made to coincide by connecting X_3 to $.X_2$, X_2 to $.X_3$, X_1 to $.X_1$.

Another example is the problem of paralleling the transformer of Fig. 23(*a*) with that of Fig. 23(*c*). By the trial-and-error process or by comparing the phase shift angles, it is evident that for any possible connection on the high-voltage side it is impossible to make the low-voltage diagrams coincide; consequently it is not possible to operate these two transformers in parallel.

As a general principle, two star-delta or delta-star transformers or banks of the same voltage ratio can always be made to coincide by proper connections between the transformers or banks, without changing the connections inside the bank.

If the phase relations are correct but the voltage ratios and impedances are different, satisfactory parallel operation of three-phase banks may be secured by the use of three-phase paralleling reactors for differences in impedance, and three-phase auto-transformers for differences in ratio or impedance. Connections similar to those for single-phase transformers, with the coils connected in series with either the high- or low-voltage windings in each phase, should be used.

13. AUTO-TRANSFORMERS

SINGLE-PHASE AUTO-TRANSFORMERS. An auto-transformer is a transformer which has a portion of its winding used in both the primary and secondary circuits. The principal advantage of auto-transformers is that they are smaller physically and cost less than two-winding transformers serving the same load. Most auto-transformers have a nameplate kva rating, based on output line current and voltage, which is the same as that of a two-winding transformer of the same rated kva output. The saving is greatest when the high-voltage and low-voltage winding voltages are nearly the same, that is, when the ratio of transformation is close to 1. The principal disadvantage of an auto-transformer is that the two windings are not insulated from each other, with the result that a fault on one winding may set up voltage stresses on the other winding. High-voltage stresses under fault conditions are most likely to occur when the auto-transformer neutral is not solidly grounded through a low-resistance ground.

Impedance. The impedance of an auto-transformer is generally less than that of a two-winding transformer of the same rating, particularly if the ratio of voltages is close to 1. A transformer with lower impedance will have better voltage regulation. The lower impedance, however, permits greater short-circuit currents to flow in the event of a fault. When the voltage ratio is close to 1, special care should be exercised to determine that the auto-transformer is designed to withstand the short-circuit currents as limited by the impedance of the circuit, including the auto-transformer. In some cases an air core current-limiting reactor is installed in series with the supply lines to the auto-transformer to limit the fault current to a safe value.

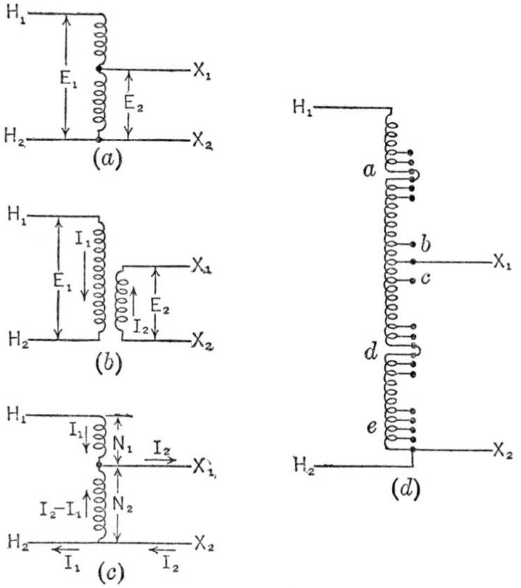

FIG. 24. Auto-transformer Connections

Particular formulas may be easily derived for calculating the currents in the windings of an auto-transformer. Another convenient method is to calculate the currents for a two-winding transformer of the same kva and voltage rating and then use the principle of superposition to determine the auto-transformer currents.

Figure 24(*a*) shows a typical auto-transformer, with voltages of E_1 and E_2. Figure 24(*b*) shows a two-winding transformer, having the same output kva and voltages. The currents I_1 and I_2 are calculated as equal to Kva rating/Winding kilovolts. If the two windings of the transformer of Fig. 24(*b*) are superimposed and connected together as in Fig. 24(*c*) the resultant current in each part of the winding is equal to the algebraic sum of the currents in that part of the winding. As shown in Fig. 24(*c*), the current in the portion of the winding connected to H_1 is I_1, the same as in Fig. 24(*b*). The current in the common portion of the winding between lead X_1 and lead X_2 is equal to the difference between the high- and low-voltage currents of Fig. 24(*b*), i.e., $I_2 - I_1$. Since the vector sum of all the load ampere-turns must be equal to zero, as for any transformer, $N_1I_1 = N_2(I_2 - I_1)$; also $(E_1 - E_2)I_1 = E_2(I_2 - I_1)$. Solving for the current in the common winding, $I_2 - I_1 = I_1\left(\frac{E_1}{E_2} - 1\right)$.

The kva parts required for an auto-transformer are less than the kva parts for a two-winding transformer of the same output and voltage rating, being equal to $K \times$ kva output, where $K = (E_1 - E_2)/E_1$. The percentage reactance equals $K \times$ (Percentage reactance, based on the current in one of the physical windings between adjacent terminals, its turns, and the voltage across it). The percentage reactance can be determined by test in the same manner as for a two-winding transformer on the basis of rated line currents and line voltages. The percentage resistance of an auto-transformer is equal to $100 \times$ Full-load copper loss in kilowatts/Rated kva output.

Auto-transformers sometimes have an additional winding, which may be called a tertiary winding, to provide power for a load at a different voltage or for any other purpose which would apply for two-winding transformers. The current in the common portion of the auto-transformer windings is generally less for combined loads on the other two windings when power is supplied to the high-voltage terminals than when it is supplied to the low-voltage terminals or tertiary-voltage terminals. The current distribution in the windings under combined loads is a factor to be considered in the design and application of three-winding auto-transformers and requires a knowledge of the load kva and power factor under all limiting conditions of operation. The current in the windings is easily obtained by use of the principle of superposition previously illustrated.

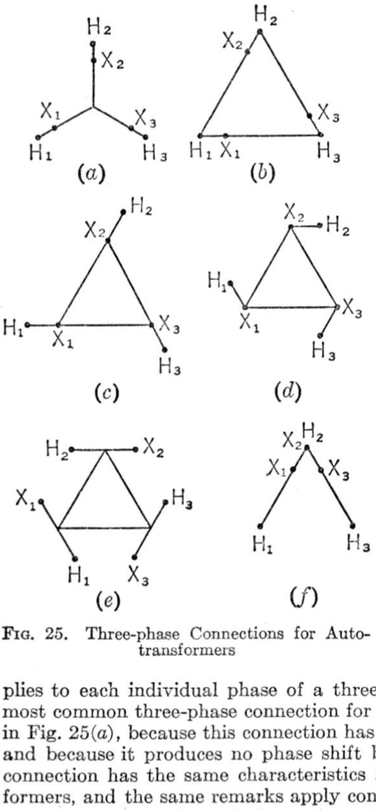

FIG. 25. Three-phase Connections for Auto-transformers

Taps. As with any transformer, the addition of taps to the windings introduces problems of mechanical and electrical stresses, which must be provided for by the designer. Although taps can be placed on the line ends of windings, a less expensive transformer results (except for some low-voltage transformers) if the taps are placed in a part of the winding remote from the line. Accordingly, in Fig. 24(*d*) taps on the high-voltage winding would be placed at *a*. Taps on the low-voltage winding could be placed at locations *b* or *c*, but because of the mechanical and electrical stresses introduced it is preferable to place them at a location such as *d*, or at *e* if $H_2 - X_2$ is grounded. Taps at *d* or *e* change not only the low-tension but also the high-tension voltage, since they change the number of turns between H_1 and H_2, so that it is necessary to provide additional taps at *a* to keep the high-tension voltage constant when the low-tension taps are changed. Because additional tap leads are necessary when taps are placed in the low-voltage winding of an auto-transformer, the voltage ratio of auto-transformers is adjusted by taps in the high-voltage winding wherever possible in preference to taps in the low-voltage winding. Of course the simplest transformer has no taps at all.

THREE-PHASE CONNECTIONS FOR AUTO-TRANSFORMERS. The information given for single-phase auto-transformers applies to each individual phase of a three-phase connection for auto-transformers. The most common three-phase connection for auto-transformers is the Y connection as shown in Fig. 25(*a*), because this connection has the smallest kva parts for a given kva output, and because it produces no phase shift between the high- and low-voltage lines. This connection has the same characteristics as the Y–Y connection for two-winding transformers, and the same remarks apply concerning triplen harmonic currents and voltages.

The delta-connected auto-transformers of Fig. 25(*b*) and the extended delta connection of Fig. 25(*c*) are used less frequently because they require larger kva parts and they produce a phase shift which varies with the ratio of transformation. The extended delta connection of Fig. 25(*d*) is used to produce a phase shift between high- and low-voltage lines with only slight change in voltage. The extended delta connection of Fig. 25(*e*) is used to produce a phase shift between lines without change in voltage. (The connections of Figs. 25(*d*) and 25(*e*) are often used with load ratio control equipment to control the flow of power between lines.) The open delta connection of Fig. 25(*f*) is frequently used for auto-transformers for motor starting because it simplifies the switching and for single-phase step voltage regulators used to regulate a three-phase three-wire (delta) line.

14. SIZE AND PERFORMANCE RELATIONS

There exist certain fundamental relations between the kva ratings of transformers and their physical size and performance; for example, larger transformers are inherently more efficient that smaller units. In order to simplify the problem of deriving the relations the following assumptions are made:

Constant physical proportions.

Constant current density in the copper.

Constant flux density in the iron.

Constant space factor of insulation. (This means that larger transformers have insulation for higher-voltage classes than the smaller transformers.)

Constant percentage eddy-current loss in the copper.

Because in practical designs these factors are only approximately constant over a limited range, actual transformers follow these relations only approximately. Where the difference in kva rating is not too great, the relations are sufficiently accurate for estimating purposes. These relations can be written by inspection between the different variables and the length l (or other linear dimensions, since the linear dimensions are proportional to each other). The equations so determined may be solved simultaneously to obtain the relations between the desired variables. In the following equations, k is a quantity which has different values in each equation. The value of k in any particular equation is constant over limited ranges.

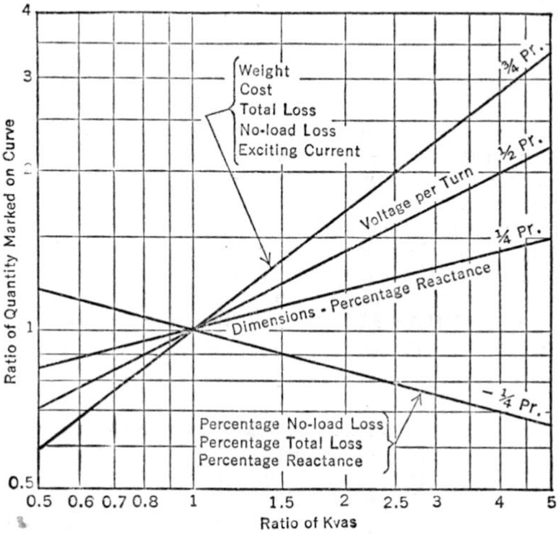

FIG. 26. Size and Performance Relations to Rating of Transformers

The most frequently used relations are:

Kva	$= kl^4$		
Weight	$= k\ \text{kva}^{3/4}$	$= kl^3$	
Cost	$= k\ \text{kva}^{3/4}$	$= kl^3$	$= k(\%\ \text{total loss})^{-3}$
Length	$= k\ \text{kva}^{1/4}$	$= kl$	
Width	$= k\ \text{kva}^{1/4}$	$= kl$	
Height	$= k\ \text{kva}^{1/4}$	$= kl$	
Total watts loss	$= k\ \text{kva}^{3/4}$	$= kl^3$	
No-load loss	$= k\ \text{kva}^{3/4}$	$= kl^3$	
Exciting current	$= k\ \text{kva}^{3/4}$	$= kl^3$	(amperes for the same voltage or kva exciting current)
Percentage total loss	$= k\ \text{kva}^{-1/4}$	$= kl^{-1}$	
Percentage no-load loss	$= k\ \text{kva}^{-1/4}$	$= kl^{-1}$	
Percentage exciting current	$= k\ \text{kva}^{-1/4}$	$= kl^{-1}$	
Percentage resistance	$= k\ \text{kva}^{-1/4}$	$= kl^{-1}$	
Percentage reactance	$= k\ \text{kva}^{1/4}$	$= kl$	
Volts per turn	$= k\ \text{kva}^{1/2}$	$= kl^2$	

For convenience in calculation, the curves of Fig. 26 have been drawn showing these relations with respect to kva rating.

One practical modification of these relations applies to the percentage reactance, which is proportional to the kva$^{1/4}$, when the insulation is increased to keep the space factor constant as the kva is increased. When the insulation class is maintained constant, the reactance naturally increases more slowly than this. Because of service requirements it is held practically constant for a wide variation in kva rating.

Another relation which may be derived in similar manner is:

$$\text{Cost} = k(\text{Percentage no-load loss} \times \text{Percentage copper loss})^{-1.5}$$

Although this relation was derived on the assumption of constant ratio of no-load loss to copper loss, it applies within limits to designs where this ratio is not the same. When applied to designs for the same kva and voltage rating, but with different losses, the exponent of the last term is usually between -1.5 and -1; in many cases a value of -1 can be used with sufficient accuracy for practical purposes. When applied to transformers of a given kva rating, the kva base is constant, and the formula may be written:

$$\text{Cost} = \left(\frac{k}{\text{No-load loss} \times \text{Copper loss}}\right)^n$$

where n is between 1 and 1.5.

15. TRANSFORMER DESIGN

Practical transformer design, reduced to its simplest terms, consists of assuming values (based on experience) of volts per turn, current density in the copper conductor, flux density in the iron core, insulating clearances, and winding and core arrangement, and calculating the resulting physical dimensions and weights, from which losses, impedance, cooling, and other values which tell the performance of the transformer can be calculated. Regardless of such theoretical helps as short cuts and specialization, design is a cut-and-try proposition because of the many variables involved. The final result is a set of data, consisting of assumed and calculated dimensions and performance, the data to be used as the starting point of the mechanical design.

FORMULAS AND DATA. The following fundamental formulas and data are used in transformer design:

1.
$$B = \frac{SV}{fA} \tag{1}$$

where B = the maximum flux density in the iron core.
V = the voltage per turn of the transformer.
f = the frequency in cycles per second.
A = the cross-section area of the iron core.
$S = 3490 \times 10^3$, if A is in square inches, and B is in gausses (lines per square centimeter).
$S = 22.5 \times 10^6$, if A is in square centimeters, and B is in gausses.
$S = 22.5 \times 10^6$, if A is in square inches, and B is in lines per square inch.

2.
$$B = \frac{KNI}{g} \tag{2}$$

is the formula for **flux density** for ampere-turns acting over a uniform field in air,

where B = the maximum flux density in the air.
NI = the number of rms ampere-turns with a sine-wave of alternating current.
g = the effective length of the leakage path.
K = 0.7 if g is in inches, and B is in gausses.
K = 1.78 if g is in centimeters, and B is in gausses.
K = 4.5 if g is in inches, and B is in lines per square inch.

3. The **reactance volts** of a transformer winding group is equal to:

$$\frac{2.01fN^2IMw10^{-7}}{g} \tag{3}$$

where f = the frequency in cycles per second.
N = the number of turns (in the base winding portion) of the winding group considered.
I = the rated rms current of the base winding.
M = the average of the mean turns of the windings.
w = the effective width of the leakage path in inches.
g = the effective length of the leakage path in inches.

(This formula may be derived by equating eq. 1 to eq. 2 and solving for the voltage.)

In determining the reactance of a transformer, one winding is chosen as the base winding, and the turns and current of this winding are used in the reactance formula with the result that the reactance voltage obtained is referred to this winding. The percentage reactance obtained will be the same regardless of which winding is used as the base winding.

For the purpose of reactance calculation a winding group consists of the windings between zero ampere-turn points on an ampere-turn curve, which is plotted as ampere-turns versus distance at the end of the core window, assuming a value of zero for the end ampere-turn value. For example, Fig. 27 represents the cross-section through a winding consisting of four winding groups, each group consisting of opposing ampere-turns of the high- and low-voltage windings, of either interleaved coils or concentric coils, shell type or core type. The curve of ampere-turns versus distance is plotted starting with an

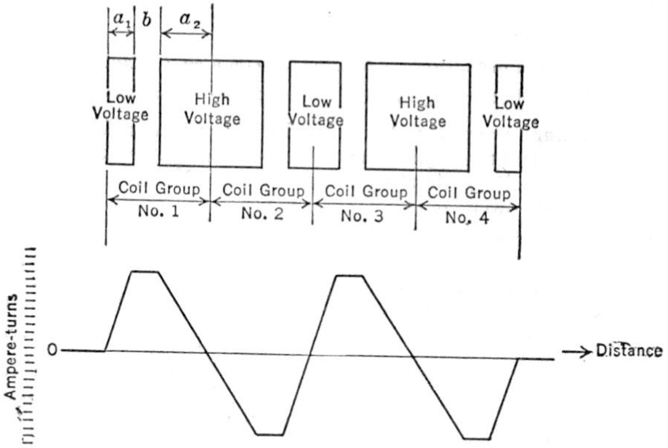

FIG. 27. Cross-section through a Winding of Four Groups with Curve of Ampere-turns

ampere-turn value of zero at one end of the core window. The coil groups may have equal or unequal numbers of ampere-turns, depending on the design, although equal winding groups are most common.

The following empirical values of w and g are applicable to determine the approximate impedance of most transformers:

$$w = b + \frac{a_1 + a_2}{3} \tag{4}$$

$$g = H + \frac{a_1 + a_2 + b}{2} \tag{5}$$

where a_1 and a_2 are the thicknesses of the low- and high-voltage windings respectively, at right angles to the leakage flux lines; b is the distance between high- and low-voltage windings, at right angles to the leakage flux lines; and H is the average height of coil parallel to the direction of leakage flux.

When the transformer consists of several winding groups, the reactance voltage of each group is calculated separately. The total reactance voltage is the sum of the reactance voltage of the separate groups, calculated individually. The percentage reactance is equal to 100 × Reactance volts/Winding voltage of the base winding.

4. The **resistance** of a copper winding at 75 deg cent is equal to

$$9.9 \times 10^{-6} \times \frac{\text{Length of conductor in feet}}{\text{Cross-section area of copper in square inches}} \tag{6}$$

The I^2R loss in one winding is equal to the rated winding current squared multiplied by the resistance. The total copper loss of the transformer is equal to the I^2R loss of both windings plus the eddy-current loss due to circulating eddy currents in the copper, which may usually be expressed as unequal current distribution in the copper of the winding. Eddy-current losses in transformers are generally kept below 20 per cent of the I^2R loss in the copper by subdividing and transposing the conductor as necessary; the larger the transformer and the heavier the current, in general, the more subdivision is required. The percentage resistance of a transformer is equal to the copper loss at rated current divided by rated kva output.

5. **Length of conductor** in feet = Mean turn of the coil in feet × Number of turns.

6. The **impedance voltage** of a transformer is the voltage required to circulate rated

current in a winding with the other winding short-circuited. Percentage impedance = 100 × Impedance voltage/Rated voltage of the winding to which voltage was applied.

Expressed in complex quantities in terms of the percentage resistance and percentage reactance, the percentage impedance is equal to: Percentage resistance + j(Percentage reactance). Consequently the impedance $z = \sqrt{r^2 + x^2}$, where r is the resistance in ohms or percentage, and x is the reactance in the same units.

7. The **voltage regulation** of a transformer is the change in secondary voltage, expressed in percentage of rated secondary voltage, between no load and full load at rated secondary voltage with the primary voltage constant.

The regulation of a transformer is commonly determined by an approximate formula, which is sufficiently accurate for practically all power and distribution transformers. It is really the first few terms of an infinite series derived from the exact formula. The commonly used formula is:

$$\text{Percentage regulation} = \text{Percentage resistance} \times \cos\theta + \text{Percentage reactance} \times \sin\theta + 0.005\,(\text{Percentage reactance} \times \cos\theta - \text{Percentage resistance} \times \sin\theta)^2$$

where $\cos\theta$ = the power factor of the load, and θ = the power factor angle. Values of $\sin\theta$ are positive for currents lagging the voltage (lagging power factor), and negative for currents leading the voltage (leading power factor). Values of $\cos\theta$ are positive for either leading or lagging power factor.

An exact formula for the regulation of a transformer is:

$$\text{Percentage regulation} = \sqrt{(\text{Percentage resistance} + 100\cos\theta)^2 + (\text{Percentage reactance} + 100\sin\theta)^2} - 100 \quad (7)$$

8. The **no-load loss** of a transformer is usually calculated by multiplying the weight of the core in pounds by the watts per pound of steel taken from a core-loss curve (Fig. 28

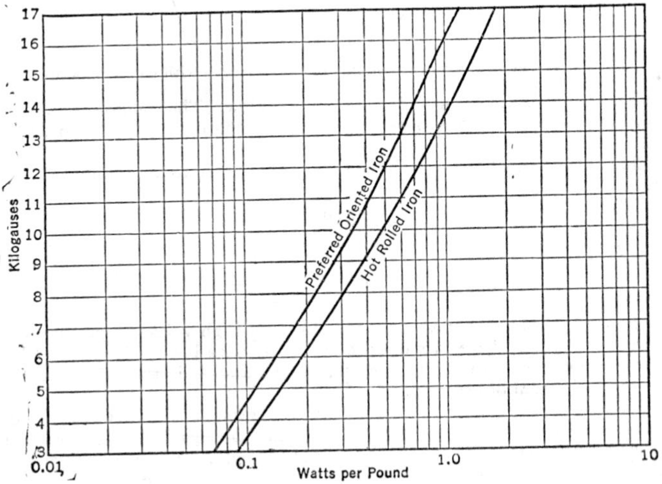

Fig. 28. Typical Core-loss Curve

is typical) for the particular kind of steel used, corresponding to the calculated flux density in the core.

9. The **exciting volt-amperes** for a transformer core may be calculated by multiplying the weight of the core in pounds by the volt-amperes per pound of steel, taken from a curve for the particular kind of steel used. If the curve is obtained from tests on similar transformers, it may be used without correction. If the curve is obtained from tests on small samples of the steel, allowance should be made for the exciting volt-amperes required for the air gaps unavoidable in building a conventional core. Although there are various methods of calculating this allowance, the only really satisfactory means of determining what this allowance should be is by tests on similar transformers. The effective length of air gap between the ends of laminations varies, and the permeability of steel differs greatly between individual sheets, so that even duplicate transformers do not have exactly the same exciting current. As a general rule, the increase in exciting current due to air gaps

and other factors will be considerably less than the exciting current required for the steel alone when the transformer is operated at normal voltage. The exciting current in amperes is of course equal to the exciting volt-amperes divided by the voltage of the winding.

10. The **efficiency** of a transformer in percentage is equal to 100 × Output/Input, and for convenience in calculation may be expressed as

$$\text{Percentage efficiency} = 100\left(1 - \frac{\text{Losses}}{\text{Output} + \text{Losses}}\right), \tag{8}$$

where the losses are equal to the sum of the no-load loss and the copper loss.

11. Design data for calculation of the **heating and cooling** of transformers are supplied in Article 16.

TYPICAL VALUES OF DESIGN FACTORS. Values for the various design factors can be selected which are typical of present-day practice. Individual transformers may have values which are different from those given, but they will seldom be greatly different. The values actually used for any given transformer are those which are required to meet the guaranteed or standard performance; they are not determined arbitrarily in the final design and may change as better materials become available and operating requirements change.

The usual starting point in the design of a transformer is the **voltage per turn,** of which typical values for 60-cycle transformers are shown in Fig. 29.

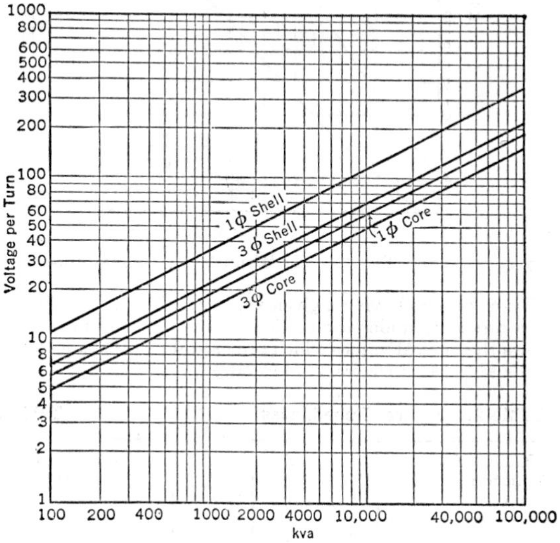

Fig. 29. Typical Curve of Volts per Turn

Typical values of **current density** in the winding copper are:

Natural-draft, dry-type transformers with class A insulation: 1000 to 1200 amp per sq in.

Distribution transformers and for dry-type transformers with class B insulation: 1200 to 1500 amp per sq in.

Oil-insulated, self- or water-cooled transformers: 1500 to 2000 amp per sq in.

Oil-insulated, forced oil-cooled transformers: 2000 to 3000 amp per sq in.

The actual current density used for any particular transformer is determined by the permissible copper loss, the permissible temperature rise in the copper, and by heating under short-circuit conditions.

Typical values of **flux density** in the steel of the core are given in Table 1 for two types of steel in common use in present-day transformers, namely, hot-rolled high-silicon steel and cold-reduced preferred oriented steel. The actual flux density used for any particular transformer is determined by the permissible no-load loss and exciting current, the inrush exciting current when the transformer is connected to the line, and the temperature rise of the core. The volt-amperes per pound for various values of flux density are shown in Fig. 30.

Table 1. Typical Flux Densities

Type of Transformer	Typical Flux Density, gausses	
	Hot-rolled High-silicon Steel	Cold-reduced Preferred Oriented Steel
Dry-type transformers.....	10,000	12,000
Distribution transformers..	12,000	15,000
Power transformers.......	13,000	16,000

Wide variations will be found in the impedances of actual transformers. In power transformers the reactance component practically always is large in comparison to the resistance component of the impedance. Typical values of impedances of power transformers of various voltage classes are given in Table 2, there being little variation in typical values with the kva rating, although considerable variation will be found among transformers of the same kva and voltage ratings.

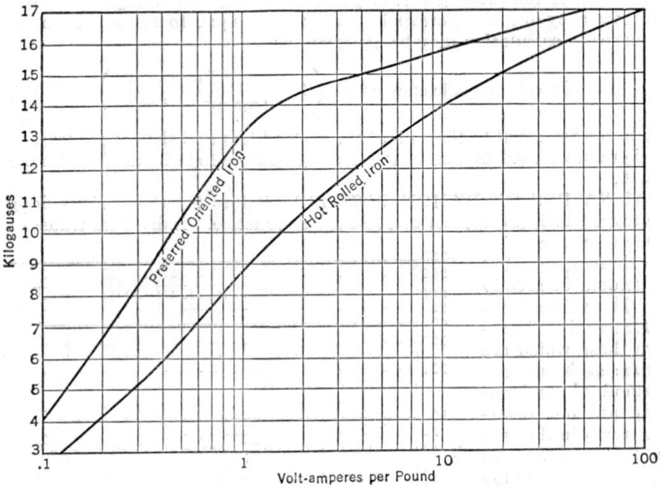

FIG. 30. Flux Densities for Various Volt-amperes per Pound

Transformers having a given voltage rating do not always receive the same test voltage, as is discussed under the section on the application of transformers. Table 3, however, shows the typical values which are used for insulation-test voltages of power transformers in the absence of contrary information.

Table 2. Typical Impedances

Kilovolt Class of Transformer, High-voltage Winding	Typical Percentage Impedance
15	4.5 to 7
25	5.5 to 8
34.5	6 to 8
46	6.5 to 9
69	7 to 10
92	7.5 to 10.5
115	8 to 12
138	8.5 to 13
161	9 to 14
196	10 to 15
230	11 to 16

Table 3. Impulse Tests

Rated Kilovolts between Terminals of Three-phase Lines	Kilovolt-60-cycle Test for 1 Min	Kilovolt-impulse Tests	
		Chopped Wave	Full Wave
1.2	10	54	45
5	19	88	75
8.66	26	110	95
15	34	130	110
25	50	175	150
34.5	70	230	200
46	95	290	250
69	140	400	350
92	185	520	450
115	230	630	550
138	275	750	650
161	325	865	750
196	395	1035	900
230	460	1210	1050
287	575	1500	1300
345	690	1785	1550

Distribution transformers of the 15-kv class and below generally receive lower tests than power transformers of the same rated kilovolts. The characteristics of the impulse-test waves are defined in the ASA standards for transformers.

16. HEATING AND COOLING

The temperature rise of a transformer is caused by iron loss in the iron core and copper loss in the winding conductors. There is also a negligible amount of heat in the insulation due to dielectric losses. In any transformer the heat is dissipated by a combination of conduction, radiation, and convection.

HEAT FLOW WITHOUT RADIATORS. In an oil-immersed transformer the heat in the core flows through the iron by conduction, both parallel to the plane of the laminations and at right angles to it, to the surface of the core, which is exposed to the oil. From the surface of the core it flows by conduction through the very thin, stationary oil film on the surface of the iron to the adjacent oil which is free to move, increasing its temperature and causing it to expand. The difference in static pressure in the oil, caused by the weight of a column of warm light oil in comparison to an adjacent column of cold heavy oil, causes the oil to circulate by convection. As the heated oil rises and flows away from the core, it is replaced by cold oil from near the bottom of the tank, which is heated in turn. The hot oil rises to the top of the tank and then flows downward along the tank wall. Because the temperature of the oil is higher than that of the tank wall, heat flows from the oil through a very thin stationary oil film on the inside of the tank wall to the tank wall by conduction. The heat passes through the tank wall by conduction, and a portion is then dissipated to the surroundings by radiation. Another portion of the heat flows by conduction through a very thin stationary air film to the adjacent air, which is free to move, increasing its temperature and causing it to expand. The difference in static pressure in the air caused by the weight of a column of warm light air, in comparison to an adjacent column of cold heavy air, causes the air to circulate by convection. As the heated air rises and flows away, it is replaced by cold air from near the bottom of the tank, which is heated in turn.

The heat generated in the copper of the windings flows by conduction through the copper, the insulation on the conductor, and the very thin stationary oil film to the adjacent oil, which is free to move. This heated oil flows upward through the oil ducts in the winding, circulating by convection in the manner described for the oil heated by the core. It joins the oil heated by the core near the top of the tank and is cooled in the same way.

HEAT FLOW WITH RADIATORS. If radiators for cooling are provided, they are connected to the tank by openings at the top and bottom, so that oil can flow through them. The hot oil enters the radiator at the top and flows by convection to the bottom, where it returns to the main tank to furnish cooled oil to the core and coils. The heat is transferred from the oil in the radiator to the surroundings by radiation and to the air by convection, in the same manner as the transfer through the tank wall to the air. In a transformer with radiators a portion of the total circulating oil stream enters the radiators, and the remainder flows down the inside of the tank wall as in a transformer without radiators, the division being determined by the relative resistances of the paths to the flow of oil and the fluid pressures acting along each path, as determined by differences in oil density due to thermal causes.

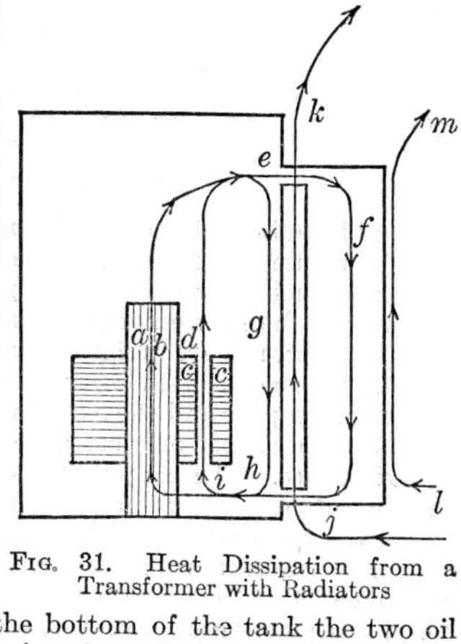

FIG. 31. Heat Dissipation from a Transformer with Radiators

The heat dissipation in an oil-insulated transformer is illustrated in Fig. 31, where the heat is transferred by conduction from the core *a* to the adjacent oil *b*, and from the coil *c* to the adjacent oil *d*. As the heated oil from *b* and *d* rises, it mixes at *e* and then divides, a portion flowing through the radiator *f* and a portion flowing along the tank wall, as at *g*. Near the bottom of the tank the two oil streams again mix at *h* and then divide, a portion entering the coil ducts at *i* and a portion flowing upward along the core to *b*. The air flow along the exterior of the tank and radiator is shown by the paths $j \cdots k$ and $l \cdots m$.

TEMPERATURE RISE AND LOAD. The load which can be placed on a transformer with safety is limited by the temperature of the windings. As a basis for rating transformers, according to the ASA standards which are in use in the United States, the average copper temperature rise by resistance above the cooling medium for a continuous rated transformer should not exceed 55 deg cent, and the hot-spot copper temperature rise should not exceed 65 deg cent. In addition, the oil rise for continuous operation should not exceed 55 deg cent for transformers with protected oil or 50 deg cent for those in which the hot oil is exposed to the air.

For convenience in calculation, the temperature rise of the copper is calculated in two parts: (1) the temperature rise of the oil, (2) the temperature rise of the copper above the oil.

Cooling Constants. Because of the complex nature of the heat transfer in a transformer as outlined and the additional factors of the multiplicity of paths of oil flow past the cooling surfaces of the core, the coils, the tank, and the radiators, and the multiplicity of paths of air flow past the external cooling surfaces, accurate calculation of temperature rises in a transformer requires empirical data on transformers having the mechanical arrangement of the various parts similar to the arrangement under consideration. However, typical cooling constants for various types of construction, which are sufficiently accurate for estimating purposes, can be given, with the caution that the values for any actual transformer may be greater or less, depending on the mechanical construction and arrangement. Such typical values are shown in Fig. 32 for a wide variety of cooling conditions and constructions.

The temperature rise of the copper above the adjacent oil may be called the copper gradient. The copper gradient plus the hot-oil temperature rise is equal to the hot-spot temperature rise of the winding. A curve of typical copper gradients versus watts per square inch of coil exposed to the oil is shown in Fig. 32, *h*. The average copper rise by resistance is lower than the hot-spot copper rise, usually by less than 10 deg cent at full load. For design purposes this rise may be determined by calculating the average temperature of the oil which is adjacent to the coils and adding the copper gradient, or it may

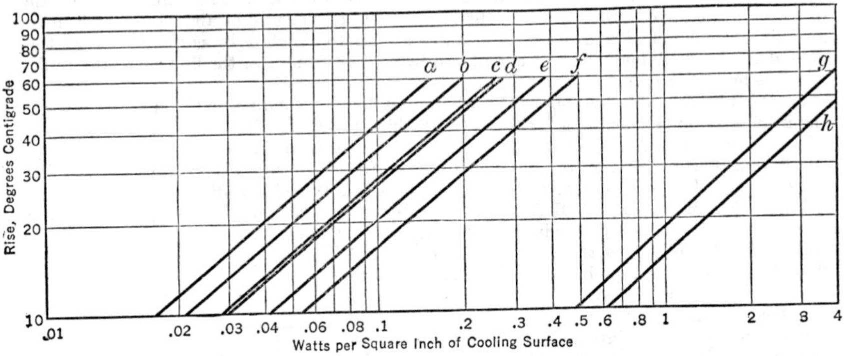

Fig. 32. Typical Cooling Constants of Transformers

be found by subtracting a correction, which is usually less than 10 deg cent, from the hot-spot copper rise. The copper gradient actually depends upon the rate of oil flow and the oil viscosity, which vary with the temperature of the oil, but the relation is so complex that accuracy can be assured only by empirical data taken on transformers of similar design. When transformers are designed for extremely low copper gradients, the rise by resistance of the transformer is often lower than the hot-oil rise, giving the effect of a negative gradient between the copper and the oil. The paradox is due to a comparison of the **average** copper rise with the **hottest** oil rise; the **hottest** copper rise will of course be greater than the **hottest** oil rise. The copper gradient varies with the loss in the windings in a complex manner but may be assumed with sufficient accuracy to vary as the 0.8 power of the loss, or the 1.6 power of the load.

The relation between oil rise and watts loss for a given construction varies for different arrangements and even with the ambient temperature, because of the change in oil viscosity. In most cases, however, sufficient accuracy is secured by assuming that the oil rise varies as the watts loss to the 0.8 power.

Cooling curves are shown in Fig. 32 for the following types of cooling surfaces commonly used for transformers:

Curve *a*: Corrugated tank.

Curve *b*: Radiators of round, oval, or flat tube construction.

Curve *c*: Round tubes welded on the tank wall.

Curve *d*: Envelope of surface with radiation only, without convection. This curve is for reference and may be used for calculations where it is desirable to separate the radiation and convection components of heat dissipation.

Curve *e*: Forced air cooling on radiators.

Curve *f*: Plain tank.

Curve *g*: Water-cooling coils for heat dissipation from water to oil through the metal of the cooling coil.

Curve *h*: Temperature gradient between copper and the adjacent oil.

Curves *a* to *f* are for dissipation from oil through a metal to air. These temperature drops may be considered to be the temperature drops due to heat flowing through the oil film and air film adhering to the solid material. By comparison with the temperature drop through the oil or air films the drop due to heat flow through copper or steel is negligible, except in some cases where the distances are large.

Typical values of temperature rise in degrees centigrade due to heat flow by conduction of 1 watt per square inch through a distance of 1 in. are as follows: core steel, 2.5; stacked laminations perpendicular to plane of individual laminations, 25; oil-treated paper tape on conductors, 250; copper, 0.1. The rate of heat flow by conduction is directly proportional to the temperature difference. For other materials than those listed, the watts per degree centigrade for a 1-in. cube can be obtained by multiplying values from available tables by 10.63 when the table value is in gram calories per second per degree centigrade for a centimeter cube, or by 0.00366 when the table value is in British thermal units per square foot per hour for 1-in. length of path per degree fahrenheit. The temperature rise due to heat flow by conduction of 1 watt per square inch through a distance of 1 in. is equal to the reciprocal of the figures thus obtained.

Effect of Sunlight. The heat dissipation from cooling surfaces of radiators exposed to the air is mostly by convection but partly by radiation. For a plain tank, however, the heat dissipation by the two means is almost equal. When any transformer is operated in the sunshine, its temperature rise will be a few degrees greater (usually not more than 10 deg cent) than when it is operated in the shade, the amount of increase depending on the intensity of the sunlight, its duration, and the design and arrangement of the cooling surfaces of the transformer. The increase in temperature due to the sun is less for tanks with radiators than for plain tanks.

Effect of Surface. The curves of Fig. 32 are for transformers painted with the usual non-metallic paints. When the cooling surfaces are painted with aluminum paint, the temperature rise of the transformer in shade will be increased from 4 to 30 per cent, the increase being least for tanks with a large number of closely spaced radiators with forced air cooling and greatest for plain tanks. When operated in sunlight during the day, a transformer with aluminum paint will run slightly cooler during part of the time when it is exposed to the sun and hotter during part of the night, with the result that the loss of life of the transformer in operation is practically the same as for a transformer with non-metallic paint.

Outdoor Operation in Shade. The curves of Fig. 32 are for indoor operation, where no wind blows on the transformers; this is the standard testing condition for transformers. The cooling is considerably better when the transformers are operated outdoors, since even a slight air movement provides some forced air cooling and a strong wind provides effective forced air cooling. The difference between curve *e* and the other curves of Fig. 32 is an indication of the improvement in cooling effect which can be expected with moderate wind velocities.

As just suggested, the cooling effectiveness of radiators can be increased by supplying these surfaces with cool moving air from any source, usually electric fans or blowers. The temperature rise in degrees centigrade of air for cooling transformers is equal to 1800 × Kilowatt loss/Cubic feet of air per minute. A typical rate of cooling air flow for forced air cooling is 200 cu ft per min, corresponding to a temperature rise in the air of 9 deg cent. The kva rating of power transformers 3333 kva and larger usually can be increased 33 per cent by forced air cooling if the winding temperature is the limiting factor. Smaller transformers can have their kva ratings increased by lesser amounts, generally not more than 25 per cent. In order to obtain effective forced air cooling, the cooling air must be properly distributed to the cooling surfaces, so that dead spots where the air moves slowly are eliminated, and a sufficient quantity of cooling air must be supplied. Where the original cooling surface is more than usually effective, as in a plain tank, the increase in capacity with forced air cooling will be less than where there is a large amount of restricted convection surface in comparison to the radiation surface.

Water Cooling. Curve *g* of Fig. 32 is typical of the cooling obtained with water-cooling coils located in the oil of the main transformer tank. The temperature rise in degrees centigrade of the water as it flows through the coils under steady-state conditions is equal to 3.8 × Kilowatt loss/Gallons of water per minute. A typical value of water rise is 10 deg cent, which means that 0.38 gal of water per min is used for each kilowatt of loss in the transformer. If the water rise is 10 deg cent, the average water temperature is 5 deg above its inlet temperature. Doubling the rate of water flow to provide increased heat dissipation will reduce the average water temperature only 2.5 deg, thereby reducing the oil rise a corresponding amount of approximately 2.5 deg, which is only 5 per cent for a transformer with a 55-deg cent rise. As this example demonstrates, increasing the water rate is not an effective means of reducing the temperature rise of a transformer, either for obtaining lower operating temperatures or increased loading.

Dry-type Transformers. The temperature rise of dry-type transformers varies approximately as the 0.8 power of the loss. A typical value of watts per square inch of cooling surface for the coils is 0.2, although considerable variation from this figure will be found, depending on design factors. For transformers with class A insulation the average rise by resistance for continuous load should not exceed 55 deg cent, and the hot-spot rise should not exceed 65 deg cent. For transformers with class B insulation the average rise by resistance should not exceed 80 deg cent. The hot-spot rise should be kept to a safe value for the insulating materials used, which often means a hot-spot rise of not more than 90 to 100 deg cent. The hot-spot rise is frequently the limitation in the design of dry-type transformers.

Transformers constructed with inorganic insulating materials, such as glass, asbestos, mica, and porcelain, and impregnated with silicone varnishes can withstand considerably higher temperatures, probably of the order of 120 deg cent average rise above 40 ambient, with 140 deg hot-spot rise, for continuous operation. The air ducts for either natural draft or forced air cooling must be large enough for sufficient air to flow through them to carry away the heat with a reasonable temperature rise, usually not more than one-third of the allowable rise for the transformer winding. The quantity of air required can be calculated from the formula previously given for forced air cooling.

Effect of Altitude. Since the watts per square inch which will be dissipated by convection to the air vary approximately as the square root of the relative air density (or barometric pressure), the cooling effect of air-cooled transformers, both dry and oil-immersed, will be less at high altitudes than at low altitudes. When transformers are operated at higher altitudes, the temperature rise will be greater because of the decreased air density. The greater the number of radiators, the greater is the effect of altitude, the effect being least for plain tanks. Figure 33 shows the variation, with the altitude, of the relative air density and of the watts dissipated by convection for the same temperature rise. The watts dissipated by radiation are not affected by the altitude. The effect of altitude for any particular transformer can be calculated by basing the dissipation by radiation on Fig. 32, curve *d*, and the surface envelope of the transformer, and correcting for the decrease in dissipation by convection in accordance with Fig. 33.

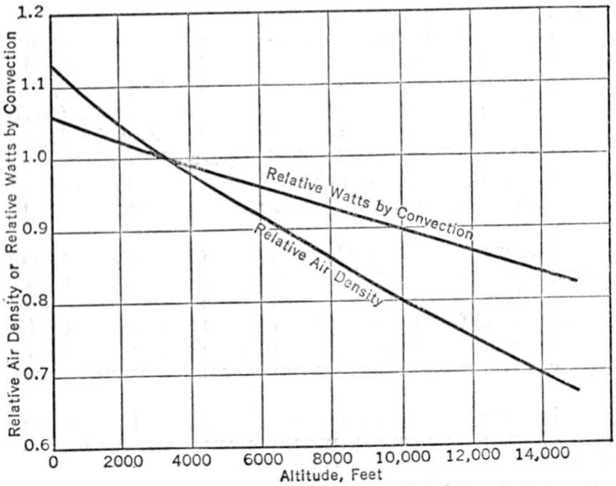

FIG. 33. Variation of Relative Density and Relative Watts Lost by Convection at Various Altitudes

TRANSIENT HEATING. Transient heating is sometimes called short-time heating of transformers. Because the materials in a transformer (the copper and insulation of the coils, the iron of the core, tank, and other parts, and the oil) have the ability to absorb heat (thermal capacity), the temperature of a transformer changes slowly when the load changes, requiring a matter of hours to reach its final temperature. Transient heating deals with the temperatures which exist for a given loading after a given time before final ultimate temperatures have been reached. The oil-temperature rise and the hot-spot copper rise above the oil may be calculated separately and added to obtain the total copper rise, the procedure being similar to that for determining the ultimate rise with continuous load. The principal use of transient heating calculations is to determine whether desired overloads can be carried for required times without the transformer exceeding safe operating temperatures.

General Method for Calculating Transient Temperature Rise. The starting point for transient heating calculations is the ultimate oil and copper rise for continuous rated load, obtained either by testing the transformer or from calculations. The core loss of a transformer operated at constant voltage is practically constant and is equal to the no-load loss. The copper loss varies as the square of the load. The total loss is the sum of the core loss and the copper loss. For all types of transformers the ultimate oil temperature for continuous operation varies approximately as the 0.8 power of the total loss

(Fig. 34). If the temperature rise of the transformer oil for a transformer with changing oil temperature is plotted against time on linear rectangular coordinate paper, the line obtained will be curved. By using suitable non-linear coordinate paper, however, a straight line may be obtained, greatly decreasing the labor required to plot the curve, since only

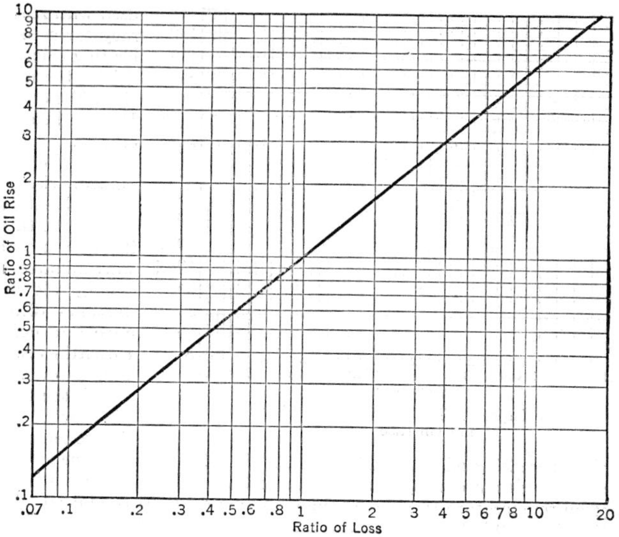

FIG. 34. Temperature Rise of Oil in Relation to Total Loss

two points are required for a straight line. Such coordinate paper is shown in Figs. 35 and 36. Figure 35 is correct for increasing oil rise; Fig. 36, for decreasing oil rise. The determination of the time scale will be described later.

The difference in temperature between the hot-spot copper temperature rise and the hot-oil rise may, for convenience, be called the copper gradient. For a particular transformer it may best be obtained from the manufacturer but can be calculated for full load approximately as 65 − Top oil temperature rise at full load, or for most transformers as 10 + (Average copper rise − Top oil rise) if the average copper rise is greater than the top oil rise. For transient calculations the ultimate copper gradient may be assumed, with reasonable accuracy, to vary as the 1.6 power of the load (Fig. 37). If the copper gradient with changing copper temperature is plotted against time on linear rectangular coordinate paper, the line obtained will be curved. If coordinate paper such as that of Figs. 35 and 36 is used, however, a straight line may be obtained. Figure 35 is correct for increasing copper gradient; Fig. 36, for decreasing copper gradient.

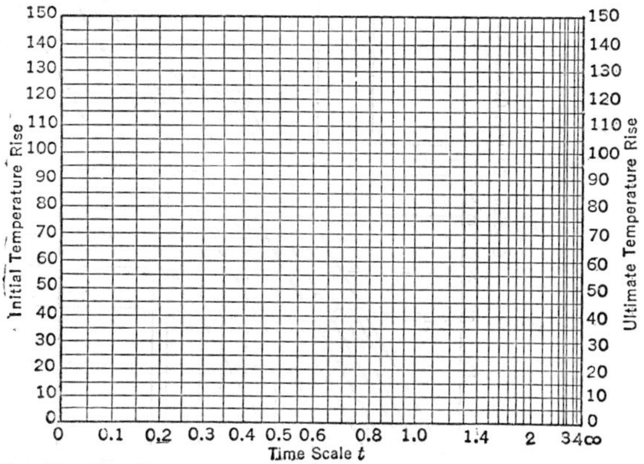

FIG. 35. Coordinate Paper for Temperature Rise of Oil for Increasing Oil Rise

For Figs. 35 and 36 t, which is the time measured on the abscissa, is equal to: Time in minutes/T, where T is a constant called the thermal time constant, which is numerically equal to the time which would be required for the transformer to reach its ultimate temperature if all heat were stored in the transformer and no heat were dissipated to the surroundings. The value of T is different for the copper and for the oil; different values of T are obtained for different loads and corresponding different ultimate oil-temperature rises. Often, however, for convenience a typical value of T may be calculated from the transformer constants and used with reasonable accuracy for a complete set of heating calculations.

The thermal time constant T is equal to UH/W. For use in calculating the oil rise:

U is the ultimate temperature rise of the oil over the ambient temperature.

H is the thermal capacity of the transformer in watt-minutes per degree centigrade = (approximately) (3.5 × Pounds core and coils + (2.4 × Pounds case) + (80 × Gallons of oil).

W is the total watts loss which will produce the ultimate temperature rise U.

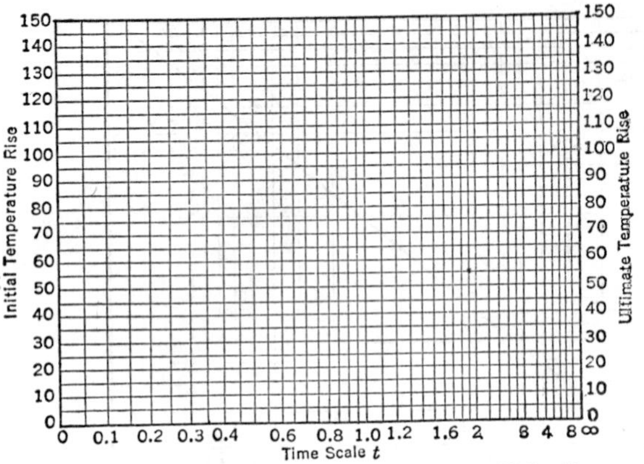

FIG. 36. Coordinate Paper for Temperature Rise of Oil for Decreasing Oil Rise

For plotting increasing oil temperature as on Fig. 35, a straight line is drawn through the origin and the ultimate temperature, infinite time point. The resulting straight line is a curve of oil-temperature rise versus time starting cold. If the initial oil rise is not 0, the initial time is not 0 but is equal to the time t, corresponding to the initial temperature

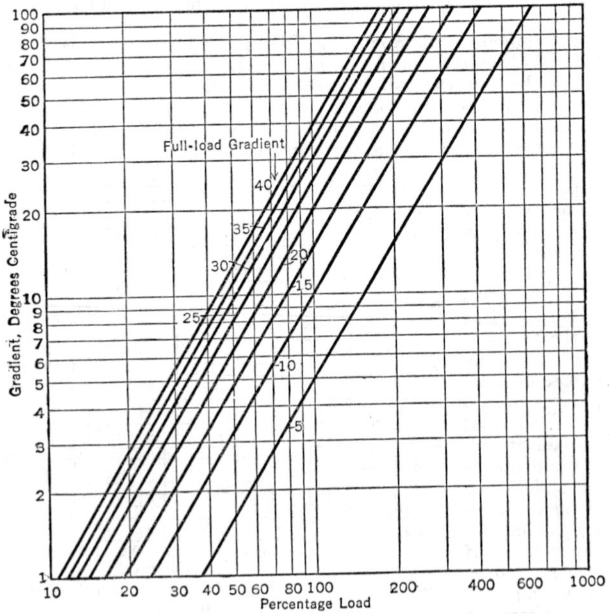

FIG. 37. Copper-temperature Gradient as Function of Load

rise. The duration of the load corresponding to the ultimate temperature rise is added to the initial time to obtain the final time, and the corresponding final temperature rise at the end of that time is read from the curve.

For plotting decreasing oil temperature as on Fig. 36, a straight line is drawn through the zero time, initial temperature-rise point and the infinite time, ultimate temperature-rise point. The resulting straight line is an approximate curve of oil-temperature rise versus time and may be used to determine the oil-temperature rise for any duration of the assumed load.

Typical values of thermal time constants for transformer oil rise are from 120 to 300 min (2 to 5 hr). The thermal time constant for copper, T_c is equal to U_cH_c/W_c, where U_c is the ultimate temperature gradient; W_c is the watts copper loss which will produce an ultimate copper gradient (rise of copper over oil temperature) of U_c deg cent; H_c is the thermal capacity of the copper in watt-minutes per degree centigrade and conventionally is equal to 2.93 × Pounds of copper, since some standard formulas for transformers are based on the premise that all heat is absorbed in the actual copper. Actually H_c is equal to 2.93 × Pounds of copper + 0.5 × Cubic inches of conductor insulation. A typical value of thermal time constant for copper of an oil-immersed transformer is 5 min. For increasing copper gradient to be plotted on Fig. 35, a straight line is drawn through the origin and the ultimate gradient, infinite time point, which may be used to determine the gradient at the end of any desired time, in the same manner as for increasing oil rise. For decreasing copper gradient to be plotted on Fig. 36, a straight line is drawn through the zero time, initial gradient point, and the infinite time, ultimate gradient point, which may be used to determine the gradient at the end of any desired time. For convenience in calculation, curves showing the relation between the ultimate gradient and the percentage load for various full-load gradients have been drawn in Fig. 37.

Table 4. Permissible Transient Temperatures

Time Rating	Permissible Final Temperature, deg cent *	Permissible Final Temperature Rise above 30 Ambient deg cent *	Typical Amperes per Square Inch in Copper	Typical Circular Mils per Ampere in Copper
1 sec.	250	220	90,000	14
2 sec.	250	220	63,000	20
5 sec †	250	220	40,000	32
6 sec.	190	160	33,000	40
10 sec †	190	160	25,000	50
11 sec.	155	125	20,000	60
1 min.	155	125	8,500	150
2 min †	155	125	6,000	200
10 min †	130	100	4,500	280
1 hr †	110	80	3,500	360
12 hr †	100	70	2,000	650
Continuous *	85	55	1,600	800

* Temperature rise for times of 2 min or less is hot-spot rise, based on all heat stored in the copper, as calculated by the formula. Temperature rise for greater times is average rise by resistance as obtained on test.

† Maximum time to which permissible temperature limits apply.

By the use of these relations the oil rise and the copper gradient for a given load carried for a given time may be calculated independently, and the copper temperature rise obtained as their sum. Because the time constant of the gradient is only a few minutes in most cases, the gradient reaches its ultimate value in less than 1 hr and, except for extremely short times, transient calculation of the gradient is not required. To calculate the temperatures for a given load cycle, a starting oil temperature is estimated and the actual load cycle is replaced by a series of constant loads carried for definite times, approximating the actual load cycle. The temperature conditions existing at the end of each block of load are calculated by the method outlined.

Method for Transients of Short Duration. For extremely short times, such as 1 min, the conventional assumption is that all the heat generated in the copper is absorbed in it, with no heat dissipated to the surroundings. A formula for the temperature of copper when carrying current with all heat absorbed by the copper is:

$$T = 2Ct\left[(1 + Ct)(T_0 + 234.5) + \frac{309.5E}{1 + Ct}\right] + T_0 \qquad (9)$$

where T = the final temperature in deg cent.
T_0 = the initial temperature in deg cent.
E = the ratio of eddy-current loss to I^2R loss at 75 C.
$C = 2.3 \times 10^{-11} \times$ (amperes per square inch of conductor)^2.
t = the time in seconds.

The formula for all heat absorbed in the copper is used mainly for short-circuit and grounding transformer calculations. For power and distribution transformers of 55 deg cent rise by resistance, an initial temperature of 95 deg cent and a final temperature not exceeding 250 deg cent are assumed, to be used with the formula. Transformers with 4 per cent impedance must withstand 25 times rated current for 2 sec; transformers with 5 per cent impedance must withstand 20 times rated current for 3 sec; transformers with 6 per cent impedance must withstand 16.6 times rated current for 4 sec; transformers with 7 per cent or more impedance must withstand full short-circuit current for 5 sec. Intermediate values are obtained by interpolation. In accordance with the formula, full-load current densities in the copper of not more than 2500 amp per sq in. (not less than 500 cir mils per amp) will in most cases meet the thermal requirements for short circuit.

Table 4 shows the permissible temperature rises above 30 deg cent average ambient and the typical current densities for grounding transformers of various time ratings. The initial temperature rise is the rise corresponding to the losses at the continuous rating, where one is given in addition to the short-time rating. Where no continuous rating is given, the initial temperature corresponds to continuous no-load loss.

17. LOSSES

The losses of transformers vary with the required characteristics. Typical total losses are shown in Fig. 38 for the usual range of voltage classes. The no-load loss is usually between 30 and 40 per cent of the total loss.

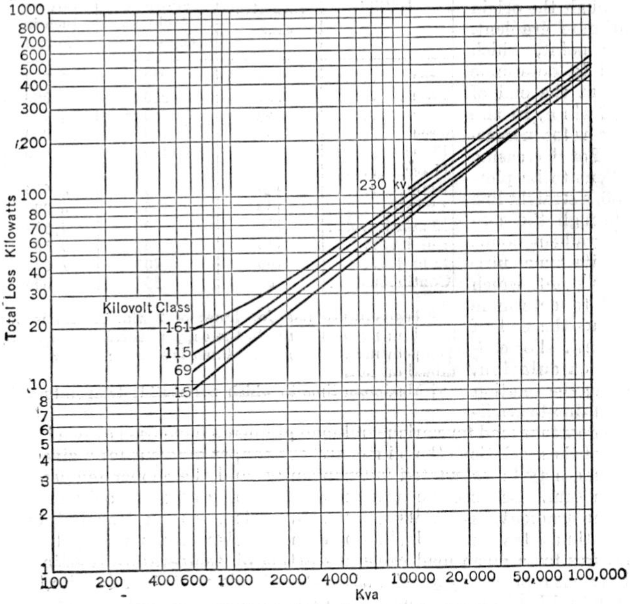

FIG. 38. Typical Total Losses of Transformers

18. IMPULSE VOLTAGE

Modern transformers, when provided with proper protection, are designed to withstand impulse voltages in excess of those usually encountered in service because of lightning, switching surges, etc. The most severe surges are due to lightning. When a transformer is subjected to a steep voltage impulse, the initial distribution of the voltage throughout the transformer is determined by the relative capacities between turns, between coils, and to ground, since only a negligible fraction of the initial surge current flows to ground by being conducted through the turns of the winding because of its high inductance. Practically all the initial surge current flows through the capacitances mentioned. If there were no capacitance to ground, the initial surge current would flow as indicated in the transformer of Fig. 39(a), in which the same surge current flows through the equal capacitances a, and consequently the voltage drop across each coil of the winding d would be the same and no concentration of voltage stress would occur in the winding.

The coils, however, do have capacity to ground, and the actual flow of initial surge current is as indicated in Fig. 39(b), in which part of the total surge current I flows to ground through the first capacitance b, the remainder flowing through the first capacitance a, after which the current divides so that part flows to ground through the second capacitance b and the remainder flows through the second capacitance a, and so on through the winding, with each capacitance a carrying a smaller current and having a proportionally smaller voltage across it. It is evident that coils close to the line end have higher voltages across them than coils farther from the line end, the relative values being determined by the ratio of the capacity between coils to the capacity to ground, and the number of coils in the winding. If the capacity between coils is large in comparison to the capacity to ground, the voltage distribution among the coils will be more uniform than if it is small.

The maximum voltage stresses near the line end are generally determined by the initial voltage distribution; the maximum voltage stresses deeper in the winding generally occur later (if a full wave is applied) as the surge current flows through the inductance of the coils, in addition to the capacity. Each part of the transformer must be provided with sufficient insulation to withstand the voltage surges to which it will be subjected.

For low-voltage transformers sufficient insulation can be provided for the voltage stresses which occur naturally in a transformer of normal design. Interleaved disk coil

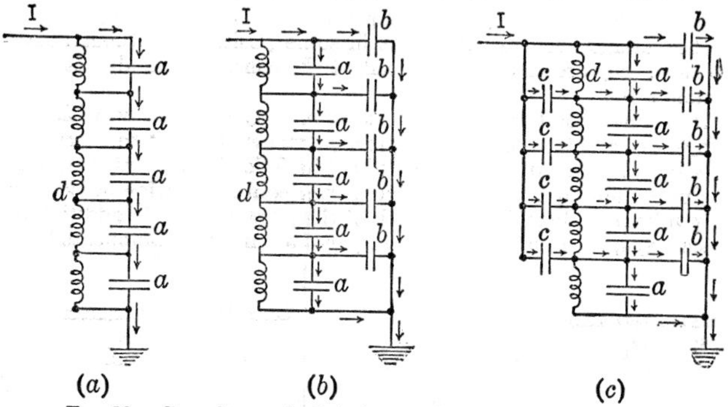

FIG. 39. Capacitances in Relation to Impulse Voltage Stresses

shell-type transformers have such proportions as to give good voltage distribution. Insulation can simply be provided for the stresses which occur in a normal design. For high-voltage shell-type transformers the addition of a shield constructed like an extra coil and connected to the line terminal of the winding helps distribute the voltage stress across the first coil and improves the naturally good voltage distribution.

Core-type transformers with disk coils have a large number of coils with a lower ratio of capacity between coils to capacity to ground, resulting in poorer voltage distribution. On low-voltage core-type transformers, sufficient insulation may be provided for the stresses which occur with normal proportions. For higher-voltage transformers these stresses may be reduced by using a shield on the line end and by using wider conductor than normal for coils near the line ends, thereby increasing the capacity between turns and reducing the voltage stress between turns to a value for which insulation can be provided.

Another means of improving the voltage distribution of core-type transformers is indicated in Fig. 39(c), in which shields are connected to the line so as to introduce capacity between the line and the coils. If the capacitances c could be made of such value that the surge currents through them were exactly equal to the surge currents through the corresponding capacitors b, uniform current would flow through the capacitances a, resulting in uniform voltage distribution in the winding. Actually, complete compensation is not practicable because of the large values of capacity required near the line end, but sufficient compensation to reduce the surge voltage stresses to reasonable values is practicable and is used on some core-type transformers.

As has been indicated, there are various means of reducing the voltage stresses which occur within a transformer subjected to surge voltages. Although there is some coordination between voltage tests which are applied to transformers, this coordination is only approximate, so that for each part of a transformer each voltage stress existing under the various test conditions must be determined and sufficient insulation provided for all the stress conditions. The insulation for a given part of a transformer may be governed by any one of such tests as the following: high potential test to ground, induced voltage test, full wave impulse test, chopped wave impulse test, steep wave impulse test. Typical values of impulse test voltages are given in Table 3, Article 15.

19. LOAD RATIO CONTROL

Load ratio control is used when it is desirable to change taps on a transformer without disconnecting it from the lines. This equipment permits changing the voltage ratio of a transformer without interrupting the load on the transformer. Transformers with load ratio control are used to regulate the voltage of lines, to control the flow of power and reactive kva between lines, and for similar purposes.

As employed in modern transformers, load ratio control equipment is made as a motor-operated device which is mounted as part of the transformer and changes taps from one position to the next in a few seconds. Many different kinds of mechanical equipment have been made for this purpose, employing various combinations of dial switches and circuit breakers or contactors.

PRINCIPLE. The general principle of load ratio control circuits is to use impedances to limit the current flow when switching from one tap to the next tap, but to select and connect these impedances so that they do not introduce excessive voltage drops in the circuit at any time. Once the circuit is selected, the current flow during a changeover position can be reduced only by increasing the current limiting impedance and thereby increasing the voltage drop in the circuit in some position. Although several different circuits are used, they are all sufficiently alike to be considered modifications of the circuit in most common use.

CIRCUITS. The circuit which is almost universally used today, with only few exceptions, employs a center tapped auto-transformer reactor to limit the current when changing from one tap to the next. This circuit has the following advantages: in most applications it produces minimum burning of the arcing contacts during switching; it provides an operating voltage midway between each of the physical taps in addition to the voltage of the physical taps; the switching means required are not too complicated for general use.

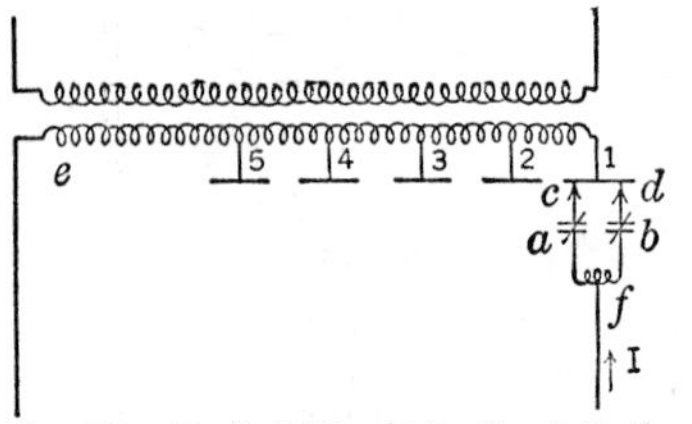

Fig. 40. Typical Circuit for Load Ratio Control: First Stage

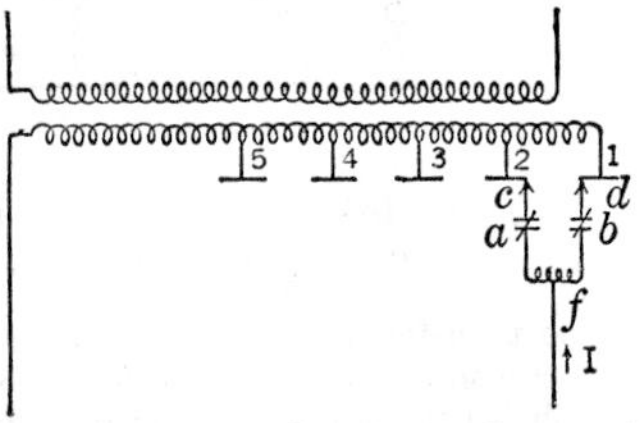

Fig. 41. Typical Circuit for Load Ratio Control: Second Stage

A common form of this circuit is shown in Fig. 40, in which *e* is the main transformer winding with taps 1, 2, 3, 4, and 5 connected to wide stationary contacts of a dial switch, as indicated. The moving contacts of the dial switch, *c* and *d*, are connected through the circuit breakers *a* and *b*, respectively, to the terminals of the mid-tapped auto-transformer reactor *f*. In the position shown, the line current *I* divides so that half flows through each half of the reactor winding. Since the two halves of the reactor winding are interlaced with each other for low impedance, the reactor introduces negligible impedance drop in the circuit. To change taps from the position shown, the mechanism opens circuit breaker *a*, causing the total line current *I* to flow through *b* and *d*, and no current to flow through *a* and *c*. In this position the line current flows through one-half the reactor *f*, causing in the circuit a voltage drop which depends upon the design of the reactor, because the total line current is magnetizing current for the reactor. Air gaps are placed in the reactor core to limit this voltage drop with full-load current to approximately half the voltage between adjacent taps. As the mechanism operates, it next moves contact *c* to stationary contact 2, after which the circuit breaker *a* is closed, resulting in the connection shown in Fig. 41, with the line voltage half-way between the voltages of taps 1 and 2. In this position each half of the reactor carries half the line current *I*, as in Fig. 40, and in addition an exciting current due to the voltage between taps 1 and 2 being applied across its terminals. (In Fig. 40, since there was no voltage across the reactor, the exciting current was equal to zero.) If the air gaps in the core are adjusted for a voltage drop in the transition position of half the tap voltage as previously mentioned, the exciting current in the reactor in Fig. 41 will be equal to one-half full-load current. The exciting current in the position of Fig. 41 can be lowered by decreasing the length of the air gaps in the reactor core, but the voltage drop in the transition position will be increased, as will the arcing voltage on the breaker contacts, when the breaker opens. The position of Fig. 41 is often used as an operating position to provide twice the number of steps.

To make the next tap change from Fig. 41 in the same direction, circuit breaker *b* is opened, after which contact *d* is moved to contact 2, and breaker *b* is closed. The line current *I* is divided equally between contacts *c* and *d* as it flows into tap 2. The other taps, such as 3 and 4, are utilized in similar fashion.

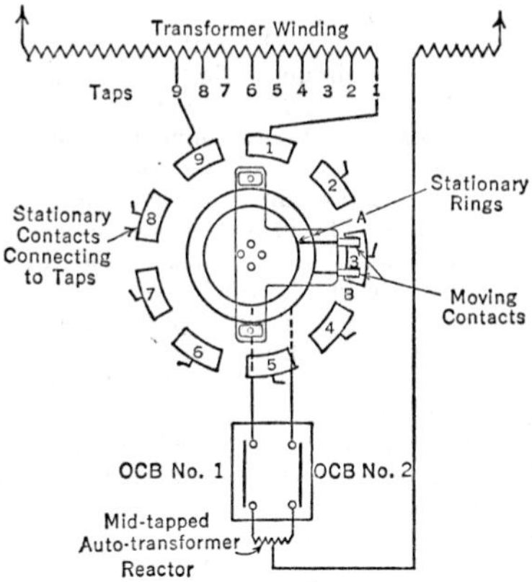

Sequence of Operations from Position No. 1 to Position No. 2, Assuming 2½% Voltage between Tap 1 and Tap 2

Position No.	Percent Winding in Circuit	Connections			
		A to Tap	B to Tap	CB No. 1	CB No. 2
1	100	1	1	Closed	Closed
		1	1	Closed	Open
		1	2	Closed	Open
		1	2	Closed	Closed
		1	2	Open	Closed
		2	2	Open	Closed
2	97½	2	2	Closed	Closed

Switch Positions for Various Tap Positions

Position No.	Percent Winding in Circuit	Moving Contacts A & B Connected To								
		1	2	3	4	5	6	7	8	9
1	100	X								
2	97½		X							
3	95			X						
4	92½				X					
5	90					X				
6	87½						X			
7	85							X		
8	82½								X	
9	80									X

FIG. 42. Mechanism of Circuit of Figs. 40 and 41

The maximum current and voltage which must be broken by the breakers *a* and *b* for a given tap voltage depend upon the load current, its power factor, and the design of the reactor. At full load the arcing kva (as the product of arcing current and voltage) is a minimum when the exciting current in the bridging position is equal to half the full-load

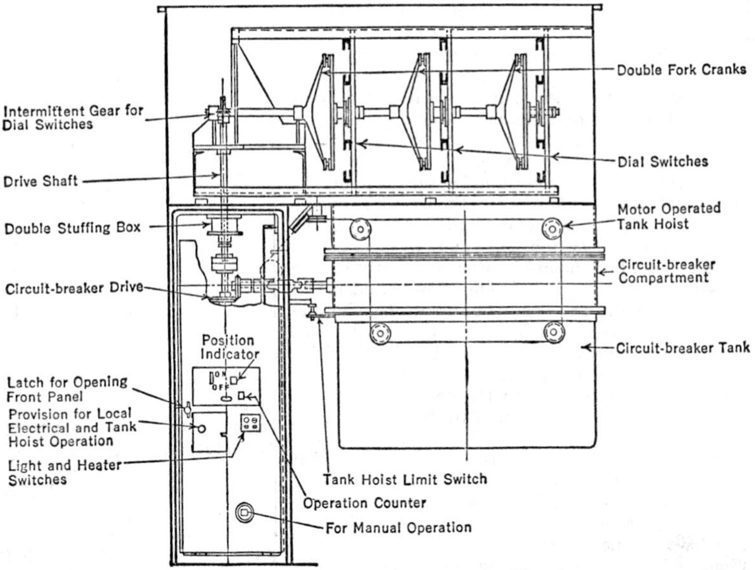

FIG. 43. Construction of Mechanism of Fig. 42

current. With this exciting current, at 100 per cent power factor, the maximum arcing voltage is equal to 141 per cent of the voltage between physical taps, and the maximum arcing current is equal to 71 per cent of full-load current; at zero power factor, the maximum arcing voltage is equal to twice the voltage between physical taps, and the maximum arcing current is equal to full-load current. The maximum reactor current at full load is equal to 71 per cent of full-load current at 100 per cent power factor, and to full-load current at zero power factor. The rated voltage of the total reactor winding is the voltage between physical taps. These relations change with a change in the exciting current, both the reactor kva parts and the arcing kva increasing with either an increase or a decrease in the exciting current of the reactor.

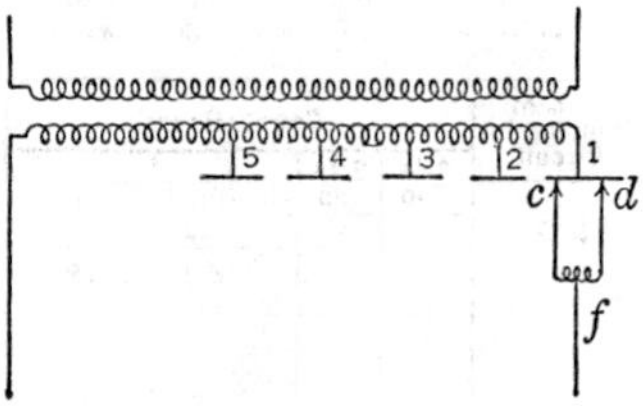

Fig. 44. Simplified Circuit for Load Ratio Control

APPARATUS. Typical physical construction of the mechanism for the circuit of Fig. 40 is indicated in Fig. 42, in which the stationary contacts are arranged in a circle; one moving contact is connected to each of two stationary slip rings by sliding contacts. The various transient and permanent connections are indicated in the table. The intermediate connections are not listed as operating positions, although they could be so used if desired. Such mechanisms are often constructed in three separate parts, each in its own enclosure, as shown in Fig. 43. In this figure the dial switch is located in the upper oil compartment, separated from the main tank and the breaker by oiltight barriers; the circuit breakers are mounted immediately below the dial switches and have their own tank and separate oil. Leads run from the breakers to the dial switches through porcelain bushings mounted on the horizontal barrier between the compartments. The motor-driven operating mechanism is located in air alongside the circuit-breaker compartment. The moving parts of the mechanism are mechanically interlocked so that they will move in proper sequence. Holding-in switches insure the completion of the sequence for one tap change, after it has been initiated, to prevent the mechanism from stopping off position.

A wide variety of modifications of the mechanical structure described will be found in load ratio control equipment for this particular circuit. Most of these modifications use dial switches in a separate oil compartment and an operating mechanism in air in its own housing. The circuit-interrupting means may consist of a conventional circuit breaker, contactors, or rotary circuit breakers. The operation may be initiated either by manual control or automatic control through a contact-making device, such as a voltage-control relay to secure constant voltage.

Where the arcing duty in changing taps is light, the circuit breakers *a* and *b* may be eliminated, simplifying the mechanical construction, and the circuit of Fig. 44 used. The dial switch construction is similar to that of Fig. 42, but the arc is broken directly on the dial switch contacts. The driving motor and mechanism of such units are often mounted in the oil of the dial switch compartment. The operating mechanisms often utilize springs, which are stretched by the driving motor and, when released, move the dial switch quickly from one position to the next, reducing the amount of contact deterioration by burning, caused by the vaporization of metal due to the arc. The control of spring-actuated mechanisms may be simplified by the elimination of holding-in switches. The circuit of Fig. 44 is fundamentally the same as that of Fig. 40; the sequence of operations is the same with the circuit-breaker operations eliminated.

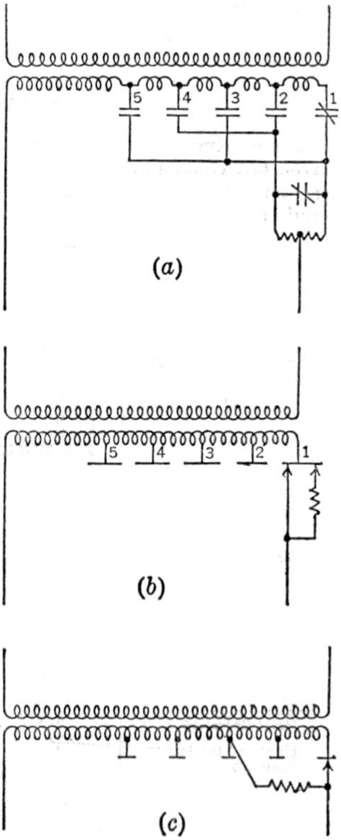

Fig. 45. Circuits for Load Ratio Control without Dial Switch

Figure 45(*a*) indicates the use of contactors or circuit breakers without a dial switch to utilize the fundamental circuit described. One contactor or breaker is provided for each physical tap, and in addition a short-circuiting contact is provided for the reactor when used to obtain the voltage of a physical tap, as in the figure.

Figure 45(*b*) shows a modification of the fundamental circuit, using an impedance, which may be a resistance or reactance, without a tap, to limit the current when changing taps. Since no mid tap is available on the reactor, voltages half-way between tap voltages are not obtainable. If a resistor is used as the impedance, it must be made very large to avoid danger that it will burn out and damage the unit if the mechanism remains in a transition position with the resistor connected between contacts.

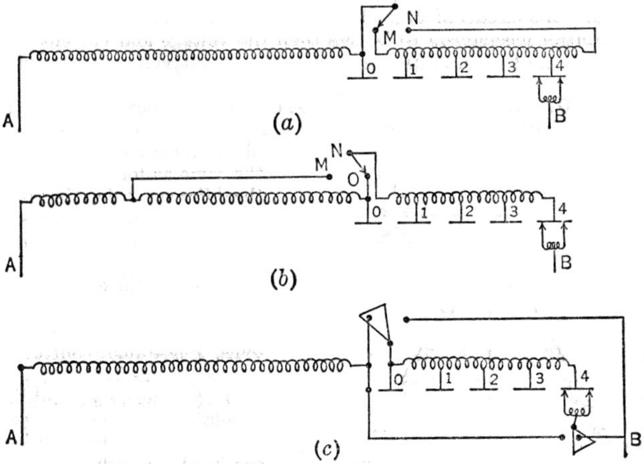

FIG. 46. Load Ratio Control with Reversing Switches

Figure 45(*c*) is really a special case of Fig. 45(*b*), used only for light duty where performance and kva parts of the impedance must be subordinated to simplicity of mechanism; the dial switch requires only narrow stationary contacts, one narrow moving contact, and only one slip ring. It imposes heavier arcing duty on the contacts, requires a larger current-limiting impedance, and causes a higher voltage dip in the circuit when changing taps.

Other connections than those shown have been used for load ratio control. The design of load ratio control circuits is closely related to the design of the mechanism with which they are used. Complicated circuits are not practical because of the complications in the mechanism which will perform the necessary switching.

Reversing switches controlled by the operating mechanism are used as a part of load ratio control circuits to double the tap range by using the same tapped portion of the winding to provide either an increase or a decrease in the winding voltage. The reversing switch of Fig. 46(*a*) performs this function by connecting lead *O* to either *M* or *N*. In this figure the leads *A* and *B* may represent the line leads of one winding of a two-winding transformer, or alternatively the leads *A* and *B* may represent the high-voltage leads, and *A* and *O* the low-voltage leads, of an auto-transformer. (This auto-transformer connection is commonly used for step voltage regulators.) In order to avoid breaking an arc on the reversing switch or interrupting the main circuit, the reversing switch is moved by the operating mechanism only when both the moving contacts of the dial switch are on *O*.

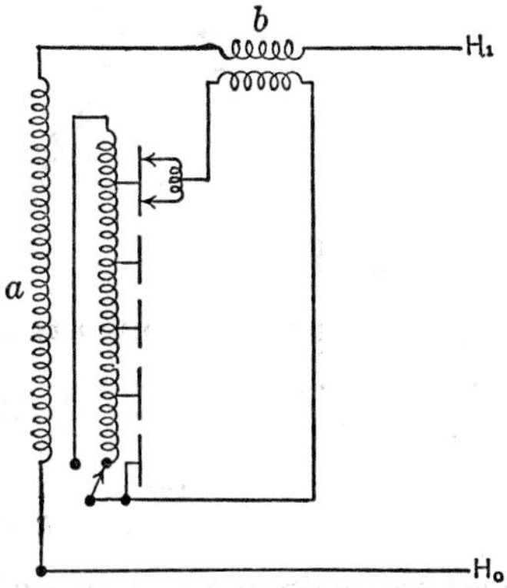

FIG. 47. Load Ratio Control with Series Transformer

Figure 46(*b*) shows another type of reversing switch connection used to obtain double the voltage range for two-winding transformers, where it is undesirable to reverse the load current direction in the tapped winding, in order to secure lower mechanical and electrical stresses in the winding. The reversing switch is moved by the operating mechanism only when both moving contacts of the dial switch are on *O*.

The circuit of Fig. 46(*c*) requires a two-pole, double-throw switch. It is used to secure twice the voltage range for applications where the other types are not applicable, as in a straight-line mechanism. During transition all four contacts are closed simultaneously while the preventive auto-transformer contacts are on 0.

Figure 47 shows a circuit in which a series transformer b is used to add a voltage to the main winding a of a transformer or to subtract it. The tapped winding of transformer a may be used to carry a separate load, it may be a tertiary winding, or the main winding of a may be tapped. Except in the third case, the transformer b may be an auto-transformer if desired. The circuit of Fig. 46 permits both the voltage and current rating of the load ratio control mechanism to be less than the voltage and current rating of the main circuit, which is of advantage when either the circuit voltage or circuit current is high.

POLYPHASE CONNECTIONS. The single-phase circuits described can be used for one phase of a variety of polyphase connections. Load ratio control equipment for phase angle control is fundamentally the same as for voltage ratio control, the difference being the way the winding terminals of the single-phase windings are connected in the polyphase circuit.

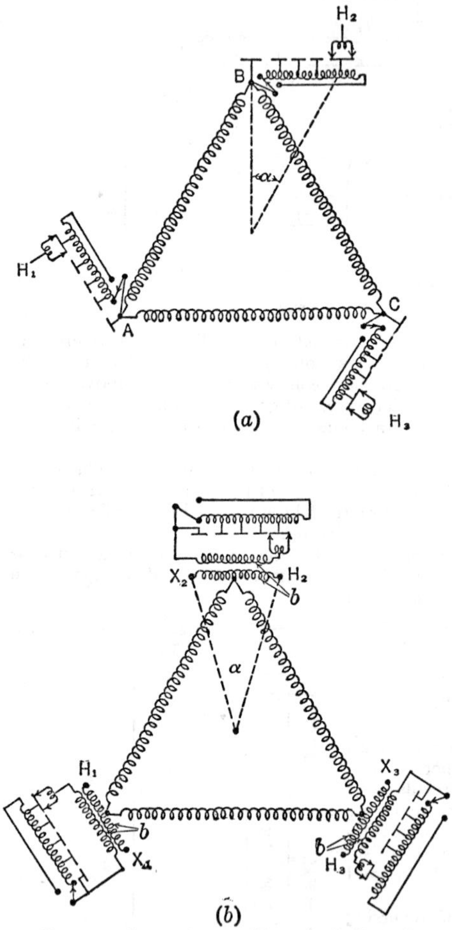

FIG. 48. Connections for Phase Angle Control

Voltage ratio control may be used to change the flow of reactive power in a loop, between lines or between systems. Phase angle control may be used to change the flow of real power in a loop or between parallel lines. (The circulating current between parallel lines caused by the addition of a voltage in phase with the line voltage to neutral, lags or leads the line voltage by an angle determined by the ratio of resistance to reactance of the line. Since the reactance is large in comparison to the resistance for practically all lines of considerable size, the circulating current will lag the voltage which is added to the circuit by nearly 90 deg. If the added voltage is in phase with the line voltage to neutral, the circulating current will be nearly 90 deg out of phase with the line voltage and hence the circulating kva is nearly all reactive kva. If the added voltage is 90 deg out of phase with the line voltage to neutral, the circulating current between paralleled lines will be in phase or 180 deg out of phase with the line voltage to neutral, and hence the circulating kva is nearly all kilowatt real power.)

A variety of connections can be used for phase angle control; typical ones are shown in Fig. 48. In Fig. 48(a) the tapped winding is placed on the line end and, by means of the reversing switch, may be used for either advancing or retarding the phase angle. Figure 48(a) may represent one winding of a two-winding transformer, or if A–B–C are connected to lines, the diagram may represent the windings of an auto-transformer. The phase angle shift produced is represented by the angle α between the dotted lines. (An insulating transformer may be interposed between the tapped winding of the main transformer and the line to insulate the mechanism and tapped winding from the line.) This connection produces a slight voltage change in addition to the phase shift.

The connection of Fig. 48(b) utilizes a series transformer b in an auto-transformer connection to produce a phase shift through the angle a between the high- and low-voltage lines. The phase angle may be either advanced or retarded by means of the reversing switch. The connection shown secures phase angle shift without voltage change.

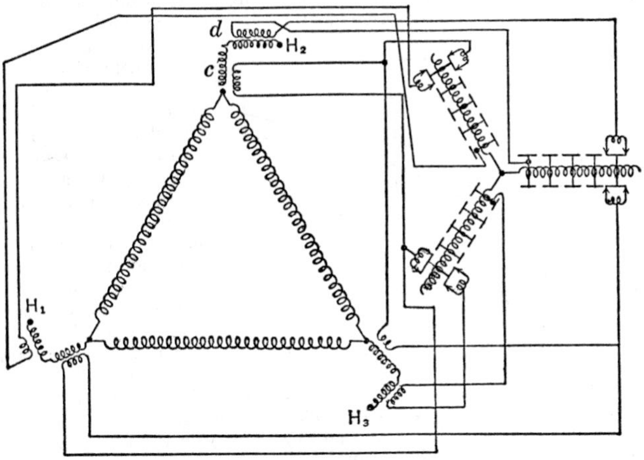

FIG. 49. Connections for Combined Load Ratio and Phase Angle Control

When control of both real and reactive power is desired, a transformer with combined ratio and phase angle control, each with its own load ratio control equipment, may be used. Figure 49 is a typical circuit utilizing the same tapped winding for both controls but two separate series insulating transformers, as shown at *c* and *d*. Figure 50 is another typical circuit, utilizing a single series insulating transformer *c* for each phase, but requiring separate tapped windings for ratio control and for phase angle control.

Polyphase circuits may utilize in each phase any of the fundamental circuits for load ratio control. Polyphase circuits for voltage control are not shown, since they usually consist of three elemental single-phase circuits connected in delta, in star, or in any of the other ways shown in Article 3. Load ratio control equipment is available for any size of transformer from the largest to the smallest but is much more expensive than a no-load tap changer.

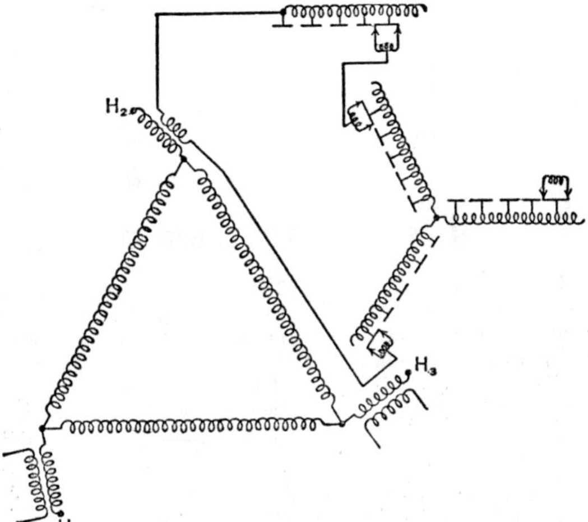

FIG. 50. Connections for Combined Load Ratio and Phase Angle Control

20. NOISE

All transformers produce noise, ranging from nearly inaudible amounts for small transformers to high levels for large transformers. Practically all the noise of transformers originates in the magnetostriction of the core steel, which changes in length when it is magnetized, the movement, although very small, producing noise. (The change in length of typical core steel when magnetized to normal density is 0.00008 per cent.) The magnetostriction and the resulting noise increase with the induction in the iron, variation such as that shown in Fig. 51 for a 10,000-kva transformer being typical. For the same induction in the iron (and the same magnetostriction) the amount of noise increases as the weight of the core is increased, tne noise level rising approximately 3 db each time the core weight is doubled.

The magnetostriction of steel is extremely variable, between not only different lots of the same grade of steel but also different portions of the same sheet of steel. Preferred oriented steels generally have low magnetostriction parallel to the direction of rolling. Some attenuation of the noise is produced by the oil and tank. If parts of the transformer are so proportioned that they vibrate in resonance, the noise level may be increased.

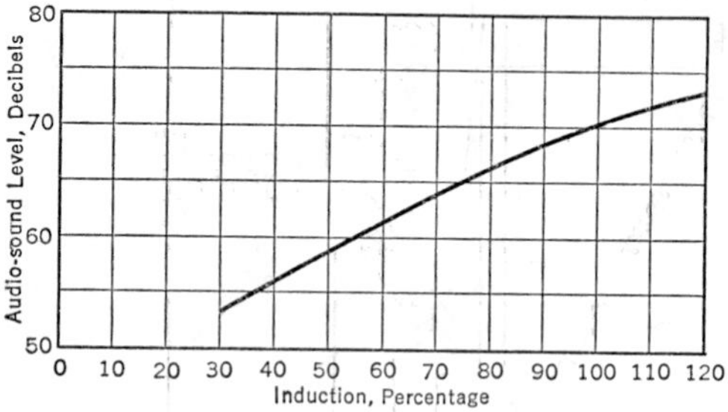

FIG. 51. Typical Relation between Noise and Induction

Noise is measured on a transformer by exciting it to full voltage and taking measurements with a sound-level meter at several locations, not over 3 ft apart, around the tank at a distance of 1 ft from the major sound-producing surfaces. The noise level of the transformer is the average of the readings thus obtained. The noise levels of typical transformers are given by the curve of Fig. 52. Transformers with less noise can be obtained by designing them with lower normal flux density in the iron, thereby decreasing the magnetostriction and the resultant noise.

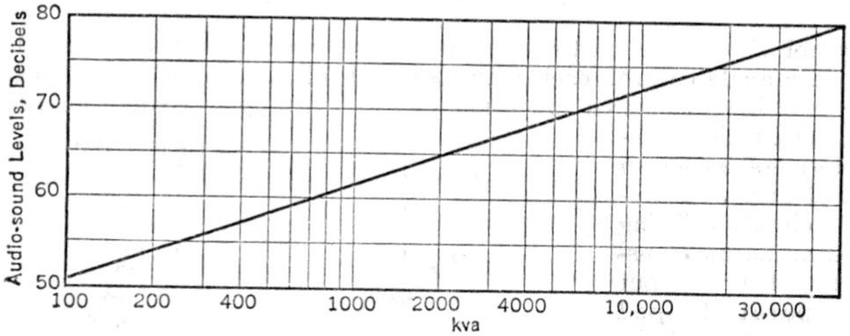

FIG. 52. Noise Levels of Typical Transformers

21. CONSTRUCTION OF TRANSFORMERS

GENERAL CONSTRUCTION. Because a transformer is a mechanical device used to secure electrical results and electrical performance, various physical arrangements of construction are possible and are used. The usual parts of a transformer, however, are core, coils, insulation, tank and cooling means, bushings, and accessories.

CORE CONSTRUCTION. On the basis of their core construction and sometimes of the associated winding arrangement, transformers are loosely classified as core type or shell type. Regardless of coil construction, the single-phase transformers of Figs. 53(*a*) and 53(*b*) and the three-phase transformer of Fig. 53(*c*) are called **core type.** The cores

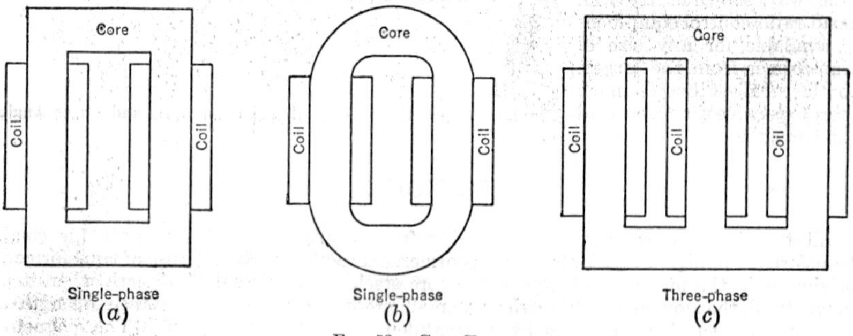

FIG. 53. Core Types

in Figs. 53(*a*) and 53(*c*) are assembled with the plane of the individual laminations parallel to the plane of the paper. For most transformers 0.014-in. thick, high-silicon steel (between 3 and 5 per cent silicon), especially made and processed to secure the desirable qualities of high permeability and low core loss at the flux densities at which it will be operated, is used. Except for wound cores like that of Fig. 53(*b*), the individual laminations are rectangular, L-shaped, or E-shaped. (The sheets are annealed after shearing to

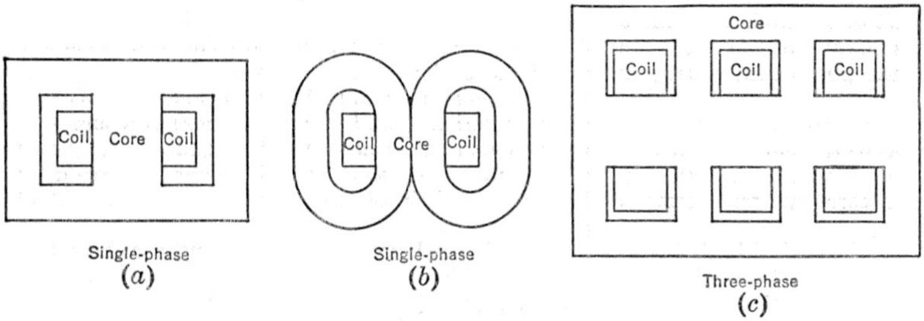

FIG. 54. Shell Types

remove mechanical stresses in the laminations.) Except in small transformers, the individual sheets are varnished or given some other coating to insulate them from each other and thus prevent circulating currents between sheets with resulting excessive core loss. The individual laminations are usually stacked with the joints in alternating layers staggered. The cross-section of the core may be rectangular or stepped to approximate the shape of the inside of the coil, which may be rectangular, oval, oblong, or round.

The core of Fig. 53(*b*) is wound of a continuous strip of steel and then annealed to remove the mechanical stresses introduced by the winding process. Sometimes the coils are wound on the core, and sometimes the core is cut across the laminations to permit slipping mold wound coils over the core. A core wound of strip steel is particularly advantageous when preferred oriented steel is used, because it permits all the flux to flow parallel to the direction of rolling of the steel, and the steel has its lowest loss and maximum permeability in this direction.

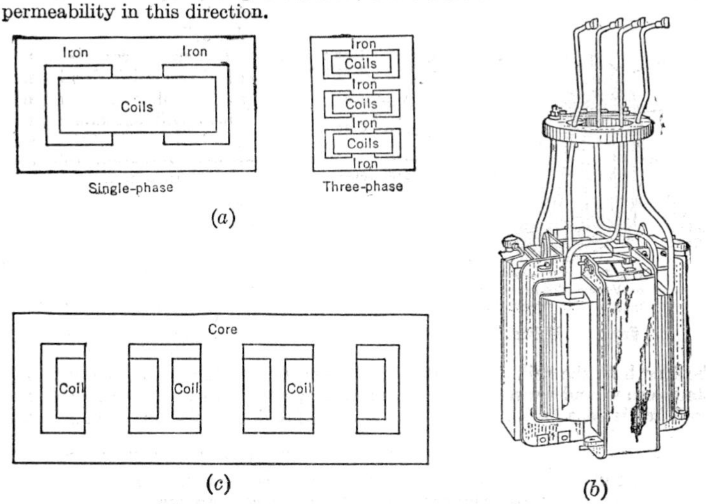

FIG. 55. Distributed Shell or Distributed Core Type

The transformers of Fig. 54 are usually called **shell type.** Sometimes, however, when the high- and low-voltage windings are concentric, as distinguished from an interleaved arrangement of disk coils, the transformers of Figs. 54(*a*) and 54(*b*) are called three-legged core type. The transformers of Figs. 54(*a*) and 54(*c*) are made with interleaved cores of flat laminations; the transformer of Fig. 54(*b*) has a wound core. The transformer of Figs. 55(*a*) and 55(*b*), which practically always has a concentric winding, is sometimes called distributed shell type and sometimes distributed core type. In the three-phase transformer of Fig. 53(*c*) the fluxes in the three legs of the core on which the windings are placed are equal and 120 elec deg apart, with positive direction assumed as upward in each leg. The flux in the yoke is the sum of two equal flux vectors 120 deg apart, or the same as the leg flux. In the three-phase transformer of Fig. 54(*c*) the flux in the portion of the yoke between the coils of adjacent phases is the difference between the fluxes of adjacent phases, when positive direction of flux is assumed to be the same for all three phases. If the center phase is not reversed, this results in 1.73 times half the phase flux in part of the iron, since the difference between two equal vectors 120 deg apart is 1.73 times one vector. If the center phase is reversed, the flux in the yoke becomes equal to half the phase flux; consequently, it is general practice to reverse the center phase of three-phase shell-type transformers. The three-phase transformer of Fig. 55(*c*) is called a five-legged core type. Its electrical characteristics are the same as those of a three-phase shell-type transformer. If the flux in the yokes is a sine-wave, it is equal to 57.7 per cent of the leg flux. The five-legged core is used when shell-type characteristics are desired with a core-type transformer and when it is desirable to decrease the core height of a core-type transformer for shipping or other reasons.

On a theoretical basis there is little to choose between shell-type and core-type transformers. Almost any given set of characteristics can be obtained equally well with either core-type or shell-type construction, the principal difference being in the cost. The practice of transformer manufacturers supports this conclusion. Core-type construction is used by some manufacturers for all types of distribution and power transformers. Other manufacturers use shell-type construction for large transformers and those of very high voltage, the remainder being core type. Where transformers have three or more windings or complications in the windings, shell-type construction is usually advantageous. Very small transformers may be either core type or shell type. The logical procedure is to decide between core type and shell type on the basis of cost for the same performance.

WINDING CONSTRUCTION. For the many variations in winding construction of transformers, three types of coils are commonly used: helical coils, disk coils of more than one turn per layer, disk coils of one turn per layer. The winding form for any of these coils may be circular, rectangular with rounded corners, oval, etc. The trend is toward the use of circular coils, because of their inherently greater mechanical strength in manufacture and in service, but other forms are also in general use. Core-type power transformers practically always use circular coils. Some shell-type power transformers use circular coils and some rectangular coils.

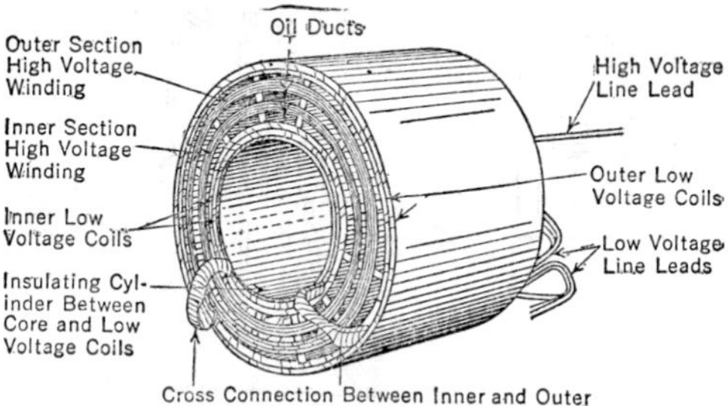

FIG. 56. Helical Coil Winding.

A helical coil is sometimes called a cylindrical coil, a straight-wound coil, or a barrel coil. A helical coil extends almost the complete length of the opening in an axial direction, the copper of the winding being wound from near one end to near the other end, as in Fig. 56. Such coils are in common use only up to 15-kv line voltage, although by using special construction they can be employed for much higher voltages. The layers are insulated from each other by oil ducts running the length of the coil, which also allow the flow of cooling oil, or by layers of paper or treated cloth wound between layers. The winding does not extend the complete length of the coil but is kept back a sufficient distance to provide insulation between layers and to ground.

A disk coil of more than one turn per layer is similar to a short helical coil, consisting of several layers of conductor wound with several turns per layer. The layers are insulated from each other by paper layer insulation. These coils are stacked together to form the complete winding, with insulating spacers, sometimes called radial spacers, between the coils. Disk coils of more than one turn per layer are used principally for small currents where the only available conductor of suitable cross-sectional area is round wire.

Disk coils of one turn per layer are wound with one turn per layer, in most cases with adjacent coils wound in opposite directions so that their starts or their finishes can be connected together. The copper is of rectangular cross-section, often with several strands in parallel. This is the most frequently used coil for power transformers of both core- and shell-type.

In order to reduce the eddy-current loss in the copper, heavy conductors required for large currents are made of many strands of rectangular conductor connected in parallel, and suitably transposed, so that each conductor has the same reactance to the other winding of the transformer.

The turn-to-turn insulation on rectangular wire and some round wire consists of untreated rope paper wound spirally on the conductor. When the paper is dried and impregnated with oil, high dielectric strength is obtained. For small transformers round wire may be insulated with a baked enamel. In larger transformers, paper tape is used, in addition, for both mechanical strength and increased dielectric strength. In some transformers cotton covering or untreated cotton tape may be added for mechanical protection.

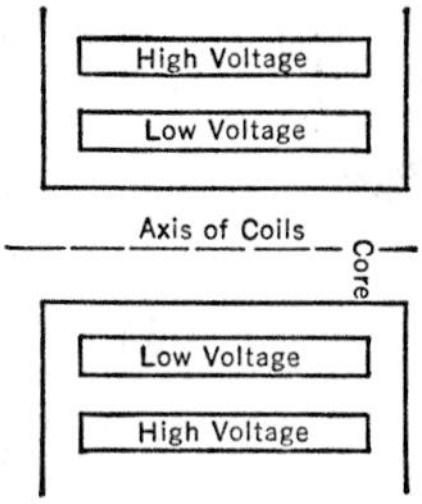

FIG. 57. Coils on Core-type Transformers

Core-type transformers usually have the coils arranged on the core concentrically, as in Fig. 57, in which the high-voltage winding is larger in diameter than the low-voltage winding and is located outside surrounding the low-voltage coil. (Some shell-type transformers which use this concentric coil arrangement are sometimes called three-legged core type.) Any of the three types of coils may be used for the high-voltage winding and for the low-voltage winding. When both the high-voltage and low-voltage windings are helical coils, they may be wound together or separately. When the windings are disk coils, they are assembled or wound in a stack, with the individual coils separated by radial spacers forming horizontal cooling and insulating oil ducts. The insulation between high-voltage and low-voltage columns consists of insulating cylinders, usually of untreated Fuller board, and, if the voltage is high, angle collars of molded Fuller board placed so as to increase the dielectric strength at the ends of the columns. If the design requires, static shields connected to the line are arranged on the line end of the coil stack and outside the coils to improve the impulse voltage distribution.

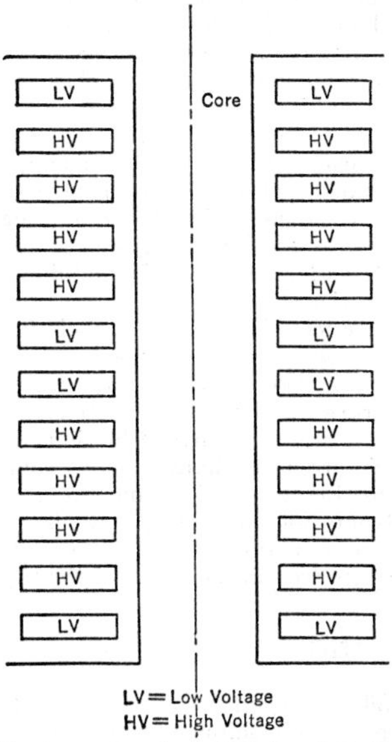

FIG. 58. Interleaved Coils of Shell-type Transformer

Most shell-type transformers use an interleaved coil arrangement of the kind shown in Fig. 58, in which all coils have approximately the same inside diameter and approximately the same outside diameter. Any desired number of coil groups can be used, depending on the design characteristics desired, each coil group consisting in most cases of two or more coils. Coil-to-coil insulation and insulation to ground consist of Fuller board washers and spacers forming oil ducts for insulation and cooling, sometimes in combination with formed angles or channels of molded Fuller board, to increase the dielectric strength. If the voltage requires, a static shield connected to the line end is placed on the line end of each line group, just like an additional coil, to improve the impulse voltage distribution.

In order to develop full dielectric strength, the cellulose material used for transformer insulation under oil must be completely dry and impregnated with oil. To accomplish this, the complete transformer after assembly is dried out, usually under vacuum, and then impregnated with hot oil, often under vacuum to eliminate the air from the insulation quickly and insure penetration of the oil. Porcelain, oil-impregnated Fuller board, or wood is used for terminal supports.

22. OIL

(See also under Article 28.)

Oil is used for insulating and cooling most power and distribution transformers. It is usually a pure unblended (or straight run) mineral oil obtained from the fractional distillation of crude petroleum. It must not contain water, inorganic acid, alkali, free sulfur, tar, asphalt, vegetable or animal oils, or other impurities. Typical characteristics of transformer oil follow.

TYPICAL INSULATING-OIL CHARACTERISTICS. Pour point not higher than minus 40 deg cent (− 40 deg fahr) (by ASTM standard test D-97). A low pour point is desirable so that the oil will not congeal in cold weather but will remain fluid.

Flash point at least 132 deg cent (270 deg fahr) (by ASTM standard test D-92). Fire point at least 149 deg cent (300 deg fahr) (by ASTM standard test D-92). High flash point and fire point are desirable to reduce fire hazard and permit high operating temperatures for overloads.

Specific gravity not more than 0.91 at 15.6 deg cent (60 deg fahr) (by pyknometer or Westphal balance). A low specific gravity aids in causing any free water which accidentally gets into the transformer to settle to the bottom of the main tank or expansion tank.

Steam-emulsion number not more than 25 sec (by ASTM standard test D-157). The steam-emulsion number is a measure of the ability of the oil to resist emulsion. A low steam-emulsion number is required for transformer oil as an indication that water will not be held in suspension but will rapidly separate out and settle at the bottom of the tank.

Viscosity not more than 63 sec Saybolt at 37.8 deg cent (100 deg fahr) (by ASTM standard test D-88). A low viscosity is desirable so that the oil will circulate freely in functioning as a cooling medium.

Neutralization number not more than 0.03 (by ASTM standard test D-188 method A, Sec. 4). Both acidity and alkalinity are undesirable; otherwise the oil may attack the materials of the transformer.

Dielectric strength of at least 26,000 volts (by ASTM standard test D-117). In addition to being required for successful operation of the transformer, high dielectric strength is an indication of freedom from moisture and impurities.

The resistance of oil to sludging, forming heavy liquid or solid compounds, is an extremely important characteristic. Some sludging is due to oxidation. Most metals, moisture, insulation, etc., act as catalysts, accelerating the rate of sludging, which occurs more rapidly at higher temperatures. The only really satisfactory test is service experience. Various short-time tests are used, however, as an indication of the chemical stability of the oil. The best correlation with service experience is obtained with tests which most closely approximate service conditions, including length of time for the tests.

INHIBITORS. It is possible to add inhibitors to transformer oils to obtain longer life. A variety of substances are employed as inhibitors, but their use is still in the experimental stage. A long period of continuous service is required to determine their operating characteristics. The use of different inhibitors complicates stocking and storing oil. Present-day oils, when used in protected transformers, have long life. So far, inhibited oils are used to a very limited extent to secure service experience.

ASKARELS. For use where inflammable oils are objectionable because of their fire hazard, non-inflammable liquids, sometimes called askarels, have been developed. Common trade names are Chlorextol, Inertol, and Pyranol. The chemical elements of which they are composed contain, among other elements, hydrogen and chlorine, so that the principal product when the liquid is decomposed by an electric arc is hydrochloric acid, which is non-inflammable. These liquids have dielectric strength characteristics comparable to those of oil. To prevent contamination and loss by evaporation, they are used in tightly sealed tanks. Since these liquids act as solvents for most of the varnishes, compounds, etc., used in oil-filled transformers, special construction for their use is required. They are usually heavier than water; consequently free water floats to the top of the liquid.

23. MECHANICAL BRACING

Mechanical bracing is provided for the core and the coils to prevent movement of parts and consequent damage to insulation, core, or mechanical parts in shipment or operation. Bracing can also prevent excessive noise in operation. The coil bracing must be designed to withstand the mechanical short-circuit stresses in accordance with the ASA standards for transformers. Transformers with 4 per cent or greater impedance are designed to withstand dead short circuit for 2 sec (or longer, if the impedance is higher) with full voltage maintained on the primary winding.

24. TANK AND COOLING MEANS

The tank of an oil-immersed transformer serves as a container for the oil, protecting it from contamination and providing cooling surface to dissipate the heat. Tanks may be round, oval, rectangular, or of other shapes as required to contain the core and coils and provide the necessary internal insulation clearances. Transformer tanks, except the pole type, are provided with bases to elevate the tank bottom for service and installation.

The principal types of transformers, classified as to cooling means, are:

Dry, natural-cooled.
Dry, forced air-cooled.
Oil-immersed, self-cooled.
Oil-immersed, forced air-cooled.
Oil-immersed, forced oil-cooled.
Oil-immersed, water-cooled.

DRY TYPE. Dry-type transformers are cooled by the circulation of air past the exposed surfaces of the core and coils. They are made only for low voltages (15 kv and below) and seldom more than 4000-kva capacity, except when forced cooling is employed. They may be installed in locations where the fire hazard of oil would require a special vault, since they are usually constructed of non-inflammable materials. The use of forced air on dry-type transformers can increase the rating to as much as 50 per cent over the self-cooled rating.

OIL-IMMERSED, SELF-COOLED TYPE. Oil-immersed, self-cooled transformers dissipate their heat by radiation and natural air circulation over the tank and by cooling

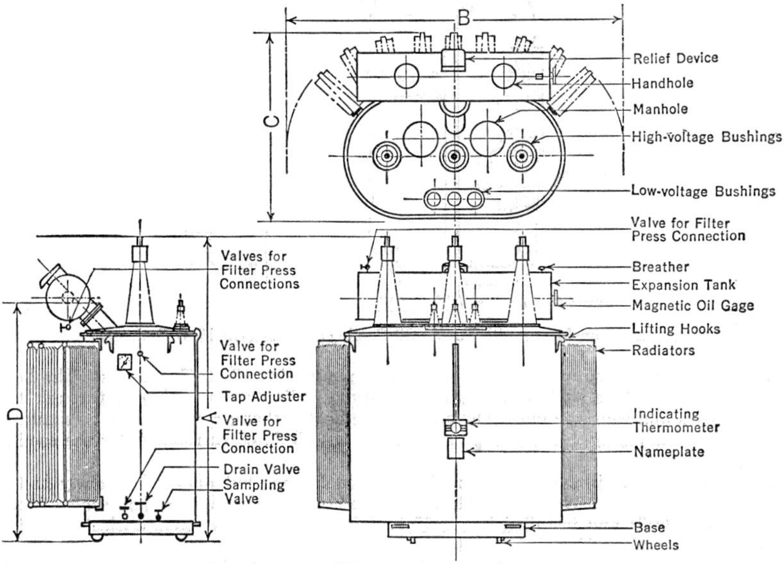

FIG. 59. Typical Large Self-cooled Transformer

tubes or radiators attached to it. Self-cooled transformers are the most efficient and the most expensive oil-immersed transformers and require the least attention. For small units ample cooling surface for heat dissipation is provided by a plain tank. At one time corrugated tanks were a popular means of increasing the cooling surface where the plain tank was not sufficient but are used less frequently today. Tubes usually 2 in. in diameter are welded in the tank wall where only a limited amount of additional cooling is required. If more than three rows of tubes are used, painting is difficult. If a large amount of additional cooling is required, as on large transformers, radiators, welded on or detachable, are used. Transformer radiators usually consist of either round or flat cooling tubes welded into headers through which the oil flows from the main tank at the top and to the main tank at the bottom. Detachable radiators are provided with radiator valves, making it unnecessary to remove the oil from the main tank to install or remove radiators. A typical large self-cooled transformer is shown in Fig. 59, in which the principal parts are marked.

OIL-IMMERSED, FORCED AIR-COOLED TYPE. Forced air-cooled transformers dissipate their heat by forced air circulation over the tank and other cooling surfaces, to which the oil flows by natural convection. Forced air-cooled transformers are self-cooled with the addition of fans or blowers, which blow air over the radiators or other cooling surfaces to increase their rate of heat dissipation. The fans may run continuously or may be automatically controlled by oil temperature or copper temperature. Special arrangements of radiators are frequently used to facilitate the application of fans. Forced air-cooled transformers are less expensive and less efficient than self-cooled transformers of the same rating. The forced air-cooled rating of transformers with combined self- and forced cooled ratings is generally 133 per cent of the self-cooled rating for transformers of 3333 kva and more, and 125 per cent of the self-cooled rating for smaller transformers.

OIL-IMMERSED, FORCED OIL-COOLED TYPE. Forced oil-cooled transformers dissipate their heat by pumping the oil to the cooling surfaces of an oil-to-air or an oil-to-water heat exchanger. Typical forced oil cooling with an oil-to-air heat exchanger is shown in Fig. 60. The hot oil is drawn from the top of the tank and circulated by a pump

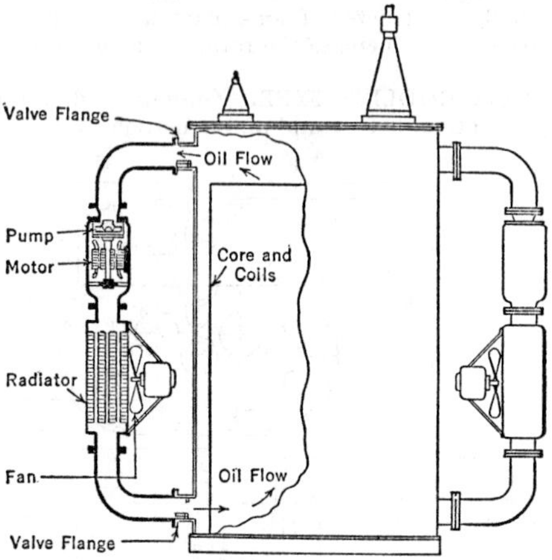

FIG. 60. Transformer with Forced Oil Cooling and Oil-to-air Heat Exchanger

unit, which has the motor immersed in the oil, eliminating all stuffing boxes. The heat exchanger is constructed like an automotive-type radiator; air is forced through the radiator by a motor-driven fan. The oil-to-water heat exchanger, which may be similarly mounted on the piping, has tubes through which the oil flows, surrounded by the cooling water, or vice versa.

Formerly, forced oil cooling was used widely in Europe but only infrequently in the United States. During World War II, however, a large number of forced oil-cooled transformers were built and installed in the United States in order to save critical material. The kva rating of a self-cooled transformer core and coils can frequently be increased as much as 67 per cent by means of forced oil cooling. Forced oil-cooled units are less efficient and less expensive in large sizes than are forced air-cooled units of the same kva rating. In order to provide for operation at reduced capacity if one cooling unit should need repairs, most forced oil-cooled transformers have at least two independent cooling systems. The cooling systems may be controlled either manually or automatically by oil or copper temperatures.

OIL-IMMERSED, WATER-COOLED TYPE. Water-cooled transformers dissipate their heat to water, which flows through coils located in the main transformer tank. Cooling coils are usually made of 1-in. copper pipe. A flow indicator, often equipped with alarm contacts to indicate failure, is used to show the water flow through the cooling coils. Sufficient parallel circuits for the cooling coils are provided to keep the required water pressure below 25 lb per sq in. Water-cooled transformers have the same efficiencies as self-cooled transformers but cost slightly less in large sizes, the cost being approximately the same as for oil-immersed, forced air-cooled transformers. They may be installed in vaults with practically no ventilation, since the water carries away the heat. Large water-cooled transformers require cooling water even with no load on the transformer, since the cooling surface of the tank is usually not sufficient to dissipate the heat due to the core loss. A typical water-cooled transformer is shown in Fig. 61, with the principal parts marked.

Transformers may be made with any desired combination of the various cooling means to fit particular applications. Combinations of self-cooling for light loads with water cooling or some form of forced cooling for heavy loads are most common.

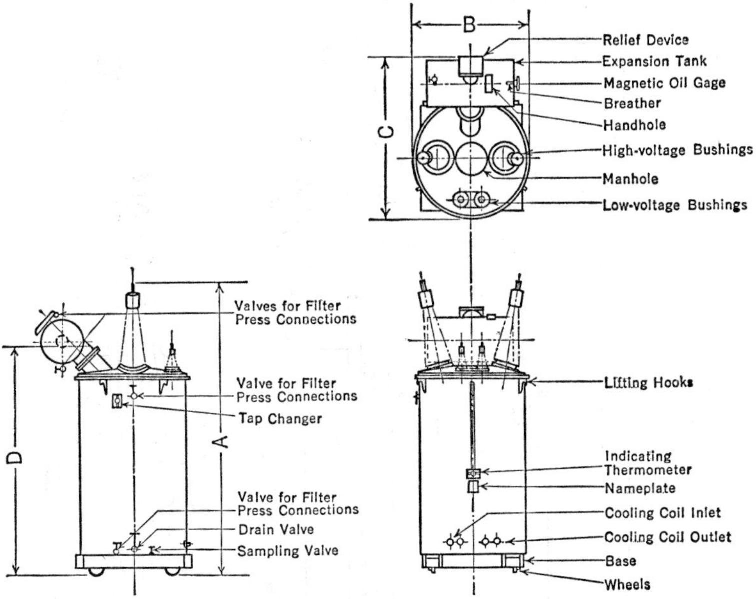

FIG. 61. Water-cooled Transformer

25. BUSHINGS

Bushings are used to insulate the leads of the transformer windings as they pass out through the case or cover. Solid porcelain bushings are commonly used for voltages up to 15,000 and less frequently for higher voltages. A typical solid porcelain bushing is shown in Fig. 62. The conductor through the center of the porcelain consists of an insulated copper cable or rod terminating in the bushing cap, where means are provided for connecting the lines running to the transformer. Where a cable is used for the conductor through the bushing, it is usually detachable at the bushing cap, from the outside of the transformer, so that the bushing can be removed or replaced without reaching inside the transformer; this construction is called "draw lead" or "fished through lead" construction. Oil- and gas-tight joints are made at the bushing cap and the flange, where the bushing is mounted on the cover by means of gaskets of such materials as composition cork or a combination of cork and oil-proof synthetic rubber. In some cases soldered joints are used, with the metal flange and bushing cap soldered to the porcelain by the use of a special glaze on the porcelain.

For voltages above 15 kv the bushings are usually oil filled in order to secure high dielectric strength in a small space with short insulating distances. (It is sometimes difficult to find room on the cover of a transformer for the necessary outlet bushings.) Typical oil-filled bushings are shown in Figs. 63 and 64, where the internal insulation consists of oil ducts separated by porcelain barriers to secure high dielectric strength. Conducting glaze may be used on porcelain surfaces to make the voltage distribution more uniform and thereby increase the dielectric strength. Oil-tight joints are secured by composition cork gaskets. In some types of oil-filled bushings the internal barriers between oil ducts are made of paper cylinders instead of porcelain.

In another type of bushing the interior cylinders are wound solidly of paper, with paper and thin layers of metal foil of graded lengths replacing the oil ducts to improve the voltage distribution and reduce the stress concentrations on the insulation. The number of layers of metal foil increases with the voltage. Bushings constructed in this manner are sometimes called condenser bushings. All these bushings have a porcelain outer shell to contain the oil and provide rain shields to increase the wet flashover of the bushing. Standard bushings used for transformers have 60-cycle dry and full wave impulse-withstand strengths at least as great as those specified for the transformer winding with which they are used, definite values for each voltage class having been standardized by the ASA.

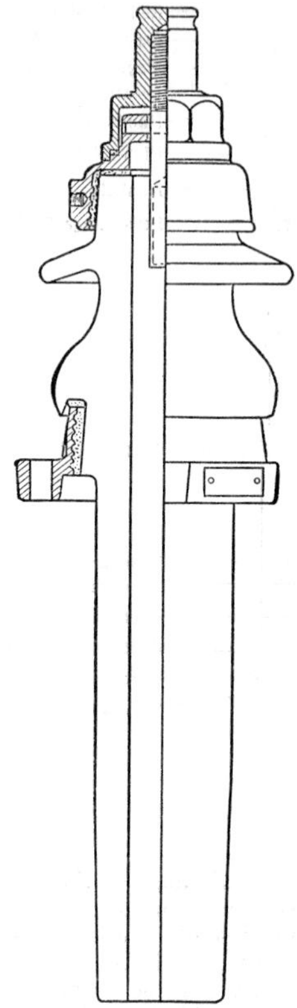

Fig. 62. Solid Porcelain Bushing

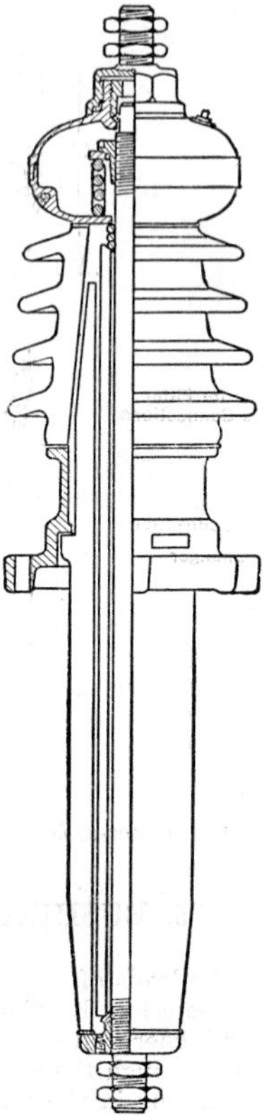

Fig. 63. Oil-filled Bushing

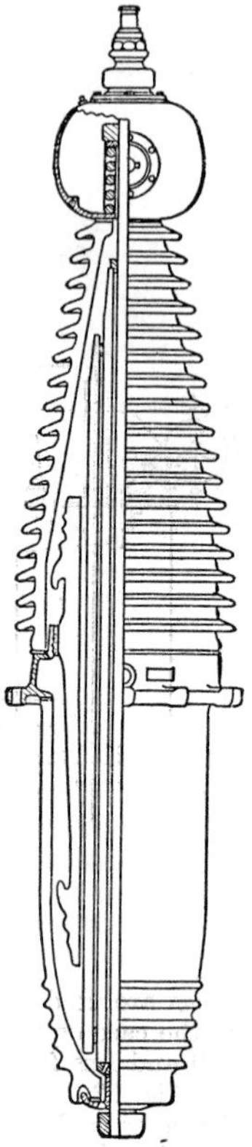

Fig. 64. Oil-filled Condenser Bushing

26. ACCESSORIES

Various accessories are used with transformers to keep them in good operating condition, to aid maintenance, or to increase their usefulness. The accessories which are in common use are described briefly in this section.

CONSERVATORS. In order to keep hot oil from coming in contact with the air, from which it may absorb moisture and oxygen, several alternative procedures are available. The tank may be sealed tight, providing an inert gas space above the oil to prevent excessive pressures as the oil expands from heating, or it may be completely filled with oil and provided with an expansion tank (sometimes called a conservator), so that only cool oil is in contact with the air. Instead, it may be provided with inert gas equipment which maintains a nitrogen atmosphere in a gas space above the oil in the main tank, by means of either an oil-seal expansion tank for the nitrogen or nitrogen cylinders connected to the gas space through mechanical reducing valves, which open to add nitrogen to the gas space when the pressure there drops to a predetermined low value. In addition to their principal function of protecting the oil from contamination, all these constructions prevent secondary explosions, which are caused by hot gases from a fault in the transformer winding rising to the surface of the oil and igniting the mixture of gas and air above the oil in an open-type transformer. All, except the expansion tank, prevent damage to the tank and cover due to primary explosions caused by a fault in the transformer below the oil level, since the gas space above the oil acts as a cushion to reduce the pressure caused by the fault. All are effective means of preventing the oil from sludging and of keeping it dry and in good operating condition.

Pressure-relief devices are used on large transformers to prevent damage to the tank or cover if excess pressures appear inside the tank. These devices usually consist of a bakelite or glass diaphragm clamped over an opening in the cover with gaskets to form a tight joint. A movable weather-proof cover should be located over the diaphragm to prevent the entrance of moisture after the diaphragm breaks because of excess pressure.

BREATHERS. When protective means such as those described are not used, the hot oil at the top of the transformer is exposed to the air, and breathers are used to prevent moisture from condensing in the air space above the oil. Open-air breathers, preferably with a separate inlet and outlet if the transformer is of any considerable size, provide for the circulation of air through the air space above the oil, so that moving air will absorb and carry away any moisture which temporarily condenses. One commonly used breather of this type has its inlet near the bottom of the tank, with a long pipe extending vertically up through the oil to the air space above, so that the air in the pipe will be heated by the oil surrounding it and will rise, causing definite air circulation through the air space above the oil and out through a breather opening in the cover. Drying breathers whereby the air is drawn in through a screen container filled with a drying agent, such as calcium chloride or silica gel, are sometimes used. Drying breathers require maintenance because the drying agent becomes saturated with moisture and must be reactivated or replaced; otherwise the drying agent may add moisture to the air when it is drawn into the transformer.

TEMPERATURE INDICATORS. When information is desired concerning the oil or copper temperatures, transformers are provided with temperature indicators or thermal relays. Power transformers are usually supplied with dial-type oil-temperature indicators with the thermometer bulb immersed in the hot oil, and the indicating instrument located at the bulb for small units or on a long tube, lower on the tank for convenience in reading, on large transformers. Oil-temperature indicators may be provided with alarm contacts to close when a predetermined oil temperature is reached. Some are provided with resettable maximum-indicating hands to show the highest oil temperature which occurred.

Winding temperature indicators, which are designed to show the hot-spot temperature of the transformer winding, reproduce the hot-spot temperature as the sum of the hot-oil temperature and the copper gradient. They may consist of a thermometer bulb, immersed in the hot oil of the transformer, to which the copper gradient is automatically added to give the hot-spot temperature, either by means of a bimetallic thermal element heated by current proportional to the winding current (supplied by the secondary of a current transformer) and mounted in the case of the indicating instrument, or alternatively by circulating this current to heat the thermometer bulb in the oil by passing the current through the bulb or through a heating coil surrounding the bulb. Where the bimetallic element is used for the gradient, the instrument may have two indicating hands, one for the oil temperature and one for the copper temperature. Winding temperature indicators of the thermometer type are usually provided with alarm contacts, which close when a predetermined hot-spot temperature is reached.

Winding temperature indicators for remote indication on switchboards (Fig. 65) are constructed similarly of a copper resistance coil immersed in the hot oil and surrounded by a heating coil, through which a current proportional to the winding current is circulated to reproduce the copper gradient. Leads from the resistance coil are connected to a d-c resistance-measuring circuit, including an instrument which is calibrated in degrees centigrade to indicate the hot-spot temperature of the transformer. In some cases a thermocouple is used instead of the resistance thermometer.

The operating principles of thermal relays are exactly the same as those of winding temperature indicators. Although they may be the same in mechanical construction, they are commonly adjusted to produce a copper gradient which is less than the actual gradient of the transformer in order to permit higher operating temperatures for short-time overloads than for long-time overloads. As commonly adjusted, they permit a higher than normal rate of loss of life of the transformer insulation for heavy short-time overloads, assuming that this rate will be compensated for by longer periods of operation at less than normal rate of loss of life or, if this compensation does not occur, by decreased length of life of the transformer. Thermal relays have alarm contacts which may be connected to sound an alarm when a certain relay temperature is reached and to trip a circuit breaker, disconnecting the load, when a certain higher relay temperature is reached. With different relay settings they may be used to control forced air-cooling equipment or forced oil-cooling equipment.

The construction of thermal relays, since they do not require an indicating instrument, may be simpler than that of winding temperature indicators, although the principle is the same. One type of thermal relay has a thermometer bulb surrounded by a heating coil immersed in the hot oil of the transformer, like a winding temperature indicator of the thermometer type. The indicating instrument is replaced by pressure switches which close the relay contacts. Another type of thermal relay consists of a bimetallic element located in the hot oil of the transformer; through this element a current proportional to the winding current is circulated, so that the temperature of the element is a function of the oil temperature and the winding current, the calibration being the same as for other

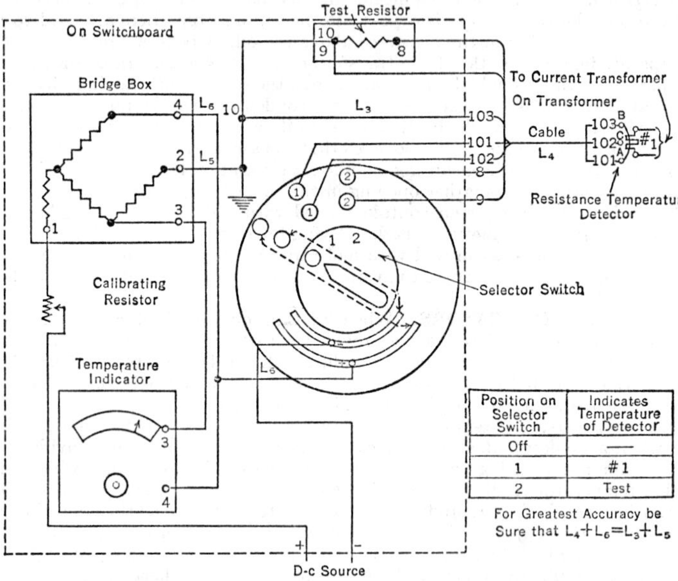

FIG. 65. Winding Temperature Indicator, Remote-control Type

types of thermal relays. The closing of the alarm contacts is controlled by the movement of the bimetallic element.

TERMINAL BOARDS. No-load tap changers or terminal boards are provided in order that the voltage ratio may be changed when the transformer is disconnected from the lines, that is, when it is de-energized. Tap changers are widely used because they reduce the time required to change connections and reduce the danger of having wrong connection, since it is unnecessary to open the tank and change the position of connecting links under the oil. The operating handle may be brought out through the cover or through the side of the tank, the latter construction having the advantage that the tap position can be observed from the ground at any time. Provision for padlocking the tap changer in any position is commonly provided. To simplify the construction, small distribution transformers have their tap-changer handles located inside the case; they are accessible by removing a cover.

Where it is desirable to change taps while the transformer is energized, the equipment described under Load Ratio Control is used.

When the lines to which a transformer is connected are in a cable or conduit, the winding leads may be brought into a terminal compartment welded on or bolted to the side of the transformer case or to the cover. The conduit may be brought into the terminal compartment through a stuffing box or other connection. Wiping sleeves or potheads are provided for lead-covered cable. The terminal compartment may be air-filled or oil-filled, depending on the voltage and other requirements. It sometimes contains a disconnecting and grounding switch for switching operation, either when the transformer is de-energized or when it is excited. One typical kind of switch has three positions: an OPEN position in which the transformer is disconnected from the line, a CLOSED position in which the transformer is connected to the lines, and a GROUND position in which the line is disconnected from the transformer and grounded for safety. Interlocks are provided to prevent grounding a live line. Another type of switch has only two positions: an OPEN position in which the switch is open, and a GROUND position in which both the line and the transformer are grounded for safety. Again, interlocks are provided to prevent grounding a live line. Other common variations in the construction of terminal compartments include the provision of test bushings for d-c voltage tests on cable and provision for several cables or potheads in one compartment.

27. TESTING

USUAL TESTS. Tests are made on transformers to determine their electrical performance and their readiness for service. The following is a list of the usual commercial tests on transformers, arranged in the order in which they are sometimes made:

Resistance.
Ratio.
Polarity and voltage vector relations.
No-load loss and exciting current.
Impedance loss and impedance voltage.
Temperature tests (heat run).
Dielectric tests.

RESISTANCE MEASUREMENTS. Resistance is used for the calculation of the I^2R loss in the copper and for determining the winding temperature at the end of the temperature tests. The resistance of all windings is measured with the transformer either in oil or out of oil; in either case all parts of the transformer should be at the same temperature, which is measured by placing the thermometer bulbs as close to the windings as possible, in contact with the coils if the transformer is out of oil, and in contact with the oil surrounding the coils, as close to the coils as possible, if the transformer is in oil.

The resistance may be measured by either the drop of potential or the bridge method. The current for the measurement should not exceed 15 per cent of the continuous rated current for the winding in order not to change the winding temperature by heating. Because of the high inductance of transformers, some time is required for the current and voltage to reach their final values.

RATIO TESTS. Ratio tests are made on all taps to determine that the transformer, as built and connected, has the specified voltage ratios and that the ratio connections are given correctly on the nameplate. Ratio tests may be made by applying a voltage not exceeding the rated voltage at rated frequency to one winding and reading the voltage of both windings simultaneously by means of a voltmeter across each winding. (Potential transformers may be used, if necessary.) Tests should be made at not less than four voltages in approximately 10 per cent steps and repeated with the voltmeters interchanged to eliminate instrument errors.

A second method of making ratio tests is by comparison with a standard transformer of known ratio. The primary of the standard transformer and that of the transformer under test are connected in parallel to a suitable a-c supply. The secondary voltages can be compared by connecting a voltmeter across each secondary winding and taking simultaneous readings. If the ratio of both transformers is nominally the same, the secondary windings may also be connected in parallel, with one of the paralleling connections completed through a voltmeter. The difference in the secondary voltages will be indicated by the voltmeter reading.

A third method is to connect a resistance potentiometer across the high-voltage winding, exciting the transformer from a suitable a-c supply. One low-voltage lead is connected to the high-voltage lead of like polarity. The slide contact of the potentiometer is connected to the other low voltage lead through a detector and moved until the detector reads zero, at which point the transformer ratio is the ratio of the potentiometer resistances.

POLARITY AND VOLTAGE VECTOR DIAGRAM TESTS. These tests are made to determine that the transformer, as built and connected, has the specified polarity and voltage vector relations. These relations must be known when two or more transformers are to be connected in parallel or in a bank. The polarity of a single-phase transformer may be determined by connecting one high-voltage lead to its adjacent low-voltage lead and applying suitable a-c voltage to the terminals of the high-voltage winding. If the voltage between the other high-voltage lead and its adjacent low-voltage lead is less than the applied voltage, the polarity is subtractive; if greater, the polarity is additive. (The voltage between the other high-voltage lead and its adjacent low-voltage lead will be either the sum or the difference of the high-voltage winding voltage and the low-voltage winding voltage.)

A second test to determine the polarity of a single-phase transformer is by comparison with a transformer of known polarity and of the same ratio. Similarly marked leads are connected together, three sets of leads connected solidly and the fourth set through a voltmeter (or a fuse or a lamp). Suitable reduced voltage is then applied to one winding. If the voltmeter reading is practically zero, both transformers have the same polarity.

A third method is by inductive kick. Direct current is applied to the high-voltage winding, and a high-voltage d-c voltmeter is connected across this winding so as to obtain a small positive deflection. Each voltmeter lead is transferred from each high-voltage bushing to its adjacent low-voltage bushing, and the d-c circuit interrupted. If the voltmeter pointer swings in a positive direction, as before, the polarity is additive; if in a negative direction, the polarity is subtractive.

Voltage vector relations of a three-phase transformer can be determined by connecting the H_1 lead to the X_1 lead and applying a suitable three-phase reduced voltage to the high-voltage winding, measuring the voltages between the other leads, and constructing a voltage vector diagram from these readings.

NO-LOAD LOSS TESTS. Tests are made to determine whether a transformer meets its no-load loss guarantees and to compare the tested loss with the calculated loss. The excitation or no-load loss of a transformer is based on the rated sine-wave voltage of rated frequency being applied to one winding of a transformer with the other winding open-circuited. Because the maximum flux density in a transformer core varies with the wave shape for a given rms voltage, the no-load loss and exciting current also vary with the voltage wave shape. A peaked voltage wave (having a form factor greater than 1.11 for a sine-wave) often results from the flow of the peaked exciting current of a transformer through the impedance of the supply circuit. (The form factor is the ratio of the rms value of the wave to the average value.) A peaked voltage wave will produce a smaller excitation loss in a transformer than a sine-wave of the same rms voltage. Consequently, if the applied voltage wave differs from a sine-wave, correction for wave shape must be made, using one of the following: average voltage voltmeter, iron loss voltmeter, or standard core, the average voltage voltmeter being preferred because it is the most accurate and most generally applicable.

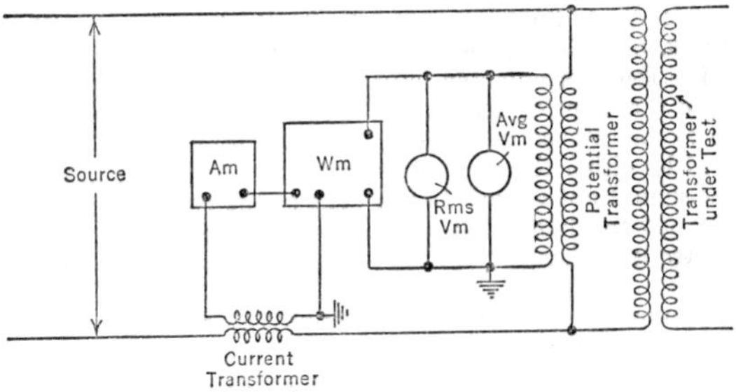

FIG. 66. Circuit for No-load Loss Tests

The circuit of Fig. 66 is used for no-load loss tests, either with or without the instrument transformers shown. When rated voltage, as read on the average voltage voltmeter, is applied to a transformer, the measured loss (corrected for tare due to instruments, etc.) will be very close to the no-load loss on a sine-wave basis. Greater accuracy may be obtained by correcting the measured loss in accordance with the following formula:

$$\text{Sine-wave no-load loss} = \frac{\text{Measured no-load loss}}{0.8 + 0.2k^2} \tag{10}$$

where k = rms test voltage/Rated voltage. (In this formula the per-unit hysteresis loss is taken as 0.8, and the per-unit eddy-current loss in the iron as 0.2; if the steel used has different relative values for the hysteresis and eddy-current loss components, they can be substituted for 0.8 and 0.2 in the formula.)

An iron loss voltmeter is used exactly like an average voltage voltmeter except that the measured loss without correction is used as the sine-wave loss. When a standard core which has been calibrated on a sine-wave of voltage is used, connected in parallel (usually through a potential transformer) with the transformer being tested, its loss is measured simultaneously with the loss of the transformer being tested, while rated voltage for the transformer is held on an rms meter. To obtain the sine-wave loss for the transformer the measured loss of the transformer under test is multiplied by the ratio of the sine-wave loss of the standard core to the measured loss of the standard core.

EXCITING CURRENT. The exciting current of the transformer is measured at the same time as the no-load loss. When rated voltage is held by the average voltage voltmeter, the exciting current for a peaked voltage wave will be slightly greater than the sine-wave voltage exciting current. In most cases this value is sufficiently accurate for practical purposes. For more accurate determination, measurements may be made with several different form factors of voltage, and a straight line drawn through points of current plotted against form factor to determine the sine-wave voltage value of exciting current at a form factor of 1.11.

An alternative method of determining exciting current is to measure the exciting current on a crest ammeter for voltages on an average voltage voltmeter of 100, 86.6, and 50 per cent of normal, designating the measured crest currents as I_1, I_2, and I_3, respectively. The exciting current corresponding to a sine-wave of voltage is equal to $\sqrt{0.333(0.5I_1^2 + I_2^2 + I_3^2)}$ for a single-phase transformer; for a three-phase transformer without large third harmonic voltages the exciting current on a sine-wave voltage basis is equal to $\sqrt{0.25I_1^2 + 0.338I_2^2}$. (Only the 100 and 86.6 per cent voltage measurements are necessary in this case.)

THE IMPEDANCE TEST. This test is made on a transformer to determine the impedance voltage and copper loss, as a check against guaranteed values, and to obtain values for efficiency and circuit calculations. The impedance test is made by short-circuiting one winding and applying sufficient reduced rated-frequency voltage to the other winding to circulate full load current in it. The circuit used for testing a single-phase transformer is shown in Fig. 67. The temperature of the windings should be measured immediately before and after the tests, and the average taken as the winding temperature. The method of measurement is the same as that described under resistance measurements. Measurements of current, voltage on an rms meter, and wattage loss are made with full-load current flowing. The voltage required to circulate full-load current is called the impedance voltage and is generally between 3 and 15 per cent of the rated voltage of the winding. Its value is corrected to 75 deg cent by a method to be described. The impedance wattage loss (copper loss) measured on this test is corrected to a standard temperature of 75 deg cent, using the following relations:

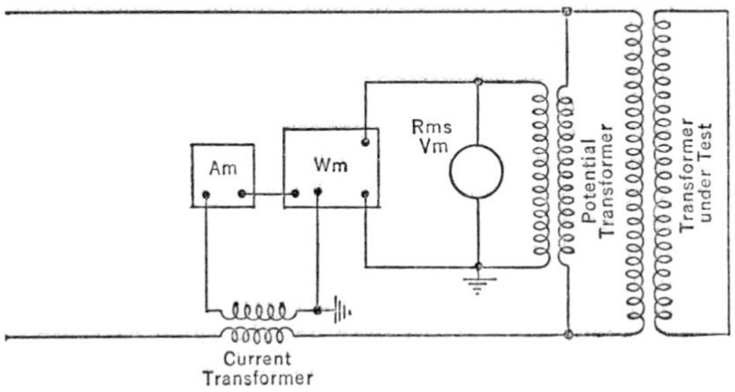

FIG. 67. Circuit for Impedance Test

The copper loss (impedance wattage) of a transformer is composed of two parts, I^2R loss due to the winding currents and the resistance of the windings, and the eddy-current loss. The I^2R loss may be calculated for any temperature by using the measured resistance and the following relation: Resistance of copper at temperature T_2 = Resistance at temperature $T_1 \times (234.5 + T_2)/(234.5 + T_1)$, or the corollary relation: I^2R loss at temperature T_2 = (I^2R loss at temperature T_1) $\times (234.5 + T_2)/(234.5 + T_1)$. The difference between the measured impedance loss and the I^2R loss is equal to the eddy-current loss when all losses are at the same temperature, which may be designated by T_1. The eddy-current loss at any other temperature T_2 is equal to: (Eddy loss at temperature T_1) $\times$ $(234.5 + T_1)/(234.5 + T_2)$. When these relations are used, both the I^2R loss and the eddy-current loss corresponding to a temperature of 75 deg cent can be calculated, and the copper loss at 75 deg cent determined as their sum.

The percentage resistance is equal to 100 $\times$ (Copper loss in kilowatts)/Rated kva. The percentage resistance and copper loss will both be at the same temperature.

The percentage impedance at any temperature T_1 equals 100 $\times$ Impedance volts at temperature T_1/Rated voltage of the winding. The impedance can be separated into its resistance and reactance components by the following formula: Percentage reactance $= \sqrt{(\text{Percentage impedance})^2 - (\text{Percentage resistance})^2}$, the impedance and resistance being at the same temperature.

The value of the reactance is unaffected by temperature; the value of the percentage resistance at 75 deg cent can be calculated and recombined with the percentage reactance to determine the percentage impedance at 75 deg cent, since Percentage impedance $= \sqrt{(\text{Percentage reactance})^2 + (\text{Percentage resistance})^2}$.

For three-phase transformers the averages of the test values of voltages and currents of the three phases are used for calculations from test data. For three-winding transformers three impedance measurements are required, one measurement for each pair of windings, with the third winding open-circuited.

ZERO-SEQUENCE IMPEDANCE. The zero-sequence impedance of transformers is used in short-circuit calculations of circuits. The zero-phase sequence impedance of a three-phase transformer or bank is the impedance to equal in-phase currents in each line or each winding. (The zero-sequence currents are not only numerically equal but are also in phase with each other.)

In many transformers the zero-sequence impedance is the same as the normal (positive-phase sequence) impedance, and separate tests to determine it are not necessary. When separate tests are made, the usual procedure is to apply single-phase power to the three line leads of the three-phase transformer, connected solidly together and the transformer neutral, circulating three times full-load current in the line to produce full-load current in each winding, and measuring the impedance wattage and voltage as in a normal impedance test. For three-winding transformers with two neutrals out, three tests are necessary, as with the normal impedance of a three-winding transformer, and a similar three-impedance circuit is necessary to represent the zero-sequence impedance of such a transformer.

The zero-sequence impedance of an auto-transformer to through zero-sequence currents in the lines is measured by connecting the three high-voltage leads together and to one terminal of the single-phase voltage source; the three low-voltage leads are connected together and to the other terminal of the source. The impedance is then measured in the usual way by circulating three times rated load current in the power supply lines to produce rated line current in each winding. Where the neutral of the auto-transformer is brought out, three measurements are required, as with a three-winding transformer, to determine the complete equivalent circuit.

TEMPERATURE TESTS. These tests are made to determine whether a transformer meets its temperature-rise guarantees. They are made with total losses equal to operating losses under the specified load conditions. A very small transformer may be tested by applying normal voltage to one winding, with an impedance connected across the secondary to give rated load on the transformer. Because of the power requirements, this method is not suitable for large transformers.

The most commonly used temperature test is the short-circuit test because it necessitates a minimum of connections and equipment and requires only one of the transformers to be tested. The connections used are the same as those for the impedance test, Fig. 67. With one winding short-circuited, sufficient a-c voltage is applied to the other winding to produce in the transformer wattage loss equal to the sum of the no-load excitation loss plus the copper loss at full load, until the oil temperature is constant. This is the ultimate oil temperature. The difference between this temperature and the ambient temperature of the cooling medium is the oil rise of the transformer. The current is then reduced to full-load current (at rated frequency) for approximately 1 hr to allow the copper-over-oil gradient to become constant; the oil temperature is then recorded, and the resistance of the windings measured. The copper temperature may be calculated from the hot resistance and the cold resistance and temperature previously measured by use of the formula: $T_2 = (234.5 + T_1) \times$ Winding resistance at temperature T_2/(Winding resistance at temperature T_1) $- 234.5$. The difference between this copper temperature and the simultaneous corresponding oil temperature is added to the ultimate oil rise previously determined to obtain the copper rise by resistance.

Since the temperature of the windings decreases as soon as the transformer is de-energized to make resistance measurements, a correction must be added to the measured temperature rise by resistance to obtain the true ultimate temperature rise. The time for taking the hot resistance must not exceed 4 min after shutdown. The correction may be made by taking a series of resistance measurements and extrapolating a curve of resistance versus time, back to zero time. In most cases a calculated correction is more convenient. A correction of 1 deg cent per min may be used for oil-immersed transformers in which the copper loss does not exceed 7 watts per pound of copper. When the wattage per pound of copper does not exceed 30, a correction in degrees centigrade, equal to the wattage per pound in the copper multiplied by the factor from Fig. 68, may be used.

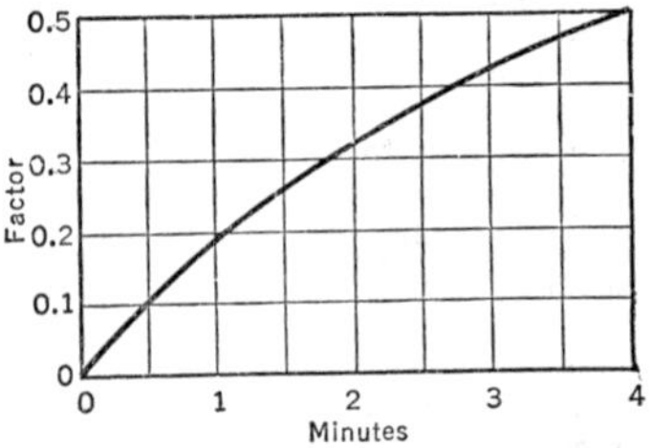

FIG. 68. Factor for Temperature Correction

When two duplicate transformers are available, they may be connected for temperature test by connecting them as for parallel operation to a source of normal voltage and frequency. The other windings are connected as for parallel operation, except that one of the paralleling connections is completed through a source of voltage for the impedance loss. Such a connection is shown in Fig. 69. The voltage of the source of impedance loss is adjusted to circulate full-load current at rated frequency or, if other than rated frequency is used, for the full-load copper loss. This connection closely approximates operating conditions, and the oil rise and copper rise are determined for both transformers after temperatures have become constant.

Among the less frequently used connections for temperature tests are the following, which apply principally to three-phase transformers but of course can also be used for a number of single-phase transformers connected as a three-phase bank:

1. Two three-phase transformers are connected as for parallel operation with normal three-phase excitation on one set of windings (high- or low-voltage of both transformers). The paralleling connections of the other windings are made through separate windings of a three-phase insulating transformer connected to a three-phase source, the voltage of which is adjusted to circulate load current.

2. The windings are connected delta high-voltage–delta low-voltage, and rated three-phase voltage and frequency are applied to one delta. The corner of the other delta is opened, and single-phase load current is circulated in the delta from a suitable source.

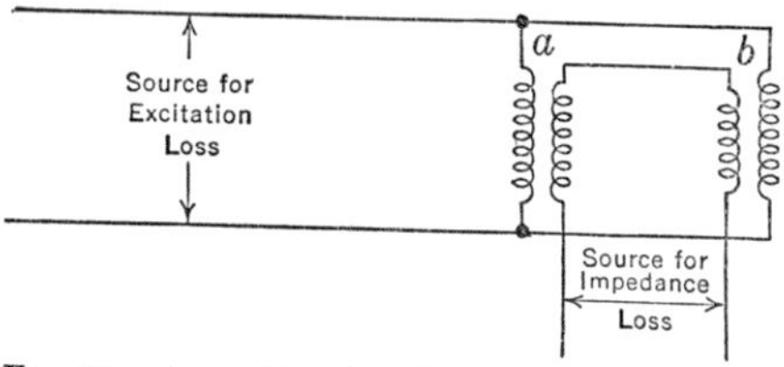

FIG. 69. Connections for Temperature Tests Using Duplicate Transformers[1]

3. Two-phase delta-Y transformers, with neutral out, are connected in parallel on the Y windings, the neutrals being connected together. The corners of the delta windings are opened, and a single-phase source of load current applied to the delta windings in series. The second and third methods generally introduce, in three-phase transformers, stray losses which do not exist in normal operation. In such cases normal load loss is used for the heat run instead of normal current in the windings.

DIELECTRIC TESTS. These tests are used to check the workmanship and insulation of transformers. The common dielectric tests applied to transformers are low-frequency applied potential test, low-frequency induced test, impulse tests.

Applied Potential Test. The applied potential test is usually made by applying 60-cycle voltage for 1 min between all terminals (connected solidly together) of the winding being tested and all other windings and ground (all terminals of the other windings are connected to ground). The voltage between the live terminals and ground is measured by a sphere spark gap connected in series with protective resistance.

Induced Potential Test. The induced potential test is normally the final dielectric test applied to a transformer. Usually it is made by applying twice normal voltage to one winding, using suitable precautions to prevent the voltage from any winding to ground from exceeding safe values. In order to limit the exciting current to a reasonable value, a frequency between 120 and 500 cycles is commonly used. The duration of the test is 1 min for 120 cycles or less and 7200 cycles for higher frequencies. Induced tests are commonly used to produce the desired test voltage from line terminals to ground on transformers with graded insulation (reduced insulation on the neutral end of the winding).

IMPULSE TESTS. When impulse tests are specified, they are made before the final induced test. Standard impulse tests consist of applying two chopped waves of specified characteristics, followed by one full wave, which reaches its specified crest value in approximately 1.5 μsec and drops to half voltage in not less than 40 μsec. In order to assist in failure detection, these tests are preceded by a reduced-voltage full wave, which is compared with the final full wave. The reduced-voltage full wave and the final full wave should have the same wave shape unless the waves are modified by the action of a protective device. Windings other than those being tested are protected by suitable means, such as grounding terminals either solidly or through resistances, gaps, or lightning arresters. Additional aids to failure detection are noise, smoke, bubbles in the oil, 60-cycle excitation, and cathode-ray oscillograph record of the current in the grounded end of the winding.

A surge generator of the type used to test transformers consists of a combination of capacitors with their associated equipment, which are charged through rectifiers to a desired potential and discharged through the transformer and series and shunt impedances which are used for wave-shape control and measurement purposes. The circuit of a typical surge generator is shown in Fig. 70, in which the individual capacitors *a*, which are insulated from each other by the sphere gaps *b*, are charged by d-c power supplied from the rectifier *c* through the resistors shown. After being charged, the capacitors are discharged through the gaps b, through the resistance R_1, to the external circuit applying voltage to the transformer being tested, the capacitance voltage divider $C_1 - C_2$ and the resistance divider $R_3 - R_4$.

The voltage dividers provide a voltage proportional to that on the transformer and suitable in magnitude to apply to the plates of a cathode-ray oscillograph, where the surge is recorded on a photographic film or observed on a fluorescent screen, usually as a curve of voltage versus time. The voltage across C_2 or R_4 is applied to the oscillograph plates. Any number of capacitors a may be used to obtain the desired voltage; the usual maximum voltage across each capacitor is 100,000 volts direct current. The resistors R_1 and R_2 are adjustable to control the shape of the voltage wave of the generator.

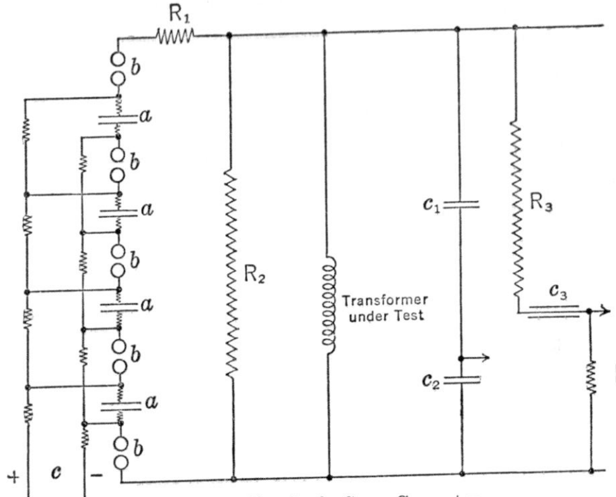

FIG. 70. Circuit of a Surge Generator

POWER FACTOR AND MEGOHMS. Miscellaneous insulation tests include power-factor tests and insulation-resistance tests. These tests are used to aid in determining the suitability of the winding for service. Windings which contain moisture generally have lower insulation resistance and higher power factor and capacity to ground. These tests are most useful when previous tests on the same and similar transformers are available for comparison.

Insulation resistance may be measured by a megger, an electronic indicator connected in series to a d-c source, or a bridge commonly constructed with an electronic indicating circuit. In order to secure comparable values of insulation resistance, voltage should be applied for the same length of time for each measurement.

Insulation power factor and capacity to ground are commonly measured with a Schering bridge from one winding, with all its terminals connected together, to all the other windings with their terminals grounded.

Since all the foregoing values vary with the temperature, readings to be compared should be taken at the same temperature. These tests have the advantage that they may be made at low voltage, thereby avoiding damage to the insulation. Moreover, compact portable equipment is available for making the tests. Consequently they are among the most common tests made in the field on equipment which has been in service or is to be placed in service.

28. INSTALLATION AND MAINTENANCE

Manufacturers furnish instructions for installing, operating, and maintaining their transformers, and reference should be made to these instructions. The following paragraphs summarize the general principles involved.

DRYOUT OF TRANSFORMERS. In order to have high dielectric strength, transformer insulation must be perfectly dry. After assembly in the factory, the core and coils are dried with hot air and vacuum in ovens used especially for drying. Drying in the field is most satisfactory and rapid when factory methods can be approximated. Transformers may be dried out either in an oven or other enclosure or in their own tanks by circulating air at approximately 90 deg cent through the coils and insulation. A desirable arrangements has ducts, leading from a fan and electrically heated resistance units, steam radiators, or other source of heat, to carry the heated air to the bottom of the core and coils. The air escapes from the enclosure through openings at the top. When the air contains considerable moisture, the rate of drying can be increased by passing the air before heating through a dryer containing a substance such as calcium chloride, silica gel, or activated alumina. When such a dryer is used, it may be desirable to recirculate the air after passing it through the dryer, instead of using fresh air.

The insulation resistance should be measured every 2 hr, starting at the beginning of the dryout and continuing until the dryout is completed. Thermocouples should be placed in contact with the coils or insulation, and a record kept of the temperatures every 2 hr. A curve of the insulation resistance and temperature versus time should be plotted to determine when the dryout can be discontinued safely. A comparison of the curve with that of similar successful dryouts is the best guide. In general, when the temperature is constant, dips in the curve of insulation resistance plotted against time indicate that moisture is coming from the interior to the surface of the insulation. The insulation resistance almost invariably decreases at first because of increase in temperature and then starts to increase. In the absence of other data, the dryout should continue until the resistance is constant for 24 hr at a value greater than that given by the formula: Megohms of one phase at 85 deg cent in air or 40 deg cent in oil = 15 × (Kilovolt test) × $\sqrt{\text{Cycles per second}}/\sqrt{\text{kva}}$, where "kilovolt test" is the low-frequency dielectric-test voltage of the winding tested in kilovolts, and "kva" is the rated kva of one phase of the transformer. The insulation resistance increases as the temperature decreases, approximately doubling for each 10-deg cent decrease in temperature.

Dryouts should be carefully supervised and constantly attended to prevent excessive temperatures or other hazards from damaging the transformer. Chemical fire extinguishers should be available in case of fire.

While part or all of the heat for drying out of oil may be supplied by circulating current in the windings, this arrangement greatly complicates the procedure without corresponding advantages; the most practicable method is by external heat as described.

When the dryout is to be made on a transformer which is nearly dry, so that only a small quantity of moisture is to be removed, dryout of the core and coils in their own tank under oil is often most convenient. Such a condition may arise when moisture leaks in during shipment, storage, or operation. To dry out a transformer in oil, the tank should be blanketed and covered to reduce heat dissipation and condensation, the top radiator valves shut off, and manholes opened or the cover raised if necessary to provide ventilation. One winding should be short-circuited, and sufficient a-c voltage applied to the other winding to circulate enough current to raise the oil temperature to 85 deg cent if the current required does not exceed 50 per cent of full-load current. If 50 per cent of full-load current does not produce an 85-deg oil temperature, the heat insulation must be increased on the tank. A less desirable alternative is to use higher current and a lower oil temperature; the dryout rate will be slower. In addition to allowing the moisture to escape from the surface of the oil at the top, the oil may be filtered to increase the dryout rate.

To determine whether the dryout can be stopped, filtering the oil is discontinued, but its temperature maintained. If the oil from the top and from the bottom of the tank tests above 26 kv with a standard 0.1-in. gap for seven consecutive tests 4 hr apart, all ventilating openings should be closed, and the run continued without filtering the oil. If, during the next 24 hr, the oil tests at 4-hr intervals show no decrease in dielectric strength, the short-circuit run may be discontinued. The transformer should be operated at two-thirds voltage with hot oil, the dielectric strength of the oil being tested and the oil filtered if necessary, until the dielectric strength is maintained at 26 kv for 24 hr without filtering. This operation should be repeated with full voltage; the dryout may then be discontinued, and the transformer placed in service. A test of the insulation resistance every 4 hr during the complete dryout is an aid in determining when the transformer is dry.

When considerable moisture is to be removed, as in a new or rebuilt transformer in which the insulation has absorbed moisture from the air during building, this method of dryout may be very slow, often requiring many weeks, in comparison to a much shorter time for dryout in air without oil.

INSPECTION. Before a transformer is filled with oil, it should be inspected to determine that it is clean and dry and that all valves, joints, etc., are tight. After its dielectric strength is checked, the oil should be pumped in through a filter press valve near the bottom of the tank (except for small units) in order to avoid filling the oil with air. After filling to the cold-oil level on the oil gage, at least 12 hr should be allowed for the oil to settle and air to escape. The dielectric strength of the oil should be checked before energizing the transformer. Some transformer tanks are made with sufficient mechanical strength to withstand the pressures incident to full vacuum on the tank, so that the transformer may be filled with oil under vacuum. For vacuum filling, full vacuum should be maintained on the tank for 2 hr to remove the air from the insulation; the oil is then allowed to flow in through a valve near the top of the tank, where the air will be removed as it flows in. Vacuum should be maintained for a short time after filling. The advantage of vacuum filling is that the air is removed from the oil and insulation at the time the transformer is filled with oil, without waiting for the oil to dissolve the air from the insulation.

INSTALLATION. Whenever feasible, transformers are shipped completely assembled in their own tanks filled with oil, since this procedure involves minimum work of installation. If necessary, the radiators, bushings, expansion tanks, or base may be removed to decrease the shipping dimensions. Sometimes shipping covers are used to meet shipping clearances, or the tank may be split horizontally, with the core and coils in the lower part, and the upper part removed for shipment.

Sometimes in order to reduce the weight to be handled, or to make it possible to ship the oil directly from the refinery, the oil used for testing is removed after test, and the core and coils shipped in their own case filled with nitrogen gas.

Exceptionally large transformers may have the core and coils shipped in a special shipping tank just large enough to hold these parts and filled with nitrogen gas. The permanent case may be split for shipment.

If a crane of sufficient capacity is available, it constitutes the easiest means of moving a transformer by the lifting hooks on the case. This is the method used at the factory. Spreaders or long slings should be used where necessary to clear bushings or avoid unnecessary strains. If a crane is not available, transformers are skidded or moved on rollers, care being taken not to damage the base or tip the transformer over. Only the lifting hooks and jack bosses provided for the purpose should be used for lifting.

Storage. When transformers are stored, they should be filled with oil and the oil-protective equipment installed and maintained just as if the transformers were in service. Indoor transformers should be stored under cover. Before transformers which have been idle for any length of time are placed in service, they should be inspected just as they were before shipping to determine that they are in operating condition.

Since self-cooled transformers depend upon air circulation to carry away the heat, adequate ventilation must be provided. These transformers should have at least 24 to 36 in. of clear air space around them for the cooling air to circulate. When placed indoors, the vaults must be well ventilated, with from 20 to 60 sq ft of air inlets near the bottom of the vault and corresponding air outlets at the top for each 1000 kva of transformer capacity.

Precautions upon Arrival. Before shipped transformers are removed from the railroad cars, they should be inspected for injury or evidence of rough handling, and a claim filed with the railroad if any damage is found.

To avoid condensation, transformers must not be opened when their temperature is below the surrounding air temperature. If brought indoors, they should be allowed to reach room temperature before the manhole covers are opened.

If the transformer is shipped in nitrogen, an analysis of the oxygen content of the gas should be made before opening the covers for inspection. If the oxygen content exceeds 3 per cent, moisture may have entered; if there are any signs of moisture, the transformer should be dried out. If after shipment it is necessary to store the transformer without filling with oil, the oxygen content of the gas should be checked at least once a week, with gas added as often as necessary to keep the oxygen content below 2 per cent and the pressure between 0 and 5 lb per sq in.

Before placing shipped transformers in service, manhole covers should be removed, and the transformer inspected for moisture, breakage, or other injury or movement of parts during shipment. All accessible bolts, nuts, and studs, including electrical connections, should be checked and tightened if necessary. The voltage connections should be checked. If the oil tests below 26 kv, it should be filtered. Sometimes transformers are untanked for a thorough inspection when they are received. Unless care is taken that moisture is not absorbed by the insulation during this inspection, dryout may be required, since insulation quickly absorbs moisture from the air. The water-cooling coils of water-cooled transformers should be tested for leaks with oil or air, preferably with a pressure of 80 to 100 lb per sq in., for approximately 1 hr. If the transformer is removed from its tank, water may be used for the pressure test of the cooling coils.

Grounding. The transformer tank should be solidly and permanently grounded, the grounding plate near the bottom of the tank being used for this purpose. Any winding leads that require grounding, as specified on the connection diagram, must be grounded.

Protection against Lightning and Surges. Protection against lightning should be provided, in the form of lightning arresters or rod gaps, connected at the transformer between live terminals of bushings and tank and ground, as short and direct leads as possible being used. Overhead ground wires on the transmission lines reduce the probability of direct strokes on the transformer.

Transformers are protected from damage due to voltage surges, such as lightning and switching surges, by connecting in parallel with the transformer winding from line terminal to the tank and ground some device such as a gap or arrester, which has a lower breakdown strength than the transformer. The dielectric strength of the transformer insulation is not a constant voltage for varying time but increases as the duration of the surge is decreased. The amount of increase is different for the various insulation arrangements used in the transformer. Unfortunately, the breakdown voltage of the available protective devices also increases as the duration of the surge voltage decreases.

Usually the breakdown voltage of the protective device increases faster than the transformer insulation strength increases. A sphere gap has the lowest rate of increase, but because the breakdown kilovoltage is reduced by dirt, moisture, or any roughness on the surface of the sphere, it is not a practical protective device. When connected directly from bushing to tank, **rod gaps** of proper length provide protection, except for the high kilovolt values which can occur with steep impulse waves. For a given steepness of wave, the shorter the gap, the less will be the kilovolt value at which the gap flashes over. Consequently, shorter gaps provide protection for steeper waves than longer gaps. If too short a rod gap were used in order to provide increased surge protection, it would flash over on switching surges which the transformer could withstand safely and thus interrupt service on the line. Table 5 gives the length of gap, which has approximately the same full wave impulse strength as the winding, in the column labeled "Test Gap," and the gap length, labeled "Protective Gap," which will seldom flash over on most lines as a result of switching surges.

Table 5. Rod Gaps

Rated Circuit Voltage, kilovolts	Test Gap	Protective Gap
1.2	0.8	0.4
5	2.2	1.1
8.66	3.3	1.7
15	4.5	2.3
25	7.1	5.3
34.5	10.2	7.6
46	13.5	10
69	20.6	15.5
92	27.5	20
115	34.7	26
138	42.1	32
161	49.0	37
196	60.0	45
230	70.4	53
287	88.0	66
345	105.6	79

Although rod gaps of the proper lengths offer considerable protection to a transformer, they do not entirely safeguard it against the high surge kilovolt values which may occur with very steep voltage waves caused by severe direct strokes of lightning close to the transformer.

Lightning arresters, which are used to protect transformers, may be thought of as air gaps in series with a varying resistance and the reactance of the leads. The gaps are constructed so that the breakdown kilovoltage increases slowly as the time is decreased. Consequently, when the surge current is small, they protect the transformer down to very short times. If the surge current is very large and increases rapidly, the voltage drop in the arrester circuit, as the surge current flows to ground through its resistance and reactance, may permit a voltage on the transformer in excess of its insulation strength. In order to minimize the reactance voltage drop due to rapidly increasing surge currents, the lead length of the arrester connections should be a minimum. In order to reduce the likelihood of direct strokes with high surge currents reaching the transformer, lines are often provided with overhead ground wires.

Arresters offer more effective protection than rod gaps because of their lower breakdown voltage, not only on long-time waves but also on short-time ones. They interrupt any power current flowing to ground, without circuit interruption, which occurs when a rod gap flashes over to protect a transformer. Rod gaps are sometimes used as back-up protection to other protective devices. Arresters are more expensive than gaps, but they sometimes permit the use on solidly grounded circuits of less expensive transformers, designed for lower test voltages, than do other protective means.

For the best protection, distribution transformers which have the center tap of the low-voltage winding grounded should have the high-voltage lightning arresters connected between the line and the tank; the center tap of the low-voltage winding should also be connected to the tank and to a good permanent ground. Gaps are usually used to protect the low-voltage winding. This connection reduces the possibility of having high surge voltages between windings and tank because of reactance drop in the leads, caused by surge currents.

In addition to arresters between lines and ground from the low-voltage terminals, auto-transformers may be protected by the use of by-pass arresters connected from the high-voltage bushing to the low-voltage bushing in each phase, with the arrester insulated from ground. Insulation from ground may be provided for a small low-voltage arrester by mounting it on the bushings; in high-voltage arresters the bypass arrester may be mounted on a separate porcelain insulator or on top of the low-voltage arrester.

Oil. Whenever feasible, the oil is shipped in the transformer. When shipped separately, it is sealed in metal cans or 55-gal steel drums; large quantities are shipped in tank cars. Drums and cans of oil should be stored indoors at uniform temperature. If outdoor storage is necessary, the containers should be covered for weather protection. Each drum should be placed on its side with the bung down at a 45-deg angle. The container should be kept tightly sealed at all times until the oil is to be used. To avoid condensation and contamination of oil by moisture, the oil temperature must not be below the air temperature when the container is opened. If brought into a warm room, the container must be allowed to stand until room temperature is reached.

Metal hose or pipe should be used for handling oil. Rubber is unsuitable because the oil will dissolve sulfur from the rubber, making the oil unfit for use.

Samples for dielectric test should consist of at least 1 pt of oil in a large-mouthed glass bottle with a cork stopper. The bottle should be cleaned with gasoline and dried before being used. Only after the oil has settled, test samples are taken from the bottom of the container or transformer by means of a sampling valve or a metal or glass thief cleaned with gasoline and dried before being used. When a sampling valve is used, a sufficient quantity must be drawn off to obtain oil from the main tank and not merely the sampling pipe. A visual inspection through the glass container should be made for water in the oil. If water is found, the cause should be determined, and the oil filtered to remove the water. Although water is not visible, it may be present in suspension, so a dielectric strength test should be made. Oil is tested for dielectric strength in a standard test cup having 1-in.-diameter disk electrodes spaced 0.1 in. apart. The oil and test cup should both be at room temperature, approximately 25 deg cent. The cleaned test cup should be rinsed out with a portion of the oil to be tested and then filled. After filling, $^1/_2$ to 1 min should be allowed for air bubbles to escape, and voltage should then be applied at about 3000 volts per second. Five breakdowns should be made on each filling, and the cup then emptied and refilled with a fresh sample. Tests should be repeated until the results on three fillings are consistent, to determine an average breakdown value, which is taken as the strength of the oil.

Oil should be filtered, if necessary, to bring its strength up to 26 kv before putting it in a transformer. If the strength of the oil in a transformer drops below 22 kv, the oil should be filtered. Preferably, oil should be drawn from one tank or container, filtered, and discharged into a clean dry tank. Another method is to draw the oil from a valve at the bottom of the tank and, after filtering, discharge it back into the tank through an upper filter press valve, taking care not to aerate it or to discharge moisture to the top of the tank, particularly if the transformer is in service.

If there is a large amount of water or dirt in the oil, the bottom oil should be filtered separately, returning it to the tank through a filter press connection near the bottom of the tank. If this cannot be done, it may be filtered into a separate tank and reconditioned before being returned to the main tank. After reconditioning the bottom oil, the remainder of the oil is filtered as described. Whatever the method, filtering should continue until the strength is well above 22 kv and preferably above 26 kv.

OPERATION. The Oil. The oil level of a transformer in service should be checked periodically to be sure that it never drops below a safe level. Samples of oil should be tested for dielectric strength before putting the transformer in service and, if the transformer is large, once each week for the first month. It is good practice to check the oil strength of transformers at 6-month intervals. If at any time the oil tests below 22,000 volts, it must be filtered. Water-cooled or forced oil-cooled transformers should not be operated, even at no load, for any extended time without the cooling equipment in operation. As a general rule, the oil temperature at no load should not exceed 80 deg cent for such operation.

Cooling Coils. Nearly all cooling water causes scale or sediment to form in cooling coils, clogging them and decreasing the rate of water flow and heat dissipation. Indications are high oil temperature and decreased water flow. If copper cooling coils of water-cooled transformers become clogged, they may usually be cleaned with a solution of hydrochloric acid with a specific gravity of 1.1, leaving it in the coils about 1 hr and repeating with a fresh solution as necessary.

Temperatures. In operation, both the load and the temperature indicators should be checked to make certain that the transformer is not operating at too high temperatures. Transformers have a nameplate kva rating, which is usually the load they will carry continuously without exceeding a 55-deg cent rise by resistance. The kva output which a transformer can deliver in service without causing undue deterioration of the insulation, however, may be more or less than the nameplate kva output, depending on operating conditions. The useful life of a transformer core and coils is the time required for the insulation to become too aged and brittle to be safe for service. Ageing of insulation takes place at all temperatures, the rate doubling for each 5- to 10-deg cent increase in operating temperature. Consequently the higher the operating temperature, the shorter will be the life of the transformer. It may be economical to use up transformer life at a faster rate in some applications than in others. Since the operating temperature depends in part upon the load carried, ageing is one of the principal factors to be considered in loading transformers.

Although the general relations between rate of ageing and temperature have been determined by tests on insulation samples, it is difficult to convert these relations into quantitative values for the length of life of transformers at various temperatures. Operating experience has shown that transformers built for 55-deg cent rise by resistance have had a satisfactory length of life of the order of 20 to 40 years. Many of these transformers, however, have been loaded as maximum rated devices so that their nameplate rating has seldom been exceeded. It is probable that continuous operation at rated kva, in a location where the average daily ambient temperature does not exceed 30 deg cent, will result in satisfactory length of life. Since the effect of ageing is cumulative, it is possible to operate at higher than normal temperatures for short times if these are balanced by longer times of operation at below normal temperatures. The amount of planned overloading of transformers can be determined by the economics of the case, balancing such factors as investment in transformers and associated equipment, rate of depreciation due to loss of life, emergency capacity, voltage regulation, and the losses in the transformer, which are higher in percentage at overloads. The most efficient operating load for a transformer is usually between three-fourths load and full load, so that it may not be profitable to operate transformers at overloads for long times, even though it is thermally permissible.

Table 6. Temperature Limits

Type of Load	Temperature (for Time Indicated), degrees centigrade		
	2 hr	8 hr	24 hr
Recurrent short-time loads (occurring once every 24 hr)	110	105	95
Short-time loads of infrequent occurrence	115	110	105

The first step in setting up safe overload values for various lengths of time for transformers is to determine safe temperatures for various lengths of time. The temperature limits in Table 6 have been suggested as maximum operating hot-spot temperatures for transformer windings for the lengths of time indicated. For overloads of very infrequent occurrence, higher temperatures are permissible. Higher temperatures are also permissible if there is no objection to sacrificing a portion of the life of the transformer.

Table 7. Oil-Temperature Limits

Type of Transformer	Permissible Load in Percentage of Full Load	Top Oil-temperature Limits for Transformers (for Time Indicated), degrees centigrade					
		Recurrent Loads			Emergency Loads		
		2 hr	8 hr	24 hr	2 hr	8 hr	24 hr
Self-cooled	100	85	80	70	90	85	80
	112	80	75	65	85	80	75
	123	75	70	60	80	75	70
	134	70	65	55	75	70	65
	144	65	60	50	70	65	60
Forced air-cooled	100	80	75	65	85	80	75
	110	75	70	60	80	75	70
	120	70	65	55	75	70	65
	129	65	60	50	70	65	60
	138	60	55	45	65	60	55

For a transformer in operation carrying a load, the hot-spot temperature for loading purposes may be obtained from the hot-spot winding temperature indicator if the transformer is equipped with one, or by reading the oil temperature on the transformer oil-temperature indicator and adding a calculated copper-over-oil gradient for the percentage load on the particular transformer.

If data on the hot-spot copper-over-oil gradient of the transformer to be overloaded are not available, the temperature limits of the oil given in Table 7 may be used to determine whether the load being carried is safe for the transformer. The oil-temperature limits for water-cooled transformers corresponding to any load are 5 deg cent lower than for self-cooled transformers.

The information given so far applies to determining whether the load a transformer is carrying is safe for the transformer; it is not directly applicable in planning for future loads, although it may be used to check whether operating loads are safe at any time.

In planning for future loads, advantage may be taken of lower ambient temperatures than 30 deg cent during a day or a year, of a load which varies during the hours of the day, and of the possibility of using up some of the useful life of the transformer at a faster than normal rate.

Effect of Ambient Temperature on Loading Transformers. A self-cooled transformer may be overloaded continuously 1 per cent of rated kva for each degree centigrade that the equivalent ambient temperature of the cooling air is below 30 deg cent (total load not to exceed 130 per cent load).

A water-cooled transformer may be overloaded continuously 1 per cent of rated kva for each degree centigrade that the equivalent ambient temperature of the cooling water is below 25 deg cent (total load not to exceed 125 per cent load). A forced air-cooled transformer may be overloaded continuously $^{3}/_{4}$ per cent of rated kva for each degree centigrade that the equivalent ambient temperature of the cooling air is below 30 deg cent (total load not to exceed 122.5 per cent load).

A forced oil-cooled transformer may be overloaded continuously 0.6 per cent of rated kva for each degree centigrade that the equivalent ambient temperature of the cooling medium is below 30 deg cent for an oil-to-air heat exchanger or below 25 deg cent for an oil-to-water heat exchanger (with maximum increase in loads of 18 and 15 per cent respectively).

For equivalent ambient temperatures above 25 deg cent for water and 30 deg cent for air, the continuous load on the transformer should be reduced 1.5 per cent of rated kva for each degree centigrade of excessive temperature for self-cooled and water-cooled transformers and 1 per cent for forced air-cooled and forced oil-cooled transformers. If the cooling air exceeds 50 deg cent or the cooling water exceeds 35 deg cent, values of load should be calculated on the basis of the characteristics of the individual transformer under consideration.

The equivalent ambient temperature for use in the foregoing rules may be on a daily, weekly, monthly, or yearly basis. On a daily basis the equivalent ambient temperature is equal to the average temperature. If other than a daily basis is used, the equivalent ambient temperature may be calculated for any locality on the basis of estimated temperatures during the time and the ageing principle that the rate of loss of life of insulation doubles for each 5- to 10-deg cent increase in temperature. The equivalent ambient temperature is the continuous temperature which would cause the same loss of life as the actual temperatures.

Effect of Load Variation. If the load on a transformer varies during the day, the maximum load may safely be in excess of the rated kva. The maximum load may be 0.5 per cent greater than rated kva for self- and water-cooled transformers and 0.4 per cent greater for forced air-cooled or forced oil-cooled transformers for each percentage that the daily load factor (ratio of average to maximum load) is below 100 per cent, with a maximum increase of 25 per cent and 20 per cent respectively, corresponding to 50 per cent load factor. The effects of ambient temperature and load factor may be calculated separately and added to determine their combined effect.

Another method of allowing for varying load is to specify the load which may be carried safely for a given time after a given continuous load. Table 8 gives typical data. Similar tables have been prepared for varying amounts of sacrifice of life and resulting life expectancy. Overloading information should be regarded with some degree of caution and checked against service experience before using extensively. Loads higher than those indicated may be used infrequently to meet emergency conditions under which it is permissible to use up some of the useful life of the transformer, on the basis either of accepting shorter total life in years or of compensating operation for long times with light loads, resulting in little loss of life.

Table 8. Loads for Short Periods

Load Duration, Hours	Previous Continuous Load for 24 Hr					
	Percentage of Full Load for Self-cooled Transformers			Percentage of Full Load for Forced Oil-cooled Transformers		
	90%	70%	50%	90%	70%	50%
0.5	159	177	189	136	147	150
1	140	154	160	124	131	134
2	124	133	137	114	118	121
4	112	117	119	109	110	110
8	106	108	108	105	106	106

When transformers are to be heavily overloaded, it should be noted that associated equipment, such as switches, circuit breakers, and current transformers, as well as bushings, leads, tap changers, or other transformer parts, may limit the safe load which a transformer can carry. When it is desired to carry heavy overloads on a transformer, consulting the manufacturer is good practice.

An effective means of providing for higher overloads is by the addition of supplemental cooling to existing transformers. Sometimes additional radiators may be added by welding new connections to the tank. Usually the increase in rating which can be secured by this means is less than 25 per cent, the increase being greatest for small transformers. Temporary or permanent fans can be mounted around the outside of a self-cooled transformer to blow air on its cooling surfaces. By supplying approximately 200 cu ft per min of cooling air per kw of loss at the overload, properly directed to the cooling surfaces, an increase in rating up to 33 per cent of rated load can be secured, the increase being greatest for large transformers and negligible for small transformers with plain tanks.

Forced oil-cooling equipment can be added to self-cooled transformers to increase the rating, sometimes as much as 67 per cent of rated load. In many cases forced oil-cooling equipment, consisting of an oil-circulating pump and a heat exchanger for cooling the oil, may be attached to the radiator valves supplied with the transformer by removing the self-cooled radiators and bolting on the forced oil-cooling equipment in their place.

Water-spray cooling may be used to increase the rating of a transformer by spraying or flowing water on the cooling surfaces of the tank or radiators. If the humidity of the air is low, spraying so that the air becomes saturated with moisture is most effective. When spraying or flowing is used, the cooling surfaces of the transformer should be kept wet. An increase in rating of as much as 60 per cent can be secured in some transformers. The disadvantages are the more rapid deterioration of the paint on the transformer and the wet conditions produced around the transformer. For these reasons water spraying is seldom used except for emergency loads.

Effect of Altitude. A transformer will carry less load for a given temperature rise at higher altitudes, because of the decreased convection cooling with the less dense air. At rated kva the temperature rise will be greater at higher altitudes. In most locations, however, the effect of the higher temperature rise at rated kva is offset by the lower ambient temperature of the higher altitudes. For example, if the transformer operating-temperature rise is 10 deg higher because of the altitude, and the equivalent ambient temperature is 20 deg (10 deg less than the normal base of 30 deg cent), no derating of the transformer because of the altitude is required.

No correction is made for altitudes up to 3300 ft inclusive. For altitudes more than 3300 feet above sea level, the correction is approximately equal to the following percentages of the temperatur- rise at the higher altitude, for each 330 feet the altitude exceeds 3300 feet.

Oil-immersed, self-cooled apparatus	0.4%
Dry-type apparatus	0.5%
Oil-immersed, forced air-cooled apparatus	0.6%
Dry-type, forced air-cooled apparatus	1 %
Oil-immersed, forced oil, forced air apparatus	0.6%

No correction is required when the cooling medium is water.

In addition to decreased cooling at higher altitudes, the effect of the decreased air density is to reduce the dielectric strength of the outer end of the bushings. The strength is proportional to the relative air density, which varies with altitude as shown in Fig. 33. Operating conditions may require the use of special high-altitude bushings on the transformer.

29. SELECTION OF TRANSFORMERS FOR A GIVEN APPLICATION

Transformers for a given application are commonly selected by preparing specifications covering the requirements. These specifications may be simple or elaborate, depending on the requirements of the application, but should include the following: frequency, phase, kva rating, high voltage, low voltage, type of cooling, reference to ASA standards, and any other mechanical or electrical requirements.

Single-phase transformers are commonly used to supply single-phase loads, where no three-phase power is required. For three-phase power either three single-phase transformers or one three-phase transformer may be used. The individual single-phase units are lighter and easier to handle, and a spare unit is less expensive. Moreover, in case of failure of one transformer, a spare can be used with the other two units or with some connections such as delta-delta. The other two units can be placed in service at reduced capacity, while the damaged unit is being repaired. Single-phase distribution transformers are lighter to handle and easier to mount on poles; usually they cost the same as a three-phase transformer of the same bank capacity.

Three-phase power transformers are less expensive than three single-phase power transformers of the same bank capacity in first cost and usually in installation expense. Because of shipping clearances, larger-bank kvas can be made with single-phase transformers than with three-phase transformers. Three-phase transformers have lower losses for the same bank kva. Because of the proved reliability of all power transformers, the tendency is toward increased use of the more complicated three-phase type.

The selection of the high-voltage and the low-voltage rating for the windings is governed by the proposed operating conditions of the circuits. The voltages must be suitable to take care of the maximum and minimum high voltages and low voltages which will occur in normal operation. Taps sufficient to keep the secondary voltage within desired limits are specified. They are usually placed on the high-voltage winding, particularly if the low-voltage winding is a heavy-current winding.

The kva rating selected for the transformer may be greater or less than the maximum kva load. Factors in the selection are present load, expected future loads, capitalization of losses, the possibility of operating the transformer above its nameplate rating in accordance with the guides for operation, and emergency capacity. Each of the several windings of a three- or four-winding transformer may have different kva ratings as determined by their loads. In some instances all three windings will carry full load simultaneously; in others the maximum sum of the loads on any two windings will not exceed the rating of the third winding. Different amounts of cooling are required for the two applications, resulting in a difference in the cost of the transformers.

The type of cooling selected for a transformer, that is, whether it is dry-type, self-cooled, water-cooled, forced air-cooled, or forced oil-cooled, depends upon an evaluation of such factors as first cost, taxes, capitalization of losses, maintenance, cost of cooling water, and the use of forced cooling for emergency or spare capacity. For purposes of comparison the total effective cost of each transformer may be calculated as the sum of: Cost price + (Kilowatt no-load loss × Capitalization per kilowatt for no-load loss) + (Kilowatt full-load loss × Capitalization per kilowatt for full-load loss) + Capitalization of water cost + Capitalization of maintenance cost + Capitalization of taxes.

Reference to ASA standards is a convenient method of taking care of standardized details, such as temperature rise and insulation tests. Practically all transformers are constructed in accordance with these standards.

Whenever the application involves other mechanical or electrical requirements, they should be specified. Typical special requirements include the following:

An auto-transformer connection.

Specification of the transformer connections such as delta high-voltage–star low-voltage, or any of the connections which are possible.

Bringing the neutral of a Y connection out of the case through an insulating bushing of specified kilovolt class.

Graded insulation on the transformer.

Insulation tests for the winding which are higher or lower than the usual test voltages.

Provision for future forced air or forced oil cooling.

Parallel operation with other transformers or impedance of certain definite values required.

No-load tap changer.

No-load tap changer with operating handle at ground level.

Motor-operated no-load tap changer.

Load ratio control equipment.

Current transformers on either or both windings for differential protection, relaying, or metering; ratios and accuracy required.

Winding temperature equipment.

Thermal relay protection.

Special arrangement of outlet bushings.

Special location of fittings.

Oil protective equipment, such as inert gas or expansion tank.

Dehydrating breather.

Wheels.

Single-circuit or multiple-circuit unit substation construction.

Terminal compartments.

Disconnecting and grounding switches.

Potheads.

Lightning arresters or provision for mounting them.

Oversized bushings for dirt or high altitudes.

Recording thermometer.

Lifting beam and slings.

30. TRANSFORMER APPLICATIONS

GENERAL PRINCIPLES. Standard transformers can be used for special applications in many cases, but often transformers built especially do the job better and more economically. For some applications standard transformers do not meet the requirements, and special transformers must be used. Special transformers are often named for the application; for example, transformers for use with mercury arc rectifiers are often called rectifier transformers. In this discussion only the characteristics which distinguish transformers for particular applications from transformers in general are included.

RECTIFIER TRANSFORMERS. In the most commonly used connections for rectifier transformers the d-c windings, which are the windings connected to the rectifier, carry current only a portion of the total time and usually in only one direction. The most common connections used for polyphase rectifier transformers are given in the section on connections. Similar connections are used for single-phase rectifiers. Rectifier transformers require special attention to the arrangement of the d-c windings on the core relative to each other and to the a-c windings in order to obtain the desired reactance, so that the core will not be magnetized with direct current, causing increased core loss and exciting current, and so that the current will divide equally among the anodes of the rectifier. In the event of an arc-back in the rectifier, short-circuit currents and resulting short-circuit stresses are much higher in the windings of rectifier transformers than in corresponding power transformers, because d-c short-circuit current flows in the d-c windings. Such short circuits occur more frequently on rectifiers than on the usual distribution or power transformers. Because of these stresses special mechanical bracing must be provided, and other precautions taken in regard to winding arrangements. Special means, such as blocking of anodes which feed into the fault and quick-opening circuit breakers, are used to clear the fault quickly and to limit the value of direct current in the d-c windings.

The temperature-load rating of a rectifier transformer is usually special to correspond to the rating of the rectifier with which it is used.

The common ratings are:

100 per cent rated load continuously: 55-deg cent rise.

200 per cent current for 1 min following 100 per cent load continuously: safe temperature.

The standard rating for industrial service:

125 per cent rated current for 2 hr following 100 per cent load continuously: 60-deg cent rise.

200 per cent rated current for 1 min following 100 per cent load continuously: safe temperature.

A standard rating for light-duty railway or mining service:

150 per cent rated current for 2 hr following 100 per cent load continuously: 60-deg cent rise.

200 per cent rated current for 1 min following 100 per cent load continuously: safe temperature.

The standard rating for medium-duty railway or mining service:

150 per cent rated current for 2 hr following 100 per cent load continuously: 60-deg cent rise.

300 per cent rated current for 1 min following 100 per cent load continuously: safe temperature.

The standard rating for heavy-duty railway service:

150 per cent rated current for 2 hr following 100 per cent load continuously: 60-deg cent rise.

300 per cent rated current for 5 min following 100 per cent load continuously: safe temperature.

SYNCHRONOUS CONVERTER TRANSFORMERS. Synchronous converter transformers are usually used to supply six-phase power to synchronous converters, one of the connections shown in the section on transformer connections being chosen. The load temperature ratings are usually special to correspond to the rating of the converter, the most common ratings being those listed under rectifier transformers. Compound-wound converters often are used with transformers constructed with higher than normal reactance, such as 15 per cent reactance, in order to obtain the desired voltage regulation on the converter as the load changes. The increase in reactance is usually obtained by placing iron laminations (with air gaps in series in the magnetic circuit) in the space between the high-voltage and low-voltage windings to increase the leakage flux. This increase in reactance is not effective in limiting the short-circuit current because the shunt iron in the high to low space saturates with high currents. The neutral connections of the transformer winding connected to the slip rings of the converter may be used as the neutral of a three-wire d-c system, for example, to provide 240 volts between the brushes on the commutator of the converter and 120 volts between either set of brushes and neutral.

D-C BALANCE COILS. Direct-current balance coils are auto-transformers used with three-wire d-c generators to provide three-wire d-c power and are connected to the slip rings provided on the d-c generator. If two slip rings are provided on the generator, the balance coil is a single-phase auto-transformer with a center tap, as shown at *a-b* of Fig. 71, and a ratio of 2 to 1. The a-c exciting current flows from *a* to *b*, since the generator through its slip rings imposes an a-c voltage of generator frequency across the winding terminals. The d-c load current flows, as indicated by the arrows, in opposite directions through the two halves of the coil. The d-c ampere-turns oppose each other and do not magnetize the iron core of the transformer. If three slip rings are provided on the generator, a three-phase zigzag-connected balance coil is used with the winding arrangement of Fig. 13(*g*) in order to avoid d-c magnetization of the core, such as would occur with a Y connection, causing increased core loss and exciting current. Except where dirt and dust conditions make the complete enclosure of an oil-immersed transformer desirable, d-c balance coils are usually of the dry type because the operating voltage is low to correspond to the voltage of the d-c generator, and the copper losses are low in order to limit the d-c voltage drop in the coil, which determines the unbalance in voltage between the two sides of the three-wire system when current flows in the neutral line.

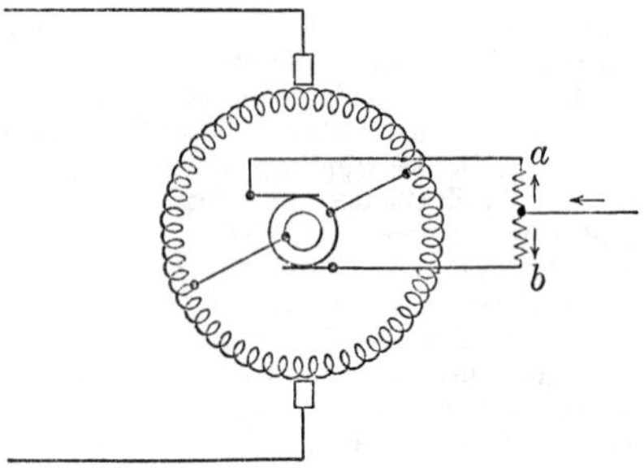

FIG. 71. Balance Coil for Three-wire System

TEST TRANSFORMERS. Test transformers are used to provide voltage and current of proper values for testing electrical apparatus. They may be constructed to furnish a multitude of voltages by taps and series-multiple connections. The voltage applied to the primary of testing transformers is usually adjusted by controlling the field of the generator which furnishes power to them or by means of step-type or induction regulators.

Testing transformers used to supply high voltages for testing, such as applied high-potential tests to transformers, are often connected in cascade in order to reduce their cost. Two transformers so connected are shown in Fig. 72. Part of the winding at the ungrounded end of transformer *a* is connected to the primary of transformer *b*, the case of which is insulated from ground by the insulating cylinders *c*. In this manner 1,000,000 volts can be obtained from two transformers, each of which is wound for 500,000 volts, one end of the winding being connected to its case and the other end insulated from the case for only 500,000 volts. The case of one unit is grounded, and the case of the other unit is insulated from ground for 500,000 volts.

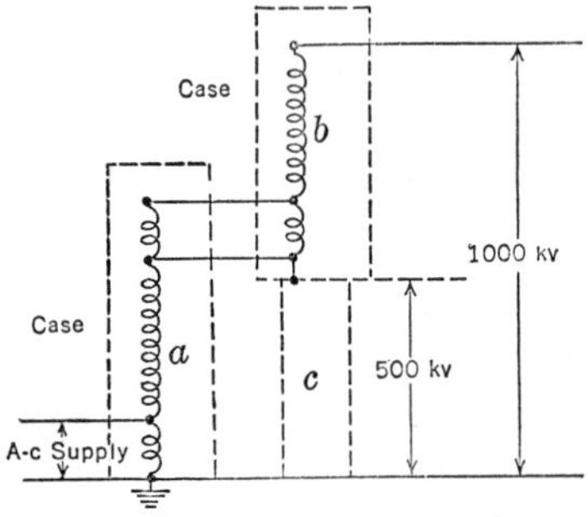

FIG. 72. Test Transformer in Cascade

NETWORK TRANSFORMERS. Network transformers have their secondary (low-voltage) windings connected in parallel to a network of low-voltage lines. A variety of circuit connections and control is used for secondary networks, some simple and others more complex. The following is a brief description of a typical network as it concerns transformers.

Typical for a three-phase four-wire secondary network is 120/208 Y low voltage, with a high-voltage supply for the network transformers of 13,800 volts. The primary windings of the transformers are usually connected to a number of radial feeders, each feeder being connected to its power source through a circuit breaker which is controlled by relays to trip out in case a fault occurs on the feeder.

The secondary windings are connected to the network through air circuit breakers, which with their controls are mounted on the transformers and called network protectors. Relays control the secondary circuit breaker so it closes only when the voltage relations are such that power will flow to the network from the feeder, and opens if the flow of power is from the network to the feeder, as it may be on a feeder fault. The secondary circuit breakers do not open on overload or a secondary fault, the intention being that the transformers in parallel will supply power to burn a secondary fault clear.

Network transformers are often made with wiping sleeves or potheads for connection to lead-covered cable. In such cases a disconnecting and grounding switch is installed inside the transformer or in a separate compartment on the side for isolating the transformer from the feeder for inspection and repair and for grounding the feeder for safety. Network transformers and their associated equipment for subway service are made submersible for satisfactory operation in water.

DISTRIBUTION TRANSFORMERS. Distribution transformers are 500 kva or less in rating. Standard kva, voltage, and current ratings have been established and are commonly used for distribution transformers. To reduce installation expense to a minimum small distribution transformers are made for pole mounting. In order to reduce size and weight, preferred oriented steel is commonly used in their construction. They are designed for low impedance so that they may carry high short-time overloads without excessive voltage drop. Connection to the primary (high-voltage) line is commonly through fuse cut-outs. The primary fuses will disconnect the transformer from the line if transformer failure occurs, and in some cases they are used for clearing secondary faults. They do not offer a practical means of protecting the transformer against overloads. For overload protection, some transformers are provided with thermal relay operated secondary breakers to interrupt the load before too high operating temperatures are reached.

GROUNDING TRANSFORMERS. Grounding transformers are transformers (usually connected zigzag or Y-delta) which establish a ground on a system to perform the following three main functions:

1. Holding the neutral shift within limits which will not permit voltages that can cause damage to connected equipment.
2. Permitting the circulation of unbalanced load current in the neutral of three-phase four-wire distribution systems.
3. Making possible the circulation of ground fault current for the operation of protective relays.

In the United States the neutral of a grounding transformer is most frequently connected solidly to ground, but it may alternatively be connected to ground through a

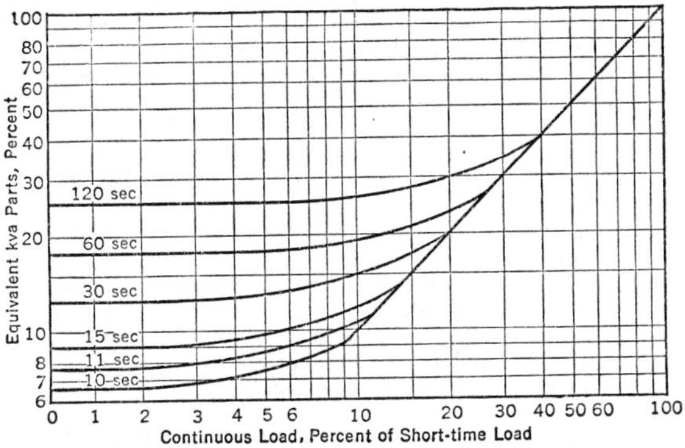

FIG. 73. Equivalent Kva Parts for a Delta-Y Grounding Transformer

resistance, a reactance, or a combination of both. By grounding the neutral of the Y winding, delta–Y-connected power transformers used to transmit power are frequently used as grounding transformers in addition. Such power transformers have a rating as a grounding transformer, with full short-circuit current flowing, of 2 to 5 sec, depending on the impedance. Circuits generally contain series impedance in addition to the transformer impedance, so that their actual permissible time of short circuit is greater than the value based on the transformer impedance alone.

Grounding transformers for apparatus fault protection usually have a short-circuit rating of 10 sec, dependence being placed on the protective equipment for the apparatus circuit to interrupt the ground current in less than 10 sec.

Grounding transformers for feeder fault protection usually have a short-circuit rating of 1 min, sometimes in addition to a continuous kva rating. Because grounding transformers carry fault current for only short times, the kva parts required are less than for continuous rating. The curves of Fig. 73 show the equivalent kva parts for a delta-Y grounding transformer in percentage of the short-time load for various amounts of continuous load expressed in percentage of the short-time load.

The equivalent kva parts for a zigzag grounding transformer are approximately 58 per cent of those for a delta-Y transformer. In most applications grounding can be secured by means of Y connections on the power-transmitting transformers on the system, but, where this is not feasible, separate grounding transformers are used.

STEP VOLTAGE REGULATORS AND REGULATING TRANSFORMERS. Step voltage regulators are auto-transformers provided with load ratio control equipment for regulating the voltage of the circuit to which they are connected, such as a feeder or bus. Typical circuits are shown in Fig. 74 for single-phase regulators. The windings are connected in star for three-phase regulators. Unlike other auto-transformers, their kva rating is not based on the total kva output as an auto-transformer but is the kva output

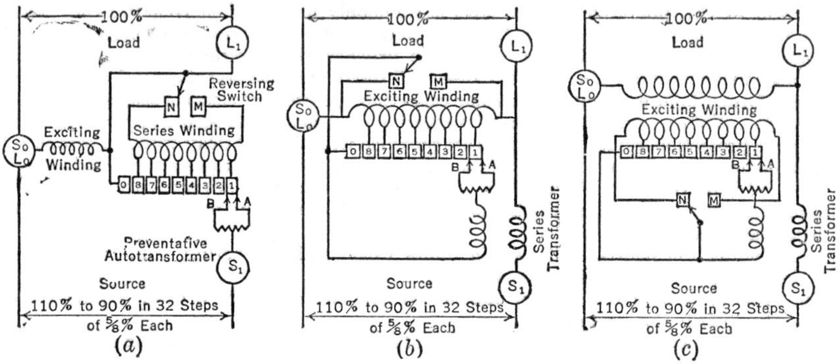

FIG. 74. Step Voltage Regulators

of the series winding based on its full-load voltage. Most regulators provide a voltage range of ± 10 per cent in thirty-two $^5/_8$ per cent steps. Typical mechanical construction is shown in Fig. 75. A typical control circuit is shown in Fig. 76.

To provide automatic operation of a voltage regulator, a voltage-control relay is excited by a voltage proportional to the output voltage, obtained from the secondary of a potential transformer connected across the output lines and mounted inside the tank of the regulator. Contacts on the voltage-control relay cause a time-delay relay or voltage integrator to operate in a "raise" direction if the output voltage is low and in a "lower" direction if the output voltage is high. After the voltage has remained high or low for approximately 1 min or less, depending on the time-delay relay setting, contacts on the time-delay relay close and cause the tap-changer motor to operate the mechanism so as to correct the voltage to the desired value. This balances the voltage-control relay, opening its contacts and causing the time-delay relay to reset. Compensation for line voltage drop (since the regulator may not be installed at the load center) is provided by passing a current (from a current transformer) proportional to the line current through an adjustable resistance and reactance proportional to the line impedance and connected in series with the voltage-control relay.

For regulating the voltage of a three-phase three-wire line, either a three-phase regulator connected in Y or two single-phase regulators connected in open delta are used. For a three-phase four-wire line, either a three-phase regulator connected in Y or three single-phase regulators connected in Y are used.

Above 250 kva per phase, transformers for regulating the voltage of circuits are called regulating transformers and, like other transformers, are given kva ratings corresponding to their output kva's. Unless specially designed for higher thermal and mechanical stresses, the short-circuit current through regulators and regulating transformers should be limited to twenty-five times the normal rated current by system impedance, including series reactors if necessary. Step voltage regulators and regulating transformers are made with winding construction like that of other transformers and are made to the same impulse-strength standards.

CONSTANT-CURRENT TRANSFORMERS. Constant-current transformers, sometimes called constant-current regulators, are used to maintain constant current in a street-lighting circuit, regardless of the number of lamps connected in series to the secondary winding. They are usually constructed for secondary currents of 6.6 amp, but current

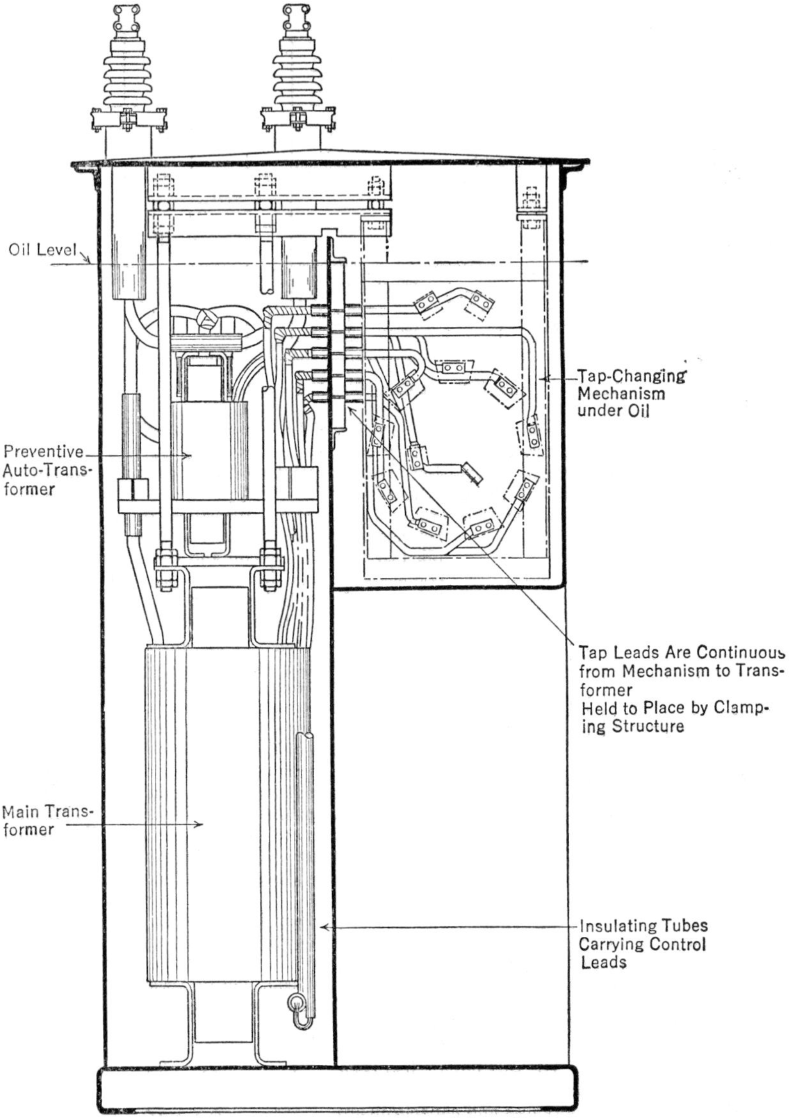

Fig. 75. Construction of Step Voltage Regulator'

ratings of 5.5, 7.5, 10, 15 and 20 amp are sometimes used, as well as kilowatt ratings from 2 to 70 kw, with corresponding secondary-load voltages up to 10,600 volts and secondary open-circuit voltages approximately 40 per cent higher. The primary voltage is usually 2400 or 4800 volts.

Two types of construction are used for constant-current regulators, the moving-coil type and the static type which employs capacitors. The moving-coil type commonly has a stationary primary coil and a moving secondary coil on a shell-type core, as shown in Fig. 77. The moving coil is suspended so that it is free to move vertically, either by chains passing over a pulley, as shown, for dry-type units, or by levers to which the coil is fastened, for oil-immersed units. In either case, counterweights are used to balance a portion of the

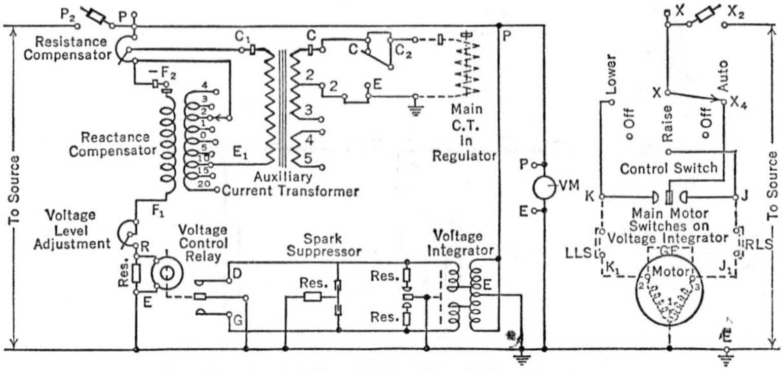

Fig. 76. Control Circuit Using Step Voltage Regulator[1]

weight of the coil, the remaining part of the coil weight being balanced by the force of repulsion between the primary and secondary coils, since they carry current in opposite directions. The counterweight is adjusted at the factory, so that mechanical balance is obtained when the rated current flows in the coils. With the type of core shown, the force of repulsion between the two coils is practically the same, regardless of the separation between the coils. When constant voltage is applied to the primary coil and the secondary coil is connected to a load, as the separation of the coils is increased, the leakage flux between the coils increases and the secondary terminal voltage decreases, decreasing the secondary current. The coil automatically adjusts its position until rated current flows, when mechanical balance obtains.

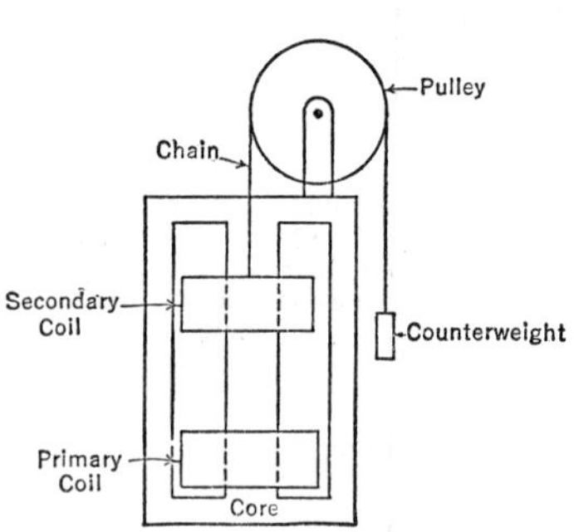

Fig. 77. Constant-current Transformer with Movable Coil

Moving-coil constant-current regulators supply loads of high power factor, usually above 99.5 per cent. Because of their high reactance, the power factor of the current drawn from the supply line is low, being lowest for partial loads, where the coil separation is greatest, and highest for full load, where the coils are closest together. The heating is greatest at low loads, and temperature run tests are made at the lowest guaranteed load, commonly 50 per cent load. These regulators are constructed so as to be capable of carrying any load between 50 per cent load and full load without exceeding their temperature-rise guarantees. Dry station-type regulators are constructed for 15-hr operation, and oil-immersed pole or subway-type regulators with class A insulation are constructed for 8-hr operation without exceeding 55-deg cent rise by resistance. Moving-coil regulators will maintain constant current in the secondary circuit even with considerable variation in the primary supply voltage.

The circuit diagram of a static constant-current regulator is shown in Fig. 78; this circuit has been called a monocyclic square circuit. The ohmic values of the individual inductances and capacitors are equal at the frequency at which the regulator is to be used; that is, they are tuned to resonance at the applied frequency. Such a circuit produces a secondary current which is proportional to the applied voltage, regardless of the ohmic value of the load on the secondary over a wide range. This circuit operates at high power factor. It requires a well-regulated supply voltage, since variations in supply voltage will cause proportional variations in the secondary current.

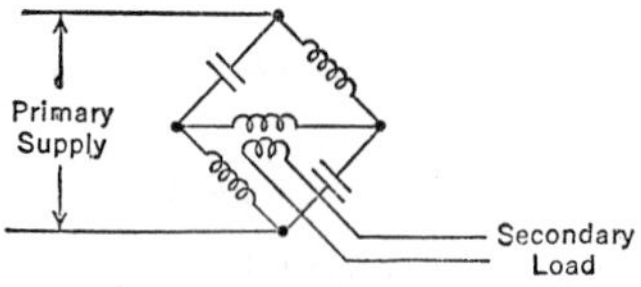

Fig. 78. Constant-current Transformer of the Static Type

SMALL DRY-TYPE TRANSFORMERS. A large number of small dry-type low-voltage transformers are used for such purposes as lighting and small power loads, bell ringing, signalling, radio transmitters and receivers, and neon signs. In the lower voltage classes, the insulation problem is one of maintaining mechanical separation between live parts and live parts and ground. The higher voltage classes are constructed like dry-type distribution transformers, although often for reduced test voltages. Dry-type transformers for installation as separate devices commonly conform to the standards of the Underwriters' Laboratories, Inc. Many small dry-type transformers are produced in large quantities as standardized articles. Small dry-type transformers are often constructed with extremely high impedance to obtain desired operating characteristics, such as ability to withstand sustained short circuit without injury.

31. BIBLIOGRAPHY

American Institute of Electrical Engineers, Neutral Grounding Devices, *Publ. No.* 32.
American Standards Association, American Standards for Transformers, Regulators, and Reactors, C 57.
Blume, L. F., G. Camilli, A. Boyajian, and V. M. Montsinger, *Transformer Engineering.* John Wiley (1938).
Dwight, H. B., *Electrical Coils and Conductors.* McGraw-Hill (1945).
Gibbs, J. B., *Transformer Principles and Practice.* McGraw-Hill (1937).
Grover, F. W., *Inductance Calculations.* Van Nostrand, N. Y. (1946).
Hill, L. H., *Transformers.* International textbooks (1937).
Hobson, J. E., and R. L. Witzke, Power Transformers in *Transmission and Distribution Reference Book,* Westinghouse Electric Manufacturing Company (1943).
Massachusetts Institute of Technology Staff, *Magnetic Circuits and Transformers.* John Wiley (1943).
National Electric Manufacturing Association, *Transformer Standards and Standards for Specialty Transformers.*
Puchstein, A. F., and T. C. Lloyd, *Alternating Current Machines.* John Wiley (1942).
Reed, E. G., *Transformer Construction and Operation.* McGraw-Hill (1927).
Rosslyn, J., *Power Transformers.* Chemical Publishing Company (1941).
Say, M. G., and E. N. Pink, *The Performance and Design of Alternating Current Machines.* Pitman, London (1936).
Stigant, S. A., and H. M. Lacey, *J. & P. Transformer Book.* Johnson & Phillips, London (1937).

See also files of *Electrical Engineering, Electric Light & Power, Electrical World, General Electric Review, Westinghouse Engineer, Allis-Chalmers Electrical Review,* and instruction books of transformer manufacturers.

SECTION 11

POWER RECTIFIERS AND INVERTERS

BY

HAROLD WINOGRAD

POWER RECTIFIERS AND INVERTERS

By Harold Winograd

1. RECTIFIER TYPES

A **rectifying device** is an elementary device which has the characteristic of conducting current effectively in only one direction. A **rectifier** is an integral assembly of one or more rectifying devices. Rectifiers are used for conversion of power from one form to another.

The following types of rectifiers have been used commercially:

1. Electronic-vacuum types.
   a. Mercury-arc, with pool cathode.
   b. Mercury-vapor, with thermionic cathode.
   c. Inert-gas, with thermionic cathode.
   d. High-vacuum, with thermionic cathode.
2. Metallic-plate types.
   a. Copper oxide.
   b. Selenium.
   c. Copper sulfide.
3. Electrolytic type.
4. Mechanical type.

From the point of view of installed kilowatt capacity and application for power-conversion purposes, the mercury-arc pool-cathode rectifier is the most important of these, and the greater part of this section will be devoted to it. The rectifier circuits and circuit theory used in connection with mercury-arc rectifiers can be applied also for the other types if their individual characteristics are taken into account.

2. ELECTRONIC-VACUUM TYPES OF RECTIFIERS

In its simplest form an electronic-vacuum type of rectifier consists of two electrodes, an **anode** and a **cathode**, which are enclosed in a vacuum envelope and are insulated from each other, as shown in Fig. 1. When connected in a circuit so that the anode is at a positive potential with respect to the cathode, and if the cathode is made to emit electrons, current is conducted through the rectifier by the flow of electrons from the cathode to the anode, because electrons are negatively charged particles and are drawn toward the positive electrode. For each ampere of current, 629 $\times 10^{16}$ electrons per second reach the anode surface. Because the direction of current flow in a circuit is considered by convention to be from the positive pole of the source to the negative pole, the direction of current flow in a rectifier is from anode to cathode, opposite to the direction of electron flow. The unidirectional conduction in a rectifier is made possible because the cathode emits electrons whereas the anode does not. If the polarity of the circuit in Fig. 1 should be reversed, the electrons would be repelled by the negative potential of the anode, and the flow of current would cease.

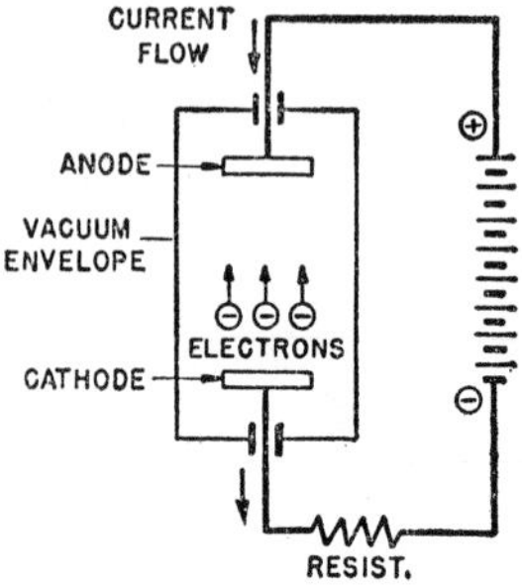

Fig. 1. Elementary Rectifying Device

In rectifiers with thermionic cathodes, electrons are emitted by heating the cathode to a high temperature with current passed through the cathode material or a heater. In a mercury-arc rectifier, electron emission takes place at **cathode spots** on the surface of the mercury-pool cathode. These are luminous spots, which have a current density of several thousand amperes per square centimeter, and move at random over the surface of the mercury. The production of electrons at a cathode spot is believed to be due to field emission, effected by a high-voltage gradient of the order of a million volts per centimeter produced by a cathode voltage drop of about 10 volts exerted over a very small space.

In a high-vacuum rectifier, containing no gas or vapor, the flow of electrons from cathode to anode is opposed by the repulsive force between electrons, called the **space charge.** A relatively high positive voltage between the anode and cathode of the order of several hundred volts is required to overcome the space charge, the voltage drop increasing with the space between electrodes and with the current density. For this reason high-vacuum rectifiers are used only for relatively small currents and high operating voltages.

In a gaseous rectifier, containing a gas or vapor at low pressure, electrons emitted from the cathode impart energy to atoms of the gas or vapor, causing some of them to lose an electron. An atom minus an electron has a positive charge and is called a **positive ion;** the process is called **ionization.** The effect of the positive ions is to neutralize the negative space charge of the electrons passing through them on the way to the anode, with the result that a relatively small voltage between the anode and cathode is required for the flow of large currents through the rectifier. This ionized path is called **plasma.**

3. MERCURY-ARC RECTIFIER

CONDUCTION OF CURRENT. A mercury-arc rectifying device, in its elementary form, is shown in Fig. 2. In addition to a mercury-pool cathode and an anode, usually made of graphite, it has an arc-starting electrode at the cathode. It may also have one or more **control grids** in front of the anode. A control grid is an electrode, which is provided with openings for the passage of electrons or ions, and to which a potential from an external source can be applied.

The starting electrode is used for establishing a cathode spot on the mercury surface. In some types of rectifiers this is accomplished by breaking contact between the starting electrode and the mercury, with current flowing from the starting electrode to the mercury. In other types of rectifiers a starting electrode made of refractory material is immersed in the mercury continuously, and a cathode spot is initiated by passing a pulse of current through the electrode to the mercury.

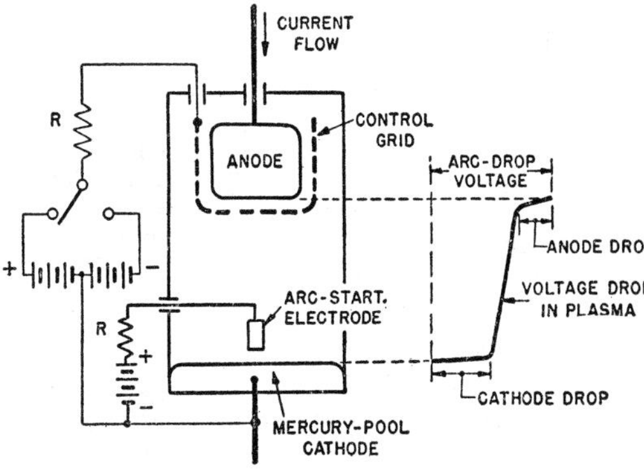

FIG. 2. Mercury-arc Rectifying Device

With a cathode spot established, the flow of current can start if the anode is connected to the cathode through a circuit which applies a positive voltage to the anode, and if the potential on the control grid (if present) permits the passage of electrons. The vacuum envelope contains mercury vapor at low pressure, which has evaporated from the cathode pool. When the flow of current starts, the mercury vapor is ionized by electrons emitted from the cathode, establishing the plasma, which provides a highly conductive path for the electrons. The plasma consists of positive ions, electrons, and neutral atoms of mercury, and these particles are in continuous random motion. The electrons move at much greater speed than the ions because of their smaller mass, about 1/370,000 of that of a mercury ion. Some positive ions combine with electrons to form neutral atoms, and new positive ions are produced by ionization of neutral atoms. There is a drift of electrons toward the anode, constituting the flow of current through the rectifier. Some of the electrons, ions, and neutral atoms strike the walls of the vacuum envelope, where the ions can become neutralized by absorbing electrons. The vapor condenses on the walls and returns to the cathode as liquid mercury, to replace mercury vaporized from the pool. In addition to vapor, small globules of mercury are ejected from the mercury pool by the concentrated heat and pressure developed at the cathode spots.

All the current at the anode consists of electrons. The greater part of the cathode current consists of electrons emitted at the cathode spots; the remainder, of positive ions entering its surface. A minimum current of 3 to 4 amp is required to maintain a stable cathode spot. A single cathode spot can carry about 15 to 30 amp. At higher current the cathode spot subdivides, creating additional cathode spots, the number depending on the magnitude of the current.

The potential distribution in the mercury arc is shown at the right of Fig. 2. The total voltage between the anode and cathode is called the **arc drop.** It consists of a drop of about 10 volts at the cathode, a drop of 0.05 to 0.2 volt per centimeter in the plasma, and a drop of approximately 5 volts at the anode. In most mercury-arc rectifiers the arc drop at rated current is in the range of 16 to 25 volts, depending on the length of the arc and other design factors and on the operating temperature. A typical arc-drop curve of a rectifying device as a function of current is shown in Fig. 3. The influence of temperature on the arc drop is caused by the effect of temperature on the mercury-vapor pressure. At low temperatures, when the vapor pressure and density are low, the density of positive ions is also lower; their compensating effect on the negative space charge of the electrons is therefore reduced, and the arc drop is increased.

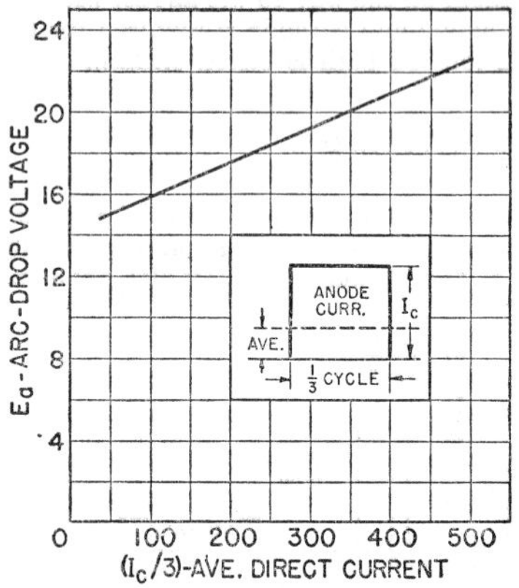

FIG. 3. Arc-drop Voltage of a Single-anode Rectifier Rated at Approximately 400 amp

When the flow of current through a rectifying device has started, it will continue as long as there is sufficient voltage in the circuit to overcome the arc drop and sufficient current at the cathode to maintain a stable cathode spot. The current is determined by the applied voltage in the circuit, less the arc drop, and by the circuit impedance. The flow of current will stop, if the net circuit voltage has dropped to zero or has become negative and the current has fallen to zero.

When the flow of current through the rectifying device has stopped, the positive ions of the plasma revert to neutral mercury atoms by combining with electrons. This process is called **deionization.** The time required for deionization depends on the ion density in the plasma and on the design of the rectifying device; it proceeds more rapidly in regions close to surfaces of conductive material. If the anode voltage becomes negative when the flow of current has stopped, positive ions will be attracted to the anode surface, resulting in a flow of a small negative current in the anode circuit. This current is called the **inverse current.**

CONTROL GRID. Before the anode starts conduction, the control grid can control the initiation of current flow through the rectifying device. If a negative voltage is applied to the grid with respect to the cathode, the electrons are repelled by the grid and are prevented from reaching the anode; no current will flow through the rectifying device, even when there is electron emission at the cathode and the anode is positive with respect to the cathode. When the grid is made positive with respect to the cathode, electrons are permitted to reach the anode, and current can flow through the rectifying device, provided there is electron emission at the cathode and the anode is positive to the cathode. The negative grid voltage required to prevent conduction depends on the design of the grid and the other elements of the rectifying device and on the magnitude of the anode-to-cathode voltage.

After the flow of current through the rectifying device has started, the grid has no influence on the current. The grid will collect electrons or ions from the plasma, depending on whether the applied grid voltage is positive or negative. In either case the resulting positive or negative current to the grid is limited by the resistance in its circuit. The applied grid voltage (minus or plus the arc drop between grid and cathode) is absorbed by the grid resistance, and the grid will be at approximately the potential of the surrounding plasma.

When the ion density in the plasma is reduced by deionization, after the anode current has stopped, a point is reached when the ion current collected by the grid drops below the value required by the circuit resistance for absorbing the applied negative voltage. The grid potential (applied voltage minus the drop in the resistance) then becomes increasingly negative with respect to the surrounding plasma, so that it repels electrons and attracts ions. The negative field of the grid is counteracted by the positive space charge of the ions flowing to it. At some distance from the grid surface, the two effects balance each other, and the negative grid potential has no effect beyond that distance. The space within this distance, which contains positive ions and is deficient in electrons, is called the **positive-ion sheath.** The thickness of the positive-ion sheath increases as the ion density is reduced, and the negative grid voltage is increased. When the positive-ion sheath has increased to such an extent that it covers the grid openings, electrons will not pass through these openings, and the grid can prevent the flow of current even when the anode becomes positive to the cathode.

A positive-ion sheath will form at the surface of a grid or any other electrode when immersed in the plasma and maintained at a negative potential, even when current is flowing between the anode and cathode. In the usual mercury-arc rectifying device, however, the magnitude of the grid resistance and the ion density during conduction are such that the positive-ion sheath is very thin or practically non-existent. It has practical significance only in connection with the ability of the grid to regain control of conduction after the current through the rectifying device has stopped.

RECTIFYING ELEMENT. From the preceding discussion of its physical properties, a mercury-arc rectifying device will permit the flow of current in a circuit to which it is connected, in the direction from anode to cathode, when there is electron emission at the cathode, when the anode is positive to the cathode, and when the grid (if present) is at such a potential that it permits the flow of electrons to the anode. The rectifying device is, in effect, a unidirectional current valve, or a switch that can be closed to permit flow of current in only one direction. Except for the effect of the arc-drop voltage, the magnitude and duration of current flow are determined by the circuit voltages and impedances. The usual methods of circuit analysis can therefore be applied to rectifier circuits. A rectifying device in an electric circuit is called a **rectifying element.**

4. POWER RECTIFIER

RECTIFICATION. The largest and best-known application of rectifying devices is for conversion of a-c power to d-c power. A rectifier unit used for that purpose is called a **power rectifier.** Its operation is illustrated in Fig. 4.

At the left of Fig. 4(*a*) a rectifying element T_1 is connected in series with a load resistance R_d to an a-c circuit having voltage e_1, represented by the sine wave at the top of Fig. 4(*a*). During the positive half-cycle of the voltage wave, current will flow through T_1 and R_d. The voltage across the resistance R_d is represented by the trace e_d, which is equal to e_1 minus the arc-drop voltage e_a. The current through the load resistance is represented by i_d, equal to e_d/R_d. Because of the valve action of the rectifying element, no current can flow during the negative half-cycle of e_1, and the d-c voltage and current traces during that part of the cycle lie along the zero lines. The average values of the d-c voltage and current, E_d and I_d, as measured by d-c permanent-magnet instruments, are represented by the straight lines in the figure.

In the rectifier circuit of Fig. 4(*b*) the power from the a-c circuit to the d-c circuit is transmitted through a **rectifier transformer.** Winding H_1-H_2 is connected to the a-c circuit and is called the **a-c winding.** Winding R_1-R_2 is connected to the d-c circuit, through the rectifying elements, and is called the **d-c winding.** The midpoint N_0 of the d-c winding is connected to the negative terminal of the d-c circuit. The a-c voltages e_1 and e_2, between the terminals of the d-c winding and its neutral point, are opposite in phase to each other and are represented by the sine waves at the top of Fig. 4(*b*). During one half-cycle, e_1 is positive, and current i_1, indicated by a solid arrow, flows from N_0 through the left side of the d-c winding, rectifying element T_1, d-c load resistance R_d, and back to N_0. The d-c voltage, e_d, is equal to e_1 minus the arc drop e_a in T_1, as shown in the second trace. During this half-cycle the anode of rectifying element T_2 is negative to its cathode, which is connected to the cathode of T_1 and is therefore at the potential e_d; consequently, no current can flow through T_2. During the second half-cycle, e_2 is positive to N_0, and current i_2 flows through the right side of the d-c winding, rectifying element T_2, and load resistance R_d, as indicated by the broken arrow. The current i_d flows in the same direction through the resistance R_d during both half-cycles. The average values of d-c voltage and current are indicated by E_d and I_d. While current is flowing in the two halves of the d-c winding, during alternate half-cycles, current is induced in the opposite direction in the a-c winding, through transformer action. This current, as indicated by the solid and broken arrows corresponding to the arrows in the d-c winding, changes direction during alternate half-cycles and is therefore an alternating current.

In Fig. 4(*c*) is shown a three-phase rectifier circuit. The three a-c voltages, e_1, e_2, e_3, are supplied by a winding, which may be the winding of a generator or the d-c winding of a transformer. The neutral N_0 is connected to the negative terminal of the load resistance R_d. The positive terminal of R_d is connected to the cathodes of the rectifying elements T_1, T_2, T_3. The a-c voltages are displaced from each other by 120 elec deg. This circuit differs from that of Fig. 4(*b*) in that the positive half-cycles of the successive voltage waves overlap by an angle of 60 deg. In this circuit the current flows from the phase which has the highest positive voltage. For example, between points *k* and *l*, e_1 has higher

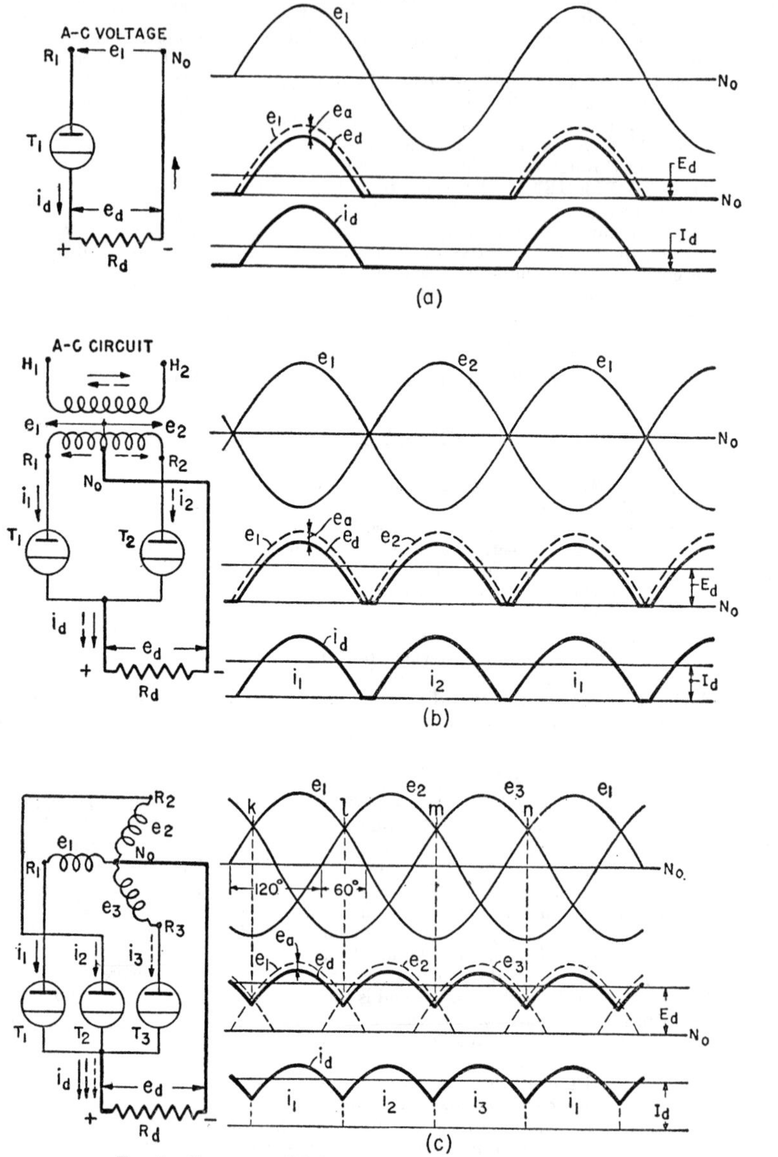

FIG. 4. Currents and Voltages of Elementary Rectifier Circuits

positive values than e_2 or e_3, and the current will flow from terminal R_1, through T_1 and R_d, to N_0. The d-c voltage e_d is equal to e_1 minus the arc drop e_a. The cathodes of T_1, T_2, and T_3, being connected together, are at the potential e_d. During this period current cannot flow through T_2 and T_3 because their anodes, which are at the potentials of e_2 and e_3, are negative with respect to their cathodes. Similarly, current will flow through T_2 between points *l* and *m*, and through T_3 between points *m* and *n*; the d-c voltage e_d is equal to the corresponding a-c voltages, e_2 and e_3, minus the arc drop. The direct current i_d follows the wave shape of the d-c voltage and consists of the currents i_1, i_2, and i_3, flowing in succession through T_1, T_2, and T_3.

The ripple in the d-c voltage or current wave, which is the difference between the instantaneous values of e_d or i_d and the average value E_d or I_d, is smaller in Fig. 4(*c*) than in Fig. 4(*b*). In general, as the number of rectifier phases is increased, the magnitude of the ripple decreases, and its frequency increases. The ripple in the current wave can be reduced by means of inductance in the d-c circuit, and if the inductance is sufficiently large, the current wave will approach the straight line of the average value.

In considering the operation of a power rectifier in this article, the reactances of the rectifier transformer and the a-c circuit were disregarded. In Article 5 the effects of these reactances are considered, and the current and voltage relations in the circuits of a power rectifier are derived.

5. RECTIFIER CIRCUIT THEORY

SYMBOLS. The following is a list of symbols used in this and later articles of this section.

α = phase control angle of retard.
$\cos\alpha$ = phase control ratio.
β = phase control angle of advance.
γ = inverter margin angle.
$\cos\delta$ = distortion component of power factor.
$\cos\phi$ = displacement component of power factor.
D_x = commutating reactance transformation constant.
e_1, e_2, etc. = instantaneous values of a-c voltages.
e_c = instantaneous values of commutating voltage.
e_d = instantaneous values of d-c voltage.
ΔE = regulation voltage, total.
E_a = arc-drop voltage.
E_d = average d-c voltage under load.
E_{d0} = theoretical d-c voltage at no load (neglecting arc drop).
E_{ii} = initial inverse voltage.
E_L = line-to-line voltage (rms) of a-c system.
E_n = line-to-neutral voltage (rms) of a-c system.
E_{pf} = peak forward voltage.
E_{pi} = peak inverse voltage.
E_r = direct-current voltage drop caused by resistance losses.
E_s = line-to-neutral voltage (rms) of transformer d-c winding.
E_x = direct-current voltage drop caused by commutating reactance.
f = frequency of a-c power system.
F_x = I_cX_c/E_s = commutating reactance factor.
i_1, i_2, etc. = instantaneous values of currents.
i_c = instantaneous values of commutating current.
I_c = direct current commutated in a set of commutating groups.
I_d = rectifier d-c load current (ave.).
I_h = total rms value of harmonic components of I_L.
I_L = alternating line current in rms amperes.
I_p = transformer a-c winding coil current (rms).
I_s = transformer d-c winding coil current (rms).
I_1 = fundamental component of I_L.
I_5, I_7, etc. = harmonic components of I_L.
I_{1P} = watt component of I_1.
I_{1Q} = reactive component of I_1.
K_L = a-c line volt-amperes.
P_d = theoretical d-c power in watts.
p = number of phases in a commutating group.
q = total number of rectifier phases.
u = commutating angle (angle of overlap).
X_c = line-to-neutral commutating reactance in ohms, referred to d-c winding of rectifier transformer.
X_{cn} = line-to-neutral commutating reactance in ohms referred to transformer a-c winding.
X_L = line-to-neutral reactance in ohms of a-c system.

Other symbols are explained when used.

COMMUTATION. In the rectifier circuit of Fig. 5(*a*) windings (which may be the d-c winding of a rectifier transformer) producing sinusoidal a-c voltages e_1, e_2, etc., which have an effective (rms) value E_s, are connected to a d-c load circuit through rectifying elements T_1, T_2, etc., similar to the circuit of Fig. 4(*c*). A reactance X_c is shown in series with each winding. This may represent the leakage reactance of a rectifier transformer and the reactance of the a-c line supplying the transformer. A large inductance L_d is connected in series with the d-c load to smooth out the ripples in the direct current, so that the d-c wave may be assumed to be a straight line. The a-c voltages constitute a

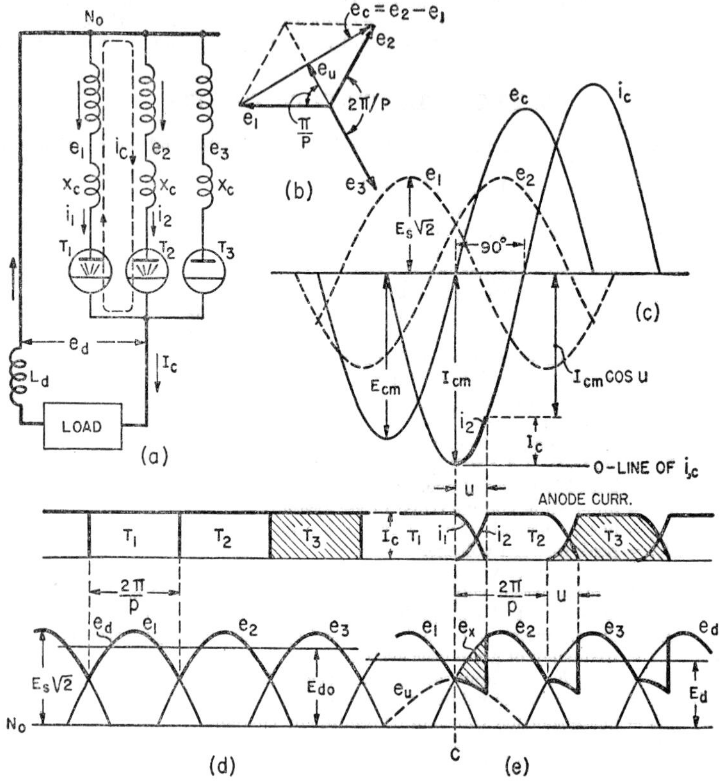

FIG. 5. Commutation of Current in a Rectifier Circuit

symmetrical polyphase system consisting of p phases, as represented in Fig. 5(*b*). The voltages are displaced from each other by an angle $2\pi/p$. Although the diagrams of Fig. 5 are shown for a three-phase system, the derivations and resultant equations are general and apply to any p-phase system in which p is any integer of 2 or larger. In these derivations the resistance losses in the rectifier transformer and the arc-drop losses in the rectifying elements will be disregarded. Their effects on the characteristics of rectifier circuits will be considered later. For the present these losses can be assumed to be included in the d-c power output to the load.

The positive half-cycles of the a-c voltages e_1, e_2, etc., are shown in Figs. 5(*d*) and 5(*e*). The direct current I_c will flow through the rectifying element having the highest a-c voltage. If the effect of the reactance X_c is disregarded, each rectifying element will conduct the current for a period $2\pi/p$, when its anode voltage is higher than the voltage of the other anodes, as shown in Fig. 5(*d*). The current I_c will be transferred from one rectifying element to the next at the point of intersection of their voltage waves. The anode current has a rectangular wave shape. Its effective (rms) value is

$$I_s = I_c/\sqrt{p} \tag{1}$$

The d-c voltage e_d is equal to the anode voltage of the conducting rectifying element, as shown in heavy outline in Fig. 5(*d*). Its average value is

$$E_{d0} = E_s\sqrt{2}\,\frac{\sin(\pi/p)}{\pi/p} \tag{2}$$

This condition, in which the effect of the transformer reactance is neglected, is approached at small direct currents, near no load. The voltage E_{d0} in eq. (2) is the theoretical no-load d-c voltage.

The effect of the reactance X_c on the transfer of current between successive rectifying elements is shown in Fig. 5(*e*). If it is assumed that rectifying element T_1 is carrying the

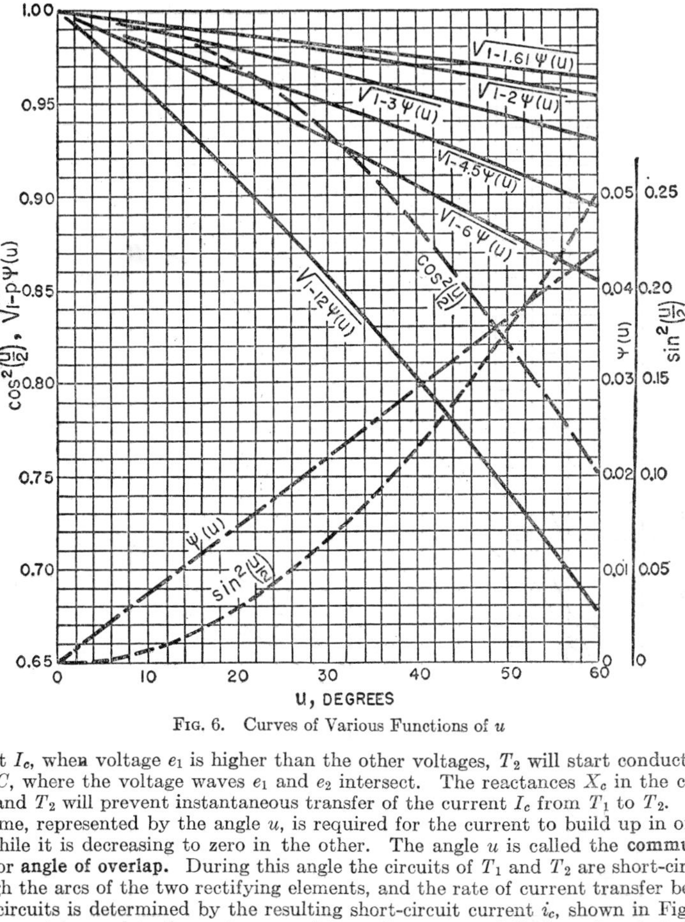

FIG. 6. Curves of Various Functions of *u*

current I_c, when voltage e_1 is higher than the other voltages, T_2 will start conducting at point *C*, where the voltage waves e_1 and e_2 intersect. The reactances X_c in the circuits of T_1 and T_2 will prevent instantaneous transfer of the current I_c from T_1 to T_2. A certain time, represented by the angle *u*, is required for the current to build up in one circuit while it is decreasing to zero in the other. The angle *u* is called the **commutating angle or angle of overlap.** During this angle the circuits of T_1 and T_2 are short-circuited through the arcs of the two rectifying elements, and the rate of current transfer between these circuits is determined by the resulting short-circuit current i_c, shown in Fig. 5(*a*). This current is equal to the vector sum of the voltages around the loop $(e_2 - e_1)$ divided by the sum of the reactances $(2X_c)$. The sum of the voltages, designated by e_c, is shown in the vector diagram of Fig. 5(*b*) and by the sine wave in Fig. 5(*c*). Since the resistances of the transformer windings are disregarded, the a-c component of i_c lags behind e_c by 90 deg. (The actual zero line for i_c is displaced from the zero line of its a-c component by its transient d-c component, which is equal to the instantaneous value of the a-c component at the start of the short-circuit.)

When T_2 starts conducting, its current i_2 is equal to i_c and follows its shape, as shown by the heavy line in Fig. 5(*c*), until it reaches the value of the direct current I_c being commutated. At that point the current i_1 of rectifying element T_1 reaches zero, and the commutation is completed. The rectifying action of T_1 will prevent i_1 from becoming negative; the flow of i_c will therefore stop, and the direct current I_c will flow through T_2 until the start of the next commutation period, at the point of intersection of e_2 and e_3.

Equation (3) for the commutating angle u can be derived readily from the circuit relations of Fig. 5.

$$\cos u = 1 - \frac{I_c X_c}{E_s \sqrt{2} \sin (\pi/p)} \tag{3}$$

Since the commutation of the direct current between two successive phases of a rectifier circuit is effected by e_c and i_c, they are called the commutating voltage and current,

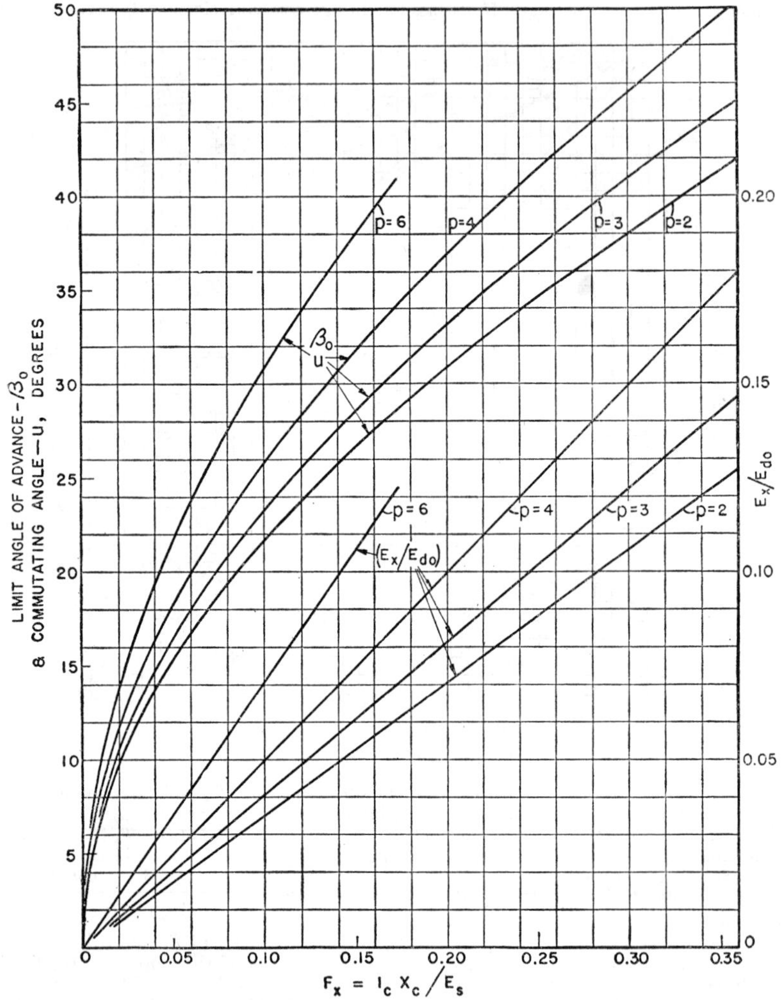

FIG. 7. Relation between Commutating Reactance Factor F_x and Other Quantities

respectively; X_c, which determines the relationship between e_c and i_c, is called the **commutating reactance.**

Taking into account the commutation period, each rectifying element conducts current during a period $(2\pi/p + u)$. The effective (rms) value of the anode current, as determined by integration, is

$$I_s = \frac{I_c}{\sqrt{p}} \sqrt{1 - p \cdot \psi(u)} \tag{4}$$

$$\psi(u) = \frac{(2 + \cos u)\sin u - (1 + 2\cos u)u}{2\pi(1 - \cos u)^2} \tag{5}$$

The function $\psi(u)$ and several factors involving it are given by curves in Fig. 6.

The d-c voltage during the commutation period is equal to the average of the voltages of the overlapping phases. The average of the voltages e_1 and e_2 is represented by the vector e_u in Fig. 5(*b*) and by its sine wave in Fig. 5(*e*). During the commutating angle u, the d-c voltage follows the trace of e_u. By comparing the d-c voltage wave in Fig. 5(*e*) with that in Fig. 5(*d*), it is seen that the effect of the commutating reactance is to reduce the area under the voltage trace by the shaded area e_x. The average value of this area, which is the drop in the d-c voltage produced by the commutating reactance, is

$$E_x = \frac{pI_cX_c}{2\pi} \tag{6}$$

The average value of the d-c voltage, taking into account this reduction, is

$$E_d = E_{d0} - E_x = E_s\sqrt{2}\,\frac{\sin(\pi/p)}{\pi/p} - \frac{pI_cX_c}{2\pi} \tag{7}$$

With the aid of eq. (3) this expression can be converted to the following:

$$E_d = E_s\sqrt{2}\,\frac{\sin(\pi/p)}{\pi/p}\cdot\cos^2\left(\frac{u}{2}\right) = E_{d0}\cdot\cos^2\left(\frac{u}{2}\right) \tag{8}$$

$$= E_{d0}\,\frac{(1 + \cos u)}{2} \tag{8a}$$

In Fig. 5(*a*) the group of rectifying elements and the alternating voltage-supply elements conductively connected to them, in which the direct current of the group is commutated between individual elements which conduct in succession, is ca led a **commutating group.** All polyphase rectifier circuits consist of one or more commutating groups, and the equations of this article can be applied by substituting for p the number of phases and for I_c the direct current of *one* commutating group.

In Fig. 7 are plotted curves of the commutating angle u and the ratio (E_x/E_{d0}) for several values of p, in function of (I_cX_c/E_s), which is designated by F_x, called the **commutating reactance factor.**

SIX-PHASE DOUBLE-Y RECTIFIER CIRCUIT. The six-phase double-Y rectifier circuit, shown in Fig. 8, has been the most generally used circuit for mercury-arc power rectifiers. The rectifier transformer windings are represented by their voltage vectors. The three-phase a-c winding may be connected in delta or Y. The d-c winding consists of two three-phase Y-connected groups, which are displaced from each other by 60 deg, with their neutral points interconnected by means of an iron-core reactor, which is called an **interphase** transformer. The midpoint of the interphase transformer constitutes the negative terminal of the rectifier unit. Each three-phase winding with the rectifier elements connected to it constitutes a commutating group. The two commutating groups operate in parallel, each carrying one-half of the total direct current.

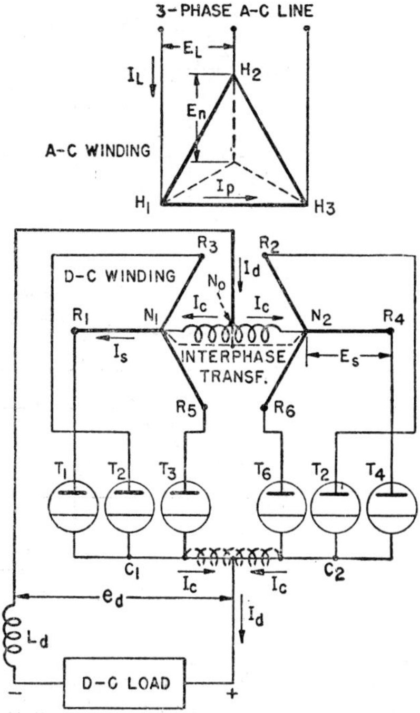

FIG. 8. Six-phase, Double-Y Rectifier Circuit

The interphase transformer absorbs the difference between the instantaneous d-c voltages of the two groups, as will be explained later. Instead of being connected between neutrals N_1 and N_2, the interphase transformer may be connected between the cathode terminals C_1 and C_2, as shown dotted in Fig. 8, without changing the operation of the circuit. The positive terminal of the load circuit will then be connected to the midpoint of the interphase transformer, and the neutrals N_1 and N_2 will be connected to the negative terminal N_0. This circuit connection will be assumed in the following analysis of the circuit.

In Fig. 9 are shown the traces of the voltages and currents in various parts of the circuit of Fig. 8, assuming a straight-line d-c wave (no ripples) and taking into account the effect of the commutating angle u.

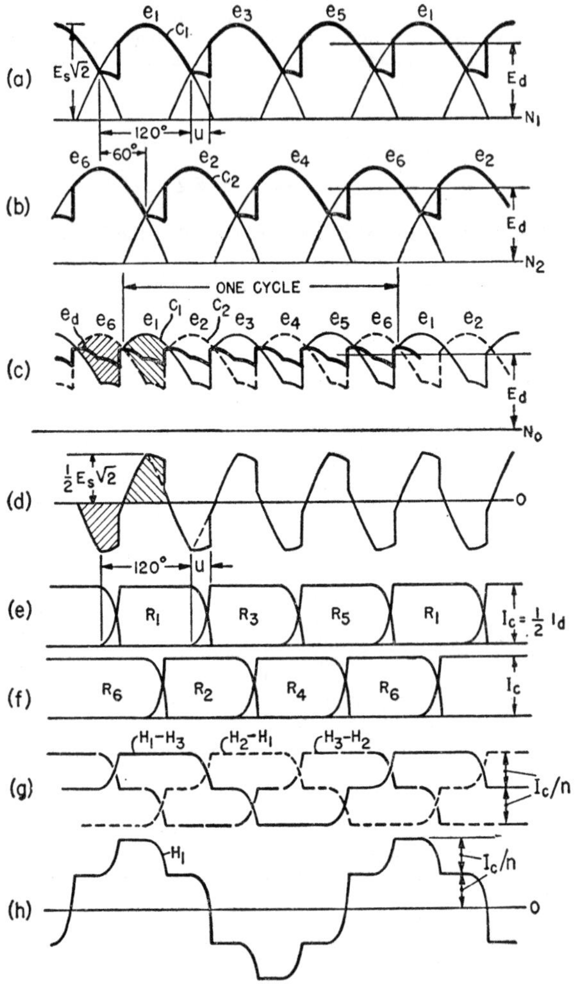

FIG. 9. Voltage and Current Waves of Circuit in Fig. 8

In Figs. 9(*a*) and (*b*) are shown the d-c voltages of the three-phase commutating groups N_1-C_1 and N_2-C_2, respectively. In each commutating group the direct current I_c is conducted by each phase in turn and is commutated between successive phases during the commutating angle u, as was explained in Fig. 5. The anode currents in the two commutating groups are shown in Figs. 9(*e*) and (*f*). The d-c voltages and anode currents of the two groups are displaced by 60 deg.

The d-c voltages of the two commutating groups are redrawn in Fig. 9(*c*) and are designated by C_1 and C_2. With the interphase transformer connected between the cathode terminals C_1 and C_2, the midpoint of the interphase transformer, which is the positive d-c terminal, is midway between the potentials of C_1 and C_2, as shown by the heavy line e_d. This is the trace of the d-c voltage and has the wave shape of a six-phase rectifier. The voltage across the interphase transformer is the difference between the potentials of points C_1 and C_2. The difference is shown shaded in Fig. 9(*c*) and is drawn in Fig. 9(*d*). The voltage across the interphase transformer is an alternating voltage having a frequency equal to three times the frequency of the a-c circuit, and its amplitude is equal to one-half the amplitude of the phase voltage of the transformer d-c winding.

The currents in the three phases of the transformer a-c winding are shown in Fig. 9(*g*). They were constructed from the anode currents shown in Figs. 9(*e*) and (*f*). The current in winding H_1-H_3, for example, is equal to the current in phase R_1 minus the current in phase R_4 (which are on the same transformer leg with winding H_1-H_3) divided by the turn ratio n between the a-c winding and the d-c winding. The a-c line current shown in Fig. 9(*h*) is constructed from the currents in the a-c winding. The line current to H_1 shown in Fig. 9(*h*) is equal to the current in winding H_1-H_3 minus the current in the winding H_2-H_1. The line currents to H_2 and H_3 have the same wave shape as that to H_1 but are displaced from it by 120 and 240 deg, respectively.

The value of p for each commutating group in Fig. 8 is equal to 3. By substituting this value for p in eq. (8), the average value of the d-c voltage for each commutating group is

$$E_d = 1.17E_s \cos^2 \frac{u}{2} \text{ volts} \tag{9}$$

This is also the average value of the six-phase d-c voltage shown in Fig. 9(*c*), which is the average of the d-c voltages of the two commutating groups. This equation takes into account the effect of the commutating angle resulting from the commutating reactance. The output d-c voltage is reduced further by the arc-drop voltage of the rectifying devices and the resistance drop in the rectifier transformer (see Article 7).

The formulas for determining the currents and other quantities of this rectifier circuit are given in Table 1B. The interphase transformer is rated in terms of the direct current (I_c) carried by its winding and the triple-frequency voltage across the winding terminals. The rms values of the interphase transformer voltage, as a function of the commutating reactance factor F_x, are given by the curves of Fig. 18. The curve for $\alpha = 0°$ corresponds with the conditions of Fig. 9.

The direct current I_c flows in opposite directions in the two halves of the interphase transformer winding and therefore produces no d-c magnetization in its core. The triple-frequency a-c voltage applied across the interphase transformer, as shown in Fig. 9(*d*), causes a triple-frequency alternating magnetizing current to flow. This current is supplied by the transformer d-c winding and flows between the two commutating groups through the rectifying elements. It must therefore flow in the forward direction (anode to cathode) through the rectifying elements of one group, and in reverse direction (cathode to anode) through the rectifying elements of the other group. The full value of the magnetizing current can flow in the reverse direction only if the direct current I_c through the rectifying elements is equal to or greater than the peak value of this a-c magnetizing current, so that the net current flow is in the forward direction. Below this value of direct current the interphase transformer is not fully magnetized; as the direct current approaches zero, the interphase transformer becomes practically ineffective, and the circuit of Fig. 8 behaves like a single commutating group of six phases, i.e., $p = 6$. The d-c load current at which the full magnetizing current of the interphase transformer can flow is called the **light transition load** and is usually of the order of 0.5 per cent to 1 per cent of the full-load current.

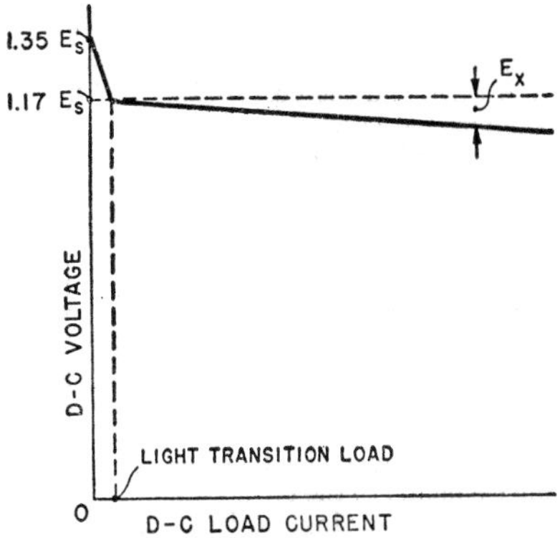

FIG. 10. No-load Voltage Rise due to Interphase Transformer

The theoretical no-load d-c voltage E_{d0} of a rectifier, as determined by eq. (2), is equal to $1.17E_s$ when p is 3, and $1.35E_s$ when p is 6. Because of the change in the mode of operation of the circuit of Fig. 8 below the light transition load, there is a steep increase in the d-c voltage between the light transition load and no load, as shown in Fig. 10.

DOUBLE-WAY CIRCUITS. In the rectifier circuits considered so far, the current between each terminal of the alternating voltage circuit, such as the d-c winding of the rectifier transformer, and the rectifying element (or elements) conductively connected to it flows only in one direction. This type of circuit is called a **single-way** rectifier circuit and is the most generally used circuit for mercury-arc power rectifiers.

There is another type of rectifier circuit, called the **double-way** circuit, in which the current between each terminal of the alternating voltage circuit and the rectifying elements conductively connected to it flows in both directions. This type of circuit has been used extensively for metallic rectifiers. It has been used also for some mercury-arc power-rectifier and power-inverter installations operating at high d-c voltages. Sometimes this circuit is called a bridge connection.

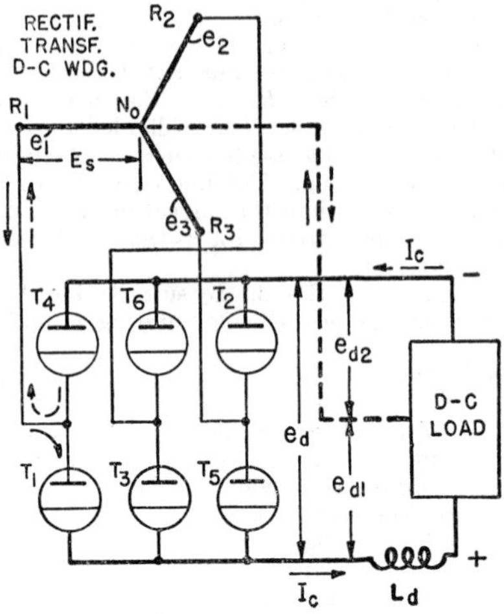

FIG. 11. Six-phase, Double-way Rectifier Circuit

In Fig. 11 is shown a six-phase double-way circuit. The three-phase Y-connected d-c winding of a rectifier transformer is connected to a group of six rectifying elements. Each terminal of the winding is connected to the anode of one rectifying element and the cathode of another. The d-c load circuit is connected between the cathodes of three rectifying elements and the anodes of the other three. To simplify the analysis of this circuit, a connection (broken line) is shown between the neutral point N_0 and the midpoint of the load circuit.

The sine waves of the three phase voltages, e_1, e_2, e_3, of the d-c winding are shown in Fig. 12(*a*). During the positive half-cycles the d-c winding and the rectifying elements T_1, T_3, and T_5 constitute a three-phase rectifier commutating group, and the current flows as indicated by the solid arrows in Fig. 11: from the d-c winding terminals, through the rectifying elements, a portion of the load circuit, to the neutral point N_0. The d-c voltage has the wave shape indicated by e_{d1} in Fig. 12(*a*). As shown in Fig. 12(*c*), each of the rectifying elements T_1, T_3, T_5 conducts the direct current I_c in turn, and the current is commutated between successive elements during the commutating angle u.

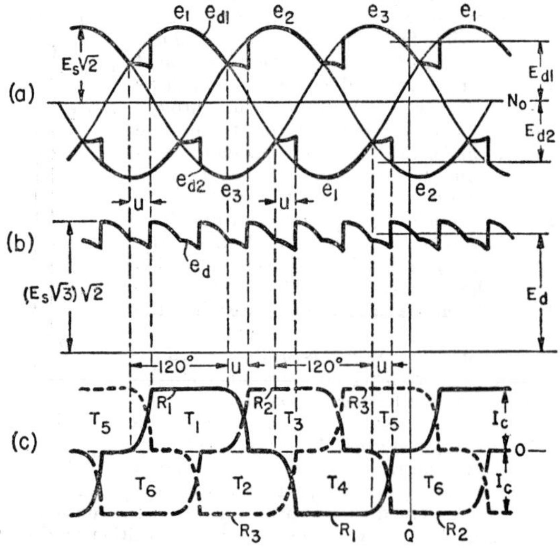

FIG. 12. Voltage and Current Waves of Circuit in Fig. 11

During the negative half-cycles of the phase voltages, the d-c winding and rectifying elements T_2, T_4, T_6 constitute another three-phase commutating group. The current in this circuit flows from N_0, through a part of the load circuit, the rectifying elements, to the terminals of the d-c winding, as indicated by the broken arrows in Fig. 11. The d-c voltage of this commutating group is designated by e_{d2} in Fig. 12(*a*). The anode currents of rectifying elements T_2, T_4, T_6 are shown in Fig. 12(*c*) below the zero line, because the direction of current in the d-c winding is opposite to that of the other rectifying elements, as indicated by the arrows in Fig. 11.

The direct currents of the two commutating groups flow in opposite directions in the neutral connection to N_0, as indicated by the arrows in Fig. 11. If the currents in the two groups are assumed equal, the current in the neutral connection will be zero, and this connection can be omitted without affecting the flow of current in the other parts of the circuit. The flow of current in the circuit of Fig. 11 at any instant selected at random, such as point Q in Fig. 12, will be as follows: $R_3 \rightarrow T_5 \rightarrow$ load circuit $\rightarrow T_6 \rightarrow R_2 \rightarrow N_0 \rightarrow R_3$. The load current flows through two phase windings and two rectifying elements in series.

The d-c voltage e_d is equal to the sum of the d-c voltages of the two commutating groups, e_{d1} and e_{d2}, and is shown in Fig. 12(b). This voltage has the wave shape of a six-phase rectifier, and its peak value is equal to the peak value of the phase-to-phase voltage of the transformer d-c winding. The value of p for each commutating group is 3. The average value of the d-c voltage of each commutating group, calculated by means of eq. (8), is

$$E_{d1} = E_{d2} = 1.17E_s \cos^2 \frac{u}{2} \tag{10}$$

The average value of the total d-c voltage is

$$E_d = E_{d1} + E_{d2} = 2.34E_s \cos^2 \frac{u}{2} \tag{11}$$

The wave shape of the current through a rectifying element is the same as for a single-way rectifier. The current at each terminal of the transformer d-c winding is equal to the sum of the currents of the rectifying elements to which it is connected and is an alternating current, as shown in Fig. 12(c). The formulas for determining the currents and other circuit quantities are given in Table 1B.

The rectifier transformer d-c winding of the six-phase double-way rectifier of Fig. 11 may be connected in delta instead of Y. The volt-ampere rating of a delta winding would be the same as for the Y-connected winding. The phase-to-neutral voltage E_s used in the equations would then be $1/\sqrt{3}$ times the delta (phase-to-phase) voltage. The transformer a-c winding may be connected in Y or delta.

RECTIFIER CIRCUITS. Tables 1A and 1B give the essential circuit relations and characteristics of the more commonly used rectifier circuits. The direct current is assumed to be a straight line. For d-c circuits in which the current departs considerably from this assumption, the values in these tables should be used with caution. This applies particularly to circuits 1 and 7. For clarity, the voltage and current waves are shown without the effect of commutation overlap. The formulas for currents and volt-ampere (VA) ratings are given on the basis of rectangular current waves but are followed by the correction factors for overlap (CFO). The equations for the d-c voltage E_d give the theoretical voltage, taking into account the voltage drop caused by the commutating reactance but disregarding the voltage drop in the rectifier and the voltage drop due to transformer resistance losses; these are considered in Article 7. The theoretical d-c power P_d is the product of E_d and I_d and is subject to the same limitations as E_d. All the values in these tables are given on the basis of zero phase control.

Circuit 3 can be made with a Y a-c winding and delta tertiary. The VA rating of the a-c winding is the same as the a-c line VA of circuit 3. The current in the tertiary winding is of triple frequency, having an amplitude and an rms value equal to $I_d/3n_t$; the CFO for the rms value is $\sqrt{1 - 12\psi(u)}$; n_t is the ratio of the tertiary winding voltage to the d-c winding phase voltage E_s. The VA rating of the tertiary winding is $0.74P_d \cdot \sqrt{1 - 12\psi(u)}/\cos^2 (u/2)$.

The direct current in the interphase transformer of circuit 5 is $I_d/2$. The voltage between terminals N_1-N_2 is of triple frequency, and its rms value is given in Fig. 18. Each of the interphase transformers N_1-N_2 and N_3-N_4 of circuit 6 has a direct current $I_d/4$, and their voltages are the same as for circuit 5; the middle interphase transformer has a direct current $I_d/2$; its voltage has an rms value of approximately $0.12E_s$; and its frequency is six times that of the a-c circuit voltage.

Data on other types of rectifier circuits are given in reference 2 of the Bibliography, Article 20.

Rectifier circuits consisting of a single commutating group, such as circuits 1, 2, 3, and 4, are called **simple** rectifier circuits. Circuits consisting of two or more commutating groups in parallel, such as circuits 5 and 6, are called **multiple** rectifier circuits.

The selection of a rectifier circuit for a specific application is governed by a number of factors, such as the circuit duty on the rectifier, the size and cost of the transformer, the efficiency and voltage regulation of the unit, and the wave shape of the currents and voltages. Circuit 5 is the most popular six-phase circuit for mercury-arc rectifiers used at d-c voltages below 3000 volts. For higher voltages circuit 4 is usually preferred. Circuit 6 has been the most widely used twelve-phase circuit for mercury-arc rectifiers. The double-way circuits 7 and 8 are used generally for metallic rectifiers. Circuit 8 has been used also for high-voltage sealed-tube rectifiers.

Table 1A. Circuit Relations and Characteristics of Rectifier Circuits

		1	2	3	4
CIRCUIT DIAGRAM		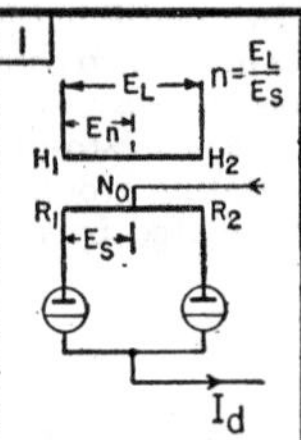	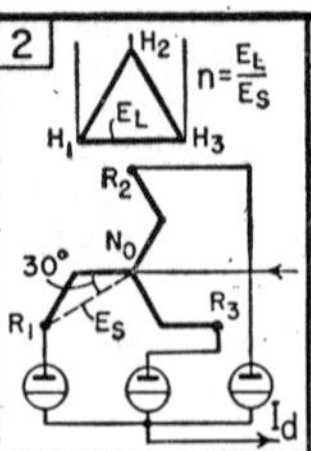	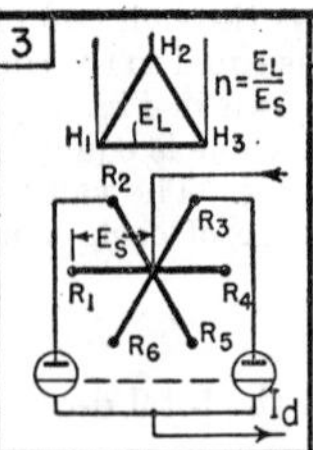	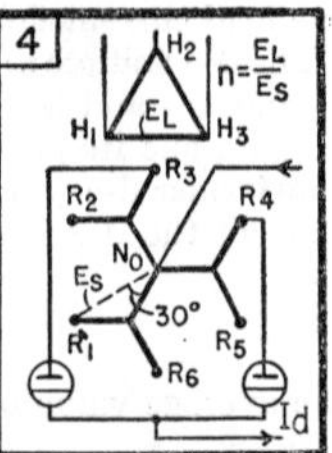
CIRCUIT NAME		DIAMETRIC (FULL WAVE)	DELTA, 3-PHASE ZIG-ZAG (WYE)	DELTA, SIX-PHASE, STAR	DELTA, SIX PHASE FORK (STAR)
CIRCUIT CHARACTERISTICS		$I_c = I_d$; $D_x = 1$ $p=2$; $q=2$	$I_c = I_d$; $D_x = 1$ $p=3$; $q=3$	$I_c = I_d$; $D_x = 3$ $p=6$; $q=6$	$I_c = I_d$; $D_x = 3$ $p=6$; $q=6$
D-C VOLTAGE	WAVE SHAPE	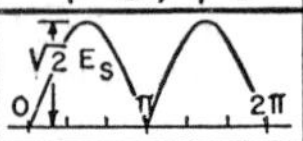	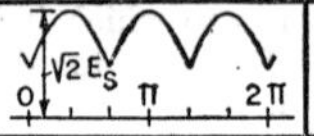	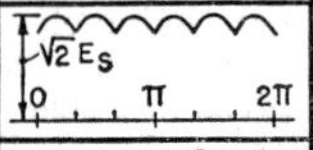	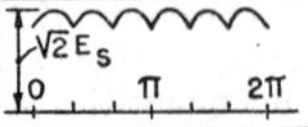
	E_d = AVE.	$0.9 E_s \cos^2 (u/2)$	$1.17 E_s \cos^2 (u/2)$	$1.35 E_s \cos^2 (u/2)$	$1.35 E_s \cos^2 (u/2)$
ANODE CURRENT	WAVE SHAPE	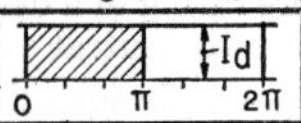	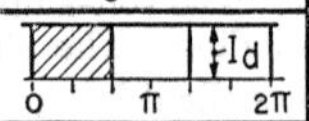	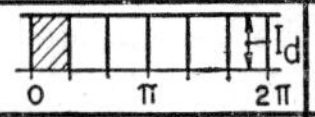	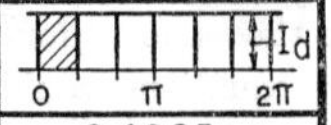
	RMS ×C FO	$\frac{0.707 I_d}{\sqrt{1-2\Psi(u)}}$	$\frac{0.577 I_d}{\sqrt{1-3\Psi(u)}}$	$\frac{0.408 I_d}{\sqrt{1-6\Psi(u)}}$	$\frac{0.408 I_d}{\sqrt{1-6\Psi(u)}}$
TRANSF. D-C WDG. CURRENT	WAVE SHAPE	SAME AS ANODE CURR.	SAME AS ANODE CURR.	SAME AS ANODE CURR.	CENTER WYE 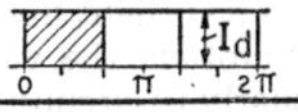
	I_s-(RMS) ×C FO	SAME AS ANODE CURR.	SAME AS ANODE CURR.	SAME AS ANODE CURR.	$\frac{0.577 I_d}{\sqrt{1-3\Psi(u)}}$
TRANSF. D-C WDG.	VA RATING ×C FO	$\frac{1.57 P_d}{\sqrt{1-2\Psi(u)}/\cos^2(u/2)}$	$\frac{1.71 P_d}{\sqrt{1-3\Psi(u)}/\cos^2(u/2)}$	$\frac{1.81 P_d}{\sqrt{1-6\Psi(u)}/\cos^2(u/2)}$	$\frac{1.79 P_d}{\sqrt{2}\sqrt{1-6\Psi(u)}+\sqrt{1-3\Psi(u)}}$ $(\sqrt{2}+1)\cos^2(u/2)$
TRANSF. A-C WDG. CURRENT	WAVE SHAPE	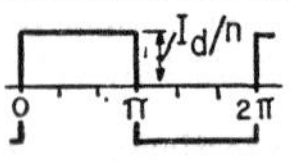	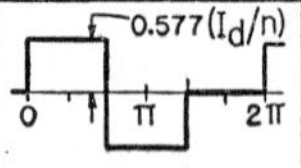	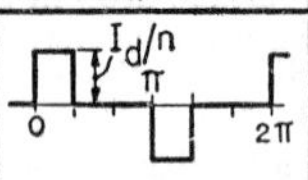	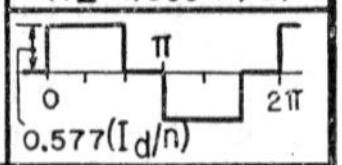
	I_p-(RMS) ×C FO	$\frac{I_d/n}{\sqrt{1-4\Psi(u)}}$	$\frac{0.471(I_d/n)}{\sqrt{1-4.5\Psi(u)}}$	$\frac{0.577(I_d/n)}{\sqrt{1-6\Psi(u)}}$	$\frac{0.471(I_d/n)}{\sqrt{1-3\Psi(u)}}$
TRANSF. A-C WDG.	VA RATING ×C FO	$\frac{1.11 P_d}{\sqrt{1-4\Psi(u)}/\cos^2(u/2)}$	$\frac{1.21 P_d}{\sqrt{1-4.5\Psi(u)}/\cos^2(u/2)}$	$\frac{1.28 P_d}{\sqrt{1-6\Psi(u)}/\cos^2(u/2)}$	$\frac{1.05 P_d}{\sqrt{1-3\Psi(u)}/\cos^2(u/2)}$
A-C LINE CURRENT	WAVE SHAPE	SAME AS A-C WINDING	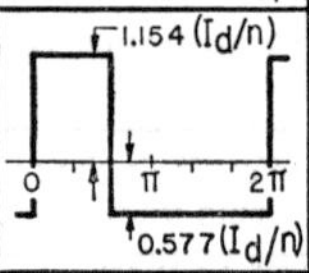	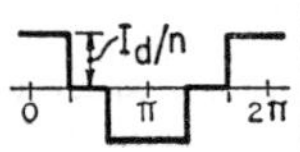	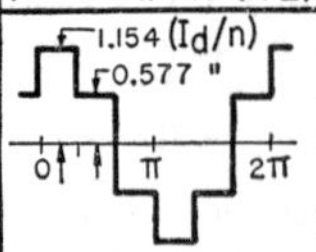
	I_L-(RMS) ×C FO	SAME AS A-C WINDING	$\frac{0.815(I_d/n)}{\sqrt{1-4.5\Psi(u)}}$	$\frac{0.815(I_d/n)}{\sqrt{1-3\Psi(u)}}$	$\frac{0.815(I_d/n)}{\sqrt{1-3\Psi(u)}}$
A-C LINE	VA RATING ×C FO	SAME AS A-C WINDING	SAME AS A-C WINDING	$\frac{1.05 P_d}{\sqrt{1-3\Psi(u)}/\cos^2(u/2)}$	SAME AS A-C WINDING
	POWER FACTOR	$\frac{0.9 \cos^2 (u/2)}{\sqrt{1-4\Psi(u)}}$	$\frac{0.826 \cos^2 (u/2)}{\sqrt{1-4.5\Psi(u)}}$	$\frac{0.955 \cos^2 (u/2)}{\sqrt{1-3\Psi(u)}}$	$\frac{0.955 \cos^2 (u/2)}{\sqrt{1-3\Psi(u)}}$

Table 1B. Circuit Relations and Characteristics of Rectifier Circuits—*Continued*

		5	6	7	8
CIRCUIT DIAGRAM		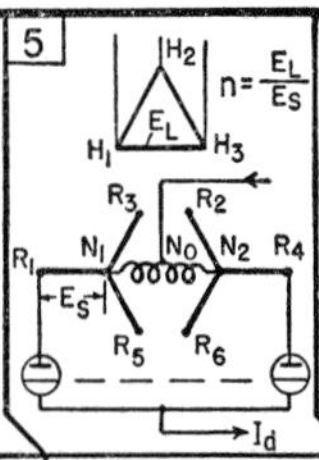	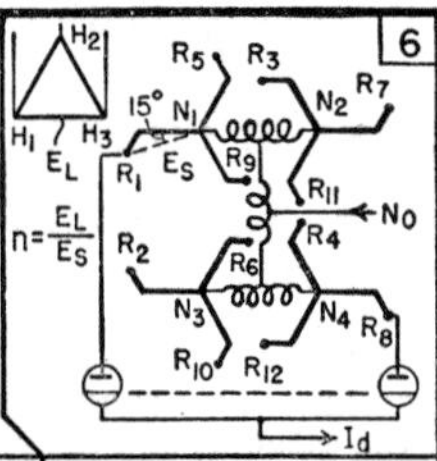	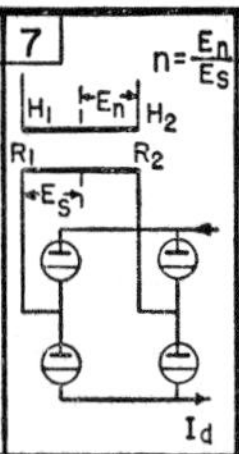	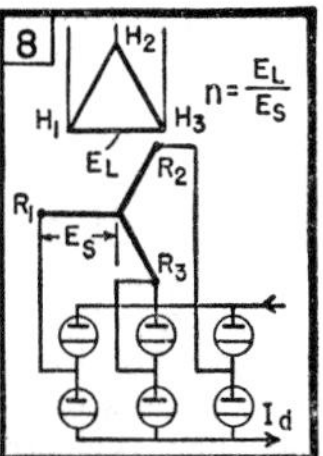
CIRCUIT NAME		DELTA, SIX-PHASE DOUBLE WYE	DELTA, TWELVE-PHASE QUADRUPLE ZIG-ZAG	DIAMETRIC DOUBLE-WAY	DELTA, SIX-PHASE WYE, DOUBLE-WAY
CIRCUIT CHARACTERISTICS		$I_c = I_d/2$; $D_x = 1$; $p = 3$; $q = 6$	$I_c = I_d/4$; $D_x = 1$; $p = 3$; $q = 12$	$I_c = I_d$; $D_x = 1$; $p = 2$; $q = 2$	$I_c = I_d$; $D_x = 3.73$; $p = 3$; $q = 6$
D-C VOLTAGE	WAVE SHAPE	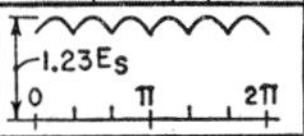	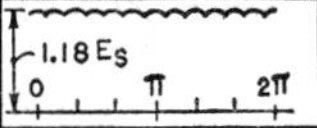	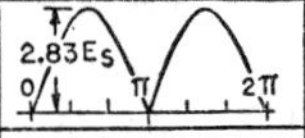	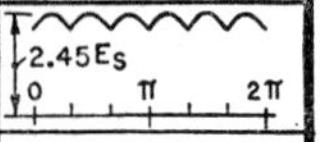
	E_d = AVE.	$1.17 E_s \cos^2(u/2)$	$1.17 E_s \cos^2(u/2)$	$1.8 E_s \cos^2(u/2)$	$2.34 E_s \cos^2(u/2)$
ANODE CURRENT	WAVE SHAPE	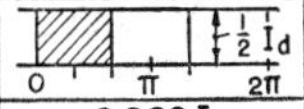	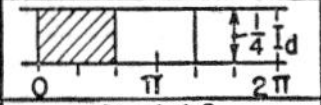	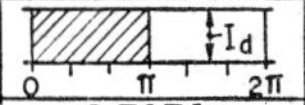	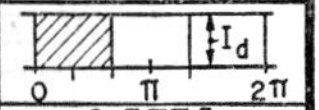
	RMS ×C F O	$0.289 I_d \sqrt{1-3\Psi(u)}$	$0.144 I_d \sqrt{1-3\Psi(u)}$	$0.707 I_d \sqrt{1-2\Psi(u)}$	$0.577 I_d \sqrt{1-3\Psi(u)}$
TRANSF. D-C WDG. CURRENT	WAVE SHAPE	SAME AS ANODE CURR.	SAME AS ANODE CURR.	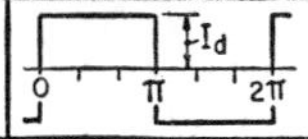	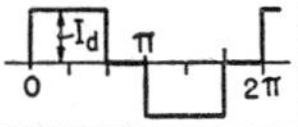
	I_s-(RMS) ×C F O	SAME AS ANODE CURR.	SAME AS ANODE CURR.	$I_d \sqrt{1-4\Psi(u)}$	$0.816 I_d \sqrt{1-3\Psi(u)}$
TRANSF. D-C WDG.	VA RATING ×C F O	$1.48 P_d \sqrt{1-3\Psi(u)}/\cos^2(u/2)$	$1.65 P_d \sqrt{1-3\Psi(u)}/\cos^2(u/2)$	$1.11 P_d \sqrt{1-\Psi(u)}/\cos^2(u/2)$	$1.05 P_d \sqrt{1-3\Psi(u)}/\cos^2(u/2)$
TRANSF. A-C WDG. CURRENT	WAVE SHAPE	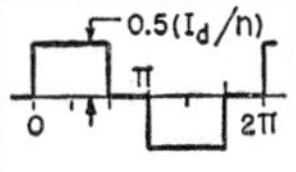	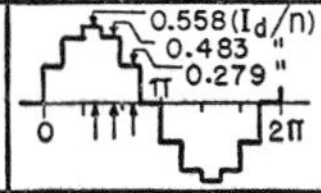	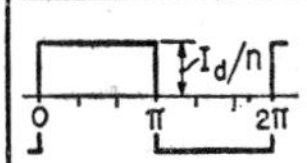	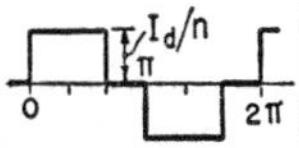
	I_p-(RMS) ×C F O	$0.408(I_d/n) \sqrt{1-3\Psi(u)}$	$0.395(I_d/n) \sqrt{1-1.61\Psi(u)}$	$I_d/n \sqrt{1-4\Psi(u)}$	$0.816(I_d/n) \sqrt{1-3\Psi(u)}$
TRANSF. A-C WDG.	VA RATING ×C F O	$1.05 P_d \sqrt{1-3\Psi(u)}/\cos^2(u/2)$	$1.01 P_d \sqrt{1-1.61\Psi(u)}/\cos^2(u/2)$	SAME AS D-C WINDING	SAME AS D-C WINDING
A-C LINE CURRENT	WAVE SHAPE	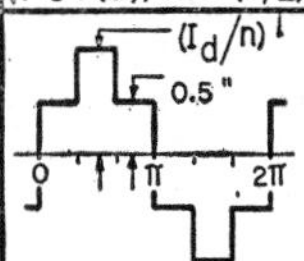	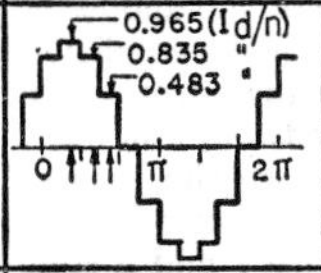	SAME AS A-C WINDING	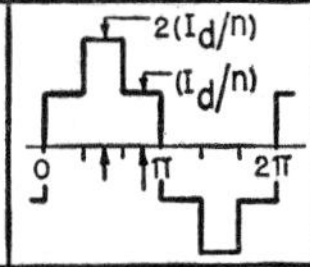
	I_L-(RMS) ×C F O	$0.707(I_d/n) \sqrt{1-3\Psi(u)}$	$0.682(I_d/n) \sqrt{1-1.61\Psi(u)}$	SAME AS A-C WINDING	$1.414(I_d/n) \sqrt{1-3\Psi(u)}$
A-C LINE	VA RATING ×C F O	SAME AS A-C WINDING	SAME AS A-C WINDING	SAME AS D-C WINDING	SAME AS D-C WINDING
	POWER FACTOR	$\frac{0.955 \cos^2(u/2)}{\sqrt{1-3\Psi(u)}}$	$\frac{0.988 \cos^2(u/2)}{\sqrt{1-1.61\Psi(u)}}$	$\frac{0.9 \cos^2(u/2)}{\sqrt{1-4\Psi(u)}}$	$\frac{0.955 \cos^2(u/2)}{\sqrt{1-3\Psi(u)}}$

Paralleling of Anodes. Two or more anodes can be paralleled on one phase of a rectifier transformer by means of anode-paralleling reactors. For paralleling two anodes on one phase, a small iron-core reactor with a center-tapped coil is used, the midpoint being connected to the transformer and the terminals to the anodes.

6. PHASE CONTROL

PRINCIPLE OF OPERATION. In Article 5 the analysis of rectifier circuits was based on the condition that each rectifying element starts conduction at the point of its alternating voltage cycle at which the anode-to-cathode voltage becomes positive. This occurs at the intersection point of the positive half-cycles of the voltage waves of successive phases in a commutating group. This condition applies if, at that point of the cycle, there is electron emission at the cathode, and the potential of the control grid (when present) is such that the anode is permitted to fire (see Article 3). By controlling either

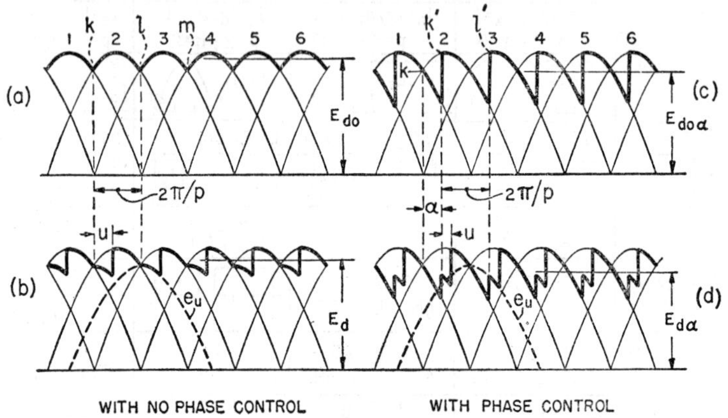

FIG. 13. Direct-current Voltage Waves, Illustrating Phase Control

of these two prerequisites, a rectifying element can be prevented from starting conduction at the point of intersection of the voltage waves and can be made to start at some chosen later point, provided the anode-to-cathode voltage at that point is positive. The process of varying the point within the cycle at which anode conduction is permitted to begin is called **phase control.**

The use of phase control for regulating the d-c voltage of power rectifiers is illustrated in Fig. 13. The positive half-cycles of the phase-to-neutral voltages applied to the anodes are shown in light lines and are designated by numerals in the order of their phase sequence. The d-c voltage, which is the voltage of the anodes when conducting current, is shown by the heavy lines. In Fig. 13(*a*) is shown the condition near no load (when the effect of the commutating reactance is negligible) for normal firing of the anodes. Anode 2 conducts from k to l, anode 3 conducts from l to m, etc. The average value of the d-c voltage is indicated by E_{d0}. In Fig. 13(*b*) is shown the normal firing of the anodes at higher loads, when the effect of the commutating reactance is taken into account. Anode 2 starts conducting at k, as in Fig. 13(*a*). Anodes 1 and 2 conduct simultaneously during the commutating angle u, while the current is transferred from anode 1 to anode 2. Similarly, anode 3 starts at point l, etc. During the commutation period u, the voltage of the firing anodes is equal to the average of the voltages of the overlapping phases and lies on the sine wave of the average voltage e_u. The average d-c voltage is indicated by E_d.

Figure 13(*c*) illustrates the effect of phase control, near no load, when the commutating reactance is neglected. By controlling the voltage of its grid or the initiation of electron emission at its cathode, anode 2 is prevented from starting at the point of intersection k of the anode voltages and is permitted to start at some later point k', displaced from k by an angle α. Anode 1 will continue to conduct the current until anode 2 starts, and the d-c voltage will follow the phase voltage of anode 1. If the start of the other anodes is delayed by the same angle α beyond their normal starting points, anode 2 will fire between k' and l', etc. The duration of the firing period of every anode is the same as in Fig. 13(*a*), but the average value of the d-c voltage is reduced below its value in Fig. 13(*a*). The greater the angle α, the lower will be the average d-c voltage. The angle α is called the **phase-control angle of retard.**

The effect of the commutating reactance on the d-c voltage wave, when phase control is used, is shown in Fig. 13(*d*). Anode 2 starts at point k' as in Fig. 13(*c*). Anodes 1 and 2 conduct simultaneously during the commutating angle u. During the commutating period the voltage of the firing anodes (and consequently the d-c voltage) is equal to the average of the phase voltages and lies on the sine wave of voltage e_u, as in Fig. 13(*b*). With retarded firing, however, the commutating angle is shorter than with normal firing, because the difference between the voltages of the overlapping anodes during commutation is greater, so that less time is required for transferring the current between anodes. The effect of the commutating angle is to reduce the average value of the d-c voltage below its value at no load, the same as for operation without phase control.

COMMUTATION. Commutation in a p-phase commutating group, when phase control is used, is shown in Fig. 14, which corresponds to Fig. 5 for normal firing of the anodes. The voltages and currents are designated by the same symbols as in Fig. 5. In Fig. 14 the start of conduction of a rectifying element, such as T_2, is delayed by an angle α beyond its starting point in Fig. 5, which was the intersection point of voltage waves e_2 and e_1 of the incoming and outgoing rectifying elements.

Element T_1 continues to carry alone the direct current I_c during the angle α, until element T_2 is permitted to start. The two rectifying elements then conduct simultaneously during the commutating angle u, while the current is increasing in T_2 and declining in T_1. The transfer of current between the two elements is effected by the commutating current i_c, produced by the commutating voltage e_c, which is the difference between the phase voltages of the two elements. During the commutation period the current i_2 of element T_2 follows the wave shape of i_c, as shown in heavy line on the sine wave of i_c. The zero line of i_c passes through the point on the wave at which the current i_2 starts flowing as in Fig. 5, because of the d-c transient component. As a result of the delayed start of T_2, the portion of i_c which effects commutation lies on a steeper portion of the sine wave, producing a steeper rise of the anode current wave and shortening the commutating angle, as compared with those in Fig. 5.

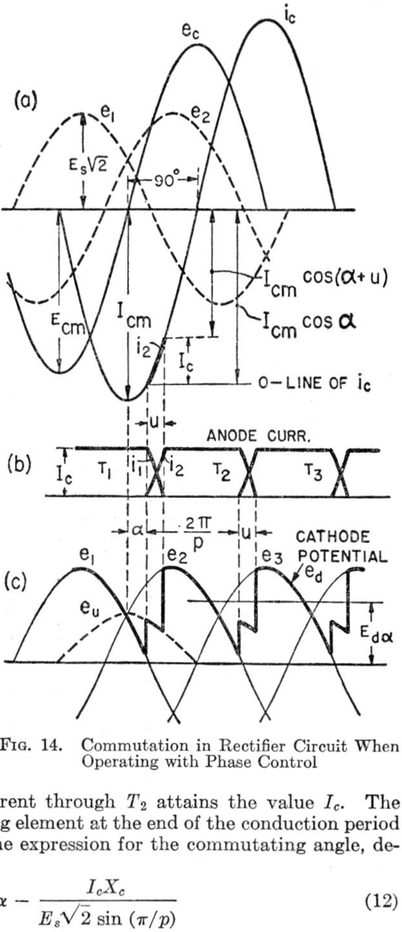

FIG. 14. Commutation in Rectifier Circuit When Operating with Phase Control

Commutation is completed when the current through T_2 attains the value I_c. The wave shape of the current through a rectifying element at the end of the conduction period follows i_1, which is equal to I_c minus i_2. The expression for the commutating angle, derived with the aid of Figs. 5 and 14, is

$$\cos(\alpha + u) = \cos\alpha - \frac{I_c X_c}{E_s\sqrt{2}\sin(\pi/p)} \tag{12}$$

The commutating angle u can be determined by means of eq. (12) for specific values of α. The relation between the commutating angle when using phase control, and the commutating angle with no phase control, is shown by the curves of Fig. 15, for several values of phase-control angle α. These curves apply for any value of p. The value of u with no phase control, for a given commutating reactance factor F_x and a specific value of p, can be obtained from the curves of Fig. 7.

The effective (rms) value of the anode current, when phase control is used, is

$$I_s = \frac{I_c}{\sqrt{p}}\sqrt{1 - p\cdot\psi(u, \alpha)} \tag{13}$$

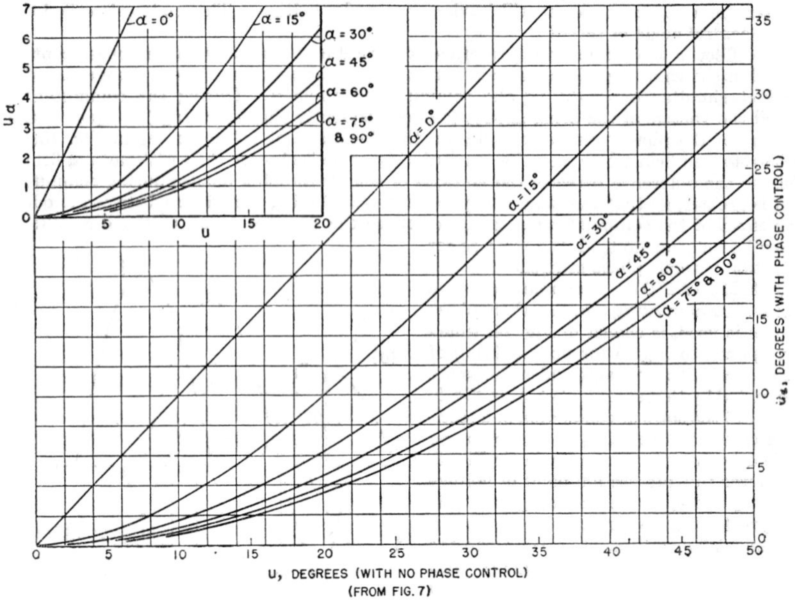

Fig. 15. Curves for Determining Commutating Angle with Phase Control

This equation is similar to eq. (4) except that the factor $\psi(u)$ has been replaced by the factor $\psi(u, \alpha)$. The approximate ratio of $\psi(u, \alpha)$ to $\psi(u)$, in function of the phase-control angle, for a given angle u, is given by the curve of Fig. 16.

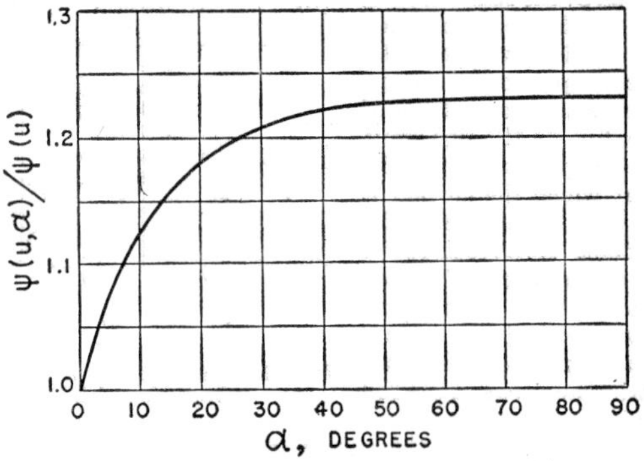

Fig. 16. Relation between $\psi(u, \alpha)$ and $\psi(u)$

The average value of the d-c voltage is

$$E_{d\alpha} = \frac{E_s\sqrt{2}\sin(\pi/p)}{\pi/p} \cdot \frac{1}{2}[\cos\alpha + \cos(\alpha + u)] \tag{14}$$

$$= E_{d0} \cdot \frac{\cos\alpha + \cos(\alpha + u)}{2} \tag{14a}$$

This equation becomes the same as eq. (8a), when $\alpha = 0$.

The average d-c voltage at no load, when u is equal to zero, is

$$E_{d0\alpha} = E_{d0}\cos\alpha \tag{15}$$

It is equal to the theoretical d-c voltage at no load, with no phase control, multiplied by the cosine of the phase-control angle.

By substituting from eq. (12) in eq. (14),

$$E_{d\alpha} = \frac{E_s\sqrt{2}\sin(\pi/p)}{\pi/p}\cos\alpha - \frac{pI_cX_c}{2\pi} = E_{d0}\cos\alpha - E_x \tag{16}$$

The first term of eq. (16) is the theoretical no-load d-c voltage with phase control. The second term, E_x, is the d-c voltage drop under load due to the commutating reactance and has the same value as given by eq. (6) and Fig. 7 for a rectifier without phase control.

The above equations for the d-c voltage apply to single-way rectifier circuits. For double-way rectifier circuits the d-c voltages have twice the value given by these equations.

7. VOLTAGE REGULATION

THEORY. The theoretical rectifier d-c voltage at no load is given by eq. (2) when no phase control is used, and by eq. (15) when phase control is used. The theoretical d-c voltage under load, without and with phase control, is given by eqs. (7) and (16). These equations do not take into account the effect of the arc-drop voltage in the rectifier and the resistance drop in the rectifier transformer. The net d-c terminal voltage, taking into account all the voltage drops, is determined by the following equations:

For single-way rectifier circuits:

$$E_d = E_s\sqrt{2}\,\frac{\sin(\pi/p)}{\pi/p}\cos\alpha - E_a - E_x - E_r \tag{17}$$

For double-way rectifier circuits:

$$E_d = 2E_s\sqrt{2}\,\frac{\sin(\pi/p)}{\pi/p}\cos\alpha - 2E_a - 2E_x - E_r \tag{18}$$

In eqs. (17) and (18) the first term is the theoretical no-load d-c voltage, which is given in Tables 1A and 1B for various transformer connections. E_a is the arc-drop voltage in a rectifying device; in eq. (18) for double-way circuits, E_a is multiplied by 2, because the current flows through two rectifying devices in series. E_x, the voltage drop caused by the commutating reactance, is given by eq. (6), in which I_c is the direct current of one commutating group and X_c the phase-to-neutral commutating reactance of the d-c winding in ohms; E_x can be obtained from Fig. 7. For double-way circuits E_x is multiplied by 2, because such circuits are equivalent to two commutating groups operating in series. The last term in the equations, E_r, is the voltage drop due to the resistance losses in the rectifier transformer.

$$E_r = \frac{P_r}{I_d} \tag{19}$$

in which P_r is the resistance (or load) loss of the rectifier transformer, in watts, and I_d is the direct current in the output circuit.

The arc-drop voltage changes only a little with changes of the direct current in the normal operating range, and the change is usually linear. The reactance and resistance drops, given by eqs. (6) and (19), vary linearly with the direct current. The voltage-regulation curve of a rectifier unit is therefore a sloping straight line.

TYPICAL REGULATION CURVE. In Fig. 17 is shown a regulation curve of a rectifier unit, and the several voltage components of eq. (17) are indicated. The heavy solid line is the **inherent** regulation curve with no phase control. The heavy broken line is a regulation curve when operating with a fixed phase-control angle α. The change in voltage from no load to rated load is indicated by ΔE, which is the same for both operating conditions. The two regulation curves are therefore parallel. The regulation of a rectifier unit is the ratio of the change in voltage, ΔE, to the rated d-c output voltage. The **rated** output voltage is the voltage specified as the basis of the rating of the unit. It may be the voltage given by the inherent regulation curve at rated current output, or it may be some lower d-c voltage obtained with a specified amount of phase control. In the latter case the d-c voltage obtained at rated current without phase control is called the **ceiling d-c voltage.**

REGULATION BY PHASE CONTROL. The rectifier d-c voltage may be regulated by phase control either manually or automatically by means of a regulator. Such regulation may be used (1) to maintain a constant output voltage over a range of load current and to compensate for variations of the a-c system voltage; (2) to provide the rectifier unit with a voltage-regulation characteristic different from its inherent regulation curve;

(3) to regulate the load current of a rectifier unit; (4) to reduce the d-c voltage for starting a motor, or for any other purpose requiring a variable d-c voltage. Phase control can be used only for reducing the d-c voltage below the inherent regulation curve. It cannot be used for obtaining voltages above that curve; this can be accomplished only by increasing the voltage of the rectifier transformer d-c winding by means of transformer taps, or by increasing the a-c voltage applied to the transformer. The voltage of the transformer d-c winding is determined by the required ceiling d-c voltage.

Phase control increases the voltage of the interphase transformer in the circuit of Fig. 8. This effect is shown by the curves of Fig. 18, which give the value of the interphase transformer voltage E_{IT} in terms of E_s for several phase control angles.

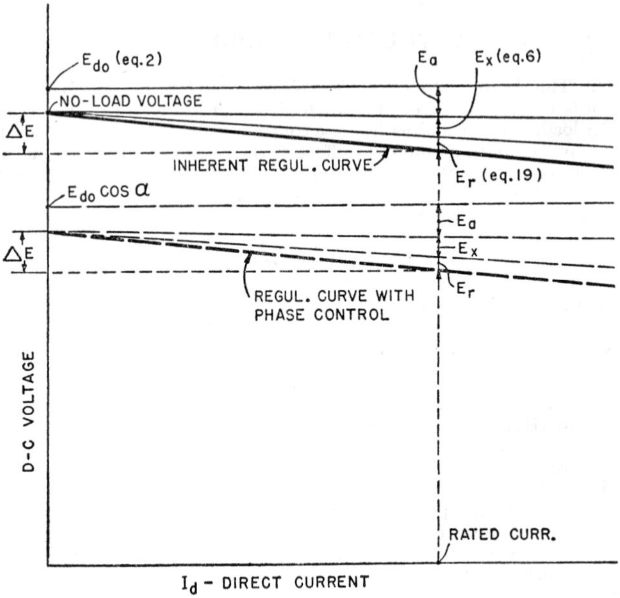

FIG. 17. Regulation Curves of Power Rectifier

EFFECT OF A-C LINE REACTANCE. Reactance in the a-c circuit to which a rectifier unit is connected has an effect similar to the commutating reactance of the rectifier transformer and will increase the component E_x of the regulation voltage ΔE. This effect can be taken into account by adding to the commutating reactance of the rectifier transformer a value given by the following equation:

$$X_c = \frac{1}{D_x}\left(\frac{E_s}{E_n}\right)^2 X_L \tag{20}$$

In eq. (20) X_c is the equivalent commutating reactance, referred to the transformer d-c winding, of the line-to-neutral a-c circuit reactance X_L, in ohms; E_s and E_n are the line-to-neutral voltages of the transformer d-c and a-c windings, respectively; D_x is the commutating reactance transformation constant, which is given in Tables IA and IB for the transformer connections shown there. Values of D_x for other transformer connections are given in reference 44 of the Bibliography, Article 20. The value of X_c obtained from eq. (20) should be added to the commutating reactance of the rectifier transformer to obtain the value of the total commutating reactance X_c used in eq. (6) for calculating E_x.

If several similar rectifier units are connected to the same a-c circuit, and if their commutating periods occur simultaneously, the effect of the a-c circuit reactance on the regulation of individual units can be determined by considering the reactance as consisting of several parallel branches through each of which flows the current of one unit. The reactance of one branch, to be used as X_L in eq. (20), will then be the total reactance multiplied by the number of branches or units.

The effect of the a-c circuit reactance is to increase the slope of the inherent regulation curve. To compensate for this effect and obtain the required d-c voltage at rated load current, a higher no-load d-c voltage is required; consequently, a higher voltage is needed for the transformer d-c winding, which increases the kva rating of the transformer.

The inherent regulation of rectifier units is usually in the range of 5 to 8 per cent. The lower values apply to rectifiers for railway and industrial service; the higher values, for electrochemical service.

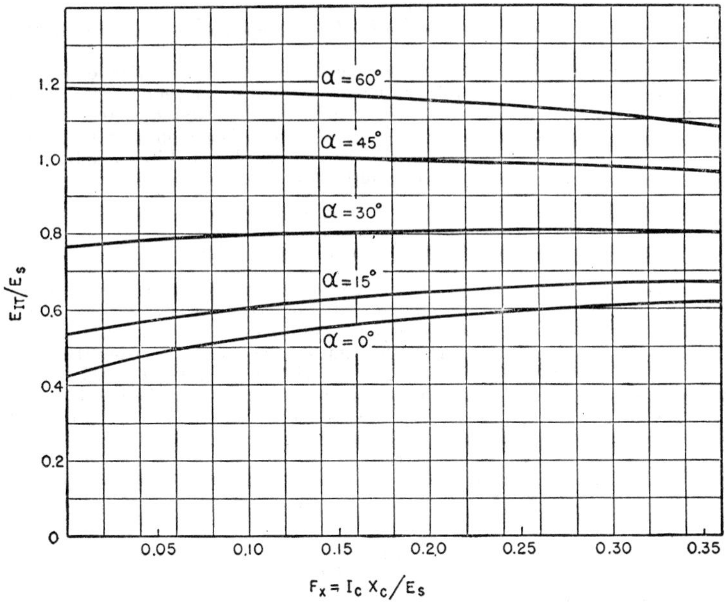

FIG. 18. Voltage (rms) of Interphase Transformer

8. POWER FACTOR

The power factor of a rectifier unit is the ratio of watts to volt-amperes at the a-c line terminals of the rectifier transformer. The a-c line current of a rectifier unit is non-sinusoidal, however, because it is derived from flat-top anode current waves [see Fig. 9(*h*)]. This current can be resolved by a Fourier series into sinusoidal components: a fundamental component (I_1) of the same frequency as the a-c voltage, and a number of harmonic components having frequencies which are multiples of the fundamental frequency. The rms value of the line current (I_L) is equal to the square root of the sum of the squares of these components:

$$I_L = \sqrt{I_1^2 + I_5^2 + I_7^2 + I_{11}^2 + \text{etc.}} = \sqrt{I_1^2 + I_H^2} \tag{21}$$

in which I_5, I_7, etc., are harmonic components having multiples of the fundamental frequency designated by their subscripts, and $I_H = \sqrt{I_5^2 + I_7^2 + I_{11}^2 + \text{etc.}}$ is the rms value of the harmonic components.

The fundamental component (I_1) can be resolved further into an in-phase or watt component (I_{1P}) and reactive component (I_{1Q}), as is done for ordinary a-c circuits. The power is produced only by the component I_{1P}. The harmonic components do not produce any power, because there are no voltages of corresponding frequencies, the a-c voltage being assumed sinusoidal. The relationship between the a-c line-to-neutral voltage E_n and the components of current I_L is shown in the diagram of Fig. 19. The relationship among I_L, I_1, and I_H is in accordance with eq. (21). From Fig. 19,

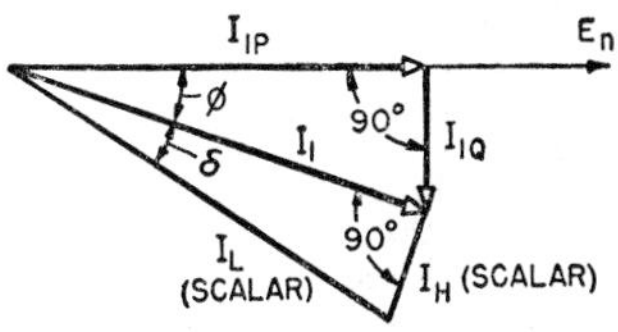

FIG. 19. Power-factor Diagram of Power Rectifier

$$\frac{I_{1P}}{I_1} = \cos\phi; \qquad \frac{I_1}{I_L} = \cos\delta \tag{22}$$

The power factor is the ratio of the watt component of the line current, to the line current:

$$\mathrm{PF} = \frac{I_{1P}}{I_L} = \frac{I_{1P}}{I_1} \cdot \frac{I_1}{I_L} = \cos\phi \cdot \cos\delta \tag{23}$$

The power factor of a rectifier unit is the product of two factors: a **displacement factor,** $\cos\phi$, and a **distortion factor,** $\cos\delta$.

An ordinary power-factor meter, which operates by the interaction of current and voltage of the same frequency, will measure only the displacement factor. The power factor of a rectifier unit can be determined by measuring the watts, current, and voltage and dividing the watts by the volt-amperes. If the power is measured by the well-known two-wattmeter method, the displacement factor can be calculated by the following equation, in which P_A and P_B are the readings of the two wattmeters:

$$\cos\phi = \frac{P_A + P_B}{2\sqrt{P_A^2 + P_B^2 - P_A P_B}} \tag{24}$$

For a rectifier unit (or a group of rectifier units) with twenty-four or more phases, the fundamental component of the a-c line current is practically the same as its rms value, and the power factor can be measured by means of a power-factor meter.

The theoretical power factor of a rectifier unit can be determined from the theoretical relations of the currents and voltages derived previously and given in Tables 1A and 1B for various rectifier circuits. The theoretical d-c power $P_d = E_d I_d$, in which I_d is the d-c output and E_d is the theoretical d-c voltage under load given by eq. (7) or (8), before subtracting the arc-drop in the rectifier and the resistance drop in the rectifier transformer. Therefore P_d includes the losses in the rectifier and the resistance (load) loss in the transformer and is the power input at the a-c line terminals of the rectifier unit, excluding the transformer core loss. The theoretical power factor is

$$\mathrm{PF} = \frac{P_d}{K_L} \tag{25}$$

in which K_L is the volt-amperes at the a-c line terminals of the rectifier transformer. The formulas for the power factor of various rectifier circuits are given in Tables 1A and 1B. Typical power-factor curves of rectifier units are shown in Fig. 38.

When the d-c voltage of a rectifier unit is reduced by phase control, the anode current, and consequently the a-c line current, is shifted in the lagging direction with respect to the a-c voltage, thus reducing the power factor. As the d-c voltage is reduced by phase control while the current is maintained at a constant value, the theoretical d-c power P_d is reduced in the same ratio as the voltage, whereas the a-c line current and volt-amperes remain practically unchanged (they actually increase slightly); the power factor is reduced by the same ratio as the d-c voltage.

9. POWER INVERTER

PRINCIPLE OF OPERATION. A **power inverter** is a rectifier unit in which the direction of average power flow is from the d-c circuit to the a-c circuit. The operation of a power inverter can be explained with the aid of Fig. 20, which shows several stages of phase control of a multiphase rectifier unit, near no load (commutating angle is zero).

In Fig. 20(*a*) is shown the d-c voltage wave with no phase control. The average value of this voltage, E_{d0}, is given by eq. (2). In Fig. 20(*b*) the d-c voltage is reduced by retarding the firing of each anode by the phase-control angle α_1 beyond the normal firing point in Fig. 20(*a*). The average value of the d-c voltage is given by eq. (15).

In Fig. 20(*c*) the phase-control angle is 90 deg, and the average value of the d-c voltage is zero, the areas of the positive and negative portions of the voltage wave being equal to each other. The average d-c power output is then also zero. It is assumed here that there is sufficient inductance in the d-c circuit to maintain a constant flow of direct current even during the negative portions of the voltage wave; the ripple voltage is then absorbed by the inductance. If there were no inductance, there would be no conduction during the negative portions of the voltage wave; the d-c voltage wave would then consist of the positive portions only, and its average value would still be positive.

In Fig. 20(*d*) is shown the d-c voltage wave when phase control is carried further, with an angle of retard α_3 greater than 90 deg. The average value of the d-c voltage is negative. If the flow of direct current is maintained through the rectifier unit into the d-c circuit, in the normal direction from anode to cathode, as for Figs. 20(*a*) to (*c*), the average d-c power output is negative, i.e., the flow of power is from the d-c circuit to the a-c circuit, and the rectifier unit then operates as a power inverter.

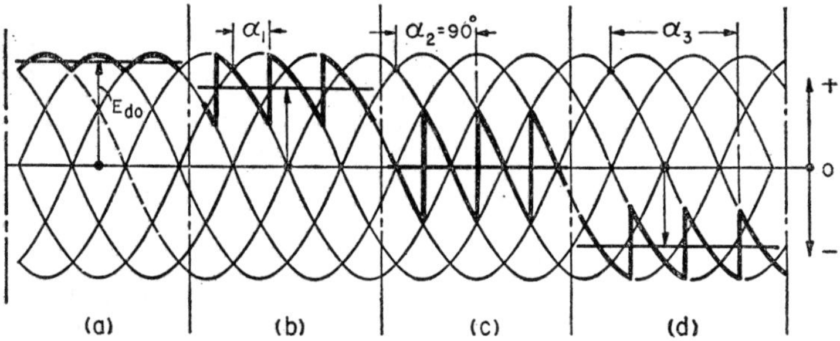

FIG. 20. Phase Control for Power Rectifier and Inverter

The circuit conditions for a power rectifier and a power inverter are shown in Fig. 21. In the power inverter the d-c voltage E_d' is in the direction from the cathode terminal to the neutral of the d-c winding, opposite to the direction of the direct current I_d. The current is made to flow through the inverter by applying to its terminals a d-c voltage from a d-c source, with the positive terminal of the d-c source connected to the neutral point of the inverter transformer d-c winding, so that the applied voltage counteracts the back voltage of the inverter. The d-c source may be a d-c generator, or a power rectifier as shown at the left of Fig. 21. The d-c voltage E_d of the power rectifier is in the same direction as the current I_d. The direction of power flow is indicated by arrows. Power from the a-c circuit *A* is converted to d-c power by the power rectifier; it is then changed to a-c power in circuit *B* by the power inverter. Power can thus be transferred between

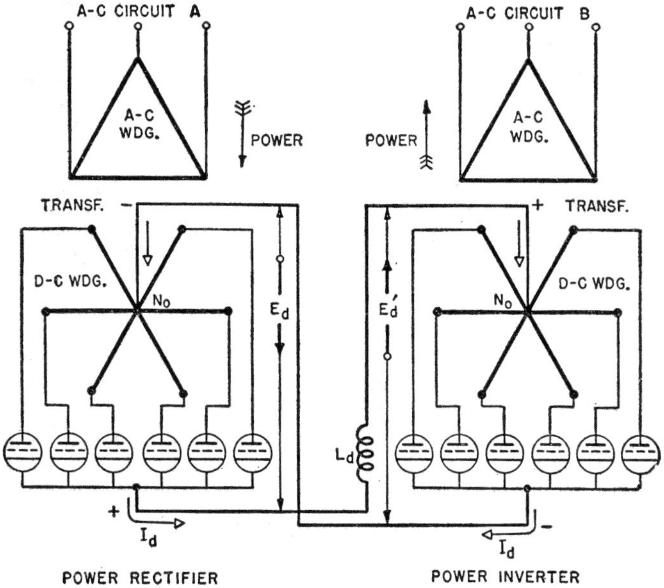

FIG. 21. Interconnection of Two A-c Circuits by Power Rectifier and Power Inverter

two independent a-c systems which operate at the same or different frequencies and voltages. The d-c voltage and current, and consequently the power flow, can be regulated by regulating the d-c voltage of the power rectifier or the power inverter or both, by means of phase control. As was shown in Fig. 20, the direction of power flow is also determined by phase control of the d-c voltage, and the power flow in Fig. 21 can be reversed by changing the starting points in the voltage cycle of the two groups of rectifying elements; this can be effected by control of either their grids or the ignition of the cathode spots, without changing the power-circuit connections.

As can be seen in Fig. 20(*a*) and (*b*), in a power rectifier current flows in the d-c winding of the transformer largely during the positive half-cycles of the phase voltages; in a power inverter [Fig. 20(*d*)] the current flows largely during the negative half-cycles of the phase voltages. This is reflected in the phase relation between the current and voltage at the a-c line terminals. For a given transformer connection the wave shapes of the anode and a-c line currents are substantially the same for a power rectifier and a power inverter. The a-c voltages of the inverter transformer, which are required for commutating the current between successive rectifying elements of a commutating group, are usually generated in the a-c circuit to which the power inverter is connected, either by a-c generators or other synchronous machines, or by oscillation between inductance and capacitance in the a-c load circuit.

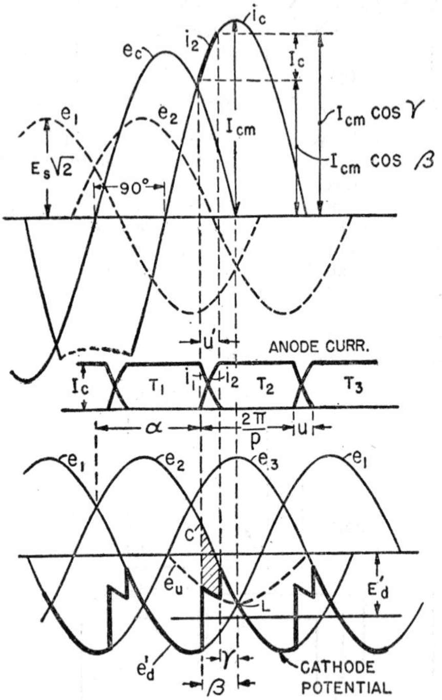

FIG. 22. Commutation of Current in Power Inverter Circuit

CIRCUIT THEORY. The voltage and current relations in a *p*-phase commutating group of a power inverter during commutation between successive rectifying elements are shown in Fig. 22, which is similar to Fig. 14. The direct current I_c is conducted in succession by rectifying elements T_1, T_2, T_3, which are connected to phases of the transformer d-c winding having a-c voltages e_1, e_2, e_3.

Rectifying element T_2 is permitted to start conduction at point *C* of its voltage wave e_2, displaced by the angle α beyond the point of intersection of the positive half-cycles of the voltages e_1 and e_2, which is the normal starting point of T_2 when operating as a power rectifier without phase control. Point *C* leads the point of intersection *L* of the negative half-cycles of voltages e_1 and e_2 by the angle β, which is called the **phase control angle of advance.** Before point *C*, the direct current I_c is conducted by rectifying element T_1, and the d-c voltage follows the voltage wave e_1, as indicated by the heavy line. After T_2 starts, T_1 and T_2 conduct the current simultaneously during the commutating angle u' while the current is transferred from T_1 to T_2. During this period the d-c voltage follows the voltage wave e_u, which is the average of e_1 and e_2. After the commutation period the current I_c is conducted by T_2, and the d-c voltage follows the voltage wave e_2 until rectifying element T_3 starts. During the commutation period the current i_2 of T_2 follows the wave shape of the commutating current i_c, as was the case in Figs. 5 and 14. The current i_1 of the outgoing rectifying element is the difference between I_c and i_2. It may be noted that the slopes of i_1 and i_2 are bent in opposite directions from those in Figs. 5 and 14 for a power rectifier, because of the shift in the position of the commutating period on the sine wave of i_c.

Commutation between T_1 and T_2 can be effected only while the voltage e_2 is more positive than the voltage e_1. Commutation must therefore be completed before the point of intersection *L* of the negative half-cycles of these voltages. Beyond that point, the voltage e_1 of rectifying element T_1 becomes positive in relation to e_2; if commutation is not completed before this point, T_1 will again take over the current and will continue to conduct during its positive half-cycle, so that the d-c voltage would become positive, and the unit would cease functioning as a power inverter. This voltage would be in the same direction as the voltage applied from the d-c source in the circuit of Fig. 21, resulting in a short circuit.

The cathode potential is the same as the potential of the conducting anode. Beyond point *L* the anode of T_1 is positive with respect to its cathode. After the current I_c has been transferred from T_1 to T_2, T_1 is prevented from starting conduction after point L by a negative voltage on its control grid or by preventing ignition of its cathode spot, depending on the type of rectifying device involved. To regain control of the conductivity of the rectifying device, in order to prevent conduction after point L, time must be allowed for deionization before point L is reached. This time is designated in Fig. 22

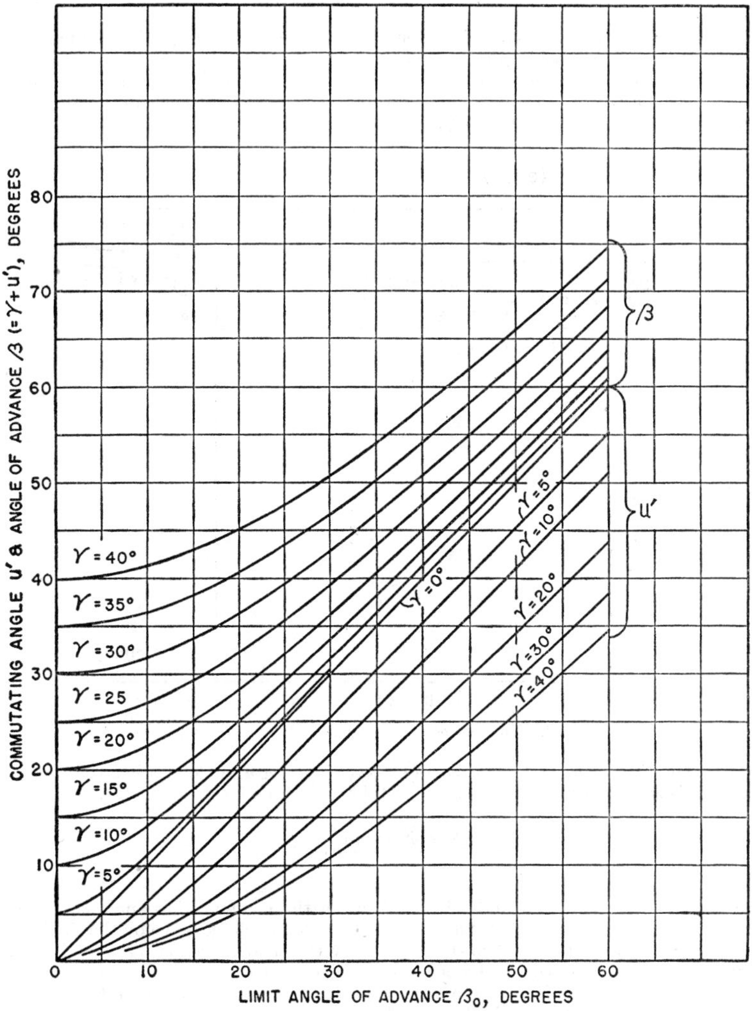

FIG. 23. Relation between Angle of Advance and Margin Angle of Power Inverter

as angle γ, which is called the **inverter margin angle** and is equal to the difference between the angle of advance β and the commutating angle u'. The relationship between angles β and γ and the circuit quantities can be derived with the aid of Fig. 5 by writing the equation for I_c from Fig. 22 and is given below.

$$\cos\gamma - \cos\beta = \frac{1}{\sqrt{2}\,\sin(\pi/p)} \cdot \frac{I_c X_c}{E_s} = \frac{1}{\sqrt{2}\,\sin(\pi/p)} F_x \qquad (26)$$

If either angle β or angle γ is given, the other angle can be determined by means of eq. (26) for a specific value of reactance factor F_x, which is proportional to the direct current I_c.

The theoretical limit for commutation in a power inverter comes when it is completed at point L in Fig. 22, i.e., when the commutating angle u' is equal to the angle of advance β, so that γ is zero. This commutation-limit angle of advance, β_0, derived from eq. (26), is given by eq. (27).

$$\cos \beta_0 = 1 - \frac{1}{\sqrt{2} \sin (\pi/p)} F_x \tag{27}$$

The curves of Fig. 7 show β_0 as a function F_x for several values of p. The actual angle of advance required to obtain a desired margin angle γ can be determined by means of eq. (26) or by means of the curves in Fig. 23, which show the relation between β and β_0 for a number of values of γ. In Fig. 23 are drawn also curves of the commutating angle u' as a function of β_0 for several values of γ.

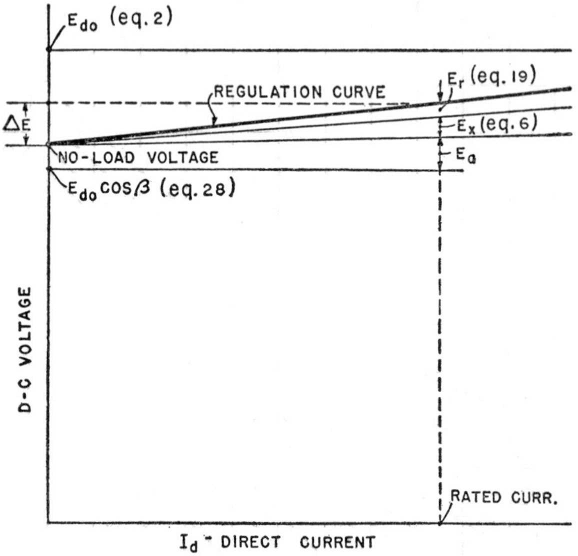

FIG. 24. Voltage-regulation Curve of Power Inverter

The theoretical average d-c voltage of a power inverter at no load, when the commutating angle is zero, is given by eq. (28), which has the same form as eq. (15).

$$E_{d0}' = \frac{E_s\sqrt{2} \sin (\pi/p)}{\pi/p} \cos \beta = E_{d0} \cos \beta \tag{28}$$

The theoretical d-c voltage under load e_d', when the commutating angle is taken into account, is shown in Fig. 22. It differs from the no-load voltage wave by the shaded area. The average value of the voltage represented by the shaded area is $E_x = pI_cX_c/2\pi$, the same value as given by eq. (6) for a power rectifier. It can be seen in Fig. 22 that the effect of the shaded area is to increase the negative d-c voltage of the inverter. The theoretical average d-c voltage under load is, therefore,

$$E_d' = E_{d0}' + E_x = E_{d0} \cos \beta + E_x \tag{29}$$

Eqs. (28) and (29) apply to single-way circuits. For double-way circuits, the equation for E_d' is

$$E_d' = 2E_{d0} \cos \beta + 2E_x \tag{30}$$

VOLTAGE REGULATION. The terminal d-c voltage of a power inverter under load is equal to the d-c voltage that must be applied to overcome the theoretical d-c voltage under load, given by eqs. (29) and (30), plus the arc-drop voltage in the rectifying devices (E_a) and the resistance drop in the transformer (E_r).

For single-way inverter circuits the terminal d-c voltage is

$$E_d' = E_{d0} \cos \beta + E_a + E_x + E_r \tag{31}$$

For double-way inverter circuits the terminal d-c voltage is

$$E_d' = 2E_{d0} \cos \beta + 2E_a + 2E_x + E_r \tag{32}$$

These equations are similar to eqs. (17) and (18) for a power rectifier, except that the voltage drops are **added** to the theoretical no-load voltage. The value of E_r is given by eq. (19), and E_x can be obtained from eq. (6) or Fig. 7.

The d-c voltage-regulation curve of a power inverter and its components as given by eq. (31) are shown in Fig. 24, assuming a constant sinusoidal voltage at the a-c line terminals. The regulation curve rises with load current, instead of dropping as shown in Fig. 17 for a power rectifier. The regulation voltage between no load and rated load is indicated by ΔE.

The effect of the a-c circuit reactance can be taken into account by **adding** to the commutating reactance of the inverter transformer an equivalent reactance given by eq. (20), the same as for a power rectifier.

POWER FACTOR. The power factor of a power inverter is the ratio of watts to volt-amperes at the a-c line terminals. As for a power rectifier, the power factor is the product of two factors, the displacement factor $\cos \phi'$ and the distortion factor $\cos \delta'$, ϕ' being the displacement angle between the fundamental component of the a-c line current and line-to-neutral voltage, and $\cos \delta'$ the ratio of the fundamental component of the line current to its total effective (rms) value. The wave shape of the a-c line current is substantially the same as for a power rectifier using the same transformer connection.

It can be seen in Fig. 22 that the center of the anode current wave of a rectifying element leads the center of the negative half-cycle of the corresponding phase voltage. This results from the requirement that the angle of advance β be sufficiently large so that commutation is completed before the intersection point L of the voltage waves. The a-c line current likewise leads the negative half-cycle of the a-c line-to-neutral voltage. The power-factor relation of the a-c line current and line-to-neutral voltage is shown by the diagram of Fig. 25, which is similar to that of Fig. 19 and in which the symbols have the same meaning. The a-c voltage $-E_n$ is opposite in phase to voltage E_n in Fig. 19 and can be considered the output a-c voltage of the inverter.

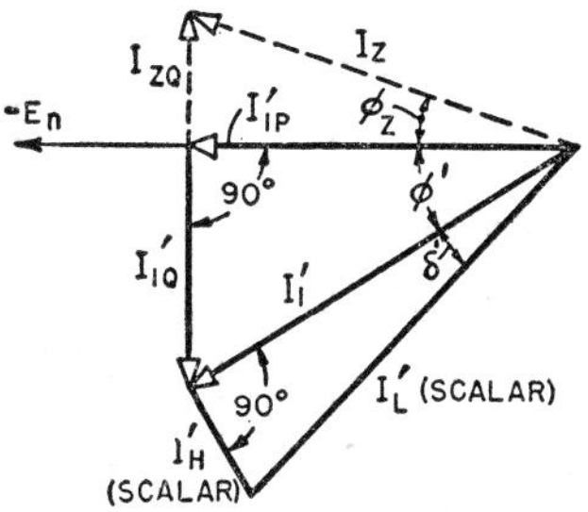

FIG. 25. Power-factor Diagram of Power Inverter

It can be seen in Fig. 25 that the fundamental component I_1' of the line current leads the voltage. One characteristic of a power inverter is that it must operate with a leading displacement factor $\cos \phi'$, which is the power factor of the fundamental component of the current I_1', as would be indicated by a power-factor meter. A power inverter can deliver power into an a-c system only if the system can receive the reactive kva of leading power factor as represented by the component I_{1Q}' of the alternating current I_L'. This can be accomplished by a-c generators or other synchronous machines or by capacitors connected to the a-c system. If the load on the system, such as induction motors, requires lagging reactive kva, the synchronous machines or capacitors have to **supply** this reactive kva and additional lagging reactive kva to offset the leading kva put out by the inverter. If the load supplied by the power inverter (see Fig. 25) draws a current I_Z, which lags the voltage by the angle ϕ_Z, the total lagging reactive current to be supplied from other sources (or the leading current to be absorbed) is equal to the sum of I_{ZQ} and I_{1Q}'.

The harmonic components of the a-c line current I_L', represented by I_H' in Fig. 25, will flow into the a-c circuit and divide among the parallel branches of the circuit in accordance with their impedances, in the same way as for a power rectifier.

The theoretical power factor of a power inverter can be calculated in the same way as for a power rectifier. It is the ratio of the power output P_d' to the volt-amperes K_L' at the a-c line terminals. The power output is equal to the power input at the d-c terminals, **minus** the losses in the rectifying devices and the resistance loss in the transformer, represented by the arc drop E_a and resistance drop E_r, respectively, in eqs. (31) and (32). The power output P_d' is therefore equal to the product of the direct current I_d and the **theoretical** d-c voltage E_d' given by eqs. (29) or (30), which does not include E_a and E_r. The a-c line volt-amperes K_L' can be determined for a given transformer connection from Tables 1A and 1B.

10. WAVE SHAPE

D-C VOLTAGE. As shown in Figs. 4, 5, 9, 13, and others, the d-c voltage of a rectifier unit contains ripples, which are made up of portions of sine waves. The fundamental frequency and magnitude of the ripple depend on the number of rectifier phases. In general, the larger the number of phases, the higher is the ripple frequency and the smaller its magnitude. For polyphase rectifier circuits, the magnitude of the ripple increases with increasing load current because of the increase of the commutating angle (see Figs. 5 and 13). It is also increased when the d-c voltage is reduced by phase control (see Fig. 13).

The ripple of the d-c voltage wave can be resolved into sinusoidal harmonic components by a Fourier series. The frequency of the first harmonic is equal to the product of the frequency of the a-c system voltage and the number of phases of the rectifier circuit (given by q in Tables 1A and 1B). The frequencies of the higher harmonics are multiples of the frequency of the first harmonic. The ratio of the frequency of an harmonic component to the frequency of the a-c system voltage is called the **order** of the harmonic. The order of the first harmonic of the voltage ripple is equal to the number of phases, q, and the orders of the higher harmonics are multiples of q. Harmonic components present in the d-c voltage, for various numbers of phases, are indicated by the letter D in Table 2. Their magnitudes are directly proportional to the magnitude of the d-c voltage (see references 9 and 36 of the Bibliography, Article 20).

Table 2. Harmonics in Rectifier Circuits

Order of Harmonic	D = Harmonic in D-c Voltage. A = Harmonic in A-c Line Current.									Frequency of Harmonic for 60-cycle A-c System
	Number of Rectifier Phases (q)									
	2	3	4	6	12	18	24	30	36	
2	D	A	......	......	......	......	......	......	......	120
3	......	D	......	......	......	......	......	......	......	180
4	D	A	D	......	......	......	......	......	......	240
5	A	A	A	A	......	......	......	......	......	300
6	D	D	......	D	......	......	......	......	......	360
7	A	A	A	A	......	......	......	......	......	420
8	D	A	D	......	......	......	......	......	......	480
9	......	D	......	......	......	......	......	......	......	540
10	D	A	......	......	......	......	......	......	......	600
11	A	A	A	A	A	......	......	......	......	660
12	D	D	D	D	D	......	......	......	......	720
13	A	A	A	A	A	......	......	......	......	780
14	D	A	......	......	......	......	......	......	......	840
15	......	D	......	......	......	......	......	......	......	900
16	D	A	D	......	......	......	......	......	......	960
17	A	A	A	A	......	A	......	......	......	1020
18	D	D	......	D	......	D	......	......	......	1080
19	A	A	A	A	......	A	......	......	......	1140
20	D	A	D	......	......	......	......	......	......	1200
21	......	D	......	......	......	......	......	......	......	1260
22	D	A	......	......	......	......	......	......	......	1320
23	A	A	A	A	A	......	A	......	......	1380
24	D	D	D	D	D	......	D	......	......	1440
25	A	A	A	A	A	......	A	......	......	1500
26	D	A	......	......	......	......	......	......	......	1560
27	......	D	......	......	......	......	......	......	......	1620
28	D	A	D	......	......	......	......	......	......	1680
29	A	A	A	A	......	......	......	A	......	1740
30	D	D	......	D	......	......	......	D	......	1800
31	A	A	A	A	......	......	......	A	......	1860
32	D	A	D	......	......	......	......	......	......	1920
33	......	D	......	......	......	......	......	......	......	1980
34	D	A	......	......	......	......	......	......	......	2040
35	A	A	A	A	A	A	......	......	A	2100
36	D	D	D	D	D	D	......	......	D	2160
37	A	A	A	A	A	A	......	......	A	2220
.	.	.	.	.	.	.	.	.	.	.
.	.	.	.	.	.	.	.	.	.	.
.	.	.	.	.	.	.	.	.	.	.

The harmonic components of the d-c voltage can cause a flow of harmonic currents in the d-c circuit, the current of any frequency being equal to the harmonic voltage divided by the circuit impedance at that frequency. In the analysis of rectifier circuits in the preceding articles, the d-c wave was assumed to be a straight line, on the assumption that there is sufficient inductance in the d-c circuit to reduce the harmonic currents to negligible values. In so far as the current and voltage relations in the rectifier circuits are concerned, this assumption leads to satisfactory results in most cases. The presence of harmonic currents and voltages in the d-c circuit, however, may have other influences. They may cause additional losses in the load, such as electrolytic cells supplied by rectifiers. If the harmonic currents are large, they may have a detrimental effect on the commutation of rotating d-c machines, although this effect is rarely experienced when multiphase rectifiers are used. The major effect of harmonic components in the d-c voltage and current has been their influence on communication circuits when they are exposed to the electric or magnetic fields of the d-c circuit, which may occur on railway systems supplied by rectifiers (see reference 2, p. 403, and reference 10 of the Bibliography, Article 20).

The harmonics of the output d-c voltage of a rectifier can be reduced to harmless values by means of a reactor in series with the d-c circuit, or by means of a series reactor in combination with resonant shunt filters as shown in Fig. 26. Each branch of the shunt filter consists of a reactor in series with a capacitor, tuned for resonance at the frequency of the harmonic which it is to suppress, the number of branches being equal to the number of harmonics to be suppressed. At resonance the impedance of the filter at the harmonic frequency is low, being equal to its effective resistance; it therefore serves to short-circuit the harmonic voltage, the filter current being limited by the impedance of the d-c reactor.

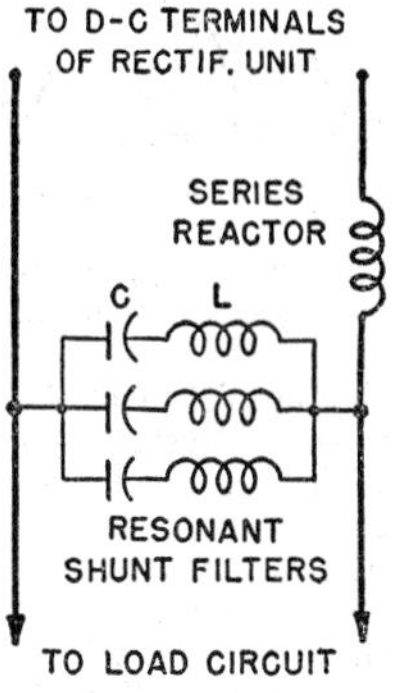

FIG. 26. Harmonic Filter for Rectifier D-c Voltage

A-C LINE CURRENT. As shown in Fig. 9 and Tables 1A and 1B, the a-c line currents of a rectifier circuit have step-type wave shapes, because they are derived from flat-top anode currents. The a-c line current can be resolved by a Fourier series into a fundamental sinusoidal component of the same frequency as the a-c line voltage, and sinusoidal harmonic components of frequencies which are multiples of the fundamental frequency. The harmonic components present depend on the number of phases of the rectifier circuit and are indicated by the letter A in Table 2 for various numbers of phases q. In general, the larger the number of phases, the higher is the frequency of the first harmonic and the fewer are the harmonics present.

The magnitudes of the harmonics in the a-c line current are given in references 14 and 36 of the Bibliography, Article 20. In general, the higher the frequency of an harmonic, the lower is its magnitude. An increase in the phase-control angle of retard (α) of a power rectifier or the phase-control angle of advance (β) of a power inverter increases the magnitude of the harmonics.

The harmonic components of the a-c line currents of rectifier circuits may cause interference in communication circuits if they are exposed to lines of the a-c system to which the rectifiers are connected. From the viewpoint of the a-c system, a rectifier installation can be considered a generator of harmonic currents which can flow into various parallel branches of the a-c system in accordance with their impedances at the harmonic frequencies. Harmonic components in the a-c line voltage are produced by the voltage drops caused by the harmonic currents in a-c circuit impedances. The possible effect of a rectifier installation on communication circuits exposed to the a-c system is reduced as the number of rectifier phases is increased. The characteristics of the a-c system have an important bearing on this effect. A comprehensive treatment of this subject is contained in reference 45 of the Bibliography, Article 20.

Filters may be used at the a-c line terminals of a rectifier unit to reduce the flow of harmonic currents into the a-c system. Such a filter usually consists of reactors connected in series with the a-c line, and capacitors or resonant shunts connected between the line terminals on the rectifier side of the series reactors. The capacitors, or the resonant filters tuned to specific harmonics, provide a low-impedance path for the harmonic currents, while the impedance of the series reactors opposes the flow of these currents into the a-c system. In large rectifier installations, consisting of a number of units, phase multiplication, by increasing the number of rectifier phases, has effectively reduced the influence of harmonics.

PHASE MULTIPLICATION. The number of phases of a rectifier unit is determined by the number of phase-voltage vectors of the rectifier transformer d-c winding. A six-phase unit has six phase voltages displaced from each other by 60 deg. A twelve-phase unit has twelve phase voltages displaced from each other by 30 deg. A twelve-phase circuit can be obtained from two six-phase rectifier units by providing the rectifier transformer of one unit with a delta-connected a-c winding and the other with a Y-connected a-c winding. This arrangement shifts the voltage vectors of the two a-c windings by 30 deg from each other and consequently shifts the voltage vectors of the d-c windings by the same angle.

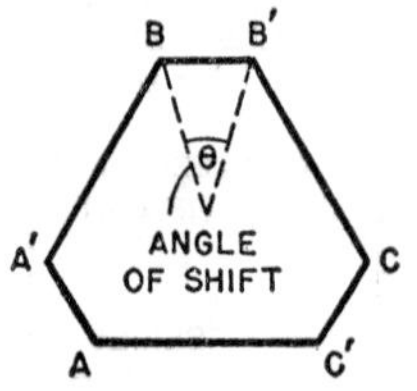

Fig. 27. Phase Shifter for Phase Multiplication

The voltage vectors of the a-c winding can be shifted by any desired angle with respect to the a-c line voltages by interposing a phase-shifting auto-transformer between the rectifier transformer and the a-c line. The voltage vectors of the d-c winding will then be shifted by the same angle. In installations having a number of rectifier units, this method has been used for shifting the voltage vectors of the individual units in order to multiply the number of rectifier phases of the installation and thus reduce the number of harmonics in the total a-c line current. The maximum number of phases obtainable is equal to the product of the number of units and the number of phases per unit. An installation consisting of six six-phase rectifier units, for example, can be converted into a thirty-six-phase system by supplying three rectifier transformers with delta-connected and three with Y-connected a-c windings and by the addition of four 10-deg phase shifters. By these means a 10-deg shift is obtained between the d-c winding voltages of successive units. The a-c line current of such an installation approaches closely a sinusoidal wave shape. The vector diagram of a phase-shifting auto-transformer used for phase multiplication is shown in Fig. 27. The angular shift is obtained without changing the magnitude of the a-c voltage. Terminals *A-B-C* may be connected to the line and terminals *A′-B′-C′* to the rectifier transformer, or vice versa, depending on the direction of shift.

Phase multiplication is most effective in reducing the harmonic content of the line current if the load is balanced among the units, which is possible if the units operate in parallel on a common d-c bus, as in electrochemical installations.

11. CIRCUIT DUTY AND FAULTS

CIRCUIT DUTY. A rectifying device must be able to carry its rated currents during the conducting period with relatively low arc drop and without overheating, and it must be able to withstand the voltage between the anode and the cathode during the scheduled non-conducting period in each cycle without failure of the solid insulation and without electric breakdown in the space between the anode and cathode. In other words, during each cycle it must serve alternately as a good conductor and a good insulator. The magnitudes and wave shapes of the anode current and anode-to-cathode voltage determine the duty imposed on a rectifying device by the circuit in which it operates. These wave shapes for a power rectifier and a power inverter, composed of three-phase commutating groups ($p = 3$), are shown in Figs. 28(*a*) and (*b*). These figures were obtained from Figs. 14 and 22, respectively. The anode-to-cathode voltages were derived from Figs. 14 and 22 by taking the difference between the voltage of one anode and the cathode potential, shown in heavy outline in these figures. During the conducting period, while the anode is carrying current, the anode-to-cathode voltage is equal to the arc drop and is therefore low. During the remainder of the cycle it is made up of portions of sine waves.

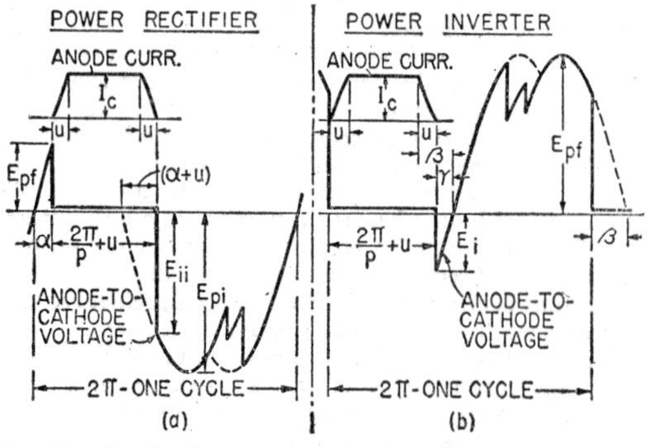

Fig. 28. Anode Currents and Anode-to-cathode Voltages of Power Rectifier and Inverter

If a power rectifier operates with retarded firing of the anodes by phase control, the anode-to-cathode voltage [Fig. 28(*a*)] becomes increasingly positive during the angle of retard α, before conduction starts. The peak forward voltage E_{pf} depends on the magnitude of α. At the conclusion of the conducting period the anode-to-cathode voltage changes suddenly from the positive value of the arc drop to a substantial negative value, which is called the **initial inverse voltage** E_{ii}. The magnitude of this voltage is a function of the angle of retard α and the commutating angle u. At lower values of these angles the anode-to-cathode voltage follows the dotted line in the figure, and the voltage E_{ii} is lower. The maximum negative voltage, called the **peak inverse voltage** E_{pi} is equal to the peak value of the maximum phase-to-phase voltage of the commutating group.

In a power inverter [Fig. 28(*b*)] the anode-to-cathode voltage becomes negative at the conclusion of the conducting period and remains negative during the margin angle γ, to permit deionization of the space about the anode and control grid, in order to keep the anode non-conductive when the anode-to-cathode voltage becomes positive. The peak forward voltage E_{pf} is equal to the peak value of the maximum phase-to-phase voltage of the commutating group. In a power inverter the anode-to-cathode voltage is positive during the greater part of the non-conducting period in a cycle, whereas in a power rectifier it is negative during most or all of the non-conducting period.

FAULTS. Two major faults can occur in a rectifying device, an arc-back and an arc-through, both signifying conduction of current by an anode during its scheduled non-conducting period in the cycle.

ARC-BACK. An arc-back is the major fault of a power rectifier. It is defined as a failure of the rectifying action which results in the flow of a principal electron stream in the reverse direction, because of the formation of a cathode spot on an anode. It can occur when the anode is at a negative potential to the cathode. When a cathode spot is established on the anode surface, current from the other anodes of the rectifier flows to the faulty anode (see Fig. 29), resulting in a short circuit on the d-c winding of the rectifier transformer, limited only by the impedance of the transformer and the a-c system and by the arc drop in the rectifying devices. The current to the faulty anode and the transformer winding to which it is connected flows in the reverse direction from normal and can attain magnitudes many times the rated current of that winding. If other sources of d-c power are connected to the same d-c circuit as the rectifier, current from the d-c circuit will also flow to the faulty anode through its cathode. The direction of current flow is from the positive side of the d-c circuit, through the faulty rectifying device and its transformer winding, to the negative side of the d-c circuit; this is the reverse direction from the normal flow of current from the rectifier to the d-c circuit. The rate of rise of the reverse current from the d-c circuit is determined by the time constant of the circuit through which it flows. Its ultimate value is limited only by the resistance of the transformer winding and the other parts of the circuit through which it flows, by the arc drop of the rectifying device, and by the voltage regulation of the d-c power source. If the d-c power source is large, such as a number of rectifier units operating in parallel with the unit which has developed an arc-back, the reverse current can reach very high values if it is not interrupted quickly.

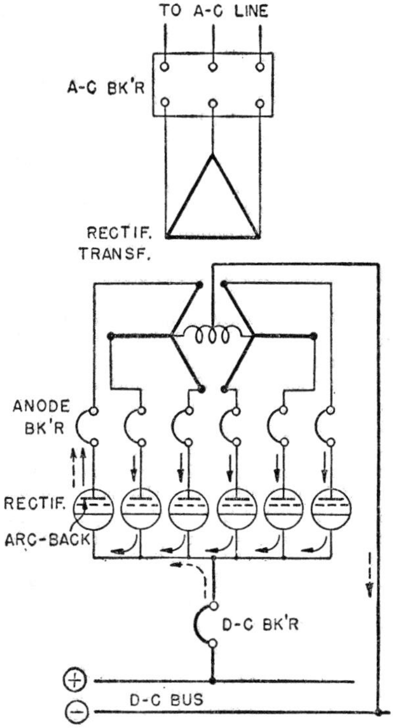

FIG. 29. Current Flow during Arc-back in Rectifier

In Fig. 29 are shown the paths of current flow during an arc-back in a power rectifier and the protective devices used for interrupting an arc-back. A six-phase double-Y rectifier circuit is shown, because it has been the most widely used power rectifier circuit. The three principal protective devices used for protection of the rectifier equipment and the a-c and d-c circuits against an arc-back are: (1) an a-c breaker between the rectifier transformer and a-c line, (2) an anode breaker between the rectifier transformer and the rectifying devices, and (3) a d-c breaker between the rectifier and the d-c load circuit. One, two, or all three of these devices may be used, depending on the specific conditions of a particular application.

The arc-back may be interrupted by the a-c and d-c breakers. Opening the a-c breaker will remove the a-c voltage from the transformer and will thus stop the flow of current to the faulty rectifying element from the other rectifying elements. The a-c breaker is tripped by means of fast overload relays connected to current transformers on the a-c side of the rectifier transformer. Opening the d-c breaker will interrupt the direct current from the d-c circuit, shown by the broken arrow in Fig. 29. The d-c breaker is tripped by means of a reverse current trip attachment. If an anode breaker is used, the arc-back can be interrupted by opening the breaker pole in the circuit of the faulty element, thus stopping the flow of current from the other rectifying elements as well as the current from the d-c circuit. The anode breaker is usually tripped by means of a reverse-current tripping device; sometimes an overload trip attachment is employed.

When an anode breaker is used for protection against arc-back, the a-c and d-c breakers may be omitted if not required for other protective or switching purposes. When a number of rectifier units in an installation are supplied from a common a-c circuit, as in an electrochemical installation, one a-c line breaker may be used for all the units to provide protection in the event of a fault in a rectifier transformer and to serve as back-up protection in the case of failure of an anode breaker. In such an installation d-c breakers will be used for protection of the rectifier unit against excessive overloads and as back-up protection for the anode breaker.

In installations with a stiff d-c system, a high-speed anode or d-c breaker having an interrupting time of about 0.015 sec or less is used to interrupt the reverse current from the d-c circuit before it reaches a high value. In installations with weak d-c systems, slower breakers may be used. With suitable protection no damage is done to a rectifier unit by an arc-back, and the unit can resume normal operation immediately.

The current supplied to a faulty rectifying element during an arc-back by the other rectifying elements of the unit can be interrupted, without opening the a-c or anode breaker, by making the other rectifying elements non-conductive. This can be accomplished by applying to the control grids of these rectifying elements a negative blocking voltage or, in single-anode rectifiers, by stopping electron emission at their cathodes. The rectifying elements which are not conducting when their conductivity is blocked are prevented from starting, and the rectifying elements which are carrying current at that time will be prevented from restarting after their current has dropped to zero. This method cannot be used to interrupt the current from the d-c circuit to the faulty rectifying element; this current has to be interrupted by a circuit breaker.

ARC-BACK CAUSES. For d-c voltages of about 500 volts and higher, the frequency of arc-back has been a major limitation in the rating of mercury-arc rectifiers. The tolerable frequency of their occurrence depends on the type of service. An average of one or two arc-backs per month for a twelve-anode rectifier has been considered acceptable for most rectifier applications. Although in some installations they have occurred more frequently, the average in many installations has been less than one per month per rectifier, and some rectifiers have operated for a year and longer without any arc-backs.

The occurrence of arc-backs is random. The service duty and the condition of a rectifier have a bearing on its tendency to arc-back. Poor vacuum or excessive temperature will increase the probability of arc-backs. It has been found from tests and operating experience that most arc-backs, particularly in the medium and lower ranges of d-c operating voltages, occur at the conclusion of the conducting period of an anode, when its voltage to the cathode suddenly reverses. It is the generally accepted theory that these arc-backs are caused by the residual ionization in the vicinity of the anode as a result of the forward current during the conducting period (see Article 3). When the anode-to-cathode voltage reverses, the positive ions are driven to the anode, producing a small negative current called the **inverse current.** The inverse current attains a peak value when the anode voltage reverses and then declines to nearly zero in a relatively short time.

When the ion density at some sensitive point on the anode surface reaches a critical value, it may produce a sufficiently high potential gradient at that point to initiate a cathode spot and result in an arc-back. An insulating particle on the anode surface, such as a small grain of sand, facilitates this process; positive ions accumulating on such a particle can produce a high potential gradient at that point of the anode surface. The greater the density of residual ions and the higher the initial inverse voltage, the higher is the peak value of the inverse current and the greater the probability of initiating a cathode spot. The residual ion density is a function of the mercury-vapor density and the magnitude of the anode current. It may also be a function of the rate of decline of the anode current during the commutation period. As the current declines, deionization reduces the ion density, the rate of deionization depending on the space between the anode and the adjacent grid or baffle surfaces, which facilitate deionization. If the current declines rapidly, there is less time for deionization, and the residual ion density is greater.

Operation of a power rectifier with retarded firing by phase control results in a higher initial inverse voltage and a steeper rate of change of the anode current at the end of the conducting period, both of which increase the inverse current and the probability of arc-back. High overloads also increase the ion density and the initial inverse voltage, because of a greater commutating angle, thus increasing the probability of arc-back.

Another vulnerable part of the inverse period, from the viewpoint of arc-back probability, is the region of the peak inverse voltage, particularly for rectifiers operating at high d-c voltages. These arc-backs are usually caused by dielectric breakdown of the mercury vapor, and high-voltage rectifiers have to be operated at a lower tank temperature than low-voltage rectifiers, in order to reduce the mercury-vapor pressure.

Arc-backs can occur also in a power inverter during the margin angle γ, when the anode-to-cathode voltage is negative [see Fig. 28(*b*)]. The resulting fault current is relatively small because of the short duration of the negative voltage, and the arc-back can extinguish itself. An arc-back in an inverter, however, may result in an arc-through when the anode voltage becomes positive.

ARC-THROUGH. An arc-through is the major fault of a power inverter. It is defined as a loss of control resulting in the flow of a principal electron stream through the rectifying element in the normal direction during a scheduled non-conducting period. An arc-through can occur during the part of the cycle when the anode-to-cathode voltage is positive [see Fig. 28(*b*)]. An arc-through may be caused by failure of the control grid to keep the anode non-conducting during this part of the cycle, either because there is a fault in the external grid-control circuit or because the blocking action of the grid is overpowered by the positive anode voltage. An arc-through will result if the margin angle γ is zero or too short, so that the space about the anode and control grid cannot be deionized before the anode-to-cathode voltage becomes positive. The anode will then continue or resume conduction during that period. An arc-through will also occur if the a-c voltage applied to the inverter transformer disappears, becomes single-phase, or drops too low, so that the current of the conducting anode cannot be commutated away by the succeeding anode.

The effect of an arc-through is that the d-c voltage of the inverter changes from its normal negative value to a positive value (or to zero, if the a-c voltage disappears). The voltage of the inverter is then in the same direction as the d-c voltage applied to it (see Fig. 21), resulting in a short circuit through the conducting rectifying element and the transformer winding connected to it. The resulting current flow is determined by the impedance of the circuit. An arc-through can be cleared by removing the d-c voltage applied to the inverter, by opening a circuit breaker in the d-c circuit, or by other means. If the d-c voltage is supplied by a power rectifier, it may be interrupted by opening the a-c breaker of the power rectifier or by making the rectifier non-conductive through grid or firing control. Under certain conditions an arc-through can clear itself without removing the d-c voltage. This is possible if the arc-through occurs late in the positive non-conducting period, or if the rate of current rise in the short-circuit is kept sufficiently low by inductance, so that the current does not reach the commutation limit (when the commutating angle u' is equal to the angle of advance β) before it can be commutated away by another rectifying element.

An arc-through can occur also on a power rectifier operating with phase control, during the angle of retard α, when the anode-to-cathode voltage is positive [see Fig. 28(*a*)]. This results in a momentary increase of the d-c output voltage but is not likely to cause a service interruption unless it recurs for many cycles.

MISFIRE. There is one other operating fault of rectifying devices, called **misfire,** which is defined as a failure to establish an arc between the main anode and cathode during a scheduled conducting period. It may be caused by failure to establish a cathode spot in rectifying devices with cyclic ignition or by failure of an anode to start conduction because of insufficient ionization. A misfire in a power rectifier is not likely to cause a service interruption unless it recurs for many cycles. In a power inverter a misfire is likely to result in an arc-through, because the failure of an anode to fire will permit the preceding anode to continue firing during the positive non-conducting period.

12. MERCURY-ARC RECTIFIER CONSTRUCTION

RECTIFIER TYPES. From the point of view of their construction there are two general classes of mercury-arc rectifiers with pool cathodes: the **multianode** rectifier and the **single-anode** rectifier. In a multianode rectifier a number of main anodes, usually six, twelve, or more, are contained in a single vacuum envelope and have a common mercury-pool cathode. In a single-anode rectifier one main anode and its mercury-pool cathode are contained in a vacuum envelope.

Until about 1940 practically all mercury-arc rectifiers were of the multianode type. The introduction of the single-anode ignitron-type rectifier changed the practice in this country, and single-anode rectifiers are now used in practically all new rectifier installations, with six or twelve single-anode tanks or tubes constituting a rectifier assembly.

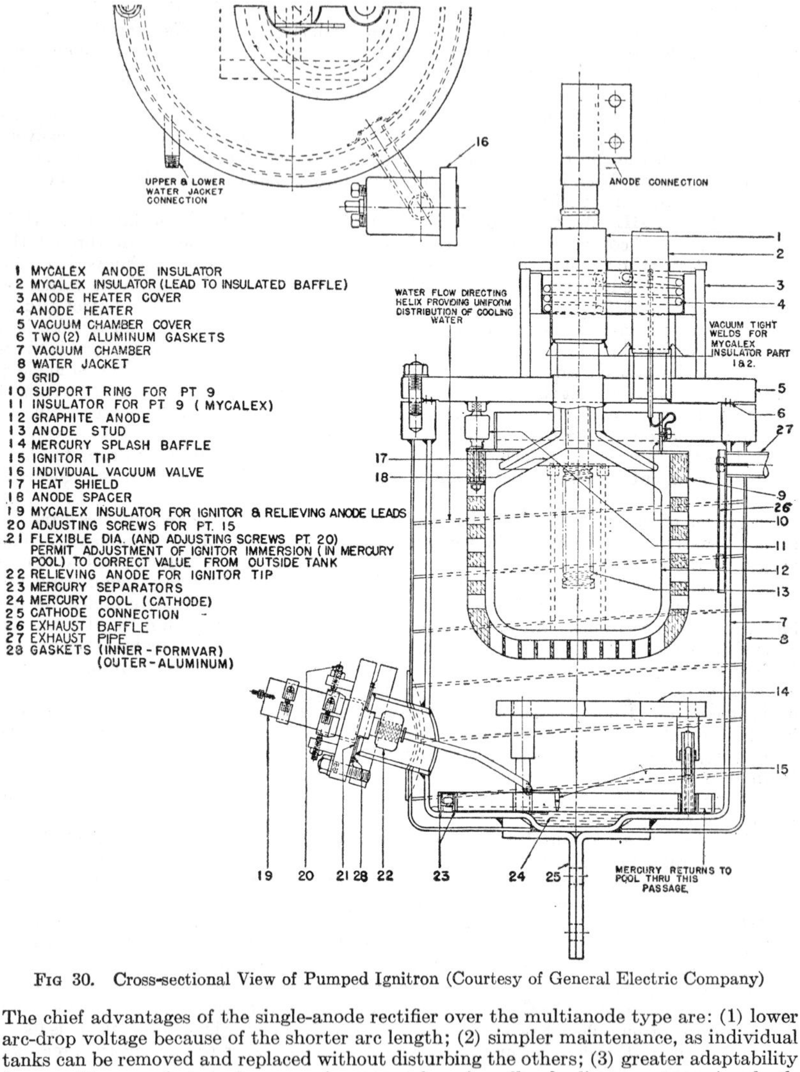

FIG 30. Cross-sectional View of Pumped Ignitron (Courtesy of General Electric Company)

The chief advantages of the single-anode rectifier over the multianode type are: (1) lower arc-drop voltage because of the shorter arc length; (2) simpler maintenance, as individual tanks can be removed and replaced without disturbing the others; (3) greater adaptability for quantity manufacture, because a larger number of smaller duplicate parts are involved; it is also possible to obtain more unit ratings from one size of tank by using different numbers of tanks per unit, thus reducing the number of tank sizes required.

There are two types of single-anode rectifiers, the **ignitron** and the **excitron.** An ignitron is a single-anode pool-cathode tube in which an ignitor is employed to initiate the cathode spot before each conducting period. An excitron is a single-anode pool-cathode tube provided with means for maintaining a cathode spot continuously.

There is still another classification of mercury-arc rectifier types: the **pumped** rectifier, which is connected continuously to evacuating equipment during operation; the **sealed** tube, which is hermetically sealed after degassing.

Most mercury-arc rectifiers used for power conversion in this country have a metal envelope and are water cooled. The anodes, grids, and baffles are usually made of graphite. Several types and makes of single-anode rectifiers are described in the following paragraphs, and their construction is shown in the accompanying figures.

IGNITRONS. The operation of an ignitron is based on the ability to initiate a cathode spot on the mercury pool by means of an ignitor at a precise and controllable instant of

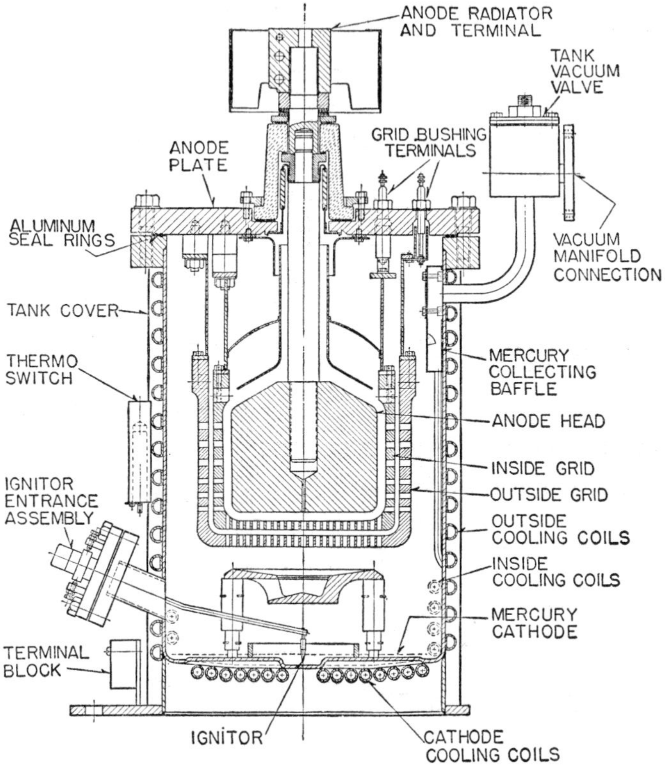

FIG. 31. Cross-sectional View of Pumped Ignitron for 3000–4000 Volts Direct Current (Courtesy of Westinghouse Electric Corporation)

each cycle. The ignitor is a stationary pencil-like electrode made of refractory material, such as boron carbide, partly immersed in the cathode mercury. When a pulse of current is passed through the ignitor into the mercury, a high-voltage gradient is established at its junction with the mercury surface, which initiates electron emission and establishes a cathode spot. The main anode then can start conduction, and cathode spots are maintained by the main-anode current during the conducting period of the cycle. When the current of the main anode drops to zero at the conclusion of its conducting period, the cathode spot disappears, and there is no electron emission at the cathode until a cathode spot is initiated by the ignitor to start the conducting period in the next cycle. Some ignitrons have an auxiliary anode to relieve the duty on the ignitor or to facilitate the "pick-up" of the main anode.

A cross-sectional view of one type of pumped ignitron for d-c operating voltages up to about 900 volts is shown in Fig. 30. The various component parts are designated in the accompanying legend. The anode plate, on which are mounted the anode assembly and grid, is bolted to the tank flange and sealed to it with a gasketed vacuum-tight joint. The tank is water cooled and is provided with a vacuum valve for connection to the evacuating system of the rectifier unit. The ignitor assembly is bolted to a side entrance of the tank with a gasketed vacuum-tight joint, and the position of the ignitor tip in the mercury is adjustable from the outside. The insulating bushings for the main anode, grid connection, and ignitor are made of Mycalex, which is an insulating material composed of mica and glass molded at high temperature and pressure together with the adjoining metal parts. The anode heater is used to prevent condensation of mercury vapor on the anode assembly at light-load currents. The grid serves as an auxiliary anode and as a deionizing member. The mercury splash baffle shields the anode against the mercury vapor and droplets generated at the cathode.

A cross-section of another type of pumped ignitron is shown in Fig. 31. This ignitron is designed for a d-c operating voltage in the range of 3000 to 4000 volts as a power rectifier and a power inverter. It is provided with two grids. Copper coils for the circulation

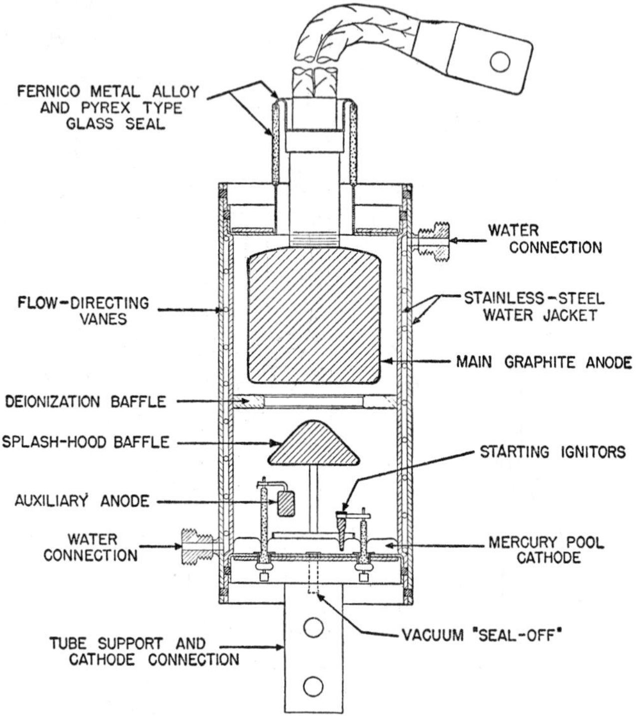

FIG. 32. Sealed Ignitron Tube (Courtesy of General Electric Company)

of cooling water are soldered to the outside surface of the tank. The anode stud is sealed to and insulated from the anode plate by means of a glass-to-metal (Kovar) seal, which is enclosed by a porcelain insulator. Lower-voltage ignitrons of the same make have a single grid, and their anode bushings are of porcelain with porcelain-to-metal soldered vacuum-tight seals. Other elements in Fig. 31 are similar to those in Fig. 30.

In Fig. 32 is shown a cross-section of a sealed ignitron tube. The tube envelope and water jacket are of stainless steel to minimize corrosion and to prevent diffusion of hydrogen from the water through the vacuum envelope. The anode bushing is made of glass fused to metal of a special alloy (called Fernico or Kovar), which has substantially the same coefficient of expansion as the glass over the range of temperature through which they pass during and after the fusion process. The same type of glass-to-metal seal is used for the inlet connections to the ignitors and auxiliary anode. The sealed tubes usually have two ignitors, one being a spare. The tube envelope, as well as the main anode and other internal parts, is degassed at an elevated temperature during the manufacture of the tube, before the evacuating connection is sealed off.

Ignitron Excitation. In Fig. 33 is shown a diagram of the ignitor excitation circuit for a pair of ignitrons which fire 180 deg apart. For a six-phase ignitron rectifier this diagram is repeated three times. The reactors L_c and L_1 and the capacitor C_1 constitute a phase-shifting network. The a-c voltage applied to the rest of the circuit can be shifted in phase by varying the direct current through the control winding of the saturable reactor L_c, which controls the phase position of the current pulse to the ignitor, in relation to the main-anode voltage, for phase control of the rectifier d-c voltage. The reactors L_2 and L_3 and the capacitor C_2 constitute a pulse-forming network. The capacitor C_2 is charged, through the current-limiting linear reactor L_2, by the a-c voltage. As the voltage across

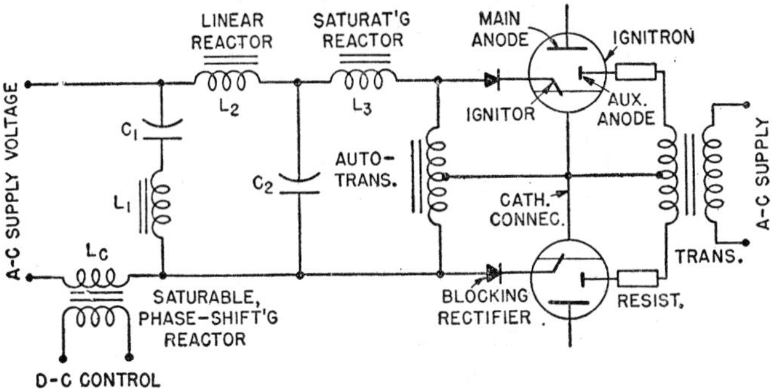

FIG. 33. Ignitron Excitation Circuit (Courtesy of General Electric Company)

C_2 increases, a small but increasing current flows to the ignitor through saturating reactor L_3, which has a high impedance below its saturation point. This current is too small to start a cathode spot on the cathode mercury. When the current through reactor L_3 reaches a predetermined critical value, it saturates, and its impedance drops to a low value. This allows capacitor C_2 to discharge through the ignitor, and its current quickly reaches the value required to initiate a cathode spot, which makes the ignitron conductive. The return circuit for the ignitor current is through a portion of the auto-transformer whose midpoint is connected to the cathode. During the second half-cycle of the a-c voltage applied to the ignitor circuit, a current pulse is applied to the ignitor of the other ignitron. A blocking rectifier, which may be of the copper oxide or selenium type, is connected in series with each ignitor to block the flow of negative current through it.

The minimum peak volt-ampere requirements for firing an ignitor, when using the excitation system of Fig. 33, are given by the curve of Fig. 34. The point on the curve at which an ignitor fires is determined by its resistance. The wave shape of the ignitor current pulse is shown in the inset of Fig. 34.

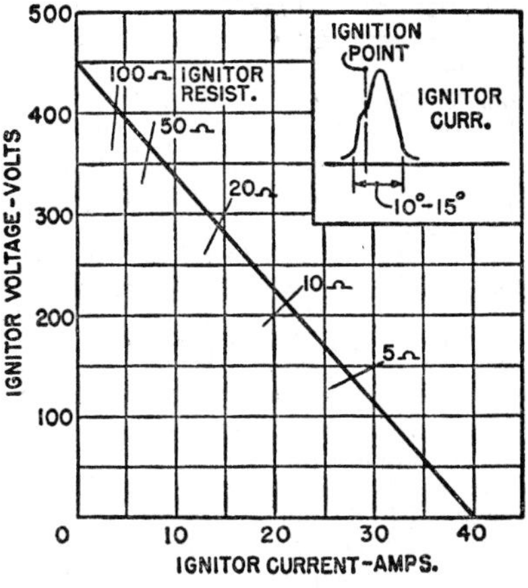

FIG. 34. Minimum Volt-ampere Requirements for Firing Ignitor

If the ignitrons have auxiliary anodes, they are usually connected to a source of a-c voltage through current-limiting resistors, as shown in Fig. 33. The a-c voltage supply for the ignitors and auxiliary anodes must be derived from the a-c system to which the rectifier is connected, in order to maintain the required phase relation to the voltages of the main anodes. The phase position of the ignitor firing point can be controlled by d-c magnetization of the saturable reactor L_c, either manually, or automatically by means of a regulator in response to the d-c voltage or some other circuit quantity.

Several other ignitor excitation systems have been used for ignitrons. In one system the ignitor is connected to the source of firing current through a thyratron tube, and the firing is controlled by grid control of the thyratron tube.

EXCITRON. In an excitron a cathode spot is maintained continuously at the cathode, when in operation, by means of an auxiliary arc from an excitation anode. The firing of the main anode is controlled by means of a control grid. With a continuous excitation arc maintained at the cathode, it is necessary to insulate the tank from the cathode or to shield the tank from the arcing space, in order to prevent establishment of a cathode spot on the tank walls, which could be maintained there by the excitation arc. Although a cathode spot may also be established on the walls of an ignitron, it is extinguished at the end of the conduction period, and preventive measures are not necessary.

A cross-section of a pumped excitron is shown in Fig. 35. The tank and the cathode plate constitute a single structure, similar to the rectifiers of Figs. 30 and 31. The tank is shielded from the arc by means of a helical cooling coil, with its turns separated by narrow gaps, which permit passage of mercury vapor and gases but prevent arc passage. The cooling coil is supported and insulated from the anode plate by means of insulating seals through which its terminals are brought out. The excitron is cooled by water circulated through the cooling coil, in series with the cathode water jacket. The main and excitation anodes, grid, and baffles are made of graphite. The anode stud is brought out through a ceramic bushing in the anode plate, with vacuum-tight seals consisting of gaskets covered by mercury. The excitation-anode and grid connections are also brought through insulating seals in the anode plate. The cathode-mercury pool is enclosed by a quartz ring.

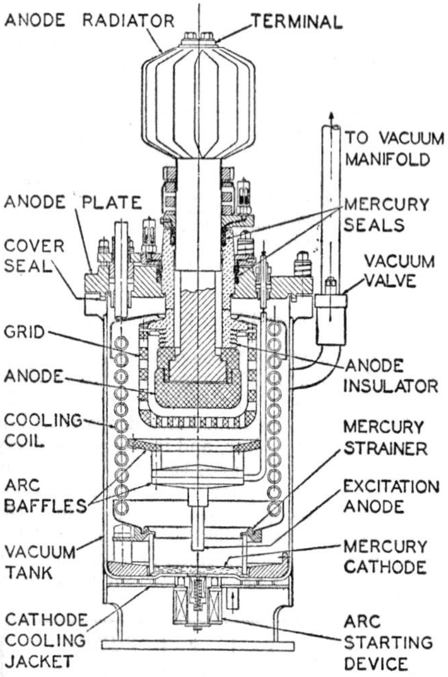

FIG. 35. Cross-sectional View of Pumped Excitron (Courtesy of Allis-Chalmers Manufacturing Company)

The excitation anode is located above the middle of the cathode and is connected to it externally in series with a source of d-c voltage. The arc-starting device consists of a nozzle-shaped steel plunger in a cylinder, which is welded to the cathode plate and filled with mercury from the cathode pool. Over the outside of the cylinder is mounted an ignition coil. The excitation arc is started by energizing the ignition coil; the plunger is pulled down, sending up a jet of mercury which momentarily short-circuits the excitation anode to the cathode. When the jet is broken, the excitation arc is established, with a cathode spot on the mercury pool, and is maintained continuously while the rectifier is in operation.

Excitron Excitation Circuit. A diagram of an excitron excitation circuit is shown in Fig. 36. The d-c excitation voltage of about 35 to 40 volts is supplied from an a-c source through an excitation transformer and a six-phase, double-way metallic rectifier (selenium or copper oxide). The excitation anode of each excitron is connected to the d-c supply in series with a reactor, a current-limiting resistor, and a relay. The ignition coil is connected to the a-c supply through a normally closed contact of the excitation relay. When a-c voltage is applied to the excitation circuit, the ignition coils are energized and initiate the excitation arcs. The resulting current in the excitation anode circuit operates the relay, which in turn opens the ignition coil circuit. The excitation current of an excitron is about 7 to 8 amp.

At the right of Fig. 36 is shown an excitation circuit for the control grids of the excitrons. The grids are supplied by a-c voltages derived from the same a-c system as the main-anode voltages and having the correct phase relation to the anode voltages. The six-phase secondary of the grid transformer is connected to the grids of a six-phase rectifier through current-limiting resistors. The grids control the starting point in the cycle of the anode conducting period. An anode is permitted to start firing when the grid-to-cathode voltage changes from a negative to a positive value. For regulating the rectifier d-c voltage by phase control, this point may be varied by shifting the phase position of the grid voltages by means of a phase shifter in the a-c supply circuit to the grid transformer. It may also be shifted by varying the d-c bias voltage between the neutral of the grid transformer and the cathode by means of a regulating rheostat, as shown in Fig. 36; this raises or lowers the sinusoidal grid voltage waves in relation to the cathode potential and thus shifts the phase position of their zero intersection points. The d-c bias supply may be obtained from the auxiliary a-c voltage supply through a transformer and a metallic rectifier, or the d-c output voltage of the rectifier may be used as the bias supply. The bias voltage may be controlled automatically, by means of a regulator, in response to the rectifier d-c voltage or some other circuit quantity.

Various other grid-control systems have been used for rectifiers. In some systems the grid voltage wave has a steep rate of rise when changing from negative to positive. Such voltages may be produced by means of saturable reactors or transformers or by electronic means.

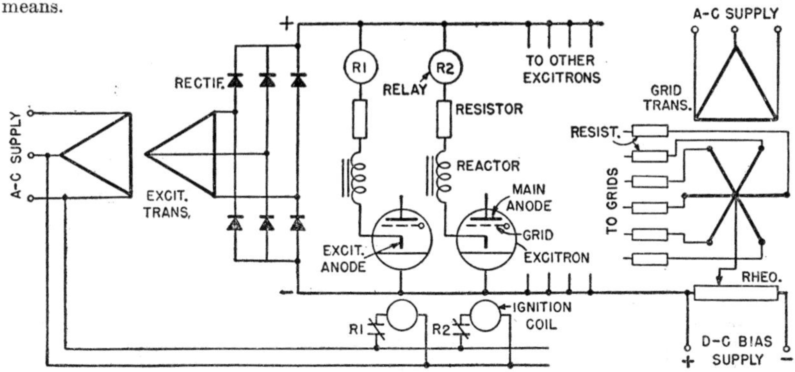

FIG. 36. Excitation and Grid-control Circuit of Excitron (Courtesy of Allis-Chalmers Manufacturing Company)

13. MERCURY-ARC RECTIFIER AUXILIARIES

In addition to the excitation equipment the major auxiliaries of mercury-arc rectifiers are the evacuating equipment for pumped rectifiers and the cooling equipment. These auxiliaries and their connection with a rectifier consisting of single-anode tanks are shown diagrammatically in Fig. 37.

EVACUATING SYSTEM. For satisfactory operation a mercury-arc rectifier must be reasonably free of gases or vapors other than mercury vapor. Some gases are given off by the rectifier materials while in operation, and minute quantities of air seep through the joints of the vacuum envelope. The function of the evacuating system is to maintain and indicate the vacuum in a rectifier. The evacuating equipment consists of vacuum pumps, vacuum gages, and their connections to the rectifier. As shown in Fig. 37, the rectifier tanks are connected to the vacuum header or manifold through individual vacuum valves, and the vacuum header is connected to the vacuum pumps and gages through a vacuum valve and piping. The purpose of the vacuum valves is to make it possible to isolate individual parts of the vacuum system for testing and servicing.

The gases from the rectifier are compressed for discharge to the atmosphere by two vacuum pumps connected in series, a mercury-condensation pump and a rotary pump. In the mercury-condensation vacuum pump, mercury is heated by means of an electric heater. The mercury vapor rising through a nozzle is deflected downward by a deflecting baffle and carries along gas molecules which have diffused from the rectifier to the gap between the baffle and the pump casing. The mercury vapor is condensed on the walls of the casing, which is cooled by water or air, and returns to the mercury pool. The gas molecules, because of the kinetic energy imparted to them by the mercury vapor, continue to move downward toward the connection to the rotary vacuum pump. Some mercury-condensation pumps are multistage, with two or three nozzles in tandem.

The rotary vacuum pump is oil sealed and is driven by a small motor. It consists of a cylinder with an eccentrically mounted rotor provided with sliding vanes. The gases received from the mercury-condensation pump are compressed in the cylinder by the rotation of the vanes and discharged to the atmosphere through a check valve and the oil. When the rotary vacuum pump stops, a vacuum valve on the suction side closes automatically to prevent oil or oil vapor from backing up into the mercury-condensation pump and the rectifier.

The pressure in vacuum systems is usually expressed in microns absolute. One micron is the pressure which will support a column of mercury $^1/_{1000}$ mm high. Two types of vacuum gages are generally used for measuring the pressure in the vacuum system of a rectifier, the **McLeod** gage and the **hot-wire** gage, which are shown in Fig. 37. The McLeod vacuum gage operates on the principle that pressure × volume of a gas is constant. Its measuring element is a glass tube of small diameter, sealed off at one end and joined at the other end to a bulb, which is connected to the vacuum system and therefore holds a sample of gas from it. The pressure of the gas sample is measured by slowly raising the mercury in the reservoir below the bulb to a preset height in the tube parallel to the measuring tube. The mercury compresses the gas sample into the measuring tube; the difference in the heights of mercury columns in the measuring tube and the parallel tube is a measure of the gas pressure and can be read on a scale alongside the measuring tube. The scale is marked in microns, and its calibration is calculated from the measured volumes of the sampling bulb and the measuring tube.

The operation of the hot-wire gage is based on the principle that the heat conductivity of a gas is a function of its pressure. The gage consists of four resistors connected in the

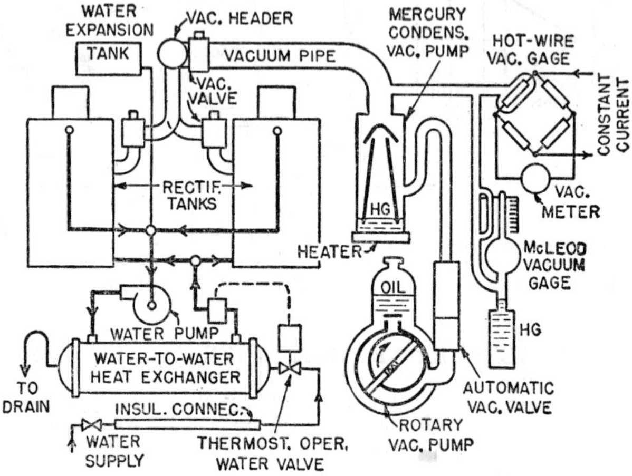

FIG. 37. Auxiliaries of Mercury-arc Rectifier

form of a Wheatstone bridge. Constant current is supplied to two opposite junction points of the bridge, and an indicating instrument is connected to the other two points. One arm (or two arms) of the bridge, made of material having a high temperature coefficient of resistivity, is sealed into a branch of the vacuum system This resistor is heated by the current; the lower the pressure, the less heat is conducted away from it, and the higher are its temperature and resistance. The change in resistance is therefore a measure of the pressure in the vacuum system and is indicated on the vacuum meter, which is calibrated in microns. The hot-wire vacuum gage, unlike the McLeod gage, provides a continuous indication of the vacuum. Contacts on the vacuum meter, or a relay connected in parallel with it, may be used for protection of the rectifier against operation with a poor vacuum or for controlling the vacuum pumps.

COOLING SYSTEM. The power loss in a rectifier, equal to the product of the arc-drop voltage and the current, is converted into heat, and most of the heat is transmitted to the cooled surfaces of the rectifier by radiation and by condensation of mercury vapor. The heat is removed from the cooled surfaces by the cooling medium, which may be water or air. The function of the cooling system is to remove the heat generated in the rectifier and to maintain its operating temperature within the optimum range. The temperature of the mercury-condensing surfaces determines the mercury-vapor pressure. Too high a temperature increases the probability of arc-back. Too low a temperature increases the arc-drop voltage, and an excessively low temperature may cause voltage surges when the mercury-vapor density is inadequate to provide the needed ionization for the current conducted by the rectifier. For rectifiers operating at d-c voltages below about 1500 volts, the normal inlet water operating temperature is usually in the range of 45 to 55 deg cent. At higher d-c voltages the operating temperature is usually lower.

In Fig. 37 is shown a typical cooling system of a water-cooled mercury-arc rectifier, using a water-to-water heat exchanger. The cooling circuits of the individual rectifier tanks are connected in parallel. The cooling water is circulated in a closed circuit between the rectifier and the heat exchanger by a motor-driven water pump. In the heat exchanger the circulated water is cooled by raw water from a water supply, which flows through the tubes of the heat exchanger and is discharged to the drain. The flow of supply water is regulated by means of a thermostatically controlled valve, in response to the temperature of the recirculated water, to maintain the desired rectifier temperature. The closed cooling system is provided with an expansion tank to permit expansion of the water when heated and to prevent entrance of air by keeping the system full. If the rectifier is insulated from ground, insulating water connections are used between the heat exchanger and the grounded water supply or between the rectifier and the heat exchanger. Rubber hose or pipes of insulating material may be used for this purpose.

The recirculated water of a rectifier may be cooled in a water-to-air heat exchanger, by air from a motor-driven fan or blower passing over the surfaces of a radiator, through which the water is circulated. The temperature is regulated by a thermostatically controlled valve which regulates the proportion of water by-passed around the radiator. The temperature may be regulated also by controlling the operation of the blower or fan.

Heat exchangers are generally used for cooling rectifiers, in order to prevent corrosion by oxygen or other corrosive agents contained in water and to prevent deposition of solids in cooling passages. The water in the closed system is usually treated with sodium chromate or other chemicals to inhibit corrosion. If the water of the cooling system is exposed to freezing temperatures, an antifreeze solution is added. If a rectifier is subjected to intermittent service or to light load for extended periods in locations with low ambient temperatures, electric heaters are sometimes installed in the cooling system to prevent starting or operating the rectifier at too low a temperature.

If the cooling surfaces of a rectifier are made of stainless steel, as they generally are in sealed-tube rectifiers, and if a supply of good-grade cooling water is available (see reference 44 of the Bibliography, Article 20), a direct raw-water cooling system is sometimes used, in which the supply water is passed directly over the cooling surfaces of the rectifier. There are two types of such cooling systems. In one type the water from the supply is passed over the cooling surfaces and is discharged to the drain. In the other type the water is recirculated by a pump, additional raw water being admitted as needed to maintain the required temperature, and the excess water is discharged to the drain. The second system permits a higher speed of water flow and better control of the operating temperature. The flow of raw water is regulated by thermostatically controlled water valves.

Water-cooled mercury-condensation vacuum pumps and vacuum pipes are usually cooled by supply water. The vacuum pipes between the rectifier tanks and the evacuating equipment are cooled to prevent transfer of mercury.

14. RECTIFIER TESTS

The tests made on the equipment of a rectifier unit and the accepted test procedures are given in reference 44 of the Bibliography, Article 20. A list of tests is given below.

The usual testing program for the equipment of a rectifier includes some or all of the following tests:

1. *Rectifier.*
   a. Degassing and vacuum test (leakage test).
   b. Dielectric-strength test.
   c. Arc-drop measurement (by means of an oscillograph).
2. *Rectifier Auxiliaries.*
   a. Dielectric-strength test.
   b. Loss measurement.
3. *Rectifier Transformer.*
   a. Loss measurement.
   b. Ratio, polarity, and phase relation check.
   c. Measurement of commutating reactance.
   d. Dielectric-strength test.
4. *Rectifier Unit.*
   a. Polarity and phase relation check.
   b. Load and temperature tests.
   c. Phase control test.
   d. Efficiency determination.
   e. Power-factor and voltage regulation.
   f. Wave-shape tests.

Degassing, for the purpose of removing the occluded gases from the internal parts of the rectifier, is part of the manufacturing process. The internal parts are heated by passing current through the rectifier at low voltage or by other means, and the liberated gases are evacuated. After installation, pumped rectifiers usually are degassed a second time before they are put into operation.

The general method for measuring the commutating reactance of a rectifier transformer is to short-circuit the line terminals of the a-c winding and apply a single-phase voltage between two terminals of the d-c winding between which commutation occurs in rectifier operation. The commutating reactance is calculated from the measured input volts, amperes, and watts.

The overall efficiency of a rectifier unit is determined by calculation from the measured segregated losses of the component parts. The power-factor and voltage regulation are determined by calculation on the basis of separately measured characteristics of the rectifier and transformer equipment.

15. APPLICATION OF MERCURY-ARC RECTIFIERS

GENERAL. It has been found generally more economical to obtain d-c power by conversion from a-c power than to generate it as d-c power by prime movers. Alternating-current power can be converted to d-c power by means of motor generators, rotary converters, or rectifiers. The commercial application of power rectifiers in this country was started about 1925. Their use has increased greatly since then, and they are now the preferred converters for most applications requiring d-c power at 250 volts or higher.

The following characteristics of mercury-arc power rectifiers have given them precedence over the rotating types of a-c to d-c converters: They are static equipment and require no special foundation and no ventilating ducts. They require no synchronizing, can be started or shut down quickly by operating circuit breakers, and are easily adaptable for automatic or remote control. They are more stable than rotating machines during disturbances on the a-c system which cause voltage or frequency fluctuations. Their maintenance and operating costs are generally lower. They have a higher efficiency than

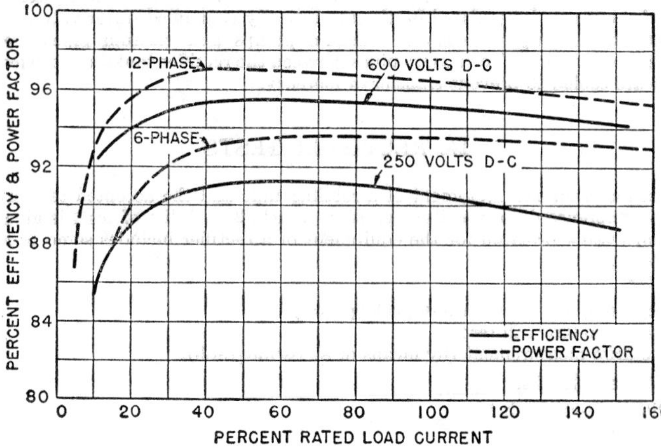

FIG. 38. Typical Efficiency and Power-factor Curves of Mercury-arc Rectifier Units (600-volt efficiency curve for 3000-kw unit; 250-volt curve for 1000-kw unit)

motor-generator sets for d-c voltages of 250 volts or higher. They have a higher full-load efficiency than rotary converters above about 400 volts, and a higher light-load efficiency even at 250 volts. The efficiency curve of a rectifier is fairly flat over the greater part of the load range and is considerably higher than that of rotating converters at light loads; this makes rectifiers particularly suitable for applications with low load factors.

Typical overall-efficiency curves of mercury-arc power-rectifier units are shown in Fig. 38. Figure 39 indicates the approximate relation between the d-c voltage and the full-load efficiency of rectifiers. As the d-c voltage is increased, the efficiency increases, because the arc-drop voltage becomes a smaller percentage of the output voltage. At higher d-c voltages the efficiency is determined largely by the efficiency of the rectifier transformer.

The application of mercury-arc rectifiers for d-c voltages below 250 volts is usually found uneconomical because they have lower efficiency and because the kilowatt rating of a rectifier at these voltages decreases in direct proportion to the voltage.

A typical mercury-arc rectifier unit consists of an a-c breaker, rectifier transformer, anode breaker (if used), rectifier and its auxiliaries, d-c breaker, and control equipment. The a-c breaker and transformer equipment may be placed outdoors or indoors, depending on local conditions. The other equipment of a unit is of the indoor type. A typical power-circuit diagram of a rectifier unit is shown in Fig. 29. The rectifier auxiliaries are described in Article 13. Several types of rectifier transformer connections are shown in Tables 1A and 1B; for most applications the six-phase double-Y connection or the twelve-phase quadruple zig-zag connection have been used. A 220- or 440-volt a-c auxiliary

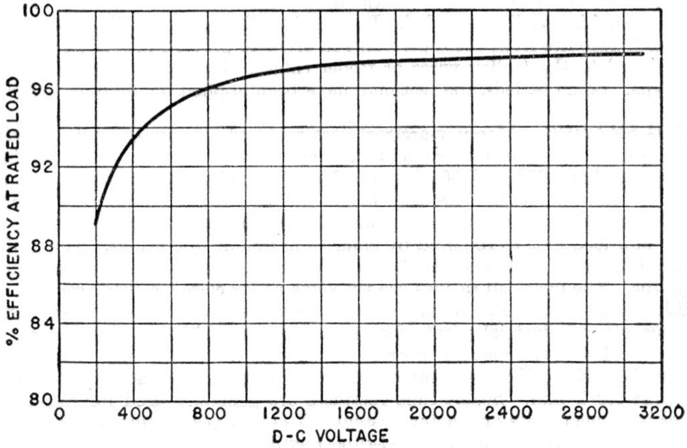

FIG. 39. Relation between D-c Voltage and Efficiency of Mercury-arc Rectifier Units

power supply is required for the auxiliary and control equipment and is usually obtained from the a-c power circuit through an auxiliary transformer. A storage battery is generally used for tripping the a-c breaker.

Various fields of application for mercury-arc rectifiers and salient factors relating to these applications are described in the following paragraphs.

ELECTROCHEMICAL SERVICE. This is the largest field of application of rectifiers, which are used for the electrolytic production of aluminum, magnesium, zinc, chlorine, hydrogen, and some chemicals. The load circuit generally consists of many cells connected in series, through which the current flows. The rectifiers are in continuous operation at approximately rated load current. Pumped-type rectifiers have been used for this service, except on some low-power applications where sealed tubes are used.

The largest total of rectifier kilowatt capacity is used for the production of aluminum. A typical aluminum-reduction circuit, called a **potline,** operates with about 55,000 amp at approximately 650 volts. This power is usually supplied by six 10,000-amp six-phase rectifier units operating in parallel. Each unit consists of two rectifiers connected to one transformer. Each rectifier is provided with a six-pole high-speed anode breaker and a semi-high-speed d-c breaker. Phase-shifting auto-transformers are connected on the a-c side of the rectifier transformers for phase multiplication, making the group of six units operate as a thirty-six-phase rectifier system; this suppresses most of the harmonic components of the a-c line current to avoid interference in telephone circuits. The rectifier units are connected to the a-c line through a common a-c breaker. The potline load current is switched on by making all the parallel rectifiers conductive simultaneously, by grid or firing control, after all the circuit breakers have been closed. The load current is switched off by simultaneous tripping of all the d-c breakers.

A typical magnesium **cell-line** is similar to an aluminum potline in rating and load characteristics. Typical chlorine cell-lines operate with 7500 to 10,000 amp at 500 to 700 volts, but there is a trend toward still higher currents with possibly lower voltages. A typical zinc cell-line operates with about 10,000 amp at about 600 volts. A hydrogen cell-line operates with about 10,000 amp at 600 to 800 volts.

The counter-emf voltages of the load circuit, as percentages of the voltages at rated current, for the various electrochemical applications, are approximately as follows: aluminum, 30 per cent; magnesium, 35 per cent; chlorine, 65 per cent; zinc, 75 per cent; hydrogen, 70 per cent.

RAILWAY SERVICE. The ratings of rectifier units for this application range between about 500 kw and about 3000 kw. The usual d-c operating voltage is about 600 volts for city railway systems, 600 or 1500 volts for suburban or interurban railways, and about 3000 volts for main-line railroad electrification. Units for railway service are rated at 150 per cent of full-load current for 2 hr and have the following shorter time overload ratings: For class I (light-duty) railway service, 200 per cent of full-load current for 1 min; for class II (medium-duty) railway service, 300 per cent of full-load current for 1 min; for class III (heavy-duty) railway service, 300 per cent of full-load current for 5 min. The rectifier substations usually are distributed over the railway system with one or two units per station, and the units are operated by automatic or supervisory control. The negative terminal is usually grounded. In some installations wave filters are required for filtering the harmonics from the d-c voltage in order to prevent interference in telephone circuits exposed to the trolley wires or feeder cables. Most of the rectifiers are of the pumped type. Some sealed-tube rectifiers have been applied for units of lower kilowatt rating.

MINING SERVICE. Mercury-arc rectifiers are used for supplying d-c power in coal mines and other mines, for haulage, and for other mine operations. For coal mines, units rated between about 200 and 500 kw, at about 275 volts direct current, are generally selected. Sealed-tube rectifiers are used for most of these applications, and the equipment is usually mounted on low-headroom mine cars which can be moved through the mine tunnels to new locations as required by changes in mining operations. The units usually are operated automatically, and the d-c voltage is regulated by phase control with a voltage regulator. The overload ratings are the same as for class I railway service.

INDUSTRIAL SERVICE. Rectifiers are used for supplying d-c power in steel mills and various other industrial plants for cranes, motors, lifting magnets, etc. Units of various ratings between about 200 kw and 1500 kw, at about 250 volts direct current, are used for such service. They are rated at 125 per cent of full-load current for 2 hr and 200 per cent for 1 min. Sealed-tube rectifiers are generally used for ratings of about 500 kw or less; such units are factory assembled in cubicles of unit-substation-type construction. For higher ratings, pumped-type rectifiers are usually applied. In installations requiring a three-wire d-c circuit, a two-machine d-c balancer set is generally used. A three-wire circuit can also be obtained with a six-phase double-way rectifier circuit.

High-voltage rectifiers can be used to supply plate power for radio-frequency oscillators used for induction or dielectric heating, when high power is required.

FREQUENCY CHANGERS. Electronic frequency changers using mercury-arc rectifiers have been applied to provide a flexible non-synchronous tie for transfer of power between a-c power systems of different frequencies. The basic circuit of such a frequency changer is shown in Fig. 21. Power taken from one a-c system is converted to direct current through a power rectifier; it is then converted to a-c power into the second power system through a power inverter. The d-c circuit serves as the flexbile link between the power systems. Power can be made to flow in either direction, as determined by the grid or firing control circuit. The amount of power transmitted is regulated by phase control. The electronic frequency changer operates with a leading power factor on the output side and can be used only if the system can receive power with a leading power factor.

Transmission of power by direct current over long distances is possible by using the transmission line as the d-c link between a power rectifier at the sending end and a power inverter at the receiving end. The power may be transmitted at a high d-c voltage by connecting in series a number of units at each end of the line. Experimental work on this method of power transmission is in progress in this country and in Europe.

Another type of electronic frequency converter using a mercury-arc rectifier has been applied for converting a-c power of low frequency (such as 60 cycles) to about 1000 cycles for induction heating applications. A power-circuit diagram of such a frequency changer is shown in Fig. 40. It consists of a polyphase low-frequency transformer, a single-phase high-frequency transformer, and six rectifying elements. The transformer windings are shown vectorially. The load circuit consists of the induction heating coil and a capacitor. The rectifying elements are connected in pairs to the three phases of the rectifier transformer d-c winding through the center-tapped d-c windings of the high-frequency inverter transformer. The direct current of the rectifier circuit is conducted by each phase of the rectifier transformer d-c winding during approximately one-third of a cycle of the low-frequency voltage. During that time, the section of the high-frequency d-c winding and the two rectifying elements connected to it operate as a single-phase power inverter, and the current flows alternately through the two parts of the winding for half-cycle periods of the high-frequency voltage. The firing of the rectifying elements is controlled by the grids, which are energized from the high-frequency output circuit through a phase shifter.

The high-frequency a-c voltage of the power inverter is generated in the load circuit by oscillation between the inductance and capacitance, and the frequency is determined by the values of these circuit constants. The load circuit operates with a leading power factor, which is required for operation of a power inverter. The load consists of the metal charge, surrounded by the induction coil and heated by induced current. The resistivity and permeability of the charge, which are among the factors influencing the power input to the induction coil, may change as the charge is heated. The power and frequency may be regulated to a limited extent by phase control of the rectifying elements. This type of frequency converter has a higher efficiency and appreciably lower no-load losses than the motor-generator type used for induction heating. Another characteristic of the electronic type frequency converter is that the frequency is not fixed and can adjust itself to the inherent frequency of the load circuit without switching capacitors, as is done when a rotating-type machine is used.

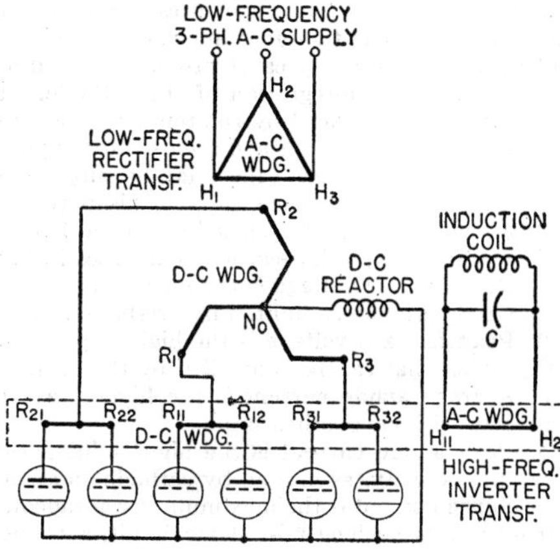

Fig. 40. Power Circuit Diagram of Electronic Converter for Conversion of Low-frequency Power to High Frequency

OTHER APPLICATIONS. Mercury-arc rectifiers have been applied for excitation of synchronous a-c generators, in place of rotating-type exciters. They have also been applied for operating variable-speed d-c motors, phase control of the d-c voltage being used for regulating the speed of the motor. A synchronous a-c motor may be operated at variable speed from an a-c supply by interposing rectifying devices between the motor and the supply and controlling their firing sequence. Mercury-arc rectifying devices are used as switches for accurate control of current in resistance welding and for other applications requiring a high-speed precision switch.

16. RECTIFIERS WITH THERMIONIC CATHODES

GAS TUBES. The electrodes of a gas tube with thermionic cathode are enclosed in a sealed glass or metal vacuum envelope which contains mercury or a gas at low pressure. There are two general types of gas-tube rectifiers. One type, called a **phanotron,** has only the main electrodes, anode and cathode. The other type, called a **thyratron,** has a control grid in addition to the main electrodes. Some tubes have a small quantity of liquid mercury, which provides mercury vapor when the tube is heated. Other tubes have an inert gas, such as argon, helium, or xenon. A combination of mercury and an inert gas is also used. The purpose of the mercury vapor or gas is to provide an ionized medium of high conductivity for the flow of electrons from the cathode to the anode, the space charge of the electrons being neutralized by the positive ions, as in the pool-cathode mercury-arc rectifier.

Electrons are emitted from the cathode when it is heated with current. In some tubes the heating filament serves as the cathode. In others the cathodes are metal cylinders which are heated indirectly by the filament. The emitting surfaces of both types of cathodes are coated with barium and strontium oxides, which are efficient electron emitters at a considerably lower temperature than is required for bare metal. The cathodes are usually surrounded by heat shields to reduce the loss of heat. The filament of the indirectly heated cathodes is enclosed by the cylindrical cathode. Hot-cathode gas tubes usually have 2.5- or 5-volt filaments. Some small thyratron tubes have 6.3-volt filaments. The filament voltage must be below the ionization voltage of the gas or vapor used, to prevent discharges between the filament terminals. The filament heating current must be applied for a specified time to heat the cathode to the emission temperature and to vaporize the mercury of mercury-vapor tubes, before a tube is permitted to carry current. The specified time varies from a few seconds to 5 min, depending on the size of tube and the cathode design. Indirectly heated cathodes require longer heating time than filamentary cathodes. The cathode voltage should not deviate from its rated value by more than 5 per cent.

The arc-drop voltage of gas tubes is in the range of 8 to 16 volts, depending on the tube design and the gas pressure. When the arc drop of a tube exceeds about 22 volts, the coating of the cathode is sputtered off by ion bombardment, which shortens materially the life of the tube. The vapor pressure of mercury-vapor tubes is a function of the tube temperature and is therefore affected by ambient temperature. Excessive temperature and vapor pressure may cause arc-backs. Too low a temperature increases the arc drop and may cause disintegration of the cathode. The control characteristics of thyratron tubes are also influenced by the temperature. Inert-gas tubes are practically unaffected by temperature variations and are used for applications involving a wide range of ambient temperatures. Inert-gas tubes may be subject to gas "clean up," which is absorption of part of the gas by the anode; this effect is more pronounced with a high initial-inverse anode voltage, which drives the ionized gas toward the anode.

The rating data of hot-cathode gas tubes include several voltage and current limits:

1. **Peak inverse voltage** is the maximum negative voltage that can be applied between anode and cathode, as limited by arc-back probability.

2. **Peak forward voltage** is the highest positive voltage between anode and cathode of a thyratron that can be controlled by the grid.

3. **Average anode current** is the highest average current that a tube can carry continuously without overheating.

4. **Peak anode current** is the highest instantaneous value of current that a tube can carry periodically, as limited by cathode emission.

5. **Surge current** is the maximum instantaneous transient current that a tube can carry during a fault, such as a d-c short circuit, without immediate failure. The duration of the surge current is usually limited to 0.15 sec. Frequent application of high currents shortens tube life.

6. The maximum voltage and current that can be applied to the grid of a thyratron tube is also specified in the tube rating. The grid-control characteristic is indicated by a curve.

Hot-cathode gas tubes with two electrodes (without grid) are available for a wide range of average-current ratings, from a fraction of an ampere up to about 50 amp, and peak inverse voltages up to about 22,000 volts. The high-voltage tubes are of the mercury-vapor type. Some low-voltage tubes have two anodes. Thyratron tubes are available for average-current ratings up to about 16 amp and peak inverse voltages up to about 12,000 volts. Most tubes have glass envelopes. Some tubes in the lower voltage range have metal envelopes and are used for applications requiring rugged equipment.

Two-electrode hot-cathode gas tubes are used as power rectifiers for many applications requiring d-c power of moderate capacity, below the ratings covered by mercury-pool rectifiers. Among such applications are battery chargers, arc welders, d-c power for magnetic chucks and similar industrial uses, plate power supply for radio transmitters and radio-frequency oscillators used for induction and dielectric heating, radio receivers, and others. Thyratron tubes are used as power rectifiers for applications requiring a variable d-c voltage and as power inverters. They supply power for electronically controlled variable-speed d-c motors; for this application they operate as power rectifiers and as power inverters for regenerative braking. They have been applied also for supplying field excitation of synchronous machines and d-c excitation for saturable-core regulating reactors. Thyratrons are also used as electronic switches for resistance welding and for firing ignitors and controlling grids of pool-cathode rectifiers.

The life of hot-cathode gas tubes is usually determined by deterioration of the cathode or impairment of the vacuum. Reasonably good tube life can be obtained if the ratings are not exceeded, and if the operating instructions are followed in regard to cathode-heating time and temperature limits. The useful life of a thyratron is sometimes limited by changes of grid characteristics with age. Hot-cathode gas tubes should be operated in a vertical position. A shock-absorbing mounting is desirable if the tubes are subjected to shock or vibration. The circuits and circuit relations used for mercury-pool type rectifiers apply also to the hot-cathode type. For tubes with a filamentary cathode the cathode connection is made at the center tap of the filament transformer winding. Tubes with indirectly heated cathodes have separate cathode terminals.

HIGH-VACUUM TUBES. A two-electrode high-vacuum tube, called a **kenotron,** consists of an anode and cathode sealed into a highly evacuated glass envelope. Conduction of current through the tube consists of a flow of electrons from cathode to anode. The cathode is a filament of pure or thoriated tungsten, which is heated with current to produce electron emission. The voltage drop from anode to cathode of various commercial tubes, at rated current, ranges from several hundred volts to about 1800 volts, depending on their current and voltage ratings. The voltage drop is due to the negative space charge of the electrons and is a function of the current. The power loss in the tube is converted into heat at the anode. The peak-current rating of a tube is limited by cathode emission; the average-current rating is limited by heating of the anode. Tubes of high current rating have water-cooled anodes. Air-cooled tubes have been made for peak-current ratings up to about 2.5 amp, and water-cooled tubes up to about 7.5 amp, for a maximum peak inverse voltage of 50,000 volts. Tubes of lower current ratings are available for maximum peak inverse voltages up to 230,000 volts.

High-vacuum two-electrode tubes are used as power rectifiers for applications requiring low currents at high d-c voltages, such as for electronic dust precipitation and high-voltage test equipment.

17. METALLIC RECTIFIERS

GENERAL. A metallic rectifier is an asymmetric electric conductor, which has a low resistance to the flow of current in one direction and a high resistance in the opposite direction; it is therefore effectively a unidirectional conductor and is used for converting alternating current into direct current. A metallic rectifier element, called a **cell,** consists of a good conductor and a semiconductor (material of high resistivity) separated by a very thin insulating barrier layer. The flow of forward current through a cell consists of a flow of electrons from the conductor, across the barrier layer, and through the semiconductor. The direction of current flow is, by convention, opposite to the electron flow and is therefore in the direction from the semiconductor to the conductor. Current is carried to the semiconductor by a metal plate in close contact with it.

Metallic rectifier cells are usually made in the form of circular disks with a hole in the center. A number of cells, with the necessary terminals, spacers, washers, and fins (when used), are assembled over an insulated stud into a tight assembly, which is called a **rectifier stack.** Cells and stacks may be connected in series or parallel, with proper polarities, to obtain the required voltage and current ratings and circuit connections for specific applications. Three types of metallic rectifiers are used commercially: copper oxide, selenium, and copper sulfide. The first two are the more widely used types.

COPPER OXIDE. A diagrammatic section through a copper oxide rectifier cell is shown at the top of Fig. 41(*a*). It consists of a copper plate on which a thin layer of cuprous oxide is formed by oxidation at a high temperature. The copper is the conductor, the cuprous oxide is the semiconductor, and the forward direction of current flow is from the cuprous oxide to the copper. A typical stack assembly is also shown. Stacks are made with and without fins, depending on the rating, and are coated with insulating varnish. Spacers are used to increase the spacing between fins and make them more effective. Disk-type cells are made up to an average-current rating per cell of about 0.25 amp, self-cooled. For higher ratings rectangular plate-type cells are used. In one design both sides of the copper plate are covered with cuprous oxide, and the outer surfaces of the oxide are coated with electroplated nickel. The nickel plating serves as one electrode, the copper plate as the other. Plate cells are made for average-current ratings of about 2.5 to about 6 amp per cell, self-cooled.

The current and voltage ratings of a copper oxide rectifier are determined mainly by thermal limitations. The losses during the conducting and the inverse periods of the cycle have to be considered. Although the negative current during the inverse period is a small percentage of the forward current, the losses are still appreciable because of the higher voltage across the cell. The inverse characteristic of a copper oxide cell is the chief cause of its voltage and temperature limitations. The resistance in the inverse direction decreases as the negative voltage is increased; i.e., the negative current and loss increase faster than the voltage. The inverse resistance has a relatively high negative temperature coefficient, so that the negative current and loss increase rapidly at higher temperatures, which may cause further increase in temperature. If the cooling is not adequate to check the temperature increase, the process can become cumulative and result in failure of the cell. The forward resistance also decreases with increased temperature, reducing the forward loss, because the forward current is determined by the load circuit. The change in the forward resistance with temperature is relatively small. The operating temperature of copper oxide cells is usually kept below 60 deg cent to reduce ageing and obtain long life.

The peak inverse voltage permitted per cell is in the range of 5 to 11 volts. The average-current density (average value of current through the cell divided by the active area) is usually in the range of about 0.08 to 0.2 amp per sq in. The values of voltage and current used are determined largely by thermal considerations and depend on the type of stack assembly, the cooling surfaces, and the method of cooling. The current rating of plate-type cells is more than doubled, and the permissible inverse voltage increased about 25 per cent, by using forced-air cooling instead of self-cooling. The type of rectifier circuit is also an important factor in determining the current and voltage ratings of a copper oxide cell, because the circuit determines the wave shapes of the forward cell current and its inverse voltage, which affect the losses. A six-cell copper oxide rectifier unit connected in a six-phase double-way (three-phase bridge) circuit, with one cell per leg, can be rated for a d-c output voltage of six volts, when adequately cooled. The rating of metallic rectifiers is based on an ambient temperature of 35 deg cent. For higher ambient temperatures the voltage rating of copper oxide rectifiers is reduced, with no reduction in

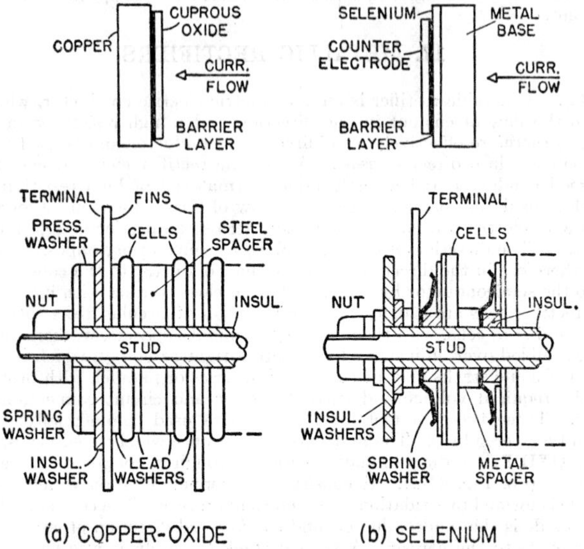

FIG. 41. Cells and Stacks of Metallic Rectifiers (*not to scale*)

the current rating. A copper oxide cell which has approximately four times the inverse resistance and about twice the forward resistance of the regular type, and can operate at twice the voltage rating given above, is now available.

The overall efficiency of a six-phase copper oxide rectifier unit, when operated at rated voltage and current per cell, is usually in the range of 65 to 75 per cent. The efficiency of a single-phase unit on a resistance load is about 15 per cent lower, if the power output is based on the average values of the d-c voltage and current rather than their rms values. If the cells are operated below their rated voltage, the efficiency is lower. The efficiency curve is fairly flat between about 20 per cent and full load, being a maximum at approximately half load. The voltage regulation between no load and full load is usually in the range of 20 to 30 per cent, depending on the type of rectifier circuit and the type of load. A copper oxide rectifier is subject to "ageing" when in service, which increases the forward resistance and thus reduces the output voltage. An average figure of 15 per cent is usually allowed for ageing, and the rectifier transformer is provided with voltage-boosting taps to compensate for this effect. Ageing is a function of time and operating temperature, being greater and more rapid at higher temperatures. At moderate temperatures, ageing is gradual over a period of several years. Ageing increases the voltage regulation of the rectifier and decreases its efficiency. The output ratings are based on the aged condition and do not have to be reduced.

SELENIUM. A diagrammatic section of a selenium rectifier cell and a typical stack assembly are shown in Fig. 41(*b*). The cell consists of a base plate of nickel-plated iron or aluminum, to one side of which is applied a thin layer of selenium. After heat treatment, the selenium is sprayed with a coat of low-melting-temperature alloy, which constitutes the counter electrode. The cell is then "formed" to produce the barrier layer, by passing current in the reverse direction for several hours. The selenium is the semiconductor and the counter-electrode is the conductor of the cell, with the barrier layer between them. The direction of the forward current flow is from the selenium to the counter-electrode. The base plate is the back electrode and does not participate in the rectification action. The spring washers of the stack serve as the connectors between the counter-electrodes of the cells and the other elements of the stack, which is usually coated with an insulating varnish. Selenium rectifier cells are made in a number of sizes up to an average-current rating of 2.5 amp per cell. Selenium rectifiers for high currents have been made with many cells mounted on a rectangular steel plate, which served as the current conductor.

A selenium rectifier is a thermally rated device, based on a maximum cell-operating temperature of 75 deg cent. As in a copper oxide rectifier, the losses during the inverse period, as well as the conducting period, have to be considered. However, the increase of the inverse current and loss with temperature are much smaller for a selenium cell than for a copper oxide cell, and a selenium rectifier is not subject to "run-away" temperature instability. Only the forward current rating is affected by the operating temperature limit. The resistance in the inverse direction decreases as the negative voltage is increased, and excessive inverse voltage will cause failure by breakdown of the barrier layer. The maximum permissible inverse voltage per cell is 18 volts rms or 25.5 volts peak. Lower values are used for large cells. Small selenium rectifiers are now available, rated for a maximum inverse voltage per cell of 26 volts rms or about 36.5 volts peak.

The average-current density of selenium cells, for an ambient temperature of 35 deg cent, is about 0.16 amp per sq in. The current may be increased 25 to 50 per cent by increasing the spaces between cells, and approximately double the normal rating may be obtained by using cooling fins. Further increases in current rating are possible by forced cooling. Increasing the current density increases the losses and lowers the efficiency. If the ambient temperature is above 35 deg cent, the current rating is reduced by approximately 15 per cent for each 5 deg increase in ambient. Above an ambient of 50 deg cent the voltage rating is also reduced.

The type of rectifier circuit and the characteristics of the load are also important factors in the current rating, because they determine the wave shape of the cell current and therefore affect the losses. A selenium rectifier may be immersed in oil for protection against a corrosive atmosphere. If a selenium rectifier is idle for a long period or operates at low inverse voltage, its inverse resistance decreases. When the full inverse voltage is applied, the inverse current is abnormally high for several minutes until the normal inverse resistance is regained. The voltage should be increased to its normal value gradually.

The overall efficiency of a six-phase selenium rectifier unit, when operated at rated voltage and current per cell, is usually between 75 and 85 per cent, and the curve is fairly flat from about 20 per cent to full load. The apparent efficiency of a single-phase selenium rectifier on a resistance load is about 20 per cent lower, because of the ratio of average to rms values of d-c voltage and current. The voltage regulation between no load and full load is usually of the order of 15 to 20 per cent, depending on the cell current density and type of rectifier circuit. The resistance of a selenium rectifier increases with ageing, which reduces the output d-c voltage by about 5 to 10 per cent, depending on the type of service. The rectifier transformer should be provided with taps to compensate for this ageing effect, which reduces the efficiency and increases the voltage regulation of the unit. Ageing usually takes place during the first 10,000 hr of continuous full-load operation.

COPPER SULFIDE. A copper sulfide rectifier cell consists of a disk of copper sulfide in contact with a disk of magnesium, with the barrier layer between them. The direction of forward current flow is from the copper sulfide (semiconductor) to the magnesium (conductor). Copper sulfide cells are assembled into stacks, as in the copper oxide rectifier, and are provided with cooling fins. The stacks are coated with a varnish as protection against humidity and corrosive fumes.

Like other metallic rectifiers, the copper sulfide rectifier is a thermally rated device with a maximum operating temperature of 85 deg cent. The maximum inverse operating voltage per cell is 3.6 volts rms or about 5 volts peak inverse. The maximum rated cell current density, when self-cooled, with an ambient temperature of 40 deg cent, ranges from less than 1 amp to about 7 amp average value per sq in., depending on the number of cells per stack and the number and size of cooling fins. To obtain long life, the operating temperature should be kept below 85 deg cent, but the rectifier can be subjected to higher temperatures without failing. The current rating can be increased considerably by forced cooling.

According to published literature, the efficiency of a copper sulfide rectifier is between 50 and 55 per cent, and the overall voltage regulation is about 20 to 25 per cent; the reduction in output voltage due to ageing is about 10 per cent and occurs within several hundred hours.

APPLICATION. Metallic rectifiers serve as power converters for numerous applications requiring direct current. Some of the more important applications are battery charging, electroplating, anodizing, cathodic protection of buried pipes, cables, and other underground installations subject to electrolysis, plate supply for radio transmitters, and operation of solenoids. These rectifiers have been applied for fly-ash and dust precipitation at voltages as high as 80 kv. They also serve as auxiliaries for the excitation and grid-control circuits of mercury-arc rectifiers. Besides such applications, metallic rectifiers are used extensively as elements in various regulator and control systems and are replacing rectifier tubes in radio receivers. For electroplating service-units rated up to 7500 amp at 12 volts d-c are in use; the rectifiers are forced cooled, and the current is regulated automatically by means of an induction regulator or saturable-core reactors in the a-c circuit.

The most generally used circuits for metallic rectifiers are the single-phase double-way (bridge) and the six-phase double-way (three-phase bridge) connections. Single-way connections are used sometimes for low output voltages, to utilize to better advantage the voltage rating of the cells. The transformer, rectifier, and other components of a unit usually are mounted in a cabinet or cubicle as integral assemblies. In view of the effect of temperature on the life of metallic rectifiers, adequate ventilation should be provided. Overvoltages in excess of maximum ratings may damage the rectifiers and should be avoided. The rating of metallic rectifiers involves many factors, such as the service duty, the type of load, and the rectifier-transformer connection. For example, a single-phase unit supplying a battery or capacitor load is usually derated about 20 per cent because of the higher peak value of the rectifier current in relation to its average value. The manufacturer should be consulted in regard to specific applications. New stacks connected in series or parallel with aged stacks may result in voltage or current unbalance between the stacks, because the forward and inverse resistances change with age.

Because of the relatively low voltage rating per cell, metallic rectifiers are particularly well suited for applications requiring low d-c voltage, such as electroplating; their efficiency, regulation, and cost per kilowatt are nearly independent of the voltage. Their rectifier action is instantaneous, no excitation or filament heating being required. They can be operated at low ambient temperatures, down to -30 or -40 deg cent. When conservatively rated, an operating life of many years can be expected. These characteristics of metallic rectifiers have led to their widespread and increasing use. There is considerable overlapping between the fields of application of the metallic and hot-cathode tube rectifiers. There is practically no overlapping between the applications of metallic and mercury-arc rectifiers; for d-c voltages above 200 volts and ratings above about 100 kw, which are approximately the lower limits for their application, mercury-arc rectifiers have higher efficiency, lower voltage regulation, and higher overload capacity, and are better able to withstand overvoltages and overtemperatures than metallic rectifiers. Metallic rectifiers cannot be used for power inversion or for any application requiring phase control; their output voltage can be varied only by varying the applied a-c voltage.

18. ELECTROLYTIC RECTIFIER

One type of electrolytic rectifier that has been used commercially is the lead-tantalum type, which consists of electrodes of lead and tantalum immersed in an electrolyte of battery-grade sulfuric acid and contained in a glass jar with a hard-rubber top, similar to a storage-battery cell. The direction of the current flow is from lead to tantalum. The rectifying action is provided by an oxide film on the tantalum electrode, formed while in operation. The flow of current is accompanied by electrolysis of the water in the electrolyte, with the liberation of oxygen and hydrogen; water has to be added about once a month. An oil film on top of the electrolyte prevents liberation of fumes and reduces evaporation of water. Special salts in the electrolyte improve electrode activity. Four rectifier cells connected in a single-phase double-way (bridge) circuit have a rated d-c output voltage of 30 volts. For higher voltages, two or more such circuits may be connected in series. The electrolytic rectifiers are used mainly for charging batteries and have been made in units rated up to 6 amp. Their overall efficiency at rated output is between 35 and 45 per cent, depending on the rating. The rectifier units also have an ampere-hour rating, which is the ampere-hours output without addition of water.

19. MECHANICAL RECTIFIER

A mechanical rectifier consists of a set of synchronously operated contacts which connect the d-c circuit in sequence to each of several phases of an a-c supply, during part of each cycle, so as to make the current flow in one direction in the d-c circuit. Each contact is in effect the equivalent of a rectifying element in rectifier circuits as shown in Figs.

8 and 11, except that the contacts are closed and opened mechanically instead of utilizing the valve action of an electronic device. Several types of mechanical rectifiers have been used, some of the rotating contact type. The most advanced mechanical rectifier design was developed in Germany about 1940. It consists of a set of silver-inlaid butt contacts arranged in a row. Each moving contact is a small plate, which is lifted about 0.1 in. from a pair of stationary contacts against a compression spring, by means of a push rod operated by a rocker arm, which is actuated by an eccentric on a synchronous-motor-driven shaft. The moving parts of the mechanism, except the contacts, are enclosed in a casing and lubricated by forced circulation of oil, which is cooled in an oil cooler; the air entering the casing is filtered. The contacts are enclosed in a removable cover provided with acoustic lining. The a-c and d-c bus-bars, which carry the stationary contacts, are cooled by a recirculating water-cooling system with a heat exchanger.

Commutation of current between successive phases of a commutating group occurs in the same way as for a mercury-arc rectifier, described in Article 5, with one important difference. In a mercury-arc rectifier or any other type of static rectifier, the current is prevented from reversing at the conclusion of the conducting period, by the valve action of the rectifying device. In a mechanical rectifier the current through the contacts can reverse if they remain closed after commutation at the end of the conducting period. If the contacts were opened before or after the zero point of current, they would deteriorate rapidly from the arcing. To avoid this, saturable commutating reactors, with a laminated core of a special alloy, are connected in series with the contacts. During the conducting period the reactor core is saturated, and its inductance is low. Near the end of the commutation period, when the current approaches zero, the core is unsaturated, and the reactor inductance is high. This limits the current resulting from the commutating voltage between phases to a fraction of an ampere for about 1 msec, which provides sufficient time for the contact to open with negligible arcing. A capacitor is connected across each contact to hold off the circuit voltage momentarily while the contact is opening. The commutating angle of a rectifier varies with the load, a-c voltage, and commutating reactance, and it is necessary to vary the instant of contact opening to correspond with the low-current step at the end of commutation. This is accomplished by mechanical and electrical regulating devices, which control the pivoting points of the rocker arms. The commutating reactor, or a separate reactor, is used also to buck the voltage across the contact when closing, in order to prevent fusing when contact is made. The output voltage of a mechanical rectifier may be regulated over a limited range by phase control, accomplished by turning the stator of the synchronous driving motor, which varies the angular position of the conducting period in the cycle. This requires larger commutating reactors.

Mechanical rectifier units of this type, with six contacts, are built for current ratings up to 5000 amp at 400 volts d-c, and dual units with twelve contacts up to 10,000 amp. According to published curves, the overall efficiency of mechanical rectifier units at rated load is 95 to 96 per cent for d-c voltages of approximately 150 volts and higher, and about 90 per cent at 100 volts. The higher efficiencies at these voltages, as compared to mercury-arc rectifiers, are due to the absence of arc-drop loss. Part of this saving, however, is offset by the losses in the commutating reactors. The voltage regulation is between 10 and 15 per cent, and the power factor is about 80 to 90 per cent. The six-phase double-way rectifier circuit (Fig. 11) has been used for the mechanical rectifiers. The rectifier is protected against arcing, during faults or system disturbances, by a high-speed contactor, which applies a short-circuit between the contacts and the commutating reactors to divert the current from the contacts until the a-c breaker is opened. Starting resistors are used in the circuit between the transformer and rectifier; these are short-circuited by a contactor when the rectifier is in operation.

20. BIBLIOGRAPHY

1. Prince, D. C., and F. B. Vogdes, *Principles of Mercury-arc Rectifiers.* McGraw-Hill (1927).
2. Marti, O. K., and H. Winograd, *Mercury-arc Power Rectifiers.* McGraw-Hill (1930).
3. Dow, W. G., *Fundamentals of Engineering Electronics.* John Wiley (1937).
4. Rissik, H., *Fundamental Theory of Arc Converters.* Chapman and Hall (London) (1939).
5. Reich, H. J., *Theory and Application of Electron Tubes.* McGraw-Hill (1939).
6. Cobine, J. D., *Gaseous Conductors.* McGraw-Hill (1941).
7. M.I.T. staff, *Applied Electronics.* John Wiley (1943).
8. Slepian, J., and L. R. Ludwig, New Method for Initiating Cathode of an Arc, *AIEE Trans.*, Vol. 52, p. 693 (1933).
9. Stebbins, F. O., and C. W. Frick, Output Wave Shape of Controlled Rectifiers, *AIEE Trans.*, p. 1259 (Sept., 1934).
10. Edison Electric Institute, Rectifier Wave Shape, *Publication* E1 (April, 1937).
11. Winograd, H., Cooling Mercury-arc Rectifiers, *Allis-Chalmers Elec. Rev.* (Sept., 1937).

12. Dortort, I. K., Interphase Transformers for Mercury-arc Rectifiers, *Allis-Chalmers Elec. Rev.* Vol. 4, p. 9 (March, 1939).
13. DeBlieux, E. V., Rectifier Transformers, *Gen. Elec. Rev.*, Vol. 40, p. 412 (Sept., 1937), p. 481 (Oct., 1937), p. 539 (Nov., 1937), p. 590 (Dec., 1937).
14. Evans, R. D., and H. N. Muller, Harmonics in A-c Circuits of Grid-controlled Rectifiers and Inverters, *AIEE Trans.*, Vol. 58, p. 861 (1939).
15. Kingdon, K. H., and E. J. Lawton, Relation of Residual Ionization to Arc-back in Thyratrons, *Gen. Elec. Rev.*, Vol. 42, p. 474 (1939).
16. Marti, O. K., and T. A. Taylor, Wave Shape of 30- and 60-phase Rectifier Groups, *AIEE Trans.*, Vol. 59, p. 218 (1940).
17. Richards, E. A., Characteristics and Applications of the Selenium Rectifier, *I.E.E.J.* (London), Vol. 88, p. 423 (1941).
18. Williams, A. L., and L. E. Thompson, Metal Rectifiers, *I.E.E. J.*, Vol. 88, p. 353 (1941).
19. Housley, J. E., and H. Winograd, Alcoa Rectifier Installation, *AIEE Trans.*, Vol. 60, p. 1266 (1941).
20. Remscheid, E. J., Water-cooled Steel-tank Rectifier Corrosion Problem, *AIEE Trans.*, Vol. 60, p. 173 (1941).
21. Hull, A. W., and F. R. Elder, High-voltage Surges in Rectifier Circuits, *J. Applied Physics*, Vol. 13, p. 372 (June, 1942).
22. Yarmack, J. E., Selenium Rectifiers and Their Design, *AIEE Trans.*, Vol. 61, p. 488 (1942).
23. Mittag, A. H., and A. Schmidt, Ignitor Excitation Circuits, etc., *AIEE Trans.*, Vol. 61, p. 574 (1942).
24. Morack, M. M., and H. C. Steiner, Sealed-tube Ignitron Rectifiers, *AIEE Trans.*, Vol. 61, p. 594 (1942).
25. Cox, J. H., and G. F. Jones, Ignitron Rectifiers in Industry, *AIEE Trans.*, Vol. 61, p. 713 (1942).
26. Rhea, T. R., and H. H. Zielinski, Electric Equipment for Large Electrochemical Installations, *AIEE Trans.*, Vol. 61, p. 733 (1942).
27. Kotterman, C. A., and E. H. Pollacek, Selenium Rectifiers, *Iron and Steel Engr.*, Vol. 20, p. 73 (Oct., 1943).
28. Kellogg, H. L., and C. C. Herskind, Testing of Mercury-arc Rectifiers, *AIEE Trans.*, Vol. 62, p. 765 (1943).
29. Vedder, E. H., and K. P. Puchlowski, Theory of Rectifiers—D-c Motor Drive, *AIEE Trans.*, Vol. 62, p. 863 (1943).
30. Herskind, C. C., Rectifier Circuit Duty, *AIEE Trans.*, Vol. 63, p. 123 (1944).
31. Steiner, H. C., J. L. Zehner, and H. E. Zuvers, Pentode Ignitrons for Electronic Power Converters, *AIEE Trans.*, Vol. 63, p. 693 (1944).
32. Smith, I. R., Copper Oxide Rectifier in Electrochemical Work, *AIEE Trans.*, Vol. 63, p. 739 (1944).
33. Cramer, F. W., L. W. Morton, and A. G. Darling, Electronic Converter for Exchange of Power, *AIEE Trans.*, Vol. 63, p. 1059 (1944).
34. Willis, C. H., R. W. Kuenning, E. F. Christensen, and B. D. Bedford, Design of Electronic Frequency Changer, *AIEE Trans.*, Vol. 63, p. 1070 (1944).
35. Winograd, H., Development of Excitron-type Rectifier, *AIEE Trans.*, Vol. 63, p. 969 (1944).
36. Christensen, E. F., C. H. Willis, and C. C. Herskind, Analysis of Rectifier Circuits, *AIEE Trans.*, Vol. 63, p. 1048 (1944).
37. Watkins, S. S., Mercury-arc Rectifiers for Railroads, *AIEE Trans.*, Vol. 64, p. 84 (1945).
38. Herskind, C. C., and H. L. Kellogg, Rectifier Fault Currents, *AIEE Trans.*, Vol. 64, p. 145 (1945).
39. Boehne, E. W., and W. A. Atwood, Anode-circuit-breaker Design and Performance Criteria, *AIEE Trans.*, Vol. 64, p. 337 (1945).
40. Evans, R. D., and A. J. Maslin, Arc-backs in Rectifier Circuits, *AIEE Trans.*, Vol. 64, p. 303 (1945).
41. Durand, S. R., Mercury-arc Heating Frequency Converter, *Electronic Industries* (June, 1945).
42. High-voltage D-c Power Transmission, *Brown Boveri Rev.* (Sept., 1945).
43. Holman, R. W., Mercury-arc Frequency Changer, *Elec. World*, Vol. 125 (April 13, 1946).
44. AIEE Report on Proposed Standards for Mercury-arc Power Converters, *AIEE Report No.* 6 (June, 1946). (NOTE: This report will be replaced at a later date by an ASA Standard.)
45. AIEE Committee Report, Inductive Coordination Aspects of Rectifier Installations, *AIEE Trans.*, Vol. 65, p. 417 (July, 1946).
46. Housley, J. E., and G. N. Hughes, Maintenance of Rectifiers for Electrochemical Installations, *AIEE Trans.*, Vol. 65, p. 436 (1946).
47. Boyer, J. L., and C. G. Hagensick, High-voltage Ignitron Rectifiers and Inverters for Railroad Service, *AIEE Trans.*, Vol. 65, p. 463 (1946).
48. Herskind, C. C., and E. J. Remscheid, Excitation, Control, and Cooling of Ignitron Tubes, *AIEE Trans.*, Vol. 65, p. 632 (1946).
49. Herskind, C. C., and H. C. Steiner, Rectifier Capacity, *AIEE Trans.*, Vol. 65, p. 667 (1946).
50. Steiner, H. C., and H. N. Price, A 400-ampere Sealed Ignitron, *AIEE Trans.*, Vol. 65, p. 680 (1946).
51. Jensen, O., A Mechanical Rectifier, *Electrochem. Soc. Trans.*, Vol. 90, p. 93 (1946).
52. Falls, W. H., Selenium and Copper Oxide Rectifiers, *Gen. Elec. Rev.*, Vol. 50, p. 34 (Feb., 1947).

SECTION 12

SWITCHGEAR AND CONTROL EQUIPMENT

BY

M. H. HOBBS, J. W. SIMPSON, H. A. TRAVERS, AND L. E. MARKLE

SWITCHGEAR AND CONTROL EQUIPMENT

By M. H. Hobbs, J. W. Simpson, H. A. Travers, and L. E. Markle

The subjects of switchgear and control are very closely related in that both are concerned with the opening and closing of circuits and frequently employ the same equipment in so doing. The distinctions between them arise from the types of circuits involved and the reasons for changing the condition of the circuit.

Switchgear is a general term covering switching and interrupting devices, also assemblies of those devices with control, metering, and protective and regulating equipment with the associated interconnections and supporting structures (NEMA Definition SG-50-1).

An electric controller is a device, or group of devices, which serves to govern in some predetermined manner the electric power delivered to the apparatus to which it is connected (ASA Definition 25.05.005).

DESIGN OF SWITCHGEAR AND CONTROL APPARATUS

1. SWITCHES

A switch is a device for making, breaking, or changing the connections in an electric circuit (NEMA Standard SG-50-620).

INSTRUMENT AND CONTROL SWITCHES. These switches are usually of the rotary type. They are very compact and are used for multicircuit switching, such as connecting one instrument to any of several circuits and for making the multipoint connections required in synchronizing generators. The instrument switches usually have two or more contact positions with a notching or star wheel device to center each contact position definitely. Control switches usually have spring return to the "off" position and must be held in the "operating" position. These switches are of two basic types. One consists of stationary fingers mounted on insulating bases, with contact being made by means of a moving contact attached to the switch rotor. The other type has both contacts mounted on stationary bases, with contact being made by means of cam segments attached to the switch rotor. These switches are of the multiposition type. The most common type of switches has a maximum of eight positions, but switches with as many as sixteen or more positions are made. They are available in varying numbers of stages, up to ten or more, with each contact usually capable of being open or closed in any position of the switch handle. Where more contacts or circuits are desired, two or more switch assemblies can be operated by a single handle through gears or links.

Instrument and control switches are usually rated at 250 volts direct current, 600 volts alternating current, or less, and have normal current-carrying capacities of 30 amp or less. The interrupting rating is normally considerably less than the current-carrying capacity.

KNIFE SWITCHES. The use of knife switches is rather limited in circuit breaking, as an arc will occur between the switch blade and the break jaw. Permissible arcing, without affecting the performance of the switch, determines its arc-breaking capacity. The more rapidly the switch is opened, the less is the damage from the arc. It is more difficult for a knife switch to open a d-c circuit than an a-c circuit of the same voltage and class of service. For this reason it has been found that a knife switch with dimensions satisfactory for opening a 250-volt d-c circuit can be safely used to open the circuit on a 600-volt a-c circuit. Hand-operated switches in d-c circuits are usually limited to 600 volts. When used to break current, quick-break attachments are recommended for all switches rated over 250 volts direct current; they must be provided for those rated at or above 200 amp and 500 volts direct current, or 800 amp and 250 volts alternating current. Figure 1 shows some of the features of a two-pole, single throw, 200-amp rear-connected switch typical of the normal American design of knife switches.

Various modifications in the standard low-voltage knife switch are made to allow for starting d-c motors, to provide auxiliary contacts for the discharge resistors of field circuits, and to take care of other special conditions.

The current-carrying capacity of a switch or circuit breaker is determined by its permissible temperature rise. The capacities usually assigned for such devices are based on the current which they can carry continuously, the temperature rise not exceeding 30 deg cent in the current-carrying parts.

The non-uniformity of current distribution in conductors of a-c circuits increases as the currents become larger and also increases with frequency. For this reason the switches of higher current capacity have different 25-cycle, 60-cycle, and d-c ratings.

Switches may be further classified, according to their principal application, as disconnecting, transfer, selector, or grounding switches.

An **interrupter switch** is a switch combining the functions of a disconnecting switch and a circuit interrupter for interrupting, at rated voltage, currents not exceeding the continuous-current rating of the switch.

Operation, as applied to a switch, is the method provided for its normal functioning. Hook operation of the switch is manual operation by means of a switch hook. Mechanical operation of the switch is operation by means of a mechanism connected to the switch by insulated mechanical linkages. Mechanically operated switches may be actuated manually, electrically, or by other suitable means.

A **quick-break switch** is a switch which has high contact-opening speed, independent of the operator. A **quick-make switch** is a switch which has high contact-closing speed, independent of the operator.

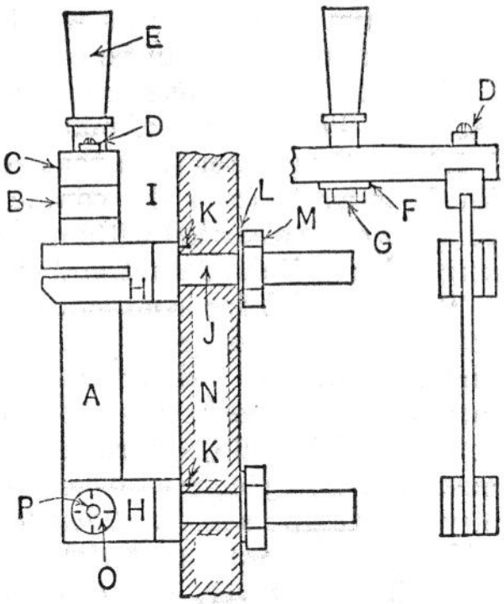

FIG. 1. 200-amp Low-jaw Knife Switch. *A*, copper blade, hard drawn, 98% conductivity; *B*, blade block pinned and sweated to *A*; *C*, Micarta cross-bar with milled slots; *D*, screw fastening cross-bar to blade block; *E*, handle bolted to cross-bar; *F*, washer for handle bolt; *G*, nut for handle bolt; *H*, jaw blades pinned and sweated in blade block; *I*, blade block with milled slots; *J*, studs electro-brazed to blade blocks; *K*, dowel pins locating switch studs; *L*, washer for switch stud; *M*, nut for switch stud; *N*, base for switch; *O*, slotted spring washer; *P*, hinge bolt.

Air switches are rated in rms amperes based on a maximum temperature rise of 30 deg cent above an ambient temperature of 40 deg cent in accordance with AIEE Standard 22. If the connections to the air switch exceed a temperature rise of 30 deg cent, the temperature rise of the switch may exceed 30 deg cent.

Air switches and bus supports are rated in rms volts in accordance with AIEE Standard 22 and are designed to meet dielectric tests in accordance with this standard. For application at altitudes above 3300 ft correction factors are used.

The standard voltage ratings of air switches are 2500, 5000, 7500, 15,000, 23,000, 34,500, 46,000, 69,000, 115,000, 161,000, 196,000, 230,000, 287,000 and 345,000 volts.

The short-time current rating of an air switch is the highest rms current, including the d-c component, that the switch is required to carry without injury for specified short-time intervals. The rating recognizes the limitations imposed by both thermal and electromagnetic effects. The standards for short-time ratings are:

1. The momentary current rating is the maximum rms total current which the switch is required to carry for any time, however small, up to 1 sec.

2. The 5-sec current rating is the rms total current, including the d-c component, which the switch is required to carry for 5 sec. For practical purposes this current is taken as the integrated heating equivalent of the 5-sec rating; the maximum test period does not exceed 10 sec.

An air switch has no overload rating because of its limited heat-storage capacity.

Air switches also have no interrupting rating. However, it is recognized that air switches must interrupt the charging current of adjacent bus supports and circuit-breaker bushings. Under certain conditions, they may interrupt other relatively low currents. For applications requiring the switch to have interrupting ability an interrupter switch should be used.

There are only two major requirements of air-switch contacts:

1. They must carry their rated current without reaching injurious temperatures, not only when new, but also after many years of service.

2. They must carry without injury the maximum short-circuit current of the system to which they are applied, until this current is interrupted by a circuit breaker. This ability must also be maintained during the life of the switch.

Low contact resistance is necessary to meet both those requirements.

The shape of the contact is important, inasmuch as it affects unit pressure and the amount of material immediately adjacent to the contacts. There are three general shapes of contacts: plane or surface contacts, as between two planes; point contacts, as between a sphere and a plane; and line contacts, as between a cylinder and a plane.

The surface contact, when new and clean, is entirely adequate for both rated and fault current. Because of low unit pressure, however, it is extremely susceptible to corrosion. Although formerly widely used on outdoor air switches, the surface contact is now used chiefly for low-current indoor knife switches.

The point contact, used to some extent in air switches, meets the requirements of high unit pressure and low contact resistance and is completely adequate for carrying rated current. Under fault conditions, however, the small amount of metal adjacent to the contact limits the rate at which heat is conducted away and thereby limits the short-circuit current that can be carried without damage.

The line contact has found increasing use in recent years and has proved very satisfactory. The amount of material adjacent to a line contact available for heat conduction is inherently greater than that adjacent to a point contact and is, therefore, able to carry the heat away from the contact more rapidly, with the result that much higher fault currents can be carried without damage. This principle is widely used in the high-pressure, single-line switch contact and the modern medium-pressure, multiple-line contact.

The material for contacts must be selected on the basis of its resistivity, chemical activity, hardness, and annealing temperature. The initial contact resistance varies directly with the resistivity of the contact material. This resistance, however, tends to increase with age because of the formation of surface film. The base metals, such as copper and its alloys, form a high-resistance oxide film when exposed to air, the rate of oxidation being comparatively low at room temperature, but becoming more rapid at elevated temperatures. The noble metals do not oxidize in air, but silver, which is a widely used contact material, reacts with sulfur in the atmosphere to form silver sulfide. This film, however, has comparatively low electrical resistance, is thin, and is easily rubbed off. The high electrical and thermal conductivity of silver makes it an ideal material for indoor or enclosed locations where a medium pressure contact is satisfactory.

HIGH-VOLTAGE SWITCHES. There are two major types of remotely operated outdoor disconnecting switches, the rotating insulator and the tilting insulator types. In a rotating insulator switch the opening and closing travel of the blade is accomplished by the rotation of one or more of the insulators supporting the conducting parts of the switch. In a tilting insulator switch the opening and closing travel of the blade is accomplished by a tilting movement of one or more of the insulators supporting the conducting parts of the switch.

A **clamp-and-release** switch has means for applying contact pressure after the blade has reached the end of the travel in the closing direction and means for releasing contact pressure before the blade starts its travel in the opening direction.

Contacts exposed to the weather must meet additional requirements, since dirt, dust, and corrosion accumulate more rapidly and since there is likelihood of ice formation. Here the contacts must have sufficient pressure to break through coatings of ice, dirt, and corrosive film. Even in enclosed locations rubbing action, which is desirable on silver, is essential on copper and its alloys in order to remove corrosion from the surface. In high-pressure rubbing contacts soft metals, such as silver and copper, are subject to excessive wear and galling, and harder alloys must be used. These alloys must not anneal when heated to temperatures of several hundred degrees centigrade, which may be reached under short-circuit conditions.

Either flexible shunts or contacts are used on the hinged end of outdoor switches at the present time. Shunts are subject to oxidation and corrosion, particularly in sulfurous industrial atmospheres, and the combination of corrosion and flexing may break the fine wires until eventually the shunt breaks completely. The same conditions that cause rapid deterioration of shunts also cause rapid deterioration of exposed contacts. At the break jaw, contacts must necessarily be left exposed. At the hinged connection, however, it is not necessary to use either a shunt or an exposed contact, since the contacts can be completely enclosed, protecting them from the elements and, to a great extent, from corrosive atmospheres.

The effort necessary to operate switches is entirely due to friction, except in the opening operation on small switches where the weight of the blade is not counterbalanced. The only essential friction is that produced by the rubbing motion in contacting the switch. This motion should be separated from the blade-elevating motion and accomplished with maximum mechanical advantage. All other friction is incidental to switch operation and should be reduced to a minimum. For this reason the number of pins and bearing points should be as small as possible. In normal service wear on a disconnecting switch can be neglected, and without maintenance its life depends on freedom from corrosion of bearings and contact surfaces. Even though bearings must be made of corrosion-resistant materials, they should still be designed with ample clearances so that corrosion will not cause seizing.

Dual-motion Switches. Dual-motion switches utilize a second motion, usually a continuation of the first (blade-closing) motion to apply contact pressure during or after entry of the blade into the break jaw. This contacting motion makes use of a much greater mechanical advantage than is available for the blade-closing motion. The dual-motion switch made possible the high-pressure switch contact as known today.

Contact pressure can be applied in a number of ways. In some switches the blade is pushed forward into the break contact. In others, contact tips are expanded by two blade members to engage the break jaw. Still others use a flat blade tip, which is rotated in a resilient break jaw to establish contact pressure.

2. CIRCUIT BREAKERS

Circuit breaking is the process of interrupting the continuity of a circuit in which an electric current is flowing. The difficulty of performing this operation is a function not only of the circuit voltage and the magnitude of the current but also of the electrical constants of the circuit. Differences in the interruption of direct current, which must be forced to zero, and alternating current, which has natural current zero periods at each reversal, justify separate consideration in describing the fundamental principles of breaking an electrical circuit.

INTERRUPTION OF DIRECT CURRENT. If the contacts of a switch carrying current are separated, an arc will be established by the high electric field and local heating at the last point of contact. The arc acts like a special kind of resistance which has a magnitude nearly proportional to its length for currents above a few hundred amperes but increases to extremely high values as the current approaches zero.

A direct-current arc will be extinguished, provided that for all current values the voltage necessary to sustain the arc is greater than the available voltage in the circuit, i.e., the electromotive force less the voltage drop in the circuit resistance.

If the circuit contains appreciable self-inductance, the switching process will usually take a longer time, since a voltage proportional to the rate of change of current will be induced in a direction tending to maintain the current. If an attempt is made to interrupt a highly inductive circuit too rapidly by opening the switch contacts and lengthening the arc at very high speed, the voltage across the switch and also across the inductance may reach dangerous values many times the normal circuit voltage.

A desirable switch characteristic is an arc voltage of approximately twice the circuit voltage maintained over most of the arcing period.

During the process of d-c interruption, all the magnetic energy stored in the inductance must be either dissipated as heat or transformed into electrostatic energy. It is evident that the arc performs the useful function of dissipating this energy and controlling the rate of decrease of current. Also the arc provides a medium which under proper conditions is capable of synchronizing the introduction of practically infinite resistance with the moment of close approach of the current to zero value.

Incandescent lamps, heating equipment, and similar resistance loads having little or no self-inductance may be switched off easily. Batteries on charge and shunt-wound motors which have a counter electromotive force independent of the current require that the switch break only the small voltage needed to force the current through the internal resistance of the battery or motor. On the other hand, field control switches and breakers for d-c series motors are subject to heavy arcing because of the relatively large amount of stored magnetic energy which must be discharged before the circuit is cleared.

INTERRUPTION OF ALTERNATING CURRENT. The periodic nature of alternating current greatly simplifies the interrupting problem, since the current in the process of reversing direction must cease flowing momentarily at half-cycle intervals. At each current zero after separating switch contacts have introduced an arc into the circuit, the arc is extinguished for an instant and must be reignited.

Since there is no stored magnetic energy in the self-inductance of the circuit at current zero, there is at least a theoretical possibility of opening a switch at just this moment with sufficient rapidity so that there will be no arcing whatsoever. In actual practice, however, such precise synchronization is extremely difficult, and it is usually more convenient to draw a low-voltage arc, parting the contacts at random in relation to the phase of the current, and then interrupt the arc at an early current zero.

When an a-c switch is opening a current at high power factor, the applied voltage is close to zero value at the instant of current zero, thus making the interruption relatively easy by limiting the available restriking voltage. On the other hand, with low power factor conditions, such as are likely to exist during a short circuit, where the current is limited largely by inductive reactance, the applied voltage will be close to peak value when the current reaches a zero pause.

Since there is always some distributed capacitance in the circuit, which must be charged up before full voltage can appear across the circuit-breaker contacts, a definite time interval is available for changing the arc space from the ionized conducting path to a deionized insulating gap. Under unfavorable conditions where the capacitance is small and the damping resistance negligible, the voltage across the contacts will rise rapidly within an interval of perhaps 100 μsec or less and overshoot to a peak value nearly twice the normal voltage crest. With such a limited time for deionization of the arc space, the interruption of a high-voltage arc may become quite difficult. The shunt capacitance of even a few hundred feet of cable, however, is sufficient to slow up the voltage rise very considerably, and the usual condition of parallel loads on the same bus to which the switch is connected further relieves the severity by damping out overswinging of the voltage transient.

The ideal a-c circuit breaker is capable of extinguishing, at rated voltage, an arc of any current value up to the interrupting rating at the first current zero, after separation of the contacts, on a circuit having the highest rate of rise of recovery voltage practically obtainable. Many circuit breakers utilize energy taken from the short-circuit current to assist in the interruption, i.e., magnetic blowout by series coils, self-generated gas blast from materials decomposed by the heat of the arc, or flow of oil under pressure provided by a separate series arc. The prompt interruption of very small currents of low power factor, such as transformer magnetizing currents or cable charging currents, however, may require supplementary extinguishing action from a special device, for instance, a fluid-driving piston, which is dependent upon energy provided by the operating mechanism.

Circuit interruption, in general, may be said to consist in substituting a highly ionized and therefore conducting gaseous medium, i.e., an arc, for a part of the metallic circuit, and then subjecting this arc to strong deionizing influences. In the d-c case the combination of arc lengthening and deionization raises the voltage required to sustain the arc to a value above the voltage available in the circuit, forcing the arc current to zero. In the a-c case there is also some forcing of the current, but particularly at the higher voltages where the arc voltage is a relatively small percentage of the circuit voltage; the deionizing processes have their primary function in interrupting the circuit by keeping the dielectric strength of the arc space, after a natural current zero pause, above the rapidly rising recovery voltage.

Deionization is accomplished by recombination of ions of opposite polarity within the arc space, diffusion from the boundaries or to solid surfaces, or actual displacement of ions by high-velocity fluid flow. In air-insulated breakers the cooling and deionizing surfaces may consist of pins, plates, barriers, or splitters of either metal or insulating material. If the arc is broken into a multiplicity of separate series arcs between metal plates, advantage may be taken of the fact that at current zero the positive space charge layer immediately adjacent to each cathode will require a minimum voltage for breakdown of approximately 250 volts crest, using copper electrodes. By directing properly a blast of compressed air into the arc stream, very effective arc extinction will be obtained as a result of the deionizing effect of the cool air particles introduced into the arc space. Under certain conditions this process may be considered a replacement of the ionized conducting medium with an unionized insulating medium.

Oil has been utilized for many years, not only as improved insulation in comparison to air for separating circuit-breaker parts of different electrical potential, but also for greater effectiveness in arc extinction. The high temperature of the arc decomposes the oil into gas, principally hydrogen, and free carbon. Rapid deionization is obtained by the cooling effect of the oil and the turbulent mixing in of fresh unionized particles from the gas-bubble boundary. By introducing various devices which serve in one way or another to keep the arc and the oil in close contact, the interrupting ability can be improved tremendously from the effect of the self-generated gas blast. Pressure developed with either a piston or a separate series pressure-generating arc may be used to provide a high-velocity oil flow into the main contact gap for still higher arc-interrupting effectiveness.

In spite of the inflammability of both oil and the gases produced by arcing, safe operation is obtained by submerging the contacts deep in oil, where oxygen is absent, and allowing the gases to cool as they rise to the surface.

MOLDED CASE CIRCUIT BREAKERS. Molded case circuit breakers are available in current ratings up to 600 amp at 250-volt direct current with 20,000-amp interrupting rating and up to 600-volt alternating current with 25,000-amp interrupting rating. They are made as 1-, 2-, 3-pole units in sizes from 10 amp up, corresponding to the ratings of the wires they are to protect. The overload characteristics of the circuit, the supply source, the wire and insulation to be protected, and other factors must be taken into account when deciding on the means of operating a circuit protective device. No one type of trip-release action is most efficient for all applications. For this reason several different methods are used. Thermal action, i.e., bimetal action, has relatively long delay on light overloads and short delay on heavy overloads. The bimetal tripping element is connected in series with the load; when an overload current flows through the bimetal and produces heat, the sensitive half of the bimetal expands while the inert half holds back. These two opposing forces bend the bimetal and operate the trip latch, thus opening the breaker.

In magnetic action the current of a heavy overload or short circuit energizes a magnet which trips the breaker instantly at a predetermined setting. Magnetic trip is not practical when used alone except on special applications, because it cannot be set low enough for safety and still handle starting inrush current. It is extremely valuable, however, when used in combination with thermal protection. Thermal-magnetic action combines the best features of both these actions for large wire sizes. Thermal elements protect on overloads where inverse time tripping is desirable, but the magnetic trip element operates the breaker instantly in case of dangerous overload or short-circuit faults.

Any thermal device such as a bimetal is inherently influenced by the temperature of the surrounding air and other ambients. The relation of this effect must be taken into consideration if proper application is to be made under extreme conditions.

Units may be rear-connected for panel-board or switchboard mounting or front-connected for use in individual enclosures. If the front panel is made removable, the stud design is such that the breaker can be removed from the front of the board without disturbing the bus connections. These breakers can be mounted either vertically or horizontally. The trip units are not usually interchangeable in the smaller sizes, but they are interchangeable in the larger sizes. These breakers can be made electrically operated and can be provided with shunt trips, undervoltage releases, auxiliary switches, and certain other accessories.

LARGE AIR CIRCUIT BREAKERS. Low-voltage air circuit breakers are those not including molded case circuit breakers, with interrupting ratings of 10,000 amp and above, as follows: standard air circuit breakers in current ratings ranging up to 10,000 amp direct current, 750 volts or less; and to 6000 amp alternating current, 600 volts or less, singlepole and multipole (not including super high-speed magnetic blowout breakers) and certain special breakers.

The published ampere interrupting rating is given in rms total amperes at the rated voltage of the breaker and is based upon the standard operating duty cycle.

The rms total amperes are determined by measuring the current flow in the test circuit as follows:

a. The circuit breakers should be short-circuited or omitted.

b. In a-c circuits the current measured should be the rms total current, including the d-c component, if any; it should be measured at an instant one-half cycle after the short circuit occurs and should be calculated in accordance with the American Standard Methods for determining the rms value of a sinusoidal current wave and a normal frequency recovery voltage. (For three-phase test circuits, the rms total current should be the average of all three phases. For a single-phase test circuit, three successive tests should be made to determine the average current in that circuit.)

c. In d-c circuits the current measured should be the maximum value.

Unit operations are of two types, namely, an "O" cycle, where the breaker opens the circuit only, and a "CO" cycle, where the breaker closes the circuit, an opening without purposely delayed action following immediately. The standard operating duty (duty cycle) of an air circuit breaker consists of an "O" unit operation, followed at a 2-min interval by a "CO" operating cycle. This is the so-called "O-CO" duty cycle and applies to trip-free breakers.

These breakers are equipped with adjustable overcurrent trips and may be of the dual magnetic type or the thermal magnetic type. A time delay feature for each overcurrent coil is provided for air circuit breakers with overcurrent tripping devices. These circuit breakers should be mechanically and electrically trip-free in any position.

Automatic tripping, by which is meant the opening of a circuit breaker under predetermined or other conditions without the intervention of an operator, may be done by means of series overcurrent, transformer overcurrent, shunt, reverse current, undervoltage, overvoltage, series undercurrent, or transformer undercurrent tripping devices.

Where air circuit breakers include pole unit frames, these frames are not insulated from the conductive parts. The use of air circuit breakers having such frames insulated from conducting parts, presumably as a safety measure, is a dangerous practice. Where greater safety is demanded than is afforded by the insulation of the handle and the tripping button of the standard breaker, remote-control operation or other means are recommended.

The dielectric test voltage for air circuit breakers is normally an alternating voltage applied continuously for a period of 60 sec and having an rms value as follows:

1. Air circuit breakers rated at 600 volts or less, twice rated voltage, plus 1000 volts.
2. Air circuit breakers rated above 600 volts, 2.25 times rated voltage, plus 2000 volts.

All poles of multipole circuit breakers must open and close simultaneously and be equipped with main and arcing contacts designed in such a manner that the arcing con-

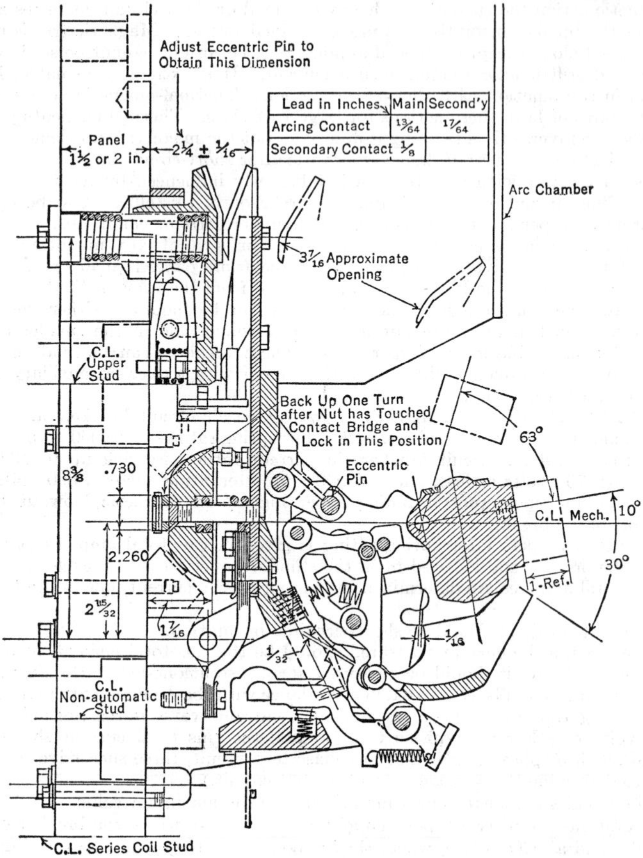

Fig. 2. Westinghouse Type DA-50 Circuit Breaker

tacts make contact before, and break contact after, the main contacts when making and breaking the circuit.

The design of circuit breakers should not be such that the operation is dependent either upon excessively precise workmanship or upon critical adjustments.

A section view of a Westinghouse type DA-50 pole unit with the "De-ion" arc chute removed is shown in Fig. 2. A similar view of an ITE circuit breaker is shown in Fig. 3.

Power Air Circuit Breakers. The breakers described herein are designed for indoor service at a-c voltages from 2300 to 15,000 volts, for continuous currents of 600, 1200, and 2000 amp, and for interrupting duty of 100,000, 150,000, 250,000, and 500,000 kva.

Contacts. Circuit breakers of this class usually have main or primary, secondary, and arcing contacts. The main contacts are of copper with contact surfaces of silver or silver-nickel. Secondary contacts, when used, are of silver-tungsten or silver-tungsten carbide with a low percentage of silver, and arcing contacts are of silver-tungsten or silver-molybdenum with a low percentage of silver.

Contact construction varies among manufacturers, but in all standard breakers the movable member is pivoted at one end, and opening is effected between the other end and the corresponding stationary member. The pivot may be of the knife switch hinge type, with the contacts either of the knife-switch or multiple-butt construction. In the multiple-butt construction several fingers on the stationary member make a short bridge between a copper block on the movable blades and the main terminal studs, a spring on each finger supplying the proper pressure. In other types the main contact consists of one or two solid copper bridges, backed by springs, which are pressed against the two main terminal

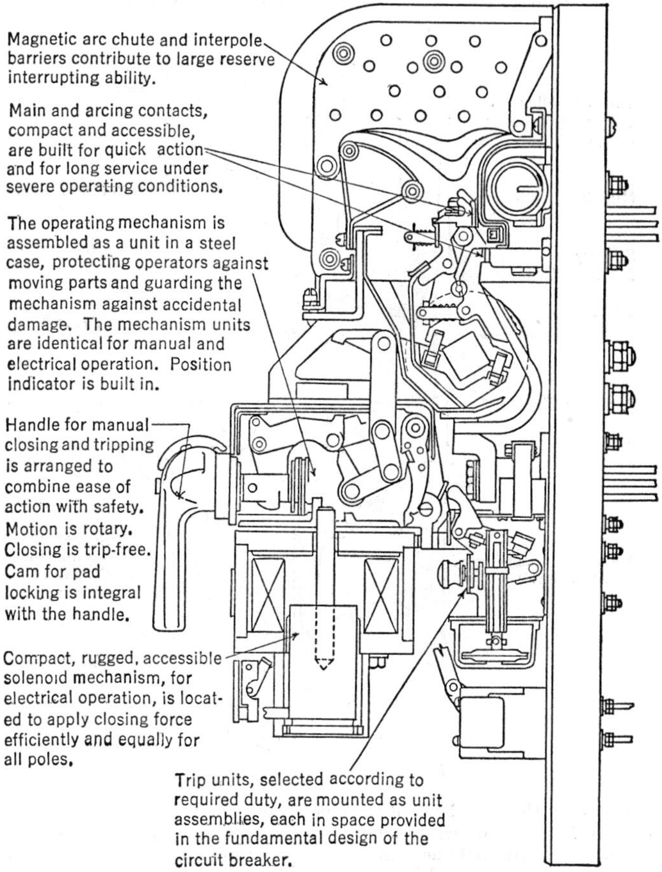
Magnetic arc chute and interpole barriers contribute to large reserve interrupting ability.
Main and arcing contacts, compact and accessible, are built for quick action and for long service under severe operating conditions.
The operating mechanism is assembled as a unit in a steel case, protecting operators against moving parts and guarding the mechanism against accidental damage. The mechanism units are identical for manual and electrical operation. Position indicator is built in.
Handle for manual closing and tripping is arranged to combine ease of action with safety. Motion is rotary. Closing is trip-free. Cam for pad locking is integral with the handle.
Compact, rugged, accessible solenoid mechanism, for electrical operation, is located to apply closing force efficiently and equally for all poles.
Trip units, selected according to required duty, are mounted as unit assemblies, each in space provided in the fundamental design of the circuit breaker.

FIG. 3. ITE Circuit Breaker

studs. The studs and springs and bridges are carried by a movable arm. The pivoted end of this arm is shunted by a small copper ribbon shunt, and the opposite end carries the secondary and arcing contacts.

Intermediate and arcing contacts are of the butt, knifeswitch, or wedge-to-finger type.

When the circuit breaker trips, the main contacts should open first, followed by the secondary and arcing contacts in that order. The secondary contacts are kept firmly pressed together until the proper time for their opening by means of springs and in some designs by the magnetic forces.

The shape of the arcing contacts is such as to accelerate the movement of the arc from the arc tips.

The stationary contacts are mounted on insulating material or on insulating bushings for mechanical support. The electrical connections are also carried to the rear through this insulating material. The circuit-breaker terminals may be of either the fixed or the draw-out type.

Standard breakers of this type are equipped for electrical closing, with provision for a hand-operated lever. Hand operation is recommended only for inspection and maintenance. The electric operating mechanism is normally of the solenoid type, in which those parts connected to the circuit-breaker contacts normally are tripped free from those parts connected to the working core of the solenoid. This is known as mechanically trip-free. The complete linkage consists essentially of the power unit, together with trip-free levers which actuate a cross-bar, which is mechanically connected to each of the moving contact arms by an insulating operating rod. The mechanism is generally isolated from live parts by a steel barrier. Inasmuch as the contact velocity is very

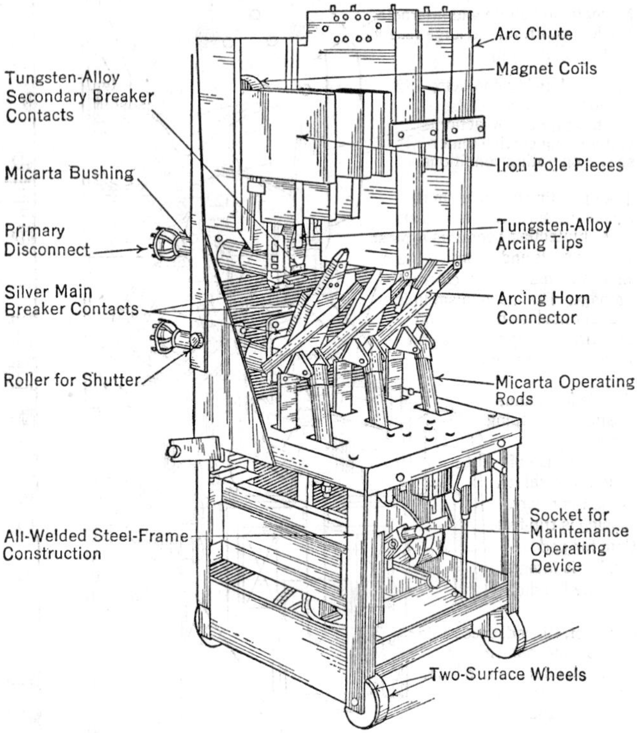

FIG. 4. De-ion Circuit Breaker

high, it is necessary that the shock of opening be cushioned. This is usually accomplished by means of a dash pot, which does not slow the opening speed during the first part of the travel but is effective only near the end of the opening stroke. The mechanism must have sufficient force during the last part of the closing stroke to insure against stalling or reversing when the breaker is closed against a short circuit.

Interrupters. The interrupter of the breaker shown in Fig. 4 is of the magnetic "De-ion" type. Briefly, the theory of operation of this type of breaker involves the use of spaced insulating plates of non-gas-forming material having tapered slots, into which the arc is moved by a magnetic field. As the arc moves into the narrow portion of the slots, electrons in the arc stream, attaining a high velocity from the magnetic field, bombard the gas particles of the arc stream and require the arc continuously to ionize fresh gas in considerable quantities. When current zero is reached, this action continues to deionize the plurality of short lengths of arc near the edges of the plates, thereby establishing sufficient dielectric strength to interrupt the circuit. It should be noted that the V-shaped slots in the insulating plates definitely limit the travel of the arc into the arc chute, thus increasing the velocity of deionizing gases, as compared to the velocity of the arc.

The arc chute of the Magne-Blast air circuit breaker is shown in Fig. 5. It consists essentially of a series of gradually interleaving fins of insulating material. The arc is formed by a multiplicity of blowout coils into this chute, and arc runners carry the arc up to the end walls of the chute. The arc length is then increased to a maximum by the interleaving fins, causing it to take a serpentine path. In addition to this lengthening of the arc the insulating material absorbs heat, thus cooling the arc. The combination of these actions results in an increase in the rate of dielectric recovery of the arc space and a decrease in the rate of rise of recovery voltage. Following a current zero after the arc is well within the chute, the rate of dielectric recovery is greater than the rate of rise of recovery voltage, the arc cannot restrike, and interruption of the circuit is thereby affected.

COMPRESSED-AIR CIRCUIT BREAKERS. Contacts. Most compressed-air circuit breakers have several sets of contacts, each designed to perform a particular function. The various contacts are designated as main or primary, secondary, arcing, and isolating contacts. The first three are used on most breakers, and the isolating contacts on a few designs. The isolating switch is in series with the interrupting contacts and opens immediately after the circuit has been interrupted to take potential off the interrupting contacts. This permits somewhat faster opening, as the inertia of the interrupting contacts is low

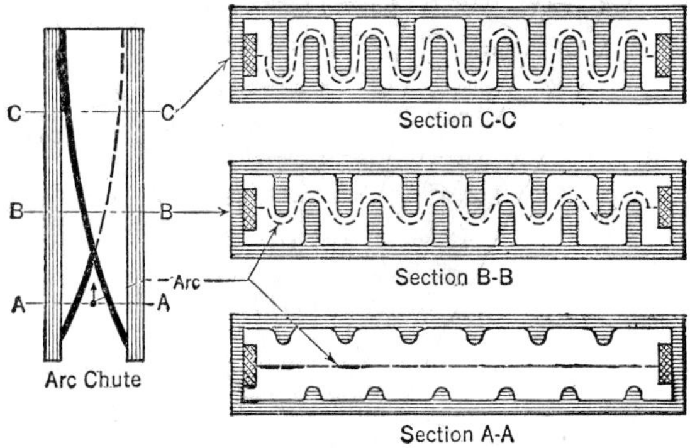

FIG. 5. Arc Chute of Magne-Blast Air Circuit Breaker

and the travel is short. With the other designs, which do not use isolating contacts, the interrupting contact travel is sufficient to perform the isolating function without the necessity for an additional switch.

The contacts are usually of copper with the contact surfaces plated or inset. The contact surfaces of the main contacts are usually silver, silver-nickel, or silver-cadmium oxide; the secondary contact surfaces are usually silver-tungsten or silver-tungsten carbide; and the arcing contact surfaces are usually silver-tungsten or silver-molybdenum, with a low percentage of silver in either alloy.

The contact designs may be butt, blade and finger, plug and socket, or rod and tulip. With the butt contact design the follow is short, causing the contacts to part very quickly after motion has begun, thus necessitating a high accelerating force at the start to insure a full blast of air at the time of contact parting. With the blade-and-finger or rod-and-tulip type, the parting of the contacts does not take place until somewhat later, and consequently a more moderate accelerating force can be used at the start to give the same contact speed at the time of parting.

The contacts at voltages above 34.5 kv are usually of the plug-and-socket or the rod-and-tulip type, with one or both contact members tubular to provide a passage for the air or the arc gases.

Up to 34.5 kv a single break is normally used; above this voltage, however, several breaks in series may be used.

The burning of the contacts of compressed-air circuit breakers is usually less than that of oil circuit breakers for a given application because of the short arcing time and the consequent limited arc energy available for burning. Although in most designs there is no blast of air on the contacts during closing, such a blast is present on some designs. The contacts are, however, usually easily renewable.

Air Supply and Operating Mechanism. The air supply is of primary importance in a compressed-air circuit breaker, inasmuch as it is used for both interruption and operation. Therefore an adequate supply of air free from moisture and oil vapor must be available at all times, and also the air blast in the interrupter must be coordinated with the contact opening. A direct mechanical connection between the operating mechanism and the air-blast valve is the best means of securing such coordination.

When an installation consists of a single breaker, the compressor and storage may be integral with the breaker. When a group of breakers are installed, however, it is normal to have two independent air-supply systems, including two compressors, two system storage tanks, and two complete sets of other necessary auxiliaries. There is an individual storage tank integral with each breaker.

Air is initially compressed and stored at pressures approximately twice the circuit-breaker-operating pressure. This procedure permits a more economic storage of usable air, and by compressing, cooling, and expanding to the breaker-operating pressure a high percentage of the moisture is removed and the dew point is depressed accordingly.

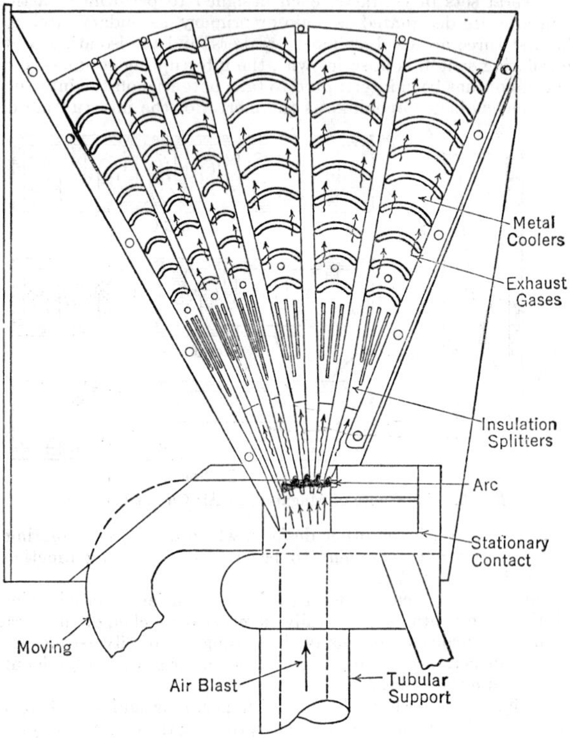

Fig. 6. Interrupter of Compressed-air Circuit Breaker

Circuit-breaker designs cover an operating range of pressure from 100 psi to 350 psi. The high-voltage breakers generally require the higher operating pressure.

The main system has a storage capacity sufficient for approximately twenty operations, and the breaker tank for two operations.

It is necessary to maintain a low oil-vapor content to prevent the possibility of explosion; this is done by using a large, slow-running compressor. It is also necessary to reduce the moisture content of the air to decrease the corrosion, protect the insulation, and prevent freezing of valves and other pneumatic devices, although the moisture in the air has no adverse effect on the interruption.

Valves are provided to control the flow of air. The main valve, which opens to supply air to the interrupter, is generally of the conventional metal-to-metal seat, poppet type mechanically operated by the breaker mechanism or an auxiliary cylinder and piston. Generally this valve is not opened during the closing operation. Valves are also necessary to control the supply of air to the operating mechanism.

An alarm is usually provided to warn the operator when the air pressure is too low. Automatic features prevent opening of the circuit breaker at a pressure below that required for a satisfactory interruption.

The operating mechanism usually consists of a piston, which is actuated by air, together with the necessary connecting rods and levers to the contacts. Air is generally supplied to one side of the piston to open the breaker and to the other side of the piston to close the breaker. The breaker is cushioned against mechanical shock during operation by either oil or air dashpots. With suitable air control the main operating piston and cylinder may be used for decelerating dashpot action.

Interrupters. Outstanding among interrupting media today is compressed air, applied in the form of a transverse blast for maximum current interruption at voltages in the powerhouse class.

An efficient circuit breaker is one in which the rate of dielectric recovery between the arcing contacts, immediately following the first current zero after contact parting, is always greater than the rate of rise of recovery voltage across the same contacts. There are two methods by which this maximum efficiency can be obtained: first, by increasing the rate of dielectric recovery sufficiently; and, second, by reducing the rate of rise of recovery voltage. This second method can be accomplished by inserting a resistor into the interrupting circuit. The transverse-air-flow type of interrupter fundamentally does not appear to have the current limitation inherent in the nozzle type of interrupting elements. The arc is drawn from back to front across the discharge end of the blast tube and is blown into the end of the slots of the splitter plates (Fig. 6). The actual interruption in an arc chamber of this type is accomplished by the air stream driving the arc against and between the lower ends of the splitter plates. When a normal current zero is reached, the arc core is in comparatively highly ionized condition, as indicated by the considerable leakage current and damping effect of the breaker on the recovery voltage after the current zero. The rate of deionization in a conducting gas column depends largely on conditions at the boundary where diffusion and recombination are most effective; consequently, it is very important to approach the current zero with a highly turbulent atmosphere surrounding the arc core. In compressed-air circuit breakers these boundary conditions are largely determined by the ways in which the air flows with respect to the arc.

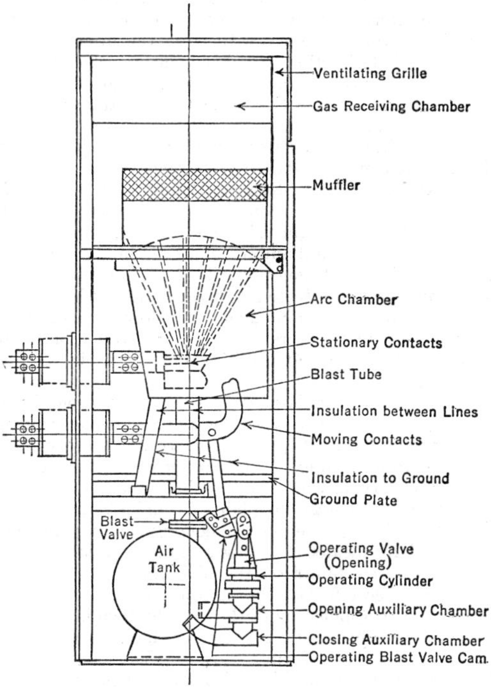

Fig. 7. Typical Compressed-air Circuit Breaker

If an arc is blown into a splitter of refractory material, it will be forced against the edge of the splitter, where it will remain more or less stationary with respect to the splitter. The flow of high-velocity air parallel to the splitter will cause one side of the arc to be highly turbulent, a condition very favorable to deionization. The other side of the arc, however, will be closely pressed to the splitter, where the air velocity is substantially low, and deionization will proceed very slowly. If, however, the splitter is made of some gas-evolving material such as fiber, the air flow parallel to the splitter presses the arc against the splitter edge, and the heat from the arc liberates gas from the fiber, which projects itself away from the splitter and into the arc stream at high velocity. This gas bombardment forces the arc stream to be highly turbulent on the side toward the splitter, and the air flow forces the arc to be highly turbulent on the side away from the splitter. Thus the entire body of the arc lends itself very readily to deionization by diffusion and recombination by reason of its high turbulence. A typical compressed-air circuit breaker is shown in Fig. 7.

Under light-current conditions the interruption problem is quite different. The air pressure and velocity required to produce satisfactory interruption is entirely a function of voltage and rate of rise of recovery voltage, and depends less upon the width of the throat.

The axial-flow type of interrupter, although its inherent limitations make it unsuitable for high kva requirements at the lower voltages, is quite satisfactory and even superior for the moderate values of current at the higher voltages. As breaker voltages are increased above 34.5 kv, the problem of insulation becomes of increasing significance, and it becomes more difficult to design a satisfactory structure which operates on the cross-blast principle. In a typical vertical-flow compressed-air interrupter, the contact members are hollow to permit a portion of the air blast to pass through them. The air blast passes through the porcelain below the interrupter, where parts of it escape through the lower working unit, and the remainder passes through the tube in the upper porcelain, from which it escapes through the upper working unit. The air approaches the arc in a radial direction from between the orifice plate and the streamliner of the contact. Upon reaching the arcing space, it turns and flows vertically along the arc stream, some of it escaping through the contact, while the greater portion escapes through the orifice plates.

In some designs the throat openings are such that the velocity of the air increases when the moving contact rods are withdrawn.

OIL CIRCUIT BREAKERS. Contacts. Contacts of oil circuit breakers have two very important duties: (1) when closed, they must carry the full-load current of the breaker with a small temperature rise; and (2) they must carry for a few seconds, without deterioration, whatever amount of short-circuit current the system can pass through them. When opening or closing under load or short-circuit conditions, the main current-carrying parts must be so protected that they will not be burned or scarred sufficiently to prevent carrying full-load current again with low temperature rise when the breaker is reclosed. On small circuit breakers the necessary protection is obtained by the special shape of the contacts. On large circuit breakers the arcing contacts are separate from the main contacts.

There are four kinds of main contacts in general use in American oil circuit breakers, namely, the solid butt type, the wedge-and-finger type, the bayonet-and-finger type, and the plug-and-socket type. Arcing contacts are generally provided to protect the main contacts from burning. They are mostly of the wedge-and-finger type, lever roll-in type, and bayonet type, with either plain-break or quick-break action.

Butt contacts may be plain or laminated. Plain butt contacts consist of properly shaped stationary members engaging moving members, one member being resiliently supported by means of springs. This type is adaptable to low-current breakers and is most frequently found on the high-voltage breakers. The laminated butt contacts are made in several forms, the elliptical form utilizing copper strip laminations being the earliest. The application of the semielliptical brush is now limited to breakers of relatively low current-interrupting capacity or short-time rating because of the tendency of electromagnetic forces to deform the brush. Various types of reversed brush contacts have been designed to overcome this difficulty. For heavier currents, non-welding alloys, plated on or inset, are used.

Wedge-and-finger contacts consist of one or more wedges, usually on the movable elements, which enter between the contact fingers when the breaker is closed. These contact fingers, which are arranged in pairs, are so constructed that they conform themselves to the wedge surfaces and present at all times a uniform contact pressure over the contact area.

The bayonet-and-finger type of contact consists of a copper rod with a special arc-resisting alloy tip, which engages a pair of fingers. The ends of the fingers may have inserts of arc-resisting alloy.

The plug-and-socket contact consists of a copper rod with an arc-resisting alloy contact tip, which engages a socket consisting of a circle of contact fingers.

Arcing contacts and contacts which perform the functions of both arcing and current carrying usually are made of silver or copper and some refractory material, such as tungsten carbide, tungsten, or molybdenum, by powder metallurgical processes.

The thermal and electrical conductivities of these contact materials are very good. Such materials are also quite strong and hard. This characteristic makes them very resistant to wear and impact. They are also resistant, to a large degree, to local surface melting under arcing conditions, a characteristic which is very desirable for arcing contacts. There is little tendency to erosion or welding, this property being directly proportional to the percentage of refractory metal. With increased refractory-metal content, however, the conductivity decreases and the contact resistance increases. Copper-tungsten contacts are particularly adaptable to use in oil break devices, as the oil prevents the oxidation of the copper and thus limits the increase in contact resistance.

A copper-tungsten composition with approximately 60 per cent tungsten and a silver-tungsten composition with about 80 per cent tungsten are used for arcing contacts. When the functions of current carrying and arcing are combined, a lower refractory-metal content is desirable, and the amount of silver is sometimes raised to 35 or even 50 per cent.

The main current-carrying contacts are usually copper or high-conductivity copper alloy. Sometimes, however, in order to increase the impact resistance, high-strength copper alloys, such as cobalt-copper-beryllium, chrome-copper, or cadmium-copper, are used. For maximum conductivity silver plating or inserts of solid silver are used.

Oil. The oil in an oil circuit breaker performs three main functions. First, as an insulating liquid it completely surrounds all potentially live contact parts, forming an insulating medium low in cost, readily applied, easily removed for inspection of the contacts, and renewed without difficulty upon any evidence of deterioration. In addition, its relatively high dielectric strength permits smaller electrical clearances than would be safe with air insulation.

Second, the oil completely surrounding the contacts forms a medium well adapted to conduct heat away from the contact surfaces, the point of greatest heat generation under normal current load. Thus contact surfaces in an oil circuit breaker may be worked at higher current densities or, conversely, lower contact pressures per unit of area than can be expected with contacts operating in still air.

Third, the oil forms a most effective interrupting medium when an arc is drawn between the contacts on opening the circuit breaker. Its action in circuit interruption was described in an earlier paragraph.

Most important for obtaining satisfactory service from an oil circuit breaker is a consideration of the characteristics of the oil used and the condition in which it must be maintained. Oil is an exceptionally good insulator when it contains neither water nor suspended particles. To insure this condition requires careful, periodic testing of the oil, especially for circuit breakers opening under heavy overload or short-circuit conditions. An oil showing insulating properties below the accepted values must be filtered to remove the moisture and suspended particles.

Tanks. The main purpose of circuit-breaker tanks is to hold the oil that submerges the contacts. Incidentally they dissipate the heat and enclose the working parts of the breaker. The type of tank is governed by considerations of the duty to be performed and of cost. Rectangular tanks are used for small breakers of moderate rupturing capacity; circular tanks with dome-shaped bottoms, for the largest-capacity oil circuit breakers, as this construction gives the strongest type of tank capable of withstanding the stresses frequently encountered when interrupting large amounts of power. It has been found experimentally that not only is the internal pressure proportional to the total arc energy, but also superimposed high-pressure impulses are present which are a function of the rate of arc energy liberated, in other words, to the instantaneous arc power. The high pressure in the arc creates a steep front pressure wave which spreads out from its origin, decreasing in proportion to the distance from the center of the arc. It is evident, therefore, that excessive speed of arc lengthening, by magnetic looping of the arc, blasting of the arc through holes or around splitter plates, or extremely high contact velocity will tend to raise the pressure peak.

Although most oil circuit breakers are of dead-tank construction so that the tank may be grounded, live-tank construction has been employed for many years on certain lines of breakers. The general design of the live-tank type has a form of inverted explosion chamber with main contacts in air. The explosion chamber consists of a steel cylindrical oil tank with insulated lining supported on a porcelain post, set in a space with clamp fitting and mounting bolts. Inside the oil tank is a system of baffle plates held in place and supported from the oil-tank top. Separation chambers above the explosion chamber contain a quantity of quartz pebbles through which the gas has to pass in escaping. The separated oil drains through the perforated disk and exhaust openings in the breaker top into the explosion chamber.

All breaker tanks are provided with insulating liners except in the largest high-voltage breakers, where the distances between the live parts and the tank are generally sufficient without them. In certain breakers a flameproof lining is also used to insure against carbonization of the insulating liners. The effectiveness of modern circuit-interrupting devices has resulted in making the tank diameters of the high-voltage breakers more dependent on insulation clearances than on interrupting capacity. Gas vented from the arc-extinguishing structure, however, momentarily substitutes a weaker dielectric for the oil usually maintained between live parts and the grounded tanks and also between the live parts on adjacent terminals, and this factor must be considered in establishing minimum tank diameters and the necessity for liners.

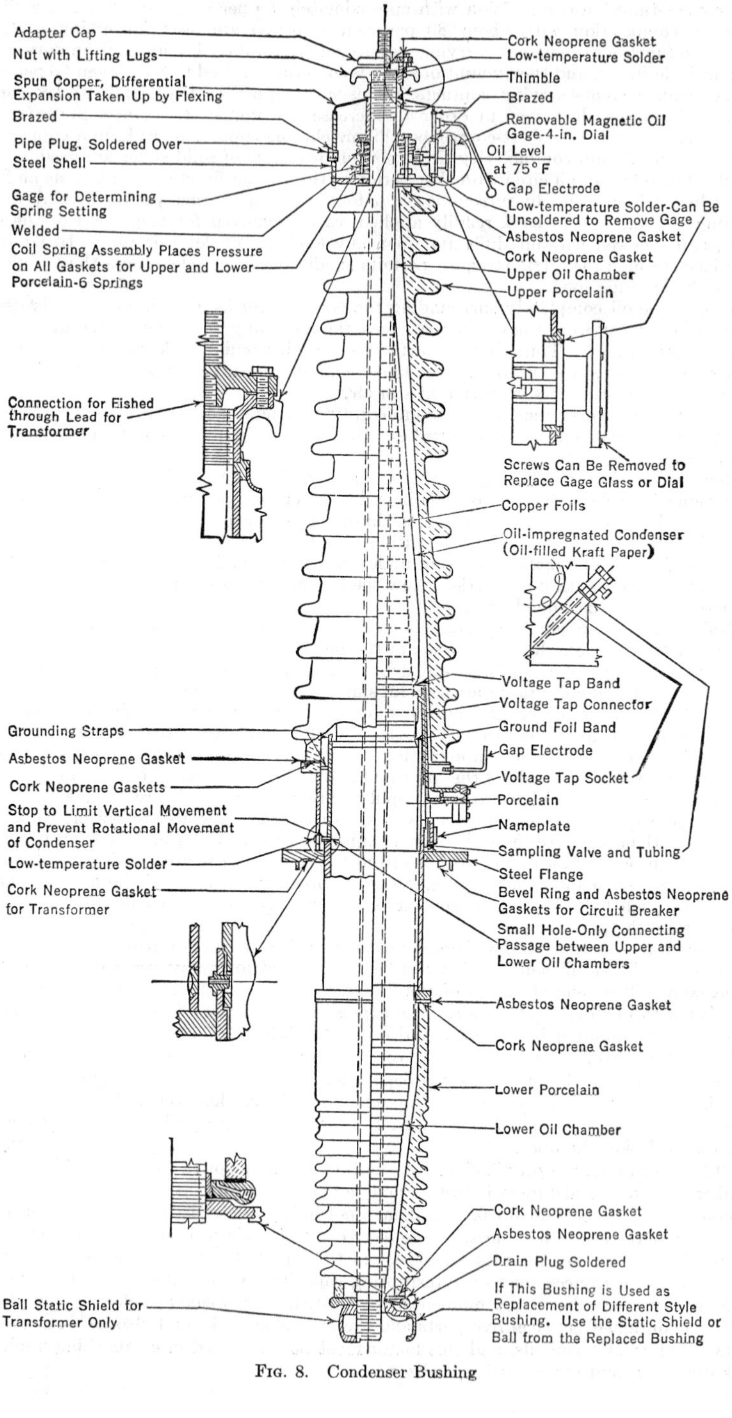
Adapter Cap
Nut with Lifting Lugs
Spun Copper, Differential Expansion Taken Up by Flexing
Brazed
Pipe Plug, Soldered Over
Steel Shell
Gage for Determining Spring Setting
Welded
Coil Spring Assembly Places Pressure on All Gaskets for Upper and Lower Porcelain-6 Springs
Connection for Fished through Lead for Transformer
Cork Neoprene Gasket
Low-temperature Solder
Thimble
Brazed
Removable Magnetic Oil Gage-4-in. Dial
Oil Level at 75° F
Gap Electrode
Low-temperature Solder-Can Be Unsoldered to Remove Gage
Asbestos Neoprene Gasket
Cork Neoprene Gasket
Upper Oil Chamber
Upper Porcelain
Screws Can Be Removed to Replace Gage Glass or Dial
Copper Foils
Oil-impregnated Condenser (Oil-filled Kraft Paper)
Voltage Tap Band
Voltage Tap Connector
Ground Foil Band
Gap Electrode
Voltage Tap Socket
Porcelain
Nameplate
Sampling Valve and Tubing
Steel Flange
Bevel Ring and Asbestos Neoprene Gaskets for Circuit Breaker
Small Hole-Only Connecting Passage between Upper and Lower Oil Chambers
Grounding Straps
Asbestos Neoprene Gasket
Cork Neoprene Gaskets
Stop to Limit Vertical Movement and Prevent Rotational Movement of Condenser
Low-temperature Solder
Cork Neoprene Gasket for Transformer
Asbestos Neoprene Gasket
Cork Neoprene Gasket
Lower Porcelain
Lower Oil Chamber
Cork Neoprene Gasket
Asbestos Neoprene Gasket
Drain Plug Soldered
If This Bushing is Used as Replacement of Different Style Bushing. Use the Static Shield or Ball from the Replaced Bushing
Ball Static Shield for Transformer Only

FIG. 8. Condenser Bushing

Mufflers are used in many breakers for scavenging the gases which are developed when the breakers are opened. The mufflers, which allow the gases to escape but prevent the throwing of oil, are usually arranged in the form of a labyrinth. Sufficient air space is also usually provided above the oil level in the larger breakers so that, even if the small exhaust-pipe openings were closed, the amount of gas generated during a maximum kva interruption would raise the static pressure by only a fraction of an atmosphere.

Tank tops are not merely covers for oil receptacles; they also carry insulating bushings for the breaker studs and contacts and usually support the mechanism for operating the moving contacts and perform other functions. Where the leads through the bushings in the top carry heavy current at 60 cycles, the resultant heating of these tops caused by hysteresis and eddy-current losses has to be investigated very carefully. To avoid these losses, breakers for the heavier currents may have non-magnetic inserts around the bushings or all-bronze tops for currents in excess of 1200 amp at 60 cycles.

In the normal design of top-connected oil circuit breakers, bushings are required for insulating, from the metal top, the copper studs that support the stationary contacts and carry the current to these contacts. These terminal bushings range in voltage requirements from 2.5 to 287 kv and in current from about 200 to 6000 amperes. Up to 23 kv for small amounts of power, porcelain bushings can be employed; for higher voltages oil-filled or condenser bushings are used.

Oil-filled bushings consist in general of metal tubes at the center and a series of short insulating cylinders of treated paper or porcelain, separated by radial disks to increase creepage distances. The cylinders and disks taper gradually from the flange toward the end. Oil is admitted through the top of the visual gage at the top of the bushing and can be removed through the oil drain at the bottom.

Condenser bushings used on the high-voltage Westinghouse breakers are usually of the design shown in Fig. 8. The condenser bushing is made by rolling, pressing, and baking onto a conductor alternate layers of treated paper and metal foil. The bushing is thus made of concentric cylinders of insulation with a layer of metal foil between them. The area of the foil and the thickness of the insulating cylinders are controlled to give suitable capacitances and uniform voltage gradient from the center to the surface. In the typical condenser bushing illustrated, the part outside the tank is covered by a weather casing of porcelain, the space between the bushings and the casing being filled with an insulating compound. Condenser bushings are also made with oil filling, using vacuum impregnation to remove all air and moisture from the paper-wound core. A portion of the bushing inside the tank under the oil and near the contact is protected by a porcelain arc shield.

Bushing-type current transformers are employed with breakers of high-voltage design. The copper conductor of the breaker bushing forms a primary winding, the toroidal secondary winding and the ring-shaped laminated iron core constituting the bushing transformer itself. Because of the voltage gradient provided by the condenser bushing, it is possible to make a connection between the last metal foil, the one nearest to ground, and the tank and to secure economically a reliable potential device to use in operating synchronizing devices and certain relay coils.

Oil circuit breakers for voltages up to 69 kv are usually frame-mounted with a single tank for all three phases at lower voltages and interrupting capacities, and with each pole in a separate tank at the higher voltages and interrupting capacities. The frames are made of pipe or angle iron and support directly the breaker tops. Tanks supported from these tops can be lowered for inspection or adjustment. Oil circuit breakers for voltages of 115 kv and above are built as large floor-mounted tanks, one per phase, with access to the inside of the tank through a manhole at the top or in the side.

Operating Mechanisms. Various methods are employed for operating oil circuit breakers. The smaller breakers are operated manually. For voltages up to 2500 volts, current up to 1200 amp, and station capacities under 3000 kva, direct control is usual, with the breakers mounted directly on the panel or panel framework. For higher voltages, heavier currents, or larger kva capacities, the breakers are mounted apart from the panel, and connection between the panels and breakers is made by rods, bell cranks, and levers. On smaller sizes of manually operated remote-control circuit breakers, the latches and their release coils, by which automatic operation is secured, are mounted directly on the cover plate of the panel. The handle is in two parts to form a trip-free feature to prevent the breaker from being held in the closed position on overload. On the larger sizes, the automatic latching details are located at the circuit breakers, thus taking the strain of the latched load off the panel and remote-control bell cranks and levers, and also reducing the mass to be accelerated at the time of automatic opening of the breaker.

Power operation in some form is used when the distance between the switchboard and circuit breaker makes the application of hand-operated breakers questionable, or when the physical size of the apparatus is too great for convenient manual operation. Electrical operation is the method generally employed in the majority of American designs. The mechanism can be mounted in any desired location with reference to the breaker itself, and the newer solenoid mechanisms include a mechanical trip-free feature which furnishes free opening of the breaker without the restraint of the inertia of the core. Because of the great simplicity of the d-c solenoid type of mechanism, it is normally employed.

Spring-operated mechanisms are also used extensively, energy being stored in the springs for both closing and opening the circuit breakers by either a-c or d-c electric motors.

Control energy for the electrically operated circuit breakers and similar apparatus in switchgear installations is a very important factor. The amount of power required for the closing of solenoid-operated breakers ranges from approximately 5 kw for the small breakers to 50 kw for the larger ones. The time demand is only $^1/_4$ to 1 sec, and usually a storage battery is furnished to supply the energy. Energy may also be supplied from an a-c source through a copper oxide rectifier. In most electrical power plants, operating voltage having a nominal value of 110 to 125 volts is used, but electrical equipment can be made suitable for almost any desired voltage.

It is usually found desirable to segregate various control circuits. Where there is a natural grouping of breakers, ordinarily a separate control circuit is provided for each group. Each circuit should be protected by its own switch and fused for not less than three times the maximum current needed for closing the largest breaker. This fusing will give protection in case of a short circuit in the control system and yet assure a supply of current under normal operating conditions. Individual circuits, usually at the breaker point, should be fused at about half the current rating of the closing coil circuits, because of the fact that, within the time of closing, the current seldom reaches the full-load amount before it is cut off by the auxiliary breaker.

A control relay is furnished with most of the larger electrically operated breakers, so that the wiring from the control switch does not have to carry the current actually needed for the closing of the breaker, but merely energizes the operating coil of the control relay or contactor. For the larger breakers, the self-induction of the closing coil is relatively high, and rather serious arcing is likely to occur in opening the closing circuit unless this is done by means of a control relay or contactor provided with blowout coils and arc chutes. This control relay is essentially a single-pole contactor with both an operating coil and release coil and is so arranged that current cannot be kept on the closing coil of the breakers if an overload has energized the release coil of the control relay.

Pneumatic operating mechanisms are used for high-speed and ultra-high-speed reclosing of oil circuit breakers of 500,000 kva and above. Their use for standard closing is increasing because of the small drain on the control battery in comparison to the current required by a solenoid mechanism. The pneumatic operating mechanism consists essentially of an air-supply system, magnetically controlled intake valve, throttle valve, cylinder, piston, latch, trip coil, and exhaust valve. Accessories include auxiliary switches, operating counter, trip-free control relays, heaters and thermostat, air interlocks, alarm switch, and safety valve, all mounted in a weatherproof housing.

The air-supply system usually consists of a motor-driven air compressor for each mechanism. A main reservoir stores air at operating pressure and is provided with a drain valve for removing condensed moisture. The combined air system stores sufficient air to provide for approximately five closing operations without recharging.

For outdoor breakers, the operating rods between the pole-unit mechanisms and the operating mechanism are enclosed in a weatherproof housing.

For manually operated circuit breakers, tripping from the secondary of current transformers is the most common method of automatic overload tripping. For some low-voltage indoor circuit breakers, series overload trip coils are used, mounted directly on the circuit breaker. Dashpots are added to these trip coils where inverse-time-limit features are wanted.

For electrically operated circuit breakers, relay tripping is most common. Current transformer trip coils rated at 5 amp or series automatic overload trip coils are also used on small electrically operated circuit breakers.

Coils for current transformer operation, with automatic overload trip, are mounted in the cover plates or on the breaker mechanism of the manually operated circuit breakers and on the operating mechanism of electrically operated circuit breakers. Ordinarily, where current transformers are used for instruments and watthour meters, the trip coils can be connected to the same transformers if high accuracy for the meters is not required. When not required for instruments or meters, lower-priced transformers of good accuracy can be connected directly to the circuit-breaker trip coils or to relays.

The shocks at the end of the opening stroke of oil circuit breakers are considerable because of the weight and inertia of the moving parts. Some form of bumper is supplied with all breakers to cushion the impact. The latest high-power breakers have a hydraulic type, such as an oil dashpot. Some of the older and smaller types employed a spring compression arrangement. These bumpers do not retard the opening of the breaker in any way, as they do not operate until the opening is almost completed.

Position indication is almost essential in connection with any breaker having an enclosed mechanism, as otherwise it is difficult to determine by inspection of the breaker whether it is closed or open. The position indicators naturally differ somewhat with various types of breakers. Usually located outside the mechanism housing, the indicator functions with the pull-rod and points to the word "Closed" or the word "Open" to show at a glance the actual position of the breaker.

A circuit-breaker-operating mechanism must be designed to provide satisfactory closing performance even when the breaker is closed on a short circuit. Under such conditions high electromagnetic forces, proportional to the square of the current and derived from the U-shaped loop, are operating during the final portion of the closing stroke to retard the closing. The closing motion of the contacts must be neither stopped nor momentarily reversed.

Interrupters. It is taken for granted that modern oil circuit breakers require some type of arc-extinguishing device in order to provide rapid low energy and reliable interruption. These devices control the arc in such a way that it is quickly extinguished by introducing high dielectric material in the path of the arc. These devices decrease the arcing time and the energy liberated, resulting in less oil deterioration, less pressure, less contact burning, and less maintenance. High effectiveness is indicated by low tank pressure and small quantities of generated gas.

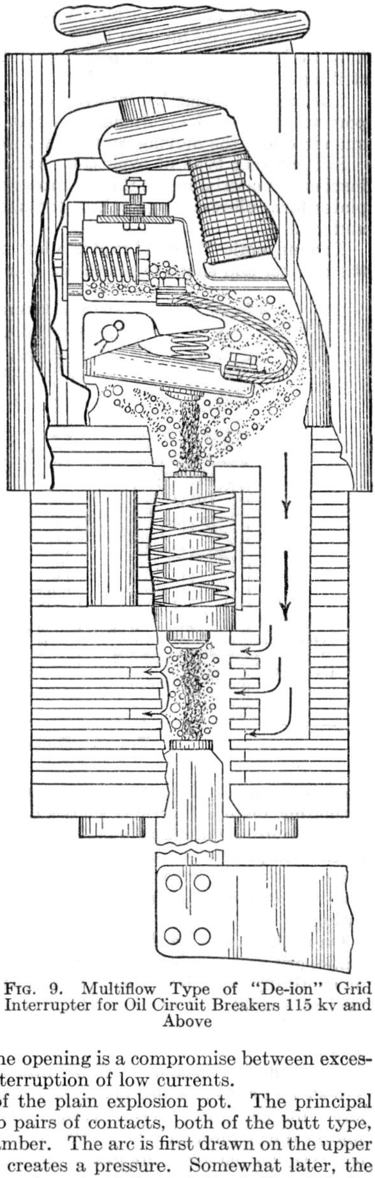

FIG. 9. Multiflow Type of "De-ion" Grid Interrupter for Oil Circuit Breakers 115 kv and Above

The earliest form of interrupter is the plain explosion pot, which consists of a chamber surrounding the stationary contact and containing oil. In the bottom of this chamber is a hole through which the moving contact rod enters to complete the circuit. To interrupt the circuit the moving contact is withdrawn from the explosion pot. The separation of the contacts draws an arc which volatilizes the oil and produces a pressure within the chamber. The moving contact is withdrawn from the throat and is followed by the arc. Partly ionized gas and atomized oil are forced into the arc at this time, causing rapid deionization and interruption of the arc at a current zero. The effectiveness is proportional to the pressure, which in turn varies with the size of the opening and the arc current. The size of the opening is a compromise between excessive pressures at high currents and ineffective interruption of low currents.

The oil-blast explosion pot is a refinement of the plain explosion pot. The principal difference consists in the fact that there are two pairs of contacts, both of the butt type, which cause two arcs to be drawn within the chamber. The arc is first drawn on the upper contacts when they separate; this immediately creates a pressure. Somewhat later, the lower contacts part. The oil and arc products are forced out through the hollow tubular moving contact. Thus the second arc is acted upon by the oil blast as soon as it is drawn, and the oil and arc products are forced across and not along the arc, as in the plain explosion pot.

Another arrangement of the oil-blast interrupter utilizes a gate-type intermediate contact, which is pushed aside as the moving contact rod enters stationary contact fingers. When interrupting fault current, the rod is withdrawn, the gate closes, and the arc above the gate creates pressure for driving oil through a channel, which directs the blast across the second arc drawn between the gate and the moving contact.

Impulse-type breakers utilize a spring-driven piston to drive oil across one or more arcs drawn in series in front of blast ports instead of generating the oil-blast pressure from a separate arc. The impulse type therefore has the advantage of providing adequate oil flow to interrupt low magnetizing or line-charging currents with little or no arc restriking.

The expulsion chamber interrupter is a type of explosion pot arranged with the exhaust port adjacent to the stationary contact finger. As the gas is vented out of this port in one direction, the moving contact is pulled down into fresh oil in the other direction.

Still another type of explosion chamber, called a ruptor, utilizes laminated plates of insulating material in the throat, through which the moving contact is withdrawn. These plates form oil-retaining pockets arranged to subject the arc to a deionizing blast of oil and gas.

The magnetic "De-ion" grid is a laminated structure made up of groups of fiber plates, the number and shape of which are determined by the circuit voltage and the current to

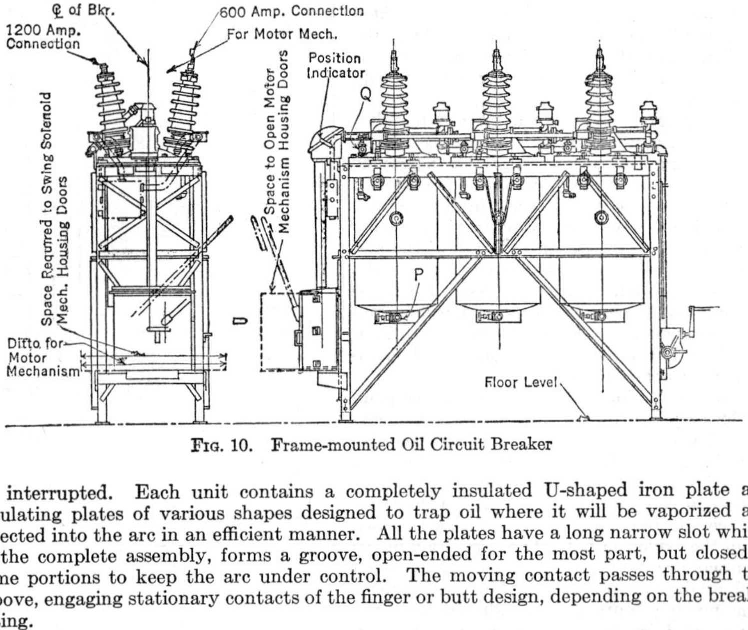

FIG. 10. Frame-mounted Oil Circuit Breaker

be interrupted. Each unit contains a completely insulated U-shaped iron plate and insulating plates of various shapes designed to trap oil where it will be vaporized and directed into the arc in an efficient manner. All the plates have a long narrow slot which, in the complete assembly, forms a groove, open-ended for the most part, but closed in some portions to keep the arc under control. The moving contact passes through this groove, engaging stationary contacts of the finger or butt design, depending on the breaker rating.

On opening, the breaker contacts separate, and an arc is drawn vertically between the stationary and moving contacts in the slot of each grid. The iron plates pull the arc magnetically back into the grid structure toward the closed end of the slot, forcing it against a solid wall of oil in the slot and trapped in the pockets. The relatively cool unionized gas formed by the heat of the arc cannot escape except through the arc space.

The multiflow type of "De-ion" grid (Fig. 9), now used on high-voltage breakers rated 115 kv and above, is also made up of laminated plates of insulating material but without any imbedded iron. Two arcs are drawn in series, the upper pressure-generating arc driving oil down through two parallel channels to pairs of horizontal inlet passages, which converge on the lower arc drawn through a succession of orifices. The oil flow deionizes the arc space as the arc products are forced out of vents at levels in the structure alternated between the inlet passages. By keeping the arc centered in the orifices without unnecessary lengthening, the arc energy is kept extremely low.

A typical high-voltage outdoor oil circuit breaker is shown in Fig. 10.

3. FUSES

The simplest fuse is the cartridge-type enclosed fuse. Certain standard dimensions and types of contacts have been adopted for various sizes up to 600 amp at 250 volts alternating or direct current and at 600 volts alternating or direct current. Up to 60 amp the ferrule type of contact is used, as shown in Fig. 11; blade contacts, shown in Fig. 12, are used for fuses from 61 to 600 amp. The rupturing capacity for cartridge fuses, as determined by test, shows that on larger systems the circuit characteristics should be such as to limit the maximum overload current passing through the fuse to approximately 10,000 amp at the rated voltage.

Two general classes of cartridge fuses have been developed: one in which a renewable element is a bare link and in which there is no powder filling, and one which has the fusible element enclosed in a powder-filled tube but which is non-renewable.

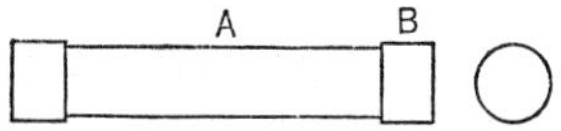

FIG. 11. Enclosed Fuse with Ferrule Contacts. *A*, fuse tube, fiber; *B*, metal ferrule contacts.

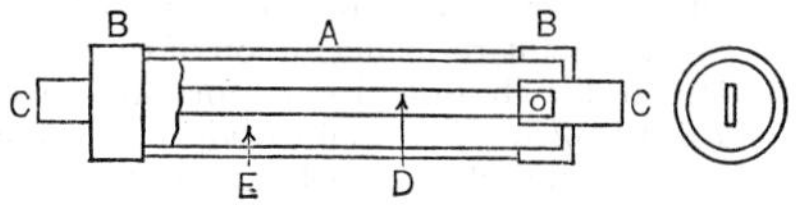

FIG. 12. Enclosed Fuse with Blade Contacts. *A*, fuse tube, fiber; *B*, metal ferrules; *C*, blade contacts; *D*, fuse strip; *E*, filler, gas-absorbing material

HIGH-VOLTAGE FUSES. These fuses are ordinarily applied as short-circuit protection for power transformers, on most transformers on distribution systems, as distribution-line sectionalizing devices, as short-circuit protectors for motor starters, for protection of high-voltage capacitors, on high-voltage d-c railway circuits, and on street-lighting transformers.

A fuse is identified by its load current, voltage, and interrupting ratings. The maximum-load current of a fuse is determined by its temperature rise, which should not exceed 30 deg cent at the fuse clips. For renewable fuses it is necessary to state the maximum current rating for both the fuse holder, which is the part used again after the fuse has operated, and for the fusible or renewable element. The time-current characteristic curve of the fusible elements must be considered when coordinating fuses with other devices.

There are two approved standards for rating high-voltage fuses. The ratings N and E have been established by EEI and NEMA. The N standard, which has been in force for many years, is applicable to the universal cable-type fuse links used in distribution-type cut-outs. Such a link will carry 100 per cent of its rated current continuously but will blow at less than 220 per cent of its rated current in 300 sec.

The recently adopted E standards, which apply to modern power fuses, require the fuse to melt at a current of 200 to 240 per cent of rating in exactly 300 sec for fuses up to and including 100 amp, and at a current of 220 to 264 per cent of rating in exactly 600 sec for fuses above 100 amp. The fuse must carry 100 per cent of the rated current continuously without exceeding a rise of 30 deg cent at the fuse clips.

The rated voltage of a fuse is the highest rms voltage at which it is designed to operate. It is expected to interrupt any current within its rating at this voltage and to maintain the circuit open with rated voltage across it after it has blown.

The interrupting rating of a high-voltage fuse is usually given as the maximum asymmetrical rms amperes that can be interrupted at the specified operating voltage. To provide a basis of rating comparable to circuit breakers and to eliminate confusion, manufacturers are beginning to list, along with the interrupting rating of power fuses, the three-phase kva rating of an equivalent circuit breaker. This rating corresponds to the symmetrical maximum fault kva for a circuit that can produce an asymmetrical current equal to the current-interrupting rating of the fuse. The kva rating is the asymmetrical rms current-interrupting rating of the fuse divided by 1.6, the assymmetry factor. The new current value is then multiplied by the circuit voltage and the three-phase factor (1.73). The asymmetry factor has been determined as representing the increase in the duty required of a power fuse when interrupting an asymmetrical instead of a symmetrical current.

In the application of distribution cut-outs the asymmetry factor is 1.2. The use of a 1.6 factor for power fuses and a 1.2 factor for distribution cut-outs results from the difference in characteristics of the circuits in which they are usually applied. A power fuse is usually close to the source of power, where there is little damping effect in the line, whereas a distribution cut-out is usually in an extended distribution system having a resistance component high enough to provide substantial damping.

The inherently fast operating characteristics of fuses require consideration of an asymmetry factor when making a fuse selection. With power circuit breakers the time of operation is usually slow enough to allow the d-c component of the circuits to decay so that the breaker interrupts only the symmetrical short-circuit current of the system. On heavy faults the fuse usually interrupts by the end of the first half-cycle of current.

CUT-OUTS. Modern fuses are divided readily into two classifications: distribution cut-outs and power fuses. The cut-out, a low-capacity fuse intended for use in a distribution system, is an assembly of a fuse support and a fuse holder of relatively low interrupting rating, which is designed to receive a universal type of cable link. Cut-outs fall into two major classifications, the open and the enclosed types. The open cut-out is used most commonly on distribution circuits above 7500 volts. The enclosed cut-out is generally applied to circuits of 7500 volts or less.

A variation of the indicating cut-out is found in the repeator type, where two or more fuse tubes are mounted so that the operation of the first connects the second into the circuit. This arrangement allows clearance of temporary faults and prompt restoration of service by the circuit-reclosing action of the second fuse tube. Cut-outs are applied almost exclusively on distribution circuits for outdoor use.

POWER FUSES. The main divisions of power fuses are expulsion fuses, boric acid fuses, liquid fuses, and current-limiting fuses. Although the boric acid fuse can be classed as an expulsion fuse, common usage has restricted the term expulsion fuse to fiber tube fuses with flexible conducting elements within the tube. Therefore the boric acid fuses and the expulsion fuses are classified separately.

Expulsion Fuses. In the expulsion fuse a tube of solid gas-generating material is mounted between the circuit terminals so that a flexible shunt can be connected through the tube with a fusible element usually at one end. Interruption is caused in this type of fuse by a blast of gas originating in the decomposition of the solid tube by the heat of the arc. This blast of gas blows the ionized arc products out of the tube and continues to do so through current zero, thus causing deionization of the arc space. Obviously, some noise and demonstration attend the interruption of high current. This factor should be considered when applying this type of fuse. Expulsion fuses are either fixed or drop-out. the trend being toward the drop-out fuse for the higher voltages.

Boric Acid Fuses. In the boric acid fuse the tube is filled with a series of blocks formed from powdered boric acid under high pressure. A hole through the center of the block forms the arcing chambers. When the fusible element melts, an arc is drawn in the arcing chamber. Steam, formed by the heat of the arc from the water of crystallization of the boric acid, deionizes the arc space, thus effecting interruption. The steam is incombustible and thus does not decrease the dielectric strength around the fuse under operating conditions. A further advantage results from the fact that the steam can be condensed to water vapor in an ordinary surface type of condenser, consisting of spaced copper plates arranged in a container fastened to the normally vented end of a fuse holder. Condensing the steam eliminates the external gas blast. As a result the fuse can be mounted with smaller clearances and will have no demonstration upon operation.

The length of the interrupting element of a boric acid fuse can be decreased by an additional small bore provided in the boric acid and by means of suitable shunts, so that the fault current is switched into the small bore, where it is interrupted if the current is small. This permits the larger bore to interrupt only the large currents, which are shunted back from the small bore. Thus the bore diameter is correlated with the magnitude of the short-circuit current. This arrangement allows efficient use of the boric acid to obtain high-voltage gradient and high interrupting capacity. These characteristics provide a totally enclosed fuse with high interrupting ability suitable for indoor application as well as vented fuses for outdoor use.

The boric acid fuse is made in both fixed and drop-out forms. The drop-out is available for load currents up to 200 amp with an interrupting ability of 7000 amp at 138 kv. The fixed type has an interrupting ability of 40,000 amp at 2400 volts, and of 20,000 amp at 33,000 volts, with a maximum load current of 400 amp. For most indoor applications the condenser-type boric acid fuse is used to provide noiseless operation.

Liquid Fuses. The liquid fuse still enjoys popularity where applicable, but its use recently has been curtailed in favor of dry fuses.

One common form of liquid fuse consists of a glass tube with suitable end caps, between which are connected the fuse elements and a flexible spring-retracted shunt. On overload the fuse link melts, and the arc is drawn into the liquid by collapse of the spring and shunt. These units are made in several current sizes up to 400 amp and in voltage sizes up to 138 kv.

Another common form of liquid fuse is the oil-filled cut-out, in which the fusible element is under oil and is commonly mounted on a rotating frame, so that the circuit can be interrupted by moving an external handle. Thus the oil cut-out acts as a load interrupter as well as a short-circuit interrupter. This type of cut-out is usually used in underground applications because it is available in the submersible type.

Current-limiting Fuses. These fuses are divided into two types: the single-element and the two-element fuses. These are dry fuses that operate on the principle of interrupting current of large magnitude by requiring the fuse element to melt far in advance of the possible peak current of the first half-cycle. The fuse is so designed that the melting of the element introduces a high arc resistance, which, in turn, holds the circuit current within the abilities of the fuse.

The single-element current-limiting fuse, which consists of small silver wires suspended in a body of fine sand, relies for its interrupting ability upon the rapid deionizing of the arc space by the condensation of the arc products on the sand. The two-element current-limiting fuse combines a circuit-clearing or clean-up fuse and a current-limiting device in series in one cartridge. The clean-up fuse is a boric acid fuse supplied with a condenser, so that it can be totally enclosed. The current-limiting element consists of silver wires mounted, in a round fiber rod, in deep longitudinal grooves of a width about equal to the diameter of the silver wire. When the silver wires melt, a controlled high arc voltage results, which prevents further rise in the fault current. Almost simultaneously with this action, the clean-up fuse element melts and clears the fault current.

Current-limiting fuses are ideal for the protection of potential and small operating transformers and are made for this service for voltages up to and including 34.5 kv. They are also made in current ratings up to 200 amp at 5 kv and 100 amp up to 15 kv. The interrupting rating is higher than for any other fuse and is 60,000 amp at 2.5 and 5 kv, 80,000 amp at 7.5 kv, and 50,000 amp at 15 kv. They are totally enclosed, eliminating all external disturbance. Their current-limiting action reduces magnetic stresses in connected equipment. They can be coordinated with motor starters to produce a combination that can be connected to a high-capacity bus. Such coordination requires extensive study of all the characteristics of both the starter and fuse. The fuse must coordinate with the thermal element of the starter, its interrupting ability, and its short-time rating, as well as with the electrical characteristics of the motor itself.

4. PROTECTIVE RELAYS

ASA Standard C37.1 defines a relay as "a device which is operated by a variation in its electrical or physical condition to effect the operation of other devices in an electrical circuit." These standards further define a protective relay as "a relay, the principal function of which is to protect service from interruption or to prevent or limit damage to apparatus."

Protective relays are used in connection with breakers that either do not have within themselves those characteristics to determine when the circuit should be opened to protect electrical service or apparatus, or in the design of which these characteristics cannot economically be incorporated.

The protective relays must be designed therefore to be responsive to electrical quantities which are different during normal and abnormal conditions and must be so arranged that those controlling the breakers in the faulted section of the system work first.

Electrical conditions which may change during faults are current, voltage, phase angle, direction, frequency, and rate of rise of current. Protective-relay designs use one or more of these conditions, together with other functions such as time, to provide a wide choice of relay characteristics to meet protective requirements of specific electrical systems.

Substantially all protective relays consist of some arrangement of one or more electromagnets with an armature or an inductive motor (disk or cup) carrying contacts which control another device, such as the trip coil of a circuit breaker. The electromagnets may have one or more current windings, potential windings, or both. These windings produce fluxes, which in turn develop torque by interaction (*a*) between fluxes from the same winding or (*b*) between fluxes from two or more windings. The reversal of current in one winding with respect to another will reverse the flux of that winding, causing a reversal of torque. This feature is utilized in relays of the power directional type. The torques thus developed are opposed by some form of mechanical restraint, either gravity or spring, to obtain a desired pickup of starting action of the relay contacts.

PLUNGER RELAYS. This is the simplest type of relay and consists of a solenoid and a movable magnetic plunger arranged as shown schematically in Fig. 13. It is used where instantaneous operation is required in a relay with an adjustable pick-up (current or voltage). The value of current necessary to raise the plunger depends upon its initial position with respect to the coil. It is customary to provide such relays with a calibrating screw to adjust the initial position of the plunger. A common range of calibration for this type of relay is the ratio of 1 to 4. There are several modifications of this type of relay, such as the use of variable weights added to the plunger, or magnetic shunts, in order to keep the ratio between pick-up and drop-out much closer. Some early designs of solenoid relays were provided with leather bellows attached to the plunger to give a certain amount of time-delay action to the relay contacts. Present-day practice favors the induction disk time-delay relays.

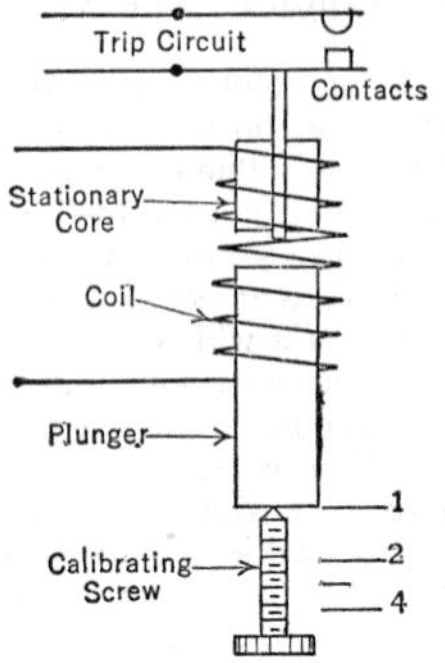

FIG. 13. Plunger-type Relay

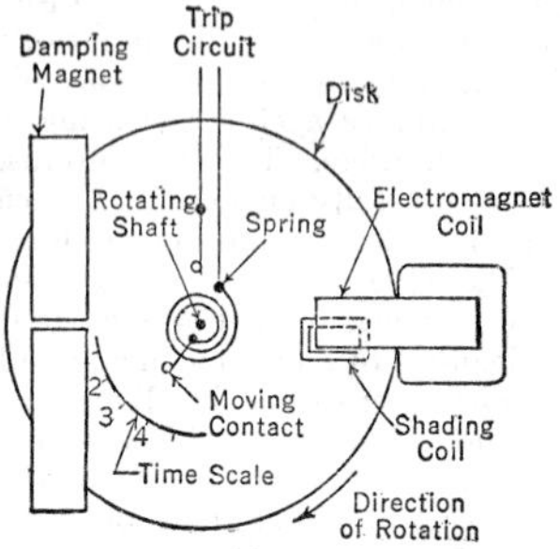

FIG. 14. Induction Disk Relay

INDUCTION DISK RELAYS. The time-delay relays most commonly used for protective purposes employ the induction disk principle. This same principle is universally used in a-c watthour meters and, when applied to relay construction, provides many varieties of time characteristic, depending on minor differences in electrical and mechanical design. The elements of an induction disk relay are shown in Fig. 14. It consists of a disk mounted on a rotating shaft restrained by a spring, with the moving contact fastened to the shaft. The operating torque on the disk is produced from fluxes developed by the coils of the electromagnet in the direction to oppose the spring restraint. The damping magnet provides restraint after the disk starts to move and is a large factor in obtaining this desired time characteristic. The time scale indicates the initial position of the moving contact when the relay is de-energized, and its setting controls the time necessary for the relay to close its contacts under any given condition.

This type of relay design lends itself most readily to obtain current, voltage, or watts by proper proportioning of the two windings on the electromagnet. The time-delay feature is controlled by initial spring tension and by separation of the moving contact from the stationary one. Two or more elements having a common shaft for the several disks provide various combinations of directional overcurrent relays, polyphase watt relays, etc.

When planning the protective-relay equipment for a modern power system, one of the first problems in selectivity is arranging the protective relays so that they will trip only when power flows in a certain direction. This is very easily provided by the use of two units: a fault detector and a directional element, with their contacts so arranged that tripping can occur only when current is flowing to a fault in the tripping direction. The same function also can be obtained by the use of a single element that combines a suitable time delay with a directional characteristic.

On two-element directional relays, there are two ways of obtaining directional discrimination. The first consists of connecting the contacts of the fault detector and the directional element in series. The second consists of supervising the operation of the fault-detector element by the contacts of the directional element.

Some power directional relays are of the three-phase type, in which case the fault-detecting elements are separate relays. They are controlled in the same way as the single-phase relay. The three-phase directional relay having only one set of contacts must control the three fault-detecting elements through the medium of an interposing auxiliary relay.

Three-phase power directional relays are sometimes equipped with a voltage-restraint element. This causes the relay to have a high current operating characteristic when the voltage is normal, but a very low current operating characteristic when the voltage is low, as is the case when there is a fault on the system. Under normal load conditions the relay will not operate, but when there is a fault, with resulting low voltage, it becomes extremely sensitive and operates very quickly.

The use of relays that operate on the magnitude of power is more often encountered for control purposes than for protection of lines or apparatus. By the proper selection of the current and voltage applied to the relay, any power relay may be used to measure reactive volt-amperes.

In an induction-type relay maximum torque is produced when the fluxes in the two magnetic circuits are 90 deg out of phase. The flux produced by the current winding is in phase with the current, but that produced by the potential coil current lags the applied voltage by an angle of approximately 85 deg and therefore the ideal 90-deg condition is not met. Correction is made by placing a metal loop, known as a **quadrature adjuster,** around the main pole. The short-circuit current flowing in this loop produces around it a local flux that lags behind a main flux.

HIGH-SPEED RELAYS. Among the many problems encountered in designing a high-speed directional element is the fact that, except at unity power factor, the direction of power flow reverses twice each cycle. If, therefore, the forces were made great, the relay would tend to follow the reversal in power flow.

Hence the design must so proportion the mass of the moving element and the operating forces that at high voltage the moving element will not follow power pulsations and yet will operate at high speed over the wide range of current and voltage encountered during faults. Another factor affecting the possible speed of operation is that no directional element can determine the average direction of power flow in less time than one alternation.

One design consists of a single loop of aluminum, free to rotate, through which flows a current that is induced by the voltage of the power circuit. Actually, the loop is the short-circuited secondary of a potential transformer, the primary of which is connected to the source of potential. The operating principle of the element is that of a conductor carrying current in a magnetic field. As the current flowing in the loop passes through the magnetic field between the two poles, it produces motion, and the operation of the element, either in the tripping or the non-tripping direction, is determined by the instantaneous relations between the applied current and voltage.

DIFFERENTIAL RELAYS. Differential protection, because of its inherent selectivity and sensitivity, is the most common form of apparatus protection. Although simple overcurrent relays have been used for such differential protection, there is always the danger, if they are made to operate on very small current magnitudes, that they may also operate on the slight difference in current that is likely to occur if the current transformers are not exactly duplicates. For this reason it is desirable to use percentage differential relays for the purpose of obtaining more sensitive protection and at the same time reducing the danger of faulty operation on external faults.

One such relay has an operating or difference coil energized by the differential current and two restraining or sum coils energized by the through current. The restraining coils produce an opening torque proportional to the sum of the input and output currents. The difference coil produces a closing torque in proportion to the difference of the two currents. When the fluxes produced by the operating and restraining coils are equal, they will circulate through the outer magnetic path without going through the disk. The winding which produces the greater flux will force the difference through the middle core, the disk, and the two upper poles. The quadrature flux, which is necessary to produce torque, is produced by two secondary coils connected in parallel to produce flux in the upper poles. The magnitude of current required for closing the contacts varies with the magnitude of the through fault current, instead of remaining constant for all load conditions. The relay is extremely sensitive to internal faults, yet it will not trip on a through short circuit.

Another type of differential relay has variable ratio characteristics. When the restraining current is low, the sensitivity is high, corresponding to a low percentage ratio. When the restraining current is high, as it is during severe faults, its sensitivity is low, which corresponds to a high percentage ratio.

To limit damage to equipment and raise the stability limit of the system, it is often desirable to reduce the time of operation below that obtainable with an induction disk-type relay. The use of high-speed differential relays is then indicated. There are several types of such relays.

In one type the operating element consists of balanced beam units, with the restraining coils wound to provide uniform restraint acting on the rear of the beam. The condensers connected in series with one restraining coil in each element provide a phase shift in the current through its associated restraining circuit. The flux created by this current, when combined with the flux from the other restraining coil, results in a uniform restraint on the balanced beam. The operating coil is wound to oppose the pull of the restraining coils and acts on the front (contact) end of the beam. Three overcurrent fault detectors are provided to prevent operation of the relay during periods of low restraint, when vibration might cause the balanced beam contact to close momentarily. The operating and restraining windings of the three elements are fed from a three-phase auxiliary saturating transformer.

Saturation of the cores of these transformers provides the variable percentage characteristic. The relay is most sensitive when the smaller restraining coil current and the operating coil current are in phase. This effect is most pronounced at high currents and is a distinct advantage, as current transformer saturation in through faults may cause phase angle errors. At small currents the phase angle characteristic is practically constant. This feature is important, as the relay may be called upon to trip a light internal fault in the presence of load current. In this case there may be legitimate phase angle difference between the operating and restraining coil currents, and it is undesirable to desensitize the relay for this condition.

The high-speed transformer relay is similar to the generator relay except that it does not have the split-phase restraining circuits but has auxiliary elements to prevent operation of the relay during heavy magnetizing-current inrushes.

The harmonic-current restraint principle is also used in percentage differential relays. The harmonic and fundamental components are separated by the use of filters. The harmonic-current components flow in the restraining winding, and the fundamental-current components in the operating winding. If the ratio of harmonic to fundamental is below a given value, the relay contacts close. Magnetizing-current inrushes develop saturation of current transformers, causing the wave form to be distorted and a large harmonic component to be present.

Some high-speed differential relays have restraining coils in which the restraint is proportional to the product of the input and output currents. The relay-sensitivity curve has a slope which increases considerably for currents above 200 per cent of normal.

DISTANCE RELAYS. Distance relays are of two general types, the first consisting of a combination of elements in which the time of operation is proportional to the distance from the relay to the point of fault, and the second of a combination of elements which balances the voltage against the currents to provide a definite cut-off point in distant measurement. The second construction is used in high-speed relaying; for back-up protection, some auxiliary timing means is utilized to provide the necessary selectivity with relays and breakers in adjacent line sections.

Either impedance or reactance can be used as the means of measuring distance in an electrical system. Resistance also indicates distance, but modern relays have not utilized this means. When using the impedance principle, it is necessary only to balance the voltage of the faulted circuit against a current in the faulted circuit to obtain an indication of distance. Fault resistance, which causes an error in impedance measurement, is usually small. In any case the fault resistance always increases the apparent impedance of the circuit, so that the relay always undertrips. The reactance principle is subject to incorrect operation at unity power factor loads, and thus a fault-detector element must be used to supervise the reactance element to prevent operation under this condition. The reactance element is also subject to considerable error if there is a difference in phase angle between the currents fed to the fault through the terminals of the line, and this error may result in either undertripping or overtripping. Therefore it is not possible to compensate for the error which may exist.

One induction-type impedance relay consists of an induction disk element with current-operating winding, through which the secondary current of the current transformer flows, and a voltage restraint winding. The slope of the time curve is adjusted by varying the voltage applied to the voltage-restraint coil. The induction disk winds up a spring, which in turn acts on the balanced beam and is opposed by the pull of the voltage-restraint coil on the beam. The time-distance curves of the relay do not pass through zero because of the inertia of the moving parts. The relay may also include a directional element to prevent operation for faults in a desired non-tripping direction.

High-speed distance relays are being used more and more in present-day interconnected systems. One such relay consists of three impedance-measuring elements, a high-speed directional element, and a synchronous timer, as well as auxiliary d-c elements and operation indicators.

The distance-measuring elements consist of a balanced beam, pivoted on jewel bearings. Two voltage coils provide restraint to hold the beam in the contact-open position, and a current coil produces operating torque which tends to move the beam in the contact-closing direction. A capacitor is in series with one of the voltage coils to give a steady restraining pull on the beam. The second and third elements are similar in construction to the first except for the contact arrangement.

The first element operates in conjunction with a high-speed directional element to trip faults within about 90 per cent of the section at high speed. The second and third elements also operate in conjunction with the directional element, but in addition must close the trip circuit through a set of contacts on the synchronous timer. The time settings of these contacts provide the necessary margin of selectivity to permit first- or second-zone relays and breakers in adjacent sections to function.

From the standpoint of back-up protection the distance relay with variable time characteristics is very desirable in that it provides an ideal way of obtaining selectivity between adjacent sections. The time of operation for close in faults is too long, however, for many applications, and as a result a relay combining a high-speed distance element and the directional element of the three-zone distance relay with the impedance element of the induction-type relay has provided a solution for many locations where full advantage cannot be taken of the characteristics of the three-zone distance relay.

Many other combinations of impedance and reactance relays are possible and available for specific applications.

PILOT-WIRE RELAYS. One such relay consists of a combination positive- and zero-sequence filter, a saturating auxiliary transformer, two Rectox units, a polar-type relay unit, and a neon lamp, all mounted in a single case. The external equipment normally supplied with the relay consists of an insulating transformer, a milliammeter, and a test switch. Current from the current-transformer secondaries is passed through a filter consisting of a three-winding iron-core reactor and two resistors. The output of this filter provides, across the primary of the saturating transformer, a voltage proportional to the positive-sequence current plus a constant times the zero-sequence current. The voltage from the filter is impressed across the tapped primary of a small saturating transformer. This transformer limits the voltage applied to the pilot wire and provides a small range of voltage for a large variation of maximum to minimum fault current. The relay-operating characteristics change from percentage differential at low current values to approximately directional characteristics at high current values. The secondary of the saturating transformer is connected to two Rectox units, the insulating transformer, and the pilot wire. The Rectox units convert the a-c output of the saturating transformers for use on the d-c polarized relay element. The polar element consists of a rectangular-shaped magnetic frame, an electromagnet, and an armature with a double set of contacts. The poles of the permanent magnet clamp directly to each side of the magnet frame. Flux from the permanent magnet divides into two paths, one across the air gap at the front of the element in which the armature is located, the other across the two gaps at the base of the frame. Two adjustable shunts, located across the rear air gaps, change the reluctance of the magnetic path so as to force some of the flux through the moving armature. Flux in the armature polarizes it and creates a magnetic bias, causing it to move toward one or the other of the poles, depending upon the adjustment of the magnetic shunts. Two concentric coils are placed around the armature and within the magnetic frame. The coils are connected in opposition, one being used as a restraining winding and the other as an operating winding.

SPECIAL-PURPOSE RELAYS. In addition to the relays described, there are many relays for special purposes. These include reverse-phase, temperature, frequency, synchronism-check, network, and carrier-current relays.

5. GENERATOR-VOLTAGE REGULATORS

To maintain proper voltage on a-c generators, regulators are available that adjust the field current of either the generator itself or its exciter. In order to maintain practically constant voltage on a-c and d-c generators, or to have these machines compound automatically to take care of feeder drop, field regulators of various kinds have been designed.

The two basic classifications are electromechanical and static. The electromechanical can further be subdivided into the vibrating and the rheostatic types, with the rheostatic type either direct acting or indirect acting. The static type can be subdivided into the network and the electronic types. The rotary amplifier regulator assembly consists of a rotating amplifier (special-design exciter), together with a control element of either the electromechanical or the static type.

VIBRATING REGULATORS. These regulators are made in two principal forms, one using a d-c vibrating magnet and the other an a-c vibrating magnet relay as the anti-hunting control device. Either design of vibrating regulator depends on the rapid opening and closing of a circuit that shunts the field rheostat and thus changes the resistance in the field circuit of the generator to be regulated. For d-c service the regulator usually works upon the main-generator field, and for a-c service upon the field of the exciter, depending upon the magnitude of the field current. In both cases the rheostat is so adjusted that when in the circuit it tends to lower the voltage considerably below normal, and when the rheostat is short-circuited the generator voltage rises. The regulator automatically closes the shunt circuit when the voltage drops to a predetermined value and opens it when the voltage rises above that value, vibrating, however, at an intermediate normal value.

The regulator with a d-c potential winding for the control of a d-c generator consists essentially of a main control magnet, whose winding is connected across the generator terminals, and one or more differentially wound relay magnets. When the pull of the potential winding increases because of a rise in generator voltage, the contact of the main control magnet is opened, and in turn one winding of the relay magnet is de-energized. Thus the relay contact is opened, and the short circuit removed from the generator rheostat. When the voltage drops, the main contact is closed, and the differentially wound relay magnet acts to short-circuit the field rheostat.

The regulator with the a-c potential winding for the control of a-c generators works on the exciter field. One design of this type of regulator employs two control magnets, one connected across the exciter bus and tending to move the main contacts farther apart as the exciter voltage rises, and the other acted on by a-c potential and current coils. When the main contact closes, it energizes the relay magnet, thus closing the relay contact, short-circuiting the exciter rheostat, and raising the voltage of the exciter and generator. The use of the exciter voltage as one of the main control elements prevents the generator voltage from overshooting, for, as the exciter voltage rises to bring up the generator a-c voltage, the d-c control tends to pull the main contacts apart and so reduce the voltage again.

Another design of the vibrating-type regulator employs an a-c vibrating magnet, which momentarily short-circuits and then reintroduces a portion of the resistance in the vibrating-magnet coil circuit, thus forcing normal vibration. When the a-c voltage drops or rises, the a-c energized solenoids cooperate to raise or lower the exciter voltages as required. The average voltage delivered by the exciter depends on the relative portion of the time that the exciter field rheostat resistance is in circuit. A compensating current winding on the a-c solenoid is provided with a dial switch to give the required amount of compensation for making generators share the reactive load or for the line drop in the feeder circuit.

DIRECT-ACTING RHEOSTATIC REGULATORS. The diagram for a typical direct-acting rheostatic-type a-c generator-voltage regulator is shown in Fig. 15. This type of regulator controls the voltage of the a-c generator by varying the resistance in the exciter or, in some cases, the generator field circuit. The regulator operates only when a correction in the voltage is necessary. The main moving part of the regulator, called the moving arm, is supported on springs with the armature centered in the air gap of the electromagnet. An assembly of spring-mounted silver buttons (or other contacts) is connected to taps on the stationary regulating resistance. The moving arm directly controls the closing and opening of the silver buttons, thus adjusting automatically the amount of regulating resistance in the exciter shunt field circuit.

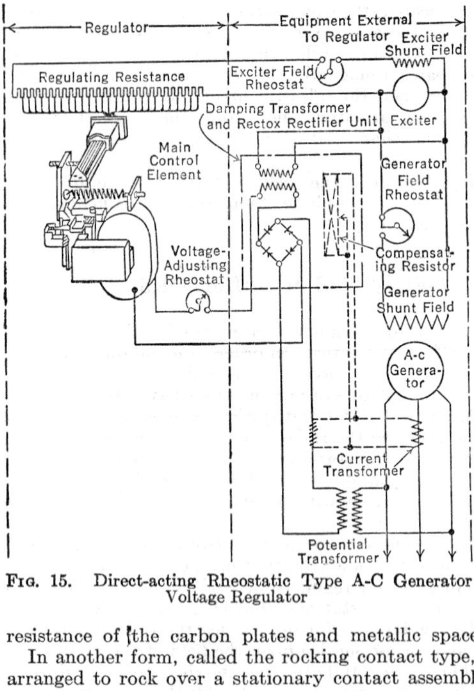

Fig. 15. Direct-acting Rheostatic Type A-C Generator Voltage Regulator

In one form of this regulator the resistance element consists of stacks of carbon plates having silver buttons in one end and insulating spacers at the other. These plates are spaced apart at the center with resistive metallic spacers. The silver-button ends are arranged to open or close like the leaves of a book, thus inserting or shorting out the resistance of the carbon plates and metallic spacers.

In another form, called the rocking contact type, a segment of conducting material is arranged to rock over a stationary contact assembly. Each contact or bar is insulated from the adjacent bar and is connected to taps on a stationary resistance, which is in series with the generator or exciter field.

The regulating action of the regulator is that of a semistatic device which operates only when a correction in voltage is necessary. For a given value of voltage and load on the generator being regulated, a corresponding value of regulating resistance is required in the exciter field circuit, and a corresponding position of the moving arm and silver buttons which will give this value of resistance. Under this condition, at the proper position of the arm, the magnetic pull on the moving arm is balanced against the pull of a coil spring. When there is a change in load, this causes a corresponding correction in the voltage.

If additional load is placed on the generator whose voltage is being regulated, the voltage drops. The drop of voltage decreases the magnetizing effect of the regulator coil and reduces the flux in the air gaps of its magnetic circuit. This in turn decreases the magnetic pull on the iron armature attached to the moving arm and allows the coiled spring to move it in a direction to begin closing in sequence more of the silver buttons. This action shunts out in small steps additional portions of the regulating resistance, which causes the field current to be increased and the exciter voltage and a-c generator voltage to be raised to their normal values. When the voltage is restored to its normal value, the moving arm of the regulator is again in a balanced state, having changed its position, however, to correspond to the change in load on the machine and the new excitation requirements. The maximum travel of the moving arm being only a fraction of an inch, the regulating resistance can be very quickly varied from maximum to practically zero when operating conditions require such control.

The damping transformer is a static device which provides for stabilizing the regulated voltage. When a change takes place in the exciter voltage as a result of the regulating action of the regulator, there is an induced transfer of energy from the primary to the secondary of the damping transformer. Thus the coil pull is reduced or increased by this impulse, and the regulator moving arm is restrained from carrying the change in regulating resistance and the consequent change in excitation too far.

A voltage-adjusting rheostat is connected in series with the regulator coil circuit. Increasing or decreasing the resistance of the rheostat in this circuit causes the regulator to change the position of its moving arm and to raise or lower the exciter voltage and the a-c generator terminal voltage in order to maintain its coil current at the normal value.

The carbon-pile regulator is another form of direct-acting rheostatic type. A coil, core stem, lever system, carbon disks, and inverted air dashpot constitute a carbon-pile regulator. The coil has a current winding for parallel operation with another regulator, or line drop compensator, and a voltage winding. Pressure on the disks is supplied by a main spring and is released by the pull of the coil on the lever system. By varying the pressure on the carbon disks the resistance is adjustable through rather wide limits, and by automatically controlling the pressure the voltage is automatically controlled.

INDIRECT-ACTING RHEOSTATIC VOLTAGE REGULATORS. The diagram for a typical indirect-acting rheostatic-type a-c generator-voltage regulator is shown in Fig. 16. This type of voltage regulator controls the voltage of the a-c generator by varying the resistance in the exciter or, in some cases, the generator field circuit. The regulator operates only when a correction in the voltage is necessary.

The rectifier voltage of potential transformers is supplied to the control element. Although the control element may be of various types, it usually has both normal- and quick-response contacts. The normal response controls the motor of the motor-operated rheostat through interposing contactors. One of the normal-response contacts closes for small variations of generator voltage from normal and causes the motor of the motor-operated rheostat to insert additional resistance in the exciter or generator field if the generator voltage is higher than normal and to remove resistance if the generator voltage is lower than normal. Large variations of the generator voltage from normal, such as are occasioned by sudden large load changes, cause one of the quick-response contacts to operate in such a manner as to give a large change in excitation in the proper direction. If the generator voltage is high, a quick-response contact opens to insert a large block of resistance in the exciter or generator field circuit. If, however, the voltage of the generator is low, a quick-response contact closes and shorts out the motor-operated field rheostat. During the operation of the quick-response contact the rheostat motor is moving the rheostat arm in the desired direction.

The motor-operated rheostat can also normally be operated from a control switch after the regulator has been disconnected by means of a regulator-control switch.

A voltage-adjusting rheostat similar to that employed with the direct-acting rheostatic regulator is also used with the indirect-acting rheostatic regulator.

Regulators of this type are also provided with means for preventing hunting. Such antihunting devices are normally associated with the control element and may take various forms.

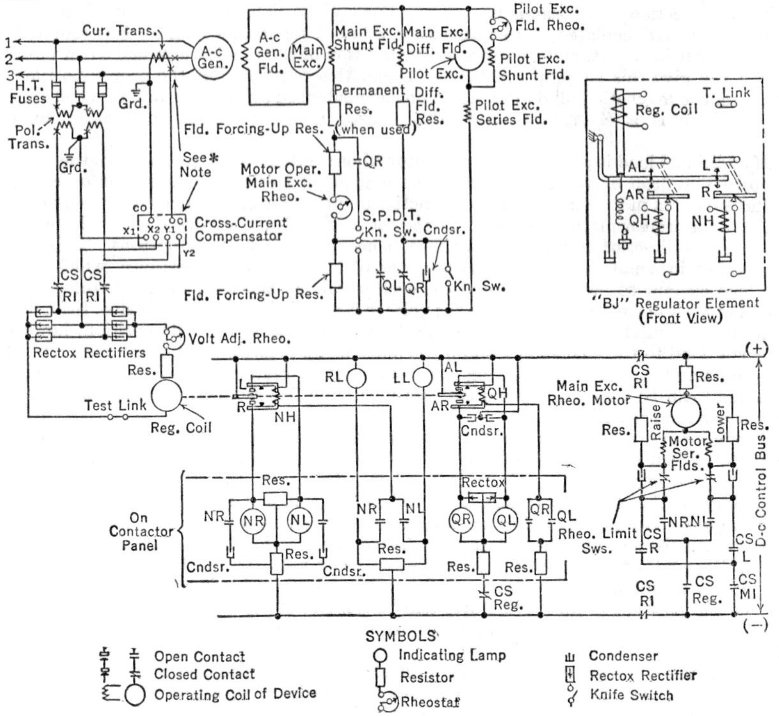

Device Nomenclature

Device	Description
R-L	Regulator Element Raise and Lower Normal-response Contacts.
AR-AL	Regulator Element Raise and Lower Quick-response Contacts.
NR-NL	Raise and Lower Rheostat Motor-control Contactors.
QR-QL	Raise and Lower High-speed Field Forcing Contactors.
QH-NH	Anti-hunt Devices for Regulator Quick- and Normal-response Contacts.
CS-R } CS-L }	Rheostat Control Switch (Manual) Raise and Lower Contacts.
CS-Reg	Transfer and Indicating Switch Contacts Closed in "Reg." Position Only.
CS-RI	Transfer and Indicating Switch Contacts Closed in "Reg." and "Ind." Positions.
CS-MI	Transfer and Indicating Switch Contacts Closed in "Man." and "Ind." Positions.
RL-LL	Raise and Lower Indicating Lamps.

* Current transformer and Cross-current compensator required when application involves parallel operation.

FIG. 16. Schematic Wiring Diagram of Type BJ Generator-Voltage Regulator for Control of an A-c Generator Excited by a Main Exciter, which is Separately Excited from a Pilot Exciter. Current Transformer and Cross-current Compensator Shown Are Required Only when Two or More A-c Generators, Under Control of Individual Regulators, Operate in Parallel

NETWORK-VOLTAGE REGULATORS. One regulator of this type utilizes the intersecting impedance characteristics of the network for the voltage references. At normal voltage two impedances are adjusted to draw equal or proportional currents; at least one impedance is non-linear. On higher voltages one impedance then draws more current than the other, and the difference current is in a direction indicating voltage above normal. This can be used for control. At lower voltages the same impedance draws less current than the other, and the difference is thus reversed, indicating voltage below normal. The currents are rectified before comparison.

Average three-phase (or positive-sequence) voltage response is usually required. Inasmuch as the static regulator volt-ampere parts are determined by the power requirements of the amplifier, these parts, particularly for the rotary amplifier type, are somewhat larger than is required by the normal burden of a conventional regulator element. Although a positive-sequence response can be obtained by the use of a positive-sequence network or filter, this device increases in size directly with the burden supplied through it. The positive-sequence voltage can also be derived by subtracting the negative-sequence voltage from the line voltage. This is accomplished by passing the secondary current of the current transformers through a network proportional to the negative-sequence impedance of the generator. The derived negative-sequence voltage is then deducted from the line voltage. The current-energized sequence filter requires only a small fraction of the volt-amperes parts, in comparison to the voltage-type sequence network.

As the frequency is varied, the static measuring circuit can be kept in balance by varying in a predetermined fashion the voltage applied to it. When it is desired to regulate for the same generator voltage with some departure from normal frequency, it is necessary to introduce, between the potential transformer and the measuring circuit, an impedance which will produce the necessary voltage variation with frequency on the measuring circuit. Network synthesis studies develop the proper combination of inductances and capacitances for this purpose. As the inductance required is a series device, it can be obtained by simply proportioning the sequence network to have this value of inductance as viewed from the output terminals.

In regulating circuits of this kind, a voltage-adjusting unit is provided for raising or lowering the "normal" voltage, thus permitting adjustment of the value which the regulator is set to maintain.

There are other static circuits; for example, a circuit in which the generator terminal voltage is balanced against the reference voltage and the differential, which appears when the generator voltage deviates from normal, is applied to the amplifier. For the reference voltage a separate source, either alternating or direct current, may be used; if none is available, a constant potential device may be employed, energized by the voltage of the generator with which the regulator is used. This device is designed to provide essentially constant a-c voltage on its output terminals, in spite of relatively large changes in its input voltage.

The output of the static network is supplied to an amplifier which may be of the rotary or other type. The output of the amplifier is then applied to the field of either the exciter or the generator.

ELECTRONIC VOLTAGE REGULATORS. It is a well-known concept that all voltage regulators are in effect amplifiers, because small deviations in regulated voltage produce relatively large changes in the generator field current to offset the internal voltage drop caused by variations in the load. In a conventional regulating system, such as a direct-acting type, the amplification is secured partly in the regulator and partly in the exciter. An electronic regulator provides a large amount of voltage amplification in the regulator circuit, with the result that the exciter needs only to provide a small amount of this amplification. The electronic regulator may be used to control the field of a d-c rotating exciter or to control an electronic exciter. This electronic exciter may be an ignitron rectifier, controlled by the thyratron firing of the ignitors. This exciter may receive its power from either the a-c generator it is exciting or from some external source. The control grids of the thyratron receive their controlling voltage direct from the regulator. The regulator thus controls the electronic exciter by raising or lowering the thyratron grid voltage in order to provide the proper exciter voltage.

The electronic voltage regulator can control the excitation of a generator by thyratron tubes which (1) supply excitation directly to the generator field, (2) supply excitation to the field of a rotating exciter, or (3) supply the firing voltage for an ignitron exciter. The ignitron exciter can be used to control the excitation on the fields of the largest ratings of generators.

One method of obtaining a voltage reference is to use the voltage drop obtained across the plate resistor of a high vacuum pentode tube. The pentode tube maintains the current flow through itself at a constant value regardless of plate voltage variation when the specific voltages are applied to its grid. There are several types of electronic-excitation control circuits using d-c amplifiers. One method uses a vacuum-tube phase shift circuit. The tube is controlled by the output of the amplifier; as the grid bias of this tube is varied, it increases or decreases the charging current on the capacitor, thus making the voltage drop across the capacitor a variable, depending upon the value of the grid voltage. The grid voltage in this case actually determines the phase shift of the voltage impressed on the grid transformers, which in turn control the angle of lag at which the thyratron tubes fire. This method gives very accurate control of the thyratron but has the disadvantage of not being able to stop the thyratron from firing because it is limited to less than 180-deg phase shift.

Another method uses variable direct current for controlling the thyratron. In this circuit an a-c voltage lagging the thyratron plate voltage by some predetermined degree is superimposed on the variable d-c voltage received from the regulator amplifier. By using this method, the a-c voltage can be made to cut the critical grid voltage of the thyratron from full delay to zero delay, thus stopping the tubes from firing. In this way all the features of field forcing, even complete removal of the excitation from the rotating machine, are made available in the regulating system.

Various types of input circuits can be used. The signal voltage representing the generator terminal voltage is balanced against a reference voltage, and the differential which appears when the generator voltage departs from normal is applied to the d-c amplifier of the regulator. For the reference voltage a separate source of either alternating or direct current may be used; if none is available, a constant potential device may be employed. This device is designed essentially to provide a constant a-c voltage on its output terminals in spite of relatively large changes in its input voltage.

The generator voltage can be rectified and then filtered and balanced against the reference voltage in order to obtain a suitable differential voltage. Another method is to use a filament controlled diode high vacuum tube. Generator voltage can be fed into the filament, and a d-c voltage, proportional to the rms value of the a-c voltage, can be obtained from the loading resistor. This voltage is then balanced against the reference voltage, and the differential voltage between the two is fed into the amplifier. This eliminates the necessity for a rectifier and at the same time offers a high degree of amplification.

Since the filament of the indicating tube can use either a-c or d-c heating voltage, a positive-sequence filter or a three-phase rectifier can be employed to provide response to polyphase voltage.

The reference voltage circuit makes use of two tubes, a high vacuum pentode and a gas-filled voltage-regulator tube. This circuit employs the theory of the pentode tube when proper grid voltages are applied and held constant to both the control grid and the screen grid. The tube then has a constant-current characteristic which is entirely independent of reasonable variations in plate voltage.

A damping or stabilizing transformer can be used for stabilizing the electronic regulator. The transformer is connected with its primary across the generator field and its secondary connected into the regulator circuit.

REGULATORS FOR OTHER QUANTITIES. The above-mentioned types of voltage regulators can be used for the regulation of other quantities than voltage. These quantities include current, speed, position, power factor, and frequency. Modifications of the regulators are required, but the basic principles remain the same.

6. INDUCTION-VOLTAGE REGULATORS

On a-c circuits containing no generating equipment or other rotating apparatus, voltage regulation is usually obtained by means of induction regulators, changing taps on transformers, or some combination of these two methods. Where two systems, or two portions of the same system, are tied together for power interchange, voltage regulation at the tie point is usually necessary to control the interchange of wattless kva. This objective has been carried out in many interesting arrangements.

The first type of regulator was a transformer with many taps and provision for connecting the feeder to any tap. This could be done by switches of various kinds, and the natural development was to arrange the contacts in the form of a ring on a suitable faceplate and to provide a movable arm for connecting the feeders with any of the taps.

Induction regulators are made for single- or three-phase service and arranged for hand operation and motor operation, for motor operation controlled from a distant point, or for completely automatic operation by means of relays.

SINGLE-PHASE REGULATOR. Figure 17 is in effect a two-winding transformer with the secondary winding arranged for connection in series and the primary winding arranged for connection directly across the line. With the transformer thus connected, a voltage will be induced in the secondary that will add to or subtract from the feeder voltage, according to the connections used. The current in the primary produces a magnetic field that induces a voltage in the secondary. The portion of this field passing through the secondary winding, and consequently the voltage induced in that winding, depend upon the angular position of the secondary with respect to the direction of the primary field. The induced voltage is a maximum when the axes of the coils coincide, zero when the coils are at right angles to each other, and maximum in the opposite direction when the axes of the coils coincide but with primary coils reversed in position. This induced voltage in the secondary therefore adds to or subtracts from the feeder voltage by a value varying from maximum regulation to zero, according to the position of the coils.

The single-phase regulator adds or subtracts its voltage directly and not vectorially in relation to the voltage from the main circuit. For this reason the use of a single-phase regulator does not cause any phase displacement.

POLYPHASE REGULATOR. This regulator resembles somewhat a vertical-shaft, phase-wound, polyphase motor, immersed in oil in a suitable tank. The regulator primary is wound with a distributed winding of the same number of phases as there are in the feeder to be regulated, and each phase of the regulator is connected across a separate phase of the feeder. The secondary winding is made up of the same number of separate windings as the primary, and each of these separate windings is connected in series with one of the feeder wires. The primary sets up a magnetic flux of constant value, which induces a constant voltage in each of the secondary windings. The induced voltage of the secondary is combined vectorially with that of the feeder. As the position of the rotor is changed, the phase angle between the feeder voltage and the secondary voltage changes, and the feeder voltage is either increased or decreased. Induction regulators can be arranged for either indoor or outdoor service and can be self-cooled or water-cooled, depending upon their size and local condition.

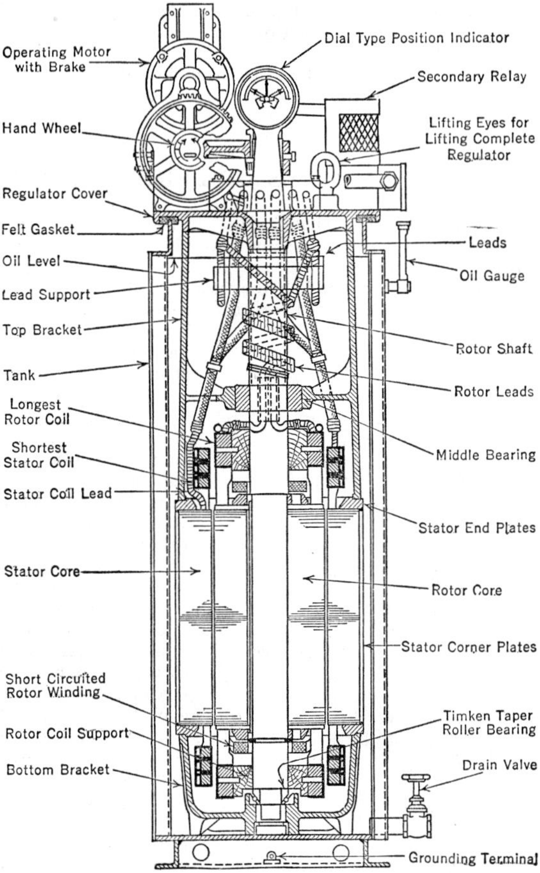

FIG. 17. Single-phase Induction Regulator, Self-cooled

CONSTANT-CURRENT REGULATORS. These are a special type of transformer with movable coils that change the constant-potential source of supply into a constant-current feeder. They are used chiefly for series lamp circuits.

AUTOMATIC OPERATION. If desired for either single- or three-phase regulation, automatic operation can be obtained by the action of a voltage relay either with or without a compensating device. This relay acts in conjunction with the motor on the regulator so that, as the load comes on or the bus voltage drops, the motor will turn the regulator in such a direction as to increase the voltage. By means of a compensator, which can be set for certain ohmic and certain inductive drops, the voltage at the point of distribution can be maintained at a constant value independent of the amount of power factor of the load, if the total drop is within the range of the regulator.

FEEDER VOLTAGE REGULATORS. These regulators of the electronic type utilize a reactor or impedance in the form of a two-winding transformer. The primary of this transformer is connected in series with the circuit to be controlled; the secondary, across the anode and the cathode of the tube. The tube varies the effective impedance of the reactor. The tube can be considered a switch to short-circuit the secondary, the time of remaining closed being varied.

STEP-INDUCTION REGULATOR. For very large capacities, either a combination of taps on a step-down transformer or an auto-transformer with an induction regulator is used, the latter being called a step-induction regulator. By designing the transformer with a voltage between the taps not greater than that of the induction regulator, a smooth voltage variation is obtained over the entire range of voltage covered by taps from the highest to the lowest.

The voltage of high-voltage transmission lines is now regulated by means of taps in the step-up or step-down transformers through which the ratio is changed without disconnecting the transformers from service or interrupting the load. (See Section 10.)

7. TYPES OF LIGHTNING ARRESTERS

There are two major classifications of lightning arresters for a-c circuits: the expulsion type and the valve type. In some fields of application the two can be used interchangeably; in others they cannot.

The **expulsion arrester** is like a fuse with the fuse links omitted. In its simplest form it is illustrated in Fig. 18. It consists essentially of a tube, usually of fiber, with electrodes in each end. This construction provides a spark gap confined within the bore of a tube of material that evolves gas under the heat of an arc. In series with this tube there is usually another spark gap to keep normal line voltage from the material of the tube, which deteriorates if subjected to voltage continuously. When a high voltage occurs, the series gap and the gap in the tube spark. The arrester thus becomes a path of low impedance, and the voltage across the terminals after sparkover drops to a low value. This voltage consists only of the arc drop in the extinction chamber. Consequently, as far as power-follow current is concerned, the device becomes nearly a short circuit, and system-fault current or something approximating system-fault current flows. When the current wave passes through zero, however, it, like the fuse, interrupts the system-fault current as a result of the gas evolved from the walls of the tube. The expulsion arrester operates repeatedly without attention. In the expulsion arrester, interruption of the power-follow current takes place in the tube. It is not interrupted by the series gap.

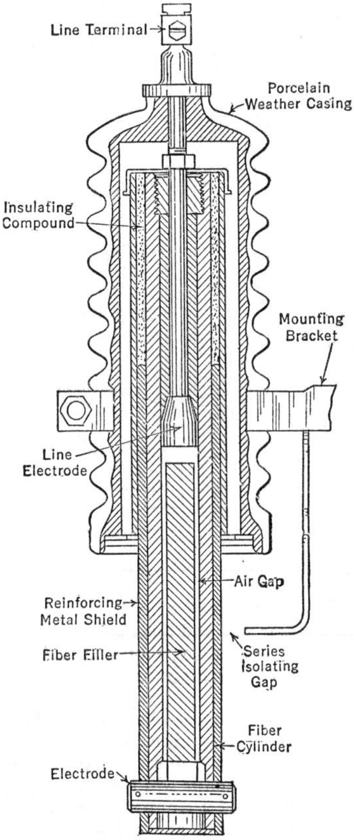

Fig. 18. Expulsion-type Lightning Arrester

The **valve arrester** has to perform the same functions as the expulsion arrester. It carries them out in a similar fashion by a somewhat different process. Its operating elements (Fig. 19) are different from those of the expulsion arresters. The valve arrester, too, has a series spark gap that normally provides insulation and acts as a switch to close by sparking, when a surge voltage appears at its terminals. In the valve arrester, it falls on the series gap also to reopen the circuit, that is, to interrupt the power-follow current when the surge has passed. The gaps in valve arresters are made to have sparkover characteristics like those of sphere gaps. Such simple air gaps are not capable of interrupting the high 60-cycle current that may be involved in system faults. Gaps such as are used in valve arresters can interrupt moderately high currents, 50 to 100 amp rms, but not much more. Therefore, to insure successful operation of the device, the power-follow current that flows after the arrester has been discharged by a lightning surge must be limited to a magnitude with which the series gap can cope. This current limitation can be provided by a resistance. However, if the resistance is the usual well-known constant or linear

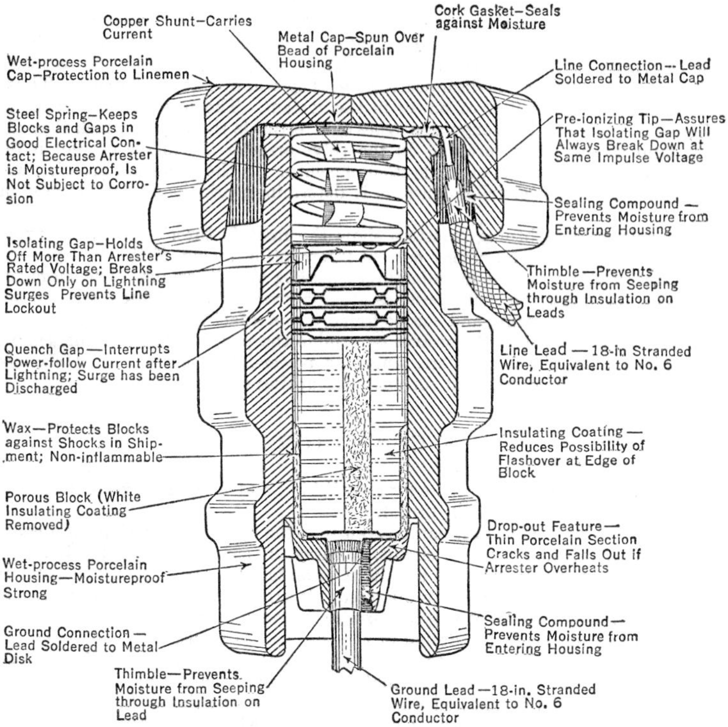

FIG. 19. Valve-type Lightning Arrester

type like a wire-wound resistor, and of such a magnitude that it will limit the power-follow current to the required small value when system voltage only is applied across the arrester terminals, the voltage drop across the resistance, when it is carrying a high lightning current, will be so high that the arrester provides no protection. For this reason, the series resistance, as it might be called, used in valve arresters is of a special kind that has a high apparent resistance at low voltage, but a low apparent resistance when high current flows (Fig. 20). It is for this reason that these materials are called valve elements, since the impedance, on the valve opening, regulates itself to the flow of current and thereby provides adequate and reliable voltage limitation. In most arresters of this type, the voltage regulation is largely a property of the electrical contacts between the particles of which the valve element consists. If the valve element is

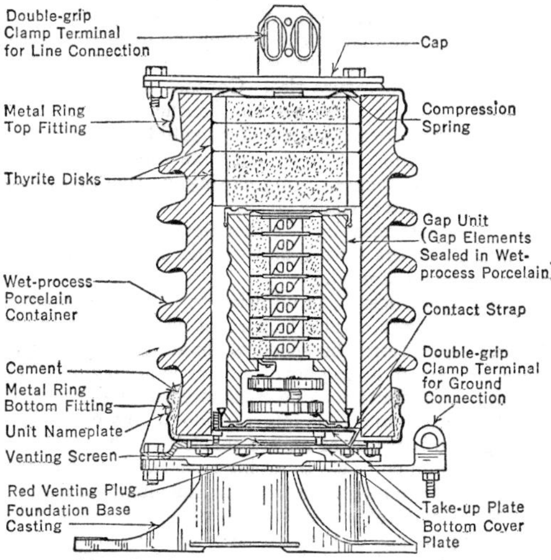

FIG. 20. Autovalve-type Lightning Arrester

viewed as a resistor, then in a 3000-volt autovalve arrester, its effective or apparent resistance at 3000 volts may be 100 or more ohms, whereas when it is passing 50,000-amp surge current its apparent resistance is less than half an ohm.

Both the expulsion and the valve types perform the function expected of lightning arresters. They are normally insulators, but they become conductors of relatively low impedance during lightning current surges. They re-establish themselves as insulators after the lightning current has disappeared, and they will operate repeatedly. Thus they do not interfere with the flow of power in the system. The principal difference in the operation of the valve and the expulsion types of arresters is the way in which they dispose of the system power-follow current. In the valve arrester the follow current is limited by the device itself regardless of the system capacity. In the expulsion arrester, with its low arc voltage, the power-follow current is determined principally by the system impedances.

The nameplates of lightning arresters show the manufacturer's name, the type, the voltage rating, the power frequency current-interrupting rating, if this is dependent on system characteristics, and the manufacturer's catalog identifications. The surge-current capacity and the protective characteristics are not given on nameplates, but are implied in the type and voltage rating and can be found in manufacturer's catalogs.

Valve arresters, since they themselves determine the power-follow current, carry only voltage ratings. Expulsion arresters, since the power-follow current is determined by the system, carry both voltage and current-interrupting ratings.

8. CONTACTORS

A contactor is a device for repeatedly establishing and interrupting an electric power circuit (ASA Standard C19.1 1943, 15–19).

From the foregoing definition it can be seen that the principal functions of a contactor are repeatedly to establish a power circuit, to carry the current during the time the contactor is closed, and to interrupt the circuit. Contactors may be of many different types. For example, they may be manually operated, electrically operated, or operated by an air cylinder, and they may be of varying sizes and details of construction.

The factors to consider in the design of the contacts to be used on contactors are the material, the pressure between contacts, the mass of the contacts, the radiating surface, and the contact surface. The contact material is usually hard-drawn or forged copper or silver. The material is selected on the basis of its resistivity, hardness, annealing temperature, and chemical activity. Copper contacts will stand more abuse and heavy arcing than silver, but when subject to high temperatures the surfaces oxidize, and the resistance at the contact increases. Silver contacts are very good for current-carrying capacity, but there is danger of their welding together when closing, if the contacts bounce or are subject to high current. Within limits the contact resistance varies inversely with the pressure and is substantially independent of the area. The mass of the contact is important, inasmuch as the greater the mass the more heat will be conducted away from the contact surfaces. The cooling surface of the contact determines the energy dissipation for a given temperature rise. Contact surfaces should be clean and, if copper, free from oxide film, which increases the contact resistance. Silver contacts become blackened because of the formation of silver sulfide, but this film does not materially increase the contact resistance. Copper contacts should be such that a slight amount of wiping or rubbing action occurs during closing, as this breaks down the copper oxide and tends to keep the contact surfaces clean.

The contacts may be of either butt, sliding, or rolling type. Silver contacts, however, are usually of the butt type to avoid excessive mechanical wear, which would result if a sliding type were used.

The contact design, in addition to being important with respect to continuous-current-carrying capacity, is also important with respect to arc-rupturing duty. In addition, the contact separation, the speed of separation, the design of the arc box, including its size, material, type of arcing horns, and type of magnetic blowout or other arc-rupturing device, are important factors to be considered in the design of a good interrupter. For more detailed information on the interruption of an a-c or a d-c arc refer to Article 2 of this section.

Magnetic blowout of an arc consists fundamentally of a means for transferring the arc from the main contacts to arcing horns and the lengthening of this arc due to the reaction between the arc magnetic field and the magnetic field of the blowout coil. There are many variations and refinements, such as the use of gas wedges to keep the ends of the arc moving by means of the resistance of the gas in the pockets. The motion of the arc end reduces the burning of the contact surface. The arc may be forced into a restricted arc chute, which facilitates the interruption of the arc by presenting a more favorable opportunity for heat transfer to the walls of the arc chute, which tend to cool and deionize the arc. In addition the vaporization of the arc-chute material because of the heat and the action of the arc on the air cause a considerable amount of energy absorption. Another commonly used method of rupturing the arc is the "De-ion" arc quencher.

A contactor normally is designed to interrupt a current equal to ten times its full load rating, but is not designed to interrupt short-circuit current of high magnitude. Fuses or high-speed breakers should be provided for such protection.

The temperature of operating coils on magnetic contactors is affected by the cooling surface and conduction of heat to and from the coil, atmospheric conditions, coil resistance, and coil current. It should be remembered that it is quite possible that other parts of the contactor are operating at a higher temperature than the coil, and consequently heat may be transferred from these parts to the coil. The important atmospheric conditions are ambient temperatures, ventilation, relative humidity. The current flowing through the coil is dependent upon the line voltage; and variation in this voltage, or the frequency on a-c circuits or the variation of the air gap of an a-c magnet will cause changes in the operating temperature.

Variation in the line voltage, in addition to possibly causing heating of the operating coil, may, if the voltage is too low, cause the contactor to fail to close.

9. MANUAL CONTROLLERS

An electric controller is a device, or group of devices, which serves to govern, in some predetermined manner, the electric power delivered to the apparatus to which it is connected (ASA Standard C19.1-1943, 15–3).

The drum controller consists of a case of iron with a cover and an operating handle or wheel connected to a square or hexagonal shaft, on which is mounted an insulating drum having copper segments secured to its outer surface. Stationary contacts are mounted on an insulating base. These contacts may consist of a single row on one side or a row on each side. The part of the stationary contact assembly which makes contact with the drum segments is flexibly mounted with a spring for obtaining contact pressure. The finger base is supported on an insulated iron base, and an arc shield may be provided between this base and the drum. The positioning of the drum is determined by a star wheel mounted on the shaft and a spring-operated roller, which drops into notches on the star wheel in the operating position of the controller. Magnetic blowouts are normally provided for the contacts, and they may have common magnetic circuit or a separate circuit for each contact. The separate-circuit type is superior but is more expensive. The copper segments attached to the drum may vary between a complete circle and a small piece making contact only in one position. The conformation of this segment depends upon the positions in which it is desired to make contact and those in which it is desired that the circuit be open. Resistors are normally mounted separately from the drum controller.

The cam controller is similar in outward appearance to the drum controller and is operated in the same manner. The rotating drum, however, is replaced by a shaft, on which cams are mounted. These cams open or close a series of contacts mounted on the stationary part of the controller. The opening or closing of the contacts in the various operating positions is determined by the contour of the cam. The design is normally so arranged that the contacts are cam-opened and spring-closed. The contacts are similar to those on a drum controller and may or may not be provided with magnetic blowouts. Because of the limited space, however, the operating duty of a cam-type controller is limited to less severe service than is a contactor. It is easier to make changes on a cam controller than on a drum controller, inasmuch as changing the sequence of contact opening and closing can be effected by changes in the cams, which may be made without completely disassembling the controller.

The faceplate controller is similar in appearance to a faceplate-type rheostat. It consists of an insulated base, upon which are mounted two or more circular rows of contacts.

The arrangement of these contacts conforms to the number of circuits to be controlled and the sequence in which the resistance of each circuit is to be changed. An arm mounted on a shaft located at the center of the contact circles carries one or more brushes bearing on the contacts. Movement of this arm changes the resistance in the various circuits. The arm is generally operated by a hand wheel or lever attached to the shaft. The resistors are mounted behind the faceplate, and the entire assembly is mounted on a framework.

The compression-type controller is an assembly consisting of an operating mechanism which, by means of cams or other similar devices, closes a contact and then, dependent upon the contour of the cam or motion of the mechanism, compresses a stack of graphite disks, thus varying the resistance. Any number of units, each consisting of a cam, contacts, and graphite disks, can be used in a single controller. Each contact can be closed in the desired position and the resistance varied as desired by proper design of the cam contour.

10. CONTROL RELAYS

A relay is a device that is operative by a variation in the condition of one electric circuit to effect the operation of other devices in the same or another electric circuit. Where relays operate in response to changes in more than one condition, all functions should be mentioned (ASA Standard C19.1-1943, 15–34).

Current relays may operate on overcurrent, undercurrent, or a combination of both. Current relays operate on several different principles: magnetic solenoid, induction, or thermal.

Voltage relays may be of either the solenoid or the induction type. They may be of the overvoltage or undervoltage type and may or may not have time delay. Voltage relays also may be of the precision-calibrated type or the auxiliary type.

Frequency relays usually operate on the principle of the variation with the frequency of the impedance of an inductive circuit. Such a relay usually consists of a coil in series with an inductance pulling on one end of a balanced arm and a coil in series with a resistor opposing this coil on the other end of the balanced arm. At the normal frequency the pulls of both coils are the same, but any variation from the normal frequency will close one or the other of the contacts.

A differential relay is a relay which functions by reason of the difference between two quantities of the same nature, such as current or voltage, etc. This term includes relays heretofore known as **ratio balance relays, biased relays,** and **percentage differential relays** (ASA Standard C19.1-1943, 15–38).

Power factor relays, which are used to indicate a power factor below a predetermined value, usually consist of an induction disk-type relay in which the torque is zero when the current and voltage are in phase. They are calibrated so that the torque will overcome the restraining spring at a predetermined lagging power factor, thus closing the contacts of the relay.

A timing relay measures a definite time after which a certain event or sequence of events in the control scheme should occur. There are many designs of timing relays, but all have the same common purpose.

One of the most familiar methods of measuring a definite time is by a clock mechanism driven by a small motor, similar to those used in electric clocks and signs. These devices are useful where a definite schedule of events is to be set up and repeated, for example, power on for 5 min, then off 1 min, with the same cycle repeated indefinitely.

A reliable and direct method of controlling a definite time sequence of events in starting and stopping motors is the motor-driven drum-type shaft, which operates a number of contacts in the control circuits. The shaft is driven by the motor through gearing, so that it turns slowly. The desired sequence is obtained by shaping the operating cams to give the correct results.

Timing relays may also be of the magnetic type. The contacts of these relays may be either open or closed when the coil is energized. The relay is built with a high-grade magnetic circuit. Although the relay can, of course, take many different forms, the basic principles are usually as follows: A copper tube surrounds the magnetic core of the relay. When the operating coil of the relay is de-energized, currents are induced in the copper tube, which constitutes a short-circuited coil, by the change in the flux of the main magnetic circuit. The induced current in the copper tube or coil resists the decay of the main flux and tends to maintain a magnetic flux. The time required for the flux decay constitutes a time delay. A delay of 5 sec is common. In order to adjust the time somewhat, an auxiliary coil, whose polarity is reversed with respect to that of the main coil, is often used to effect a neutralization of the main flux after the main coil is de-energized. By using a resistance to control the strength of the auxiliary coil, the time required to accomplish the neutralization of the main flux can be controlled. Timing can also be controlled by using thin spacers of non-magnetic material in the magnetic circuit to introduce an air gap. The timing is then regulated by the length of the air gap. The greater the air gap, the shorter the timing will be. Additional adjustments are possible by changing the spring pressure that opposes the pull of the main magnetic circuit.

Other types of timing relays are the induction disk and gear train type, the escapement type, and the racheting type with reciprocating drive. In addition to the above-described relays there are a great many relays for special purposes.

11. MASTER SWITCHES

A master switch is a switch which dominates the operation of contactors, relays, or other magnetically operated devices (ASA Standard C19.1-1943, 15–17).

Associated with every magnetic controller is some form of master switch through which the operation of the entire equipment is controlled. Some of these master switches, such as float switches, pressure switches, vacuum switches, and thermostats, are entirely automatic in operation and require no attention by an operator.

Some master switches are operated by the motion of the driven machine itself and are, therefore, automatically operated. Such devices, called limit switches, are mounted close to a machine, so that some part of the machine strikes and operates them at the limit of travel. They are sometimes geared to a machine and measure motor revolutions to determine the limits of travel. A common example of limit-switch application is the planer, which is automatically reversed at the end of each stroke.

Other forms of master switches are the manually operated ones, which always require an operator. Perhaps the most generally used one is the push-button station in its various combinations. The push buttons carry only control circuits, and close or open circuits to the operating coils of magnetic contactors.

The cam-type master switch is another manually operated device. With a master switch and a suitable magnetic controller, a motor may be operated in either direction of rotation at different speeds by moving the master switch handle to the proper position. This combination is used for severe operating conditions where operations are frequent and long life is important.

Sometimes a combination drum-type controller that combines the duty of master-switch service with that of a field rheostat or speed regulation of a d-c motor is used. This type of master station is most frequently used on machine-tool controllers.

12. ELECTRICAL INTERLOCKS

Electrical interlocks are switches mounted on contactors or other devices and operated by rods or levers. These interlocks open or close, depending upon the open or closed position of the contactor or device with which they are associated. They are used to govern succeeding operations of the same or allied devices.

APPLICATION OF SWITCHGEAR AND CONTROL APPARATUS

13. CIRCUIT BREAKERS

MOLDED CASE CIRCUIT BREAKERS. Molded case circuit breakers are designed to protect insulated conductors from the effect of overcurrent. These breakers may be used on any low-voltage circuits, such as the interior network of a building for light and power, extending from the incoming service to the final branch circuits. With a few exceptions, overcurrent protection is required at every point where the wire size is reduced.

For example, in a typical building the service conductors are protected at the point of entrance, and the main distribution center protects the light and power feeders, which in turn terminate in distribution centers for the final branch circuits, or the branch circuits may be individually tapped from the feeder at convenient points. The distribution centers may be switchboards, panel boards, or groups of individual circuit breakers. Regardless of the type of enclosure, circuit-breaker mechanisms and characteristics are identical.

Molded case breakers are designed for conductor protection, and their tripping characteristics are coordinated with the thermal capacity of the various wires they protect. When a circuit breaker is applied for motor branch circuit protection, its purpose is to protect the circuit only from short-circuit faults. The motor starter or overload protective relays prevent overloading of the circuits. When properly selected, the circuit breakers permit motor starting yet trip automatically to protect against faults in conductors, motors, or starting equipment.

Requirements for coordinating the tripping characteristics of molded case breakers with those of the motor-running protective device or devices vary according to whether one motor or more than one is in the circuit. (1) For a single-motor circuit the motor-running protective device, properly selected, will permit the use of a circuit breaker of higher rating than that required by the conductor. The result is that, with overloads up to approximately the locked-rotor current of the motor, the motor-running protective device will open before the circuit breaker. With heavier overloads and short circuits, the circuit breaker will open first. (2) If there are two or more motors grouped on one branch circuit, protected by one thermal circuit breaker, proper selection of each motor-running protective device permits, as with a single motor, the use of a circuit breaker of higher than wire-size rating. In this case, however, the rating for the circuit breaker is determined under the provisions of the National Electric Code.

LOW-VOLTAGE AIR CIRCUIT BREAKERS. Low-voltage air circuit breakers should have an ampere rating at least as great as the maximum rated 1-hr (or more) overload current of the apparatus that the breakers will be required to control. Air circuit breakers should have a current-interrupting rating at least as great as the total obtainable rms current, calculated as follows:

1. In a-c circuits, the average value of the rms total current, including the d-c component at an instant one-half cycle after short circuit occurs. (For practical purposes, this average may be taken as 1.25 times the average symmetrical value.)
2. In d-c circuits, the maximum value of current.

Circuit Breakers in Cascade. Properly selected air circuit breakers may be applied in cascade. The following requirements, however, should be observed.

1. Cascading should be limited to three steps of interrupting capacity:

a. The interrupting rating of the breaker or breakers nearest the source of power should be equal to at least 100 per cent of the short-circuit current, as calculated in accordance with the paragraph above. The breaker or breakers in this step should also be equipped with instantaneous trip features set at not more than 80 per cent of the lowest interrupting rating of the breakers in the next lower step.

b. The breaker or breakers in the second step should be selected so that the calculated short-circuit current through this step will not exceed 200 per cent of their interrupting rating. The breaker or breakers in this step should also be equipped with instantaneous trip features set at not more than 80 per cent of the lowest interrupting rating of the breakers in the next lower step.

c. The breaker or breakers in the third step should be selected so that the calculated short-circuit current through this step will not exceed 300 per cent of their interrupting rating.

Where cascading is proposed, recommendations should be obtained from the manufacturer in order to insure proper coordination between successive steps.

Rating. The operation of breakers in excess of their interrupting rating is limited to one interruption, after which inspection and maintenance may be required (NEMA Standard Sg7-61).

The rating of low-voltage air circuit breakers includes the following items: (1) rated voltage, (2) rated frequency, (3) rated continuous current, (4) rated short-time current, (5) rated interrupting current.

Circuit breakers should not be applied on circuits whose voltage exceeds their rated voltage.

Air circuit breakers provided with magnetic tripping devices have an ampere rating based on a temperature rise at the terminal connections not exceeding 30 deg cent. When air circuit breakers are provided with thermal tripping devices, they should have an ampere rating based on a temperature rise at the terminal connections not exceeding 40 deg cent. The temperature rise of contacts will increase because of oxidation of the contact surfaces. The rating of breakers is, therefore, based on sufficient maintenance to keep the temperature rise within the specified limits. The standard ambient temperature is considered to be 40 deg cent.

Momentary ratings apply to air circuit breakers which are inherently instantaneous in tripping operations. Each rating shall be expressed in amperes (if for a-c circuits, in rms total amperes, including the d-c component) and should be numerically equal to the interrupting rating.

Circuit breakers are not intended for applications requiring highly repetitive duty. The number of operations with normal maintenance at rated load or no load varies from 50,000 for 15,000-amp interrupting rating circuit breakers to 5000 for 100,000-amp interrupting rating circuit breakers with a frequency of operation of not more than 20 in 10 min.

Standard breakers are available in one-, two-, three-, and four-pole units. They may have an overcurrent trip on each pole, or they may have some poles with, and other poles without, such trips, to suit the requirements of the service.

A-c Circuit Protection. A-c generators in attended stations are usually provided with short-circuit trips of the instantaneous type and are not provided with time-overcurrent trips. In unattended substations, however, it is necessary to provide time-overcurrent protection in order to prevent damage due to excessive current values.

For a-c circuits in general, one circuit-breaker pole is provided for each ungrounded conductor, and these poles are arranged to trip all phases simultaneously. Time-overcurrent trips are normally provided; these may be of either the dual-magnetic or the thermal-magnetic type. Time-overcurrent protection is not used on breakers, however, when they are applied in cascade.

Low-voltage air circuit breakers may be used on motor-starting circuits up to 1600 amp. These breakers are frequently supplied with shunt trip or undervoltage trip operated by various types of protective relays. The starting may be across-the-line at reduced voltage by the auto-transformer, or the line reactor, method. The use of these circuit breakers for motor-starting duty, however, is limited to a-c motors of modern design which have essentially normal starting conditions and are not subject to highly repetitive duty, jogging, etc. These breakers also should not be applied where the starting inrush current exceeds 600 per cent of the normal value.

D-c Generator Protection. With d-c generators it is necessary to take precautions against the burning of the commutator and brushes and against flashover. For this purpose the circuit breakers usually have an instantaneous trip adjustable from 100 to 200 per cent but set at 200 per cent to effect high-speed opening on short circuits. For marine applications or other uses where continuity of service is of paramount importance, time-overcurrent protection, at approximately 125 per cent of full load with instantaneous short-circuit trip set at from 600 to 700 per cent, is frequently provided.

Three-wire d-c generators require protection on both sides of the armature. When there are equalizer connections, algebraic sum overload protection is normally used to obviate the necessity of running extra leads from the machine to the circuit breaker. Each overload device functions on the algebraic sum of the current in the armature and equalizer load.

No overcurrent protection is normally supplied for exciter breakers. If the exciters are to operate in parallel, however, directional protection should be used. The setting of such protection should be above the current caused by inductive action between the generator armature and its field circuit. Inasmuch as this may be above the continuous current, it is frequently necessary to use the reverse-current relays rather than direct-acting reverse-current devices.

Although knife switches can be used for the equalizer connection of compound-wound machines, the advantages afforded by the use of multipole circuit breakers are frequently sufficient to justify their use. There is no necessity for overcurrent protection of the equalizer leads.

When d-c generators and synchronous converters are operated in parallel or in parallel with another source of power, current-directional or reverse-current protection should be provided to take care of reversals of power flow and to provide fast tripping for internal fault protection. The setting of such protective devices should be as low as possible but yet high enough to obviate the possibility of tripping on the normal magnitude of regenerated load or on a small interchange of current and light load.

When d-c generators and synchronous converters are operated in parallel with another source, undervoltage trips should be provided to disconnect them from the system after a shut-down to prevent their being subsequently subjected to voltage. Undervoltage trips are also used in connection with generator overspeed or other such devices.

Shunt trips are used on low-voltage air circuit breakers in connection with protective relays or in conjunction with remote manual operation.

POWER CIRCUIT BREAKERS. The successful concentration of large blocks of electrical energy and the interconnection of heavy power systems are possible only because adequate circuit breakers are available which will not only transmit the desired normal power flow at its particular voltage, but also, of utmost importance, have the ability to disconnect a faulted circuit from an electrical system promptly with minimum disturbance to the system and without material depreciation of the breaker.

Since the limitation of any particular breaker design is fixed, and as there is such a wide choice available, it becomes necessary for the prospective user to assure himself that the particular system on which he desires to use a breaker will not cause any of the breaker's limits to be exceeded. Otherwise, unsatisfactory service may be experienced.

For 2500- to 15,000-volt service, indoor air circuit breakers have the advantages of freedom from oil-fire hazard and the ability to withstand reasonable repetitive duty without excessive maintenance. At the present time these self-contained air circuit breakers are available in interrupting ratings of from 50,000 to 500,000 kva three-phase Manufacturer's catalog ratings should be consulted for specific applications.

Both indoor and outdoor oil breakers are available for 2.5- to 34.5-kv service in interrupting ratings from 25,000 to 2,500,000 kva.

Compressed-air circuit breakers have recently been developed for indoor and outdoor 15- and 34.5-kv service with interrupting ratings of from 500,000 to 2,500,000 kva. Compressed air is employed both for breaker operation and arc extinction. These breakers offer freedom from oil-fire hazard without sacrifice in space requirements. For either air or oil breakers the compressed-air operating mechanisms, which were developed with the compressed-air breakers, permit faster reclosing speed than electric mechanisms.

For 46 kv and higher, oil circuit breakers are used almost exclusively at present. In the highest voltages, 115 kv and up, each pole unit is mounted separately on the foundation.

The several ratings of power circuit breakers are defined in AIEE Standards No. 19 and 20 and NEMA Standards 4167 and 4271. It is the intention here, not to review these ratings in detail, but only to discuss the principal factors involved in the selection of a circuit breaker for a particular application.

Rated Voltage. The standard rms voltage rating of a particular breaker is based on its use at an altitude of 3300 ft or less. For operation at higher altitude the voltage rating is reduced 1 per cent for each 330 ft of altitude above 3300 ft. Voltage ratings up to and including 15,000 volts are maximum values, and no breaker should be operated above its rating under any conditions. Breakers rated above 15,000 volts are rated on the basis of normal voltage; 5 per cent tolerance above rated voltage is allowable for emergency operation.

Rated Current. Continuous-current rating is based on an ambient temperature of 40 deg cent and an altitude of 3300 ft or less. For operation with a higher ambient temperature, the current rating must be so reduced that the total temperature of any part will not exceed that specified in the AIEE Standards.

The frequency of the current affects the heating of magnetic parts because of induction, and of current-carrying parts because of a change in current distribution. The continuous-current ratings of each breaker for 60 and 25 cycles and for direct current are published by the manufacturers of the breakers.

Short-time current ratings are based on mechanical stress, thermal characteristics, or both. Two values are given for all circuit breakers, one of which is for any period from 1 to 5 sec. These current ratings are based on rms total amperes, as determined from the envelope of the fault-current curve, and include the d-c component. The 5-sec current rating is based on the value of rms total current flowing at the end of 1 sec. Only one short-time current rating is given for air circuit breakers, and it is based on the maximum rms total amperes at any time.

The **rated making current** is the maximum rms total current against which the breaker must be capable of closing without welding or undue damage to the circuit breaker or contacts with rated control voltage at the closing mechanism. This rating should be at least as great as the maximum fault current which can be obtained during the first cycle. The rated latching current is the maximum rms total current against which the breaker must be capable of closing and latching with rated control voltage at the closing mechanism.

The **rated interrupting current** at rated voltage is an important factor in the application of a breaker; it is the maximum rms total current that the breaker should be required to interrupt in a circuit operating at the rated voltage of the breaker. This rating is also usually given approximately as a kva interrupting capacity for three-phase breakers by multiplying the current by 1.73 times the rated voltage. No additional significance concerning the effect of the value of current in the second and third poles is implied. The greatest possible rms total current (including the d-c component) in any pole governs the selection. The interrupting current is that flowing when the breaker contacts part.

When a power circuit breaker is used in a circuit operating at a voltage lower than the voltage rating of the breaker, it can be used to interrupt larger currents than at rated voltage. The current that may be interrupted increases as the voltage is lowered until the interrupting-capacity current limitation of the breaker is reached, which is the maximum permissible interrupting current, regardless of how low the circuit voltage is. For intermediate values of voltage, the interrupting current is equal to rated voltage divided by circuit voltage times rated interrupting current at rated voltage.

The conditions assumed in rating power circuit breakers include consideration of the stored electrostatic and magnetic energy of the system, re-establishment of an arc under transient voltage conditions, decrement of the system, and other variables. These influences are considered as not differing widely in average systems and are to be taken into account in the factor of safety employed in the rating of the breakers.

The **interrupting time** of a circuit breaker is the maximum interval from the time the trip coil is energized at normal control voltage until the arc is extinguished. This time is published for standard breakers for the interruption of currents from 25 to 100 per cent of the rated value.

The rated interrupting current of a breaker is based on a specified operating duty (duty cycle). This duty is defined in terms of a given number of unit operations, each consisting of a closing operation followed without purposeful delay by the opening of the breaker. This is designated by the letters "CO."

The standard operating duty for power oiltight and air circuit breakers consists of two such unit operations with an interval of 15 sec between them. Oil circuit breakers having standard interrupting ratings of 50,000 kva and above are oiltight. For non-oiltight breakers (those having standard interrupting ratings below 50,000 kva) the standard interval is 2 min. When breakers are subjected to several successive unit operations or to duty cycles more severe than the standard, derating factors must be used.

For reclosing and other special-duty cycles, the interrupting rating of the standard breakers may be obtained by applying factors to the standard interrupting rating for the particular breaker.

The standard reclosing time in cycles for 60-cycle outdoor reclosing oil circuit breakers up to and including 1200 rated amp and 1,500,000 kva interrupting rating varies from 30 cycles on a 60-cycle basis for 7500-volt breakers to 45 cycles for higher-voltage breakers. For fast-reclosing breakers the time varies from 20 to 30 cycles. Special operating mechanisms are available to provide faster reclosing.

The standard ratings of most power circuit breakers are given in terms of three-pole breakers for three-phase systems. These voltage ratings are based on the line-to-line voltage of the circuit, and the interrupting ratings are given in amperes and approximate three-phase kva. In order to select the proper one-, two-, three-, or four-pole circuit breaker for special services on three-phase circuits and for use on two-phase and single-phase circuits, the equivalent three-phase breaker rating must be determined.

DETERMINATION OF THE SHORT-CIRCUIT CURRENT. Within the first complete cycle following the short circuit, the current usually rises to a maximum value. If, now, the generator field circuit is assumed to remain unchanged by the use of automatic generator-voltage regulators or similar devices, then, in succeeding cycles, the current will decrease with time until a sustained value exists on the system. The current wave during the transient period, as shown by oscillograms, may be symmetrical or asymmetrical with respect to the line representing zero current, depending upon the point of the voltage wave at which the short circuit occurred. The transient current is said to be symmetrical when the assumed line connecting points on the current wave midway between the peaks coincides with the line of zero current. Similarly, a transient current is said to be asymmetrical when the assumed line connecting points on the current wave midway between the peaks does not coincide with the line of zero current.

The transient of an electrical system at a point remote from the source of power supply will be different from that at the terminals of an a-c generator. Its amplitude and duration will depend upon the electrical constants of the system and the distance between the point of short circuit and the source of power supply. In general, the greater the distance between the source of power and the point of short circuit, the less the amplitude of the transient and the shorter its duration.

The current flowing in a short circuit in a simple a-c system can be expressed as a function of time in the form of three exponentially decaying components and a term which is substantially constant. The components are termed the subtransient, transient, d-c, and steady-state components.

The greatest transient disturbance of the system which can occur at the time of contact separation of the circuit breaker when the system is short-circuited governs the selection of a suitable breaker.

The rms of the total short-circuit current wave at any instant is the square root of the sum of the squares of the values of the direct component and the rms value of the alternating component at that instant. In actual systems all these terms are affected to some extent by the number of machines, the phase angles between them, and also their excitation. The phase relations between machines may change during the period of short circuits and thus add further complications. Automatic voltage regulators introduce another factor, which tends to increase the short-circuit current a short time after the initial rush, since the regulator action causes an increase in machine excitation.

The rigorous determination of short-circuit currents as a function of time involves too laborious a calculation to be practical. Fortunately, such rigorous calculations are not generally required to determine the proper application of interrupting devices and relay settings, for many other factors having greater proportional effects than the variations cited contribute to the determination of the limits for a rational choice of circuit breakers. It is, therefore, perfectly proper and logical to make certain approximations in the calculation of short-circuit current to permit simple solutions of electrical systems for the purpose of application of equipment. Such assumptions have been in general use for a considerable time and are recognized by the electrical engineering profession.

The modern trend of practice is to speed up materially the clearing of short circuits in order to reduce damage and maintain continuity of service. In general, however, the more quickly a short circuit is cleared, the greater is the current to be interrupted. Thus it is quite easy, by speeding up the relay system, to jeopardize the breakers by requiring them to interrupt more current than that for which they were originally chosen. Also, in the past there was sometimes a tendency to save investment in breakers by the use of slow-speed relays in order to take advantage of the decrease in short-circuit current due to its decrement with time. There is no guarantee, even with slow-speed relays, that a short circuit will not change its character and bring about a high current flow in a breaker while it is opening. Furthermore, there is always the danger of shortening the time settings of slow-speed relays, thus tending to increase the interrupting duty on the breakers.

This being the case, the problem of interrupting-device application resolved itself into a study of what happens in a system during the first few cycles of short circuit.

Short-circuit Conditions. There are two short-circuit conditions that must be considered.

a. The momentary rating of the breaker must be equal to or greater than the highest rms value of currents that the system can produce through the breaker in question at any time. The multiplying factor for the general case is 1.6. At 5000 volts and below, however, unless current is fed predominantly by directly connected synchronous machines or through current-limiting reactors only, the multiplying factor is 1.4.

b. The interrupting rating of the breaker must be equal to or greater than the short-circuit current at the time of contact separation. The proper multiplying factors, which depend on the speed of opening of the breaker, vary from 1.0 for an eight-cycle breaker to 1.4 for a two-cycle breaker in the general case. In special cases the factor varies from 1.1 for an eight-cycle breaker to 1.5 for a two-cycle breaker, for breakers at generator voltage, where the rating is more than 500,000 kva three-phase, if either fed predominantly from generators directly or through current-limiting reactors only.

Factors Affecting Short-circuit Current. The rms current at any point of a system under short-circuit conditions is affected by several factors:

1. The reactance of generators and meters (subtransient or transient), as well as all other elements of the circuit to the point of assumed fault, is used in the calculation, based on arrangement as shown in detail on the circuit diagram in question.

2. Calculations made for three-phase faults and for single-phase-to-ground, if the system neutral is grounded, will be adequate for other types of faults, thus simplifying the scope of necessary calculations.

3. Generally, resistance at the point of fault or in the circuit itself can be neglected, as it is rarely of sufficient magnitude to affect the short-circuit calculation appreciably. Impedances in the neutrals of generators or transformers, however, should be included in single-wire line-to-ground calculations. Obviously, such neutral impedances have no influence in three-phase or wire-to-wire fault calculations.

4. The multiplying factors previously given are based on the assumption of fully loaded generators, 80 per cent power factor for turbogenerators, and 90 per cent power factor for hydrogenerators.

5. The short circuit is assumed to be established at the point of voltage wave giving the maximum total current.

6. The use of high-speed relays is assumed.

7. Power circuit breakers are applied on the maximum current in any phase. For air breakers at 600 volts or less, these interrupting ratings are based on the average current in the three phases. The multiplying factors for such breakers take this basis of rating into account.

8. The use of automatic generator-voltage regulators rarely affects the interrupting duty of breakers, because of time lag in increasing the generator voltage and hence the magnitude of fault current.

Calculation Procedure. The following is a summary of the procedure for calculating the highest values of short circuit to determine suitable breaker application.

Determine the required interrupting capacity of any power circuit breaker by the following procedure:

1. Calculate the highest value of rms symmetrical current for any type of fault on the basis of no load, normal voltage, using:

a. Subtransient reactance for synchronous generators.

b. Transient reactance for synchronous motors.
c. No recognition of induction motors.
d. Reactance of all other circuit elements to point of fault.
2. Multiply this highest value of current by the proper multiplying factor.

Determine the required momentary current rating for any power circuit breaker by the following procedure:

1. Calculate the highest value of rms symmetrical current for any type of fault on the basis of no load, normal voltage, using:
a. Subtransient reactance of all rotating machines, i.e., generators, synchronous motors, and induction motors.
b. Reactance of all other circuit elements to point of fault.
2. Multiply this highest value of current by the proper multiplying factor.

In order to determine short-circuit current values for breaker applications it is necessary to obtain the requisite data covering machines, transformer, reactors, and lines. When the problem involves existing equipment, such data may be available or can be obtained and should be used whenever possible. For problems involving new projects, either isolated or in combination with existing systems, average values of reactance may have to be used. In the determination of this current, the following symbols are used:

E = Line-to-neutral voltage.
X_1 = Positive-sequence reactance viewed from the point of fault, including transient or subtransient direct axis reactance of machines in ohms per phase.
X_0 = Zero-sequence reactance.
R_0 = Zero-sequence resistance.

The "highest value of rms symmetrical current for any type of fault" is determined from the formula E/X_1 or $3E/(2X_1 + X_0)$, whichever is greater. When R_0 is greater than $5X_1$, no consideration need be given to the latter term. This value should be taken for the maximum connected synchronous capacity. This current is then multiplied by the proper factors. The resulting interrupting and momentary current should be used to select the circuit breaker. The factors represent the ratio between the rms total current at the instant of contact parting and the initial value of rms symmetrical current. In determining these factors, it was assumed that circuit breakers would be installed which would permit the use of high-speed relays at some later date, and the time of contact parting was selected on this basis. Contact parting times of 4, 3, 2, and 1 cycles were assumed for 8-, 5-, 3-, and 2-cycle breakers.

Note that the total fault current calculated above may in some cases divide between two or more circuits. It is necessary to determine the maximum fault current that must be interrupted by each breaker under any circuit condition.

For most apparatus and circuits the resistance may be neglected as a justifiable approximation. For underground cables and very light aerial lines the resistance may be as great as the reactance. For these elements the impedance should be used instead of the reactance. Unless it constitutes a major part of the total current impedance, this impedance may be added arithmetically to the reactance of the rest of the circuit without appreciable error.

In ascertaining short-circuit currents for preliminary relay settings, conditions should be determined for each class of relay:

1. For maximum connected synchronous capacity: The maximum initial symmetrical current is determined by E/X_1 or $3E/(2X_1 + X_0)$, whichever is greater, except that, when R_0 is greater than $5X_1$, no consideration need be given the latter term.

2. For minimum connected synchronous capacity: The minimum symmetrical current is determined by $1.866E/X_1$ or $3E/\sqrt{(2X_1 + X_0)^2 + R_0^2}$, whichever is smaller. For each of these conditions a multiplying factor of 1 is chosen, and subtransient reactance is used for high-speed relays and transient reactance for time overcurrent relays.

14. FUSES

The interrupting requirements may be low if the fuse is located some distance from the substation on a rural distribution line, or they may be extremely high if the fuse is connected to the generating station bus. In applying a distribution cut-out, the symmetrical fault current of the system at the location under consideration is determined and a cut-out chosen with a current-interrupting rating at least equal to 1.2 times the symmetrical fault current.

The power fuse is chosen so that its kva rating equals or exceeds the calculated symmetrical fault kva possible at that point. This is consistent with the practice followed in applying circuit breakers. If the symmetrical fault current of the circuit is known, a fuse should be selected with a current-interrupting ability equal to at least 1.6 times the system symmetrical fault current.

In general, the continuous-current rating of a fuse should exceed by about 50 per cent the full-load current of the circuit in which it is applied. If the fuse is applied too closely, that is, with the load current equal or nearly equal to the continuous-current rating of the fuse, unnecessary operation may result because the fuse link will be heavily loaded and so will be easily blown by a momentary overcurrent. Such an overcurrent might be caused by the starting of a large motor, the charging current of a line or bank of capacitors, or external disturbances, such as lightning.

Exceptions to the 50 per cent margin occur where the fuse must be coordinated with other protective devices. The fuse is then applied from consideration of its time-current characteristics, although its continuous-current rating must not be exceeded. Sometimes the small size of the fuse, indicated by load-current considerations, makes it desirable to choose a larger fuse to prevent unnecessary fuse blowings by lightning currents, such as appear on distribution systems. A typical case is a 1.5-kva transformer on a 6900-volt system, where a $^1/_2$-amp fuse would be ample for the load current but a 5- or 10-amp fuse is used to prevent unnecessary outages.

The fuse must meet the requirements of insulation to ground and insulation across the blown fuse or the fuse in its disconnected position. Occasionally, special conditions warrant the use of the next higher voltage class of insulation, e.g., in steel mills, where the dirt necessitates 15-kv insulation for 6900 volts. A special case is found where two different supply sources are connected together through fuses, and either source is ungrounded. Under fault conditions two of the three fuses might blow, and the third unblown fuse would tie one phase of the two systems together. Under these circumstances, if there were no stabilizing force in the two systems tending to keep them in synchronism, a part of the time one system would be diametrically opposed in phase to the other system, thus producing twice the system voltage across the blown fuses. Under such conditions adequate voltage ratings must be chosen.

Although most outdoor fuse equipment can be used indoors, indoor equipment is not suitable for outdoor use. Special precautions are usually taken with outdoor equipment to prevent the placing of voltage across exposed organic insulation, which might be broken down by leakage current. At times the choice of a fuse is affected by the space requirements. Another consideration is the noise produced by the operation of the ordinary expulsion fuse, which consequently might not be suitable for indoor use. Instead, a totally enclosed fuse would be chosen.

The type of load and the frequency of operation of the fuse affect the choice. For a motor-starting load, such as in a rubber mill, where the motor is "inched" and draws its full starting current repeatedly during brief periods, a somewhat larger current rating of fuse is used than would be necessary for a motor under continuous operation, such as a blower. A fuse would be undesirable at a location where a large number of fuse operations would be expected, such as a long line exposed to lightning, falling branches, and other hazards. Here the cost of outages and replacement of blown fuses would be high enough to dictate the use of a circuit breaker.

A careful consideration of design and performance factors will reveal that there is less overlapping of fuses than is generally believed. When the economic factors are included, it is seen that the overlapping of fuse units is almost non-existent. Although there are often several fuses that will carry the load and meet specific interrupting requirements, investigation of all the factors will indicate clearly the best fuse for a given application.

15. PROTECTIVE RELAYS

The function of relays and circuit breakers in the operation of a power system is to prevent or limit damage during faults or overloads and to minimize their effect on the remainder of the system. This is accomplished by dividing the system into protective zones separated by circuit breakers. These zones may be divided into four classes: (1) generator, (2) bus, (3) transformer, and (4) line. During a fault the zone which includes the faulted apparatus is de-energized and disconnected from the system.

The application problem consists of selecting a relay scheme which will recognize the existence of a fault within a given protective zone and initiate proper circuit-breaker operation. After the general plan of the system has been established, the location of circuit breakers tentatively determined, and the system divided into protective zones, the protective-relay scheme for each zone must be selected.

Selection of the relay scheme involves a consideration of the protective requirements of the system and the characteristics of available methods of protection. Usually a general consideration of the main features alone either determines the scheme or narrows the selection to two or three schemes that can be considered in greater detail. Specific information, such as burdens, connection diagrams, and method of setting, is given in the manufacturer's catalogs and instruction books pertaining to the particular relays.

A-C GENERATORS. Experience has shown that most of the internal faults in rotating electrical apparatus develop originally as a ground in one of the phase windings. The higher the value of impedance in the neutral, the smaller the ground fault current in the apparatus; the voltage stress and the difficulty of protecting the apparatus and the rest of the system, however, increases.

Although overcurrent relays, which protect against ground faults only, may have very sensitive current settings, selectivity requirements may prevent the use of low time-settings, and it is still necessary to protect against the faults which start as phase-to-phase faults and also those which develop into phase-to-phase faults before the ground relays can operate.

Phase-to-phase overcurrent protection has decided limitations from the standpoint of sensitivity, selectivity, and speed of operation, since overcurrent relays must be set above maximum load current and must also have time-settings which select with other relays on the system. For these reasons ordinary overcurrent and phase-to-ground relaying are seldom satisfactory for protection against internal faults.

Differential schemes of protection are almost universally used in the protection of generators. These schemes are based on the principle of balancing the secondary currents of the current transformers at the terminals of each phase winding so that, under normal conditions of through current flow, the current will circulate in the transformer secondaries. When a protective relay is connected to this balanced circuit, it will receive current only when a fault occurs in the equipment. Differential schemes are inherently selective and consequently can be operated without intentional time delay.

The use of overcurrent relays in a differential connection permits sensitive protection only when used with ideal current transformers. Hence, if overcurrent relays are used, they have to be set so as not to respond to the maximum error current which can flow in the relay on a through fault.

This disadvantage is overcome by the use of a percentage-differential relay. The relative number of turns on the restraining and the operating windings are so proportioned and the connections so made that the relay tripping current is proportional to the total current for through faults. The relay is very sensitive on internal faults because of the large difference in magnitude and phase relation of the currents in the restraining coils.

In installations where some scheme of overload protection is necessary, overcurrent or overpower relays can be used. It is usually desirable to have a time delay in the relay, especially if the protection is only against overloads, as the relay must otherwise be set above the maximum momentary peak loads.

In the application of overcurrent relays to generator protection, consideration must be given to the fact that the time delay necessary for selectivity in the relay may cause the synchronous reactance of the machine to control the current received by the relay. Since the short-circuit current, as determined by the synchronous reactance, may be in the order of magnitude of full load, the overcurrent protection of generators cannot be as satisfactory as the same scheme for motors.

If the power supplied by a generator, rather than the magnitude of current, is used as the means of detecting overloads, an overpower relay should be used. If the power of the various phases is unequal, a relay in each phase will be required. If the load is balanced, however, two of the relays may be replaced by reactors.

Since the overheating of the windings in electrical apparatus is closely related to overcurrent protection, temperature protection is not always necessary. In some cases, however, it is desirable to provide relays whose operation is affected by the temperature of the windings, as well as the overload, or by the temperature of the windings alone.

TRANSFORMERS. Although the principles of differential protection can be applied to the protection of transformers as advantageously as to the protection of generators and motors, the conditions are not exactly the same. Because of the ratio of transformation, the actual current entering the transformer will differ from that leaving the transformer, and frequently it is impossible to obtain standard current transformers having ratios which will produce equal secondary currents. Also, if a transformer bank is delta-star connected, there will be a phase displacement between the currents on the two sides of the transformer. The protection of Scott-connected transformers and of three-winding transformers must be given special consideration.

The secondary currents can be equalized by means of current-balancing auto-transformers and can be brought into phase by connecting the current transformers on the delta side in star and on the star side in delta. It is usually simpler, however, to use a differential relay provided with taps which permit normally unequal secondary currents to be balanced within the relay. As the current transformers have different ratios, the differential current caused by dissimilar characteristics will be considerably greater than in generator protection. For this reason ordinary overcurrent relays cannot be given sensitive settings, and induction-type transformer differential relays are provided with a characteristic requiring a 50 per cent unbalance in current to operate.

The protection of a three-winding transformer does not differ in principle from that of a two-winding transformer. A summation of the secondary currents in each of the three windings is made.

A few power transformers have very high values of magnetizing-current inrushes when the breaker is closed, and this magnetizing inrush may be high enough in magnitude and duration to cause the relay to operate. To prevent this, a magnetizing-inrush tripping suppressor is necessary. In many cases increasing the contact travel, increasing the damping on the relay, or both will allow induction-type differential relays to be used on transformer banks having a high magnetizing-inrush current, without resorting to auxiliary devices to block tripping.

There are many installations where the use of high-speed differential protection of transformer banks is necessary to keep damage to a minimum and to fit in with high-speed relays and circuit breakers for transmission-line protection.

BUSES. The method of construction of generating-station and substation buses adopted by operating companies has resulted in the buses being one of the components of modern power systems least subject to faults. Since faults rarely occur on buses, it was considered for a long time that they were practically faultproof and that installation of protective relays introduced a hazard greater than no protection at all, because of the possibility of incorrect operation. Contrary to this belief, however, faults have occurred on buses and, where concentrations of power are high, have resulted in severe damage to the equipment and long outages.

The major differences between bus protection and generator or transformer protection are found in the number of circuits in the protected zone and in the values of currents involved in the various circuits. Current transformers of the same ratio are inserted in the leads of all circuits entering and leaving the bus section, and the vector sum of all the secondary currents is impressed upon the operating winding of an overcurrent relay for each phase. A ground relay can be used to measure the vector sum of the residual or ground current, since this gives more sensitive protection on ground faults.

Since the currents in the various branches are different during external faults, varying degrees of saturation are produced in the current transformers. Induction-type relays, because of their time delay and because of their tendency not to operate on the distorted current wave form resulting from saturation, can be given a lower setting than instantaneous relays. In either case, however, the current transformers must be accurate enough for the relays to operate on minimum internal faults and not to operate on maximum external faults.

When the current transformers are not sufficiently accurate, and there is a large ratio between maximum and minimum fault currents, or when the d-c time constant is high, it is necessary to use differential relays with restraint windings in addition to the overcurrent winding.

A differential relay with two restraint windings may be used if the circuits can be combined into two groups. This, however, causes difficulty when all the fault current comes from circuits in the same group. In such cases there is no current in one of the restraint windings, and the error current flows in one of the restraint windings and the operating winding. The relay may not operate correctly, as it then has the characteristics of an overcurrent relay with a low current pick-up.

A relay having three restraining elements permits a more flexible grouping of the source circuits. To take full advantage of the percentage differential-relay characteristics a restraining winding for each circuit would be required. The use of two or three three-restraint windings per phase with the operating windings in series permits the ideal scheme to be approached. This is generally sufficient, as a careful analysis will usually show that even with more than six circuits the windings can be grouped so that there will be sufficient restraint for any external fault.

A variable percentage relay inherently compensates for errors introduced by current transformer saturation and at the same time is very sensitive for low fault currents. This characteristic is particularly desirable on internal ground faults, where the ground current is limited by a neutral impedance.

Differential protection for buses is usually desirable but is not always feasible in existing stations. This situation may be due to lack of current transformers on some circuits or to the existing transformers already having a burden so great that it prevents satisfactory differential protection of the bus. Also, in many double-bus installations it is found that the existing current transformers are improperly located to allow the individual protection of each circuit of the double bus.

For such cases modified protection schemes must be devised. Most of these schemes do not give the full degree of protection possible but provide protection for ground faults, which are the most common, particularly on outdoor buses or indoor buses with segregated phases. A simplified bus differential scheme, giving protection only against ground faults, can be used without the addition of the equipment necessary for complete differential protection. An auxiliary current transformer is connected in the neutral of the current transformers in each circuit. The vector sum of all the residual currents in the secondaries of the auxiliary transformers is impressed on a ground differential relay. To prevent operation on an open circuit in a main current transformer secondary, the relay is interlocked with a ground overcurrent relay connected in a generator or transformer bank neutral.

A modification of the differential scheme, using impedance relays, can be applied to bus sections having reactors in the bus tie and outgoing feeder circuits. The vector sum of the currents in all sources is applied to the current coil, and the voltage coil is connected to receive line-to-neutral voltage on the bus. This becomes in reality an overcurrent relay with voltage restraint.

Another method of protection, which is more adaptable for new stations, is the fault bus. This involves the insulation of all circuit-breaker structures and the bases of all insulators and switch supports from ground, all these being connected together by means of a fault bus, and then the fault bus being connected to ground through a current transformer. With this arrangement, ground faults cause a current to flow through some part of the fault bus and then to ground through the current transformer. A relay connected in the secondary of the current transformer trips the breaker connected to the bus.

LINES. Transmission circuits, unlike apparatus or buses, have their terminals some distance apart, and the application of differential protection to them introduces many problems, which become more difficult as the circuits grow longer. The expense of differential protection, therefore, cannot always be justified.

There are wide differences in transmission circuits, and as a result many different types of relays are available for their protection, each of which has its field of application. The deciding factors in selecting a scheme are relative cost of the schemes, the amount and arrangement of existing station equipment, and the degree of protection desired.

Radial System. The simplest system is a radial system having a single source of power with a number of feeders leaving the generator bus, each feeder in turn being subdivided into a small number of feeders. The protection of such a system against short circuits may be secured by overcurrent protective relays. The most remote branches may be disconnected automatically from the rest of the system by the blowing of a fuse or the operation of instantaneous circuit breakers. The circuit breakers near the generator, as well as the intermediate breakers, must be equipped with time delay relays. The settings of the relays nearest the generator are highest, and each successive relay is lower by a sufficient time interval to insure selective operation.

Instantaneous overcurrent relays may also be used if the variation in fault current permits settings which will cover a large portion of the section, but not cause tripping for a fault beyond the next bus section. In applying instantaneous relays, they must be considered supplemental to timed relays, and consideration must be given to maximum and minimum fault currents in determining the relay settings and the protection afforded.

The time delay necessary between relays in adjacent sections is largely determined by the time required for circuit-breaker operation.

The number of radial circuits in series which can be protected by using timed overcurrent relays is limited by the maximum permissible tripping time. Where such conditions exist, impedance relays provide a solution. The time of operation of these relays, being proportional to the distance from the relay location to the point of fault, provides the same minimum operating time at all locations for close-in faults, with an increasing time of operation for remote faults, in which the voltage on the relay is higher and the current usually lower. Instantaneous phase protection can be provided by means of high-speed impedance relays. As they can be set to cover only 90 per cent of the protected section, however, they must be used in conjunction with relays utilizing time as the means of obtaining selectivity.

Loop System. In the loop system, fault current may flow in either direction through any substation, depending on the location of the fault. The direction of current flow, therefore, can be used as an aid in obtaining a selective relaying system. When fault current flows into the line section which the relay is protecting, the directional-element contacts close to permit tripping through the contacts of a fault-detector element, but if fault current flows out of the protected section, the directional-element contacts block tripping. A directional element may be used in conjunction with either an overcurrent or impedance-type fault detector.

The number of circuits which can be included in the loop is subject to the same limitations with the directional overcurrent relay as with the straight overcurrent relay.

A typical loop system is shown in Fig. 1. The arrows indicate the direction in which the overcurrent must flow to trip the relay, and the figures show the definite minimum time settings of the relays. Overcurrent relays only are used at the generating station or source of power, as any fault on the system can cause current flow in only one direction from this point.

A fault on any section causes only the breakers at the ends of that section to trip, since the time delay on other breakers is successively higher as the generating station is approached. If, however, these breakers should fail to operate, back-up protection is provided by the breakers at the far end of the next adjacent sections.

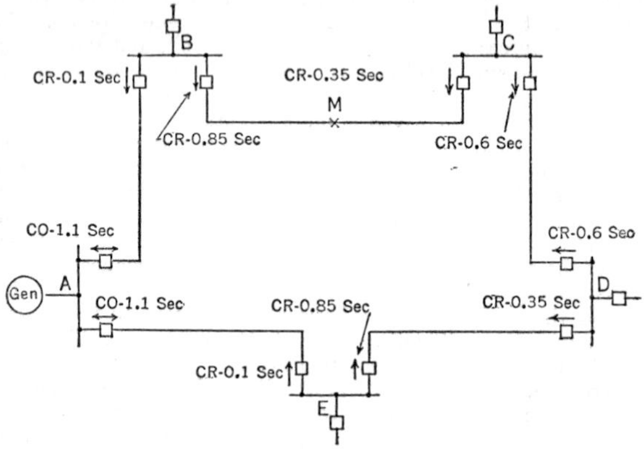

FIG. 1. Protection of Typical Loop System

Where a loop system has more than one source of power, the use of directional overcurrent relays becomes more complicated. If all sources are connected to the loop at all times, it may be possible, by utilizing the inverse portion of the overcurrent relay characteristic, to apply directional overcurrent relays and to use the same system of successive time settings as in the loop with a single source of supply. If one source of power is disconnected, it is neither practicable nor desirable to change relay settings when the system set-up changes.

A better solution, therefore, for relaying a loop system is the use of impedance or distance relays. Each relay is set to trip its own section of the line in a time corresponding to the distance to the fault, the maximum time required for a fault at the far end of protected circuit usually being from 0.75 to 1.0 sec. Regardless of the number of sources of power or the point on the system where the fault occurs, the proper relay will trip.

Systems with a Wide Range of Generating Capacity. On systems where the range between minimum and maximum connected generating capacity is very large, the three-phase and phase-to-phase fault currents, when minimum capacity is connected, may be less than the maximum-load current which the lines carry with maximum capacity connected. Since the relays must be set higher than the maximum-load current, they will be inoperative for minimum fault current unless means are employed to permit setting the relays for lower current values. One method is to connect the impedance relay so that it is inoperative until a voltage relay drops out. The voltage relays are set to drop out at a voltage which is less than normal line voltage but above the voltage that the relays will receive with minimum fault current flowing to a fault in the next section of the line.

Quick reversals of power flow often occur after faults are cleared, and instantaneous overcurrent or distance elements cannot be used satisfactorily with slow-speed directional elements when directional discrimination must be used for selectivity.

Figure 2 shows the tripping characteristics of a relay having an instantaneous impedance element and a directional element combined with a time-delay impedance element.

Distance protection, using instantaneous impedance elements, a high-speed directional element, and a timer combined, gives the tripping characteristic of Fig. 3. Zone 1 covers

90 per cent of the first line section; zone 2, the remaining 10 per cent of the first line section and 50 per cent into the second line section; and zone 3, the remaining 50 per cent of the second line section and 25 per cent into the third line section.

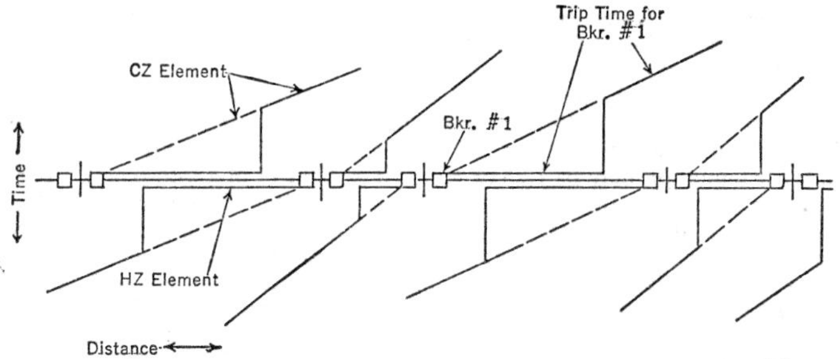

FIG. 2. Tripping Characteristics of a Relay Having an Instantaneous Impedance Element and a Directional Element Combined with a Time-delay Impedance Element

Pilot Relaying. Pilot relaying provides simultaneous tripping of the breakers at the line terminals for any internal fault. Pilot-relaying schemes compare conditions at the terminals of a transmission line to determine whether a fault is internal or external to the protected section. This comparison is made by means of a communication channel of either pilot wires or high-frequency carrier current superimposed on the transmission line. Two general methods of comparison are used: (1) a comparison of magnitude and direction of current flow, and (2) a comparison of the direction of power flow. The principles involved in the current-comparison schemes are the same as those for generator or transformer differential protection and are limited to use over a pilot-wire channel. The distance between the line terminals, however, introduces into the problem factors which make differential schemes of pilot protection more difficult than differential schemes applied to apparatus. For example, a set of relays for each line terminal is usually required, and the length of the pilot-wire circuit imposes a burden on the current transformers. Other factors which enter the problem are induced voltages during faults and the possibility of short circuits, grounds, or open circuits on the pilot wires.

The directional-comparison scheme is applicable to a d-c pilot-wire channel as well as to a carrier-current pilot channel. In this scheme electrical quantities at one line terminal are not compared directly with those at the others, but their effects on the operation of directional relays are compared. Carrier-pilot schemes are not subject to the limitation of induced voltages, grounds, short circuits, or open circuits on the pilot wires, and can be applied to transmission circuits of any voltage and over distances much longer than can be covered with pilot wires. The practical distance over which a carrier channel can be obtained, consistent with reasonable power output from the carrier transmitter, is largely determined by the losses in the transmission line at the frequency being used.

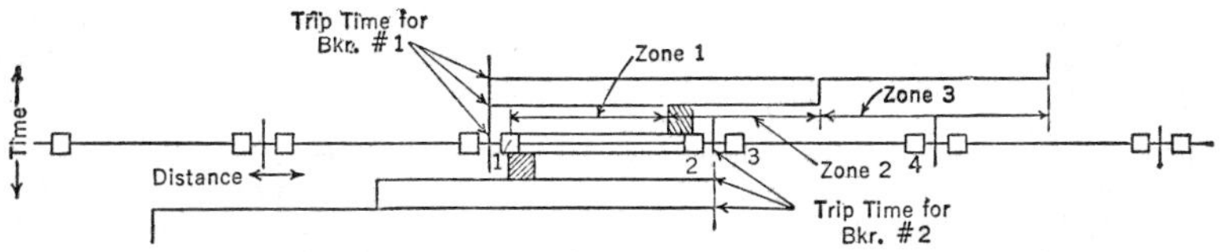

FIG. 3. Tripping Characteristics of a Relay Having Instantaneous Impedance Elements, a High-speed Directional Element, and a Timer

The choice between a carrier or a pilot-wire channel is an economic one, and the dividing line is in the neighborhood of 7 miles for most types of pilot-wire schemes. The cost of installation and maintenance of the pilot wires must be balanced against the cost of the terminal equipment for a carrier channel. The distance over which current-comparison schemes can be applied economically has been greatly increased. Recent improvements in pilot-wire relaying make it necessary to use only two wires. The burden on current transformers is extremely low with the new methods, and telephone circuits may be used for pilot wires.

Carrier-current System. For many lines the system known as carrier current is the most practical and reliable medium for comparing the conditions at the two ends of the line. Carrier current is a term applied to currents of 50- to 150-kilocycle frequency superimposed to the wire lines and not radiated into space, as is common in radio broadcasting. This results in greater efficiency and makes it possible to transmit greater distances with less high-frequency energy.

With carrier-current transmission the efficiency is very low, being in the order of 10 per cent, but as the energy losses do not involve large amounts of power, they do not represent an appreciable economic loss.

Carrier frequency may be impressed on a circuit between one conductor and ground or between any two phases. Although phase-to-ground transmission is usually less expensive, since only one set of coupling units is necessary at each end of the transmission channel, the attenuation may be 1.25 or more times that of phase-to-phase. The interference level is greater with phase-to-ground carrier circuits. When two- or three-lines-to-ground coupling is used, the attenuation is increased, since the resistance of the earth return is greater than the combination earth and phase return. In general, relaying and supervisory control will usually employ a phase-to-ground carrier channel because (*a*) the distances involved are seldom greater than 100 miles, and (*b*) the interference level or interference with signals is usually not serious for these applications. For other types of transmission, especially communication, the interphase circuit is preferable.

If the presence of branch lines, taps, or spurs offers interference from the reflection or absorption of certain carrier frequencies, resonant choke coils may be used to isolate a particular section of the transmission line. The choice of a carrier frequency with which the transmission characteristics are good is important.

Carrier frequency is introduced on the transmission line by means of a capacitor and a drain coil connected from the phase conductor to ground for phase-to-ground transmission. The capacitor offers an impedance of several million ohms to power frequency current. The drain coil offers high impedance to the carrier frequencies but low impedance to the power frequency current. Thus the ungrounded terminal is at a potential of less than 100 volts above ground, with the 60-cycle charging current flowing through it.

The carrier frequency is impressed directly across the drain coil. The carrier voltage is applied to the transmission line through the capacitor. The capacitor has a low impedance to carrier frequencies, so that in effect the carrier voltage is impressed directly on the transmission conductors without resorting to a high-voltage connection. For phase-to-phase transmission, this same connection is used on each phase conductor, so that half the carrier voltage appears between each phase and ground.

Several different types of relays can be used in a carrier-current relaying system. One scheme utilizes the time-distance characteristics of an impedance relay to provide high-speed simultaneous tripping with carrier in service, and step-type distance protection with carrier either in or out of service. The first element operates independently of the carrier current. The second element trips at high speed for faults in the section because carrier tripping contacts short around the synchronous timer. These tripping contacts close immediately if the fault is within the section but are held open by the carrier-current signals to block tripping if the fault is beyond the section being protected. This arrangement provides simultaneous tripping over the entire line section. The synchronous timer is used in connection with the second impedance element to provide back-up protection for the second zone section. The tripping circuit of the third element is independent of carrier current and operates with time delay for overall back-up protection. The directional element, supervised by the second impedance element, controls the transmission of carrier current. Additional interlocks can be included to prevent tripping of any of the elements due to out-of-synchronism surges.

Either receiver may receive a signal from its own transmitter or from the transmitter at the opposite end of the section. The correct functioning of the carrier current is not affected by internal transmission-line faults because it is used to block tripping in unfaulted sections and therefore is not required to transmit a signal over a faulted section.

For internal faults the relays at each end will provide simultaneous tripping, as the blocking carrier will be stopped at both ends. For external faults the fault current will be in such a direction as to cause one relay to desire to trip its breakers and the other to block tripping. Since the fault current, as seen by the relay at the other end, is in the wrong direction, a blocking carrier current will be transmitted to the first relay.

Another scheme utilizes relays which operate on symmetrical components of current. This obviates the necessity for using separate relays in each phase and reduces the relay requirements to one per terminal to handle all phase and ground faults.

Basically the relays determine whether a fault is internal or external to the protected line section by comparing, over the carrier channel, the relative directions of the currents at the two ends of the line. This comparison takes place in the grid circuit of a vacuum tube.

The necessity for the ground-preference feature in carrier relay systems depends to a large extent on the type of current connections used in the phase relays.

Ground-fault Protection. When star currents are used in the phase relays, the ground-preference feature definitely is needed during the flow of fault current and load current.

If delta currents are used in the phase relays, the ground-preference feature is needed only where the fault current is all zero-sequence and extreme load currents are present.

Phase preference is required on systems subject to simultaneous grounds in different line sections, and, if delta currents are used, the phase-preference connection is feasible.

Most transmission systems are operated with the neutral of transformer banks grounded either solidly or through an impedance. Since the majority of faults involve ground, ground relays are widely used in addition to phase relays for the protection of transmission lines. Residual or zero-sequence current and voltage are used either alone or in combination to obtain relaying quantities for the operation of ground relays. For normal balanced system operation, neither residual current nor voltage is present, and consequently ground relays can be made more sensitive than phase relays. Selectivity also is usually simpler to obtain because of the presence of many sources of ground current on most systems.

Overcurrent protection for ground faults, using time as a means of selectivity, follows the same principles as outlined for phase protection. Settings can be made only after the magnitudes of short-circuit current are known.

The directional overcurrent relay can be used for ground-fault protection where current may flow in either direction to a fault through the relay location. High-speed ground relays may be used, but they can only supplement slow-speed relays, since it is impossible to protect at high speeds for faults at the far end of the line and still maintain selectivity.

Impedance relays generally are not recommended for ground protection because of possible wide variations in the ground-circuit impedance. Relays with reactance characteristics can be used to give step-type distance protection for ground faults. Three relays are required for complete ground protection and must be connected to receive line-to-neutral voltage and compensated line current. Also, since reactance elements are subject to operation on in-phase load currents, a fault-detecting impedance element must be used to control the reactance elements. The impedance element should be given a setting so that it will operate for any fault within the zone of the reactance elements regardless of the circuit impedance.

Ground faults do not usually result in as severe a shock to a system as do phase faults, but if they are not rapidly removed they may develop into phase-to-phase faults. High-speed ground protection, therefore, is desirable.

Parallel Lines. In radial, loop, or network systems identical or similar lines connecting two or more of the stations are often encountered. Such parallel lines provide a convenient means of increasing transmission capacity, form a more flexible system for the economical operation of lines, and lend themselves very readily to relay protection.

To supplement single-line protection, it is desirable to protect parallel feeders by making use of the fact that under normal conditions electrically similar parallel feeders will have equal currents in them.

Relays which function on the balanced current in the lines may be used at the source end of any number of parallel lines. A system having a power source at both ends or at the receiving end of three or more parallel feeders in a radial system may also use balanced-current relaying. It cannot be used at the receiving end of a single pair of radial parallel feeders, for in the case of a fault on one line there would be an equal and opposite current in the two relay elements, and the relay would not operate. On new systems high-speed relays should be used in preference to low-speed relays. Normally, during single-line operation auxiliary switches on the breakers prevent the operation of balanced-current relays.

Another method of obtaining balanced-current protection is by the cross-connection of directional relays. This scheme may be applied to any system, no matter how complex, if its feeders are run parallel between the switching points. Since cross-connected schemes are inherently selective, high-speed relays can be used. In addition, the relays can be connected to give long-time operation for correct selectivity during single-line operation when a combination of high- and slow-speed relays is used.

Direct-current Systems. Direct-current power systems have been almost wholly replaced by the a-c system and at present are confined to the supply of power to transportation systems and to a few isolated sections of metropolitan areas. It must be remembered, however, that there are still many applications where alternating current cannot be used, or, at least, where no means have been devised for using it instead of direct current. On such installations, therefore, it is important to have dependable relay protection for times when trouble occurs.

The inherent characteristics of d-c systems prevent the use of many of the means employed on a-c systems to obtain selectivity. For example, differential schemes and graded time settings cannot readily be used. Selectivity in most cases is obtained by graded current settings, but satisfactory operation cannot always be secured because of the wide variation in short-circuit currents for various fault locations. In many cases the protection for the circuit is built into the circuit breaker in the form of a series trip coil. This may be supplemented by a time-delay feature, usually a dashpot attached to the tripping mechanism, to increase the margin of selectivity.

The protection of d-c apparatus may be classed as protection against overcurrent, reverse current, and overtemperature. Usually such protection on motors and generators is secured by means of circuit breakers equipped with the necessary overcurrent or reverse-current coils or by fuses of suitable capacity. In many installations, however, it is found necessary to use a protective relay in order to secure the needed accuracy and reliability of protection. Such relays may be of the d'Arsonval movement type, the solenoid type, or the thermal type.

Flashover of commutators on rotary converters has been the source of much trouble in the past, and, although improvement in design has reduced the probability of such occurrences, the difficulty still exists. The machine should be instantly disconnected upon the occurrence of a flashover. As flashovers usually occur to the frame of the machine, a relay whose coil is connected directly between the frame of the machine and ground can give high-speed protection.

16. GENERATOR-VOLTAGE REGULATORS

The application of generator-voltage regulators depends on several factors entirely independent of the size and design of the regulator itself. It is not only necessary that the regulator be properly designed, but it is also essential that the exciters, generators, and prime mover have characteristics that will harmonize with each other and assist in keeping the voltage at the desired value under rapidly changing load conditions.

Only sensitivity is normally specified by the manufacturers in connection with the performance of generator-voltage regulators. Sensitivity represents the band or zone of voltage, expressed in terms of percentage of the normal value of regulated voltage, within which the regulator will normally hold the voltage under steady-load conditions. This does not mean that the regulated voltage will not vary outside the sensitivity zone. It does mean, however, that, when the regulated voltage varies more than the percentage sensitivity from the regulator setting, because of sudden changes in load or other conditions, the regulator will immediately apply corrective action to restore the voltage to the sensitivity zone.

Regulator sensitivity must not be confused with overall regulation, which involves not only regulator sensitivity but also the time constants of the machines and the character and magnitude of load changes. The magnitude and rate of load change determine how far the voltage will vary outside the regulator-sensitivity zone, and the time constant of the machine chiefly determines the time required to restore the voltage to the sensitivity zone of the regulator.

Although on many special applications speed of response is of considerable importance, it is well to keep in mind that the time constant of the machines have a large effect in determining the speed of response of the regulation system to changes in voltage. With the increased use of high-speed circuit breakers the necessity for quick response of voltage regulators in maintaining system stability has been lessened to a considerable extent.

After the general type of regulator has been selected, it is necessary to determine the specific size required. This can usually be determined from the manufacturer's application tables if the a-c generator kva and the average speed of the a-c machine and the exciter are known. The limiting characteristic which governs the size of regulators and of associated apparatus, such as a motor-operated rheostat, is normally the current-carrying and watt-dissipating capacity of the resistance element in the exciter or generator field circuit.

VIBRATING REGULATOR. The vibrating regulator can be used economically in central-station, municipal, and industrial generating plants of moderate capacity, office- or apartment-building power plants, d-c power plants of small capacity, and synchronous condenser stations of moderate capacity.

An outstanding advantage of the vibrating regulator is its suitability to control a number of exciters simultaneously. Although widely used, this method of control imposes certain restrictions that are often ignored. Obviously, a single regulator cannot be adjusted to suit the peculiarities of several generating units at the same time if these units are of dissimilar characteristics, such as might be caused by a difference in manufacture, age, or types of machine.

The extended-broad-range regulator is, in a sense, a combination of a vibrating regulator and a rheostatic regulator. The vibrating element, however, is the more rapid and always makes instant correction of excitation upon a change of load. The operation of the rheostat in the lower ranges of excitation merely follows the action of the vibrating element in order to maintain the exciter voltage within the limits of rapid response. An application of extended-broad-range regulators requires that synchronous machines have individual excitors and regulators.

RHEOSTATIC REGULATOR. The direct-acting rheostatic regulator is designed for the automatic voltage control of small and medium-sized a-c and d-c machines. These types are suitable for the automatic voltage control of constant-speed one-, two-, or three-phase a-c generators with individual self-excited exciters. The exciter must be designed for shunt field control and self-excited operation, with its minimum operating voltage not less than 30 per cent of its rated voltage. Each regulator is designed for and limited to the control of one exciter.

These regulators are also suitable for the automatic voltage control of constant-speed d-c generators, designed for shunt field control, and either self-excited or excited from a separate constant source. They may also be applied to d-c generators separately excited by individual variable-voltage exciters, in which case the regulator operates in the exciter shunt field circuit.

The indirectly acting exciter rheostatic generator-voltage regulator is applicable to the automatic voltage control of medium- and large-sized a-c generators, such as are used in central stations and in large municipal and industrial generating stations. Synchronous condensers and synchronous motors are also controlled by this regulator.

This regulator is particularly applicable on systems requiring quick-response excitation of a-c machines, as it has a high-speed operation, the regulator operating within 3 cycles after a voltage change. The regulator can be readily adjusted to match the individual characteristics of various kinds of a-c machines, such as slow-speed waterwheel-driven a-c generators, high-speed turbine-driven a-c generators, synchronous converters, and synchronous motors. This adjustment is normally built in as a standard part of the regulator, and it is not necessary to modify or change regulator parts in order to make such adjustments. These regulating equipments can be obtained in ample capacity to handle the largest generating units.

NETWORK-TYPE REGULATOR. The network-type control element, used with a static or a rotary amplifier, and the electronic voltage regulators provide quicker response and somewhat better sensitivity than the other types of regulators. It must be remembered, however, that these regulators may be more expensive than other types and should be used only where increased sensitivity and speed of response are required.

The division of kilowatt load among paralleled a-c generators is dependent upon the power supply to each generator and is controlled by the governor of its prime mover. Thus the division of the kilowatt load is practically independent of the excitation. Changes in the field excitation of paralleled a-c generators, however, affect the reactive kva or wattless component of the output. Since the voltage regulator acts directly on the field excitation, it will be seen that, although the division of kilowatt load between generators is not influenced by the voltage regulator, the division of wattless current is directly affected by the operation of the regulator. To secure stability in parallel operation it is, therefore, necessary to give the regulated voltage of each generator a droop with an increase in the wattless component of the generator current. This is accomplished by the use of either a current-energized compensator or an equalizer-reactor.

These devices are normally adjustable from zero to approximately 12 per cent. Compensation of from 3 to 6 per cent at full load and zero power factor is ample for the usual application. It is more important to set the percentage compensation relatively high and thus provide for stable operation of generators in parallel, with resultant good equalizing or reactive kva, than it is to have minimum droop in the regulated voltage when the reactive load is increasing. Under the usual operating conditions, with the a-c generators carrying the customary load, the operator maintains the normal value of regulated voltage by means of the voltage-adjusting rheostat.

17. LIGHTNING ARRESTERS

The purpose of lightning arresters is to assure, as far as possible, the maintenance of an uninterrupted supply of electric power to users regardless of the disturbing effects of electrical storms.

LIGHTNING AND ITS EFFECTS. Lightning is an electrical discharge in air, one terminal of which is a cloud. By certain atmospheric processes that occur during thunderstorms, charges are accumulated in clouds or portions of clouds, and equal charges of opposite polarity are formed in the earth beneath. As these charges increase, the voltage gradient in the air adjacent to the charge center in the cloud increases. When the gradient exceeds the insulation strength of the air, a low current streamer starts downward from the cloud and continues to grow. When the streamer makes contact with the earth, it is like closing a switch between the two charges of opposite polarity, one in the earth, the other in the streamer channel and in the cloud. Consequently large impulse currents flow. The crest magnitudes of these currents vary between about 1000 and more than 160,000 amp. Their duration is measured in microseconds. The currents and voltages are preponderantly of a given polarity; i.e., in any one stroke, the discharge is usually unidirectional and does not reverse in polarity.

Lightning discharges produce voltages in structures such as masts, trees, buildings, and transmission lines. If lightning strikes a structure, the stroke current flows in the structure and produces a voltage. In an object of high impedance, such as a tree or a masonry structure, high voltage is developed. Damage such as physical rupture or fire usually occurs. In conductors of low impedance the voltages between electrically adjacent points are low, and damage is avoided. On transmission lines high voltages result from the flow of lightning current into the distributed inductance and capacitance of the line. The current and voltage are related by a quantity called surge impedance, Z, in ohms. In its simplest terms Z equals $\sqrt{L/C}$, where L and C are the inductance and capacity per unit length of the lines. The voltage to ground, assuming no flashover, is $E = IZ$.

The other way in which lightning produces voltage in a line is by induction, principally electrostatic, from lightning that strikes near the line without actually striking it. Such voltages are limited in magnitude to less than 1000 kv and usually to less than 400 or 500 kv. They are less important than the voltages produced by direct stroke on the line and are usually not hazardous on high-voltage systems insulated for operation at more than 69,000 volts. Lightning protection is designed to cope with the most severe conditions.

METHODS OF LIGHTNING PROTECTION. Several methods are used for lightning protection. For objects such as oil tanks, buildings, munition dumps, and valuable trees, wires, rods, masts, or overhead grounded wires are used to catch the lightning stroke and to take the lightning current to ground without giving rise to high voltage in the conductors or to grounded objects adjacent to the lightning conductor.

LIGHTNING ARRESTERS. On electrical systems the most widely used means of protection is the lightning arrester. Overhead ground wires, masts, and rods are used to a considerable extent to protect transmission lines and substation structures against direct strokes, and for this purpose they are very effective if properly installed. This is called shielding. Even with good direct-stroke protection, however, voltages can appear in the line conductors by coupling, and for the protection of electrical apparatus the lightning arrester is the best and most widely used device.

Simple spark gaps, such as rod, needle, horn, or sphere gaps, are not lightning arresters, because, although they can be made to provide a certain degree of protection against overvoltage, they are unable to interrupt the power current path. It is well known that, even though a high voltage may be required to break down an insulating wall of air, or any gas for that matter, it takes very little voltage to keep an arc going after it has once been started. Thus the power-follow currents of a plain gap must be cleared by the operation of a circuit breaker or fuse somewhere else on the system.

Principle of Operation. The lightning arrester operates on voltage and then copes with the current which the disturbance discharges through the arrester. It should discharge these currents without permitting high voltage across its terminals while the current is flowing. These voltages are of the nature of impulses. Lightning voltages on power lines have been recorded by means of the electronic oscillograph, and considerable data have been accumulated on the nature of lightning currents, both in direct strokes and through lightning arresters, by means of various types of specially developed recording instruments. The current and voltage magnitudes and wave shapes vary over wide ranges, and there appears to be no typical lightning current or voltage. Current peaks in strokes of from less than 1000 to more than 160,000 amp have been measured. The maximum voltage on lines exceeds the flashover of the highest insulation and will reach several million volts. The current or voltage peaks may be reached in from 1 to 10 or more μsec, so that the rate of voltage rise may be from a few thousand to several million volts per microsecond. After the crest of the discharge has been reached, the current and voltage decay relatively slowly on a more or less exponential curve that drops to half the crest in from 10 to 100 μsec. The total duration of the voltage to its complete exhaustion may be from 100 to many thousand microseconds.

Some strokes have high current peaks and not much of a drawn-out low-current tail. Others have low peaks and long tails, and some have both. Also, some strokes have a number of peaks; as many as 17 have been recorded. These are called multiple or repetitive strokes.

It is evident that lightning arresters, if they are to protect other apparatus, must become conducting very quickly. Also it is obvious that there must be a margin between the impulse voltage that the apparatus will withstand and the voltage permitted by the arrester.

The strength of insulation against impulse voltages differs from its 60-cycle strength. The impulse strength is generally higher as a result of the short time for which impulse voltages are applied. Furthermore, the insulation strength in volts will vary with the shape of the applied voltage. Generally speaking, the voltage required to flashover or break down insulation becomes higher the shorter the time of application of voltage or the steeper the rate of rise of the applied voltage.

CHARACTERISTICS OF ARRESTERS. Four principal characteristics must be known about a lightning arrester if it is to be used properly.

1. Its voltage rating, expressed in terms of power frequency voltage, to enable the user to choose a lightning arrester suitable for the voltage of his system.

2. Its power frequency current-interrupting rating, expressed in terms of system short-circuit current, to enable the user to choose the arrester that will satisfactorily interrupt the power-follow current and thus restore itself to an insulator.

3. Its surge-current capacity, expressed in amperes crest of surge current, to give the user information on the lightning current that the arrester will handle without being damaged.

4. Its protective or impulse characteristics, to enable the user to correlate the arrester with the equipment he intends to protect.

The voltage rating, together with the type designation of the arrester, indirectly indicates the impulse of protective characteristics of the arrester. The standard ratings are listed in manufacturer's catalogs, and with the ratings are stated the protective characteristics in terms of impulse sparkover and discharge voltages. When the ratings and the types are known, the degree of correlation between the arrester and the apparatus to be protected can be estimated. Thus, to get the most efficient and economical results from a lightning arrester, the impulse-withstand strength of the apparatus and the system voltage conditions should be known, and the meaning of arrester voltage rating should be understood.

If the system voltage that is impressed across the arrester terminals exceeds its rating, the arrester is discharged by a surge. Two conditions exist that make it difficult, if not impossible, for the arrester to restore itself to an insulator. The first condition is that the power frequency voltage against which the arrester tries to clear is higher than the arrester is designed for, which makes it hard for the series gaps to recover their insulation strength when the power-follow current passes through zero. The other condition is that the magnitude of power current will be considerably greater, because of the increased voltage, than it would be if the voltage were within the lightning-arrester rating. For these two reasons, it is quite likely that the power-follow current will not be interrupted, but will continue to flow. This will produce heat in the arrester, which will probably become damaged beyond repair. It must be remembered that a lightning arrester is a short-time operating device. Several cycles are a long time for a lightning arrester. Furthermore, such a continuous flow of power-follow current will cause a system outage, which is just what the arrester is expected to prevent.

Since the lightning arrester is sensitive to line-to-ground voltage, it is necessary to give consideration to such voltages on systems rather than to line-to-line voltages. To avoid all risk of damage to arresters, the rating of the arresters chosen for a given system should be sufficiently high so that it will not be exceeded under normal or abnormal conditions.

The foregoing are the general aspects of the application of arresters, from the standpoint of system voltages that may appear across the arrester terminals. The manufacturer's catalog usually gives some general suggestions for the ratings of arresters that are suitable for standard system voltages and neutral grounding. These arresters are satisfactory for average use, but sometimes unusual considerations or conditions arise that require special treatment. This is a subject of considerable scope and is not covered herein. It is the intention here merely to call attention to the essential factors and to emphasize the importance of the voltage to ground.

There are also some application factors that should be borne in mind from the standpoint of the protection to apparatus. It is naturally desirable to limit the voltages as much as possible at the terminals of the protected equipment. Certain conditions may permit higher voltages at the equipment than at the arrester, and these will now be briefly discussed.

Arresters generally have some lengths of line between them and ground. These lines should be kept as short as possible because wires have inductance, in which rapidly changing currents produce voltage drops. Such voltage drops in wires that are in series with arresters add to the voltage across the arrester terminals, causing the voltage across the transformer insulation to be higher than the arrester voltage by the inductive drop.

Another factor external to the arrester is ground resistance. The electrodes buried or driven into the ground to make electrical contact with it have some resistance to earth, depending on the extent of the electrodes and the characteristics of the soil. Ground resistances may differ from a small fraction of an ohm to hundreds of ohms. The discharge current that flows through the arrester produces an *IR* drop in this ground resistance, which adds to the arrester voltage and the lead drop, and further increases the voltage impressed on the protective equipment. The effect of ground resistance can be eliminated by connecting the ground terminal of the arrester directly to the tank or frame of the protected apparatus.

Ground wires and arresters are economically competitive, and in general the choice is determined by relative cost. For this reason the arresters must be inexpensive. The lightning arresters used on transmission lines are usually of the expulsion type.

Motors and generators require protective equipment that is different from and more elaborate than that required by static apparatus, such as transformers or circuit breakers.

The lightning arresters for air-insulated dry-type transformers must be of the same type as those used for the protection of rotating machines, because the impulse-withstand strength of these transformers is of the same order as that of rotating machines.

On arresters for use on d-c circuits the problem of interrupting the power-follow current is much more difficult than on a-c circuits because the current does not go through zero. Direct-current arresters therefore are generally of different types from those used on a-c systems. The principal use of d-c arresters is on d-c street cars, trolley buses, and railway systems, operating principally at 500 to 550 volts direct current, although some operate at considerably higher voltages. There are three types in general use: the magnetic blow-out type, the resistance block type, and the capacitor type. The electrolytic types were in use to a large extent until the development of suitable capacitor types but have become almost obsolete.

18. REACTORS

(See also Section 6)

Some of the principal applications of current-limiting reactors are as follows:

GENERATOR REACTORS. When several generators are in parallel in one station, it is often necessary to limit the fault current they can feed into the bus by connecting current-limiting reactors in series with the generator leads. When this is done, there is a voltage drop through the reactor, caused by normal current flow. Consequently it is necessary to operate the generator at a higher voltage than name plate rating if normal voltage is to be maintained on the load bus. Ten per cent above normal voltage is usually the maximum at which a generator can be operated continuously; this limits the value of reactance that can be connected in series with the leads.

BUS SECTIONALIZING AND SYNCHRONIZING BUS REACTORS. In order to reduce the fault current that can flow from one bus section into another, sectionalizing reactors are connected between them. A particular and much-used form of bus sectionalizing is the synchronizing or star bus. In this form each generator or power-source bus supplies a group of feeders directly, while being connected through a reactor to a common bus, called a synchronizing or star bus. Each bus is then connected to the other through the synchronizing bus, but the fault current that can flow from one to the other is limited by the synchronizing bus reactors. The extent that fault current can be limited and the maximum value of reactance that can be used in a current-limiting reactor depend on the voltage drop that can be tolerated under full-load conditions.

FEEDER REACTORS. Current-limiting reactors are used to reduce the fault current on feeder circuits, thereby making it possible to use breakers of lower kva rupturing ability than would otherwise be necessary. Cable feeder circuits are sometimes connected in series with reactors in order to limit the explosive and fusing effect of otherwise high short-circuit currents.

NEUTRAL GROUNDING REACTORS. Reactors are, under some conditions, connected between the neutrals of generators or star-connected transformer banks and ground for the purpose of limiting the flow of ground-fault current. Limitations to this type of application, which comes under the subject System Grounding, are covered in detail in Section 14, Article 93. (1) Reactors in this category are not required to carry load current and are therefore only short-time rated per AIEE Standard No. 32.

MISCELLANEOUS USES OF REACTORS. Wherever it is desired to limit the flow of alternating current, as in series with low reactance auto-transformers and induction regulators, reactors can be used. Arc furnace circuits sometimes require reactors in series with the transformer bank to stabilize the influence on the arc and to limit the flow of current when the electrodes become short-circuited within the furnace.

19. CAPACITORS

(See also Section 6)

Abnormal capacitor currents due to excess fundamental or harmonic voltage may occasionally be encountered. Blowing of fuses and/or overtemperature operation may be the first noticeable indication of such an overload condition.

The inrush current drawn by a capacitor is larger than is encountered with other apparatus, because of its impedance frequency characteristics. Momentary ratings of breakers should be well above the possible inrush current. Contact or switch maintenance is affected especially in repetitive switching. The inrush current is substantially reduced in automatically switched capacitors by the use of discharge devices, which discharge the capacitor in a predetermined time after it is de-energized. Fuse ratings must be liberal, never less than 165 per cent of the normal current. The inrush current depends upon the size of the capacitor, the line voltage, and the short-circuit current the line is capable of producing without the capacitor. In parallel banks, the inrush is governed by the impedance between banks and can be easily reduced to, or below, the value of line inrush for the same size of bank, by introducing a very small amount of inductance between banks.

20. CONTROL APPLICATIONS

An electric controller is intimately connected with the motor and should be considered with it. The functions of control in connection with motors are starting, changing the direction of rotation, limiting the load, regulating the speed, and stopping. All or any combination of these functions may be either manual or automatic. The starting may be manual, as determined by the operator, or it may be automatic at a fixed point in the cycle of operation when certain predetermined conditions occur. During starting it is the function of the controller to limit the current and in some instances the torque. The speed of a motor may be regulated by means of armature control, field control, voltage control, or combinations depending upon the type of motor. Motors may be stopped by friction or magnetic brakes, used alone or in combination with dynamic braking or plugging. The motor may be stopped as determined by the operator, or it may be stopped automatically at a fixed point in the cycle, when predetermined conditions occur, on overload or on voltage or field failure.

SQUIRREL-CAGE MOTORS. Full-voltage starters are referred to as across-the-line starters. When a squirrel-cage motor is connected directly to the power lines, the inrush of current to the motor is approximately six times the normal full-load current. Power companies sometimes object to such heavy line currents because of the temporary reduction in line voltage, with consequent "blinking" of lights connected to the same line. A marked improvement in power lines, generating equipment, and motors has, however, eliminated many objections to across-the-line starters for squirrel-cage motors, so that they are one of the most active items of industrial control for motors.

The requirements of manually operated across-the-line starters are such that the design is a self-contained unit, except on high-voltage installations, where circuit breakers are used. Low-voltage designs require additional fusible safety switches or circuit breakers for disconnecting purposes and for protecting the branch circuit wiring and the overload heaters in the starters when short circuits occur. The time required for the overload device to operate permits momentary overload current without tripping the starter. This feature is essential in order to prevent tripping by the unavoidable current peaks that occur when a motor is starting.

Figure 4 shows a common form of control scheme for a reversing across-the-line starter. A non-reversing starter consists of a contactor, a push-button switch, and an overload relay. The equipment for a reversing starter is similar except that an additional three-pole motor contactor is required for reversed operation, and a mechanical interlock is used between the two contactors to prevent both from closing simultaneously to cause a short circuit between phases. Electrical interlocking is also usually provided.

Reduced-voltage starters are used on squirrel-cage motors to reduce the inrush-current peaks to satisfactory values. Reduced voltage is obtained either by auto-transformers with taps for various voltages or by resistors or reactors in series with the motor windings.

The starting current of a squirrel-cage motor varies almost directly with the voltage. The starting torque, however, varies as the square of the applied voltage. Any reduced-voltage starter, therefore, must consider both the torque required to start the motor and the current taken from the line.

The **auto-transformer type of starter** consists of either two auto-transformers connected in open delta or a three-coil auto-transformer with one coil connected in each phase. For the reduced-voltage starting conditions the motor is connected to taps on the transformers. Taps for 50, 65, and 60 per cent voltage are commonly used. When the motor has accelerated sufficiently, the motor is disconnected from the taps of the transformers and connected directly to the power line.

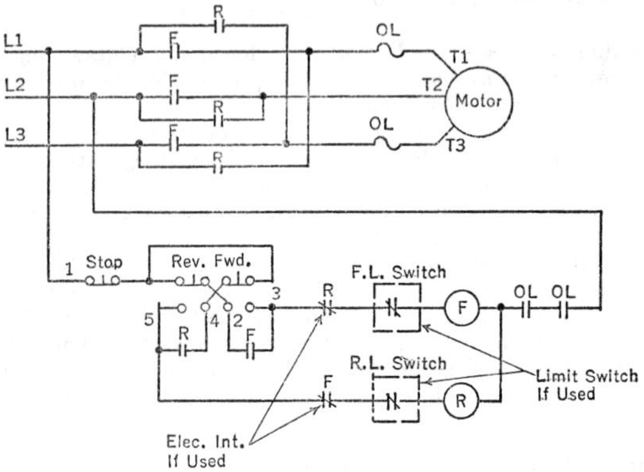

FIG. 4. Control Scheme for Across-the-line Reversing Starter for Induction Motor

A disadvantage of this type of reduced-voltage starter is the complete loss of power on the motor terminals when the motor is disconnected from the transformer taps and transferred to the power line. The open circuit during the transition results in a second current peak when the motor is connected to the line, and this second peak may be objectionable.

The objectionable open-circuit transfer from reduced- to full-voltage connections on auto-transformers can be eliminated by circuit arrangements that maintain power on the motor during the transition period. Additional apparatus is required, and except for special cases the need is not considered of enough importance to justify the additional expense.

With the primary-resistance type of starter, the motor is connected to the line through a resistor. When the motor is started, all the current to the motor passes through the starting resistor, which is designed to reduce the motor terminal voltage to the desired value. The starting resistor is finally short-circuited to connect the motor directly to the line.

As the motor accelerates, the current decreases; the voltage drop on the resistor decreases; and the motor terminal voltage increases. This automatic increase in voltage at the motor terminals is a great aid in accelerating the motor and is a decided advantage not provided by the auto-transformer starter. The resistance type of starter does not have any open-circuit period during transfer from reduced to full voltage and consequently provides smoother acceleration than the auto-transformer type of starter.

The circuits for reactance-type starters are similar to those for the resistance type. The reactance type has the same advantages and disadvantages as the resistance type. It is somewhat more expensive but has the advantage of smaller size, especially for large motors, for which it is used most frequently.

Squirrel-cage motors are not usually considered when speed control is a requirement of the installation. However, they can be operated at different speeds by reconnecting the windings to form a different number of poles on the primary winding. Sufficient leads are brought out of the motor to make the proper connections external to the motor to form the different pole combinations.

A squirrel-cage motor may be stopped by disconnecting it from the power line and then applying direct current to one phase to provide dynamic braking action. If no separate source of d-c power is available, it can be provided from the a-c system by a rectifier. After the motor has stopped, the direct current is disconnected by a timing relay to prevent overheating of the windings.

Squirrel-cage motors may be stopped in a very short time by connecting them to the power system with two phases reversed to provide a torque in the opposite direction of rotation. If actual reversal of operation is not desired, the operator must disconnect all power when the motor comes to rest, or an automatic device, such as a zero-speed switch, must be used to open the contactor just before reversal occurs. Plugging resistors may be used in the circuit to control the motor torque for the plugging condition.

SYNCHRONOUS MOTORS. Except for the d-c field circuit, a synchronous motor is very similar to a squirrel-cage motor. The primary or stator circuits of the two are comparable; hence they require practically the same primary control equipment for either full-voltage or reduced-voltage starting.

Synchronous motors run at constant speed, depending upon the frequency of the power system and the number of poles in the d-c field.

For the primary circuits any of the starting schemes that are used for squirrel-cage induction motors can be applied to synchronous motors, and the control circuits are essentially similar; i.e., the control can be semimagnetic or magnetic, and the motor can be started on full voltage or reduced voltage, as the load requirements dictate. Reduced-voltage starters may be of the auto-transformer, series resistor, or series reactor types. When the need for extremely smooth acceleration warrants the added cost, a d-c controlled "saturable" reactor is sometimes used.

When power is applied to the stator, the field circuit is not energized but is connected to a starting resistor through the closed contact of a field control relay. This resistor improves the starting torque of the motor and provides a discharge circuit to prevent excessive induced voltages in the field circuits during acceleration or at times when the field supply circuit is opened.

As the rotor approaches synchronous speed, the field circuit must be disconnected from the starting resistor and connected directly to the d-c power supply. The torque that is provided when the field is energized is known as the "pull-in" torque and serves to lock the rotor electrically to the rotating field of the stator. The motor then operates at synchronous speed. The field circuit is never completely opened during the transfer period.

Definite time relays are often used to determine when the d-c excitation should be applied to the motor field. With this scheme it is necessary only to set the timing relays for a period long enough to permit the motor to accelerate to a speed at which the pull-in torque will be ample to pull the rotor into synchronism with the rotating stator field.

Field frequency relays may also be used to control the closing of the field contactor. When the motor starts, the rotor is at relatively low speed, and the rotating field of the stator is at synchronous speed. The difference in speed is a maximum; hence the frequency and value of the induced voltage and current in the field circuit are at a maximum. As the rotor speed increases, the induced voltage and current in the field winding decrease until, at a speed near synchronism, where "pull-in" should occur, they are low and may be used to control relays, which in turn cause the field contactor to close.

As the rotor approaches synchronous speed, field excitation may also be applied by relays responsive to current and frequency conditions in the stator circuit.

If a synchronous motor is sufficiently overloaded, or if the line voltage gets too low, the rotor is dragged out of synchronism with the rotating field of the stator, and the motor stalls. The control must then operate to disconnect the power from the primary of the stator until another start can be made. It sometimes happens that the overload or low-voltage condition exists for only a short time. Under such conditions the control is often arranged to remove field excitation, to reconnect the field to the starting or discharge resistor, or to permit the motor to reaccelerate and resynchronize without any interruption in the motor operation. This scheme requires that an overload sufficient to cause a pull-out must disappear or be removed immediately, because the motor cannot reaccelerate a load that causes a pull-out.

Power systems and synchronous motors are so constructed today that, as a rule, motors can be started and field excitation applied without consideration of line disturbances or pull-in torque conditions. On some installations, however, it is necessary to apply field excitation at a certain definite position of the d-c field poles with respect to the rotating stator poles of the primary in order to obtain desirable pull-in conditions. This practice is known as angle switching.

If d-c excitation is applied when the field poles are in certain definite angular positions with respect to the rotating stator poles, the rotor "locks" into step with the rotating field of the primary with minimum line disturbance and maximum pull-in torque. If the d-c poles are behind the "lock-in" position, the rotor is jerked forward when d-c excitation is applied, and certain line disturbances occur. If the d-c rotor poles happen to be ahead of the lock-in position, the rotor "falls back," with a consequent momentary reversal of torque on the motor shaft.

An a-c line ammeter and a d-c field ammeter are usually provided with the control for a synchronous motor. Power factor meters with the necessary individual current and potential transformers, as well as watthour meters and wattmeters, are sometimes furnished.

Synchronous-motor controllers provide undervoltage protection and overload protection for the motor windings. For voltages above 600, time-delay undervoltage protection frequently is provided and the contactors or circuit breakers are of the latched-in type or have other equally effective means. The time delay is from 1/2 to 3 sec, although it does not have to be adjustable. Overload protection should be provided in two phases.

The starting current for reduced-voltage starting should not exceed 400 per cent of the motor full-load current, and for full-voltage starting should not normally exceed 600 per cent. Certain applications require quick stopping of the synchronous motor under emergency conditions. For such stopping either a magnetic clutch and magnetic brake or dynamic braking may be used. In the magnetic type of stopping, the magnetic clutch disconnects the motor shaft from the load, and the magnetic brake stops the motion of the machinery. Dynamic braking consists of opening the motor contactor, thus disconnecting the motor from the line and connecting the motor windings to Y-connected resistors. The motor then acts as a generator and is quickly stopped. Stopping can usually be effected as quickly with dynamic braking as with the magnetic clutch and brake.

WOUND-ROTOR MOTORS. The primary connections to wound-rotor motors and overload protection for them are similar to across-the-line starter connections to a squirrel-cage motor. The primary circuit can be closed by a manual device, but the overload protection will require an overload relay and a magnetic contactor. A magnetic type of across-the-line starter, controlled by a manually operated secondary controller, is therefore often used. The manual controller may be of either the faceplate or the drum type.

The three-phase secondary resistor of the faceplate type is mounted behind the faceplate within the enclosure. Each phase of the resistor is connected to a group of contacts on the faceplate. The primary circuit is closed by the full-voltage magnetic starter controlled by auxiliary contacts on the faceplate. These contacts are closed at the first motion of the three-phase arm. The arm makes the auxiliary contact only on the first position; consequently, if power is interrupted, restarting is prevented without returning to the zero position.

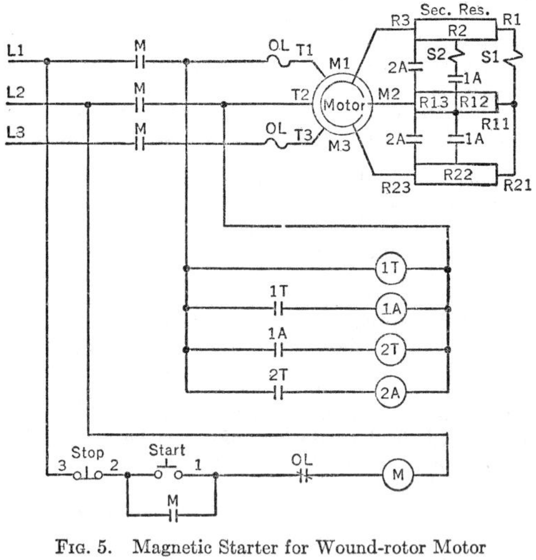

FIG. 5. Magnetic Starter for Wound-rotor Motor

The resistor may be designed for either continuous duty or starting duty only. Faceplate controllers are not normally used for high ratings.

Drum controllers can be made more rugged than faceplate controllers. They also permit better design for installation requirements and can be entirely enclosed more satisfactorily. Since the space requirements are less, they are more easily mounted in available spaces. They can be designed so that the handle can be moved in either direction from an "off" position, and the circuits can be arranged to reverse primary connections to obtain either forward or reverse motor operation. Mechanical arrangement and insulation problems do not readily permit primary connections for reversing service on a faceplate controller. Since low-voltage primary circuits can be used on a drum controller, a primary line starter is not absolutely necessary. The line starter is required, however, with the drum controller if overload and low-voltage protection are needed.

Neither faceplate nor drum controllers are satisfactory for high-voltage service. For such service some suitable high-voltage primary switch must be used. This unit can be manually operated or magnetically controlled from the secondary faceplate or drum.

The connections for a magnetic starter for a wound-rotor motor are shown in Fig. 5. To short-circuit the secondary resistor, the first contactor to close is $1A$, that next to the star connection. Two-pole contactors can then be used, and the resistor short-circuited across phases. With the scheme illustrated in Fig. 5 two timing relays $1T$ and $2T$ and two accelerating contactors $1A$ and $2A$ are used. When the line contactor M closes, the first timing relay $1T$ starts and eventually closes the first accelerating contactor $1A$. After this contactor closes, the second timing relay $2T$ starts and in proper time will close the second accelerating contactor $2A$.

Current-limit acceleration could be furnished instead of time acceleration by using two current relays with coils indicated by $S1$ and $S2$ instead of the two timing relays.

Wound-rotor motors are sometimes used on applications that require rather fine speed adjustments, which, of course, necessitate a large number of operating points. Furthermore, the rating of the motor is sometimes large, and for such cases the fixed number of points on a master switch and the grid resistor units are neither satisfactory nor practical. On such applications liquid rheostats are generally used for secondary control.

Dynamic braking and plugging can be applied to wound-rotor motors in the same manner as to squirrel-cage motors. Since the secondary resistor can be used as a plugging resistor, a separate plugging resistor for the primary circuit need not be provided unless the design will not permit one resistor to serve both purposes.

DIRECT-CURRENT MOTORS. Principles of Speed Control. Shunt-wound motors operate at almost constant speed at all loads within their capacity. Their speed may be increased, within limits, by inserting resistance in series with the shunt field. The permissible increase in speed of standard motors varies from 10 to 100 per cent, depending upon their size and rated speed.

Compound-wound motors are provided with a series field in addition to the shunt field. This provides additional field strength when the loads are above normal, as when starting and during acceleration. It has the disadvantage of causing a greater drop in speed with increasing load than occurs with a shunt motor. The speed of compound-wound may be increased in the same manner as that of shunt motors.

Adjustable-speed motors usually employ both shunt and series field windings and are generally provided with commutating poles to insure good commutation at the higher speeds. The series field winding is provided to stabilize the speed and has fewer turns than are used on a compound-wound motor and consequently has very little effect on the change of speed with change of load. The speed may be increased to as much as 400 per cent of the base speed by inserting resistance in series with the shunt field.

Series motors have only a series field, through which the motor-load current flows. At no load the field is therefore very weak, and the motor may attain runaway speeds. The motor must, therefore, always operate with some load or precaution taken to prevent runaway speeds.

The series motor is well adapted to crane installations, where, when hoisting heavy loads, it develops ample torque and operates at reduced speed. When hoisting light loads, it operates at higher speeds and therefore handles more material. To prevent runaway speeds when lowering loads, the motor connections are made so that the series field, in combination with resistors, is connected to the power supply like a shunt field to give the motor characteristics similar to those of a shunt motor. Speed control is obtained by using resistors in the armature and series field circuit.

Methods of Speed Control. In order to limit the current peaks in a d-c motor to safe values, most motors are started with resistance in the armature circuit. This resistance is reduced in value, by suitable amounts of one or more values, called points, until it is entirely removed from the circuit. The motor then runs at its rated full-field speed. The full-field speed is that for which the motor was designed to run with its rated voltage on the armature and maximum field strength.

When it is necessary to operate a d-c motor at less than its full-field speed, the voltage on the armature circuit must be reduced by changing the supply voltage or by using resistance in the armature circuit. The first method requires an individual generator for the power supply to the motor; otherwise the change in supply voltage will affect other motors. The second method involves a rather large waste of power because the reduction in voltage must be absorbed in the resistor. The loss in the resistor changes as the square of the armature current and as the value of the resistance. Since armature currents are relatively high, the losses in the resistor must be considered.

Control at speeds higher than full-field speed on shunt or compound motors is usually obtained by a resistor or rheostat connected in series with the shunt field. The shunt field currents are comparatively small; hence the power lost in the rheostat is relatively negligible. The rheostat itself is compact and may be mounted in a convenient location. To raise the speed the resistance of the rheostat is increased, thus decreasing the shunt field current.

If extremely low speeds are attempted by using only series resistors, the armature voltage falls so low that the motor becomes unstable as load changes occur. If additional resistors are used in parallel with the armature, the load changes have less influence on the armature at low speeds, and the motor operation is much better. Various combinations of series and parallel resistors are frequently used, therefore, to obtain different operating speeds of relatively low value.

Shunt fields sometimes will overheat if energized while the motor is not running. Therefore it may be necessary to disconnect the field circuit or to insert resistance to reduce the field current to a safe value while the motor is at rest.

A d-c motor exerts maximum torque when its field strength is at maximum value. Since motors may be stopped while running at maximum speed (minimum field strength), it is often necessary to short-circuit the field resistor to obtain maximum torque for starting. In like manner, the maximum dynamic braking effect is obtained with maximum field strength. The relays used to provide full-field strength for starting and dynamic braking duty are known as full-field relays.

Controllers and Relays. Controllers are sometimes built to permit automatic accelerating to a previously determined operating speed. These designs are called preset speed controllers. The preset speed usually represents a reduced field strength as determined by a rheostat setting.

On preset speed controllers the acceleration in one step from full field (maximum field strength) to the preset speed point (weakened field strength) would cause excessive armature current peaks unless a field accelerating relay was used. This relay uses a series coil which, when the current becomes excessive, will short-circuit the field resistor to strengthen the motor field and provide more torque to accelerate the load without higher armature current. As the acceleration progresses and the motor current decreases, the relay will reinsert the field resistor to increase the speed further. A number of operations of the relay are often necessary to change the motor speed from full-field value to preset value. This relay is called a field accelerating relay.

Field decelerating relays are used to prevent motors from regenerating when an attempt is made to reduce the speed too rapidly, and a load of high inertia drives the motor as a generator. Under such conditions the regenerated voltage, unless restricted in value, might rise to a value sufficient to cause flashing at commutator and brushes. When the regenerated voltage and reversed current attain dangerous values, the decelerating relay inserts resistance in the field circuit and decreases the generating ability of the motor.

If the shunt field should accidentally open while the armature circuit remains closed, the motor is likely to attain a runaway speed and be damaged. Field-loss or field-failure relays are used to prevent such damage. The coil of the relay is connected in series with the shunt field circuit. Its contacts are connected in the control circuit like a stop button. Then, if the field circuit opens, the relay acts to stop the motor.

Manual methods for controlling d-c motors use either faceplate designs or drum controllers. Faceplate starters can be made for various services such as starting duty only, starting duty with speed regulation by armature resistance, starting duty with speed regulation by field resistance, and starting duty with speed regulation by both armature and field resistance. Drum controllers and resistors for d-c use are similar to those for wound-rotor motors. Magnetic starters and controllers are also used for d-c motors.

Acceleration. The acceleration of a motor is generally accomplished by the use of resistors in the motor circuit. The number of speed or resistor points used to accelerate a motor will depend entirely upon the current peak permitted and the load to be accelerated. The resistors should be so proportioned that the motor will accelerate from zero speed to full-load speed in the least number of points, with current peaks consistent with good motor operation.

The simplest and least expensive form of acceleration for magnetic starters on d-c motors uses the counter-electromotive force principle. With this scheme the acceleration depends upon the increasing value of armature voltage or counter emf as the motor accelerates.

The counter-emf scheme of acceleration is limited to small motors with relatively few points of acceleration. A starter with more than three points of acceleration would require setting the accelerating contactors to operate at voltages relatively close together. Large contactors also close too sluggishly on the rising counter-emf force.

If the line voltage is low or if heavy overloads are to be accelerated, the motor may not accelerate to a point where the counter-emf is high enough to close the accelerating contactors. Under such conditions the starting resistors may remain in the circuit long enough to be overheated.

Current-limit acceleration requires that the current relays be adjusted to close the contacts at a value of current high enough to accelerate the load. If the current needed to accelerate the load is higher than the relay setting, the relays will not operate; the starting resistance will remain in circuit and become overheated. If the relay settings are much higher than the load to be accelerated, the accelerating contactors close in rapid succession and are practically useless. The relays, therefore, must always be set to accelerate the maximum load. For lighter load conditions a different relay setting would be desirable but usually is not made.

Acceleration by the definite-time principle is the most positive. If the motor is heavily overloaded and stalled on the first point, the starting and acceleration are forced to a later point. The overload relay will protect the motor against exceedingly high current, as well as normal overload. Definite-time acceleration is generally used on both a-c and d-c applications.

Shunt Field Rheostat. The shunt field rheostat is a convenient means of obtaining speed control on d-c motors. The relatively small field current and high resistance requirements permit the resistor units to be of small cross-section and light weight. The losses in the rheostat are not objectionably high. A large number of speed points are easily provided by connecting the buttons or points on the faceplates to taps on the resistor units. Sixty points of speed control are common.

When it is necessary to provide a large number of accurately adjusted, low operating speeds, the variable-voltage system of control will give the best results. It is practically impossible to obtain a large number of low operating speeds by the armature resistor scheme on a constant-voltage system.

Dynamic Braking and Lowering. Reversing control with dynamic braking or with plugging can be provided. Plugging is quite commonly used on steel mill or other applications where quick reversal is required. Plugging service is used most with series motors or very heavily compounded motors, because it is difficult to apply dynamic braking to them effectively. They have very little field strength for dynamic braking unless the series field is included in the dynamic braking circuit. This is difficult to accomplish, since the connections to the series field would have to be reversed. To reverse the series field adds serious complications to the control equipment.

When a motor is running in one direction and is quickly connected to the power line in the opposite direction, the voltage on the armature is double the line voltage for a moment, because the counter voltage of the armature and the line voltage are in the same direction. Under such conditions the armature current is exceedingly high. A plugging resistor is connected into the circuit by means of a plugging contactor and plugging relays, to limit the armature currents to safe values when the motor is plugged.

Dynamic lowering is a term applied to the control circuits used in connection with a series motor for crane applications. When hoisting, it is necessary to take up slack in the chain slowly and carefully. Loads must be hoisted smoothly and at varying speeds. For these conditions the armature and series fields are connected in series with each other. Various combinations of resistors in series or parallel with the armature serve to permit operation at different speeds. Four speeds in either direction are usually satisfactory.

When lowering the loads, the conditions are different. If the load is just heavy enough to counteract the friction load, the no-load condition exists. If the hook is empty, the motor must drive the hook down. On the other hand, the load may be very heavy and therefore able to overhaul the motor and drive it above normal speed as a generator. To avoid runaway speeds when lowering, the series field is reconnected in parallel with the armature, but with resistance in each circuit. For lowering, therefore, the motor acts as a shunt motor. Either the armature or the field may be reversed to reverse the motor for hoisting or lowering. Different speeds are obtained by various combinations of the resistors in both the armature and the field circuits.

When lowering a light load at high speeds, the master switch and controller will change the value of resistance to provide a weak field and remove all resistance from the armature circuit. To lower a heavy load the master switch must be moved to a different position for resistor combinations that strengthen the field and retard the armature to give complete control of the load.

When the master switch is moved to the "off" position the motor is stopped by both a series coil type of magnetic brake and a dynamic brake. The magnetic brake prevents the load from falling.

Regeneration. Regeneration is a term usually applied for the condition on a system during which the motors are delivering negative torque or are acting as generators and delivering electrical energy to the power line. Regeneration takes place when the induced voltage of the motor is higher than the power line voltage, and such a condition obtains when the motor is driven above its normal speed. A regenerative system must be provided with current control, as otherwise rapid changes of line voltage might cause damage to the motor if the control apparatus did not insert armature resistance rapidly enough. Mechanical means of braking should be provided in order to stop the motor if the electric circuit should become disconnected. Regeneration may be used with either shunt d-c motors or induction motors, and with induction motors the inherent characteristics of such motors make it possible to obtain regeneration with little or no additional complication. Regeneration with d-c motors may be obtained by either the field-control or the voltage-control method. The voltage-control method may take the form of a motor-generator set, a booster, or a combination motor-generator set and booster.

ASSEMBLIES OF APPARATUS

21. POWER SWITCHBOARDS

DEFINITIONS. A power switchboard is that part of switchgear which consists of one or more panels upon which are mounted the switching control, meter, and protective and regulator equipment (AIEE Standard 27–89). The panels or panel supports may also carry the main switching and interrupting devices, together with their connections.

Switchboards may be either live-front, having live parts on the front of the panels, or dead-front, having no live parts on the front of the panels.

A distribution switchboard is a power switchboard used for the distribution of electrical energy at the voltages common for such distribution within a building. Knife switches, air circuit breakers, and fuses are generally used for circuit interruption on distribution switchboards, and voltages seldom exceed 600. Such switchboards, however, often include switchboard equipment for a high-tension incoming supply circuit and a step-down transformer (AIEE Standard 27–85).

A vertical switchboard is a switchboard composed of vertical panels (AIEE Standard 27–90). It may be either of the live-front or the dead-front type and may be either self-supporting by means of an angle iron or pipe framework or supported by means of braces to the wall.

MATERIALS. Although the most generally used panel material is steel, slate, marble, ebony asbestos, or laminated phenolic material may be used. Steel is almost always used for dead-front switchboard construction. Each panel normally consists of a single sheet formed to provide angle sections at the top and bottom and channel sections at the side. The angle sections provide a ready means of bolting the bottom of the panel to a channel sill and the top of the panel to a steel top member. The channel-shaped sides provide all necessary vertical framing and also serve as raceways or wiring gutters for the vertical wiring. The corners are welded to increase rigidity and are ground smooth to insure a good appearance.

EQUIPMENT. Instruments, meters, relays, voltage regulators, and similar equipment may be either semiflush- or projection-mounted on the steel panel, the former being preferred on most modern switchboards. The handles of instrument and control switches project through the panel. Rheostats and circuit breakers are mounted on brackets behind the panel and have their operating handles extending through the steel front panels. The equipment and wiring mounted on the rear of the panel should be so supported that welds or boltheads will not be visible from the front of the panel. Several general principles govern the relative location of apparatus on the front panels; no specific rules which are always applicable, however, can be given. The two most important considerations governing locations are facilitation of operation and pleasing appearance. Instruments should be at approximately eye level. Apparatus having operating handles should be located below the instruments and should be kept up some distance from the floor. Apparatus such as relays and voltage regulators, which are not indicating and do not have operating handles, may be placed at the bottom of the panels. Instruments should be located in a standard manner, as it is important that the locations be consistent on the same switchboard so as not to confuse the operator.

Instrument switches should be located in a logical position with respect to the instruments associated with them, preferably directly below the instruments and in the same order from left to right. There is seldom any need for locating instrument and control switches in any special relationship with each other, except in the case of synchroscope and associated breaker-control switches. Considerable thought should be given, however, to the proper location of associated control switches, with respect to both other control switches and other apparatus, such as field rheostat hand wheels and voltage-adjusting rheostats. Apparatus that requires frequent operation or attention should be located in the most convenient places. It will seldom be possible to have the ideal location for every piece of apparatus, and consequently the locations chosen must necessarily be the result of compromise. On live-front switchboards most of the apparatus is mounted on the front surface of the panel, with studs projecting through the panel to the rear of the switchboard for wiring and bus connections. In general, the rules for the location of apparatus on dead-front switchboards apply also to live-front switchboards, except that air circuit breakers are normally located at the top with the instruments and instrument and control switches located beneath them.

A switchboard ground bus should be provided, to which all equipment requiring grounding should be connected, except for instruments and relays which are grounded to the steel panel by their mounting studs. An adequate copper ground bus should tie steel panels together and to ground.

A swinging bracket, on which a synchroscope, voltmeter, and frequency meter are mounted, is frequently added at the end of a switchboard.

WIRING. The wiring of switchboards deserves careful consideration, as it is an important factor in the satisfactory performance of the equipment. Wires going in the same direction are grouped, even though grouping may involve slightly longer runs for many of them. Frequently both horizontal and vertical wiring troughs are provided for these wiring groups. These troughs usually take the form of a channel and cover, although in detail of construction there are numerous variations. The channel may be a structural member of the switchboard as well as a wiring trough, it may be a separate channel, or the equivalent result may be secured by means of a Z-shaped member, which together with the cover and the switchboard members forms a box section. The covers may have perforations with each wire emanating from an individual hole, a series of slots with several wires emanating from the same slot, or a narrow gap along one or both sides, so that wires may be brought out at any point. Care must be exercised in the formation of entrances to the trough to prevent chafing of wire insulation.

It is common practice to use only right-angle bends for small wiring on switchboards. Bends shaper than 90 deg or of less radius than the overall diameter of the wire should not be used, as they tend to crack the insulation.

Wires or groups of wires should normally be supported shortly before they enter or after they leave wiring troughs. Wires or groups of wires not in wiring troughs should be supported at frequent intervals, with proper precautions taken to prevent chafing of wire insulation. Wire groups may be round or rectangular and made up of precise layers of several wires each. In any event the wires in the groups must be adequately held together by means of clips, metal bands, cord, or other such devices. Wires are subject to considerable vibration on moving freight cars, and this movement has been known to cause fracture, particularly close to the terminals. Flexible wire would be better from this viewpoint but is usually undesirable because it cannot be easily formed and does not stay in position as well as solid conductors.

Each wiring terminal or connection should be marked with the wiring-diagram designation. Terminal blocks should be provided for all small wiring leaving the panels and for cross-panel wiring. When a switchboard is shipped in more than one section, it should be possible to disconnect all cross-panel wiring easily at these terminal blocks.

The switchboard wire should be No. 12 AWG solid copper, soft annealed, and tinned. The insulation should consist of varnished cloth, a non-hardening compound applied between the layers to exclude moisture and air, and a layer of felted asbestos thoroughly impregnated with a flame-resisting and moistureproof compound. The outer finish should be a closely woven cotton braid, also completely saturated with a flame-resisting and moistureproof compound. New synthetic insulations are also coming into use and undoubtedly will prove satisfactory. The outer wire cover is normally gray unless color coding is used. It should be approved by the Underwriters' Laboratories for the application.

DUPLEX SWITCHBOARD. A duplex switchboard is a structure with front and rear panels of metal or insulation material, separated a comparatively short distance by an aisle, and enclosed at both ends. No primary switching devices are located between front and rear panels. The rear panels may be hinged for access to panel wiring (AIEE Standard 27–86).

These switchboards are extensively used for central-station, industrial, and other applications where it is desired to have centralized control of power-operated equipment. Any number of sections, including front and rear panels and tie members, may be bolted together in a horizontal line to comprise a complete duplex switchboard. The instruments, meters, and control switches are normally mounted on the front panels, as they require the constant attention of the operator. Such equipment as relays and recording meters, which require only infrequent attention, are normally mounted on the rear panel. Semi-flush instruments and relays are used almost exclusively. There are tie members between the front and rear panels at the top, and often a top sheet covers the entire switchboard if there are several panels. There are no tie members between front and rear panels at the bottom, each panel being secured to a channel sill.

Access to the aisleway between front and rear panels is gained through a hinged door at the end. The location, mounting, and wiring of apparatus are similiar to those of the vertical switchboard previously described.

The front and rear panels are sometimes brought close together, and the aisle eliminated. Access to the space between the boards is then obtained by hinging one row of panels, either the front or the rear panel. The panel with sensitive protective devices and with more wires should be kept stationary.

Flexible wire is always used for connections to hinged panels, the standard being No. 12 gage wire made of 65 strands of 0.010-diameter conductors. These are usually clamped near the hinged side of the panel and on the fixed structure, allowing a loop with sufficient slack to give full panel opening. The proper flexing of this loop when the panel moves is very important to avoid ultimate breakage. To be suitable a clamping arrangement should have a maximum bend of about 75 deg around a $^3/_8$-in. radius. Another method is to bundle the wires and run them parallel to the hinged side of the panels, securing one end of the group on the moving panel and the other to the fixed structure. The bundle twists as the panel moves, and if the individual wires are allowed some freedom of movement and at least 18 in. are provided between clamps, no undue flexure occurs.

Draw-out relays with knife-blade or other types of test switches in the same units are frequently used. They are for panel mounting, and their construction conserves switchboard space. The relay elements are mounted on a removable unit chassis permitting convenient testing and maintenance. Maximum flexibility is provided, since test connections can be made by either clip leads or test plug.

Switchboards and control desks for generating stations and primary substations frequently have miniature bus layouts which are replicas of the power circuits. They are usually of metal with different finishes or colors for circuits of different voltages.

BENCHBOARDS. A benchboard, sometimes designated as a control desk, is a switchboard having a horizontal or slightly inclined section for mounting control switches, indicating lamps, and instrument switches, and constructed with or without vertical instrument sections (AIEE Standard 27–80).

The continuous type of benchboard has no opening between the vertical instrument section and the bench. The duplex type of benchboard is a combination structure of a continuous-type board with hinged rear panels of metal or insulation material, separated at comparatively short distance. Grille or solid end enclosures are provided, and the hinged panels provide access to panel wiring and also serve to carry part of the equipment. The open type of benchboard has a space between the bench and the vertical instrument sections. The separate type of benchboard consists of a bench or desk without vertical instrument sections.

The front apron is frequently inclined inward from the front edge to provide toe room at the bottom of the board. Often the front edge of the board is also provided with a wearing strip. Rear hinged panels or doors are sometimes used for the mounting of watthour meters, relays, and other devices.

It is recommended that the benchboard be provided with an under walk-way from which complete access can be gained to the devices and wiring on the bottom of the bench panel. The side sheets can be constructed with an opening for headroom, which permits free movement on the walk-way along the length of the structure. Wiring troughs extending the length of the board are provided for interpanel wiring and potential buses. Circuits between front and rear also are enclosed. Additional troughs can be provided where required so that a minimum of wiring will be exposed. All troughs should be equipped with removable covers, permitting easy addition or removal of panel wiring.

The separate type of benchboard is normally used where no instruments or relays are associated with the control switches mounted on the benchboard. A separate benchboard is also desirable where the operator must see the machines or operations which he controls.

Additions may sometime dictate the use of the open type of benchboard with the instruments, relays, mimic bus, etc., located on a separate vertical instrument section.

The duplex type of benchboard is the most commonly used and provides a very desirable method of mounting instruments, instrument and control switches, relays, watthour meters, and other control apparatus.

Control benchboards may be either straight or curved. The general construction features are similar to those of the previously described duplex switchboard.

LOW-VOLTAGE ENCLOSED AIR CIRCUIT-BREAKER SWITCHBOARDS. An enclosed air circuit-breaker switchboard is a switching structure containing dead-front air circuit breakers, buses, and connections, with enclosures on the ends, back, and top. The air circuit breakers are contained in individual compartments and are controlled remotely from the front panels. The structures are of two kinds, stationary type and draw-out type. The stationary type contains air circuit breakers which are mounted rigidly on bases and have no special arrangement for quick removal from the structure. The draw-out type contains air circuit breakers so arranged that they may be easily drawn out of the structure and disconnected from the buses by means of self-coupling disconnecting devices.

The stationary structure of the draw-out type of metal-enclosed switchgear is the fixed portion which receives and supports the removable elements. This structure also includes the breaker compartments, buses, connections, stationary disconnecting contacts, supporting rails, and instrument transformers where used. The removable element of metal-enclosed air-circuit-breaker switchgear of the draw-out type is the portion which comprises and supports the air circuit breaker and its associated removable apparatus (AIEE Standard 27–100). There are three removable element positions: operating, test, and disconnect (AIEE Standards 27-120, 27-121, and 27-122).

METAL-ENCLOSED SWITCHGEAR. Metal-enclosed switchgear is primarily intended for indoor operation. It can, however, be furnished in weatherproof houses suitable for outdoor operation.

The switchgear is suitable for 600 volts maximum service. The cubicles forming part of the stationary structure may be fabricated from stretcher levelled steel forming a self-contained housing which has one or more individual compartments for the removable air-circuit-breaker units, with a full-height rear compartment for the bare buses, instrument transformers, and outgoing connections. Instead, the entire structure may be fabricated from steel angles and shapes, covered with sheet metal as required.

The individual circuit-breaker compartments are equipped with primary and secondary contacts, rails, stationary disconnecting mechanism parts, and the cell interlocks. Grille openings are provided in the front panels to provide ventilation. These grilles are usually provided with baffles to protect the operator from the direct flash or the products of a circuit interruption. The structure is normally so designed that future additions may readily be made at either end at any time.

Air Circuit Breaker. The removable element consists of an air circuit breaker on an insulating base mounted in a formed and welded steel frame. Removable elements requiring a cell height of 45 in. or more are floor wheeled. There are several different types of mechanism for moving the circuit breaker from the operating to the disconnect position. One type utilizes easy-rolling wheels mounted on the removable element frame and running in jig-welded rails secured to the stationary structure. Extension rails are provided for supporting the breaker in the disconnect position.

Another method of construction uses a C-shaped formed metal guide secured in the center at the bottom of the breaker stationary cubicle and has a T-section secured to the removable element. Sliding supports are also provided at the side underneath the removable element to stabilize the element while it is being withdrawn.

A third method of construction utilizes a pantograph mechanism for holding the breaker in a vertical position during connecting and disconnecting. The draw-out mechanism is operated by a screw provided with a removable crank handle. When fully withdrawn, the breaker may be lifted off the pantograph frame. Pantograph-mounted circuit breakers are provided with feet and are free-standing when removed from their compartment. Another method of construction utilizes telescoping rails.

Breakers of 2000-amp or greater capacity and some smaller electrically operated breakers are normally mounted on a truck, which is provided with wheels at the bottom. Guide members are located within the stationary structure.

When the breakers are in the disconnect position, they may be lifted off the rails, suitable lifting provisions normally being made.

For the large breakers mechanisms offering considerable mechanical advantage are provided for connecting and disconnecting the fixed and movable contacts.

Primary Disconnecting Devices. There are also several types of primary disconnecting devices (AIEE Standard 27-126), one of which will be described. The stationary part of the primary disconnecting devices for each circuit breaker consists of a set of cylindrical contacts mounted on an insulating base. Buses and outgoing cable connections are directly connected to them. The corresponding moving contacts consist of a set of contact fingers equally spaced around the periphery of the round circuit-breaker studs. In the operating position these contact fingers will engage with the stationary contacts, forming a current-carrying bridge. The assembly will provide a multitude of silver-to-silver high-pressure point contacts. High pressure on each finger is maintained by individual short leaf springs. The entire assembly is fully floating and is provided with ample flexibility between the stationary and moving elements. Contact engagement is maintained only in the connected position.

Secondary Disconnecting Devices. The secondary disconnecting devices consist of floating fingers mounted on removable units and engaging with flat contact segments located at the rear of the equipment. The secondary disconnecting devices are silver-plated to insure permanence of contacts. Contact pressure is provided by helical springs. Contact engagement is maintained in the connected and test positions.

Buses. The buses usually consist of high-conductivity bare copper bars. The main bus joints and all tap connections are silver-plated and tightly clamped with through bolts to insure maximum conductivity. Molded terminal blocks with integral-type barriers are provided for the secondary circuits. The terminal blocks are mounted at the rear of the unit and are accessible through a removable cover. They can be mounted at either the top or the bottom as required.

Standard Units. Most manufacturers of this type of switchgear have a considerable number of standard units which can be combined to meet specific application requirements in regard to number and rating of circuit breakers and selection of attachments, instruments, and relays. The basic units and a selection of standard instrument panels are available as completely engineered, completely tooled designs. From these, switchgear can usually be assembled to meet any requirement of 600-volt main and feeder switching and protective equipment. With this method of construction and assembly, future additions are easily made at either end of the structure at any time. In addition to the basic units and instrument panels, bus transition units are available. Panel widths vary from 16 to 60 in., depending upon the circuit breaker involved. Although the normally recommended depth is 54 in., the depth may vary from 42 to 72 in., depending upon specific requirements.

Advantages of Draw-out Construction. Draw-out construction offers three distinctive advantages, all of which are important for continuity of service. Inspection is easier and can be more thorough with the breakers drawn out of the operating position. The removable element of the breaker serves as a disconnect, and any breaker can be replaced with a reserve breaker of the same type and rating or with a breaker from a less important circuit.

Access to buses, cable connections, instrument transformers, and secondary terminals is provided by removing the rear sheets of the switchgear. For ease of handling, two separate removable back sheets are usually supplied for each unit.

Full-height steel barriers separate breaker enclosures from bus and connection compartments. Current transformers for use with instruments, watthour meters, or relays are supplied. Cable cleats are provided for cable supports. Wiring troughs for cross-panel connections are provided as required. The cells for the larger breakers may be equipped with shutters, which cover the stationary contacts. These shutters close automatically when the breaker is removed and open automatically when the breaker is moved back into operating position.

Potential transformers are generally located on the rear barrier of instrument sections. Gang-operated disconnect switches can also be provided when required.

When breakers are used in the second or third tier, it is recommended that a monorail overhead crane, which provides the simplest method of lifting circuit breakers in or out of the stationary structure, be installed. Where height of ceiling or other restrictions prohibit a monorail installation, however, breakers can be elevated by a handling carriage. This carriage can be provided with a roller base for the breakers which do not have wheels and with a platform for mobile-type breakers.

Metal-enclosed bus runs are often furnished with switchgear installations. These low-voltage runs comprising a metal housing supporting high-conductivity bare bus-bars are of course designed and built to meet installation requirements. They assure the same high level of insulation and the same safety features as the switchgear itself.

The preferred method for anchoring metal-enclosed switchgear to the floor is to imbed steel channels in a concrete floor with the top surfaces of the channels flush with the finished floor. Housings are welded or bolted at the mounting bolt holes to the channels.

The main circuits are given a dielectric test of 2200 volts for 1 min between live parts and ground and between opposite polarities. The wiring and control circuits are given a dielectric test of 1500 volts for 1 min between live parts and ground.

Instruments, meters, and relays are normally mounted on a front hinged panel. The relays may be of either the fixed or the draw-out construction.

Instruments, meters, relays, and associated devices, such as instrument switches, test blocks, and indicating lamps, cannot be mounted on the doors of circuit-breaker compartments, but may be mounted on the rear panels or in the front doors of bus transition units or on special instrument panels.

An enclosed air-circuit-breaker switchboard may also be of the stationary type. There is considerably more variation in the construction of this class of switchgear than in that of the draw-out type. Such switchboards may be almost identical with draw-out switchboards except that the air circuit breakers are mounted rigidly on bases and have no special arrangement for quick removal from the structure.

22. METAL-CLAD SWITCHGEAR

DEFINITION. Metal-clad switchgear consists of a metal structure containing a circuit breaker and other associated equipment, such as instrument transformers, buses, and connections. The transformers, insulated buses, and connections are placed in separate grounded metal compartments, which may be either unfilled* (dry type) or may contain fluid, semifluid, or other insulating mediums. The circuit breaker is equipped with self-coupling primary and secondary contacts and is arranged with a disconnecting mechanism for moving it physically (through vertical travel or horizontal travel) from the connected to the disconnected position, after which it may be removed from the stationary structure. Interlocks are provided to insure proper sequence and safe operation. With circuit breakers of great weight or volume an alternative construction is used, whereby disconnection

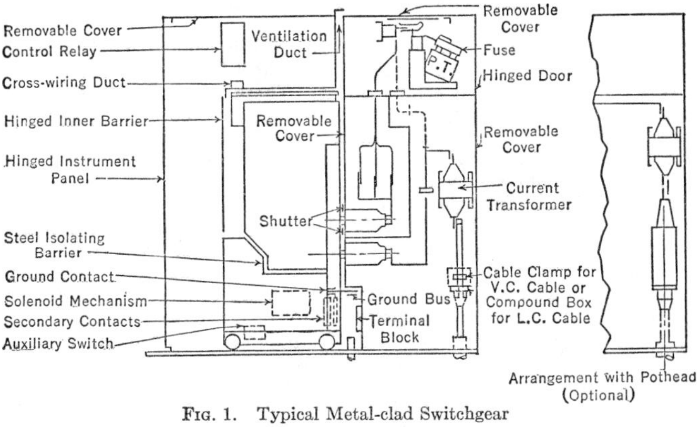

FIG. 1. Typical Metal-clad Switchgear

is accomplished by disconnecting switches which automatically ground the breaker† when the switch is opened (AIEE Standard 27-102). A typical metal-clad switchgear assembly is shown in Fig. 1.

TYPES OF METAL-CLAD SWITCHGEAR. Metal-clad switchgear is divided into light- and heavy-duty classifications, each of which is further divided into indoor and outdoor types. Any of these types can be furnished with either oil or air circuit breakers. Light-duty metal-clad switchgear is built for service up to 5000 volts, is rated at 600 or 1200 amp, and has an interrupting capacity of 50,000 kva.

Heavy-duty metal-clad switchgear is built for service up to 15,000 volts, is rated at 600, 1200, or 2000 amp, and has an interrupting capacity of 500,000 kva. Common bus capacities are 1200, 1600, and 2000 amp, but capacities of 3000 and 4000 amp are occasionally used. There is very little fundamental difference in the construction of light- and heavy-duty metal-clad switchgear. The principal differences are the result of the increased voltage and capacity of the circuit breakers associated with each class.

Most manufacturers have a considerable number of standard units from which the complete switchgear may be assembled according to the individual circuits for which units are required. These units are grouped so as to meet the specific requirements of the application. There are many advantages to this type of switchgear. The installed cost is reduced because such switchgear permits greater use of standardization and repetitive factory-production methods. Since each unit is self-contained, self-supporting, and generally interchangeable with the others, complete flexibility is provided for future changes.

The entire structure is completely assembled, adjusted, wired, and tested at the factory. Shipment is made in completely assembled groups, depending in size upon handling facilities in transit and the destination. Installation consists merely of properly anchoring the

* The unfilled or dry type is generally used in American installations. The filled types are common in European applications.

† In equipment of this type the grounding features on the switches are omitted, as they have little value and complicate the structure.

gear to the foundation and making primary and secondary connections. Obviously, these methods save expensive assembly, wiring, and testing of unassembled apparatus, as well as the cost of separate field-built cell structures.

Five bus and connection arrangements are considered standard for metal-clad switchgear: (1) a single bus with single circuit breaker; (2) main and transfer bus, the transfer bus connected to the feeder by means of single throw disconnecting switches; (3) bus sectionalizing arrangement with single-circuit breaker; (4) bus tie arrangement with single-circuit breaker; and (5) double bus arrangement with one or two circuit breakers.

STATIONARY STRUCTURE. The stationary structures are usually fabricated from sheet steel not less than 1/8 in. thick or No. 11 U.S.S. gage supported on a suitable structural-steel framework. It is the usual practice to utilize a common side sheet between units. This practice results in a material saving in sheet steel and also permits the assembling of a large portion of the detailed equipment in the housings before they are assembled together, access being gained through the open side. Rigidly accurate jigs and fixtures are used throughout for punching, bending, welding, and assembling the cells and details, thus insuring exact duplication of essential parts. In this way interchangeability of circuit-breaker units of comparable type and rating and duplication for future extensions are assured. The interchangeability may not, however, include electrical interchangeability of secondary and control circuits.

Each unit may be divided into as many as five separate compartments, each separated from all adjacent compartments by means of sheet-steel barriers to confine any possible disturbance within an individual compartment. These five compartments are for the circuit breaker, bus, current transformers and connections, potential transformers, and instruments, relays, and other control devices. Removable covers or doors are provided so that access may be gained to any compartment for maintenance or inspection. By virtue of this arrangement an operator working in one compartment is not exposed to live circuits in adjacent compartments.

CIRCUIT-BREAKER COMPARTMENT. The design and construction of the circuit-breaker compartment depends on the type and rating of circuit breaker used. Either air or oil circuit breakers may be of the lift (vertical-travel) or draw-out (horizontal-travel) type. Both lift and draw-out types of construction are commonly used for air circuit breakers, but lift-type construction is customary for oil circuit breakers.

Primary disconnecting devices consisting of separable contacts, usually of the plug-and-socket type, are provided to connect or disconnect the main circuits between the removable element and its stationary structure. The plug is usually mounted on the stationary structure, and the socket on the removable element. The details of the design of primary disconnecting devices differ widely among manufacturers; usually, however, the socket consists of a number of contact fingers equally spaced around the periphery of the circuit-breaker stud and the stud which is connected to the bus. High-pressure contact is maintained by means of coiled garter or leaf springs. The contact-finger assembly is locked in some manner to either the circuit-breaker or the stationary-structure stud. When the circuit breaker is in the connected position, the entire primary-disconnecting-device assembly is encased in an insulating housing of either wet-process porcelain or molded phenolic material, which is mounted on the stationary structure. On switchgear rated at 15,000 volts automatic shutters are provided in the stationary structure to prevent accidental contact with the current-carrying parts when the removable element is in the disconnect position or has been removed. Such shutters are also frequently provided on lower-voltage switchgear. The shutters may be fully automatic; that is, they may close when the breaker is moved from the operating position and open when the breaker again approaches the connected position, by means of a mechanism actuated by the motion of the circuit breaker. In some designs, however, the shutters close automatically but must be opened manually before the circuit breaker can be moved to the connected position. Provision is also made for manually opening the shutters for inspection or maintenance.

Secondary disconnecting devices are separable contacts which connect or disconnect the auxiliary or control circuits between the removable element and its stationary structure. These secondary disconnecting devices automatically disconnect the small wiring connections when the removable element is moved from the operating position, unless provision is made for the secondary contacts to be closed in the test position. In some designs test jumpers are provided for completing all secondary connections for checking the control circuits in the disconnect position.

A mechanism is provided for moving the removable element from the connected to the test and disconnect positions. The connected position is that position of the removable element in which both primary and secondary disconnecting devices are in full contact. The disconnect position is that position of the removable element in which the primary disconnecting devices are separated by a safe distance for isolation of the breaker element, although the secondary disconnecting devices may still be in contact. The test position is that position in which the primary disconnecting devices of the removable element are separated by a safe distance from those in the stationary structure, and the secondary disconnecting devices are in operating contact for test operations. This position may correspond with the disconnect position.

Mechanical interlocks are provided to prevent the circuit breaker from being moved to or from the operating position with the circuit breaker in the closed position and to prevent the circuit breaker from being closed unless the primary disconnecting devices are in full contact or separated by a safe distance.

This mechanism may be operated either by hand or by motor. Motor-operated mechanisms are generally used with the vertical-lift type of construction for larger and heavier circuit breakers.

Suitable guides must be provided to secure accurate alignment of contacts and to prevent jamming of the removable element during travel to or from the connected position. Position-limit stops are also provided on this class of equipment.

Adequate provision must be made for securing the circuit breaker in the connected position.

Transfer trucks are usually provided for the vertical-lift type of construction and are sometimes provided for the draw-out type of construction. Many circuit breakers of the draw-out type, however, have wheels on the removable element and are therefore individually mobile.

On all units accommodating oil circuit breakers a pipe header is provided, into which the breaker muffler or separating chamber will exhaust. The connection to this header is automatic when the breaker is inserted into the operating position, and a valve automatically closes the opening to the header when the breaker is removed.

Provisions also must be made for venting the gases generated during the interruption of an air circuit breaker.

INSTRUMENT, RELAY, AND CONTROL-SWITCH COMPARTMENT. All metal-clad units can be furnished with a steel-front enclosure having a hinged-front instrument panel. On this panel are mounted the various secondary control devices which are furnished for each particular type of control. If desired, swinging brackets can be added at the end for mounting the synchronizing equipment. Also the instruments may be located on a panel opposite the breaker side.

All instrument and control wiring in circuit-breaker compartments is shielded from primary circuits by metal barriers.

BUS COMPARTMENT. The bus-bars are made of high-conductivity copper bar with, usually, rounded edges. The copper at each main bus joint and each tap joint is silver-plated and tightly clamped with through bolts to insure maximum conductivity. The buses are supported by insulating material, frequently consisting of heavy laminated phenolic plates. All buses and bus taps are completely insulated by preformed laminated phenolic-material boxes and varnished cambric tape. In some cases, means are provided to isolate the bus compartment of each unit to prevent the free passage of gases between units of a group.

A substantial ground contact is usually provided between the circuit-breaker units and the housing, and positive contact is automatically made in the operating position. The ground bus has a capacity at least 25 per cent of that of the largest bus.

All buses are so arranged that they can be readily extended when additional units are added. In some designs screened ventilating openings are provided in the front and top of the bus compartment.

CURRENT-TRANSFORMER AND CONNECTION COMPARTMENT. This compartment is arranged for the power conductors to enter from above or below the housing, as required. In each compartment wiping sleeves are provided for lead-covered power cables, or supports for braid-covered power cables. Sealed-type potheads for lead-covered cables can be supplied as optional equipment, if desired. Connections for external control wiring are provided at terminal blocks, conveniently located with respect to the incoming control cable.

Current transformers for instrumentation and relaying are coordinated with the circuit breaker to assure adequate thermal and mechanical capacity under short-circuit conditions.

POTENTIAL-TRANSFORMER COMPARTMENT. The potential transformers may be of either the fixed or the disconnecting type and may be provided with current-limiting fuses.

If the primaries of the potential transformers have standard fuses, resistors should be used in series with them if the interrupting rating of the fuse is not adequate to take care of the short-circuit current that the system can deliver.

If the potential transformers are of the disconnecting type, opening the compartment doors automatically disconnects them, grounds the disconnecting contacts, and withdraws the transformer and fuse to a completely safe position. Secondary circuits are short-circuited and grounded in the disconnect position.

BUS-ENTRANCE COMPARTMENT. Where an isolated installation of any one unit is made, such as a feeder unit or a motor unit, and no additional auxiliary compartment is included, it is necessary to provide a suitable means for joining the incoming cables to the bus of the base unit. This is accomplished by means of a so-called bus-entrance compartment, which consists of a steel entrance box secured to the base unit immediately adjacent and external to the unit bus. This box, or compartment, is equipped with either a wiping sleeve or bushing to receive the incoming cables, and the necessary bus extension. This compartment also may be used where cables join the bus of any one group of units with the bus of any other separate group of units, or other installed equipment.

DIELECTRIC TESTS. Standard insulation tests at 60 cycles are made on the completely assembled switchgear. A standard full-wave impulse-test rating is also being considered. The equipment must withstand for 1 min the a-c voltage test prescribed for its voltage class; potential or operating transformers having inherently lower insulation strength, however, may be disconnected for this test. For the impulse ratings, a positive or negative (whichever has the lower value), 1.5×40-μsec wave will be used. The impulse values are phase-to-phase and phase-to-ground. The impulse-test voltages will probably vary from 45 kv for equipment rated at 2.5 kv to 95 kv for equipment rated at 15 kv. The standard insulation-test voltage for equipment rated at 2.5 kv is 15 kv; for 5 kv, 19 kv; and for 15 kv, 36 kv.

OUTDOOR METAL-CLAD SWITCHGEAR. Both light- and heavy-duty metal-clad switchgear can be mounted in suitable weatherproof enclosures for outdoor installations. The base units are the same for both indoor and outdoor applications. The weatherproof housing is constructed integrally with the basic structure and is not merely a steel enclosure. The outdoor enclosure is often wider than the narrower breaker units in order to provide reasonable accessibility for inspection and maintenance. This is not a disadvantage, inasmuch as space is not at the same premium for outdoor installations as it is for indoor installations.

The basic structure, including the mounting details and withdrawal mechanism for the circuit breakers, bus compartments, transformer compartments, etc., is the same as that of indoor metal-clad switchgear. Interunit barriers of sheet steel are omitted, and this basic structure is provided with a weatherproof housing, one side of which acts as an interunit barrier. Front and rear doors are provided. Any number of units are assembled in line to make up the complete assembly. Heaters and ventilation are provided to control condensation. The instruments and relays are normally mounted on a panel on the opposite side from the circuit breaker so as not to interfere with its withdrawal when the panel opening is restricted by the outer, weatherproof door.

23. CUBICLE SWITCHGEAR

Station cubicle switchgear is primarily intended for heavy-duty service in central stations and more important outlying feeder substations. It is used on feeder, generator, sectionalizing, and similar segregated phase and high-level insulation circuits where continuity of service is extremely important. Basically, a unit of station cubicle switchgear is a large-capacity, heavy-duty, oil or air circuit breaker completely enclosed within a metal cubicle with the phases segregated by metal barriers.

Circuit breakers have fixed mountings. Being large, they are not built to be removed by trucks or by draw-out arrangements, but are made easily and safely accessible for inspection and servicing. Besides the circuit breaker, the cubicle includes disconnect switches for isolating the breaker and current transformers. Because of the grounded heavy-gage metal enclosure of the cubicle, phase-to-phase faults are practically impossible. A typical example of station-type cubicle switchgear is shown in Fig. 2.

Operation is safe for personnel because all live parts are enclosed by grounded metal barriers. Simple and dependable mechanical key- or lever-type interlocks positively prevent improper operating sequence. Although fully enclosed, the circuit breaker and all other parts are safely and quickly accessible for inspection and maintenance.

The cubicles consist of rigid, welded structural frames, completely enclosed by metal sheets. Metal barriers are provided between phases in all units which have primary conductors. The barriers, which isolate breaker pole units, are removable where necessary to provide access to equipment for maintenance. All individual compartments are sufficiently gastight to prevent passage of gas from one compartment to another without its undergoing complete deionization.

Hinged doors provide access to the front of the circuit-breaker compartment and the breaker-mechanism compartment. Bolted covers with inspection windows allow access to the disconnecting switch pole units. Other removable covers allow access to each phase of the bus.

The design of station cubicles provides a completely grounded metal enclosure for the entire unit. The enclosure is subdivided into three major compartments: circuit-breaker compartment, main bus and disconnect-switch compartment, and the line disconnect-switch and connection compartment, which are isolated from each other by tightly fitted metal barriers through which the circuits are carried in bushings. Each major compartment is subdivided by metal barriers into segregated phase compartments. Thus each breaker unit comprises nine isolated compartments. The purpose of the three major subdivisions is primarily to provide safety for operating and maintenance personnel, as the equipment in any one of the three compartments can be isolated from the others so that maintenance operations can be performed in safety. The purpose of phase segregation is to reduce damage to equipment and circuits because faults will be limited to those between phase and ground, which can be cleared by fast-acting ground relays.

The general construction of all types and ratings of breakers is similar. Only the details of construction change with the voltage, ampere, and interrupting capacity ratings. As voltage ratings increase, greater insulation distances are required, and the units become larger. Increases in ampere ratings require increased use of non-magnetic metals for enclosures. Ratings up to 1200 amp will operate satisfactorily in magnetic steel enclosures when spaced for high-level 15,000-volt service; 2000-amp ratings are equipped with non-magnetic interphase barriers and sheets through which bushings pass; and ratings of 3000 amp and above have complete non-magnetic enclosures for the bus and disconnecting-switch compartments. The strength and number of bus supports are coordinated with the different interrupting ratings, which determine the highest current to which the equipment may be subjected.

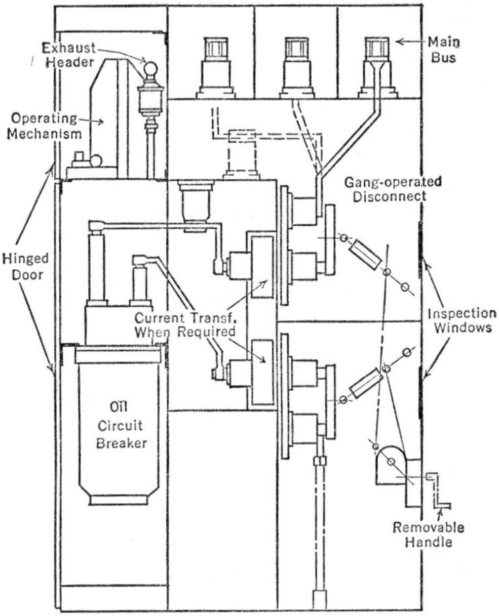

FIG. 2. Station-type Cubicle Switchgear

Auxiliary compartments using the same type of segregated phase construction are provided for such auxiliary equipment as potential transformers, bus connections, and separate disconnecting switches. No equipment is located in the main structure in a manner to give phase-to-phase exposure. Potential transformers are applied as three transformers, star-connected with grounded neutral to maintain the phase segregation. They are usually of the disconnecting type. Circuits leaving the units in three conductor potheads terminate in a compartment external to the switchgear unit, so that the conductors can enter their respective phase compartments properly separated.

Each breaker unit has provision for four sets of bushing-type current transformers on the disconnecting-switch bushings, two sets on either side of the breaker. The breaker-control relay panel is located in the operating-mechanism compartment of the breaker, with terminal blocks for control and instrument transformer wiring. Secondary circuits may enter either the top or the bottom of the units. There is no provision for control or protective devices on this class of switchgear, as it is planned for operation from a remotely located control board.

Access to the breaker-mechanism compartment can be had at all times by means of hinged doors. Access to the breaker primary compartment is by means of hinged doors, which are interlocked with the disconnecting switches so that they can open only when the switches are open, thus isolating the circuits in that compartment. The main bus is provided with bolted-on cover plates which can be removed to expose the bus for inspection. The disconnecting-switch compartments are enclosed by bolted covers provided with inspection windows for each phase switch.

The disconnecting-switch compartments are located at the rear of the cubicle. Each cubicle includes two three-pole disconnecting switches for isolating the circuit breaker from the bus and from the incoming line. The disconnecting switches are operated in a six-pole group. All the disconnecting-switch pole units in the cubicle are operated simultaneously by a geared mechanism. The switches are operated from the outside of the cubicle by means of a removable handle, which is mounted at a convenient distance above the floor level. The disconnecting-switch blades are connected to the operating shaft by means of insulating operating rods. When the blades are closed, switch mechanisms have an over-center position to prevent accidental opening of the blades under short-circuit conditions. Position indicators are provided to show the open and closed positions of the switches. Interlocks are provided between disconnecting switches and the circuit breaker to prevent operation of the switches unless the breaker is in the open position. They are also provided between the switches and the breaker-compartment doors to prevent opening the doors unless the switches are open and to prevent closing the switches as long as the doors are open. The interlocking may be either key type or mechanical.

Isolated disconnecting-switch compartments cannot be interlocked to provide safe entrance, as either or both of the line and bus terminals of the switches may be alive. It is therefore necessary to enclose these compartments in a manner which necessitates the authorization and supervision of competent personnel before entrance may be made.

The main buses are located at the top of the cubicle. The buses and main connections are fabricated of bare copper bars. The buses are mounted on heavy-duty porcelain supports. The buses and connections are capable of withstanding mechanical stresses produced by short-circuit current corresponding to the current-interrupting rating of the circuit breaker at the service voltage.

A heavy copper ground bus extends the entire length of the cubicle structure. Each circuit-breaker frame is connected to the ground bus by copper strap.

Many of the higher-ampere circuits connecting with the switchgear, such as generating connections, bus-tie connections, and synchronizing buses, are economical bus-run applications. These bus runs have the same features of insulation, phase segregation, bus, and bus supports as the bus in the switchgear to which they connect. The application of bus runs requires complete and accurate information concerning the building in which they are to be located, so that proper lengths, bends, and methods of support can be designed.

Bus protection is usually applied to this class of equipment. The common protective schemes are ground bus-fault and bus-differential protection with either current transformers or linear couplers.

The insulation levels of station cubicle switchgear are those specified by AIEE and NEMA rules. The 60-cycle insulation test for service up to 15 kv, which is 50 kv for 1 min, is now referred to as high-level 15-kv insulation. For service voltages above 15 kv, such as 23 kv, 25 kv, 26.4 kv, and other values up to 34.5 kv, equipment suitable for an 80-kv 1-min test is available. The basic impulse levels soon to be effective for this class of equipment are 110 kv for the high-level 15-kv class and 150 kv for the 34.5-kv class. The application of the 34.5-kv equipment to circuits above 28.5 kv is on the basis of grounded neutral.

Available interrupting ratings for this class of switchgear are 500,000, 1,000,000 and 1,500,000 kva at 15 kv and 1,000,000 and 1,500,000 kva at 34.5 kv for oil circuit breakers, and also 2,500,000 kva at 15 kv for air circuit breakers. The current capacity varies from 1200 to 5000 amp.

24. UNIT SUBSTATIONS

Closely coupled combinations of transformers and factory-assembled switchgear, which supply moderate-capacity blocks of power directly at centers of load from high-voltage sources stepped down to distribution voltage, are conventionally known as unit substations, power centers, or load centers. Their purpose is to provide economy in the use of material and improved voltage regulation by transmitting power at high voltage to the center of load, where it is transformed to service voltage for distribution over a minimum of high-current low-voltage circuits. These assemblies may be of either outdoor or indoor design.

The indoor assemblies usually are applied in industrial plants having large building areas in which space can be assigned for the substation. In industries using small buildings widely spaced, or for secondary utility-power distribution, outdoor units are found economical. The economy of the outdoor application decreases with the number of switchgear sections involved.

The general usage of these assemblies is to supply secondary distribution at 2300 or 4000 volts from 15,000-volt or higher primary circuits and to supply concentrated centers of load with power at service voltage of 440 volts or lower. There is also a trend to use such compact assemblies to supply 15-kv primary distribution from high-tension service of 33 kv or more.

Indoor designs use standard metal-clad switchgear assemblies associated with air-insulated, oil-insulated, or non-inflammable-liquid-filled transformers. The outdoor designs are similar to the indoor designs except that liquid-filled transformers only are used, and the switchgear is of the outdoor metal-clad construction. Standard arrangements of transformer and switchgear provide the high-voltage terminal and switchgear on the right and the low-voltage switchgear on the left of the transformer (when facing the front or panel side of the assembly).

The switchgear high- or low-voltage design for indoor installations consists of standard metal-enclosed switchgear units arranged for throat connections with the associated transformer. The largest proportion of installations have metal-clad high-voltage switchgear units and draw-out low-voltage units. The combination is throat-connected between the transformer and the switchgear to keep the floor-space requirements and connections to a minimum. The usual length of throat connections is 36 in. or 18 in. on both the transformer and the switchgear. Some types of transformers, however, are arranged adjacent to switchgear without throats. The throat connection between the transformer and switchgear is usually made with provision for a $^1/_2$-in. misalignment of parts in any direction. The primary circuits are completed with flexible shunts, and the secondary with copper bus-bars with slotted holes to compensate for misalignment.

The switchgear for outdoor unit substations and power centers consists of indoor metal-clad base units with weatherproof enclosures. The base unit may have the breaker removable from the side opposite the panel. Indoor and outdoor removable units of the same rating can be interchangeable. The outdoor construction includes ventilation of cell compartments through screened ventilators, baffled to exclude weather. The doors are made weatherproof by the use of recessed door jambs to carry any water which may get through to the bottom of the house. Each breaker section is equipped with space heaters in the breaker compartments and the panel compartment to permit warming the air above the outside temperature, thus preventing moisture condensation. These heaters are controlled by a small low-voltage breaker in the auxiliary compartment so that they can be turned on during severe weather conditions.

Unit substation switchgear groups seldom exceed five circuit-breaker sections and an auxiliary compartment, which can be assembled and shipped as a complete unit for air or oil breakers up to 250,000-kva interrupting rating. The 500,000-kva breaker enclosures are considerably larger and are not recommended for shipment in groups exceeding four units.

The functions of transformation, regulation, protection, control, and metering of electric power, performed by separate pieces of apparatus, in the open structural type of substation, are combined in a completely enclosed factory assembled, wired and tested unit substation including all necessary auxiliary equipment for its operation.

Complete portable substations can be mounted on a trailer or on a railway car. Those mounted on a trailer are normally limited to a maximum of 69-kv and up to 6000-kva capacity of transformer. The sizes are limited, however, only by the maximum size and weight of trailer permitted by various state laws. Railway-car-mounted portable substations, for single- or three-phase service at all voltages up to 138 kv and in ratings up to 50,000 kva, are entirely practicable where railway clearances are the only limitation.

25. CONTROL ASSEMBLIES

Control panels are generally of insulating materials, such as slate, marble, laminated phenolic, or asbestos composition. Slate and marble panels must be clear of metallic veins that are conductors. Some control contactors and relays are so designed that the panel upon which they are finally assembled need not be of insulating material. Steel panels are therefore sometimes used.

The arrangement of apparatus on the panel is important for many reasons. If the apparatus is located in a symmetrical arrangement, with similar apparatus placed in straight lines and with some thought for balance, the entire assembly will have a more pleasing appearance than when arranged carelessly. Cost and space usually make it imperative that panels be as small as possible. There are, however, certain clearance distances that must be observed for electrical and sometimes for mechanical reasons. Certain standards have been created for creepage and arcing distances under different conditions. On clean dry surfaces less distance is required between parts of opposite polarities than on moist and dirty surfaces. Insulating distances under oil need not be as great as in air. No apparatus should be so located that an arc will contact other apparatus or enclosures.

Any relays or parts that require adjustments or attention from time to time should be so mounted that they will be readily accessible. Knife switches should be mounted in such a way that gravity will not tend to close them.

Thermal devices should be mounted where they will be least affected by heat. This precaution is especially important if the panel is enclosed in a small cabinet. Resistors and other heat-producing parts should be as near the top as possible, so that other apparatus will not be subjected to unnecessary heat. This precaution is most important when the panel is enclosed.

Apparatus should be so arranged on the panel that the heavy copper connections needed to carry the full-load motor currents will be as short and straight as possible. Terminal points, to which the external connections from power lines and motor and power devices will be brought, must be made accessible.

Panels may be mounted on open frames or enclosed in cabinets. Open frames may be either angle iron or pipe. The angle-iron frame is most commonly used because it provides a better support for the panel. Enclosures may be of the general-purpose type, which is primarily designed to protect against accidental contact with the apparatus and is suitable for general-purpose applications indoors, where atmospheric conditions are normal. It gives some protection against dust and indirect splashing but is not dust-tight or drip-proof. Enclosures may also be dust-tight, drip-tight, weather-resistant, watertight, or submersible, or they may be designed especially for application in hazardous locations.

All enclosures should be so designed and assembled that they will withstand handling during shipment and installation. There should be sufficient space within the enclosure to permit uninsulated parts of wire terminals to be separated so as to prevent their coming in contact with each other. Enclosures should also be such as to permit proper wire connections to be made with adequate spacing of the terminals and ends of conductors from adjacent parts of the enclosures. Exposed non-arcing current-carrying parts within the enclosure should have an air space between them and the uninsulated walls of the enclosure, including conduit fittings, of at least $^1/_2$ in. for 600 volts or less. Enclosures of such size, material, or shape as not to have adequate rigidity need greater spacing. A suitable lining of insulating material not less than $^1/_{32}$ in. in thickness may be considered acceptable where the spacing referred to above is less than $^1/_2$ in.

All enclosures and parts of enclosures, such as doors, covers, or tanks, should be provided with means for firmly securing them in place. Among the available means are locks, interlocks, screws, and seals. A flush-type enclosure for mounting in a wall should be provided with adjustments to align the device with the flushplate and to compensate for differences in thickness of the wall finish.

Some controllers are so designed that the complete unit is a single assembly. Others are composed of a number of units that must be assembled on a common structure, and the various parts must then be interconnected.

Controllers of the first type are well illustrated by the manually operated faceplate and across-the-line starters. Those of the second type are represented by magnetic controllers, consisting of a number of contactors, relays, and other devices, mounted on a panel properly interconnected and assembled on a framework or in a cabinet. In such cases the main or motor circuits are made of copper strap. For small motors a properly insulated wire may be used. The control-circuit connections are made of insulated wire of relatively small capacity.

The magnetic contactors, relays, and similar parts are the units that make possible magnetic controllers for the many varied and complicated operations in modern industry. When used with suitable resistors, limit switches, master switches, etc., they control motors to accomplish any desired results.

MISCELLANEOUS. The rating of control apparatus, in general, should be expressed in volts, amperes, horsepower, or kilowatts, as may be appropriate. For convenience in application, starters are frequently rated in horsepower. In such cases the starter rating is the horsepower rating of the largest motor for which the starter is designed to be used.

Standard voltage ratings for industrial control are as follows: (1) a-c multiphase: 110, 220, 440, and 550; (2) a-c single-phase: 115 and 230; (3) d-c: 115, 230, and 550. It is considered that 208-volt a-c is adequately taken care of by 220-volt equipment.

The various kinds of ratings recognized are continuous, periodic, and 8-hour.

General-purpose industrial-control equipment is classified in accordance with its ability to interrupt current. There are two classes of general-purpose control equipment, defined as follows:

Class A: Alternating-current, air-break and oil-immersed controllers for service on 600 volts or less, and controllers for service on 2300 and 4600 volts with capacity to interrupt 10 times normal motor rating.

Class B: Direct-current controllers for service on 600 volts or less with capacity to interrupt 10 times normal motor rating.

The 1-sec thermal capacity of class A and class B controllers is 15 times the current corresponding to the horsepower rating.

The temperature rise of contacts above the temperature of the cooling air, when tested in accordance with the rating, should not exceed the following values: (1) laminated contacts, 50 deg cent, (2) solid contacts, 65 deg cent.

Mechanical parts not in contact with insulation may reach such temperatures as will not be injurious in any respect. The temperature rise of buses, connecting straps, and terminals above the temperature of the cooling air when the test is made in accordance with the rating should not exceed 50 deg cent. The temperature rise of the connectors to a source of heat, for example, resistors, thermal heaters, and power-tube anodes, should not be subject to these limitations.

Inasmuch as disconnect switches are considered unit pieces of apparatus and are usually mounted separately from the control, it is recommended that such disconnecting means be not mounted within the controller enclosure. An exception is recognized in the combination of a-c magnetic across-the-line starting switches and disconnecting means.

Where furnished, overload protection should be provided by a contactor with an overload relay or some sort of circuit breaker which will respond to excessive current in one side of a d-c and a single-phase a-c circuit and to excessive current in two sides of a polyphase circuit.

26. AUTOMATIC OPERATION

Automatic control of generating stations and substations has, in general, been applied to practically all new installations, as well as many existing ones. Unattended operation, with its increased protection and operating efficiency, is often remotely supervised and controlled from a distant control office by supervisory control.

The field of application for automatic substations and hydroelectric generating stations is rapidly broadening. In general, it may be said that any substation which would require the presence of an operator if manually operated, can be made automatic. If there are not more than two or three machines, savings in operating cost will usually result.

EQUIPMENT FOR SUBSTATIONS. Automatic substation equipment has been designed to duplicate in every way the manual operation of substation apparatus, without the attention of an operator. Starting and shutting down the station are functions of the load demand. In addition, many protective devices uncommon to the average substation give absolute protection which is free from the human element.

Automatic equipment for railway service has usually been designed for fully automatic operation; that is, the starting impulse is received automatically when the trolley voltage drops below a predetermined value. The station also stops automatically when the load demand falls below a predetermined minimum for a continuous period of time, which can be adjusted from 1 to 60 min.

Protective equipment is exceptionally complete. Every effort has been expended to make the substation equipment simple and compact without the sacrifice of any protective or automatic feature. Each device of an automatic station has its identifying number marked directly on the switchboard and on the wiring diagram. The American companies furnishing automatic substation equipment have agreed on a system of numbering so that the same devices or similar devices performing certain functions are assigned definite numbers, and an operating engineer familiar with the connections and equipment of one manufacturer can readily understand the corresponding equipment of another manufacturer.

Synchronous condensers are used for power-factor correction and for voltage maintenance in terminal substations supplied from power-transmission circuits. Automatic equipment is available for starting such condensers, connecting them to the power system, and adjusting their field circuits to obtain the maximum correction benefit from the use of the synchronous condenser. When conditions are such that the condenser is no longer needed, it is automatically disconnected from the circuit.

TYPES OF STATIONS. For railway service, automatic substations have been built for a-c self-starting synchronous converters up to 3000 kw, 600 volts direct current.

Equipment also has been supplied for motor-started converters and for 1200- or 1500-volt d-c service with one 1200- or 1500-volt machine or two 600- or 750-volt machines in series. Automatic equipment has also been furnished for use on 3000-volt railway service and for mercury arc rectifier stations.

The Ignitron rectifier is being supplied on all new installations, as well as replacing many existing rotating machines, because of its higher operating efficiency, simplicity, small spatial requirement, and reduction in weight.

Automatic equipment is frequently used in mine service. This equipment has been standardized for motor-generator sets and Ignitron rectifiers up to 300-kw, 25- or 60-cycles, 2300-volt alternating current, 275- or 550-volt direct current. This capacity is usually the maximum required, but larger capacities can be readily taken care of by substituting larger circuit breakers and contactors. Automatic synchronous converter substation equipment of like capacities can also be supplied. Three-wire automatic stations have been provided for d-c systems, usually supplied by motor-generator sets or booster converters to allow for proper variation in the d-c voltage by means of suitable voltage regulators.

Where automatic operation is desired in a-c substations without attendants, the main feature to be desired is that all transformer and feeder circuits should be normally closed ready for the delivery of power. In case of a feeder overload, the feeder breaker should trip after a short time interval and should try several times to have service restored on that feeder. For this automatic reclosing, relays are used. These relays are made in various forms and are arranged to energize one or more circuits several times in succession with regular time intervals between each energization. Also multiunit transformers are sometimes used to switch in and out automatically on load demand.

HYDROELECTRIC STATIONS. Hydroelectric stations can be furnished with automatic switching equipment for the control of waterwheels and generators. These stations may be arranged to start in any of the following ways:

1. Push-button control from a distant station.
2. Reduction of the frequency of the line.
3. The waterhead exceeding a certain height.
4. The load on other generators exceeding their rating.
5. Supervisory control equipment, operated by the dispatcher, located at some distant point.

Automatic control equipment can be applied to almost any hydroelectric generating station regardless of size, type, or head of water. Either induction generators or synchronous generators may be used, the synchronous type being preferable on most systems. With synchronous machines either the self-synchronizing method or the automatic synchronizing method of starting is used. The self-synchronizing method is employed if the relative size of the generator is small in comparison to the adjacent sections of the power system. When the waterwheel is within 3 per cent of normal speed, the generator, provided with a damper winding, can be thrown on the system and the field circuit closed. It pulls into step any self-starting synchronous motor under similar conditions.

If the generator is large in comparison to the adjacent parts of the system, or if it does not have a damper winding, an automatic synchronizing equipment is provided, which regulates the waterwheel speed and closes the breaker at the first favorable point of synchronism. Generators up to 82,500 kva have been controlled by this method. Protection against troubles in bearings, governor, exciter, etc., or overspeed, overvoltage, and similar conditions is provided by suitable relays.

Shutting-down operation in an automatically controlled hydroelectric station is controlled by protective devices operating on either the lockout relay or a master relay, or the station is shut down by hand by opening a switch. The only immediate action resulting is the de-energization of the governor solenoid. This causes the gates to close, thus shifting the load from this unit to other parallel units of the system. The generator is still connected to the line with normal excitation-delivering load in proportion to the diminishing gate opening. As the gates reach a point slightly below the no-load running position, the shunt trip coil on the oil circuit breaker opens the breaker. As the gates reach the zero gate-opening position, brakes, if used, are applied to stop the machine.

27. LOAD-DISPATCHER SYSTEMS

In any large power system with several generating stations, many substations, and possible interconnections with other systems, it is customary to center the responsibility and authority for the best operation of the system in the hands of the system operator, or load dispatcher. It is necessary for him to know at all times what stations are connected to the line, what machines are in operation, what load is being carried, what power resources are available, and what is likely to be the power demand at any hour of the day and any day of the year.

Supervisory control equipment has been created to meet a demand for equipment which automatically gives the load dispatcher visual indication of the conditions of his power equipment and at the same time provides him with a means of controlling his power apparatus. The automatic station equipment protects itself during starting, running, and stopping conditions. There are certain emergency conditions, however, over which the automatic station equipment has no control, such as a fire in a certain section of the city where the electrical circuits are supplied from automatic stations, so that it would be necessary for the dispatcher to open feeder circuits supplying this district.

An example of development along these lines is the Visicode Supervisory Control. This system combines simplicity of design with minimum line-wire cost, as all the controlling and answer-back signals are handled over two wires. The entire equipment is normally at rest, and any number of apparatus units can be controlled and supervised with a very compact equipment. Up to 60 points, the dispatcher equipment can be mounted on a single panel 16 by 90 in., and the substation equipment placed on a similar panel. From 61 to 140 points two panels are required. Greater numbers can be handled, if desired, by using more panel space.

At the substation there are a number of interposing relays, which serve to relay the signals from the relay panel to the power equipment. These are usually mounted on the supervisory switchboard.

Selective remote metering may be accomplished over the wires used for the supervisory control equipment by merely assigning a position for that function and by operating the selection key associated with that position, so that the drive circuits will be stopped. In this way the remote metering transmitters are connected with the supervisory lines at the substation, and at the same time the indicating instruments of the proper type are connected to the lines at the dispatching station and the dispatcher is given an indication of the current or voltage or other quantities to be metered.

28. BIBLIOGRAPHY

Ainsworth, C. D., The Ruptor—A Modern High-efficiency Interrupting Device for Oil Circuit Breakers, *Allis-Chalmers Elec. Rev.*, Vol. 1, pp. 9–12 (September, 1936).

Boehne, E. W., and W. A. Atwood, Anode-circuit Breaker Design and Performance Criteria, *AIEE Trans.*, Vol. 64, pp. 337–345 (1945).

Boehne, E. W., and L. J. Linde, "Magne Blast" Air Circuit Breaker for 5000-volt Service, *AIEE Trans.*, Vol. 59, pp. 202–208 (1940).

Bower, J. L., Fundamentals of the Amplidyne Generator, *AIEE Trans.*, Vol. 64, pp. 873–881 (1945).

Bowie, A. J., and C. P. Garman, 287-kv Boulder Dam Disconnecting Switches, *AIEE Trans.*, Vol. 55, pp. 582–589 (1936).

Byrd, H. L., and M. F. Beall, A Three-cycle 3500-megavolt-ampere Air-blast Circuit Breaker for 138,000-volt Service, *AIEE Trans.*, Vol. 64, pp. 229–232 (1945).

Crabbs, H. J., Control and Instrument Switches, *Elec. J.*, Vol. 30, pp. 62–65 (1933).

Edsall, W. S., and S. R. Stubbs, Circuit Interruption by Air Blast, *AIEE Trans.*, Vol. 59, pp. 503–509 (1940).

Fahnoe, H. H., A New High-interrupting-capacity Fuse, for Voltages through 138 kv, *AIEE Trans.*, Vol. 62, p. 630 (1943).

Graybill, H. W., and J. S. Ferguson, A New Outdoor Air Switch and the Principles Involved in Its Design, *AIEE Trans.*, Vol. 64, pp. 583–586 (1945).

Graybill, H. W., and J. S. Ferguson, General Electric Rotating-cam Switches, *Gen. Elec. Rev.*, Vol 40, p. 164 (1937).

Harder, E. L., and C. E. Valentine, Static Voltage Regulator for Rototrol Exciter, *AIEE Trans.*, Vol. 64, pp. 601–606 (1945).

Hill, A. W., and W. M. Leeds, High-voltage Oil Circuit Breakers for Rapid-reclosing Duty, *AIEE Trans.*, Vol. 63, pp. 113–118 (1944).

Hill, A. W., and W. M. Leeds, The Next Step in Interrupting Capacity—5,000,000 kva, *AIEE Trans.*, Vol. 64, pp. 317–323 (1945).

James, H. D., and L. E. Markle, *Controllers for Electric Motors*, McGraw-Hill Book Co. (1945).

Lapple H., New High-voltage High-capacity S.S.W. Fuse, *Siemens Review*, Vol. 7, No. 5, pp. 106–110 (1931).

Linde, L. J., and B. W. Wyman, Magnetic-type Air Circuit Breaker for 15,000-volt Services, *AIEE Trans.*, Vol. 63, pp. 140–144 (1944).

Lingal, H. J., H. L. Cole, and T. R. Watts, Oil-impregnated-paper High-voltage Condenser Bushings for Circuit Breakers and Transformers, *AIEE Trans.*, Vol. 62, pp. 269–275 (1943).

Lohansen, R. A., Recent Developments in High-voltage Fuses, *A.E.C. Progress* No. 2, p. 28 (1936).

Ludwig, L. R., and R. H. Nan, Magnetic "De-ion" Air Breaker for 2500–5000 Volts, *AIEE Trans.*, Vol. 59, pp. 518–522 (1940).

Ludwig, L. R., H. L. Rawlins, and B. P. Baker, New 15-kv Pneumatic Circuit Interrupter, *AIEE Trans.*, Vol. 59, pp. 528–533 (1940).

MacNeill, J. B., and A. W. Hill, Multiple-grid Breakers for High-voltage Service, *AIEE Trans.*, Vol. 58, pp. 427–431 (1939).

Monseth, I. T., and P. H. Robinson, *Relay Systems—Theory and Application.* McGraw-Hill Book Co (1935).

Paxton, R., and H. E. Strang, Design of Current-carrying Contacts in Modern Switchgear, *Gen. Elec. Rev.*, Vol. 36, pp. 524–528 (1933).
Prince, D. C., Theory of Oil-blast Circuit Breakers, *AIEE Trans.*, Vol. 51, pp. 166–170 (1933).
Prince, D. C., and W. F. Skeats, Oil-blast Circuit Breaker, *AIEE Trans.*, Vol. 50, pp. 506–512 (1931).
Prince, D. C., and E. A. Williams, Jr., The Current-limiting Power Fuse, *AIEE Trans.*, Vol. 58, pp. 11–18 (1939).
Rawlins, H. L., and J. M. Wallace, Modern High-voltage Fuses, *Westinghouse Engineer*, pp. 153–157 (September, 1944).
Rawlins, H. L., and H. H. Fahnoe, A New Three-element Current-Limiting Power Fuse, *AIEE Trans.*, Vol. 63, p. 156 (1944).
Schaelchlin, W., 150,000-Ampere Contractor, *Elec. J.*, Vol. 33, pp. 363–366 (August, 1936).
Seaman, J. W., Modern Trends of Low-voltage Air Circuit Breakers, *AIEE Trans.*, Vol. 59, pp. 24–30 (1940, January Section).
Slepian, J., Extinction of an A-c Arc, *AIEE Trans.*, Vol. 47, pp. 1398–1407 (1928).
Slepian, J., Theory of the "De-ion" Circuit Breaker, *AIEE Trans.*, Vol. 48, pp. 523–527 (1929).
Slepian, J., Extinction of a Long A-c Arc, *AIEE Trans.*, Vol. 49, pp. 421–430 (1930).
Slepian, J., and C. L. Denault, The Expulsion Fuse, *AIEE Trans.*, Vol. 51, pp. 157–165 (1932).
Spurck, R. M., and H. E. Strang, New Multibreak Interrupter for Fast-clearing Circuit Breakers, *AIEE Trans.*, Vol. 57, pp. 705–710 (1938).
Strang, H. E., and A. C. Boisseau, Design and Construction of High-capacity Air-blast Circuit Breakers, *AIEE Trans.*, Vol. 59, pp. 522–527 (1940).
Strom, A. P., and H. L. Rawlins, The Boric Acid Fuse, *AIEE Trans.*, Vol. 51, pp. 1020–1025 (1932).
Todd, R. W., and W. H. Thompson, *Outdoor High-voltage Switchgear.* Sir Isaac Pitman and Sons, Ltd., London (1937).
Van Sickle, R. C., and W. M. Leeds, *Recent Developments in Arc-rupturing Devices, AIEE Trans.*, Vol. 51, pp. 177–184 (1932).
Wagner, C. F., and G. D. McCann, *Lightning Phenomena, Elec. Engineering*, Vol. 60, pp. 374–384 (1941, August Section); pp. 438–443 (September Section); pp. 483–500 (October Section).
Wagner, C. F., and R. D. Evans, *Symmetrical Components.* McGraw-Hill Book Co. (1933).
Westinghouse Electric and Manufacturing Company, *Electrical Transmission and Distribution Reference Book* (1942).
Westinghouse Electric and Manufacturing Company, Industrial Control, *Extension Course* 15 (November, 1940).
Westinghouse Electric and Manufacturing Company, *Silent Sentinels, Protective Relays* (1940).
Wyman, B. W., and J. H. Keagy, New Solenoid Mechanism for the "Magne-blast" Breaker, *AIEE Trans.*, Vol. 64, pp. 268–274 (1945, May Section).

SECTION 13

POWER STATIONS AND SUBSTATIONS

STEAM-ELECTRIC POWER STATIONS

By F. S. Bennett

HYDROELECTRIC POWER STATIONS

By R. A. Hopkins

INTERNAL-COMBUSTION POWER STATIONS

By W. A. Sloan

POWER-STATION CIRCUITS

By R. A. Hopkins

SUBSTATIONS

By W. F. Wetmore

Substations for Light and Power Systems

Railway Substations

POWER STATIONS AND SUBSTATIONS

STEAM-ELECTRIC POWER STATIONS

By F. S. Bennett

1. PLANNING STEAM-ELECTRIC STATIONS

The factors affecting the planning are: location, capacity, size of units, and steam pressure and temperature.

Data on power-station steam equipment will be found in Kent's *Handbook for Mechanical Engineers*, John Wiley; on structural details, in Merriman's *American Civil Engineers' Handbook*, John Wiley; and on buildings in Kidder-Parker's *Architects' and Builders' Handbook*, John Wiley.

LOCATION. This factor may be subdivided as follows:

1. Real estate values.
2. Location of market.
3. Water supply (quantity and temperature).
4. Cost of intake and outlets for water.
5. Prevailing direction of winds (should be away from residential sections).
6. Railroad and water transportation facilities.
7. Local restrictions on transmission lines.
8. Area of real estate available for coal storage and ash disposal.

CAPACITY. The capacity is determined by:

1. Water supply.
2. Market, present and future.
3. Interconnections to existing systems.

SIZE OF UNITS. The steam and electric units should be carefully studied for the following:

1. Present requirements.
2. Future growth.
3. Type of service (industrial or utility).
4. Expected load curve of station.
5. Spare capacity.

The industrial power plant requires careful study, as small units generally prevent expansion to larger units because of building limitations. One solution is to build in units with little or no building to hamper the next unit, which may be larger or smaller.

Central stations are now so interconnected electrically that each one is a potential source of emergency service for the others. The size of units, therefore, need not be determined by the spare capacity or the load curve, which may be adjusted by making certain stations for base load. The other factors are more important.

STEAM PRESSURE AND TEMPERATURE. These factors affect the fuel economy only. The contributing items that may increase or decrease the value of the fuel economy are:

1. Cost of fuel.
2. Load curve (total output and daily output).
3. Reliable apparatus available.
4. Water supply.

Figure 1 shows the maximum values possible to design turbine generating equipment by increase in steam pressures. The limitations in application are restricted by the ability of the manufacturer to make reliable apparatus for these values.

Figure 2 gives data on actual performances of generating stations using turbines and regenerative feedwater heating. The binary cycle of mercury-vapor use is in service and is being extended. Its development along the lines that are expected may well supplement or replace the higher steam pressures.

The present commercial limit of steam pressure and temperature is 1200 to 1800 lb per sq in. and 1050 deg fahr total temperature. Machines which have been built for 2500 lb and 950 deg fahr are in successful operation.

The central generating station or utility plant has been, in general, the user of high steam pressure and temperature. The industrial plant which has coincidental demands of relatively large sizes for steam and power could economically employ the high steam pressures and temperatures. There are, however, several detrimental factors which have prevented this type of development in industrial plants.

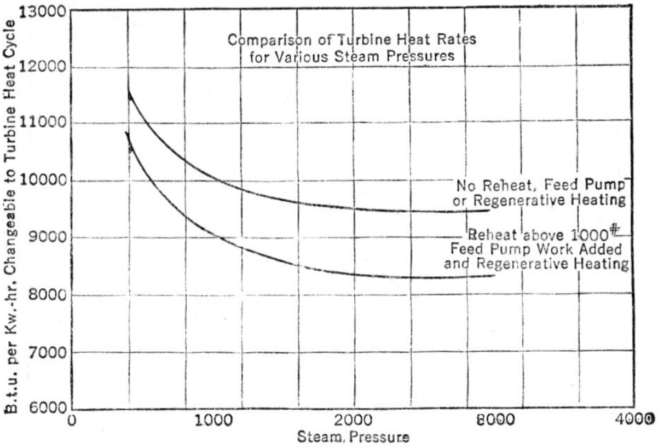

FIG. 1. Heat Rates for Various Steam Pressures

The first factor is cost of spare generating capacity for protection of continuity of service.

The second factor relates to boiler operation. It is necessary to have pure scale-free water. This is difficult and costly to obtain in the industrial plant, which usually loses from 25 to 50 per cent of the steam generated, so that this loss must be made up from some other source, such as a feed treating system. Such a treating system is expensive and many times is entirely impracticable.

The third factor relates to operation. The industrial plant does not have the technical help necessary for the operation of the more complicated system required by the high

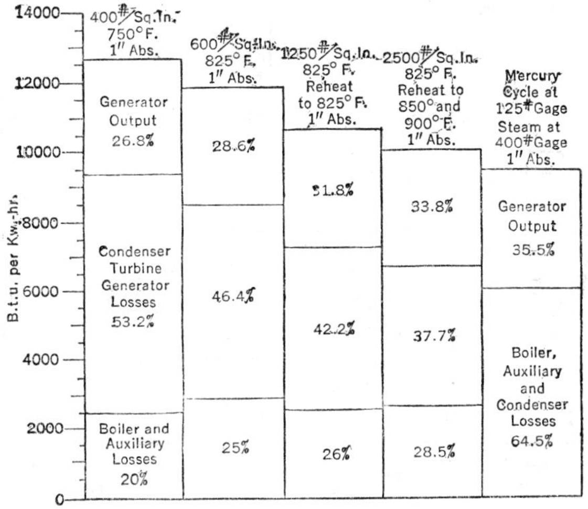

FIG. 2. Distribution of Losses in Steam Power Stations

temperature and pressures. The general results are service interruptions and high maintenance costs.

These factors have held industrial pressures to 400 to 600 lb and 750 to 800 deg fahr total temperature. The general tendency, however, is to increase both pressures and temperatures in new plants and in extensive additions to old plants.

2. HEAT BALANCE

The distribution of heat in a power station must be known to fix the requirements for apparatus.

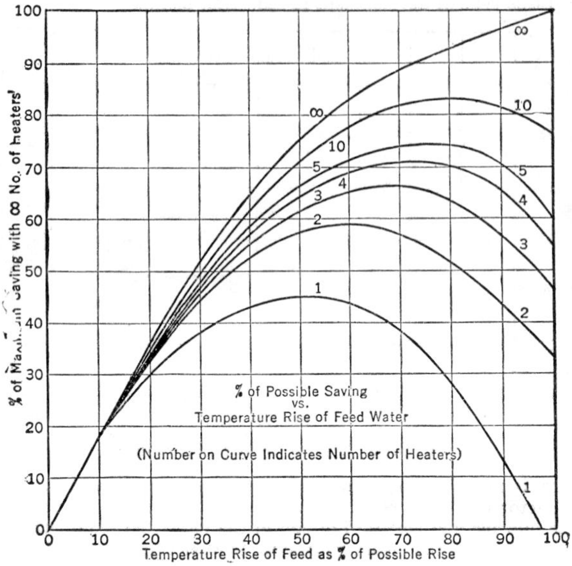

Fig. 3. Basis for Selecting Number of Feedwater Heaters

The distribution of the heat delivered to the turbine has become a very complicated affair because of regenerative feedwater heating, deaeration of feedwater and evaporation of boiler make-up water.

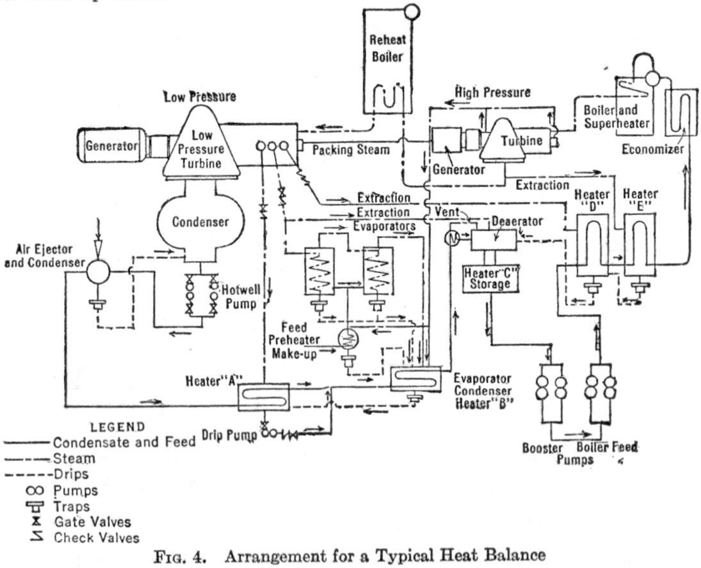

Fig. 4. Arrangement for a Typical Heat Balance

FEEDWATER HEATERS. Figure 3 gives a basis for selecting the number of feedwater heaters and their location in the steam pressure range of the machine. It should

Table 1. Calculations Turbine Heat Cycle: 80,000-kw unit

Conditions: 1422 lb gage, 825 deg fahr total temperature, 1 in. mercury absolute boiling point.

Kw output	48,985	81,330
Throttle pressure, lb gage	1,422	1,422
Throttle superheat, deg fahr	235	235
Btu per lb at throttle	1,387	1,387
Throttle flow, lb per hr	362,000	652,000
Throttle w.r	7,390	8,016
Btu per kwhr charged to cycle	9,010	9,000
Condenser flow, lb per hr	271,700	460,920
Condenser w.r	5,543	5,669
Btu per lb in exhaust l.p.t	1,024.0	1,011.4
H.p.t. exhaust pressure, lb abs	231	405
Btu per lb in exhaust of h.p.t	1,239.6	1,277.8
Flow through reheater, lb per hr	328,180	582,900
L.p.t. inlet pressure, lb abs	215.0	377.5
Btu per lb at l.p.t	1,438.0	1,431.9
Superheat at l.p.t inlet	437	386
Generator losses, h.p. unit	576 kw	640 kw
Generator losses, l.p. unit	1,362 kw	1,618 kw
Total packing leakage, lb per hr	6,000	10,700
Btu per lb in packing steam	1,333.9	1,359.9
Evaporator condenser pressure, lb abs	7.27	11.77
Pounds per hour steam to air ejector	2,000	2,000
Temperature condensate from hotwell	79.3	79.2
Temperature condensate from air ejector cond	87.4	84.2
HEATER A		
Shell pressure, lb abs	4.56	7.56
Heater pressure, lb abs	4.07	6.80
Btu per lb in extracted steam	1,123.0	1,113.0
Saturated steam temperature at heater	153.8	175.5
Condensate flow through heater	273,700	462,920
Pounds per hour steam extracted	15,200	38,360
Temperature condensate leaving	143.8	165.5
Temperature rise due to return of drips	0.7	1.0
EVAPORATOR CONDENSER HEATER B		
Pressure in heater, lb abs	7.27	11.77
Saturated temperature in heater	178.5	201.0
Condensate flow, lb per hr	294,900	511,980
Temperature entering heater	144.3	166.3
Temperature leaving heater	168.5	191.0
DEAERATOR HEATER C		
Shell pressure, lb abs	25.9	44.3
Heater pressure, lb abs	23.3	39.9
Btu per lb in extracted steam	1,238.4	1,231.0
Saturated temperature at heater	236.3	267.2
Condensate flow to heater	294,900	511,980
Extracted steam, lb per hr	14,680	27,820
Temperature leaving heater	235.3	266.2
Condensate leaving heater	362,000	652,000
HEATER D		
Shell pressure, lb abs	100.3	173.1
Heater pressure, lb abs	90.2	155.8
Saturated temperature steam at heater	320.5	361.5
Btu per lb in extracted steam	1,359.8	1,354.0
Condensate flow through heater	362,000	652,000
Pounds per hour steam extracted	24,300	51,700
Temperature leaving heater	310.5	351.5
Evaporator condenser pressure, lb abs	7.27	11.77
Pounds per hour steam to air ejector	2,000	2,000
Temperature condensate from hotwell	79.3	79.2
Temperature condensate from air ejector cond	87.4	84.2
HEATER E		
Shell pressure, lb abs	231	405
Heater pressure, lb abs	207.9	364.5
Btu per lb in extracted steam	1,239.6	1,277.8
Saturated temperature steam at heater	385.2	435.8
Condensate flow through heater	362,000	652,000
Pounds per hour steam extracted	26,120	60,500
Temperature leaving heater / Final feedwater temperature	375.2	425.8

be noted that practical difficulties in building a high-pressure machine to extract steam at high initial pressures may be encountered. A simple calculation will show the greater gain from regenerative feedwater heating that can be made by 1400-lb initial steam pressure as compared to 400-lb initial pressure.

HEAT BALANCE. Figure 4 shows an arrangement for a typical heat balance. Sample calculations based on this arrangement are given in Table 1.

To make these calculations the following are required: an up-to-date steam table, Mollier diagram with expansion lines of the turbine, pressure flow curves for the extraction points, exhaust loss of the machine, efficiency of the generator, and packing leakage. The machine designer must furnish all these data but the steam table. (See Kent's *Handbook for Mechanical Engineers.*)

The heat distribution from the fuel to the turbine is generally very simple so far as station design is concerned. The cost and load factor fix the heat recovery beyond the boiler. For pressures up to 600 lb, the boiler takes out 75 to 80 per cent of the heat, and an air preheater gets an additional 10 per cent. The preheater limit is to keep the exit-flue gases above the dew point, which varies from 250 to 300 deg fahr. Exit gases below these values form corrosive acids, which rapidly destroy the heat-recovery equipment.

Pressures above 600 lb require **economizers** to keep the preheated air temperature to a safe value.

3. SELECTION AND TYPE OF APPARATUS

The selection of suitable apparatus is largely a matter of experience. The following general rules should be applied:

1. Make a diagram showing services and connections of other apparatus.
2. Write a specification defining services and capacities.
3. Determine the most suitable manufacturers by records of past performance in other plants or by other suitable methods.
4. Interview the manufacturers' representatives, requesting data on a definite set of requirements.

DIAGRAM. The preparation of a diagram allows the engineer to complete the picture of the requirements and relations of the various pieces of apparatus and paves the way for a more complete specification.

SPECIFICATIONS. It is generally undesirable to try to make a specification cover the manufacturing details. The writer usually knows less than the manufacturer and quite often radically increases the cost without any gain in utility or reliability.

The specifications, however, should define the services required of the equipment, the capacities (normal and maximum), the auxiliaries to be furnished, the foundations, erection, and other work to be performed by the contractor.

The types of suitable apparatus are governed by so many factors that only a general discussion of the principal apparatus can be given with a few definite rules.

BOILER TYPES. Steam boilers 250 lb operating pressure and above are all water tube type. Pressures below this value are, in general, fire tube types. The multiple-drum boiler has in general succeeded the header-type boiler, although this type is still manufactured to a limited extent.

BOILER FURNACES. Boiler furnaces are of two general types, those with refractory walls and those with water-cooled walls or combinations of water-cooled and refractory walls.

The furnace with complete refractory walls is limited in capacity by the requirement for temperatures such that the walls do not melt or spall.

A common method of rating furnace capacity is heat released per cubic foot of furnace volume by the burning fuel. The maximum value for solid refractory walls is 25,000 to 15,000 Btu per cu ft.

The furnace with completely water-cooled walls has a maximum heat capacity of 60,000 Btu per cu ft and an average value of 30,000. The furnace for burning pulverized coal is usually water-cooled, and the heat release is regulated by the coal to be burned, primarily its ash content and melting point, if the ash is to be removed in a dry state. The high values of heat release in general give liquid ash. In some cases the walls are even shielded by refractory or metal shields.

Stokers can be successfully operated with solid refractory furnaces and very seldom have completely water-cooled furnaces.

TYPES OF FUEL BURNING. There are in general four types of fuel used for steam generation: coal, oil, gas, and industrial waste. The burning of coal is usually accomplished on grates, hand or stoker, and in pulverized form. Oil is burned through burners, pressure-atomized, steam- or air-atomized, and natural draft. Gas is burned in the same type of burner as oil and sometimes in combination with it. Industrial waste is burned on grates, a special design being generally required for the particular type of fuel available.

The selection of the type of fuel burning is affected by the characteristics of the coal. High ash content and low melting point give difficult operating conditions for pulverized coal and most types of stokers. The spreader type of stoker has been developed to handle this coal and is the most successful so far available. Coals of medium ash content, 10 per cent or less, with melting points of 2300 deg fahr and above give good performance with almost any type of equipment.

The type of coal burning selected is also affected by the operating characteristics of the station. Higher efficiency can be obtained with pulverized coal, and greater instantaneous load swings can be handled than with stokers. Boilers below 75,000 lb per hr have better operating characteristics with stokers than with pulverized fuel.

ASH SYSTEMS AND CINDER ELIMINATION. The disposal of ashes has become a very considerable problem for powdered-fuel coal burning. The ash from powdered fuel is of no practical use for fill unless it is mixed with some material that will retain it against winds, particularly if the ash is obtained from cinder-elimination systems. In many cases it is necessary to keep the ash wet until it can be covered with a retaining material.

The removal of ashes from the furnace is generally accomplished by either a water-jet system or an air-transport system. Both systems are used extensively, and their application depends in general upon local conditions for the disposal of the ash.

The elimination of ashes from the stack gases of a power station has become a necessity. People living near industrial and central-station generating plants have become conscious of the ash discharged from the stacks and spread over the surrounding countryside, and they demand its elimination. There are three systems for accomplishing this result: special draft fan, centrifugal or mechanical precipitator, and Cotrell or electric precipitator. (See Section 18, Articles 24 to 31.)

The fan type has been used the least, although it is economical in space and cost, since it uses the same fan as is used for boiler draft. This type, however, has not been very effective and requires further development.

The mechanical or centrifugal type has been most widely installed. It requires a considerable cost for maintenance and power for operation.

The **Cotrell** or electric process has been installed to a considerable extent. It is the most efficient type but has some serious disadvantages. It has a very high first cost for apparatus and requires extensive foundations and building space. The material that it catches is very fine and is difficult to retain and dispose of.

The general problem of ash elimination has not been very well solved. Considerable study is still needed, for there will undoubtedly be adverse legislation in the future.

BOILER AUXILIARIES. Most stations have electric drive for auxiliaries. An exception is the spare boiler-feed pump, which is usually a steam-driven unit. Feed pumps are generally of the centrifugal type above 50 gal per min and of constant speed. Fans for draft are constant in motor speed with vane inlet control for forced draft. The induced-draft fans are usually variable in speed. Three types of speed control are available: adjustable-speed motors, hydraulic coupling between motor and pump, and induction coupling between motor and pump.

The provision of scale-free boiler feedwater is very important. The method of providing this water is dependent upon the chemical constituents of the supply and the volume requirements for water. Plants that have requirements of 10 per cent or less generally use evaporators. Plants having 10 to 100 per cent make-up use lime soda or Zeloite exchangers and add chemicals in the boiler drums for final conditioning. The selection of the Zeloite or lime soda depends upon the chemicals in the raw make-up and is beyond the scope of this article.

INSTRUMENTS. The modern station requires indications of apparatus performance. The instrument has taken the place of the guess of the experienced operator. The following list of instruments is about the minimum required to operate one modern boiler.

1. Meter to indicate and record steam pressure, steam temperature. steam flow (integrate), and air flow.
2. Meter to indicate and record feedwater pressure, temperature, and flow (integrate).
3. Meter to indicate and record gas temperatures (usually four pens).
4. Meter to indicate and record water level in boiler drum.
5. Meter to indicate air and gas pressures (usually six to eight separate indicators).
6. Meter to indicate and record temperature of coal and air mixture and mill differential pressure (pulverized fuel only).
7. Meter to record boiler gas and air temperatures, including preheater and economizer.
8. Indicating gages for special requirement of boiler, such as spare pump and head tank pressure.

The turbine room also requires instruments for measuring the following quantities:

1. Inlet steam pressure.
2. Vacuum.
3. Barometric pressure.
4. Generator air temperature.
5. Turbine stage pressure.
6. Turbine speed.
7. Generator output.

These instruments are the minimum; generally several special ones are required for the particular installation. In addition, a signal outfit is necessary.

TURBINE GENERATOR. The selection of a turbine generator requires application studies for capacity, average condenser vacuum, and steam pressure and temperature. The capacity is determined by the service requirements. The vacuum is a function of circulating-water temperatures; data on this point are usually available from local, state, or federal sources. The steam pressure and temperature are a function of fuel economy that can be justified by calculations of output load curves and fuel costs.

There has been a gradual increase in turbine capacity for 60-cycle generation at 3600 rpm. The maximum capacity at present is 100,000 kw at 3600 rpm for 1250 lb per sq in., 1050 deg fahr total temperature. Machines are available for higher pressures, up to 2500 lb. They are, however, special, requiring justification for their application, and hence are not likely to come into general use.

CONDENSERS. The features that must be decided are tube length, size of tubes, material, single or double water pass, and divided water boxes.

The tube length is often a factor of available space. Generally speaking, it is desirable to keep the tube short, say 12 to 14 ft for 5000- to 10,000-kw machines, 16 to 18 ft for 15,000- to 25,000-kw machines, and 20 to 30 ft for larger machines. One-inch tubes are desirable where the water is dirty, and $^3/_4$- or $^5/_8$-in. tubes are used only where the water is clean. The merits of single or double water-pass arrangements are dependent upon the amount of water available and the pumping head, exclusive of the condenser.

The condenser auxiliaries have become very nearly standard: steam-jet air-removal apparatus with surface coolers and constant-speed motor-driven centrifugal circulating pumps. The large units have at least two pumps, one of which may be used with cold water, thus saving power input. The condensed steam is handled by constant-speed motor-driven centrifugal pumps.

4. PIPING AND VALVES

The increase in pressure and temperature of steam power plants has made it necessary to study very carefully the expansion stresses and flange construction. The general arrangement tends toward unit systems, particularly in large plants. This trend is due to increased pressure and temperature, which make it difficult to handle the large headers required.

PIPING ARRANGEMENTS. The parallel system of piping is shown in Fig. 5. This system has many merits and should be used where the capacities and pressure do not require headers beyond the limits of safety.

The unit system (Fig. 6) is used for larger systems and particularly for high pressures, such as 400 to 1200 lb. The majority of operators like to interconnect these units to obtain more flexibility. This interconnection does not provide for header capacity to carry the capacity of both units connected.

EXPANSION PROVISIONS. The theory underlying the standard forms of pipe bends is well covered by the ASME paper of W. H. Shipman, 1930. See Kent's *Mechanical Engineers' Handbook*, Vol. 2, Section 5, of this series.

FLANGES AND WELDED JOINTS. The electric-arc welding of pipe has practically replaced flange construction for steam lines and valve connections. Many types of apparatus to which the pipe can be welded are also available. In general, however, most apparatus still has flange connections.

The welding of pipe lines has been extended to include low-pressure and water lines and has been found to eliminate the considerable maintenance cost that was incurred with flange connections.

PIPE SIZES. The pipe size is fixed by the allowable drop in pressure. Steam pipes usually have **vapor velocities** from 7000 to 11,000 ft per min. Lower velocities generally represent too large an investment. The usual velocity for a water pipe is 5 to 8 ft per sec. Condenser hot-well suction pipe may be run at 4 ft per sec to obtain suitable pump suctions and syphon effects. Gas and air pipes are usually run near the lower limit for steam, 5000 to 7000 ft per min.

VALVES. Steel valves, either cast or forged, are required at temperatures above 600 deg fahr. Globe valves should be used wherever the medium will be throttled. Gate valves should be used for stop valves because their pressure loss is low. The material of

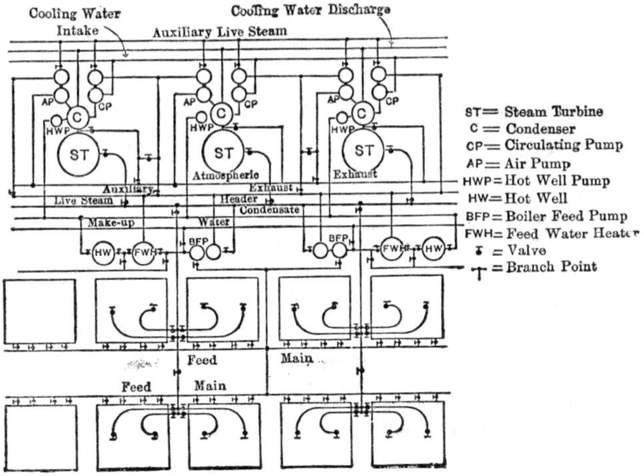

FIG. 5. Parallel Method of Connection

the seats and disks of a valve is very important. Very hard non-abrasive surfaces free from warping and cracks are required. These seats and disks should be renewable and capable of refitting.

Motor operation allows valves to be installed in otherwise inaccessible places. The motor operation is a valuable feature for relatively quick operation and for shutting off the lines in case of accident.

A safe rule is to by-pass all valves 6 in. and larger for any pressures above 100 lb. Stop valves should be placed on all pipe lines where they leave the header. Drains are very

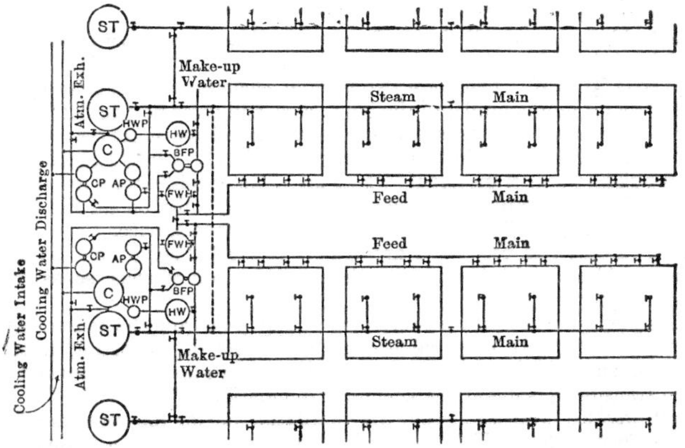

FIG. 6. Unit Method of Connection

important on steam lines. They should take care of both the live and dead conditions of the pipe lines.

PIPE INSULATION. Steam and hot-water lines require insulation from heat loss; cold-water lines, from condensation. The general type of insulation is 85 per cent magnesia for temperatures up to 500 deg fahr, and special high-temperature type for higher values. There are many other types, such as air cell, hair felt, rock wool, and aluminum foil. All these have a place, but require special study to justify their application. The thickness of insulation should be the subject of a cost study. Quite often the facts are not available for this study, and insulation is arbitrarily selected. Table 2 can serve as a guide.

Table 2. Thicknesses of Pipe Coverings

Pipe Size	Thickness Higher Temperature Type, Inches	Thickness 85 Per Cent Magnesia, Inches	Thickness Cement, Inches
Temperatures 800 to 500 Deg Fahr			
18 to 10 in. inc.	1 1/2	2	1/2
9 to 6 in. inc.	1 1/2	1 3/4	1/2
5 to 4 in. inc.	1	2	1/2
3 in. and smaller	2	....	1/2
Temperatures 500 to 300 Deg Fahr			
18 to 10 in. inc.	....	2 1/2	1/2
9 to 6 in. inc.	....	2	1/2
5 to 4 in. inc.	....	1 1/2	....
3 in. and smaller	....	1 1/2	....

The covering of the insulation depends upon the location. Indoors, where no water or high-moisture air is present, canvas can be used, sewed on where appearance must be considered and pasted on where appearance is not important. Wet places should have two layers of roofing felt wired on and painted with asphalt between layers.

5. ARRANGEMENT OF APPARATUS

The arrangement of power-station apparatus is largely dependent upon the kind of service, water conditions, soil conditions, and individual ideas. A few typical considerations will be given.

1. Natural light, as far as possible, should be provided for.
2. Ventilation should leave no dead pockets.
3. Apparatus should be accessible for repairs.
4. Piping should be laid out before the major apparatus is fixed in location.
5. Picture drawings should be made so that the plant can be visualized.
6. Future expansion of the plant should be provided for by drawing studies of the largest and smallest units.
7. It is desirable to group all the feed pumps and service water pumps so that one operator can control them.

GENERATING ROOM. There are two general schemes for arranging a turbine electric generating room: units parallel to the boiler axis, and units perpendicular to the boiler axis. The parallel units fit better where there is a multiplicity of boilers. The perpendicular units are arranged to better advantage for a single boiler or unit scheme. The parallel units save on building and crane spans, which are a factor with large, long turbine units.

The well-designed generator room sets the generating units on an island foundation and leaves the basement well lighted and accessible to crane service.

The modern turbine and condenser are usually connected solidly on the exhaust, and expansion is provided for by supporting the condenser on springs, so as to give very little uplift on the turbine.

The switching equipment and transformers have all gone from the generating room and are arranged outdoors, with only cable connections and disconnecting switches at the machine. A convenient and general arrangement is to have these switches and the current and potential transformers under the generator foundation.

The general practice is to water-cool the ventilating air for the generator by means of finned tube surface coolers. The ventilating air is circulated in a closed-duct system through these coolers and the generator by means of fans on the rotor of the generator or separate motor-driven fans. This system excludes dirt and limits the fire hazard.

The use of steam temperatures from 500 to 1000 deg fahr has created a considerable fire hazard from the lubricating oil. Storage tanks, filtering systems, circulating pipes, and everything containing oil should be kept as far as possible from the steam lines. Flanges should be welded on, and as few joints as possible made. If flanges are used, they should have not less than four bolts, and in general it is desirable to construct them to the 400-lb steam standard.

BOILER ROOM. The modern boiler room should be well lighted, well ventilated, and free from dust and smoke. These features are obtained by making the fuel-handling equipment dust-tight and placing the operating floor at the lowest level with no intervening floors, access to the upper part being by galleries and stairs. Good ventilation can be obtained where there are no floors and usually natural light. The general tendency is to make boilers large and to expect continuous service with few interruptions for repairs. Making boiler rooms clean and cool has removed the reason for walling off the generator room. The present tendency is to combine the boiler and the turbine room. This arrangement saves space and cost and tends to reduce the number of operators required. Another step that has followed the combination of the turbine and the boiler room is partial elimination of the building structure. A notable example is the General Electric Company Outdoor Station at Schenectady, N. Y.

AUXILIARY EQUIPMENT. The auxiliary equipment, such as boiler-feed pumps, house-service pumps, and feedwater treating, should be concentrated as much as possible, leaving the piping simple and compact. This concentration helps to keep the number of operators small, as the control of this equipment can well be combined with that of the condensing equipment by locating the two adjacently. The general practice is to make the normal units motor-driven and have a few spare steam units for emergency and starting.

6. POWER-STATION COSTS AND ECONOMICS

The general division of costs among parts, such as buildings, apparatus, facilities, and construction labor, remains constant. Figures 7 and 8 show overall cost and division of

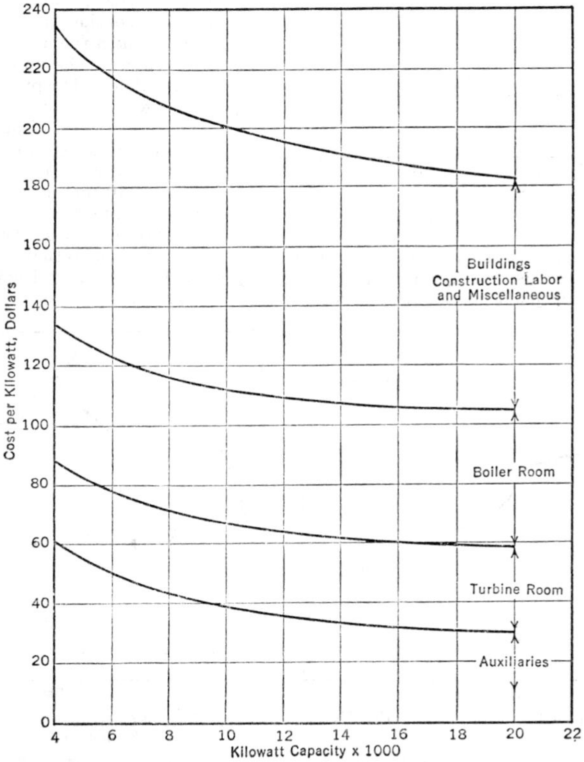

FIG. 7. Power-station Costs

costs for 1946. These figures should be corrected for the cost indexes that are in use at the moment. The slope of the curve for size also does not change within the range for which a general figure can be calculated.

The possible improvements that can be made to modernize an existing central station are: higher steam pressure and temperature, better boiler efficiency, and new topping turbines or new condensing turbines.

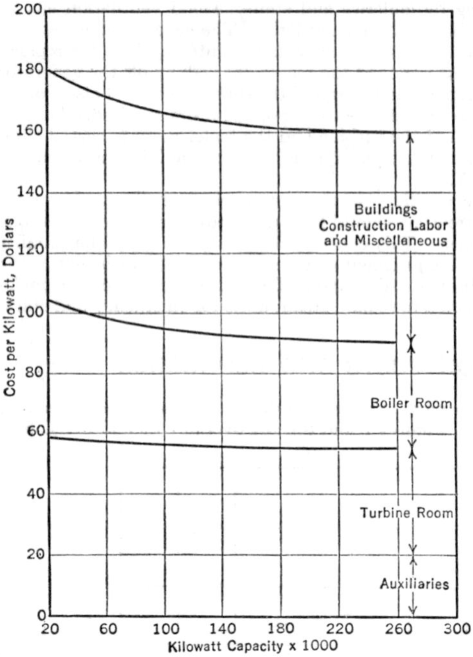

FIG. 8. Power-station Costs

The need and extent of such changes depend upon the balance between cost and saving in electric production costs. Extensive studies usually precede such changes. Diagrams such as those shown should be used to determine sizes and costs. The exact diagram is a function of type of changes. For instance, if the old plant is to be retained, turbines will be used as reducing valves to supply steam for the existing units. New boiler and facilities will be required. This is called topping a plant. If, on the other hand, the old plant is to be replaced by a new plant retaining only location, coal, water, and building, entirely new turbines, boilers, and facilities will be required.

Figure 9 defines the use factor of the overall station, providing the data for the calculation of heat consumption of the station. From it can also be taken the data needed to decide a size for the unit or units topping a plant. The preliminary selection is generally for not less than a 50 per cent use factor. The diagram also provides data to calculate the fuel costs for the station.

Figure 10 provides the data for a selection of machine sizes for greatest economy and lowest cost for adequate spare capacity. The usual plant has considerable daily variation, so that it is necessary to prepare diagrams for a high, an average, and a low day. It is usual to select the machines for the topping plant so that the existing units carry the peaks and emergency load.

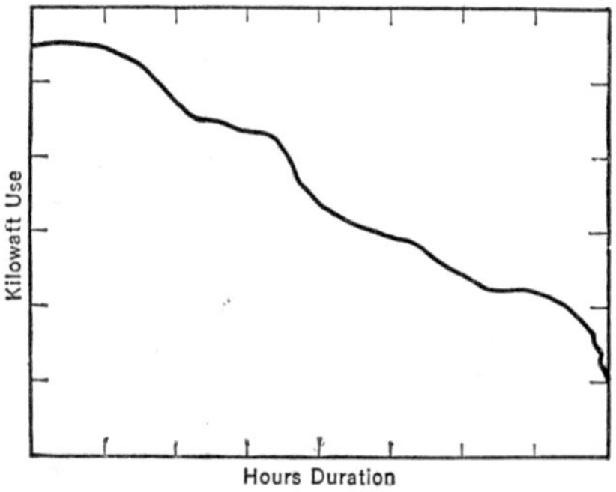

FIG. 9. Form of Chart for Defining Use Factor

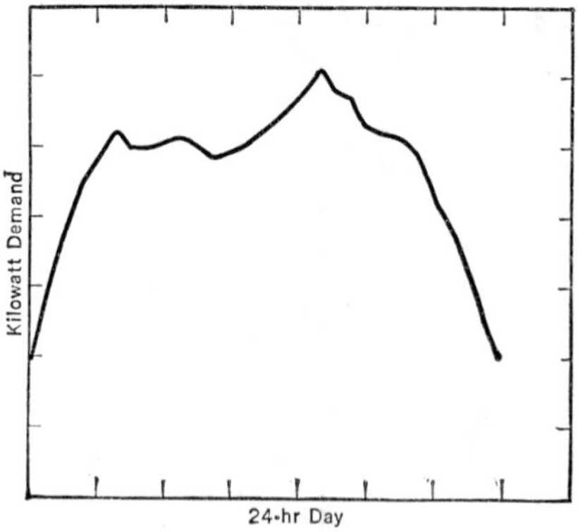

FIG. 10. Form of Chart Giving Data for Selection of Machine Size

Figure 11 generally has to be prepared for more than one steam pressure and temperature. Data can be taken from this figure to determine the fuel consumption of the plant for the actual or assumed conditions of operation.

Correlation of the data from these diagrams provides a means of determining the most economical cost, both investment and operating, for a plant.

The industrial power station generally has to provide large quantities of steam for heating and manufacturing and variable quantities of electric power. The plant which has to provide large quantities of steam can economically generate its electric power.

The heating and manufacturing steam can be provided by generating steam at a higher pressure than is required and reducing to the required pressure through a turbine generator. This procedure provides electric power at a heat rate equivalent to a kilowatt-

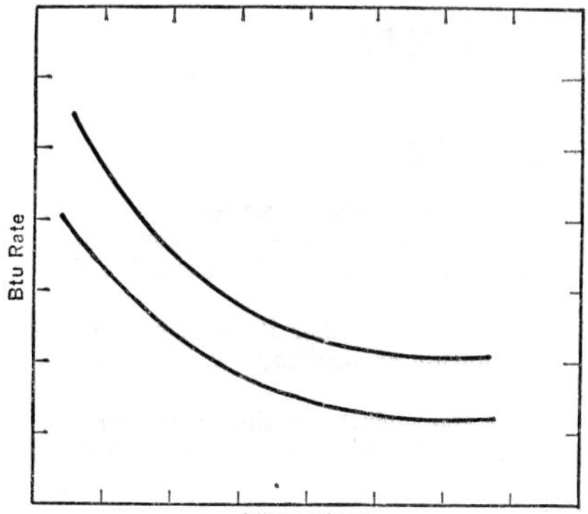

Fig. 11. Form of Chart for Use in Determining Fuel Consumption

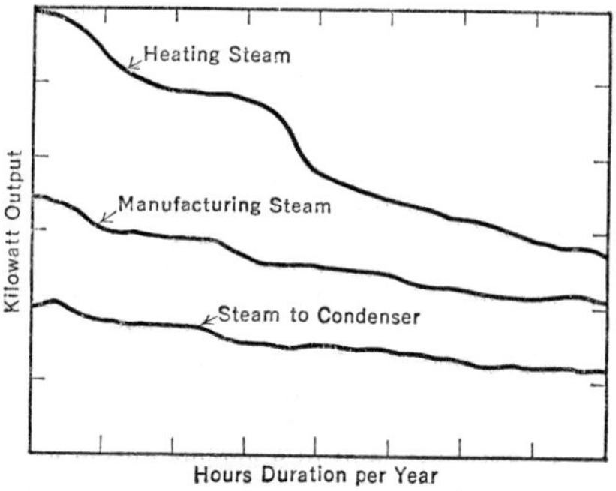

Fig. 12. Steam for By-product Power

hour plus the external losses, such as bearings and electric generator losses. This rate is about 4000 Btu per kwhr.

The diagrams required to make a study application are shown in Figs. 12, 13, and 14. Figures 9 and 10 are also required for the study. Figure 12 is similar to Fig. 9 and defines the available steam for by-product power production. The size of the machine or machines is determined from Fig. 10. Figures 13 and 14 then determine what electric power can be produced by by-product; the rest of the power can be made condensing or purchased. The steam pressure and temperature generally will not be less than 400 lb and 750 deg fahr nor higher than 800 to 850 deg fahr. The data required for selection include investment costs and fuel costs for a year's operation. These costs are generally compared to costs of purchasing power from a central station company.

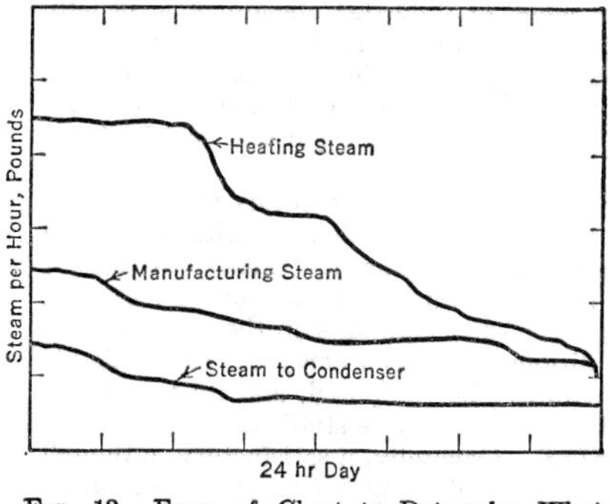

Fig. 13. Form of Chart to Determine What Electric Power Can Be Produced by By-product

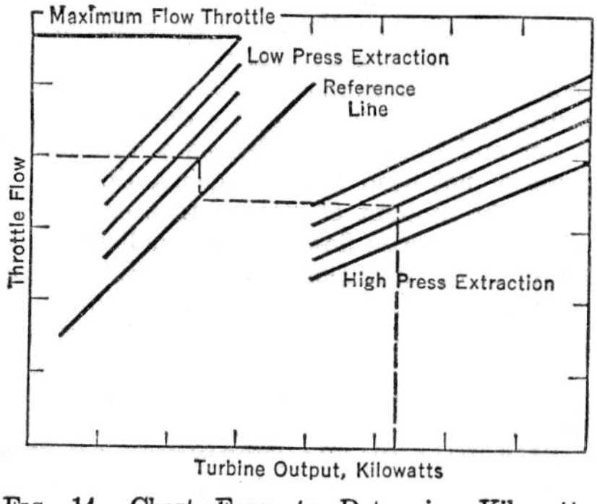

Fig. 14. Chart Form to Determine Kilowatt Output for Steam Extraction

The preparation of diagrams, as indicated, requires accurate records of past performance and reliable future predictions. Considerable technical knowledge and information that must be obtained from equipment manufacturers are also necessary.

7. BIBLIOGRAPHY

Bailey, E. G., Modern Boiler Furnaces, *Trans. ASME*, Vol. 61, p. 561 (1939).
Fernald, R. H., and G. A. Orrok, *Engineering of Power Plants*. McGraw-Hill (1927).

Fieldner, A. C., A. H. Emery, and M. W. Von Bernwitz, Bibliography of U. S. Bureau of Mines Investigation on Coal and Its Products 1910–1935, *Bur. Mines Tech. Paper* 576.
Harding, L. A., *Steam Power Plant Engineering.* John Wiley (1932).
McAdams, W. H., *Heat Transmission.* McGraw-Hill (1933).
Moore, H., *Liquid Fuels.* Van Nostrand (1935).
Newman, L. E., A. Keller, J. M. Lyons, L. B. Wales, *Modern Turbines.* John Wiley (1944).
Stodola, A., *Steam and Gas Turbines.* McGraw-Hill (1927).

HYDROELECTRIC POWER STATIONS

By R. A. Hopkins

8. SELECTION OF SITE

AVAILABLE POWER AND ENERGY. Analysis of the available power and energy includes the study of hydrological conditions (rainfall, runoff, and stream flow), head limitations, facilities for storage and pondage, and the conversion of the available flow and head into electrical energy.

The United States Geological Survey, the United States Weather Bureau, and many of the individual states publish much pertinent information on **stream flow, climatology, geology, flood flows,** and similar subjects.

When records of stream flow are not available for the vicinity of the particular site under consideration, they may be obtained or approximated by one or more of the following methods:

1. Establishment of a gaging station at the site.
2. Deduction from local rainfall statistics.
3. Comparison with flow of adjacent or similar watersheds where records are available.

Stream flow data usually are summarized by the preparation of **hydrographs,** for which daily averaged flows are plotted chronologically, and **flow duration curves,** for which flow quantities are plotted against percentage of time of occurrence. Where storage is a factor, use frequently is made of the **mass diagram,** on which the summation of daily or monthly flows is plotted chronologically for the entire period under study. The slope of the mass curve at any point represents the rate of flow, and from this curve the amount of storage required to regulate the flow to a definite minimum may be determined graphically. Particular attention should be given to data for years of maximum and minimum stream flow. (See Article 18, Reference 11.)

The **head** to be developed is subject to headwater and tailwater limitations. Headwater elevation is usually controlled by the cost of reservoir lands, hydraulic structures, and water rights, and by limitations due to the backwater influence of water impounded by the dam. The range of **headwater elevation** is influenced by the type of spillway, the reservoir operation, and flood flows. The range of **tailwater elevations** is usually established by the natural river levels at the lower end of the tailrace. In the case of overlapping developments the backwater influence of the lower development controls the tailwater levels of the upper development. From the various combinations of headwater and tailwater conditions a **duration of head curve** may be plotted, which shows gross or net head plotted against time.

On the amount of **storage** that can be economically provided depends the extent to which the stream flow can be equalized, and the amount of installed capacity which can be justified. A run-of-river plant has no storage, and the stream flow must be used as it occurs. Sometimes storage can be provided economically in a separate storage reservoir, particularly if the same storage would serve a series of developments on the same stream.

After the stream flow, head, and storage conditions have been analyzed, the available power and energy for dry, average, and wet years can be computed, and its conversion into electrical energy and absorption by available markets can be studied.

PHYSICAL CHARACTERISTICS OF THE SITE. Generally, a satisfactory, although not necessarily economical, type of dam can be constructed on almost any kind of foundation, but it is imperative that the engineer have the most complete knowledge possible of all foundation conditions. Data on foundations are usually obtained from core or wash borings or from test pits.

It is also necessary to investigate the geology and topography of the entire reservoir area in order to discover possible future sources of leakage to adjoining watersheds or to the stream below the dam.

Each site should also be considered for its practical advantages for construction, including particularly such items as accessibility and transportation facilities, possibilities for stream diversion during construction, and local supplies of materials, such as sand, gravel, and lumber.

9. TYPES OF DEVELOPMENTS

The **general type** of a development may be classified in several ways, although no classification can be complete and exact. With respect to head, a classification is given in Article 11 according to the type of waterwheel employed for various heads. Developments using impulse wheels are usually considered high; those using reaction waterwheels, medium, and those using propeller waterwheels, low.

DIVERSION DEVELOPMENT. With respect to the development of head, a system may be classed as a **diversion** development where the water is diverted from the stream by means of a small diversion dam, carried for a considerable distance through a canal flume, pipe, or tunnel without much loss in elevation, and suddenly dropped through penstocks, which develop most of the head, to the powerhouse located at a lower elevation than the dam and on the same stream or on another stream. Instead, it may be classed as a **direct** development if the powerhouse is close to the dam, which itself develops practically the entire head.

RUN-OF-RIVER DEVELOPMENT. With respect to continuity of generation, a development may be classed as a **run-of-river** development if only slight pondage is provided, and except for abnormal flows the entire river is passed through the turbines continuously, furnishing principally kilowatthours for base load. It may be classed as a **storage** development if a large amount of storage capacity is provided, and the plant is shut down part of the time, particularly during the filling season, and can be operated intermittently to furnish principally kilowatts for peak load.

With respect to character of terrain and navigation requirements, a development may be classified as **main-river,** if the terrain is sometimes flat, the river is broad, and a navigation lock is required. It may be classed as a **tributary** development if the terrain is steep and rugged, and navigation facilities are not required.

TYPICAL DEVELOPMENTS. Three **typical developments** of recent construction are sketched in Figs. 1, 2, and 3 to illustrate some of the principal types. Figure 1 shows a

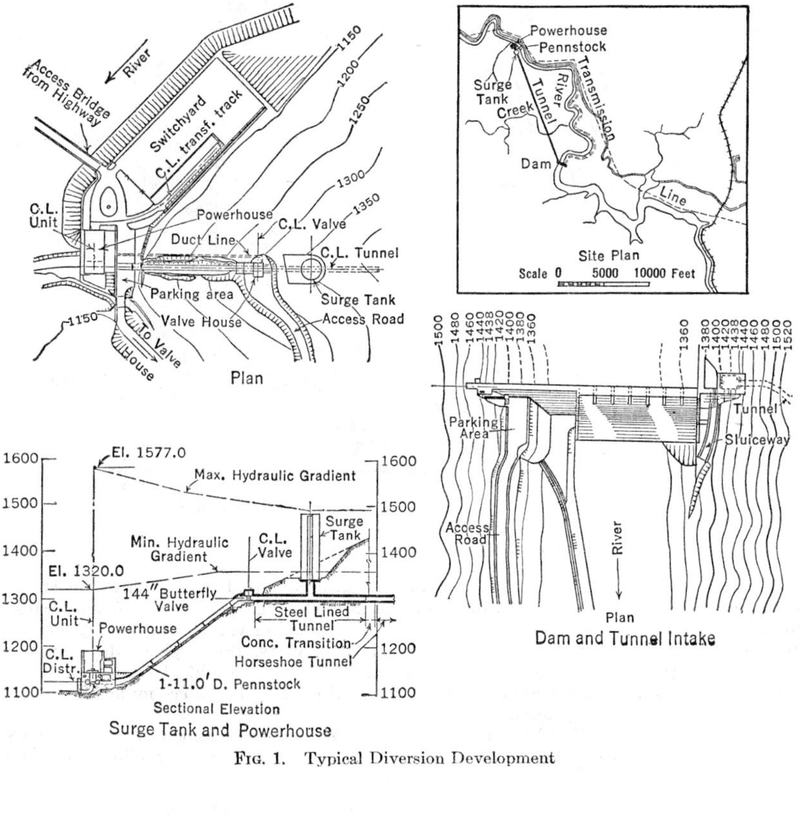

FIG. 1. Typical Diversion Development

diversion development at a site having a potential head of 294 to 316 ft and a stream flow of 125 cu ft per sec minimum, 1150 average, and 133,000 probable maximum. It has an 80,000-cu yd concrete diversion dam 110 ft high, impounding a total volume of 14,440 acre-feet, a 100,000-cu ft per sec open spillway controlled by seven radial gates, a lift-type

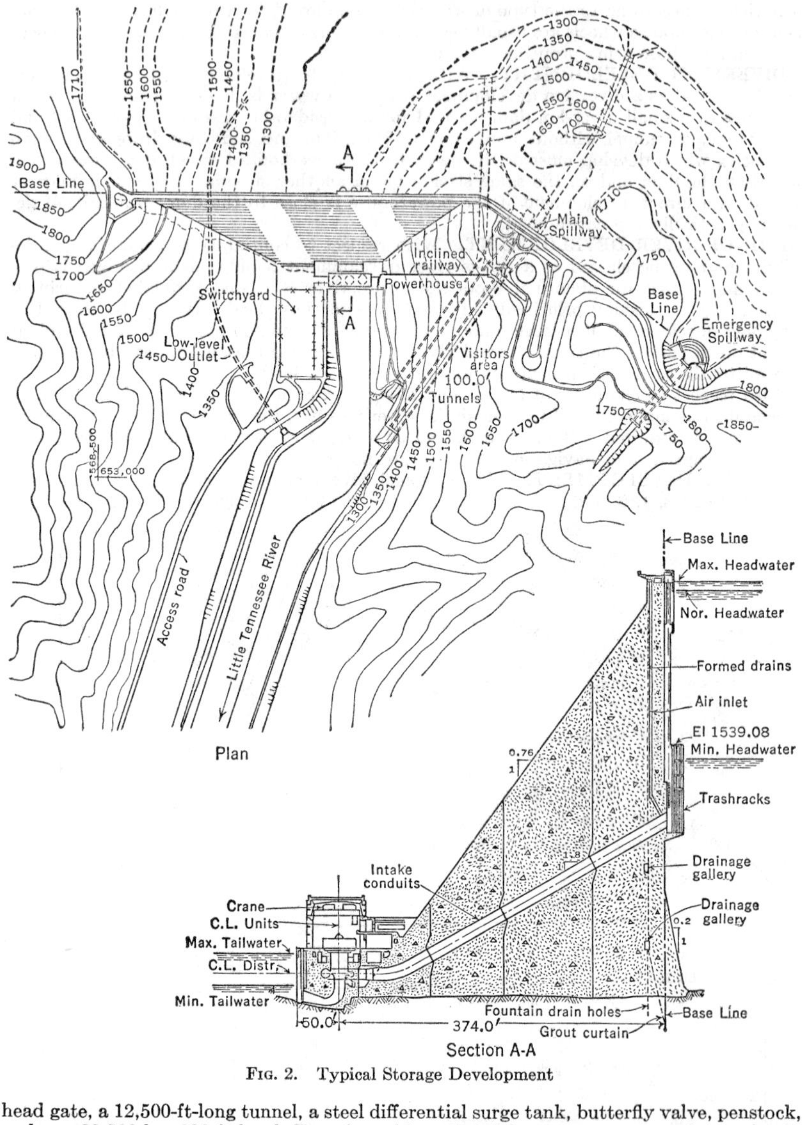

FIG. 2. Typical Storage Development

head gate, a 12,500-ft-long tunnel, a steel differential surge tank, butterfly valve, penstock, and one 33,500-hp, 280-ft head, Francis turbine, 27,000-kw generator, transformer bank, and switchyard for three outgoing lines at two transmission voltage levels. To absorb sudden fluctuations in water flow caused by load changes, this type of development requires a surge tank or some form of by-pass and sometimes both. It also usually requires fairly slow governor action and large flywheel effect. To be economically sound, the head must be relatively high.

STORAGE DEVELOPMENT. Figure 2 shows a **storage** development at a site having a potential head of 248 to 433 ft, and a stream flow of 420 cu ft per sec minimum, 3700 average, and 199,000 probable maximum. It has a 2,812,000-cu yd dam 480 ft high, impounding a total volume of 1,450,000 acre-feet, two 79,000-cu ft per sec spillway tunnels controlled by four Taintor gates and six sluice gates, lift-type head gates, steel-lined penstocks, two initial and three ultimate 91,500-hp, 330-ft-head Francis turbines, 67,500-kw generators, transformers, and switchyard for several outgoing transmission lines. With this type of development the governor action can be as fast as desired for good load and speed control.

MAIN-RIVER DEVELOPMENT. Figure 3 shows a **main-river** development at a site having a potential head of 40 to 60.5 ft, and a flow of 2900 cu ft per sec min, 26,400 average, and 658,000 probable maximum. It has a navigation lock, a 480,000-cu yd, 112-ft-high dam, impounding a total volume of 1,132,000 acre-feet, a 550,000-cu ft per sec spillway

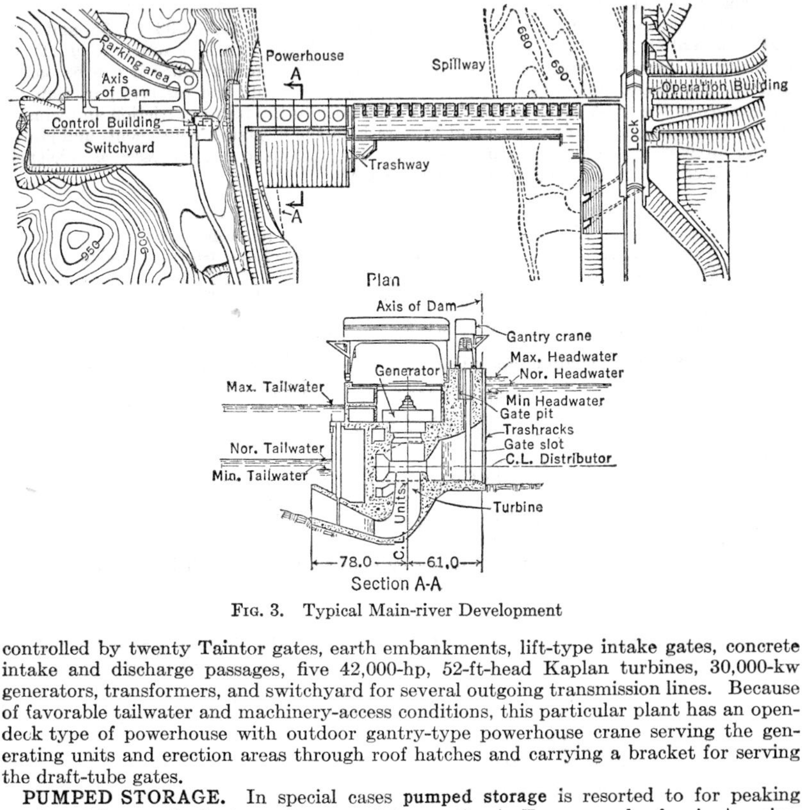

FIG. 3. Typical Main-river Development

controlled by twenty Taintor gates, earth embankments, lift-type intake gates, concrete intake and discharge passages, five 42,000-hp, 52-ft-head Kaplan turbines, 30,000-kw generators, transformers, and switchyard for several outgoing transmission lines. Because of favorable tailwater and machinery-access conditions, this particular plant has an open-deck type of powerhouse with outdoor gantry-type powerhouse crane serving the generating units and erection areas through roof hatches and carrying a bracket for serving the draft-tube gates.

PUMPED STORAGE. In special cases **pumped storage** is resorted to for peaking service. Many such plants are in successful operation in Europe, and a few in America. An outstanding development of this type, the Rocky River Development of the Connecticut Light and Power Company, was completed in 1929. In this type of development, tidewater or water from a low-level stream or reservoir is pumped to a high-level artificial or natural reservoir during off-peak hours, the stored water being used to provide system peak capacity. Off-peak energy available from other plants in the system is used for pumping. The maximum overall efficiency of a pumped storage plant from bus-bar ingoing to bus-bar outgoing will vary from 60 to 70 per cent. In Germany the Herdecke plant on the Ruhr River, with an installation of four 48,000-hp units under 534-ft head, develops an average overall efficiency of 65 per cent and a maximum of 68 per cent.

UNATTENDED STATIONS. These have been developed to a considerable extent during the past decade. The saving in cost of operation is considerable on a per kilowatt-hour basis, and the cost of maintenance has sometimes proved to be even less than for an equivalent attended station. At one modern supervisory-controlled station the total cost of operation and maintenance is reported as less than 11 cents per 1000 kilowatthours of gross generation.

Many methods of control are in use. The station can be started, loaded, unloaded, and stopped by an operator in another station, who also receives indications and annunciations. This rather complete control can be handled by a multipair cable for short distances or by carrier channels for longer distances. A simpler control sometimes provides only for starting and stopping by an operator in another station, by a float switch in the forebay, or by other means. In the simplest arrangement the station can be started and stopped by sending an operator to the station for the purpose. The two last described types of control often are found very useful in reducing operating costs of old plants, where complete remote control would be difficult and expensive to apply. With any form of control the station usually is provided with protective relays, which shut it down and disconnect it in case of trouble. A maintenance crew usually visits the station at regular intervals and performs scheduled checks and adjustments.

For an extension of this subject, see Article 18, References 31, 32, 33, 34, 35, and 36.

The **potential power** that can be developed at a hydro site is a function of (1) the quantity of water available in the stream, (2) the head through which it can be passed, and (3) the losses, including evaporation and seepage from the reservoir, friction and turbulence in the trash racks and waterways, velocity in the tailrace, friction and windage in the generating units, resistance and hysteresis in the electrical equipment, and station use for auxiliaries, pumping, heating, lighting. The various losses, expressed as efficiencies and grouped to facilitate the usual calculations, are:

Item	Range	Based on
Reservoir	0.95 to 0.98	Inflow to reservoir
Waterways	0.90 to 0.99	Gross head from reservoir to tailrace
Turbines	0.88 to 0.94	Rated head on turbine
Generators	0.94 to 0.98	Input to generator
Electrical equipment	0.97 to 0.99	Output from generator
Station use	0.92 to 0.98	Output from generator

The **fundamental equations** for calculating electric power from falling water are:

$$hp = 0.1136 \times cfs \times h \times e \tag{1}$$

$$kw = 0.746 \times hp \tag{2}$$

where hp = power in horsepower.
cfs = water flow in cubic feet per second.
h = head in feet.
e = product of all efficiencies involved.
kw = power in kilowatts.
0.1136 = constant based on 62.5 lb per cu ft of water and 550 ft-lb per sec equivalent of 1 hp.
0.746 = ratio of kw to hp.

The **energy** available is a function of the power and the time the power is available, which depends upon the duration of stream flow and the amount of storage capacity. The energy equation is:

$$kwhr = kw \times t \tag{3}$$

where $kwhr$ = energy in kilowatthours.
kw = power in kilowatts.
t = time duration in hours.

See Article 18, Reference 10.

10. ELEMENTS OF A DEVELOPMENT

THE RESERVOIR. The chief function of the reservoir is to store water during periods of excessive stream flow in order that the surplus water may be used to advantage during periods of deficient flow.

The construction and operation of a reservoir are subject to local and federal regulations concerning matters which involve public health and the conservation of natural resources, such as reservoir clearing, navigation facilities, fish propagation, and mosquito control.

Water stored in reservoirs is subject to losses from natural causes, such as evaporation and leakage.

THE DAM. The choice of type of dam to be used in connection with the development of a particular site depends on many conditions, chief of which are the character of the foundation, the topography of the site, the height of the dam, and the availability of materials for construction. The various types of dams in common use are:

1. **Solid Masonry Dam.** Usually of mass or cyclopean concrete, but sometimes of stone masonry. May be of gravity or arch type or a combination. Requires rock foundation of unquestioned soundness and durability.

2. **Hollow Masonry Dam.** Usually of reinforced concrete, but occasionally of precast concrete slabs. Includes such special types as slab and buttress, multiple arch, and multiple dome. Ordinarily suitable for rock foundation of poorer quality than is required for a solid masonry dam, although if high bearing stresses are used, the best rock foundation is necessary.

3. **Earth Dam.** Built of selected soils with or without core walls of concrete, wood piling, or steel-sheet piling. Cores of sluiced material or puddled clay sometimes used. Suitable for soft foundations where adequate materials are available.

4. **Rock-fill Dam.** Built of rock fill either with concrete or timber deck or with a core wall. Suitable for rock foundation, where material is available, at less cost than concrete.

5. **Timber Dam.** May be of timber frame and deck construction or of rock-filled timber cribs. Seldom practical for heights exceeding 20 ft.

The **Spillway section** of the dam, and sometimes all the dam, is designed to discharge surplus and flood water. That portion not designed as a spillway is usually called a retaining or impounding section. The discharge capacity of the spillway is established so as to provide a margin of safety over the largest recorded flood. The spillway may be built with a free crest or with a controlled crest. The discharge over a free-crest spillway is controlled only by the water level in the reservoir. The discharge over a controlled-crest spillway is regulated by flashboards or crest gates.

Crest gates of the following types are commonly used for gate-controlled spillways: sliding, roller (fixed wheels), Stoney (roller trains), radial or Taintor, drum, tilting, bear trap, and rolling. Gates may be operated by individual hoists or by traveling hoists. In some cases, particularly where gates are partially wet most of the time, it is necessary to provide upstream slots and either a spare gate or a set of stop logs to close off the water for maintenance of the main gates.

Water may be discharged past the dam also through siphons or various types of sluice gates and valves.

Accessories sometimes required in connection with the construction of a dam include such structures as navigation locks, fish ladders, and log chutes.

WATERWAYS. Under this heading are classified all structures and equipment between the reservoir and the tailrace with the exception of the scroll case and the draft tube, which are closely associated with the turbine, of the reaction and the propeller type, and usually are designed by the turbine builder.

The **forebay** is the pool adjacent to the intake. Many developments have no definite forebay, but when water is brought to the intake by canal or flume it is usually necessary to expand the section at the intake in order to reduce the velocity.

The **intake** is located at the upstream end of the conduits leading to the turbines and contains structures and equipment necessary to control the flow entering the conduits. The intake is usually provided with racks or screens to keep trash and ice from entering the conduits and with gates or valves to close off the waterways. Where the dam is of earth or rock fill, intake towers isolated from the dam are generally used. In cold climates the intake is frequently housed to prevent freezing.

Trash racks, in front of the intake, are usually constructed of round-edged vertical bars spaced from 1 1/2 to 8 in. on centers, depending on the size and type of waterwheel runner. Water velocity at the bars is usually between 1.5 and 2.5 ft per sec. A motor-operated rake and a sluiceway generally are required to dispose of trash. In cold climates, gate-rack heating generally is required. (See Article 16.)

Intake gates of the following types are in common use: sliding, Stoney (roller trains), roller (fixed wheels), Sirnit (fixed wheels unequally spaced, resting in recesses in guide rail), caterpillar, radial or Taintor, cylinder (usually for circular intake towers). Intake valves of the following types are also used: rectangular butterfly (horizontal shaft), circular butterfly (horizontal or vertical shaft), needle or Johnson, rotary. The circular butterfly, the needle, and the rotary valves and the cylinder gates are suitable for use under high heads; the remaining types are rarely used for heads exceeding 100 ft. Permissible velocities at the headgate openings vary from 2 to 10 ft per sec. Upstream slots and either a spare gate or a set of stop logs are required to close off the water for maintenance of the main gates.

Artificial water conduits may be classified as: (1) open conduits; canals or flumes; (2) closed conduits; tunnels or pipes. Frequently a combination of two or more types of conduit will be found practical as well as economical. Long conduits are usually constructed in two portions: (1) the **high-level conduit,** which is designed for relatively low pressure and follows the topography or hydraulic gradient to a suitable location near the powerhouse, from which point (2) the **penstock** conveys the water to the turbine, being designed for the necessary higher pressures. A **surge tank** or regulator is located at the junction of the high-level conduit with the penstock to prevent excessive changes in pressure during fluctuations in the flow.

Canals may be lined or unlined, depending upon the character of the ground through which they pass, the shape of cross-section, and the water velocity. Ordinary soils will stand velocities up to 2.5 ft per sec without erosion. Canals may be lined or paved with wood, rock, brick, concrete, or gunite.

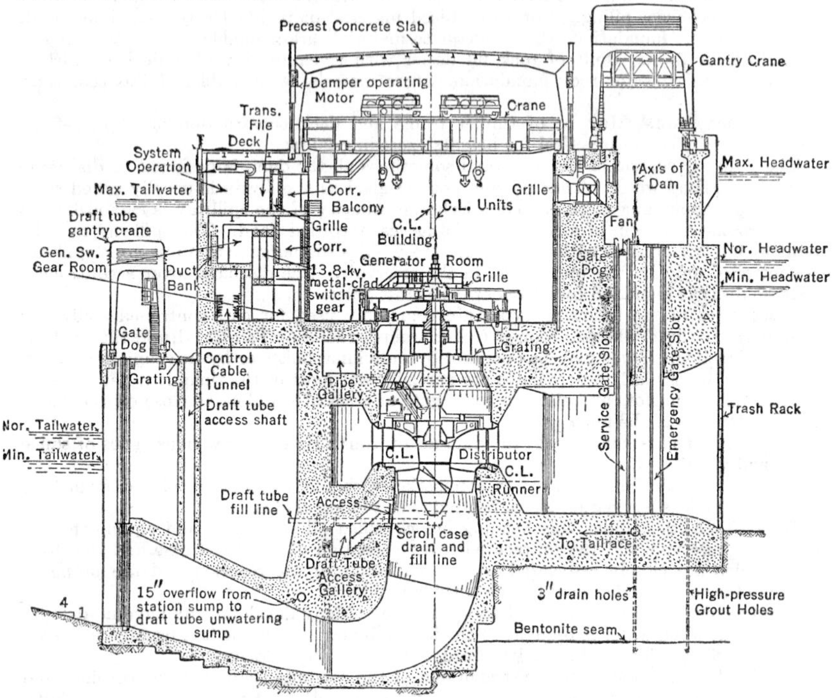

FIG. 4. Conventional Enclosed Generator Room

Flumes may be of wood, concrete, or steel. In section they may be rectangular or semicircular, or they may be semicircular or semioval with vertical sides. The wooden flume is seldom used for permanent construction on account of its short life and high maintenance cost. In cold climates long flumes or canals are subject to ice troubles.

The **tunnel** is the most permanent, but usually the most expensive, form of conduit. Whether designed as a flow or a pressure tunnel, it is usually lined to provide additional structural safety, to minimize head losses by providing a smooth-surfaced and uniform section, and to reduce losses of water by leakage. Tunnel linings are usually of reinforced concrete, although brick linings have occasionally been used.

Pipe lines may be of wood, concrete, or steel, although concrete pipe lines are rarely used except for very short sections.

Wood-stave pipe is used extensively for conduits under moderate heads in locations where the topography is favorable. It is relatively cheap in first cost, and if continuously filled has fairly long life. It is particularly adaptable for inaccessible locations. Two types of wood-stave pipe are in general use: (1) machine-banded pipe, and (2) continuous wood-stave pipe. Machine-banded pipe is used for heads up to 500 ft and in diameters up to 42 in. Continuous wood-stave pipe is built up in place and has been constructed up to a diameter of 16 ft.

Steel-pipe conduit is usually constructed with either riveted or welded joints, certain advantages being inherent in each type. For penstocks under very high heads, forged-steel pipe has been used successfully. Steel pipe, when properly painted and maintained, has a long life. In the design of a steel pipe line or penstock, careful attention must be paid to such details as supporting saddles, expansion joints, and anchorages. Permissible velocities in steel pipe are as high as 20 ft per sec for high heads, but for ordinary heads usually range from 8 to 12 ft per sec. Maximum economical velocity will generally depend on the length of conduit and the load and head conditions.

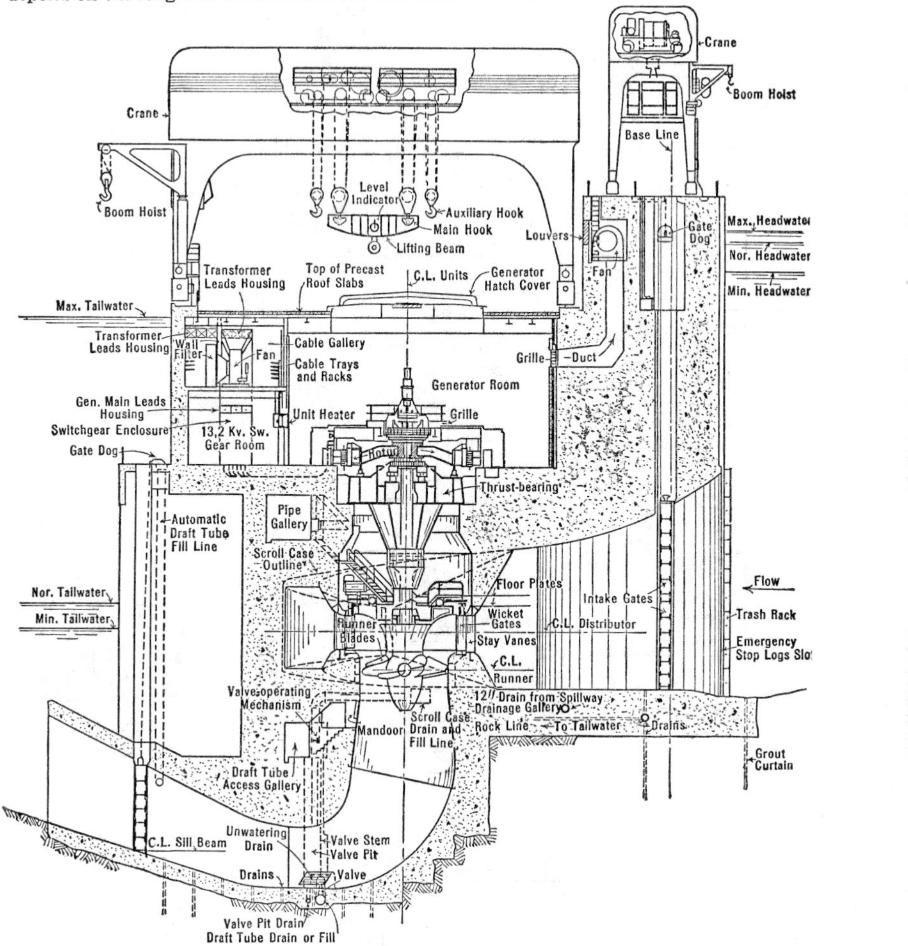

Fig. 5. Semiopen Powerhouse

The **tailrace** is the channel or waterway that conducts the water from the discharge end of the turbine draft tubes back to the natural river channel. Except under very favorable topographical conditions it is rarely economical to obtain additional head by extensive tailrace excavation. Permissible tailrace velocities usually vary from 3 to 5 ft per sec, although somewhat higher velocities are permissible for the higher head plants.

Impulse wheels discharge above tailwater level, and the tailrace therefore has no important effect on their operation other than to reduce maximum water level and permit lower setting of the waterwheel.

POWERHOUSE. The powerhouse is the structure that houses the power facilities, such as turbines, generators, low-voltage switchgear, control boards, auxiliaries, and offices. The powerhouse may be far from the dam or a part of it, depending upon the type of development. The **arrangement** of the powerhouse and its equipment requires a great deal of detailed study to provide low cost and convenient operation. (See Article 18, Reference 7.)

The **generating units** are located primarily by the hydraulic requirements of the penstocks, scroll cases, and draft tubes. The height of the generator room is determined by the height required to assemble the generating units and to lift major parts over completed units. The conventional enclosed generator room with indoor cranes (Fig. 4) affords

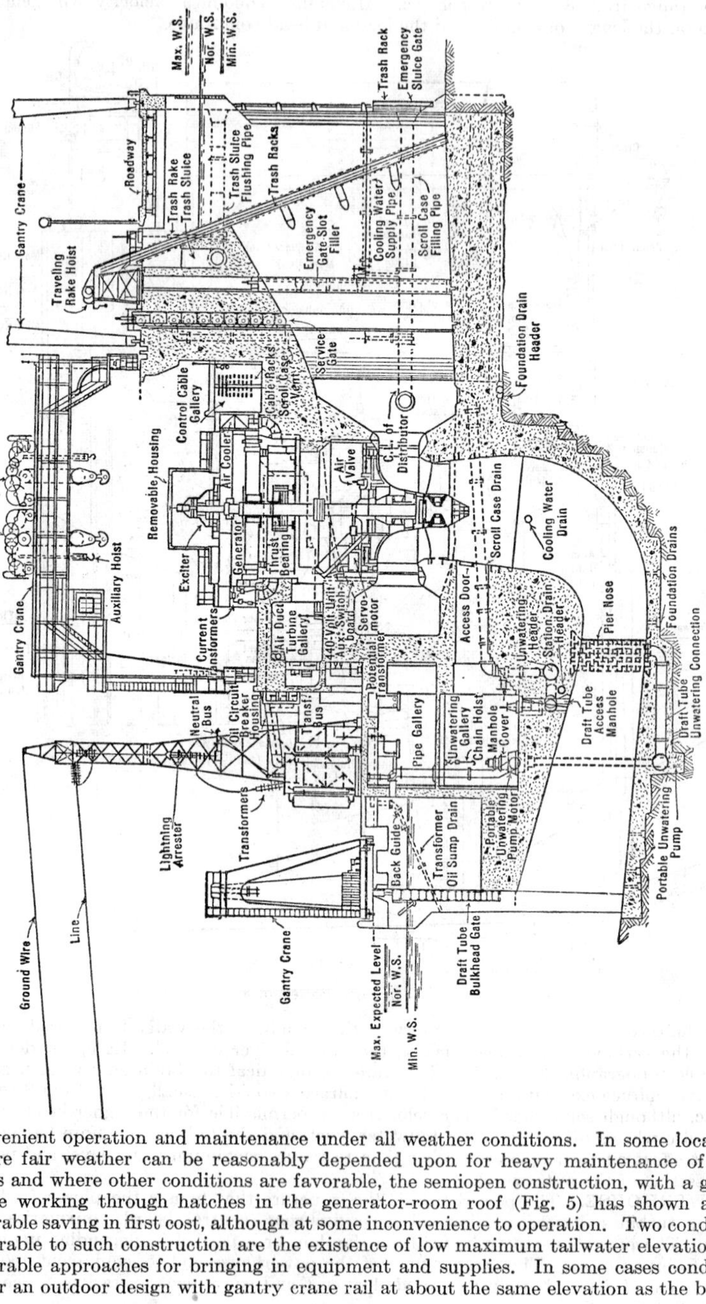

FIG. 6. Typical Outdoor Powerhouse[1]

convenient operation and maintenance under all weather conditions. In some localities, where fair weather can be reasonably depended upon for heavy maintenance of main units and where other conditions are favorable, the semiopen construction, with a gantry crane working through hatches in the generator-room roof (Fig. 5) has shown a considerable saving in first cost, although at some inconvenience to operation. Two conditions favorable to such construction are the existence of low maximum tailwater elevation and favorable approaches for bringing in equipment and supplies. In some cases conditions favor an outdoor design with gantry crane rail at about the same elevation as the bottom of the generator, and weatherproof steel housings over the generators (Fig. 6). It is essential that first cost be weighed very carefully against operating convenience in laying out the general arrangement of the powerhouse.

Erection space must be provided under the crane that serves the generating units. It is good practice to locate the space at the shore end of the station, where it conveniently can receive incoming equipment, materials, and supplies. Space must be allowed for setting down a completed generator rotor and turbine runner. Other large items, such as main bearing bracket and turbine cover plate, also may require setting-down space. With many station arrangements the main transformers are brought under the powerhouse crane for untanking, and both floor space and untanking height must be provided. The station **machine shop** should be located adjacent to the erection space and should contain sufficient tools for all maintenance intended to be done at the plant.

Governors and their pressure tanks are located close to the turbines so as to reduce to a practical minimum the length of the oil piping for operating the servomotors. In modern designs the front of the governor cabinet is used as a control board for the turbine. The governor is located on the operating floor, which usually is at the elevation of the base of the generator. It is important to have short piping between the governor and the servomotors and to have reasonably good access to the turbines below and the direct-connected exciters above.

The **electrical bay** usually extends the full length of the generator room, either upstream or downstream from it. The electrical bay contains the generator-voltage switchgear, generator buses if any, excitation and neutral grounding cubicles, main conductors to the outdoor transformers, unit auxiliary power switchboards, control-cable gallery, generator fire-protection apparatus, piping gallery for water, oil, and air, and all such services to the generating units. If station service power is taken from the generator-transformer connections, the transformers and main distribution board also may be located in the electrical bay.

The **service bay** usually is located at the shore end of the powerhouse and extends upstream and downstream the full width of the generator room and electrical bay. It contains equipment common to all generating units and used by the station as a whole, such as sump pumps, air compressors, oil-purifying equipment and storage tanks, machine shops, electrical shop, erection space, storage space, emergency gasoline generator if any, water-purifying plant, laboratory, offices, and public spaces.

The **control room** and its associated equipment, such as battery chargers and telephone apparatus, usually are located as a group at a strategic position in the electrical bay or the service bay. Being the control center for the generating units, the generator-voltage switchgear, the main transformers, and the transmission-voltage switchgear, this group should be located centrally to all these facilities for an economical control-cable layout and for convenience of operation. Control cables must be routed from the switchboards to all equipment controlled, and in an important station it is usual to provide a cable room below the control room and control-cable galleries throughout the length of the powerhouse and sometimes to the switchyard to accommodate the cables on open shelves. The battery room and charging equipment should be located close to the control room. The communications room can be located in the electrical bay or service bay. Since it should be well ventilated and air-conditioned, it usually is located close to the control room, offices, or public spaces where these services are available.

SWITCHYARD. An important hydroelectric generating station usually is connected to a high-voltage power system and requires step-up transformers and **high-voltage**

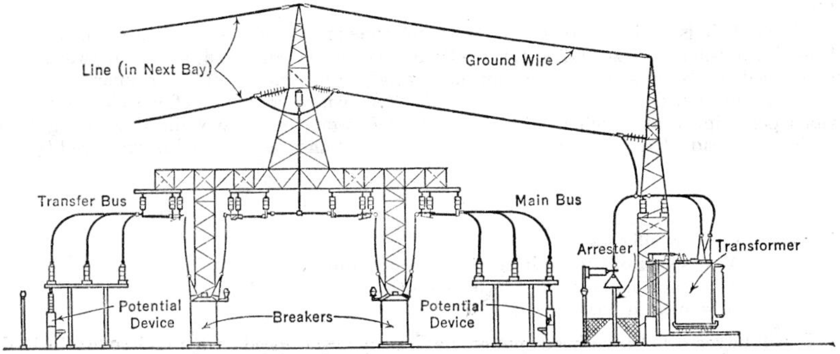

Fig. 7. Typical Switchyard Section

switching. Many arrangements are possible. The transformers should be reasonably close to the generators to reduce the lengths of the low-voltage conductors, but must also be arranged to provide adequate spacing for the high-voltage conductors. In some cases the transformers have been placed on the intake deck or the draft-tube deck, and the switching on the powerhouse roof, but only rarely is sufficient space available on the roof or is it possible to get the lines away satisfactorily. A further objection is that the impact loads imposed by high-capacity breakers necessitate an expensive roof structure. Occasionally the transformers can be located on the draft-tube deck, and the switching on shore, but this arrangement usually results in long spans from transformers to switchyard. The most common arrangement is to locate the transformers alongside the switchyard and use insulated cables in ducts, or bare conductors on insulators in a tunnel between the generators and the transformers. Such arrangements are indicated in Figs. 1, 2, and 3. A cross-section through a typical 161-kv switchyard is shown in Fig. 7.

MISCELLANEOUS PERMANENT STRUCTURES AND FACILITIES. Various miscellaneous structures and facilities which may be required in connection with a hydroelectric development include:

1. Permanent railroad connection for handling equipment and supplies.
2. Permanent highway connection for easy access to plant.
3. Operators' quarters (houses, clubhouse, water supply, sewage system, lighting).
4. Parking space for visitors and garages for employees.
5. Office building (occasionally provided for important stations).
6. Storage building.
7. Boats and boathouse for reservoir patrol.
8. Gage houses (for automatic recording of water levels).

The extent of these facilities will vary with the size and importance of the development and with its distance from settled communities. (See Article 18, References 2, 3, and 4.)

11. WATERWHEELS

Waterwheels are classified broadly in three general types: impulse or Pelton, reaction or Francis, and propeller or Kaplan. Ordinarily impulse wheels are used for high heads, Francis wheels for intermediate heads, and propeller wheels for low heads. There are no well-defined limits separating the head ranges for these three types of wheels. Generally, impulse wheels are used for heads above 850 ft, Francis wheels for heads from 900 to 40 ft, and propeller wheels for heads below 75 ft. Figure 8 shows a cross-sectional view through a typical 42,000-hp, Kaplan wheel for 52-ft head coupled to a single-bearing 30,000-kw generator.

Relative **efficiencies** and characteristic shapes of the load-efficiency curves of the three classes of large modern waterwheels are indicated in Fig. 9. For the propeller wheel, the solid-line curve is for an adjustable-blade (Kaplan) wheel, on the assumption that the blade tilt is controlled along with the wicket-gate opening to give best efficiency at each load; the dotted curves indicate the efficiencies through small load ranges with the blades fixed at various angles of tilt.

Generally, it is economical to run the smallest **number of units** which can be operated to use the water efficiently with respect to the load requirements. At run-of-river plants, however, it may be desirable to install units of different capacities in order to use the water most economically under varying conditions of flow and head. At low-head plants a similar result may be obtained by the use of propeller runners with adjustable blades.

Although it is desirable to use **waterwheel speeds** as high as practicable in order to keep down generator size and cost, it is also necessary to maintain a balance between head and speed to obtain smooth operation and avoid serious **pitting of the runners.** An important consideration, therefore, is **specific speed,** which may be defined as the revolutions per minute at which a homologous runner would operate while developing 1 hp under 1-ft head. The relation between specific speed and actual speed is expressed by the formula:

$$N_s = \frac{N\sqrt{P}}{H^{5/4}}$$

in which N_s = specific speed.
N = speed of runner in revolutions per minute.
P = capacity of runner in horsepower.
H = head in feet.

It will be seen that high-head runners have low specific speeds, whereas low-head wheels have high specific speeds.

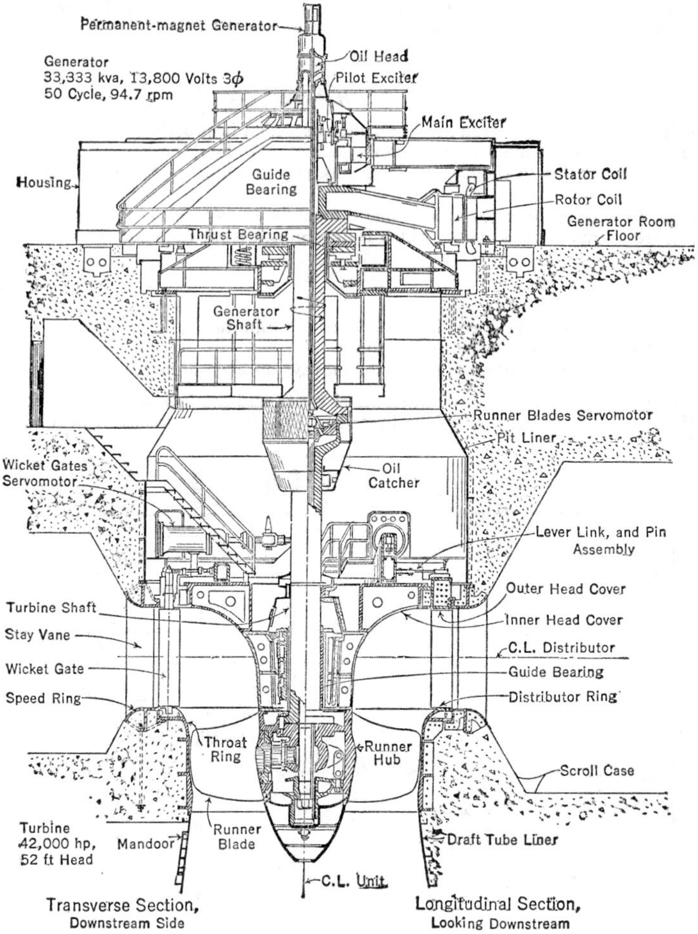
Permanent-magnet Generator
Generator
33,333 kva, 13,800 Volts 3ϕ
50 Cycle, 94.7 rpm
Oil Head
Pilot Exciter
Main Exciter
Housing
Guide Bearing
Stator Coil
Rotor Coil
Generator Room Floor
Thrust Bearing
Generator Shaft
Runner Blades Servomotor
Pit Liner
Wicket Gates Servomotor
Oil Catcher
Lever Link, and Pin Assembly
Outer Head Cover
Turbine Shaft
Inner Head Cover
Stay Vane
C.L. Distributor
Wicket Gate
Guide Bearing
Speed Ring
Distributor Ring
Throat Ring
Runner Hub
Scroll Case
Turbine
42,000 hp,
52 ft Head
Mandoor
Runner Blade
Draft Tube Liner
C.L. Unit
Transverse Section,
Downstream Side
Longitudinal Section,
Looking Downstream

FIG. 8. Cross-section of Kaplan Wheel

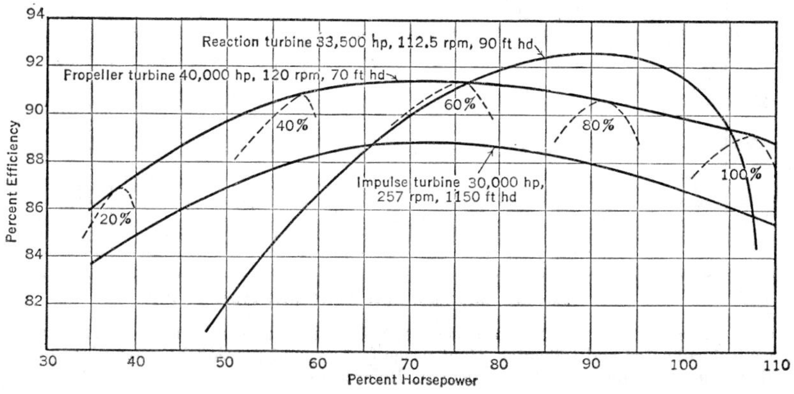
Reaction turbine 33,500 hp, 112.5 rpm, 90 ft hd
Propeller turbine 40,000 hp, 120 rpm, 70 ft hd
Impulse turbine 30,000 hp,
257 rpm, 1150 ft hd
20%
40%
60%
80%
100%
Percent Efficiency
Percent Horsepower
94
92
90
88
86
84
82
30
40
50
60
70
80
90
100
110

FIG. 9. Efficiencies of Waterwheels

In *Hydro-electric Handbook*, John Wiley, Creager and Justin give the following empirical formula for determining the maximum safe specific speed to avoid pitting of the runners of Francis wheels for various heads:

$$\text{Max. } N_s = \frac{5050}{H + 32} + 19$$

Speeds based on this formula are frequently exceeded in practice, since a small amount of pitting is not objectionable, and even in cases of considerable pitting the saving in first cost of equipment due to the higher speed may more than offset the cost of repairing the pitted runner.

The speeds derived from the specific speed formula are subject to correction to conform to the nearest **synchronous speed,** which is determined from the equation:

$$S = \frac{120f}{p}$$

in which S = synchronous speed in revolutions per minute.
f = frequency in cycles per second.
p = number of poles of generator (always an even number).

A very general classification of waterwheel types and settings appropriate for various conditions of head and capacity is indicated in Fig. 10. The selection of a waterwheel

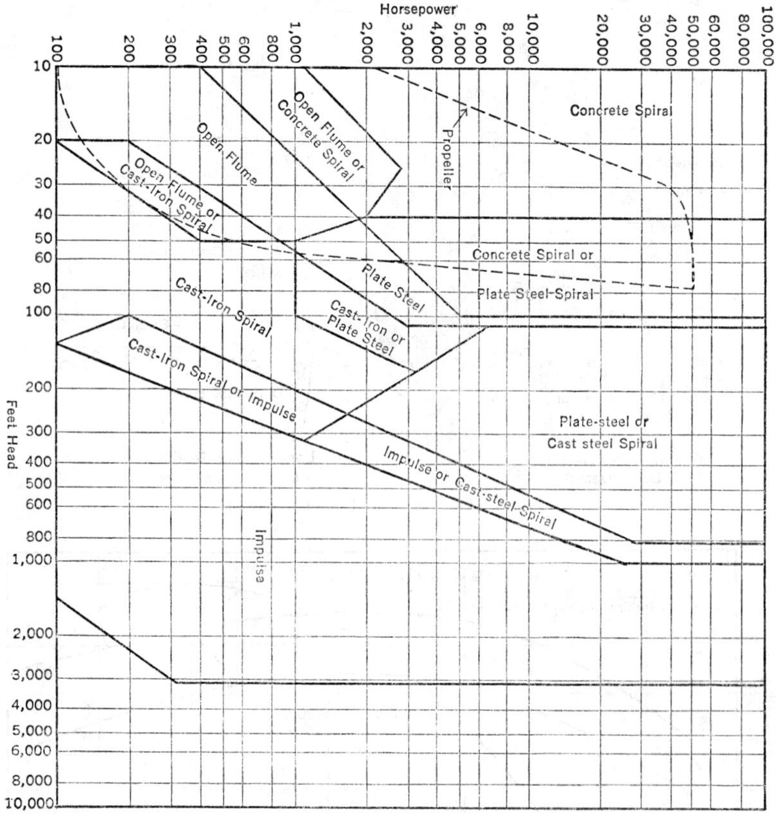

FIG. 10. Classification of Waterwheel Types

involves a judicious correlation of specific speed, peripheral coefficient, cavitation, variations in headwater and tailwater, plant usage factors, and relation between plant output and system requirements. (See Article 18, References 8, 9, 12, 15, 16, 17, 19, 21, 22, and 23.)

12. GOVERNORS

A **hydraulic governor** is a device for controlling the speed of the waterwheel. A modern hydraulic governor for a **Francis** or **propeller** turbine consists essentially of a sensitive flyball mechanism driven in exact speed relation with the turbine, a pilot valve the position of which is controlled by the flyballs, a main valve operated by the pilot valve, and one or more servomotors, which receive oil pressure through the main valve and in turn open or close the wicket gates as needed to maintain closely constant speed regardless of load. Oil pressure of 150 to 300 lb for operating the governor and gates is supplied by one or more motor-driven pumps and stored in a pressure tank under a cushion of air. The return oil is collected in a sump and brought back to the suction of the pumps. In order to prevent overtravel and hunting, a restoring mechanism, which consists of a mechanical linkage between the servomotor piston and the pilot valve, is provided. The oil sump, pumps, flyballs, valves, and mechanisms usually are housed in a cabinet, the front face of which is used as the turbine-control board.

For **Kaplan** turbines, a second main valve and servomotor are provided to operate the runner blades in correct relation with the wicket gates. For **Pelton** wheels the governor operates on one or more of the needle valves, the jet deflector, or the by-pass valve.

The **flyballs** sometimes are driven by the turbine through a belt or gear, but more frequently by a synchronous motor connected either to a transformer on the generator leads or to a permanent-magnet synchronous generator coupled to the generator shaft.

A **speed-level control** located on the flyball-restraining spring and operated by a small motor allows speed change by the switchboard operator when synchronizing. It also allows the connection of a **frequency-control device** or a time-correcting device to maintain automatically standard frequency and standard time on the power system. A **speed-droop control,** usually adjustable from zero to six per cent of rated speed at rated load, provides the desired sharing of load between generating units operating in parallel. When adjusted for zero speed droop, isochronous operation, a unit will accept all fluctuations of load. When adjusted for a speed droop greater than zero, it will tend to reject load fluctuations. A **gate limit control,** which blocks the movement of the pilot valve, places any desired load limit on the turbine. An **overspeed trip,** operated by a centrifugal switch on the generator shaft, shuts down the unit, usually to no-load speed. A **brake control** provides manual and sometimes automatic application of the generator brakes when shutting down. A **manually operated control valve** permits manual operation of the wicket gates. Most of these controls, along with their indicators, are provided both at the governor for use by the turbine operator and also in the main control room for the switchboard operator.

The **rating** of a governor refers to the energy it can exert to operate the gates or blades of the turbine. The approximate energy required, in foot-pounds, is:

$$\text{Foot-pounds} = \frac{\text{Maximum horsepower} \times C}{\sqrt{\text{Maximum head}}}$$

An average value for the coefficient C is 45. For an open-flume turbine with submerged gate mechanism it may be increased to 50. For a high-head reaction turbine the value may be reduced to 35, and for the deflector on an impulse wheel it may be as low as 20. In addition to specifying the rating, some engineers specify:

Pressure-tank volume to be 20 times the servomotor volume.

Sump-tank volume to be one-half the pressure-tank volume.

Oil-pump capacity in gallons per minute to be 2.5 times the servomotor volume in gallons.

Speed regulation can be calculated by any of several formulas, all based upon the fundamental principle that the change in kinetic energy of the moving column of water must be balanced by the change in kinetic energy of the rotating mass of the generating unit. All formulas assume the unit to be operating in isolation from a power system and therefore cover runaway conditions. Under normal operation on a power system, the actual speed regulation is assisted by the kinetic energy of the other rotating units, and the speed regulation is better than is indicated by the formulas. (See Article 18, Reference 18.)

The **sensitivity** of a governor refers to the amount of speed change necessary for the governor to detect it and start corrective gate operation. Modern governors are sensitive to a speed change of $^1/_{25}$ to $^1/_{100}$ of 1 per cent of rated speed.

The **time constant** of a governor is the interval of time between the occurrence of a speed change of a specified percentage of normal speed and the starting of the corrective movement of the gates. Modern governors have a time constant as low as 10 cycles or $^1/_6$ sec.

The **gate-operating time** refers to the time required for a complete movement, either opening or closing, of the gates. A modern waterwheel governor usually is capable of causing a complete gate travel within 2 sec. This, however, is much quicker than can be allowed in most cases without causing excessive penstock pressure when closing or vacuum when opening. Usually the gate-operating time is specified to be adjustable from 4 to 8 sec. With long penstocks, tunnels, or pipes, the inertia of the water column may be very great, and careful analysis must be made of permissible pressure and vacuum limits and of gate-operating time. For many plants an operating time of 3 to 6 sec is satisfactory, but results of penstock failure are too important to risk any rule-of-thumb solution. (See Article 18, References 6 and 13.)

13. GENERATORS

Generators directly driven by hydraulic turbines must have certain design and construction features not required by other prime movers. These will be outlined briefly. For the usual generator design and construction features, reference should be made to Section 9.

OUTPUT AND CAPACITY. These are derived from and matched to the turbine capability. The generator rating usually is matched to the turbine output at normal head, and the generator is specified to have class B insulation and to operate at this rated load continuously without exceeding a 60-deg cent temperature rise. It then has available a 15 per cent continuous overload capacity without exceeding an 80-deg cent rise for use during infrequent high-head conditions. It must have a short-time overload capacity, usually involving mechanical stresses rather than heating, to match the turbine maximum output under abnormal conditions.

EFFICIENCIES. As defined by AIEE Standards, efficiencies take into account (1) core losses of armature and field, (2) resistance losses of armature and field (and field

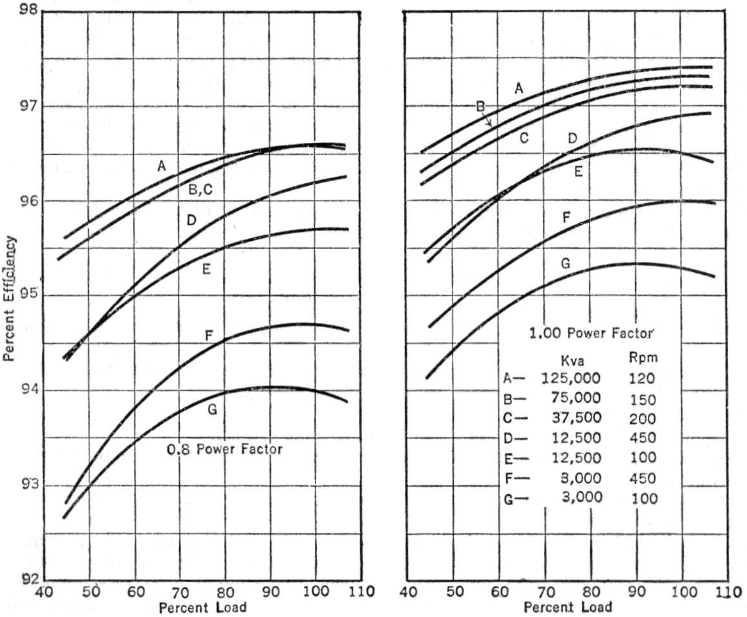

Fig. 11. Typical Efficiencies of Hydroelectric Generators

rheostat if used), (3) stray-load losses (eddy-current, iron, and other minor losses resulting from load current), (4) windage, bearing-friction, and brush-friction losses, (5) brush-contact losses (often neglected), and (6) ventilating-fan input if separately driven fans are used. Typical efficiencies of medium and large modern hydroelectric generators are given in Fig. 11.

SPEEDS. Normal speeds of medium and large hydroelectric units range from 50 to 720 rpm and are dictated largely by the turbines. The relation between speed in revolutions per minute (*rpm*), frequency in cycles per second (*cps*), and number of pairs of field poles (p) is expressed by the equation: $rpm \times p = 60 \times cps$. **Runaway speeds,** resulting from loss of load with turbine gates fully open under highest head, must be provided for. Such speeds usually amount to as much as 180 per cent of normal for impulse wheels, 200 per cent for reaction wheels, and 220 per cent for propeller wheels.

FLYWHEEL EFFECT. Flywheel effect (WR^2) of the rotating parts of the generating unit must be coordinated with the tunnel, penstock, surge tank, governor, and gates supplying energy to the unit and with the transformers, breakers, relays, and transmission system receiving energy from the unit in order to insure adequate power stability under normal and abnormal conditions. The generator rotor contains a large portion of the total WR^2 of the unit. A reasonable increase in WR^2 above that inherent in a normal generator design can be obtained by increasing the weight of the rotor rim at reasonable cost. Excessive increase involves considerable expense and sometimes necessitates a separate flywheel. (See Article 18, References 5 and 14.)

DAMPER WINDINGS. These usually are required to improve power stability and especially to reduce voltage distortion resulting from single-phase faults on the power system. Sufficient damping effect to reduce the ratio of quadrature to direct axis subtransient reactances to not more than 1.35 usually can be obtained without the use of damper connections between poles, which constitute a mechanical hazard, and this amount of damping usually is found to be adequate.

CHARGING CAPACITY. Line-charging capacity up to 80 per cent of the rated kilovolt-amperes is usually inherent in this type of generator. With long transmission lines additional capacity usually is required by special design of the generator, the excitation system, or both.

SYNCHRONOUS-CONDENSER OPERATION. This is important, since during the filling season or during low flows a hydroelectric plant often is required to operate condensing to assist movement of energy into its locality from other parts of the power system. If the turbine runner is set above tailwater, as is usually the case with reaction wheels, water must be supplied to lubricate its sealing glands; if it is set below tailwater, as with a propeller wheel, the water level must be lowered by compressed air or other means to avoid excessive losses. A normal generator inherently has condenser capacity equal to 55 per cent of its rated kilovolt-amperes.

SHAFTS. Vertical shafts are used for the great majority of reaction wheels and with all propeller wheels, and this situation dictates a vertical-shaft generator. For large, low-speed units the vertical shaft is favorable to the generator design as well as the turbine design. The rotating field is built up with a hub, spider, floating laminated rim, and salient field poles. The weights of the turbine runner, generator rotor, shafts, and exciters and also the downward hydraulic thrust all are carried on a single thrust bearing built into the generator either above or below the rotor.

BEARINGS. When the bearing is located *above the rotor*, a heavy bearing bracket is required above the stator frame. The rotor usually is shrunk onto the shaft, and the bearing carries the rotating load through a removable ring key around the top end of the shaft. The bearing runner is a continuous washer, and all bearing parts can be removed by the crane after the exciters are removed. The rotor cannot be removed without removing the bearing.

When the bearing is located *below the rotor*, the heavy bearing bracket is located below the stator frame, where the span is shorter and the sole plates can be embedded in the substructure. Usually the overall height of the generator is reduced. The shaft can be flanged at the top to rest upon the bearing and to receive the rotor, which is bolted and diametrically keyed to the top of the flange. The bearing runner is split, and the bearing parts can be removed from below without disturbing the rotor, which then rests upon its brake-jacks. The rotor can be removed without disturbing the bearing. If desired, the rotor arms can be sloped, and the guide bearing placed just above the thrust bearing in the same oil reservoir. In this position the guide bearing is so close to the horizontal plane of magnetic pull that only the one generator guide bearing is required. This design is known as the single-bearing, or overhung, or umbrella design.

COUPLING. The **coupling** between the generator and turbine shafts assumes major importance because the two shafts which are flanged to form the coupling usually are fabricated and finished by different manufacturers. It is the usual practice to place with the generator manufacturer the responsibility of final finishing of the two flanges, fitting the coupling bolts, coupling the two shafts together, and checking them by rotation both in a lathe at the shop and in operating position at the site. A check of major importance is the perpendicularity of the thrust-bearing runner surface with the common axis of the guide-bearing journals.

FIELD ASSEMBLY. This becomes a major operation because of the disassembly necessary to meet shipping requirements. The stator usually must be shipped in several sections, and the joint coils and end connections assembled in the field. In some instances the entire stator has been wound in the field. The rotor spider usually must be partially assembled, and the rim and poles completely assembled in the field. These operations are so akin to manufacture that both usually are included in the same contract.

Another important field-assembly feature is the handling of turbine parts through the generator stator after it is in position. This involves coordination of design between the generator and the turbine manufacturers and coordination of erection schedules to avoid conflict in assembly.

The windings must be dried out after assembly and before the final dielectric tests. The method usually preferred is by rotation under short circuit, but dry-out has sometimes been accomplished by applying external heat.

Acceptance tests of field-assembled machines must be made after installation. The efficiency usually is obtained from a careful measurement of losses and often is used later in conducting turbine-efficiency tests. The friction and windage losses, which cannot be separated accurately as between the generator and the turbine, usually are agreed to by the two manufacturers on the basis of estimates.

14. GENERATOR EXCITATION

The generator-excitation system, including exciters, field rheostats, voltage regulators, and the associated main and control wiring, is the most vital auxiliary service in the hydroelectric generating station. Upon its unfailing operation depends the uninterrupted station output, and upon its accurate performance depends the correct generator voltage and power factor.

The **power required** for hydroelectric synchronous generator excitation is stated approximately as follows:

$$kw = 10\sqrt{\frac{K}{S}}$$

where kw = required excitation in kilowatts.
K = generator rating in kilovolt-amperes.
S = generator speed in revolutions per minute.

The **exciter rating** in kilowatts should be not less than 110 per cent of the excitation required. Exciter rated voltages in common use are 125 and 250 volts. Nominal ceiling voltage of the exciter should be not less than 120 per cent, and nominal collector-ring voltage of the generator not more than 90 per cent of the exciter rated voltage.

The **excitation system** may be centralized, having a small group of exciters feeding an excitation bus, which serves all generators of the station, or it may be unitized, having an individual exciter for each generator. In some cases the **central system** may require a smaller number of larger exciters, thus possibly reducing maintenance and increasing efficiency very slightly. The exciters can be serviced without shutting down the generators. The central system has the disadvantages that the main wiring of the exciters must be carried greater distances and thereby subjected to more hazards, and that a fault may involve more than one generator. A further disadvantage is that rheostats are required in the generator fields, and the voltage regulators must operate on these rheostats. These limitations practically restrict the use of the central system to small stations. The **unit system,** with an individual exciter for each generator, is used in almost all modern major hydroelectric stations. Each generator is independent of disturbances in the excitation of the others. The connections from each exciter to its generator are short. Generator field rheostats are not required.

A **spare exciter** can be bused to all generators for emergency use, but this procedure seldom is justified, particularly with direct-connected rotating exciters, which are necessarily of slow-speed, rugged design. Although having one exciter of a station large enough for two generators would provide spare capacity and would be convenient in conducting acceptance tests on generators after installation, a set usually can be borrowed for testing purposes, and a permanent installation of this capacity is hardly justified. Spare parts for the main exciter and a complete spare pilot exciter often are kept in stock.

Drive for the rotating-type exciter can be from the generator shaft, by motor, or by prime mover. **Direct drive** from the generator shaft is by far the most commonly used drive in hydroelectric practice. The principal advantage is ability to maintain excitation during system disturbances. The efficiency of the large prime mover is better than that of any small motor or prime mover for exciter drive. Performance is very dependable. Main connections between exciter brushes and generator slip rings are very short and rugged. The physical arrangement is compact and convenient. **Motor drive** often is used. Being independent of the generator, the exciter can be serviced without shutting down the generator, provided that an alternate source of excitation is available. Motor drive may have space advantage under some unusual condition of station arrangement. Because of its higher speed the exciter itself costs less; but the total cost of exciter, motor, control, and wiring may be more. **Prime-mover drive** is one of the least used drives. If the prime mover is a small waterwheel, it is subject to clogging by trash and ice, and the first cost is high. It has the unique advantages, however, that its continuous operation and its speed are independent of the generator and the transmission system, and that it can be used when no other sources are available, as may be required in some unusual case when starting up a station. Very rarely a dual drive by **motor and prime mover** has been used and provided with automatic controls, so that normally it will be driven by either source as selected, and upon failure of that source will be sustained by the other. The station **battery** has sometimes been used to back up a pilot exciter or even a main exciter for a short time.

The **electronic exciter** has recently come into limited use in place of the conventional rotating exciter. It consists of a group of water-cooled Ignitron power rectifier tubes usually arranged so that a tube may be replaced readily without interrupting the excitation. Like the motor-driven exciter, it is no more dependable than its source of a-c supply but can be serviced without shutting down the generator, provided that an alternate source of excitation is available.

Selection of the particular scheme of excitation equipment to meet most successfully the local requirements involves a review of generator and transmission system characteristics, the proposed operating practice, and the various types of equipment available. For a small, simple generating station, a self-excited main exciter and a direct-acting voltage regulator may be sufficient.

For a more important station, particularly where quick response is required for improving power-system stability, one choice, using equipment now regarded as conventional, might be a direct-connected main exciter to supply controlled excitation to the generator field without a generator-field rheostat, a pilot exciter to supply constant voltage for excitation of the main exciter, and an indirect-acting voltage regulator with fast raise and lower contactors to operate upon a rheostat in the main exciter field circuit. A recent development accomplishing similar results consists of an electronic exciter and an electronic regulator. Such equipment has given outstanding performance in limited commercial use for several years, but as yet its initial cost and maintenance are considerably higher than those of the conventional equipment. Another recent development consists of a direct-connected, self-excited main exciter, a rotating amplifier (amplidyne) to control the excitation of the main exciter, and a static (impedance or electronic) voltage regulator acting upon the rotating amplifier. This equipment has proved superior in some ways to the conventional equipment at a somewhat comparable cost. (See Article 18, References 25, 27, 28 and 29, and Section 12, Article 16.)

15. AUXILIARY POWER SUPPLY

The auxiliaries of a hydroelectric station are, as a rule, driven by electric motors, except that the exciters generally are driven directly by the main shafts. The power demand usually amounts to 0.5 to 3.0 per cent of the station rating. Normal and emergency power commonly are obtained from several sources.

A **transformer on the generator bus** has low maintenance and operating costs but usually requires a heavy-duty circuit breaker, which constitutes an important item of initial cost. Although in rare cases a transformer is provided on the high-voltage bus, the initial cost of the transformer is considerable, and the high cost of a breaker usually results in the use of a high-voltage outdoor fuse. Supply from a bus is dependable during normal operation but is not independent of power-system disturbances.

A **transformer on the generator-transformer leads** is particularly desirable in a station without a generator bus. It can take its supply from either the generator or the transmission system. Two such auxiliary power transformers, each of sufficient capacity for the entire station service load, provide auxiliary power from either of two generators or from the system. Disconnects usually are provided in the primary connections of the auxiliary power transformer, and the transformers are included within the differential relay zone of the main transformer. The supply is dependable during normal operation but is not entirely independent of power-system disturbances.

An **outside source,** such as a low-voltage distribution line in the vicinity, is particularly useful as a source of auxiliary power during periods when the generators may be shut down. With the main connection schemes of some stations a small source of this kind is essential for starting.

An **auxiliary generator driven by a main waterwheel (shaft generator)** is fairly simple and efficient, reliable, and independent of electrical disturbances on the main system except speed changes. It is quite expensive, however, in case of a slow-speed main shaft and is also more expensive to maintain and operate than are transformers.

An **auxiliary generator driven by an auxiliary waterwheel (house generator)** is the most expensive in first cost, maintenance, and operation and is subject to disturbance from trash and ice. Its efficiency is less than that of a transformer or a shaft generator. It is definitely free from disturbances on the main system and is available when the main units are shut down. It sometimes is used as an emergency source to supplement transformers.

An **auxiliary generator driven by fuel** sometimes is used to start the station and to supply the necessary lighting and pumping when the station is shut down. A small gasoline-driven generator set is often very desirable as an emergency source to operate spillway gates and other vital auxiliaries in case the generators are shut down and all outside sources are disconnected.

Voltages commonly used for auxiliary power systems are 2200, 440, and 220 volts. The most economical voltages are indicated by an analysis of the wiring and the control equipment, taking into account the locations of the transformers, buses, and motors. In many larger stations it is found economical to use 2200 volts for the large motors and one of the lower voltages for the smaller motors.

Among the many possible **auxiliaries** which may require electric power supply are the following: head gates, sluice gates, trash-rack rakes, navigation locks, forebay ice prevention, trash-rack heating, gate guide heating, filler gates, fish protection; lubricating oil pumps, governor oil pumps, lubricating- and insulating-oil handling and filtering pumps, house-service water pumps, generator-cooling water pumps and fans, air compressors, transformer oil and water pumps and blowers; elevators, incline railway, car pullers, turbine-room crane, gantry cranes, transformer hoists, building-ventilating fans, water coolers, water heaters, space heaters, drying ovens, machine shop; exciters, battery charging, telephones, signals, clocks, water-level recorders and indicators, power receptacles, transmission-line ice melting, local lines, village light and power; lighting for powerhouse, switchyard, grounds, storehouse, locks, employees' cottages, display signs.

16. ICE PREVENTION

The operation of hydroelectric plants located in cold climates requires definite provision for combating ice during the winter seasons. Sometimes ice is encountered in such large quantities that it must be cut and sluiced away, involving considerable expense and waste of water. Except in extreme cases, however, a plant can be kept operative by a small expenditure of energy properly applied at vital points if adequate provision is made in the original design.

Gates should be protected from excessive pressure of sheet ice, and their sills, guides, and seals should be kept sufficiently free of ice to allow operation. This has been accomplished by applying small amounts of steam or electric heat inside the gates and at embedded parts of the guides, which are made hollow for this purpose. To secure economical heating it is very important that the downstream side of the gates, exposed to the cold air, be thoroughly housed with heat-insulating material and that the heat inside the gates be evenly distributed by fans, chimneys, and other means. The data in Table 1 are from NELA reports of successful installations.

Table 1. Gate-heating Installations

Installation	1	2	3	4
Size of gate	50 × 19.5	50 × 33	40 × 24	30 × 36
Kw in gate proper	35	36	45	36
Kw at guides and seals	10	16	16	16
Total kw	45	52	61	52
Watts per sq ft	46	31.5	64	48

Flashboards should be protected against surface ice. The most effective method appears to be to maintain an open channel about 3 or 4 ft wide clear of ice along the upstream face. This has been accomplished by placing electric heaters or low-efficiency lamps in reflectors just above the water surface, covering them with canvas or heavy paper. Installations of this type have been reported using from 6 to 17 kw per 100 ft of flashboards.

Air bubbler systems have been used to maintain an open channel clear of ice along the upstream faces of gates and flashboards. A bubbler system consists of perforated pipes or groups of small nozzles installed along the lower, upstream face of the gates or flashboards, and supplied with air under pressure, which rises and creates a circulation of warmer water from the bottom of the pond to the surface. The data in Table 2 are taken from NELA reports of installations of this type.

Table 2. Air Bubbler Installations

Installation	1	2	3	4	5	6	7
Diameter of orifices, in.	0.0314	0.0314	0.062	0.0156	0.08	0.062	0.0156
Spacing of orifices, ft	5	5	10	3	4	3	3
Depth of orifices, ft	5	15	18	3	12	8.5	12
Air pressure, lb per sq in.	80	80	10	60	6.5	22	20
Cu ft per min of free air per 100 ft of channel	34	34	6	7	17.5	25	7
Kilowatts per 100 ft of channel including compressor and motor losses	3.4	3.4	0.25	0.55	0.7	1.44	0.5

Surge tanks of small size are likely to freeze, especially during periods of steady load. Satisfactory performance in cold weather has been obtained by insulation alone or by insulation with steam or electric heat. When heat is applied in the air space between tank and insulating sheathing, part of it is obviously lost through the sheathing; when applied directly to the water, part is obviously lost by exchange of water between tank and penstock. The most economical application, therefore, depends upon local conditions. Some conditions favor applying the heat directly to the water. In a few cases, freezing has been prevented by anchoring a wood pole in the tank in such position that its motion breaks up any surface ice which may start to form.

Penstocks, if exposed to the weather and designed for low velocity, may partially freeze, causing loss of head. In thawing again, the ice may pass on in sufficient quantities to clog the turbine gates. The usual precaution against this trouble is to insulate or bury the penstock. It is an advantage, in this connection, to take the water from the lower part of the reservoir so as to secure warmer water.

Trash racks must be protected from sheet ice, the pressure of which may injure the structure, and from **frazil ice,** which tends to adhere to the bars and clog the passages. Protection from sheet ice is afforded in some designs by a curtain wall in front of the rack extending down well below the surface of the water. **Frazil ice** is usually the greatest of all ice hazards. Although it generally appears for only a short time in early winter and occasionally in the spring, it is likely to clog the racks at these times sufficiently to curtail the station output seriously. At one plant not more than 25 per cent of the available water could be used during winter months before rack heating was installed. Only where the pond or forebay is of large size and the intake is deep, can complete immunity from frazil ice be assured. Trouble from frazil ice is prevented in some cases where the waterwheels have large passages, such as with the propeller-type wheel, by raising the trash racks during the frazil ice flow, allowing the ice to pass through the plant. Air bubbler systems have been installed in front of the racks, designed to carry the frazil ice up to the surface, where it can be sluiced off. In one plant a mechanical rake of special design is used successfully to clear racks of frazil ice. The method most extensively used consists of heating the trash-rack bars. It is found that the bars need be raised only a fraction of a degree above the freezing temperature to prevent the frazil from sticking to them. If the top of the rack extends above the surface of the water and is enclosed, heat can be applied at this point by means of warm-air fans or steam coils, and some heat will be carried down in the rack by conduction through the bars. This method has proved successful in moderately cold locations. In a number of installations, some of which are located in places where the climate is severe, the racks are heated by passing electric current through them. The energy required to keep the racks clear, in watts per square inch of bar surface exposed to the water, has been reported as 0.87, 0.88, and 0.89, respectively, from three installations which have successfully passed through one or more ice seasons. The power factor has been found to vary from 0.72 to 0.83. The impedance of the rack is subject to wide variation, depending upon the quality of steel, size, and shape of bars, spacing of bars, and arrangement of bars in series parallel circuits. It is also probably dependent to a considerable extent upon the voltage, the frequency, and the water resistance. Tests upon trash racks installed in 1930 and 1931, respectively, provide the data in Table 3.

Table 3. Rack-heating Installations

Installation	1		2	
	Volts	Ohms	Volts	Ohms
A-c 60-cycle tests	15	0.078	44	0.264
	29	0.068	121	0.197
	34	0.059	160	0.163
D-c test	....	0.018	.....	0.042

17. COSTS

The cost of hydroelectric developments probably varies more widely than that of any other class of projects. Such items as land requirements for reservoir and plant, water rights, reservoir clearing, relocation of roads and railroads, length and height and type of dam, and necessity for and layout of tunnels, pipe lines, forebays, and tailraces vary greatly for each development. The cost per kilowatt of the initial installation will usually be higher than the cost per kilowatt of the ultimate installation, where the plant capacity is increased over a period of years, since such items as the reservoir, dam, and headworks must be completed in connection with the initial development.

Accurate cost figures of hydroelectric developments are difficult to obtain for publication, since the complete cost of a development includes not only the costs of the physical structures and machinery, but also the costs of preliminary investigations, surveys, foundation exploration, land acquirement, federal and state licenses, financing, and other, less tangible items. For these reasons no attempt will be made to give overall costs of developments beyond the broad statement that the cost per kilowatt will range from about $90 under exceptionally favorable conditions or $125 for more general conditions for a large plant to $250 for a small plant constructed under disadvantageous conditions. Unit costs for individual items of construction will likewise vary over a wide range, depending on location, distance from material supply sources, prevailing wage scales, and other factors.

The procedure suggested for obtaining a preliminary cost estimate for the development of a proposed site is to prepare preliminary design of the project and base the estimate on this preliminary design and the best information available on local unit costs of structures and equipment.

A suggested outline for the preparation of a complete project estimate is as follows:

1. Preliminary and General Expense. Includes cost of preliminary investigation and exploration, organization and legal expenses, franchises, federal and state licenses, financing, and interest and taxes during construction.

2. Engineering. Includes studies of economical development of the site, preparation of general and detailed plans, specifications for structures and equipment, purchase, inspection and expediting of equipment and material, and general consulting services.

3. Hydraulic Work. Includes land and water rights, surveys and borings, reservoir-basin clearing, relocation of railroads and highways, removal of buildings, river diversion, cofferdams, dams, dikes, headworks, forebay structures, waterwheel settings, intake gates, crest gates, stop logs, hoists and handling equipment, canals, flumes, tunnels, pipe lines, penstocks, surge tanks, turbine equipment, and tailrace.

4. Electrical Work. Includes generators, exciters, low-voltage switchgear, step-up transformers, high-voltage switchgear, auxiliary power supply, conduits, ducts, wiring, grounding, controls, switchboards, arresters, neutral grounding devices, annunciators, communication equipment.

5. General. Includes power-station building, station yard, private roads, railroad serving development, auxiliary equipment, shop equipment, auxiliary buildings, permanent operators' quarters, preliminary operation and tests.

6. Substations. Includes land, substation buildings, equipment foundations, switching structures, electrical equipment switchgear, wiring, auxiliary equipment, preliminary operation and tests.

7. Overhead Transmission Lines. Includes land, right-of-way, clearing, fencing, foundations, steel and wood towers, conductors, insulators, communication and maintenance facilities, preliminary operation and tests.

18. BIBLIOGRAPHY

1. Progress in Power Generation, *Trans. AIEE*, p. 748 (1927). Extensive bibliography, 33 columns.
2. A Survey of Hydroelectric Developments, *Trans. AIEE*, Part I, p. 988; Part II, p. 1086 (1934).
3. A. C. Clogher, Hydroelectric Practice in the United States, *Trans. ASME*, p. 65 (1937). Bibliography.
4. T. H. Hogg, Hydroelectric Practice in Canada, *Trans. ASME*, p. 79 (1937).
5. First Report of Power-system Stability, by Subcommittee on Interconnection and Stability Factors, *Trans. AIEE*, p. 261 (1937). Bibliography.
6. E. B. Strowger, Relation of Relief-valve and Turbine Characteristics in the Determination of Water Hammer, *Trans. ASME*, p. 701 (1937).
7. C. C. Whelchel, Trends in the Design and Arrangement of Electrical Equipment in Hydraulic Power Plants, *Trans. AIEE*, p. 78 (1939).
8. P. L. Heslop and G. A. Jessop, The Kaplan Turbines at Bonneville, *Trans. ASME*, p. 97 (1939).
9. F. H. Rogers, Hydro-generation of Energy, *Trans. ASCE*, p. 952 (1939). Part of a symposium on Economic Aspects of Energy Generation.
10. H. K. Barrows, Hydro-generated Energy, *Trans. ASCE*, p. 1088 (1939). Part of a symposium on Cost of Energy Generation.
11. J. W. Hackney, Energy-mass Diagrams for Power Studies, *Trans. ASCE*, p. 1644 (1939).
12. R. S. Quick, Problems Encountered in the Design and Operation of Impulse Turbines, *Trans. ASME*, p. 15 (1940).
13. Symposium on Governing, *Trans. ASME*, pp. 167–240 (1940). Bibliography.
14. S. H. Wright, Generator Characteristics as Affecting System Operation, *Trans. ASME*, p. 185 (1940). Bibliography.
15. R. V. Terry, Development of the Automatic Adjustable-blade-type Propeller Turbine, *Trans. ASME*, p. 395 (1941).
16. R. E. B. Sharp, Cavitation of Hydraulic-turbine Runners, *Trans. ASME*, p. 567 (1940).
17. G. R. Rich and J. F. Roberts, Kaplan-turbine Installations of Tennessee Valley Authority, *Trans ASME*, p. 309 (1941).
18. J. D. Scoville, Speed Regulation of Kaplan Turbines, *Trans. ASME*, p. 385 (1941).
19. G. R. Rich and J. F. Roberts, Francis-turbine Installations of the Norris and Hiwassee Projects, *Trans. ASME*, p. 19 (1942).
20. Central Station Engineers of the Westinghouse Electric and Manufacturing Company, *Electrical Transmission and Distribution Reference Book* (1942).
21. G. H. Bragg, Maintenance of Hydroelectric Generating Units, *Trans. ASME*, p. 329 (1944).
22. Arnold Pfau, Mechanical Features of the Glenville Impulse Turbines, *Trans. ASME*, p. 513 (1944).
23. Robert Lowy, Efficiency Analysis of Pelton Wheels, *Trans. ASME*, p. 527 (1944).
24. G. W. Spaulding, The Co-ordinated Operation of Hydro and Steam Capacity in Electric Power Systems, *Trans. ASME*, p. 545 (1944).
25. S. B. Crary and J. B. McClure, Excitation Systems for Synchronous Machines, *AIEE Misc. Paper* 46–78 (December, 1945).
26. A. H. Frampton, Hydroelectric Power Plants, *Elec. Engineering*, p. 151 (April, 1946).
27. R. A. Hopkins and H. J. Petersen, Tennessee Valley Authority Hydroelectric Stations—Electrical and Mechanical Design, *Trans. AIEE*, p. 920 (1946).
28. F. M. Porter and J. H. Kinghorn, The Development of Modern Excitation Systems for Synchronous Condensers and Generators, *Trans. AIEE*, p. 1020 (1946).
29. J. B. McClure, S. I. Whittlesey, and M. E. Hartman, Modern Excitation Systems for Large Synchronous Machines, *Trans. AIEE*, p. 939 (1946).
30. W. P. Creager and J. D. Justin, *Hydro-electric Handbook*, John Wiley (Second Edition in preparation).
31. P. Sporn and E. L. Peterson, Design and Operating Features of Winfield Hydro Development, *Power Plant Engineering*, p. 36 (February, 1940).
32. Bibliography on Automatic Stations, 1930–1941, *Trans. AIEE*, p. 1111 (1942).
33. T. J. Corwin and D. P. Dinapol, Dutch Flat Designed as Fully Automatic, *Elec. West*, p. 60 (September, 1943).
34. P. Peterson, Automatic and Remote Control of Substations and Hydroelectric Power Plants, *Elec. News and Engineering*, pp. 28, 46, 50 (December 15, 1943).
35. M. Ebersberger, Economic Operation of Medium-size Power Stations through Remote Control and Automatic Switchgear, *Brown-Boveri Rev.*, p. 222 (July 7, 1944).
36. C. W. Bohner and A. P. Maness, Supervisory Control of 30,000-kva Hydro Plant, *AIEE Tech. Paper* 46–71 (December, 1946).

INTERNAL-COMBUSTION POWER STATIONS

By W. A. Sloan

19. INTERNAL-COMBUSTION ENGINES

APPLICATIONS. The Diesel engine is an excellent prime mover for electrical generation in plants up to about 10,000 kw. It has the advantages of low fuel cost, short periods for warming up, and no standby losses, and it requires little water. Small sizes are about as efficient to operate as large ones, whereas with steam engines or turbines the steam rates are often twice as great in small units as in large. However, unlike the steam turbine, the plant floor area and cost increase nearly proportionally with capacity, which fixes an economic limit of 5000 to 10,000 kw for Diesel generating stations at present.

Except where a large supply of by-product gas is available, as in steel mills, the Diesel engine is the present choice in the small power field. It is used by public utilities, municipalities, hotels, and factories.

PRINCIPLE. The internal-combustion engine is a prime mover whose action depends upon the heating of air by burning fuel within the cylinder. This fundamental principle is the same regardless of the fuel, which may be a combustible gas, vapor, or oil. Pulverized coal has been used successfully but is in the experimental state. The combustion causes a rapid rise in temperature and pressure in the products of combustion, which expand behind a piston, moving it forward. In its simplest form the internal-combustion engine is similar to a reciprocating single-acting steam engine, the reciprocation of the piston in the cylinder being changed to rotation of the shaft by means of a crank and connecting rod.

CLASSIFICATION. There are two distinct methods of applying the fundamental principle of operation of internal-combustion engines, so that, with reference to the thermodynamic cycle, engines may be classified as either spark-ignition (explosion) or compression-ignition (Diesel).

In the **spark-ignition engine,** air which has been mixed with a suitable quantity of fuel in gaseous or vaporized form is moderately compressed in the cylinder and is then ignited, giving a rise in pressure which is almost instantaneous. The ignition is accomplished by an electric spark. In this class belong all gas and gasoline engines and some oil engines. The compression must be so regulated that the temperature of ignition is not reached before the spark jumps, and the mixture of fuel and air must be of such proportions that it will burn rapidly. (See Section 17, Article 15.)

The **compression-ignition** or **Diesel engine** compresses only air, to a predetermined temperature above the ignition point of the fuel. The fuel is then forced into the cylinder in a finely atomized state and ignites spontaneously without explosion, by reason of the heat resulting from the compression of the air. In the pure Diesel engine the fuel is injected into the cylinder by means of a blast of high-pressure air, and the timing of injection is such that there is no rise in pressure during combustion. The development of engines using mechanical means of injecting fuel into the cylinder (commonly called solid-injection Diesel) gives rise to a cycle somewhat different from the constant-pressure combustion of the pure Diesel. The recognized definition of a Diesel now is: an engine in which the fuel, injected after compression is practically completed, is ignited solely by the heat resulting from the compression of the air supplied for combustion.

There is one type of solid-injection engine, commonly called **semi-Diesel,** which has lower compression than the Diesel, so that the temperature is below that required for ignition, and in which sufficient additional heat is supplied by a hot plate or bulb forming part of the combustion chamber. The plate or bulb must be heated for starting, after which the heat of combustion will keep it at the required temperature unless the load is too light. These engines are less efficient than those having compression ignition and are rough-running in multicylinders because of the uncertainty of ignition.

Any of the thermodynamic cycles for internal-combustion engines may be mechanically performed in an engine giving an impulse in one end of each cylinder in either one or two revolutions of the crankshaft.

An engine requiring two revolutions of the crankshaft to complete its working cycle is called a four-cycle engine, with reference to the four piston strokes in two revolutions. The following sequence of operations takes place during four consecutive strokes: (*a*) inspiration of a mixture of gas and air during an entire stroke, or air only in the case of a Diesel; (*b*) compression during the second (return) stroke; (*c*) ignition or injection, causing combustion at or near the dead center and expansion during the third stroke; (*d*) expulsion or exhaust of the products of combustion during the fourth (return) stroke.

In the two-cycle engine the working cycle is completed in one revolution of the crankshaft or two strokes of the piston. Starting with the combustion, at or near dead center, the piston moves toward the crankshaft with the gases expanding behind it. Toward the end of this first stroke the piston uncovers an exhaust port, and the burned gases escape. Shortly after, an inlet port opens, admitting a mixture of fuel and air (air only in a Diesel) from a reservoir in which it has been slightly compressed. The inlet and exhaust ports close early in the return stroke, and during the remainder of this second stroke compression occurs. In some designs admission is accomplished through mechanically operated valves in the cylinder head. Other engines have inlet ports at the bottom of the stroke and exhaust valves in the cylinder head. Some engines use blowers for forcing in the air or mixture, but many small engines have a closed crankcase, which is used as a reservoir, and the piston displacement itself furnishes the pressure.

When the design is such that combustion occurs on only one side of the piston, the engine is single-acting; if on both sides, double-acting. The majority of internal-combustion engines are single-acting, the double-acting principle being restricted mostly to very large sizes. The choice between single- and double-acting engines depends on the cost of production. When plain uncooled pistons can be used, the single-acting engine is cheaper;

but when it becomes necessary to water-cool the pistons, it is generally desirable to use the double-acting principle. In some single-acting engines the pistons are cooled by a spray of lubricating oil on the crankcase side. Such an engine requires a lubricating-oil cooler.

POWER. In contrast to steam prime movers, the internal-combustion engine has a definite limit of power. The power depends upon the weight of air that can be passed through the engine in a given time and upon the efficiency with which it can be burned. This fact is evident, since, however much air is present in the cylinder, sufficient fuel can always be admitted to combine with it, but further addition of fuel will be useless unless there is sufficient oxygen for its combustion.

The power developed in one end of a single cylinder of an internal-combustion engine may be calculated by the following formula:

$$ihp = \frac{PLAN}{33{,}000}$$

where ihp = indicated horsepower.
P = indicated mean effective pressure in pounds per square inch.
L = length of stroke in feet.
A = area of cylinder bore in square inches.
N = number of impulses per minute;
= revolutions per minute for two-cycle engines;
= revolutions per minute/2 for four-cycle engines.

The mean effective pressure can be obtained from an indicator diagram and varies with the load. At full load it depends on the kind of fuel and the amount of compression.

A portion of the power developed in the cylinder is absorbed in engine friction, so that the available power at the shaft is less than that developed in the cylinder and is called the brake horsepower, being that power measured by a dynamometer or prony brake. Mechanical efficiency is the ratio of brake to indicated horsepower and varies from 70 to 90 per cent at full load, depending on the size and type of the engine.

RATING. Internal-combustion engines are rated on their brake horsepower, usually with an allowance for overload. This overload allowance may vary from 10 to 20 per cent of rating, and the characteristic of limitation of power in such an engine makes it desirable to know either the maximum power or the overload capacity when selecting an engine for a definite service. A good criterion for the selection of an engine is the brake mean effective pressure (bmep) on which it was rated. The brake mean effective pressure is really the product of indicated mean effective pressure and mechanical efficiency and may run as high as 100 lb per sq in. or above in ordinary engines. For continuous service it is well to select four-cycle engines rated on about 70 to 80 lb per sq in. brake mean effective pressure. Two-cycle engines with blowers and high-pressure lubrication may also be rated on about 70 to 80 lb per sq in. brake mean effective pressure, but two-cycle crankcase compression engines must use low-pressure lubrication, and the brake mean effective pressure for continuous rating should be limited to 40 to 50 lb per sq in.

The brake mean effective pressure may be calculated, when the engine dimensions are known, by putting the brake horsepower in the formula for indicated horsepower, thus:

$$bhp = \frac{P_1LAN}{33{,}000}$$

where bhp = brake horsepower.
P_1 = brake mean effective pressure in pounds per square inch.

Other terms are the same as in the formula for indicated horsepower.

Although internal-combustion engines are mechanically somewhat limited as to size, engines have been built up to 10,000 brake horsepower, the majority being of less than 3000 brake horsepower. A few years ago piston speeds of 700 to 800 ft per min were considered high for stationary engines, but recent designs are utilizing speeds as high as 1500 ft per min. This increase naturally helps to reduce the weight per brake horsepower.

Since the power of an internal-combustion engine is limited by the weight of air taken into the cylinder in a given time, altitude will have an effect on the power. Manufacturers rate their engines for sea-level operation, and allowance should be made for a decrease in capacity at altitudes above 2000 ft. Altitudes of 10,000 ft will reduce the brake horsepower about 30 per cent, and other altitudes have proportional effect. Supercharging by means of a blower may be used to offset the effect of altitude on power output.

EFFICIENCY AND ECONOMY. The internal-combustion engine uses a cycle theoretically more efficient than that of the steam engine or turbine and consequently should be a more economical engine. Unlike the steam engine, the internal-combustion engine does not become appreciably more economical as the size is increased. Although in small powers it is far more economical than the steam engine, in very large powers it does not retain this advantage to anything like the same extent, particularly in view of recent developments of the regenerative and reheat steam cycles. Full-load test figures show thermal efficiencies above 20 per cent, referred to brake horsepower, for gas and gasoline engines, and above 30 per cent for Diesel engines.

In a spark-ignition engine the efficiency is largely dependent upon the amount of compression, but in the Diesel engine the length of time required for combustion has a decided influence. Although theoretically the spark-ignition engine gives a higher efficiency for equal compressions, this advantage cannot be obtained in practice, since a spark-ignition engine compresses a mixture which would pre-ignite if the compression were too high, whereas the Diesel compresses only air.

The fuel consumption of internal-combustion engines may be expected to vary somewhat with the kind of engine and the fuel, but Table 1 gives an idea of what might be expected, on the average, in terms of British thermal units in the fuel.

Table 1. Average Consumption in British Thermal Units per Brake Horsepower-hour

Per cent of rated load	100	75	50	25
Gas and gasoline engines	10,500	12,000	14,700	20,000
Diesel engines	8,550	8,800	9,500	11,500

The losses in an internal-combustion engine are the heat carried away in cooling water and that carried away in the exhaust gases, friction, and radiation. For spark-ignition engines the amounts vary considerably with compression and somewhat with speed. For a Diesel engine the full-load heat balance will not vary much from that given in Table 2.

Table 2. Average Full-load Heat Balance in Diesel Engine

Brake horsepower	30%
Cooling water loss	32
Exhaust gas loss	28
Friction, compressor, radiation	10
In fuel	100

LUBRICATION. Because of the high temperatures that prevail in the cylinder of the internal-combustion engine, the question of proper lubrication is a serious one. Cylinder oil should be of a high grade, free from acids, and composed of hydrocarbons that leave no residue after combustion. Only mineral oils, therefore, are suitable for the purpose.

The amount of oil required per horsepower-hour varies with the character of the installation and the method of operation. With a properly designed and operated lubricating system the average consumption of lubricating oil will be about 0.0005 gal per hr per rated bhp for four-cycle engines and possibly twice as much for two-cycle.

COOLING. Internal-combustion engines must be cooled by circulating water through jackets surrounding the heated parts. The quantity of water which must be circulated depends upon the initial and final temperatures.

Inlet water temperatures are usually from 80 to 90 deg fahr. In general the outlet temperature should not be above 160 deg fahr, or there may be danger of breaking down the film of lubricating oil on the cylinder walls or warping of valves. In order to avoid excessive heat stresses in large engines, the temperature is usually not allowed to go above 120 deg fahr, but in engines of medium size it may run as high as 130 to 140 deg fahr. If the water contains foreign matter, the temperature must be kept low to avoid precipitation and formation of scale, or a water-softening plant must be used.

Excessive circulation resulting in low final temperatures increases the fuel consumption. It is particularly undesirable with gasoline engines, as it may cause precipitation of part of the fuel. Very low cooling-water temperatures increase the viscosity of the lubricating oil, resulting in an increase of piston friction.

Sometimes a desirable grade of water is available in sufficient amount to allow it to be wasted after circulation through the jackets, but where water is scarce or expensive, cooling towers or spray ponds are generally employed for cooling before recirculation. A heat exchanger may be used, so that distilled water is circulated through the engine jacket and raw water used for cooling it in the exchanger.

The quantity of water to be circulated depends on the thermal efficiency of the engine, the proportion of heat loss to the jackets, and the rise in temperature through the engine.

Table 3 gives the quantity required in gallons per hour per brake horsepower for a wide range of conditions.

Table 3. Gallons of Cooling Water per Hour per Brake Horsepower

Thermal Efficiency, %	Jacket Loss, %	Temperature Rise, deg fahr				
		30	35	40	45	50
15	30	20.4	17.4	15.3	13.6	12.2
	35	23.7	20.3	17.8	15.8	14.2
	40	27.1	23.2	20.3	18.0	16.2
20	30	15.3	13.1	11.5	10.2	9.1
	35	17.8	15.3	13.4	11.9	10.7
	40	20.4	17.4	15.3	13.6	12.2
25	30	12.2	10.5	9.1	8.2	7.3
	35	14.2	12.2	10.7	9.5	8.5
	40	16.3	13.9	12.2	10.9	9.7
30	30	10.2	8.8	7.6	6.8	6.1
	35	11.9	10.2	8.9	7.9	7.1
	40	13.5	11.7	10.2	9.1	8.2

20. POWER PLANTS

SELECTION OF UNITS. Division of the capacity into two or more units allows the operator to meet the load curve with units loaded to somewhere near the point of best efficiency. In addition it makes the plant more reliable, as at least partial service can be maintained in case of accident to one of the units.

High rating for a given engine lowers the cost per horsepower but shortens the life. It is better, therefore, to select engines of rather conservative rating in speed and brake mean effective pressure. The design should be as simple as possible. Details of design, workmanship, and materials, as proved by the engine's record and the manufacturer's reputation, are more important than the type of engine. The specifications should be drawn up by experts and should contain only points which are essential to the purchaser, such as weight and space limitations, giving the manufacturer the choice of type and details of design. It should be remembered that, although bids may be tabulated, quality cannot be tabulated, and the cheapest engine may not give the most economical service.

PLANT ARRANGEMENT. It is common practice to set the units on parallel lines. When so arranged, the average plant, having from two to four units, is a nearly square building. When steam engines are replaced by Diesels and the old building is used, the arrangement will naturally depend on the shape of the old plant, and engines may have to be placed with their center lines in line. Ample clearance must be allowed for the dismantling of engine, generator, and exciter.

FOUNDATIONS are very important in the installation of Diesel engines, as the mass must be sufficient to absorb the vibration; engines located in basement of hotels, department stores, and similar places should be insulated from the rest of the building. In certain localities silencers are required in place of ordinary mufflers for the exhaust. The air for combustion should be free of dust and dirt.

The **oil storage** should be located outside the plant, either above or below ground, depending on the local conditions. The capacity of the oil storage is determined by the maximum rate of fuel consumption of the plant and the longest expected time between deliveries. The oil may be handled directly from the storage tank to the engine, or an overhead day tank may be placed inside the building.

21. ECONOMICS

The costs given in this section are representative only, and it must be remembered that unusual market or local conditions may change these figures materially. The data given refer only to slow-speed units, but the large majority of generating plants belong in this class. High-speed generating sets are treated in Article 22.

INVESTMENT COSTS AND FIXED CHARGES. The average unit cost of representative Diesel electric generating plants in installed capacity up to 10,000 kw will be found in Figs. 1 and 2. The costs are divided into four different accounts, land being

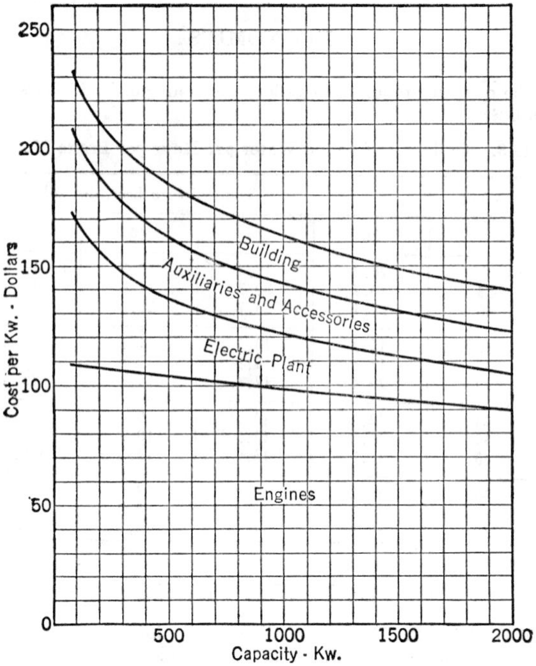

Fig. 1. Unit Costs of Diesel Electric Generating Plants

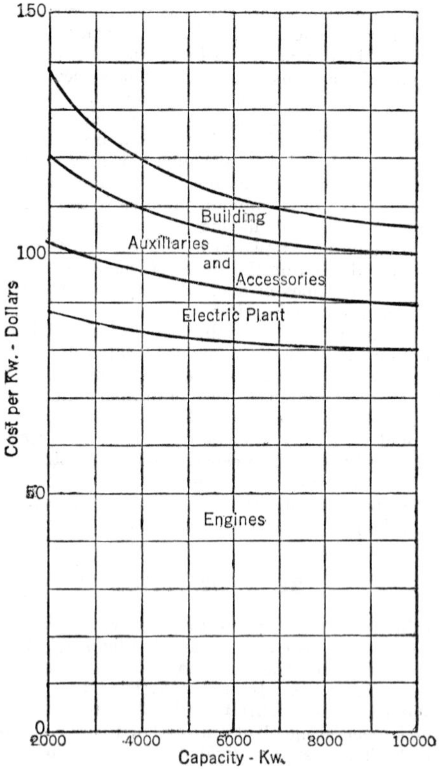

Fig. 2. Average Cost of Diesel Electric Generating Plants

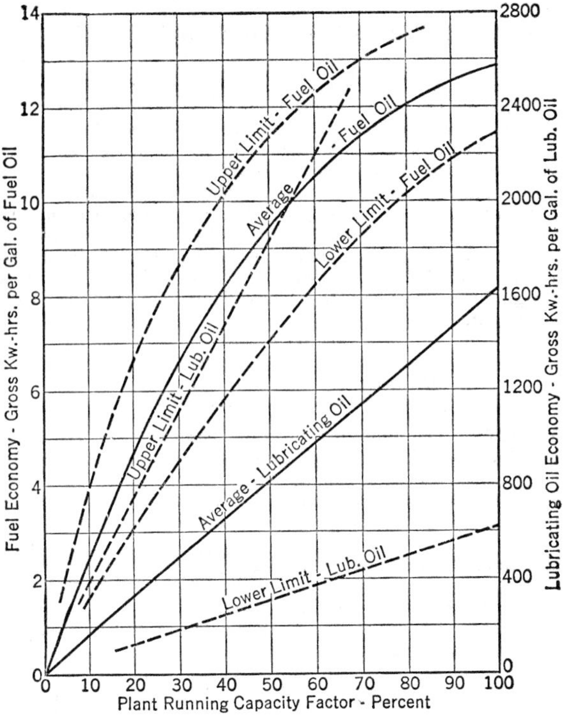

Fig. 3. Fuel Economy of Diesel Electric Generating Plants

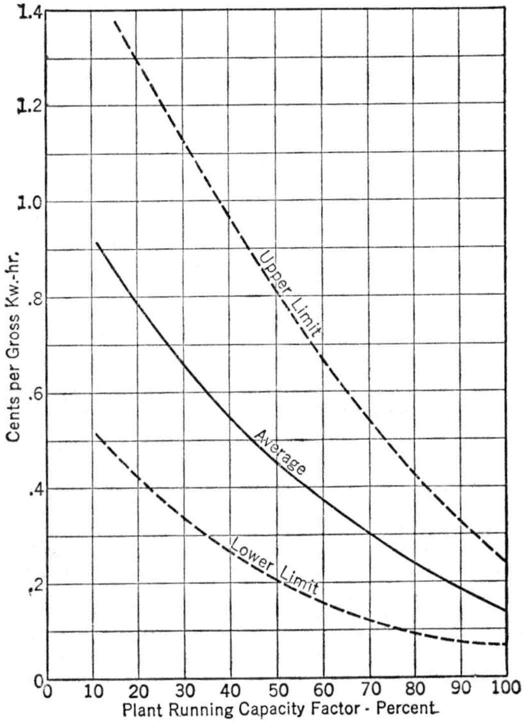

Fig. 4. Cost of Attendance of Diesel Electric Generating Plants

omitted, since it is recognized that location would have more effect on this item than the others. The building costs are for substantial but not elaborate structures. The figures given are for each item erected, including its foundations, but no allowance has been made for overhead. Overhead costs would cover administrative expense, financing, engineering, etc., and usually run about 10 to 15 per cent of the costs as shown.

The first costs of operating a generating plant which must be considered are the fixed charges on the investment. These charges include interest, depreciation, insurance, and taxes. Interest may be considered constant for all the individual parts which make up the capital investment, but there is a difference in depreciation, insurance, and taxes among the individual parts of the plant. For the purpose of a rough estimate, however, the following average figures may be used. These figures are the yearly cost in per cent of the capital investment.

Fixed Charges

Interest	5%
Depreciation	6%
Insurance and taxes	2%

OPERATING COSTS. The operating costs of Diesel electric generating plants will include the cost of fuel and lubricating oil, attendance, engine and other plant repairs, supplies, and miscellaneous expenses, including the cost of water. These items will vary with the character of load and method of operation, and the factor which they appear to follow is the plant running capacity factor, which is defined as follows:

$$\text{Plant running capacity factor, per cent} = \frac{\text{Plant output in gross kwhr} \times 100}{\text{Total rated kwhr of individual units}}$$

The expression rated kwhr refers to the kilowatt rating of an engine-generator set multiplied by the number of hours operated. For example, if a unit having a rating of

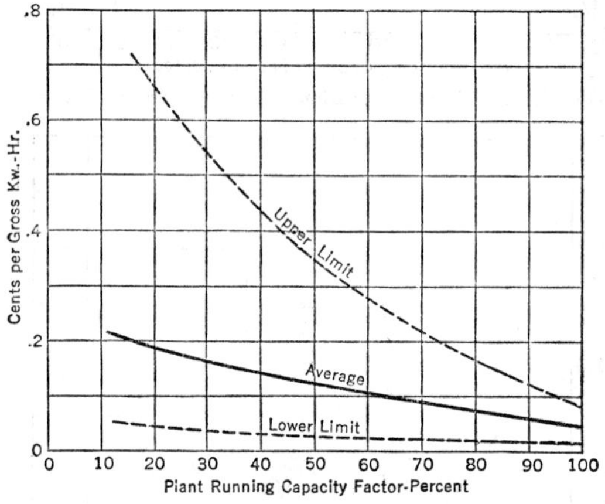

Fig. 5. Combined Operating Costs of Diesel Electric Plant

All data in Figs. 3, 4 and 5 are based on gross kilowatthours generated. The net kilowatthour output is found by subtracting the power used for plant auxiliaries and station lights from the total gross output of the plant. From 2 to 10 per cent of the gross output may be used in the plant, from 4 to 6 per cent representing a good average.

200 kw is operated 4000 hr, the rated kwhr equals 800,000, no matter what the actual output may have been. Total rated kilowatthours of individual units may be estimated by fitting the units to the expected load curve and calculating the hours of operation required for each unit.

Estimating the cost of fuel and lubricating oil involves a knowledge of the quantities to be used and the price per gallon. Figure 3 gives the fuel and lubricating oil economics of Diesel electric generating plants in gross kilowatthours per gallon of each oil, with respect to the plant running capacity factor. These curves were taken from yearly reports of the ASME Committee on Oil-engine Power Cost and represent actual performance of several hundred Diesel plants. These same reports show that the cost of fuel oil used in these plants varied from 2.2 to 7.7 cents per gallon, the most common cost being from 5 to 6 cents. The cost of lubricating oil varied from 27 to 66 cents per gallon, the most common cost being from 45 to 55 cents.

These reports were analysed for the other operating costs, which are shown in Figs. 4 and 5, in cents per gross kilowatt-hour generated, against plant running capacity factor. Figure 4 gives the cost of attendance, including superintendence, for low-speed units. Figure 5 gives the combined cost of all repairs and supplies and miscellaneous costs, including water.

22. HIGH-SPEED GENERATING SETS

High-speed generating sets in sizes up to 100 kw are useful in small industrial plants in locations where electric rates are high, and as peak-load or emergency-standby sets for places where failure of electric service is serious, as in hospitals. These sets are usually self-contained with engine, generator, switchboard, radiator, and fan all mounted on one base, requiring little or no foundation.

The cost of such sets may be $100 per kilowatt or less. The interest item of the fixed charges is therefore lower than for slow-speed engine sets, but the depreciation is much higher.

Few plants of this type have kept accurate records of operating costs. Figure 3 may be used to determine the fuel oil required, as small engines are almost as efficient as large ones. These engines have a tendency to use more lubricating oil than the large slow-speed engines, and in using Fig. 3 to estimate this item the median line should be taken as the upper limit. Figures 4 and 5 must not be used for estimating attendance and maintenance costs for high-speed sets. Attendance costs for these sets is usually almost negligible, but maintenance expenses may be much higher than for slow-speed engines.

23. GAS-TURBINE GENERATING SETS

Much research has been done and many developments have taken place recently in connection with the gas turbine, but in spite of some industrial applications it may be said to be still in the experimental state. For electric power generation it is at present useful only where fuel is cheap or water lacking, or as a standby with low fixed charges and small space requirements.

Securing efficiencies high enough to compare at all favorably with Diesel plants depends on the temperature at which the gases may be used in the turbine. At present the limit seems to be about 1000 deg fahr, which is too low for high efficiency. Higher operating temperatures depend on metallurgical development.

24. BIBLIOGRAPHY

American Society of Mechanical Engineers, Yearly Reports on Oil-Engine Power Costs.
Degler, H. E., *Internal-combustion Engines.* John Wiley (1938).
Jennings, B. H., and E. F. Obert, *Internal Combustion Engines.* International Text-Book (1944).
Lichty, *Internal Combustion Engines.* McGraw-Hill (1939).
Maleev, V. L., *Internal Combustion Engines.* McGraw-Hill (1945).
Morrison, L. H., *Diesel Engines.* McGraw-Hill (1923).
Morrison, L. H., *Diesel Engineering Handbook.* Diesel Publishers (New York) (1923).
Polson, J. A., *Internal Combustion Engines.* John Wiley (1942).
Riccardo, H. R., *The Internal Combustion Engine.* Blackie (London) (1941).
Taylor, C. F., and E. S. Taylor, *Internal Combustion Engines.* International Text-Book (1938).

POWER-STATION CIRCUITS

By R. A. Hopkins

25. MAIN ELECTRICAL CONNECTIONS

FUNDAMENTAL REQUIREMENTS. The scheme of electrical connections to be used for generators, transformers, transmission lines, and station auxiliaries largely determines the physical arrangement of the electrical bay, the equipment, and the wiring. (See Section 12.)

Efficient operation of the station at all loads should be provided for by suitable grouping and switching of generators, transformers, and lines. The efficiency of a steam turbine-generator varies considerably with the load, and the efficiency of a waterwheel generator varies considerably with both the load and the head. For this reason, if the load is not constant, it may be necessary to switch the units from hour to hour to secure greatest economy. Switching of transformers is less important because transformer efficiency is relatively high at all loads. The lines have small loss at light load, and for this reason, as well as to safeguard continuity of service, they are usually left in service at all loads. An exception to this rule is found in long, multiple-circuit lines, where one or more circuits are usually disconnected at light load to avoid overloading transformers and generators with reactive kilovolt-amperes.

Retirement of generators, circuit breakers, and other moving equipment at regular intervals for inspection and maintenance should be provided for. Where interruption of station output cannot be permitted, the connections must provide for the substitution of spare equipment during this routine maintenance.

Faults occurring in any part of the equipment or connections should be promptly detected and isolated with minimum disturbance to other circuits and minimum loss of load. This requires a certain amount of busing of generators and lines, so that the portion of load carried by the faulted part is automatically transferred without interruption. The relay scheme should be so arranged that the retirement of a breaker for inspection does not impair any of the relay protection.

Synchronizing must be provided for. Low-voltage circuit breakers were formerly preferred for synchronizing on account of their quicker action in comparison to high-voltage breakers. High-voltage breakers inherently required somewhat longer closing time on account of their longer contact travel. The time of closing of modern breakers is so well controlled and of such consistent duration, however, that synchronizing is now feasible at all voltages up to 220 kv.

Short circuits and system stability must be carefully investigated. (See Section 3.) The possibility of severe short circuits necessitates the use of heavy bus supports, current transformers, and other equipment to withstand the mechanical stresses and of heavy circuit breakers to interrupt the current, all of which contribute materially to the cost of the installation. Low stability may cause serious rejection of load during system disturbances. Reactance in generators, transformers, and circuits reduces short-circuit severity but also lowers the stability. The amount of reactance to be used in generators, transformers, and main circuits, therefore, must be a compromise. **Sectionalizing** of the station, as opposed to busing, greatly reduces short-circuit severity, but some degree of busing is usually demanded in order to equalize the load among the transformers and generators and to transfer load from a faulted machine. For the small and the medium-sized station the busing is usually provided as required, and the breakers and other devices are designed for the resulting short-circuit currents. In very large stations, solid low-voltage busing might result in short-circuit currents beyond the ratings of available breakers and other equipment, and in this case some form of high-reactance busing may be used, such as bus reactors, double-primary transformers, and double-circuit generators. These devices insert reactance between main circuits, but not directly in them, thus limiting short-circuit current without impeding the delivery of normal station output. The use of high-speed relays and fast breakers to improve stability increases breaker duty but does not affect mechanical stresses. Some of the other means of increasing stability. such as flywheel effect, damper windings, quick-response excitation, quick-acting governors, and sensitive voltage regulators, may at the same time increase both mechanical stresses and breaker duty. Therefore the studies of short circuits and stability should be conducted jointly.

Charging current for long transmission lines at light load must be provided. Under extreme conditions this may require the connecting of two generators to one line at light load, although one generator is used per line for normal load conditions. It has been found much more satisfactory, however, to be able to charge a line with a single generator.

ELEMENTARY CONNECTIONS. A few elementary connections, which may be modified and combined to meet local requirements most completely, form the basis of all connection schemes. The more useful of these elementary connections are illustrated in Fig. 1 and analyzed as follows:

Figure 1 (*a*) shows the **single, low-voltage bus.** Any number of generators may be operated to secure best station efficiency at all loads, but the transformers cannot be operated this way unless all lines are in parallel to the same substation bus. Only the bus sectionalizing breakers can be retired for inspection without curtailing the output, and this can be done only by sectionalizing the station. A bus fault will shut down at least one generator, transformer, and line. A generator fault will not cause immediate loss of load if the bus is operated united, that is, with sectionalizing breakers closed. A transformer fault will cause immediate loss of a line. Short-circuit stresses and breaker duty are high when the bus is united. Auxiliary power may be taken from two or more sections of the bus. A large number of lines cannot be accommodated without an expensive transformer layout. Air break switches may be used in the lines, since interrupting duty can be handled by the transformer low-voltage breakers. The scheme is simple, compact, and inexpensive.

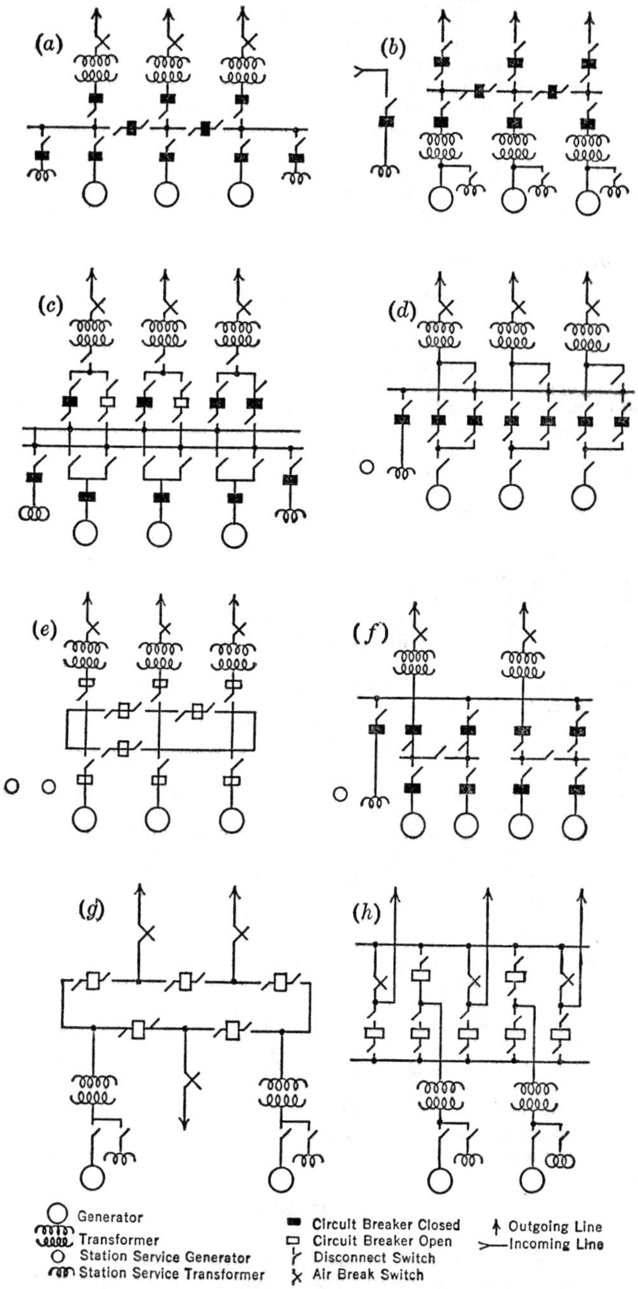

FIG. 1. Elementary Connections for Power-station Circuits

Figure 1 (*b*) shows the **single, high-voltage bus.** Generators and transformers can be operated for best station efficiency provided that their individual best efficiencies are at the same load. Only the bus sectionalizing breakers can be retired without curtailing the output, and this can be done only by sectionalizing the station. A bus fault will shut down at least one generator, transformer, and line. A generator or transformer fault will not cause immediate loss of load if the bus is operated united. Short-circuit stresses and breaker duty are high when the bus is united. Auxiliary power cannot be taken from the bus without using expensive high-voltage transformers. Any number of lines may be accommodated. Air break switches cannot be used, as every breaker shown must be used for interrupting. Generator breakers may be omitted. The scheme is simple but, in comparison to the single low-voltage bus, requires more space, and the breakers are more expensive.

Figure 1 (*c*), **the double bus,** is shown for low voltage, but may be used also for high voltage. Some stations have both. Selector breakers are shown for the transformers, and selector switches are shown for the generators, but either construction may be used for either location. The principal advantages of the double bus are that either bus with its breaker may be retired for inspection without curtailing the output or sectionalizing the station, and that both buses may be operated isolated from each other with any desired grouping of generators and lines on each bus. If selector breakers are used throughout and both buses are operated in parallel, with all breakers closed, the double bus has the further decided advantage that a bus fault will not interrupt station output. Bus sectionalizing breakers are sometimes used to localize bus faults. The two buses may be tied together by closing both breakers of any two-breaker circuit, and they must be so tied together while transferring any of the circuits not having selector breakers.

Figure 1 (*d*), the **straight transfer bus,** sometimes called the **synchronizing bus,** may be used either for low voltage, as shown, or for high voltage. The breakers may be used either on the source side of the bus, as shown, or on the load side. The bus or any breaker can be retired without curtailing the output if station service has a separate source. A bus fault does not cause loss of any load except that which is being transferred. If current-limiting reactors must be used, this arrangement provides a place for them between each generator and the transfer bus. With the reactors thus located, the station output normally is not delivered through the reactors, but short-circuit current from one main circuit to another must pass through two reactors in series. The scheme may be made simple, compact, and relatively inexpensive.

Figure 1 (*e*), the **ring transfer bus,** is shown for low voltage, but may be used also for high voltage. The principal advantage of this type of transfer bus is greater flexibility in sectionalizing.

Figure 1 (*f*), the **group bus,** is of particular value where several generators or several feeders are to be connected to a transformer.

Figure 1 (*g*), the **simple ring bus,** is shown for high voltage but may be used also for low voltage. Bus relaying is not required because each section of the bus is a part of and is relayed with a circuit. The relaying of any circuit trips two breakers and opens the ring without separating any other circuit. The air break switches in the lines allow a faulty line to be disconnected and the ring to be reclosed. Sometimes such air break switches are used in the transformer circuits also. The scheme is inexpensive, since it requires only one breaker per circuit, yet any breaker can be retired without interrupting the station output. Usually not more than four to six circuits are used in a ring, but sometimes two or more rings are used and connected together by means of tie breakers.

Figure 1 (*h*), the **main-and-transfer bus,** is probably in wider use than any of the other schemes. By judicious physical arrangement of apparatus and structures and by omitting the air break switches in the lines, the switching scheme can be started in the initial development of a station as a simple ring bus up to four circuits; as the station is expanded, it can be converted to the main-and-transfer form with a minimum of changes. It then can be extended to any number of circuits, and bus sectionalizing breakers can be used if desired. By adding a second breaker for each line, the scheme can be converted to a double bus (*c*).

26. STATION WIRING

Station wiring should be as direct as possible to reduce initial cost, operating losses, and exposure to hazards; of generous capacity to avoid excessive temperature rise, power loss, and voltage drop; adequately insulated against abnormal as well as normal voltages; adequately supported to resist mechanical stresses resulting from short circuits; completely protected against mechanical or other injury; isolated to prevent spread of trouble from one circuit to another; enclosed to safeguard attendants but reasonably accessible for inspection and maintenance. These requirements demand thorough coordination of the wiring with the connected equipment and planning in the initial stages of station design so that the wiring may be built into the building structures.

LOW-VOLTAGE POWER WIRING. This wiring consists of the generator connections through the switching and busing to the step-up transformers, the field and excitation circuits, generator-voltage feeders, and auxiliary power circuits. Current is the controlling factor in the design of low-voltage circuits. Normal current determines the conductor size from the standpoint of either temperature rise or voltage drop. Short-circuit current determines the mechanical strength of the supports. Two constructions are in common use: (1) bare conductors supported on insulators and (2) insulated cables pulled into ducts, or conduits.

Buses. Bare conductors supported on insulators are generally preferred for buses because they can be tapered to suit the loading along the bus and because the many splices and tap connections can be constructed easily. They are preferred also for short connections between breakers, disconnects, current transformers, and other equipment because they are easily joined to the apparatus terminals.

The **conductor** may consist of copper or aluminum flat bars, tubes, hollow squares, channels or angles, used singly or in multiple and in various sizes and weights. **Flat bars** are most convenient for making connections and for tapering a bus. They have good mechanical rigidity edgewise and flexibility sideways, both features being very useful in bus design. They have excellent radiating surface per unit of cross-sectional area when used singly; but when used in multiple, mutual heating materially reduces the current capacity per unit of cross-sectional area. When used for alternating current, the skin effect further reduces the capacity. **Tubes** are more difficult to connect and tap but have equal rigidity in any axis. Although they have less outside radiating surface than single bars of equal sectional area, they have less skin effect. **Hollow squares** have rigidity and outside radiating surface slightly less than tubes, but their flat sides facilitate making connections and taps. Perforations generally improve the current capacity by providing ventilation, even at the expense of some cross-sectional area. **Channels and angles** offer probably the greatest net advantages for the usual design, particularly when used in pairs to form a hollow, ventilated conductor. Such a conductor is easy to splice and tap; it can be tapered to some extent; it is fairly rigid in all axes; and it has good radiating surface. By slightly separating the two channels or two angles horizontally to provide vertical ventilation for the full length of the conductor, excellent current capacity is obtained.

Carrying Capacity of Bare Conductor Bars in Open Air. With as many as four parallel vertical strip bars per conductor, spaced for ventilation, the distribution of current between the bars is fairly uniform, but with five or more in parallel planes, the outer bars carry considerably more current than the inner ones.

Let I = current in amperes, per conductor.
R = resistance in ohms per foot, all bars in parallel. For alternating current, R will be the d-c resistance multiplied by the skin-effect factor, for which consult Section 3, Article 24.

$$W_r = \text{watts per square inch dissipated by radiation}$$
$$= 36.6 \times 10^{-12} e[(273 + T_1)^4 - (273 + T_2)^4] \quad (1)$$

where T_1 = temperature of bars in degrees centigrade.
T_2 = temperature of walls, to which heat is radiated, in degrees centigrade.
e = emissivity factor, which is about 0.8 to 0.9 for well-lacquered copper and about 0.3 for tarnished copper.

W_{c1} = watts per square inch dissipated by convection from outside surfaces of bars

$$= \frac{0.0022 P^{0.5} \theta^{1.25}}{h^{0.25}} \quad (2)$$

W_{c2} = watts per square inch dissipated by convection from the in-between surfaces

$$= \frac{0.0017 P^{0.5} \theta^{1.25}}{h^{0.25}} \quad (3)$$

If the bars are more than 7 in. high,

$$W_{c1} = 0.0014 P^{0.5} \theta^{1.25} \quad (4)$$

where P = pressure of air in atmospheres.
θ = temperature of bars above ambient air, unheated by the bars, in degrees centigrade.
h = vertical height of bars in inches.

Let A_1 = exposed outside area in square inches of all bars, per foot length.
A_2 = inside area in square inches of all bars, per foot length.

Then

$$I = \frac{(W_r A_1 + W_{c1} A_1 + W_{c2} A_2)}{R} \tag{5}$$

The above formulas are adapted from Dwight, *Electrical Coils and Conductors*, and McAdams, *Heat Transmission*, McGraw-Hill. Additional data on this subject are given in *Copper for Bus-bars*, Copper Development Association, London, where the statement is made that the temperature of copper bars should not exceed 80 deg cent, since above that temperature the rate of oxidation increases rapidly and may give rise to excessive local heating at joints and contacts.

The carrying capacity of any bare conductor depends considerably upon ventilation, which itself depends upon the position of the bars and shapes. Relative 60-cycle current capacities of various structural shapes are as follows:

	Percentages
Four copper bars, each 1/4 in. thick by 4 in. wide, on edge, with 1/4-in. vertical ventilation between bars	100
Square tubing, not ventilated	111
Round tubing	118
Square tubing, ventilated	133
Two channels, ventilated	133
Two angles, ventilated	137

Insulators. Insulators for bare conductors usually consist of glazed porcelain bodies with cemented bases and caps. Both dielectric and mechanical strengths are important. A flashover test from conductor to ground equal to about 2 1/2 times working voltage between phases and a minimum mechanical breaking strength equal to 2 or 3 times the maximum computed short-circuit stress usually are specified. Where mechanical stresses are high, it is customary to use two or more insulators at each support in order to utilize the compressive strength of the porcelain to best advantage.

Repulsive Force between Bars. The repulsive force, in pounds per foot of circuit, between two long, parallel, straight wires equals

$$\frac{5.40 I_1 I_2 \cos\theta}{s \times 10^7} \tag{6}$$

where I_1 and I_2 = the currents in amperes.
θ = phase angle between them.
s = axial spacing in inches.

For flat bars of dimensions $a \times b$ inches and spaced axially s inches, the force will be at least 90 per cent of the value given by eq. (6) for values of $(s - a)/(a + b)$ greater than 1, a being the dimension parallel with s.

For values of $(s - a)/(a + b)$ equal to 1 or less, the force calculated by eq. (6) must be reduced by the following factors.

$(s - a)/(a + b)$	a/b			
	0	0.1	0.25	0.50
0.1	0.24	0.42	0.64	0.81
0.2	0.42	0.56	0.73	0.86
0.4	0.64	0.72	0.83	0.91
0.6	0.76	0.82	0.88	0.95
0.8	0.83	0.88	0.92	0.96
1.0	0.88	0.91	0.94	0.98

Note: This method neglects skin effect and thereby makes the force slightly less for low values of $(s - a)/(a + b)$ and slightly more for high values. For experimental data, see C. J. Barrows, *Trans. AIEE*, p. 392 (1911).

This subject is treated at length by O. R. Shurig and M. F. Sayre, *Trans. AIEE*, Vol. 44, p. 217 (1925).

Enclosures. Bare conductors need to be enclosed to protect the conductors from injury, to protect personnel from contact, and to isolate any flashover that may occur. Materials used are masonry, Transite, steel, copper, and aluminum. Any enclosure obviously reduces the current capacity by restricting the ventilation. In the case of alternating current a metal enclosure further reduces the capacity by introducing eddy-current loss if of non-magnetic metal and both eddy-current and hysteresis losses if of magnetic metal. These alternating-current losses are greatly intensified and usually accompanied by vibration and noise in any single-phase sections of the enclosure. Where enclosures are of masonry or Transite, these alternating-current losses may appear in any structural or re-enforcing steel if located close to a conductor, particularly if it surrounds a single phase of the circuit. (See Article 27, References 1 and 5.)

Joints. Joints in a conductor should have no more resistance than the conductor itself. Bolted joints must have a large area of contact or heavy contact pressure. Since oxidation of the copper surfaces of a joint increases the resistance, which raises the temperature and accelerates the oxidation, good practice consists of silver-plating joints that must be bolted or clamped and welding or brazing all others.

Expansion and contraction of the conductor resulting from load cycles must be provided for either by a flexible layout or by flexible joints at frequent intervals.

Effect of Enclosures on Carrying Capacity. Most of the published data are based upon bare conductors in still air without enclosing structures. Experience proves that bus enclosures reduce the allowable carrying capacity by 15 to 25 per cent. The standards of the National Electrical Manufacturers' Association allow a temperature rise of 15 deg cent from air outside to air inside the enclosure and permit a copper temperature of 70 deg cent with bolted or clamped joints and 85 deg cent with silver-plated, brazed, or welded joints. (See Article 27, References 4, 6, and 8.) For a further investigation of the many factors involved in circuit design, reference should be made to the very extensive bibliography contained in *Trans. AIEE*, p. 391 (1944).

Cables. Insulated cables pulled into ducts or conduits are preferred, particularly for long, continuous circuits. The conductor is continuous from end to end, and the circuit is well protected within the conduit or duct, which usually is buried in the building structure.

The **conductor** of the insulated cable is practically without exception annealed copper and generally is stranded to make the cable more flexible and to avoid breakage by vibration or bending. Standard concentric stranding is generally used, but some advantage in pulling around bends can be gained by using one of the more flexible strandings. For important heavy-capacity circuits which require single-conductor cables, it is usual to consider special strandings. These include annular stranding with hemp or hollow metal cores to reduce skin effect and increase heat radiation, segmental stranding to reduce skin effect, and compact stranding to reduce diameter.

Insulations in common use at present are rubber, varnished cambric, asbestos, and impregnated paper. **Oil-base rubber** and butyl rubber are particularly satisfactory for power-station use since they resist ozone that may originate within the cable or come from other sources in the station. They also resist water, oil, and most acids. Such cables are comparatively easy to splice and can be terminated without compound-filled terminals. **Varnished cambric** is more sensitive to moisture, although it resists ozone and oil. Being applied to the conductor in the form of tapes, it is likely to be more concentric than rubber, particularly rubber that is extruded onto the conductor. Being firmer, it is less likely to decentralize after installation. The tapes, however, sometimes become displaced when the cable is pulled around bends, especially if the outer covering is not firm. Varnished glass, which is less sensitive to moisture, is coming into use. **Asbestos** combined with varnished cambric (type AVC) often is used as insulation in steam plants where ambient temperatures are high. **Oil-impregnated paper** is most compact and least expensive and therefore is generally preferred for long runs of underground cables. It must be covered with lead and sealed at the ends with compound-filled terminals to exclude all moisture. Being saturated with oil, it has a tendency to build up pressure sufficient to stretch the lead at low points of the circuit, particularly with the ratcheting action of load cycles. The runs must therefore be limited to not more than 20 to 30 ft difference in elevation, or special lead or re-enforcement of the lead sheath must be provided. **Relative advantages** of the three types of insulation are indicated in the following tabulation, which applies to ungrounded neutrals and conductors of over 1,000,000-cir-mil cross-section:

	5 kv	10 kv	15 kv
Required insulation thickness, 64th inch:			
Rubber	12	19	28
Cambric	10	15	21
Paper	7	11	16
Permissible conductor temperature, degrees centigrade:			
Rubber	75	70	70
Cambric	85	81	77
Paper	85	83	81

Shielding by means of conducting metallic or non-metallic tapes or braids applied directly over the insulation generally is used for cables operating at over 5000 volts between conductors. In steam plants where insulation is likely to become especially dry from high temperatures, it is good practice to use shielding for voltages as low as 2300 volts between conductors.

Outer coverings for power-station cables consist of tapes, braids, rubber jackets, and lead sheaths. **Tapes and braids** are satisfactory for rubber and varnished-cambric cables in dry locations, but rubber jackets usually are preferred because they afford better protection, are easier to install, and last longer. **Jackets** of synthetic material, such as Neoprene, are highly resistant to moisture, oil, alkalies, and most acids. For large varnished-cambric-insulated cables, the rubber jacket, if used, should be re-enforced with fabric like a rubber hose to provide firmer support during pulling. **Lead sheaths** are desirable for varnished-cambric insulation and are necessary for paper insulation. The composition and thickness of the lead must be adequate to withstand the ratcheting action of load cycling where differences of static head exist along the circuit. Lead is subject to corrosion by alkaline environment and also to d-c and a-c electrolysis. Unless other means can be used, it is often necessary to protect the lead by means of heavily impregnated tapes or plastic.

Grounding. Grounding is required for lead sheaths and static shielding tapes to safeguard employees and to control voltage gradients along the insulation. Mutual induction between the conductor and the lead sheath of single-conductor cables may result in excessive sheath current if the sheath is grounded at both ends, or in excessive sheath voltage if grounded at only one end. It is usual practice therefore to sectionalize the sheaths of long, single-conductor cables by means of insulating sleeves and to ground each section at one point only. (See Article 27, Reference 9.)

Design of Insulated-cable Systems. This is a very important item of generating station and substation design, involving the installation features discussed in the preceding paragraphs and many cable design features. Reference should be made to Section 14, Articles 63 to 77, inclusive.

The **configuration** of single-phase cables in a circuit with several cables in parallel in each phase, in order to balance the impedances and equalize the parallel currents, requires careful analysis. (See Article 27, Reference 3.)

Conduits. Conduits for the low-voltage wiring may be of rigid steel, brass, aluminum, fiber, Transite, or tile, or of flexible steel or bronze. Rigid conduits are used as far as possible, the flexible types being used only where required by vibration or expansion joints in the building or by connections to movable machinery. Steel conduit is the most universally accepted for both concealed and exposed construction on account of its superior strength and low cost. For sizes larger than $1^1/_2$ in., however, fiber conduit is generally less expensive for concealed construction, but it cannot be used exposed. For circuits of such capacity that each phase must occupy an individual conduit, fiber is generally used for concealed construction and brass or aluminum for exposed work, since steel conduit would cause hysteresis losses under these conditions. Heavy-current-capacity circuits that must be run exposed are sometimes made up of a number of paralleled smaller circuits, each in a steel conduit. This construction has the advantage that skin effect and proximity effect are practically eliminated, but great care must be taken to make all the paralleled circuits of the same length and to equalize them thoroughly with heavy buses and tight connections at each end. **Tile duct** is used occasionally for power-station wiring but has the disadvantage that bends are difficult to make. Tile finds its greatest usefulness in straight duct lines. Both tile and fiber must be embedded in concrete for strength and protection.

Installation of the Conduit System in most cases must proceed with the building structure. Careful planning is necessary to avoid unnecessary bends and offsets. It generally costs less to embed the conduits in the structure, even at the expense of slightly thicker floors and walls in certain panels, than to support the conduits exposed after the structure is in place. Conduits embedded in concrete structures should have at least 2 in. of clear space between them to allow the concrete to flow into place. Large groups of conduits should be installed in floor fills or wall fills, independent of the building structure. Wherever steel conduits are embedded in cinders or cinder concrete, they should be protected from the acid of the cinders by a coating of cement or other suitable material. Joints in conduits should be made watertight, and each run should be drained and vented, if possible, to prevent the accumulation of condensed moisture. All metal conduits should be thoroughly grounded. Before cables are pulled, the conduits should be thoroughly cleaned and dried.

Pull boxes should be provided for long conduit runs or runs with difficult bends. A safe rule for the average case is to allow between pull boxes not over 50 ft with bends equivalent to two 90-deg bends, or not over 100 ft with bends equivalent to one 90-deg bend, or not over 200 ft without bends. Pull boxes should be of substantial, fireproof construction and of such size and shape that the cables will not have to be bent during installation to less inside radius than 6 times their outside diameters if rubber insulated or 7 times if cambric insulated or 16 times if paper insulated. The cables of these important main low-voltage circuits should be supported in the boxes by porcelain insulators, and where they enter the boxes they should be protected by fiber sleeves inside the conduit ends. Figure 2 indicates conduit sizes appropriate for various cable sizes and groupings under usual conditions. These do not all comply with NE Code requirements. (See Section 14, Article 81.)

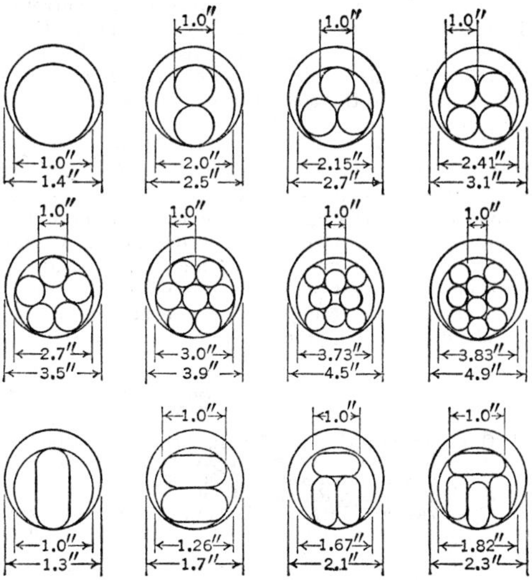

FIG. 2. Conduit Sizes for Various Cable Sizes and Groupings

HIGH-VOLTAGE POWER WIRING. This wiring consists of the connections from the step-up transformers through the high-voltage switching and busing to the outgoing lines. Voltage is the controlling factor of design. Rated and short-time currents should be investigated, but they seldom control the design of the wiring. Present practice is to locate all high-voltage equipment and wiring outdoors.

Buses. **Two general types of** buses and connections are used: the **rigid type,** consisting of copper tube or steel pipe spanned between porcelain insulators; and the **flexible type,** consisting of stranded copper or aluminum cable supported on strain insulators. Combinations of the two types sometimes meet the conditions to best advantage. The general arrangement of the wiring and structures is materially affected by the mounting of the disconnecting switches, that is, whether the insulator stacks are vertical or horizontal. Vertical stacks are generally preferred for the higher voltages on account of their greater stability; but for lower voltages, where either vertical or horizontal stacks are permitted, a more compact station can usually be arranged with disconnecting switches mounted with stacks horizontal. (See Article 27, References 2 and 7.)

Structures. **Steel structures and supports** for the flexible type of wiring are of two general types: trussed structures, suitable for the lower voltages, where spans are comparatively short; and dead-end towers, used generally for higher voltages, where the bays are too large for economical trussing. For the trussed structures, the steel members may be solid rolled sections for short spans but are usually latticed. The structures should be designed for the stresses of the pull-off lines as well as those of the station buses and connections. Steel structures should be galvanized after fabrication and assembled with non-rusting bolts; or, if plain steel and riveted connections are used, the structure should be designed to facilitate routine painting.

CONTROL WIRING. These important circuits should be of careful design and high quality. It is particularly important that the wiring be accessible and flexible to facilitate tracing circuits and making extensions. Most of the circuits originate at the switchboards and extend throughout the powerhouse and switchyard to the equipment controlled. It is very important that provision for these circuits be made early in the design of the station.

Cables are generally multiple-conductor, with each cable carrying the conductors necessary for a particular circuit, such as the control of a circuit breaker or the secondaries of an instrument transformer. Each cable usually carries one or more spare conductors to bring the total up to one of a few cable sizes selected as standard stock sizes. Conductors have often been made up of 19 strands of No. 25 wire or 19 strands of No. 22 wire, but there is a trend toward the use of standard 7-strand No. 12 or 7-strand No. 9 wire. Although some users prefer varnished-cambric insulation, oil-base rubber and polyethylene insulations are coming into extensive use. Braids, rubber or plastic jackets, and lead sheaths are all used for various conditions; but the Neoprene jacket is coming into general use as a single, all-purpose covering, except for polyethylene insulation.

Conduits of rigid steel commonly enclose control cables, particularly where they terminate at equipment. A separate conduit is used for each cable. Duct lines of Transite or fiber ducts with several cables per duct often are used to convey cables to remote distribution points, as from the control room to the switchyard. In important stations it is considered good practice at present to run the control cables from the switchboards on open Transite trays in a control gallery throughout the length of the powerhouse and sometimes to the switchyard also. Trays of $^3/_8$-in. Transite with edges rounded upward are available for this specific purpose. When the tray system is used, the individual cables are carried from the trays through individual steel conduits to their terminations at the control equipment.

STATION GROUNDING. Grounding of all equipment frames, metal structures, instrument transformer secondaries, piping, fences, and all non-live metal parts in the station and in the switchyard is provided to safeguard personnel. A ground grid usually is provided below the surface of the transformer and switchyard to reduce voltage gradient along the ground during arrester discharge. Certain neutral points of circuits, such as generator neutrals and transformer neutrals, are grounded solidly or through suitable impedances to assist relaying and to limit voltage rise under conditions of fault. Lightning arresters and gaps are grounded as a part of the protection against voltage surges. For details of grounding practice see Section 14, Articles 90 to 98.

27. BIBLIOGRAPHY

1. O. R. Schurig and H. P. Kuehni, Temperature Rise and Losses in Solid Structural Steel Exposed to the Magnetic Fields from A-C Conductors, *Trans. AIEE*, p. 184 (1926). Bibliography.
2. C. W. Frick, Current-carrying Capacity of Bare Cylindrical Conductors for Indoor and Outdoor Service, *Gen. Elec. Rev.*, p. 464 (1931).
3. C. F. Wagner and H. N. Muller, Jr., Unbalanced Currents in Cable Groups, *Elec. J.*, p. 390 (1938).
4. *Alcoa Aluminum Bus Conductors.* Aluminum Company of America, Pittsburgh, Pennsylvania (1940).
5. S. C. Killian, Temperature Rise in Steel Adjacent to A-C Conductors, *Delta-Star Magazine* (July, August, 1941).
6. *Mechanical and Electrical Data on Conductors and other Electrical Materials.* Chase Brass & Copper Co., Waterbury, Connecticut (1941).
7. S. C. Killian, Evaluate Steel Pipe as A-C Conductor, *Elec. World*, p. 67 (April 18, 1942).
8. Current Ratings for Electrical Conductors; Physical Properties and Technical Data, *Publication* C-51, Anaconda Wire and Cable Co. (1942).
9. R. C. Waldron, The Effect of Grounding Metallic Shielding and Sheaths on Insulated Power Cables, *Elec. World* (September 30 to November 25, 1944).
10. *Carrying Capacity of Insulated Wires and Cables.* Insulated Power Cable Engineers' Association (1945).

SUBSTATIONS

By W. F. Wetmore

28. CLASSIFICATION OF SUBSTATIONS

A substation is an aggregation of electrical apparatus for the purpose of control, regulation, subdivision, and transformation or conversion of electrical energy. It is the connecting link between two or more sections of a transmission or distribution system.

Most substations perform two or more of these functions.

FUNCTIONAL CLASSIFICATION. Substations may be classified according to their functions as follows:

1. Interconnecting two or more transmission lines for tying together two or more sources of power. This may be a simple "switching station," or it may involve the transformation of voltage if the lines operate at different voltages. The flow of energy may be either way.

2. Transforming voltage by means of transformers from some higher voltage to a lower one, usually interconnecting the high-voltage transmission line with a primary or secondary distribution system. Such a station is called a distribution substation, and the energy usually flows only one way.

3. Regulating the voltage or power factor by means of regulating transformers, step-voltage regulators, induction regulators, capacitors, or synchronous condensers.

4. Converting from alternating current to direct current to supply direct current to railways or customers.

5. Converting from alternating current of one frequency to alternating current of another frequency.

6. An industrial substation receiving power at some high voltage and transforming to some convenient voltage or converting to some other form to suit the particular needs of an industrial establishment.

CLASSIFICATION BY INDOOR AND OUTDOOR TYPES. Substations may be of the indoor or outdoor type, depending upon the degree of protection from the weather. The tendency is to put the transformers, regulators, high-voltage circuit breakers, capacitors, synchronous condensers, and frequency changers outdoors, and the control equipment indoors. Circuit breakers of 15 kv and below are usually of the indoor type but in small substations are frequently located outdoors in metal cubicles. Converters, rectifiers, and storage batteries are usually placed indoors.

CLASSIFICATION BY CONTROL. Substations may be of the manually controlled type, in which operators are always in attendance; of the automatic type, in which all operations are performed by automatic features; or of the remote-control or supervisory type, in which most operations are initiated by an operator at a distant station.

SUBSTATIONS FOR LIGHT AND POWER SYSTEMS

By W. F. Wetmore

29. LOCATION OF SUBSTATIONS

SWITCHING SUBSTATION. The location of a switching station is dictated by engineering features of the system rather than by economics. Generally it is at one or both ends of a transmission line: at the generating end to interconnect the various generators to several outgoing lines, or at the receiving end to tie one or more transmission lines to branches serving various parts of a city or district. Usually a switching station is also a transforming station, and the physical and mechanical construction of the two are similar.

TRANSFORMER SUBSTATIONS. The location of a transformer substation is usually within the load area to be served or adjacent to it. Right-of-way for high-tension lines, particularly in the vicinity of an urban community, will often have an important bearing upon the location of such a station. Transformer substations serve to interconnect high-voltage transmission lines with secondary transmission lines. In systems involving long lines (hydro) the transformation is usually from 230 kv to 115 or 69 kv, whereas in systems involving shorter distances (steam) the transformation is usually from 138 or 115 kv to 34.5, 23, or 13.2 kv, the lower secondary voltages being more suitable where cable circuits are required. Transforming substations usually require rather large areas, and railroad facilities are desirable if large units are to be installed.

DISTRIBUTION SUBSTATIONS. A distribution substation is a particular type of transformer station designed to transform from a medium transmission voltage (13.2 to 46 kv) to a distribution voltage (2400 to 7200 volts). The location is governed to some extent by economics. The greater the number and the closer together the stations, the higher will be the first and the maintenance costs of the stations, but the investment in the distributing mains will be less and the voltage regulation at the customers' premises will be better. The number, location, and capacity of distribution stations are judged theoretically by the maximum load demand of the customers per square mile, the diversity factor of the load, and the length and cost of the distribution feeders. In the practical case of a growing community the substations are rather widely spaced initially and intermediate substations provided as required by load growth and voltage conditions.

SELECTION OF SUBSTATION SITE. In selecting the site for a substation, care should be exercised in regard to the following factors: right-of-way for lines and cable outlets to provide for both present and future requirements, proximity to center of load and to existing transmission systems, cost of grading and drainage, assurance of satisfactory foundation conditions as indicated by soil borings, availability of water and sewer facilities, if required, space for future as well as present facilities, land cost and tax rate, access to the site from railways, heavy-duty highways, or secondary highways, depending upon the size of equipment to be handled, building restrictions, plans for future road widening or grade separations, proximity to gas mains and electric transit systems, and the possibility of noise complaints by residents in the vicinity.

30. DESIGN OF SUBSTATIONS

SWITCHING SUBSTATION. A switching station may be of either the indoor or the outdoor type. Small stations are usually outdoor, but in large stations with voltages up to 23 kv it is advantageous to install the equipment indoors. The tendency is toward the use of circuit breakers of low oil content or air-blast circuit breakers installed in metal enclosures, which minimizes the hazard of fire and permits a compact arrangement suitable for indoor installation.

Indoor metal-enclosed switchgear lends itself to the use of ground-fault relaying. Either this method or bus-differential relays are generally used at major stations to insure rapid clearing of faults. Switching stations at the receiving end of transmission lines in this voltage class are usually transforming stations and are discussed under that heading.

For voltages of 34.5 kv and above the switching equipment is usually located outdoors. Since most of the circuit breakers are of the oil type, allowance should be made for a fire of considerable magnitude. This requirement, together with the increased electrical clearances required for the higher voltages, favors the use of an outdoor design. The tendency is to provide ample spacing between equipment on one bus section and that on another, but if sufficient space is not available, barrier walls are installed to limit the spread of fire. The hazard of a failure in a section circuit breaker is common to both adjacent sections, and either means should exist to isolate quickly a fault at that point or double section breakers should be provided. The advent of airblast breakers will reduce the need for segregation of breakers, but the hazard of transformer fires will remain. To permit maintenance of the station equipment without a line shut-down, double-bus designs are sometimes used with either two breakers per line or one breaker connected to the main bus and a disconnecting switch connected to the reserve bus. This practice, however, is somewhat less warranted with modern equipment, since lines can usually be removed from service occasionally without objectionable loss to the system.

TRANSFORMER SUBSTATIONS. Transformer substations may be of either the outdoor or indoor type, but where the secondary voltage is in excess of 15 kv the outdoor type is more usual. This type of station is generally the receiving end of transmission circuits from a source of generation and serves to reduce the voltage to a value suitable for the secondary transmission system of a load area, such as a city or a rural district (Fig. 1). The total transforming capacity may vary over a wide range, but the tendency is toward high ratings (30,000 to 300,000 kva) for supply voltages of 115 kv and above. Steel structures are almost universally employed for supporting line terminals, buses, disconnecting switches, and lightning arresters. Circuit breakers and transformers are provided with independent concrete foundations. Crushed stone is placed to a suitable depth around oil-filled equipment to serve as a reservoir in the event of rupture of the tanks. For moderate voltages (up to 69 kv) the equipment other than breakers and transformers is usually mounted on a common steel structure, but for higher voltages the disconnecting switches and lightning arresters are often independently mounted. At 230 kv the required spacing between conductors is so great that it is sometimes economical to provide tubular-type rigid bus conductors supported from the ground, with rigid taps to the associated equipment. In such a design the only elevated structures required are those for the line terminals. The use of steel is minimized, but the space requirement is increased.

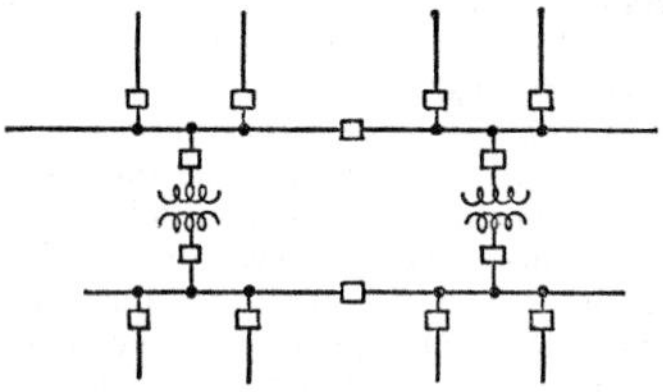

Fig. 1. Typical Substation Circuits

The presence of oil-filled equipment constitutes a serious fire hazard. Space separation of major equipment or the use of fire barriers is desirable. Sometimes the equipment associated with a bus section is grouped under one structure and separated by a considerable distance (100 ft or more) from the equipment associated with a second bus section (Fig. 2). This "island" type of design provides space separation with a minimum of steel. The circuit breaker between sections may be considered part of one section and mounted adjacently to that structure, or it may be provided with an independent structure midway between sections with disconnecting switches located at each section group.

The design of the secondary-voltage switching structure will vary somewhat with the voltage and station rating. In large stations with secondary voltages of 23 kv or below the outgoing lines are usually carried away from the station in cables. Some lines continue in cable, whereas others are brought up to overhead lines at a convenient distance from the station. In small stations, or with high voltages, it is practicable to employ overhead entrances if the lines are overhead.

Transformer substations sometimes are supplied by taps from one or more transmission lines. In a single-tap station having no high-voltage breaker, a failure of either the line or the station equipment will cause a service interruption. A single breaker in the tap protects the line from station failures but does not protect the station from line outage.

Under normal operating conditions breakers located in the line on both sides of the tap will prevent a station outage due to line failure. Two transformers tapped to the same line, with a circuit breaker connected in the line between the two taps, minimizes service

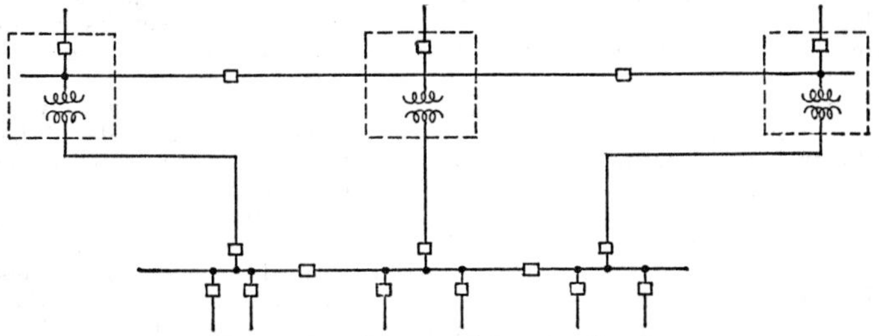

Fig. 2. Island Type of Substation Circuit

interruptions due to either line or transformer failures and permits transformer shut-downs for maintenance purposes.

If two transmission lines are available, transformers may be tapped from both lines, and the need for high-voltage switching is materially reduced in comparison to the single-line design. The connection of tap stations to transmission lines tends to make the relaying of the lines rather difficult. **Carrier-current relaying** is particularly useful for this purpose.

Voltage-regulating equipment is sometimes provided at transformer substations. This may consist of automatic tap-changing equipment on the transformer banks or of synchronous condensers or of capacitors. Synchronous condensers or capacitors are connected to either the low-voltage buses or to tertiary windings in the transformers.

DISTRIBUTION SUBSTATIONS. Distribution substations usually transform from a transmission voltage of from 13.2 to 46 kv three-phase to a distribution voltage of 2400, 4160, 4800, or 7200 volts three-phase. Some systems employ higher voltages, such as 4800/8320 Y grounded or 7200/12470 Y grounded volts, particularly for rural distribution. Distribution substations vary in rating from the order of 50,000 kva in heavy industrial areas to a few hundred kva in rural communities. The distribution feeders may be of the radial type or may connect to a primary network, the fundamental distinction being that the radial type requires a station design which is self-reliant for most types of equipment failures, whereas the primary-network type can suffer a long station outage without loss of service to the customers. A station feeding a radial system usually is provided with two or more transformers, whereas one feeding a network may consist of only one transformer.

Distribution substations may be of either the indoor or the outdoor type. The tendency, however, is to locate transformers and regulators outdoors and at least the secondary switching equipment indoors in the larger stations. Increasing building costs, however, tend to decrease the number of indoor installations. If the station is located in dense urban territory and the voltage does not exceed 23 kv, the high-voltage switching equipment, if any, is usually installed indoors. In suburban stations, or if the voltage exceeds 23 kv, this equipment is usually located outdoors.

Single-tap Substation. The simplest type of distribution substation is the single-tap station, consisting of a single three-phase transformer equipped with a horn-gap switch, fuses, and lightning arresters on the high-voltage side, and a fault-interrupting device and lightning arresters on the low-voltage side (Fig. 3). The interrupting device is usually a repeater fuse or a reclosing circuit breaker providing at least two reclosures before locking out on sustained fault. Voltage regulation is provided by tap-changing equipment in the transformer or by a regulator of the induction or step-voltage type. A station of this kind is usually fed by a tap from a transmission circuit and may either feed a radial load or tie into a primary network. If the load is radial, service must be restored after a transformer failure by replacing the unit or transferring the load either to a near-by substation or to a portable substation.

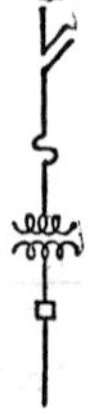

Fig. 3. Single-tap Substation

For substations feeding radial loads it is more usual to provide a bank of three single-phase transformers, which can be operated in an emergency with only two units connected in open delta. An alternate is to provide the fourth single-phase unit with means for readily connecting it in place of a defective unit.

This second method may be required if the transformers are star connected, and with either type of connection will often increase the firm load-carrying capacity of the station.

Single-tap substations, which usually feed rural districts or small suburban communities, seldom exceed a rating of 3000 kva. They may feed more than one distribution circuit but are usually limited to three or four. Each circuit may have its own reclosing circuit breaker, or one reclosing circuit breaker may feed a bus to which two or more fused circuits are connected.

Service to a single-tap substation is materially improved if automatically operated horn-gap switches are placed in the high-voltage line on either or both sides of the tap.

Double-tap Substation. In the larger suburban communities the improved reliability of a double-tap type of substation is justified if a second feed is available at reasonable cost. A simple form of this type of substation consists of two transformers, each normally feeding one distribution feeder, with automatic throw-over equipment provided to connect both feeders to one transformer or its supply line (Fig. 4). The firm load-carrying capacity of the station is the emergency rating of one transformer and its associated equipment. Since the transformers are not operated in parallel, each of the transmission lines feeding a station of this type can come from different switching centers with a resultant improvement in reliability. Three circuit breakers are required, but in other respects the equipment associated with each transformer is similar to that required for a single-tap station. Either single-phase or three-phase transformers may be used, but the trend is toward the three-phase design. The equipment is usually either of the outdoor type or housed in metal cubicles. A small building is generally provided to house a telephone, maintenance equipment, and sometimes toilet facilities.

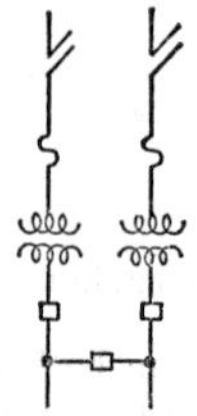

FIG. 4. Double-tap Substation

The double-tap design can be applied to a single transmission line by installing a high-voltage circuit breaker in the line between the two transformer taps. This arrangement is employed if its cost is less than that required for a tap from a second line. The high-voltage circuit breaker is usually provided with automatic reclosing relay equipment. In a modified version of this design a horn-gap switch is substituted for the high-voltage circuit breaker. There is only a minor reduction in service reliability if the switch is automatically operated. For the initial installation of a station of this type the horn-gap switch may be manually operated and the low-voltage automatic throw-over equipment omitted. The reserve advantages of the two-transformer design are obtained, and by adding supplementary equipment the service continuity can be improved as warranted by the service conditions.

If more than two distribution feeders are required, the switching arrangement becomes somewhat more elaborate; for example, a station supplying four feeders will require two transformer breakers, a bus tie breaker, and four feeder breakers (Fig. 5). With this number of circuit breakers it becomes advantageous to provide a building to house the circuit breakers, control and metering equipment, and tripping battery, as well as telephone and toilet facilities. Some form of heating equipment is usually provided for use during maintenance work or at times when an operator is in attendance. Normally a station of this type is unattended, but an alarm circuit to some central point is provided to call attention to station trouble.

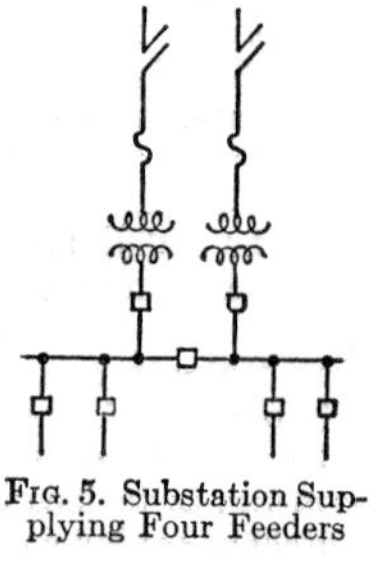

FIG. 5. Substation Supplying Four Feeders

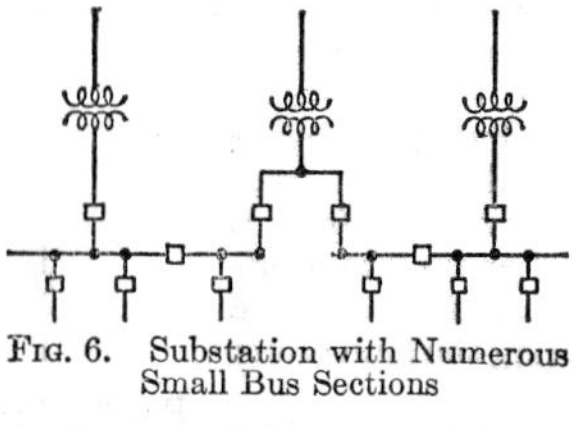

FIG. 6. Substation with Numerous Small Bus Sections

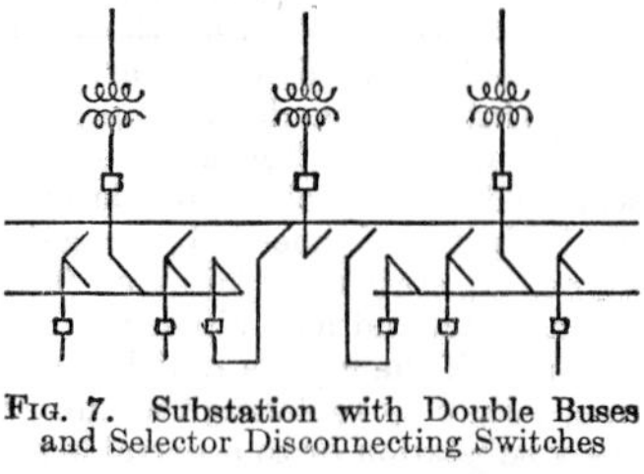

FIG. 7. Substation with Double Buses and Selector Disconnecting Switches

The double-tap design can be carried up to rather large capacities, although the difficulty of interrupting the transformer exciting current with horn-gap switches tends to limit the transformer size, unless circuit breakers are provided or a line shut-down is permissible for this purpose. The larger units (5000 to 10,000 kva) are usually supplied by radial lines from one or more major switching centers, with the transformer rating matched to that of the line or cable feed. The number of distribution feeders required is determined by the average feeder load and tends to vary inversely with the distribution voltage.

The principle employed in the double-tap design can be extended to any number of units provided that either the throw-over transformer is normally lightly loaded or means are provided to transfer some of its load to other units in the event of an emergency operation. A multiplicity of small bus sections (Fig. 6) or double buses with selector disconnecting switches (Fig. 7) are sometimes employed to obtain the desired division of load under emergency conditions. The double-bus arrangement is also of value in balancing loads during normal operation and in permitting bus shut-downs without interruption of load.

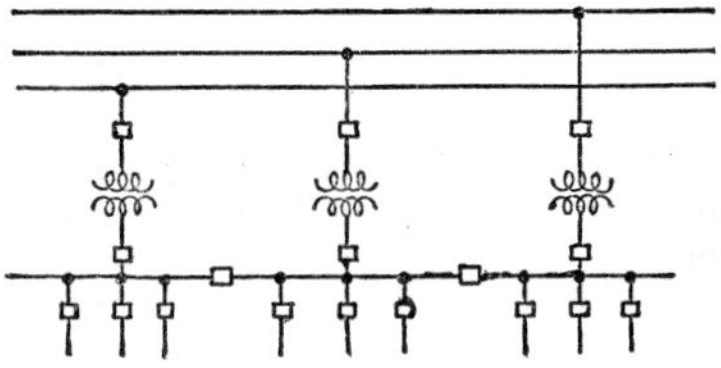

FIG. 8. Substation with Transformer Secondaries in Parallel

Parallel Operation of Transformer Secondaries. If the transmission lines feeding a substation are from the same source, the transformer secondary buses can be operated in parallel through section circuit breakers (Fig. 8). In comparison with non-parallel operation, the relaying is somewhat more complicated and breakers of higher interrupting capacity are required, but the loss of a transmission line and its associated transformer causes no interruption to the distribution feeders, and the load will remain balanced on the remaining transformers. Reasonable interrupting duties are obtained by the use of small transformers or by interposing reactors between sections or groups of sections. Large substations (50,000 kva), wherein the several transformer sections are linked together through reactors in the form of a synchronizing bus, are in service. It is advantageous to employ stations of this type to supply loads for which a continuous power supply is essential.

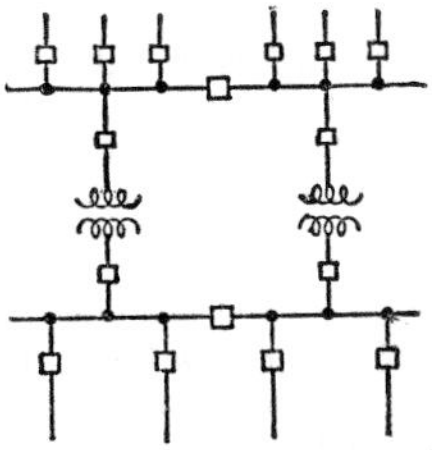

FIG. 9. Distribution Substation with Several Incoming Lines

Substations with High-voltage Bus. Distribution substations equipped with a high-voltage bus system fed by two or more incoming lines provide a high degree of service continuity to the distribution feeders (Fig. 9). In addition to serving as a source for the transformers within the station, this bus may also serve as a switching center for the transmission system and nearby radially fed substations. In a station of this type the transformer secondary buses are seldom provided with automatic throw-over facilities, since transformer failures are too infrequent to justify the cost and maintenance of the required relay equipment. Section breakers are provided to permit maintenance shut-downs, but they normally stand open to avoid high short-circuit duties. Stations of this type vary in size from those required for small suburban commmunities to those serving large urban loads. The need for the high-voltage bus, however, is more frequently determined by the requirements of the transmission system than by those of the distribution system.

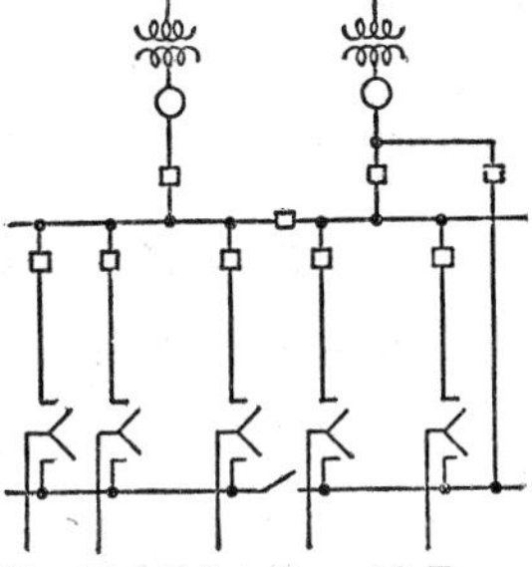

FIG. 10. Substation with Transfer Bus

Transfer Buses. It is considered good practice to maintain service to a distribution feeder under all conditions other than a fault on the feeder. Since, however, the station equipment associated with a feeder must be removed from service for periodic maintenance or equipment changes, means must be provided to transfer the feeder to another source, either by transfer to another feeder outside the station or by connection to a transfer bus within the station. For the latter method the transfer bus is energized either by connection to one of the other feeders in the station or by a special transfer bus feed (Fig. 10). Transfer buses are usually provided in medium and large substations, particularly if they are of the indoor type.

Voltage Regulation. Voltage control is provided in distribution substations by bus regulation, individual feeder regulation, or both. Bus regulation is obtained by tap-changing equipment in the transformers or by regulators of either the induction or the step type connected in series with the transformer secondaries. If all feeders have similar load and impedance characteristics, the bus voltage can be varied with the load cycle to satisfy the needs of all feeders. If, however, a few of the feeders have materially different characteristics, it becomes necessary to provide individual supplementary regulation for those lines. In some cases it is advantageous to regulate all feeders individually, but the tendency is toward group regulation. Capacitors or synchronous condensers are sometimes installed in distribution substations to assist in maintaining satisfactory voltage and to reduce transformer loading. The reduction in cost of capacitors in recent years has resulted in a material increase in their use for this purpose. They are usually switched in as large blocks as is permissible without objectionable voltage flicker. The switching may be controlled by one or more of several variables, such as voltage, transformer load, power factor, or time.

31. EQUIPMENT

Because of the great variety of requirements to be satisfied in substation design, no one arrangement can be considered typical. The equipment required to satisfy most designs can be summarized as follows:

TRANSFORMERS. Transformers are usually of the oil-insulated, self-cooled type, except where prohibited by local fire regulations. Auxiliary fans are sometimes provided to obtain increased ratings for emergency use. In large units forced-oil designs are sometimes employed to decrease first cost and reduce space requirements, the oil being cooled by radiators with motor-operated fans or by water-cooled heat exchangers. Conventional water-cooled transformers are sometimes installed at generating stations.

Transformers may be of the single-phase type with three in a bank or of the three-phase type. The tendency is toward the use of the three-phase design except for very large high-voltage units or for substations having only one bank of transformers.

Equipment for changing taps under load is frequently included in transformers. The control may be either automatic or manual.

HIGH-VOLTAGE EQUIPMENT. The high-voltage equipment consists of lightning arresters, horn-gap switches, fuses, buses of either the rigid or strain type, disconnecting switches, potential and current transformers, and circuit breakers. Oil circuit breakers predominate, but development is toward air circuit breakers employing compressed air for arc extinction.

LOW-VOLTAGE EQUIPMENT. The equipment required for the low-voltage section of a substation is of the same general type as that required for the high-voltage section, with the addition of voltage regulators of either the induction or step type. Outdoor circuit breakers are usually of the oil type, but for indoor installations below 15 kv the tendency is toward air circuit breakers of the magnetic blow-out design.

RELAYS AND METERS. Relays are provided to operate circuit breakers or to energize an alarm in the event of short circuit, overload, undervoltage, reverse power, high temperature, grounds, and sometimes low oil level in oil-filled equipment.

The metering equipment will consist of one or more indicating, recording, or integrating instruments measuring volts, amperes, watts, vars, watthours, varhours, or power factor.

STATION SERVICE EQUIPMENT. One or more station service feeds are required to supply lights, battery chargers, heaters, oil burners, regulator controls, pumps, and miscellaneous power devices. If two feeds are provided, automatic throw-over equipment is usually included.

COMMUNICATION. Telephone or alarm circuits to the main powerhouse, power dispatcher, and other substations may be provided in several ways: leased lines, private lines, carrier-current system over power lines, or radio.

32. SYNCHRONOUS-CONDENSER SUBSTATIONS

Synchronous condensers are usually installed at stations required for other purposes, such as switching, transformation, or distribution. Units up to 10,000 kva may be connected to distribution buses, but larger ratings are usually associated with the major transmission system.

These condensers are synchronous motors with a very generously designed field, so that they may be greatly overexcited. Overexcitation causes the condenser to draw a leading (anti-inductive) component of line current, which, in passing through the inductive reactance of the transmission line or added lumped inductance, causes a rise in voltage at the receiving end. Conversely, underexcitation causes a lagging current and a lowering of voltage. By operating underexcited at light loads and overexcited at heavy loads, the delivered voltage may be kept at a specified value in spite of line loss. Thus

$$E = e + (i + ji_1)(r + jx)$$

where E = voltage per Y phase at the sending end.
e = voltage per Y phase at the receiving end.
i = component of load current in phase with e.
i_1 = reactive component of current delivered.
$r + jx$ = impedance of one line of transmission.

E, e, and the power, per phase, delivered, may be set, and i_1 will be the reactive current required of the condenser, positive for leading and negative for lagging. If the line voltage is very high, that is, greater than 100,000, the distributed capacity of the line may cause a charging current of sufficient magnitude to help considerably.

Synchronous condensers are usually provided with complete automatic voltage control and reduced voltage-starting equipment.

The tendency is toward the use of hydrogen-cooled units and outdoor installations.

33. FREQUENCY-CHANGER SUBSTATIONS

Frequency-changer stations are usually designed for converting from 60 cycles to 25 cycles for railway or industrial uses or for the interconnection of two utility systems. The sets are usually of the rotating type, either synchronous-synchronous or one of several types of synchronous-induction designs. Some units are hydrogen cooled, and a few electronic installations are in service.

34. CONVERSION SUBSTATIONS

Substations for converting from alternating to direct current may employ motor-generators, synchronous converters, or rectifiers. For 120- to 240-volt distribution service motor-generators and synchronous converters are in most common usage. For railway service synchronous converters and rectifiers are usually employed, and are discussed under that heading.

The use of motor-generators protects the d-c system from voltage disturbances on the a-c system. The generators are of either the shunt type or the differential compound type, the latter being particularly adapted to re-energizing the d-c network in the event of a complete shut-down. Voltage is regulated by shunt field control.

Synchronous converters tend to have a lower first cost and are more efficient than motor-generators. Voltage regulation is obtained by change of power factor resulting from shunt field control or by means of series boosters on the same shaft as the converter. A series booster is a three-phase generator with its phase windings independent and each in series with one of the three leads which supply the collector rings of the converter. The field of the booster is separately excited and controlled to provide the voltage required.

Rectifiers are particularly adapted to the higher d-c voltages and capacities. In recent years units up to 36,000 kw have been installed by the chemical industry. The rectifier tanks may be of either the single-anode or multianode design.

35. PORTABLE SUBSTATIONS

A portable substation is a complete distribution station mounted on a truck and designed to assume the load of a distribution station during periods of scheduled maintenance or emergencies. It consists of a transformer with high-voltage horn-gap switch, fuses and lightning arresters, and low-voltage circuit breaker and lightning arresters. Voltage regulation is usually provided by a regulator or tap-changing equipment in the transformer. In the higher ratings the transformer is usually forced-oil cooled to keep the total weight within highway load limits.

RAILWAY SUBSTATIONS

By W. F. Wetmore

(See also Synchronous Converters, Section 8, Article 30; Energy Requirements for Railways, Section 17, Article 2; Switchboards, Section 12; Switchgear Equipment, Section 12; Transformers, Section 10.) Substations are used for electric railways when the length of the road is so great that the whole road cannot be supplied by one power station at the voltage required by the motors without either an excessive drop in voltage or a prohibitive amount of copper or both. Practically all electric railways require substations. In practice there are two types of substations: (1) those for transforming from alternating to direct current, and (2) those for transforming from high-voltage to low-voltage alternating current.

36. LOCATION AND CAPACITY

The location, capacity, and number of substations involve not only the cost of the copper required for distribution and the cost of the substations, but also the distribution of traffic and special local conditions. The fundamental economics of the subject are expressed by the general theorem that the cost of operating the substations plus the cost of interest and fixed charges on the investment in substations and line copper should be a minimum. This is explained by the fact that if to any given arrangement an additional substation is added with the proper rearrangement of the spacing, the amount of copper used in the distributing system may be considerably decreased on account of the lesser distance between substations. If the saving in interest on the value of the copper is greater than the cost of operation of this new substation plus the interest and fixed charges on the first cost of the substation, the change is warranted.

ALLOWABLE VOLTAGE DROP. The distance between substations depends directly upon the allowable loss in voltage and the amount of copper in the trolley and varies inversely as the load. The allowable loss of voltage is a matter of the special conditions of each road. In city roads a drop in voltage of 8 per cent with average load and 15 per cent with maximum load is frequently observed. In interurban roads a drop of 12 per cent with average load and 30 per cent with maximum load is customary.

The two conditions which determine the allowable drop in voltage are the effect on the lights and the effect on the control circuit of the car, each factor requiring a smaller drop in voltage than the actual operating characteristics of the motors.

DISTANCE BETWEEN SUBSTATIONS. For the service usually found in practice, Table 1 shows the average distance between substations.

Table 1. Usual Distances between Railway Substations

Volts	Type	Miles between Substations	
		Single-track Road	Double-track Road
600	Direct-current, trolley......	10	15
600	Direct-current, third rail....	13	19
3,300	Single-phase, trolley........	17	23
1,200	Direct-current, trolley......	19	25
1,200	Direct-current, third rail....	38	43
6,600	Single-phase, trolley........	45	50
11,000	Single-phase, trolley........		70

Where the length of track receives its power from only one direction, the allowable distance for a given drop in voltage varies between one-half and one-quarter the distance allowable between substations, depending on the degree of concentration of the load.

37. CONVERTER AND MOTOR-GENERATOR SUBSTATIONS

Substations may convert alternating current into direct current by means of synchronous converters, mercury rectifiers, or motor-generator sets. Converters have been in vogue the longest and are the most common. They are particularly adapted for operation on 25-cycle alternating current and delivering 600 volts direct current. At 60 cycles they do not operate as well as the other types but are still satisfactory. For delivering 1200 volts or more they are not as desirable as the motor-generator or the rectifier because of commutation difficulties. The converter has the advantage, however, that it readily lends itself to power-factor and voltage control. For 1200 to 3000 volts, the rectifier is more efficient, because the principal loss is due to the counter emf of the arc, which is about 20 to 30 volts, irrespective of the delivered voltage. The rectifier has a higher all-day efficiency than the other types because of the small losses at light load.

EQUIPMENT. A standard a-c to d-c railway substation usually contains the following pieces of apparatus: converters or motor-generators, transformers, reactances, cables, lightning arresters, and switching equipment.

Converters. (See Rotary Converters for data regarding converters.) Compound-wound commutating-pole converters are used where the load is variable to provide automatically any desired voltage regulation. The converter, by means of its series winding, is made to take a leading current, which, in passing through the reactance of the line, the transformers, or an additional reactance introduced for the purpose, tends to neutralize the line drop. In city service, where a large number of cars are operating on one section, the load is fairly constant, and shunt-wound converters are generally used.

Both three-phase and six-phase converters are used, the three-phase for the smaller and the six-phase for the larger capacities. In a three-phase converter for a direct emf voltage of 600 at no load, the transformers should supply the converter with 370 volts between rings or between lines. In the six-phase converter it is customary to use the diametrical connection, in which each transformer secondary supplies 430 volts to the converter, giving 600 volts at the commutator. Each converter for railway work is customarily supplied with a speed-limiting device on one end of the shaft and an end-play device on the other end.

Transformers. (See Section 10, Transformers.) The transformers used in railway substations may be of the oil-cooled, air-blast, or water-cooled type. The oil-cooled type is used where the expense of the complications for air blast or water cooling is not warranted. Air-blast transformers may be used for voltages up to and including 33,000; the objections are the necessity of providing a pit, air ducts, and blower to supply the ventilation. Water-cooled transformers, which are oil-insulated transformers with water circulating in a special coil submerged in the oil, are built in sizes from 500 kw upward and for all voltages. Their use depends upon the availability of water for cooling purposes. The usual aggregate capacity of a transformer for a particular size of synchronous converter is about the same as the converter capacity.

The transformers may be of either the single-phase or three-phase type. For small or moderate installations the single-phase type in banks of three is preferable on account of the economy of maintaining only one single-phase transformer as a spare for a whole station. In railway work it is customary to connect the secondary of the transformers in delta for three-phase, because of the possibility of operating at reduced output on open delta in case of failure of one transformer. It is common practice to provide transformers for railway work with four 2 1/2 per cent taps on the high-potential winding, in order to use similar transformers in all substations and yet make allowances for the difference in the line drop between the power station and the various substations. Reduced-voltage taps are provided on the low-potential side for starting the converters.

Reactances for Voltage Regulation. The voltage at the d-c bus-bars is regulated automatically by means of line compounding, which consists in adjusting the shunt field excitation of each converter so that the converter takes lagging current at no load, operates at unity power factor at about three-fourths load, and takes leading current at all loads greater than three-fourths. To accomplish this compounding, it is necessary that there be a certain amount of reactance between the power-station bus-bars and the converters. (See Rotary Energy Converters, Section 8.) There is seldom enough reactance in the transmission line for this purpose, so that additional reactance is inserted either by the use of special transformers having considerable leakage reactance or by means of reactance coils.

Switchboards. (See also Section 12.) The following switchboard panels are standard for converter substations:

1. Incoming a-c line panel.
2. Outgoing a-c line panel.
3. High-tension a-c converter or rectifier panel.
4. Direct-current converter or rectifier panel.
5. Direct-current feeder panel.
6. Equalizer and negative panel (on the converter).

Where a substation is tapped off a transmission line at an intermediate point, it is good practice to bring the transmission line into the substation, interpose control switches, and then carry the circuit out of the substation on to the next substation. For this reason, in all but terminal substations, it is customary to provide both an incoming and an outgoing a-c line panel. In connection with the a-c panel of the switchboard there are a line switch, lightning arrester, current transformers, and main oil switch, by means of which potential may be removed from all transformers. Between the transformers and the converter are the starting switches, the reactance coil, and possibly measuring devices.

Single-pole switchboard panels are used for the direct current, the positive main bus-bar being the only one on the board. The negative terminals of the converters are connected with switches to the negative or ground return bus-bar, which is frequently located beneath the converter. The series field is connected on the negative side, and the equalizer, series field shunt, and field break-up switches are frequently placed on the machine itself.

Crane. Where ground space is limited, it is good economy to provide a crane in order that the various pieces of apparatus may be lifted over each other when they are taken apart for repairs; otherwise considerable space must be left to move them about to and from the entrance.

Methods of Starting Converters. There are several methods of starting converters, as is explained under Synchronous Converters. Starting from the a-c end as an induction motor is most desirable for railway work, as it avoids the necessity of synchronizing and requires less time. The ability to start a machine quickly and get it on the line in the shortest possible time is very important in railway work and is an advantage inherent in this method of starting. Three-phase converters are started, by means of suitable starting switches, from one-half-voltage taps on the transformer secondaries and take approximately full-load current from the line. Six-phase converters are started from one-half-voltage taps and take three-fourths full-load line current. Since 60-cycle converters generally take a greater starting current than 25-cycle converters and are usually operated on a system supplying power for other purposes, where voltage disturbances are objectionable, special means of starting 60-cycle converters are frequently employed.

LOAD FACTOR AND EFFICIENCY. The load factor (q.v.) of a converter substation is usually low, from 30 to 50 per cent; i.e., the load on the station is relatively light except during the morning and afternoon rush hours, 7 to 9 A.M. and 5 to 7 P.M. respectively. The all-day efficiency of the station itself, or the ratio of the kilowatthours output to the kilowatthours input, is less than the efficiency at maximum or rated load. The overall efficiency between the a-c generators in the powerhouse and the cars, however, is about constant throughout the day, since the efficiency of a transmission line increases with decrease of load, thus offsetting the low light-load efficiencies of the transformers and converters. In general practice in interurban 600-volt railways the maximum and all-day efficiencies of the various apparatus are approximately as given in Table 2.

Table 2. Efficiencies of Various Kinds of Apparatus

Apparatus	Full-load Efficiency, Per Cent	All-day, Efficiency, Per Cent
Step-up transformers	98	97
High-tension line	95	98
Step-down transformers	97	94
Converters	90	88
Low-tension distribution	85	88
Overall, a-c generator to motors	69	69

AUTOMATIC SUBSTATIONS. To decrease the cost of operation by eliminating the need for attendants, substations are now equipped so that the entire control and operation are accomplished mechanically and automatically. When the voltage on the adjacent section of the trolley falls below a specified value (450 volts), a contact-making voltmeter closes a control circuit, which starts a motor-driven controller operating a large gang of switches and contactors. These switches automatically and consecutively accomplish the various starting operations. The converters are started from the a-c side, so that there is no synchronizing, but the polarity of the d-c end is regulated, the field excitation is adjusted at the proper time, and the several converters are connected in parallel. Protection against overload, low voltage, and overspeed is provided.

When the load falls below a specified value, a contact-making ammeter closes a circuit which opens all switches and shuts the converters down. A large number of these automatic substations have been installed containing up to three converters per station and ranging in size up to 2000-kw units of converters.

38. RECTIFIER SUBSTATIONS

APPLICATION. The great majority of the older 600-volt railroads use synchronous converters, but many of the newer 600-volt roads and the 1500-volt roads use rectifiers. Some roads using 3000 volts, e.g., St. Paul, have motor-generator sets with two 1500-volt generators in series electrically, both driven by one synchronous motor; others, e.g., Lackawanna, use rectifiers for 3000 volts direct current.

TRANSFORMERS. Mercury-arc rectifiers are usually constructed with six anodes and are supplied by three-phase transformers with double secondaries giving six phases. Larger sizes may be constructed with twelve anodes and supplied from three-phase transformers with specially designed secondaries. Ideally the voltage on the d-c side is equal to the peak voltage of the a-c side ($\sqrt{2} \times$ the rms value), but because of the drop in the arc and the transformers, the d-c voltage obtained is less by some 30 or 40 volts.

EFFICIENCY. Six-hundred volt rectifiers have an overall efficiency at rated load of about 94 per cent, including the losses in the auxiliaries. For 1500 volts the efficiency is about 96 per cent, and for 3000 volts it is 97 per cent.

The efficiency at one-fourth load is usually above 92 per cent, as compared to 88 per cent in a converter. The power factor is from 92 to 95 per cent (higher at overloads) within the working range. The cause of the low value is primarily wave distortion; it is not due to a lagging current of fundamental frequency. This wave distortion, common to all rectifiers, may cause interference in near-by telephone circuits, in which case the offending harmonies may be suppressed by a special wave filter.

AUXILIARY APPARATUS. Each rectifier requires the following auxiliary apparatus; transformers, vacuum control and pumps, temperature control and water circulation,

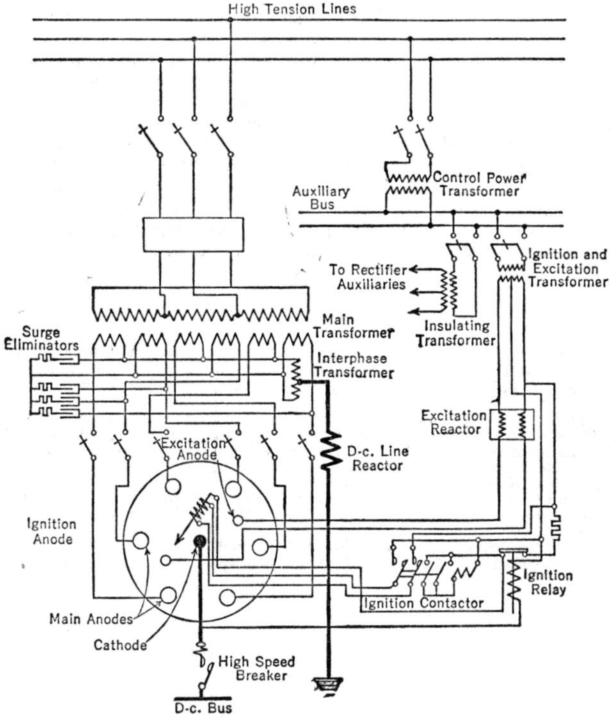

Fig. 11. Rectifier Substation Diagram

ignition and excitation equipment, heater for starting, voltage control for compounding if required, usual overload protection, and protection against arc-back (Fig. 11).

Many rectifier substations, such as those in the New York Independent Subway, are entirely automatic in their operation and require no attendants.

SPACE. A rectifier substation usually requires slightly more space for a given power than a converter substation, but as there are no moving parts very light foundations may be used, and there is no vibration or noise. They may be adapted for regeneration on the train by a special arrangement to provide for the flow of energy from the d-c to the a-c side. Rectifier substations cost about 30 per cent more than converter substations.

39. ALTERNATING-CURRENT SINGLE-PHASE SUBSTATIONS

For a single-phase railway operating at a high voltage (11,000 or 12,000) on the trolley it is not necessary to place substations as near together as in d-c systems. Even with 12,000 volts, however, there is a limit to the length of road which may be supplied from one feeding point. For instance, on the New Haven or Pennsylvania railroads, with heavy traffic on four tracks, it is necessary to step up the voltage at the powerhouse to 44,000 or 88,000 and step down again to trolley voltage at intervals of 10 to 20 miles. A substation for this purpose need contain only transformers and protective devices and usually has neither a building nor attendants, the circuit breakers being reset automatically or by remote control from a supervisory station. Some roads use a three-wire system with auto-transformers by which the trolley wire may serve also as one of the transmission wires. (See Fig. 12.)

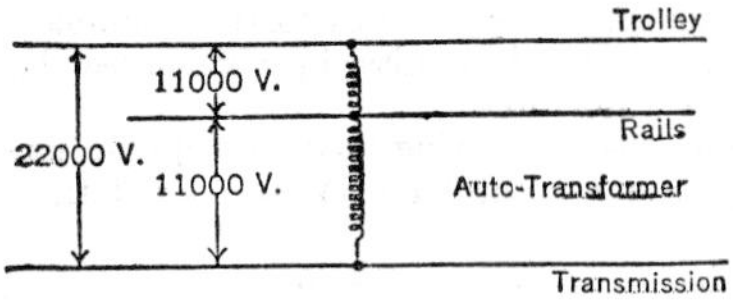

Fig. 12. Single-phase Railway Distribution by Auto-transformers

If power is purchased by the railway from existing 60-cycle commercial systems, a frequency-converter substation is used, as the motors on the cars and locomotives will not operate with 60 cycles but must have 25 cycles, and the commercial power companies will not allow a single-phase load to be connected to the same lines as supply the other customers. A frequency converter consists of a 60-cycle, three-phase synchronous motor driving a 25-cycle single-phase (in this case) generator, the usual combination being a 24-pole 60-cycle machine and a 10-pole 25-cycle machine, both running at 300 rpm. A synchronous frequency changer ties the two systems together rigidly, so that the slightest change in frequency or phase in one system reacts upon the other system. One of the two machines has its stationary part (stator) in a cradle, so that it may be rotated a few degrees in order to adjust the load. In modern practice these frequency converters are totally enclosed and self-ventilated, and they require no building or attendant.

40. BIBLIOGRAPHY

Ambrose, J. S., Substations That Grow with Loads, *Elec. Light & Power*, Vol. 19, No. 5, p. 42.
Farnham, S. B., Recent Trends in Substation Design, *Gen. Elec. Rev.*, Vol. 48, p. 14.
Hentz, R. A., and J. A. Thielman, Power Supply for Suburban Areas, *Elec. Engineering*, Vol. 59, p. 234 (1940).
Johnson, J. A., and R. T. Henry, Energy Delivery Systems, *Trans. AIEE*, Vol. 53, p. 1704 (1934).
Monteith, A. C., Modern Equipment Aids System Savings, *E.E.I. Bulletin*, Vol. 12, p. 253.
Payne, B., Standardizing Substation Designs, *Elec. World*, Vol. 116, p. 1047 (1941).
Sanderson, C. H., *Electrical Systems Handbook*. McGraw-Hill.
Seelye, H. P., *Electric Distribution Engineering*, New York.
Stanley, R. M., and C. T. Sinclair, Primary Network, *Trans. AIEE*, Vol. 50, p. 871 (1931).
Westinghouse Electric and Manufacturing Company, *Electrical Transmission and Distribution Reference Book* (1944).
Worth, D. G., J. D. Ross Substation, *Elec. West*, Vol. 89, p. 41.
Publications of the National Electric Light Association and the Edison Electric Institute, such as the *Relay Handbook*.

SECTION 14

POWER TRANSMISSION AND DISTRIBUTION

POWER TRANSMISSION AND DISTRIBUTION

SYSTEMS

Revised by W. A. Del Mar
Previous Revision by R. A. Philip

Circuits designed for transmitting relatively large amounts of power from one fixed point to another are called transmission lines; those for delivering small amounts at numerous points are called distribution circuits. Transmission lines usually have no or few branches, whereas it is characteristic of distribution circuits to have many branches. The various systems of transmitting and distributing electric energy for light and power may be classified under two general heads, viz., constant-potential or multiple systems and constant-current or series systems, and each of these may be subdivided into d-c and a-c systems.

1. CONSTANT-POTENTIAL OR MULTIPLE SYSTEM

In this system, which is the one principally used for electrical distribution, the voltage between conductors is kept as constant as practicable, and the current varies as the load changes.

Direct or alternating current can be used equally well for certain purposes, principally those for which the heating effect of the current is used, including the lighting of incandescent lamps, cooking, and heating. For certain purposes, where the current effects chemical or physical changes, such as charging storage batteries and electroplating, direct current is essential; for other purposes, such as operating arc lamps, it is better. For motive power, the direct current is most favorable where acceleration, variable speed, and adjustable speed are desirable, whereas alternating current gives best results where uniform unvarying speed is desired. Although the fields of the two kinds overlap to such a large extent that either type can be used for general distribution, it is often found more advantageous to use both. For low-voltage, underground distribution, direct current has the advantages that heavy currents can best be carried on single-conductor cables and that no subway transformers are required. Developments in a-c networks, however, have led to gradual replacement of d-c systems by more economical a-c systems.

Alternating current may be supplied either directly from the generators in the power station or from the secondaries of transformers in substations. Direct current may be supplied either directly from d-c generators or from converter substations supplied with high-voltage alternating current.

LIGHT AND POWER CIRCUITS. For a load consisting of both electric lamps and electric motors the same circuit may be used throughout for both classes of service, or entirely or partially independent circuits may be employed. The use of the same circuit throughout for light and power service has the advantages of reduced number of wires, transformers, and meters, reduced weight of wire, and reduced capacity of transformers and meters; the use of separate circuits for light and power has the advantage of requiring less capacity of feeder regulators and, sometimes, of less weight of wire for the same perfection of regulation on the lighting circuits. It also simplifies the problem of balancing the phases in polyphase distribution. In the business districts of many cities the Edison three-wire d-c system is used, in which case light and power are equally supplied from the same circuit, the motors as a rule being connected to the outside wires. In business districts where the lighting is done by alternating current the lighting is frequently on single-phase circuits, and the power on either separate polyphase circuits or on 500-volt d-c circuits. In residential districts, where the lighting is done by alternating current, small motors are usually put on the same services as the lights, whereas larger motors are put on separate meters, services, and transformers, but the same primaries are used for both services. In factory districts when the power is the predominating load the incidental lighting is sometimes taken off the same services, but generally there are separate meters and transformers. Usually the same polyphase primaries are used for power and light with separate transformers and secondaries; in these cases the lighting transformers are commonly distributed as equally as convenient between the several phases, in order to balance the load. (See Section 15, Article 3, for effects of voltage variations on lamp performance.)

TWO-WIRE SYSTEM. The simplest multiple system is the two-wire system, where all devices are connected directly in multiple. This system is used very extensively for d-c light and power from isolated plants for power circuits (usually 500 volts), and railway circuits from central stations. It is also used for single-phase a-c distribution for both primary and secondary circuits.

THREE-WIRE SYSTEM. The three-wire system (Fig. 1) is obtained by replacing the outgoing wire of one two-wire system and the return wire of a second two-wire system by a single wire, called the neutral. The voltage between the outside wires is then double the voltage between the neutral and either outside wire. For example, 110-volt lamps may be connected between the neutral and either outside wire, and 220-volt motors may be connected between the two outside wires, and both the lamps and motors be supplied with their rated voltage. The neutral wire carries a current which depends only upon the *difference* in the loads on the two sides of the system and their distribution. As a rule the neutral of a three-wire main is made equal in cross-section to each outside wire. With perfectly balanced load the three-wire system with all three wires of the same size results in a saving in copper of 62.5 per cent, as compared with a two-wire system supplying the same load at the same regulation.

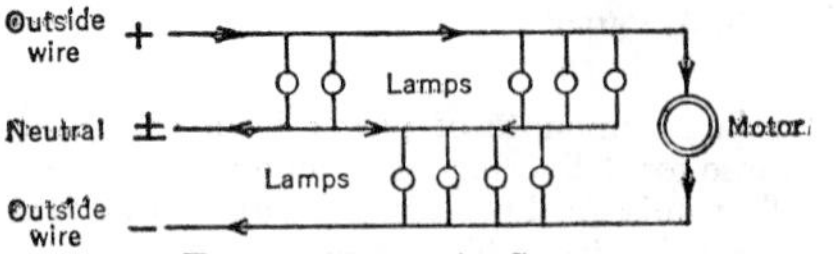

FIG. 1. Three-wire System

The three-wire system is used very extensively for d-c light and power distribution from central stations, also for large isolated plants and for alternating current for lighting on the secondary circuit. Three-wire systems are usually 110 volts on each side of neutral, or 220 volts between outsides, although there are several systems using 220 volts on each side and 440 volts between outsides.

The Edison three-wire distribution system, as used in the business sections of large cities, consists of a set of interconnected three-wire mains supplied by two-conductor or three-conductor feeders from one or more powerhouses or substations. The different feeders feed into the same set of mains at different points. At the powerhouse or substation end the feeders are often all supplied from a common bus. Where there is a great difference in the length of the feeders, however, there are sometimes two buses, which are run in multiple when the load is light but separated when it is heavy, the short feeders on one called the low bus, and the long feeders on the other called the high bus, because of the relative voltages. As the feeders from the several substations connect to interconnected mains, the bus voltage in each is raised or lowered in accordance with the drop in its own feeders. All mains which connect feeding points supplied from a given bus are made large enough to allow large equalizing currents to flow through them, without requiring any great drop in the mains, in order to equalize the voltages at these points.

2. CONSTANT-CURRENT OR SERIES SYSTEM

The lamps or other devices are connected in series, and the current through them is kept constant, the voltages varying automatically to increase or decrease the energy delivered. The constant-current system is the principal one used in the United States for street lighting, especially in the less congested areas, but it is now little used for any other purpose. The current used for street lighting is usually 6.6 amp at voltages from 2200 to 11,000.

SOURCE OF CURRENT. Alternating current is usually obtained from a constant-current transformer on a constant-potential a-c circuit.

CUT-OUTS, BY-PASSES, AND TRANSFORMERS. Switching out of lamps on series circuits is accomplished by short-circuiting them; this leaves the lamps charged to the potential of that point in the circuit. To make them safe to handle, "absolute cut-outs" are used which also disconnect both conductors to the lamp, leaving the circuit closed.

To avoid the excessively high voltage which would occur when the circuit opens, because of lamps burning out, an automatic by-pass is provided in multiple with the lamps, sometimes consisting of a piece of paper, which punctures on a moderate rise of voltage, or of a choke coil, which takes but little current at normal lamp voltage. Sockets for series incandescent lamps are arranged to close the circuit automatically when the lamp is withdrawn.

Transformers, one at each lamp, are sometimes used with their primaries in series to insulate the secondary circuit containing the lamp from the primary or for reducing the current for low-current lamps used in series with lamps taking higher current.

ADVANTAGES AND DISADVANTAGES OF SERIES LIGHTING. Constant-current or series systems have the advantage that low-voltage lamps may be used on high-voltage circuits without the expense, losses, or complication of transformation. They have the disadvantage that, unless series transformers are used, the lamps are dangerous to handle, the efficiency is low at light loads, and it is impracticable to distribute any large amount of power on a single circuit. In the series system the current and consequently the loss in the conductor are the same irrespective of the load; in the multiple system the current is proportional to the load, and the watts lost in the line therefore vary as the square of the load and the percentage loss directly as the load. For this reason the series system is not an economical one where the load varies and averages much below full load, as with most commercial loads. For street lighting, where all the lamps are turned on and off at once, the efficiency at partial loads is of no importance. In constant-current systems the resistance of the circuit does not affect the uniformity of the voltage on the lamps at different parts of the circuit; with constant-potential systems it does. For a scattering load of lamps, such as street lamps, a uniform light can therefore be obtained from the several lamps, with a much smaller weight of copper and fewer wires, by using the series than by using the multiple system.

THURY SYSTEM. In Europe the constant-current or series system, with direct current of extra-high voltage, has been used for long-distance power transmission under the name Thury system. The current is obtained by connecting several generators in series and is utilized by a number of motors also in series. The advantages are simple switchboards, no transformers, and minimum strain on line insulators, the last due first to the fact that in direct current the effective voltage is as high as the maximum voltage, and second, that in a constant-current system the working voltage remains at its maximum value only during the short period when the load is also a maximum. Among the disadvantages are the necessity for insulating frames of generators and motors, the need of speed governors on motors, and the necessity of converting the current by moving machinery in every case where it is used for lighting and in most cases for power. The Thury system is not used in the United States.

3. DIRECT-CURRENT SYSTEMS

Direct current has the following advantages for lighting: (1) safety, since the lines are in no way associated with high-voltage conductors: (2) freedom from power factor, reactance, and skin effect, which results in superior voltage regulation in heavily loaded low-voltage circuits; (3) the direct availability of the storage battery as a reserve and a load regulator; (4) the self-exciting and self-regulating features of d-c generators; (5) the superiority of d-c motors for adjustable-speed service and for the operation of elevators and cranes; and (6) the marked superiority of d-c arc lamps. Direct current is generally used in isolated plants and in congested city districts because of the greater ease with which good voltage regulation is maintained.

4. ALTERNATING-CURRENT SYSTEMS

The simplest form of a-c system is the single-phase system, for which the connections are exactly the same as for d-c systems. Both two-wire and three-wire circuits are used, the former for primary circuits and small secondary circuits and branches, and the latter for large secondary circuits. The principal use of single-phase circuits is for electric lighting and auxiliary uses, such as heating or cooking, and fan motors. For delivering large amounts of power a two-phase or three-phase system is generally used.

TWO-PHASE SYSTEMS. In this system there are two single-phase currents having a difference in phase of 90 deg, or a quarter of a cycle. These currents may be distributed on the three-wire, four-wire, or five-wire system.

Three-wire Two-phase System. Each single-phase current has a separate outgoing wire but unites in a common return wire. Each two-phase motor has two circuits, each connected between an outside wire and the return wire. The voltage between the two outside wires is 41 per cent greater than between outside wire and return wire, and the current in the return wire is 41 per cent greater than in each outside wire. This is an unsymmetrical system and has the disadvantage that even a balanced load will cause a distortion and unbalancing of the delivered voltage of the two phases because of the unsymmetrical drop in the common return wire.

Four-wire Two-phase System. Each of the two single-phase circuits has a complete, independent two-wire circuit. There are two variations: first where the circuits are insulated from each other, in which case a cross between either wire of one circuit with either wire of the other will change the voltage stress to ground, but will not affect the delivered voltage or cause a short circuit; second, when the neutrals of the two circuits are connected. In the second case, from each wire of one circuit to either wire of the other circuit, the voltage is 71 per cent of the voltage between wires of the same phase. The four-wire system with insulated phases is probably the most extensively used of the two-phase systems, but even this system is largely a relic of the past.

THREE-PHASE SYSTEMS. In this system there are three single-phase alternating currents with a phase difference of 120 deg, or of one-third of a cycle. These currents may be distributed on the three-, four- or six-wire systems.

Three-wire Three-phase System. Each single-phase current has a separate outgoing wire; the three return currents neutralize so that no return wire is required. The three wires are necessarily interconnected; the voltages are usually the same between any two, and the currents are equal in each of the three conductors, provided the loads on the three phases are equal, i.e., provided the load is balanced. When equally loaded, the voltage drops in the three conductors are equal and symmetrical. This is the most extensively used of the three-phase systems.

Four-wire Three-phase System. This is a modification of the three-wire system, in which a neutral wire is extended as a fourth wire. Lamps or transformers may be connected from each of the three wires to the neutral, which carries only the unbalanced current, due to the differences in loading of the three phases. The voltage between the three outside wires is 73 per cent greater than from each outside wire to neutral.

Six-wire Three-phase System. If to a three-wire, three-phase system, three wires are added, one with voltage midway between that of each pair of outside wires, lamps may be divided into six groups, between the three outside wires and the three adjacent middle wires. The result is the same as though there were three single-phase three-wire circuits, one for each phase, with the six outside wires combined in pairs, giving three common outside wires in place of the three pairs. When connected in this way, the three middle wires cease to be neutrals, as between the three there is a three-phase voltage equal to one-half of that between the three outside wires.

SIX-PHASE SYSTEM. This is used for circuits in the interior of substations (see Sections 8 and 13), such as from transformers to rotary converters, but is not used for distribution.

FREQUENCIES IN USE IN THE UNITED STATES. At present the two frequencies in most general use and adopted as standards in new work are:

60 cycles per second, used by the majority of companies operating a-c lighting systems.

25 cycles per second, generally used for a-c railway work or where the alternating current is to be converted into direct current before final use.

USE OF TWO FREQUENCIES. When the bulk of the load is direct current, but a small though important part is a-c lighting, two frequencies, 25 and 60 cycles, are sometimes used. Sometimes the two frequencies are generated by separate prime movers, but all current may be generated at 25 cycles and the 60-cycle current obtained from frequency changers. (See Section 8, Article 28, Motor-Generators.)

5. COMBINED D-C AND A-C SYSTEMS

Three-phase transmission and d-c distribution are readily combined but require either rotary or electronic conversion equipment. (See Section 11.) Electric railways, in particular, have commonly used synchronous converters for 500- to 600-volt systems, but the demand for higher voltages on overhead trolley systems led to the use of motor-generators and electronic converters. (See Section 13, Article 37.)

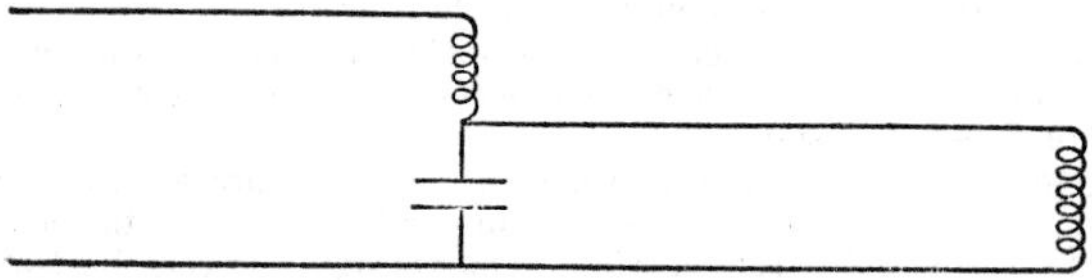

Fig. 2. Boucherot Circuit

A system of transmission is being developed comprising conversion of constant-potential a-c power to constant-current d-c power, transmission of constant-current d-c power, and reconversion to constant-potential a-c power.

A constant potential applied to a reactor and condenser in series may be made to yield constant current to a circuit derived from the condenser as shown in Fig. 2. This is known as the **Boucherot circuit** (Steinmetz, *Alternating-current Phenomena*, McGraw-Hill).

Constant alternating current is obtained from such a circuit, transformed up, and rectified by means of inverters with grid-controlled vapor-discharge tubes. At the receiving end the operations are reversed.

The system has the advantage that the d-c transmission line may be short-circuited without damage to line or equipment.

A description of an experimental line for 10 amp at 15,000 volts is given by C. H. Willis and E. Bedford, *Trans. AIEE*, Vol. 54, p. 102 (1935).

6. RELATIVE VOLTAGES AND WEIGHTS OF COPPER REQUIRED FOR VARIOUS SYSTEMS

The comparison of the various systems with respect to the weight of copper required, as shown in Table 1, is based upon the assumption that (1) the energy delivered, (2) the energy loss, and (3) the maximum voltage strain on insulation between any wire and ground are respectively *the same* for all systems.

When the neutral or middle point of the system is grounded, then the maximum voltage strain on the insulation to ground is the maximum instantaneous value of the voltage between any outside wire and neutral. With an alternating sine wave of voltage the maximum instantaneous value is equal to $\sqrt{2}$ times the effective value determined by a voltmeter. When the neutral is not grounded, an accidental ground on any leg will throw full line voltage across the insulation between any other leg and ground.

On this basis of comparison, assuming a sine wave of voltage and 100 per cent power factor for the a-c systems, the relative voltages between outside wires and the relative weights of copper are as follows:

Table 1. Relative Voltages and Relative Weights of Copper

System *	Relative Voltages between Outers		Relative Weights of Copper *	
	Grounded neutral	Insulated neutral	Grounded neutral	Insulated neutral
	Per cent	Per cent	Per cent	Per cent
Direct-current (2-wire)	100	50	100	400
Single-phase (2-wire)	71	35.5	200	800
Two-phase (4-wire)	71	35.5	200	800
Three-phase (3-wire)	61	35.5	200	600

* Neutral wire not included; when neutral is added increase the figures given in the ratio of weight of neutral to combined weight of the outside wires for the system in question. For example, the addition of a neutral wire equal in size to either outside wire to a d-c 2-wire system with grounded neutral gives a relative weight of copper of $100 \times 1.5 = 150$ per cent.

EFFECT OF POWER FACTOR. To correct Table 1 for a-c systems where the power factor is less than unity, the relative weight of copper given is divided by the square of the power factor (e.g., for a single-phase two-wire system with insulated neutral and 70 per cent power factor, the weight of copper will be $800/0.70^2 = 1630$).

7. STANDARD VOLTAGES

Unless otherwise specified, the "voltage" of a polyphase system refers to the potential difference between phases.

VOLTAGES COMMONLY USED. The voltages most used are:

Direct current:

100 to 125 volts	Lighting, small power, and field excitation.
200 to 240 volts	Power.
500 to 600 volts	Power, electric urban railways.
1200 to 1500 volts	Interurban electric railways.
2400 to 3000 volts	Electric railway trunk lines.

Alternating current:

120/240	12,000	138,000
120/208 Y	7,200/12,470 Y	161,000
240	7,620/13,200 Y	230,000
480	13,200	Higher voltages up to 500,000 are projected experimentally.
600	14,400	
2400	23,000	
2400/4160 Y	34,500	
4800	46,000	
7200	69,000	
4800/8320 Y	115,000	

The foregoing voltages, which are only approximated by the actual voltages in use, are the standard ones for "general apparatus." See also Articles 46 and 66.

8. EFFICIENCY OF DISTRIBUTION

The ratio of the energy which is registered by the customers' meters, or which would be so registered if all customers had meters, to the energy supplied by the generator to the bus-bar in the generating station may be called the overall efficiency of distribution, the word distribution being used in a broad sense to include transmission. The losses may be divided into several kinds: line loss, transformer loss, converter loss, meter loss and error, leakage, and loss unaccounted for. For some purposes it is useful to consider each loss from two points of view: (1) as a loss of energy, and (2) as a loss of power at full load; the first may be called the energy loss, and the second the capacity loss. The corresponding efficiencies are usually designated as all-day or energy efficiency and full-load or capacity efficiency, respectively.

FIXED AND VARIABLE LOSSES. The total energy loss consists of two components: (1) a fixed loss independent of load, including the core loss of transformers, the core loss, excitation, friction, and windage of rotating converting apparatus, the loss in the shunt coils of meters, dielectric loss, the copper loss in constant-current circuits, and the loss in the arc of mercury-arc rectifiers for constant-current circuits; (2) a variable loss proportional to the square of the current, including the copper loss of constant-potential circuits and of transformers, the armature copper loss of rotating converting apparatus. The effect of the fixed loss on the per cent efficiency depends on the load factor of the load; that of the variable loss depends on both the load factor and the shape of the load curve.

REPRESENTATIVE LOSSES. For a lighting system, the full-load losses in primary feeders, primary mains, transformers, secondary mains, services, and meters may be expected to be as much as 17.5 per cent of the power generated, and the daily energy loss 33.3 per cent of the energy generated, giving 82.5 per cent capacity efficiency and 66.7 per cent energy efficiency, respectively.

EFFECT OF NATURE OF LOAD. In making estimates of the efficiencies of particular systems, the effect of the following items should be considered: (1) relation of transformer capacity to maximum loads, (2) load factor, (3) shape of load curve, (4) power factor, (5) diversity factor.

9. REGULATION

An ideal constant-potential distribution would have one uniform, unvarying voltage and would be said to have perfect regulation. The greater the variation from such constancy the poorer, i.e., the greater numerically, the regulation. The regulation is usually specified in per cent variation, either above or below a standard mentioned. The standard is usually either the nominal voltage desired or the actual average voltage obtained.

EVIL EFFECTS OF POOR REGULATION. The evil effects of high voltage are short life of electric lamps, excessive speed of d-c motors, excessive exciting current of induction motors, and burning out of motors and other devices; on the other hand, low voltage greatly diminishes both the candlepower and efficiency of electric lamps, decreases the maximum power of motors, and increases the current which a motor will take for a fixed horsepower output. As electric lamps are much more sensitive to change of voltage than motors, separate circuits are often used for lighting and power, the former having devices for regulation which are omitted from the latter.

The following figures give roughly the quantitative effect of voltage variation between 5 per cent below and 5 per cent above normal:

Each per cent decrease in voltage decreases candlepower of tungsten incandescent lamps	3.5 per cent
decreases torque of induction motor	2 per cent
Each per cent increase in voltage decreases the life of tungsten incandescent lamps *	10 per cent
increases the magnetizing current of induction motors	2 per cent

ORDINARY LIMITS OF REGULATION. Roughly, the maximum voltage variation at the lamps on a lighting system should never exceed 5 per cent; i.e., the regulation above or below normal should never be greater than 2.5 per cent, and should be as much less as is economically feasible. The voltage variation on power systems is usually about 10 per cent (5 per cent above or below normal) and is sometimes considerably more.

CALCULATION OF REGULATION. In order to calculate the variation in voltage at any receiving device or group of such devices from no load to full load, the voltage at the generating or substation or feeding point being assumed constant, the impedance drops in all parts of the distribution system must be calculated and properly combined. The various parts of the system to be considered are the house wiring, service wires (or leads from street mains to the house), secondaries, distribution transformers, primary mains, primary feeders, substation transformers, transmission lines, and power station transformers.

In making such calculations, account should be taken of the fact that the loads or currents in the several parts of the system seldom have their maximum values at the same time. The maximum drop in house wiring will not occur simultaneously in all houses, nor will the maximum service drop occur together on all services, etc. Furthermore, the maximum house wiring drop for a given house may not occur at the same time as the maximum drop on the secondary mains, transformer or primary mains to which it is connected.

Effect of Line Reactance. The regulation of an a-c system of unity power factor will be poorer than that of a d-c system of the same copper efficiency (see foregoing tabulation) because of additional drop due to line reactance.

Effect of Lagging Power Factor. A lagging power factor usually makes the regulation of an a-c system worse than it would be for unity power factor and therefore much worse than for a d-c system of the same copper efficiency.

Effect of Leading Power Factor. A leading power factor usually makes the regulation better than it would be for unity power factor and may give even better regulation than can be obtained from a d-c system of the same copper efficiency.

Effect of Currents in Neutral. Any current in the neutral wire of a balanced system produces a drop which tends to unbalance the voltage of the two sides. In the case of a three-wire d-c or single-phase system *1 per cent drop in neutral produces 2 per cent difference in the voltages on the two sides.* Voltage drop in the neutral therefore affects the regulation more seriously than an equal amount in the outside wires. If all the currents in the neutral have the same direction, say toward the station, the voltage drops in the various parts of it will be cumulative, and though the drop in each section may be small and no current may actually reach the station, the aggregate effect may be serious. The individual loads on the two sides of the neutral should be connected so that the unavoidable neutral currents flow alternately in each direction, thus causing drops in alternate directions and thereby neutralizing each other over the total length of the circuit.

FEEDER NEUTRAL IN THREE-WIRE SYSTEMS. When the "feeder and main" system is used with a three-wire system, two-wire feeders should be used only on the outside wires, and only a single feeder neutral be used. That is, with respect to the neutral, the "tree" system gives better regulation for the same weight of copper than the feeder and main system, but with respect to the outside wires the latter system gives the better results.

10. VOLTAGE CONTROL

The more common methods of controlling the voltage at feeding points when the feeder and main system of distribution are used are the single bus system, the high and low bus system, and feeder regulators.

SINGLE BUS SYSTEM. All the feeders may be connected to a single bus and the bus voltage be raised as load increases so as to compensate for average drop of all the

* Average for 5 per cent increase in voltage; the first per cent increase in voltage decreases the life 13 per cent. (See Section 15, Article 3.)

feeders on it. This method gives excessive voltage on such feeders as are comparatively short, and low voltage on those that are long. Usually maximum voltage is carried in the evening during the lighting peak and lower voltage during the day, giving very poor regulation on power circuits having maximum load during the day and low load at night. This method is used on most small d-c and a-c distribution systems.

HIGH AND LOW BUS SYSTEM. The bus may consist of two parts which may be separated and operated at different voltages, feeders of greater average drop connected to the "high" (voltage) bus and the others to the "low" (voltage) bus. The voltage of each bus is raised or lowered in accordance with average drop of feeders on it, as in the single bus system. This method is used extensively on the Edison three-wire d-c systems, as noted above. The unequal drop of the several feeders on the same bus is equalized by heavy interconnection through mains.

FEEDER REGULATORS. A feeder may have a separate regulator adjustable to compensate for its own drop. This method is sometimes used on d-c railway feeders and is employed very extensively on a-c lighting feeders. A feeder or other distribution circuit which contains a voltage regulator is frequently referred to as a "boosted circuit." Boosted circuits are not usually interconnected but may be if the boosting is of the same nature in each. This system is the more flexible and economical for extensive distribution systems, since each feeder is independently regulated and may be proportioned in cross-section with regard to economy rather than inherent voltage regulation. Single-phase regulators are preferred for close control if the loads are subject to unbalancing.

CONTROL OF POLYPHASE SYSTEMS. In a polyphase system the load may be as equally divided among the phases as possible, and the voltage regulated with respect to any one phase taken as representing the average of all; or the lighting load may be connected on a single phase, and the voltage regulated for this phase alone. The former method is more common, as it permits of full output of all the phases of the generator being used and gives a more equal voltage on the several phases of polyphase motors.

TRANSFORMERS AS OUTSIDE BOOSTERS. An auto-transformer or an ordinary transformer may be connected to a feeder at any point and used as a "booster" to raise or lower the voltage. In an ordinary transformer the primary and secondary are connected in series, thus converting the transformer into an auto-transformer; see Auto-transformers in Section 10. Such boosters are not adjustable and have a bad effect on the regulation, as they give excessive voltage on the boosted part of line at light load.

To cut such a booster out of service without taking it off the circuit the primary coil must be open-circuited and the secondary short-circuited. *Caution:* The main circuit must be open before cutting out booster, because if the primary is opened while current is flowing in the secondary, the booster becomes a step-up transformer and may give a dangerously high voltage on the primary. If, on the other hand, the secondary is short-circuited first, a destructive short-circuit current may flow through it and the primary, and when the primary is opened a dangerous arc will form.

USE OF TRANSFORMERS OF DIFFERENT RATIO. Transformers are also made with taps, so that a uniform secondary voltage can be obtained with a varying primary voltage. The plan is not a good one, because if the difference in ratio is correct for uniform voltage at full load, it gives unequal voltages at light load. It also has the very serious disadvantage that the haphazard changing of transformer ratios by ignorant linemen to compensate for dim light, due perhaps to lamps already blackened by excessive voltage, makes it impossible to carry out any systematic plan for securing the best average regulation for the system as a whole.

11. BALANCING OF LOAD

In three-wire d-c or single-phase systems and in all polyphase systems supplying single-phase load it is necessary to balance the load approximately between the two sides or the several phases, as the case may be. Unbalanced load has two bad effects: (1) it loads the two sides or phases of the system unequally, making it impossible to get full output out of the lightly loaded side or phase without overloading the other; (2) it makes the regulation of the system worse by causing high voltage on the lightly loaded side or phase and low voltage on the other.

The first difficulty is not serious, because through the conduction of heat from the loaded to the unloaded coils of transformers and generators the machine capacity is not reduced in proportion to the unbalance; for moderate unbalancing, say up to 10 per cent greater load on one side or phase than on the other, it is doubtful if any appreciable effect could be discovered. As the total load usually consists of a great number of small parts, a very little foresight in dividing the load in the first place between the sides or phases, and in suitably distributing subsequently connected load, will give a balance good enough for all practical purposes. When polyphase alternators were first installed for supplying existing single-phase lighting circuits, it was supposed to be important to have a close balance of load between the phases, and early switchboards were therefore provided with transfer switches for throwing single-phase feeders from phase to phase. It was later found that, instead of being necessary to transfer the load from phase to phase, following the diurnal or annual variations of load, the circuits could stay on the same phase indefinitely and that transfer switches were unnecessary and undesirable. The time when rebalancing of this kind is necessary is usually when new circuits are established, at which time the changes of connections are best made by changing the taps to the bus-bars.

MOTOR-GENERATOR BALANCER. On d-c three-wire systems the difference in current on the two sides may be balanced without taking the neutral current to the generator by a balancer consisting of two similar machines mechanically coupled and electrically connected in series. Each machine is wound for the voltage of one side of the system (110 volts for a 110/220-volt system); the common connection between the machines is connected to the neutral and the other two terminals to the two outsides; see Fig. 3 (the field windings are omitted for clearness). The unit acts as a motor-generator (see Section 8, Article 28), whichever machine happens to be on the light side being the motor for the time being and the other the generator. By strengthening the field of the one on the side where the voltage is low and weakening that of the other, so as to keep the same total voltage, the voltages on the two sides, as well as the currents, may be balanced if necessary.

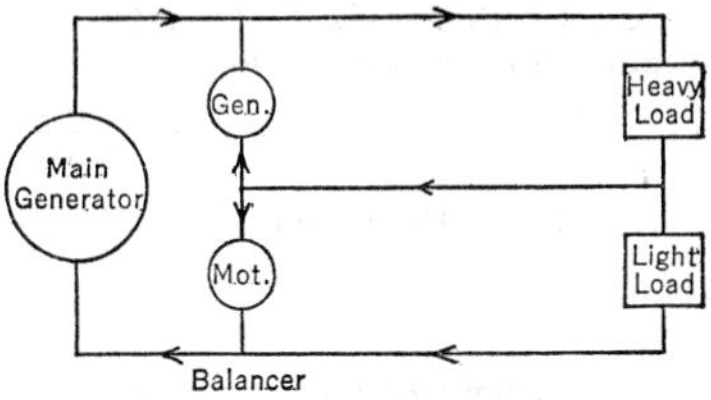

FIG. 3. Motor-generator Balancer

The output of the balancer generator and the input of the balancer motor are practically equal to each other, neglecting the losses in the machines, and each is equal to

$$\frac{P_1 - P_2}{2}$$

where P_1 is the load on the heavily loaded side of the system, and P_2 the load on the lightly loaded side. For example, if the load on one side is 110 kw and that on the other side 90 kw, the load on each unit of the balancer is 10 kw.

It would, however, be unsafe to use a balancer as small as such calculations for the normal unbalancing would indicate as correct, because in case of a short circuit on one side of the system, or of loss of a large amount of load on one side, caused, say, by the blowing of fuses, the balancer would be dangerously overloaded. This might not only destroy the balancer, but might also burn out many lamps and motors on one side because of excessive voltage. In small systems, say up to 100-kw total capacity, the balancer should be able to operate momentarily with any loading up to the capacity of the main generator, and in large systems should have capacity sufficient to burn off a short circuit on one side of the system without creating an unduly high voltage on the lightly loaded side.

DYNAMOTOR AS A BALANCER. A smaller and cheaper balancer is obtained by using a dynamotor (see Section 8, Article 29), which is a single machine with two windings on one armature and two commutators. The armatures are connected in series and balance the currents, but as there is only one field the voltages cannot be balanced.

A-C BALANCING COIL. On a-c single-phase three-wire systems it is usual to obtain the neutral from the middle point of the coil of the transformer supplying the current. It can, however, be obtained at any point of the two-wire circuit without going back to the transformer by using a balance coil. A balance coil is a transformer with two similar windings connected in series across the circuit and with the neutral wire connected to the common point between the windings. Whichever coil is on the lightly loaded side acts as the primary and the other as the secondary, thereby balancing the circuit by transferring one-half of the difference in load from the heavily loaded side to the lightly loaded side. Such coils, which are essentially auto-transformers with a 2 : 1 ratio, may be used in various ways: to obtain the neutral for an unbalanced three-wire circuit from a two-wire circuit; to supply a 110-volt load from a 220-volt circuit; to supply a large 110-volt load from a 110/220-volt circuit without connecting to the neutral. In practice balance coils are not much used, as the neutral can more cheaply be obtained from the transformer.

12. BIBLIOGRAPHY

Reports of Edison Electrical Institute Committees.
Reports of Transmission and Distribution Committee of A.I.E.E.
See also Articles 19, 44, 53, and 93.

ELECTRICAL DESIGN OF TRANSMISSION AND DISTRIBUTION SYSTEMS

Revised by W. A. Del Mar

13. DEFINITIONS AND FUNDAMENTAL RELATIONS

The following definitions and relations apply to all types of transmission and distribution lines.

GENERATOR END AND LOAD END. By the generator end of the line is meant the end which is connected to the source of power (either directly or through transformers), and by the load end is meant the end which is connected to the load or substation which is supplied with power over the line.

PER CENT POWER LOSS (Q). By per cent power loss, as used in this chapter, is meant the percentage ratio

$$Q = 100 \cdot \frac{\text{Total power lost in the line}}{\text{Total power delivered at load end}} \tag{1}$$

Hence, if P is the power delivered, then the total power supplied to the line and load is

$$P_0 = P\left(\frac{100 + Q}{100}\right) \tag{2}$$

PER CENT VOLTAGE LOSS (D). By per cent voltage loss, as used in this chapter, is meant the percentage ratio *

$$D = 100 \cdot \frac{(\text{Voltage at generator end}) - (\text{Voltage at load end})}{\text{Voltage at load end}} \tag{3}$$

Hence, calling E the voltage at the load end, the voltage at the generator end is

$$E_0 = E\left(\frac{100 + D}{100}\right) \tag{4}$$

The per cent voltage loss allowed under various conditions is discussed in detail in Article 9; the allowable voltage loss is usually between 2 and 20 per cent, the most common figure, for transmission lines, being 10 per cent.

In a d-c line with a single load at its far end, the per cent power loss and the per cent voltage loss are always equal, but in an a-c line the per cent voltage loss may be either greater or less than the per cent power loss, depending upon the constants of the line and the power factor of the load, or may even be negative, i.e., there may be an actual rise of voltage at the load end above the voltage at the generator end (see below).

EFFICIENCY OF TRANSMISSION. By the efficiency of transmission is meant the percentage ratio

$$100 \cdot \frac{\text{Power output of line at load end}}{\text{Total power input to line at generator end}} \tag{5}$$

The per cent efficiency is related to the per cent power loss Q as follows:

$$\text{Per cent efficiency} = \frac{10{,}000}{100 + Q} \tag{6}$$

14. ELECTRICAL DESIGN OF D-C LINES

Two types of problems arise: (1) given a definite line with known constants, what is the power loss and voltage loss for a given load? and (2) to transmit a given amount of power a given distance with a given allowable loss, what will be the size and weight of the conductor required? In the following paragraphs are given the necessary formulas for the several cases. (See also Article 17.)

TWO-WIRE LINE; CONCENTRATED LOAD AT FAR END. Let

E = volts between wires at the load end of the line.

$P = \dfrac{EI}{1000}$ = kilowatts taken by load.

* In an a-c line the difference in the numerator of this ratio is the algebraic difference between rms values of the two voltages; not the vector difference.

$I = \dfrac{1000P}{E}$ = amperes taken by load.

l = length of each line wire in feet.

r = ohms per 1000 feet of conductor; see tables in Articles 54 to 62, Bare Wires and Cables.

$R = \dfrac{rl}{500}$ = total resistance of line (both conductors) in ohms.

The following relations then hold:

$$\text{Total kilowatts lost} = p = \frac{RI^2}{1000} = \frac{rlI^2}{500{,}000} = \frac{2rlP^2}{E^2} \quad (7)$$

$$\text{Total volts lost} = v = RI = \frac{rlI}{500} = \frac{2rlP}{E} \quad (8)$$

$$\text{Per cent power loss} = Q = \frac{100p}{P} = \frac{rlI^2}{5000P} = \frac{200rlP}{E^2} \quad (9)$$

$$\text{Per cent voltage loss} = D = \frac{100v}{E} = \frac{rlI}{5E} = \frac{200rlP}{E^2} \quad (10)$$

$$\left.\begin{array}{l}\text{Resistance of each con-}\\ \text{ductor per 1000 ft}\end{array}\right\} = r = \frac{500v}{lI} = \frac{QE^2}{200lP} = \frac{DE^2}{200lP} \quad (11)$$

CALCULATION OF SIZE AND WEIGHT OF CONDUCTOR FOR CONCENTRATED LOAD. From the value of r calculated from any one of the relations given in eq. (11), the size of wire may be found from the tables in Articles 54 to 62, Bare Wires and Cables; the next larger size of wire (next smaller gage number) should usually be chosen when the calculated resistance lies between that of two commercial sizes. The wire selected must also have sufficient current-carrying capacity; see Articles 61 and 69. For outside lines, however, the current-carrying capacity will in general be ample unless the allowable voltage loss is excessive. For an outside overhead line a wire smaller than No. 6 A.W.G. (or B. & S.) gage is seldom used, chiefly on account of its lack of mechanical strength.

Let w = weight per 1000 ft of the wire finally selected; then

$$\text{Total weight of conductor in pounds} = W = \frac{wl}{500} \quad (12)$$

DIRECT CALCULATION OF TOTAL WEIGHT OF CONDUCTOR (W); TWO-WIRE LINE. For preliminary estimates it is sometimes convenient to calculate the total weight of conductor directly, without reference to a wire table. The total weight of conductor for a two-wire line with concentrated load at its end is given by the formula

$$W = \frac{KP}{Q}\left(\frac{l}{E}\right)^2 \text{ pounds} \quad (13)$$

where P is the power taken by the load, l the length of the line (length of each wire), E the voltage at the load, Q the per cent power loss (= per cent voltage drop for two-wire d-c line), and K a constant depending upon the material of the conductor and the units in which P, l, and E are expressed, as given in Table 1.

Table 1. Values of K in Equation (13)

Material	E in volts, l in feet, P in kilowatts	E in kilovolts, l in miles, P in kilowatts
Copper (98 per cent conductivity)	13.5	380
Aluminum (61 per cent conductivity)	6.5	185
Any material of specific gravity δ having a conductivity of c per cent at 20° C	$\dfrac{141\,\delta}{c}$	$\dfrac{3940\,\delta}{c}$

NOTE: The values of K given for copper and aluminum are about 5 per cent greater than their theoretical values to allow for stranding, higher working temperature, etc.; 100 per cent conductivity corresponds to 1.724 microhms per centimeter cube at 20 deg cent.

Example: Two-wire D-c Line, Concentrated Load. A load of 100 kw is to be transmitted over a two-wire line to a motor operating at 230 volts, the motor being 1000 ft from the power-house switchboard. For a 10 per cent power loss or voltage drop in the line, the approximate total weight of copper required is, from eq. (13),

$$W = \frac{13.5 \times 100}{10}\left(\frac{1000}{230}\right)^2 = 2550 \text{ lb}$$

From eq. (11) the resistance per 1000 ft is

$$r = \frac{10 \times (230)^2}{200 \times 1000 \times 100} = 0.0264 \text{ ohm per 1000 ft}$$

The nearest even circular mil size is 400,000 cir mils (stranded), which has a resistance ot 0.0270 ohm per 1000 ft at 77 deg fahr (see Articles 54 and 55, Bare Wires and Cables), and a weight of 1240 lb per 1000 ft. From eq. (12) the total weight of conductor is then

$$W = \frac{1240 \times 1000}{500} = 2480 \text{ lb}$$

This wire, if bare, weatherproofed, or insulated with paper or varnished cambric, will safely carry the required current of 100,000/230 = 435 amp, but if rubber insulated and mounted indoors, a larger wire should be required, viz., 600,000 cir mils, according to the National Electric Code (see Articles 61 and 82).

CALCULATION OF TWO-WIRE D-C LINE IN TERMS OF VOLTAGE AT GENERATOR END. When the volts E_0 at the generator end are given instead of the volts E at the load end, the calculations for a line of given total resistance of R ohms with a concentrated load of P kilowatts at the load end may be made in the same manner as above by first finding the volts E at the load by the formula

$$E = \frac{E_0}{2}\left[1 + \sqrt{1 - \frac{4000RP}{E_0^2}}\right] \tag{14}$$

For an efficiency of transmission of less than 50 per cent, the sign before the radical should be − instead of +, but an efficiency of less than 50 per cent practically never occurs in power transmission. It is of interest to note that for an efficiency of 50 per cent $P = E_0^2 \div 4000R$, which is the maximum power which can be delivered at the far end of the line for a given impressed voltage at the generator end.

When E has been calculated by eq. (14), eqs. (7) to (13) may be applied directly.

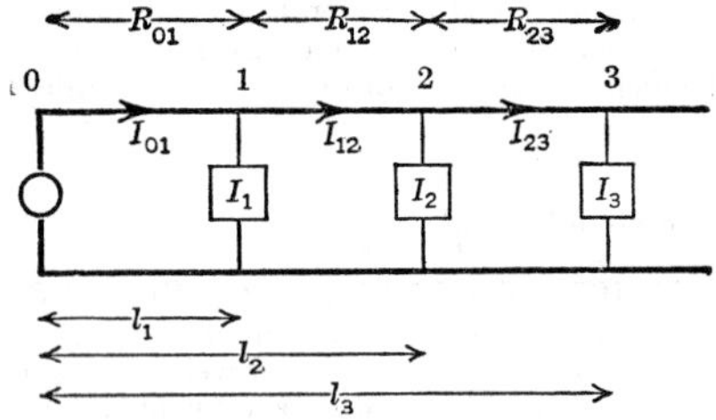

FIG. 1. Distributed Load

TWO-WIRE LINE; DISTRIBUTED LOAD. When a line supplies a number of loads at different distances from the generator end, the voltage loss to the far end of the line is the same as would be produced by a load concentrated at the "center of gravity" of the line and taking a current equal to the total current taken by all the loads. The center of gravity of the line is defined as follows: Let I_1, I_2, I_3, etc., be the currents taken from the line by the various loads (Fig. 1) and let R_1, R_2, R_3, etc., be the total line resistances (both wires) from the generator end to the respective loads, and put

$$I = I_1 + I_2 + I_3 + \cdots \tag{15}$$

Then the center of gravity is that point between which and the generator end of the line the total line resistance is

$$R_g = \frac{R_1I_1 + R_2I_2 + R_3I_3 + \cdots}{I} \tag{16}$$

When the line conductor has the same cross-section throughout its length, then the center of gravity is at the distance

$$l_g = \frac{l_1I_1 + l_2I_2 + l_3I_3 + \cdots}{I} \tag{16a}$$

from the generator end, where l_1, l_2, l_3, etc., are the distances of the respective loads from the generator end.

The total voltage loss to the far end of the line is then

$$v = R_gI = \frac{rl_gI}{500} \tag{17}$$

where the distance l_g is in feet and r is the resistance of the conductor per 1000 ft, the second relation in (17) holding only when the conductor has the same cross-section throughout its length.

The total kilowatts lost in the line are

$$p = \frac{1}{1000}(R_{01}I_{01}^2 + R_{12}I_{12}^2 + R_{23}I_{23}^2 + \cdots) \tag{18}$$

where, referring to Fig. 1, R_{01} = total resistance (both wires) from 0 to 1, R_{12} = total resistance (both wires) from 1 to 2, etc., and $I_{01} = I_1 + I_2 + I_3 + \cdots$ = the current in the line from 0 to 1, $I_{12} = I_2 + I_3 + \cdots$ = the current in the line from 1 to 2, etc. When the cross-section of the line conductor is the same throughout its length, eq. (18) may be also written

$$p = \frac{r}{500{,}000}(l_{01}I_{01}^2 + l_{12}I_{12}^2 + l_{23}I_{23}^2 + \cdots) \tag{18a}$$

where the distances l_{01}, l_{12}, l_{23}, etc., are as shown in Fig. 1 and are measured in feet, and r is the resistance of the line conductor per 1000 ft.

Calculation of Size and Weight of Conductor for Distributed Load. For a conductor of the same cross-section throughout, the required resistance per 1000 ft for a given voltage loss of v volts to the end of the line may be calculated from the formula

$$r = \frac{500v}{l_g I} \tag{19}$$

where l_g, expressed in feet, and I are given by eqs. (16a) and (15). The size and weight of conductor can then be found by reference to the wire tables Articles 54 to 62, Bare Wires and Cables.

When the loads are far apart and the smaller loads are farthest from the generator, it is sometimes advisable to use different sizes of conductors for the various portions of the line. For a given voltage loss v to the end of the line, the minimum weight of conductor is obtained when the voltage lost per unit length of conductor in each section of the line is proportional to the square root of the current in this portion of the line. For minimum total weight of conductor then, referring to Fig. 1, the resistance per 1000 ft *of wire* for the section between 1 and 2, say, must be

$$r_{12} = \frac{1}{\sqrt{I_{12}}} \cdot \frac{500v}{l_{01}\sqrt{I_{01}} + l_{12}\sqrt{I_{12}} + l_{23}\sqrt{I_{23}} + \cdots} \tag{20}$$

where the lengths are in feet, and similarly for the other sections. The weight of wire for each section may then be found by reference to the wire tables in Articles 54 to 62, and the total weight W can then be computed; vice versa, for a line proportioned in this manner the voltage loss to the end of the line for a given total weight of copper will be a minimum.

THREE-WIRE D-C LINE. When a three-wire circuit is exactly balanced, i.e., when the loads between each of the two outer wires and the neutral are the same and are connected to the neutral at the same point or points, no current flows in the neutral wire. The foregoing equations for a two-wire line then apply directly to a balanced three-wire line, noting, however, that the E (= volts between wires) in these formulas is to be taken as the volts between the *outer* wires, and that the weight as calculated by the foregoing equations is the weight of the two outer wires. The neutral wire is usually made equal in size to each outer wire, but when only slight unbalancing is expected it is sometimes made smaller. When the neutral is made equal in size to the outer wire, the total weight of the three conductors will be 50 per cent more than that given by eq. (13), when E in this formula is taken equal to the volts between the outer wires.

15. ELECTRICAL DESIGN OF A-C LINES

As a rough guide in fixing upon a preliminary design, the following facts should be noted; complete formulas for the various calculations required are given later.

1. A power loss of approximately 10 per cent of the delivered power is usually allowed.
2. A line voltage of approximately 1000 volts per mile of line is common practice for long-distance lines not over 150 miles in length; that is, for a 10-mile line a line voltage of at least 10,000 volts would be employed; for a 100-mile line a line voltage of at least 100,000 volts would be used. The maximum line voltage now employed is 238,000 volts, and the maximum distance of transmission is 350 miles.
3. On the basis of 1000 volts per mile of line, unity power factor at the load, a 10 per cent power loss, and copper at 10 cents per pound, the cost of the copper required for a

three-phase line is $2.66 per kilowatt delivered, and for a single-phase or two-phase four-wire line $3.45 per kilowatt delivered.

TOTAL WEIGHT OF CONDUCTOR FOR A-C LINES. The size and total weight of the conductor required for any conditions † of length, power delivered, power factor, line voltage, and power loss may be calculated as follows:

Let E = voltage between wires at the load end of the line.
P = total power taken by all phases of the load.
$\cos \phi$ = power factor of load, as a decimal.
l = length of line (= length of each line wire).
Q = allowable total power loss in per cent of delivered power.

Then the total weight of all conductors is given by the formula

$$W = \frac{KP}{Q}\left(\frac{l}{E \cos \phi}\right)^2 \text{ pounds} \tag{21}$$

where K is a constant depending upon the number of phases and wires, the material of the conductor, and the units in which the various quantities are expressed, as given in Table 2.

Table 2. Values of K in Equation (21)

Material and Units	Single-phase or Balanced 4-wire 2-phase *	Balanced 3-wire 3-phase
Copper (98 per cent conductivity):		
E in volts, l in feet, P in kilowatts.............	13.5	10
E in kilovolts, l in miles, P in kilowatts........	380	280
Aluminum (61 per cent conductivity):		
E in volts, l in feet, P in kilowatts	6.5	4.9
E in kilovolts, l in miles, P in kilowatts........	185	140
Any material of specific gravity δ having a conductivity of c per cent at 20° C:		
E in volts, l in feet, P in kilowatts.............	$\frac{141\ \delta}{c}$	$\frac{106\ \delta}{c}$
E in kilovolts, l in miles, P in kilowatts........	$\frac{3940\ \delta}{c}$	$\frac{2950\ \delta}{c}$

NOTE: The values of K given for copper and aluminum are taken about 5 per cent greater than their theoretical values to allow for stranding, higher working temperatures, etc.; 100 per cent conductivity corresponds to 1.724 microhms per centimeter cube at 20 deg cent.

* For a 3-wire 2-phase system with the middle conductor having a cross-section equal to $\sqrt{2}$ times the cross-section of either outer wire multiply these constants by 0.85, taking for E the voltage (volts or kilovolts) between the middle wire and either outer wire.

COMMERCIAL SIZE OF CONDUCTOR AND CORRESPONDING TOTAL WEIGHT. Equation (21) takes no account of the available commercial sizes of wire. These sizes differ successively by approximately 25 per cent in cross-section. The weight can also be determined by calculating the resistance of the required conductor per 1000 ft or per mile, and taking from the wire tables in Articles 54 to 62 the nearest commercial size. Neglecting the charging current, the required resistance per unit length of wire is

$$r = \frac{K_1 Q (E \cos \phi)^2}{lP} \text{ ohms} \tag{22}$$

where K_1 is a constant depending on the number of phases and wires and the units in which the other quantities are expressed, as given in Table 3.

From the wire table the corresponding weight (w) per unit length of conductor having a resistance nearest to that calculated by formula (22) is obtained; the total weight of conductor, including all wires, is then

$$W = K_2 wl \text{ pounds} \tag{23}$$

where K_2 is a constant depending upon the number of phases and wires and the units in which w and l are expressed, as given in Table 3.

† These formulas are based on the assumption that the charging current is negligible in comparison with the load current, which condition is practically realized in all but the longest high-voltage lines; formulas for power loss taking the charging current into account are given later.

Table 3. Values of K_1 and K_2 in Equations (22) and (23)

Units	Single-phase	Balanced 4-wire 2-phase *	Balanced 3-wire 3-phase
E in volts, l in feet, P in kilowatts, r in ohms per 1000 feet and w in pounds per 1000 ft...	$K_1 = 0.005$ $K_2 = 0.002$	$K_1 = 0.01$ $K_2 = 0.004$	$K_1 = 0.01$ $K_2 = 0.003$
E in kilovolts, l in miles, P in kilowatts, r in ohms per mile and w in pounds per mile....	$K_1 = 5$ $K_2 = 2$	$K_1 = 10$ $K_2 = 4$	$K_1 = 10$ $K_2 = 3$

* The values of K_1 given in this column when used in eq. (22) will give the resistance per 1000 ft or per mile of either outer wire in a 3-wire 2-phase system; the middle wire should, for the same energy loss per pound, have a cross-section 41 per cent greater than either outer, but when commercial sizes (A. W. G.) are used, either a wire one gage number smaller (25 per cent greater cross-section) or two gage numbers smaller (60 per cent greater cross-section) may be used. In the first case the corresponding value of K_2 is 0.81 times the values given in this column, and in the second case 0.90 times the values given in this column.

Current per Wire; Heating of Line Conductors. The size of wire, as determined from eq. (22), must be ample to carry the required current without overheating. Heating of the line conductors is seldom a limitation in outside overhead lines, but for inside wiring or underground cables the temperature rise may set a limit to the size of wire which may be used. It is therefore always wise to determine the current which the conductor must carry and make sure that the wire is sufficiently large not to overheat; see Articles 61 on Bare Wires and Cables, and Article 69 on Insulated Wires and Cables, for tables of current-carrying capacity under various conditions.

The current per line wire in amperes may be calculated from the following equations, in which *E is the kilovolts* between wires at the load end, *P the total kilowatts* (all phases) delivered to the load, and cos ϕ the power factor of the load as a fraction.

$$\left.\begin{aligned} &\text{Single-phase:} && I = \frac{P}{E\cos\phi} \\ &\text{Two-phase,‡ 4-wire, balanced:} && I = \frac{P}{2E\cos\phi} \\ &\text{Three-phase, 3-wire, balanced:} && I = \frac{P}{\sqrt{3}E\cos\phi} \end{aligned}\right\} \quad (24)$$

Example: Calculation of Weight and Size of Conductor for a Three-phase Line. A load of 20,000 kw is to be transmitted by means of an overhead three-phase line of copper wire to a substation 50 miles away operating at 60,000 volts between wires, the frequency being 25 cycles per second, and the power factor of the load 80 per cent with the current lagging; a power loss of 10 per cent of the delivered power is to be allowed. From eq. (21) the required total weight of copper is

$$W = \frac{280 \times 20{,}000}{10}\left(\frac{50}{60 \times 0.8}\right)^2 = 607{,}000 \text{ lb}$$

From eq. (22) the required resistance per mile of conductor is

$$r = \frac{10 \times 10(60 \times 0.8)^2}{50 \times 20{,}000} = 0.231 \text{ ohm per mile}$$

The nearest commercial size is 250,000 cir mils (stranded), which has a resistance of 0.228 ohm per mile at 77 deg fahr (see Bare Wires and Cables), and a weight of 4080 lb per mile. From eq. (23) the total weight is then

$$W = 3 \times 4080 \times 50 = 612{,}000 \text{ lb}$$

The current corresponding to the given load is, from eq. (24),

$$I = \frac{20{,}000}{\sqrt{3} \times 60 \times 0.8} = 241 \text{ amp}$$

which will give a negligible temperature rise in the wire. See Article 61, Current-carrying Capacity, in Bare Wires and Cables.

‡ E is here the volts between the two wires of the same phase. This formula also gives the current in each outer wire of a balanced 2-phase 3-wire line, E being the kilovolts between either outer and the middle wire; the current in the middle is $\sqrt{2}$ times the current in each outer wire.

CALCULATION OF SIZE AND WEIGHT OF CONDUCTORS FOR A GIVEN PER CENT VOLTAGE LOSS. The voltage loss in an a-c line depends not only upon the resistance of the line, but also upon the line reactance, and in long lines upon the capacitance of the line. It is therefore impossible to express directly in a simple formula the size or weight of the wire in terms of the voltage loss. The most practical method of making such calculations is to assume first that the per cent power loss is equal to the given per cent voltage loss, and to calculate the size by formula (22); then, using this size of wire, to calculate the per cent voltage loss by the equations given below. If this calculated voltage loss differs appreciably from the given voltage loss, the next larger or smaller size of wire (accordingly as the calculated loss is greater or less than the given loss) is chosen, and the voltage loss is recalculated, and so on, until the proper size of wire has been found.

16. FACTORS WHICH AFFECT THE VOLTAGE AND POWER LOSS IN A-C LINES

Because of the inductance and capacitance the per cent voltage loss in an a-c line is not so easily calculated as the voltage loss in a d-c line and is in general different from the per cent power loss by an amount dependent upon the inductance and capacity of the line, the frequency, and the power factor of the load.

DETERMINATION OF LINE CONSTANTS. The four fundamental line constants are the resistance (r) and inductance (L) of the line conductors per unit length and the capacitance (C) and leakage conductance (G) per unit length. For all but the shortest transmission lines the mile is usually the most convenient unit of length, and this unit will be used throughout the remainder of this article unless it is distinctly stated otherwise. Tables of resistance, inductance, and capacitance are given respectively in Articles 20 to 23. From the inductance and capacitance per mile or per 1000 ft may be calculated, for any given frequency, the reactance x ($= 2\pi fL$) and the capacitance susceptance § b ($= 2\pi fC$) for the corresponding unit of length.

SEQUENCE IMPEDANCES. The method of calculation of circuit characteristics described in Section 3, Article 37, under the heading "Symmetrical Components of Currents and Voltages" requires a knowledge of positive, negative, and zero-phase-sequence impedances.

Positive-phase-sequence impedances are balanced three-phase impedances as used in balanced line calculations. These are the impedances used in this chapter. Negative-phase-sequence impedances are identical with positive-sequence-impedances for lines, transformers, and other static equipment. For synchronous equipment, however, negative-phase-sequence voltage is a balanced three-phase voltage with direction of rotation opposite to the positive-sequence rotation.

Zero-phase-sequence impedance is that which determines the flow of current in the earth when unbalanced conditions exist. Zero-phase capacitance is also a factor when the system is not grounded or is grounded through a high neutral resistance or reactance. (See Section 3, Article 47.)

ALLOWANCE FOR SKIN EFFECT IN CONDUCTORS. The skin effect is practically negligible at 25 cycles for all copper conductors smaller than 1,000,000 cir mils, and at 60 cycles for all copper conductors smaller than 450,000 cir mils. The corresponding limiting sizes for aluminum are about 30 per cent larger. The skin effect is quite appreciable in copper or aluminum cables with a steel core; it is usual to neglect the conductivity of the steel core entirely in calculating the resistance of such cables. (See Article 59 and Section 3, Article 25.)

APPARENT RESISTANCE AND REACTANCE IN UNSYMMETRICAL ARRANGEMENTS OF WIRES. When the three wires of a three-phase line are so arranged that they form the three edges of an equilateral prism, the reactance of each wire is the same as for one wire of a two-wire line. However, when the wires are arranged all in one plane, as is frequently done, the unequal mutual induction sets up a reactive electromotive force in each outer wire which is not in quadrature with the current in this wire. As a result, both the apparent resistance and the apparent reactance of each outer wire are different from its true resistance and reactance. Let r = the true resistance per mile of each wire in ohms; x = reactance per mile of each wire in ohms (Articles 20 to 23); and f = frequency

§ The capacitance susceptance to neutral, in micromhos per mile, is equal to the charging current in amperes per mile per million volts between wire and neutral; the charging current for any other voltage to neutral is in proportion.

in cycles per second. Then the apparent resistances and reactances per mile of the three wires, No. 2 being the middle wire, are approximately:

$$\left.\begin{array}{lll} r_1 = r - 0.00121f & r_2 = r & r_3 = r - 0.00121f \\ x_1 = x + 0.00070f & x_2 = x & x_3 = x + 0.00070f \end{array}\right\} \qquad (25)$$

The changes in the apparent resistances indicate, not any change in the power dissipated as heat in the wires, but a transfer of energy from one wire to the other by the magnetic field surrounding the wires. These relations assume sine-wave currents equal in effective value and differing in phase by exactly 120 deg. The assumption that the currents are exactly balanced cannot be strictly true, since the inequality in the apparent resistances and reactances of the three wires tends to unbalance the system, but the values just given may be taken as a fair approximation when the voltage loss in the line is not over 10 per cent, say. An exact treatment in terms of Bessel functions is given in H. B. Dwight's *Electrical Coils and Conductors*, p. 196. When the line wires are transposed, these mutual-inductance effects are eliminated from the line as a whole, although the apparent impedances of the three wires in any one "exposure" of the transposition will be different; the transpositions, however, keep the currents balanced.

LEAKAGE CONDUCTANCE. The leakage current, even at very high voltages, is usually negligible in power transmission lines, but for telephone lines the leakage is much greater, on account of the large number of small insulators used, and has a very appreciable effect on both the attenuation and distortion of the voice currents.

When the voltage is sufficiently high on a power line to cause the formation of corona, an appreciable leakage current passes from one wire to the other. Even for a sine-wave voltage this leakage current is by no means sinusoidal, since its instantaneous values are practically zero except during the peak of the voltage wave, and consequently the corona loss cannot be accurately represented by a constant leakage conductance. Roughly, however, calling p_c the average value of the corona loss from each wire in watts per mile, corresponding to the given line voltage, the leakage conductance to neutral in micromhos per mile due to the corona may be taken equal to

$$g_c = \frac{p_c}{V^2} \qquad (26)$$

where V is the effective (rms) kilovolts to neutral.

RISE OF VOLTAGE AT LOAD END OF LINE ON OPEN CIRCUIT. In every a-c transmission line the voltage at the load end when this end is open is higher than at the generator end, although in short low-frequency lines this rise is inappreciable. In overhead lines for which the product

$$(\text{Cycles per second}) \times (\text{Length of line} \,\|\, \text{in miles}) < 10{,}000 \qquad (27)$$

this no-load rise as a percentage of the delivered voltage is, to a close approximation when the resistance of the wire is less than that of a No. 0 B. & S. copper wire, equal to $\left(\frac{fl}{4000}\right)^2$, where f is the frequency in cycles per second, and l the length of the line in miles. For example, in a 25-cycle line 160 miles long this no-load rise is 1 per cent of the delivered voltage; in a 60-cycle line of the same length the no-load rise in voltage is 5.8 per cent of the delivered voltage. The relation expressed by the above formula under the conditions stated is independent of the value of the delivered voltage and of the size and spacing of the wires, at least for all practical cases.

This rise, which is due to the charging current taken by the line, may be looked upon as present at all loads, but when the load is appreciable the voltage drop due to the load current, unless leading, more than offsets this voltage rise. A leading current may increase the rise in voltage at the load end as the load comes on.

SYNCHRONOUS CONDENSERS (OR PHASE MODIFIERS) TO MAINTAIN CONSTANT VOLTAGE AT LOAD END. By making the line current at the load end of the line lead the line voltage by the proper phase angle, it is possible to compensate entirely for the change in the load voltage which normally takes place as the load current increases. An overexcited synchronous motor connected in parallel with the load is sometimes used for this purpose, as described in detail in Section 9, Article 15. A synchronous motor so used is commonly called a synchronous condenser, since the current taken by it leads the voltage impressed on its terminals.

REACTIVE POWER. Because of the fact that transmission and distribution lines must have sufficient capacity to carry the magnetizing current of the load, many utilities

‖ Length of each conductor.

make a charge for the kilovar load, i.e., the reactive component of the kilovolt-amperes. The term **vector power** is used to designate the vector sum of the real power and the reactive power. It may also be defined as the product of the impedance, treated as a complex quantity, and the square of the magnitude of the current.

Reactive power which is absorbed by an inductive load is customarily given the positive (counterclockwise) direction of rotation, although, according to AIEE and IEC definitions current in 1947, the direction is negative. There is however, a movement to change the definition to accord with present practice.

17. METHODS OF CALCULATING VOLTAGE AND POWER LOSS IN A-C LINES

The absolutely rigorous calculation of an a-c line requires that the distributed nature of the inductance and capacity be considered, i.e., that the line be considered as made up of an infinite number of sections such as shown in Fig. 4. However, simpler approximate methods may be employed for nearly all power lines such as are now used, with results sufficiently accurate for all practical purposes. The accuracy of these approximate methods depends primarily upon the frequency and length of the line; the less the value of the product of these two quantities, the simpler is the method which may be employed. Accurate calculation of even a short a-c line with distributed load can be effected only by rather complicated network equations; see under Kirchhoff's laws in Section 3.

LIMITATIONS OF APPROXIMATE METHODS OF A-C LINE CALCULATIONS. The usual approximate methods of calculating a-c lines, in the order of their simplicity, may be designated as (1) the simple impedance method; and (2) the single end-condenser method. For short low-voltage lines, such as distribution lines, the simple impedance method, which entirely neglects the charging current, is usually sufficiently accurate. The single end-condenser method takes the charging current into account in a manner usually sufficiently accurate for all but the longest high-voltage lines. For exact calculations the rigorous method given below should be used. In fact, in border cases it is well to check an approximate solution by this method.

FUNDAMENTAL ASSUMPTIONS ON WHICH FORMULAS GIVEN BELOW ARE BASED. All the formulas given below are based on the assumptions of pure sine-wave currents and voltages and a perfectly balanced system. By using the voltage to neutral instead of the voltage between wires, the formulas are also put in such shape that they may be applied directly to a single-phase, a two-phase four-wire line, or a three-phase three-wire line, a two-phase four-wire line being considered two separate single-phase lines. The fundamental idea in this method of treatment is that each line wire is considered a separate circuit. The return wire shown in the various diagrams is therefore to be considered as having no impedance. This method of treatment of *balanced* polyphase circuits is strictly accurate (see Alternating Currents, Section 3), but when the system is not balanced the circuit must be treated by the more general methods of network calculations; see Section 3. Also, when the voltage and current waves are not sinusoidal, each harmonic must be treated separately as described in Section 3, Article 8.

SIMPLE IMPEDANCE METHOD. This method is based upon the assumption that the capacitance of the line may be neglected entirely, and that the line may be considered simply as an impedance in series with the load, this impedance having a resistance equal to the total resistance of the line conductor and a reactance equal to the total

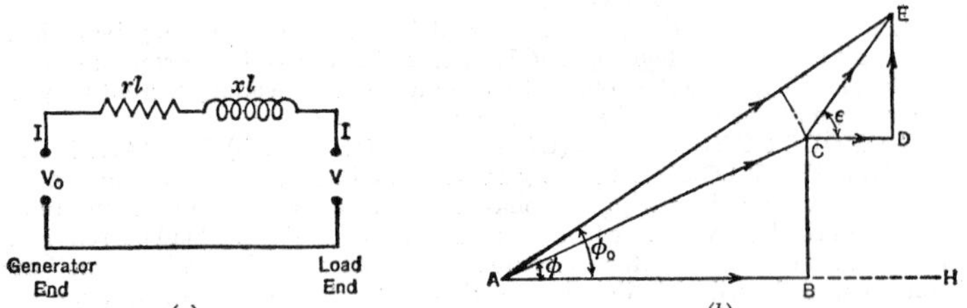

(a) (b)

FIG. 2. Circuit and Vector Diagrams of Simple Impedance Circuit

inductive reactance of the line conductor. Figure 2(*a*) is a diagram of the circuit, and Fig. 2(*b*) is a complete vector diagram of the current and voltage. When the wires are unsymmetrically arranged, either transpositions are assumed or, in the case of a three-phase line, the reactance is based on a conductor spacing equal to the cube root of the product of the three actual spacings.

$$\overline{AH} = I \qquad \overline{BC} = V\sin\phi \qquad \overline{CE} = zlI$$
$$\overline{AC} = V \qquad \overline{CD} = rlI \qquad \overline{AE} = V_0$$
$$\overline{AB} = V\cos\phi \qquad \overline{DE} = xlI \qquad \overline{FE} = V_0 - V$$

where $\overline{AB}$ is the power component of V.
$\overline{BC}$ is its reactive component.
$\overline{AD'}$ is the power component of V_0.
$\overline{D'E}$ is its reactive component.

Let V = volts to neutral at load end of the line (= volts between wires divided by 2 in the case of a single-phase line, and volts between wires divided by $\sqrt{3}$ in the case of a three-phase line).
I = amperes per wire; see eq. (24).
l = length of each line wire in miles.
$Z = \dfrac{V}{lI}$ = "equivalent" impedance of the load per mile of line.
$\cos\phi$ = power factor of the load at end of line.
$\cos\phi_0$ = power factor of line and load.
$\sin\phi = \sqrt{1 - \cos^2\phi}$ = the reactive factor of the load; $\sin\phi$ is to be taken positive for a lagging and negative for a leading current.
V_0 = volts to neutral at the generator end of the line.
r = conductor resistance per mile of line in ohms; see tables in Articles 54 and 55 on Bare Wires and Cables.
x = conductor reactance per mile of line in ohms; see tables in Articles 20 to 23.
$z = \sqrt{r^2 + x^2} = r\cos\epsilon + x\sin\epsilon$ = conductor impedance ¶ per mile of line in ohms.
Q = per cent power loss, as a percentage of delivered power.
D = per cent voltage loss, as a percentage of delivered voltage.

From the vector diagram it is evident that the voltage at the generator end is

$$\overline{V_0} = \overline{V} + \overline{ZlI} \tag{28}$$

or

$$V_0 = \sqrt{(V\cos\phi + rlI)^2 + (V\sin\phi + xlI)^2} \tag{29}$$

which may also be written

$$V_0 = V\sqrt{\left(\cos\phi + \frac{r}{Z}\right)^2 + \left(\sin\phi + \frac{x}{Z}\right)^2} \tag{30}$$

where r and x are per mile of line.

The current at the generator end is the same as at the load end.

The per cent power loss is

$$Q = \frac{100\,rlI}{V\cos\phi} = \frac{100\,r}{Z\cos\phi} \tag{31}$$

and the per cent voltage loss is

$$D = \frac{100(V_0 - V)}{V} = 100\left[\sqrt{\left(\cos\phi + \frac{r}{Z}\right)^2 + \left(\sin\phi + \frac{x}{Z}\right)^2} - 1\right] \tag{32}$$

The power factor at the generator end is

$$\cos\phi_0 = \left(\frac{100 + Q}{100 + D}\right)\cos\phi \tag{33}$$

RELATION BETWEEN IMPEDANCE DROP AND VOLTAGE LOSS. The total impedance drop, which is zlI volts, should be carefully distinguished from the voltage loss, which is $v = V_0 - V$ volts. The vector diagram, 2(*b*), will make the difference clear.

¶ A convenient way of calculating an expression of the form $\sqrt{a^2 + b^2}$ is to write it $a\sqrt{1 + (b/a)^2}$ or $b\sqrt{1 + (a/b)^2}$, according as a is greater or less than b; the expression under the radical will then always lie between the numbers 1 and 2, and no difficulty will be experienced with decimal points. When b/a is less than 0.3, then the expression $\sqrt{a^2 + b^2} = a + (b^2/2a)$ with an error of less than 0.1 per cent; and when a/b is less than 0.3, then $\sqrt{a^2 + b^2} = b + (a^2/2b)$ with an error of less than 0.1 per cent. The error in the approximate expressions diminishes very rapidly as the ratio of b/a or a/b, as the case may be, decreases, being only 0.02 per cent when the ratio is 0.2.

For a given impedance drop of, say, A per cent, the voltage at the load end of the line may be anything from A per cent less than the voltage at the generator end to A per cent greater than the voltage at the generator end. The determining factor is the difference between the power-factor angle (ϕ) of the load and the power-factor angle $\left(\epsilon = \tan^{-1}\frac{x}{r}\right)$ of the line; only when $\epsilon - \phi = 0$ are the voltage loss and impedance drop the same. When $\epsilon - \phi$ is greater than 90 deg (which may occur for a leading current, since ϕ is then negative), the voltage at the load end will in general be higher than at the generator end, although the impedance drop in the line may be very large. As a fair approximation, when the impedance drop is less than 20 per cent, that is, when z/Z is less than 0.2, the percentage voltage loss may be written

$$D = \frac{100z}{Z}\cos(\epsilon - \phi) \tag{34}$$

z and Z being the impedances per mile of line of the line and load respectively, and ϵ and ϕ the power-factor angles of the line and load respectively.

Example of Calculation by Simple Impedance Method. Take the case of a three-phase, 60-cycle line 50 miles long, the wires being No. 0000 A.W.G. (or B. & S.) stranded copper spaced symmetrically with 6 ft between centers, and let the load be 15,000 kw at 60,000 volts between wires and at a power factor of 80 per cent with lagging current. The voltage to neutral is then $60{,}000 \div \sqrt{3} = 34{,}600$, and the current per wire $15{,}000 \div (\sqrt{3} \times 60 \times 0.8) = 180$ amp. The resistance of each wire per mile is 0.269 ohm at 77 deg fahr, and the reactance 0.728 ohm. The equivalent impedance of the load per mile of line is $Z = 34{,}600 \div (50 \times 180) = 3.84$. $\cos\phi = 0.8$, and $\sin\phi = \sqrt{1 - 0.8^2} = 0.6$. Whence

$$\text{Per cent power loss} = Q = \frac{100 \times 0.269}{3.84 \times 0.8} = 8.76 \text{ per cent}$$

$$\text{Per cent voltage loss} = D = 100\left[\sqrt{\left(0.8 + \frac{0.269}{3.84}\right)^2 + \left(0.6 + \frac{0.728}{3.84}\right)^2} - 1\right] = 17.5\%$$

$$\text{Power factor at generator end} = \cos\phi_0 = \left(\frac{100 + 8.76}{100 + 17.5}\right) \times 0.80 = 74.1 \text{ per cent}$$

$$\text{Per cent impedance drop} = \frac{100\sqrt{0.269^2 + 0.728^2}}{3.84} = 20.2 \text{ per cent}$$

Using eq. 34,

$$D = \frac{100 \times 0.775 \times .84}{3.84} = 17.0$$

SIMPLE IMPEDANCE METHOD APPLIED TO DISTRIBUTION CIRCUITS. In the case of distribution circuits consisting of a number of differently loaded sections in series, it is necessary to know the voltage loss in each section in order to obtain, by summation, the total voltage loss in the circuit. Where the loads on the different sections have widely different power factors, the $(V\cos\phi + rlI)$ and the $(V\sin\phi + xlI)$ must be calculated separately for each section, the power components and the reactive components added separately, and the square root of the sum of their squares taken to obtain the station voltage V_0. In practice, however, this is seldom found necessary, and it is customary to make use of the approximate eq. (34), modified as follows to give the voltage loss v:

$$v = lIz(\cos\epsilon - \phi) \tag{35}$$

$$v = lz\frac{kva}{kv}(\cos\epsilon - \phi) \tag{36}$$

In eq. 36 the kva delivered by each section is usually estimated from the product of the connected transformer kva and the demand factor. To be exact, the kv should be the line to neutral kv at the load, but, as that is unknown when V_0 is assumed, the station voltage may be used without appreciable error. The v for the most distant section is first calculated, then the next nearer section, using for it the total accumulated kva, and so on for any number of sections.

Example of Calculation. Assume a circuit consisting of No. 4/0 conductors spaced symmetrically on 6 ft equivalent spacing. Assume that at 15 miles from the substation there is a 50-kva transformer, and at 5 miles there is a 25-kva transformer. Assume, further, that the demand factor on both sections will be 0.6, and the substation circuit voltage 4.5 kv or 2.6 kv to neutral.

Then, as in the previous example,

$$r = 0.269 \qquad x = 0.728 \qquad z = 0.775$$

$$\frac{x}{r} = 2.706 \qquad \tan^{-1} 2.706 = 69.72°$$

$$\phi = \cos^{-1} 0.80 = 36.87° \qquad \epsilon - \phi = 32.85° \qquad \cos 32.85° = 0.84$$

For the most distant section

$$v = 10 \times 0.775 \times \frac{50 \times 0.6}{2.6} \times 0.84 = 75$$

For the near section

$$v = 5 \times 0.775 \times \frac{75 \times 0.6}{2.6} \times 0.84 = 57$$

Total drop per phase 132 volts or 5 per cent

If the transformer ratio is 35 to 1, the equivalent secondary voltage drop will be 3 3/4 volts at the most distant point, between each phase and neutral.

SINGLE END-CONDENSER METHOD.** This method assumes that the total current at the load end is equal to the actual load current plus (vectorially) the charging current which would be taken by a single condenser shunted across the line at the load end, the capacitance of this condenser being taken equal to the total capacitance of the line. This method gives too low a voltage at the generator end by approximately the same amount that the straight impedance method gives it too high, and also gives the power loss too low by approximately the same amount that the straight impedance method gives it too high. By averaging the losses obtained by the two methods a close approximation to their true values is obtained.

Figure 3(a) is a diagram of the circuit, and Fig. 3(b) is a complete vector diagram of the voltage and current; voltages are shown by full lines and currents by dotted lines. The

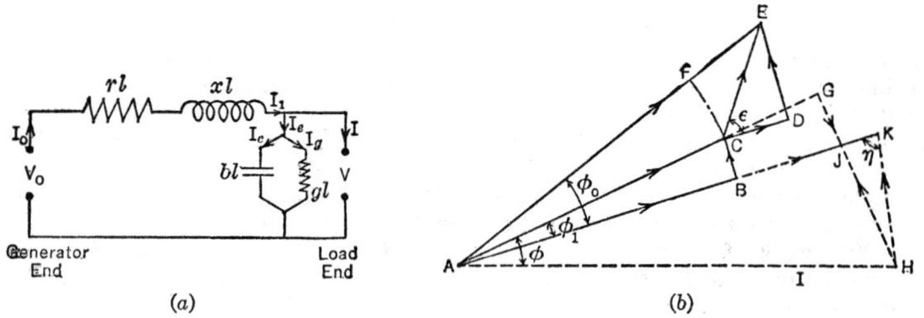

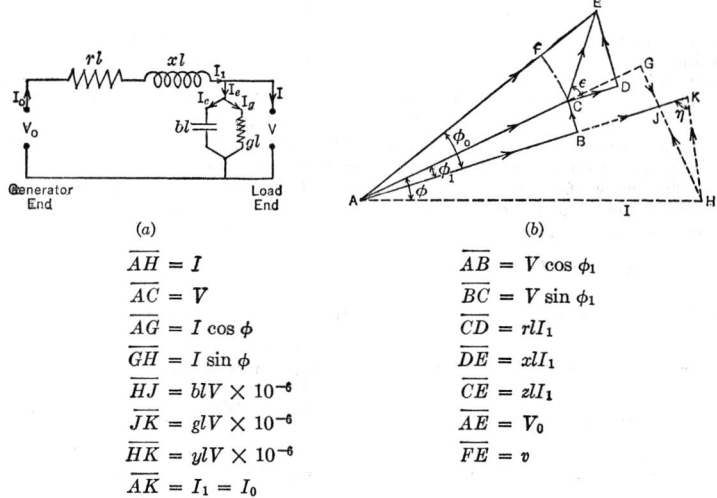

$$\overline{AH} = I \qquad \overline{AB} = V\cos\phi_1$$

$$\overline{AC} = V \qquad \overline{BC} = V\sin\phi_1$$

$$\overline{AG} = I\cos\phi \qquad \overline{CD} = rlI_1$$

$$\overline{GH} = I\sin\phi \qquad \overline{DE} = xlI_1$$

$$\overline{HJ} = blV \times 10^{-6} \qquad \overline{CE} = zlI_1$$

$$\overline{JK} = glV \times 10^{-6} \qquad \overline{AE} = V_0$$

$$\overline{HK} = ylV \times 10^{-6} \qquad \overline{FE} = v$$

$$\overline{AK} = I_1 = I_0$$

FIG. 3. Circuit and Vector Diagrams of Circuit with End Condenser

diagrams and formulas are for the general case of a line with leakage, but for nearly all practical cases the leakage may be neglected.

The effect of the electrostatic capacitance of the line is to change both the numerical value and the phase angle of the line current, or the condenser and the load may be

** See Section 3, Article 18 for equivalent T and π circuits.

looked upon as forming together an equivalent load taking a current I_1 at a power factor $\cos\phi_1$ differing from the actual current and power factor of the load. Let

V = volts to neutral at the load end of the line.
I = actual amperes per wire at the load end.
$\cos\phi$ = actual power factor at the load end.
$\sin\phi = \sqrt{1-\cos^2\phi}$ = actual reactive factor at the load end.
l = length of each line wire in miles.
b = capacitive susceptance to neutral per mile of line, in micromhos; see tables in Article 20.
g = leakage conductance to neutral per mile of line in micromhos, usually taken equal to zero in power lines, as explained above.
$y = \sqrt{g^2+b^2}$ = dielectric admittance to neutral per mile of conductor, in micromhos. Note that for no leakage $y = b$.

The total leakage current, total charging current, and total exciting current of the line are then, respectively,

$$I_g = glV \times 10^{-6} \qquad I_c = blV \times 10^{-6} \qquad I_e = \sqrt{I_g^2 + I_c^2} \tag{37}$$

The total line current, i.e., the resultant of the actual load current and the exciting current, is

$$I_1 = I\sqrt{\left(\cos\phi + \frac{I_g}{I}\right)^2 + \left(\sin\phi - \frac{I_c}{I}\right)^2} \tag{38}$$

On the assumptions of this method of calculation, I_1 is also the current at the generator end.

The power factor of the equivalent load formed by the actual load and the condenser is then

$$\cos\phi_1 = \frac{I\cos\phi + I_g}{I_1} \tag{39}$$

and the reactance factor of this equivalent load is

$$\sin\phi_1 = \frac{I\sin\phi - I_c}{I_1} \tag{40}$$

The equations given above for the straight impedance method are then directly applicable, using for I, $\cos\phi$, and $\sin\phi$ in those equations the values of I_1, $\cos\phi_1$, and $\sin\phi_1$ just calculated; i.e., the straight impedance method is to be applied not to the actual load but to the equivalent load formed by the actual load and a condenser having an admittance equal to the total admittance of the line.

Example of Calculation by Single End-condenser Method. Assume a 3-phase line, 50 miles long, No. 0000 A.W.G. stranded copper, 6 ft between centers, frequency 60 cycles, load 15,000 kw, 60,000 volts between wires at load end, 80 per cent power factor at load, as in the example used for the straight impedance method. Then $V = 34{,}600$, $I = 180$, $\cos\phi = 0.8$, $\sin\phi = 0.6$, $l = 50$, $b = 6.03$, $g = 0$, $r = 0.269$, $x = 0.728$. Then,

Charging current $= I_c = 6.03 \times 50 \times 34{,}600 \times 10^{-6} = 10.4$ amp

Resultant current $= I_1 = 180\sqrt{(0.8)^2 + \left(0.6 - \frac{10.4}{180}\right)^2} = 174$ amp

Power factor of equivalent load $= \cos\phi_1 = \dfrac{180 \times 0.8}{174} = 0.828$

Reactive factor of equivalent load $= \sin\phi_1 = \dfrac{(180 \times 0.6) - 10.4}{174} = 0.561$

Impedance of equivalent load per mile of line $= Z_1 = \dfrac{34{,}600}{50 \times 174} = 3.98$

Per cent power loss $= Q = \dfrac{100 \times 0.269}{3.98 \times 0.828} = 8.16$ per cent

Per cent voltage loss $= D$

$$= 100\left[\sqrt{\left(0.828 + \frac{0.269}{3.98}\right)^2 + \left(0.561 + \frac{0.728}{3.98}\right)^2} - 1\right] = 16.30 \text{ per cent}$$

Power factor at generator end $= \cos\phi_0 = \left(\dfrac{100 + 8.16}{100 + 16.3}\right) \times 0.822 = 76.4$ per cent

The per cent power loss and voltage loss obtained by the straight impedance method neglecting the line capacity are 8.76 and 17.6 per cent, respectively. As noted above, the single end-condenser method gives these losses too low (for inductive loads) by approximately the same amount that the straight impedance method gives them too high, whence closer approximations to the true losses are: per cent power loss = (8.16 + 8.76) ÷ 2 = 8.46 and per cent voltage loss = (16.3 + 17.6) ÷ 2 = 17.0.

CALCULATION OF EFFECT OF SYNCHRONOUS CONDENSER. Equations (38) to (40) apply directly to the calculation of the effect of a synchronous condenser at the end of the line, taking a current having an in-phase or energy component equal to I_g and a quadrature leading component equal to I_c. Figure 3(*b*) then represents the vector relations of the currents and voltages, the vector JK being the in-phase component of the current taken by the synchronous condenser and HJ the quadrature component.

EXACT CALCULATION OF A-C LINES OF ANY LENGTH AND FREQUENCY.†† The charging current (and also the leakage current) for any element of a transmission line passes through only that portion of the line conductor between the given element and the generator end. The exact determination of the line current, voltage, and power at any point therefore requires that this fact be taken into account; in other words, the capacity and leakage of the line are distributed and not lumped, as assumed in the foregoing methods of calculation. In Fig. 4(*a*) are shown three successive elements of one wire of a line, a return or neutral of zero impedance being also shown to complete the circuit. This method of treating separately each wire of a line is fully explained above.

$\overline{MN} = I \qquad \overline{MK_0} = A\epsilon^{\alpha l}$

$\overline{HK} = YV \qquad \overline{MH_0} = B\epsilon^{-\alpha l}$

$\overline{MK} = A \qquad \overline{MN_0} = I_0$

$\overline{MH} = B \qquad \overline{H_0K_0} = YV_0$

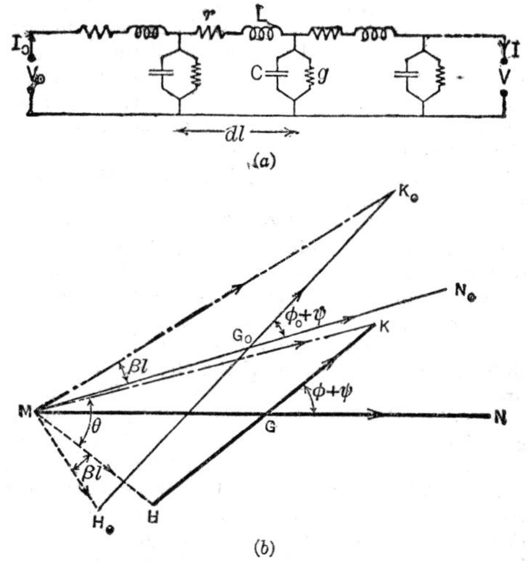

FIG. 4. Circuit and Vector Diagrams of Circuit with Distributed Capacitance and Leakance

In order to make clear the physical meaning of the various terms employed, the general solution is first given in terms of instantaneous values. The working formulas are then given (1) in terms of exponentials, viz., ϵ^x and ϵ^{-x}, and (2) in terms of real hyperbolic functions.

GENERAL EQUATIONS OF TRANSMISSION LINE. Let l = the distance in miles of any point along a transmission line, measured from any arbitrarily chosen point (say, from the load end); let i = the instantaneous current in amperes in the conductor at this point, taken positive in the direction *opposite* to that in which l is measured (i.e., positive when *toward* the load end); let v = the potential drop in volts between this point and the neutral; and let r = the conductor resistance in ohms per mile, L = the inductance of the conductor in henrys per mile, C = the capacity to neutral in microfarads per mile, and g = the leakage conductance to neutral in micromhos per mile. The following relations then hold at any point along the line, t being time in seconds measured from any arbitrarily chosen instant:

$$\frac{dv}{dl} = ri + L\frac{di}{dt} \tag{41}$$

$$10^6\frac{di}{dl} = gv + C\frac{dv}{dt} \tag{42}$$

If the circuit is composed of two or more sections of different constants (e.g., an overhead section and an underground section, or a circuit formed by a step-up transformer, transmission line, and a step-down transformer), then a similar set of equations holds for each section of the circuit, the constants r, L, g, and C being in general different for the several sections.

†† This division is abstracted from the lecture notes of Dr. H. Pender.

The complete solution of any two equations of the form given by (41) and (42) consists of an infinite series of terms for both v and i, corresponding terms in the two series having the following values:

$$i = \sqrt{2}\epsilon^{-(u-s)t}\,[A\epsilon^{\alpha l}\sin(\omega t + \beta l + \theta_1) + B\epsilon^{-\alpha l}\sin(\omega t - \beta l + \theta_2)] \tag{43}$$

$$v = \frac{\sqrt{2}}{Y}\epsilon^{-(u-s)t}\,[A\epsilon^{\alpha l}\sin(\omega t + \beta l + \theta_1 - \psi) - B\epsilon^{-\alpha l}\sin(\omega t - \beta l + \theta_2 - \psi)] \tag{44}$$

PHYSICAL INTERPRETATION AND NAMES GIVEN TO THE VARIOUS CONSTANTS. The constant ω in eqs. (43) and (44) is equal to $2\pi f$, where f is the frequency of the oscillation represented by these equations. In the most general case any change in the circuit conditions, such as closing or opening a switch, or a lightning stroke in the vicinity of the circuit, may set up an infinite number of oscillations of different frequencies, the frequencies being determined by the initial conditions at the instant the change is made. The current and voltage set up by each oscillation are represented by a set of terms of the form given by (43) and (44), and the resultant current and voltage will be the sum of all the current terms and the sum of all the voltage terms, respectively. The oscillation of any given frequency, however, may be considered separately, as it is uninfluenced by the presence of the other oscillations. Moreover, in the case of a composite circuit, consisting say of a step-up transformer, transmission line, and step-down transformer, if an oscillation of frequency f is set up in one part of the circuit, it will also appear in all other sections of the circuit, though it may be greatly damped in these sections and therefore produce no appreciable effect.

Attenuation Constant (α), Wavelength Constant (β), Wavelength (λ), and Velocity of Propagation (U). Referring to eqs. (43) and (44), each oscillation sets up in each section of the circuit two waves, each of which has a wavelength $\lambda = 2\pi/\beta$; the constant β is therefore called the wavelength constant; it is a function of the frequency and of the constants r, L, g, and C of the circuit. The two waves travel along the line in opposite directions, each with a velocity $U = \omega/\beta$; in a composite circuit this velocity U is in general different for each section of the circuit. One wave may be looked upon as the incident and the other as the reflected wave. The amplitude of each wave diminishes by the factor ϵ^{α} as the wave travels unit distance; the factor ϵ^{α} is called the attenuation factor, and the constant α is called the attenuation constant. The attenuation constant is a function of the frequency and the constants r, L, g, and C of the circuit; see below.

Surge Admittance (Y) and Its Power-factor Angle (Ψ). The constant Y, which is equal to the quotient of the amplitude (or rms value) of the incident current wave by the incident voltage wave, is called the surge admittance, and its reciprocal is called the surge impedance; it is a function of the frequency and the constants r, L, g, and C of the circuit; see below. The constant ψ, which is equal to the angle by which the incident current wave leads the incident voltage wave, is called the power factor angle of the surge admittance; it is a function of the frequency and the constants r, L, g, and C; see below.

Amplitude Constants (A and B) and Phase-angle Constants (θ_1 and θ_2). The constants are equal to the amplitudes of the incident and reflected current waves at the point from which the distance l is measured, and the constants θ_1 and θ_2 give the phase of these two waves at this point ($l = 0$) at the instant from which time is measured ($t = 0$). Note that the incident current wave at $l = 0$ leads the reflected current wave by the angle $\theta = \theta_1 - \theta_2$. The determination of these constants for steady-state conditions is given below.

Natural Damping Constant (u), Energy Transfer Constant (s), and Composite Damping Constant ($u - s$). In the general case of a natural oscillation in a composite circuit the amplitude of each wave diminishes in unit time by the factor $\epsilon^{(u-s)}$; this factor is called the composite damping factor, and the constant $(u - s)$ is called the composite damping constant. The composite damping constant, like the frequency f, is the same for all sections of a composite circuit. In the case of a line of uniform constants throughout, not connected to any terminal apparatus, it can readily be shown that $s = 0$, in which case the amplitude of the oscillations diminishes in unit time by the factor ϵ^{u}; this factor is therefore called the natural damping factor, and the constant u, which for a section having the constants r, L, g, and C per unit length is equal to

$$u = \frac{1}{2}\left(\frac{r}{L} + \frac{g}{C}\right) \tag{45}$$

is called the natural damping constant. Any section of a composite circuit for which the actual damping $\epsilon^{(u-s)}$ is less than the natural damping ϵ^{u} must receive energy from some other section; consequently a positive value of s for a given section means that energy is transferred into this section from some other section of the circuit. Similarly, a negative value of s for a given section means that energy is transferred from this section to some other section. The constant s may therefore be called the energy transfer constant. Since the voltage and current in the circuit cannot increase indefinitely, the energy transfer constant s can never have a positive value greater than u.

Since the composite damping constant $(u - s)$ is the same for all sections of a circuit, it follows that the transfer of energy from one section to another by oscillations in a composite circuit will always be into the section in which u is the larger from the section in which u is the smaller. Neglecting the leakage conductance g, this means that energy will be transferred into section 1 from section 2, when r_1/L_1 is greater than r_2/L_2; i.e., energy is transferred from the section of the larger time constant (L_2/r_2) to that of the smaller time constant (L_1/r_1). When the resistances are small, this means that in the limiting case all the energy $(= {}^1/_2\, L_2 I^2)$ of the magnetic field of the second section may go into electrostatic energy $(= {}^1/_2\, C_1 V^2)$ in the first section, producing therefore a very high voltage at the junction point when the inductance L_2 of the second section is large compared with the capacity C_1 of the first section. This accounts for the very high voltages sometimes set up during switching operations at the junction point of an overhead line with an underground line, or in a transformer connected to a long overhead line.

STEADY-STATE CONDITIONS IN A TRANSMISSION LINE. From the above discussion it is evident that, when a sufficient time (usually a small fraction of a second) has elapsed after any change in the circuit conditions, the only terms left in the general equations of a transmission line for a given impressed sine-wave voltage of frequency f are those for which $s = u$, viz.,‡‡

$$i = \sqrt{2}[A\epsilon^{\alpha l} \sin (\omega t + \beta l) + B\epsilon^{-\alpha l} \sin (\omega t - \beta l - \theta)] \tag{46}$$

$$v = \frac{\sqrt{2}}{Y} [A\epsilon^{\alpha l} \sin (\omega t + \beta l - \psi) - B\epsilon^{-\alpha l} \sin (\omega t - \beta l - \theta - \psi)] \tag{47}$$

The effective value of the current at any point is then equal to the sum of two vectors having the lengths $A\epsilon^{\alpha l}$ and $B\epsilon^{-\alpha l}$, the former leading the latter by the angle $(2\beta l + \theta)$, and the effective value of the voltage is equal to the difference of these same two vectors divided by Y. The phase angle between the voltage and current is equal to the phase angle between the sum and difference of the A and B vectors less the angle ψ.

Notation for Steady-state Conditions. These relations are clearly shown in the vector diagram, Fig. 4(b), which is a complete vector diagram of a transmission line with distributed capacity and leakage. The four constants α, β, Y, and ψ are constants of the line, independent of the load, and are expressed in terms of the ordinary line constants as follows: Let

f = frequency in cycles per second.

r = conductor resistance per mile, in ohms; see tables in Articles 54 and 55 of this section.

$x = 2\pi f L$ = conductor reactance per mile, in ohms, corresponding to the impressed frequency f; see tables in Articles 20 to 23.

$z = \sqrt{r^2 + x^2}$ = conductor impedance per mile, in ohms.

g = leakage conductance to neutral per mile of line, in micromhos. For power lines g is usually taken equal to zero.

$b = 2\pi f C$ = capacity susceptance to neutral per mile of line, in micromhos, corresponding to the impressed frequency f; see tables in Article 20 of this section.

$y = \sqrt{g^2 + b^2}$ = dielectric admittance per mile, in micromhos; when $g = 0$, then $y = b$.

$\alpha = 10^{-3}\sqrt{\dfrac{yz - bx + gr}{2}}$ = the attenuation constant; for r and g small compared with x and b, respectively, this reduces to $\dfrac{10^{-3}}{2}\left(r\sqrt{\dfrac{C}{L}} + g\sqrt{\dfrac{L}{C}}\right)$.

$\beta = 10^{-3}\sqrt{\dfrac{yz + bx - gr}{2}}$ = the wavelength constant; for r and g small compared with x and b, respectively, this reduces to $2\pi f \times 10^{-3}\sqrt{LC}$, which for an overhead line equals approximately $\dfrac{2\pi f}{180{,}000}$.

‡‡ For steady-state conditions time may be counted from any arbitrarily chosen interval, i.e., θ_2 in eqs. (43) and (44) may be put equal to zero, and for convenience $-\theta$ may be used for θ_2.

$Y = 10^{-3}\sqrt{\frac{y}{z}}$ = surge admittance; for r and g small compared with x and b, respectively, this reduces to $10^{-3}\sqrt{\frac{C}{L}}$. The reciprocal of the surge admittance is called the surge impedance.

$\psi = \tan^{-1}\sqrt{\frac{yz - bx - rg}{yz + bx + rg}}$ = the power factor angle of the surge admittance, taken positive for $gx < br$ and negative for $gx > br$. For r and g small compared with x and b, respectively, then $\psi = 28.7\left(\frac{r}{x} - \frac{g}{b}\right)$ deg.

$U = \frac{2\pi f}{\beta}$ = velocity of propagation in miles per second; for a frequency f sufficiently high to make r negligible compared with x, and g negligible compared with b, this reduces to $\frac{10^3}{\sqrt{LC}}$, which for an overhead line with wires far apart is equal to the velocity of light in air, viz., 180,000 miles per second, approximately.

$\lambda = \frac{2\pi}{\beta} = \frac{U}{f}$ = wavelength of each wave in miles; for a frequency f sufficiently high to make r negligible compared with x, and g negligible compared with b, this reduces to $\frac{10^3}{f\sqrt{LC}}$, which for an overhead line is equal approximately to $\frac{180{,}000}{f}$ miles.§§

In the vector diagram and in the formulas given below, let

l = length of the line in miles.
I = effective (rms) value of the amperes per wire at the load end.
V = effective (rms) value of the volts to neutral at the load end.
ϕ = the power-factor angle at the load end; i.e., $\cos\phi$ is the power factor at the load end. ϕ is taken positive for a lagging and negative for a leading current.

I_0, V_0, ϕ_0 = corresponding quantities at the generator end.

Solution by Vector Diagram [Fig. 4(*b*)]. Having calculated the constants α, β, Y, and ψ, and knowing the current I, voltage V, and power-factor angle ϕ of the load, lay off $\overline{MN} = I$ as the base line, and bisect it at G.

At the angle $(\phi + \psi)$ ahead of $\overline{MN}$ lay off the line $\overline{HK}$ equal in length to YV, so that it is also bisected by G.

Then measure off $\overline{MK} = A$ and $\overline{MH} = B$.

Lay off at the angle 57.3 βl deg ahead of A the line $\overline{MK_0}$ equal in length to $A\epsilon^{\alpha l}$. (See Section 1 for tables of ϵ^x and ϵ^{-x}.)

Lay off at the angle 57.3 βl deg behind B the line $\overline{MH_0}$ equal in length to $B\epsilon^{-\alpha l}$.

Bisect H_0K_0 at G_0.

Then the line $\overline{MN_0} = 2\overline{MG_0}$ is equal to the current at the generator end; the line $\overline{H_0K_0}$ divided by Y is equal to the voltage at the generator end; the angle between $\overline{G_0N_0}$ and $\overline{G_0K_0}$ less the angle ψ is the power-factor angle at the generator end.

Note that the voltage at the load end, if drawn in the diagram, would be at the angle ψ behind the vector $\overline{GK}$, and the generator end would be at the angle ψ behind $\overline{G_0K_0}$.

Algebraic Solution for Steady-state Conditions. The vector diagram may be solved algebraically as follows: Calculate first the constants A, B, and θ from the formulas:

$$A = \tfrac{1}{2}\sqrt{I^2 + (YV)^2 + 2YVI\cos(\phi + \psi)} \tag{48}$$

$$B = \tfrac{1}{2}\sqrt{I^2 + (YV)^2 - 2YVI\cos(\phi + \psi)} \tag{49}$$

$$\theta = \tan^{-1}\left[\frac{2YVI\sin(\phi + \psi)}{I^2 - (YV)^2}\right] \tag{50}$$

§§ The above formulas for z, y, α, β, Y, ψ, U, and λ also hold for the transient or free oscillations in a single circuit, and also for each section of a composite circuit, provided in these formulas $r_1 = r(u - s)L$ is substituted for r, and $g_1 = g \div (u - s)C$ is substituted for g.

Note that $\sin \theta$ has the same algebraic sign as the numerator of this fraction, and $\cos \theta$ has the same algebraic sign as the denominator of this fraction; this fixes the quadrant in which θ lies.

Put

$$A_0 = A\epsilon^{\alpha l}, \quad \text{and} \quad B_0 = B\epsilon^{-\alpha l} \tag{51}$$

The current, voltage, and power-factor angle at the generator end are, then, expressing all angles in degrees,

$$I_0 = \sqrt{A_0^2 + B_0^2 + 2AB \cos (114.6\beta l + \theta)} \tag{52}$$

$$V_0 = \frac{1}{Y}\sqrt{A_0^2 + B_0^2 - 2AB \cos (114.6\beta l + \theta)} \tag{53}$$

$$\phi_0 = \tan^{-1}\left[\frac{2AB \sin (114.6\beta l + \theta)}{A_0^2 - B_0^2}\right] - \psi \tag{54}$$

Note that the quadrant in which $(\phi_0 + \psi)$ lies is determined by the algebraic signs of the numerator and denominator of the fraction in the brackets, just as in the case of the angle θ.

Solution of Steady-state Conditions in Terms of Hyperbolic Functions. (See Section 1 for tables.) The above expressions for the current, voltage, and power factor at the generator may also be put in the form,

$$I_0 = I\sqrt{\frac{\cosh (2\alpha l + \gamma) + \cos (114.6\beta l + \theta)}{\cosh \gamma + \cos \theta}} \tag{55}$$

$$V_0 = V\sqrt{\frac{\cosh (2\psi l + \gamma) - \cos (114.6\beta l + \theta)}{\cosh \gamma - \cos \theta}} \tag{56}$$

$$\phi_0 = \tan^{-1}\left(\frac{\sin (114.6\beta l + \theta)}{\sinh (2\alpha l + \gamma)}\right) - \psi \tag{57}$$

where θ and γ are given by the formulas

$$\gamma = \tanh^{-1}\left(\frac{YVI \cos (\phi + \psi)}{I^2 + (YV)^2}\right) \tag{58}$$

$$\theta = \tan^{-1}\left(\frac{2YVI \sin (\phi + \psi)}{I^2 - (YV)^2}\right) \tag{59}$$

The other quantities are as above defined. Note that θ is the same angle as given by eq. (50), and the quadrant in which it lies is to be determined as described in the note under (50). Also note that the constant γ given by (58) may be expressed in terms of A and B, given by (48) and (49), by means of the formula

$$\gamma = \log_\epsilon\left(\frac{A}{B}\right) = 2.302 \log_{10}\left(\frac{A}{B}\right) \tag{60}$$

Formulas for Open Circuit and Short Circuit at Load End. When the line is open at the load end, $I = 0$, whence, from eqs. (48) to (50), $A = {}^1/_2 YV$, $B = {}^1/_2 YV$, and $\theta = 180$ deg (since the denominator of the fraction is negative). When the line is short-circuited at the load end, $V = 0$, and $A = {}^1/_2 I$, $B = {}^1/_2 I$, and $\theta = 0$ deg (since the denominator of the fraction is positive). The current, voltage, and power-factor angle at any point along the line may then be found in either case by substituting these values in eqs. (52) to (54), which reduce to the simple hyperbolic forms:

On Open Circuit	On Short Circuit
$I_0 = YV\sqrt{\sinh^2 (\alpha l) + \sin^2 (57.3\ \beta l)}$	$I_0 = I\sqrt{1 + \sinh^2 (\alpha l) - \sin^2 (57.3\ \beta l)}$
$V_0 = V\sqrt{1 + \sinh^2 (\alpha l) - \sin^2 (57.3\ \beta l)}$	$V_0 = \frac{I}{Y}\sqrt{\sinh^2 (\alpha l) + \sin^2 (57.3\ \beta l)}$
$\phi_0 = -\tan^{-1}\left[\frac{\sin (114.6\ \beta l)}{\sinh (2\ \alpha l)}\right] - \psi$	$\phi_0 = +\tan^{-1}\left[\frac{\sin (114.6\ \beta l)}{\sinh (2\ \alpha l)}\right] - \psi$

18. CORONA

Original Author, F. W. Peek
Revisions by W. A. Del Mar

If a potential difference is established between smooth parallel wires, or between concentric cylinders, and gradually increased, a voltage is finally reached at which a hissing noise is heard. If it is dark, a pale glow, called the electric "corona," will be seen to surround the wires. When wattmeters are inserted in the line, a loss is noticed to start at this critical voltage. The characteristic odor of ozone is also noted, and the air around the wires becomes ionized. An oscilloscope will often reveal high-frequency oscillations of the charging current. These phenomena are referred to as "corona effects."

The corona starts at the conductor surface, because the voltage stress per unit distance, or "voltage gradient," is highest there. The breakdown or corona extends out to a point where the stress is below the breakdown point of air. If the wires are close together, so that the stress is fairly uniform, the breakdown immediately extends from conductor to conductor without corona, and the phenomenon is called a "sparkover." The corona does not extend to the other conductor when the separation is large compared to the wire diameter, because it reaches a point in space where the gradient is below the breakdown gradient of air. Increasing the voltage after the corona point is reached causes the corona to extend until finally a spark occurs from conductor to conductor. The power loss increases very rapidly with increase in voltage above the critical point.

Corona is caused by either a-c or d-c voltages and starts at approximately the same maximum stress. At the critical voltage, corona occurs only at the crest of the alternating wave. As the voltage is increased above the critical point, the loss extends over a greater portion of the wave. In the a-c corona the eye sees a superposition of the corona caused by the plus and minus half-cycles of the a-c waves. If the effects of the half-waves are viewed separately, it is noticed that a reddish haze surrounds the wire while it is negative, whereas the surface of the wire glows bluish-white while it is positive. Positive and negative d-c corona have exactly the same appearance as the positive and negative corona of the corresponding a-c half-waves. When a negative wire becomes corroded, reddish tufts appear. If there are rough spots on the wire, the corona starts at these points at a lower voltage. At points the positive corona extends out as a bluish-white spray; the negative corona appears as a red tuft.

Corona is also caused by transient voltages. Corona produced by voltages lasting less than a millionth of a second can be seen by the eye, and distinction can be made between a positive and a negative half-wave.

LAWS OF CORONA FORMATION. The laws of corona formation have been quite definitely worked out. The chief factors affecting corona formation are:

For a given spacing and air density the corona starting voltage is lowered by decreasing the wire diameter.

Increasing the spacing increases the corona starting voltage, but the effect of changing the diameter is relatively much greater.

Decreasing barometric pressure decreases air density and decreases the corona starting voltage.

Increasing temperature decreases air density and decreases the corona starting voltage.

Dirt, water, etc., on the conductor surfaces lower the corona starting voltage by increasing the stress.

The *apparent strength* of air is not constant in irregular fields but is greater at the surface of small conductors than large ones. The strength at the start of corona is always constant, however, at a distance from the conductor which is a known function of the radius, known as the **energy distance.** This corresponds to the strength in a uniform field of 76 kv per in. (maximum value in case of a-c voltage), which seems to mean that the actual strength of air is 76 kv per in. In order, however, that a finite thickness only may be brought to this stress in an irregular field, the stress at the conductor surface must be higher. At a given spacing the corona always starts at a lower voltage on small wires than large ones, because, although the *apparent strength* of air at the small conductor surface is greater, the stress produced by a given voltage is relatively much higher for the smaller wires.

Corona does not start at exactly the same voltages when the wires are alternately plus and minus. The difference is at most a few per cent and is greater for small wires than large ones.

For perfectly clean, smooth wires of uniform diameter, there is no loss until the *visual critical voltage* is reached, when the loss assumes a *definite value* and increases as a function of the ratio of the applied voltage and the *disruptive critical voltage.* The *visual critical voltage* takes into account the *apparent strength* of the air, which is a function of the wire diameter; the *disruptive critical* voltage corresponds to the constant strength of air for uniform fields, which is about 76 kv per in., maximum value. The *visual critical voltage* is always higher than the *disruptive critical voltage. Because of irregularities there is always a loss below the visual critical voltage.*

Corona loss increases with increasing frequency, being proportional to the frequency, between 25 and 60 cycles.

The formulas for calculating corona on transmission lines are given below.

CORONA ON TRANSMISSION LINES. It is of great importance in the design of high-voltage transmission lines to know the various factors that affect corona formation and to be able to estimate accurately the starting voltage and loss for a given line.

The various characteristics may be calculated from the following formulas, based on the work of F. W. Peek, as modified by William S. Peterson, A. Nuttall, V. Siegfried, J. S. Carroll, and B. Cozzens, *Trans. AIEE*, Vol. 52, pp. 55–63 (1933).

Let e = rms value of the kilovolts to neutral ($= 1/\sqrt{3}$ kv between wires for three-phase; $= 1/2$ kv between wires for single-phase).

e_0 = disruptive voltage in kilovolts to neutral, at which the voltage gradient at the surface of the conductor is equal to the air breakdown value.

t = temperature of the conductor in degrees fahrenheit.

b = barometric pressure in inches.

$\delta = \dfrac{17.9b}{459 + t}$ = specific gravity of air, referred to air at 77 deg fahr and 29.9 in. barometric pressure.

r = radius of conductor in inches.

s = distance between conductor centers in inches.

f = frequency in cycles per second.

$g_0 = 53.6$ = disruptive critical gradient, kilovolts per inch, rms value.

$$g_v = g_0\,\delta\left(1 + \frac{0.189}{\sqrt{\delta r}}\right) = \text{visual critical gradient, kilovolts per inch, rms value.} \quad (61)$$

Then, for smooth round conductors, the **disruptive critical voltage**, rms value of kilovolts to neutral, is

$$e_0 = 2.302 m g_0 r \delta^{2/3} \log_{10}\left(\frac{s}{r}\right) \quad (62)$$

where m is the **roughness factor** and has the following values:

Condition of Conductor	m
New, unwashed conductors	0.67 to 0.74
Washed with a grease solvent *	0.912 to 0.93
Scratch-brushed	0.885
Buffed	1.00
Dragged and dusty	0.72 to 0.75
Weathered (5 months)	0.945
Weathered wire at low humidity	0.92
Same, at night, as low as	0.78
For general design	0.87 to 0.90

* It is found that die grease greatly increases corona loss.

For **stranded conductors** having 12 or more strands in the outer layer, the disruptive critical voltage, rms value of kilovolts to neutral, is

$$e_0 = \frac{123.4\delta^{2/3} m\left[\log_{10}\dfrac{s}{cr_1} + (n-1)\log_{10}\dfrac{2s}{(d-2cr_1)}\right]}{\dfrac{1}{cr_1} + (n-1)/(d-2cr_1)} \quad (63)$$

where r_1 is the radius of an individual strand, n is the number of strands in the outside layer, d is the overall diameter of the conductor, and c is given by

$$c = \left[\frac{1 - \sin\left(\dfrac{\pi}{2} + \dfrac{\pi}{n}\right)}{\left(\dfrac{\pi}{2} + \dfrac{\pi}{n}\right)}\right] \quad (64)$$

For **stranded conductors** having 6 outside strands,

$$e_0 = \frac{123.4\delta^{2/3} rm\left[\left(\log_{10}\dfrac{s}{r}\right) + 0.0677\right]}{1.37} \quad (65)$$

The **visual critical voltage,** rms value of kilovolts to neutral, is given approximately by

$$e_v = 2.302 m_v g_v r \log_{10}\left(\frac{s}{r}\right) \quad (66)$$

where m_v, for polished wires, = 0.72 for local corona and = 0.82 for decided corona all along cable.

The **power loss in fair weather,** in kilowatts per mile of single conductor, is

$$p = \frac{0.0000337fe^2F}{\left(\log_{10}\frac{s}{r}\right)^2} \tag{67}$$

where F is a function of e/e_0, having the values given in Table 4.

Table 4. Empirical Corona Loss Function, *F*

$\frac{e}{e_0}$	F	$\frac{e}{e_0}$	F	$\frac{e}{e_0}$	F	$\frac{e}{e_0}$	F	$\frac{e}{e_0}$	F
0.6	0.011	1.4	0.32	2.2	8.4	6.0	22	12	28
0.8	0.018	1.5	0.92	2.4	9.8	7.0	24	14	29
1.0	0.036	1.6	2.70	3.0	10.5	8.0	26	16	29
1.2	0.083	1.8	5.1	4.0	16.5	9.0	27	18	29
1.3	0.140	2.0	7.0	5.0	19.0	10.0	28	20	30

Methods of Increasing Size of Conductors. For equal conductivity an aluminum conductor has about a 25 per cent greater diameter than a copper conductor and, therefore, approximately 25 per cent higher critical voltage. The advantage of aluminum may be still further increased by the addition of a steel cable core. Aluminum, however, is more easily nicked or scratched than copper during installation, thereby losing some of this advantage.

Copper cables of large diameter are now made with tubular strands, or with ordinary strands on a twisted I-beam core, or in annular form with interlocked or keyed strands. (See Article 59.)

ARRANGEMENT OF CONDUCTORS. In the foregoing equations it has been assumed for three-phase lines that the conductors are so arranged that they form the edges of an equilateral prism. When the conductors are not so arranged, but are placed symmetrically in a plane, corona will start at a lower voltage on the center conductor than on the outside conductors. The actual critical voltage on the center conductor will be approximately 4 per cent lower, and on the outside conductors 6 per cent higher, than for the equilateral prism arrangement with the same spacing.

VOLTAGE VARIATION ALONG LINE. In practice, because of the drop in voltage along the line, the corona loss will be different at different points on the line. This may be allowed for by calculating the loss per mile at various points, and plotting a curve with loss per mile as ordinates and length in miles as abscissas. The area of this curve then represents the total loss.

If an insulated system is operating near the critical corona voltage, and one conductor becomes grounded, the corona loss will be quite high.

THE CORONA LIMIT OF HIGH-VOLTAGE TRANSMISSION LINES. SAFE AND ECONOMICAL VOLTAGES. It will generally be found that it is safe and economical to operate a line up to, but not above, the fair-weather value of the disruptive critical voltage, i.e., up to the value of e_0 given by eq. (62), (63) or (65), for average barometer and summer temperatures. This will give loss during storms, but as storms do not extend over the whole line at one time, this loss generally will not be serious. During cold weather the critical voltage will be higher, and the storm loss less.

In addition to the loss of energy, corona loss may be undesirable from another standpoint: since the loss occurs only on part of the wave, it may introduce harmonics if it is excessive.

Table 5. Approximate Altitude Correction Factors at 25 Deg Cent

Altitude		Correction Factor	Altitude		Correction Factor
Feet	Meters		Feet	Meters	
0	0	1.00	5,000	1525	0.87
500	152	0.99	6,000	1830	0.86
1000	305	0.97	7,000	2135	0.84
1500	459	0.96	8,000	2400	0.82
2000	610	0.95	9,000	2745	0.80
2500	765	0.94	10,000	3050	0.77
3000	915	0.93	12,000	3660	0.74
4000	1220	0.91	14,000	4270	0.70

Table 5 gives the factors by which these voltages must be multiplied to give the corresponding disruptive critical voltage at various elevations This correction factor is equal to the ratio of the barometric pressure at the given altitude to the barometric pressure at sea level.

Special Calculations. For small conductors the loss, in kilowatts per mile of conductor, is given by the expression,

$$p = \frac{390}{\delta}(f + 25)\sqrt{\frac{r + \frac{0.93}{s} + 0.016}{s}}(e - e_d)^2 \times 10^{-5} \qquad (68)$$

where $$e_d = 2.302 m g_d r \log_{10}\left(\frac{s}{r}\right)$$

and $$g_d = g_0\delta\left[1 + \frac{0.189}{\sqrt{\delta r}}\frac{1}{(1 + 1480r^2)}\right]$$

19. BIBLIOGRAPHY

Calculations

Clarke, E., *Circuit Analysis of A-c Power Systems.* John Wiley (1943).
MIT Staff, *Electric Circuits.* John Wiley (1943).
Nesbit, W., *Electrical Characteristics of Transmission Lines*, E. Pittsburgh (1926).
Pernot, F. E., *Electrical Phenomena in Parallel Conductors.* John Wiley (1918).
Richardson, D. E., *Electrical Network Calculations.* Van Nostrand (1946).
Russell, A., *Alternating Currents.* Cambridge Univ. Press (1914).
Waddicor, H., *Principles of Electric Power Transmission and Distribution.* John Wiley (1935).
Westinghouse Engineers, *Electrical Transmission and Distribution Reference Book.* E. Pittsburgh (1944).
Woodruff, L. F., *Principles of Electric Power Transmission.* John Wiley (1938).

Corona

Peek, F. W., *Dielectric Phenomena in High Voltage Engineering.*
Lewis, W. W., Some corona tests, *Gen. Elec. Rev.*, Vol. 23, pp. 419–426 (May, 1920).

Also numerous articles by F. W. Peek, J. B. Whitehead, H. J. Ryan, H. H. Henline, R. Wilkins, W. S. Peterson, W. D. Weidlein, J. C. Clark, F. F. Evenson, C. F. Harding, and others in the *Trans. AIEE*, 1908 to date. There was an important group of papers in 1924, dealing especially with 220,000-volt lines, and another in 1933.

CAPACITANCE, INDUCTANCE, AND SEQUENCE IMPEDANCE

By W. A. Del Mar

20. CAPACITANCE

See also section on electricity in Eshbach, *Fundamentals of Engineering.*

OVERHEAD WIRES. Single Round Wire Parallel to the Ground.

H = height of wire above ground.
d = diameter of wire, in same unit as H.

$$\text{The capacitance} = \frac{7.354 \times 10^{-3}}{\log_{10}\frac{4H}{d}} \text{ microfarads per 1000 ft} \qquad (1)$$

Two Parallel Round * Wires.

D = distance apart, center to center.
d = diameter of wire, in same unit as D.

The exact capacitance †

$$= \frac{8.467 \times 10^{-3}}{\cosh^{-1}\frac{D}{d}} \text{ microfarads per 1000 ft} \qquad (2)$$

* When the wires are far apart compared with the linear dimensions of their cross-section, eqs. 2 and 3 also apply approximately to wires of any shape of cross-section, provided d is taken equal to the perimeter of the cross-section divided by π, i.e., equal to the "equivalent" diameter of the cross-section.

† Taking into account the non-uniform distribution of the charge on each wire; see Pender and Osborne, *Elec. World*, Vol. 56, p. 667 (1910).

When D is greater than $10d$, the following formula for the capacitance *between wires* may be used instead of the above with an error of less than 0.1 per cent:

$$= \frac{3.677 \times 10^{-3}}{\log_{10} \dfrac{2D}{d}} \text{ microfarads per 1000 ft} \qquad (3)$$

The *capacitance to neutral* in all cases is $C_0 = 2C$. The capacitances to neutral for various sizes of wires and various spacings are given in Tables 1 and 2. Note that these tables and the above formulas are strictly applicable to ordinary overhead lines only when the distance from the wires to other conductors, particularly the earth, is large compared with their distance apart. However, the effect of the earth is usually small in most practical cases (see below), and the formulas and tables give a very fair approximation to the actual capacitances.

The capacitances of standard strands given in the following tables are calculated by the same formula as for smooth round wires, using for the diameter d the diameter of the stranded conductor; see Articles 54 to 62 on Bare Wires and Cables. The values as thus calculated are therefore not exact, but the error is probably less than 3 per cent for all practical cases.

Effect of the Earth on the Capacity of Two Overhead Wires. If both wires are at the same height, the effect of the earth is to increase the capacitance by an amount equal to

Table 1. Capacitance to Neutral * of Smooth Round Wires

Microfarads per 1000 FEET of each wire of a single-phase or of a symmetrical three-phase line

Size of Wire, A.W.G.	Diam. of Wire, inches	Inches between Wires, center to center							
		1	3	6	9	12	18	24	30
0000	0.4600	0.01199	0.006608	0.005192	0.004618	0.004282	0.003884	0.003643	0.003477
000	0.4096	0.01099	0.006317	0.005013	0.004477	0.004161	0.003783	0.003555	0.003396
00	0.3648	0.01016	0.006055	0.004847	0.004344	0.004045	0.003688	0.003470	0.003319
0	0.3249	0.009458	0.005812	0.004692	0.004218	0.003936	0.003597	0.003390	0.003245
1	0.2893	0.008855	0.005587	0.004546	0.004100	0.003833	0.003511	0.003313	0.003174
2	0.2576	0.008332	0.005381	0.004408	0.003988	0.003735	0.003428	0.003239	0.003107
4	0.2043	0.007455	0.005010	0.004157	0.003781	0.003553	0.003274	0.003102	0.002980
6	0.1620	0.006753	0.004688	0.003933	0.003595	0.003388	0.003134	0.002975	0.002863
8	0.1285	0.006177	0.004406	0.003732	0.003426	0.003238	0.003005	0.002859	0.002755
10	0.1019	0.005693	0.004155	0.003551	0.003273	0.003100	0.002886	0.002751	0.002655
12	0.08081	0.005277	0.003931	0.003386	0.003132	0.002974	0.002776	0.002651	0.002562
14	0.06408	0.004921	0.003730	0.003235	0.003003	0.002858	0.002675	0.002558	0.002475

Size of Wire, A.W.G.	Feet between Wires, center to center								
	3	4	5	6	8	10	15	20	25
0000	0.003351	0.003171	0.003043	0.002947	0.002806	0.002706	0.002542	0.002436	0.002361
000	0.003276	0.003103	0.002981	0.002889	0.002753	0.002657	0.002498	0.002396	0.002323
00	0.003204	0.003039	0.002922	0.002833	0.002702	0.002610	0.002456	0.002358	0.002287
0	0.003135	0.002977	0.002864	0.002779	0.002653	0.002564	0.002416	0.002320	0.002251
1	0.003069	0.002917	0.002809	0.002727	0.002606	0.002520	0.002376	0.002284	0.002217
2	0.003006	0.002860	0.002756	0.002677	0.002560	0.002477	0.002338	0.002249	0.002184
4	0.002887	0.002752	0.002656	0.002582	0.002474	0.002396	0.002266	0.002182	0.002121
6	0.002777	0.002652	0.002563	0.002494	0.002392	0.002319	0.002197	0.002118	0.002061
8	0.002676	0.002559	0.002476	0.002412	0.002317	0.002248	0.002133	0.002059	0.002004
10	0.002581	0.002473	0.002395	0.002335	0.002245	0.002181	0.002073	0.002002	0.001951
12	0.002493	0.002392	0.002319	0.002262	0.002178	0.902118	0.002016	0.001949	0.001900
14	0.002411	0.002316	0.002247	0.002194	0.002115	0.002058	0.001961	0.001898	0.001852

* The capacitance *between* wires equals one-half the values given in this table.

the increase in capacitance which would result from decreasing the distance between the wires from the actual distance D to the "equivalent" distance D'.

$$D' = \frac{D}{\sqrt{1 + \left(\frac{D}{2H}\right)^2}} = \text{"equivalent" distance apart} \qquad (4)$$

Three Parallel Wires.
Equilateral Triangle Arrangement.

D = distance apart, center to center.
d = diameter of wire in same units as D.

The normal capacitance between any two of the wires with the third wire insulated is

$$\frac{3.677 \times 10^{-3}}{\log_{10} \frac{2D}{d}} \text{ microfarads per 1000 ft} \qquad (5)$$

which is the same as the capacitance between two parallel wires by themselves; see above.

Equilateral Triangle Arrangement with Balanced Three-phase Voltages. For sine-wave voltages between the wires equal in effective value to V and differing in phase by

Table 2. Capacitance to Neutral * of Standard Stranded Conductors

Microfarads per 1000 FEET of each conductor of a single-phase or of a symmetrical three-phase line

Size of Cable, C.M. or A.W.G.	Diam. of Strand, inches	Inches between Conductors, center to center							
		1	3	6	9	12	18	24	30
1,000,000	1.152	....	0.0105	0.00725	0.00617	0.00558	0.00492	0.00454	0.00428
750,000	0.998	....	0.00959	0.00683	0.00586	0.00533	0.00472	0.00437	0.00414
500,000	0.814	0.0245	0.00856	0.00630	0.00547	0.00501	0.00447	0.00415	0.00394
350,000	0.681	0.0181	0.00783	0.00591	0.00517	0.00476	0.00427	0.00398	0.00378
250,000	0.575	0.0147	0.00725	0.00558	0.00492	0.00454	0.00409	0.00383	0.00364
0 000	0.528	0.0135	0.00699	0.00542	0.00480	0.00444	0.00401	0.00376	0.00358
000	0.470	0.0122	0.00666	0.00523	0.00465	0.00431	0.00390	0.00366	0.00349
00	0.418	0.0112	0.00637	0.00504	0.00450	0.00418	0.00380	0.00357	0.00341
0	0.373	0.0103	0.00610	0.00488	0.00437	0.00407	0.00371	0.00349	0.00333
1	0.332	0.00958	0.00586	0.00472	0.00424	0.00396	0.00361	0.00341	0.00326
2	0.292	0.00891	0.00561	0.00456	0.00411	0.00384	0.00352	0.00332	0.00318
4	4.232	0.00790	0.00520	0.00429	0.00389	0.00365	0.00336	0.00318	0.00305
6	0.184	0.00712	0.00486	0.00405	0.00369	0.00348	0.00321	0.00304	0.00293

Size of Cable, C.M. or A.W.G.	Feet between Conductors, center to center								
	3	4	5	6	8	10	15	20	25
1,000,000	0.00410	0.00383	0.00365	0.00351	0.00331	0.00317	0.00295	0.00281	0.00271
750,000	0.00396	0.00371	0.00354	0.00341	0.00322	0.00309	0.00288	0.00274	0.00265
500,000	0.00378	0.00355	0.00339	0.00327	0.00310	0.00298	0.00278	0.00266	0.00257
350,000	0.00363	0.00342	0.00328	0.00316	0.00300	0.00289	0.00270	0.00258	0.00250
250,000	0.00351	0.00331	0.00317	0.00307	0.00292	0.00281	0.00263	0.00252	0.00244
0 000	0.00345	0.00326	0.00312	0.00302	0.00287	0.00277	0.00260	0.00249	0.00240
000	0.00337	0.00318	0.00306	0.00296	0.00282	0.00272	0.00255	0.00245	0.00237
00	0.00329	0.00312	0.00299	0.00290	0.00276	0.00267	0.00251	0.00240	0.00233
0	0.00322	0.00305	0.00293	0.00284	0.00271	0.00262	0.00247	0.00237	0.00229
1	0.00315	0.00299	0.00288	0.00279	0.00266	0.00257	0.00242	0.00233	0.00226
2	0.00308	0.00292	0.00281	0.00273	0.00261	0.00252	0.00238	0.00229	0.00222
4	0.00295	0.00281	0.00271	0.00263	0.00252	0.00244	0.00230	0.00222	0.00215
6	0.00284	0.00271	0.00261	0.00254	0.00244	0.00236	0.00223	0.00215	0.00209

* The capacitance *between* conductors equals one-half the values given in this table.

120 deg, the following relations hold between the *effective* values of the voltages and charging currents for each of the three wires.

$$I = 2\pi f C_0 V_0 \tag{6}$$

where $V_0 = \dfrac{V}{\sqrt{3}}$ = voltage to neutral.

$$C_0 = 2C_{12} = \text{capacity to neutral}$$

when the C's are in microfarads and the charging current in microamperes.

The charging current for any wire is 90 deg ahead of the voltage drop from that wire to the neutral. For the same voltage V between wires in a single-phase system as in a three-phase balanced system, the charging current per wire in the three-phase system with the equilateral triangle arrangement of wires is $2/\sqrt{3} = 1.155$ times the charging current per wire in the single-phase system.

For any other arrangement of wires and for an unbalanced three-phase system, the general equations in Eshbach, *Fundamentals of Engineering*, may be used.

Effect of the Earth on the Capacity of Three Overhead Wires. Ordinarily the formulas given above, which neglect the effect of the earth, are sufficiently accurate for all practical purposes; compare with the effect of the earth on a two-wire line. See below under Zero-sequence Capacitance, and Table 1 on p. 14-34.

CABLES. Single-conductor Cable.

Round Wire in Concentric Sheath.

D = inside diameter of sheath.
d = diameter of conductor (both in same units).
K = specific inductive capacity of dielectric.

The capacitance of a length of cable long compared to its diameter

$$= \frac{7.354 \times 10^{-3} K}{\log_{10} \dfrac{D}{d}} \text{ microfarads per 1000 ft} \tag{7}$$

Round Wire in Eccentric Sheath.

a = radius of wire.
b = inside radius of sheath.
c = distance between center of wire and center of sheath (all in same units).

The ratio of the capacitance, eccentric, to the capacitance, concentric,

$$= \frac{\cosh^{-1}\left(\dfrac{b^2 + a^2}{2ab}\right)}{\cosh^{-1}\left(\dfrac{b^2 + a^2 - c^2}{2ab}\right)} \tag{8}$$

Two-conductor Cable.

D = inside diameter of sheath.
a = distance between centers of wires.
d = diameter of wires (all in same units).

Assuming the entire space between conductors and sheath to have uniform specific inductive capacity K, the capacitance from conductor to conductor

$$= \frac{3.677 \times 10^{-3} K}{\log_{10}\left(\dfrac{2a}{d} \cdot \dfrac{D^2 - a^2}{D^2 + a^2}\right)} \text{ microfarads per 1000 ft} \tag{9}$$

Three-conductor Cable.

D = inside diameter of sheath.
a = distance between centers of wires.
d = diameter of wires (all in same units).

Assuming the entire space between conductors and sheath to have uniform specific inductive capacity K, the capacitance between one conductor and any other one

$$= \frac{3.677 \times 10^{-3} K}{\log_{10} \dfrac{2a\rho}{d}} \text{ microfarads per 1000 ft} \tag{10}$$

where

$$\rho = \sqrt{\frac{(3D^2 - 4a^2)^3}{(3D^2)^3 - (4a^2)^3}}$$

That is, the capacitance is the same as that of two parallel wires by themselves, but at a distance ρa between centers instead of the actual distance a.

Relations between capacitances of various combinations of conductors and sheath will be found in *Alternating Currents*, by Alexander Russell, and in *Electric Cables*, by W. A. Del Mar.

ZERO-SEQUENCE CAPACITANCE. The zero-sequence capacitance is useful in the case of ungrounded circuits or circuits grounded through a high impedance. Formulas for calculating zero-sequence capacitance are given in Table 3.

Table 3. Formulas for Zero-sequence Capacitance of Three-phase Lines

Number of Ground Wires	Zero-sequence Capacitance in Microfarads per 1000 Ft	
	Single Circuit	Twin Circuit
0	$\dfrac{0.0339}{A_{11} + 2A_{12}}$	$\dfrac{0.0678}{A_{11} + 2A_{12} + 3A_{14}}$
1	$\dfrac{0.0339}{A_{11} + 2A_{12} - 3\dfrac{A_{17}^2}{A_{77}}}$	$\dfrac{0.0678}{A_{11} + 2A_{12} + 3A_{14} - 6\dfrac{A_{17}^2}{A_{77}}}$
n	$\dfrac{0.0339}{A_{11} + 2A_{12} - 3\dfrac{nA_{17}^2}{A_{77} + (n-1)A_{78}}}$	$\dfrac{0.0678}{A_{11} + 2A_{12} + 3A_{14} - 6\dfrac{nA_{17}^2}{A_{77} + (n-1)A_{78}}}$

$A_{11} = 4.604 \log \dfrac{4h_m}{d_c}$ $\quad A_{14} = 4.604 \log \sqrt{\dfrac{4h_m^2}{S_{14}^2} + 1}$ $\quad A_{77} = 4.604 \log \dfrac{4h_g}{d_g}$

$A_{12} = 4.604 \log \sqrt{\dfrac{4h_m^2}{S_{12}^2} + 1}$ $\quad A_{17} = 4.604 \log \sqrt{\dfrac{4h_m h_g}{S_{17}^2} + 1}$ $\quad A_{78} = 4.604 \log \sqrt{\dfrac{4h_g^2}{S_{78}^2} + 1}$

where d_c = diameter of conductors.
d_g = diameter of ground wires.
h_m = mean height of conductors above ground.
h_g = mean height of ground wires above ground.

S_{12}, S_{14}, S_{17}, and S_{78} are the mean spacings between conductor groups and are found in the same manner as the mean spacings for computing the zero-sequence impedances (see Article 22). Logarithms are to the base ten.

21. INDUCTANCE

POSITIVE- AND NEGATIVE-SEQUENCE INDUCTANCES. These are the inductances between phases of a system when currents are balanced. The positive and negative values are identical for transmission lines and other static equipment. Unless otherwise designated, the term inductance will be used for self-inductance.

INDUCTANCE OF A TWO-WIRE TRANSMISSION LINE. For the usual case, where the length is so great that the distance apart of the conductors is negligible in comparison,

$$L = 0.01524 + 0.14037 \log_{10} \frac{2D}{d} \quad \text{millihenrys per 1000 ft} \tag{11}$$

where D = distance between centers of wires, and d = the diameter of the wires. They may be in any unit provided they are in the same unit. The total inductance of both wires is obtained by multiplying by twice the length of the line. Table 1 is based on eq. (11) for one wire. See below for the effect of stranding.

INDUCTANCE OF A THREE-PHASE THREE-WIRE LINE. The inductance per wire is as given in eq. (11) or Table 4, but the effective inductance for a pair of wires of an equilateral three-phase line is $\sqrt{3}$ times those values. Usually, however, the inductance per wire is used, and the effect of phase difference is taken into account in combining impedances or voltage drops. When the line is not equilateral, the geometric mean separation, as shown in eq. (18) is used. Equation (11) is not applicable to conductors at close spacing when the frequency is high. The effect of close spacing is to introduce proximity effect, which, by bringing the centers of the currents closer than the centers of the wires, reduces the inductance. The calculation of proximity effect is quite involved, but the following examples will give some idea of its magnitude.

Conductor Size, Cir Mils	Spacing, Inches, Center to Center	Inductance as Fraction of Value without Proximity Effect
750,000	1.32	0.94
1,500,000	1.70	0.88
2,500,000	2.17	0.81

The inductance of three cables in a steel pipe or conduit may be estimated by eq. (11), adding 30 per cent to the actual spacing, D.

See F. B. Silsbee, *Elec. World*, Vol. 68, p. 125 (July 15, 1916), or H. B. Dwight, *Electrical Coils and Conductors*, Chapter 23, *Trans. AIEE*, Vol. 42, p. 850 (1923), and *G.E. Review*, Vol. 3, p. 531 (1927).

EFFECT OF STRANDING. If d in eq. (11) is used to represent the overall diameter of a stranded conductor, the first term of that equation changes from 0.01524 to the following values:

Three wires:	0.02371	Thirty-seven wires:	0.01611
Seven wires:	0.01957	Sixty-one wires:	0 01569
Nineteen wires:	0.01690	Ninety-one wires:	0.01560

If, however, d in eq. (11) is made the diameter of a solid wire of the same cross-section as the stranded conductor, the logarithm will be less than if d were made the overall diameter, and this practically offsets the foregoing increase in the first term. Using this latter procedure, the error will be of the order of 1 1/2 per cent at 18-in. spacing, 1 per cent at 36-in. spacing, etc.

EFFECT OF CONDUCTORS OF MAGNETIC METAL. For conductors of magnetic metal of permeability μ, the first term of eq. (11) changes from 0.01524 to 0.01524 μ. Table 3 gives the inductance in millihenrys per mile of A.C.S.R. cable, i.e., aluminum cable with a steel core.

ZERO-SEQUENCE INDUCTANCES. These are the inductances between conductors and ground-return circuits. Because of the multiplicity of conductors in the ground elements of the circuit, ground wires, and the earth itself with overhead wires and the sheath or sheaths with cables in ducts, special methods of treatment have been developed, which are given in Section 14, Article 22.

INDUCTANCE OF A SOLID WIRE IN A CYLINDRICAL SHEATH. The following formula gives correct values whether the cylinders are or are not coaxial.

Let r_1 = radius of wire in centimeters.

r_2 = inside radius of sheath in centimeters.

r_3 = outside radius of sheath in centimeters.

$$G = \left[\frac{r_3^2 \log_\epsilon r_3 - r_2^2 \log_\epsilon r_2}{r_3^2 - r_2^2} - \frac{1}{2}\right]$$

$$L = 0.06096\,[G - (\log_\epsilon r_1) + 1/4] \text{ millihenrys per 1000 ft.} \quad (12)$$

INDUCTANCE OF TWO TUBULAR CONDUCTORS, ONE INSIDE THE OTHER. Here, again, the formula is correct for both coaxial and excentric cylinders.

Let r_1 = inner radius of inner tube in centimeters.

r_2 = outer radius of inner tube in centimeters.

r_3 = inner radius of outer tube in centimeters.

r_4 = outer radius of outer tube in centimeters.

G_1 = value of G from following table for radial ratio r_3/r_4.

G_2 = value of G from following table for radial ratio r_2/r_1.

$$L = 0.06096\left[\log_\epsilon \frac{r_4}{r_2} + \frac{2\left(\frac{r_3}{r_4}\right)^2}{1 - \left(\frac{r_3}{r_4}\right)^2}\log_\epsilon \frac{r_4}{r_3} - 1 + G_1 + G_2\right] \text{ millihenrys per 1000 ft} \quad (13)$$

Values of G for Eq. (13)

Radial Ratio	G	Radial Ratio	G
0	0.25	0.80	0.0663
0.5	0.1603	0.85	0.0499
0.6	0.1304	0.90	0.0333
0.7	0.0989	0.95	0.0167
0.75	0.0827	1.00	0

MUTUAL INDUCTANCE OF TWO EQUAL, PARALLEL STRAIGHT WIRES. Let D be the distance between centers of the two wires, and l the length of each wire, both in centimeters. Then the mutual inductance, millihenrys, between the two wires is

$$M = 2\left[l \log_\epsilon \left(\frac{l + \sqrt{l^2 + D^2}}{D}\right) - \sqrt{l^2 + D^2} + D\right] \times 10^{-6} \text{ millihenrys} \quad (14)$$

$$= 2l\left[\log_\epsilon \frac{2l}{D} - 1 + \frac{D}{l}\right] \times 10^{-6} \text{ (approximately) millihenrys} \quad (15)$$

when the length l is great in comparison with D.

Equation (14), which is an exact expression when the wires have no appreciable cross-section, is not an exact expression for the mutual inductance of two parallel cylindrical wires, but is not appreciably in error even when the section is large and D is small, if l is great compared with D. See p. 14-41 for rectangular bars.

Table 4. Self-inductance of Solid Non-magnetic Wires *

Millihenrys per 1000 FEET of each wire of a single-phase or of a symmetrical three-phase line

Size of Wire, cir mils or A.W.G.	Diam. of Wire, inches	Inches between Wires, center to center							
		1	3	6	9	12	18	24	30
1,000,000	1.0000	0.05750	0.1245	0.1667	0.1915	0.2090	0.2337	0.2512	0.2648
750,000	0.8660	0.06627	0.1332	0.1755	0.2002	0.2178	0.2425	0.2600	0.2736
500,000	0.7071	0.07863	0.1456	0.1879	0.2126	0.2301	0.2548	0.2724	0.2860
350,000	0.5916	0.08950	0.1565	0.1987	0.2235	0.2410	0.2657	0.2832	0.2968
250,000	0.5000	0.09976	0.1667	0.2090	0.2337	0.2512	0.2760	0.2935	0.3071
0000	0.4600	0.1048	0.1718	0.2141	0.2388	0.2563	0.2810	0.2986	0.3122
000	0.4096	0.1119	0.1789	0.2211	0.2459	0.2634	0.2881	0.3057	0.3193
00	0.3648	0.1190	0.1860	0.2282	0.2529	0.2705	0.2952	0.3127	0.3263
0	0.3249	0.1260	0.1930	0.2353	0.2600	0.2775	0.3022	0.3198	0.3334
1	0.2893	0.1331	0.2001	0.2423	0.2671	0.2846	0.3093	0.3269	0.3405
2	0.2576	0.1402	0.2072	0.2494	0.2741	0.2917	0.3164	0.3339	0.3475
4	0.2043	0.1543	0.2213	0.2635	0.2883	0.3058	0.3305	0.3481	0.3617
6	0.1620	0.1685	0.2354	0.2777	0.3024	0.3199	0.3447	0.3622	0.3758
8	0.1285	0.1826	0.2496	0.2918	0.3165	0.3341	0.3588	0.3763	0.3899
10	0.1019	0.1967	0.2637	0.3060	0.3307	0.3482	0.3729	0.3905	0.4041
12	0.08081	0.2109	0.2778	0.3201	0.3448	0.3623	0.3871	0.4046	0.4182
14	0.06408	0.2250	0.2920	0.3342	0.3590	0.3765	0.4012	0.4187	0.4323
16	0.05082	0.2391	0.3061	0.3484	0.3731	0.3906	0.4153	0.4329	0.4465

Size of Wire, cir mils or A.W.G.	Feet between Wires, center to center								
	3	4	5	6	8	10	15	20	25
1,000,000	0.2760	0.2935	0.3071	0.3182	0.3358	0.3494	0.3741	0.3916	0.4052
750,000	0.2847	0.3023	0.3159	0.3270	0.3445	0.3581	0.3828	0.4004	0.4140
500,000	0.2971	0.3146	0.3282	0.3393	0.3569	0.3705	0.3952	0.4127	0.4263
350,000	0.3080	0.3255	0.3391	0.3502	0.3678	0.3814	0.4061	0.4236	0.4372
250,000	0.3182	0.3358	0.3494	0.3605	0.3780	0.3916	0.4163	0.4339	0.4475
0000	0.3233	0.3408	0.3544	0.3656	0.3831	0.3967	0.4214	0.4390	0.4526
000	0.3304	0.3479	0.3615	0.3726	0.3902	0.4038	0.4285	0.4460	0.4596
00	0.3374	0.3550	0.3686	0.3797	0.3972	0.4108	0.4356	0.4531	0.4667
0	0.3445	0.3620	0.3756	0.3867	0.4043	0.4179	0.4426	0.4601	0.4737
1	0.3516	0.3691	0.3827	0.3938	0.4114	0.4250	0.4497	0.4672	0.4808
2	0.3586	0.3762	0.3898	0.4009	0.4184	0.4320	0.4568	0.4743	0.4879
4	0.3728	0.3903	0.4039	0.4150	0.4326	0.4462	0.4709	0.4884	0.5020
6	0.3869	0.4045	0.4181	0.4292	0.4467	0.4603	0.4850	0.5026	0.5162
8	0.4011	0.4186	0.4322	0.4433	0.4608	0.4744	0.4992	0.5167	0.5303
10	0.4152	0.4327	0.4463	0.4574	0.4750	0.4886	0.5133	0.5308	0.5444
12	0.4293	0.4469	0.4605	0.4716	0.4891	0.5027	0.5274	0.5450	0.5586
14	0.4435	0.4610	0.4746	0.4857	0.5033	0.5169	0.5416	0.5591	0.5727
16	0.4576	0.4751	0.4887	0.4998	0.5174	0.5310	0.5557	0.5732	0.5868

* The inductances given in this table also apply, with a practically negligible error (about 1 per cent), to ordinary stranded wires of the *same cross-section.*

Table 5. Self-inductance * of Aluminum Steel Core (A.C.S.R.) Cable, per Mile of Single Conductor

(Multiple-layer Conductors†—All Current Densities)

Circular Mils or A.W.G. (B. & S.) Aluminum	Number of Wires		Copper Equivalent CM or A.W.G. Based on Copper 97% Alum. 61%	Inductance L in Millihenries per Mile of Each Conductor of a Single-phase, Two-phase, or Three-phase Circuit, Distance D between Centers of Conductors,‡ in feet													
	Al.	St.		3	4	5	6	8	11	13	15	17	19	23	25	30	35
1,590,000	54	19	1,000,000	1.30	1.40	1.47	1.53	1.62	1.72	1.78	1.82	1.86	1.90	1.96	1.99	2.05	2.10
1,510,500	54	19	950,000	1.31	1.41	1.48	1.54	1.63	1.73	1.79	1.83	1.87	1.91	1.97	2.00	2.06	2.10
1,431,000	54	19	900,000	1.32	1.42	1.49	1.55	1.64	1.74	1.79	1.84	1.88	1.92	1.98	2.00	2.06	2.11
1,351,500	54	19	850,000	1.33	1.42	1.50	1.56	1.65	1.75	1.80	1.85	1.89	1.93	1.99	2.01	2.07	2.12
1,272,000	54	19	800,000	1.34	1.43	1.51	1.56	1.66	1.76	1.81	1.86	1.90	1.94	2.00	2.02	2.08	2.13
1,192,500	54	19	750,000	1.35	1.44	1.52	1.57	1.67	1.77	1.82	1.87	1.91	1.95	2.01	2.03	2.09	2.14
1,113,000	54	19	700,000	1.36	1.45	1.53	1.59	1.68	1.78	1.84	1.88	1.92	1.96	2.02	2.05	2.10	2.15
1,033,500	54	7	650,000	1.38	1.47	1.54	1.60	1.69	1.79	1.85	1.89	1.93	1.97	2.03	2.06	2.12	2.17
954,000	54	7	600,000	1.39	1.48	1.55	1.61	1.70	1.81	1.86	1.91	1.95	1.98	2.04	2.07	2.13	2.18
900,000	54	7	566,000	1.40	1.49	1.56	1.62	1.71	1.82	1.87	1.91	1.96	1.99	2.05	2.08	2.14	2.19
874,000	54	7	550,000	1.40	1.49	1.57	1.62	1.72	1.82	1.87	1.92	1.96	2.00	2.06	2.08	2.14	2.19
795,000	54	7	500,000	1.42	1.51	1.58	1.64	1.73	1.84	1.89	1.94	1.98	2.01	2.07	2.10	2.16	2.21
795,000	26	7	500,000	1.41	1.50	1.58	1.63	1.73	1.83	1.88	1.93	1.97	2.01	2.07	2.09	2.15	2.20
795,000	30	19	500,000	1.40	1.49	1.56	1.62	1.71	1.81	1.87	1.91	1.95	1.99	2.05	2.08	2.14	2.19
715,500	54	7	450,000	1.43	1.53	1.60	1.66	1.75	1.85	1.91	1.95	1.99	2.03	2.09	2.12	2.18	2.23
715,500	26	7	450,000	1.43	1.52	1.59	1.65	1.74	1.85	1.90	1.95	1.99	2.02	2.08	2.11	2.17	2.22
715,500	30	19	450,000	1.41	1.51	1.58	1.64	1.73	1.83	1.89	1.93	1.97	2.01	2.07	2.10	2.15	2.20
666,600	54	7	419,000	1.45	1.54	1.61	1.67	1.76	1.86	1.92	1.96	2.00	2.04	2.10	2.13	2.19	2.24
636,000	54	7	400,000	1.45	1.55	1.62	1.68	1.77	1.87	1.92	1.97	2.01	2.05	2.11	2.14	2.19	2.24
636,000	26	7	400,000	1.45	1.54	1.61	1.67	1.76	1.87	1.92	1.97	2.01	2.04	2.10	2.13	2.19	2.24
636,000	30	19	400,000	1.43	1.53	1.60	1.66	1.75	1.85	1.90	1.95	1.99	2.03	2.09	2.12	2.17	2.22
605,000	54	7	380,500	1.46	1.55	1.63	1.68	1.78	1.88	1.93	1.98	2.02	2.06	2.12	2.14	2.20	2.25
605,000	26	7	380,500	1.45	1.55	1.62	1.68	1.77	1.87	1.93	1.97	2.01	2.05	2.11	2.14	2.20	2.25
605,000	30	19	380,500	1.44	1.53	1.60	1.66	1.76	1.86	1.91	1.96	2.00	2.03	2.10	2.12	2.18	2.23
556,500	26	7	350,000	1.47	1.56	1.63	1.69	1.78	1.89	1.94	1.99	2.03	2.06	2.12	2.15	2.21	2.26
556,500	30	7	350,000	1.45	1.55	1.62	1.68	1.77	1.87	1.93	1.97	2.01	2.05	2.11	2.14	2.19	2.24
500,000	30	7	314,500	1.47	1.56	1.63	1.69	1.79	1.89	1.94	1.99	2.03	2.06	2.13	2.15	2.21	2.26
477,000	26	7	300,000	1.49	1.59	1.66	1.72	1.81	1.91	1.97	2.01	2.05	2.09	2.15	2.18	2.23	2.28
477,000	30	7	300,000	1.48	1.57	1.64	1.70	1.79	1.90	1.95	2.00	2.04	2.07	2.13	2.16	2.22	2.27
397,500	26	7	250,000	1.52	1.62	1.69	1.75	1.84	1.94	1.99	2.04	2.08	2.12	2.18	2.20	2.26	2.31
397,500	30	7	250,000	1.51	1.60	1.67	1.73	1.82	1.93	1.98	2.03	2.07	2.10	2.16	2.19	2.25	2.30
336,400	26	7	0000	1.55	1.64	1.71	1.77	1.86	1.97	2.02	2.07	2.11	2.14	2.20	2.23	2.29	2.34
336,400	30	7	0000	1.53	1.63	1.70	1.76	1.85	1.95	2.01	2.05	2.09	2.13	2.19	2.22	2.28	2.33
300,000	26	7	188,700	1.57	1.66	1.73	1.79	1.88	1.99	2.04	2.09	2.13	2.16	2.22	2.25	2.31	2.36
300,000	30	7	188,700	1.55	1.65	1.72	1.78	1.87	1.97	2.02	2.07	2.11	2.15	2.21	2.23	2.29	2.34
266,800	26	7	000	1.59	1.68	1.75	1.81	1.90	2.00	2.06	2.10	2.14	2.18	2.24	2.27	2.33	2.38

Self-inductance* of Aluminum Steel Core Cable, per Mile of Single Conductor

(Single-layer Conductors § —Current Density, 0 amp per sq in.)

266,800	6	7	000	1.59	1.68	1.76	1.81	1.91	2.01	2.06	2.11	2.15	2.19	2.25	2.27	2.33	2.38
0,000	6	1	00	1.74	1.84	1.91	1.97	2.06	2.16	2.22	2.26	2.30	2.34	2.40	2.43	2.48	2.53
000	6	1	0	1.79	1.88	1.95	2.01	2.10	2.20	2.26	2.30	2.34	2.38	2.44	2.47	2.53	2.58
00	6	1	1	1.82	1.92	1.99	2.05	2.14	2.24	2.30	2.34	2.38	2.42	2.48	2.51	2.57	2.61
0	6	1	2	1.86	1.95	2.02	2.08	2.18	2.28	2.33	2.38	2.42	.245	2.52	2.54	2.60	2.65
1	6	1	3	1.89	1.99	2.06	2.12	2.21	2.31	2.37	2.41	2.45	2.49	2.55	2.58	2.63	2.68
2	6	1	4	1.93	2.02	2.09	2.15	2.24	2.34	2.40	2.44	2.48	2.52	2.58	2.61	2.67	2.72
2	7	1	4	1.90	2.00	2.07	2.13	2.22	2.32	2.38	2.42	2.46	2.50	2.56	2.59	2.64	2.69
3	6	1	5	1.96	2.05	2.12	2.18	2.27	2.38	2.43	2.48	2.52	2.55	2.61	2.64	2.70	2.75
4	6	1	6	1.99	2.08	2.16	2.21	2.31	2.41	2.46	2.51	2.55	2.59	2.65	2.67	2.73	2.78
4	7	1	6	1.97	2.06	2.14	2.20	2.30	2.39	2.45	2.49	2.53	2.57	2.63	2.66	2.72	2.76
5	6	1	7	2.02	2.12	2.19	2.25	2.34	2.44	2.50	2.54	2.58	2.62	2.68	2.71	2.77	2.81
6	6	1	8	2.06	2.15	2.22	2.28	2.38	2.48	2.53	2.58	2.62	2.65	2.72	2.74	2.80	2.85

Table 5. Self-inductance* of Aluminum Steel Core Cable, per Mile of Single Conductor—*Continued*

(Single-layer Conductors §—Current Density, 600 amp per sq in.)

Circular Mils or A.W.G. (B. & S.) Aluminum	Number of Wires		Copper Equivalent CM or A.W.G. Based on Copper 97% Alum. 61%	Inductance L in Millihenries per Mile of Each Conductor of a Single-phase, Two-phase, or Three-phase Circuit, Distance D between Centers of Conductors,‡ in feet													
	Al.	St.		3	4	5	6	8	11	13	15	17	19	23	25	30	35
266,800	6	7	000	1.64	1.73	1.80	1.86	1.95	2.06	2.11	2.16	2.20	2.23	2.29	2.32	2.38	2.43
0,000	6	1	00	1.80	1.89	1.96	2.02	2.11	2.21	2.27	2.31	2.35	2.39	2.45	2.48	2.54	2.59
000	6	1	0	1.82	1.92	1.99	2.05	2.14	2.24	2.30	2.34	2.38	2.42	2.48	2.51	2.56	2.61
00	6	1	1	1.85	1.94	2.02	2.07	2.17	2.27	2.32	2.37	2.41	2.45	2.51	2.53	2.59	2.64
0	6	1	2	1.88	1.97	2.04	2.10	2.19	2.30	2.35	2.40	2.44	2.47	2.53	2.56	2.62	2.67
1	6	1	3	1.91	2.00	2.07	2.13	2.22	2.33	2.38	2.43	2.47	2.50	2.56	2.59	2.65	2.70
2	6	1	4	1.94	2.03	2.10	2.16	2.25	2.36	2.41	2.46	2.50	2.53	2.59	2.62	2.68	2.73
2	7	1	4	1.92	2.01	2.08	2.14	2.24	2.34	2.39	2.44	2.48	2.51	2.58	2.60	2.66	2.71
3	6	1	5	1.97	2.06	2.13	2.19	2.28	2.38	2.44	2.48	2.52	2.56	2.62	2.65	2.71	2.76
4	6	1	6	1.99	2.09	2.16	2.22	2.31	2.41	2.47	2.51	2.55	2.59	2.65	2.68	2.74	2.78
4	7	1	6	1.98	2.07	2.15	2.20	2.30	2.40	2.45	2.50	2.54	2.59	2.64	2.66	2.72	2.77
5	6	1	7	2.03	2.12	2.19	2.25	2.34	2.45	2.50	2.55	2.59	2.62	2.68	2.71	2.77	2.82
6	6	1	8	2.07	2.16	2.23	2.29	2.38	2.49	2.54	2.59	2.63	2.66	2.72	2.75	2.81	2.86
(Single-layer Conductors §—Current Density 1200 amp per sq in.)																	
266,800	6	7	000	1.85	1.94	2.01	2.07	2.16	2.27	2.32	2.37	2.41	2.44	2.50	2.53	2.59	2.64
0,000	6	1	00	1.87	1.96	2.03	2.09	2.18	2.29	2.34	2.39	2.43	2.46	2.52	2.55	2.61	2.66
000	6	1	0	1.90	2.00	2.07	2.13	2.22	2.32	2.38	2.42	2.46	2.50	2.56	2.59	2.64	2.69
00	6	1	1	1.93	2.02	2.09	2.15	2.24	2.35	2.40	2.45	2.49	2.52	2.58	2.61	2.67	2.72
0	6	1	2	1.95	2.04	2.11	2.17	2.26	2.36	2.42	2.46	2.50	2.54	2.60	2.63	2.69	2.74
1	6	1	3	1.96	2.05	2.13	2.19	2.28	2.38	2.43	2.48	2.52	2.56	2.62	2.64	2.70	2.75
2	6	1	4	1.97	2.07	2.14	2.20	2.29	2.39	2.45	2.49	2.53	2.57	2.63	2.66	2.72	2.76
2	7	1	4	1.95	2.04	2.11	2.17	2.26	2.36	2.42	2.46	2.51	2.54	2.60	2.63	2.69	2.74
3	6	1	5	1.99	2.08	2.15	2.21	2.30	2.41	2.46	2.51	2.55	2.58	2.64	2.67	2.73	2.78
4	6	1	6	2.00	2.10	2.17	2.23	2.32	2.42	2.48	2.52	2.56	2.60	2.66	2.69	2.74	2.79
4	7	1	4	1.99	2.08	2.16	2.21	2.31	2.41	2.46	2.51	2.55	2.60	2.65	2.67	2.73	2.78
5	6	1	7	2.05	2.14	2.21	2.27	2.36	2.46	2.52	2.56	2.60	2.64	2.70	2.73	2.79	2.84
6	6	1	8	2.08	2.17	2.25	2.30	2.40	2.50	2.55	2.60	2.64	2.68	2.74	2.76	2.82	2.87

* The inductance values of the table are based upon actual tests on various sizes of cable at various current densities at 1-ft spacing. The inductances at other spacings were calculated from those at 1-ft spacing by means of the fundamental inductance formula.

The inductance L' for any spacing D' not given in the table is equal to the inductance L at the next smaller spacing D given in the table, plus the quantity $0.74113 \log_{10} D'/D$. Thus $L' = L + 0.74113 \log_{10} D'/D$. Or the inductance in millihenries to be added to that at the next smaller spacing may be taken from the following tabulation:

D'/D........	1.05	1.10	1.15	1.20	1.25	1.30	1.35	1.40	1.45	1.50
$L+$.........	0.016	0.031	0.045	0.059	0.072	0.084	0.097	0.108	0.120	0.131
D'/D........	1.55	1.60	1.65	1.70	1.75	1.80	1.85	1.90	1.95	2.00
$L+$.........	0.141	0.151	0.161	0.171	0.180	0.189	0.198	0.207	0.215	0.223

† By multiple-layer conductors is meant conductors with two or more layers of aluminum over the steel core.

‡ For any three-phase arrangement of conductors $D = \sqrt[3]{ABC}$. This resolves itself into $D = A$, B, or C for symmetrical triangular spacing, and into $D = 1.26A$ or B for regular flat spacing, it being immaterial whether the conductors are in a horizontal or vertical plane.

§ By single-layer conductors is meant conductors with one layer of aluminum over the steel core.

Equation (14) is also applicable, with a practically negligible error, to bars of rectangular section, and in fact to the mutual inductance between any two parallel conductors of any section and external to each other, e.g., between an overhead wire and a rail, the distance D being the geometric mean distance between the two sections.

MUTUAL INDUCTANCE BETWEEN A TUBE AND AN INTERIOR WIRE. (For mutual inductance between a tube and an exterior wire, see Woodruff, *Principles of Electric Power Transmission.*)

Let r_1 = radius of interior wire in centimeters.

r_2 = inner radius of sheath in centimeters.

r_3 = outer radius of sheath in centimeters.

l = length of wire in centimeters.

$$M = 2l\left[\log_\epsilon 2l - \frac{r_3^2 \log_\epsilon r_3 - r_2^2 \log_\epsilon r_2}{r_3^2 - r_2^2} - \frac{1}{2}\right] \times 10^6 \quad \text{millihenrys,} \tag{16}$$

or approximately

$$M = 2l\left[\log_\epsilon \frac{2l}{r_2} - \frac{1}{2}\right] \times 10^6 \quad \text{millihenrys} \tag{17}$$

Equation 17 holds only for thin sheaths. In general, when the cross-sectional dimensions of the conductors are small compared with their distance apart, it suffices, in calculating mutual inductances, to assume that their mutual inductance is the same as that of the filaments along their axes. Where the cross-section is too large to justify this assumption, it is necessary to average the mutual inductances of all the filaments of which the conductors may be supposed to consist. This is done by averaging the logarithms of the distances between all the pairs of filaments. This average is the **geometric mean distance** (GMD), and with it we can use the filament formulas if we replace the axial distance between conductors by their GMD.

INTERNAL AND EXTERNAL INDUCTANCES. In eq. (11) the constant 0.01524 or 0.01524 μ represents the inductance due to the magnetic field within the wire, and the logarithmic quantity represents that due to the external field. In some calculations, especially where the conductor is of magnetic material, it is convenient to separate these two elements of the inductance. As noted above, the internal inductance is influenced by the stranding, so that, if the external inductance is calculated on the basis of actual conductor diameter, the internal inductance appropriate to the stranding should be used. The reactances in Tables 8 and 9 are calculated on this basis.

22. ZERO-SEQUENCE INDUCTANCES AND IMPEDANCES OF POWER LINES AND CABLES

The computation of short-circuit currents in power networks by the method of symmetrical components requires, in the case of faults to ground, the use of zero-sequence impedances of the power lines. (See Section 3, Article 37.) For a three-phase circuit, this impedance is defined as three times the impedance of the phase conductors in parallel, with ground return. The presence of ground wires in the case of overhead lines, or of cable sheaths in the case of power cables, modifies the zero-sequence impedance, as defined above, and it is necessary to take the ground wires and cable sheaths into account in computations.

FORMULAS AND CHARTS. Tables 6 to 9 and Figs. 1 and 2 are reproduced or abstracted, by permission, from Computation of Zero-sequence Impedances of Power Lines and Cables, *Report No.* 37, Joint Committee on Development and Research, Edison Electric Institute and Bell Telephone Laboratories, 1936. This report is quite comprehensive, including tables for 25-cycle systems and formulas for the case where the distributed conductance to ground has to be considered. Additional data are given in *Electrical Transmission and Distribution Reference Book*, Westinghouse Electric and Manufacturing Company. The self- and mutual impedances to be used in the formulas for zero-sequence impedance may be taken from the charts and formulas of Figs. 1 and 2. The values of X_{11} are the external reactances, and the internal reactances must be added, as indicated by the formula on the chart. These internal reactances are given in Tables 8 and 9. The resistances to be used are the 60-cycle a-c resistances of each wire, as given in Article 59, expressed in ohms per mile.

GEOMETRIC MEAN DISTANCE. The geometric mean separations (logarithmic averages) can be used as the equivalent separations in determining the mutual impedances involved in the formulas. As an example, let wires 1, 2, and 3 have separations S_{12}, S_{13}, and S_{23}; then the geometric mean spacing S_m is:

$$S_m = \sqrt[3]{S_{12} \cdot S_{13} \cdot S_{23}} \tag{18}$$

To obtain the mean spacing between three phase wires and one ground wire, let S_{17}, S_{27}, S_{37} be the distances from each phase wire to the ground wire; then

$$S = \sqrt[3]{S_{17} \cdot S_{27} \cdot S_{37}} \tag{19}$$

For two ground wires, the mean spacing of each is computed from the phase wires as above. If they are not the same, the geometric mean of the two is used.

With twin circuits, the simplest method to arrive at the mean separation between wires of the two circuits is to get first the geometric mean separation between each wire of one circuit and the three wires of the other circuit. The geometric mean of these results is then taken for the final value. For example:

$$
\begin{aligned}
S_{m1} &= \sqrt[3]{S_{14} \cdot S_{15} \cdot S_{16}} \\
S_{m2} &= \sqrt[3]{S_{24} \cdot S_{25} \cdot S_{26}} \\
S_{m3} &= \sqrt[3]{S_{34} \cdot S_{35} \cdot S_{36}} \\
S_m &= \sqrt[3]{S_{m1} \cdot S_{m2} \cdot S_{m3}}
\end{aligned} \qquad (20)
$$

A similar process would be followed in getting the mean spacing from phase wires to ground wires, the mean spacing being first obtained from each circuit to each ground wire and the mean of these distances then obtained.

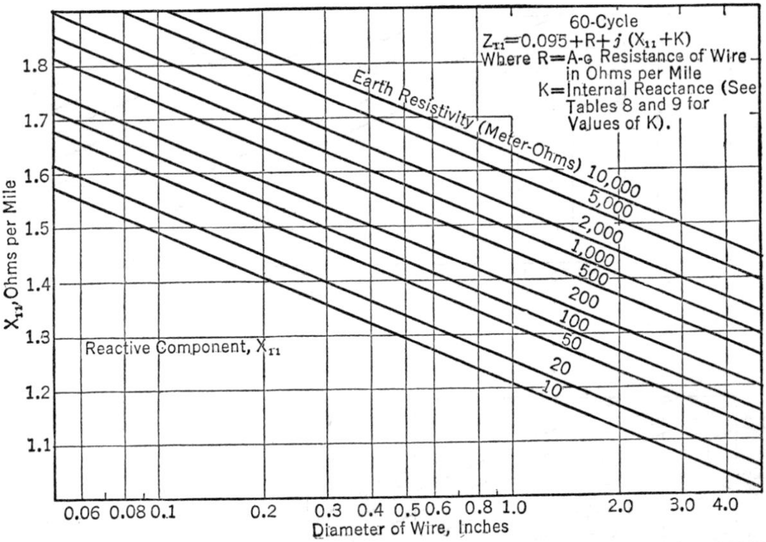

FIG. 1. Self Reactance and Impedance of Wire with Earth Return for Use in Formulas of Table 6

OPEN-WIRE LINES. Formulas for the calculation of zero-sequence impedance and ground-wire current of single- and twin-wire circuits, with any number of ground wires, are given in Table 6. These formulas apply where, as is usually the case, the ground wire may be considered to be perfectly grounded.

UNDERGROUND CABLES. The ground-return impedance of an insulated wire buried in the earth is substantially independent of the depth of the wire below the earth's surface. Also, as mentioned earlier, the variation of self- and mutual impedances with height of conductor above ground, within the range of power line configurations and ground resistivities usually encountered, is small. Therefore, within ordinary engineering accuracy, the self- and mutual impedances involved in the formulas for zero-sequence impedances of cables, either overhead or underground, may be obtained from Figs. 1 and 2.

Available data on the sheath-to-ground leakance of a cable installed in a duct indicate that, for cables more than 2 miles in length, the sheath generally may be treated as though perfectly grounded. Consequently, within this limitation, the formulas in Table 7 may be used for computing the zero-sequence impedance of underground cable as well as aerial cable.

If the cable sheath is paralleled throughout its length by conductors whose dimensions, resistances, and locations are known, e.g., cables in the same duct run, their effect can be included in the calculation of zero-sequence impedance by setting up equations similar to those developed for overhead lines with more than one ground wire. In the more usual case the cable is paralleled for various portions of its length by other cables in the same and other duct banks, by water and gas mains, and by other miscellaneous structures, and an accurate calculation of the zero-sequence impedance is impossible.

SHIELDED CABLE WITH STEEL BINDER TAPE. Within a certain range the zero-sequence impedance of shielded cables with steel binding tapes varies with the cable

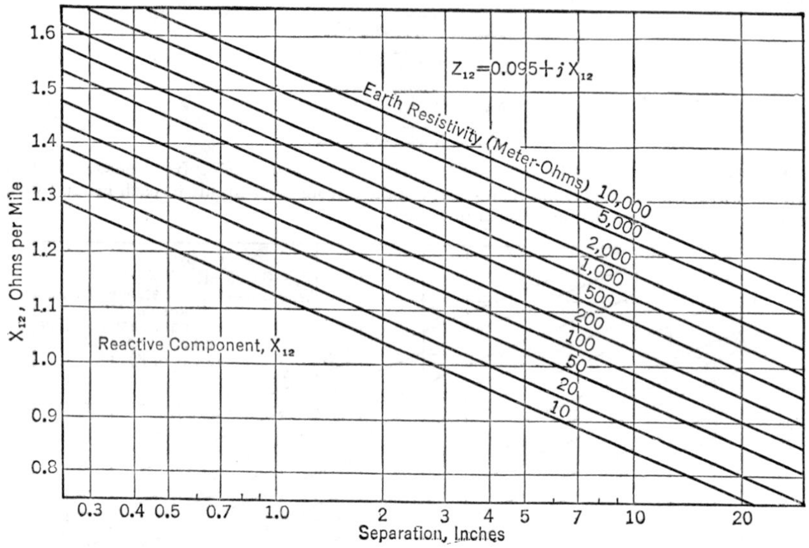

(a)

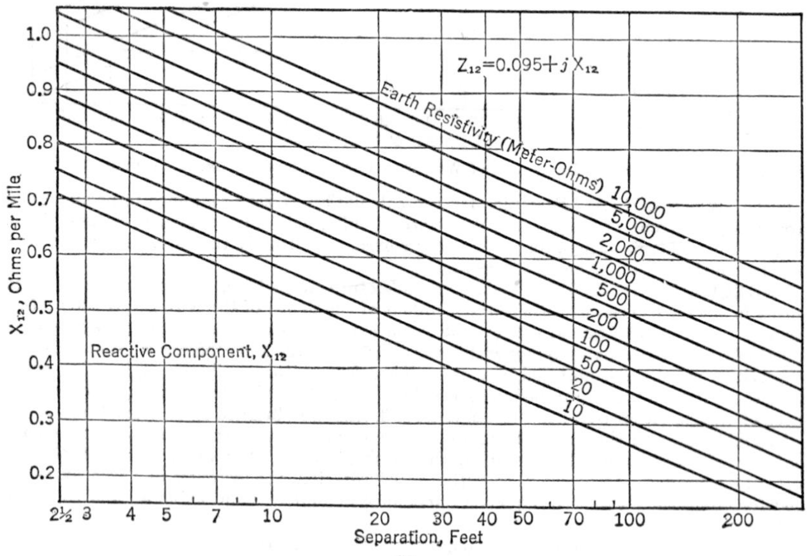

(b)

FIG. 2. Mutual Reactance and Impedance of Wires with Earth Return for Use in Formulas of Table 6 residual current. Both field and laboratory tests on a three-conductor 350,000-cir mil shielded cable have indicated that, for circuit residual currents above about 600 amp, the zero-sequence impedance is practically independent of current. For most cable systems fault currents will be in excess of 600 amp, and, although the zero-sequence impedance at such currents is slightly higher than if no magnetic material were present, this increment may usually be neglected in view of the other uncertainties entering into such computations.

AERIAL CABLES ON MESSENGERS. The formulas for open-wire lines may be used, the reactances of cable sheaths and messengers being substituted for those of ground wires.

ARMORED CABLES. Because of magnetic material being outside the lead sheath and in greater bulk than in a shielded cable with steel binder, the internal components of the reactance are not susceptible of accurate predetermination and should be determined experimentally. A procedure for this is given in Report No. 37, cited below Table 6.

Table 6. Formulas for Zero-sequence Impedances and Ground-wire Currents of Open-wire Lines—Ground Wires Solidly Grounded *

(See Section 1, Article 2, for Laws of Combination of Vectors)

No. of Ground Wires	Zero-sequence Impedance	Total Ground-wire Current
	Single Three-phase Circuits	
0	$Z_0 = Z_{11} + 2Z_{12}$	
1	$Z_0 = Z_{11} + 2Z_{12} - 3\dfrac{Z_{17}^2}{Z_{77}}$	$I_{GW} = -\dfrac{Z_{17}}{Z_{77}} I_F$
2	$Z_0 = Z_{11} + 2Z_{12} - 3\dfrac{2Z_{17}^2}{Z_{77} + Z_{78}}$	$I_{GW} = -\dfrac{2Z_{17}}{Z_{77} + Z_{78}} I_F$
3	$Z_0 = Z_{11} + 2Z_{12} - 3\dfrac{3Z_{17}^2}{Z_{77} + 2Z_{78}}$	$I_{GW} = -\dfrac{3Z_{17}}{Z_{77} + 2Z_{78}} I_F$
n	$Z_0 = Z_{11} + 2Z_{12} - 3\dfrac{nZ_{17}^2}{Z_{77} + (n-1)Z_{78}}$	$I_{GW} = -\dfrac{nZ_{17}}{Z_{77} + (n-1)Z_{78}} I_F$
	Twin Three-phase Circuits	
0	$Z_0 = 1/2(Z_{11} + 2Z_{12} + 3Z_{14})$	
1	$Z_0 = \dfrac{1}{2}\left(Z_{11} + 2Z_{12} + 3Z_{14} - 6\dfrac{Z_{17}^2}{Z_{77}}\right)$	$I_{GW} = -\dfrac{Z_{17}}{Z_{77}} I_F$
2	$Z_0 = \dfrac{1}{2}\left(Z_{11} + 2Z_{12} + 3Z_{14} - 6\dfrac{2Z_{17}^2}{Z_{77} + Z_{78}}\right)$	$I_{GW} = -\dfrac{2Z_{17}}{Z_{77} + Z_{78}} I_F$
3	$Z_0 = \dfrac{1}{2}\left(Z_{11} + 2Z_{12} + 3Z_{14} - 6\dfrac{3Z_{17}^2}{Z_{77} + 2Z_{78}}\right)$	$I_{GW} = -\dfrac{3Z_{17}}{Z_{77} + 2Z_{78}} I_F$
n	$Z_0 = \dfrac{1}{2}\left(Z_{11} + 2Z_{12} + 3Z_{14} - 6\dfrac{nZ_{17}^2}{Z_{77} + (n-1)Z_{78}}\right)$	$I_{GW} = -\dfrac{nZ_{17}}{Z_{77} + (n-1)Z_{78}} I_F$

I_F = total fault current through section of line considered. I_{GW} = sum of ground-wire currents.

* From Computation of Zero-sequence Impedances of Power Lines and Cables, *Report No.* 37, Joint Committee on Development and Research, Edison Electric Institute and Bell Telephone Laboratories (1936).

NOTE 1. Although positive- (or negative-) sequence impedance is usually computed from tables of inductance and resistance, it may be obtained from the ground-return self- and mutual impedances of the phase conductors by using the following formula:

For single-circuit line: $Z_1 = Z_2 = Z_{11} - Z_{12}$
For twin-circuit line: $Z_1 = Z_2 = 1/2(Z_{11} - Z_{12})$

NOTE 2. EXPLANATION OF SUBSCRIPT NOTATION. To avoid complicated notation, subscript numerals signify the group of wires for which the specified impedance is to be obtained. Usually there are not more than three groups to be considered: phase-wire group 1, 2, 3, phase-wire group 4, 5, 6, and ground-wire group 7, 8, . . . n. The lowest numbered wires of each group are used as the group subscripts; thus Z_{11} signifies the self-impedance for the group containing phase wires 1, 2, 3, Z_{12} the mean mutual impedance between phase wires of this group, Z_{14} the mean mutual impedance between phase-wire groups, etc.

Table 7. Formulas for Zero-sequence Impedances and Cable-sheath Currents of Cable Circuits, Sheath Considered Grounded at Ends of Section through Negligible Resistance *

(See Table 6 for explanation of subscript notation and Section 1, Article 2, for addition and division of complex quantities.)

No. of Cables with Phase Conductors in Parallel	No. and Type of Additional Shielding Conductors in Parallel	Cables with Non-magnetic Sheaths: Zero-sequence Impedance	Cables with Non-magnetic Sheaths: Total Cable-sheath Current	Cable Sheaths Containing Magnetic Material: Zero-sequence Impedance	Cable Sheaths Containing Magnetic Material: Total Cable-sheath Current
		Three-phase Cables		Three-phase Cables with Steel Binding Tape	
1	None	$Z_0=Z_{11}+2Z_{12}-3\frac{Z_{17}^2}{Z_{77}}$	$I_{sh}=-\frac{Z_{17}}{Z_{77}}I_F$	$Z_0=Z_{11}'+2Z_{12}'+3\frac{R_{dc}Z_{ex}}{R_{dc}+Z_{ex}}$	$I_{sh}=-\frac{Z_{ex}}{R_{dc}+Z_{ex}}I_F$
1	One sheath †	$Z_0=Z_{11}+2Z_{12}-3\frac{2Z_{17}^2}{Z_{77}+Z_{78}}$	$I_{sh}\dagger=-\frac{2Z_{17}}{Z_{77}+Z_{78}}I_F$	$Z_0=Z_{11}'+2Z_{12}'+3\frac{(R_{dc}-Z_{ex}+Z_{78})Z_{ex}}{R_{dc}+Z_{ex}+Z_{78}}$	$I_{sh}\ddagger=-\frac{2Z_{ex}}{R_{dc}+Z_{ex}+Z_{78}}I_F$
1	Two sheaths †	$Z_0=Z_{11}+2Z_{12}-3\frac{3Z_{17}^2}{Z_{77}+2Z_{78}}$	$I_{sh}\ddagger=-\frac{3Z_{17}}{Z_{77}+2Z_{78}}I_F$	$Z_0=Z_{11}'+2Z_{12}'+3\frac{(R_{dc}-2Z_{ex}+2Z_{78})Z_{ex}}{R_{dc}+Z_{ex}+2Z_{78}}$	$I_{sh}\ddagger=-\frac{3Z_{ex}}{R_{dc}+Z_{ex}+2Z_{78}}I_F$
1	Three sheaths †	$Z_0=Z_{11}+2Z_{12}-3\frac{4Z_{17}^2}{Z_{77}+3Z_{78}}$	$I_{sh}\ddagger=-\frac{4Z_{17}}{Z_{77}+3Z_{78}}I_F$	$Z_0=Z_{11}'+2Z_{12}'+3\frac{(R_{dc}-3Z_{ex}+3Z_{78})Z_{ex}}{R_{dc}+Z_{ex}+3Z_{78}}$	$I_{sh}\ddagger=-\frac{4Z_{ex}}{R_{dc}+Z_{ex}+3Z_{78}}I_F$
1	One messenger	$Z_0=Z_{11}+2Z_{12}-3\frac{(Z_{77}+Z_{mm}-2Z_{7m})Z_{17}^2}{Z_{77}Z_{mm}-Z_{7m}^2}$	$I_{sh}\ddagger=-\frac{(Z_{77}+Z_{mm}-2Z_{7m})Z_{17}}{Z_{77}Z_{mm}-Z_{7m}^2}I_F$	One Three-phase Steel Armored Cable	
2	None	$Z_0=\frac{1}{2}\left(Z_{11}+2Z_{12}+3Z_{14}-6\frac{2Z_{17}^2}{Z_{77}+Z_{78}}\right)$	$I_{sh}\ddagger=-\frac{2Z_{17}}{Z_{77}+Z_{78}}I_F$	$Z_0=Z_{11}'+2Z_{12}'-3Z_{17}'+3\frac{R_{dc}(Z_{17}'+Z_{ex})}{R_{dc}+Z_{17}'+Z_{ex}}$	$I_{sh}=-\frac{Z_{17}'+Z_{ex}}{R_{dc}+Z_{17}'+Z_{ex}}I_F$
n	None	$Z_0=\frac{1}{n}\left[Z_{11}+2Z_{12}+3(n-1)Z_{14}-3n\frac{nZ_{17}^2}{Z_{77}+(n-1)Z_{78}}\right]$	$I_{sh}\ddagger=-\frac{nZ_{17}}{Z_{77}+(n-1)Z_{78}}I_F$		
		Three Single-phase Cables			
3	None	$Z_0=Z_{11}+2Z_{12}-3\frac{3Z_{17}^2}{Z_{77}+2Z_{78}}$	$I_{sh}\dagger=-\frac{3Z_{17}}{Z_{77}+2Z_{78}}I_F$		

Z_{11}, Z_{77}, and Z_{mm} are the self impedances of conductors, sheaths and messengers, respectively.
Z_{12} is the mean mutual impedance between the conductors of one three-phase circuit.
Z_{14} is the mean mutual impedance between the conductors of one three-phase circuit and the conductors of other circuits.
Z_{17} is the mean mutual impedance between the conductors of one cable and the sheaths of the n cables.
Z_{78} is the mean mutual impedance between the n sheaths.
Z_{7m} is the mean mutual impedance between the sheath and the messenger.

Z_{11}', Z_{12}', and Z_{17}' are the internal components of impedances Z_{11}, Z_{12}, and Z_{17}.
Z_{ex} is the external impedance component.
R_{dc} is the d-c resistance of the sheaths.
Z_{78} is the same as defined under cables with non-magnetic sheaths.
I_F is the total fault current through section of line considered.

* From Computation of Zero-sequence Impedances of Power Lines and Cables, *Report No.* 37, Joint Committee on Development and Research, Edison Electric Institute and Bell Telephone Laboratories (1936). † Sheath of same dimension as sheath of cable for which zero-sequence impedance is sought. ‡ Sum of currents in all sheaths or other conductors considered in formula.

Table 8. Internal Reactance (*K*) of Copper, Copper-weld, and Steel Conductors, 60 Cycles *

Diameter, Inches	Stranding	Current Amperes	Internal Reactance (*K*), Ohms per Mile
		Siemens-Martin Steel	
1/4	....	5	0.42
1/4	....	15	0.48
1/4	....	25	0.54
3/8	....	5	0.40
3/8	....	15	0.46
3/8	....	25	0.50
1/2	....	5	0.41
1/2	....	15	0.44
1/2	....	25	0.47
		High-strength Steel	
3/8	....	5	0.39
3/8	....	15	0.43
3/8	....	25	0.47
		Copper-weld—40% Conductivity	
3/8	7	10	0.17
3/8	7	50	0.18
3/8	7	100	0.21
3/8	7	200	0.17
1/2	7	10	0.15
1/2	7	50	0.15
1/2	7	100	0.16
1/2	7	200	0.16
		Copper-weld—30% Conductivity	
5/16	3	5–120	0.18
5/16	7	5–120	0.20
3/8	7	5–160	0.19
1/2	7	5–200	0.18
21/32	19	5–215	0.19

Copper

The internal reactance (*K*) of copper conductors depends on the stranding. Values applicable to the usual strandings for the wire sizes above are:

Size of Conductor, Circular Mils	No. of Strands	Internal Reactance (*K*), Ohms per Mile, 60 Cycles
25,251 to 211,600	7	0.039
211,600 to 400,000	19	0.033
450,000 to 600,000	37	0.032
650,000 to 1,000,000	61	0.031
All sizes	Solid	0.031

* From Computation of Zero-sequence Impedances of Power Lines and Cables, *Report No.* 37, Joint Committee on Development and Research, Edison Electric Institute and Bell Telephone Laboratories (1936).

Table 9. Internal Reactance of A.C.S.R. Conductors *

Size of Conductor, Circular Mils or A.W.G.	Number of Wires		Outside Diameter of Cable, Inches	Internal Reactance (*K*) Ohms per Mile, 60 Cycles	Size of Conductor, Circular Mils or M.W.G.	Number of Wires		Outside Diameter of Cable, Inches	Internal Reactance (*K*) Ohms per Mile, 60 Cycles
	Alum.	Steel				Alum.	Steel		
1,590,000	54	19	1.545	0.026	500,000	30	7	0.904	0.023
1,510,500	54	19	1.506	0.026	477,000	30	7	0.883	0.025
1,431,000	54	19	1.465	0.026	477,000	26	7	0.858	0.023
1,351,500	54	19	1.424	0.026	397,500	30	7	0.806	0.025
1,272,000	54	19	1.382	0.026	397,500	26	7	0.783	0.023
1,192,500	54	19	1.338	0.026	336,400	30	7	0.741	0.025
1,113,000	54	19	1.293	0.026	336,400	26	7	0.721	0.023
1,033,500	54	7	1.246	0.026	300,000	30	7	0.700	0.025
954,000	54	7	1.196	0.026	300,000	26	7	0.680	0.023
900,000	54	7	1.162	0.026	266,800	26	7	0.642	0.025
874,500	54	7	1.146	0.026	0000	6	1	0.563	0.128
795,000	54	7	1.093	0.026	000	6	1	0.502	0.152
795,000	30	19	1.140	0.025	00	6	1	0.447	0.157
795,000	26	7	1.108	0.023	0	6	1	0.398	0.159
715,500	54	7	1.036	0.026	1	6	1	0.355	0.153
715,500	30	19	1.081	0.025	2	6	1	0.316	0.139
715,500	26	7	1.051	0.023	3	6	1	0.281	0.121
666,600	54	7	1.000	0.026	4	6	1	0.250	0.105
636,000	54	7	0.977	0.026	5	6	1	0.223	0.097
636,000	30	19	1.019	0.025	6	6	1	0.198	0.090
636,000	26	7	0.990	0.023					
605,000	54	7	0.953	0.026	For other frequencies the internal reactance may be derived proportionately.				
556,500	30	7	0.953	0.026					
556,500	26	7	0.927	0.023					

* From Computation of Zero-sequence Impedances of Power Lines and Cables, *Report No.* 37, Joint Committee on Development and Research, Edison Electric Institute and Bell Telephone Laboratories (1936).

23. BIBLIOGRAPHY

Carson, J. R., Ground Return Impedance, *Bell System Tech. J.*, Vol. 8, p. 94 (1929).

Clarke, E., *Circuit Analysis of A-c Power Systems*, Vol. I. John Wiley (1943).

Clem, J. E., Reactance of Transmission Lines with Ground Return, *Trans. AIEE*, Vol. 50, p. 901 (1931).

Deutsch, W., Graphical Method of Calculating the Capacity of Cylindrical Combinations, *Arch. Elektrotechnik*, Vol. 2, p. 435 (1914).

Dwight, H. B., *Electrical Coils and Conductors*, McGraw-Hill (1945); also several papers in *Trans. AIEE*.

Edison Electric Institute and Bell Telephone Laboratories, Shielding of Ground-return Circuits at Low Frequencies, *Eng. Report No.* 26 of Joint Subcommittee on Development and Research; Computation of Zero-sequence Impedance of Power Lines and Cables, *Report No.* 37.

Grover, F. W., *Inductance Calculations*. Van Nostrand (1946). *See also* Rosa and Grover.

Jacobs, A. M., Mutual Inductance of Two Straight Wires Which Are Not Parallel, *J. AIEE*, Vol. 39, p. 244 (March, 1920).

Linder, F. W., Graphical Method of Calculating Fault Currents on Rural Distribution System, *Trans. AIEE*, Vol. 64, p. 16 (1945).

Massachusetts Institute of Technology Staff, *Electrical Currents*. John Wiley (1943).

Muller, H. N., Electrical Characteristics of Cables, *Electrical Transmission and Distribution Reference Book*. Westinghouse Electric and Manufacturing Company (1944).

National Electric Light Association, *Underground Systems Reference Book* (1931).

Railroad Commission of State of California, *Inductive Interference of Electric Power and Communications Circuits* (1919).

Rosa and Grover, *Scientific Paper No.* 169, National Bureau of Standards (1916).

Russel, A., *Alternating Currents*. Cambridge, England (1914).

Simmons, D. M., Calculation of the Electrical Problems of Transmission by Underground Cables, *Elec. J.*, Vol. 22, p. 366 (1925); Calculation of the Electrical Problems of Underground Cables: Geometric Factor, Capacity, and Leakage, *Elec. J.*, Vol. 29, p. 336 (1932).

Starr, F. M., Equivalent Circuits, *Trans. AIEE*, Vol. 51, p. 287 (1932).

Waddicor, H., *The Principles of Electric Power Transmission*. John Wiley (1933).

Wagner and Evans, *Symmetrical Components*. McGraw-Hill (1933).

Webb, R. L., and O. W. Manz, Impedance Measurements on Underground Cables, *Elec. Engineering*, p 359 (April, 1936).

Westinghouse Electric and Manufacturing Company, *Electrical Transmission and Distribution Reference Book* (1944).
Woodruff, L. F., *Electric Power Transmission*. John Wiley (1938).
Wright, S. H., and C. F. Hall, Characteristics of Aerial Lines, *Electrical Transmission and Distribution Reference Book*. Westinghouse Electric and Manufacturing Company (1944).

POLE AND TOWER LINES

By Howard A. Enos

(Some of the original material by R. A. PHILIP and C. M. SPOFFORD has been used.)

24. BASIC FACTORS IN MECHANICAL DESIGN

In designing a pole or tower line the following factors, in addition to the purely electrical characteristics, must be considered: (1) character of the route, (2) clearances, (3) grade of construction, (4) type of supporting structures, (5) type of insulators, (6) conductors, (7) mechanical loading, (8) joint use by other utilities, (9) right-of-way. The requirements imposed by these several conditions are noted below, together with tables, formulas, and curves useful in making the necessary calculations. Rural distribution pole lines are treated in Section 19.

CHARACTER OF THE ROUTE. Usually the lower-voltage lines are run along streets and highways, wherever possible, in order to reach customers more easily and to make the lines accessible for maintenance. The higher-voltage transmission lines are more often run across country on private right-of-way in order to obtain the most direct route, as well as to avoid buildings and low-voltage lines, and to obtain adequate space for towers. Lines which are run along streets and highways must be designed to give proper clearances from buildings, other lines, railroad tracks, and trees. Poles must be so located as not to obstruct the public use of streets and driveways or to be subject to damage from traffic.

Urban Distribution Lines usually require close spacing of poles on account of the necessity of keeping the span length of service wires from pole to house as short as possible. Poles are usually set on property lines to avoid obstructing the front of each lot and to avoid overhanging, with the service wires, property adjacent to the building served. Normal **span lengths for urban distribution lines** vary from about 100 to about 175 ft, the average length being about 125 ft. Poles are usually set from 6 in. to 1 ft inside the curb when along streets, but in alleys it is often necessary to set the poles within the traveled space.

The restrictions mentioned do not exist, for the most part, in respect to **rural distribution lines.** Much longer spans are possible, and the length of span is determined largely by economic considerations. Spans of 200 to 300 ft are common, and many lines are built with spans as long as 600 ft.

Higher-voltage transmission lines on private right-of-way are usually built with long spans, and the type of terrain covered by the line largely determines the kind of construction to be used. In fairly level country the height of towers is determined mainly by the length of span, the ground clearances required, and economic considerations. In mountainous territory the valleys may often be crossed by single spans of more than a mile in length, using fairly low towers. Under such conditions the towers must be specially designed for wide spacings of conductors and for heavy transverse and longitudinal stresses. Lines on private right-of-way often require special consideration of side-hill construction.

CLEARANCES. The following clearances for conductors must be considered: to ground, tracks, buildings, trees, conductors and structures of another line, other conductors on the same structure, the structure itself, guy wires and other equipment on the structure, and the edge of the right-of-way.

Space does not permit the tabulation here of clearances for all these conditions for all voltages used for overhead lines. The National Electrical Safety Code gives clearances which are considered good practice. For higher-voltage transmission lines on private right-of-way the clearance to ground should be 20 ft or more, depending on the voltage. The clearance to any grounded portion of the structure, to trees, or other structures, or to the edge of the right-of-way should exceed by a safe margin the arcing distance across the insulator string when the string is deflected the maximum amount by wind pressure. Consideration should also be given to the behavior of lightning flashovers as affected by the presence of grading shields or rings.

Horizontal separations between conductors for high-voltage transmission lines depend on the conductor size and material, the span length, the voltage, and whether pin-type or suspension insulators are used. Opinions of engineers vary as to exact values of spacing to be used for a given set of conditions. The following spacings represent average practice. for 33-kv lines on pin-type insulators, 4 to 6 ft for spans of 150 to 500 ft, respectively; for 33-kv lines on suspension insulators, 6 to 10 ft for spans of 300 to 1000 ft, respectively; for 132-kv lines, 10 to 17 ft for spans from 500 to 1500 ft, respectively; for 220-kv lines, 20 to 26 ft for spans from 800 to 1500 ft, respectively.

P. H. Thomas gives the following formula for computing the minimum horizontal spacing for conductors [*AIEE Trans.* (1928)]:

$$S = Cd\frac{D}{w} + A + \frac{L}{2} \qquad (1)$$

where S = horizontal spacing in feet; C = a constant depending on conductor material (C = 4 for copper and C = 3.5 for A.C.S.R. usually) and character of terrain; d = per cent sag; D = diameter of conductor in inches; w = horizontal loading in pounds per foot; A = arcing distance of the line voltage in feet; L = length of suspension insulator string in feet.

Vertical separations should be about the same as those stated for horizontal separations, but consideration should be given to unequal sags due to unequal ice loading. Also the conductors of a vertical configuration should be offset away from the vertical planes through the other two conductors from 1 to 3 ft, depending upon the span length and the voltage. Horizontal separations between conductors of different circuits should be somewhat greater than between conductors of the same circuit to avoid involving both circuits in the same flashover. Horizontal spacings for conductors on steel towers determined by the required clearances to the tower are usually adequate.

TELEPHONE CIRCUITS FOR POWER LINES. When the line voltage does not exceed 66,000 volts, the telephone circuits are usually carried on the same poles or towers as the power circuits, being placed below the power conductors. A separate line of wooden poles on the same right-of-way is occasionally employed when the voltage is higher than 66,000.

Where the telephone wires are on the same supporting structure as the power wires, sufficient clearance between the power and telephone circuits must be allowed to make the telephone line accessible for repairs and also to prevent the two circuits touching under abnormal conditions. On wood-pole lines of short span (100 to 125 ft) the vertical clearance at the poles or towers ranges from 4 ft for 22,000 volts to 6 ft for 66,000 volts. On long-span lines greater spacing is necessary to allow for safe clearance in the middle of the longest span, because of the change in the sag of the power and telephone wires under all conditions of unequal ice loading and all variations of side deflection due to wind.

Telephone wires are ordinarily of copper, although copper-clad steel or A.C.S.R. is sometimes used for long spans. For spans up to 125 ft No. 10 B. & S. copper may be used (although No. 8 is preferable); for longer spans larger sizes (usually No. 8 or No. 6 or even No. 4) are necessary to allow for sleet load. A spacing of 12 in. between wires may be used for 125-ft spans, but a wider spacing is necessary for longer spans. Where inadequate spacing is used, the telephone lines will frequently become crossed by the wind, unless they are strung with little sag, in which case they are overloaded and broken by sleet. With wide spacing and large sag higher poles must be used.

Carrier Current Communication. It is impracticable to carry telephone circuits on the structures of long-span high-voltage transmission lines, and it is very expensive to provide separate wood-pole telephone lines. As a result, many power companies now use the power circuit conductors themselves as the telephone circuit. This is accomplished by modulating a high-frequency voltage by the amplitude or frequency of the voice current and impressing the modulated carrier voltage upon the transmission circuit by capacity coupling. At the receiving end the transmission line is capacity coupled to the receiving apparatus, which demodulates the carrier voltage, and the resultant is a reproduction of the original voice current, which operates an ordinary telephone receiver or a loud speaker. The apparatus is quite similar to ordinary radio transmitting and receiving equipment, and provision can be made for either one-way or two-way conversations.

For additional data on this subject see the volume on Communication Engineering.

GROUND WIRES. Grounded cables or wires are placed above the transmission line circuits to protect the circuits from lightning discharges (see Lightning Protection, Section 12). They are usually grounded at each supporting structure except where short spans are used. The same care must be exercised to obtain clearances between conductors and ground cables or wires, as outlined under telephone circuits. For tower lines having flexible towers and single-circuit towers having conductors arranged in a horizontal plane, two ground cables are preferable, but for double-circuit tower lines either one or two may be used. As a general rule a line drawn through the ground cable and any conductor should not make an angle of more than 45 deg with the vertical. See Articles 91 to 99.

GRADE OF CONSTRUCTION. The criterion for the strength requirements of a line is known as the "grade of construction." In the National Electrical Safety Code the grades are designated by the letters B, C, D, E, and N, grade B being the highest and requiring the greatest strength. The grade to be used depends upon the type of circuit, the voltage, and the surroundings of the line. For example, a power line of any voltage crossing over a main track of a steam railroad requires grade B construction, but under certain other conditions may be as low as grade N. Regulatory bodies of various states have set up similar requirements.

TYPE OF SUPPORTING STRUCTURES. Wood poles are used to support lines of any voltage. Steel structures of various types are used to a limited extent to support low-voltage lines but are largely used for higher-voltage transmission lines. Reinforced-concrete poles are also used to a limited extent where good appearance is a factor and where the expense can be justified.

Wood poles have the advantage of increasing the insulation value of a line against lightning and therefore are coming into considerable use for high-voltage transmission lines as an economical means of minimizing flashovers. Wood poles are used with pin-type insulator construction for low-voltage distribution lines and low-voltage transmission lines. Single wood poles are used with suspension-type insulator construction up to about 66 kv. Single-circuit lines of higher voltage are common, two-pole H-frame construction being used. Such construction has been used even for extremely long spans. With wood-pole structures the stresses which cannot be sustained by the poles themselves are sustained by proper guying. (See Article 28 on Wood Poles.)

Steel structures are used in various forms, such as tubular poles, square latticed poles, and wide-base towers. Tubular steel poles are used primarily for low-voltage distribution lines and as trolley supports where good appearance is a factor, although they are being largely superseded by smoothly finished wood poles. Square latticed steel poles are used to some extent for medium-voltage transmission lines on narrow right-of-way and where high lateral strength, which cannot be obtained by the use of wood poles without guying, is required.

Wide-base steel towers are used for long-span lines on private right-of-way where considerable height is necessary to provide ground clearance for the conductors. Such towers usually have square or rectangular bases, although triangular-base towers have been used to some extent for single-circuit medium-voltage transmission lines. Wide-base towers are usually classified as suspension towers, angle towers, and strain or dead-end towers. Suspension towers are used in the straight sections of a line and are designed mainly to withstand stresses in a transverse direction resulting from wind pressure on ice-coated conductors and the tower itself. In a longitudinal direction they are designed to carry a certain amount of unbalanced stress due to broken conductors and wind loading. Certain torsional stresses due to unequal conductor loadings must also be taken into account. Angle towers are designed to withstand both the balanced stresses due to the angle in the line and also any unbalanced stresses due to wind pressures and unequal conductor loading as well as possible broken conductors. Strain towers are used at dead-end points in the line and must be designed for full lateral stresses the same as suspension towers, also for the full longitudinal stresses of all the loaded conductors and for torsional stresses due to broken conductors. Semistrain towers are sometimes used for situations where the longitudinal stresses are intermediate between those of suspension and strain towers.

For economic reasons the great majority of towers for lines up to 154 kv are double circuit. Even though the second circuit may not be added until later, the additional cost for double-circuit towers above that for single-circuit towers is so small as usually to be justified. Such towers commonly have each circuit arranged in a vertical configuration on opposite sides of the tower. For lines of 220 kv or higher, single-circuit towers are nearly always used, although double-circuit towers may be economically desirable under certain conditions. The cost of double-circuit towers for such voltages is not far different from the cost of separate lines. Single-circuit towers for the extremely high voltages invariably have the conductors arranged in a horizontal configuration.

After all the mechanical and electrical features of the line have been considered, the final choice of the type of structure to be used is determined by the overall economics of the problem. The selection should be made on the basis of minimum annual cost, modified by the other factors involved. (See Article 29 on Steel Towers.)

TYPE OF INSULATORS. Pin-type or post-type insulators are used almost exclusively for straight line work for voltages up to 15 kv and quite generally for that purpose up to 22 or 33 kv. Suspension or disk-type insulators are used for dead-ending lines of any voltage, although small conductors of low-voltage lines are often dead-ended on double arm construction, using pin-type insulators. Suspension insulators are used for tangent and angle construction for practically all lines above 33 kv and sometimes for 22- and 33-kv lines. Pin-type insulators are unsatisfactory and expensive for the higher voltages. (See also Article 31 on Insulators.)

CONDUCTORS. Conductor materials for overhead lines include copper, bronze, aluminum, steel, copper-clad steel, stranded copper combined with copper-clad steel, and steel-reinforced aluminum. All-aluminum and all-steel conductors have a limited application in modern overhead construction because of the low strength of aluminum and the tendency of steel to corrode. High strength for long-span construction is obtained by the use of bronze, hard-drawn copper, copper-clad steel, or steel-reinforced aluminum. Conductors are practically always stranded in sizes larger than No. 2 A. W. G. Tubular copper conductors have been used on high-voltage lines. Steel-reinforced aluminum conductors are always stranded. The steel reinforcement forms the core and may be either solid or stranded. The aluminum wires are concentrically stranded around the steel core. Where large-diameter conductors are required to raise the corona limit of the line, they may be obtained by the use of annular or multitubular copper conductors or aluminum cables with steel-core reinforcement. The mechanical characteristics of conductor materials are given in Table 1 below. (See Articles 56 to 58.)

Table 1. Mechanical Characteristics of Conductor Materials

Item	H. D. Copper	Aluminum	Steel
Ultimate strength, in lb per sq in.	60,000–65,000	25,000–50,000	60,000–80,000
Yield point, in lb per sq in.	30,000–35,000	11,000–14,000	35,000–40,000
Modulus of elasticity, in lb in. units	16×10^6	9×10^6	22×10^6–28×10^6
Coefficient of linear expansion per deg fahr.	9.6×10^{-6}	12.8×10^{-6}	6.6×10^{-6}
Weight in pounds of a 1-ft length having a cross-section of 1,000,000 cir mils (stranded)	3.09	0.92	2.67

NOTE. See also Articles 54–62, Bare Wires and Cables.

The maximum stress in conductors should not exceed the yield point (usually called "elastic limit"). The elastic limit and modulus of elasticity of a conductor are not, however, fixed under all conditions. They are reduced by fluctuating stresses. Probably the working limit for fluctuating stress should not be over 50 to 70 per cent of the ultimate strength for uniformly increased stress, depending upon the material.

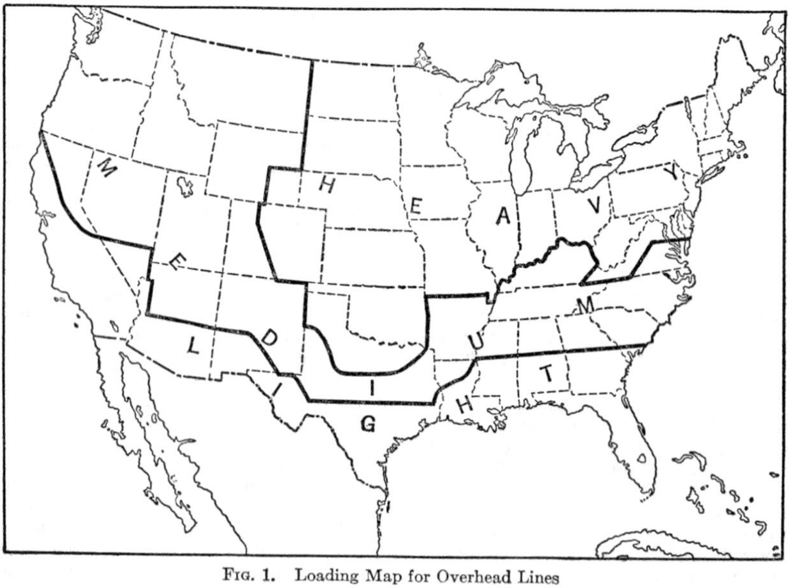

FIG. 1. Loading Map for Overhead Lines

MECHANICAL LOADING. The term mechanical loading refers to the external conditions which produce mechanical stresses in the line conductors and supports. It also includes the weight of the conductors and structures themselves. The external conditions referred to include the temperature range to be expected, the collection of ice on the conductors and structures, and the variations in wind pressure on conductors and structures. The map shown in Fig. 1 is usually taken as a basis for determining the heaviest loading conditions for a line in a certain locality.

When local conditions are known to vary from the general conditions in the surrounding area indicated on this map, such variations should be taken into account. The loading on conductors is assumed to be the resultant loading per foot equivalent to the vertical loading per foot of the conductor, including the ice covering, if any, combined with the transverse loading per foot due to the transverse horizontal wind pressure upon the projected area of the conductor, including the ice covering, if any, to which equivalent resultant is added a constant. In Table 2 are given the vertical and transverse loadings for each loading area, together with the proper constants to be added to the resultant loadings to obtain the final total loading.

Table 2. Conductor Loading

	Loading District		
	Heavy	Medium	Light
Radial thickness of ice, in inches	0.50	0.25	0
Horizontal wind pressure, in pounds per square foot	4	4	9
Temperature, in degrees fahrenheit	0	+15	+30
Constant to be added to the resultant, in pounds per foot:			
For bare conductors of copper, steel, copper alloy, copper-covered steel, and combinations thereof	0.29	0.19	0.05
For bare conductors of aluminum (with or without steel reinforcement)	0.31	0.22	0.05
For weatherproof and similar covered conductors (all materials)	0.31	0.22	0.05

TEMPERATURE RANGE. The maximum and minimum air temperatures which have been observed in any locality for a period of years can be obtained from the records of the United States Weather Bureau or from similar records for other countries. The minimum air temperature recorded (which may be as low as −40 deg fahr in some of the northern states) will be the minimum temperature which the conductor may be expected to reach. However, since the conductors are exposed to the direct rays of the sun, they will reach a maximum temperature in the summer considerably in excess of the Weather Bureau records, which give the temperatures in the shade.

Another important temperature which should be determined is that of the wire when coated with ice. As noted below, a sleet storm is usually followed by a fall in temperature, and although the ice forms at 32 deg fahr, the wire may reach a much lower temperature while the ice is on it.

The following temperature ranges have been used in the design of certain lines:

	Maximum	Minimum
Eastern Canada	+120 deg fahr	−40 deg fahr
Mississippi Valley	+120 deg fahr	−20 deg fahr
Southern California	+140 deg fahr	+10 deg fahr

COLLECTION OF ICE ON WIRES. Records of the Weather Bureau lead to the conclusion that sleet and ice storms are generally followed by falling temperatures and high winds, and transmission lines should be designed to meet these conditions. Records indicate that under favorable conditions ice and sleet will collect on wires and cables to the same amount in any climate where freezing temperatures are obtained. Mild, moderate, and cold climates differ in the frequency with which conditions are favorable. In general, sleet storms are most frequent in the moderate climates, since precipitation takes place more often at freezing temperatures. Destructive sleet storms occur in the eastern part of the United States at least as far south as Atlanta. One-half inch thickness of solid ice on wires and cables is generally assumed in designing transmission lines, but thicknesses of one-quarter and three-quarters inch are also assumed in the more favorable and unfavorable localities, respectively.

Ice and sleet generally collect quite uniformly on wires throughout their length. The collection is sometimes in the form of icicles but more often is egg-shaped in cross-section, with the wire in the small end of the section. Ice frequently falls off non-uniformly in sections.

Clear solid ice weighs 57 lb per cu ft or 0.033 lb per cu in., but sleet or frozen snow, such as often collects on wires, weighs much less, sometimes as little as 8 lb per cu ft.

WIND PRESSURE. Wind pressure is a subject upon which little exact information exists, although many experiments have been made and much study has been given to the subject by engineers and scientists. Among the unsettled questions are:

a. The relation between pressure and velocity.

b. The variation of pressure with size and shape of exposed plane surfaces.

c. The direction and intensity of pressure upon non-vertical surfaces.

d. The intensity of pressure upon non-planar surfaces.

e. The total pressure upon a number of parallel bars or other members placed side by side.

f. The decrease of pressure upon leeward surfaces.

g. The lifting power of the wind.

Relation between Indicated (United States Weather Bureau) Wind Velocity and Actual Velocity. The indications of the anemometers used by the United States Weather Bureau give, not the *actual* wind velocity, but values considerably higher than the actual velocities, as shown in Table 3.

Table 3. Relation between Indicated and Actual Wind Velocity

Indicated Velocity, mi per hr	Actual Velocity, mi per hr	Indicated Velocity, mi per hr	Actual Velocity, mi per hr
10	9.6	60	48.0
20	17.8	70	55.2
30	25.7	80	62.2
40	33.3	90	69.2
50	40.8	100	76.2

In the United States Weather Bureau reports the indicated and not the actual wind velocities are given. However, as the anemometers used give the *average* velocity for several minutes, the instantaneous velocities due to sudden gusts may be considerably greater than the indicated velocities; the indicated velocity probably more nearly represents the "gust" velocity than the actual average velocity. In all calculations of maximum wind pressure it is therefore recommended that the *indicated* velocity be used.

The Weather Bureau records give no indication of the "gust" velocities which may occur during the 5-min periods, and which may greatly exceed the average velocity. Tests with a Dines pressure tube anemometer have shown that the extreme maximum is about 50 per cent greater than the average for short periods.

The extreme maximum wind velocity observed in Chicago in the whole 36 year period from 1873 to 1910 was 84 miles per hour (uncorrected) in February, 1894. A velocity of 76 miles per hour (uncorrected) was observed once in November, 1898, and a velocity of 72 miles per hour (uncorrected) was observed seven times. During the 10-year period from 1894 to 1903 the maximum wind velocities in a few other representative localities were as follows, all velocities being the observed or uncorrected velocities: Bismarck, N.D., 72; Eastport, Me., 78; Buffalo, N. Y., 90; New York City, N. Y., 78; Galveston, Tex., 84; Savannah, Ga., 76; Salt Lake City, Utah, 60. All the maxima range between 60 miles and 90 miles per hour. (See N.E.L.A. *Overhead Systems Reference Book* for more complete data.)

Relation between Pressure and Velocity. The pressure varies about as the square of the velocity, the results given by different experimenters for the pressure due to a *normal wind on a plane surface* ranging from

$$P = 0.004V^2 \quad \text{to} \quad P = 0.0032V^2 \tag{2}$$

where P = pressure in pounds per square foot.
V = actual wind velocity in miles per hour.

The former of these values is for small flat surfaces; the latter represents the results of unusually careful experiments by Stanton (see *Minutes of Proc. Inst. Civil Engineers,*

Vols. 156 and 171) upon the intensity of pressure on plates varying in size from 25 to 100 sq ft and is probably more nearly correct than the higher value. In the Stanton formula the values are reduced to correspond to a temperature of 60 deg fahr and an atmospheric pressure of 14.7 lb per sq in., i.e., barometric pressure of 30 in.

The influence of size and shape of exposed surface is an important question and is not well understood, although it is known that the resultant pressure on a large surface may be taken as less per square foot than that on a small surface, since the maximum intensity of the wind is due to gusts of comparatively small cross-section.

Formulas for Pressure on Plane Surfaces When Wind Is Not Normal. The pressure upon vertical plane surfaces may be taken as normal to the surface and equal in intensity to the assumed wind pressure. Upon surfaces which are not vertical, the pressure is usually considered to be normal to the surface but lower in intensity than upon vertical surfaces. The variation in pressure with respect to the slope is not well understood, and a number of empirical formulas are in use, among which are the Duchemin formula

$$P_n = P\frac{2 \sin i}{1 + \sin^2 i} \tag{3}$$

and the Hutton formula

$$P_n = P(\sin i)^{(1.84 \cos i - 1)} \tag{4}$$

where P = intensity of normal pressure upon the vertical surface.
P_n = intensity of normal pressure upon the given surface.
i = angle made by surface with the horizontal.

The following theoretical formula results from the assumption that the wind always blows in horizontal lines, and that if the pressure is resolved into normal and tangential components, the tangential component may be neglected:

$$P_n = P \sin^2 i \tag{5}$$

This formula gives lower values than the empirical formulas and probably gives too low results, since it makes no allowance for the reduction in pressure on the leeward side which is known to occur, and which may in part be attributed to the influence of the tangential component. It should also be noted that the wind does not blow uniformly in horizontal lines but may deviate considerably from the horizontal.

The values given by these three formulas are tabulated for comparison in Table 4, an assumed value of 30 lb per sq ft being used for P. In the absence of further experience upon this phase of wind pressure it would seem wise to use one of the empirical formulas instead of the theoretical one. The Hutton formula is used quite generally by structural engineers in England and the United States.

Pressure on Non-planar Surfaces. The pressure upon non-planar surfaces is important in the case of chimneys, standpipes, and other similar objects.

Upon the same assumptions as were made in the preceding paragraph, it may be demonstrated that theoretically the pressure on a cylinder is two-thirds of the total pressure on a plane diametrical section. This value is quite generally used. The pressures thus obtained lack experimental proof but are probably more nearly correct than the pressure obtained by the same method upon plane surfaces.

Effect of Reduction of Pressure on Leeward Side. The pressure upon the windward side of an exposed surface is a function of the density and velocity of the air currents. The pressure on the leeward side is also a function of the shape of the surface and has been shown by numerous experiments to be less than the static pressure of the air current. The resultant total pressure upon a surface is in consequence a function not only of the direct pressure on the windward side, but also of the pressure on the leeward side, which in turn is a function of the form of the surface. No algebraic formula can be given which will indicate the pressure on surfaces of varying shape with any considerable degree of precision.

Wind Pressure on Wires. H. W. Buck [*Trans. Int. Elec. Cong., St. Louis,* Vol. 2, p. 318 (1904)] gives the following formula for the pressure due to a normal wind on a stranded wire:

$$P = 0.0025V^2 \tag{6}$$

where P is the pressure in pounds per square foot of projected area of wire and sleet, if any (length times diameter), and V is the velocity in miles per hour. This formula is based upon tests made on a 950-ft span at Niagara Falls; the wind velocities were measured by a United States Weather Bureau anemometer corrected to give actual average velocities.

Table 4. Wind Pressure in Pounds per Square Foot

(P = 30 lb. per sq. ft)

Angle i, degrees	Theoretical $P \sin^2 i$	Duchemin $P \frac{2 \sin i}{1 + \sin^2 i}$	Hutton $P (\sin i)^{(1.84 \cos i - 1)}$
5	0.0	5.2	3.9
10	0.9	10.1	7.3
15	2.0	14.6	10.5
20	3.5	18.4	13.7
25	5.3	21.5	16.9
30	7.5	24.0	19.9
35	9.9	25.8	22.6
40	12.4	27.3	25.1
45	15.0	28.3	27.0
50	17.6	29.0	28.6
55	20.1	29.4	29.7
60	22.5	....	....
65	24.6	Above 60 deg	Above 60 deg
70	26.4	use 30 lb	use 30 ib
75	28.0		
80	29.1		
85	29.7		
90	30.0		

Practical Rules for Wind-pressure Allowance. The many uncertainties connected with wind pressure make worthless the attempts to specify with precision its magnitude and direction. In the lack of additional information and further theoretical studies there seems to be no reason for deviating from the common rules which have been used for many years with satisfactory results.

For wind pressure on roofs and buildings it is common practice to allow 30 lb per sq ft acting horizontally on the sides and ends of buildings or on the vertical projection of roofs. It is also very important to figure the wind stresses on the steel frame, considering it as an independent structure without walls, floors, or partitions, since failures often occur in erection.

Steel Towers. The National Electrical Safety Code requires the stated wind pressure for a given loading area to be increased 60 per cent for flat surfaces. For latticed structures the wind pressure is taken on 150 per cent of the actual exposed area of one lateral face to allow for pressure on the opposite face. For grade A, B, and C construction the wind pressure is taken as three times the specified wind pressure for the loading area in computing the strength of the tower without conductors.

Conductor Vibration and Dancing. Both these phenomena are caused by wind action, but they occur under quite different circumstances and should not be confused. In "vibration" the conductor is set into vibration by light breezes, which produce periodically varying eddy turbulence on the leeward side of the conductor. In short spans the amplitude is usually extremely small, and the vibration is evident only by the humming sound produced. As a result of this musical tone the phenomenon has been called aeolian vibration.

The conductor vibrates in loops between nodal points. These loops travel back and forth across the span as the wind pressure varies, but occasionally stand still if the wind blows absolutely steadily for a short period. The frequency and amplitude of the vibration are functions of the wind velocity, the span length, the distance between nodes, the tension in the conductor, the diameter of the conductor, and its weight per foot. These functions have been expressed in fundamental equations. (See Bibliography.)

Such vibrations cause fatigue of the conductor at the point of support, which eventually results in breakage at that point. The remedies consist of preventing the vibration or absorbing the energy which produces it. To prevent the vibration from starting it is necessary to use a different size or type of conductor or to reduce the tension below the critical value. Such remedies are seldom feasible economically; therefore it is necessary to use damping devices. The most commonly used and the most successful device is the Stockbridge damper shown in Fig. 2. Torsional dampers have also been useful in some cases. One or more of these devices is required on the conductor at each side of the support.

Armor rods absorb only about 10 per cent as much energy as the Stockbridge damper, but on short spans with small conductors they are quite effective. A layer of small armor rods surrounding the conductor and extending each side of the support is twisted in the same direction as the lay of the outer strands of the conductor and clamped at each end to the conductor. On short spans special insulator ties, with the tie wire twisted around the conductor and extending out on it for 6 or 8 in., serve the same purpose.

Dancing is caused by gusty winds acting upon ice-coated conductors when the ice coating is irregular in shape. Traveling waves having large amplitudes are produced. No means of damping these vibrations has yet been found, and the forces developed are often sufficient to break conductors and wreck supporting structures. The only way known to

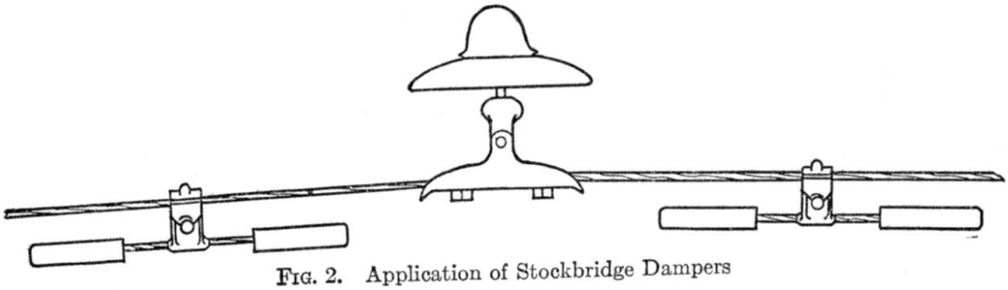

FIG. 2. Application of Stockbridge Dampers

prevent the phenomenon is to keep the ice coat from forming or to melt it off quickly enough to avoid damage. Most operators in glaze areas have established routines for increasing the current in the conductors sufficiently to melt the ice or to prevent its formation.

VERTICAL AND TRANSVERSE FORCES ON A SUSPENDED WIRE. The resultant force acting on one foot of a suspended wire is in general made up of three components:

c = weight of the conductor (including insulation, if any) per foot length, in pounds.
i = weight of the ice coating per foot length of the conductor, in pounds.
h = wind pressure per foot length of the conductor, in pounds.

The weight of the conductor per foot length may be taken directly from the tables in Articles 54–62, Bare Wires and Cables. Let d be the diameter in inches, and let t be the thickness of the ice coating; then the weight of the ice coating per foot length of the conductor is

$$i = 1.24t(d + t) \tag{7}$$

Let p be the wind pressure per square foot of projected area; then the wind pressure per foot length of the conductor, i.e., the horizontal component of the resultant force, is

$$h = \frac{p(d + 2t)}{12} \tag{8}$$

The vertical component of the resultant force per foot length of conductor, which is equal to the resultant force for no wind, is

$$v = c + i \tag{9}$$

Values of v and h for various ice thicknesses and various sizes of wires are given in Table 5. When v and h are known, the resultant force w for any combination of wind and ice loads is readily determined by the formula

$$w = \sqrt{v^2 + h^2} + \text{Constant (if required)} \tag{10}$$

Table 5. Strength; Vertical, Transverse, and Total Loadings

Conductor size, Circular Mils or A.W.G.	Overall Diameter, Inches	Cross-sectional Area, Square Inch	Ultimate Strength,* Pounds			Loading, Pounds per Linear Foot of Conductor								
						Light Loading			Medium Loading			Heavy Loading		
			Hard Drawn (Min.)	Medium Hard Drawn (Min.)	Annealed (Max.)	Vertical (Weight of Conductor Without Ice)	Transverse (9 Lb Wind Per Sq Ft on Conductor Without Ice)	Total Conductor Loading †	Vertical (Conductor Plus 1/4 In. Radial Ice)	Transverse (4 Lb Wind Per Sq Ft on Ice-covered Conductor)	Total Conductor Loading †	Vertical (Conductor Plus 1/2 In. Radial Ice)	Transverse (4 Lb Wind Per Sq Ft on Ice-covered Conductor)	Total Conductor Loading †
Copper Wire, Solid—Bare														
4/0	0.4600	0.1662	8,143	6,980	5,983	0.6405	0.3450	0.7775	0.8613	0.3200	1.1038	1.2376	0.4867	1.6199
3/0	0.4096	0.1318	6,722	5,667	4,745	0.5079	0.3072	0.6436	0.7130	0.3032	0.9648	1.0737	0.4699	1.4620
2/0	0.3648	0.1045	5,519	4,599	3,763	0.4028	0.2736	0.5869	0.5940	0.2883	0.9503	0.9407	0.4549	1.3349
1/0	0.3249	0.08289	4,517	3,730	2,984	0.3195	0.2437	0.4518	0.4983	0.2750	0.7592	0.8326	0.4416	1.2325
1	0.2893	0.06573	3,688	3,024	2,432	0.2533	0.2170	0.3835	0.4210	0.2631	0.6865	0.7441	0.4298	1.1493
2	0.2576	0.05213	3,003	2,450	1,929	0.2009	0.1932	0.3287	0.3587	0.2525	0.6287	0.6720	0.4192	1.0820
3	0.2294	0.04134	2,439	1,984	1,530	0.1593	0.1721	0.2845	0.3083	0.2431	0.5826	0.6128	0.4098	1.0272
4	0.2043	0.03278	1,970	1,584	1,213	0.1264	0.1532	0.2486	0.2676	0.2348	0.5460	0.5643	0.4014	0.9825
5	0.1819	0.02600	1,591	1,265	961.9	0.1002	0.1364	0.2193	0.2345	0.2273	0.5166	0.5242	0.3940	0.9458
6	0.1620	0.02062	1,280	1,010	762.9	0.07946	0.1215	0.1952	0.2075	0.2207	0.4929	0.4911	0.3873	0.9155
7	0.1443	0.01635	1,030	806.6	605.0	0.06302	0.1082	0.1752	0.1856	0.2148	0.4739	0.4636	0.3814	0.8903
8	0.1285	0.01297	826.0	643.9	479.8	0.04997	0.09638	0.1586	0.1676	0.2095	0.4583	0.4408	0.3762	0.8695
9	0.1144	0.01028	661.2	514.2	380.5	0.03963	0.08580	0.1445	0.1530	0.2048	0.4456	0.4218	0.3715	0.8521
10	0.1019	0.008155	529.2	410.4	314.0	0.03143	0.07643	0.1326	0.1409	0.2006	0.4351	0.4058	0.3673	0.8373
11	0.09074	0.006467	422.9	327.6	249.0	0.02492	0.06806	0.1225	0.1309	0.1969	0.4264	0.3924	0.3636	0.8250
12	0.08081	0.005129	337.0	261.6	197.5	0.01977	0.06061	0.1138	0.1227	0.1936	0.4192	0.3810	0.3603	0.8144
Copper Cable—Bare														
1,000,000														
61 strand	1.152	0.7854	45,030	35,100	29,060	3.088	0.8640	3.2566	3.5240	0.5504	3.7567	4.1156	0.7173	4.4676
37 strand	1.151	0.7854	43,830	34,350	29,060	3.088	0.8633	3.2564	3.5237	0.5504	3.7564	4.1149	0.7170	4.4669
750,000														
61 strand	0.998	0.5890	34,090	26,510	21,790	2.316	0.7485	2.4840	2.7041	0.4993	2.9398	3.2478	0.6660	3.6054
37 strand	0.997	0.5890	33,400	26,150	21,790	2.316	0.7478	2.4837	2.7038	0.4990	2.9395	3.2471	0.6657	3.6046
500,000														
37 strand	0.814	0.3927	22,510	17,550	14,530	1.544	0.6105	1.7103	1.8749	0.4380	2.1154	2.3613	0.6047	2.7275
19 strand	0.811	0.3927	21,950	17,320	14,530	1.544	0.6083	1.7095	1.8740	0.4370	2.1143	2.3594	0.6037	2.7254

400,000 19 strand	0.726	0.3142	17,560	13,850	11,620	1.235	0.5445	1.3997	1.5385	0.4087	1.7819	1.9976	0.5753	2.3688
350,000														
19 strand	0.679	0.2749	15,590	12,200	10,170	1.081	0.5093	1.2450	1.3699	0.3930	1.6152	1.9143	0.5597	2.1887
12 strand	0.710	0.2749	15,140	12,020	10,170	1.081	0.5325	1.2550	1.3796	0.4033	1.6274	1.9336	0.5700	2.2102
300,000														
19 strand	0.629	0.2356	13,510	10,530	8,718	0.9263	0.4718	1.0895	1.1997	0.3763	1.4473	1.6285	0.5430	2.0066
12 strand	0.657	0.2356	13,170	10,390	8,718	0.9263	0.4928	1.0992	1.2084	0.3857	1.4585	1.6460	0.5523	2.0262
250,000														
19 strand	0.574	0.1964	11,360	8,836	7,265	0.7719	0.4305	0.9338	1.02816	0.3580	1.2787	1.4398	0.5247	1.8224
12 strand	0.600	0.1964	11,130	8,717	7,265	0.7719	0.4500	0.9435	1.03625	0.3667	1.2892	1.4561	0.5333	1.8407
4/0														
12 strand	0.552	0.1662	9,483	7,378	6,149	0.6533	0.4140	0.8234	0.9027	0.3507	1.1584	1.3076	0.5173	1.6962
7 strand	0.522	0.1662	9,154	7,269	6,149	0.6533	0.3915	0.8116	0.8934	0.3407	1.1462	1.2890	0.5073	1.6752
3/0														
12 strand	0.492	0.1318	7,556	5,890	4,876	0.5181	0.3690	0.6861	0.7489	0.3307	1.0087	1.1351	0.4973	1.5293
7 strand	0.464	0.1318	7,366	5,812	4,876	0.5181	0.3480	0.6741	0.7402	0.3213	0.9969	1.1177	0.4880	1.5096
2/0 7 strand	0.414	0.1045	5,926	4,641	3,868	0.4109	0.3105	0.5650	0.6174	0.3047	0.8785	0.9794	0.4713	1.3769
1/0 7 strand	0.368	0.08289	4,752	3,703	3,066	0.3257	0.2760	0.4769	0.5179	0.2893	0.7832	0.8656	0.4560	1.2684
1														
7 strand	0.328	0.06573	3,804	2,958	2,432	0.2584	0.2460	0.4068	0.4382	0.2760	0.7079	0.7734	0.4427	1.1811
3 strand	0.360	0.06573	3,620	2,875	2,432	0.2559	0.2700	0.4220	0.4456	0.2867	0.7199	0.7908	0.4533	1.2015
2														
7 strand	0.292	0.05213	3,045	2,361	2,007	0.2049	0.2190	0.3499	0.3735	0.2640	0.6474	0.6975	0.4307	1.1098
3 strand	0.320	0.05213	2,913	2,299	1,929	0.2029	0.2400	0.3643	0.3802	0.2733	0.6582	0.7129	0.4400	1.1278
3 3 strand	0.285	0.04134	2,359	1,835	1,529	0.1609	0.2138	0.3176	0.3273	0.2617	0.6091	0.6492	0.4283	1.0678
4 3 strand	0.254	0.03278	1,879	1,465	1,213	0.1276	0.1905	0.2793	0.2843	0.2513	0.5694	0.5966	0.4180	1.0185
6 3 strand	0.201	0.02062	1,203	933.9		0.08026	0.1508	0.2035	0.2205	0.2337	0.5113	0.5163	0.4003	0.9433

* The values in this column are called Breaking Strength by some wire manufacturers.

† The values in this column give the resultant conductor loading per foot as specified in Rule 251, National Electrical Safety Code, Fifth Edition.

Table 5. Strength; Vertical, Transverse, and Total Loadings—*Continued*

Conductor Size, Circular Mils or A.W.G.	Ultimate Strength,* Pounds			Loading, Pounds per Linear Foot of Conductor								
				Light Loading			Medium Loading			Heavy Loading		
	Hard Drawn (Min.)	Medium Hard Drawn (Min.)	Annealed Max.	Vertical (Weight of Conductor Without Ice)	Transverse (9 Lb Wind Per Sq Ft on Conductor Without Ice)	Total Conductor Loading †	Vertical (Conductor Plus 1/4 In. Radial Ice)	Transverse (4 Lb Wind Per Sq Ft on Ice-covered Conductor)	Total Conductor Loading †	Vertical (Conductor Plus 1/2 In. Radial Ice)	Transverse (4 Lb Wind Per Sq Ft on Ice-covered Conductor)	Total Conductor Loading †
Copper, Stranded—Triple Braid Weatherproof												
1,000,000	41,610	32,630	30,510	3.670	1.275	3.935	4.276	0.733	4.559	5.038	0.900	5.429
750,000	31,730	24,840	22,880	2.822	1.125	3.088	3.366	0.667	3.652	4.066	0.833	4.460
500,000	21,390	16,450	15,260	1.894	0.938	2.163	2.361	0.583	2.651	2.983	0.750	3.385
350,000	14,380	11,590	10,680	1.345	0.788	1.609	1.749	0.517	2.053	2.309	0.683	2.718
250,000	10,790	8,394	7,628	0.984	0.683	1.248	1.346	0.470	1.646	1.862	0.637	2.278
4/0	8,696	7,105	6,456	0.800	0.623	1.064	1.136	0.443	1.440	1.627	0.610	2.048
3/0	6,998	5,521	5,119	0.653	0.563	0.912	0.964	0.417	1.270	1.431	0.583	1.855
2/0	5,630	4,409	4,061	0.522	0.503	0.775	0.808	0.390	1.117	1.250	0.557	1.678
1/0	4,514	3,518	3,219	0.424	0.465	0.679	0.695	0.373	1.009	1.121	0.540	1.554
1	3,614	2,810	2,554	0.328	0.413	0.577	0.578	0.350	0.896	0.981	0.517	1.419
2	2,893	2,243	2,107	0.270	0.383	0.518	0.507	0.337	0.829	0.898	0.503	1.339
Copper, Solid—Triple Braid Weatherproof												
4/0	7,740	6,630	6,280	0.767	0.578	1.010	1.084	0.423	1.384	1.557	0.590	1.975
3/0	6,390	5,380	4,980	0.629	0.533	0.875	0.927	0.403	1.231	1.381	0.570	1.804
2/0	5,240	4,370	3,950	0.502	0.473	0.740	0.776	0.377	1.083	1.205	0.543	1.632
1/0	4,290	3,540	3,130	0.407	0.428	0.641	0.662	0.357	0.972	1.072	0.523	1.503
1	3,503	2,870	2,550	0.316	0.390	0.552	0.555	0.340	0.872	0.950	0.507	1.387
2	2,850	2,330	2,030	0.260	0.356	0.492	0.485	0.325	0.803	0.866	0.492	1.306
3	2,320	1,885	1,606	0.199	0.315	0.423	0.407	0.307	0.730	0.771	0.473	1.215
4	1,872	1,505	1,274	0.164	0.293	0.386	0.363	0.297	0.689	0.717	0.463	1.164
6	1,216	960	801	0.112	0.251	0.325	0.294	0.278	0.625	0.631	0.445	1.082
8	785	612	504	0.075	0.218	0.281	0.243	0.263	0.578	0.566	0.430	1.021
10	503	390	330	0.053	0.188	0.245	0.208	0.250	0.545	0.519	0.417	0.976
12	402	249	207	0.035	0.158	0.212	0.176	0.237	0.516	0.476	0.403	0.934

Aluminum Cable, Steel Reinforced

Conductor Size, Circular Mils or A.W.G. and Stranding	Overall Diameter, Inches	Cross-sectional Area, Square Inches	Ultimate Strength,* Pounds	Loading, Pounds per Linear Foot of Conductor								
				Light Loading			Medium Loading			Heavy Loading		
				Vertical (Weight of Conductor Without Ice)	Transverse (9 Lb Wind Per Sq Ft on Conductor Without Ice)	Total Conductor Loading †	Vertical (Conductor Plus 1/4 In. Radial Ice)	Transverse (4 Lb Wind Per Sq Ft on Ice-covered Conductor)	Total Conductor Loading †	Vertical (Conductor Plus 1/2 In. Radial Ice)	Transverse (4 Lb Wind Per Sq Ft on Ice-covered Conductor)	Total Conductor Loading †
1,590,000 54A/19St	1.545	1.4070	56,000	2.045	1.159	2.401	2.603	0.6817	2.911	3.317	0.8483	3.734
1,510,000 54A/19St	1.506	1.3366	53,200	1.943	1.130	2.298	2.489	0.6687	2.797	3.191	0.8353	3.609
1,431,000 54A/19St	1.465	1.2663	50,400	1.840	1.099	2.193	2.373	0.6550	2.682	3.062	0.8217	3.480
1,351,500 54A/19St	1.424	1.959	47,600	1.738	1.068	2.091	2.259	0.6413	2.568	2.935	0.8080	3.354
1,272,000 54A/19St	1.382	1.1256	44,800	1.636	1.037	1.987	2.144	0.6273	2.454	2.807	0.7940	3.227
1,192,500 54A/19St	1.338	1.0552	43,100	1.534	1.006	1.884	2.028	0.6127	2.339	2.677	0.7793	3.095
1,113,000 54A/19St	1.293	0.9849	40,200	1.429	0.970	1.777	1.909	0.5977	2.220	2.544	0.7643	2.966
1,033,500 54A/7St	1.246	0.9169	37,100	1.330	0.935	1.676	1.795	0.5820	2.107	2.416	0.7487	2.839
954,000 54A/7St	1.196	0.8464	34,200	1.227	0.897	1.570	1.677	0.5653	1.990	2.282	0.7320	2.707
900,000 54A/7St	1.162	0.7985	32,300	1.158	0.8715	1.499	1.597	0.5540	1.910	2.192	0.7207	2.617
874,500 54A/7St	1.146	0.7759	31,400	1.125	0.8595	1.466	1.559	0.5487	1.873	2.149	0.7153	2.575
795,000 30A/19St	1.140	0.7668	38,400	1.237	0.8550	1.554	1.669	0.5467	1.976	2.257	0.7133	2.677
26A/7St	1.108	0.7261	31,200	1.098	0.8310	1.427	1.520	0.5360	1.832	2.098	0.7027	2.523
54A/7St	1.093	0.7053	28,500	1.026	0.8198	1.363	1.444	0.5310	1.759	2.017	0.6977	2.444
715,500 30A/19St	1.081	0.6901	34,600	1.111	0.8108	1.425	1.525	0.5270	1.833	2.094	0.6937	2.516
26A/7St	1.051	0.6535	28,100	0.987	0.7883	1.313	1.392	0.5170	1.705	1.952	0.6837	2.378
54A/7St	1.036	0.6348	26,300	0.922	0.7770	1.256	1.322	0.5120	1.638	1.877	0.6787	2.306

* The values in this column are called Breaking Strength by some wire manufacturers.
† The values in this column give the resultant conductor loading per foot as specified in Rule 251, National Electrical Safety Code, Fifth Edition.

Table 5. Strength; Vertical, Transverse and Total Loadings—*Continued*

Aluminum Cable, Steel Reinforced

Conductor Size, Circular Mils or A.W.G. and Stranding	Overall Diameter, Inch	Cross-sectional Area, Square Inch	Ultimate Strength,* Pounds	Loading, Pounds per Linear Foot of Conductor								
				Light Loading			Medium Loading			Heavy Loading		
				Vertical (Weight of Conductor Without Ice)	Transverse (9 Lb Wind Per Sq Ft on Conductor Without Ice)	Total Conductor Loading †	Vertical (Conductor Plus 1/4 In. Radial Ice)	Transverse (4 Lb Wind Per Sq Ft on Ice-covered Conductor)	Total Conductor Loading †	Vertical (Conductor Plus 1/2 In. Radial Ice)	Transverse (4 Lb Wind Per Sq Ft on Ice-covered Conductor)	Total Conductor Loading †
666,600												
54A/7St	1.000	0.5914	24,500	0.858	0.7500	1.190	1.247	0.5000	1.564	1.791	0.6667	2.221
636,000												
30A/19St	1.019	0.6134	31,500	0.986	0.7643	1.297	1.381	0.5063	1.691	1.931	0.6730	2.355
26A/7St	0.990	0.5809	25,000	0.874	0.7425	1.197	1.262	0.4967	1.576	1.801	0.6633	2.229
54A/7St	0.977	0.5643	23,600	0.818	0.7328	1.148	1.199	0.4923	1.516	1.737	0.6590	2.168
605,000												
30A/19St	0.994	0.5835	30,000	0.938	0.7455	1.248	1.325	0.4980	1.635	1.867	0.6647	2.292
26A/7St	0.966	0.5526	24,100	0.832	0.7275	1.155	1.211	0.4900	1.526	1.746	0.6567	2.175
54A/7St	0.953	0.5368	22,500	0.779	0.7148	1.107	1.153	0.4843	1.471	1.683	0.6510	2.115
556,500												
30A/7St	0.953	0.5391	27,200	0.870	0.7148	1.176	1.244	0.4843	1.555	1.774	0.6510	2.200
26A/7St	0.927	0.5083	22,400	0.765	0.6953	1.084	1.131	0.4757	1.447	1.653	0.6423	2.083
500,000												
30A/7St	0.904	0.4843	24,400	0.7819	0.6780	1.085	1.140	0.4680	1.452	1.654	0.6347	2.082
477,000												
30A/7St	0.883	0.4620	23,300	0.7459	0.6623	1.0475	1.098	0.4610	1.411	1.606	0.6277	2.034
26A/7St	0.858	0.4356	19,430	0.6557	0.6435	0.9685	1.000	0.4527	1.318	1.500	0.6193	1.933
397,500												
30A/7St	0.806	0.3850	19,980	0.6216	0.6045	0.9171	0.9500	0.4353	1.2650	1.434	0.6020	1.865
26A/7St	0.783	0.3630	16,190	0.5464	0.5873	0.8522	0.8677	0.4277	1.1874	1.344	0.5943	1.780
336,400												
30A/7St	0.741	0.3259	17,040	0.5261	0.5558	0.8153	0.8343	0.4137	1.1512	1.298	0.5803	1.732
26A/7St	0.721	0.3072	14,050	0.4624	0.5408	0.7615	0.7644	0.4070	1.0860	1.222	0.5737	1.660
300,000												
30A/7St	0.700	0.2906	15,430	0.4691	0,5250	0.7540	0.7646	0.4000	1.0829	1.216	0.5667	1.652
26A/7St	0.680	0.2740	12,650	0.4110	0.5100	0.7050	0.7002	0.3933	1.0231	1.145	0.5600	1.585
266,800												
26A/7St	0.642	0.2436	11,250	0.3656	0.4815	0.6546	0.6430	0.3807	0.9673	1.076	0.5473	1.517
6A/7St	0.633	0.2367	9,645	0.3413	0.4748	0.6347	0.6159	0.3777	0.9425	1.046	0.5443	1.489
203,000												
8A/7St	0.584	0.2025	11,140	0.3356	0.4380	0.6018	0.5950	0.3613	0.9161	1.010	0.5280	1.444

203,200 16A/19St	0.714	0.3020	27,500	0.6749	0.5355	0.9115	0.9747	0.4047	1.275	1.430	0.5713	1.850
211,300 12A/7St	0.663	0.2628	19,640	0.5263	0.4973	0.7741	0.8099	0.3877	1.118	1.250	0.5543	1.677
190,800 12A/7St	0.631	0.2373	17,730	0.4752	0.4733	0.7207	0.7492	0.3770	1.058	1.179	0.5437	1.608
176,900 12A/7St	0.607	0.2200	16,440	0.4406	0.4553	0.6836	0.7071	0.3690	1.018	1.129	0.5357	1.560
159,000 12A/7St	0.576	0.1977	15,200	0.3960	0.4320	0.6360	0.6529	0.3587	0.9650	1.065	0.5253	1.498
134,600 12A/7St	0.530	0.1674	12,920	0.3352	0.3975	0.5700	0.5778	0.3433	0.8921	0.9759	0.5100	1.411
110,800 12A/7St	0.481	0.1378	10,730	0.2760	0.3608	0.5043	0.5033	0.3270	0.8202	0.8862	0.4937	1.324
101,800 12A/7St	0.461	0.1266	9,860	0.2536	0.3458	0.4788	0.4747	0.3203	0.7927	0.8513	0.4870	1.291
80,000 8A/1St	0.367	0.0847	5,200	0.1498	0.2753	0.3634	0.3417	0.2890	0.6675	0.6891	0.4557	1.136
4/0 6A/1St	0.563	0.1939	8,420	0.2921	0.4223	0.5634	0.5449	0.3543	0.8700	0.9533	0.5210	1.396
3/0 6A/1St	0.502	0.1538	6,675	0.2316	0.3765	0.4920	0.4655	0.3340	0.7929	0.8548	0.5007	1.301
2/0 6A/1St	0.447	0.1219	5,345	0.1837	0.3353	0.4323	0.4005	0.3157	0.7300	0.7727	0.4823	1.221
1/0 6A/1St	0.398	0.0967	4,280	0.1456	0.2985	0.3821	0.3471	0.2993	0.6783	0.7042	0.4660	1.154
1 6A/1St	0.355	0.0767	3,480	0.1155	0.2663	0.3403	0.3037	0.2850	0.6365	0.6473	0.4517	1.099
2 7A/1St 6A/1St	0.325 0.316	0.0653 0.0608	3,525 2,790	0.1072 0.0916	0.2438 0.2370	0.3163 0.3041	0.2860 0.2676	0.2750 0.2720	0.6168 0.6016	0.6204 0.5992	0.4417 0.4387	1.072 1.053
3 6A/1St	0.281	0.0482	2,250	0.0727	0.2108	0.2730	0.2378	0.2603	0.5726	0.5585	0.4270	1.013
4 7A/1St 6A/1St	0.257 0.250	0.0411 0.0383	2,288 1,830	0.0674 0.0576	0.1928 0.1875	0.2542 0.2461	0.2251 0.2131	0.2523 0.2500	0.5581 0.5485	0.5382 0.5241	0.4190 0.4167	0.9921 0.9796
5 6A/1St	0.223	0.0303	1,460	0.0457	0.1673	0.2234	0.1928	0.2410	0.5286	0.4954	0.4077	0.9516
6 6A/1St	0.198	0.0240	1,170	0.0362	0.1485	0.2029	0.1755	0.2327	0.5115	0.4704	0.3993	0.9270

* The values in this column are called Breaking Strength by some wire manufacturers.
† The values in this column give the resultant conductor loading per foot as specified in Rule 251, National Electrical Safety Code, Fifth Edition.

Table 5. Strength; Vertical, Transverse and Total Loadings—*Continued*

Conductor Size,	Overall Diameter, Inch	Cross-sectional Area, Square Inch	Ultimate Strength,* Pounds	Loading, Pounds per Linear Foot of Conductor: Light Loading: Vertical (Weight of Conductor Without Ice)	Light Loading: Transverse (9 Lb Wind Per Sq Ft on Conductor Without Ice)	Light Loading: Total Conductor Loading †	Medium Loading: Vertical (Conductor Plus 1/4 In. Radial Ice)	Medium Loading: Transverse (4 Lb Wind Per Sq Ft on Ice-covered Conductor)	Medium Loading: Total Conductor Loading †	Heavy Loading: Vertical (Conductor Plus 1/2 In. Radial Ice)	Heavy Loading: Transverse (4 Lb Wind Per Sq Ft on Ice-covered Conductor)	Heavy Loading: Total Conductor Loading †
Copperweld–Copper and Copperweld—Bare (Copperweld Steel Company)												
2A	0.366	0.06799	5,876	0.2568	0.2745	0.4258	0.4483	0.2887	0.7232	0.7953	0.4553	1.2065
3A	0.326	0.05392	4,810	0.2036	0.2445	0.3682	0.3827	0.2753	0.6615	0.7172	0.4420	1.1325
4A	0.290	0.04276	3,938	0.1615	0.2175	0.3209	0.3294	0.2633	0.6118	0.6527	0.4300	1.0716
5A	0.258	0.03391	3,193	0.1281	0.1935	0.2821	0.2860	0.2527	0.5716	0.5994	0.4193	1.0216
6A	0.230	0.02689	2,585	0.1016	0.1725	0.2502	0.2508	0.2433	0.5395	0.5555	0.4100	0.9804
7A	0.223	0.02516	2,754	0.09366	0.1673	0.2417	0.2407	0.2410	0.5306	0.5432	0.4077	0.9692
8A	0.199	0.01995	2,233	0.07427	0.1493	0.2168	0.2139	0.2330	0.5063	0.5089	0.3997	0.9371
1/0F	0.388	0.09207	6,536	0.3541	0.2910	0.5083	0.5524	0.2960	0.8167	0.9062	0.4627	1.3075
1F	0.346	0.07303	5,266	0.2809	0.2595	0.4324	0.4662	0.2820	0.7349	0.8069	0.4487	1.2133
2F	0.308	0.05792	4,233	0.2228	0.2310	0.3709	0.3963	0.2693	0.6691	0.7252	0.4360	1.1362
4C	0.284	0.04098	3,231	0.1548	0.2130	0.3133	0.3208	0.2613	0.6038	0.6423	0.4280	1.0618
6C	0.225	0.02577	2,143	0.0973	0.1688	0.2448	0.2450	0.2417	0.5342	0.5481	0.4083	0.9735
8C	0.179	0.01604	1,362	0.06067	0.1343	0.1974	0.1940	0.2263	0.4881	0.4829	0.3930	0.9126
9 1/2D	0.174	0.01539	1,743	0.05646	0.1305	0.1922	0.1883	0.2247	0.4832	0.4755	0.3913	0.9059
3 No. 10	0.220	0.02446	2,882	0.08713	0.1650	0.2366	0.2332	0.2400	0.5246	0.5348	0.4067	0.9619
3 No. 11	0.196	0.01940	2,286	0.06910	0.1470	0.2124	0.2078	0.2320	0.5015	0.5019	0.3983	0.9310
3 No. 12	0.174	0.01539	2,040	0.05480	0.1305	0.1915	0.1866	0.2247	0.4821	0.4739	0.3913	0.9046
Amerductor—SCP and SCG (American Steel and Wire Company)												
2	0.392	0.0781	6,378	0.291	0.294	0.464	0.491	0.297	0.764	0.845	0.464	1.254
4	0.310	0.0489	4,486	0.182	0.233	0.346	0.356	0.270	0.637	0.685	0.437	1.103
6	0.248	0.0312	3,060	0.116	0.186	0.269	0.271	0.249	0.558	0.581	0.416	1.005
8	0.196	0.0195	2,112	0.073	0.147	0.214	0.212	0.232	0.504	0.506	0.399	0.934
8X	0.216	0.0234	2,700	0.086	0.162	0.233	0.231	0.239	0.522	0.531	0.405	0.958
9X	0.193	0.0186	2,346	0.069	0.145	0.211	0.207	0.231	0.500	0.500	0.398	0.929
10	0.220	0.0245	3,853	0.088	0.165	0.237	0.235	0.240	0.526	0.535	0.407	0.962
12	0.174	0.0154	2,426	0.055	0.131	0.192	0.187	0.225	0.483	0.474	0.391	0.904
Galvanized Steel Conductors—3 Wire (Bethlehem Steel Company)												
4R3-A	0.297	0.04487	5,610	0.156	0.223	0.322	0.326	0.266	0.611	0.652	0.432	1.072
6R3-A	0.252	0.03225	4,295	0.112	0.189	0.270	0.269	0.251	0.558	0.579	0.417	1.004
8R3-A	0.207	0.02171	2,915	0.075	0.155	0.222	0.217	0.236	0.511	0.515	0.402	0.943

Galvanized Steel Strand

Nominal Diameter, Inch	Overall Diameter, Inch	Cross-sectional Area, Square Inch	Ultimate Strength,* Pounds					Loading, Pounds per Linear Foot of Conductor								
								Light Loading			Medium Loading			Heavy Loading		
			Utilities Grade	Common Grade	Siemens-Martin Grade	High-strength Grade	Extra-high-Strength Grade	Vertical (Weight of Conductor Without Ice)	Transverse (9 Lb Wind Per Sq Ft on Conductor Without Ice)	Total Conductor Loading †	Vertical (Conductor Plus 1/4 In. Radial Ice)	Transverse (4 Lb Wind Per Sq Ft on Ice-covered Conductor)	Total Conductor Loading †	Vertical (Conductor Plus 1/2 In. Radial Ice)	Transverse (4 Lb Wind Per Sq Ft on Ice-covered Conductor)	Total Conductor Loading †
1/8 7 strand	0.123	0.00924	......	540	910	1,330	1,830	0.0318	0.092	0.147	0.147	0.208	0.445	0.420	0.374	0.852
5/32 7 strand	0.156	0.01487	......	870	1,470	2,140	2,940	0.0513	0.117	0.178	0.178	0.219	0.472	0.459	0.385	0.889
3/16 7 strand	0.186	0.02113	......	1,150	1,900	2,850	3,990	0.0729	0.140	0.208	0.208	0.229	0.499	0.499	0.395	0.926
7 strand	0.195	0.02323	2,400	......	......	......	......	0.0803	0.146	0.216	0.218	0.232	0.508	0.513	0.398	0.939
7/32 7 strand	0.216	0.02850	......	1,540	2,560	3,850	5,400	0.0983	0.162	0.239	0.244	0.239	0.532	0.543	0.405	0.967
1/4 3 strand	0.259	0.03393	3,150	......	......	......	......	0.1167	0.194	0.277	0.275	0.253	0.564	0.589	0.420	1.013
3 strand	0.259	0.03393	4,500	......	......	......	......	0.1167	0.194	0.277	0.275	0.253	0.564	0.589	0.420	1.013
7 strand	0.240	0.03519	......	1,900	3,150	4,750	6,650	0.1210	0.180	0.267	0.274	0.247	0.559	0.581	0.413	1.003
9/32 7 strand	0.279	0.04755	4,600	2,570	4,250	6,400	8,950	0.1640	0.209	0.316	0.328	0.260	0.609	0.649	0.426	1.066
5/16 3 strand	0.312	0.04954	6,500	......	......	......	......	0.1706	0.234	0.340	0.346	0.271	0.629	0.675	0.437	1.094
7 strand	0.312	0.05946	......	3,200	5,350	8,000	11,200	0.2050	0.234	0.361	0.380	0.271	0.657	0.710	0.437	1.124
7 strand	0.327	0.06532	6,000	......	......	......	......	0.2250	0.245	0.383	0.404	0.276	0.679	0.740	0.442	1.152
3/8 3 strand	0.356	0.06415	8,500	......	......	......	......	0.2203	0.267	0.396	0.409	0.285	0.689	0.752	0.452	1.167
7 strand	0.360	0.07917	11,500	4,250	6,950	10,800	15,400	0.2730	0.270	0.434	0.463	0.287	0.735	0.807	0.453	1.215
7/16 7 strand	0.436	0.1156	18,000	5,700	9,350	14,500	20,800	0.3990	0.326	0.565	0.612	0.312	0.577	0.981	0.478	1.381
1/2 7 strand	0.495	0.1497	25,000	7,400	12,100	18,800	26,900	0.5170	0.371	0.686	0.748	0.332	1.008	1.136	0.498	1.530
19 strand	0.500	0.1492	......	7,620	12,700	19,100	26,700	0.5040	0.375	0.678	0.738	0.333	1.000	1.126	0.500	1.522
9/16 7 strand	0.564	0.1943	......	9,600	15,700	24,500	35,000	0.6710	0.423	0.843	0.925	0.355	1.181	1.332	0.521	1.720
19 strand	0.565	0.1905	......	9,640	16,100	24,100	33,700	0.6370	0.424	0.815	0.891	0.355	1.149	1.300	0.522	1.691
5/8 7 strand	0.621	0.2356	......	11,600	19,100	29,600	42,400	0.813	0.466	0.987	1.084	0.374	1.337	1.510	0.540	1.894
19 strand	0.625	0.2332	......	11,000	18,100	28,100	40,200	0.796	0.469	0.974	1.068	0.375	1.322	1.496	0.542	1.881
3/4 19 strand	0.750	0.3358	......	16,000	26,200	40,800	58,300	1.155	0.563	1.335	1.466	0.417	1.714	1.932	0.583	2.308

* The values in this column are called Breaking Strength by some wire manufacturers.
† The values in this column give the resultant conductor loading per foot as specified in Rule 251, National Electrical Safety Code, Fifth Edition.

25. CALCULATION OF SAG AND TENSION

FUNDAMENTAL EQUATIONS. A wire or cable suspended so as to hang freely between two points of support conforms very closely to a catenary curve. The curve would be a perfect catenary if the wire were perfectly flexible and its weight were uniformly distributed along its length. For all practical purposes the curve is closer to a catenary than to any other curve. In the case of short spans with small sags the curve can be considered a parabola. If the sag is less than 6 per cent of the span length, the error in sag computed by the parabolic equations is less than 1/2 per cent. The flatness of the curve allows of some further simplifications in even the parabolic formulas, viz.: (1) the tension is considered uniform throughout the span, the slight excess of tension at the ends over that in the middle being neglected; (2) the change in length of the wire due to elastic stretch or temperature expansion is taken as equal to the change of length of a wire equal in length to the horizontal distance between the points of support.

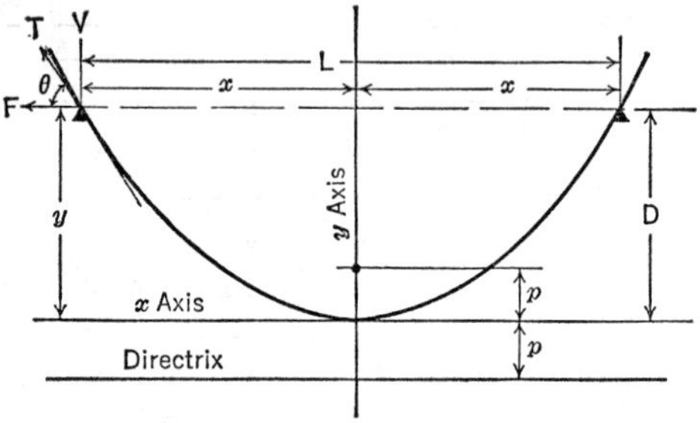

FIG. 3. Parabola

PARABOLIC EQUATIONS. The fundamental mathematical relations with respect to the parabola shown in Fig. 3 are as follows:

$$x^2 = 4py \tag{11}$$

$$\tan\theta = \frac{x}{2p} = \frac{2y}{x} = \frac{4D}{L} \tag{12}$$

$$\tan\theta = \frac{wx}{F} = \frac{wL}{2F} \tag{13}$$

From the fundamental relations the following working equations for wire spans have been derived. In these equations the notation indicated is used throughout.

A = cross-section of the conductor (actual metal cross-section), in square inches = circular mils ÷ 1,273,000.

a = coefficient of linear expansion of the conductor per degree fahrenheit; see Table 1 above.

D = deflection, in feet, of the lowest point of the conductor from the line through supports when suspended from two points of support at the same elevation and at a distance L apart. (D is measured in the direction of the resultant transverse force.)

e = difference in elevation of the two points of support, in feet.

F = longitudinal horizontal component of the tension in the conductor, in pounds. (The resultant tension T in the wire at the insulator is equal to $\sqrt{F^2 + H^2 + V^2}$, where V is the weight of the conductor and ice from the insulator to the lowest point of the span, and H is the total wind pressure on half the length of span; H and V in this expression are usually negligible compared with F.)

h = wind pressure, in pounds per foot length of conductor, assumed perpendicular to the vertical plane through the two points of support; see eq. (8) and Tables 4 and 5.

L = length of span, in feet, i.e., the horizontal distance between the two points of support, in feet.

l = length, in feet, of the arc of the curve in which the conductor hangs, i.e., the length of stretched conductor between the two points of support.

M = modulus of elasticity of the conductor in pound-inch units; see Table 1 above.

$S = \frac{vD}{w}$ = sag of the lowest point of the conductor below the horizontal line through the points of support; for no wind $S = D$.

T_0 = maximum allowable tension in the conductor, in pounds.

v = vertical force, in pounds, on a 1-ft length of the conductor, including the weight of conductor and the weight of the ice, if any, on it; see eq. (9) and Table 5.

$w = \sqrt{v^2 + h^2}$ = resultant load, in pounds, on a 1-ft length of the conductor.

$Z = \frac{hD}{w}$ = side swing, in feet, of the middle point of the conductor, measured perpendicularly to the vertical plane through the two points of support.

The symbols with the subscript "0" will be used to designate the values of the various quantities under the conditions of maximum assumed loading.

Deflection, Sag, and Side Swing. For a given length of span L, loading w, and stress F, the deflection D for the points of support at the same elevation is given by the relation

$$D = \frac{wL^2}{8F} \tag{14}$$

When there is no wind, this is also equal to the vertical sag, i.e., $S = D$. When there is wind, w is greater than the vertical loading v, and the vertical sag for the points of support at the same elevation is

$$S = \frac{vD}{w} = \frac{vL^2}{8F} \tag{15}$$

D in eq. (15) has the value given by eq. (14). When one point of support is at an elevation e above the other, then the vertical sag of the lowest point of the conductor below the lower point of support is

$$S' = S\left(1 - \frac{e}{4S}\right)^2 \tag{16}$$

where S is given by eq. (15). The horizontal distance of the lowest point of the conductor from the lower point of support is

$$L' = \frac{L}{2}\left(1 - \frac{e}{4S}\right) \tag{17}$$

The side swing Z of the middle point of the conductor, which is the point deflected the maximum distance from the vertical plane through the two points of support, is

$$Z = \frac{hD}{w} = \frac{hL^2}{8F} \tag{18}$$

D in eq. (18) has the value given by eq. (14).

Length of Stretched Conductor. The length of conductor between the two points of support for a given length of span L, loading w, stress F, and difference of elevation e is

$$l = L + \frac{8}{3}\frac{D^2}{\mathrm{L}} + \frac{e^2}{2L} \tag{19}$$

where D has the value given by eq. (14), i.e., D is the deflection for the same length of span, loading, and tension, but for the points of support at the same elevation.

Effect of Changes of Temperature and Loading. If the wire is unstressed or the stress does not change while the temperature undergoes a change, then the change in length of the wire is

$$l_{et} = la(t_2 - t_1) \tag{20}$$

If the wire is subjected to a change of stress (or loading) at constant temperature, the change in length of the wire is

$$l_{el} = \frac{l(T_2 - T_1)}{MA} \tag{21}$$

When both the loading and the temperature change, the stress in the wire will change to some new value, say F_0, and the deflection will change to some new value, say D_0. Let the new loading be w_0 and the new temperature t_0, the initial temperature being t; also let a be the coefficient of linear expansion, M the modulus of elasticity, and A the cross-section of the conductor in square inches. Then, when the points of support remain fixed, the following relation must hold

$$\frac{8}{3L^2}(D^2 - D_0^2) = a(t - t_0) + \frac{1}{MA}(F - F_0) \tag{22}$$

The D's in this equation are the same as given by eq. (14) for a loading of w and w_0, respectively, and stresses of F and F_0, respectively. Note that eq. (22) is independent of the difference in elevation of the two points of support; also that the two sets of symbols, with and without the subscripts, refer to *any* two sets of conditions.

In order to apply the foregoing equations to the calculation of deflections and stresses under various temperature and loading conditions, eqs. (14) and (22) may be written as follows:

$$D = \frac{wL^2}{8F} \tag{23}$$

$$D^3 = D\left[\frac{L^2a}{2.66}(t - t_0) + D_0^2 - \frac{F_0L^2}{2.66MA}\right] + \frac{3wL^4}{64MA} \tag{24}$$

From solutions of these equations under various conditions, curves showing the relation between deflection and temperature and between stress and temperature may be plotted. Such curves provide stringing charts and show maximum and minimum sags from which clearance templates may be constructed.

CATENARY EQUATIONS. The fundamental mathematical relations with respect to the catenary shown in Fig. 4 are as follows:

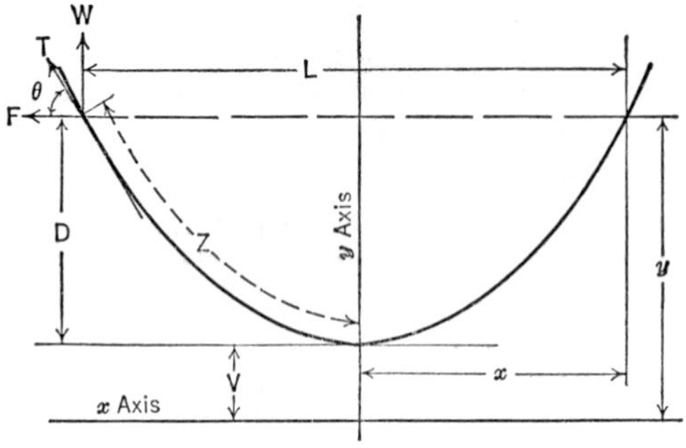

FIG. 4. Catenary

$$y = \frac{V}{2}\left(e^{x/V} + e^{-x/V}\right) \tag{25}$$

$$y = V\left(\cosh \frac{x}{V}\right) \tag{26}$$

$$Z = \frac{V}{2}\left(e^{x/V} - e^{-x/V}\right) \tag{27}$$

$$Z = V\left(\sinh \frac{x}{V}\right) \tag{28}$$

$$D = y - V \tag{29}$$

$$D = V\left(\cosh \frac{x}{V} - 1\right) \tag{30}$$

From the fundamental relations are derived the following additional working equations, in which these additional notations are used:

Z_e = change in length Z due to change in loading.
Z_t = change in length Z due to temperature change.
K_u = permanent set per unit length of cable.
K_t = total permanent set of cable due to loading.

$$L = 2X \text{ (supports at same elevation)} \tag{31}$$

$$l = 2Z \text{ (supports at same elevation)} \tag{32}$$

$$T = Yw \tag{33}$$

$$T = F + Dw \tag{34}$$

$$F = Vw \tag{35}$$

$$Z_e = \frac{w}{2AM}(XV + YZ) \tag{36}$$

$$Z_e = \frac{FX + TZ}{2AM} \tag{37}$$

$$Z_t = Za(t - t_0) \tag{38}$$

$$K_t = K_u Z \tag{39}$$

BEHAVIOR OF A WIRE UNDER VARIABLE STRESS. In order to understand clearly the application of the equations given above to practical problems, it is desirable to analyze the behavior of a wire or cable when it is strung on supports and subjected to changes in stress as a result of changes in loading or temperature.

When a previously unstressed wire is suspended between two supports, the weight of the wire produces a stress, which in turn causes the wire to elongate. A complete removal of the stress will, because of the elasticity of the wire, result in a reduction in its length, but will not cause it to return to its original length. The difference between the original and subsequent unstressed lengths is the amount of permanent set produced by the amount of stress in the wire when suspended from the supports.

When a wire is first strung, an initial tension is produced in it, but after a time that tension is reduced slightly because the wire continues to elongate until the tension becomes just balanced by the elastic resistance of the wire. If the tension could be maintained constant for a considerable length of time, say 1 hour or more, the elongation would be greater than under the previous condition. In practice, the tension would be held constant at the maximum value only under such circumstances as the accumulation of ice loading at a sufficiently rapid rate or a reduction of temperature at an equivalent rate.

Figure 5 illustrates graphically the behavior of a solid wire in a testing machine under a series of changes in stress.

If we assume that a platform or trough is placed under the wire between the jaws of the machine, so that the wire cannot sag, and constantly increasing tension is applied until, say, 60 or 70 per cent of the ultimate strength of the wire is reached and then the tension is quickly reduced to zero, the elongation will follow the line *o . . . a . . . h . . . i*. The slope of line *h . . . i* is approximately the final modulus of elasticity. If, instead, the stress is raised to point *a* and the machine is stopped for, say, 15 minutes, the tension and elongation will arrive at some such point as *b*. If then the machine is reversed to remove all stress, the elongation becomes $o - c$, which is the permanent set for stress *a*.

If now the machine is started up and the stress increased to *d*, the elongation follows the solid line from *c* to *d*. If the machine is now kept running just fast enough to maintain the tension *d* for a period of, say, 1 hour, the elongation will reach *e*. A reduction of stress to zero brings the elongation to *f*. Increasing the stress to *g* and holding it for another hour brings the elongation to *h*, and a reduction of stress to zero brings the elongation to point *i*.

Plotting the permanent set values against the maximum stresses responsible for them produces the permanent set curve shown. Such a curve is essential to the solution of

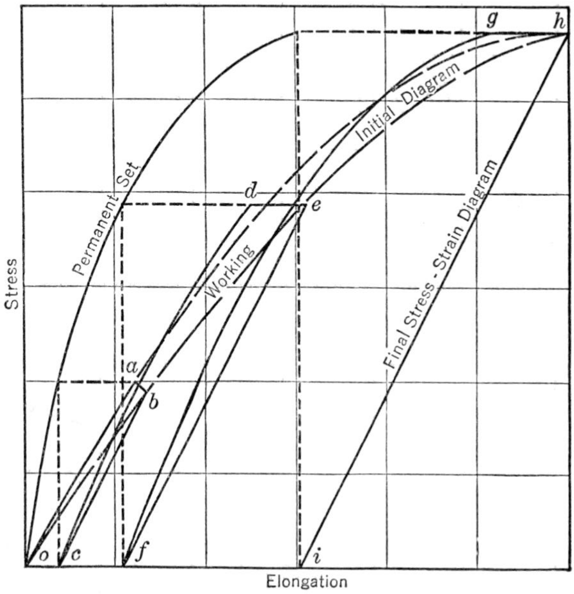

FIG. 5. Behavior of a Wire under Changes in Stress

actual problems. The long dash curve, *o . . . b . . . e . . . h*, is used as the working initial stress-strain diagram. It is evident that the working stress-strain diagram depends upon the loading cycles used in developing it. These are probably never the same as the loading cycles experienced by an actual line between the time when it is first strung and the time when it experiences full design loading, which may never occur. Variations in conductors, in handling, and in testing will obviously introduce errors which cannot be predicted. The discrepancies between the performance in the line and in the testing machine are not serious, however, if the stress-strain diagrams used in calculation of the line are the result of averaging a large number of carefully made tests.

Figure 5 shows typical behavior of solid wires or stranded cables made up of similar wires. If a stranded cable is made up of more than one kind of wire, such as A.C.S.R. or copperweld-copper, the behavior is somewhat different because the temperature coefficients and moduli of the steel and aluminum or copper strands are not the same. At low values of stress the load is entirely transferred to the material of higher strength. This is shown graphically by Fig. 6, which illustrates the behavior of an A.C.S.R. cable under variations of stress. The tension was increased, then held constant for about an hour, then decreased to zero. This was followed by increasing the stress to a new maximum value, holding it there, then reducing to zero again, then again increasing. The dotted line is the initial working diagram.

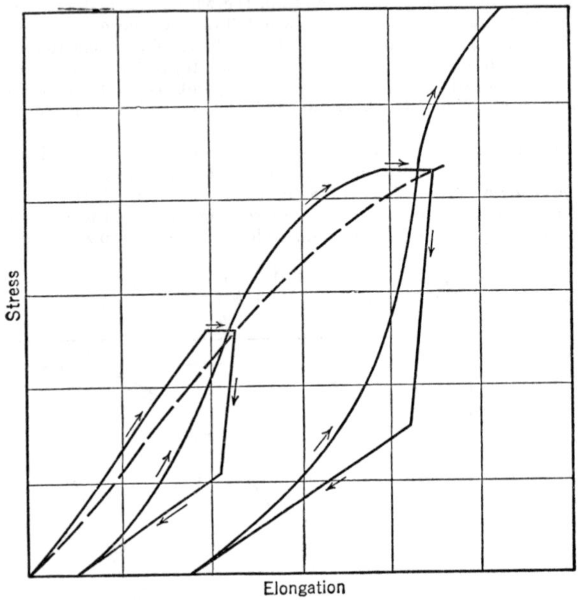

FIG. 6. Behavior of ACSR Conductor under Changes in Stress

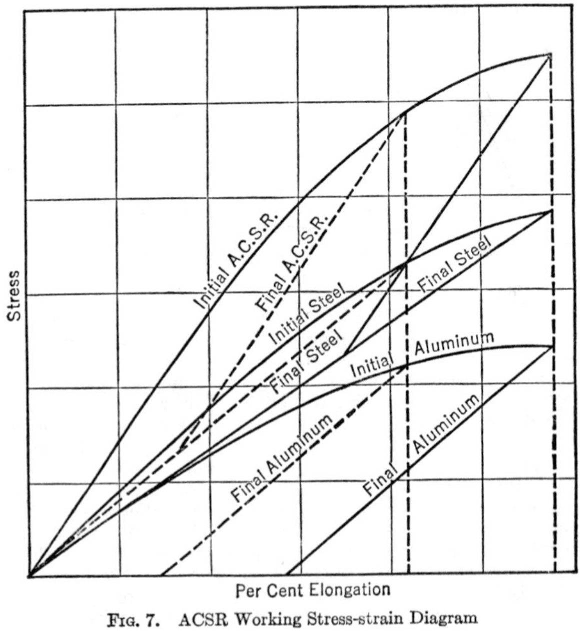

FIG. 7. ACSR Working Stress-strain Diagram

Figure 7 shows a complete working stress-strain diagram for A.C.S.R., not only for the complete cable but also for the components. In a composite cable such as A.C.S.R. the virtual modulus of elasticity and the virtual coefficient of expansion are not the same as for either the steel or the aluminum. Approximate values may be calculated from the following equations:

$$M_v = M_a R_a + M_s R_s \tag{40}$$

$$C_v = \frac{C_a M_a R_a}{M_v} + \frac{C_s M_s R_s}{M_v} \tag{41}$$

where M is modulus of elasticity, C is coefficient of expansion, R is fraction of total cross-section represented by each material. Subscripts a and s indicate aluminum and steel, respectively, and subscript v indicates the virtual value for the composite cable.

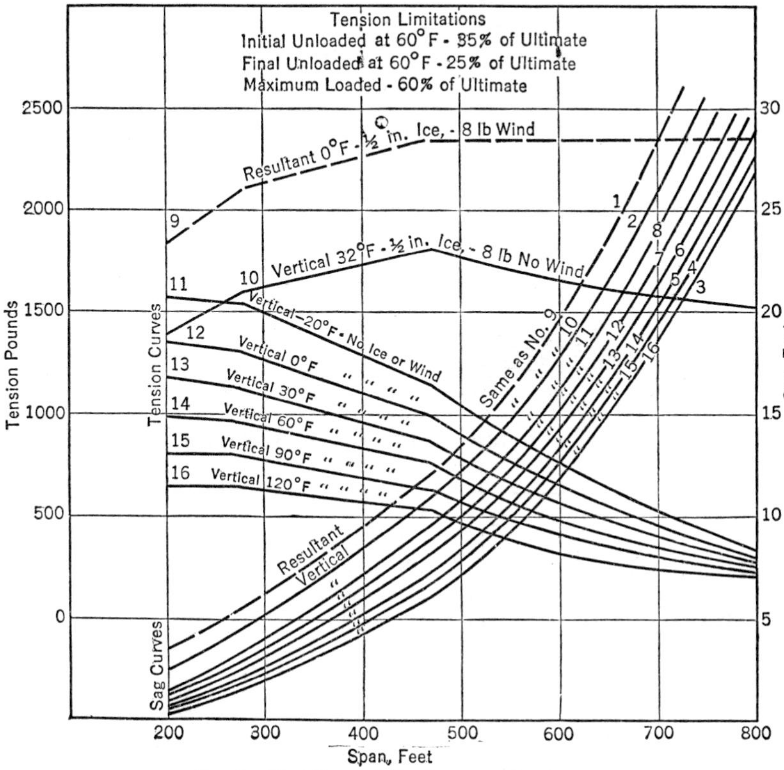

Fig. 8. Final Sags and Tensions—Dead-end Spans

SAG-TENSION CHARTS. For the practical design and layout of lines it is necessary to obtain or prepare charts on which are plotted curves giving sags and tensions for various span lengths of each kind and size of conductor which might be used. The kinds of charts needed are:

1. Final Sag-tension—Dead-end Spans.
2. Initial Sag-tension—Dead-end Spans.
3. Final Sag-tension—Ruling Span.
4. Stringing Sag-tension—Ruling Span.

Chart No. 1 is used for preliminary design and calculation of ruling span final curves and also for stringing prestretched conductors on dead-end spans. Chart No. 2 is used for calculation of ruling span stringing curves or for stringing unstretched conductors on dead-end spans. Chart No. 3 is used for checking ground clearances, locating structures, or stringing prestretched conductors in spans related to the ruling span. Chart No. 4 is used for stringing unstretched conductors in spans related to the ruling span. Figures 8, 9, 10, and 11 are examples of the four types of charts.

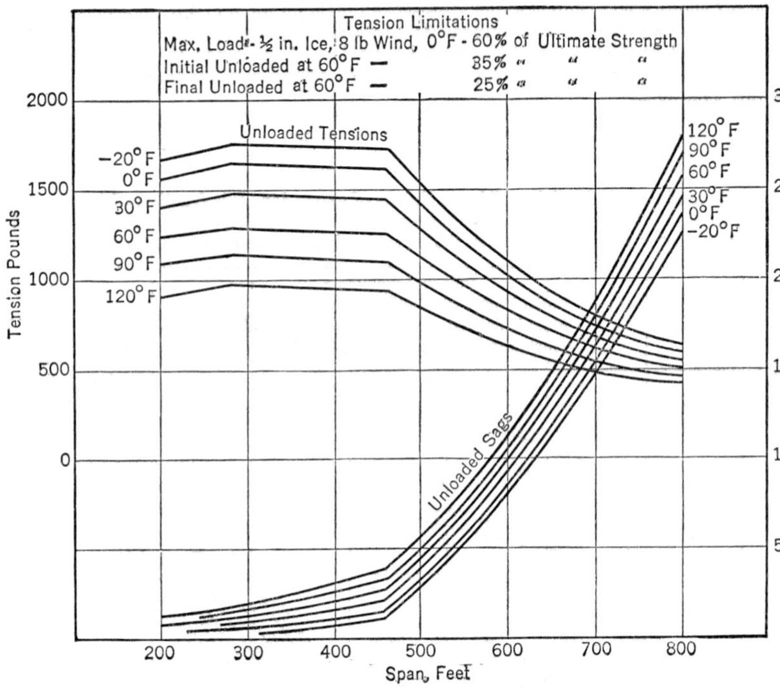

FIG. 9. Initial Sags and Tensions—Dead-end Spans

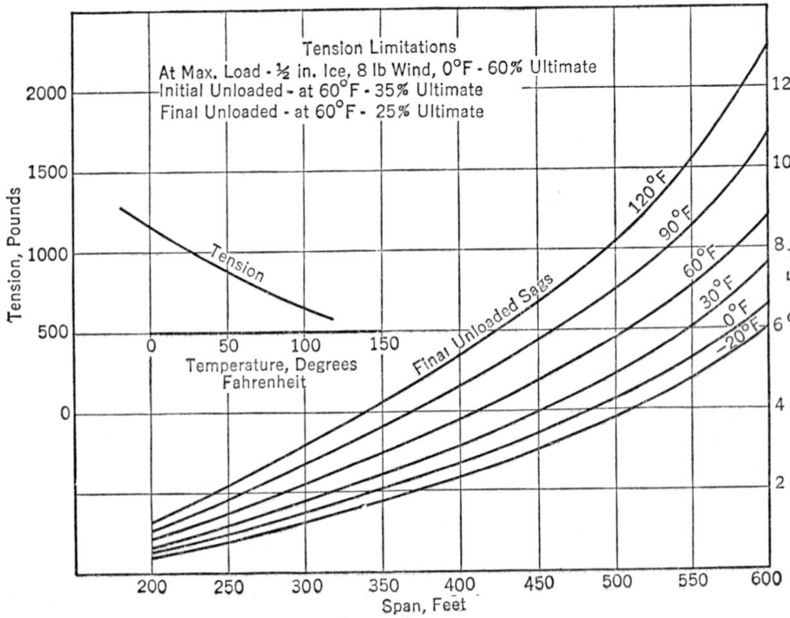

FIG. 10. Final Sags and Tensions—400-ft Ruling Span

SAGS AND TENSIONS IN UNEQUAL SPANS. Where the spans between dead-end structures in an actual line are unequal, the stringing sags for each different span length might be obtained from a Chart No. 2 such as described above. In that case the same tension would be reached in all spans under the maximum design loading, but under any other condition the tensions would be different in spans whose lengths were not the same.

The unequal stresses on the two sides of an insulator will tend to deflect some part of the structure if the conductor is rigidly attached to it. If suspension insulators are used, the insulator string will swing. Any motion of the point of support will tend to equalize the unbalanced stresses, and the amount of movement necessary to equalize the stresses is usually very small, especially if there is little inequality in length of spans.

In order to minimize the required movement all spans should be strung to a tension corresponding to a span length approximating a weighted average of the actual span

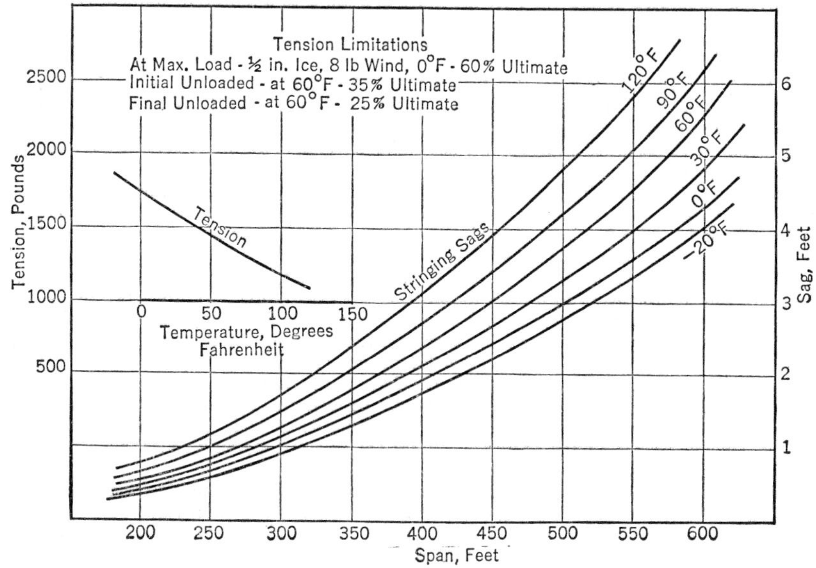

FIG. 11. Stringing Sags and Tensions—400-ft Ruling Span

lengths between dead-end structures. Such a span length is called the **ruling span.** Its length is usually taken as the average span plus two-thirds of the difference between the average and the maximum span.

It should be noted that in the ruling span method it is necessary to equalize the horizontal component of the tension in the conductor throughout the several spans, and not the total tension. Therefore the horizontal component of the tension corresponding to the ruling span length under each loading condition is used as a basis for computing the sags for the various span lengths related to the ruling span.

When the conductor is thus strung, the tension under minimum temperature and maximum loading will usually exceed the assumed tension in the shorter spans and be less than the assumed tension in the longer spans. Similarly, under maximum temperature conditions, the longer spans will have a sag in excess of the value calculated on the assumption of full design tension being reached with maximum design loading, and the shorter spans will have a sag less than the calculated value. An approximate method of calculating the actual stresses and sags will be found in the paragraph on calculation of stresses in unequally loaded spans (p. 14-78).

CALCULATIONS FOR LEVEL SPANS. For short spans or for approximate calculations the parabolic equations (14) to (24) inclusive may be used. If the wires have been prestretched, eq. (24) may be used for a direct calculation of sag after a change of loading and temperature. An accurate method of solving the cubic equation was given by Dr. Pender in an article in the *Electrical World*, Vol. 66, p. 344 (1915). The following is a fairly quick slide-rule method of solving eq. (24). Although limited by slide-rule accuracy, it may serve a useful purpose for many calculations. The combination of eqs. (23) and (24) may be put in the following form:

$$D^3 = D\frac{Q}{K} + \frac{wL^2}{8K} \tag{42}$$

where

$$\frac{Q}{K} = D_0^2 + \frac{L^2a}{2.66}(t - t_0) - \frac{F_0}{K} \tag{43}$$

$$K = \frac{2.66MA}{L^2} \tag{44}$$

and are constant for a given set of conditions.

Having calculated $\frac{Q}{K}$ and $\frac{wL^2}{8K}$, set the index of scale C to a tentative value of D on the D scale of the slide rule. Calculate D^3, using the A and B scales, and $D\frac{Q}{K}$, using the C and D scales. Usually this can be done without changing the setting. Mental addition of the right-hand side of the equation will determine how closely the equation balances. Two or three trial values for D will result in a very close approximation. One or two more trials with accurate addition will determine a value of D which is accurate enough for many purposes.

Analytical Methods. To calculate sags and tensions for various initial and final conditions for a wire or cable which is unstretched before stringing, it is necessary to use analytical methods in order to allow properly for the permanent set produced by stressing the wire. These might also be called step-by-step methods.

The steps involved are few and quite simple. First, select a certain condition of loading, tensions, and temperature as a basis for starting the calculation. Limitations are placed not only on the final maximum loaded tension, but also on the stringing tension and the final unloaded tension at 60 deg fahr; therefore the starting condition chosen should be that which results in neither of the other limitations on tension being exceeded. Which of the three conditions governs is not always clear, but in heavy-loading areas the 60-deg fahr tensions govern in the short span lengths. In medium- and light-loading areas the 60-deg fahr tensions govern in the medium span lengths as well. In the long spans the loaded tension nearly always governs. Under the selected starting condition, calculate the length of the wire between supports, and the sag.

Second, calculate the length of the wire which would result if all tension were removed from it, the temperature remaining constant at the starting condition value.

Third, calculate the length of the wire while still unstressed but as modified by changing its temperature to as many values as needed for stringing purposes.

Fourth, for each of the temperatures used in the third step, calculate the length of the wire when stressed by adding the tension due to the weight of the wire plus any desired external loading of ice or wind.

Fifth, correct the lengths obtained in the fourth step by the difference between the amounts of permanent set produced by the tensions at the loadings involved in the fourth step and in the starting condition. If the starting condition was a stringing condition, add the permanent-set correction to the lengths obtained in the fourth step. If the starting condition was that of maximum loading, subtract the correction. In determining permanent-set values use the average tension in the wire, i.e., one-half the sum of the tension at the support and the tension at the lowest point of the wire.

Sixth, calculate the sags and tensions corresponding to the wire lengths determined in the fourth and fifth steps.

The foregoing procedure is applicable to dead-end spans only, and it should be pointed out that the method cannot be applied accurately to composite cables, i.e., those having some of the strands composed of material different from the rest. To apply the method to composite cables it is necessary to use "virtual" moduli of elasticity. Since these are not true constants, but are weighted averages, they cannot result in accurate calculations. Graphic methods, described later, are much more desirable. The foreging procedure is not applicable to spans with supports at different elevations without modifications.

To apply the analytical method to the calculation of sags and tensions for other span lengths related to the ruling span, use the following procedure, remembering that the horizontal component of the wire tension is the same in all related spans for a given loading condition. Having calculated the tension in the ruling span for each initial and each final condition, calculate the horizontal component of the total tension for each such condition for the ruling span. Using these horizontal components, calculate the total tension for each related span length at each loading and temperature condition. From these tensions the corresponding sags may be calculated.

When all the values necessary have been calculated by using the foregoing methods, it is possible to plot sag and tension charts for all the conditions needed. It is seldom necessary, however, to perform these calculations except for unusual conditions, because very complete sets of charts are available on request from the manufacturers of various types of conductors. Even if the only charts available are the initial and final dead-end span charts, sometimes called loci charts, the ruling span charts may be easily and quickly prepared from them.

Calculation Methods. The parabolic equations are not easily adaptable to the step-by-step method because it is always necessary when using them to solve cubic equations. On the other hand, the catenary equations lend themselves very well to this method, since it is possible to provide many short cuts and helps for rapid calculation. In 1911 P. H. Thomas presented a method of solving catenary problems in which he reduced the actual span of wire to a unit catenary by dividing each of the dimensions of the catenary curve of the actual span by the span length. This device resulted in a considerable simplification of the catenary calculations. A similar principle has been made use of by others who have proposed simplified methods of calculation. (See Section 19.)

Other helps, such as tables and charts, have been developed by various authors. J. S. Martin presented a set of tables and a method for their use before the Engineers Society of Western Pennsylvania in 1922. In 1931 these tables were republished by the Copperweld Steel Company. In 1943 the same company published a table and a set of curve charts, plotting on a large scale the factors covered by Martin's tables. The use of these charts obviates the necessity of interpolating between values in the Martin tables. These new charts, the table, and the blank calculating forms recommended for use with them provide a method of deriving initial values and ruling span values which was lacking in the original Martin method.

The Martin tables give the values of four functions corresponding to the value of the function $\frac{\text{Span } (S) \times \text{Weight } (W)}{\text{Tension } (T)}$. The other four functions given by the tables are:

1. $$\text{Sag factor} = \frac{\text{Sag}}{\text{Span } (S)} \tag{45}$$

2. $$\text{Length factor} = \frac{\text{Length of wire}}{\text{Span } (S)} \tag{46}$$

3. Stretch factor

4. $\frac{SW}{T} \times$ Stretch factor

Other equations used are:

Change in length factor of a wire caused by adding or removing weight (W) = Elongation factor × Stretch factor (47)

Change in length factor of an unstressed wire caused by a change in temperature of t deg = Coefficient of expansion × t × Unstressed length factor (48)

$$\text{Elongation factor} = \frac{\text{Weight } (W) \times \text{Span } (S)}{\text{Area } (a) \times \text{Modulus } (e)} \tag{49}$$

$$\text{Tension } (T) = \frac{SW}{SW/T} \tag{50}$$

It will be noted that all these functions are either unit values or are inherent functions of the catenary, except the function SW/T, which is calculated from known factors. The calculations are therefore carried out on a unit span basis by the step-by-step method. The calculation begins with the determination of SW/T for the starting condition of load and temperature. The other factors are then obtained from the tables or charts. The change of length factor due to removing all stress is calculated by eq. (47). Subtracting this change from the original length factor gives the unstressed length factor at the original temperature. Adding the change of length factor calculated from eq. (48) produces the unstressed length factor at the new temperature.

The next step is to calculate the elongation factor for the new loading. This, multiplied by the final stretch factor, gives the change in length factor. This change, added to the last unstressed length factor, gives the final length factor. Since the final stretch factor and the final length factor must correspond exactly in the tables, it is necessary to calculate them together by trial. The final SW/T is then calculated from the last column of the tables, and the final sag factor obtained also from the tables. The final unloaded sag and tension are simply calculated from eqs. (45) and (50).

To calculate initial sags and tensions, start with the final unstressed length factor at the minimum temperature, as determined by the foregoing method. Subtract the difference between the permanent set per unit length at maximum tension and the permanent set at the estimated initial tension to obtain the initial unstressed length factor at the same temperature. Using the coefficient of expansion, obtain, by eq. (48), the change of length factor to add or subtract to obtain the unstressed length factors for other temperatures. Calculate elongation factor SW/ae, using final modulus. From the tables or charts determine SW/T and the sag factor, using the initial unstressed length factor. Calculate the initial sags and tensions by Martin's eqs. (45) and (50).

It should be remembered that the average tension in the wire should be used in determining permanent set values, and the "final" modulus of elasticity should be used instead of the so-called initial modulus. The so-called initial modulus is only the slope at the origin of the initial stress-strain curve for a previously unstressed wire. It is not a true modulus of elasticity, since it represents a combination of elastic and non-elastic elongation. Only that portion representing the elastic elongation is the true modulus. The latter is practically the same regardless of the amount of stress applied, so long as it is below the yield point.

When the maximum loaded condition is used as the starting condition for the calculation of initial or stringing sags and tensions, it is not difficult to determine the values of permanent set necessary to complete the calculations. Although it is necessary to estimate the average initial tension, the estimate does not need to be extremely close because the rate of change of permanent set at the low tension values is very small. In the case of starting with initial conditions, however, to obtain final loaded sags and tensions it is necessary to estimate the average final loaded tension. Since the rate of change of permanent set is relatively large at the loaded tensions, the estimate needs to be very accurate. Several trials with different estimated values of tension and their corresponding values of permanent set may be required. After adding each value of permanent set and computing from the new unstressed length the loaded lengths and tensions, a value of loaded tension will be found which corresponds closely with the value of permanent set used to calculate that particular loaded tension.

Another analytical method has been developed by J. F. Nash and J. F. Nash, Jr., and described in a paper in *Electrical Engineering* (November, 1945). This method is also based upon a unit catenary and unit values of catenary functions. When the catenary functions are divided by X, which is equal to one-half the span length, the following relations arise:

$$\frac{Y}{X} = \frac{V}{X} \cosh \frac{X}{V} \quad \text{(see eq. 26)}$$

$$\frac{Z}{X} = \frac{V}{X} \sinh \frac{X}{V} \quad \text{(see eq. 28)}$$

$$\frac{S}{X} = \frac{Y}{X} - \frac{V}{X} \quad \text{(see eq. 29)}$$

A table is provided which gives the related values of V/X, Y/X, Z/X, and S/X. When V, Y, Z or S is known for any span, the other values are quickly determined from the table by interpolation.

This method uses exactly the same step-by-step procedure as the Martin method but a different table and different equations. The Nash method eliminates the trial-and-error procedure necessary in the Martin method in determining the final stretch factor and the final length factor.

Graphic Methods. Many graphic methods have been devised. The Thomas method and the Pender and Thompson charts are covered in detail in the N.E.L.A. *Overhead Systems Reference Book*. A method developed by Theodore Varney was described by the Aluminum Company of America in a booklet entitled "A.C.S.R. Graphic Method for Sag-Tension Calculations." Varney's method is particularly adapted to calculations involving composite conductors, although it can be used for ordinary conductors as well.

This method requires the use of a set of charts plotted for a particular conductor. The so-called catenary charts are plotted, one for each span length considered. Each catenary chart has a curve relating sag to per cent increase of arc length over span length, the latter scale being the horizontal axis. Each chart also has a family of curves relating stress for various loadings to per cent increase of arc length over span length.

A set of stress-strain diagrams is also required, each diagram applying to a particular temperature for the conductor under consideration. These charts are plotted on transparent paper and show the initial and final stress-strain diagram for the complete cable, as well as the division of stress between the aluminum and the steel by separate stress-strain diagrams for each material. The stress in pounds per square inches is plotted against elongation in per cent. The scale of these charts is the same as that of the "catenary charts" and is arranged the same. See Fig. 7 for a typical diagram of this kind.

To use, place the stress-strain diagram for the temperature of maximum loading over the proper catenary chart with the horizontal axes of elongation coinciding, but with the initial stress-strain diagram of the complete cable intersecting the maximum loading curve at the stress of maximum loading. Now, at the intersection of the initial stress-strain curve with the bare cable loading curve, read the initial stringing tension at the temperature of this particular chart. Immediately above this point read the corresponding sag on the sag curve. Draw a line parallel with the final stress-strain curve through the point on the initial stress-strain curve corresponding to the maximum stress. Such a line is the final stress-strain curve for this particular maximum stress. Draw a similar line for each of the component stress-strain diagrams. At the intersection of this line for the complete cable with the bare loading curve, read the final unloaded tension for the complete cable. Directly below this point read the component tensions on their respective final stress-strain curves. Directly above these points read the final unloaded sag on the sag curve. Read the final loaded sag, of course, on the sag curve above the point on the initial stress-strain curve corresponding to the maximum tension.

All these values have been at the temperature corresponding to the maximum loading. Before separating the two charts used, mark on the horizontal scale of the catenary chart the position of the zero point of the horizontal scale of elongation of the stress-strain diagram. To obtain initial and final sags and tensions at some other temperature take the stress-strain diagram for that temperature and lay it over the catenary chart, with the zero of the elongation scale of the new stress-strain diagram located at the reference point previously marked on the catenary chart horizontal scale. Now all values needed at the new temperature may be read off as before.

CALCULATIONS FOR SIDE-HILL SPANS. Figure 12 shows the configuration of a typical span with supports at different elevations, together with its significant dimensions.

The vertical sag S measured between the straight line between points of support and the tangent to the conductor parallel to that line is very nearly equal to the sag in a level span with the distance between supports equal to the inclined distance L''. For short spans the error in calculating sag on that basis is negligible. For longer spans a more accurate method has been worked out by J. S. Martin, using an equivalent span length equal to $(2L'' - L)$ and employing the ordinary catenary methods for calculating the sag as though it were a level span. This method does not, however, provide an accurate determination of the wire tensions at the two supports.

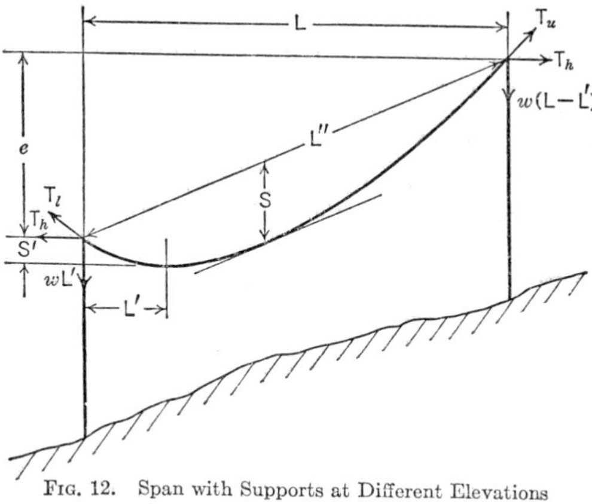

Fig. 12. Span with Supports at Different Elevations

Martin has developed a more accurate method, and the Nashes, as reported in their paper in *Electrical Engineering* (November, 1945), have also developed an extremely accurate and complete method. For usual purposes it is entirely satisfactory to make the calculations by using the parabolic eqs. (14) to (19) inclusive. Since the horizontal component of the total tension at each support is required to be the same as that in the ruling span, it is known, and the tensions at the supports can be quickly calculated. The vertical component of tension at each support is equal to the total loading per foot times the distance from that support to the low point of the wire. The vector sum of the vertical and horizontal components at a support is the total tension in the wire at that support.

In practice it is seldom necessary to make any calculations of side-hill spans except to determine total tensions or conditions which might develop uplift. Clearance conditions are best handled by means of the templates described later. The same template is used for both level and side-hill spans once it has been prepared. Even for a dead-end side-hill span it is best to prepare a template so that no undesirable clearance condition will be overlooked.

CALCULATION OF STRESSES IN UNEQUALLY LOADED SPANS.* When suspension insulators are used, any tendency of the stresses in two adjacent spans to become unequal will produce such a deflection of the insulator, in the direction of the span with greater stress, as will establish equilibrium in the line. This state of affairs will occur (1) when adjacent spans carry unequal ice loads, (2) when the wire on one side of the insulator breaks, and (3) to a slight extent with changes in temperature when the adjacent spans are unequal in length, as noted above. The following method of calculating the "equilibrium" stresses in the wires and corresponding sags is applicable to all cases of initially unbalanced stresses, irrespective of their cause. The method may also be used to calculate the stress and sag in spans supported on pin insulators, provided the moment of bending of the pin and of the pole or tower is known or can be calculated.

Change of Stress Due to Change in Length of Span. When the length L of the span (i.e., distance between points of support) increases by λ inches, because of a horizontal displacement of the insulators (without slipping of the wire), the stress in the wire is increased by the same amount as would be produced by a fall in temperature of

$$t - t' = \frac{\lambda}{12aL}, \quad \text{or} \quad t' = t - \frac{\lambda}{12aL} \text{ deg fahr} \tag{51}$$

where t is the actual temperature, and t' may be called the "equivalent" temperature corresponding to the change in length λ. In this equation λ is the actual increase in the distance between the points of support in inches, a the temperature coefficient of linear expansion per degree fahrenheit, and L the original length of the span in feet. For example, in an 800-ft span of copper wire an increase of 1 in. in L corresponds to a drop of temperature of $1 \div (12 \times 9.6 \times 10^{-6} \times 800) = 10.85$ deg, and an increase in length of λ inches corresponds to a drop of temperature of $t - t' = 10.85\,\lambda$ deg.

Hence a stress-temperature chart for any given length of span may be used directly to determine the stress in the wire after any change in the length of the span, due to the deflection of the insulator. For example, consider an 800-ft span of 300,000-cir mil bare copper conductor, at 32 deg fahr, without ice or wind, initially stressed to 4170 lb, and let the length of the span be increased 4 in. as the result of the deflection of the insulators by this amount. This increase in length of span will then give the same stress in the wire as would be produced if the temperature fell from 32 deg to $t' = 32 - 4 \times 10.85 = -11.4$ deg. The point on the stress-temperature curve corresponding to a temperature of -11.4 deg gives the new stress, viz., 4700 lb.

Horizontal Pull of Insulator. Referring to Fig. 13, let m = the horizontal distance in inches (measured along the span) which any insulator is deflected from the vertical, taken positive when to the right, say, and negative when to the left. Let x = the length of the insulator string in inches, i.e., distance from point of attachment to tower to point of attachment to wire; V = total weight of wire and ice between the lowest point of the wire in the span to the left of the insulator and the lowest point of the wire in the span to the right of the insulator, plus one-half the weight of the insulator; H = total wind pressure on the length of wire between the middle points of the two adjacent spans, plus half the wind pressure on the insulator; and put $W = \sqrt{V^2 + H^2}$. Then the horizontal component of the pull of the insulator toward the left along the line of the span is

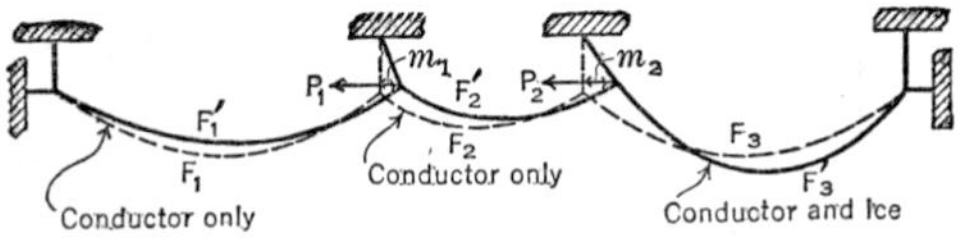

FIG. 13. Unequally Loaded Spans

$$P = \frac{m}{\sqrt{x^2 - m^2}} \cdot W \,\dagger \tag{52}$$

For example, consider two adjacent spans of 300,000-cir mil copper, each 800 ft long and with points of support at the same elevation. Let the span to the left be free of ice, and let the one to the right have a 1/4-in. ice coating; assume the insulator to be 60 in long and to weigh 100 lb. Then for no wind $H = 0$, $V = 1.19 \times 400 + 0.915 \times 400 + 100/2 = 892$ lb; hence for deflections of the insulator of less than 12 in. the horizontal pull of the insulator is $P = (892 \times m) \div 60 = 14.9m$, or 14.9 lb per in. deflection.

* From lecture notes by Dr. H. Pender.

† When m is less than 20 per cent of x, this may be written, with an error of less than 2 per cent,

$$P = \frac{m}{x} \cdot W \tag{53}$$

STRESSES IN A SERIES OF SPANS WHEN POINTS OF SUPPORT ARE NOT FIXED. Referring to Fig. 13, let the left-hand end of span 1 be anchored, and assume the insulator at the right-hand end to be deflected a horizontal distance of m_1 inches, due, for example, to a change in the loading on the succeeding spans (or to a change in temperature when the spans are of unequal length). From eq. (51) calculate the "equivalent" temperature t'_1 corresponding to this change in length, and from a stress-temperature curve corresponding to the assumed loading w_1 of this span find the stress on this curve corresponding to a temperature of t'_1 degrees; call this stress F'_1.

Next calculate the transverse and vertical loads on the insulator, viz., H_1 and V_1, and the resultant load $W_1 = \sqrt{V_1^2 + H_1^2}$, as explained above. Then from eq. (52) or (53) calculate the horizontal pull P_1 of the insulator. The stress in the second span, assuming the value of m_1 chosen at the start is correct, must then be

$$F'_2 = F'_1 + P_1 \tag{54}$$

t'_2 is then determined from F'_2 on the stress-temperature curve. From eq. (51) the corresponding increase in the length of span 2 must then be

$$\lambda_2 = 12aL_2(t - t'_2) \tag{55}$$

where L_2 is the length of the second span. The corresponding deflection of the insulator at the right-hand end of span 2 must then be

$$m_2 = m_1 + \lambda_2 \tag{56}$$

always taking the insulator deflection positive when to the right, say.

Using the values of λ_2 and m_2 thus found, calculate λ_3 and m_3 in exactly the same manner as λ_2 and m_2 were calculated, and similarly for the succeeding spans until the next anchor tower is reached. For the anchor tower at the right-hand end of the nth span, say, the deflection of the insulator must be zero, viz.,

$$m_n = 0 \tag{57}$$

If m_n as calculated comes out greater than zero, then the assumed value of m_1 is too great; if m_n comes out less than zero, the assumed value of m_1 is too small. By calculating m_n for two or three assumed values of m_1, and plotting m_n as ordinates against m_1 as abscissas, the correct value of m_1 will be where this curve crosses the axis of abscissas. By using this correct value of m_1, the stresses and deflections in each span may then be accurately calculated by the process just given. The complete process is best shown by an example.

Example. Consider three spans between anchor towers (Fig. 13), all of the same length, 800 ft, and all supports at the same elevation, 300,000-cir mil copper being used for the conductor. Let the temperature be 32 deg fahr, and let the middle span have a 1/4-in. ice coating, but the other two spans have no ice on them; also assume no wind. By applying eq. (51) for an 800-ft span of copper wire, it is determined that an increase of 1 in. in L corresponds to a drop in temperature of 10.85 deg. Assume that each insulator weighs 100 lb and has a length of 60 in. Then, for an increase of λ inches in the length of any span, the "equivalent" temperature is, from eq. (51)

$$t' = 32 - 10.85\lambda$$

or, if the equivalent temperature rise t' is known,

$$\lambda = 0.092(32 - t')$$

The horizontal pull of any insulator for a deflection of m inches (small compared with the length of the insulator) is, from eq. (53),

$$P = 14.9m$$

In the following table are given the calculations for assumed values of m_1 of 1, 2, 3, and 4 in., and in Fig. 14 are plotted the corresponding calculated values of m_3 against m_1. It is seen that the relation between m_3 and m_1 is practically a straight line cutting the horizontal axis at $m_1 = 2.4$, which is therefore the correct value of m_1. The calculations for $m_1 = 2.4$ in. are given in the last column of the table. Hence the stresses and deflections in the two end spans (without ice) are $F'_1 = F'_3 = 4460$ lb and $D'_1 = D'_3 = 16.3$ ft, respectively, and the stress and deflection in the middle span loaded with 1/4 in. of ice are $F'_2 = 4496$ lb and $D'_2 = 21.1$ ft, respectively.

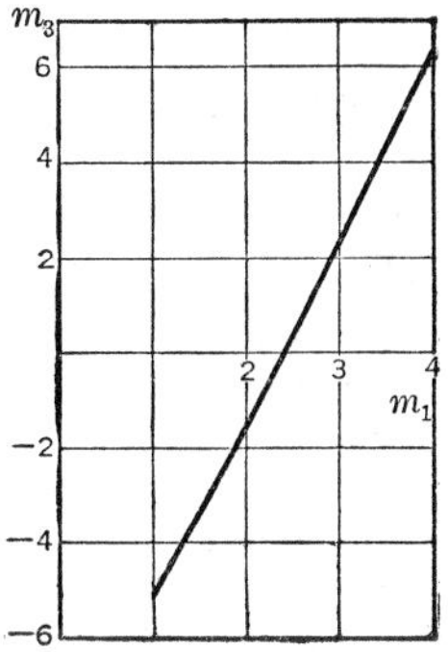

Fig. 14

These calculations will be facilitated, if many are to be made, by plotting stress-temperature curves for each loading assumed. The equivalent temperatures corresponding to calculated stresses are then taken directly from the curves. If only a few calculations are necessary, eqs. (23) and (24) may be employed.

m_1	= assumed	1	2	3	4	2.4
t'_1	$= 32 - 10.85 m_1$	21.1	10.3	−0.6	−11.4	6
F'_1	From Fig. 3 ($w = 0.915$)	4300	4410	4550	4700	4460
P_1	$= 14.9 m_1$	15	30	45	60	36
F'_2	$= F'_1 + P_1$	4315	4440	4595	4760	4496
t'_2	($w = 1.19$)	105	90	74	55	84
λ_2	$= 0.092(32 - t'_2)$	−6.72	−5.33	−3.86	−2.12	−4.78
m_2	$= m_1 + \lambda_2$	−5.72	−3.33	−0.86	1.88	−2.38
P_2	$= 14.9 m_2$	−85	−50	−13	28	−36
F'_3	$= F'_2 + P_2$	4230	4390	4582	4788	4460
t'_3	($w = 0.915$)	26	13	−3	−16	6
λ_3	$= 0.092(32 - t'_3)$	0.55	1.75	3.22	4.42	2.39
m_3	$= m_2 + \lambda_3$	−5.17	−1.58	2.36	6.30	0.01

LOCATION OF STRUCTURES AND DETERMINATION OF CLEARANCES. The height of structures is determined so as to give some specified minimum clearance from conductor to ground for some length of span chosen as a nominal standard on the basis of level ground. In practice the ground is rarely level, and the structures are actually located to conform to the irregularities of the ground. In locating structures of a given height the spans are made as long as is consistent with maintaining the ground clearance. The irregularities of the ground are ordinarily advantageous and permit of slightly longer spans on the average than could be obtained with the same height of structures on level ground.

Profile and Plan of Right-of-way. In order to locate structures properly it is necessary to have a profile of the right-of-way. Profiles are conveniently plotted on standard ruled profile section paper to a vertical scale of 20 ft to the inch and a horizontal scale of 200 ft to the inch. Three profiles are desirable, one along the center of the line and one on each side, say at each edge of right-of-way, as shown on Fig. 15 at *A*, *B*, and *C*. The two side profiles indicate the amount and direction of the slope of the ground across the line which must be allowed for in determining ground clearance and foundation or tower extensions.

A plan of the right-of-way is of course also necessary for determining the construction at angles in the line, and the clearances from the conductor to the edge of the right-of-way when the conductor is deflected horizontally by the wind. Such a plan is shown at the bottom of Fig. 15.

TEMPLATES FOR LOCATING STRUCTURES. Three templates are required, one for ground clearance with maximum sag, marked *M* in Fig. 15, one for uplift at times of minimum sag, marked *N*, and one for maximum side swing, marked *Z*. These are cut from thin celluloid and are to the same horizontal and vertical scales as used for the profile and plan of the right-of-way.

Since the curvature of the catenary or parabola in which the wire hangs depends only on the tension and loading and not on the length of the span or on the difference in elevation of the points of support, all spans having the same tension and loading can be drawn (for any one predetermined scale) from a single template, irrespective of their lengths or of the differences in elevation of the points of support. When the elevations of the points of support are not the same, however, the lowest point of the curve is shifted from the middle of the span toward the lower support, but the axis of the curve remains vertical.

Construction of Maximum Sag Template *M*. From the final sag-tension chart for the ruling span of the line (Fig. 10), select the curve which shows the maximum vertical sag. This is usually the curve for maximum temperature. By the use of the first part of eq. (15) check this by determining the vertical sag below supports of the curve of the wire under maximum wind and ice loading. Determine the vertical sag for the longest span shown on the chart. This span should be nearly as long as twice the longest hill-side span in the line. Using this sag as the sag at the center of the corresponding span, plot the remainder of the catenary curve on each side of the center by the use of the multipliers in Table 6. Each multiplier gives the percentage of center-span sag for a particular percentage of the span length. Level supports are assumed.

If it is necessary to extend the template to greater span lengths than are shown by the sag-tension chart, this can be accomplished by the use of Table 6. Divide the longest span length shown on the chart by the longest span length needed, and express this ratio as a percentage. From Table 6 obtain the corresponding percentage of center-span sag. This is the percentage that the sag from the chart is of the sag for the required maximum span for the template. Intermediate sags to plot the extension of the template are calculated from the table as before.

On the same piece of celluloid should be plotted two other curves. One is the ground-clearance line. This is the same as the template *M* but is dropped vertically a distance equal to the minimum ground clearance. The other curve is the structure base line. It

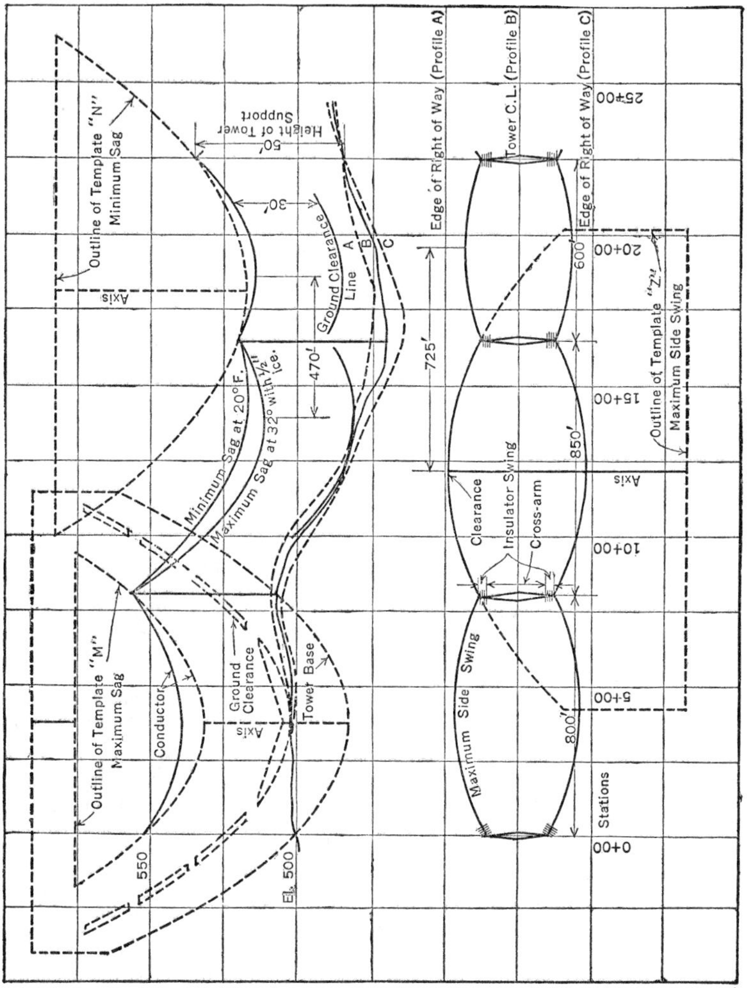

FIG. 15. Use of Templates

is also the same as the template *M* but is dropped vertically a distance equal to the standard height above ground of the lowest conductor at the point of support.

Construction of Minimum Sag Template *N*. From the initial (stringing) sag-tension chart for the ruling span of the line (Fig. 11) select the curve for the minimum temperature which may be expected to occur in the region where the line is constructed. Determine the vertical sag for the longest span shown on the chart. Using this sag as the sag at the center of the corresponding span, plot the catenary curve by means of Table 6, as was done in the maximum sag template.

Construction of Maximum Side-swing Template *Z*. The maximum side swing occurs at time of maximum wind pressure and may be at maximum temperature or at 32 deg fahr when covered with ice. In the latter case the side swing depends on the shape (circular or elliptical) of the ice covering and its specific gravity. For a circular covering of solid ice one particular thickness (usually but not necessarily the maximum thickness) gives the greatest side swing. From the final sag-tension chart for the ruling span of the line (Fig. 10) determine the total deflection of the wire in the plane of the resultant for the maximum span shown on the chart. By means of the first part of eq. (18) calculate the side swing of the mid-point of the wire. The projection of the curve of the wire on a horizontal plane is also a catenary; therefore it can be plotted by means of Table 6, as for the other templates, considering the side swing of the wire at the middle of the span as the center-span sag.

Table 6. Sags in Terms of Spans

Percentage of Span Length	Percentage of Sag at Center of Span	Percentage of Span Length	Percentage of Sag at Center of Span
5 or 95	19.5	30 or 70	84.4
10 — 90	36.5	35 — 65	91.3
15 — 85	51.0	40 — 60	96.0
20 — 80	64.0	45 — 55	98.9
25 — 75	75.0	50 — 50	100.0

LOCATING TOWERS BY MEANS OF TEMPLATE *M*. Choose a starting point, as shown for example in Fig. 15, at station 0 + 00, elevation 500.0 ft for the first tower location. The template *M* is then placed over the profile and shifted until its axis is vertical, the lower curve is at station 0 + 00, elevation 500.0 ft, and the middle curve is tangent to the ground profile as shown. The proper location for the second tower is at the point where the lower curve again intersects the ground profile, or at station 8 + 00, elevation 506.0 ft, in the example. The operation is then repeated for the next tower. Adjustments in length of span are usually necessary to meet local conditions, in order to avoid locating towers in roads or swamps and to bring towers at angle points. Adjustments which increase the ground clearance are of course allowable.

The position of the conductors with maximum sag may be drawn on the profile from the top curve of the template.

UPLIFT ON INSULATOR; USE OF TEMPLATE *N*. An insulator sustains the weight of the lengths of conductor from the insulator to the lowest point of the span on each side. If the conductor leaves the insulator horizontally on one side, the lowest point of that span is at the insulator, which then sustains no weight due to that span. If the conductor has an upward inclination where it leaves an insulator, it is exerting an uplift equal to the weight of a length of conductor extending from the insulator along the span produced in the reverse direction to the lowest point of the catenary. Where the conductor has a downward inclination on one side and an upward one on the other side of the insulator, there will be a weight or uplift on the insulator equal to the difference between the weight of conductor on one side and uplift on the other.

Suspension insulators, when used hanging downward to sustain weight, are incapable of resisting uplift. Where uplift occurs, the conductor may be dead ended or may be tied down or weighted down.

The method used for locating towers ordinarily precludes uplift under the loading which gives maximum sag, but uplift may occur when the loading is less. To determine this the minimum sag is drawn on the profile with template *N*. The minimum sag curve is drawn between points of support, keeping the axis of the catenary vertical as before.

SIDE SWING OF SUSPENSION INSULATORS. Let l_1 = the length of conductor between the lowest point in the span to the left of the insulator and the lowest point in the span to the right of the insulator, and let l_2 = the distance between the middle points of these two spans, both in feet. Also let v = the weight of the conductor and ice per foot length, and h = the wind pressure per foot length (see Table 5). Then the vertical pull on the insulator is vl_1, and the transverse horizontal force is hl_2. Also let v_1 = the weight of the insulator, and h_1 = the total wind pressure on it. Then the insulator is deflected sidewise from the vertical by approximately the angle

$$\theta = \tan^{-1}\left[\frac{hl_2 + 0.5h_1}{vl_1 + 0.5v_1}\right] \qquad (58)$$

Usually the weight of the insulator and the wind pressure on it are negligible compared with the weight and wind pressure on the conductor, in which case

$$\theta = \tan^{-1}\left(\frac{hl_2}{vl_1}\right) \qquad (59)$$

When the points of support are at the same elevation, $l_1 = l_2$ and

$$\theta = \tan^{-1}\frac{h}{v} \qquad (60)$$

Calling X the length of the insulator in feet, then the transverse horizontal deflection of the insulator is $X \sin \theta$ ft.

For example, consider the side swing of the third insulator (from the left) in Fig. 15. Then $l_1 = 470$, $l_2 = 850/2 + 600/2 = 725$, $v = 1.61$ (for 300,000-cir mil conductor with $1/2$ in. of ice), and $h = 0.82$ (for wind pressure of 6 lb per sq ft). Hence, neglecting the weight of the insulator and the wind pressure on it,

$$\theta = \tan^{-1} \frac{0.82 \times 725}{1.61 \times 470} = 38 \text{ deg}$$

If the insulator is 5 ft long, the transverse horizontal deflection is then 5 sin 38 deg = 3.1 ft.

If the angle of swing as thus determined is excessive, the cables and insulators will be lifted up into the cross-arms at times of low temperatures and high winds. The remedy is the same as in direct uplift.

SIDE CLEARANCE; USE OF TEMPLATE *Z*. Where a right-of-way of definite width is obtained, it is necessary to determine whether the conductor will swing beyond the edge of the right-of-way. Therefore, after the towers have been located by the use of the profile, they should be marked on the plan and the side swing marked in from template Z, as shown in Fig. 15. In determining side swing, the swing of the insulator (if suspension type) must be allowed for, as well as the side swing of the conductor. Adequate margin should be allowed between the extreme position of conductor and the edge of the right-of-way, so that a safe clearance will be preserved from any structures erected adjacently. Where extraordinarily long spans must be used, an adequate extra width of right-of-way should be obtained in the first place.

LOSS OF CLEARANCE BETWEEN CONDUCTORS BECAUSE OF UNEQUAL ICE LOADING. Where one span is loaded with ice and the one immediately below it is not, the clearance is reduced. This condition may sometimes arise because of the ice falling off the lower wire before it falls off the upper wire. Where the wires are directly over each other, the normal clearance must be great enough to prevent the crossing of wires under these conditions. For ice loading without wind, clearance under unequal loading is most easily obtained by offsetting the wires horizontally for the required clearance instead of increasing the vertical clearance. To prevent crossing of the unequally loaded wires when deflected by wind pressure, however, this horizontal offset must be considerable, as the clearance must then be obtained between the wires in their inclined positions.

If 300,000-cir mil bare copper conductors initially stressed to 4170 lb at 32 deg fahr are normally 10 ft apart vertically on an 800-ft span, the sag would be 17.5 ft without ice at 32 deg fahr, 19.0 ft with $1/4$ in. ice, and 20.9 ft with $1/2$ in. ice at 32 deg fahr. Consequently, if two cables are used, one above the other, and ice forms on the upper but not on the lower, then, assuming fixed points of support (pin insulators), the clearance will be reduced by 3.4 ft for $1/2$ in. ice, and 1.5 ft for $1/4$ in. ice, making the clearances 6.6 ft and 8.5 ft, respectively, instead of 10 ft.

Where suspension insulators are used, the reduction of clearance from unequal ice loading is greater. If one span is loaded with ice and the adjacent spans of the same wire are not loaded, the sag of the loaded span will be increased because the insulators will swing toward it. Similarly, if one span is unloaded and adjacent spans are loaded, the sag will be decreased. The minimum clearance occurs where only one span of the upper wire is loaded and is immediately over the only unloaded span of the lower wire. The actual reduction is readily calculated by the method given in the paragraph on Calculation of Stresses in Unequally Loaded Spans (p. 14-78). The amount of reduction depends on the number of spans between anchor towers, and the distance from the anchor towers at which the unbalanced loading occurs. For a 300,000-cir mil copper cable on 800-ft spans at 32 deg fahr with unequal loadings on sections of one, two, three, and five spans (see Fig. 16) the assumed conditions of unequal loading and the loss of clearance are as follows:

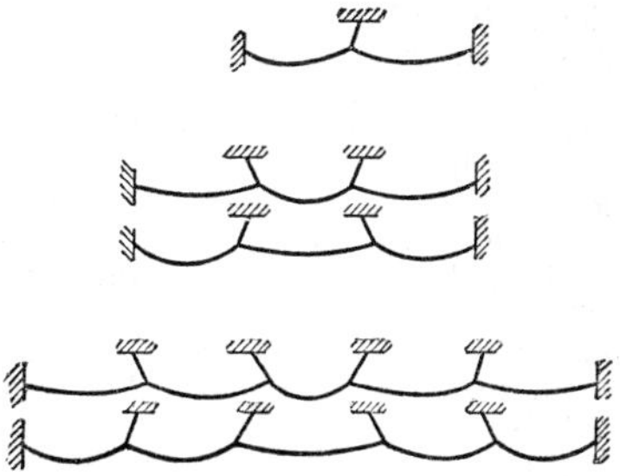

FIG. 16. Unequally Loaded Spans

Number of Spans between Anchor Towers	1	2	3	5
Upper conductor; 1/4 in. ice on spans Nos.:*	1	1	2	3
Lower conductor; 1/4 in. ice on spans Nos.:*	....	2	1, 3	1, 2, 4, 5
Sag of middle span of upper conductor, feet	19.0	20.5	21.1	21.6
Sag of middle span of lower conductor, feet	17.5	15.9	15.5	15.2
Loss of clearance in middle span, feet	1.5	4.6	5.6	6.4

* The other spans assumed to have no ice load.

STRESSES AND DEFLECTIONS DUE TO BROKEN CONDUCTORS. When a conductor breaks in a span supported by suspension insulators, the insulators adjacent to the broken span swing up into line with the cable, throwing increased slack into the unbroken part of the cable equal to the length of the insulator. This slack divides between the unbroken spans, increasing the deflection of each. The stresses and deflections of the unbroken spans may be determined by the method given for Calculation of Stresses in Unequally Loaded Spans (p. 14-78), calling the span in which the break occurs span 1.

26. JOINT USE OF POLES WITH OTHER UTILITIES

When supporting structures of power lines are used jointly by other utilities, such as telephone or other communication systems, additional factors are introduced into the problem of line design beside those needing consideration in the case of power lines alone. Often a higher grade of construction is necessary, and attention must be given to the required separations between the conductors and equipment of the two utilities. When the poles are used jointly by power and communication utilities, the separations required are generally greater than those necessary between attachments of the same or different power utilities. Usually contracts which set forth in detail the construction specifications to be followed are entered into between the joint-use parties. The Edison Electrical Institute and the Bell system have agreed upon and published complete specifications for joint-pole practices.

27. RIGHT-OF-WAY

The right-of-way should be as short and straight as practicable. Its width will depend largely on the type of line to be constructed. A width of 50 ft is generally considered adequate for a single pole line with moderately short spans. For a double-circuit tower line, even with fairly long spans, a width of 100 ft is generally adequate. With exceptionally long spans and extremely high voltages widths of as much as 200 ft may be necessary. Where two parallel pole or tower lines are to be constructed on the same right-of-way, the distance between the center lines of the two lines should be added to the width of right-of-way required for a single line. The towers for the two lines should be placed directly opposite each other in order to reduce the width of right-of-way required.

The direction of the line will often need to be shifted to avoid buildings or other obstructions. In such a case consideration should be given to the maximum side swing of the conductors in the wind, and adequate clearance should be provided between the conductors and such obstructions under the worst condition of side swing. In computing the side swing, allowance should be made for swing of suspension insulators.

When it is necessary that the right-of-way cross railroads, roads, or other lines, the length of crossing should be reduced to a minimum by making the crossing as nearly at right angles as is practicable. Rights-of-way through swamps often require expensive road building and expensive tower foundations, although small swamps (up to about 1000 ft across) can often be crossed in a single span. Steep side hills require extra expense for foundations and tower extensions and introduce a hazard of injury to tower from sliding earth, rocks, trees, or snow. A right-of-way through forests requires expensive clearing.

It is sometimes advantageous to own and fence the right-of-way. When the right-of-way passes through farm lands, however, it is usually advantageous not to fence it in, but to have it cultivated and kept free from brush. Instead of purchasing a right-of-way, it is often sufficient to obtain easements covering the location of towers and suspension of wires. Easements and right-of-way agreements should include the right to remove and trim trees under and adjacent to the line. The right-of-way should be passable (or at least accessible) for patrolling as well as for construction.

The first step in selecting a route for a line is to lay out the possible routes on accurate maps. Topographic maps are of considerable assistance in this work. Rough reconnaissance surveys are made to determine the most suitable route. Aerial photographic surveys are very valuable, especially in mountainous country. The next step is to obtain the right-of-way either by outright purchase or by obtaining easements. This is followed by sending out a party to make an accurate survey, including plans and profiles showing buildings, obstructions, natural features, and property lines. (See Location of Structures and Determination of Clearances, p. 14-80.)

28. WOOD POLES FOR OVERHEAD LINES

BASIS FOR SELECTION OF POLES. The length of pole required for a particular location is determined by the following factors:

1. Amount of vertical space required for wires and equipment.
2. Clearance required above ground or obstructions for wires and equipment.
3. Sag of conductors.
4. Depth pole is to be set in the ground.

The size (class) of pole is determined by the strength required to sustain the mechanical loading imposed upon it. The strength of an unguyed pole is usually determined by its circumference at the ground line. This dimension determines the resisting moment of the pole when bending as a cantilever. For a guyed pole the resisting moment at the point of guy attachment must be sufficient to withstand the bending stresses imposed at that point. The top of the pole must also be of adequate diameter to permit the attachment of cross-arms without unduly weakening the pole near the top. Four and one-half inches is about the minimum top diameter suitable for cross-arm construction.

Long life is very desirable, provided that it can be obtained at a low annual cost and provided that the line is expected to be more or less permanent. Long life for poles is obtained by the use of durable wood species and usually by preservative treatment applied to both durable and non-durable species. In order to obtain the quality of pole timber desired, purchase specifications are required, and the specifications should be enforced by careful inspection of the poles before, during, and after the preservative treatment.

METHODS OF SPECIFYING POLE DIMENSIONS. A wood pole is usually specified by its total or "nominal" length and by its class. The word class refers to the dimensional classifications set up by the specifications of the American Standards Association. In those specifications classes numbered from 1 to 10 inclusive are provided for poles of the wood species principally used. Classes 1 to 7 inclusive specify the minimum top circumference for each class and the minimum circumference at 6 ft from the butt end for each nominal length in each class. Class 1 provides the largest ground line circumference, and Class 7 the smallest. Classes 8 to 10 inclusive specify minimum top circumferences only. Table 7 shows dimensions for most wood species used for poles.

The dimensions specified for each class of Classes 1 to 7 inclusive are so selected that for any one wood species any pole of a certain class, regardless of the length, will support the same conductors located in the same manner relative to the top of the pole (including wind and ice loading on conductors and pole) with the same fiber stress (within close limits) in the pole. This permits the same class pole to be used throughout a line if the size and arrangement of the conductors do not change, but a different class may be necessary on account of changes in grade of construction, angles, etc. Poles of different species but of the same class will support the same mechanical loading.

Taper of Poles. The taper of various kinds of poles, specified as the difference, measured in inches, between two circumferences 10 ft apart, is given as follows in *Forest Service Bull.* 84: chestnut (Maryland), 3.8 to 4.0; northern white cedar (Michigan), 5.2; western yellow pine (California), 4.0; lodgepole pine (Montana), 3.0; loblolly pine (Texas), 2.4; western red cedar (Washington), 3.5. Trees grown upon a high elevation have a greater taper in the trunk than trees grown lower down.

Timber, to be desirable for use as poles, should have the following qualities: straightness, small taper, low weight, few knots, good resistance to decay, lack of serious defects, high strength, and softness enough to permit the spikes of a climber to enter readily. No pole timber is perfectly resistant to decay, but any timber can have its life greatly prolonged by preservative treatment. Treatment methods are discussed later in this article.

Uncertainties in Names of Timber Trees. Each of the terms cedar, pine, etc., used in describing poles and cross-arms, and each of even the apparently more exact terms, such as white cedar, yellow pine, etc., cover several kinds of trees and have different meanings in different localities.

At least eight pines (of the thirty-five native ones) are in the market, some of which so closely resemble each other in their minute structure that they can hardly be told apart; yet they differ in quality and should be used separately, although they are often mixed or confounded in the trade.

Table 7. American Standards Association Standards for Dimensions of Wood Poles

Southern Yellow Pine, Douglas Fir, Eastern Hemlock, Western Hemlock, Eastern Larch, and Western Larch Poles
(Fiber stress, 7400 lb per sq in.)

Class		1	2	3	4	5	6	7	8	9	10
Minimum Circumference at Top, Inches		27	25	23	21	19	17	15	18	15	12
Length of Pole, Feet	Ground Line Dist. from Butt, Feet	Minimum Circumference at 6 ft from Butt, Inches									
16	3 1/2					21.5	19.5	18.0			
18	3 1/2			26.5	24.5	22.5	21.0	19.0			
20	4	31.5	29.5	27.5	25.5	23.5	22.0	20.0			
22	4	33.0	31.0	29.0	26.5	24.5	23.0	21.0			
25	5	34.5	32.5	30.0	28.0	26.0	24.0	22.0			
30	5 1/2	37.5	35.0	32.5	30.0	28.0	26.0	24.0			No butt requirements
35	6	40.0	37.5	35.0	32.0	30.0	27.5	25.5		No butt requirements	
40	6	42.0	39.5	37.0	34.0	31.5	29.0	27.0	No butt requirements		
45	6 1/2	44.0	41.5	38.5	36.0	33.0	30.5	28.5			
50	7	46.0	43.0	40.0	37.5	34.5	32.0	29.5			
55	7 1/2	47.5	44.5	41.5	39.0	36.0	33.5				
60	8	49.5	46.0	43.0	40.0	37.0	34.5				
65	8 1/2	51.0	47.5	44.5	41.5	38.5					
70	9	52.5	49.0	46.0	42.5	39.5					
75	9 1/2	54.0	50.5	47.0	44.0						
80	10	55.0	51.5	48.5	45.0						
85	10 1/2	56.5	53.0	49.5							
90	11	57.5	54.0	50.5							

Lodgepole Pine, Jack Pine, Red (Norway) Pine, and White Fir Poles
(Fiber stress, 6600 lb per sq in.)

Class		1	2	3	4	5	6	7	8	9	10
Minimum Circumference at Top, Inches		27	25	23	21	19	17	15	18	15	12
Length of Pole, Feet	Ground Line Dist. from Butt, Feet	Minimum Circumference at 6 ft from Butt, Inches									
16	3 1/2					22.0	20.5	19.0			
18	3 1/2			27.5	25.5	23.5	21.5	20.0			
20	4	32.5	30.5	28.5	26.5	24.5	22.5	21.0			
22	4	34.0	32.0	30.0	27.5	25.5	23.5	22.0			
25	5	36.0	33.5	31.0	29.0	27.0	25.0	23.0			
30	5 1/2	39.0	36.5	34.0	31.5	29.0	27.0	25.0			No butt requirements
35	6	41.5	38.5	36.0	33.5	31.0	28.5	26.5		No butt requirements	
40	6	44.0	41.0	38.0	35.5	33.0	30.5	28.0	No butt requirements		
45	6 1/2	46.0	43.0	40.0	37.0	34.5	32.0	29.5			
50	7	48.0	45.0	42.0	39.0	36.0	33.5	31.0			
55	7 1/2	49.5	46.5	43.5	40.5	37.5	34.5				
60	8	51.5	48.0	45.0	42.0	38.5					
65	8 1/2	53.0	49.5	46.0	43.0						
70	9	54.5	51.0	47.5							
75	9 1/2	56.0	52.5								

Forestry Bulletin 10 states: " 'Yellow pine,' is applied in the trade to all the southern lumber pines; in the Northeast it is also applied to the pitch pine; in the West it refers mostly to bull pine. 'Yellow longleaf pine,' 'Georgia pine,' chiefly used in advertisement, refers to longleaf pine."

TIMBERS ORDINARILY USED FOR POLES AND CROSS-ARMS. The principal timber trees used for poles and cross-arms are briefly described below in accordance with the names used in the trade.

Chestnut grows throughout the Appalachian Mountain region. A blight has reduced the supply of satisfactory chestnut timber to a negligible amount. The sapwood is very

Table 7. American Standards Association Standards for Dimensions of Wood Poles —*Continued*

Northern White Pine, Ponderosa Pine, Sugar Pine, Chestnut, and Western White Pine Poles
(Fiber stress, 6000 lb per sq in.)

Class		1	2	3	4	5	6	7	8	9	10
Minimum Circumference at Top, Inches		27	25	23	21	19	17	15	18	15	12
Length of Pole, Feet	Ground Line Dist. from Butt, Feet	Minimum Circumference at 6 ft from Butt, Inches									
16	3 1/2					22.5	21.0	19.5			
18	3 1/2			28.0	26.0	24.0	22.0	20.5			
20	4	33.5	31.5	29.5	27.0	25.0	23.0	21.5			
22	4	35.0	33.0	30.5	28.5	26.5	24.5	22.5			No butt requirements
25	5	37.0	34.5	32.5	30.0	28.0	25.5	24.0		No butt requirements	
30	5 1/2	40.0	37.5	35.0	32.5	30.0	28.0	26.0	No butt requirements		
35	6	42.5	40.0	37.5	34.5	32.0	30.0	27.5			
40	6	45.0	42.5	39.5	36.5	34.0	31.5	29.5			
45	6 1/2	47.5	44.5	41.5	38.5	36.0	33.0	31.0			
50	7	49.5	46.5	43.5	40.0	37.5	34.5	32.0			
55	7 1/2	51.5	48.5	45.0	42.0	39.0	36.0				
60	8	53.5	50.0	46.5	43.5						

Western Red Cedar, Southern White Cedar, and Spruce Poles
(Fiber stress, 5600 lb per sq in.)

Class		1	2	3	4	5	6	7	8	9	10
Minimum Circumference at Top, Inches		27	25	23	21	19	17	15	18	15	12
Length of Pole, Feet	Ground Line Dist. from Butt, Feet	Minimum Circumference at 6 ft from Butt, Inches									
16	3 1/2					23.0	21.5	19.5			
18	3 1/2			28.5	26.5	24.5	22.5	21.0			
20	4	34.5	32.0	30.0	28.0	25.5	23.5	22.0			
22	4	36.0	33.5	31.5	29.0	27.0	25.0	23.0			No butt requirements
25	5	38.0	35.5	33.0	30.5	28.5	26.0	24.5		No butt requirements	
30	5 1/2	41.0	38.5	35.5	33.0	30.5	28.5	26.5	No butt requirements		
35	6	43.5	41.0	38.0	35.5	32.5	30.5	28.0			
40	6	46.0	43.5	40.5	37.5	34.5	32.0				
45	6 1/2	48.5	45.5	42.5	39.5	36.5					
50	7	50.5	47.5	44.5	41.0	38.0					
55	7 1/2	52.5	49.5	46.0	42.5	39.5					
60	8	54.5	51.0	47.5	44.0						

thin, usually from about 1/8 to 3/8 in. in thickness. It is not so straight as cedar and is likely to be knotty. It is slightly stronger and heavier than cedar.

Northern White Cedar has its principal source of supply in the region of the Great Lakes. It is used mainly in the northeast quarter of the United States for small poles. Its large taper and spreading butts make it undesirable for long poles. The sapwood varies from 1/2 to 1 in. in thickness and is usually thicker at the top than at the butt. Frequently the butts are decayed at the center. It is very slow in growth, requiring about 190 years for a 30-ft pole.

Southern White Cedar grows mainly in the southern swamps and is somewhat less durable than the northern cedar. It is used very little for poles, as its sapwood decays very quickly.

Red Cedar is a small to medium-sized tree scattered through the forests or, in the West, sparsely covering extensive areas (cedar brakes). The red cedar is the most widely distributed conifer of the United States, occurring from the Atlantic to the Pacific and from Florida to Minnesota, but it attains a suitable size for lumber only in the South, more especially in the Gulf States, and is seldom used for poles.

The term **juniper** is commonly used by telephone men for southern white cedar; the term also is applied to red cedar. Juniper poles come from Virginia, the Carolinas, and other South Atlantic States.

Western Red Cedar is light, straight, and durable and is one of the principal pole timbers used at present. The main sources of supply are northern Idaho, western Washington, and British Columbia. Its durability is greatly increased by treating the butts to a point about 1 ft above the ground line with creosote. (See Preservative Treatments, p. 14-91.) It does not have as great strength as southern pine and does not stand impact very well. The sapwood is somewhat subject to dry rot in certain localities.

Cypress is a large deciduous tree, occupying much of the swamp and overflow land along the coast and rivers of the southern states. Cypress is usually considered a durable wood, and the heartwood is, in fact, one of the most durable of our native species. The sapwood, however, decays quickly, seriously weakening the pole. The width of the sapwood on pole-size trees is from $^3/_4$ to $1\,^1/_4$ in. Cypress frequently is too large for use as a pole and has greater value for lumber. Even when its general diameter is small enough, the butt will often be so big that it adds too much weight.

Southern Yellow Pine is a trade term covering the longleaf, shortleaf, loblolly, slash, and pond pines. These species are strong, straight, and symmetrical but are subject to quick decay as poles unless treated with preservative throughout their length. There is a plentiful supply of these pines, and they easily reproduce, with rapid growth.

Longleaf Pine is a large tree which forms extensive forests and furnishes the hardest and strongest pine lumber in the market. It is obtained from the coast region from North Carolina to Texas. The longleaf pine is strikingly heavy, hard, and resinous, and usually very regular and narrow ringed, showing little sapwood, and differing in this respect from the shortleaf pine and loblolly pine, which generally have wider rings and more sapwood, the loblolly excelling in that respect.

Shortleaf Pine resembles loblolly pine and often approaches in its wood the Norway pine. It is the common lumber pine of Missouri and Arkansas, and North Carolina to Texas and Missouri.

Loblolly Pine is a large tree; it forms extensive forests. It is wider-ringed, coarser, lighter, and softer, with more sapwood than the longleaf pine, but the two are often confounded. This is the common lumber pine from Virginia to South Carolina and is found extensively in the southern states from Virginia to Texas.

Norway Pine is a large tree; it never forms forests, being usually scattered or in small groves, together with white pine. It is largely sapwood and hence not durable. It grows from Minnesota to Michigan, also in New England to Pennsylvania. The Norway pine, which may be confounded with the shortleaf pine, can be distinguished by being much lighter and softer. It may also, but more rarely, be confounded with heavier white pine, but it has sharper definition of the annual ring, weight, and hardness.

Western Yellow Pine is used for poles to a limited extent in certain parts of the Southwest, where the high cost of more durable pole timbers makes it necessary to find a cheaper substitute. The life of this timber, untreated, is very short. In the upper part of the San Joaquin Valley of California, where a study of this species was made, untreated pine poles last only 2 or 3 years; but, since the wood when not exposed to the soil is fairly durable, it is believed that a butt treatment with a good wood preservative will result in a pole that will give good service. A butt-treated pine pole costs considerably less than an untreated cedar pole in this locality.

Lodgepole Pine is cut to a limited extent for poles. It grows at high altitudes in the Rocky Mountains. It decays quickly in contact with the soil, but is durable when not so exposed. The tree grows tall and straight, with very little taper, and makes a well-shaped pole. In certain parts of the West, where large bodies of fire-killed lodgepole remain standing for many years, sound and thoroughly seasoned, conditions for effective treatment are excellent. If given a butt treatment, this dead timber makes a durable pole, and in many localities the cost of the pine pole, plus the cost of the treatment, is less than that of the Idaho cedar untreated. The sapwood of pole-sized timber may be an inch or an inch and a quarter thick.

Douglas Fir is a conifer found largely in the North Pacific coast region and principally in Oregon. It is light, strong, and straight-grained. It is very durable when not in contact with the soil and therefore particularly adapted for use as cross-arms.

TERMS DESCRIBING PART OF A POLE. Knots are the heartwood of branches extending transversely through the sapwood of the trunk outwardly from the central heartwood.

Annular Rings are the concentric rings added yearly under the bark as the tree grows. Each ring is composed of two rings, one called spring wood and the other summer wood.

Spring Wood and Summer Wood are differentiated by their density and color. The spring wood is relatively porous and usually is light in color; the summer wood is denser and darker in color. In conifers the summer wood is a reddish yellow on account of its resin content. In most woods the summer wood is thinner than the spring wood.

Pith. The pith of a tree is the central core about which the annual rings are formed. It goes through the tree from top to bottom and branches into the limbs. The pith is quite thick, usually $1/8$ to $1/5$ in. in Norway pine and in the southern species, although much thinner in white pine and very thin, $1/15$ to $1/25$ in., in cypress, cedar, and larch. The pith of the tree is the weakest part on account of the many knots which it invariably and necessarily contains.

Sapwood. The sapwood of a tree is a zone of wood next to the bark, 1 to 3 or more in. wide and containing 30 to 50 or more annular rings (in coniferous trees). It is of lighter color than the inner, darker part of the log, which is the heartwood. Sapwood changes to heartwood as the tree grows.

The width of the sapwood is small for longleaf and white pine and great for loblolly and Norway pines. In old trees of longleaf pine the sapwood forms about 40 per cent of the merchantable log; in the loblolly and in all young (coniferous) trees the bulk of the wood is sapwood.

Sapwood, being the normal condition of the outer rings of a tree, is not a "defect" in poles, where the whole cross-section of the tree (except bark) is used. Being weaker and more liable to decay, it is considered a defect in pins and cross-arms, which are better if made from the heartwood only.

DEFECTS IN WOOD USED FOR POLES AND CROSS-ARMS. The following are the defects in timber which are frequently referred to in specifications for poles and cross-arms. The first six definitions are from *Carpentry and Joinery*, by Paul N. Hasluck. The next twelve definitions are those used in the timber-test work of the Forest Service in describing defects (*Forest Service Circular* 38, Revised).

Cup-shakes. These are cracks extending circumferentially at one or more places, caused by the separation of the annual rings.

Dote. This is a speckled stain found in beech, American oak, and other timber, due to incipient decay. It is produced by imperfect seasoning or by exposure for a long period to a stagnant atmosphere.

Heart-shakes. These are splits or clefts occurring in the center of the tree. They are common in nearly every variety of timber and are very serious when they twist in the length, as they interfere with the conversion of the tree into boards or scantlings. They sometimes divide the log in two for a few feet from the end.

Star-shakes. When several heart-shakes occur in one tree, they are called star-shakes from the appearance produced by their radiation from the center.

Wind Cracks. Shakes or splits on the sides of a balk (a log which has been squared off) of timber, caused by shrinkage of the exterior surface, are called wind cracks.

Dry Rot. Dry rot is a special form of decay in timber caused by the growth of a fungus, which spreads over the surface like a close network of threads, white, yellow, or brown, and causes the inside to perish and crumble. Causes which render timber favorable to the growth of this fungus are: large proportion of sapwood; felled at wrong season when full of sap; cutting down in the spring or fall of the year instead of in midwinter or midsummer, when the sap is at rest; stacked for seasoning without sufficient air spaces being left; fixed before thoroughly seasoned; painted or varnished while containing moisture.

Sound Knot. A sound knot is one which is solid across its face and as hard as the wood surrounding it; it may be either red or black, and is so fixed by growth or position that it will retain its place in the piece.

Loose Knot. A loose knot is one not firmly held in place by growth or position.

Pith Knot. A pith knot is a sound knot with a pith hole not more than $1/4$ in. in diameter at the center.

Encased Knot. An encased knot is one which is surrounded wholly or in part by bark or pitch. Where the encasement is less than $1/2$ in. in width on both sides, not exceeding one-half the circumference of the knot, it shall be considered a sound knot.

Rotten Knot. A rotten knot is one not as hard as the wood it is in.

Pin Knot. A pin knot is a sound knot not over 2 in. in diameter.

Spike Knot. A spike knot is one sawn in a lengthwise direction. The mean or average width shall be considered in measuring these knots.

Pitch Pocket. A pitch pocket is an opening between the grain of the wood containing more or less pitch or bark.

Pitch Streak. A pitch streak is a well-defined accumulation of pitch at one point in the piece. When not sufficient to develop a well-defined streak, or where the fiber between grains—that is, the coarse-grained fiber, usually termed "spring wood"—is not saturated with pitch, it shall not be considered a defect.

Wane. Wane is bark, or lack of wood from any cause, on edges of timber.

Shakes. Shakes are splits in timber which usually cause a separation of the wood between annual rings.

Checks. Checks are splits in timber which usually cause a separation of the wood across annual rings.

Wind-shake. This is a crack or incoherence in timber produced by violent winds while the timber was growing.

Wind. A turn or bend; a piece of timber is out of wind when it is perfectly straight or flat.

Warped. Wood twisted out of shape by seasoning is warped.

Cat-faces. These are old wounds, partially overgrown, leaving a long, narrow, dead surface exposed.

Insect Damage. This damage is caused by the boring of various insects and is indicated by small holes or bumps on the pole surface.

Crook. An offset in the axis of the pole is called crook. The axis above the crook may be parallel to or coincident with the axis below the crook, or it may be curved.

Sweep. This is produced by a curving axis. It may be in one plane and one direction, or in two planes (double sweep), or in two directions in one plane (reverse sweep).

VOLUME AND WEIGHT OF POLES. A quick way to find the approximate volume of a pole is to multiply the area of the circle at the center of gravity by the length of the pole. The formula for the volume, considering a pole as a frustum of a cone, is

$$v = \frac{\pi}{1728}(d_1^2 + d_1 d_2 + d_2^2)h \tag{61}$$

where v = volume in cubic feet; d_1 = diameter at butt in inches; d_2 = diameter at top in inches; h = length of pole in feet.

The weight of a pole may be found by multiplying its volume in cubic feet by its weight per cubic foot. Table 8 gives the weight per cubic foot.

Table 8. Average Weight per Cubic Foot of Poles *

Kind of Pole	When Cut		When Seasoned	
	Weight, Pounds per Cubic Foot	Moisture, Per Cent of Oven-Dry Weight	Weight, Pounds per Cubic Foot	Moisture, Per Cent of Oven-Dry Weight
Cedar				
Western red	27	37	23	12
Northern white	28	55	22	12
Cypress, southern	51	91	32	12
Douglas fir				
Coast	38	36	34	12
Intermediate	38	48	31	12
Rocky Mountain	35	38	30	12
Hemlock				
Western	41	74	29	12
Eastern	50	111	28	12
Pine				
Lodgepole	39	65	29	12
Southern yellow	52	81	36	12
Ponderosa	45	91	28	12
Redwood	43	146	21	12

* Department of Agriculture, *Tech. Bull.* 479.

SEASONING. Timber which is not to be treated with a preservative by a pressure method should be thoroughly seasoned. For poles this should be done by air seasoning. For cross-arms and other sawn timbers kiln drying should be applied also. Timbers which are to be pressure-treated are improved by at least partial air seasoning, even though it is followed by steaming before the preservative is applied. For timbers which are very susceptible to decay, the steaming process should be applied before preservative treatment to sterilize the wood and stop incipient decay.

Green timber contains a large amount of moisture, as indicated by Table 8. The purpose of seasoning is to reduce this moisture content to an approximate equilibrium with the humidity of the air. The moisture content at equilibrium varies considerably, depending upon the average humidity of the air at the seasoning location. In arid regions it may be as low as 5 per cent and in very humid locations as high as 20 per cent. The general average in the United States is about 12 per cent of oven-dry weight.

Contrary to general opinion, the moisture content of green wood does not vary a great deal from season to season throughout the year. The length of time required to season wood which is cut at different times during the year depends, not on the moisture content at the time of cutting, but almost entirely upon the temperature and humidity of the air during the seasoning period. Even in the same general area the practice of different timber suppliers as to length of seasoning period varies considerably. Table 9 gives an idea of the range of seasoning periods in general use for different species and in different localities. Small pieces season more quickly than large pieces, short pieces more quickly than long ones, and sawn pieces more quickly than round ones.

Table 9. Range of Air-seasoning Periods in Use for Various Wood Species in the United States

Species	Location of Seasoning	Length of Period, Months
Pine		
Southern yellow	Southeast	2 to 4
	Northeast	5 to 6
Lodgepole	Rocky Mountain	3 to 12
Ponderosa	Rocky Mountain	3 to 12
Douglas fir		
Coastal	North Pacific	3 to 12
Mountain	Rocky Mountain	3 to 6
Hemlock		
Western	Rocky Mountain	6 to 12
Eastern	North Central	8 to 12

Large timbers naturally dry out first at the outside, and it is not absolutely necessary that timber which is to be given a full preservative treatment be dry all the way through. Only the area to receive the preservative, i.e., the outside layers, needs to be dry. Too prolonged or too rapid drying is not desirable. The seasoning period should not continue until decay starts. This is particularly true of timber such as southern yellow pine, which is very susceptible to decay. Too rapid drying at high temperatures may harden the surface and prevent adequate drying of the interior and the proper penetration of the preservative.

ROOFING. Poles are sometimes "roofed" by cutting the top to form an inclined surface. Either a single cut is made at an angle of 15 to 30 deg, or two such planes at an angle of 30 to 45 deg are cut to meet in a horizontal ridge. Roofs are sometimes painted, or coated with a bituminous substance, or covered with zinc, lead, or copper caps.

PRESERVATIVE TREATMENTS. Practically all wood species, when used as poles, are subject to decay, either where in contact with the earth or throughout the length of the pole. Untreated cypress, pine, and juniper poles last only from 5 to 10 years; untreated cedar poles last from 15 to 20 years under favorable conditions. Very few reliable data are available to show the full value of preservative treatments in prolonging the life of poles. Pole service records available do show, however, that increases in life of at least 10 years, on the average, may be expected as a result of the best preservative treatments. From an economic standpoint such treatments are justifiable, since they result in:

1. Increase of pole life.
2. Possibility of using smaller poles, as less allowance need be made for decay.
3. Possibility of using wood species not naturally durable.

Cedar poles decay mainly in contact with soil, and therefore the butt-treatment method is principally used for that species. Most conifers are subject to decay throughout their length and therefore should be full-length treated. Poles should not be painted with oil paints, as these prevent evaporation of moisture and promote decay.

Cause of the Decay of Timber. Decay of wood is due to low forms of plant life called fungi. The germs of decay are not inherent in the wood. The wood-destroying fungi start from the outside, either from adjacent rotten wood or by spores, which correspond to seeds, being carried by the wind and deposited on the surface. Although the fungi from these spores begin at the "surface" of the wood, this surface must be understood to include all holes or cracks which the spores may enter.

For their growth and development fungi require air, heat, moisture, and food. Warmth, preferably between 60 and 100 deg fahr, favors decay. Cold retards it, and temperatures above 150 deg fahr prevent it. Under water or deep under the surface of the ground where the air is excluded, decay does not take place. Ordinarily wood which is seasoned until it is air-dry does not contain sufficient moisture to support the growth of fungi.

Preservatives. The best method of checking the growth of fungi is to poison their food, which is the wood itself. Such poisonous substances are called preservatives and are injected into the wood by various methods. In order to be effective over a long period of time the preservative used for pole-line timber must be chemically stable, must remain in the wood, and must not evaporate or leach out because of the action of the elements. These requirements do not apply so strictly to timber for building construction which is not exposed to the weather. Very few preservatives meet the requirements for pole-line timber. Of these, coal-tar creosote oil has been used the longest, and the results obtained with it have been given long and careful study. When properly applied, it protects the wood which it penetrates for 25 years or more. Water-gas tar, coal tar, and wood-tar creosote mixed with coal-tar creosote or petroleum have been used to a limited extent as emergency preservatives for poles. Petroleum solutions of copper naphthanate are receiving increasing attention as possible substitutes for creosote.

Coal-tar Creosote Oil is a distillation by-product of coal tar and is produced in the manufacture of coke and illuminating gas. It is sometimes known as dead oil of coal tar. The American Wood-Preservers' Association has specifications for two grades of creosote oil. That known as Grade 1 is mainly used in the preservation of poles. Creosote is a mixture of chemical compounds, consisting principally of liquid and solid aromatic hydrocarbons, tar acids, and tar bases. It has a specific gravity of not less than 1.03 and a continuous boiling range from 200 to 325 deg cent. In Grade 1 creosote oil the amount of residue at several distillation temperatures is strictly limited. It is believed that oils boiling below 200 deg cent will not remain in the wood to preserve it, but that the fractions above 200 deg cent are admirably suited to the purpose. In order to produce treated poles and cross-arms which are clean and not tarry on the surface, it is desirable to place stricter limitations on the characteristics of creosote oil than are required by the specification for standard Grade 1 oil. Used oil should not contain matter insoluble in benzol in excess of 0.75 per cent. After distillation up to 355 deg cent the residue should not be more than 20 per cent for new oil or 22 per cent for used oil.

Pentachlorphenol, a chlorinated phenol of high toxicity, which is coming into general use as a preservative, is practically insoluble in water but quite soluble in oils. A 5 per cent solution, by weight, in medium-heavy fuel oil as commonly used, appears to have a toxicity equal to the ordinary coal-tar creosote or greater. Its stability, volatility, and non-leaching qualities seem to be satisfactory. Solvents should be used which do not sludge when heated or reheated, but oils which are too light are not entirely satisfactory. Oils of medium viscosity seem to give the best results in that respect, as well as providing good penetrating qualities. It has been suggested that the flash point should not be less than 190 deg fahr. Mixtures containing one-half coal-tar creosote and one-half pentachlorphenol-oil solution have been tried with very satisfactory results as to penetration, and it seems logical to expect that the life of timber so treated should be as long as that obtained by using creosote alone.

Water-gas Tar is a by-product in the manufacture of water gas and is somewhat similar to the tar obtained in the manufacture of coal gas. The oil obtained by the distillation of water-gas tar is known as **water-gas tar distillate** and is quite similar to creosote oil, although its toxic properties are not considered as great as those of creosote. A mixture of 60 per cent of the distillate with 40 per cent of refined or filtered water-gas tar is known as **water-gas tar solution.**

METHODS OF TREATMENT. The methods of applying the preservatives to the pole are the brush treatment, open-tank treatment, and pressure-tank treatment.

The brush treatment is applied to a part of the butt at the ground line, the open-tank treatment to the whole butt, and the pressure-tank treatment to the whole pole.

The brush treatment is least expensive and gives the least protection, the pressure tank is most expensive and gives the most protection.

Open-tank Treatment is used for poles which are durable except where in contact with the soil. It is used mainly for cedar and chestnut poles. In order to promote the penetration of the preservative the section of the pole to be treated is usually incised. The wood is punctured by machines in a definite pattern of short, narrow incisions to a depth of $^3/_8$ to $^1/_2$ in. The butts of the poles are immersed in creosote oil at a temperature of about 230 deg fahr for not less than 6 hours. The depth of immersion is at least 1 ft more than the height of the ground line above the butt end. After the hot bath the butts are immersed in a cold creosote bath to the same depth. The temperature of the cold bath is not allowed to exceed 150 deg fahr. The length of time in the cold bath must not be less than 2 hours. All the outer bark and practically all the inner bark must be removed before the treatment

It is now recognized that poles with butt treatment alone are subject to relatively early decay in the above-ground portion when installed in areas other than those of very low humidity and low annual rainfall. Therefore full-length treatments are increasingly used for poles of species formerly considered durable above ground. Open-tank full-length treatments have been devised which provide excellent penetration of the sapwood in species such as lodgepole pine, western cedar, western hemlock, and Douglas fir. The most promising of these treatments consists of a hot creosote soaking for 5 hours, followed by a cold pentachlorphenol-oil solution bath for 5 hours. If necessary, a second hot creosote bath for 2 hours is employed. Another method used for cedar poles employs first a standard butt treatment with creosote, followed by a full-length bath for 2 hours in cold pentachlorphenol solution.

Pressure Treatment is used for all poles, such as pine, which decay quickly even in the above-ground section. It is a full-length treatment, in which small cars are loaded with poles and run into horizontal cylinders which may be made pressure-tight. Pressure treatments are classified as full-cell and empty-cell processes. In the full-cell process the wood cells are left practically full of preservative at the end of the treatment. In the empty-cell process the wood cells are practically emptied of preservative before the completion of the treatment.

In the full-cell process the poles are first subjected to a vacuum to draw out as much moisture as is practicable; then, without breaking the vacuum, the preservative is introduced into the tank at a temperature between 165 and 200 deg fahr. The pressure is then raised to 100 lb per sq in. or more and maintained until the maximum penetration and injection which are practicable have been obtained. Lastly the tank is quickly drained of preservative, and a quick high vacuum applied to remove the preservative from the surface of the poles. The net final retention of preservative is usually specified at 10 to 20 lb per cu ft of pole volume. To insure the long life of the pole, the entire sapwood content should be penetrated by preservative.

In the Reuping empty-cell process, air pressure of sufficient amount to assist properly in the final emptying of the wood cells is applied. After the application of the air pressure, and without reducing it, the preservative is introduced into the tank at a temperature between 165 and 200 deg fahr and under sufficient pressure to obtain the desired penetration. After the pressure application is completed, the cylinder is quickly drained and a vacuum is applied and maintained until the net retention of preservative is reduced to the specified amount. In the Lowry empty-cell process the application of initial air pressure is omitted, thus permitting lower pressures while the preservative is in the cylinder; however, higher vacuum is needed at the end to remove the excess preservative. The net final retention of preservative is usually specified as 6 to 10 lb per cu ft of pole volume. Complete penetration of the sapwood is also desirable in this process. It should be noted that in all these pressure processes excessive temperatures, pressures, and vacuums tend to injure the timber.

SPECIFICATIONS FOR POLES AND INSPECTION. The American Standards Association has prepared specifications covering dimensions and quality of timber for pole use. These specifications are widely used, although individual purchasers often prepare their own.

Inspection. The purchaser's inspector should examine each pole in the white for dimensions, shape, and defects and reject those which do not conform to the specification. During the treating process the inspector should observe the procedure and read all gages to determine the amount of preservative retained in the wood. He should maintain a frequent check upon the analysis of the preservative to make sure that it conforms to the specification. After the treatment is completed, the inspector should take borings from the poles in each charge to determine the depth of penetration of the preservative. Some purchasers require that all poles be bored, and those not conforming must be rejected. Generally, however, a sampling method is used. Usually twenty poles in each charge are bored. The specification of the American Wood-Preservers' Association permits the whole charge to be accepted if one or two of the twenty borings are non-conforming. Stricter specifications in general use require that every pole in the charge be bored if any of the twenty borings are non-conforming, with the acceptance only of those poles which are found to be conforming. Usually poles found to be insufficiently penetrated the first time are allowed to be retreated and again offered for inspection.

Since the value of the preservative treatment is dependent upon the extent to which the vulnerable portion of the pole is treated, the depth of penetration is the most important factor in satisfactory wood preservation. The sapwood of most species is the most vulnerable to decay, but the heartwood is by no means invulnerable in any of the usual pole timbers. To get the most protection the entire sapwood and as much as possible of the heartwood should be penetrated by the preservative. With modern treating methods there is very little difficulty in producing full sapwood penetration and even considerable heartwood penetration in southern pine poles. Therefore, purchasers desiring high-grade poles should specify 100 per cent sapwood penetration for southern pine poles with less than 3 in. of sapwood and at least 95 per cent sapwood penetration for southern pine poles with greater sapwood thickness. Under normal production conditions there should be no difficulty in obtaining such poles at no premium in price. The insistence of several purchasers on such exacting specifications has been largely responsible for the great improvement in treating methods in recent years.

FORCES ACTING ON A POLE. A pole is subject to the following forces:

1. Vertical forces due to weight of pole, wires, sleet, etc., and to downward pull of guys.
2. Lateral horizontal forces due to wind across line on pole, wire, sleet, etc.
3. Longitudinal horizontal forces due to unbalanced pull of wires.
4. Torsional forces due to unbalanced pull of wires.

A pole is strong in respect to the vertical forces but weak for horizontal forces, and the cross-arms are weak for the torsional forces. The theory of good line work is, therefore, first to reduce the horizontal and torsional forces as much as possible by balancing the stresses, and second to convert remaining unbalanced horizontal stresses into vertical stresses on the pole by the use of guys.

In practice the lateral horizontal force of the wind cannot ordinarily be provided for by guys. Calculations for strength of poles, when made, are ordinarily limited to the effect of side wind.

BREAKING OF POLE BY CROSS WIND. The principal forces tending to break a pole are wind pressures on pole and conductors when the wind blows transversely. These tend to break it by cross bending.

Let M_1 = moment of the wind on the pole.
M_2 = moment of the wind on the wires.
M = moment of resistance of the pole.

Then the condition that the pole shall not break is that

$$M_1 + M_2 < M \tag{62}$$

The calculation of M_1, M_2, and M is given below.

Moment of Wind on Pole (M_1). Moment at ground level due to wind pressure on pole is

$$M_1 = \frac{P_1 H_1^2 (D_1 + 2D_2)}{72} \tag{63}$$

where M_1 = moment at the ground in pound-feet.
P_1 = wind pressure in pounds per square foot of projected area of pole.
H_1 = height of pole in feet.
D_1 = diameter of pole at ground in inches.
D_2 = diameter of pole at top in inches.

The maximum bending moment due to horizontal forces at the top of the pole is ordinarily assumed to be at the ground level; it is really a little below ground level and opposite the center of pressure of resistance furnished by the ground.

Moment of Wind on Wires (M_2). Moment at ground level due to wind pressure on the wires is

$$M_2 = \frac{P_2 H_2 nd(S_1 + S_2)}{24} \tag{64}$$

where M_2 = moment at the ground in pound-feet.
P_2 = wind pressure in pounds per square foot of projected area of wires.
H_2 = height of wires above ground in feet.
n = number of wires.
d = diameter of wires (including ice) in inches.
S_1 and S_2 = lengths of adjacent spans in feet.

Where wires are of different diameters or at different levels, the formula is to be applied to each size and each level separately, and the moments are to be summed.

Moment of Resistance (M). The moment of resistance or strength of a circular pole for bending as a cantilever is

$$M = 0.000264fC^3 \tag{65}$$

where M = maximum allowable moment of resistance at the ground line in pound-feet.
f = maximum allowable fiber stress in pounds per square inch.
C = circumference of the pole at the ground line in inches.

Table 10. Fiber Stresses and Breaking Loads

Kind of Timber	Ultimate Fiber Stresses, N.E.S. Code, Pounds per Square Inch	Test Results		
		Fiber Stress at Rupture, Pounds per Square Inch	Breaking Load, Pounds	Elastic Limit, Pounds per Square Inch
Cedar				
Western red	5600	6065 *	1310 †	3200 ‡
Western red	....	....	2215 §	....
Western red	....	....	1930 §	....
Oregon	....	....	3040 §	....
Northern white	3600	3621 *	2650 ‖	2600 ‡
Chestnut	6000	6480 *	3240 ‖	3100 ‡
Cypress	5000	7110 ‡	....	4200 ‡
Douglas fir	7400	....	....	3800 ‡
Western hemlock	....	7400 *	....	3400 ‡
Pine				
Lodgepole	6600	5500 ‡	1430 †	3000 ‡
Southern yellow	7400	8026	....	....
Longleaf	7400	8700 ‡	....	5200 ‡
Shortleaf	7400	7300 ‡	....	3900 ‡
Ponderosa	....	5000	....	3100 ‡
Redwood	3600	6100 ‡	....	3600 ‡
Engelmann spruce	....	4200 ‡	1405 †	2500 ‡

* American Standards Association.
† *Forest Service Cir.* 204; 25-ft poles, 7 in. top, force applied at the top.
‡ Department of Agriculture, *Tech. Bull.* 479.
§ Pacific Telephone & Telegraph Company; 25- to 35-ft poles, 6 to 9 in. top, force applied at the top.
‖ L. W. Winchester, *Elec. World* (March 16, 1911); 29- to 31.5-ft poles set in ground 4 to 6 ft.

Fiber Stress (f) and Actual Tests of Strength. Table 10 gives the value of the fiber stress for various kinds of timber and the actual breaking load from tests of a number of poles.

Weakest Point of a Pole. A pole is approximately a truncated cone in shape. For a bending force applied at one end, such a cone is weakest at the point where the diameter is $^3/_2$ the diameter at the point (near the small end) where the force is applied. A pole with 8-in. diameter at the cross-arm is, therefore, weakest where it is 12 in. in diameter and may be expected to break at this point provided this point is above the place where maximum bending occurs. If it is less than 12 in. in diameter at the point of maximum bending, then the break may be expected there. This rule must be considered approximate, as it neglects the fact that the pole is not homogeneous; i.e., outer annual rings are sapwood and inner are heartwood, and also neglects effect of knots, etc. In practical working calculations the weakest section is taken at the ground line, since that point tends to become weaker than any point above ground as a result of its greater moisture content and its greater tendency to decay.

STRESS DUE TO ANGLE IN LINE. If there is an angle in the line, an additional stress is imposed upon the supporting structure at the angle point because of the tensions in the conductors. If the conductors in the adjacent spans have equal tensions (t) and the angle of departure of the line is (a), the resultant force on the structure will be:

$$F = 2t \sin \frac{a}{2} \text{ pounds} \tag{66}$$

and the angle between the direction of the resultant and the direction of either span will be: $90 \text{ deg} - \frac{a}{2}$. If the tensions in the adjacent spans are not equal, the resultant force will be:

$$F = \sqrt{t_1^2 + t_2^2 - 2t_1t_2 \cos a} \tag{67}$$

and the angle (b) between the resultant and the span in which the tension is (t) is obtained from:

$$\cos b = \frac{F^2 + t_1^2 - t_2^2}{2Ft_1} \tag{68}$$

ATTACHMENT OF CROSS-ARMS TO POLES. Wooden cross-arms are attached to wooden poles:

1. By gaining the pole, see below and Fig. 17.
2. By one or two bolts.
3. By one or two cross-arm braces.

The forces at the point of attachment which these fastenings must resist are:

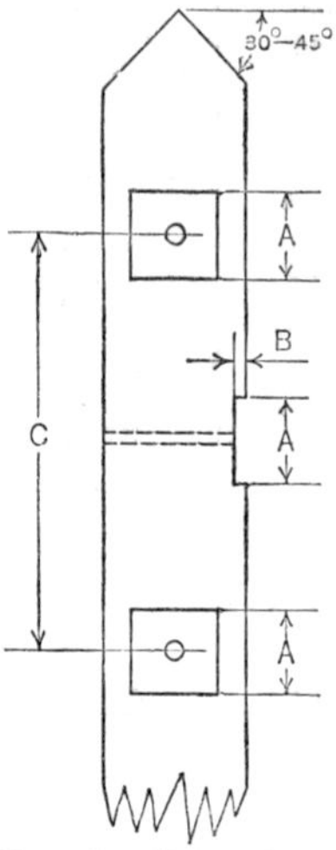

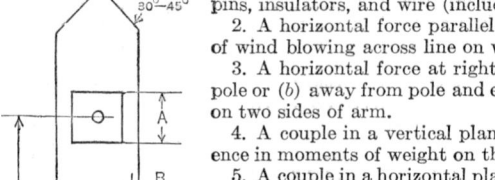

FIG. 17. Gaining and Roofing

1. A force vertically downward, equal to weight of cross-arm, pins, insulators, and wire (including sleet).
2. A horizontal force parallel to axis of arm, equal to pressure of wind blowing across line on wires.
3. A horizontal force at right angles to axis of arm: (*a*) toward pole or (*b*) away from pole and equal to difference in pull of wires on two sides of arm.
4. A couple in a vertical plane parallel to arm, equal to difference in moments of weight on the two ends of arm.
5. A couple in a horizontal plane parallel to arm, equal to difference in moments of wire pull on the two ends of arm.
6. A couple in a vertical plane at right angles to arm, equal to difference in moments of wire pull (caused by pin leverage) in the two directions.

GAINING. A gain is a notch cut in the side of a pole to receive a cross-arm. The width (vertical dimension) of the gain should be just large enough for the cross-arm. The depth of gain varies from $^1/_2$ to 1 in. With gains shallower than $^1/_2$ in. the cross-arm has insufficient support below, and the flat bearing surface at the back is inadequate unless the pole is of larger diameter than usual. Deep gains greatly weaken the top of the pole, especially when double arms are used. Another type of gain, called a **slab gain,** is coming into use. It is formed by flattening the side of the pole from the top down to a point below the lowest cross-arm. The depth to which the wood is removed in making such a gain is also from $^1/_2$ to 1 in. In most cases a $^1/_2$-in. depth is sufficient to provide the required width of flat surface. Poles which require full-length preservative treatment should be drilled, gained, and roofed completely before treatment.

CROSS-ARM CONSTRUCTION. On ordinary straight-line construction single cross-arms are generally used. If conductors are carried at more than one level, a single arm is used at each level, and all crossarms are placed on the same side of the pole. If the pole has any sweep, the arms are placed on the concave side to leave the convex side clear for climbing. At four-way corners single arms are used with alternate arms at right angles to each other. At two-way corners, dead-ends, angles, and places where the line stresses are not balanced longitudinally, double arms are used. In alleys and at other places to avoid obstructions, alley arms are often used.

The size of cross-arms depends on the size of conductors carried and upon the voltage of the circuits. The spacing of arms depends upon the voltage of the circuits carried and upon the length of spans.

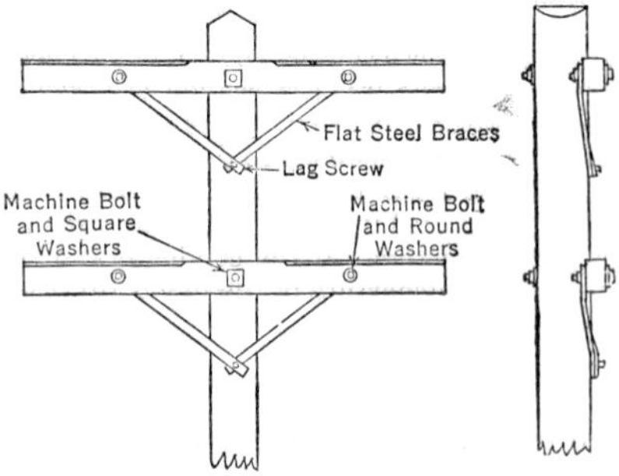

FIG. 18. Single Cross-arm Construction

Figures 18 to 24 show various types of wood-pole construction. Figure 18 shows single-arm construction for low- and medium-voltage lines. Figure 19 is an example of double-arm construction at a dead-end in the line. Figure 20 shows typical construction for a four-way corner. Figure 21 is an example of alley-arm construction. Figure 22 shows a typical medium-voltage transmission line pole for straight line construction using suspension insulators. Figure 23 is the type of construction sometimes employed for corners on single-circuit transmission lines. Two poles of this type are sometimes used for corners in double-circuit lines of vertical configuration. Figure 24 is an example of construction largely used with single-circuit wood-pole transmission lines of high voltage.

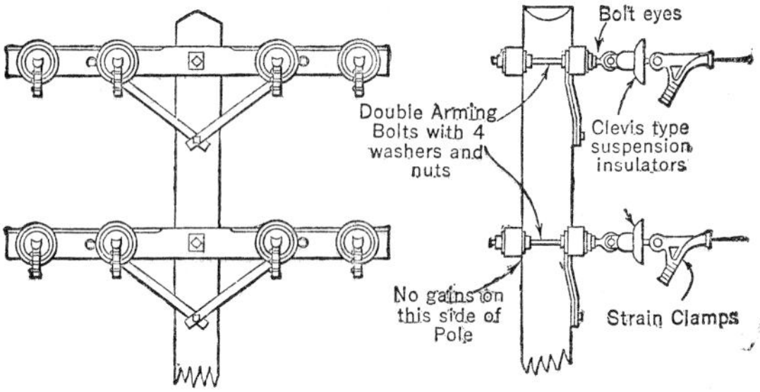

Fig. 19. Double Arm Construction

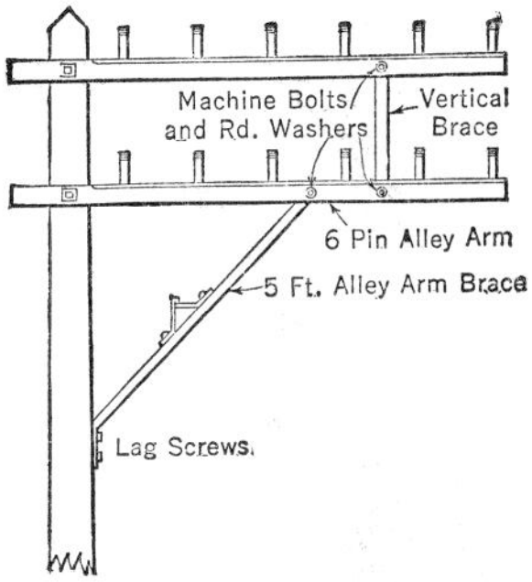

Fig. 21. Alley Arm Construction

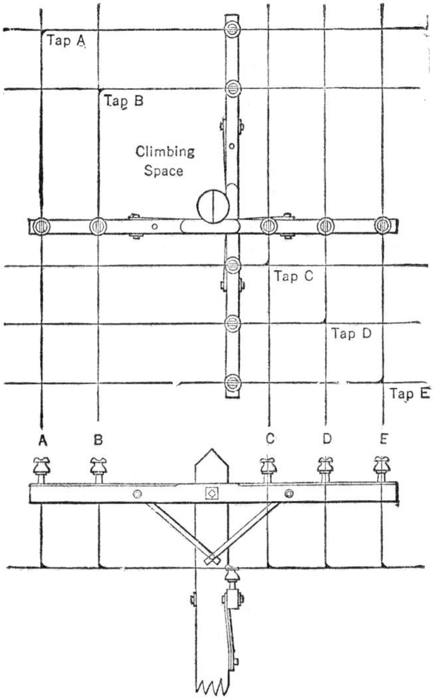

Fig. 20. Four-way Corner

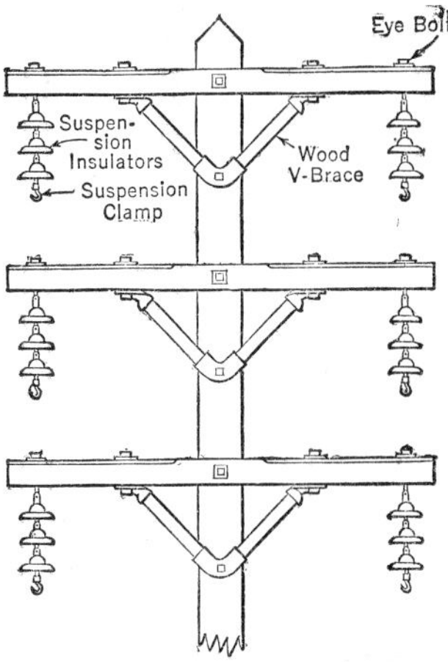

Fig. 22. Straight Line Suspension

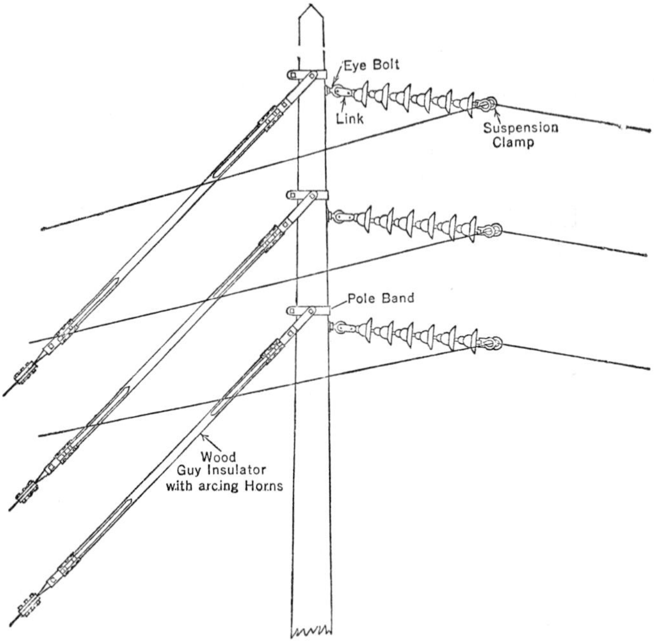

FIG. 23. Single-circuit Flying Corner

HARDWARE. Hardware for wood-pole lines is usually of steel which is hot-dip galvanized. The Edison Electrical Institute has prepared specifications covering material, dimensions, and galvanizing for the commonest items of hardware. Specifications of the ASTM also cover material and galvanizing. Hardware items include: machine bolts, double-arming bolts, lag screws, square and round washers, eye bolts, clevises, guy plates, hooks and shims, hub plates, pole steps, and cross-arm braces. Most items are manufactured in a wide range of sizes to cover the varied conditions to be met in wood-pole construction. Wood braces with steel connectors are often used on transmission lines to minimize phase-to-phase flashovers.

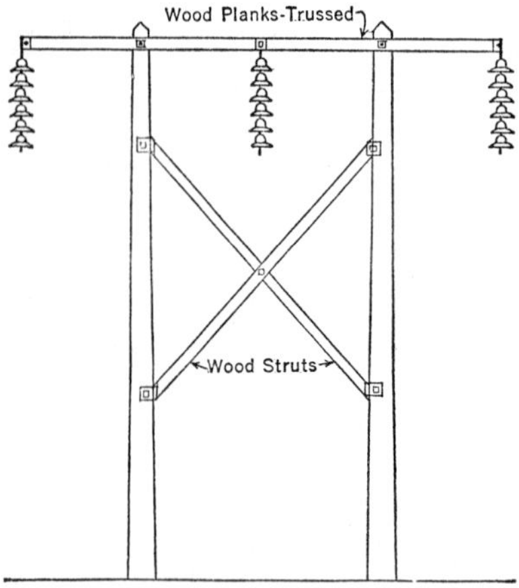

FIG. 24. Single-circuit H-frame Construction

SETTING POLES. Table 11 gives the average depth of setting which is generally considered good practice. In soft or swampy soil, poles should be braced by cribs or swamp braces and guyed if necessary. Unguyed poles at angles in the line should be set from 6 in. to 1 ft deeper.

In ordinary firm soil, pole holes are dug by hand, using digging bars and long-handled spoon-shaped shovels, or by hole-digging machines. Such digging machines are earth augers driven by gasoline engines and mounted on trucks, trailers, or tractors The motive power of the truck or tractor is generally used to drive the auger. Caterpillar tractors are especially suitable for this work in rural territory, as they are able to travel over uneven or soft ground without difficulty. Pole holes should be the same diameter at the top as at the bottom and should be only enough larger than the pole butt to allow the use of tamping bars to tamp the backfill thoroughly around the pole.

In swampy soil or quicksand a barrel or split caisson is required to prevent the side of the hole from caving in. As the hole is dug, the barrel or caisson is gradually pushed down; the barrel may be left in the hole after the pole is set, but the caisson may be withdrawn to be used again. Another successful method employs a high-pressure water jet. In this method the hole is started with shovels, and the pole is set up in position and held by hand lines. A hose with a long pipe nozzle is laid alongside the pole, and the water under high pressure is started. The jet loosens the soil and brings it to the surface in suspension. By reason of its weight the pole drops slowly into the hole. Eighty-foot poles have been set 40 ft deep by this method in extremely deep marshes.

Table 11. Average Depth of Pole Setting

Length of Pole, Feet	Feet in Soil	Feet in Rock	Length of Pole, Feet	Feet in Soil	Feet in Rock
20	5.0	3.0	55	7.0	5.0
25	5.0	3.5	60	7.5	5.0
30	5.5	3.5	65	8.0	6.0
35	6.0	4.0	70	8.0	6.0
40	6.0	4.0	75	8.5	6.0
45	6.5	4.5	80	9.0	6.5
50	7.0	4.5	....	....	....

Explosives are often used for setting poles in soft or sandy soils or in rock. In soft soils a charge of dynamite is laid at the bottom of a small pipe driven to the required depth of hole. The pipe is then withdrawn, and the pole is set on the surface directly over the charge and held upright by hand lines. When the charge is exploded, the pole drops into the hole and the soil drops back around the pole. When the pole is to be set in rock, a small hole is drilled to the required depth, and the charge laid at the bottom. The small hole is backfilled and tamped. Firing the charge opens a hole to take the pole. Experience in such work is necessary to determine the size of charge and the proper depth to open a hole no larger than necessary for the pole.

A pole may be raised and set in the hole by a hoisting rig, such as a gin pole or a derrick mounted on a truck or tractor. Poles may be set by three to five men using pike poles. After the pole is set in the hole, it is supported in proper position by pike poles while the backfill is made and tamped.

GUYING OF POLES. Whenever a pole is not strong enough to withstand the bending stresses imposed on it by unbalanced forces, it should be guyed. Guys should be strong enough to take the entire stress in the direction in which they act, the pole acting only as a strut. Guys consist usually of stranded steel wires, together with means of attaching to poles and anchors, and also include strain insulators where they are required.

Stresses in Guys and Anchors. To compute the tension in a guy wire use the formula:

$$T = \frac{M}{H \sin a} \text{ pounds tension} \qquad (69)$$

where M is the bending moment applied to the pole in the plane containing the pole and the guy, in pound-feet; H is the height above ground of the point of attachment of the guy to the pole, in feet; and a is the angle which the guy makes with the vertical.

Guy Wire. Stranded steel wires used for guys are usually heavily galvanized, but in locations where corrosion is excessive, stranded copper-clad steel wires are often used. Mild steel is common for lightly loaded lines, but Siemens-Martin or high-strength steel is used for many important or heavily loaded lines. Table 12 gives the characteristics of copperweld stranded cables used for guy and messenger wires. Table 13 gives similar data for stainless steel cables. The characteristics of galvanized stranded steel cables may be found in Table 5.

Attachment to Poles and Anchors. Guys are attached to poles by wrapping the end of the wire twice or more around the pole and clamping the free end to the main part of the guy by means of one or more guy clamps. Steel shims or plates are placed between the wrappings of the guy wire and the pole. Steel hooks are also used to prevent the guy from slipping down the pole. Hooks, shims, and plates are attached to the pole by nails or lag screws. An alternative method is to attach the guy wire to an eye-bolt or patented attaching device, which is bolted through the pole. Anchors sunk in the ground are provided with long anchor rods extending slightly above the ground surface and terminating in an eye. The free end of the guy is passed through the eye of the anchor rod, bent back parallel to the main portion of the guy, and clamped to it. The loop is protected by a guy thimble where it bears on the anchor rod eye.

Table 12. Mechanical Characteristics of Copperweld Guy Strand

Nominal Diameter, Inch, and Size, A.W.G.	Actual Diameter, Inch	Area, Square Inch	Weight, Pounds per 1000 ft	Breaking Load, Pounds	
				High Strength	Extra High Strength
1/2 (7 No. 6)	0.486	0.1443	515	16,880	20,410
7/16 (7 No. 7)	0.432	0.1140	407	13,860	16,820
3/8 (7 No. 8)	0.384	0.0901	322	11,340	13,800
11/32 (7 No. 9)	0.342	0.0715	255	9,320	11,280
5/16 (7 No. 10)	0.306	0.0572	204	7,750	9,200
3 No. 6	0.349	0.0618	220	6,830	8,260
3 No. 7	0.310	0.0489	174	5,610	6,800
3 No. 8	0.276	0.0386	138	4,590	5,580
3 No. 9	0.246	0.0306	109	3,770	4,560
3 No. 10	0.220	0.0245	87	3,140	3,720

Table 13. Mechanical Characteristics of Page Stainless-steel Strand

Nominal Diameter, Inch	Number and Diameter of Strands	Weight, Pounds per 1000 ft	Breaking Strength, Pounds	
			235,000 lb per sq in.	165,000 lb per sq in.
7/32	7 × 0.072	100	6,300	4,500
1/4	7 × 0.083	124	8,500	6,400
5/16	7 × 0.104	200	13,200	9,200
3/8	7 × 0.120	275	18,000	12,500
7/16	7 × 0.145	400	26,000	18,200
1/2	7 × 0.165	500	33,700	23,600
13/64	3 × 0.093	70	4,500	3,150
7/32	3 × 0.104	90	5,650	3,950
1/4	3 × 0.120	120	7,550	5,300
5/16	3 × 0.145	175	11,000	7,700
3/8	3 × 0.165	225	14,300	10,000
3/8	19 × 0.075	300	16,800	12,800
7/16	19 × 0.087	400	22,500	15,800
1/2	19 × 0.100	505	30,000	21,000
9/16	19 × 0.110	640	36,200	25,400
5/8	19 × 0.125	800	47,000	33,000
3/4	19 × 0.150	1160	67,500	47,500
7/8	19 × 0.175	1585	91,400	64,000

Strain Insulators. Strain insulators are placed in guys to prevent the lower part from becoming electrically energized by contact of the upper part with conductors or by leakage. Two strain insulators are used whenever a single insulator will not effectively prevent a hazard to the public or to linemen. No guy insulator should be located less than 8 ft from the ground. Insulators are not necessary where no part of the guy is within 8 ft from the ground or where the guy is permanently and effectively grounded.

Two types of strain insulators are used. One type is made of porcelain, and the other is of wood with steel connecting parts. Porcelain insulators are usually designed so that the porcelain is in compression with the guy wire interlinked. In this type the porcelain may fail and allow the guy wire loops to come together, making the insulation ineffective without a mechanical failure of the guy. Such insulators are used only in connection with low- and medium-voltage lines. The National Electrical Safety Code requires guy insulators to have a mechanical strength equal to that of the guy in which they are installed, a dry flashover value of twice the line voltage, and a wet flashover value not less than the line voltage. Wood insulators are now used extensively with high-voltage wood-pole transmission lines. They are made of treated wood with steel connecting parts at each end and are provided with arcing horns to prevent flashovers from burning the wood. Generally the wood portion is long compared with its cross-sectional dimensions, and usually the cross-section is rectangular. The wood is used in tension, and of course a failure of the insulator causes a mechanical failure of the guy.

Guy Anchors and Stubs. Guys may be anchored to the earth, to other poles, and occasionally to trees or buildings. When guys are anchored to the earth, it is necessary to bury in the earth some object to which the guy is to be connected. Such an object, known as an anchor, must be so designed as to press against a sufficient volume and weight of earth to resist completely the force due to the tension in the guy. A common form of anchor is the so-called "dead-man" or log anchor. It usually consists of a log or section of a pole buried in a trench. An anchor rod is inserted through a diametrical hole bored through the middle of the log and is held in place by a plate and by nuts on the end of the rod. A transverse trench is provided running from the surface of the ground, at the proper angle with the ground surface, down to the bottom of the trench dug for the log. After the anchor is in the proper position the trenches are, of course, backfilled and tamped. A large volume of undisturbed earth is provided for the anchorage by this method, and it is therefore useful for anchoring heavy guys.

Many forms of patented anchors are available. These are usually made of steel or malleable iron. Some are designed to screw into the earth, and some are placed or driven into small holes and afterwards expanded to increase the bearing area. Such patented anchors are made in a wide range of sizes to be used with small or large guy wires and light or heavy loads.

Guy stubs are short poles, to the tops of which guys from line poles are attached when the guy is required to clear the ground by a considerable distance. An anchor guy is usually run from the top of the guy stub to the ground, in which case the stub is raked a sufficient amount so that it will act only as a strut. Sometimes guy stubs are made self-supporting by giving them considerable rake and by burying a log just below the surface of the ground and another log at the bottom of the stub on the opposite side for the stub to bear against. Figure 25 shows a self-supporting guy stub and a common form of log anchor.

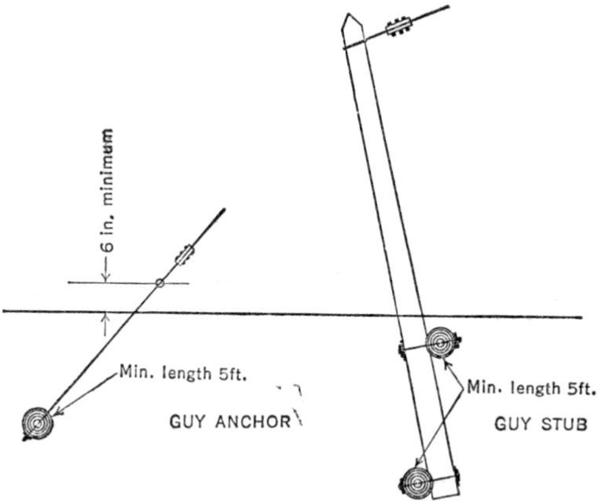

FIG. 25. Guy Anchor and Stub

REPAIRING DECAYED POLES. If poles have been weakened at the ground line by decay, the strength may be restored by cutting off the decayed butt and resetting the pole, thus reducing its height by 6 to 8 ft. If reduction of height is not permissible, the pole may be stubbed by setting alongside of it a short pole or stub extending a few feet above ground, to which the old pole or the undecayed part above ground is bolted or otherwise fastened. This does not look well and is unsuitable for city distribution lines but has been used for transmission lines. Another method is to reinforce the decayed pole by a sleeve of concrete (usually reinforced) extending above and below the decayed portion.

Several patented devices are also available. Some act as sockets for the upper portion of the pole, which has been cut off at or above the ground line. Another method leaves the pole intact, but T-shaped steel bars are driven into the ground close beside the pole and are clamped tightly to it by curved steel rods.

29. STEEL TOWERS AND POLES

Steel structures are often used instead of wood poles for supporting electrical circuits. They have the advantages over wood structures that they can be made as large as necessary and that they may be made self-supporting without the use of guys. They have the disadvantage of not having an inherent insulating value, and, therefore, when steel structures are used, the line insulators must be selected with that fact in mind. They do, however, have the advantage of being less susceptible to damage by grass, brush, or forest fires.

GENERAL FEATURES OF DESIGN. Towers are composed of two parts, the tower proper and the foundation. Towers are usually built of standard structural-steel shapes. Angles are used for most members. Channels are used for the larger members of some towers, usually the cross-arms, or for posts of flexible towers. Flat pieces are sometimes used for the minimum-sized bracing of light towers. Round rods are used for tension members in some types. The principal members of a tower are the corner posts or legs, which are vertical or approximately vertical, and are usually the heaviest members of the tower proper, and the horizontal and diagonal web members which connect the posts together in vertical planes which constitute the sides or faces of the tower.

The spread of a tower at the base is generally between one-fourth and one-fifth of the height. The greatest economy in cost of tower plus foundation usually requires a little wider base than that which gives the least cost for the tower alone.

Towers are usually designed for either one or two three-wire circuits, usually with one or two ground wires above and sometimes with a telephone circuit below. Where two circuits are on one tower, they are generally located on opposite sides to reduce the hazard of repairing the line. (See Fig. 26.) The three wires of a circuit are occasionally arranged to form an equilateral triangular prism but frequently lie in a single plane, which is usually horizontal for single-circuit towers. (See Figs. 27 and 28.)

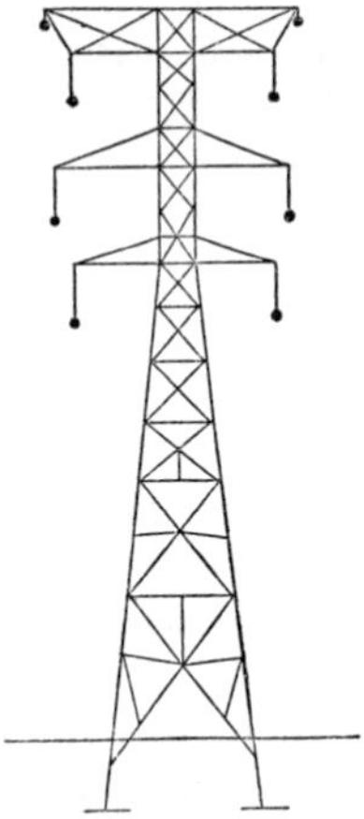

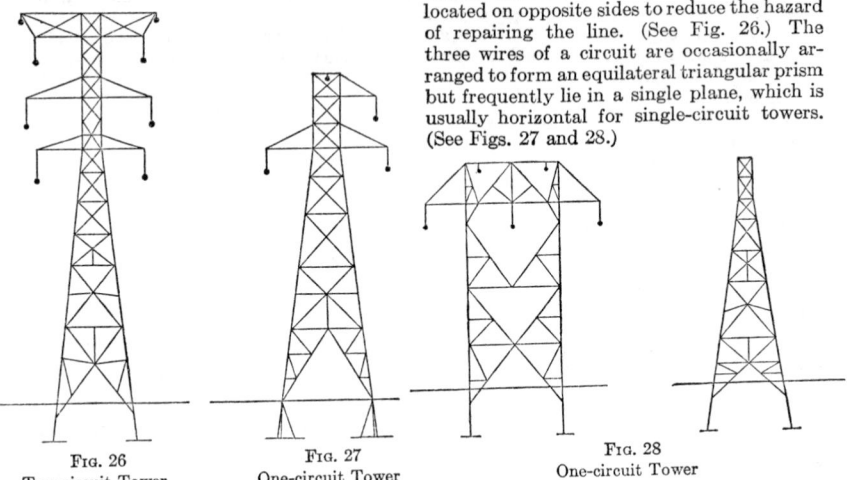

Fig. 26
Two-circuit Tower

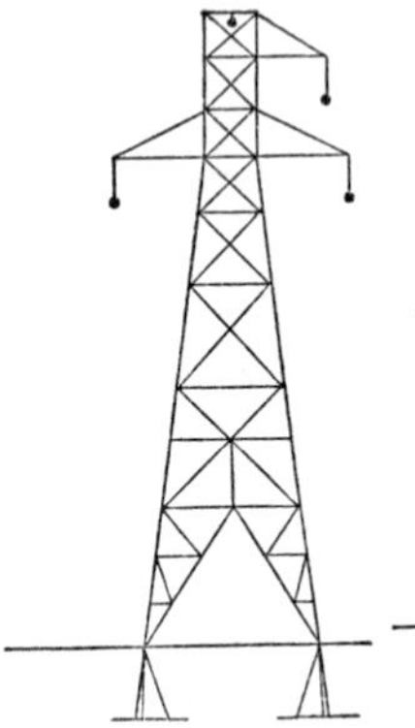

Fig. 27
One-circuit Tower

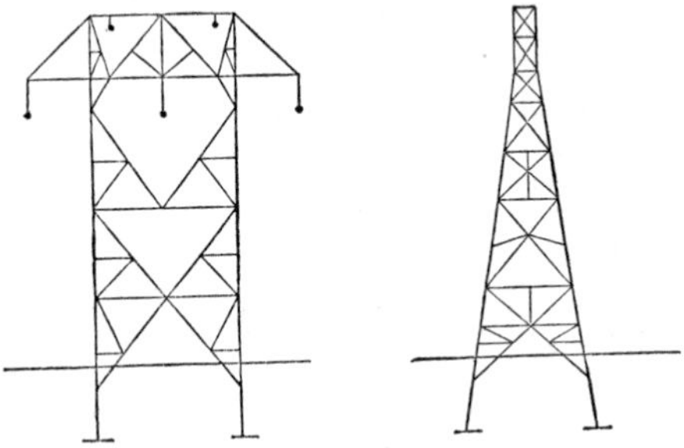

Fig. 28
One-circuit Tower

Steel poles are commonly of the tubular type or of the square latticed type. Tubular poles serve mainly to support trolley conductors and feeders on city streets, and latticed poles are used to support distribution lines and low- or medium-voltage transmission lines. Latticed steel poles have narrow bases; i.e., the taper of the pole is small. The vertical members are usually angles, and the lacing on each face is either flat steel bars or small angles with one leg turned inward. They may be designed for one or more low-voltage circuits arranged in horizontal or triangular configuration. Single-circuit high-voltage lines may be arranged in triangular configuration, but double-circuit high-voltage lines are always arranged in vertical configuration with one circuit on each side of the pole.

TYPES OF TOWERS. Two general types of towers have been used, the flexible and the rigid. **Flexible Towers** are more or less obsolete at the present time. The general form of such a tower is a flat A-frame having good lateral strength but none in the direction of the line. Strength in that direction is usually attained by attaching a ground wire rigidly to the top of each A-frame. Such structures are not adequate for important or high-voltage lines, and they have been largely superseded by wood-pole structures.

Rigid Towers are usually triangular, square, or rectangular. The square tower is probably the most common. Square towers usually have the four faces framed with the same size members (even though the stresses in the longitudinal and lateral faces rarely figure the same), because of the economy of manufacture and erection which results from the simplicity. This feature has an advantage in design in that the torsional stresses are more simply determined. Rectangular (including square) towers have the disadvantage that the unequal settlement of the foundations may produce high internal stresses not allowed for in the design. Triangular towers avoid internal stresses from unequal foundation settlement but present difficulties in the joining of standard structural shapes, and stresses in them are difficult to calculate.

CONNECTIONS OF MEMBERS. The members of a tower are usually connected by bolts. By using no rivets, the members may be fabricated in quantities, compactly bundled, easily handled even in rough country, and erected by less skillful labor; also the galvanizing can be done after all shop work is completed. All the bolts of a tower should be of one diameter ($^5/_8$ in. is suitable for the members generally used) and of as few different lengths as possible. Bolt holes should be slightly larger than the bolts ($^1/_{16}$ in. is a usual amount). By designing bolted connections so that friction between the surfaces develops the full compressive strength of members, the play in the bolt holes with changing compression and tension in the members is eliminated.

If the unit shearing stress is made two-thirds of the unit tensile stress in the members connected, and the unit bearing stress is made twice the unit shearing stress, a bolted connection will be not less than 10 per cent stronger than the members connected.

CLEARANCE BETWEEN CONDUCTORS AND TOWER. Clearance from conductor to tower is usually based on that existing at maximum side swing of the insulator string toward the tower members, due to wind pressure on the conductors. The side swing is usually taken at 60 deg, but sometimes, under favorable conditions, at only 45 deg from the vertical. The side swing should be calculated for maximum wind pressure both with and without ice. Reduced clearance of the conductor itself from the corners of the tower should also be considered. The angle of swing will be greater for small or light-weight conductors than for large or heavy ones. Clearance from the cross-arm should be considered, as well as from the vertical portion of the tower. If arcing rings or arc-suppressing devices are used on insulator strings, clearances from their live parts should be determined. Flashovers ordinarily are due to lightning potentials impressed on the conductors; therefore arcing distances should be selected on the basis of the characteristics of impulse flashovers. The flashover distance to the tower, through air, should be at least 10 per cent greater than that over the insulator string, with its protective equipment, if any.

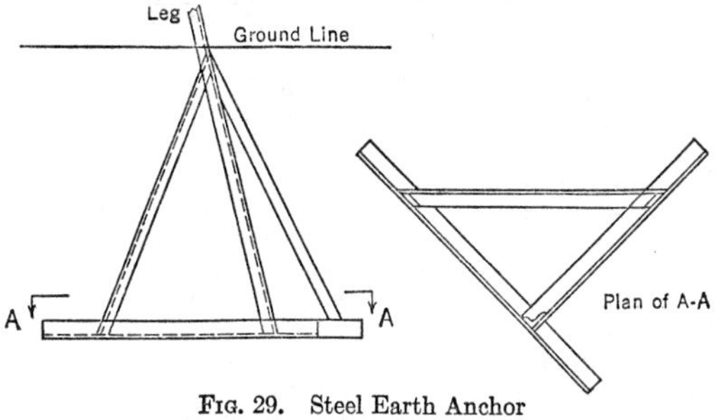

FIG. 29. Steel Earth Anchor

FOUNDATIONS FOR TOWERS. Structural steel, mass (unreinforced) concrete, reinforced concrete, or piles may be used for tower foundations. Rock footings are also used in special locations.

Structural-steel Foundations are cheap and easily transported. They usually consist of grillages of structural-steel I-beams with tie members consisting of I-beams or channels. (See Figs. 29 and 30.) These grillages are either buried at a proper depth in the earth or surrounded by a mass of concrete. They should have sufficient horizontal area so that they will have adequate bearing surface to withstand downward pressures and carry sufficient earth to withstand possible uplift forces. The corner leg members are usually carried down to the grillage and connected to it in a proper manner to transmit the vertical and horizontal stresses involved.

Concrete Foundations have an advantage over structural steel in that they can more easily be varied in depth, spread, etc., to accommodate themselves to local conditions of

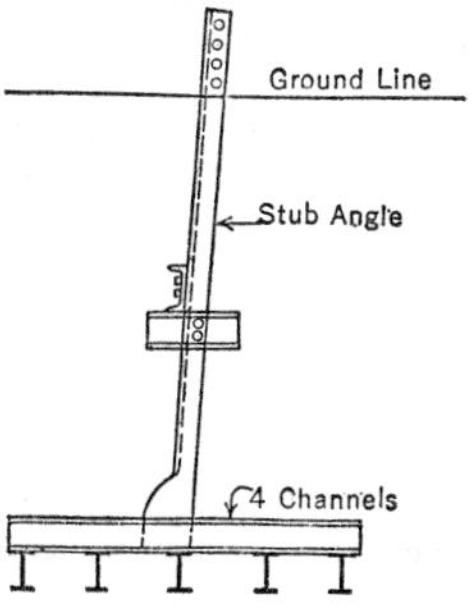

FIG. 30. Steel Grillage Foundation

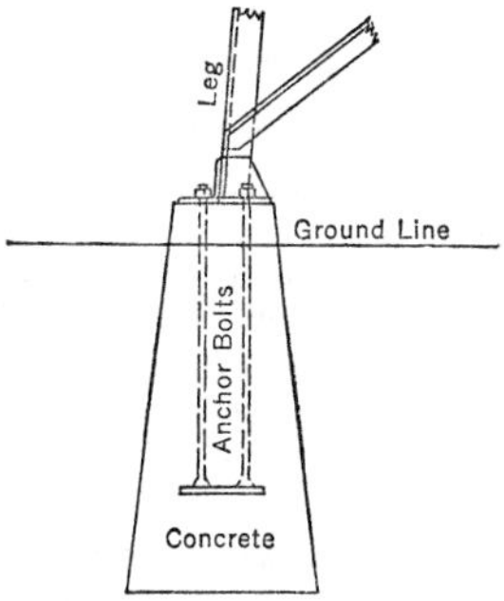

FIG. 31. Concrete Foundation

soil. This is especially advantageous where boulders or irregular ledges interfere with the use of a standard-sized foundation. (See Fig. 31.)

Mass-concrete foundations are advantageous where it is necessary or desirable to have a foundation of such weight as to withstand much uplift with little reliance on the holding power of the earth. The towers may be conveniently attached to anchor bolts imbedded in the foundation, or the leg member may extend down to the bottom of the concrete. To avoid tension in the concrete the bolts must extend to the bottom, with proper plates for distributing the stress. The anchor bolts and plates then become a crude system of reinforcement.

Reinforced-concrete foundations are durable and require less material than mass concrete, thereby facilitating transportation.

Piles are used under or for foundations in very marshy ground where the holding power of other foundations is unreliable.

Rock Footings for towers standing on ledges may consist of anchor bolts grouted into holes drilled in rock and extending through level bearing plates grouted to the rough rock surface at the proper elevation.

FORCES ACTING ON TOWERS. The stresses in towers are caused by: (1) the weight of tower, insulators, clamps, cables (conductors, ground wires, telephone wires), and ice loads on them; (2) the wind pressure on above; * (3) the unbalanced tension in cables when dead ended or broken on one side; (4) the unbalanced resultant due to cable tension at angles in the line; (5) the loads imposed when erecting towers, stringing wire, or repairing line.

A careful study should be made of all the combinations of these loads which are possible or probable. Often no single combination can be found which will produce the maximum stress in all tower members, and therefore several combinations must be used to determine the design.

In a square anchor tower carrying six wires, three on each side, the maximum stress may be expected in the corner posts when all six wires are pulling in the same direction; the maximum stress in the web members will probably be produced by three wires pulling on one side in one direction and the three on the other side pulling in the opposite direction. In the first case the tower is subject to a bending stress, and in the second to a torsional stress. In each case the stresses due to weight and wind are to be superimposed. The wind may act as a force along the line or across the line, but generally its longitudinal effect is negligible while its lateral effect is important.

STRESSES IN TOWER MEMBERS. The stresses in the several members of a tower are usually determined graphically from the assumed loadings by means of stress diagrams; see section on trusses in Kent's *Mechanical Engineers' Handbook*. In most designs the distribution of stress is not fully determinate.

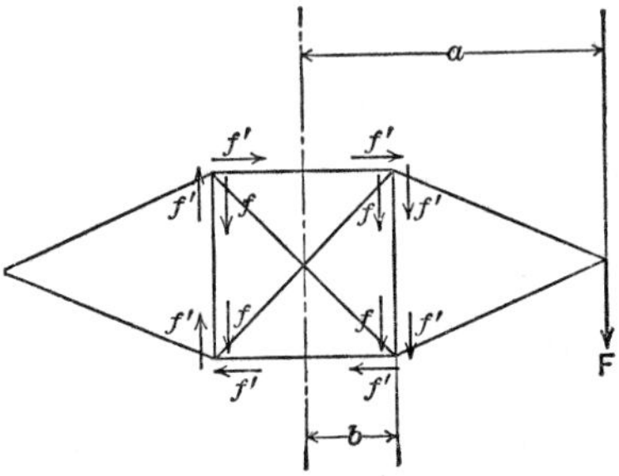

FIG. 32. Forces Acting on a Tower

Certain assumptions are, however, commonly made which give a determinate distribution for the purposes of design. Among these assumptions are:

1. An unbalanced stress on the tower (say a broken wire pulling on one side) can be resolved into an equal stress at the axis of the tower and a torsional moment.

2. The equivalent stress at the axis of a rectangular tower can be considered as balanced between the two faces parallel with it, each face taking one-half the stress.

3. The torsional moment can be considered as divided among all four faces of a rectangular tower.

4. If the tower is square, each face takes one-fourth of the torsion.

The above relations may be expressed as follows (see Fig. 32):

Let F = unbalanced force, in pounds, applied at end of cross-arm.

a = distance, in feet, from *end of cross-arm* to axis of tower.

b = distance, in feet, from *side of tower body* to axis of tower.

Then

$f = \dfrac{F}{4}$ = balanced force, in pounds, applied at each corner post equivalent to F in bending effect on body of tower. (70)

* The wind pressure on a tower is assumed to be uniformly distributed per square foot of surface against which the wind blows, one-half of it consequently being on the windward side of the windwind face and the other half on the windward side (inside) of the leeward face. For simplicity in calculation this uniformly distributed force is replaced by a series of concentrated forces, one at each panel point equivalent to the total distributed force extending over a half panel above and a half panel below the panel point. By panel point is meant a point of intersection of principal members, for example, of horizontal members with vertical members; and by panel is meant the section of a side between panel points, a panel usually being bounded by two vertical and two horizontal members.

$f' = \frac{Fa}{8b}$ = torsional force, in pounds, applied at each corner post equivalent to F in twisting effect on body of tower. **(71)**

5. In a tower framed with a double system (i.e., diagonals in duplicate and suitable for compression as well as tension) each system may be considered as taking one-half of the stress as far as possible.

Approximations Made in Calculating Tower Stresses. The stress diagrams are usually simplified by employing certain approximations:

a. Faces of towers are usually battered so that they deviate slightly from a true vertical plane, but the stress diagrams generally neglect this inclination and are based on the vertical projection of the face.

b. Where the face of a tower does not lie in one plane (i.e., has a change of batter, as occurs frequently at bottom cross-arm where a prismatic cage joins a pyramidal base), the change of inclination is neglected, and the diagrams are based on a single vertical projection as before.

Subject to the limitations of the assumptions and approximations given above, the four faces of the tower can be regarded as four cantilevers, supported at the base and loaded at the top, which are independent except that each of the four corner posts is common to two faces and must contain the resultant of both stresses.

Where a face of a tower or any part of a face has any considerable inclination, the above approximations may not be used without danger of serious error.

Unstressed Members. A tower usually contains members no stress in which is shown by the stress diagram, viz.:

1. Diagonal members in a horizontal plane do not usually appear in the stress diagram when located below the lowest cross-arm. These members play an important part in the distribution of torsion among the faces. In a rectangular tower the torsion will usually redistribute among the four faces at each level where there are horizontal diagonals; therefore the failure of the stress diagram to show stresses in them may be taken to indicate that the assumed distribution of torsion is not quite correct rather than a true absence of stress.

2. Redundant members are braces which carry no determinate stress but perform the important function of supporting the compression members which do carry stress. The unit stress allowable in a compression member diminishes as the unsupported length increases. The weight of compression members is therefore diminished by dividing their unsupported length by braces applied at one or more intermediate points.

Unit Stresses in Towers. Towers are designed for certain combinations of loads. A tower designed to withstand all possible combinations of loads with each load at its maximum possible value would be uneconomical, as its overload capacity under conditions which would occur on the average would be out of reason. Therefore, towers are designed for average loading conditions, and unit stresses are used which will take care of a reasonable overload. The term factor of safety is no longer used in reference to tower design, since it is misleading; the term overload capacity factor is used in its place.

Stresses in Steel. If tests of a particular tower are lacking, and if steel conforming to ASTM Specification A7-42 for structural-grade steel is available, it is recommended that the following values for the yield point of steel members be used for the purpose of design:

Tension:	33,000 lb per sq in.
Compression:	$33{,}000 - 130L/R$ lb per sq in.
Shear:	30,000 lb per sq in.
Bearing:	60,000 lb per sq in.

L = unsupported length in inches.

R = least radius of gyration in inches.

Where the yield point of the steel used is different from that in Specification A7-42, the foregoing unit stresses should be changed in proportion to the ratio of the actual yield point to the yield point of the specification.

The values of L/R should not exceed the following for compression members:

For leg members:	150
For other members having figured stresses:	200
For secondary members without figured stresses:	250

The use of the foregoing stresses results in an overload capacity factor of 1.9 to 1.96 for ordinary transmission towers and about 2.2 for river-crossing towers.

Eccentricity in Stresses at Joints. As tower members are ordinarily connected together, the stresses in them are slightly eccentric, thereby preventing the full strength of the members from being developed. The eccentricity should be eliminated or reduced as much as possible by having the center of gravity of the several members at each connection meet at one point as exactly as possible.

FORCES ACTING ON TOWER FOUNDATIONS. Foundation stresses are of two classes: first, the foundations resist the tendency of the tower to slide and overturn because of the external forces on it, considering the tower as a self-contained structure; second, the foundations resist certain stresses which would be internal tower stresses were the tower framed as a complete self-contained structure, but which become external stresses because the ground is depended on for the function of certain omitted members. The weight of the tower can evidently be reduced by thus substituting the ground for certain members, but the size of foundation is thereby increased. The amount of these latter stresses depends on the outline and framing of the tower, and their effect should not be overlooked in determining loadings on foundations.

The magnitude and direction of the forces acting on the tower foundations may be illustrated by taking the case of a rectangular tower and considering a transmission line which runs north and south.

Let a = width in feet of base of tower (east and west).
b = length in feet of base of tower (north and south).
W = total weight in pounds of tower, insulators, fittings, and one span of all the wires, including ice load, if any.
W' = total weight in pounds of any unbalanced load, such as a wire c feet off center.
F = resultant force in pounds of the wind on the tower and a complete span of all the wires (with ice coat, if any), acting at a distance of d feet above the foundation, wind assumed blowing across line from west to east.
P = pull in pounds of any unbalanced force toward the south applied at a distance of e feet above the foundation and f feet to the west of axis of tower, e.g., a dead-ended wire or a broken wire on the north side of the tower.

Then, assuming that the forces divide equally among the four foundations and that the torsional forces are in a circumferential direction, the relations given in Table 14 hold. These assumptions are reasonably correct for a tower with the four legs joined at the bottom with a horizontal strut in each of the four faces and with horizontal ties across the diagonals, unless the framing which is usually provided in the other faces is inadequate. Probably few towers in use fully meet these requirements. Therefore, there are usually additional stresses of large magnitude due to the foundations performing the function of missing or inadequate members, as pointed out in the notes appended to Table 14.

From the above relations the **resultant force** on each of the four foundations may be found. In general there will be on each foundation: (1) a downward pressure, (2) a direct uplift, and (3) a horizontal overturning force, producing a tendency to slide and an uplift on one side of the foundation and a downward pressure on the other side.

Downward Pressure. The downward pressure usually is of little importance in determining the size of the foundation, as a foundation large enough for uplift and overturning is unnecessarily safe against downward pressure.

Direct Uplift. The uplift is very important, as the weight of tower and foundation is rarely sufficient to provide more than a small fraction of the holding-down power required. The excess uplift is usually resisted by the earth in which the foundation is buried. Not only is the weight of the earth directly over the foundation effective, but also there is an additional resistance due to friction or cohesion of the earth which may be several times greater. These forces are usually computed on the assumption that they are equivalent to the weight of the earth in a frustum of an inverted cone or pyramid covering the foundation and extending to the surface of the earth. The face of this cone is generally taken as making an angle of 30 deg with the vertical.

Horizontal Overturning Force, Sliding, and Indirect Uplift. The horizontal overturning force is also important. Its effect on the base may be resolved into two components: a horizontal force tending to slide the foundation, and a moment tending to rotate the base about a horizontal axis. The resistance of the earth to these forces is an obscure subject, especially if the foundation is of irregular shape. The following discussion, which neglects several favorable elements, may be considered a conservative view of the earth resistance.

The resistance to sliding may be considered due to the friction of the bottom of the base on the earth, and it may further be assumed that any base large enough to resist uplift will also furnish sufficient friction to prevent sliding. The arm of the overturning force is then the same as the height of the foundation (bottom of base to top where tower is attached), and the overturning moment is equal to the horizontal force multiplied by this arm. The resisting moment may be considered due entirely to the vertical reaction of the earth on the top and bottom of the foundation. These vertical pressures may be taken as varying uniformly from zero at a horizontal neutral axis through the middle of the base, to a maximum at the edges of the base most remote from the axis. The moment of resistance is calculated as for a beam subject to bending and having a cross-section identical with the area of the base. It may be assumed that any unit pressure allowable on the uplift edge will be amply safe on the opposite edge, where the pressure is downward, so that calculations for uplift only are necessary. The maximum allowable unit stress on the uplift edge may be taken as equal to the average unit resistance to uplift of the whole foundation, determined as described under Direct Uplift, p. 14-82.

Table 14. Forces on Tower Foundations

Magnitude of Force on Each of the Four Foundations	Direction at Each Foundation			
	N. E. corner	S. E. corner	S. W. corner	N. W. corner
$\frac{W}{4}$	Down	Down	Down	Down
$\frac{cW'}{a}$	Down	Down	Up	Up
$\frac{F}{4}$ (Note 1)	East	East	East	East
$\frac{dF}{a}$	Down	Down	Up	Up
$\frac{P}{4}$ (Note 2)	South	South	South	South
$\frac{eP}{b}$	Up	Down	Down	Up
$\frac{afP}{4(a^2+b^2)}$ (Note 3)	North	North	South	South
$\frac{bfP}{4(a^2+b^2)}$ (Note 3)	West	East	East	West

NOTES: Where there are no struts between the bottoms of the legs, and especially where the bottom panel is framed on the single system, both of which conditions are usual:

1. The force of the wind F will give a greater force than $F/4$ in an easterly direction on the two west foundations and a correspondingly less force on the other two, and will in addition produce four new forces tending to force the legs apart on the compression side and draw them together on the tension side.

2. Similarly the pull of the wire P will give a greater force than $P/4$ in a southerly direction on the two north foundations and a correspondingly less force on the other two, and will in addition produce a westerly force at the N. E. and S. W. corners and an easterly force at the S. E. and N. W. corners.

3. Similarly, the torsional forces due to P will increase in magnitude and change in direction, and the unbalanced pull P may develop new forces tending to raise two diagonally opposite legs and depress the other two.

Limiting Conditions. Usually the most severe condition that a tower foundation is required to meet consists of a combination of uplift and overturning. For this condition the unit stress of uplift proper must be added to the maximum unit stress of uplift due to overturning, and the sum must be within the allowable average unit resistance to uplift.

STRENGTH OF FOUNDATIONS. (See also Kent, on Strength and Elasticity.) A foundation is subject to stresses from the tower tending to move it and from the resistance of the earth preventing motion. The foundation should, of course, be strong enough not to break when subject to these opposing forces. As the points of application of the resistance of the earth and the magnitude of the unit stresses transmitted by the earth at any point are subject to great uncertainty, the foundation should be designed for strength for the distribution of earth resistance which is most severe, considering, for example, that, while the holding power is calculated on a uniformly distributed earth resistance, it may be developed in practice by concentrated pressure from stones or timber located near the outer edge of the base.

IMPORTANT POINTS REGARDING DESIGN OF FOUNDATIONS. The following important conclusions follow from the foregoing discussion:

1. The inverted cone theory of resistance to uplift gives a calculated resistance which increases at a rapid rate with the depth (eventually increasing approximately as the cube of the depth). It would be unsafe, however, to apply the theory for foundations differing

much from those of usual dimensions, say for depths much exceeding 6 ft, and for foundations where the spread of base is much less than the depth to which it is buried.

2. The foot of the tower (top of foundation) should be brought as close to the surface of the ground as possible to reduce the overturning moment.

3. The tower diagonals of the bottom panel of the tower should intersect the corner posts as low as possible, because this is the actual point of application of the overturning force to the foundation, and if this intersection is above the foundation, this extra length must be added to the arm of the overturning moment.

4. By inclining the axis of the foundation approximately in line with the inclined tower leg (i.e., to bring it as near as possible into line with the resultant of the horizontal overturning force and the vertical pressure or uplift), a more economical use of material may be made to resist the combined uplift and overturning.

TESTING OF TOWERS. Since towers are generally indeterminate structures, new designs are usually tested. Test loads proportional to the loads specified for the design are applied until the required factor of safety has been proved or failure occurs. Towers are usually mounted on a rigid base for testing. This gives the strength which the tower would develop on an "ideal" foundation. In practice the tower must be expected to develop somewhat less strength, as unequal movement of the foundations will ordinarily overstrain certain members. Test loads may be applied by means of weights suspended directly at proper points for vertical loads and applied by means of pulleys attached to a tower-testing structure ("test tower") for horizontal loads. This method of application makes the determination of the test loads easy, but it is inconvenient to have the weights fall any considerable distance if the tower is tested to failure.

TESTING OF FOUNDATIONS. Foundations are occasionally tested, but the test is usually for determining the holding power of the soil rather than the strength of the foundation. For the former purpose the test result will depend largely on the character of the soil and its condition (dry, wet, or frozen). Tests for holding power are necessary only for uplift and overturning forces. In testing it is important that the testing machine should not press down on the surface of the soil near the foundation. The machine should rest on the ground outside the base of an inverted cone of angle of 45 deg from the vertical and enveloping the base of the foundation under test.

SPECIFICATIONS, CONTRACTS, AND PROPOSALS. Different manufacturers of towers use different details of construction, so that specifications written for the purpose of obtaining proposals should contain only conditions and requirements. These should state the loadings, possible overload, and maximum allowable stresses and should show outline dimensions of the tower as determined from clearances required.

The specifications should state whether towers are bolted or riveted, galvanized or painted, shipped assembled or partly assembled, tested or not tested, etc., besides containing the usual structural-steel specifications.

The proposals should show the arrangement and sizes of members and typical details of connections.

Contracts are often let on the basis of furnishing an approximate number of towers when the exact number required cannot be determined in advance, and a unit price per pound is included to cover extensions, special foundations, and modifications that may be required.

The galvanizing of towers and parts is usually specified to pass the ASTM specification for galvanized steel. All members of towers that are not so heavy that there is danger of their buckling in the process may be specified to be galvanized by the hot-dip process.

INSTALLATION AND ERECTION. Towers are generally shipped disassembled, with each piece marked to show the type of tower of which it is a part, also with a part number to show its location in the tower. Similar parts are usually bundled together, with the proper number of each for an individual tower. A tower schedule should be drawn up, showing, for each tower number, the type of structure; shipping point; extensions, if any; adjacent spans; and angle in line, if any. A bill of material for each different kind of tower in the line should be provided. These schedules will aid materially in the distribution and checking of material.

The staking gang follows the right-of-way clearing gang, staking the tower locations accurately according to the survey. Closely following the staking gang comes the material distribution crew, and after them the foundation gang.

PREPARATION OF FOUNDATIONS. Foundations should be set with their bases below the frost line. The hole should not be excavated deeper than necessary, so that the foundation may rest on undisturbed earth. If any of the several foundations of a tower rest on loose backfill, unequal settlement, which may greatly weaken the tower, may be expected. The hole should not be larger than necessary, as the backfill will be less effective than undisturbed earth in resisting uplift. The resistance of concrete foundations can sometimes be increased by digging a hole smaller in diameter than the base and undercutting it at the bottom; with structural-steel foundations large stones can sometimes be placed against the steel to increase its resistance to motion, either vertical or lateral. The backfill should be well tamped in place, and especial care should be used if it is probable that the foundation will be subject to heavy stress before the earth has had time to settle. On sloping ground, filling may be required on the low side to give the designed weight of earth against uplift. In water or mud the floating power of hydrostatic pressure beneath the foundation must be allowed for. Where towers are raised on extensions, the increased foundation stresses due to increased moment must not be overlooked.

The several foundations of a tower should be set accurately by template, in respect to both spacing and elevation, so that towers stand truly vertical and have no initial stresses due to distortion.

ERECTION OF SUPERSTRUCTURE. The method of erection depends largely upon the type of tower and its location. Under favorable conditions, short, light towers may be completely assembled lying on the ground. Such a tower may be raised into position by the use of an A-frame or gin pole. A line is connected to the tower slightly above its center of gravity and run over the top of the gin pole or A-frame and through proper rigging to the source of power. Sufficient guys and hand lines should be provided to control the position of the tower while being raised.

Heavy towers must be assembled piece by piece, although panel sections may be assembled on the ground and raised into position by means of booms, floating gin poles, or other suitable rigging. Individual pieces are raised to position by hand lines or tackle attached to the leg members.

With bolted connections the nuts should be prevented from working loose by checking threads on bolts, by riveting over the end of the bolt, or by lock-nut or washer.

Each tower member should have a number, which should be shown on the erection drawings and marked on the corresponding member of each tower.

The members should be cut, bent, and punched to template, so that the parts will be interchangeable and the assembled tower will fit the foundation prepared for it.

A clean-up gang should follow the wire-stringing crew to inspect and tighten bolts, pick up left-over material, and make sure everything is in proper condition. This gang should work with the engineer who makes the final inspection.

MAINTENANCE OF TOWERS. All bolted connections on towers should be carefully watched and kept tight, and the galvanizing or paint should be inspected regularly, as towers must be repainted before any deterioration from rusting occurs. Foundations are the most likely source of trouble in operation. These should be kept properly backfilled and should be watched for unequal settlement of legs.

DIMENSIONS, WEIGHTS. Data on these items for a number of towers are given in Table 15.

Table 15. Data on Typical Steel Towers

Use	Conductors		Number of Ground wires	Voltage, kv	Height, feet		Base, feet		Normal Span	Maximum Angle, degrees	Weight, pounds
	Number	Size, circular mils			To Lower Cross-arm	Overall	Across Line	Along Line			
T	6	336,400 ACSR	1	132	64.0	95.0	18.6	18.6	1060	5	10,900
L	6	336,400 ACSR	1	132	64.0	96.6	22.0	22.0	1060	15	13,600
A	6	336,400 ACSR	1	132	64.0	100.0	22.0	22.0	1060	45	20,000
T, L, A	3	336,400 ACSR	1	132	45.0	53.0	23.0	17.0	3000	45	13,000
T, L, A	3	397,500 ACSR	1	132	45.0	55.0	23.0	17.0	4000	45	15,000
T	3	795,000 ACSR	2	220	71.1	79.6	33.0	33.0	1100	1 1/2	13,110
A	3	795,000 ACSR	2	220	71.1	79.6	33.0	33.0	1100	60	30,910
*	*	*	*	*	*	55.0	2.5	2.5	*	*	2,640

* Latticed steel pole, not including arms, max. bending moment = 165,000 lb-ft, weight includes 6-ft length in ground. T-suspension tower; L-angle tower; A-dead-end tower.

Data on Foundations for Above Towers

Type	Number per Tower	Depth below Ground, Feet	Area, Inches	Weight Each
Steel........................	4	9	45 × 60	406
Steel........................	4	10	45 × 60	403
Steel........................	4	12	72 × 76	1082
Steel........................	4	9	72 × 74	665
Steel........................	4	11	60 × 82	843
Steel........................	4	7	....	392
Steel........................	4	7	....	2675

30. CROSS-ARMS

Cross-arms are usually of wood, although sometimes of steel. "Buck-arms" or "reverse arms" are cross-arms attached to a pole at right angles to the principal arms, and are used for taking off wires at right angles to the line, either at the junction of intersecting lines or at services. "Double arms" are pairs of cross-arms attached to opposite sides of a pole so as to act as one compound arm. Double arms are used to increase the strength of an arm and to permit the use of two pins and two insulators for supporting a single wire where additional strength is required. (See Article 28.)

Cross-arms are used principally for supporting pins, insulators, and wires, although lightning arresters, transformers, switches, and other miscellaneous appliances are often mounted on them, usually for the purpose of keeping the pole free of incumbrances so that it will be more easy to climb.

When city distribution lines are located in alleys, it is common to locate poles next to the property line. Where it is not permissible to let arms overhang private property, special arms which extend on one side of the pole only may be used. These must be well braced and should not be used for dead-ending wires. Another type of construction is obtained by locating two poles on opposite sides of the alley and putting special cross-arms across the alley between them.

FORCES ON AND STRESSES IN CROSS-ARMS. The forces which a cross-arm resists are:

a. Vertical forces due to weight of pins, insulators, wires (with sleet), and accidental loads due to linemen standing on arms, etc.

b. Transverse horizontal forces due to wind pressure on the wires (with sleet) at right angles to the line.

c. Longitudinal horizontal forces due to the pull of the wires where the pull is unbalanced. Unbalanced pull is usually due to an angle in the line, the ending of the wire at the arm, a change in the size of the wire at arm, and an unequal tension in the spans on the two sides of the arm.

The principal internal stresses produced in an arm from these forces are:

1. A bending force in a vertical plane due to vertical forces (*a*).
2. A bending force in a horizontal plane due to horizontal forces (*c*).
3. A twisting force about the longest axis of the arm due to the "pin leverage" of the horizontal forces (*c*). The pin leverage is the distance from the center of the wire to the axis of the arm.

Of these stresses the most destructive is probably the twisting stress, which tends to split the arm in a vertical plane through the pinholes and along the grain of the wood. On this account the pin and insulator should be no taller than necessary, and the pin should extend completely through the arm to give the best distribution of bearing pressure.

The vertical and horizontal bending stresses are of some importance and may be computed by the usual beam formula. Data of tests on strength of cross-arms for these stresses are given below.

STRENGTH TESTS OF CROSS-ARMS. (*Forest Service Cir.* 204.) Tests made on 3 1/4 in. by 4 1/4 in. by 6 ft, 6-pin air-dried cross-arms with vertical load distributed equally at each pin hole gave the results in Table 16.

Table 16. Strength Tests of Cross-arms

Kind of Wood	Average Maximum Load, lb	Maximum Crushing Strength, lb per sq in.
Longleaf pine, 75 per cent heart	10,180	8950
Longleaf pine, 100 per cent heart	9,780	8940
Shortleaf pine	9,260	7300
Longleaf pine, 50 per cent heart	8,980	5425
Shortleaf pine, creosoted	7,650	5770
Douglas fir	7,590	7080
White cedar	5,200	4700

The tests showed that for ordinary use the strength, for vertical loads of 6-pin arms, need not be considered in calculations of line construction, except in the rare case of abrupt change of the grade of the line.

CROSS-ARM CLASSIFICATIONS. Cross-arms are classified generally as (1) telephone arms, (2) power distribution arms, and (3) transmission arms. They are more specifically designated by stating the number of pin positions, overall length, cross-sectional dimensions, and type—whether standard arms, alley arms, or other special kinds.

REQUIREMENTS FOR CROSS-ARMS. The dimensions of cross-arms standardized by the EEI are shown in Table 17.

Table 17. Dimensions of Cross-arms

Use of Arms	Number of Pins	Width, in.	Depth, in.	Length, in.
Distribution—Std	2	3 1/4	4 1/4	36
Distribution—Std	4	3 1/2	4 1/2	67
Distribution—Std	6	3 1/2	4 1/2	96
Distribution—Std	8	3 3/4	4 3/4	120
Distribution—Alley	6	3 1/2	4 1/2	96
Transmission—Std	2	3 3/4	4 3/4	84
Transmission—Std	4	3 3/4	4 3/4	120

Transmission cross-arms suitable for use with suspension insulators are usually special and have cross-sectional dimensions 4 in. by 5 in., 5 in. by 7 in., or larger if necessary.

Figure 33 shows dimensions and drilling for E.E.I. standard cross-arms. The strength of double arms, with spacing bolts, in the direction of the line should not be considered as more than 30 per cent greater than that of a single arm.

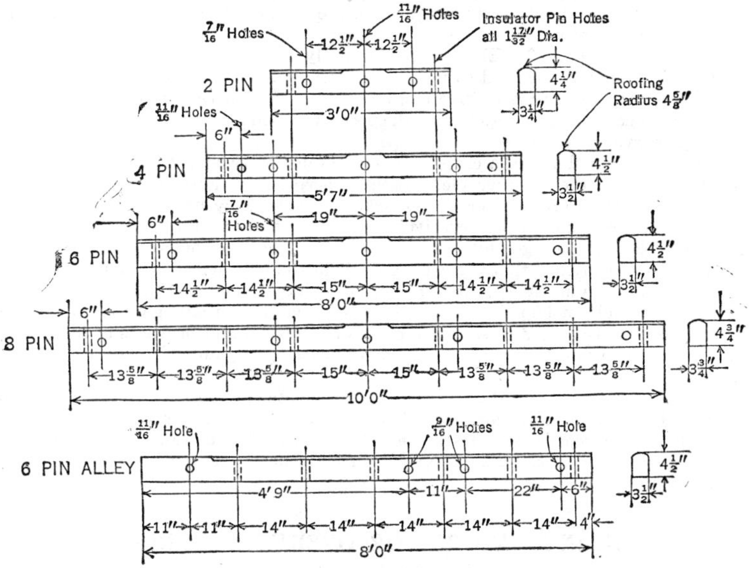

Fig. 33. Standard Cross-arms

Distance between Pole Pins. The requirements for climbing space determine this dimension. If telephone cross-arms are used on joint poles with power cross-arms above them, the pole-pin spacing must be the same as that required for the power cross-arms. On power cross-arms the pole-pin spacing is required to be: 24 in. for voltages up to 300, 30 in. for voltages up to 7500, 36 in. for voltages from 7500 to 15,000, and more than 36 in. for voltages above 15,000.

Distance between Other Pins. For low-voltage power distribution cross-arms the pin spacing called for by the EEI Standards is 14 1/2 in. The 4-pin medium-voltage transmission cross-arm standardized by the EEI has a pin spacing of 36 in. On long-span construction, even on low-voltage lines, it may be necessary to have greater conductor spacings. This is often accomplished by using longer arms with more pin positions, but omitting intermediate pins. Pins not used should be left out of the arm.

Distance from End Pin to End of Arm. This distance is 4 in. for low-voltage power arms and 5 in. for medium-voltage transmission arms. These distances are chosen to minimize splitting of the arm by pin leverage.

Pinholes. Pinholes are specified to be of a diameter which will give a close fit for the pin. EEI standard wood pins have a maximum diameter at the shank of 1 1/2 in. Pinholes for such pins are to have nominal diameters of 1 $^{17}/_{32}$ in. and are tested with a steel gage having plugs of 1 $^{1}/_{2}$ in. and 1 $^{9}/_{16}$ in. The hole must take the 1 $^{1}/_{2}$-in. plug without forcing, but not the 1 $^{9}/_{16}$-in. plug. Holes for steel pins must take a testing plug of diameter equal to the diameter of the pin shank without forcing, but not $^{1}/_{8}$ in. larger.

Bolt Holes. For fastening the arm to the pole one $^{11}/_{16}$-in. hole is placed at the center of the arm for a $^{5}/_{8}$-in. machine bolt. For fastening flat cross-arm braces to the arm two $^{7}/_{16}$-in. holes are placed each side of the middle of the arm 38 in. apart. These holes are drilled to take $^{1}/_{2}$-in. machine bolts. Bolt holes are tested by steel gages and must take a test plug $^{1}/_{16}$ in. smaller than the bolt diameter without forcing.

SPECIFICATIONS. The points to be covered in specifications for cross-arms are as follows:

Material. Usually Douglas fir or creosoted yellow pine.

Quality of Material. Should be free from loose or unsound knots, loose heart, rot, shakes, wane, or wormholes. Checks should not exceed 12 in. in length, $^{3}/_{4}$ in. in depth, or $^{1}/_{16}$ in. in width. Grain should be parallel with the axis of the arm within 5 deg. Knots larger than $^{1}/_{4}$-in. diameter should be avoided. Large pitch pockets are not desirable. Sapwood should not be allowed in yellow pine arms, nor sapwood in excess of 25 per cent in Douglas fir arms. Excessive warp should not be allowed.

Dimensions, Pinholes, etc., as noted in previous paragraphs.

Seasoning. Cross-arms should be thoroughly air-dried or kiln-dried before manufacture. When arms are dried after manufacture, the pinholes become elliptical because of shrinkage across the grain.

PRESERVATIVE TREATMENT. Until recently it was considered that cross-arms made of Douglas fir did not need to be treated with preservatives, but the increased life of poles due to improved preservative methods has made it necessary to prolong the life of cross-arms to correspond. Practically all species of wood suitable for cross-arms, therefore, need preservative treatment.

Since most of the volume of a cross-arm consists of heartwood, it is difficult to obtain much penetration of the preservative, especially perpendicular to the grain, unless low-viscosity preservatives are used or the viscosity is lowered by heating. Pressure treatments using the best types of preservatives for poles give satisfactory results. Open-tank treatments with hot creosote, followed by cold pentachlorphenol oil, have also been quite satisfactory, provided that the arms are left in the bath for a sufficiently long time. Two to four hours are necessary. (See Preservative Treatment, p. 14-91.)

Cross-arms should not be painted, especially with paints containing oil, although creosote stains may be used. Paint seals the pores of the wood and prevents evaporation of moisture. Retained moisture promotes fungus growth, and painted arms soon rot internally.

STORAGE. Cross-arms held in storage should be stacked on creosoted skids. Each layer should be placed at right angles to that below it. Plenty of air space should be left around each arm for ventilation. Each stack should be roofed to drain off rain and should be protected from the direct sun. Cross-arms should not be stored in heated buildings; open sheds are best.

31. INSULATORS FOR OVERHEAD LINES

Insulators for overhead lines may be classified as pin-type, suspension, and strain insulators. Pin-type insulators are used for low- and medium-voltage lines; suspension insulators, for all voltage lines; strain insulators, in guys and for dead-ending low-voltage lines. The general features of design common to all classes of line insulators will be first considered.

DESIGN OF LINE INSULATORS. The insulating materials principally used for line insulators are: wet-process porcelain, dry-process porcelain, and glass. (See Table 18). Wet-process porcelain is used for this purpose to a far greater extent than dry-process porcelain. Wet-process porcelain has greater resistance to impact and is practically impervious to moisture without glazing; dry-process porcelain is not. However, dry-process porcelain, if well made, has a somewhat higher crushing strength. Dry-process porcelain is therefore used only for the lowest-voltage lines, and even there to only a limited extent.

Prior to the development of high-grade wet-process porcelain, glass was used extensively for all lines up to about 44 kv. The glass available at that time was fragile and subject to breakage due to temperature changes combined with the effect of internal strains. Improved porcelain quickly eliminated glass from line use except for low-voltage signal lines. Recently, however, great advances have been made in glass manufacture, resulting in the development of glass insulators which are quite tough and have low internal strains. The use of these insulators is therefore increasing for low- and medium-voltage lines.

An insulator must be designed to stand extreme and sudden temperature changes, sleet, and rain, as well as smoke, dust, and often special conditions, such as salt fogs, salt-water sprays, and chemical fumes, without deterioration from chemical action, breakage from mechanical stresses, or electrical failure. The design of high-tension insulators is a process of compromise between requirements often antagonistic in nature.

Table 18. Properties of Porcelain and Glass

Property	Glass	Porcelain	Pyrex
Tensile strength, lb per sq in.	7600–12,100	6000–8500	3000
Crushing strength, lb per sq in.	12,100–50,000	44,000–60,000	120,000
Modulus of elasticity, lb per sq in.	7–12 ($\times 10^6$)	10–15 ($\times 10^6$)	9×10^6
Coefficient of expansion per deg F	4.4–4.92 ($\times 10^{-6}$)	1.82–3.7 ($\times 10^{-6}$)	1.8×10^{-6}
Coefficient of expansion per deg C	7.9–8.83 ($\times 10^{-6}$)	3.3–6.6 ($\times 10^{-6}$)	3.2×10^{-6}
Weight, lb per cu in.	0.09–0.125	0.08–0.085	0.0815
Puncture strength, kv per in.	1800–3000	316–695	2800+
Dielectric constant	6.0–8.0	6.15	4.48

Shape and Thickness. The insulating surfaces should conform to the flow lines of the electrostatic field; the surfaces of the rain sheds or petticoats should conform to equipotential surfaces. The leakage resistance per shell for multipart pin-type insulators should be about equal; the plane of mechanical rupture should not coincide with the plane of electrical stress, and the unit should have approximately equal capacitances per shell.

The insulating material must be thick enough to resist puncture by the combined working voltage of the line and any probable transient whose time lag to sparkover is great. If this thickness is greater than desirable from a manufacturing standpoint, two or more pieces are used to give the proper aggregate thickness. The thickness of a porcelain part must be so related to the distance around it that it will arc over before it will puncture. The ratio of puncture strength to arc-over voltage is the factor of safety of the part or of the insulator against puncture.

This ratio should be high to give sufficient margin to protect the insulator from puncture by the transients before mentioned.

Leakage and Arcing Distances. The leakage and the dry and wet arcing distances are criteria of the effectiveness of insulators in their function. These terms are defined in the following paragraphs and are illustrated in Fig. 34. Table 19 gives values for these factors for representative insulators.

Leakage distance is the shortest distance between conductor and pin or between cap and pin, measured along the surface of the insulating material. This distance is roughly proportional to the amount of voltage required to cause leakage current to flow over the surface of the insulator. It does not, however, take into account the varying width of leakage path nor the conductivity of the dirt or moisture film which may be deposited on the surface.

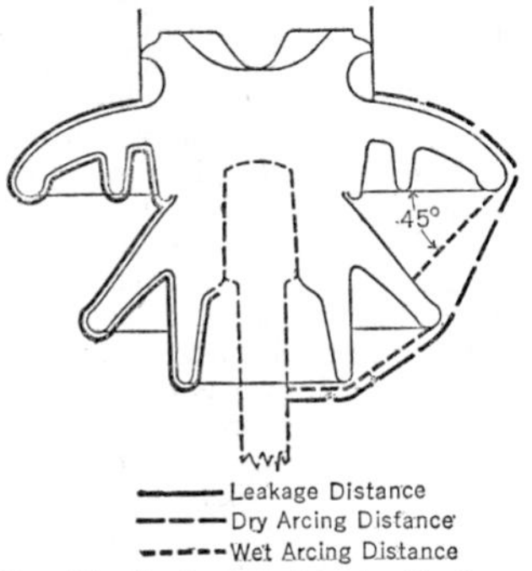

Fig. 34. Leakage and Arcing Paths of Insulator

Wet arcing distance is the shortest distance between conductor or cap, and pin, measured partly over the surface of the insulating material and partly through the air. Where, in any portion of the path to be measured, the striking distance through air between any

Table 19. Characteristics of Typical Insulators

Type	Dry Flashover, kv	Leakage Distance, in.	Dry Arcing Distance, in.	Wet Arcing Distance, in.
Low-voltage pin-type	50	4.75	3.12	1.37
Low-voltage pin-type	70	9.00	4.50	2.25
Medium-voltage pin-type	95	12.12	6.25	4.00
Medium-voltage pin-type	170	35.50	14.37	9.50
Suspension 10 in. disc	80	12.00	8.00	4.12
Suspension 12 in. disc	95	12.50	9.20	4.30

two points is less than the length of path over the surface between those points, the air striking distance is measured for that portion of the wet arcing distance in place of the distance over the surface. The wet arcing distance is a fairly good measure of the voltage required to arc over the insulator when exposed to rain.

Dry arcing distance is the shortest distance between conductor or cap, and pin. It is measured entirely through the air and is a fairly good measure of the voltage necessary to arc over the insulator when it is perfectly dry.

Free Arcing. The porcelain must extend beyond the charged conducting connections (i.e., tie wire or cap at the top and pin at the bottom) sufficiently so that the distance between the connections through the air around the porcelain is greater than the arcing distance of the maximum voltage to be carried. The arcing distance required for a given voltage may be determined roughly (but only very roughly) from the tables of arcing distances between needle points; see Spark Gap, Section 5, Article 16. The greater radius of curvature (compared with needle points) of such metal parts as the insulator pin decreases the potential gradient at the terminals. Also the porcelain has a much greater specific inductive capacity than the air, and its proximity to the arcing path disturbs the electrostatic field through the air. Surface charges on the porcelain because of surface leakage or corona also modify the field.

Free arcing is the property of arcing over along a line which does not touch the porcelain body from the point where the arc leaves the metal cap to where it strikes the metal pin. Where the arc touches any part of the porcelain, the great heat fractures the porcelain in a few seconds; hence the desirability of designing the insulator so that it is free arcing. A properly designed insulator will arc over as a whole before any individual part (i.e., shell or unit) arcs over. In many defective designs the insulator will fail by some parts arcing over, thereby increasing the voltage on others, which then fail by puncture or arcing over.

Spread of Petticoats of Pin Insulators. In two concentric shells the two surfaces which lie opposite to each other are at different potentials except where they are cemented together. The difference in potential between two points on opposite surfaces is greater the further they are removed from the joint. Unless the shells diverge correspondingly so as to increase the distance between the shells as the potential increases, the air will break down and part of the leakage surface will be short-circuited by a corona discharge. This divergence is shown in Fig. 39, where the top is a disk made slightly convex to shed water, and the inner shells are cone shaped.

Minimum Height of Pin. Pin insulators mounted on metal cross-arms should be provided with metal pins which have sufficient length above the cross-arm to insure that flashover will take place to the pin rather than the arm; otherwise the full flashover value of the insulator will not be obtained. The distance to the arm from the lowest skirt should never be less than the shortest distance from the skirt to the pin, even with wood arms. On the other hand, the pin should not be longer than necessary on account of the increase in bending moment on the pin with increased height.

Color and Glazing. Brown, slate, and white are the common colors used in glazing porcelain. Brown is the most common color, since it is more of an aid in determining faults. Slate-colored glazing of the same color as galvanizing on towers makes insulators a less conspicuous target for malicious destruction.

Glaze is somewhat similar in composition to porcelain. The ingredients are mixed with water and held in suspension while the unfired insulator is dipped in the mixture. Upon firing, the glaze takes on a glass-like consistency which retards the collection of dirt and allows it to be washed off easily by rain.

Cementing of Insulators. Porcelain insulator parts are cemented together with neat Portland cement which is selected for quality, strength, fineness, and absence of metallic particles. After cementing, the insulators are placed in compartments or tanks and slowly cured under automatically controlled humidity and temperature conditions. The surfaces to be cemented are left unglazed and are either corrugated or sanded to obtain good bond.

Pinholes. The pinholes in pin-type insulators are made in various forms to accommodate different types of pins. Insulators for pins with full-size threads are made with the threads molded in the porcelain or have threaded sheet-metal thimbles cemented into unthreaded sockets. Some pins have malleable-iron thimbles, and insulators for use with these pins have the pin sockets unglazed, but with either plain or corrugated surfaces for cementing to the thimbles. Some insulators for this purpose have the sockets glazed but with sharp porcelain grains held to the surface by the glaze. Special flexible cements are often used for thermal relief.

Faults in Insulators. The more common faults in porcelain are porosity, folds and flaws in molding, the development of checks and hair cracks in the process of drying, incomplete and non-uniform glazing, warping, air bubbles, conducting impurities, under- and overfiring, and chipping of edges. Only 50 to 75 per cent of molded shapes ordinarily pass final test, and even fewer of the more difficult shapes. Inspection and testing are essential to eliminate faults in both design and manufacture.

DISTRIBUTION-LINE INSULATORS. Insulators for power lines from 115 to 17,000 volts are included in this classification. Both pin-type and suspension insulators are used for lines within this voltage range. For secondary lines carried on racks, spool insulators are used.

Pin-type Insulators. The smallest insulator in this class is the side-groove rounded-top type. The side groove is 3/8 in. in radius, the overall height is 3 9/16 in., and the skirt diameter is 3 1/4 in. The dry flashover voltage is about 50 kv, and the wet flashover voltage about 20 kv. This insulator is suitable for voltages up to 5 kv, although it has some disadvantages in not being able to accommodate large wires and in not having a top groove.

All the other insulators in this class are provided with both top and side grooves. The most-used insulator of this class is suitable for lines up to 7500 volts. It has top and side grooves of 7/16-in. radius, accommodating weatherproof wires up to and including No. 4/0 stranded. It is 3 1/4 in. high and has a skirt diameter of 3 3/4 in. The dry flashover voltage is about 50 kv, and the wet flashover voltage about 30 kv. The outline of this insulator is shown in Fig. 35.

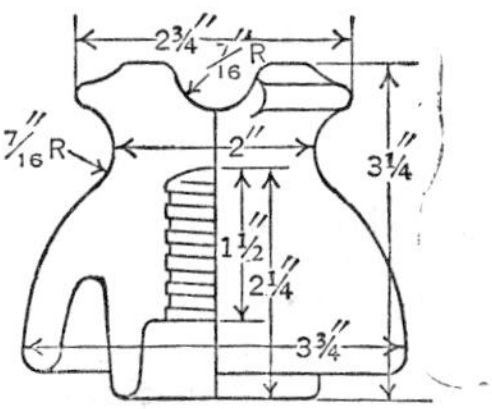

FIG. 35. Small One-piece Insulator

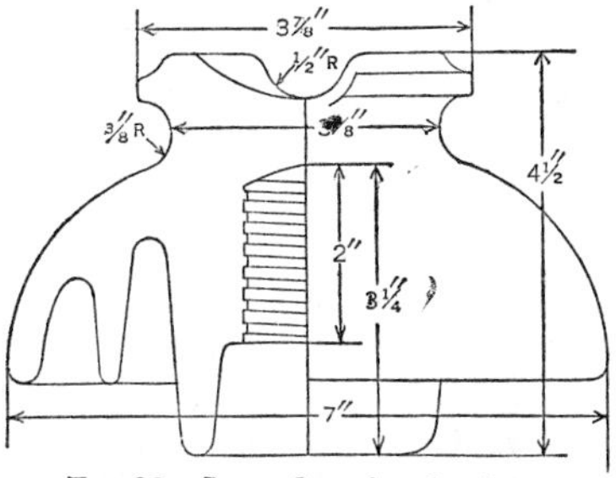

FIG. 36. Large One-piece Insulator

The larger pin-type insulators of this class are similar in form to the insulator described in the preceding paragraph. All dimensions are increased to provide the increased flashover values and leakage distances required for the higher voltages. An example of an insulator suitable for 17-kv lines is shown in Fig. 36. This insulator has a dry flashover of 90 kv and a wet flashover of 50 kv.

All the pin-type insulators of this class are made for pins with 1-in. diameter tops, but the larger sizes are also made for 1 3/8 pins.

Suspension Insulators. In this voltage class, suspension insulators are used mainly for dead-ending lines or for angle construction, although a few lines of 15 to 17 kv have been built, using suspension insulators throughout. The ones most used are of the cap and pin (or disk) type. The insulator disks have a diameter of 6 to 10 in. The caps are malleable iron or steel and are provided with clevis tops for connecting to the cross-arm. The pin takes the form of a hook, eye, or ball, depending upon the method of connection desired. Still other variations of cap and pin designs are available for special purposes.

The top of the insulator disk is provided with a hollow projection in the shape of a truncated cone. This projection is cemented into the hollow cap, and the pin is cemented into the hollow portion of the cone. Figure 37 shows a typical suspension insulator for low-voltage lines. It has a dry flashover of 50 kv and a wet flashover of 30 kv and is suitable for lines up to 7500 volts.

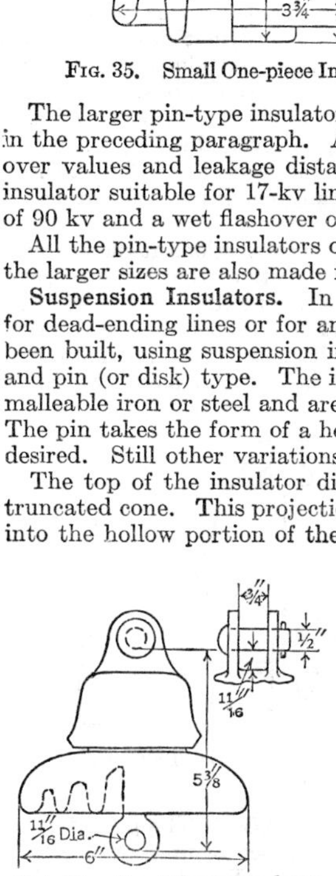

FIG. 37. Small Disk Insulator

Spool Insulators. Insulators to be used on secondary racks having vertical rods for insulator supports are of the spool type and are available in two standard sizes, one for light racks and one for heavy racks. Figure 38 shows a typical insulator for heavy racks. The smaller size spool is similar in shape with a height of 2 1/8 in. and a diameter of 2 1/4 in. The groove radius is 1/2 in.

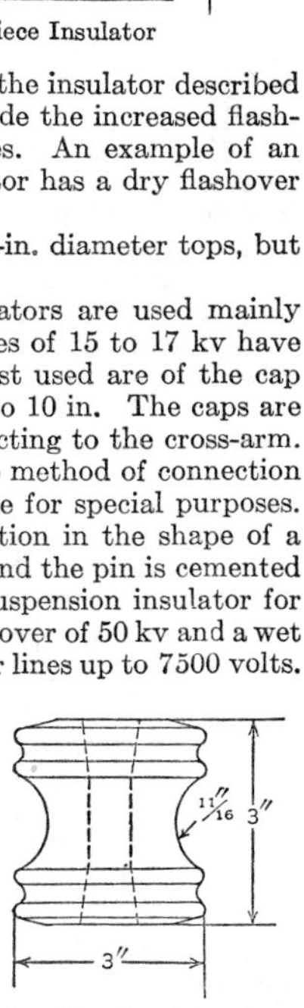

FIG. 38. Spool Insulator

TRANSMISSION-LINE INSULATORS. For transmission lines above 17 kv both suspension insulators and pin-type insulators are used, although the pin-type insulators are now mainly confined to lines below 45 kv.

Pin-type Insulators. Pin-type insulators for high-voltage transmission lines must necessarily be made up of two or more parts cemented together. This construction has often resulted in radio interference, due to lack of a good bond between the cemented parts, and in early failure, due to the effect of temperature changes. It is not now considered good practice to use pin-type insulators having more than two parts. Figure 39 shows a pin-type insulator suitable for lines of 35 kv. The dry flashover of this insulator is about 125 kv, and the wet flashover about 85 kv.

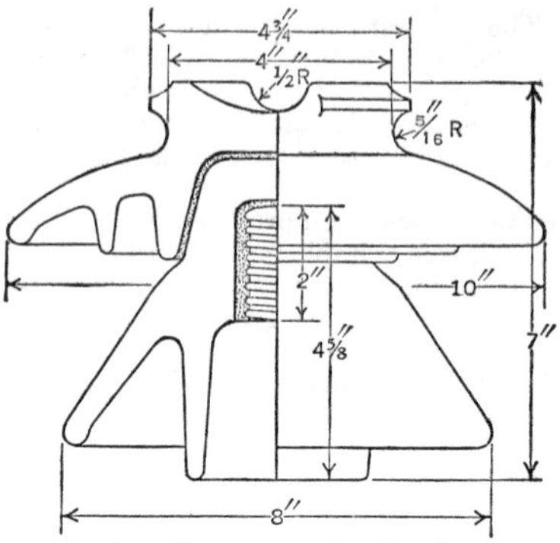

FIG. 39. Large Two-piece Insulator

Suspension Insulators. Suspension insulator units (Figs. 40 and 41) are usually connected together in strings of two or more for high-voltage transmission lines to provide sufficient insulation. The construction of these units is similar to that described under Distribution Line Insulators, p. 14-115. Ten-inch disks are used almost exclusively, although high-strength 12-in. disks have been used under special conditions. Suspension insulators are designed to give spacings between individual units of 4 3/4 in. to 7 in., to meet varying insulation requirements. For most purposes, however, the 5 3/4-in. spacing is satisfactory. Suspension insulators are made with different mechanical strengths. It is therefore possible to select insulators to meet almost any conditions, including the economic requirements of a given line.

The suspension-type insulator is always used in tension, the connections at the two ends being made so that the insulator is free to swing in any direction; the insulator takes such a position that its axis coincides with the direction of the mechanical stress. This type is used hanging below the cross-arm with axis vertical as a suspension insulator for sustaining the weight of cable at points where there is little horizontal force, and also with axis approximately horizontal as a "strain" or "dead-end" insulator at points where the horizontal force predominates.

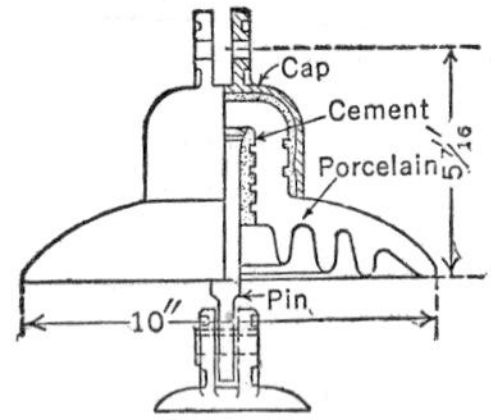

FIG. 40. Large Disk Insulator—Clevis Type

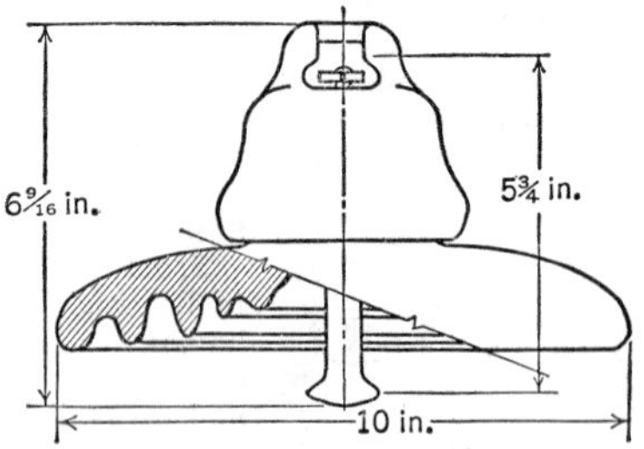

FIG. 41. Large Disk Insulator—Ball-socket Type

ELECTRICAL CHARACTERISTICS OF INSULATOR STRINGS. The potential per unit required to flash over a suspension insulator string composed of units of the same design decreases with an increasing number of units, as shown in Fig. 42, because of an unequal potential gradient. The potential gradient for various insulator strings is shown in Fig. 43. If grading shields or arcing rings are used in connection with an insulator string, the voltage gradient is changed and made more uniform. A typical effect of adding grading rings is shown in Fig. 44.

The impulse flashover characteristics of typical insulator strings is shown in Fig. 42. Curve *A* gives the crest voltage for flashover of a positive wave of 1 × 5 μsec (i.e., reaching crest value in 1 μsec and dropping to $^1/_2$ crest value in 5 μsec from the start of the wave). Curve *B* gives flashover values for a positive wave of 1 $^1/_2$ × 40 μsec.

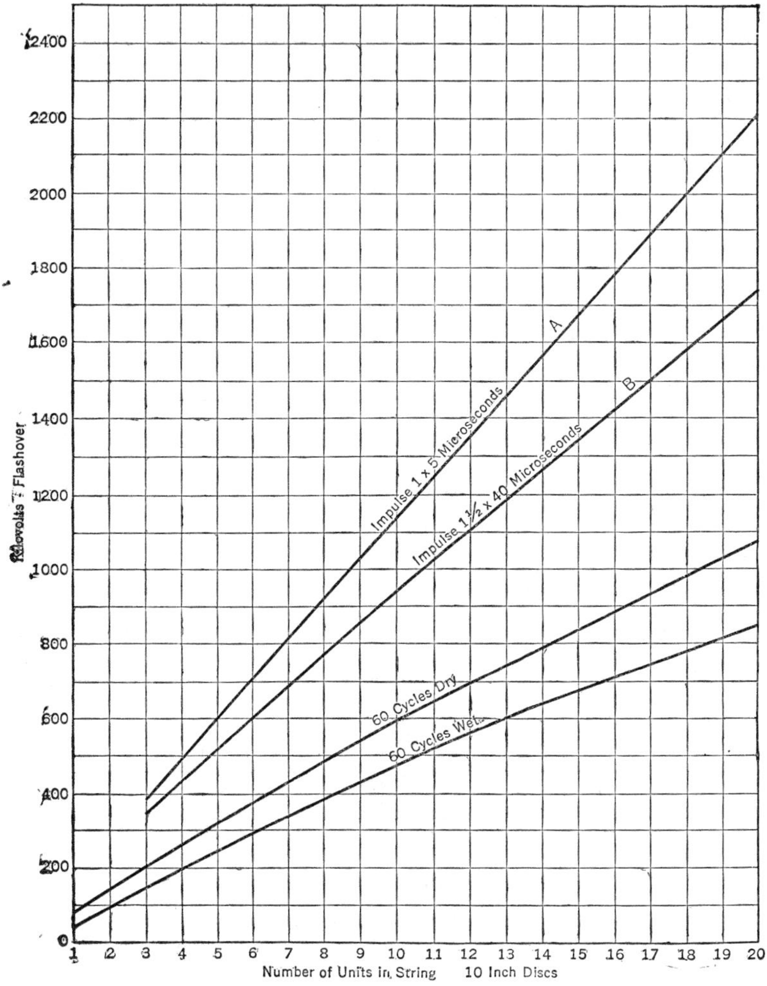

FIG. 42. Flashover Characteristics, Suspension Insulators

INSULATION COORDINATION. It would be desirable, of course, to make a transmission line proof against all flashovers, if that were possible. It is doubtful whether it is possible to so insulate a line as to achieve this result. Lightning voltages are so great that any amount of insulation that is economically feasible would probably fail to eliminate flashovers. Increasing the insulation of the transmission line itself only transfers the flashovers to the terminal equipment, where they cause greater damage than they would on the line. Therefore, the insulation of terminal equipment should be greater and better protected than the line insulation. Lightning arresters, shielding, and other methods are feasible and economical at substations but not on the line. For line insulation, protection arcing and grading rings, ground wires and counterpoises, and expulsion protective gaps have been used with fairly good results. (See Bibliography, p. 14-127.)

STRAIN INSULATORS. Insulators of this type are used primarily for insertion in guy wires. Most porcelain strain insulators are designed to place the porcelain under compression and are of cylindrical form with grooves for the loops of the guy wires. The design is such as to allow the loops of the two sections of guy wire to be interlaced with porcelain between them. Insulators for the higher voltages are provided with deep fluting or fins to provide additional leakage surface. This design prevents mechanical failure of the guy even if breakage of the porcelain takes place. Insulators of this type are also used

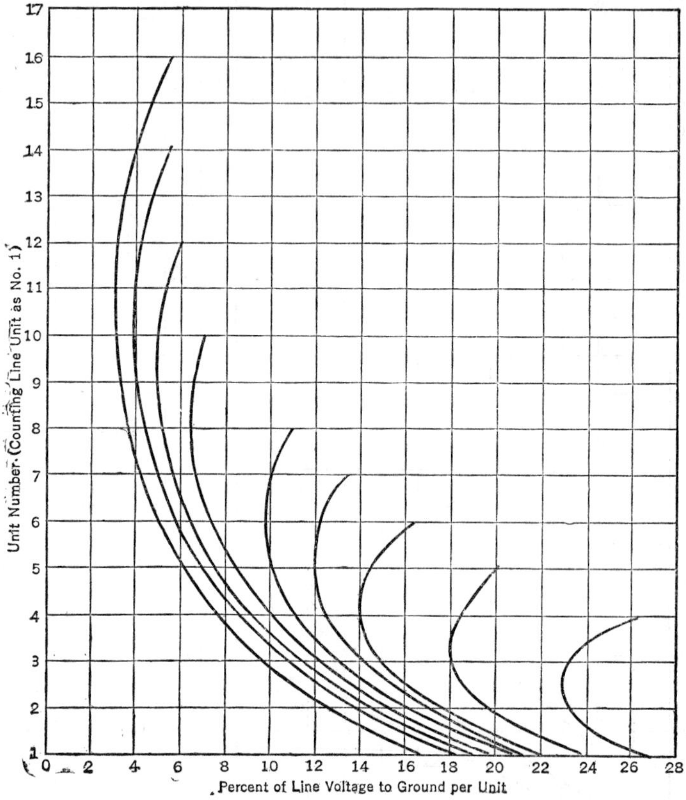

FIG. 43. Potential Gradient, Suspension Insulators

for dead-ending circuits up to about 2500 volts. Several styles of metal clevises are available for attaching them to poles or cross-arms when used for dead-ending purposes. Figure 45 shows a typical strain insulator.

Wood strain insulators are coming into greater use for guys on wood transmission line structures in order to increase the overall insulating value of the wood structure against lightning voltages. The flashover value of insulation under lightning voltages depends primarily on its length between live parts and grounded parts. Wood provides a long flashover path at lower cost than any other material of the same mechanical strength. The wood is used in tension and is connected to the guy wires by metal end-pieces specially designed to develop the full mechanical strength of the wood. Arcing horns are provided to make sure that any possible arc across the insulator will be kept away from the wood.

TESTS OF INSULATORS. Tests of insulators may be classified as: (1) design tests, made on a few insulators to determine the electrical and mechanical characteristics of each different design; and (2) routine tests, made on all or a certain percentage of each lot of insulators purchased, to detect defects of material or workmanship. Standard 41 of the AIEE gives complete test specifications for porcelain insulators, which are generally accepted. These specifications, slightly modified, may also be used for glass insulators. Space does not permit the incorporation of these specifications here. They are reprinted in several manufacturers' catalogs.

Design Tests include: wet and dry flashover, corona, and puncture tests. For suspension insulators a combined mechanical and electrical test is required.

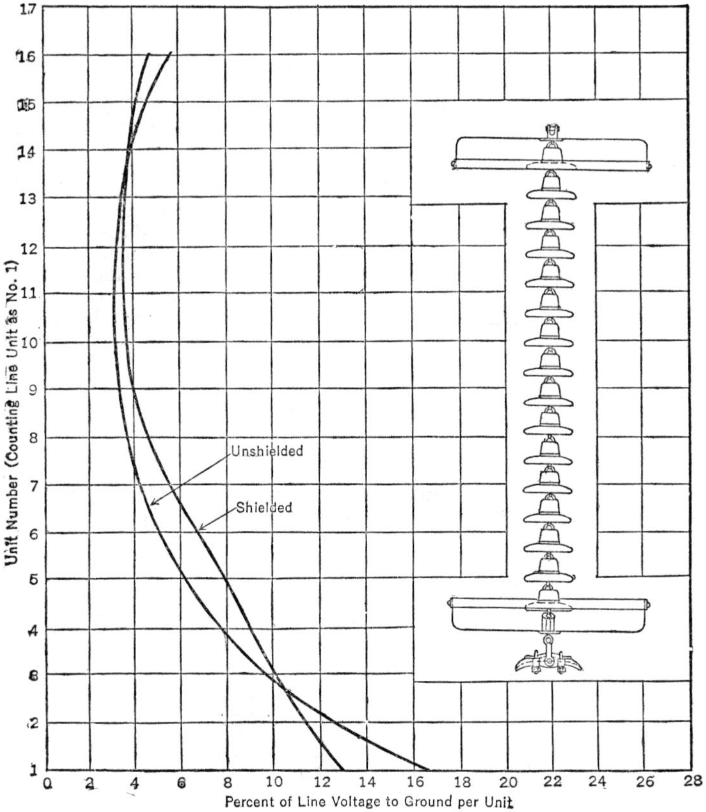

FIG. 44. Effect of Grading Rings

Routine Tests include: (1) Continuous dry flashover for 3 minutes for pin-type insulators, both on individual parts and on the assembled insulator. For suspension insulators the flashover is continued for 5 minutes. (2) Thermal change tests on the assembled suspension insulators and on individual parts as well as assembled pin-type insulators. (3) Porosity tests. (4) Puncture tests. (5) High-frequency and combined mechanical and electrical tests on suspension insulators only.

Testing on the Line. Insulators on the line should be tested periodically to eliminate deteriorating units before they fail and interrupt service. Various patented methods are available for line testing of insulators. Generally these methods use the voltage gradient method. Measurement of the dielectric power factor is an excellent method for testing off the line.

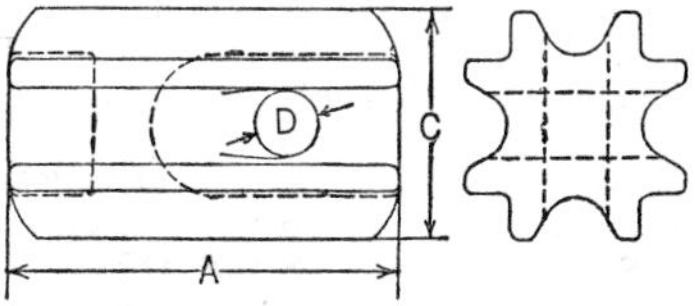

FIG. 45. Strain Insulator

SELECTION OF INSULATORS. The first step in the selection of insulators is to determine the insulation level of the line relative to that of the terminal equipment. Both normal-frequency flashover voltages and lightning voltages should be considered. The type of supporting structures (wood or steel) has a considerable effect on the insulation efficiency of the line. The possibility of overvoltages due to switching surges and to arcing grounds on isolated neutral systems should be taken into account. The mechanical strength requirements should also be evaluated. Manufacturers' catalogs now give very complete and reliable data to aid in insulator selection.

32. INSULATOR PINS AND HARDWARE

Insulator pins are made of wood or steel. Insulator hardware is made of steel or malleable iron.

WOOD PINS. All-wood pins are now used only for lines up to 7500 volts. They are unsatisfactory on higher voltages, for they deteriorate rapidly because of the combination of moisture and leakage current. Black and yellow locust are the most satisfactory woods for the purpose, although eucalyptus, oak, birch, osage orange, gum, and other woods have been used successfully. Figure 46 shows the EEI standard wood pin. It should be noted that the 4 1/4-in. shank will not develop the full strength of cross-arms 4 1/2 in. or more in depth.

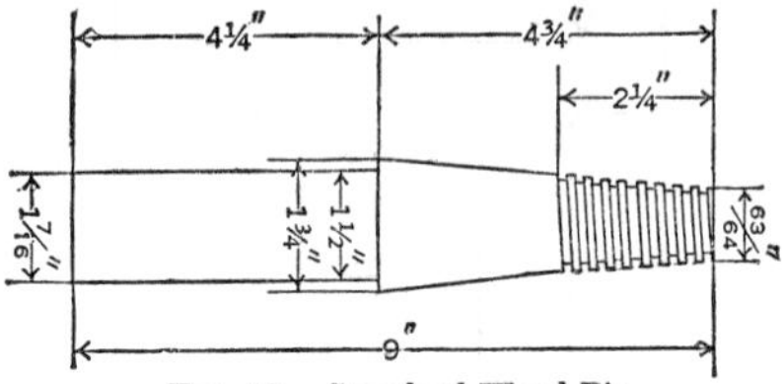

Fig. 46. Standard Wood Pin

STEEL PINS. Steel pins are available in many designs for any voltage class of insulators. They are made of forged steel and are usually galvanized. Steel pins for low-voltage insulators are made with long and short bolt shanks and also with lag screw shanks. The threaded head is made of cast lead, pressed steel, or a helix of spring steel wire. The lead head has the advantage of giving a tight fit without tending to burst the insulator under temperature changes. Figure 47 shows a typical pin of this type.

Steel pins for higher-voltage insulators are similar to those for low voltages, except in size and strength. In addition to being made with lead heads, they are also made with special threaded tops for screwing into thimbles, which are cemented into the insulators before placing on the pins. Figure 48 shows a pin of this general type.

Since steel pins may be designed to have greater strength than is obtainable with wood pins, they are more desirable for heavy conductors and long spans. Because of the smaller size of shank, steel pins require less wood to be removed from the cross-arms and therefore do not reduce the cross-arm strength as much as wood pins do. Cross-arms rot around the pinholes rather quickly when wood pins are used, but not when steel pins are used.

ATTACHMENT OF PINS TO ARMS. Pinholes in cross-arms for wood pins are bored 1 17/32 in., and the diameter at the top of the shank of pins is required to be 1 1/2 in. On account of the swelling or shrinking of the wood, the pins may fit tightly or loosely. In order to make sure that the pins will not pull out of the arm, a 4 or 6 penny galvanized nail is driven through the arm and the pin. Pinholes for steel pins are bored 1/16 in. larger than the shank diameter. The end of the pin shank is threaded, and in addition to the nut the pin is furnished with a locking washer. Drawing the nut tight insures a rigid connection and seals the pinhole against moisture, thus retarding decay.

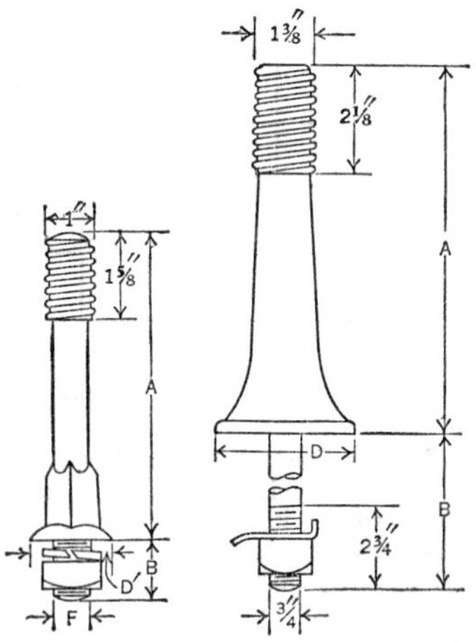

Fig. 47 Fig. 48
Steel Pins

SUSPENSION INSULATOR HARDWARE. The suspension insulator type of construction requires a wide variety of hardware fittings. For attaching insulators to structures, use is made of eyes, shackles, clevises, or hooks. These are usually designed to permit free swing of the insulator string in every direction. For any particular situation fittings should be selected which will provide the simplest form of connection. Eyes of some form are usually provided on tower cross-arms for attaching insulators. Eye bolts or specially designed fittings are used to attach insulator strings to wood cross-arms. With parallel strings of insulators for extra strength, yokes provide a common connection for the strings and space them apart properly. These yokes are, of course, used at each end of the multiple string.

Suspension clamps or strain clamps serve to attach conductors to insulator strings. Suspension clamps are used on insulator strings at all structures at which the conductor is not dead-ended, including tangent sections, slight angles, and angles which are turned with flying corner construction. Such clamps are generally made of galvanized malleable iron with steel U-bolts or J-bolts, although cast or forged steel has been employed in some cases. These clamps should have smooth, well-rounded seats and corners. They should be quite light in weight and therefore of small inertia to reduce conductor vibration. They are provided with aluminum liners, if desired, for use with aluminum cables. Figure 49 shows a typical suspension clamp. Recent developments at Boulder Dam and elsewhere have led to the development of clamps for extra-long spans, with special features to prevent concentration of vibration on the cable near the clamp.

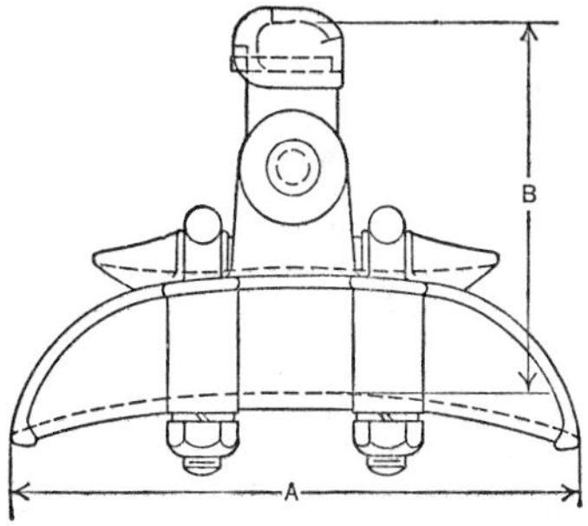

FIG. 49. Suspension Clamp

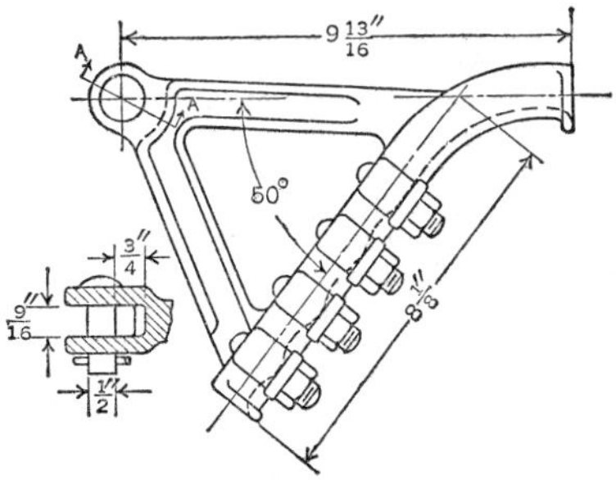

FIG. 50. Strain Clamp

Strain clamps, used for dead-ending conductors, are made in a wide variety of styles and sizes to meet the tension requirements imposed by the line conductors. They are commonly made in such form as to put a bend in the conductor to provide some snubbing action and are also provided with steel U-bolts or J-bolts, and pressure pieces to hold the conductor by friction. Slight waves are often made in the surfaces of pressure pieces and conductor seats to increase the snubbing effect. Strain clamps are designed to hold without slippage the largest conductor they will fit. They are also provided with liners for aluminum cable, if desired. Figure 50 shows a typical strain clamp. Clamp bodies are usually made of galvanized malleable iron or steel. Another form of strain clamp which is coming into general use provides nearly 360-deg snubbing effect combined with clamping. This type is used with the smaller conductors on distribution lines.

33. AERIAL-CABLE CONSTRUCTION

Certain conditions, such as the following, make open-wire construction undesirable and are best suited to aerial-cable construction. These conditions are:

1. Thick growth of trees and branches, preventing adequate tree trimming.
2. Inadequate clearance from other structures.
3. Inadequate isolation from contact by the public.
4. Need for improving appearance of lines.
5. Need for reducing reactance of circuits.

Three important phases of the aerial-cable problem require consideration: (1) types of cable, (2) cable supports, and (3) installation methods. Both single- and multiconductor cables are in use, as well as groups of single-conductor cables, hung on a single messenger strand, operating as a multiconductor circuit. Since aerial cables should have adequate flexibility, the conductors should be stranded in all sizes down to No. 8 A.W.G. Insulation for aerial cables need be no different than that for underground cables, except in its ability to withstand flexing at the lowest temperature likely to be encountered at the location where it is installed. Underground cables only need to provide this flexibility at the time of installation, but aerial cables are subject to movement a great deal of the time because of variable wind pressure.

THE CABLE. The atmospheric conditions to which an aerial cable is subjected make the requirements for coverings somewhat different from those for underground cables. The action of sunlight, snow, ice, and rain, as well as possible chemical fumes in the air, necessitate rather exacting requirements. Lead sheath satisfies most of the requirements, although some trouble has been experienced with it because of cable flexure and also the gnawing of squirrels. A sheath consisting of spirally applied nickel-bronze tape has been used with good success because it has none of the disadvantages of lead sheath. **Neoprene** jackets have been found satisfactory where there is no possibility of severe chafing. Asbestos braids are quite permanent, but the low-melting-point asphalt impregnants used soon leach out and leave the light-colored asbestos very conspicuous. Other fibrous coverings are unsatisfactory because early rotting of the braids results in unsightly festooning. Some of the newly developed plastics show considerable promise as cable coverings. In many places **polyethylene** has been used as insulation, without outer covering.

MESSENGERS. Galvanized steel and copperweld messengers are mainly used in construction where the cable is suspended from the messenger strand. Hard-drawn copper cable is sometimes used for messengers, and recently a stainless-steel strand has been developed. A so-called self-supporting type of cable has been developed, in which the supporting strand is cabled in a layer around the insulated cable like the armor on a submarine cable. One form of this cable has copper wire for every other strand of the supporting covering. This added conductivity allows the use of the supporting covering as a return or neutral conductor. In the suspended type of cable the messenger strand is also used for this purpose, especially in overhead network cables, where a hard-drawn copper cable is used for the messenger.

INSTALLATION. For suspending the cable from the messenger a wide variety of hangers has been used. At one time the ordinary marline hangers were used with cables without metallic sheaths. Their short life resulted in their displacement by copper, zinc, or bronze strap-type hangers. Of these the bronze is undoubtedly the best but is also the most expensive. A very satisfactory type of hanger has been developed recently and is coming into general use. This is the continuous spiral type, which can be applied in the field, by means of a simple machine, or in the factory. When applied in the factory, it is possible to ship the combination of cable and messenger, already bound together, on an ordinary cable reel ready for installation as a unit. The binding may be wire or tape, the wire being copper or galvanized steel, and the tape being copper, bronze, or zinc. The choice of metal depends upon the metals used for messenger or cable sheath. The pitch of the spiral varies from 2 to 4 in., depending upon the size and type of cable. Tape is preferred to wire for cables with non-metallic coverings.

The method of attaching aerial power cable messengers to poles follows generally the same practice as has been developed by long experience for communication aerial cables. Some special messenger hangers and dead-end attachments have been developed, particularly for power cable use.

Where cables are installed separately from the messengers, the messengers are installed first and sagged to a predetermined amount calculated by the usual methods outlined on pp. 14-66 to 14-84 for calculation of sag and tension. The fully loaded condition should be considered as including a coating of the necessary thickness of ice around the messenger and cable assembly. The ice should be considered as filling all the interstices between cables and between them and the messenger. The unloaded condition should be taken as the weight of the messenger strand alone. When the messenger and cable are installed as a unit, the unloaded condition should be taken as the weight of the messenger and cable combined, without ice or wind.

When a cable or group of cables is installed separately from the messenger, it generally is pulled into place through roller sheaves hung from the messenger. At the point where the cable first reaches the messenger after leaving the reel, a special feeding sheave should be used, having multiple rollers arranged along a framework in an ascending curve. This appliance prevents any sharp bends in the cable at this point. After pulling in the cable section, the hangers should be placed on the cable and messenger, starting from one end and meanwhile taking up the slack by tension at the other end.

Self-supporting Cables. "Preassembled" or "self-supporting" cables are installed in a similar manner, but such cables are pulled in through sheaves hung from each pole. When the cables have been pulled up to proper sag, they are transferred from the sheaves to the permanent supporting attachments on the poles. For self-supporting types of cables the support attachments are specially designed saddles or dead-end clamps. For preassembled cables the messenger is clamped into the usual type of messenger bracket or clamp. In order to permit this it is necessary to separate the cable from the messenger for a short distance each side of the pole. The binding tape, therefore, must be cut and unwound for the desired distance, and the loose ends wrapped securely around the assembly. If taps into the cable are to be made, the distance within which the binding is to be removed must be great enough to allow considerable separation of the conductors themselves. Self-supporting cables must be dead-ended at tap points in order to permit separation of the live conductor and the surrounding messenger strand.

34. CONDUCTOR INSTALLATION

Stringing conductors is the final major job in erecting overhead lines. For city distribution lines the process is somewhat different from that for transmission lines, although the underlying principle is the same.

ON DISTRIBUTION LINES. In stringing conductors on lines on city streets the conductors are either coiled on pay-off reels or left on shipping reels, which are supported on reel jacks. The reels may either be mounted on the line truck or set up on the ground underneath the line. On account of the short spans the conductors may be pulled in by manila hand-lines running over the cross-arms. Power winches mounted on the line trucks are often employed for pulling heavy conductors. Secondary conductors to be run on vertical racks are often laid out along the line and lifted by hand-lines to the rack points, where they are held loosely in place behind the rack rods. Where conductors are being strung on poles already carrying energized circuits, every possible precaution should be taken to prevent injury to the linemen. Throughout the process of stringing, sagging, and tying-in, the live conductors should be covered at the poles with rubber blankets, line hose, or similar protective devices, and the linemen should be required to wear their rubber gloves and glove protectors. They should be required to have their safety belts fastened around the pole at all times when in working position. Hand-lines should be perfectly dry and clean.

After pulling the wires over the arms and securing them temporarily at the dead-end arms, they are tied in permanently at one dead-end arm. They are then pulled up at the other end to proper sag for each span. It is not feasible to measure tension in the conductors for such construction, and therefore the sag is measured by sighting across targets fixed in proper position on adjacent poles. After completion of the sagging operation the other dead end is tied in permanently, and then the wires are lifted upon the insulators at each intermediate pole and tied to them by tie wires. Long-span rural line conductors on wood poles are strung in much the same manner.

ON TRANSMISSION LINES. Conductors for long-span transmission lines are shipped in long lengths on wooden reels. These reels of wire are distributed along the line in accordance with the length of wire on each. The reels are set up on reel jacks which allow the reels to revolve when the wire is payed out. Stringing sheaves which have very free-running rollers are hung on the towers in place of the insulator strings. The conductors are placed in the sheaves on the first tower away from the reel set-up and pulled to a point beyond the next tower. The pulling is stopped until the conductors are lifted to the sheaves on the second tower, and the pulling process is then continued as before until the next reel set-up is reached. When the first set of reels are emptied, the conductors are tied temporarily to the adjacent dead-end tower. The conductors of the second set of reels are then pulled into their section of the line, and when the reels are emptied, the conductors of both sections are spliced together. Pulling of conductors is usually done with horses. Under favorable conditions tractors provide excellent power for the purpose. In mountainous country oxen have been used successfully when horses or tractors were out of the question. Locomotives have sometimes been used for lines along railroads. In pulling conductors care should be exercised not to injure them by sharp bends, faulty cable grips, or dragging them over sharp stones.

This general procedure is followed until the next dead-end tower is reached, when the entire section between dead-ends is pulled up to proper sag and tension for the prevailing temperatures. If prestretching is done in the field, the conductors should be pulled up to nearly the elastic limit before adjusting to normal sag and tension. (See Article 24, Mechanical Design.) Dynamometers are used to measure the tension in conductors for prestretching for approximate sagging. Since dynamometers are not sufficiently accurate, the final determination of correct sag and tension should be made with surveying instruments. Wind loads on conductors at the time of stringing should be taken into account if they are of sufficient magnitude to affect the tension more than a negligible amount. On long-span transmission lines "walkie-talkie" radio outfits facilitate communication between sections of the stringing crews.

Where ground wires or telephone wires are also on the tower, it is equally important that they be strung at the proper tension; otherwise they may cross and ground the conductors.

Suspension towers (i.e., those intermediate between the dead-end towers) are ordinarily not strong enough to stand the strain of dead-ended cables during high winds and heavy sleet storms; consequently care must be used if cables are temporarily dead-ended on them during construction.

After the conductors are properly sagged, the insulators are put in position and connected to the conductors. Care should be taken to connect the insulators at proper points on the conductors so that the tension will not be changed and so that suspension insulators will hang in a vertical plane. At dead-end towers the jumpers must be clamped together and bent to shape, not left so that they may ground to the tower because of twist of cable or wind pressure.

CONDUCTOR SPLICING. Conductors are spliced by the use of twisting sleeves or compression sleeves or by twisting the conductors together without sleeves. Twisting without sleeves is used only with small conductors and is being superseded by the other methods to some extent. Twisting sleeves are used quite generally for splicing conductors which are not too large to be twisted and which have less strength than the twisted splice, by a safe margin. Figure 51 shows a typical twisting sleeve and a twisted joint. Twisting sleeves are made with a figure-eight cross-section, as well as elliptical. It is claimed that

FIG. 51. Twisted Splice

greater strength is obtained with the figure-eight shape. Twisting-sleeve joints are twisted by hand wrenches, which are of various sizes, depending on the conductor size. Rachet wrenches are quite popular for this purpose.

Compression sleeves are made in various forms and for various methods of application. A common form used for steel-reinforced aluminum cable really consists of two sleeves, a steel sleeve for the steel core and an aluminum sleeve covering the whole splice. In making the splice a machine known as a compressor applies the pressure hydraulically to properly shaped dies. The aluminum sleeve is placed on the cable and slid back some distance from the joint. The aluminum strands are cut off to expose the steel core a little more than one-half the length of the steel sleeve. Both the aluminum and the steel strands are wrapped near the ends with wire to prevent spreading. Both ends of cable are dipped into red lead paint to cover slightly more than the total length of the joint. The steel sleeve is slipped over the ends of the steel cores and centered and then compressed with the machine. The steel sleeve is wrapped with aluminum wire to the diameter of the outer aluminum strands, after which the aluminum sleeve is slipped over the joint, centered, and compressed. Figure 52 shows such a joint. Item 1 of the figure is the steel sleeve, 2 the aluminum sleeve, and 3 the filler wire.

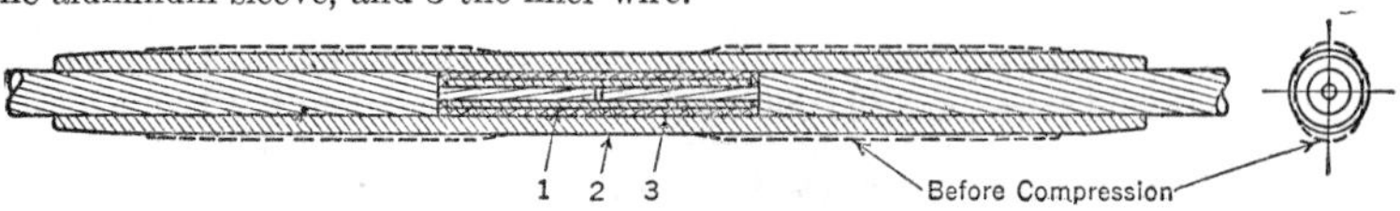

FIG. 52. Compression Splice, ACSR

Compression joints for copper conductors are quite similar, and many patented methods have been devised to make them up in the field. One method for small conductors involves a die which rolls from one end of the sleeve to the other. Another method employs stationary dies, with the pressure applied by exploding gunpowder cartridges. Hydraulic compressors are also used. In general the strength of sleeve joints depends on the length of the sleeve to give adequate bond and on its thickness to give adequate tensile strength.

TRANSPOSITION OF CONDUCTORS. Communication circuits closely paralleling a-c power circuits are susceptible to inductive interference. The interference is brought about by the linking together of the two separate systems by means of their own external electromagnetic or electrostatic fields. Interference may also arise because of the fact that the earth forms a common part of the circuits of both systems. Interference is proportional to the closeness of the two circuits, the length of parallelism or exposure, and the magnitude of the interfering voltage or current.

Certain frequencies produce more serious effects than others. In telephone circuits, frequencies from 800 to 1500 cycles per second are most troublesome from the standpoint of noise. Eleven hundred cycles is by far the worst noise frequency. Telegraph circuits are susceptible to frequencies of 60, 180, 300, and 420 cycles. The triplen harmonics and their odd multiples, together with the odd non-triplen harmonics, are those which may be present and which require attention from an interference standpoint.

Perfectly balanced symmetrical three-phase systems may have the fundamental frequency and all odd harmonics present, but there will be no single-phase or residual components of current or voltage. Unbalance in voltage between phases, currents, or capacity to ground produces a residual component. The third harmonic and its odd multiples may be produced by certain connections of transformers or by the generator wave shape.

The third harmonic and its multiples are best taken care of by suppressing them; this may be accomplished by providing a delta-connected winding across the phases. (See Transformer Connections, Section 10.) Residuals are best taken care of by repressing them either by delta-Y grounding banks or by wave traps. The inductive effects of balanced voltages and currents may be overcome by a coordinated system of transpositions of both power and communication lines.

Transposition of power lines is necessary only within the distance in which the lines are closely parallel. This distance is called the length of exposure. A transposition scheme requires that each conductor of the power line have equal coupling to the communication line. This is accomplished in three-phase lines by successively bringing each of the three conductors into the position previously occupied by one of the other conductors. At each transposition structure this replacement of conductors takes place in such a manner as to rotate the phases spirally along the line. The length of line in which each conductor occupies each position for an equal distance is called a barrel. The number of barrels necessary in a given exposure depends upon the amount of inductive coupling between circuits.

In order to neutralize the voltages induced between the two sides of the communication circuit it is also necessary to transpose the communication circuit. The two transposition schemes must, therefore, be coordinated. Figure 53 shows a typical coordinated scheme.

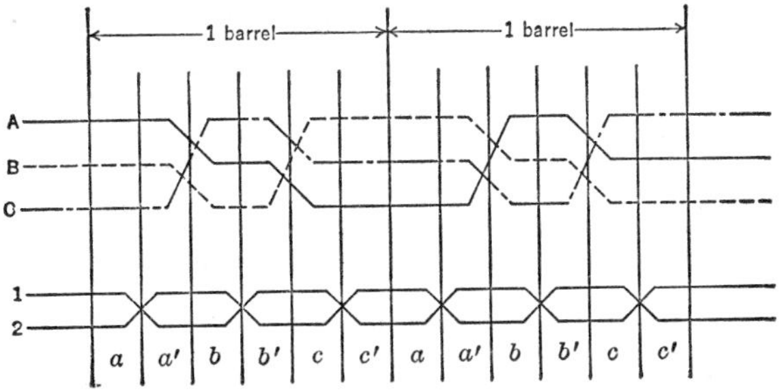

FIG. 53. Transposition Scheme

The conditions required for balance for such a coordinated scheme are shown in the following table.

Component of Voltage or Current in Power Line	Voltage Induced in Communication Circuit	Condition Required for Balance
Balanced	Between wires	$a - a' = b - b' = c - c'$
Balanced	To ground	$a + a' = b - b' = c + c'$
Residual	Between wires	$a + b' + c = a' + b + c'$

Specially designed structures are usually necessary at transposition points to provide proper clearances between conductors and between conductors and the structures. For a more complete discussion of inductive interference and remedial measures see Pender and McIlwain's *Handbook of Electric Communication and Electronics*.

35. INSPECTION OF POLE AND TOWER LINES

All parts entering into a line are usually tested and inspected before shipment; therefore, in general, the final inspection of a line before putting it into service consists in inspecting the final assembly, which is ordinarily done by an inspection crew under the supervision of an engineer. This crew gives a visual inspection of all parts to detect breakage or other defects which can be seen. Tightness of bolts, position of insulators, alignment of structures, clearances of jumpers, conductor connections, rust spots on steel, and condition of right-of-way are among the items checked. Phasing-out of the lines is usually done by the crews working on the electrical construction of the substations. After the line is in operation, periodic inspections similar to the final construction inspection are made, and the inspection crews usually carry out the necessary maintenance work.

The proper maintenance of a line includes resetting foundations that have settled; covering foundations with earth to proper depth after heavy rains; repainting towers before they are affected by rust; renewal of rusted ground cables; replacement of cracked or partially defective insulators that have not failed; and correcting sag of any cable where sag has changed as a result of stretch of cable, change of length during emergency repairs, etc.

In addition to the periodic inspections which are made 1 or 2 years apart, the lines should be patrolled every few days by regular patrolmen who should note and report the condition of foundations, poles, towers, insulators, and conductors. They should keep weeds, brush, and inflammable material away from the structures and also make minor and emergency repairs. Arrangements should be made for means of communication for patrolmen. If a private telephone line is constructed along the power-line right-of-way, telephones for patrolmen should be placed at convenient intervals. Emergency repair storerooms should also be provided at strategic points along the line.

36. BIBLIOGRAPHY

General

DeWeese, Fred, *Transmission Lines.* McGraw-Hill (1945).
Gear, H. B., and P. F. Williams, *Electric Service Distribution Systems.* D. Van Nostrand (1926).
Imlay, L. E., *Mechanical Characteristics of Transmission Lines.* Westinghouse Technical Night School Press, East Pittsburgh, Pa. (1928).
Jones, R. E., Rural Line Construction in Ontario, *J. AIEE* (July, 1930).
Kurtz, E., *The Lineman's Handbook.* McGraw-Hill (1928).
Loew, E. A., *Electric Power Transmission.* McGraw-Hill (1928).
Miller, M. C., Bare Wire in Overhead Distribution Practice, *Trans. AIEE* (July, 1929).
National Electrical Safety Code, Fifth Edition. U. S. Bureau of Standards (1941).
N.E.L.A., *Overhead Systems Reference Book.* New York (1927).
Pannell, E. V., *High Tension Line Practice.* D. Van Nostrand (1926).
Seelye, Howard P., *Electrical Distribution Engineering.* McGraw-Hill (1930).
Still, Alfred, *Electric Power Transmission.* McGraw-Hill.
Waddicor, H., *Principles of Electric Power Transmission.* John Wiley (1935).
Westinghouse Electric and Manufacturing Company, *Electrical Transmission and Distribution Reference Book* (1944).
Woodruff, L. F., *Principles of Electric Power Transmission.* John Wiley (1946).

Mechanical Design

Buchanan, W. B., Vibration Analysis—Transmission Conductors, *Elec. Engg.* (November, 1934).
Bullard, W. R., and D. H. Keyes, Joint Use of Poles with 6900-volt Lines, *Trans. AIEE* (1934).
Carlson, C. B., and Battey, W. R., Mechanical and Electrical Construction of Modern High Voltage Lines, *Trans. AIEE* (1923).

See also papers by W. Dreyer, M. T. Crawford, H. R. Wakeman, and W. H. Lines; also L. J. Corbett in *Trans. AIEE* (1923).

Davison, A. E., J. A. Ingles, and V. M. Martinoff, Vibration and Fatigue in Electrical Conductors, *Trans. AIEE* (December, 1932).
Joint Use of Poles—Telephone Circuits and 6.6- and 13.2-kv Power Circuits, E.E.I. *Publication* E-4 (1937).
Joint-Pole Practices for Supply and Communication Circuits, E.E.I. *Publication* M-12 (1945).
Ehrenburg, D. O., Transmission Line Catenary Calculations, *Trans. AIEE* (1935).
Fowle, F. F., A Study of Sleet Loads and Wind Velocities, *Elec. World*, Vol. 56, p. 995 (1910).
Greisser, V. H., Effects of Ice Loading on Transmission Lines, *Trans. AIEE* (1913).
Martin, J. S., Sag Calculations, *Proc. Engineers' Soc. Western Pa.* (November, 1922).
Monroe, R. A., and R. L. Templin, Vibration of Overhead Transmission Lines, *Trans. AIEE* (1932).
Neff, G. C., Rural Electrification, *Trans. AIEE* (1926).
Pender, Harold, and H. F. Thomson, The Mechanical and Electrical Characteristics of Transmission Lines, *Trans. AIEE* (1911).
Pipes, L. A., Cable and Damper Vibration Studies, *Trans. AIEE* (1936).
Scattergood, Engineering Features of Boulder Dam—Los Angeles Lines, *Trans. AIEE* (1936).
Seelye, H. P., and Myron Zucker, Pole Flexibility as a Factor in Line Design, *Trans. AIEE* (1937).
Stickley, G. W., Stress-strain Studies of Transmission Line Conductors, *Trans. AIEE* (December, 1932).
Stockbridge, G. H., Overcoming Vibration in Transmission Cables, *Elec. World* (Dec. 26, 1925).
Sturm, R. G., Cable and Damper Vibration Studies, *Trans. AIEE* (1936).
Thomas, Percy H., Sag Calculations for Suspended Wires, *Trans. AIEE* (1911), *Elec. World* (Nov. 16, 1912).
Thomas, Percy H., Formula for Minimum Horizontal Spacing, *Trans. AIEE* (1928).
Varney, Theodore, Notes on Vibration of Transmission Line Conductors, *Trans. AIEE* (1926).
Varney, Theodore, *A.C.S.R. Graphic Method for Sag-Tension Calculations.* Aluminum Company of America (1927).
Varney, Theodore, Vibration of Transmission Line Conductors, *Trans. AIEE* (1928).
Wright, E. M., and J. Mini, Jr., Field Tests on Conductor Vibration, *Elec. Engg.* (July, 1934).

Wood Poles and Cross-arms

American Standards Association, Specifications for Wood Poles, No. 05b1, 05c1, 05d1, 05e1; Dimensions of Wood Poles, No. 05b2, 05c2, 05d2, 05e2; Ultimate Fiber Stresses of Wood Poles, No. 05a1.
American Wood-Preservers' Association, Standard Specifications.
Bell Telephone Laboratories, *Monograph* B-615, Ultimate Fiber Stresses for Wood Poles, by R. H. Colley; B-616, Knot Sizes in Wood Poles, by R. H. Colley and R. C. Eggleston; B-617, Standard Dimensions and Uniform Classification for Wood Poles, by R. C. Eggleston and A. P. Jahn.

Betts, H. S., *Timber: Its Strength, Seasoning, and Grading.* McGraw-Hill (1919).
Brown, H. P., and A. J. Panshin, *Commercial Timbers of the United States.* McGraw-Hill (1940).
Brown, Nelson C., *Timber Products and Industries.* John Wiley (1937).
Brown, Nelson C., *Lumber.* John Wiley (1947).
Colley, R. H., Poles and Pole Treatment, *Trans. AIEE* (1942).
Garratt, George A., *Mechanical Properties of Wood.* John Wiley (1931).
Grondal, B. L., Strength Tests of Solid and Laminated Cross Arms, University of Washington Engineering Experiment Station, *Bulletin* 47.
Hansen, Howard J., *Modern Timber Design.* John Wiley (1943).
Hunt, G. M., and G. A. Garratt, *Wood Preservation.* McGraw-Hill (1938).
Koehler, Arthur, *Properties and Uses of Wood.* McGraw-Hill (1924).
United States Department of Agriculture, *Wood Handbook* (1935).
United States Forest Products Laboratory, *Bulletins* and *Circulars.*
Weiss, H. F., *The Preservation of Structural Timber.* McGraw-Hill (1916).
Young, C. R., *Elementary Structural Problems in Steel and Timber.* John Wiley (1935).

Towers

American Bridge Company, *Transmission Towers* (1925).
Scholes, D. R., Fundamental Considerations Governing the Design of Transmission Line Structures, *Trans. AIEE* (1908).
Silver, A. E., Problems of 220-kv Power Transmission, *Trans. AIEE* (1919).
Still, Alfred, Flexible Supports for Overhead Transmission Lines. *Elec. World*, Vol. 60, p. 97 (1912).
Thomas, P. H., New Transmission Line Construction—Post Type Towers, *Trans. AIEE* (July, 1930).
Walls, J. A., J. B. Leeper, W. E. Mitchell, P. M. Downing, F. C. Connery, Symposium on Foundations for Transmission Line Towers and Tower Erection, *Trans. AIEE*, Vol. 34 (1915).

Insulation

Brand, High Tension Insulator Tests: A Study of Design Factors, *Gen. Elec. Rev.*, Vol. 14 (1913).
Farr, C. C., and H. E. Philpott, Test and Investigations on Extra-high Tension Insulators, *Trans. AIEE* (1922).
Feder, Surface Leakage as a Factor in Insulator Design, *J. AIEE* (September, 1920).
Fortesque and Fransworth, Air as an Insulator When in the Presence of Insulating Bodies of Higher Specific Inductive Capacity, *AIEE*, Vol. 33, p. 893.
Gilchrist and Klinefelter, Applications of Theory and Practice to the Design of Transmission Line Insulators, *Elec. J.*, Vol. 16, p. 8; Experimental Investigation of Porcelain Mixes, Vol. 15, p. 77; Design of Porcelain Insulators from a Theoretical Standpoint, Vol. 15, p. 443; Design of Porcelain Insulators from a Ceramic Standpoint, Vol. 15, p. 489; A Comparison of Different Methods of Testing Electrical Porcelain, Vol. 12, p. 282.
Gillam, G. H., Stabilized Insulators. An Account of the Properties and Applications of Semi-conducting Ceramic Glazes, *Elect. Times*, Vol. 112, p. 289 (1947).
Hawley, K. A., High Voltage Insulator Development, *Elec. World* (1925).
Hawley, K. A., Development of the Porcelain Insulator, *Trans. AIEE* (1931).
Iler, G. A., Locating Defective Insulators, *Elec. World* (1925).
Lapp, G. W., Overpotential Test for Insulators, *Trans. AIEE* (1922).
Lloyd, W. L., Jr., J. E. Clem, V. M. Montsinger, Coordination of Insulation, *Trans. AIEE* (June, 1933).
McEachron, K. B., I. W. Gross, H. L. Melvin, A. M. Opsahl, J. J. Torok, Expulsion Protective Gaps, *Trans. AIEE* (December, 1933).
Melvin, H. L., Impulse Insulation Characteristics of Wood Pole Lines, *Trans. AIEE* (January, 1930).
Peaslee, Factors Controlling the Design and Selection of Suspension Insulators, *J. AIEE*, Vol. 76 (June, 1920).
Peek, F. W., Jr., *Dielectric Phenomena in High Voltage Engineering.* McGraw-Hill (1929).
Ryan, Unit Voltage Duties in Long Suspension Insulators, *J. AIEE* (July, 1920).
Ryan, Ceramics in Relation to Durability of Porcelain Suspension Insulators, *AIEE*, Vol. 35, p. 1437.
Smith, H. B., Development of a Suspension-Type Insulator, *Trans. AIEE* (1924).
Sothman, Comparative Tests on High Tension Insulators, *AIEE*, Vol. 31, p. 2169.
Sporn, Philip, Rationalization of Transmission System Insulation Strength, *Trans. AIEE* (1928).
Sporn, P., E. L. Peterson, and V. A. Mulford, Selection and Performance of Suspension Insulators, *Elec. Engg.* (June, 1934).
Torok, J. J., and W. Ramberg, Impulse Flashover of Insulators, *J. AIEE* (December, 1928).
Torok, J. J., Surge Characteristics of Insulators and Gaps, *Trans. AIEE* (July, 1930).

Line Erection

Elliott, V. D., Location and Right-of-way, *Trans. AIEE* (1924).
Idail, M. J., Stringing Suspension Insulator Lines in Rough Country, *Elec. World* (April 9, 1921).
Pendleton, T. P., Map Compilation from Aerial Photographs, *U. S. Geol. Surv. Bull.* 788-F.
Woodruff, W. W., and G. I. Wright, Utilization of R. R. Rights-of-way for Electric Power Transmission, *Trans. AIEE* (March, 1931).

Inductive Coordination

Barnett, S. J., A Report on Electromagnetic Induction, *Trans. AIEE* (1919).
Trueblood and Cone, Power Distribution and Telephone Circuits—Inductive and Physical Relations, *Trans. AIEE* (1925).

UNDERGROUND CONDUITS

By Robert E. Morse

37. TERMINOLOGY

The following definitions are generally accepted in the industry.

Duct. A pipe or tube designed or used for the accommodation of electric wire or cable.

Service Duct. A duct entering a consumer's premises.

Duct Bank. A number of ducts joined together in a group.

Duct Bell. A flared termination for a duct.

Manhole. A chamber in which one or more duct banks terminate, thus giving access to the cables contained therein.

Service Box or **Distribution Manhole.** A small manhole placed in a duct bank in which not necessarily all ducts in the bank are exposed and in which service ducts terminate.

Transformer Vault or **Transformer Manhole.** A large manhole designed to accommodate one or more transformers and associated apparatus.

Handhole. A small pull box cut into a service duct.

Manhole Frame. A casting surrounding the entrance to a manhole and containing a seat for a manhole cover.

Manhole Cover. A removable casting fitting the seat in the manhole frame and closing the entrance to a manhole.

Conduit. Usually synonymous with duct but, especially in the plural, sometimes used to designate an electrical subway system consisting of duct banks, manholes, service boxes, and transformer vaults.

38. USE OF CONDUITS

Any conduit system designed for electric power cables must have ducts ample in size for the accommodation of any cables to be placed therein, must have manholes so designed that cables can be properly racked both for splicing and for taking up expansion and contraction, and also so that ample room is allowed for working on them, and must fit the requirements of present circuits and those that may reasonably be expected for many years in the future.

TRANSMISSION UNDERGROUND CONDUITS. Manhole sections may be long, up to 600 ft or more, but the limiting feature in manhole-section lengths is pulling stresses on the cables as they are drawn in. Economic factors also enter into consideration, as long manhole sections require the keeping in stock of long and expensive replacement lengths.

DISTRIBUTION UNDERGROUND CONDUITS. These conduits are primarily designed for distribution primary and secondary cables, and section lengths are determined principally by distribution requirements. These systems also include service boxes and transformer vaults. It is often desirable that such conduit systems be so designed that they may be used at some future date for transmission cables, at least at moderate voltages (up to 44 kv).

SERVICE CONDUITS. These are ducts leading either from a manhole or service box of a distribution underground conduit system into customers' premises, or to street-lighting fixtures. They often follow the route of the main duct bank, turning into their destination at the nearest point to it.

39. TYPES OF DUCTS

Ducts made of fiber, tile, asbestos cement, iron, concrete, wood, and soapstone are available.

Of these materials fiber duct is the most widely used by the power industry on account of its lightness, moderate cost, and ease of handling. It is always set in a monolithic bank of concrete. Single tile and asbestos cement (transite), both of which are heavier, follow in order. Tile and the heavier grades of asbestos cement (transite) do not always require a concrete envelope, however. Iron is often used for street light and service laterals without a concrete envelope.

FIBER DUCT. Fiber duct is made of wood pulp formed on a mandrel, dried in an oven, and then impregnated under pressure with hot asphalt. Ends are machined either with a taper for a sleeve coupling ("Harrington joint") or for a socket joint. The former type is the more commonly used. Table 1 shows this coupling, together with the standard sizes of conduit. Conduit made for socket-joint connection has a tenon machined on one end of the length and a mortise in the other, into which the tenon fits. This type of coupling cannot be used readily on short lengths in the field and cannot be made as watertight as the sleeve type. Factory-made bends are available in fiber conduit of various radii and degrees of curvature, as shown in Table 2. Special bends can be made on special order.

Table 1. Dimensions of Harrington Joint Fiber Conduit and Couplings

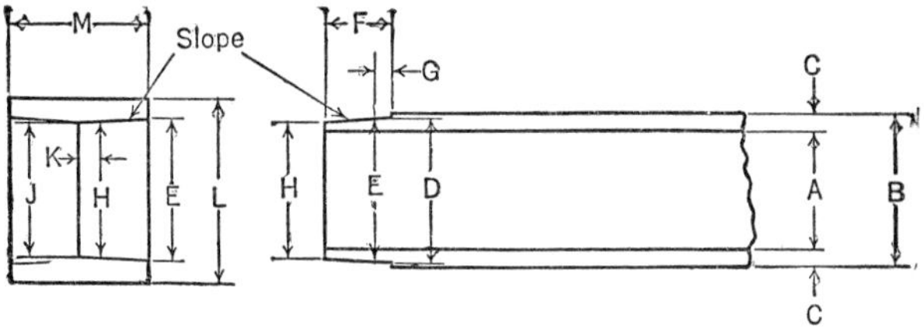

Harrington Joint Conduit and Coupling

	Size of Conduit, Inches							
	1 1/2	2	2 1/2	3	3 1/2	4	4 1/2	5
	Dimensions, Inches							
A	1.594	2.063	2.563	3.030	3.563	4.046	4.602	5.138
B	2.094	2.563	3.063	3.530	4.063	4.566	5.142	5.738
C	0.250	0.250	0.250	0.250	0.250	0.260	0.270	0.300
D	2.012	2.490	3.000	3.468	4.015	4.513	5.092	5.688
E	1.992	2.470	2.980	3.448	3.995	4.493	5.072	5.668
F	1.313	1.438	1.438	1.688	1.688	1.938	1.938	1.938
G	0.287	0.287	0.287	0.287	0.287	0.287	0.287	0.287
H	1.920	2.390	2.900	3.350	3.897	4.378	4.957	5.553
J	1.896	2.365	2.875	3.326	3.873	4.353	4.932	5.528
K	0.350	0.350	0.350	0.350	0.350	0.350	0.350	0.350
L	2.469	2.969	3.469	3.969	4.531	5.000	5.656	6.371
M	2.750	3.000	3.000	3.500	3.500	4.000	4.000	4.000
	Slope, Degrees and Minutes							
	2°-0″	2°-0″	2°-0″	2°-0″	2°-0″	2°-0″	2°-0″	2°-0″

Table 2. Dimension of Standard 90-deg and 45-deg Bends Fiber Conduit

Dimensions, Inches

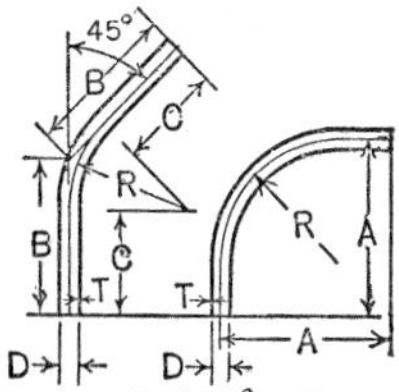

Standard 90° and 45° Bends

Size of Conduit, Inches	Thickness	For 18-in. Radius			For 24-in. Radius			For 36-in Radius		
		A	B	C	A	B	C	A	B	C
1 1/2	0.25	34	31	23 1/2	35	31	21	38	31	16
2	0.25	34	31	23 1/2	35	31	21	38	31	16
2 1/2	0.25	....	....	......	35	31	21	38	31	16
3	0.25	....	....	......	35	31	21	38	31	16
3 1/2	0.25	....	....	......	....	....	....	38	31	16
4	0.26	....	....	......	....	....	....	38	31	16
4 1/2	0.27	....	....	......	....	....	....	38	31	16
5	0.30	....	....	......	....	....	....	38	31	16

TILE DUCT. Tile duct is available as either single ducts or multiple ducts consisting of two to nine raceways in each piece. The multiple type is seldom used now by the power industry.

Single tile duct consists of a single clay tube with either a round or a square bore. Round-bore ducts are octagonal in outside shape. Single lengths are usually 18 in. long and are made with a mortise about 1 1/2 in. long on one end and a tenon on the other. Square duct, which is also square outside, is laid with dowels set in a hole provided in each corner of the end, so that the square bore will always line up. Since it is impossible to cut and fit this type of duct in the field, many intermediate lengths, together with short mitered sections, are available for making sweeps. Bent sections of various radii from 12 to 96 in. are also provided. Standard sizes and shapes of single tile duct are given in Table 3.

Table 3. Standard Sizes and Shapes of Single Tile Duct

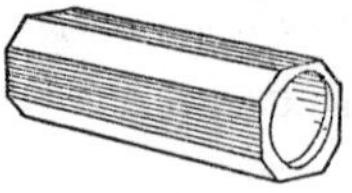

Standard Single Duct Round Bore

3 1/4″ bore—18″ long
3 1/2″ bore—18″ long
4″ bore—18″ long
4 1/4″ bore—18″ long
5 1/4″ bore—24″ long
(also in shorts)

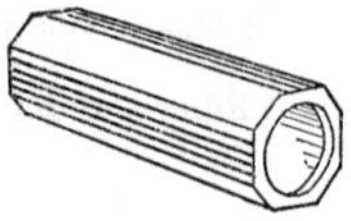

N. Y. T. Single Duct Round Bore

3 1/2″ bore—18″ long
(also in shorts)

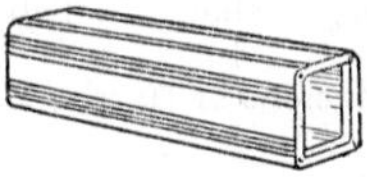

Standard Single Duct Square Bore with Dowel Holes

3 1/4″ bore—18″ long
4 1/4″ bore—18″ long
(also in shorts)

Standard Single Duct Square Bore without Dowel Holes

3 1/2″ square bore—18″ long (also in shorts)

Nominal Bore, Inches	No. of Dowel Holes	Standard Length, Inches	Duct Feet per Piece	Actual Size of Duct Hole, Inches	Approx. Outside Dimension, Inches
3 1/4 round	0	18	1 1/2	3 3/8	4 1/2 × 4 1/2
3 1/2 round	0	18	1 1/2	3 5/8	4 7/8 × 4 7/8
3 1/2 round N.Y.T.	0	18	1 1/2	3 3/4	5 × 5
4 round	0	18	1 1/2	4 1/8	5 3/8 × 5 3/8
4 1/4 round	0	18	1 1/2	4 3/8	5 5/8 × 5 5/8
5 1/4 round	0	24	2	5 3/8	6 7/8 × 6 7/8
3 1/4 square	4	18	1 1/2	3 3/8	4 3/4 × 4 3/4
3 1/2 square	0	18	1 1/2	3 5/8	5 × 5
4 1/4 square	4	18	1 1/2	4 3/8	5 7/8 × 5 7/8

NOTE. Short lengths of 3, 4, 6, 9, 12 and 15 in. are available for each of the above shapes.

ASBESTOS-CEMENT DUCT. Asbestos cement (transite) is made from a mixture of Portland cement and asbestos fiber formed on a mandrel under pressure and cured in a controlled atmosphere. It is made to the same dimensions as fiber duct and uses the same (Harrington joint) type of coupling. The material is fireproof, mechanically strong, and inert to most chemicals. Several weights, of which two are particularly used for underground conduits, are available. One has a thin wall for installation in a concrete envelope, and the other a heavier wall for burial directly in the ground. The latter can also be run exposed and is often used as riser conduit and other exposed runs, either horizontal or vertical. Either type can be cut and machined with special tools in the field. Bends are available, just as in fiber duct. Standard sizes are shown in Table 4.

IRON CONDUITS. Iron conduits used as part of an underground conduit system are usually standard galvanized steel conduit with threaded couplings, laid directly in earth. When used as service laterals, they are often pushed under sidewalks into buildings by means of a hydraulic or pneumatic pipe pusher.

CONCRETE CONDUITS. Concrete conduits are square outside with mortise and tenon joints and are made of a sand and cement mixture, sometimes with the addition of asbestos fibers for a binder. The mixture is assembled on a mandrel and, in one process, placed in a mold and rotated at high speed to compress the material by centrifugal force. After curing, the resulting conduit is mechanically strong, has a smooth, round bore, and can be laid directly in earth. It is very heavy, however, and has so far found only a limited field for use.

PUMP LOG. Wood conduit or pump log usually comes in 5-ft lengths with mortise and tenon joints. It is ordinarily made from southern pine, is always square on the outside with a round bore, and is thoroughly impregnated with creosote oil. It is always laid directly in the ground without an envelope. Its use in the power field has been limited largely to service laterals, for although it is inexpensive and easy to install, it is very vulnerable to the effects of an electric arc.

SOAPSTONE DUCT. Soapstone duct is made by an extrusion process from a mixture of powdered soapstone, clay, and Portland cement. In order to attain a sufficient mechanical strength the walls have to be made very thick, so that the finished product is quite heavy. Its principal advantage is that the coefficient of friction between it and cables being drawn into it is quite low in comparison to other types of conduit, and pulling stresses are therefore reduced. The soapstone duct has found only limited acceptance.

Table 4. Dimensions of Standard and Thin-wall Transite Conduit

Size (I. D.), Inches	Wall * Thickness, Inches	Standard † Length, Feet	Weight,* Pounds per Feet
Standard Conduit			
2	0.35	5	2.0
3	0.37	10	3.0
4	0.37	10	4.0
5	0.40	10	5.5
6	0.40	10	6.5
Thin-wall Conduit			
2	0.25	5	1.5
3	0.27	10	2.1
4	0.27	10	2.6
4 1/2	0.27	10	3.0
5	0.30	10	3.5
6	0.30	10	4.8

Standard Conduit Tapered for Harrington Coupling

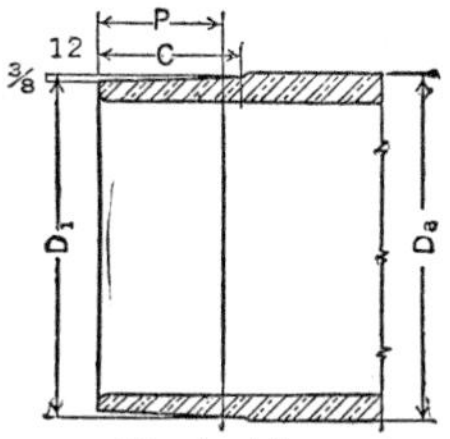

Standard Taper

Size (I. D.), Inches	D_1	D_3	C	P
2	2.58	2.70	1.75	1.50
3	3.62	3.74	1.75	1.50
4	4.62	4.74	2.00	1.75
5	5.68	5.80	2.00	1.75
6	6.68	6.80	2.00	1.75

Thin-wall Conduit Tapered for Harrington Coupling

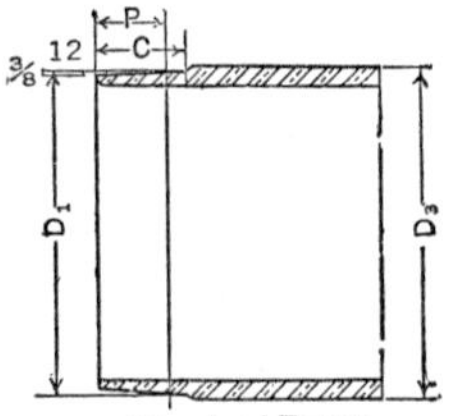

Standard Taper

Size (I. D.), Inches	D_1	D_3	C	P
2	2.36	2.50	1.00	0.75
3	3.39	3.54	1.13	0.88
4	4.39	4.54	1.25	1.00
4 1/2	4.89	5.04	1.25	1.00
5	5.45	5.60	1.38	1.13
6	6.46	6.60	1.50	1.25

* Tolerances for wall thickness and weight are as follows: Wall thickness, ±0.05-in.; weight, ±15 per cent. Weights shown include one Harrington coupling per length.

† Right is reserved to ship lengths shorter than standard, not to exceed 15 per cent of the order. All such short lengths will be in multiples of 6 in. and no length will be shorter than 3 ft.

40. MANHOLES

Manholes serve a dual purpose. They are primarily points in an underground conduit system where all conduits in the system at that point are exposed, so that cables may be drawn into them and spliced after installation. They also serve as expansion and contraction points for the cables passing through them, a function which must never be lost sight of during the initial design and the subsequent racking of cables in the manhole itself.

STRUCTURES. Manhole structures are usually boxes of various shapes with reinforced-concrete or brick walls and a reinforced-concrete slab roof. Until 1925 or 1926 manholes were usually built with brick walls and often a brick arch roof resting on railroad iron spanning the manhole walls. This type of structure is not recommended for modern construction to withstand present-day highway loads. Modern highway bridges are designed for wheel loads of up to 20,000 lb with a 50 per cent additional allowance for impact, and manhole walls and roofs must be equally strong. This requirement necessitates reinforced walls at least 8 in. thick and a heavily reinforced roof slab at least that thick for an average manhole. A minimum clear height of 6 1/2 ft between floor and roof should be maintained.

Manholes in general should be rectangular in shape. Short, straight walls or curved walls should be avoided, as they hinder splicing. A typical design is given in Fig. 1.

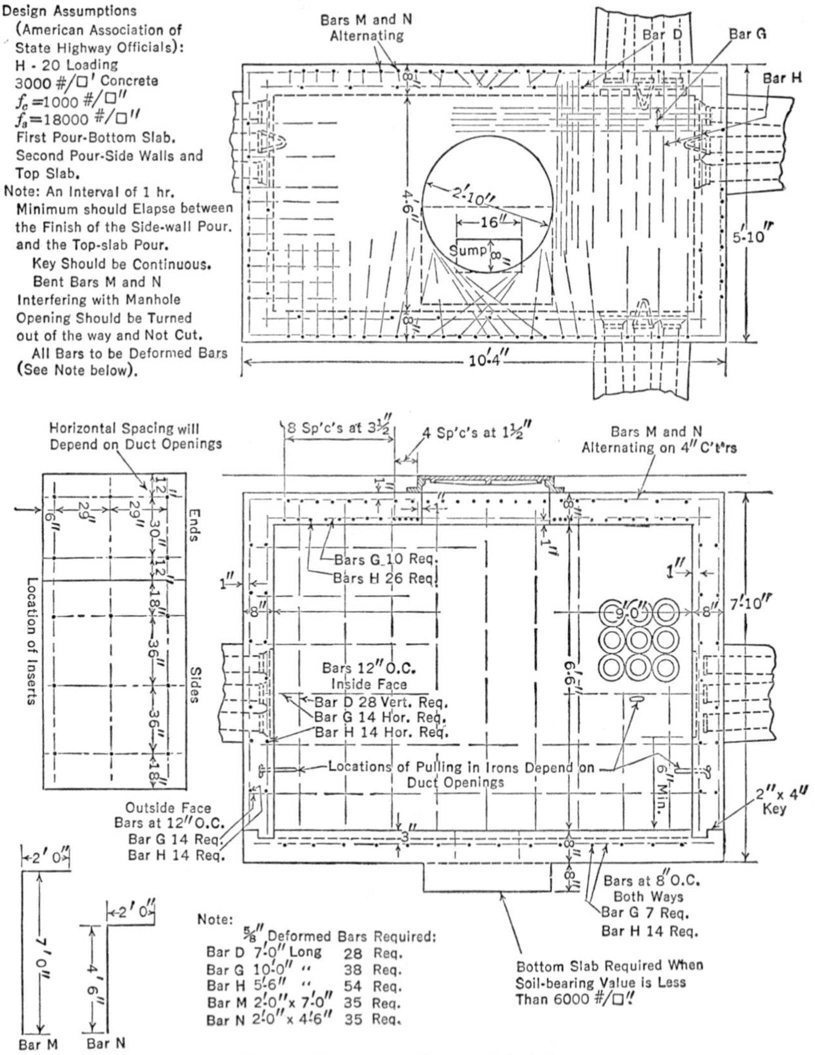

Fig. 1. Rectangular Concrete Manhole

DUCT ENTRANCES. Duct entrances into a manhole should be so located that sharp bends of cable, at the duct mouth especially, but also anywhere else in the manhole, will be unnecessary. For this reason they should be placed near one edge of the wall in which they terminate, so that there will be either no bend in cables emerging from ducts or plenty of room for a reverse bend before the cables straighten out on the wall on which they are to be racked.

Ducts terminating in manholes, service boxes, and transformer vaults should be provided with a flared end flush with the manhole wall. Space for this is attained by spreading the ducts in the last 10 ft as they approach the manhole. Manufactured flared ends, known as duct bells, are available in fiber, transite, or glazed porcelain. Often molds are placed in the concrete form so that the flare, either straight or rounded, is cast in the concrete. These flares permit the bending of the cable, if it is necessary on entering a manhole, to begin several inches back from the manhole wall, thus saving space, and they also permit fireproofing, if used, to be carried well beyond the manhole wall into the duct. Additional space for cable bends at duct entrances is often obtained by recessing the whole window 3 to 6 in. in the wall. Glazed porcelain duct bells and, to a less extent, fiber or transite duct bells provide a smooth, rounded surface over which cable sheaths can ride during expansion and contraction without scoring of the lead. Some engineers believe that duct bells at duct mouths make cable shields at these points unnecessary.

LENGTH OF MANHOLES. Manhole walls must be long enough to permit the splicing of any cable which may ever be installed in the manhole. The largest cable which can be drawn into a 4 in. duct, the usual size used in distribution conduits, is slightly over 3 in. in diameter. Such a cable, made for the highest voltage practicable for installation in distribution conduits, requires a sleeve about 36 in. long. A wall 8 ft long is the minimum on which such a cable can be conveniently spliced, and hence ordinary junction manholes from which duct lines go out in three or more directions should have two and preferably four walls at least 8 ft long. Fewer than this number will limit the size and voltage of cables which can be run through the manhole.

The cover must be located so that no cables will be directly under the opening after they are racked and also so that cables can be fed directly into any duct with a minimum of bending during pulling.

FLOORS. In good firm soil providing natural drainage into it of water which may accumulate in a manhole, the floor may be of earth, firmly tamped. Where the bottom of the manhole is below the natural water table or the earth is of such a nature that it will not support the manhole structure on the wall footings alone, a concrete floor must be provided. This may require steel reinforcement as well. The floor slab should be provided with a sump located near, but not necessarily directly under, the manhole opening. The surface should be floated when the slab is poured and drained from all directions toward this sump. The sump should be at least large and deep enough to submerge the foot valve of a suction hose below the surface of the manhole floor. It is sometimes possible to connect this sump directly to a storm sewer through a trap. These connections, however, tend to become clogged and useless after a few years.

WATERPROOFING. In very wet soil it is desirable that manhole and vault floors, walls, and roofs be made as impervious to water seepage as possible by the use of any of the recognized means of waterproofing concrete, such as integral admixture. Concrete made from high early strength cement will often give concrete dense enough for this purpose and in addition reduce materially the time when traffic can be permitted over the structure.

PULLING EYES AND INSERTS. Pulling eyes (Fig. 2) are heavy galvanized steel loops, which should be set in concrete manhole walls opposite every duct bank entrance. They provide an eye of great strength for fastening a snatch block, through which the pulling rope passes when cables are being pulled. They should also be installed in the roofs of transformer vaults directly over the contemplated location of transformers or network protectors to aid in placing or servicing equipment.

FIG. 2. Pulling Eye

Other inserts, by means of which cable racks or apparatus can be bolted directly to the manhole wall, should be provided. Many types of these inserts, designed for different-sized bolts, are available. Any type is simple to place, by boring a hole in the inside form at the point desired and fastening the insert to the form by means of a bolt through the hole. The bolt is removed when the form is stripped, leaving the insert ready for use. Inserts not immediately used should be well greased with heavy grease, and a bolt run into the threads to protect them from corrosion.

CHIMNEYS. Manhole roofs are seldom near enough to the surface so that the manhole frame can be set directly on the roof and be just flush with the finished paving. In order to provide a seat for the frame at the proper distance from the paving surface, a brick chimney is built around the opening in the manhole roof. This chimney should have walls at least 9 in. thick laid in cement mortar. Chimneys over 42 in. in length should be of reinforced concrete.

MANHOLES FOR HIGH-VOLTAGE CABLES. Manholes for conduit systems designed for high-voltage transmission cables, 66 kv and above, have to be extra long to accommodate the splices and apparatus associated with them that are necessary with this type of cable. These manholes usually require a minimum length of 10 ft. Essentially they are the same as manholes built for lower-voltage cables, except in size. Conduit systems intended for these high-voltage cables should not be used at the same time for cables of any other class.

SERVICE BOXES. Service boxes are usually placed above the duct structure, and only the ducts in the top layer of the duct bank are exposed in them. The remaining ducts, if deep enough, are carried underneath the floor, which need be no more than 4 ft deep, or even less. Only secondary, street-lighting, and traffic-control cables are ever exposed in these boxes. Joints on these cables take little room and are not difficult to make, provided the cover of the box has a large enough opening for a man to work in the box without being restricted by the roof. The covers usually used on junction manholes are large enough for this purpose. Thus it is unnecessary to keep a special design of manhole ring and cover for service boxes in stock, provided the boxes are of such shape that they can accommodate the standard cover. These considerations dictate a box about 4 ft square and 3 to 4 ft deep inside. Sometimes it is desirable to offset a service box to one side of the main duct run in order to find space for a standard unit. This can easily be done by diverting the top layer of ducts to one side into the box and back again. A sketch of a typical service box is given in Fig. 3.

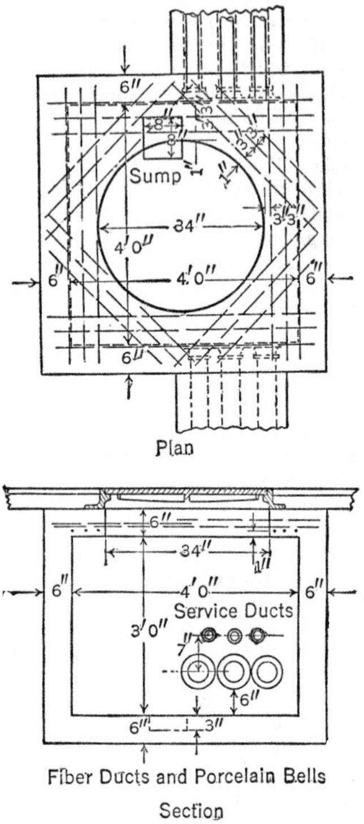

FIG. 3. Service Box

TRANSFORMER VAULTS. Transformer vaults are manholes specially designed to accommodate transformers and associated equipment. Usually they can be designed for some latitude in the choice of equipment they are to accommodate, but in any given design there is always a maximum in the total kva of transformer capacity which it will accommodate. This maximum depends primarily on the ventilation provided, for although some of the heat developed in the transformers is radiated to the walls and thence to the surrounding earth, most of it is carried away by the air in the vault, and plenty of area should be provided for the escape of heated air and the entry of cooler air from outside. The more ventilating area provided, the less volume of air in the vault around the transformers is required. Ordinarily a vault should be at least large enough so that a man can inspect equipment in it and have room enough to make cable connections to transformers and associated equipment, such as network protectors, primary cut-outs, and oil switches.

The equipment which is installed in transformer vaults is necessarily large, and openings in the roof for the entry of this apparatus must be provided. Once installed, the equipment usually need not be disturbed for many years. For this reason vault roofs are often built with the roof containing one or more removable sections, which, being replaced after the setting of transformers and other large apparatus inside the vault, are paved over with street or sidewalk paving. Entrance for personnel into the vault is obtained through an ordinary manhole cover conveniently placed in a permanent part of the roof. The same result can be obtained by means of a large cast-iron frame set over the manhole opening, which is correspondingly large, with two covers set in this frame, the larger fitting the frame, and the smaller fitting an opening in the larger. The larger cover may or may not be paved over. This type of vault opening can be used only in sidewalks, as it is impracticable to design such a cover for the loads imposed by street traffic. The same considerations govern vault design as govern manhole design, but concrete floors should always be provided to support the concentrated load of transformers.

Ventilation. Ventilation of transformer vaults is usually provided by means of openings in opposite walls of the vault, one near the floor, and the other near the roof. These openings lead to a vertical shaft reaching to the surface and covered by a grating. Theoretically cool air is drawn onto the floor of the vault through the lower opening and discharged through the upper after picking up heat from flowing over the transformer cases. Actually almost as good results can be obtained from two large grated openings in the roof, as cool air will flow down and warm air up through the same grating. A minimum figure of 2 sq in. per kva of maximum transformer capacity in a vault is suggested as an adequate area for the sum of all ventilating openings and ducts, in and out, of the vault.

It is often desirable to include special inserts in the walls of transformer vaults for the attachment of associated equipment, such as network protectors, primary oil switches, or

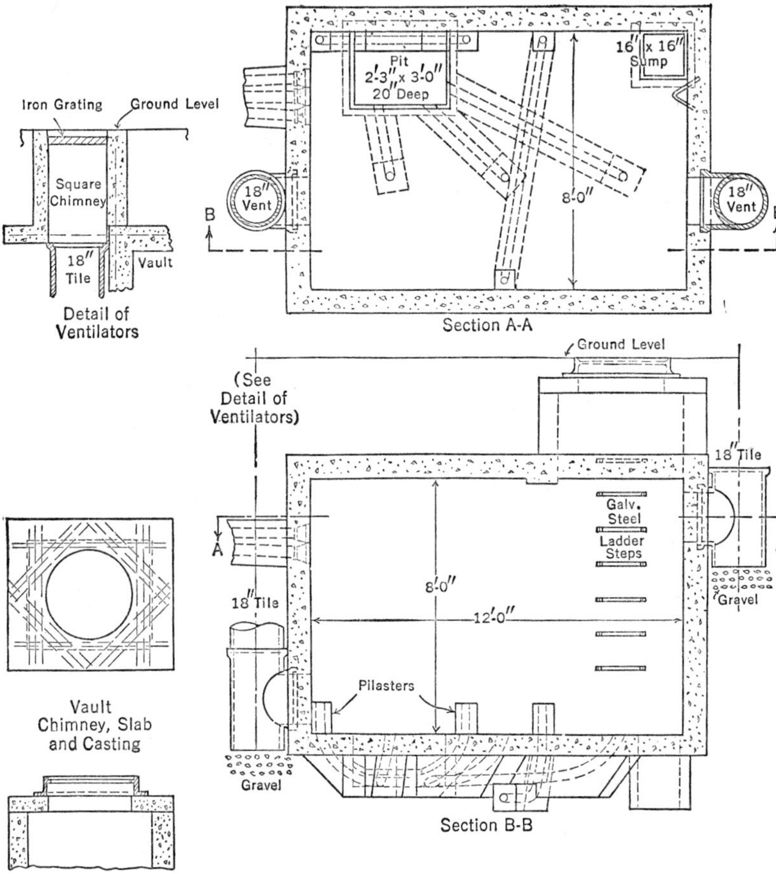

FIG. 4. Transformer Vault

cut-outs. It may also be advantageous to install ducts in the floor for the accommodation of primary cables between an oil switch or cut-out and the transformers, or for secondary cables between the transformers and a network protector.

A sketch of a transformer vault designed to accommodate a maximum of 450 kva in transformers is shown in Fig. 4. For a three-phase transformer part of the roof slab would have to be made removable. Floor ducts are shown for primary and secondary cables.

MANHOLE FRAMES AND COVERS. The entrance to a manhole, service box, or vault is through a removable cover set in a frame imbedded in the street or sidewalk paving and flush with it. The frame must be large enough to permit the easy passage of a man's body, in addition to a ladder set at a safe angle, and also large enough to permit the setting up of pulling rigging sometimes required. The cover itself, while strong enough to carry traffic, must also be light enough for one man to "pull" it. A manhole frame and cover with a clear opening 32 in. in diameter can be designed to meet all these requirements. Most subway equipment, even including single-phase transformers up to 200 kva, can also be designed to go through such an opening. Circular covers should be used, because they cannot slip through the opening under any circumstances and because the seats can be accurately machined on both frame and cover, so that there is uniform bearing all around and rocking is avoided.

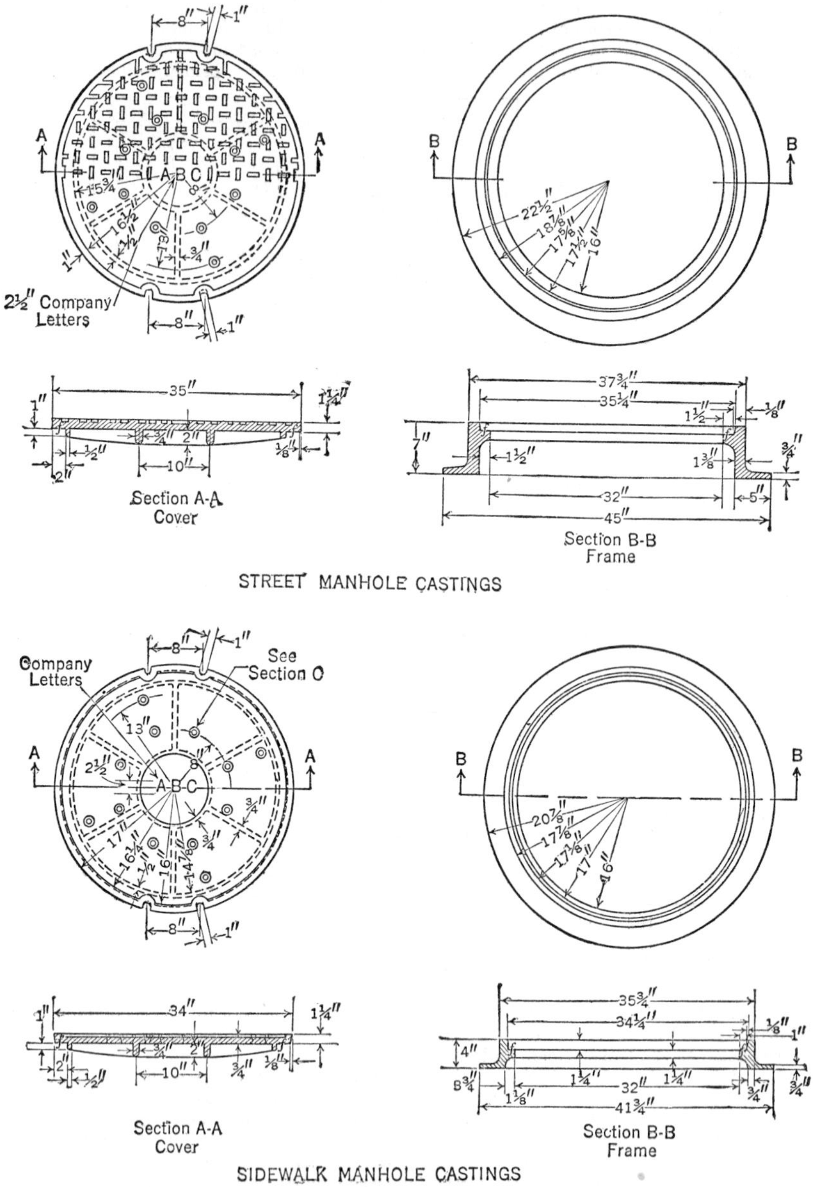

Fig. 5

Openings in the cover should be provided for ventilation and also for the insertion of tools for lifting the cover but should be small enough so that pedestrians' heels will not catch in them. A ratio of 1 sq in. of ventilating holes for every 30 sq in. of cover area is suggested as adequate without weakening the cover.

Inner covers designed to prevent the entrance of surface water into manholes are not recommended. They interfere with ventilation and permit pressures from an explosion due to the ignition of an accumulation of explosive gases in a manhole to rise to destructive proportions. A single cover not locked down will lift and relieve these pressures before they reach such heights.

The surface of a cover intended for use in the street is usually deeply indented to proside traction for wheels passing over it. This type of surface is unsuitable for sidewalk vue, as cast iron wears smooth and becomes slippery. It is customary to use a lighter frame and cover for sidewalk manhole openings and sometimes to provide means of coating the surface of the cover with a concrete-like material which is practically the same to walk on as granolithic sidewalk. Typical designs of manhole frames and covers for both street and sidewalk use are shown in Fig. 5.

HANDHOLES. These are small pull boxes cut into a service duct. They consist of a concrete box with an earth floor and a removable round or rectangular cover over the entire box. They have a total area of seldom more than 1 1/2 to 2 sq ft and a depth of only a few inches, so that work on cables in them can be done from the surface. They are very convenient for cutting an existing underground service originating on a pole over to a new underground system. A handhole is built over the existing service pipe, which is cut off at the inside edges of the handhole. A service duct is run from the handhole to the nearest service box or manhole. At the proper time the service can thus be transferred to the underground system with a minimum of interruption to the customer. Handholes can often be precast and installed complete and ready for use as soon as the excavation has been made.

41. DESIGN OF CONDUIT SYSTEMS

GENERAL FEATURES. As it must be determined in advance just what provision, if any, must be made for future requirements, a study of business trends in the area under consideration should be made, so that further loads can be assumed with a minimum of additional conduit construction. A study of soil conditions should be made, and existing underground structures located as accurately as possible.

MAPS. Maps on a large scale, approximately 20 ft to 1 in., should be drawn, showing the location and dimensions of all other underground structures and utilities, together with the depth to the top and bottom of each. Records of these structures are often nonexistent, but a great deal of information can be obtained from a careful survey of the streets, noting all evidences that appear on the surface, such as manhole covers, water valves and hydrants, catch basins, and gas drips. Electric utilities are usually the last to go underground in any given area, and consequently it is often difficult to find convenient locations for the structures proposed.

The proposed construction is drawn to scale on these maps. If a distribution underground system is being laid out, these maps should also show every building in the area and the location of the service which will be brought into it. Service ducts are also drawn on the maps. After construction is completed, these maps, suitably corrected, will constitute a permanent record of the installation.

A careful load survey should be made, and transformer vaults located. Because of the cost of these vaults, transformers should be of fairly large rating so as to limit the number of vaults required.

MANHOLES. Manholes are usually located at street intersections, but they are also required wherever it is necessary to have access to primary cables, such as at transformer vaults. Transformers should never be located in line manholes but always in a separate vault placed as close as possible to a line manhole and containing nothing but transformers and associated equipment. It is usually not advisable to distribute services from transformer vaults, but one or two large services may be taken directly out of a vault in special cases.

SERVICE BOXES. Service boxes will be located between manholes as may be necessary, depending on the number of services to be provided for. One service box should not be required to provide for more than eight services, as a rule, and service ducts should be limited to a maximum length of about 125 ft. These ducts usually contain at least one right-angle bend, which increases the difficulty of pulling service cables, and, as the cost of these boxes is comparatively small, it is better to space them at fairly close intervals.

DESIGN OF DUCT STRUCTURES. Conduits are placed underground solely for the accommodation of electric cables, and every feature of design must be considered primarily for its effect on them.

Mechanical damage is caused by pulling stresses imposed on cables during installation because of too long sections between manholes or frequent and sharp bends in the conduit or both, friction at duct mouths due to rubbing during expansion and contraction, too sharp bends in manholes, crowding of cables in manholes as a result of restricted space for racking, and duct of too small a bore for the cable. Although it is possible to pull a cable through a straight duct only a little smaller in outside diameter than the bore of the duct, this does not permit "snaking" of the cable in the duct to take up expansion and contraction, which must then all be taken up in the manholes.

Chemical effects are those which cause corrosion of lead sheaths or destruction of nonmetallic cable coverings. Incompletely cured concrete will corrode lead which is in contact with it. For this reason it is advisable to delay the installation of lead-covered cables in a new duct structure for at least 6 weeks after the concrete has been poured. Some types of cement will continue to corrode lead, so that this matter must be considered in the choice of cement.

GRADING. The grading must be such that there is no possibility of water standing anywhere in the duct line if the manhole at each end is dry. For this reason the ducts should slope, either from one manhole continuously to the next, or from a high point between them down to each manhole. A slope as low as 1 in. per 100 ft will prevent water from standing in any part of the duct if the remainder is dry. For purposes of identification each duct in a continuous section between two manholes should terminate in each manhole in the same relative location.

CONFIGURATION FOR HEAT DISSIPATION. In order to facilitate dissipation of heat, as many ducts as possible must be located on the periphery of the duct bank next to the earth. A duct bank in which the ducts are arranged two high or wide, as the case may be, fulfills this condition for all ducts, but it is also the most expensive, increasingly so as the number of ducts rises beyond eight.

It has frequently been found that the removal of cables from interior ducts has actually increased the power-carrying capacity of the whole duct bank because of the fact that the remaining cables can carry more load.

SEPARATION. The wall between ducts should be heavy enough to withstand the stresses set up during a cable fault, so that damage will not be communicated to unfaulted cables, and should also be thick enough to retard transmission of heat from one cable to the other. These considerations have led to the general abandonment by the electric-power industry of multiple tile duct. For single tile ducts this separation need not be great, as the two duct walls fulfill most of this requirement. For fiber and thin-wall transite a separation of 1 1/2 to 2 in. of concrete is usually specified.

Considerations of cable arrangement in manholes, as well as of heat dissipation, make it impracticable to use for power cables a duct bank containing more than sixteen ducts, and even this number is apt to cause excessive congestion in manholes. A study of present and future requirements will generally disclose that from six to twelve ducts in a single duct bank will suffice. If more than twelve are needed, it is usually better to consider the construction of another duct system, including manholes, sufficiently removed from the first so that heat generated in one will not affect the other. Experience indicates that a minimum separation of 2 ft in earth will accomplish this purpose.

CURVES. Changes in the direction of conduits, whether vertical or horizontal, should, wherever possible, be made in sweeps of at least 25-ft radius. With fiber or transite duct this is done with short lengths of straight conduit with a slight angle (not over 7 deg between them at each coupling. With single tile duct short mitered sections are available for placing between straight lengths. The use of manufactured bends of short radius (18 to 36 in.) should be avoided wherever possible except on short runs, and then only at or close to the end of the run. Bends of less than 18-in. radius, while available, are not recommended for any point in a conduit run. Whenever it is necessary to use manufactured bends, those of 36-in. radius or more, if possible, should be used. Smaller radii should be limited to 2- and 3-in. conduit containing small cables.

42. CONSTRUCTION

Unless accurate and reliable information is already available, it may be desirable to make test borings throughout the area where it is proposed to install underground conduits.

Excavation in sandy soil, particularly if it contains water, requires tight sheathing heavily braced and may even necessitate steel sheet piling to prevent settlement of adjacent structures. Pumping may also be required. In average soil the bracing and pumping required will be much less or may be dispensed with altogether. Some degree of light sheathing is almost always required. Mechanical means of excavating for manholes and vaults are sometimes advantageous. The use of mechanical equipment in excavating city streets, however, is apt to be hazardous, as it is difficult to avoid damage to other buried utilities.

TEST TRENCHES. At the beginning of construction the first step in actually cutting the surface is the digging of test trenches. These are short, narrow trenches dug at the proposed site of a manhole or at right angles to the axis of the proposed duct bank and to a sufficient depth to disclose any unforeseen obstructions. By judicious selection of the points at which to dig these test trenches the engineer can save a great deal of fruitless excavation.

Water and gas pipes, which are ordinarily the most difficult to locate in advance of excavation and which may be in the way of proposed manholes or duct banks, can usually be detoured out of the way, provided they are not too large. This is often less expensive than radically changing the whole layout of duct construction in order to avoid them. Sewers, steam pipes, and other duct lines and manholes are usually impossible to detour, but fortunately they can ordinarily be located fairly accurately, even to depth, from surface surveys.

CONDUIT CONSTRUCTION. Trenches for duct lines should be excavated completely from one manhole to the next so that the proper grade can be established for the completed duct bank to drain into one manhole or both of them. It may be necessary to raise or lower the grade of the whole trench in order to pass over or under some obstruction. The depth must be such that at the highest point of the duct line it will still have the minimum cover specified, usually 30 in. The trench will be wide enough to give adequate working space for the laying of ducts, and this width will ordinarily be more than just that required for the duct bank. A trench 24 in. wide is about the minimum which affords working space for laying duct, and even this is hardly adequate. Some excavations require continuous pumping as well. In localities where the water table is high and the ground porous, excavations can be kept dry by the use of well points temporarily sunk alongside the excavation.

Sometimes, in excavating trenches in paved streets, the paving and sub-base are left intact across the trench for a short distance every 20 ft or so. This is done to keep the pavement from slipping and cracking in the vicinity of the excavation and may be required by municipal authorities.

After the grade of the trench is established, it should be tamped smooth and hard, preparatory to the laying of conduits. If fiber or transite conduits are to be laid in concrete, a bed of concrete 3 in. thick is placed in the bottom of the trench, and the laying of duct begun. Most engineers require the placing of aprons at the sides along the duct bank to retain the concrete while setting. If the trench is close sheeted, this sheeting may be used as the form for one side. Without these aprons, and with the sides of the trench used as forms, the thickness of the concrete envelope at the sides is indeterminate, and it is also difficult to maintain accurate separation between ducts.

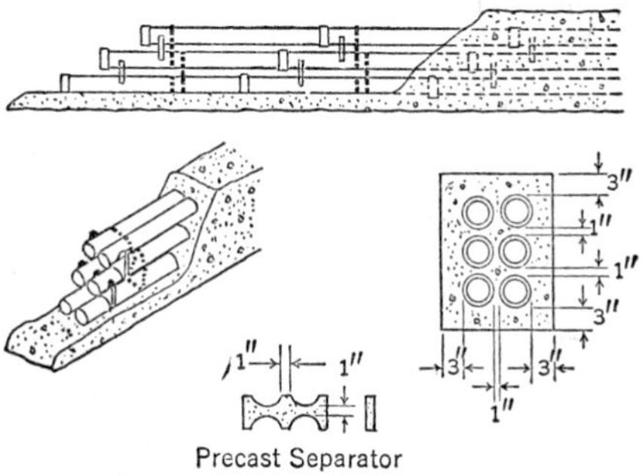

FIG. 6. Fiber Conduit Construction

If **fiber** or **transite** conduits are being used, there are two methods of laying them from this point on. In one method, known as the **group method,** the whole bank of conduits is assembled on cinder-concrete spacer blocks (Fig. 6). The conduits are laid in the first tier on the initial concrete bed, spacer blocks are placed on top of them, the second tier is laid on those, and so on until all conduits in the bank are laid. Joints are staggered in such a way that no two are adjacent horizontally or vertically in the bank. Some companies make a practice of painting the tapered end of each length with a thick asphalt paint just before it is inserted in the sleeve in order to decrease the permeability of the joint. A sleeve is applied to one end of each length before it is passed into the trench, so that the open end of every duct in the unfinished bank has a sleeve on it. Concreting follows immediately after all conduits in the bank are assembled in place.

This method is open to the objection that it is very difficult to ram concrete uniformly between all ducts, especially in the lower tiers. A more common method, known as the **tier-by-tier method,** consists in laying the ducts in the same way except that concrete is poured in each tier before the ducts in the tier above it are laid. In this method it is customary to proceed immediately with the second tier after the concrete in the first has been poured. If work on a section of duct bank is interrupted, the unfinished end of each tier should extend as close as possible to the end of the tier below it. All open ends of conduits should have sleeves on them, and each should be left with a tapered wooden plug in it to be removed when construction is renewed.

For the tier-by-tier method spacers to hold the conduits in line may be wooden or metal "combs," which are removed as soon as the concrete is poured, their space being immediately filled with concrete. Precast cinder-concrete spacers may be used instead, and these are left in. It is claimed that wooden or metal combs which are removed insure a homogeneous concrete mass more impervious to water seepage.

Concrete for duct banks should be slightly stiff when poured in order to reduce the tendency of ducts to float out of line. A water-cement ratio of 5 gal per sack is suggested. A mix no richer than 1 : 2 : 4 nor leaner than 1 : 3 : 6 is suggested for duct banks, depending on conditions. Aggregate should be reasonably fine, not over $^1/_2$ in., so that it can be thoroughly spaded between ducts. Spading should be carefully done so as not to injure ducts. The use of mechanical vibrators should be avoided, as it tends to make the ducts get out of line in wet concrete. Concrete should cure for 24 hours before backfilling is commenced. Backfill should be flushed and firmly tamped. The replacement of paving over it should be postponed as long as possible to allow the backfill ample time to settle.

Single tile ducts are laid by a mason in cement mortar on a concrete base. The base, usually 3 in. thick, is laid on the tamped floor of the trench. The ducts are laid one at a time and one tier at a time. All joints, which are of the mortise and tenon type, are wrapped with wide muslin tape dipped in grout to prevent the entrance of any concrete materials through the joint. In order to keep the bores of these ducts in line as they are laid, a mandrel (Fig. 7) about three times the length of a single section of tile

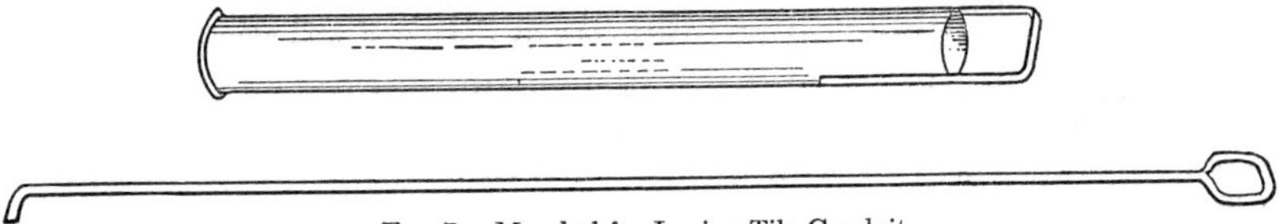

FIG. 7. Mandrel for Laying Tile Conduit

duct is drawn through each duct as it is laid. The foreward end of this mandrel has a rubber washer, which fits the duct snugly and wipes off any foreign matter inside the duct. The tile is trowelled in position, with the mandrel projecting a short length from its open end. As each length is finished, the mandrel is pulled forward, and a new length is slipped over it and fitted into the socket of the previous length. The joint is wrapped with muslin tape, and seams between the new length and those alongside it are filled with cement mortar. A separate mandrel is necessary for each duct in the bank. Square single tile is laid in the same manner, except that the mandrel is square and small steel dowels may be used in holes provided in each corner of the end of each duct. After the ducts are built up to the desired number and the bank is complete, 3 in. of concrete is placed around the two sides and the top to complete the envelope around the ducts. This concrete should be formed by a continuous apron on each side.

Multiple tile ducts are laid in the same manner on a concrete bed. Joints are doweled with steel dowels (Fig. 8) inserted into holes in the four corners of each end of each piece. These are necessary to line up the various ducts correctly. It is not necessary to draw a mandrel through these ducts until the whole bank is completed. Joints on multiple tile duct are first wrapped, after the ends have been fitted all around the joint, to a thickness of about $^1/_2$ in. A concrete envelope is desirable.

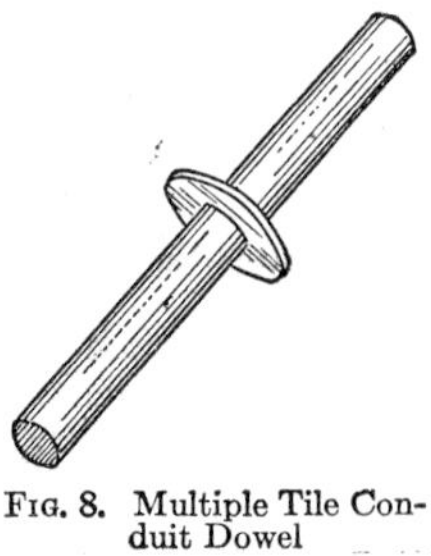

FIG. 8. Multiple Tile Conduit Dowel

Service ducts may be cast integrally in a concrete duct bank but are more often laid on top of the completed main bank while the trench is still open. They are usually smaller in diameter than the main ducts and made of a material which can be buried directly in earth, such as iron or thick-walled transite. If more fragile materials are used, it is advisable first to cover them with 2 or 3 in. of screened earth and then to lay a wooden plank on this earth before backfilling in order to protect them from future excavations. Separate trenches for service duct are required only when an open main line trench is not available. These trenches are shallow and only wide enough to accommodate the necessary number of ducts. They are often run in sidewalks. When service ducts enter buildings, every precaution should be taken to prevent them from becoming a channel for the entrance of water into the premises.

MANHOLE AND VAULT CONSTRUCTION. A light-wood template should be made in the form of the outside perimeter of the manhole. By placing this over the test trenches, the most convenient position for the manhole can be determined, as well as the location of the cover. Decision will then be made whether to detour such foreign structures as may be disclosed or to go below them. Perhaps an oddly shaped manhole can avoid doing either and still serve the purpose. If immovable structures are not too deep, it may be advantageous to build the manhole under them, with access to it through a chimney. The presence of a gas or water pipe or any other foreign utility in a power manhole should not be tolerated. If this arrangement is absolutely unavoidable, such pipe should be covered with concrete and benched through. Transformer vaults, although larger than junction manholes, can usually be located under sidewalks, thus avoiding large obstructions.

After the manhole or vault is located, the earth is excavated to the required depth, sheeting, bracing, or pumping being used as needed. At the proper depth the earth is tamped and the floor poured, if one is to be used, the entire area of the excavation to and against the sheathing on all sides being covered. Mortises for construction joints are left in this slab properly located for the side walls. The floor is float-finished directly from the slab. Eight to twenty-four hours later the building of the forms is started, and when these are completed, including forms for the roof slab, concrete for walls and roof should be poured in one operation.

Duct entrances into manholes or vaults should be completed and poured at the same time that the manhole walls are poured. Leaving windows for the later construction of duct entrances, and especially of knockouts for the future entry of duct banks, should be discouraged. It is very difficult to make either of these watertight. Collapsible steel forms for the inside forms of manholes are advantageous if a great many manholes of the same size and shape are to be constructed. These forms can be used repeatedly, and although the first cost is high, it is amortized in a very few uses. They are limited to one size and shape. Reusable outside forms are not usually feasible.

Concrete for manhole and vault construction is usually specified as 1 : 2 : 4 mix with $^3/_4$-in. crushed stone or gravel as aggregate. Waterproofing is added if needed. Considerable economy can be realized by the use of ready-mixed concrete if this is available near by.

Ground rods are frequently driven in the excavation of manholes, or vaults and copper leads from these are brought out into the manhole. It is desirable that this should be done before any concrete is poured. The leads should consist of soft-drawn, solid, bare copper wire brought up through the wall and out into the manhole 1 to 2 ft above the floor. This arrangement will prevent any water seepage along the leads into the manhole.

Inserts, chimneys, floor ducts, ventilators, and duct entrances in the manholes and vaults have been discussed in detail.

RODDING, CLEANING, AND WIRING OF DUCTS. In order to test the continuity of conduits after completion it is customary to draw a mandrel through them followed by a wire brush. There are three methods of getting a rope, to which the mandrel and brush are to be attached, through a duct in the first place. The oldest and most reliable method is by means of hickory rods, 1 in. in diameter and 3 to 4 ft long, fitted with special couplers at each end (Fig. 9). These rods are pushed into the duct, each rod being coupled to the one in advance of it as the latter disappears into the duct. The pulling rope is attached to the last rod and pulled through, the rods being uncoupled as they emerge. This method is slow, but if there are obstructions in the duct, it is the only way of getting through. Special tools are available (Fig. 10) which can be attached to these rods for cutting obstructions.

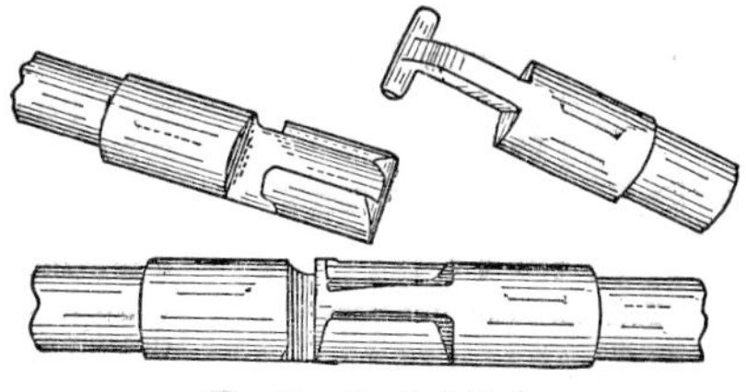

Fig. 9. Conduit Rods

Heavy fish wires which can be pushed through a duct that is reasonably clear are also available and can be pushed through in a much shorter time. A single fish wire, however, can be pushed a maximum of only about 200 ft. There is another type on the market, made of round spring steel, which comes in two sections, one 400 and the other 300 ft long. One section is fitted with a special head so designed that when the other section is pushed into the duct from the other end it will pick it up and be locked to it. With this device a 500- to 600-ft section, provided it is reasonably clear, can be rodded in a comparatively short time.

The quickest method of getting a rope through a clear duct is by blowing it with compressed air. A traveller (Fig. 11) is attached to a length of light sash cord on a freely turning reel and inserted in the duct. A tapered plug, with a pipe through the center and valve and air-hose connection at the larger end, is then inserted in the duct mouth back of the traveller. A slot in the side of the plug permits the cord to run freely past the plug when air pressure is turned on. This method saves a great deal of time if many

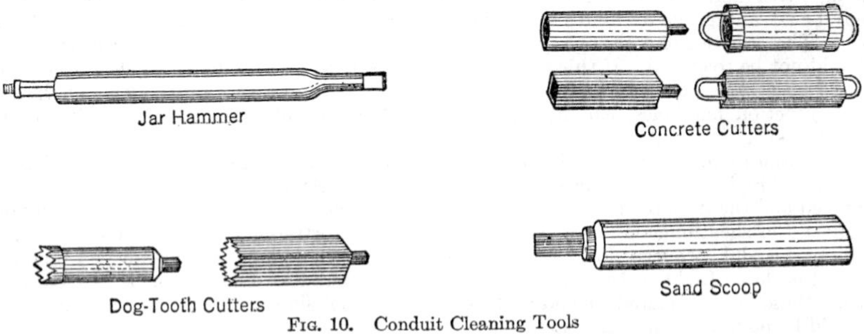

FIG. 10. Conduit Cleaning Tools

ducts in a single bank are to be rodded, as on completion of construction. For rodding a single duct, more time is usually lost in setting up for it than is saved in actually running the rope through.

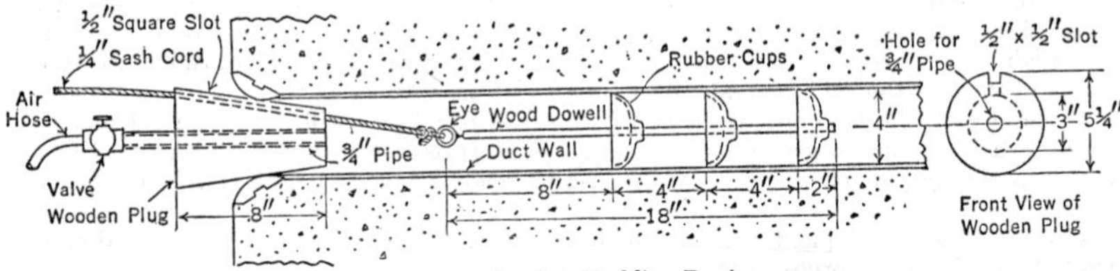

FIG. 11. Conduit Rodding Device

With any of these methods a heavy rope must be pulled through after the first rods, fish wire, or sash cord has passed through. A flexible mandrel, 1/4 in. less in diameter than the bore of the duct (Fig. 12), is then attached to this rope, and another rope to the other end of the mandrel. This second rope is important, as it provides a means of withdrawing the mandrel if it encounters obstructions which it cannot get by. It also leaves a rope in the duct after the mandrel has gone through. A wire brush, 1/4 in. larger in diameter than the bore of the duct (Fig. 13), is then pulled through the duct in the same way. The pulling rope is attached to this brush if cable is to be immediately drawn into the duct. If the duct is not to be used immediately, but use is contemplated for it within a few months, a galvanized steel or copperweld wire is drawn into the duct and left there to be used later for drawing a pulling rope through. This wire should be solid and from No. 8 to No. 10 AWG in size. These wires should not be left more than 6 months in ducts, however, as they are liable to corrode and break inside the duct.

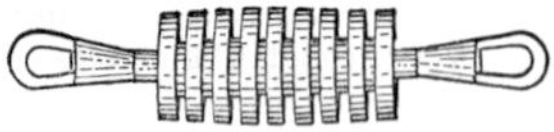

FIG. 12. Conduit Mandrel

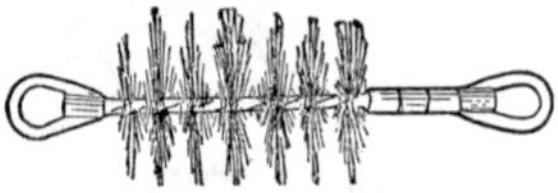

FIG. 13. Conduit Brush

43. MAINTENANCE

A properly built underground system requires very little maintenance of the structures themselves. Regular maintenance consists in the removal of water from the system and in the removal of mud and debris which collect in manholes, particularly under grating covers such as those placed over ventilating shafts of transformer vaults. As most, if not all, cables and apparatus which are installed in underground conduits are designed for submerged operation, the presence of water need not be disturbing, but means must be provided for removing water from manholes and vaults in a minimum of time for the entrance of employees in emergencies or for routine inspection.

WATER PUMP. In localities where the water table is close to the surface and manholes tend to be wholly or partly submerged the greater part of the time, it is well to provide a truck with a pump driven through a power take-off by the truck engine. This pump should have a capacity of 60 to 100 gal per min. Localities which experience flooded manholes only occasionally can be served by a small gasoline-engine-driven pump weighing about 100 lb and able to deliver about 30 gal per min.

VENTILATING EQUIPMENT. Special ventilating equipment for manholes and vaults, consisting of a small gasoline blower and a length of canvas hose about 10 in. in diameter should be provided. This outfit is placed beside the manhole opening and delivers a large volume of air through the canvas hose to the bottom of the manhole or vault. This device will clear a manhole or vault of smoke and noxious gases in a very few minutes, making it safe for repair crews to enter quickly after trouble develops. The device will also keep a manhole cool in hot weather while men are working in it.

TEMPERATURE. Overloaded transformers and cables in an underground conduit system are sometimes kept within safe temperature limits by the deliberate introduction of water into the system to submerge the cables and equipment. A great deal of additional capacity can be obtained from apparatus cooled in this way, but since water in an underground system is always a problem, its deliberate introduction should be avoided except in extreme cases.

Some reduction in temperatures can be realized by the installation of fans or blowers in manholes and vaults. For adequate results a considerable amount of power has to be consumed, and the fans and the motors driving them require a great deal of attention. If the overloaded condition continues, larger equipment or facilities for additional circuits separated a reasonable distance from existing ones should be considered.

Other methods of cooling duct lines have been tried with varying success. One method is to keep the earth surrounding the duct banks moist by means of a perforated water pipe laid on each side of the bank. This arrangement is effective and not expensive if installed with the duct bank. Its application is limited, however, to short, heavily loaded banks such as those carrying generator leads. Another method is to provide an air passage along either side of the duct line. Neither of these methods is applicable to ducts in city streets.

44. BIBLIOGRAPHY

Edison Electric Institute, *Reports of Transmission and Distribution Committee.*
N.E.L.A., *Underground Systems Reference Book* (1931).
Ruhling, T. C., *Underground Systems for Electric Light and Power* (1927).

Much of the material for this section has also been obtained from manufacturers' catalogs and bulletins.

DISTRIBUTION

By C. T. Hatcher

Electrical distribution includes the distributing of energy for light and power from its point of generation or conversion to points of utilization.

Distribution of energy is accomplished by the use of either an a-c system or a d-c system. The d-c systems are, in general, those which were used originally for supplying the centers of large cities and operate at 120 to 240 volts. The a-c systems followed the d-c systems and are supplementing them because of the fact that they are more economical.

A distribution system can normally be said to consist of two parts: the primary distribution, which extends either from the generating stations or substations to distribution transformers, in the case of an a-c system, or from the generating station to points of conversion, in the case of a d-c system; and the secondary distribution, which extends from these transformers or conversion points to the points of utilization.

Overhead construction, which is normally used where the areas are not thickly populated, has the following advantages: (1) the first cost is lower, and (2) it is easier to repair. Underground construction, normally used in large cities and congested areas, has the following advantages: (1) it is less unsightly, and (2) it is less subject to damage by external agencies.

45. DIRECT-CURRENT DISTRIBUTION CIRCUITS

PRIMARY SYSTEM. Alternating-current feeders are normally used to transfer energy from the generating station to substations, where it is converted to direct current. Utilizing alternating current enables the transfer of bulk power at lower losses. These feeders are normally operated at 25 cycles and either 6600 or 11,000 volts.

SECONDARY SYSTEM. From the conversion substation d-c feeders operating at 120 to 240 volts (three-wire system) are used to supply a mains system which is joined together at every street intersection to form a grid. The grid is fed by the feeders through junction

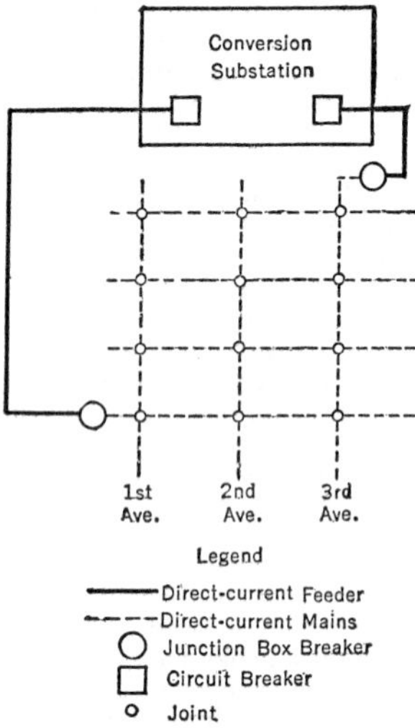

Fig. 1. Direct-current Secondary System

boxes equipped with fuses or circuit breakers, which are used to clear a faulted feeder from the grid (Fig. 1). Individual services are taken off the mains to points of utilization. In some cases direct feeders are used to supply customers' bulk loads.

46. ALTERNATING-CURRENT DISTRIBUTION CIRCUITS

PRIMARY SYSTEMS. Alternating-current feeders are used to transfer energy from the generating station or substation to distribution transformers.

Primary distribution systems are operated at various voltages and are generally three-phase circuits with grounded neutrals. The 2400/4160-volt system is most generally used. The following are approved systems:

Nominal System Voltage

2,400	Delta	12,000	Delta
2,400/4,160	Y	7,200/12,470	Y
4,800	Delta	7,620/13,200	Y
7,200	Delta	13,200	Delta
4,800/8,320	Y	14,400	Delta

The prime consideration in any distribution system is to provide constant potential at the utilization point. A number of methods are used to accomplish this purpose; the outstanding ones are the following.

Radial Distribution. Radial distribution is a system in which independent feeders branch out radially from a common source of supply, such as a generating station or substation, without intermediate connection between feeders. Distribution transformers are connected to the radial feeders. The transformers are connected either in the tree system or to mains which are connected to a center of distribution.

The tree system, as the name implies, has branches or taps which are taken off and connected to transformers along the length of the feeder (Fig. 2). In such a system it is possible to grade the conductor size down for the radial feeder in accordance with the connected transformer capacity.

Where the tree system cannot maintain the required voltage, it is necessary to utilize the feeder and mains system. In this case the feeder is run to a point near the load center of a district, and mains branch therefrom to the distribution transformers (Fig. 3). Normally sectionalizing devices are provided at tie points between the feeder and mains so that in case of trouble a connection can be made temporarily to another radial feeder.

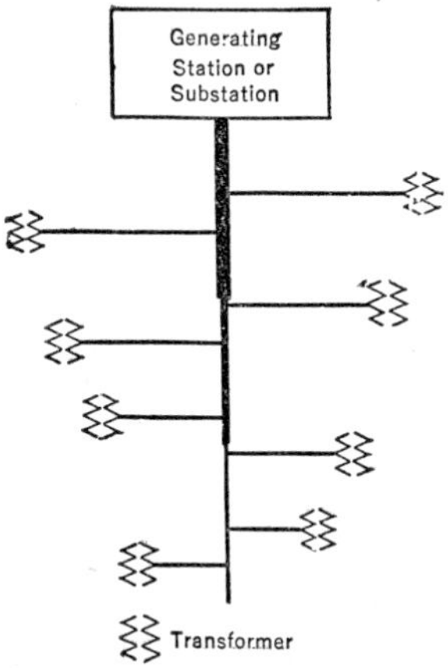

FIG. 2. Radial Distribution, Tree System

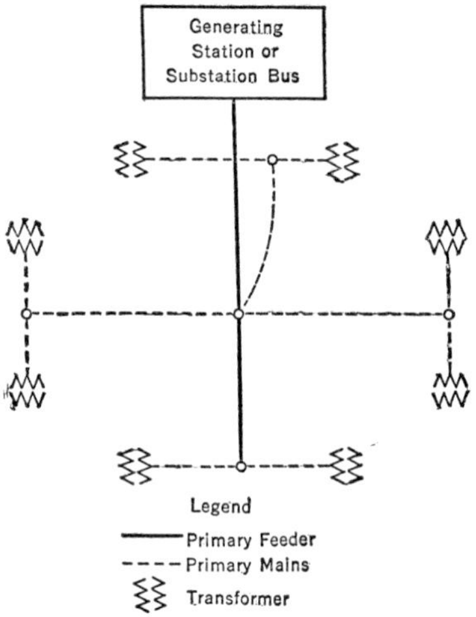

FIG. 3. Radial Distribution, Feeder and Mains

Network Distribution. Network distribution is a system in which a secondary mains grid is fed through distribution transformers by multiple feeders from a generating station or from substations supplied from the same generating station. This system is applicable to districts having high load densities and those where continuity of service is important. The mains grid is of such a capacity as to insure good regulation.

The distribution transformers are connected to the secondary mains by means of automatic network switches which disconnect the transformer from the grid when there is a fault on the primary feeder or when the feeder is de-energized and which close automatically when the feeder is re-energized (Fig. 4).

The network system permits the feeding of current into a load from several directions. Proper location of transformers adjacent to heavy load centers and regulation of the feeders at the station bus provide for adequate voltage at utilization points.

Vertical Distribution. Vertical distribution is a specialized application of network distribution. In tall buildings, where there is concentration of load, considerable amounts of copper would be required to obtain satisfactory voltage on secondary runs. In order to avoid this, primary feeders are run vertically in the buildings and supply distribution transformers at various floor levels. The secondary cables are tied together in the same manner as a secondary street network.

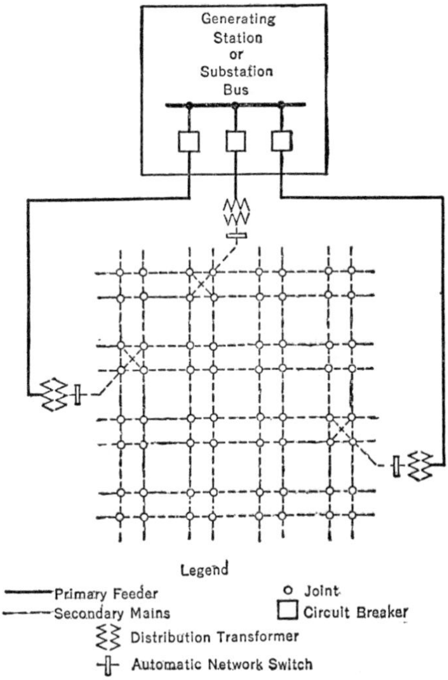

FIG. 4. Alternating-current Network Distribution

Loop Distribution. The loop-distribution system is used for supplying bulk loads where they are too large and important to connect to feeders used for miscellaneous radial distribution and where a network system does not exist. Normally a feeder extends from a substation or generating station to a group of customers connected in series and either back to the same station or to one adjacent (Fig. 5). In some cases sectionalizing switches are non-automatic trip and hand-operated, whereas in others directional relays are used for switch operation.

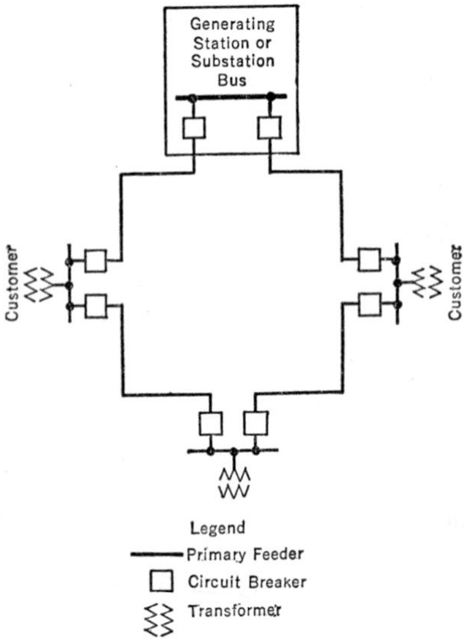

FIG. 5. Alternating-current Loop Distribution

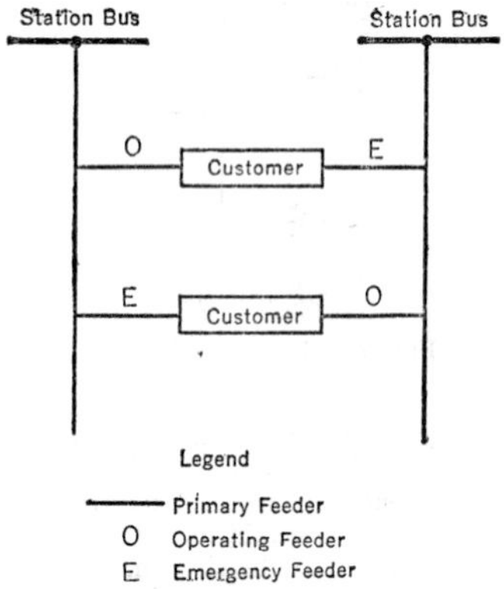

FIG. 6. Alternating Current, Duplicate Service

Duplicate Service. Duplicate service is a modification of the radial system, used where additional reliability is required for bulk load being supplied from radial feeders. Two feeders are used for the load supply, one normally energizing the transformer and the other held for emergency (Fig. 6). In case of failure of the normally loaded feeder the second feeder is connected to the transformer.

SECONDARY SYSTEM. Secondary a-c distribution systems are generally operated at 120/208 volts or 120/240 volts, 60 cycles.

The systems may be either radial or network.

Radial. Radial secondary distribution is a system of supplying power from distribution transformers to customers where each of a number of isolated secondaries is supplied by one transformer or transformer bank.

Network. Network secondary distribution is a system which is completely joined together so as to form a grid and is fed from the secondary side of a number of distribution transformers and feeders. A secondary network must be designed so that any conductors which become short-circuited with a low resistance path will receive enough current from the network to establish an arcing fault and burn themselves clear. In obtaining this requirement consideration must be given to the size of secondary cable and the size and spacing of transformers for specific load densities. A secondary network is normally fed with multiple feeders from synchronized sources.

Network voltage regulation methods include bus regulation and feeder regulation.

The 120/208-volt system is the most generally used, but the following are also used: 115/199, 115/230, 117/202, and 120/240 volts.

Network Protectors. Network protectors are automatic air circuit breakers between the distribution transformers and the secondaries. The network protectors are used to disconnect the secondary network when a fault occurs in a distribution transformer or high-voltage cable supplying the transformer. The protector opens on a reversal of energy when failure occurs in the primary supply and recloses when the feeder is again energized, this being accomplished by the use of relays. Protectors are made in submersible and non-submersible types to suit manhole and vault conditions.

Cables. Secondary-network cable are of sizes which depend upon the design of the system, varying from 4/0 to 500,000 cir mils, and are normally of the single conductor type. The cables are usually either paper insulated or rubber insulated. Lead covering is used for the paper-insulated cable; lead, cotton braids, or a synthetic jacket, for the rubber-insulated cable. Rubber-insulated cables with a lead sheath will probably give the longest life, but it may be economically justifiable to use other coverings because of their lower initial cost.

A large percentage of secondary cable faults in a network will burn clear rapidly. In these faults the short-circuit current does not exist for a sufficiently long period of time to raise the temperature of the cable conductor on either side of the fault to a high temperature that will fuse or damage the cable insulation. Where the faults will not burn clear rapidly, considerable damage may be done to adjacent cables before the fault is cleared manually by cutting, unless it is cleared automatically. Automatic clearing can be accomplished by installing fuses or limiters at selected points so that they will blow at certain current values which will not cause damage to the particular insulation being used (Fig. 7). The fuses are so designed as to form an integral part of the joint or cable terminal and to facilitate ease of installation.

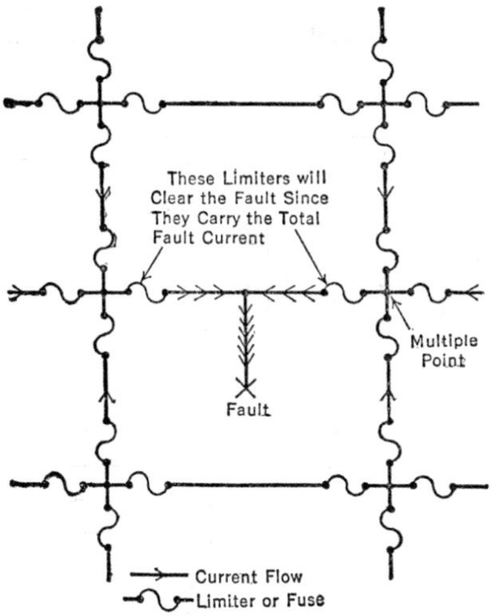

FIG. 7. Limiter or Fuse Installation

Transformers. Network transformers are of either the submersible or the non-submersible type. They may be either single-phase or three-phase units. Transformers may be of the oil type, air cooled, or askarel filled. For high-density load areas transformers of various sizes up to three-phase, 1000-kva units may be used.

47. OVERHEAD DISTRIBUTION

DESIGN OF LINES. Overhead distribution systems should be designed so as to obtain the maximum use of the poles in serving all buildings, both present and future.

New pole lines should be designed to care for initial installation plus future additional wires, cross-arms, etc., to provide facilities for the expected service life of the line.

Pole lines may be located on main streets and thoroughfares, side streets, and rear alleys, depending upon the specific requirements.

In order to obtain maximum use of poles it is often desirable to have joint usage of them with other utilities, such as telephone and telegraph. Where joint usage is desirable, arrangements are required for obtaining proper allocation of space for wires and equipment, so that safe and satisfactory working conditions are available for all occupants. These requirements are obtained by joint cooperation between the parties occupying the poles.

Overhead distribution is accomplished by means of open-span construction or of insulated conductors supported by a messenger. In open-span construction it is necessary to obtain physical separation between phases and insulation from ground, whereas with messenger-supported cable the conductors are insulated and can be in physical contact with ground and each other.

MATERIALS FOR OPEN-SPAN CONSTRUCTION. Poles. Poles for distribution lines are generally wood. Cedar, creosoted pine, chestnut, and wallaba poles are some of the types which have been found to be satisfactory. Poles range in length from 25 to 65 ft, depending on the locations where they are to be used or the equipment which they are to support.

Poles are spaced normally in equal span lengths varying from 100 to 125 ft so as to provide safe sag limits and points for service connections.

Conductors. Primary and secondary conductors are normally of the weatherproof type, although some bare conductors have been used in recent years. Copper wire is generally used, either solid or stranded, depending upon the size, and is covered with saturated braids, papers, or felts having weather-resisting characteristics. Sizes from No. 6 A.W.G. to 4/0 are generally satisfactory, but larger sizes up to 500,000 cir mils are often used. Since the No. 6 size is likely to stretch when used for span lengths and to cause undue breakage, it is recommended that careful consideration be given to its selection for main runs. Medium hard-drawn copper is preferred by most companies for sizes up to and including 2/0, and annealed copper for the larger sizes.

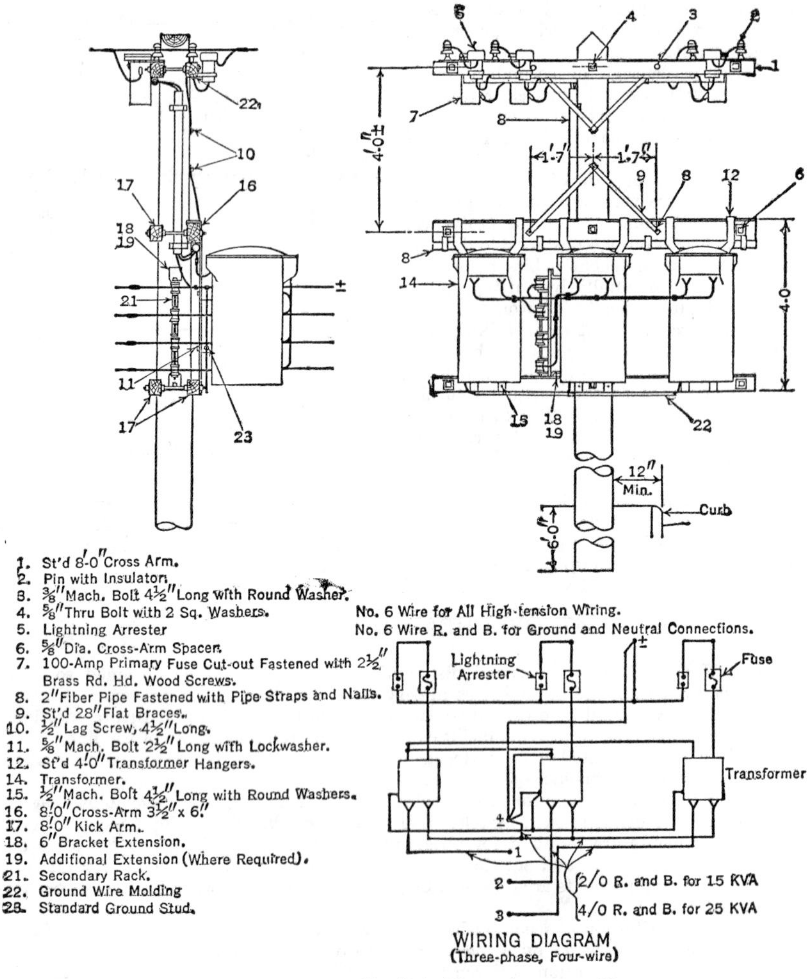

FIG. 8. Transformer Line Pole, Three-phase Assembly

Where it is necessary for lines to pass through trees, rubber-insulated conductors covered with special protective braids or tapes are used for protection against contact grounding. These cables are known as tree wire. In place of the tree wire, protection can be obtained by wood or plastic guards over the weatherproof wire.

Covering on Overhead Wires. The weatherproof covering on overhead wires, which normally consists of three saturated cotton braids, is solely for the purpose of limiting the short-circuit current due to an accidental cross or grounding. The wires are installed on insulators, and the normal insulation of the line thereby maintained without any consideration being given to the wire covering. Although weatherproof braid is an imperfect insulator, it serves to eliminate some of the short circuits and arcs which would occur, because of momentary contact, if bare wires were used. Wires with a thin coating of neoprene or polyethylene are being used for the same purpose.

Insulators. Insulators are normally either porcelain or glass. For primary wires the pin type is used, and these insulators are of either side-groove or top-groove design. For secondary wires porcelain spools are normally used, although pin-type insulators on cross-arms are satisfactory.

The conductor is attached to the insulator by a tie wire of the same material as the conductor, although soft wire is usually employed even for hard-drawn conductors. The tie wire can be either bare or covered to correspond to the conductor. Ties should be relatively simple in design and easy to apply. They should bind the conductor securely to the insulators and should reinforce the conductor on both sides of the insulator.

Arrangement of Wires on Poles. Primary wires are normally supported on cross-arms. Secondary wires are supported on either cross-arms or secondary racks mounted on the side of the pole (Fig. 8).

The arrangement of wires is governed by mechanical, electrical, and practical considerations. In planning any system, it is quite desirable that a phase sequence be established for the location of the operating phases and neutral on the cross-arms. A standard arrangement enables workmen to determine quickly the location of a trouble on a particular phase. Wires should be arranged symmetrically so as to present a satisfactory appearance.

In order to provide safe working conditions for the workmen it is desirable to install the highest-voltage wires on the top cross-arms and the lower-voltage wires on the lower cross-arms.

Trees. Trees constitute a serious obstacle to the proper construction and operation of open-span overhead lines. This difficulty can be overcome by:

1. Using tall poles for carrying the lines over the trees.
2. Selecting streets without trees.
3. Tree trimming.

The tree-trimming method involves obtaining the permission of city authorities or individual owners for the cutting of limbs. Moreover this method entails considerable maintenance expense, since the trees will continue to grow and will require yearly trimming.

MATERIALS FOR MESSENGER-SUPPORTED CONDUCTORS. Conductors. Where tree conditions are especially bad or where a more attractive appearance is desired, insulated conductors supported by a messenger are used. This construction may be a messenger, fastened to poles and carrying rings, wherein the insulated conductors are pulled in the field, or a completed cable, assembled in the factory, wherein the messenger is assembled with the conductors and is termed self-supporting. The messenger is either of copperweld construction or of copper and steel strands and serves as the neutral.

Self-supporting cable is fastened directly to the poles and does not require cross-arms. Since some protection is provided against falling limbs and trees, self-supporting cable may provide an increased factor of safety in operation.

The insulated conductors can be either shielded or non-shielded, depending upon the requirements of the system. Tees or service connections can be taken off a line either at a pole or in midspan. The self-supporting type of cable can be used for either primary or secondary circuits.

Where self-supporting cables are used, tree-trimming costs are practically eliminated, and pole heights can be reduced in heavily wooded areas.

POLE TRANSFORMERS. The pole-mounted transformer provides the connection between the primary system and the secondary system. The overhead transformers are normally mounted on poles, and connections are run from the primary feeder to the transformer. The secondary side of the transformer is connected to the mains supplying load in that area (Fig. 8).

Pole transformers range in size normally from 1.5 to 25 kva, although transformers of larger capacity are used for heavy industrial loads.

In determining the loading limits of transformers, consideration should be given to the type of load, namely, residential or industrial, being supplied. Loading limits should be based on an analysis of the thermal characteristics of the transformers, the ambient temperatures, including the effect of the sun, and the daily load cycles under which the transformers operate.

AUTO-TRANSFORMERS. When it is necessary to correct excessive flicker caused by the operation of 120-volt motors or other 120-volt utilization devices on single-phase, 120/240-volt, three-wire circuits, balancing transformers should be used if they will provide the necessary degree of flicker correction and at the same time will permit a substantial saving over the cost of alternative methods.

48. SERVICES

SERVICE CABLES. Service cables connect the overhead street mains with the building wiring. Usually these cables extend in a single span from the nearest pole to the house or in some cases from the street main at its nearest point to the house (Fig. 9).

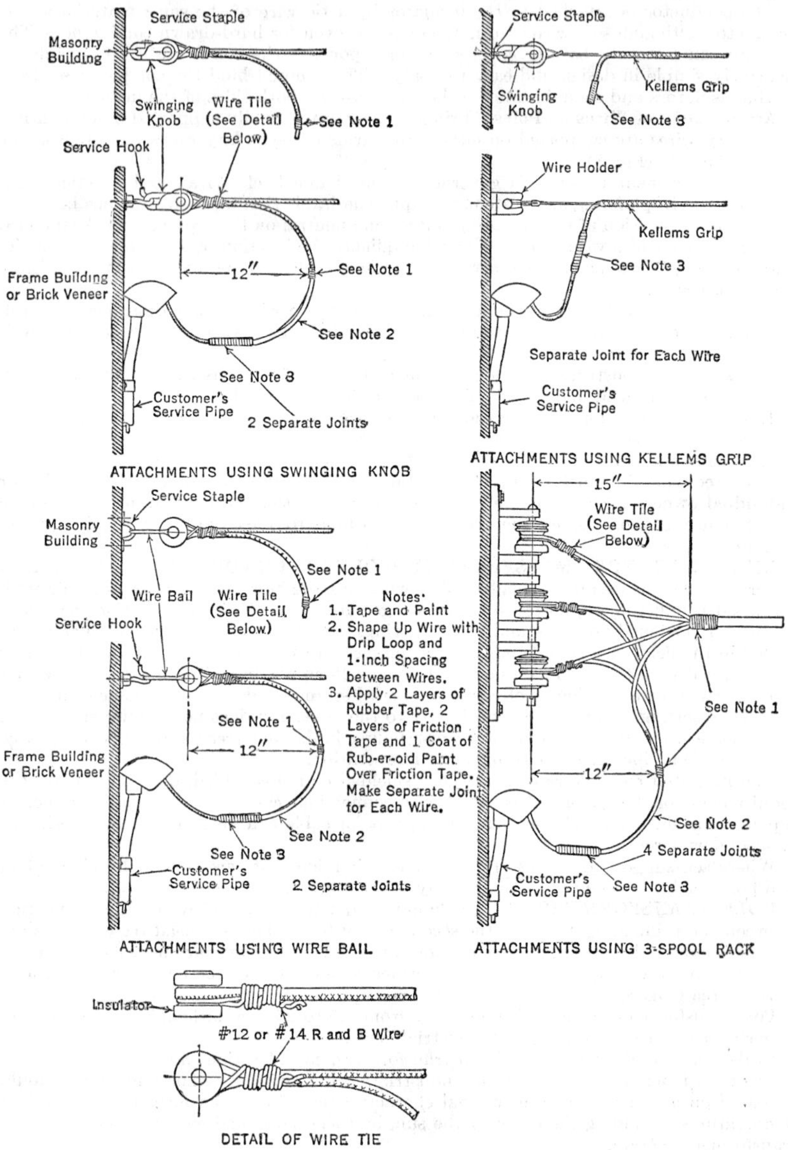

FIG. 9. Overhead Service Attachments to Buildings

Service cables should embody the following features: (1) light weight, (2) flexibility to fit corner contours, (3) small diameters, and (4) simple fittings for house connections. These cables are divided into three main classes:

1. Cables between overhead mains and weatherhead on the building.

2. Cables between the building weatherhead and the meter, switch, or service equipment.

3. Cables continuing without a break between the overhead mains and the meter switch or service equipment.

In all three classes the service cable has to be fastened to the building for support. This is accomplished by fastening the cable or cables to spool insulators or by the use of a clamp on the cable which is secured to a hook on the building. The spool insulators or hook is so designed and located on the building wall as to take the strain of the span, so that, where the service cable passes into the building, it will not be under strain.

Service wires are attached directly to the mains that supply them at the nearest pole or, in some of the most recent designs, at any point where the mains span is nearest to the building.

Single-conductor and multiple-conductor cables are used for services. Present-day tendencies are toward the use of multiple-conductor cables because they present a more satisfactory appearance in the span length and at the building, where the number of spool insulators can be decreased.

Service cables between the mains and the weatherhead and between the weatherhead and the meter or switch are normally rubber insulated and covered with a single heavy cotton braid saturated with a weather-resisting, flame-retarding compound. For the connection from the mains to the weatherhead a cable, wherein the phase conductors are rubber insulated and a concentric uninsulated neutral is stranded around the assembled conductors, can be used.

Service cables for continuous run between the mains and the meter or switch are normally rubber-insulated phase conductors with a concentric uninsulated neutral. It is necessary to install this type of cable in conduit on the building and to the meter or switch. Where it is desired to install the cable without conduit, the concentric-type cable covered with either a heavy cotton braid saturated with weather-resisting compound or a galvanized steel tape and heavy cotton braid is used. This protection enables the cable to be fastened to the building and also provides safety against tampering.

Where service cables enter the building through the attic, a drip loop is installed between the supporting bracket and the point of entrance, so that water will not follow the cable into the building.

The weatherhead is a metal hood with its opening covered with a porcelain bushing plate, the opening and plate facing downward so that rain drip will not enter.

FRONT AND REAR SERVICES. Buildings are supplied from either the front or the rear, depending upon the location of the pole lines. When supplied from the rear, the mains are run in rear alleys, and the supply is to individual houses or groups of houses.

SWITCHES, POTHEADS, AND CUT-OUTS. An overhead system is protected and sectionalized (1) at the station, either generating station or substation, by automatic circuit breakers and disconnecting switches, (2) at the poles, where the circuit branches out by oil switches or fused cut-outs, (3) at transformer primaries by fused cut-outs.

In order to sectionalize 2.4-kv and 4-kv circuits it is necessary to install switches or fused cut-outs. Under some operating conditions two positive breaks are required before working on a circuit, and in these cases oil switches and disconnecting potheads are used in series. This combination is used where underground primary and secondary circuits are connected on poles to overhead circuits.

Radial transformers installed on poles generally have fused cut-outs at the transformer.

Network transformers fed from overhead mains may have oil switches and disconnecting potheads or fused cut-outs on the riser poles.

49. LIGHTNING PROTECTION

Lightning strokes cause damage to poles, burning of transformers or switches, puncturing of insulators or underground cable, and blowing of fuses. Damage may also occur to meters, switches, or appliances on the premises of consumers. The protection principally used includes: (1) lightning arresters and choke coils at the station, (2) lightning arresters at intervals on the lines, (3) ground wires over the lines. and (4) grounding of the circuit.

The following general principle applies to grounding. At each lightning-arrester location, the lightning-arrester ground leads should be connected to an adequate ground. Adequate grounds are as follows: (1) a secondary neutral, either overhead or underground, at the pole on which the lightning arrester is located, provided there are at least two water-pipe grounds at services supplied from the secondary system involved; (2) an overhead secondary neutral if one exists at the pole on which the lightning arrester is located, provided the overhead secondary involved is connected to underground network mains; (3) a driven ground at the lightning arrester location, supplemented if possible by one water-pipe ground if an adequate ground is impracticable to obtain.

Ground wires are used principally over transmission lines, but in some cases they may be desirable over distribution wires in exposed places. (See Article 96.)

50. UNDERGROUND DISTRIBUTION

Underground distribution is normally accomplished by installing both primary and secondary cables in conduits, manholes, and service boxes. Primary and secondary cables may be installed in the same conduit and manhole system or in separate systems, depending upon the degree of protection desired. Primary circuits should not be installed in service boxes. In some cases the circuits are buried directly in the ground, and the cables protected with metallic coverings on the cables or slabs of concrete or treated planking over them.

Alternating-current primary circuits are installed from generating stations or substations to distribution transformers, which may be installed in either manholes or building vaults. From the transformers secondary mains passing through manholes and service boxes are installed throughout the area to be supplied. From the service boxes service cables are run into the buildings to meters and service switches.

CONDUCTORS. Primary cables are either three-conductor or single-conductor and may be insulated with paper, rubber, or varnished cambric. Cables of the 15-kv and 27-kv class are generally paper insulated and lead covered. Rubber insulation is ordinarily used for cables of the 5-kv class and may be either lead covered or non-leaded, depending upon the type of rubber used. The usual sizes are No. 4/0 A.W.G. to 500,000 cir mils, although the range extends from No. 6 A.W.G. to 2,500,000 cir mils.

Secondary cables are generally single conductor and may be insulated with rubber or paper. Paper has the advantage of greater carrying capacity but is susceptible to water. Rubber is a satisfactory insulation because of its moisture-resisting characteristics and is therefore widely used. Rubber cables are either lead covered or non-leaded. The insulation of non-leaded cables is covered by a heavy braid or a synthetic jacket. In selecting a leaded or non-leaded cable the choice is generally made on economic considerations based on the assumed life of the cables. Selection of conductor size is based upon carrying capacity or voltage regulation. Sizes for mains and services range from No. 6 A.W.G. to 500,000 cir mils, with 4/0 A.W.G. being commonly used for mains.

Rubber cable is better suited for services because of the fact that it is easier to terminate in the consumer's premises. Paper requires a terminal which will prevent leakage of oil or admittance of water.

Synthetic insulations, such as polyvinyl compounds and polyethylene, are now being considered because of their excellent ageing and moisture-resisting characteristics. In using these compounds careful consideration must be given to the flow temperature characteristics when they are subjected to temperatures under fault conditions.

CONDUIT LINES. See Underground Conduits, Articles 37 to 44.

TRANSFORMERS. Underground distribution transformers are installed in manholes or vaults under the street or sidewalk. In some cases they are installed in buildings, either in basements or, in large buildings, on upper floors. Transformers installed in manholes are of the submersible type, being equipped with waterproof tanks and a welded or gasketed cover.

Where installations are made in manholes in streets and sidewalk vaults, gratings are used for ventilation and access for removal of the transformers. (See Article 40.)

BONDING. Underground lead-covered cables are bonded in manholes and boxes and connected to primary or secondary neutrals in order to minimize troubles arising from faults. Bonding can be accomplished by the use of a bond tree or series strip bonding. The bond tree is more rugged and in general provides better protection than the series method. Bonding is installed in one end of a manhole or box, all cables are connected to the tree or loop, and the tree or loop is then connected to the neutral if it exists in the manhole or box (Fig. 10). When bonding a circuit composed of three single-conductor cables installed in separate ducts care must be taken not to tie the cables together and cause circulating currents. In these cases special bonding systems are used, such as cross-bonding or bonding transformers, where insulated sleeves break the continuity of the sheath. Stanchions, rack plates, and isolated hangers should not be bonded to the neutral or common bonding system.

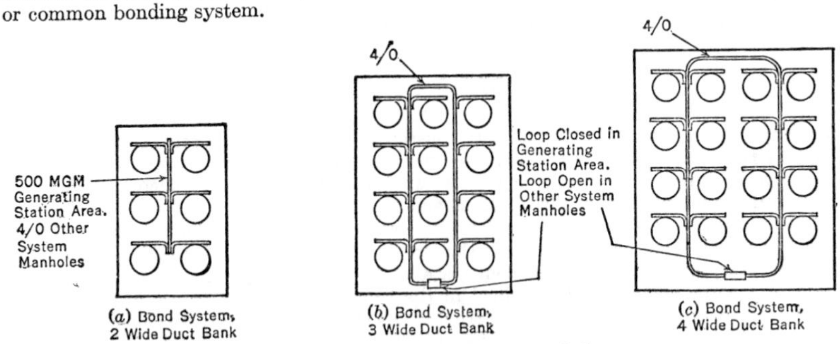

FIG. 10. Methods of Bonding in Manholes

CABLE ARCPROOFING. Primary cables operating on circuits 2400 volts and above, where exposed with cables of other circuits, should be covered with arcproof materials to protect them against arcs generated by electrical failures in cables, joints, or apparatus. This covering may be sand and cement, asbestos, or materials of a similar nature.

SUBMARINE CABLES. Where a distributing system is divided by a waterway, the connection is made by submarine cables laid either on the surface of the bottom or below the surface in trenches dredged for the purpose. Cables may be single conductor or multiple conductor, the latter being used whenever possible for alternating currents in order to avoid reactance due to the steel-wire armor of the cable.

Submarine cables are normally made in one continuous length so as to avoid joints in the waterway bottom. Paper and rubber are the insulations generally used. For the lower voltages, rubber is an excellent material for underwater burial because of its moisture-resisting characteristics. When paper is used, it is desirable to maintain a positive internal pressure by using an oil- or gas-filled type of cable.

Submarine cables are terminated either in manholes on each side of the water if they are jointed to cables in conduits or on poles if they are connected to an overhead system.

51. STREET LIGHTING

Lighting of streets is accomplished by two methods: (1) series circuits, and (2) multiple circuits.

SERIES CIRCUITS. In a series circuit all lamps are connected in series, and the same current passes through each lamp in completing its path to the source of supply. Circuits are normally operated at 6.6 amp, and the terminal voltage is a function of the total wattage of the lamps on the circuit. Individual lamp sockets are constructed with disk film cut-outs, so that a failure in a lamp will not extinguish the entire circuit but will only "short out" the individual lamp and continue the circuit operative.

MULTIPLE CIRCUITS. In a multiple circuit all lamps are connected in multiple, and the voltage is maintained constant at 120 volts. A failure in an individual lamp will only cause an outage on that lamp.

DESIGN. Street lighting is provided by either (1) lamp posts fed by underground services, or (2) lamp brackets or swinging lamps fed either from series circuits on primary cross-arms or multiple circuits on secondary racks.

Street lights are required to be turned on and off in accordance with the daily hours of darkness. In order to follow such a schedule it is necessary to control directly the source of supply. This is accomplished by a direct supply circuit from a substation controlled by a time switch or an attendant, a control circuit or pilot wire from a substation, a photo cell, or an astronomical time switch.

Astronomical time switches are used to control individual lamps, series circuits, and groups of multiple lamps.

LIGHTNING ARRESTERS. Lightning arresters are installed on series and multiple circuits to protect the circuits from disturbances due to lightning and induced surges.

STREET-LIGHTING TRANSFORMERS. Constant-current transformers are used in various ratings from 5 to 20 kw. They are used to supply and regulate a series circuit from a constant potential source.

Series insulating transformers are used where it is not desirable to bring the high-voltage series circuit into the lamp socket. They provide a low-voltage secondary circuit which may be grounded separately. Ratings vary from $^1/_{10}$ to 3 kw.

Multiple-series transformers are used to supply small series circuits from a 120-volt source.

52. RECORDS OF CIRCUITS

Underground and overhead circuits are constantly being changed because of extensions additions, removals, and interferences. In order that operations may be conducted quickly and correctly during emergencies or routine work it is necessary that adequate and correct records be maintained of the locations and type of all facilities.

Maps of circuits and plates of secondary grids should be maintained at all times. These maps and plates should show routes of the circuits, points of ties, size and type of cable and apparatus, and type of replacement material.

Master records should be maintained in the central office, and prints of these maps and plates kept in field offices and field trucks. As changes are made in the field, the field prints should be corrected, and the information transmitted immediately to the master records so that changes can be made and new maps and plates issued.

53. BIBLIOGRAPHY

General Reference Books

Edison Electric Institute, *Overhead Systems Reference Book* (1927).
Edison Electric Institute, *Underground Systems Reference Book* (1931).
Gear, H. B., and P. F. William, *Electric Service Distribution Systems*. Van Nostrand (1926).
Parsons, J. S., and H. G. Barnett in *Electrical Transmission and Distribution Reference Book*. Westinghouse Electric and Manufacturing Company (1943).
Ruhling, T. C., *Underground Systems for Electric Light and Power*. McGraw-Hill (1927).
Sanderson, C. H., *Electric System Handbook*. McGraw-Hill (1930).
Seelye, Howard P., *Electrical Distribution Engineering*. McGraw-Hill (1930).

Transactions of American Institute of Electrical Engineers

Chase, P. H., Two-phase, Five-wire Distribution, Vol. 44, p. 737 (1925).
Holben, W. P., A Review of Overhead Secondary Distribution, Vol. 56, pp. 114 and 189 (1937).
Johnson, J. A., and R. T. Henry, Fundamentals of Design of Electric Energy Delivery Systems, Vol. 53, p. 1704 (1934).
Kehoe, A. H., Underground A-c Network Distribution for Central Station Systems. Vol. 43, p. 844 (1924).
Kehoe, A. H., Vertical Networks in Office Buildings, Vol. 50, p. 1159 (1931).
Miller, M. C., Bare Wire Overhead Distribution Practice, Vol. 48, p. 988 (1929).
Richter, H., Combined Light and Power Systems for A-c Networks, Vol. 46, p. 216 (1927).
Sutherland, G., and D. S. MacCorkle, Burn-off Characteristics of A-c Low-voltage Network Cables, Vol. 50, p. 831 (1931).
Wulfing, H. E., Radial and Primary Network Distribution, Vol. 53, p. 38 (1934).
Xenis, C. P., Short-circuit Protection of Networks by the Use of Limiters, Vol. 56, p. 1191 (1937).

Other Periodicals

Beard, J. R., and T. G. N. Haldane, Design of City Distribution Systems and the Problem of Standardization, *J. Amer. Inst. Elec. Engineers*, Vol. 65, p. 97 (January, 1927).
Blake, D. K., Low-voltage A-c Networks, *Gen. Elec. Rev.*, pp. 82, 140, 186 (February, March and April, 1928).
Parsons, J. S., and H. G. Barnett, Distribution Systems and Primary and Secondary Network Distribution Systems, *Electrical Transmission and Distribution Reference Book* (1943).

Bulletins of Edison Electric Institute, covering Transmission and Distribution Committee reports.

BARE WIRES AND CABLES

By W. A. Del Mar

54. DIMENSIONS, WEIGHTS, AND D-C RESISTANCES OF SOLID WIRES

A wire may be either solid or stranded, i.e., made up of a number of smaller wires twisted or braided together. A large bare stranded wire is usually called a bare cable. Data on the insulation and protection of wires and cables will be found in Articles 63 to 79. Data on **resistance wires** will be found in Section 2, Article 14.

CONDUCTIVITY. Per cent conductivity refers to the **International Annealed Copper Standard.** On the assumption of a resistivity temperature coefficient of 0.00393 at 20 deg cent this per cent conductivity is 0.283 per cent higher than the conductivity referred to Matthiessen's Standard. If the length of a given wire is L cm, its cross-section is A sq cm, and its resistance at 20 deg cent is R_{20} ohms, then the conductivity of this wire is

$$C = \frac{15.328L}{88{,}900AR_{20}} \text{ per cent}$$

Annealed copper usually has a conductivity of about 100 per cent; hard-drawn copper, a conductivity of about 97 per cent. Ordinarily hard-drawn aluminum has a conductivity of 61 per cent. The conductivity of iron or steel wire ranges from 8 to 16 per cent.

The resistances given in this article are in International (U.S.A.) ohms which are greater than the present Legal Absolute ohm by about 5 parts in 10,000. See Section 1, Article 14. This may affect the fourth significant figure. As this book goes to press, it is not known whether the standard tables or the standard of conductivity will be altered to conform to the revised value of the ohm. See Section 1, Article 14.

Table 1 gives the maximum resistivities of copper permitted by ASTM Specifications.

RESISTANCE AT 20 DEG CENT. If the cross-section of a wire is A cir mils and its conductivity C per cent, its resistance, in ohms per 1000 ft, will be

$$\frac{1.0372 \times 10^6}{AC}$$

RESISTANCE AT ANY TEMPERATURE. The tables give resistances at 20 deg cent. These may be converted to resistances at other temperatures by the following formulas.

Let C = per cent conductivity.
R_{20} = resistance of 100 per cent conductivity wire at 20 deg cent (from table)
R_t = resistance of wire of conductivity C at temperature t deg cent.

For copper, $$R_t = \frac{100}{C} R_{20}[1 + 0.00393(t - 20)]$$

aluminum, $$R_t = \frac{61}{C} R_{20}[1 + 0.004(t - 20)]$$

steel, see Table 7.

TABLES FOR SOLID WIRES. Tables 2 to 8 for copper and aluminum are compiled from tables in Circular 31 of the Bureau of Standards. Table 9 is compiled from data published by the American Steel and Wire Company.

Table 1. Maximum Resistivities of Copper Wire as Permitted by ASTM Specifications

Diameters, Inches	Soft-annealed	Medium Hard-drawn	Hard-drawn
	Ohms, Mile, Pound, 20° C		
0.460–0.325	875.20	896.15	900.77
0.324–0.0403	875.20	905.44	910.15
	Per Cent Conductivity		
0.460–0.325	100	97.66	97.16
0.324–0.0403	100	96.66	96.16
	Ohms, Mil, Foot *		
0.046–0.325	10.371	10.619	10.674
0.324–0.0403	10.371	10.729	10.785

* The resistance, in ohms per foot, at 20 deg cent, of a copper conductor of cross-sectional area A cir mils, is equal to this factor divided by A.

Table 2. Solid Copper Wire

A.W.G. or B. & S. Gage; English Units

100 per cent conductivity; density 8.89 at 20 deg cent

Gage No.	Diameter in Mils	Cross-section		Resistance at 20° C or 68° F		Weight, Pounds		Feet per Pound
		Circular Mils	Square Inches	Ohms per 1000 ft	Ohms per Mile	Per 1000 ft	Per Mile	
0000	460.0	211,600	0.1662	0.04901	0.259	640.5	3380	1.561
000	409.6	167,800	0.1318	0.06180	0.326	507.9	2680	1.968
00	364.8	133,100	0.1045	0.07793	0.411	402.8	2130	2.482
0	324.9	105,500	0.08289	0.09827	0.519	319.5	1680	3.130
1	289.3	83,690	0.06573	0.1239	0.654	253.3	1340	3.947
2	257.6	66,370	0.05213	0.1563	0.825	200.9	1060	4.977
3	229.4	52,640	0.04134	0.1970	1.04	159.3	841	6.276
4	204.3	41,740	0.03278	0.2485	1.31	126.4	667	7.914
5	181.9	33,100	0.02600	0.3133	1.65	100.2	529	9.980
6	162.0	26,250	0.02062	0.3951	2.09	79.46	420	12.58
7	144.3	20,820	0.01635	0.4982	2.63	63.02	333	15.87
8	128.5	16,510	0.01297	0.6282	3.32	49.98	264	20.01
9	114.4	13,090	0.01028	0.7919	4.181	39.63	209	25.23
10	101.9	10,380	0.008155	0.9989	5.28	31.43	166	31.82
11	90.74	8,234	0.006467	1.259	6.649	24.92	132	40.12
12	80.81	6,530	0.005129	1.588	8.38	19.77	104	50.59
13	71.96	5,178	0.004067	2.002	10.57	15.68	82.8	63.79
14	64.08	4,107	0.003225	2.525	13.3	12.43	63.3	80.44
15	57.07	3,257	0.002558	3.184	16.8	9.858	52.0	101.4
16	50.82	2,583	0.002028	4.015	21.2	7.818	41.3	127.9
17	45.26	2,048	0.001609	5.064	26.7	6.200	32.7	161.3
18	40.30	1,624	0.001276	6.385	33.7	4.917	26.0	203.4
19	35.89	1,288	0.001012	8.051	42.5	3.899	20.6	256.5
20	31.96	1,022	0.0008023	10.15	53.6	3.092	16.3	323.4
21	28.46	810.1	0.0006363	12.80	67.6	2.452	12.9	407.8
22	25.35	642.4	0.0005046	16.14	85.2	1.945	10.3	514.2
23	22.57	509.5	0.0004002	20.36	108	1.542	8.14	648.4
24	20.10	404.0	0.0003173	25.67	135	1.223	6.46	817.7
25	17.90	320.4	0.0002517	32.37	171	0.9699	5.12	1,031
26	15.94	254.1	0.0001996	40.82	216	0.7692	4.06	1,300
27	14.20	201.5	0.0001583	51.46	272	0.6100	3.22	1,639
28	12.64	159.8	0.0001255	64.90	343	0.4837	2.55	2,067
29	11.26	126.7	0.00009953	81.84	432	0.3836	2.03	2,607
30	10.03	100.5	0.00007894	103.2	545	0.3042	1.61	3,287
31	8.928	79.70	0.00006260	130.1	687	0.2413	1.27	4,145
32	7.950	63.21	0.00004964	164.1	866	0.1913	1.01	5,227
33	7.080	50.13	0.00003937	206.9	1,090	0.1517	0.814	6,591
34	6,305	39.75	0.00003122	260.9	1,380	0.1203	0.635	8,310
35	5.615	31.52	0.00002476	329.0	1,740	0.09542	0.504	10,480
36	5.000	25.00	0.00001964	414.8	2,190	0.07568	0.400	13,210
37	4.453	19.83	0.00001557	523.1	2,762	0.06001	0.317	16,660
38	3.965	15.72	0.00001235	659.6	3,480	0.04759	0.251	21,010
39	3.531	12.47	0.000009793	831.8	4,392	0.03774	0.199	26,500
40	3.145	9.888	0.000007766	1049	5,540	0.02993	0.158	33,410
41	2.800	7,842	0.000006159	1323	6,983	0.02374	0.125	42,130
42	2.494	6.219	0.000004884	1668	8,806	0.01882	0.0994	53,120
43	2.221	4.932	0.000003873	2103	11,100	0.01493	0.0788	66,990
44	1.978	3.911	0.000003072	2652	14,000	0.01184	0.0625	84,470

Table 3. Solid Copper Wire

A.W.G. or B. & S. Gage in Metric Units

100 per cent conductivity; density 8.89 at 20 deg cent

Gage No.	Diameter, mm	Cross-section, sq mm	Ohms per Kilometer 20° C	Kilograms per Kilometer
0000	11.68	107.2	0.1608	953.2
000	10.40	85.03	0.2028	755.9
00	9.266	67.43	0.2557	599.5
0	8.252	53.48	0.3224	475.4
1	7.348	42.41	0.4066	377.0
2	6.544	33.63	0.5126	299.0
3	5.827	26.67	0.6464	237.1
4	5.189	21.15	0.8152	188.0
5	4.621	16.77	1.028	149.1
6	4.115	13.30	1.296	118.2
7	3.665	10.55	1.634	93.78
8	3.264	8.366	2.061	74.37
9	2.906	6.632	2.598	58.98
10	2.588	5.261	3.277	46.77
11	2.305	4.192	4.131	37.10
12	2.053	3.309	5.211	29.42
13	1.828	2.624	6.569	23.34
14	1.628	2.081	8.285	18.50
15	1.450	1.650	10.45	14.67
16	1.291	1.309	13.18	11.63
17	1.150	1.038	16.61	9.226
18	1.024	0.8231	20.95	7.317
19	0.9116	0.6527	26.42	5.803
20	0.8118	0.5176	33.31	4.602
21	0.7230	0.4105	42.00	3.649
22	0.6438	0.3255	52.96	2.894
23	0.5733	0.2582	66.79	2.295
24	0.5106	0.2047	84.22	1.820
25	0.4547	0.1624	106.2	1.443
26	0.4049	0.1288	133.9	1.145
27	0.3606	0.1021	168.8	0.9078
28	0.3211	0.08098	212.9	0.7199
29	0.2859	0.06422	268.5	0.5709
30	0.2546	0.05093	338.6	0.4527
31	0.2268	0.04039	426.9	0.3590
32	0.2019	0.03203	538.3	0.2847
33	0.1798	0.02540	678.8	0.2258
34	0.1601	0.02014	856.0	0.1791
35	0.1426	0.01597	1079	0.1420
36	0.1270	0.01267	1361	0.1126
37	0.1131	0.01005	1716	0.08931
38	0.1007	0.007967	2164	0.07083
39	0.08969	0.006318	2729	0.05617
40	0.07987	0.005010	3441	0.04454
41	0.07113	0.003973	4339	0.03532
42	0.06334	0.003151	5472	0.02801
43	0.05641	0.002499	6900	0.02222
44	0.05023	0.001982	8700	0.01762

Table 4. Solid Copper Wire

British Standard Wire Gage; English Units

100 per cent conductivity; density 8.89 at 20 deg cent

Gage No.	Diameter, mils	Cross-section		Ohms per 1000 ft, 15.6° C or 60° F *	Pounds per 1000 ft
		Circular Mils	Square Inches		
7–0	500	250,000	0.1964	0.04077	756.8
6–0	464	215,300	0.1691	0.04734	651.7
5–0	432	186,600	0.1466	0.05461	564.9
4–0	400	160,000	0.1257	0.06370	484.3
3–0	372	138,400	0.1087	0.07365	418.9
2–0	348	121,100	0.09512	0.08416	366.6
0	324	105,000	0.08245	0.09709	317.8
1	300	90,000	0.07069	0.1132	272.4
2	276	76,180	0.05983	0.1338	230.6
3	252	63,500	0.04988	0.1605	192.2
4	232	53,820	0.04227	0.1894	162.9
5	212	44,940	0.03530	0.2268	136.0
6	192	36,860	0.02895	0.2765	111.6
7	176	30,980	0.02433	0.3290	93.76
8	160	25,600	0.02011	0.3981	77.49
9	144	20,740	0.01629	0.4915	62.77
10	128	16,380	0.01287	0.6221	49.59
11	116	13,460	0.01057	0.7574	40.73
12	104	10,820	0.008495	0.9423	32.74
13	92	8,464	0.006648	1.204	25.62
14	80	6,400	0.005027	1.592	19.37
15	72	5,184	0.004072	1.966	15.69
16	64	4,096	0.003217	2.488	12.40
17	56	3,136	0.002463	3.250	9.493
18	48	2,304	0.001810	4.424	6.974
19	40	1,600	0.001257	6.370	4.843
20	36	1,296	0.001018	7.864	3.923
22	28	784.0	0.0006158	13.00	2.373
24	22	484.0	0.0003801	21.06	1.465
26	18	324.0	0.0002545	31.46	0.9807
28	14.8	219.0	0.0001720	46.54	0.6630
30	12.4	153.8	0.0001208	66.28	0.4654
32	10.8	116.6	0.00009161	87.38	0.3531
34	9.2	84.64	0.00006648	120.4	0.2562
36	7.6	57.76	0.00004536	176.5	0.1748
38	6.0	36.00	0.00002827	283.1	0.1090
40	4.8	23.04	0.00001810	442.4	0.06974
42	4.0	16.00	0.00001257	637.0	0.04843
44	3.2	10.24	0.000008042	995.3	0.03100
46	2.4	5.760	0.000004524	1,769	0.01744
48	1.6	2.560	0.000002011	3,981	0.007749
50	1.0	1.000	0.0000007854	10,190	0.003027

* Let C = per cent conductivity, R_{60} = resistance of 100 per cent conductivity wire at 60 deg fahr (from table), R_t = resistance of wire of conductivity C at any temperature t deg fahr; then

$$R_t = \frac{100}{C} R_{60}[1 + 0.00223(t - 60)]$$

Table 5 (*a*). Solid Copper Wire

"Millimeter Gage"; Metric Units and Circular Mils

100 per cent conductivity; density 8.89 at 20 deg cent

Diameter, mm	Cross-section, sq mm	Ohms per Kilometer, 20° C	Kilograms per Kilometer	Cross-section, cir mils*
10.0	78.54	0.2195	698.2	155,000
9.0	63.62	0.2710	565.6	125,550
8.0	50.27	0.3430	446.9	99,200
7.0	38.48	0.4480	342.1	75,950
6.0	28.27	0.6098	251.4	55,800
5.0	19.64	0.8781	174.6	38,750
4.5	15.90	1.084	141.4	31,380
4.0	12.57	1.372	111.7	24,800
3.5	9.621	1.792	85.53	18,990
3.0	7.069	2.439	62.84	13,950
2.5	4.909	3.512	43.64	9,690
2.0	3.142	5.488	27.93	6,200
1.8	2.545	6.775	22.62	5,010
1.6	2.011	8.575	17.87	3,970
1.4	1.539	11.20	13.69	3,040
1.2	1.131	15.24	10.05	2,230
1.0	0.7854	21.95	6.982	1,550
0.90	0.6362	27.10	5.656	....
0.80	0.5027	34.30	4.469	....
0.70	0.3848	44.80	3.421	....
0.60	0.2827	60.98	2.514	....
0.50	0.1964	87.81	1.746	....
0.45	0.1590	108.4	1.414	....
0.40	0.1257	137.2	1.117	....
0.35	0.09621	179.2	0.8553	....
0.30	0.07069	243.9	0.6284	....
0.25	0.04909	351.2	0.4364	....
0.20	0.03142	548.8	0.2793	....
0.15	0.01767	975.6	0.1571	....
0.10	0.007854	2195	0.06982	....
0.05	0.001964	8781	0.01746	....

* One square millimeter equals 1973.52 circular mils.

Table 5 (*b*). Large Metric Conductors

Square-millimeter Gage; Metric Units, Circular Mils, and Square Inches

Cross-section, sq mm	Cross-section		Cross-section, sq mm	Cross-section	
	Cir Mils	Sq in		Cir Mils	Sq in
1.5	2,960	0.002325	150	296,000	0.2325
2.5	4,934	0.003875	185	365,100	0.2867
4.0	7,894	0.006200	240	473,600	0.3720
6.0	11,840	0.009300	300	592,100	0.4650
10.0	19,740	0.01550	400	789,400	0.6200
16.0	31,580	0.02480	500	986,800	0.7750
25	49,340	0.03875	625	1,233,000	0.9687
35	69,070	0.05425	800	1,579,000	1.240
50	98,680	0.07750	1000	1,974,000	1.550
70	138,100	0.1085			
95	187,500	0.1472			
120	236,800	0.1860			

Table 6. Solid Aluminum Wire

A.W.G. or B. & S. Gage; English Units

61 per cent conductivity; density 2.70

Gage No.	Diameter in Mils	Cross-section		Resistance at 20° C or 68° F *		Weight, Pounds		Feet per Pound
		Circular Mils	Square Inches	Ohms per 1000 ft	Ohms per Mile	Per 1000 ft	Per Mile	
0000	460.0	211,600	0.1662	0.0804	0.424	195	1027	5.14
000	409.6	167,800	0.1318	0.101	0.535	154	815	6.48
00	364.8	133,100	0.1045	0.128	0.675	122	646	8.17
0	324.9	105,500	0.08289	0.161	0.851	97.0	512	10.31
1	289.3	83,690	0.06573	0.203	1.073	76.9	406	13.00
2	257.6	66,370	0.05213	0.256	1.353	61.0	322	16.39
3	229.4	52,630	0.04134	0.323	1.706	48.4	255	20.7
4	204.3	41,740	0.03278	0.408	2.15	38.4	203	26.1
5	181.9	33,100	0.02600	0.514	2.71	30.4	160.7	32.9
6	162.0	26,250	0.02062	0.648	3.42	24.1	127.4	41.4
7	144.3	20,820	0.01635	0.817	4.31	19.1	101.0	52.3
8	128.5	16,510	0.01297	1.03	5.44	15.2	80.2	65.9
9	114.4	13,090	0.01028	1.30	6.86	12.0	63.5	83.1
10	101.9	10,380	0.008155	1.64	8.65	9.55	50.4	104.8
11	90.74	8,234	0.006467	2.06	10.90	7.57	39.8	132
12	80.81	6,530	0.005129	2.61	13.76	6.00	31.7	167
13	71.96	5,178	0.004067	3.28	17.34	4.76	25.1	210
14	64.08	4,107	0.003225	4.14	21.9	3.78	19.93	265
15	57.07	3,257	0.002558	5.22	27.6	2.99	15.81	334
16	50.82	2,583	0.002029	6.59	34.8	2.37	12.54	421
17	45.26	2,048	0.001609	8.31	43.8	1.88	9.94	531
18	40.30	1,624	0.001276	10.5	55.3	1.49	7.89	670
19	35.89	1,288	0.001012	13.2	69.7	1.18	6.25	844
20	31.96	1,022	0.0008023	16.7	87.9	0.939	4.96	1,065
21	28.46	810.1	0.0006363	21.0	110.9	0.745	3.93	1,343
22	25.35	642.4	0.0005046	26.5	139.8	0.591	3.12	1,693
23	22.57	509.5	0.0004002	33.4	176.3	0.468	2.47	2,130
24	20.10	404.0	0.0003173	42.1	222	0.371	1.961	2,690
25	17.90	320.4	0.0002517	53.1	280	0.295	1.556	3,390
26	15.94	254.1	0.0001996	67.0	353	0.234	1.233	4,280
27	14.20	201.5	0.0001583	84.4	446	0.185	0.978	5,400
28	12.64	159.8	0.0001255	106	562	0.147	0.776	6,810
29	11.26	126.7	0.00009953	134	709	0.117	0.615	8,580
30	10.03	100.5	0.00007894	169	894	0.0924	0.488	10,820
31	8.928	79.70	0.00006260	213	1127	0.0733	0.387	13,650
32	7.950	63.21	0.00004964	269	1421	0.0581	0.307	17,210
33	7.080	50.13	0.00003937	339	1792	0.0461	0.243	21,700
34	6.305	39.75	0.00003122	428	2260	0.0365	0.1929	27,400
35	5.615	31.52	0.00002476	540	2850	0.0290	0.1530	34,510

* Let C = per cent conductivity, R_{20} = resistance of 61 per cent conductivity wire at 20 deg cent (from table). R_t = resistance of wire of conductivity C at any temperature t deg cent; then

$$R_t = \frac{61 R_{20}}{C} [1 + 0.004(t - 20)]$$

Pounds per 1000 ft = 0.9145 × cir mils × 10^{-3}

Table 7. Solid Aluminum Wire

A.W.G. or B. & S. Gage in Metric Units

61 per cent conductivity; density 2.70; temperature 20 deg cent or 68 deg fahr *

Gage No.	Diameter, mm	Cross-section, sq mm	Ohms per Kilometer	Kilograms per Kilometer
0000	11.68	107.2	0.264	289
000	10.40	85.03	0.333	230
00	9.266	67.43	0.419	182
0	8.252	53.48	0.529	144
1	7.348	42.41	0.667	114
2	6.544	33.63	0.841	90.8
3	5.827	26.67	1.06	72.0
4	5.189	21.15	1.34	57.1
5	4.621	16.77	1.69	45.3
6	4.115	13.30	2.13	35.9
7	3.665	10.55	2.68	28.5
8	3.264	8.366	3.38	22.6
9	2.906	6.632	4.26	17.91
10	2.588	5.261	5.38	14.2
11	2.305	4.192	6.74	11.32
12	2.053	3.309	8.55	8.93
13	1.828	2.624	10.77	7.08
14	1.628	2.081	13.6	5.62
15	1.450	1.650	17.1	4.46
16	1.291	1.309	21.6	3.53
17	1.150	1.038	27.3	2.80
18	1.024	0.8231	34.4	2.22
19	0.9116	0.6527	43.3	1.76
20	0.8118	0.5176	54.6	1.40
21	0.7230	0.4105	68.9	1.11
22	0.6438	0.3255	86.9	0.879
23	0.5733	0.2582	110	0.697
24	0.5106	0.2047	138	0.553
25	0.4547	0.1624	174	0.438
26	0.4049	0.1288	220	0.348
27	0.3606	0.1021	277	0.276
28	0.3211	0.08098	349	0.219
29	0.2859	0.06422	440	0.173
30	0.2546	0.05093	555	0.138
31	0.2268	0.04039	700	0.109
32	0.2019	0.03203	883	0.0865
33	0.1798	0.02540	1110	0.0686
34	0.1601	0.02014	1400	0.0544
35	0.1426	0.01597	1770	0.0431

* Let C = per cent conductivity, R_{20} = resistance of 61 per cent conductivity wire at 20 deg cent (from table), R_t = resistance of wire of conductivity C at any temperature t deg cent; then

$$R_t = \frac{61R_{20}}{C}[1 + 0.004(t - 20)]$$

The temperature coefficient is approximate only.

Table 8. Solid Steel Wire

American Steel Wire Gage; English Units

12.5 per cent conductivity; density 7.78

Am. Steel Wire Gage No.	Diameter		Cross-section		Resistance at 20° C or 68° F *		Weight in Pounds		Feet per Pound
	In.	Mils	Circular Mils	Square Inches	Ohms per 1000 ft	Ohms per Mile	per 1000 ft	per Mile	
	1/2	500.0	250,000	0.1964	0.332	1.752	662.5	3499	1.51
7-0		490.0	240,100	0.1886	0.346	1.825	636.3	3360	1.57
	15/32	468.8	219,800	0.1726	0.378	1.993	582.4	3075	1.72
6-0		460.0	211,600	0.1662	0.392	2.07	560.8	2961	1.78
	7/16	437.5	191,400	0.1503	0.433	2.29	507.2	2678	1.97
5-0		430.0	184,900	0.1452	0.449	2.37	490.0	2587	2.04
	13/32	406.3	165,000	0.1296	0.503	2.65	436.8	2306	2.28
4-0		393.8	155,100	0.1218	0.535	2.82	411.9	2175	2.42
	3/8	375.0	140,600	0.1104	0.590	3.12	372.6	1967	2.68
3-0		362.5	131,400	0.1032	0.631	3.33	348.2	1839	2.87
	11/32	343.8	118,200	0.09280	0.702	3.71	313.1	1653	3.19
2-0		331.0	109,600	0.08605	0.757	4.00	290.3	1533	3.44
	5/16	312.5	97,660	0.07670	0.850	4.49	258.8	1366	3.86
0		306.5	93,940	0.07378	0.883	4.66	249.0	1315	4.02
1		283.0	80,090	0.06290	1.036	5.47	212.2	1121	4.71
	9/32	281.3	79,100	0.06213	1.049	5.54	209.6	1107	4.77
2		262.5	68,910	0.05412	1.204	6.36	182.6	964.1	5.48
	1/4	250.0	62,500	0.04909	1.328	7.01	165.6	874.5	6.04
3		243.7	59,490	0.04665	1.397	7.38	157.4	831.0	6.35
4		225.3	50,760	0.03987	1.635	8.63	134.5	710.2	7.43
	7/32	218.8	47,850	0.03758	1.734	9.15	126.8	669.5	7.89
5		207.0	42,850	0.03365	1.936	10.22	113.6	599.5	8.81
6		192.0	36,860	0.02895	2.25	11.88	97.7	515.8	10.23
	3/16	187.5	35,160	0.02761	2.36	12.46	93.2	491.9	10.73
7		177.0	31,330	0.02461	2.65	13.98	83.0	438.4	12.04
8		162.0	26,240	0.02061	3.16	16.69	69.6	367.2	14.38
	5/32	156.3	24,410	0.01917	3.40	17.95	64.7	341.6	15.46
9		148.3	21,990	0.01727	3.77	19.92	58.3	307.8	17.16
10		135.0	18,200	0.01431	4.55	24.0	48.3	255.0	20.70
	1/8	125.0	15,630	0.01227	5.31	28.0	41.4	218.6	24.15
11		120.5	14,520	0.01140	5.71	30.2	38.5	203.2	25.98
12		105.5	11,130	0.00874	7.45	39.4	29.5	155.7	33.90
	3/32	93.8	8,789	0.00690	9.44	49.8	23.3	123.0	42.94
13		91.5	8,372	0.00658	9.91	52.3	22.1	117.2	45.16
14		80.0	6,400	0.00503	12.96	68.5	17.0	89.55	58.97
15		72.0	5,184	0.00407	16.01	84.5	13.7	72.53	72.80
16		62.5	3,906	0.00307	21.2	112.1	10.4	54.66	96.60
	1/16	62.5	3,906	0.00307	21.2	112.1	10.4	54.66	96.60
17		54.0	2,916	0.00229	28.5	150.2	7.73	40.80	129.5
18		47.5	2,256	0.00177	36.8	194.2	5.98	31.57	167.2
19		41.0	1,681	0.00132	49.4	261	4.45	23.52	224.4
20		34.8	1,211	0.00095	68.5	362	3.21	16.95	311.5
21		31.8	1,008	0.00079	82.3	435	2.67	14.11	374.4
	1/32	31.3	977	0.00076	85.0	449	2.59	13.66	386.5
22		28.6	818	0.00064	101.4	536	2.17	11.45	461.1
23		25.8	666	0.00052	124.6	658	1.76	9.31	567.0
24		23.0	529	0.00042	156.8	828	1.40	7.40	713.5
25		20.4	416	0.00033	199.4	1053	1.10	5.82	907.0

* Let C = per cent conductivity,
R_{20} = resistance of 12.5 per cent conductivity wire at 20 deg cent (from table),
R_t = resistance of wire of conductivity C at any temperature t deg cent; then

$$R_t = \frac{12.5\ R_{20}}{C}[1 + 0.006(t - 20)]$$

The temperature coefficient is approximate only.

Table 8. Solid Steel Wire—*Continued*

American Steel Wire Gage; English Units

12.5 per cent conductivity; density 7.78

Am. Steel Wire Gage No.	Diameter		Cross-section		Resistance at 20° C or 68° F *		Weight in Pounds		Feet per Pound
	In.	Mils	Circular Mils	Square Inches	Ohms per 1000 ft	Ohms per Mile	per 1000 ft	per Mile	
26		18.1	328	0.00026	253	1337	0.87	4.58	1152
27		17.3	299	0.00024	277	1464	0.79	4.19	1261
28		16.2	262	0.00021	316	1669	0.70	3.67	1438
29		15.0	225	0.00018	469	1947	0.60	3.15	1677
30		14.0	196	0.00015	424	2240	0.52	2.74	1925
31		13.2	174	0.00014	476	2510	0.46	2.44	2166
32		12.8	164	0.00013	506	2670	0.43	2.30	2303
33		11.8	139	0.00011	596	3150	0.37	1.95	2710
34		10.4	108	0.00008	767	4050	0.29	1.51	3489
35		9.5	90	0.00007	919	4850	0.24	1.26	4193
36		9.0	81	0.00006	1023	5410	0.21	1.13	4659

* Let C = per cent conductivity,
R_{20} = resistance of 12.5 per cent conductivity wire at 20 deg cent (from table),
R_t = resistance of wire of conductivity C at any temperature t deg cent; then

$$R_t = \frac{12.5\,R_{20}}{C}[1 + 0.006(t - 20)]$$

The temperature coefficient is approximate only.

COPPERWELD WIRE. This wire consists of a steel core and a concentric coat of copper permanently welded to it. It is used chiefly for long-span transmission and messenger cable. It is made in several grades, which differ in the relative amounts of steel and copper. The grades are designated by the corresponding conductivity expressed as percentages of the Annealed Copper Standard: e.g., 40 per cent grade has a conductivity of 40 per cent. Properties of this wire are given in Table 9.

Table 9. Copperweld Wire

A. W. G. or B. & S. Gage; English Units

40 per cent conductivity; density 8.26

Gage No.	Diameter, mils	Cross-section		Resistance at 23.9° C or 75° F *		Weight in Pounds		Feet per Pound
		Circular Mils	Square Inches	Ohms per 1000 ft	Ohms per Mile	per 1000 ft	per Mile	
0000	460.0	211,600	0.1662	0.123	0.649	595	3140	1.68
000	409.6	167,800	0.1318	0.154	0.813	471	2490	2.12
00	364.8	133,100	0.1045	0.195	1.03	374	1970	2.67
0	324.9	105,500	0.08289	0.246	1.30	297	1570	3.37
1	289.3	83,690	0.06573	0.310	1.64	235	1240	4.26
2	257.6	66,370	0.05213	0.390	2.06	186	982	5.38
3	229.4	52,630	0.04134	0.492	2.60	148	781	6.76
4	204.3	41,740	0.03278	0.622	3.28	117	618	8.55
5	181.9	33,100	0.02600	0.782	4.13	92.9	491	10.76
6	162.0	26,250	0.02062	0.987	5.21	73.7	389	13.57
7	144.3	20,820	0.01635	1.25	6.60	58.5	309	17.09
8	128.5	16,510	0.01297	1.57	8.29	46.4	245	21.6
9	114.4	13,090	0.01028	1.98	10.5	36.8	194	27.2
10	101.9	10,380	0.008155	2.50	13.2	29.2	154	34.2
11	90.74	8,234	0.006467	3.15	16.6	23.1	122	43.3
12	80.81	6,530	0.005129	3.97	21.0	18.3	96.6	54.6
13	71.96	5,178	0.004067	5.00	26.4	14.6	77.1	68.5
14	64.08	4,107	0.003225	6.31	33.3	11.5	60.7	87.0

* Let C = per cent conductivity,
$R_{23.9}$ = resistance of 40 per cent conductivity wire at 23.9 deg cent (from table),
R_t = resistance of wire of conductivity C at temperature t deg cent; then

$$R_t = \frac{40R_{23.9}}{C}[1 + 0.00432(t - 23.9)]$$

The temperature coefficient is approximate only.

ALLOY WIRES OF HIGH TENSILE STRENGTH. Copper alloys having a low conductivity but a tensile strength from 50 per cent to 100 per cent greater than that of copper are sometimes used where strength or hardness is a primary requisite. as in long spans of small wires or for trolley wires. See Article 56.

55. DIMENSIONS, WEIGHTS, AND D-C RESISTANCES OF STRANDED CONDUCTORS

FACTORS AFFECTING DIMENSIONS, WEIGHTS, AND RESISTANCES OF STRANDED WIRES. Individual stranded wires or cables are of four different types, namely: (*a*) bunched wire; (*b*) wire braids; (*c*) concentric-lay cables; and (*d*) rope-lay cables.

Bunched Wires. Bunched wires are used especially for those extra-flexible cables known as cords, wherein the individual wires are so small that concentric stranding is not necessary to keep them together. The wires are assembled parallel and then generally given a slight twist. Sometimes they are kept together by being wound with soft cotton thread, which also serves to prevent adhesion between the insulation and wires.

Wire Braids. In the flat form, wire braids are used for potential leads, etc., in lighting cables, where a flexible flat conductor is necessary. Tubular wire braids, known as **basket weave,** are also frequently formed over the insulation of cables in order to afford mechanical protection. Cables for naval or marine purposes and for automobile work are frequently thus protected.

Concentric-lay Cables. A concentric-lay cable is a stranded conductor composed of a central core surrounded by one or more layers of helically laid wires. A **rope-lay cable** is a stranded wire made up in the same manner by using stranded wires instead of individual solid wires for the core and layers. The cores of concentric-lay cables may be composed of one, two, three, or four wires of equal diameter. A five- or six-wire core would not be symmetrical, and seven wires would themselves constitute a core and a layer.

Number of Wires in Concentric-lay Cables. Table 10 gives all the possible concentric-lay cables with eight or fewer layers of equal size wires and formulas for calculating the number of wires with any number of layers. (See Table 14.)

Table 10. Number of Wires in Concentric-lay Cables

(All wires of same diameter)

Number of Layers over Core	Number of Wires in Core			
	1	2	3	4
0	1	2	3	4
1	7	10	12	14
2	19	24	27	30
3	37	44	48	52
4	61	70	75	80
5	91	102	108	114
6	127	140	147	154
7	169	184	192	200
8	217	234	243	252
n	$3n^2 + 3n + 1$	$3n^2 + 5n + 2$	$3n^2 + 6n + 3$	$3n^2 + 7n + 4$

The number of wires per layer increases by six for each successive layer when the core has one wire, the first layer over the core having six. With cores having more than one wire, the increment per layer is not constant.

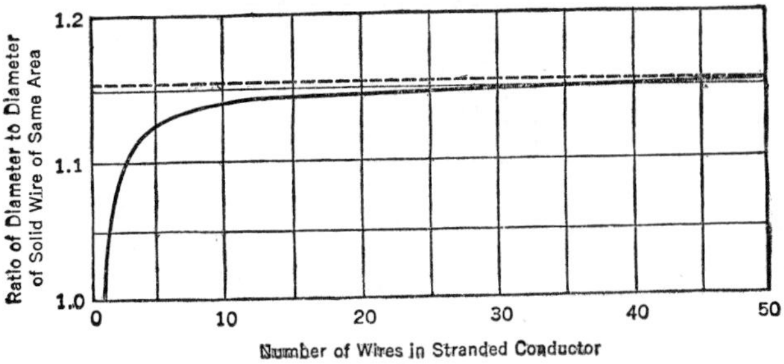

FIG. 1

Diameter of Concentric-lay Cables. The diameter of the circumscribing circle of any of the above cables is equal to $(2n + b)$ times the diameter of each wire, where n is the number of layers over the core and b has the following values: 1 wire in core, $b = 1$; 2 wires in core, $b = 2$; 3 wires in core, $b = 2.155$; 4 wires in core, $b = 2.414$.

The relation between the number of component wires and the diameter of the cable is shown in Fig. 1.

Table 11. Wires in Rope-lay Cables

Number of Layers over Core	Number of Strands *	Total Number of Wires — Wires per Strand * 7	19	37	61	91
0	1	7	19	37	61	91
1	7	49	133	259	427	637
2	19	133	361	703	1159	1,729
3	37	259	703	1369	2257	3,367
4	61	427	1159	2257	3721	5,551
5	91	637	1729	3367	5551	8,281
6	127	889	2413	4699	7747	11,557
n	$3(n^2 + n) + 1$	$21(n^2 + n) + 7$	$57(n^2 + n) + 19$	$111(n^2 + n) + 37$	$183(n^2 + n) + 61$	$273(n^2 + n) + 91$

* By "strand" is here meant the group of stranded wires of which the rope is built up.

The number of wires in a rope-lay cable is frequently designated by a product; thus, 7 × 19 indicates a conductor made up of 7 strands, each strand containing 19 wires.

Table 12. Properties of Concentric-lay Cables

		Regular; $d_c = d$	Special; $d_c \neq d$
I.	Number of wires in terms of number of layers (and core diameter)	$N = 3\,(n^2 + n) + 1$ (including core)	$N = 3\left(n\frac{d_c}{d} + n^2\right)$ (excluding core)
II.	Diameter of cable in terms of diameter of wires and number of layers	$D = d\,(1 + 2n)$	$D = d_c + 2nd$
III.	Diameter of cable in terms of total area and number of wires	$D = 10^{-3}\sqrt{\frac{1}{3}\left(4 - \frac{1}{N}\right)}\cdot\sqrt{A}$	
IV.	Ratio of wire area to area of circle circumscribing the outside of cable	$R = \frac{3(n^2 + n) + 1}{(2n + 1)^2}$	
V.	Weight of cable in terms of weight of wire, number of layers and pitch factors	$W = w\,(1 + 6p_6 + 12p_{12} + \text{etc.})$	$W = w_c + w\,(6p_6 + 12p_{12} + \text{etc.})$
VI.	Strength of cable in terms of strength of the component wires and the pitch factors	$T = \frac{\pi}{4}d^2\left[s + t\left(\frac{6}{p_6} + \frac{12}{p_{12}} + \text{etc.}\right)\right]$	$T = \frac{\pi}{4}\left[sd_c^2 + d^2t\left(\frac{6}{p_6} + \frac{12}{p_{12}} + \text{etc.}\right)\right]$
VII.	Minimum pitch in terms of wire diameter and core diameter	$\frac{3\,\pi dp}{\sqrt{(\pi + 3)(\pi - 3)}} = 10.1$ times pitch diameter	$\frac{\pi d_p ld}{\sqrt{(\pi d_p)^2 - (ld)^2}}$
VIII.	Diameter of wires in terms of total conductor area and number of wires	$d = \frac{1}{1000}\sqrt{\frac{A}{N}}$	
IX.	Volume of space, not occupied by wires, within circumscribed cylinder, cubic inches per 1000 ft	kd^2, where k = 16,366 for 7 strands; 31,356 " 19 "; 49,765 " 37 "; 71,741 " 61 "; 97,022 " 91 "	

Rope-lay Cables. As already noted, a rope-lay cable is made up in the same way as a concentric-lay cable except that stranded wires are used for the core and layers instead of individual solid wires. Rope strands are used for large conductors which would be too stiff if stranded concentrically. The formulas for regular concentric-lay cables may be readily modified to apply to rope-lay cables, as each stranded wire bears the same relation to the rope as each individual solid wire does to the concentric-lay cable. Table 11 gives the principal forms of rope-lay cables.

Diameters of Component Wires in Commercial Stranded Conductors. Table 16 gives the diameters of the component wires in the types of stranded conductors ordinarily used.

Effect of Lay on Resistance and Weight. In Tables 14, 15, and 17, for stranded cables, the values given for "ohms per unit length" and "weight per unit length" are 2 per cent greater than for a solid rod of cross-section equal to the total cross-section of the wires of the cable. This increment of 2 per cent means that the values are correct for cables having a lay of 1 in 15.7. For any other lay, equal to 1 in n, resistance or mass may be calculated by increasing the above tabulated values by

$$\left(\frac{484}{n^2} - 2\right) \text{ per cent}$$

Table 13. Number of Wires for Concentric Standing

(IPCEA)

Class A for bare, weatherproof, and slow-burning cables.
Class B for rubber, varnished-cambric, and paper cables for normal use.
Class C for use where greater degrees of flexibility are desired and for rubber-insulated cables for over 1000 volts.
Class D for cables under class C where extra flexible stranding is desired.

M C M or A.W.G.	Class A	Class B	Class C	Class D	M C M or A.W.G.	Class A	Class B	Class C	Class D
5000	169	217	271	271	500	37	37	61	91
4500	169	217	271	271	450	37	37	61	91
4000	169	217	271	271	400	19	37	61	91
3500	127	169	217	271	350	19	37	61	91
3000	127	169	217	271	300	19	37	61	91
2500	91	127	169	217	250	19	37	61	91
2000	91	127	169	217	4/0	7–19	19	37	61
1900	91	127	169	217	3/0	7–19	19	37	61
1800	91	127	169	217	2/0	7	19	37	61
1750	91	127	169	217	1/0	7	19	37	61
1700	91	127	169	217	1	7	19	37	61
1600	91	127	169	217	2	7	7	19	37
1500	61	91	127	169	3	7	7	19	37
1400	61	91	127	169	4	7	7	19	37
1300	61	91	127	169	5	7	7	19	37
1250	61	91	127	169	6	1	1	19	37
1200	61	91	127	169	7	1	1	19	37
1100	61	91	127	169	8	1	1	19	37
1000	61	61	91	127	9	1	1	7	19
900	61	61	91	127	10	1	1	7	19
800	61	61	91	127	12	1	1	7	19
750	61	61	91	127	14	1	1	7	19
700	61	61	91	127	16	1	1	7	7
650	61	61	91	127	18	1	1	7	7
600	37	61 *	91	127	20	1	1	7	7
550	37	61 *	91	127					

* Thirty-seven wires permissible for rubber-insulated cables.

General Formulas for Properties of Cables in Terms of the Properties of the Constituent Wires. Table 12 gives the principal formulas for concentric-lay cable having a core of one wire.

A = total area, in circular mils, of the component wires measured at right angles to their axes, when laid out straight.
D = diameter of cable overall, in inches.
d = diameter of each of the component wires, in inches
d_c = diameter of core, in inches.
d_p = pitch diameter, in inches, of any layer (= mean diameter of the helix made by any layer).
e = elongation, per cent, at which the wires (other than the core) break.

l = number of wires in any layer having pitch diameter d_p.
N = total number of wires except where the core is of special size, in which case N is the number of wires exclusive of the core.
n = number of layers of wire over the core.
P = pitch of any layer of wires = distance, in inches, measured along the axis of the cable for one complete turn of the helix formed by any wire of this layer.
p = pitch-factor of any layer of wires = ratio of the actual length of a wire to the corresponding axial length of the cable.
R = ratio of wire area to the total area of the circle circumscribing the outside of the conductor.
s = stress, in pounds per square inch, in the core when the elongation is e.
t = tensile strength of each outer wire, in pounds per square inch.
T = tensile strength of conductor, in pounds.
W = weight of conductor, in pounds per foot.
w = weight of each wire of the cable, in pounds per foot.
w_c = weight of the core of the cable, in pounds per foot.

Table 14. Copper Cables, Concentric-lay

Circular Mils and A.W.G. or B. & S. Gage; English Units

100 per cent conductivity; density 8.89 at 20 deg cent

Circular Mils and A.W.G.	Resistance at 25° C or 77° F		Weight in Pounds, Bare		Class B Stranding			Class C Stranding		
	Ohms per 1000 ft	Ohms per Mile	per 1000 ft	per Mile	Number of Wires	Diameter of Wires, mils	Outside Diameter, mils	Number of Wires	Diameter of Wires, mils	Outside Diameter, mils
2,000,000	0.00539	0.0285	6180	32600	127	125.5	1631	169	108.8	1632
1,900,000	0.00568	0.0300	5870	31000	127	122.3	1590	169	106.0	1590
1,800,000	0.00599	0.0316	5560	29300	127	119.1	1548	169	103.2	1548
1,700,000	0.00634	0.0335	5250	27700	127	115.7	1504	169	100.3	1504
1,600,000	0.00674	0.0356	4940	26100	127	112.2	1459	169	97.3	1460
1,500,000	0.00719	0.0380	4630	24500	91	128.4	1412	127	108.7	1413
1,400,000	0.00770	0.0407	4320	22800	91	124.0	1364	127	105.0	1365
1,300,000	0.00830	0.0438	4010	21200	91	119.5	1315	127	101.2	1315
1,200,000	0.00899	0.0475	3710	19600	91	114.8	1263	127	97.2	1264
1,100,000	0.00981	0.0518	3400	17900	91	109.9	1209	127	93.1	1210
1,000,000	0.0108	0.0570	3090	16300	61	128.0	1152	91	104.8	1153
950,000	0.0114	0.0600	2930	15490	61	124.8	1123	91	102.2	1124
900,000	0.0120	0.0633	2780	14670	61	121.5	1093	91	99.4	1094
850,000	0.0127	0.0670	2620	13860	61	118.0	1062	91	96.6	1063
800,000	0.0135	0.0712	2470	13040	61	114.5	1031	91	93.8	1031
750,000	0.0144	0.0759	2320	12230	61	110.9	998	91	90.8	999
700,000	0.0154	0.0814	2160	11410	61	107.1	964	91	87.7	965
650,000	0.0166	0.0876	2010	10600	61	103.2	929	91	84.5	930
600,000	0.0180	0.0949	1850	9780	61	99.2	893	91	81.2	893
550,000	0.0196	0.1036	1700	8970	61	95.0	855	91	77.7	855
500,000	0.0216	0.1139	1540	8150	37	116.2	814	61	90.5	815
450,000	0.0240	0.1266	1390	7340	37	110.3	772	61	85.9	773
400,000	0.0270	0.1424	1240	6520	37	104.0	728	61	81.0	729
350,000	0.0308	0.1627	1080	5710	37	97.3	681	61	75.7	682
300,000	0.0360	0.1899	926	4890	37	90.0	630	61	70.1	631
250,000	0.0431	0.228	772	4080	37	82.2	575	61	64.0	576
0000	0.0509	0.269	653	3450	19	105.5	528	37	75.6	533
000	0.0642	0.339	518	2735	19	94.0	470	37	67.3	471
00	0.0811	0.428	411	2170	19	83.7	418	37	60.0	420
0	0.102	0.540	326	1720	19	74.5	373	37	53.4	374
1	0.129	0.681	258	1364	19	66.4	332	37	47.6	333
2	0.162	0.858	205	1082	7	97.4	292	19	59.1	296
3	0.205	1.082	163	858	7	86.7	260	19	52.6	263
4	0.259	1.365	129	680	7	77.2	232	19	46.9	234
5	0.326	1.721	102	540	7	68.8	206	19	41.7	209
6	0.410	2.170	81.0	428	7	61.2	184	19	37.2	186
7	0.519	2.74	64.3	339	7	54.5	164	19	33.1	166
8	0.654	3.45	51.0	269	7	48.6	146	19	29.5	147

Table 15. Copper Cables, Concentric-lay

Circular Mils and A. W. G. or B. & S. Gage in Metric Units

100 per cent conductivity; density 8.89 at 20 deg cent

Circular Mils and A.W.G.	Total Cross-section, sq mm	Ohms per Kilometer at 25° C	Kilograms per Kilometer, Bare	Class B Stranding			Class C Stranding		
				Number of Wires	Diameter of Wires, mm	Outside Diameter, mm	Number of Wires	Diameter of Wires, mm	Outside Diameter, mm
2,000,000	1013	0.0177	9190	127	3.19	41.4	169	2.76	41.4
1,900,000	963	0.0186	8730	127	3.11	40.4	169	2.69	40.4
1,800,000	912	0.0197	8270	127	3.02	39.3	169	2.62	39.3
1,700,000	861	0.0208	7810	127	2.94	38.2	169	2.55	38.2
1,600,000	811	0.0221	7350	127	2.85	37.1	169	2.47	37.1
1,500,000	760	0.0236	6890	91	3.26	35.9	127	2.76	35.9
1,400,000	709	0.0253	6430	91	3.15	34.7	127	2.67	34.7
1,300,000	659	0.0272	5970	91	3.04	33.4	127	2.57	33.4
1,200,000	608	0.0295	5510	91	2.92	32.1	127	2.47	32.1
1,100,000	557	0.0322	5050	91	2.79	30.7	127	2.36	30.7
1,000,000	507	0.0354	4590	61	3.25	29.3	91	2.66	29.3
950,000	481	0.0373	4370	61	3.17	28.5	91	2.60	28.5
900,000	456	0.0393	4140	61	3.09	27.8	91	2.53	27.8
850,000	431	0.0416	3910	61	3.00	27.0	91	2.45	27.0
800,000	405	0.0442	3680	61	2.91	26.2	91	2.38	26.2
750,000	380	0.0472	3450	61	2.82	25.3	91	2.31	25.4
700,000	355	0.0506	3220	61	2.72	24.5	91	2.23	24.5
650,000	329	0.0544	2990	61	2.62	23.6	91	2.15	23.6
600,000	304	0.0590	2760	61	2.52	22.7	91	2.06	22.7
550,000	279	0.0643	2530	61	2.41	21.7	91	1.97	21.7
500,000	253	0.0708	2300	37	2.95	20.7	61	2.30	20.7
450,000	228	0.0786	2070	37	2.80	19.6	61	2.18	19.6
400,000	203	0.0885	1840	37	2.64	18.5	61	2.06	18.5
350,000	177	0.101	1610	37	2.47	17.3	61	1.92	17.3
300,000	152	0.118	1380	37	2.29	16.0	61	1.78	16.0
250,000	127	0.142	1150	37	2.09	14.6	61	1.63	14.6
0000	107	0.167	972	19	2.68	13.4	37	1.93	13.5
000	85	0.211	771	19	2.39	11.9	37	1.71	12.0
00	67.4	0.266	611	19	2.13	10.6	37	1.52	10.7
0	53.5	0.334	485	19	1.89	9.46	37	1.36	9.50
1	42.4	0.423	385	19	1.69	8.43	37	1.21	8.46
2	33.6	0.533	305	7	2.47	7.42	19	1.50	7.51
3	26.7	0.673	242	7	2.20	6.61	19	1.34	6.68
4	21.2	0.849	192	7	1.96	5.88	19	1.19	5.95
5	16.8	1.07	152	7	1.75	5.24	19	1.06	5.30
6	13.3	1.35	121	7	1.56	4.67	19	0.944	4.72
7	10.5	1.70	95.7	7	1.39	4.16	19	0.841	4.20
8	8.37	2.14	75.9	7	1.23	3.70	19	0.749	3.74

Table 16. Commercial Stranded Conductors

Area of Conductor, cir mils	Number of Wires in the Stranded Conductor								
	7	19	37	7×7=49	61	91	127	169	217
	Diameter, in Inches, of Each Wire in the Cable								
2,000,000	0.5345	0.3244	0.2325	0.202	0.181	0.1482	0.1255	0.1086	0.096
1,750,000	0.5000	0.3035	0.2175	0.189	0.169	0.1387	0.1174	0.1020	0.090
1,500,000	0.4629	0.2810	0.2013	0.175	0.157	0.1285	0.1087	0.0940	0.083
1,250,000	0.4226	0.2565	0.1838	0.1507	0.143	0.1174	0.0992	0.0860	0.076
1,000,000	0.3779	0.2294	0.1644	0.1429	0.1285	0.1048	0.0887	0.0769	0.0678
950,000	0.3684	0.2236	0.1602	0.1392	0.1247	0.1021	0.0864	0.0749	0.0661
900,000	0.3585	0.2176	0.1559	0.1355	0.1214	0.0995	0.0841	0.0729	0.0644
850,000	0.3484	0.2115	0.1515	0.1317	0.1180	0.0966	0.0818	0.0709	0.0625
800,000	0.3380	0.2050	0.1470	0.1278	0.1145	0.0937	0.0793	0.0687	0.0607
750,000	0.3273	0.1986	0.1423	0.1237	0.1108	0.0907	0.0769	0.0666	0.0588
700,000	0.3163	0.1919	0.1375	0.1195	0.1071	0.0887	0.0742	0.0643	0.0567
650,000	0.3047	0.1849	0.1325	0.1152	0.1032	0.0845	0.0715	0.0620	0.0547
600,000	0.2927	0.1776	0.1273	0.1107	0.0991	0.0812	0.0687	0.0595	0.0525
550,000	0.2803	0.1701	0.1219	0.1060	0.0949	0.0777	0.0658	0.0571	0.0503
500,000	0.2673	0.1622	0.1162	0.0110	0.0905	0.0741	0.0628	0.0543	0.0480
450,000	0.2535	0.1538	0.1103	0.0958	0.0858	0.0703	0.0595	0.0516	0.0455
400,000	0.2390	0.1457	0.1039	0.0904	0.0809	0.0663	0.0561	0.0486	0.0429
350,000	0.2236	0.1357	0.0972	0.0845	0.0757	0.0620	0.0526	0.0455	0.0401
300,000	0.2070	0.1256	0.0903	0.0783	0.0701	0.0574	0.0486	0.0421	0.0371
250,000	0.1889	0.1147	0.0824	0.0714	0.0640	0.0524	0.0443	0.0384	0.0339
Size A.W.G.									
0000	0.1739	0.1055	0.0756	0.0657	0.0589	....	....	....	....
000	0.1548	0.0940	0.0674	0.0586	0.0525	....	....	....	....
00	0.1379	0.0837	0.0600	0.0521	0.0467	....	....	....	....
0	0.1228	0.0745	0.0534	0.0464	0.0416	....	....	....	....
1	0.1094	0.0664	0.0475	0.0413	....	....	....	....	....
2	0.0974	0.0591	0.0424	0.0369	....	....	....	....	....
3	0.0867	0.0525	0.0377	0.0327	....	....	....	....	....
4	0.0772	0.0468	0.0335	0.0291	....	....	....	....	....
5	0.0688	0.0418	0.0299	0.0260	....	....	....	....	....
6	0.0612	0.0372	0.0266	0.0231	....	....	....	....	....
7	0.0545	0.0331	0.0237	0.0206	....	....	....	....	....
8	0.0484	0.0294	0.0211	0.0184	....	....	....	....	....
9	0.0432	0.0263	0.0188	0.0164	....	....	....	....	....
10	0.0386	0.0233	0.0168	....	....	....	....	....	....
12	0.0306	0.0185	0.0133	....	....	....	....	....	....
14	0.0242	0.0148	0.0105	....	....	....	....	....	....

Weights in pounds per 1000 ft of all bare copper cables are computed by multiplying the circular mils by 0.00309.

Table 17. D-c Resistance of Copper Wires and Cables at Various Temperatures *

(Based on the Standards of the AIEE)

Resistance, Ohms of Wire or Cable which is 1000 ft Long at 20 deg cent

(Stranded except for sizes smaller than No. 6 A.W.G.)

Size A.W.G. or Cir In.	20° C 68° F	25° C 77° F	30° C 86° F	35° C 95° F	40° C 104° F	45° C 113° F	50° C 122° F
14	2.525	2.574	2.624	2.674	2.723	2.773	2.822
12	1.588	1.619	1.650	1.682	1.713	1.744	1.775
10	0.9989	1.018	1.038	1.058	1.077	1.097	1.116
8	0.6282	0.6404	0.6527	0.6651	0.6774	0.6898	0.7021
6	0.403	0.410	0.419	0.427	0.435	0.442	0.450
4	0.253	0.259	0.263	0.268	0.273	0.278	0.283
2	0.159	0.162	0.166	0.169	0.172	0.175	0.178
1	0.126	0.129	0.131	0.134	0.136	0.139	0.141
0	0.100	0.102	0.104	0.106	0.108	0.110	0.112
00	0.0795	0.0811	0.0826	0.0842	0.0857	0.0873	0.0888
000	0.0630	0.0642	0.0655	0.0667	0.0680	0.0692	0.0705
0000	0.0500	0.0509	0.0519	0.0529	0.0539	0.0549	0.0559
Cir In.							
0.25	0.0423	0.0431	0.0440	0.0448	0.0456	0.0465	0.0473
0.35	0.0302	0.0308	0.0314	0.0320	0.0326	0.0332	0.0338
0.50	0.0211	0.0216	0.0220	0.0224	0.0228	0.0232	0.0236
0.75	0.0141	0.0144	0.0147	0.0149	0.0152	0.0155	0.0158
1.00	0.0106	0.0108	0.0110	0.0122	0.0114	0.0116	0.0118
1.25	0.00846	0.00863	0.00879	0.00896	0.00913	0.00929	0.00946
1.50	0.00705	0.00719	0.00733	0.00747	0.00760	0.00774	0.00788
1.75	0.00604	0.00616	0.00628	0.00640	0.00652	0.00664	0.00676
2.00	0.00529	0.00539	0.00550	0.00560	0.00570	0.00580	0.00591

Size A.W.G. or Cir In.	55° C 131° F	60° C 140° F	65° C 149° F	70° C 158° F	75° C 167° F	80° C 176° F	85° C 185° F
14	2.872	2.922	2.971	3.021	3.071	3.120	3.170
12	1.806	1.838	1.869	1.900	1.931	1.962	1.994
10	1.136	1.156	1.175	1.195	1.215	1.234	1.254
8	0.7144	0.7268	0.7391	0.7515	0.7638	0.7762	0.7885
6	0.458	0.466	0.474	0.482	0.490	0.498	0.506
4	0.288	0.293	0.298	0.303	0.308	0.313	0.318
2	0.181	0.184	0.188	0.191	0.194	0.197	0.200
1	0.144	0.146	0.149	0.151	0.154	0.156	0.158
0	0.114	0.116	0.118	0.120	0.122	0.124	0.126
00	0.0904	0.0920	0.0935	0.0951	0.0967	0.0982	0.0998
000	0.0717	0.0729	0.0742	0.0754	0.0767	0.0779	0.0791
0000	0.0569	0.0578	0.0588	0.0598	0.0608	0.0618	0.0628
Cir In.							
0.25	0.0481	0.0490	0.0498	0.0505	0.0514	0.0523	0.0531
0.35	0.0344	0.0350	0.0356	0.0362	0.0368	0.0378	0.0379
0.50	0.0241	0.0245	0.0249	0.0253	0.0257	0.0261	0.0266
0.75	0.0160	0.0163	0.0166	0.0169	0.0171	0.0174	0.0177
1.00	0.0120	0.0122	0.0125	0.0127	0.0129	0.0131	0.0133
1.25	0.00962	0.00979	0.00996	0.0101	0.0103	0.0105	0.0106
1.50	0.00802	0.00816	0.00830	0.00844	0.00857	0.00871	0.00885
1.75	0.00687	0.00699	0.00711	0.00723	0.00735	0.00747	0.00759
2.00	0.00602	0.00612	0.00622	0.00633	0.00643	0.00654	0.00664

* The resistances to alternating currents are obtained by multiplying these values by the skin-effect ratio and, in the case of conductors close together, by the combined skin and proximity effect ratio. See Article 59.

56. STRENGTH, ELASTICITY, AND EXPANSION COEFFICIENTS OF WIRES

The strength and elasticity of a wire of any material depend to a considerable extent upon the method of manufacture, heat treatment, etc. The tensile strength of soft copper is between 25,000 and 35,000 lb per sq in., as against 60,000 lb per sq in. for hard-drawn copper. Again, because of the greater relative thickness of the hard "skin" and comparatively soft "core" of small hard-drawn copper wires as compared with large wires, the tensile strength, in pounds per square inch, of a small hard-drawn copper wire is greater than that of a large hard-drawn wire. For example, a No. 0000 A.W.G. hard-drawn copper wire has a tensile strength of about 50,000 lb per sq in. as against approximately 65,000 lb per sq in. for a No. 18. A similar but smaller variation holds for soft annealed wires. The tensile strength of steel wire depends to a very great extent upon the composition of the steel.

METALS AVAILABLE FOR CONDUCTORS. Table 18 gives, for a No. 0 A.W.G. wire, representative values of the various quantities stated for representative metals and alloys used for wires and cables. These values do not hold, except to a rough approximation, for other sizes of wire.

Table 18. Strength, Elasticity, and Coefficient of Expansion

Of a No. 0 A. W. G. Wire

Kind of Wire	Tensile Strength, lb per sq in.	Elastic Limit, lb per sq in.*	Modulus of Elasticity, lb-in. units	Coefficient of Linear Expansion	
				per °F	per °C
Copper, soft-annealed	36,000	....	....	9.6×10^{-6}	17×10^{-6}
Copper, hard-drawn	57,330	30,000	16×10^6	9.6×10^{-6}	17×10^{-6}
Aluminum, soft-annealed	16,000	....	....	12.8×10^{-6}	23×10^{-6}
Aluminum, hard-drawn	25,000	25,000	9×10^6	12.8×10^{-6}	23×10^{-6}
Copper-clad steel, 40% grade	60,000	51,000	22×10^6	6.7×10^{-6}	12×10^{-6}
P. M. G. (Phelps-Dodge) bronzes	60,000 to 140,000	27,000 to 63,000	15×10^6	9.6×10^{-6}	17×10^{-6}
Anaconda copper alloys	69,000 to 135,000	....	13×10^6 to 15×10^6	....	....
Phonoelectric bronze	75,800	55,000	18×10^6	8.3×10^{-6}	14.9×10^{-6}
Steel, ordinary	68,000	40,000	24×10^6 to 30×10^6 (all steels)	7.0×10^{-6}	12.6×10^{-6}
Steel, Siemens-Martin	90,000	45,000		7.0×10^{-6}	12.6×10^{-6}
Steel, high-strength	150,000	82,000		7.0×10^{-6}	12.6×10^{-6}
Steel, extra high-strength	225,000	135,000		7.0×10^{-6}	12.6×10^{-6}

* There is no elastic limit for soft annealed copper, and the elastic limits of hard-drawn copper and aluminum are not precise quantities.

PHYSICAL CHARACTERISTICS OF COPPER WIRE. The maximum and minimum tensile strengths and elongations of soft annealed, medium hard-drawn, and hard-drawn copper wires, as specified by ASTM, are given in Table 19.

TENSILE BREAKING LOAD. The tensile strengths in pounds for solid wires of various metals from $^1/_{16}$ to $^1/_2$ in. in diameter are given in Table 20.

Table 19. Physical Characteristics of Copper Wire

As Required by ASTM Specifications

Diameter, Inches	Soft annealed		Medium Hard-drawn			Hard-drawn	
	Maximum Tensile Strength, psi	Minimum Elongation, Per Cent in 10 in.	Tensile Strength, psi		Minimum Elongation, Per Cent	Minimum Tensile Strength, psi	Maximum Elongation, Per Cent in 10 in.
			Maximum	Minimum			
0.4600	36,000	35	49,000	42,000	3.75	49,000	3.75
0.4096	36,000	35	50,000	43,000	3.60	51,000	3.25
0.3648	36,000	35	51,000	44,000	3.25		
0.3249	36,000	35	52,000	45,000	3.00	52,800	2.80
0.2893	37,000	30	53,000	46,000	2.75	56,100	2.40
0.2576	37,000	30	54,000	47,000	2.50	57,600	2.17
0.2294	37,000	30	55,000	48,000	2.25	59,000	1.98
0.2043	37,000	30	55,330	48,330	1.25	60,100	1.79
0.1819	37,000	30	55,660	48,660	1.20	61,200	1.24
0.1620	37,000	30	56,000	49,000	1.15	62,100	1.14
0.1443	37,000	30	56,330	49,330	1.11	63,000	1.14
0.1285	37,000	30	56,660	49,660	1.08	63,700	1.07
0.1144	37,000	30	57,000	50,000	1.06	64,300	1.06
0.1019	38,500	25	57,330	50,330	1.04	64,900	1.00
0.09074	38,500	25	57,660	50,660	1.02	65,400	0.97
0.08081	38,500	25	58,000	51,000	1.00	65,700	0.95
0.07196	38,500	25	58,330	51,330	0.98	65,900	0.92
0.06408	38,500	25	58,660	51,660	0.96	66,200	0.90
0.05707	38,500	25	59,000	52,000	0.94	66,400	0.89
0.05082	38,500	25	59,330	52,330	0.92	66,600	0.87
0.04526	38,500	25	59,660	52,660	0.90	66,800	0.86
0.04030–			60,000	53,000	0.88	67,000	0.85
0.02257	38,500	25	....	....	....	....	....
0.02010 and smaller	....	20	....	....	....	....	....

Table 20. Breaking Load for Solid Wires in Pounds per Wire

Gage No. A.W.G. or B. & S.	Diameter In.	Diameter Mils	Hard-drawn Copper (A.S.T.M.) *	Hard-drawn Aluminum (23,000 to 33,300 lb per sq in.)	Copper-clad Steel, 40 per cent Grade	Steel (100,000 lb per sq in.) †
	1/2	500	9310	4520	11,400	19,640
0000		460	8140	3820	10,000	16,620
	7/16	437	7500	3460	9,250	15,030
000		410	6720	3030	8,300	13,180
	3/8	375	5800	2540	7,150	11,040
00		365	5540	2400	6,850	10,450
0		325	4520	1910	5,700	8,289
	5/16	312	4220	1770	5,400	7,670
1		289	3680	1530	4,800	6,573
2		258	3000	1240	4,000	5,213
	1/4	250	2830	1170	3,780	4,909
3		229	2420	1000	3,200	4,134
4		204	1950	810	2,600	3,278
	3/16	187	1680	693	2,300	2,761
5		182	1570	655	2,200	2,600
6		162	1270	532	1,800	2,062
7		144	1020	432	1,450	1,635
8		129	822	351	1,200	1,297
	1/8	125	780	335	1,150	1,227
9		114	660	287	975	1,028
10		102	528	234	800	816
11		91	423	191	650	647
12		81	337	155	510	513
13		72	268	126	410	407
14		64	213	103	330	323
	1/16	62	203	98	310	307

* Tensile strength in pounds per square inch ranging from 49,000 for No. 0000 to 66,200 for No. 14; see below.

† For wires having a tensile strength of S pounds per square inch, multiply by $S/100{,}000$. The tensile strength of steel varies from 60,000 to 225,000 lb per sq in.

Table 21. Physical Properties of Hard-drawn Copper Transmission Cables

Size A.W.G. or B. & S.	Area, cir mils	Number of Wires	Diameter of Wires, nominal mils	Diameter of Cable, nominal inch	Elastic Limit Cable lb	Elastic Limit Lb per sq in.	Breaking Strength Nominal lb	Tensile Strength, Minimum, lb per sq in.	Weight, Nominal, per 1000 ft, lb
4	41,740	3	118.0	0.254	1128	34,400	1879	57,330	128.9
3	52,640	3	132.5	0.285	1415	34,240	2359	57,060	162.5
2	66,370	3	148.7	0.320	1748	33,530	2913	55,890	204.9
2	66,370	7	97.4	0.292	1827	35,050	3045	58,410	204.9
1	83,690	3	167.0	0.360	2172	33,050	3620	55,080	258.4
1	83,690	7	109.3	0.328	2282	34,720	3804	57,870	258.4
1/0	105,500	7	122.8	0.368	2851	34,400	4752	57,330	325.8
2/0	133,100	7	137.9	0.414	3556	34,020	5926	56,700	410.9
3/0	167,800	7	154.8	0.464	4420	33,530	7366	55,890	518.1
4/0	211,600	7	173.9	0.522	5492	33,050	9154	55,080	653.3
4/0	211,600	12	132.8	0.552	5690	34,240	9483	57,060	653.3
4/0	211,600	19	105.5	0.528	5770	34,720	9617	57,870	653.3
Circular Mils									
250,000		12	144.3	0.600	6680	34,020	11130	56,700	771.9
250,000		19	114.7	0.574	6754	34,400	11260	57,330	771.9
300,000		12	158.1	0.657	7900	33,530	13170	55,890	926.3
300,000		19	125.7	0.629	8105	34,400	13510	57,330	926.3
350,000		19	135.7	0.679	9352	34,020	15590	56,700	1081
400,000		19	145.1	0.726	10540	33,530	17560	55,890	1235
450,000		19	153.9	0.770	11850	33,530	19750	55,890	1389
500,000		19	162.2	0.811	13170	33,530	21950	55,890	1544
500,000		37	116.2	0.813	13510	34,400	22510	57,330	1544

HARD-DRAWN COPPER CABLES. Table 21 gives the mechanical properties of hard-drawn copper transmission cables.

MODULUS. Tests by B. Welbourne (*J.I.E.E.*, Vol. 56, p. 53 (1917)) indicate that the elastic modulus of copper cables varies with the number of strands, as shown in Table 22.

Table 22

Strands	Modulus, lb-in. units
7	20.0×10^6
19	17.5×10^6
37	15.5×10^6
61	14.0×10^6
91	12.5×10^6

ALUMINUM CABLES. Table 23 gives characteristics of aluminum cables.

Table 23. Aluminum Cables

A. W. G. and Circular Mils; English Units

A. W. G. Gage or Circular Mils		Usual Number of Strands	Diameter of Bare Cable, in.	Resistance at 25° C or 77° F		Weight in Pounds	
Copper (97 per cent) Equivalent	Aluminum 61 per cent			Ohms per 1000 ft	Ohms per Mile	per 1000 ft	per Mile
1,000,000	1,590,000	61	1.454	0.0111	0.0587	1462	7872
950,000	1,515,500	61	1.416	0.0118	0.0618	1393	7482
900,000	1,431,000	61	1.380	0.0124	0.0652	1317	7086
850,000	1,351,500	61	1.341	0.0131	0.0691	1243	6695
800,000	1,272,000	61	1.301	0.0139	0.0734	1171	6299
750,000	1,192,500	61	1.257	0.0148	0.0783	1098	5908
700,000	1,113,000	61	1.215	0.0159	0.0839	1025	5512
650,000	1,033,500	37	1.170	0.0171	0.0903	950	5116
600,000	954,000	37	1.124	0.0186	0.0979	877	4726
550,000	874,500	37	1.077	0.0202	0.107	805	4330
500,000	795,000	37	1.026	0.0223	0.117	732	3939
450,000	715,500	37	0.974	0.0247	0.131	658	3543
400,000	636,000	37	0.918	0.0278	0.147	585	3152
350,000	556,500	19	0.856	0.0318	0.168	512	2756
300,000	477,000	19	0.793	0.0371	0.196	439	2362
250,000	397,500	19	0.724	0.0444	0.235	365	1969
0000	336,400	19	0.657	0.0525	0.278	310.2	1666
000	266,800	7	0.586	0.0662	0.350	245.7	1322
00	211,950	7	0.522	0.0836	0.441	195	1048
0	167,800	7	0.464	0.105	0.556	155	831
1	133,077	7	0.414	0.133	0.702	122.6	659
2	105,535	7	0.368	0.167	0.885	97.2	523
3	83,693	7	0.328	0.211	1.12	77	414
4	66,371	7	0.293	0.267	1.41	61.2	329
5	52,635	7	0.258	0.336	1.77	48.5	261
6	41,741	7	0.232	0.423	2.24	38.5	207

57. CABLES FOR HIGH VOLTAGES AND LONG SPANS

LARGE-DIAMETER COPPER CABLES. Aerial cables for high voltages should be of sufficiently great diameter to keep the corona loss down to an economic minimum. The diameter required may be calculated from the formulas in Article 18. As the cross-section is generally determined by the current, conductivity, and span, high-voltage lines often require conductors whose cross-section and diameter must be independent of each other, if minimum weight and cost are to be attained. Such conductors are made in a variety of forms, as follows:

P.D.C.P. Hollow Core Cable. This consists of a thin-walled copper core tube which is concentrically stranded with one or more layers of copper wires. These wires may be either solid or hollow, or a combination of solid and hollow wires, as shown in Fig. 2.

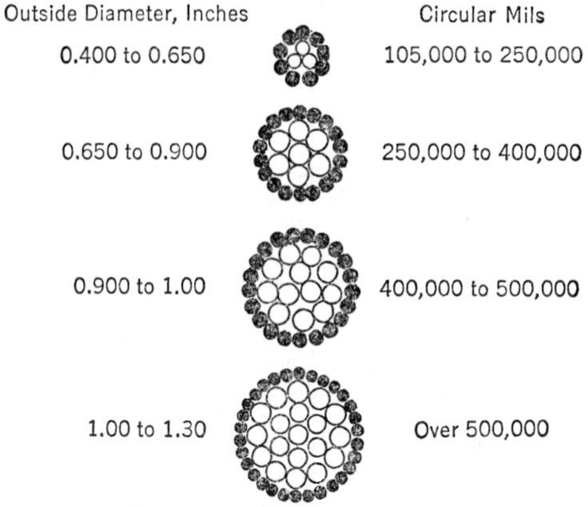

FIG. 2. Type PDCP Conductors

Anaconda I-beam Cable. A copper section like a thin I-beam is given a helical twist and used as the core on which one or more layers of wire are stranded, as shown in Fig. 3.

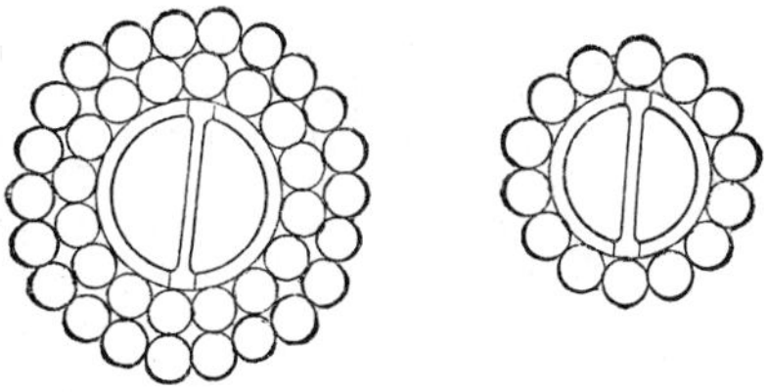

FIG. 3. I-Beam Type Conductors

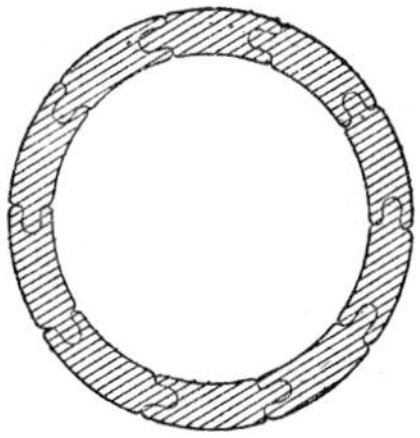

FIG. 4. Type HH Conductor

Type HH (or Heddernheim) Cable. Interlocking segments of copper are stranded together as shown in Fig. 4, forming a single flexible tube.

A. S. & W. Type Cable. This is similar to Type HH except that the segments are grooved at both ends and held together by flat elliptical key wires, which fit into opposing grooves.

These various types differ somewhat in diameter, corona suppression, a-c resistance, ease of installation and splicing, ability to resist vibration, and general mechanical stability in tension with and without lubrication, all of which properties should be given due weight. On long spans, consideration must be given, especially during installation, to proper balance of turning moments of the helically laid wires. Where members have substantially different lays, the distribution of load between members must also be considered.

Aluminum-steel Cables. Cables made of a central core of steel, solid or stranded, surrounded by one or more layers of aluminum wire are in extensive use for aerial lines. Their great advantages are that they may be erected not only with less sag than aluminum but with from 60 to 70 per cent of that of hard-drawn copper and that their diameter is often sufficiently great, except at the highest voltages, to prevent corona formation.

Table 24. Aluminum-steel (A.C.S.R.) Cable *

Size of Conductor		Cross-section, sq in.		Copper Equivalent Based upon Copper 97% Aluminum 61 %	Stranding		Outside Diameter, in.	Per Cent Aluminum		Ultimate Strength lb	Weight per 1000 ft, lb.	D-c Resistance in Ohms per Mile at 0 Amp and 25° C†
Circular Mils (Aluminum)	A.W.G. or B.&S.	Aluminum	Total		Aluminum	Steel		By Weight	By Area			
1,590,000	....	1.249	1.4071	1,000,000	54x.1716	19x.1030	1.545	73.4	88.5	56,000	2033	0.0587
1,510,500	....	1.186	1.3367	950,000	54x.1673	19x.1004	1.506	73.4	88.5	53,200	1932	0.0618
1,431,000	....	1.124	1.2664	900,000	54x.1628	19x.0977	1.465	73.4	88.5	50,400	1830	0.0652
1,351,500	....	1.062	1.1959	850,000	54x.1582	19x.0949	1.424	73.4	88.5	47,600	1728	0.0691
1,272,000	....	0.9990	1.1256	800,000	54x.1535	19x.0921	1.382	73.4	88.5	44,800	1627	0.0734
1,192,500	....	0.9366	1.0553	750,000	54x.1486	19x.0892	1.338	73.4	88.5	43,100	1526	0.0783
1,113,000	....	0.8741	0.9850	700,000	54x.1436	19x.0862	1.293	73.4	88.5	40,200	1424	0.0839
1,033,500	....	0.8117	0.9170	650,000	54x.1384	7x.1384	1.246	72.9	88.5	37,100	1330	0.0903
954,000	....	0.7493	0.8464	600,000	54x.1329	7x.1329	1.196	72.9	88.5	34,200	1227	0.0979
900,000	....	0.7069	0.7985	566,000	54x.1291	7x.1291	1.162	72.9	88.5	32,300	1158	0.104
874,000	....	0.6868	0.7759	550,000	54x.1273	7x.1273	1.146	72.9	88.5	31,400	1125	0.107
795,000	....	0.6244	0.7054	500,000	54x.1214	7x.1214	1.093	72.9	88.5	28,500	1024	0.117
795,000	....	0.6244	0.7261	500,000	26x.1749	7x.1360	1.108	68.2	86.0	31,200	1094	0.117
795,000	....	0.6244	0.7668	500,000	30x.1628	19x.0977	1.140	60.5	81.5	38,400	1234	0.117
715,500	....	0.5620	0.6348	450,000	54x.1151	7x.1151	1.036	72.9	88.5	26,300	920	0.131
715,500	....	0.5620	0.6535	450,000	26x.1659	7x.1290	1.051	68.2	86.0	28,100	984	0.131
715,500	....	0.5620	0.6900	450,000	30x.1544	19x.0926	1.081	60.5	81.5	34,600	1109	0.131
666,600	....	0.5235	0.5914	419,000	54x.1111	7x.1111	1.000	72.9	88.5	24,500	857	0.140
636,000	....	0.4995	0.5642	400,000	54x.1085	7x.1085	0.977	72.9	88.5	23,600	819	0.147
636,000	....	0.4995	0.5808	400,000	26x.1564	7x.1216	0.990	68.2	86.0	25,000	875	0.147
636,000	....	0.4995	0.6135	400,000	30x.1456	19x.0874	1.019	60.5	81.5	31,500	987	0.147
605,000	....	0.4752	0.5369	380,500	54x.1059	7x.1059	0.953	72.9	88.5	22,500	779	0.154
605,000	....	0.4752	0.5525	380,500	26x.1525	7x.1186	0.966	68.2	86.0	24,100	833	0.154
556,500	....	0.4371	0.5083	350,000	26x.1463	7x.1138	0.927	68.2	86.0	22,400	766	0.168
556,500	....	0.4371	0.5391	350,000	30x.1362	7x.1362	0.953	59.9	81.1	27,200	871	0.168
500,000	....	0.3927	0.4843	314,500	30x.1291	7x.1291	0.904	59.9	81.1	24,400	782.8	0.187
477,000	....	0.3746	0.4357	300,000	26x.1355	7x.1054	0.858	68.2	86.0	19,430	656.6	0.196
477,000	....	0.3746	0.4620	300,000	30x.1261	7x.1261	0.883	59.9	81.1	23,300	746.8	0.196
397,500	....	0.3122	0.3630	250,000	26x.1236	7x.0961	0.783	68.2	86.0	16,190	546.8	0.235
397,500	....	0.3122	0.3850	250,000	30x.1151	7x.1151	0.806	59.9	81.1	19,980	622.4	0.235
336,400	....	0.2642	0.3073	0 000	26x.1138	7x.0885	0.721	68.2	86.0	14,050	643.0	0.278
336,400	....	0.2642	0.3259	0 000	30x.1059	7x.1059	0.741	59.9	81.1	17,040	526.7	0.278
300,000	....	0.2356	0.2739	188,700	26x.1074	7x.0835	0.680	68.2	86.0	12,650	412.7	0.311
300,000	....	0.2356	0.2906	188,700	30x.1000	7x.1000	0.700	59.9	81.1	15,430	469.7	0.311
266,800	....	0.2095	0.2367	000	6x.2109	7x.0703	0.633	72.9	88.5	9,645	343.3	0.350
266,800	....	0.2095	0.2436	000	26x.1013	7x.0788	0.642	68.2	86.0	11,250	367.2	0.350
211,600	0000	0.1662	0.1939	00	6x.1878	1x.1878	0.563	67.6	85.7	8,420	293.4	0.441
167,806	000	0.1318	0.1537	0	6x.1672	1x.1672	0.502	67.6	85.7	6,675	232.4	0.556
133,077	00	0.1045	0.1219	1	6x.1490	1x.1490	0.447	67.6	85.7	5,345	184.5	0.702
105,535	0	0.0829	0.0967	2	6x.1327	1x.1327	0.398	67.6	85.7	4,280	146.4	0.885
83,693	1	0.0657	0.0767	3	6x.1182	1x.1182	0.355	67.6	85.7	3,480	116.1	1.12
66,371	2	0.0521	0.0608	4	6x.1052	1x.1052	0.316	67.6	85.7	2,790	92.1	1.41
52,635	3	0.0413	0.0482	5	6x.0937	1x.0937	0.281	67.6	85.7	2,250	73.0	1.78
41,741	4	0.0328	0.0383	6	6x.0834	1x.0834	0.250	67.6	85.7	1,830	57.9	2.24
33,102	5	0.0260	0.0308	7	6x.0743	1x.0743	0.223	67.6	85.7	1,460	45.8	2.82
26,251	6	0.0206	0.0240	8	6x.0661	1x.0661	0.198	67.6	85.7	1,170	36.4	3.56

* From Electrical Characteristics of A.C.S.R. Cable, issued by Aluminum Co. of America.

† A-c resistances are higher, owing to skin effect, especially at high current densities. Values for various current densities are given in Electrical Characteristics of A.C.S.R. Cable, which also gives tables of inductance, reactance, and capacitance and a chart for calculating carrying capacity.

The modulus of elasticity and the temperature coefficient of expansion depend upon the relative cross-section of aluminum and steel. The anchor clamps hold both the aluminum and steel, so that at low temperatures the load is divided between the two metals. As the temperature rises, the aluminum expands faster than the steel and the load gradually shifts over to the steel, which in turn stretches more to keep pace with the expanded length of aluminum, tending to eliminate relative motion between the two metals. Eventually the steel carries the total load, and the expansion follows the coefficient for steel only.

Let H_a = fraction of area covered by aluminum.
H_s = fraction of area covered by steel.
M_a = modulus of elasticity of aluminum (9,000,000).
M_s = modulus of elasticity of steel (30,000,000).
M = modulus of elasticity of aluminum-steel cable.
G_a = coefficient of expansion of aluminum (0.0000128).
G_s = coefficient of expansion of steel (0.0000064064).
G = coefficient of expansion of cable.

$$M = M_aH_a + M_sH_s = (9 \times H_a + 30H_s)10^6,$$

$$G = \left(\frac{1}{1 + \dfrac{M_aH_a}{M_sH_s}}\right)G_s - \left(\frac{1}{1 + \dfrac{M_aH_a}{M_sH_s}} - 1\right)G_a$$

$$= 0.0000064\left(\frac{1}{1 + 0.3\dfrac{H_a}{H_s}}\right) - 0.0000128\left(\frac{1}{1 + 0.3\dfrac{H_a}{H_s}} - 1\right)$$

Table 24 gives the principal mechanical characteristics of aluminum-steel cables, and Table 32 their resistances.

58. MESSENGER CABLES

STEEL CABLE. A modulus of 22×10^6 lb-in. units is representative of ordinary steel messenger cable.

Table 25 is compiled from tables published by the General Electric Company (*Bulletin* 4538). "High" and "extra-high" strength steel should be used only where absolutely necessary, as, on account of its stiffness, it requires special mechanical fastenings.

Table 25. Steel Cable for Catenary Suspension

Extra-galvanized Siemens-Martin Steel Strand, 90,000 lb per sq in.

Diameter, in.	Tensile Strength, lb	Elastic Limit, lb	Elongation, per cent	Lay, in.
1/4	3,060	1,830	6 to 9	3
5/16	4,860	2,910	6 to 9	3 1/2
3/8	6,800	4,080	5 to 8	4
7/16	9,000	5,300	5 to 8	4 1/2
1/2	11,000	6,600	5 to 8	4 1/2
5/8	19,000	11,400	4 to 6	5
Extra-galvanized High-strength Crucible Steel Strand				
1/4	5,100	3,315	3 to 5	3 1/2
5/16	8,100	5,265	3 to 5	4
3/8	11,500	7,475	3 to 5	4 1/2
7/16	15,000	9,500	3 to 5	5
1/2	18,000	11,700	3 to 5	5
5/8	25,000	16,250	2 to 4	5 1/2
Extra-galvanized Extra-high-strength Plow Steel Strand				
1/4	7,600	5,700	2 1/2 to 4	4
5/16	12,100	9,075	2 1/2 to 4	4 1/2
3/8	17,250	12,930	2 1/2 to 4	5
7/16	22,500	16,800	2 1/2 to 4	5 1/2
1/2	27,000	20,250	2 1/2 to 4	5 1/2
5/8	42,000	31,500	1 1/2 to 3	6

COPPERWELD MESSENGER CABLE. Table 26 gives the physical properties of standard sizes of copperweld aerial cable. Tables 27 and 28 give the approximate sizes of such cable for various spans and cable weights. These tables are based on heavy loading conditions of 1/2-in. ice and 8 lb wind at 0 deg fahr with the maximum tension not exceeding 50 per cent of rated messenger strength, and with the normal tension not exceeding 35 per cent of messenger strength. The tables are for estimating purposes and indicate the messenger size for the average type of installation. Where increased safety is desired, or where smaller sags are to be used, a larger messenger will be required. Smaller messengers should be used only when larger sags than indicated are permissible. The exact size of messenger will depend upon the requirement for the particular installation.

Table 26. Physical Properties of Standard Sizes of Copperweld Aerial Cable Messenger *

Nominal Diameter, Inch and Size, A.W.G.	Actual Diameter, Inch †	Breaking Load,‡ Pounds		Weight		Cross-section	
		High Strength	Extra-high Strength	Pounds per 1000 ft	Pounds per Mile	Circular Mils	Square Inches
9/16 (7 No. 5)	0.546	20,470	24,650	649.4	3,429	231,700	0.1820
1/2 (7 No. 6)	0.486	16,890	20,460	515.0	2,719	183,800	0.1443
7/16 (7 No. 7)	0.433	13,910	16,890	408.4	2,157	145,700	0.1145
3/8 (7 No. 8)	0.385	11,440	13,890	323.9	1,710	115,600	0.09077
11/32 (7 No. 9)	0.343	9,393	11,280	256.9	1,356	91,650	0.07198
5/16 (7 No. 10)	0.306	7,758	9,196	203.7	1,076	72,680	0.05708
3 No. 6	0.349	6,835	8,281	220.3	1,163	78,750	0.06185
3 No. 7	0.311	5,629	6,838	174.7	922.4	62,450	0.04905
3 No. 8	0.277	4,629	5,621	138.5	731.5	49,530	0.03890

Modulus of elasticity: Strand, 23,000,000. Coefficient of linear expansion: 0.0000072 per degree fahrenheit.

* Courtesy of Copperweld Steel Company.

† Diameter of circumscribing circle.

‡ Breaking load of 7-wire strand is taken as 90 per cent of the sum of the breaking loads of the individual wires; breaking load of 3-wire strand is taken as 85 per cent of the sum of the breaking loads of the individual wires; these strengths are the same as for 30 per cent conductivity cooperweld strands.

Table 27. Construction with Usual Sags for Light- and Medium-weight Cables *

Weight of Cable per Foot, Pounds	Length of Span, Feet									
	40	60	80	100	120	140	160	180	200	220
	Approximate Sag of Messenger with Cable at 60 deg fahr, Inches									
	6	8	12	18	24	30	36	42	51	60
	Approximate Normal Tension at 60 deg fahr and Approximate Size of Messenger for Standard Construction									
1	515	870	1030	1075	1160	1260	1375	1490	1510	1550
	3—#7 EHS or 5/16 in. HS									
2	915	1540	1830	1900	2060	2250	2510	2725	2770	2850
	3—#7 EHS or 5/16 in. HS						11/32 in. HS			
3	1320	2230	2685	2800	3020	3350	3650	3960	4030	4140
	5/16 in. HS		11/32 in. HS			3/8 in. HS		3/8 in. EHS		
4	1740	2940	3540	3690	3980	4330	4800	5210	5300	5450
	11/32 in. HS		3/8 in. HS		3/8 in. EHS		7/16 in. EHS			
5	2170	3660	4340	4520	4950	5400	6000	6500	6600	6800
	3/8 in. HS		3/8 in. EHS		7/16 in. EHS		1/2 in. EHS			
6	2570	4340	5200	5420	5950	6500	7200	7800	7940	8170
	3/8 in. EHS		7/16 in. EHS		1/2 in. EHS		9/16 in. EHS			
7	3000	5070	6100	6350	6970	7600	8260			
	7/16 in. EHS		1/2 in. EHS		9/16 in. EHS					
8	3450	5800	7000	7300	7880					
	1/2 in. EHS		9/16 in. EHS							

HS = High Strength; EHS = Extra-high Strength.

* Courtesy of Copperweld Steel Company.

Table 28. Construction with Increased Sags for Heavy-cable Installation *

<table>
<tr><th rowspan="5">Weight of Cable per Foot, Pounds</th><th colspan="8">Length of Span, Feet</th></tr>
<tr><th>40</th><th>60</th><th>80</th><th>100</th><th>120</th><th>140</th><th>160</th><th>180</th></tr>
<tr><th colspan="8">Approximate Sag of Messenger with Cable at 60 deg fahr, Inches</th></tr>
<tr><th>9</th><th>13</th><th>20</th><th>30</th><th>40</th><th>50</th><th>60</th><th>70</th></tr>
<tr><th colspan="8">Approximate Normal Tension at 60 deg fahr and Approximate Size of Messenger for Standard Construction</th></tr>
<tr><td rowspan="2">6</td><td>1700</td><td>2640</td><td>3050</td><td>3210</td><td>3470</td><td>3780</td><td>4110</td><td>4460</td></tr>
<tr><td colspan="3">11/32 in. HS</td><td colspan="3">3/8 in. HS</td><td colspan="2">3/8 in. EHS</td></tr>
<tr><td rowspan="2">7</td><td>1960</td><td>3050</td><td>3560</td><td>3710</td><td>4000</td><td>4370</td><td>4800</td><td>5210</td></tr>
<tr><td colspan="2">11/32 in. HS</td><td colspan="2">3/8 in. HS</td><td colspan="2">3/8 in. EHS</td><td colspan="2">7/16 in. EHS</td></tr>
<tr><td rowspan="2">8</td><td>2250</td><td>3500</td><td>4050</td><td>4210</td><td>4550</td><td>5000</td><td>5440</td><td>5900</td></tr>
<tr><td colspan="2">3/8 in. HS</td><td colspan="3">3/8 in. EHS</td><td colspan="2">7/16 in. EHS</td><td>1/2 in. EHS</td></tr>
<tr><td rowspan="2">9</td><td>2510</td><td>3915</td><td>4525</td><td>4750</td><td>5130</td><td>5600</td><td>6150</td><td>6670</td></tr>
<tr><td colspan="3">3/8 in. EHS</td><td colspan="3">7/16 in. EHS</td><td colspan="2">1/2 in. EHS</td></tr>
<tr><td rowspan="2">10</td><td>2780</td><td>4330</td><td>5040</td><td>5250</td><td>5730</td><td>6240</td><td>6790</td><td>7460</td></tr>
<tr><td colspan="2">3/8 in. EHS</td><td colspan="2">7/16 in. EHS</td><td colspan="3">1/2 in. EHS</td><td>9/16 in. EHS</td></tr>
<tr><td rowspan="2">11</td><td>3050</td><td>4780</td><td>5520</td><td>5800</td><td>6270</td><td>6800</td><td>7520</td><td>8150</td></tr>
<tr><td>3/8 in. EHS</td><td colspan="2">7/16 in. EHS</td><td colspan="3">1/2 in. EHS</td><td colspan="2">9/16 in. EHS</td></tr>
<tr><td rowspan="2">12</td><td>3335</td><td>5200</td><td>6050</td><td>6300</td><td>6800</td><td>7500</td><td>8160</td><td></td></tr>
<tr><td colspan="2">7/16 in. EHS</td><td colspan="3">1/2 in. EHS</td><td colspan="2">9/16 in. EHS</td><td></td></tr>
</table>

HS = High Strength; EHS = Extra-high Strength.
* Courtesy of Copperweld Steel Company.

59. SKIN EFFECT AND A-C RESISTANCE

The a-c resistance of conductors is greater than the d-c resistance given in the preceding tables, because of skin effect. (See Section 3, Article 22.) The skin-effect ratio is the ratio of the a-c resistance to the d-c resistance. Table 29 gives these ratios for concentric-lay conductors, Table 30 for segmental conductors, and Table 31 for annular conductors, such as are used in insulated cables. The skin-effect ratio for annular conductors for aerial transmission should be obtained from the manufacturers. The a-c resistance of A.C.S.R. conductors is given in Table 32.

Table 29. Skin-effect Ratios

Concentric-lay Copper Conductors; Class B Stranding

Size, Circular Mils	Skin-effect Ratio	
	25 Cycles	60 Cycles
5,000,000	1.237	1.765
4,500,000	1.200	1.685
4,000,000	1.165	1.605
3,500,000	1.130	1.513
3,000,000	1.100	1.424
2,500,000	1.071	1.326
2,250,000	1.059	1.276
2,000,000	1.048	1.233
1,750,000	1.037	1.185
1,500,000	1.027	1.142
1,250,000	1.019	1.102
1,000,000	1.012	1.067
950,000	1.011	1.061
900,000	1.010	1.055
850,000	1.009	1.049
800,000	1.008	1.044
750,000	1.007	1.039
700,000	1.006	1.034
650,000	1.005	1.029
600,000	1.005	1.025
550,000		1.021
500,000		1.018
450,000		1.014
400,000		1.011
350,000		1.009
300,000		1.006
250,000		1.005

The following skin-effect ratios apply to tubular stranded conductors at 60 cycles and 65 deg cent.

Circular Mils	Inside Diameter, Inches						
	0.25	0.50	0.75	1.00	1.25	1.50	2.00
3,000,000	1.39	1.36	1.29	1.23	1.19	1.15	1.08
2,500,000	1.28	1.24	1.20	1.16	1.12	1.09	1.05
2,000,000	1.20	1.17	1.12	1.09	1.06	1.05	1.02
1,500,000	1.12	1.09	1.06	1.04	1.03	1.02	1.01
1,000,000	1.05	1.03	1.02	1.01	1.01	....	....

Table 30. Resistance of Segmental Copper Conductors

(IPCEA)

Circular Mils	D-c Resistance, Ohms per 1000 ft		Skin-effect Ratio, 60 Cycles	
	25° C	65° C	25° C	65° C
1,000,000	0.0108	0.0125	1.018	1.013
1,250,000	0.00867	0.0100	1.028	1.021
1,500,000	0.00723	0.00834	1.039	1.030
1,750,000	0.00619	0.00714	1.053	1.040
2,000,000	0.00542	0.00626	1.068	1.052
2,500,000	0.00434	0.00501	1.102	1.078
3,000,000	0.00361	0.00417	1.141	1.109
3,500,000	0.00310	0.00358	1.182	1.141
4,000,000	0.00271	0.00313	1.225	1.178

Table 31. Annular Copper Conductor Cables

(IPCEA)

Nominal Size, cir mils	Approximate Rope Size, in.	Diameter Wires, in.	Number of Strands in Each Layer				Actual Cir-mil Area	Maximum O.D., in.	Approximate Weight Copper, lb per 1000 ft	Approximate D-c Resistance ohms per 1000 ft.		Skin Effect Ratio at 65° C	
			1st	2nd	3rd	Total				25° C	65° C	25 Cycles	60 Cycles
750,000	0.375	0.1172	12	18	24	54	741,735	1.108	2,312	0.0147	0.0169	1.004	1.021
800,000	0.468	0.1110	16	21	28	65	800,865	1.164	2,497	0.0136	0.0157	1.004	1.021
900,000	0.500	0.1172	16	22	28	66	906,565	1.234	2,826	0.0120	0.0139	1.005	1.025
1,000,000	0.563	0.1255	16	21	28	65	1,023,766	1.346	3,192	0.0106	0.0123	1.006	1.031
1,250,000	0.750	0.1255	21	26	33	80	1,260,020	1.533	3,928	0.00864	0.0100	1.007	1.034
1,500,000	1.000	0.1255	26	32	38	96	1,512,024	1.783	4,714	0.00720	0.00831	1.007	1.037
1,750,000	1.125	0.1280	30	35	42	107	1,753,088	1.923	5,466	0.00621	0.00717	1.008	1.043
2,000,000	1.3125	0.1284	34	40	46	120	1,978,387	2.114	6,168	0.00550	0.00635	1.009	1.045
2,500,000	1.500	0.1440	34	40	46	120	2,488,320	2.394	7,758	0.00438	0.00505	1.012	1.066
3,000,000	1.625	0.1620	33	38	45	116	3,044,304	2.627	9,492	0.00358	0.00413	1.019	1.105
3,500,000	2.000	0.1620	40	45	52	137	3,595,428	3.007	11,319	0.00306	0.00353	1.020	1.110
4,000,000	2.250	0.1620	45	51	57	153	4,015,332	3.262	12,641	0.00274	0.00316	1.021	1.116
4,500,000	2.500	0.1620	50	56	62	168	4,408,992	3.517	14,013	0.00252	0.00291	1.022	1.118
5,000,000	2.875	0.1620	57	63	69	189	4,960,116	3.897	15,765	0.00224	0.00258	1.023	1.121

Table 32. A-c Resistance per Mile of A.C.S.R. Cable *

Circular Mils or A.W.G. (B.&S.) Aluminum	Number of Layers of Aluminum Over Steel Core	Number of Wires		Copper Equivalent Circular Mils or A.W.G., Based on Copper 97%, Aluminum 61%	D-c Resistance, Ohms per Mile at 0 amp and 25° C	Ohms per Mile of Single Conductor at 25° C																							
						0 amp per sq in.			200 amp per sq in.			400 amp per sq in.			600 amp per sq in.			800 amp per sq in.			1000 amp per sq in.			1200 amp per sq in.			1400 amp per sq in.		
		Aluminum	Steel			25 Cycles	50 Cycles	60 Cycles	25 Cycles	50 Cycles	60 Cycles	25 Cycles	50 Cycles	60 Cycles	25 Cycles	50 Cycles	60 Cycles	25 Cycles	50 Cycles	60 Cycles	25 Cycles	50 Cycles	60 Cycles	25 Cycles	50 Cycles	60 Cycles	25 Cycles	50 Cycles	60 Cycles
1,590,000	3	54	19	1,000,000	0.0587	0.0588	0.0590	0.0591	0.0589	0.0592	0.0594	0.0590	0.0595	0.0598	0.0592	0.0602	0.0607	0.0595	0.0611	0.0619	0.0600	0.0625	0.0638	0.0606	0.0643	0.0662	0.0615	0.0670	0.0698
1,510,500	3	54	19	950,000	0.0618	0.0619	0.0621	0.0622	0.0620	0.0623	0.0625	0.0621	0.0626	0.0629	0.0623	0.0632	0.0637	0.0626	0.0642	0.0650	0.0630	0.0655	0.0668	0.0636	0.0672	0.0691	0.0545	0.0698	0.0726
1,431,000	3	54	19	900,000	0.0652	0.0653	0.0655	0.0656	0.0654	0.0658	0.0659	0.0655	0.0661	0.0663	0.0657	0.0666	0.0671	0.0660	0.0676	0.0684	0.0665	0.0690	0.0703	0.0670	0.0706	0.0724	0.0678	0.0730	0.0756
1,351,500	3	54	19	850,000	0.0691	0.0692	0.0694	0.0695	0.0693	0.0696	0.0698	0.0694	0.0699	0.0702	0.0695	0.0705	0.0709	0.0698	0.0713	0.0721	0.0703	0.0728	0.0740	0.0708	0.0743	0.0761	0.0715	0.0765	0.0790
1,272,000	3	54	19	800,000	0.0734	0.0735	0.0737	0.0738	0.0736	0.0740	0.0742	0.0737	0.0743	0.0746	0.0738	0.0748	0.0752	0.0741	0.0756	0.0764	0.0746	0.0770	0.0782	0.0751	0.0785	0.0802	0.0757	0.0805	0.0829
1,192,500	3	54	19	750,000	0.0783	0.0784	0.0786	0.0788	0.0785	0.0789	0.0791	0.0786	0.0792	0.0795	0.0787	0.0796	0.0801	0.0790	0.0804	0.0812	0.0794	0.0818	0.0830	0.0799	0.0832	0.0849	0.0805	0.0851	0.0874
1,113,000	3	54	19	700,000	0.0839	0.0840	0.0842	0.0844	0.0841	0.0845	0.0848	0.0842	0.0849	0.0852	0.0843	0.0853	0.0857	0.0846	0.0860	0.0867	0.0850	0.0873	0.0885	0.0855	0.0887	0.0903	0.0860	0.0904	0.0927
1,033,500	3	54	7	650,000	0.0903	0.0905	0.0907	0.0909	0.0906	0.0910	0.0913	0.0907	0.0913	0.0917	0.0908	0.0918	0.0922	0.0910	0.0924	0.0931	0.0914	0.0935	0.0945	0.0919	0.0950	0.0965	0.0924	0.0967	0.0988
954,000	3	54	7	600,000	0.0979	0.0980	0.0981	0.0982	0.0980	0.0983	0.0985	0.0981	0.0986	0.0989	0.0983	0.0992	0.0997	0.0986	0.100	0.101	0.0990	0.101	0.102	0.0994	0.102	0.104	0.0998	0.104	0.106
900,000	3	54	7	566,000	0.104	0.104	0.104	0.104	0.104	0.104	0.105	0.104	0.105	0.105	0.104	0.105	0.106	0.104	0.106	0.107	0.105	0.107	0.108	0.105	0.108	0.110	0.106	0.110	0.112
874,500	3	54	7	550,000	0.107	0.107	0.107	0.108	0.107	0.107	0.108	0.107	0.108	0.109	0.107	0.108	0.109	0.107	0.109	0.110	0.108	0.110	0.111	0.108	0.111	0.113	0.109	0.113	0.115
795,000	3	54	7	500,000	0.117	0.118	0.118	0.119	0.118	0.119	0.119	0.118	0.119	0.119	0.118	0.119	0.119	0.118	0.119	0.120	0.118	0.120	0.121	0.119	0.121	0.123	0.119	0.123	0.125
795,000	2	26	7	500,000	0.117	0.117	0.117	0.117	0.117	0.117	0.117	0.117	0.117	0.117	0.117	0.117	0.117	0.117	0.117	0.117	0.117	0.117	0.117	0.117	0.117	0.117	0.117	0.117	0.117
795,000	2	30	19	500,000	0.117	0.117	0.117	0.117	0.117	0.117	0.117	0.117	0.117	0.117	0.117	0.117	0.117	0.117	0.117	0.117	0.117	0.117	0.117	0.117	0.117	0.117	0.117	0.117	0.117
715,500	3	54	7	450,000	0.131	0.131	0.131	0.132	0.131	0.132	0.133	0.131	0.132	0.133	0.131	0.132	0.133	0.131	0.133	0.133	0.131	0.133	0.134	0.132	0.135	0.136	0.133	0.137	0.139
715,500	2	26	7	450,000	0.131	0.131	0.131	0.131	0.131	0.131	0.131	0.131	0.131	0.131	0.131	0.131	0.131	0.131	0.131	0.131	0.131	0.131	0.131	0.131	0.131	0.131	0.131	0.131	0.131
715,500	2	30	19	450,000	0.131	0.131	0.131	0.131	0.131	0.131	0.131	0.131	0.131	0.131	0.131	0.131	0.131	0.131	0.131	0.131	0.131	0.131	0.131	0.131	0.131	0.131	0.131	0.131	0.131
666,600	3	54	7	419,000	0.140	0.140	0.141	0.141	0.140	0.141	0.142	0.140	0.141	0.142	0.141	0.141	0.142	0.141	0.143	0.143	0.141	0.143	0.144	0.142	0.145	0.146	0.142	0.146	0.148
636,000	3	54	7	400,000	0.147	0.147	0.148	0.148	0.147	0.148	0.149	0.147	0.148	0.149	0.147	0.149	0.149	0.148	0.149	0.150	0.148	0.151	0.152	0.149	0.152	0.154	0.149	0.153	0.156
636,000	2	26	7	400,000	0.147	0.147	0.147	0.147	0.147	0.147	0.147	0.147	0.147	0.147	0.147	0.147	0.147	0.147	0.147	0.147	0.147	0.147	0.147	0.147	0.147	0.147	0.147	0.147	0.147
636,000	2	30	19	400,006	0.147	0.147	0.147	0.147	0.147	0.147	0.147	0.147	0.147	0.147	0.147	0.147	0.147	0.147	0.147	0.147	0.147	0.147	0.147	0.147	0.147	0.147	0.147	0.147	0.147
605,000	3	54	7	380,500	0.154	0.155	0.155	0.155	0.155	0.155	0.156	0.155	0.156	0.156	0.155	0.156	0.157	0.155	0.157	0.158	0.156	0.159	0.160	0.156	0.160	0.162	0.157	0.161	0.163
605,000	2	26	7	380,500	0.154	0.154	0.154	0.154	0.154	0.154	0.154	0.154	0.154	0.154	0.154	0.154	0.154	0.154	0.154	0.154	0.154	0.154	0.154	0.154	0.154	0.154	0.154	0.154	0.154
605,000	2	30	19	380,500	0.154	0.154	0.154	0.154	0.154	0.154	0.154	0.154	0.154	0.154	0.154	0.154	0.154	0.154	0.154	0.154	0.154	0.154	0.154	0.154	0.154	0.154	0.154	0.154	0.154
556,500	2	26	7	350,000	0.168	0.168	0.168	0.168	0.168	0.168	0.168	0.168	0.168	0.168	0.168	0.168	0.168	0.168	0.168	0.168	0.168	0.168	0.168	0.168	0.168	0.168	0.168	0.168	0.168
556,500	2	30	7	350,000	0.168	0.168	0.168	0.168	0.168	0.168	0.168	0.168	0.168	0.168	0.168	0.168	0.168	0.168	0.168	0.168	0.168	0.168	0.168	0.168	0.168	0.168	0.168	0.168	0.168
500,000	2	30	7	314,500	0.187	0.187	0.187	0.187	0.187	0.187	0.187	0.187	0.187	0.187	0.187	0.187	0.187	0.187	0.187	0.187	0.187	0.187	0.187	0.187	0.187	0.187	0.187	0.187	0.187
477,000	2	26	7	300,000	0.196	0.196	0.196	0.196	0.196	0.196	0.196	0.196	0.196	0.196	0.196	0.196	0.196	0.196	0.196	0.196	0.196	0.196	0.196	0.196	0.196	0.196	0.196	0.196	0.196
477,000	2	30	7	300,000	0.196	0.196	0.196	0.196	0.196	0.196	0.196	0.196	0.196	0.196	0.196	0.196	0.196	0.196	0.196	0.196	0.196	0.196	0.196	0.196	0.196	0.196	0.196	0.196	0.196
397,500	2	26	7	250,000	0.235	0.235	0.235	0.235	0.235	0.235	0.235	0.235	0.235	0.235	0.235	0.235	0.235	0.235	0.235	0.235	0.235	0.235	0.235	0.235	0.235	0.235	0.235	0.235	0.235
397,500	2	30	7	250,000	0.235	0.235	0.235	0.235	0.235	0.235	0.235	0.235	0.235	0.235	0.235	0.235	0.235	0.235	0.235	0.235	0.235	0.235	0.235	0.235	0.253	0.235	0.235	0.235	0.235

336,400	2	26	7	4/0	0.278	0.278	0.278	0.278	0.278	0.278	0.278	0.278	0.278	0.278	0.278	0.278	0.278	0.278	0.278	0.278	0.278	0.278	0.278	0.278	0.278	0.278	0.278	0.278	0.278
336,400	2	30	7	4/0	0.278	0.278	0.278	0.278	0.278	0.278	0.278	0.278	0.278	0.278	0.278	0.278	0.278	0.278	0.278	0.278	0.278	0.278	0.278	0.278	0.278	0.278	0.278	0.278	0.278
300,000	2	26	7	188,700	0.311	0.311	0.311	0.311	0.311	0.311	0.311	0.311	0.311	0.311	0.311	0.311	0.311	0.311	0.311	0.311	0.311	0.311	0.311	0.311	0.311	0.311	0.311	0.311	0.311
300,000	2	30	7	188,700	0.311	0.311	0.311	0.311	0.311	0.311	0.311	0.311	0.311	0.311	0.311	0.311	0.311	0.311	0.311	0.311	0.311	0.311	0.311	0.311	0.311	0.311	0.311	0.311	0.311
266,800	2	26	7	3/0	0.350	0.350	0.350	0.350	0.350	0.350	0.350	0.350	0.350	0.350	0.350	0.350	0.350	0.350	0.350	0.350	0.350	0.350	0.350	0.350	0.350	0.350	0.350	0.350	0.350
266,800	1	6	7	3/0	0.350	0.350	0.351	0.351	0.350	0.351	0.352	0.351	0.353	0.355	0.353	0.360	0.363	0.359	0.378	0.388	0.368	0.406	0.425	0.378	0.434	0.463	0.384	0.454	0.489
4/0	1	6	1	2/0	0.441	0.442	0.444	0.445	0.443	0.445	0.446	0.444	0.450	0.453	0.447	0.458	0.464	0.452	0.474	0.485	0.458	0.493	0.510	0.463	0.508	0.530	0.468	0.521	0.548
3/0	1	6	1	1/0	0.556	0.557	0.559	0.560	0.557	0.560	0.561	0.560	0.566	0.569	0.562	0.574	0.579	0.566	0.584	0.594	0.570	0.597	0.611	0.574	0.610	0.628	0.579	0.624	0.647
2/0	1	6	1	1	0.702	0.702	0.704	0.706	0.703	0.705	0.707	0.704	0.709	0.712	0.706	0.714	0.718	0.708	0.721	0.727	0.711	0.730	0.739	0.715	0.742	0.756	0.720	0.758	0.777
1/0	1	6	1	2	0.885	0.885	0.887	0.888	0.885	0.888	0.889	0.886	0.890	0.892	0.887	0.891	0.893	0.888	0.895	0.899	0.890	0.901	0.907	0.893	0.911	0.920	0.898	0.927	0.941
1	1	6	1	3	1.12	1.12	1.12	1.12	1.12	1.12	1.12	1.12	1.12	1.12	1.12	1.12	1.12	1.12	1.12	1.12	1.12	1.13	1.13	1.12	1.14	1.15	1.13	1.16	1.17
2	1	6	1	4	1.41	1.41	1.41	1.41	1.41	1.41	1.41	1.41	1.41	1.41	1.41	1.41	1.41	1.41	1.41	1.41	1.41	1.42	1.42	1.41	1.43	1.43	1.42	1.44	1.45
2	1	7	1	4	1.41	1.41	1.41	1.41	1.41	1.41	1.41	1.41	1.41	1.41	1.41	1.41	1.41	1.41	1.41	1.41	1.41	1.42	1.42	1.41	1.42	1.42	1.41	1.43	1.43
3	1	6	1	5	1.78	1.78	1.78	1.78	1.78	1.78	1.78	1.78	1.78	1.78	1.78	1.78	1.78	1.78	1.78	1.78	1.78	1.78	1.78	1.78	1.79	1.79	1.78	1.80	1.80
4	1	6	1	6	2.24	2.24	2.24	2.24	2.24	2.24	2.24	2.24	2.24	2.24	2.24	2.24	2.24	2.24	2.24	2.24	2.24	2.24	2.24	2.24	2.24	2.24	2.24	2.24	2.24
4	1	7	1	6	2.24	2.24	2.24	2.24	2.24	2.24	2.24	2.24	2.24	2.24	2.24	2.24	2.24	2.24	2.24	2.24	2.24	2.24	2.24	2.24	2.24	2.24	2.24	2.24	2.24
5	1	6	1	7	2.82	2.82	2.82	2.82	2.82	2.82	2.82	2.82	2.82	2.82	2.82	2.82	2.82	2.82	2.82	2.82	2.82	2.82	2.82	2.82	2.82	2.82	2.82	2.82	2.82
6	1	6	1	8	3.56	3.56	3.56	3.56	3.56	3.56	3.56	3.56	3.56	3.56	3.56	3.56	3.56	3.56	3.56	3.56	3.56	3.56	3.56	3.56	3.56	3.56	3.56	3.56	3.56
203,000	1	8	7	127,700	0.460	0.460	0.461	0.461	0.460	0.461	0.461	0.461	0.462	0.463	0.462	0.465	0.466	0.463	0.469	0.472	0.466	0.477	0.483	0.471	0.490	0.500	0.477	0.509	0.524
203,200	1	16	19	127,800	0.460	0.460	0.461	0.461	0.460	0.461	0.461	0.461	0.462	0.463	0.462	0.465	0.466	0.463	0.469	0.472	0.466	0.477	0.483	0.471	0.490	0.500	0.477	0.509	0.524
211,300	1	12	7	132,900	0.442	0.442	0.443	0.443	0.442	0.443	0.443	0.443	0.444	0.445	0.444	0.447	0.448	0.445	0.451	0.453	0.448	0.459	0.464	0.453	0.472	0.481	0.458	0.487	0.503
190,800	1	12	7	120,000	0.490	0.490	0.491	0.491	0.490	0.491	0.491	0.491	0.492	0.493	0.492	0.495	0.496	0.493	0.499	0.502	0.497	0.509	0.515	0.502	0.523	0.533	0.508	0.542	0.558
176,900	1	12	7	111,200	0.528	0.528	0.529	0.529	0.528	0.529	0.529	0.529	0.530	0.531	0.530	0.533	0.535	0.532	0.538	0.541	0.535	0.548	0.555	0.540	0.563	0.574	0.548	0.584	0.601
159,000	1	12	7	100,000	0.587	0.587	0.588	0.588	0.587	0.588	0.588	0.588	0.589	0.590	0.589	0.592	0.594	0.591	0.598	0.602	0.595	0.610	0.617	0.601	0.627	0.639	0.609	0.649	0.669
134,600	1	12	7	84,600	0.694	0.694	0.695	0.695	0.694	0.695	0.695	0.695	0.698	0.699	0.696	0.701	0.703	0.699	0.707	0.711	0.703	0.721	0.729	0.710	0.740	0.755	0.720	0.767	0.790
110,800	1	12	7	69,700	0.843	0.843	0.845	0.845	0.843	0.845	0.845	0.845	0.848	0.850	0.846	0.852	0.854	0.849	0.859	0.864	0.855	0.876	0.886	0.863	0.899	0.917	0.874	0.932	0.960
101,800	1	12	7	64,160	0.917	0.917	0.918	0.918	0.918	0.919	0.919	0.919	0.922	0.923	0.920	0.926	0.929	0.923	0.934	0.940	0.929	0.952	0.963	0.939	0.979	0.998	0.951	1.008	1.044
80,000	1	8	1	50,310	1.170	1.170	1.170	1.170	1.170	1.170	1.170	1.170	1.170	1.170	1.170	1.170	1.170	1.171	1.172	1.172	1.176	1.178	1.181	1.177	1.191	1.197	1.184	1.214	1.223

* Courtesy of Aluminum Company of America.

NOTE 1. The values of d-c resistance given in the table above are based on the following resistance values:
a. The International Annealed Copper Standard is defined as follows: "At a temperature of 20 deg C (68 deg F), the resistance of a wire of standard annealed copper 1 meter long and having a uniform section of 1 sq mm is 1/58 (=0.017241) ohm."
b. Commercial annealed copper wire at standard temperature is taken at 100% conductivity.
c. Commercial hard-drawn copper wire is taken at 97% conductivity.
d. Commercial hard-drawn aluminum wire at standard temperature is taken at 61% (AIEE Standard 60.97%) conductivity.
e. On this basis the resistance of 1 circular mil-foot of annealed copper is 10.371 ohms at 20 deg C.
f. The resistance of 1 circular mil-foot of aluminum is 17.002 ohms at 20 deg C.
g. For stranded conductors, the resistances are 2% greater than for the equivalent solid conductor. This allows for increase of length due to an average length of lay, as recommended by the American Institute of Electrical Engineers.
h. In computing the resistances of aluminum cable steel reinforced no deduction is made for the conductance of the steel core.
i. The effect of temperature on the unit of resistance is defined as follows: "At a temperature of 20 deg C (68 deg F), the constant mass temperature coefficient of resistance of standard annealed copper, measured between two points rigidly fixed to the wire, is 0.393% per degree Centigrade."
j. The effect of temperature on the resistance of aluminum wire is to increase it by 0.403% per degree at 20 deg C.
NOTE 2. The a-c resistances are based on calculations developed from actual tests.
NOTE 3. No allowance has been made for increased length because of sag when the conductors are suspended.

60. TESTS OF BARE WIRES AND CABLES

The usual tests on bare wires are gaging diameter and measuring tensile strength, elongation, modulus, elastic limit, and electrical conductivity.

GAGING DIAMETER. The usual type of gage for measuring wire diameters is shown in Fig. 5. The wire is placed between the measuring surfaces, and the screw adjusted until a click occurs. The number of large divisions exposed on the axis is multiplied by 100; the number of small divisions exposed on the axis is multiplied by 25; and the sum of these two items is added to the number indicated on the revolving scale. The sum will be the diameter of the wire in mils.

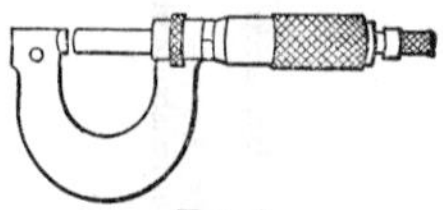

FIG. 5

TENSILE STRENGTH, ELONGATION, MODULUS, AND ELASTIC LIMIT. The essential features of a wire-testing machine are a means of applying a measurable pulling force to the wire, and a means of taking up the elongation. Accordingly, the usual testing machine consists of two pairs of jaws for gripping the wire, one pair being connected to a balance lever and the other to a power-driven mechanism which draws it in the direction of the axis of the wire. A typical machine is shown diagrammatically in Fig. 6 where *A* and *B* are the two pairs of jaws between which the wire is stretched. The machine is operated by setting in motion the mechanism which makes the jaw *A* move steadily in the direction indicated. The operator then moves the counterpoise *C* by hand, in the direction indicated, so as to keep the beam balanced. This operation is continued until the wire breaks, when the elongation of the sample is measured by the travel of the jaw *A* and its breaking strength by the weight indicated on the balance beam at the counterpoise *C*.

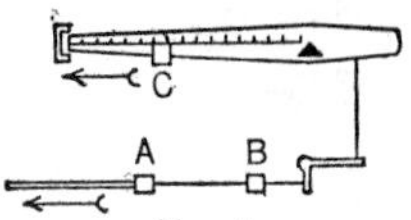

FIG. 6

Measurement of Strain. The amount by which the wire is stretched is measured by means of an extensometer, which consists of a pair of clamps to grip the wire at points a definite distance apart, and a magnifying scale for measuring the increase of distance between these clamps as the wire stretches. The stress-strain curve obtained by plotting the elongations thus measured against the stresses measured by the machine described above is not a true one, as there is initially an abnormal elongation due to the straightening of the wire, as shown by curve *OA* in Fig. 7. The standard method of overcoming this is described in the ASTM Specification for Hard-drawn Copper Wire.

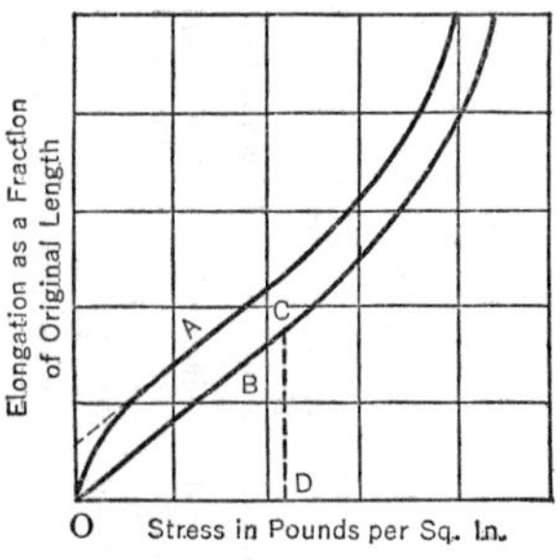

FIG. 7

Modulus of Elasticity. The modulus of elasticity is obtained from the slope of the straight part of the corrected curve. In Fig. 7 the modulus of elasticity is *OD–CD* pound-inch units.

Elastic Limit. The elastic limit can be obtained only by applying a series of increasing loads, releasing the load (leaving, however, a sufficient load to keep the wire straight), and measuring the elongation between successive loads. The load at which a permanent elongation begins is the elastic limit.

CONDUCTIVITY. In order to maintain a wire at a uniform and known temperature, it must be short. Unless the wire is very small, the test sample will therefore have a very low resistance, and an ordinary Wheatstone bridge will not be sufficiently accurate to measure it. This difficulty is avoided by using a bridge of the Kelvin, Hoop, Willyoung, or Reeves type (see Section 5, Article 7).

61. CURRENT-CARRYING CAPACITY OF BARE AND WEATHERPROOF WIRES AND CABLES

Table 33 gives the current-carrying capacities of bare and weatherproof wires and cables on the basis of the conditions listed at the end of the table. In some cases the copper temperature is limited, by the joints or other hardware, to about 70 deg cent (L. M. Olmstead, *Elec. World*, Vol. 127, p. 42 (1947)).

Table 33. Current-carrying Capacities of Bare and Weatherproof Covered Copper Wires and Cables in Normal Operation *

(Current in Amperes)

Wire Size, A.W.G. or Circular Mils		Bare		Weatherproof	
		Summer	Winter	Summer	Winter
Solid:	8	77	104	87	125
	6	104	142	114	165
	4	137	186	151	218
	2	184	250	202	292
Stranded:	2	188	256	204	296
	1	220	300	235	341
	0	252	346	274	390
	00	292	400	315	458
	000	341	469	366	533
	0000	391	540	424	619
	250,000	436	604	476	694
	300,000	489	679	532	776
	350,000	535	750	589	860
	400,000	585	815	641	935
	500,000	674	940	743	1083

* From *Calculating of Current-carrying Capacity of Overhead Conductors* by H. A. Enos, EEI (February, 1943, revised November 1946.)

Table based on air temperatures 60 deg fahr (16 deg cent) in winter and 110 deg fahr (43 deg cent) in summer.

Maximum copper temperature 80 deg cent, and maximum temperature of braid surface 70 deg cent.

Do not use "Winter" values if there is any possibility of the air temperature being above 60 deg fahr at time of maximum load.

Wind velocity assumed at 1 mile per hour crosswise to the conductor.

62. BIBLIOGRAPHY

American Institute of Electrical Engineers, Reports of Transmission and Distribution Committee.
American Society for Testing Materials, Specifications and Committee Reports.
Copper Wire Engineering Association, Publications (Washington, D. C.).
Edison Electric Institute, *Overhead Systems Handbook.*
International Congress on H. T. Electrical Systems, Reports (Paris).
National Bureau of Standards, Bulletins and Circulars (Washington).

See also publications of aluminum- and copper-wire manufacturers and the Bibliography of Section 19.

INSULATED WIRES AND CABLES

By W. A. Del Mar

63. TERMINOLOGY

Wire. A wire is a slender rod or filament of drawn metal. The definition restricts the term wire to what would ordinarily be understood by the term solid wire. In the definition the word slender is used in the sense that the length is great in comparison with the diameter. If a wire is covered with insulation, it is properly called an insulated wire. Although primarily the term wire refers to the metal, nevertheless when the context shows that the wire is insulated, the term wire will be understood to include the insulation.

Conductor. A conductor is a wire or combination of wires, not insulated from one another, suitable for carrying an electric current.

Stranded Conductor. A stranded conductor is a conductor composed of a group of wires, or of any combination of groups of wires. The wires in a stranded conductor are usually twisted together.

Strand. A strand is one of the wires, or groups of wires, of any stranded conductor.

Cable. A cable is either a stranded conductor (single-conductor cable) or a combination of conductors insulated from one another (multiple-conductor cable).

The first kind of cable is a single conductor; the second kind is a group of several conductors. The component conductors of the second kind of cable may be either solid or stranded, and this kind of cable may or may not have a common insulating covering.

The term cable is applied by some manufacturers to a solid wire heavily insulated and lead-covered; this usage arises from the manner of the insulation, but such a conductor is not included under this definition of cable. The term cable is a general one, and in practice it is usually applied only to the larger sizes. A small cable is called a stranded wire or a cord. Cables may be bare or insulated, and insulated cables may be sheathed with lead, or armored with wires or bands.

Cord. A cord is a small, very flexible insulated cable. There is no sharp dividing line in respect to size between a cord and a cable.

Concentric-lay Conductor. A concentric-lay conductor is a conductor composed of a central core surrounded by one or more layers of helically laid wires. In the most common type of concentric-lay conductor, all wires are of the same size, and the central core is a single wire.

Rope-lay Conductor or Cable. A rope-lay cable is a cable composed of a central core surrounded by one or more layers of helically laid groups of wires. This kind of cable differs from the preceding in that the main strands are themselves stranded. In the most common type of rope-lay conductor or cable, all wires are of the same size, and the central core is a concentric-lay conductor.

Concentric-lay Cable. A concentric-lay cable is either: (*a*) A concentric-lay conductor as defined. (*b*) A multiple-conductor cable composed of a central core surrounded by one or more layers of helically laid insulated conductors.

***N*-conductor Cable.** An N-conductor cable is a combination of N conductors insulated from one another. It is not intended that the name as here given be actually used. One would instead speak of a three-conductor cable, a twelve-conductor cable, etc. In referring to the general case, one may speak of a multiple-conductor cable.

***N*-conductor Concentric Cable.** An N-conductor concentric cable is a cable composed of an insulated central conductor with $(N - 1)$ tubular stranded conductors laid over it concentrically and separated by layers of insulation. This kind of cable usually has only two or three conductors. Such cables are used particularly for alternating currents. The remark on the expression N-conductor given for the preceding definition also applies here.

Duplex Cable. A duplex cable is a cable composed of two insulated stranded conductors twisted together. They may or may not have a common insulating covering.

Twin Cable. A twin cable is a cable composed of two insulated stranded conductors laid parallel and having a common covering.

Twin Wire. A twin wire is a cable composed of two small insulated conductors laid parallel and having a common covering.

Twisted Pair. A twisted pair is a cable composed of two small insulated conductors, twisted together, without a common covering. The two conductors of a twisted pair are usually substantially insulated, so that the combination is a special case of a cord.

Triplex Cable. A triplex cable is a cable composed of three insulated single-conductor cables twisted together. They may or may not have a common insulating covering.

Sector Cable. A sector cable is a multiple-conductor cable in which the cross-section of each conductor is substantially a sector, an ellipse, or a figure intermediate between them. Sector cables are used in order to obtain decreased overall diameter and thus permit the use of larger conductors in a cable of given diameter.

Round Conductor. A round conductor is either a solid or a stranded conductor of which the cross-section is substantially circular.

Split Conductor Cable. A split conductor cable is one in which each conductor is composed of two or more insulated conductors normally connected in parallel.

Shielded Conductor Cable. A shielded conductor cable is a cable in which the insulated conductor or conductors are enclosed in a conducting envelope or envelopes.

Factor of Assurance. The factor of assurance of wire or cable insulation is the ratio of the voltage at which it is tested to that at which it is used.

Insulation Resistance. The insulation resistance of an insulated conductor is the resistance offered by its insulation to an impressed direct voltage tending to produce a leakage of current through the insulation.

Mil. A mil is the one-thousandth part of an inch. There are 1974 circular mils in a square millimeter.

Circular Mil. A circular mil is a unit of area equal to $\pi/4$ (= 0.7854 . . .) of a square mil. The cross-sectional area of a circle in circular mils is therefore equal to the square of its diameter in mils. A circular inch is equal to 1,000,000 circular mils.

Lay. The lay or pitch of any helical element of a cable is the axial length of a turn of the helix of that element. Among the helical elements of a cable may be each strand in a concentric-lay cable, or each insulated conductor in a multiple-conductor cable.

Direction of Lay. The direction of lay is the lateral direction in which the strands of a cable run over the top of the cable as they recede from an observer looking along the axis of the cable.

64. MATERIALS

(See also Section 2)

CONDUCTORS. Round copper wire, solid or stranded, is almost invariably used for insulated wires and cables. Aluminum, requiring a larger cross-section for the same conductance per unit length, requires more insulating material for the same thickness of insulation.

Stranding. Conductors are stranded in order to make them more flexible. The flexibility is approximately proportional to the square root of the number of strands. In large cables, flexibility is required, first, in order to permit them to be put on reels; second, to permit easy installation; and third, to permit them to be bent around the walls of splicing chambers. Cords, elevator cables, mining machine cables, etc., must have considerable flexibility in order to permit them to be readily shifted from place to place in service.

For conduit work it is found unnecessary to strand conductors smaller than No. 6 A.W.G., although some contractors use stranded conductors as small as No. 12.

Sector-shaped Conductors. Two-, three-, and four-conductor cables with sector-shaped conductors are in general use, the three-conductor type being most common. The principal advantage of this type of cable is the greater carrying capacity for a given outside diameter.

PREPARATION OF CONDUCTORS FOR INSULATION. Where rubber insulation is used, it is necessary to cover the conductor with a thin film of tin or lead alloy, or with a separator, as described below.

Tinning. Copper and rubber, when brought into contact, react upon one another chemically. It was formerly thought that this action was due entirely to the sulfur in the rubber compound combining with the copper. Pure rubber, however, also reacts with the copper, the rubber breaking down into a gluey, sticky mass. Coating the copper with either tin or an alloy of tin and lead affords the necessary protection.

Separators. Small stranded conductors are usually covered with either a winding of soft cotton threads or a thin film of a plastic for the following purposes:

1. To protect the copper and rubber from mutual chemical action where a coating of tin on the copper cannot be used.
2. To hold together a group of fine wires so that individual wires will not stand up and penetrate the insulation during the manufacturing process.
3. To prevent adhesion between the rubber and copper in order that the copper may be easily bared for making connections.

INSULATION. The materials used for wire and cable insulation are vulcanized rubber, varnished cloth, impregnated paper, vinyl plastics, polyethylene, asbestos, cotton and silk thread, and enamel.

Rubber Insulation. This is composed of either natural or synthetic rubber mixed with modifying agents to give it the desired mechanical and electrical properties. The synthetic rubbers commonly used for cable insulation and sheaths are as follows:

GR-S, formerly known as Buna-S.
GR-I, or butyl rubber.
GR-M, or Neoprene.

There is also GR-A, or Perbunan, which, however, has comparatively little application in this field.

There are two processes by which rubber compound is ordinarily put on the wire, the **strip** and the **extrusion** or **seamless processes.** In the strip process the compound is first made into long narrow strips and then pressed around the wire; in the extrusion process the wire is run through a die through which the compound is pressed onto the wire. Insulation made by the strip process shows a seam or ridge where the sides of the strip have united, which is often partially obliterated by a tape applied before vulcanization; that made by the **extrusion** process is seamless.

Rubber insulating compounds are made by mixing new rubber with some or all of the following ingredients:

Vulcanizing agent: For natural rubber or GR-S, either sulfur or a sulfur-bearing organic compound, capable of liberating sulfur at vulcanizing temperatures. For GR-M, either zinc oxide or litharge modified by magnesia. For GR-A, the same as for natural rubber, or litharge. For GR-I, the same as for natural rubber, but with stronger acceleration.

Accelerator: Either litharge or an organic compound, usually of the basic nitrogenous type.

Activator: Usually zinc oxide, when sulfur or sulfur-bearing compound is used as vulcanizing agent.

Fillers: Inert mineral matter, such as talc or clay.

Hardener: Usually carbon black or clay.

Softener: Paraffin wax, reclaimed rubber, resins, peptizing agents, coumarone-indine resins, hydrocarbons, and hydrocarbon polymers.

Peptizing agents: For natural rubber, aromatic mercaptans. For GR-M, thiurams, mercapto thiazoles, and guanidines. For GR-S, the same as for natural rubber, but in greater concentration.

Antioxidant: Organic compounds, usually of nitrogenous type.

Ozone-resistant: Vulcanized oil, ceresin, GR-I, Vistanex.

Tackifier: Materials like Koresin, a condensation product of acetylene with an alkalated phenol.

Sun-checking inhibitor: Amorphous waxes, used with natural rubber, GR-S, and GR-A. GR-M is highly sunlight resistant.

The types of rubber compounds in common use are listed in Table 1.

In addition to the processes mentioned, there are several processes for the deposition of rubber from latex, whereby thin walls are obtained for multiple-conductor cables for communication and control.

VINYL PLASTICS (PVC). Polyvinyl chloride with a plasticizer is the most popular plastomer for low-voltage wires, being used without outer covering for house wiring. It is highly flame resistant, tough, and pliable, but not elastic like rubber. It is extruded hot on the wire, like rubber, and hardens without vulcanization.

POLYTHENE OR POLYETHYLENE. Polyethylene is a derivative of ethylene which possesses remarkably low power factor, low S.I.C., and virtually zero water absorption. It is pliable and tough and, like the vinyl plastics, is extruded on the wire hot and hardens upon cooling.

VARNISHED CAMBRIC. Cotton fabric coated with multiple layers of varnish is applied to the conductor in the form of tape, the successive turns being staggered. A thin layer of oil or petrolatum is applied between layers. Where the cable is to be used for vertical risers, only a small quantity of this "slipper compound" should be used.

IMPREGNATED PAPER. Tapes of special wood pulp paper, 5 to 8 mils thick, are applied to the conductor and then dried and thoroughly impregnated with heavy oily compound, sometimes containing about 15 to 20 per cent of rosin. The utmost care is required in the drying, evacuating, and impregnating to keep out every trace of air and moisture, especially for high-voltage cables. The lead is applied after impregnation. Oil-filled cable is similar except that a thin oil is used, and impregnation is effected after application of the sheath.

ASBESTOS. Asbestos may be applied in felted form, as a tape, or as a braid. It is usually impregnated with either synthetic chlorine compound or asphaltic material, and it is often associated with varnished cambric, as its own electrical properties are very poor. This combination is known as the AVC type.

COTTON AND SILK. Cotton or silk insulation consists of one to three layers of thread spun on to the wire. It is usually paraffined or varnished.

ENAMEL. Wire is passed through successive baths of quick-drying enamel until the requisite thickness is attained.

FILLERS. Fillers, used to round out multiple-conductor cables, are usually made of crinkled paper for paper-insulated and varnished-cambric-insulated cables and of jute for rubber- or varnished-cambric-insulated cables. The jute may be paraffined, tarred, or vacuum-impregnated with asphaltic compound, depending on the degree of waterproofness required. Dry jute may be "cutched," a processing which preserves it against decay.

PROTECTIVE COVERINGS. Coverings of lead, steel wire, steel tape, treated cotton, asphalted jute, rubberized cotton tape, reinforced rubber, hardened rubber, vulcanized oil, treated paper, etc., are used to protect cable insulation from mechanical injury or water.

Cotton Braid. The usual covering for rubber-insulated conductors is a cotton braid saturated with asphaltic material and filled with stearine pitch. The larger sizes of cables have a rubberized cotton tape under the braid.

Weatherproofing. Saturated cotton braid is also put on uninsulated hard-drawn copper wire for overhead service, in order to protect the wire from destructive arcing due to accidental contact with tree branches and other foreign bodies. This type of wire is being largely superseded by wire with a thin wall of either neoprene or polyethylene.

Lead Sheath. Lead sheath is put over the insulation by passing the cable through a die, while hot lead is pressed hydraulically around it through an annular die, forming a continuous and close-fitting pipe. Were it not for electrolytic corrosion and a tendency to crack when repeatedly bent, this sheath would be practically permanent. Unfortunately, lead is subject to electrolytic and even chemical corrosion. It is also rendered brittle and eventually breaks into pieces when exposed to vibration, this effect being due to separation along intercrystalline surfaces. Unalloyed lead is suitable for cables which

Table 1. Voltage Classification and Grades of Insulation

(IPCEA)

Circuit Voltage	Single-conductor Cables		Multiple-conductor Cables	
	Fibrous-covered	Lead-covered	Fibrous-covered	Lead-covered
0–2000	All grades	All grades	All grades	All grades
2001–5000	RP-RH-MR Ozone-resistant	All grades	RP-RH-MR Ozone-resistant	RP-RH-MR Ozone-resistant
5001 and over	Ozone-resistant or manufacturers' recommendations for all constructions.			

NOTE: 0–600-volt Type T may be used without other coverings.

Code grade (Type R) or other grades lower than Type RH, RP, or Ozone-resistant are not recommended for service over 2000 volts.

Grades of insulation referred to in this table are as follows:

Code grade (Type R)	NEMA Standard for Code insulation. Underwriters' Laboratories, Inc., Type R.
Performance grade (Type RP)	NEMA Standard for Performance grade insulation. ASTM Performance grade, Spec. D-353 and D-755.
Heat-resisting grade (Type RH)	NEMA Standard for Heat-resisting insulation. ASTM Heat-resisting insulation, Spec. D-469 and D-754. Underwriters' Laboratories, Inc., Type RH.
Moisture-resistant grade	Underwriters' Laboratories, Inc., Type RW.
Ozone-resistant compound	IPCEA Standard for Ozone-resistant Compound. ASTM Ozone-resistant insulation, Spec. D-574.
Manufacturers' recommendation (Type MR)	Higher-grade special compounds not listed above but recommended by the manufacturer. Physical and electrical characteristics must be obtained from the manufacturer.
Thermoplastic synthetic (Type T)	IPCEA Standard for Thermoplastic insulation. Underwriters' Laboratories, Inc., Type T.

Where type letters are used to indicate the type of construction of rubber-covered wires and cables in accordance with the requirements of the National Electrical Code and the Underwriters' Laboratories, Inc., Standards, the following type letters shall be used:

R: Indicates a single conductor having rubber insulation of Code grade and a fibrous covering.

RP: Indicates a single conductor having rubber insulation of Performance grade and a fibrous covering.

RH: Indicates a single conductor having rubber insulation of Heat-resisting grade and a fibrous covering.

D: Used as a suffix, indicates a twin wire having two insulated fibrous-covered conductors laid parallel under an outer fibrous covering.

M: Used as a suffix, indicates a cable having two or more rubber-insulated fibrous-covered conductors twisted together under an outer fibrous covering.

L: Used as a suffix, indicates that an outer covering of lead has been applied.

Type letters as given above, when used alone, indicate conductors for use at not more than 600 volts. Conductors for use at higher voltages shall be indicated by adding numerical suffixes to the type letters, as follows:

10—for use at not more than 1000 volts
20—for use at not more than 2000 volts
30—for use at not more than 3000 volts
40—for use at not more than 4000 volts
50—for use at not more than 5000 volts

T: Indicates a single conductor having thermoplastic insulation without covering.

are hard and compact. Lead containing about 0.06 per cent of copper or 0.1 per cent arsenic is to be preferred. Other cables require a small quantity of tin or antimony to harden the lead. For a given thickness pure lead is cheaper, but for a given tensile strength alloy is cheaper. The usual proportions for alloy sheath are either 2 per cent of tin or $^3/_4$ per cent of antimony. Ternary alloys of lead, tin, and cadmium are also used. Alloyed lead is more resistant to intercrystalline cracking, at atmospheric temperatures, than pure lead.

Non-metallic Sheaths. Sheaths of Neoprene, PVC, polyethylene, vulcanized oil, asphalt-saturated asbestos, and other materials water-resistant in various degrees are being largely used for rubber-insulated cables which are to be buried underground.

Armor Wire and Tape. Submarine cables are usually covered with steel wire armor, and cables to be buried directly in the ground are generally covered with galvanized steel tape armor. In either case the armor is usually covered with asphalted jute.

65. APPLICATIONS OF VARIOUS TYPES

The type of insulation and protection to employ in any instance depends upon the purpose for which the conductor is to be used and the place in which it is to be installed.

Power Wires and Cables, that is, those for the transmission or distribution of electric energy, may be installed in buildings, in cars, in underground conduits, on pole lines, buried in the ground, or under water.

Wires and cables for **buildings** are usually rubber-insulated and covered with a saturated braid. Vinyl plastics are also used without covering. Varnished cambric with saturated braid is often used for the larger sizes.

Wires and cables for **powerhouses** are usually varnished-cambric insulated with a flameproof braid.

Wires and cables for **underground conduit lines** are generally paper-insulated, but rubber or varnished cambric is sometimes used for distribution.

Wires for **mining** machinery and for **railway signals** are rubber-insulated.

Submarine Power Cables are usually insulated with rubber, sometimes sheathed in lead, and armored with steel wire. Paper-insulated lead-sheathed and armored submarines are in extensive use for high voltages.

Insulated Conductors for Instrument and Machine Windings: Enameled wire or wire insulated with cotton or silk is used for the former; varnished cambric, mica, and asbestos compounds for the latter.

66. DESIGN

INSULATION THICKNESS. The thickness of insulation required for low-voltage cables is merely that necessitated by inevitable irregularities of manufacture and roughness in handling. At voltages above 1000, the dielectric stress begins to have an influence and, when 2000 volts are reached, becomes the dominant factor.

Potential Gradient in Insulation. Let

F = potential gradient, in volts per mil, at any point P in the dielectric at a distance x mils from the center of the wire (Fig. 1).
E = volts between wire and sheath (or outside surface of insulation).
r = radius of wire, in mils.
R = outside radius of insulation, in mils.
F_s = dielectric strength of the insulation, in volts per mil.

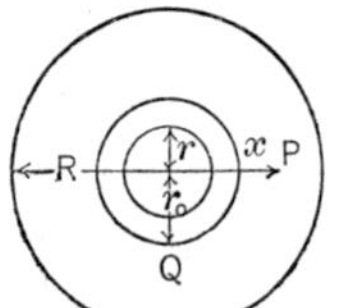

Fig. 1

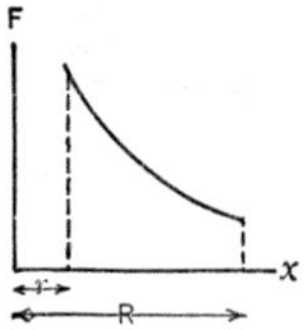

Fig. 2

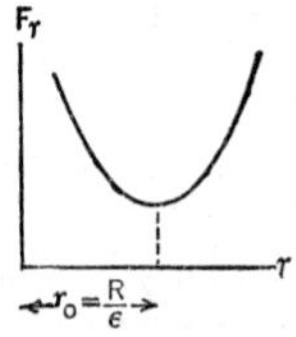

Fig. 3

Then, on the assumption of a perfectly homogeneous dielectric and perfect symmetry between conductor and insulation,

$$F = \frac{E}{x \log_\epsilon \left(\frac{R}{r}\right)}$$

That is, the potential gradient is the greatest at the surface of the conductor and decreases toward the outer surface of the insulation, as shown in Fig. 2. The potential gradient at the surface of the conductor is

$$F_{\max} = \frac{E}{r \log_\epsilon \left(\frac{R}{r}\right)}$$

For the same outside diameter of the insulation this stress at the surface varies with the radius of the conductor as shown in Fig. 3 and has a *minimum* value theoretically when $r = R/2.72$, but actually when R/r is somewhat greater than 2.72.

The volt-ampere characteristics of insulation are generally of the form shown in Fig. 4. Because of the form of the curve at high stresses, it is evident that, when the critical stress F_c is reached, the current will rise indefinitely, i.e., the insulation will fail, even if the stress is lowered. If, however, in cylindrical insulation, the inner layers are stressed to the value F_c, the remainder may act as a ballast resistor to keep the current down. Then the inner insulation will be overstressed but uninjured. The above formulas for stress, therefore, are not necessarily applicable at or near the breakdown point.

In impregnated-paper insulation, a potential gradient of about 50 volts per mil at the conductor surface will start ionization of occluded gas, but ionization is likely to die out unless the stress is over 60 volts per mil. At atmospheric pressure oil between paper tapes ionizes at an overall stress of about 230 volts per mil. (Inge and Walther.)

Rubber insulation is as likely to be injured by excessive minimum as by excessive maximum stress, as air of the outer surface of the insulation may be ionized and converted into ozone, which rapidly destroys the rubber.

Tables 3 to 12 inclusive give the standard thicknesses of insulation in American practice.

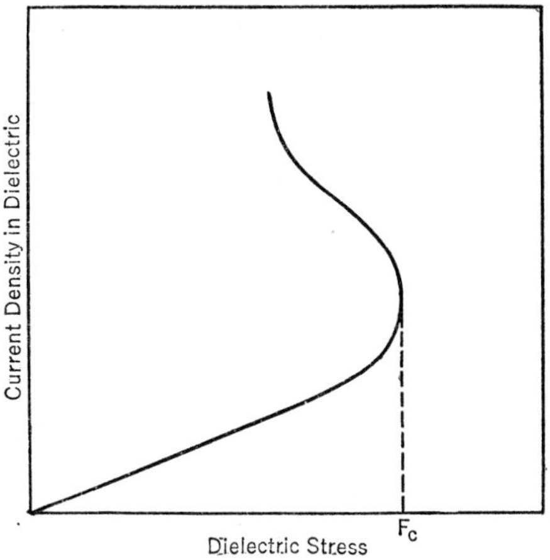

Fig. 4. Stress-current Density Curve of Dielectric

PREFERRED VOLTAGE RATINGS. Table 2 lists preferred voltage ratings for insulated power cables corresponding to preferred system voltages. These preferred voltage ratings designate the maximum continuous system voltages at which the cable should be operated.

Rated voltages for either single- or three-conductor cables intended for use on three-phase systems are expressed in terms of phase-to-phase voltages.

All values in this tabulation assume use on grounded neutral systems, except that this limitation does not apply for cables to be operated on systems having nominal system voltages of 600 volts or less. Cables to be used on non-grounded systems operating at voltages in excess of 600 volts should be especially designed for this application.

Table 2. Preferred Cable-voltage Ratings

Nominal System Voltage	Preferred Maximum Voltage Rating	Nominal System Voltage	Preferred Maximum Voltage Rating
120/240	600	23,000	25,800
120/208Y	600	34,500	37,500
240	600	46,000	48,300
480	600	69,000	72,500
600	1,000	115,000	121,000
2,400	3,000	138,000	145,000
2,400/4,160Y	5,000	161,000	169,000
4,800	6,000	230,000	242,000
7,200	9,000		
4,800/8,320Y	10,000		
12,000	15,000		
7,200/12,470Y	15,000		
7,620/13,200Y	15,000		
13,200	15,000		
14,400	15,000		

Note. For system voltages above 14,400 volts, cable installations are usually made in consideration of conditions and circumstances of the specific application, and cable voltage ratings selected accordingly. For example, if it is known that the continuous operating voltage on a particular 115-kv system will not exceed 115 kv, then a cable rated 115 kv can be used rather than a cable rated 121 kv as shown in this table.

Table 3. Insulation Thicknesses for Rubber-insulated Wires and Cables

(See Table 1 for voltage limitations of various grades)

Code, Performance, Heat-resisting and Ozone-resistant Grade Compounds

Rated Three-phase Circuit Voltage,* Volts	Size of Conductor, A.W.G. or MCM	Insulation Thickness on Each Conductor, 64th Inch		Rated Three-phase Circuit Voltage,* Volts	Size of Conductor, A.W.G. or MCM	Insulation Thickness on Each Conductor, 64th Inch	
		Grounded	Ungrounded			Grounded	Ungrounded
0–300	18–16	2	2	12,001–13,000	6–4/0	17	23
0–600 †	14–9	3	3		225–1,000	17	23
	8–2	4	4		Over 1,000	18	24
	1–4/0	5	5	13,001–14,000	6–4/0	18	25
	225–500	6	6		225–1,000	18	25
	525–1,000	7	7		Over 1,000	19	26
	Over 1,000	8	8	14,001–15,000	6–4/0	19	27
601–1,000	14–8	4	4		225–1,000	19	27
	7–2	5	5		Over 1,000	20	28
	1–4/0	6	6	15,001–16,000	4–4/0	20	
	225–500	7	7		225–1,000	20	
	525–1,000	8	8		Over 1,000	21	
	Over 1,000	9	9	16,001–17,000	4–4/0	21	
1,001–2,000	14–8	5	5		225–1,000	21	
	7–2	6	6		Over 1,000	22	
	1–4/0	7	7	17,001–18,000	4–4/0	22	
	225–500	8	8		225–1,000	22	
	525–1,000	9	9		Over 1,000	23	
	Over 1,000	9	9	18,001–19,000	4–4/0	23	
2,001–3,000	10–8	7	7		225–1,000	23	
	7–4/0	8	8		Over 1,000	24	
	225–1,000	9	9	19,001–20,000	2–4/0	24	
	Over 1,000	10	10		225–1,000	24	
3,001–4,000	10–4/0	9	9		Over 1,000	25	
	225–1,000	10	10	20,001–21,000	2–4/0	25	
	Over 1,000	11	11		225–1,000	25	
4,001–5,000	8–4/0	10	10		Over 1,000	26	
	225–1,000	11	11	21,001–22,000	2–4/0	26	
	Over 1,000	12	12		225–1,000	26	
5,001–6,000	8–4/0	10	12		Over 1,000	27	
	225–1,000	11	12	22,001–23,000	2–4/0	27	
	Over 1,000	12	13		225–1,000	27	
6,001–7,000	8–4/0	11	14		Over 1,000	28	
	225–1,000	11	14	23,001–24,000	2–4/0	28	
	Over 1,000	12	15		225–1,000	28	
7,001–8,000	8–4/0	12	16		Over 1,000	29	
	225–1,000	12	16	24,001–25,000	2–4/0	29	
	Over 1,000	13	17		225–4/0	29	
8,001–9,000	6–4/0	13	17		Over 1,000	30	
	225–1,000	13	17	25,001–26,000	2–4/0	30	
	Over 1,000	14	18		225–1,000	30	
9,001–10,000	6–4/0	14	18		Over 1,000	31	
	225–1,000	14	18	26,001–27,000	1–4/0	31	
	Over 1,000	15	19		225–1,000	31	
10,001–11,000	6–4/0	15	20		Over 1,000	32	
	225–1,000	15	20	27,001–28,000	1–4/0	32	
	Over 1,000	16	21		225–1,000	32	
11,001–12,000	6–4/0	16	22		Over 1,000	33	
	225–1,000	16	22				
	Over 1,000	17	23				

* For single- or two-phase systems, operating at over 5,000 volts with the center grounded, multiply the circuit voltage (phase to phase) by 0.866 ($=1/2\sqrt{3}$), and use the resulting voltage value to select the corresponding insulation thickness in the grounded neutral column.

For single- or two-phase, ungrounded systems operating at over 5,000 volts, multiply the circuit voltage (phase to phase) by 0.866 ($=1/2\sqrt{3}$), and use the resulting voltage value to select the corresponding insulation thickness in the ungrounded neutral column.

For d-c systems up to and including 2,000 volts, use thickness values in accordance with the grounded neutral column.

† For building wire the National Electrical Code permits the use of a 2/64-in. wall of Types R and RH insulations in sizes 14 and 12 A.W.G. for 600-volt service.

A circuit is considered to be grounded for application of cables, if the neutral is permanently connected to earth and if facilities are provided to insure prompt isolation of a faulty element of the circuit.

RUBBER INSULATION. The thicknesses of insulation in Table 3 are those recommended by the Insulated Power Cables Engineers Association and apply to single-conductor cable and the individual conductors of multiple-conductor cables, leaded or braided, except for some special appplications, such as non-leaded submarines, vertical risers, railway signals, and series lighting.

Non-leaded submarine cables should have $^1/_{32}$ in. more insulation than specified in Table 3.

Table 4 gives insulation thicknesses for series lighting cables. Table 5 gives the recommendations of the Association of American Railroads for railway signal wire. Constructions for other special types are given in *General Specifications for Wire and Cable with Rubber and Rubber-like Insulations* of the Insulated Power Cable Engineers' Association (68 pp.).

Table 4. Insulation Thicknesses for Series Lighting Cables with Rubber Insulation

Open Circuit Voltage	Conductor Size, A.W.G.	Insulation Thickness, 64th Inch: Without Circuit Protectors	Insulation Thickness, 64th Inch: With Circuit Protectors
Less than 1,000	10	3	3
	8–4	4	4
601– 1,000	10–8	4	4
	7–4	5	4
1,001– 2,000	10–8	5	4
	7–4	6	5
2,001– 3,000	10–8	7	5
	7–4	8	6
3,001– 4,000	10–8	9	7
	7–4	9	8
4,001– 5,000	10–4	10	9
5,001– 6,000	8–4	10	9
6,001– 7,000	"	11	10
7,001– 8,000	"	12	11
8,001– 9,000	"	13	11
9,001–10,000	"	14	12
10,001–11,000	"	15	12
11,001–12,000	"	16	14
12,001–13,000	"	17	14
13,001–14,000	"	18	15
14,001–15,000	"	19	15

Table 5. Insulation Thicknesses for Railway Signal Wire with Rubber Insulation*

Conductor Size, A.W.G.	Insulation Thickness, 64th Inch: 0 to 600 Volts			601 to 2500 Volts		
	A	B	C	D	E	F
14–8	5	4	3	10	9	7
17–2	6	5	4	10	9	8
1–4/0	7	6	5	10	9	8

* From Specifications of Signal Section of the American Railroad Association.

NOTE. Classes A, B, C, D, E, and F are specified by the signal engineer in accordance with his estimate of the importance of the circuits and the severity of the service.

VARNISHED-CAMBRIC INSULATION. Tables 6 and 7 give the IPCEA recommendations for varnished-cambric-insulated cables for various three-phase line-to-line voltages. The insulation thicknesses for various systems may be determined as indicated below:

System	Grounded (G) or Ungrounded (U)	Circuit Voltage, Volts	Insulation Thickness
3-Phase Δ or Y——...	G or U	Any	Use Table 6 or 7.
3-Phase Δ..........	One phase permanently grounded	Any	Multiply circuit voltage by 1.73, and use resulting voltage to select thickness in the grounded neutral column of Table 5 or 6.
1-Phase or 2-phase...	U	Up to and 3000	Use thickness from ungrounded neutral column of Table 6 or 7.
1-Phase or 2-phase...	One phase grounded	Over 600	Multiply circuit voltage by 1.73, and use resulting voltage to select thickness in grounded neutral column of Table 6 or 7.
1-Phase or 2-phase...	Center grounded	Over 600	Multiply circuit voltage by 0.866, and use resulting voltage to select thickness in grounded neutral column of Table 6 or 7.
1-Phase or 2-phase...	U	Over 3000	Multiply circuit voltage by 0.866, and use resulting voltage to select thickness in ungrounded neutral column of Table 6 or 7.
D-c................		Up to and including 2000	Use thickness from column for grounded neutral in Table 6 or 7.
Series lighting.......	G or U	Up to 15,000	Use Table 8.

Table 6. Recommended Thickness of Varnished-cambric Insulation
(IPCEA)
Single-conductor Cable and Multiple-conductor Shielded Cable

Rated Voltage, Volts, Phase to Phase	Size, A.W.G. or 1,000 cir mil	Ungrounded Neutral Wall of V. C. in 64ths inch (and mils)	Grounded Neutral Wall of V. C. in 64ths inch (and mils)	N.E. Code 1947
0–600	14–8	3 (47)	3 (47)	3 (47)
	7–2	4 (63)	4 (63)	4 (63)
	1–4/0	5 (78)	5 (78)	5 (78)
	213–500	6 (94)	6 (94)	6 (94)
	501–1,000	7 (109)	7 (109)	7 (109)
	1,001 and larger	8 (125)	8 (125)	8 (125)
601–1,000 *	14–2	4 (63)	4 (63)	4 (63)
	1–4/0	5 (78)	5 (78)	5 (78)
	213–500	6 (94)	6 (94)	6 (94)
	501–1,000	7 (109)	7 (109)	7 (109)
	1,001 and larger	8 (125)	8 (125)	8 (125)
1,001–2,000	12–2	5 (78)	5 (78)	5 (78)
	1–4/0	6 (94)	6 (94)	6 (94)
	213–500	6 (94)	6 (94)	6 (94)
	501–1,000	7 (109)	7 (109)	7 (109)
	1,001 and larger	8 (125)	8 (125)	8 (125)
2,001–3,000 (incl. 2,500 *)	10–2	6 (94)	6 (94)	6 (94)
	1–4/0	6 (94)	6 (94)	6 (94)
	213–500	7 (109)	7 (109)	7 (109)
	501–1,000	7 (109)	7 (109)	7 (109)
	1,001 and larger	8 (125)	8 (125)	8 (125)
3,001–4,000	8–4/0	7 (109)	7 (109)	7 (109)
	213–500	8 (125)	8 (125)	8 (125)
	501–1,000	8 (125)	8 (125)	(125)
	1,001 and larger	9 (141)	9 (141)	9 (141)
4,001–5,000 (incl. 4,500 *)	8–4/0	9 (141)	9 (141)	9 (141)
	213–1,000	10 (156)	10 (156)	10 (156)
	1,001 and larger	10 (156)	10 (156)	10 (156)
5,001–6,000	8–4/0	10 (156)	9 (141)	
	213–1,000	11 (172)	10 (156)	
	1,001 and larger	11 (172)	10 (156)	
6,001–7,000	8 and larger	11 (172)	10 (156)	
7,001–8,000 (incl. 7,500 *)	6 and larger	12 (188)	11 (172)	
8,001–9,000	6 and larger	13 (203)	12 (188)	
9,001–10,000	6 and larger	15 (234)	12 (188)	
10,001–11,000	6 and larger	16 (250)	13 (203)	
11,001–12,000	6 and larger	16 (250)	14 (219)	
12,001–13,000	6 and larger	18 (281)	15 (234)	
13,001–14,000	6 and larger	19 (296)	15 (234)	
14,001–15,000 *	6 and larger	21 (328)	16 (250)	
15,001–16,000	4 and larger	22 (344)	17 (266)	
16,001–17,000	4 and larger	23 (359)	18 (281)	
17,001–18,000	4 and larger		19 (296)	
18.001–19,000	4 and larger		20 (313)	
19,001–20,000	2 and larger		21 (328)	
20,001–21,000	2 and larger		22 (344)	
21,001–22,000	2 and larger		23 (359)	
22,001–23,000 *	2 and larger		24 (375)	
23,001–24,000	2 and larger		25 (391)	
24,001–25,000	2 and larger		26 (406)	
25,001–26,000	2 and larger		27 (422)	
26,001–27,000	2 and larger		28 (438)	
27,001–28,000	1 and larger		29 (453)	

* These ratings are recommended as standard. The highest rated voltage at each step represents maximum normal operating voltage. All cables have an emergency operating tolerance of 5 per cent above the rated voltage. Emergency overvoltage should not exceed a total time of more than 300 hours in any one year. In addition to the foregoing emergency overvoltage tolerance, a short-time overvoltage tolerance of 10 per cent of the rated voltage for not more than 15 min is allowed to take care of such contingencies as load switching, sudden loss of load, and the like. (Subject to future revision.)

Unless otherwise specified, two-conductor shielded cable will be of the round type.

A circuit is considered to have a grounded neutral if the neutral is permanently connected to earth and if facilities are provided to insure prompt isolation of a faulty element of the circuit.

Table 7. Recommended Thickness of Varnished-cambric Insulation

Multiple-conductor Cable without Individually Shielded Conductors

Rated Voltage Volts Phase to Phase	Size A. W. G. or 1000 cir mil	Ungrounded Neutral Wall of V. C. in 64ths inch (and mils)		Grounded Neutral Wall of V. C. in 64ths inch (and mils)	
		On Condrs. (C)	On Belt (B)	On Condrs. (C)	On Belt (B)
0–600	14–8	3 (47)	0 (0)	3 (47)	0 (0)
	7–2	4 (63)	0 (0)	4 (63)	0 (0)
	1–4/0	5 (78)	0 (0)	5 (78)	0 (0)
	213–500	6 (94)	0 (0)	6 (94)	0 (0)
	501–1000	6 (94)	2 (31)	6 (94)	2 (31)
	1001 and larger	7 (109)	2 (31)	7 (109)	2 (31)
601–1,000*	14–2	4 (63)	0 (0)	4 (63)	0 (0)
	1–4/0	5 (78)	0 (0)	5 (78)	0 (0)
	213–500	6 (94)	0 (0)	6 (94)	0 (0)
	501–1000	6 (94)	2 (31)	6 (94)	2 (31)
	1001 and larger	7 (109)	2 (31)	7 (109)	2 (31)
1,001–2,000	12–2	5 (78)	0 (0)	5 (78)	0 (0)
	1–4/0	6 (94)	0 (0)	6 (94)	0 (0)
	213–500	6 (94)	0 (0)	6 (94)	0 (0)
	501–1000	6 (94)	2 (31)	6 (94)	2 (31)
	1001 and larger	7 (109)	2 (31)	7 (109)	2 (31)
2,001–3,000 (incl. 2,500*)	10–2	5 (78)	2 (31)	5 (78)	2 (31)
	1–4/0	6 (94)	2 (31)	6 (94)	2 (31)
	213–500	6 (94)	2 (31)	6 (94)	2 (31)
	501–1000	6 (94)	3 (47)	6 (94)	3 (47)
	1001 and larger	7 (109)	3 (47)	7 (109)	3 (47)
3,001–4,000	8–4/0	6 (94)	3 (47)	6 (94)	3 (47)
	213–500	6 (94)	3 (47)	6 (94)	3 (47)
	501–1000	6 (94)	4 (63)	6 (94)	4 (63)
	1001 and larger	7 (109)	4 (63)	7 (109)	4 (63)
4,001–5,000 (incl. 4,500*)	8–4/0	6 (94)	4 (63)	6 (94)	4 (63)
	213–1000	7 (109)	4 (63)	7 (109)	4 (63)
	1001 and larger	7 (109)	5 (78)	7 (109)	5 (78)
5,001–6,000	8–4/0	6 (94)	5 (78)	6 (94)	5 (78)
	213–1000	7 (109)	5 (78)	7 (109)	5 (78)
	1001 and larger	7 (109)	5 (78)	7 (109)	5 (78)
6,001–7,000	8 and larger	7 (109)	6 (94)	7 (109)	5 (78)
7,001–8,000 (incl. 7,500*)	6 and larger	7 (109)	7 (109)	7 (109)	6 (94)
8,001–9,000	6 and larger	8 (125)	8 (125)	8 (125)	6 (94)
9,001–10,000	6 and larger	9 (141)	9 (141)	9 (141)	6 (94)
10,001–11,000	6 and larger	10 (156)	10 (156)	10 (156)	6 (94)
11,001–12,000	6 and larger	10 (156)	10 (156)	10 (156)	7 (109)
12,001–13,000	6 and larger	11 (172)	11 (172)	11 (172)	7 (109)
13,001–14,000	6 and larger	12 (188)	12 (188)	12 (188)	7 (109)
14,001–15,000*	6 and larger	13 (203)	13 (203)	13 (203)	7 (109)
15,001–16,000	4 and larger	14 (219)	14 (219)	14 (219)	7 (109)
16,001–17,000	4 and larger	14 (219)	14 (219)	14 (219)	7 (109)

* These ratings are recommended as standard. The highest rated voltage at each step represents maximum normal operating voltage. All cables have an emergency operating tolerance of 5 per cent above the rated voltage. Emergency overvoltage should not exceed a total time of more than 300 hours in any one year. A short-time overvoltage tolerance of 10 per cent of the rated voltage for not more than 15 min is allowed to take care of such contingencies as load switching, sudden loss of load, and the like. (Subject to future revision.)

Unless otherwise specified, two-conductor cable will be of the round type; when laid parallel, there shall be no belt; and the thickness of the insulation on each conductor shall be equal to the total thickness of conductor and belt insulation given above.

For special service applications requiring a multiple-conductor non-belted cable, the thickness of insulation on each conductor shall be equal to the total thickness of conductor and belt insulation given above.

A circuit is considered to have a grounded neutral if the neutral is permanently connected to earth and if facilities are provided to insure prompt isolation of a faulty element of the circuit.

Table 8. Varnished-cambric Insulation for Series Lighting Circuits

Secondary Circuit Voltage at 6.6 amp	Without Protectors		With Protectors	
Open, Volts	Conductor Size, A.W.G.	Recommended Wall of V.C. in 64ths Inch	Conductor Size, A.W.G.	Recommended Wall of V.C. in 64ths Inch
225	8	3	8	3
	7–4	4	7–4	4
430	8	3	8	3
	7–4	4	7–4	4
660	8	4	8	3
	7–4	4	7–4	4
1,080	8–4	5	8–4	4
1,600	8–4	5	8–4	5
2,090	8–4	6	8–4	5
3,090	8–4	7	8–4	6
4,115	8–4	8	8–4	7
5,110	8–4	9	8–4	7
6,130	8–4	10	8–4	8
8,260	6–4	12	8–4	10
10,350	6–4	13	6–4	11
12,400	6–4	15	6–4	12
14,500	6–4	16	6–4	13

IMPREGNATED PAPER INSULATION, SOLID TYPE. The insulation thicknesses most commonly used by the larger utilities are those specified by the Association of Edison Illuminating Companies and are given in the following formulas, which are taken from the Seventh Edition of their Specifications. Many companies, especially the smaller ones, use thicker insulation.

Single-conductor Cables. The thicknesses of insulation for single-conductor cables intended for use on grounded neutral systems are calculated by the following equations:

For rated voltages of 1–4 kv, inclusive: $T = 50 + 8.3E$.

For rated voltages of 5–30 kv, inclusive: $T = (50 + 8.3E)K$.

For rated voltages of 31–69 kv, inclusive: $T = (30 + 9.0E)K$.

where T (thickness in mils) is adjusted to the nearest 5-mil value.

E is rated voltage in kilovolts.

K is a factor (Table 9) depending on conductor size.

Table 9. Multiplying Factors (K) for Conductor Size

Single-conductor Cable

Conductor Size, Circular Mils or A.W.G.	Factor, K
300,000 and larger	1.00
250,000	1.01
4/0	1.03
3/0	1.07
2/0	1.11
1/0	1.15
1	1.19
2	1.22
3	1.25
4	1.27
5	1.29
6	1.30
7	1.31
8	1.32

Table 10. Multiplying Factors (K') for Conductor Size

Three-conductor Cable

Conductor Size, A.W.G.	Factor, K'
2/0 and larger	1.00
1/0	1.03
1	1.05
2	1.11
3	1.14
4	1.18
5	1.21
6	1.24
7	1.27
8	1.29

Three-conductor Shielded-type Cables. The thicknesses of insulation for three-conductor, shielded-type cables intended for use on grounded neutral systems are calculated by the equations:

For rated voltages of 9–30 kv, inclusive: $T = (50 + 8.3E)K'$.

For rated voltages of 31–46 kv, inclusive: $T = (30 + 9.0E)K'$.

where T (thickness in mils) is adjusted to the nearest 5-mil value.

E is rated voltage in kilovolts.

K' is a factor (Table 10) depending on conductor size.

Three-conductor Belted Cables. The thicknesses of insulation for multiconductor, belted-type cables intended for use on grounded neutral systems are calculated by the equations:

For rated voltages of 1–4 kv, inclusive: $T = 50 + 7.0E.$

$t = 30 + 3.0E.$

For rated voltages of 5–15 kv, inclusive: $T = (50 + 7.0E)K'.$

$t = (30 + 3.0E)K'.$

where T (thickness of conductor insulation in mils) and t (thickness of belt insulation in mils) are adjusted to the nearest 5-mil value.

E is rated voltage in kilovolts.

K' is a factor (Table 10) depending on conductor size.

OIL-FILLED AND OTHER IONIZATION-FREE CABLES. The usual insulation thicknesses for oil-filled, oilostatic and compression cables are given in Table 11.

Table 11. Recommended Insulation Thicknesses for Oil-filled Cable

(1945)

Rated Circuit Voltage, Kilowatts	Range of Conductor Sizes, A.W.G. and thousands of circular mils		Insulation Thickness,† mils (Grounded Neutral)
	Single-conductor	Three-conductor, Round *	
15	1/0–1000	1–750	110
20	1/0–2500	"	130
23	"	"	145
25	"	"	155
30	"	"	170
34.5	"	0–750	190
35	"	"	190
40	2/0–2500	"	210
45	"	0–1000	225
46	"	"	225
50	"	"	245
55	"	"	255
60	"	"	280
65	"	2/0–1000	295
69	"	"	315
70	"		320
75	3/0–2500	3/0–1000	340
80	"	"	355
90	"		390
100	"		430
110	"		465
115	4/0–2500		480
120	"		500
130	"		535
138	"		560
140	"		570
150	"		610
160	250–2500		645
230	750–2500		925

* For sector conductors the minimum sizes must be somewhat larger to avoid excessive stresses in the insulation at corners having small radii.

† For conductors larger than 750,000 cir mils, up to 1,000,000 cir mils, the thickness of insulation should be increased as follows:

Rated Voltage, Kilovolts	Increase, Mils
15–33	15
34–38	10
39 and 40	5

INSULATION THICKNESSES FOR LOW-GAS-PRESSURE CABLES. Thicknesses usual for this type of cable are given in Table 12.

INSULATION THICKNESSES FOR SYSTEMS WITH UNGROUNDED NEUTRAL. It is usual to add one-third to the line voltage and use the thickness of insulation appropriate for grounded neutral at this higher voltage.

Table 12. Insulation Thicknesses for Low-gas-pressure Cable

Grounded Neutral

Maximum Voltage Rating, Volts	Minimum Size Conductor		Insulation Thickness, Mils
	Round	Sector	
15,000	6 A.W.G.	0 A.W.G.	120 *
25,800	2 A.W.G.	3/0 A.W.G.	220
37,500	2/0 A.W.G.	300 MCM	310
48,300	3/0 A.W.G.	350 MCM	400

* For sizes through 650 MCM, 120; for sizes from 651 through 850 MCM, 130; for sizes from 851 MCM and up, 140 mils.

LEAD SHEATHS. The IPCEA recommended thicknesses for lead sheaths for rubber-, varnished-cambric-, and asbestos-insulated cables are given in Table 13. Those for paper-insulated cables usually follow a formula recommended by the Association of Edison Illuminating Companies.

Table 13. Recommended Lead Thicknesses for Rubber-, Varnished-cambric-, and Asbestos-insulated Cables

Diameter of Core, Inches	Thickness of Lead Sheath	
	64th Inch	Mils
0 –0.425	3	47
0.426–0.700	4	63
0.701–1.050	5	78
1.051–1.500	6	94
1.501–2.000	7	109
2.001–3.000	8	125
3.001 and larger	9	141

The average thickness for flat twin cable shall be that corresponding to the major core diameter. For submarine cables the thickness for the first two core diameters shall be 5/64 in. (78 mils).

Lead Thickness (A.E.I.C.) for Solid-type Paper Cables. Where the value inside the parenthesis of the equation for core diameter is less than 2 in. use D'; where it is greater than 2 in. use D in the following formula for the thickness of sheath:

$$S = 65 + 0.027D$$

where D is the calculated core diameter in mils.

S is the thickness of the sheath, adjusted to the nearest 5-mil value.

Core Diameter. The core diameters are calculated by use of the following equations For single-conductor cables:

$$D = 1.03(d + 2T)$$

$$D' = (d + 2T) + 60$$

For three-conductor cables with round conductors:

$$D = 1.03(2.155d + 4.31T + 4.31S + 2t + 2b)$$

$$D' = (2.155d + 4.31T + 4.31S + 2t + 2b) + 60$$

For three-conductor cables, with sector conductors:

$$D = 1.03(2.06V + 4.45T + 4.45S + 2t + 2b)$$

$$D = (2.06V + 4.45T + 4.45S + 2t + 2b) + 60$$

In these equations d is diameter in mils of round conductor.

V is V-gage depth in mils of sector conductor. (Values of V are found in Table 14.)

T is thickness in mils of conductor insulation.

t is thickness in mils of belt insulation for belted cable and is zero for shielded cable.

s is thickness in mils of outer shielding tape and intercalated paper tape for shielded-type cable (generally taken as 8 mils) and is zero for belted cable.

b is thickness in mils of binder for shielded-type cable (generally taken as 10 mils) and is zero for belted cable.

Table 14. V-gage Depth for Estimating Diameters of Sector Cables

Compact Type

Size, A.W.G. or MCM	V-gage Depth, Mils	
	For 200 Mils or Less of Insulation	For Over 2000 Mils of Insulation
0	288	288
00	323	323
000	364	364
0000	417	410
250	455	447
300	497	490
350	539	532
400	572	566
500	606	600
600	700	690
700	754	742
750	780	767
1000	900	898

Table 15. Armor Dimensions and Thickness of Jute Bedding for Armored Cable

(IPCEA, 1940)

64th Inch

(a) Jute Bedding				(b) Width and Thickness of Steel Tapes			(c) Size of Galvanized Steel Armor Wire	
Diameter of Cable under Jute, in.	Minimum Thickness of Jute			Diameter of Cable under Jute Bedding, in.	Minimum Steel Tape Thickness, mils	Maximum Width of Steel Tape, in.	Diameter of Cable under Jute Bedding, in.	Minimum Size of Armor Wire, B. W. G.
	Steel Tape Parkway Cable	Round Wire Armor Cables						
		Lead Sheathed	Non-leaded					
0 to 0.450	2	3	5	0 to 0.450	20	0.75	0 to 0.750	12
0.451–0.750	3	3	5	0.451–0.750	20	1.00	0.751–1.000	10
0.751–1.000	3	4	6	0.751–1.000	20	1.00	1.001–1.700	8
1.001–2.500	4	5	7	1.001–1.400	30	1.25	1.701–2.500	6
2.501 and larger	4	6	8	1.401–2.000	30	1.50	2.501 and larger	4
				2.001 and larger	30	2.00		

Table 16. NEMA Class A Braids

Diameter under Braid, in.	Minimum Thickness of Braid, in.	Corresponding Minimum Size and Ply of Cotton Yarn
0–0.200	0.015	30/2 or 14/1
0.201–0.350	0.017	26/2 or 12/1
0.351–0.800	0.020	20/2 or 10/1
0.801–1.500	0.025	12/2
1.501–3.000	0.031	8/2

ARMOR. Standard armor for ordinary installation and service conditions is made in accordance with the dimensions given in Table 15. The IPCEA Specifications for Metallic Coverings give special designs for borehole, dredge, and vertical riser cables.

BRAIDS. Three IPCEA grades of braid are in use, as follows:

a. Class A (NEMA) Braid Specifications, which are to be used only for Code grade wires and cables. The thickness shall be as specified in Table 16.

b. Class B (ASTM "Standard") Braid Specifications, which are designed for indoor service, installation in conduits, or as the inner braids for multiple-conductor cables, both for rubber and varnished cambric. The thickness is given in Table 17.

c. Class C (ASTM "Heavy") Braid Specifications, which are designed for outdoor and rough service. The thickness is given in Table 18.

d. Unless otherwise specified, Class B (ASTM "Standard") Braid Specifications are used for 30 per cent grade.

RUBBERIZED TAPE. Rubberized tape has a thickness of not less than 10 mils, and is applied with a lap of not less than 10 per cent of its width. The base sheeting of the tape has not less than 56 × 60 picks per inch.

Table 17. ASTM Standard or Class B Braids

Diameter under Braid, in.	Minimum Thickness of Braid, in.	Corresponding Minimum Size and Ply of Cotton Yarn
0–0.200	0.016	30/2
0.201–0.350	0.017	26/2
0.351–0.800	0.020	20/2
0.801–1.500	0.025	12/2
1.501–3.000	0.031	8/2

Table 18. ASTM Heavy Braids

Diameter under Braid, in.	Minimum Thickness of Braid, in.	Corresponding Minimum Size and Ply of Cotton Yarn
0–0.200	0.020	20/2
0.201–0.300	0.022	16/2
0.301–0.600	0.025	12/2
0.601–1.000	0.031	8/2
1.001–1.500	0.037	6/2
1.501–2.000	0.044	4/2
2.001–3.000	0.053	4/3

67. SHIELDING

Shielding is tentatively defined by IPCEA as follows: "Shielding of an electric power cable is the practice of confining its dielectric field to the inside of the cable insulation or insulated conductor assembly by surrounding the insulation or assembly with a grounded conducting medium called a shield."

There are three principal purposes for shielding insulated electric power cables:

1. To obtain symmetrical radial stress distribution within the insulation and to control tangential and longitudinal stresses or discharges on the surfaces of the insulation.
2. To protect cables connected to overhead lines or otherwise subject to induced potentials.
3. To provide increased safety to human life.

Cables for permanent installation, which require shielding, are usually provided with a metallic tape wound tightly around the insulation. Some rubber-insulated cables have a sheath of semi-conducting material over the insulation. In American practice paper-insulated cables are shielded with perforated copper tape, but in Europe a perforated metallized paper tape is preferred. Shielding, of whatever nature, should be grounded, and its terminations should have stress cones.

Where the shielding is in substantially continuous contact with a ground conductor or metallic sheath capable of carrying fault currents, the metallic tapes may be replaced with non-metallic conducting tapes of sufficiently low resistance to maintain the surface of the insulation at substantially ground potential under normal operating conditions.

The foregoing discussion refers to shielding over the insulation; it is, however, not uncommon to shield the interior surface of the insulation to protect it against ionization of air spaces between it and the conductor. In rubber insulation such "conductor shielding" may be a layer of rubber containing carbon black, and in paper cable, either an aluminum coated paper or a paper containing carbon black.

RUBBER-INSULATED CABLES. Table 19 gives the recommended practice for the shielding of rubber-insulated cables, as proposed by IPCEA.

PAPER-INSULATED CABLES. Multiple-conductor cables are almost invariably shielded for voltages of 15 kv and over, and present practice favors a lower limit of 9 kv. Single-conductor cable is seldom shielded for voltages below 60 kv except in the case of low-gas-pressure cable, for which shielding is essential because it assists in forming the necessary gas channel.

Table 19. Recommended Shielding Practice for Rubber-insulated Cables (IPCEA)

(Cables shall be shielded at three-phase operating voltages (line to line) above the values listed below. See Note *h*.)

	Ozone or Non-ozone-resistant Type of Insulation	Single and Multi-conductor		
		Basic Limits		Special Limits
		Grounded Neutral, Kilovolts	Ungrounded Neutral, Kilovolts	
Leaded Cables				
All except submarine installations	Ozone	10	8	....
All except submarine installations	Non-ozone	10	6	....
Submarine installations	Ozone	10	8	*a*
Submarine installations	Non-ozone	10	6	*a*
Non-leaded, Metallic Armored Cables				
Ordinary conditions	Both	10	6	*b*
Submarine installations	Both	10	6	*c*
Directly in earth	Both	6	3	....
Connected to overhead lines	Both	See Note *g*		
Non-leaded, Rubber-jacketed, and/or Fibrous-covered cables				
Ordinary conditions	Both	6	3	*b*
Directly in earth	Both	6	3	*d*
In ducts or metal conduits, moist	Both	2	2	....
In ducts or metal conduits, permanently dry	Both	6	3	*e*
On insulators	Both	6	3	*f*
Aerially in metal rings	Both	2	2	....
Aerially in marlin ties	Ozone	6	3	....
Aerially in marlin ties	Non-ozone	3	2	....
Connected to overhead lines	Both	See Note *g*		

NOTES. *a*. Single-conductor leaded submarine cables are shielded at shore ends above the voltages specified. Above 23 kv grounded neutral and 17 kv ungrounded neutral, they should be shielded throughout.

b. Ordinary conditions are defined as those permitting a cable, throughout its length, to operate without being subjected to interruption of its uniform leakage and charging current flow or change in its normal stress concentration.

c. Non-leaded submarine cables should be shielded at shore ends above the voltages specified and may be shielded throughout. The latter practice is preferable, as without shielding the application of an a-c test voltage might cause corona cutting.

d. Cables installed in exceptionally dry earth or in earth that changes from dry to moist should either be shielded above 2 kv or provided with a bare ground wire laid alongside it.

e. Supposedly dry metallic conduits may contain sufficient moisture to augment surface leakage to the point where surface discharges occur. For conservative practice it may be advisable to shield below the limits given in the table.

f. Single-conductor cables on insulators do not require shielding within the voltage limits usual for rubber-insulated cables.

g. Cables connected directly to overhead lines subject to lightning surges require shielding at voltages lower than the basic limits, especially if the cable is a short link in the overhead system.

h. For systems other than three-phase, the equivalent voltages for use in the table are as follows: Single-phase and two-phase systems for over 5000 volts, multiply the line-to-line voltage by the following figures to obtain the equivalent three-phase voltage: one side grounded, 1.73; midpoint grounded, 0.866; ungrounded, 0.866.

VARNISHED-CAMBRIC-INSULATED CABLES.* Shielding by the use of metallic tapes is recommended for cables rated above the three-phase line voltages specified for the following classes.

Class A: Single-conductor non-leaded cables.

Installed on individual full-voltage insulators: No shielding is necessary at any voltage.

In contact with conducting or semi-conducting surfaces: Shielding is desirable for voltages from 4000 to 8000 inclusive and is necessary above 8000 volts, except for series street lighting cable, on which shielding is necessary above 9000 volts.

* Courtesy, IPCEA.

Class B: Multiple-conductor non-leaded and unarmored cables.

Shielding is desirable for voltages from 4000 to 8000 inclusive and is necessary under any condition of installation above 8000 volts. If all the insulation is on the conductors, each conductor should be shielded; if part of the insulation is in the belt, the shielding tape should be applied over the belt.

Class C: Multiple-conductor lead-sheathed cables and/or metallic armored cables.

Non-belted cable with individually shielded conductors is recommended under any condition of installation above 10,000 volts. Where belted cables are specified, shielding is unnecessary.†

68. TESTS

GENERAL. Rubber-insulated wires and cables are tested for imperfections by a high-voltage test and for dryness by a megohm test. General quality is tested by tensile strength, elongation, modulus, and permanent set measurements. Life expectancy is estimated by accelerated aging tests, in which oxidation is promoted by heat or concentration of oxygen.

High-voltage rubber insulation is tested for corona suppression by noting either the voltage at which visual corona appears, when viewed in the dark, or the voltage at which high-frequency oscillations appear in the charging current, using an oscilloscope as detector. Ozone resistance is determined by exposure to ozone of 0.02 per cent concentration.

Water resistance is tested by determining the rise of S.I.C. when immersed in water or the increase in weight due to water absorption. The former test is usually made at 50 deg cent, and the critical factor is the rise of S.I.C. between the seventh and fourteenth days. The latter test is usually made at 70 deg cent for 7 days, and results expressed in milligrams per square inch of exposed surface.

Varnished-cambric-insulated wires and cables are given high-voltage and megohm tests for the reasons given above for rubber. They are sometimes tested for power factor and ionization. See IPCEA specifications and Table 21(*a*).

Impregnated-paper-insulated cables are given high-voltage and megohm tests, although recent specifications do not have any megohm requirements. Cables for 8000 volts and over are tested for power factor and ionization, both as an economic measure and as a guard against reduction of carrying capacity which might result from dielectric loss. The usual limiting power factors are given in Table 48. Voltage-time tests are made on short samples to determine the ability of the cable to withstand overvoltages and to measure, however inadequately, their stability.

TEST VOLTAGES. Factory Tests. Factory-test voltages for rubber-insulated cables are given in Table 20. The duration of test for Code grade is 1 min; for other grades, 5 min. Factory-test voltages for varnished-cambric-insulated cables are given in Table 21. The duration of test is 5 min. Factory-test voltages for "solid-type" impregnated-paper-insulated cables are given in Table 21(*b*), which also gives the duration.

Oil-filled Cable. Single-conductor and three-conductor cables are tested at 300 volts per mil of specified insulation thickness for 15 min.

For all types of cables, a test voltage applied after armoring should be 80 per cent of that applied before armoring.

Tests after Installation and Proof Tests. Tests after installation and proof tests at intervals are made at voltages not exceeding the following values, the percentages being taken on the full reel factory-test voltage unless otherwise stated.

Test	Rubber	V.C.	Paper
After installation	80% for same period, but not over 2000 volts for 600-volt wire	60% for 60 min or 80% for 5 min	60% for 4 hr or 80% for 1/4 hr
Proof	75% for same period	150% of rated voltage or 60% for 5 min in either case	175% of rated voltage or 60% for 5 min in either case
D-c tests equivalent to above a-c tests *	2.2 for non-ozone resistant and 3 for ozone resistant	2	2.4

* If the cable temperature exceeds 25 deg cent, the ratio shall be reduced by 0.013 for each degree over 25 deg.

† The above shielding recommendations are based only on considerations of continuity of cable operation under ordinary conditions. Where considerations of safety or unusual operating conditions assume importance, lower limits may be required for non-leaded cables.

IONIZATION. Cables for circuit voltages of 8000 and above are tested for ionization, i.e., rise of power factor from 20 to 100 volts per mil with paper insulation and from 20 to 80 volts per mil with varnished-cambric insulation. The purpose of this test is to detect entrained gas. The usual ionization limits are as follows:

Number of Conductors	Rated Kilovolts	Ionization Factor (Rise of Power Factor)	
		Paper	Varnished Cambric
Single.	8–20	0.3	3.5
	21–35	0.2	3.5
	36 and over	0.1	3.5
Multiple, shielded.	8–20	0.3	3.5
	21 and over	0.2	3.5
Multiple, belted.	. . . .	No requirement	5.0

If the test is made between 10 and 24 deg cent, the test value is reduced from the test value 2 per cent of that value for each degree below 25 deg, with paper insulation, and 4 per cent, with varnished-cambric insulation.

BENDING TESTS. These tests are made to ensure flexibility of paper and varnished-cambric cables. In paper cables they are made at a diameter of 12 times the diameter of the cable, and in varnished cambric at a diameter ratio of 8.

INSULATION RESISTANCE. The insulation resistance of a cable is usually expressed in megohm-miles, sometimes erroneously called megohms per mile. The total insulation resistance of a cable varies inversely as its length; e.g., a cable 2 miles long has half the resistance between conductor and sheath that 1 mile of this cable has. The formulas for insulation resistance of various types of cables are given below.

The insulation resistance of a single-conductor cable of length l centimeters is

$$R' = \frac{\rho}{2\pi l} \log_\epsilon \frac{D}{d}$$

where d = diameter of conductor.
D = outside diameter of insulation.
R' = insulation resistance.
l = axial length centimeter.
ρ = specific resistance.

From the above formula the megohm-miles are

$$R = K \log_{10} \frac{D}{d}$$

where R = the insulation resistance in megohms for a specified unit length.
K = megohms constant. When the insulation resistance is to be determined in megohm-mile units, K equals 0.00000228 times the resistivity of the insulation expressed in megohm-centimeter units.
D = outside diameter of insulation.
d = diameter of conductor.

The value of K for various types of insulation at 60 deg fahr, after an electrification of approximately 1 min under a constant d-c voltage, is given in the accompanying table. K varies with the time of electrification, the temperature, and the humidity.

Insulation	K at 60° F (15.5° C)	
	Limits	Usual Values
Vulcanized rubber	780 to 10,000	2,000 to 6,000
Gutta-percha	500 to 4,000	2,500
Varnished cloth	400 to 2,000	1,000
Impregnated paper	500 to 3,000	1,000
Commercial Grades of Rubber		
Code		3,800 } for 1,000 ft
Intermediate		8,000 } for 1,000 ft
30% and performance type		21,120 } for 1,000 ft

Table 20. Test Voltages, in Kilovolts, for Rubber-insulated Wires and Cables

Where the insulated conductor or conductors are covered by rubber or rubber-like impervious sheaths, either integral with the insulation or separate from it, or where the thickness of the insulation is increased for mechanical reasons, such as the insulation between conductors of a concentric-type cable, the test voltage shall be determined by the size of the conductor and rated voltage of the cable and not by the apparent thickness of the insulation.

Rated Circuit Voltage	Size A.W.G. or MCM	Insulation Thickness, 64th Inch and Mils		Grade of Insulation					
				Code (R)		Performance (RP) Heat-resisting (RH)		Ozone Resistant	
		Neutral Grounded	Neutral Ungrounded	Grounded	Ungrounded	Grounded	Ungrounded	Grounded	Ungrounded
0–600	14–9	3 (47)	3 (47)	1.5	1.5	3.0	3.0 *	4.5	4.5
	8	4 (63)	4 (63)	1.5	1.5	3.5	3.5	6.0	6.0
	7–2	4 (63)	4 (63)	2.0	2.0	3.5	3.5	6.0	6.0
	1–4/0	5 (78)	5 (78)	2.5	2.5	4.0	4.0	7.5	7.5
	225–500	6 (94)	6 (94)	3.0	3.0	5.0	5.0	8.5	8.5
	525–1,000	7 (109)	7 (109)	3.5	3.5	6.0	6.0	10.0	10.0
	Over 1,000	8 (125)	8 (125)	3.5	3.5	7.0	7.0	11.5	11.5
601–1,000	14–8	4 (63)	4 (63)	4.0	4.0	5.0	5.0	6.0	6.0
	7–2	5 (78)	5 (78)	5.0	5.0	6.0	6.0	7.5	7.5
	1–4/0	6 (94)	6 (94)	6.0	6.0	7.5	7.5	8.5	8.5
	225–500	7 (109)	7 (109)	7.0	7.0	9.0	9.0	10.0	10.0
	525–1,000	8 (125)	8 (125)	8.0	8.0	10.0	10.0	11.5	11.5
	Over 1,000	9 (141)	9 (141)	9.0	9.0	11.0	11.0	13.0	13.0
1,001–2,000	14–8	5 (78)	5 (78)	5.0	5.0	6.0	6.0	7.5	7.5
	7–2	6 (94)	6 (94)	6.0	6.0	7.5	7.5	8.5	8.5
	1–4/0	7 (109)	7 (109)	7.0	7.0	9.0	9.0	10.0	10.0
	225–500	8 (125)	8 (125)	8.0	8.0	10.0	10.0	11.5	11.5
	525–1,000	9 (141)	9 (141)	9.0	9.0	11.0	11.0	13.0	13.0
	Over 1,000	9 (141)	9 (141)	9.0	9.0	11.0	11.0	13.0	13.0
2,001–3,000	10–8	7 (109)	7 (109)	7.0	7.0	9.0	9.0	10.0	10.0
	7–4/0	8 (125)	8 (125)	8.0	8.0	10.0	10.0	11.5	11.5
	225–1,000	9 (141)	9 (141)	9.0	9.0	11.0	11.0	13.0	13.0
	Over 1,000	10 (156)	10 (156)	10.0	10.0	12.5	12.5	14.0	14.0
3,001–4,000	10–4/0	9 (141)	9 (141)	9.0	9.0	11.0	11.0	13.0	13.0
	225–1,000	10 (156)	10 (156)	10.0	10.0	12.5	12.5	14.0	14.0
	Over 1,000	11 (172)	11 (172)	11.5	11.5	13.5	13.5	15.5	15.5
4,001–5,000	8–4/0	10 (156)	10 (156)	10.0	10.0	12.5	12.5	14.0	14.0
	225–1,000	11 (172)	11 (172)	11.5	11.5	13.5	13.5	15.5	15.5
	Over 1,000	12 (188)	12 (188)	13.0	13.0	15.0	15.0	17.0	17.0
5,001–6,000	8–4/0	10 (156)	12 (188)			12.5	15.0	14.0	17.0
	225–1,000	11 (172)	12 (188)			13.5	15.0	15.5	17.0
	Over 1,000	12 (188)	13 (203)			15.0	16.5	17.0	18.5
6,001–7,000	8–1,000	11 (172)	14 (219)			13.5	17.5	15.5	20.0
	Over 1,000	12 (188)	15 (234)			15.0	19.0	17.0	21.0
7,001–8,000	8–1,000	12 (188)	16 (250)			15.0	20.0	17.0	22.5
	Over 1,000	13 (203)	17 (266)			16.5	21.0	18.5	24.0
8,001–9,000	6–1,000	13 (203)	17 (266)			16.5	21.0	18.5	24.0
	Over 1,000	14 (219)	18 (281)			17.5	22.5	20.0	25.0
9,001–10,000	6–1,000	14 (219)	18 (281)			17.5	22.5	20.0	25.0
	Over 1,000	15 (234)	19 (297)			19.0	24.0	21.0	26.5
10,001–11,000	6–1,000	15 (234)	20 (313)			19.0	25.0	21.0	28.0
	Over 1,000	16 (250)	21 (328)			20.0	26.5	22.5	29.5
11,001–12,000	6–1,000	16 (250)	22 (344)			20.0	28.0	22.5	31.0
	Over 1,000	17 (266)	23 (359)			21.0	29.5	24.0	32.0
12,001–13,000	6–1,000	17 (266)	23 (359)			21.0	29.5	24.0	32.0
	Over 1,000	18 (281)	24 (375)			22.5	31.0	25.0	33.5
13,001–14,000	6–1,000	18 (281)	25 (391)			22.5	32.5	25.0	35.0
	Over 1,000	19 (297)	26 (406)			24.0	34.0	26.5	36.0

* Building wire sizes 14 and 12 with $^2/_{64}$ in. thickness of insulation shall be tested at 1.5 kv.

Table 20. Test Voltages, in Kilovolts, for Rubber-insulated Wires and Cables—*Continued*

Rated Circuit Voltage	Size A.W.G. or MCM	Insulation Thickness, 64th Inch and Mils		Grade of Insulation					
				Code (R)		Performance (RP) Heat-resisting (RH)		Ozone Resistant	
		Neutral Grounded	Neutral Ungrounded	Grounded	Ungrounded	Grounded	Ungrounded	Grounded	Ungrounded
14,001–15,000	6–1,000	19 (297)	27 (422)			24.0	35.5	26.5	37.5
	Over 1,000	20 (313)	28 (438)			25.0	37.0	28.0	39.0
15,001–16,000	4–1,000	20 (313)	28 (438)					28.0	39.0
	Over 1,000	21 (328)	29 (453)					29.5	40.5
16,001–17,000	4–1,000	21 (328)	30 (469)					29.5	41.5
	Over 1,000	22 (344)	31 (484)					31.0	43.0
17,001–18,000	4–1,000	22 (344)						31.0	
	Over 1,000	23 (359)						32.0	
18,001–19,000	4–1,000	23 (359)						32.0	
	Over 1,000	24 (375)						33.5	
19,001–20,000	2–1,000	24 (375)						33.5	
	Over 1,000	25 (391)						35.0	
20,001–21,000	2–1,000	25 (391)						35.0	
	Over 1,000	26 (406)						36.0	
21,001–22,000	2–1,000	26 (406)						36.0	
	Over 1,000	27 (422)						37.5	
22,001–23,000	2–1,000	27 (422)						37.5	
	Over 1,000	28 (438)						39.0	
23,001–24,000	2–1,000	28 (438)						39.0	
	Over 1,000	29 (453)						40.5	
24,001–25,000	2–1,000	29 (453)						40.5	
	Over 1,000	30 (469)						41.5	
25,001–26,000	2–1,000	30 (469)						41.5	
	Over 1,000	31 (484)						43.0	
26,001–27,000	1–1,000	31 (484)						43.0	
	Over 1,000	32 (500)						44.5	
27,001–28,000	1–1,000	32 (500)						44.5	
	Over 1,000	33 (516)						46.0	

Table 22 gives the value of $\log_{10} \frac{D}{d}$ for various sizes of wire and thicknesses of insulation.

If the insulation resistance is measured with alternating current, it will appear to be much less than with direct current because of the dielectric loss.

Multiple-conductor Cables, Concentric. The insulation resistance between the inner conductor and the adjacent conductor of a concentric cable is calculated as for a single-conductor cable. The insulation resistance between any other conductor and its adjacent conductors, or conductor and sheath (if the sheath is adjacent), of a concentric cable, is the product divided by the sum of the resistances of the layers of insulation adjacent to such conductor, each being calculated separately as for a single-conductor cable.

Multiple-conductor Cables, Non-concentric, Round or Sector Type. The insulation resistance of each conductor to all other conductors connected to the sheath or water shall be calculated assuming the multiple-conductor cable to be replaced by a single-conductor cable having a round conductor of the same cross-sectional area and an outside diameter over insulation of

$$D = \frac{(d + 3c + 2b)}{d}$$

where d is the diameter of a round conductor of the same cross-sectional area, c is the thickness of conductor insulation, and b is the thickness of belt insulation.

Where the braid or outer surface of the insulation acts as a conducting surface (as it sometimes does in multiple-conductor, rubber-insulated cables), the values of D are as in single-conductor cables (AIEE Standard No. 30).

Table 21(*a*). Test Voltages, in Kilovolts, for Varnished-cambric-insulated Wires and Cables

Size of Conductor, A.W.G. or Circular Mils	Thickness of Insulation, 64th Inch (and Mils)								
	3 (47)	4 (63)	5 (78)	6 (94)	7 (109)	8 (125)	9 (141)	10 (156)	11 (172)
14–13	2.5	3.5							
12–11	2.5	3.5	4.5						
10–9	2.5	3.5	4.5	6.5					
8	2.5	3.5	4.5	6.5	9.0	11.0	13.5	15.5	17.5
7		3.5	4.5	7.0	9.5	11.5	14.0	16.0	18.0
6–5		4.0	5.0	7.5	10.0	12.0	14.5	16.5	18.5
4–3		5.0	5.5	8.0	10.5	12.5	15.0	17.0	19.0
2		5.0	5.5	8.0	10.5	12.5	15.0	17.0	19.0
1–4/0			6.0	8.5	11.0	13.5	15.5	18.0	19.5
213,000–500,000				9.0	11.5	14.0	16.0	18.5	20.0
501,000 and larger				9.0	11.5	14.0	16.0	18.5	20.0
	12 (188)	13 (203)	14 (219)	15 (234)	16 (250)	17 (266)	18 (281)	19 (297)	20 (313)
6–5	20.0	22.0	23.5	25.0	27.0	29.0	30.5	32.5	35.0
4–3	20.5	22.5	24.0	26.0	27.5	29.5	31.0	33.0	35.5
2	20.5	22.5	24.0	26.0	27.5	29.5	31.0	33.0	35.5
1–4/0	21.0	23.0	24.5	26.5	28.0	30.0	31.5	33.5	36.0
213,000–500,000	22.0	23.5	25.5	27.0	29.0	30.5	32.5	34.5	37.0
501,000 and larger	22.0	24.0	26.0	27.5	29.5	31.0	33.0	35.0	37.5
	21 (328)	22 (344)	23 (359)	24 (375)	25 (391)	26 (407)	27 (422)	28 (438)	29 (453)
6–5	37.5								
4–3	38.0	40.5	43.0						
2	38.0	40.5	43.0	45.0	47.5	49.5	50.5	52.0	
1–4/0	38.5	41.0	43.5	45.5	48.0	50.0	51.0	52.5	54.0
213,000–500,000	39.0	41.5	44.0	46.0	48.5	51.0	52.0	53.5	55.0
501,000 and larger	39.5	42.0	44.5	46.5	49.0	51.5	52.5	54.0	55.5

NOTES: (*a*) Single-conductor cables and multiple-conductor non-belted cables with individually shielded conductor shall be tested from conductors to sheath or ground at a voltage specified in the above table for the thickness of insulation from conductor to ground.

(*b*) Multiple-conductor cables with a belt of varnished cambric applied over the assembled conductors shall be tested between conductors at twice the voltage given in the above tables for single-conductor cables having the same thickness of insulation as applied to the individual conductors. The test voltage to sheath or to ground shall be 58 per cent of this value, except that for ungrounded circuits over 6000 volts the test voltage to sheath or ground shall be 80 per cent of the test voltage between conductors.

(*c*) Multiple-conductor non-belted, non-shielded cables or cables with a belt of material other than varnished cambric shall be tested between conductors and between conductors and ground at the voltage given in the above tables for a single-conductor cable having the same thickness of insulation as applied to the individual conductors.

(*d*) Where three-phase testing equipment is available, it is permissible, at the discretion of the manufacturer, to apply three-phase testing voltage to three-conductor cable, the voltage to ground being that above specified.

(*e*) Cables having a metallic covering over the insulation shall be tested with this covering grounded. A ground shall be provided for cables which do not have a metallic covering by immersing the cable in water for the minimum time required to make the tests, except that this immersion test shall not be applied to cables having an insulating wall thickness on the conductor, in case of single-conductor cables, or on the belt, in case of multiple-conductor belted cables, of less than 5/64 in. For single-conductor cables a 5-ft length on one end of the cable shall be wrapped with metal foil and given the voltage test in Table 21, applied between conductor and metal foil, the latter being at high potential and the conductor grounded. This test may be omitted if 100 per cent of the cable is acceptably spark tested. The insulation, in that case, shall be capable of withstanding, without breakdown, a test potential of 7000 volts for insulation 4/64 in., and 5000 volts for insulation 3/64 in., in thickness. For multiple-conductor cables with belt less than 5/64 in., the test voltage specified in Table 21 and footnotes shall be applied only between conductors.

Table 21(*b*). Test Voltages, in Kilovolts, for Impregnated-paper-insulated Cables

Conductors to sheath, $5 + 2.1E$; conductor to conductor, $8.7 + 3.65E$.

E is phase-to-phase voltage in kilovolts. Duration of test is 5 minutes for E less than 8 kv, and 15 minutes for E of 8 kv or more. Test voltage is reduced 3 per cent for each AWG size under No.1 AWG.

Table 22. Values of Log D/d

Equivalent Insulation, Thickness

Cond. Size A.W.G. or M. Cir. Mils.	64th Inches													
	3	4	5	6	7	8	9	10	12	14	16	18	20	22
	Decimals													
	(0.047)	(0.063)	(0.078)	(0.094)	(0.109)	(0.125)	(0.141)	(0.156)	(0.188)	(0.219)	(0.250)	(0.281)	(0.313)	(0.344)
14*	0.392	0.470	0.537	0.594	0.645	0.691	0.732	0.770	0.836	...	...	...	...	...
12*	.334	.405	.467	.520	.568	.611	.651	.686	.751	0.806	0.856	...	...	...
10*	.283	.348	.404	.453	.498	.538	.575	.609	.670	.723	.771	0.814	0.853	0.889
8*	.239	.296	.347	.392	.432	.470	.505	.537	.594	.645	.691	.731	.770	.804
6	...	.225	.267	.305	.340	.373	.403	.431	.483	.529	.570	.608	.643	.676
5	...	.206	.245	.281	.314	.346	.373	.401	.450	.495	.535	.572	.606	.637
4	...	.187	.224	.257	.289	.318	.345	.371	.418	.460	.500	.535	.568	.598
3	...	.171	.204	.236	.265	.293	.318	.343	.388	.429	.466	.500	.532	.562
2	...	.155	.186	.215	.243	.269	.293	.316	.359	.398	.433	.466	.497	.526
1	...	...	.168	.195	.220	.244	.267	.288	.328	.365	.399	.431	.461	.487
1/0	...	...	.152	.177	.201	.223	.244	.264	.302	.337	.369	.399	.428	.454
2/0	...	...	.138	.161	.183	.204	.223	.242	.278	.311	.342	.370	.397	.422
3/0	...	...	.125	.146	.166	.185	.204	.221	.255	.286	.315	.342	.367	.392
4/0	...	...	.113	.132	.151	.168	.187	.202	.233	.262	.289	.315	.339	.362
250	...	...	...	.123	.140	.157	.173	.189	.218	.246	.272	.296	.320	.342
300	...	...	...	.113	.130	.145	.160	.175	.203	.229	.254	.278	.300	.321
350	...	...	...	.106	.121	.136	.150	.164	.190	.215	.239	.262	.283	.303
400	...	...	...	.0995	.114	.128	.142	.155	.181	.204	.227	.249	.269	.289
450	...	...	...	.0944	.108	.122	.136	.148	.172	.195	.217	.238	.258	.277
500	...	...	...	.0901	.103	.116	.129	.141	.165	.187	.208	.228	.248	.266
600	...	...	...	...	.0952	.107	.119	.130	.152	.173	.193	.212	.230	.248
650	...	...	...	...	.0919	.104	.115	.126	.147	.168	.187	.206	.223	.241
700	...	...	...	...	.0888	.100	.111	.122	.143	.163	.181	.199	.217	.234
750	...	...	...	...	.0861	.0971	.108	.118	.139	.157	.176	.194	.211	.228
800	...	...	...	...	.0836	.0943	.105	.115	.135	.154	.172	.189	.206	.222
900	...	...	...	...	.0793	.0895	.0994	.108	.128	.146	.164	.180	.196	.212
1000	...	...	...	...	.0755	.0851	.0948	.104	.122	.140	.157	.173	.189	.203
1250	...	...	...	...	...	.0770	.0856	.0943	.111	.127	.142	.157	.172	.186
1500	...	...	...	...	...	.0708	.0789	.0870	.102	.116	.132	.146	.159	.172
1750	...	...	...	...	...	.0658	.0734	.0810	.0954	.110	.123	.136	.149	.162
2000	...	...	...	...	...	.0619	.0691	.0761	.0898	.103	.116	.128	.141	.153
2500	...	...	...	...	...	.0558	.0623	.0687	.0812	.0934	.105	.117	.128	.139

* Solid conductor. Conductors larger than No. 8 are assumed to be stranded. For intermediate thickness interpolate.

Table 22. Values of Log *D/d*--*Continued*

Equivalent Insulation, Thickness

Cond. Size A.W.G. or M. Cir. Mils.	64th Inches													
	24	26	28	30	32	34	36	38	40	44	48	52	56	60
	Decimals													
	(0.375)	(0.406)	(0.438)	(0.469)	(0.500)	(0.531)	(0.562)	(0.594)	(0.625)	(0.687)	(0.750)	(0.812)	(0.875)	(0.937)
10*	0.922	....	....	....	....	....	....	....	....	....	....	....	....	....
8*	.836	0.866	0.894	....	....	....	....	....	....	....	....	....	....	....
6	.706	.734	.760	0.786	....	....	....	....	....	....	....	....	....	....
5	.667	.694	.720	.744	....	....	....	....	....	....	....	....	....	....
4	.625	.653	.678	.703	0.725	....	....	....	....	....	....	....	....	....
3	.589	.615	.640	.663	.685	0.706	....	....	....	....	....	....	....	....
2	.553	.578	.602	.625	.646	.666	0.686	....	....	....	....	....	....	....
1	.513	.538	.561	.583	.603	.623	.642	0.661	....	....	....	....	....	....
1/0	.479	.502	.525	.546	.566	.585	.604	.622	0.639	0.671	....	....	....	....
2/0	.446	.469	.490	.511	.531	.549	.567	.584	.601	.632	0.662	....	....	....
3/0	.414	.436	.457	.476	.495	.513	.531	.547	.563	.594	.622	0.649	....	....
4/0	.384	.405	.425	.443	.462	.480	.491	.512	.527	.557	.585	.610	0.635	0.658
250	.363	.383	.402	.420	.438	.455	.471	.487	.502	.530	.557	.583	.607	.630
300	.341	.360	.379	.396	.413	.430	.445	.460	.475	.503	.529	.554	.578	.600
350	.323	.341	.359	.376	.393	.408	.424	.438	.453	.480	.506	.530	.553	.575
400	.308	.326	.343	.360	.376	.391	.406	.420	.434	.461	.486	.510	.532	.553
450	.295	.312	.329	.345	.361	.376	.391	.405	.418	.444	.469	.492	.514	.535
500	.284	.301	.317	.333	.348	.363	.377	.391	.404	.430	.454	.477	.498	.519
600	.265	.281	.297	.312	.326	.340	.354	.367	.380	.405	.428	.450	.471	.491
650	.257	.273	.288	.303	.317	.331	.345	.357	.370	.395	.417	.439	.460	.480
700	.250	.266	.281	.295	.309	.323	.336	.349	.361	.385	.408	.429	.450	.469
750	.243	.259	.273	.288	.302	.315	.328	.341	.353	.376	.399	.420	.440	.459
800	.237	.252	.267	.281	.294	.308	.320	.333	.345	.368	.390	.411	.431	....
900	.227	.242	.255	.269	.282	.295	.307	.319	.331	.354	.375	.396	....	....
1000	.218	.232	.245	.259	.271	.284	.296	.308	.319	.341	.362	....	....	....
1250	.199	.212	.225	.237	.249	.261	.272	.284	.294	....	....	....	....	....
1500	.185	.197	.210	.221	.233	.244	.255	.265	....	....	....	....	....	....
1750	.174	.185	.197	.208	.219	.229	.240	....	....	....	....	....	....	....
2000	.164	.176	.187	.197	.207	.218	.228	....	....	....	....	....	....	....
2500	.150	.160	.170	.180	.190	.199	....	....	....	....	....	....	....	....

* Solid conductor. Conductors larger than No. 8 are assumed to be stranded. For intermediate thickness interpolate.

69. CURRENT-CARRYING CAPACITY *

The current-carrying capacity of insulated conductors is limited by electrical and chemical deterioration of the insulation at high temperatures and by the strains due to longitudinal thermal expansion in lead sheaths near joints.

TABLES OF CARRYING CAPACITY. Tables of carrying capacity for cables insulated with various kinds of rubber, impregnated paper, and varnished cambric and installed in various groupings in air, in conduits, and in underground ducts are published by the Insulated Power Cable Engineers' Association. Tables 23 to 32 are typical examples from the IPCEA publication. The permissible current-carrying capacities of building wire are set by the National Electrical Code and by municipal codes in certain cities. The N.E.C. values are given in Section 14, Article 82.

REDUCTION OF CARRYING CAPACITY BY GROUPING. It will be noted from the IPCEA tables that the current-carrying capacity of cables in a duct bank is reduced by the proximity of other loaded cables. This may be regarded as due to the higher group ambient temperature set up by the neighboring cables, but for purposes of calculation it is more convenient to regard it as due to the greater temperature drop to base ambient temperature in the external elements of thermal resistance. The reduction is greater for large cables than for small, and greater for high-voltage than for low-voltage cables. The following examples illustrate this for a three-conductor cable (75 per cent load factor), assuming six equally loaded cables in a duct bank, as compared with a single cable.

Conductors	Percentage Reduction, Amperes	
	0–7,500 Volts	34,000 Volts
0 A.W.G..........	15	20
500,000 cir. mils...	20	25

Table 23. Impregnated-paper-insulated Cable in Underground Ducts: Single-conductor Cable, 69,000 Volts; Copper Temperature, 60° C; 75 and 100 Per Cent Load Factors

Conductor Size, 1000 Cir Mils	Number of Equally Loaded Cables in Duct Bank							
	Three		Six		Nine		Twelve	
	Per Cent Load Factor							
	75	100	75	100	75	100	75	100
	Amperes per Conductor							
350	360	336	333	305	312	279	293	259
400	389	362	358	328	335	300	315	278
500	441	409	406	370	379	337	354	312
600	490	454	450	409	419	371	391	343
700	536	495	490	444	455	403	425	372
750	556	514	508	460	472	417	439	384
800	575	531	525	475	487	430	453	396
1000	652	599	592	533	547	481	508	442
1250	734	672	664	595	610	535	564	489
1500	804	733	724	647	664	580	612	529
1750	865	788	776	692	711	618	653	563
2000	924	840	822	732	750	651	688	592
2500	1001	903	892	791	811	700	741	635

* Written with the assistance of E. H. Kirkham.

Table 24. Impregnated-paper-insulated Cable in Underground Ducts; Single-conductor Cable, 69,000 Volts; Copper Temperature, 60° C; 30 and 50 Per Cent Load Factors *

Conductor Size, 1000 Cir Mils	Number of Equally Loaded Cables in Duct Bank							
	Three		Six		Nine		Twelve	
	Per Cent Load Factor							
	30	50	30	50	30	50	30	50
	Amperes per Conductor							
350	395	382	387	364	375	348	365	332
400	428	413	418	393	405	375	394	358
500	489	470	477	446	461	425	447	405
600	545	524	532	496	513	471	497	448
700	599	573	582	543	561	514	542	489
750	623	597	605	562	583	533	563	506
800	644	617	626	582	603	554	582	523
1000	736	702	713	660	685	622	660	589
1250	832	792	806	742	772	698	741	659
1500	918	872	886	814	848	763	812	718
1750	994	942	957	876	913	818	873	770
2000	1066	1008	1020	931	972	868	927	814
2500	1163	1096	1115	1013	1060	942	1007	880

Correction Factors for Various Ambient Earth Temperatures for Tables 23 and 24

10° C	1.13	1.13	1.13	1.14
20	1.00	1.00	1.00	1.00
30	0.85	0.85	0.84	0.84
40	0.67	0.66	0.65	0.64
50	0.42	0.40	0.36	0.32

* The ratings in Table 24 are for load factors lower than those ordinarily met except for railway loads.

Table 25. Impregnated-paper-insulated Cable in Underground Ducts: Three-conductor Cable, 15,000 Volts, Shielded; Copper Temperature, 81° C; 75 and 100 Per Cent Load Factors

Conductor Size, A.W.G. or 1000 Cir Mils	Number of Equally Loaded Cables in Duct Bank									
	One		Three		Six		Nine		Twelve	
	Per Cent Load Factor									
	75	100	75	100	75	100	75	100	75	100
	Amperes per Conductor									
6	88	83	81	75	74	66	69	60	64	56
4	115	107	104	95	95	85	89	77	83	72
2	146	137	133	121	120	107	112	97	104	90
1	166	156	149	136	136	121	125	109	117	100
0	182	176	169	154	154	137	141	122	131	112
00	215	202	193	175	174	156	162	139	148	127
000	245	230	220	198	198	174	182	157	168	144
0000	281	261	250	223	224	196	205	176	189	162
250	310	290	276	246	245	215	224	193	207	177
300	344	320	305	272	271	236	246	211	227	194
350	375	346	330	293	293	255	267	227	245	208
400	403	373	354	314	313	273	285	242	262	222
500	450	418	399	350	350	303	318	269	292	247
600	501	460	437	385	384	330	346	293	317	269
700	543	497	472	414	410	354	372	313	339	285
750	562	514	485	426	423	365	383	323	348	293

Table 26. Impregnated-paper-insulated Cable in Underground Ducts: Three-conductor Cable, 15,000 Volts, Shielded; Copper Temperature, 81° C; 30 and 50 Per Cent Load Factors *

Conductor Size, A.W.G. or 1000 Cir Mils	Number of Equally Loaded Cables in Duct Bank									
	One		Three		Six		Nine		Twelve	
	Per Cent Load Factor									
	30	50	30	50	30	50	30	50	30	50
	Amperes per Conductor									
6	94	91	91	87	89	83	87	78	84	75
4	123	120	119	114	116	108	113	102	109	96
2	159	154	153	144	149	136	144	129	139	123
1	179	174	172	163	168	153	162	145	158	138
0	203	195	196	185	190	173	183	164	178	156
00	234	224	225	212	218	198	211	187	203	177
000	270	258	258	242	249	225	241	212	232	202
0000	308	295	295	276	285	257	275	241	265	227
250	341	327	325	305	315	283	303	265	291	250
300	383	365	364	339	351	313	337	293	322	276
350	417	397	397	369	383	340	366	318	350	301
400	453	428	429	396	413	366	394	340	376	320
500	513	487	483	446	467	410	444	381	419	358
600	567	537	534	491	513	450	488	416	465	390
700	618	586	583	533	558	486	528	449	503	419
750	643	606	602	551	576	502	545	464	519	432

Correction Factors for Various Ambient Earth Temperatures

10° C	1.08	1.08	1.08	1.08	1.08
20	1.00	1.00	1.00	1.00	1.00
30	0.91	0.91	0.91	0.91	0.91
40	0.82	0.82	0.82	0.82	0.81
50	0.71	0.71	0.71	0.71	0.70

* The ratings in Table 26 are for load factors lower than those ordinarily met except for railway loads.

Table 27. Rubber-insulated Cables in Air: Single-conductor Cable; 30–100 Per Cent Load Factors; Copper Temperatures, 50, 60, 70, and 75° C

Conductor Size, A.W.G. or 1000 Cir Mils	Copper Temperature										
	50° C	60° C					70° C			75° C	
	Voltage Range										
	0–5000	0–5000	5001–8000	8001–15,000	15,001–20,000	20,001–26,000	8001–15,000	15,001–20,000	20,001–26,000	0–5000	5001–8000
	Amperes per Conductor										
14	14	20	......	......	......	......	......	......	......	26	......
12	19	26	......	......	......	......	......	......	......	33	......
10	25	35	......	......	......	......	......	......	......	44	......
8	34	47	54	......	......	......	......	......	......	61	70
6	46	64	72	73	......	......	88	......	......	83	93
4	62	86	94	94	93	......	116	113	......	110	121
2	84	117	126	127	122	120	153	148	147	150	163
1	97	135	145	147	142	139	178	173	170	172	187
0	114	158	166	167	163	158	201	198	193	202	214
00	131	183	190	191	188	180	230	228	220	235	245
000	153	212	220	217	214	208	261	258	251	273	284
0000	176	245	255	251	246	238	302	299	291	315	329
250	197	275	280	275	270	263	333	328	322	352	362
300	220	306	312	310	302	293	373	367	359	393	403
350	249	346	345	342	330	323	410	401	395	443	445
400	270	375	377	370	361	347	446	435	425	481	486
450	287	399	402	397	384	377	478	467	461	516	519
500	305	425	430	422	412	397	508	497	493	546	554
600	341	474	480	473	458	444	570	554	550	608	620
700	373	518	529	520	504	490	625	611	602	665	685
750	387	538	555	545	527	517	659	640	636	691	716
800	401	557	575	565	547	532	678	662	652	717	742
900	430	597	618	608	585	573	730	712	700	767	798
1000	461	642	660	649	624	610	778	760	745	824	852
1250	526	731	760	745	716	700	897	870	857	938	980
1500	578	805	840	823	786	766	991	954	939	1032	1084
1750	632	879	920	905	870	840	1090	1055	1030	1130	1187
2000	682	948	990	973	932	904	1172	1130	1116	1220	1290

Correction Factors for Various Ambient Air Temperatures

	0–5000 Volts	0–8000 Volts	8001–26,000 Volts	8001–26,000 Volts	0–8000 Volts
10° C	2.00	1.58	1.62	1.43	1.36
20	1.73	1.41	1.44	1.30	1.25
30	1.41	1.22	1.24	1.16	1.13
40	1.00	1.00	1.00	1.00	1.00
50	0	0.71	0.68	0.80	0.85

Table 28. Rubber-insulated Cables in Air: Three-conductor Cables; 30–100 Per Cent Load Factors; Non-shielded and Shielded; Copper Temperatures, 50, 60, 70, and 75° C

Conductor Size, A.W.G. or 1,000 Cir Mils	Copper Temperature												
	50° C	60° C		70° C	75° C		60° C			70° C		75° C	
	Voltage Range												
	0–5,000	0–6,000	6,001–10,000	0–6,000	0–6,000	6,001–10,000	0–6,000	6,001–10,000	10,001–15,000	6,001–10,000	10,001–15,000	0–6,000	6,001–8,000
	Non-Shielded						Shielded						
	Amperes per Conductor												
14	13	18	....	21	23	....	19	....	....	....	....	25	....
12	16	22	....	27	29	....	25	....	....	....	....	32	....
10	21	29	....	35	38	....	33	....	....	....	....	43	....
8	28	39	42	47	50	54	43	45	....	53	....	55	58
6	36	50	55	60	65	71	56	58	59	70	71	72	75
4	48	67	72	80	86	93	72	76	79	92	96	93	99
2	63	88	94	105	113	121	96	98	100	118	122	124	127
1	72	100	108	120	129	140	110	111	115	134	140	142	144
0	82	114	122	137	147	157	126	129	132	156	160	163	168
00	95	132	140	159	171	181	148	150	152	181	185	191	195
000	108	150	161	181	194	208	166	169	172	205	209	214	220
0000	123	171	183	206	221	236	190	192	196	234	238	245	250
250	136	190	203	227	245	264	211	214	216	260	262	271	277
300	152	211	229	255	273	296	235	239	239	289	290	300	311
350	169	235	249	280	303	322	258	263	263	318	318	333	342
400	181	252	270	303	325	349	279	283	281	343	341	360	368
450	193	268	289	323	347	373	298	304	300	368	364	385	395
500	208	289	307	348	374	396	316	325	321	393	390	408	423
600	231	321	341	387	415	440	350	359	353	435	432	452	467
700	255	355	373	428	457	482	384	393	387	476	470	495	512
750	267	371	389	447	480	502	400	410	403	496	490	517	533

Correction Factors for Various Ambient Air Temperatures

10° C	2.00	1.58	1.41	1.36	1.58	1.59	1.61	1.42	1.43	1.36
20	1.73	1.41	1.29	1.25	1.41	1.42	1.43	1.30	1.31	1.25
30	1.41	1.22	1.15	1.13	1.22	1.23	1.24	1.16	1.17	1.13
40	1.00	1.00	1.00	1.00	1.00	1.00	1.00	1.00	1.00	1.00
50	0	0.71	0.82	0.85	0.71	0.70	0.69	0.81	0.80	0.85

Table 29. Varnished-cambric-insulated Cables in Underground Ducts: Single-conductor Cable, 7500 Volts; Copper Temperature, 84° C; 75 and 100 Per Cent Load Factors

Conductor Size, A.W.G. or 1000 Cir Mils	Number of Equally Loaded Cables in Duct Bank					
	Three		Six		Nine	
	Per Cent Load Factor					
	75	100	75	100	75	100
	Amperes per Conductor					
6	109	103	103	96	98	90
4	142	135	134	124	128	115
2	186	175	174	161	166	150
1	212	200	199	183	189	170
0	243	228	227	208	215	193
00	279	260	259	237	246	220
000	323	300	299	272	282	251
0000	370	340	340	310	323	285
250	405	375	374	340	352	311
300	452	416	415	377	392	345
350	493	456	455	402	427	375
400	535	492	490	444	460	405
500	612	560	558	500	521	455
600	676	619	616	550	573	500
700	735	672	667	596	620	540
750	763	695	693	619	641	558
800	792	720	717	640	664	578
1000	893	810	805	715	745	645
1250	1005	910	900	798	834	720
1500	1101	992	985	869	907	780
1750	1184	1064	1056	929	970	832
2000	1262	1130	1125	987	1029	879

Table 30. Varnished-cambric-insulated Cables in Underground Ducts: Single-conductor Cable, 7500 Volts; Copper Temperature, 84° C; 30 and 50 Per Cent Load Factor *

Conductor Size, A.W.G. or 1000 Cir Mils	Number of Equally Loaded Cables in Duct Bank					
	Three		Six		Nine	
	Per Cent Load Factor					
	30	50	30	50	30	50
	Amperes per Conductor					
6	116	113	114	110	112	107
4	154	149	152	144	149	140
2	202	196	199	189	196	183
1	232	224	227	216	224	208
0	268	258	262	248	257	239
00	308	296	301	288	295	274
000	356	344	350	328	342	316
0000	412	393	402	377	393	360
250	455	434	443	413	433	396
300	510	491	499	463	486	442
350	560	537	546	506	532	483
400	607	580	593	548	576	522
500	698	660	675	625	659	595
600	772	735	749	692	727	658
700	842	798	817	752	791	713
750	878	829	849	781	821	738
800	912	858	882	810	852	765
1000	1032	974	995	913	965	863
1250	1172	1102	1127	1027	1092	969
1500	1300	1212	1246	1129	1200	1061
1750	1408	1308	1350	1216	1296	1140
2000	1512	1402	1449	1300	1388	1213

Correction Factors for Various Ambient Earth Temperatures

	Three	Six	Nine
10° C	1.08	1.08	1.08
15	1.04	1.04	1.04
20	1.00	1.00	1.00
25	0.96	0.96	0.96
30	0.92	0.92	0.92
35	0.87	0.87	0.87

* The ratings in Table 30 are for load factors lower than those ordinarily met except for railway loads.

NOTE. Ratings in Tables 23–30 are based on:

a. 60-cycle alternating current and an ambient earth temperature of 20° C.

b. Round standard strand conductors or sectors in the case of three-conductor paper-insulated cables, sizes 0 and larger.

c. Open-circuited sheath operation, i.e., sheaths (if any) bonded and grounded at one point only.

d. One cable per duct, all cables equally loaded and in outside ducts only.

e. Ratings include dielectric loss and skin effect.

THERMAL UNITS USED IN CALCULATIONS. **A Thermal Ohm** is a thermal resistance expressed in degrees centigrade per watt rate of heat flow.

Thermal Volume Resistivity (ρ) is the thermal resistance of a unit cube expressed in thermal ohms.

Thermal Surface Resistivity (B) is expressed in degrees centigrade of temperature drop caused by heat flowing at the rate of 1 watt from each square centimeter of surface. It is not a constant, but a function of the cable diameter and the nature of its surface.

A Thermal Resistance (R_{th}), expressed in thermal ohms, is represented by the following typical equations:

$$R_{th} = \rho G, \quad \text{or} \quad R_{th} = \frac{B}{A}$$

where G is a **Geometric Factor** dependent on the dimensions of the cable, and A is the area of a surface in square centimeters. The following formulas use the metric values of ρ but express R_{th} per foot of cable.

Table 31. Varnished-cambric-insulated Cables in Underground Ducts: Three-conductor Cable, 15,000 Volts; Shielded; Copper Temperature, 77° C; 75 and 100 Per Cent Load Factors

Conductor Size, A.W.G. or 1000 Cir Mils	Number of Equally Loaded Cables in Duct Bank							
	One		Three		Six		Nine	
	Per Cent Load Factor							
	75	100	75	100	75	100	75	100
	Amperes per Conductor							
8	65	62	59	54	54	48	50	44
6	84	80	76	70	70	63	65	56
4	108	103	98	90	89	79	82	71
2	142	135	128	115	115	99	102	90
1	161	152	145	131	130	111	116	101
0	183	172	163	146	146	124	130	113
00	211	197	185	167	165	143	149	127
000	240	224	210	188	185	162	167	142
0000	272	253	236	212	209	181	188	159
250	299	278	259	231	229	198	204	172
300	330	306	285	254	250	217	222	188
350	357	332	309	273	270	233	238	201
400	384	356	330	281	288	248	253	219
500	431	398	369	324	319	274	280	236
600	469	433	400	350	344	294	301	252
700	512	470	432	378	370	315	321	270
750	528	486	445	390	382	323	330	278

The basic equation used in continuous-carrying-capacity calculations is the equality of the sum of the I^2R and dielectric losses with the heat dissipated, which is the temperature rise divided by the thermal resistance, as defined above, or

$$I^2R_c + W_d = \frac{T}{R_{th}}$$

FORMULAS FOR CALCULATING CURRENT-CARRYING CAPACITY. The following symbols and formulas apply to **all types of cables:**

T_c = maximum allowable copper temperature in degrees Centigrade (Table 33).
T_d = copper temperature with dielectric loss, but no load, in degrees Centigrade.
R_c = effective a-c resistance of each conductor, in ohms per foot, at temperature T_c (Articles 59 and 69 and Table 47).
I = current per conductor.
w_c = copper loss per conductor in watts per foot.
n = number of conductors per cable.
w_c = copper loss in watts per foot of cable.
W_d = dielectric loss in watts per foot of cable.
W_t = total watts lost per foot of cable.
ρ = thermal resistivity of insulation (Table 34).
G = geometric factor of cable (Tables 35 and 36).
B = surface thermal resistivity of cable (Table 37).
d = overall diameter of cable in inches.
R_{thi} = thermal resistance of insulation per foot of cable, all conductors thermally in parallel.
R_{ths} = surface thermal resistance of cable.
R_{th} = total thermal resistance from all conductors of cable to the base ambient.

$$W_c = nw_c = nI^2R_c \tag{1}$$

$$W_t = W_c + W_d \tag{2}$$

$$R_{thi} = \frac{0.00522\rho G}{n} \tag{3}$$

$$R_{ths} = \frac{0.00411B}{d} \tag{4}$$

Table 32. Varnished-cambric-insulated Cables in Underground Ducts: Three-conductor Cable, 15,000 Volts; Shielded; Copper Temperature, 77° C; 30 and 50 Per Cent Load Factors *

Conductor Size, A.W.G. or 1000 Cir Mils	Number of Equally Loaded Cables in Duct Bank							
	One		Three		Six		Nine	
	Per Cent Load Factor							
	30	50	30	50	30	50	30	50
	Amperes per Conductor							
8	69	67	67	64	65	59	62	57
6	92	87	88	83	84	78	83	73
4	118	112	114	106	109	100	105	94
2	155	149	149	140	142	130	134	120
1	176	169	169	159	161	147	152	136
0	200	192	192	179	182	165	171	153
00	230	221	220	205	210	188	198	175
000	264	252	250	232	237	213	225	197
0000	300	288	284	264	268	240	253	221
250	330	317	312	290	294	263	278	240
300	368	352	347	319	323	289	305	263
350	400	383	377	347	351	313	330	285
400	433	412	406	372	377	335	354	303
500	488	464	456	415	422	373	394	337
600	533	505	498	452	459	404	425	363
700	584	552	547	491	499	435	459	390
750	605	572	561	508	514	450	472	402

Correction Factors for Various Ambient Earth Temperatures

	One	Three	Six	Nine
10° C	1.09	1.09	1.10	1.10
15	1.05	1.05	1.05	1.05
20	1.00	1.00	1.00	1.00
25	0.95	0.95	0.95	0.95
30	0.90	0.90	0.89	0.89
35	0.85	0.84	0.83	0.83

* The ratings in Table 32 are for load factors lower than those ordinarily met except for railway loads.

NOTE. Ratings in Tables 31 and 32 are based on:
a. 60-cycle alternating current and an ambient earth temperature of 20° C.
b. Round standard strand conductors.
c. One cable per duct, all cables equally loaded and in outside ducts only.
d. Ratings include dielectric loss and all induced a-c losses.

The following additional symbols and formulas apply to **cables in still air:**

T_a = base ambient air temperature in degrees Centigrade (Table 38).

$$R_{th} = R_{thi} + R_{ths} \tag{5}$$

$$T_d - T_a = W_d R_{th} \tag{6}$$

$$(T_c - T_a) - (T_d - T_a) = W_c R_{th} \tag{7}$$

$$I = \sqrt{\frac{(T_c - T_a) - (T_d - T_a)}{nR_c R_{th}}} \tag{8}$$

The above applies to one cable; for a group of cables the current is reduced by the factor K_a from Table 39.

The following symbols and formulas apply to **cables in underground ducts:**

D = duct correction factor which, added to R_{th} for cables in still air, gives the thermal resistance of each of any number of loaded cables in a duct bank (Table 40).

T_e = base ambient earth temperature in degrees centigrade.

$$T_d - T_e = W_d(R_{th} + D) \tag{9}$$

In formula (9) a value of D is used for 100 per cent load factor.

$$(T_c - T_e) - (T_d - T_e) = W_c(R_{th} + D) \tag{10}$$

In formula (10) a value of D is used for the actual load factor being considered.

$$I = \sqrt{\frac{(T_c - T_e) - (T_d - T_e)}{nR_c(R_{th} + D)}} \quad (11)$$

F = loss factor (as a fraction).

The following symbols and formulas apply to **cables in conduits in buildings:**

T_a = base ambient air temperature in degrees centigrade.
Q = conduit correction factor which, added to R_{th} for cables in still air, gives the thermal resistance of each of any number of loaded cables in a conduit (Table 41).
K_c = a group correction factor which may be used to reduce the current calculated for the cables in one conduit to the value where the conduit is one of a group (Table 42).
C = group correction factor which, multiplied by Q, takes into account the number of conduits in a group (Table 43).

Both K_c and C accomplish the same thing, the C method being more accurate.

$$I = \sqrt{\frac{T_c - T_a}{nR_c(R_{th} + QC)}} \quad (12)$$

or

$$I = K_c\sqrt{\frac{T_c - T_a}{nR_c}} \quad (13)$$

The following symbols and formulas apply to **cables buried directly in the ground:**

L = axial depth of burial, ft.
d = overall diameter of cable, in.
R_{the} = thermal resistance from cable surface to base ambient.
g = thermal resistivity of soil (a commonly used value of which is 90) * (Table 44).

$$R_{the} = 0.012g \log_{10} \frac{48L}{d} \quad (14)$$

$$I = \sqrt{\frac{(T_c - T_e) - (T_d - T_e)}{nR_c(R_{thi} + FR_{the})}} \quad (15)$$

Table 46 gives some examples for buried cables.

The following symbols and formulas apply to **cables in buried pipe with gas or oil pressure medium.** Where water is the pressure medium, R_{thp} may be taken as 1 deg cent.

d_p = outside diameter of pipe or pipe covering in inches.
L = axial depth of pipe in feet.
d_i = inside diameter of pipe in inches.
d_c = diameter of circle circumscribing the complete cables.
ρ_1 = apparent thermal resistivity from surface of cable to pipe (Table 45).
R_{thp} = thermal resistance from surface of cable to pipe.†

$$R_{thp} = 0.0361\rho_1 \log_{10} \frac{d_i}{d_c} \quad (16)$$

$$R_{the} = 0.012g \log_{10} \frac{48L}{d_p} \quad (17)$$

In the case of cables in steel pipes, it is not practicable to use the value of R_c derived from ordinary skin-effect factors because the magnetism induced in the pipe creates an extra loss in the conductor, as well as a loss in the pipe itself. It is therefore usual to have two values of R_c, designated as R_{c1} and R_{c2}, to be derived by multiplying the d-c resistance by the factors indicated for this purpose in Table 47.

If there is a pipe covering of thermal resistivity ρ_2, add to R_{the} $0.012\rho_2 \times \log_{10}$ (ratio of outside to inside diameter of covering). The usual value of ρ_2 is 215.

$$I = \sqrt{\frac{(T_c - T_e) - (T_d - T_e)}{n[R_{c1}(R_{thi} + R_{thp}) + R_{c2}FR_{the}]}} \quad (18)$$

* The factor 0.012 is used if g is measured *in situ,* i.e., in undisturbed earth and with heat flowing as from a buried cylinder. If g is measured on earth samples in a laboratory set-up, the factor should be 0.008.

† At pressures under 200 psi it makes little difference whether the pressure medium is oil or nitrogen gas.

Table 33. Maximum Permissible Copper Temperatures for Continuous Operation

a. Impregnated-paper-insulated Cables, Solid and Low-gas-pressure Types

Voltage Rating, Kilovolts, Phase to Phase	Single-conductor and Three-conductor, Shielded, Degrees Centigrade	Three-conductor, Belted, Degrees Centigrade
4.5	85	85
7.5	85	85
15	81	75
23	77	67
34.5	70	60
46	63	......
69	60	......

b. Impregnated-paper-insulated Cables, Ionization-free Types

Voltage Rating, Kilovolts, Phase to Phase	Single- and Three-conductor, Degrees Centigrade
15	81
25	76
26–75	75
76–162	70
163–230	65

c. Varnished-cambric-insulated Cables

Voltage Rating, Kilovolts, Phase to Phase	Single-conductor and Three-conductor, Shielded, Degrees Centigrade	Three-conductor, Belted, Degrees Centigrade	Voltage Rating, Kilovolts, Phase to Phase	Single-conductor and Three-conductor, Shielded, Degrees Centigrade	Three-conductor, Belted, Degrees Centigrade
0–5	85	85	14	78	70
6	85	83	15	77	70
7	84	82	16	76	70
7.5	84	81	17	75	70
8	83	80	18	75	......
9	82	78	19	74	......
10	81	76	20	73	......
11	80	75	21	72	......
12	80	73	22	71	......
13	79	71	over 22 to 28	70	......

d. Rubber- and Vinyl-plastic-insulated Cables

Type of Compound	Voltage, Volts	Maximum Copper Temperature, Degrees Centigrade	Type of Compound	Voltage, Volts	Maximum Copper Temperature, Degrees Centigrade
Code (Type R).....	Up to 5000	60	Ozone-resistant....	Up to 8000	75
Performance type (Type RP).......	Up to 8000	60	Thermoplastic synthetic (Type T).......	1000	60
			Ozone-resisting....	Above 8000	70
Heat-resisting (Type RH)......	Up to 8000	75	Polyethylene......	All	75

Table 34. Thermal Resistivities of Cable Insulations (Thermal Ohms per Centimeter Cube)

Type of Insulation	Value of ρ Commonly Assumed in Calculations
Asbestos, impregnated	500
Dry paper	770
Impregnated paper, solid type	700 *
Impregnated paper, fluid-filled type	550
Oil	330
Polyethylene	350
Rubber compounds	500
Varnished cambric	600
Vinyl plastic	600

* New cable has a resistivity of 500 or less. The conservative value given allows for loosening of the sheath and other forms of deterioration in service. The other values given are conservative for reasons specific to each material.

Table 35. Geometric Factors, G, for Single-conductor and Three-conductor Belted Cables

T = Conductor-insulation thickness; t = belt-insulation thickness; d = conductor diameter.*

Ratio: $\frac{T+t}{d}$, Three-conductor; $\frac{T}{d}$, Single-conductor	Single-conductor	Three-conductor, Belted			
		Round Conductors		Sector Conductors	
		$t/T = 1$	$t/T = 1/2$	$t/T = 1$	$t/T = 1/2$
0.1	0.19	0.57	0.57	....	....
0.2	0.34	0.85	0.85	....	....
0.3	0.47	1.08	1.08	0.76	0.76
0.4	0.59	1.29	1.28	0.99	0.98
0.5	0.70	1.46	1.43	1.18	1.16
0.6	0.78	1.61	1.57	1.35	1.32
0.7	0.88	1.74	1.69	1.51	1.47
0.8	0.95	1.86	1.80	1.64	1.58
0.9	1.03	1.98	1.90	1.76	1.69
1.0	1.10	2.08	1.99	1.87	1.79
1.1	1.16	2.15	2.06	1.95	1.88
1.2	1.22	2.23	2.13	2.05	1.96
1.4	1.34	2.36	2.27	2.20	2.11
1.6	1.44	2.50	2.39	2.35	2.25
1.8	1.52	2.61	2.50	2.48	2.37
2.0	1.61	2.72	2.60	2.58	2.47
2.2	1.69	2.81	2.68	2.70	2.57
2.4	1.75	2.90	2.76	2.78	2.65

* With sector cables use diameter d of round conductor of same cross-sectional area.

More complete data on geometric factors are given by D. M. Simmons in *Elec. J.* (1932) and in *Trans. AIEE*, Vol. XLII, p. 600 (1923).

Table 36. Geometric Factor G for Three-conductor Shielded Cables

t = Insulation thickness;
d = conductor diameter.*

$\frac{d+2t}{d}$	Sector Conductors	Round Conductors
1.2	0.19	0.21
1.4	0.35	0.39
1.6	0.49	0.54
1.8	0.61	0.68
2.0	0.72	0.80
2.2	0.82	0.91
2.4	0.91	1.01
2.6	0.99	1.10
2.8	1.07	1.19
3.0	1.14	1.27
3.2	1.21	1.34

* With sector cables use diameter d of round conductor of same cross-sectional area.

Table 37. Surface Thermal Resistivity

Outside Diameter of Cable, Inches	Surface Thermal Resistivity, C°/Watt/cm
0.5	805
1.0	952
1.5	1117
1.75 and over	1200

Table 38. Ambient Temperatures

a. Base ambient temperature is the no-load temperature of a cable, duct, or conduit in a group without load on any cable, duct, or conduit in the group.
b. Group ambient temperature is the no-load temperature of a cable, duct, or conduit in a group with all other cables, ducts, or conduits in the group loaded.

Representative values are:	Degrees Centigrade
a. In well-ventilated buildings without other sources of heat	28–30
b. In buildings with other sources of heat, such as power stations and industrial plants	35–40
c. In poorly ventilated enclosures, such as attics and sun-exposed runways	40–45
d. Furnace and boiler rooms (higher temperatures in close proximity)	40–60
e. In the shade outdoors	35–40
f. Indirect exposures to hot sun	50
g. Earth	20

Table 39. Values of K_a for Cables in Air

Number Vertically	Number Horizontally					
	1	2	3	4	5	6
1	1.00	0.93	0.87	0.84	0.83	0.82
2	0.89	0.83	0.79	0.76	0.75	0.74
3	0.80	0.76	0.72	0.70	0.69	0.68
4	0.77	0.72	0.68	0.67	0.66	0.65
5	0.75	0.70	0.66	0.65	0.64	0.63
6	0.74	0.69	0.64	0.63	0.62	0.61

NOTE. These correction factors apply only when the spacing between cable or conduit surfaces is not greater than cable or conduit diameter or less than one-quarter of cable or conduit diameter.

DAILY LOAD FACTORS. The daily load factor represents the percentage ratio of the average 24-hr load current to the 1-hr average of the maximum peak load current occurring in a 24-hr day. The heating of cables, however, is proportional, not to the average current, but to the rms current. Representative loss factors, based on this principle and derived from an average of many daily load curves, are given below.

Load Factor, Per Cent	Loss Factor, Per Cent
30	15
40	24
50	33
60	44
70	57
75	62.5
80	70
90	85
100	100

For cables in free air the calculations are made at 100 per cent load factor but are applicable, without appreciable error, to any load factor down to 30 per cent.

For cables in conduit the calculations are made at 100 per cent load factor because of lack of information from tests at other load factors. For enclosed conduit the correction for load factor undoubtedly lies somewhere between that for cables in free air and that for cables in underground ducts.

For cables in underground ducts or pipes or for cables buried in the ground, the heat generated in the conductors during the 1-hr peak is influenced almost entirely by the peak load rms current, whereas the heat generated in the duct bank and contiguous soil is a function of the daily rms current because of the greater heat capacity involved. It is, therefore, usual to assume that all the heat generated in the peak hour passes from the conductor to the duct, but that only the daily rms current is involved in the heat that passes from the duct to the ambient earth, an approximation that errs on the conservative side. As a practical consideration it is more convenient to use only the peak load current in the calculations and to reduce the thermal resistance from the duct to ambient earth by multiplying it by the loss factor.

DUCT CORRECTION FACTOR. For convenience it is usual to base the carrying capacities of cables in ducts on those for cables in air by adding to the thermal resistance of a cable in air a factor to take care of the greater thermal resistance offered by a duct bank and its surrounding earth. Its value is a function of the number of loaded ducts in the bank and of the load factor. Typical values are given in Table 40.

Table 40. Duct Correction Factors

Number of Loaded Ducts in Duct Bank	Per Cent Daily Load Factor			
	30	50	75	100
	Per Cent Loss Factor			
	15	33	62.5	100
	Duct Correction Factor, D			
1	0.2	0.5	0.9	1.5
3	0.5	1.0	1.9	3.1
6	0.7	1.6	3.1	4.9
9	1.0	2.2	4.2	6.8
12	1.3	2.8	5.4	8.6

Table 41. Conduit Correction Factor, Q

Nominal Conduit Diameter, Inches	100% Load Factor		
	Number of Either Single- or Multiple-conductor Cables per Conduit		
	One	Two	Three
0.50	3.9	9.5	15.4
0.75	3.6	8.6	13.9
1.00	3.2	7.6	12.4
1.25	2.8	6.7	11.0
1.50	2.6	6.2	10.2
2.00	2.3	5.5	8.9
2.50	2.1	4.8	7.6
3.00	1.9	4.2	6.6
3.50	1.8	3.8	6.0
4.00	1.7	3.5	5.5
4.50	1.6	3.3	5.2

NOTES. *a.* These values of Q are for a single loaded circuit, corresponding to a single loaded duct underground. For a group of conduits in air and spaced closer than the diameter of a conduit, it is necessary to modify these values of Q, because of the effect of mutual heating, by multiplying the current value for a single loaded conduit by the correction factor K_c given in Table 42 for the particular grouping of conduits.

b. The above values refer to number of cables in a conduit, not to number of conductors in a conduit. EXAMPLE: The correction factor for one three-conductor cable in a 2-in. conduit is 2.3, whereas for three single-conductor cables in the same-sized conduit it is 8.9.

c. Tables do not extend beyond 4.5 in. nominal O.D. because of unknown losses when very large conductor sizes are used.

Table 42. Values of K_c for Cables in Conduits

Number Vertically	Number Horizontally					
	1	2	3	4	5	6
1	1.00	0.94	0.91	0.88	0.87	0.86
2	0.92	0.87	0.84	0.81	0.80	0.79
3	0.85	0.81	0.78	0.76	0.75	0.74
4	0.82	0.78	0.74	0.73	0.72	0.72
5	0.80	0.76	0.72	0.71	0.70	0.70
6	0.79	0.75	0.71	0.70	0.69	0.68

Table 43. Group Correction Factor C for Various Groupings of Conduits

Number Vertically	Number of Conduits Horizontally					
	1	2	3	4	5	6
1	1.00	1.20	1.40	1.51	1.58	1.60
2	1.33	1.54	1.75	1.90	1.99	2.03
3	1.67	1.88	2.10	2.25	2.35	2.40
4	1.85	2.11	2.40	2.52	2.59	2.65
5	1.94	2.24	2.59	2.71	2.79	2.85
6	2.00	2.30	2.70	2.83	2.92	3.00

Table 44. Thermal Resistivity of Soil

Thermal Ohms per Centimeter Cube*

g	$H_2O\%$		
	Sandy Loam	Heavy Clay	Chalky
340	0	1	2
180	5	17	10
120	10	....	16
90	15	....	20

* From tests on laboratory samples.

Table 45. Apparent Thermal Resistivity ρ_1 from Surface of One Cable to Pipe

Copper Temperature T_c, Degrees Centigrade	ρ_1
40	180
50	150
60	130
70	110
75	100
80	95

Table 46. Approximate Carrying Capacity of Cables Buried in the Earth

600-volt cables 18 in. deep; 15° to 65° = 50° C temperature rise
11,000-volt cables 36 in. deep; ρ = 180 watt-cm-C°

A.W.G. or Cir Mils	Amperes			A.W.G. or Cir Mils	Amperes		
	One-conductor	Three-conductor			One-conductor	Three-conductor	
	600 Volts	600 Volts	11,000 Volts		600 Volts	600 Volts	11,000 Volts
4	180	120	100	350,000	590	380	350
2	230	150	130	400,000	640	400	370
1	270	170	150	500,000	730	450	420
0	300	190	160	600,000	820	500	460
00	330	210	190	750,000	930	570	520
000	390	250	220	1,000,000	1100	...	...
0000	440	280	250	1,250,000	1270	...	...
250,000	490	300	280	1,500,000	1420	...	...
300,000	540	340	310				

Table 47. A-c/d-c Ratio for Cables in Steel Pipes *

60 Cycles

Because some of the loss caused by the magnetism of the pipe is in the pipe and some in the conductor, two ratios are used in eq. (18).

Conductor Size, Circular Mils or A.W.G.	Ratio for Losses in Conductor Factor for R_{c1}		Ratio for All Losses Factor for R_{c2}	
	Concentric	Segmental	Concentric	Segmental
2,000,000	1.84	1.47	2.05	1.59
1,750,000	1.70	1.41	1.89	1.51
1,500,000	1.57	1.35	1.72	1.43
1,250,000	1.46	1.27	1.57	1.34
1,000,000	1.33	1.20	1.41	1.25
750,000	1.22	....	1.27	....
600,000	1.16	....	1.20	....
500,000	1.11	....	1.14	....
400,000	1.10	....	1.11	....
350,000	1.08	....	1.10	....
300,000	1.07	....	1.09	....
250,000	1.05	....	1.06	....
4/0	1.00	....	1.05	....
3/0	1.00	....	1.04	....
2/0	1.00	....	1.03	....
1/0	1.00	....	1.03	....
1	1.00	....	1.02	....

* Based on IPCEA test data.

DIELECTRIC LOSS. The 60-cycle dielectric loss of a cable may be calculated by the following formula:

$$W_d = \frac{0.00636E^2n^2k\cos\theta}{G_2}$$

where E = voltage to ground in kilovolts.
n = number of conductors in cable.
k = S.I.C. of insulation.
G_2 = geometric factor (Table 49).
$\cos\theta$ = power factor (Table 48).

Table 48. Usual Maximum Power Factors (Per Cent) of Cable Insulation

Temperature of Cable, Degrees Centigrade	Impregnated Paper		Varnished Cambric		Rubber ozone-resistant Type
	Solid Type	Ionization-free Types	Single-conductor and Multiple-conductor, Shielded	Multiple-conductor, Belted	
60	0.9	0.60	....	....	5.0
70	1.5	0.75	8.0	12.5	5.0
75	1.8	0.83	8.7	14.0	5.0
80	2.1	0.90	9.6	16.0	5.0
85	2.5	0.97	11.0	18.0	5.0
90	3.0	1.05	....	....	....

Geometric Factor for Three-phase Dielectric Loss G_2. Table 49 gives the geometric factors for three-conductor belted cables. The geometric factor G_2 for shielded (Type H) cables may be taken as $3G$, where G is the geometric factor for a single-conductor cable of the same size and insulation thickness, as given in Table 35.

Table 49. Geometric Factors, G_2 for Three-conductor Belted Cables

T = thickness of conductor insulation; t = thickness of belt insulation; d = diameter of conductor.

$(T + t)/d$	Round Conductors		Sector Conductors	
	$t/T = 0$	$t/T = 1$	$t/T = 0$	$t/T = 1$
0.1	0.85	0.75	....	....
0.2	1.40	1.20	....	....
0.3	1.85	1.55	1.30	1.09
0.4	2.25	1.85	1.73	1.42
0.5	2.60	2.10	2.11	1.70
0.6	2.95	2.35	2.51	2.00
0.7	3.20	2.55	2.78	2.22
0.8	3.45	2.75	3.04	2.42
0.9	3.70	2.95	3.31	2.64
1.0	3.90	3.15	3.54	2.86
1.1	4.05	3.30	3.71	3.02
1.2	4.25	3.45	3.92	3.18
1.4	4.60	3.75	4.30	3.51
1.6	4.90	4.05	4.61	3.81
1.8	5.15	4.30	4.89	4.09
2.0	5.40	4.55	5.16	4.35
2.2	5.65	4.75	5.42	4.56
2.4	5.85	4.95	5.62	4.76

SHEATH LOSSES AND CARRYING CAPACITY. The effect of sheath currents on the carrying capacity may be calculated by substituting for the thermal resistance from conductor to ground, $R_1 + R_2$, the larger quantity

$$R_1 + R_2\left(1 + \frac{W_s}{W_c}\right)$$

where R_1 = thermal resistance from conductor to sheath.
R_2 = thermal resistance from sheath to ambient earth or air.
W_c = watts loss in conductor plus dielectric loss.
W_s = watts loss in sheath.

TEMPERATURE RISE OF CABLE WITH VARIABLE LOAD.

$$T = T_m(1 - \epsilon^{-t/t_1})$$

where T = temperature rise at time t hr in degrees centigrade.
T_m = ultimate temperature rise in degrees centigrade.
ϵ = base of Naperian logarithm.
t_1 = time constant, or time in hours required to reach 63 per cent of final temperature rise with a constant power loss.
= SH
where H = heating factor or thermal resistance in degrees centigrade per watt per foot.
S = thermal storage factor in watthours to raise the copper in 1 ft of cable 1 deg cent.

This may be calculated from the following factors:

Material	Watthours Stored per Pound per Degree Centigrade	Material	Watthours Stored per Pound per Degree Centigrade
Copper	0.048	Impregnated paper	0.24
Paper	0.137	Saturated asbestos	0.10
Oil	0.237	Vinylite	0.18
Lead	0.016	Steel	0.058
Water	0.51	Polyethylene	0.35

The ratio T/T_m or $(1 - \epsilon^{-t/t_1})$ for a period t = 1 hr is called the hourly attainment factor A. Its usual value is between 0.4 and 0.8.

This factor may be used to calculate the copper temperature curve for any load curve which may be divided for convenience into hourly steps. If the copper temperature rise at the beginning of any hour is represented by T_0 and the ultimate temperature rise by

T_m, corresponding to the average watts lost during that hour, the change in temperature rise during the hour will be $A(T_m - T_0)$. The copper temperature rise at the end of the hour (T_t) is then equal to $T_0 + A(T_m - T_0)$, and the copper temperature is equal to $T_t + T_d$, where T_d is the temperature of the air in an adjacent empty duct, which may be assumed constant. By evaluating $T_0 + A(T_m - T_0)$ for each hourly period, the entire temperature curve may be plotted step by step.

The heating of cables buried in the ground, either directly or in pipes, is quite slow because of the heat capacity of the soil. The same is true, although in a less degree, of cables in ducts. The above exponential equation is, therefore, liable to give unduly conservative overload heating. Tests have shown that about 75 per cent of the temperature rise in the ground occurs within five diameters of the cable or pipe and 85 per cent within ten diameters. With a knowledge of the specific heat of the soil, it is theoretically possible to set up a few equations for zones of soil around the cable, but the calculations are too complicated for ordinary treatment and must be worked out on a calculating machine on which the equations may be set up.

DIRECT-CURRENT-CARRYING CAPACITIES. The preceding methods of calculation are based on 60-cycle alternating current. Some of the formulas may be used for direct current, using the d-c resistance of the conductor and neglecting dielectric loss. If, however, the a-c current is known, the d-c-carrying capacity may be obtained by multiplying the a-c value by the factors in Table 50.

Table 50. Conversion Factors for Determining Direct-current Ratings

Conductor Size, A.W.G. or 1000 Cir Mils	Single-conductor Cable in Air or Separate Non-metallic Conduit	Multiple-conductor Cable or 2 or 3 Single-conductor Cables in Same Metallic Conduit
Up to No. 1	1.00	1.00
0	1.00	1.01
00	1.00	1.01
000	1.00	1.02
0000	1.00	1.02
250	1.00	1.03
300	1.00	1.03
350	1.00	1.04
400	1.00	1.05
500	1.01	1.06
600	1.01	1.08
700	1.02	1.09
750	1.02	1.10
800	1.02	....
1000	1.03	....
1250	1.05	....
1500	1.07	....
1750	1.09	....
2000	1.11	....

NOTE. To obtain d-c ratings multiply the alternating current by the above conversion factors.

CONTROL CABLE. The thermal resistance of control cables and other multiple-conductor cables with seven or more conductors may be estimated from the following formula:

$$R_{thi} = FR_u$$

where R_{thi} is the thermal resistance from the central conductor to the periphery of the group of insulated conductors, R_u is the thermal resistance of each insulated conductor, and F is a factor from Table 51.

Table 51. Factor *F*

Number of Conductors	F	Number of Conductors	F
7	3.7	80	29
10	5	100	35
20	8	120	42
40	15	140	49
60	22		

70. ALTERNATING-CURRENT RESISTANCE AND REACTANCE

RATIO OF A-C RESISTANCE TO D-C RESISTANCE. Table 52 gives the ratios of a-c to d-c resistance for single- and three-conductor cables up to 2,000,000 cir mils. Values for larger single conductors and special constructions are given in Article 57 and in Table 47 of Article 69.

Table 52. A-c/D-c Resistance Ratio

60 Cycles

Conductor Size, A.W.G. or 1000 Cir Mils	Single Conductors in Air or Separate Non-metallic Conduit	Multiple-conductor Cable or 2 or 3 Single-conductor Cables in Same Metallic Conduit
Up to 3	1.00	1.00
2 and 1	1.00	1.01
0	1.00	1.02
00	1.00	1.03
000	1.00	1.04
0000	1.00	1.05
250	1.005	1.06
300	1.006	1.07
350	1.009	1.08
400	1.011	1.10
500	1.018	1.13
600	1.025	1.16
700	1.034	1.19
750	1.039	1.21
800	1.044	1.23
1000	1.067	....
1250	1.102	....
1500	1.142	....
1750	1.185	....
2000	1.233	....

REACTANCE OF TRIPLEX CABLES. The reactance of three-conductor cables without magnetic binder tapes or armor may be calculated by the ordinary formulas for reactance of parallel conductors. See Article 21. This reactance is of the order of 0.027 to 0.035 ohm per 1000 ft at 60 cycles. The effect of steel binder tape is to increase the reactance about 25 per cent. If the conductors are large in relation to their distance apart, the reactance will be reduced by proximity effect. See Article 21.

Table 53 gives approximate values of the reactance of three single-conductor cables, as installed for compression or oilostatic systems, in terms of the reactance of similar cables in the open air.

Data on zero-sequence impedance of cables are given in Article 22.

Table 53. Approximate Reactance of Three Single-conductor Cables in a Steel Pipe

Area of Each of Three Conductors, Circular Mils	Ratio of Reactance in Steel Pipe to That in Air
250,000	1.50
500,000	1.40
750,000	1.30
1,000,000	1.25
1,250,000	1.20
1,500,000	1.15
2,000,000	1.14
Larger	1.13

This table is based on the usual diameter ratio of cable to pipe.

71. SINGLE-CONDUCTOR CABLES FOR THREE-PHASE SYSTEMS

It is common practice, for transmission lines operating at 40,000 volts or over, to use three single-conductor, paper-insulated, lead-sheathed cables for three-phase systems. Voltages are induced along the sheaths, and therefore considerations of safety make it necessary to ground them. If any sheath is grounded or connected to another sheath at more than one point, current will flow, which will lower the carrying capacity of the cable and cause expensive energy loss. Means for keeping down the sheath voltages and currents have been devised, such as the transposition of sheaths. [See Halperin and Miller, *Trans. AIEE*, Vol. 48, p. 399 (1929).]

The following formulas are approximate, the exact ones being unwieldy for practical purposes:

Let b = radius of conductors in inches, carrying I_c amperes each.
e = inner radius of lead sheath, in inches.
f = outer radius of lead sheath, in inches.
D = distance between centers of conductors, in inches, assuming them to be set in an equilateral triangle.
M = mutual inductance of conductor and sheath, in henrys per 1000 ft.
W_s = power loss in sheath, in watts per 1000 ft of each cable.
E_s = emf, in volts, induced per 1000 ft of sheath. The potential difference between sheaths at 1000 ft from any bond will be $1.73E_s$.
$\omega = 2\pi f$, where f = frequency.
L_0 = inductance of each cable, with sheath open-circuited, in henrys per 1000 ft.
L_c = inductance of each cable with sheath short-circuited, in henrys per 1000 ft.
L_s = inductance of the sheath of each cable, in henrys per 1000 ft.
I_s = sheath current, in amperes when bonds of negligible resistance are used.
R_s = resistance of sheath, in ohms per 1000 ft.

$$M = 1.405 \times 10^{-4} \log_{10} \frac{2D}{e + f}$$

$$E_s = \omega M I_c$$

$$L_0 = \left(0.152 + 1.4 \log_{10} \frac{D}{b}\right) \times 10^{-4}$$

$$L_c = L_0 - M$$

$$L_s = \frac{1.4}{10^4} \log_{10} \frac{D}{f}$$

$$R_s = \frac{0.036}{f^2 - e^2} \text{ at 40 deg cent}$$

$$I_s = \frac{E_s}{\sqrt{R_s^2 + (\omega L_s)^2}}$$

$$W_s = I_s^2 R_s$$

An important group of papers on this subject will be found in *J.I.E.E.* (London), 1928 and 1929. Halperin and Miller, *Trans. AIEE*, p. 399 (April, 1929) give formulas for groupings other than equilateral. Figure 5 enables approximate values of sheath voltages to be obtained for these various groupings, if the ratio S/R of the axial spacing to the mean radius of the sheath is known.

EFFECT OF MAGNETIC ARMOR.

Let μ = permeability of armor.

$$a = \frac{1 - \mu}{\mu}.$$

k = ratio of total flux in magnetic armor to flux that would exist if armor were replaced by air, usually from 20 to 40.

Then

$$k = \frac{1}{a}\left(\frac{\pi}{2} - \frac{2}{\sqrt{1 - a^2}} \tan^{-1} \frac{1 - a}{1 + a}\right)$$

Let L = inductance in millihenrys per mile.
d_1 = diameter over armor.
d_2 = diameter under armor.
D = distance between conductors.
d = radius of conductor.

$$L = 0.0805 + 0.741 \left(\log_{10} \frac{D}{d} + k \log_{10} \frac{d_1}{d_2}\right)$$

This assumes perfect magnetic contact between armor wires. In practice, because of imperfect contact, the inductance will be about 0.7 of the above value. This formula requires some assumption to be made regarding the permeability of the wire, a factor which depends both on the material of the armor and the current in the cable. Numerical calculations require the permeability to be deduced from the magnetic flux density in the armor, a quantity which may be estimated from k and the conductor current.

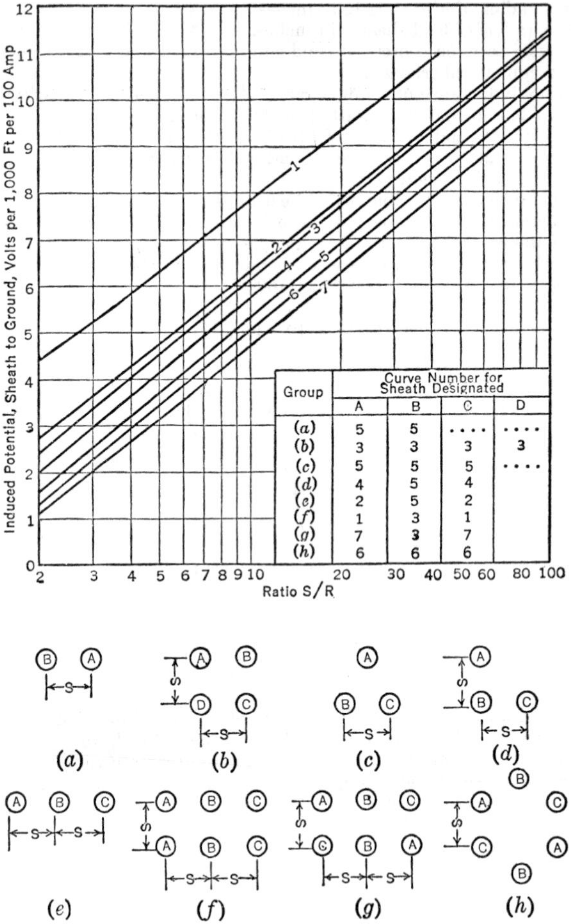

Group	Curve Number for Sheath Designated			
	A	B	C	D
(a)	5	5	....	....
(b)	3	3	3	3
(c)	5	5	5	....
(d)	4	5	4	
(e)	2	5	2	
(f)	1	3	1	
(g)	7	3	7	
(h)	6	6	6	

FIG. 5. Diagram for Estimation of Voltages Induced in Sheaths of Single-conductor Cables

72. IONIZATION-FREE CABLES FOR HIGH VOLTAGES

EXPANSION AND CONTRACTION IN IMPREGNATED-PAPER CABLES. The ordinary (solid-type) impregnated-paper cable is characterized by high resistance to fluid flow both radially and longitudinally, because of the density of the paper, the compactness of construction, and the viscosity of the oil. As the oil has a large coefficient of thermal expansion, compared to the other materials of the cable, the effect of thermal cycles is to create, alternately, local high pressures and local vacua. The high pressures may cause stretching of the sheath and thus accentuate the vacua in subsequent cycles. The local vacua are weak spots because they ionize at low voltage, causing hot spots. Therefore, either the insulation thickness must be great enough to overcome this weakness, or means must be adopted to prevent the formation of voids.

In ordinary cables the insulation is made thick enough to reduce the stresses to a point at which destructive ionization, in minute voids, does not occur. The elimination of voids may be accomplished in either of two ways, to be described.

OIL-FILLED CABLE. The formation of voids may be prevented by the use of oil of low viscosity, not over 20 minutes Saybolt, and preferably much less. Such oil flows radially through the insulation and longitudinally through oil ducts furnished to carry the oil along the cable from reservoir-equipped joints. The insulation thickness is about one-half the ordinary value, permitting the use of cables of practicable diameter for voltages as high as 250,000, three-phase. Single-conductor cables have their oil duct at the center of the conductor, whereas three-conductor cables have theirs in the grooves be-between insulated conductors. The ducts in either case are made by tubes of helically wound metal strip. The diameters of hollow core conductors are given in Table 54.

Approximate Formulas for Oil-filled Cable.

Let Z = viscosity of oil, in centipoises.
d = effective channel diameter, in inches.

$$b = \frac{0.0000283Z}{d^4}.$$

If the resistance is to be measured by forcing external oil through the hollow core, let

p = pressure, in pounds per square inch.
q = flow of oil, in cubic inches per second.
l = length of cable, in feet.

$$b = 0.4\frac{p}{ql}.$$

The usual order of value of b is about 0.05.

The maximum rate of oil demand of a cable may be approximated on the assumption that, at the beginning of heating or cooling, the entire heating or cooling depends on the thermal capacity of the copper and oil contained therein.

Let a = oil demand, in cubic inches per foot per second.
α = coefficient of expansion of oil per degree centigrade, usually about 0.00074.
v = volume of oil in conductor (including hollow core), in cubic inches per foot of cable.
W = watts per foot lost in cable.
$h = 1899(0.511W_0 + 0.101W_c)$
= watt-seconds per foot per degree Centigrade.
W_0 = weight of oil in conductor, in pounds per foot.
W_c = weight of copper, in pounds per foot.

$$a = \frac{\alpha v W}{h}$$

= the rate of volume expansion in cubic inches per foot per second.

The usual order of value of a is about 50×10^{-6}.

The pressure drop in the hollow core, due to oil demand, may be calculated as follows:

Let p = pressure drop, pounds per square inch to distance x from oil-feeding point, at moment of application of load.
a = oil demand, defined above.
b = resistance factor, defined above.
l = total length of section, in feet.

$$p = 2ab\left(lx - \frac{x^2}{2}\right).$$

At the end of the line, where $x = {}^1/_2\, l$,

$$p = {}^1/_2\, abl^2$$

It is usual to design the line so that the pressure will not exceed 15 lb per sq in. at any point or fall below atmospheric pressure. The maximum pressure, however, does not occur at the moment of application of load, but some 5 or 10 minutes later, when some of the oil in the insulation has expanded. It is usually 10 to 20 percent greater than the initial pressure, p. For a complete treatment see *Trans. AIEE*, Vol. 52, p. 98 (1933).

Reservoirs. Reservoirs for oil-filled cable consist of cylindrical tanks of either stainless steel or zinc-flashed steel containing a stack of cylindrical cells, capable of expanding or contracting in volume. The Type AC reservoir contains oil in the tank and gas at a fixed pressure in the cells, the tank being connected with the cable joint. Types CC and DC have the oil, which connects with the cable oil, in their cells. Type DC has gas (CO_2) under pressure in the tank, whereas Type CC contains oil which is open to the atmosphere.

Type CC is used where there is suitable elevation to give the requisite pressure; otherwise the other types are used.

OILOSTATIC SYSTEM. Single-conductor paper-insulated cables without impervious sheath are operated in oil of medium viscosity, maintained under a pressure of 12 to 15 atm in a steel pipe by means of a pumping system. Expansion and contraction of the oil under load cycles are taken up in terminal reservoirs. The cable is protected during shipment by a temporary sheath, which is stripped off as the cable is drawn into the pipe.

Table 54. Diameters of Hollow Core Conductors for Oil-filled Cables

Size, A.W.G. or Circular Mils	Overall Diameter of Conductor, Inches	
	Inside Diameter of Spring Core = 0.500 in.	Inside Diameter of Spring Core = 0.690 in.
2/0	0.736	....
3/0	0.768	0.924
4/0	0.807	0.956
250,000	0.837	0.983
300,000	0.880	1.017
350,000	0.917	1.049
400,000	0.953	1.082
450,000	0.989	1.112
500,000	1.028	1.145
600,000	1.084	1.201
650,000	1.121	1.228
700,000	1.151	1.256
750,000	1.180	1.286
800,000	1.212	1.309
850,000	1.242	1.341
900,000	1.261	1.365
1,000,000	1.310	1.416
1,250,000	1.434	1.524
1,500,000	1.547	1.635
1,750,000	1.650	1.730
2,000,000	1.760	1.833

COMPRESSION SYSTEM. Single-conductor paper-insulated cables with impervious sheath, either lead or polyethylene, are operated in either nitrogen gas or oil under a pressure of 12 to 15 atm in a steel pipe. In a gas-pressure medium, the pressure is maintained by gas cylinders at the terminals, whereas with an oil-pressure medium pressure is maintained and expansion allowed for, as in the oilostatic system.

LOW-GAS-PRESSURE CABLE SYSTEM. Lead sheaths will withstand normal internal pressures of 15 to 20 psi. Advantage is taken of this fact to make ionization-free cables for voltages up to about 40,000 volts by equipping the cables with longitudinal ducts, draining the ducts of oil, and replacing the oil by nitrogen gas maintained at a pressure of 15 to 20 psi by means of terminal gas tanks. Three-conductor cables have two ducts formed by helices of bronze tape and one duct consisting of a solid copper tube, open only at the joints, which serves as a through passage to carry the gas pressure past any obstruction, such as slugs of oil, which may occur in the open-sided ducts. Insulation thickness is about midway between that for solid type and that for oil-filled cable.

MEDIUM-GAS-PRESSURE CABLE SYSTEM. This is similar to the low-pressure system except that the gas pressure is about 40 psi, and the sheath is either thickened or reinforced to withstand this pressure. By this means the three-phase voltage limit is raised to about 69,000 volts.

73. INSTALLATION

For the installation of wires and cables in buildings see Wiring of Buildings, Articles 78 to 89. For the construction of the pole lines carrying the messenger wire see Articles 28 and 33. For the construction of conduit systems see chapter on Underground Conduits, Articles 37 to 44. Below is given a brief description of modern practice in installing insulated wires and cables on messenger wires and in underground conduit systems. See also Distribution Lines, Articles 45 to 53, and Transmission Lines, Articles 24 to 36.

MESSENGER CONSTRUCTION. The messenger wire is erected in the same manner as an ordinary line wire; see Article 34. A "leading-up" wire is stretched from the bottom of one pole to the messenger wire on the starting pole, forming an incline on which to pull up the cable. A pulling rope is then fastened to the end of the cable by means of a cable grip and carried alongside the messenger to the point where the cable is to reach, thence through a snatch-block down to the terminal pole to a second block at the bottom and thence to a capstan, winch, locomotive, or whatever is to be used for pulling. Either temporary rollers should be provided on the poles over which the rope runs, or the rope should be suspended from the messenger by wire hooks. The cable is then slowly drawn up the inclined wire and along the messenger, attaching temporary carriers to the cable as it is payed out and hooking them over the messenger to carry the weight of the cable. Linemen must be stationed on each pole to pass these carriers around the messenger clamp or insulator. The final suspension of the cable may be accomplished in either of the following ways: (1) When the end of the cable arrives at the beginning of the last span, the lineman on each pole replaces the temporary carriers by permanent hangers, spacing them regularly along the cable, so that when the last span is pulled, all the hangers will be in place. (2) When the cable has been pulled all the way, a lineman rides along the messenger wire in a carriage, replacing each temporary carrier by a hanger. This plan is preferable, as the hangers may be attached more tightly to the messenger wire and are less likely to slip on the cable.

INSTALLATION OF CABLES IN DUCTS. The conduit system having been constructed, it must be prepared for the reception of the cable by being cleaned out or rodded as described in Article 42 on Underground Conduits.

If several ducts are available, the choice of the particular duct to be used should be governed by the following considerations: (1) avoiding unnecessary crossings of cable in the splicing chambers, substations, etc.; (2) avoiding the obstructing of empty ducts; (3) keeping the cables cool; and (4) keeping d-c cables away from others. The coolest and best heat-radiating ducts are those located at the lower corners of the system; next are those nearest the outside of the system; and lastly come the middle and top ducts, which not only take up heat from the lower cables, but must also dissipate heat through adjoining ducts.

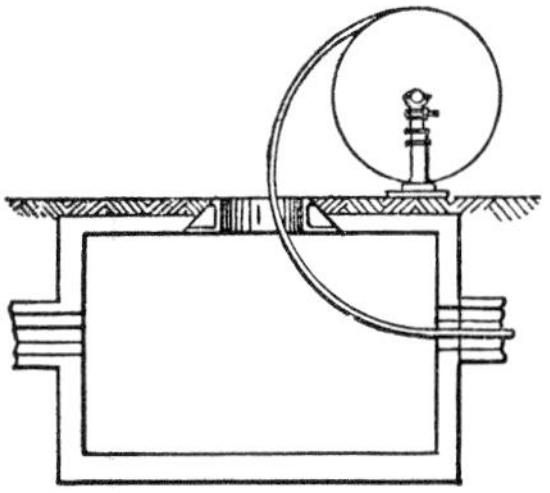

FIG. 6. Feeding Cable into Duct

The next step is to set the cable reel on the shaft of a pair of wheels of slightly greater diameter than the reel itself or on jacks and raise it slightly above the ground, taking care to locate it as shown in Fig. 6, so that the cable will unreel into the manhole without making a reverse bend.

The pulling rope having been left in the duct after rodding, the cable is unreeled sufficiently to bring its end close to the mouth of the duct, and either a wire pulling grip (Fig. 7) is drawn over its end or attachment is made to a pulling eye soldered to the conductors and sheath of the cable. The end of the grip or pulling eye is hooked to the rope, and the rope pulled from the other end. The pulling may be done by capstan, winch, motor truck, horse, or hand, depending upon the size and amount of cable to be pulled and upon local conditions. The cable should be carefully guided into the duct so as to avoid sharp bends and abrasions, and a small quantity of grease, about 100 lb per mile, may with advantage be spread over the cable as it enters the duct. It is also necessary to cover the edges of the duct with pieces of lead or other suitable material to prevent abrasion of the cable.

FIG. 7. Kellem Grip

Pulling Forces. Where a pulling eye is used, it is customary to allow a maximum (kinetic) pulling stress of 0.008 lb per total circular mil area of copper, with a maximum of 6000 lb.

The maximum permissible pull in pounds is therefore $S_m = 0.008 \times$ Total circular mil area of copper.

A maximum stress of 1500 lb per sq in. in the lead sheath is recommended for pulling by woven grip, except that a somewhat greater stress is permissible to overcome static friction.

On this basis the maximum permissible (kinetic) pull in pounds is

$$S_m = 4712t(D - t)$$

where t = thickness of sheath, in inches.
D = outside diameter of sheath, in inches.

The maximum permissible pulling length L will be

$$L = \frac{S_m}{Wf}$$

where W = weight of cable, in pounds per foot.
S_m = maximum permissible pull, in pounds.
f = coefficient of kinetic friction.

The coefficient of friction recommended by *Underground Systems Handbook* is 0.4 for paper-lead cables. For cables with skid wires, 0.3 is a more usual value.

Where the conduit is continuously curved, the following factors, given in *Underground Systems Handbook*, may be used as multipliers for the pull calculated for straight runs:

Radius of Curvature, Feet	Multiplier
75	2.0
100	1.65
150	1.34
200	1.19
250	1.10
300	1.07

Bends in Installation. In order to avoid injury to the insulation, cables should not be installed with bends of less radius than those indicated in Tables 55, 56, and 57.

Table 55. Minimum Bending Radii for Rubber-insulated Cables

Thickness of Conductor Insulation	Minimum Bending Radius as a Multiple of Overall Diameter	
	Up to 500,000 Cir Mils	500,000 Cir Mils and over
Up to 8/64 in.	3	4
9/64 to 12/64 in.	4	5
13/64 to 20/64 in.	5	6
21/64 in. and over	6	6

Table 56. Minimum Bending Radii for Paper-insulated Cables

D is the overall diameter of the cable.

Cable Diameter, Inches	Total Area of All Conductors, Circular Mils	Minimum Bending Radii, Inches	
		Up to 20 kv	Over 20 kv
Up to 1	....	8	....
Over 1	Up to 1,500,000	$8D$	$10D$
	Over 1,500,000	$10D$	$12D$

Table 57. Minimum Bending Radii for Varnished-cambric-insulated Cables

Single-conductor and Shielded or Non-shielded Non-belted Cables *

(IPCEA)

Conductor Insulation Thickness, 64ths Inch	Outside Diameter of Cable, Inches		
	0–0.750	0.751–1.500	Over 1.500
	Minimum Bending Radii As Multiple of Cable Diameter		
10 and less	5	6	7
11 to 20	6	6 1/2	7
Over 20	6 1/2	7	7

Multiple-conductor Belted Cables

Outside Diameter, Inches	Minimum Bending Radii As Multiple of Cable Diameter
1.000 and less	5
1.001 to 1.500	6
Over 1.500	7

* Flat twin cables shall be bent on the minor axis only, and the minor diameter shall be used in computing the bending radius.

PROTECTION OF CABLES IN SPLICING CHAMBERS. Wherever there are several large cables in a splicing chamber, there is always danger that a burn-out of one cable will involve some or all of the remainder. Hence it is usual to protect such cables by means of one or more of the following methods:

1. Cement coating with 1/4-in. rope bond.
2. Asbestos tape saturated with silicate of soda.
3. Asbestos tape covered with soft steel-tape armor.
4. Asbestos rope.
5. Special trade-marked products designed for this purpose.

74. CABLE JOINTS

The problem of cable splicing may be resolved into the following elements:

a. Joining the conductors to retain their full carrying capacity and requisite tensile strength.

b. Insulating the joint to resist both the normal radial stresses and the tangential stresses caused by the termination of the sheaths.

c. Reducing tangential stresses as far as practicable in high-tension joints.

d. Keeping out air and moisture.

e. In oil-filled cables, providing (1) an oil barrier, (2) a continuous channel for the oil, (3) suitable oil passages for connecting the cable oil channels to an external reservoir.

CONDUCTORS. The **joining of conductors** is a standard procedure, involving the cleaning and "tinning" of the conductors with stearine flux and 50/50 solder. The conductor ends thus prepared are inserted into standard NELA connectors of the dimensions shown in Table 58. Solder, 50/50, is then ladled over the connector until all interstices are filled, and then all surplus solder is wiped or filed off, so as to leave the connector and abutting conductors free of roughness or projections. With sector conductors, it is usual

Table 58. Nominal Dimensions Straight Tinned Copper Connectors

Conductor Size	Inside Diameter, mils	Outside Diameter, mils	Wall Thickness, mils	Slot Width, mils	Length, inches
8	151	201	25	30	1 1/2
7	169	225	28	30	1 1/2
6	189	251	31	30	1 1/2
5	211	281	35	30	1 1/2
4	237	315	39	30	2
3	265	353	44	30	2
2	297	395	49	30	2
1	337	449	56	70	2
0	378	504	63	70	2
00	423	565	71	70	2
000	475	635	80	70	2
0000	533	713	90	70	2 1/2
250 M	581	777	98	120	2 1/2
300 M	635	849	107	120	2 1/2
350 M	690	920	115	120	2 1/2
400 M	740	986	123	120	3
450 M	784	1046	131	120	3
500 M	826	1102	138	120	3
550 M	868	1154	143	175	3
600 M	906	1206	150	175	3 1/2
650 M	948	1260	156	175	3 1/2
700 M	983	1307	162	175	3 1/2
750 M	1018	1356	169	175	3 1/2
800 M	1052	1400	174	175	4
850 M	1083	1441	179	220	4
900 M	1115	1483	184	220	4
950 M	1145	1525	190	220	4
1000 M	1175	1565	195	220	4 1/2
1250 M	1320	1754	217	220	4 1/2
1500 M	1440	1912	236	280	5
1750 M	1560	2074	257	280	5 1/2
2000 M	1664	2214	275	280	6
2500 M	1855	2455	300	280	6 1/2

to hammer the ends into cylindrical form and use standard connectors of the next larger size in the table. Soldered connectors are being largely superseded by mechanical pressure connectors.

INSULATION. The next step is to prepare the cable insulation surface to receive the joint insulation. Except in certain special designs the insulation is penciled down from its full diameter to approximately the diameter of the connector, and the short spaces (1/4 to 1/2 in.) between the penciled ends and the connector are filled with insulating tape of the same material, but much narrower, as is used for the joint insulation. The length of the bared insulation commonly used is shown in Fig. 8. The penciling is usually about

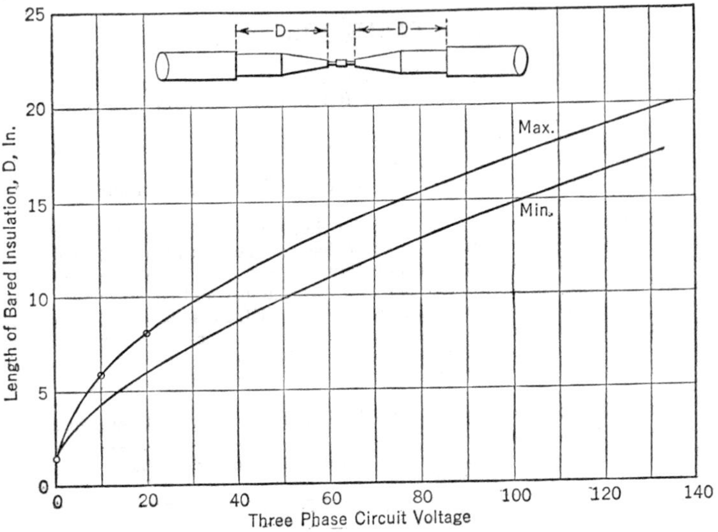

FIG. 8. Bared Insulation Required for Cable Joints

0.2 in. in length per 64ths inch of thickness. Some engineers prefer to have the insulation stepped, rather than penciled gradually, in which case the slope may be somewhat steeper, as the unequal stretching of the tape does not have to be considered.

The main insulation of a joint consists of tape applied lapped, in cigar form, from beyond the ends of the penciling on one side, to a corresponding point on the other, with a maximum thickness over the connector. This maximum is usually about 40 per cent more than the thickness of cable insulation.

For rubber-insulated cables, rubber tape is used. For both varnished-cambric and impregnated-paper insulation it is common to use bias-cut varnished-cambric tape. For high-voltage cables, say 11,000 volts and over, the varnished-cambric tape is generally of the low dielectric-loss type. The usual thicknesses of varnished-cambric tape are 7, 10, or 12 mils. Rubber tape up to 25 mils thick is used. Care must be taken not to stretch the tape unduly, lest its dielectric strength be impaired.

Some power companies use impregnated-paper tape, usually 5 or 6 mils thick, and a few use wide rolls of paper with converging sides, the purpose of the convergence being to make the paper roll up with conical ends.

The higher-voltage cables generally begin the slope of the tape at the end of the lead; the slope is covered with a flared metallic shield, usually of copper gauze tape.

Where the cable insulation is shielded, it is usual to carry the shield over the entire joint insulation. The proper slope of shield has much to do with the reduction of tangential stresses.

The design of joints from the standpoint of stress is based on the formula for the capacity between concentric cylinders and on the fact that the total voltage divides inversely as the capacities. By means of these relations the voltages at the various layers may be calculated, and the axial potential drops calculated therefrom.

It is necessary to keep air and moisture out of a joint, and to this end, it is customary, with either varnished cambric or paper tape, to apply a liberal amount of oil or compound as each layer of tape is being added. The insulation is finally "boiled out" by pouring hot compound over it.

COMPOUND. When the joint is equipped with its sleeve, it is filled with compound of one or other of the following types, usually in accordance with Table 59:

A, insulating oils, that remain liquid over the operating temperature range encountered and which migrate into cable.

B, insulating greases and viscous compounds (not an asphaltic compound) which have melting points within the operating temperature range and which migrate (more or less) into cable.

C, plastic and hard compounds which have melting points above the maximum operating temperature encountered and which do not migrate into cable.

Table 59. Usual Types of Joint-filling Compounds

Standard Walls of Insulation and Lead

	0–18 kv	19–23 kv	24–34.5 kv (Note 3)	Above 34.5 kv (Note 3)
Belted 3-conductor cable....	*C*	......	....	....
Shielded 3-conductor cable..	*C*	*C*	*A*, for underground *B*, for underground (Note 1) *C*, for submarine or aerial	*A*
Single conductor cable......	*C*	*C*	*A*, for underground *B*, for underground (Note 1) *C*, for submarine or aerial	*A*, for underground *C*, for submarine (Note 2)

NOTES: For heavier walls of insulation than standard the limits of Class *C* compound may be raised 10 per cent.

(1) With proper servicing.

(2) For submarine use, a full length of cable without joints is obviously first choice. Where piers for bringing ends out of water are feasible, oil-filled joints are recommended. Where this is not possible, solid filled joints (fully shielded) can be used.

(3) There has been successful experience with joints filled with Class *C* compounds as follows:
Shielded, 3-conductor cable up to 34.5 kv.
Single-conductor cable, up to 46 kv.

Oil-filled Cables present special problems of their own, which are treated in *Trans. AIEE*, p. 200 (1928) and *Elec. World*, p. 97 (1929).

The success of joint making for high voltages depends on careful design with respect to potential distribution, mechanical features, and care in workmanship to avoid leaving dirt, air pockets, sheath leaks, and rough edges on conducting parts.

COVERINGS. Lead-sheathed cable joints are usually enclosed in lead sheaths, but very long joints sometimes have brass or copper sleeves. The sleeve diameter is generally based on a clearance of about 1/4 to 1/2 in. over the joint insulation.

The lead sleeve is slipped over the cable before the conductors are joined and is brought into position when the structure of insulation and shielding has been completed. The ends are beaten down to meet the cable sheath and joined to it by wiping with 45 tin/55 lead solder.

Rubber cables with braid covering do not have lead sleeves on their splices but are finished either with "anhydrous tape" or friction tape, over which asphaltic compound is poured. Low-voltage rubber joints are often made without other compound, but high-voltage rubber-lead cables always have their sleeves filled with plastic compound.

75. FAULT LOCATION

Faults may be classified as follows:

a. Conductors not burned apart.

1. Solid ground on one or more conductors with one clear conductor.
2. Solid ground on all conductors.
3. Short or cross between two or more conductors with one clear conductor.
4. Short or cross between two conductors with one conductor grounded.
5. Short or cross between all conductors and ungrounded.

b. One or more conductors burned apart.

1–5. Conditions 1–5 inclusive (see above) from either end of cable.

6. All conductors burned off and clear of grounds.

The general procedure followed in locating a fault varies somewhat with local conditions but is typified by the following:

Failure
|
Sectionalization as Far as Possible
|
Test Analysis by Above Classification
|
Fault Reduction, as Required
|

1	2	3	4	5
Inspection	Cut and Test Until Localized	Tone Test	Loop Test	Capacity Test
		Final Cut and Test	Tone Test	Cut and Test for Approximate Location
			Final Cut and Test	Final Cut and Test

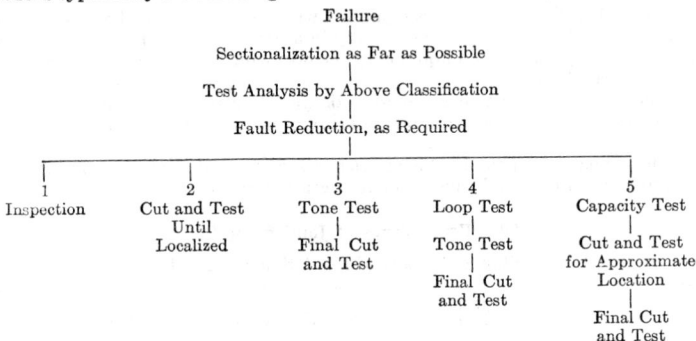

In short cables, the fault is usually located by inspection, i.e., looking for smoking manholes or listening for crackling sounds when the kenotron is applied to the faulty cable.

The final step in every fault hunt is a cut and try, although ordinarily the fault is found at the first cut.

The two major methods of locating circuit faults are the **tracing current** or **tone-test method** and the **bridge method.** The tone-test method is more generally useful and certainly more accurate, although the bridge method is quite indispensable for long transmission lines.

TONE-TEST METHOD. The tracing current method depends on the Lundin or some similar fault locator, which includes three main elements: the interrupter, the analyzer, and the pick-up or listening equipment.

The interrupter uses a 110-volt a-c source of 50-amp capacity and applies current of suitable voltage to the cable and analyzer in characteristic periodic impulses, which may be picked up along the cable by the listening equipment.

The analyzer, with its reactive loading coils, indicating lamp, and step-up transformers, enables an experienced operator to obtain a description of the fault.

The procedure of analysis is as follows:

The inductance coils of the analyzer are connected in series with the conductor to be studied. If this conductor is open and clear of ground, the current, in passing through the inductive loading coils, will produce across the capacitative conductor a voltage drop which is greater than the applied voltage. If, however, there is a high-resistance fault, the voltage may be equal to or less than the applied voltage. Where a solid ground exists, the circuit voltage is reduced nearly to zero. These voltage conditions are tested by means of a lamp which is bright, dim, or extinguished, according to which of the conditions exists.

A partial ground is usually reduced by the application of sufficient voltage, using either a kenotron or a Thyratron tube, in order to facilitate the locating process.

The locating depends upon an abrupt decrease or change in the signal at a certain point along the cable, which indicates that the fault has been passed. A pick-up coil with telephone receiver is used for detecting "grounds."

In triplex cables, if the detector is moved circumferentially around the cable, it will sometimes be near and sometimes away from the conductor carrying the tracing current. The resultant variation in loudness, called the "hump effect," always indicates that the fault is farther on. Sheath current produces no hump effect. The change from humped to uniform sound around the cable gives the precise location of the trouble. If the trouble is in a duct, it is usual to check the location by "ringing" the lead at the near joint and checking again with the earphone and by trying for a spark across the cut.

Another device is a bar listening coil, which is used on the surface of the ground, and which, up to the fault, gives a louder sound with the bar vertical than horizontal, and beyond the fault gives a louder sound with the bar horizontal than vertical [J. A. Vahey, E.E.I. (Feb. 1939)].

In some cases, a large listening coil is used on an automobile which is driven along the route of the cable until the sound changes.

In single-conductor cables, where there is no "hump effect," the stray sheath currents beyond the fault will make locations uncertain. In such cases a heavy copper strap is used to divert current from the sheath locally; the exploring coil being applied between the sheath and strap will, in effect, be between the sheath and conductor currents, so that, when the fault is beyond, the signal will be strengthened, and when it is behind, it will be weakened, as the shunt is applied.

The usual cable fault is at its lowest resistance when a current of 1 to 5 amp is flowing through it. Less current will allow the fault to cool, and more, to burn up, either tending to increase the resistance. If the pilot lamp flickers when starting to reduce the fault at 1100 or 2200 volts, it is probable that there is moisture at the break. In such cases it is usual to resort to the bridge method.

BRIDGE METHOD. The bridge method, moreover, is much used for transmission lines because the tracing current method is necessarily slow on long lines and because high-voltage cable faults are often hard to reduce to low enough resistance to be located by a tracing current.

The bridge uses the Murray loop as shown in Fig. 9. It consists essentially of a resistance wire graduated into two equal scales, each reading from 0 to 100, the zero points being at the ends of the wire. A galvanometer is connected across the zero points, and a flexible lead connected to the positive terminal of a source of direct current is put in sliding contact with the wire. The negative terminal of the d-c source is grounded. A protective gap is often placed across the galvanometer terminals to protect against surges which occur when a sudden change in fault resistance takes place. The usual source of d-c supply for a bridge test is a motor-generator set, usually giving about 1500 volts. However, a 40-kv slide-wire bridge, which both breaks down the fault and then locates it, is now in successful use.

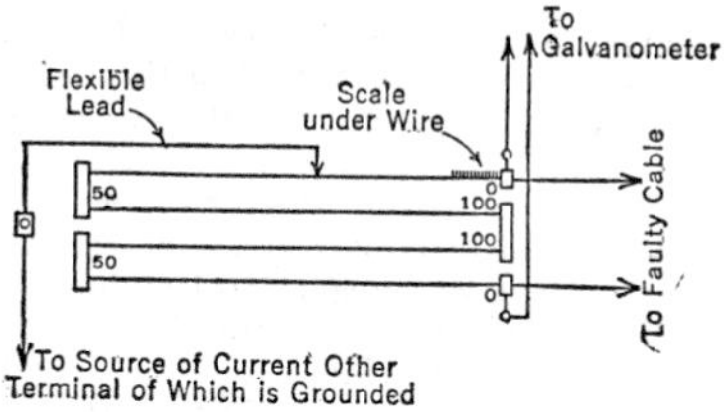

FIG. 9. Murray Loop

Capacity tests for open-circuited conductors without ground are usually made by measuring the charging current with an a-c milliameter.

Trouble in buried series lighting cable is usually located by tests at the lamps. The section having been found in this way, the location in the section is determined by a bridge test, using a return conductor above ground.

The usual time required to locate a service fault in a cable in a duct line is about 2 to 3 hours. Test failures may take about an hour longer.

CONDENSER DISCHARGE METHOD. A steep front wave is impressed on the cable from the discharge of a capacitor through a sphere gap. This wave jumps the fault and discharges the condenser with a loud snap, which can be heard, above ground, in most cases, without the aid of an amplifier.

LOCATION OF FAULTS IN NON-METALLIC SHEATH-TYPE CABLES FOR SERIES LIGHTING. Such cables are usually buried back of the curb or under pavements, making inspection impossible.

Grounds can be located in a section between adjacent lights by noticing which lamps do not burn properly. After the ends of the faulty section of cable are exposed at the lamp posts, a loop test with portable battery energy can be used for approximate localization. It is necessary to lay a temporary return wire along the ground between the exposed cable ends. Then the cut-and-test method must be used for final localization. Faults can generally be found within 10 ft of loop test locations.

In an open circuit, the cables being shielded by the ground, the electrostatic field will not extend outside the sheath. The lamp filaments, however, being unshielded, may still be used as sources of sound for a Lundin detector. By this means opens may be located between lamps. The exploring coil may occasionally be used to trace opens, but here, the entire current from the station being charging current, the signal dies out uniformly to the point where the open exists. Usually, the trouble hunter can get to within 200 ft of the open before the signal becomes inaudible.

In some recent types of series circuits with buried cables and buried insulating transformers at the base of the lighting standards, the electrostatic field does not extend to the secondaries and lamp filaments. In this case, however, the charging current of the cable up to the fault is transformed and circulated through the lamp filaments. For this condition a special coil, vertically mounted, is attached to the end of a fish pole and connected to the earphones or amplifier, which picks up the sound at the lamp. This satisfactorily locates the fault between two lamps. More precise location may sometimes be made with triangle and amplifier from the surface of the ground if the fault is solid; otherwise digging is necessary.

If the circuits do not enter the stations, it is necessary to move a portable set to the location to analyze and apply tracing current to the faulty circuit.

76. WEIGHTS OF INSULATED WIRES AND CABLES

So many variables enter into the weight of an insulated wire or cable, and so many forms are in use, that it is impossible to give comprehensive tables here. The weight of a cable may be calculated from its dimensions by finding the sum of the weights of the conductors and the weights of the insulation, braid, tape, sheath, etc. The weight of the conductor may be found in the wire tables under Bare Wires and Cables, Articles 54 and 55. The weight of the insulation, braid, tape, or sheath may be found by calculating the cross-section of each of these materials and multiplying by the length of the cable and a factor proportional to the density of the material; see Table 60.

The cross-section of a tube having an internal diameter d and thickness t is $\pi t(d + t)$. When the diameter and thickness are in inches and the specific gravity of the material forming the tube is δ, then the weight in pounds per 1000 ft of tube is

$$W = 1362\ \delta t(d + t)$$

The values of the specific gravity δ, the product 1362 δ, and the weight per cubic inch for the common materials used in the construction of cables are given in Table 60. From this relation the weight per 1000 ft of any tubular (circular cross-section) layer of insulation, braid, tape, or sheath may be readily calculated. The formula is, therefore, directly applicable to single-conductor tables.

The weight of duplex or triplex cables is calculated as for a group of single-conductor cables, except with regard to the fillers. The cross-section of the filling material is most readily calculated by subtracting from the cross-section of the entire cable the cross-sections of the individual conductors and the tubular insulation, sheath, etc. The diameter of the circle circumscribing three round insulated conductors of diameter d is $2.15d$.

The weight of separators in pounds per 1000 ft is usually between 5 and 10 times the mean diameter, expressed in inches, of the conductor and separator.

The weight of saturated braids per 1000 ft is from 12 to 30 times the diameter over the insulation, expressed in inches. The corresponding multiplier for 12-mil rubber-filled cotton tape with 1/4 lap is 24.

Table 60. Weights of Cable Materials

Material	Specific Gravity, δ	Pounds per Cubic Inch, 0.03613 δ	1362 δ
Rubber compound, 30 per cent rubber:			
Organic base	1.2	0.0434	1,635
Mineral base	1.5 to 2.0	0.054 to 0.072	2,043 to 2,724
Varnished cloth	1.11	0.0401	1,515
Impregnated paper	1.17	0.0423	1,594
Lead	11.37	0.411	15,530
Untreated braid	1.11	0.0402	1,515
Saturated braid	1.33	0.0480	1,809
Tarred jute, in cable	0.63	0.0228	858
Dry jute, in cable	0.267	0.00965	364
Dry hemp, in cable	0.267	0.00965	364
Gutta-percha	1.0	0.0361	1,362
Rubber-filled tape	1.0	0.0361	1,362
Copper	8.89	0.3212	12,108
Polyethylene	0.92	0.033	1,255
Polyvinyl chloride compound	1.30	0.047	1,770
Saturated asbestos	1.02	0.037	1,390

77. BIBLIOGRAPHY

SPECIFICATIONS. Practically all wires and cables are purchased under standard specifications. The principal ones are as follows:

Subject	Standardizing Organization	Standard or Specification Number
Definitions, Methods of Test, etc.	AIEE ASA	30 C–8–1
Bare Soft or Annealed Copper Wire	ASTM ASA	B–3 H–4.1
Tinned Soft or Annealed Copper Wire	ASTM ASA	B–33 H–4.4
Lead- or Lead-alloy-coated Soft Copper Wire	ASTM	B–189
Rubber-insulated and Plastic-insulated House Wire and Cord	UL	Various
Rubber-insulated Wire, 30 per cent Hevea, Class AO	ASTM ASA	D–27 C–8–17
Rubber-insulated Wire, Heat-resistant Type	ASTM	D–460
Rubber-insulated Wire, Performance Type	NEMA ASTM	...... D–353
Rubber-insulated Wire, Moisture-resisting Type	UL	Type RW
Rubber-insulated Wire, Ozone-resisting Type	ASTM	D–574
Synthetic Rubber-insulated Wire, Heat-resistant Type	ASTM	D–754
Synthetic Rubber-insulated Wire, Performance Type	ASTM	D–755
Polyvinal-insulated Wire	ASTM	D–734
Impregnated-paper-insulated Cables, Solid Type	AEIC	7th Ed.
All Types Rubber and Rubber-like Insulated Wires	IPCEA	S–19–81
Impregnated-paper Cables, Oil-filled Type	AEIC	4th Ed. (1945)
Varnished-cambric Wires and Cables	IPCEA	S–2–313
Magnet Wire	NEMA ASA	36–34 and 41–69 8.5, 8.6, and 8.7
Cable Shipping Reels	IPCEA	A–2–234 and A–6–9
Braid Colors for Control Cable	IPCEA	S–19–81
Marine Wires and Cables	AIEE	45

a. The braces indicate that the specifications included by them are identical in substance.

b. AIEE = American Institute of Electrical Engineers.
ASA = American Standards Association.
ASTM = American Society for Testing Materials.
IPCEA = Insulated Power Cable Engineers Association.
UL = Underwriters' Laboratories.
NEMA = National Electrical Manufacturers Association.

BOOKS

Apt, R., *Isolierte Leitungen und Kabel.* J. Springer, Berlin (1928).
Beaver, C. J., *Insulated Electric Cables.* Van Nostrand (1926).
Coyle, D., and F. J. O. Howe. *Electric Cables.* Spon, London (1909).
Del Mar, W. A., *Electric Cables.* McGraw-Hill (1924).
Dunsheath, P., *High Voltage Cables.* Pitman, London (1929).
Emanueli, L., *High Voltage Cables.* John Wiley (1932).
Gemant, A., *Liquid Dielectrics.* John Wiley (1933).
Main, F. W., *Electric Cables.* London (1930).
National Electric Light Association, *Underground Systems Reference Book* (1931). Contains extensive classified bibliography.
Pyne, A. P., and N. A. Allen, *Power Cables.* Pitman, London (1929).
Robinson, D. M., *Dielectric Phenomena in High Voltage Cables.* Van Nostrand (1936).
Ruhling, T. C., *Underground Systems.* McGraw-Hill (1927).
Russel, A., *The Theory of Electric Cables and Networks.* Van Nostrand (1925).
Simmons, D. M., Calculation of Electrical Problems of Underground Cables, Reprint by General Cable Corp. from *Elect. J.* (May–Nov. 1932). Contains an extensive bibliography.
Stubbings, G. W., *Underground Cable Systems.* Chapman & Hall, London (1929).
Waddicor, H., *Principles of Electric Power Transmission.* Chapman & Hall, London (1935).
Westinghouse Electric Corporation, *Electrical Transmission and Distribution Reference Book* (1944).
Whitehead, J. B., *Impregnated Paper Insulation.* John Wiley (1935).

WIRING OF BUILDINGS

By George W. Zink

78. GENERAL REQUIREMENTS

Wires and fittings designed to conduct electricity in a building should be selected as to size and insulation and installed in such a manner that: (1) the entire wiring system shall conform to the rules and regulations of any authority having jurisdiction over the building in question; (2) the attending fire risk and the possibility of an electric shock to the inhabitants shall be a minimum; (3) the electric power efficiency of the system shall be reasonable; (4) the voltage at the receiver shall approximate the rated voltage of the receivers and shall remain sensibly constant; (5) the mechanical arrangement of the system shall be simple and convenient for inspection and use; (6) the conductors shall be mechanically protected from external injury; (7) the service shall not be interrupted under normal load; and (8) the cost of the materials, labor of installation, and replacement due to depreciation shall not be excessive.

Items (1) and (2) are governed by one or more of the regulatory bodies described in this article. Items (3) to (8) are largely matters of design and depend upon the type of building, available supply of energy, and economic considerations. The design features are covered by Articles 85 to 89.

The National Electrical Code is a compilation of rules and recommendations which represent acceptable standards of safety in the construction and installation of electrical equipment. It is the electricians' installation manual throughout the United States.

The Code rules are drawn up primarily with regard to fire hazard but also give consideration to accident prevention. Safety rules, however, are more specifically covered by municipal regulations and safety codes, such as the National Safety Code.

Revisions of the Code are made periodically, usually every 3 years, by the Electrical Committee of the National Fire Protection Association, which sponsors the Code under the rules of procedure of the American Standards Association. It is published and distributed by the National Board of Fire Underwriters.

It should be noted that the Code represents minimum accepted standards and insures, not an adequate or efficient wiring system, but only a reasonably safe system in relation to fire hazard and insurance risk.

References to the Code in this article are to the 1947 edition.

UNDERWRITERS' LABORATORIES, INC. It is to be noted that the Code does not specify test requirements for equipment, but refers to "approved" devices or apparatus. This means approved by the Underwriters' Laboratories, Inc., an organization established to examine and test materials and equipment to determine their fitness for use under the Code. Detailed specifications for the construction and performance under test and in service of approved electrical fittings and material will be found in the Standards of the Underwriters' Laboratories.

The List of Inspected Electrical Appliances is published semiannually by the Underwriters' Laboratories, and all "listed" materials bear some evidence of their approval.

Copies of the National Electrical Code and the List of Inspected Electrical Appliances may be obtained from most local inspection bureaus or from either the National Board of Fire Underwriters, 85 John Street, New York, N. Y., or the Underwriters' Laboratories, 161 Sixth Avenue, New York, N. Y. (Chicago address: 207 East Ohio Street).

MUNICIPAL REGULATION. Most states and municipalities have laws concerning the installation of electric wiring, and these laws usually include the National Electrical Code with such modifications as are considered advisable to suit local conditions. These modified rules are often more rigid than the Code in that they prohibit certain practices recognized by the Code. For example, concealed knob and tube wiring is one method of wiring recognized by the Code but prohibited by most municipalities. Local regulations are often broader than the Code in that they cover accident risk, inspection, licensing of electricians, and penalties for violations. Municipal authorities should be consulted for copies of local regulations.

REGULATIONS OF THE UTILITY COMPANIES. Most lighting and power companies have rules which must be observed by consumers obtaining service from their lines. These rules are made necessary by conditions pertaining to the operation of their particular systems and usually concern the location of meters, the type and location of service entrances, and, in the case of power loads, the type and size of motors, etc.

ENFORCEMENT AND INSPECTION. The National Electrical Code, as issued by the National Board of Fire Underwriters, has in itself no legal status. However, inasmuch as the Code rules form part of ordinances of most cities throughout the country, it has almost universal enforcement by the police power of such municipalities. Where not so enforced, insurance companies require the Code to be observed if the buildings in question are to be insured. In order to ensure the observance of the Code, the Fire Underwriters and most municipalities have inspectors to check the electrical work in buildings. These two inspection agencies may act independently, but often they cooperate and have one office handle both inspections. Before a new building is occupied, and when the occupancy of an old building is changed, it is advisable to be sure that proper inspection has been made.

79. SYSTEMS OF WIRING FOR INTERIOR DISTRIBUTION

There are several systems of wiring suitable for interior distribution, the more usual ones being given below. The selection of a wiring system, where power is purchased, is influenced or may be definitely controlled by the distribution system used on the lines of the utility company supplying the service.

SINGLE-PHASE AND D-C DISTRIBUTION. **Direct-current or single-phase a-c two-wire systems** are used for small, isolated generating plants or low-wattage loads on utility lines. They are uneconomical in the use of copper when heavy loads are involved and are seldom chosen for a complete distribution system. Many utility companies limit the amount of power that will be applied to such a system. See Fig. 1.

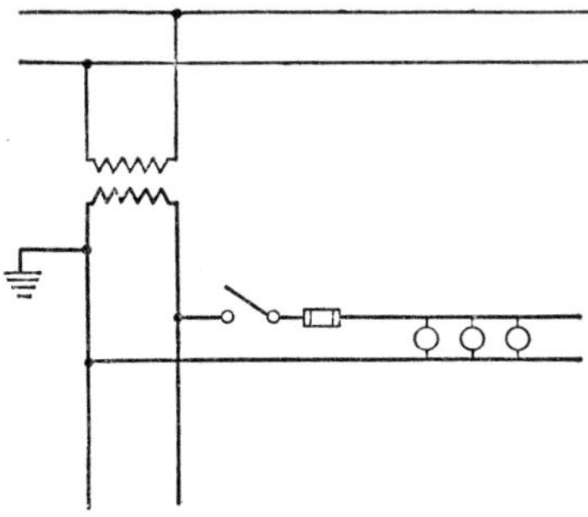

FIG. 1. Single-phase Two-wire System

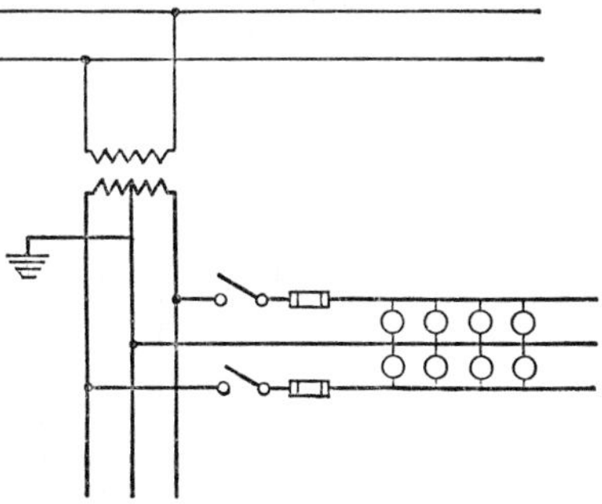

FIG. 2. Single-phase Three-wire System

Direct-current or single-phase a-c three-wire systems may be considered alike and are used mostly for 115/230-volt distribution on lighting and power circuits when the motor load does not exceed 4 or 5 hp. The branch circuits may be taken off as 115 volts or 230 volts as desired. In laying out the circuits, care must be taken to keep the load on both legs approximately the same; otherwise there may be considerable difference in the voltage of the two legs, resulting in poor illumination on the low side and shortened lamp life on the high side.

The neutral wire is grounded and therefore should not be fused. Although the neutral wire carries only the unbalanced current, the Code requires that the same size of wire be used as in the ungrounded conductors for loads up to 200 amp. For unbalanced loads above 200 amp 70 per cent of the excess is used in calculating the size of the neutral.

Branch circuits may be two-wire 115 or 230 volts or three-wire 115/230 volts. See Fig. 2.

TWO-PHASE DISTRIBUTION. **Two-phase systems** are seldom used in modern installations, although there are some older systems in operation. For distributing two-phase alternating currents, either three or four wires are employed (Fig. 3). A four-wire two-phase system may be treated as two separate two-wire systems, which cannot in any case be connected in parallel. A single wire 41 per cent larger than either of the wires it displaces may be substituted for any two of the wires of the four-wire two-phase system, thus making a three-wire two-phase system. Either three or four wires may be used for two-phase lighting or power loads.

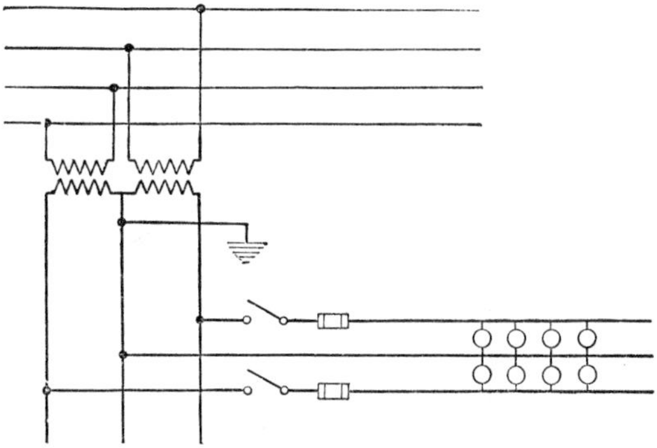

FIG. 3. Three-wire Two-phase System

All the three or four wires are used for power loads, the motors being connected to both phases. In the three-wire system lighting loads are connected to each phase, one wire being common to both phases. It is important that the phases are kept in balance by maintaining equal loads on each side of the circuit.

THREE-PHASE DISTRIBUTION. **The three-wire system** is the most usual one for power distribution and for mixed power and small lighting loads in small industrial plants. Where power is the only load, a three-wire system is used. If a small lighting load is required on a three-phase, 230-volt system, one phase may have a neutral tap and a three-wire lighting circuit obtained. This, however, will tend to throw the system out of balance and can be used only for small loads. See Fig. 4.

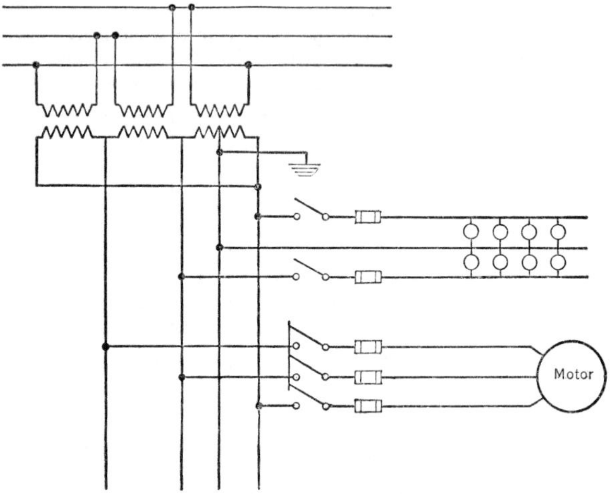

FIG. 4. Three-phase Three-wire System

The three-phase four-wire system is in general use where the power load is subordinate to the lighting load. In this case, the fourth wire is the neutral and usually grounded, and there should be no fuse in this wire in any part of the system. Branch lighting circuits may be run two-wire, using one outside wire and the neutral; three-wire, using two

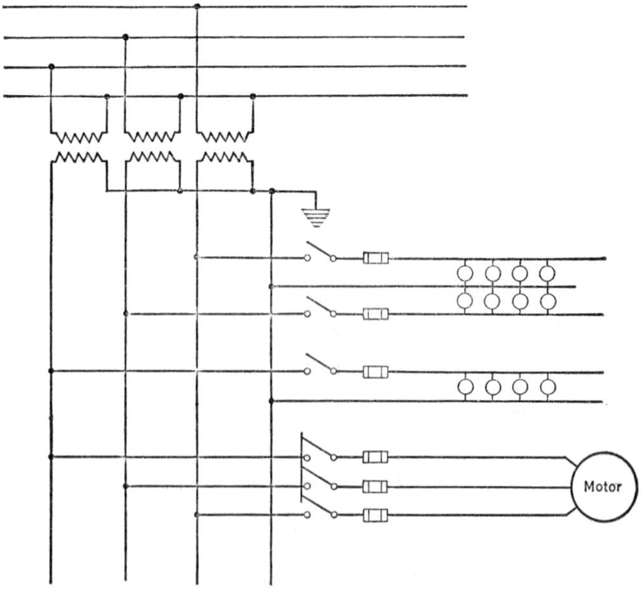

FIG. 5. Three-phase Four-wire System

outside wires and the neutral, or three-wire, using the three-phase wires for three-phase motor circuits. Care must be taken to distribute the lighting load so as to keep all phases of the system in balance. This system is usually 120–208 volts, that is, 120 volts from any outside wire to neutral and 208 volts between the outside wires. See Fig. 5.

80. STANDARD VOLTAGES FOR INTERIOR DISTRIBUTION

GENERAL. The most popular voltages are 115, 230, 440, 550, or 2300 volts. Before a voltage is selected for any installation, the utility company furnishing the power should be consulted as to the types of service available, particularly if small capacity loads are to be taken directly from utility lines at the service voltage. If the loads are sufficiently large to permit the use of step-down transformers on the premises, any suitable secondary voltage can be used.

The National Electrical Code does not permit interior wiring above 7500 volts, and special rules apply for systems over 600 volts.

RESIDENTIAL BUILDINGS. Single- or multiple-family dwellings are usually supplied by a 115/230-volt single-phase three-wire system. The lights are supplied by 115-volt branch circuits, and ranges, water heaters, and other heavy-wattage equipment are supplied at 230 volts.

COMMERCIAL BUILDINGS. This classification includes office buildings, warehouses, and other structures where the power load is not large and consists mostly of small motors. In larger cities where the utility company maintains a low-voltage network system these buildings are usually supplied with a 120/208-volt three-phase four-wire system in which the lighting load is taken off at 120 volts and the motor load as three-phase 208 volts. Where a network system is not available, the lighting load is generally supplied by a 115/230 single-phase three-wire system, and a separate three-phase three-wire system is used to supply the power load at 230 volts.

INDUSTRIAL PLANTS. The distribution systems in industrial plants vary widely, depending upon the size of plant, type of load, and size of motors.

Small plants can frequently be treated in the same manner as commercial buildings and supplied directly from the utility mains. If the plant is large or composed of many buildings, however, a high-voltage primary distribution system may be required in addition to the secondary distribution system supplying the load within the building.

Such plants require an extended study of load conditions, both for initial requirements and future expansion. (See *Electric Power Distribution for Industrial Plants*, published by AIEE.)

81. METHODS OF WIRING

The National Electrical Code recognizes a number of methods of wiring, one or more of which must be selected for any wiring layout. Inasmuch as certain types of wiring recognized by the Code are not permitted in some municipalities, this point should be checked before deciding upon any particular method. The methods listed by the 1947 Edition of the Code are as follows:

1. Installed on insulators.
   - *a.* Open wiring.
   - *b.* Concealed knob and tube.
2. The "pull-in and pull-out" or raceway systems.
   - *a.* Conduit work—rigid or flexible metallic conduit.
   - *b.* Electrical metallic tubing.
   - *c.* Wireways.
   - *d.* Busways.
   - *e.* Underfloor raceways.
   - *f.* Cellular metal floor raceways.
3. Cable assemblies.
   - *a.* Armored cable.
   - *b.* Non-metallic sheath cable.
   - *c.* Non-metallic waterproof wiring.
4. Extension materials.
   - *a.* Surface metal raceways.
   - *b.* Under plaster extensions.
   - *c.* Non-metallic surface extensions.
5. Special feeders.
   - *a.* Bare bus-bars and risers.
   - *b.* Service entrance conductors.

All methods listed, unless otherwise limited in the descriptions given below, are suitable for 600 volts. Where the voltage between conductors exceeds 600 volts in locations "accessible to other than qualified persons," conductors shall be in rigid metal conduit, duct, or armored cable especially approved for the purpose.

OPEN WIRING ON INSULATORS. This method is used extensively where the appearance of exposed wiring is not objectionable, as in industrial buildings, and where the wiring is not exposed to mechanical injury. It is a satisfactory method in buildings or rooms subject to moisture, corrosive vapors, or heat, such as paper mills, dye houses, or dry kilns. It is often used in hot, dry locations, such as boiler rooms, inasmuch as one of the high-temperature insulations, such as Type AVC or AI, may be used. This permits much heavier current loading of the conductor than is possible for any of the so-called raceway systems under the same temperature conditions.

Open wiring is not permitted in (1) commercial garages, (2) theaters, (3) motion-picture studios, (4) hoistways and elevator shafts, and (5) hazardous locations.

In installing, single-conductor wires are supported on knobs or cleats. Wires must never be fastened with staples. For voltages not exceeding 300 volts, wires must be separated 2 1/2 in. from each other and 1/2 in. from the surface wired over. For voltages from 301 to 600, the wires shall be separated 4 in., and at least 1 in. from the surface wired over. In damp places, 1-in. separation from the surface wired over must be maintained for all voltages. Wires must be supported at least every 4 1/2 ft, except that in buildings of mill construction wires No. 8 A.W.G. and larger may be separated 6 in. and run from timber to timber with supports on each timber.

Where exposed to mechanical injury, the wires must be protected by suitable guard strips—boxing or other means approved by the National Electrical Code. Wires passing through floors, walls, timbers, or partitions must be protected by insulating tubes or bushings. Insulating tubes must be used over the wire where within 2 in. of piping or other conducting material.

Most approved types of insulated wire may be used for open wiring in dry locations (see Column F, Table 7). The most popular constructions are Types SB, SBW, R, AVC, and AIA. For moist locations one of the rubber-covered or thermoplastic types should be used: Types R, RW, RH, T, or TW.

For the installation of open wiring the National Electrical Code specifies minimum sizes of insulators and cleats to be used with any given size of wire.

CONCEALED KNOB AND TUBE. This type of wiring is used in buildings of frame construction where economical concealed wiring is desired. Most cities and towns prohibit this method of wiring, and its use therefore is restricted largely to isolated rural dwellings.

The method of installing consists of running single insulated wires in the hollow spaces of walls and between floors. Where run parallel to timbers or floor, the wires are supported on single wire insulators or knobs, and where passing through timbers or floors, by insulating tubes. Wires must be separated at least 5 in., and where this spacing cannot be maintained, as in entering outlet boxes, it shall be encased in flexible tubing, which must extend from the last knob into and be secured to such boxes. The requirements as to supporting, clearances, etc., which are specified for open wiring apply also to knob and tube.

Approved single-conductor rubber-covered wire Types R, RH, RW, and RU and thermoplastic Types T and TW are suitable for knob and tube wiring.

RIGID METAL CONDUIT. The best and most extensively used type of wiring consists of wires enclosed in metallic conduit, either rigid or flexible. Rigid conduit is standard for general work; flexible conduit is frequently used for short runs, such as from the end of a rigid conduit to a motor mounted on a machine frame. Conduit is employed in all types of buildings and may be run exposed or concealed and may be used for all voltages. Rigid conduit, unless of corrosion-resistant material, cannot be used under cinder fill when subject to moisture unless protected by at least 2 in. of non-cinder concrete or unless the conduit is at least 18 in. under the fill.

Rigid conduit is standard steel pipe, protected against corrosion by galvanizing, sherardizing, enameling, singly or combined; or pipe of non-corrosive metal may be used. It is regularly furnished in 10-ft lengths and cannot be used in sizes smaller

Table 1. Radius of Conduit Bends

Size of Conduit, Inches	Wires without Lead Sheath, Inches	Wires with Lead Sheath, Inches
1/2	3.7	6.2
3/4	4.9	8.3
1	6.3	10.5
1 1/4	8.3	13.8
1 1/2	9.6	16.1
2	12.4	20.6
2 1/2	14.8	24.6
3	18.4	30.6
3 1/2	21.3	35.5
4	24.1	40.2
4 1/2	27.0	45.0
5	30.3	50.4
6	36.4	60.6

than 1/2 in. The system must be installed complete from outlet to outlet and be securely fastened in place. The wires shall not be pulled in until all mechanical work on the building is completed.

Numerous designs of fittings are made for joining or terminating the conduit. These may be of the threaded or threadless type, and in either case must be installed to give adequate electrical continuity to the system, which must be grounded, as provided for by the Code.

A run of conduit between outlets or fittings must not contain more than four quarter bends. Bends of rigid conduit shall be so made that the internal diameter shall not be reduced and the radius of the curve of the inner edge of any field bend shall not be less than shown in Table 1.

When long vertical feeders are required, as in tall buildings, special vertical riser cables designed to be self supporting may be used.

The types of wire commonly used for conduit work under various conditions are as follows:

a. Dry locations: rubber-covered wire Type R, RH, or RU (sizes 14 to 6 A.W.G.), thermoplastic Type T (sizes 14 to 4/0 A.W.G.), and varnished-cambric Type V or AVB.

b. Wet locations: rubber-covered wire Type RW, thermoplastic Type TW (sizes 14 to 4/0), lead-covered cable of all types.

The number of wires which may be installed in a conduit is limited by certain "conduit fill" rules and tables in the National Electrical Code.

Tables 2, 3, and 4 give the number of wires per conduit when all wires are the same size and rated at not over 600 volt.

Table 5 is used for groups of conductors not included in Tables 2, 3, and 4 and for rewiring existing raceways.

Table 2. Number of Conductors in Conduit or Tubing *

Rubber-covered Types RF-32, R, RH, RW, RU; Thermoplastic Types TF, T, TW
One to Nine Conductors

Size, A.W.G. MCM	Number of Conductors in One Conduit								
	1	2	3	4	5	6	7	8	9
18	1/2	1/2	1/2	1/2	1/2	1/2	1/2	3/4	3/4
16	1/2	1/2	1/2	1/2	1/2	1/2	3/4	3/4	3/4
14	1/2	1/2	1/2	1/2	3/4	3/4	1	1	1
12	1/2	1/2	1/2	3/4	3/4	1	1	1	1 1/4
10	1/2	3/4	3/4	3/4	1	1	1	1 1/4	1 1/4
8	1/2	3/4	3/4	1	1 1/4	1 1/4	1 1/4	1 1/2	1 1/2
6	1/2	1	1	1 1/4	1 1/2	1 1/2	2	2	2
4	1/2	1 1/4	1 1/4	1 1/2	1 1/2	2	2	2	2 1/2
3	3/4	1 1/4	1 1/4	1 1/2	2	2	2	2 1/2	2 1/2
2	3/4	1 1/4	1 1/4	2	2	2	2 1/2	2 1/2	2 1/2
1	3/4	1 1/2	1 1/2	2	2 1/2	2 1/2	2 1/2	3	3
1/0	1	1 1/2	2	2	2 1/2	2 1/2	3	3	3
2/0	1	2	2	2 1/2	2 1/2	3	3	3	3 1/2
3/0	1	2	2	2 1/2	3	3	3	3 1/2	3 1/2
4/0	1 1/4	2	2 1/2	3	3	3	3 1/2	3 1/2	4
250	1 1/4	2 1/2	2 1/2	3	3	3 1/2	4	4	4 1/2
300	1 1/4	2 1/2	2 1/2	3	3 1/2	4	4	4 1/2	4 1/2
350	1 1/4	3	3	3 1/2	3 1/2	4	4 1/2	4 1/2	5
400	1 1/2	3	3	3 1/2	4	4	4 1/2	5	5
500	1 1/2	3	3	3 1/2	4	4 1/2	5	5	6
600	2	3 1/2	3 1/2	4	4 1/2	5	6	6	6
700	2	3 1/2	3 1/2	4 1/2	5	5	6	6	....
750	2	3 1/2	3 1/2	4 1/2	5	6	6	6	....
800	2	3 1/2	4	4 1/2	5	6	6	....	....
900	2	4	4	5	6	6	6	....	....
1000	2	4	4	5	6	6	....	....	....
1250	2 1/2	4 1/2	4 1/2	6	6	....	....	....	....
1500	3	5	5	6	....	....	....	....	....
1750	3	5	6	6	....	....	....	....	....
2000	3	6	6	....	....	....	....	....	....

* National Electrical Code.

Table 6 gives nominal sizes and areas of conduits or tubing, and Tables 11, 12, and 13 give nominal size and areas of the more commonly used conductors recommended for computing size of conduit and tubing for various combinations of conductors in accordance with the limitations of Table 5. These dimensions represent average conditions, and although variations will be found in the dimensions of conductors and conduit of different manufacturers, these variations will not seriously affect the computation.

Varnished-cambric-insulated conductors, except for sizes 14, 12, and 8, have the same insulation thickness as Type R conductors; therefore Type R tables and computation may be used for determining the number of varnished-cambric-insulated conductors in conduit or tubing.

Table 3. Number of Conductors in Conduit or Tubing *

More Than Nine Conductors

Rubber-covered Types RF-32, R, RH, RW, RU; Thermoplastic Types TE, T, and TW

Size, A.W.G.	Maximum Number of Conductors in Conduit or Tubing						
	3/4 in.	1 in.	1 1/4 in.	1 1/2 in.	2 in.	2 1/2 in.	3 in.
18	12	20	35	49	80	115	176
16	10	17	30	41	68	97	150
14	....	10	18	25	40	59	90
12	....	....	15	21	35	50	77
10	....	....	13	17	29	41	64
8	....	....	....	10	17	25	38
6	....	....	....	....	....	15	23

* National Electrical Code.

Table 4. Number of Lead-covered Wires and Cables in Conduit—600 Volts *

Types RL and RHL

Size of Wire	Size of Conduit to Contain Not More than Four Cables											
	Single-conductor Cable				Two-conductor Cable				Three-conductor Cable			
	1	2	3	4	1	2	3	4	1	2	3	4
	Cables in One Conduit				Cables in One Conduit				Cables in One Conduit			
14	1/2	3/4	3/4	1	3/4	1	1	1 1/4	3/4	1 1/4	1 1/2	1 1/2
12	1/2	3/4	3/4	1	3/4	1	1 1/4	1 1/4	1	1 1/4	1 1/2	2
10	1/2	3/4	1	1	3/4	1 1/4	1 1/4	1 1/2	1	1 1/2	2	2
8	1/2	1	1 1/4	1 1/2	1	1 1/4	1 1/2	2	1	2	2	2 1/2
6	3/4	1 1/4	1 1/2	1 1/2	1 1/4	1 1/2	2	2 1/2	1 1/4	2 1/2	3	3
4	3/4	1 1/4	1 1/2	1 1/2	1 1/4	2	2 1/2	2 1/2	1 1/2	3	3	3 1/2
3	3/4	1 1/4	1 1/2	2	1 1/4	2	2 1/2	3	1 1/2	3	3	3 1/2
2	1	1 1/4	1 1/2	2	1 1/4	2	2 1/2	3	1 1/2	3	3 1/2	4
1	1	1 1/2	2	2	1 1/2	2 1/2	3	3 1/2	2	3 1/2	4	4 1/2
0	1	2	2	2 1/2	2	2 1/2	3	3 1/2	2	4	4 1/2	5
00	1	2	2	2 1/2	2	3	3 1/2	4	2 1/2	4	4 1/2	5
000	1 1/4	2	2 1/2	2 1/2	2	3	3 1/2	4	2 1/2	4 1/2	4 1/2	6
0000	1 1/4	2 1/2	2 1/2	3	2 1/2	3	3 1/2	4 1/2	3	5	6	6
250,000	1 1/4	2 1/2	3	3	......	......	......	......	3	6	6	......
300,000	1 1/2	3	3	3 1/2	......	......	......	......	3 1/2	6	6	......
350,000	1 1/2	3	3	3 1/2	......	......	......	......	3 1/2	6	6	......
400,000	1 1/2	3	3	3 1/2	......	......	......	......	3 1/2	6	6	......
500,000	1 1/2	3	3 1/2	4	......	......	......	......	4	6	......	......
600,000	2	3 1/2	4	4 1/2	......	......	......	......	......	......	......	......
700,000	2	4	4	5	......	......	......	......	......	......	......	......
750,000	2	4	4	5	......	......	......	......	......	......	......	......
800,000	2	4	4 1/2	5	......	......	......	......	......	......	......	......
900,000	2 1/2	4	4 1/2	5	......	......	......	......	......	......	......	......
1,000,000	2 1/2	4 1/2	4 1/2	6	......	......	......	......	......	......	......	......
1,250,000	3	5	5	6	......	......	......	......	......	......	......	......
1,500,000	3	5	6	6	......	......	......	......	......	......	......	......
1,750,000	3	6	6	......	......	......	......	......	......	......	......	......
2,000,000	3 1/2	6	6	......	......	......	......	......	......	......	......	......

The above sizes apply to straight runs or with nominal offsets equivalent to not more than two quarter-bends.

* National Electrical Code.

Table 5. Combination of Conductors

For groups or combinations of conductors not included in Tables 3 and 4, it is recommended that the conduit or tubing be of such size that the sum of the cross-sectional areas of the individual conductors will not be more than the percentage of the interior cross-sectional area of the conduit or tubing shown in the following table:

Per Cent Area of Conduit or Tubing

	Number of Conductors				
	1	2	3	4	Over 4
Conductors (not lead covered)	53	31	43	40	40
Lead-covered conductors	55	30	40	38	35
For rewiring existing raceways for increased load where it is impracticable to increase the size of the raceway because of structural conditions	60	40	50	50	50

Table 6. Dimensions and Per Cent Area of Conduit and Tubing *

Areas of Conduit or Tubing for the Combinations of Wires Permitted by Table 5

Trade Size	Internal Diameter, Inches	Area, Square Inches									
			Not Lead Covered				Lead Covered				
		Total 100%	1 Cond. 53%	2 Cond. 31%	3 Cond. 43%	4 Cond. and Over 40%	1 Cond. 55%	2 Cond. 30%	3 Cond. 40%	4 Cond. 38%	Over 4 Cond. 35%
1/2	0.622	0.30	0.16	0.09	0.13	0.12	0.17	0.09	0.12	0.11	0.11
3/4	0.824	0.53	0.28	0.16	0.23	0.21	0.29	0.16	0.21	0.20	0.19
1	1.049	0.86	0.46	0.27	0.37	0.34	0.47	0.26	0.34	0.33	0.30
1 1/4	1.380	1.50	0.80	0.47	0.65	0.60	0.83	0.45	0.60	0.57	0.53
1 1/2	1.610	2.04	1.08	0.63	0.88	0.82	1.12	0.61	0.82	0.78	0.71
2	2.067	3.36	1.78	1.04	1.44	1.34	1.85	1.01	1.34	1.28	1.18
2 1/2	2.469	4.79	2.54	1.48	2.06	1.92	2.63	1.44	1.92	1.82	1.68
3	3.068	7.38	3.91	2.29	3.17	2.95	4.06	2.21	2.95	2.80	2.58
3 1/2	3.548	9.90	5.25	3.07	4.26	3.96	5.44	2.97	3.96	3.76	3.47
4	4.026	12.72	6.74	3.94	5.47	5.09	7.00	3.82	5.09	4.83	4.45
4 1/2	4.506	15.95	8.45	4.94	6.86	6.38	8.77	4.78	6.38	6.06	5.57
5	5.047	20.00	10.60	6.20	8.60	8.00	11.00	6.00	8.00	7.60	7.00
6	6.065	28.89	15.31	8.96	12.42	11.56	15.89	8.67	11.56	10.98	10.11

* National Electrical Code.

ELECTRICAL METALLIC TUBING. A form of rigid conduit having the same internal diameter as standard conduit, but with a much thinner wall and referred to as electrical metallic tubing, may be used in the same manner as rigid conduit with certain restrictions imposed because of the lighter material. The tubing, unless of non-corrodible metal, must be protected with a rust-resisting coating. It is installed under the same rules as rigid conduit, except that it shall not be used (1) when subject to severe mechanical injury during or after installation, (2) in cinder concrete or fill where subject to permanent moisture, unless protected on all sides by a layer of non-cinder concrete at least 2 in. thick, or unless tubing is at least 18 in. under fill, (3) in hazardous locations, (4) when exposed to corrosive vapor unless corrosion-resistant material is used.

No tubing smaller than 1/2 in. electrical trade size may be used except for under plaster extensions. The maximum size of tubing permitted is 2 in. electrical trade size.

WIREWAYS. In industrial plants where exposed work is not objectionable and a readily accessible low-voltage wiring system is desired, branch circuits are often run part of their length in wireways. These consist of sheet-metal troughs having hinged covers and made in convenient length so designed as to be securely fastened together. Convenient knockouts are provided in the sides and bottom so that extensions of conduit, tubing, armored cable, etc., may be run from any point. Rigid supports must be provided at least every 5 ft. Wireways may be used only in dry locations and for voltages not exceeding 600 volts.

After the system is installed, the cover is raised and the wires are laid in place. Not more than 30 wires may be placed in one wireway unless the conductors are signal or control wires. No conductor larger than 500,000 cir mils shall be used, and the cross-sectional area of all cables shall not exceed 20 per cent of the area of the wireway.

Rubber-covered wire Type R and RH, thermoplastic insulated wire Type T, and varnished-cambric-insulated wire Type V are most frequently used. For high ambient temperature location asbestos-varnished-cambric Type AVB may be used.

BUSWAYS. Busways consist of sheet-metal troughs similar to wireways, and their installation is governed by essentially the same rules. Instead of the conductors being insulated, however, they are bare copper, usually rectangular in shape, mounted on suitable insulators.

Although this type of wiring is usually classed as a raceway system, it is in effect a specially designed wiring assembly. It is becoming increasingly popular for use in machine shops and light manufacturing when frequent taps are required to feed a large number of individually motor-driven machines.

UNDERFLOOR RACEWAYS. In buildings of fire-resisting construction, metal or fiber raceways laid in concrete floors are extensively used for running branch circuits. This method is particularly adaptable to office buildings, where the location of partitions is frequently varied with a change of occupancy.

Raceways are made in several styles, the metal ones usually being of rectangular section and the fiber ones of semicircular or oval section, and have conveniently spaced outlets formed in the duct. They are usually installed in the form of a grid covering an entire floor in such a manner that, no matter what the office layout may be, there will be an outlet convenient to any desk or other equipment.

The ducts are often run parallel in pairs, one for light and power wire and one for telephone or signal systems.

Raceways may be placed in the concrete fill between the rough and finished floor or in the slab, provided raceways of half-round section or of flat-top section not over 4 in. in width shall have at least $^3/_4$ in. of concrete above the raceway, except that in office occupancies, metal flat-top raceways not over 2 in. in width may be laid flush with the concrete if covered with substantial linoleum or equivalent floor covering.

Underfloor raceways are restricted to dry locations. The cross-sectional area of all conductors in the raceway must not exceed 40 per cent of the area of the raceway, except when armored cable or non-metallic sheathed cable only is used, in which case there is no restriction on area of fill.

Any general-purpose wire may be used, but no conductor shall be larger than No. 4 A.W.G.

CELLULAR METAL FLOOR RACEWAYS. A cellular metal floor raceway is the hollow spaces of cellular metal floors, together with suitable fittings approved as enclosures for electrical conductors, and may be considered as a special type of underfloor raceway.

The same general rules for area of fill and type of wire apply as in the case of underfloor raceways, except that the largest permissible size of conductor is No. 1/0 A.W.G.

ARMORED CABLE. Armored cable, sometimes known as BX or ABC cable, is employed in residence wiring and is also a satisfactory method of installing concealed wiring in old buildings. It is frequently used for branch circuits in installations where the feeders are run in conduit. It is permitted for all voltage up to 600 volts.

The cable consists of one or more suitably insulated conductors, protected by an interlocking metal strip, usually galvanized steel, wound helically to form a tube similar to flexible conduit. Where of more than one conductor, the insulated conductors are cabled together and enclosed in a braided or impregnated fiber or paper covering before the armor is applied.

Armored cable may be installed exposed or concealed, may be run in the hollow spaces of walls or between floors, or may be "fished" through such spaces in completed buildings.

Special fittings or outlet boxes are available to fit cables of various sizes. All cable must be continuous from outlet to outlet and must be connected both electrically and mechanically thereto, and the armor shall be properly grounded.

Where exposed to weather or continuous moisture, or where embedded in masonry or otherwise exposed to materials likely to affect the rubber insulation, a type of cable having a lead covering under the armor shall be used.

Armored cable cannot be used in (1) theaters, (2) motion-picture studios, (3) hazardous locations, (4) when exposed to corrosion fumes or vapors, (5) in storage battery rooms, (6) on cranes, hoists, or elevators, or in hoistways except under special conditions designated by the code. It may be used under plaster extensions and embedded in plaster finish on brick or masonry walls except in damp locations.

NON-METALLIC SHEATHED CABLE. This material is used in a manner similar to armored cable but requires more protection where exposed to mechanical injury. It is prohibited by the Code for certain types of buildings. It may be used for voltages up to 600 volts in sizes No. 14 to No. 4 A.W.G. inclusive.

The cable consists of suitably insulated conductors in two- and three-wire assemblies, protected by paper wrappings and cotton braids thoroughly impregnated with moisture-resisting compounds and with the outer covering treated to make it fire-resisting. It is furnished with or without a bare ground wire in the assembly.

Non-metallic sheathed cable cannot be embedded in masonry, concrete, fill, or plaster. It cannot be used (1) in commercial garages, (2) in theaters, (3) in motion-picture studios, (4) in storage battery rooms, (5) in hoistways, (6) in any hazardous location, (7) in breweries, ice plants, and similar wet locations.

NON-METALLIC WATERPROOF WIRING. Subject to special approval of the authority enforcing the National Electrical Code in the location involved, non-metallic waterproof wiring may be used for exposed work in breweries, ice plants, and similar wet locations provided that the voltage does not exceed 300 volts between conductors or 150 volts to ground.

The material consists of rubber-insulated conductors, not smaller than No. 12 A.W.G., cabled together, with or without a grounding conductor and covered with a rubber sheath. The construction is similar to Type S portable cord except that flexible conductors are not required.

SURFACE METAL RACEWAYS. **Metal Raceways,** sometimes called metal molding, are used for exposed surface wiring, usually in completed buildings for the extension of existing systems where a neat and inconspicuous metallic wire enclosure is desired.

Raceways are made in various designs and sizes of substantially rectangular cross-section, with removable covers or backs. They are made in various sizes, together with the necessary fittings, to accommodate from two to ten wires. The strips are fastened in place by screws or bolts so placed as not to interfere with the wire space.

The same rules regarding electrical and mechanical continuity between outlets or fittings and the same grounding rules apply as for rigid conduit.

Surface raceways are not permitted for wire sizes larger than No. 6. They must not be used in damp locations or where exposed to corrosive fumes or subject to severe mechanical injury.

UNDERPLASTER EXTENSIONS. Where the absence of hollow spaces in the walls and floors of finished buildings does not permit fishing between outlets, extensions may be made to existing systems by means of underplaster extensions. This method is permitted in fire-resisting buildings only and used mostly in short runs. Such extensions must not extend beyond the limits of the floor where they originate.

The installation is made by channeling the plaster and fastening the raceway, cable, or conduit to the wall, then plastering over flush with the original finish.

Special small sizes of conduit, tubing, and raceways are permitted for these extensions, which are not permitted for other wiring, either open or concealed. Standard "all metal" wiring systems are permitted for underplaster extensions, but usually they are too large to be properly covered.

NON-METALLIC SURFACE EXTENSIONS. For short extensions from existing convenience outlets, a non-metallic surface extension material may be used for short runs not extending beyond the room in which they originate. The inclusion in the National Electrical Code of properly supervised material of this nature is intended to discourage the dangerous practice of using lamp cord for this purpose.

The material consists of two insulated conductors enclosed in a suitable fabric or other covering so designed that it may be tacked or otherwise fastened to walls or trim. It can be used only for circuits up to 150 volts in residences or offices.

BARE-CONDUCTOR FEEDERS. Conductors serving as main feeders in buildings of fire-resisting construction may be run without insulating coverings when suitable provisions are made in the design of the building. Such installations are treated as special cases and are subject to review by the authority enforcing the Code in the locality where the installation is made.

Such installations may not be used for circuits in excess of 600 volts. The conductors are supported on insulators in properly guarded shafts or channels, provided with fire cut-offs at each floor.

The conductors may be bars of round or rectangular section, or they may be of tubing. Copper tubing made in standard iron-pipe sizes is frequently used. Having a tubular section, this has the advantage in the larger sizes on a-c systems of reducing the skin-effect losses, which may be quite appreciable on conductors having an area above 500,000 cir mils.

The maximum current loading shall not exceed 1000 amp per sq in. of cross-sectional area of conductor in unventilated enclosures and 1200 amp per sq in. in ventilated enclosures.

Table 7. Types of Building Wire Recognized by the National Electrical Code

Name	Type Letter	Maximum Operating Temperature	Insulation	Thickness of Insulation		Outer Covering	Special Provisions and Use
Rubber-covered Fixture Wire, Solid or Stranded	RF-64	60° C 140° F	Code Rubber	18	1/64 in.	Cotton	Fixture wiring limited to 300 volts
	RF-32	60° C 140° F	Code Rubber	18–16	2/64 "	Cotton	Fixture wiring and as permitted in Sect. 3103 NEC
Rubber-covered Fixture Wire, Flexible Stranding	FF-64	60° C 140° F	Code Rubber	18	1/64 "	Cotton	Fixture wiring limited to 300 volts
	FF-32	60° C 140° F	Code Rubber	18–16	2/64 "	Cotton	Fixture wiring
Thermoplastic-covered Fixture Wire, Solid or Stranded	TF	60° C 140° F	Thermoplastic	18–16	2/64 "	None	Fixture wiring and as permitted in Sect. 3103 NEC
Thermoplastic-covered Fixture Wire, Flexible Stranding	TFF	60° C 140° F	Thermoplastic	18–16	2/64 "	None	Fixture wiring
Cotton-covered Heat-resistant Fixture Wire	CF	90° C 140° F	Impregnated cotton	18–14	2/64 "	None	Fixture wiring limited to 300 volts
Asbestos-covered Heat-resistant Fixture Wire	AF	125° C 275° F	Impregnated asbestos	18–14	2/64 "	None	Fixture wiring limited to 300 volts
Code Rubber	R	60° C	Code Rubber	14–12 10 8–2 1–4/0 213–500 501–1000 1001–2000	2/64 " 3/64 " 4/64 " 5/64 " 6/64 " 7/64 " 8/64 "	Moisture-resistant flame-retardant fibrous covering	General use

Heat-resistant Rubber	RH	75° C 167° F	Heat-resistant rubber	14–12 10 8–2 1–4/0 213–500 501–1000 1001–2000	2/64 " 3/64 " 4/64 " 5/64 " 6/64 " 7/64 " 8/64 "	Moisture-resistant flame-retardant fibrous covering	General use
Moisture-resistant Rubber	RW	60° C 140° F	Moisture-resistant rubber	14–12 8–2 1–4/0 213–500 501–1000 1001–2000	3/64 " 4/64 " 5/64 " 6/64 " 7/64 " 8/64 "	Moisture-resistant flame-retardant fibrous covering	General use and in wet locations
Latex Rubber	RU	60° C 140° F	90% unmilled grainless rubber	14–12 8–6	18 mils 25 "	Moisture-resistant flame-retardant fibrous covering	General use
Thermoplastic	T	60° C 140° F	Flame-retardant thermoplastic	14–10 8 6–2 1–4/0 213–500 501–1000 1001–2000	2/64 in. 3/64 " 4/64 " 5/64 " 6/64 " 7/64 " 8/64 "	None	General use, No. 14 to 4/0 inclusive, Open work No. 14 to 2,000,000 CM
Moisture-resistant Thermoplastic	TW	60° C 140° F	Flame-retardant moisture-resisting thermoplastic	14–10 8 6–2 1–4/0 213–500 501–1000 1001–2000	2/64 " 3/64 " 4/64 " 5/64 " 6/64 " 7/64 " 8/64 "	None	General use and in wet locations, No. 14 to 4/0 inclusive, open work No. 14 to 2,000,000 CM
Thermoplastic Asbestos	TA	90° C 194° F	Thermoplastic and asbestos	 14–8 6–2 1–4/0	*Thermoplastic* / *Asbestos* 20 mils / 20 mils 30 " / 25 " 40 " / 30 "	Flame-retardant cotton braid	Switchboard wiring only

Table 7. Types of Building Wire Recognized by the National Electrical Code—*Continued*

Name	Type Letter	Maximum Operating Temperature	Insulation	Thickness of Insulation	Outer Covering	Special Provisions and Use
Varnished Cambric	V	85° C 185° F	Varnished cambric	14–8 3/64 in. 6–2 4/64 " 1–4/0 5/64 " 213–500 6/64 " 501–1000 7/64 " 1001–2000 8/64 "	Fibrous covering or lead sheath	Dry locations only, smaller than No. 6 by special permission
Asbestos Varnished Cambric	AVA and AVL	110° C 230° F	Impregnated asbestos and varnished cambric	(dimensions in mils) VC / AVA Asb / AVL Asb 14–8 sol. only 30 20 25	AVA-asbestos braid	AVA dry locations only
				1st Asb / VC / AVA 2d Asb / AVL 2d Asb 14–8 10 30 15 25 6–2 15 30 20 25 1–4/0 20 30 30 30 213–500 25 40 40 40 501–1000 30 40 40 40 1001–2000 30 50 50 50	AVL-asbestos braid and lead sheath	AVL wet locations
Asbestos Varnished Cambric	AVB	90° C 194° F	Impregnated asbestos and varnished cambric	VC / Asb 14–8 sol. only 30 20 6–2 " " 40 30 1–4/0 " " 40 40	Flame-retardant cotton braid (Switchboard wiring)	Dry locations only
				Asb / VC / 2d Asb 14–8 10 30 15 6–2 15 30 20 1–4/0 20 30 30 213–500 25 40 40 501–1000 30 40 40 1001–2000 30 50 50	Flame-retardant cotton braid	
Asbestos	A	200° C 392° F	Asbestos	14 30 mils 12–8	Without asbestos braid	Dry locations only, not for general use, in raceways only for leads to or within apparatus limited to 300 volts

Asbestos	AA	200° C	Asbestos	14–8 6–2 1–4/0	30 mils 30 " 60 "	With asbestos braid	Dry locations only, open wiring, not for general use, in raceways only for leads to or within apparatus limited to 300 volts
Asbestos	AI	125° C 257° F	Impregnated asbestos	14 12–8	30 " 40 "	Without asbestos braid	Dry locations only, not for general use, in raceways only for leads to or within apparatus limited to 300 volts
Asbestos	AIA	125° C 257° F	Impregnated asbestos	14 12–8 6–2 1–4/0 213–500 501–1000	Sol. / Str. 30 mils / 30 mils 30 " / 40 " 40 " / 60 " 60 " / 75 " .. " / 90 " .. " / 105 "	With asbestos braid	Dry locations only, open wiring, not for general use, in raceways only for leads to or within apparatus
Paper		85° C 185° F	Paper			Lead sheath	For underground service conductors or by special permission
Slow-burning	SB	90° C	3 braids of impregnated fire-resistant cotton thread	14–10 8–2 1–4/0 213–500 501–1000 1001–2000	3/64 in. 4/64 " 5/64 " 6/64 " 7/64 " 8/64 "	Outer cover finished smooth and hard	Dry locations only, open wiring, in raceways where temperature will exceed those permitted for rubber- or varnished-cambric covered conductors
Slow-burning Weatherproof	SBW	90° C 194° F	2 layers impregnated cotton braid	14–10 8–2 1–4/0 213–500 501–1000 1001–2000	3/64 " 4/64 " 5/64 " 6/64 " 7/64 " 8/64 "	Outer fire-retardant coating	Dry locations only, open wiring only
Weatherproof	WP	80° C 176° F	At least 3 impregnated cotton braids or equivalent				Open wiring by special permission or when other insulations not suitable for existing conditions

SERVICE ENTRANCE CABLE. This is a wire assembly or cable designed to be used as a feeder running from the service company's lines outside the building to the service equipment within the building. It is used in place of conduit and is fastened to the side of the building in a similar manner. The most frequent application is for residence services, where it is less conspicuous than conduit. There are no restrictions, however, regarding its use for any type of service up to 600 volts.

Entrance cable is made in protected and unprotected types and may have a bare neutral wire. The protected type has a heavy interlocked armor over the assembly, and the unprotected type usually has the bare neutral, composed of several small wires, wound helically about the insulated conductors, over which may be placed a metal or fiber tape and protecting braids. The unprotected type is approved only for use where not subject to mechanical injury.

Underground service entrance cables are also made in several types, with both metallic and non-metallic coverings.

Service entrance cables (Types SE and ASE) may be used in interior wiring if all the conductors of the cable are of the rubber-covered or thermoplastic type; but if without an individually insulated grounded conductor, these cables may be used only for range or domestic water-heater circuits, or as feeders from master service cabinets to supply other buildings, provided the following conditions are met: (*a*) the cable has a final non-metallic covering, such as a braid or sheath; (*b*) the supply is alternating current not exceeding 150 volts to ground; (*c*) no domestic water-heater is supplied through a conductor without individual insulation.

VERTICAL RISER CABLES. In tall buildings it is common practice to locate the service transformers on a floor about half way up the building and distribute low-voltage current from this point. The service is then carried from the basement, as an extension of the street mains, up through the building to the transformer vault by means of a vertical riser cable. The voltage most frequently used is 13,200 volts.

Vertical riser cables are armored with steel wires wound helically about the insulated conductors and so designed that the entire cable may be suspended from its upper end by means of the armor wire. Inasmuch as long cables suspended from one end tend to untwist and cause damage to the insulation, the design of the complete cable must be such as to prevent untwisting.

The insulation may be rubber, varnished cambric, or paper, but is usually either rubber or varnished cambric so made as to prevent "bleeding" of the compound used between the tapes as a slipper.

82. WIRES AND CABLES

A large number of types of wire and cable are used for interior wiring, and as new insulating and covering materials are developed, new types are being added. Some of these types are special-purpose wires, and others are suitable for general use in any approved wiring method. (See also Article 81.)

Table 7 lists the approved types of wire recognized by the 1947 Code with their more important characteristics, such as permitted maximum operating temperature and limitations of use. The insulation thicknesses in this table are given only for voltages up to 600 volts; for higher voltages in the types permitted for interior wiring see Article 66 and Table 8.

Carrying capacities are given in Tables 14, 15, and 16.

Lead-covered cables, for use in wet locations or other places specified in the Code, consist of one or more conductors insulated with one of the standard types of insulation and covered with a lead sheath. The type letter designation of lead-covered cable is determined by adding L after the type letter of the wire or cable over which the lead sheath is placed, e.g., Type RL for rubber lead, Type VL for varnished-cambric lead.

Flexible cords are not classed as a wiring material by the National Electrical Code but are considered as equipment and may be used only for (1) pendants, (2) connection of portable lamps or appliances, (3) elevator cables, (4) wiring of cranes or hoists, (5) connection of stationary equipment to facilitate their interchange, or (6) prevention of the transmission of noise or vibration. Flexible cords should not be used (1) as a substitute for fixed wiring, (2) where run through holes in walls, ceilings, or floors, (3) where run through doorways, windows, or similar openings, (4) where attached to building surfaces, or (5) where concealed behind building walls, ceilings, or floors.

The various types of flexible cords and their uses are given in Table 9. The allowable current-carrying capacities are given in Table 16.

SHIELDING OF RUBBER-INSULATED CONDUCTORS OVER 600 VOLTS. If non-leaded, fibrous-covered, rubber-insulated conductors for permanent installation operate at voltages higher than those indicated in Table 10 and under the conditions men-

Table 8. Thickness of Asbestos and Varnished-cambric Insulation for Single-conductor Cable, Types AVA, AVB, and AVL, in Mils

Conductor Size, A.W.G. or MCM	First Wall Asbestos	Varnished Cambric					Second Wall Asbestos
	For Voltages Not Exceeding						
	1000–5000	1000	2000	3000	4000	5000	1000–5000
14–2	15	45	60	80	100	120	25
1–4/0	20	45	60	80	100	120	30
213–500	25	45	60	80	100	120	40
501–1000	30	45	60	80	100	120	40
1001–2000	30	55	75	95	115	140	50

Table 9. Flexible Cords *

Trade Name	Type Letter	Size A.W.G.	Number of Conductors	Insulation	Braid on Each Conductor	Outer Covering	Use		
Asbestos-covered Tinsel Cord	AT See Note 3	27	2 or 3	Rubber and asbestos	Cotton	None	Attached to a device	Dry places	Not hard usage
					None	Cotton or rayon			
Cotton-covered Tinsel Cord	CT See Note 3	27	2 or 3	Rubber	Cotton	Cotton or rayon	Attached to a device	Dry places	Not hard usage
Rubber-jacketed Tinsel Cord	ATJ See Note 3	27	2 or 3	Rubber and asbestos	None	Rubber	Attached to a device	Damp places	Not hard usage
	CTJ See Note 3			Rubber	Cotton				
Asbestos-covered-Heat-resistant Cord	AFC	18, 16, 14	2 or 3	Impregnated asbestos	Cotton or rayon	None	Pendant	Dry places	Not hard usage
	AFPO		2						
	AFPD		2 or 3		None	Cotton, rayon, or saturated asbestos			
Cotton-covered-Heat-resistant Cord	CFC	18, 16, 14	2 or 3	Impregnated cotton	Cotton or rayon	None	Pendant	Dry places	Not hard usage
	CFPO		2						
	CFPD		2 or 3		None	Cotton or rayon			
Parallel Cord	PO-64	18	2	Rubber	Cotton	Cotton or rayon	See Note 2	Dry places	Not hard usage
	PO-32	18, 16					Pendant or portable		
	PO	14 and over							
All Rubber Parallel Cord	POSJ-64	18	2	Rubber	None	Rubber	Pendant or portable	Damp places	Not hard usage
	POSJ-32	18, 16							
All Plastic Parallel Cord	POT-64	18	2	Thermoplastic	None	Thermoplastic	Pendant or portable	Damp places	Not hard usage
	POT-32	18, 16							
Lamp Cord	C	18 and over	2 or more	Rubber	Cotton	None	Pendant or portable	Dry places	Not hard usage
Armored Cord	CA	18, 16, 14	2	Rubber	Cotton	Fibrous and metal armor	Pendant or portable	Dry places	Hard usage
Twisted Portable Cord	PD	18 and over	2 or more	Rubber	Cotton	Cotton or rayon	Pendant or portable	Dry places	Not hard usage
Reinforced Cord	P-64	18	2 or more	Rubber	Cotton	Cotton over rubber filler	Pendant or portable	Dry places	Not hard usage
	P-32	18, 16							Hard usage
	P	14 and over							

Table 9. Flexible Cords *—*Continued*

Trade Name	Type Letter	Size A.W.G.	Number of Conductors	Insulation	Braid on Each Conductors	Outer Covering	Use		
Moisture-proof Reinforced Cord	PWP-64	18	2 or more	Rubber	Cotton	Cotton, moisture-resistant finish over rubber filter	Pendant or portable	Damp places	Not hard usage
	PWP-32	18, 16							
	PWP	14 and over							Hard usage
Braided Heavy-duty Cord	K See Note 4	18 and over	2 or more	Rubber	Cotton	Two cotton, moisture-resistant finish See Note 5	Pendant or portable	Damp places	Hard usage
Vacuum Cleaner Cord	SV	18	2	Rubber	None	Rubber	Pendant or portable	Damp places	Hard usage
	SVT			Thermoplastic		Thermoplastic			
Junior Hard-service Cord	SJ	18, 16	2, 3, or 4	Rubber	None	Rubber	Pendant or portable	Damp places	Hard usage
	SJO			Thermoplastic		Oil-resistant compound			
	SJT					Thermoplastic			
Hard-service Cord	S See Note 6	18 to 10 incl.	2 or more	Rubber	None	Rubber	Pendant or portable	Damp places	Extra-hard usage
	SO					Oil-resistant compound			
	ST			Thermoplastic		Thermoplastic			
Rubber-jacketed Heat-resistant Cord	AFSJ	18, 16	2 or 3	Impregnated asbestos	None	Rubber	Portable	Damp places	Portable heaters
	AFS	18, 16, 14							
Heater Cord	HC	18, 17, 16, 15, 14	2 or more	Rubber and asbestos	Cotton	None	Portable	Dry places	Portable heaters
	HPD				None	Cotton or rayon			
Rubber-jacketed Heater Cord	HSJ	18, 17, 16, 15, 14	2 or more	Rubber and asbestos	None	Cotton and rubber	Portable	Damp places	Portable heaters
Heat- and Moisture-resistant Cord	AVPO	18 to 10 incl.	2	Asbestos and varnished cambric	None	Asbestos, Flame-retardant, moisture-resistant	Pendant or portable	Damp places	Not hard usage
	AVPD		2 or 3						
Elevator Cable	E See Note 7	18 and over	2 or more	Rubber	Cotton	Three cotton, outer one flame-retardant and moisture-resistant See Note 5	Elevator lighting and control	Non-hazardous locations	
	EO See Note 7					One cotton and a Neoprene jacket See Note 5		Hazardous locations	

* National Electrical Code.

NOTES 1. Except for types AFPO, CFPO, PO-64, PO-32, PO, POSJ-64 and POSJ-32, and POT-64, POT-32, and AVPO individual conductors are twisted together.

2. Type PO-64 is for use only with portable lamps, portable radio-receiving appliances, portable clocks, and similar appliances which are not likely to be moved frequently and where appearance is a consideration.

3. Types AT, CT, ATJ, and CTJ are suitable for use in lengths not exceeding 8 ft when attached directly, or by means of a special type of plug, to a portable appliance rated at 50 watts or less and of such a nature that extreme flexibility of the cord is essential. Types AT and ATJ are for use only with heating appliances.

4. Type K is suitable for use on theater stages.

5. Rubber-filled or varnished-cambric tapes may be substituted for the inner braids.

6. Types S, SO, and ST are suitable for use on theater stages, in garages, and elsewhere where flexible cords are permitted by this code.

7. Types E and EO may have a composite assembly of steel and copper strands in the make-up of the individual conductors or may have one or more supporting fillers of cotton or hemp rope, or of cotton-covered or rubber-covered steel steel wire laid up with the conductors under the outer covering of the cable. In cables containing six or more conductors the steel supporting strands shall run straight through and not be cabled with the conductors.

tioned, they shall be of a type having a metallic or semiconducting shielding for the purpose of confining the electric field. If a semiconducting shield is used, it must be a type approved by the Underwriters' Laboratories. This rule in the Code is necessary because of the susceptibility of rubber-covered conductors to damage from ozone which is generated in the vicinity of cables operating at the higher voltages.

Table 10. Shielding

Method of Installation	Voltage, in Kilovolts, above Which Shielding is Required	
	Neutral Grounded	Neutral Ungrounded
In ducts or metal conduit in wet or damp locations	2	2
In ducts or metal conduit in permanently dry locations	6 *	3 *
On insulators, only if multiple-conductor	6	3

* In some cases supposedly dry metallic conduits may contain sufficient moisture to allow formation of ozone and thus damage non-shielded cable at voltages above 2000. For conservation practice it may be desirable to shield below the limits given.

Grounding of Shielding. The shielding of shielded cable, whether of metallic or semiconducting type, must be grounded, preferably at two or more points. Stress cones must be made at all terminations of shielding, as in potheads and joints. Ungrounded shielding normally has a potential above ground and should be regarded as a live conductor.

For grounding practice of interior wiring systems see Article 87.

Table 11. Dimensions of Rubber-covered and Thermoplastic-covered Conductors *

Size, A.W.G. MCM	Types RF-32, R, RH, RW		Types TF, T, TW		Type RU	
	Approx. Diam., Inches	Approx. Area, Square Inches	Approx. Diam., Inches	Approx. Area, Square Inches	Approx. Diam., Inches	Approx. Area, Square Inches
18	0.146	0.0167	0.106	0.0088	....	....
16	0.158	0.0196	0.118	0.0109	....	....
14	0.171	0.0230 †	0.131	0.0135	0.146	0.0167
12	0.188	0.0278 †	0.148	0.0172	0.163	0.0208
10	0.242	0.0460	0.169	0.0224	0.184	0.0266
8	0.311	0.0760	0.228	0.0408	0.228	0.0406
6	0.397	0.1238	0.323	0.0819	0.317	0.0787
4	0.452	0.1605	0.372	0.1087	....	....
3	0.481	0.1817	0.401	0.1263	....	....
2	0.513	0.2067	0.433	0.1473	....	....
1	0.588	0.2715	0.508	0.2027	....	....
0	0.629	0.3107	0.549	0.2367	....	....
00	0.675	0.3578	0.595	0.2781	....	....
000	0.727	0.4151	0.647	0.3288	....	....
0000	0.785	0.4840	0.705	0.3904	....	....
250	0.868	0.5917	0.788	0.4877	....	....
300	0.933	0.6837	0.843	0.5581	....	....
350	0.985	0.7620	0.895	0.6291	....	....
400	1.032	0.8365	0.942	0.6969	....	....
500	1.119	0.9834	1.029	0.8316	....	....
600	1.233	1.1940	1.143	1.0261	....	....
700	1.304	1.3355	1.214	1.1575	....	....
750	1.339	1.4082	1.249	1.2252	....	....
800	1.372	1.4784	1.282	1.2908	....	....
900	1.435	1.6173	1.345	1.4208	....	....
1000	1.494	1.7531	1.404	1.5482	....	....
1250	1.676	2.2062	1.577	1.9532	....	....
1500	1.801	2.5475	1.702	2.2748	....	....
1750	1.916	2.8895	1.817	2.5930	....	....
2000	2.021	3.2079	1.922	2.9013	....	....

* National Electrical Code.

† The diameters of Type RW in Nos. 14 and 12 are 0.204 and 0.221, respectively, and the areas are 0.0327 and 0.0384, respectively.

NOTE. No. 18 to No. 8, solid; No. 6 and larger, stranded.

Table 12. Dimensions of Lead-covered Conductors *

Types RL and RHL

Size, A.W.G. MCM	Single-conductor		Two-conductor		Three-conductor	
	Diam., Inches	Area, Square Inches	Diam., Inches	Area, Square Inches	Diam., Inches	Area, Square Inches
14 (2/64)	0.25	0.049	0.25 × 0.41	0.089	0.52	0.212
12 (2/64)	0.26	0.053	0.26 × 0.47	0.108	0.55	0.238
14 (3/64)	0.28	0.062	0.28 × 0.47	0.115	0.59	0.273
12 (3/64)	0.29	0.066	0.31 × 0.54	0.146	0.62	0.301
10 (3/64)	0.35	0.096	0.35 × 0.59	0.180	0.68	0.363
8	0.41	0.132	0.41 × 0.71	0.255	0.82	0.528
6	0.49	0.188	0.49 × 0.86	0.369	0.97	0.738
4	0.55	0.237	0.54 × 0.96	0.457	1.08	0.916
2	0.60	0.283	0.61 × 1.08	0.578	1.21	1.146
1	0.67	0.352	0.70 × 1.23	0.756	1.38	1.49
0	0.71	0.396	0.74 × 1.32	0.859	1.47	1.70
00	0.76	0.454	0.79 × 1.41	0.980	1.57	1.94
000	0.81	0.515	0.84 × 1.52	1.123	1.69	2.24
0000	0.87	0.593	0.90 × 1.64	1.302	1.85	2.68
250	0.98	0.754	....	....	2.02	3.20
300	1.04	0.85	....	....	2.15	3.62
350	1.10	0.95	....	....	2.26	4.02
400	1.14	1.02	....	....	2.40	4.52
500	1.23	1.18	....	....	2.59	5.28

NOTE. No. 14 to No. 8, solid conductors; No. 6 and larger, stranded conductors.
* National Electrical Code.

Table 13. Dimensions of Asbestos–varnished-cambric-insulated Conductors *

Size, A.W.G. MCM	Type AVA		Type AVB		Type AVL	
	Approx. Diam., Inches	Approx. Area, Square Inches	Approx. Diam., Inches	Approx. Area, Square Inches	Approx. Diam., Inches	Approx. Area, Square Inches
14	0.245	0.047	0.205	0.033	0.320	0.080
12	0.265	0.055	0.225	0.040	0.340	0.091
10	0.285	0.064	0.245	0.047	0.360	0.102
8	0.310	0.075	0.270	0.057	0.390	0.119
6	0.395	0.122	0.345	0.094	0.430	0.145
4	0.445	0.155	0.395	0.123	0.480	0.181
2	0.505	0.200	0.460	0.166	0.570	0.255
1	0.585	0.268	0.540	0.229	0.620	0.300
0	0.625	0.307	0.580	0.264	0.660	0.341
00	0.670	0.353	0.625	0.307	0.705	0.390
000	0.720	0.406	0.675	0.358	0.755	0.447
0000	0.780	0.478	0.735	0.425	0.815	0.521
250	0.885	0.616	0.855	0.572	0.955	0.715
300	0.940	0.692	0.910	0.649	1.010	0.800
350	0.995	0.778	0.965	0.731	1.060	0.885
400	1.040	0.850	1.010	0.800	1.105	0.960
500	1.125	0.995	1.095	0.945	1.190	1.118
550	1.165	1.065	1.135	1.01	1.265	1.26
600	1.205	1.140	1.175	1.09	1.305	1.34
650	1.240	1.21	1.210	1.15	1.340	1.41
700	1.275	1.28	1.245	1.22	1.375	1.49
750	1.310	1.35	1.280	1.29	1.410	1.57
800	1.345	1.42	1.315	1.36	1.440	1.63
850	1.375	1.49	1.345	1.43	1.470	1.70
900	1.405	1.55	1.375	1.49	1.505	1.78
950	1.435	1.62	1.405	1.55	1.535	1.85
1000	1.465	1.69	1.435	1.62	1.565	1.93

NOTE. No. 14 to No. 8, solid, No. 6 and larger, stranded; except AVL where all sizes are stranded.
* National Electrical Code.

Table 14. Allowable Current-carrying Capacities of Conductors, in Amperes *

Not More Than Three Conductors in Raceway† or Cable

(Based on Room Temperature of 30° C, 86° F)

Size, A.W.G. MCM	Rubber Type R, Type RW, Type RU, (14–6); Thermoplastic Type T, (14–410), Type TW (14–410)	Rubber Type RH	Paper; Thermoplastic-asbestos Type TA; Varnished-cambric Type V; Asbestos-varnished-cambric Type AVB	Asbestos-varnished-cambric Type AVA, Type AVL	Impregnated Asbestos Type AI (14–8), Type AIA	Asbestos Type A (14–8), Type AA
14	15	15	25	30	30	30
12	20	20	30	35	40	40
10	30	30	40	45	50	55
8	40	45	50	60	65	70
6	55	65	70	80	85	95
4	70	85	90	105	115	120
3	80	100	105	120	130	145
2	95	115	120	135	145	165
1	110	130	140	160	170	190
0	125	150	155	190	200	225
00	145	175	185	215	230	250
000	165	200	210	245	265	285
0000	195	230	235	275	310	340
250	215	255	270	315	335	...
300	240	285	300	345	380	...
350	260	310	325	390	420	...
400	280	335	360	420	450	...
500	320	380	405	470	500	...
600	355	420	455	525	545	...
700	385	460	490	560	600	...
750	400	475	500	580	620	...
800	410	490	515	600	640	...
900	435	520	555	...	...	...
1000	455	545	585	680	730	...
1250	495	590	645	...	...	...
1500	520	625	700	785	...	...
1750	545	650	735	...	...	...
2000	560	665	775	840	...	...

Correction Factor for Room Temperatures over 30° C, 86° F

C. F.						
40 104	0.82	0.88	0.90	0.94	0.95	...
45 113	0.71	0.82	0.85	0.90	0.92	...
50 122	0.58	0.75	0.80	0.87	0.89	...
55 131	0.41	0.67	0.74	0.83	0.86	...
60 140	...	0.58	0.67	0.79	0.83	0.91
70 158	...	0.35	0.52	0.71	0.76	0.87
75 167	...	...	0.43	0.66	0.72	0.86
80 176	...	...	0.30	0.61	0.69	0.84
90 194	...	...	...	0.50	0.61	0.80
100 212	...	...	...	...	0.51	0.77
120 248	...	...	...	...	...	0.69
140 284	...	...	...	...	...	0.59

* National Electrical Code. † All "pull-in" and "pull-out" systems.

Table 15. Allowable Current-carrying Capacities of Conductors, in Amperes *

Single Conductor in Free Air

(Based on Room Temperature of 30° C, 86° F)

Size, A.W.G. MCM	Rubber Type R, Type RW, Type RU (14–6) / Thermoplastic Type T, Type TW	Rubber Type RH	Thermoplastic-asbestos Type TA / Varnished-cambric Type V / Asbestos-varnished-cambric Type AVB	Asbestos-varnished-cambric Type AVA, Type AVL	Impregnated Asbestos Type AI (14–8), Type AIA	Asbestos Type A (14–8), Type AA	Slow-burning Type SB / Weatherproof Type WP, Type SBW
14	20	20	30	40	40	45	30
12	25	25	40	50	50	55	40
10	40	40	55	65	70	75	55
8	55	65	70	85	90	100	70
6	80	95	100	120	125	135	100
4	105	125	135	160	170	180	130
3	120	145	155	180	195	210	150
2	140	170	180	210	225	240	175
1	165	195	210	245	265	280	205
0	195	230	245	285	305	325	235
00	225	265	285	330	355	370	275
000	260	310	330	385	410	430	320
0000	300	360	385	445	475	510	370
250	340	405	425	495	530	....	410
300	375	445	480	555	590	....	460
350	420	505	530	610	655	....	510
400	455	545	575	665	710	....	555
500	515	620	660	765	815	....	630
600	575	690	740	855	910	....	710
700	630	755	815	940	1005	....	780
750	655	785	845	980	1045	....	810
800	680	815	880	1020	1085	....	845
900	730	870	940	....	....	....	905
1000	780	935	1000	1165	1240	....	965
1250	890	1065	1130	....	....	....	....
1500	980	1175	1260	1450	....	....	1215
1750	1070	1280	1370	....	....	....	....
2000	1155	1385	1470	1715	....	....	1405

Correction Factor for Room Temperatures over 30° C, 86° F

C. F.							
40 104	0.82	0.88	0.90	0.94	0.95	....	....
45 113	0.71	0.82	0.85	0.90	0.92	....	....
50 122	0.58	0.75	0.80	0.87	0.89	....	....
55 131	0.41	0.67	0.74	0.83	0.86	....	....
60 140	....	0.58	0.67	0.79	0.83	0.91	....
70 158	....	0.35	.052	0.71	0.76	0.87	....
75 167	....	....	0.43	0.66	0.72	0.86	....
80 176	....	....	0.30	0.61	0.69	0.84	....
90 194	....	....	....	0.50	0.61	0.80	....
100 212	....	....	....	....	0.51	0.77	....
120 248	....	....	....	....	....	0.69	....
140 284	....	....	....	....	....	0.59	....

* National Electrical Codes.

NOTES to Tables 14 and 15. 1. *Aluminum Conductors.* For aluminum conductors, the allowable current-carrying capacities shall be taken as 84 per cent of those given in the table for the respective sizes of copper conductor with the same kind of insulation.

2. *Bare Conductors.* If bare conductors are used with insulated conductors, their allowable current-carrying capacity shall be limited to that permitted for the insulated conductor with which they are used.

3. *Application of Table.* For open wiring on insulators and for concealed knob and tube work, the allowable current-carrying capacities of Table 15 shall be used. For all other recognized wiring methods, the allowable current-carrying capacities of Table 14 shall be used.

4. *More Than Three Conductors in a Raceway.* Table 14 gives the allowable current-carrying capacity for not more than 3 conductors in a raceway or cable. If the number of conductors in a raceway or cable is from 4 to 6, the allowable current-carrying capacity of each conductor shall be reduced to 80 per cent of the values in Table 14. If the number of conductors in a raceway or cable is from 7 to 9, the allowable current-carrying capacity of each conductor shall be reduced to 70 per cent of the values in Table 14.

5. *Neutral Conductor.* A neutral conductor which carries only the unbalanced current from other conductors, as in the case of normally balanced circuits of three or more conductors, shall not be counted in determining current-carrying capacities as provided for in the preceding paragraph.

In a 3-wire circuit consisting of two phase wires and the neutral of a 4-wire, 3-phase system, a common conductor carries approximately the same current as the other conductors and is not therefore considered a neutral conductor.

Table 16. Allowable Current-carrying Capacity of Flexible Cord and Fixture Wire, in Amperes *

(Based on Room Temperature of 30° C, 86° F)

Size, A.W.G.	Flexible Cord						Fixture Wire	
	Rubber and Cotton Types CT, CTJ / Rubber and Asbestos Types AT, ATJ	Rubber Types PO, C, PD, P, PWP, K, E, EH / Armored Type CA	Rubber Types S, SO, SJ, SJO, SV, POSJ / Thermoplastic Types ST, SJT, SVT, POT	Types AFS, AFSJ, HC, HPD, HSJ	Types AVPO, AVPD	Cotton Types CFC,† CFPO,† CFPD † / Asbestos Types AFC, † AFPO,† AFPD †	Rubber / Types RF-64, RF-32, FF-64, FF-32	Thermoplastic Types TF, TFF / Cotton Type CF † / Asbestos Type AF †
27 ‡	0.5	....	....	....	....	....	....	....
18	....	5	7	10	17	6	5	6
17	....	....	....	12	....	....	....	....
16	....	7	10	15	22	8	7	8
15	....	....	....	17	....	....	....	....
14	....	15	15	20	28	17	....	17
12	....	20	20	....	36	....	....	....
10	....	25	25	....	47	....	....	....
8	....	35	....	....	....	....	....	....
6	....	45	....	....	....	....	....	....
4	....	60	....	....	....	....	....	....
2	....	80	....	....	....	....	....	....

* National Electrical Code.

† These types are used almost exclusively in fixtures where they are exposed to high temperatures and ampere ratings are assigned accordingly.

‡ Tinsel cord.

MORE THAN THREE CONDUCTORS IN A CORD

Table 16 gives the allowable current-carrying capacities for not more than three current-carrying conductors in a cord. If the number of current-carrying conductors in a cord is from four to six the allowable current-carrying capacity of each conductor shall be reduced to 80 per cent of the values in the table.

83. WIRING ACCESSORIES

OUTLET, SWITCH, AND JUNCTION BOXES. Standard boxes of various designs and sizes are available for use in connection with the systems of wiring permitted by the National Electrical Code. They must be used for housing switches, as outlets for attaching fixtures, or as junction boxes for housing joints or splices. Metal boxes are most commonly used and are required for all raceway or armored cable systems. Non-metallic boxes may be used with open wiring, knob and tube work, non-metallic sheathed cable, and non-metallic waterproof wiring.

In wet or hazardous locations special water- or vaportight fittings are required.

SWITCHBOARDS AND PANELBOARDS. There is no clearly defined distinction between switchboards and panelboards. In general, however, switching equipment and protection equipment housed in metal cabinets are considered panelboards and are used almost exclusively for interior wiring, except where large blocks of power are being used. For large distribution centers heavy switchgear is mounted on "switchboards" or in special cabinets usually located in rooms or enclosures accessible only to authorized persons.

Panelboards contain switches, fuses, or other protection devices and are available for any required number and types of circuits within the limits of the Code requirements. The switches are mounted so as to be available for operation, but fuses and live parts are enclosed so that they are accessible only to authorized persons.

SWITCHES. A wide variety of switches are available for controlling any of the common types of circuits. They are made in combination with overload protection devices such as fuses or circuit breakers and may be had either for manual operation or for operation by remote control.

The more important requirements regarding the use of switches are as follows: (1) any switch shall open all ungrounded conductors of the circuit; (2) no switch shall disconnect the grounded conductor of a circuit unless the switch simultaneously opens all ungrounded conductors; (3) a single pole switch must never be used in a grounded conductor.

Circuit breakers, operated directly by a hand lever, will serve as a switch and are commonly used for this purpose on power circuits and feeders. Recent developments in small air circuit breakers suitable for mounting on panelboards make these devices popular for controlling lighting and appliance branch circuits.

OVERCURRENT PROTECTION DEVICES. Every circuit must be protected from overloads at one or more points. This is accomplished by means of fuses, circuit breakers, or thermal cut-outs, all of which are designed to open the circuit when the current reaches a predetermined value.

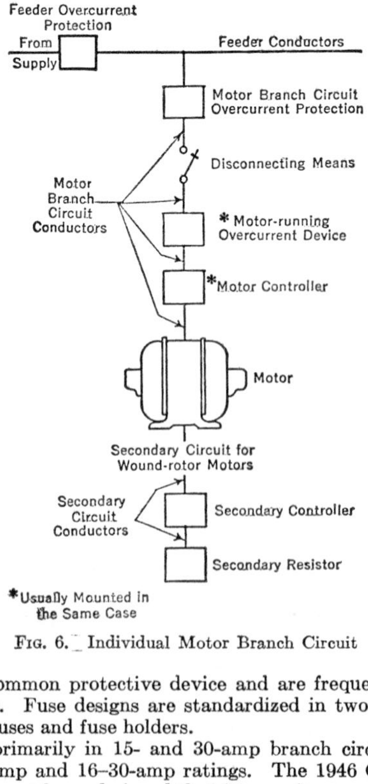

FIG. 6. Individual Motor Branch Circuit

Fuses are the most common protective device and are frequently used in series with other protective devices. Fuse designs are standardized in two classifications: (1) plug fuses and (2) cartridge fuses and fuse holders.

Plug fuses are used primarily in 15- and 30-amp branch circuits and are limited to 125 volts and to 0–15-amp and 16–30-amp ratings. The 1946 Code requires that in all new construction, fuses must be so designed that 30-amp fuses cannot be used in a 15-amp receptacle and be of the so-called non-tamperable type.

Cartridge fuses and fuse holders are made in the following classification:

Not Over 250 Volts, Amperes	Not Over 600 Volts, Amperes
0–30	0–30
31–60	31–60
61–100	61–100
101–200	101–200
201–400	201–400
401–600	401–600

Circuit breakers are magnetically operated and are available in a wide variety of designs and current ratings.

Thermal cut-outs are operated by means of a heating unit, which opens the circuit through a bimetallic element or a fusible link. These devices are particularly valuable for protection against small overloads of long duration.

The proper use of overcurrent protection equipment is one of the most important features of any wiring installation, and for interior wiring is covered fully by Article 240 of the National Electrical Code.

In general, overcurrent protective devices are located at the point where the conductors to be protected receive their supply and in the conductors designated in Table 17.

The rating or setting of a protective device is determined by the rating of the branch circuit or feeder. See Table 21 for lighting branch circuits and Tables 18, 19, and 20, and Fig. 6 for individual motor branch circuits. For two or more motors on one branch circuit special rules apply. See Article 430 of the National Electrical Code.

Table 17. Number of Overcurrent Units, Such as Trip Coils or Relays, for Protection of Circuits *

Systems	Number and Location of Overcurrent Units †
2-wire, single-phase a-c or d-c, ungrounded	Two (one in each conductor)
2-wire, single-phase a-c or d-c, one wire grounded	One in ungrounded conductor
2-wire, single-phase a-c or d-c, mid-point grounded	Two (one in each conductor)
2-wire, single-phase a-c derived from 3-phase, with ungrounded neutral	Two (one in each conductor)
2-wire, single-phase derived from 3-phase, grounded neutral system by using outside wires of 3-phase circuit	Two (one in each conductor)
3-wire, single-phase a-c or d-c, ungrounded neutral	Three (one in each conductor)
3-wire, single-phase a-c or d-c, grounded neutral	Two (one in each conductor except neutral conductor)
3-wire, 2-phase, a-c, common wire ungrounded	Three (one in each conductor)
3-wire, 2-phase, a-c, common wire grounded	Two (one in each conductor except common conductor)
4-wire, 2-phase, ungrounded, phases separate	Four (one in each conductor)
4-wire, 2-phase, grounded neutral, or 5-wire, 2-phase, grounded neutral	Four (one in each conductor except neutral conductor)
3-wire, 3-phase, ungrounded	Three (one in each conductor) ‡
3-wire, 3-phase, 1 wire grounded	Two (one in each ungrounded conductor)
3-wire, 3-phase, grounded neutral	Three (one in each conductor) ‡
3-wire, 3-phase, mid-point of one phase grounded	Three (one in each conductor) ‡
4-wire, 3-phase, grounded neutral	Three (one in each ungrounded conductor) ‡
4-wire, 3-phase, ungrounded neutral	Four (one in each conductor)

* National Electrical Code.

† An overcurrent unit may consist of a series overcurrent tripping device or the combination of a current transformer and a secondary overcurrent tripping device. Either two or three secondary overcurrent tripping devices may be used with three current transformers on a 3-phase system.

‡ When three current transformers are used instead of three series overcurrent tripping devices, the secondary tripping devices may consist of three secondary overcurrent tripping devices or two secondary overcurrent tripping devices with a residual current tripping device of a lower range.

Where standard devices are not available with three or four overcurrent units as required in the table, it is permissible to substitute two overcurrent units and one fuse where three overcurrent units are called for, two overcurrent units and two fuses where four overcurrent units are called for. The fuse or fuses are to be placed in the conductors not containing an overcurrent unit. This practice of substituting fuses for overcurrent units, however, is to be discouraged for obvious reasons.

Table 18. Maximum Rating or Setting of Motor-branch-circuit Protective Devices for Motors Marked with a Code Letter Indicating Locked Rotor Kva *

Type of Motor	Percentage of Full-load Current		
	Fuse Rating (See also Table 20, Columns 6, 7, 8, 9)	Circuit-breaker Setting	
		Instantaneous Type	Time-limit Type
All a-c single-phase and polyphase squirrel-cage and synchronous motors with full-voltage, resistor or reactor starting:			
Code Letter A	150	....	150
Code Letter B to E	250	....	200
Code Letter F to R	300	....	250
All a-c squirrel-cage and synchronous motors with auto-transformer starting:			
Code Letter A	150	....	150
Code Letter B to E	200	....	200
Code Letter F to R	250	....	200

* National Electrical Code.

The values given in the last column also cover the ratings of non-adjustable, time-limit types of circuit-breakers which may also be modified as in Section 4342 of NEC.

Synchronous motors of the low-torque, low-speed type (usually 450 rpm or lower), such as are used to drive reciprocating compressors, pumps, etc., which start up unloaded, do not require a fuse rating or circuit-breaker setting in excess of 200 per cent of full-load current.

For motors not marked with a Code Letter, see Table 19.

Table 19. Maximum Rating or Setting of Motor-branch-circuit Protective Devices for Motors Not Marked with a Code Letter Indicating Locked Rotor Kva *

Type of Motor	Percentage of Full-load Current		
	Fuse Rating (See also Table 20, Columns 6, 7, 8, 9)	Circuit-breaker Setting	
		Instantaneous Type	Time-limit Type
Single-phase, all types	300	....	250
Squirrel-cage and synchronous (full-voltage, resistor and reactor starting)	300	....	250
Squirrel-cage and synchronous (auto-transformer starting)			
Not more than 30 amp	250	....	200
More than 30 amp	200	....	200
High-reactance squirrel-cage			
Not more than 30 amp	250	....	250
More than 30 amp	200	....	200
Wound-rotor	150	....	150
Direct-current			
Not more than 50 hp	150	250	150
More than 50 hp	150	175	150

* National Electrical Code.

The values given in the last column also cover the ratings of non-adjustable, time-limit types of circuit-breakers which may also be modified as in Section 4342 of NEC.

Synchronous motors of the low-torque low-speed type (usually 450 rpm or lower), such as are used to drive reciprocating compressors, pumps, etc., which start up unloaded, do not require a fuse rating or circuit-breaker setting in excess of 200 per cent of full-load current.

For motors marked with a Code Letter, see Table 18.

Table 20. Conductor Sizes and Overcurrent Protection for Motors *

Full-load Current Rating of Motor, amperes	Minimum Size Conductor in Raceways † A.W.G. and MCM: Type R, Type T	Minimum Size Conductor in Raceways † A.W.G. and MCM: Type RH	For Running Protection of Motors: Maximum Rating of NEC Fuses, amperes	For Running Protection of Motors: Maximum Setting of Time-limit Protective Device, amperes	Maximum Allowable Rating of Branch Circuit Fuses: *With Code Letters* Single-phase and Squirrel-cage and Synchronous. Full Voltage, Resistor and Reactor Starting, Code Letters F to R inc. *Without Code Letters* Same as above	Maximum Allowable Rating of Branch Circuit Fuses: *With Code Letters* Single-phase and Squirrel-cage and Synchronous. Full Voltage, Resistor or Reactor Starting, Code Letters B to E inc. Autotransformer Starting, Code Letters F to R inc. *Without Code Letters* Squirrel-cage and Synchronous, Autotransformer Starting, High-reactance Squirrel-cage.‡ Both not more than 30 amperes	Maximum Allowable Rating of Branch Circuit Fuses: *With Code Letters* Squirrel-cage and Synchronous Autotransformer Starting, Code Letters B to E inc. *Without Code Letters* Squirrel-cage and Synchronous, Autotransformer Starting, High-reactance Squirrel-cage.‡ Both more than 30 amperes	Maximum Allowable Rating of Branch Circuit Fuses: *With Code Letters* All Motors, Code Letter A *Without Code Letters* D-c and Wound-rotor Motors
1	14	14	2	1.25	15	15	15	15
2	14	14	3	2.50	15	15	15	15
3	14	14	4	3.75	15	15	15	15
4	14	14	6	5.0	15	15	15	15
5	14	14	8	6.25	15	15	15	15
6	14	14	8	7.50	20	15	15	15
7	14	14	10	8.75	25	20	15	15
8	14	14	10	10.00	25	20	20	15
9	14	14	12	11.25	30	25	20	15
10	14	14	15	12.50	30	25	20	15
11	14	14	15	13.75	35	30	25	20
12	14	14	15	15.00	40	30	25	20
13	12	12	20	16.25	40	35	30	20
14	12	12	20	17.50	45	35	30	25
15	12	12	20	18.75	45	40	30	25
16	12	12	20	20.00	50	40	35	25
17	10	10	25	21.25	60	45	35	30
18	10	10	25	22.50	60	45	40	30
19	10	10	25	23.75	60	50	40	30
20	10	10	25	25.00	60	50	40	30
22	10	10	30	27.50	70	60	45	35
24	10	10	30	30.00	80	60	50	40
26	8	10	35	32.50	80	70	60	40
28	8	10	35	35.00	90	70	60	45
30	8	8	40	37.50	90	70	60	45
32	8	8	40	40.00	100	80	70	50
34	6	8	45	42.50	110	90	70	60
36	6	8	45	45.00	110	90	80	60
38	6	6	50	47.50	125	100	80	60
40	6	6	50	50.00	125	100	80	60
42	6	6	50	52.50	125	110	90	70
44	6	6	60	55.00	125	110	90	70
46	4	6	60	57.50	150	125	100	70
48	4	6	60	60.00	150	125	100	80
50	4	6	60	62.50	150	125	100	80
52	4	6	70	65.00	175	150	110	80
54	4	4	70	67.50	175	150	110	90
56	4	4	70	70.00	175	150	120	90
58	3	4	70	72.50	175	150	120	90
60	3	4	80	75.00	200	150	120	90
62	3	4	80	77.50	200	175	125	100
64	3	4	80	80.00	200	175	150	100
66	2	4	80	82.50	200	175	150	100
68	2	4	90	85.00	225	175	150	110
70	2	3	90	87.50	225	175	150	110
72	2	3	90	90.00	225	200	150	110
74	2	3	90	92.50	225	200	150	125
76	2	3	100	95.00	250	200	175	125
78	1	3	100	97.50	250	200	175	125
80	1	3	100	100.00	250	200	175	125
82	1	2	110	102.50	250	225	175	125
84	1	2	110	105.00	250	225	175	150
86	1	2	110	107.50	300	225	175	150

Table 20. Conductor Sizes and Overcurrent Protection for Motors*—*Continued*

Full-load Current Rating of Motor, amperes	Minimum Size Conductor in Raceways † A.W.G. and MCM		For Running Protection of Motors		Maximum Allowable Rating of Branch Circuit Fuses			
	Type R, Type T	Type RH	Maximum Plating of NEC Fuses amperes	Maximum Setting of Time-limit Protective Device, amperes	*With Code Letters* Single-phase and Squirrel-cage and Synchronous. Full Voltage, Resistor and Reactor Starting, Code Letters F to R inc. *Without Code Letters* Shown as above	*With Code Letters* Single-phase and Squirrel-cage and Synchronous. Full Voltage, Resistor or Reactor Starting, Code Letters B to E inc. Autotransformer Starting, Code Letters F to R inc. *Without Code Letters* Squirrel-cage and Synchronous, Autotransformer Starting, High-reactance Squirrel-cage.‡ Both not more than 30 amperes	*With Code Letters* Squirrel-cage and Synchronous Autotransformer Starting, Code Letters B to E inc. *Without Code Letters* Squirrel-cage and Synchronous, Autotransformer Starting, High-reactance Squirrel-cage.‡ Both more than 30 amperes	*With Code Letters* All Motors, Code Letter A *Without Code Letters* D-c and Wound-rotor Motors
88	1	2	110	110.00	300	225	200	150
90	0	2	110	112.50	300	225	200	150
92	0	2	125	115.00	300	250	200	150
94	0	1	125	117.50	300	250	200	150
96	0	1	125	120.00	300	250	200	150
98	0	1	125	122.50	300	250	200	150
100	0	1	125	125.00	300	250	200	150
105	00	1	150	131.5	350	300	225	175
110	00	0	150	137.5	350	300	225	175
115	00	0	150	144.0	350	300	250	175
120	000	0	150	150.0	400	300	250	200
125	000	00	175	156.5	400	350	250	200
130	000	00	175	162.5	400	350	300	200
135	0000	00	175	169.0	450	350	300	225
140	0000	00	175	175.0	450	350	300	225
145	0000	000	200	181.5	450	400	300	225
150	0000	000	200	187.5	450	400	300	225
155	0000	000	200	194.0	500	400	350	250
160	250	000	200	200.0	500	400	350	250
165	250	0000	225	206.	500	450	350	250
170	250	0000	225	213.	500	450	350	300
175	300	0000	225	219.	600	450	350	300
180	300	0000	225	225.	600	450	400	300
185	300	0000	250	231.	600	500	400	300
190	300	250	250	238.	600	500	400	300
195	350	250	250	244.	600	500	400	300
200	350	250	250	250.	600	500	400	300
210	400	300	250	263.	....	600	450	350
220	400	300	300	275.	....	600	450	350
230	500	300	300	288.	....	600	500	350
240	500	350	300	300.	....	600	500	400
250	500	350	300	313.	....	....	500	400
260	600	400	350	325.	....	....	600	400
270	600	400	350	338.	....	....	600	450
280	600	500	350	350.	....	....	600	450
290	700	500	350	363.	....	....	600	450
300	700	500	400	375.	....	....	600	450
320	750	600	400	400.	....	....	....	500
340	900	600	450	425.	....	....	....	600
360	1000	700	450	450.	....	....	....	600
380	1250	750	500	475.	....	....	....	600
400	1500	900	500	500.	....	....	....	600
420	1750	1000	600	525.	....	....	....	....
440	2000	1250	600	550.	....	....	....	....
460	....	1250	600	575.	....	....	....	....
480	....	1500	600	600.	....	....	....	....
500	....	1500	....	625.	....	....	....	....

* National Electrical Code.

† For conductors in air or for other insulations see Tables 14 and 15.

‡ High-reactance squirrel-cage motors are those designed to limit the starting current by means of deep-slot secondaries or double-wound secondaries and are generally started on full voltage.

84. ADEQUATE WIRING

In laying out a wiring system, in order to ensure its being adequate, safe, and efficient, consideration should be given certain general principles and accepted practices. The more important of these follow.

In designing new buildings, provision should be made for channeling and pocketing of buildings to accommodate the various wiring systems, keeping in mind that telephone, fire alarm, or other systems should be provided for but must be kept separate from light and power wiring.

Distribution centers should be centrally located in easily accessible places where switches and fuses for branch circuits can be grouped for convenient operation.

Branch circuit loads should be evenly distributed and complicated wiring avoided, so that the mechanical execution of the work will be as simple as possible.

With the increasing use of electrical devices, consideration should be given to possible future loads, and provision made for ample capacity in branch circuits and mains. This applies particularly to residences and apartments.

The Industry Committee on Interior Wiring Design has published *A Handbook of Interior Wiring Design*, which includes recommendations for adequate wiring, with particular reference to residential and commercial buildings. For industrial plants the AIEE publication, *Electrical Power Distribution for Industrial Plants*, has much valuable information.

85. BRANCH CIRCUITS

GENERAL-PURPOSE BRANCH CIRCUITS. The National Electrical Code recognizes four general-purpose branch circuits, in addition to motor branch circuits, which are treated as a separate class. The principal use and limitations of these circuits having two or more outlets are given in Table 21.

Table 21. Branch Circuit Requirements

Circuit rating	15 amp	20 amp	30 amp	50 amp
Conductors, min size				
Circuit wires	14	12	10	6
Taps	14	14	14	12
Overcurrent protection	15 amp	20 amp	30 amp	50 amp
Outlet devices				
Lamp holders permitted	Any type	Heavy duty	Heavy duty	Heavy duty
Receptacle rating	15 amp max	20 amp	20 or 30 amp	30 amp
Maximum load	15 amp	20 amp	30 amp	50 amp
Maximum load if motor-operated appliances are supplied	12 amp	16 amp	24 amp	Not permitted
Permissible load	Notes *a–d*	Notes *a–d*	Note *b*	Note *c*

a. 15- and 20-amp Branch Circuits. Lighting units and/or appliances. The rating of any one portable appliance shall not exceed 80 per cent of the branch circuit rating. The total rating of fixed appliances shall not exceed 50 per cent of the branch circuit rating, if lighting units or portable appliances are also applied.

b. 30-amp Branch Circuits. Fixed lighting units, in other than dwelling occupancies, or appliances, in any occupancy. The rating of any one portable appliance shall not exceed 24 amp.

c. 50-amp Branch Circuits. Fixed lighting units in other than dwelling occupancies, fixed cooling appliances, fixed range and water heater or infra-red lamp industrial heating appliances.

If motor-operated appliances are supplied by any of the above branch circuits, the total load shall not exceed 80 per cent of the circuit rating. If the circuit supplies only motor-operated appliances, then the requirements of motor branch circuits apply.

d. In dwelling-type occupancies the voltage between conductors shall not exceed 150 volts, except for permanently connected appliances or portable appliances of more than 1650 watts.

For a small appliance load in kitchen, laundry, pantry, dining room, and breakfast room one or more branch circuits shall be provided for all receptacle outlets in these rooms, and such circuits shall have no other outlet. The conductor size shall be not smaller than No. 12. On these circuits 15-amp receptacles may be used.

At least one outlet shall be for laundry appliances, and this shall be of the 3-pole type designed for grounding.

The number of receptacles per circuit is not specified for dwellings, but at least one receptacle is required for every 20 linear feet of wall space or fraction thereof in every room except bathrooms.

INDIVIDUAL BRANCH CIRCUITS. A branch circuit may supply individual pieces of equipment, but in this case the overcurrent-protective device must not exceed the carrying capacity of the conductor nor exceed 150 per cent of the rating of the appliance.

MOTOR BRANCH CIRCUITS. There are several types of layouts for motor branch circuits, and the choice is governed by the size and number of motors, the location, the importance of uninterrupted service, etc.

For maximum flexibility and reliability of service individual motor branch circuits may be run from a branch circuit distribution center. This is always desirable for large motors. Where the motors are small or intermediate in size, it is more economical to supply several motors through a sub feeder from the distribution center. In such circuits an effort should be made to keep the motors in the group of approximately the same size.

The National Electrical Code does not limit the capacity of branch circuits for power; therefore the number and capacity of such circuits are matters of engineering design to meet the requirements of each installation. The Code in Article 430, however, has other extensive regulations governing motor branch circuits. The fundamental requirements for continuous-duty motors only are given below.

The conductors of branch circuits supplying only one motor must have a current-carrying capacity of not less than 125 per cent of the full-load rating of the motor (see Table 20).

The conductors of branch circuits supplying two or more motors must have a carrying capacity of not less than 125 per cent of the full-load current of the largest motor of the group plus the full-load current of all the other motors in the group. If the motors are all the same size, 25 per cent of the full-load current of one motor is added to the total full-load current of all the motors. For full-load currents of the more usual sizes of motors see Tables 27 to 30.

Motor controllers should be in sight of the motors they control. When remote-control devices are used, special rules apply.

Overcurrent protection devices should be provided in accordance with Tables 18 and 19 and Fig. 6 for individual motor circuits. Where two or more motors are on one branch circuit, each overcurrent device and motor controller must be of a type approved for group installation.

Circuit Identification. For the purpose of distinguishing circuits during and after installation the outer surfaces of insulated conductors are colored. This color may be continuous throughout the length of the conductor or, for sizes larger than No. 6 A.W.G., may consist of painting the exposed ends.

A white conductor in any circuit is the grounded conductor and can serve no other purpose except for certain limited use in cable assemblies.

A green conductor in portable cords and cables is a conductor for grounding exposed parts of equipment. This color is also often used for the grounding conductor on fixed equipment where such a means of grounding is necessary.

Conductors installed in raceways or conduit, as open wiring or as concealed knob and tube work, and the conductors of multiwire branch circuits and two-wire branch circuits connected to the same system are identified by the following color code: two-wire circuits, one black, one white; three-wire circuits, one black, one white, one red; four-wire circuits, one black, one white, one red, one blue; five-wire circuits, one black, one white, one red, one blue, one yellow. If more than one multiwire branch circuit is carried through a single raceway, the ungrounded conductors of the additional circuit may be of colors other than those specified.

In cable assemblies, such as non-metallic sheathed cable, two-conductor cables have one white conductor, and either a red or a black conductor. Three-conductor cables have one white, one red, and one black conductor. In four-conductor cables the fourth conductor may be any color.

Table 22. Standard Loads for Illumination in Commercial Interiors

(From Industry Committee on Interior Wiring Design, *A Handbook of Interior Wiring Design*)

Occupancy	Electrical Demands, Watts per Square Foot	Occupancy	Electrical Demands, Watts per Square Foot
Armories		Hospitals—*Continued*	
Drill sheds and exhibition halls	2.5	Private rooms	3.5
This does not include lighting circuits for demonstration booths, special exhibit spaces, etc.		Including allowance for convenience outlets for local illumination.	
Art galleries		Operating room	4.0
General	1.0	Operating tables or chairs	....
On paintings	....	Major Surgeries—3000 watts per area.	
100 watts per running foot of usable wall area.		Minor Surgeries—1500 watts per area.	
Auditoriums	1.5	The above two figures include allowance for directional control. Special wiring for emergency systems must also be considered.	
Automobile showrooms	4.0		
Banks			
Lobby	2.5		
Counters	....		
75 watts per running foot, including service for signs, small motor applications, etc.		Laboratories	4.0
		Hotels	
		Lobby	2.5
Offices and cages	4.0	Not including provision for conventions, exhibits.	
Barber shop and beauty parlors	4.0		
This does not include circuits for special equipment.		Dining room	2.5
		Kitchen	2.5
Bowling		Bedrooms	3.0
Alley runway and seats	2.5	Including allowance for convenience outlets.	
Pins	....		
300 watts per set of pins.		Corridors	....
Billiards		10 watts per running foot.	
General	2.5	Writing room	4.0
Tables	....	This includes allowance for convenience outlets.	
450 watts per table.			
Churches		Library	
Auditoriums	1.5	Reading rooms	6.0
Sunday school rooms	2.5	This includes allowance for convenience outlets.	
Pulpit or rostrum	3.5		
Clubrooms		Stack room	....
Lounge	1.5	12 watts per running foot of facing stacks.	
Reading rooms	4.0		
The above two uses are so often combined that the higher figure is advisable. It includes provision for convenience outlets.		Moving-picture theater	
		During intermission	1.5
		During pictures	0.5
		These figures do not include auxiliary circuits for color or other spectacular effects.	
Courtrooms	2.5		
Dance halls	1.5	Museum	
No allowance has been included for spectacular lighting, spots, etc.		General	2.5
		Local illumination of special exhibits	....
Drafting rooms	7.0	Allow wattage for local illumination equal to total calculated for general lighting.	
Fire-engine houses	2.0		
Gymnasiums			
Main floor	3.0	Office buildings	
Shower rooms	2.5	Private offices	3.0
Locker rooms	1.5	No close work.	
Fencing, boxing, etc.	4.0	Private offices	5.0
Handball, squash, etc.	5.0	With close work.	
Halls and interior passageways	....	General offices	3.0
15 watts per running foot.		No close work.	
Hospitals		General offices	4.5
Lobby, reception room	2.5	With close work.	
Corridors	....	File room, vault, etc.	2.5
8 watts per running foot		Reception room	1.5
Wards	2.5	Post office	
Including allowance for convenience outlets for local illumination.		Lobby	2.5
		Sorting, mailing, etc.	4.0
		Storage, file room, etc.	2.0

Table 22. Standard Loads for Illumination in Commercial Interiors—*Continued*

(From Industry Committee on Interior Wiring Design, *A Handbook of Interior Wiring Design*)

Occupancy	Electrical Demands, Watts per Square Foot	Occupancy	Electrical Demands, Watts per Square Foot
Professional offices		Schools	
Waiting rooms	2.5	Auditoriums	2.5
Consultation rooms	4.5	If used as a study hall	4.0
Operating offices	4.5	Class and study rooms	4.0
Dental chairs	....	Drawing room	7.0
600 watts per chair.		Laboratories	3.0
Railway		Manual training	4.0
Depot—waiting room	2.5	Sewing room	7.0
Ticket offices—general	2.5	Sight-saving classes	7.0
Ticket counters	....	Stores—department, specialty, and miscellaneous large	
75 watts per running foot.			
Rest room, smoking room	2.5	Main floor	5.0
Baggage room	2.0	Other floors	3.5
Concourse	1.5	Stores in outlying districts	3.0
Train platform	1.0	Theaters	
Restaurants, lunchrooms, and cafeterias		Auditoriums	1.5
		This figure does not include auxiliary circuits for color or other spectacular effects.	
Dining area	2.5		
Food displays	....		
100 watts per running foot of counter (including service aisle).		Foyer	2.5
		Lobby	3.5
		Wall cases	....
		50 to 75 watts per running foot, depending on height and depth.	

Table 23. Standard Loads for General Illumination in Industrial Occupancies

(From Industry Committee on Interior Wiring Design, *A Handbook of Interior Wiring Design*)

In many cases the desirable level of illumination is much higher than that obtainable by using prevailing methods of general illumination. In such instances, designated by (*), the watts per square foot values specified are intended to provide only for the general illumination needed. Supplementary illumination must then be provided by local or localized general methods. These additional load considerations are entirely dependent on specific studies of machine spacing, actual size of areas requiring high intensities, color control, special glare or directional features, etc. Where it is not possible to make this preliminary study, the watts per square foot allowed for the supplementary illumination must be at least equivalent to the watts per square foot provided for general illumination.

Occupancy	Electrical Demands, Watts per Square Foot	Occupancy	Electrical Demands, Watts per Square Foot
Aisles, stairways, passageways	....	Candy making	4.0
10 watts per running foot.		Canning and preserving	4.0
Assembly		Chemical works	
Rough	2.0	Hand furnaces, stationary driers and crystallizers	1.0
Medium	4.0		
Fine	* 3.0	Mechanical driers and crystallizers, filtrations, evaporators, bleaching	2.0
Extra-fine	* 3.0		
Automobile manufacturing			
Assembly line	* 2.0	Tanks for cooking, extractors, percolators, nitrators, electrolytic cells	3.0
Frame assembly	3.0		
Body assembly	4.0		
Body finishing and inspecting	* 3.0	Clay products and cements	
Bakeries	4.0	Grinding, filter presses, kiln rooms	1.0
Book-binding			
Folding, assembling, pasting	2.0	Moldings, pressing, cleaning, trimming	2.0
Cutting, punching, stitching, embossing	4.0	Enameling	3.0
Breweries		Glazing	4.0
Brew house	1.0	Cloth products	
Boiling, keg-washing, etc.	2.0	Cutting, inspecting, sewing	
Bottling	3.0	Light goods	4.0
		Dark goods	* 3.0

Table 23. Standard Loads for General Illumination in Industrial Occupancies—*Continued*
(From Industry Committee on Interior Wiring Design, *A Handbook of Interior Wiring Design*)

Occupancy	Electrical Demands, Watts per Square Foot
Cloth products—*Continued*	
Pressing, cloth treating (oil cloth, etc.)	
Light goods	2.0
Dark goods	4.0
Coal breaking, washing, screening	1.0
Dairy products	4.0
Engraving	* 2.0
Forge shops	
Welding	2.0
Foundries	
Charging floor, tumbling, cleaning, pouring, shaking out	1.0
Rough molding and core making	2.0
Fine molding and core making	4.0
Garages	
Storage	
Live	2.0
Dead	1.0
Repair and washing	* 3.0
Glass works	
Mixing and furnace rooms, pressing and lehr glass-blowing machines	2.0
Grinding, cutting glass to size, silvering	4.0
Fine grinding, polishing, beveling, etching, inspection, etc.	* 3.0
Glove manufacturing	
Light goods	
Cutting, pressing, knitting, sorting	2.0
Stitching, trimming, inspecting	4.0
Dark goods	
Cutting, pressing, etc.	4.0
Stitching, trimming, etc.	* 4.0
Hangars—aeroplane	
Storage—live	2.0
Repair department	* 3.0
Hat manufacturing	
Dyeing, stiffening, cleaning, and refining	
Light	2.0
Dark	4.0
Forming, sizing, pouncing, flanging, finishing, and ironing	
Light	3.0
Dark	6.0
Sewing	
Light	4.0
Dark	* 3.0
Ice making	
Engine and compressor room	2.0
Inspection	
Rough	2.0
Medium	4.0
Fine	* 3.0
Extra-fine	* 4.0
Jewelry and watch manufacturing	* 2.0
Laundries and dry cleaning	4.0
Leather manufacturing	
Vats	1.0
Cleaning, tanning, and stretching	2.0
Leather manufacturing—*Continued*	
Cutting, fleshing, and stuffing	3.0
Finishing and scarfing	4.0
Leather working	
Pressing, winding, and glazing	
Light	2.0
Dark	4.0
Grading, matching, cutting, scarfing, sewing	
Light	4.0
Dark	* 3.0
Locker rooms	1.0
Machine shops	
Rough bench and machine work	2.0
Medium bench and machine work, ordinary automatic machines, rough grinding, medium buffing, and polishing	4.0
Fine bench and machine work, fine automatic machines, medium grinding, fine buffing and polishing	* 4.0
Extra-fine bench and machine work, grinding	* 4.0
Meat packing	
Slaughtering	2.0
Cleaning, cutting, cooking, grinding, canning, packing	4.0
Milling—grain foods	
Cleaning, grinding, and rolling	2.0
Baking or roasting	4.0
Flour grading	6.0
Offices	
Private and general	
No close work	2.0
Close work	4.0
Drafting rooms	7.0
Packing and boxing	2.0
Paint manufacturing	2.0
Paint shops	
Dipping, spraying, firing, rubbing, ordinary hand painting and finishing	4.0
Fine hand painting and finishing	* 3.0
Extra-fine hand painting and finishing (automobile bodies, piano cases, etc.)	* 3.0
Paper-box manufacturing	
Light	2.0
Dark	4.0
Storage of stock	1.0
Paper manufacturing	
Beaters, grinding, calendering	2.0
Finishing, cutting, trimming	4.0
Plating	2.0
Polishing and burnishing	3.0
Power Plants	
Boiler, coal and ash handling, storage battery rooms	1.0
Auxiliary equipment, oil switches and transformers	2.0
Switchboards, engines, generators, blowers, compressors	3.0

Table 23. Standard Loads for General Illumination in Industrial Occupancies—*Continued*

(From Industry Committee on Interior Wiring Design, *A Handbook of Interior Wiring Design*)

Occupancy	Electrical Demands, Watts per Square Foot
Printing industries	
Matrixing and casting	2.0
Miscellaneous machines	3.0
Presses and electrotyping	4.0
Lithographing	* 4.0
Linotype, monotype, typesetting, imposing stone, engraving	* 4.0
Proof reading	* 3.0
Receiving and shipping	2.0
Rubber manufacturing and products	
Calenders, compounding mills, fabric preparation, stock cutting, tubing machines, solid tire operations, mechanical goods building, vulcanizing	2.0
Bead building, pneumatic tire building and finishing, inner tube operation, mechanical goods trimming, treading	4.0
Sheet-metal works	
Miscellaneous machines, ordinary bench work	3.0
Punches, presses, shears, stamps, welders, spinning, medium bench work	4.0
Tin plate inspection	* 3.0
Shoe manufacturing	
Hand turning, miscellaneous bench and machine work	2.0
Inspecting and sorting raw material, cutting and stitching	
Light	4.0
Dark	* 4.0
Lasting and welting	4.0
Soap manufacturing	
Kettle houses, cutting, soap chip and powder	2.0
Stamping, wrapping and packing, filling and packing soap powder	4.0
Steel and iron mills, bar, sheet and wire products	
Soaking pits and reheating furnaces	1.0
Charging and casting floors	2.0
Muck and heavy rolling, shearing (rough by gage), pickling, and cleaning	2.0
Plate inspection, chipping	* 4.0
Automatic machines, light and cold rolling, wire drawing, shearing (fine by line)	3.0
Stone crushing and screening	
Belt conveyor tubes, main line shafting spaces, chute rooms, inside of bins	1.0
Stone crushing and screening—*Continued*	
Primary breaker room, auxiliary breakers under bins	1.0
Screens	2.0
Storage-battery manufacturing	
Molding of grids	2.0
Store and stock rooms	
Rough bulky material	1.0
Medium or fine material requiring care	2.0
Structural-steel fabrication	2.0
Sugar grading	5.0
Testing	
Rough	2.0
Fine	4.0
Extra-fine instruments, scales, etc.	* 3.0
Textile mills	
Cotton	
Opening and lapping, carding, drawing, roving, dyeing	2.0
Spooling, spinning, drawing, warping, weaving, quilling, inspecting, knitting, slashing (over beam end)	4.0
Silk	
Winding, throwing, dyeing	3.0
Quilling, warping, weaving, finishing	
Light goods	3.0
Dark goods	5.0
Wool	
Carding, picking, washing, combing	2.0
Twisting, dyeing	2.0
Drawing-in, warping	
Light goods	3.0
Dark goods	5.0
Weaving	
Light goods	3.0
Dark goods	5.0
Knitting machines	4.0
Tobacco products	
Drying, stripping, general	2.0
Grading and sorting	* 3.0
Toilets and washrooms	1.0
Upholstering	
Automobile, coach, furniture	4.0
Warehouse	1.0
Woodworking	
Rough sawing and bench work	2.0
Sizing, planing, rough sanding, medium machine and bench work, gluing, veneering, cooperage	4.0
Fine bench and machine work, fine standing and finishing	6.0

86. WIRING DESIGN

GENERAL. Before proceeding with any wiring layout it is necessary (1) to determine the anticipated load, (2) from the load requirements to establish the number and type of branch circuits, (3) to determine feeder loads, (4) to determine conductor sizes for branch circuits and feeders on the basis of current-carrying capacity and to check for excessive voltage drop.

For residential and commercial buildings and small industrial plants standards have been set up which can be used in these calculations. This information is included in the following paragraphs. For large industrial plants the procedure is more involved, particularly in connection with power loads. It often includes a rather extensive primary distribution system, as well as the secondary distribution supplying power to the load, and can be covered only briefly in this article. The booklet of the American Institute of Electrical Engineers, *Electrical Power Distribution for Industrial Plants*, covers the subject in considerable detail and should be consulted for further information.

LOAD CALCULATIONS. The load on which feeder and branch circuit conductor sizes are based may be determined as outlined below. Certain demand factors may be applied to their loads for determining the sizes of feeder conductors and are covered in the discussion of feeder calculations, p. 14-278.

Lighting Loads. In most types of buildings lighting loads are calculated on the basis of the area to be illuminated. The National Electrical Code has established minimum requirements (Table 25) for various types of occupancies. These values, however, are seldom adequate for an efficient or satisfactory layout. Tables 22 and 23 give more complete recommendations and should be followed in preference to the minimum values of the Code.

To determine the number and distribution of lighting branch circuits is relatively simple, inasmuch as the circuits fall into fixed rating classifications, and with the load computed the number of circuits becomes a matter of dividing the load by the circuit rating. The location of the circuits will depend chiefly upon the physical layout of the building, but it is good practice to have every room or area supplied by at least two circuits, so that the loss of one circuit will not isolate any area.

Appliance loads are of importance mainly in residential occupancies, where the cable requires that at least 1500 watts be allowed for small appliance loads in each dwelling or apartment. This does not include special appliance loads, such as ranges, water heaters, clothes driers, oil burners, or deep freeze units, which should be allowed for on the basis of natural load where such load is known. Otherwise, if heavy appliance loads are anticipated, generous estimates should be made, and branch circuit and feeder capacity should be installed accordingly. Table 24 gives the approximate wattage of commonly used appliances. See also table in Section 15, Article 16.

Table 24. Approximate Wattage of Appliances

Flat irons	500–1,200	Waffle iron	500– 750
Electric grill	500–1,000	Immersion heater	300– 500
Coffee maker	500	Curling iron	25
Toaster	500	Domestic range	1,750–21,000
Vacuum cleaner	150	Bake ovens	5,000 up
Heating pad	60– 150	Dish washer	200
Radiant heater	500–1,500	Washing machine	200
Sewing machine	50	Ironing machine	1,500– 6,000
Sun lamp	250–1,500	Soldering iron	100– 500
Refrigerator	100– 350	Office fans	30– 60
Deep freeze unit	200– 750	Water heaters	600– 5,000
Poultry picker	500–1,200	Radio	60– 150
Domestic water pumps	250– 750		

Table 25. Unit Loads and Feeder Demand Factors

The unit values and the demand factors herein are based on minimum load conditions, 100 per cent power factor, and may not provide sufficient capacity for the installation contemplated.

In view of the trend toward higher intensity lighting systems and increased loads due to more general use of fixed and portable appliances, each installation should be considered as to the load likely to be imposed and the capacity increased to ensure safe operation.

Where electric discharge lighting systems are to be installed, high power-factor type should be used or the conductor capacity may need to be increased.

Type of Occupancy	Unit Load Per Square Foot, Watts	Load to Which Demand Factor Applies, Watts	Demand Factor, Per Cent
Armories and auditoriums	1	Total wattage	100
Banks	2	Total wattage	100
Barber shops and beauty parlors	3	Total wattage	100
Churches	1	Total wattage	100
Clubs	2	Total wattage	100
Court rooms	2	Total wattage	100
Dwellings—Single-family	2	2,500 or less Over 2,500	100 30
Dwellings—Multifamily (other than hotels)	2	3,000 or less Next 117,000 Over 120,000	100 35 25
Garages—commercial (storage)	1/2	Total wattage	100
Hospitals	2	50,000 or less Over 50,000	40 * 20
Hotels, including apartment houses without provisions for cooking by tenants	2	20,000 or less Next 80,000 Over 100,000	50 * 40 30
Industrial commercial (loft) buildings	2	Total wattage	100
Lodge rooms	1 1/2	Total wattage	100
Office buildings	2	20,000 or less Over 20,000	100 70
Restaurants	2	Total wattage	100
Schools	3	15,000 or less Over 15,000	100 50
Stores	3	Total wattage	100
Warehouses, storage	1/4	12,500 or less Over 12,500	100 50
In any of above occupancies except single-family dwellings and individual apartments of multifamily dwellings:			
Assembly halls and auditoriums	1	Total wattage as specified for the specific occupancy	
Halls, corridors	1/2		
Closets, storage spaces	1/4		

* For subfeeders to areas in hospitals and hotels where entire lighting is likely to be used at one time, as in operating rooms, ballrooms, and dining rooms, a demand factor of 100 per cent shall be used.

Table 26. Demand Loads for Household Electric Ranges and Other Cooking Appliances over 1 3/4 kw Rating

Column A to be used in all cases except as otherwise permitted in Note 3 below.

Number of Ranges	Maximum Demand, Kilowatts (See Notes)	Demand Factors, Per Cent (See Note 3)	
	A (Not over 12 kw Rating)	*B* (Less than 3 1/2 kw Rating)	*C* (3 1/2 kw to 8 3/4 kw Rating)
1	8	80	80
2	11	75	65
3	14	70	55
4	17	66	50
5	20	62	45
6	21	59	43
7	22	56	40
8	23	53	36
9	24	51	35
10	25	49	34
11	26	47	32
12	27	45	32
13	28	43	32
14	29	41	32
15	30	40	32
16	31	39	28
17	32	38	28
18	33	37	28
19	34	36	28
20	35	35	28
21	36	34	26
22	37	33	26
23	38	32	26
24	39	31	26
25	40	30	26
26–30	15 plus 1 for each range	30	24
31–40		30	22
41–50	25 plus 3/4 for each range	30	20
51–60		30	18
61 & over		30	16

Notes. 1. Over 12 kw to 21 kw ranges. For ranges individually rated more than 12 kw but not more than 21 kw, 5 per cent shall be added to the above maximum demand (Column *A*) for each additional kw of rating or major fraction thereof by which the individual range rating exceeds 12 kw.

2. Over 21 kw ranges. Ranges individually rated more than 21 kw are not considered as household electric ranges, and the demand should be determined on the basis of rating and use. Generally, the demand for commercial ranges should be based on the maximum nameplate rating.

3. Over 1 3/4 kw to 8 3/4 kw. In lieu of the method provided in Column *A*, the load for ranges individually rated more than 1 3/4 kw but not more than 8 3/4 kw may be considered as the sum of the nameplate ratings of all the ranges, multiplied by the demand factors specified in Columns *B* or *C* for the given number of ranges.

4. Branch circuit load. Branch circuit load for one range may be computed in accordance with the above table.

POWER LOADS. Motors constitute the bulk of most power loads, and the current requirements may be obtained from Tables 27 to 30 inclusive, which give approximate values for standard motors. The manufacturer's rating should be used for special or very large motors. Where heating, electroplating, or other heavy current-consuming equipment is involved, the load must be determined on the basis of the manufacturer's recommendation.

The distribution of power loads in relation to the source of supply is governed entirely by the machine and equipment layouts rather than by consideration of wiring efficiency, and therefore the branch circuit capacities and locations require a careful study which is possible only after the equipment layout has been made.

Table 27. Full-load Current, in Amperes, of D-c Motors

Horsepower	115 Volts	230 Volts	550 Volts
1/2	4.6	2.3	....
3/4	6.6	3.3	1.4
1	8.6	4.3	1.8
1 1/2	12.6	6.3	2.6
2	16.4	8.2	3.4
3	24	12	5.0
5	40	20	8.3
7 1/2	58	29	12
10	76	38	16
15	112	56	23
20	148	74	31
25	184	92	38
30	220	110	46
40	292	146	61
50	360	180	75
60	430	215	90
75	536	268	111
100		335	148
125		443	184
150		534	220
200		712	295

NOTE. These values are average for all speeds.

Table 28. Full-load Current, in Amperes, of Single-phase A-c Motors

Horsepower	115 Volts	230 Volts *	440 Volts
1/6	3.2	1.6	
1/4	4.6	2.3	
1/2	7.4	3.7	
3/4	10.2	5.1	
1	13.0	6.5	
1 1/2	18.4	9.2	
2	24	12	
3	34	17	
5	56	28	
7 1/2	80	40	21
10	100	50	26

NOTE. The above values of full-load current are for motors running at speed usual for belted motors and motors with normal torque characteristics. Motors built for especially low speeds or high torques may require more running current, in which case the nameplate current rating should be used.

* For full-load currents of 208- and 200-volt motors, increase corresponding 230-volt motor full-load current by 10 and 15 per cent, respectively.

CALCULATION OF CURRENT TO BE CARRIED BY FEEDERS. The computed load of a feeder shall be not less than the sum of all branch circuit loads supplied by the feeder, subject to special provisions for certain types of loads. The most important of these provisions are as follows:

1. The demand factors given in Table 25 may be applied to compute feeder circuit loads for general illumination.

2. In single-family dwellings or individual apartments of multifamily dwellings having provision for cooking, a feeder load of not less than 1500 watts shall be included for small appliances in dining room, kitchen, and laundry.

3. For electric ranges of more than 1 3/4 kw rating, the feeder load shall be calculated in accordance with Table 26. Where a number of ranges are supplied by a three-phase four-wire feeder, the current shall be computed on the basis of the demand of twice the maximum number of ranges connected between any two phase wires.

4. Motor loads computed in accordance with the discussion of motor branch circuits, p. 14-270, shall be included in feeder calculations.

5. The neutral feeder load shall be the maximum unbalance of the load determined as above. The maximum unbalanced load shall be the maximum connected load between the neutral and any one ungrounded conductor, except that the load thus obtained shall be multiplied by 140 per cent for a five-wire two-phase system.

Table 29. Full-load Current, in Amperes, of Two-phase A-c Motors (Four wire)

Horsepower	Induction Type, Squirrel-cage and Wound-rotor					Synchronous Type,* Unity Power Factor			
	110 Volts	220 Volts	440 Volts	550 Volts	2300 Volts	220 Volts	440 Volts	550 Volts	2300 Volts
1/2	4	2.0	1.0	0.8					
3/4	6.8	2.4	1.2	1.0					
1	6.4	3.2	1.6	1.3					
1 1/2	8.8	4.4	2.2	1.8					
2	11.2	5.6	2.8	2.2					
3		8	4	3.2					
5		13	7	6					
7 1/2		19	9	8					
10		24	12	10					
15		34	17	14					
20		45	23	18					
25		55	28	22	6	47	24	19	4.7
30		67	34	27	7.5	56	29	23	5.7
40		88	44	35	9	75	37	31	7.0
50		108	54	43	11	94	47	38	9
60		129	65	52	13	111	56	44	11
75		158	79	63	16	140	70	57	13
100		212	106	85	21	182	93	74	17
125		268	134	108	26	228	114	93	22
150		311	155	124	31	....	137	110	26
200		415	208	166	41	....	182	145	35

NOTE. These values of full-load current are for motors running at speeds usual for belted motors and motors with normal torque characteristics. In the case of motors built for especially low speeds or high current, the nameplate current rating should be used. Current in the common conductor of a two-phase three-wire system will be 1.41 times the value given.

* For 90 and 80 per cent power factor the above figures should be multiplied by 1.1 and 1.25, respectively.

Table 30. Full-load Current, in Amperes, of Three-phase A-c Motors

Horsepower	Induction Type, Squirrel-cage and Wound-rotor					Synchronous Type,* Unity Power Factor			
	110 Volts	220 Volts	440 Volts	550 Volts	2300 Volts	220 Volts	440 Volts	550 Volts	2300 Volts
1/2	4.0	2.0	1.0	0.8					
3/4	5.6	2.8	1.4	1.1					
1	7	3.5	1.8	1.4					
1 1/2	10	5.0	2.5	2.0					
2	13	6.5	3.3	2.6					
3		9	4.5	4					
5		15	7.5	6					
7 1/2		22	11	9					
10		27	14	11					
15		40	20	16					
20		52	26	21					
25		64	32	26	7.0	54	27	22	5.4
30		78	39	31	8.5	65	33	26	6.5
40		104	52	41	10.5	86	43	35	8
50		125	63	50	13	108	54	44	10
60		150	75	60	16	128	64	51	12
75		185	93	74	19	161	81	65	15
100		246	123	98	25	211	106	85	20
125		310	155	124	31	264	132	106	25
150		360	180	144	37	....	158	127	30
200		480	240	192	48	....	210	168	40

NOTE. For full-load currents of 208- and 200-volt motors, increase the corresponding 220-volt motor full-load current by 6 and 10 per cent, respectively.

* For 90 and 80 per cent power factor the above figures should be multiplied by 1.1 and 1.25, respectively.

For range feeders the maximum unbalance shall be considered 70 per cent of the load on the ungrounded conductors as determined according to Table 26. For three-wire d-c or single-phase a-c, four-wire 3-phase, and five-wire, two-phase systems a further demand factor of 70 per cent may be applied to that portion of the unbalanced load in excess of 200 amp.

Conductor Size. The minimum size of conductors required for branch circuits, based only upon the safe current-carrying capacity of the conductors, is given in Table 21 for the four special branch circuits recognized by the Code and in Table 20 for individual motor branch circuits. The minimum size of conductor for feeders and subfeeders can be determined from Tables 14 and 15 after the maximum loads have been calculated.

Conductor sizes determined on the basis of safe carrying capacity, as given above, are usually adequate for short runs or for lighting branch circuits where the full rating of the circuit is seldom needed, for example, in small residences. For long runs and fully loaded circuits the conductor size is often determined by the allowable voltage drop between the source of supply and the load.

The size of the feeder conductors should be such that the voltage drop up to the final distribution point for the load as computed above will not be more than 3 per cent for power loads and not more than 1 per cent for lighting loads. The drop in the branch circuit beyond the feeder distribution center for motor loads should be such that the total drop to any motor will not exceed 5 per cent under full load; i.e., if the feeder drop is 3 per cent, the branch circuit drop should not be more than 2 per cent. The drop in lighting branch circuits should not exceed 2 per cent or a total of 3 per cent from outlet to source of supply.

It is almost always desirable, particularly in industrial plants, to install feeder conductors larger than the initial computed load. This not only provides for future additional loads but also reduces the copper loss (I^2R loss), which may run to a considerable amount when the loads are large.

Dimensions of wires and cables most commonly used for interior wiring are given in Tables 11, 12, and 13.

Properties of bare copper conductors are given in Articles 54 and 70.

Current-carrying capacities of conductors, as designated by the National Electrical Code, are given in Tables 14, 15, and 16.

For information on wires and cables of types other than those used for building wiring see Section 14, Articles 63–77.

EXAMPLES OF BRANCH CIRCUIT AND FEEDER CALCULATIONS. Single-family Dwelling with Floor Area of 2500 sq ft Exclusive of Unoccupied Cellar and Unfinished Attic or Open Porches. Minimum requirements only considered.

COMPUTED LOAD

General lighting load (Table 22): *

2500 sq ft at 2 watts per sq ft	5000 watts
Small appliance load	1500 watts
	6500 watts

MINIMUM NUMBER OF BRANCH CIRCUITS REQUIRED

General lighting load: 5000 ÷ 115 = 43 amp or three 15-amp 2-wire circuits.

Small appliance load: 1500 ÷ 115 = 13 amp or one 2-wire circuit of No. 12 wire, the minimum size allowed for appliance circuit.

MINIMUM SIZE FEEDER (in this case service conductors)

Total load: 6500 watts

2500 watts at 100%	2500 watts
4000 watts at 30%	1200 watts
Net load	3700 watts

For 115-volt 2-wire system: 3700 ÷ 115 = 32 amp or a No. 8 Type R Conductor (Table 15)

MINIMUM NUMBER OF RECEPTACLES

Living room: 16 ft × 22 ft; 76 ft total distance around room: 76 ÷ 20 = 3.8 or 4 receptacles.

Each bedroom: 14 ft × 15 ft; 58 ft total distance around room: 58 ÷ 20 = 2.9 or 3 receptacles.

Dining room: 16 ft × 16 ft; 64 ft total distance around room: 64 ÷ 20 = 3.2 or 3 receptacles.

Breakfast room: 10 ft × 12 ft; 44 ft total distance around room: 44 ÷ 20 = 2.2 or 2 receptacles.

Kitchen: 14 ft × 14 ft; 56 ft total distance around room: 56 ÷ 20 = 2.8 or 3 receptacles.

NOTE: Dining room, breakfast room, kitchen, and laundry must be supplied by a circuit or circuits of not less than No. 12 wire.

* The above examples assume that the entire general lighting load is likely to be used for long periods of time, and the load is therefore increased by 25 per cent. The 25 per cent increase is not applicable to any portion of the load not used for long periods.

Store Building. A store 50 ft by 60 ft, or 3000 sq ft, has 30 ft of show window.

COMPUTED LOAD:

General lighting load:

3,000 sq ft at 3 watts per sq ft × 1.25 .... 11,250 watts

Show-window lighting load:

30 ft at 200 watts per ft .... 6,000 watts

MINIMUM NUMBER OF BRANCH CIRCUITS REQUIRED:

General lighting load: 11,250 ÷ 230 = 49 amp for 3-wire, 115-230 volts; or 98 amp for 2-wire, 115 volts:

Three 30-amp, 2-wire, and one 15-amp, 2-wire circuits; or
Five 20-amp, 2-wire circuits; or
Three 20-amp, 2-wire, and three 15-amp, 2-wire circuits; or
Seven 15-amp, 2-wire circuits; or
Three 15-amp, 3-wire, and one 15-amp, 2-wire circuits.

Special lighting load (show window): 6000 ÷ 230 = 26 amp for 3-wire, 115–230 volts; or 52 amp for 2-wire, 115 volts:

Four 15-amp, 2-wire circuits; or
Three 20-amp, 2-wire circuits; or
Two 15-amp, 3-wire circuits.

MINIMUM SIZE FEEDERS (or SERVICE CONDUCTORS) REQUIRED:

For 115–230-volt, 3-wire system:

Ampere load: 49 + 26 = 75 amp.
Size of each feeder, No. 3.

For 115-volt system:

Ampere load: 98 + 52 = 150 amp.
Size of each feeder, No. 3/0.

VOLTAGE-DROP CALCULATIONS. In d-c circuits where the voltage drop depends only upon the known values of current and conductor resistance, the calculations are simple and accurate. For a-c circuits, however, reactance, power factor, and the a-c resistance of the conductor must be considered, and, although exact numerical values can seldom be determined for these characteristics, approximate values based upon experimental data and experience may be used as noted in connection with the formulas given below.

Notation used in connection with voltage-drop formulas:

W = watts load.
I = line current, in amperes.
E_0 = voltage to neutral, sending end.
e_0 = voltage to neutral, receiving end.
E_l = voltage between lines, sending ends.
e_l = voltage between lines, receiving end.
$V_0 = E_0 - e_0$ = voltage drop to neutral, in volts.
$V_l = E_l - e_l$ = voltage drop between lines, in volts.
$V_p = E_p - e_p$ = voltage drop between phases, in volts.
R = d-c or a-c resistance, in ohms per 1000 ft per conductor. (For d-c resistance see Article 54; for a-c resistance see Article 70.)
X = reactance = $2\pi f(0.140 \log_{10} S/A + 0.0153) \times 10^{-3}$ ohms per 1000 ft. For 60 cycles $X = 0.0529 \log_{10} S/A + 0.00574$ ohms per 1000 ft. If conductors are in iron conduit, increase results by 50 per cent for magnetic effect and random lay.
f = frequency.
S = spacing between centers of conductors. For conductors in conduit, S is usually taken as diameter of overinsulated conductors plus 10 per cent.
A = radius of conductor, in same unit as S.
Z = impedance = $\sqrt{R^2 + X^2}$ ohms.
L = length of line, in feet. (Supply to load, one way.)
θ = power-factor angle of load.
pf = $\cos\theta$ = power factor.
$\sin\theta = \sqrt{1 - \cos^2\theta}$.

D-c Two-wire Circuit.

$$I = \frac{W}{e_0} \text{ amp}$$

$$V_0 = E_0 - e_0 = \frac{IR2L}{1000} \text{ volts drop}$$

D-c Three-wire Circuit (Balanced Load).

$$I = \frac{W}{2e_0} = \frac{W}{e_l} \text{ amp}$$

$$V_0 = E_0 - e_0 = \frac{IRL}{1000} \text{ volts drop to neutral}$$

$$V_l = E_l - e_l = \frac{IR2L}{1000} \text{ volts drop between lines}$$

Basic Equation for A-c Circuits. The basic equation for volts drop to neutral on all a-c wiring circuits is

$$V_0 = E_0 - e_0 = [\sqrt{(e_0 \cos\theta + IR)^2 + (e_0 \sin\theta + IX)^2} - e_0] \times \frac{L}{1000} \text{ volts drop to neutral}$$

and represents the volts drop per conductor for balanced load conditions.

The voltage drop between lines and between phases, together with related current and voltage values for the commonly used circuits, is given below.

A-c Single-phase Two-wire Circuit.

$$I = \frac{W}{e_l \times \text{pf}} \text{ amp}$$

$$V_l = 2V_0 \text{ volts drop between lines}$$

$$= \frac{IZ2L}{1000} \text{ volts drop between lines (approx.)}$$

A-c Single-phase Three-wire Circuit (Balanced Load).

$$e_0 = \frac{e_l}{2}$$

$$I = \frac{W}{2e_0 \times \text{pf}} = \frac{W}{e_l \times \text{pf}}$$

$$V_0 = E_0 - e_0 = \text{volts drop to neutral}$$

$$= \frac{IZL}{1000} \text{ volts drop to neutral (approx.)}$$

$$V_l = E_l - e_l = 2V_0 \text{ volts drop between lines}$$

$$= \frac{IZ2L}{1000} \text{ volts drop between lines (approx.)}$$

A-c Two-phase Four- or Five-wire (Balanced Load).

$$e_0 = \frac{e_l}{2}$$

$$I = \frac{W}{4e_0 \times \text{pf}} = \frac{W}{2e_l \times \text{pf}} \text{ amp}$$

$$V_0 = E_0 - e_0 = \text{volts drop to neutral}$$

$$= \frac{IZL}{1000} \text{ volts drop to neutral (approx.)}$$

$$V_l = E_l - e_l = 2V_0$$

$$= \frac{IZ2L}{1000} \text{ volts drop between lines (approx.)}$$

The formulas for two-phase four- and five-wire circuits are the same provided that the load is balanced. The neutral (fifth wire) carries no current.

A-c Three-phase Three- or Four-wire (Balanced Load).

$$e_0 = \frac{e_l}{\sqrt{3}} \qquad I = \frac{W}{3 \times e_0 \times \text{pf}} = \frac{W}{\sqrt{3} \times e_l \times \text{pf}}$$

$$V_0 = E_0 - e_0 = \text{volts drop to neutral}$$

$$= \frac{IZL}{1000} = \text{volts drop to neutral (approx.)}$$

$$V_l = E_l - e_l = \frac{IZ\sqrt{3}L}{1000} = \text{volts drop between lines (approx.)}$$

With a balanced load the neutral carries no current; therefore the formulas are the same for three- and four-wire circuits.

Table 31 gives values which may be used for checking voltage drop in the most commonly used types of circuits. Values are given for the various conductor temperatures permitted by the standard types of insulations. The 60 deg column is used for Types R, RW, RU, T, and TW; the 75 deg column, for Type RH; the 85 deg column, for varnished-cambric Type V and Type AVB.

Table 31. Factors for Determining Voltage Drop, Conductor Size, Length, and Current for Single- and Three-phase Circuits in Iron Conduit

Based on 80 Per Cent Lagging Power Factor and Various Conductor Temperatures

Size, A.W.G. or MCM	Factors for Single-phase Circuits			Factors for Three-phase Circuits		
	60° C	75° C	85° C	60° C	75° C	85° C
14	207.0	195.8	190.9	239.4	226.6	220.8
12	328.9	310.5	301.6	379.1	359.1	350.9
10	514.6	492.1	478.8	595.2	566.1	551.0
8	807.7	769.3	745.0	930.7	996.9	860.0
6	1,258	1,203	1,163	1,447	1,385	1,346
4	1,922	1,849	1,795	2,225	2,133	2,068
2	2,928	2,806	2,714	3,379	3,232	3,139
1	3,553	3,406	3,340	4,092	3,943	3,850
0	3,950	3,790	3,737	4,547	4,379	4,318
00	4,713	4,533	4,416	5,438	5,225	5,097
000	5,528	5,313	5,225	6,391	6,149	6,040
0,000	6,485	6,256	6,138	7,469	7,222	7,086
250	6,923	6,758	6,610	7,988	7,778	7,616
300	7,735	7,496	7,355	8,925	8,664	8,492
350	8,450	8,242	8,093	9,750	9,486	9,328
400	8,936	8,736	8,592	10,341	10,114	9,928
500	9,985	9,741	9,615	11,516	11,231	11,110
600	10,555	10,399	10,193	12,214	11,943	11,778
700	11,242	11,018	10,882	12,975	12,724	12,590
750	11,502	11,258	11,144	13,293	12,968	12,851
1,000	12,175	12,019	11,867	14,058	13,878	13,703

To find the maximum permissible length of a given cable that may be used with a given current and voltage drop:

$$\text{Maximum permissible length, in feet} = \frac{\text{Factor} \times \text{Actual volts drop}}{\text{Current}}.$$

To find the drop in volts that will be obtained with a length of given cable carrying a certain current:

$$\text{Volts drop} = \frac{\text{Length of cable} \times \text{Current in amperes}}{\text{Factor}}.$$

To find the maximum current that can be carried over a length of given cable without exceeding a definite value of volts drop.

$$\text{Maximum current, in amperes} = \frac{\text{Factor} \times \text{Actual volts drop}}{\text{Length, in feet}}.$$

NOTE: Care must be taken to select the proper factor for either single- or three-phase and the one corresponding to the maximum permissible operating temperature for the particular size of conductor and grade of insulation under consideration.

87. GROUNDING PRACTICE FOR INTERIOR WIRING SYSTEMS

Grounding in interior wiring systems is done primarily as a safety measure, and two classes of grounds are concerned:

In a **system ground,** one of the current-carrying conductors of a system is connected to ground.

In an **equipment ground** non-current-carrying parts are connected to ground. This covers such things as metallic conduit, cable armor, motor frames, switch boxes, and the like.

The purpose of equipment grounds is to keep all exposed metal parts at as near ground potential as possible by providing a low-resistance path to ground, so that in case of a breakdown in the insulation, the non-current-carrying parts of the equipment cannot be raised above ground potential sufficiently to be dangerous to a person coming in contact with them. In ungrounded equipment it is also possible to have a high-resistance path to ground through some part of the building structure, such as metal lath, which would introduce a fire hazard as well as the accident risk, due to shock.

Equipment Grounds are required on exposed metal parts of practically all fixed equipment, including all metal duct and raceway systems. There are a few exceptions to this rule, the most important being that grounding of conduit or armored cable runs under 25 ft is not required if free from metallic contact with the ground or grounded metal parts and protected from contact by persons.

The exposed metal parts of portable equipment operating at more than 150 volts to ground must be grounded, and, when possible, other portable equipment should be grounded. The attachment cords for grounded portable equipment are made with an extra conductor for this purpose.

The sizes of grounding conductors for equipment grounds are determined by the class of equipment being grounded and by the size of the overcurrent device limiting the current.

System Grounds on interior wiring systems protect the low-voltage wiring in case of accidental contact with higher-voltage lines. Where the neutrals of both systems are grounded, a low-resistance path to ground is provided through the neutral of the low-voltage system, and in the event of a cross between the two systems a fuse should blow and isolate that part of the system where the trouble occurred.

If such a cross occurs with an ungrounded system, the low-voltage system will be brought up to the potential above ground of the high-voltage line. Under these conditions there is danger of shock to persons coming in contact with the low-voltage system, as well as probable damage to the insulation of the low-voltage system and the equipment connected to it.

In order to be effective, a ground connection must be of low resistance. Therefore the grounding electrode must make good contact with moist earth. Continuous water-piping systems afford the most convenient and best grounds. Where such a ground is not available, the metal frame of a building, a gas-piping system, or a properly constructed "artificial ground" consisting of a buried plate or drain pipe may be used. Artificial grounds must not have a resistance of over 25 ohms and should be less. Water-pipe grounds are of the order of 3 ohms or less.

The size of a grounding conductor for a system ground for a d-c system shall not be less than the largest conductor supplied by the system and in no case less than No. 8 A.W.G. copper. The size of the grounding conductor for an a-c system, a common grounding conductor, or a grounding conductor for service equipment shall be not less than that given in Table 32, except that, when connected to a buried plate electrode or other "made" grounds, the grounding conductor need not be larger than No. 6 or its equivalent.

Table 32. Size of Grounding Conductor

Size of Largest Service Conductor or Equivalent for Paralleled Conductors	For Wiring System and/or Service Equipment, Copper Wire No.	For Service Equipment Only, Conduit or Pipe, Inch	For Service Equipment Only, Electrical Metallic Tubing, Inches
2 or smaller	8	1/2	1/2
1 or 0	6	1/2	1
00 or 000	4	3/4	1 1/4
Over 000 to 350,000 cir mils	2	3/4	1 1/4
Over 350,000 to 600,000 cir mils	0	1	2
Over 600,000 to 1,100,000 cir mils	00	1	2
Over 1,100,000 cir mils	000	1	2

Conduit, pipe, or electrical metallic tubing cannot be used alone as the grounding conductor for a wiring system. Wire sizes apply to both bare and insulated conductors.

88. WIRING DIAGRAMS AND LAYOUTS

GENERAL. After the load and the number and type of branch circuits have been determined, it is usually necessary to make a wiring diagram showing the location of all outlets, switch equipment, motors, etc. These diagrams are generally laid out directly on a floor plan of the building with a separate schematic diagram of risers and feeders when necessary. For small residences the architect will often locate the desired outlets on the construction plans and allow the electrical contractor to locate the wiring in the most convenient manner.

The American Standards Association electrical symbols for use in architectural plans are given in Fig. 7.

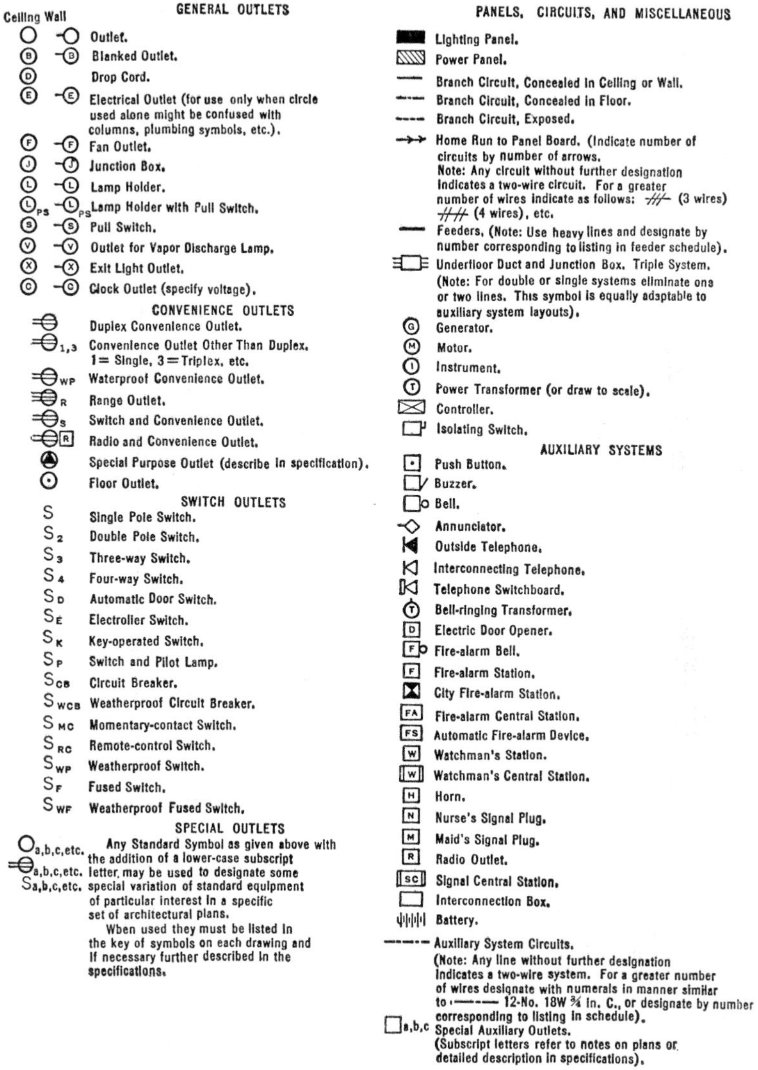

FIG. 7. Electrical Symbols for Architectural Plans (ASA)

Every wiring layout should be supplemented by a detailed specification which will control the quality of material, the wiring method, and, usually, the make or design of the wiring devices used, together with such other information as cannot be clearly indicated on a wiring layout.

It is not possible in an article of this length to cover the details of a wiring layout, and only the more important considerations are noted below. For further information see the references in the Bibliography, Article 90.

RESIDENCE WIRING. The service entrance must first be located with reference to the type of dwelling and the position of the power company's lines. In general, the power company will supervise this part of the work and quite frequently install the complete service through the meter and service switch, up to the point where the branch circuits are connected. Service conductors are brought in overhead or underground, usually through conduit or by means of service entrance cables.

The service equipment located within the building should be easy of access, and provision should be made on the panel for extra branch circuits to take care of circuits that may be added in the future.

The branch circuits should be so distributed that the blowing of a fuse in one circuit will not leave an entire floor without light. Basement, stair, and hall lights should be so wired as to operate from more than one location, and it is desirable to have several conveniently located lights operated from a switch at the main entrance.

In locating wall switches care should be taken to have them near the most-used entrance to the room, on the latch side of the door.

Consideration should be given to the use of remote control of both interior and outside lighting circuits. It is possible, by means of relays operated by low-voltage control circuits, to control lighting throughout the building from one or more central points. For example, a set of push buttons in the entrance or master bedroom can be made to operate the light switch on any circuit throughout the house. This makes it unnecessary to carry the feeders to such circuits to the central point and therefore reduces the amount and size of wire required.

In larger residences provision should also be made for telephone service, call bells, radio, and other convenience services which are not part of this lighting system.

WIRING OF COMMERCIAL BUILDINGS. The first step in laying out a wiring system for buildings other than residences is to locate on a floor plan the lighting and appliance outlets and, if possible, the probable location and capacity of power equipment. Panelboard locations should then be determined in such a manner that the branch circuits serviced from them are of as nearly the same length as possible and preferably no longer than 100 ft.

In multistory buildings the distributing boxes, from which mains run to the various panelboards, should be placed in the same locations on each floor. It is customary to provide pockets or closets in the building structure for such boxes and channels for the feeders to them.

The load for each panel should be calculated for use in determining the capacity of feeders. (See Wiring Calculations in Article 86.)

The Number of Feeders depends upon the occupancy of the building and also upon (1) the power taken by the receiving devices, (2) the degree of control required at the main distributing center or switchboard, (3) the character of the receiving devices, and (4) the uniformity of voltage required.

In multiple-family dwellings one feeder is required for each apartment, the main distributing center being a metering panelboard where the meters for the entire building, or a large section of it, are grouped.

In other buildings, if it is not required to control each floor separately, one set of feeders may be used for the entire building or a large section of it.

It is advisable to sectionalize public buildings, or buildings occupied by a large number of people, in such a manner that not all the lights will fail in the event of an open circuit on any one set of feeders. It is not good practice to supply lighting loads and large fluctuating loads from the same feeders, as the voltage will vary with the load and cause undesirable flickering of the lamps. This means that power and lighting loads should not be supplied from the same feeders.

The **maximum size** of feeders should not exceed 500,000 cir mils, as it is not practicable to run larger conductors. Where larger capacity is required, several conductors of equal size having the necessary combined carrying capacity should be used. Large conductors on a-c systems are also undesirable because of loss due to skin effect.

WIRING OF INDUSTRIAL PLANTS. If the plant is small and consumes only a nominal amount of power at low voltages, the wiring problem is not far different from that of commercial and loft buildings. If the plant is large or consumes power in large blocks, however, then the problem may become quite involved, and usually experience with the particular industry involved is essential.

In making the preliminary load survey the demand factors and power factors of the various units should be known and used in calculating feeders. Demand factors for power loads such as those established for lighting loads by the National Electrical Code (Table 25) are not available for the wide variety of industries encountered. For the continuous-operation type of industry, such as textile mills, furnaces, or pumping stations, the demand

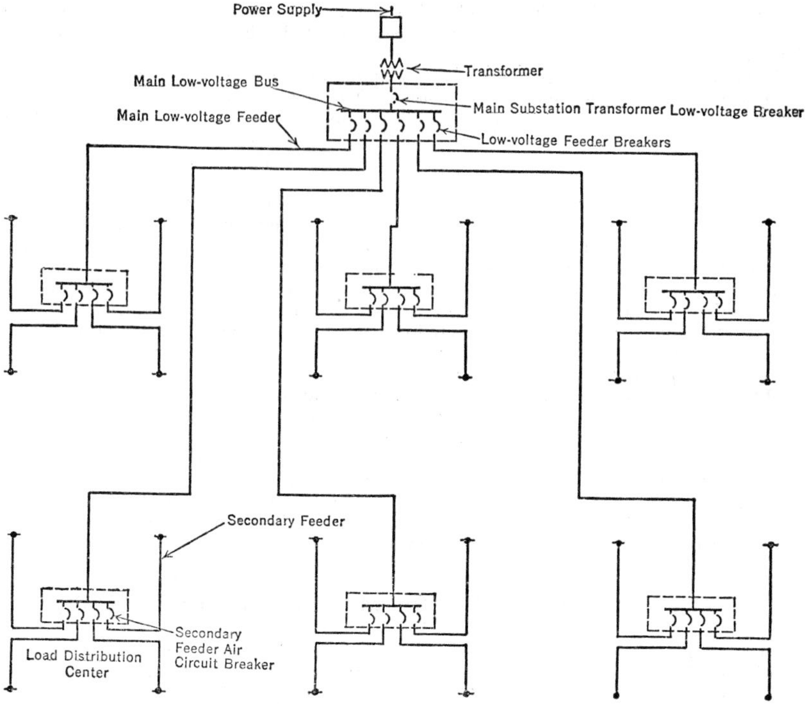

FIG. 8. Radial System of Distribution

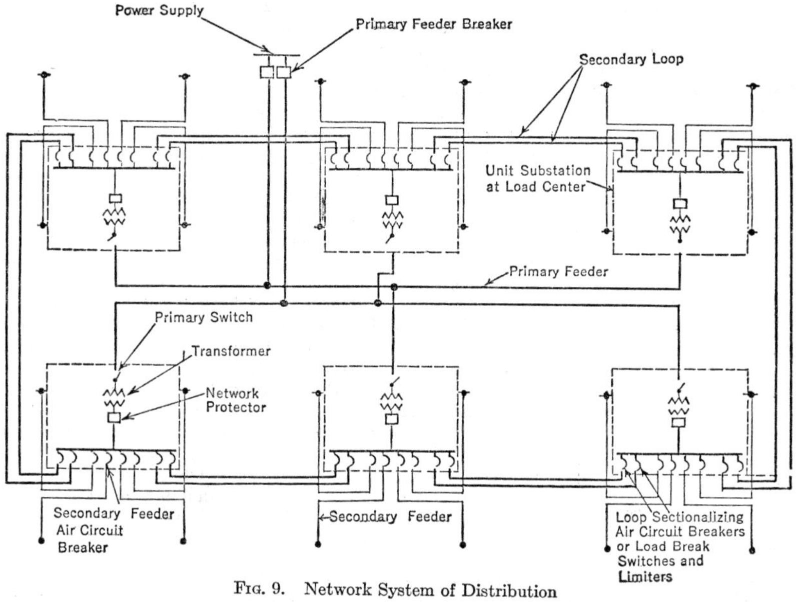

FIG. 9. Network System of Distribution

factor may run more than 90 per cent of the installed capacity; welding equipment may not exceed 25 per cent; and general manufacturing will fall somewhere between these limits.

In large plants it is also probable that an extensive feeder system is involved to take care of a number of load distribution centers. The feeder system will be some form of (1) a "radial system," where the conductors supplying the load centers receive their supply from one main supply point (Fig. 8), (2) a "network system," where the load centers are interconnected and fed from several unit substations (Fig. 9), or (3) combinations of the above.

The radial system is economical in small plants and those where the load is concentrated in a small area. When the plant is large, it has the disadvantage that the load is transmitted over considerable distance at low voltages. Radial systems may be modified in several ways, such as a number of secondary systems fed by a loop primary feeder.

In network systems the supply voltage is transformed to the utilization voltage as near the loads as practicable, the transformers being fed by two or more primary feeders and the transformer secondaries connected to secondary bus or loop. Each primary and its transformers form a unit, and the system must be so designed that when one primary goes out the remainder will carry the load.

When only two or three primary feeders are available, considerable excess conductor and transformer capacity is required to carry emergency loads due to the loss of a feeder. When several feeders are used, the spare capacity required decreases as the number of feeders increases. Where continuity of service is important, the reliability of a network system is usually worth the additional capacity required.

A complete discussion of the various types of radial and network systems is given in the AIEE publication, *Electric Power Distribution for Industrial Plants*, from which Figs. 8 and 9 were taken to illustrate the basic principle of each system.

89. WIRING BUILDINGS FOR MISCELLANEOUS DEVICES

In addition to the wires designed to conduct electrical energy to lamps, motors, and heating devices in a building, wires may be installed in connection with telephone, telegraph, district messenger and call-bell circuits, fire and burglar alarms, door-opening devices, watchman's clocks, and electric clocks. Since all these devices are operated at low voltage, it is unnecessary to use the same care in selecting and installing the wires as in the higher-voltage systems, except that in all low-voltage systems care must be exercised that the conductors do not become crossed with light and power circuits.

PROTECTION OF LOW-VOLTAGE WIRING. When the conductors of any low-voltage system are brought into a building from the outside, an approved protective device must be located as near as possible to the entrance of the wires to the building. With the exception of instrument circuits of telegraph systems, where cut-outs only are required, protective devices must contain a lightning arrester with a ground connection and a cut-out or heat coil (see Pender and McIlwain, *Communication and Electronics Handbook*). The conductors beyond the protective device in low-voltage systems need be insulated only sufficiently to prevent short circuits and the consequent interruption of service. When bunched together in vertical runs, the wires must be inclosed in a fire-resisting covering to prevent the wires carrying fire from floor to floor. Low-voltage circuits may be run in the same shaft with light and power circuits, provided the two classes of wires are separated by at least 2 in. or one of the classes is run in a non-combustible tubing. Low-voltage wires may not be run in any case in the same conduit or raceway with lighting or power wires.

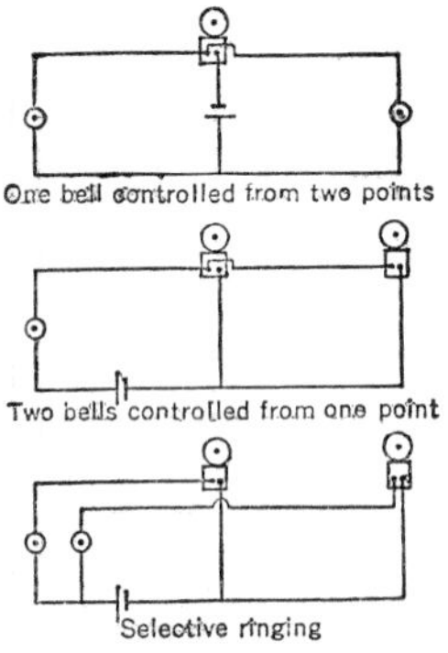

Fig. 10. Electric Bell Circuits

BELL WIRING. In its simplest form a bell-wiring system consists of a battery (or a bell-ringing transformer connected to the lighting circuit), a bell push, a bell or buzzer, and the connecting wire. Paraffin-impregnated double-cotton covered wires, or rubber-covered fixture wire in damp locations, ranging in size from No. 16 to No. 22 A.W.G. gage, are usually employed, either singly or in the form of twin wires. The wires may be fished, concealed in molding, or run exposed along the finish of the room. When a large number of bell wires must be run together through a building, as in a hotel annunciator system, the wires are frequently enclosed in a cotton braid or lead sheath. In many buildings supplied with an a-c source of power a small step-down transformer is used in place of the batteries, thus effecting a saving in the cost of the installation and eliminating the possibility of run-down batteries. In Fig. 10 are shown several systems of bell wiring; these diagrams are self-explanatory.

BURGLAR ALARMS. In place of the bell push used in bell-wiring systems, the bell circuit may be closed by some circuit-closing device attached to windows, doors, etc., so that

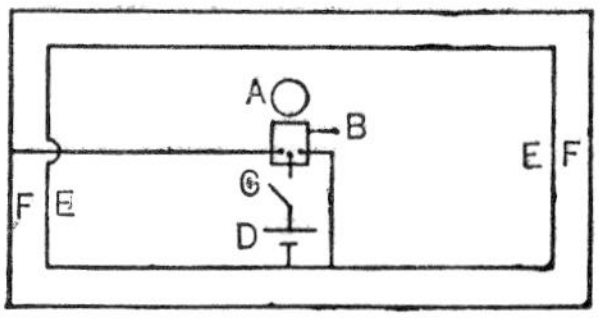

A Bell B Constant Ringing Drop
C Day Switch D Battery
E and F Wires to which closing springs are connected

FIG. 11. Open-circuit Burglar-alarm System

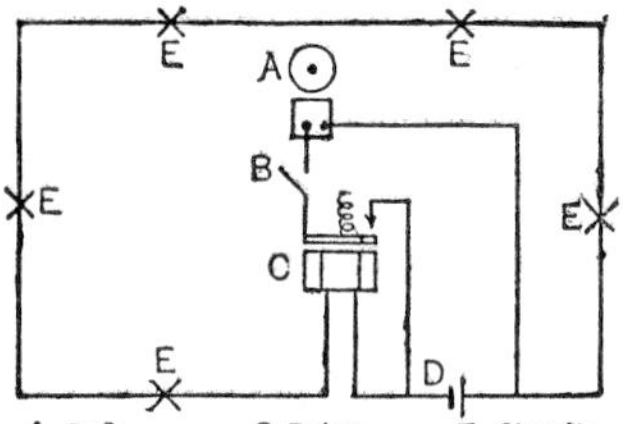

A Bell C Relay E Circuit-opening Springs
B Day Switch D Battery

FIG. 12. Closed-circuit Burglar-alarm System

the opening of any window or door in the building will be made known by the ringing of a bell. As such a system has the objection that it will not operate if the battery is run down or if the wires are cut, a closed-circuit system is preferable, in which the opening of a window or door opens the circuit and a relay in turn closes the bell circuit. In Fig. 11 is shown a typical open-circuit and in Fig. 12 a typical closed-circuit system. In Fig. 13 are shown a door spring *A* and a window spring *B*. Photoelectric cells are also used for actuating alarms.

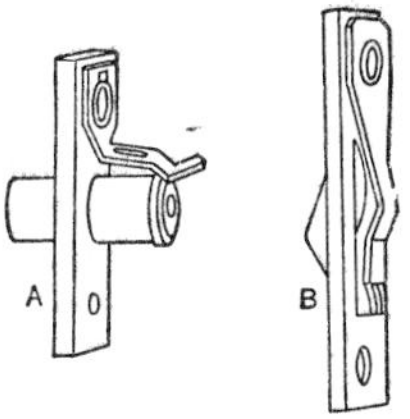

FIG. 13. Door (*A*) and Window (*B*) Springs

COMMUNICATION SYSTEMS. This classification includes telephone, telegraph (except radio), district messenger, fire and burglar alarms, and similar central station systems. The National Electrical Code, which governs other wiring systems in detail, concerns itself only with protective measures essential to safeguard the operation of communication systems. Detailed service requirements for fire alarm, sprinkler supervisory systems, etc., are covered by Vol. V of the National Fire Codes, "Extinguishing and Alarm Equipment," published by the National Fire Protection Association. The general requirements of low-voltage wiring as given above apply to these systems.

RADIO EQUIPMENT. Antenna systems, counterpoise, and lead-in conductors are required to be of hard-drawn copper, bronze, or other corrosion-resistant high-strength material. Outdoor antenna and lead-in conductors shall be securely supported, shall not be attached to poles or structures carrying power wire of more than 250 volts, and shall be kept well away from such wires to avoid the possibility of accidental contact.

The sizes of antenna and counterpoise required for various spaces are as follows:

Material	Minimum Size of Conductors, A.W.G.		
	0–35 ft Span	35–150 ft Span	Over 150 ft
Receiving Stations			
Hard-drawn copper	19	14	12
Copper-clad steel or bronze	20	17	14
Transmitting Stations			
Hard-drawn copper	14	14	10
Copper-clad steel or bronze	14	14	12

Protective grounding conductors are required to be not smaller than No. 14 copper or No. 17 bronze or copper-clad steel for receiving stations. For transmitting stations they must be as large as the lead-in, but in no case smaller than No. 14. A single conductor may be used for both protective and operating ground.

All noise suppressers connected to power-supply leads must have the approval of the Underwriters' Laboratories.

90. BIBLIOGRAPHY

Abbott, A. L., *National Electrical Code Handbook.* McGraw-Hill.
American Institute of Electrical Engineers, *Electric Power Distribution for Industrial Plants.*
Cook, Arthur L., *Electric Wiring for Lighting and Power Installations,* Third Ed. John Wiley (1933).
Croft, T., *American Electricians' Handbook.* McGraw-Hill (1941).
Industry Committee on Interior Wiring, *Handbook of Interior Wiring Design* (1937), *Handbook of Residential Wiring Design* (1946), *Handbook of Farmstead Wiring Design* (1946).
McKinley, R. B., L. D. Madsen, Voltage-drop Calculating Charts for Cable, *Gen. Elec. Rev.* (Sept., 1941).
National Board of Fire Underwriters, *National Electrical Code.* New York (1947).
National Bureau of Standards, *National Electrical Safety Code.* Washington, D. C.
Rosch, S. J., The Current-carrying Capacities of Rubber-insulated Conductors, *Trans. AIEE,* Vol. 57, p. 155 (1938).
See also articles in *Electrical World, Electrical Contracting, Qualified Contractor.*

GROUNDING OF ELECTRIC CIRCUITS

By A. S. Brookes

91. FUNDAMENTAL THEORY

The subject of grounding covers the problems relating to the conduction of electric currents to the earth and through the ground. The earth now rarely serves as a part of the return circuit, being used mainly for fixing the potential of circuit neutrals. The ground connection improves service continuity and protects lives and equipment.

RESISTANCE OF THE SOIL. The electrical conductivity of the materials constituting the earth's surface is very low compared with the high conductivity of metals, since the main constituents of the earth, silicon dioxide and aluminum oxide, are excellent insulators. The conductivity of the ground is due largely to salts and moisture. Even such a semiconductor may carry a considerable amount of current if the cross section is large enough.

Because of the high resistivity all currents flowing through the ground suffer a considerable voltage drop. These potential gradients will be considered in two different zones, namely, the spaces in proximity to the ground electrodes and the long paths between such electrodes.

A quantitative analysis is necessary to permit numerical calculations and to draw definite conclusions.

The resistivity of the soil depends on its type and dryness and varies with distance as well as depth. The resistivity is much lower below the subsoil water level than above it. In frozen soil, as in the surface layer in winter, it is particularly high. For good ground electrodes it is important, therefore, to avoid the frost limit and to reach the subsoil water plane.

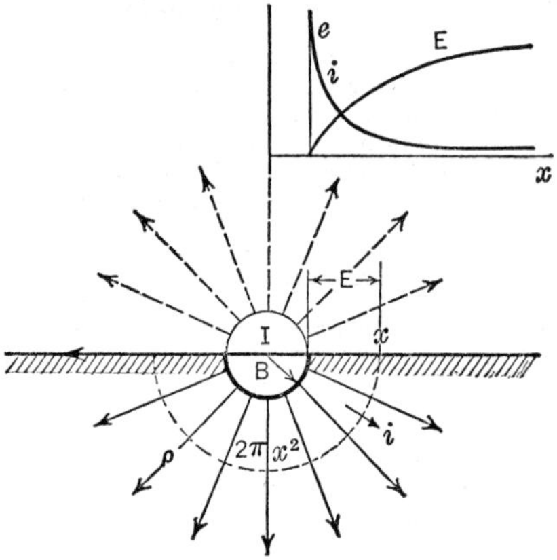

Fig. 1. Radial Flow of Current from Hemispherical Electrode to Ground (From R. Rudenberg, *Elec. Eng.*, Jan. 1945)

The greater the geological age of the strata, the higher the resistivity, but there are exceptions. In the strata under the surface, loam, clay, and limestone usually have lower resistivity; sandy and rocky materials have higher resistivity. For power networks and for radio waves the resistivity near the surface is more important, but for telephone interference by action at a distance the resistivity of the deeper layers has greater significance.

SPHERICAL GROUND ELECTRODES. The simplest electrode is a sphere in ground symmetrical in all directions, as shown in Fig. 1. The lower hemisphere embedded under the surface of the earth may be taken as an example. If a current I flows from the electrode, spreading out radially in the ground, the current density at distance x from the center of the hemisphere is

$$i = \frac{I}{2\pi x^2} \tag{1}$$

Such a current produces in soil of resistivity ρ an electric field strength

$$e = \rho i = \frac{\rho I}{2\pi x^2} \tag{2}$$

Developing these equations to determine the resistance encountered by the current from the hemisphere gives

$$R = \frac{E}{I} = \frac{\rho}{2\pi B} \tag{3}$$

where B is the radius of the sphere.

FIELD STRENGTH ON SURFACE. If a man is walking through a surface field of electric potential, such as that shown in Fig. 2, near a faulty transmission tower, the body diverts some current from the earth and may suffer damage. With a tower footing current of I amp, and the footing represented by an equivalent sphere of radius B, the field strength is given by eq. (2) and is a maximum (e_B) at distance B. Then

$$e_B = \frac{\rho I}{2\pi B^2} \tag{4}$$

If the man's internal body resistance is low in comparison with his foot resistance, and b is his equivalent foot radius (an average figure is 7 cm), the maximum body current, with a step length s, is:

$$i_s = \frac{sb}{2x^2} I \tag{5}$$

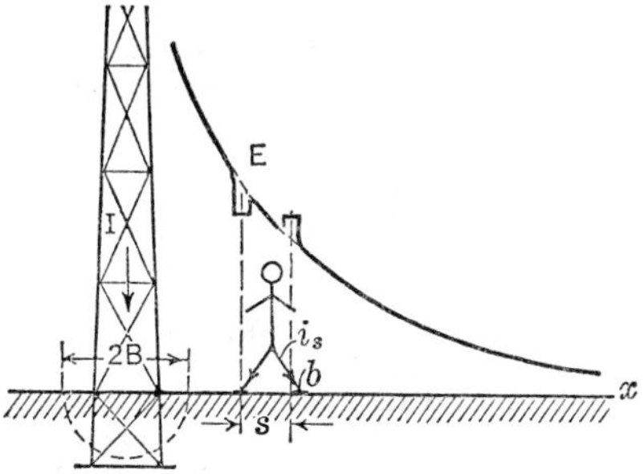

FIG. 2. Potential at the Surface of the Earth near a Current-carrying Electrode (From R. Rudenberg, *Elec. Eng.*, Jan. 1945)

MEASUREMENTS BY PROBES. Ground resistivity may be measured by passing current through two short ground rod probes of radii B and b, at a spacing x, much greater than B and b.

$$\frac{E}{I} = \frac{\rho}{2\pi}\left[\frac{1}{B} + \frac{1}{b} - \frac{2}{x}\right] \tag{6}$$

If the electrode radius b is made small in comparison to B and x, the resistivity

$$\rho = 2\pi b \frac{E}{I} \tag{7}$$

If no specific resistivity measurements for a definite spot of ground are made, the values in Table 1 may be used as typical. (See Article 22.)

Table 1. Average Resistivity of the Ground

Type of Ground	Resistivity ρ, Ohmmeters
Wet organic soil..........	10
Moist soil...............	10^2
Dry soil.................	10^3
Bedrock..................	10^4

TWO PARALLEL GROUND ELECTRODES. The resistance of two spheres in the ground at a spacing of $2Z$ and carrying together the current $2I$ is

$$R = \frac{E}{2I} = \frac{\rho}{4\pi b} \cdot \frac{1}{2}\left(1 + \frac{b}{2Z}\right) \tag{8}$$

The last factor for far-distant electrodes with $Z = \infty$ becomes $^1/_2$; but for close electrodes, for example, with $Z = b$, it becomes $^1/_2(1 + {}^1/_2) = {}^3/_4$. Thus distant spheres are mutually independent in their resistance, and their parallel resistance can be computed according to ordinary rules. Close electrodes, however, experience an increase in their ground resistance by mutual interference.

DEPTH ELECTRODE. With a spherical electrode buried at a depth Z, the resistance is given by the equation:

$$R = \frac{E}{I} = \frac{\rho}{4\pi b}\left(1 + \frac{b}{2Z}\right) \tag{9}$$

With a depth electrode the space of high current concentration is not accessible to persons; thus the danger of stepping near the electrode decreases. The maximum field strength is

$$e = \frac{\rho I}{3\sqrt{3}\pi Z^2} \tag{10}$$

The ratio of this value to maximum value of eq. (4) for a surface electrode is

$$\frac{2}{3\sqrt{3}}\left(\frac{b}{Z}\right)^2 = 0.39\left(\frac{b}{Z}\right)^2$$

DRIVEN ROD AT EARTH SURFACE. The most common form of ground electrode is a rod having a relatively small radius a in comparison with the length l. The potential at a remote part is

$$V_x = \frac{\rho I}{2\pi x} \tag{11}$$

The resistance

$$R = \frac{E}{I} = \frac{\rho}{2\pi l}\left(\log_\epsilon \frac{4l}{a} - 1\right) \tag{12}$$

The field strength, which is much larger than for the hemisphere because of the high concentration of current around the small rod, is

$$e_0 = \frac{\rho I}{2\pi l x} \tag{13}$$

MULTIPLE ROD ELECTRODE. The analysis of the resistance of several rods in parallel cannot be undertaken here. In general, the mutual effect increases as the rods are more closely spaced. In other words, the greater the spacing, the more effective the individual rod. [See R. Rudenberg, *Elec. Eng.* (Jan., 1945) or H. B. Dwight, *Trans. AIEE*, Vol. 55, p. 1319.]

IMPULSE CHARACTERISTICS. The performance of a ground electrode under impulse currents must be known if lightning-protection problems are to be accurately analyzed. The capacity and inductance of a rod are too small to influence the performance of a driven rod, even with lightning surges. However, lightning discharges generally produce such high voltages at the rod that the ground breaks down electrically, thus increasing the effective diameter of the rod. Hence, in general, the effective resistance under impulse conditions is less than under low-frequency currents.

92. TYPES OF GROUNDS

WATER PIPES. Because of the great extent of water-piping systems, they offer good conductance to earth. The actual resistance is usually a fraction of an ohm. The low resistance of such systems is due to their size and the fact that the pipes are in contact with a great volume of earth below the frost line. They are the most widely used means of grounding for electric and telephone services.

The resistance of a water-piping system, as measured at a specific point, may vary greatly, depending upon whether near-by pipe joints are electrically conductive or are the high-resistance cement or "leadite" type.

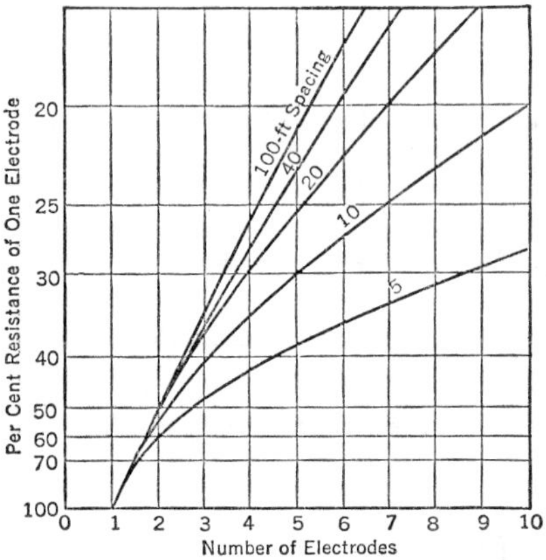

FIG. 3. Efficiency of Driven Electrodes in Multiple at Various Spacings (From C. Jensen, *Elec. Eng.*, Feb. 1945)

OTHER STRUCTURES. Structural-metal frameworks of buildings which make direct contact with the soil are permitted by various codes as an alternate method of grounding when a water-piping system is not available. The steel pipe of a well casing is sometimes used as an effective ground in rural areas.

DRIVEN ELECTRODES. The most widely used form of installed ground electrode is the driven ground rod. These rods are generally 8 ft long, but sectional rods are used to obtain greater lengths and thus secure greater conductivity in high-resistance soils. Multiple rods are also used when the resistance of a single rod is too high. Such rods must be well spaced to obtain any benefit, as is shown in Fig. 3, which gives the relative resistances of multiple rods at various spacings.

BURIED ELECTRODES. Buried metal strips or cable are frequently used in areas where underlying rock makes driven grounds impractical. The extent of such a system depends upon the resistance and type of protection desired. For substation protection it is common to use a grid of buried conductors to provide an equipotential area over the entire station plot. Typical dimensions are a depth of 6 in. and cable spacings to form squares, 10 or 12 ft on a side.

CHEMICAL TREATMENT. Chemical treatment of grounds to secure lower resistance is sometimes used. A typical method is to place the chemical in a circular trench around but not touching the electrode. Common salt and magnesium sulfate are most commonly used. The treatment is effective but is not permanent, as the chemicals are dissipated by the natural drainage of the soil and must be replaced every few years. A further disadvantage is that corrosion of the rod is accelerated by the chemical treatment.

93. MEASUREMENT OF GROUND RESISTANCE

Instruments for measuring ground resistance generally require two auxiliary grounds with resistance comparable in magnitude to the one being measured. The grounds must be placed at distances sufficient to eliminate proximity effect.

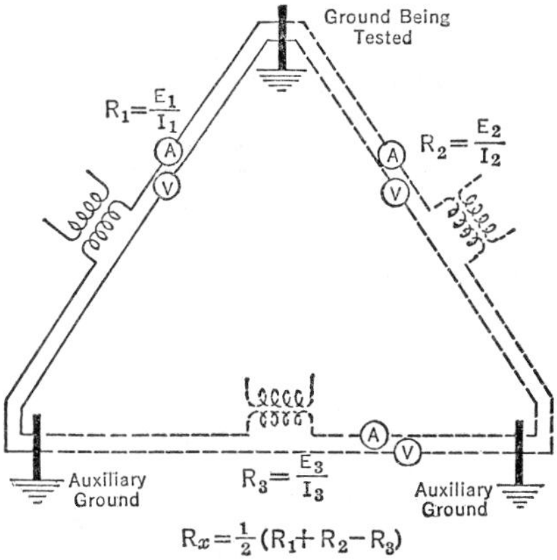

FIG. 4. Connections for Three-point Method of Measuring Ground Resistance, Using A-c Ammeter and Voltmeter (From C. Jensen, *Elec. Eng.*, Feb. 1945)

The well-known three-point method of measurement shown in Fig. 4 requires ammeter, voltmeter, and power source, a-c or d-c. After the three series resistance measurements have been successively made, the unknown resistance can be determined from the formula

$$R_x = {}^1/_2(R_1 + R_2 - R_3) \qquad (14)$$

If the current is direct, the average of readings made with reversed polarity should be used to eliminate polarization effects.

94. GENERATOR NEUTRAL GROUNDING DEVICES

In the grounding of a generator, theoretical considerations, as well as experience and operating records, should be given considerable weight. A generator is a highly insulated, rugged piece of equipment, and a failure on a particular unit is not expected. The importance and high cost of a generator make it essential to guard against any such failure. Protective relays are installed, phase-to-ground winding fault currents are limited to safe values, lightning-surge equipment is applied, and other precautions are taken.

The method of grounding the generator neutral has an important bearing on the magnitude of the transient voltages which the generator must withstand. The usual methods of grounding a generator neutral follow:

REACTOR. The reactor chosen must limit the fault currents to a safe value and yet not have such a high reactance that transient voltages are excessive. It is usual to choose a reactor of such size that the single phase-to-ground fault current, under the worst operating condition, is limited to that value which would result from a three-phase fault.

RESISTOR. The current through a neutral resistor is usually limited for large generators to a minimum of about 100 amp and to a maximum of about 1.5 times the normal rated generator current. Relaying requirements and inductive coordination needs will determine the value chosen.

TRANSFORMER WITH SECONDARY RESISTOR. A generator may be grounded through a small power or distribution transformer, the secondary of which is shunted by a resistor to provide safe and economical phase-to-ground fault protection to the generator circuit. The transformer serves only as a means of changing a low ohm resistor, in effect, into a high resistance in the neutral.

POTENTIAL TRANSFORMER. A generator with a potential transformer in the neutral is grounded only through the capacitance to ground of the winding and associated circuits. The potential transformer is only a device for measuring the voltage between neutral and ground and actuating an alarm or tripping relay. The machine is essentially ungrounded.

INFINITE IMPEDANCE (UNGROUNDED). The disadvantages of ungrounded generators are:

1. It is not possible to relay for ground faults.
2. Arcing ground faults or circuit-breaker operation may develop transient voltages high enough to cause generator failure.
3. Phase-to-phase faults are more probable than on a grounded generator.
4. Feeders supplied at generator voltage will not trip to remove the hazard of a broken wire on the ground.

GENERATOR-SURGE PROTECTION. For 11.5-kv and 13.8-kv grounded generators, one standard 0.25 μf capacitor unit on the terminals is recommended, whereas for generators in the ungrounded classification two units such as these are recommended. Any generator with a solid ground or neutral resistor of less than 50 ohms is considered grounded, and all others are considered ungrounded.

95. SUBSTATION GROUNDING

In this discussion of grounding, the term substation refers to switching stations, distribution substations, and the larger industrial substations. Smaller industrial stations have grounding requirements similar to those of distribution transformers, which are discussed in Article 96. The principles apply to all types of stations but are most applicable to those with overhead transmission and distribution lines.

OBJECTIVES. Grounding is an important feature of substation design. Its effectiveness is important in maintaining service continuity, protecting costly equipment, and safeguarding operating and maintenance personnel. To accomplish these objectives, the following specific purposes must be provided for:

1. Grounding for lightning and surge protection.
2. System grounding to stabilize circuit potentials to ground and to permit circuit relaying for the clearing of ground faults.
3. Grounding of non-current-carrying structures and auxiliary low-voltage circuits for safety to persons.

DESIGN. Since the interconnection of all grounds results in maximum effectiveness, modern substation grounding practice is to use a single all-purpose ground bus, or two or more interconnected buses if the layout makes this arrangement more convenient.

Since it is important to have a nearly uniform potential of earth surfaces over the area under consideration, distributed ground beds of driven rods are economical in areas free of rock. If practicable, the rods should be driven to levels of permanent moisture. Where pile foundations are needed for structural reasons, grounds have been made by attaching copper wire in grooves on one side of the piles. These may be used in combination with other types of electrodes.

Where ground beds of driven rods are used, the most efficient arrangement is to have a fair amount of separation between rods. This leads to the common practice of driving a series of rods around the outer perimeter of the station, supplemented by additional rods scattered through the station area at points of connection to lightning protective devices or other important equipment. The conductor connecting the rods forms a portion (sometimes the major portion) of the ground bus, and in outdoor stations it may be buried several inches under the surface of the earth.

In areas of high-resistance soil, chemical treatment is sometimes used. In some very dry, sandy soils, deep wells have been sunk, and the metal casing or immersed plates used for the ground electrodes. In other cases of difficult grounding at the substation site, better facilities have been found near by, such as a stream or other body of water.

This requires extending the ground bus out to the ground bed. Care must be taken that, under fault conditions, the ground bed and associated connections do not constitute a high-potential hazard to persons.

In some thickly populated areas the water-piping system is used to supplement the substation ground. Conditions should be investigated to ensure that the water pipes have adequate current-carrying capacity and that the local resistance is low enough to avoid high-potential hazards to water consumers in the vicinity and to prevent local high-potential gradients at the ground surface in the substation area and vicinity.

RESISTANCE. The need for a specific value of ground resistance at a station depends upon the purposes for which the grounding is used and other related factors, such as the voltage of station circuits, the fault current required for relaying, and the impedance of circuits and station equipment. Therefore the requirements for conductivity to ground may be determined by the needs for system neutral grounding. Unless the ground resistance is low enough to prevent hazardous potentials on the ground bus under fault conditions, however, special precautions must be taken to equalize potentials between exposed objects. This means solidly connecting all such structures, usually including the metal fence, to the ground bus and also providing that any surface on which a person may stand is at substantially the same potential as surrounding objects. In outdoor stations this purpose sometimes is accomplished by placing steel plates well connected to surrounding objects, such as switch handles, for personnel to stand upon in locations involving specific operating duties.

Another means of equalizing surface potentials is the use of a grid of light conductors buried a few inches. It may be desirable to extend the ground grid outside the fence for a few feet, if the fence is of metal, for personnel protection.

SIZE OF GROUND BUS. Liberal size of ground bus is a good investment. All connections should be brazed or clamped. It is good practice to provide two or more ground paths for equipment that may receive heavy fault currents. Figure 5 gives the calculated

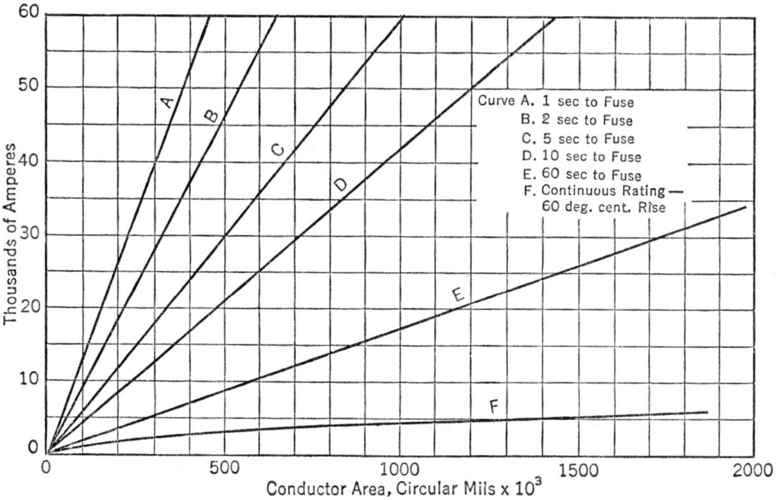

FIG. 5. Calculated Current Required to Fuse Copper Conductors for Various Melting Times (Radiation Neglected)

burn-off currents for various fault durations, plotted against conductor size. These values and the calculated fault currents, with a liberal added factor of safety, may be used in ground-bus design. Low resistance along the bus is essential to prevent potential differences which may cause insulation breakdown on connected low-voltage apparatus, such as relays and instrument transformers. Safety to personnel is another consideration requiring liberal conductivity to reduce voltage differences.

MATERIAL. Because of its high conductivity and long life copper is the usual material for a buried ground bus. The question of possible corrosion caused by the electrolytic couple between the copper and connected structures, such as footings, steel pipes, and cables, is being raised at the present time. Some users are considering cathodic protection to minimize such corrosion.

96. TRANSMISSION-SYSTEM GROUNDING

Grounding of overhead transmission lines is done almost entirely for lightning protection. Ground resistance, in this article, refers to impulse resistance, which is usually somewhat lower than resistance at power frequencies.

PURPOSE. The primary purpose of lightning protection is to prevent circuit tripouts by minimizing flashovers. Practically any degree of immunity from service interruption due to lightning can be secured, the only limitation being the justifiable cost. The two principal methods of protection for transmission lines are the drainage of surges from the conductors by expulsion gaps and the shielding of conductors by overhead ground wires.

EXPULSION GAPS. Expulsion gaps may be classed as a special form of lightning arrester and are connected (with external air gap) so as to bridge the insulation to be protected. They limit surge potentials to values below insulation flashover and provide a path to ground for the surge current. If gaps are installed at every structure, low-resistance ground connections are not required. This is a useful feature in areas of high-resistivity soil. Low-cost types of grounds are used, such as the normal footing resistance of steel towers, pole-butt-wrapped grounds, or single driven rods for wood pole structures. In the simplest form of connection one terminal of each gap is connected to the ground lead or tower. The other terminal forms one electrode of an air gap to the conductor. Sets of expulsion gaps may be installed at every structure or at greater spacings, depending upon the degree of flashover immunity desired. As the spacing is increased, however, ground resistance plays a more important part.

A special form of expulsion-gap application uses one of the line conductors as an overhead ground wire and requires only one gap per structure. For this purpose the line conductor so used must be elevated above the other conductors. Lightning striking the upper conductor passes through an air gap to the expulsion gap and thence to ground. In this type of protection the ground resistance is important and is governed by the principles of ground wire protection.

GROUND WIRE PROTECTION. The most widely used form of protection is that of shielding by overhead ground wires. Such ground wires, paralleling the line conductors, are supported above them and arranged so as to shield the lines and take the strokes. Rules have been worked out for such arrangements.

The ground wires are connected to earth, usually at each structure. When a direct lightning stroke hits the ground wire, the current flows to ground. The resulting potential on the ground wire can be found approximately by multiplying the ground resistance in ohms by the amperes in the stroke, which may be estimated as 50,000 for a moderate stroke or 100,000 for a top practical value. The resulting tower-top potential may be millions of volts. Since the line conductors, relatively speaking, stay at zero potential, the chances are great that there will be flashovers from tower to line conductors, followed by a flow of power current and line tripping. Since the tower potential is limited by the resistance of the ground, a low resistance reduces the amount of insulation required to prevent flashover. It is usually more economical to reduce ground resistances than to increase insulation.

Actual calculations require consideration of the surge impedance of tower and ground wire, the distance between grounding points, the successive reflections of the surge waves along the ground wire at adjacent towers, and the surge coupling between ground wires and conductors. Practical formulas exist for these calculations, as, for example, in L. W. Bewley's *Travelling Waves on Transmission Lines*. If the frequency of storms in a given area and the approximate percentage of strokes having various degrees of severity are known, a line can be designed for protection against any desired percentage of expected strokes, and the cost of such protection evaluated.

The footings of steel towers are in contact with the soil, but the resistance is frequently too high for adequate lightning protection. Supplementary grounds may take the form of rods driven to permanent moisture level. For maximum effectiveness the rods should be well spaced from the tower legs.

COUNTERPOISE. Where ground rods are ineffective because of dry soil, rock, or other factors, the counterpoise may be used. The counterpoise consists of one or more bare conductors buried a few inches and extending away from the tower footing in two or more directions. Maximum efficiency is obtained from four radial wires equally spaced, but practical right-of-way considerations and safety factors usually lead to the use of one or two conductors parallel to the line conductors and extending from tower to tower. Counterpoise wires may be of steel or copper. Their effectiveness is largely independent of diameter.

WOOD POLE LINES. Ground wire protection on wood pole lines follows the same general principles as that for steel tower lines, except that the wood may be used effectively to augment the line insulation. Roughly speaking, for surge voltages, 1 ft of wood is equivalent to one suspension-insulator unit. Butt-wrapped grounds on wood poles are approximately equivalent to a ground rod driven to the same depth.

97. DISTRIBUTION-SYSTEM GROUNDING

Grounding of primary distribution circuits has two principal purposes: stabilizing circuit potentials to ground to avoid overstressing insulation, and providing a path for ground faults to aid in the operation of overcurrent devices and secure prompt disconnecting under fault conditions. This second function is important in providing safety to the public.

PURPOSES. Distribution-system insulation levels are inherently low, and it is not economical to provide shielding from direct lightning strokes. A certain number of line trip-outs are expected. Service continuity is secured by automatically reclosing circuit breakers, duplicate lines in some cases, and other measures to minimize the effect of trip-outs. Lightning protection is largely limited to prevention of insulation failures in transformers and other equipment where repairs are costly and involve lengthy service interruptions.

MULTIGROUNDING. Multigrounding of secondary-circuit neutrals became common early in the development of grounding a-c distribution systems. Individual grounds at transformers and at services, interconnected through the circuit neutral, and sometimes supplemented by grounds at other locations, constitute the basis of multigrounded secondary neutrals. Multigrounding of primary neutrals developed more slowly but came into common use with the development of the 2400/4160-volt four-wire grounded neutral system. As the next logical step in the trend toward common grounding facilities, the **common primary and secondary neutral** is coming into general use. It aids grounding efficiency and requires only one neutral wire instead of two. On overhead lines the common neutral is located at the same general level on the poles as the secondary conductors. The insulator may be omitted.

Whereas in urban areas multigrounded networks may be continuous over the area and may involve thousands of grounds, in rural areas connection in networks is not often practicable, and grounds are spaced more widely. Also, water pipes, so effective as a ground means in urban areas, generally are lacking in rural areas. Secondary mains are infrequent, customers being served by individual transformers. In rural areas, therefore, all available means of making the grounding system effective are needed. Since low-resistance grounds are uncommon in rural areas, there was early opposition to the common neutral. It is now realized, however, that the common neutral increases safety in rural areas, and its use is widespread. Where secondary mains are not present, the term common neutral means that the grounded secondary neutrals of transformers are connected to the multigrounded primary neutral.

Multigrounding is also effective in reducing the impedance of the neutral conductor and improving voltage regulation in a circuit in which the neutral carries some current. Data are available for calculating the magnitude of this effect.

GUYS. Connection of effectively grounded guys to the common neutral results in reducing the impedance of neutral and guy. This is often done, instead of using insulators in the guy, where permitted by the National Electric Code.

TRANSFORMER PROTECTION. The distribution transformer is the piece of overhead equipment most frequently requiring lightning protection. It has been conclusively shown that, if the ground lead of a lightning arrester protecting the primary side of a transformer is connected to the secondary neutral (or better still to the common neutral), the quality of the protection is greatly improved, because the arrester then bridges directly the insulation to be protected. In addition, grounding is improved by the addition of the arrester ground to the common grounding facilities. The most effective lightning protection requires connecting together the primary and secondary neutrals, the lightning arrester ground lead, and the transformer tanks. Some organizations wish to use wood insulation as a measure of safety to workmen and prefer not to ground the tank directly. In such cases a gap may be used between the tank and ground connection.

CABLE SHEATHS. See Articles 71 and 73.

NEUTRAL GROUNDS AT SERVICES. In addition to the grounds found at transformers and other equipment to provide circuit neutral grounding and lightning protection, there are neutral grounds at services. Ground electrodes may be of various kinds; continuous water-piping systems are preferred, and driven grounds and pole-butt grounds are probably next in order of extent of use. The aim is maximum safety provided by using the best available means at hand for making individual grounds and making full use of the principle of pooling ground facilities.

UTILIZATION EQUIPMENT. On the consumer's premises, exposed metallic parts of wiring systems and some types of utilization equipment are grounded for safety. The National Electrical Code deals in detail with such grounding (see Article 87). Grounding is an important safety measure, but at times insulation may be a superior alternative. For example, the metallic shell on lamps on flexible cords can never be considered permanently and adequately grounded. An insulated cover provides safety in this case, which is merely one of the numerous uses where grounding cannot be the only answer.

98. STATIC ELECTRICITY IN INDUSTRY

Static electricity in industry causes many fires and explosions resulting in property damage and injuries and deaths to workers. Static electricity is generated when two dissimilar substances are brought into contact. Voltages appear when the substances are separated. The charge per unit of contact area of any two such substances in contact is proportional to the difference of the two dielectric constants.

Some examples of cases where static electricity is generated in quantities sufficient to form sparks, which may ignite flammable gases or mixtures, are the following:

1. Induced charges may be formed on ungrounded metallic objects by passing clouds.
2. Individuals may be charged by walking over a wide variety of surfaces, such as dry concrete, carpets, linoleum.
3. Power and conveyor belts are very active producers of static electricity.
4. The continuous flow of paper or textiles through manufacturing operations may generate dangerous voltages.
5. The flow of hydrocarbon liquids, such as gasoline, through pipe and hose lines may build up dangerous charges on the hose or tank, if either is insulated.
6. Rubber-tired vehicles travelling at high speed over highways may develop 15,000 to 25,000 volts or more.

There are many examples in addition to these. No one remedy will apply to all cases, and remedies should be chosen only after careful study. Incorrect application may accentuate the danger instead of removing it. A brief summary of the general principles of grounding as a means of mitigating static electricity follows:

1. All stationary parts of machinery which are employed in processes where static electricity is generated should be permanently grounded.
2. Rotating parts of such machinery should also be grounded, a suitable form of grounding brush being used to bear on the rotor or shaft. The insulating film of lubricating oil in the bearings makes this necessary.
3. Where explosive atmospheres pervade wooden buildings or surround outdoor structures, all metal objects, such as roofs, tanks, fences, railway sidings, and truck bodies, should be grounded.
4. Where static electricity is generated on rapidly moving paper, textiles, power belts, etc., the processing parts of the machine should be grounded, but other grounded objects should be removed to safe distances to prevent jumping of sparks.
5. The temporary grounding of portable equipment during certain operations in flammable atmospheres may constitute a highly dangerous practice, unless approved grounding methods are employed.

99. BIBLIOGRAPHY

Grounding Principles and Practices (in five parts), *Elec. Engineering* (Jan. to May, 1945).
Peters, O. S., Ground Connections for Electrical Systems, *Nat. Bur. Standards Tech. Paper* 108 (1918).
American Institute of Electrical Engineers, *Lightning Reference Book* (1918–1935).
AIEE Standard 32, Neutral Grounding Devices (May, 1947).
National Electrical Code.
Edison Electric Institute, Principles and Practices in Grounding, *Pub.* D-9 (Oct., 1936).
National Electrical Safety Code.

See also articles on various aspects of grounding in *Elec. Engineering* (Aug., 1934, Nov., 1935, Dec., 1936, and Sept., 1942) and in *AIEE Trans.* (1936, 1939, 1941, 1942, 1943, and 1944).

ELECTROLYSIS OF UNDERGROUND STRUCTURES

By L. J. Gorman

100. CORROSION PROCESSES

CAUSES OF CORROSION. The underground corrosion of metal is caused principally by electrochemical reactions, which are accompanied by a flow of electric current between a portion of the metal and the moisture in its environment. The electric currents associated with the electrolytic processes of corrosion may originate externally, as in stray current electrolysis, or they may result from local electrolytic cells which form on the surface of the metal or in its environment. Aside from the source of current, the corrosion processes are essentially the same, and they may be treated under the general subject of electrolysis. Other causes of underground corrosion include the chemical reactions of corrosive ingredients in the environment and the action of certain anaerobic bacteria.

ELECTROLYTIC CELL. Essentially, an electrolytic cell comprises two electrodes, the anode and the cathode, and a conducting solution which constitutes the electrolyte. In the corrosion process metal ions pass into solution at the anode, and hydrogen ions are displaced at the cathode. The anode and cathode reactions are accompanied by an electric current which flows from the anode, through the electrolyte, to the cathode. Simultaneously, various counter reactions occur which retard the corrosion processes. These counter reactions are usually grouped under the general term polarization. The rate of corrosion may be influenced by certain chemical reactions which modify the behavior of the electrolytic cell.

When electric currents flow through an electrolyte, the positive ions, metal and hydrogen, travel in the direction of the current to the cathode; and the acid, or negative, ions move in a direction opposite to that of the current and appear at the anode. In the usual corrosion, very little, if any, of the metal reaches the cathode. Most of the metal ions combine with the oxygen or some other element in the electrolyte and form insoluble compounds which appear as corrosion products. The hydrogen may combine with oxygen, or it may be evolved as hydrogen gas. The electrolytic cells responsible for corrosion are frequently called corrosion cells. Corrosion normally occurs at the anode, where the current enters the solution. Under certain conditions, however, cathode corrosion may occur as the result of secondary cathode reactions.

RATE OF CORROSION. The amount of corrosion by electrolysis depends on the current and the electrochemical equivalent of the metal. The rate of penetration will vary as the current density. With the current densities and the other conditions usually found in practice, the corrosion of metal may be calculated, approximately, by Faraday's law and a coefficient of corrosion (see Section 6). This coefficient will vary between 0.2 and 1.5, depending on the various local action and passivity effects. In practice, the more usual values of the coefficient vary between 0.5 and 1.1.

When the coefficient is unity, iron dissolves at the rate of 1.042 grams per ampere-hour. Accordingly, a current density of 0.1 ma per sq cm will penetrate iron at the rate of 0.104 cm (0.041 in.) per year. Lead corrodes at the theoretical rate of 3.866 grams per ampere-hour. Therefore the rate of penetration for 0.1 ma per sq cm is 0.268 cm (0.105 in.) per year. These rates of penetration are seldom realized because the coefficient is usually less than 1.0, and it varies during the progress of corrosion. Because of the uncertainty of the coefficient and the other variables, the calculation of the rate of penetration is not generally practicable.

APPEARANCE OF CORROSION. Corrosion by electrolysis is usually localized in the characteristic pitting observed in the electrolysis of lead cable sheath or iron pipe. The pitting is characteristic of both stray current electrolysis and the local cell types of corrosion. The type of action cannot be determined with certainty from the appearance of the metal surface.

101. CONTROLLING FACTORS

Of the several controlling factors in underground corrosion the most important appear to be polarization, film formation, circuit resistance, and relative anode and cathode areas. Other factors, such as passivity and overvoltage, will influence underground corrosion in some cases. The oxygen dissolved in the electrolyte has a wide influence on the behavior of the corrosion cell.

POLARIZATION. This is a general term used to group the various factors which influence the potential of the electrolytic cell with the passage of current. Polarization may include such factors as the effects of ion concentration, resistance changes resulting from the electrolytic processes, and certain electrode effects which depend on the composition and character of the electrodes. Polarization occurs, in varying degree, at both the anode and the cathode; it is usually most pronounced at the cathode. When cathode polarization is predominant, the corrosion process is said to be cathode controlled. Anode control occurs when polarization is predominant at the anode. Depolarization of hydrogen is effected by its combination with oxygen, or it may be removed as hydrogen gas. Any agency which increases the availability of oxygen or assists in the removal of the polarizing gases will accelerate anode corrosion.

FILM FORMATIONS. Films increase the resistance at the surface of the cable or pipe. Certain chemical inhibitors, such as silicates, phosphates, and chromates, are effective in the formation of protective films. Soil waters which contain sulfates or silicates tend to form protective films on lead sheath. Nitrates, on the other hand, prevent the film formation and thereby increase the rate of corrosion. Iron oxides, which form under alkaline conditions, are usually dense; they adhere closely to the pipe surface and have more protective value than the oxides formed in acid soils.

CIRCUIT RESISTANCE. This is the component of resistance which is independent of the current. It includes the resistance of any films, a component of the electrolyte resistance, and the resistance of the external circuit. In the usual soil corrosion, the major component of the resistance exists at the surface of the electrodes, and any film formation that increases the surface resistance is an important factor in controlling the corroding current. In practice, the resistivity of the environment in contact with a pipe or cable may vary between wide limits in a single duct section or over a short length of pipe. The severity of the corrosion will be most pronounced at the points of low resistivity.

THE RELATIVE AREAS OF THE ANODE AND THE CATHODE. This will influence the polarization of the corrosion cell. When the cathode area is relatively large, cathode polarization will be correspondingly weak, and the anode will corrode at a higher rate than under conditions where the cathode area is equal to or less than the anode. This factor is of particular importance in selecting metals to be used together in sea water or soil. For example, the metal of bolts or rivets should be slightly cathodic to that of the structure to which they are attached. Likewise, the solder used in cable splices should have, as nearly as possible, the same electrode potential as the lead sheath. The relative area effects are most pronounced where attempts are made to protect cable or pipe by the application of coatings. Any imperfections in the coating applied to anodic areas will restrict these areas and weaken the polarization at the corresponding cathodes. This will have the effect of concentrating the corrosion at the exposed anodic areas.

PASSIVITY. This is the property exhibited by some metals whereby they become abnormally inactive toward certain chemical reactions. Passivity is also manifested in the phenomenon of current flowing through an electrolyte without producing the full amount of anodic corrosion which would occur under normal conditions. Experimental work has associated passivity with invisible films formed by the reaction between the metal and some element in its environment.

OVERVOLTAGE. This is the voltage required to overcome the resistance introduced by the evolution of hydrogen or some other element as a gas. Hydrogen, for example, is plated out on the cathode as atomic hydrogen at its electrode potential. If hydrogen is to be removed as a gas, an additional voltage must be available to form molecular hydrogen. The overvoltage varies with different metals and with the condition of the metal surface. For example, the hydrogen overvoltage for zinc, at the upper end of the overvoltage scale, is 0.75 volt. Platinum, at the lower end of the scale, has an overvoltage of 0.12 volt. Likewise a certain overvoltage is required for the removal of oxygen, chlorine, or the other gases that polarize at the anode.

102. CORROSION CELLS

TYPES OF CORROSION CELLS. Corrosion cells may be grouped under three general electrolytic types: (1) bimetal, or galvanic, cells in which metals having different electrode potentials are in contact in the same electrolyte, (2) environmental, or concentration, cells in which the same metal is exposed to different environments, and (3) electrolytic cells in which the source of current is external to the cell, as in stray current electrolysis. In practice, environmental, or concentration, cells are the cause of widespread corrosion and present more complex corrosion problems than the bimetal types or stray current electrolysis.

BIMETAL CELLS. These are set up when two metals having different electrode potentials are electrically connected and exposed in the same electrolyte. For example, the static (open-circuit) potential difference of lead and copper is approximately 0.4 volt in a solution of sodium chloride. Any other salt solution, acid or alkaline, can be used with some variations in the performance of the cell. Lead is anodic to copper, and on closed circuit the flow of current through the electrolyte is from the lead to the copper. In this case the lead is corroded, and the copper is protected. Similar cell activity is exhibited by other metal combinations when they are exposed in solutions of the various salts. In general, the cell behavior depends on the relative electrode potentials of the metals, the salt solution constituting the electrolyte, and the various factors relating to polarization, relative electrode areas, etc., as explained in Article 101. Impurities on the metal surface have the effect of bimetal cells, and, if the impurity is anodic to the base metal, pitting of the surface will result from local action. In the zinc-iron cell, for example, the zinc is anodic to the iron; and a flow of current from the zinc will neutralize the galvanic currents caused by the impurities in the iron and protect the entire surface of the iron. This is the basic principle of cathodic protection.

ELECTRODE POTENTIALS. **Standard electrode potentials** for the various metals are given in the well-known electrochemical series. The order of nobility in which the metals appear in the series and their electrode potentials are determined in standard solutions and under conditions which are not usually obtainable in practice. Electrode potentials vary considerably in soil water, and frequently the order of nobility is reversed. This is particularly true of metals which are near together in the series. For this reason the standard emf series is not directly applicable to the corrosion of metals in contact with soil moisture. A more practical arrangement for the common metals is given in Table 1, which was developed for sea water by La Que and Cox [*Proceedings, American Society for Testing Materials*, Vol. 40, pp. 670–687 (1940)].

ENVIRONMENTAL CELLS. These are caused by the reaction of different solutions on the same metal; i.e., the electrode potential of a given metal varies with different salt solutions or with different concentrations of the same solution. For this reason corrosion cells will form on the surface of a cable or pipe when it is in contact with different soil types or with different salt concentrations in the soil moisture. The conditions most favorable for the environmental types of corrosion are usually found in areas where the soil is of a heterogeneous nature. For example, a pipe which passes from a natural soil to filled-in ground may be subjected to corrosion at or near the soil boundary line; a pipe normally laid in a light sandy soil will become anodic where it passes through a stratum of dense clay or wet soil. Coal cinders are extremely corrosive, and no unprotected cable or pipe should be laid in a cinder-filled soil.

Table 1. Galvanic Series for Sea Water

Anodic (Corroded) End

Magnesium	Lead
Magnesium alloys	Tin
Zinc	Muntz metal
Galvanized steel or galvanized wrought iron	Manganese bronze
Aluminum 52SH	Naval brass
Aluminum 4S	Nickel (active)
Aluminum 3S	Inconel (active)
Aluminum 2S	Yellow brass
Aluminum 53ST	Admiralty brass
Alclad	Aluminum bronze
Cadmium	Red brass
Aluminum A17ST	Copper
Aluminum 17ST	Silicon bronze
Aluminum 24ST	Ambrac
Mild steel	70–30 copper nickel
Wrought iron	Comp. G bronze
Cast iron	Comp. M bronze
Ni-resist	Nickel (passive)
13% chromium stainless steel, Type 410 (active)	Inconel (passive)
50–50 lead tin solder	Monel
18–8 stainless steel, Type 304 (active)	18–8 stainless steel, Type 304 (passive)
18–8–3 stainless steel, Type 316 (active)	18–8–3 stainless steel, Type 316 (passive)

Cathodic (Protected) End

DIFFERENTIAL-AERATION CELLS. These occur where there is a variation in the supply of oxygen to adjacent parts of the same metal. The part to which there is a deficiency of oxygen will be anodic to the part which is better aerated. This type of cell occurs, for example, in cable ducts where the degree of aeration varies from point to point because of the accumulation of mud or sediment in the duct, or in a congested cable system, where the supply of air to the cables in the bottom ducts is more or less restricted.

A pipe may be subjected to differential aeration where it is in contact with both well-aerated and poorly aerated soils.

CORROSION CELL POTENTIALS. Potentials resulting from such causes as differential aeration, concentration cells, and other local environmental conditions are usually small. They may vary from a few hundredths of a volt to 0.1 volt, or more in severe cases. On the other hand, the potentials resulting from variations in relatively large masses of the soil along the route of the cable or pipe may be considerable, depending upon the relative soil characteristics. For example, anodic potentials of 0.8 to 1.0 volt have been obtained in cases where the duct bank or pipe passed through cinder-filled soil. These potentials may be local, or they may extend over a considerable area and be the cause of severe corrosion. In this respect it is important to distinguish between the purely local type of cell and the type resulting from differences in the characteristics of relatively large masses of soil.

Electrolytic cells of all types are most active under conditions of low resistance. Local cell potentials of the environmental types may be expected in any duct or on any pipe. The resistivity of the soil or of the duct contents may vary from less than a hundred ohm-centimeters to several thousand ohm-centimeters within short distances. Consequently, the resistivity of the environment may frequently be the principal factor controlling the corroding current.

103. STRAY CURRENT ELECTROLYSIS

ORIGIN OF STRAY CURRENTS. Electric railways, which operate on direct current and use the track rails as the return circuit, are the principal sources of stray current. The rails of street-car lines are usually embedded in the street pavement and make close contact with the earth. This results in relatively low rail-to-earth resistances. Electrified railways, which operate on private right-of-way, usually, have better rail ballast and higher rail-to-earth resistance, with a corresponding reduction in the amount of leakage current. Railways electrified with alternating current do not ordinarily set up electrolysis conditions. However, the rails may act as carriers of direct current from their connections with direct-current systems. Some direct-current power distribution systems, which operate with grounded neutrals, may set up local electrolysis conditions. These conditions, however, are usually caused by local faults which can be readily corrected.

EFFECTS OF STRAY CURRENTS. The general principle of stray current electrolysis is shown in Fig. 1, which illustrates the return flow of current on the rails of a trolley system, with a portion of the current escaping into the earth. At points remote from the power bus the rails are usually several volts positive with respect to the earth. This causes a portion of the return current to flow from the rail into the earth, where it is collected by any cable, pipe, or other subsurface structure which is in close proximity to the track rails. Normally, this current flows on the structure toward the railway substation, near which the cable or pipe becomes positive, and the current is discharged into the earth, thereby causing electrolysis. In practice, the cable or pipe subjected to electrolysis is a part of an extensive network in which the electrolysis problem may present many ramifications and complexities.

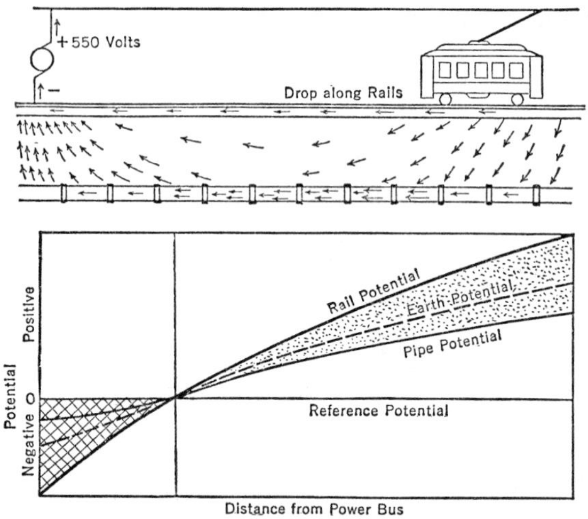

FIG. 1. Stray Current Electrolysis. (*Top*) Distribution of stray current in rails and earth. (*Bottom*) Distribution of potential.

The area in which the current is discharged from the cable or pipe is commonly referred to as the positive area; and that in which current is collected, as the negative area. A section will be found between the positive and negative areas where the pipe-to-earth potential is reversing in polarity and where the current is reversing in its direction of flow. Normally, corrosion occurs in the positive areas where the current is discharged into the earth. In many cases, however, corrosion is experienced on lead cable sheath in the negative areas because of the corrosive salts, such as sodium hydroxide, which are deposited on the cable sheath through secondary cathodic reactions. Iron pipe is not subject to the usual types of cathodic corrosion.

The amount of leakage current entering the earth depends upon the track potentials, the condition of the road bed, and the existence of metallic connections between the track structure and the subsurface structures. Excessive track potentials are usually caused by defective rail bonds, rails of insufficient weight for the service, and other failure to provide adequate negative returns. Insulated negative-return feeders, connected to the rails at suitable intervals, are frequently used to supplement the rail conductivity and to reduce the voltage drop in the rails.

TRACK RAIL RESISTANCE. In computing the resistance of track rails, an average resistance value of 0.0003 ohm per pound-foot of rail is generally used. This value may vary between 0.00027 and 0.00033 ohm, depending upon the composition of the rail. The resistance of rail bonds in good condition amounts to the equivalence of 3 to 6 ft of rail. These values may be used in determining the rail conductivity and the efficiency of the rail bonding.

The contact resistance between the track and the earth varies between wide limits. In street railways, where the rails are embedded in the pavement, the rail-to-earth resistance may amount to 0.5 to 1.5 ohms for 1000 ft of single track. On the other hand, where good rail ballast is provided, the dry-weather resistance may be from 10 to 15 ohms or even more per 1000 ft of single track. In wet weather this figure may drop to 3 to 5 ohms. These values are approximate and serve only to indicate the range of rail-bed resistance that may be expected (American Committee on Electrolysis).

CONTROL. Although stray current electrolysis can be reduced to minor proportions through the application of the appropriate measures, the problem will continue to exist potentially and must be controlled through a systematic program of supervision and maintenance of the measures installed for electrolysis mitigation. This requires the continued cooperative efforts of the electric transit companies and the cable- and pipe-operating utilities concerned.

104. ELECTROLYSIS SURVEYS AND TESTS

CLASSIFICATION OF SURVEYS AND TESTS. The supervision and control of electrolytic corrosion are best accomplished through systematically planned electrolysis surveys and tests, which can be conveniently grouped into (1) maintenance tests, (2) special investigations, and (3) general surveys.

Maintenance tests are scheduled on a periodic basis to cover the inspection of existing installations for electrolysis protection. Usually maintenance tests are made annually. Some installations, such as those involving drainage switches, rectifiers, and other special equipment, require more frequent inspection. In other cases, where the conditions are well established, less frequent inspection schedules may be found sufficient.

General electrolysis surveys are made to obtain overall information on electrolysis conditions affecting the system as a whole. The surveys should cover systematically the various parts of the system and should be supplemented by special tests to explore in detail the corrosive areas, so that appropriate protective measures can be determined. The general surveys should be repeated at intervals to check the conditions and to keep up to date the electrolysis information on the system.

Special tests are made in situations where a more or less extended investigation is required to supplement the general surveys and to obtain information concerning specific locations. The special work includes investigation of the conditions incidental to corrosion failures and the surveys made in connection with new cable or pipe installations.

SURVEY PLANNING. The survey plan should include all the information available from the system records and that obtainable through consultations with the electric transit companies and the cable and pipe utilities operating in the territory. Suitable maps showing the physical relationship between the railway and subsurface structures are essential and should be prepared preliminary to the survey. All pertinent information concerning electric railways, cable, and piping systems should be obtained and shown on the maps in so far as practicable. The maps should also include the principal terrain features, such as streams, swamps, and filled-in ground, that may have a bearing on the electrolysis conditions. The data later obtained from the survey should be plotted on these maps or on map overlays in a form convenient for study.

In general, the electrolysis tests should be planned to cover three types of electrolytic action: (1) stray currents from electric railways or other sources, (2) electrolytic corrosion caused by conditions in the environment, and (3) galvanic potentials between different metals in contact with the soil moisture. The procedures for making electrolysis surveys are described in detail in the 1921 Report of the American Committee on Electrolysis and in *Technologic Paper No.* 355 of the National Bureau of Standards. The subject of soil corrosion is presented in *Soil Corrosion and Pipe Line Protection*, by Scott Ewing, American Gas Association.

TEST PROCEDURE. An electrolysis survey comprises a series of tests made at selected test locations to determine where the structure is collecting or discharging current and to obtain data on the probable cause and magnitude of the electrolysis conditions. The essential elements of the tests include the measurement of the potential and current on the structure and the determination of the resistivity and other controlling factors in the environment.

Potential Measurements. The potential differences between the structure and the adjacent earth are the most significant. They give qualitative data on the areas in which current is collected or discharged and indicate the locations at which corrosions may be expected. In the areas where the structure is negative with respect to the earth the current tends to flow to the structure; where the structure is positive, there is a tendency for current to discharge into the earth. However, the potential does not indicate the magnitude of the current collected or discharged at any specific point. For any given structure-to-earth potential, the magnitude of this current is determined by the resistance to earth, the polarization effects, and the other controlling factors discussed in Article 101.

The survey should also include the potentials between the structure under survey and the adjacent trolley rails, cables, pipes, or other grounded metallic structures. These data assist in evaluating the electrolysis conditions and furnish significant data concerning the control of the earth potentials.

Earth Potentials. In the measurement of earth potentials, difficulties are presented by the galvanic potentials of the electrodes. The value of the measurement depends upon the electrode used to make contact with the earth and its location with respect to the structure. For lead-covered cable, a section of lead cable sheath, or a lead plate, is used extensively by cable companies. The electrode is usually placed in an empty duct adjacent to the cable or, when a duct is not available, on the bottom of the manhole. Sufficient moisture should be present to provide a suitable earth contact. For iron or steel pipe, it is desirable to use a non-polarizing electrode, such as the copper sulfate half cell shown in Fig. 2. Frequently an iron or steel rod of material similar to that in the pipe is used. The electrode should be placed in the earth, directly over the pipe. If the pipe is exposed, the measurements should be taken at two or more positions around the pipe so as to obtain the most representative reading. The earth-potential measurements should be made with high-resistance voltmeters (200,000 ohms or more), potentiometers, or electronic voltmeters, so as to minimize the errors caused by high-resistance contacts, polarization, etc.

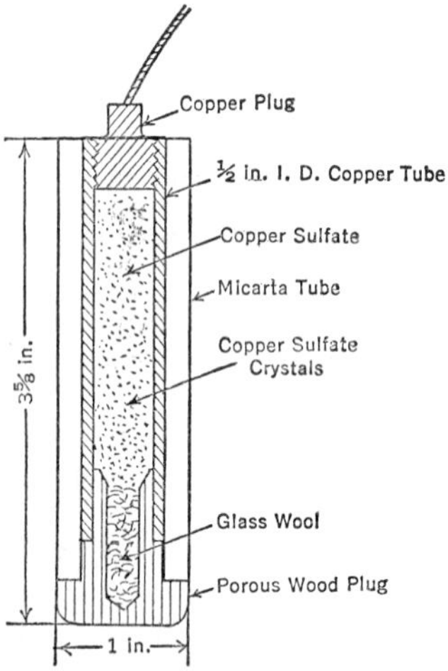

FIG. 2. Copper Sulfate Electrode Used in the Measurement of Earth Potential

The galvanic potential of the earth electrode may introduce unpredictable errors into the earth potentials indicated. The potential of the copper sulfate electrode, for example, is approximately 0.55 volt cathodic with respect to clean iron pipe in the soil. This potential varies somewhat in different soils, and with the condition of the pipe. It is customary, however, to use the electrode as a reference and to record the reading as indicated. For example, if the copper sulfate reading is 0.75 volt, the pipe is approximately 0.20 volt negative with respect to the earth. A reading of 0.55 indicates a potential difference of zero between the pipe metal and the earth. Although iron and lead electrodes are used extensively, their potentials are very uncertain; and, when used as earth electrodes, they may introduce errors amounting to ±0.20 volt or more, depending upon their condition. When used with caution, however, they are convenient for making routine surveys and serve to indicate the existence of unusual potential conditions.

Potential Characteristics. The earth potentials caused by stray railway current will vary between wide limits, and they may reverse in polarity. For this reason it is necessary to observe the readings over periods of approximately 5 min in the zones where the rail traffic is relatively heavy, and for 15 min or more in zones where the car or train schedules are less frequent. When readings are taken with indicating instruments, they should be observed to obtain the maximum, minimum, and average values. Where reversals in polarity occur, the average positive reading and the average negative reading should be recorded separately, with the corresponding maximum readings. Also the relative time for each polarity should be noted. The characteristic of the potential can be used to identify the source of the electrolysis current. For example, the potentials associated with stray railway current fluctuate with the railway load, whereas the galvanic potentials, resulting from soil conditions, have a steady battery characteristic.

Current Measurements. There is no direct relation between the total amount of current flowing on a cable or pipe and the discharge currents that cause corrosion. The current measurements supplement the potential data, however, and assist in identifying the areas in which current is collected or discharged. Also they furnish other data which are pertinent to an evaluation of the electrolysis conditions, such as the location of contacts, the heating effects on cable sheath, the danger of pipe-joint corrosion, and the possible fire hazards resulting from breaks in gas mains carrying current.

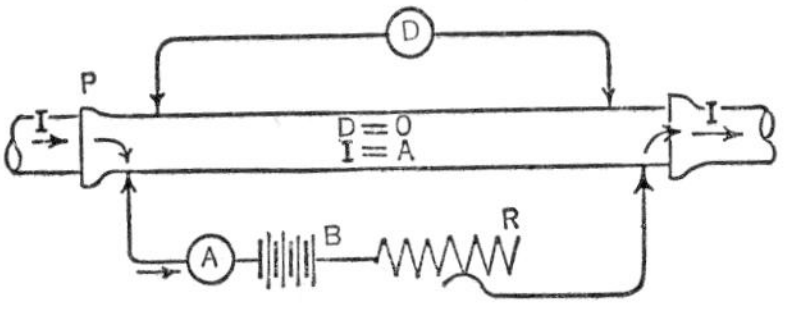

FIG. 3. Divided Circuit Method of Measuring Current in Pipes. When the indication on the galvanometer, D, is zero, the current in the ammeter, A, is equal to the current in the pipe

The current flowing on a cable or pipe is most conveniently measured by the drop-in-potential method. With this method the millivolt drop is measured between two points on the structure taken at a measured distance apart. The current is then computed from the millivolt readings and the resistance of the structure between the points of contact. There must be no splices, joints, or connections between the points of contact to interrupt the uniformity of the structure. The resistance and the current constant (amperes per millivolt per foot) may be computed from the physical dimensions of the structure and the resistivity data given in Table 2, or the constants may be obtained from the tables published in the references given earlier in this article. The drop-in-potential method gives results which are sufficiently accurate for most purposes. Where more accurate

Table 2. Resistance and Current Data—Average Values at 20 Deg Cent

Metal	Weight, lb per cu in.	Resistivity, Microhms per Centimeter Cube	Resistivity, Microhms per ft-in.^{2}	Resistivity, Microhms per lb-ft	Current per Millivolt Drop		Temperature Coefficient of Resistivity
					per ft-in.^{2}, Amperes	per lb-ft, Amperes	
Steel pipe	0.280	13.6	64.5	215.8	15.5	4.65	0.003
Wrought-iron pipe	0.283	13.05	61.7	209.3	16.2	4.76	0.003
Cast-iron pipe	0.256	84.5	400.0	1225.0	2.5	0.815	....
Lead sheath	0.410	22.0	104.0	510.0	9.6	1.96	0.0039
Copper wire	0.321	1.724	8.15	....	122.0	....	0.0039
Steel rail	....	....	....	300.0	....	3.33	....

Formulas for calculating current:

$$I = \frac{EAK_a}{L} \text{ (sixth column)} \quad \text{or} \quad I = \frac{EWK_w}{L} \text{ (seventh column)}$$

$$A = 3.1416\,(D - r) \times r$$

$$K_{at} = K_a[(1 - 0.003)(t - 20)] \quad \text{or} \quad K_{wt} = K_w[(1 - 0.003)(t - 20)]$$

where I = current, in amperes.
E = potential drop, in millivolts, between points of contact.
L = distance between points of contact.
A = cross-sectional area of metal, in square inches.
W = weight of pipe, in pounds per foot.
K_a = current for 1 mv drop per foot on 1 sq in. of metal.
K_w = current for 1 mv drop on 1 ft of metal weighing 1 lb. (Subscript t refers to values at deg cent.)
D = the outside diameter, in inches.
r = the thickness of metal, in inches.
t = temperature, in degrees centigrade.

data are desired or where the current constant cannot be computed, the cable or pipe can be calibrated by the divided circuit method shown in Fig. 3. Other methods for measuring the flow of current have been described, but they are seldom used.

It is frequently desired to measure the current and potential simultaneously for the purpose of identifying the current with the potential. Simultaneous readings of current are also taken at two or more test points when it is desired to determine the loss or gain of current in the section between the test points. Unless the loss or gain amounts to 10 or 15 per cent of the total current, the test is not particularly significant. In making simultaneous tests, the same type of instrument should be used at each test point so as to avoid the errors resulting from different instrument periods.

Special Tests. The tests made in cable manholes and at test points on pipe lines do not accurately reflect the local conditions existing between test locations. Furthermore, the conductivity of the soil and the contents of cable ducts will vary widely from point to point, and variables exist that are unpredictable from the usual potential and current tests, particularly where the corroding currents are of a local galvanic origin. For this reason tests of a special nature are often required to obtain complete information on the causes of corrosion. This special work includes the soil surveys and the tests described by Ewing in *Soil Corrosion and Pipe Line Protection.* For lead-covered cables valuable information can be obtained from the duct survey described below.

SOIL SURVEYS. Essentially, a soil survey consists of mapping the soil types and texture along the route of the structure. The survey should include any special features of the terrain, such as stream crossings, marsh land, and filled-in ground, that may have a bearing on the causes of corrosion. The principal tests include the measurement of soil resistivity, *p*H determinations, and the various other tests which have been devised to measure the corrosiveness of soil. In some cases a chemical analysis of the soil at selected locations may be useful. Where a high degree of accuracy is required, the soil resistivity can be determined by laboratory methods. For most practical purposes, however, the field measurements made with the Shepard resistivity meter are satisfactory. When it is desired to measure the resistivity of relatively large masses of soil at varying depths, the measurements can be made by an arrangement of the McCollum earth current meter or by a standard ground ohmmeter, using the four-electrode method. The resistivity meter and the earth current meter are described in Ewing, *Soil Corrosion and Pipe Line Protection.*

THE DUCT SURVEY. The duct survey is used by cable companies to explore the electrolysis conditions existing in the duct sections between manholes and, particularly, to investigate the conditions incidental to cable-sheath corrosion. The survey is also used to advantage where it is desired to explore the electrolysis conditions in ducts before the installation of new cable.

The survey is made by pulling a test electrode through a vacant duct and measuring at frequent intervals (5 to 10 ft) the potential, current, and resistance by means of instruments connected between the test electrode and the cables occupying the other ducts in the bank. The arrangements and connections for the survey are shown in Fig. 4. The test electrode most generally used consists of a piece of lead-covered cable, 12 in. long and 2.5 to 2.75 in. in diameter, fitted with a pulling eye and test leads.

The potential between the cable sheath and the test electrode, in contact with the duct wall, is measured with a high-resistance voltmeter (200,000 ohms per volt or more) so as to minimize the contact errors and the polarization effects. The readings thus obtained give an approximation of the potential of the cable sheath with respect to the duct wall from point to point through the duct. The magnitude of the current discharged to the duct wall, per foot of cable, is estimated by connecting the test electrode to the cable sheaths through a low-resistance milliammeter. In the measurement of both potential and current at points within the duct, it is necessary to correct for the IR drop due to any stray current flowing on the cable sheath. Where the cables are carrying appreciable stray current, a reading taken at some midposition in the duct and referred to the sheath in the manhole will not give an accurate reading of the potential and current at the point of test. To correct for sheath current, the usual practice is to record the magnitude and direction of the current flowing on the sheath and to compute the IR drop to the point of test.

The resistance per foot of cable sheath, through the duct wall, furnishes valuable data on conditions in the duct. It is conveniently measured by connecting a low-voltage battery (1.5 volts) in series with a milliammeter between the test electrode and the cable sheath; the resistance is computed from the battery voltage and the measured current. Where cable sheath is not available in unoccupied banks, another grounded structure, such as a water pipe, is frequently used. A reversing switch is required to reverse the potential on the test electrode so as to minimize errors introduced by sheath potentials, polarization, and other effects. A low-range ohmmeter, operating on alternating current, is a satisfactory instrument for measuring duct resistance, and its use will eliminate the question of electrode polarization.

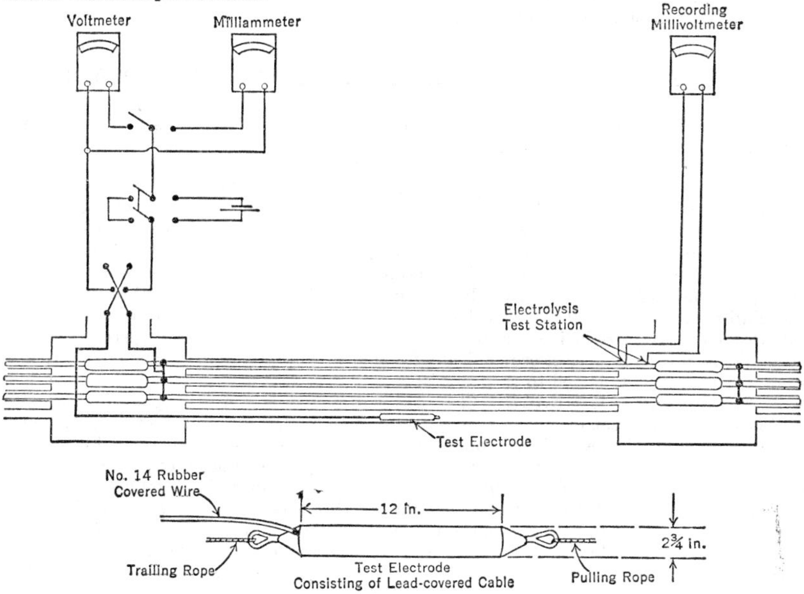

FIG. 4. Equipment and Method for Making Duct Surveys

IDENTIFYING THE CAUSES OF CORROSION. A guide for distinguishing between electrolytic corrosion and chemical attack on lead cable sheath is suggested in *The Corrosion of Underground Cables* by Radley and Richards (see Bibliography, Article 106). The outstanding characteristic of electrolytic corrosion is the pitting and furrowing caused by this type of attack on both lead and iron. Chemical corrosion, on the other hand, is indicated by a more uniform attack, with shallow, saucer-like pits or corroded patches. In lead sheath, the attack may be intercrystalline, whereas chemical corrosion does not subject the crystal boundaries to preferential damage. In the electrolysis of cast-iron pipe the surface of the pipe becomes graphitized; i.e., the metallic iron is removed from the pipe surface, leaving pits filled with a soft material rich in carbon. Usually, the amount of this product, resulting from electrolytic corrosion, is small. The products of lead corrosion are rich in lead chloride or lead carbonate. Lead oxides may be formed in severe cases. The corrosion products from iron are usually the ferrous or ferric oxides, sulfides, or sulfates.

A large amount of iron-pipe corrosion is caused by certain **anaerobic bacteria.** The anaerobic type of attack occurs in the form of pitting, which is very similar to the electrolytic types of attack. The graphitization of cast iron is also a characteristic of anaerobic corrosion. For this reason the anaerobic attack cannot be distinguished from electrolytic corrosion by the appearance of the pipe metal. The products of anaerobic corrosion are, however, invariably rich in iron sulfide, which can be detected by the evolution of hydrogen sulfide gas when the product is treated with hydrochloride acid. Other and more reliable means of identifying this type of corrosion are in the course of development (see Bibliography, Article 106).

The causes of electrolytic corrosion in a particular case cannot be identified with certainty from the appearance of the metal surface. Therefore, to complete the diagnosis, the inspection should be followed by field surveys to determine (1) the presence of stray current potentials, (2) the existence of electrolytic cell potentials resulting from the soil conditions in the earth surrounding the duct bank or pipe, and (3) the resistivity of the soil and moisture in the cable ducts. For iron pipe the tests for stray current and soil potentials should be supplemented by tests to identify the anaerobic-bacteria type of attack.

Where chemical corrosion is suspected, an analysis of the soil or duct contents, together with pH determinations, should be made with a view to identifying the chemical cause of the attack. For example, an alkalinity of pH 8.5 or higher is detrimental to lead cable sheath. Lead is also attacked by organic acids, such as acetic acid. Acid soils are more corrosive to iron pipe than neutral or alkaline soils.

105. THE MITIGATION OF ELECTROLYTIC CORROSION

STRAY CURRENT. In the mitigation of stray current electrolysis a survey should be made of the rail system for the purpose of locating defective rail bonding and any metallic contacts that may exist between the track structure and such underground structures as lead-covered cables, water pipes, gas pipes, and metallic bridges. The defective rail bonds should be repaired, and the contacts with other structures should be cleared. Good rail bonds will lower the rail potentials, and the removal of contacts will greatly reduce the amount of leakage current. Where practicable, the road bed and the footings supporting railway structures should be reconditioned so as to increase the rail-to-earth resistance and to further reduce the leakage current. If necessary, the rail returns should be reinforced by insulated copper cables, connected to the rails at points where they will be most effective in reducing the overall rail potentials. After the rail system has been reconditioned in so far as it is economically feasible, attention should be given to the underground structures subjected to electrolysis.

ELECTROLYSIS DRAINAGE. Electrolysis drainage is used principally for the mitigation of stray current electrolysis on cable systems and, to a limited extent, on piping systems. Essentially, a drainage connection consists of a low-resistance bond connected between the cable sheaths and the railway negative system so as to complete a metallic circuit for the stray current flowing on the cables. The drainage bond reduces the potential of the cable sheath to approximately that of the negative return to which it is connected, the structure-to-earth potential is made negative, and the discharge of stray current into the earth is eliminated. When cables or pipes are maintained at a negative potential, they are also protected by cathodic action against the galvanic types of corrosion. Drainage bonds are installed in the positive areas at or near the railway substation, or they may be connected to the railway negative returns at other points where their use will improve the electrolysis conditions.

Because of the low resistance in the bond connection, the amount of current flowing on the drained structure is greatly increased. This increased current does not endanger the cable sheath and does not interfere with the operation of the cable, provided the current is not sufficient to cause appreciable heating of the cable sheath. With power cables a sheath current density amounting to 50 amp per sq in. section of lead can be permitted under the usual conditions without appreciably derating the cable.

In the drainage of pipes there is danger of developing potentials at high-resistance pipe couplings and thereby setting up electrolysis conditions at these joints. For this reason it is necessary to bond all pipe couplings to insure electrical continuity where pipe drainage is used. Moreover, with gas pipe, a fire hazard is introduced by any current flowing on the pipe; it is difficult and frequently impracticable to drain a large bare pipe because of its intimate contact with the earth and the low pipe-to-earth resistance.

The unrestricted use of drainage is likely to cause potential differences between the drained structure and other structures that are not included in the drainage system. This condition frequently occurs at locations remote from the drainage point, where the other structure is subjected to an electrolysis condition that otherwise would not exist. Furthermore, the influence of drainage on cables that are fairly well isolated, such as communication cables or isolated power feeders, is greater than on power cables that are bonded into a low-resistance distribution system. Consequently, where isolated cables run parallel to or intersect with distribution feeders, there may be a considerable difference of potential between the cable systems. To minimize these potential differences, controlled drainage should be used where practicable.

DRAINAGE SWITCHES. In many drainage installations an automatic switch is required to open the drainage bond during certain periods when the railway current tends to flow from the railway into the cable system. This occurs, for example, when a substation shuts down or during light load periods, at which time the potential of the negative bus becomes positive with respect to the cable sheath. In other cases, variations in the railway load cause wide fluctuations in the negative bus potential, and at frequent intervals the bus becomes positive with respect to the power cables. When this happens, the drainage switch is required to open automatically on a reversal in the flow of current and to close the circuit again when the cable-sheath potential becomes positive with respect to the bus.

INSULATING COUPLINGS. Insulating couplings are an effective means of reducing stray current electrolysis on the piping systems. They are used to break the electrical continuity of the pipe and to insulate it from sources of stray current.

There are no specific rules for locating the insulating couplings. The object is to break the continuity of the pipe without developing serious potentials across the couplings. The allowable potential will depend on the soil conditions. The locations are selected by surveying the route of the proposed installation, utilizing, so far as possible, the existing structures to determine the potential gradients. The information thus obtained is used as a basis for estimating the locations where insulating joints will be the most effective. As a general rule, they are installed most frequently where the potential gradients are the highest.

The insulating joints used on cast-iron pipe consist usually of flange couplings with insulating gaskets and washers. The Dresser insulating coupling is used on wrought-iron and steel pipe under 20 in., and the Macallen insulating coupling for small pipes and services. The couplings should be given a protective coating of pitch or coal-tar enamel to protect them against moisture and soil action.

Although insulating couplings, properly installed, are an effective means for reducing stray current electrolysis on pipe lines, they do not protect against the local action resulting from unfavorable soil conditions. In such cases some form of cathodic protection or protective covering is required.

CATHODIC PROTECTION. The cathodic method is used to protect both cables and pipes against electrolytic corrosion where electrolysis drainage is not feasible. Essentially, cathodic protection comprises a system of earth electrodes, suitably arranged, to which a positive potential is applied for the purpose of raising the earth potential sufficiently to render the protected structure negative with respect to the earth. The earth electrodes are the anodes, and the protected cable or pipe is made the cathode, as shown in Fig. 5. The current flowing from the anode, through the earth, to the structure tends to neutralize any positive potentials likely to cause corrosion. Current at the required potential is supplied by motor generators or suitable rectifiers.

The earth electrode system must be designed to meet the conditions encountered in particular cases. In general, three types of earth anodes are commonly used: (1) the single ground bed, located at considerable distance from the protected structure as shown in Fig. 5(*a*), (2) the distributed anodes shown in Fig. 5(*b*), (3) a continuous pipe, rail, or metallic covered cable running parallel to the structure under protection, as in Fig. 5(*c*). The ground bed may consist of pipe or rail buried horizontally or driven vertically into the ground. Coke breeze is frequently used to lower the resistance of the ground bed and to preserve the life of the metal. The distributed anodes may consist of steel rails embedded in coke breeze, or the special carbon electrodes, which have been designed for this purpose, can be used.

The cathodic method of protection has met with considerable success on long pipe lines and cables where other structures are not seriously involved and where there is sufficient right-of-way for the proper installation of the earth electrodes. Under the congested conditions found in large cities, the cathode method has but limited application. Where used, precautions must be taken to avoid discharging current to other structures which are not included in the protective scheme. Moreover, the current discharged from individual electrodes or from sections of a continuous electrode will vary between wide limits from point to point, depending on the soil conditions. There is danger in supplying too

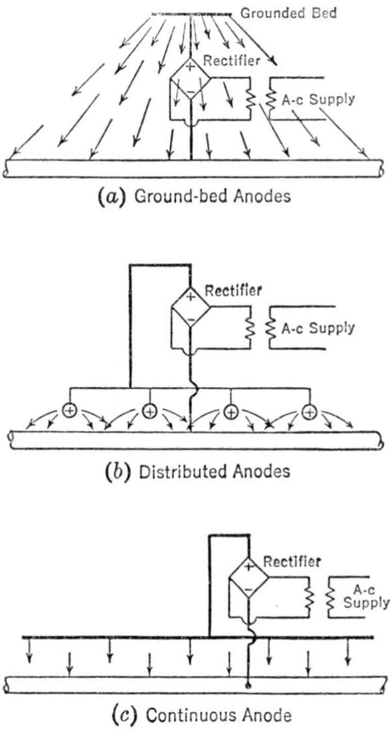

FIG. 5. Anode Systems Used in Cathodic Protection. (*a*) Ground-bed anode. (*b*) Distributed anodes. (*c*) Continuous anode.

much current in some sections in order to obtain protection in others. With lead-covered cables, there is grave danger of cathodic corrosion in the sections receiving an excessive amount of current. To avoid the possibility of cathodic reactions, the current should be limited to that required to give protection.

Cathodic protection is most effective on pipe lines when used in connection with pipe coverings. In this application the cathodic currents reach the pipe through any defects in the covering. A few amperes, depending on the condition of the covering, may be sufficient to protect several miles of pipe. Where bare pipe is in direct contact with the soil, it is often desirable to sectionalize the pipe with insulating joints and to apply cathodic protection to the sections subjected to corrosion. The current requirement for bare pipe may be 50 to 100 times that required for covered pipe in good condition.

ANODIC METALS FOR CATHODIC PROTECTION. A considerable amount of work is being done in connection with the use of anodic metals, such as zinc, magnesium, and aluminum, with a view to their use in the cathodic protection of underground cables and pipes. In the application, advantage is taken of the difference between the electrode potential of the anode metal and the potential of iron or lead when the metals are in contact with the soil. For example, magnesium is approximately 1.0 volt anodic to iron, aluminum is 0.7 volt, and zinc is 0.45 volt, under the average soil conditions. The anodic metal, in suitable electrode form, is buried in the soil and connected to the pipe or cable for which protection is required. Under the potential difference between the electrodes and the pipe, galvanic current will flow from the electrode to the pipe and protect the pipe cathodically. The amount of anodic metal required, its location with respect to the pipe, and other design features will depend on the soil conditions and the allowable current density on the surface of the anodic metal used. The efficiency of the metal is influenced by its polarization, the formation of corrosion products, and other factors resulting from soil effects.

PROTECTIVE COVERINGS. Protective coverings can be used to advantage on cable sheaths and pipe lines in corrosive areas where drainage or other electrical methods of protection are impractical. In corrosive areas practically all important steel pipe lines should be installed with protective coverings, supplemented by cathodic protection where necessary.

The principal protective coverings used on cable sheaths consist of fabric tapes saturated with asphalt or tar compounds and special rubber-faced tapes, vulcanized over the cable sheath. Some of the newer synthetic and plastic compounds, such as Neoprene and polyethylene, have properties which make them attractive as cable-sheath coverings. Cable-sheath coverings must be sufficiently strong to withstand pulling into the cable ducts without damage, and the material must withstand the action of the duct contents without serious deterioration. The fabric of fabric tapes may disintegrate in a few years, leaving the compound to protect the sheath. Considerable difficulty may be experienced in pulling this type of cable into and out of the cable ducts, because of the increased overall diameter of the cables and the tendency of the compound to cement to the duct walls.

Heavy petroleum cable grease and various emulsions containing corrosion-inhibiting compounds have been successfully used for cable protection under moderately corrosive conditions. The grease is applied to the cable sheath by hand as it is pulled into the duct, care being taken to have the grease rubbed on all parts of the cable. In the use of grease it is important that the amount be ample and that the cable be installed in clean ducts. Any sand or mud in the duct will mix with the grease and leave a flaky coating without protective value.

106. BIBLIOGRAPHY

American Institute of Electrical Engineers, *Report of the American Committee on Electrolysis* (1921).
Beckwith, T. D., The Bacterial Corrosion of Iron and Steel, *J. Amer. Water Works Assoc.*, Vol. 33, No. 1 (1941).
Brown, R. H., and R. B. Mears, The Electrochemistry of Corrosion, *Trans. Electrochem. Soc.*, Vol. 74, p. 495 (1938).
Burns, R. M., Corrosion of Metals, I. Mechanism of Corrosion Processes, *Bell System Tech. J.*, Vol. 15, No. 1, p. 20 (1936).
Ewing, Scott, *Soil Corrosion and Pipe Line Protection.* Headquarters, American Gas Association.
Gorman, L. J., Electrolysis and Corrosion of Underground Power System Cables, *Trans. AIEE*, Vol. 64, p. 329 (1945).
Hadley, Raymond F., *Studies in Microbiological Anaerobic Corrosion.* Headquarters, American Gas Association (1940).
LaQue, F. L., and G. L. Cox, Some Observations of the Potentials of Metals and Alloys in Sea Water, *Proc. ASTM*, Vol. 40, p. 670 (1940).
McCullum, Burton, and G. H. Ahlborn, The Influence of Frequency of Alternating or Infrequently Reversed Current on Electrolytic Corrosion, *Trans. AIEE*, Vol. XXXV, Part I, p. 301 (1916).
McCullum, Burton, and R. H. Logan, Electrolysis Testing, *Tech. Papers of Bur. of Standards No.* 355 (1935).

Mears, R. B., and R. H. Brown, A Theory of Cathodic Protection, *Trans. Electrochem. Soc.*, Vol. 74 p. 519 (1938).
Miller, M. C., Cathodic Protection of Buried Metals, *Elec. Light and Power*, Vol. 25, No. 6, p. 92 (1944).
Radley, W. G., and C. E. Richards, The Corrosion of Underground Cables, *J. Inst. Elec. Eng.*, Vol. 85, p. 685 (1939).
Smith, A. V., Cathodic Protection Interference, *Amer. Gas Assoc. Monthly*, Vol. 25, p. 421 (1943).
Speller, Frank N., *Corrosion, Causes and Prevention*, 2nd Ed. McGraw-Hill (1935).
Starkey, R. L., and K. M. Wight, *Anaerobic Corrosion of Iron in Soil.* Headquarters, American Gas Association (1945).
Uhlig, H. H., *Corrosion Handbook.* John Wiley (1948).
Wahlquist, H. W., Use of Zinc for Cathodic Protection, *Proc. Nat. Assoc. Corrosion Eng.*, p. 61 (1944).

SECTION 15

LIGHTING AND HEATING

* Contributed by E. R. Ambrose.

LIGHTING AND HEATING

LAMPS AND ILLUMINATION

By C. E. Weitz

1. TERMS AND DEFINITIONS

The terms and definitions in this section represent only a few of those related to the subject. The Illuminating Engineering Society's bulletin on nomenclature and standards contains some 200 items of nomenclature, together with related standards data, as approved by the American Standards Association. Presently under consideration are new terms and modified definitions, one purpose of which is the terminology of various technical and scientific groups on a common basis. The few terms and definitions included herewith seem to be of most practical importance, and some liberties have been taken with the precise language or wording of official and approved definitions. The present edition of the IES nomenclature and standards bulletin represents the tenth revision since the first set of provisional definitions first published in 1910. The address of the Illuminating Engineering Society is 51 Madison Ave., New York 10, New York. Other related technical societies, such as the Optical Society of America and the Intersociety Color Council, have in the past formulated definitions of equivalent terms, and intersociety committees are working for standardization of fundamental concepts and their expression in a common terminology.

GENERAL RADIATION TERMS. Radiant Energy, U. Radiant energy is energy traveling in the form of electromagnetic waves. It is measured in units of energy, such as ergs, joules, calories, or kilowatthours.

Radiant Flux, $\phi = dU/dt$; alternate symbol, P. Radiant flux is the time rate of flow of radiant energy. It is expressed preferably in watts, or in ergs per second.

Radiance. The radiance of a radiator is the total radiant flux emitted by it.

Irradiancy, $H = d\Phi/dA$. Irradiancy is the radiant flux density upon an irradiated surface.

Radiant Intensity, $J = d\phi/dw$. The radiant intensity of a source is the energy emitted per unit time per unit solid angle about the direction considered; e.g., watts per steradian.

Radiant Flux Density, $W = d\Phi/dA$. Radiant flux density at an element of surface is the ratio of radiant flux at that element of surface to the area of that element; e.g., watts per square centimeter.

Steradiance, $J = d\Phi/dw$. The steradiance (radiant intensity) of a radiator, in a given direction, is the solid-angular radiant flux density in the direction in question.

Steradiancy, $N = d^2\Phi, (d\ dA \cos\theta) = dJ/(dA \cos\theta)$. Steradiancy, in a given direction, is the steradiance of any surface per unit of projected area of the surface as viewed from that direction.

θ, in the defining equation, is the angle between the direction of observation and the normal to the surface.

ILLUMINATION TERMS. International Candle, c. The candle is the unit of luminous intensity. The unit used in the United States is a specified fraction of the average horizontal candlepower of a group of 45 carbon-filament lamps preserved at the National Bureau of Standards, when the lamps are operated at specified voltages. This unit is identical, within the limits of uncertainty of measurement, with the International Candle established in 1909 by agreement between the national standardizing laboratories of France, Great Britain, and the United States, and adopted in 1921 by the International Commission on Illumination.

The international agreement of 1909 fixed only the unit at low color temperatures as represented by carbon-filament lamps. In rating lamps at higher temperatures, differences developed between the units used in different countries. The International Committee on Weights and Measures adopted in 1937 a new system of units based upon (1) assigning 60 candles per square centimeter as the brightness of a black body at the temperature of freezing platinum, and (2) deriving values for standards having other spectral distributions by use of the accepted luminosity factors. It was planned to introduce the new units into use January 1, 1940, but because of the war this step was deferred and put into effect by the National Bureau of Standards, January 1, 1948.

Candlepower, cp. Candlepower is luminous intensity expressed in candles.

Lumen. The lumen is the unit of luminous flux. It is equal to the flux through a unit solid angle (steradian) from a uniform point source of one candle, or to the flux on a unit surface all points of which are at unit distance from a uniform point source of one candle.

Illumination, $E = dF/dA$. Illumination is the density of the luminous flux on a surface; it is the quotient of the flux by the area of the surface when the latter is uniformly illuminated.

The term illumination is also commonly used in a qualitative or general sense to designate the act of illuminating or the state of being illuminated. Usually the context will indicate which meaning is intended, but occasionally it is desirable to use the expression **amount of illumination** to indicate that the quantitative meaning is intended.

Footcandle, ft-c. The footcandle is the unit of illumination when the foot is taken as the unit of length. It is the illumination on a surface one square foot in area on which there is a uniformly distributed flux of one lumen, or the illumination produced at a surface all points of which are at a distance of one foot from a uniform point source of one candle. (**Lux:** one lumen per square meter; **phot:** one lumen per square centimeter; footcandle = 10.764 lux = 0.001076 phot.)

Brightness, $B = dI/dA \cos\theta$. Brightness is the luminous intensity of any surface in a given direction per unit of projected area of the surface as viewed from that direction.

In the defining equation, θ is the angle between the direction of observation and the normal to the surface.

In practice no surface follows exactly the cosine formula of emission or reflection; hence the brightness of a surface generally is not uniform but varies with the angle at which it is viewed. Brightness can be measured not only for sources and illuminated surfaces, but also for virtual surfaces such as the sky.

In common usage the term brightness usually refers to the intensity of **sensation** resulting from viewing surfaces or spaces from which light comes to the eye. This sensation is determined in part by the definitely measurable "brightness" defined above and in part by conditions of observation, such as the state of adaptation of the eye.

Units of Brightness. The practice recognized internationally is to express brightness in candles per unit area of surface. The brightness of **any** surface, in a specified direction, can also be expressed in terms of the lumens per unit area from a perfectly diffusing surface of equal brightness.

For numerical factors by which values of brightness expressed in one unit can be converted to other units, see Section 1, Equivalents and Conversion Factors.

Footlambert, ft-L. The footlambert is a unit of brightness equal to $\frac{1}{\pi}$ candle per square foot, or to the uniform brightness of a perfectly diffusing surface emitting or reflecting light at the rate of one lumen per square foot, or to the average brightness of any surface emitting or reflecting light at that rate.

The average brightness of any reflecting surface in footlamberts is, therefore, the product of the illumination in footcandles by the reflection factor of the surface.

The footlambert is the same as the "apparent footcandle."

(Lambert: one reflected or emitted lumen per square centimeter; millilambert = 0.001 lambert; one footlambert = 1.076 millilambert = 0.001076 lambert. One candle per square inch = 452 footlamberts.)

2. LAMPS AND THEIR RADIATION

Illuminating engineering is that branch of applied science dealing with the production and application of light. By extension, illuminating engineering also deals with the production and application of radiant energy in the adjoining regions of spectrum. It is concerned principally with three main aspects:

1. Production of light or radiation.
2. Control of light or radiation.
3. Specification in practical service.

LAMPS AND THE RADIATION SPECTRUM. The term **lamp** is a generic one meaning an artificial source of light. Its meaning, in popular language, has been extended to include devices (radiators) which generate radiant energy adjacent to the visible spectrum; e.g., germicidal lamps, sunlamps, "black light" lamps, photochemical lamps, and infrared lamps. Most of these "lamps" serve primary purposes other than the supplying of light; i.e., their light output is of secondary importance, and in some may be a distinct disadvantage in their application. On the other hand, lighting installations may use lamps which combine all types of radiation from such sources.

DEFINITIONS. The definitions given in Article 1 which deal with radiant energy in general are applicable to the entire domain of energy output of lamps; the term spectral radiant energy refers to the radiant flux isolated for the purpose of considering any wavelength or wavelength interval. Within the visible spectrum the evaluation of radiant energy has long been agreed upon, and special terms and units—lumens, candlepower, footcandles, etc.—have been standardized and accepted in everyday language to describe radiant energy within the visible spectrum. This is not yet true of the radiant energy in the ultraviolet and infrared regions. Although there are several terms (ergs, joules, calories, watts, or kilowatthours) that may be used to measure spectral radiant energy in total output, outside the scientific laboratory and in practical usage the terms watts, milliwatts, or microwatts are most commonly used. Thus a germicidal lamp (which for its special spectral radiant energy qualities) radiates most of its energy at 2537 A, may be described as radiating so many watts, milliwatts, or microwatts of 2537 A energy; such a description allows comparison between sources as to output and efficiency. Alternative descriptions of performance may be watts (milliwatts or microwatts) per unit solid angle or steradian; and since most sources do not radiate or distribute their energy in spherical uniformity, some explanation is needed to describe the direction with respect to the axis of the lamp or other angular descriptions. Similarly, flux density of spectral radiant energy is often reported as so many watts (milliwatts or microwatts) per unit area (square inch or square centimeter) at a given distance (inches, feet, meters, etc.), and these may be related to the bare source itself or may be used to describe the performance of a combination of lamp and reflector.

SPECTRAL DOMAIN OF LAMPS. Figure 1(*a*) shows the usual spectral boundaries of electric lamps as regards ultraviolet, visible, and infrared radiation. Superimposed is a curve showing the radiation from the sun reaching the earth. The distribution is approximately 54 per cent in the ultraviolet region, 40 per cent in the visible, and 5 per cent in the infrared; these values vary with the season, atmospheric conditions, and latitude.

ENERGY DISTRIBUTION OF LAMPS. Figure 1(*b*) shows the energy distribution of typical incandescent and electric discharge lamps, as well as the location of the principal lines of mercury and sodium lamps. The curves are all plotted to peak at 100 on a relative scale and therefore offer no quantitative comparison with respect to the energy output of the sources.

Electric discharge lamps, such as mercury and sodium, radiate principally at a few specific points represented by the vertical arrows on the chart; other minor lines also are generally present. The intensity or amount of energy of the various lines will vary with the design of the lamp, particularly with respect to the vapor pressure at which the lamp is operated. The extremely low vapor pressure (1/100,000) in fluorescent and germicidal lamps tends to concentrate the mercury radiation at the single 2537 A line; in the Type H-6 mercury lamp, operated at 110 times atmospheric pressure, the distribution of energy shifts to longer wavelengths, and the lines tend to widen into bands which become quite continuous.

Filament lamps produce relatively little ultraviolet, and, as indicated by the curves, most of the energy is in the longer wavelengths of visible and infrared. The higher the filament temperature, the more efficiently light is produced, and the light becomes whiter as more radiation is shifted to the shorter visible wavelengths. Conversely, as the temperature is decreased, less light is produced, and the radiation peak is at longer wavelengths in infrared, as indicated in the 375 deg K and 700 deg K radiators shown.

Fluorescent phosphors transform 2537 A radiation into longer wavelengths in both the ultraviolet and the visible portions. Thus with a choice and blending of phosphors, continuous radiation through these regions can be produced. Germicidal, sunlamp, and "black light" lamps require special ultraviolet transmitting bulbs. The curve shown for the daylight fluorescent lamp is that from the phosphor alone; the four mercury lines within the visible spectrum add to this phosphor curve to give the final spectral output.

APPRAISAL OF RADIANT ENERGY OF LAMPS. Eye Sensitivity and Color. The spectral content or emissivity of a lamp or radiant energy source, expressed in watts or other units, is of little practical value in itself. The value of such measurements and data needs interpretation in terms of the effect or response that such radiation has in fact. For instance, the spectral radiant energy of a source manifest within the visible spectrum and expressed in watts has little meaning unless it is related to the spectral response of the eye. This is variously called the luminosity (visibility or eye-sensitivity) curve. This eye-sensitivity curve, shown in Fig. 1(*c*), is the average response in tests on many eyes and indicates how the eye appraises radiant energy. The numerical values

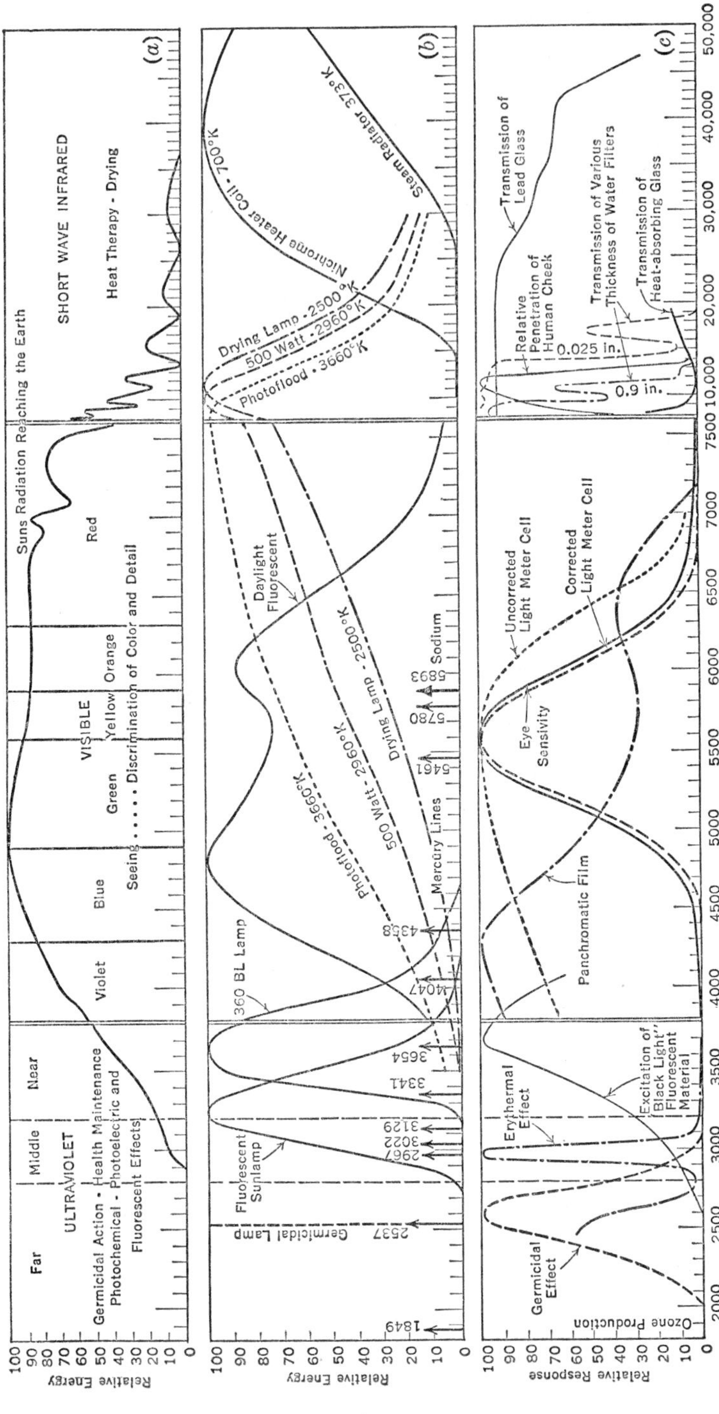

FIG. 1. Spectral Data. (*a*) Range of radiation output of lamps; (*b*) radiation characteristics of specific lamps and other radiators; (*c*) response characteristics of various organisms and materials to radiation from lamps

standardized for each wavelength between 3800 A and 7600 A are to be found in the IES and other Standards publications. It can be shown that under certain conditions of adaptation, the eye responds to wider wavelength limits on both ends of the "standardized" spectrum. Many of these aspects become highly important, and much new data were established by researches during the past war. It has long been known that the luminosity or eye-sensitivity curve is not fixed but shifts to the left under decreasingly low levels of illumination. This is known as the **Purkinje effect.** Many wartime researches on blackout specifications, light and color sensitivity of dark-adapted eyes, and other aspects were of paramount importance in concealment tactics but have lesser value in ordinary times and in ordinary illumination problems.

Eye-sensitivity data are complicated by the fact that each wavelength or, of more practicable significance, each wavelength interval registers its own color sensation, popularly grouped in the 6 colors of the rainbow—violet, blue, green, yellow, orange, and red—though there are more than 150 perceptible color differences. Although white light (by definition) results when a radiator emits equal energy (wattage) throughout the entire visible spectral range, there is a wide range of "white light" as appraised by the eye, and many misconceptions can result unless all samples are viewed at one time. A description of the color locus of any light source can be determined by reference to the ICI (International Commission on Illumination) system of color coordinates. See Deane B. Judd, the 1931 I.C.I. Standard Observer and Coordinate System for Colorimetry, *J. Optical Soc. America*, Vol. 23, pp. 359–374 (1933).

RESPONSE TO ULTRAVIOLET RADIATION. The effectiveness of radiation within the ultraviolet region is represented by the three response curves (1) the germicidal shortwave ultraviolet—this curve peaks at about 2600 A; (2) the erythemal (sun-tanning effect) curve, which peaks at 2967 A; and (3) the fluorescent or black-light response of fluorescent paints and lacquers, many of which have maximum response in the 3650 A region. All three of these curves overlap to some extent. The most efficient source for each of these uses is the one which concentrates the greatest amount of radiation at the point of greatest response, in per cent of the total wattage input. Low-wattage germicidal lamps are practical because their radiation is concentrated at 2537 A, which is near the point of maximum effectiveness. Use is made of special phosphors to produce longer wavelengths of ultraviolet, of erythemal, and of black-light radiation; examples are fluorescent sunlamps, which have their maximum energy emission at 3250 A, and the 360BL lamps, which have their peak response at 3600 A in the black-light region.

Figure 2 shows the ultraviolet transmission curves for various standard glasses used in lamps for ultraviolet radiation. These are for glass 1 mm in thickness, except for two

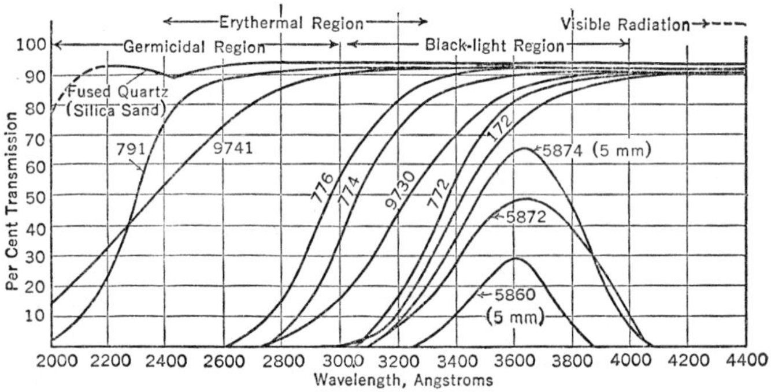

FIG. 2. Transmission or Absorption Characteristics of Various Glasses Related to Ultraviolet and Visible Radiation Wavelengths

samples (5874 and 5860) which are 5 mm, commonly used as black-light filters. Ultraviolet transmission is reduced as thickness is increased.

TRANSMISSION OF INFRARED RADIATION. From 75 to 85 per cent of the wattage input to a filament lamp is dissipated as heat through infrared radiation. The greater portion of this energy is emitted from the limit of the visible spectrum at 7600 A out to 50,000 A. Wavelengths longer than 50,000 A are absorbed by the glass bulb. The heat thus absorbed by the bulb is either carried away by conduction or convection or is reradiated at wavelengths corresponding to those for a radiating body at the temperature of hot glass. The use of lamps to supply radiant energy for heat therapy and industrial-drying applications has grown tremendously in the last few years. Any ordinary incandescent lamp is a most efficient generator of infrared radiation. Special lamps for heat and drying applications are merely ordinary lamps designed for low efficiency in light output and long life.

SPECTRAL DATA ON LAMPS. The curves of Fig. 1 were all plotted to peak at 100 per cent relative energy, allowing no quantitative comparison in output between the lamps of different construction and wattage. Table I gives a summary of the radiated watts and the percentage of input watts for a wide range of lamp sizes and types. Within the visible range, the output is expressed in lumens; for ultraviolet radiation the data are explained as follows:

Sunburn, Suntan Radiation, the E-viton. Familiarity with the energy to produce lumens can be applied to understanding measurements of ultraviolet energy and its relative effects. For example, in sunlamp radiation the skin instead of the eye is the receiver, and its response within its range is shown by the erythemal curve, which indicates how effective different wavelengths are in producing temporary reddening of untanned skin. For rating purposes these values have been standardized for the range between 2400 and 3600 A. The erythemal flux column in Table 1 summarizes the effectiveness of the sources listed; corresponding to the lumen in concept, the unit used is that quantity of radiant energy which produces as much temporary reddening as 10 microwatts of energy at 2967 A —the wavelength of maximum effectiveness.

The effects of ultraviolet radiation on other biological functions can also be portrayed by response curves, although they differ from the erythemal curve. The production of Vitamin D and the antirachitic effectiveness of wavelengths down to 2800 A, however, are known to coincide closely with the erythemal effectiveness. The unit E-viton has been employed to express the health and erythema value of radiations above 2800 A, which are present in natural sunlight. Antirachitic benefits of shorter wavelengths have also been demonstrated in laboratory experiments. Except for germicidal lamps, sunlamps, and the A-H6 mercury lamp (in quartz water jacket), the erythemal flux column could have been called E-vitons; for those sources the erythemal flux below and above 2800 A is given in the following table:

Lamp	Erythemal Flux	
	Above 2800 A, E-vitons	Below 2800 A, Equivalent E-vitons
8-watt germicidal	750	84,250
15-watt germicidal	1,400	158,600
30-watt germicidal	3,500	396,500
A-H6 mercury—quartz jacket	2,550,000	950,000
S-1 sunlamps	67,720	280
S-4 sunlamps	49,925	75
RS sunlamps	25,000	50

The shorter wavelengths under the erythemal curve seem to create quick, temporary reddening or burning of the skin, whereas the longer wavelengths appear to penetrate deeper to cause a tanning of the skin without burn, unless exposure is excessive. The use of suntan lotions to prevent sunburn is effective because these create a film which absorbs the shorter, burning rays and allows the longer, tanning rays to penetrate to the skin, causing it to tan without burn.

Germicidal Radiation. The radiation most effective in its lethal effect on airborne bacteria is around 2600 A, and the relative potency of short-wave ultraviolet in this vicinity is indicated by the germicidal curve. Germicidal lamps radiate most of their energy at 2537 A, and comparative efficiencies of such sources are stated in terms of equivalent milliwatts of 2600 A.

Black-light Radiation. The term black light has, by popular usage, come to mean the radiant energy between 3200 A and 4000 A, and the term **fluoren** has been proposed to represent 1 milliwatt of energy emitted within this region. The effectiveness of a fluoren in creating fluorescent effects will depend on the product of the incident energy (fluorens) and the response characteristics of the fluorescent material, integrated over all the wavelengths in the range from 3200 to 4000 A. The fluoren column in Table 1 shows the relative fluoren content of energy for the various lamps; however, these values are likely to be misleading, since most applications require dark-glass filters to absorb the visible light, and at the same time these filters absorb considerable of the black-light radiation. The B-H4 lamp incorporates a red-purple bulb which absorbs the lumens.

Table 1. Spectral Data on Lamps

Sources	Bulb	Below 2800 A		2800–3200 A		3200–3800 A		Total Ultraviolet below 3800 A		Total Visible 3800–7600 A		Germicidal Effective-ness (Equivalent Milliwatts of 2600 A)	Units of Erythemal Flux	"Black Light," Fluorens	Light, Lumens
		Watts	% of Lamp Watts	Watts	% of Lamp Watts	Watts	% of Lamp Watts	Watts	% of Lamp Watts	Watts	% of Lamp Watts				
Mercury Lamps															
Germicidal															
8-watt	#9,741	1.5	19	0.03	0.41	0.03	0.33	1.6	20	0.14	1.7	1,470	85,000	28	27
15-watt	#9,741	2.9	19	0.06	0.42	0.05	0.34	3	20	0.26	1.75	2,840	160,000	54	53
30-watt	#9,741	7.2	24	0.16	0.53	0.13	0.42	7.5	25	0.65	2.15	7,050	400,000	135	132
Sunlamps															
S-1 (400-watt)	#776	0.01	0.002	3.2	0.8	4.5	1.1	7.7	1.95	45	11	66	68,000	4,850	7,200
S-4 (100-watt)	#721	0.01	0.01	2.1	2.1	3.6	3.6	5.7	5.7	13.2	13.2	71	50,000	3,450	3,300
RS-4 (100-watt)	#776	0.01	0.01	1.5	1.5	2.9	2.9	4.4	4.4	9.6	9.6	59	35,000	2,950	2,550
RS (275-watt)	#776			1.2	0.46	2.9	1.1	4.2	1.5	8.1	3.0	32	25,000	2,950	2,700
Black Light															
B-H4 (100-watt)	#5,872	0	0	0.001	0.001	0.83	0.83	0.83	0.83	0.06	0.06	0	18	845	...
E-H4 (100-watt)	#776	...	...	0.05	0.05	2.2	2.2	2.2	2.2	7.8	7.8	0	200	2,300	2,450
A-H5 (250-watt)	#774	0	0	0.43	0.17	7.6	3.1	8.1	3.2	31	12.5	2.1	3,850	8,350	10,000
6-watt 360BL	...	0	0	...	...	0.31	5.2	0.31	5.2	0.2	3.25	...	40	400	14
15-watt 360BL	...	0	0	...	...	1.2	7.8	1.2	7.8	0.73	4.9	...	150	1,500	53
30-watt 360BL	...	0	0	...	...	2.6	8.8	2.6	8.8	1.65	5.5	...	335	3,400	120
40-watt 360BL	...	0	0	...	...	3.8	9.5	3.8	9.5	2.4	6.0	...	480	4,900	175
Type H															
A-H1 (400-watt)	#772	0	0	0.001	0.0003	4.3	1.1	4.3	1.1	44	11	0	55	4,500	16,000
A-H4 (100-watt)	#772	0	0	0.03	0.03	2.3	2.3	2.4	2.4	12	12	0.03	180	2,450	3,000
A-H6 (1000-watt)	#774	0	0	6.8	0.68	62	6.2	69	6.9	290	29	68	90,000	78,500	65,000
A-H6 (1000-watt)	Quartz	31	3.1	75	7.5	90	9.0	195	19.5	290	29	32,000	3,500,000	110,000	65,000
A-H9 (3000-watt)	#772	0	0	0	0	22	0.72	22	0.72	370	12.5	0	58	24,500	120,000

Lamp	Color Temp.														
Sodium Lamp NA-9 (180-watt)		0	0	0	0	0.03	0.02	0.03	0.02	24.5	13.5	0	17	33	10,000
Fluorescent Lamps															
20-watt white		0	0	0.005	0.03	0.04	0.19	0.04	0.22	3.3	17	...	13	41	920
30-watt daylight		0	0	0.01	0.03	0.23	0.77	0.24	0.81	5.7	19	...	32	280	1,350
30-watt soft white										4.8	16	...			1,170
30-watt blue		0	0	0.01	0.03	0.30	1.0	0.31	1.0	5.3	17.5	...	13	610	780
30-watt green		0	0	0.01	0.03	0.10	0.32	0.11	0.35	5.3	17.5	...	49	100	2,250
30-watt pink		0	0	0.01	0.03	0.05	0.16	0.06	0.19	3.6	12	...	20	48	750
30-watt red		0	0	0	0	...	0.0003	...	...	1.35	4.5	0	0	...	120
30-watt gold		0	0	0	0	...	0.0005	...	...	2.9	9.6	0	0	...	930
40-watt white		0	0	0.01	0.03	0.10	0.24	0.11	0.27	8.0	20	...	31	100	2,300
40-watt 4500° white										7.6	19	...			2,100
40-watt daylight		0	0	0.01	0.03	0.34	0.85	0.35	0.88	8.0	20	...	46	400	1,920
100-watt white		0	0	0.03	0.03	0.19	0.19	0.22	0.22	14.5	14.5	...	62	200	4,200
100-watt daylight		0	0	0.03	0.03	0.70	0.70	0.73	0.73	16.3	16.3	...	96	835	3,900
	Color Temp.							Total Infrared 7600-Infinity							
Filament Lamps															
40-watt standard	2750	0	0	0	0	0.01	0.03	24.5	61	2.9	7.2	0	2	25	465
100-watt standard	2865	0	0	0	0	0.06	0.06	65	65	9.8	9.8	0	7	115	1,620
500-watt standard	2960	0	0	0.006	0.001	0.46	0.09	360	72	59	12	...	67	870	9,960
1000-watt standard	2990	0	0	0.03	0.003	0.96	0.096	740	74	125	12.5	...	195	1,800	21,500
1000-watt photoflood	3360	0	0	0.17	0.017	3.1	0.31	710	71	185	18.5	...	1,000	5,200	33,000
250-watt R-40 drying	2500	0	0	0	0	0.01	0.004	177	71	10	4	0	...	35	1,450
250-watt R-40 heat	2000	0	0	0	0	0	0	185	74	1.7	0.69	0	0	0	180
250-watt purple X	3100	0	0	0.006	0.003	0.28	0.11			0.24	0.09	...	100	470	0.1

3. LAMPS AS LIGHT SOURCES

The ever-expanding fields of electric lamps derive from two important developments: (1) the incandescent or filament source of Edison, and (2) the electric discharge or arc source of Sir Humphry Davy.

THE INCANDESCENT LAMP. Aside from a few specialized services, such as in resistances, infrared radiators, and indicators, carbon-filament lamps have practically disappeared from the market, having been replaced by the tungsten-filament type, the commercial introduction of which dates back to 1907. No substance known today that can be used as a lamp filament is as efficient as tungsten in converting electricity into light on the basis of practical life—with no reservations as to cost, difficulty of manufacture, or any other consideration. Although carbon has a higher melting point (6510 deg fahr) than tungsten (6120 deg fahr), it had to be operated at a much lower temperature because of the high rate of evaporation.

Lamp-efficiency Range. Theoretically, a tungsten bar at the melting point would yield an efficiency as high as 52 lumens per watt, but in practical lamps the highest efficiency of a standard lamp listed in the lamp catalog is 35.8 lumens per watt (the 250-watt photoflood, 3-hour life), and the highest for general lighting service is 22 lumens per watt (1500-watt, 115-volt, 1000-hour life). Incidentally, the lowest efficiency standard lamp is the 6-watt, 115-volt, 1500-hour sign lamp at 6.6 lumens per watt. Between these extremes lies the complete range of today's operating efficiencies of tungsten-filament lamps.

The first consideration is that a source produce light most economically for the service intended, i.e., that the best balance of overall lighting cost in terms of lighting results be secured. To realize this objective most fully, however, requires the highest skill in lamp design—a definite specification of filament length, diameter and form, coil spacing, mandrel size (the mandrel is the form on which the filament is wound), lead wires, number of filament supports, method of mounting, proper gas, gas pressure, bulb size, shape, and temperature.

Gas-filled Lamps. The purpose of gas inside the bulb is to introduce pressure on the filament to retard evaporation. Although the gas conducts some heat away from the filament, this is more than offset by the higher temperatures at which the filament may be operated. Inert gases, i.e., those that do not combine chemically with tungsten, must be used, and the best gas is the one with lowest heat conductivity. Nitrogen was first used exclusively because of its lower cost; argon was recognized as better than nitrogen but was scarce and relatively expensive. Present-day lamps have an atmosphere of argon and nitrogen in varying proportions depending on the wattage. Argon alone ionizes at normal circuit voltages and tends to arc between the leads.

The rate of evaporation of a metal when surrounded by a gas varies with the size of molecule of the gas. Krypton gas has a lower heat conductivity than either nitrogen or argon and, if used for lamps, would indicate a slight gain in efficiency. This gain would be less for higher-voltage lamps. Krypton is too scarce and expensive, however, to be used ordinarily for general-service lamps, since it would double the present cost of the lamp. It is practical today only in special types of lamps, such as the small-bulb miner's cap lamps, where highest efficiency is desirable to prevent excessive drain on the battery.

Hydrogen has high heat conductivity and is inefficient for general lamps but has been employed in special lamps used for signaling purposes, where quick flashing is desired.

In manufacture, gas is introduced at about 80 per cent of atmospheric pressure. When the lamp is operated under normal conditions, the pressure rises to about atmospheric pressure. A lamp operated at more than normal temperature will develop higher than atmospheric pressure within the bulb, causing it to blister and bulge if the temperature becomes sufficient to soften the glass. When hard glass bulbs are used or where bulbs may be cooled by artificial ventilation, such as in projection housings, greater efficiency may be obtained by increasing the internal gas pressure.

Operating Data on Standard Lamps. Data are given in Table 2 for twenty-two standard types, comprising a considerable range in both wattage size and design characteristics. These values were obtained by averaging the test results of thousands of lamps. The effect of design voltage on luminous efficiency, filament length, and filament diameter for lamps of the same wattage should be noted. The maximum bare bulb temperature is measured with the lamp operating vertically base up; the base temperature is measured at the junction of the base and bulb. The ambient temperature is 77 deg fahr.

Luminous and Thermal Characteristics. Table 3 shows the distribution of input energy and other thermal and luminous characteristics of three vacuum and seven gas-filled lamps. The filament dissipates its energy by radiation beyond the bulb, by conduction and convection of the surrounding gas, by conduction of the leads and supports, and by absorption by the bulb and base. The table gives the percentage of the input energy dissipated by each method.

Table 2. Operating Data on Standard Lamps

Watts	Bulb Size	Volts	Amperes	Approximate Initial Lumens	Rated Initial Lumens per Watt	Rated Average Life, Hours	Uncoiled Filament Length, Inches	Filament Diameter, Inches	Filament Temperature, °F	Maximum Bare Bulb Temperature, °F	Base Temperature, °F
6 *	S-14	120	0.050	40	6.6	1,500	14.5	0.00047	3,860	93	88
10 *	S-14	120	0.083	78	7.8	1,500	17.0	0.00064	3,900	106	106
25 *	A-19	120	0.21	260	10.4	1,000	22.2	0.0012	4,190	110	108
40	A-19	120	0.34	465	11.6	1,000	15.0	0.0013	4,490	260	221
60 †	A-19	120	0.50	835	13.9	1,000	21.0	0.0018	4,530	252	195
100 †	A-21	120	0.83	1,620	16.2	750	22.7	0.0025	4,670	261	201
100	A-23	240	0.42	1,240	12.4	1,000	35.5	0.0016	4,470	285	228
100	A-23	30	3.12	1,780	17.8	1,000	7.4	0.0061	4,660	285	228
100 (Proj.)	T-8	120	0.83	1,920	19.2	50	19.0	0.0025	4,890	....	....
150	PS-25	120	1.25	2,600	17.3	750	23.0	0.0032	4,710	290	209
200	PS-30	120	1.67	3,650	18.2	750	23.9	0.0038	4,750	307	212
300	PS-35	120	2.50	5,850	19.5	750	27.9	0.0050	4,825	374	173
500	PS-40	120	4.17	9,850	19.7	1,000	32.1	0.0071	4,840	389	213
1,000	PS-52	120	8.3	21,000	21.0	1,000	38.3	0.0111	4,930	475	235
1,000	PS-52	240	4.2	19,100	19.1	1,000	66.8	0.0072	4,760	475	235
1,000 (Proj.)	T-20	120	8.3	28,000	28.0	50	32.6	0.0108	5,590	....	....
1,000 (Spot) ‡	G-40	120	8.3	22,500	22.5	200	38.3	0.0114	5,200	756	192
1,500	PS-52	120	12.5	33,500	22.3	1,000	43.3	0.014	5,010	505	265
2,000	PS-52	120	16.7	44,000	22.0	1,000	46.3	0.018	5,030	855	....
3,000	T-32	32	93.8	88,500	29.5	100	13.6	0.048	5,390	....	....
5,000	T-64	120	41.7	16,400	32.7	75	44.4	0.029	5,360	860	....
10,000	G-96	120	83.4	325,000	32.7	75	54.4	0.046	5,540	....	....

* Vacuum. † Coiled-coil filament. ‡ Vertical base down.

Cooling Effects of Lead Wires and Supports. Lead wires and filament supports reduce light output because they drain heat from the filament by conduction. The curves of Fig. 3 show the distribution of temperature, resistance, radiation intensity, and brightness near a cooling junction for a long tungsten wire heated to a central maximum temperature of 3860 deg fahr. The number of support wires required depends on the service; lead wires are indispensable. Coiled filaments require less supports than straight filaments; coiled-coil filaments even fewer. Lamps with many supports, rough-service lamps, for instance, will have higher heat-conduction losses and lower efficiencies. These curves show only general relationships, which vary with each specific lamp size, type, and voltage. Because high-voltage lamps require longer filaments, they must have more supports. The cooling junction effects will therefore be more pronounced.

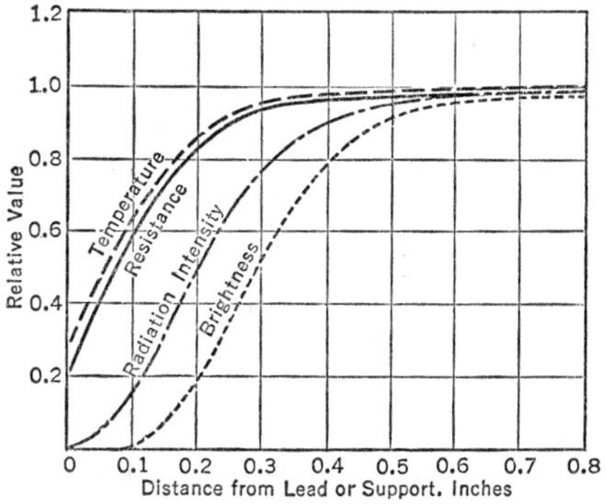

FIG. 3. Curves Showing the Effect of Filament Cooling by Lead and Support Wires

Gas Loss and Wattage. The percentage of heat conducted away from the filament is less as the wattage is increased. This results largely because the filament, regardless of its size, is surrounded by a fairly stationary sheath of hot gas of nearly constant thickness.

The percentage heat loss due to gas conduction from the thicker filaments of large lamps is therefore less than with filaments of smaller diameter. Lamps below 40 watts are still of vacuum type because the additional heat loss by the gas offsets the advantage gained by the lower rate of filament evaporation. Coiled filaments made gas-filled lamps practicable; double coiling gives even more advantage (the 60-watt coiled-coil filament has only 13.5 per cent gas loss) and may extend the gas-filled advantage to lower-wattage lamps. (See Table 3.)

Table 3. Luminous and Thermal Characteristics

Watts	Filament Radiation beyond Bulb			Losses by Absorption			Filament Heat Content, Joules	Heating Time to 90% Lumens, Seconds	Cooling Time to 10% Lumens, Seconds	Variation of Light Output from Mean, Per Cent	
	As Light, %	As Infrared, %	Total, %	Gas Loss, %	End Loss, %	Bulb and Base Loss, %				60 Cycles	25 Cycles
6 *	6.0	87.0	93.0	....	1.5	5.5	0.25	0.04	0.01	29	69
10 *	7.1	86.4	93.5	....	1.5	5.0	0.62	0.06	0.02	17	40
25 *	8.7	85.3	94.0	....	1.5	4.5	2.8	0.10	0.03	10	28
40	7.5	63.8	71.3	20.0	1.6	7.1	2.5	0.07	0.03	13	29
60 †	7.5	73.3	80.8	13.5	1.2	4.5	5.5	0.10	0.04	8	19
100 †	10.0	72.0	82.0	11.5	1.3	5.2	14.1	0.13	0.06	5	14
200	10.2	67.2	77.4	13.7	1.7	7.2	39.5	0.22	0.09	4	11
300	11.1	68.7	79.8	11.6	1.8	6.8	80	0.27	0.13	3	8
500	12.0	70.3	82.3	8.8	1.8	7.1	182	0.38	0.19	2	6
1000	12.1	65.3	87.4	6.0	1.9	4.7	568	0.67	0.30	1	4

* Vacuum. † Coiled-coil filament.

Lamp Efficiency and Design Voltage. Because of the less proportionate gas loss with larger filaments, the efficiency is higher as the wattage is increased. On the other hand, as the design voltage is decreased for the same wattage, the filament must be larger to handle the extra current. Lamps for 115-volt circuits are thus more efficient than 230-volt lamps; 60- and 30-volt lamps used for train lighting are still more efficient.

Lamp Efficiency and Lamp Wattage. The higher the wattage, the higher will be the lumen-per-watt efficiency. This holds only for lamps of essentially the same construction, voltage class, and life. The initial efficiency of standard filament construction, general-

Table 4. Standard Large Lamps

(115-, 120-, and 125-volt Circuits)

Watts	Bulb	Filament Construction	Rated Average Life, Hours	Maximum Overall Length, Inches	Average Light Center Length, Inches	Approximate Initial Lumens	Rated Initial Lumens Per Watt
6	S-6	C-7A	1,500	1 7/8	....	40	6.6
7	C-7	C-7A	2,000	2 1/8	....	60	7.1
10	S-11	C-7A	1,500	2 5/16	1 5/8	80	8.0
25	A-19	C-9	1,000	3 5/16	2 1/2	260	10.4 ‡
40	A-19	C-9	1,000	4 1/4	2 7/8	465	11.7 ‡
50	A-19	CC-6	1,000	4 7/16	3 1/8	665	13.3 ‡
60	A-19	CC-6	1,000	4 7/16	3 1/8	835	13.9 ‡
75	A-21	C-9	750	5 5/16	3 7/8	1,120	14.9 ‡
100	A-21	CC-6	750	5 5/16	3 7/8	1,630	16.3 ‡
150	PS-25	C-9	750	6 15/16	5 1/4	2,600	17.2 ‡
200	PS-30	C-9	750	8 1/16	6	3,700	18.5 ‡
300	PS-35	C-9	750	8 3/4	6	5,900	19.7 ‡
500	PS-40	C-7A	1,000	9 3/4	7	9,950	19.9 ‡
750	PS-52	C-7A	1,000	13 1/16	9 1/2	1,500	20.6
1,000	PS-52	C-7A	1,000	13 1/16	9 1/2	21,500	21.6
1,500	PS-52	C-7A	1,000	13 1/16	9 1/2	33,000	22.0
5,000 *	T-64	C-13	75	13 3/8	6 1/2 †	164,000	32.7
10,000 *	G-96	C-13	75	17 3/8	10 †	325,000	32.7

* For motion picture, color photography, and airport floodlight applications. Approximate color temperature, 3350 deg K. Special heat-resisting glass bulb.

† The light center length of this lamp is exclusive of base prongs; base prong being the small diameter part of metal post.

‡ Lumens per watt listed are for 120-volt lamps only.

purpose 115-volt, 1000-hour lamps ranges from 6.6 lumens per watt for the 6-watt size up to 22.2 lumens per watt for the 1500-watt size. Technical data on the more common sizes and types of standard lamps are given in Table 4. Lamps designed for lower voltage, such as those used in train or plane lighting, or for shorter life, as represented by photo and projection lamps, have higher efficiency. Lamps designed for higher voltage, 230–250-volt circuits, or for longer life, such as sign-service, rough-service, and lumiline lamps, have efficiencies lower than standard lamps.

Lamp Efficiency and Filament Temperature. Lamp efficiency increases as filament temperature is raised. This is true for a lamp of any given design and construction. The theoretical efficiency of tungsten metal as such decreases when this same metal is put into a lamp as a filament because of (1) end loss, (2) poorer radiating properties of the filament as a result of coiling, (3) bulb, base, and lead loss, and (4) gas loss. The present 40-watt lamp has an efficiency 64.7 per cent of the theoretical efficiency for tungsten metal; the 500-watt lamp, 80.6 per cent of theoretical efficiency. The decrease in gas loss with higher-wattage lamps accounts for most of the approach toward the theoretical maximum efficiency. At the melting temperature of tungsten the theoretical maximum efficiency is about 52 lumens per watt; the practical limit is about 40 lumens per watt.

Table 5. Data on Current Inrush for a Given Set of Laboratory Conditions

Lamp Wattage	120-volt Normal Current, Amperes	Theoretical Inrush—Basis Hot to Cold Resistance, Amperes	Actual Maximum Current Inrush by Test, Amperes	Time for Current to Reach Maximum Value, Seconds	Time for Current to Fall to Normal Value, Seconds
75	0.625	9.38	7.2	0.0004	0.07
100	0.835	13.0	9.0	0.0007	0.10
200	1.67	26.2	17.2	0.0008	0.10
300	2.50	40.0	26.2	0.0011	0.13
500	4.17	67.9	45.7	0.0014	0.15
750	6.25	101.9	51.7	0.0021	0.17
1000	8.33	142.4	65.2	0.0031	0.23

Resistance and Temperature. Tungsten has a positive resistance characteristic; i.e., the resistance increases with the temperature. Carbon, used in the earlier lamps, had a negative resistance characteristic. Since the cold resistance of the tungsten filament is much less than the hot resistance—one-fifteenth to one-seventeenth of it—the initial current will be much greater than the final value. This momentary current surge is of short duration but must sometimes be considered in fuse and circuit-breaker operation and may have some significance in the design of switch contacts (Table 5). Recent measurements indicate that standard voltage lamps do not initially heat to a temperature higher than the final average filament temperature. The boxed areas on the graph of Fig. 4 indicate temperature ranges of present vacuum and gas-filled lamps.

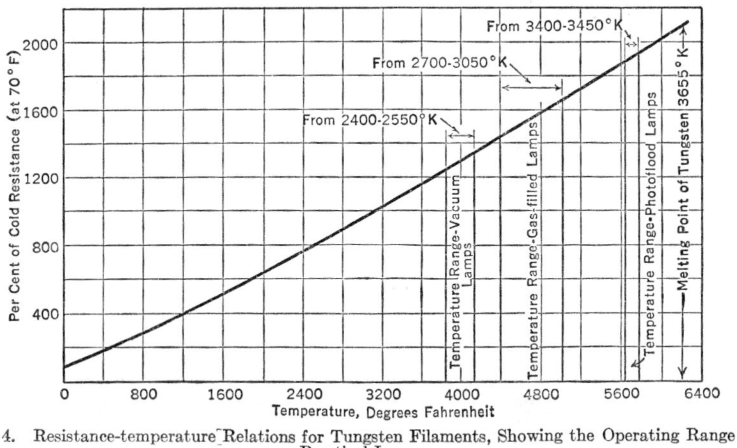

FIG. 4. Resistance-temperature Relations for Tungsten Filaments, Showing the Operating Range for Practical Lamps

Voltage Characteristics. The most important aspect of lamp operation is the effect of departure under operating conditions from lamp-design voltage. These relations are shown in Fig. 5 for large gas-filled lamps; vacuum-type lamps show slightly different values.

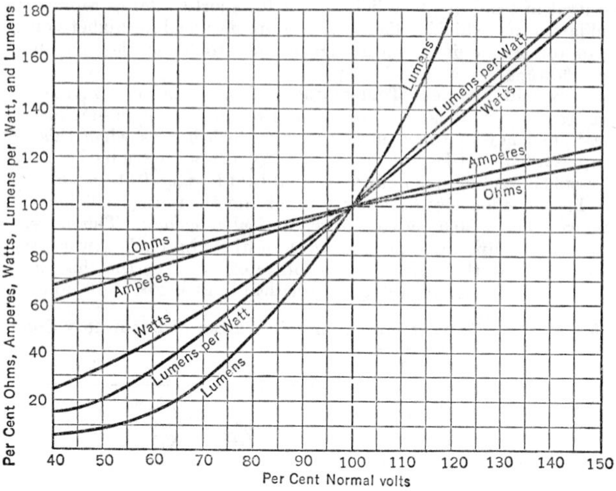

FIG. 5. Tungsten-filament Lamp Characteristics as a Function of Normal Design Voltage

These differences are given in Table 6 for the various exponents as they apply to the equations of lamp characteristics. It will be noted from these exponents that lamp life is most importantly affected by actual operating voltage.

Table 6. Exponential Functions

$$\frac{\text{Life}}{\text{LIFE}} = \left(\frac{\text{LUMENS}}{\text{lumens}}\right)^a = \left(\frac{\text{LUMENS/WATT}}{\text{lumens/watt}}\right)^b = \left(\frac{\text{VOLTS}}{\text{volts}}\right)^d = \left(\frac{\text{AMPS}}{\text{amps}}\right)^u$$

$$\frac{\text{lumens}}{\text{LUMENS}} = \left(\frac{\text{volts}}{\text{VOLTS}}\right)^k = \left(\frac{\text{lumens/watt}}{\text{LUMENS/WATT}}\right)^h = \left(\frac{\text{watts}}{\text{WATTS}}\right)^s = \left(\frac{\text{amps}}{\text{AMPS}}\right)^y = \left(\frac{\text{ohms}}{\text{OHMS}}\right)^z$$

$$\frac{\text{LUMENS/WATT}}{\text{lumens/watt}} = \left(\frac{\text{LUMENS}}{\text{lumens}}\right)^f = \left(\frac{\text{VOLTS}}{\text{volts}}\right)^g = \left(\frac{\text{AMPS}}{\text{amps}}\right)^j$$

$$\frac{\text{amps}}{\text{AMPS}} = \left(\frac{\text{volts}}{\text{VOLTS}}\right)^t \quad \text{and} \quad \frac{\text{watts}}{\text{WATTS}} = \left(\frac{\text{volts}}{\text{VOLTS}}\right)^n$$

Capitals represent normal rated values.

The exponents are as follows:

	a	b	d	u	h	k	s	y
Gas-filled lamps.....	3.86	7.1	13.1	24.1	1.84	3.38	2.19	6.25
Vacuum lamps......	3.85	7.0	13.5	23.3	1.82	3.51	2.22	6.05

	z	t	n	f	g	j
Gas-filled lamps.....	7.36	0.541	1.54	0.544	1.84	3.40
Vacuum lamps......	8.36	0.580	1.58	0.550	1.93	3.33

Exponents d, k, and t are taken as fundamentals, and other exponents are derived from them. Values given apply to lamps operated at efficiencies near normal and are accurate enough for calculations in the voltage range normally encountered.

The theoretical life of lamps calculated by the exponential relationship of life and voltage is seldom realized in practical installations in the case of excessive "undervoltage" burning, since handling, cleaning, vibration, etc., introduce breakage factors which tend to reduce lamp life.

Lamp Life and Economics. These are matters of design, as is evident from the fact that incandescent lamps ranging in rated life from 3 hours (photoflood) to 3000 hours (street-lighting service) and extending to many thousands of hours for heat and drying applications are available as standard catalog items. The manufacturer must select, on the basis of the most satisfactory and economical service under average conditions, some definite operating efficiency for lamp design. Individual users may present departures from average conditions, but even in such instances it is generally advisable to take advantage of low-cost, mass-production types wherever possible.

In cost-of-light calculations it is convenient to use a million lumenhours as the quantity of light. This is roughly equivalent to the total quantity of light produced by a 100-watt lamp during its normal life when operated at normal voltage. A standard 750-hour, 1600-lumen, 100-watt lamp produces 1,047,500 lumenhours during its normal life, taking into account normal depreciation. The formula for the **unit cost of light** is given in the following convenient form:

$$\text{Cost per million lumenhours, in dollars} = \frac{10}{\text{Efficiency}}\left(\frac{\text{Lamp cost}}{\text{Watts} \times \text{Life}} + \text{Kilowatt-hour cost}\right)$$

or

$$= \frac{10}{E}\left(\frac{P}{WL} + R\right)$$

where E = average lumens per watt throughout life.
W = average watts consumed throughout life.
L = average lamp life, in thousands of hours.
P = net cost of lamp delivered in socket, in cents.
R = cost of electrical energy, in cents per kilowatt-hour.

Depreciation and Useful Life. In Fig. 6 are combined several important aspects of lamp economics. Curve (*a*) shows the typical depreciation in light output of a filament lamp throughout life. This depreciation is due to blackening of the bulb as the tungsten gradually evaporates and deposits on it. As the filament material evaporates, the filament decreases in diameter, the resistance increases, and the current and wattage decrease in accordance with Ohm's law. This applies to multiple- or constant-voltage circuits. On constant current, as employed in series street-lighting circuits, the watts and volts show a slight increase above rated values as the lamps age and the initial efficiency is well maintained throughout life, dropping off slightly in the latter part as the increase of bulb blackening more than offsets the higher efficiency of light generation. Curve (*b*) shows the normal mortality pattern of filament lamps, emphasizing the fact that rated lamp life represents at least an average of life tests on thousands of lamps under precise operating conditions in the laboratory life-testing racks. Despite all the research in lamp design and precision manufacture, some lamps fail before their expected life, whereas others last longer.

Group Replacements. In large installations the economics of group replacement is introduced. Curve (*c*) combines the normal depreciation and mortality curve and represents by the area under the curve the total lumenhours produced by any considerable installation of lamps. This suggests the logic of group replacement of lamps, i.e., replacement of all lamps in an installation at one time and at an overall lower cost than is the case if lamps are replaced individually as failures occur. The fact is that lamps that continue to burn after rated life will fail soon and are operating at low efficiency and beyond the point where the balance between lamp cost and current cost dictates lamp replacement in order to get the lowest cost of light. This fact, which is indicated on the graph, is the

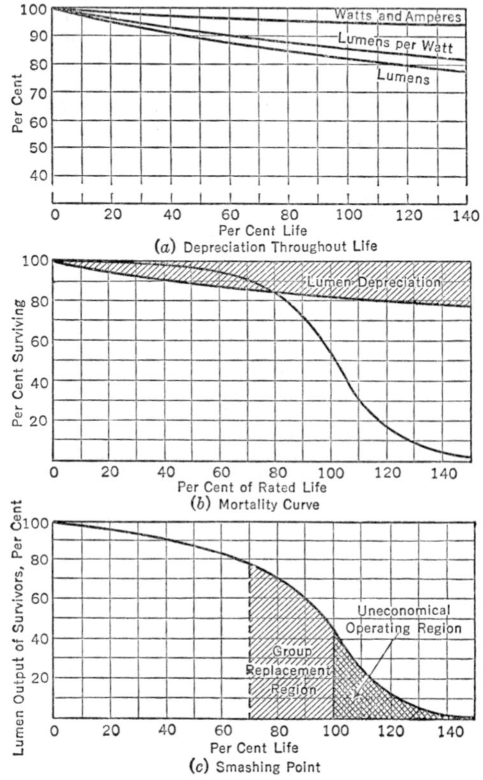

FIG. 6. Performance Curves of Typical Tungsten-filament Lamps throughout Life

basis for group-replacement practices in street-lighting maintenance, as well as in large industrial and commercial installations. On the same basis any individual lamp, such as is used in the home, that has become excessively blackened by extra-long life may well be replaced before actual burnout if true economies are observed.

Bulb, Base, and Filament Designations. In the many hundreds of lamps listed in the catalogs of manufacturers, the three most important characteristics are the bulb size and shape, the base, and the filament form. These are offered in various combinations, which, coupled with a variety of bulb finishes and voltage ratings, multiply into the thousands the different types of lamps that are regularly being supplied. The principal types of each are illustrated in Fig. 7.

Bulb Designations. Bulbs are described in shape and size by a letter and number, the letter indicating the shape and the number representing the diameter at the largest part in eighths of an inch; e.g., the R-40 is a reflector-type bulb 40/8ths or 5 in. in its largest diameter. Originally the letter designation attempted to describe the shape, thus the S for straight-side, G for globular or round, PS for pear-shaped, T for tubular, but with some of the newer shapes this procedure has been hard to follow strictly, though the idea of a descriptive letter has been kept in mind.

Standard bases for filament lamps have increased considerably with the development of the bayonet, prefocus, and bipost types, all of which serve the needs of progress. The value of prudent standardization is evident from the fact that in the early days of Edison about 150 different designs of lamp bases and socket were in existence, and through the logic of standardization this number was reduced to six common types that have endured through the years.

Filament Forms and Designations. Filament forms are designated by a letter and number combination, the letter S denoting straight, uncoiled wire, C single-coiled wire, and CC coiled-coil. The various forms of filament are based on considerations of burning position, minimum supports for efficiency, and ruggedness, or high concentration for optical control.

Lamp Temperatures. Operating a lamp at high bulb temperature will usually not affect the life unless this temperature is sufficient to produce blistering of the bulb. In Table 2 maximum bare bulb temperatures are shown; these are within safe limits unless the lamp is operated in enclosed equipments having insufficient ventilating facilities. The maximum safe temperature at which most lamps having lime or "soft" glass bulbs may be operated is about 700 deg fahr. The maximum safe temperature for hard glass bulbs is 885 deg fahr, and for Pyrex bulbs 975 deg fahr. Bulb blackening increases bulb temperature about 8 per cent. Gas-filled lamps operate at much higher bulb temperatures than vacuum types, and the bulbs are likely to crack if exposed to rain or snow. Where hot bulbs come in contact with fixture parts, thermal cracks are also likely to occur because of strains set up as a result of localized cooling.

Street Series Lamps. Street series lamps are designed to operate in series on constant-current circuits. The most common circuit employs 6.6 amp automatically regulated to maintain this current flow, regardless of the number or size of lamps used on the circuit. Lamps in the larger sizes are also designed for 15- and 20-amp operation, this higher current being usually obtained for each lamp by an individual step-up current transformer connected to a standard series circuit.

Lamps are designated by their rated initial lumen output and ampere rating, e.g., the 6000-lumen, 6.6-amp lamp or the 25,000-lumen, 20-amp lamp. Although series lamps as small in output as 250 lumens are available, the standard sizes range from 1000 to 25,000 lumens. Wattage and voltage ratings, as used to designate multiple lamps, are not commonly employed. Multiple lamps are designed for a definite wattage at a definite voltage, and changes in efficiency are shown by changes in lumen output; the lumen outputs of series lamps, on the other hand, remain fixed because the lumen output is generally specified in street-lighting contracts, and changes in efficiency due to improvements are reflected by changes in wattage or voltage. This usually results in odd numbers and fractions; e.g., the present 6000-lumen 6.6-amp lamp has an average rating of 46.9 volts and 310.0 watts. On a constant-current circuit the filaments for all sizes of lamps of a given current rating are of approximately the same diameter but of different length. The lamp voltage will vary with the lumen output, ranging from a few volts in the smaller sizes to 40 or 50 volts for the lamps of high lumen rating.

The common sizes of series lamps are listed in Table 7. Series circuits should be closely regulated, as fluctuations from normal current will cause considerable variation in lamp performance. The effect of current variation in series operation is considerably greater than voltage variation on multiple operation. Roughly, a 1 per cent change in amperes (0.066 amp on a 6.6-amp circuit) will produce about 1 $^3/_4$ per cent change in volts, about

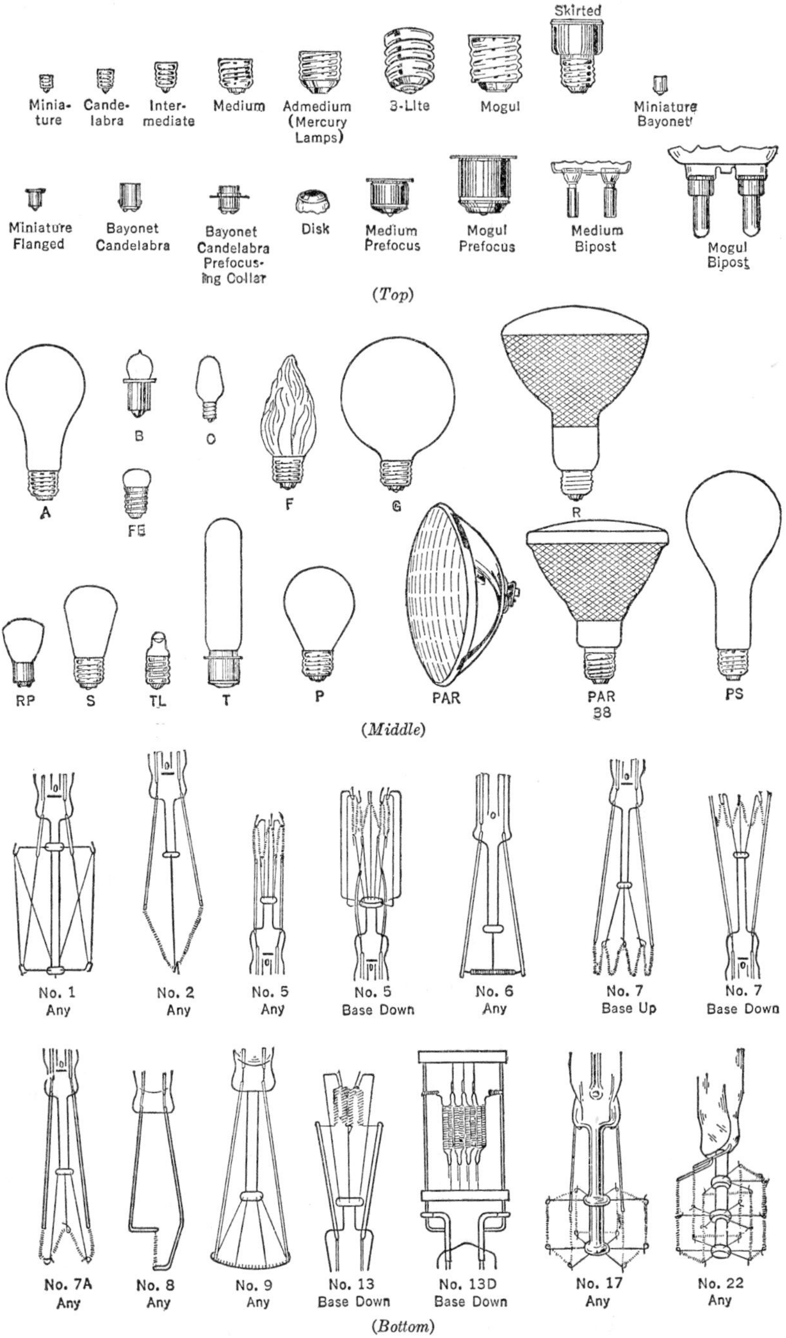

FIG. 7. Base, Bulb, and Filament Designations for Incandescent Lamps. (*Top*) standard types of bases; (*middle*) bulb shapes; (*bottom*) filament forms

2 3/4 per cent change in watts, about 3 1/2 per cent change in efficiency, and about 7 per cent change in light output.

The voltage and wattage of series lamps increase with life, amounting to about 4 per cent above their initial values at the end of rated life, and averaging about 2 per cent during life. Provision should be made in the capacity of constant-current transformers for this increase in voltage.

Table 7. Lamps for Street Series Circuits

Rated Initial Lumens	Bulb	Amperes	Filament Construction	Maximum Overall Length, Inches	Average Light Center Length, Inches	Approximate Initial Volts	Approximate Initial Watts	Rated Initial Lumens per Watt	Lumens at 70% of Rated Life, Per Cent
1,000	PS-25	6.6	C-8	7 1/8	5 3/8	9.4	61.5	16.2	100
2,500	PS-35	6.6	C-2V	9 3/8	7	21.7	143.0	17.5	100
4,000	PS-35	6.6	C-2V	9 3/8	7	31.9	210.0	19.0	98
4,000	PS-35	15	C-2V	9 3/8	6 1/4	13.5	205.0	19.7	95
6,000	PS-40	6.6	C-2V	9 3/4	7	46.9	310.0	19.4	95
6,000	PS-40	20	C-2V	9 3/4	6 1/4	14.7	295.0	20.4	94
10,000	PS-40	20	C-7	9 3/4	6 1/4	24.3	485.0	20.6	91
15,000	PS-40	20	C-7	9 3/4	6 1/4	35.7	715.0	21.0	85

Bulb Finishes and Colors. Inside frosting is widely applied to all types and sizes of bulbs. Frosting gives moderate diffusion of the light, thus reducing glare when lamps are used exposed, and eliminating striations and shadows when used in most types of equipment. Frosting inside the bulb leaves the outer surface smooth and easy to clean; furthermore, the inside frosting absorbs little light—perhaps 1 to 1 1/2 per cent. Although white glass or white-coated bulbs give greater diffusion, the loss of light is of the order of 15 per cent.

White-bowl lamps have a white diffusing enamel on the inside of the bowl and are applicable principally in open direct lighting reflectors. This coating redirects about 80 per cent of the incident light upward, 20 per cent being transmitted diffusely through the bowl. This control in redirection and diffusion of the light greatly lessens glare and softens shadows.

Daylight lamps have blue-green glass bulbs which absorb some of the red and yellow rays, producing a whiter light. The color correction falls about midway between unmodified tungsten-filament light and standard natural daylight. The color temperature of daylight lamps ranges from 3500 to 4000 deg K.

Colored lamps in diffusing bulbs are available in three different types of finishes: (1) outside spray coated, (2) inside coated, and (3) ceramic glazed or enameled. Outside-coated lamps are suitable for indoor use where not exposed to the weather. Their surfaces collect dirt readily and are not easily cleaned. Inside-coated bulbs have smooth outside surfaces that are easily cleaned. The pigments are not subjected to weather and therefore have some advantage in permanence of color. Ceramic glazed or enameled finish is a recent development which gives a permanent finish to the bulb, with the ceramic pigments fused into the glass. Purity and uniformity of color are only moderately realized, and efficiency of light transmission is perhaps 20 per cent lower than in equivalent lamps of clear or natural-colored glass. **Natural-colored glass lamps** find application where purity, efficiency, and permanence of color are desired and are used more for specialized applications than for ordinary sign and decorative color effects. These lamps cost somewhat more than coated lamps, and only a few colors are regularly available. Because of their greater efficiency of light transmission, the overall cost of producing colored light with natural-colored lamps is about the same as with coated lamps.

Integral Reflector Lamps. Many types of lamps are available for specialized applications where accurate beam control is necessary. Such applications include spotlights, floodlights, locomotive headlights, and stereopticon and motion-picture projection. In addition to these types there has been an ever-increasing use of integral reflector lamps, i.e., lamps in which light control is built into the lamp itself by applying reflecting materials (silver or aluminum) to either the outside or the inside surface of the bulb. This type of lamp employing the "sealed beam" principle has simplified both equipment design and maintenance problems for widely diverse services.

The silvered-bowl lamp has a finish of pure silver deposited on the bulb and sealed with an electrolytic coating of copper; over these two metallic coatings is applied an aluminum or bronze finish. The reflecting surface is thus protected from all dust, dirt, and deterioration. Light control is achieved with an initial loss of only 6 to 10 per cent in light output. Since lamps of this type perform the function of light control, economic comparisons with ordinary lamps should be made from photometric and maintenance tests of such lamps in combination with auxiliary reflecting equipment of such a design as to be comparable in light control and distribution. This process has also been applied in neck silvering, and such processed lamps are being used to provide specialized light distribution as required for street-lighting service. The semi-silvered-bowl lamp, along with the reflector showcase lamp, has many applications in store-lighting service.

Projector and Reflector Lamps. Any desired pattern of light beam can be incorporated in a lamp by coordinating filament positioning with respect to special bulb reflecting contours. One of the most significant advances in lamp construction is employed in the projector flood and spot lamps. Instead of the conventional blown bulb the lamp is constructed of two molded glass sections. A bowl-shaped section of parabolic or other suitable contour, on which a highly efficient reflecting film of aluminum has been vaporized, serves as the reflector. This section incorporates the base and filament. A molded glass cover plate, either clear or configurated by any desirable lens pattern, is then fused to the reflector section. Made of hard glass, this type of lamp may be used out of doors without danger of thermal cracks, and louvers, shields, and color filter fittings may be supported by the bulb.

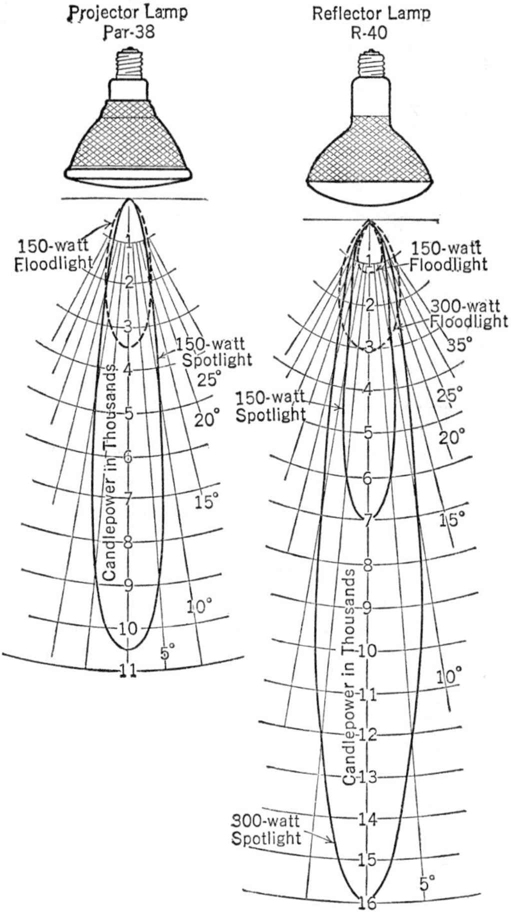

Fig. 8. Candlepower-distribution Curves of Several Sizes of Reflector and Projector Lamps for Spot and Floodlight Service

The reflector lamp has a blown rather than molded glass bulb of special reflector contour and inside aluminized surface. This construction is less expensive than the PAR bulbs (Fig. 7) and is suitable for all manner of interior spotlighting and floodlighting purposes.

The beam pattern and candlepower distribution from the several types of projector and reflector lamps are represented by the curves in Fig. 8.

Automotive or "Sealed Beam" Types. The advantages of the integral reflector lamp led to the development of the so-called "sealed beam" system of automobile headlighting, which was introduced by the automobile industry on most 1940 car models. Since then this principle has been extended to many different applications, such as airplane landing lamps, tractor lamps, fog lamps, and various specialized types for signalling, spotlamps, etc. The merit of such lamps lies in the precise contour into which glass can be molded and which is not subject to denting or springing out of shape during processing and handling, as in the case of metal reflectors. Furthermore, filament positioning may be done with optical accuracy, and since the filament, reflector, and lens elements are combined as a single manufacturing procedure, the result is a precise optical device of greater efficiency and uniformity.

Projection Lamps. Light sources used in picture-projection systems represent extreme precision in lamp making. Light must be accurately controlled in the interest of efficiency and compactness. Filaments must be precisely located at the focal points of the optical system by extreme care in positioning the filament with respect to prefocusing bases. Highly concentrated monoplane and biplane filaments accurately regulated as to horizontal and vertical dimensions are employed. Tubular bulbs allow the spherical reflector and condensing lens to be placed close to the filament to intercept all possible light. A lamp with a 750-watt filament operating at a temperature of 5300 deg fahr and only $^3/_4$ in. away from the glass envelope, which has a maximum safe operating temperature of 975 deg fahr, must dissipate as much energy as a 1-hp motor. Forced ventilation of lamp housings is necessary in many cases to keep glass from softening.

Photocell exciter lamps are small, precision-made lamps of special design for sound-reproduction systems. They are used to illuminate the sound track on motion-picture films. The sound track regulates the amount of light transmitted through the film, the variations of which are picked up by a photosensitive cell and in turn are transformed into electric impulses which produce the sound.

LAMPS FOR PHOTOGRAPHY. The design of lamps specifically for photographic service is concerned with actinic quality, i.e., providing sources which are best adapted to the response or sensitivity of several classes of film emulsions. Some lamps are specified in terms of color temperature, which serves as a basic rating for film-exposure data. Thus several lines of lamps are available for the requirements of commercial studios and portraiture as well as for taking pictures in color. Photographic efficiency and dependability with respect to unvarying spectral quality are of major concern to the photographer. Comparatively, life is of little importance, lamps of various sizes being matched for color temperature. The rated life varies as necessary with the wattage size to achieve the specified color temperature.

Photoflash lamps, physically patterned after the regular incandescent lamps, are actually "combustion" sources. Instead of burning fats or oils as in the historical candle and oil lamps, they burn aluminum foil or wire or, in the SM, solid material on the lead wires. This combustible material ignites or oxidizes readily in an atmosphere of pure oxygen. The lamp bulb is simply a container for the inflammable metal and the oxygen; the familiar filament serves simply as the match or spark plug which ignites the fire and can be designed to function at any voltage. The bulbs are coated with clear lacquer inside and out to safeguard against shattering the glass. Aluminum is used because it produces a continuous spectrum of intense white light with a color temperature of approximately 3800 deg K, favorable to almost all film emulsions. Special blue lacquer may be used on the bulb to raise the color temperature to approximately 6000 deg K.

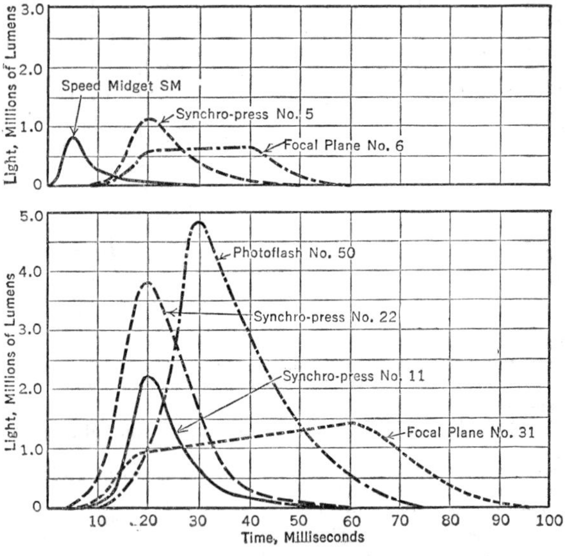

Fig. 9. Light-time Characteristics of Typical Photoflash Lamps

Photoflash lamps have a burning life of only a few hundredths of a second, and to study their flash characteristics an entirely new precise technique of photometric measurements had to be devised. Photoflash-lamp design is predicated on the service, the type of camera, and the necessary synchronism of flash and shutter opening. Photoflash lamps are rated in lumen-seconds, which is a measure of their photographic effectiveness. This effectiveness is the product of luminous intensity (lumens) and time.

The burning of metallic foil or wire will produce lumens of light in proportion to its total mass; the rapidity or time of burning will be dependent on the thickness in the same way that wood shavings burn faster than logs. Thus, by choice of amount and thickness of material within the bulb, the light outputs and time characteristics of photoflash lamps can be regulated as desired. The flash characteristics of the common types of photoflash lamps are shown in Fig. 9.

Photoflood lamps are merely high-efficiency sources of the same character as other photo lamps, the color temperature ranging from 3380 to 3450 deg K. These lamps, because of their high filament temperature, produce, in general, about twice the lumens (three times the photographic effectiveness) produced by similar wattages of general-service lamps. Relatively small bulb sizes are employed (the 250-watt No. 1 photoflood is in the same size bulb as was formerly used for the 60-watt general-service lamp), so that these lamps may be conveniently used in less bulky reflecting equipments or for certain effects in ordinary residential or commercial fixtures.

Flashtubes represent a relatively new practical light source, especially developed and greatly accelerated for wartime photographic purposes. Of the score or two types developed for specialized purposes, several are likely to find their way into future standardization as photographic light sources. Flashtubes are mercury-arc sources and depend upon the characteristics of an unballasted electric discharge for their principal function, which, in turn, calls for rather complicated systems of control and timing of the charge and discharge of electricity through the flashtubes. These tubes may be straight or elongated, or, for purposes of concentrating the source, may be a spiral of glass tubing. At present this is a new field with many possibilities.

4. LAMPS AS HEAT SOURCES

INFRARED DRYING AND HEAT LAMPS. The special considerations in the design of lamps for drying and heating service are to obtain long life at a sacrifice of light output and, in the high-wattage lamps, to use specialized filament forms to obtain better control for high concentration of energy on the work. Vacuum lamps emit, as radiant energy, a greater proportion of the filament heat than do gas-filled lamps, yet because of lack of pressure on the filament the bulbs blacken more quickly, which in turn reduces radiation efficiency and interferes with proper control. Carbon filaments are rugged and are efficient generators of infrared radiation but, like the old carbon lamps, also depreciate quite rapidly because of bulb blackening. The theoretical design life of drying lamps is many thousands of hours in terms of normal filament evaporation, and failure from this cause is unusual; life experienced in service is very long (rated around 5000 hr) and failure is determined to a greater extent by handling and conditions of installation.

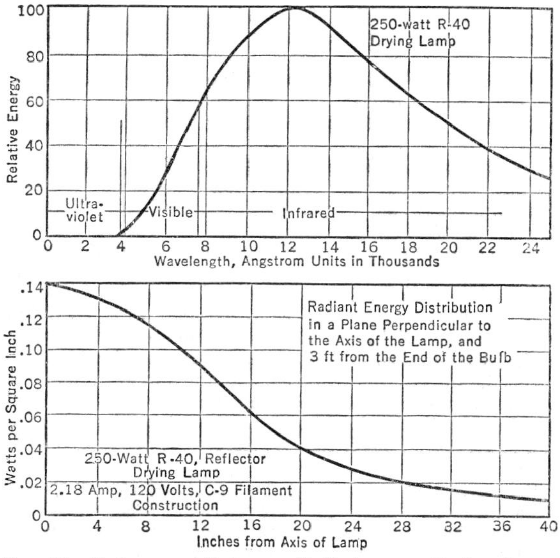

FIG. 10. Performance Data on Radiant-energy Drying Lamps. (*Top*) Spectral distribution of energy of the 250-watt R-40 drying lamp; (*bottom*) radiant-energy distribution for the same lamp for a specific distance

The advantage of infrared drying is the simplicity and flexibility of its use, since banks or tunnels of lamps are easily constructed as a part of the organized flow of materials on a production line. No warm-up time is required, and there is no waste heat as in a hot oven, where drying by air convection takes a much longer time than is required by radiant energy, which produces a rather quick transfer of heat and in most instances reduces drying time to a few minutes.

Drying lamps operate at a color temperature of 2500 deg K at the low efficiency of light production of 6 lumens per watt Color temperature is unimportant as a measure of radiation effectiveness for heating and drying, as there is but a slight difference in sources within the range of 1200 to 3000 deg K for equal concentrations of energy. Heat lamps are a modification of drying lamps and are operated at 2000 deg K in order to reduce light output to approximately 1 lumen per watt. Figure 10 shows the energy distribution of a 250-watt reflector-type drying lamp. Electrolytic gold, silver, Alzak aluminum, and copper are all good reflectors of infrared radiation.

The reflector-type lamp has gained the most popularity, since it incorporates the advantages of sealed reflecting surfaces common to such lamps for other purposes because of their compactness, efficiency, and control, when mounted close together in flat banks, these lamps will provide up to about 6 watts of radiant energy per square inch of work surface. With 375-watt reflector lamps a concentration up to 10 watts per square inch is possible.

5. ARC LAMPS

THE CARBON ARC. The first commercial type of electric lamp, the arc lamp, was introduced in 1878 for street lighting and later was used for general industrial lighting. The principal uses today are for blueprinting, photoengraving, photography, motion-picture projection, high-intensity searchlights and floodlights, and ultraviolet-light production for therapy and irradiation. Many types of carbons are employed, depending on the application and the spectral energy distribution required. There are three basic types of carbon arc lamps: (1) the low-intensity arc between solid or neutral-cored electrodes, (2) the flame arc, and (3) the high-intensity arc. (See the National Carbon Company's Catalog A-4000.)

In the **low-intensity arc** the principal light source is incandescent solid carbon at or near its sublimation temperature. Solid or neutral-cored carbons are used for this type of arc. The composition of the neutral core is predominantly carbon, with a small percentage of an arc-supporting material, such as a potassium salt, which does not contribute significantly to the light. Most low-intensity arcs are operated on direct current, and the crater of the positive electrodes provides a steady concentrated light of the greatest brilliancy and whiteness available from a solid incandescent source, attaining a peak brilliancy of 175 to 180 candles per sq mm (17,500 to 18,000 candles per sq cm) and a color temperature of about 3550 deg K. In comparison with the snow-white quality of some other types of arcs, however, the light from the low-intensity carbon arc has a yellowish tint. The usual color composition is about 18 per cent violet and blue, 32 per cent green and yellow, and 50 per cent orange and red. Some light is emitted from the incandescent tip of the negative carbon and from the arc stream, but the crater face of the positive carbon is the source of about 90 per cent of all the light from the arc.

When operated on alternating current, the low-intensity neutral-cored carbon arc is less efficient than on direct current. Current density in the positive carbon of the d-c low-intensity arc ranges from 50 to 200 amp per sq in. (7.75 to 31 amp per sq cm) for the familiar commercial types of lamps. The area of the crater or anode spot adjusts itself to the current value, so that a temperature near the sublimation temperature of carbon is attained. Increase of current enlarges the area of the anode spot but never raises the temperature above the sublimation temperature of carbon. Therefore, increasing arc current may increase the total light emission from the arc, if size of carbon permits the formation of a larger crater, but it cannot increase the brilliancy above the limiting value of 175 to 180 candles per sq mm.

It is important to note that the term low-intensity arc does not imply low emission of radiant energy. It merely distinguishes one basic type of arc from two other basic types. Some low-intensity arcs, such as those used in motion-picture projection, blueprinting, accelerated testing, and searchlights, have very high radiant energy emission.

THE FLAME ARC. Most flexible of all carbon arcs in respect to quality of radiation is the flame arc. The flame arc is a natural development from the low-intensity arc, obtained by enlarging the core in the electrodes and replacing part of the carbon with chemical compounds capable of radiating efficiently when in a highly heated gaseous form. These compounds are vaporized along with the carbon and diffuse throughout the arc flame, rendering it luminescent. The high concentration of flame materials in the core reduces the arc and brilliance of the anode spot, so that, at the low current densities used in flame arcs, the contribution of the electrode incandescence to the total light becomes unimportant. The evaporation of flame materials is slow in comparison to that obtained in a high-intensity arc, and the resulting concentration of flame elements in the arc stream is low, so that a high brilliance does not result. In the yellow flame arc the brilliancy is on the order of 8 candles per sq mm (800 candles per sq cm). Since the whole flame is made luminous, however, the light source is one of large area, and the luminous efficiency is high.

The radiation emitted by the flame arc consists chiefly of the characteristic line spectra of the elements in the flame material and the band spectra of the compounds formed. Rare earth metals of the cerium group are used as flame materials where a white light is desired; calcium salts, for a yellow light; and strontium salts, for a reddish tint. Other metallic compounds emphasize specific bands of ultraviolet.

Flame-type carbons cored with compounds of the cerium group are known as Sunshine carbons. They produce the closest approach to the radiation of natural sunlight available from any light source of suitable power for industrial purposes. This type of flame arc is used on many applications where it is desired to duplicate the color quality of sunlight or to reproduce the photochemical effects of both the visible and ultraviolet rays of natural sunlight. The core of the white-flame photographic carbon is similar in composition to that of the Sunshine carbon, whereas the panchromatic, or orange-flame, photographic carbon contains a mixture of metallic compounds in the core. Various metallic core compositions are used for industrial irradiation and photochemical processes and for light therapy, the specific bands of ultraviolet at which most effective reaction is obtained being emphasized by selection of core materials.

HIGH-INTENSITY ARCS. The high-intensity carbon arc is one in which, in addition to the light from the incandescent crater surface, a large amount of light originates in the gaseous region immediately in front of the carbon as the result of the combination of a high current density and an atmosphere rich in flame materials.

To produce a d-c high-intensity arc, the positive carbon must be cored with chemical compounds similar to those used in flame-arc electrodes. The current density, however, is much higher, so that the anode spot spreads over the entire tip of the carbon, resulting in the rapid evaporation of flame material as well as carbon from the core. Since the flame material is more easily ionized than carbon, its presence in the anode layer results in a lower anode drop at the core area than at the shell of the carbon. This tends to concentrate the current at the core surface, resulting in the hollowing out of a crater as the current is increased. The rapid evaporation of the flame material produces a high concentration of this efficiently radiating gas in the crater and immediately in front of it. There is a close correlation between the crater depth and the brilliancy of the arc gas within and immediately in front of the crater, and for a given type of positive carbon there is a linear relationship between the crater depth and the excess brightness over that of a low-intensity arc.

The increase in brilliancy of a high-intensity over a low-intensity arc is produced by radiation from the high concentration of flame materials within the confines of the crater. A rise of current in a high-intensity arc increases the crater area only slightly but produces a marked increase in brilliancy. The maximum brilliancy of the crater obtained in various types of d-c high-intensity arcs used in common commercial lamps ranges from 350 to 1200 candles per sq mm (35,000 to 120,000 candles per sq cm) with current densities in the positive carbon ranging from 400 to well over 1000 amp per sq in. (62 to 186 amp per sq cm). Experimental carbons have been produced with brilliancies as high as 2000 candles per sq mm (200,000 candles per sq cm).

A high-intensity arc can also be produced with alternating current. This is a true high-intensity arc within the meaning of the definition previously given. Carbons of the same diameter are used in both holders, each being cored with flame materials. High current density and high concentration of flame materials combine to produce light from both the incandescent electrode tips and also from the gaseous region immediately adjacent, as in the d-c high-intensity arc. The a-c high-intensity arc is operated with a shorter arc gap and at lower arc voltage than the d-c arc of corresponding current. Under correct operating conditions there is a highly luminous area adjacent to each electrode face, and the tail flames from the two electrodes merge into a luminous ball of lower brilliancy, with two well-defined tips at the top. A crater brilliancy of about 280 candles per sq mm (28,000 candles per sq cm) is obtained with this arc.

Except for some high-intensity carbons used in light therapy, the cores of practically all high-intensity positive carbons contain compounds of the cerium group, similar to those used in the white-flame arc. These carbons produce a snow-white light with a color temperature in the vicinity of 5800 deg K and approximately equal energy emission in the violet-blue, green-yellow, and orange-red bands of the spectrum. The principal use of the high-intensity carbon arc is in applications such as searchlights and motion-picture production or projection, where a light source of very high brilliancy is required to provide the needed volume of light from the limited area of source permitted by the optical system.

6. PRODUCTION OF LIGHT BY GASEOUS CONDUCTION

Mercury, sodium, and neon are the elements most widely used for the production of light by gaseous conduction methods. The temperature, pressure, voltage, and other considerations necessary to produce light by this method make these three elements most economically and practically feasible. Different metals may be used for electrodes, often with a coating of high electron-emissive barium or strontium oxide.

Light is produced by gaseous conduction methods when proper energy transitions result from electron displacement within the atomic structure of the gas involved. Voltage applied at the electrodes gives acceleration to free electrons which, in the course of their travel, strike atoms and displace electrons from their normal atomic positions. Radiations of a particular wavelength result as the displaced electrons return to their normal position in the atomic structure; this wavelength depends on the gas used, its pressure, and the degree of electron displacement.

MERCURY VAPOR LAMPS. In mercury lamps the vapor pressure at which a lamp operates accounts in a large measure for the difference in spectral distribution of energy between the several types. In general, higher operating pressure tends to shift a larger proportion of the emitted radiation to longer wavelengths, and at extremely high pressures there is the tendency to broaden the line spectra into wider bands. Within the visible region the mercury spectrum, or energy radiated, consists of four principal wavelengths, which result in a greenish-blue light at efficiencies of 30 to 65 lumens per watt. Although the light source itself appears to be a bluish white, there is an absence of red radiation, and most colored objects appear distorted in color value. Blue, green, and yellow in objects are emphasized, whereas orange and red appear brownish or black For this reason mercury lamps are often combined with filament lamps in installations where good color appearance is important. Sodium vapor lamps produce only yellow light and similarly distort the appearance of all other colors; the color, however, is nearly monochromatic and near the point of maximum theoretical efficiency, although practical lamps today operate at about 50 lumens per watt.

Because electric arcs have an inherent negative resistance characteristic, suitable current ballast, as well as starting equipment, must be employed with electric discharge lamps.

Mercury-lamp Types. The principal types of mercury lamps for various applications, together with their operating characteristics, are shown in Table 8. The general purpose of the several types may be divided as follows.

Sunlamps. The S-1 was the earliest form of Type S sunlamp and generously provided light, radiant heat, and suntan ultraviolet radiation. It incorporates a filament across the mercury-arc electrodes. The S-4 and the reflector type RS-4 employ a small arc tube similar in design to that used in Type A-H4, the difference being in the bulb shape and in the ultraviolet-transmitting glass used for the bulbs. The RS incorporates, in addition, a filament ballast and a bimetallic starting switch within the reflector bulb, so that no auxiliary ballast is required. The filament ballast of the RS, of course, provides additional light and heat. The glass bulbs used for sunlamps transmit practically no radiation shorter than 2800 A.

Black-light Lamps. Most mercury lamps generate considerable ultraviolet radiation in the region between 3300 and 4000 A, the principal line being 3650 A, which is high in effectiveness for the usual fluorescent materials. So-called black-light lamps differ in that a type of glass is used for the bulbs which transmits this black-light radiation. For most black-light applications the absence of visible light is essential; this visible light may be absorbed by a red-purple filter, either as an outer bulb, as in the B-H4, or as accessories attached to the unit.

General Lighting Types. The 400-watt H-1 lamp is by far the most widely used of all mercury lamps, because of its general application in factory lighting and for occasional exterior floodlighting and street lighting. The 3000-watt A-H9 meets the demand for a high-wattage lamp for high-bay industrial lighting. The A-H4 and A-H5 fill the need for lower-wattage mercury lamps for various uses. The 1000-watt H-6 has been employed for searchlights, television-studio lighting, and similar specialized applications where water or air cooling is practicable.

Type H Lamps. The Type H lamp consists essentially of two main electrodes located at opposite ends of a glass tube in which the mercury maintaining the arc is vaporized Figure 11 shows the construction features of the 100-watt H-4 lamp, which in general are the same for the 400-watt H-1 and the 250-watt H-5 lamps. The electrodes are of tungsten wire, coiled and covered with barium-strontium oxide, which makes it possible for them to function satisfactorily at a correct temperature and for a long useful life.

Table 8. Technical Data on Mercury Lamps

Designation	Sunlight Lamps				Black-light Lamps			General Lighting Lamps				
	S-1	S-4	RS-4	RS	B-H4	C-H4 (Spot) E-H4 (Flood)	A-H5	A-H1 B-H1	A-H4	A-H5	A-H6 *	A-H9
Lamp watts (rated)	400	100	100	275	100	100	250	400	100	250	1,000	3,000
Watts, with single-lamp transformer	500	123	123	...	123	123	290	452	120	290	1,095	3,220
Watts, with Tulamp transformer	...	...	...	...	...	...	286/lamp	440/lamp	...	286/lamp	...	...
Lumens at 100 hours	...	3,000	Reflector-type lamps not rated in lumens	Reflector-type lamps not rated in lumens	...	Not rated in lumens	11,200	16,000	3,300	11,200	...	120,000
Lumens (approximate initial)	7,200 †	...			Black-light bulb		...	...	...	...	65,000	...
Lumens per watt at 100 hours	...	30					40	40	30	40	...	40
Initial lumens per watt	18 †	...					...	...	...	...	65	...
Overall lumens per watt (single-lamp trans.)	14.4	24.4	...	...	...	...	34.5	35.4	24.4	34.5	59.4	37.3
Rated life, hours (see Note 1)	400	Approx.	400 app-	lications	1,000	1,000	2,500	4,000	1,000	2,500	75	3,000
Bulb	PS-22	A-21	R-40	R-40	T-16	PAR-38	T-18	T-16	T-10	T-18	T-2	T-9 1/2
Finish	I. F.	Clear	I. F. reflector type	I. F. reflector type	Natural, red-purple	Alum. reflector and clear lens	Clear	Clear	Clear	Clear	Clear	Clear
Base	Mogul	Admed.	Admed.	Medium	Admed.	Admed. skt.	Mogul	Mogul	Admed.	Mogul	3/16" sleeve	S. C. term.
Burning position	Base up	Any	Any	Any	Any	Any	Any	See Note 2	Any	Any	Horiz.	Any
Maximum overall length, inches	6 7/16	5 1/4	6 1/2	6 7/8	5 1/2	5 7/16	8	13	5 5/8	8	3 1/4	54 7/8
Light center length, inches	5	3 7/16	...	...	3 7/16	...	5	7 3/4	3 7/16	5	...	...
Pressure, atmospheres	0.9	8	8	2	8	8	4	1.2	8	4	110	0.7
Number of electrodes	2	3	3	3	3	3	3	3	3	3	2	2
Lamp operating volts	14	130	130	110–125 (50–60 cycles a-c)	130	130	140	137	130	140	840	535
Lamp starting current, amperes	9.5	1.3	1.3	3.2	1.3	1.3	2.9	5	1.3	2.9	2.5	9.3
Lamp operating current, amperes	30	0.9	0.9	2.5	0.9	0.9	2.1	3.2	0.9	2.1	1.4	6.1
Supply voltage (primary volts) †	115	115, 230	115, 230	...	115, 230	115, 230	115, 230	115, 230	115, 230	115, 230	115, 230	230, 460, 575
Transformer secondary open circuit voltage	33	245	245	No Trans.	245	245	250	220	245	250	1,200	850
Power factor, per cent (See Note 3)	50	50, 90	50, 90	90	50, 90	50, 90	50, 90, 95	60, 90, 95	50, 90	50, 90, 95	...	90
Starting time to full output	5 min	3 min	3 min	3 min	3 min	3 to 8 min	4 min	7 min ‡	3 min	4 min	4 sec	7 min
Restarting time	0	3 min	3 min	5 min	3 min	3 to 8 min	4 min	7 min ‡	3 min	4 min	2 sec	8 min

* B-H6 is air-cooled and rated at 900 watts. Characteristics are similar to A-H6.
† Nominal voltage—lamp design is centered for the range of standard voltage circuits.
‡ On lag circuits.

NOTES: 1. In A-H1, B-H1, A-H4, B-H4, C-H4, E-H4, A-H5, C-H5, and A-H9 lamps, the rated life is based on specified test conditions with the lamps turned off and restarted no oftener than once every 5 burning hours. The life rating of the C-H5 is 3000 hours and of the A-H1 is 5000 hours when burned 10 hours per start. The life of S-4, RS-4, and RS sunlamps cannot be adequately expressed in hours for ordinary household service because comparatively short burning periods are employed. The A-H6 life rating is based on tests employing 25-min burning periods, and the life may not be more than one-third as much on very short burning periods such as 3 to 5 min. An approved type water cooling jacket must be used with A-H6 lamps.

2. Although the S-1 lamp can be operated in any position from base up to horizontal, the maximum ultraviolet output is obtained when vertically base up. The life of the A-H5 may be somewhat impaired if the lamp is burned in a horizontal position. A-H1 is for base-up burning; B-H1 for base-down burning. Both types must be operated within 10 deg of vertical. The E-H1 lamp has an inner quartz bulb and may be operated horizontally.

3. The higher power factor is obtained with transformer incorporating integral correction. Two-lamp transformers have an overall power factor of 95 per cent.

The arc tube contains a small amount of pure argon gas, which is used as a conducting medium to facilitate the starting of the arc before the mercury is vaporized. Near the upper end of the tube is a starting electrode which is electrically connected to the lower electrode; hence, when current is applied, an electric field is set up between the starting electrode and the upper main electrode, causing an emission of electrons from the active surface of the main electrode. This imparts energy to the gas in the arc tube so that it becomes conducting.

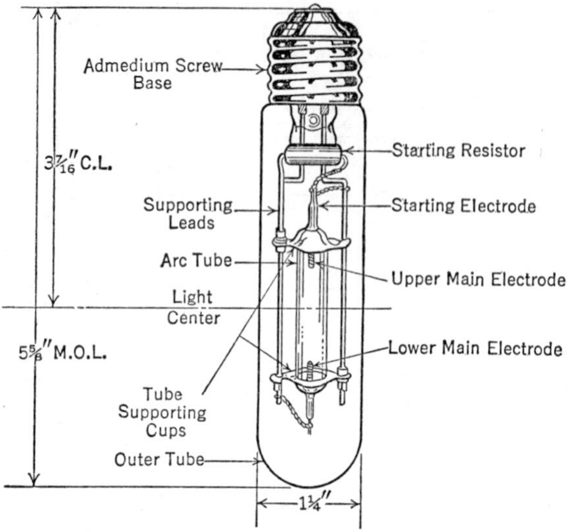

Fig. 11. Construction and Parts of the 100-watt H-4 Mercury Lamp

The life rating for Type H lamps is based on specified test conditions with the lamps burning on the average of 5 burning hours per start. The life is increased as the burning hours per start are increased; e.g., the life of the H-1 lamp rated at 3000 hours on a 5-hour cycle is increased to 6000 hours on a 10-hour burning period per start.

It will be noted from Table 8 that several minutes elapse from the time a lamp is turned on until it comes up to full output. Although each type has its own characteristics, the curves in Fig. 12 (*a*) and (*b*) for the 400-watt H-1 lamp are representative of starting characteristics. Figure 12 (*b*) shows the line characteristics using a 220-volt high-power-factor transformer, the dotted curve indicating the line current taken with an uncorrected (60 per cent) power factor transformer.

Since Type H lamps must be operated within rather close voltage limits, transformer taps are provided for satisfactory operation over a wide range of line voltages. It is recommended that mercury lamps not be operated from line voltages more than 5 per cent above or more than 2 $^1/_2$ per cent below the rated tap voltage of the transformer involved.

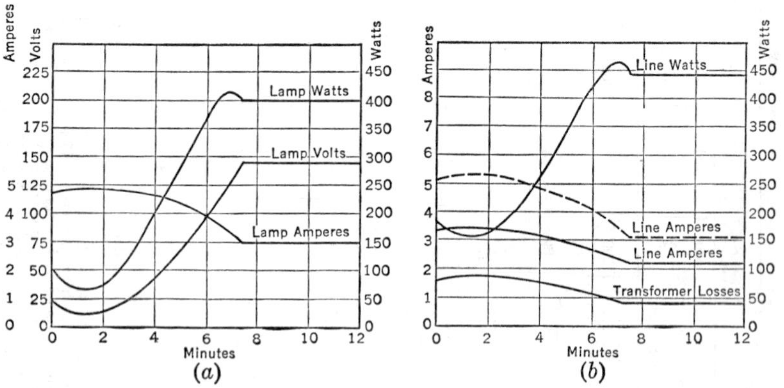

Fig. 12. Starting Characteristics of the 400-watt H-1 Mercury Lamp Operating from a 220-volt High-power-factor Transformer

If the line voltage is consistently above 105 per cent of the tap setting, the increased wattage delivered to the lamp may boost the temperature of glass or other lamp parts to the trouble point and may overheat and damage the transformer (ambient temperatures for transformers of all mercury lamps should not exceed 110 deg fahr). These possibilities arise because both lamp current and wattage increase above 2 per cent for every per cent increase in line voltage over the nominal tap setting. At line voltages lower than 97 1/2 per cent of the tap rating, lamp-starting reliability is reduced, and because of the greater difficulty in starting, cathode deterioration may be more rapid. In fact, lamp life may be shortened by either excessive under- or overvoltage operation.

The curves of Fig. 13 show how the lamp volts, current, watts, and light output vary with line voltage for the H-1 lamp. These curves refer to lag circuits only; on lead circuits of two-lamp transformers the variations will be approximately half of those shown in the curves. These characteristics vary in amount for the different sizes and types of lamps, but the general shape of the curves is similar.

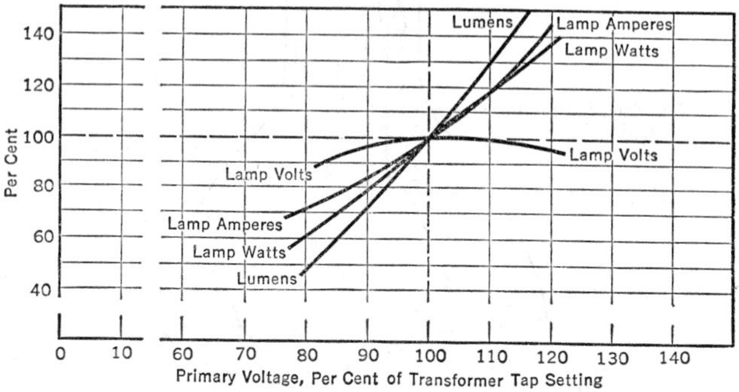

FIG. 13. Lamp Characteristics of the 400-watt H-1 Mercury Lamp as Related to Line Voltage

Type H-6 Water-cooled Lamp. The 1000-watt Type A-H6 mercury lamp consists of a capillary quartz tube about 1 1/2 in. long, having an outside diameter of 1/4 in. and a bore of 3/32 in. Sealed into each end is a tungsten wire which serves as both electrode and lead. The tips of these wires project just through the surface of a small mercury pool in each end of the lamp. The pressure when not lighted is about 1/15 atm, which is the pressure of the argon gas with which the lamp is filled. The lamp reaches its full brilliancy in 1 or 2 sec after power is applied, the heat from the arc quickly vaporizing the mercury and building up the pressure to about 110 atm or 1620 lb per sq in.

Because of the high wattage in such a small volume, it is necessary that water be passed over the tube fast enough to prevent the formation of steam bubbles on the surface of the quartz tube.

The lamp produces 65,000 lumens with a maximum surface brightness of 195,000 candles per sq in., one-fifth the brightness of the sun. Because the heat storage is small and cooling rapid, H-6 lamps may be restarted at once after the current has been turned off. During life the lamp voltage gradually increases, and the current and wattage decrease. Either fracture of the quartz bulb or failure to start terminates the useful life of the lamp. It is advisable to include a pressure switch in the control circuit so that the power is turned off automatically if the water jacket breaks or if the water supply is interrupted. The life is dependent upon the number of times the lamp is started and the type of service.

Mercury Vapor Sunlamps. The bulbs of these lamps are made of special glass which transmits ultraviolet radiations having wavelengths of more than 2800 A and absorbs the shorter wavelengths that may be harmful to the eyes. Of the four types of sunlamps listed, Type S-1 differs considerably in design from the S-4, RS-4, and RS, the last three employing a similar design of 100-watt arc tube. The S-1 sunlamp consists essentially of a mercury arc between tungsten electrodes, with a tungsten filament in parallel with the electrodes to facilitate starting. The bulb contains a small pool of mercury which is vaporized by the heat of the filament upon starting; the ionized vapor becomes conducting, and the arc strikes between the electrodes.

The S-4 and RS-4 lamps differ only in that the RS-4 employs a reflector-type R-40 bulb, both requiring separate ballasts. In the RS lamp a 175-watt filament resistance ballast is incorporated in series with the 100-watt arc tube. The construction of the RS lamp is shown in Fig. 14.

E-viton Rating of Sunlamps. The unit E-viton (erythemal viton) has been developed and is now used to measure the energy which produces sun-tanning of the skin. The E-viton is that amount of radiant energy which will produce the same erythemal effect as 10 microwatts of 2967 A wavelength. (Refer to Table 1 for erythemal flux output data of lamps.)

Exposure times vary with individuals, and it is difficult to give recommended values which are effective and satisfactory for all people. Too, the exposure can be increased as a person becomes accustomed to the radiation. A value of approximately 225 E-viton

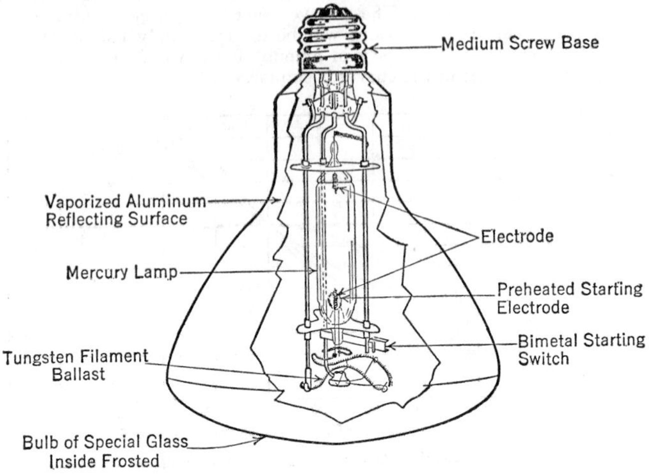

FIG. 14. Construction and Parts of the Internal Ballasted RS Sunlamp

min per sq in. is considered the average value to produce a minimum perceptible erythema of untanned human skin. In general, an exposure of 5–10 min at a distance of 30 in. from new lamps will produce a mild sunburn on untanned skin.

The ultraviolet output of sunlamps is significantly a function of primary voltage, as shown by the curves of Fig. 15. The relative E-viton output increases rapidly if S-1 lamps are operated overvoltage; lamp life will, however, be reduced.

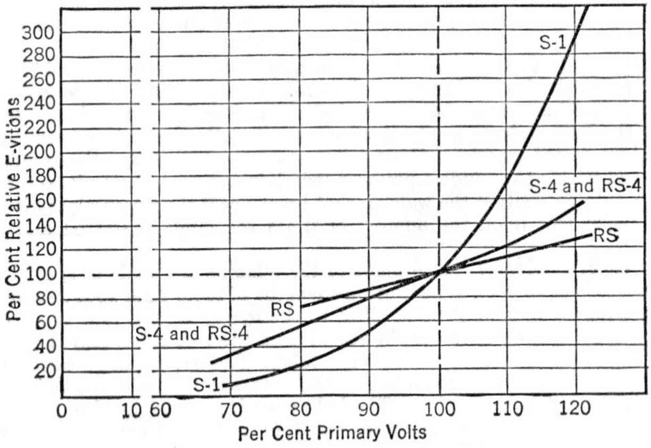

FIG. 15. Relation of Voltage to Ultraviolet Output of Various Mercury-type Sunlamps

SODIUM LAMPS. Electric discharge lamps using sodium vapor possess inherent possibilities for luminous efficiency because the radiation from such a discharge is very close to the point of maximum luminosity. Efficiencies of 100 lumens per watt have been obtained with laboratory sodium lamps and 50 lumens per watt are secured in practice.

Several sizes of sodium lamps have been made, but only two are in commercial use at present. These are the 180-watt 10,000-lumen lamp, which is the most widely used sodium lamp, and the 145-watt 6000-lumen lamp. They are applied principally to street and highway lighting and can be used on either series or multiple circuits.

The 10,000-lumen lamp consists of a tubular inner bulb about 12 in. long and about 3 in. in diameter placed within a double-walled vacuum flask to maintain the proper temperature. The inner bulb contains a small quantity of sodium and some neon gas to facilitate starting. Coiled filaments at either end serve as cathodes, with one side of each filament connected to molybdenum anodes, so that only four base contacts are required.

On starting, a time-delay relay in the luminaire housing allows the cathodes to heat. Then the circuit is broken, and the inductive kick of the transformer starts a discharge of neon with its characteristic red color. As the temperature rises, the sodium evaporates, and gradually the sodium vapor reaches its full brilliancy and normal color. This requires about 30 min. The 10,000-lumen sodium lamp has an average life of 3000 hours under normal street or highway lighting service. It has a starting voltage of 50 and a normal operating voltage of 30. Its current rating is 6.6 amp.

GLOW LAMPS. Various sizes and types of glow lamps are now available, ranging in power consumption from about 1/25 watt to 3 watts.

Neon gas, which is most generally used in glow lamps, produces an efficiency of about 0.3 lumen per watt. Since the light output of glow lamps is not great, they find only limited use as sources of illumination; they are valuable, however, as signals, pilots, night lights, and indicators of live circuits and by intensity of glow give some hint of the applied voltage.

The low current consumption is indicated by the current range, which is from only 0.0004 amp for the smallest to 0.030 amp for the largest lamp now manufactured. These lamps have a useful life of approximately 3000 hours.

Glow lamps, like all electric discharge lamps, have a negative or "run away" characteristic. Because of this characteristic, if the lamp were connected directly across a source of voltage sufficiently high to ionize the gas, the current would immediately rise to such proportions as to destroy the lamp. It is therefore essential that a limiting resistance be used in series with the lamp. In screw-base lamps this resistor is incorporated in the base; in bayonet-base lamps, which are manufactured without a resistor, the proper resistance must be supplied externally in the circuit by the user. Common types of glow lamps are listed in Table 9.

Table 9. Glow Lamps (105–125 Volts)

Order Designation	Watts (Nominal)	Bulb	Base	Maximum Overall Length, Inches	Approximate Starting Voltage A-c	Approximate Starting Voltage D-c	Approximate Series Resistance, Ohms
NE-2	1/25	T-2	Unbased (wire term.)	1 1/16	65	90	200,000 ex.
NE-51	1/25	T-3 1/4	S. C. bay. min.	1 3/16	65	90	200,000 ex.
NE-48	1/4	T-4 1/2	D. C. bay. cand.	1 1/2	65	90	30,000 ex.
NE-45	1/4	T-4 1/2	Cand. screw	1 5/8	65	90	30,000 in.
NE-27	1/2	G-10	Medium screw	2 1/16	105		3,500 in.
NE-30	1	G-10	Medium screw	2 1/16	60	85	4,800 in.
NE-32	1	G-10	D. C. bay. cand.	2	60	85	4,800 ex.
NE-34	2	S-14	Medium screw	3 5/16	60	85	3,500 in.
NE-36	2	S-14	Sk. D. C. bay. cand.	3 3/4	60	85	3,500 ex.
NE-40	3	S-14	Medium screw	3 5/16	60	85	2,200 in.
NE-42	3	S-14	Sk. D. C. bay. cand.	3 3/4	60	85	2,200 ex.

Ex. refers to external resistor; in., to internal resistor

FLUORESCENT LAMPS. The fluorescent lamp is a form of electric discharge source based on extremely low mercury vapor pressure. This pressure under normal conditions is of the order of 6 to 10 μ (roughly one-one hundred thousandth of normal atmospheric pressure) and accounts for the high transformation (60 per cent) of electrical input watts into radiation at 2537 A. (See Fig. 16.) By coating the inside of the glass bulb or tubing with fluorescent chemicals, or phosphors as they are called, the 2537 A ultraviolet radiation is converted into longer wavelengths of ultraviolet in fluorescent sunlamps and 360BL lamps, and into visible light in standard fluorescent lamps.

Table 10 gives a list of various phosphors used in respect to color of light and the general range of wavelengths in Angstrom units for exciting range, sensitivity peak, emitted range, and emitted peak.

Table 10. Fluorescent Chemicals

Phosphor	Lamp Color	Exciting Range *	Sensitivity Peak	Emitted Range	Emitted Peak
Calcium tungstate.........	Blue	2200–3000	2720	3100–7000	4400
Magnesium tungstate	Blue-white	2200–3200	2850	3600–7200	4800
Zinc silicate..............	Green	2200–2960	2537	4600–6400	5250
Zinc beryllium silicate.....	Yellow-white	2200–3000	2537	4800–7500	5950
Cadmium silicate.........	Yellow-pink	2200–3200	2400	4800–7400	5950
Cadmium borate..........	Pink	2200–3600	2500	5200–7500	6150
360 BL phosphor	Blue ultra	2200–3200	2500–2800	3200–4500	3600
"E" phosphor............	Blue ultra	2200–2650	2475	2700–4000	3250

* 2200 A is lower limit of measurements.

Conversion Losses and Efficiency. Figure 16 shows the approximate distribution of energy in the 40-watt fluorescent lamp. The top bar indicates the electrical energy input, the middle bar shows the conversion of energy within the lamp, and the third bar gives the ultimate nature of the energy output. Skillful lamp design and proper operating conditions result in three-fifths of the input energy being converted into exciting radiation, practically all of which is in a single line (2536.7 A) less than 1 A in width. A little over 2 per cent of the energy is represented in the four principal mercury lines within the visible spectrum: 4047, 4358, 5461, and 5780 A. The rest of the input, plus the conversion loss in the phosphor coating, is emitted as infrared radiation or dissipated by conduction and convection. Part of this unavoidable loss keeps the cathodes hot, an essential condition for the free emission of electrons and highest efficiency at low operation voltages.

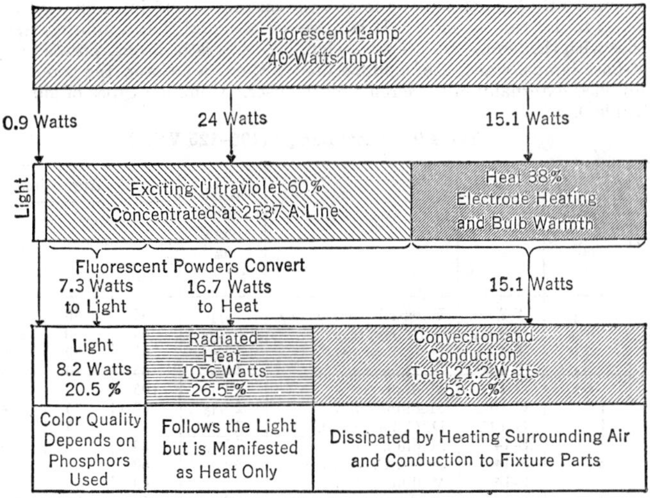

Fig. 16. Energy Input and Energy Conversion for the 40-watt Standard Fluorescent Lamp

The regions marked "light" represent the radiant energy emitted in the visible portion of the spectrum. The phosphor determines the distribution of this energy and thus the lumen output and color quality.

Figure 17 indicates why the rated efficiency of a 40-watt white fluorescent lamp is but a small fraction of the theoretical maximum—621 lumens per watt if all the input wattage could be radiated as yellow-green light, to which the eyes are most responsive. The values given are for the present 40-watt lamp; the values are different for lamps of other wattage and design.

Because it converts the shortwave ultraviolet radiation to visible light, the fluorescent chemicals are, in effect, the heart of the lamp. This coating must be subjected to close manufacturing control; careful blending and heat-treating of the chemicals themselves and close tolerances on the coating thickness are essential.

Voltage and Current-efficiency Factors. The choice of lamp dimensions and electrical values is determined not only by the maximum luminous efficiency, but by numerous other factors, such as brightness, lumen output, lumen maintenance, reliable starting and

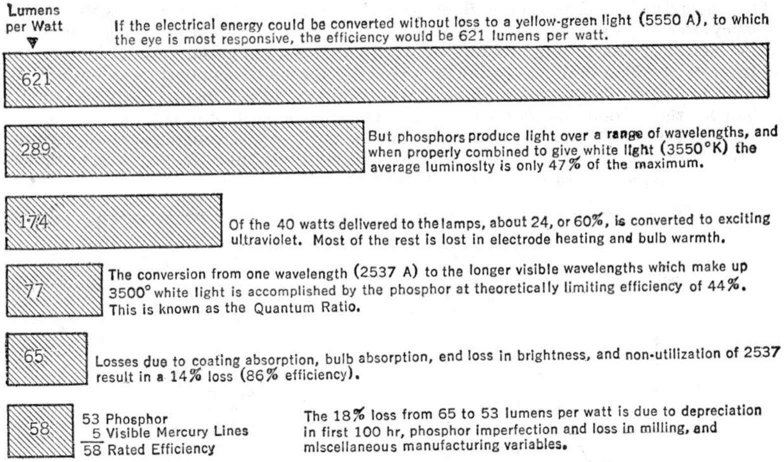
Lumens per Watt
If the electrical energy could be converted without loss to a yellow-green light (5550 A), to which the eye is most responsive, the efficiency would be 621 lumens per watt.
621
289
But phosphors produce light over a range of wavelengths, and when properly combined to give white light (3550°K) the average luminosity is only 47% of the maximum.
174
Of the 40 watts delivered to the lamps, about 24, or 60%, is converted to exciting ultraviolet. Most of the rest is lost in electrode heating and bulb warmth.
77
The conversion from one wavelength (2537 A) to the longer visible wavelengths which make up 3500° white light is accomplished by the phosphor at theoretically limiting efficiency of 44%. This is known as the Quantum Ratio.
65
Losses due to coating absorption, bulb absorption, end loss in brightness, and non-utilization of 2537 result in a 14% loss (86% efficiency).
58
53 Phosphor
5 Visible Mercury Lines
58 Rated Efficiency
The 18% loss from 65 to 53 lumens per watt is due to depreciation in first 100 hr, phosphor imperfection and loss in milling, and miscellaneous manufacturing variables.

FIG. 17. Efficiency Losses in a 40-watt White Fluorescent Lamp

satisfactory regulation, preferably without step-up or step-down from suitable line voltages, minimum wattage loss in ballast equipment, and commercial adaptability to manufacture, shipment, and use.

The chief determinants of lamp voltage are arc length, bulb diameter, and lamp current. The type of cathode is also a factor; hot-cathode lamps have lower voltages than lamps of corresponding size operated cold-cathode at the same current, because the hot-cathode lamps commonly have 70–100 volts less fixed voltage drop at the electrodes. For a given current and bulb diameter the lamp voltage rises as the length is increased, and for a given current and length falls as the diameter is increased. For fixed dimensions the lamp

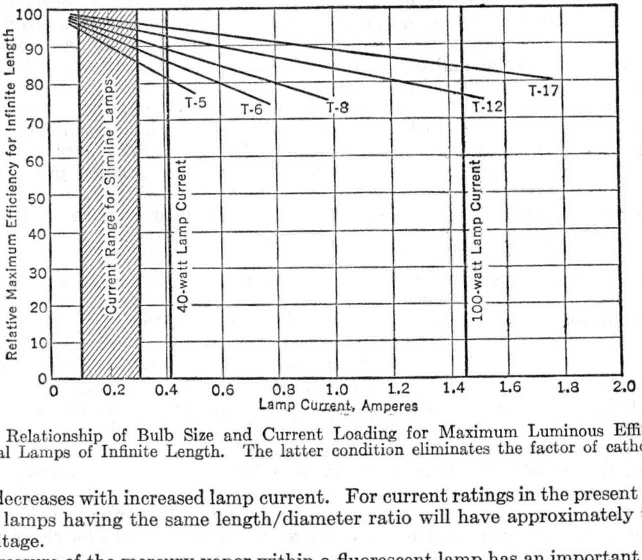

FIG. 18. Relationship of Bulb Size and Current Loading for Maximum Luminous Efficiency for Theoretical Lamps of Infinite Length. The latter condition eliminates the factor of cathode losses.

voltage decreases with increased lamp current. For current ratings in the present standard line, the lamps having the same length/diameter ratio will have approximately the same lamp voltage.

The pressure of the mercury vapor within a fluorescent lamp has an important effect on the electrical characteristics of the lamp. The normal pressure of a given size of lamp depends on the bulb-wall temperature, which in turn is determined by the input wattage and the area (a function of length and diameter) available to dissipate the heat. The optimum pressure is that which produces most efficiently the ultraviolet radiation that excites the phosphor.

The theoretical efficiency of a fluorescent lamp of infinite length (positive-column efficiency) decreases as the lamp current is increased because conditions are less favorable for conversion of input energy to 2537 and other exciting radiation. For a given current the theoretical efficiency is improved by increasing the bulb diameter; in an actual lamp of ordinary length efficiencies are somewhat less because of end losses. Larger diameters are best for the currents necessary to secure high lumens per foot of lamp (Fig. 18).

Effect of Bulb-wall Temperature. At low temperatures the mercury condenses out, and the internal vapor pressure drops below the point at which the exciting ultraviolet radiation is produced most efficiently. This is shown in Fig. 19.

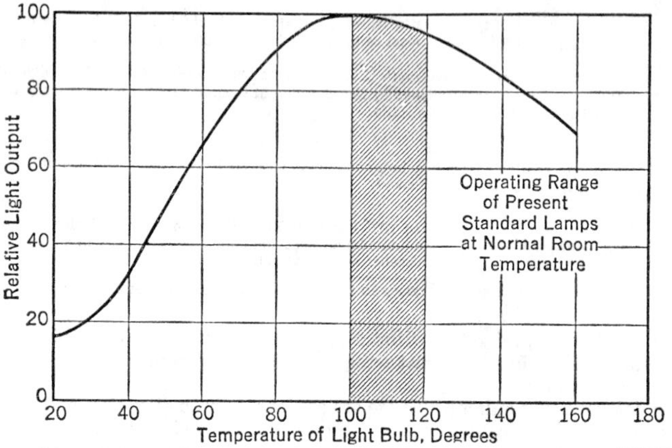

FIG. 19. General Effect of Bulb-wall Temperature on the Efficiency of Fluorescent Lamps

If lamps are sufficiently loaded or if enclosures raise the bulb-wall temperature to the extent that the mercury vapor pressure exceeds the value for optimum production of the exciting radiation, such lamps or units may produce more light below normal room temperature than under ordinary conditions. Because of the difference in watts per square inch of surface area, the 100-watt lamp will be better able to maintain a favorable bulb-wall temperature at lower ambient temperatures than the 40-watt size. On the other hand, the 100-watt lamp will fall off more in light output at high ambient temperatures.

Fluorescent Lamp Types. The principles of all fluorescent lamps are the same regardless of size, design, or type, and whether mass-produced as stock catalog items or custom-built and shaped to order in letter or pattern form. All make use of phosphor-coated glass tubing, excited by the predominant 2537 A line ultraviolet radiation produced by an electrical discharge through a mercury vapor medium of very low pressure. As previously discussed, the problem of lamp design is one of composing many variables. Among these variables are the phosphor efficiency, proper control of mercury vapor pressure and lamp-filling pressure, lamp loading or current density, and bulb-wall area, which is controlled by lamp length and diameter; these factors, in turn, affect the loss or conservation of heat, which in turn affects the internal vapor pressure. These design elements are common to every type of fluorescent lamp. Any further differences in lamp design will depend on the external circuits, with respect to the method of starting the lamp and developing the proper current and voltage relations. This statement applies whether the lamp is to be operated in series or multiple, or with instantaneous or preheat starting.

Table 11 lists the various sizes and types of standard fluorescent lamps. In addition to the various color temperatures of "white" light, the 15-, 20-, and 30-watt sizes of the standard line are available to furnish blue, green, pink, gold, and red light.

Cathode Characteristics. When a discharge or flow of electricity takes place through a gas or vapor medium, electrons flow or, in a sense, are pulled away by the difference of electrical pressure or voltage between one end of the lamp and the other. This requires an abundance of free electrons, and although tungsten and iron, as used for electrodes, have free electrons, such cathodes are usually fortified by coating with barium or strontium oxides for greater electron emission.

Table 11. Technical Data on Fluorescent Lamps

Standard Line Lamps

Bipin base construction for preheat starting circuits

	Miniature Bipin Base			Medium Bipin Base						Mogul Bipin	
Wattage size	6	8	13	14	15 (T-8)	15 (T-12)	20	30	40 *	40 *	100
Nominal length, inches	9	12	21	15	18	18	24	36	48	60	60
Diameter, inches	5/8	5/8	5/8	1 1/2	1	1 1/2	1 1/2	1	1 1/2	2 1/8	2 1/8
Bulb	T-5	T-5	T-5	T-12	T-8	T-12	T-12	T-8	T-12	T-17	T-17
Approximate lamp amperes	0.145	0.16	0.16	0.37	0.30	0.33	0.35	0.34	0.41	0.40	1.45
Approximate lamp volts	48	57	100	41	56	48	62	103	108	110	72
Circuit voltages	Depends on ballast types available for various lamps										
Rated average life †	2500	2500	2500	2500	2500	2500	2500	2500	2500	2500	3000
Lumen output											
White	210	330	585	490	615	600	920	1470	2300	2100 ‡	4200
Daylight	186	295	520	435	585	540	800	1350	1920	......	3900
Footlamberts											
White	2620	2950	2700	1410	2080	1360	1450	2260	1750	920 ‡	1850
Daylight	2330	2640	2400	1260	1980	1230	1260	2080	1470	......	1710

Slimline Lamps

All Slimline lamps have a single pin base for instant-start hot-cathode operation

Nominal length, inches, and bulb diameter	42–T-6			64–T-6			72–T-8			96–T-8		
Maximum lamp length, inches	40			62			70			94		
Diameter, inches	3/4			3/4			1			1		
Minimum starting voltage	450			600			600			750		
Rated life †	2500			2500			2500			2500		
Lamp watts (add auxiliary watts for total)	16	25	33	24	39	51	22	38	51	29	51	69
Lamp current, milliamperes	100	200	300	100	200	300	100	200	300	100	200	300
Approximate lamp volts	180	150	130	285	230	200	250	220	200	335	295	265
Lumen output—white	930	1400	1710	1440	2250	2700	1410	2350	3000	1890	3200	4150
Footlamberts—white	1700	2600	3150	1700	2600	3150	1050	1750	2250	1050	1750	2250

Circline Lamps

Four-prong, connector-type base

Lamp watts	32	Light output (white), lumens	1600
Outside diameter of circle, inches	12 ± 1/4	Brightness, footlamberts	2040
Diameter of tube, inches	T-10 (1 1/4)	Circline lamps in 8- and 16-in. diameters (approx. 20 and 40 watts) will also be available as production facilities permit	
Lamp amperes (operating)	0.43		
Lamp volts (operating)	84		

* The 40-watt T-12 instant-start lamp has a medium bipin base with pins short-circuited inside end caps and will not operate on preheat ballast circuits; the 40-watt T-17 mogul bipin is of the same construction.

† Based on 3 burning hours per start; the 2500-hour lamps are rated as 4000 hours life at 6 burning hours per start, 6000 hours at 12 burning hours per start. Equivalent ratings for 100-watt lamps are 4500 and 6500 hours.

‡ 4500 deg white.

Many hundreds of different cathode designs have been tried experimentally or used in practice, but these narrow down to two principal classes, each adapted to the electrical circuit that may be used for starting and operating. The two types of cathodes shown in Fig. 20 have become known as cold and hot cathodes.

The cold-cathode type in a variety of forms has been used for neon tubing ever since its development over 30 years ago. Its use developed around long lengths of tubing or a series of lengths fabricated into letter or pattern form, and special high-voltage circuits were favorable to this application because instant start is essential for sign service, where many displays demand flashing and sometimes dimming sequences for attention purposes. High voltage dictates that the cathode be of fairly large size in order to provide a large electron-bearing surface. High-voltage cold-starting introduces quite a shock to the cathode and drives off an excessive number of electrons, but with the large-area iron thimble-type cathode an abundance of electrons are provided to insure long life, even when used on flashing circuits. During operation the cathode attains a temperature of about 150 deg cent but does not go higher because of the low current and large metal area, which dissipates the heat. Because of the higher voltage required at the cathode to maintain the electron emission, there will be a higher voltage drop at this point, which, for a given current, actually means a greater wattage loss at the cathode, and the bulb area at each end will be 40 to 60 deg higher in temperature than the same bulb area of hot-cathode lamps.

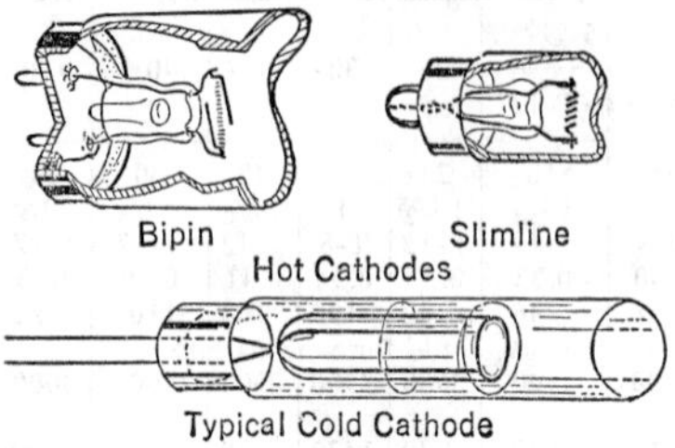

FIG. 20. Examples of "Hot-" and "Cold-" Cathode Construction

The hot-cathode lamps employ coiled tungsten filaments coated with electron-bearing materials. By passing a current through the filament at starting, it is heated to around 950 deg cent, and the arc is established at a lower voltage. Because of the small size of the cathode the normal current flow maintains a high temperature at a small portion of the cathode, although the voltage drop and consequently the wattage loss at the cathode are relatively low; this in turn makes for greater lamp efficiency and actually a cooler operation than with a so-called cold cathode. Hot-cathode lamps can be started cold-cathode, i.e., without preheating, by sufficient starting voltage; they will operate hot-cathode, however, by virtue of the impinging arc heating a few segments of the small filament wire to red-hot temperature.

Although lamps designed for preheat circuits have bipin bases, these terminals may be connected together, and the lamps can be started by high voltage and operated as hot-cathode lamps. Slimline lamps have only one terminal at each end, since they are designed for instant starting, yet because of the filament-type cathode, they operate as hot-cathode lamps. The higher efficiency of this type of construction, due to lower electrode losses, results in considerable advantage, particularly in lamps of shorter length.

Fluorescent Ballasts. A fluorescent lamp ballast may be simply a coil of insulated copper wire wound on an iron core made up of layers of thin iron stampings. The ballast is placed in series with the lamp and, if properly designed, will limit the current to value for which the lamp is designed.

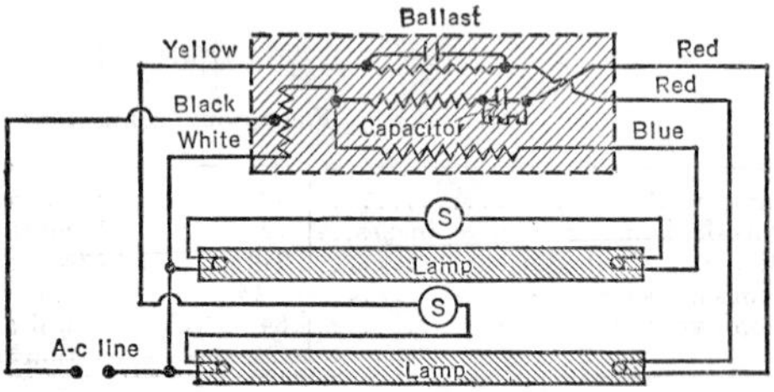

FIG. 21. Transformer Elements and Typical Wiring Diagram for a Two-lamp Fluorescent-lamp Circuit

It should be remembered that the length, diameter, or wattage of any fluorescent lamp requires a specific lamp voltage and current, and it is the function of the ballast to deliver these essentials. Therefore, ballast design must be predicated (1) on the circuit or distribution voltage on the user's premises, and (2) on the frequency of the system. The first may involve four voltage classifications: 118, 208, 240, and 260 volts as nominal circuit ratings, and the latter involves principally three frequencies: 60, 50, and 25 cycles. Although the lamp remains the same for each condition, this means that ballasts must be properly specified for the circuit on which fluorescent lamps are to be used.

Ballasts are available to operate single lamps and two, three, and four lamps. The most common type is the two-lamp, which uses the "split-phase" principle, with one of the lamps ballasted by inductive reactance only and the other by inductive and capacitive reactances in series. The result is an overall power factor of 95 per cent or more, and at the same time the stroboscopic effect is reduced because of the 120-deg phase displacement in the two branches of the circuit. Two-lamp ballasts for 118-volt operation of 30-, 40-, and 100-watt lamps consist of an autotransformer winding and two reactor windings on a single core. A typical two-lamp ballast circuit is shown in Fig. 21. The characteristic curves of a two-lamp ballast are shown in Fig. 22. The four-lamp ballast for the 100-watt size takes advantage of the voltage rating of the 100-watt lamp to operate two-in-series on each leg of a modified two-lamp ballast. Its application is confined to a Y-connected network distribution rated at 254/440, 265/460, and 277/480 volts.

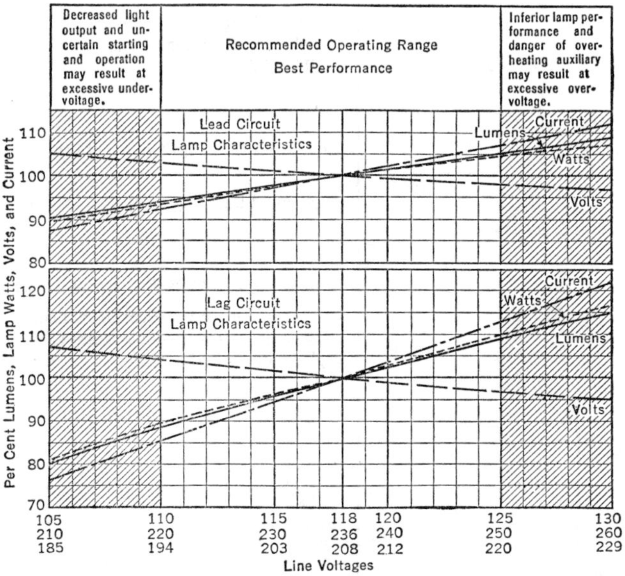

FIG. 22. Ballast and Lamp Performance Characteristics for Two-lamp Fluorescent Operation on Lead and Lag Circuits

Direct-current Operation. Although fluorescent lamps are designed for a-c operation, they may be used on d-c circuits if the proper series resistance is provided in connection with a suitable inductance coil and starting switch. Failure to provide the necessary resistor equipment will result in failure of the lamp and/or accessories. The series resistance consumes about as much wattage as the lamp itself, and therefore the overall efficiency of light production will be much less than in normal a-c operation.

Because of somewhat more difficult starting conditions on direct current, useful lamp life may be impaired because of failure to start. Lumen maintenance may likewise be somewhat less favorable. On the other hand, color quality and total light output of lamps compare favorably with a-c operation, and problems of power factor and stroboscopic effect are, of course, eliminated.

Because direct current flows in only one direction through the tube, lamps of 36 in. or longer may develop a considerable dim region at one end of the tube. This may be corrected by equipping installations with line reversing switches. A switch suitable for 240-volt inductive circuits and of a type which opens the circuit for an instant before reversing it should be used.

Lamp Starting. Lamps are designed for operation either on preheat starting circuits or so-called high-voltage instant-start circuits. The preheat starting circuit makes use of a replaceable starter, the function of which is to complete a separate circuit, so that a preheat current can flow through the filament cathodes and heat them momentarily, after which the switch opens automatically and the lamp starts. A small 0.006 μf condenser across the switch contacts aids in starting but is primarily useful to shunt out line-lead harmonics, which may cause radio interference. Several types of starter switches are available. Instant-start circuits require higher starting voltage, which is a function of ballast design. A characteristic of the instant-start circuit is the difficulty of starting under conditions of high humidity, and to solve this problem instant-start lamps are provided with a thin metallic strip running lengthwise of the lamp on the outside of the bulb.

Lamp Colors and Color Temperature. With filament lamps colored light was produced by means of absorption filters; although amber, orange, and red could be produced with

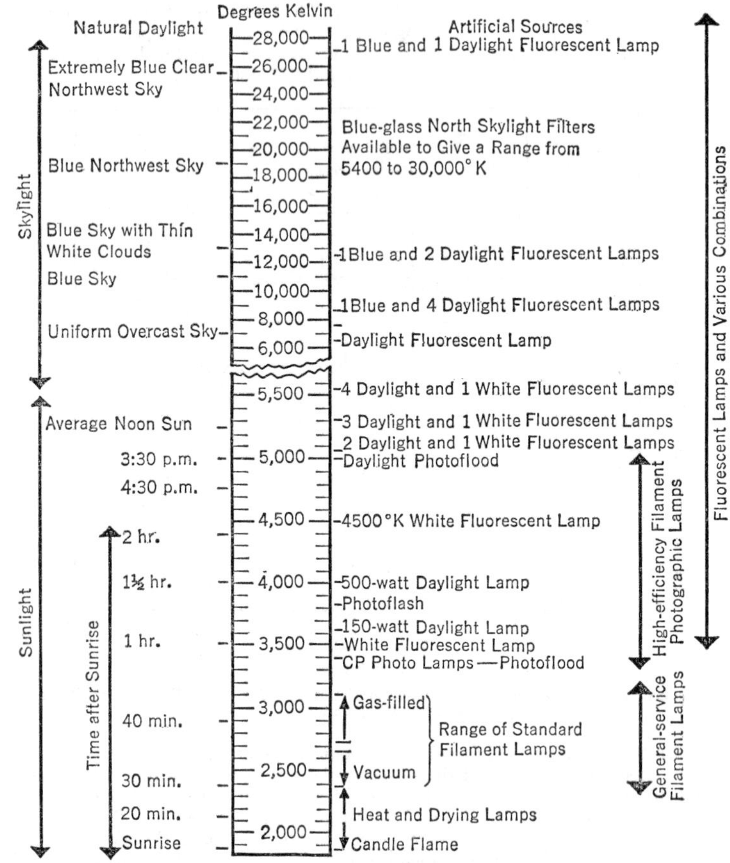

FIG. 23. Color-temperature Scale, Showing Various Conditions of Natural Daylight and Related Values for Various Artificial Sources

fair efficiency, blues and greens required an absorption of 85 to 99 per cent of the light from the less efficient filament lamp. An efficiency of 1 or 2 lumens per watt was accepted. With the fluorescent lamp colored light is produced directly by choice of phosphor; the green phosphor, for example, produces green light at an efficiency of 70 lumens per watt. Blue, pink, and amber have an efficiency of 25 to 30 lumens per watt; at present red is low, at about 4 or 5 lumens per watt. White light is a mixture of all colors and is obtained by blending in proper proportion the phosphors that in themselves produce colored light. Obviously, therefore, almost any degree of "whiteness" or almost any tint of saturation may be obtained.

In the interest of standardization and manufacturing economy, four different "white lamps" are listed—namely, "daylight" (6500 deg K), the 4500 deg K white, the 3500 deg K white, and the "soft white," a warmer-toned white. Figure 23 shows a color-temperature scale for natural daylight, together with the color temperatures of filament lamps and various combinations of fluorescent lamps.

7. LIGHTING SYSTEMS

CLASSIFICATION OF LIGHTING SYSTEMS. The Illuminating Engineering Society classifies general lighting systems into five fundamental types, which will be described. These are based on the light distribution, as indicated in Fig. 24.

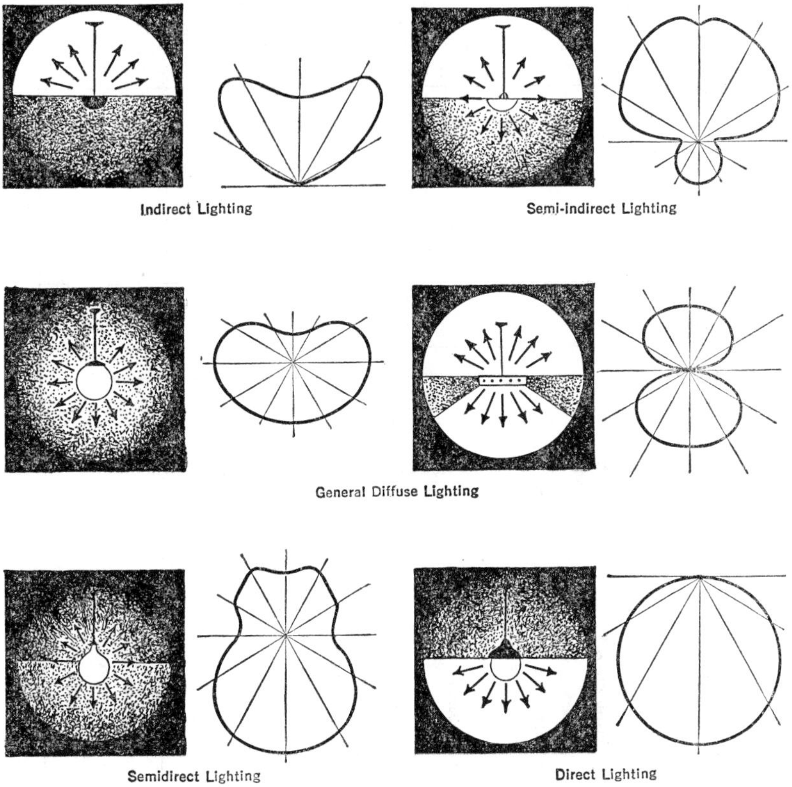

FIG. 24. General Classifications of Lighting Equipment in Terms of Lighting Distribution in Upper and Lower Hemispheres

INDIRECT LIGHTING. With this system 90 to 100 per cent of the light from the luminaires is first directed to the ceiling and upper side walls, from which it is diffusely reflected to all parts of the room. In effect, the entire ceiling becomes the light source, and shadows and reflected glare are minimized. Since the light source is the entire ceiling, it cannot be avoided by the observer; therefore care must be taken to keep the brightness low enough to prevent it being a source of glare. Because the ceiling constitutes an important part of such a lighting system, it should be as light in color as possible, being given a matte finish with a high reflection factor, and should be maintained in good condition. Figure 25 shows the relative reflection efficiencies of various materials.

The lighting units, which may have either opaque or luminous bottoms, should be easy to clean, because a layer of dust and dirt can absorb one-fourth or more of the light output of the unit.

Indirect lighting produces a quality of lighting highly desirable for such visual tasks as those in drafting rooms, general offices, and private offices.

Semi-indirect Lighting is defined as any system in which 60 to 90 per cent of the luminaire output is emitted upward toward the ceiling and upper side walls, while the rest is directed downward. As this system also utilizes the ceiling as the main source of light, the same considerations of ceiling finish and good maintenance should be observed as

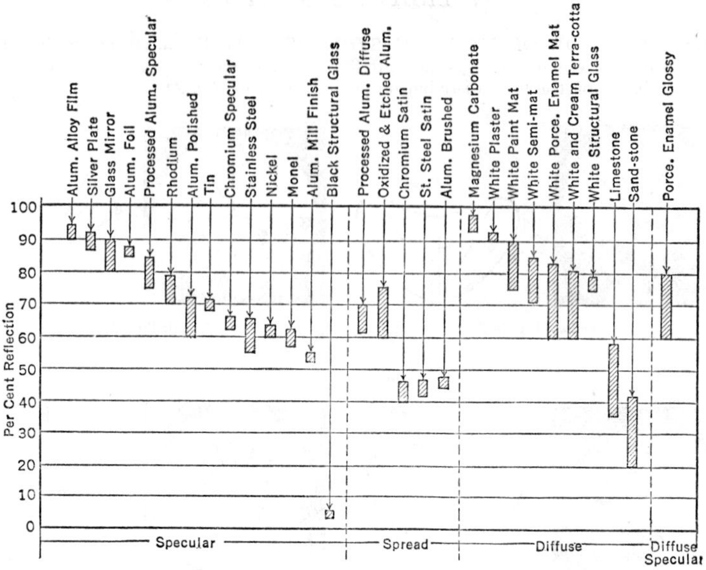

Fig. 25. Relative Reflecting Efficiency of Various Materials for White Light

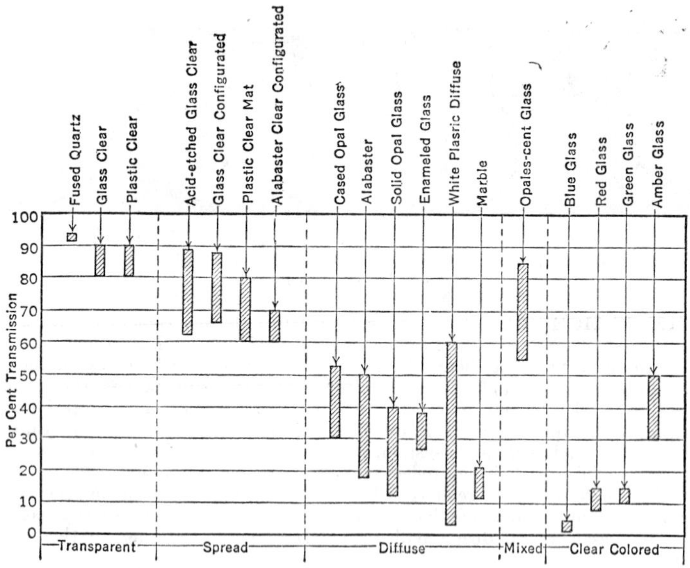

Fig. 26. Relative Transmission Efficiencies of Various Translucent Materials for White Light

for indirect lighting. In general, semi-indirect units give a little more light for the same wattage than do indirect units, but more attention may have to be given to the factors of direct and reflected glare. Figure 26 shows the transmission characteristics of translucent materials.

GENERAL DIFFUSE LIGHTING. This classification refers to systems where the predominant illumination on horizontal working surfaces (40 to 60 per cent) comes directly from the lighting units, but where there is also a considerable contribution from upward light, reflected from ceiling and upper wall areas.

One type of unit of this classification is the glass diffusing, enclosing globe. Diffusing globes should be of sufficient opacity to conceal the lamp completely. Diffusing globes should not be used to provide the higher illuminations, since they are likely to become too bright from the standpoint of both direct and reflected glare.

Another general type of unit in this classification is represented by the direct-indirect, where little or no light is emitted in the normal fields of view.

Although general diffuse lighting systems give more illumination for a specified wattage than do indirect or semi-indirect systems, shadows are more noticeable, and some difficulty may be experienced with both direct and reflected glare.

SEMIDIRECT LIGHTING. In this classification 60 to 90 per cent of the output of the luminaire is directed downward to the working surface. There is, therefore, some contribution to the illumination at the working plane from light which is directed upward and reflected by the ceiling and upper wall areas. For the most part luminaires in this glass are of the open-bottom type, although some have closed bottoms of glass or plastic material. If used for general office lighting, the comments in the immediately preceding and following classifications should be noted.

DIRECT LIGHTING. Units in this classification can be defined as those which direct practically all (90 to 100 per cent) of the light of the luminaires in angles below the horizontal, i.e., directly toward the usual working areas. Although, in general, such systems provide most effective illumination on the working surfaces, this efficiency may be obtained at the sacrifice of other factors. For example, disturbing shadows may result unless the area of the lighting units is relatively large. Direct and reflected glare may be distressing if the relative brightnesses are not kept within the recommended limits. An appearance of dinginess may result unless some special means of brightening the ceiling is provided, such as upward light from the luminaires or light-colored floorings, desk tops, and furniture finishes.

The general characteristics of these systems prevail, even though the details of equipment design and installation differ widely. The advent of fluorescent lighting has stimulated built-in or architectural lighting, using troffer* systems in continuous rows or in pattern design. Units of this character, either open louvered or with glass or plastic cover plates, are set flush with the ceiling, and, in fact, the construction may be such that the lighting units support the finished ceiling. Built-in coves and coffers are also finding more favor, and better designs are possible with fluorescent lamps or high-voltage tubing, which avoids the scallop effect often noted when filament lamps are used. The long life of fluorescent lamps is also an advantage in such places, where frequent lamp replacement is particularly difficult. Flush ceiling-mounted spotlights and lens plate units, often referred to as downlights, are likewise growing in popularity.

LIGHTING-DESIGN CONSIDERATIONS. The problem of lighting design in its simplest terms is to deliver lumens from the source to the surface to be lighted. The lumens per square foot or footcandle result will depend on many factors which cause some of the lumens generated to be lost before reaching the surface to be lighted. Design calculations merely account for the light losses in the fixture or reflector and light absorbed by ceiling and walls.

The point-by-point method of calculation is useful for determining the footcandles produced by a direct lighting reflector or a floodlight or searchlight projector. The point-by-point method of lighting calculation is based on the "inverse square law," i.e., that the intensity of light varies inversely as the square of the distance from the light source to the point of measurement. From the candlepower distribution curve of a reflector, the footcandles at any given point may be computed from the formula:

$$\text{Footcandles} = \frac{CP \text{ (candlepower)}}{D^2 \text{ (distance in feet)}} \qquad \text{(Normal to the beam)} \tag{1}$$

$$\text{Footcandles} = \frac{CP}{D^2} \times \text{Cosine of angle } X \qquad \text{(On horizontal plane)} \tag{2}$$

* A long channel- or gutter-shaped container for fluorescent lamps, usually installed in ceilings with opening flush with the ceiling.

This method is long and tedious where any significant number of sources is involved, since the contribution of each source to the lighting at any one point must be individually computed. The method is not applicable in rooms where light reflection from ceiling and wall surfaces is substantial. Neither is it generally applicable for extended sources, such as large fluorescent luminaires, since the inverse square law is inoperative at distances closer than five times the maximum dimension of the source.

THE LUMEN METHOD OF DESIGN. In planning a general lighting system for a working area the aim is to provide substantially a uniform level of illumination throughout the room. This eliminates spottiness and dark corners and makes the entire area equally suitable as a work space or for display, sales, or other general purpose.

The number of outlets to provide for any given area is determined by the maximum allowable spacing between lighting units and is in turn regulated by their height above the floor.

Strictly speaking, the spacing for uniform illumination on the work depends upon the height of the light source above the surface to be illuminated, but since most work surfaces are from 2 1/2 to 3 1/2 ft above the floor, the spacing may for practical purposes be considered a function of the mounting height of lamps above the floor. In general, a spacing in feet which does not substantially exceed this mounting height will result in reasonably uniform illumination. The ceiling height, or rather the height at which units may be mounted clear of obstructions, therefore limits the maximum permissible spacing. The spacing of lighting units is not influenced by the size or type of lamp used but is regulated by the distribution characteristics of the reflector.

The maximum permissible spacing for uniformity of illumination for most general lighting units is determined by multiplying the distance between the light source and the work plane by 1.5. For concentrating- or focusing-type units the spacing should be not more than their mounting height above the work, and often should be less, depending on the actual distribution characteristics of the reflector.

UTILIZATION FACTORS. In order to specify the lamp size necessary to provide the footcandles desired, the first step is to determine the percentage of light emitted by the lamp that actually gets down to and is useful on the working level. This percentage, called the utilization factor for the particular installation, is given in Table 12.

Reflector Characteristics. The selection of reflecting equipment depends not only upon its efficiency but also upon suitable distribution of light. Utilization factors are computed from the candlepower-distribution curves in accordance with basic experimental data [Harrison and Anderson, *Trans. IES*, Vol. 15 (1920)]. It will be quite evident that a narrow or concentrating distribution will direct the light strongly downward, keeping less of it from striking the walls and ceiling than will a broad distribution. In a narrow distribution the influence of the size of room or the color of walls and ceiling on the utilization is much less than in a broad distribution, where a large proportion of light strikes the walls and ceiling, only a part being rereflected to working surfaces.

Interior Finish. Utilization factors also coordinate the effect of interior finish on lighting results, and Table 12 embraces a range of general ceiling- and wall-reflection conditions. Note in Table 12 that the influence of the interior finish is least important with direct reflectors, becoming increasingly influential with semidirect lighting and constituting a major factor in lighting efficiency with semi-indirect and indirect lighting. The net reflection value of even light walls seldom exceeds 50 per cent when allowance is made for wall furnishings and door and window openings. In glass-enclosed rooms or buildings the effective wall reflection is practically negligible.

Room Index. In general, large rooms use light more efficiently than do small rooms because there is less wall area to absorb light in proportion to the floor space. For the same reason rooms with high ceilings are less efficiently lighted than low-ceilinged rooms with the same arrangement of floor space.

Table 13 classifies room proportions into ten classes, as indicated by letters A to J. This serves merely as a reference index to be applied in Table 12 for the particular type of reflecting equipment used. Note that this factor of room size and proportion may influence the utilization factor from 100 to nearly 300 per cent, depending on the type of reflecting equipment.

Maintenance Factor. Allowance must always be made for depreciation of lamps, reflectors, and reflecting surfaces, so that desired footcandle levels may be maintained in service, as contrasted to initial values. Lamps average about 80 to 90 per cent of their initial lumen output, and the inevitable film of dust that collects quickly on reflecting surfaces accounts for another 15 to 25 per cent of normal depreciation even with a reasonable cleaning schedule. The average illumination maintained in service will, under good conditions, be of the order of 70 per cent of the initial value, or 0.70 when expressed as a **maintenance factor.** In some instances, particularly with direct lighting equipment where

Table 12. Utilization Factors for Typical Luminaires of the Five Classifications of Lighting Systems

Luminaire Type	Luminaire Light Distribution	Ceiling	70%			50%			30%	
		Walls	50%	30%	10%	50%	30%	10%	30%	10%
		Room Index	Utilization Factors							
Direct		J	0.37	0.31	0.27	0.36	0.31	0.27	0.31	0.27
Maintenance Factor		I	0.45	0.41	0.38	0.45	0.40	0.37	0.40	0.37
Good........ 0.75	↑	H	0.49	0.45	0.42	0.49	0.45	0.42	0.45	0.42
Average...... 0.65		G	0.53	0.49	0.46	0.53	0.49	0.46	0.48	0.46
Poor......... 0.55	0/79	F	0.56	0.53	0.49	0.55	0.52	0.49	0.51	0.49
		E	0.61	0.58	0.55	0.60	0.57	0.55	0.56	0.55
	↓	D	0.66	0.63	0.60	0.64	0.62	0.60	0.61	0.60
		C	0.67	0.65	0.62	0.66	0.64	0.62	0.63	0.61
		B	0.71	0.68	0.66	0.69	0.67	0.65	0.66	0.64
		A	0.72	0.70	0.67	0.71	0.68	0.67	0.67	0.66
Semidirect		J	0.27	0.25	0.19	0.26	0.22	0.19	0.20	0.18
Maintenance Factor		I	0.35	0.29	0.26	0.33	0.28	0.25	0.27	0.24
Good........ 0.75	↑	H	0.38	0.34	0.30	0.36	0.32	0.29	0.30	0.28
Average...... 0.65		G	0.43	0.38	0.34	0.40	0.36	0.32	0.33	0.31
Poor......... 0.55	25/60	F	0.46	0.41	0.37	0.43	0.39	0.35	0.37	0.33
	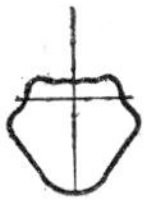	E	0.50	0.46	0.42	0.47	0.43	0.40	0.40	0.38
	↓	D	0.55	0.50	0.46	0.51	0.47	0.44	0.44	0.42
		C	0.58	0.53	0.49	0.53	0.49	0.46	0.46	0.44
		B	0.62	0.57	0.53	0.57	0.53	0.51	0.50	0.48
		A	0.64	0.60	0.56	0.59	0.55	0.52	0.51	0.49
General Diffuse		J	0.24	0.19	0.16	0.22	0.18	0.15	0.16	0.14
Maintenance Factor		I	0.29	0.25	0.22	0.27	0.23	0.20	0.21	0.19
Good........ 0.75	↑	H	0.33	0.28	0.26	0.30	0.26	0.24	0.24	0.21
Average...... 0.70		G	0.37	0.32	0.29	0.33	0.29	0.26	0.26	0.24
Poor......... 0.65	39/45	F	0.40	0.36	0.31	0.36	0.32	0.29	0.29	0.26
		E	0.45	0.40	0.36	0.40	0.36	0.33	0.32	0.29
	↓	D	0.48	0.43	0.39	0.43	0.39	0.36	0.34	0.33
		C	0.51	0.46	0.42	0.45	0.41	0.38	0.37	0.34
		B	0.55	0.50	0.47	0.49	0.45	0.42	0.40	0.38
		A	0.57	0.53	0.49	0.51	0.47	0.44	0.41	0.40
Semi-indirect		J	0.20	0.16	0.13	0.16	0.13	0.11	0.10	0.09
Maintenance Factor		I	0.24	0.20	0.18	0.20	0.17	0.15	0.13	0.12
Good........ 0.70	↑	H	0.28	0.24	0.21	0.23	0.19	0.17	0.15	0.13
Average...... 0.65		G	0.31	0.27	0.24	0.26	0.22	0.20	0.17	0.15
Poor......... 0.60	66/20	F	0.34	0.30	0.27	0.28	0.24	0.22	0.19	0.17
	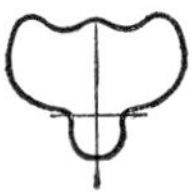	E	0.38	0.34	0.31	0.31	0.27	0.25	0.21	0.19
	↓	D	0.42	0.38	0.35	0.34	0.30	0.28	0.23	0.22
		C	0.45	0.41	0.37	0.36	0.32	0.30	0.25	0.23
		B	0.49	0.45	0.42	0.39	0.36	0.34	0.27	0.25
		A	0.51	0.47	0.44	0.41	0.38	0.36	0.28	0.27
Indirect		J	0.15	0.11	0.10	0.09	0.08	0.06	0.04	0.03
Maintenance Factor		I	0.19	0.15	0.13	0.12	0.10	0.09	0.06	0.04
Good........ 0.60	↑	H	0.22	0.19	0.16	0.14	0.12	0.10	0.07	0.05
Average...... 0.50		G	0.26	0.22	0.19	0.17	0.14	0.13	0.08	0.07
Poor......... 0.40	80/0	F	0.28	0.24	0.21	0.19	0.16	0.14	0.09	0.08
		E	0.32	0.28	0.25	0.21	0.18	0.17	0.11	0.10
	↓	D	0.35	0.31	0.29	0.23	0.21	0.19	0.12	0.11
		C	0.38	0.34	0.31	0.25	0.22	0.21	0.13	0.12
		B	0.42	0.39	0.36	0.27	0.25	0.24	0.15	0.14
		A	0.43	0.41	0.38	0.29	0.27	0.25	0.16	0.15

Table 13. Room Index

		Ceiling Height, Feet										
For Semi-indirect and Indirect Lighting		9 and 9 1/2	10 to 11 1/2	12 to 13 1/2	14 to 16 1/2	17 to 20	21 to 24	25 to 30	31 to 36	37 to 50		
		Mounting Height above Floor, Feet										
For Direct and Semidirect Lighting		7 and 7 1/2	8 and 8 1/2	9 and 9 1/2	10 to 11 1/2	12 to 13 1/2	14 to 16 1/2	17 to 20	21 to 24	25 to 30	31 to 36	37 to 50
Room Width, Feet	Room Length, Feet	Room Index										
9 (8 1/2–9)	8– 10	H	I	J	J							
	10– 14	H	I	I	J							
	14– 20	G	H	I	J	J						
	20– 30	G	G	H	I	J	J					
	30– 42	F	G	H	I	J	J	J				
	42 up	E	F	G	H	I	J	J				
10 (9 1/2–10 1/2)	10– 14	G	H	I	J	J						
	14– 20	G	H	I	J	J	J					
	20– 30	F	G	H	I	J	J					
	30– 42	F	G	G	H	I	J	J				
	42– 60	E	F	G	H	I	J	J				
	60 up	E	F	F	H	H	I	J				
12 (11–12 1/2)	10– 14	G	H	I	I	J	J					
	14– 20	F	G	H	I	J	J					
	20– 30	F	G	G	H	I	J	J				
	30– 42	E	F	G	H	I	J	J				
	42– 60	E	F	F	G	H	I	J				
	60 up	E	E	F	G	H	I	J				
14 (13–15 1/2)	14– 20	F	G	H	H	I	J	J				
	20– 30	E	F	G	H	I	J	J				
	30– 42	E	F	F	G	H	I	J	J			
	42– 60	E	E	F	F	H	I	J	J	J		
	60– 90	D	E	E	F	G	H	J	J	J		
	90 up	D	E	E	F	F	G	I	J	J		
17 (16–18 1/2)	14– 20	E	F	G	H	I	J	J				
	20– 30	E	F	F	G	H	I	J				
	30– 42	D	E	F	G	H	H	J	J	J		
	42– 60	D	E	E	F	G	G	I	J	J	J	
	60–110	D	E	E	F	G	G	I	J	J	J	
	110 up	C	D	E	E	F	G	H	I	J	J	
20 (19–21 1/2)	20– 30	D	E	F	G	H	I	J	J			
	30– 42	D	E	E	F	G	H	I	J	J		
	42– 60	D	D	E	E	F	G	I	J	J	J	
	60– 90	C	D	E	E	F	G	H	J	J	J	
	90–140	C	D	D	E	F	F	H	I	I	J	J
	140 up	C	D	D	E	F	F	H	H	I	J	J
24 (22–26)	20– 30	D	E	E	F	G	H	I	J	J		
	30– 42	C	D	E	F	G	G	I	J	J		
	42– 60	C	D	D	E	F	G	H	I	J	J	
	60– 90	C	D	D	E	F	F	H	I	J	J	J
	90–140	C	C	D	E	E	F	G	H	I	J	J
	140 up	C	C	D	E	E	F	G	H	I	I	J
30 (27–33)	30– 42	C	D	D	E	F	G	H	I	J	J	
	42– 60	C	C	D	D	F	F	H	H	I	J	
	60– 90	B	C	C	D	E	F	G	H	I	J	J
	90–140	B	C	C	D	E	E	F	G	H	I	J
	140–180	B	C	C	D	E	E	F	G	H	I	J
	180 up	B	C	C	D	E	E	F	G	H	I	J

Table 13. Room Index—*Continued*

		Ceiling Height, Feet										
For Semi-indirect and Indirect Lighting		9 and 9 1/2	10 to 11 1/2	12 to 13 1/2	14 to 16 1/2	17 to 20	21 to 24	25 to 30	31 to 36	37 to 50		
		Mounting Height above Floor, Feet										
For Direct and Semidirect Lighting		7 and 7 1/2	8 and 8 1/2	9 and 9 1/2	10 to 11 1/2	12 to 13 1/2	14 to 16 1/2	17 to 20	21 to 24	25 to 30	31 to 36	37 to 50
Room Width, Feet	Room Length, Feet	Room Index										
36 (34–39)	30– 42	B	C	D	E	F	F	H	I	I	J	
	42– 60	B	C	C	D	E	F	G	H	I	J	J
	60– 90	A	C	C	C	E	E	F	H	H	J	J
	90–140	A	B	C	C	D	E	F	G	H	I	J
	140–200	A	B	C	C	D	E	F	F	G	H	I
	200 up	A	B	C	C	D	E	F	F	G	H	I
42 (40–45)	42– 60	A	B	C	C	E	F	G	H	I	I	J
	60– 90	A	B	B	C	D	E	F	G	H	I	J
	90–140	A	B	B	C	D	D	E	F	G	H	J
	140–200	A	A	B	C	D	D	E	F	G	H	I
	200 up	A	A	B	C	D	D	E	F	F	G	I
50 (46–55)	42– 60	A	A	B	C	D	E	F	G	H	I	J
	60– 90	A	A	B	C	C	D	F	F	G	H	J
	90–140	A	A	A	C	C	D	E	F	F	G	I
	140–200	A	A	A	C	C	D	E	E	F	G	I
	200 up	A	A	A	C	C	D	E	E	F	G	H
60 (56–67)	60– 90	A	A	A	B	C	D	E	F	G	H	I
	90–140	A	A	A	B	C	C	D	E	F	G	H
	140–200	A	A	A	B	C	C	D	E	E	F	H
	200 up	A	A	A	B	C	C	D	E	E	F	H
75 (68–90)	60– 90	A	A	A	A	B	C	D	E	F	G	I
	90–140	A	A	A	A	B	C	D	E	F	F	H
	140–200	A	A	A	A	B	B	C	D	E	F	G
	200 up	A	A	A	A	B	B	C	D	E	F	G

Room index is the classification of a room according to its proportions; large and small rooms of the same proportion have the same index. Hence, for large rooms of dimensions greater than those shown, divide each dimension by the same number and use the index determined for the smaller room. Example: A room 200′ × 600′ × 40′ would have the same room index as a room 50′ × 150′ × 10′.

there is little dust and smoke in the atmosphere, a higher maintenance value may be obtained, but in open indirect equipment, cove lighting, skylights, and similar types of installations hard to reach and likely to be neglected, a considerably lower maintenance factor should be assumed.

COMPUTATIONS. The lamp lumens required to light a room are computed from the following formulas:

$$\text{Lamp lumens required} = \frac{\text{Footcandles} \times \text{Area of room}}{\text{Utilization factor} \times \text{Maintenance factor}}$$

$$\text{Lamp lumens per luminaire} = \frac{\text{Total lumens required}}{\text{Number of luminaires to be installed}}$$

or, conversely,

$$\text{Footcandles} = \frac{\text{Total lamp lumens} \times \text{Utilization factor} \times \text{Maintenance factor}}{\text{Total area in square feet}}$$

8. SPECIFICATION OF LIGHTING

There is an almost infinite variety of lamp and lighting applications, which account for the several thousand different types and sizes of lamps, both large and miniature. The specifications of lamps for projection, optical systems, signalling, and indicators, and a host of specialized uses of light, such as automotive and aviation requirements, are generally handled by specialist engineers. The broader fields of specification involve the lighting of factories, stores, offices, schools, public buildings, and streets and are the concern of people whose talents and activities cover the range from the research engineer, the lamp and equipment manufacturer, the distributor and salesman, the electrical contractor, and finally the user.

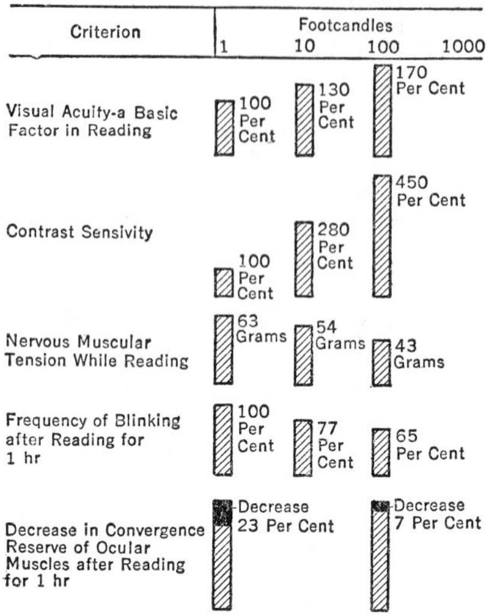

FIG. 27. Summary of Research Results of the Influence of Illumination on Visual Performance and Effects of Seeing

The fundamentals to be observed in every lighting application are simply that the effect should be comfortable to the eyes and that the quantity of light should be adequate for the visual task. This is true regardless of the application or the technique of installation, and in this respect the approaches to the lighting of a factory, a store, and an office may not differ greatly. In addition to its function of aiding the eye to see quickly, accurately, and comfortably, light has potentialities for luminous decoration, color, and atmosphere which require the skill and talent of architectural designers, decorators, and color experts. These attributes are being utilized today more generously, since new lamps and modern materials represent tools which can be used with better results and greater economy than could filament lamps alone.

LIGHTING RESEARCH. During the past twenty-five years research on light and vision has established a background of knowledge on the partnership relation of light and the eyes, and the specification of lighting can proceed today on the basis of science rather than by the guesswork or hit-or-miss methods so commonly employed in the past. Figure 27 is a summary of results of extensive research, which revealed the influence of illumination on important aspects of eye performance. Figure 28 illustrates the scale of footcandle effectiveness. Numerous and diverse researches have established the fact that the illumination (footcandles) must be approximately doubled to produce equal and significant improvements in seeing. Obviously, the reflectance of the illuminated surface must be considered in specifying recommended levels of illumination. For example, a surface having a reflectance of 80 per cent and illuminated with 10 ft-c will be equal in brightness to a surface having a reflectance of 8 per cent and illuminated with 100 ft-c.

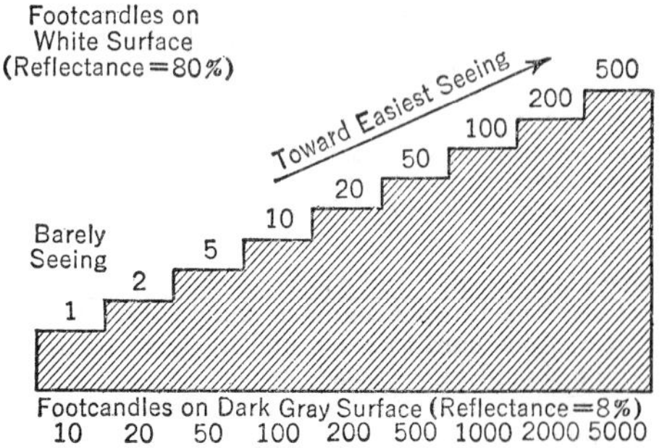

FIG. 28. Logarithmic Scale of Footcandles for Equal Steps in Seeing Effectiveness. Scales show the relative footcandles to create the same brightness for objects of 80 per cent reflectance as compared with 8 per cent reflectance

Reports of these researches are contained in the volumes of *Transactions* of the IES throughout the past quarter century; among the books covering these researches are *The Science of Seeing* (1937), Luckiesh and Moss, *Reading as a Visual Task* (1942), Luckiesh and Moss, *Light, Vision, and Seeing* (1944), Van Nostrand Company.

VISIBILITY AND COMFORT. There are two ideals toward which the design of any lighting installation should strive: (1) maximum visibility of the objects to be seen or the tasks to be performed, and (2) maximum comfort of the observer at all times and, more broadly, maximum ease of seeing when critical seeing is being done. Lighting practice is often a necessary compromise between practical considerations and ultimate ideals, but lighting progress through the years has been the record of improvements in visibility and comfort [Harrison and Luckiesh, Comfortable Lighting, *IES Trans.* (Dec., 1941)].

Ideal standards of artificial illumination, as far as best visibility is concerned, cannot generally be achieved in practice. The general order of footcandles recommended is as follows:

100 ft-c or more: For very severe and prolonged tasks, such as fine needlework, fine engravings, fine penwork, fine assembly, sewing on dark goods, and discrimination of fine details of low contrast, as in inspection.

50 to 100 ft-c: For severe and prolonged tasks, such as proofreading, drafting, difficult reading, watch repairing, fine machine work, average sewing, and other needlework.

20 to 50 ft-c: For moderately critical and prolonged tasks, such as clerical work, ordinary reading, common benchwork, and average sewing and other needlework on light goods.

10 to 20 ft-c: For moderate and prolonged tasks of office and factory and, when not prolonged, ordinary reading and sewing on light goods.

5 to 10 ft-c: For visually controlled work in which seeing is important, but more or less interrupted or casual, and does not involve discrimination of fine details or low contrasts.

0 to 5 ft-c: The danger zone for severe visual tasks and for quick and certain seeing. Satisfactory for perceiving larger objects and for casual seeing.

The best examples of modern practice employ around 100 ft-c of general lighting; supplementary localized lighting is employed where higher levels of lighting are required. The **visibility meter** is a new instrument developed to aid in the study of the severity of visual tasks and to lead to a rational basis for footcandle recommendations. It consists essentially of two circular gradient filters which may be rotated in front of the eyes while looking at any object or actually performing a seeing task. These filters incidentally reduce the brightness of the object in the visual field, but primarily they are slightly diffusing for the purpose of altering the contrast between the object and its background. Thus threshold conditions are obtained. The meter has two scales, one of which is calibrated to read relative footcandles, and the other, relative visibility. The relative footcandle scale extends from 1 to 1000. When the circular filters are adjusted simultaneously to the threshold of visibility for any object, the scale reading indicates the footcandles required in order that the object will be as visible as 8-point Bodoni type when viewed from a distance of 14 in. under 10 ft-c of illumination. The relative visibility also is read when the object being viewed is reduced to the threshold of visibility. The scale readings obtained in this manner for various seeing tasks represent a range of visibility levels from 1 to 20 as established with the parallel-bar test standards.

COMFORT FACTORS. In general, the factors which affect visibility also influence eye comfort. The most obvious aspects of discomfort concern direct glare and reflected glare. The measurement and appraisal of glare have long defied numerical rating, since glare involves many elements, such as size and brightness of the source, distance, angle of view, adaptation of the eye, length of viewing time, and overall psychological reaction or sensitivity of the individual. [See report of the IES Committee, Quality and Quantity for Interior Illumination, *Trans. IES* (Dec. 1944), and Crouch, Relation Between Illumination and Vision, *IES Trans.* (Nov. 1945)]. For prolonged visual tasks or exposure, such as in offices or schools, source brightness of luminaires should not greatly exceed 200 ft-L.

In brightness engineering much emphasis is being given to the proper ratio between the brightness of the immediate work and that of surroundings. Much can be gained in the way of visual comfort and pleasing atmosphere by attention to this point. According to the IES, the brightness ratio for best seeing conditions is obtained when the ratio of the brightness of the visual task to the brightness of its immediate surroundings is unity; for good seeing conditions, the ratio should be no greater than three. This had led to the advocacy of lighter finishes for industrial machines, desk tops, school desks, floor coverings, filing cabinets, etc., in order to reduce brightness contrasts.

Highly disturbing brightness ratios require that the eyes go through the process of adapting for the different brightnesses as the eyes follow a natural course throughout the field of view. The effort involved undoubtedly leads to fatigue and in no wise contributes to the better performance of the visual task. If the eyes are normal, they can stand abuse surprisingly well. On the other hand, if the eyes are subnormal and the seeing task is unnecessarily severe, the added eye fatigue may reasonably result in a more serious condition, leading to general fatigue, which is the source of many industrial accidents.

RECOMMENDED PRACTICE. The Illuminating Engineering Society publishes a series of Recommended Practice bulletins, some of which are approved by the American Standards Association. The series comprises publications on industrial, office, school, home, and street and highway lighting. In addition, publications are available on test standards and specifications on the testing of lighting equipment and on the standard method for measuring and reporting illumination from artificial sources in building interiors. Part of the discussion of several lighting fields which follows is abstracted from the publications mentioned.

OFFICE AND SCHOOL LIGHTING. The visual requirements of these applications are comparable, and the same technique of lighting is applicable to both. For offices involving difficult seeing tasks, such as auditing, accounting, business-machine operation, drafting, and designing, a minimum of 50 ft-c of general lighting is recommended. For ordinary seeing tasks, such as in file rooms, mail rooms, and general correspondence, a minimum of 25 ft-c is desirable; for reception rooms, washrooms, and service areas, 5–10 ft-c. Continuous rows of fluorescent troffers have been widely installed, and for applied fixtures indirect or semi-indirect lighting is preferred. With semi-indirect fixtures the density of the glass or plastic is most agreeable when the brightness of these parts is of the same order as the ceiling brightness against which the fixture is viewed.

The relations between brightness and illumination stated below for an illumination of 50 ft-c sustained in service are indicative of the trend of brightness researches, and it is suggested that they be taken as a desirable objective toward which to work. It should be recognized that many present-day installations which are considered good practice will not meet these limitations for the direct glare zone, commonly considered as the zone from the horizontal to approximately 45 deg below the horizontal.

1. The brightness of individually mounted luminaires, spaced in accordance with the conventional spacing, mounting-height relationship, should not exceed 800 ft-L (1.8 candles per sq in.).

2. The brightness of luminaires mounted in continuous rows, or nearly so, should not exceed 400 ft-L (0.9 candle per sq in.).

3. The brightness of very large sources (such as indirectly lighted ceilings or completely luminous ceilings) should not exceed an average of 175 ft-L (0.4 candle per sq in.).

For lower levels of illumination these values should be reduced. Brightness above these values will result in an increasing degree of discomfort, and, conversely, decreasing brightness may be expected to produce a greater degree of comfort, a desirable condition for work areas.

It is recommended that light-colored desk tops having a simple pattern and with a reflection-factor of 20 to 30 per cent be used.

LIGHT-REFLECTION VALUE AND FINISH OF CEILINGS AND WALLS. Light-colored surfaces are of particular value in providing a high utilization of light by reflecting a large part of it toward the working surfaces. The fact that bright window areas and artificial light sources are less uncomfortable to the eye when viewed against light backgrounds is an additional reason for the use of light colors on the ceiling and upper walls.

The ceiling should reflect at least 75 per cent of the light which strikes it; higher values are desirable. Although white is the most desirable finish, bluish white, cream white, or ivory white may be used.

The walls are directly in the line of vision, and finish with a reflection factor of 50 to 60 per cent should be used for the upper walls. Lighter-colored walls may appear annoyingly bright; darker walls decrease the illumination and introduce uncomfortable brightness contrasts.

Much is to be gained by an understanding of the psychological effect of color. For example, a room finished in tones such as cream or buff seems warmer, and the color has the effect of making the room seem smaller. On the other hand, light green or blue seems to recede, causing the room to appear larger and cooler.

The dado, or wall area below the window sills, may be of moderately dark color. To insure against annoying brightness in the normal line of vision, white plaster or very light colors are to be strictly avoided on the walls, particularly below eye level.

DRAFTING ROOMS AND CLASSROOMS. Drafting makes very serious demands upon the eyes, since it involves the accurate discrimination of fine details, frequently over long periods of time. The contrast between the work and the background may be very poor, e.g., when tracing a faint blueprint or a worn pencil drawing. Reflected glare from a specular drawing surface, as well as from the polished T-square, celluloid triangles, or scales, may be particularly annoying and should be avoided. Care must also be taken to eliminate shadows along the drawing edge of the T-square or triangles, as well as multiple shadows from the drawing instruments or the draftsmen's hands.

Horizontal boards present a problem, because with boards in this position any ceiling or luminaire brightness may be reflected by the work to the eyes of the draftsman, and also the T-squares, triangles, and curves may cast shadows. With the boards in a vertical position, specular reflections for all practical purposes do not exist, and shadows are minimized. Furthermore, the boards may be high enough to shield the eyes of the draftsmen from any brightnesses of lighting units otherwise in the field of view.

For drafting the minimum sustained illumination recommended on the work is 50 ft-c. The design of lighting systems for such areas should consider the desirability of using a source of light which can be so controlled as to present a low surface brightness to the eye.

Where the boards are horizontal or nearly so, straight edges such as T-squares, positioned parallel with fluorescent sources of illumination, may cast sharp shadows unless the edges are beveled. It is usually particularly desirable that such shadows be eliminated when the straight edges are used parallel to the sides of the drafting table, as they generally are. It is recommended, therefore, that, where line sources of illumination are used, the boards be placed at an angle with the lines of lighting equipment, or vice versa. In many cases an angle of 45 deg has been found to offer a practical solution to this problem.

A drafting table with a frosted- or white-glass top illuminated from below is recommended for tracing. Indirect lighting with concentric louvered silvered bowl lamps is favored for school classrooms because of its efficiency and simplicity of maintenance. In standard classrooms the common practice is to provide six units. Five-hundred-watt lamps provide about 15 to 20 ft-c; 750-watt lamps, 20 to 30 ft-c. The preferred system of fluorescent lamps makes use of three continuous rows, seven or eight in a row, of two-lamp 40-watt units; these provide around 40 ft-c.

INDUSTRIAL LIGHTING. This field involves many specialized applications in the lighting of the varied industrial processes, in specialized lighting for production machinery, and in many inspection lighting situations. Fundamental and common to all work areas is the requirement for a good general lighting system to provide from 20 to 50 ft-c, depending on the nature of the work. For severe seeing tasks and critical discernment, higher levels of 100 to 200 or more ft-c are generally provided by localized supplementary lighting.

General lighting has customarily been provided by direct lighting, porcelain-enamelled reflectors using filament, mercury, or fluorescent lamps, or combination systems. Aside from providing light on the work, the general system illuminates the surroundings, creates a bright, cheerful appearance, and reduces contrasts, thus lessening eyestrain and fatigue. This reduction of contrasts is so important that direct lighting with opaque reflectors, although most efficient in directing light to the work, is subject to the criticism that more light is needed above the units, so that ceilings and roof structures should be painted in light colors to minimize glare by reducing contrasts of light sources with the background. Occasional installations have been planned with some units devoted specifically to lighting the upper portions of the room. Where conditions are favorable, indirect lighting has been used.

The use of continuous rows of two-lamp fluorescent units has been widely accepted in newer plants, with rows 8 to 10 ft apart and providing 40 to 50 ft-c. The 400-watt H-1 mercury lamp, often in combination with 750-watt filament lamps in pairs or on alternate spacing, has been used fairly widely, especially in the metal-working industries. The 3-kw mercury lamp is used in high-bay installations in foundries, steel mills, erecting shops, and the like; because of the high lumen output and high brightness, its application is best limited to mounting heights of 35 ft or greater. The overall cost per footcandle, however, is the lowest of any system of general factory lighting, because of the high efficiency, the lower investment cost, and the lower maintenance cost brought about by the relatively few units required.

SPECIALIZED APPLICATIONS. These are encountered in paint shops, spray booths, and powder and chemical plants, where explosion dusts or vapors may be present. The requirements call for dust-tight, moisture- or vaporproof units and special fittings. Such hazardous locations are classified by the Underwriters' Laboratories, and manufacturers' equipment must be approved for installation in accordance with the various requirements.

Color identification and discrimination require special attention to the spectral quality of the light and often necessitate high intensities of the order of 500 ft-c. Such applications formerly resorted to high-wattage filament units equipped with deep blue "north skylight" glass filters, but combinations of fluorescent lamps to provide the proper color temperature have proven adequate for many applications and with manyfold increase in efficiency. These problems are encountered in wool and cotton grading, tobacco sorting, paint mixing, tile and enamel sorting and grading, and the like.

Special inspection processes involve many ingenious lighting devices, such as polarized light for detecting strains in glass, or ultraviolet light, as in the "Magnaglo" process to detect flaws in castings. Various projection devices are also used to secure magnification, as in the inspection of lamp filaments, or to direct light into difficult places, as in gun-barrel inspection. For some applications highly diffuse or transmitted light from large surfaces is required; for others, concentrated directional lighting is needed. Several hundred footcandles may be provided for benchwork assembly and inspection by a continuous trough of fluorescent lamps, preferably two-lamp units to reduce stroboscopic effect, mounted a foot or two above the table or bench top. Infrared industrial lamps for heating and drying and germicidal lamps to protect against airborne bacteria and for sterilizing purposes are also widely used in industry.

STORE LIGHTING. The application of light in stores goes far beyond the purely utilitarian purpose of merely seeing, although the more modern stores employ general lighting of the order of 20 to 50 ft-c. A major application of lamps and lighting, supplementary to the general lighting, is found in the lighting of show and display cases, niches, and shelves, and above counters, in spot- and floodlighting of special displays, and in the use of colored light for decorative effects and advertising. The object of brightly lighted displays—three to five times as bright as the general interior lighting—is to provide interest and attraction, to create atmosphere, and to accent special merchandise displays. This use of light proves profitable to the merchant in attracting customers and in giving emphasis to featured products. The fluorescent lamp and cold-cathode tubing have introduced new concepts in store lighting, as far as architectural design and theatrical effects are concerned; the line of reflector and projector lamps simplifies the use of spot- and floodlighting in show windows and on interior displays, and most modern systems combine both fluorescent and filament lighting. Reflector-type spot and flood lamps are adaptable to color and louvering accessories, and various types of lampholders are available for flexible control of the light.

HOME LIGHTING. This field represents a wide range of individual tastes in decoration and appointments which defy much standardization of fixtures. However, the engineering approach to the varied applications concerns itself again with the fundamentals of comfortable seeing in terms of an adequate amount of light and in the design of fixtures of comfortable brightness. To aid the hundreds of manufacturers and designers the Illuminating Engineering Society offers a guide in the form of a Recommended Practice bulletin [*Trans. IES*, (June, 1945)], and groups of fixture manufacturers have developed their own standards, testing, and approval specifications for advancement and progress in fixture design. Although filament lamps will probably long dominate home-lighting applications, fluorescent lighting for kitchens, bathroom mirrors, bedlamps, desk lamps, window valances, etc., is rapidly coming into use. Many new types of portable lamps are designed around the circular fluorescent lamp.

STREET LIGHTING. The Institute of Traffic Engineers classifies streets with respect to volume of vehicular traffic as follows: very light traffic, under 150 per hour in both directions; light, 150–500; medium, 500–1200; heavy, 1200–2400; very heavy, 2400–4000; heaviest, over 4000. These classifications are related to the recommended practice of street lighting as developed by the IES [*Trans. IES* (Feb., 1946)]. The recommended average horizontal footcandles (lumens per square foot) for urban streets are as follows:

Pedestrian Traffic	Vehicular Traffic Classification			
	Very Light (Under 150)	Light (150–500)	Medium (500–1200)	Heavy to Heaviest (1200 up)
	Lumens per Square Foot			
Heavy........	*	0.8	1.0	1.2
Medium.......	*	0.6	0.8	1.0
Light or none..	0.2	0.4	0.6	0.8

* This condition is unusual, but if it should occur, the footcandle figures appearing in the column to the right may be used.

Light Distribution and Luminaire Types. The choice of light distribution of a luminaire is determined by mounting height, spacing, and transverse location. Good practice requires that light emitted from the luminaire be largely directed toward the street and properly distributed to insure efficiency and the recommended minimum average illumination. Some light should be directed back of the curb line to provide adequate illumination on the sidewalk and adjacent areas.

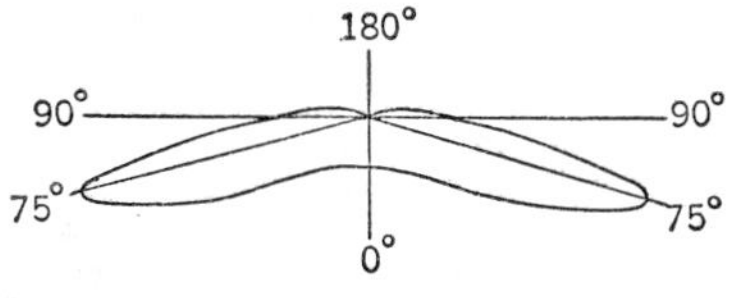

FIG. 29. Typical Recommended Vertical Light Distribution for Street-lighting Units

The type of vertical light distribution which is generally recommended is shown in Fig. 29. Distributions of this type have maximum candlepower and maximum flux of light between the angles of 70 deg and 80 deg from nadir or 20 deg and 10 deg below the horizontal.

Table 14. Typical Arrangement of Luminaires, Mounting Height, and Spacing for Various Initial Footcandle Values

Footcandles	Street Width, Feet	Lamp Lumens	Type Distribution	Luminaire Arrangement	Mounting Height	Approximate Spacing, Feet
0.2	30	2,500	I	Center	25	170
	30	2,500	II	Staggered	20	165
	30	4,000	I	Center	25	200
	40	4,000	II	Staggered	25	200
	40	4,000	III	Staggered	25	200
0.3	30	4,000	I	Center	25	180
	30	4,000	II	Staggered	25	150
	40	6,000	II	Staggered	25	210
	40	6,000	III	Staggered	25	210
0.4	40	6,000	II	Staggered	25	155
	50	6,000	II	Staggered	25	135
	50	6,000	III	Staggered	25	140
	50	6,000	IV	Staggered	25	110
0.5	40	6,000	II	Staggered	25	125
	50	6,000	II	Staggered	25	110
	50	6,000	III	Staggered	25	110
	50	6,000	IV	Staggered	25	90
0.6	50	10,000	III	Staggered	30	140
	60	10,000	III	Staggered	30	125
	60	10,000	IV	Staggered	25	115
0.7	50	10,000	III	Staggered	30	120
	60	10,000	III	Staggered	30	110
	60	10,000	IV	Staggered	25	95
0.8	50	10,000	III	Staggered	30	105
	60	10,000	III	Staggered	30	95
	60	10,000	IV	Staggered	25	85
	70	10,000	III	Staggered	30	85
	70	10,000	IV	Staggered	25	75
1.0	60	10,000	III	Staggered	30	75
	60	10,000	IV	Staggered	25	70
	70	10,000	IV	Opposite	25	120
	80	10,000	IV	Opposite	25	110
1.2	60	15,000	IV	Staggered	30	75
	70	15,000	IV	Opposite	30	130
	80	15,000	IV	Opposite	30	120
1.4	60	15,000	IV	Opposite	30	120
	70	15,000	IV	Opposite	30	110
	80	15,000	IV	Opposite	30	100
1.6	60	15,000	IV	Opposite	30	105
	70	15,000	IV	Opposite	30	95
	80	15,000	IV	Opposite	30	90
1.8	60	15,000	IV	Opposite	30	95
	70	15,000	IV	Opposite	30	85
	80	15,000	IV	Opposite	30	80
2.0	60	15,000	IV	Opposite	30	85
	70	15,000	IV	Opposite	30	75
	80	15,000	IV	Opposite	30	70

In sideways or lateral directions the light distribution may be symmetric or asymmetric. Since the street area to be illuminated by each luminaire is generally rectangular, the asymmetric light distributions are usually more efficient.

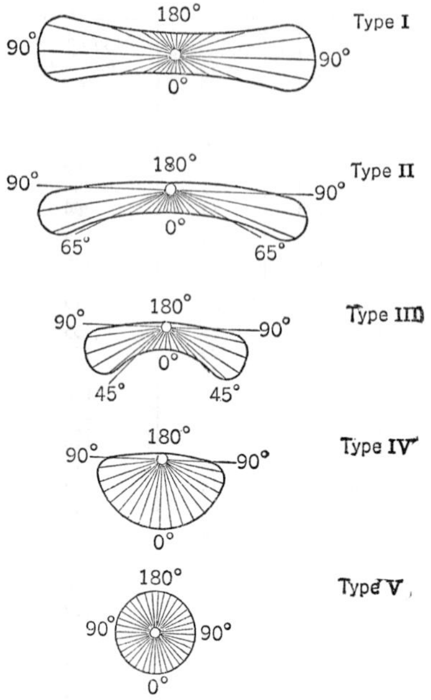

FIG. 30. Five Types of Lateral Distribution Which Define Five Standard Classes of Street-lighting Luminaires

Five different types of lateral distribution are shown in Fig. 30 and may be described as follows:

Type I Luminaire—Two-way Distribution: Intended for luminaire mounting approximately over the center of the street. It projects two beams of light in opposite directions along the street, their axes being parallel with the curb line.

Type II Luminaire—Narrow Asymmetric Distribution: Intended for luminaire mounting at or near the side of the street. It is narrow distribution, having a lateral width up to approximately 25 deg in the cone of maximum candlepower at approximately 75 deg.

Type III Luminaire—Medium-width Asymmetric Distribution: Intended for luminaire mounting at or near the side of the street but having a lateral width up to approximately 45 deg in the cone of maximum candlepower at approximately 75 deg. It is intended for wide streets.

Type IV Luminaire—Wide Asymmetric Distribution: Still wider laterally than Type III. The width is approximately 90 deg in the cone of maximum candlepower at approximately 75 deg.

Type V Luminaire—Symmetric Distribution: Candlepower in the 75-deg cone is the same throughout 360 deg. It is useful where lighting must be installed in center parkways and to some extent for intersections.

Luminaires which approximate the first four of the foregoing light-distribution characteristics will meet a large portion of all street- and highway-lighting applications. The recommended mounting heights for luminaires having the distribution characteristics described are given below. Where practicable, higher mounting may often be preferable.

Lamp Size, Lumens	Mounting Heights, Feet			
	Type I	Type II	Type III	Type IV and V
2,500	25	20	20	20
4,000	25	25	25	25
6,000	25	25	25	25
10,000	....	30	30	25
15,000	....	....	30	30

Street-lighting Design and Recommendations. The desirable light distribution, lamp size, spacing, and arrangement of lighting units to provide the required illumination for any street-lighting project may be determined with accuracy and convenience by methods described in technical literature. The data in Table 14 give typical lighting arrangements for various footcandle levels for several street widths.

9. BIBLIOGRAPHY

Radiant Energy—General

Forsythe, W. E., *Measurement of Radiant Energy.* McGraw-Hill (1937).
Luckiesh, Matthew, *Applications of Germicidal, Erythemal, and Infrared Energy.* D. Van Nostrand (1946).

Incandescent Lamps

Forsythe, W. E., and Adams, *The Tungsten-filament Incandescent Lamp.* Dennison University Bulletin (1937).
Forsythe, W. E., Adams, and Cargill, *Some Factors Affecting the Operation of Incandescent Lamps.* Dennison University Bulletin (1939).
National Bureau of Standards Circular C459, May 15, 1947.
Schroeder, Henry, *History of the Incandescent Lamp.* Smithsonian Institution (1923).
Weitz, C. E., *G-E Lamp Bulletin,* LD-1 (1946).

Electric Discharge Sources

Buttolph, L. J., *A Review of Gaseous Discharge Lamps.* IES (1932).
Ferree, H. M., *Some Characteristics and Applications of Negative-glow Lamps.* AIEE (1940).
Fonda, G. R., *The Fundamental Principles of Fluorescence.* AIEE (1938).
Forsythe, W. E., Adams, and Barnes, *Mercury Vapor Lamps.* Dennison University Bulletin (1942).
Forsythe, W. E., Adams, and Barnes, *Fluorescence and Fluorescent Lamps.* Dennison University Bulletin (1941).
Hays, R. F., Jr., *Development of the Glow Switch.* AIEE (1941).
Inman and Thayer, *Low-voltage Fluorescent Lamps.* AIEE (1938).
Inman and Thayer, *Characteristics of Fluorescent Lamps.* IES (1938).
Marden, Beese, and Meisten, *Cadmium and Zinc Vapor Lamps.* IES (1936).
Modern Carbon Arcs, Catalog A-4000. National Carbon Company (1944).
Noel and Farnham, *A Water-cooled Quartz Mercury Arc.* SMPE (1938).
Thayer, *Ballasting Requirements for Fluorescent Lamps.* Electrochemical Society (1943).
Townsend, Mark A., *Electronics of the Fluorescent Lamp.* AIEE (1942).
Weitz, C. E., *G-E Lamp Bulletin,* LD-1 (1946).

Light Measurement—Control and Specification

Amick, *Fluorescent Lighting Manual.* McGraw-Hill (1947—new edition in process).
Higbie, H. H., *Lighting Calculations.* John Wiley (1934).
Illuminating Engineering Society, *Lighting Handbook* (1947).
Illuminating Engineering Society, *Nomenclature and Standards* (1942).
Illuminating Engineering Society, Publications on Recommended Practice: (1) Home, (2) Office, (3) School, (4) Industrial, (5) Street.
Kraehenbuehl, J. O., *Electrical Illumination.* John Wiley (1942).
Luckiesh, Matthew, *Light, Vision, and Seeing.* D. Van Nostrand (1944).
Luckiesh, Matthew, and Moss, *The Science of Seeing.* D. Van Nostrand (1937).
Moon, *Scientific Basis of Illuminating Engineering.* McGraw-Hill (1936).

ELECTRIC HEATING AND AIR CONDITIONING

By E. R. Ambrose

10. HEAT-LOSS CALCULATIONS

To obtain the accurate heating requirement of a structure, it is necessary to make a heat-loss calculation based on:

1. Design temperature, outdoor and indoor.
2. Heat transmission through the external surfaces.
3. Infiltration.
4. Humidification requirements.
5. Internal heat gain.

DESIGN TEMPERATURE. The outdoor design temperature for all localities has not been definitely determined because of lack of climatological data. In most large cities, however, sufficient weather data are available to permit the selection of a proper temperature. If weather data are not available, the best plan is to have available data for the nearest city corrected, on the basis of experience and judgment, for the locality under consideration. Even when the daily dry-bulb temperature is available, there is quite a difference of opinion on what design temperature should be used. It is safe to say that the extremely low temperature, occurring rather infrequently and of short duration, should not be chosen, but, on the other hand, the use of any average minimum-maximum temperature is debatable. The ASHVE Technical Advisory Board on Weather Design Conditions has recommended the adoption of an outdoor design temperature which is equalled or exceeded during 97 $^1/_2$ per cent of the hours in December, January, February, and March. Unfortunately, design temperature on this basis is not yet available. The design temperatures shown in Table 1 are representative of the practices in various parts of the country.

The indoor temperature is usually referred to as the temperature at the breathing line, which is taken at 5 feet above the floor and not less than 3 feet from an outside wall. The inside condition to be maintained for comfort is dependent upon the dry-bulb temperature, relative humidity, air motion, and radiation effect. An index or scale, called "effective temperature," is now generally used to designate the proper inside temperature. A com-

Table 1. Inside and Outside Design Temperatures

(Summer and Winter)

State and City	Winter			Summer		
	Lowest * Recorded Temp., °F	Design † Temp. Actually Used, °F	Average Wind Vel. and Direction, mph	Design ‡ Dry-bulb Temp., °F	Design ‡ Wet-bulb Temp., °F	Average Wind Vel. and Direction, mph
Alabama						
Birmingham	-10	10	8.6 N	93	77	5.2 S
Mobile	-1	20	8.3 N	94	78	8.6 SW
Montgomery	-5	15				
Arkansas						
Fort Smith	-15	-15	8.0 E			
Little Rock	-12	0	9.9 NW	95	77	7.0 NE
California						
Los Angeles	28	30	6.1 NE	88	70	6.0 SW
Sacramento	17	30				
San Francisco	27	30	7.5 N	85	68	11.0 SW
Connecticut						
New Haven	-15	-5	9.3 N	88	74	7.3 S
Dist. of Columbia						
Washington	-15	0	7.3 NW	95	78	6.2 S
Florida						
Jacksonville	10	27	8.2 NE	94	78	8.7 SW
Pensacola	7	20				
Georgia						
Atlanta	-8	10	11.8 NW	91	75	7.3 NW
Macon	7	10				
Savannah	8	20	8.3 NW	95	79	7.8 SW
Illinois						
Champaign	-25	-10				
Chicago	-23	-10	17 SW	95	75	10.2 NE
Rockford	-26	-10				
Indiana						
Ft. Wayne	-24	-10				
Indianapolis	-25	-5	11.8 S	90	73	9.0 SW
Terre Haute	-18	-10				
Iowa						
Davenport	-27	-10				
Des Moines	-30	-15		92	74	6.6 SW
Sioux City	-35	-18	12.2 NW			
Kansas						
Wichita	-25	-5		95	77	
Topeka	-25	-5				
Kentucky						
Louisville	-20	-5	9.3 SW	94	75	8.0 SW
Louisiana						
Baton Rouge	2	20				
New Orleans	7	25	9.6 N	94	79	7.0 SW
Maryland						
Baltimore	-7	0	7.2 NW	93	76	6 9 SW
Massachusetts						
Amherst	-22	0				
Boston	-18	0	11.7 W	88	73	9.2 SW
Fitchburg	-17	0				
Nantucket	-6	0				
Michigan						
Detroit	-24	0	13.1 SW	93	73	10.3 SW
Saginaw	-23	0				
St. Joseph	-21	-10				
Minnesota						
Minneapolis	-33	-15	11.5 NW	84	72	8.4 SE
Rochester	-42	-20				
Missouri						
Kansas City	-22	-10		92	75	9.5 S
St. Louis	-22	-5	11.8 NW	95	78	9.4 SW
Nebraska						
Omaha	-32	-20		94	75	
New Hampshire						
Concord	-35	-10	6.0 NW			
New Jersey						
Atlantic City	-9	0	10.6 NW			
Dover	-21	0				
Jersey City	-12	0				
Newark	-13	0				
Paterson	-13	0				
Trenton	-14	0		95	76	10.0 SW
New Mexico						
Albuquerque	-10	0				

Table 1. Inside and Outside Design Temperatures—*Continued*

(Summer and Winter)

State and City	Winter: Lowest * Recorded Temp., °F	Winter: Design † Temp. Actually Used, °F	Winter: Average Wind Vel. and Direction, mph	Summer: Design ‡ Dry-bulb Temp., °F	Summer: Design ‡ Wet-bulb Temp., °F	Summer: Average Wind Vel. and Direction, mph
New York						
Albany	−24	−10	7.9 S	90	74	7.1 S
Binghamton	−28	−5				
Buffalo	−21	0	17.7 W	83	72	12.2 SW
Elmira	−24	−10				
New York	−14	0	17.1 NW	95	75	12.9 SW
Syracuse	−24	−2				
North Carolina						
Greensboro	−2	20				
North Dakota						
Bismarck	−45	−30	− NW	88	69	8.8 NW
Fargo	−48	−30				
Ohio						
Cincinnati	−17	−5		95	78	6.6 SW
Cleveland	−18	0	14.5 SW	95	73	9.9 S
Dayton	−16	0				
Oklahoma						
Oklahoma City	−17	5	12.0 N	96	76	10.1 S
Oregon						
Portland	−2	5	6.5 S	83	65	6.6 NW
Pennsylvania						
Erie	−16	5				
Harrisburg	−14	0				
Lancaster	−27	0				
Philadelphia	−11	0	11.0 NW	95	78	9.7 SW
Pittsburgh	−20	0	13.7 NW	91	73	9.0 NW
Scranton	−19	−5				
Uniontown	−22	0				
York	−17	0				
Rhode Island						
Providence	−17	0	14.6 NW	85	73	10.0 NW
South Carolina						
Charleston	7	20	11.0 N	94	80	9.9 SW
Columbia	−2	10	8.0 NE	100	78	
South Dakota						
Aberdeen	−46	−20				
Tennessee						
Chattanooga	−10	0		94	76	6.5 SW
Nashville	−13	0				
Texas						
Dallas	−10	10		99	76	9.4 S
Ft. Worth	−24	14	11.0 NW	95	75	
Houston	−5	20		93	79	7.7 S
San Antonio	4	20	8.2 N	100	78	7.4 SE
Utah						
Salt Lake City	−20	0	4.9 SE	95	67	8.2 SE
Vermont						
Burlington	−30	−10	12.9 S	85	71	8.9 S
Virginia						
Lynchburg	−7	0	5.2 NW			
Richmond	−3	10	7.4 S	95	78	6.2 SW
Washington						
Seattle	3	10	9.1 SE	83	61	7.9 S
Spokane	−30	−10	5.2 SW	89	63	6.5 SW
Tacoma	7	10				
Walla Walla	−29	−10				
West Virginia						
Wheeling	−25	0				
Wisconsin						
Madison	−29	−15		89	73	8.1 SW
Milwaukee	−25	−17	11.7 W	93	74	10.4 S
Sheboygan	−38	−10				
Wyoming						
Laramie	−42	−10				
Sheridan	−45	−30	5.3 NW			

* As reported in publications of U. S. Dept. of Agriculture, Weather Bureau, covering long periods, 20 to 50 years, including winter of 1933–34.

† As reported by readers of *Heating and Ventilating*.

‡ Collected from various published sources.

Reprinted by kind permission of *Heating and Ventilating*.

fort chart, using effective temperatures, has been devised by the ASHVE. Studies indicate that an effective temperature of 66 to 67 deg fahr is satisfactory for 98 per cent of the subject. A dry-bulb temperature of 70 to 72 deg fahr, with a relative humidity of 30 to 40 per cent and an air motion of 15 to 25 ft per min, will result in an effective temperature of 66 to 67 deg fahr. Other recommended effective temperatures are found in the ASHVE Guide.

Radiation between occupants and the windows, outside walls, and cold objects of the structure has an important bearing on the feeling of warmth and may change the required effective temperature.

Table 2. Heat Transmission Coefficients *

A. Exterior Walls

Exterior Walls, Frame Construction		Type of Sheathing			
Exterior Finish	Interior Finish	Gypsum 1/2 in. Thick	Plywood 5/16 in.	Wood 25/32 in. Paper	Insulating 25/32 in. Thick
Wood siding (clapboard)	Metal lath and plaster	0.33	0.32	0.26	0.20
	Gypsum board	0.32	0.32	0.25	0.20
	Wood or gypsum lath and plaster	0.31	0.31	0.25	0.19
	Plywood	0.30	0.30	0.24	0.19
	Insulating board lath 1/2 in. and plaster	0.22	0.22	0.19	0.15
Wood shingles	Metal lath and plaster	0.25	0.25	0.26	0.17
	Gypsum board	0.25	0.25	0.25	0.17
	Wood or gypsum lath and plaster	0.24	0.24	0.25	0.16
	Plywood	0.24	0.24	0.24	0.16
	Insulating board lath 1/2 in. and plaster	0.19	0.18	0.19	0.13
Brick veneer	Metal lath and plaster	0.37	0.36	0.28	0.21
	Gypsum board	0.36	0.36	0.28	0.21
	Wood or gypsum lath and plaster	0.35	0.34	0.27	0.20
	Plywood	0.34	0.33	0.27	0.20
	Insulating board lath 1/2 in. and plaster	0.24	0.24	0.20	0.16

Coefficients are expressed in Btu per hour per square foot, per degree fahrenheit difference in temperature between the air on the two sides and with a wind velocity of 15 mph.

For coefficients with insulation between framing, see Part *D*.

B. Floors and Ceilings, Frame Construction

Floors and Ceilings, Frame Construction	Insulation between or on Joists—No Flooring Area												With Flooring	
Type of Ceiling	None	Insulating Board on Joists		Blanket or Bat Insulation between Joists			Vermiculite Insulation between Joists			Mineralwool Insulation between Joists			Single Wood Floor	Double Wood Floor
		1/2 in.	1 in.	1 in.	2 in.	3 in.	2 in.	3 in.	4 in.	2 in.	3 in.	4 in.		
No ceiling		0.37	0.24										0.45	0.34
Metal lath and plaster	0.69	0.26	0.19	0.19	0.12	0.093	0.18	0.14	0.11	0.12	0.092	0.066	0.30	0.25
Gypsum board (3/8 in.)	0.67	0.26	0.18	0.19	0.12	0.092	0.18	0.14	0.10	0.12	0.092	0.066	0.30	0.24
Wood lath and plaster	0.62	0.25	0.18	0.19	0.12	0.092	0.17	0.14	0.10	0.12	0.091	0.065	0.29	0.24
Gypsum lath and plaster	0.61	0.25	0.18	0.19	0.12	0.092	0.17	0.14	0.10	0.12	0.091	0.065	0.28	0.24
Plywood (3/8 in.)	0.59	0.24	0.18	0.19	0.12	0.09	0.17	0.14	0.10	0.12	0.09	0.065	0.28	0.23
Insulating board (1/2 in.)	0.37	0.19	0.15	0.16	0.11	0.082	0.15	0.11	0.092	0.11	0.082	0.060	0.22	0.19
1/2-in. insulating board lath and plaster	0.35	0.19	0.15	0.15	0.10	0.081	0.14	0.11	0.091	0.10	0.082	0.060	0.21	0.18
1-in. insulating board lath and plaster	0.23	0.15	0.12	0.12	0.089	0.072	0.12	0.097	0.080	0.089	0.072	0.055	0.16	0.14

Coefficients are expressed in Btu per hour per square foot per degree fahrenheit difference in temperature between the air on the two sides and with still air on both sides (no wind).

For coefficients of last two columns with insulation between framing, see Part *D*.

* Reprinted by permission of the American Society of Heating and Ventilating Engineers, ASHVE *Guide* (1944).

Table 2. Heat Transmission Coefficients—*Continued*

C. Pitched Roofs

Pitched Roofs, Frame Construction	Type of Roofing and Sheathing											
	Wood Shingles on 1 in. × 4 in. Wood Strips Spaced 2 in.				Asphalt Shingles or Roll Roofing Solid Wood Sheathing				Slate or Tile on Solid Wood Sheathing			
Type of Ceiling (Applied to Roof Rafters)	Insulation between Rafters				Insulation between Rafters				Insulation between Rafters			
	None	Blanket or Bat			None	Blanket or Bat			None	Blanket or Bat		
		1 in.	2 in.	3 in.		1 in.	2 in.	3 in.		1 in.	2 in.	3 in.
No ceiling	0.48	0.15	0.11	0.083	0.53	0.15	0.11	0.085	0.55	0.16	0.11	0.085
Metal lath and plaster	0.31	0.14	0.10	0.082	0.33	0.15	0.10	0.083	0.34	0.15	0.11	0.083
Gypsum board 3/8 in.	0.31	0.14	0.10	0.082	0.32	0.15	0.10	0.082	0.34	0.15	0.11	0.083
Wood lath and plaster	0.30	0.14	0.10	0.080	0.31	0.14	0.10	0.082	0.32	0.15	0.10	0.082
Gypsum lath and plaster	0.29	0.14	0.10	0.080	0.31	0.14	0.10	0.082	0.32	0.15	0.10	0.082
Plywood 3/8 in.	0.29	0.14	0.10	0.080	0.30	0.14	0.10	0.080	0.31	0.14	0.10	0.082
Insulating board 1/2 in.	0.22	0.12	0.09	0.073	0.23	0.12	0.12	0.093	0.24	0.13	0.094	0.076
Insulating board lath 1/2 in. plastered	0.22	0.12	0.09	0.073	0.22	0.12	0.090	0.073	0.23	0.12	0.093	0.074
Insulating board lath 1 in. plastered	0.16	0.10	0.077	0.065	0.17	0.10	0.077	0.065	0.17	0.10	0.080	0.066

Coefficients are expressed in Btu per hour per square foot per degree fahrenheit difference in temperature between the air on the two sides and with a wind velocity of 15 mph.

D. Frame Construction

Frame Construction Insulation between Framing	With Insulation between Framing			
	Blanket or Bat Insulation between Framing			3 5/8 in. Loose Mineral Wool between Framing
Coefficient With No Insulation between Framing	1 in.	2 in.	3 in.	
0.12	0.083	0.067	0.057	0.054
0.14	0.092	0.073	0.061	0.058
0.16	0.10	0.077	0.065	0.060
0.18	0.11	0.082	0.068	0.063
0.20	0.12	0.087	0.070	0.066
0.22	0.12	0.090	0.073	0.069
0.24	0.12	0.094	0.076	0.070
0.26	0.13	0.096	0.077	0.072
0.28	0.14	0.098	0.078	0.073
0.30	0.14	0.10	0.080	0.075
0.32	0.15	0.10	0.082	0.076
0.34	0.15	0.11	0.083	0.078
0.36	0.16	0.11	0.085	0.079
0.38	0.16	0.11	0.087	0.080
0.40	0.16	0.11	0.088	0.082
0.42	0.16	0.11	0.088	0.082
0.44	0.17	0.12	0.090	0.083

This table is to be used to determine the coefficients when insulation is added between framing. For example, refer to Part *A*. The coefficient for a frame wall with wood siding, metal lath, and plaster is .20. In Part *D* find .20 in the first column. If a 3-in. blanket is used, the overall coefficient will be .070.

Table 2. Heat Transmission Coefficients—*Continued*

E. Windows and Skylights

Description	Coefficient
Single	1.13
Double	0.45
Triple	0.281

F. Hollow Glass Block Walls

Description	Coefficient
Smooth surface blocks 7 3/4 in. × 7 3/4 in. × 3 7/8 in.	0.49
Ribbed-surface Blocks 7 3/4 in. × 7 3/4 in. × 3 7/8 in.	0.46

G. Solid Wood Doors

Nominal Thickness	Coefficient	
	Exposed Door	Glass Storm Door
1	0.69	0.42
1 1/4	0.59	0.38
1 1/2	0.52	0.35
1 3/4	0.51	0.35
2	0.46	0.32
2 1/2	0.38	0.28
3	0.33	0.25

Coefficients are expressed in Btu per hour per square foot per degree fahrenheit difference in temperature between the air on the two sides and with an outside wind velocity of 15 mph.

H. Masonry Side Walls

Masonry Side Walls	Uninsulated				Insulated		
	Thickness, Inches	Plain Walls, No Finish	1/2 in. Plaster on Walls	Plaster on Gypsum Lath Furred	1/2 in. Plaster 1/2 in. Rigid Insulation Furred	1/2 in. Plaster 1 in. Rigid Insulation Furred	1/2 in. Plaster Gypsum Lath 1 in. Flexible Insulation Furred
Brick, 4-in. hard, remainder common	8	0.50	0.46	0.30	0.22	0.16	0.14
	12	0.35	0.34	0.24	0.19	0.14	0.13
	16	0.28	0.27	0.20	0.16	0.13	0.12
Hollow tile, stucco exterior	8	0.40	0.37	0.26	0.20	0.15	0.13
	12	0.30	0.29	0.21	0.17	0.13	0.12
	16	0.25	0.24	0.19	0.15	0.12	0.11
Limestone or sandstone	8	0.70	0.64	0.36	0.25	0.18	0.16
	12	0.57	0.53	0.33	0.23	0.17	0.15
	16	0.49	0.45	0.30	0.22	0.16	0.14
Concrete monolithic	6	0.79	0.71	0.39	0.26	0.19	0.16
	10	0.63	0.58	0.34	0.24	0.18	0.15
	12	0.58	0.53	0.33	0.23	0.17	0.15
Hollow concrete blocks gravel aggregate	8	0.56	0.52	0.32	0.22	0.17	0.15
	12	0.50	0.46	0.30	0.19	0.16	0.14
4-in. brick veneer hollow tile backing	6 †	0.35	0.34	0.24	0.18	0.14	0.13
	8 †	0.34	0.32	0.23	0.18	0.14	0.13
4-in. brick veneer concrete backing	6 †	0.59	0.54	0.33	9.23	0.17	0.15
	8 †	0.54	0.50	0.31	0.23	0.17	0.15
4-in. cut stone veneer hollow tile backing	6 †	0.37	0.35	0.25	0.19	0.15	0.13
	8 †	0.36	0.34	0.24	0.19	0.15	0.13
4-in. cut stone veneer concrete backing	6 †	0.63	0.58	0.34	0.24	0.18	0.15
	8 †	0.57	0.53	0.33	0.23	0.17	0.15

Coefficients are expressed in Btu per hour per square foot per degree fahrenheit difference in temperature between the air on the two sides and with a wind velocity of 15 mph.

† Thickness of backing.

Table 2. Heat Transmission Coefficients—*Continued*

I. Floors and Ceilings, Concrete Construction

Floors and Ceilings, Concrete Construction		Type of Flooring				
Type of Ceiling	Thickness of Concrete	No Flooring (Bare Concrete)	Tile or Terrazzo on Concrete	1/4 in. Battleship Linoleum on Concrete	Parquet Flooring in Mastic on Concrete	Double Wood Floor in Sleepers
No ceiling	3	0.69	0.65	0.45	0.45	0.25
	6	0.59	0.56	0.41	0.41	0.23
	10	0.50	0.48	0.36	0.36	0.22
1/2 in. plaster applied directly to underside of concrete	3	0.62	0.59	0.43	0.43	0.24
	6	0.54	0.52	0.39	0.39	0.22
	10	0.46	0.44	0.34	0.34	0.21
Suspended or furred metal lath and plaster	3	0.38	0.37	0.30	0.30	0.19
	6	0.35	0.34	0.28	0.28	0.18
	10	0.32	0.31	0.26	0.26	0.17
Suspended or furred gypsum board and plaster	3	0.36	0.35	0.28	0.28	0.19
	6	0.33	0.32	0.27	0.27	0.18
	10	0.30	0.29	0.24	0.24	0.17
Suspended or furred insulating board lath and plaster	3	0.25	0.24	0.21	0.21	0.15
	6	0.23	0.23	0.20	0.20	0.15
	10	0.22	0.21	0.19	0.19	0.14

Coefficients are expressed in Btu per square foot per degree fahrenheit difference in temperature between the air on the two sides, and with still air on both sides (no wind).

Until more complete data are available, it is recommended that a coefficient of .10 be used for all types of concrete floors on the ground, with or without insulation. For basement walls below grade, use the same average coefficient (.10).

For basement walls below grade, a ground temperature of 32 deg fahr is recommended. For basement floors below grade, a ground temperature of from 40 to 55 deg fahr, depending on geographic location, should be used.

J. Flat Roofs

Flat Roofs Covered with Built-up Roofing		Without Ceiling-underside Roof Exposed			With Metal Lath and Plaster Ceilings		
Type of Roof Deck	Thickness of Roof Deck, Inches	No Insulation	1/2 in. Rigid Insulation	1 in. Rigid Insulation	No Insulation	1/2 in. Rigid Insulation	1 in. Rigid Insulation
Precast cement tile	1 5/8	0.84	0.37	0.24	0.43	0.26	0.19
Concrete	2	0.82	0.37	0.24	0.42	0.26	0.19
	4	0.72	0.34	0.23	0.40	0.25	0.18
	6	0.65	0.33	0.22	0.37	0.24	0.18
Wood	1	0.49	0.28	0.20	0.32	0.21	0.16
	1 1/2	0.37	0.24	0.18	0.26	0.19	0.15
	2	0.32	0.22	0.16	0.24	0.17	0.14
	4	0.23	0.17	0.14	0.18	0.14	0.12
Flat metal roofs	....	0.94	0.39	0.24	0.46	0.27	0.19

NOTE: Bare corrugated iron without roofing has a coefficient of 1.50 based on the projected area.

Coefficients are expressed in Btu per hour per square foot per degree fahrenheit difference in temperature between the air on the two sides and with a wind velocity of 15 mph.

Table 2. Heat Transmission Coefficients—*Continued*

K. Infiltration through Walls

Type of Wall	Wind Velocity, Miles per Hour					
	5	10	15	20	25	30
8 1/2-in. brick wall						
Plain	1.75	4.20	7.85	12.2	18.6	22.9
Plastered	0.017	0.037	0.066	0.107	0.161	0.236
13-in. brick wall						
Plain	1.44	3.92	7.48	11.6	16.3	21.2
Plastered	0.005	0.013	0.025	0.043	0.067	0.097
Frame wall, lath and plaster	0.03	0.07	0.13	0.18	0.23	0.26

Expressed in cubic feet of air per hour through 1 sq ft of wall.

TRANSMISSION. The heat loss due to transmission can be obtained by the formula:

$$H = AU(t_i - t_o) \tag{1}$$

where H = heat loss transmitted through wall, roof, ceiling, floor, etc., in Btu per hr.
A = area surface, in square feet.
U = overall transmission coefficient from Table 2.
t_i = inside dry-bulb temperature, in degrees fahrenheit.
t_o = outside dry-bulb temperature, in degrees fahrenheit.

It can be noticed from Table 2 that insulation on the wall, roof, ceiling, and floor, and weather stripping and double glaze windows and doors result in a considerable reduction of the transmission coefficient U.

The extent to which insulation can be economically justified is shown for a typical two-story, unattached residence. The design heat loss is approximately 150,000 Btu per hr with a 70 deg fahr temperature difference between inside and outside. The distribution of the heat loss is approximately:

30 per cent through side walls.
26 per cent through windows and doors.
20 per cent through infiltration.
15 per cent through ceiling and roof.
9 per cent through floor.

Recent investigation indicates that the following heat-loss reductions can be made:

1. Window weather stripping: 10 per cent.
2. 2-in. roof insulation: 13 per cent.
3. 2-in. wall insulation: 17 per cent.
4. Storm windows and doors with resulting infiltration reduction: 28 per cent.

The installation of storm windows and doors causes a material reduction in the infiltration of air, as well as a reduction of the heat loss through the structure. Weather stripping is not usually required when storm windows and doors are used, since they should be made to be tight fitting. If items 2, 3, and 4 are provided, a 58 per cent reduction in heat loss results. If 3 5/8 in. of insulation is used on the roof and walls instead of 2 in., an additional 5 per cent reduction results.

INFILTRATION. It can be noticed that the transmission coefficient U in Table 2 is based on 15-mph wind, which is considered good practice and above the average wind velocity for most locations. Air leakage into a space through windows and door cracks is a direct function of the wind pressure and can be calculated by either the crack method or the air change method.

The crack method is the generally accepted method of computing the infiltration for a structure. It consists of using the total amount of crack around the windows and doors on the windward side of the structure, plus the infiltration through the walls themselves. In most cases, because of the construction, the infiltration through the walls is negligible. The length of cracks for a double-hung window is equal to three times the width plus two times the height. The amount of infiltration in cubic feet per hour for various types of windows, with various wind velocities, is given in Table 3.

Table 3. Infiltration through Windows *

Expressed in Cubic Feet per Foot of Crack per Hour

Type of Window	Remarks	Wind Velocity, Miles per Hour					
		5	10	15	20	25	30
Double-hung wood-sash windows (unlocked)	Around frame in masonry wall—not calked	3	8	14	20	27	35
	Around frame in masonry wall—calked...	1	2	3	4	5	6
	Around frame in wood-frame construction	2	6	11	17	23	30
	Total for average window, non-weatherstripped, 1/16-in. crack and 3/64-in. clearance. Includes wood-frame leakage....	7	21	39	59	80	104
	Ditto, weatherstripped................	4	13	24	36	49	63
	Total for poorly fitted window, non-weatherstripped, 3/32-in. crack and 3/32-in. clearance. Includes wood-frame leakage..............................	27	69	111	154	199	249
	Ditto, weatherstripped................	6	19	34	51	71	92
Double-hung metal windows	Non-weatherstripped, locked...........	20	45	70	96	125	154
	Non-weatherstripped, unlocked..........	20	47	74	104	137	170
	Weatherstripped, unlocked.............	6	19	32	46	60	76
Rolled section steel-sash windows	Industrial pivoted, 1/16-in. crack.........	52	108	176	244	304	372
	Architectural projected, 1/32-in. crack.....	15	36	62	86	112	139
	Architectural projected, 3/64-in. crack.....	20	52	88	116	152	182
	Residential casement, 1/64-in. crack.......	6	18	33	47	60	74
	Residential casement, 1/32-in. crack.......	14	32	52	76	100	128
	Heavy casement section, projected, 1/64-in. crack..............................	3	10	18	26	36	48
	Heavy casement section, projected 1/32-in. crack..............................	8	24	38	54	72	92
Hollow metal, vertically pivoted window.....................		30	88	145	186	221	242

* Reproduced by kind permission of ASHVE from their 1946 *Guide.*

The infiltration can also be roughly estimated by using a certain number of air changes per hour for each room, based on the type, exposure, use, and location of the room. Tables giving the recommended air changes by this method are found in the ASHVE *Guide.*

On some installations, especially if there are a large number of occupants, it may be necessary to supply more ventilation air than will result by natural infiltration. The exact quantity to be used depends to a great extent on the judgment of the design engineer. Some influencing factors to be considered are the quantity of smoke present, the number and activity of the people, and the amount and kind of odors. Helpful recommendations concerning the amount of ventilation air normally used under various conditions are found in most heating and ventilating handbooks.

The sensible heat required to raise the dry-bulb temperature of the infiltration air is given by the formula:

$$H_s = Q_1(\text{Density of air})(\text{Specific heat})(t_i - t_o) \tag{2}$$

For 70 deg fahr the specific heat is 0.24 Btu per lb and the density is 0.075 lb per cu ft. These values can be substituted in eq. (2) to obtain:

$$H_s = 0.018Q_1(t_i - t_o) \tag{2a}$$

where Q_1 = quantity of infiltration air from Table 3, cubic feet per hour.
t_i = inside dry bulb temperature, degrees fahrenheit.
t_o = outside dry bulb temperature, degrees fahrenheit.
H_s = sensible heat required to raise dry-bulb temperature of infiltration air, Btu per hr.

HUMIDIFICATION. If it is desired to maintain a given relative humidity in the space to be heated, the amount of heat required to evaporate the water may be determined by the formula:

$$H_H = \frac{Q_1 d\,(h_i - h_o)}{7000}\,h_l \tag{3}$$

For standard conditions of 70 deg fahr dry bulb and density of 0.075 lb per cu ft, with latent heat of 1060 Btu per lb, eq. (3) becomes:

$$H_H = 0.0114Q_1(h_i - h_o) \tag{3a}$$

where Q_1 = quantity of infiltration air from Table 3, cubic feet per hour.
d = density of inside air, pounds per cubic feet.
h_i = vapor density of inside air, grains per pound of dry air.
h_o = vapor density of outside air, grains per pound of dry air.
h_l = latent heat of moisture in inside air, Btu per pound.

INTERNAL HEAT GAIN OF STRUCTURE. The sensible and latent loads produced by people, electric lights, motors, operating equipment, and other miscellaneous heat sources may be considered useful heat. Generally, in residences and office buildings, these heat sources are so small that they can be neglected. In industrial plants, theaters, assembly halls, and the like, however, auxiliary heat sources should be investigated and taken into consideration.

Motors and Machinery. Motors and operating equipment convert all the electrical energy supplied them into heat, which is retained in the room, if the product being manufactured is not removed until its temperature is the same as the room temperature. In this case:

$$\text{British thermal units supplied per hour} = \frac{2546 \times \text{Horsepower output of motor}}{\text{Efficiency of motor}} \tag{4}$$

If power is transmitted to the machinery from the outside, then only the heat equivalent of the brake horsepower supplied is used:

$$\text{Btu supplied per hr} = 2546 \times \text{Brake horsepower} \tag{4a}$$

In many industrial plants the heat given out by the machinery is the chief source of heating and is frequently sufficient to overheat the building, even in zero weather, thus requiring cooling by ventilation or mechanical refrigeration the year round.

In a transformer substation or similar space from which electrical energy is transmitted, the heat supplied by the electrical apparatus is due to the losses in the apparatus. The loss is equivalent to 3415 Btu per kilowatthour of loss.

People seated at rest dissipate 290 Btu per hr, of which 145 Btu is sensible and 145 Btu latent heat. An office worker, moderately active, dissipates 490 Btu per hr, of which 225 is sensible and 265 latent heat. A metal worker at a bench dissipates 862 Btu per hr, of which 277 is sensible and 515 latent heat. Sensible and latent heats given off by people in other activities and under other working conditions can be found in air-conditioning handbooks.

Lamps. The heat (Btu) supplied by the electric lamps in a room is equal to:

$$3.413(\text{Watts per lamp})(\text{Number of lamps}) \tag{5}$$

The heat supplied by gas lamps may be calculated from the following data:

1 cu ft producer gas: 150 Btu.
1 cu ft illuminating gas: 530 Btu.
1 cu ft natural gas: 1000 Btu.

A Welsbach burner averages 3 cu ft of gas per hr, and a fishtail burner 5 cu ft per hr.

Sun Effect. The internal heat gain caused by solar radiation is entirely sensible heat. This heat gain is always taken into account in determining the cooling load but seldom considered in the heating load calculations, because it is not dependable as useful heat. Some engineers consider solar radiation a safety factor during the heating cycle. In many sections of the country, however, it is an important item, and its influence should be given full weight.

11. ESTIMATING ELECTRIC CONSUMPTION FOR HEATING

Either the calculated heat-loss method or the degree-day method can be used to estimate the fuel consumption of a heating system. The degree-day method is not as theoretically accurate as the calculated heat-loss method, but is more generally used.

CALCULATED HEAT-LOSS METHOD. In predicting fuel consumption for heating a space, the general formula is:

$$F = \frac{H(t_i - t_a)N}{E(t_d - t_o)C} \tag{6}$$

where F = amount of fuel (same unit as C).
H = Heat loss, Btu per hour.
t_i = average inside temperature maintained during heating season, degrees fahrenheit.
t_a = average outside temperature through heating season (see U. S. Weather Data), degrees fahrenheit.
t_d = inside design temperature, degrees fahrenheit.
t_o = outside design temperature, degrees fahrenheit.
N = number of heating hours in heating period (for October 1 to May 1 season is 5088 hr).
E = efficiency of utilization of fuel over the period expressed as a decimal.
C = heating value of one unit of fuel.

The estimated fuel consumption by the calculated heat-loss method is theoretical. To be worth while, compensation must be made for fluctuation experienced in the heating of buildings, poor heating systems, winter heat gains, sun gains, and similar factors.

DEGREE-DAY METHOD. This method is based on computing the actual consumption on a degree-day basis.

A degree day is the difference between 65 deg fahr and the mean outdoor temperature.

$$\text{Degree days} = 65 \text{ deg fahr} - \frac{(t_{\max} + t_{\min})}{2} \tag{7}$$

The average number of degree days per month (on a 65 deg fahr basis) which have occurred over a period of a year can be found in heating and ventilating handbooks.

The general equation for calculating the probable fuel consumption by the degree-day method is:

$$F = UND \tag{8}$$

where F = fuel consumption for estimated period.
U = quantity of fuel used per degree day per thousand Btu hourly calculated heat loss.
N = heat loss, thousand Btu per hr.
D = number of degree days for the estimated period.

12. TYPES OF ELECTRIC COMFORT HEATING SYSTEMS

From the standpoint of the heating customer, advantages offered by electric heating are: (1) elimination of dirt, soot, and fumes, (2) elimination of firing and other furnace labor, (3) elimination of fuel and combustion products from the premises, (4) accurate and reliable room-temperature regulation, (5) minimum radiation losses in the basement and other parts of the building where heat is not required, (6) minimum maintenance and reconditioning. From the standpoint of the power company, the heating load may offer possibilities of improving the system load factor and power factor, and at the same time greatly increasing the kilowatthour output. The offpeak load especially may be desirable if it fills in the valleys in the load curve and does not increase the generating or distribution capacity required.

Three different types of electric heating systems are in use today:

1. Direct radiator and space heaters, which take electric energy coincidental with the heating demand.
2. Thermal storage systems, which take electric energy only at offpeak hours when the power system is operating sufficiently below rated load to allow carrying the added heating load without additional plant or distribution capacity.
3. Heat-pump systems.

DIRECT RADIATORS AND SPACE HEATERS. This class of equipment includes radiant heaters, convection heaters, unit heaters, and electric hot-water and electric steam radiators using immersion-type heating units. Each has certain characteristics and certain advantages.

The chief advantages of direct radiators and space heaters are their low first cost and their portability, which permits them to be moved from place to place where needed. Small electric heaters may be connected to appliance outlets, but larger ones involve the installation and cost of electric wiring of the proper capacity. Care must be exercised to comply with the National Electrical Code of the National Board of Fire Underwriters, which specifies: (1) portable electric heaters rated at 6 amp or 600 watts or less may be used on lighting branch circuits or on combination lighting and appliance branch circuits; (2) fixed or portable electric heaters of not over 1320 watts may be supplied by an ordinary appliance branch circuit; (3) up to 1650-watt equipment may be supplied by a medium-duty or heavy-duty appliance branch circuit. Equipment of higher rating can be supplied by power circuits of proper capacity.

Radiant Heaters. The radiant electric heaters deliver most of their heat by radiation. Reflectors direct the heat rays in the direction desired. This type of heat gives an immediate and pleasant sensation of warmth when the heat rays strike the body. These heaters have a high efficiency of utilization but are not satisfactory for general heating, as radiant rays do not warm the air through which they pass. The location of the heaters is important, since they must be directed toward the objects to be heated. They must not face a window, because any radiation toward the window passes through and is lost. Radiant heaters are built in portable types and in wall types for both flush and recessed mounting.

Convection Heaters. Convection types of electric heaters are built in a number of different forms. They are designed to deliver heat by convection air currents set up by the tendency of the heated air to rise. The heating units are made of coils of resistance wire, metal ribbons, grids, and other forms and shapes of resistance materials. The heating elements have comparatively large area and operate at moderately low temperatures (temperatures below that which makes the element visibly red). The enclosures should give proper stack effect in order to draw cold air from the floor and direct it into the room.

Unit Heaters. Electric unit heaters are similar to steam or hot-water unit heaters in that they combine the heater with a fan or blower. They are made in portable, wall-mounting, and built-in types in sizes ranging from 1 to 50 kw or larger, if required. This is the best type of electric heater for general applications. The warm air can be directed toward the floor or in any direction desired. The positive air circulation insured by the fan lowers stratification of the air and reduces the possibility of high ceiling temperatures. Because of the forced air circulation, the area of the heating elements may be made smaller, and the weight and cost of the resistors reduced to a minimum. None of the electrical energy taken by the fan is lost, since it is all delivered to the air as heat. Portable units of artistic design, free from radio interference, are available for home use. Larger units may be purchased for industrial plants, power plants, and substations and for temporary use during construction work, repairs, etc.

Hot-water and Steam Heaters. Electric hot-water and electric steam radiators have heating elements to heat water in the lower part of the radiator. The temperature of the water or the pressure is controlled by a thermostat or by a thermostat and a pressure switch, which automatically turns the heating element off and on as required. The cost of this type of equipment is high because of the relatively expensive construction required. The main advantage is the larger area of low-temperature direct radiating surface. The water and steam do not increase the efficiency of the heating elements.

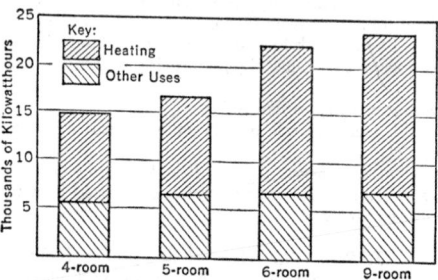

Energy Consumption. The kilowatthours of electrical energy used per year, including electric resistance heating, for well-insulated four-, five-, six-, and nine-room houses, located in the vicinity of Knoxville and Chattanooga, Tennessee, are shown in Fig. 1. The four-room houses have an approximate heat loss of 35,000 Btu per hr based on a 70 deg fahr outside-inside temperature differential. The normal winter averages 3200 degree days. The average kilowatthours for heating, per heating season for 10 four-room houses, as shown in Fig. 1, is 9300, for 10 five-room houses is 10,200, for 5 six-room houses is 15,500, and for 5 nine-room houses is 17,760. The average for heating the 30 houses is 1.35 kwh per cu ft per season. It must be kept in mind that the data for houses with different types of construction, located in different geographic areas, will vary materially from those given in Fig. 1.

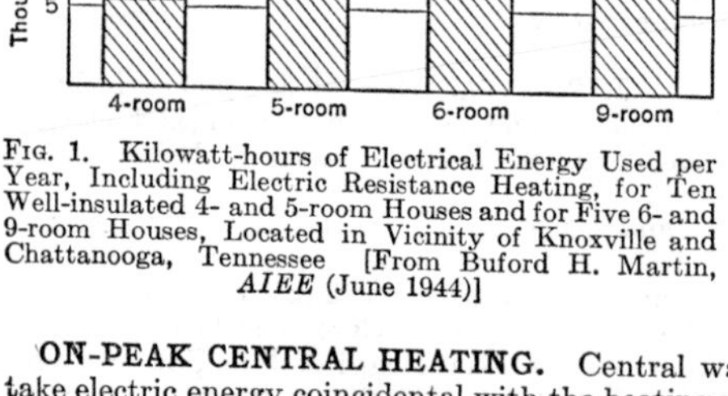

FIG. 1. Kilowatt-hours of Electrical Energy Used per Year, Including Electric Resistance Heating, for Ten Well-insulated 4- and 5-room Houses and for Five 6- and 9-room Houses, Located in Vicinity of Knoxville and Chattanooga, Tennessee [From Buford H. Martin, *AIEE* (June 1944)]

ON-PEAK CENTRAL HEATING. Central warm-air electric heating systems, which take electric energy coincidental with the heating requirements, have the heating elements installed in a plenum chamber connected to the warm-air-duct system. A fan or blower circulates the air past the heating units and through the ducts. In this system it is important to circulate the proper quantity of air in order to give the desired supply air temperature. The electric energy used in this type of system is given by eq. (9):

$$E = \frac{Q_1(\text{Density of air})(\text{Specific heat of air})(t_2 - t_1)}{3.413} \tag{9}$$

For 70 deg fahr air, the specific heat is 0.24 Btu per lb, and density is 0.075 lb per cu ft. These values can be substituted in eq. (9) to obtain:

$$E = 0.00527Q_1(t_2 - t_1) \tag{9a}$$

where E = electric energy, watts.
Q_1 = quantity of air, cubic feet per hour.
t_1 = temperature of air entering heating coil, degrees fahrenheit.
t_2 = temperature of air leaving heating coil, degrees fahrenheit.

Central electric hot-water or steam systems differ only in the use of electric boilers to supply the hot water or steam. For hot water, small, heavily insulated boilers with immersion-type heating units are used. The same type of boiler can also be used for steam. The hot water or steam leaving the boiler is delivered to the heating surfaces in the same manner as in conventional heating systems.

OFF-PEAK THERMAL STORAGE SYSTEMS. A number of different types of thermal storage electric heaters have been developed.

One type of thermal storage system uses a well-insulated central storage tank filled with water. The tank is heated to a temperature of 250 to 300 deg fahr during hours designated by the power company. Charging periods ranging from 12 to 20 hours are used, depending upon the characteristics of the utility's load curve. During the charging period the storage tank absorbs enough heat to meet the demands of the 24-hour period. Heat is delivered under thermostatic control as required to maintain the desired room temperature. Several different arrangements of the equipment have been developed, depending on the type of heating surfaces used to distribute the heat to the building. These may be listed as follows:

1. Hot water with gravity circulation through standard hot-water radiation.
2. Hot water with pump circulation through standard hot-water radiation.
3. Steam circulated through standard steam radiators.
4. Warm air utilizing a central hot-water- or steam-heating coil.
5. Warm air utilizing hot-water or steam unit heaters.
6. Warm air utilizing stored heat in connection with waste heat from synchronous condensers or other equipment.

The essential parts of all the different arrangements of this equipment are: (*a*) heavily insulated storage tank, (*b*) immersion-type heating elements, (*c*) automatic charging control, (*d*) heating surfaces, and (*e*) automatic room-temperature-control system.

Heating Capacity Required. The heating capacity in kilowatts required is found by dividing the Btu requirement of the building for the outdoor design temperature by the product of 3413 times the number of hours that charging is permitted. This should be increased to allow for a small tank loss, irregularity in heating-element rating, and low voltage, remembering that the heating-element capacity varies as the square of the voltage. Where the voltage remains practically constant, 10 to 15 per cent is usually sufficient allowance to cover these factors.

The size of the storage tank depends upon the following factors:

1. The number and allocation of the hours during which heat must be supplied from storage.
2. The temperature range over which the storage water is worked.
3. The Btu requirements of the building during the time that heat is supplied from storage.

Charging Schedules. Various charging schedules have been used, depending on the times that peak loads occur on the electric system. A number of schedules have permitted charging the tank at any time except for 4 to 7 hours during the evening lighting peak. In other places, charging has been prohibited for an additional 2 hours in the morning to care for a morning peak. Charging schedules as short as 12 hours have been used successfully by power companies having a large industrial load and a high load factor. Where the hours of heating from storage are broken up, the tank will recover or partly recover its charge between discharge periods, so that the size of the tank is materially reduced.

Heating Cycle. It has been found that with standard hot-water radiation a storage temperature of 250 deg fahr, giving a pressure of 15 lb per sq in., may be used, and that the water may be returned from the radiators at a temperature of about 150 deg fahr, giving a working range of 100 deg fahr. The temperature of the water to the radiators is varied by means of a mixing valve controlled automatically by the heat requirements of the building. The return water stratifies at the bottom of the storage tank, so that the

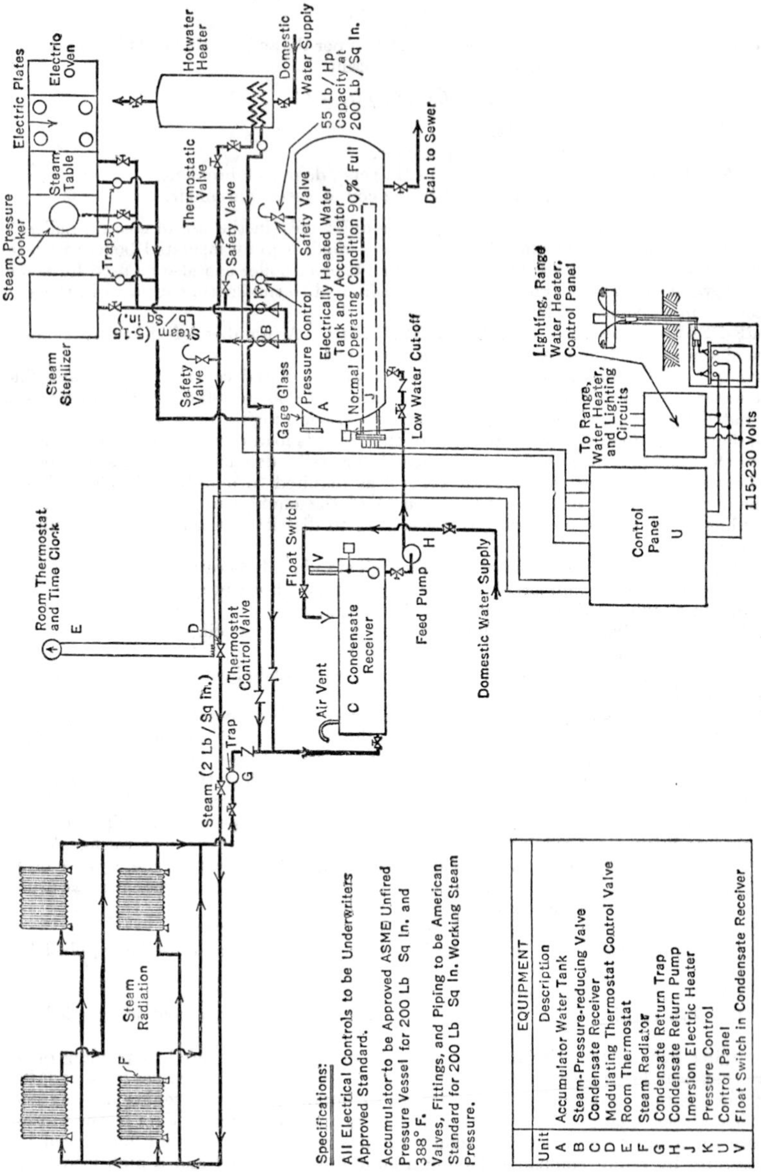

FIG. 2. Schematic Diagram for an Electrothermal Low-pressure Heating System [From Krueger, *Heating and Ventilating* (Jan. 1945)]

hot water above remains at almost a constant temperature until the whole tank is nearly discharged. In warm-air systems the storage tank is used to supply hot water or steam vapor to unit heaters or other heating surfaces, in which case it is found that a working range of 300 to 200 deg fahr can be obtained. The working temperature range can be increased by increasing the amount of heating surface, which lowers the return water temperature, or by raising the storage-tank temperature to the limit that the pressure will permit. Where the heating from storage is confined to a single period, the capacity of the storage tank in gallons equals the maximum Btu storage required, divided by the product obtained by multiplying the working temperature range of the storage water by the weight of a gallon of water at the fully charged temperature by the storage-tank efficiency, allowing for a small tank radiation loss. Where the hours of heating from storage are broken or short, and the heat required is low, the storage tank acts like a storage battery floating on the line. The effect is to smooth out the charging curve and reduce the connected load to an average value over the whole charging period. Since in this case the heat stored in the tank fluctuates up and down, it is necessary to calculate the input and output over the whole 24-hour cycle in order to determine the maximum Btu storage at any one time and the size of the tank.

Tank Construction. In new buildings, where openings to the basement permit, factory-fabricated tanks are used; otherwise the tanks are electrically welded in the basement. As yet there has been no definite specifications of storage tanks with reference to the boiler code, but this has been under consideration by the Hartford Boiler Code Committee. The present tendency is to limit electric welding to tanks not over 4 1/2 ft in diameter and not over 15 lb per sq in. in gage pressure (250 deg fahr) and to use a single V butt weld on both circumferential and longitudinal seams, all edges being level planed. Tanks of larger diameter and higher pressures are usually riveted in accordance with the American Society of Mechanical Engineers' code for unfired pressure vessels. All tanks should be provided with manholes for inspection.

Heating Elements and Control. Standard immersion-type heating elements, usually 5 or 10 kw in size, are inserted directly into the tank in a group assembly in a nozzle welded into the tank, or they are installed singly through smaller nozzles. The tank construction, piping, and the connections to the tank are made quite simple, so that the whole unit lends itself to efficient insulation. The automatic charging control is obtained by a synchronous time switch and a suitable tank thermostatic switch, which open and close the main magnetic contactors supplying current to the heating elements.

Examples of Installations. During the past 10 years a great number of off-peak thermal storage installations have been made in residences, office buildings, substations, and other buildings. The connected load for these installations ranges from 30 to about 300 kw. The first cost of this equipment varies greatly, but in general it is only slightly higher than that of modern gas-, oil-, or coal-heating equipment with fully automatic control.

Figure 2 shows adaptability of an off-peak thermal storage system to existing dwellings having low-pressure steam-heating installations. The main thermostat E responds to the room temperature and regulates the steam flow through the modulating thermostat control valve D. By this method of control the steam flow to the radiators varies with the demand for heat as indicated by the thermostat. The condensate from the radiators is returned through the common drain header, back to the condensate receiver C, through the thermostatic trap G. The condensate is continually reused, with only slight make-up water being required. With this low-pressure steam-heating system, no control of relative humidity or air conditioning is possible, unless additional ventilating ducts and equipment are installed. In the design of the off-peak thermal storage system given in Fig. 2, it is possible to incorporate a steam sterilizer, steam pressure cooker, and steam table, as illustrated.

HEATING BY A HEAT PUMP. The heat pump is a compression-type refrigeration system, consisting of an evaporator, condenser, compressor, and expansion valve. When operated as a refrigerating or cooling system in the conventional way, it pumps heat from an enclosed space and rejects it to the outside. When used for heating, the system takes low-temperature heat from an external heat source, raises it to a higher level, and delivers it to the space to be heated.

A simple illustration of the heat pump is shown in Fig. 3. The enthalpies, in Btu per lb, are from "Freon" (F-12) tables. It can be noticed that the heat absorbed by the evaporator is 45.6 Btu per lb of refrigerant; the heat equivalent of the work of compression is 12.7 Btu per lb; the heat rejected by the condenser is $91.5 - 33.2 = 58.3$ Btu per lb. In this manner a low-temperature heat at the evaporator is raised to a sufficiently higher temperature level at the condenser (approximately 108 deg fahr in the figure) to become useful for heating a space.

Efficiency. By definition, the efficiency or coefficient of performance is the total useful heat output divided by the heat equivalent of the work done in producing the effect. The ideal coefficient of performance obtainable during the heating cycle, between the two temperatures, is represented by the formula:

$$cp = \frac{T_2}{T_2 - T_1} \qquad (10)$$

T_1 is the absolute temperature at the low temperature level, and T_2 the absolute temperature at the high temperature level. The coefficient of performance on actual installations is lower than ideal because of inherent inefficiencies in the compressor cycle. In the example illustrated in Fig. 3, $cp = 58.3/12.7 = 4.58$.

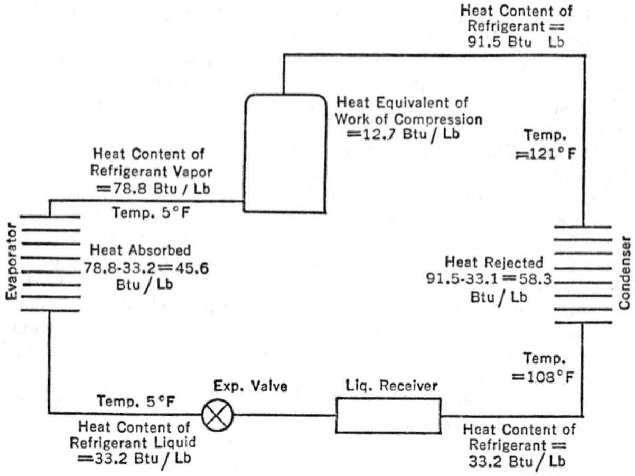

FIG. 3. Schematic Diagram of Heat-pump Machine to Show How a Low-temperature Heat at the Evaporator Is Raised to Sufficiently Higher Temperature at the Condenser to Be Useful [Sporn and Ambrose, *Heating and Ventilating* (Jan. 1944)]

Heat Cycle. The heat absorbed by the evaporator of a heat-pump system during the heating cycle can be taken from the air, water, ground, or any other suitable source, as illustrated by Figs. 4 and 5. Figure 4, usually referred to as air-to-air design, uses air as the source of heat and air to deliver the heat from the conditioner coil to the conditioned space. During the heating cycle the gas refrigerant goes from the refrigerating compressor, through valve A, path 1–2, to the conditioner coil, where it is liquefied, giving up its heat to the air being delivered to the conditioned space. From the conditioner coil the liquid refrigerant goes through the check valve, to the liquid receiver, the expansion valve, and

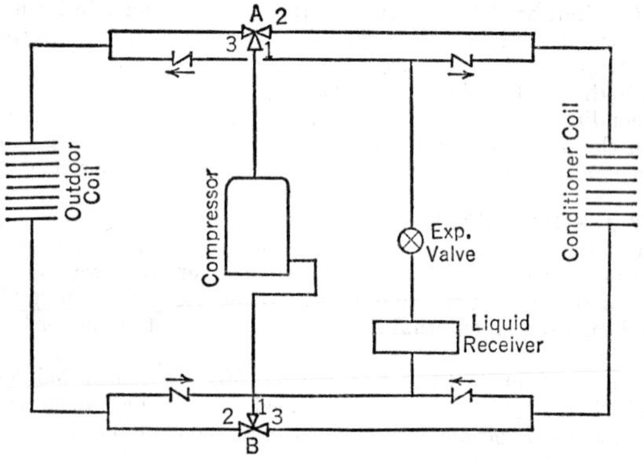

FIG. 4. Air-to-air Design Heat Pump. Air is used as heat source and to transfer heat to conditioned space

the check valve, to the outdoor coil, where it is gasified, absorbing heat from air outside the conditioned space. From the outdoor coil the gas returns to the compressor, through valve B, path 2–1. During the cooling cycle, the gas refrigerant goes from the refrigerating compressor, through valve A, path 1–3, to the outdoor coil, where it is liquefied, giving up heat to the air outside the conditioned space. From the outdoor coil the liquid refrigerant goes through the check valve, to the liquid receiver, through the expansion valve and check valve, to the conditioner coil, where it is gasified, absorbing heat from the air being delivered to the conditioned space. From the conditioner coil the gas refrigerant returns to the compressor, through valve *B*, path 3–1.

Figure 5, usually referred to as liquid-to-air design, uses water or ground as the heat source and air to deliver the heat from the condenser to the conditioned space. During the heating cycle, the gas refrigerant goes from the compressor, through valve *A*, path 1–2, to the conditioner coil, where it is liquefied, giving up its heat to the air being delivered to the conditioned space. From the conditioner coil the liquid refrigerant goes through the check valve, to the liquid receiver, the expansion valve, and the check valve, to the condenser-cooler, where it is gasified, absorbing heat from the liquid being circulated by the pump. From the condenser-cooler the gas returns to the compressor, through valve *B*, path 2–1. During the cooling cycle the gas refrigerant goes from the refrigerating compressor, through valve *A*, path 1–3, to the condenser-cooler, where it is liquefied, giving up heat to the liquid being circulated by the pump. From the condenser-cooler the liquid refrigerant goes through the check valve, to the liquid receiver, through the expansion valve and check valve, to the conditioner coil, where it is gasified, absorbing heat from the air being delivered to the conditioned space. From the conditioner coil the gas refrigerant returns to the compressor, through valve *B*, path 3–1.

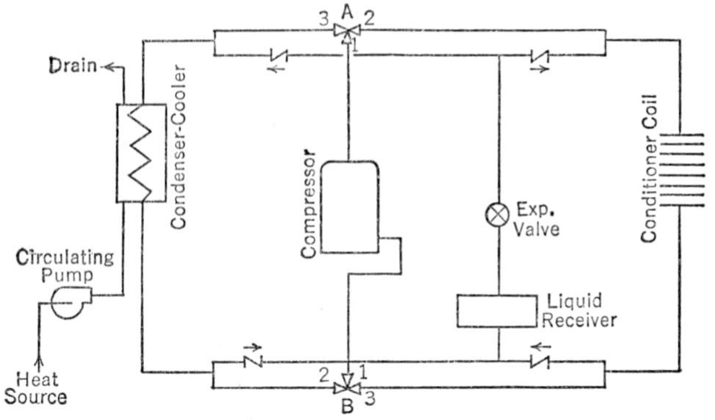

FIG. 5. Liquid-to-air Design Heat Pump. Water or ground is used as heat source, and air to deliver heat from condenser to the conditioned space.

Examples of Installations. A number of heat-pump installations have been made in the past 10 years in many parts of the United States and in several sections of Europe. An air-to-air system and a water-to-air system have recently been installed in the Ohio Power Company's office buildings at Portsmouth and Coshocton, Ohio. The Portsmouth building, erected in 1940, has four stories and is 104 ft long by 45 ft wide by 45 $^1/_2$ ft high. The Coshocton building, erected in 1940, has two stories and a basement and is 88 ft long by 55 ft wide by 35 ft high. Both buildings are of non-combustible construction, well-insulated with double glazed windows throughout.

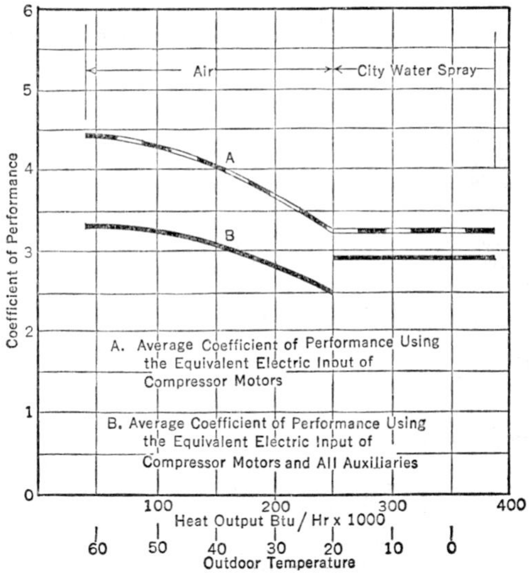

FIG. 6. Coefficient of Performance at Various Outdoor Temperatures of Heat-pump System, the Ohio Power Company, Portsmouth, Ohio [From *ASHVE Trans.*, Vol. 50 (1944), by permission]

Table 4 gives the actual performance data of the Portsmouth system, and Fig. 6 the average coefficient of performance throughout the heating season. Table 5 gives the actual performance data of the Coshocton system, and Fig. 7 the average coefficient of performance throughout the year.

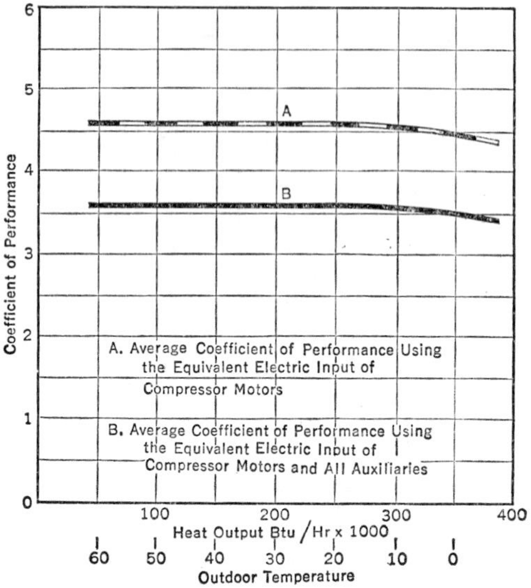

FIG. 7. Coefficient of Performance at Various Outdoor Temperatures of Heat-pump System, the Ohio Power Company, Coshocton, Ohio [From *ASHVE Trans.*, Vol. 50 (1944), by permission]

Table 4. Actual Performance Data for Portsmouth System

[By permission of the American Society of Heating and Ventilating Engineers, *ASHVE Trans.*, Vol. 50 (1944)]

	Test 1	Test 2	Test 3	Test 4
Supply air to conditioned space				
Cubic feet per minute	18,000	18,000	18,000	18,000
Temperature, degrees fahrenheit	77.1	85.1	78.9	87.2
Outside air				
Cubic feet per minute	2,000	2,000	2,000	2,000
Temperature, degrees fahrenheit	32	32	32	32
Conditioner heating coil				
Entering liquid temperature, degrees fahrenheit	81.8	90	83	92
Leaving liquid temperature, degrees fahrenheit	85	97.2	87.5	100
Gallons per minute	95	95	95	95
Cooler				
Entering liquid temperature, degrees fahrenheit	21.2	19	28.5	26
Leaving liquid temperature, degrees fahrenheit	18	13.8	25	20
Gallons per minute	95	95	95	95
Electric consumption				
Compressors, kilowatts	16.6	32.4	17.8	35.1
Auxiliaries, kilowatts	9.2	11.1	4.8	4.8
Total kilowatts *	25.8	43.5	22.6	39.9
Capacity				
Refrigeration, Btu per hour	140,000	250,000	168,000	284,000
Heat output, Btu per hour	187,200	342,000	218,500	383,500
Coefficient of performance				
Using kilowatt input to compressor	3.3	3.1	3.6	3.2
Using total kilowatt input †	2.2	2.4	3.0	2.9

Test 1—1 25-hp compressor: 135 psi head press, 11 psi suction.
Test 2—2 25-hp compressors: 160 psi head press, 10 psi suction.
Test 3—1 25-hp compressor using water spray: 140 psi head press, 16 psi suction.
Test 4—2 25-hp compressors using water spray: 170 psi head press, 12 psi suction.

* Kilowatts to conditioner supply fan not included.

† 60 per cent of kilowatt input to circulating pumps included as useful work.

The heat-pump system for heating, cooling, and year-round air conditioning is theoretically and practically sound. Its principal drawback, high installation cost, has been greatly reduced by the development of improved equipment. Further developments and design improvement, combined with quantity production and improved production technique, are sure to widen the field of application of this type of system.

Table 5. Actual Performance Data for Coshocton System

[By permission of the American Society of Heating and Ventilating Engineers, *ASHVE Trans.*, Vol 50 (1944)]

	Test 1	Test 2	Test 3
Supply air to conditioned space			
Cubic feet per minute	11,000	11,000	11,000
Temperature, degrees fahrenheit	78.6	80.5	94.3
Outside air			
Cubic feet per minute	2,500	2,500	2,500
Temperature, degrees fahrenheit	37	35	35
Conditioner heating coil			
Entering water temperature, degrees fahrenheit	88	92.8	103
Leaving water temperature, degrees fahrenheit	83	87.0	113.5
Gpm	70	70	70
Cooler			
Entering water temperature, degrees fahrenheit	55	55	55
Leaving water temperature, degrees fahrenheit	47	46.5	47.3
Gpm	37	38	75
Electric consumption			
Compressors, kilowatts	10.2	15.5	27.5
Auxiliaries, kilowatts	5.2	5.2	5.2
Total kilowatts *	15.4	20.7	32.7
Capacity			
Refrigeration, Btu per hour	148,000	161,500	288,600
Heat output, Btu per hour	175,000	203,800	366,600
Coefficient of performance			
Using kilowatt input to compressors	5.0	3.85	3.9
Using total kilowatt input †	3.5	3.0	3.3

Test 1—10-hp compressor: 120 psi head pressure, 40 1/2 psi suction.
Test 2—15-hp compressor: 140 psi head pressure, 35 psi suction.
Test 3—10-hp compressor: 150 psi head pressure, 42 1/2 psi suction.
15-hp compressor: 175 psi head pressure, 38 1/2 psi suction.

* Kilowatts to conditioner supply fan not included.
† 60 per cent of auxiliaries kilowatt input was considered useful heat.

13. COMPARATIVE FUEL COSTS

It can be seen from Fig. 8 and from the examples of heating cost which have been given, that, even with the many inherent advantages of electric resistance heating, there are very few areas in the United States where it can be justified at the present time when the relative cost of delivered energy is compared with the corresponding cost of coal, oil, or gas. The heat pump, however, offers possibilities because it eliminates the handicap of electric energy on a straight energy-conversion basis by furnishing 3 to 5 kwhr of equivalent heating for every kwhr of energy used, and at the same time maintains all the advantages of electric resistance heating.

A comparison of the cost of heating by electricity with that by coal, oil, or gas is readily obtained from eqs. (11), (12), and (13) by assuming the expected efficiencies and calculating the equivalent prices for these fuels to give the same cost for a given amount of heat delivered.

The equivalent rate for coal is given by:

$$R_c = \frac{R_e}{E_e} 5.85 B_c E_c \times 10^{-3} \qquad (11)$$

where R_c = cost of coal, dollars per ton (2000 lb).
R_e = cost of electricity, cents per kilowatthour.
E_c = overall efficiency for electric heating, per cent.
B_c = Btu per pound of coal.
E_e = overall efficiency for coal heating, per cent.

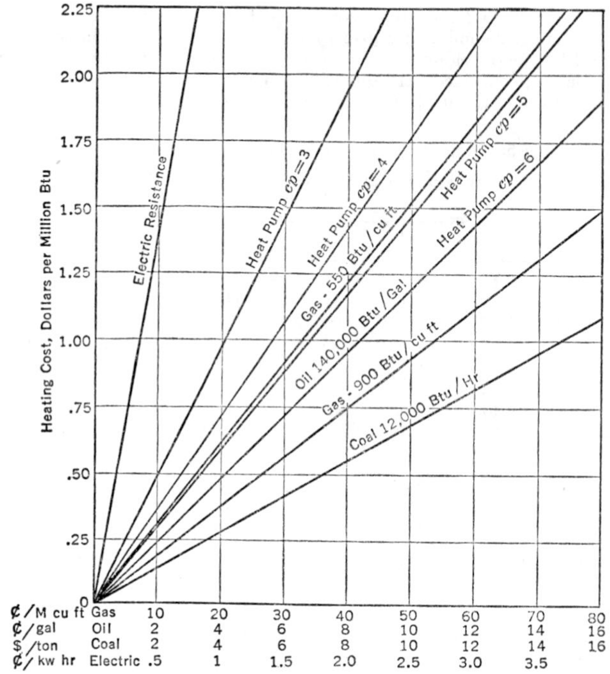

FIG. 8. Comparative Heating Cost in Dollars per Million Btu of Electric Resistance Heating, Heat-pump Systems, and Coal, Oil, and Gas. Based on an assumed burning efficiency of 60 per cent for coal, oil, and gas.

The equivalent rate for oil is given by:

$$R_o = \frac{R_e}{E_e} 2.93 B_o E_o \times 10^{-4} \tag{12}$$

where R_o = cost of oil, cents per gallon.
R_e = cost of electricity, cents per kilowatthour.
E_e = overall efficiency for electric heating, per cent.
B_o = Btu per gallon of oil.
E_o = overall efficiency for oil heating, per cent.

The equivalent rate for gas is given by:

$$R_g = \frac{R_e}{E_e} 0.293 B_g E_g \tag{13}$$

where R_g = cost of gas, cents per 1000 cu ft.
R_e = cost of electricity, cents per kilowatthour.
E_e = overall efficiency for electric heating, per cent.
B_g = Btu per cubic foot of gas.
E_g = overall efficiency for gas heating, per cent.

Figure 8 gives the comparative heating cost in dollars per million Btu of electric resistance heaters, heat-pump systems, and coal, oil, and gas. The burning efficiency of 60 per cent used for coal, oil, and gas in preparing Fig. 8 is representative but will vary considerably with conditions and the design of the equipment. From Fig. 8 the required cost of energy or fuel to obtain a heating cost of $1.25 per million Btu will be as follows:

1. Electric energy for resistance heater	0.43 cents per kwhr
2. Electric energy for heat pump (cp of 3)	1.29 cents per kwhr
3. Electric energy for heat pump (cp of 4)	1.72 cents per kwhr
4. Electric energy for heat pump (cp of 5)	2.14 cents per kwhr
5. Electric energy for heat pump (cp of 6)	2.57 cents per kwhr
6. Gas (550 Btu per cu ft heating value)	41.2 cents per M cu ft
7. Gas (900 Btu per cu ft heating value)	67.8 cents per M cu ft
8. Oil (140,000 Btu per gal heating value)	10.5 cents per gal
9. Coal (12,000 Btu per lb heating value)	18.5 dollars per ton

To obtain a heating cost of 75 cents per million Btu, the required cost of energy and fuel will be:

1. Electric energy for resistance heater	0.26 cents per kwhr
2. Electric energy for heat pump (cp of 3)	0.77 cents per kwhr
3. Electric energy for heat pump (cp of 4)	1.03 cents per kwhr
4. Electric energy for heat pump (cp of 5)	1.28 cents per kwhr
5. Electric energy for heat pump (cp of 6)	1.54 cents per kwhr
6. Gas (550 Btu per cu ft heating value)	27.7 cents per M cu ft
7. Gas (900 Btu per cu ft heating value)	40.5 cents per M cu ft
8. Oil (140,000 Btu per gal heating value)	6.3 cents per gal
9. Coal (12,000 Btu per lb heating value)	10.80 dollars per ton

14. AIR CONDITIONING

A modern air-conditioning system provides for control of the temperature, humidity, purity, odor, and movement of air during both the heating and cooling cycles.

The temperature sensation of the human body depends not only upon the dry-bulb temperature of the surrounding air, but also upon the humidity (or wet-bulb temperature) and upon air movement, because of the effect of humidity and air movement on the dissipation of body heat. The best representation of the combined effect of these factors on comfort is given by comfort charts, which are found in the ASHVE *Guide.*

In addition to heating, winter air conditioning consists of humidification, as well as filtering and sometimes deodorizing. Relative humidities between 40 per cent and 60 per cent are usually desirable in winter, which generally requires the addition of moisture to the air. In warm-air systems this is accomplished by passing the air through sprays or over surfaces wetted by a continuous flow of water. With steam or hot-water heating systems, the moisture is added by unit humidifiers, or, where more adequate equipment cannot be supplied, some improvement in the humidity may be obtained by the use of radiator pans.

Complete summer air conditioning consists of lowering the effective temperature by cooling and by reducing the relative humidity of, filtering, deodorizing, and circulating the conditioned air. The temperature must not be lowered too far below that of the outside, however, and the relative humidity must be kept low enough (50 per cent or less) to induce a rate of evaporation which will keep the clothing and skin dry. The recommended indoor-outdoor temperature differentials for different kinds and types of structures are found in the ASHVE manual and other air-conditioning guides.

METHODS FOR LOWERING TEMPERATURE. There are four methods of lowering the effective temperature:

1. Using some form of refrigeration to lower the dry-bulb and wet-bulb temperatures by the removal of sensible and latent heat.
2. Dehumidifying and lowering of the dewpoint temperature without lowering the dry-bulb temperature.
3. Evaporative cooling, which lowers the dry-bulb temperature by evaporation of moisture without addition or subtraction of heat.
4. Increasing the air movement, which results in higher evaporation from the skin.

The most satisfactory method depends on climatic and other conditions. The first two methods are, as a general rule, the most satisfactory. Satisfactory cooling by evaporation and air motion alone requires a low relative humidity and thus is applicable only in restricted sections of the country.

DEHUMIDIFICATION. Dehumidification may be obtained in three ways:

1. Cooling the air below the dewpoint by refrigeration.
2. Extraction by liquid absorbents.
3. Extraction by solid adsorbents.

Dehumidification by the first method is based on the fact that, as the temperature of air is lowered, the amount of moisture that can be retained is decreased. For a given moisture content (grains per cubic foot) there is a definite temperature called the "dewpoint temperature," for which the air will be saturated (100 per cent humidity). If the temperature is dropped below the dewpoint temperature, a definite amount of moisture condenses out until the amount left corresponds to that required to give 100 per cent relative humidity for the new temperature. Thus, by lowering the temperature to the proper point, reducing the moisture, and reheating by means of the internal heat gain or an external source, air at any desired temperature and relative humidity can be obtained in the conditioned space.

In extraction by liquid absorption moisture is taken up by a substance which undergoes a change in chemical or physical structure. This type of absorbent is usually a water solution of lithium chloride, calcium chloride, lithium bromide, or a similar solution.

When liquid absorbents are used, the solution is brought into intimate contact with the air by means of sprays. Since the aqueous brine solution has a lower vapor pressure, moisture in the air will be absorbed by the solution, resulting in a conversion of latent heat to sensible heat, which raises the solution temperature and consequently the air temperature. In extraction by adsorption a substance is used which does not undergo a change in chemical or physical structure. Solid adsorbents include activated alumina, silica gel, and activated carbon. When solid adsorbents are used, the water vapor condenses as it passes through the material, giving up its latent heat of condensation, plus the so-called heat of mixing, to the air.

Both the absorbents and adsorbents have to be reactivated from time to time by applying an external source of heat to drive off the moisture which has been collected.

COOLING LOAD. The cooling load for any building or part of a building is made up of four parts:

1. Heat transmitted through walls, roofs, and glass, with allowances for sun-exposed surfaces and heat capacities.
2. Solar radiation transmitted through windows.
3. Heat and moisture from infiltration of outside air.
4. Heat and moisture from occupants, lights, and machinery.

The transmission for surfaces not exposed to the sun is calculated by using the heat-transmission constants and temperature difference in the same manner as for a heat loss. For surfaces exposed to the sun, the heat transmission will be greater.

One hundred per cent of the sun's solar energy is transmitted into a building through an opening which is without glass or other obstructions and which is normal to the sun's rays. The amount of solar radiation transmitted to the conditioned space for the different hours of the day and for different angles of incidence of the sun's rays is found in the ASHVE and other air-conditioning guides.

AIR-CONDITIONING EQUIPMENT. Air-conditioning systems may be classified as central air-conditioning systems and unit air-conditioners. The first is used principally in theaters, restaurants, office buildings, and manufacturing plants. The second is used in homes and separate offices and for other applications where the required capacity is not too great. In the central system the fans, heating and cooling surfaces, humidifiers, filters, etc., as well as the refrigeration compressor, are located in a suitable apparatus room, from which the conditioned air is supplied, usually by means of supply and return ducts. In the unit conditioner the fans, heating and cooling surfaces, humidifier, compressor, controls, and other equipment form a complete unit, which is assembled at the factory.

Unit air conditioners usually have capacities from 6,000 to 120,000 Btu per hr for cooling and up to 300,000 Btu per hr for heating. The different types may be classified as year-round units, summer units, and winter units. The year-round units perform all the functions of air conditioning, namely, cooling dehumidification, air circulation, air cleaning, heating, and humidification. Summer units provide cooling, dehumidification, air cleaning, and air circulation. The winter units provide heating, humidification, air cleaning, and air circulation. Units which do not provide simultaneous control of at least four of the recognized functions of an air conditioner are usually more properly classified as unit heaters, unit ventilators, window-type ventilators, humidifiers, coolers, or dehumidifiers.

Cooling is usually accomplished by mechanical refrigerators using low-temperature evaporation or a liquid to absorb the heat. Air or water is cooled by passing it over the evaporator. The evaporated liquid is compressed and then cooled by passing air or water over the condensing coils. The apparatus, exclusive of the evaporator or cooling coils, is called the condensing unit. There are two methods of applying the refrigeration. In the direct-expansion system the evaporator is in direct contact with the air in the conditioned space. In the indirect-expansion system the evaporator cools the water or some other liquid, which in turn cools the air in the conditioned space. For both the direct and indirect systems, odorless, non-toxic refrigerant fluids such as "Freon" are normally used. Compressors are of multicylinder or rotary designs and are driven by electric motors. The compressor assembly is carefully mounted to give quiet operation. Much of the success in comfort air conditioning is due to improvements in the design, efficiency, and dependability of the compressor. The heat pump, as has been pointed out, is an all-electric year-round heating and cooling system. Since the same equipment is used for both cycles of operation, it holds great promise in air conditioning for homes, office buildings, and commercial establishments.

Filters. Filters or air cleaners are of three general types: (1) dry-air filters, (2) air washers, (3) viscous filters.

Dry-air filters are made in the form of a screen from felt, cloth, wool, and other materials. The filter material may be designed for vacuum cleaning or for replacement as it becomes filled with dust and dirt. The air-washer type cleans the air by passing it through sprays. In the viscous type of filter the dust and dirt are trapped by allowing them to impinge on viscous sheets placed in the air stream from the fan. A great many different arrangements and substances are used. The viscous screens are washed, and the viscous coating is renewed either manually or automatically.

Controls. The automatic controls for heating and air conditioning constitute a very important part of the equipment. This apparatus is made up of devices sensitive to temperature, devices sensitive to relative humidity, and devices sensitive to pressure, together with relays, contactors, valves, dampers, time switches, and other auxiliary equipment. Both compressed air and electricity are used for operating the auxiliary equipment. The temperature-sensitive devices, called thermostats, actuate electric or pneumatic switches. Depending on their intended use, thermostats may be classified, according to the type of temperature-sensitive element, as straight, curved, or spiral bimetal strip thermostats, diaphragm type filled with either volatile liquid or a gas, and liquid-bulb type. Devices which are sensitive to changes in humidity cause the humidifying apparatus to increase or decrease the supply of moisture. These devices, which are called humidistats or hydrostats, operate on several different principles. One type, which is directly sensitive to humidity, uses wooden blocks, air, fiber, paper, and membranes. A second type uses a wet-bulb and a dry-bulb thermostat.

15. BIBLIOGRAPHY

American Society of Heating and Ventilating Engineers, *Guide* and *Heating and Ventilating Handbook.*

Krueger, George H., Blueprint of Post-war Realities—Electrothermal Space Heating and Air Conditioning, *Heating and Ventilating* (Jan., 1945).

Martin, Buford H., Electrically Heated Homes in the Tennessee Valley, *Trans. AIEE* (July, 1944).

Roberson, L. N., Radiant Heating by Electricity, *Heating and Ventilating* (Sept., 1946).

Sharp, W. L., Interim Report on Electric House-heating Test, *Electric Light and Power* (Oct., 1946).

Sporn, Philip, The Sixth Ingredient—the Heat Pump, *EEI Bull.* (August, 1944).

Sporn, Philip, and E. R. Ambrose, The Heat Pump, an All-electric Year-round Air-conditioning System, *Heating and Ventilating* (Jan., 1944).

Sporn, Philip, and E. R. Ambrose, Description and Performance of Two Heat-Pump Air-conditioning Systems, *Heating, Piping, and Air Conditioning* (June, 1944).

Sporn, Philip, E. R. Ambrose, and T. Baumeister, *Heat Pumps,* John Wiley (1947).

Electric Power Markets in Washington, *Univ. Washington Bull. No.* 93 (1937).

HOUSEHOLD HEATING AND COOKING APPLIANCES

By E. K. Clark

16. POWER RATING OF APPLIANCES

Ranges and water heaters are the major household heating appliances counted upon by utilities for load building, the other heating appliances occupying a very secondary role, as may be seen from the following table:

Appliances	Power Consumption	
	Input, Watts	Annual Consumption, Kilowatthours (Average)
Hair curler	20	....
Waffle iron	600– 1,000	12
Toaster	450– 1,200	25
Percolator	350– 600	72
Pressing iron	550– 1,000	100
Roaster	1,000– 1,320	300
Hot plate	500– 1,650	....
Warming pad	60	....
Electric blanket	180– 225	150
Portable air heater	800– 1,320	45
Sandwich grill	600– 1,000	....
Sun lamp	200– 500	....
Radiant heater	1,000– 1,500	45
Range	7,000–13,600	1,000
Water heater	1,000– 6,500	3,500
Ironing machine (heater only)	1,000– 1,500	100
Clothes dryer (heater only)	4,600– 5,000	150

See Section 2, Article 6, and Section 18, Article 21, for resistor materials; also Section 2, Article 10, and Section 18, Article 22, for insulating materials.

17. SMALL APPLIANCES

The one universal appliance is the pressing iron, commonly called the flat iron. These irons range in weight from 2 1/2 to 6 lb, are highly polished, and are either nickel or chromium plated. Non-automatic irons are rated 500 to 600 watts. The automatic iron contains a compact, built-in thermostat to regulate the temperature, thereby eliminating the necessity of watching the temperature to avoid scorching. The temperature is adjusted by a knob on top of the iron, within ready reach of the operator's finger. Some temperature positions are indicated in degrees, but the majority are marked with the names of the various commercial fabrics used in clothes, draperies, etc. The maximum temperature to which the thermostat can be set is some point below 600 degree fahr ironing surface temperature. This temperature is above the usual ironing temperature but is not high enough to start a fire if inadvertently left standing too long in contact with a combustible material.

Automatic irons with inputs up to 1000 watts are now available. This wattage provides sufficient heat to avoid any delay due to wet cloth or fast work by the operator.

A relatively new development in household pressing irons, as distinguished from tailor irons, is the steam iron. These irons are heated electrically, much the same as flat irons, are automatic with adjustable temperature settings, and are somewhat heavier and larger due to the built-in tank for containing the water supply. One type of steam iron utilizes a flash-boiler in which drops of water are converted into steam as required, the frequency of the drops being controlled by the operator by the manipulation of a control valve. Another type contains a secondary chamber which acts as an ordinary boiler, with the flow of steam from the boiler being controlled by an adjustable valve. In all types of steam irons, the steam emits from the iron onto the cloth being ironed, through small ports in the toe of the ironing surface of the iron.

Waffle irons are also available with adjustable automatic temperature regulation and with pilot lights to indicate the proper baking temperature.

Toasters are manufactured with the simple, turnover principle, with time-control features, or with fully automatic, pop-up, time-temperature brownness control.

Percolators have protective thermostats or fuses to prevent overheating when operated dry.

Roasters have adjustable, automatic temperature regulation, and, when an accessory time control turns the roaster on and off at desired times, this appliance approximates the utility and convenience of a fully automatic electric range.

Electric hot plates, having one or two burners of the same or different sizes, are available. The burners may be of the open-coil refractory brick type, or of the tubular or solid metal-encased refractory oxide insulated type.

Warming pads are of two types. One type, termed "three-speed," is set for "low," "medium," or "high" by a switch, which changes the circuit for inputs of 20, 40, or 60 watts respectively. Thermostats limit the top temperature. "Multiple-heat" pads give positive control at different temperatures.

Electric blankets are full double-bed size and are the only covering required for room temperatures as low as 35 deg fahr. The heating element is a waterproof insulated wire. Two hundred feet or more is used to give uniform distribution. An adjustable temperature control increases the input to the blanket as the room temperature lowers.

There is a wide variety of small appliances of many different types of construction. Nickel-chromium wire or ribbon is used in practically all of them. The principal electrical insulations are natural or built-up mica, refractory porcelain, or a refractory oxide powder or cement. When mica is used, the resistance element is usually wound around a mica form and then is electrically insulated, top and bottom, with additional mica sheets, the whole being clamped between metal sheets or castings. In the toaster elements, the element ribbon is left exposed. When a refractory porcelain is used, the porcelain serves merely as a support for exposed high-temperature-operating radiant resistance coils. When a refractory oxide powder or cement is used, the resistance coil is embedded in or encircled by the powder or cement, which, in turn, is encircled by or enclosed in a metal sheath. Usually the unit is swaged or pressed to compact the oxide and thus improve its thermal conduction.

Because the heater must be designed as a part of the completed appliance, there are no standardized heating elements for use in electric heating appliances, although attempts have been made by several concerns to supply universal renewal heaters for electric ranges.

18. RANGES

Present-day household ranges are rather standardized in sizes and vary mainly in appearance and accessories. They all consist of one or two ovens and two or more surface or top burners. The majority include storage space for pots and pans.

Practically all ovens are 16 in. wide, with the heights varying somewhat in different models and different makes. They are well insulated with mineral wool or glass wool to reduce loss of heat. Today, practically all ovens are below the level of the cooking top or platform. This gives a working surface of much greater extent.

Most surface burners are standardized at approximately 6 and 8 in. in diameter. The following table gives the range of watts input found in ovens and surface burners:

Heaters in Electric Ranges

Heater	Wattage
Oven: for preheating	3700–4700
Oven: for baking or roasting	2200–3700
Oven: for broiling	2400–4000
Surface unit: 6 in. in diameter	1200–1300
Surface unit: 8 in. in diameter	1500–2200
Warming compartment (when provided)	250– 300

Oven heaters are of several different designs. One type consists of a skeleton metal frame, provided with refractory porcelain bushings through which exposed helical resistance coils are threaded. Another type consists of practically self-supporting metal tubes, enclosing helical resistance coils compacted in refractory oxide insulating powder.

The top oven heater is used for broiling, and two broiling speeds or wattages are sometimes provided. In most cases, to provide a balanced heat for baking, a low-wattage coil in the top oven heater gives just enough heat, in addition to that in the lower or baking heater, to equalize the temperature in the entire oven.

Hand control is provided for each heater, although both heaters of the oven are usually controlled by one knob or dial. The surface burners are controlled by switches which provide five, seven, or, in some designs, an infinite number of different heats. The latter design works on the principle of varying the length of time the heating coil is on the circuit.

Practically all ovens are provided with automatic temperature control, usually of the hydraulic or expanding-liquid type, although some designs utilize a bimetal expansion element. Most ranges either are provided with means for starting and stopping oven operation at predetermined times or can be so equipped through the addition of a standard accessory.

Surface burners are principally of two designs, both of the enclosed type. One, the most universally used, consists of one or two spirally coiled tubular elements, welded or mechanically secured to a metal spider or frame and mounted in a metal reflector pan. The tubular element consists of a helical resistance coil surrounded by compacted refractory oxide insulating powder and encased in a heat-resisting metal tube. The other design consists of a spirally wound and partly flattened resistance coil embedded in a refractory insulating cement and encased in a flat, circular, heat-resisting metal trough.

The total capacity of an electric range may be from 7000 to 13,600 watts, depending on the size and number of burners, but the average is about 10,000 watts. Since all the burners are seldom, if ever, used simultaneously at full heat, however, the maximum demand is usually much less than this value.

19. CLOTHES DRYERS

A recent development in home appliances is the electric dryer for the home laundry. This device eliminates the necessity for clothes to be hung outside for drying and greatly decreases the waiting time between washing and ironing. The damp clothes are slowly tumbled in a metal basket, while warm air is circulated through them. Because of the requirement of a high-wattage electric heater in the dryer (approximately 4800 watts), a special electric circuit is needed.

COMMERCIAL COOKING

By J. L. Shroyer

20. COUNTER APPLIANCES AND INSTALLATION EQUIPMENT

Electricity became a recognized practical source of heat for commercial cooking during the period between World Wars I and II. Before the introduction of electricity, most cooking was done on the top surface or in the ovens of ranges. The ease with which electric heat can be directly applied in the cooking of various types of food led to the development of numerous specialized cooking devices. Designed for a particular cooking operation, they provide better cooking and permit more efficient use of electricity than a general-purpose range. Specialized cooking devices also permit sectionalizing the various cooking operations, thereby eliminating the confusion which results when several cooks use the same device.

The wide difference in types of food-service establishments, varying from the small lunch counter to large state and federal institutions, has resulted in the development of two distinct lines of equipment, commonly known as counter appliances and installation equipment. Counter appliances are used by lunchrooms and small food-service establishments and include such items as griddles, grills, toasters, waffle bakers, fry kettles, coffee makers, and hot plates. Installation equipment is used by larger establishments, such as restaurants, hotels, hospitals, and governmental institutions, and consists of such items as sectional roasting and baking ovens, ranges, cooking tops, fry kettles, broilers, stock kettles.

Different manufacturers vary widely both the size and the watts per square inch of the heated surfaces. Tables 1 and 2 show typical wattage ratings for various items:

Table 1. Counter Appliances

Item	Wattage
Griddle (18 × 24 in.)	4000
Grill (9 × 11 in.)	1650
Toaster (4 slice)	2450
Waffle baker (2 section)	2000
Fry kettle (basket 9 × 9 in.)	4000
Coffee maker (per unit)	660
Hot plate (8 in.)	1500

Table 2. Installation Equipment

Item	Wattage
Range (3 ft)	
Oven (22 × 26 in.)	6,000
Top plates, 3 (12 × 24 in.)	15,900
Sectional roasting ovens	
Compartment (28 × 37 × 12 in.)	5,000
Fry kettle	
Fat container (20 × 20 in.)	18,000
Broiler	
Grid (25 × 22 in.)	10,500
Stock kettle (20 gal.)	5,000

The energy consumption per meal may vary from approximately 25 to 1000 whr, depending on the type of food service and the extent to which steam is used for such items as coffee and tea urns, kettles, steamers, storage tables, and dishwashers. Some large establishments, such as state and federal institutions, may use steam for practically all cooking except baked products and even purchase part of their bakery goods. Other establishments catering to a high-class trade may use very little steam and cook most of their food to order. This kind of service necessitates keeping the equipment hot over long periods during which very little food is cooked. The energy consumption per meal in this case is relatively high.

21. ENERGY CONSUMPTION

Data in Table 3 may be used as a basis for estimating the watthour consumption per meal for institutional food-service establishments, such as state and federal institutions.

This table is based on the average use of steam, and allowances should be made for variations from the stipulated usage. "Main kitchen food service" implies bulk cooking, and "special food service" implies separate cooking of individual orders or small quantities of food.

Electric ovens, fry kettles, and stock kettles are very efficient in the use of electric energy. Range tops and broilers are relatively less efficient. Therefore the watthour consumption per meal is substantially affected by the relative amount of usage of these groups of equipment.

Close estimates of the required kilowatthour consumption for any type of food-service establishment can usually be obtained from cooking-equipment manufacturers if they are provided with a full knowledge of the food-service plan, the kind of equipment to be used, and the size of the project.

Table 4 provides data that may be used as a basis for comparing load characteristics for different types of food-service establishments.

Table 3. Typical Consumption Allowances for Electrically Heated Installation Equipment in Institutional Service

Assumptions on which allowances are based:
Hot-water requirements per day per person
Beverages: 1.56 lb, 150 deg fahr rise
Dishwashers: 8.1 lb, 125 deg fahr rise
Four lb of cooked food per person per day

	Average Watthour Per Meal			
	Main Kitchen Food Service		Special Food Service †	
	All Electric Equipment	Steam and Electric *	All Electric Equipment	Steam and Electric
Group 1 (electric or steam)				
Coffee and tea urns	31	0	31	0
Kettles and steamers	51	0	51	0
Dishwasher	125	0	125	0
Storage tables	55	55	55	55
Group 2 (electric)				
Roasting, frying, and boiling equipment	49	49	210	210
Toasting and miscellaneous equipment	6	6	28	28
Baking	31	31	50	50
Total	348	141	550	343

* Per cent of food by weight cooked with steam when steam is used for kettles and steamers

Cereals:	100 per cent
Vegetables:	90 "
Meat and fish:	20 "

† Vegetables, meats, and fish cooked by Group 2 equipment and in individual or small quantities.

Table 4. Characteristics of Four Representative Cooking Loads Selected to Bracket the Major Portion of the Food-service Industry

(Based on available test data)

Case designation	A	B	C	D
Type of establishment	Lunch counters	Small restaurants	Large restaurants	Institutions
Cooking equipment used	Counter	Counter and installation	Installation	Installation
Relative number, per cent	79.8	15.90	3.98	0.32
Avg. continued load for cooking, kilowatts	7.50	25.00	60.00	125.00
Avg. demand, kilowatts	2.83	10.00	27.00	56.00
Avg. incremental demand, kilowatts *	2.29	8.75	20.20	35.00
Consumption per year per kilowatt connected, kilowatthours	675	1200	1500	1500

Large cafeterias, etc., serving 3000 to 10,000 meals per day generally require about 0.04 kw continued of cooking and baking load for each meal served per day.

* The amount by which the lighting and power kilowatt demand is increased by the addition of the cooking load.

DOMESTIC WATER HEATING

By J. H. Reifenberg

22. PRINCIPLES

The most popular method of heating water for household use is by means of an insulated, resistance-type heating element immersed in the water contained in a storage tank. Some water heaters are built with resistance heaters clamped to the outside of the storage tank; very rarely induction-type heaters are used. Various sizes of heating elements and storage tanks are used, depending upon the amount of hot water required and the length of time the heater is permitted to be connected to the power circuit.

1 Btu will heat 1 lb of water 1 deg fahr.

1 gal of water weighs 8.253 lb at 120 deg fahr.

1 kwhr equals 3412 Btu.

2.42 watts will raise 1 gal of water 1 deg fahr in 1 hr.

$$\text{Size of heater in kilowatts} = \frac{\text{Gallons of water} \times \text{Degrees fahrenheit rise desired} \times 2.42}{1000 \times \text{Time, hours}}$$

1 kwhr will heat 4.1 gal of water 100 deg fahr at 100 per cent efficiency.

The following table gives the approximate quantities of hot water required for various household uses and the temperatures at which the water will be used.

Use	Quantity, gallons	Temperature, degrees fahrenheit
Bathtub	10–15	95–100
Shower	12–20	95–100
Dishwashing		
Hand	2– 5	115–145
Mechanical	3– 7	140–160
Washing face and hands	1/2– 1	95–105
Washing machine	36	125
Rinsing		
First	30	105
Second	30	95
Automatic washing machine (per cycle avg.)	28	150

Experience shows that in lower-priced homes the daily hot water used per person is from 7 to 10 gal; in medium-priced homes this figure increases to 10 to 15 gal, and in larger homes may run as high as 20 to 30 gal per person.

The wattage of the heating element to be installed depends upon the amount of water to be heated and the number of hours during which electricity is available for water heating. The various electric utility companies determine the changing hours for water heating in order to level off their system load curve and in this way are able to establish rates of 1 cent per kwhr or less. Some electric companies are able to furnish this low-cost current 24 hr per day, whereas other utilities having high peak loads at certain hours cut off the water heaters during these peaks. This latter method is known as controlled or off-peak service.

In some cases semi-off-peak service is furnished, the lower heating element of a two-element water heater being time controlled, and the top element allowed to be on the line at any time during the 24 hr. In case of abnormal hot-water demands the top element is available to supply normal requirements and prevent complete exhaustion of the hot water. Power for this top element is usually charged at a higher rate than that for the bottom or off-peak element.

In order to reduce the connected load of the water heater and thus lower the demand on the distribution system many utilities use limited demand wiring. In this case the bottom-heater circuit is wired through back contacts on a double-throw top thermostat, so that, when the top thermostat closes the circuit to the top heating element, it opens the circuit to the bottom element. Thus at no time are both heating elements on the line simultaneously.

23. STANDARDS

In cooperation with the electric utility companies the electric water-heater manufacturers have through their trade associations set up a number of standards for electric water heaters, in order to establish a uniform testing procedure to make possible the determination of the performance in terms of safety, effectiveness, durability, and convenience. This work is still in progress, but to date the following standards have been adopted:

1. Voltage rating: The voltage rating of electric water heaters shall be 230 to 240 volts. The design voltage for this range shall be 236 volts.

2. Pressure rating for tanks: The minimum standard test pressure rating shall be 300 lb. per sq in.

3. Element wattage rating: The standard wattage rating for 230 to 240-volt water-heater elements is as follows:

600	2000
750	2500
1000	3000
1250	4000
1500	5000

4. Standard wattages for single-element water heaters:

Tank Size, Gallons	Wattage Rating
30	1500
40	2000
52	2500
66	3000
80	3000

The above ratings are based on 50 watts of heating unit per gallon of tank volume.

5. Standard wattages for two-unit water heaters:

Tank Size, Gallons	Wattage	
	Upper Unit	Lower Unit
30	1000	600
40	1250	750
52	1500	1000
66	2000	1250
80	2500	1500
110	3000	2000
120	4000	2500
140	4000	3000

The foregoing ratings are based on 30 watts of heating unit per gallon of tank volume in the upper unit and 20 watts per gallon in the lower unit. All wattages are rounded off to the nearest standard element wattage.

The standards for element wattages are under discussion, and it is possible that they will be revised upward. This rise is indicated by the increasing use of hot water in the home, particularly with the introduction of automatic clothes washer and dishwashers.

6. Temperature setting of thermostat shall be at the 150 deg fahr point.

7. Temperature adjustment: The thermostats shall be adjustable and provide a low cut-off temperature not greater than 120 deg fahr and a high cut-off temperature not less than 170 deg fahr.

8. Color coding: The wiring shall be color coded as red, black, blue, and yellow. The wiring diagrams (Fig. 1) show the application of this color coding to three popular methods of wiring.

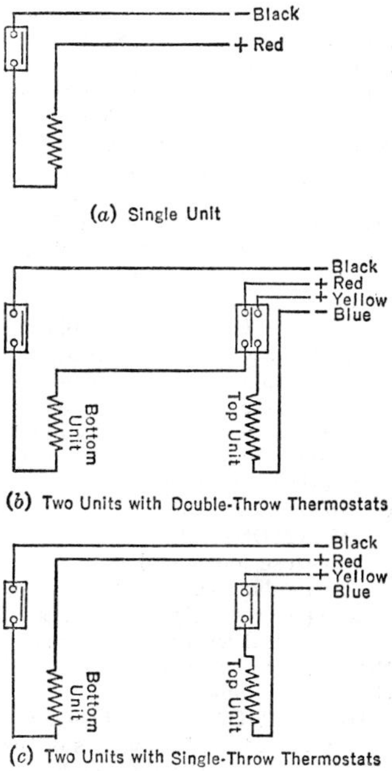

FIG. 1. Wiring Diagrams

24. DESIGN

The water heater consists of a storage tank completely surrounded by heat insulation, with the heating element mounted through the side wall of the tank and the assembly enclosed in a sheet-metal casing mounted on legs. Thermostats for controlling the water temperature are mounted on the side wall of the tank or installed in tubes extending into the tank.

STORAGE TANKS. Galvanized tanks are the type most widely used in electric water heaters. The shells are rolled and electric welded, and the die-formed heads inserted and electric welded to the shell. The tank is thoroughly cleaned and hot-dip galvanized. The gage of steel used in side wall and heads is determined by the test pressure specified. The galvanized tank is the most economical in first cost and is suitable in most localities. In the so-called soft-water areas, particularly New England, the South Atlantic States, and the Pacific Northwest, however, galvanized tanks have a very short life. In these areas copper, copper-alloy, Monel, or porcelain-enameled steel tanks are used. On copper tanks the seams are usually riveted and soldered, whereas copper-alloy and Monel tanks have welded seams. Copper and copper-alloy tanks are block tin lined to prevent the discoloration of some waters which react with the copper. On the porcelain-enamel-lined tanks the shell and heads are enameled separately and inspected for defects in the enamel. The three parts are then assembled by arc welding. Heavier-gage steel is used in these tanks, primarily to resist flexing of the steel under pressure, with the consequent danger of damage to the enamel.

HEATING ELEMENTS. Heating elements are of either the immersion or the strap-on type. Immersion heaters are of two kinds: (1) helical coils of nickel-chromium resistance wire surrounded by magnesium oxide, or a similar insulating material, within a copper tube or casing; and (2) helical coils of resistance wire mounted in or around blocks of porcelain or other similar refractory material. Immersion heaters of the first type are mounted either singly or in multiple in a head which is threaded with standard pipe thread for screwing into the tank or arranged for bolting to flanged adapters welded to the tank wall. Immersion heaters of the second type are installed in water-tight tubular wells welded to the tank. They may be inserted horizontally through the tank wall or vertically through the top or bottom head. When heating elements are mounted horizontally, the usual practice is to install one near the bottom of the tank and a second approximately one-fourth of the distance from the top of the tank. Each heating element is controlled by a thermostat responsive to the water temperature near the level of the heating element. Since the cold water is admitted at or near the bottom of the tank, the lower thermostat is the first affected when hot water is drawn. The top heating element is not called into service until approximately three-fourths of the hot water is withdrawn from the tank. When its thermostat closes, the top element heats only the water above its own level and, being of a higher wattage than the lower element, is on a relatively short time. Thus it acts as an emergency or stand-by element to provide smaller quantities of hot water quickly, leaving the larger part of the heating load to the lower element.

Strap-on or belt-type heating elements are designed for clamping around the outside of the tank and are of three general types: (1) a series of porcelain blocks with the resistance wire laid in grooves in the blocks; (2) tubular elements mounted in channels clamped to the tank wall; or (3) mica-insulated laminated elements enclosed in a metal sheath. It is customary to use two elements, one near the bottom and one near the top, as in the immersion type. The strap-on element has a large heat-dissipating area, which is claimed as an advantage, particularly in extremely hard-water areas. It is not necessary to drain the water from the tank to replace elements, as in the direct-immersion type. However, it is desirable to apply more insulation around the outside of the tank, particularly at the heating element, to prevent undue heat losses.

THERMOSTATS. Thermostats are designed to maintain the water temperature within close limits and are adjustable to provide water temperatures most suitable for the particular use. The heat-responsive means may be bimetal, either of the snap-acting disk type or a strip operating a toggle mechanism, gas-filled or liquid-filled bellows, or rods of metal with wide difference in coefficient of expansion. Depending upon design, the thermostats are either mounted on the wall of the tank or installed in tubes inserted through the tank wall.

INSULATION. The efficiency of operation of an electric water heater depends primarily upon the effectiveness of the heat insulation in reducing the radiation and convection losses from the tank and its fittings. Fiberglas, mineral wool, balsam wool, Palco bark, and granulated cork are the heat insulations generally used. The thickness of insulation varies from 2 1/2 to 5 in. For equal thickness the insulating values of these materials differ only slightly. The insulation is retained in an outside casing, which usually is made of sheet steel with japanned or synthetic enameled finish.

25. TESTING

Electric water heaters are tested for (1) rate of heating, (2) stand-by loss, and (3) draw-off. The water temperatures are measured by thermocouples and a potentiometer. The thermocouples are attached to a copper tube closed at one end, with the other end open but soldered into a drilled pipe plug, which is inserted in the top of the tank. The thermocouples are located so that one is at the center of each of a given number of equal volumes from top to bottom of the tank. Additional thermocouples or thermometers are placed at the inlet and outlet of the tank to measure temperatures of water drawn into and from the tank. A calibrated watthour meter measures the power consumption, and a water meter determines the water draw-off.

RATE OF HEATING. In this test the tank is filled with water at 50 deg Fahr (±2 deg fahr), and the heating element energized to deliver rated wattage. At the first cut-off of the thermostat the kilowatthour input is recorded. The efficiency of heating is determined by the following formula:

$$\text{Per cent efficiency} = \frac{C \times (T_2 - T_1) \times 0.00242 \times 100}{\text{Kilowatthours consumed by heater}} \qquad (1)$$

where C = actual capacity of tank in gallons.
T_1 = average water temperature at start.
T_2 = average water temperature at end.

STAND-BY LOSS. Stand-by loss is determined by permitting the water heater to operate under the control of its thermostat without any water being withdrawn, and measuring the wattage consumed. The water heater is allowed to operate approximately 24 hr to reach a steady condition, and the test started at the next thermostat cut-off. Water temperatures and watthour readings are taken, and the heater is allowed to operate for at least 48 hr. Water temperatures are taken at each cut-on and cut-off of the thermostat during the test, and the watthour reading is taken at the end of the test. Room-temperature measurements are taken at intervals frequent enough to assure an average reading. The stand-by loss in watts, corrected to an 80 deg fahr rise, is determined as follows:

$$\text{Watts loss} = \frac{2.42C(T_1 - T_2) + Wh}{N} \times \frac{80}{(T_t - T_r)} \qquad (2)$$

where C = actual capacity of tank in gallons.
T_1 = average water temperature at start, degrees fahrenheit.
T_2 = average water temperature at end, degrees fahrenheit.
Wh = measured watthour consumption.
N = number of hours of test.
T_r = average room temperature.
T_t = average water temperature throughout test.

DRAW-OFF TESTS. Draw-off tests are used as a basis for comparison and for determining the ability of the water heater to furnish an adequate hot-water service. Because of variation in customer habits and utility system practices and conditions it is impractical to establish standard tests to meet all conditions. The three schedules shown in Table 1 have been selected as covering an adequate range of conditions to serve as bases for comparison and to indicate probable delivery performance under other schedules.

Table 1. Draw-off Test Schedules

Indicated Time, A.M.	Total Withdrawal, Per Cent	Schedules for Withdrawals, Gallons per Day		
6:30	11	6	8	9
7:30	5	3	4	4
8:00	17	9	13	15
9:00	14	8	10	12
10:00	13	7	10	11
11:00	4	2	3	3
12:00	4	2	3	4
1:00	7	4	5	6
5:00	3	2	2	2
6:00	4	2	3	3
7:30	5	3	4	5
10:00	13	7	10	11
Totals	100	55	75	85

Schedules are based on testing a 50-gal water heater; for other capacities the schedules are adjusted proportionately. The selected conditions for the three schedules are:

55-gal day: Controlled 10-hr heating with a 1000-watt bottom element and a 1500-watt top element. The bottom-element circuit is controlled by a time switch, which turns it off between 7:00 A.M. and 9:00 P.M., whereas the top thermostat is free to operate at any time. The wiring is for limited demand.

75-gal day: Uncontrolled 24-hr heating with a single 1500-watt bottom element.

85-gal day: Controlled 18-hr heating with a 1000-watt bottom element and 1500-watt top element connected for limited demand. A time switch cuts the bottom element off at 10:30 A.M. and on at 1:30 P.M., then off at 4:30 P.M. and on at 7:30 P.M., whereas the top element is free to operate at any time.

In conducting the draw-off tests the water heater is preheated and idling under control of its thermostat. The first withdrawal is made within a half hour after the thermostat cut-out. Records are made of the following: (1) average temperature of water in tank at beginning of test, (2) average temperature of water in tank at end of test, (3) temperature of each gallon of hot water withdrawn, (4) temperature of each gallon of cold water, (5) average room temperature, (6) kilowatthour input, (7) average temperature of water in tank for the 24-hr period.

The data are corrected to standard conditions of 70 deg fahr room temperature, 50 deg fahr cold water, and 150 deg fahr hot water, and the following calculations made:

$$\text{Tank losses} = \text{Input} - \text{Output}$$

$$\text{Efficiency} = \frac{\text{Output}}{\text{Input}} \tag{3}$$

$$\text{Gallons of water of 100 deg fahr rise per kilowatthour} = \frac{\text{Gallons drawn}}{\text{Input}} \tag{4}$$

$$\text{Daily load factor} = \frac{\text{Input}}{24 \times \text{Connected load}} \tag{5}$$

In eq. (3), the input is the corrected observed kilowatthours, and the output is measured in gallons of water delivered and the temperature of the water, corrected to standard conditions and converted to kilowatthours by the factor given in Article 22. In eq. (5), the connected load is the wattage rating of the heating element. Where two element heaters are connected for limited demand, the wattage rating of the top element only is used.

26. INSTALLATION

Electric water heaters should be installed as close as possible to the point of most frequent use of hot water in order to reduce pipe losses. After each withdrawal from the water heater, hot water is left in the pipe to cool, resulting in heat loss. Covering the pipes with heat insulation will minimize these losses. Heat losses from bare pipe are shown in Table 2. The values therein are based on the assumption that the water in the pipes is at 150 deg fahr and the surrounding air at 70 deg fahr.

Table 2. Heat Loss from Bare Pipe

Nominal Pipe Size, Inches	Heat Loss	Btu Lost per Foot Length per Hour	
		Galvanized Steel Pipe	Copper Tubing
3/8	Convection	15.81	12.41
	Radiation	9.60	5.78
	Total	25.41	18.19
1/2	Convection	18.83	14.85
	Radiation	11.93	7.22
	Total	30.76	22.07
3/4	Convection	22.55	19.40
	Radiation	14.94	10.10
	Total	37.49	29.50
1	Convection	27.00	23.82
	Radiation	18.71	13.01
	Total	45.71	36.83

Return circulation piping systems should be avoided wherever possible. These systems assure hot water at the faucet at all times without the need of drawing the cold water from the line, but they do so by providing a constant circulation of hot water throughout the system. This results in a great amount of heat being lost by radiation and convection from the pipes.

27. HEAT-PUMP SYSTEM

HEAT-PUMP WATER HEATER. A self-contained heat-pump unit, designed especially for heating water and similar to an arrangement shown in Fig. 2, is feasible and

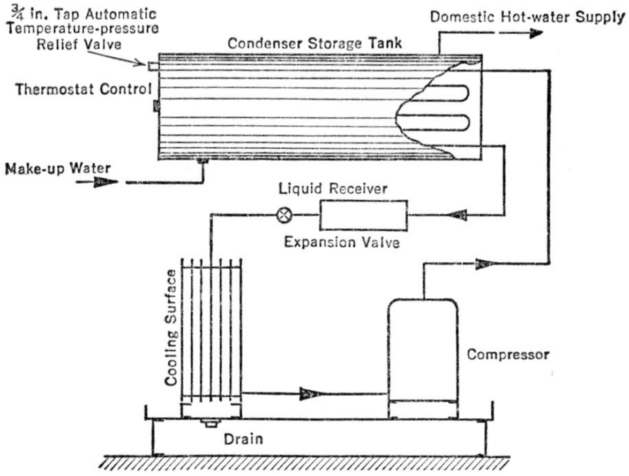

FIG. 2 Diagram of Heat-pump Water Heater

practical. With delivered hot-water temperatures of 130 to 150 deg fahr, average coefficients of performance of 2 $^1/_2$ to 4 are possible. The heat-pump water heater maintains the advantage of resistance electric heaters and reduces to a considerable degree the disadvantages of high operating cost when compared with other fuels.

The heat source for a heat-pump water heater can be ambient air or any other source which has been mentioned in Article 12 as suitable for the year-round heat-pump unit.

The equipment for a heat-pump water heater consists of a refrigerating compressor, cooling surface, expansion valve, condenser-storage tank, and connecting piping.

28. BIBLIOGRAPHY

American Society of Heating and Ventilating Engineers, *Guide* (1941).
National Electrical Manufacturers Association, *Pub. No.* 103 and *Pub. No.* 45-104.
Heating, Piping, and Air Conditioning (Nov., 1936).
Elec. World (Sept. 11, 1937, Oct. 22, 1938, and April 8, 1939).
Elec. Light and Power (July, 1938, and Oct., 1938).

SECTION 16

INDUSTRIAL APPLICATIONS OF MOTORS AND SERVOMECHANISMS

INDUSTRIAL APPLICATIONS OF MOTORS AND SERVOMECHANISMS

INDUSTRIAL APPLICATIONS OF MOTORS

By Francis A. Westbrook

1. ADVANTAGES OF ELECTRIC DRIVE

In comparison to old-fashioned line shafting, the more important advantages of motorization are:

1. Flexibility in placing machines and the use of portable tools.
2. Elimination of overhead belting, providing better illumination and ventilation and making space for overhead handling equipment.
3. Confinement of breakdowns, in most cases, to a single machine or group.
4. Use of recording or indicating meters to determine the performance of any machine, making it possible to take whatever maintenance steps are indicated to secure efficient operation. Accurate records of power consumption for all operations are thus easily obtained.

2. GROUP VERSUS INDIVIDUAL DRIVE

There are two general systems of drive, group and individual. When line shafting is to be changed to motor drive, a simple way of making the change is to split the shafting up into such sections as would be most convenient and to drive each by a comparatively large motor. On the other hand, it frequently happens that it is necessary to operate only one machine of the group for a considerable time, as in overtime work, and to do this it would be necessary to keep the motor and the line shafting of the whole group running. Since the efficiency of the motor at this light load would be small, and the friction losses of the entire drive would have to be supplied, it is evident that such operation would be most inefficient.

If, on the other hand, a large number of similar machines which run continuously, if at all, as with much textile machinery, are involved, they are often driven in comparatively small groups. If the plant is not on full production, certain groups are shut down, whereas others are operated without any individual being idle. In this way the investment in motors is reduced, but the advantages of simplified shafting, and protection against complete paralysis of operations due to the breakdown of one motor, are realized. Modern installations, therefore, use both group and individual drive.

INFLUENCE OF NATURE OF LOAD. It is generally agreed that all large tools or other machinery should have individual motors, especially if their service is intermittent. The present tendency is certainly in that direction. With group drives there are two distinct loads, the friction of the line shafting and belting and the variable load of the machines.

In many process machines it is necessary to perform multiple operations, each requiring variable-speed control in addition to the basic variable-speed control of the whole machine. For such conditions it is customary to use a fixed-speed motor and a mechanical or hydraulic variable-speed transmission, or an adjustable-speed motor with suitable power-transmission media to the various operations, which in turn are equipped with variable-speed transmissions. Thus speed control of the process as a whole is provided for, and at the same time it is easily possible to adjust the speed relations of the different operations as required by conditions which are not constant. For example, a process machine handles various kinds of cloth, involving unwinding from a roll, washing, chemical treatment, drying, and rewinding. Each step may require individual speed adjustment to control stretch or shrinkage and tension on the unwinding and winding operations, and the whole machine requires variable-speed control to assure proper processing of different kinds of cloth. Such control may be exercised by means of mechanical variable units, such as those made by the Reeves Pulley Company and the Link-Belt Company, for example, or by electrical equipment, such as the Westinghouse Rototrol.

Occasionally where connecting media are difficult to install or maintain, and fixed process speeds are desired, synchronous motors and variable-speed transmissions are used at the various operating points. Where variable process speeds are required, an a-c motor-generator set may be used to produce variable frequency for the synchronous motors. The latter are suitable only when the machine starts and stops without processing material.

Specific recommendations for the driving equipment and the type of variable-speed unit or units employed must be determined by the process and its requirements, by the limitations of the transmission media to perform within the predetermined specifications, and the cost of the process. Almost any set of driving requirements may be provided if the cost can be justified on the basis of improved quality of product, increased production, reduced losses, etc.

INFLUENCE OF SPEED. Wide ranges of control and variations of speed are often sufficient reason for individual drive. With group drive the methods of speed control for the individual machines are slower when accomplished by shifting belts on cone pulleys or changing gears. The increasing use of variable-speed pulleys and mechanical variable-speed power-transmission units for individual machines, however, makes group drives less limited in applicability.

INFLUENCE OF RELATIVE COSTS. Whether group or individual drive is to be installed is an economic question, and each case must be properly analyzed. Individual drive necessarily means a larger investment (forty 1-hp motors cost over five times as much as one 40-hp motor), but in nearly all cases a much greater percentage income will be realized than with line-shaft drive. With individual drive the size of the motor is set by the maximum requirements of the machine, whereas with group drive its size is determined by the average load requirement of the group, which generally is roughly one-third less.

The smaller group-drive motors mean a lower connected load and lower installation and operating costs than motors for individual drives, but the maximum power demand of the plant should be about the same in either case. If power is purchased, the price should be based on the actual maximum power demand and not, as sometimes is required, on the total connected horsepower of the motors.

POWER TRANSMISSION COUNCIL. In the introduction to one of its booklets the Power Transmission Council, 41 Park Row, New York 7, N. Y., states that it is "Founded on the accepted premise that no one type of drive can economically meet the varied operating conditions and production requirements of industry. . . . An extensive survey of American industrial plants shows that nearly a billion dollars' worth of power is being wasted annually as a result of inefficient and unsuitable power transmission methods. . . . Inquiries from manufacturers concerning power transmission problems are welcomed. . . . Consultation with experienced and qualified power transmission engineers involves neither charge nor obligation and cannot fail to suggest improvements which will result in power savings and increased production." Studies on the relative merits of group and individual drives are among the problems to which detailed consideration has been given. Industrial plant engineers are more than likely to receive from this authority very competent assistance with almost any phase of their power transmission problems.

3. SELECTION OF MOTORS

GENERAL CONSIDERATIONS. The installation of a motor having too large a capacity should be avoided, unless an increase in the load is expected in the near future, because the efficiency of a motor is usually a maximum at its normal rated output. With a-c motors the effect on the power factor must be considered. The power factor decreases rapidly if the motors are operated below normal load, and because of the low power-factor penalty exacted by many public utilities, and the bad effect of low power factor on the regulation of the system, it should be kept as high as practicable. Ordinarily it is possible to group the machinery so that the motors may be operated near their rated output at all times. Too small a motor is naturally also very undesirable, as it will in all probability be subject to overloads, which may result in overheating and a burn-out of the motor, causing a shutdown of the machinery. Furthermore, the operating conditions of the plant may be such, as in steel mills, that the failure of a single motor may necessitate shutting down the entire mill.

USE OF STANDARD MOTORS. The American Institute of Electrical Engineers has devised and promulgated standards on which the ratings of motors used by the most competent manufacturers are based. These ratings are the best to use because of the standardized basis of comparison. In addition, many of the better motor manufacturers have changed their motor frames to conform with standardized motor-mounting dimensions adopted under the guidance of the National Electrical Manufacturers Association (NEMA). The interchangeability thus realized is a great advantage because, among other benefits, the user is not committed to one make of motor. The Standards of the NEMA also cover ratings, performance, and some dimensional features.

TORQUE AND SPEED. Some machines require motors with very heavy starting torque, although they run under light load when up to speed; for others the requirements may be just the opposite. With a varying-speed motor the torque-speed characteristics should agree as nearly as possible with the load. For example, to start and accelerate the bridge of a crane requires a motor capable of developing a high starting torque, but after the bridge is accelerated much less power is needed to keep it in motion.

The maximum torque must also be given consideration. A motor driving a heavy punch may, in spite of the flywheel, develop insufficient torque to keep up the speed. As a rule, however, where the motor is large enough for starting and normal operation, but not large enough for the maximum overload required for a second or two, a suitable flywheel will cut down the maximum torque required.

AVERAGE LOAD. The average load usually should be the determining factor, although under many conditions the starting load controls the size of the motor. In some instances the insertion of a clutch, which can be slipped in gradually when starting a machine, permits using a motor of smaller size than otherwise is possible. The frequency of starting, the duration of the starting period, and unusual starting current conditions are important in selecting the right motor. The amount of power required to drive a given machine should be known as accurately as possible. See Section 14, Article 86, for current and voltage calculations.

HIGH-SPEED MOTORS. In general, high-speed motors cost less and are more efficient from the operating standpoint than low-speed motors. It is, therefore, best to employ motors having a high normal speed unless expensive mechanical transmission equipment will be required where slow-moving machines are to be driven. Close speed regulation is required for machine tools and textile machinery and under other conditions requiring substantially constant speed. On the other hand, where there are peak loads likely to damage the motor if it does not slow down automatically, or where the machine speed is partially governed by a flywheel, as with punch presses, wide speed regulation is necessary. For still other kinds of work calling for variable speed control by an operator it may be necessary to use wound-rotor motors or d-c motors unless mechanical variable speed transmission units are applicable. In other words, the nature of the drive must be fully considered in selecting a motor for any particular service.

GENERAL-PURPOSE MOTORS. Motors not designed for some special application are usually referred to as general-purpose motors. They are constant-speed motors but are designed to carry reasonably varying loads above or below normal. These motors are made by most manufacturers in sizes up to 200 hp and with speeds of 450 rpm and higher. They meet most requirements, and their use is consequently so wide that the National Electrical Manufacturers Association has adopted a **service factor** for all general-purpose motors, which is applied to the normal horsepower rating and gives the overload at which a motor may be run with safety. The service factor, which appears on the nameplates of new motors, reads: "Service factor 1.15 at rated volts and cycles." In other words, if the rating of a motor is 50 hp, the permissible loading will be $1.15 \times 50 = 57.5$ hp, provided the line voltage and frequency do not change. This means that, if the load requirement is 55 hp, a 50-hp motor will be large enough.

INFLUENCE OF ENVIRONMENT. The environment in which the motor has to function should be carefully considered. Many manufacturers make an effort to design motors for operation under almost every imaginable kind of working conditions, such as heat, cold, moisture, dust, or inflammable or corrosive gases. The purchaser should get the benefit of this by consulting with reliable makers, since their advice costs nothing. The right motor may cost a little more, but it is poor economy to save at the sacrifice of quality, suitability, and production.

A-C VERSUS D-C SYSTEMS. As a rule the a-c system should be selected if possible, even for the isolated industrial plant, as it permits of throwing over to a central-station service if it is found that power can be more economically purchased than generated on the premises, or if emergency central-station current is needed.

4. DIRECT-CURRENT MOTORS

Except in parts of cities having only d-c distribution or in small plants where such a service predominates, there is no reason for installing d-c motors unless an adjustable speed service is required. If direct current is required, it can be obtained with a motor-generator set.

Table 1 lists the types and some typical applications of d-c motors.

Table 1. Types of Direct-current Motors for Various Kinds of Loads

Functional Character of Load	Applications	Type	Usual Horsepower Range	Usual Rpm Range, Full Load	Speeds			Starting Torque	Methods of Mounting						Protective Enclosures				
									Sleeve or Ball Bearing		Ball Bearing Only								
									Horizontal		Horizontal		Vertical						
					Constant Speed	Multispeed	Adjustable Speed		Rigid Base	Cushion Base	Face Type, Bracket Mounting	Flange Type, Bracket Mounting	Face Type, Bracket Mounting	Flange Type, Bracket Mounting	Open Type Protected	Splashproof	Totally Enclosed Nonventilated	Totally Enclosed Fan-cooled	Explosionproof
Constant-speed characteristics from no load to full load or adjustable speed and all but extreme starting torque	Blowers—Fans, machine tools, centrifugal pumps	Shunt	1/6 to 300	690 to 1750	Yes	No	Yes	Limited only by commutation	Yes	Up to 1 1/2 hp	Yes	Yes	Yes	Yes	Yes	Yes	Yes	No	No
The full load must be started from rest and smoothly accelerated to full-load speed. Has varying speed characteristics from no load to full load.	Refrigeration compressors, reciprocating pumps, air compressors, shears, draw presses, stokers	Compound	"	"	"	"	"	"	"	"	"	"	"	"	"	"	"	"	"
Widely varying speed characteristics from no load to full load, very high starting torque or periodic torque, and motor cannot be operated without load.	Cranes, hoists, traction	Series	"	"	No	"	"	"	"	"	"	"	"	"	"	"	"	"	"

Table 2. Types of Induction Motors for Various Kinds of Loads

Functional Character of Load to Be Driven	Applications	Type	Horsepower Range	Rpm Range, Full Load	Electrical Forms Available					Mechanical Forms Available										
					Speeds			Starting Torque	Starting Current	Methods of Mounting						Protective Enclosures				
										Sleeve or Ball Bearing, Horizontal		Ball Bearing Only, Horizontal		Ball Bearing Only, Vertical						
					Constant Speed	Multispeeds	Adjustable Speed			Rigid Base	Cushion Base	Face Type, Bracket Mounting	Flange Type, Bracket Mounting	Face Type, Bracket Mounting	Flange Type, Bracket Mounting	Open Type Protected	Splashproof	Totally Enclosed Non-ventilated	Totally Enclosed Fan-cooled	Explosionproof
Polyphase Motors																				
For Moderately Easy-to-start Loads For moderately easy-to-start loads or where the load is applied as the motor approaches full speed. This motor has general purpose characteristics and will cover about 90% of industrial applications.	Blowers—fans, centrifugal pumps, machine tools, drill presses—grinders, milling machines	Squirrel Cage Induction	1/6 to 600	290 to 3500	Yes	Yes	No	Normal	Normal	Yes	1/6 through 3 hp only	Yes	Yes	Yes	Yes	Yes	Up to 100 hp, yes	Yes	From 2 to 15 hp	From 2 to 15 hp
For Hard-to-start Loads Where the full load must be started from rest and smoothly accelerated to full-load speed.	Refrigeration compressors, reciprocating pumps, air compressors		5 to 600	830 to 1750				High	Low											
For Hard-to-start Intermittent Loads and Shock Loads Two types of high-torque, high-slip motors—one for shock loads, like hoists, cranes, elevators or similar intermittent duty. The second for fluctuating loads susceptible of absorption by inertia of rotating parts, like shears, brakes, draw presses.	Hoists, elevators, cranes, shears, brakes, deep well pumps, draw presses, oil-well pumping		1/2 to 75	500 to 1500				High torque, high slip	Low											
For Applications Where Load-starting Current Is Paramount and Load Easy to Start A saving in otherwise expensive control is also often possible.	Larger sizes centrifugal pumps, unloaded compressors		40 to 600	1160 to 1750				Low	Extremely low											

For Hard-to-start Loads with Low Starting Current Where extremely low starting current is required in starting heavy loads and bringing them up to speed smoothly. Also applications requiring adjustable varying speed.	Blowers—fans, hoists, conveyors, gates, bridges, elevators, cranes	Slip Ring	1 to 350	570 to 1750	Yes	No	Yes Varying	Range of torques	Range of starting currents	Yes	No	Yes	Yes	Yes	Yes	Yes	Yes	Yes	No	No
Single-phase Motors																				
For Easy-to-start Loads Where the load builds up gradually with speed or where load is applied after full speed is attained.	Small drill presses, small tools, light fans—blowers	Split Phase	1/20 to 1/4	850 to 3500	Yes	Yes	No	Low	High	Yes	Yes	Yes	Yes	Yes	Yes	Yes	Yes	Yes	No	No
For Hard-to-start Loads Where the full load must be started from rest and smoothly accelerated to full-load speed.	Refrigeration compressors, air compressors, reciprocating pumps, stokers	Capacitor	1/6 to 20	850 to 3500				High	Normal	Yes	Yes	Yes	Yes	Yes	Yes	Yes	Yes	Yes	1 1/2 hp and larger	1 1/2 hp and larger
For Moderately Easy-to-start Loads For moderately easy-to-start loads or where the load is applied as the motor approaches full-load speed. This motor has general-purpose characteristics. Use Type CSH or RS for hard-to-start loads.	Blowers—fans, machine tools		1 to 20	850 to 3500				Normal	Low											
For Hard-to-start Loads This motor is for the same type of loads as the CSH capacitor motor but has extremely low starting current, which is very important where line voltage regulation is poor, as in farm application.	Stokers, refrigeration compressors, air compressors, reciprocating pumps	Repulsion Start Induction	1/2 to 15	850 to 1750		No		High	Extremely low	Yes	Yes	Yes	Yes	Yes	Yes	Yes	Yes	Yes	No	No

5. INDUCTION MOTORS

The induction motor is the least expensive, the most simple and rugged, and the most economical to maintain.

If varying speed is required only intermittently, the phase-wound induction motor may be used to advantage; but if the motor must run at reduced speed a considerable portion of the time, the varying-speed, brush-shifting motor should be considered. As a rule, a considerable part of industrial machinery requires a constant-speed service, for which the a-c motor is admirably adapted. The constant-speed induction motor is adaptable to such a wide range of applications that this type suits the great majority of uses.

Table 2, from data prepared by the Century Electric Company, gives a good indication of the types of induction motors available, their electrical characteristics, and some typical applications. Comparison with the operating characteristics of loads not listed will help in choosing the type of motor which should be used. (See Section 9.)

6. SYNCHRONOUS MOTORS

Synchronous motors have two outstanding advantages from the standpoint of the industrial plant engineer. First, the efficiency is higher than that of induction motors of the same rating, especially with large, low-speed motors. Second, in addition to driving machinery more economically, in many instances synchronous motors serve to correct the plant power factor. They are available in horizontal- and vertical-shaft types and in ratings up to several thousand horsepower. High- and low-speed types are available. (See Section 9.)

HIGH-SPEED TYPE. Power-factor correction is the usual reason for installing high-speed synchronous motors, the driven machines generally being started without load. The principal other advantages are high efficiency, constant speed irrespective of load conditions, and starting torque comparable with that of squirrel-cage motors. Unless power-factor correction is the controlling consideration, however, the high-speed induction motor is usually more economical, especially if it operates at close to full load, as its power factor and efficiency are then high.

If the advantages of high-speed induction motors are to be retained for low-speed drives, it is necessary to install speed-reducing equipment of some kind, such as belts, V-belts, gears, gear speed-reducing units, or chains. All these increase the first cost of a given drive, its maintenance costs, and its power-transmission losses and usually take up more floor space.

LOW-SPEED TYPE. This type has greatly improved starting and operating characteristics and greater simplicity and reliability. Its control is also more simple and flexible.

The costs of induction and synchronous motors become about equal at 250-hp 600-rpm ratings, and as the horsepower increases and the speed decreases, the costs favor the synchronous motors.

Low-speed synchronous motors, with their good starting characteristics, may be direct-connected to low-speed machinery with the following advantages: power-factor correction, higher efficiency, constant-speed irrespective of load conditions, suitable torque available for the driven load, automatic starting with simple, inexpensive, and reliable across-the-line type of full-voltage starters by push-button control, direct connection to the driven machine, minimum floor space, cost comparable to that of high-speed induction motors, and lower maintenance costs.

Low-speed, synchronous motors of 450 rpm and less are normally furnished in the engine type, which permits mounting directly on the shaft extension of the driven machine, thereby eliminating much driving equipment, such as the motor shaft, bearings, base, and speed-reducing units. Two-bearing pedestal-type motors are also available for coupling to the load or for using gears, chains, or belts. The advantages of the low-speed type have resulted in its very satisfactory application in steel, cotton, rubber, cement, paper and flour mills, refineries, hotels, office buildings, drainage, pumping, sewage disposal, irrigation, etc., where constant-speed operation is called for. These motors are also used for direct connection to reciprocating compressors and in other applications where the load involves large power pulsations. Under these conditions it is desirable to add flywheel effect to the motor in order to limit these pulsations. The low-speed synchronous motor is also very adaptable to varying demands of starting and running torque, making it applicable to a diversity of types of drives.

From Table 3, which gives the types of synchronous motors, their speeds and horsepower ratings, and some typical applications, it is possible to get a reasonably good idea of the requirements of other applications.

Table 3. Synchronous Motor Applications
(Courtesy, Westinghouse Electric Corporation)

Typical Loads		Motor Types	Speed, in Rpm, and Hp Ratings	Construction
Coupled	Fans and blowers Compressors Pumps	Bracket-type bearings. Includes general-purpose motors. Have starting torque comparable to squirrel-cage induction motors and are generally used for the same applications. (Type G).	1800 and 1200—up to 800 900 and 720—up to 600 600 and 514—up to 500 450 and 360—up to 400	Cast frame Brackets with sealed-sleeve bearings Laminated or fabricated rotor Poles dovetailed or bolted to spider
Belted	Fans Line shafts Grinding and crushing machinery Beaters (pulp)			
Geared	Line shafts Grinding and crushing machinery			
Coupled	Fans, blowers, and pumps Grinding and crushing machinery Jordans (pulp)	Same as above.	1800 and 1200— 400 and above 900— 800 and above 720 and 600—1500 and above 514—1000 and above 450—above 2500 400—above 4000	Fabricated frame Pedestal bearings—self-aligning Laminated rotor Poles dovetailed to spider With or without fabricated bedplate
Belted	Fans and compressors Grinding and crushing machinery Beaters (pulp) Line shafts			
Geared	Grinding and crushing machinery Tube and ball mills Rubber mill line Plasticators and banbury mixers Steel and metal-rolling mills Line shafts			
Coupled	Jordans (pulp) Chippers and wood hogs Fans and compressors Pumps Plasticators and banbury mixers Tube and ball mills Steel and metal-rolling mills Line shafts	Built for wide range of applications, usually with fabricated bedplates but sometimes furnished without bedplates so as to be mounted on the driven machine base, especially with centrifugal pumps. Favored for coupled or geared drives. (Type HG).	720—500 to 1500 600 to 514—400 to 900 450—up to 2500 400 to 360—up to 4000 327 and below—all ratings	Fabricated frame Fabricated rotor Pedestal bearing Poles bolted to spider With or without fabricated bedplate
Direct-connected	Compressors (air, gas, and refrigeration) Vacuum pumps Centrifugal pumps Chippers Direct-current generators	For use with reciprocating compressors, etc., where flywheel effect is necessary. Design is adaptable to changing normal WR- where necessary to be furnished by the motor rotor. Low and medium speeds. (Type HR).	450 and below—40 and up	Fabricated frame Fabricated rotor Split rotor hub or completely split spider Variable flywheel effect
Coupled	Line shafts (flour mills) Tube and ball mills (cement and mining) Rock and ore crushers Cold-roll steel mills	High-torque applications with limited maximum kva. Some control of motor torques is possible. Similar to wound-rotor induction motor in starting (Type HS, Simplex). Where the motor rating is small compared to the total connected load, high-torque motors with somewhat higher starting kva than the above may be used (Type HM).	Type HS Simplex 450 to 150—300 and above Type HM 720 to 150—200 and above	Fabricated frame Fabricated rotor Pedestal bearings Special damper winding With bedplate

7. MECHANICAL FORMS OF MOTORS AND THEIR APPLICATIONS

Motors of practically all the different electrical forms and characteristics, except synchronous motors, suitable for the various kinds of loads encountered in practical applications are also available with a variety of protective enclosures. These enclosures have been designed with a view to simplifying maintenance, reducing out-of-service interruptions to a minimum, and increasing safety. A brief explanation will indicate their significance. All of these types have been standardized by the National Electrical Manufacturers Association (NEMA), and their technical definitions are given in Section 9.

The following discussion relates particularly to a-c motors, but the basic functional constructions are also available in d-c motors.

STANDARD OPEN-TYPE MOTORS. These are for service in dry locations where no corrosive or abrasive agents are present to attack the working parts, and where the atmosphere is not unduly laden with dust, lint, etc., which will clog the air spaces. The absence of excessive oil, which tends to damage insulation, is also assumed. This form is suitable for probably 90 per cent of all industrial motor applications.

PROTECTED (OPEN-TYPE) MOTORS. These motors are designed for protection against dripping liquids, metal chips, and other falling materials. The end-plates act as shields, having only one opening, which is below the center line and protected on the top and sides, so that falling material cannot enter the motor. A ventilating fan is provided.

SPLASHPROOF MOTORS (sometimes referred to as weatherproof motors). This type is essentially similar to the protected type, except that the end-plates are so designed that, with the frame, they totally enclose the motor except on the bottom, which has baffled openings to provide ventilation. The baffles prevent the entrance of liquids into the motor. This type is for use in dairies, food-processing plants of many kinds, breweries, paper mills, etc., where the splashing of liquids occurs, as when ceilings, walls, and floors are washed down with a hose.

TOTALLY ENCLOSED FAN-COOLED MOTORS. These are used where dust, fumes, oil, coolant spray, moisture, and other damaging materials are present in such quantities as to necessitate frequent rewinding and other repairs to open-type motors.

EXPLOSIONPROOF MOTORS. This type is designed for situations where the atmosphere contains explosive elements due to the presence of natural gases, lacquer solvent vapors, acetone, petroleum, etc., and explosive dust. The vital parts of the motor are completely enclosed to prevent a spark from igniting the explosively laden atmosphere. Cooling air is passed through the space between an inner and outer shell by means of a fan.

TOTALLY ENCLOSED FAN-COOLED EXPLOSION-RESISTING MOTORS. The enclosures of these motors are built to withstand possible internal explosions within the motor in especially hazardous locations. The seals around the shaft are designed to prevent the escape of flame which might set off an explosion in the surrounding atmosphere. This type is available in a limited number of ratings and may be used outdoors unless snow is liable to block the ventilating openings. For corrosive atmospheres both indoors and outdoors, the exposed parts may be specially treated or plated or be made of cast iron or alloy. Bearings are carefully sealed.

ENCLOSED MOTORS—NON-VENTILATED. This type is for use where there is so much dust in the atmosphere or surroundings or where the dust is of such a nature that it will close the air passages of ventilated motors.

CHEMICAL PLANT MOTORS. These are of the totally enclosed, fan-cooled type, with special emphasis on the impregnation and insulation of the windings, moistureproof and dustproof conduit box, and special drain plugs in the end-plates, conduit box, etc. They are designed for environments which are likely to cause excessively rapid deterioration of motors without such protection.

TEXTILE MOTORS. These motors are designed primarily for the individual drive of textile machinery, such as roving frames, pickers, winders, and spinning and twisting frames, which produce lint and dust. They may, of course, be used for other machines when conditions of service are more or less similar. These motors have streamlined, highly polished stator windings and smooth internal surfaces, which offer no foothold for accumulations of lint and dust. They have skeleton frame construction with free air paths and are easily accessible for ordinary cleaning and inspection. They are constant-speed, squirrel-cage motors.

SLEEVE AND BALL-BEARING MOTORS. Practically all types may be had with either sleeve or ball bearings. The advantage of ball bearings is that, although the first cost is higher, maintenance costs are lower, because the motors require much less frequent lubrication, they are constructed so that dirt is far less likely to get in from outside, and there is practically no leakage of lubricant. Friction losses are also greatly reduced, and the problem of overheated bearings, with resultant shutdowns and fire hazard, seldom occurs with ball bearings.

GEAR MOTORS. These are available for a-c and d-c operation and in the various mechanical forms outlined above. They consist of a high-speed motor and speed-reducing gears in a self-contained unit, designed for economical power transmission. Three types are listed as available in standard ratings from 1 to 75 hp. The speed-reduction unit is manufactured in accordance with the recommendations of the American Gear Manufacturers' Association (AGMA). Their purpose is to provide a compact, economical, slow-speed drive while retaining the advantages of the high-speed motor. The three types referred to give a variety of speeds, as shown in Table 4. Typical gear-motor applications are indicated in Table 5.

Table 4. Three Types of Gear Units Giving Wide Variety of Speeds

(Courtesy, Westinghouse Electric Corporation)

Specifications: Type A

Single Reduction Unit Available in the Following Ratios:

Gear Ratios	AGMA Full-load Output Speeds (Motor Rpm, 1750)
1.22	1430
1.50	1170
1.84	950
2.24	780
2.73	640
3.37	520
4.17	420
5.00	350
6.25	280

Available with the following motor types:
Single-phase: 1 to 5 hp—Type CR
1 to 7 1/2 hp—Type CU

Polyphase (2- and 3-phase):
Squirrel-cage—1 to 75 hp—Type CS
Wound-rotor—1 to 75 hp—Type CW
Direct-current: 1 to 7 1/2 hp—Type SK

Specifications: Type C

Double Reduction Unit Available in the Following Ratios:

Gear Ratios	AGMA Full-load Output Speeds (Motor Rpm, 1750)
7.61	230
9.21	190
11.3	155
14.0	125
17.5	100
20.8	84
25.7	68

Available with the following motor types:
Single-phase: 1 to 5 hp—Type CR
1 to 7 1/2 hp—Type CU

Polyphase (2- and 3-phase):
Squirrel-cage—1 to 15 hp—Type CS
Wound-rotor—1 to 15 hp—Type CW
Direct-current: 1 to 7 1/2 hp—Type SK

Specifications: Type E

Double Reduction Unit Available in the Following Ratios:

Gear Ratios	AGMA Full-load Output Speeds (Motor Rpm)		
	1750	1165	870
31.2	56	....	....
38.9	45	....	....
47.3	37	25	....
52.7	....	....	16.5
58.3	30	20	....

Available with the following motor types:
Single-phase: 1 to 5 hp—Type CR
1 to 7 1/2 hp—Type CU

Polyphase (2- and 3-phase):
Squirrel-cage—1 to 75 hp—Type CS
Wound-rotor—1 to 75 hp—Type CW
Direct-current—1 to 7 1/2 hp—Type SK

MOTORIZED VARIABLE-SPEED UNITS. Motorized variable-speed mechanical power-transmission units are also available. They consist of an a-c motor assembled with a mechanical variable-speed power transmission as a single enclosed unit and are available for a considerable choice of ratings. Such units have the advantage of compactness and neatness and are sometimes employed as drives by machinery manufacturers. They are also sometimes used by plant engineers and master mechanics when applying individual drive to equipment formerly driven from shafting, or for something like a conveyor drive installed by the plant mechanical crew. In other words, they are particularly helpful where variable speed is called for and the space is congested.

Table 5. Typical Gear-motor Applications *

Three AGMA classes give wide variety of service. AGMA standard practice recognizes three classes of gear motors, based on load conditions and service required. Table 5 illustrates the difference between these classes. For load conditions heavier than those classified, such as 24-hour heavy shock loads, refer to manufacturer.

Class I: For steady loads not exceeding normal rating of motor and 8-hr-a-day service. Moderate shock loads where service is intermittent.

Class II: For steady loads not exceeding normal rating of motor and 24-hr-a-day service. Moderate shock loads for 8 hr a day.

Class III: Moderate shock loads for 24 hr a day. Heavy shock loads for 8 hr a day.

Application	8- to 10-hr Service	24-hr Service
Agitators		
Pure liquid	I	I
Variable density	II	II
Blowers		
Centrifugal	I	II
Lobe	II	II
Brewing and distilling		
Bottling machinery	I	II
Brew kettles, continuous duty	II	II
Cookers, continuous duty	II	II
Mash tubs, continuous duty	II	II
Scale hopper (frequent starting peaks)	II	II
Car dumpers	II	III
Car pullers	II	III
Clarifiers	I	II
Classifiers	II	II
Clay-working Machinery		
Brick press	III	III
Briquette machine	III	III
Clay-working machinery	II	II
Pug mill	II	II
Compressors		
Centrifugal	I	II
Lobe	II	II
Reciprocating—multicylinder—adequately flywheeled (within 3% cyclic variation)	II	III
Reciprocating—single cylinder	Refer to Factory	

Application	8- to 10-hr Service	24-hr Service
Conveyors		
(Uniformly loaded or fed)		
Apron	I	II
Assembly	I	II
Belt	I	II
Flight	I	II
Oven	I	II
Screw	I	II
(Heavy-duty or dual drive—not uniformly fed)		
Apron	II	II
Assembly	II	II
Belt	II	II
Flight	II	II
Oven	II	II
Screw	II	II
Reciprocating	III	III
Shaker	III	III
Cranes and hoists		
Main hoists—medium duty	II	II
Main hoists—heavy duty	III	III
Skip hoists	II	II
Travel motion	II	II
Trolley motion	II	II
Dredges		
Cable reels	II	II
Conveyors	II	II
Cutter head drives	III	III

Application	8- to 10-hr Service	24-hr Service
Dredges—*Continued*		
Jig drive	III	III
Screen drive	III	III
Stackers	II	II
Utility winches	II	II
Elevators		
(Conveyor type—same as conveyors)		
Freight	II	II
Passenger	III	III
Fans		
Centrifugal	I	II
Cooling towers	II	I
Large (mine, etc.)	II	III
Light, small diameter	I	II
Food industry		
Beet slicers	II	II
Cereal cookers	I	II
Dough mixers	II	II
Meat grinders	II	II
Hoists—See Cranes		
Laundry washers	II	II
Laundry tumblers	II	II
Line shafts		
Driving processing equipment	II	II
Other line shafts	I	II
Machine tools		
Punch press (gear-connected to load) and shears	III	III

Application	8- to 10-hr Service	24-hr Service
Machine tools—*Continued*		
Notching press (belt-driven)	I	II
Plate planers	III	III
Other machine tools—main drives	II	II
Auxiliary drives (feed-traverse, etc.)	I	II
Metal mills		
Draw bench carriage and main drives	III	III
Forming machines	III	III
Pinch, dryer, and scrubber rolls (reversing)	Refer to Factory	
Slitters	III	III
Small rolling-mill drives	III	III
Table conveyors (nonreversing)	II	II
Table conveyors (reversing)	Refer to Factory	
Wire drawing and flattening machines	III	III
Mills (rotary type)		
Ball	III	III
Cement kilns		III
Dryers and coolers		II
Hammer	III	III
Kilns	II	III
Pebble	II	III
Rod	III	III
Tumbling barrels	III	III
Mixers		
Concrete mixers—continuous duty	II	II
Concrete mixers—intermittent duty	I	
Constant density	I	II
Irregular density	II	II

Application	8- to 10-hr Service	24-hr Service
Oil industry		
Oil-well pumping (not over 150% peak torque)	III	III
Refineries		
Chillers	II	II
Paraffin filter press	II	II
Rotary kiln	II	II
Paper mills		
Agitators (mixers)		II
Bleachers		II
Beaters and pulpers		III
Calenders		III
Conveyors		II
Couch		III
Cylinders		II
Dryers		III
Felt stretchers		II
Jordans		III
Presses		III
Stock chests		III
Suction roll		III
Winders		II
Rubber industry		
Mixers	III	III
Calenders	II	III
Mills	III	III
Sheeters	III	III
Tire-building machines	II	II
Tire and tube press openers	I	I
Tubers or strainers	II	II

Application	8- to 10-hr Service	24-hr Service
Pumps		
Centrifugal—with surge tanks or equivalent	I	II
Centrifugal—without surge tanks	II	II
Gear and rotary—constant-density fluid	I	II
Gear and rotary—variable-density fluid	II	II
Proportioning pumps	II	III
Reciprocating—with open discharge	I	II
Reciprocating—multicylinder, double acting	II	II
Reciprocating—single-cylinder	Refer to Factory	
Sewage-disposal equipment		
Inside service	I	II
Screens		
Rotary—stone or gravel	II	II
Traveling water intake	I	I
Textile industry		
Batchers	II	II
Calenders	II	II
Card machines	II	III
Dry cans	II	II
Dyeing machinery	II	II
Looms	Refer to Factory	
Mangles	II	II
Nappers	II	II
Soapers	II	II
Spinners	II	II
Tenter frames	II	II

* All data quoted from *AGMA Recommended Practice*, Section 8, pp. 14–16.

8. BIBLIOGRAPHY

Fox, Gordon, *Electric Drive Practice.*
Harwood, P. B., *Control of Electric Motors.* John Wiley (1936).
Johnson, T. C., ed., *Electric Motors in Industry.* John Wiley (1942).
Shoults, D. R., and C. J. Rife (ed. by T. C. Johnson), *Electric Motors in Industry.* John Wiley (1942)
Recent issues of *Elec. J.*, *Elec. Eng.*, *Gen. Elec. Rev.*, *J. Inst. Iron and Steel Elec. Eng.*, and others.
Transactions of World Power Conference, Berlin, Vol. 1, Section 2 (1930).
Transactions of World Power Conference, Scandinavia, Vol. 5 (1933).
Standards of National Electrical Manufacturers' Association.
Reports of Edison Electrical Institute (EEI).
Trade journals of the various industries to which electric drive is applied.
Bulletins and reports of Power Transmission Council, Boston.
Bulletins and reports of motor manufacturers.

See also Bibliographies of Sections 8 and 9.

SERVOMECHANISMS

By S. W. Herwald

9. DEFINITIONS, FUNDAMENTALS, AND COMPONENTS

DEFINITION. Servomechanisms are a class of automatic regulators that perform the basic function of keeping a regulated quantity matched to a reference quantity. Many of the regulators that have been used for years in electric-power engineering are servomechanisms. These include voltage, speed, sectional paper mill, and steel mill regulators. The name servomechanism originated with the use of the term servo or slave in dealing with torque amplifiers or position regulators and was extended to include all closed-cycle regulators when it was found that identical theory applied.

BASIC ELEMENTS. The elements contained in a servomechanism are shown in the schematic diagram, Fig. 1. The three basic elements are:

1. Error measurer.
2. Amplifier.
3. Error corrector.

As shown in Fig. 1, each is essential in matching the regulated quantity to the reference quantity. The error measurer determines when the regulated quantity is different from

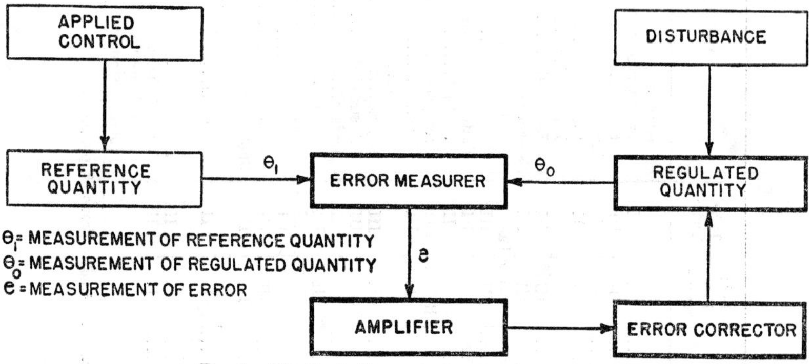

FIG. 1. The Basic Elements of a Servomechanism

the reference quantity. It then sends an error signal to the amplifier, which in turn supplies power to the error corrector. With this power the error corrector changes the regulated quantity so that it matches the reference quantity. The closed loop in Fig. 1, comprised of the error measurer, the amplifier, the error corrector, and the regulated quantity, is characteristic of all servomechanisms and accounts for both their high accuracy and their tendency to "hunt." Four important characteristics that can be obtained with servomechanism control are:

1. High accuracy.
2. Fast response.
3. Unattended control.
4. Remote operation.

Although there are many simple servomechanism applications, such as a thermostat controlling a furnace for home heating, the merit of this type of control increases with the difficulty of the problem.

Any quantity, such as voltage, speed, temperature, position, direction, or torque, can be servomechanism regulated. In all cases an error measurer, an amplifier, and an error corrector must be used. Each of these is in all probability a familiar device. Therefore a servomechanism, which is a combination of these devices into the closed loop of Fig. 1, should be considered a system rather than a particular device.

The disturbance shown in Fig. 1 represents load variations affecting the regulated quantity; for example, the generator voltage drop with load for a voltage regulator. Applied control (Fig. 1) is the controlled variation of the reference quantity in order to obtain desired changes in the regulated quantity. An example is changing the voltage base of a voltage regulator. Both the applied control and the disturbance to the regulated quantity are independent. They upset the previously existing balance between the reference quantity and the regulated quantity. Servomechanisms restore this balance between reference and regulated quantities continuously and automatically.

ERROR MEASURERS. All error measurers, regardless of whether they are electrical, mechanical, hydraulic, or any combination thereof, perform the following three functions:

1. Measure the reference quantity (θ_1, Fig. 1).
2. Measure the regulated quantity (θ_0, Fig. 1).
3. Produce an error signal (e, Fig. 1).

The points in which a designer is generally interested when selecting an error measurer are:

1. Energy required to measure reference quantity.
2. Accuracy of error measurement.
3. Size and weight.
4. Reliability.

Most electrical error measurers involve some sort of bridge circuit. Two of the most familiar are:

1. An a-c or d-c resistance bridge formed by two potentiometers across a voltage source. Error voltage e appears between the movable arms of the potentiometers whenever their positions do not match.

2. A tachometer bridge, in which two tachometer generators are connected bucking, and error voltage e appears whenever the tachometer speeds are not identical.

In addition, magnetic bridge circuits, including synchro-generator and control transformer combinations, are used to measure positional error; frequency-sensitive networks are used in a bridge to detect frequency error; and thermocouples and phototubes are used in bridge circuits to detect temperature and light-intensity errors, respectively. Numerous other devices, such as a mechanical differential to measure position error, a flyball to measure speed error, a float to measure liquid-level error, or a bimetal to measure temperature error, are all error measurers.

AMPLIFIERS. The broad similarity of all amplifiers is apparent, since each contains a "gate" element that controls the flow of power from the power source to the load. In a generator the field is that gate. Each of these gates is a low power or small force element when compared to the power it controls. Some of the most common amplifiers are contacts, relays, generators, electronic tubes, saturable reactors, silverstats, valves, throttles, and clutches. The amplification factor of such devices ranges from slightly above unity to a million or more. Huge amplifications can be obtained by cascading amplifiers. Such a cascade, composed of an electronic tube, a relay, and a motor-operated throttle, provides tremendous amplification. Microwatts control thousands of horsepower.

In servomechanism applications the load on the amplifier is always the error corrector. The difference between the power available at the error measurer and that required to correct the regulated quantity determines the amount of amplification necessary. The faster it is desired to correct the regulated quantity, the greater is the power required for the error corrector, and consequently the greater is the amplification necessary.

ERROR CORRECTORS. Error correctors are power devices and include electric motors and solenoids as well as hydraulic motors and pistons, steam or gas prime movers, and fuel burners. In the past these were the devices around which servomechanisms were built. The present tendency is to design special error correctors for servomechanism applications so that they may be coordinated into the overall system. In the case of electric motors this may require low-inertia rotors or special speed-torque characteristics, as well as low electrical time delays. These time delays, caused by energy-storage elements, such as springs or inertias in a mechanical system, or inductance or capacitance in an electrical system, are of utmost importance in the selection of an error corrector. The smaller the delay, the faster the response of the servomechanism system, with a consequent reduction in transient overshoot and system oscillation.

SPEED SERVOMECHANISMS. Figure 2 shows the combination of common elements, such as an electric motor, an electronic tube, a potentiometer and battery, and a tachometer, into a speed-servomechanism system. Each element is labeled in accordance with the notation of the basic schematic Fig. 1. Similar diagrams can be drawn for all servomechanisms, and in most cases the individual devices used in the servomechanism system will also be well known.

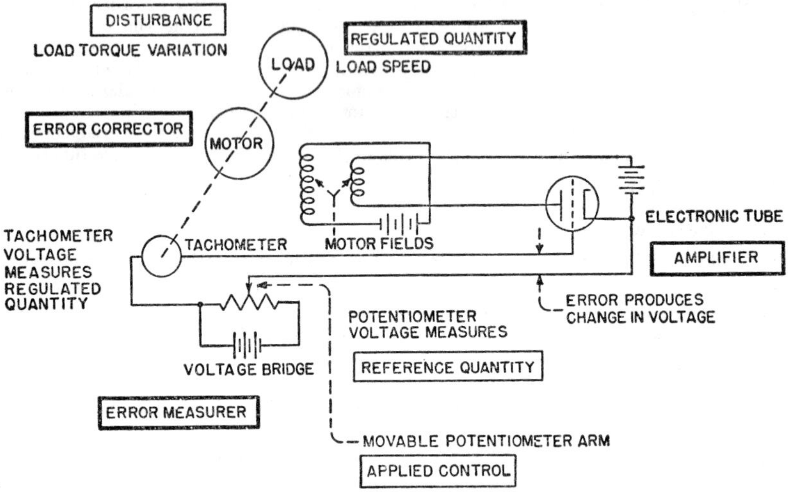

FIG. 2. A Speed-servomechanism System

Speed servomechanisms such as that shown in Fig. 2 are used in paper mills to control the speed of a section and to keep each section running at the proper speed in relation to the others. Similar applications can be found in regulating motor-generator sets, steel mill drives, and machine tools.

10. CALCULATION OF SERVOMECHANISM PERFORMANCE

The foregoing discussion of servomechanisms was limited to fundamentals, definitions, and components. Little mention was made of the problem of stability. This problem becomes more difficult as the desire grows for higher accuracy and better transient response in servomechanisms. In general, systems that are stable because of their inherent damping are no longer satisfactory, and additional "antihunt" or "stability" means must be used.

GENERAL EQUATIONS. If a signal is applied to the reference quantity of a perfect servomechanism, then θ_1, the measurement of the reference quantity, is a time function, and θ_0, the measurement of the regulated quantity, appears as an identical time function exactly 180 deg out of phase, because e, the error measurement, always tends to remain zero. In a real servomechanism energy-storage elements (inertia, inductance, capacitance, springs) and energy-absorbing elements (dashpots, resistors, orifices) introduce phase and amplitude distortion. If, for example, θ_1 is sinusoidal with time, at a certain frequency these energy elements could cause θ_0 to have a 180-deg phase shift. Then θ_1 and θ_0 are exactly in phase, and if θ_0 is of sufficiently great magnitude, the servomechanism system will be unstable and "hunt." Several methods may be used for determining whether a servomechanism is stable. In most cases, however, particularly in electric-power applications, more than the knowledge of whether a system is stable or unstable is desired. Frequently transient response is desired, so that such items as initial voltage drop in a voltage regulator for a suddenly applied load, or speed droop in a steel-mill motor when an ingot hits the rolls, can be determined.

The basic servomechanism equation can be written as

$$e = \frac{\theta_1}{1 + Kf(p)} \tag{1}$$

where K = system amplification constant.
$f(p)$ = system time-response function.
p = differential operator equivalent to d/dt if all quantities are zero at $t = 0$.
θ_1 = measurement of reference quantity.
θ_0 = measurement of regulated quantity.
e = measurement of error.

Also, since

$$e = \theta_1 - \theta_0 \tag{2}$$

$$\frac{\theta_0}{\theta_1} = \frac{Kf(p)}{1 + Kf(p)} \tag{3}$$

and

$$\frac{\theta_0}{e} = Kf(p) \tag{4}$$

In a series circuit composed of an inductance L and two resistances R and R_0, with voltage E applied across the circuit, E can be considered as the input, and the voltage E_0 across the resistor R_0 as the output. Then, when we compare with eq. (4), E_0, E, $\frac{R_0}{R_0 + R}$, and $\frac{1}{1 + Lp(R_0 + R)}$ correspond to θ_0, e, K, and $f(p)$, respectively. In a servomechanism K is the product of amplification as well as attenuation factors, and $f(p)$ contains terms for all the energy-storage elements in the system, mechanical as well as electrical. The differential equations of any servomechanism can be thrown into the form of (1) or (3). Some of the more complicated systems have energy-storage elements in the feedback loop. In those cases (3) becomes

$$\frac{\theta_0}{\theta_1} = \frac{Kf(p)}{1 + Kf(p)K_0f_0(p)} \tag{5}$$

where K_0 = amplification constant of the feedback loop.
$f_0(p)$ = response function of the feedback loop.

It is to be noted that, if $K_0f_0(p) = 1$ in (5), (3) results.

STABILITY. Three general methods are used to determine servomechanism stability. The first consists of either totally or partially solving (1) by classical methods. The second is to use a variation of the Nyquist method for feedback amplifiers [(5) is identical to the standard feedback amplifier equation]. This second method generally consists of plotting the amplitude and phase of $Kf(jw)$, where jw has replaced p for the range of angular frequency w from zero to infinity. This steady-state plot of amplitude and phase can then be interpreted for stability and, to some extent, for damping and transient response. The third method is the analogy method. Usually an electrical circuit is formed that has the same eq. (1) as the real servomechanism. Then, by the use of a transients analyzer or similar device, the response of the equivalent electrical circuit to the desired transient is obtained, and thus the exact response of the real servomechanism is known.

The classical methods using operational processes to solve (1) involve the laborious procedure of finding roots of equations and then obtaining the residue at each pole. Even when the results are obtained, they are good only for a particular set of numerical values. Each time the numerical value of a parameter is changed, the entire calculation has to be repeated. This makes it difficult to obtain optimum values.

The **Nyquist or frequency plot method,** as applied to servomechanisms, has been developed to the point where good predictions of system damping and transient behavior can be made. In general, amplitude peaks occur at important natural frequencies, and the magnitudes of these peaks are measures of the damping. The frequency band over which the amplitude is flat is a measure of the speed of response. Although the frequency method is excellent for initial design stages, it still requires experience and interpretation to predict the exact response of a servomechanism to a transient. The frequency response of any element can be obtained experimentally by varying its input sinusoidally at frequencies from zero to infinity and recording the amplitude and phase of the output.

In the analogy method the equivalent system could be mechanical, hydraulic, or electrical. However, the ease of coupling electrical elements, together with the low power levels at which electric tubes will transmit intelligence, usually makes the choice electrical. A transients analyzer provides a means of viewing and measuring on the screen of an oscilloscope the actual transient response of the electrical network equivalent to the servomechanism system. The effect of varying any parameter can be seen immediately, and consequently optimum design points can be obtained quickly.

SIMPLE STABILIZING MEANS. In order to increase the stiffness of a servomechanism system so that the transient response may be improved and the steady-state errors reduced, special stabilizing means are necessary. All devices used have no effect at frequencies approaching zero. Consequently, the steady-state error is not affected. Nevertheless at the higher frequencies, where "hunting" is present, powerful "antihunt" factors are introduced. The analysis of servomechanism performance to find optimum performance with different means and amounts of antihunt can be carried out by any one of the three methods outlined.

Some of the most common stabilizing means used in electric-power applications are phase advance or anticipatory networks, resistor-capacitor (RC) feedback from the armature of a variable-voltage generator, RC feedback from the output of a d-c tachometer, damping transformers, and gyroscope anticipators. The simplest RC feedback circuit consists of a capacitor and resistor in series. This network is placed across any voltage, such as generator armature or d-c tachometer, that is to be fed back. The voltage that appears across the resistor is then physically brought back to the closed servomechanism loop (Fig. 1) and fed into a point somewhere near the input of the amplifier. Hence, because a voltage obtained with a resistor-capacitor combination is taken from the servomechanism loop and brought back in a clockwise direction, as contrasted to the normal counterclockwise direction of the closed servomechanism loop, comes the name RC feedback. More complicated RC feedback circuits involve the use of more than just a simple series resistor-capacitor network.

The anticipatory type of stabilizing means raises the natural frequency of the servomechanism system, materially increases the damping, and tends to diminish the initial overshoot. Any RC feedback or damping transformer circuit, on the other hand, tends to lower the natural frequency and increase the initial overshoot of the system, although it provides excellent damping. The balancing factor is that RC feedback can be simply obtained, whereas getting sufficient anticipation provides a more difficult problem. Therefore, in systems having an adequate margin in transient performance, RC feedback or damping transformers can be used to advantage. In systems where the best possible transient performance is desired, it is usually worth the extra difficulties and complications to provide anticipatory control. In some cases optimum control is obtained by using both anticipation and RC feedback.

11. BIBLIOGRAPHY

Brown, G. S., and A. C. Hall, Dynamic Behavior and Design of Servomechanisms, *ASME Paper No.* 45-A-20.

Brown, G. S., and D. P. Campbell, *Principles of Servomechanisms.* John Wiley (1948).

Gardner, M. F., and J. L. Barnes, *Transients in Linear Systems.* John Wiley (1942).

Hanna, C. R., K. A. Oplinger, and S. J. Mikina, Recent Developments in Speed Regulation, *Trans. AIEE,* Vol. 59, pp. 692–700.

Hanna, C. R., K. A. Oplinger, and C. E. Valentine, Recent Developments in Generator Voltage Regulation, *Trans. AIEE,* Vol. 58, pp. 838–844.

Herwald, S. W., Considerations in Servomechanism Design, *Trans. AIEE,* pp. 871–877 (Dec. 1944).

MacColl, L. A., *Theory of Servomechanisms.* Van Nostrand (1945).

McCann, G. D., S. W. Herwald, and H. S. Kirschbaum, Electrical Analogy Methods Applied to Servomechanism Problems, *Trans. AIEE,* pp. 91–96 (Feb. 1946).

SECTION 17

TRANSPORTATION

ELECTRIC TRACTION

Revision by Cassius M. Davis of Article by H. A. Currie and Robert S. Rhodes

ELECTRICAL EQUIPMENT OF INTERNAL-COMBUSTION AUTOMOBILES

By Edmund B. Neil

ELECTRICAL EQUIPMENT IN AIRCRAFT

By T. B. Holliday

ELECTRIC PROPULSION OF SHIPS

By W. N. Zippler

RAILWAY SIGNALING

By Robert B. Elsworth

* Revision by Cassius M. Davis of article by W. A. Del Mar.

ELECTRIC TRACTION

Revision by Cassius M. Davis of article by H. A. Currie and Robert S. Rhodes

1. TRACTION SYSTEMS

APPLICATIONS OF ELECTRIC TRACTION. There are certain well-defined fields in which electric traction is superior to other methods, the most important being the following:

1. Where the frequency of stops is so great as to require a high rate of acceleration in order to make a good schedule speed. Motor buses, however, are replacing trolley lines to an increasing extent for such service. If the service requires trains of several cars, any desired number of these cars can be made motor cars, and thus a sufficient weight on the driving wheels can be economically secured to give the required adhesion and tractive effort.

2. Where local conditions prohibit the nuisance of the smoke, exhaust gases, and noise of steam locomotives, as in cities, tunnels, and mines.

3. In heavy trunk-line service, where the density of traffic is so great that a high load factor can be obtained with respect to both the power house and distribution system. The *operating* cost of an electric train is always less than the operating cost of a steam train, but the *fixed* charges (interest on investment and depreciation) of the electric system are high, for the first cost for the electric equipment, viz., locomotives, motor-car equipment, distribution system, and power house, is much greater than the first cost of steam locomotives and their accessory equipment. If there are sufficient trains in a system so that the pro rata share of the fixed charges for each train is less than its share of the difference between the operating cost of an equivalent steam equipment and the operating cost of the total electric equipment, then electric traction is advantageous, even in the absence of the secondary and intangible advantages which usually exist.

4. Where cheap electric power is available and fuel is expensive.

SYSTEMS. Three different types of motors are in use for electric traction, the d-c motor, the single-phase commutator motor, and the three-phase induction motor; see Section 9 on Motors. The operating voltages (i.e., volts between trolley and track rails or between third rail and track rails or, in double-trolley roads, between the two contact wires) and the motor voltages employed are as follows:

Direct-current Systems. Trolley or third-rail voltages of 600, 750, 1200, 1500, 2400, and 3000 are in use. The motors for the 600- and 750-volt systems are designed for operation at full trolley voltage. For all the other systems, two motors are usually connected permanently in series electrically, each designed for half the trolley voltage, but insulated for the full voltage.

Single-phase Systems. Trolley voltages (third rails are not used) of from 3000 to 11,000 volts and frequencies of 15 and 25 cycles per second are in use. In the United States 11,000-volt single-phase 25 cycles is standard, and in Europe 15,000-volt 15 or 16 $^2/_3$ cycles. The trolley voltage is stepped down to values suitable for direct application to the motors, from 200 to 1150 volts.

Three-phase Systems. Two trolley wires for each track are required, the two wires and the track forming the necessary three conductors for the three-phase distribution. This system is used to some extent in Europe but not in the United States.

The voltages used between the two contact wires and between each contact wire and track range from 3700 to 10,000, and the frequency is 15, 16 $^2/_3$, and 45 cycles.

Comparison of the Various Systems. For ordinary street railway service the 600-volt d-c system is almost universally employed, but for interurban and trunk-line service there has been much difference of opinion as to which of the various systems is the most economical when all the factors are taken into account. The factors which must be considered in comparing the three systems in any particular case are the following:

1. For a given weight and length of trolley or third rail the per cent power loss for a given amount of power transmitted varies inversely as the square of the trolley or third-rail voltage.

2. The higher the trolley or third-rail voltage, the smaller is the number of substations required for the same efficiency of distribution and weight of conductor.

3. The higher the trolley or third-rail voltage, the more costly is the insulation and supporting structure, and also the greater is the cost of maintenance of the distribution system.

4. Both the first cost and the annual expense of the substations are less for the a-c systems than for the d-c systems, since for the former static transformers only are required, whereas for the latter mercury-arc rectifiers or rotating machinery must be used for conversion.

5. The less-than-unity power factor of a-c motors (90 to 95 per cent) as well as of the line (because of the reactance of the trolley wire and track return) gives rise to a greater power loss in the a-c distribution system for the same power delivered than in the d-c system, and makes necessary generating apparatus of greater kva capacity.

6. The high-voltage d-c motor, for the same horsepower rating and speed, costs more, weighs more, and occupies more space than the 600-volt d-c type or the modern a-c motor.

7. With the a-c motors, transformers are required on the locomotive or car, which adds to the cost and weight of the equipment.

8. The 600-volt d-c motor costs less to maintain and is liable to fewer operating troubles than any of the other motors.

9. With the commutating type of a-c motor the power lost in the control equipment is practically negligible, since the "potential" type of control can be used. For both the d-c motor and the induction motor a resistance control is necessary, with consequent loss in power (see Article 5, Control Systems for Railway Motors).

10. The induction motor is inherently a constant-speed machine, and consequently the power input varies directly as the opposing resistance. The d-c motor and the a-c commutator motor are inherently variable-speed machines, and the power input varies approximately as the square root of the opposing resistance, the speed at the same time falling off.

11. The three-phase induction motor, when kept connected electrically to the source of power, automatically operates as a generator when the train is going down grade at a speed greater than the synchronous speed of the motor, the motor thus returning power to the line and at the same time acting as a brake to prevent any considerable increase in speed. "Regeneration," as this action is called, can also be obtained with the other types of motors but only at increased expense for the additional control equipment required.

Examples of the Use of the Various Systems. Tables 1 and 2 list the most notable examples of electrified steam railroads.

Table 1. The Most Notable Examples of Electrified Steam Roads in the United States and Canada (1938)

Name of Road	Date Installed	Miles of Single Track	Main Line, Tunnel, or Terminal	Trolley or Third Rail	Trolley Voltage	System	Locomotive (L.) or Motor Cars (M.C.)
Baltimore & Ohio	1895	9.1	Tun.	3rd R.	650	D-c	L.
B. & O. Staten Island R. T.	1925	44.7	M.L.	3rd R.	650	D-c	M.C.
Butte, Anaconda & Pacific	1913	144.7	M.L.	T.	2,400	D-c	L.
Canadian National R. R.	1908	15.0	M.L.	T.	3,300	A-c	L.
C., M., St. P. & Pacific	1915	892.1	M.L.	T.	3,000	D-c	L.
Cleveland Union Terminals Co.	1930	56.0	Term.	T.	3,000	D-c	L.
Delaware, Lackawanna & Western R. R.	1930	160.0	Term.	T.	3,000	D-c	Both
Great Northern Railway	1927	87.5	M.L.	T.	11,000	S-p	L.
Illinois Central R. R.	1926	156.6	Term.	T.	1,500	D-c	Both
Long Island R. R.	1905	448.1	Term.	3rd R.	600	D-c	Both
Long Island R. R., Bay Ridge	1927	84.4	M.L.	T.	11,000	S-p	L.
Michigan Central R. R.	1910	28.5	Term.	3rd R.	600	D-c	L.
New York Central R. R.	1906	381.6	Term.	3rd R.	650	D-c	Both
N. Y., N. H. & H. R. R.—A-c.	1907	597.5	M.L.	T.	11,000	S-p	Both
N. Y., N. H. & H. R. R.—D-c.	1895	65.1	Term.	T. & 3rd R.	600	D-c	M.C.
Norfolk & Western Ry.	1915	208.4	M.L.	T.	11,000	S-p	L.
Pennsylvania Railroad Company	1930–1938	2,196	M.L.	T.	11,000	S-p	Both
Reading Company (Philadelphia)	1931	197.8	Term.	T.	11,000	S-p	M.C.
Virginian Railway	1925	231.0	M.L.	T.	11,000	S-p	L.

Table 2. Selected Examples of Foreign Railroad Electrifications *

Name of Road	Date Installed	Miles of Single Track	Main Line, Tunnel, or Terminal	Trolley or Third Rail	Trolley Voltage	System	Locomotives (L.) or Motor Cars (M.C.)
Americas							
Canada							
Canadian National Ry....	1918	32.2	Tun.	T.	2,400	D-c	Both
Montreal Harbor........	1922	63.0	Term.	T.	2,400	D-c	L.
London & Port Stanley...	1915	44.9	Term.	T.	1,500	D-c	Both
Canadian Nat'l—Grand Trunk...............	1908	15.2	Tun.	T.	3,300	S-p	L.
Cuba							
Hershey Cuban..........	1920	110.0	M.L.	T.	1,200	D-c	Both
Mexico							
Mexican Ry. Co., Ltd....	1924	70.23	M.L.	T.	3,000	D-c	L.
South America							
Bethlehem Chile Iron Mines Co.............	1916	24.0	M.L.	T.	2,400	D-c	L.
Anglo-Chilean Consol. Nitrate Corporation......	1927	33.47	M.L.	T.	1,500	D-c	L.
Chilean State Railways...	1924	244.6	M.L.	T.	3,000	D-c	L.
Paulista Railway.........	1921	252.5	M.L.	T.	3,000	D-c	L.
Transandine Railway.....	1927	23.7	M.L.	T.	3,000	D-c	L.
West of Minas...........	1930	46.0	M.L.	T.	1,500	D-c	Both
Central Railway of Brazil	1937	82.0	Term.	T.	3,000	D-c	Both
Sorocabana Railway......	1945	174.0	M.L.	T.	3,000	D-c	Both
Buenos Aires & Western Railway.............	1927	82.1	M.L.	3rd R. & T.	800	D-c	Both
Central Argentine Railway.................	1931	111.2	Term.	3rd R.	800	D-c	M.C
Australasia							
Victorian Railways.........	1919	439.0	Term.	T.	1,500	D-c	M.C.
New South Wales Railways.	1926	278.4	Term.	T.	1,500	D-c	M.C.
New Zealand Govt. Railways	1924	13.4	Tun.	T.	1,500	D-c	L.
New Zealand Govt. Railways	1929	18.6	Tun.	T.	1,500	D-c	L.
New Zealand Govt. Railways	1938	65.0	Tun.	T.	1,500	D-c	Both
Austria							
Austrian Federal...........	1912	820.0	M.L.	T.	15,000	S-p	L.
Belgium							
Belgian State Railways.....	1935	54.0	Term.	T.	3,000	D-c	M.C.
China							
South Manchurian Railway.	1914	157.8	M.L.	T.	1,200	D-c	Both
France							
Midi Railway.............	1929	1,502.5	M.L.	T.	1,500	D-c	L.
P. L. M. Railway..........	1925	165.1	M.L.	3rd R.	1,500	D-c	L.
Paris-Orleans.............	1924	1,931.2	M.L.	T.	1,500	D-c	Both
State.....................	1924	154.5	Term.	3rd R.	650	D-c	M.C.
Germany							
German State Rys.—16 2/3 cy.....................	1914–28	627.6	M.L.	T.	15,000	S-p	Both
German State Rys.—15 cy..	1913	63.0	M.L.	T.	15,000	S-p	L.
German State Railways—25 cy.	1908–24	68.5	Term.	T.	3,000–6,000	S-p	M.C.
German State Railways.....	1924	365.6	Term.	3rd R.	800	D-c	M.C.
Italy							
Italian State Railways......	1901–27	127.0	Term.	3rd R.	650	D-c	M.C.
Italian State—3-ph 15 cy...	1902	77.0	M.L.	T.	3,300	3-ph	L.
Italian State—3-ph 16 2/3 cy.....................	1911–29	1,366.8	M.L.	T.	3,700	3-ph	L.
Italian State—3-ph 45 cy...	1928	150.0	M.L.	T.	10,000	3-ph	L.
Italian State—3,000 V. D-c.	1927	87.0	M.L.	T.	3,000	D-c	Both
North of Milan...........	1929	62.0	M.L.	T.	3,000	D-c	Both
Valli Di Lanzo............	1921	26.1	M.L.	T.	4,000	D-c	Both

Table 2. Selected Examples of Foreign Railroad Electrifications *—*Continued*

Name of Road	Date Installed	Miles of Single Track	Main Line, Tunnel, or Terminal	Trolley or Third Rail	Trolley Voltage	System	Locomotives (L.) or Motor Cars (M.C.)
India							
Bombay, Baroda & Central India	1928	75.1	Term.	T.	1,500	D-c	Both
Great Indian Peninsula Ry.	1926	571.4	M.L.	T.	1,500	D-c	Both
So. Indian Railway	1931	43.1	Term.	T.	1,500	D-c	Both
Japan							
Imperial Government Ry.	1908–28	602.0	M.L.	T.	1,500	D-c	Both
Morocco							
Moroccan Railway	1927	237.4	M.L.	T.	3,000	D-c	Both
Netherlands							
Netherlands Ry.	1927	292.1	M.L.	T.	1,500	D-c	M.C.
Netherlands East Indies							
Java State Railways	1925	165.0	Term.	T.	1,500	D-c	Both
Norway							
Norwegian State—15 cy.	1923	40.4	M.L.	T.	16,000	S-p	L.
Norwegian State—16 2/3 cy.	1922–7	159.0	M.L.	T.	15,000	S-p	L.
Norwegian State—16 2/3 cy.	1911	22.4	M.L.	T.	10,000	S-p	L.
Poland							
Polish State Railways	1935	134.0	Term.	T.	3,000	D-c	Both
Russia							
Transcaucasian	1932	47.0	M.L.	T.	3,000	D-c	L.
Spain							
Santander-Bilbao	1928	30.6	M.L.	T.	1,650	D-c	L.
Spanish Northern	1924	49.7	M.L.	T.	3,000	D-c	L.
Vascongados Railway	1929	153.6	M.L.	T.	1,650	D-c	L.
Spanish Northern	1929	367.5	M.L.	T.	1,500	D-c	Both
Madrid-Avila-Segovia	1941	500	M.L.	T.	1,500	D-c	Both
Sweden							
Nordmark Klaralvens Ry.—25 cy.	1921	123.3	M.L.	T.	16,000	S-p	L.
Swedish State Rys.—15 cy.	1915	365.0	M.L.	T.	16,000	S-p	L.
Swedish State Rys.—16 2/3 cy.	1925	488.0	M.L.	T.	16,000	S-p	L.
Switzerland							
Swiss Federal—16 2/3 & 25 cy.	1906–28	2,408.5	M.L.	T.	3,300 5,500 15,000	S-p	Both
Bernese Alps—16 2/3 cy.	1910–27	110.0	M.L.	T.	15,000	S-p	L.
Montreux-Oberland	1901	50.5	M.L.	T.	1,000	D-c	L.
Rhaetian Rys.—16 2/3 cy.	1913–22	198.5	M.L.	T.	11,000	S-p	L.
United Kingdom							
London, Midland & Scottish	1904	195.7	Term.	3rd R.	600	D-c	M.C.
London & Northeastern	1914	48.25	M.L.	T.	1,500	D-c	L.
Southern Railway	1909–29	7.99	Term.	3rd R.	650	D-c	M.C.
London & Northeastern	1904	80.5	M.L.	3rd R.	600	D-c	Both
Africa							
Algerian State Railways	1932	87.5	M.L.	T.	3,000	D-c	L.
South African Railways-Reef	1938	402.0	Tun. M.L.	T.	3,000	D-c	M.C.

* Because of unsettled conditions in Europe it is not possible to be certain that this table is complete and up to date.

ANALYSIS OF AN ELECTRIC RAILWAY PROJECT. In the analysis of a particular problem the general line of procedure indicated below is followed. Methods of calculation will be found in Article 2.

1. Determine the number and capacity of cars to supply the service desired.
2. Determine the power and energy required to propel these cars at the schedule speed desired.
3. Select the motors to correspond to the power determined in 2.
4. Lay out distribution of cars, by train diagrams if necessary.
5. Calculate the capacity of the low-potential distribution system and of the substations.

6. Determine the capacity of the generating stations and transmission system.
7. Estimate the first cost of the system.
8. Estimate the cost of operation.
9. Estimate the earning power of the system.

A. For a new road the earning power must exceed the sum of the operating cost and fixed charges by an amount sufficient to pay dividends.

B. For the electrification of a steam road it must be possible to show either: (1) that the result of electrification will be to reduce the operating charges by an amount more than sufficient to pay the fixed charges on the electrical apparatus, or (2) that the result of electrification will increase the capacity of the road or attract sufficient new business so that the increased earning power will more than balance the increased fixed charges.

C. The study should include other traction systems, such as internal-combustion engine systems.

2. ENERGY REQUIREMENTS AND MOTOR EQUIPMENT

Revision by Cassius M. Davis of article by W. A. Del Mar

From a consideration of the forces acting on a moving train it is possible to determine the motor capacity and energy required to operate it when the profile and contour of the road, the time table, and the characteristics of the available motors are known.

UNITS AND ABBREVIATIONS. Throughout this article the various quantities employed will be expressed in the following units unless specifically stated otherwise: distances in feet, weights in tons of 2000 lb, forces in pounds, speeds in miles per hour (abbreviated mph), accelerations in miles per hour per second (abbreviated mphps), mechanical power in horsepower, energy in watthours.

FORCES ACTING ON A TRAIN. The forces tending to accelerate a train are the tractive effort developed by the motors and the component of the weight along the track on down grades. The forces which retard the motion of the train are the various frictional forces and the component of the weight along the track on up grades, as well as, in braking, the frictional force due to the brakes. All the various frictional forces, except the braking resistance, such as track friction, journal friction, and air friction, which oppose the motion of a train on a straight track are usually considered together and are referred to as the train resistance. The extra friction due to track curvature is usually considered an equivalent up grade.

Tractive Effort and Draw-bar Pull. The tractive effort of a motor is the force exerted by the motor at the rim of the driving wheel to which the motor is geared. The tractive effort of a locomotive is the force exerted by the locomotive at the rim of the drivers. The draw-bar pull of a locomotive is the force transmitted through the draw-bar of the locomotive,* and is less than its tractive effort by an amount equal to the resistance due to the rolling friction of the locomotive wheels on the track and the air resistance of the locomotive.

Train Resistance. The total train resistance may be expressed as the sum of four terms, described as follows by W. J. Davis, Jr., in the *General Electric Review* (1926).

The first two terms of the equations represent journal friction almost entirely. They have been derived from dynamometer and coasting tests on standard freight and passenger cars and electric locomotives and are based on oil lubrication with average temperature conditions. Journal friction may be increased 20 to 40 per cent at temperatures below freezing.

The third term comprises resistances due to flange friction, concussion, swaying, and miscellaneous frictions proportional to the speed. The factor for this element is decreased by increase in length of truck wheel base and increased by poor roadbed conditions and inferior riding qualities of motor cars.

The last term gives air resistance for average weight of car or locomotive in pounds per ton for standard types of equipment. No allowance is made for head winds or strong side winds.

Locomotive resistance represents tractive effort delivered to driving axles and does not include friction losses in gears, motor bearings, or other parts of the driving equipment, as these are usually covered in the motive power efficiency.

The formulas in Table 3 are based on tests taken under mild weather conditions. Values obtained from them may be used as modified above in calculations relating to electric

---

*For a steam locomotive, the draw-bar pull usually refers to the force transmitted through the coupling between the tender and train; i.e., the tender is considered a part of the locomotive.

distributing systems, substations, energy consumption, and power demand. In the determination of electric motor characteristics and gear reductions to meet particular speed requirements, however, it may be desirable to add a small percentage to the required speed as a protection against unusual conditions.

Train Resistance at Starting. The formulas given are not applicable to speeds below about 10 mph. The New York Central tests on electric trains show that the train resistance decreases with decrease in speed to 10 mph, but that, as the speed still further decreases, the resistance per ton increases. The resistance at starting may be from 6 to 18 lb per ton, depending upon the condition of the bearings, track, etc., and upon the duration of the stop preceding the starting. These figures also apply to freight trains. Tests on the Rock Island system showed that in a train which had stood overnight in cold weather (i.e., which had become "frozen up"), the starting resistance was 30 lb per ton. The slack in the car couplings, however, renders it unnecessary for a locomotive to exert sufficient effort to start all the cars at once.

Table 3

Symbols

R = Tractive resistance in pounds per ton (2000 lb) on tangent level track.
A = Area in square feet of cross-section of locomotive or car body and trucks.
V = Speed in miles per hour.
n = Number of axles per car.
w = Average weight per axle in tons.
wn = Average weight of locomotive or car.

Values of A

Locomotives: 50 tons	105 sq ft
" 70 tons	110 "
" 100 tons and over	120 "
Freight cars	85–90 "
Passenger cars	120 "
Multiple-unit cars	100–110 "
Motor cars: 2 trucks	80–100 "
" " 1 truck	70–75 "

Where Used	Usual Formulas Recommended for convenience in calculation. Approved for axle weights in excess of 5 tons.	General Formulas Applicable to all axle weights. To be used when axle weights are less than 5 tons
Locomotives	$R = 1.3 + \frac{29}{w} + 0.03\,V + \frac{0.0024\,AV^2}{wn}$	$R = \frac{9.4}{\sqrt{w}} + \frac{12.5}{w} + 0.03V + \frac{0.0024AV^2}{wn}$
Freight cars	$R = 1.3 + \frac{29}{w} + 0.045\,V + \frac{0.0005\,AV^2}{wn}$	$R = \frac{9.4}{\sqrt{w}} + \frac{12.5}{w} + 0.045V + \frac{0.0005AV^2}{wn}$
Passenger cars (vestibuled)	$R = 1.3 + \frac{29}{w} + 0.03\,V + \frac{0.00034AV^2}{wn}$	$R = \frac{9.4}{\sqrt{w}} + \frac{12.5}{w} + 0.03V + \frac{0.00034\,AV^2}{wn}$
Multiple-unit trains: Leading car (vestibuled)	$R = 1.3 + \frac{29}{w} + 0.045\,V + \frac{0.0024\,AV^2}{wn}$	$R = \frac{9.4}{\sqrt{w}} + \frac{12.5}{w} + 0.045V + \frac{0.0024AV^2}{wn}$
Multiple-unit trains: trailing cars	$R = 1.3 + \frac{29}{w} + 0.045\,V + \frac{0.00034AV^2}{wn}$	$R = \frac{9.4}{\sqrt{w}} + \frac{12.5}{w} + 0.045V + \frac{0.00034\,AV^2}{wn}$
Motor cars	$R = 1.3 + \frac{29}{w} + 0.09\,V + \frac{0.0024\,AV^2}{wn}$	$R = \frac{9.4}{\sqrt{w}} + \frac{12.5}{w} + 0.09V + \frac{0.0024AV^2}{wn}$

Grades and Curvature. An actual up grade of G per cent produces a retarding force of $20G$ pounds per ton, and a down grade of G per cent produces an accelerating force of $20G$ pounds per ton. A curve gives rise to a retarding force which is small in comparison to the total train resistance. It is usually taken as 0.8 lb per ton for each degree of curvature. Track curvature is expressed by the formula:

$$\text{Degree of curvature} = \frac{5730}{\text{Radius of curve, in feet}} \tag{1}$$

ACCELERATION AND BRAKING. The permissible rate of acceleration depends upon a number of factors.

1. The rating of the motors: the larger the motors, the higher the tractive effort they can develop and therefore the greater the acceleration.
2. The weight on the driving wheels of the car or locomotive.
3. The comfort of the passengers. This also depends to some extent upon the uniformity of the acceleration.
4. To make a given schedule speed with the least amount of energy the acceleration rate should be as high as possible. Very high rates of acceleration, however, are not in general justified on this score, as the increase in the size of the motors required may more than offset the saving of energy.

Table 4. Acceleration Rates

Service	Miles per Hour per Second
Steam locomotive, freight service	0.1 to 0.2
Steam locomotive, passenger service	0.2 to 0.5
Electric locomotive, passenger service	0.3 to 0.6
Electric motor cars, interurban service	1.0 to 1.5
Electric motor cars, city service	2.0 to 4.75
Electric motor cars, rapid transit service	1.5 to 2.0
Highest practical rate, service conditions	2.5 to 4.75
Highest practical rate, emergency	7 to 8

The rates of acceleration in Table 4 represent common practice.

Braking. The maximum retardation in braking is limited by the discomfort of the passengers, and injury to equipment, a retardation of 1.5 mphps being the usual practical limit for electric or steam passenger trains, although 2.5 mphps is sometimes attained. For freight trains the braking retardation is from 0.7 to 1.0 mphps. The higher the rate of braking, the less is the energy consumption for a given schedule speed.

The tractive effort required to give to 1 ton (2000 lb) a linear acceleration of 1 mphps is 91.2 lb. To accelerate a train of W tons requires a tractive effort of 91.2 aW lb to produce a linear acceleration of a miles per hour per second, but on account of the accompanying angular acceleration of the rotating parts an additional force is required. This additional force is proportional to the linear acceleration a and also depends upon the radius of gyration (see Eshbach, *Handbook of Engineering Fundamentals*, Vol. I) of all rotating parts, and upon the gear ratio of the motors (i.e., ratio of number of teeth in gear to number of teeth in pinion). The effect of the moment of inertia may be looked upon either as increasing the effective weight W or as increasing the acceleration constant, the acceleration constant being defined as the quotient of total accelerating force divided by the product of weight and linear acceleration.

The acceleration constant is raised by the flywheel effect discussed above by about 5 per cent for heavy cars and locomotives, and between 5 per cent and 10 per cent for light low-speed cars, 8 per cent being an average figure. However, the acceleration constant is usually taken as 100, corresponding to an increase in effective weight of about 10 per cent. A given linear acceleration of a miles per hour per second then requires an accelerating force of $100a$ pounds per ton.

TRACTIVE EFFORT REQUIRED.

Let F = tractive effort, in pounds per ton, exerted by motors.
G = per cent actual grade (+ for up grade).
g = degrees of curvature.
r = train resistance, in pounds per ton.
a = acceleration in miles per hour per second (− for retardation).

Then the tractive effort required per ton of total train weight is

$$F = 100a + r + 20G + g \tag{2}$$

Example. Given a train of three 45-ton cars moving with a speed of 20 mph and accelerating at a rate of 1.5 mphps up a 1 per cent grade on a straight track; what is the total tractive effort required, assuming a train resistance of 7.8 lb per ton?

Answer: $(100 \times 1.5 + 7.8 + 20 \times 1) \times 3 \times 45 = 24{,}000$ lb

GEAR RATIO AND SPEED. By gear ratio is meant the ratio of the number of teeth in the gear on the wheel axle to the number of teeth in the pinion on the motor shaft. A gear ratio greater than 6 : 1 is seldom used for railway motors. For a given torque developed by the driving motor, the tractive effort at the wheel rim and the linear speed for a given current depend upon the gear ratio and wheel diameter. Let D = the diameter of the wheel in inches, K = the gear ratio, F = the tractive effort for a given current input; then the tractive effort F_1 for this same current input but for a wheel diameter D_1 and gear ratio K_1 is

$$F_1 = \frac{DK_1}{D_1K} F$$

If V is the speed corresponding to the tractive effort F for a given motor voltage, then the speed V_1 corresponding to the tractive effort F_1 for the same motor voltage and current is

$$V_1 = \frac{D_1K}{DK_1} V$$

If the gear ratio is low, the maximum speed will be high and the rate of acceleration low; if the gear ratio is high, the maximum speed will be low and the rate of acceleration high;

For a given motor equipment, train weight, schedule, and profile the energy consumption and temperature rise of the motors depend upon the gear ratio selected, since this in turn determines the amount of coasting. The proper gear ratio can be found only by trial calculation, plotting speed-time and distance-time curves from motor curves based upon different gear ratios, and calculating the energy consumption and temperature rise in the motors as described below.

MAXIMUM POSSIBLE TRACTIVE EFFORT—ADHESION COEFFICIENT. The adhesion or "tractive" coefficient is the quotient (expressed usually as per cent) of the tractive effort in pounds which will slip the drivers, divided by the weight in pounds on the drivers. Burch gives the values in Table 5. The maximum possible tractive effort is the product of the adhesion coefficient (as a decimal fraction) and the weight (in pounds) on the drivers.

Table 5. Adhesion Coefficients

Condition of Track	Without Sand	With Sand
Most favorable condition	35	40
Clean, dry rail	28	30
Thoroughly wet rail	18	24
Greasy moist rail	15	25
Sleet-covered rail	15	20
Dry-snow-covered rail	11	15

Maximum Grade Train Can Ascend.

Let W = total weight of train in tons.
W_d = total weight on all drivers in tons.
p = adhesion coefficient in per cent.
r = train resistance in pounds per ton of total weight.
G = per cent grade.
g = degree of curvature.
a = acceleration in miles per hour per second.

Then the maximum tractive effort which the drivers can exert is 20 pW_d pounds, and therefore $(r + 20G + g + 100a)W$ must be less than 20 pW_d, or the maximum per cent grade which the train can ascend is *

$$G = \frac{pW_d}{W} - \frac{(r + g + 100a)}{20} \qquad (3)$$

This grade is greater the less the acceleration, or the greater the retardation. The greater the speed before the train strikes the grade, the greater may the retardation be without bringing the train to rest on the grade, and therefore the steeper the grade it may ascend.

Example. Assume no acceleration or retardation and no curvature, a train resistance of 8 lb per ton, an adhesive coefficient of 15 per cent, and 25 per cent of total weight of train on drivers. Then the maximum grade the train can ascend is $G = 15 \times 0.25 - 8/20 = 3.35$ per cent.

The highest value of G is reached when all the weight is on the drivers, e.g., single cars or trains of motor cars with all axles equipped with motors. On steam freight roads the maximum grade seldom exceeds 2 per cent and is usually considerably less, except in very mountainous country.

Weight of Locomotive. The weight of locomotive required to accelerate a train weighing W tons at the rate of a miles per hour per second up a grade of G per cent on a g degree curve against a frictional resistance of r pounds per ton, when the q per cent of the weight is on the drivers and the coefficient of adhesion is p per cent, is given by the following formula:

$$\text{Weight of locomotive} = \frac{5W}{pq}(100a + r + 20G + g)$$

Example. What weight of locomotive is required to accelerate a 400-ton train at the rate of 0.5 mphps up a 0.1 per cent grade against a frictional resistance of 8 lb per ton, when 80 per cent of the weight is on the drivers and the coefficient of adhesion is 20 per cent?

$$\text{Weight of locomotive} = \frac{5 \times 400}{20 \times 80}(50 + 8 + 2) = 75 \text{ tons}$$

* To be exact W_d should be multiplied by $\sqrt{1 - (G/100)^2}$, but except for very heavy grades this correction is negligible.

POWER REQUIRED AT GIVEN SPEED.

Let r = train resistance in pounds per ton of total train weight.
G = per cent grade.
g = degree of curvature.
a = acceleration in miles per hour per second.
v = speed in miles per hour.
W = total weight of train in tons.

Then the power required *at the rims of the drivers* is 1.99 $v(r + 20G + g + 100a)$ watts per ton, or

$$p_0 = 2.67 \times 10^{-3}\, vW(r + 20G + g + 100a) \text{ horsepower, total.} \quad (4)$$

The *power input* p_i to the car or locomotive is equal to the power at the rims of the drivers divided by the overall efficiency ϵ of the controller, motors, and gears, i.e.,

$$p_i = \frac{1.99\, Wv(r + 20G + g + 100a)}{1000\epsilon} \text{ kilowatts} \quad (5)$$

Efficiency of Motors. The overall efficiency of the motors and gears when the motor is operating at full line voltage does not vary considerably for loads ranging from 50 to 150 per cent of rated load. The maximum efficiency is usually at about rated load and has the values given in Table 6. At 50 and 150 per cent load, the efficiency may be from 3 to 10 per cent less, depending upon the design and the type of motor. The variation in efficiency with load is usually greater with a-c series than with d-c motors.

Table 6. Maximum Overall Efficiency of Motors and Gears at Rated Voltage

Horsepower, 1-hr rating	Kind of Motor	Maximum Efficiency per cent
30–100	D-c single-reduction spur geared...	83–88
30–100	D-c single reduction worm geared..	79–83
30–100	D-c double reduction spur geared..	82–86
100–250	D-c geared....................	88–89
250–500	D-c gearless..................	91–93
50–200	A-c series geared...............	70–80*
200–500	3-phase induction geared........	85–89

* Including step-down transformers.

SPEED-TIME AND DISTANCE-TIME CURVES. To determine the energy required to propel a car or train a given distance over a given track in a given time requires consideration of a number of factors which can best be taken into account by the construction of various kinds of time curves. Such curves may be constructed with practically any degree of accuracy desired when the profile of the track, the weight of the train, the various resistances, schedule speed, time of stops, etc., are accurately known. Such data are, however, seldom known with any great precision, and consequently elaborate methods of plotting and calculation are seldom justified. Below will be given some results of actual tests and a rough but simple method of approximating the energy requirements.

The following terminology will be employed:

Speed-time Curve. A curve showing the speed (in miles per hour) plotted as ordinates against elapsed time (in seconds) as abscissas; e.g., the curve *ABCEF* in Fig. 1. A speed-time curve may be conveniently divided into four parts, namely:

Controller Period. The period from starting until full line voltage is established across each motor, i.e., the portion *AB* of the curve in Fig. 1.

Motor-curve Period. The period during which the motor is operating on full line voltage, i.e., the portion *BC* in Fig. 1. The relation between speed and tractive effort during this portion of the run is fixed by the motor characteristics, specifically by the speed-torque curve of the motor and the gear ratio.

Coasting Period. The period during which the car or train is coasting, i.e., the portion *CE* in Fig. 1.

Braking Period. The period during which the brakes are applied, i.e., the portion *EF* in Fig. 1.

Distance-time Curve. A curve showing the distance covered (in feet) plotted as ordinates against elapsed time (in seconds) as abscissas, e.g., the curve *AG* in Fig. 1.

Current-time Curve. A curve showing the line current (in amperes) plotted as ordinates against elapsed time (in seconds) as abscissas, e.g., the curve marked "Amperes" in Fig. 1.

Voltage-time and Power-time Curves. Curves showing respectively the voltage per motor (or the line voltage) and the power input to the motor (or train) plotted as ordinates against elapsed time (in seconds) as abscissas.

Average Speed and Schedule Speed. The average speed V is the total distance run L' (in miles) divided by the time (in hours) the train is actually running. The schedule speed S is the total distance run (in miles) divided by the total time (in hours) of the run from one end of the road to the other, including time of all stops at intermediate stations. If the *total* time of all the *intermediate* stops is T_s' seconds, then

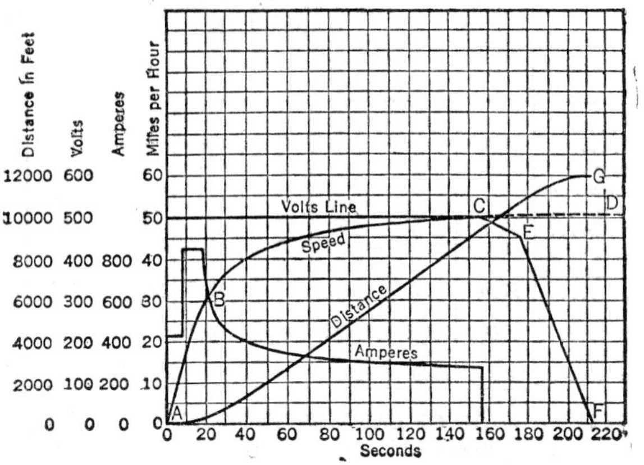

FIG. 1. Speed, Distance, and Current Curves

$$V = \frac{S}{1 - (T_s' S/3600 L')} \tag{6}$$

where V and S are in miles per hour and L' is the *total* length of route in *miles*.

Duration and Frequency of Station Stops. The duration of each stop for surface cars ranges from 5 to 10 sec, for elevated and subway trains from 10 to 30 sec, for interurban trains from 10 to 40 sec. The stops per mile are the reciprocal of the average distance in miles between stops. For the six most important elevated and subway lines in Europe and America, the stops per mile average 2.5.

Average Equivalent Grade (G). Grades may be taken into consideration by calculating the sum H_1 of all the rises on up grades and the sum H_2 of all the drops on down grades, and taking for the average "equivalent" up grade in percentage

$$G = \frac{100(H_1 - 0.5\, H_2)}{L} \tag{7}$$

where H_1 and H_2 are in feet, and L is the total length of the route in feet. On a round trip $H_1 = H_2$, and the "equivalent" grade in per cent is

$$G = \frac{50H_1}{L}$$

This method of dealing with grades is equivalent to assuming that half the kinetic energy stored in the train on down grades is utilized in taking the train up the following up grade. The amount of energy thus rendered available of course will depend upon the amount of braking necessary on down grades to prevent excessive speeds and also upon the location of the stops with respect to the grades. The figure $1/2$ is taken as an approximate average; it may be varied as seems reasonable in view of the actual profile.

Average Angle of Curvature (g). Curves may be taken into account by finding the average curvature, i.e., finding for each curve the product of the degree of curvature by the length of the curve, adding all these products, and dividing by the length of the route.

ENERGY CONSUMPTION FROM TESTS. Table 7 gives the energy consumption, as found by tests, for a number of typical services. This table will be found useful as a rough check on any calculations made for a specific service. Methods of making such calculations are given below.

Table 7. Electric Trains: Energy at Collector, from Actual Tests

	New York Central R. R.	New York, New Haven & Hartford R. R.					Interborough Subway, N. Y.		Urban Trolley Lines
							Express	Local	
Service	P.	F.	F.	P.	P.	P.	P.	P.	P.
A-c or D-c	D-c	A-c	A-c	A-c	A-c	A-c	D-c	D-c	D-c
Locomotive or cars	L.	L.	L.	L.	L.	C.	C.	C.	C.
Train weight, including locomotive, tons	1226	3060	4049	909	519	529	355	226	4–12
Length of run, miles	32.7	69	87	60	60	21	10.2	15.2	....
Average grade, per cent	0	0.033	−0.055	−0.02	−0.02	−0.033	0	0	....
Average speed, including stops	37.4	21.7	16.1	47.4	38.7	23	24.4	16.5	....
Stops per mile, including one terminal	0.092	0.043	0.034	0.017	0.18	0.67	0.59	2.6	2–10
Energy consumption, watt hours per ton-mile	28.3	17.0	22.2	28.3	48.8	72.0	43.6	104.2	100 to 160

APPROXIMATE METHOD OF CALCULATING ENERGY CONSUMPTION. The following method is based upon simple kinetic principles and, if certain characteristics of the run are known, gives the actual energy output at the wheel rims. This fact makes the method useful, not only for rough calculations, but also to check calculations made by the step-by-step method.

When the method is applied to checking purposes, the column of Table 8 headed "Actual Energy Output" should be used, and the input calculated from the known efficiencies. When applied to rough calculations, the column headed "Approximate Electrical Energy Input" should be used. In the latter case the maximum speed and length of run with power on are not known, but it is possible to assume certain values, based upon experience, which will give a rough approximation of the energy required.

Table 8. Output at Wheel Rim and Input to Car in Watt-hours per Ton-mile

Energy for	Actual Energy Output at Wheel Rims of Cars	Approximate Electrical Energy Input to Cars
Acceleration	$\dfrac{V^2_m}{36.2L}$	$\dfrac{K^2 n V^2}{25}$
Train resistance	$\dfrac{1.99\, rL_p}{L}$	$\dfrac{2.9\, r}{Q}$
Grades	$\dfrac{39.8\, GL_p}{L}$	$\dfrac{57\, G}{Q}$
Curves	$\dfrac{1.99\, gL_p}{L}$	$\dfrac{2.9\, g}{Q}$
Total	Sum	Sum

Let V = average running speed in miles per hour.
V_m = maximum speed in miles per hour.
L = length of run in miles.
L_p = distance traveled, with power on, in miles.
$n = 1/L$ = number of stops per mile including one terminus.
r = average train resistance in pounds per ton (say that corresponding to a speed from 10 to 20 per cent greater than the average speed).
G = average equivalent grade in per cent.
g = average curvature in degrees.

$$K = \frac{V_m}{V} = \text{ratio of maximum to average speed; see Table 9.}$$

$$Q = \frac{L}{L_p} = \text{ratio of length of run to distance traveled with power on; see Table 9.}$$

NOTE. $25 = 36.2\,\epsilon$, $57 = 39.8 \div \epsilon$, and $2.9 = 1.99 \div \epsilon$, where ϵ is the efficiency, taken as 0.7 The formula for energy due to curves assumes each degree of curvature to be equivalent to a train resistance of 1 lb per ton, which is probably high.

Table 9. Values of K and Q

(D. C. Woodbury and W. A. Del Mar)

Stops per Mile n	K: Locomotive Passenger Trains	K: Single Cars, Multiple-unit Trains, and Freight Trains	Q: All Trains	Stops per Mile n	K: Locomotive Passenger Trains	K: Single Cars, Multiple-unit Trains, and Freight Trains	Q: All Trains
0	1.00	1.00	1.00	1.4	1.93	1.59	2.34
0.1	1.18	1.10	1.11	1.6	1.94	1.62	2.44
0.2	1.35	1.18	1.24	1.8	1.94	1.65	2.52
0.3	1.48	1.25	1.38	2.0	1.95	1.68	2.58
0.4	1.60	1.31	1.52	2.5	1.95	1.75	2.71
0.5	1.68	1.36	1.67	3.0	1.96	1.80	2.81
0.6	1.75	1.40	1.78	3.5	1.96	1.85	2.87
0.7	1.82	1.44	1.89	4.0	1.97	1.90	2.91
0.8	1.86	1.47	1.99	4.5	1.97	1.94	2.95
0.9	1.90	1.50	2.07	5.0	1.98	1.97	3.00
1.0	1.93	1.52	2.15	over 5.0	2.00	2.00	3.00
1.2	1.93	1.56	2.24				

Example. A multiple-unit train has a speed (excluding stops) of 25 mph and makes 0.8 stop per mile. It ascends an average grade of 0.143 per cent. What will be its energy consumption in watthours per ton-mile?

From Table 9 we find, for $n = 0.8$, that $K = 1.47$ and $Q = 1.99$. Then, using the formulas in Table 8, the results in Table 10 are obtained, assuming 6.5 lb per ton for friction.

Efficiency of Run. The formulas given above enable one to judge the effect, upon the energy consumption, of altering any of the principal physical elements upon which the run is based.

From the formulas for kinetic energy it is obvious that a low value of K means a low energy consumption. A low value of K, however, also means a "square" speed-time curve; i.e., for low energy consumption the controller, acceleration, and braking periods should be as short as possible, or, in other words, the rate of acceleration and braking should be as great as practicable.

The quantitative effect of changing any of these variables may be estimated by the analytical method outlined below.

Table 10

Energy for	Watthours per Ton-mile	
Acceleration	$\frac{1.47^2 \times 0.8 \times 25^2}{25} =$	43.2
Train resistance	$\frac{2.9 \times 6.5}{1.99} =$	9.5
Grades	$\frac{57 \times 0.143}{1.99} =$	4.1
Total		56.8

This example, worked out by the step-by-step method, gives 60.5 whr per ton-mile.

STEP-BY-STEP METHOD OF PLOTTING SPEED-TIME CURVES. There is no way of exactly predetermining a speed-time curve except by a number of successive trials. That is to say, the time the current is kept on, the time of coasting, and the time of braking must each be guessed, and it is usually necessary to make a number of trials, varying the proportion of motor run, coasting, and braking, before the given distance is traversed in the desired time.

If the characteristics of the train and its equipment are expressed numerically, the principles of mechanics enable such trial runs to be plotted on paper, and the proper proportion of motor run, coasting, and braking can be selected to make the train travel the desired distance in the given time. For a given motor equipment on a given route it is possible to plot by a step-by-step method these speed-time curves, and then from these curves and the characteristic curves of the motors the various characteristics, such as energy consumption and root-mean-square current, may be determined. The accuracy of this method depends solely upon the accuracy with which the assumed data are known. The necessary data are:

Profile and alignment of road.
Characteristic curves of the motors.
Total weight of train in tons.
Time of run in seconds between successive stops.
Permissible starting current, or acceleration in miles per hour per second during the controller period.
Braking rate in miles per hour per second.
Train resistance in pounds per ton at any speed.

A number of schemes for working out the step-by-step method have been devised. For some of these see textbooks and articles mentioned under Bibliography, Article 14.

MOTOR CAPACITY. Quoting from and amplifying the AIEE No. 11, 1943, American Standards for Rotating Electrical Machinery on Railway Locomotives, and Rail Cars, and Trolley, Gasoline-electric and Oil-electric Coaches, we should note that the earliest streetcar motors were of the open type like ordinary industrial motors but were soon followed by fully enclosed motors, which were the rule for about 15 years. These motors were rated on the 1-hr basis, starting cold, and most of them had no continuous rating on full voltage. Motors with separate ventilation were then introduced on heavy locomotives and were followed by the development of self-ventilated motors for streetcars. Now non-ventilated motors are practically obsolete.

Motors are applied to cars and locomotives with much higher continuous ratings than formerly, and at the same time there has been a material reduction in motor weights. Whereas the continuous rating current was 40 per cent of the 1-hr rating for enclosed non-ventilated motors, the continuous rating of the self-ventilated motor is about 60 tc 80 per cent and of separately ventilated motors 80 to 100 per cent of the 1-hr rating.

The above standards state that, in the absence of any specification as to the kind of rating, the continuous horsepower rating shall be understood. The current densities in the windings of the non-ventilated motors at the continuous ratings are so low that the windings have relatively large thermal capacity and can carry very heavy overloads based on the continuous current rating. The self-ventilated motor, however, usually has a much higher proportional current density at its continuous rating and correspondingly less thermal capacity and less overload capacity. The separately ventilated motor has still less relative thermal and overload capacity. The thermal capacity is usually the determining factor for the overloads the motor can carry.

Modern railway motors are overloaded primarily during the acceleration portion of the duty cycle, which is a matter of seconds in urban transit applications and a very few minutes in locomotives starting heavy trains.

The commutation of modern motors is so good that overloads approximating twice the continuous rating are easily handled, and, with rapid acceleration, overload thermal capacity is generally ample. Modern high-horsepower light-weight motors can be applied with sufficient continuous capacity so that sustained overloads are not necessary.

When the motor reaches a constant temperature under a constant load, the temperature of any part will then be proportional (approximately) to the total power (kilowatts) lost in the windings and core. Similarly, under a fluctuating load continuing over a long period during which there are no excessively long breaks or excessive overloads, the temperature becomes fairly constant, and the rise is proportional to the average power (kilowatts) lost during this period. There will be times at which the temperature rise will exceed this average and times at which it will be less, but on account of the heat-storage capacity (or thermal capacity) of the materials of which the motor is made the fluctuations in temperature will be very much less than the fluctuations in the load.

Size of Motor Limited by Average Temperature Rise. The manufacturers supply information as to the current which any motor will carry continuously (on stand test) without overheating, at various voltages from one-half to full voltage; see Table 11. From this information, making the assumptions noted in the preceding paragraph, it is possible to determine the approximate temperature of the motor for any given run or series of runs. The process is to calculate the root-mean-square current per motor and the average motor voltage for the particular service contemplated. Call these values of the rms service current and average motor voltage I_e and E_m respectively. Let I_c be the continuous-current capacity at a given voltage E_c, as given by the manufacturer (see Table 11), and let T_c be temperature rise corresponding to this continuous rating. [Motors having ordinary fibrous insulation are rated on the basis of a 75 deg cent temperature rise on stand test, which corresponds to about 65 deg cent rise in actual service, because of better ventilation (see Standards of the AIEE); hence for ordinary motors T_c is 65 deg cent]. Let J_c be the corresponding core loss, K_c the corresponding copper loss, and $L_c = J_c + K_c$

Table 11a. Rated and Continuous Capacity of Westinghouse D-c Railway Motors

Type Westinghouse	Rated Horsepower	Rated Voltage	Rated Amperes	Continuous Current Capacity		Weight, Pounds
				At 300 Volts	At 450 Volts	
Surface Cars—Subway—Elevated						
336-A1	125	600	174	120	127	4100
570-D5	190	600	255	160	165	5300
559-DR3	210	600	285	....	260	4650
1442-A1	140	600	194	....	....	1360
1447-A	100	300	280	250	....	1780
1432-J	55	300	156	135	....	695
1433	90	300	260	224	....	1155
Trolley Locomotives						
562-D5	100	600	140	85	87	4900
582-D5	140	600	195	115	120	6000
369	325	600	435	370	370	7700
Diesel Electric Locomotives						
362-D	....	....	....	520	500	6170
370-F	....	....	....	930	930	7380
1443-A	125	300	350	317	....	1425
1444-A	140	300	387	340	....	1430

Table 11b. Rated and Continuous Capacity of General Electric Company Railway Motors

Type	Application	Continuous Horsepower	Volts	Continuous Tractive Effort	Max. Speed, Miles per Hour	Continuous Amperes	Weight, Pounds
GE-1240	Fast schedule subway car	59	300/600	....	....	250	1,620 *
GE-1220	P.C.C. streetcar	48	300/600	....	....	135	720 *
GE-1234	Trolley coach	123	600	....	....	170	1,440 *
GE-729	Electric road locomotive	665	1,500/3,000	....	....	360	12,200
GEA-627	Electric road locomotive	780	Transformer †	....	....	1,500	14,900
GEA-626	Electric suburban car	360	Transformer †	....	....	1,480	6,800
GE-747	Diesel-elec. loco. 550-hp eng.	‡	Generator §	6,250	30	285	2,900
GE-731	Diesel-elec. loco. 1,000-hp eng.	‡	Generator §	8,500	60	850	6,700
GE-752	Diesel-elec. loco. 1,500-hp eng.	‡	Generator §	9,350	75	900	7,800

* These motors have special gearing included in the driving axles; hence weight is for motor only. Weights of all other GE motors listed include weight of gears.

† Voltage applied varies, depending upon the transformer secondary taps to which the motor is connected. These motors are 25-cycle single-phase series commutator-type machines.

‡ These motors, together with the generators which supply power to them, are designed to transmit the full net power delivered by the Diesel engine over a wide range of locomotive operating conditions; they do not have a definite horsepower rating. Roughly, their horsepower output, for four-motor equipments, is about 20 per cent of the Diesel engine rating.

§ The voltage applied varies with the current load on the generator; the voltage falls off rapidly with increasing current so as to maintain constant Diesel engine output over a wide range of generator currents.

the corresponding total electrical losses. Then the total electrical losses corresponding to the average load are

$$L_a = \frac{E_m}{E_c} J_c + \left(\frac{I_e}{I_c}\right)^2 K_c$$

and the average temperature attained by the motor in service will be approximately

$$T_a = \frac{L_a}{L_c} \cdot T_c$$

For safe operation the average temperature rise T_a should never exceed the value T_c, which for motors with ordinary fibrous insulation is 65 deg cent.

Approximate Values of J_c and K_c. When the core loss and copper loss are not given separately, a rough estimate of J_c and K_c may be made by assuming that at rated load (1-hr rating and line voltage) the core loss is, say, one-fourth of the total electrical losses. The total electrical losses L_r in kilowatts at rated load may be found from the characteristic curves of the motor by using the formula

$$L_r = P\left(\frac{0.97}{\epsilon} - 1\right)$$

where P is the 1-hr rating in kilowatts, and ϵ the efficiency of the motor with gears; the 0.97 takes into account the frictional losses in the motor and the gears. Let I_r be the rated current and E_r the rated voltage; then

$$J_c = \frac{L_r E_c}{4 E_r} \quad \text{and} \quad K_c = \frac{3 L_r}{4}\left(\frac{I_c}{I_r}\right)^2$$

Size of Motor Limited by Short-time Temperature Rise. When the service is such that the motor must take a heavy current for a comparatively long interval (e.g., a long starting period or a heavy grade for a considerable distance), followed by a like period of light load or no load, the average temperature for the run, as calculated above, may be within the required limits, but the short-time temperature rise may be excessive. This short-time temperature rise depends upon the heat-storage capacity of the motor, i.e., upon the energy loss (number of kilowatthours of heat developed in it) required to raise its temperature 1 deg, say, assuming no radiation of heat from its surface.

Final Choice of Motor. No motor should be employed for a given service which does not meet the above requirements regarding the maximum current and heating limits. A larger motor than that fixed by these requirements may prove cheaper in the long run if, by using such a motor, the energy consumption can be materially reduced by increasing the amount of coasting during the run. In any event the motor should be of sufficient capacity to permit a reasonable amount of coasting under normal conditions, so that there will be a sufficient margin in which to make up for lost time due to unexpected slowdowns or extra stops.

ANALYTICAL METHOD OF PREDETERMINING ENERGY AND MOTOR EQUIPMENT. A useful analytical method was developed by Cary T. Hutchinson and described in two papers in *AIEE Trans.*, Vol. 19, p. 129 (1902) and Vol. 22, p. 657 (1903). In the method a speed-time curve similar in shape to the curve *ABCEF* in Fig. 1 is assumed. That is, the acceleration during the controller period, the train resistance, and the braking retardation are all assumed constant, but a "motor-curve" period (*BC* in Fig. 1) is also taken into account, this last constituting the essential difference between this method and the "straight-line" speed-time curve method frequently employed for approximate calculations. The introduction of this motor-curve period in the calculations enables one to approximate much more closely actual working conditions, and the results are much more accurate. In addition this method enables one to predetermine, without choosing any particular equipment, the effect of rate of acceleration, rate of braking, per cent of coasting, etc. Where extensive calculations are to be carried out, it is well worth while to consult the above references and apply this method.

TRAIN AND LOAD DIAGRAMS. The current-time curve for a train making a number of stops may be represented as shown by (*a*) in Fig. 2. On a railway line where there are several trains, the total current may be obtained by placing the current curve for each train at its proper place in the time scale, and adding the ordinates of the curves. Such a process is very tedious and unnecessary where there are a large number of trains. In such cases the high and low parts of the curves become staggered with respect to one another more or less according to the laws of chance, so that each current curve may be replaced by a rectangle of the same area but with a base extending over the entire running time as shown by (*b*) in Fig. 2. When this is done, the kilowatts and amperes per train are derived from the watthours per ton-mile by the following formulas:

$$\text{Kilowatts} = \frac{WV}{1000} \times (\text{Watthours per ton-mile})$$

$$\text{Amperes} = \frac{WV}{E} \times (\text{Watthours per ton-mile})$$

where W = weight of train in tons.
V = average running speed (excluding stops) in miles per hour.
E = line voltage.

In Fig. 2 the time when the current is cut off is indicated by 1, and that when the train stops by 2.

Another approximation, which is even more often used, is to replace the series of rectangles shown at (*b*) by a single rectangle as shown at (*c*) in Fig. 2. The area of this rectangle will be equal to the sum of the areas of the smaller rectangles or current curves.

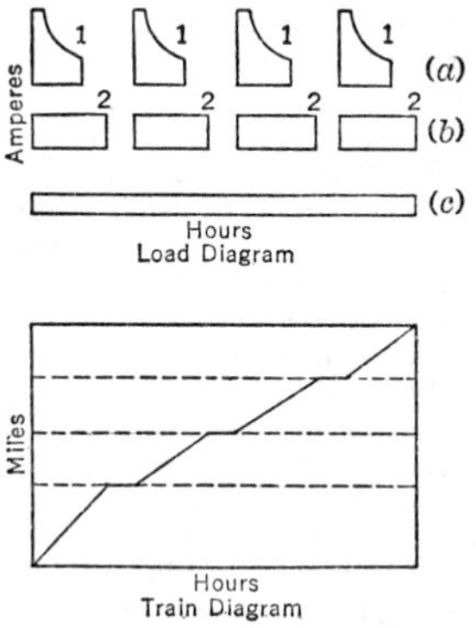

FIG. 2

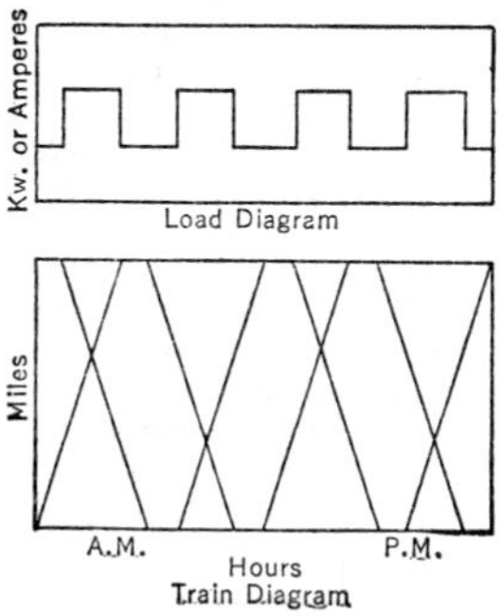

FIG. 3

Using this approximation, kilowatts or amperes may be obtained from the above formulas, taking, however, V to be the schedule speed, i.e., speed including stops. The procedure is to plot a train diagram showing when each train comes on and off the line, neglecting intermediate stops, as shown for a simple case in Fig. 3. Each time a train comes on or off, the corresponding kilowatts or amperes are added to, or subtracted from, the load diagram.

Power Required for Car Heating and Lighting. In addition to the energy required for propelling the cars, a very appreciable amount is also required in the winter for heating them, and a small amount at night for lighting. In making up a load diagram this energy should be included.

The average power for car heating varies, of course, with the climate and time of year. The figures in Table 12 represent usual requirements in the northern parts of the United States.

Table 12. Heating and Lighting of Cars

Length of Car, ft	Average* Kilowatts for Lighting	Average† Kilowatts for Heating	
		Average Conditions	Severe Conditions
14–20	0.25	3.5	4
20–28	0.35	4.5	5.5
28–34	0.55	5.5	7.5
34–40	0.70	7.5	10.5

* During the hours lights are on, using tungsten lamps.
† During the time car is in service.

Substation and Power-station Loads. The load diagram obtained as described above gives the total load at the trains. To obtain the load at a substation, the kilowatts or amperes must be increased by a suitable amount to allow for the losses in the distribution system. The load diagram of the power station should allow for all transmission and distribution losses between the power house and substation and also for all auxiliary power, such as that required for station lighting or shop machinery.

3. ELECTRIC LOCOMOTIVES

For many years a new type of electric locomotive was designed for nearly every new proposition. This was due in part to the wide diversity in the conditions to be fulfilled. However, with the very numerous applications of electric locomotives of 30 to 80 tons weight for slow freight and switching service on interurban roads and in terminals, the double-truck bogie type finally demonstrated its superior fitness and came to be adopted almost universally in America for all work involving speeds less than 45 mph. For higher speeds special provision must be made for guiding the locomotive around curves by the addition of guiding trucks or axles for placing as much of the weight of the motors as possible on springs, and for raising the center of gravity to a reasonable height (5 ft or over). With a low center of gravity every sidewise movement of the mass of the locomotive strikes a blow sidewise on the track, but with a high center of gravity a side swaying is transformed into a downward thrust on the track. As the track is not usually designed to withstand great side thrusts, it is better to avoid a low center of gravity in high-speed locomotives. For higher capacities, the cab underframe must be relieved of the draft forces by articulating the trucks with a hinge joint and mounting the draft gear on the truck frame.

The coefficient of adhesion which can be relied upon in electric locomotives at running speeds is from 18 to 22 per cent, tending to decrease at higher speeds. It is higher for electric than for steam locomotives on account of the uniform torque of the electric motor, while the steam locomotive has the advantage of coupled drivers, which minimize the effect of local slippery spots on the rail. The coefficient at starting is usually taken at 25 per cent, and with clean dry rails may at low speeds be as high as 30 to 40 per cent.

CLASSIFICATION. Locomotives are usually classified by the arrangement of their wheels and the subdivision of the wheels into driving wheels and guiding wheels. The Whyte classification for steam locomotives has been applied to electric locomotives to some extent, but a distinct system, which distinguishes more clearly between driving and idle axles, has come into use both in the United States and abroad. Numerals represent the number of either idle axles in a guiding truck or simple weight-bearing axles in a driving truck. The number of adjacent drivers in one truck or driving wheel-base is indicated by the letters A, B, C, D, etc., for one, two, three, and four axles respectively. For example, the symbol 2 – D – 2 represents an electric locomotive which, in the Whyte classification, would be designated 4 – 8 – 4. If two driving wheel-bases are used, two letters are required, separated by a minus sign for a non-articulated unit (transmitting the tractive force through the truck center pins and the cab underframe) or a plus sign for an articulated arrangement. For example, 2 – C + C – 2 would represent two "ten wheeler" or 4 – 6 – 0 wheel arrangements coupled back to back with an articulated joint or hinge.

TYPES OF MOTORS. Locomotives are built with various types of motors and operate from various systems of electrical distribution, e.g.,

a. Direct-current system, with d-c series motors operating from 600-, 750-, 1200-, 1500-, 2400-, or 3000-volt trolleys. In all cases each motor must be insulated for full trolley voltage, but on the higher voltages two or more motors are operated in series.

b. Alternating-current system.

1. Alternating-current series motors fed from transformer secondaries at approximately 250 volts at 25 cycles. The contact line may be 3300, 6600, 11,000, or 22,000 volts, though 11,000 is by far the most common. Freedom from the restriction of inconvenient high trolley potential on the motor commutator is a strong point in favor of the a-c system.

2. Direct-current series motors fed from a motor generator carried in the locomotive cab. This makes available the advantages of the high-voltage, a-c distribution and the ruggedness of the d-c series motor. Economical use of this type of unit is confined to heavy-drag, slow-speed service.

3. Three-phase induction motors fed from a single-phase trolley through a Scott-connected transformer and phase converter. Acceleration is secured by inserting resistance, usually by means of a liquid rheostat, in the rotor circuits.

4. Three-phase induction motors fed through a transformer, from a three-phase trolley system. This system is widely used in Italy, but the only installation of its sort in the United States has been superseded. The complexity of the three-phase distribution is a great handicap to this system.

CONTROL SYSTEMS. The control of all modern electric locomotives is by some form of remotely operated power control, as the currents required are too large or the voltage too high for a hand-operated control. Many locomotives are equipped with multiple-unit control., one engineman operating and controlling two or three locomotive units at the same time.

Table 13. Electric Locomotives

	Cleveland Union Terminals—Passenger	New York, New Haven & Hartford R. R.—Passenger	New York Central Railroad—Passenger	Chicago, Milwaukee, St. Paul & Pacific Railroad—Gearless Passenger	Chicago, Milwaukee, St. Paul & Pacific Railroad—Geared Quill Pass.	Great Northern Ry. Passenger and Freight M. G. Type	Great Northern Ry. Freight M. G. Type
Date installed	1929	1931	1926	1919	1920	1930	1926
System	D-c	Single-phase	D-c	D-c	D-c	Single-phase	Single-phase
Trolley voltage	3,000	11,000	600	3,000	3,000	11,000	11,000
Service	Passenger	Passenger	Passenger	Passenger	Passenger	Freight & pass.	Freight
Weight in working order (lb) Total	419,000	403,000	285,000	528,000	620,000	539,000	368,600
Weight in working order (lb) drivers	312,000	272,000	285,000	464,000	420,000	426,200	282,700
Weight of equipment	144,220	173,000	110,800	234,000	270,000	278,500	176,800
Number of motors	6	6 (twin)	8	12	6 (twin)	6	4
Diameter, drivers	48 in.	56 in.	36 in.	44 in.	68 in.	55 in.	56 in.
Total wheel-base	69 ft 0 in.	66 ft 0 in.	46 ft 5 in	67 ft 0 in.	79 ft 10 in.	58 ft 8 in.	62 ft 10 in.
Rigid wheel-base	15 ft 0 in.	13 ft 8 in.	6 ft 6 in.	13 ft 9 in.	16 ft 9 in.	15 ft 4 in.	16 ft 9 in.
Total horsepower—1-hr rating	2,900	3,440	2,488	4,020	4,680	3,260	2,165
Tractive effort—1-hr rating	29,200	25,200	18,440	59,500	66,000	67,200	56,250
Speed, miles per hour—1-hr rating	37.3	51.2	50.6	25.3	26.7	18.2	14.0
Classification	2–C+C–2	2–C+C–2	B–B+B–B	1B+D+D+B1	2–C–1+1–C–2	1C+C1	1–D–1

	New York Central Railroad—Freight	Mexican Railway Co.—Freight and Passenger	Virginian Railway—Freight	New York Central Railroad—Switcher (3-power)	Illinois Central Railroad—Switcher	Pennsylvania Railroad—Freight	Pennsylvania Railroad—Passenger
Date installed	1930–1	1925	1925	1930	1930	1932	1935–1943
System	D-c	D-c	Single-phase	D-c	D-c	Single-phase	Single-phase
Trolley voltage	600	3,000	11,000	600	1,500	11,000	11,000
Service	Freight	Freight	Freight	Switching	Switching	Freight	Passenger
Weight in working order (lb) Total	266,000	308,000	429,530	257,000	200,000	392,000	460,000
Weight in working order (lb) drivers	266,000	308,000	315,100	257,000	200,000	220,000	300,000
Weight of equipment*	106,200	135,000	157,400	132,600	73,000	160,000	164,000
Number of motors	4	6	2	4	4	3 (twin)	6 (twin)
Diameter, drivers	44 in.	46 in.	62 in.	44 in.	45 in.	72 in.	57 in.
Total wheel-base	41 ft 0 in.	40 ft 6 in.	37 ft 6 in.	34 ft 1 in.	27 ft 3 in.	49 ft 10 in.	69 ft 0 in.
Rigid wheel-base	14 ft 6 in.	9 ft 2 in.	16 ft 6 in.	8 ft 3 in.	8 ft 7 in.	20 ft 0 in.	13 ft 8 in.
Total horsepower—continuous	2,490	3,030	2,030	1,665	1,550	3,750	4,620
Tractive effort—continuous	41,800	58,000	54,000	34,100	20,400	22,300	19,140
Speed, miles per hour—continuous	22.3	19.6	14.1	18.3	28.5	63.0	90
Classification	C+C	B+B+B	1–D–1†	B–B	B–B	2–C–2	2–C+C–2

* Includes all electrical equipment, and air brake parts except foundation brake rigging and air brake cylinders.
† Three cabs normally operated as a single locomotive.

TYPES OF ELECTRIC LOCOMOTIVES. The simplest form of locomotive is a two-truck car ballasted to provide greater adhesion. Freight and baggage cars are frequently so used on interurban roads. In general, the ordinary car motors are not suited to locomotive service, being wound for too high speed.

The type of locomotive used for light haulage and terminal switching consists of a cab containing the control equipment and supported by two four-wheeled trucks, each with two axle-hung motors. The tractive forces are transmitted through the cab underframe. Such units range in weight up to 100 tons and in horsepower up to 1850. They are well adapted to switching purposes in terminal freight yards, and in smaller units to hauling freight trains on interurban electric railways.

A further development of this type of unit is exemplified in the Detroit River Tunnel locomotives of the Michigan Central Railroad. The heaviest of these units weighs 126 tons. They are similar to the type described above except that the two trucks are fastened together by an articulated coupling. The draft gear is carried at the ends of the truck frames, thereby relieving the cab underframe of all tractive and buffing forces.

A still further step is illustrated by the West Side freight locomotives of the New York Central Railroad, which have two three-axle articulated trucks to obtain greater motor capacity and to subdivide the load sufficiently to keep the axle loading within limits.

The Cleveland Union Terminal passenger locomotives have a two-axle truck for guiding purposes at each end of a wheel-base, like the New York Central units just described.

While the above types have been evolving by steps from the streetcar prototype, another type has been developing, designed especially for high-speed railway service. Several earlier attempts have contributed features. In 1906 the New York, New Haven, and Hartford Railroad purchased 41 passenger locomotives having a number of new features. In addition to being the first a-c installation of considerable size, these locomotives introduced the quill drive. To relieve the dead weight on the axle, attendant upon an axle-hung motor, the motor armature was mounted on a quill which was frame-supported while surrounding the axle and driving through flexible spring elements. In 1912 the New Haven made further progress by introducing the twin motor. A gear was mounted on the quill, which was supported in bearings below a frame-mounted twin motor, both pinions meshing with the gear.

Meanwhile the Pennsylvania was using locomotives patterned closely after the steam locomotive, driving from one or two large motors by means of side rods. Their most recent development, in connection with the most comprehensive electrification yet attempted, abandons the side rod in favor of the individually powered quill-drive axle mounted in a frame with integral cab, like a steam locomotive. Their latest passenger unit has a $2 - C + C - 2$ wheel arrangement, each of the six driving axles having a twin motor and quill drive.

Table 13 gives data on a number of electric locomotives.

4. ELECTRIC CARS AND TROLLEY COACHES

PCC TYPE CARS. Practically all the streetcars built since 1936 are the PCC (Presidents Conference Committee) type. This represents a radical departure from previous designs, with its light weight, high-speed motors, hypoid gearing, resilient wheels, all-steel body, and dynamic motor braking for the main service brake.

CONSTRUCTION OF TRUCKS. By liberal use of rubber throughout the truck, noise and transmission of shock at crossings have been largely eliminated, resulting in much quieter and smoother operation than with the older trucks with axle-hung motors. The truck frame is built up of steel castings and structural shapes welded together, the two motors in each truck being clamped rigidly to the upper portion of the frame. A large conical bearing in the center is provided for the kingpin which keeps the body in position on the trucks. Wheels are the so-called "resilient" type, including rubber sandwiches held between plates, this construction being effective in reducing noise and providing a smoother ride. Flexible shunts carry the motor current around the insulated portions of the wheel.

MOTORS. Four 300-volt motors are used per car, the two on each truck being connected permanently in series for operation on 600 volts. Each motor has an hourly rating of 55 hp and weighs 715 lb. This corresponds to 13 lb per hp, compared to an average figure of 25 lb per hp for axle-hung motors for city service. With the standard gear ratio of 7.17 to 1 and 25-in.-diameter wheels, the motor makes approximately 100 revolutions per minute for each mile per hour of the car. This relatively high speed, combined with careful design, accounts for the light weight per horsepower. For several years these motors were self-ventilated, but in the present design the fan has been removed to permit installation of a drum brake on the drive end, and the motor is cooled by forced ventilation at a rate of 250 cu ft per min.

CONTROL. Smooth operation is obtained during acceleration and dynamic braking by use of an automatic multipoint control, which provides about 100 series resistance steps, as compared to 8 formerly used with drum-type controllers. This permits accelerating and braking rates up to 4 mphps with comfort, making possible schedule speeds of 14 mph with 8 stops per mile. Controls are foot operated for most applications, a master controller governing the action of a pilot motor, which in turn operates the motor controller.

BRAKE EQUIPMENT. The major portion of the braking duty is performed by the traction motors acting as generators, down to a car speed of about 1 mph, at which point the friction brake is applied automatically to make the final stop and hold the car at rest. Air brakes acting on the car wheels were used for this purpose until a few years ago. Present designs include a drum brake on the drive end of the motor controlled by a solenoid type of actuator, energized from the 32-volt battery. With the elimination of the air brake the compressor is no longer required, and all functions previously performed by air pressure are now electrically accomplished. A supplementary brake is provided in the form of four electric track brakes suspended by springs over the track between the wheels on each truck. These brakes are pulled to the rail by their own magnetism when energized from the car battery and are particularly useful in making stops under slippery rail conditions, as they present a separate friction surface to the rail and are not dependent upon wheel-to-rail adhesion for their effectiveness.

CAR BODY. All steel construction is used in the car body except for the floor, which is wood. The majority of cars are single end for one-man operation. Train operation has been provided in some instances up to three cars in multiple unit.

TROLLEY COACHES. The trolley coach or trackless trolley has been in fleet service many years and has been installed in cities all over the world. The use of modern trolley coaches in cities in the United States began in 1928, and they are now operating in quantity in some 60 cities, including Seattle, Shreveport, Cleveland, Boston, Kansas City, and Pittsburgh.

The modern trolley coach is constructed along standard automotive lines, including conventional wheels, tires, steering mechanism, air brakes, bodies, and seats. The most common size now being put into service has a seating capacity of 44 passengers; wide aisles facilitate heavy rush-hour loading.

Electrical traction equipment consists of one 140-hp motor and suitable multipoint automatic control to assure smooth accelerating, running, and dynamic braking characteristics. Electrical power is taken from two 600-volt overhead wires by two flexibly mounted poles on top of the coach. The flexible mounting of these poles, coupled with the standard type of motor-vehicle steering, enables this vehicle to maneuver easily in traffic and to pass normal road obstacles.

Table 14. Trolley Coach

Item	Dimension
Length	34 ft
Width	100 in.
Wheel-base	239 1/2 in.
Seating capacity	44
Weight (empty)	21,000 lb
Weight (seated load)	27,160 lb
Maximum speed (seated load)	37.5 mph
Schedule speed (8 stops/mile)	13 mph
Number motors	1
Motor capacity	140 hp

The characteristics of a typical trolley coach are given in Table 14.

DOUBLE-TRUCK CARS. Electric railway cars are now almost entirely of the double-truck type and range in size from 42 ft 0 in. overall length, with seating capacity of 42 passengers and maximum speed of 35 mph for city service, to 80 ft 0 in. overall length, seating capacity of 121 passengers, and maximum speed of 75 mph for interurban and electrified steam railroad suburban service.

The PCC car for city service, which has been described, falls in this class. It is rapidly superseding the older style of streetcar. The following brief descriptions refer chiefly to interurban, rapid transit, and suburban cars.

The body of these cars is supported by two bolsters, each having a lubricated pivot bearing, and two bearing plates located about 2 ft 6 in. each side of longitudinal center line of car. These bearings rest upon similar bearings on the truck, body and trucks being held together by a large kingpin. This arrangement permits the truck to swivel through a large angle in running on curves as sharp as 30-ft radius.

Construction of Trucks. The truck consists of a rigid frame of either structural steel, pressed steel, or cast steel, or a combination of all three, having vertical guides at all four corners, in which slide journal boxes equipped with either plain babbitt-lined bronze bear-

Table 15. Data on Electric Cars and Trolley Coaches

	Trolley Coach	Street-Car	PCC City Car	Inter-urban	Subway, N.Y. City	Subway, Phila-delphia	Suburban, Lacka-wanna R.R.	Suburban, N.Y., N.H., & H. R.R.
Seating capacity	44	44	54	38	60	75	84	120
Length overall, ft.	34	41 1/2	46 1/2	43 3/4	60 1/2	67 1/2	70 1/4	79 1/2
Number of motors	1	4	4	4	2	2	4	4
Horsepower of each motor	140	35	55	100	190	210	255	260
Weight of body, less propulsion equip., lb	18,190	15,350	18,300	23,100	45,400	64,000	69,800	84,400
Weight of trucks, less motors, lb	......	8,350	13,140	12,300	25,400	31,800	35,700	43,200
Weight of electrical equipment, less motors, lb	1,350	1,200	1,700	1,740	2,500	2,800	19,800	23,720
Total weight of motors, lb	1,460	6,100	2,860	10,660	10,700	11,400	21,900	26,680
Total weight of vehicle, lb	21,000	31,000	36,000	47,800	84,000	110,000	147,200	178,000
Weight of seated passenger load, at 140 each, lb	6,160	6,160	7,560	5,320	8,400	10,500	11,750	16,800
Weight per seated passenger, lb	480	706	667	1,260	1,400	1,467	1,752	1,483

ings or antifriction bearings. The truck frames are supported on springs resting on the journal boxes, or on equalizer beams spanning both journal boxes.

Suspension of Motors. Motors have been almost universally mounted by the nose suspension method, with motors located between truck transoms and axles. Motor frames have two split bronze bushed bearings which bear on the axle, and a nose which bears on the transom, with wear plates and springs between nose and transom. Power is transmitted by single reduction spur gears, sometimes with helical teeth. Some experimental special drives with completely spring-borne high-speed motors have been applied, one embodying an application of automotive type axle and worm drive with armature shafts longitudinal, others using double reduction spur gears.

CAR BODIES. The arrangement and dimensions of car bodies have become fairly well standardized according to the service for which they are designed: interurban, rapid transit (subway and elevated), and suburban.

Interurban car bodies are usually provided with front entrance and rear exit doors on both sides, but some have been built with a single entrance door at front for single-end operation by one man. Seating capacity is from 45 to 50, with practically all transverse seats of the reclining type upholstered in various fabrics.

Rapid-transit car bodies are built with two, three, or four doors per side to facilitate loading and unloading from station platforms level with the car floor. Seats are provided for 50 to 65 passengers and are usually nearly all longitudinal and upholstered in rattan.

Suburban car bodies for electrified steam railroad service are generally duplicates of steam road practice, modified by restrictions of weight and clearance. Enclosed vestibules are provided, with covered connecting doorways in some cases, for passing between cars. Seating capacity is from 82 to 120. Nearly all seats are transverse, upholstered in rattan or imitation leather.

DIMENSIONS AND WEIGHTS OF TYPICAL EQUIPMENTS. Table 15 gives the principal data for electric cars operated in various typical services.

5. CONTROL SYSTEMS FOR RAILWAY MOTORS

The function of the control equipment is to regulate the speed and direction of the motors by certain definite systematic changes in connections. The speed of d-c railway motors is controlled in three ways: (1) by connecting suitable resistors in series with the motors, which will reduce the voltage across the motors and thereby the current which they will take; (2) by changing the connection of the motors so that they will be connected at first in series, thereby applying half the line voltage to each motor, and then in parallel across full line voltage, (3) by field shunting or field tapping after full line voltage has been applied. This reduces the motor field strength, thereby increasing the speed of the motor in order to produce the same counter-emf as with full field.

In most control equipments a combination of all three methods is used. For speed control of a-c railway motors an auto-transformer, or preferably a two-winding transformer with taps, is used instead of the resistors.

The direction of rotation of the motors, d-c or a-c, is changed by changing the direction of the current in either the fields or the armatures; it is customary to connect the terminals of each field coil to a reversing switch in order to accomplish this effect.

TERMINOLOGY. The following terms are in general use.

Cylinder or Drum Control or Direct Control are names commonly applied to an equipment in which all the connections are made by contacts on a cylinder or drum, which is manually operated by the motorman and located on the platform. This may, therefore, be called direct control.

Multiple-unit, Indirect, Remote Control, or Train Control are names applied to an equipment in which the changes in connection of the main power circuit are made by switches called "contactors," usually located underneath the floor of the car, and controlled by electric circuits coming from a small master controller on the platform. Two systems of multiple-unit control are in use in this country:

Magnetic Contactor System. In this system the contactors are closed by electromagnets.

Pneumatic Contactor System. In this system the contactors are closed by compressed air from the air-brake system, the air valves at the switches being controlled electrically from the master controller.

Non-automatic Control is a term applied to that method of control in which the motorman has it in his power to regulate the current to any value he pleases by moving the controller handle, the change in connections depending only upon the motion of the handle.

Automatic Control, as distinguished from non-automatic, is a type of control in which certain automatic devices prevent the motorman from causing the motors to take a current greater than a predetermined value. With this method the motors start with a definite current, and as soon as the current has decreased to a specified value, a change in the connections is automatically made. Thus the rate of acceleration and the current are kept practically uniform throughout the period of control. It is nearly always used in connection with multiple-unit control.

Rheostatic Control consists in connecting a resistor in series with the motor and short-circuiting consecutively parts of this resistor. It is seldom used at present, except on mining locomotives and for single-motor operation.

Series-parallel Control, which is used on practically all railway equipments, includes the feature of connecting two motors and their resistors in series on the first step, then short-circuiting portions of the resistor consecutively until all resistance is cut out, under which condition the motors will operate efficiently at approximately half speed. On the next step of the controller the two motors with resistor in series are connected in parallel and subjected to full line voltage. There are two methods of accomplishing the change from series to parallel.

Transition with Series Resistance. During the transition from series to parallel a resistor is placed in series with one motor and the other motor is first short-circuited, then disconnected from the main circuit, and finally placed in parallel with the first motor. This method is in general use in equipments of small motors with the so-called type K controller.

Bridge Transition. The so-called "bridge" method consists in grouping the motors and their resistors like the arms of a Wheatstone bridge, so that after the two motors are in full-series position the resistors may be placed in circuit again in parallel with the motors, without opening the circuit; the two motors are then connected in parallel with each other and each in series with its own resistor.

Series-parallel Control with Four-motor Equipment. Whereas the two methods just described apply particularly to two-motor equipments, they are equally applicable to four-motor equipments by connecting two motors permanently in series or parallel and treating them as a unit.

DRUM CONTROLLERS. The type of control, as well as the construction of the controllers for ordinary single-car equipments, has been practically standardized in this country, the chief manufacturers supplying control equipments which are practically identical. Railway drum controllers are designated by the type letters K, B, and R with appropriate subnumbers and subletters to distinguish smaller variations. The most common form is the K controller. On the first notch this controller connects the motors in series with each other and with the starting resistor. The resistor is then cut out by steps until the motors alone are in series. Transition is then effected from series to parallel connection of motors with resistance in the circuit. The resistor is then again cut out in steps until the motors are connected directly across the line. B controllers, in addition to the starting steps, include a certain number of braking steps. They find little use in the United States, where air brakes are nearly universal. R controllers provide resistance steps only, without series-parallel connection of motors. Applications of R controllers are largely confined to single-motor equipments and to very small locomotives where simplicity is desirable.

Railway drum controllers generally include two drums held in a cast frame. A cover of insulating material encloses the front and sides and, when removed, gives access to all working parts. The main drum, which makes the resistor and transition connections, consists of an insulated shaft carrying a set of contacts. Fingers press on these contacts to make the proper connections. Between fingers are barriers of arc-resisting material. A magnetic blowout operates between these barriers to assist in extinguishing arcs. The second drum is used to reverse the direction of motion of the motors. It consists of a wooden drum with copper contacts with fingers pressing on the contacts. It is mechanically interlocked with the main drum so that it cannot be moved with power applied.

Railway drum controllers find their widest application on cars used on city streets. Except in special cases they are not applied to equipments of an aggregate motor capacity of more than about 260 hp.

Where type K controllers are used for motors aggregating more than about 100 hp, they are sometimes adapted with a modification of the remote control by the addition of an electrically operated line breaker, placed underneath the floor of the car, the function of which is to open the main power circuit whenever necessary and thus remove all flashing and arcing from the controller. This expedient makes it possible to use a smaller controller for a given capacity of motors and obviates all danger to the passengers from fire and fright. The scheme is accomplished by substituting for the main power circuit on the controller an auxiliary circuit carrying only a fraction of an ampere; every time this auxiliary circuit is opened in the main controller, the line breaker underneath the car opens the power circuit. When the auxiliary circuit is closed, the line breaker closes the main circuit. By means of an overload trip operated by a coil in the main circuit, this breaker is also used as a circuit breaker, and if the current taken by the car exceeds a certain value, a relay opens the auxiliary circuit, which in turn causes the line breaker to open.

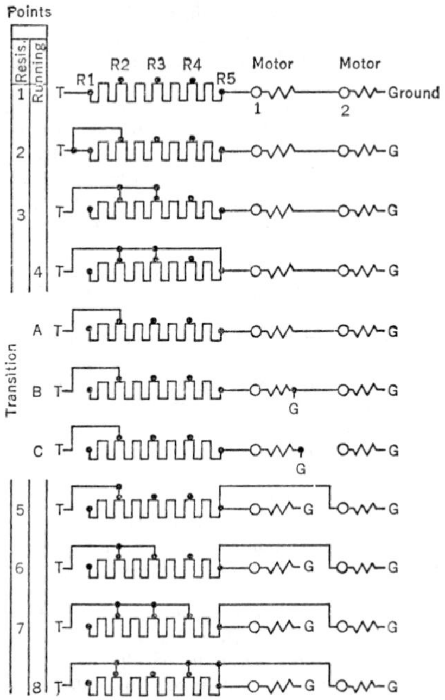

Fig. 4. Type K Control

Method of Operation of the Type K Control. The principle of the type K control for small and moderate-size motors is shown in Fig. 4. The controller has an operating handle which moves the main cylinder and thereby changes the connections, and also a reversing handle which moves the reversing cylinder. The latter merely changes the direction of the current through the fields of all the motors with respect to the armature. These two handles or cylinders are interlocked, so that the reversing handle can be moved only when the operating handle is in the off-position, thus preventing reversal with voltage on the motors.

One terminal of one of the motors is grounded throughout. The first three points are known as accelerating steps. As resistance is in circuit for each of these points, the controller should not be left on any of them for a considerable length of time, as there is a considerable power loss in the rheostats and they are not designed for continuous operation. The fourth step, full series, is an efficient "running point," giving about half normal speed. The next two steps are transition steps and are not marked as points on the controller, as they must be passed over rapidly. During this period one terminal of the second motor is grounded, thus short-circuiting the motor, which has one terminal grounded initially; the connection between the two motors is then opened; finally the two motors are connected in parallel but in series with a part of the rheostat. Points 5, 6, and 7 are accelerating steps with motors in parallel and resistors in series and are therefore not to be used continuously. At the last point the motors are in parallel, and all resistance is cut out; it is therefore an efficient high-speed running point. The two terminals of the armatures of each motor are led to fingers on the reversing cylinder, as are also one terminal of each field and the two points of the main circuit to which the motors are connected. Thus, in reversing, the current is reversed in the armatures, but not in the fields.

In all controllers a magnetic blowout and an "arc chute" are employed to interrupt the current quickly and direct it away from the contacts, in order to prevent short-circuiting other contacts. In the older types of controllers this was obtained from one large magnet coil and a large iron pole piece covering all the contacts. In the later forms each contact has an independent blowout coil and small pole pieces. This gives a more powerful effect and more accurately directs the arc in the proper direction.

REASONS FOR MULTIPLE-UNIT CONTROL. When the total capacity of the motors on a car or locomotive exceeds 300 hp, it is advisable, and when the capacity exceeds about 450 hp, it is necessary, to use the indirect or multiple-unit control, for the cylinder type of controllers required to handle the large currents becomes too bulky and dangerous to place on the platforms of passenger cars. The cylinder control is also inadequate when it is desired to control simultaneously the motors on the several cars in a train, which is necessary in order to obtain the high tractive effort essential in high-speed service on elevated and underground railways.

GENERAL ELECTRIC MULTIPLE-UNIT CONTROL. Two types are in use, the M and the MA. The chief difference is that the type M is non-automatic, whereas the

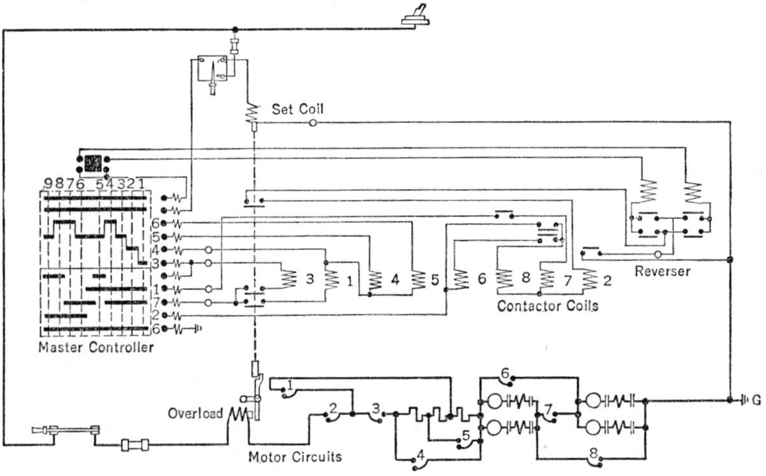

Fig. 5. General Electric Type M Control

type MA is provided with a current-limiting relay. Figure 5 is a diagram of the type M, showing the control circuits in light lines and the motor circuits in heavy lines.

The material included in the control equipment of a motor car consists of:

2 master controllers.
1 motor controller containing 8 or 10 contactors and a reverser.
3 master control switches.
1 main switch.
1 main fuse box.
1 set of rheostats.
Cables, train couplers.

Master Controller. This is very similar to an ordinary railway drum-type controller but is much smaller, as the current carried by it is small. Each master controller is equipped with an operating handle, reversing handle, individual magnetic blowouts, and (optionally) a "deadman's handle," which automatically interrupts the current and applies the brakes when the motorman's hand is removed from the button located in the top of the handle.

Motor Controller. This consists of an iron box lined with asbestos, in which the several contactors and the reverser are placed, and mounted under the car.

Contactors. Each contactor consists of a powerful magnet operating an arm by means of a toggle joint against a spring pressure. This arm closes and opens the circuit in a strong magnetic field which acts as a blowout. As one contactor can carry and break currents of several thousand amperes, no extra circuit breakers are required. Each contactor is provided with interlocks, so that contactor 2 cannot be operated until after contactor 1 is closed, thus providing the proper sequence of operation under all conditions.

Reverser. This is a switch with several circuits and contacts and is comparable to the reverser cylinder in an ordinary controller. These contacts are mounted on a rocker arm actuated by two electromagnets, one for moving the switch to the forward and the other to the reverse position. The electromagnets receive their current from the master controller and are interlocked, so that only one can be operated at a time.

Switches and Fuses in Control Circuit. Motor cut-out switches are located on the reverser to permit cutting out a disabled motor. In the control circuit there is one main control switch and fuse to protect the control circuits, and near each master controller is located a "control and reset" switch, which in one position closes the circuit to the resetting coil of the overload relay in the main controller and in the other position closes the supply circuit for the master controller.

Main Switch. A knife-blade switch is placed in the power circuit and is intended to disconnect the motor circuits from the trolley when it is desired to test the motor controller.

Train Line Couplers are provided where cars are to be operated in trains. These couplers and jumpers provide a means of connecting similar control circuits of the different cars. They contain from 8 to 12 wires and are so designed that it is impossible to couple the cars together improperly.

Current-limiting Relay. For automatic control, or acceleration at a predetermined current, a current-limit relay is provided on each car, and this prevents each successive contactor from operating until the current in the motors has decreased to a predetermined value. On roads having a fairly level profile and operating with frequent stops this refinement is desirable, as it makes it possible for the motorman to accelerate the train every time at the maximum allowable rate and yet never exceed that rate except on a down grade. When this relay is provided, the control equipment is known commercially as the type MA control.

PC Control. This type was first introduced in 1914 and is a simplification of the older multiple-unit systems. It uses air from the air-brake system to operate the switches in groups, thus requiring only two or three air cylinders and valves instead of many, which is the cause of its simplicity and improved reliability. On account of its cheapness it is quite generally used for single cars as well as trains.

The PC control (manufactured by the General Electric Company) uses a combination of electrically controlled pneumatic cylinders to rotate, notch by notch, a main camshaft, and the cams on this shaft open and close spring-actuated switches or "contactors." These cam-operated switches cut out the starting resistors and change the motor connections from series to parallel. A separate air cylinder operates the line breaker, while other cylinders actuate the reversing drum. The line breaker has suitable magnetic blowout coils and serves as a circuit breaker by means of a series trip coil.

The advantages claimed for both forms of this type of control are:

1. All power circuits and the weight of the main control apparatus are removed from the car platform; i.e., the system is operated by remote control.
2. The main line current is always broken by large, suitably designed breakers.
3. Multiple-unit operation may be secured with less apparatus and therefore adapted to smaller equipments.
4. The operation of the contactors, being mechanical, is more positive, and therefore the time element is definite.
5. Automatic or selective acceleration may be provided with less complexity.
6. A greater number of motor cars may be operated in multiple in one train (16 motor cars in one train is claimed).
7. The current from the control circuits is so small that storage batteries may be used, thus making this control easily adaptable for operation on a-c or high-voltage d-c systems.

WESTINGHOUSE "UNIT SWITCH" SYSTEM. The Westinghouse "unit switch control" is designed for either multiple-unit operation of several motor cars in a train or operation of single cars or locomotives using either large currents or high voltage. Unit switch control equipments are classified according to whether they are arranged for hand or automatic acceleration, whether they employ energy from the line or from a battery, and whether they have certain other features, such as field control. The standard designations are as follows: HB, ABF, HL, HLF, AL, etc. Thus it will be noted that "H" signifies hand acceleration, "A" automatic, "B" battery control, "L" line control, "F" field control of motors, etc.

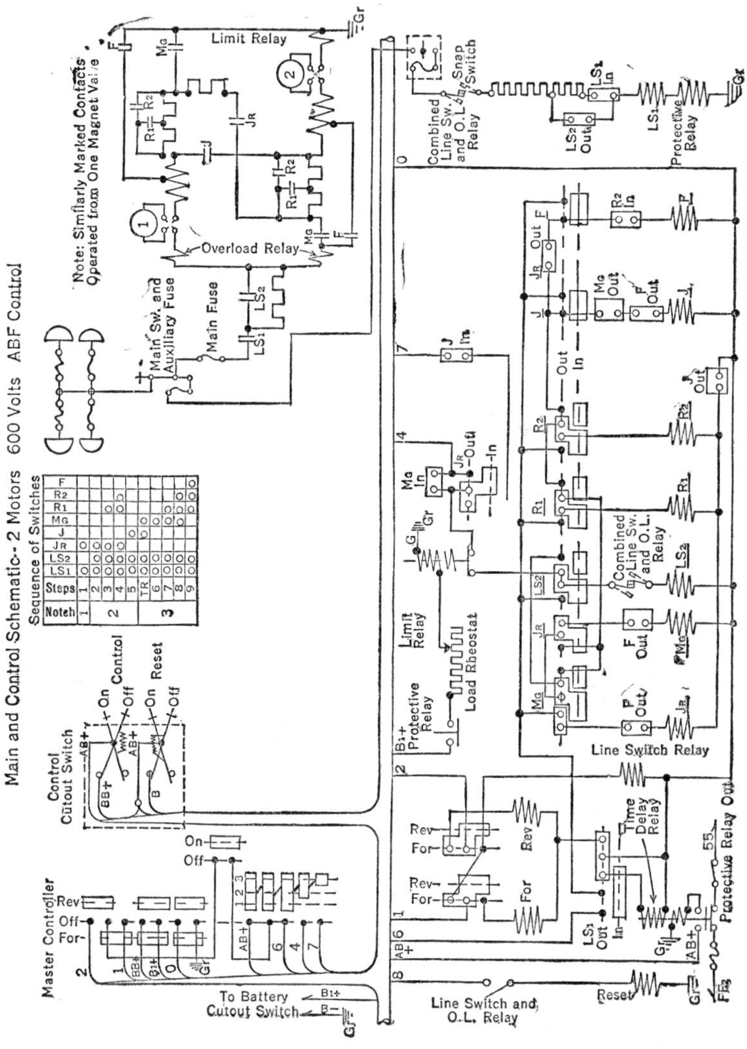

FIG. 6. Westinghouse Unit Switch Control

Rapid-transit Control. ABF control is automatic in its progression, being under control of a limit relay which permits each additional step to be taken when motor current has dropped to a predetermined value. It provides series-parallel operation of motors, using bridging transition, which ensures smooth, uniform acceleration. The distinctive features of ABF control are the three-wire system, which makes it necessary to complete three circuits at the controller before power can be applied to the motor, inherent rail gap protection, and the use of low battery voltage.

The apparatus included in ABF control includes a master controller, line switch, a group of unit switches, and the necessary protective and accelerating relays. The unit switches are electropneumatically operated and are used to make the motor and resistor connections.

ABF control is suitable for all classes of rapid-transit service, such as subway, elevated, and interurban lines.

Operation. The interlocking shown in Fig. 6 provides a simple and effective means for automatic progression. The unit switches are provided with interlocks, which are electrically connected with the valve magnet in such a manner that the closing of one switch energizes the magnet of the next, thus producing automatic progression of the unit switches under the control of a limit relay. This relay controls the rate at which resistance is cut out of the circuit so as to give uniform accelerating current. The relay is of the solenoid type, being operated by one motor current. When this current exceeds a predetermined value, for which the relay is adjusted, the control feed circuit is opened, thus halting the switch progression. When the current falls below the predetermined value, the control feed circuit is closed and allows the unit switches to continue their progression.

The air required for operating the unit switches is taken from the air-brake system through a reducing valve at a constant pressure of 70 psi.

HL Control. HL control provides series-parallel operation of motors with resistor steps in both series and parallel like the K controller. The changes in motor and resistor connections are made by a number of independent pneumatic switches, each provided with a strong magnetic blowout and held normally open by a powerful spring. The switches are controlled by a magnet valve, which admits or exhausts air from the switch cylinder. A pneumatically operated reverser controls the direction of motion of the car. This reverser is operated by two pistons similar to those used in the unit switches.

Energy to operate the magnet valves is taken from the trolley and reduced to a low voltage by taking taps across a resistor, which is connected from trolley to ground. This low-voltage energy is distributed by a master controller to the various magnet valves of the unit switches and reverser. These same wires supplying energy to the magnet valves also run to each end of the car, from where they may be run to other cars by means of jumpers. In this way multiple operation of a number of cars may be obtained from a single master controller. The number of master controller notches is the same as the number of steps of operation, each step being under direct control of the operator.

HL control is generally used on equipments with aggregate motor capacity of more than 200 or 250 hp where stops are relatively infrequent. It is also used on smaller equipments where train operation is required.

CONTROL FOR PCC CARS. The control for these cars is quite unlike that applied to former streetcars. Service requirements demand accelerating rates up to 4.75 mphps and braking rates of the same order and higher in emergencies. Such high accelerating rates necessitate a very smooth and rapid building up of line current from zero to the maximum. This is accomplished by cutting out the accelerating resistor in a large number of steps in an accurately timed sequence. The four 55-hp, 300-volt traction motors are permanently connected two in series and the two groups in parallel; thus there is no transition from series to parallel, as in most earlier streetcars. There are no air brakes on the most modern type of PCC cars. Instead, retardation is accomplished by a combination of dynamic, friction, and electromagnetic track brakes. The dynamic brake provides the principal braking function in controlling the retardation of the car; the friction brake is held inoperative but in readiness, until the dynamic brake "fades out" just before the car comes to rest; and the track brake functions only when very high or emergency rates of retardation are required, and the operator depresses the brake pedal to its extreme position. The friction brake also acts as a self-applied parking brake when the car is standing unattended. There are four track brakes, one over each rail on each side of both trucks. There are also four friction brakes, each mounted on the propeller shaft connecting each motor to the corresponding car axle. The control is so designed that the three types of brakes are properly "blended" in action so as to produce a positive and smooth retardation under all service conditions.

PCC control equipment is manufactured by both General Electric and Westinghouse. Basically both function in the same manner, differing only in details and design of the component parts. The greatest difference appears in the motor controller or accelerator.

The General Electric motor controller consists of a stationary commutator (similar to a motor commutator), with the accelerating resistor sections connected between adjacent segments. A revolving brush arm, driven by a small electric motor, cuts out the resistors in small increments as it passes from segment to segment. One continuous movement of the brush arm over the commutator provides 136 steps of acceleration. The same device controls the dynamic braking at any rate up to 3.5 mphps. Depending upon the car speed at which dynamic braking is applied, it can provide up to 272 steps for braking.

The Westinghouse accelerator consists of 99 spring fingers mounted on an insulated barrel and connected to resistors mounted on the outside of the barrel. The fingers are successively pressed against a bus ring by two spring-loaded rollers, which are driven by a pilot motor. There are 77 accelerating resistor steps and 4 field shunting steps. This control also provides dynamic braking with 95 steps. The dynamic braking is entirely automatic, with rates ranging from 1 to 3.5 mphps.

The other control apparatus on the car includes master and braking controller, line switch, contactor assembly, and relays. A motor generator supplies low-voltage control power and drives two fans to ventilate the control and traction motors.

CONTROL FOR TROLLEY COACHES. Westinghouse has developed Electrocam Control for trolley coaches equipped with one 140-hp series motor. A total of 15 notches are provided in acceleration, including 2 field shunting positions. The maximum accelerating rate of 4 mphps is attained for an average loaded vehicle. A dynamic braking rate of 2 mphps is available.

The control consists of a group of cam-operated switches actuated by a balanced air engine, together with a control panel, forward and reverse selector switch, set of resistors, master controller, and brake controller. Power for operating the system is supplied from a 12-volt battery.

The master and brake controller are located beneath the floor at the operator's position and connected by means of a simple linkage to the operator's control pedals. Movement of the master or brake controller sets up electric circuits to permit the cam controller to advance under the control of a limit switch as traffic conditions dictate.

This type of control provides the maximum amount of simplicity in the way of operating rods, a minimum of 600-volt cable, and the minimum of fatigue for the operator of the vehicle.

The General Electric MRC control is used with a 140-hp compound-wound motor and provides for acceleration, speed control, and dynamic braking. All the control equipment, except the accelerating and braking resistors, is mounted on a single framework. This consists of a mechanically operated master cam accelerating and braking controller, the electrically operated accelerating contactors, the braking contactor, line breaker, relays, etc. The master controller is operated by rods connected to pedals at the driver's position. The master controller governs the operation and sequence of the contactors and relays for applying power, cutting out the accelerating resistor sections, and varying the strength of the motor shunt field. There are 10 steps of series acceleration and 4 steps of motor field strength for speed control of the vehicle above about 17 mph. Dynamic braking is controlled by varying the shunt field excitation of the traction motor operating as a separately excited generator. Control power is taken from the trolley wires at 600 volts.

The use of the compound motor and its control makes it possible, between coach speeds from 17 to 40 mph, to increase speed without rheostatic loss or to decrease speed by returning power to the line, by the regenerative effect, with the minimum use of air brakes.

Both Electrocam and MRC control provide up to 2-mphps dynamic braking effort in making service stops. This feature greatly relieves the duty on brake blocks and drums. For higher rates of braking and for bringing the vehicle to a standstill, air-operated automotive-type brakes, with which dynamic braking is coordinated, are provided.

CONTROL FOR MINING AND INDUSTRIAL LOCOMOTIVES. On small mining and industrial locomotives, where space limitations will not permit mounting full magnetic or pneumatic control, a semimagnetic or semipneumatic system is often used.

With this type of control all accelerating connections of the motors and main resistors are made with switches, magnetically or electropneumatically operated. The controller consists of a master drum which controls the sequence of operation of the switches, and a reverse drum, through which the motor connections are made for the desired direction of operation.

The term "permissible," as applied to control equipment, implies that, when this equipment is mounted and wired with the apparatus necessary to the operation of a storage-battery locomotive or other mining machinery, the complete locomotive or machine is approved by the U. S. Bureau of Mines for operation in gaseous mines. This control may be of the drum or semimagnetic type. All apparatus is enclosed in metal enclosures with metal-to-metal joints between cover and container, the joints being of such a width or construction that no sparks or flame will pass from the interior of the enclosure to the outside in case of an explosion on the inside.

The term explosion-tested or explosion-proof, as applied to control equipment, implies that the enclosures for this equipment have passed the explosion tests prescribed by the

U. S. Bureau of Mines for permissible control but, because of some method of operation or condition of installation, the complete locomotive or mining machine does not have the approval of the Bureau of Mines.

CONTROL OF HIGH-VOLTAGE D-C MOTORS. The motors operating on systems of 1200 to 3000 volts are usually designed to operate two in series; thus each receives normally 600 to 1500 volts. However, the insulation of each motor must be designed to withstand the whole line potential, and each motor must be able to withstand momentarily the line voltage across its commutator, for if one motor slips the voltage will be unevenly divided between them. For operation on high voltage two motors in series are normally treated as a unit, and the series and parallel connection made with these double units. The multiple-unit control is preferable with these high voltages, and contactors or unit switches similar to those for 600 volts are used. To operate the control it is customary to supply either a self-starting dynamotor which provides 600 volts or a motor generator supplying 32 volts, for this purpose as well as for the lights and other auxiliary apparatus.

Provision for 600-volt Operation. Where these equipments operate also over 600-volt sections of road, provision has to be made to change the connection of the dynamotor or motor generator when the transfer is made. If the cars are to operate at reduced speed on the lower voltages, as is usually the case on entering the city districts, no change need be made in the motor connections. If the cars must operate at 600 volts at high speed over an interurban section, however, then provision must be made to separate the pairs of motors so that all motors will be in parallel for full-speed operation on 600 volts. This requires a commutating switch with automatic protection in order to provide that it is always changed when the car passes from one section to another.

CONTROL OF A-C COMMUTATOR MOTORS. To transform the line voltage (3000, 6000, or 11,000 volts) to a voltage suitable for the motors, which is usually from 200 to 300 volts per armature, early equipments used an auto-transformer. Most of the later equipments use a two-winding transformer which, though it is slightly heavier for a given capacity, has the advantage of operating with the secondary ungrounded. Taps on the auto-transformer or the two-winding transformer provide the various voltages necessary to start and control the motors, and thus there is no need of series-parallel control or rheostats, and the energy loss in the rheostats is obviated. Early transformers for cars were oil-insulated. The modern transformer for both cars and locomotives is insulated with askarel, a synthetic liquid. The use of oil, especially in passenger service, is objectionable as a fire hazard. Fewer steps are required for the control of a-c motors on account of the reactance of the circuits. To avoid open-circuiting the connection to the motors in changing from tap to tap or short-circuiting the portion of the transformer between the taps, a "preventive" resistor or reactor is connected in the circuit momentarily during the transition. A reverser is provided to reverse the connection of the series fields or exciting windings.

Provision for D-c Operation. For operation on direct current as well as alternating current, provision must be made to perform the following operations: (1) cut the transformer out of circuit, (2) connect the motors for series-parallel control, (3) connect rheostats in circuit, (4) change the field connection of the motors, (5) change the connections of the compressor motors, (6) change the connections of the lighting circuits. All this is done by a "commutating switch," which is thrown over at the instant the change is made. This switch is so arranged that it can be moved only when the controller is at the off-position. The commutating switch is usually operated by the motorman when the car reaches a dead section of the trolley between the a-c and d-c sections.

6. CURRENT COLLECTORS

Current collectors for electric cars or locomotives are divided into three classes in accordance with the form of the working conductor from which they collect current, as follows:

a. Overhead collector, which may be of the pole type with wheel or sliding shoe contact, the pantograph type with sliding or roller contact, or the bow type.

b. Third-rail shoes, which may be of the over-running or under-running type.

c. Underground conduit plow.

WHEEL TROLLEY. The wheel trolley consists of a grooved brass or copper wheel held in bearings in a prong called a "harp" at the end of a steel pole, which is pressed upward by a system of springs and levers carried by a spring-equipped base, with provision for vertical movement and horizontal rotation, together with adjustment for pressure against the contact wire.

The contact wire is generally from 17 to 18 ft above rails on urban and interurban roads and 22 ft on electrified steam roads with this type of collector. The pressure between wheel and contact wire varies between 20 lb and 35 lb, depending upon speed and current to be collected. The current-collecting capacity of the wheel collector, though governed by speed and by the condition of the overhead system, is generally given as 400 to 450 amp for average urban and interurban operating conditions. As much as 1200 amp can be collected during the brief period of acceleration.

The wheel collector is generally used on street railways and is subject to objections for heavier work on account of limited capacity, wear on rotating parts, liability of dewirements, and necessity for reversal in case of reverse movements, unless they are limited in extent.

Sliding Shoes. The slider shoe is a grooved block of bronze or steel which is used in the "harp" in place of the wheel. It gives sliding contact between the collector and the contact wire. The sliding shoe is replacing the wheel on many interurban and some city lines and very generally on trolley buses. With lubricated contact wires, longer shoe and wire life is obtained, and radio interference is reduced.

PANTOGRAPH COLLECTOR. The pantograph collector, as generally used, consists of one or two flat sliding shoes, mounted on a collapsible frame of pantograph form. Adjustable pressure is provided by springs, compressed air, or both. A roller has been used to a limited extent instead of the shoe or shoes, but, except in one case, has been replaced by the shoe on account of undesirable weight, limited capacity, and rotating parts.

With relatively small current values, 150 to 200 amp, a plain steel shoe is used, without lubrication. For higher values, steel shoes are fitted with renewable wearing strips of copper, copper and steel, or copper alloy and are lubricated. It is practicable to collect from 1000 to 2000 amp, with one collector, with shoes so fitted and lubricated and with suitable overhead construction. Pantographs have been manufactured with operating ranges as high as 129 in. Pantograph pressures vary from 10 to 20 lb per shoe.

BOW COLLECTOR. The bow collector consists of a contact member of bow shape, generally having a renewable part of aluminum or copper alloy, mounted on one or two flexible poles carried by a spring-equipped base. For high-speed, heavy traction work, the bow is sometimes mounted on a pantograph frame. It is suitable for only small currents. The bow is little used in the United States but is popular in Europe.

THIRD-RAIL COLLECTORS. A third-rail collector consists of a shoe or slipper, made of cast iron or steel, which makes contact with the third rail, together with the necessary frame and attachments. The gravity type, for over-running rail, suspended by links, has been replaced largely by a type employing springs or compressed air for pressure. Collectors are of the over-running, under-running, or side contact type, to suit the kind of third rail involved.

Third-rail shoes have relatively high current-collection capacity—about 2000 amp at speeds of 30 to 35 mph and from 500 to 600 amp at 60 mph. (See Article 12 on Third Rails.) As they are self-adjusting, two or three may be placed on a locomotive or car just as well as one, so that there is practically no limit to the current that can be collected in this way. In fact it is customary to put two on each side of each locomotive or car in order to prevent a cessation of current when passing over breaks in the third rail due to switches or crossings.

UNDERGROUND CONDUIT PLOW. The underground plow consists of an insulated steel plate hung from a movable structure on the car. On the two sides of this plate and thoroughly insulated from it are two shoes pressed outward from the plate by springs. These shoes press against the two working conductors, which are usually steel tees separated from each other by about 6 in. and supported on some form of ceramic insulator. The current is led from the shoes to the car body by flexible insulated conductors.

7. BRAKES AND BRAKING SYSTEMS

(See also Electric Cars, Article 4, and Energy Requirements, Article 2.) In order to stop a car or train a torque must be applied to the wheels in a direction opposite to the direction of motion of the car. This may be accomplished by applying a frictional retarding force to the wheel rims, by applying a retarding force to the axle by means of a magnetically operated friction clutch, by applying a reverse torque to the axles by operating the motors as generators, or by applying a frictional force to the rails directly by a "track brake." The method of applying a retarding frictional force to the wheel rims is the most general one and lends itself most readily to the system of manipulating the brakes by compressed air. This system has done much to improve the safety of travel on railroads. The other methods have all been tried and some have been put into practical operation, either to meet special local conditions or for special car designs.

FRICTIONAL RESISTANCES IN BRAKE-SHOE SYSTEM. The application of the usual brake-shoe system makes use of the frictional adhesion between the wheels and the track and between the brake shoes and the wheels. Both these quantities vary throughout a considerable range, and it is therefore necessary to adjust the pressure between the various members so that it is possible to rely on a definite minimum value.

Adhesion between Wheel Rim and Rails. The coefficient of adhesion between the wheel rims and rails varies from less than 15 per cent to over 30 per cent, depending upon the condition of the track and the relative motion between the track and wheel rim. (See Energy Requirements, Article 2.) An adhesion of 15 per cent can usually be depended upon with normal track and can be increased to 25 per cent by the use of sand, but these values obtain only while the wheels are rolling on the track. If they begin to slide, the coefficient decreases considerably. For this reason the braking effort must always be controlled so that the wheels do not slip.

Adhesion between Brake Shoe and Wheel Rim. When the brakes are applied, the retarding force is applied below the center of gravity of the car body. The latter is therefore subjected to a couple and tends to press downward at the forward end and upward at the rear end, thus changing the distribution of weight on the axles and decreasing the adhesion on some axles or trucks. It is therefore not possible to figure on using for braking purposes the same weight per axle as exists at standstill. For this reason the brake-shoe adhesion must be less than the track adhesion. The coefficient of adhesion between the customary cast-iron brake shoe and the steel tire of the wheel varies with the speed and decreases as the time of application increases. As the speed increases, the coefficient drops off, being a maximum of 30 to 25 per cent at speeds from 0 to 5 mph, 20 per cent at 20 mph, 14 per cent at 40 mph, and 7.5 per cent at 60 mph. Thus at high speeds a heavy pressure may be applied without stopping the wheels, but as the speed of the car diminishes, the pressure on the brake shoes must be decreased in order to prevent gripping the wheels and causing them to slide on the track.

Effect of Angular Momentum. In addition to overcoming the linear momentum of the cars the brakes must overcome the angular momentum of the gears and motor armatures. The effect of the latter is to introduce a tendency of the whole motor to rotate around the car axle and add further strains on the gears and on the trucks. For this reason brake shoes hung between the wheels of a truck are better than those hung on the outside of the wheels.

HAND BRAKES are always provided on cars and electric locomotives whether power brakes are employed or not, as they are necessary to hold a car left out of service on a grade, because the air brakes will not hold a car standing idle for any length of time. In hand braking equipment the "foundation" brakes (see next paragraph) are actuated through a drum or lever system which is hand operated.

The "foundation" brakes are that part of the brake equipment usually furnished separately from the power-braking equipment and consist of the brake shoes, hangers, equalizers, levers, etc., back to the brake cylinder. To this is attached the desired form or make of power brake. Of the various forms of power brakes in use, viz., air, electric, regenerative, and electropneumatic, the air brake is the most common.

AIR BRAKES. In electric railway practice three systems of air brakes are in use, each of which is best suited to a definite type of service and has its particular field: (1) straight air-brake system for cars always operated singly; (2) semiautomatic air-brake system for cars operated in trains of two or three but never more than three cars; (3) automatic air-brake system for cars operated in trains of any number of cars; (4) combined automatic and straight air-brake system for cars operated singly or in trains of any number of cars.

Straight Air-brake System. The equipment for the straight air-brake system consists essentially of a motor-driven compressor, a reservoir, a brake cylinder, a motorman's valve, a train pipe, and the foundation brakes. The brakes are applied by direct pressure, i.e., by admitting air from the reservoir directly to the cylinder. The advantages of the system are that it is quick-acting and the braking effort is easily controlled to any value desired. Its disadvantage is that in trains a break in the train pipe renders the brakes ineffective, so that there is no means of applying the brakes on the trail-cars if the train should break apart.

Semiautomatic Air-brake System. This system involves the use of an emergency valve and an emergency pipe on each car, so arranged that if the air pressure in the emergency pipe drops because of cars breaking apart, this valve automatically operates to admit pressure into the brake cylinders and apply the brakes. The usual operation in service is like that of the straight air brake.

Automatic Air-brake System. This system involves an auxiliary reservoir and a "triple valve" on each car. Whenever the air pressure in the train pipe is reduced, either intentionally or accidentally, this triple valve admits the air pressure of the auxiliary reservoir on each car into the brake cylinders. The motorman applies the brakes by opening the train pipe to the atmosphere by means of the engineer's valve. The engineer's valve has at least four positions: "release," "lap," "service," and "emergency." When the handle is turned to the service position, the train pipe is opened to the atmosphere, and the air continuously escapes. This applies the brakes with a continuously increasing pressure. When the desired pressure has been reached, the handle is turned to the lap position, and the pressure of air in the train pipe and the pressure of the brakes on the wheels remain constant. When the handle is turned to the emergency position, a very rapid escape of air from the train pipe causes a quick, heavy application. Turning the handle to the release position restores pressure in the train pipe, releases the brakes, and recharges the reservoirs.

Combined Automatic and Straight Air-brake System. This system combines the advantages of the flexibility and promptness of straight air operation on a single car, and the safety features of the automatic air brake.

Electropneumatic Brakes. This system is now in extensive use in subway and suburban service. It involves controlling the application and release of the brakes on each car by means of electromagnets receiving current from the locomotive cab. By energizing these circuits the motorman can apply or release the air brakes on all cars practically simultaneously. Each car is equipped with the usual automatic brake mechanism, so that the brakes operate pneumatically in case of failure of the electric circuits.

ELECTRIC BRAKES. Electric brakes in most cases are used to supplement the standard air-brake equipment. Three types are in common use: rheostatic brakes, electromagnetic track brakes, and regenerative braking.

Rheostatic Brakes. With this type of equipment the motors are arranged to operate as generators supplying current to rheostats, which absorb the energy. This type of braking is effective, of course, only when the car is in motion. It is possible to operate rheostatic brakes without the refinements in control that are required with regenerative braking.

Electromagnetic Track Brakes. For this type of braking equipment electromagnetic brake shoes energized by trolley current are spring-suspended from the car trucks, normally clear of the rails. When a brake application is desired, the shoes are moved down to the rails by means of small air cylinders, or, with modern PCC cars, the shoes are automatically pulled down by the magnetic attraction between the shoes and the rails, There is no mechanical connection between the electric brake shoes and the wheel brake, and the air control is also separate. By exciting the magnet coils in the brake shoe, the friction between the shoe and the rail acts to retard the motion of the car. When the brake is applied, no weight is taken from the car wheels and full braking is therefore available with the air brakes. Since the excitation current for the shoe is taken from the trolley, there is no regenerative load on the motors. Braking is not dependent on the motion of the car.

Regenerative Braking. The term regenerative braking is normally used to apply to equipments where the motors act as generators and return current to the trolley line, thus effecting a retardation of the locomotive by making use of the stored energy in the moving vehicle. This system has been most extensively used for heavy locomotive duty. It was first used by the Chicago, Milwaukee, St. Paul, and Pacific Railroad on large d-c freight locomotives and subsequently on similar locomotives for various American-equipped foreign d-c electrified roads. The principal feature of these equipments is the provision of an accurately controlled exciter which separately energizes the fields of the motor, thus giving a definite control of the voltage generated. On some locomotives a separate motor-generator set, and on others a mechanically driven exciter mounted on the idle trucks, supplies this current. A further variation is the use of one or more of the motor armatures as the source of excitation for the other motors.

There are three installations in this country using regenerative braking with single-phase a-c supply. On the Norfolk and Western and the Virginian Railway polyphase induction motors are used, and three-phase regeneration occurs automatically when the locomotive exceeds synchronous speed. On the Great Northern Railway, with motor-generator type locomotives converting the single-phase supply to direct current for driving the traction motors, an unusually wide range of regeneration is obtained by controlling the fields of the d-c generators.

With d-c motors and with single-phase commutator motors it is possible to secure regenerative braking over quite a wide range of speeds. With the three-phase motor regenerative braking is limited to a range fairly close to the synchronous speed of the motor. Half speed on three-phase motor equipments is usually obtained by a change in the number of field poles, permitting half-speed regeneration as well as motoring.

Regenerative braking not only relieves the air-brake equipment but also effects a saving of an appreciable amount of energy and a reduction in wear on the wheel tires and brake shoes. Savings in power for the Chicago, Milwaukee, St. Paul, and Pacific have been reported as averaging 11 per cent of the power consumed by the motors. On other installations the savings have been even greater.

8. THE USE OF ELECTRICITY ON TRAINS

Before 1930 electricity was used on railroad passenger cars to provide a relatively low level of illumination and to operate small ventilating fans. Generators having outputs of 1 to 5 kw were adequate for the axle generator-battery system that was generally used. In the late 1920's and early 1930's air conditioning came into general use on main line railroads. Railroad cars with electromechanical air-conditioning systems used axle-driven generators with capacities up to 20 kw. Those using ice systems required generator capacities up to 10 kw; those using engine-driven compressors (ice engines), generator capacities up to 7.5 kw.

At the end of World War II the railroads of the United States embarked upon a program of building new cars and modernizing old ones to provide luxurious appointments for coach passengers and for Pullman travel. The new coaches require generator outputs of 20 to 40 kw, and the new diners require generator outputs of 30 to 60 kw.

The introduction of front-to-rear-end communication on freight trains and the desire to provide greater comforts for freight-train crews require the use of a generator on cabooses. At this date generator capacities between 1 and 3 kw are being considered.

This discussion will be confined to modern railroad passenger cars.

SYSTEMS USED. The following systems are generally used for supplying electric power on modern passenger cars:

a. Axle-driven generator.
b. Axle-driven motor-generator.
c. Engine-driven generator.

There have been experimental installations of head-end power-supply systems, but operational disadvantages have prevented their adoption. Their obvious advantages will continue to keep them in the picture for "feature" trains. A discussion of head-end power is therefore included.

AXLE-DRIVEN GENERATOR SYSTEM. This method of generating electric power and the axle-driven motor-generator system are used by the great majority of railroad cars operating east of the Mississippi. A large number of western railroads use engine-driven generators. Axle generators or axle-driven motor generators are used on cars which enter Grand Central Terminal or Pennsylvania Terminal, New York. Existing regulations prevent the use of engine-driven equipment under passenger cars that enter the New York tunnels.

Axle generators are used with storage batteries of the lead-acid or nickel-iron-alkali type. The generator supplies the load and charges the battery when the car is running above the "cut-in speed" of the generator. The generator "cuts in" when the d-c generated voltage is slightly higher than the battery voltage. The generator "cuts out" when the battery reverses into the generator. A reverse current relay opens the line contactor disconnecting the generator from the battery, leaving the load connected to the battery. In some generators a field reversing switch reverses the polarity of the generator, depending on the direction of motion of the car. This method has been superseded, however, by the armature reversing switch. This is a switch mounted on the shaft of the generator which reverses on the first revolution of the shaft of the generator if the direction of rotation is opposite to that of the previous operation. The switch is pulled closed in firm contact at a speed well below the cut-in speed.

The field of the generator is controlled by a regulator or voltage-control relay which adjusts the field strength to hold constant voltage with varying load and with car speeds varying, in some cases, from 20 to 120 mph. A device is included in the control to limit the current to a predetermined value. The more simple form of "current limit" reduces the voltage of the generator by reducing its field strength if a given current is exceeded. This reduces the voltage applied to the load and transfers some of the load current to the battery if the load is predominantly motors. A more complicated form of current limit allows the current output to increase beyond the current limit as the voltage is reduced. A further refinement may permit the maximum current output at a given voltage to increase as the speed of the generator increases, to take advantage of increased cooling at higher speeds of self-ventilated generators.

The voltage at which axle-driven generators operate depends upon the type of battery and the number of cells in the battery. More cells of a nickel-iron-alkali battery than of a lead-acid battery are required for the same "nominal" voltage. The discharge and charging characteristics are also quite different. There are three different "nominal" voltages in use on railroad cars in the United States: 32, 64, and 114 volts. All modern axle-driven generators use the constant voltage system of charging. Most railroads depend on rising volt-ampere charging characteristics to prevent overcharge. As these characteristics change with temperature, a lower charging voltage is used in summer than in winter by some railroads. One railroad employs a temperature relay located in the battery box to change the charging voltage as the temperature changes. This gives protection against overcharge under widely varying conditions. An older method of protection against overcharge consists of an undercurrent relay that reduces the generator voltage when the current input into the battery falls below a certain value. Lead-acid batteries use a charging voltage approximately 16 per cent higher than the nominal battery voltage in summer and up to 25 per cent higher than the nominal battery voltage in winter. Nickel-iron-alkali batteries use a charging voltage approximately 35 per cent higher than the nominal battery voltage. Axle-driven generators will deliver their loads at voltages up to 140 per cent of the nominal battery voltages.

Except where considerations of standardization predominate, the trend is toward the use of 64 volts or 114 volts on modern cars. The higher voltages reduce the weight of cables and permit a more economical selection of power-consuming devices. The hazard of accidental grounds, on the other hand, is increased by the higher voltages. The d-c circuit on a passenger car is not normally grounded.

The required ampere-hour capacity of the battery depends on the load and operating conditions. Modern cars use batteries of 1000 to 1200 ampere-hours for 32-volt systems and batteries of 600 ampere-hours for 64-volt systems. The ampere-hour capacity of 114-volt batteries is usually between 450 and 600. The capacity of the axle generator depends upon the ampere-hour capacity of the battery, the load, and the operating conditions. Generators are selected on the basis of cut-in speed and ampere capacity, to reduce the discharge from the battery to a minimum under average operating conditions. These conditions vary on different railroads.

Axle-driven generators are customarily rated in kilowatts, but the suitability of a generator for a given application depends on the amperes that it can deliver. The voltages on which the kilowatt ratings of axle generators are based have not been standardized by the manufacturers.

The majority of axle-driven generators are body mounted and are belted or geared by a drive shaft to the axle of the car. There are some axle-mounted generators geared to the axle like a traction motor. One railroad uses this generator with a shear disk to prevent locking the wheels of the car in case of mechanical failure of the generator. Modern body-mounted generators use a drive consisting of three major units: gear unit, propeller shaft, and clutch. Power is transmitted from the axle through the axle mounting to the hypoid gear and pinion unit, then through the propeller shaft to the clutch and generator. In any extreme overload on the generator the safety clutch starts slipping at the friction surfaces, generating heat in the safety plate. This heat is transferred to the safety plugs and causes them to melt. The clutch then runs free on the bearings.

AXLE-DRIVEN MOTOR-GENERATOR SYSTEM. This is essentially the same as the straight axle-driven generator, except that an induction motor is mounted on the same shaft with the d-c generator. The induction motor is plugged in, in yards and terminals, to carry part or all of the load or to charge the battery. This has the advantage of a regulated source of voltage for battery charging, as compared with the variable voltage from d-c yard charging outlets. Provision is, however, usually made on cars to permit charging from d-c outlets if necessary. A 15-hp, 220-volt, three-phase motor permits a d-c output of between 9 and 10 kw on plug-in power. A 25-hp motor permits a d-c output of between 15 and 16 kw. Alternating-current outlets have been standardized at 220 volts, three phase, 60 cycles. They are of limited capacity and are sometimes fused close to the rated output so that special time-delay fuses are required to stand the inrush current on starting. The control of the d-c generator is arranged to change the "current limit" on plug-in power, so that the capacity of the a-c outlet is not exceeded.

The drive of the motor generator differs from that of the simple axle generator in that an automatic clutch keeps the drive from engaging with the axle of the car when the generator is being driven by the motor. This clutch engages by centrifugal action when the car reaches a speed corresponding to 280 rpm of the drive shaft; it includes a safety feature to protect against excessive overload.

ENGINE-DRIVEN GENERATOR. A number of railroads that do not operate cars into the New York terminals use engine-driven compressors for air conditioning and a separate engine-driven generator to charge the battery and to operate fans, lights, etc. Propane gas is used for fuel in the Hesselman type of engines. The gas is carried in steel bottles in racks under the car, and the fuel lines are equipped with safety devices to shut off the gas in case of a break in the fuel line. The engine generator is mounted on a suitable carriage under the car, so that it may be drawn out and inspected or removed without having to take the car over a pit. This system has the advantage of not taking any power from the locomotive at the head of the train. It has the disadvantages inherent in small engines.

In recent years a number of manufacturers have undertaken the development of a Diesel-engine-driven generator for undercar mounting. The output of the generator is of the order of 25 kw. Because of clearance limitations for equipment mounted at the side of the car only vertical engines of limited stroke can be used. Pancake engines may permit greater capacity when they are developed to the degree of reliability required for this service. The engine runs at constant speed, and the generator may be either d-c or a-c. A 220-volt, three-phase, 60-cycle generator provides a source on the car of the same voltage and frequency as is available on plug-in power, permits the use of hermetically sealed compressors for air conditioning, and makes it possible to use small motors, fluorescent lighting, etc., which are produced in volume for other applications. Direct-current generators have the advantage of simplicity of wiring and control but retain the maintenance problems of axle-driven generators.

HEAD-END SYSTEMS. A few railroads make limited use of d-c turbogenerators mounted on the locomotive with a train line to the cars on the train. These and other similar systems are of limited application. In one variation, Diesel-engine-driven generators are mounted in a power car with all the necessary control. Three-phase power is transmitted by means of couplers between cars. The voltage used depends on the load and on the length of the train.

Although experimental installations of this method of power supply were made several years ago, it has not yet been adopted for general use, principally because only cars equipped for the system can be used. When a car is shopped, it can be replaced only by a car that is equipped for head-end power. Another disadvantage is that the same power is used for an 8-car train as for a 20-car train.

ILLUMINATION. The following types of lighting installations are in general use:

a. Incandescent lamps for use on 30 volts d-c, 60 volts d-c, and 110 volts d-c. These require a lamp regulator to keep the voltage at the rated value when the battery is being charged. A carbon pile regulator is generally used.

b. Fluorescent lamps with resistance ballasts for use on 60 volts d-c. These require a lamp regulator.

c. Fluorescent lamps for use on 110 volts or 220 volts, 60 cycles, in three-phase or single-phase banks. These require a conversion unit to convert the d-c generator or battery voltage to a-c. The conversion unit is provided with an inherent speed- and voltage-regulating system.

General lighting design is a procedure which determines the lumens per square foot (foot-candles) delivered in a horizontal work plane. The proportions of the car, color of the walls and ceiling, fixture efficiency, and light distribution all enter into the lumens required at the source to secure a required foot-candle standard in the reading plane. The lamp efficiency, expressed in lumens per watt, determines the power input for a given lumen output. The losses in auxiliaries and in conversion or regulating devices determine the watts taken from the generator or battery to deliver a given number of watts to the lamps. A combination of these factors determines the power required to secure a given level of illumination in the reading plane.

Fluorescent lighting, because of its high efficiency, low surface brightness, novelty appeal, and quality of light, has come into general use on railroad cars. A variety of lamp lengths are available for 110 volts and 220 volts, 60 cycles. The Slimline lamps extend the range of length above 48 in. for decorative assemblies. The majority of railroads which use fluorescent lighting have adopted the a-c type, but one railroad has successfully applied the 60-volt d-c fluorescent lamp. The 60-volt lamp is limited in length and is not in general use for commercial and residence lighting. It avoids a conversion unit but requires a lamp regulator. The lighting load on a modern car varies between 1.5 kw and 3 kw, depending upon the level of illumination, use of decorative lighting, efficiencies of conversion units, etc.

OTHER ELECTRICAL LOADS. The principal electrical load on a railroad car is the air-conditioning system. (See Section 15.) Other electrical loads include water coolers, electrical dust- and smoke-precipitating devices, announcer systems, wire recorders, radios, refrigeration. and electric cooking devices in dining cars. It is not unusual to have a connected load in excess of 20 kw on a modern coach and in excess of 25 kw on a diner. An all-electric diner will have a connected load considerably in excess of this. The peak load may be estimated by assuming a diversity factor of 80 per cent of the connected load if more accurate data are not available.

CONVERSION UNITS. Three types of converter units are in general use. One consists of a motor alternator, with a special arrangement of field connections to maintain constant voltage and frequency on the a-c side with a wide variation in d-c voltage. The second consists of an amplidyne mounted on the same shaft as a synchronous inverter. The amplidyne regulates the d-c voltage applied to the inverter as the d-c generator or battery voltage and the a-c output varies. An amplidyne is used because of the small amount of power (less than 1 watt) required to control its output and because of its rapid response to change in d-c voltage and d-c load. The third is the vibrating type, used for loads up to 500 watts but more generally used to supply power for razor outlets on cars not equipped with rotating-type inverters or supplied with a-c power from engine-driven generators.

GENERAL. From an operating standpoint the axle-driven motor-generator system is more flexible than the others. A car so equipped may be operated on any division of any railroad independent of charging facilities at the division terminal points, availability of fuel, or availability of train line connections between cars. Because of its battery, it may be cut out of one train at a junction point for connection to another train without special arrangements being made to supply light from an outside source. It may operate in tunnels and terminals where internal-combustion engines are not permitted to operate.

9. THIRD-RAIL OR CONTACT-RAIL SYSTEMS

The contact rail, or third rail, is a conductor supported on insulators near the ground and presenting a continuous contact surface to a collector or shoe attached to the rolling stock. Electrical continuity between adjacent lengths is obtained by copper bonds across the joints.

TYPES OF CONSTRUCTION. Third-rail construction may be classified into the top-contact and under-contact types, each of which is susceptible of important variations in design, especially with reference to the type of protection.

Interborough Top-contact Type (Fig. 7). One of the most commonly used types is often called the Interborough type on account of its use in the subways of the Interborough Rapid Transit System of New York. The rail rests on porcelain insulators. A board protection is attached to the rail itself by means of clamps and uprights and is thereby kept in perfect alignment.

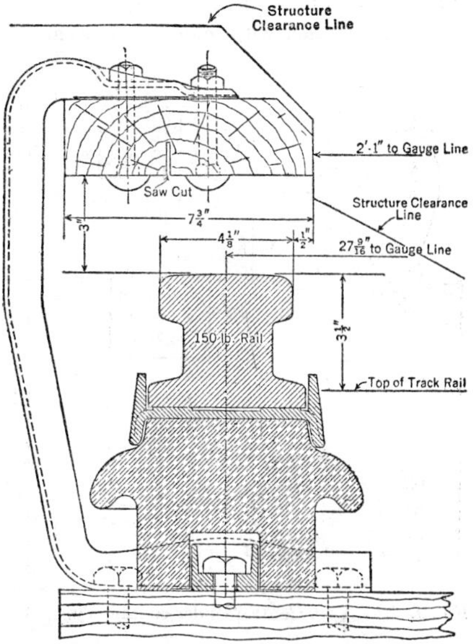

FIG. 7. Interborough Top-contact Type

Pennsylvania Top-contact Type (Fig. 8). Another type has the protection supported on separate brackets independent of the third rail itself. Although this reduces the amount of labor which has to be done on the live rail when repairs are being made, it cannot be relied upon as well as the Interborough type to keep the rail and protection in perfect alignment.

New York Central Under-contact Type (Fig. 9). Although the top-contact types have given first-class service, they are considered to have certain disadvantages for exposed locations, as they cannot be wholly protected from snow, ice, and sleet. The lower part is only a few inches from the ties, and holding clips generally reduce this clearance, increasing the danger of grounding from accumulation of wet snow and ashes and from flooding. The occasional suspension of traffic during sleet and snowstorms and floods, on railroads using the top-contact type of third rail, led to the idea of an undercontact third rail loosely clasped in insulators by hook bolts hung from brackets, with the top and

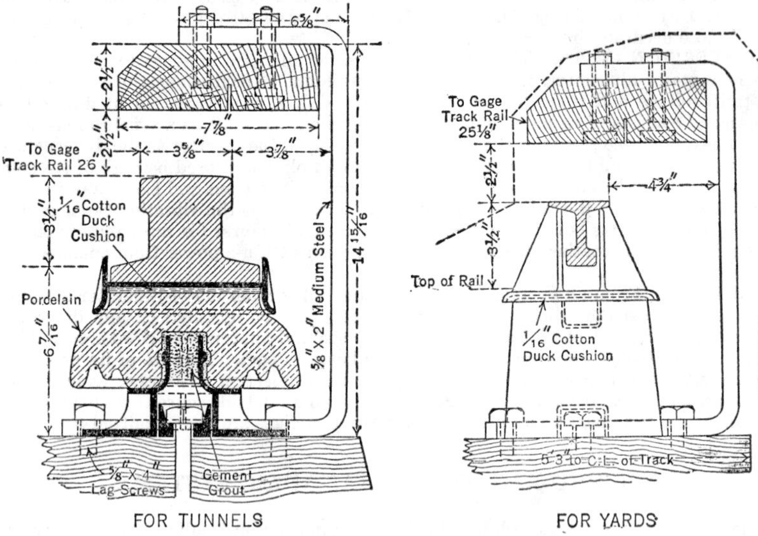

FIG. 8. Pennsylvania Top-contact Type

sides of the rail completely sheathed in an insulating material for protecting the rail from accidental contact with man and beast and from sleet, snow, and spray. With this type of rail the protection is of such character that there is no packing of snow between the protection and the contact rail, as in some other forms, and in sleet storms no ice forms on the contact surface; some icicles may form at the edge of the petticoats but are easily broken off by the passing shoe.

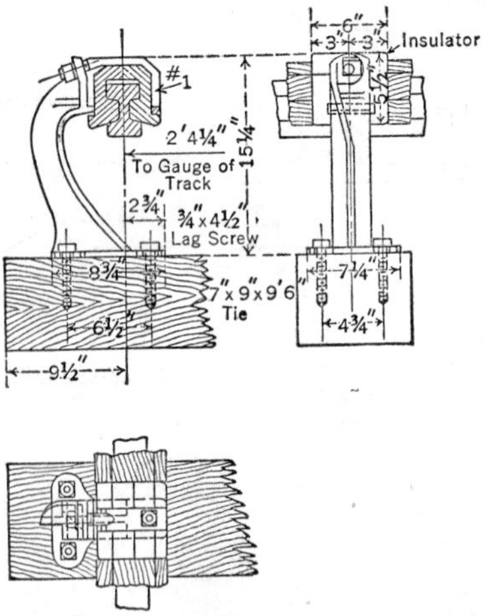

FIG. 9. New York Central Under-contact Type

Where the rail is buried in snow, the passage of the contact shoe breaks the snow away, leaving the rail surface clear, instead of ironing the snow down on the rail, as may happen with the top-contact type.

Protection and Special Work. The protection between the insulator blocks, depending upon local conditions and the price of materials, as well as the potential used, is usually formed of three wooden strips, one grooved on the underside and enclosing the head of the rail, and the other two, attached to and dependent from it, reaching in toward the web of the rail. Soft rubber has sometimes been used in places where brine and manure drippings have tended to reduce the insulation resistance of wood sheathing.

10. OVERHEAD CONTACT SYSTEMS

CONTACT WIRE. The contact wire is usually made of hard-drawn copper. Where a harder material is required to withstand greater wear and give longer life, a bronze alloy is used. The conductivity of alloy contact wire is usually between 55 and 80 per cent of that of copper, depending on the particular materials used in the alloy and on the degree of wear resistance desired.

SUSPENSION. The contact wire is usually suspended at heights varying from 15 to 25 ft above the top of the running rails. In any one particular installation, however, it is unusual to find a variation in the height of the contact wire of more than 6 ft. The contact wire presents a continuous contact surface (except for section insulating gaps in some types of construction) to the trolley wheel, bow, or pantograph mounted on the rolling stock.

There are two general classes of contact-wire suspensions, direct and catenary.

Direct Suspension. In the simple direct-suspension class the contact wire is suspended by insulators from span wires stretched across the tracks between poles or building walls, or from bracket arms attached to the poles. A combination of the two types is sometimes made by using a short-span wire connecting two bracket arms on the same pole. In this single bracket type of construction the poles are usually placed from 90 to 120 ft apart on tangent track, and the contact wire sags considerably between supports, allowing a vertical motion between supports with a very decided "hard spot" at bracket-arm supports. This construction is unsuited for high-speed service, because of the considerable collector motion with resultant arcing and decreased life of the collector and contact wire.

Catenary Suspension. The catenary suspension was designed to overcome the difficulties encountered with the previous type of construction. In general it consists, in its simplest form, of a messenger wire sagging between supporting structures with the contact wire supported from the messenger wire by hangers spaced at equal intervals. The hanger spacing varies from 15 to 30 ft. This method of support results in the contact wire being in a plane approximately parallel to the plane of the running rails by using hangers of varying length at the different points in the span. The bracket arm, span wire, or cross-bridge type of support may be used with the catenary system, with supporting structures placed usually 150 to 300 ft apart on tangent track.

APPLICATIONS OF VARIOUS TYPES OF CONSTRUCTION. The direct-suspension type of contact-wire construction is generally used on urban street railways and trolley coach systems, where it is not objected to on the grounds of unsightliness or danger due to the exposed construction. It is also used on some of the older interurban installations. Center-pole construction with bracket arms is usually chosen for street systems for double-track work where it is not objectionable so to locate the poles. The alternate method is the use of cross-span wires with the poles located near the curb or attached to buildings. This latter method is used quite extensively in Europe.

The catenary-type suspension is generally used for all other overhead contact systems, except where the current required by the rolling stock is too great to be economically carried by copper wires or to be satisfactorily collected without undue arcing. This is illustrated by the overhead third-rail construction for special conditions.

ELECTRICAL DESIGN OF CONDUCTORS. The electrical design of railway distribution systems involves the consideration of potential drop, economy in use of copper, and the heating of the conductors. The heating of conductors is seldom an important feature in railway distribution system design, as it is generally necessary to use a low current density in order to keep the potential within the limits required. A discussion of the heating of conductors will be found in Section 14, Articles 61 and 69. The feeder system being sufficient to meet the conditions imposed by the allowable potential drop, it will be economical to install more copper if the resulting saving in the cost of energy is greater than the increase in interest and other charges on the additional investment, plus any additional maintenance charges that may be encountered because of the new facility.

Allowable Potential Drop. The maximum allowable potential drop is limited by various features, such as:

1. The necessity of running the trains or cars at certain speeds and meeting schedule requirements.

2. The necessity of keeping the illumination in street railway or multiple-unit cars at a suitable intensity. This varies with the requirements of various localities.

3. The danger of electrolysis in the grounded portion of the return circuit. The maximum drop in grounded circuits, under maximum load conditions, is limited by law in certain localities. In Great Britain the limit is 7 volts drop between any two points in the negative system. In Germany it is generally 1 volt per kilometer (1.61 volts per mile). A usual limit for some cities in the United States is 1 volt per 1000 ft. In trolley-coach systems, using positive and negative contact wires, the negative may or may not be grounded at the substations. If grounded at the substations, it may be permanently grounded or grounded only when the substation is supplying power. When the negative contact wire is not grounded or is grounded only when the substation is supplying power, the above potential-drop limitation does not apply.

4. The necessity of maintaining voltage at the rolling stock to such a value that satisfactory operation of the control circuits will be obtained.

5. The necessity of tripping-out feeder breakers on remote faults or short circuits.

CALCULATION OF POTENTIAL DROP. The method of calculation of potential drop depends upon the following conditions: (1) whether the current is direct or alternating, (2) whether the load is concentrated at one point, sparsely distributed, or evenly distributed, (3) the distribution of metal in the feeder circuits, and (4) whether the section is being fed by one or by two or more substations. Calculations for a-c lines differ from those for d-c lines only in taking into account the inductance, as described below. On city railways it is usual to assume the load to be evenly distributed, it being stated as a given number of amperes per foot. (See Article 2.) If the load is actually concentrated at n equidistant points, the drop will exceed that calculated on the assumption of uniform distribution by about $100/n$ per cent. On interurban and trunk lines, the cars are usually concentrated at one or two points between substations, making the assumption of uniform distribution impracticable. In such cases the loads should be located so as to give the worst conditions, and calculations made as for any network.

Where electrolytic damage is to be guarded against, the drop of potential in the track rails themselves has to be calculated, as well as the total drop in the rails and feeders.

Values of the **resistance of trolley wires** to direct current will be found in Section 14, Article 54, and values of the **resistance of rails** to direct current will be found in Article 12. It should be noted that the resistances of the trolley and positive feeders are in parallel and that the track rails and negative feeders are in parallel. Also note that in high-voltage systems a considerable portion of the current returns through the earth and not through the rails, and consequently the drop in the rails is due only to that part of the current which returns through them. For preliminary calculations, however, the full current may be assumed as returning through the rails.

FORMULAS FOR D-C TROLLEY CIRCUITS. The following formulas apply to certain typical circuits which frequently occur in practice.

Let I = total current in amperes taken by all cars on section considered.
L = total length of section in 1000 ft.
V_p = total drop in volts, in positive conductors between substation bus and far end of line.
V_n = total drop in volts in negative conductors between substation bus and far end of line.
$V = V_p + V_n$ = total drop in volts in both positive and negative conductors.
r_p = resistance in ohms of all the positive conductors in multiple per 1000 ft of line.
r_n = resistance in ohms of all the negative conductors in multiple per 1000 ft of line.
$r = r_p + r_n$ = total resistance in ohms per 1000 ft of line.
l = distance in 1000 ft, from far end of line to any point P.
v = drop to the point P, subscripts used as for V.

Uniformly Distributed Load, Uniform Conductor, Fed from One Substation. Then

$$v = \frac{rIl^2}{2L}$$

$$V = \frac{rIL}{2}$$

These formulas are applicable to either the positive or negative conductors considered separately or to both in series.

Uniformly Distributed Load, Conductor Tapered to Give Minimum Weight of Metal, Fed from One Substation. For minimum weight the tapering must be such that at any point P

$$r = \frac{3V}{2I\sqrt{L}\sqrt{l}}$$

i.e., the cross-section, if all the conductors are of the same metal, must increase directly as the square root of l. The drop to the point P is

$$v = \frac{V\sqrt{l^3}}{L^3}$$

These formulas also apply to either the positive or the negative conductors separately or to both in series.

Uniformly Distributed Load, Conductor Divided into Sections (Fig. 10); Each Section of Constant Resistance, Fed from One Substation. The drop from P_n to the far end of line is

$$\frac{I}{2L}\sum_{n=1}^{n=n} r_n[l^2{}_n - l^2{}_{(n-1)}]$$

This formula is applicable to either the positive or the negative conductors. The subscript n is here the general form of the subscript corresponding to each section, represented by 1, 2, 3, and 4 in Fig. 10.

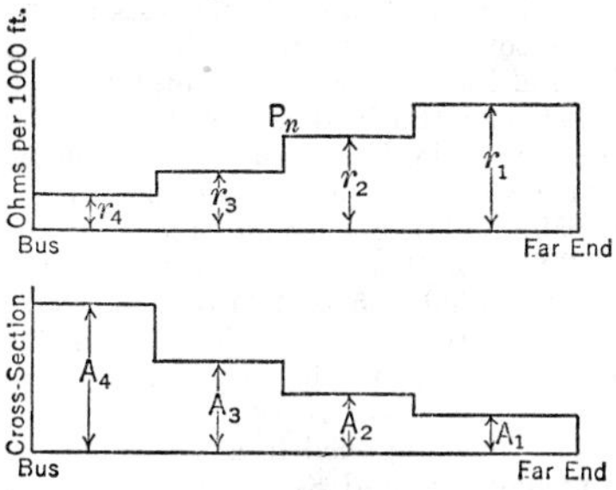

FIG. 10

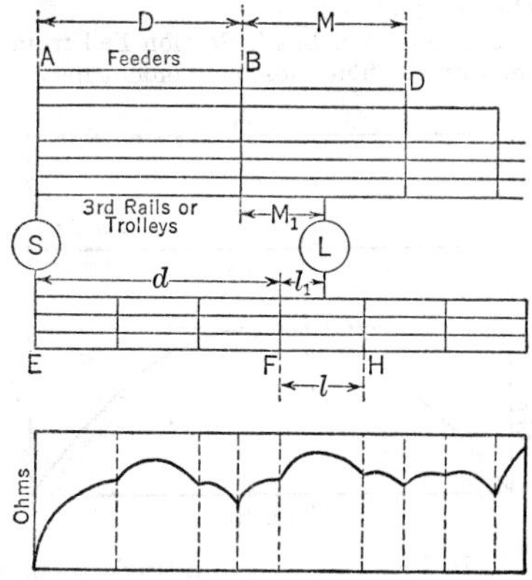

FIG. 11

Concentrated Load, Section Fed from One End. Figure 11 shows a 4-track road with 4 trolley wires and 3 feeders with the tracks cross-bonded at intervals. The solution given below is a general one, and may be applied to any case from 1 track and 1 trolley wire up. Let all distances be expressed in 1000 ft and let

N_f = number of feeders in section considered, e.g., for the section AB, $N_f = 3$, and for the section BD, $N_f = 2$.
N_c = number of contact conductors, trolley wires, or third rails in section considered, e.g., 4 are shown in Fig. 11.
N_t = number of tracks in section considered.
R_f = resistance of each feeder per 1000 ft.
R_c = resistance of each contact conductor per 1000 ft.
R_t = resistance of each track per 1000 ft (1 rail or 2 rails in multiple, depending upon whether 1 or 2 rails are used for return conductor).
$n = N_c + \frac{R_c}{R_f} N_f$ for the section considered.

Then for the section in which the load may be, the resistance of the positive conductors from the load to the end of that section in the direction of the substation, e.g., the resistance from L to B, is

$$R_cM_1\left[1 - \frac{(n-1)M_1}{nM}\right]$$

The resistance of the positive conductors in any section, such as BA, is

$$\frac{R_cD}{n}$$

(Note that the value of n for this section is not the same as for the section BD.)

The resistance of the negative conductors from the load to the first cross-bond in the direction of the substation, e.g., the resistance from L to F, is

$$R_tl_1\left[1 - \frac{(N_t-1)l_1}{N_tl}\right]$$

The resistance of the remaining portion of the negative conductors, e.g., from F to E, is

$$\frac{R_t d}{N_t}$$

(Note that, if negative feeders are used, each negative feeder having a resistance of R_f' per 1000 ft, then for N_t in the last two formulas substitute $n' = N_t + \frac{R_t}{R_f'} N_f'$, where N_f' is the number of negative feeders for that section.)

The total resistance from the load to the substation is the sum of the resistances as above calculated.

Concentrated Load, Section Fed from Both Ends, Substation Voltage at the Two Ends the Same. The most convenient method of treating such problems is to plot an "equivalent-resistance-distance" curve, such as those shown in Figs. 11 to 15. By "equivalent resistance" is here meant that resistance by which the total current taken by the load must be multiplied to give the total drop in voltage between the load and either substation. For example, if the substation voltage is 600 at each end, the voltage across the load is 550, and the current taken by the load is 200 amp, then the equivalent resistance is $R = (600 - 550)/200 = 0.25$. This method avoids the determination of the distribution of the current in the various parts of the network, and the resistance when once determined can be applied to any load.

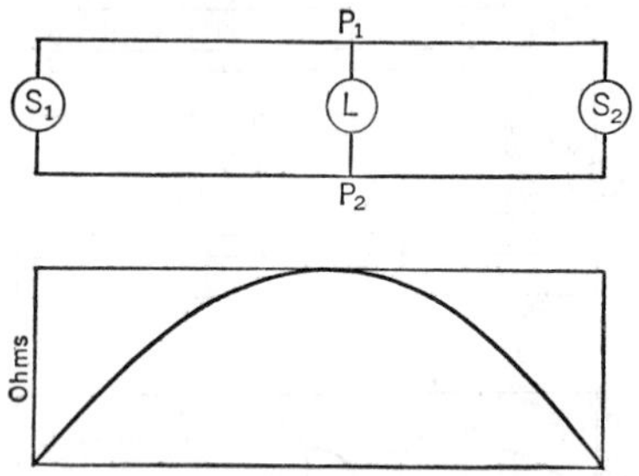

FIG. 12

1. In Fig. 12 is shown a single track and single trolley. S_1 and S_2 are substations; L is a load placed arbitrarily between corresponding points P_1 and P_2 on the positive and negative conductors respectively.

Let a = resistance of the conductors between the points $S_1\ P_1\ P_2\ S_1$;
b = resistance of the conductors between the points $S_2\ P_1\ P_2\ S_2$.

These resistances are the resistances of the transmitting conductors and do not include the internal resistances of the substations and load. Then the equivalent resistance is

$$R = \frac{ab}{a + b}$$

2. In Fig. 13 is shown a 4-track road with 4 trolleys, both track and trolley cross-bonded. Figure 14 is a simplified diagram of Fig. 13, corresponding points being designated by identical letters.

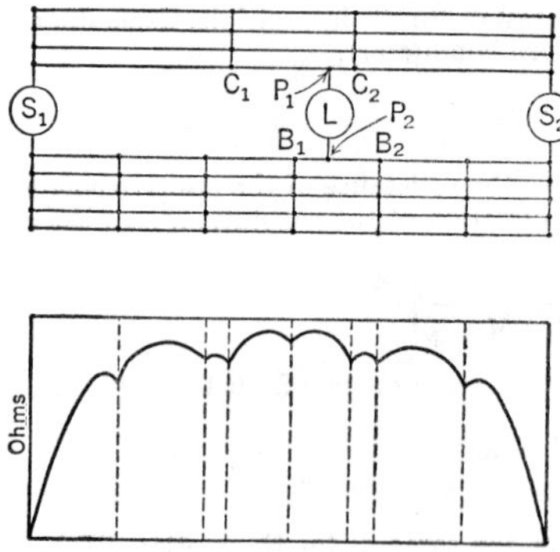

FIG. 13

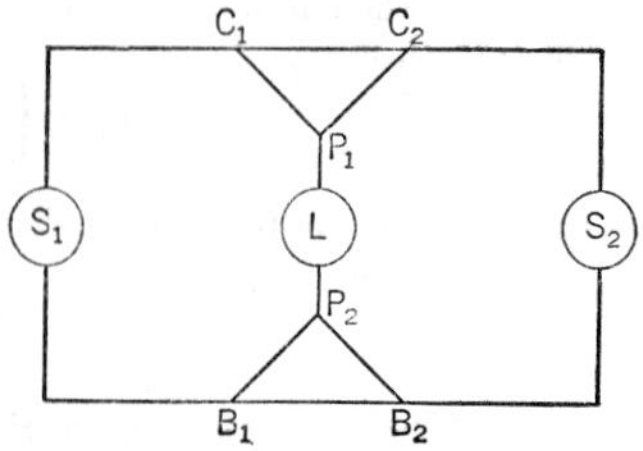

FIG. 14

S_1 and S_2 are substations; L is a load placed arbitrarily between points P_1 on the positive system and P_2 on the negative system; C_1 and C_2 are ties between positive conductors, and B_1 and B_2 ties between negative conductors.

Let a = resistance of loop $S_1\ C_1\ P_1\ LP_2\ B_1\ S_1$.
b = resistance of loop $C_1\ P_1\ C_2\ C_1$.
c = resistance of contact conductor $C_1\ P_1$.
d = resistance of contact conductor $C_2\ P_1$.
e = resistance of loop $B_1\ P_2\ B_2\ B_1$.
f = resistance of track $B_1\ P_2$.
g = resistance of track $B_2\ P_2$.
h = resistance $\dfrac{d^2}{b} + \dfrac{g^2}{e}$.
i = resistance $\dfrac{c^2}{b} + \dfrac{f^2}{e}$.
j = resistance $S_2\ C_2\ P_1\ LP_2\ B_2\ S_2$.

These resistances are the resistances of the transmitting conductors and do not include the internal resistances of the substations and load. The equivalent resistance is

$$R = \frac{aj - ah - ji - 2\left(\dfrac{fg}{e} \times \dfrac{cd}{b}\right) + \dfrac{f^2d^2 + c^2g^2}{eb}}{a + j - h - i - 2\left(\dfrac{fg}{e} + \dfrac{cd}{b}\right)}$$

Concentrated Load, Section Fed by Feeders from Both Ends. Figure 15 shows a simple case with feeders for the positive conductors only. No general formula is available for such a circuit. Any such network can, however, be calculated by Kirchhoff's laws (Section 3, Article 15), using the numerical values of the resistances for the various elements. The equivalent resistance from the substations to the load is then the drop in voltage for a load of 1 amp.

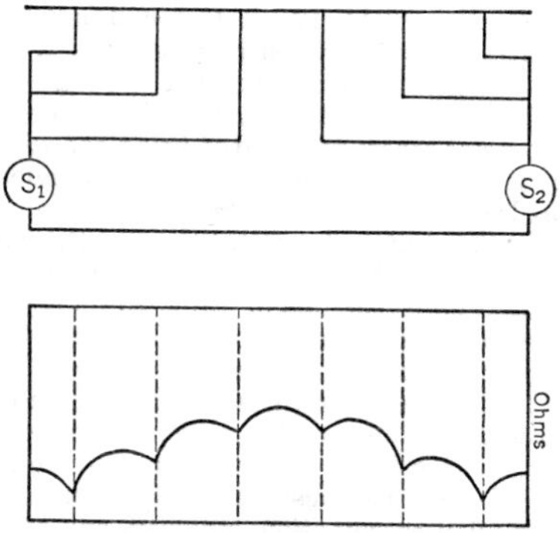

FIG. 15

METHODS OF REDUCING TOTAL DROP. Three methods are employed for reducing the total drop between the substation and the load. The one in most general use is the installation of additional copper as feeders connected to the contact wire, third rail, or track at frequent intervals. These feeders are usually bare and carried on insulators on the pole line. The resistance of the positive or contact wire side of the circuit is usually high compared to the negative or rail side of the circuit, and the feeders are therefore added usually on the positive side of the circuit only. Feeders on the negative side of the circuit are usually installed only as an electrolysis protective measure, but are sometimes required in order to permit track repairs on a single-track line to prevent the danger of a high voltage existing across the gap when track rails are removed at the time that a car or train may be using power. Figures 11 and 15 indicate two methods of making feeder connections.

The second method is to connect a feeder to the contact wire or third rail at one point only, sometimes using a booster in the feeder circuit at the substation. The principle of this method, when a booster is used, is to make the feeder carry more current than it would on the basis of resistance ratios. The first cost and operating costs of a booster are usually prohibitive from an economical standpoint, when measured by the improvement obtained. This method is, however, successful and economical on the negative side of the circuit as a means of electrolysis protection, in which case, however, boosters are usually omitted.

A third method consists of the use of an additional substation, either fixed or portable, at or near the point of low voltage

A-C RAILWAY CIRCUITS. When alternating current is used, the *impedance* of the trolley-rail circuit becomes the determining factor in calculating the circuit constants, as contrasted to the *resistance* when direct current is used. The problem becomes a little involved in several respects because of:

a. Skin effect in the rails which causes a non-uniform distribution of current within the rails.

b. Proportion of current returning in the earth. For a railroad having the contact wire approximately 20 ft above the rails it may be assumed that the current divides between the rails and ground approximately as follows:

Single track—40 per cent in rails, 60 per cent in ground
Double track—60 per cent in rails, 40 per cent in ground
Four track—70 per cent in rails, 30 per cent in ground

c. Number of rails in parallel.

d. Number of conductors in the overhead circuit.

These items affect the a-c resistance as well as the internal and external reactance of the conductors. The a-c resistance and internal reactance of rails have been experimentally determined and reduced to empirical formulas. See *Journal of Franklin Institute*, Vol. 182, pp. 135–190 (1916) (Kennelly, Archard, and Dana) and *Transactions AIEE* (1934), pp. 1771–1780 (Trueblood and Wascheck).

Of the reactance components of the trolley-track circuit, that of the trolley is simple to calculate by the usual methods (see Sections 3 and 14); however, that of the track and earth return is rather complicated. For extensive analyses of this problem the reader is referred to *Bell System Technical Journal* (October 1926), pp. 539–554 (Carson); *Electric Journal* (April 1931), pp. 239–244 (Wagner and Evans), and *Circuit Analysis of A-C Power Systems* by Edith Clarke (John Wiley & Son).

There are two practical methods of estimating the constants of a-c railway circuits. One is to use the constants measured on a road which has been electrified. This method is most useful when extending a system already in operation or in making calculations on a new system having the same size of rails and the same overhead circuits. The other method is to use certain approximations which have been found satisfactory. In this connection, Table 17 will be found useful.

The current division in a two-wire or three-wire system may be calculated most easily by using a calculating board.

The more laborious method is to work back to the source of supply, by starting from the point where the load is being considered, and then expand the circuit as the current divisions and voltage drops are determined. In general a separate calculation must be made for each load point and for each source of power. A unit current load at zero angle is assumed at each substation, and the current division throughout the network is determined. The load between substations is assumed to be equivalent to a load at the substation on either side of the load. The equivalent load at a substation is assumed to be inversely proportional to the distance of the substation from the load. The principle of superposition is then applied to determine the current division for the entire system.

High-voltage transmission lines for a-c railway systems are usually single phase. In two-wire systems and in some three-wire systems the transmission line must act also as a distribution line. The constants of the line may be calculated by the methods used for calculating the constants of three-phase lines. The problems of wave form, relaying, insulation coordination, stability, and other problems of high-voltage engineering require special methods of attack, but the same fundamental principles may be used as in considering three-phase systems.

Inductive coordination must be considered in all railway systems, whether d-c or a-c. In a-c systems acoustic shock is probably the most serious problem. For railroads the application of the a-c high-speed breaker has done much to reduce this trouble, in addition to protecting overhead lines, current collectors, and apparatus. In laying out a power system for a railroad all interferences are considered, and the system is so designed as to reduce such interferences to the minimum consistent with economical design.

The three distinct types of a-c railway distribution systems are illustrated in Figs. 16, 17, and 18.

The Two-wire System (Fig. 16). In this system the power, after being transformed to a suitable transmission voltage, is carried by single-phase lines to step-down transformer stations at suitable locations, stepped down to the catenary voltage, and connected to the catenary and rail. Examples of this type of system are:

Railroad	Transmission Voltage, Kilovolts	Catenary Voltage, Kilovolts
Pennsylvania	132 and 44	12
Norfolk and Western	88 and 44	11
Great Northern	44	11.5

The Three-wire System (Fig. 17). For systems of limited extent (25 to 30 miles either way from the power-supply source), the transmission voltage may be fairly low,

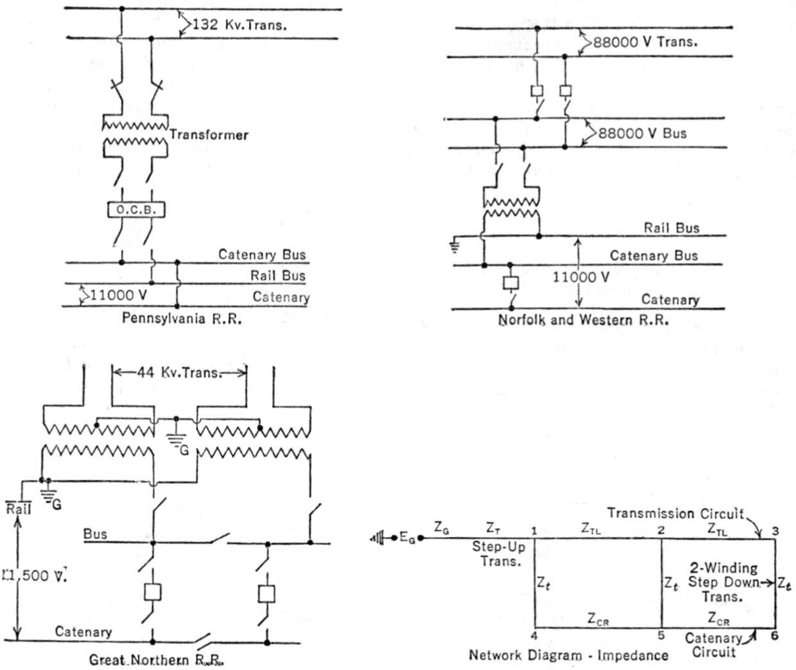

FIG. 16. Two-wire A-c Railway Distribution System

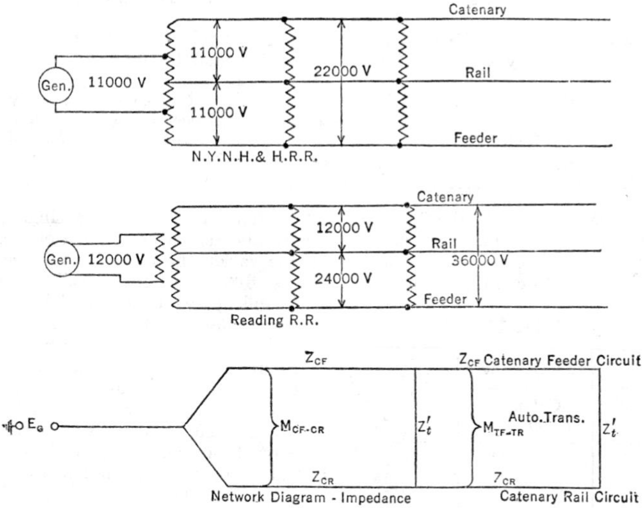

FIG. 17. Three-wire A-c Railway Distribution System

and the catenary is used as one side of the transmission circuit. The other wire is called the feeder. Three-winding transformers are generally used at the power source. Power is transformed to a suitable transmission voltage, usually 24 or 36 kv, and fed over the transmission circuit, consisting of the catenary and feeder, to auto-transformer stations, where it is transformed to catenary-rail voltage. The three-winding transformer at the power source also supplies power directly to the catenary-rail circuit. By varying the reactance of the three-winding transformer the power flow may be proportioned between catenary-rail and catenary-feeder circuits to reduce interference in communication circuits. The following are examples of this system:

Railroad	Catenary-feeder Voltage, Kilovolts	Catenary-rail Voltage, Kilovolts
New York, New Haven & Hartford........	22	11
Reading (initial installation)..............	36	12

The Three-wire System with Transmission Circuit (Fig. 18). Economies may sometimes be effected by a combination of the two previously described systems. The transmission circuit may be run on a shorter route cross-country, meeting the railroad right-of-way only at infrequent intervals with the three-wire system between these points.

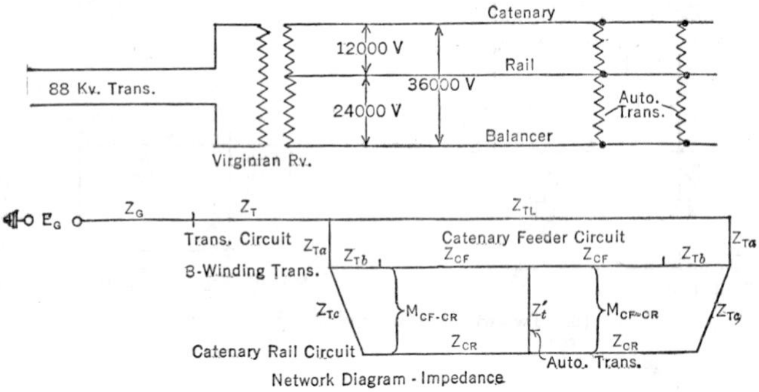

Fig. 18. Three-wire A-c Distribution System with Transmission

The three-wire system may be used for the initial electrification, and the transmission system added in later years when increased power demand requires it. Examples of this system are:

Railroad	Transmission Voltage, Kilovolts	Catenary-feeder Voltage, Kilovolts	Catenary-rail Voltage, Kilovolts
Virginian......................	88	36	12
Reading.......................	66	36	12

Impedance Diagrams. In the figure for each system there is shown a single line impedance diagram for calculating short-circuit currents, regulation, induction, load division, relaying problems, etc. The constants of the various branches having been determined, these networks can then be solved by the usual "network methods." Impedances for all apparatus and aerial circuits can be calculated as for power systems. Impedances of circuits having ground and rail return, such as a catenary-rail circuit, can be calculated as for ground return power circuits in which ground wires are accounted for. A larger proportion of the current, however, generally returns in the rails than in the ground wires. Average values for 100-lb rails are given in Table 16. Since a rail is made of steel, its internal reactance is high, so that a fictitiously low equivalent radius must be used to account for this. The radius for a 100-lb rail has been found experimentally to be 0.35 in. On the basis of this equivalent radius for 1 rail, geometric mean radii for 1, 2, 3, and 4 tracks are given in Table 16.

Table 16

No. of Tracks (Pairs of Rails)	Rail Current in Percentage of Catenary Current	Geometric Mean Radius, ft
1	40	0.37
2	60	2.16
3	65	4.54
4	70	7.17

Because of the large skin effect the ratio of a-c to d-c resistance of rails is high, the 25-cycle resistance per pair of rails being 0.092 ohm per mile.

On the basis of the proportions of rail current given in Table 16 total impedances of typical catenary circuits are given in Table 17. Variations in earth conductivity or catenary height encountered in practice would not ordinarily cause variations of more than 10 per cent in the values given in this table.

Table 17.* Impedance of Catenary Circuits

Catenary-rail Circuit	Cate-naries	Tracks	Catenary Circuit	Return Circuit	Total Impedance
3/0 55% 1/0 copper 100-lb rails	1	1	0.316 + j 0.306	0.037 + j 0.084	0.353 + j 0.390 = 0.526
	1	2	0.316 + j 0.310	0.028 + j 0.073	0.344 + j 0.383 = 0.514
	1	4	0.316 + j 0.322	0.016 + j 0.051	0.332 + j 0.373 = 0.498
	2	2	0.158 + j 0.170	0.028 + j 0.073	0.186 + j 0.243 = 0.305
	2	4	0.158 + j 0.180	0.016 + j 0.049	0.174 + j 0.229 = 0.289
	4	4	0.079 + j 0.094	0.016 + j 0.049	0.095 + j 0.143 = 0.171
3/0 55% 2/0 copper 100-lb rails	1	1	0.270 + j 0.309	0.037 + j 0.084	0.307 + j 0.393 = 0.498
	1	2	0.270 + j 0.313	0.028 + j 0.073	0.298 + j 0.386 = 0.488
	1	4	0.270 + j 0.325	0.016 + j 0.051	0.286 + j 0.376 = 0.472
	2	2	0.135 + j 0.172	0.028 + j 0.073	0.163 + j 0.245 = 0.294
	2	4	0.135 + j 0.182	0.016 + j 0.049	0.151 + j 0.231 = 0.275
	4	4	0.068 + j 0.095	0.016 + j 0.049	0.084 + j 0.144 = 0.166
3/0 55% 3/0 copper 100-lb rails	1	1	0.232 + j 0.312	0.037 + j 0.084	0.269 + j 0.396 = 0.478
	1	2	0.232 + j 0.316	0.028 + j 0.073	0.260 + j 0.389 = 0.467
	1	4	0.232 + j 0.327	0.016 + j 0.051	0.248 + j 0.378 = 0.450
	2	2	0.116 + j 0.173	0.028 + j 0.073	0.144 + j 0.246 = 0.282
	2	4	0.116 + j 0.183	0.016 + j 0.049	0.132 + j 0.232 = 0.266
	4	4	0.058 + j 0.096	0.016 + j 0.049	0.074 + j 0.145 = 0.162
4/0 55% 4/0 copper 100-lb rails	1	1	0.189 + j 0.309	0.037 + j 0.084	0.226 + j 0.393 = 0.452
	1	2	0.189 + j 0.313	0.028 + j 0.073	0.217 + j 0.386 = 0.442
	1	4	0.189 + j 0.325	0.016 + j 0.051	0.205 + j 0.376 = 0.428
	2	2	0.094 + j 0.172	0.028 + j 0.073	0.122 + j 0.245 = 0.273
	2	4	0.094 + j 0.182	0.016 + j 0.049	0.110 + j 0.231 = 0.256
	4	4	0.047 + j 0.095	0.016 + j 0.049	0.063 + j 0.144 = 0.157
4/0 copper 100-lb rails	1	1	0.260 + j 0.368	0.037 + j 0.084	0.297 + j 0.452 = 0.540
	1	2	0.260 + j 0.372	0.028 + j 0.073	0.288 + j 0.445 = 0.528
	1	4	0.260 + j 0.384	0.016 + j 0.051	0.276 + j 0.435 = 0.513
	2	2	0.130 + j 0.200	0.028 + j 0.073	0.158 + j 0.273 = 0.315
	2	4	0.130 + j 0.210	0.016 + j 0.049	0.146 + j 0.259 = 0.296
	4	4	0.065 + j 0.109	0.016 + j 0.049	0.081 + j 0.158 = 0.177

* From Regulation and inductive effects in single phase railway circuits, by A. W. Copley, in *Elec. J.*, August, 1920.

MECHANICAL DESIGN OF OVERHEAD SYSTEMS. The American Transit Association has adopted specifications for overhead line material and construction which cover most of the mechanical details pertaining to ordinary contact wire suspension systems.

Direct Suspension Construction. The contact wire is secured to an "ear" by peening, clamps, or other means; the ear is bolted to a "suspension" which may or may not be provided with an insulating portion. These suspensions may be fastened directly to the bracket arm or carried by span wires, which in turn are fastened to the poles or bracket arms. If the ear is not insulated, strain insulators are inserted in the span wire between the ear and the point of attachment to the poles or bracket arms. A slightly different construction, called "pull-off," is used on curves. In a system designed for use with trolley wheels, "trolley-frogs" must be used at turnouts to guide the trolley wheels. Complete information and dimensions of the various parts referred to are contained in the ATA specifications previously cited. A few of the more important items of construction are covered briefly in the following paragraphs.

Length of Span. In general this type of construction should not be used for spans greater than 135 ft, 90 to 110 ft being most generally used.

Height of Contact Wire. The height of contact wire above the top of the rail varies between 15 and 22 ft, with a recommended height of 18 ft in streets and general locations, except where special conditions govern; the usual height over steam railroad crossings should be 22 ft. These values are for conditions of maximum sag.

Rake of Poles. Wood poles for bracket-arm construction should in general have a backward rake from the track of 6 in. in 24 ft. For cross-span construction wood poles should have a rake of 12 in. in 24 ft. Steel poles should have a rake half that of wood. Center poles (between tracks) should be set without rake.

Anchorage. A permanent anchorage for the contact wire may be provided at intervals of 1/4 to 3/4 mile and at both ends of every grade and curve, depending upon conditions.

CURVES. On curves of less than 500-ft radius a pole spacing of 50 ft should be used; for 500- to 800-ft radius, 75-ft spacing; and over 800-ft radius, the same spacing as for tangent track but not over 100 ft.

At curves the contact wire is made to follow the curvature of the track by pull-offs

Table 18. Overhead Catenary Construction

Railroad	Trolley Voltage	Equivalent Conductivity, cir mils	Span Length Tangent, ft	Type of Curve Construction	Main Messenger Sag	Main Messenger Kind	Main Messenger Diameter, in.	Auxiliary Messenger Material	Auxiliary Messenger Diameter	Grooved Contact Wires	
										Material and Size	No. Used
Chi. M. St. P. & P.	3,000	443,200	150	Chord	20 in.	SM	0.500	.....	....	Cu 4/0	2
Black River....	3,000	450,000	150	"	20 in.	(1)	0.375	.....	....	Cu 4/0	2
Cleve. Union Term	3,000	793,800	300	"	4 ft 4 in.	(3)	0.878	Cu	0.528 in.	Bz 4/0	2
Illinois Central....	1,500	838,900	300	"	4 ft 9 in.	(2)	0.810	Cu	0.512 in.	Bz 3/0	2
" "	1,500	898,700	300	"	4 ft 9 in.	(2)	0.810	Cu	0.375 in.	Cu 4/0	2
" "	1,500	840,000	300	"	4 ft 9 in.	(1)	0.810	Cu	0.512 in.	Bz 3/0	2
D.L. & W.R.R.....	3,000	792,000	300	"	5 ft 0 in.	(3)	0.821	Cu	0.414 in.	Bz 4/0	2
N.Y. N.H. & H. Woodln.-Stamford..........	11,000	335,360	300	"	6 ft 5 in.	(4)	9/16	Cu *	0.482 in.	Bz 4/0	1
Har. Stam.-N. Haven	11,000	320,640	{300 150}	Inclined	15 in. for 150-ft span	(5)	5/8	Cu *	0.482 in.	Bz 4/0	1
Danbury.......	11,000	315,800	250	"	4 ft 9 1/2 in.	(5)	9/16	Cu *	0.482 in.	Bz 4/0	1
Norfolk & Western	11,000	216,700	300	"	4 ft 6 in.	(6)	1/2	Cu *	0.392 in.	Bz 3/0	1
Penn. R.R. "Paoli"	11,000	335,600	285	"	5 ft 5 in.	(6)	0.625	Cu	0.482 in.	Bz 4/0	1
" Wilmington.	11,000	334,200	300	"	5 ft 6 in.	(6)	0.625	Cu	0.460 in.	Bz 4/0	1
" Trenton....	11,000	373,370	300	"	5 ft 6 in.	(6)	0.625	Cu	0.460 in.	Bz 4/0	1
N. Y., N. H. & H., Yds., Sidings	11,000	61,760	300	{Mostly chord}	5 ft 3 in.	(6)	3/8	Cu	....	Bz 2/0	1
Virginian Railway.	11,000	714,600	320	Inclined	5 ft 0 in.	(3)	0.725	Cu	2–3/0	Bz 3/0	1
" " .	11,000	645,000	320	"	in	(3)	0.725	Cu	2–2/0	Bz 3/0	1
" " .	11,000	512,000	320	"	300 ft	(3)	0.555	Cu	2/0	Bz 3/0	1
Reading R.R.....	11,000	328,000	250	Inclined	3 ft 0 in.	(1) (7)	0.636	.....	....	Bz 4/0	1
" "	11,000	337,600	300	Inclined	5 ft 6 in.	(8)	0.625	Cu	0.460 in.	Bz 4/0	1

SM—Siemens Martin hi-strength steel.

1. Hi-strength red brass, 7 strand.
2. Hi-strength copper-covered steel, 7 strand steel, 12 strand copper.
3. Hi-strength bronze, 19 strand bronze, 16 strand copper.
4. Hi-strength steel gal. ex. hi. 7 strand 2 cables.
5. Hi-strength steel gal. ex. hi. 19 strand.
6. Hi-strength steel gal. ex. hi. 7 strand.
7. Copper H.D.
8. 19-wire, bronze.

* Grooved auxiliary messenger.

Chi. M. St. P. & P.: Modified simple catenary, 885.21 miles, wood pole, bracket construction with cross-span at sidings and passing tracks. Feeder tap every 1000 ft. Contact trolleys side by side.

Cleve. Union Terminal: Compound catenary cross-bridge construction, 56 miles. Two trolleys; on main, side by side; in yards, one above the other.

Illinois Central: Compound catenary cross-bridge and center pole construction, 154.5 miles. Cross-span construction in yards.

D. L. & W. R. R.: Compound catenary cross-bridge construction on main line, 160 miles. Simple catenary on side tracks and yards. Bracket construction over single-track branch.

N. Y., N. H. & H. R. R.: 576.8 miles. Three point compound catenary, regular compound catenary, special double compound catenary, and simple catenary in yards. Cross-bridge on main line, cross-span in yards, and pole and bracket on the Danbury branch.

Norfolk & Western: 208.63 miles of cross-span compound catenary construction. Poles and brackets on single-track sections.

Penn. R. R.: 2200 miles. Cross-span compound catenary construction. H poles and tubular poles are used which also carry feeders.

Virginian Railway: 228.54 miles. Cross-span compound catenary construction. Some sections have two auxiliary messengers in vertical plane with contact wire. Single track, pole and bracket construction.

(wires pulling the contact wire toward the outside of the curve). Pull-offs should be radial to the contact wire where possible and not too expensive.

The number of pull-offs should be such that the contact wire is kept within about $2\ ^1/_2$ in. from the theoretical curve and is given by the formula

$$L = \sqrt{\frac{2aR}{3} - \left(\frac{a}{6}\right)^2}$$

where L = distance between pull-offs in feet.

R = radius of curve in feet.

a = offset of contact wire in inches from the theoretical curve, midway between pull-offs, the ears being assumed to be on the theoretical curve.

Or

$$L = 0.815\sqrt{aR}$$

with an error less than $^1/_4$ per cent for all radii greater than 40 ft. If a is to be $2\ ^1/_2$ in., then

$$L = 1.29\sqrt{R}, \text{ approximately.}$$

Catenary Suspension Construction. In general, catenary construction may be divided into simple and compound. In the former the contact wire or wires are supported by hangers directly from the messenger wire. In the latter another wire, called the auxiliary messenger, supports the contact wires by clips or very short hangers, being itself supported by hangers from the main or primary messenger wire.

The compound catenary is more flexible than the simple catenary unless some form of flexible or loop hanger is used, which may necessitate shunts in each span in order to prevent arcing at the hangers. Flexibility in the overhead construction is very important, especially where high-speed operation is encountered, as a "hard spot" in the overhead results in arcing with decreased life of contact wire and collector device. An aid to flexibility is the use of two contact wires in approximately the same horizontal plane, each wire being fastened to alternate hangers or clips. The choice between simple and compound construction also depends on the total conductivity required and the train speed.

The same general requirements for pull-offs, anchorages, etc., are encountered with catenary construction as with the direct suspension type.

At curves the catenary construction may be handled by either the "tangent chord" or the "inclined" construction. This latter type does not require pull-offs but does necessitate more detailed attention to hanger design.

It is almost the universal custom to provide "steadies" in tangent track at each supporting structure to prevent side sway caused by wind or resulting from rolling stock motion. "Steadies" are very similar in construction to pull-offs. See Table 18.

11. UNDERGROUND TROLLEY SYSTEMS

This system is little used on account of its cost, the installations at Budapest, New York, Washington, and London being the only notable ones. The essential feature is the underground conductor which is reached from the car by a "plow" extending through a continuous slot parallel to the tracks.

NEW YORK SYSTEM. The New York type, in successful use for many years, will be described briefly.

General Description. The street is excavated and cleared of obstructions for a width of about 5 ft 6 in. and a depth of about 3 ft, and yokes (Fig. 19) are set in the excavation about 5 ft apart. The track and slot rails are supported on these yokes, and the whole system is made solid with concrete, which fills the excavation from the foundation to near the top of the rails, leaving only a tunnel under the slot free from masonry. The conductors, of a special rolled-steel flanged tee section, are suspended in this tunnel by special strain insulators, no part of the electrical system being grounded. The use of two insulated conductors avoids trouble in case of accidental grounding.

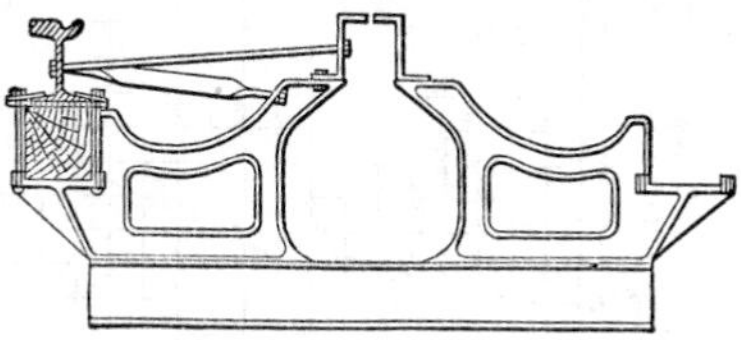

FIG. 19

Yoke. The yoke, shown in Fig. 19, is made in three parts: an I-beam, which rests on the floor of the excavation, and two castings riveted to it, each supporting one track rail and one slot rail. The track rail is carried on a timber stringer, which extends from yoke to yoke along the whole line. The stringers are held to the yokes by bolts, which serve also to fasten the rails to the stringers. The rails are further secured by long bolts running to the center of the yoke. The slot rails are bolted directly to the yoke and are connected to the track rails by long bolts every 30 in.

Insulator Boxes. Cast-iron boxes are laid across from the slot rails to the track rails every 15 ft and bolted to them. These contain the insulators, which support the contact conductors. In each box is a pair of shelves; resting on these shelves is a cast-iron bridge which holds the insulator. A cast-iron cover is flush with the street surface.

Concrete Work. The tunnel which is to protect the conductors and do the draining is made with collapsible forms. Concrete is then poured into the excavation, completely filling every part but the tunnel and its offsets at the insulator boxes to within the height of a paving block of the rail tops. The depth of the completed tunnel is 18 in. from the base of the slot rails.

THE LONDON SYSTEM has alternate long and short yokes, the long ones fulfilling the same function as the New York ones, the short ones serving merely to give additional support to the slot rails. No stringers are used, the rails resting on hardwood blocks at the yokes only.

THE BUDAPEST CONSTRUCTION has the slot in the track rail, thus saving considerable iron, but necessitating an excessively wide slot to accommodate the wheel flanges.

12. TRACK AND THIRD RAILS

(See also Railway Track Bonds, Article 13; Traction Systems, Article 1; Third-rail or Contact-rail Systems, Article 9; and Underground Trolley Systems, Article 11.)

The American Transit Association has standardized the rail sections summarized in Table 19, which are described fully in its Engineering Manual.

Table 19. Rail Sections

Designation of Rail	Height, in.	Weight, lb per yd	Designation of Rail	Height, in.	Weight, lb per yd
Standard section.....	5	80	Girder grooved.....	7	103, 122
" "	5 5/8	90	" "	9	134
" "	6	100	" guard.......	7	140
Plain girder..........	7	82, 92, 102	" "	9	152

The standard ("tee") and plain girder sections are listed for use on private right-of-way and in macadam or other shallow paving, and the girder grooved and guard sections for paved streets.

Standard lengths are as follows: plain girder and girder grooved rail, 60 and 62 ft; girder guard rail, 30 and 32 ft; standard section ("tee") rail, 33 or 39 ft as ordered, 62 ft when so specified.

RESISTANCE AND CHEMICAL COMPOSITION. The chemical composition of steel rails with respect to the impurities or elements other than iron (chiefly carbon, manganese, phosphorus, sulfur, and silicon) varies over a considerable range, depending

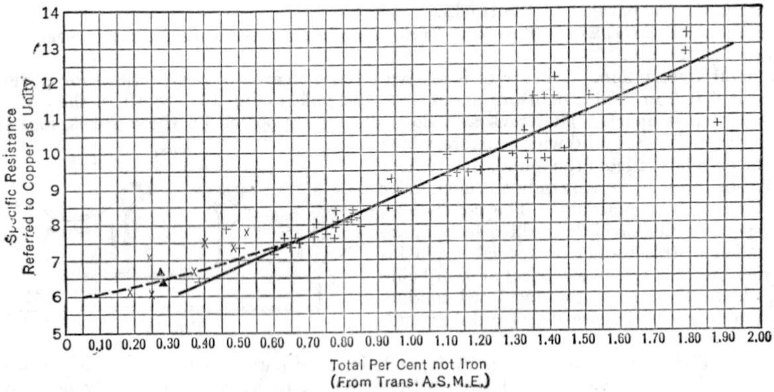

Fig. 20

upon the process of manufacture. According to J. A. Capp, the specific resistance of an ordinary * steel rail may be taken as a rough indication of the total impurities present. Figure 20 shows the results of tests on a number of samples of different makes, ranging in total impurities from 0.1 per cent to 1.9 per cent, the specific resistance referred to copper as unity ranging from 6.1 to 13.3. The greater hardness caused by the presence of impurities is an advantage in track rails which offsets the disadvantage of low conductivity, but it is usually economical to employ for the third or contact rail a rail of fairly high conductivity.

Table 20. Chemical Composition, Track Rails

Designation of Rail		Carbon, %	Manganese, %	Phosphorus, %	Silicon, %
	Weight, lb per yd				
Standard Section Rails	50–69	0.50–0.63	0.60–0.90	< 0.04	> 0.15
	70–84	0.53–0.70			
	85–100	0.62–0.77			
	101–120	0.67–0.83	0.50–0.90		
	121–140	0.72–0.89			
Girder Rails	A *	0.60–0.75	0.60–0.90	< 0.04	0.15–0.40
	B *	0.70–0.85			
	C *	0.75–0.90			

* Girder guard rails are class A. Plain or grooved girder rails 135 lb or over are class C; under 135 lb may be specified A or B.

The compositions given in Table 20 are recommended by the American Transit Association. The specific resistance of track rails, referred to copper as unity, ranges between 10 and 12.5.

As shown by specification requirements in Table 21, current practice favors copper in third rails and low content of carbon and sulfur, to reduce both corrosion and electrical losses.

Table 21. Chemical Compositions and Resistances, Third Rails

	New York Central R. R.	Pennsylvania R. R.	Long Island R. R.	Baltimore & Ohio R. R. (Staten Island)	Board of Transportation N.Y. City	Manhattan Elevated R. R.
Weight per yard, lb	70	150	150	....	150	100
Process	O.H.	O.H.	O.H.	O.H.	O.H.	Mfr's option
Length, ft	33	39	39	33	33	60
Specific resistance (ratio to copper) not to exceed:						
No. 1 rails	7.	7.	7.	7.	6.85	7.
No. 2 rails	7.35	7.35	7.35	7.35	7.	7.3
Bonus below	...	6.85	6.85	....	....	....
Chemical composition, %:						
Carbon (not to exceed)	0.20	0.06,*0.10	0.06 *	0.06*	Note A	Note B
Sulfur (not to exceed)	.06	.06	.06 *	.05*	Left to discretion of manufacturer	
Manganese (not to exceed)	.30	.20	.15 *	.15*		
Phosphorus (not to exceed)	.04	.015	.008*	.025*		
Silicon (not to exceed)	.05	.05	.04 *	.04*		
Copper from	.20	.15	.15 *	.20		
to	.30	.30	.30 *	.30		
Total, other than iron Not to exceed	...	.70	.70	.70		

* " Desired " (not positive requirement.)

A " May be low carbon, but should have sufficient stiffness to retain shape during shipment and installation."

B Desired not less than 0.15; must not be less than 0.12.

ALLOWANCE FOR WEAR IN RESISTANCE CALCULATIONS. Resistance calculations should be made for rails worn down to the weight at which they will be scrapped. It is usual to scrap rails when they have lost from 10 to 20 per cent of the original weight,

* This does not apply to special forms of rails submitted in the process of rolling to extra heavy pressures

depending upon the importance of the line. The resistances in Table 22 should therefore be increased from 10 to 20 per cent. They are d-c resistances. See p. 17-44 for a-c resistances.

Table 22. Resistance of Steel Rails

(Full Cross-section)

Weight, lb per yd	Cross-section, sq in.	Area, millions of circular mils	Specific Resistance 12.5 times that of copper		Specific Resistance 8 times that of copper *	
			Ohms per 1000 ft	Ohms per mils	Ohms per 1000 ft	Ohms per mils
40	3.90	4.95	0.0261	0.138	0.0167	0.0882
45	4.40	5.60	0.0231	0.122	0.0148	0.0782
50	4.90	6.23	0.0208	0.110	0.0133	0.0702
55	5.40	6.86	0.0189	0.0996	0.0121	0.0637
60	5.90	7.50	0.0173	0.0911	0.0110	0.0583
65	6.40	8.14	0.0159	0.0840	0.0102	0.0538
70	6.90	8.77	0.0148	0.0779	0.00944	0.0499
75	7.437	9.45	0.0138	0.0729	0.00884	0.0467
80	7.80	9.9	0.0131	0.0689	0.00835	0.0441
85	8.34	10.5	0.0122	0.0645	0.00781	0.0413
90	8.83	11.2	0.0115	0.0609	0.00738	0.0390
95	9.30	11.8	0.0109	0.0570	0.00701	0.0370
100	9.82	12.5	0.0104	0.0547	0.00664	0.0350
105	10.31	13.1	0.00991	0.0521	0.00632	0.0333
110	10.80	13.8	0.00945	0.0497	0.00604	0.0318
115	11.29	14.4	0.00904	0.0476	0.00577	0.0304
120	11.78	15.0	0.00867	0.0456	0.00553	0.0292
125	12.27	15.6	0.00832	0.0438	0.00531	0.0280
130	12.77	16.2	0.00800	0.0421	0.00511	0.0269
135	13.26	16.9	0.00770	0.0405	0.00492	0.0259
140	13.75	17.5	0.00743	0.0391	0.00474	0.0250
145	14.24	18.1	0.00717	0.0377	0.00458	0.0241
150	14.73	18.7	0.00693	0.0365	0.00443	0.0233
155	15.22	19.4	0.00671	0.0353	0.00428	0.0226
160	15.71	20.0	0.00650	0.0342	0.00415	0.0219

* To find the resistance of rails of any specific resistance x, referred to copper as unity, multiply these resistances by x and divide by 8.

Alternating-current Resistance and Reactance. See Overhead Contact Systems, Article 10, and Railway Signaling, Article 49.

13. RAILWAY TRACK BONDS

(See also Third-rail or Contact-rail Systems, Article 9; Track and Third Rails, Article 12; Wires and Cables.) Rail bonds are electrical conductors for bridging the joints of rails. They consist of stranded or laminated copper conductors welded or pressed into copper or steel terminals. Sometimes the conductors are made up of copper and steel wires stranded together.

TYPES OF BONDS. Bonds may be classified, according to the method of fastening them to the rail, as welded bonds, brazed bonds, and bonds applied by mechanical pressure.

Welded Bonds (Figs. 21, 22, and 23) are designed for oxyacetylene flame welding, steel-electrode arc welding, copper-electrode arc welding, and carbon-electrode arc welding.

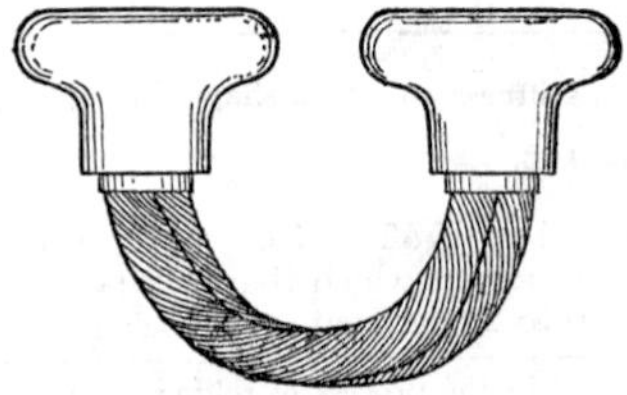

Fig. 21

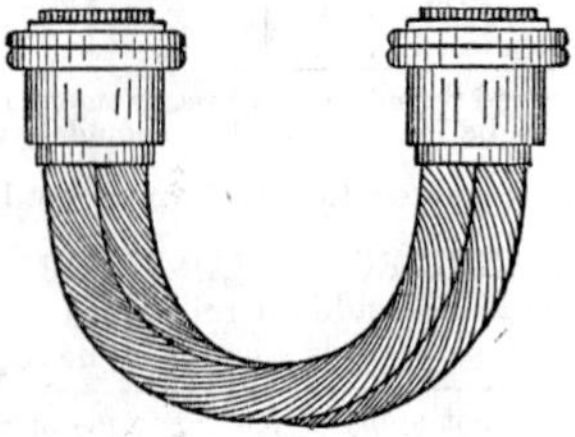

Fig. 22

They usually consist of all-copper stranded conductors with mechanically attached steel or copper terminals or with steel terminals butt-welded to the conductors. The flame-welded bonds are usually installed with a flux-containing copper-alloy welding metal.

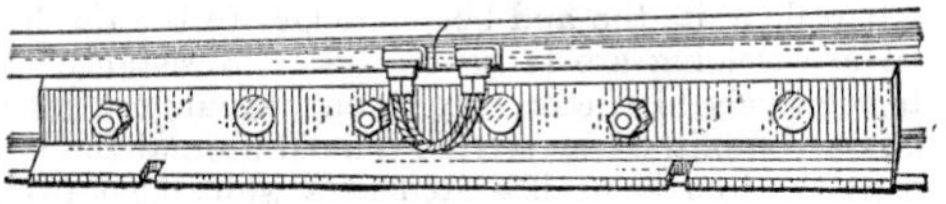

FIG. 23

The arc weld bonds are installed with steel electrode or flux-containing copper-alloy electrode, or with carbon electrode in conjunction with a carbon mold with flux-containing copper as the welding metal.

The designs of welded bonds are many and are changing so rapidly that no details of construction are given.

Brazed Bonds (Fig. 24) are brazed or welded to the rail by heat generated electrically in a carbon which constitutes the clamp holding the bond against the rail.

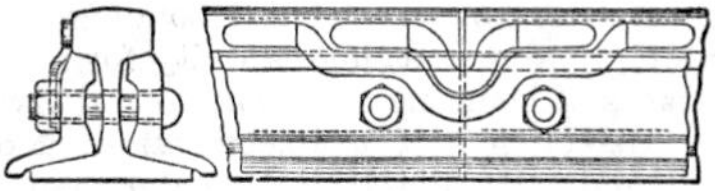

FIG. 24

Expanded and Compressed Terminal Bonds. Bonds fastened to the rail by mechanical pressure may be divided into two general classes, expanded terminal and compressed terminal bonds. Both kinds are called stud terminal bonds.

Pin-expanded Terminal Bonds (Fig. 25) have their heads drilled with an axial hole, through which a tapered steel pin (*d*) is driven, forcing the copper outward and against the steel. This type of bond is fastened to the web of the rail.

Compressed Terminal Bonds have solid copper terminals and are installed with pressure applied at both ends of the stud. The bond is applied to the web of the rail by means of a heavy screw or hydraulic press which engages the bond head and causes it

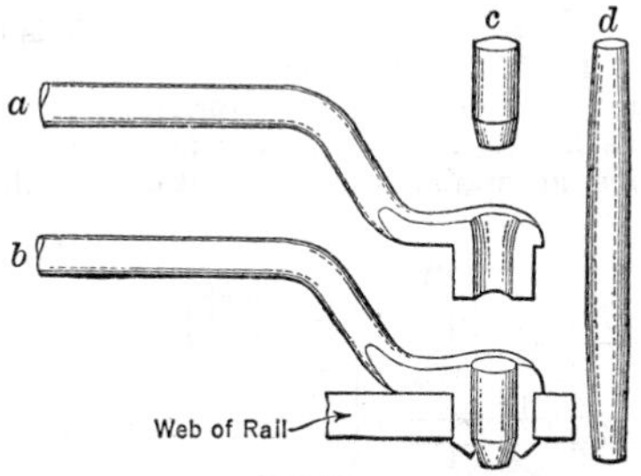

FIG. 25

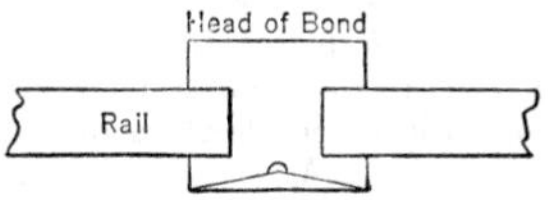

FIG. 26

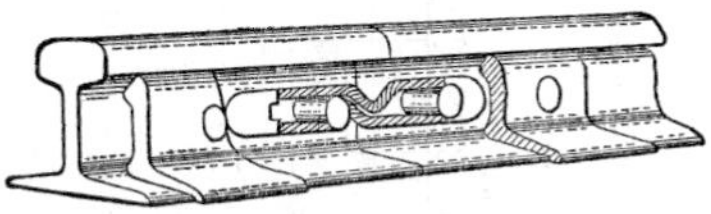

FIG. 27

to compress longitudinally and expand laterally as the pressure is applied, bringing the copper into firm contact with the steel and spreading the projecting end of the terminal into a button-shaped rivet-head, as shown in Fig. 26.

Exposed versus Concealed Bonds. Bonds may be either exposed or concealed (Fig. 27) under the fish-plates. The exposed condition is preferable, if there is no likelihood of theft, as it permits inspection to be easily made.

Welded-type bonds are always of the exposed type and are applied either to the head or the top of the flange of the rail. Stud terminal bonds (pin or compressed) may be of the concealed or exposed type. They are always applied in holes drilled in the web of the rail. The choice of type is largely a matter of economics. Although concealed bonds are necessarily applied to the web of the rail, exposed bonds may be applied to the foot or head. Head bonds have the advantage of being short. Web or base bonds, unless concealed, have to be long in order to span the splice bars.

USES OF VARIOUS TYPES OF BONDS. Bonds are used for track rails, third rails, and girders of elevated and subway lines. Expanded or compressed terminal bonds, especially of the concealed type, are excellent for heavy traction work, where space underneath the splice bars is adequate. Welded bonds are nearly always used on the head of the rail, within the limits of the splice bars; they are easy to inspect.

SUBSTITUTES FOR BONDING. Several efficient substitutes for bonding are now in use, such as electrical welding, cast welding, and thermit welding. Welded joints are used almost exclusively in paved streets.

Electric Welding is almost exclusively of the arc welding type. Various methods of "sewing" the splice bar along its top and bottom edges to the rail head and rail base by different methods of arc welding are in general use. Resistance welding of the splice plate, because of its relatively high cost, has given way almost entirely to arc welding.

Thermit Welding is accomplished by setting a mold around the rail joint and pouring molten iron around it, the iron being liberated at a white heat from a mixture of iron oxide and aluminum, which is ignited in a crucible.

Flame-welding by oxyacetylene torch is also in successful use.

SELECTION OF TYPE AND SIZE OF BOND. Considerations determining the choice between concealed and exposed bonds are liability to theft, electrolytic corrosion, facility of inspection, and injury to bonds in service.

The choice between mechanically attached and welded bonds depends largely upon the importance of rapid installation, the mechanical stresses to be withstood in service, the type of labor available, and the facilities for the use of drills, presses, etc.

Single versus Double Bonding. Joints are sometimes bonded with one and sometimes with two bonds. Double bonding has the advantages of less chance of complete failure and greater carrying capacity for a given cross-section of copper. It has the disadvantages of being more expensive and giving uncertain results in testing, however.

Selection of Cross-sectional Area of Bond. The cross-sectional area of a rail bond should, as a rule, be not greater than is necessary to keep its temperature at a safe working value, unless greater area is required for mechanical strength. The resistance of the bonded joint is of secondary importance unless very high, because the resistance of the joints is usually a mere fraction of the total track-rail resistance (see below).

Carrying Capacity of Bonds. The excellent heat conductivity of copper and the large heat-storage capacity of steel rails tend to make the carrying capacity of bonds in cold weather considerably greater than that of free wire of the same size, especially if the bonds are short. In hot weather, however, the rails and consequently the bonds are likely to become hot from exposure to the sun's rays, thereby reducing the effective carrying capacity of the bonds.

It is not safe to set up a formula for carrying capacity. A very conservative rule is to use 1 amp per 500 cir mil cross-sectional area. If 250 circ mils per amp is used, there will be no injurious heating of the bond, but the length of bond conductor is a very important factor. Very short bonds will carry considerably more current than long bonds.

Resistance of Bonds. The ohmic resistance of stud terminal rail bonds, at 68 deg fahr, including rail contact resistance, is given in Fig. 28.

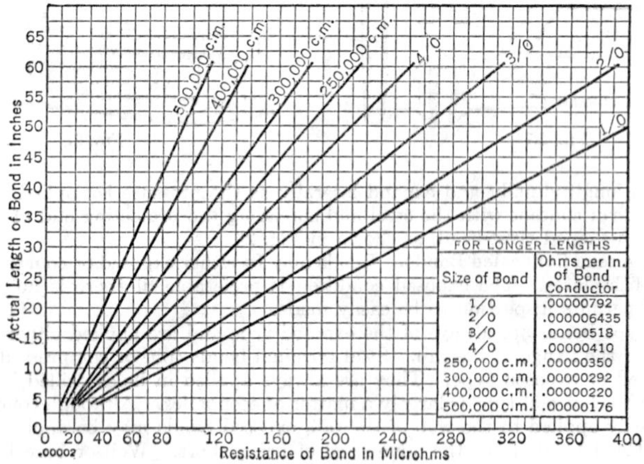

Fig. 28

Although the resistance values of welded bonds are in general higher, the stud terminal bond-resistance values can be used with fair accuracy for welded bonds of all types. There is very little difference between the resistance of one form of welded bond and another when properly applied.

Resistance of Bonded Joint. The conductivity of the splice bars may be neglected. The resistance of a bonded joint can be computed as follows:

Let b = resistance of joint, ohms.
a, a_1, a_2 = resistance of bond A, A_1, A_2 (extended length) ohms, from Fig. 28.
d = resistance of length D of rail, ohms.

Then, for a single bonded joint, $b = a$. For a joint bonded with two equal bonds, lapped, Fig. 29(*a*), $b = {}^1/_2\,(a + d)$, and for a joint bonded with two unequal bonds, Fig. 29(*b*) $b = \dfrac{a_2(a_1 + 2d)}{a_2 + a_1 + 2d}$.

Resistance of Bonded Rail. The resistance of bonded rail can be computed from the following formula:

$$x = nb + r\frac{(5280 - ns)}{5280}$$

where x = resistance of bonded rail per mile, ohms.
b = resistance of joint, ohms, from preceding paragraph.
s = length between outside bond terminals, feet (Fig. 29).
n = number of joints per mile.
r = resistance of rail per mile, ohms. (See Rails, Article 12.)

Length of Bond. The length of concealed bonds is necessarily determined by the spacing of the bolt holes. The bonds for attachment to the head of the rail are usually 7 or 7 ½ in. long for single bonding and about 8 ½ in. long for double bonding where U-shaped bonds are used. Bonds to span the splice bar must be of sufficient length to have the terminals beyond the ends of the splice bar and far enough away to permit the bond conductor to clear the splice bar properly.

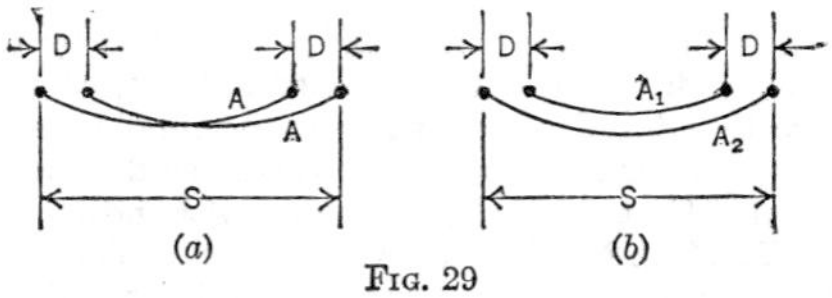

FIG. 29

SPECIFICATIONS. (See also article on Specifications and Contracts in Eshbach, *Fundamentals of Engineering.*) Specifications for rail bonds should state the exact service conditions under which the bonds are to be used, the style of attachment desired, the part of the rail they are to be applied to, the style of conductor (ribbon, solid wire, or cable), the cross-sectional area, the contact area of the stud, the formed length between centers of terminals, and the fish-plate and bolt layout, if the bonds are to be concealed.

INSTALLATION. An important consideration in the installation of bonds is the cleanliness of the bonds and bond holes, or other adhesion surface. Unless this is secured, the bonds will be electrically defective, whatever their mechanical strength may be.

Welded Bonds. For the correct installation of any type of welded bond it is essential that the proper welding apparatus and welding metal for applying the bonds be selected. Flame-welding requires the use of flux-containing copper alloy of the proper composition, which the leading manufacturers have developed. Steel-electrode welding requires the use of low-carbon steel. Coated electrodes are very desirable. Copper-electrode welding requires the use of a flux-containing copper-alloy electrode and high amperage in order to "puddle" the metal as much as possible. Carbon-electrode welding requires the use of a carbon mold to surround the bond terminal. Flux-containing copper alloy (the same as copper-alloy electrode) is melted in the carbon arc until the mold cavity is filled. Proper polarity of electrode is essential. Steel electrodes may be used either positive or negative. Copper electrodes should always be positive. Carbon electrodes must be negative.

Brazed Bonds. The rail surface is first ground clean. The bond terminal, being covered with a brass plate, is then placed against the clean surface. A carbon block is pressed against the surface of the bond terminal opposite the rail. The carbon block is then heated to incandescence by means of an electric arc, which does not come in contact with the bond. Borax flux is introduced between the bond terminal and the rail. When the terminal reaches the proper temperature, a braze is effected between the bond terminal and the rail.

Pin-expanded Bonds. The rail is drilled through the web, with or without lubricant. Drilling without lubricant has the advantage of giving a perfectly clean hole. Holes are drilled to the same diameter as the bond studs, and rust or scale on the web of the rail is removed around the hole. The bond stud is then inserted into the hole, and a long taper punch lubricated with grease is driven entirely through the terminal. Then a short drift pin is driven home, as shown in Fig. 25. About 100 bonds may be installed with one taper punch.

This type of bond requires a smaller equipment in tools and materials than most other types and does not necessitate any apparatus which obstructs the track and thereby endangers traffic.

Compressed-terminal Bonds. The drilling having been performed as for a pin-expanded bond, and the bond studs inserted into their respective holes, a screw or hydraulic compressor is applied at both ends of the bond head, the conical point of the press fitting into the conical depression of the bond. Pressure is applied, either until a collar on the ram touches the rail, or until the head of the bond acquires the proper shape. Where no collar is used, the point of the press (if of the screw type) sometimes cuts into the bond head; this may be avoided by placing a small amount of flake graphite mixed with oil in the depression of the bond head.

For durable contact the bond terminal stud should be installed with a direct pressure of at least 25 tons per square inch of contact area. Actual contact resistances of stud terminals at 15 tons per square inch contact pressure are given in Table 23.

Table 23. Contact Resistance of Stud Terminals

Diameter of Terminal Stud, in.	Area of Contact, sq in.	Contact Resistance, ohm	Diameter of Terminal Stud, in.	Area of Contact, sq in.	Contact Resistance, ohm
1	1.77	0.00000040	5/8	1.10	0.00000064
7/8	1.55	0.00000045	1/2	0.88	0.00000080
3/4	1.33	0.00000053	2 twin terminal studs	2.00	0.00000035

The copper of a bond head is hardened by the pressure it is subjected to, and, like the steel, is distorted within its elastic limit, causing the surfaces to adhere even if the pressure is reduced to one-third its original value, say 10,000 lb per sq in. Between these two pressures, the electrical resistance does not vary. Expansion due to heat, therefore, has no effect upon the resistance of bonds.

TESTS AFTER INSTALLATION. The usual method of testing is to measure the drop of potential across the bonded joint and find simultaneously the length of continuous rail in which the same drop occurs, i.e., the "equivalent" length of the bonded joint. Several ingenious instruments have been devised for making this comparison with ease and accuracy.

Voltmeter Method. The simplest form of bond tester consists of a center zero millivoltmeter with three connections to the rail. (See Fig. 30.) With current flowing through the rail in the direction from A to C, the current flows through the millivoltmeter movable coil in the direction A to B to indicate the difference of potential between A and B. The drop between B and C is indicated by the current flowing through the same coil in the direction B to C, thus opposing the drop between A and B. The millivoltmeter gives the difference. By varying the distance BC, the drop across the joint and the drop from the solid rail can be equalized. The distance between B and C is then measured and represents the length of rail equivalent to the resistance across the joint.

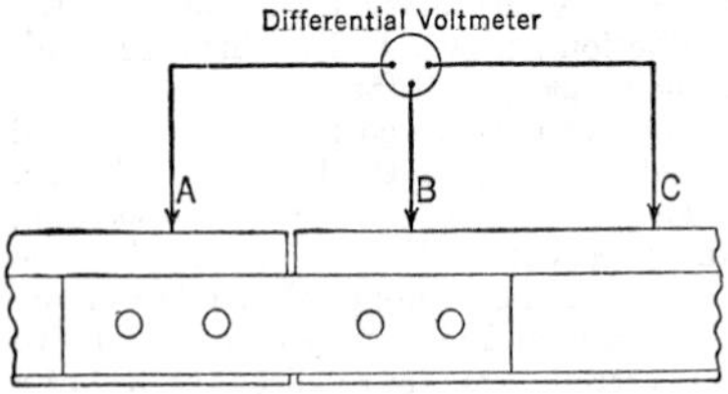

FIG. 30

Many commercial bonding sets have rail contact devices with a fixed spacing of the contacts, which span the rail joints, of 3 ft. In order to obviate correction of the readings of the rail joints as tested, the resistance of 3 ft of joint may be calculated and used for comparison.

14. BIBLIOGRAPHY

General

Carter, F. W., *Railway Electric Traction.* London (1922).
Dickinson, R. E., *Electric Trains.* London (1927).
Dover, A. T., *Electric Traction.* London and New York (1917).
Harding, C. F., and D. D. Ewing, *Electric Railway Engineering.* New York (1926).
Healy, K. T., *Electrification of Steam Railroads.* New York (1929).
Manson, A. J., *Railroad Electrification.* New York (1923).
Richey, A. S., *Electric Railway Handbook.* New York (1924).
Sheldon S., and E. Hausmann, *Electric Traction and Transmission Engineering.* New York (1920).
Copper and Brass Research Association, *Survey of Electrification of Steam Railroads.* New York (1929).
Railway Age (weekly): *Transit Journal,* formerly *Electric Railway Journal* (monthly); *Railway Electrical Engineer* (monthly); *Railway Mechanical Engineer* (monthly); *Railway Gazette* (weekly,

London; particularly *Diesel Railway* Traction Supplement, and *Electric Railway* Traction Supplement, each once in four weeks.) *Electric Journal* (monthly); *Electric Railway Bus and Transportation Journal* (weekly, London); *General Electric Review* (monthly).
Association of American Railroads, Proceedings and Manuals of Mechanical and Engineering Divisions.
American Railway Engineering Association, Proceedings and Manual.
American Transit Association, Proceedings.
American Transit Engineering Association, Proceedings and Engineering Manual.
International Railway Congress Association, Bulletin (monthly).
American Institute of Electrical Engineers, Transactions and Standards.
American Society of Mechanical Engineers, Proceedings.

Energy Requirements and Motor Equipments

Anderson, G. H., cited by F. W. Carter, *Trans. AIEE*, Vol. 36, p. 240 (1906).
Carter, F. W., Predetermination in Railway Work, *Trans. AIEE*, Vol. 22, p. 133 (1903).
Carter, F. W., The Mechanics of Train Movement, *JIEE*, Vol. 50, p. 434 (1913).
Davis, W. J., Jr., The Tractive Resistance of Electric Locomotives and Cars, *Gen. Elec. Rev.*, Vol. 29, pp. 685–707 (1926).
Frendenberger, L. A., Plotting of Speed-time Curve from the Acceleration-speed Curve, *Elec. World*, Vol. 42, p. 96 (1903).
Hutchinson, C. T., The Conditions Governing the Rise of Temperature of Electric Railway Motors in Service, *Trans. AIEE*, Vol. 22, p. 657 (1903).
Kimball, E. E., Notes on the Determination of Power Station Capacities, *Gen. Elec. Rev.*, Vol. 10, pp. 77 and 120 (1908).
Mailloux, C. O., Notes on the Plotting of Speed-time Curves, *Trans. AIEE*, Vol. 19, p. 901 (1902).
Mailloux, C. O., Method of Determining the Continuous Current Having the Same Heating Effect as a Variable Current, Int. Elect. Congress, Turin (1911).
Schmidt, Ed. C., Train Resistance, *Bull.* 43, 59, 110, and 167, University of Illinois, 1910, 1912, 1918, and 1927; Train Tests in the Cambridge Subway, *Elec. Ry. J.*, Vol. 40, p. 280 (1912).
Totten, A. I., Resistance of Lightweight Passenger Trains, *Trans. ASME*, Vol. 59, No. 4, p. 329 (May 1937).
Valentine, W. S., A Graphical Method of Making Time-speed Curves, *St. Ry. J.*, Vol. 20, p. 303 (1920).

Locomotives and Cars

Jackson, W., *Electric Car Maintenance.* New York, 1914.
Locomotive Cyclopedia. New York.
Car Builders' Cyclopedia. New York.
Association of American Railroads, Mechanical Division, *Manual of Standard and Recommended Practice.*
American Transit Association, *Engineering Manual.*
Association of Railway Electrical Engineers, Proceedings; *Manual of Recommended Practice.*

Car Lighting

Bender, C. W., *Railway Electrical Engineers' Handbook.* Cleveland, National Lamp Works.
Stuart, C. W. T., *Car Lighting by Electricity.* New York (1923).
Association of Railway Electrical Engineers, *Handbook, Electric Car Lighting Equipment;* Manual; and Proceedings.
Bulletins of Gould Coupler Co., Safety Car Heating and Lighting Co., U. S. Light and Heat Co.

Bonds

Brown, H. F., Railbonding Practice and Experience, *Trans. AIEE*, Vol. 49, p. 1299 (1930). (The discussion covers European practice.)
Shepard, E. R., Modern Practice in the Construction and Maintenance of Rail Joints and Bonds in Electric Railways, *Tech. Paper* 62, *Bur. Standards* (1920).
Track Circuit Handbook. Ohio Brass Co., Mansfield, O.
Publications of American Steel & Wire Co.

Overhead Systems

Manuals cited above.
Viele, S. M., Current Collection from an Overhead Contact System; Brown, H. F., Catenary Design for Overhead Contact System; Thorp, J. S., Catenary Construction for Chicago Terminal, I.C.R.R.; Wade, R. E., and Linebaugh, J. J., The Collection of Current from Overhead Contact Wires; Jorstad, O. M., Railway Inclined Catenary Standardized Design; all in *Trans. AIEE.*, Vol. 46, pp. 1072–1138 (1927).
Copley, A. W., Regulation and Inductive Effects in Single-phase Railway Circuits, *Elec. J.*, Vol. 17, p. 326 (August, 1920).

Recent Papers in *Trans. AIEE*

Chapman, R. L., and O. K. Kjolseth, Modern Motive Power for the Sorocabana Railway, Vol. 63, p. 558 (1944).
Craton, F. H., and F. M. Turner, Electric Braking for Railroad and Urban Transit Equipment, Vol. 59, p. 489 (1940).
Gage, S. S., Electrification and Signalling of the Canadian National Railways Terminal, Montreal, Vol. 64, p. 41 (1945).
Gardner, J. E., The Application of Electricity for the Auxiliaries of Railroad Trains, Vol. 60, p. 34 (1941).
Green, H. W., Application of Electricity to Railways, Vol. 90, part I, p. 71 (1943).
Harder, E. L., Pennsylvania Railroad—New York–Washington–Harrisburg Electrification—Relay Protection of Power Supply System, Vol. 58, p. 266 (1939).
Monroe, W. P., Selection and Design of the Electrification of the San Francisco–Oakland Bay Bridge Railway, Vol. 57, p. 791 (1938).
Muylaert, D., The Sorocabana Railway Electrification, Vol. 62, p. 804 (1943).
Thomas, B. F., Energy Consumption on Street Railways, Vol. 53, p. 326 (1934).
Torchio, P., Railway Power Supply, Vol. 59, p. 550 (1940).
Trueblood, H. M., and G. Wascheck, Investigation of Rail Impedances, Vol. 53, p. 1771 (1934).
Wilcox, H. C., and A. G. Oehler, Factors Involved in the Selection of Railroad Motive Power, Vol. 62, p. 235 (1943).
Wright, A. M., Modern Rail Transport, Vol. 59, p. 1306 (1940).

ELECTRICAL EQUIPMENT OF INTERNAL-COMBUSTION AUTOMOBILES

By Edmund B. Neil

15. ELECTRIC IGNITION

FUNCTION. It is the function of the ignition system of an internal-combustion engine to initiate the burning of the charge of fuel and air within the engine cylinder at the proper instant in the cycle; i.e., when the explosive mixture of gas and air is at or near the point of maximum compression. The typical system (see Fig. 1) comprises (1) a source of electrical energy; (2) a timer for correlating the production of the spark with the rotation of the engine (this usually is a "breaker" which interrupts the primary circuit of an induction coil); (3) an induction coil for transforming the energy to a high voltage; (4) a

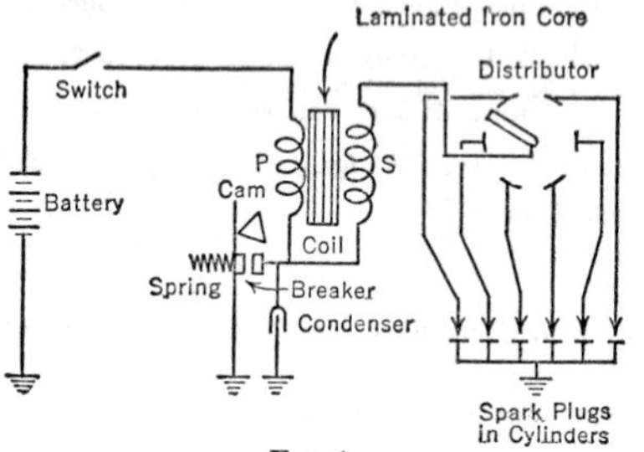

FIG. 1

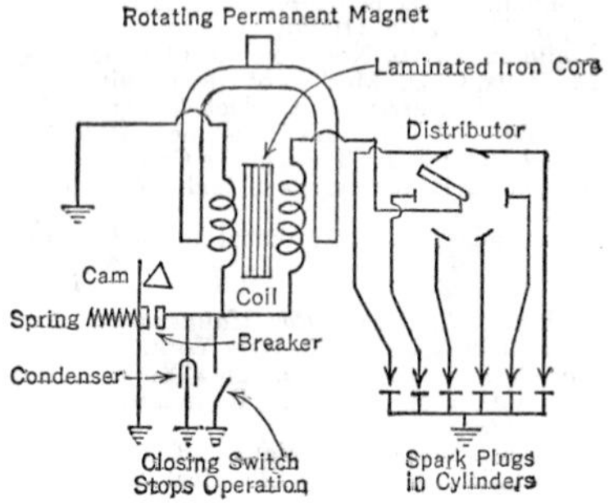

FIG. 2

distributor for leading the high voltage discharge to the successive cylinders in the proper sequence; and (5) a spark plug in each cylinder to provide a fixed insulated gap at which the igniting spark is produced.

The energy source (1) above may be either a battery or magneto generator. In the latter case the circuit is as shown in Fig. 2. Notice that the switch in the battery circuit (Fig. 1) is in the open position to prevent current flow and operation of the ignition system, whereas in the magneto circuit it is in the closed or grounded position.

IGNITION OF GASES. When combustible mixtures of air and gas or finely atomized liquid fuel are ignited, a wave of flame spreads through the mixture. Inflammation of the mixture is not instantaneous. In general, the rate of flame propagation is relatively slow, but under some conditions the charge is said to **detonate.** Detonation is the autoignition of the remaining or unburned portion of the charge of fuel and air in the cylinder. While the exact cause has not yet been determined, it is probably due in part to the rise in pressure of the unburned gases accompanying initially normal burning of the first portion of the charge. Two opposing pressure waves are thus set up, which cause a violent increase in cylinder pressure, accompanied by an abnormal increase in temperature. It is desirable to avoid detonation not alone because of the high temperatures that accompany it but also because the rapid pressure rise may result in mechanical injury. The rate of flame propagation, before detonation, varies with different gases, with different proportions of air and fuel in otherwise similar mixtures, with the density and the temperature of the charge, and with other factors, such as the chemical structure of the fuel.

When severe detonation occurs, the power developed usually decreases but, if the compression pressure and consequent density of the gases compressed can be increased without detonation, power output can be materially increased. Knowledge of this fact and of the destructive effects of detonation have led to the extensive use of fuels less subject to detonation and to the use of tetraethyl lead. Even a few cubic centimeters of this liquid per gallon of fuel will suppress detonation and permit marked increase in the allowable compression pressure and in power output. Changes in the molecular grouping of hydrocarbon fuels, especially of petroleum derivatives, such as can be effected by variations in the refining process, are capable of producing fuels that are not subject to detonation even at the high compressions ordinarily encountered in automobile engines. The availability of these types of fuels has led in recent years to the use of much higher compression pressures and to obtaining much higher power outputs from engines of a given displacement. This in turn has necessitated important changes in ignition equipment, some of which have been made necessary by the increase in voltage required at the spark plug to break down (ionize) the gap between spark-plug electrodes.

Detonation is much more likely to occur if ignition takes place at a point in the cylinder remote from a highly heated area, so that the flame wave is propagated toward the hot area, than if the reverse condition is established. For this reason, the spark plug is sometimes located over the exhaust valve instead of over the inlet valve, where it was formerly customary to place it. Regardless of the location of the ignition source, however, there is a measurable time lag between the instant of combustion and that at which the entire charge is inflamed. It is necessary, therefore, in order that all the charge may be ignited at or very near the top of the stroke, to ignite the charge well in advance of top center, and usually to increase this advance by some means as the engine speed increases. To prevent back kicks or sudden reversal of the direction of rotation during the cranking period, it is usually necessary to provide means for retarding the spark during this period. Although both hand-operated and automatic means of effecting the required advance and retard of spark timing or a combination of both were formerly used, automatic control is now virtually universal for automobiles. As timing of the spark has a marked effect both upon power output and upon specific fuel consumption, as well as upon the general responsiveness of the engine to demands for change in speed and power output, increasing attention has been paid to means for securing automatic control by centrifugally operated mechanisms and devices actuated by the pressure difference between the inlet manifold and outside air.

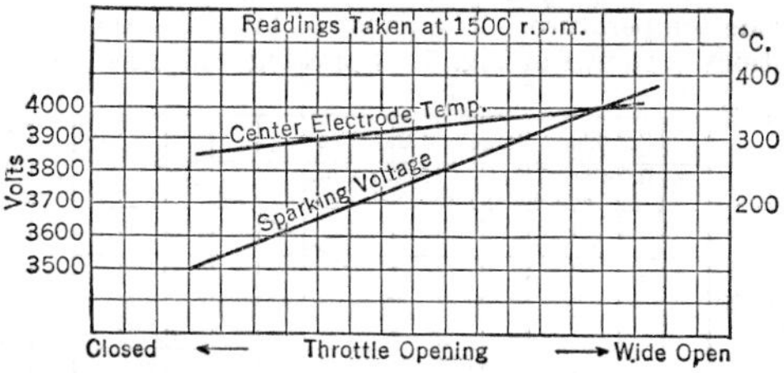

FIG. 3

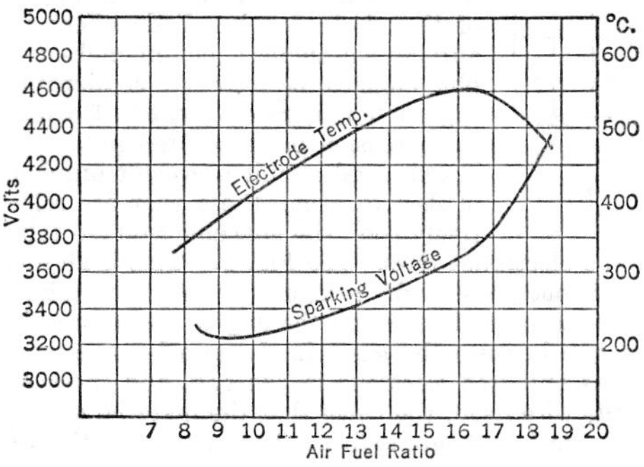

FIG. 4

The voltage required to cause a spark to jump the gap in a spark plug, when the gap measures less than 0.035 in., is nearly proportional to the width of the gap and to the density of the gases in the gap. Under apparently identical conditions in the engine, however, consecutive sparks may have as much as 50 per cent variation in the breakdown or sparking voltage at the plug. In addition to the size of the gap and the density of the gases in the gap, the sparking voltage is affected mainly by the shape, temperature, material, and deterioration of the electrode material. It is also affected materially by electrically conducting deposits on the insulator, especially when, as is often the case, these deposits are not composed entirely of carbon but contain metallic materials, such as lead from the antiknock fluid in the fuel. The lower the resistance of the shunt which this deposit forms, the lower is the voltage which the spark coil is capable of delivering to the gap.

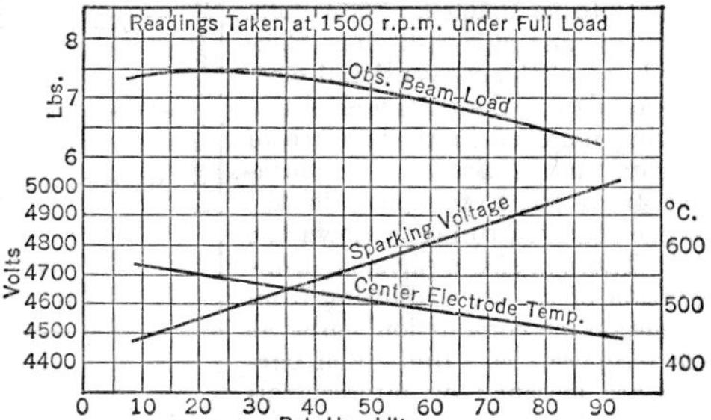

FIG. 5

For a given engine speed, the actual compression pressure and resulting density of the charge at the time of ignition increase as the throttle is opened. The secondary or sparking voltage rises accordingly, even though the temperature of the center electrode also rises. See Fig. 3. The sparking voltage and the temperature of the electrode also vary with changes in the air-to-fuel ratio, as shown in Fig. 4. Atmospheric humidity, as well as barometric pressure and turbulence of the gases in the cylinder at the time of ignition, have an effect upon combustion temperature, and also upon sparking voltage and electrode temperature. The effect of humidity on these factors under full-load engine conditions is shown in Fig. 5. Sparking voltage increases as the gap between electrodes increases, but not in a straight-line ratio, as shown in Fig. 6, which applies to an engine of rather low compression. A small increase in conductance (resulting from fouling of the insulator) or in the size of the spark-plug gap may result in missing. This is shown by reference to Fig. 7, which shows the maximum allowable fouling conductance on the insulator for a given gap size.

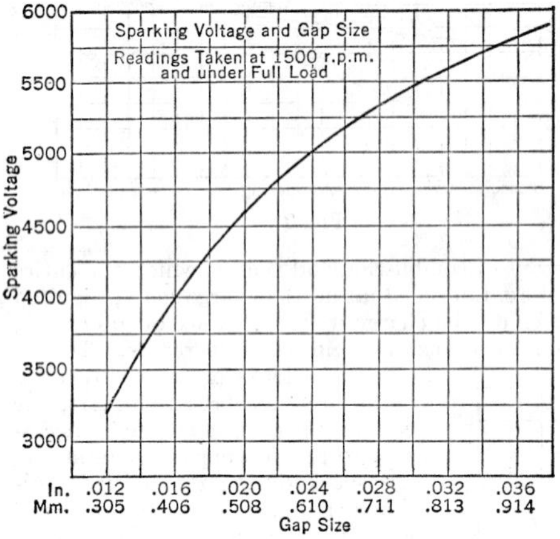

Fig. 6

Sharp edges on electrodes may slightly lower the sparking voltage, but such edges wear away after a few hours of engine operation. The shape of the electrodes has little effect in automobile engines, but in high-output engines it may be significant. Figure 8 illustrates the effect when the explosion pressures are high. As hot electrodes emit a greater number of electrons than those that are relatively cool, the sparking voltage falls off quite rapidly as the temperature of the electrodes rises. See Fig. 9. This is an important effect frequently overlooked and is one reason why a higher voltage is required in starting a cold engine than is needed to cause sparking when the engine and electrode become heated. Since a spark jumps a gap because the gases in the gap become conductors as a result of the presence of ions and free electrons in the gap, and most of the electrons are furnished by the electrode wire itself, the belief that the material of the electrode has no effect upon sparking voltage is not admissible. The characteristics of the ordinary commercial electrode alloy are such that the emission of electrons decreases as the plug is used and requires a proportional increase in the sparking voltage necessary to break down the gap. The increased voltage required sometimes amounts to as much as 3500 volts under the same gap conditions. Continued use of the conventional electrode tends to exhaust the electron-emitting material in the surface of the wire and results in wide variations from spark to spark in the voltage required to break down the gap. There is also a physical disintegration of the wire resulting from the action of combustion gases at high temperature. Individual particles are loosened by intercrystalline corrosion and become detached with resultant widening of the gap.

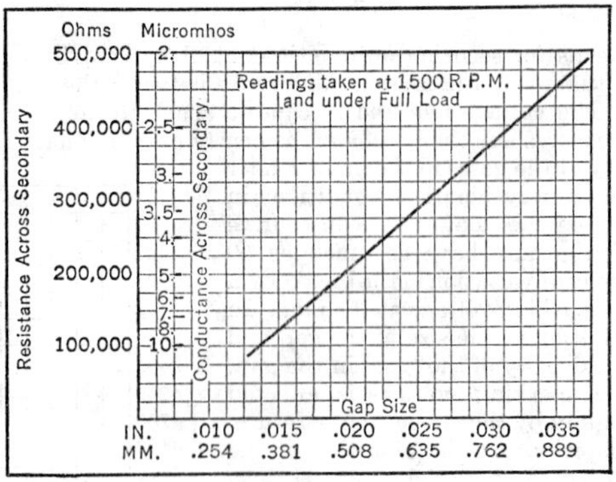

Fig. 7

As a result of the fundamental work of Rabezzana and Randolph, the following recommendations are made to minimize spark-plug missing: (1) correct spark-plug gap, modified to conform with latest practice; (2) electrode shape least affected by corrosion; no sharp edges or points at gap; (3) good electrode wire; (4) correct carburetion; (5) proper position of plug in combustion chamber; (6) correct breaker point gaps; (7) clean and square breaker points; (8) correct plugs for conditions under which engine is used (see p. 17-68); (9) avoidance of metal shielding on secondary leads and minimum length of said leads.

TYPES OF IGNITION SYSTEMS. All automobile ignition systems in use today are high-tension jump-spark types in which the initial or primary current is generated at a comparatively low voltage, which is stepped up sufficiently to break down the gap at the spark plug and produce an ignition arc or spark at the gap.

Make-and-break Low-tension Systems were extensively used at one time but are now obsolete. This type of system employed a battery or low-tension magneto, the current from which was stepped up through a coil of high inductance connected to a pair of contacts (one mechanically actuated) within the combustion chamber of the engine. These contacts were arranged so that mechanical separation occurred at a predetermined point in the cycle, resulting in an arc which ignited the charge. This type of system was not well suited for high-compression engines nor for those operating at high speed, and it also suffered the disadvantage that, if one breaker within a cylinder of a multicylinder engine became grounded (short-circuited) by carbon or maladjustment, ignition current for the remaining cylinders could not be supplied by the current source.

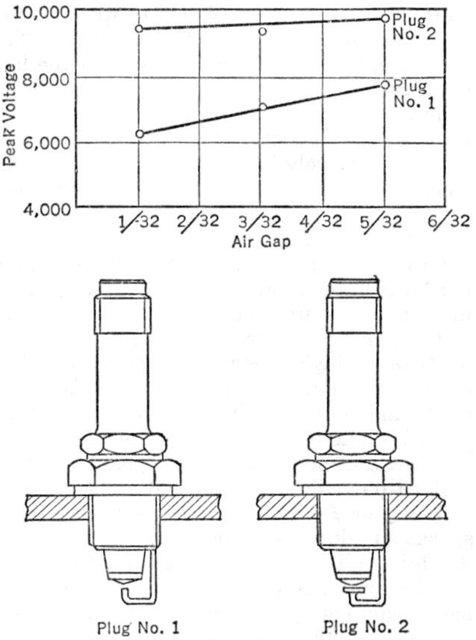

Fig. 8

Jump-spark or High-tension Systems may be classified as follows:

Vibrator Systems, in which the primary circuit is broken by vibrating contacts, as on Model T Ford cars, are now obsolete.

Single-spark Systems, in which a single spark is produced per spark plug (two plugs may be used) by one interruption of the primary circuit for each power stroke of the engine, are now universal on American passenger-car engines and are extensively used on many other types of engines. The source of electrical energy is now almost invariably a storage battery, usually floating on the line with a charging generator, but it may be a set of dry cells or a magneto. The primary circuit remains closed during a considerable proportion of the engine cycle in order that the flux in the induction coil may build up to a maximum value. The field thus created collapses rapidly, of course, when the primary circuit is broken by separating the cam-actuated contacts, the lines of force cutting the secondary winding and building up a voltage high enough to jump the gap of the spark plug. Because of the long time that the primary circuit of this type of system remains closed, as compared to so-called open-circuit types previously sometimes used, the current flow is greater, and the energy in the spark is correspondingly increased.

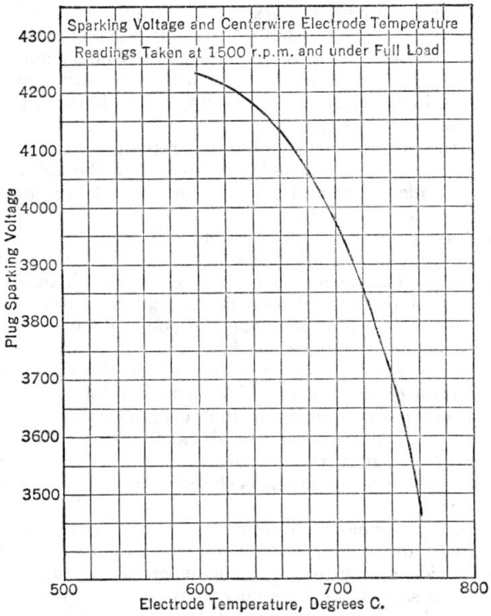

Fig. 9

Electric sheets of high-silicon or other special steel or annealed core wire are used for the core of the battery "powered" induction coil in order to break up eddy currents, to avoid residual magnetism, and to ensure the most rapid collapse of the magnetic field when the primary circuit is broken. In general, the primary winding has from 150 to 225 turns and is wound on the core, the size of wire running from No. 19 to No. 22 A.W.G. From two to four layers are used, and enameled wire is generally employed. Most coils either are designed to withstand continuous current flow or are protected by a resistance unit with a high temperature coefficient, so arranged that, if the current becomes excessive, the increasing resistance of the heated resistance unit is sufficient to prevent an excessive flow of current. Another benefit derived from a resistance unit in series with the primary is that it permits of a higher current flow when the engine is being started and the unit is still cool. During operation at low speed, when the primary circuit is closed for comparatively longer intervals, the resistance unit heats and reduces the current flow, whereas at higher speed the reverse effect is secured. This tends to equalize the current and resulting sparks at all speeds.

Secondary windings of No. 38 plain enameled wire are commonly used. About 1 volt per turn is generated; hence, if there are 250 turns per layer, the difference in voltage per layer approximates 500. Insulating paper is therefore applied between layers, which in many coils are wound outside the primary. If the insulation between layers of the coil becomes punctured, missing of the engine is likely to occur, especially under heavy loads or at high speeds. In some coils the secondary is wound next to the core, with the primary outside. Although this is a less favorable arrangement from the standpoint of the magnetic circuit, it makes it possible to eliminate a resistance unit, since the primary is made to act in place of such a unit in addition to performing its normal function. This arrangement also makes possible a less expensive case (stamped metal in place of molded phenolic material) and a cheaper secondary winding.

Two-plug Ignition. Detonation may be avoided by producing two (or more) ignition sparks simultaneously or nearly at the same time some distance apart in each cylinder. Smoother idling and low-speed (part throttle) operation are also obtained. In this manner the combustible mixture (charge), by being ignited at two points, burns more rapidly, hence completely at a normal rate. Combustion is thus completed before the rise in pressure is sufficient to cause auto-ignition of any unburned portion of the charge. If the two sparks do not occur simultaneously, the plug nearest the hotter or exhaust area of the cylinder is timed to fire from 4 to 7 deg ahead of the plug nearest the cooler or intake area of the cylinder.

COIL AND CONDENSER. In the ordinary ignition system using a coil connected as shown in Fig. 1, when the switch is closed, the current flows through the resistance unit, through the primary winding, through the closed contact points, and back to the battery. Opening of the contact points by the cam provided for this purpose and driven, of course, in timed relation to the engine, causes a surge of current in the primary and a sudden increase in voltage which, except for a condenser in parallel with the breaker gap, would tend to continue the flow of current in arcing at the gap and result in burning the breaker points. The condenser temporarily absorbs the surge, reduces the arcing, and reverses the direction of current flow in the primary, thus aiding the rapidity of collapse of the magnetic field in and around the core and correspondingly increasing the voltage induced in the secondary winding. The condenser should be of proper capacity to minimize arcing at the contact points.

Some observers have contended that so-called hot sparks (presumably those in which a relatively large amount of energy is dissipated) are of little advantage, so long as a spark occurs at all at the spark-plug gap. This belief appears to be doubtful, partly because coils of low inductance which build up rapidly and give adequate voltage to jump the plug gap have not proved conducive to easy starting and good operation, especially when the engine is cold. Coils of higher inductance, in combination with cam and breaker designs such as to increase the *dwell* or period during which the contacts remain closed, and permit the magnetization of the coil to build up to high values, are said to improve running conditions and be more economical of fuel, especially in engines that are inclined to run too cold in service. In any case, some makers of ignition systems have found it advantageous to use coils of high inductance along with breaker and cam designs such as to increase the period during which the circuit is closed, especially in high-speed engines.

Because of the fact that most automobile engines have four or more cylinders, some having as many as twelve or even sixteen, and engine speeds may be as high as 4000 rpm or more, it becomes difficult to provide a sufficient number of sparks per minute from a single coil. Therefore two coils and sets of breakers may be required for engines of eight or more cylinders. The two breakers may be operated from a single cam by mounting them on opposite sides of it, while still retaining a single distributor with the required number of plug leads or outlets. An important exception to the use of two coils is the system developed for the Ford V-8 engine (Fig. 10). In this system a single coil is used with two sets of breaker points, which are in parallel in the primary circuit. One set of points performs the usual function of breaking the primary circuit in timed relation to the crankshaft of the engine. The other set is so timed that the points close immediately afterward. Thus the time which would ordinarily be required to bring the breaker set of points back to the closed position is eliminated, and the time or "dwell" during which the circuit is closed is increased. More time is therefore allowed for the primary current to charge the coil and condenser. Obviously the exact relationship between the timing of the two sets of points is important to proper action of the ignition system. The manufacturer therefore provides concerns engaged in ignition service with both mechanical and electrical equipment for point timing and gap setting.

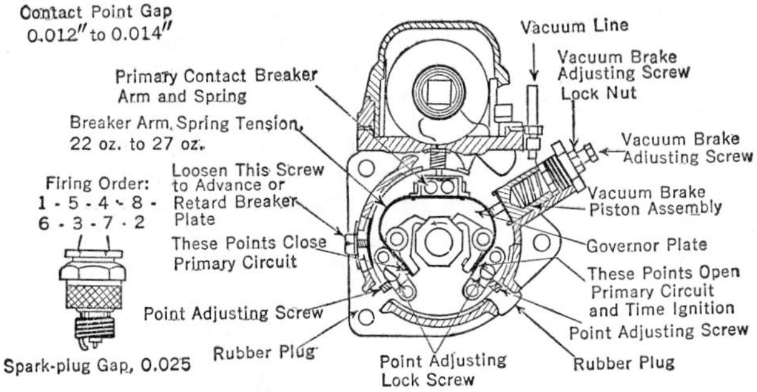

Fig. 10

MAGNETOS. Highly successful ignition from battery sources has resulted in eliminating magneto ignition on American passenger cars and largely on trucks and on many other types of automotive equipment in which a battery is required anyway for lighting and/or starting purposes. This is in no sense an indication that magneto ignition is less successful than formerly, but rather that the higher cost of magneto equipment is not justified when less expensive and satisfactory ignition means are available. For airplane engines, many forms of industrial engines, tractor engines, and some smaller marine engines, magnetos are extensively used with excellent results.

Magnetos are electric generators in which the field is produced by permanent steel magnets rather than by excitation through a field winding. In some cases there is a single primary winding on the rotating armature, which produces a low-tension current that is afterward stepped up to sparking voltage in a separate induction coil. More frequently, however, the armature has a high-tension as well as a primary winding, and then the magneto is termed a high-tension type. This type is fitted with fixed U-shaped magnets, between the ends of which the armature turns. An alternative arrangement, however, and the one most used today, involves a primary and a secondary winding that are fixed while the magnet or magnets are rotated or oscillated.

Magnetos are also classified as single-spark plug and two-spark plug types, the former being the more common, and having one end of the secondary grounded through the primary while the other end is connected to a distributor. So-called two-spark magnetos have both ends of the secondary insulated and connected through a double distributor to two spark plugs in the same cylinder. With this arrangement the charge is ignited at two points simultaneously. Magnetos are driven in timed relation to the engine and are fitted with a breaker mechanism by which the primary circuit is opened at the desired instant. Multicylinder types also incorporate a distributing device for connecting spark plugs successively to the terminal or terminals of the secondary winding.

A further classification of magnetos follows:

High-tension Shuttle Type. This type, which is little used today, has an H-shaped shuttle or armature on which both primary and secondary windings are placed. The armature rotates in ball bearings with a minimum gap between the soft-iron laminated pole pieces that are attached to the ends of the U-shaped magnets. The primary usually has about 150 turns of No. 22 wire and is next to the laminated shuttle. Secondary turns may approximate 10,000, the layers being insulated with oiled paper and/or oiled silk. One end of the secondary is usually grounded to the primary, and the other is connected to a collector ring, which in turn is connected to the central point of the distributor. The latter has a rotating brush which is driven by gearing from the armature shaft within a distributor head having one metal segment for each plug to which the high-tension current is to be led. The rotating brush makes contact with or comes close to the metal segments, so that the current may be led successively to various spark-plug wires, either by jumping the gap between the rotating arm and the segments or by actual contact of a carbon brush on the arm with them. The rubbing surface between contact segments against which the brush bears may be made from vulcanized rubber to minimize the effect of arcing, but this and the remainder of the distributor are usually molded from phenolic materials which have excellent insulating qualities and are not softened by heat. To operate the breaker mechanism two cam surfaces are provided on the breaker cover, which is coaxial with the armature, and a plate carrying the breaker contacts. One of the latter is on an L-shaped lever, one end of which strikes the cams as the breaker plate rotates with the armature. This mechanically opens the breaker points. The breaker cover and the cams attached to it can be turned by a lever, which is attached to an advance and retard linkage. With this type of magneto there are two sparks per revolution of the armature, hence for a four-cylinder four-cycle engine the magneto is driven at crankshaft speed and for a six-cylinder engine at one and one-half times crankshaft speed.

Inductor Types. Inductor-type magnetos differ from the shuttle type in that the windings are stationary and the magnets themselves or their pole pieces rotate. A diagram of one inductor-type magneto that has seen extensive use is shown in Fig. 11. In this

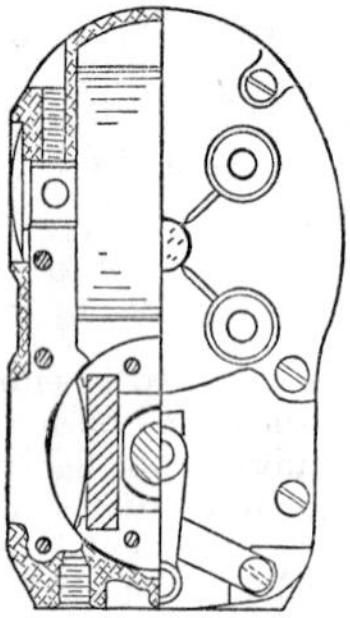

End Elevation and Half-section at Axis of Coil

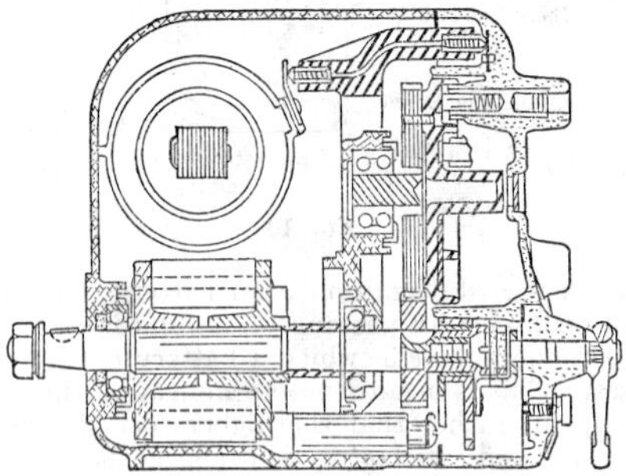

Longitudinal Section through Bosch Inductor-type Magneto

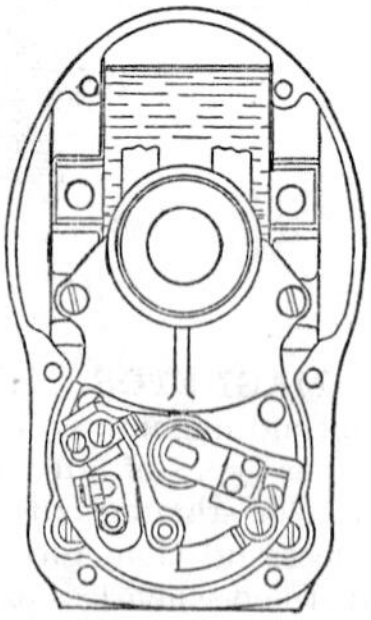

End Elevation with Cover Removed, Showing Details of Interrupter

Fig. 11

type, the rotating magnets form a bell-shaped structure with two or more pairs of poles. The rotor is made from a special chrome magnet steel, and the poles have laminated extremities which are held in place by an end plate on which a breaker cam is mounted. The rotating magnet, in the design diagrammed in Fig. 11, has four poles. The pole extremities of the magnets rotate in close proximity to the two pole shoes that are fastened to the core of the coil. The latter has an inner primary winding and an outer secondary winding. The paths of the magnetic flux are reversed each time that a pole piece passes through the gap between pole shoes. One breaker point is carried on a rocker lever, the other end of which has a shoe riding on cam. The cage which carries the pivot for this rocker lever carries the second contact and is pivoted so as to turn about the armature axis in advancing and retarding the spark. In this, as in some other magnetos, a tap is sometimes provided whereby an external source of high-tension current can be applied during the starting period.

Other Inductor-type Magnetos of several designs are used. In one type, known as the "polar," the coils are wound on a laminated core having extended pole pieces, between which rotate a pair of soft-iron blocks mounted on opposite sides of a non-magnetic shaft. The outer ends of these blocks form disks which rotate close to the poles of a permanent U-shaped magnet. Four or more pole pieces can be used, the number of sparks produced per revolution being equal to the number of poles. In the **sleeve-inductor-type** magneto there is a fixed shuttle-type armature. Between this and fixed permanent magnet poles rotates a sleeve of non-magnetic material having segments of soft iron. As the sleeve rotates, there is a reversal of flux through the armature core. With two segments four sparks per revolution can be produced.

Other Types of Magnetos, of which there are many, include the **oscillating type,** in which the armature is rotated through part of a revolution against springs and is then released so that it is snapped back across the position of flux reversal, thereby generating enough energy to produce a spark.

Impulse Couplings are often fitted between the magneto shaft and the engine shaft which drives the magneto. Their purpose is to snap the rotating part of the magneto through a fraction of a revolution at a relatively high speed at the instant when a spark is required during the cranking or starting period of the engine, instead of allowing these parts to turn at the slow uniform speed which a positive coupling would impose. This sudden acceleration produces a sufficiently high rotational speed to cause the spark to occur. The coupling includes a spring and latch device so arranged that the spring is wound up during a part of the revolution and is then suddenly released, imparting to the rotating part of the magneto a momentary angular velocity equivalent to that at about 800 rpm uniform speed. When the engine starts and the magneto attains a speed of 150 to 200 rpm, a centrifugal device renders the latch inoperative, and the coupling then functions at a normal uniform angular velocity.

Combination Ignition Systems, involving two or more systems that are more or less independent but to some extent interconnected, were widely employed, chiefly on automobile engines, at one time. The terms dual, duplex, double, and two-spark have been used in different senses by different makers who have sought to increase dependability or provide more convenient starting. Today such combinations are not common, largely because of their added expense and because the degree of dependability of a single system is sufficient to meet most requirements. Combinations of magneto and battery ignition, once very popular, also are seldom employed today.

CYCLE OF OPERATION. There are four periods in the cycle of operation of all jump-spark systems, each marked by an abrupt change in conditions. During the *first period*, the impressed voltage from the battery or magneto builds up the primary current and consequent storage of magnetic energy. With a battery system the build-up is exponential and gradually approaches the closed-circuit value. In the magneto the emf wave is peaked, the short-circuit current wave at moderate and high speeds being flat-topped but of sufficient value to produce a spark if the break occurs over a 90-deg electrical half wave. At slower speeds the current wave is more peaked, and the spark can be produced only over a narrower range of timing. The normal current at break ranges from 3 to 5 amp. The inductance of the magneto is usually greater; hence the heat energy in the spark is greater. Discharge of the battery on the closed circuit or the operation of the magneto as a short-circuited a-c generator is suddenly interrupted by the opening of the breaker contacts, introducing the second period.

The *second period* is of extremely short duration (of the order of 0.00005 sec) and extends from the opening of the breaker contacts to the initiation of the spark at the spark-plug gap. At the start of this period the primary current continues to flow unabated, charges the condenser, and introduces a back-emf into the primary circuit. This decreases the primary current and the resultant flux in the core and induces in the secondary a voltage which is higher than that of the primary in approximately the ratio of the number of turns, or from about 50 or 60 to 1. The secondary is thus charged as a condenser, one "plate" of which is the secondary winding, distributor, and spark-plug leads, and the other the engine and other grounded metal parts of the assembly. When the charge is sufficient to jump the gap at the spark plug, the discharge occurs, often when only a small fraction of the magnetic energy available has been transformed.

In the *third period*, immediately after attainment of the breakdown voltage of the gap in the second period, the gap becomes conducting and the secondary "condenser" discharges through it. After completion of this condenser discharge there follows the *fourth period*, during which the magnetic energy still left in the coil continues to deliver an inductive spark for several thousandths of a second until the entire energy is dissipated. The spark delivered by a magneto is usually of longer duration than that delivered by a battery system, both because of the greater amount of stored magnetic energy and because the voltage generated in the secondary by continued rotation (if the spark is sufficiently advanced) may force an additional flow of current across the gap.

A spark of low amperage and high voltage may jump the gap at the plug and still not ignite the mixture because the heat generated is insufficient. This is true, especially when the engine is cold. A lack of sufficient heat in the spark may result from too wide a gap at the spark plug, a gap in some other portion of the secondary circuit, an inadequate or imperfect ignition coil, too short a dwell, leakage (or so-called "corona" effect) resulting from deterioration of the insulating qualities of secondary wiring, and several other factors. As engine speed increases, the length of time that the primary circuit remains closed decreases, with a corresponding decrease in the time available for building up the flux in the core. The result is a weaker spark with more or less tendency toward missing at very high speed.

COIL PERFORMANCE. Figure 12 shows the current output of three coils over wide ranges of speed. Curve *A* indicates good performance at engine speed up to about 1800 rpm, but above this speed the output is low and becomes zero at a little over 2200 rpm. At low speed and in starting, coil *A* is much better than coil *B*, but at high speeds, as shown by curve *B*, this coil continues to give a spark up to speeds approximating 4500 rpm. Its maximum output, however, never exceeds 1.3 ma. Coil *C*, as shown by curve *C*, has a higher output than either *A* or *B* through most of its range and a maximum more than twice as great as coil *B*. A coil such as *C*, if made with one secondary winding, would be costly, bulky, and difficult to insulate, as there would be about 1000 volts pressure difference between layers. Outer layers would be far from the core, and the length of winding so great as to increase the resistance to a point where it would affect performance. To avoid such drawbacks, some coils have been made with two secondary windings (on an elongated core) connected in series. In this case insulation problems are less difficult, the diameter is reduced, despite the use of larger wire of lower resistance, and a total of 30,000 turns, or 15,000 per coil, is feasible. Such coils give good starting sparks on only 2 volts, and with a properly designed circuit breaker and distributor meet the requirements of even an eight-cylinder engine at speeds up to 5000 rpm.

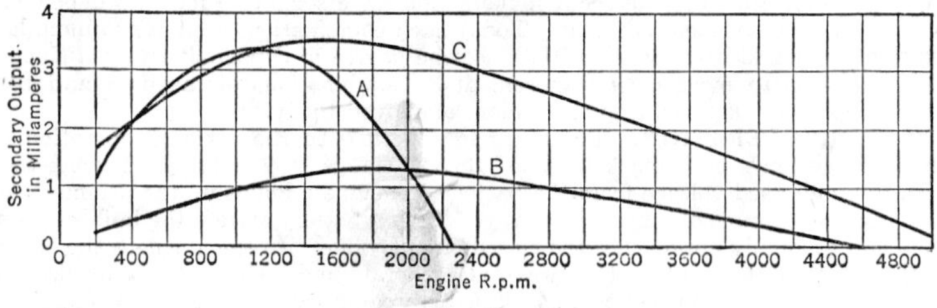

FIG. 12

During the starting period of engines that are cranked by a motor driven by energy from a storage battery, the heavy drain on the battery may drop its voltage as low as 2 volts, especially when the engine is cold. As ignition current is taken from the same battery during this period, only 2 volts may be available at the primary. This imposes a difficult problem upon the designer of ignition coils, and has led to the development of so-called automatic coils, designed to give a hot spark at low voltage and automatically accommodating themselves to the higher voltage which obtains when the engine picks up speed and the generator raises the voltage to about 8 volts.

SPARK TIMING. To obtain maximum power output and greatest fuel economy from an engine depending upon electric ignition, it is necessary to vary the timing of the spark with changes in speed and in throttle opening. This is partly because the time interval for completion of the engine cycle decreases in direct proportion to increase in speed and partly because the rate of combustion is influenced by changes in compression pressure, which varies in turn (in the four-stroke throttling type of engine) with changes in throttle opening, and by the octane rating of the fuel. For many years, hand setting of the spark timing device was depended upon to provide the optimum setting, but is now virtually obsolete. The chief disadvantage of manual control lay in dependence upon the human element, as its success, if best results were to be secured, depended upon the skill and willingness of the operator in varying the setting with changes in conditions.

Automatic Spark Timing Controls. Two general types of automatic spark advance and retard mechanisms are now quite generally employed, usually in combination and without an added manual control. The first type is operated by a centrifugal governor designed to advance the spark with increase in speed but not necessarily in direct proportion thereto, although some designs have this objective. It should also be either capable of adjustment or especially designed for the particular engine, as engines vary greatly in the degree of advance required, some needing only 10 deg or less and others requiring as much as 50 deg. Centrifugal governors are generally considered a decided improvement over hand advance mechanisms and are now extensively used. When used alone, however, they have the disadvantage that they are controlled only by variations in speed and not by changes in load.

A vacuum control, in addition to centrifugal control, has been used to overcome the disadvantage of a centrifugal device alone. The difference in pressure which actuates the device is that which exists between the interior of the inlet manifold of the engine and the atmosphere. Manifold depression decreases as the throttle is opened to meet increases in load and increases again when the throttle is closed with decrease in load. Since spark advance should be greater when the throttle is closed, because the compression pressure is then lowered and flame propagation is less rapid, the vacuum control is so arranged that the advance is increased as the load falls off, and vice versa. This type of control is superimposed upon that effected with a centrifugal governor, so that the combined effect is one of speed and load, as is desired.

Sectional views of such a distributor are shown in Fig. 13. The two governor weights *G*, which revolve with the distributor shaft, are eccentrically mounted, so that the centrifugal force tends to cause them to turn on their mounting studs. The governor weights carry pins *J*, by which the motion around their mounting studs is communicated to the cam-advance plate *H*, which is rigidly secured to the hollow shaft of cam *I*.

Most centrifugal governors are provided with a spring which counteracts the centrifugal force on the governor weights and pulls these weights toward the axis of rotation. In the centrifugal-vacuum governor this retracting force is fractional and is due to the pressure of the brake shoe *F* against the circumference of the cam-advance plate *H* under the influence of spring *C* and of the vacuum in cylinder *B*. Brake shoe *F* is mounted on piston *A*, which is adapted to slide in cylinder *B*. This cylinder communicates with the inlet manifold of the engine through a tube. Piston *A* therefore is under the combined influence of the spring *C* and the vacuum within the cylinder.

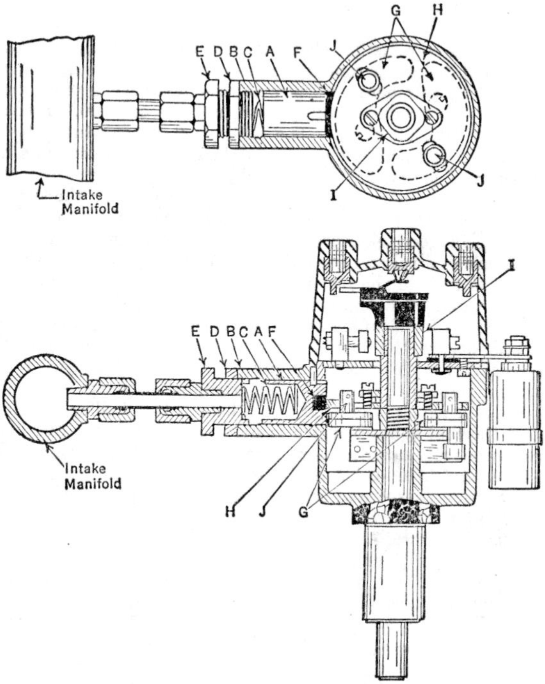

FIG. 13

If the operator gradually opens the throttle while the engine is idling or disconnected from its load, the engine naturally increases its speed. The greater speed causes the centrifugal weights to move further from the axis of rotation, thereby advancing the spark. At the same time, however, the vacuum in the inlet manifold is decreased, and the reduced suction in cylinder *B* enables the spring *C* to press the brake shoe *F* harder against the cam-advance plate *H*, with the result that the increased friction balances the increased centrifugal force and the spark is maintained in a definite, more advanced position.

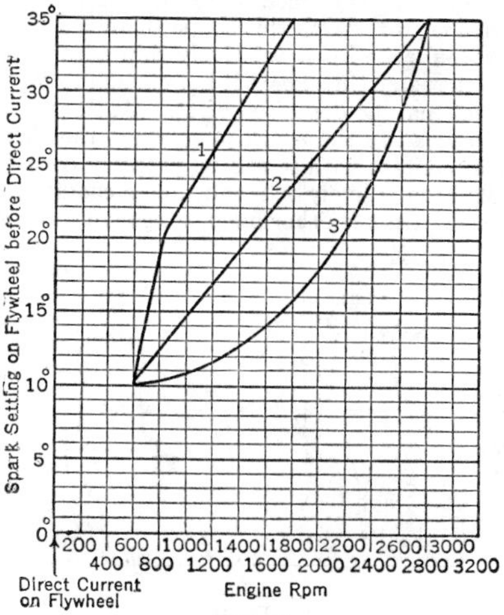

FIG. 14

If, when the engine is running at a low speed, the throttle is suddenly opened for acceleration, the vacuum in cylinder *B* is greatly reduced; hence piston *A* presses brake shoe *F* more forcibly against cam-advance plate *H*, and the greater friction thus produced draws the governor weights *G* closer to the axis of rotation, thereby retarding the spark. Closing the throttle naturally has the opposite effect. Figure 14 shows the effect of a centrifugal governor alone, and when combined with vacuum control on the spark advance curve.

SPARK PLUGS. The function of the spark plug is to provide a gap having a fixed breakdown voltage between two electrodes, one of which is almost always grounded to the shell of the plug, which has a male thread and is screwed into a threaded hole in the wall of the combustion chamber, and the other (the central electrode) is insulated and connected to the secondary of the ignition coil. Among the requirements sometimes imposed by purchasers of automobile-engine spark plugs are the following: The voltage necessary to produce a spark shall not exceed 6000. Insulation resistance shall not be less-than 100,000 ohms. The plug shall be substantially gastight. In practice these conditions sometimes must be fulfilled under pressures as high as 1000 psi. and in the presence of gases that vary rapidly in temperature from atmospheric to 3000 deg fahr or above and frequently tend to deposit carbon on the surface of the insulator.

The electrode passes through the center of a vitreous ceramic insulator and is cemented in place in good electrical contact with a brass cap for receiving the secondary lead. Shells are of steel and are either crimped or spun over a shoulder on the insulator in such a way as to hold the latter permanently in gastight contact with a seat, or are provided with a threaded bushing which performs the same function but enables the insulator to be removed for cleaning and inspection. Gastight joints are essential, for if appreciable leakage occurs, the flow of hot gases between the shell and the insulator is likely to cause overheating with consequent preignition of the charge. Most modern spark plugs have straight central electrodes. The grounded electrode is usually an L-shaped piece of wire, one end of which is pressed into a hole in the shell or is welded to the shell to ensure the good thermal contact which is necessary to prevent overheating. Electrodes are made from relatively non-corrosive wire, usually containing a considerable percentage of nickel.

SPARK-PLUG GAP SETTINGS. The optimum setting depends partly upon the compression pressure of the engine, partly upon the engine temperature, and partly upon the voltage which the ignition system is capable of generating under the operating conditions that obtain. One authority recommends a gap of 0.015 to 0.018 in. for high-compression engines and of 0.018 to 0.022 in. for low-compression engines. A setting of 0.025 to 0.030 in. is most frequently used, but the specific recommendations of the engine manufacturer or maker of the ignition equipment should be followed. Gaps in excess of 0.035 in. are likely to result in leakage in the external secondary circuit. Gaps that are too close are likely to cause rough engine performance, especially under part-throttle operating conditions. In setting the gap it is recommended that only the grounded electrode be bent, as bending of the center electrode is likely to result in breakage of the insulator at the time or later when it is heated in use.

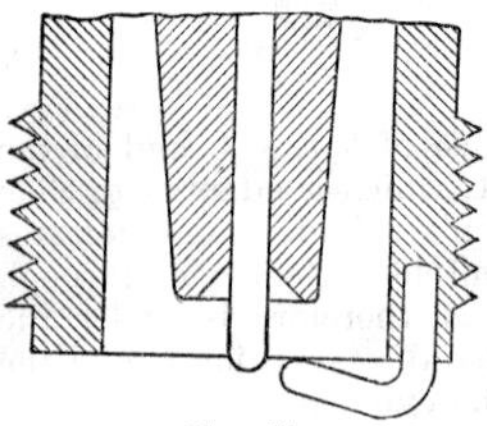

Fig. 15

SPARK-PLUG INSULATORS. Insulators in most modern spark plugs have what is termed a "semipetticoat" tip, such as is shown in Fig. 15. With this shape there is a heat gradient which prevents the lip or relatively thin edge from overheating but at the same time keeps it hot enough to burn off oil or carbon that may be deposited on it. Much care is usually exercised in locating the spark plug where the shell and the insulator in contact with it will be properly cooled, as otherwise overheating with consequent pre-ignition is likely to result unless conditions are such that the engine normally runs rather cool. On the other hand, excessive cooling of the plug is likely to result in fouling of the insulator with consequent formation of a high-resistance shunt across the gap. For these reasons many plugs are specially designed to meet the particular cooling conditions that exist in engines of various types. Thus there are hot-running plugs for cool-running engines and cool-running plugs for hot-running engines, and it is important that the right plug for the particular conditions encountered be selected. Plugs designed to run hot in cold engines have an insulator that is relatively long from its firing end to the shoulder where it seats against the shell. Plugs designed to run cool in hot engines have an insulator which is relatively short between the points mentioned, so that the heat is conducted away from the hottest end more rapidly. (See Fig. 16.)

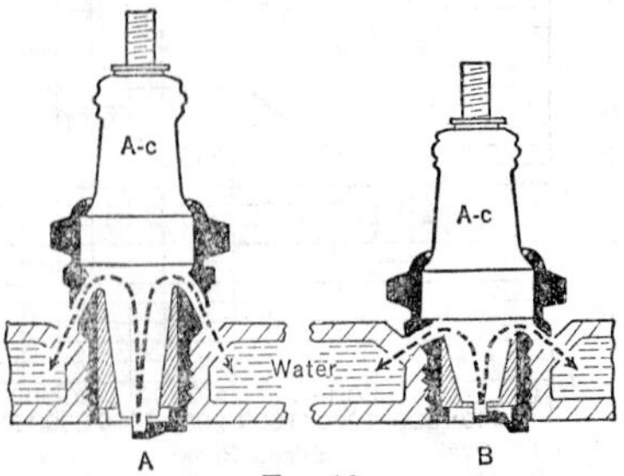

Fig. 16

Spark-plug shells may be varied in length so as to fit into heads of differing thicknesses and in some cases to bring the spark gap well away from the inner wall of the combustion chamber, so that the gases, which are rendered turbulent as a result of velocity imparted during the intake and compression stroke, may sweep across the gap during the ignition period. This practice, however, is no longer common, and cylinder heads are so designed that standard plugs can be used without placing the plug points away from the turbulent gas. Many engine manufacturers have made extensive tests to determine the best type, location, and size of spark plug to be used, and when new plugs are installed it is usually desirable to make sure that they duplicate the type selected as a result of such tests.

Spark-plug Standards. Three sizes of spark plugs are in rather general use in this country, namely, the 18 mm, the 14 mm, and the 10 mm. Formerly, the SAE 7/8-in. 18 thd was much used, but it has largely been superseded by the later SAE standard 18-mm size. The 14-mm size is extensively used, and the 10-mm plug has recently been adopted by several automobile manufacturers. Neither of these sizes, however, has been standardized by the SAE. The 1/2-in. pipe size plug is now completely obsolete. Although two sizes of threads have been used for many years for the electrode terminal nut, the external dimensions of the terminal post or nut are the same for all plug sizes. The 7/8-in. 18 thd and 18-mm plug dimensions are given in Fig. 17. The 14-mm plug has a thread pitch of 1.5 mm, and the 10-mm plug a pitch of 1.0 mm. All plugs have straight threads and depend upon seating against a copper-asbestos gasket to insure a gastight joint.

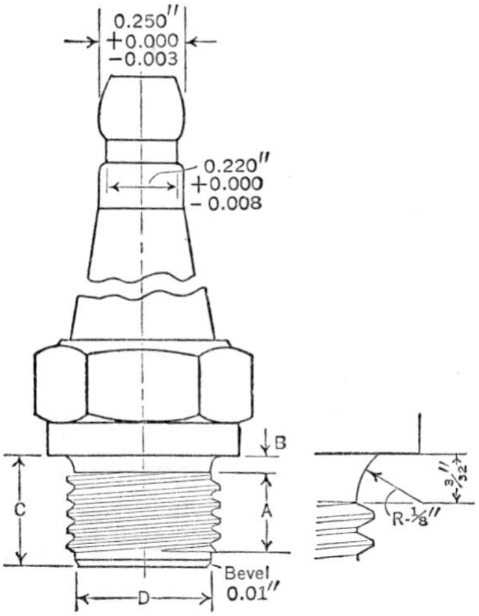

FIG. 17

Spark-plug Dimensions

Spark Plug	Hexagon Size, Inches, ±0.005	Thread Length A ± 1/64	Neck Length B	Skirt Length C	Max. Pilot Diameter D
7/8 in.–18	15/16 or 1 1/8	3/8	1/8	5/8 or 15/16	25/32
18–1.5 mm *	1	13/32	3/32	1/2	5/8

* Form of thread is International Standard, which is the same as the American Standard (B1.1-1935) except that the truncation at the root of the thread is half as much as in the American Standard Thread.

Spark-plug Threads

Spark Plug	Major Diameter		Pitch Diameter		Max. Minor Diam.
	Max.	Min.	Max.	Min.	
7/8 in.–18	0.8750	0.8668	0.8384	0.8343	0.8068
18–1.5 mm	0.7077 in. 17.975 mm	0.7028 in. 17.850 mm	0.6693 in. 17.001 mm	0.6644 in. 16.876 mm	0.6246 in. 15.864 mm

Although no limitations are placed upon spark-plug manufacturers as to internal features of design, types of plugs having a removable insulator are possibly less used than those in which the insulator is machine-crimped into permanent place in the shell. It is claimed that a more gastight joint is thus obtained when the shell is tightly crimped over the gaskets generally used above and below the insulator.

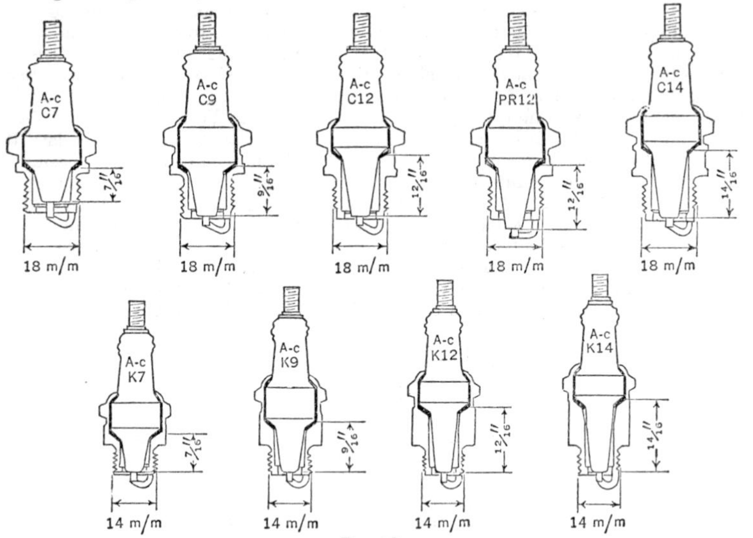

Fig. 18

Trend toward Smaller Spark Plugs. The use of successively higher compression pressures in automobile engines in recent years and the wide range of operating loads on engines in automobile driving have imposed severe conditions upon spark plugs.

It is necessary to keep the spark-plug insulator at a temperature above the range within which carbon can form on it, as a coating of carbon will by-pass some or all of the high-

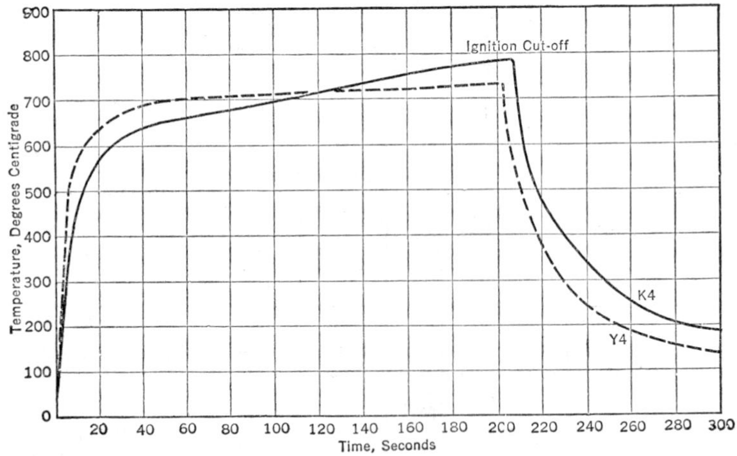

Fig. 19

tension current. On the other hand, too high a temperature of the insulator is the cause of rapid deterioration of the spark plug and of preignition.

It is important that the time which elapses between the starting of the engine and the moment when the insulator gets hot enough to burn away any carbon that may be deposited upon it be as short as possible. Also, the insulator should not become hot enough when the engine is operating at full throttle to cause preignition.

If the heat range of the plug is narrower than that required by the engine, the plug will fail at high speed or it will fail in city driving in cold weather, or perhaps in both cases.

Engines of today require a wider heat range in spark plugs than did earlier designs. The increase in the number of cylinders has resulted in greater engine speed and higher compression. These developments have increased greatly the rate of heat flow which the spark plug must handle and necessitate the selection of spark plugs of correct heat range. (See Fig. 18.)

Another factor which increases the spark-plug heat range required in most engines is the difference in the combustion temperature attained in different cylinders, because of inequalities of mixture distribution.

Figure 19 shows that the maximum temperature reached by the new 10-mm plugs is lower than that reached by the 14-mm plugs, with the result that electrode wear and incrustation formed over the insulator by combustion materials is reduced in the 10-mm plug. One plug manufacturer claims that extensive road tests have proved the ability of the 10-mm plug to keep its insulator cleaner from carbon, as well as from high-temperature combustion residuals, than the various plugs with threads of larger diameter.

Spark-plug Suppressors. When radio receiving and/or transmitting sets are installed in automobiles, it is frequently necessary to suppress the normal tendency of the secondary or high-tension part of the ignition system, as well as the primary, to emit waves of any frequency that can disturb the radio set. This is accomplished by the use of so-called "spark-plug suppressors" for the secondary, and condensers at suitable points in the primary circuit. Shielding of the secondary wiring or harness, i.e., completely surrounding the secondary wires by well-grounded metal conduits, is usually not recommended because of other effects on the spark characteristics, although preferred and often necessary in aviation engine ignition systems.

16. STARTING AND LIGHTING SYSTEMS

The items comprising the conventional d-c electric starting and lighting equipment of a gasoline automobile are (1) generator, (2) storage battery, (3) starting motor, (4)

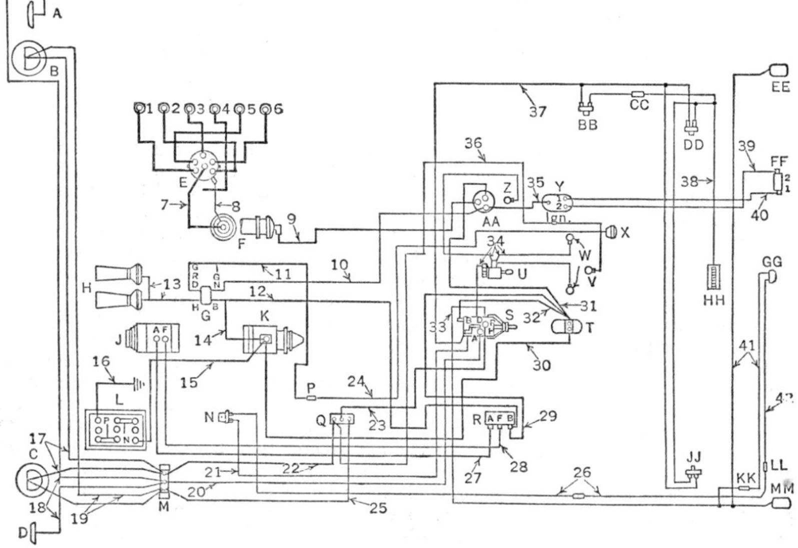

FIG. 20

lamps, (5) wiring, (6) switches, cut-outs, and fuses, (7) ammeter. The battery is used not only to supply energy for starting and lighting but also for ignition and often for operating supplementary apparatus, such as windshield wipers, cigar lighters, radio receiving sets and in many recent makes of vehicles several gages, instrument panel lamps, etc. Although most electrical systems are nominally of 6 volts, many vehicles now use systems operating at 12, 24, and 32 volts. A 30-volt system is also used under certain conditions.

In addition to the conventional d-c systems commonly used, a system of a-c–d-c type, comprising an alternator, rectifier, and output control unit, has been introduced for passenger cars, motor coaches, and motor trucks. Complete systems largely interchangeable with those of conventional type are available for several makes of cars. A typical wiring diagram is shown in Fig. 20.

GENERATOR. Practically all generators now employed on passenger cars are shunt wound. They are driven by the engine, sometimes through gears or silent chains, but more often by belts which as a rule are V types. The primary function of the generator is to keep the storage battery, which floats on the line, charged, but in some installations, especially on buses, the generator is designed to carry the lighting and ignition load even though the battery becomes disconnected. Since with a shunt-wound generator the voltage normally varies in substantially direct proportion to the speed, and the speed constantly varies with changes in engine and vehicle speed, means for maintaining a nearly constant voltage have to be applied.

BATTERY CUT-OUT. To prevent the battery from discharging through the generator when the battery voltage exceeds that of the generator, it is usually necessary to

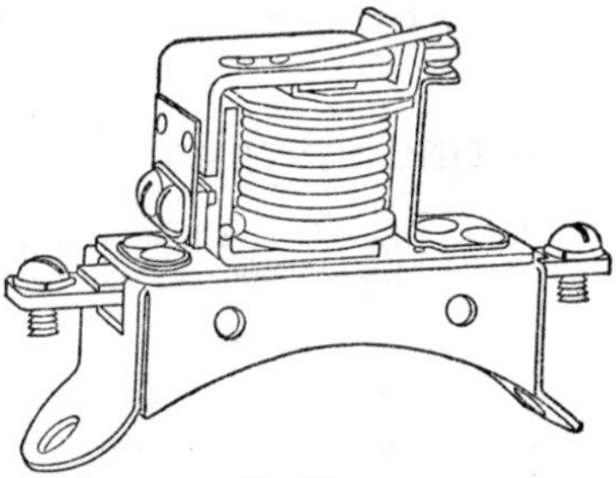

FIG. 21

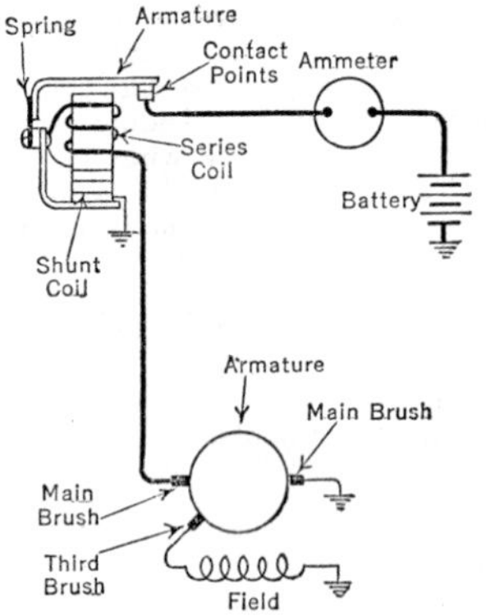

FIG. 22

provide an automatic switch, variously termed a cut-out, a circuit breaker, and a reverse-current relay, designed to open the circuit when the speed of the generator falls below that at which the battery is charged. A relay of this type is shown in Fig. 21; the wiring diagram, Fig. 22, shows the windings of the relay. Essentially the relay consists of an iron core with a shunt and a series winding. The shunt winding is connected permanently across the main generator terminals, one of which is usually grounded. It serves to magnetize the core and attract a hinged armature as soon as the voltage is high enough and the magnetizing force great enough to overcome the tension of a spring which tends to hold the armature away from the core. When the armature is attracted, it closes a pair of contact points, thereby closing the battery circuit and also the series winding circuit of the cut-out. The additional flux added by this winding presses the contacts more firmly together. Both the spring and the stops that limit the travel of the armature of the relay are usually so set that the contacts close at 7 to 7.5 volts, equivalent to a charging rate of about 3 amp, with an ordinary automobile-type relay for a 6- to 8-volt system. Contacts open again at a discharge current not greater than 2.5 amp, when the series coil bucks the shunt coil and decreases the pull on the armature enough for the spring to pull it away from the core. In general this actuation of the relay takes place at an engine speed of about 600 rpm, equivalent to a car speed, in high gear, of around 10 mph.

TYPES OF VOLTAGE CONTROL. Although the so-called "third-brush" system of generator regulation was in almost general use for many years, the need for more precise control of generator output and voltage, due to the widely varying demand for current in modern automotive vehicles, has resulted in the development of numerous methods of regulation in conjunction with generators of the two-brush type but also applied to those of the older three-brush type. In all of these the primary purpose is to proportion the discharge rate of the generator to the demand for current for battery charging, lamp load, especially for night driving, and heaters, radio set, and other auxiliary equipment when these are in use. Operating requirements may vary from a few amperes for ignition only, up to a total of approximately 35 amp for continuous operation of all equipment, and momentary delivery of several hundred amperes for engine starting, all of which must be met by the generator working in combination with the battery. The desire of the vehicle owner to add electrical equipment other than that originally supplied has further complicated the problem of providing adequate generator capacity and of obtaining precise control.

Third-brush Regulation. In this system, use is made of the distorting effect of armature reaction. When current flows through the armature, the resultant flux is skewed around in the direction of rotation (see Fig. 23), with the result that the number of lines of force entering the armature from the leading pole tip is reduced and a corresponding increase in the number of lines of force takes place at the trailing pole tip. Consequently, referring to Fig. 23, it is evident that the electromotive force generated between the brush *A* and the third brush *C*, located as shown, will decrease as the current through the armature increases. If the shunt field is connected between the brushes *A* and *C*, the current through the shunt field will therefore decrease as the current in the main circuit increases, which in turn will reduce the electromotive force between *A* and *C* still further. Consequently, if the battery is connected across the main brushes *A* and *B*, any tendency for the current to the battery to increase when the speed of the generator increases is counteracted by the decrease in the field current. Such a generator, when connected to a storage battery, tends to act as a constant-current machine, irrespective of the speed. As a matter of fact, because of the effect of the short-circuit current in the coil passing under the brush *C*, which current is practically proportional to the speed and which produces a demagnetizing action, the current to the battery actually reaches a maximum value at a definite speed and then falls off. The maximum current supplied to the generator is readily controlled merely by shifting the position of the intermediate brush.

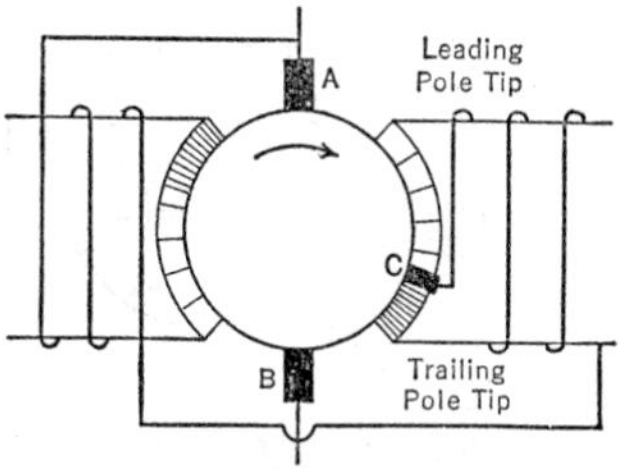

Fig. 23

Among the advantages of the third-brush control system are its simplicity and the fact that it requires, in most instances, no supplementary apparatus outside the generator. It also permits of ready variation of the current output by the simple expedient of shifting the brush. Shifting the brush in the direction of armature rotation increases the output, and shifting it in the reverse direction decreases the output. The output rises slightly as the battery becomes charged, and it is this characteristic that gives rise to overcharging of the battery when the output of the generator continues to progress much more rapidly than the battery is being discharged. It is partly to avoid this disadvantage that supplementary regulators are used. Another reason for the use of voltage regulators is that, without them or some other protective device, an open circuit or high resistance in the charging circuit of a third-brush generator results in a rapid rise in voltage which is likely to burn out all the lights or cause a sufficiently heavy overload on the generator to damage its windings. A protective fuse in the field circuit of the generator is frequently provided.

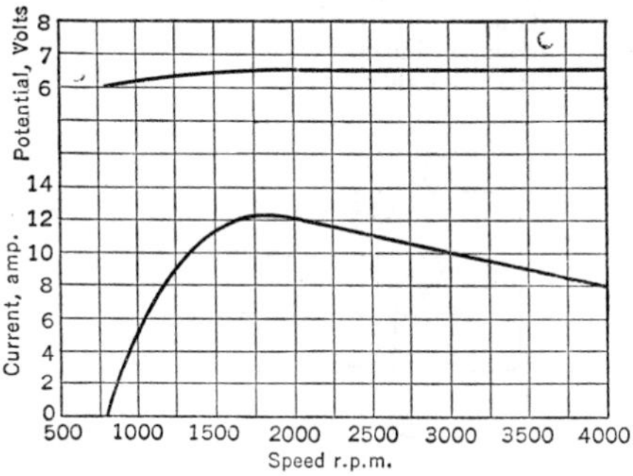

Fig. 24

In Fig. 24 are given the characteristic potential and current-speed curves of a third-brush 6- to 8-volt generator with a battery load. Ordinarily the setting of the third brush is such that the peak of the curve comes at or near the average speed at which the vehicle is operated, but since this speed often varies widely, even for the same vehicle, the chances of overcharging or undercharging are considerable.

Thermostatic Control. One means of reducing the overcharging characteristics of three-brush generators, and one also used in some two-brush systems, consists of a simple bimetal thermostat and resistance, as shown in Fig. 26, connected in the field circuit of the generator. This thermostat is placed inside the field frame of the generator, where the bimetallic element (the lower one in Fig. 25) is subjected to variations in the temperature of the generator itself. When the temperature rises to about 165 deg fahr, the contacts open and cut the resistance into the generator field circuit. The effect of this change is to decrease the output of the generator and thereby to lower its temperature. The resistance element, which is made easily replaceable, also acts as a fuse to protect the generator in case an open circuit occurs between the generator and the battery.

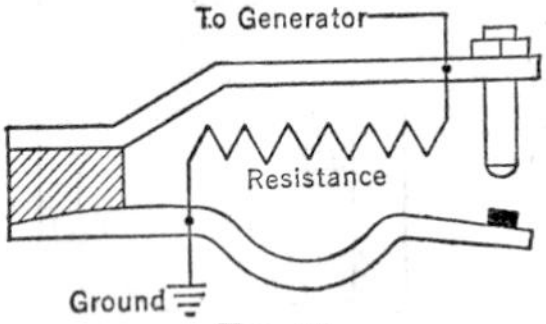

Fig. 25

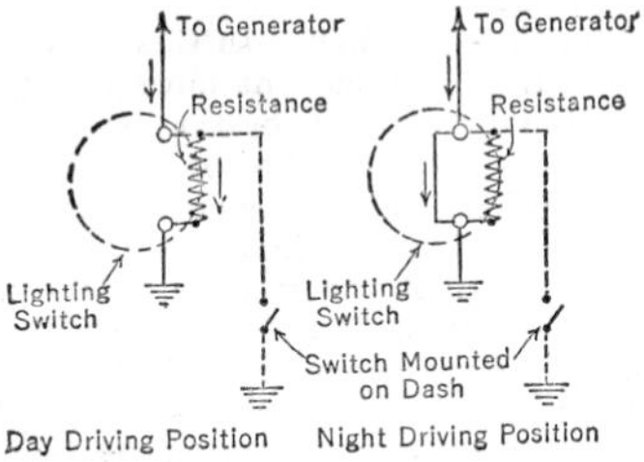

Fig. 26

Switch Resistance Control. With this method of control the field winding is brought out to an external insulated terminal in the generator frame, rather than being grounded within the machine. The terminal is then connected to the headlamp switch, as shown in Fig. 26. When the lighting switch is in its "off" or "parking lamp" position, the ground is through a resistance in the headlamp switch. When the headlamps are turned on, the resistance is cut out by a direct ground at the switch. This allows a higher output to be taken from the generator when driving at night and affords protection to the battery from overcharging during extensive daylight driving. A 1-ohm resistance is the standard used in the switches, but $^1/_2$-, $^3/_4$-, and 1 $^1/_2$-ohm resistances are available. The 1-ohm resistance is suitable for average driving. An excessive amount of any one type of driving will warrant changing to another size of resistance.

Step Voltage Regulator. The step voltage regulator, as shown in Fig. 27, is composed of a voltage-control unit and a cut-out relay in conjunction with a third-brush type of generator. The purpose of the unit is to increase or decrease the generator output in accordance with the battery requirements and the connected electrical load. When the battery becomes fully charged, the voltage-control unit operates and automatically inserts a resistance in series with the generator field circuit, and the generator output is reduced to a safe value. The generator continues to charge at a lower rate as long as the battery is fully charged and the electrical load is small.

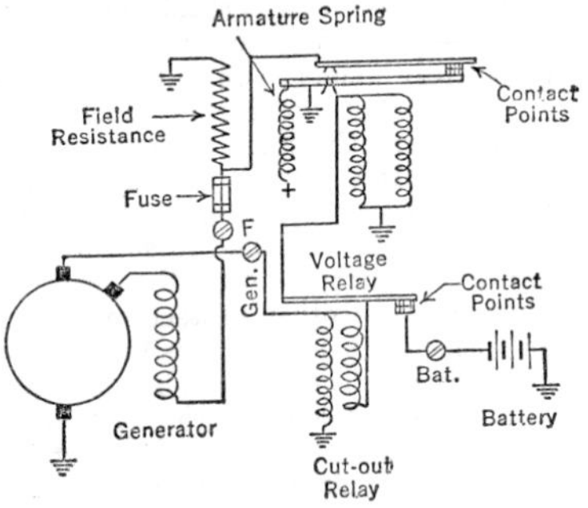

Fig. 27

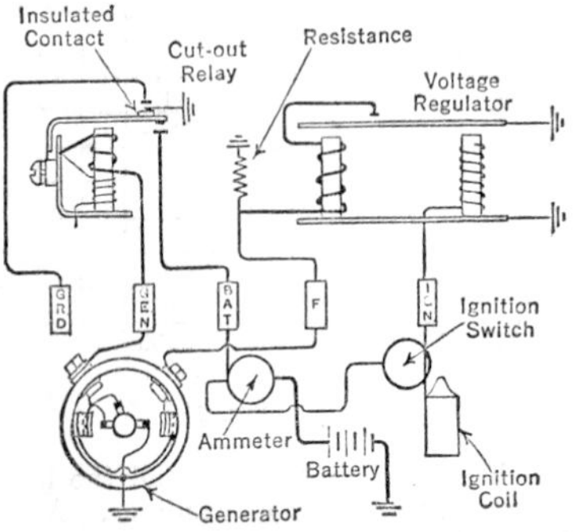

Fig. 28

If the battery becomes partially discharged, the voltage-control unit ceases to operate and allows the generator output to increase to its maximum. The voltage-control unit does not increase the maximum output of the generator, as this is dependent upon the position of the third brush.

When a sufficient electrical load is being used, such as lamps, radio, and heater, to require a higher generator output, the voltage-control unit stops operating, and the resistance is automatically removed from the generator field circuit.

Vibrating Voltage Regulator. The vibrating-voltage-type regulator has two units and two functions to perform. The cut-out relay opens and closes the circuit between the generator and the battery. The regulator prevents the voltage from exceeding a predetermined value, regardless of the generator output. The current is limited by the conventional third-brush action.

The regulator (Fig. 28) has a shunt winding consisting of many turns of fine wire, which is connected so that the battery voltage controls its operation. In addition to the shunt winding, a series winding of a few turns of heavy wire acts as an accelerator winding to speed up the action of the armature, while the shunt winding is the governing winding.

When the battery is low, the armature contact points are closed. The generator field current flows to the ground through the contact points and the series winding. When the battery voltage reaches a predetermined value, it is forcing sufficient current through the shunt winding to increase the magnetic force of the coil to a point where it overcomes armature spring tension and pulls the armature down toward the core, opening the contact points. The generator field current is now forced to flow through a resistance to the ground, reducing the generator output.

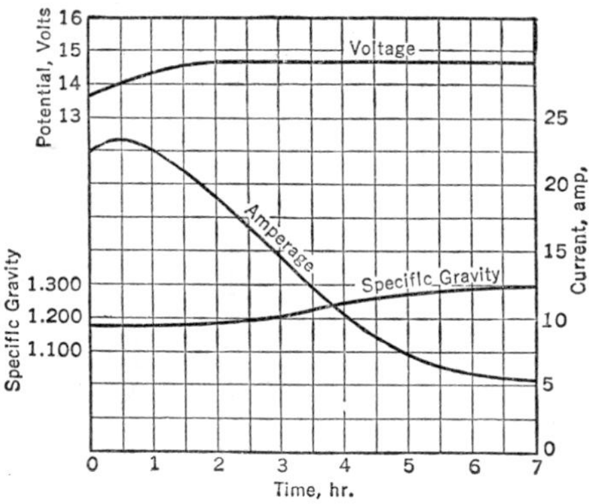

FIG. 29

When the points separate, the magnetic field of the series winding collapses completely. The reduced voltage of the generator as the points separate, as well as the collapse of the series winding magnetic field, causes the shunt winding magnetic field to drop until it can no longer hold the armature down, and the points close quickly. This shorts out the resistance and the voltage rises again, thus completing one cycle of operation. These cycles may take place 150 to 250 times a second, to maintain the voltage at its correct value, as long as the voltage is high enough to keep the regulator in operation. When a current load is added which is great enough to lower the battery voltage below the operating voltage of the regulator, the points will remain closed, and the generator will maintain its maximum output.

The regulator is compensated for temperature variations, so that a higher voltage is required to vibrate the contact points under cold operating conditions than under hot ones. As a higher voltage is required to charge a battery when it is cold than when it is hot, this feature maintains the generator voltage to meet battery requirements at all times without changing the voltage setting.

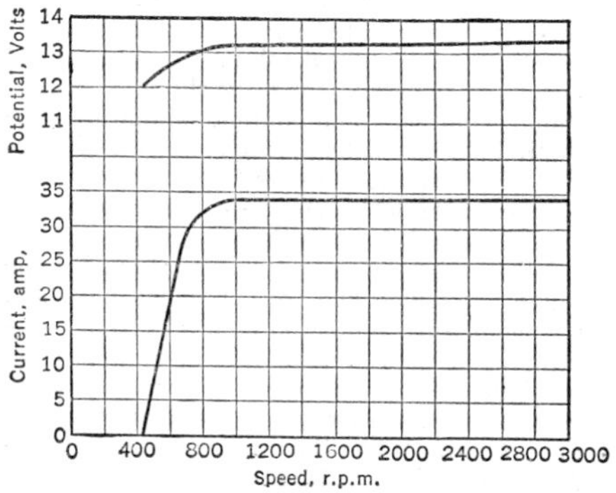

FIG. 30

Since the regulator holds the voltage substantially constant and that of the battery gradually increases as charging progresses, the voltage differential, and therefore the rate of charging, become steadily less as charging proceeds. In Fig. 29 is given a characteristic curve showing how the charging current of a voltage-controlled generator falls off as the battery becomes charged, until the current delivered assumes a low and almost constant value which is only slightly in excess of that normally consumed by the ignition system of the engine which drives the generator. Figure 30 shows typical performance curves of a voltage-controlled generator operating with a battery load. It will be seen that both the voltage and current remain substantially constant over a 3-to-1 speed range from about 900 to 3000 rpm.

Regulation of Two-brush Generators. As noted previously, the third-brush type of generator has been supplanted by generators using but two brushes, namely, those of straight shunt type without modification in so far as the internal construction of the generator is concerned. With generators of this type both regulators, namely those of voltage and of current, operate independently in order to provide adequate control of output characteristics. Control is by the adjustment of resistances in the field circuit of the generator. Various types are sometimes designated by the terms single core and double core, depending upon the design of the regulator.

Current and Voltage Regulator (Double Core) (Fig. 31). When the requirements of the connected load are large, and the battery is low, the current regulator operates to prevent

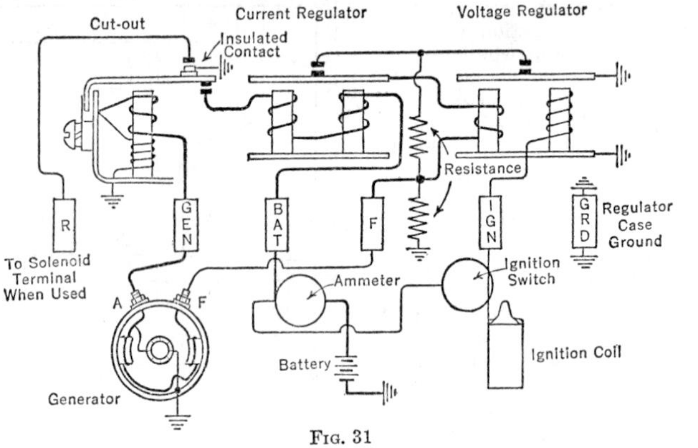

Fig. 31

the output from exceeding the rated output of the generator, at which time the voltage regulator is inoperative. If the requirements of the connected load are reduced and the battery comes up to charge, the voltage regulator operates to prevent high voltage at the battery and in the circuit. The increasing resistance of the battery, as it becomes charged, tapers down the generator output.

Two resistors are used, a low resistance for the current regulator, when the two resistors are in parallel, and a higher resistance for the voltage regulator where a single resistor is used.

The operation of the cut-out relay and voltage-control part of the double-core regulator is the same as that described for third-brush regulation. The current regulator is also an electromagnet and is composed of a few turns of heavy wire of low resistance connected in series between the generator and the series winding of the cut-out relay. Thus the entire output of the generator flows through the current regulator. The contact points are normally held closed by spring tension, which retains the armature in its "up" position, and the generator field current is conducted directly to the ground through these points.

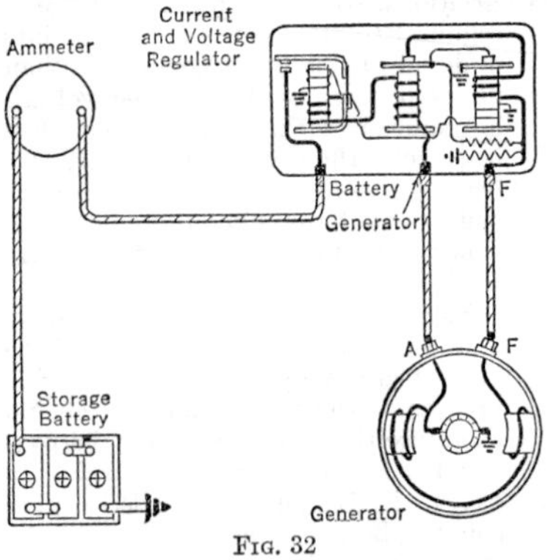

Fig. 32

When the total generator output reaches the value for which the regulator has been set, sufficient force is created by the current regulator winding to overcome the armature spring tension. The armature is thus attracted downward toward the core, and the contact points are opened. When the points open, the resistance shown in Fig. 32 is inserted in the field circuit of the generator, reducing the current output of the generator. The magnetic field in the regulator winding therefore becomes weaker, and when it decreases to a point where it can no longer hold the armature down against the tension of the spring, the points again close, and the generator output rises, completing one cycle of operation of the regulator. The cycles are repeated at such high frequency that the generator output is limited to its predetermined maximum. Movement of the armature is so rapid that it cannot be observed. The current regulator operates only at the value for which it has been set, so that, if it is used with a generator having, say, a 28-amp output, it will operate only when the generator attempts to exceed this output. The action is so sensitive that the regulator must be mounted in order that engine vibration will not affect proper opening and closing of the contact points.

In some motor vehicles an auxiliary set of contact points is mounted on the armature of the cut-out relay. These are connected into the starting motor solenoid relay circuit as a safety device. (See Starting Motor Drives and Controls, p. 17-91.) A generator-charging light, used in place of an ammeter, may also be connected to this set of contact points, as a means of indicating whether the generator is charging. When such a light is used, the relay armature contact point must be insulated from the armature.

Single-core Regulator. Although differing in design from the double-core regulator, operation of the single-core unit (Fig. 32) is the same. The wiring diagram of another

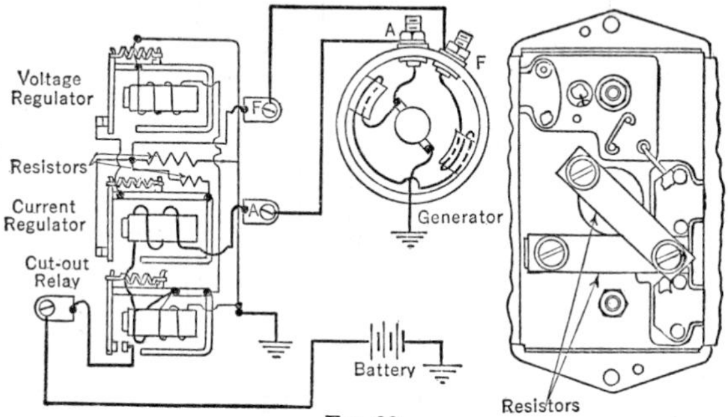

FIG. 33

make of regulator in which the cores are arranged differently is shown in Fig. 33. In this type the resistors are mounted in the back of the assembly, as shown.

Still another modification of regulator design (Fig. 34) is used on various makes of cars produced by a well-known manufacturer. Notice the use of two instead of three sets of contact points in this regulator. Operation of the regulator is as follows:

The regulator contact points are normally closed, and the generator fields are subjected to full generator voltage. Field current with the regulator points closed is controlled by the generator voltage, which in turn is controlled by the battery voltage, the resistance of the circuit, and the load. The opening voltage varies with different temperatures but, when the points open at a given temperature, the field current must pass through a series of resistances, thus reducing the field current. With the field current reduced, the generator voltage drops and the regulator points again close, the voltage again rises and the points open, thus completing a cycle of operation. The greater the load or the lower the state of charge of the battery, the longer the points will remain closed for each cycle, and the greater the charging current. When the battery becomes fully charged or the load is reduced, the regulator points open sooner, and the length of the cycle is reduced, thus decreasing the charging current. With a fully charged battery the generator output may be as low as 6 or 7 amp when only the ignition is on. It may be as high as 30 to 35 amp with a low battery and maximum load.

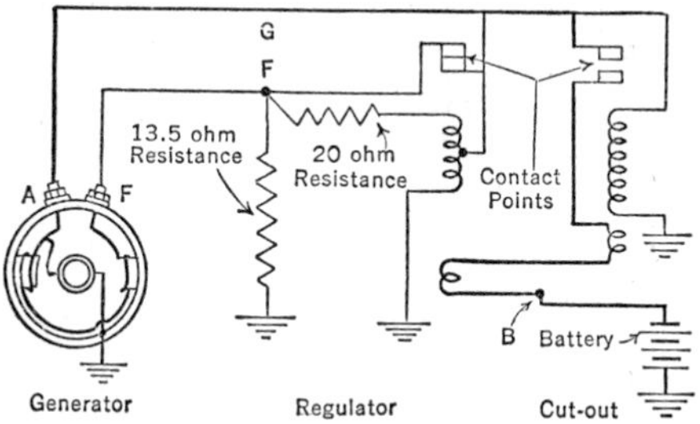

FIG. 34

ALTERNATING-CURRENT–DIRECT-CURRENT GENERATING SYSTEM. A generating system for motor vehicles which combines an alternator with a rectifier became available during 1946 and is applicable on new cars, motor coaches, trucks, etc. High current output at low engine speeds, safe operation at high maximum engine speeds, and rapid rise in output within a narrow engine speed range are given as principal advantages for the system. Suggested applications include passenger cars and transport trucks equipped with mobile radio-telephones, police cruisers and fire-department vehicles which are driven either slowly or only at intermittent periods, and other services where frequent battery charging is required with conventional systems to provide necessary current.

The alternator (Fig. 35) is of the revolving field type, so that the conductors in which the output current and voltage are generated are stationary. The stator winding is three-phase. The field coil in the rotor is energized from the battery through one brush riding on a small slip ring, and another brush making contact with the end of the rotor shaft. Since the polarity of the field does not affect the operation of the alternator, nor

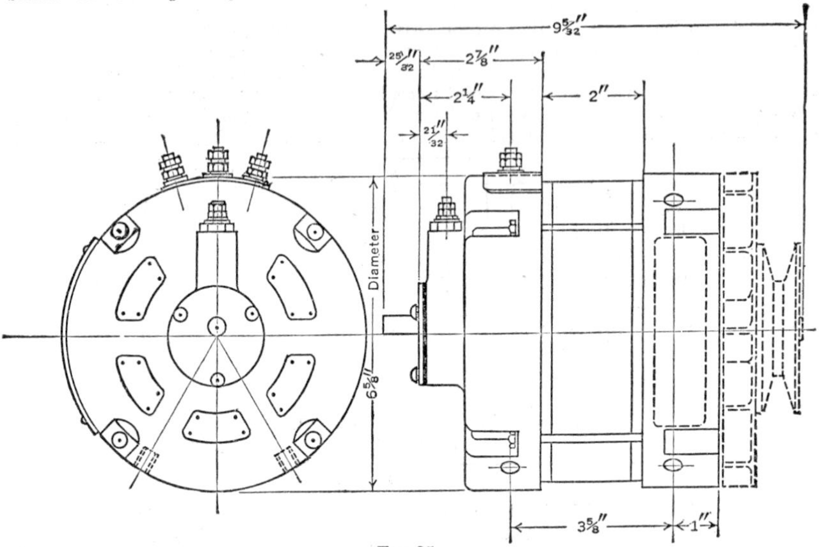

Fig. 35

does the direction of rotation, the same unit can be used for either clockwise or counterclockwise rotation. No changes other than in connections to the rectifier are necessary for either positively or negatively grounded batteries. The alternators are of nominal

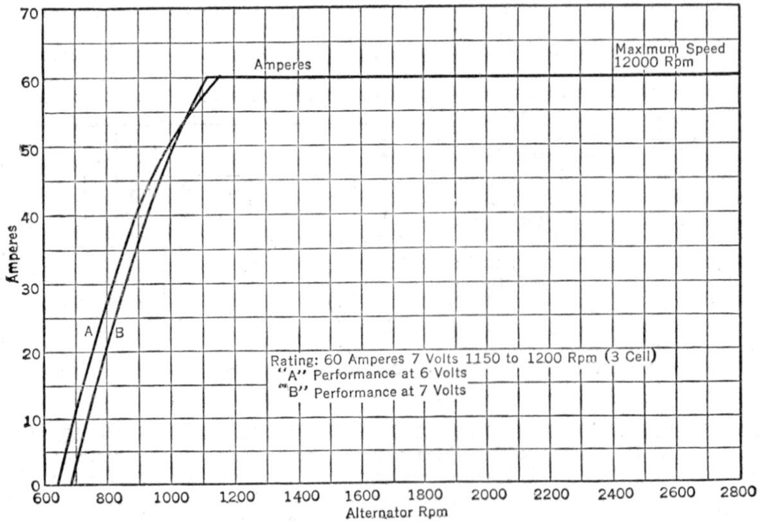

Fig. 36

6- or 12-volt type. With the 6-volt type current output ranges from 25 to 35 amp at the idling speed of most engines when a 3-to-1 speed ratio is used between engine drive shaft and alternator shaft. Current generation begins at 230 engine rpm and rises to 60 amp maximum output at 375 rpm (Fig. 36). This output is maintained at 7.5 volts to maximum engine speed (4000 rpm), corresponding to a rotor speed of 12,000 rpm. The performance of production generators is said to be within ± 5 per cent of these values.

The output curve for the 12-volt alternator is shown in Fig. 37. Maximum continuous output is 100 amp at 14 volts from 1660 to 12,000 rpm of the rotor shaft. The size

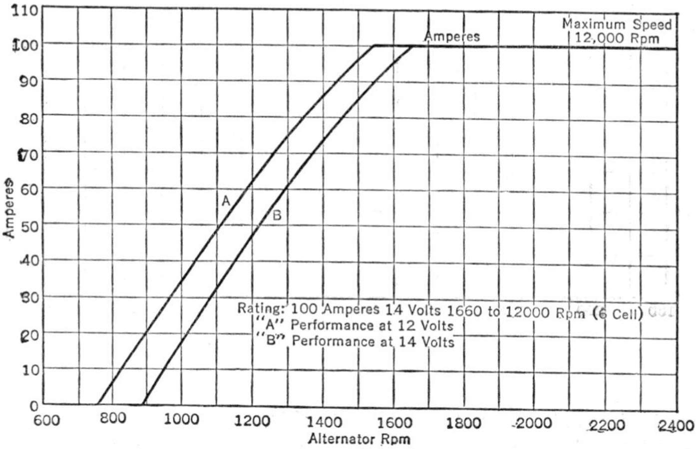

FIG. 37

of the alternator for both 6- and 12-volt systems is practically the same. The units are much smaller than for d-c generators of equivalent output, so that weight is reduced and problems in mounting the alternators on engines are eliminated. The larger and heavier d-c generators necessary for high output have caused difficulties in mounting because of both weight and limited space available under the vehicle hood.

The conversion from alternating to direct current is made by the use of a three-phase, full-wave, bridge-type dry-plate copper sulfide or selenium rectifier (Fig. 38), connected into the circuit as shown in Fig. 39. By locating the rectifier close to the engine-cooling

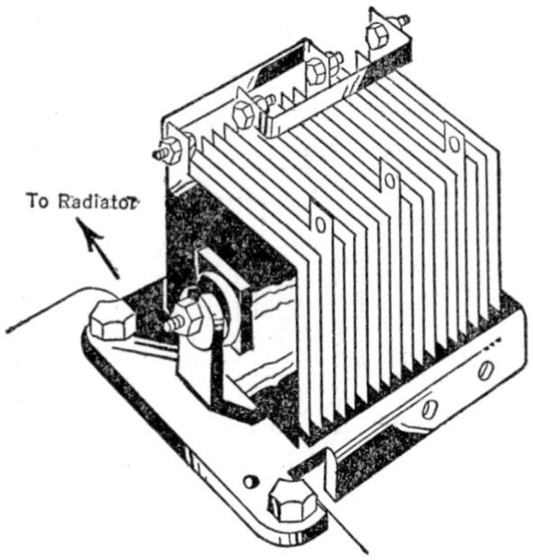

FIG. 38

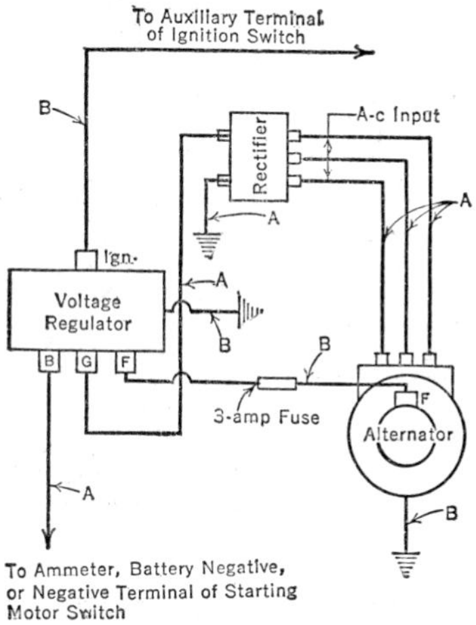

FIG. 39

fan, proper ventilation to assure a wide margin of safety against damage of the unit is obtained. For continuously satisfactory rectifier performance the manufacturer has stipulated minimum ambient air velocities and maximum air temperatures for various current outputs. These conditions are shown in Fig. 40.

After rectification the output of the system is controlled by a voltage regulator connected on the d-c side of the rectifier in a manner similar to that of the conventional d-c generator, but with the exception that the conventional cut-out relay is replaced by a load relay. The load relay is used to complete the excitation of the field coil of the alternator from the battery and is connected into the circuit at the time the ignition switch

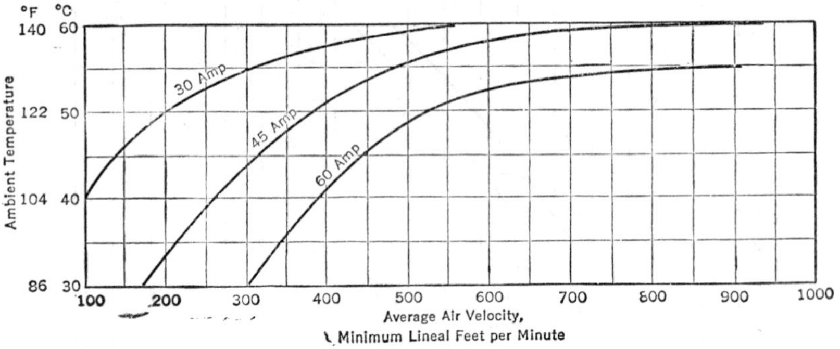

Fig. 40

of the engine is closed. Figure 41 is a complete wiring harness diagram, showing how the alternator, rectifier, and voltage-control units are connected and attached to the engine of a popular-priced automobile.

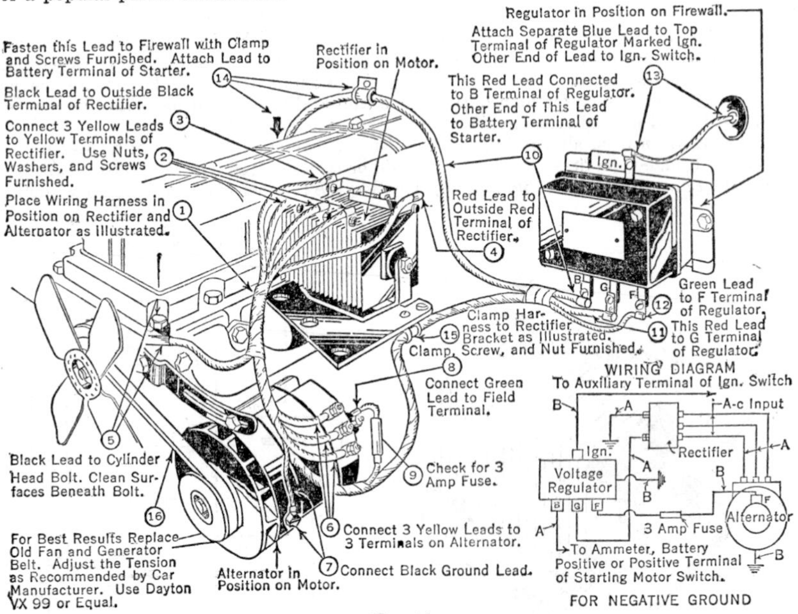

Fig. 41

GENERATOR MOUNTINGS. Generator mountings for automotive lighting systems of four types have been standardized by the Society of Automotive Engineers. Where the generator is to be driven directly from an accessory drive gear of the engine, a flange-type mounting may be used. The flange has slotted holes for adjustment of a driving chain if used. The generator also may be mounted in a cradle or on a base. Much used today is the hinge-type mounting, wherein flanges at both ends of the generator are hinged and then the generator is swung in or out for driving belt adjustment. A locking pin or rod is provided to retain any given adjustment indefinitely. Generator overall diameters, lengths, and shaft end and pulley dimensions, as well as terminal posts, have also been standardized, so that generators are readily replaceable.

17. BATTERY

Storage batteries of the lead-acid type are universally employed. All up-to-date American passenger cars are fitted with three-cell (nominally 6-volt) batteries, but motor coaches often use the six-cell type, and still higher voltages are required in some forms of automotive applications. Ampere-hour capacities for passenger cars vary from about 80 up to 140, but considerably larger batteries are used in some buses and for certain other automotive applications, notably in motor boats.

STANDARDS. A number of standards covering dimensions and ratings for storage batteries have been adopted by the Society of Automotive Engineers. A summary of the more important follows:

Ratings. Batteries for combined starting and lighting service shall have two ratings except as noted. The first rating shall indicate the lighting ability and shall be the capacity in ampere-hours of the battery when it is discharged continuously to an average final terminal voltage equivalent to 1.75 per cell at the 20-hr rate for passenger-car and motor-truck service and at the 4-hr rate for motor-coach service. The second rating shall apply only to batteries used in passenger-car and motor-truck starting and lighting service. This rating shall indicate the cranking ability of the battery at low temperatures and shall be (1) the time in minutes when the battery is discharged continuously at 300 amp to a final average terminal voltage equivalent to 1.0 volt per cell, the temperature of the battery at the beginning of such discharge being 0 deg fahr; and (2) the terminal battery voltage 5 sec after beginning such discharge.

Location of Battery Parts. The location and polarity of the terminal posts and the position of handles, when used, shall be as shown in Fig. 42 and Table 1.

Type Designations and Markings. Type letters, numbers, or symbols, which shall enable the user to determine ratings from the manufacturers' catalogs, shall be stamped or molded on the case or cell connectors or on a nameplate permanently attached to the end or side of the battery, or the ampere-hour capacity at the 20-hr rate for passenger-car and motor-truck batteries and at the 4-hr rate for motor-coach batteries shall be molded on the side or end of the container.

Grounding. The positive side of the storage battery should be securely and adequately grounded to the car frame or a substantial part thereof.* All contact surfaces in the grounding circuit should be clean, tinned, and free from oxide or paint. The ground connection from the battery should be readily accessible for servicing.

Terminal Posts. Polarity shall be plainly marked as follows: The positive post shall be marked Pos. or P or +. When taper posts are used for terminals of lead-acid storage batteries, the dimensions in inches shall be

Small diam. of neg. post	5/8	Minimum length of taper	5/8
Small diam. of pos. post	11/16	Taper per foot, incl. angle	1 1/3

In addition to the foregoing requirements for battery dimensions, discharge rates, etc., elaborate test apparatus and testing procedures have been standardized by the Society of Automotive Engineers. For methods of conducting these tests, refer to SAE Handbook of Standards.

Diesel Electrical Systems-Recommended Practice. Generators, storage batteries, starting motors, lighting and auxiliary electrical equipment shall be for nominal system ratings of 12, 24, or 32 volts, as determined by the power requirements of the application. It is recommended that no intermediate voltages be considered, except that a 30-volt system may be used when cranking requirements permit and no lighting or auxiliary electrical equipment is involved.

The combination of a 24-volt starting motor and two 12-volt batteries connected in series for cranking, and reconnected in parallel for charging from a 12-volt generator and operating lights and other auxiliary equipment, is considered practical where it can be adapted to the installation.

* The negative pole of the battery is occasionally grounded in preference to the positive.

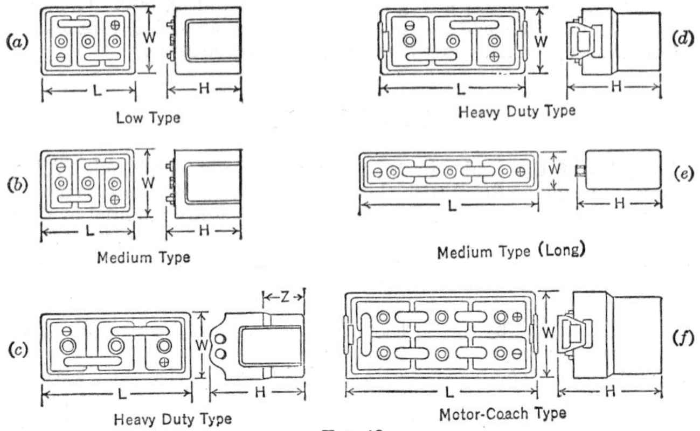

FIG. 42

Table 1. Battery Classification, Ratings, and Dimensions

Battery Type	SAE Battery No.	Assembly Fig. 42	No. of Cells	Minimum Capacity at 20-hr Rate, amp-hr	Minimum Time * at 300 amp, 0° F, Minutes	Minimum 5 sec * Volts at 300 amp, 0° F.	Life Cycles,* Minimum	Max. Overall Dimen., Inches Length *L*	Width *W*	Height *H*
Low	2-L	*a*	3	100	3.20	4.20	234	10 5/8	7 1/4	8
	3-L	*a*	3	110	4.00	4.30	277	11 7/8	7 1/4	8
Medium	1-M	*b*	3	90	3.00	4.10	234	9 1/8	7 1/8	9 1/8
	1-ME	*e*	3	96	3.10	4.15	234	19 3/8	4 1/8	9 1/8 †
	2-M	*b*	3	105	3.70	4.20	277	10 3/8	7 1/8	9 1/8
	2-ME	*e*	3	110	4.00	4.25	277	19 3/8	4 1/8	9 1/8 †
	3-M	*b*	3	120	4.40	4.30	316	11 3/4	7 1/8	9 1/8
	3-ME	*e*	3	122	4.90	4.35	316	19 3/8	4 1/8	9 1/8 †
	4-M	*b*	3	135	5.20	4.40	358	13 1/8	7 1/8	9 1/8
	5-M	*b*	3	150	6.10	4.50	398	14 1/2	7 1/8	9 1/8
High	1-H	*b*	3	100	3.30	4.15	269	9 1/8	7 1/8	9 3/8
	2-H	*b*	3	116	4.30	4.25	314	10 3/8	7 1/8	9 3/8
	3-H	*b*	3	133	5.30	4.40	358	11 3/4	7 1/8	9 3/8
	4-H	*b*	3	150	6.30	4.50	403	13 1/8	7 1/8	9 3/8
	5-H	*b*	3	166	7.50	4.60	448	13 3/4	7 1/8	9 3/8
	7-H	*b*	3	200	9.60	4.80	538	16 1/4	7 1/8	9 3/8
Heavy duty	1-T	*c* ‡ or *d*	3	118	1.50	3.45	300	13 3/8	7 5/8	10 1/2
	2-T	*c* ‡ or *d*	3	137	2.75	3.65	350	15 1/8	7 5/8	10 1/2
	3-T	*c* ‡ or *d*	3	157	4.00	3.80	398	16 7/8	7 5/8	10 1/2
	5-T	*c* ‡ or *d*	3	196	6.40	4.05	490	20 7/8	7 5/8	10 1/2
	7-T		3	236	8.60	4.30	560	23 3/8	7 5/8	10 1/2
Motor-coach	1-B	*d*	3	112 §	....	....	418	19 1/2	7 5/8	10 7/8
	2-B	*d*	3	160 §	....	....	595	26 1/4	7 5/8	10 7/8
	4-B	*d*	6 ‖	73 §	....	....	249	21 1/2	7 5/8	10 7/8
	5-B	*f*	6	87 §	....	....	300	21 1/4	8 3/4	10 7/8
	6-B	*f*	6	102 §	....	....	350	21 1/4	10	10 7/8
	7-B	*f*	6	117 §	....	....	398	21 1/4	11 1/8	10 7/8

* Ratings for Type T and B batteries are for double insulation. When double insulation is used in other types, deduct 10 per cent from the time and voltage ratings and add 15 per cent to the life cycles.

† This dimension is to the top of the terminal post.

‡ Dimension Z shall not exceed 3 1/2 in.

§ For motor-coach batteries, the minimum ampere-hour capacity is at the 4-hr rate.

‖ Side-to-side assembly of cells.

18. LIGHTING EQUIPMENT

Lighting equipment on most automotive vehicles includes not only equipment legally required in most states, such as head-, tail-, and license-plate lamps, but numerous supplementary lamps for the instrument panel, car interior (dome), and others. The principal improvement in automotive lighting in recent years was the adoption (1940) of the so-called "Sealed Beam" headlamp, which has made obsolete practically all other lamps for road lighting. Most states have stringent regulations covering the aiming or adjustment of headlamps, and in some cases the position and candlepower of other lamps used on the vehicle.

HEADLAMPS. The following is a summary of SAE recommended practice pertaining to the construction and installation of Sealed Beam headlamps (Fig. 43):

Definitions. *Sealed Beam Unit:* An integral and indivisible optical assembly with the name Sealed Beam branded on the lens.

Country or Upper Beam: A clear road beam intended for distant illumination and for use on the open highway when not meeting other vehicles.

Traffic or Lower Beam: A beam low enough on the left to avoid glare in the eyes of oncoming drivers and intended for use in congested areas and on highways when meeting other vehicles within a distance of 500 ft.

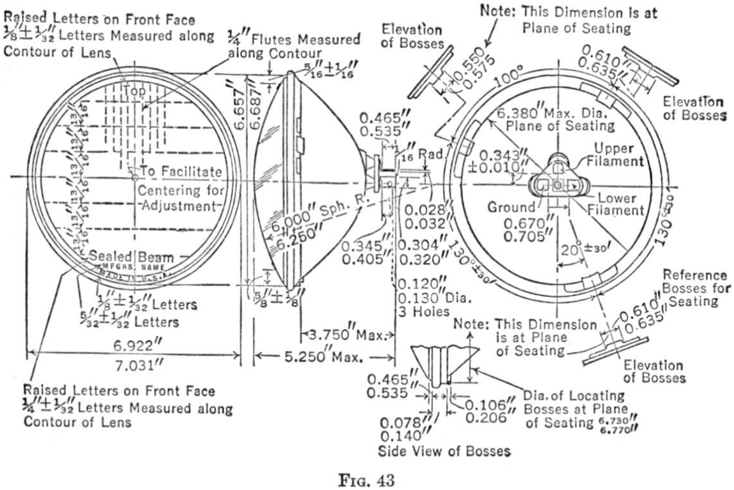

Fig. 43

Photometric Test. Photometric tests shall be made with the photometer at a distance of 60 ft from the lamps. Units shall be operated at their rated voltage during the tests.

The country or upper beam shall have a sufficiently well-defined high-intensity area or hot spot to permit the aiming of both beams from the center of this area.

The upper beam from each lamp shall be aimed visually so that the zones of maximum intensity superimpose at the photometric test plate and so that the geometric center of the zone of highest intensity falls 0.6 deg vertically below the photometer axis.

The combined beams from the two lamps shall meet the following specification:

Maximum Beam Intensity. The maximum intensity of the beam shall not exceed 75,000 cp.

In locating the test points shown in Figs. 44 and 45 the following nomenclature shall apply. The line formed by the intersection of the median vertical plane parallel to the lamp axis and the test screen is designated as *V*. The line formed by the intersection of the horizontal plane through the lamp centers and the test screen is designated as *H*. The point at the intersection of these two lines is designated as *H–V*. The other points on the screen are designated by similar symbols to indicate the number of degrees of arc above or below *H* and the number of degrees of arc to the left or right of *V*, e.g.: 4D–3L is a point 4 deg below *H* and 3 deg to the left of *V*, and 1U–V is a point 1 deg above *H* in the median vertical plane.

Table 2. Motor-vehicle Electric Lamp Bulbs* (for New Designs or Original Equipment) (SAE)

Type of Service	Trade No.	Design: Cp	Design: Volts	Rated Average Lab. Life at Design Volts	Max. Amperes at Design Volts	Filament: Type	Filament: Center Length, in.	Filament: Focal Center Length Tol., + or −	Filament: Axial Align. Tol., + or −	Bulb: Type	Bulb: Max. Diam., in.	Max. Overall Length, in.	Base: Type	Base: Design	Base: Contacts	Base: Plane of Pins or Major Locking Eyelet with Respect to Filament, deg
						6- to 8-volt Circuits										
1, 2	51	1	7.5	1000	0.25	C–2R	1/2	3/32	3/32	G–3 1/2	0.463	15/16	A–1	Min. bay.	S.C.	90
1, P	55	2	7.0	500	0.45	C–2R	1/2	3/32	3/32	G–4 1/2	0.588	1 1/16	A–1	Min. bay.	S.C.	90
L,M,P,T	63	3	7.0	1000	0.73	C–2R	3/4	3/32	3/32	G–6	0.744	1 7/16	B–1	Cand. bay.	S.C.	90
3	81	6	6.5	500	1.14	C–2R	3/4	3/32	3/32	G–6	0.744	1 7/16	B–1	Cand. bay.	S.C.	90
3, D	1129	21	6.4	200	2.91	C–2V	1 1/4	0.04	0.03	S–8	1.041	2	B–1	Cand. bay.	S.C.	90
4	1209S	32	6.1	125	4.30	C–6 Short	7/8	0.010	0.010	RP–11	1.420	2 1/4	F–1	Cand. pref.	S.C.	90
7, F	1323	32	6.2	200	4.52	C–2V	7/8	0.010	0.010	RP–11	1.420	2 1/4	F–1	Cand. pref.	S.C.	90
D,L,S,T	1154	{ 21	6.4	200	2.97	C–6	1 1/4	0.04	0.03	S–8	1.041	2	C–2	Cand. bay. indexing	D.C.	90
		{ 3	7.0	1000	0.73	C–2R										
6	2220	{ 21	6.4	200	3.05	C–2V	1 1/8	0.010	0.010	RP–11	1.420	2 1/4	F–2	Cand. pref.	D.C.	90
		{ 21	6.4	200	3.05	C–2V										
5	2320	{ 32	6.2	200	4.52	C–2V	1 1/8	0.010	0.010	RP–11	1.420	2 1/4	F–2	Cand. pref.	D.C.	90
		{ 21	6.4	400	3.30	C–2V										
						12- to 16-volt Circuits										
1, 2	57	1 1/2 (nom)	14.0	1000	0.22	C–2V	1/2	3/32	3/32	G–4 1/2	0.588	1 1/16	A–1	Min. bay.	S.C.	90
1, M,T	67	3	13.0	1000	0.41	C–2R	13/16	3/32	3/32	G–6	0.744	1 7/16	B–1	Cand. bay.	S.C.	90
D,M	89	6	13.0	750	0.63	C–2R	3/4	3/32	3/32	G–6	0.744	1 7/16	B–1	Cand. bay.	S.C.	90
D	1141	21	12.8	500	1.50	C–2R	1 1/4	0.04	0.03	S–8	1.041	2	B–1	Cand. bay.	S.C.	90
3	1142	21	12.8	500	1.50	C–2R	1 1/4	0.04	0.03	S–8	1.041	2	B–2	Cand. bay.	D.C.	90
4, 7, F	1327	32	12.75	300	2.33	C–2V	7/8	0.010	0.010	RP–11	1.420	2 1/4	F–1	Cand. pref.	S.C.	90
D,P,S,T	1016	{ 21	12.8	300	1.47	C–2R	1 1/4	0.04	0.03	S–8	1.041	2	C–2	Cand. bay. indexing	D.C.	0
		{ 6	13.0	1000	0.63	C–12										
3	F42T6/W					Fluorescent				T–6		40				

* Complete lists of available bulbs, including specials and replacements, are issued by the bulb manufacturers.

1—Instrument. 2—Indicator. 3—Interior. 4—Spot. 5—Motorcycle Headlamp. 6—City Delivery-truck Headlamp. 7—Farm Tractor. D—Turn Signal. F—Fog. L—License. M—Marker. P—Parking. S—Stop. T—Tail.

The terms S.C. and D.C., as used in this table, refer to single- and double-contact bulb bases and not to the type of current for which the bulbs are intended.

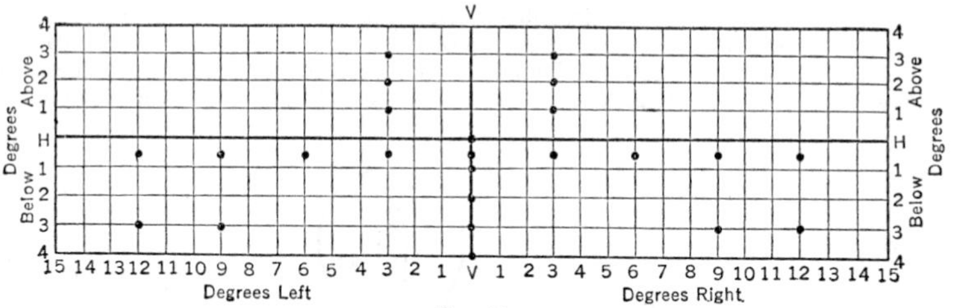

FIG. 44

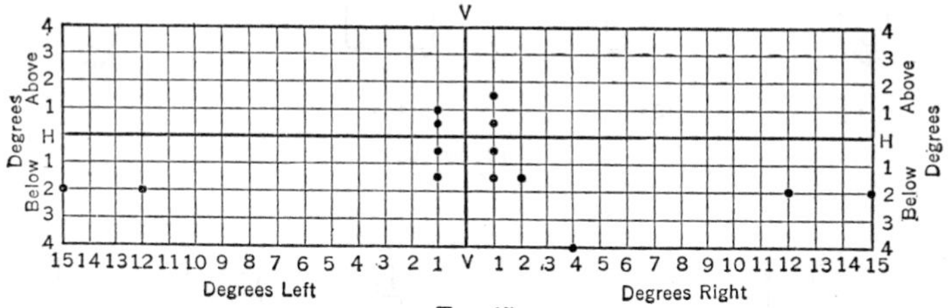

FIG. 45

Table 3. Sealed Units and Bulbs Used in Units (for New Designs or Original Equipment) (SAE)

Type of Service	Trade No.	Design		Rated Average Lab. Life at Design Volts	Max. Amperes at Design Volts	Filament Type	Bulb		Max. Overall Length, in.	Base		
		Watts	Volts				Type or Trade No.	Max. Diam., in.		Type	Design	Contacts
					6- to 8-volt Circuits							
H	Unit	45W	6.4	300	7.38	C-6	2400	7.031	5 1/4	H-3	Lug term.	3
		35W	6.4	500	5.74	C-6						
H	4030	40W	6.4	300	6.56	C-6	PAR-56	7.031	5 1/4	H-3	Lug term.	3
		30W	6.4	500	4.93	C-6						
F	Unit	35W	6.2	300	5.93	C-6	2404	5.70	4 1/8	G-1	Screw term.	S.C.
F	4012A	30W	6.2	300	5.08	C-6	PAR-46	5.70	4 1/8	G-2	Screw term.	D.C.
4	4535	30W	6.2	300	5.08	C-6 Short	PAR-46	5.70	4 1/8	G-2	Screw term.	D.C.
7	Unit							5.70	4 1/8	G-2	Screw term.	D.C.
7	4013	25W	6.4	300	4.1	C-6	PAR-46	5.70	4 1/8	G-2	Screw term.	D.C.
F	4015A	30W	6.2	300	5.08	C-6	PAR-36	4.46	2 5/8	G-2	Screw term.	D.C.
					12- to 16-volt Circuits							
H	Unit	45W	12.8	300	3.76	C-6	2412	7.031	5 1/4	H-3	Lug term.	3
		35W	12.8	500	2.93	C-6						
H	4430	45W	12.8	300	3.76	C-6	PAR-56	7.031	5 1/4	H-3	Lug term.	3
		35W	12.8	500	2.93	C-6						
F	Unit	35W	12.8	300	2.93	C-6	2414	5.70	4 1/8	G-1	Screw term.	S.C.
F		35W	12.8			C-6	PAR-46	5.70	4 1/8	G-2	Screw term.	D.C.

4—Spot. F—Fog. 7—Tractor. H—Sealed Beam Headlamp.
The terms S.C. and D.C., as used in this table, refer to single- and double-contact bulb bases and not to the type of current for which the bulbs are intended.

SUPPLEMENTARY LAMPS. In addition to headlamps, various supplementary driving and passing lamps, foglamps, direction-signal lamps and others have been standardized by the SAE. Tests covering the effects of moisture, dust, corrosion, and vibration have been established, as well as tests to determine photometric characteristics. These standards have been accepted by many state authorities as well as by the automotive industry.

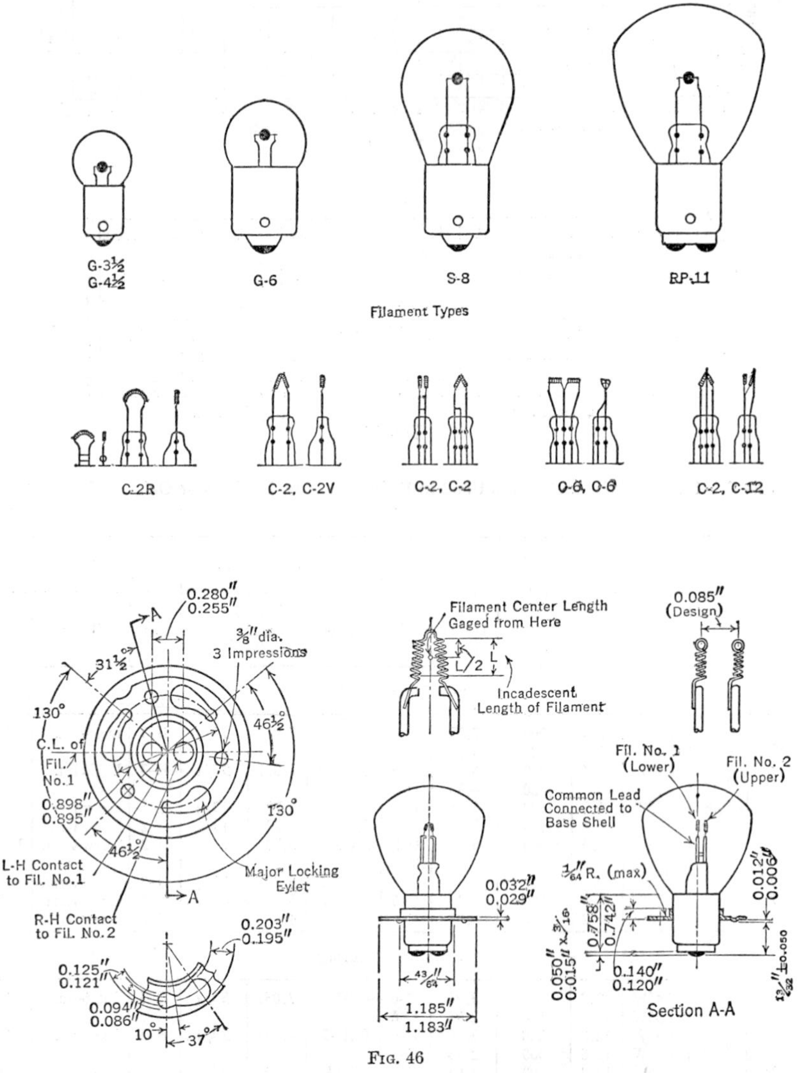

Fig. 46

Lamp bulbs for older-type prefocused and asymmetric-type headlamps also are covered by these standards.

LAMP BULBS, BASES, AND SOCKETS. In Fig. 46 some of the types of lamps and filaments in former and current use are illustrated. Lamps are made with both single- and double-contact bases, depending upon whether the return circuit from lamp to battery is grounded. The bases also may be flanged to locate the bulb in a fixed position if the lamp is of prefocused type, so as to require no adjustment forward or backward in the reflector to obtain the required beam characteristics. Lamp bases for all except the

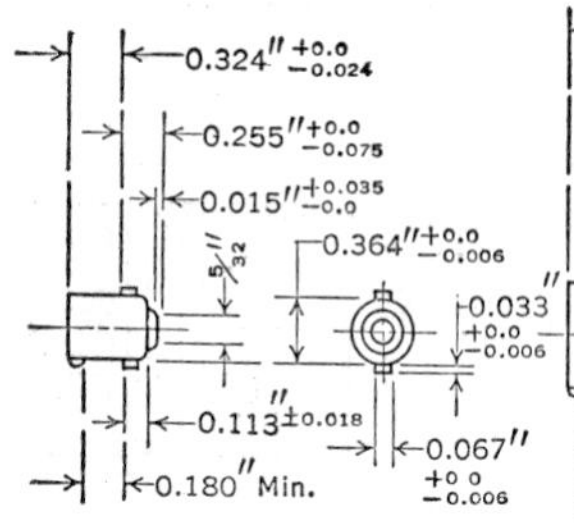

Type A-1, Min. Bay, S.C.

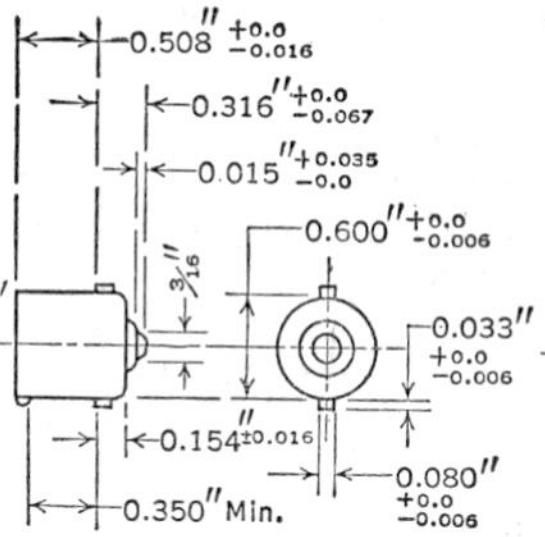

Type B-1, Cand. Bay, S.C.

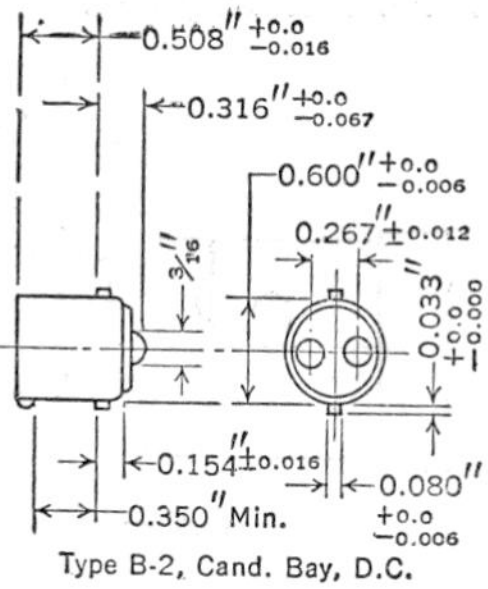

Type B-2, Cand. Bay, D.C.

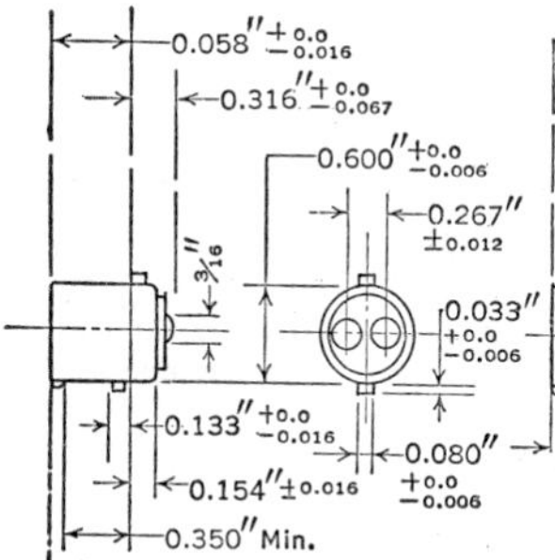

Type C-2, Index. Cand. Bay, D.C.

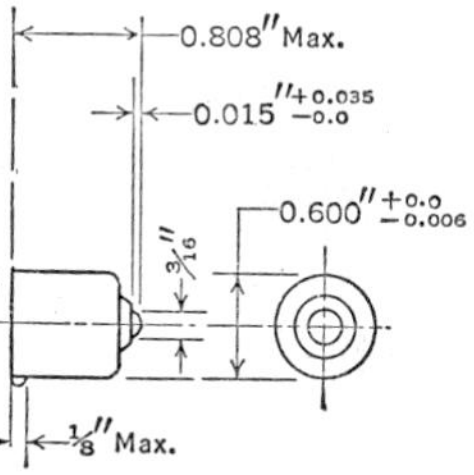

Type D-1, Pinless Cand., S.C.

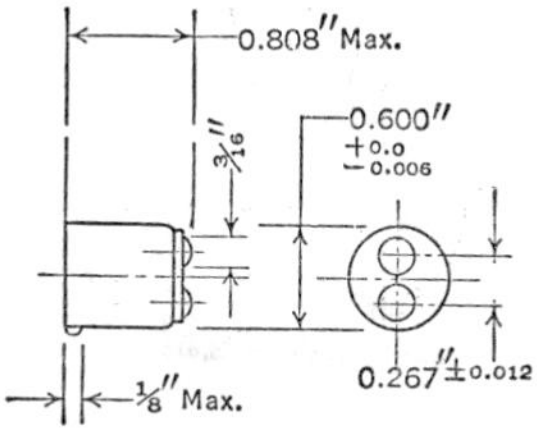

Type D-2, Pinless Cand., D.C.

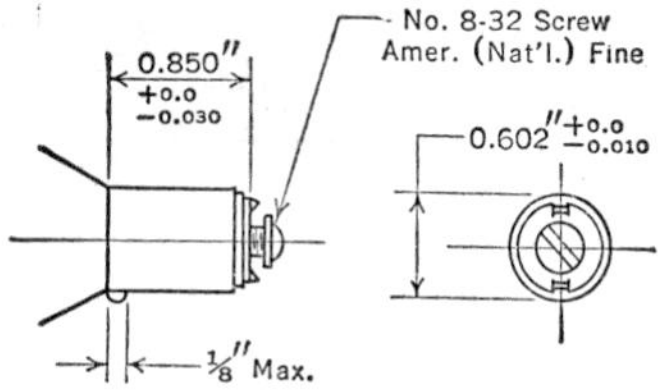

Type G-1, Screw Terminal

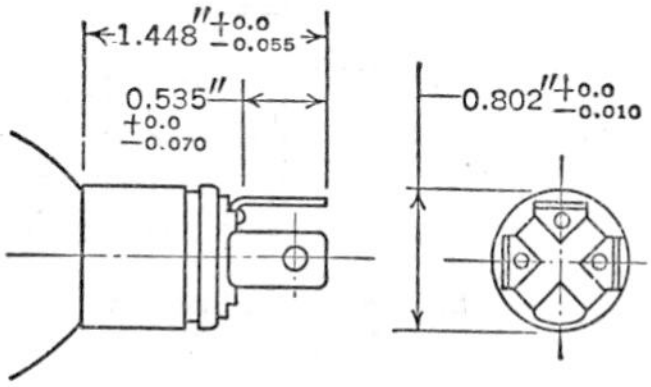

Type H-3, Three Contact Lugs

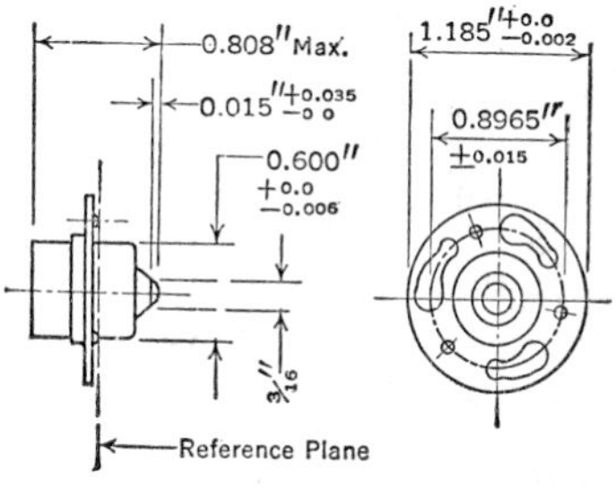

Type F-1 Cand. Prefocus S.C.

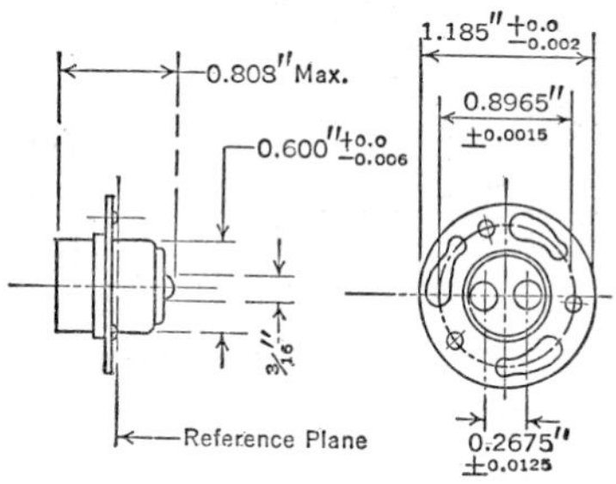

Type F-2, Cand. Prefocus D.C.

Fig. 47

Sealed Beam lamp, which has no base, are shown in Fig. 47. A few of the socket types, including the connector for the Sealed Beam headlamp, are shown in Fig. 48. Tables 2 and 3 list only the sizes and types of lamp bulbs recommended for vehicles.

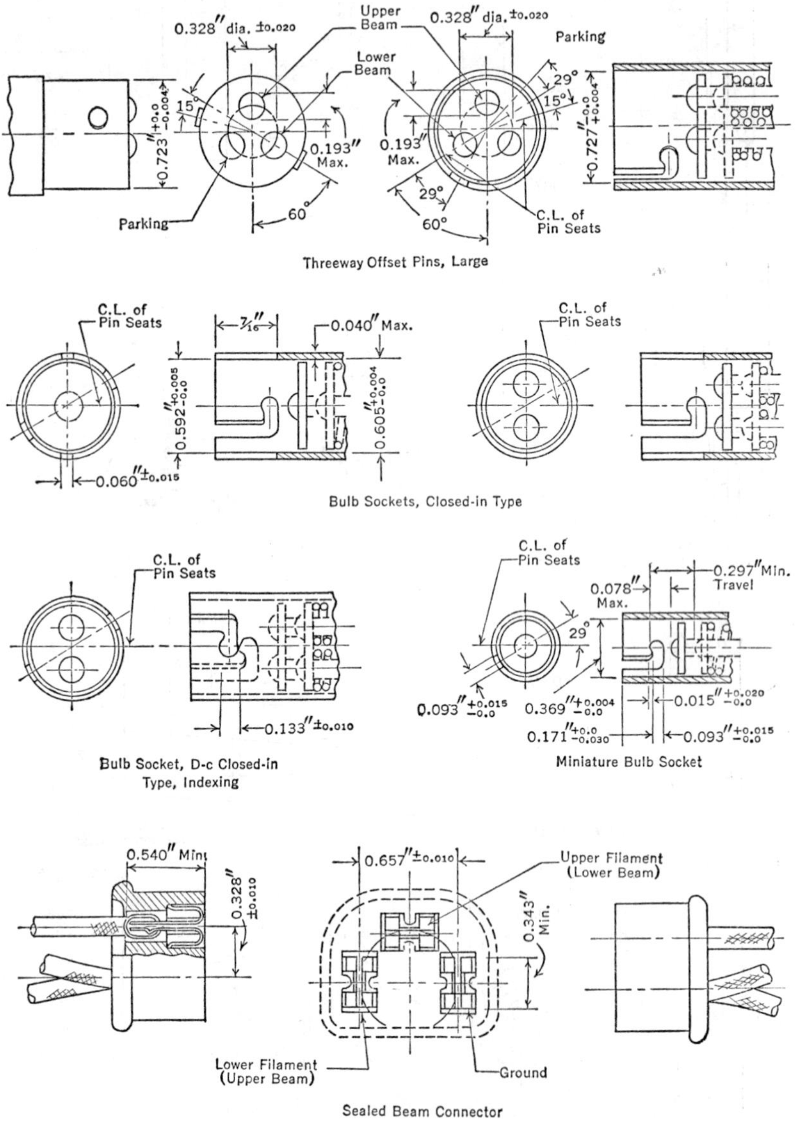

FIG. 48

FUSES. The SAE standards and specifications for automotive fuses stipulate, in addition to their dimensions and capacities, as given in Fig. 49, that the caps be of nickel or cadmium-plated brass or copper. Only clear glass shall be used, so that the fuse strip can readily be seen. The rated ampere capacity of each fuse shall be marked on one of the caps. A fuse of other than standard rating must be of the same dimensions as the next larger standard size. All fuses shall carry their rated current continuously, but shall blow at 125 per cent of load in less than 30 min.

CIRCUIT BREAKERS. To avoid the use of fuses and to provide a means of knowing instantly that a short circuit exists in the lighting system of the vehicle, circuit breakers are frequently employed. Those in current use are of two types, the so-called "vibrator"

Fuse Capacities and Dimensions

SAE Capacity, Amperes	*L* Nom. Length, Inches	*C* Min. Inches	*T* Min., Cap Thickness, Inches
4	5/8	1/4	0.010
6	3/4	1/4	0.010
9	7/8	1/4	0.010
14	1 1/16	1/4	0.010
20	1 1/4	1/4	0.010
30	1 7/16	3/8	0.015

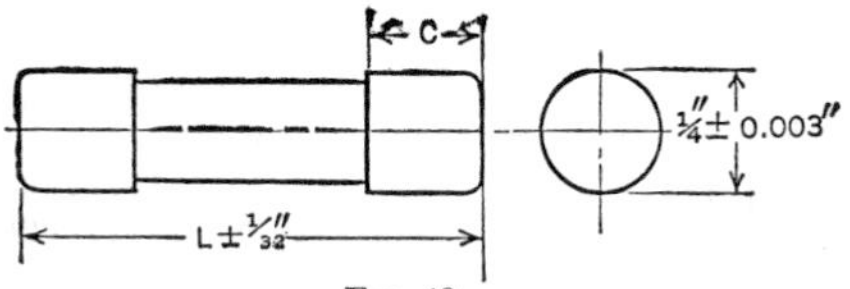

FIG. 49

type, and a combination thermostatic and magnetic breaker. The vibrator type has an electromagnet in series with a pair of contact points, the unit placed in series with the lamp load in a wire carrying the entire load. When the current becomes excessive, as it does if a short circuit exists, the magnet is sufficiently energized to open the circuit. Breaking the circuit then causes the points to close. This cycle continues to repeat rapidly and maintain a current flow of sufficiently reduced value to prevent overheating of the lamp wiring. The rapid vibration of the points produces a buzzing sound to warn the vehicle operator that a short circuit exists. With the thermostatic type of breaker a bimetallic bar is used in place of a spring for mounting one of the contact points. A magnetic coil is connected in parallel with the pair of points, instead of in series with the points, as with the vibrating type of breaker (Fig. 50). When the current flow through the bimetallic bar exceeds a predetermined value, usually about 50 amp, the bar becomes sufficiently hot to open the points. Current then continues to flow in reduced amount through the coil. If the bar cools sufficiently to close the points again, they will be reopened, completing the cycle. If the circuit has been cleared of the short circuit, however, the points will remain closed. This type of breaker may also be combined with one of vibrator type, so that the audible warning of a short is produced. By permitting some current to flow even in the presence of a short circuit, the lamps may continue to burn at reduced brilliance instead of going completely "out," as when a main fuse is used.

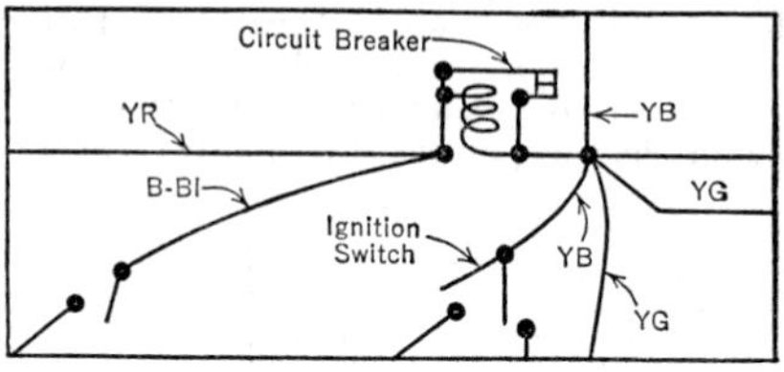

FIG. 50

19. STARTING MOTOR

Starting motors for automotive engines are series wound machines, designed to produce high torque at comparatively low speed and for use during only short time intervals. Consequently the motors are smaller and lighter than series motors for most other applications. Characteristic curves of a starting motor are given in Fig. 51. As this figure indicates, the torque is a maximum at zero speed, where, of course, the current drawn is a maximum. Both torque and current fall off rapidly with increase in speed. The power output increases with the speed to a maximum and then decreases. Since motors of this class are rarely used for more than 1 min of continuous running, high efficiencies are not of special importance, and ratings can be considerably higher than for series motors designed for continuous use.

For the same reasons the allowable rate of temperature rise may be much greater in the starting motor. High torque output is required, not alone to overcome the inertia of the parts put into motion, but also to break down the resistance or shear of the oil films in engine bearings and cylinders and to overcome the load imposed by compression of the charge in engine cylinders. Although a part of the energy used in compressing the charge is returned on the expansion stroke even before ignition occurs, much of it is lost through wire-drawing of the gases and dissipation of the heat of compression to the cylinder jackets. Torque requirements vary during each revolution of the engine, the variation increasing as the number of cylinders in the engine decreases. Other factors also affect the torque required, one of the most important being the viscosity of the engine lubricant, which varies in turn with engine temperature and is highest with a given oil when it is new and undiluted by fuel which passes the pistons. Chiefly because of higher oil viscosities in cold weather, low-temperature starting imposes some of the most difficult limiting conditions in starter design. In recent years the development of lubricants which flow at cold temperatures and the practice of disengaging the clutch when starting an automobile engine in cold weather have somewhat reduced the torque requirements of the starting motor. In any case the current drain on the battery by the starting motor must not be so great that sufficient current is not available for ignition, or the engine cannot be started under any conditions.

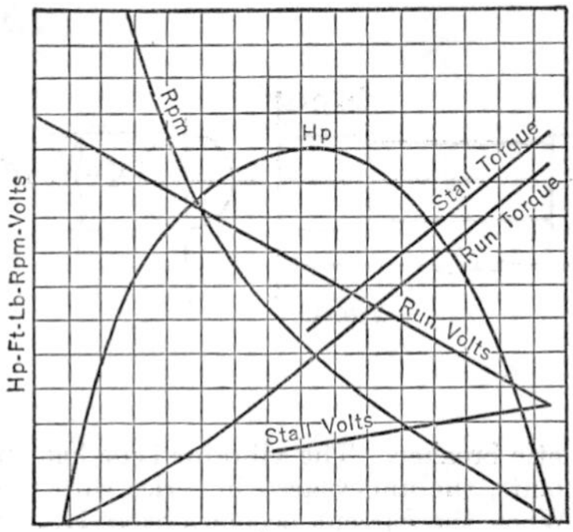

FIG. 51

Depending upon the size and type of engine, with compression ratio a factor, a minimum cranking speed approximating 40 rpm may be stipulated for the engine. The starting motor is so geared to the engine that this engine speed or higher will be obtained without raising the speed of the motor above safe limits for the armature coil windings. Figure 52 gives the relationship between the starting-motor voltage and the current flow for 6-, 12-, and 24-volt motors as recommended by the SAE. As will be noticed from this series of curves, the starting current may attain as high a value as 800 amp for the 6-volt system and 1000 amp for a heavy-duty 24-volt system, while the voltage may drop to 2, 4, and 8 volts for 6-, 12-, and 24-volt systems respectively. Because of these high battery-discharge rates, the wire sizes between the battery and motor must be sufficient to prevent overheating and too high a voltage drop between the two units. The SAE stipulates that the difference between the voltage at the battery terminals and the voltage at the starting motor shall not exceed 0.12 volt per 100 amp.

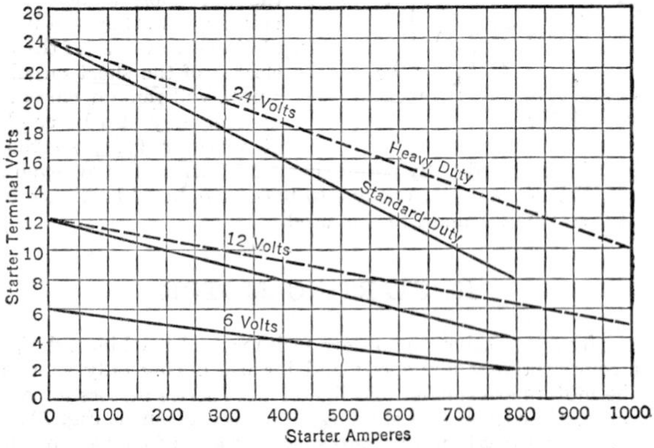

FIG. 52

with the circuit at a temperature of 68 deg fahr. Sufficient current to provide these discharge rates will be available if SAE recommended practice is followed by the battery and the vehicle manufacturer.

In Fig. 53 the curves show the actual relation between the piston displacement and the cranking speed for five different six-cylinder engines of similar design and approximately the same compression ratio, at 0 deg fahr, the same size of battery being used in each case. Both motors are of the direct-geared type, the gear ratio being 9 to 115. Starting motor *A* had four poles wound, and motor *B* had field coils on two poles. The following general conclusions may be drawn from the foregoing data: (1) The larger the engine, the more difficult is winter cranking. (2) For any given starting motor, increasing the size of the battery does not add greatly to cranking speed but materially increases cranking capacity. (3) Cranking speeds of an engine may vary between wide limits depending upon the type of starting motor, the gear reduction employed, and the viscosity of the oil.

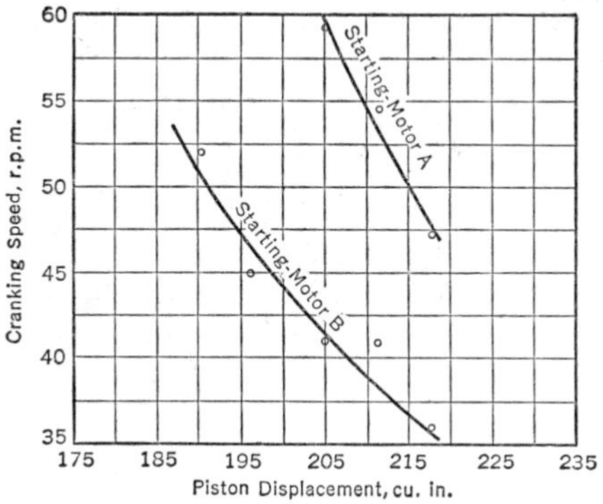

Fig. 53

STARTING MOTOR DRIVES AND CONTROLS. Several methods of engaging the starting motor with the engine are in use: (1) automatic engagement and disengagement of the motor driving-pinion with a flywheel ring gear, by means of the so-called "Bendix pinion"; (2) engagement and disengagement of the pinion by means of a separate pedal, the movement of which also closes the starting motor circuit; (3) engagement by means of a solenoid, sometimes in conjunction with a relay; and (4) engagement by means of the accelerator or clutch pedal in conjunction with a relay and disengagement by engine vacuum cylinder. Combinations and variations of these methods also have been used. In addition an automatic engine-restarting unit known as the "Startix" was used on some cars until about 1936. With this unit the starting motor began to crank the engine by turning on the ignition switch, and as long as this switch was turned on, the engine would again be cranked if it stalled at any time. Two solenoids operating in conjunction with a Bendix pinion were used to obtain the necessary control. Circuits and methods of operating the various control systems in recent and current use are as follows:

Bendix Pinion. This type of engagement is used on many automotive engines. A sleeve having screw threads (usually a triple thread) with stops at each end to limit the lengthwise travel of a driving pinion is mounted on a threaded extension of the motor shaft. The pinion, which is unbalanced by means of an integral counterweight, has internal threads for mounting on the sleeve. The sleeve is connected to the motor shaft by a drive spring attached to a collar pinned to the motor shaft. The pinion, when engaged, meshes with a ring gear attached to the outside of the flywheel of the engine. When the starting motor is not running, the pinion is out of mesh with the ring gear. When the starting motor circuit is closed, the shaft starts to rotate immediately at high speed. Since the pinion is unbalanced, it does not rotate immediately with the shaft, but, because of its inertia, moves toward engagement (forward or backward, depending upon the design used) on the revolving threaded sleeve until it meshes with the ring gear. If the teeth of the pinion and ring gear meet instead of meshing, the drive spring permits the pinion to revolve and force it into engagement. When the pinion is fully meshed, it is then driven by the motor through the drive spring and cranks the engine. When the engine starts, the ring gear on the flywheel drives the pinion at a higher speed than does the starting motor, causing the pinion to turn in the opposite direction on the threaded sleeve and automatically become disengaged from the ring gear. The drive spring acts as a cushion against variations in engine-cranking torque and also reduces the shock on the teeth and motor shaft when the pinion and gear are meshed or when the engine may backfire and rotate in the opposite direction. Principal advantages of the Bendix pinion are its simplicity, automatic action, and high starting motor speed before cranking begins, the latter being also a disadvantage in that the quick impulse given the pinion may cause nicking of the gear teeth and possible breakage of the driving spring because of the high torque loading imposed thereon. Closing of the starting motor circuit may be by means of a foot switch carrying the full starting motor current, or by means of a solenoid switch having its auxiliary circuit connected to a hand or foot button, or by other means. In any case only the circuit is closed, meshing and demeshing of the pinion occurring automatically as described. The type of solenoid used with an auxiliary circuit to a convenient button is

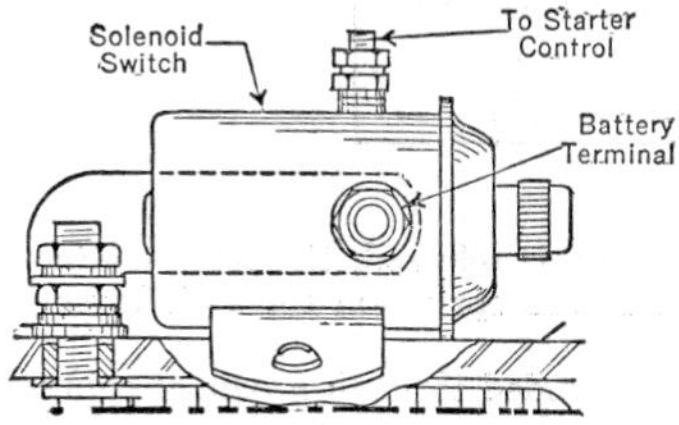

Fig. 54

shown in Fig. 54. This type of switch can be operated manually in case of failure of the auxiliary circuit to the button by depressing the plunger shown at the right end of the unit.

Engagement by Solenoid. With controls of this type a powerful solenoid is used to pull the pinion into engagement with the flywheel ring gear, as shown in Fig. 55. The circuit diagram is shown in Fig. 56. The solenoid is controlled by a push-button switch on the instrument panel. When the push button is depressed, current flows from the

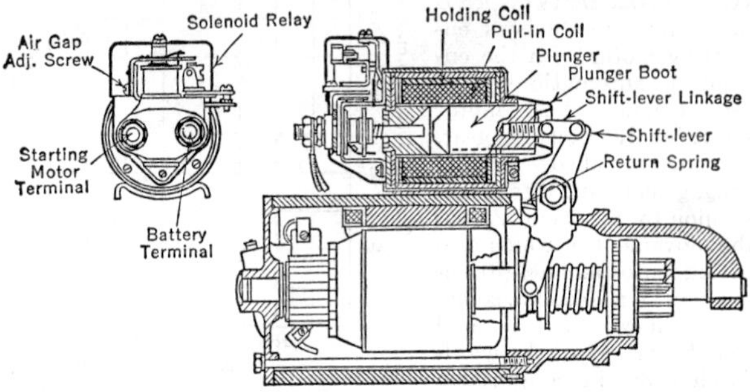

Fig. 55

ignition switch through the push button and the solenoid relay coil to ground. This causes the relay armature to close the contact points. The current then flows from the battery connections at the solenoid through the relay points and coils of the solenoid.

The solenoid winding is made up of two coils: a series winding connected from the relay stationary contact to the main switch terminal connecting with the starting motor, and a shunt winding connecting from the relay stationary contact to ground. When the relay points are first closed, current flows through both coils, immediately pulling in the solenoid plunger and shift lever and engaging the pinion of the starting motor with the flywheel. When the plunger reaches the end of travel, the starter switch disk is closed, and the motor cranks the engine. As the starter switch disk is closed, one coil of the solenoid (series) is shorted out. The shunt coil (grounded) remains in the circuit with sufficient current to hold the pinion in engagement with the flywheel while the engine is

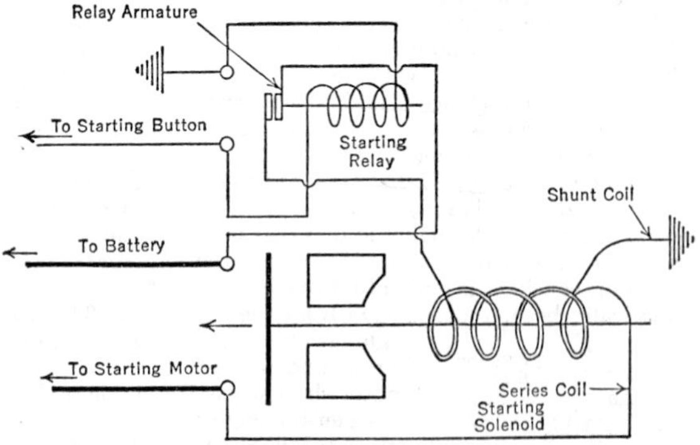

Fig. 56

being cranked. As soon as pressure on the push button is released, the relay contacts open, breaking the solenoid circuit and allowing the return spring on the shift lever to disengage the pinion.

In a variation of this control the relay solenoid is grounded through a pair of contact points mounted on the generator output relay. When the generator starts to charge the battery, these contacts open and stop the flow of current through the starting motor solenoid winding. The starter is thus prevented from accidental engagement when the engine is running. A torsional spring on the starter engaging lever pulls the solenoid plunger back, allowing the starter switch to open and pulling the pinion out of mesh with the ring gear.

Vacuum Controls. In its simplest form this type of control is in reality operated manually through the clutch pedal for engagement and disengagement of the pinion, combined with a vacuum release and vacuum-release lock effective while the engine is running. (See Fig. 57.)

When the clutch pedal is depressed with the engine stopped, the switch cam is turned counterclockwise. This pulls the horizontal switch lever to the left, and the vertical

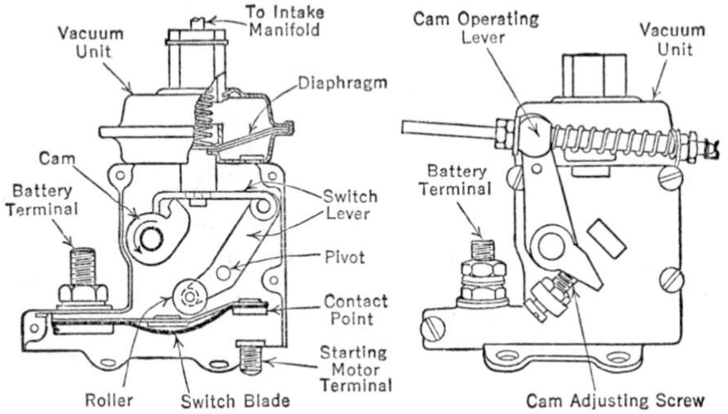

Fig. 57

switch lever forces the switch blade down until its contact point makes contact with the starting motor terminal. The diaphragm is then in its lowest position. As soon as the engine fires, the reduced pressure in the intake manifold of the engine raises the diaphragm, lifting the horizontal switch lever and breaking the contact at the starting motor terminal. The cam can then rotate without touching the horizontal switch lever as long as the engine is running. As soon as the engine stops, the vacuum diaphragm and the horizontal switch arm drop, ready to make contact again when the clutch pedal is depressed.

A vacuum control switch is sometimes used in the starting motor circuit in addition to the solenoid switch, described previously. The solenoid may also be equipped with a relay or auxiliary switch circuit. Such a system of control is shown in Fig. 58. With

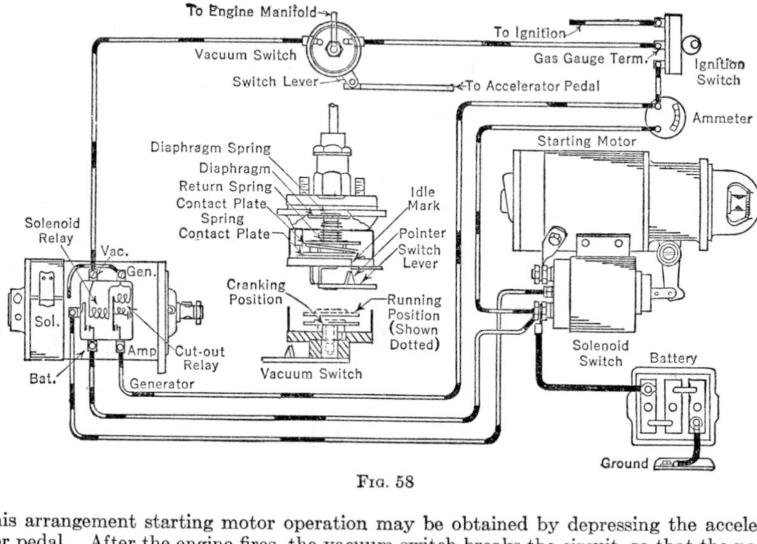

Fig. 58

this arrangement starting motor operation may be obtained by depressing the accelerator pedal. After the engine fires, the vacuum switch breaks the circuit, so that the pedal can operate as usual as long as the engine is running without closing the circuit through the vacuum switch.

Vacuum and carburetor controls of more complex type than those described previously are used on recent passenger automobiles. In one of them, after turning on the ignition switch, the starter can be operated by pulling out the hand throttle about two-thirds of its travel or by depressing the accelerator a small distance. The starter circuit is opened and the gears are automatically disengaged as soon as the engine starts. The starter control mechanism consists of a switch mounted on the automatic choke (see Fig. 59), which is operated by engine vacuum and linkage connected to the throttle valve and a solenoid with relay. Auxiliary contacts on the cut-out relay in the voltage regulator interrupt the control circuit as soon as the engine picks up. The generator windings are used for completing the control circuit to the ground.

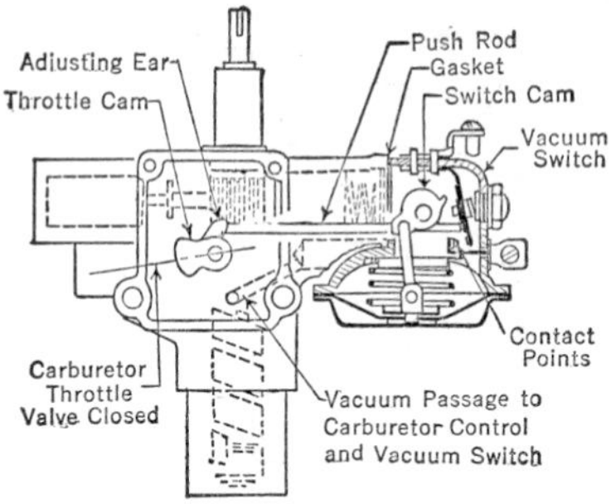

FIG. 59

Movement of either the hand throttle or the accelerator pedal causes the throttle to open and the vacuum switch contact points to close. This allows current to flow from the battery through the ignition switch, vacuum switch, solenoid relay windings, upper contacts on the cut-out relay, and generator to the ground. Completion of this circuit causes the solenoid and starting motor to operate as described.

Normally, as soon as the engine is running, the vacuum switch will be opened by manifold vacuum, causing the solenoid relay contacts to open and thus breaking the solenoid circuit. As there are certain conditions when the engine vacuum is not sufficient to open the contacts in the vacuum switch, this circuit is also opened by the solenoid relay current passing through the generator and an auxiliary set of contacts on the cut-out relay when the generator is turning. After the engine fires, the increase in speed results in generating a voltage which prevents the current passing through the magnet coil of the solenoid relay from continuing its flow through the generator to the ground. This causes the contact points of the solenoid relay to open at a lower voltage than that required to operate the cut-out relay. In completing the circuit through the generator, the connection is made to the insulated main brush, which provides a positive circuit through the field coils, as well as the generator brushes, so that the operation is not dependent upon the circuit through the generator brushes alone. The auxiliary contacts on the cut-out relay are retained to prevent completion of the control circuit through the ignition coil, should the ignition switch be turned off and again turned on before the engine stops running.

With this hook-up there are the following means of preventing engagement of the gears while the engine is running: manifold vacuum acting on the vacuum switch; auxiliary contacts on the cut-out relay; blocking effect of the generator voltage and mechanical lockout in the vacuum switch.

In a recent type the switch is incorporated in the carburetor. When the accelerator is depressed, with the engine stopped, a steel ball which rests on a flat spot on the throttle shaft is forced against a plunger, which raises a copper contact spring until it makes an electrical connection between two brass blocks in the bakelite top of the switch. This closes the solenoid relay circuit.

When the engine starts, the manifold vacuum raises the steel ball away from the shaft and plunger to a seat in the casting, where it remains as long as the engine runs. As soon as the ball is raised, the coil spring pushes down on the contact, forcing the contact and plunger down, which breaks the connection, opening the solenoid relay circuit. The ball cannot return to the starting position until the engine stops and the throttle is returned to the idle position.

20. ENGINE AUXILIARIES

The electrical system of the modern motor vehicle not only encompasses the principal units and controls previously described, but a number of auxiliaries, including gages for determining the engine oil pressure, fuel supply, and cooling-water temperature. Bimetallic heating elements, heating coils, rheostats, and similar means are used for operation of these auxiliaries. They are usually connected through the ignition switch of the engine, so that they function only when this switch is turned on.

OIL-PRESSURE GAGES (Fig. 60). This type of gage consists of two units, the engine unit, which is attached to an oil-pressure tube of the lubrication system, and the gage unit, which is located on the instrument panel of the vehicle. A single wire connects these two units, and the circuit is completed through the ignition switch, car battery, and ground. The engine unit contains a diaphragm, which deflects varying amounts in proportion to the oil pressure. When the diaphragm is deflected, the circuit is closed through a contact point on a bimetal arm, allowing current to pass through a heating coil formed around the arm. Heat from this coil causes the bimetal to distort and break the contact. The coil then cools and recloses the contact, thus completing the cycle of operation. This cycle is repeated continuously at frequent intervals, the rate being approximately 120 cycles per minute for an oil pressure of 25 psi.

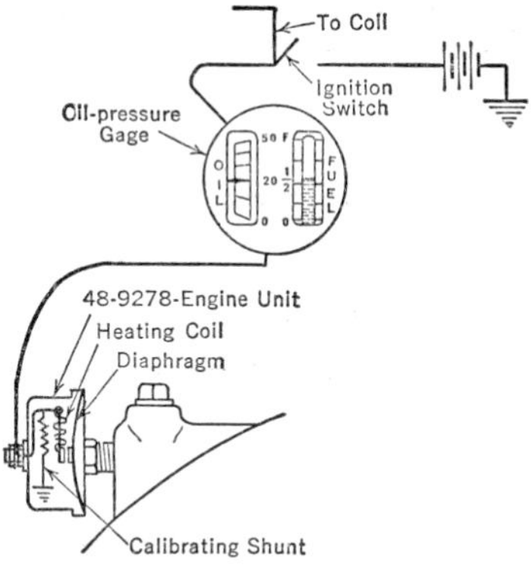

Fig. 60

Connected in series with the heating coil of the engine unit is a similar coil formed around a bimetal arm in the gage unit. Heating of the engine unit coil simultaneously causes heating of the gage coil, causing the bimetal arms to distort simultaneously. By linking the gage pointer to the gage unit bimetal arm, the amount of distortion and thus the amount of oil pressure are shown by the gage unit. As the oil pressure increases, causing greater deflection of the diaphragm of the engine unit, a greater distortion of the bimetal arm is required before the contacts open and break the heating coil circuit. Both heating coils must therefore operate for longer intervals than at a lower oil pressure. This in turn causes greater distortion of the bimetal arm in the gage unit, which registers a higher oil pressure on the calibrated dial. Accuracy of the device is assured by the use of temperature-compensating shunts, which are calibrated at the time of initial installation in the vehicle.

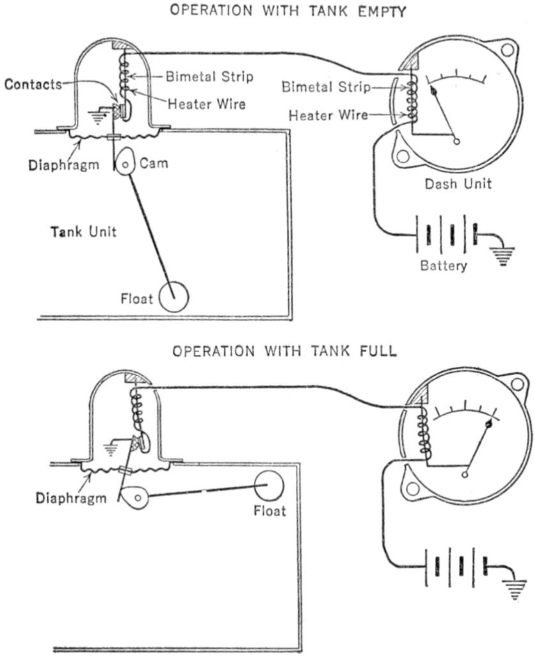

Fig. 61

FUEL-SUPPLY GAGES. Electric gages for registering the amount of fuel in the gasoline tank of a motor vehicle are of two general types: (1) bimetallic elements in the instrument panel and tank units, somewhat similar to those used for oil pressure, as above, and (2) balanced coil type. In both types compensation for differences in outdoor temperatures may be provided. Actuation of the tank unit is by means of a float on the surface of the fuel, connected by a lever or lever and cam to the tank element, whether a bimetal arm or resistance coil, similar to a rheostat.

An example of the bimetal-bar and heating-coil type is shown in Fig. 61. Since the gage depends entirely upon temperature for its operation, a change in voltage in the system will not affect the reading. A higher voltage will show a change in fuel level faster, but the final reading will be the same. Because the strips heat and cool slowly, any sudden change in gasoline level caused by rough roads, etc., is damped out, and a steady reading of the average level in the tank is given.

A gage with balanced coils operated by a rheostat and used on many makes of cars is illustrated in Fig. 62. The instrument panel unit consists principally of two coils spaced

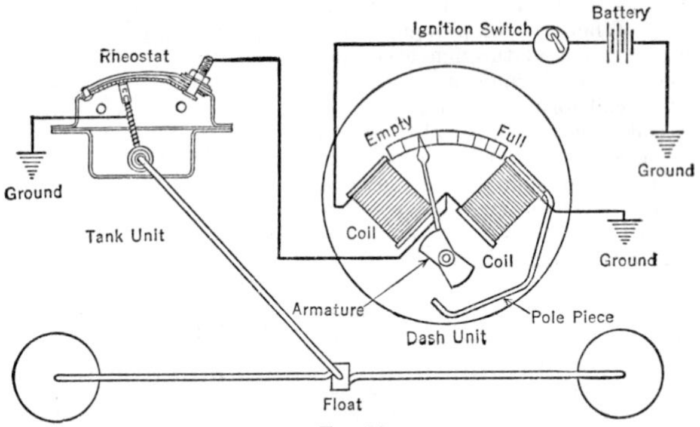

Fig. 62

90 deg apart, with an armature and pointer assembly mounted at the intersection of the coil axes. An inertia damper is provided on the armature assembly to prevent vibration of the pointer on rough roads.

The tank unit is essentially a rheostat, the movable contact being actuated by a float that rests on the surface of the gasoline in the tank. Movement of the float is transmitted to the rheostat contact arm by a set of gears in larger tanks and through a lever in smaller tanks.

When the gasoline tank is empty, the float assembly is at its lowest position, where the rheostat in the tank unit is completely grounded. All current through the dash unit therefore flows through the coil at the empty side of the indicator, and the pointer is pulled to the empty mark. As fuel is added in the gasoline tank, the float assembly rises. This moves the contact brush in the rheostat, introducing resistance into the circuit that grounds the full coil in the dash unit, so that part of the current flows through this coil. The pointer is attracted away from "empty" to a position of balance between the two coils, its point of rest depending upon the amount of resistance, which in turn is governed by the amount of gasoline in the tank.

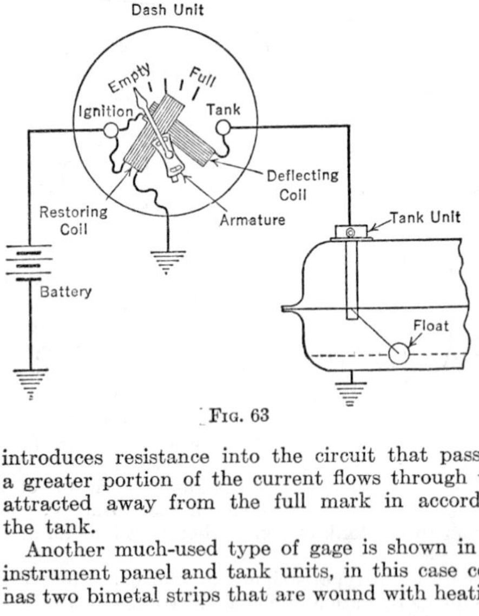

Fig. 63

A gage similar in construction to the one shown in Fig. 62, except for the arrangement of the two coils in the instrument panel unit, is shown in Fig. 63. When the gasoline tank is filled, the float assembly is at its highest position where the rheostat in the tank unit is completely grounded. All current through the instrument panel unit is therefore equally divided between the two coils, and the pointer is pulled to the full mark. As fuel is used from the tank, the float sinks, moving the contact brush on the rheostat. This introduces resistance into the circuit that passes through the deflecting coil, so that a greater portion of the current flows through the restoring coil. The pointer is then attracted away from the full mark in accordance with the amount of gasoline in the tank.

Another much-used type of gage is shown in Fig. 64. This gage includes the usual instrument panel and tank units, in this case connected by two wires. The dash unit has two bimetal strips that are wound with heating coils, and two bimetal strips without coils that compensate for external temperatures and also protect the coils from overheating. The heating coils are wound around the strips that actuate the pointer and are welded to the strip at the lower end. The other end of each coil is connected to terminals at the rear of the gage. At the lower end of one unheated strip is a set of contact points held together by a coil spring. There are no contact points at the lower end of the other unheated strip, but there are two stops. One is an insulating button, and the other is a flat spring which is insulated from the frame.

The tank unit consists of a resistance and a contact arm, which are grounded to the case. The position of the float in the gasoline tank governs the movement of the arm on the resistance, which in turn controls the current in the heating circuits of the dash unit. The amount which the heated strips in the dash unit bend depends upon current in the heating coils. As the strips bend, the lugs at the lower end of the strips grip the point and move it into position on the dial to indicate the amount of fuel in the tank.

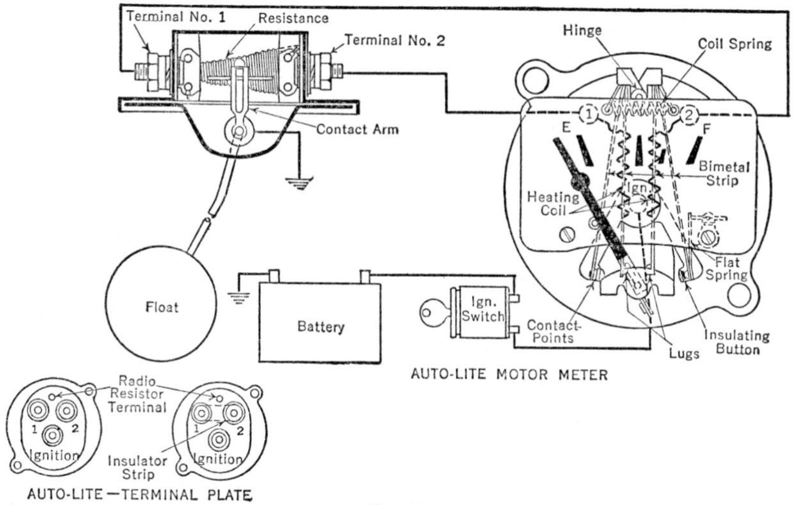

FIG. 64

The unheated strips are assembled so that any external temperature change causes the hinge to rotate slightly and thus move the strips that actuate the pointer, so that the gripping lugs remain stationary. If the coils become overheated, the excessive bending of the heated strips rotates the hinge and opens the contact points. Cooling begins immediately, and the points swing together before the position of the pointer is affected.

COOLING-WATER-TEMPERATURE GAGES. In general, gages for registering cooling-water temperature are similar to those previously described for oil pressure and fuel supply, but with one essential difference. Since motor vehicles operate under normal temperatures most of the time and the cooling water accordingly registers normal temperatures, gages for this purpose are so arranged that the smaller the current consumption the higher is the instrument-panel-gage reading. When the ignition switch is turned off, the bimetal strip or bar is straight, as shown at the right in Fig. 65(*b*). This does not mean that the cooling water has reached boiling temperature but indicates only that the circuit is open. When the switch is turned on, the strip takes the position shown at left in Fig. 65(*a*) if the engine is cold, or some intermediate position depending upon the actual temperature of the cooling water.

ELECTRIC FUEL PUMPS. Although standard American practice is to use a mechanically operated fuel pump to draw fuel from the tank and force it to the carburetor, occasions arise when an auxiliary pump is desirable. Motor coaches, and especially motor trucks and road tractors operated at night over long distances, vehicles requiring special fuels obtainable only in certain localities, and engines used for stationary purposes where the tank is some distance from the engine may require units of this type. The electric pump may be used in addition to the mechanical pump, drawing fuel from one or more auxiliary tanks carried in addition to the regular fuel-supply tank of the vehicle. Several pumps may be used to avoid opening a supply valve from one tank and closing that of another, which has been emptied of fuel. Large engines also may require several pumps, since the capacity of single pumps varies from approximately 8 to 15 gal per hr.

The fuel-pumping mechanism in units of the electric type is similar to that used in a mechanical pump, namely, a movable diaphragm or flexible disk, or may be of the piston or plunger type. The diaphragm is actuated by means of an electromagnet, the core

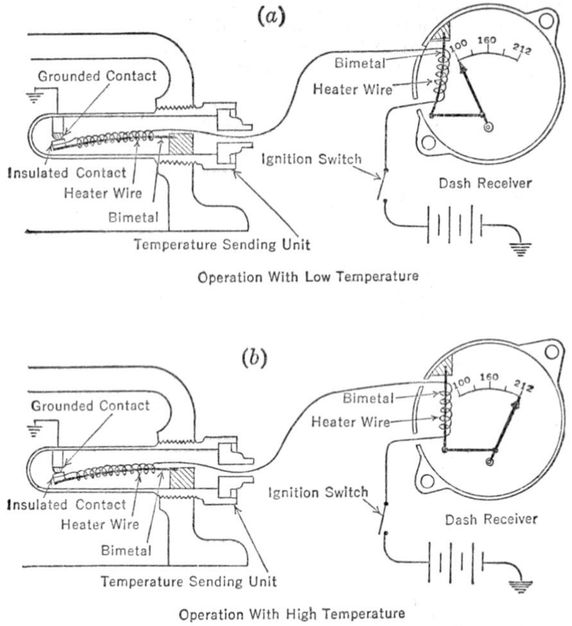

FIG. 65

of which is free to move in and out of the coil and which is caused to vibrate rapidly by opening and closing contact points connected into the coil circuit. The end of the core is attached to the diaphragm or plunger so that movement of the core is transferred to it. Although only a small movement of the diaphragm takes place, so that merely a small amount of fuel is pumped with each stroke, these vibrations or pulsations take place so rapidly that fuel is supplied in virtually a continuous stream.

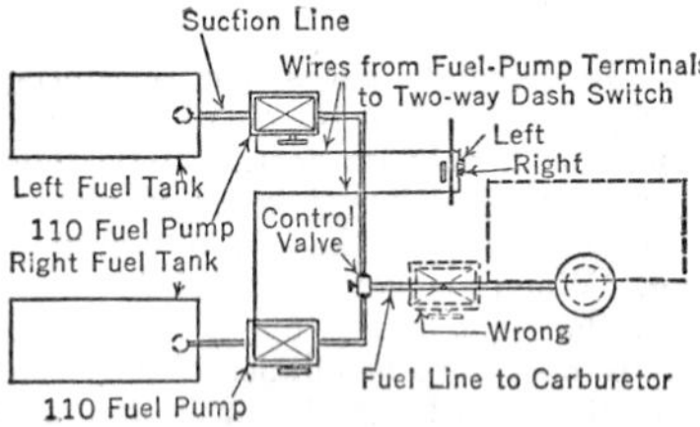

FIG. 66

A particular advantage of an electric pump is that it may be located close to a fuel tank, thus placing the entire fuel line to the carburetor under pressure. This assists in avoiding vapor-locking (vaporization) of the fuel in the supply line in hot weather, as may be possible when the line between the tank and pump is below atmospheric pressure. Electric pumps supply fuel at 2 to 3 psi pressure and are available with coils for either 6- or 12-volt electrical systems. Figure 66 shows how two pumps should properly be connected to supply fuel from two auxiliary tanks.

BIMETAL-TYPE VOLTMETERS. Many years ago voltmeters were extensively used to determine battery condition and whether the generator was charging. Although today their use is restricted to special installations where the driver can understand the value of such an instrument, several makes of cars were equipped with a so-called "battery gage" a few years ago, the gage actually being a voltmeter of bimetal-bar-and-coil type, as shown in Fig. 67. Dials on these meters were marked in colors, as indicated. The voltages corresponding to the color-separation lines are also shown. The face of the actual "battery gage" is shown in Fig. 68. The current flow through the meter is

$^1/_{10}$ amp at 6 volts, but even this small current drain on the battery occurs only when the ignition switch is closed. The production use of the meters was discontinued and ammeters again installed after about 2 years, since the average driver could not interpret the meter readings correctly.

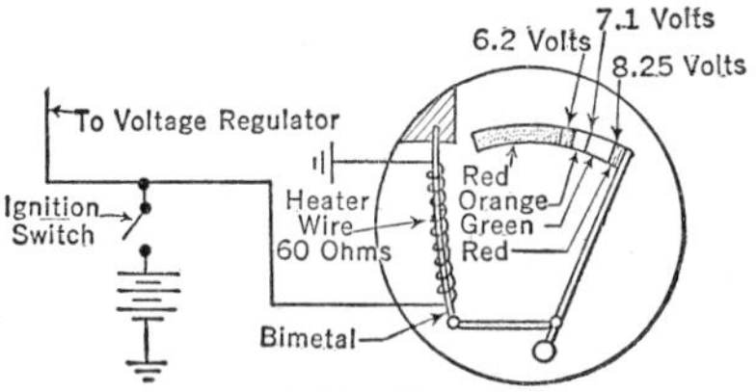

Fig. 67

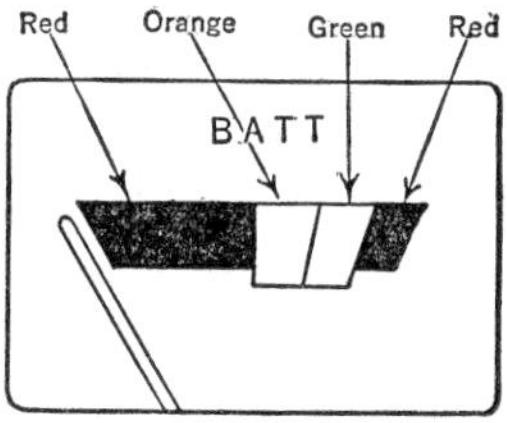

Fig. 68

AUTOMATIC CHOKES. Several makes of automobiles have been equipped with automatic and semiautomatic chokes of electromagnetic type. A rod operated by the starting motor closes these chokes when the starting motor circuit is closed. The choke is then held closed by a solenoid combined with a thermostatic control until a predetermined temperature, usually about 70 deg fahr, is reached, when the choke opens. This type may also be fitted with a manual control, so that the operator may utilize either automatic or manual operation. The arrangement of a choke of this type is shown in Fig. 69. Another electrically controlled choke consists of a bimetal thermostat, near which an electric heating coil is placed. When the ignition switch is closed, current flows through the heating coil, the temperature of which controls the tension of the thermostat, and thus the amount the choke is moved toward its fully closed position. A vacuum piston opens the choke when the engine begins to fire. When a 70 deg temperature is reached, the choke is fully opened, and the current to the heating coil is cut off.

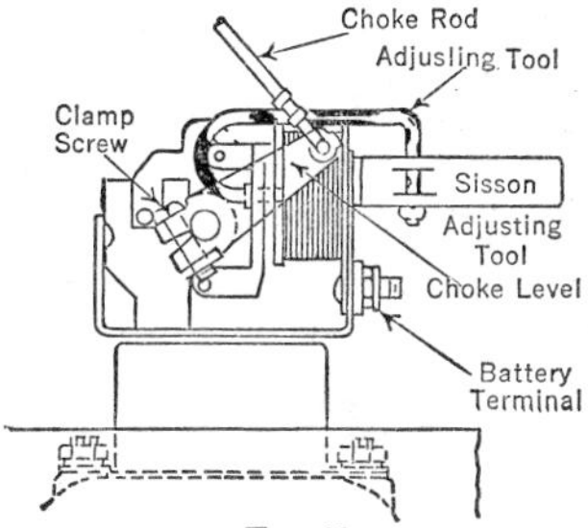

Fig. 69

INSTALLATION OF RADIO SETS. The satisfactory operation of radio receiving and transmitting sets in a motor vehicle presents many problems, largely because of the fact that the ignition system of the vehicle is in a sense a radio transmitting system, in fact has been so used in extreme emergencies. In aircraft engines the ignition system is often completely shielded to minimize interference. In a motor vehicle, however, this practice would materially increase cost and make the ignition system rather inaccessible for service. Suppressors (resistors) are therefore usually installed on each spark plug, and several condensers (capacitors) placed in the primary circuit of the ignition system, between the charging line of the generator and ground (to reduce commutator effect), and at the voltage-control unit or at any other point in the system where a low-tension current may be opened and closed intermittently at a sufficiently rapid rate to cause objectionable radio interference. View at lower left of Fig. 65 shows the location of a resistor used in conjunction with a fuel-supply gage.

Interference may be caused not only directly by the electrical system of the vehicle but also by the building up of static charges from the friction of operating parts of the vehicle and between the tires and road surface. Various methods of direct grounding and dissipation of charges of this kind have been used, including the grounding of wheel hubs to spindles or axles. Low-resistance materials also have been incorporated by tire manufacturers between the bead and tread of the tire to provide a means of carrying any static potential built up within the entire vehicle directly to the road surface or ground.

21. WIRING

WIRING SYSTEMS. Although double-wire (non-grounded) starting and lighting systems were used for many years, the single-wire (grounded) system is now generally employed. Only where the operation of some device or instrument requires it are two wires used. In the single-wire system the cylinder block of the engine, car frame, body, and other parts of the vehicle form the return circuit for virtually all equipment comprising the electrical system. Where a firm contact between one part of the vehicle and another cannot be obtained, as when the entire engine is mounted on rubber vibration-absorbing blocks, a special ground strap or connector is used between the two units, so that the electrical resistance of the return will be minimized.

The SAE recommended practice for grounded return systems not only includes preferred specifications for wire sizes, thickness of insulation, etc., but also outlines tests for conformance to these requirements. A summary of the more important recommendations follows:

Insulated Cable. Terminals on other than starting motor cables should be clamped to the insulation and soldered or welded to the conductors. The use of crimped terminals should be restricted to applications in the lighting circuits. SAE Standard terminals or lugs should be used on starting motor cables. Conductor cross-sections should be such that the potential loss at normal load will not exceed 10 per cent.

The insulated cables comprising the automobile wiring circuits should, where possible, be grouped together and protected by non-metallic tape or braid covering capable of withstanding severe abrasion except where otherwise protected or not in contact with metal surfaces. This wiring assembly should be cleated at intervals so as to be properly supported and should be located so that no portion is closer than 1 in. to the carburetor, gasoline line, and moving parts, and at least 2 in. from the exhaust pipe. The edges of all holes in metal members through which the cable passes should be rolled or bushed with rubber grommets. For extra protection double or triple tape or braid covering should be supplied.

Grounding. Wherever a conductor is connected to ground, it should be accessible for repair. Ground return connections should be made to the chassis frame or engine. The surface on which the terminals make contact should be clean and free from oxide or paint.

Battery Installation. The storage-battery compartment, or metal parts near the storage battery, should be painted with an acid-resisting paint and should have openings to provide ample battery ventilation and drainage. At the point where the live line passes through a metal compartment, the cable should be protected against grounding by an acid and waterproof insulated bushing.

Overload Protective Devices. The current to all low-tension circuits, except starting motor and ignition circuits, should pass through protective devices connected to the battery feed side of switches. The circuits should preferably be arranged so that the opening of a protective device will not extinguish all the lights.

Specification for Lighting and Starting Motor Cables. Conductors shall be insulated in one of the following ways to be specified by purchaser:

Type 1. Insulated without braid.

Type 2. Insulated and single braided.

Type 3. Varnished cambric tape and single braid.

The braid shall be lacquered, varnished, or weatherproof. Construction of these cables shall be as shown in Table 5.

Specification for Double-braid Lighting Cables. The conductor, insulation, and first braid shall be as shown in Table 5. The second braid covering shall be made up of one-half paper twine and one-half yarn, the paper twine wound in opposite direction to the yarn.

The braid shall be lacquered, varnished, or weatherproof, as specified in paragraph under General Specifications. Construction of these cables shall be as shown in Table 4.

Table 4. Double-braid Lighting Cables

Nom. Size, AWG	Size, Ply, and Ends of Cotton	Size and Ends of Paper Yarn	Picks Per Inch	Max. O.D.
16	30/2/4	0.028/1	13	0.185
14	30/2/4	0.028/1	13	0.211
12	30/2/4	0.028/1	16	0.245
10	30/2/4	0.028/1	20	0.282

Specification for High-tension Cable. This specification covers braid-covered insulated cable for use as high-tension leads in motor-vehicle and tractor-engine ignition systems.

Conductor. The conductor shall comprise 19 copper wires of No. 29 AWG (0.0112 to 0.0114 in. diameter), cabled to form a concentric strand with a left-hand lay. Each wire shall be thoroughly annealed and tinned.

Insulation. A covering of insulating compound shall be applied directly over the conductor. The insulating compound shall adhere closely to, but shall strip readily from, the wires of the conductor, leaving them reasonably clean.

Braid. A protective braid shall be applied over the insulating compound and so protected as to be resistant to heat, oil, water, and gasoline.

Adjacent layers of cable, when wound on a reel or when packed in sets, shall not stick to one another at any temperature under 105 deg fahr.

Size. The outside diameter of the cable shall be within 0.270 to 0.285 in.

Table 5. Lighting and Starting Motor Cables

SAE Wire Size	Number of Wires	Nom. Size of Strand AWG	Nom. Size of Strand Inches	Approx. Diam. of Stranded Conductor, Inches	Min. Area of Conductor, Cir Mils	Nom. Thick. of Insulated Wall	Max. O. D. of Finished Cable
20	10	30	0.0100	0.040	980	0.022	0.115
18	16	30	0.0100	0.050	1,568	0.022	0.135
16	19	29	0.0113	0.060	2,340	0.022	0.145
	16	28	0.0126				
14	26	28	0.0126	0.075	3,777	0.022	0.165
	19	27	0.0142				
12	41	28	0.0126	0.090	5,947	0.027	0.195
	19	25	0.0179				
10	65	28	0.0126	0.115	9,443	0.031	0.230
	49	27	0.0142				
	19	23	0.0226				
8	103	28	0.0126	0.160	15,105	0.037	0.301
	49	25	0.0179				
	19	21	0.0285				
6	172	28	0.0126	0.210	26,440	0.047	0.360
	133	27	0.0142				
	37	21	0.0285				
4	133	25	0.0179	0.275	38,430	0.047	0.437
	61	22	0.0253				
	49	21	0.0285				
2	259	26	0.0159	0.335	63,119	0.047	0.505
	127	23	0.0226	0.294			0.480
1	259	25	0.0179	0.375	80,010	0.047	0.557
	127	22	0.0253	0.330			0.503
0	259	24	0.0201	0.420	100,965	0.047	0.600
	127	21	0.0285	0.370			0.545
00	259	23	0.0226	0.475	126,822	0.047	0.655
	127	20	0.0320	0.415			0.590
000	259	22	0.0253	0.535	163,170	0.078	0.750
0000	418	23	0.0226	0.595	207,746	0.078	0.810

NOTE. Because of the special care needed in processing small conductors, it is recommended that nothing smaller than No. 16 wire be used in the wiring harnesses.

The conductors may be of bare copper or of tin-coated copper as required. A fibrous covering may be used as a separator between the bare conductors and the insulation.

When bunched stranding is used, the length of core lay for sizes No. 20 to No. 12 inclusive shall be 2 in. max. For No. 10 to No. 4 inclusive, the lay shall be 3 in. max. For larger cable the stranding shall be such that, when the cable is bent around a 5-in.-diameter mandrel, the cable shall not be injured.

The cross-sectional area of stranded conductors shall be not less than 98 per cent of the values specified in the table, which allow for the effects of tinning and stranding.

On varnished-cambric cables the tape must be wrapped sufficiently tightly to prevent the conductor being pulled from a 3-in. length of cable under a load of 10 lb. Varnished-cambric tape shall have a minimum overlap of 25 per cent.

Coloring of Cables. Both high (secondary) and low (primary) cables may have the insulation in a single color or have one or more threads or strands of the insulating material in colors differing from the base color, but for two different reasons. High-tension cable may be colored when specified by the vehicle manufacturer to designate the year and quarter of the year when the cable was manufactured. The SAE recommended practice for this method of designation is as follows:

To identify the date of manufacture (when specified), three distinguishable colored threads shall be used in the braid. Two of these threads shall indicate the year of manufacture, the colors being:

1944—Yellow	1946—Red
1945—Green	1947—Blue

For succeeding years this cycle will repeat.

The third thread shall indicate the quarter of the year of manufacture, the colors being

First quarter—Red	Third quarter—Green
Second quarter—Blue	Fourth quarter—Yellow

To identify the manufacturer, two distinguishable colored threads shall be used in the braid. Colors assigned to manufacturers by the National Board of Fire Underwriters for rubber-insulated wire and cable shall be employed if practicable.

Location and direction of colored threads in the braid used for date marking shall not conflict with those used for manufacturer's identification.

Manufacturers identification and a coded date shall be marked on the outside of the cable at 2-ft intervals.

Low-tension cable is colored at the discretion of the manufacturer to simplify the installation and connection of various circuits and to provide a means of tracing a given circuit from one point in the vehicle to another in case of cable breakage or short circuit. In various wiring diagrams used by vehicle manufacturers for instruction of assembly personnel and in servicing, the color designations are frequently abbreviated by letters, such as YB for yellow with black tracer and BR for black with red tracer. A color system that has been used for some time is as follows:

Passenger car wiring, where cable is bought in coils: Red for unprotected live wires; yellow for protected live wires.

Passenger car wiring where cables are bought in the form of a harness: Red for unprotected live wires; red with yellow tracer for low-tension or primary ignition; red with black tracer for ammeter to battery; yellow for protected live wire; brown with black tracer from lighting switch to junction block (parking lamp) and for all ground connections except battery ground; black for lighting switch to tail-lamp; black with red tracer to bright headlamps (or upper beam); green to dim headlamps (or lower beam) and from switch to signal lamp.

Table 6. Flexible Galvanized Conduit—Unpacked

Nom. Inside Diam., Inches	Actual Inside Diam., Inches		Max. Outside Diam., Inches	Min. Thickness of Strip, Inches	Approx. Weight, Pounds per 100 ft Extended	Min. Tension Load, Pounds	Min. U Bend Load, Pounds	Min. Bending Radius, Inches
	Min.	Max.						
3/16	0.184	0.196	0.278	0.010	3.8	75	30	1
1/4	0.245	0.260	0.345	0.010	4.5	100	40	1
5/16	0.308	0.323	0.413	0.010	5.4	110	40	1 1/8
3/8	0.368	0.388	0.478	0.010	6.2	125	40	1 1/4
7/16	0.431	0.451	0.551	0.011	8.2	150	45	1 1/4
1/2	0.493	0.513	0.633	0.011	9.0	175	60	1 3/8
9/16	0.556	0.576	0.696	0.011	10.4	185	75	1 5/8
5/8	0.618	0.638	0.758	0.011	11.3	200	75	1 3/4

Motor-coach and truck wiring (assumed to be bought in bulk coils): Red for unprotected live wires; yellow for protected live wires; brown with black tracer for generator cut-out or regulator to ground and all ground connections except that of battery; black for bright headlamps (or upper beam) and for bodylamp feed wires, switch to lamp; black with red tracer for dim headlamps (or lower beam); green from switch to signal-lamp and to signal-lamp indicator or pilot.

Table 7. Dimensions of Non-metallic Conduit

Nominal Size, Inches	Inside Diam., Inches		Outside Diam., Inches	
	Min.	Max.	Min.	Max.
3/16 *	0.187	0.207	0.277	0.297
3/16	0.187	0.207	0.287	0.307
1/4 *	0.250	0.270	0.340	0.360
1/4	0.250	0.270	0.350	0.370
5/16	0.312	0.332	0.412	0.432
3/8 *	0.375	0.395	0.475	0.495
3/8	0.375	0.395	0.505	0.525
7/16	0.437	0.457	0.567	0.587
1/2	0.500	0.520	0.630	0.650
9/16	0.562	0.582	0.722	0.742
5/8	0.625	0.645	0.785	0.805
11/16	0.687	0.707	0.847	0.867
3/4	0.750	0.770	0.934	0.954
13/16	0.812	0.832	0.996	1.016
7/8	0.875	0.895	1.079	1.099
15/16	0.937	0.957	1.141	1.161
1	1.000	1.020	1.204	1.224

For use with standard ferrules.

STEEL CONDUIT. Flexible steel conduit is made to the dimensions shown in Table 6.

NON-METALLIC CONDUIT. Non-metallic flexible conduit or loom is recommended for use as an insulated covering giving mechanical protection over insulated wire, metal tubing, or other parts requiring a water, oil, and acid-proof covering resistant to fire or abrasion. It is also recommended for use as a covering for copper or other metal tubing to prevent crystallization and to eliminate rattles. The dimensions of the standard sizes are listed in Table 7.

Alternate dimensions for the three sizes indicated in the table are for loom that can be used with standard ferrules.

22. BIBLIOGRAPHY

Books

Handbook of the Society of Automotive Engineers, New York (1946).
Factory Shop Manual, Motor Magazine, New York (1942 and 1946).
Crouse, W. H., *Automotive Mechanics*. McGraw-Hill, (1946).
Heitner, Shidle, and Bissell, *Elements of Automotive Mechanics*. Van Nostrand (1942).
Heldt, P. M., *Electric Equipment of Gasoline Automobiles*. P. M. Heldt, Nyack, N. Y.
Ricardo and Glyde, *The High Speed Internal Combustion Engine*. Science Publ., N. Y. (1941).
Stone, P. McD., *Electricity and Its Application to Automotive Vehicles*. Van Nostrand (1923).

Publications

Kent, P. J., Low-temperature-starting Developments of Automobile Engines, *J. SAE*, p. 141 (Aug. 1931).
Rabezzana, H., and D. W. Randolph, Sparking Plugs: Some Factors Affecting Their Electrical Performance, *Automotive Eng.*, p. 224 (June 1930).
Rabezzana, H., The 10-mm Spark Plug Added to AC's Metric Family, *Automotive Industries*, p. 434 (Apr. 7, 1934).
Rabezzana, H., Some Factors Controlling Part-load Economy, *J. SAE*, pp. 511, 528 (Dec. 1938).
Tognola, T. and D. W. DeChard, Analysis of Improvements in Aviation Spark Plugs, SAE Annual Meeting (Jan. 10–14, 1938).

ELECTRICAL EQUIPMENT IN AIRCRAFT

By T. B. Holliday

23. SPECIAL PROBLEMS PRESENTED BY AVIATION

ACCESSORY POWER. The main engine or engines of an airplane, in addition to delivering propulsive power, usually furnish power for the operation of accessories within the airplane, because such engines represent the minimum weight in equipment and in fuel consumption.

Accessory power must be converted into a form which facilitates distribution throughout the airplane, distributed, and utilized at remote points. Three types of power have been used: electric, hydraulic, and pneumatic. In each type will be found a generator to convert mechanical power into another form, a distribution network, and special devices to utilize the accessory power.

WEIGHT. Reduction of weight is the chief difference between designing aircraft equipment and designing other types of equipment. Weight limits the performance of the airplane in speed, range, and payload. An increase of weight empty of an airplane increases the cost of operation by a ratio of approximately 4 to 1; i.e., an increase of 1 per cent in weight empty will cause an increase of 4 per cent in operating cost. In long-range aircraft the addition of 1 lb will require an added 2 lb of fuel or even more. Weight reduction is least important in personal aircraft, which are designed for comparatively short-range flights at moderate performance levels.

EFFICIENCY. The requirement for high efficiency is second only to that for light weight, because efficiency also determines weight. Since the accessory power comes from the fuel carried by the airplane, the efficiency of a piece of equipment will determine the amount of fuel it consumes. Efficiency is most important in continuous-duty equipment. For example, a continuous-duty electric motor will consume its own weight of fuel in 5 to 10 hr. The fixed weight of the motor and its efficiency must be balanced to obtain the least total of fixed and fuel weight. Equipment having a low power rating and intermittent duty can have a low efficiency without penalizing weight. If such intermittent-duty equipment has a high power rating, however, efficiency becomes important, because the power-input requirement may exceed the limited capacity of the accesory power system. This is very important in electric motors, which have a very high power inrush and low efficiency at starting.

RELIABILITY. Aircraft equipment must be reliable despite the need for minimum weight. The only concession made to weight reduction is a lowering of useful life. If the main engines are overhauled after perhaps 1000 hr of service, an opportunity is then provided for inspection or overhaul of some other pieces of equipment. Between such overhaul periods the equipment must be utterly reliable. The service life of equipment is thus dependent on the life of engines and the airplane. Continuous-duty electrical equipment should be able to operate without attention for 5000 hr if possible. Lower values will increase the number of inspections and replacements of equipment.

MAINTENANCE. Consideration of maintenance problems is a design requirement for aircraft equipment because aircraft operators practice preventive maintenance. Frequent inspections and replacement of items to prevent failure are the rule in aviation. The engineering design should be such that these inspections are made easily or are unnecessary.

COST. In aircraft the cost of equipment has less relative importance than in any other form of transportation. Any requirement for cost comes after those which have already been discussed. Only in personal aircraft is cost important. In any type of aircraft cost of equipment will be considered when all other factors are equal.

ENVIRONMENT. The conditions under which aircraft equipment must operate have led to the development of many environment-free types of electrical equipment.

The most obvious special operating condition for aircraft equipment is altitude. The density and temperature of air decrease with increasing altitude. The mass of air will decrease in proportion to density, but this lowering of cooling effectiveness is partially compensated for by the lower ambient temperature surrounding most pieces of equipment. Some items will be so located within the airplane that the temperature of the surrounding air does not decline as fast as does that of the outside air. Examples include equipment installed in heated passenger compartments and in engine nacelles. In the latter, the ambient temperature surrounding a piece of equipment may actually rise with altitude, because the decreased cooling effectiveness of the upper air does not adequately carry off the losses of the engine. In still air an average value for the cooling effectiveness of air at 40,000 ft is 40 per cent of that at sea level. If the ambient temperature is higher than normal for that altitude, cooling will be correspondingly reduced.

Advantage can be taken of the speed of the aircraft to use forced cooling, thereby utilizing a greater mass of air. By this means the effectiveness of cooling can be increased to approximately 80 per cent of the sea-level value. It must be realized that this method gives added cooling by adding to the aerodynamic drag of the airplane. The ram pressure available by this method varies from more than 20 in. of water to less than 1 in., depending on speed, the length of the air duct, the point from which the air is taken, and the design of the duct.

During climb and descent of an airplane, with resulting change in air density, a piece of equipment will "breathe" unless it is hermetically sealed. This breathing carries moisture and contamination into all parts of a mechanism. Therefore all parts of the design must be carefully protected against corrosion.

The ionization potential of air decreases very rapidly with altitude. Surface-leakage-insulating values decrease in proportion, and it becomes necessary to use higher spacing or creepage distances between points of opposite polairty in aircraft equipment than is necessary for ground equipment. Figure 1 includes several charts which show the effect of altitude on the factors discussed.

A fourth effect of altitude is that some types of carbon brushes will dust away in a very few hours or even minutes of service at altitudes above 35,000 ft. It has been necessary to add special ingredients to the brushes to obtain normal brush life. To procure suitable brush material the electrical designer should specify operating conditions.

Commercial aircraft can operate efficiently at altitudes to 40,000 ft, and military aircraft at more than 60,000 ft; pilotless aircraft can be expected to achieve ultimately altitudes of 500,000 ft.

Aircraft equipment must face extreme limits of temperature, both high and low. Specifications for aircraft equipment usually require satisfactory performance at 160 deg fahr. This value was determined by adding the highest temperature encountered on the ground, approximately 130 deg, and the rise of temperature due to absorbed solar radiation. Within the continental United States the maximum temperature will be approximately 20 deg less. In flight still higher temperatures can be encountered in the engine nacelles. Ambient temperatures therein can reach values of 250 deg fahr or even higher in high-performance military aircraft, where a maximum of power is compressed into the smallest possible volume.

Extremely low temperatures can be encountered on the ground and in flight. Normal specifications require performance at 65 deg fahr below zero. Still lower temperatures have been encountered on the earth's surface, but the limit as stated is adequate for practically all aircraft operations. In flight over the tropics temperatures lower than 110 deg fahr below zero have been recorded. With continuous-duty electrical machinery, the ground minimum is the worst operating condition. Since the equipment operates in flight, its own losses nullify the further lowering of temperature. Intermittently operated equipment which may be called on to function at extreme altitudes should be designed to start and operate at the lower temperature. Low temperatures seldom interfere directly with the operation of electrical machinery. Troubles such as unequal contraction of different metals should be anticipated in the design. Electrical items can be adversely affected, however, by lowered performance of associated equipment. For example, electric motors are often used to drive screw shafts and oil pumps. In either of these, low temperatures will increase the viscosity of lubricants and oil until the torque required may exceed the capabilities of the motor.

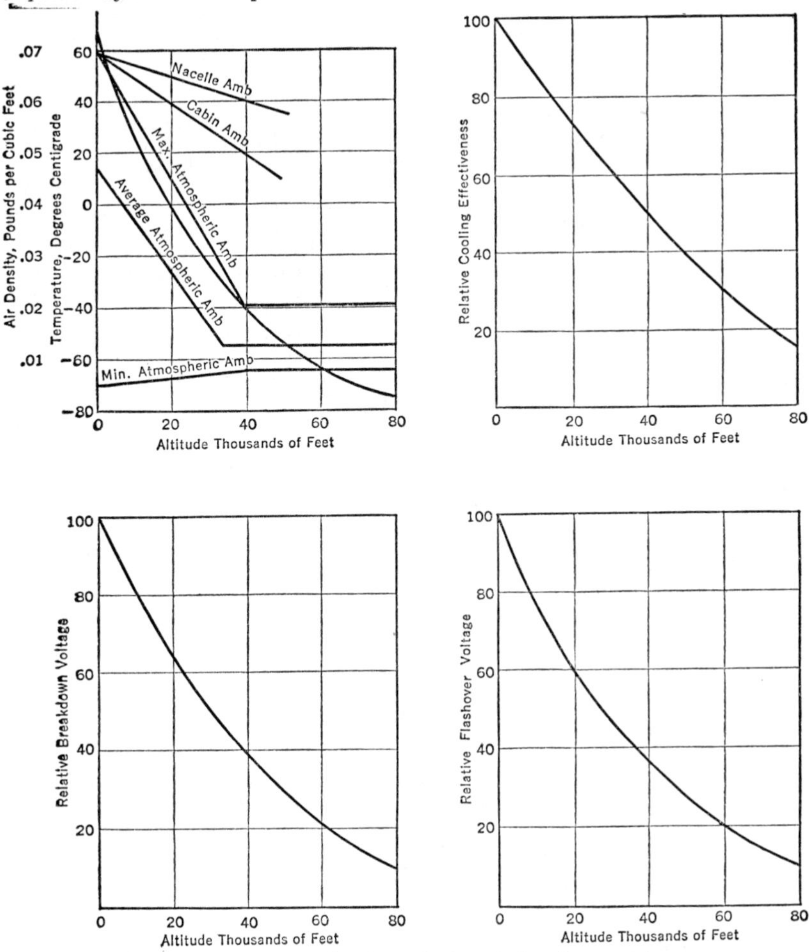

FIG. 1. Effect of Altitude on Cooling and Insulation Characteristics

A fourth environmental condition is the contamination which follows exposure to sand, dust, dirt, and special chemicals which are used to clean the airplane and engine. Undoubtedly, this exposure of aircraft equipment is not as severe as for similar items in other forms of transportation. Careful design to reduce weight may, however, eliminate needed protection from dust and dirt. For example, the reliability of sensitive relays could be jeopardized by open construction, which would admit dust.

Another environment, which occurs especially in the tropics, is excessive humidity. Accelerated corrosion is caused by humidity. The combination of breathing with ascent and descent and of humidity will place moisture in the innermost parts of mechanisms and create contamination and friction in sensitive pivots, critical bearings, and control devices. Fungicidal growths are a by-product of humidity. Fungi will corrode textile materials and metals, and some forms of algae will etch glass. Hermetic sealing will not necessarily prevent fungicidal growth, since many types of fungi are resident but dormant at the point where equipment is manufactured. Such fungi will become active when the equipment is used in areas where local temperatures foster their growth.

The sixth environmental condition is localized in seacoast areas and concerns exposure to salt air. Moisture and salt produce a most corrosive combination, and furthermore a deposit of salt will conduct electricity, thereby impairing insulation. Salt will cause corrosion between dissimilar metals in proportion to the difference in electrolytic potential between them. In aircraft electrical equipment, copper and aluminum are often used in the same piece. To prevent corrosion such parts must be separated by an intermediate metal. The surface creepage distance across insulating materials must be increased to prevent failure by arc-over or leakage currents. Hermetic sealing will protect against salt corrosion in some types of accessories, such as sensitive relays, but larger parts, such as generators, cannot be sealed. In fact, the generator, one of the most rugged of aircraft electrical accessories, is very seriously affected by salt corrosion. It cannot pass as severe a test as can most other parts. Neither can sealing be applied to junction panels and connectors, where the other weakness, insulation breakdown, can occur. Adequate inspection and cleaning are the best solution for these parts.

Vibration is another condition that characterizes almost any part of the airplane. It is most severe on the engine but is present on any portion of the structure. Frequencies ranging from a few cycles to 3000 per second can be found, and accelerations will be as high as 10 g on structure. On the engine-accessory section, the accelerating force will sometimes reach values of 75 g on some types of engines. Most accessories should be designed either to avoid or to withstand critical frequencies from 10 to 3000 cycles per second. They should also withstand accelerating forces of 15 g at those critical frequencies. Engine-mounted accessories should withstand accelerating forces of 75 g, and any critical frequency must lie above or below the operating range. The generator is driven by the engine and is subject to torsional vibration in addition. It is necessary to add a damper within the generator to absorb such torsional vibration and the accelerating forces of a critical frequency in case the operating period goes through such a critical frequency.

STANDARDIZATION. Standardized requirements are published in Army-Navy specifications and standard sheets, in Civil Aeronautics Administration specifications, and in the SAE handbook. Such data must be used by the aircraft electrical designer and should be employed to simplify the work of the airplane user.

24. POWER SYSTEMS

TYPES. Electrical power systems in aircraft include both d-c and a-c types. The d-c system has one or more generators as the basic source of power, with a storage battery floating on the line to absorb peak surges of power and to serve as an emergency source of power. The system voltage varies with the size and use of the airplane to give a minimum weight to the electrical system.

Developments in a-c systems have had two results. A small power system has become a normal part of the electrical circuits in d-c aircraft. As much as 25 per cent of the d-c power normally generated in the airplane has been supplied to converters, which delivered a-c power at frequencies from 60 to 800 cycles per second, depending on the needs of special apparatus. As these special loads were standardized, it became possible to supply their power needs with a miniature a-c network in which the converter or inverter became the source of power. Most large aircraft have a primary power system which is d-c and a secondary system that is a-c.

In very large aircraft it has been possible to consider a-c power as the primary power. Experimental work has explored the advantages of several operating frequencies from 200 to 800 cycles per second and has shown that 400 cycles per second is a desirable compromise. Apparatus designed for the industrial frequency of 60 cycles per second is too heavy for use in aircraft.

DIRECT-CURRENT SYSTEM. In practically all d-c systems, the source of power is a variable-speed generator that is driven by and mounted on the accessory section of the main engine. The rating of the generator must equal or exceed the highest load under the worst operating condition. Most designs of multiengine aircraft can operate safely with only half the engines. Therefore the generators must have such capacities that half their number can supply the maximum load. In normal service the generators will be used at a small fraction of their capacity, and the demand factor will range from 25 to 50 per cent.

The generator is geared to the crankshaft by step-up gearing, whose ratio is such that a standard speed is produced at the rated speed of the engine. The operating speed of the generator at the rated speed of the engine is 3750 rpm or 7500 rpm, depending on the choice of the engine designer. The lower value was used before 1940, and the higher value has been used in most large engines since 1943. The change in speed was brought about by the increasing demand for electrical power and an equally insistent restriction on the space that was available on the accessory section of the engine. Generators having ratings as high as 12 kw can be mounted on the engine. Larger units are mounted on aircraft structure and are driven by a flexible shaft.

The generator is rated at the lowest operating speed which is expected in normal service of the aircraft. This will be the minimum cruising speed, and it will be 55 to 60 per cent of the engine design speed discussed. The rated speed of generators has dropped steadily with improvements in aerodynamic efficiency, since these improvements have made it possible for the airplane to fly at lower engine power ratings. For example, the rated speed of the "low"-speed-range generators was 2600 rpm in 1938 and 2200 rpm in 1943.

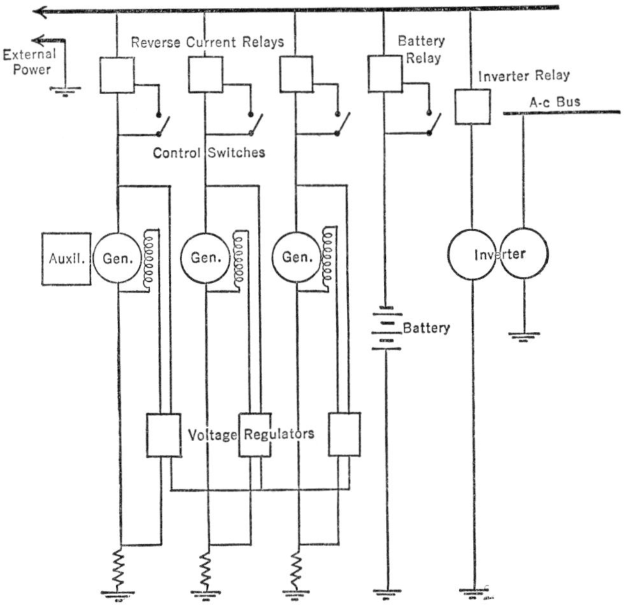

Fig. 2. Multigenerator D-c Power System

The trend toward lower operating speeds is still more pronounced in large transport-type aircraft. Aerodynamic efficiency is so high in these aircraft that the engines can be reduced to 40 per cent of rated speed. The trend toward lower speeds will be altered by the internal-combustion turbine, whose power output has an exponential ratio of three instead of a linear ratio with respect to speed. Generators driven by these engines will seldom be operated at less than 75 per cent of rated speed.

The generator must also withstand a take-off speed of approximately 125 per cent of rated speed. To this must be added a "growth" or improvement factor in the engines. As experience is accumulated on a new engine, weaknesses are eliminated, and it is found possible to raise the rated speed and thereby increase the power rating. Such increases also affect the generator, since it is inadvisable to change the gear ratio. Therefore a "high"-speed generator may be required to deliver rated power at 4000 rpm and partial power at 3000 rpm in one airplane, and to withstand a diving speed of 11,000 rpm in a military fighter airplane. The speed range of aircraft generators is not as wide as that of automotive types, but it is steadily broadening.

The system voltage will depend on the size and purpose of the airplane. The 6- to 8-volt automotive system is quite suitable for small personal aircraft. Larger aircraft, up to multiengine types of approximately 20,000 lb, can use a 12- to 15-volt system. A 24- to 30-volt system will satisfy the needs of larger airplanes up to four-engine types perhaps 100,000 lb in gross weight. The reference to two voltages is made to show the rated voltage of the battery-generator combination. The next step is the 110- to 120-volt system. At this voltage the battery is impractical, and its place is taken by an auxiliary engine-generator power unit. One hundred and fifteen volts is the regulated or network voltage and corresponds to the 28-volt regulator setting of the 24- to 30-volt system. System voltage determines the weight of distribution wiring. Therefore, as the size of the airplane increases, both the power loads and length of the circuits increase. These cause a disproportionate increase in wiring weight unless the system voltage is increased. The weight of wiring should not exceed 10 per cent of the weight of the electrical installation. It has been as high as 25 per cent in some aircraft, and this is an unnecessary weight penalty.

The system voltage is controlled by regulators for each generator. In d-c systems having a battery and a generator, the voltage setting was usually about 15 per cent above

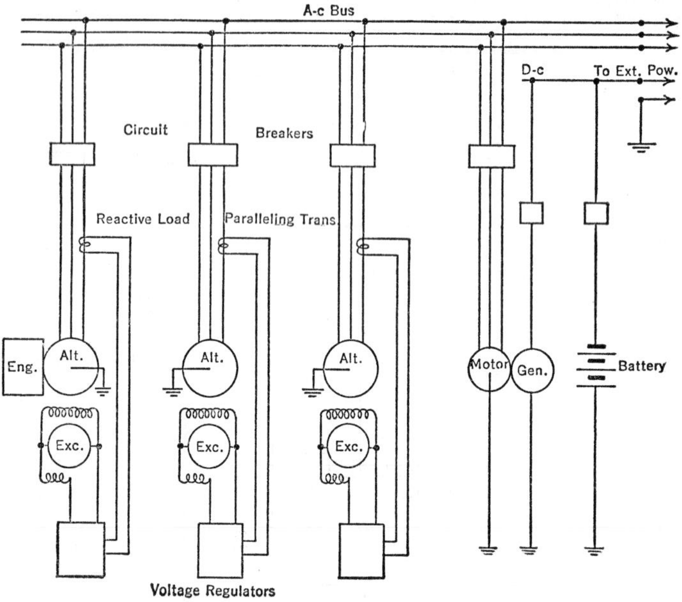

Fig. 3. Three-phase A-c Power System

the battery voltage. In turn, the rated voltage of the generator is approximately 7 per cent above the regulated voltage to allow for line drop in the supply bus and to give some margin of performance during emergency or overload.

Regulators also include provision for division of load between the units in a multigenerator system. Such parallel operation is important in aircraft because it provides one source of power for all the electrical system, and because it permits the use of smaller generators in multiengine aircraft.

To permit operation of essential accessories while the airplane is on the ground, a connection from the power bus wiring to an external plug is provided on all but personal aircraft. Use of ground power for such service avoids the necessity for a large and heavy battery within the airplane.

Power is distributed throughout the airplane by a bus network, which connects all sources of power to the distribution circuits. Basically the bus is quite similar in all types of aircraft.

Figure 2 gives the schematic arrangement of a multigenerator d-c system.

ALTERNATING-CURRENT SYSTEM. The need for a-c power grew with the increasing application of electrical power to radio and control problems. Radio equipment needed higher voltages, which were not easy to attain through the medium of d-c motor-generator sets but were easily reached with a transformer. The accurate control needed for effective automatic pilots and remote operation of guns in military aircraft was easily supplied by a-c circuits. In aircraft whose primary system was d-c, the a-c power needed for a given piece of apparatus was supplied by an inverter. As the number of the latter increased they were merged into two or more units that supplied an a-c network.

In very large aircraft the source of power was engine-driven alternators instead of inverters, but the problems of the a-c system were quite similar. Although separate engines were studied and tested as prime movers for the alternators, they could not compete with the larger main engines for weight or fuel economy. Variable ratio couplings were developed to connect the main engine to a constant-speed alternator. The combined drive and alternator then became a source of power comparable to the d-c generator discussed.

The speed range through which the combination must supply power is higher than for a d-c generator, because a battery cannot be used as a reservoir of standby power. The drive should deliver rated and constant output speed with a variation of input speed of approximately 5 to 1, to include all speeds from idling to maximum. The minimum requirement is a spread of 3 to 1.

The place of the battery in the system is taken by an auxiliary engine-generator unit, which serves as a source of ground power and of emergency power in flight. Its rating must be such that it alone can supply the largest single load to be encountered during ground operation. A plug for connection to a source of external power is also used with a-c systems.

Systems proven by flight tests include 120-volt, single-phase, 800-cycle; 120-volt, three-phase, 400-cycle; and 208/120-volt, three-phase, four-wire, 400-cycle systems. The last-named has shown greatest promise. The fourth and neutral wire is structural ground in metal aircraft. British developments along parallel lines have included a three-phase, 250-cycle system.

The power rating of individual alternators increased from 6 kw to 50 kw in the period from 1937 to 1945. System capacities of 120 kw have been installed in single aircraft. Still higher capacities will be needed to meet steadily increasing demands for more accessories and more comfort in large passenger transports.

To provide proper operation in parallel it is necessary to control both voltage and speed. Governors for the variable ratio drives include circuits responsive to load and to the share of the load carried by the individual alternator. The governors, in response to load, will lower the operating speed from 105 per cent of rated at no-load to 95 per cent of rated at 150 per cent rated load. This drop in speed facilitates accurate division of load. The performance of the governor controls division of real load.

Division of reactive load is the function of the voltage regulator. It is accomplished by a circuit that is responsive to the current flowing from its alternator, and the proportion of that current to the total load on the system.

Both loop and radial types of networks have been studied and tested. The radial type has many advantages for aircraft. The loop type may have advantages of less vulnerability in military aircraft, but a radial type can be so installed as to best use the structural features and protection of the airplane and thereby overcome some of this advantage.

A schematic arrangement of an a-c system is given in Fig. 3.

25. SOURCES OF POWER

GENERATOR. The weight of generators is plotted against power rating in Fig. 4. Data given in this illustration are based on minimum speed ratings of 2500 and 4400 rpm, which correspond to the speeds at rated power of 3750 and 7500 rpm. The maximum speeds at which these generators will be expected to operate are 5000 and 10,000 rpm respectively. Voltage rating has little effect on weight, since the length of the commutator and brushes is the only dimension that is directly proportional to voltage.

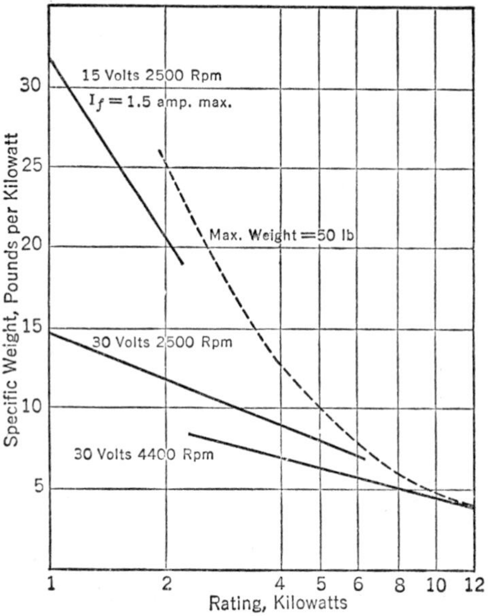

Fig. 4. Weight of Engine-driven D-c Generator

All generators in common use have a shunt field with regulator control. At current ratings of 100 amp and higher it becomes necessary to add interpoles for control of armature reaction. Very little series field can be used to avoid the danger of reversed polarity in case the circuit to the battery is accidentally closed. A comparison of the output constant for aircraft-generator designs and for industrial practice is given in Fig. 5. The output constant is a valuable measure of design, since space is so limited on the accessory section that the maximum output per cubic inch is essential.

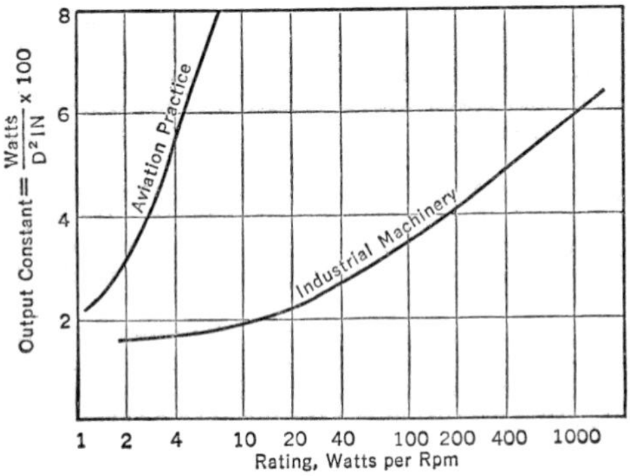

FIG. 5. Comparative Design Practice

The greater output constant of aircraft machines is partially explained by the higher densities used in conductors and brushes, as shown in Fig. 6. The higher current densities increase the losses per pound, but the weight of active material decreases at a rate that compensates for much of this loss, and the overall efficiency of aircraft designs is relatively high. The output constant is also improved by working the magnetic sections at the highest possible values. Armature laminations are made from transformer grade steels to minimize iron losses at high flux densities and to reduce the hysteresis and core losses at the relatively high internal frequencies used in the d-c generator designs. At the higher ratings the most successful low-speed designs have six or eight poles, whereas the high-speed units have four or six poles. The frequency of magnetic reversals in the armature varies from 80 to 300 cycles per second, depending on the design and speed.

The mechanical design is very important at three points. The first is the mounting flange. If the weight of the generator exceeds approximately 35 lb, the mounting flange should be of cast or forged steel, and the design should be such that high local stresses are avoided. The commutator end housing is equally important, in that it must absorb the forces of acceleration that are applied to the weight of the armature. Finally, the

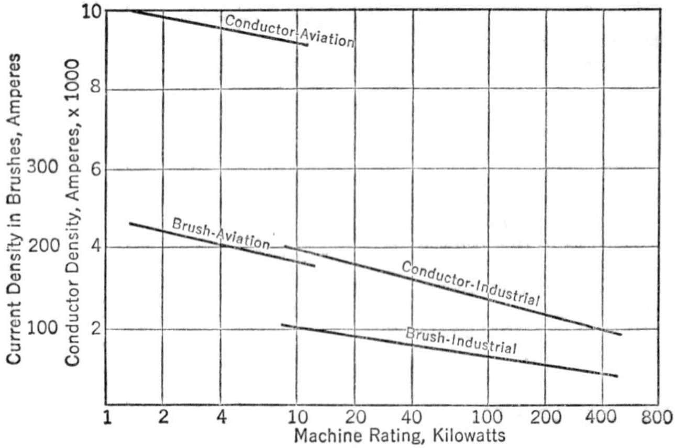

FIG. 6. Comparative Design Practice Regarding Current Density

method of absorbing torsional vibration in the driving member and the method of damping critical vibration are very important. In some designs a quill shaft is used to absorb the vibration. The armature is driven by the shaft at the commutator end. Damping is obtained by a braking effect between the shaft and an asbestos wrapping between the shaft and the inner shell of the armature. Other methods include two plates coupled with rubber and two plates coupled by a non-resonant combination of springs.

Forced or ram cooling of the generator is necessary if it is to operate at altitudes above

20,000 ft. The coupling for a 2-in. air duct is provided at the commutator end. The cooling air leaves the generator through apertures in or near the mounting flange. The generator must be designed for a minimum ram of 6 in. of water pressure. Although pressures of 20 in. and more are usually available, approximately half is lost in the ducts to the generator, and another 4 in. are lost in back pressure within the engine nacelle, into which the generator cooling air must exhaust.

The performance of typical generators of the higher ratings is shown by the curves of Fig. 7. The overload capacity of compensated designs is a considerable advantage, in that storage-battery capacity can be reduced.

An installation detail which must be mentioned concerns uniform tightening of the mounting studs. The six studs on which the generator flange are bolted are not accessible, and on some engines one or two of them may be virtually impossible to reach without removal of another accessory. It is a temptation under such circumstances to forget to tighten those stud nuts or to tighten them improperly. All six nuts must be tightened to a uniform and correct amount with the aid of a torque wrench.

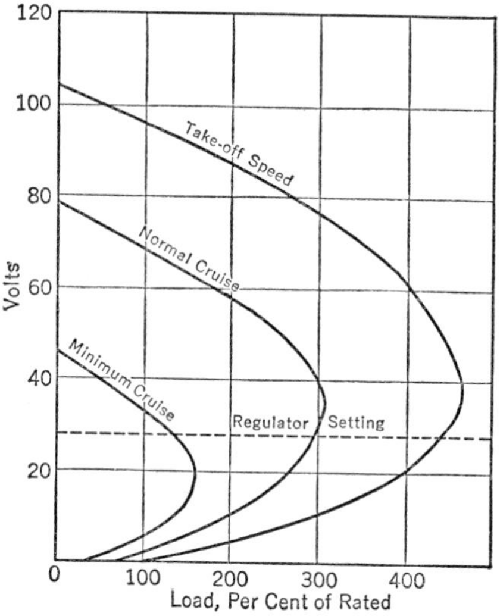

FIG. 7. Characteristic of D-c Generator

VOLTAGE REGULATOR. Accurate and reliable control of generator voltage gives unexpected weight savings in other items of equipment. For example, it has been possible to design a better storage battery, which has more power capacity per pound, because accurate control of voltage is given by the regulator. Lamps have a design voltage very close to actual operating voltage because the regulator maintains close limits. Control apparatus can be designed to function within narrower voltage limits and thereby improve accuracy as well as reduce weight.

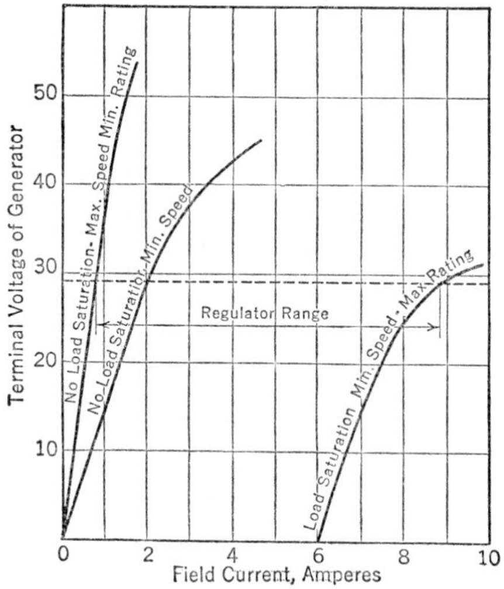

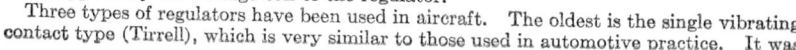

FIG. 8. Generator Characteristics Which Define Voltage Regulator Requirements

The voltage regulator functions by inserting a resistance in the shunt field circuit of the generator. A preferred design would be one whose resistor is continuously variable from zero to its maximum resistance. The maximum value will depend on the generator design. The saturation curves of two generators at no load and rated load are given by the curves of Fig. 8. The two generators selected represent low and high ratings. A regulator which will serve both must be able to limit field current to the minimum required by the small rating at no-load and high speed and yet pass the current needed by the larger generator when supplying its rated load at its minimum speed.

A second function of the regulator is control of the division of load. This is accomplished by a paralleling coil, which responds to a signal in proportion to the load being delivered by its own generator, as compared to that being delivered by others, and which supplements the control being applied by the voltage coil of the regulator.

Three types of regulators have been used in aircraft. The oldest is the single vibrating contact type (Tirrell), which is very similar to those used in automotive practice. It was reasonably satisfactory when the generator-voltage rating did not exceed 15, and the maximum field current did not exceed 1.5 amp. The best available alloys of precious metals would seldom give a life without attention of more than 100 hr. When the generator-voltage rating was increased to 30, it became virtually impossible to design a satisfactory vibrating contact type of regulator. Units having three successive contacts were not satisfactory. At the same time it became necessary to increase the field current of the generator to achieve satisfactory reduction in weight.

The multicontact design was the first answer to this need. As many as twenty contacts were used in either a parallel or a series arrangement. Instead of interrupting a field voltage of approximately 30 volts, each contact now opened only one-twentieth of this amount, and soft silver contacts were quite adequate. Such contacts could also control field currents as high as 8 amp. This regulator can be described as a step type.

The carbon pile design is the only continuously variable type that is used in aircraft. The resistance element is a stack of carbon disks, which are compressed to a point of minimum resistance by a leaf spring. As the generator voltage increases, the solenoid coil exerts a magnetic force contrary to the spring and by reducing the spring pressure loosens the carbon stack and increases resistance. The finned housing around the stack radiates the heat lost in the resistance element. The second coil of the solenoid and the one having the larger wire is the paralleling or load-equalizing coil. The major disadvantage of the carbon-pile type is its high minimum resistance. It is difficult to obtain a minimum value as low as 0.5 ohm while retaining a maximum value of more than 50 ohms.

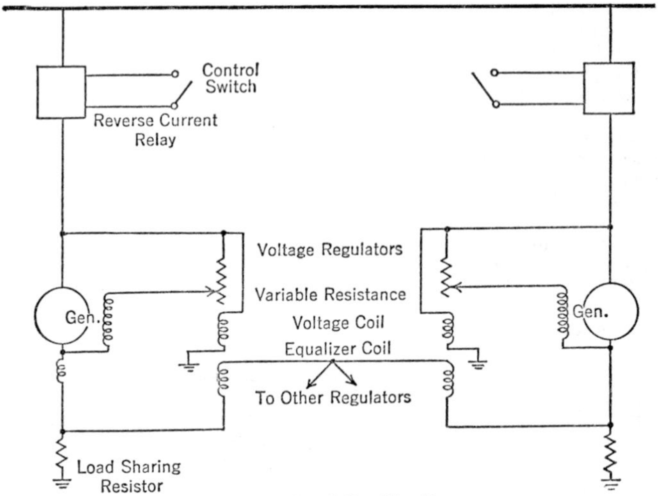

FIG. 9. Paralleling Circuits

A schematic arrangement of the regulator, showing its relation to its own and other generators, is given in Fig. 9. The paralleling connection should be noted particularly. The signal for load division must come from a shunt connection in the generator output circuit. The arrangement shown has a separate resistor in the negative circuit from the generator. The resistor can be the series field of the generator and can be located in the positive side of the line. The principle of operation will be the same. When the division of load is not equal, current will flow in the equalizing coils in such manner as to neutralize the difference. Generators of different ratings can be operated in parallel, with proportionate distribution of load, by selecting values for the series resistances such that the voltage loss therein is the same at the rated load of each generator.

REVERSE CURRENT CUT-OUT RELAY. The basic function of the reverse current cut-out is the protection of the generator. Two additional duties have been added in some designs. To protect the generator, the cut-out will not close until the terminal voltage of the generator exceeds the line voltage, and it opens when the generator voltage falls below the line voltage. The opening of the cut-out is caused by the flow of "reverse" current from the line into the generator when the generator voltage is low. A third function is remote control of the generator circuit. This is easily provided by placing a switch in the coil circuit of the magnetic contactor. Some designs have also been polarized, so that the cut-out refuses to close if the generator voltage is reversed in polarity.

When the generator rating does not exceed 50 amp, the cut-out relay can be a relatively simple single unit similar to those used in automotive engineering. If the generator rating is 100 amp or more, the weight of the wiring assumes considerable importance, and it is advantageous to design the cut-out relay as a remote-control device also. Then it can be located in the circuit at a position which will reduce the amount of heavy cable that is required. With this additional requirement a better design can be achieved by dividing the cut-out into two relays, a pilot control relay and a magnetic contactor.

The pilot relay is the control element of the cut-out. It has two coils, the voltage coil and the reverse current coil. In one form the voltage coil measures the generator voltage, and in another it measures the difference between generator and line voltage. The first is called a voltage type, and the coil acts to close the relay contacts against the opposing force of a spring. The spring tension is usually set so that the closing voltage is approximately 110 per cent of battery voltage. The reverse current coil has relatively few turns, only two in some designs, and it acts to assist the voltage coil when current is flowing from the generator to the line. When the speed of the generator is reduced, a point will be reached where it will not be able to maintain line voltage. Current will then flow in a reverse direction from the line into the generator. If the relay has been designed for a particular generator rating, the reverse current need not exceed 5 per cent of this rating. However, one design of cut-out relay is suitable for several ratings of generators, and experience has shown that the reverse current required for opening the cut-out can be as high as 20 per cent of the generator rating. This situation arises only when a relay suitable for a large generator is used with a small one.

The contacts of the pilot relay close the circuit of the coil in the magnetic contactor. Polarization of the cut-out can be achieved by placing a rectifier in the contactor coil circuit or in the pilot relay voltage coil circuit, or by magnetic polarization of the pilot relay structure.

The differential type of pilot relay is similar in appearance to the voltage type. Since it operates on a difference in voltage between the generator and line, the coil must be designed for a much smaller voltage. A common value is 0.5 volt. The pilot relay will close when the generator voltage exceeds line voltage by 0.5 volt. In a multigenerator system, the first generator will come on the line at 0.5 volt above battery voltage. The second generator will come on the line when its voltage exceeds that of the first generator by the same amount. This would seem to keep the second generator off the line until the first generator was carrying more than rated load. This does not occur, because the paralleling circuit acts to equalize the load, and the amount of its control exceeds the differential control. The combination of the differential type of cut-out relay and the paralleling type of voltage regulator provides a stable and trouble-free system. The differential design does away with the need for adjustment, since a considerable variation in the 0.5 closing voltage will be negligible in comparison with the rated voltage of the generator.

To protect the low-voltage coil, a variable ballast resistance, such as a lamp, is needed in series with it. When the generator is delivering rated voltage, with the cut-out relay open, and the line voltage is zero, full generator voltage will be applied across the differential coil. The lamp will absorb much of the excess voltage.

The magnetic contactor is the power relay and its contacts open the generator circuit. At high current ratings the design of the contacts is most important. Bridging contacts have proved best. It has already been noted that the ionization potential at 40,000 ft is a fraction of the sea-level value. Experience has shown that a single contact will not open a 30-volt circuit at that altitude, and that a current of 200 amp will flow across a $^3/_8$-in. gap. The bridging type has two gaps in series and will open 30-volt inductive circuits.

To obtain maximum performance from the cut-out relay, two types of contact materials are needed. The first should be a good conductor to minimize voltage loss across the cut-out relay. The second should be a tough material to serve as the arcing contact. Both are used in the design outlined in Fig. 9.

Two aspects of mechanical design are quite important. The cut-out relay is subject to severe vibration, since it is installed in the engine nacelle in many cases. The armatures of both the pilot relay and the contactor must be statically and dynamically balanced to avoid accidental closure or opening. Bouncing and chattering of contacts are also avoided by balanced design. Since the pilot relay is quite sensitive, sand or dirt can easily impair its operation. For this reason a sealed enclosure should be placed around the cut-out. The design should be such that the position of the relay is not restrictive.

The cut-out relay in the generator circuit can be located at any point between the generator and the point at which the first load is taken from the positive bus. In single-engine aircraft the relay will usually be found in the engine nacelle because the battery and the starter circuits are connected to the bus at that point. The same practice is usually followed in multiengine aircraft, but a revision of this practice is desirable when the number of generators exceeds four. To gain access to the relays in flight, it is advantageous to bring the bus into the fuselage and feed the loads through a radial network. By concentrating the point of load distribution at a small loop in the center of the airplane, the voltage regulation can be improved considerably. Since the voltage loss from the generator to the point of voltage regulation does not affect the voltage delivered to the loads, it is possible to use somewhat smaller conductors for this section of the wiring and thereby compensate for part of the weight penalty which results from the radial feeder network. Of course, this gain in weight is obtained at the expense of the increase of generated voltage and a loss in overall efficiency.

BATTERY. Storage batteries or accumulators commonly used in aircraft are of the lead-acid type. They are connected in parallel with the engine-driven generator. The battery absorbs load surges, acts as a suppressor of radio noise, and can serve as a source of power in emergency. The time through which it can supply power will depend on the load and on temperature.

Aircraft batteries are made up of 3, 6, or 12 cells for use with 6- to 8-, 12- to 16-, or 24- to 28-volt systems, respectively. Characteristic data for several batteries are shown in Table 1.

Table 1. Dimensions and Weights of Aircraft Batteries

Voltage	Ampere-hour 5-hr Rate	Max. Weight, Pounds	Approximate Dimensions			
			Length	Width	Width plus Terminal	Height
12	17	27	7 3/4	5 1/8	8 1/2	9
	34	40	12	5 1/8	8 1/2	10 3/8
	68	71	17	6 1/8	9 3/8	10 3/8
	88	78	14	7 1/4	10 1/2	11
	105	91	14	7 1/4	10 1/2	11
24	11	34	9 3/4	7 5/8	11	7 3/4
	17	52	7 3/4	10	13 1/4	8 3/4
	34	76	9 3/4	10	13 1/4	10 1/4

In construction, aircraft batteries differ from automotive and other types only in the effort to reduce weight. A secondary emphasis on the ability to deliver current in large amount has become more important with the development of the jet engine and electric starters for use thereon. To meet both requirements the plates are made as thin as possible. Similarly, to reduce the space occupied by the battery, the separators are also made as thin as possible. In the high-capacity battery used with the jet-engine starter, these separators are porous glass cloth. A battery designed to meet such requirements for performance is more fragile than industrial counterparts and cannot withstand electrical abuse to the same extent.

Few batteries listed in Table 1 will have a specific weight of less than 100 lb per kwhr. This ratio is more favorable at shorter time intervals and is theoretically less than 1.5 lb per kw-min for the larger batteries. It has already been explained that the specific weight of engine-driven generators varies from 20 to 5 lb per kw, depending on rating. Therefore, when the load duration exceeds 2 to 6 min, it is advisable to supply the load from additional capacity in the generators, rather than from batteries. The point at which a minimum total weight of generators and batteries will be achieved can be determined by careful analysis of loads, generator capacity, and battery characteristics.

The acid used in batteries is a serious hazard because it will destroy and weaken structure. Batteries installed in aircraft which perform aerobatics must have a special device in the cell-vent plugs which will automatically seal when the battery is inverted.

It is necessary to convey gases from the battery to an exhaust point, where they will not damage the aircraft. To do this, the battery is enclosed in a case which is vented at two points. One of these openings is connected to a point which has a slight pressure during flight. The second vent is connected either to an exhaust point which clears the airplane both in flight or at rest or to an alkaline solution which neutralizes the acid and releases harmless vapors.

The performance of a typical cell, as used in aircraft batteries, is portrayed by the curves of Fig. 10. The ability to deliver high discharge current is of primary importance. It will be noted that the cell will deliver little current at any voltage above 2.0, and it is

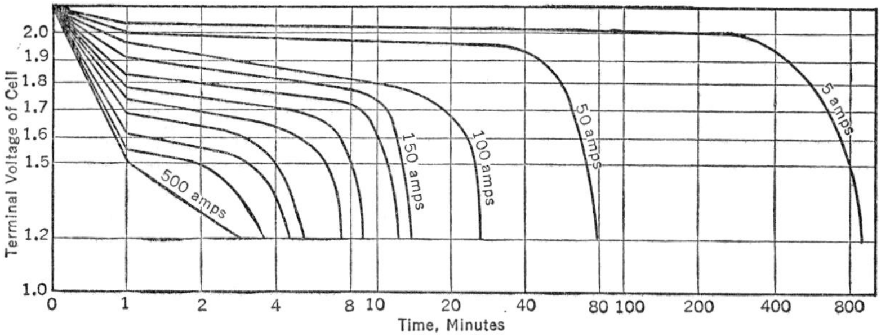

FIG. 10. Characteristics of Lead-acid Battery Cell

at the end of its usefulness when the terminal voltage has dropped to 1.5 volts. The battery will not contribute to any load until the generator voltage falls to about 75 per cent of its normal rating.

The effect of temperature (Fig. 11) is to decrease capacity as temperature is lowered. Batteries are normally rated at a temperature of 80 deg fahr. A battery has very little capacity at a temperature of −40 deg, although it will recover normal capacity when warmed to normal temperatures. The specific gravity of the acid solution should be adjusted for the climate in which the airplane is normally operated. A reading of 1.300 is suitable for the continental United States. In tropical climates a value of 1.275 is satisfactory, and in arctic winters a reading of 1.330 will improve the performance of the battery.

The battery should be located near the electrical load center of the system and near the heavy feeder lines. Since it is a very heavy unit that requires frequent replacement or inspection, its location must be accessible and must include space for measuring specific gravity and adding liquid to the electrolyte.

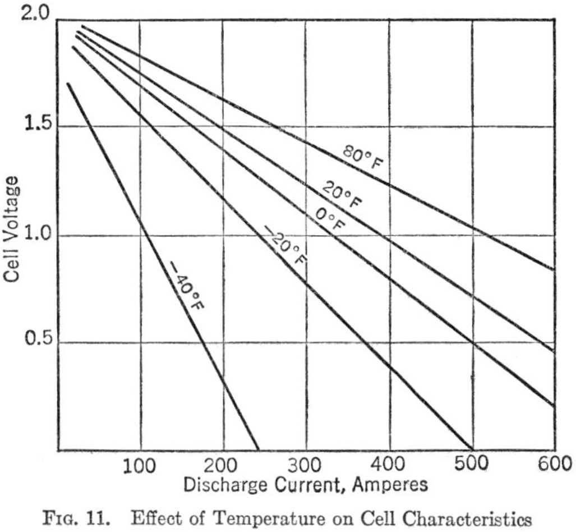

FIG. 11. Effect of Temperature on Cell Characteristics

AUXILIARY POWER UNIT. Since the battery is a heavy source of power for any load whose duration exceeds a few minutes, much research has been devoted to the design of engine-generator units that may be a substitute for or supplement to batteries. The power unit has several advantages. Within its capacity it will maintain the system voltage at the value normally supplied by the main engine-driven generators. All items in the electrical system can then be designed for a much narrower range of voltage than when the system may be supplied by either a generator or battery.

The engine-generator unit will deliver its rated load indefinitely, as it does not have a limited capacity with respect to time. The weight of the engine-generator unit is considerably less than that of equivalent batteries, unless the comparison is based on load duration of a very few minutes. Units having a continuous rating of 5 kw and a 5-min rating of 7.5 kw weigh less than 125 lb and have been used extensively. Refinements in design can reduce this weight to little more than half the above value.

The engines used with the auxiliary units must include features of aircraft-engine construction. To reduce weight it is imperative to select rather high operating speeds.

Air-cooled designs are favored for their simplicity and lower weight. In practically all cases the engine is designed for constant-speed operation. Variable-speed operation would permit the design of a much lighter engine, but it is difficult to obtain a governor which will accelerate the engine before the generator has demanded too much torque. A constant-speed design does not match the torque characteristics of the generator, and it is possible for the generator to overpower the engine.

The auxiliary unit is not suitable as a source of primary power. The rating described above weighs approximately 25 lb per kwhr. A generator mounted on the main engine with its proportionate share of the engine weight will weigh about 8 lb per kwhr, less than one-third as much as the auxiliary unit. Furthermore, the fuel consumed by the auxiliary unit will be 1.0 lb per hp-hr. The corresponding value for main engines is little more than 0.4 lb per hp-hr. Of course, higher power ratings will reduce the specific weights given, but those weights are not likely to become less than that of the main engines. The only power unit which promises competitive weights is the internal-combustion turbine, and this is due to the possibility of using very high operating speeds.

INVERTER. Applications for alternating current have increased until it is necessary to convert a portion of the aircraft electrical power from d-c to a-c. Such applications include radio, radar, precision controls, and instruments for remote indication. During initial development of the equipment it was common practice to include an inverter as an integral part. Fortunately, efforts toward standardized ratings of the local sources made it possible to merge them into one system, thereby making a secondary a-c system within the primary d-c system.

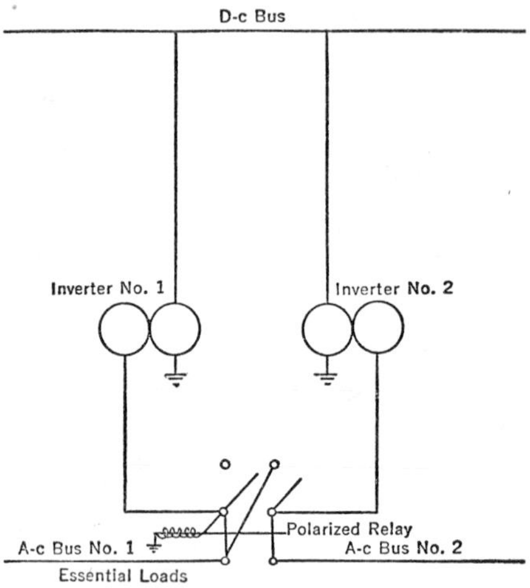

Fig. 12. Inverter Power System

Trends of standardization have been toward a 115-volt, 400-cycle, three-phase system. Initial steps of this standardization did not include parallel operation of inverters, but such usage is entirely practical. It is likely that a choice of 208/120-volt, four-wire, three-phase system will be an improvement, since it will give a greater choice of single-phase lines, opportunities for improved control, and reduced vulnerability.

The term inverter, as used here, includes any device which will convert d-c to a-c power. It includes mechanical, electrical, and electronic devices. Mechanical designs include the vibrating-switch type, which is limited to power ratings of less than 200 volt-amp and to frequencies of less than 200 cycles. Developments of electronic designs have been handicapped by the low voltage of the primary system. If the system voltage were 115 volts instead of 28 volts, the electronic inverter would be more suitable.

Therefore, commonly used inverters are motor-generator sets of some form. The types include the dynamotor and two types of motor alternators, the rotating field, and inductor. To meet requirements of load-consuming equipment, controls for both voltage and frequency are required. To reduce radio noise it is usually necessary to add filter elements to both input and output circuits.

The inverter should be installed near the point of voltage regulation to minimize the fluctuation of input voltage. Furthermore, it should be accessible for inspection and should have sufficient free space to permit adequate cooling, particularly during operation at altitude.

The schematic circuit for two units serving divided loads is shown in Fig. 12. The arrangement is a substitute for parallel operation. Essential loads are connected to one bus and are supplied by one inverter. In case of failure of this inverter, a transfer switch disconnects the second inverter from the other loads and connects it to the essential loads.

26. POWER DISTRIBUTION

WIRING SYSTEM. The wiring system consists of bus and feeder networks, and in large aircraft the system problems are very similar to those in industrial networks. It is common practice to use a ground return circuit in all-metal aircraft. The negative side of d-c circuits is connected to ground, i.e., to the structure. When it is necessary to use a two-wire system, as in fabric-covered aircraft, the negative side is connected to the metal framework. With a-c systems one phase or the neutral of a four-wire system is connected to ground. Figure 13 shows the schematic arrangement of the power bus in several types of aircraft.

CABLE. The wiring in aircraft has been designed to meet unusual requirements. Cable for power and lighting circuits is restricted to sizes from A.W.G. 4/0 to 22. Ignition cable is 5, 7, or 9 mm in diameter, and the conductor is either stranded copper or steel.

Cable must be strong enough to withstand pulling through conduit and rough handling in aircraft which do not have conduit. This is the reason that the smallest size is limited to A.W.G. 22, despite the fact that many loads could use a still smaller conductor. In addition the cable must be flexible and able to withstand considerable abrasion.

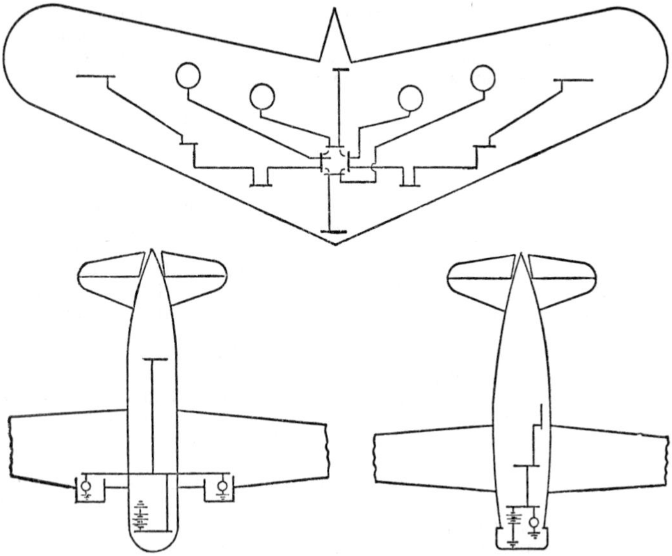

FIG. 13. Power-bus System in Various Types of Aircraft

The cable must be non-combustible. In addition, it should not emit smoke or fumes during severe overload (although this is a difficult requirement), lest it give a false alarm of danger to the airplane crew.

The cable insulation must resist moisture and fungicidal growth. The insulation should not slide on the conductor and yet should be capable of being easily stripped off for the connection of terminals. The ability to resist high temperatures is a steadily growing requirement. Special types of insulation which use asbestos, vitreous materials, or synthetic enamels have been developed for engine nacelles and for wiring within internal-combustion turbines.

The conductor consists of many strands of fine soft copper wire to achieve the needed flexibility. The conductor is insulated with a coating such as rubber or polyvinyl chloride plastic. Rubber is being superseded by members of the plastic family because the latter have better aging characteristics and less tendency to burn. Over this insulation is placed an outer cover of cotton, rayon, glass, asbestos, or other material to meet special requirements. Cotton is most resistant to abrasion and is most commonly used. The outer covering is impregnated with varnish or lacquer.

At a system voltage of 28, the weight of the wiring in an airplane is approximately 25 per cent of the total weight of the electrical system. Efforts to reduce weight have caused the development of aluminum conductors in the larger sizes. Soft aluminum has 60 per cent of the conductivity and 30 per cent of the weight of copper wire. Since the amount of insulation is increased by the larger size of an equivalent conductor, the theoretical gain in weight is not achieved. Instead of being only one-half the weight, the aluminum conductor will be approximately 75 per cent the weight of a copper conductor. Aluminum is not used in small sizes because the cable is weak. The primary problem in using aluminum conductors is the installation of terminals. Mechanical connections are most practical and have been found quite satisfactory when the pressure of application is sufficient to break down the oxide film on conductor strands and so seal the clamped joint that the oxides cannot reform. Fused connections would be superior, but the greater difficulty of making such connections offsets this superiority.

TERMINALS. Stranded conductors require terminals at the end of each section. Besides holding the strands together, the terminals make it possible to connect adjacent sections of the circuit.

Soldering has been the most popular method of attaching the terminal to the conductor, but mechanically crimped terminals are superseding the soldered connection. Solder can seldom be applied properly when working in the cramped and poorly lit portions of aircraft structure. Therefore cable assemblies are usually fabricated in the shop, so that only installation is required in the airplane. Solder flowing into the strands of the conductor will stiffen the cable and increase the likelihood of vibrational failure.

Mechanically applied terminals have their shank merged with the strands of the conductor by applying pressure on one side of the shank to force it into a U-shaped indenture. A specially designed tool prevents the terminal from expanding beyond the desired diameter.

The shank of the terminal and a short portion of the cable are covered with plastic tubing which has the dual purpose of insulating the former and strengthening the conductor at the sharp edge of the terminal.

JUNCTION PANELS. When terminals have been installed on the ends of two sections of cable, the two can be electrically connected by bolting them together. It is the function of the junction panel to facilitate such connections for many circuits. The panel is used at structural breaks in the aircraft structure and at points where the circuits divide. The panel is a strip of insulating material through which project a number of terminal posts. A second sheet of insulating material separates the lower end of the posts from structural ground. It is the function of the posts to provide means for clamping the terminals together.

Special designs of panels have been molded to include separating insulators, which prevent accidental contact between adjacent terminals. When the wiring system is shielded, the panel is enclosed in a metal box, which serves the dual purpose of shielding and protection. With unshielded systems it is still necessary to protect the exposed terminals. This has been done quite effectively with a heavy canvas cover, which hangs over the panel and is fastened on the lower edge.

CONNECTORS. The combination of plugs and receptacles is described as a connector, and it is the function of the connector to duplicate those of the junction panel but in much less space. Connectors are ideal for quickly and accurately connecting a large number of circuits. They are most suitable for connecting circuits into pieces of equipment which must be removed for inspection or replacement at frequent intervals.

Considerable work has been done on the problem of standardizing connectors to increase the interchangeability of equipment and circuits. The contact elements in common use are the equivalent of conductors from size 20 A.W.G. to 0 A.W.G. The number of such contacts in a single connector assembly range from 1 to more than 100. The connector is the ideal means for connecting circuits to pieces of equipment, but it should seldom be used for other applications.

The contact is the heart of the connector. The pin element is a simple rod and is the basis for the dimensional standards of the contact. Its diameter has been selected to match that of copper conductors, and the terminal cup on one end will receive the stranded conductor of the corresponding size. The terminal has been constructed in both the solder cup and mechanical types.

The socket half of the contact must match the pin dimensions, keep within limiting overall dimensions, and provide the specified electrical performance. This performance is given in terms of a continuous-current rating at a maximum drop in voltage. The contact must carry the required current with the same voltage loss (or less) that would occur in the equivalent conductor having four times the length of the contact. Pressure between socket and pin determines the voltage drop more than any other factor. To provide adequate pressure many arrangements of springs have been tried. It is also possible to use a split type of socket, but this requires much closer tolerances on the pin diameter than is possible to maintain in production. The connector must be so designed that any type of plug can be used with any receptacle.

The connector shell is not important electrically or structurally, but its dimensions must be controlled quite accurately to assure interchangeability. The receptacle is defined as the fixed element, and the plug as the movable one, since either pin or socket contacts can be put in either half. The receptacle usually has a flanged base for attachment to equipment. The shell has an external thread of extra-fine pitch, which matches a coupling nut on the plug shell. This nut provides a mechanical advantage during insertion or removal of the plug. Polarization of the contacts is achieved by a key in the receptacle shell and a keyway in the plug shell. The outer end of the plug shell is usually threaded to permit the attachment of conduit.

The inserts which carry the contact elements are of plastic materials which can withstand the several operating conditions, which do not char during arcing, and which meet other special requirements. These inserts usually rest against shoulders within their respective shells.

Weaknesses of connectors which make their use in other applications inadvisable are numerous. The fact that many circuits are taken through an extremely small space makes maintenance very difficult. If space is available, other solutions, such as connector panels, should be used. Again the closeness of terminals leads to short circuits and current leakage across the surface of the inserts. The contact elements must use special alloys and silver plating to meet the severe performance requirements, and these materials will corrode in service.

Since connectors will receive little maintenance, they must be designed accordingly.

CONTROL OF CIRCUITS. The common types of control equipment include switches and relays. Switches are operated manually or mechanically, and most of the latter type are precision units which are operated by a very short travel of the actuator. Relays can be divided into two types, power and circuit. The former is often described as a magnetic contactor and is used to reduce the amount of heavy wiring. The latter reduces the amount of wiring by using one cable to control several circuits. In many designs protective devices have been added to manually operated switches and to relays making local and remote circuit breakers. These become both control and protective devices.

First on the list of requirements for switches is small size, particularly in panel space. In addition, they must carry quite high currents with low loss in voltage. The mechanism should have a snap action, since most of the circuits are inductive. Special requirements are added for pressure-actuated switches, thermally operated units, and critical control units. For example, the limit switches in landing-gear circuits must be located on the gear, where they are exposed to dirt and moisture. Such units must be protected from this abuse by a sealed construction. Special designs of handles are helpful to the operator who must find and use the switches in darkness.

Weight is of first importance in relays, since the only reason for using them is saving weight in conductors. Functioning of the relay must not be disturbed by vibration or acceleration. Moving elements should be balanced or should be held by a force stronger than any acceleration up to 15 g. Relays should operate correctly at any voltage from 75 to 125 per cent of normal, and they should remain closed at voltages of less than 10 per cent of normal. They should be capable of installation in any position. Designs that are to be used in the presence of gasoline vapors should be explosionproof. Contacts should be non-welding and should have a wiping action. Those having high current ratings generally require two types of contacts, one for current conductivity and the other for arcing. To interrupt arcs at high altitudes a bridging type of contact is superior. To prevent damage by corrosion, sand, and dirt, a sealed container is advisable.

Outline dimensions and ratings of several switches are given in Fig. 14; Fig. 15 gives similar data for relays.

In the design of switches or relays the contacts are the important part. Pressure between contacts must be as high as practicable, as this determines current-carrying ability. To avoid welding, the design must minimize heat losses and provide for the conduction of heat from the contacts. Contacts are often made of a good conductor faced with a thin layer of material which will withstand arcing. Contacts should not bounce when the circuit is closed, particularly in lamp or motor circuits, since the inrush current is very high in both of these.

The effect of altitude on gap breakdown values is given in Fig. 1. These curves emphasize the need for the bridging type of contact design, which doubles the effective gap and makes each half more effective.

Switches and relays, like connectors, are not likely to receive maintenance. Both should be designed with this fact in view. Often cleaning of relay contacts results in more damage than good.

Manually operated switches should be installed with full consideration of the problems of the crew member who will use them. The end-product of such an installation will not be the neat and regular panel that the electrical engineer desires, but it will produce a safer airplane. Operation of the switch lever should be convenient and natural. The

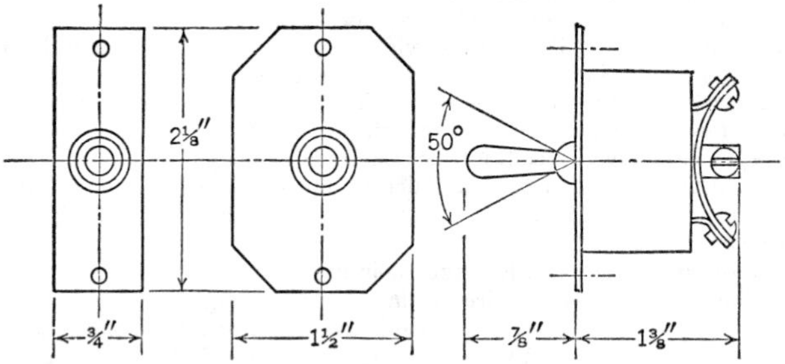

Arrange.	Contacts	Weight, Pounds	Current Rating Amperes at 26 Volts			
			Cont.	Lamp	Res.	Induct.
2-pole	On-off moment	0.20	55 55	7 5	40 30	20 18
1-pole	On-off moment	0.10	55 55	7 5	25 20	15 10
1-pole mounting hole	On-off moment	0.08	40 40	5 4	20 15	15 10

FIG. 14. Toggle Switch

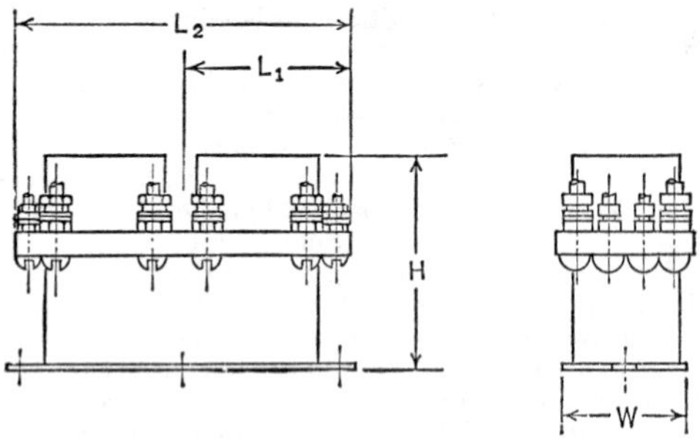

Rating Res./Ind.	Type	Weight, Pounds	Duty	Coil Data			Dimensions, Inches			
				Current, Amperes	Voltage					
					Pick-up	Drop	*L*1	*L*2	*W*	*H*
400/100	SPST	2.6	Int.	5.5	7.5	3.0	5 1/2	....	2 1/4	4
10/6	SPDT	0.4	Cont.	0.2	18	7.0	2 1/8	....	1 3/4	2 1/4
25/20		0.7	Cont.	0.25	18	7.0	3 1/4	....	2 5/8	2
50		0.75	Cont.	0.5	20	7.0	2 1/4	....	2	2 7/8
50/40	2PST	0.75	Cont.	0.5	20	7.0	3	....	2 5/8	2 3/4
100/80		1.5	Cont.	0.6	20	7.0	3	....	2 3/4	3 1/2
10	2PDT	0.5	Cont.	0.5	18	7.0	2 3/4	....	1 3/4	2 3/16
50		1.85	Int.	2.5	7.5	3.0	....	5 3/16	2 1/8	3
100/80		3.75	Int.	3.5	7.5	3.0	....	6 3/4	2 3/4	3 3/4
200/100		6.0	Int.	4.5	7.5	3.0	....	7 3/4	3 1/2	4 1/8

FIG. 15. Relay Switch

requirement for convenience will scatter the switches all over the pilot's cabin. For example, the landing light switch will be near the throttle, the microphone switch will be in the control wheel, and all starting switches will be grouped together. Other switches that are seldom used may be grouped together on a panel. It is customary to mount these switches so that the "off" position is down or to the right.

Mechanically operated switches should be located above any leakage of water, oil, or other fluid and where they will not be damaged during the maintenance of other equipment. Relays may be located in any position that will best serve their primary purpose of saving weight.

PROTECTION. The power-distribution network can be divided into four parts for analysis of the protection problem: generator, bus, circuit, and equipment, each of which requires a special type of protection for maintaining continuity of electrical service. Failure of any one part must not affect normal operation of any other part.

Equipment which is available to give this protection includes three types: relay, fusible link, and circuit breaker. The last can be divided into two types, magnetic and thermal. A combination of the two types occurs in a thermal type of circuit breaker, which is equipped with a magnetic reset device. The fusible link or fuse has been a very popular form of protective device but has one serious weakness, and that is the matter of supply. Normal functioning of the fuse causes its destruction and requires a replacement. Since circuit breakers can be reset, they are rapidly superseding the fuse in most applications. The thermal type of breaker more nearly matches the thermal characteristics of the equipment and is much more widely used than is the magnetic breaker.

The generator is protected by the reverse current relay, but experience has shown that a sudden fault between the generator and relay can cause the relay to misfunction. There is little lag in the rise of reverse current in such a fault, and as a result the current will reach a magnitude sufficient to hold the relay closed by the influence of the current coil, and the current will reach this value before the relay can open normally. Therefore, it becomes necessary to add a second protective device at the relay. Fusible links, thermally operated circuit breakers, and a combined thermal and magnetic breaker have been used. The last is the most successful design. It has a magnetic trip that is delayed by the thermal element until the reverse current relay has had time to function. This avoids nuisance faults in the generator circuit.

Protection of the bus is more difficult. To be effective the sections of the bus should be divided into three or more channels. Protective devices can then be placed at the ends of each channel section, requiring six protective devices. A fault on any one channel will be fed through two devices in other channels and one device in the faulted channel. The one protective device, being loaded twice as much or more than any other, will fail first and clear the fault. The protective device can be either a fusible link or a relay type. The fusible link has the advantage of weight, but it does not give satisfactory protection throughout all possible combinations of generator and line faults. For example, if half the generators are off the line, the remaining half will not deliver enough current to rupture the link. Again, two faults in certain places can so isolate channels that normal load current will cause the remaining channel to open. Relay types of protective devices can offer complete protection by making the control coils sensitive to differential and reverse currents. This method is almost out of the question for aircraft because the weight of the great number of relays that would be needed is excessive.

At one time it was thought that enough electrical power was being installed in the larger aircraft to burn any fault clear. Experience has shown that this is not true. It is known that a conductor which has been ruptured will not weld to structure and cause a permanent fault, but permanent faults can occur when a rigid bus-bar conductor is forced into contact with structure or when a metal shielding box is forced into the posts of a junction panel.

Circuit protection is simply protecting the cable from the bus to the piece of equipment served by the cable. The protective device should match the thermal characteristics of the cable as accurately as possible. It should be located as near the bus as convenient, since the short length of cable from the bus to the protective device is a vulnerable spot. Here the functions of the control switch and protective device can be combined if the bus and the circuit end meet at or very near the point of control. If not, the protective device should be located at the bus, and the switch connected in series with it at the point of control. The protective device can be either a fusible link or a thermal type of circuit breaker. The low-melting-point type of fusible links do not match the cable characteristics as well as do those of copper or higher-melting-point alloys. Thermal-type circuit breakers are available which quite closely match the cable characteristics.

The device which protects a piece of equipment must be integral with it. A remotely located device cannot be sensitive to all the factors which determine the likelihood of the equipment to fail.

Similarly, electric motors present a difficult problem in protection. Three motors of equal rating, one in an engine nacelle, one in a heated cabin, and the third in the unheated wing, present quite different requirements if the protective device is located at the power-supply bus. If the protective device is located within the motor, however, the same type will serve all three. When it is integral with the motor, it is sensitive to the load on the motor as well as the ambient conditions. The device must respond to both load current and frame temperature to protect the motor properly against all conditions of overload to and including stalled rotor. A thermal circuit breaker with a supplementary heater proportional to load current is a suitable device for motor protection.

CONDUIT. A fourfold purpose is served by conduit. Metallic conduit is a shield to reduce the propagation of electromagnetic waves that would cause interference to

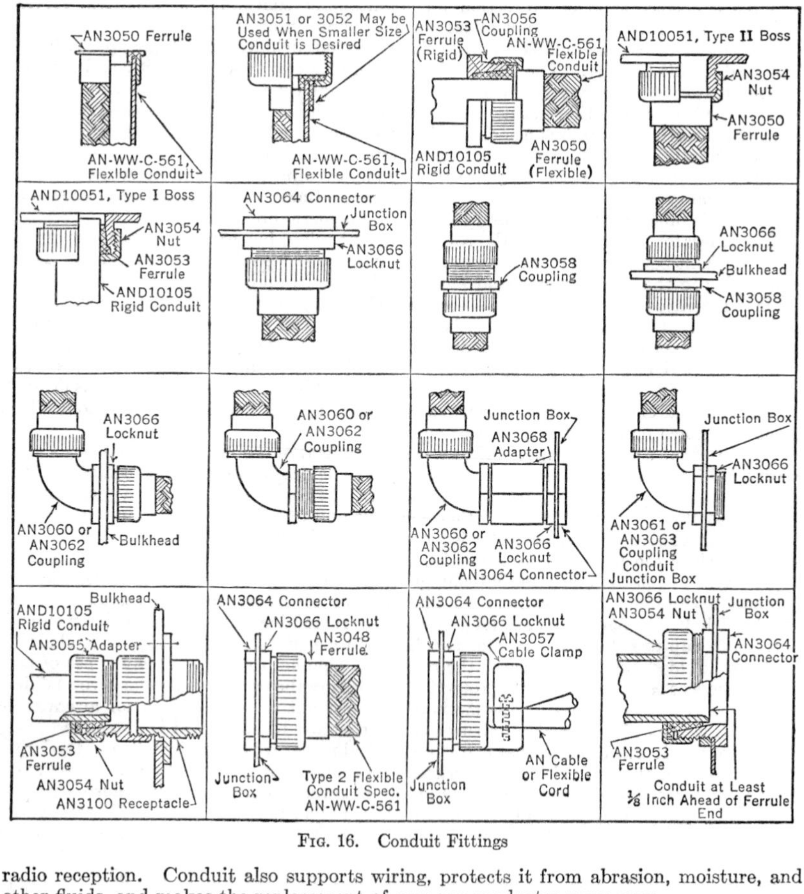

Fig. 16. Conduit Fittings

radio reception. Conduit also supports wiring, protects it from abrasion, moisture, and other fluids, and makes the replacement of any one conductor very easy.

Metallic conduit is usually aluminum or aluminum-alloy tubing. Half-hard 2-S alloy is a good material because it has the necessary strength combined with good conductivity. Flexible conduit is constructed by helically winding a strip of metal in such manner that the edges interlock. The tube is then covered with a woven metal braid. Materials are usually aluminum alloys, although stainless steel has been used for ignition harness shields where considerable resistance to high temperatures is needed. Flexible conduit is twice as heavy and five times as costly as rigid conduit, and for that reason its use is restricted to lengths where only it can be employed. For example, it is used with the plugs of connectors, since the plug is removable. It is also used in the engine nacelle to connect wiring to generator and starter. As already noted, the terminal end of these pieces can move nearly 2 in. during low-speed operation of the engine.

Fittings used with either type of conduit usually depend on a clamping action to hold the conduit end. The clamp is energized by a nut which works over a tapered thread on the fitting. Various forms of the fitting are available, some of which are shown in Fig. 16. These fittings are one reason that soft metal cannot be used in conduit, as it would crush and become loose. Basic requirements for the fitting include a positive clamp to the conduit, ready attachment to boxes, and a complete absence of sharp edges.

With the development of methods to reduce radio noise, other than by shielding, the need for metallic conduit has been greatly reduced. Other needs for conduit persist, however, and there has been considerable use of plastic materials to support and protect wiring.

When designing and laying out a conduit run, every effort should be made to use direct routes. Such routes make it easier to pull wiring through the conduit and will reduce the weight of conduit and wiring. Conduit supports can be more than 3 ft apart in protected installations, but if the conduit is so located that it can be a convenient handhold, supports at closer intervals will be advisable. Conduit should have large bends to facilitate the installation of wiring. The cross-sectional area of all the wiring within a conduit should not exceed 80 per cent of the internal area of the conduit. This value can be exceeded if the run is short and straight, but, if there is more than 180 deg of bend in the run, it will be found difficult to install wiring with 20 per cent excess space. Wiring such as ignition cable, which is located near hot cylinders, should be protected with special conduit, e.g., the stainless-steel flexible conduit described above.

Most of the wiring in the airplane is open or unshielded. This has been made possible by the development of other and better means of reducing radio-noise influence voltages. Other benefits are obtained by open wiring. In military aircraft the vulnerability of the wiring and the electrical system to enemy gunfire is greatly reduced. Quantity production of aircraft is made easier. Repairs to open wiring are made much more easily in flight and on the ground. Finally, the weight of the wiring is reduced.

In designing an open-wiring installation, most of the rules given above for conduit runs apply, except that it is particularly important to take every advantage of the protection given by aircraft structure. All the wiring going to one piece of equipment should be kept in a group if possible. Open wiring must be supported at more frequent intervals than conduit, and this interval should not exceed 24 in. Almost the only objection to be raised against open wiring is poor appearance, but in commercial aircraft the wiring can often be concealed behind cabin lining and floors.

During installation of open wiring great care must be taken to avoid sharp edges and corners. Runs around doors should be protected so that personnel cannot use the cable group as a handhold. Cable groups should not be routed through holes in structural bulkheads, as more time will be required for installation and still more time for replacement. A properly designed installation will permit the removal and replacement of a cable group in less time than would be required to make repairs within the airplane. This should be the goal of the aircraft electrical engineer when designing the installation.

WIRING DIAGRAM. When preparing the estimate of the load, that required for each piece of equipment and the duration in time must be tabulated. From the totals, the maximum demand load and its duration are determined. The rating of generators and batteries to supply the loads under all anticipated conditions of operation can then be selected.

The next step is the preparation of a schematic diagram showing all circuits in the simplest possible form. The schematic diagram will prove the correctness of all circuits very quickly. The second diagram is a block diagram, the only one on which any attempt to scale the wiring dimensionally is made. With this diagram, the routing and length of circuits and the size of cables are determined. It is common practice in the 24- to 30-volt system of aircraft wiring to allow a loss of 1 volt in the wiring from the point of voltage regulation to the load, which represents a loss of little more than 3 per cent. This relatively low permissible loss means that many conductors will be larger than necessary, so that at this voltage it is the loss of voltage and not the current that determines wire size.

The chart of Fig. 17 gives the relation of load current, voltage loss, and length of circuit to cable size. The resistance of the return circuit through the structure of an all-metal airplane can be neglected. In other types the length of circuit will be doubled, although the size of the negative return cable may be much greater than that of the positive cable. The chart shows that the weight of the cable is inversely proportional to the square of voltage, and this fact is the basis for continuous studies of higher-voltage systems and of methods to tolerate a higher loss in the cable.

The next diagram to be prepared is a composite diagram. Here an attempt is made to portray all the wiring of the airplane on one drawing. This is practical for aircraft having a gross weight of less than 25,000 lb. For larger aircraft it is necessary to sectionalize the drawing. The next step is to portray individual circuits on drawings. These circuit diagrams are the ones most used by the electrician, since they isolate the circuits. It is likely that the composite diagram will become a collection of circuit diagrams. This collection can be assembled in book form, and the composite diagram will be a simple series of line diagrams, with references to the detailed circuit diagrams. Within the book will be lists of equipment items and their identification, as well as lists of cables with numbers, sizes, and lengths.

Special diagrams are needed by the aircraft manufacturer during production to facilitate wiring jobs that are done on the bench or installations within the airplane. Finally,

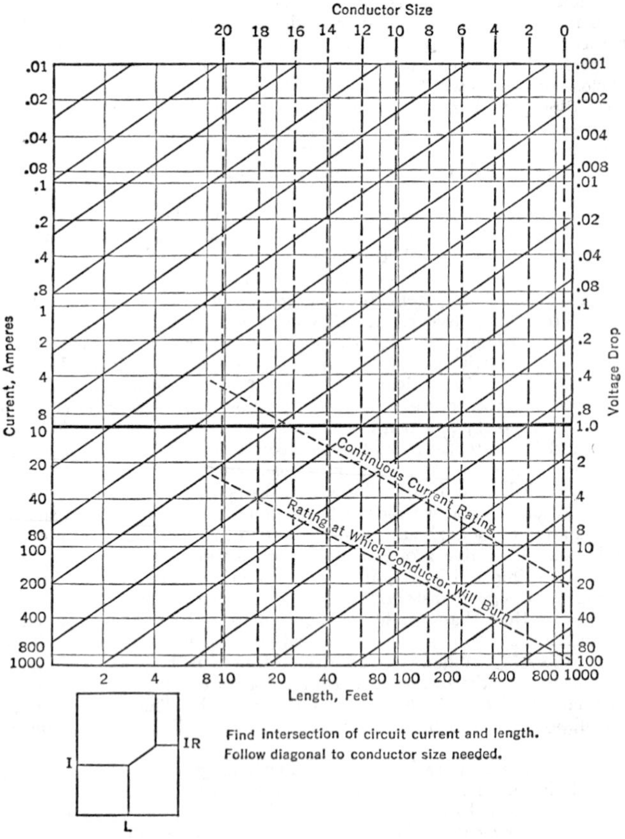

FIG. 17. Nomograph for Selection of Wire Size

diagrams of the connections at junction panels are very helpful during both production and maintenance. If the panel is protected by a box, the diagram can be located in the lid of the box.

The industry has developed a system of wiring identification in which one or two prefix letters are used to designate the equipment item with which the circuit is associated, and a number represents the circuit. A suffix letter shows the position of a given cable in the circuit. The total is a part number, characteristic of only one length of cable. Separated from the part number by a short distance or by a symbol, such as an arrow, is a second number which is the size of the cable. The whole number can be set up in a machine which will imprint it in the cable insulation at intervals of approximately 1 ft. The numbers, less wire size, are referenced on the wiring diagram.

Some conventions that have been established in the preparation of diagrams include use of 1/4-in. spacing between lines of the diagram and a standard set of symbols, which has been coordinated by the armed services and has been submitted to the American Standards Association. Crossing lines do not indicate a connection. A terminal black dot shows a connection. Detailed wiring is shown only at panels. The cable group that connects panels is shown as a heavy line. Such practices greatly reduce the space required to show the complete wiring of an airplane.

27. ELECTRIC MOTORS

APPLICATIONS. Before World War II scarcely any motors were used in aircraft, with the single exception of the engine starter. Before aircraft developments initiated and accelerated by that war were complete, one type of four-engine bombardment airplane used 140 motors. The applications for motors can be classified as power, control, and instrument. Examples of power motors are those used for engine starters, retraction of the landing gear, and operation of the wing-flap mechanism. Control motors are those which position propeller blades, rotate gun turrets, and adjust trim tabs. Instrument motors are used in small gyroscopes and remote indicating instruments.

TYPES AND CONSTRUCTION. Aircraft motors can be classified by the type of power: d-c or a-c. Direct-current motors can be further classified by describing the duty cycle, frame, and field winding. Continuous-duty motors are expected to carry rated load continuously under any altitudes and ambient temperatures described in the operating conditions of the applicable specification. Intermittent-duty ratings are given to motors which must perform for only a short time. In most cases this is a specific rating, since the weight of the motor will be proportional to the length of the duty cycle. At one time much consideration was given to a standardized rating of 3-min operation, followed by 17-min cooling. Since the motor cooled only half the temperature rise during the cooling period allotted, it was necessary to add additional cooling provisions at a penalty in weight. With the development of protective devices which reduced the likelihood of failure during abuse in service it became possible to design the motors more closely to operating conditions. For example, it is necessary to feather a propeller in normal service only once during a flight, and the motor designer is tempted to base his calculations on a total duty of only 15 sec. In practice, however, it was found that the motor would be required to do as many as four complete cycles during the course of ground tests or instruction. The motor must protect itself against such abuse.

The frame construction is usually either ventilated or explosionproof. The ventilated type is used with most continuous-duty motors. Intermittent-duty motors are of closed construction and usually meet aircraft requirements for explosionproofing. These requirements are not the equivalent of industrial requirements, since only one explosive vapor, gasoline, is present. The motor is required to demonstrate that normal operation in an explosive mixture of gasoline vapor will not ignite the vapor, and that explosion of such vapor within the frame of the motor will not cause the external vapor to ignite.

Direct-current motors are also classified by the field windings. Most motors are of the series type, since starting torque is very important. Reversible motors will often have dual series field windings, as this construction is lighter than a single field with an external relay. The shunt field type of winding is seldom used, and then only in low-torque applications, such as instruments. Compound windings will find more favor in motors of high power rating, since compensating fields must be added to neutralize armature reaction.

Alternating-current motors are usually of the induction type. Both three- and single-phase designs are used, although the latter is found in only small power applications. The a-c induction motor is almost sufficiently explosionproof for normal use in aircraft without special enclosure, since there are no conducting surfaces for electric current. The a-c motor must be designed for duty cycles similar to those already described for the d-c motor.

CHARACTERISTICS. The curves of Fig. 18 give estimated potential weight values for various speeds and power ratings of both d-c and a-c motors. Increase in speed is an effective method for reducing the weight of motors. Practical limits for continuous-duty motors appear to be approximately 12,000 rpm in both d-c and a-c designs. With d-c intermittent-duty motors further gains can be obtained at speeds of more than 20,000 rpm. At some point near this value the penalties of adding mechanical strength offset the gains in electrical structural weight.

The torque curves for both types of motors, as given in Fig. 19, are similar to industrial counterparts. The d-c motor delivers more torque per ampere, and this is a powerful reason for the retention of d-c systems in aircraft. At low temperature the performance of either type improves, since the losses are dissipated more easily.

Both d-c and a-c motors have similar weaknesses. The starting current is six to ten times normal running current at rated load. This penalizes the system by requiring more capacity in the power source and larger conductors in the power circuit. Under starting conditions the voltage at the motor terminal will be only 80 per cent of regulated voltage instead of 97 per cent, and if the power source is overloaded by the starting current the voltage will be even less.

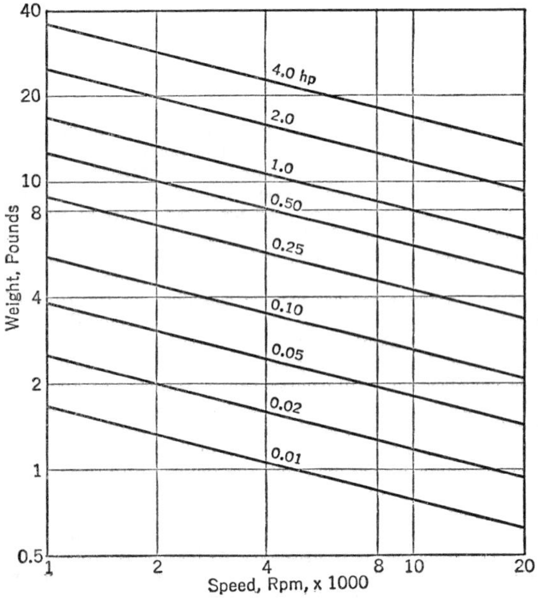

FIG. 18. Weight Characteristics of Electric Motors

The motor cannot long endure the heating that results either from starting current or from load current during operation at low speed and high load. If the rotor is locked, the motor will fail in a short time, varying from a few seconds to several minutes, depending on the conservatism of the design. A properly designed motor of minimum weight will fail under such conditions in approximately 30 sec. Therefore, to achieve a design that will best serve the airplane, the motor must have integral protection against locked rotor and overload that will prevent failure due to burnout. The condition of locked rotor is usually caused by malfunction of the driven device, which is serious enough, but failure of the motor must not be added to the list of troubles.

The greatest weakness of electric motors is the lack of variation in speed. Direct-current motors are better than a-c motors, since variation of either shunt field or armature voltage can give a variation in speed of approximately 2 to 1. To obtain an infinite range of speed it is necessary to add an auxiliary device.

In aircraft motors the d-c type has its greatest advantage in its ability to deliver torque. The a-c motor is lighter, is simpler in construction, is more efficient, and has no arcing contacts, but it has the disadvantage of power factor.

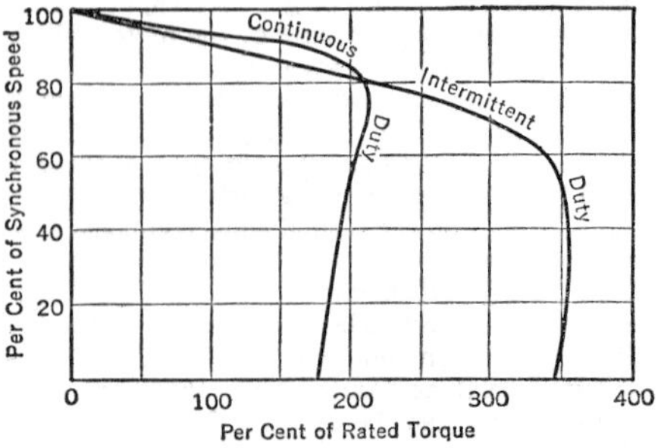

FIG. 19. Torque Characteristics of Electric Motors

SPEED CONTROL. To vary the output speed of an electric motor continuously, it is necessary to add an auxiliary device. Even in the simplest applications of motors it is often necessary to protect the driven device from the inertia of the high-speed rotor. Brakes and clutches are used to perform both these functions. A clutch which is engaged with the motor is useful in positioning a device, because it will disengage the rotor at the limit of travel and allow the rotor to coast to a stop without regard to the over-running time. The driven device is positioned accurately and released from the motor. A brake is also useful in holding the driven device with a very small amount of torque within the motor, whereas a large brake might be needed at the other side of a gear reduction.

Clutches have also been used to give speed control. The driving motor is operated at constant speed and is connected to the load through a solenoid operated clutch. The frequency and duration of clutch engagements determine the speed of the driven member.

AMPLIDYNE. One solution to the need for variable speed has been an adaptation of the **Ward-Leonard control system,** consisting of a motor-generator set and a driven motor. The generator is a special design giving a high degree of amplification and is known in one form as the **amplidyne.** A shunt field supplies excitation for a short-circuited armature. The current in the armature sets up a cross-magnetization much higher than the normal field. A second set of brushes is located 90 deg from the normal and short-circuited brushes to deliver power that is proportional to the cross-magnetization. Reaction resulting from the load current would oppose the original field and reduce the amplification. Therefore, a series field opposes the reaction field. In some designs the amplification from field circuit input to armature output is 10,000 to 1.

The driven motor receives only armature power from the generator, and field power is taken from the constant-voltage electrical system. Therefore the complete system delivers torque in an amount proportional to and in the direction given by the input signal to the generator field.

The system is an excellent solution to any problem requiring wide variations in speed combined with precision control. It has found wide application in the remote control of gun turrets in military aircraft and in automatic pilots for all types of aircraft. The accuracy of the system is limited only by the inertia of the motor and the characteristics of the follow-up system.

ELECTRIC VERSUS HYDRAULIC MOTORS. Hydraulic power is widely used for many aircraft accessories, particularly those which require a large amount of power for a short time. Although hydraulic power cannot operate many accessories, such as radio and lights, it offers enough saving in weight to justify itself, despite the fact that its use makes it necessary to design and install another complete accessory system. Electric motors must meet the challenge of the hydraulic system before there can be any thought of a single accessory system in the airplane.

Electric motors are better suited to rotary motion than to linear; a hydraulic piston and cylinder are an ideal solution to the need for linear motion. The electrical solution to this problem is a motor, gear reduction, screw shaft, and protective devices such as limit switches and clutches. In many installations a gear reduction and crank with linkage would be an equally effective solution and would be far more favorable to the electric motor. Most comparisons are made, however, on the basis of a linear actuator alone.

At low temperatures the electric motor has all the advantage. No hydraulic fluid retains its low viscosity at extremely low temperatures, and for most of them the viscosity doubles with each drop in temperature of about 15 deg fahr from 20 to −40 deg. Since most hydraulic lines are deliberately operated with a high loss to decrease the weight of tubing, it is evident that the losses at low temperature absorb practically all the hydraulic power.

The overall efficiency of the electrical system is higher than that of the hydraulic. That of generator and motor is less than that of pump and motor, but the higher losses in hydraulic lines lower the overall efficiency of the hydraulic system. The pump will deliver power at any speed, whereas the generator must be operated at its rated minimum speed before it will deliver rated voltage and load current.

The hydraulic pump and motor are much lighter than the electric equivalents, the generator and motor. In many cases, however, the saving in weight of wiring compensates for the heavier weight of the power units. Operating pressures of 3000 psi and a loss of 35 per cent in lines will reduce the weight of hydraulic lines to values that are competitive with the 24-30-volt wiring system. The latter value is a relatively low-pressure electrical system, however, and even the conventional 110-volt system would lower the weight of wiring to a fractional part of that used with the low-voltage system.

The hydraulic system has one important advantage in that manual power can be used in emergency to pump pressure into a hydraulic reservoir and thence into the system.

Hydraulic power is better suited to applications for boost power because a proportionate resistance or feel can be fed back into the control mechanism. This is important when power boost is used to help the pilot during operation of very large aircraft. If power boost is supplied electrically, it is necessary to provide an artificial simulation of resistance.

The electric solution is ideal for most automatic control problems and has been used to the advantage of the hydraulic system, since in many aircraft designs merging the two systems will give the lightest and most efficient accessory system. Long runs and frequent bends lower the efficiency of the hydraulic system. If the electric system is used to control remotely located valves, hydraulic lines can be shortened to minimum lengths of straight runs. Most advantageous use of the best characteristics of both systems is obtained.

It may be possible to use only an electrical accessory system in many types of aircraft, since the time of operation of most hydraulic loads comes at the off-peak periods of the electrical system. If the load in flight is heavy, there will be an excess of electrical capacity at take-off and landing, which can be used for operation of landing gear and wing flaps. This fact makes the overall weight of the electrical system competitive with that of the combined electric and hydraulic system.

ENGINE STARTER. Methods of starting aircraft engines have included four manual, four electrical, and one combustion. The four manual methods include turning the propeller itself, turning the engine through a handcrank and gear box, energizing an impulse coupling which, when released, turns the engine, and finally storing a great amount of manual energy in a rotating flywheel which, when released, in a short time is able to start

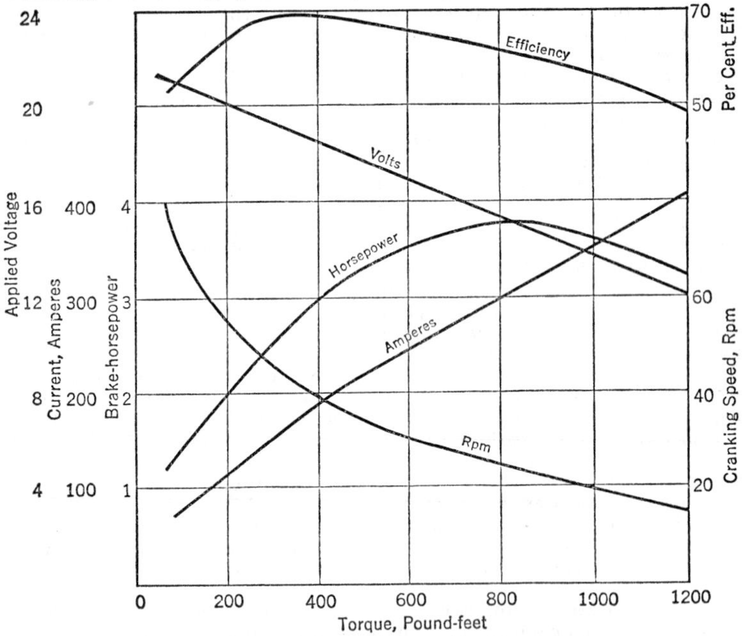

FIG. 20. Characteristics of Direct-cranking Engine Starter

rather large engines. The last method is described as a hand inertia starter, because it uses the inertia of the spinning flywheel. The combustion method uses the chemical energy stored in nitrous powder. Expansion of the gases from the burning powder rotates the engine for approximately one revolution but at rather high speed, approximately 5 per cent of rated speed.

In the remaining and electrical methods, an electric motor gives the necessary power. The first application of a motor was to an inertia starter. The motor was used to energize the flywheel, and the power for starting was taken from the kinetic energy so stored. Logically, the next step was to use the motor during the cranking operation also, and this gave the combined inertia and direct cranking starter. The kinetic energy in the flywheel was used to accelerate the engine, while the electric motor kept the engine turning.

With the advent of more reserve power, either in the airplane or connected thereto from an external source, it became possible to use direct cranking motors. These motors had approximately 10 hp, about three times that in the inertia type of starters. It is necessary to consider the limitations of the power source very carefully when designing or selecting starters. The aircraft type of storage battery can supply a 3-hp starter motor, but the 10-hp design required an auxiliary power unit in addition to the battery. Another application for a motor of high power rating is the propeller-feathering pump unit. In one design of starter one motor was used for both purposes, with a resultant weight saving.

The internal-combustion turbine type of engine has quite different starting characteristics, since there is little drag, but the engine must be rotated at more than 25 per cent of rated speed to start. The power rating of a starter to meet this need is nearly 20 hp. Starters of this rating have been merged with generators of 400-amp capacity. The combined starter-generator weighs about 5 lb more than would either machine separately, approximately 60 lb.

The inertia and direct cranking starters use a high-speed electric motor, a gear reduction, and a clutch. The motors have been designed for operating speeds of 12,000, 16,000, and 24,000 rpm. Since torque is the most important characteristic, the field windings are of the series type. The clutch softens the engagement of the inertia type of starter with the engine and protects the starter in case of backfire from the engine. The starter-generator design needs no gear reduction, since it is connected directly to the shaft of the turbine. Generator needs determine the rated speed, which is usually about 7500 rpm, but this will vary as the turbine type of engine is developed further.

Characteristics of a typical direct cranking starter are given in Fig. 20.

Several accessories are needed to assist the starter or engine during starting. A solenoid engages the starter with the engine crankshaft. A brush lifting mechanism was added to one design of starter to free the inertia type of the drag of brushes and yet allow the motor to be reconnected for the direct cranking operation. An ignition booster coil was used to provide a high-voltage spark during the low-speed operation of starting when the magneto was unable to generate an adequate voltage. This has been superseded by a simple interrupter, which feeds pulses of direct current into the magneto, thereby saving the weight of the secondary winding in a booster coil.

Another electrical accessory which facilitates engine starting is a solenoid operated valve. Opening this valve dilutes the engine oil with gasoline and reduces viscosity to levels where starting in cold weather is much easier. The diluting operation is performed just before stopping the engine. The gasoline is quickly driven out of the oil by the engine heat when the engine is started again.

PROPELLER CONTROL. To obtain maximum performance from the engine and propeller, it is necessary to control the pitch of the blades. The pitch is varied by rotating the blades within the hub, and this rotation has been accomplished both electrically and hydraulically. Variation of blade angle allows the engine to be run at constant speed and at the best condition to achieve desired performance, whether it be maximum range, maximum speed, maximum climb, or minimum fuel consumption. It also protects the engine from excessive speed during take-off and descent.

One type of hydraulic control uses engine oil for normal control. An electric motor-driven governor furnishes the reference signal. Hydraulic pressure is valved in accordance with signals from the reference to rotate the propeller blades. In case of engine failure it is desirable to feather the propeller blades of the dead engine in a multiengine airplane. Feathering is the rotation of the blades to a position of minimum drag or to a position where the width of the blade is parallel to the line of flight. In the hydraulic control it is necessary to overpower the engine oil pressure. An electric motor-driven oil pump is used to supply the higher pressure. In one design, as mentioned above, the motor was that of the engine starter.

Another propeller design uses an electric motor as the control element. It is located in the hub of the propeller and actuates the blades through gearing having a ratio of approximately 27,000 to 1. Feathering is more simply accomplished, since the motor can drive the blades to the limit of travel. To accomplish the feathering operation in less time, a motor generator has been used to deliver approximately four times rated voltage to the motor and thereby speed its operation. As the size of propeller blades increases to absorb more power, it is likely that balanced designs will be sought to reduce the amount of power needed in the control motor.

LANDING-GEAR AND WING-FLAP RETRACTION. The operation of landing gear and wing flaps has been accomplished successfully by both electric and hydraulic means. Since both operations are for short periods and require large amounts of power, hydraulic solutions are well suited. Therefore the majority of such mechanisms in large airplanes are hydraulic, and in every case a piston and cylinder are used. As the demand for electric power has increased, however, generator capacity has increased also, until electric operation has become practical and economical.

The inefficiency of the screw shaft required with electric motor operation added to the power requirements and the weight of the electric actuator until it could not be competitive with the hydraulic unit. The ball-bearing type of screw shaft tends to reduce this handicap by delivering efficiencies of nearly 90 per cent, even at low temperatures. Rotary-link designs would be still more compact and efficient.

Since torque is the dominating requirement, the motor is always of the series type. This motor should always have integral protection. In operation it is protected by limit switches at either end of its travel, but experience has shown that these switches are vulnerable to dirt, moisture, corrosion, and maladjustments. The motor should not be destroyed by improper functioning of the limit switch. One design protects the motor with a clutch assembly, but failure of the limit switch results in failure of the clutch, motor, or both.

28. LIGHTING

EQUIPMENT. Three groups of problems are encountered when lighting an airplane. These may be classed as external, instrument, and internal lighting. The automotive type of lamp is used with few exceptions. The 6- and 12-volt lamps were adopted directly from automotive practices, and 24-volt designs were added, using the same base and bulb. When needed, special types have been developed. Practically all these lamps use the bayonet type or a similar locking type of base. A base design which might loosen during vibration cannot be considered. The vibration which occurs in an airplane also limits the voltage which can be used for lighting. Since most of the fixtures must be small, the size of bulb and also the power requirement are small. The filament for a 24-volt lamp in the 3 candlepower rating is very fragile. Designs for 110 volts are probably impractical at ratings below 50 candlepower, and these include most of the lamps in an airplane.

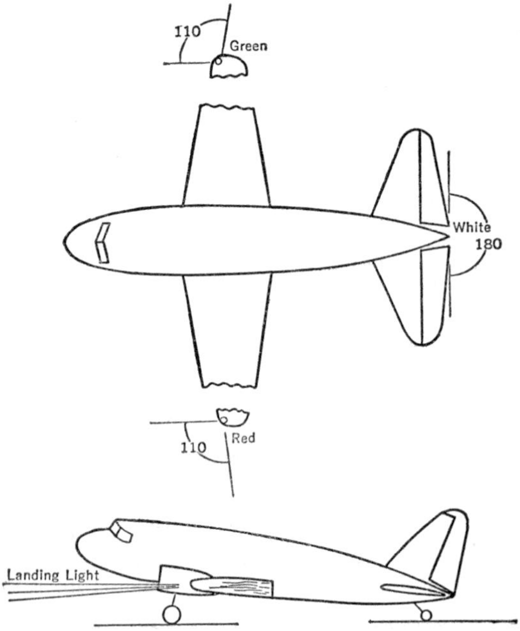

FIG. 21. Position of External Lights

EXTERIOR LIGHTING. The lights which mark the position of the airplane are the only lights which aid in controlling traffic and in preventing aerial collision. The location, color, and angles of visibility of each unit, in conformance with international agreements, are shown in Fig. 21. The color assignment and angular cut-off were adopted directly from marine practice. The angular cut-off is designed so that an observer can see only one light at a time.

Night flight was actively and commercially entered about 1930. Almost immediately a high-intensity spotlight with a red cover glass was added to the position lights. The "passing" light projected an intense but narrow beam parallel to the line of flight and thereby gave warning to approaching aircraft at much greater distances. As traffic increased, the hazards of collision during the approach to the airport became serious. It was found that the position lights were easily confused with city lights on the ground. To circumvent this new danger, a brighter tail light was used, and a flashing switch was added to the position light circuit. The flashing lights are quite distinctive and adequate.

It should be recognized, however, that any type of lighting cannot adequately meet all the conditions of flight. The range of visibility decreases rapidly when moisture in any form is present in the atmosphere, and the range cannot be increased to safe values with any power of lighting equipment. Therefore, the task of giving warning and showing the location of approaching aircraft must ultimately be relegated to radio and electronic devices. Lighting will then provide only short-range warning for local traffic conditions.

Formation lights were added to military aircraft to overcome another shortcoming of position lights. It is difficult to judge the position of an airplane by observation of only one light, and protracted observation of one light leads to vertigo. For many years the military services used position lights which were visible through 180 deg for night formation flight. The tail light was also divided and so positioned that it illuminated the tail surfaces. This arrangement was reasonably satisfactory, since the position of the lead airplane could be judged by the perspective of the tail surfaces and the position of the wing-tip light. Finally a separate set of lights was added to the airplane. It consisted of a line of low-intensity blue lights spaced across the tail surface, along the fuselage, and across the wing. Five to seven units outline the airplane very well. Blue was chosen because it is least visible at a distance.

The landing light is a powerful projector that is used only during the landing and take-off periods. The rating of the lamp varies from 100 to 750 watts or larger in proportion to the size of the airplane. The focal lengths of reflectors range from $1/2$ to $1\ 1/2$ in. The peak candlepower of units that are designed for multiengine aircraft ranges from 300,000 to 1,000,000. Clear cover glasses are almost universally used, as there is little attempt to spread the beam.

The location of the lighting unit and the direction of the beam depend on the design of the airplane. It is desirable to locate the beam as far as possible from the pilot's position to avoid side glare and looking through the beam. In single-engine aircraft, the units are usually located in the wings and are directed outward enough to parallel the pilot's line of vision over and beside the engine cowl. In multiengine aircraft the pilot has clear vision ahead, and the light beams are directed parallel to the line of flight. In some instances the beams have been directed across each other, because the changing pattern of the intersecting beams gave valuable information regarding elevation and attitude.

The beam direction in elevation is shown by Fig. 21. At least the lower portion of the beam should lie on the ground during the taxiing period. Yet the beam cannot be tilted downward too much, or it will lose its effectiveness during the approach. The best position is one in which the lighting is barely adequate for ground taxiing.

The design of the lighting unit has undergone some wide variations. Initially, a reflector approximately 12 in. in diameter was attached beneath the wing tips. The reflector was streamlined, with a 36-in. cone. The unit was heavy, unsightly, and inefficient. The next step was to partially bury a smaller unit in the leading edge of the wing. As the wing increased in size, it became possible to completely enclose the lighting unit in the leading edge and to cover it with a cover glass conforming exactly to the contour of the leading edge. This is still a favorite type of installation but often cannot be used because it will interfere with de-icing provisions, whether by rubber boots or by thermal methods. Again, in very high-speed aircraft the tendency is toward a thin wing having a sharp leading edge. This does not give adequate space for the landing light.

Retractable units located in the wing at the thickest section satisfy such objections. The only other piece of equipment which can interfere with the retractable light is the fuel cell. Retractable lights are usually an alternate choice because they are heavier, more costly, and less reliable than the simple fixed unit.

INSTRUMENT LIGHTING. The function of instrument lighting is to give maximum visibility to characters on the dial with minimum interference to vision outside the airplane. The eye accommodates itself to darkness very slowly, requiring approximately 20 min, and yet that accommodation can be lost in a second of exposure to bright light. General lighting within the cockpit is not satisfactory as instrument lighting, because the eye can never gain any night vision. Indirect lighting of instruments, using a reflecting cover, was the first effort to improve this lighting. Three-candlepower lamps were located between groups of four instruments, and the intensity of the light was controlled by rheostats.

A more efficient and effective lighting was achieved by inserting a small (grain of wheat) lamp into the instrument between the dial and its cover glass. This design was costly, and since the lamp was tiny it could be designed for only 3 volts. A series resistor as well as a variable rheostat was required for control of lamp voltage.

Since the start of night flying fluorescent paint has been used to mark the important characters and the needle of instruments. To increase the fluorescence, ultraviolet lighting was used in later designs. This gave ideal lighting, since the characters became the source of light and there was absolutely no glare. Some pilots, however, reported that the ultraviolet light irritated their eyes.

The next step was to use general floodlighting of the cabin, but with red light. Adaptation to darkness and night vision are not affected by red light. An excellent solution to the problem of lighting the pilot's cockpit may be a combination of red floodlighting and ultraviolet light.

INTERIOR LIGHTING. General illumination of cabins and interiors is usually achieved with simple fixtures described as dome lights. A 6-candlepower lamp in a dome reflector is adequate. Individual seat lighting, particularly to achieve a satisfactory light intensity for reading, has been troublesome. It is usually necessary to control the light distribution very carefully so that none reaches the adjacent seat. Candlepower ratings as high as 50 have been used, but since space is very limited, the fixtures are inefficient and subject to much improvement.

Lighting is also widely used in special applications, such as signals, warning signs, calling devices, and instruments. In the last application the lamp is again a signal, but it is actuated by an instrument element such as pressure or fire detection. Since the ever-present possibility of lamp failure is a constant worry, a "push to test" socket was developed for aircraft. Pushing the fixture closes the lamp circuit and tests the lamp.

29. INSTRUMENTS

Instruments can be divided into three groups: flight, engine, and control. The last type is increasing in importance and will soon be a greater problem than the others.

FLIGHT GROUP. Instruments used during the flight and navigation of the airplane show speed, direction, altitude, attitude, and time. The airspeed indicator, altimeter, and rate-of-climb indicator are actuated by changes in atmospheric or impact air pressure. The electrical system is not concerned with them. Directional instruments include the magnetic compass and the directional gyroscope. Attitude instruments are also gyroscopic.

Electrical wiring must be kept away from the magnetic compass to avoid errors of deviation. The directional gyro augments the compass, because it is rigid in space and gives a true indication of direction during a turn. The two instruments have been merged in two different ways. One uses a gyroscope to give the compass a stable platform and thereby avoid turning errors. The second method uses a compass to correct constantly the precession of the gyro, which provides an ideal directional instrument.

Attitude instruments include the turn-and-bank indicator and the attitude gyro, flight indicator, or artificial horizon. The first accurately indicates the rate of turn; the second shows the attitude of the airplane in bank and pitch.

Many successful gyroscopic instruments receive power from air suction. Disadvantages of this power include introduction of moisture and dirt into high-speed mechanisms and a decreasing effectiveness with altitude. A closed pressure system corrects these difficulties, but electrical designs have proved lighter and more satisfactory. Both d-c and a-c power have been used. Dust from the brushes of d-c designs is objectionable, and for that reason the a-c type is gaining favor. This type usually receives power from a 115-volt, 400-cycle inverter.

ENGINE GROUP. Engine instruments indicate pressure, temperature, quantity, and speed. A pressure indication may be given by a signal light, as noted previously, or by a pressure-indicating element in combination with a means for transmitting the signal electrically. For example, the pressure can be converted to rotary motion with a Bourdon tube element. A self-synchronous motor is attached to the pressure unit and in turn is connected electrically to another and similar motor at the instrument panel. The second motor is attached to the needle of the indicating instrument. The indicator motor keeps synchronized with the transmitter and thereby delivers the indication.

Both a-c and d-c types of self-synchronous indicators have been developed. The a-c design is a miniature replica of industrial units. Direct-current selsyns use at the transmitter a potentiometer that is tapped at 120-deg intervals. Movement of the instrument element moves a pair of contacts diametrically opposite around the potentiometer. At the indicator a toroidal coil is correspondingly tapped at 120-deg intervals. A permanent magnet to which is attached the indicating needle follows the resultant magnetic field around the coil and effectively transmits the position of the potentiometer contacts.

Temperature measurements are made and transmitted electrically in practically all cases. Two devices commonly used are thermocouples and resistance bulbs. The former device is well suited to the measurement of high temperatures, such as cylinder heads or jet-engine tail pipe. The junctions are usually iron-constantan or copper-constantan.

The resistance-bulb method is suited to the measurement of low temperatures, such as oil and coolant values. The bulb is a resistance element located at the point where temperature is to be measured and connected into a Wheatstone bridge as one leg. The indicator is the galvanometer of the bridge circuit.

Quantity measurements are needed particularly for fuel. A float-actuated potentiometer has been widely and successfully used in this application. The indicator is simply a voltmeter. A later design measures the capacitance between a pair of cylindrical plates standing in the fuel cell. The circuit is balanced to zero when the cell is empty. Since the dielectric constant of gasoline is more than twice that of air, a measurement of the plate capacity can be calibrated in terms of fuel quantity. As the size of airplanes increases, the need for greater accuracy in the measurement of fuel quantity and flow also increases.

Rotational speed is measured electrically by two methods. First, a permanent magnet field type of generator is used as a source of the signal. The indicator is a voltmeter, since the generated voltage will be directly proportional to speed. The second method uses an a-c generator as the transmitter and a synchronous motor as the indicator. The latter drives a mechanical type of tachometer, such as the magnetic drag type, and the combination is the indicator.

CONTROLS. The type of control to be discussed here is exemplified in the automatic pilot. It has been classed with instruments because the reference signal is taken from gyroscopes. The signal is amplified and given to servos, which move the control surfaces and maneuver the airplane. One of the first successful designs was electric, but it was succeeded by a vacuum-hydraulic type that was widely used for many years. Vacuum drove the gyroscopes, and hydraulic power was used in the servos.

The first widely used electric design was operated entirely by d-c power. The signal was picked off the gyros by potentiometers, amplified through electronic circuits and delivered to relays. The servos were continuously running d-c motors with two clutches and a planetary gear system. Closing a relay energized a clutch and connected the servomotor to a control surface. This was a jogging control.

Later designs use a-c gyro motors, and the reference signal is picked off magnetically. The signal is again amplified through electronic circuits and delivered to servomotors, which have used both the amplidyne system previously described and a-c motors. The vacuum hydraulic system weighed 140 lb and was suitable for a 30,000-lb airplane. The electric system weighs less than 50 lb and will control the largest aircraft. Use of the gyro instruments as the reference will give still further weight saving.

Linking the aircraft radio equipment into the automatic pilot makes possible automatic take-off, flight, and landing. **Pilotless aircraft** are also derived from this combination. When automatic navigation is added, all the elements for guided missiles are at hand. More prosaic duties include continuous adjustment of trim tabs, temperature control of engines, and power boosters for the control surfaces. Electrical devices are ideal for most of these applications.

30. RADIO

POWER. The aircraft electrical system is concerned with only two aspects of radio equipment. Power of proper characteristics must be delivered to the radio receivers and transmitters, and the electrical system must create no interference with proper functioning of the radio equipment. For many years the primary power source was a 12-volt d-c system, and the other voltages needed for radio equipment were taken from a dynamotor. This was a heavy and inefficient solution, and the adoption of a 24-volt system did not alter the practice.

The introduction of high-power transmitters and radar equipment forced a revision of these practices. An inverter was substituted for a dynamotor, and alternating current was used in the radio equipment. With a-c power the design of radio equipment was made much more flexible, because higher voltages and any number of voltages could be used through transformers. The inverter was more efficient than the dynamotor, although some of this gain must be attributed to the higher power ratings.

Although both 400- and 800-cycle constant frequency and variable frequencies from 400 to 2400 cycles were tried, the trend has been to a constant frequency of 400 cycles.

Elimination of radio noise is obtained through the combined efforts of the equipment designer and the aircraft electrical engineer. The former must design the equipment so that it produces a minimum of noise in the frequency spectrum from 20 to 200 megacycles. The aircraft engineer then installs the equipment and its wiring so that the noise which is produced does no damage. Experience has shown that an absence of radio noise is proof that a given design has other desirable qualities, such as efficiency and reliability. The design problem can be expected to become more severe in the future, since the tendency is for the frequency spectrum to widen to values of 500 megacycles or higher.

SOURCES OF INTERFERENCE. The two sources of radio noise are the ignition and electrical systems. The ignition system for internal-combustion aircraft engines usually consists of dual magnetos on the accessory section of the engine, connected to the spark plugs with shielded ignition cable. Interference can be conducted into the electrical system through the booster coil and ignition switch. Radiation of interference through space is minimized by the shielding harness, the shielding effect of the engine cowling, and the shielding of radio equipment by the fuselage. The shielding harness consists of flexible metal tubing, which encloses each ignition cable from the point where it leaves the magneto until it enters the spark-plug enclosure. The metal tubing is grounded or bonded to the engine as often as feasible and usually at intervals of about 12 in.

Most of the radio interference that comes from the electrical system originates in commutator-type machines and arcing contacts. The noise voltage can enter the radio receiver through the power leads or through the antenna circuits. A machine having good characteristics will produce less noise than an inefficient and cheap design. The noise can be further reduced by judicious filtering within the frame of the machine itself.

The final step in reducing noise must be made in the installation of equipment and wiring within the airplane. Noise-producing machines must be separated from radio equipment, and wiring must be carefully routed. These are the functions of the aircraft engineer, who is ultimately responsible for producing a noise-free installation. The final test is made with the aircraft receiver and all electrical equipment functioning. The noise produced by electrical equipment must not exceed the threshold noise level in the receiver itself because, regardless of airplane type, the flight mission is often dependent on radio equipment. Radar has only added to the complexity of the problem, since noise will give false indications on the radar screen.

PREVENTION OF NOISE. Noise is reduced or prevented by full utilization of four methods: shielding, filtering, routing, and bonding. Shielding is the placing of a conducting metal wall between the source of noise and the receiver. Aircraft structure is used as a shield as often as possible. Aluminum conduit, rigid or flexible, is used when other shielding is not available. In some types of aircraft all the radio equipment can be isolated in one compartment, which shields it from the rest of the airplane. When this is possible, the remainder of the shielding problem is reduced to a negligible amount. In military aircraft this procedure cannot always be used, because radio equipment is scattered throughout the airplane and because changes in installation are required in the combat zone.

The routing of wiring and antenna lead-ins can do much to reduce noise. Wiring should be kept as far from the receiver, its antenna, and its power leads as possible. Whenever possible, the wiring should be shielded by structure. The routing should be such that the linkage of magnetic fields between wiring and antenna is reduced to a minimum.

Filtering is a last-resort method of reducing noise because it increases weight. Filtering can be effectively accomplished at either the source of noise or at the receiver. If done at the source, the wiring to the electrical equipment is free of noise, and the routing and shielding of that wiring are much less critical. Source filtering is more effective but will add considerable weight. Another objection to source filtering is that aircraft designs differ, and the same piece of equipment will need quite different filters for different airplanes. A filter that is needed for the most critical airplane becomes useless weight in another, where routing and shielding will reduce the noise.

Receiver filtering has many advantages. Since most receivers consume little power, a good filter will have little weight. A filter, combined with routing and shielding of electrical wiring and equipment, will permit almost unlimited noise levels from the electrical equipment itself. Receiver filtering will not be satisfactory when the installation of wiring and equipment cannot be controlled, as in military aircraft. Here source filtering is better because the entire electrical system is quiet, and either electrical or radio equipment can be shifted at will.

A compromise requirement that might be suited to all aircraft is to require equipment designers to meet a minimum noise level which is consistent with good design. If the machine exceeds this level, the designer must add enough source filtering to meet the minimum requirements, or revise the design. The radio receiver will also have its filter. Such requirements will divide the responsibility for noise elimination but will lead to the installation of lightest weight.

Atmospheric noise cannot be controlled completely by the methods outlined. Experience has shown that portions of the aircraft can reach high differences of potential, and that the breakdown of this voltage will cause noise. Bonding is a method of reducing this noise. A bond is a good electrical connection and is applied between any two structural parts between which such differences of potential may rise. For example, the control surfaces are surprisingly well insulated from the wing and fuselage by ball bearings. It is necessary to connect a flexible lead, or bond, across one or more of these bearings. Shielding must be bonded to structure at intervals of 2 or 3 ft. Long control cables and rods which are insulated from structure should be bonded at one end. The amount of bonding required is reduced by good shielding and filtering. When there is no noise, there is no need for bonds.

31. IGNITION

TYPES. The battery ignition system depends for its operation on the electrical system. Although this type has been widely used in automotive practice, it has not gained such popularity in aviation. It is generally considered to be less reliable, since the two systems must function correctly, and it may also need more care. Since the battery ignition type is thoroughly discussed in the automotive section, the remainder of this discussion will be devoted to the magneto types.

The magneto is the source of electrical energy and replaces the generator, battery, and induction coil of the battery system. The remainder of the system, consisting of distribution wiring and spark plugs, is similar to the same equipment in the battery system. Magnetos have been constructed in two-, four-, and eight-pole types. A nine-pole inductor type has also been used. For aircraft service the rotating field design is most popular. Dual ignition is used on aircraft engines larger than 100 hp to improve efficiency. A dual magneto is used as a source of power. Some mechanical driving parts are common to the two magnetos.

To shorten high-voltage distribution leads, the magneto has been divided into two parts in some designs. The low-voltage or primary side of the magneto is left on the accessory side of the engine. The secondary coil and distributor are located on the nose section. This variation can be carried a step further by locating individual transformers at each spark plug, an idea which has been developed experimentally in combination with a resonant circuit. It gives a single hot spark of sufficient voltage to jump any gap normally encountered. Since current across the gap is reduced, so is erosion, and the reduced wear, combined with higher voltages at the plug, should give much longer life to plugs.

REQUIREMENTS. The most important requirement of the ignition system is reliability. The second is efficiency, including efficiency in the ignition equipment and also in the performance of the engine. The engine must deliver ignition at the right instant in each cylinder to give smooth operation. As in any part intended for aircraft, weight should be reduced to a minimum. Installation should be easy, servicing simple, and maintenance reduced. Ignition is a constant source of radio noise, and the propagation of such noise-influence voltages should be neutralized or confined to the ignition equipment to a maximum extent.

The ignition system is subject to more extreme environmental conditions than most pieces of electrical equipment. It must be serviced at temperatures of -40 to -60 deg fahr, just as other pieces of electrical equipment. At the other extreme, however, the magneto is exposed to 250 deg fahr, cable to 350 deg fahr, and plugs to 500 deg fahr.

Since the system is mounted on the engine, all parts are subject to severe vibration at frequencies below 200 cycles per second, with forces as high as 75 g.

The reduction of dielectric strength and ionization potential which accompanies increase in altitude is very damaging. Corona discharge forms ozone and nitric oxides. The oxides combine with moisture to make nitric acid, which corrodes all metal parts. Both corona leakage and weakened dielectrics combat attempts to increase ignition voltage and thereby gain more effective ignition. To avoid the detrimental effects of altitude, the ignition system has been pressurized in some designs and in others it has been filled with a sealing compound. Both are expensive and heavy remedies for the environmental troubles of altitude.

MAGNETO. The magneto (Fig. 22) generates sufficient voltage to break down the spark-plug gap at the right time. One spark is produced for each flux reversal, or 2 per cycle. The voltage required is dependent on the plug gap, fuel-air ratio, and density of the gaseous mixture within the cylinder. If gaps do not exceed 0.025 in., the maximum design voltage of the magneto is about 12,000 volts. The plug will fire at less than half this value under normal service. The voltage required will rise with increase in capacitance of the distribution system.

The magneto must also deliver a small amount of power. The power delivered at the plug is smaller than the losses in the remainder of the system. Since ignition occurs with the first spark, any subsequent discharge across the plug gap is wasted and should be minimized. Losses occurring in the low-voltage or primary side of the magneto are the copper and iron losses of a simple generator. The capacitor which protects the breaker points will absorb some charging loss. Losses in the secondary side are due to leakage, capacity, and coil losses. Leakage occurs across the surface of distributor insulation, through the cable, and across fouled plugs. The capacity of the shielded distribution system causes a loss. Coil losses are due to the resistance of the winding and to iron losses not included with those calculated with the primary side. Mechanical losses include friction and windage.

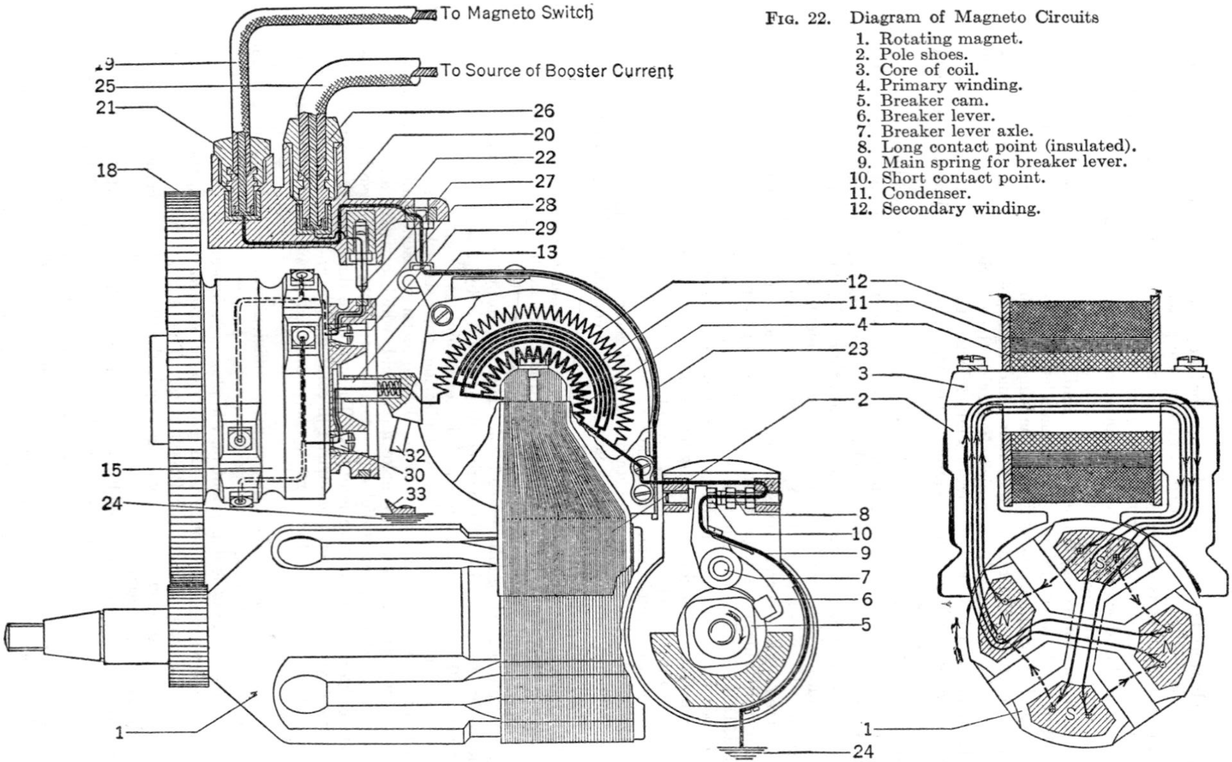

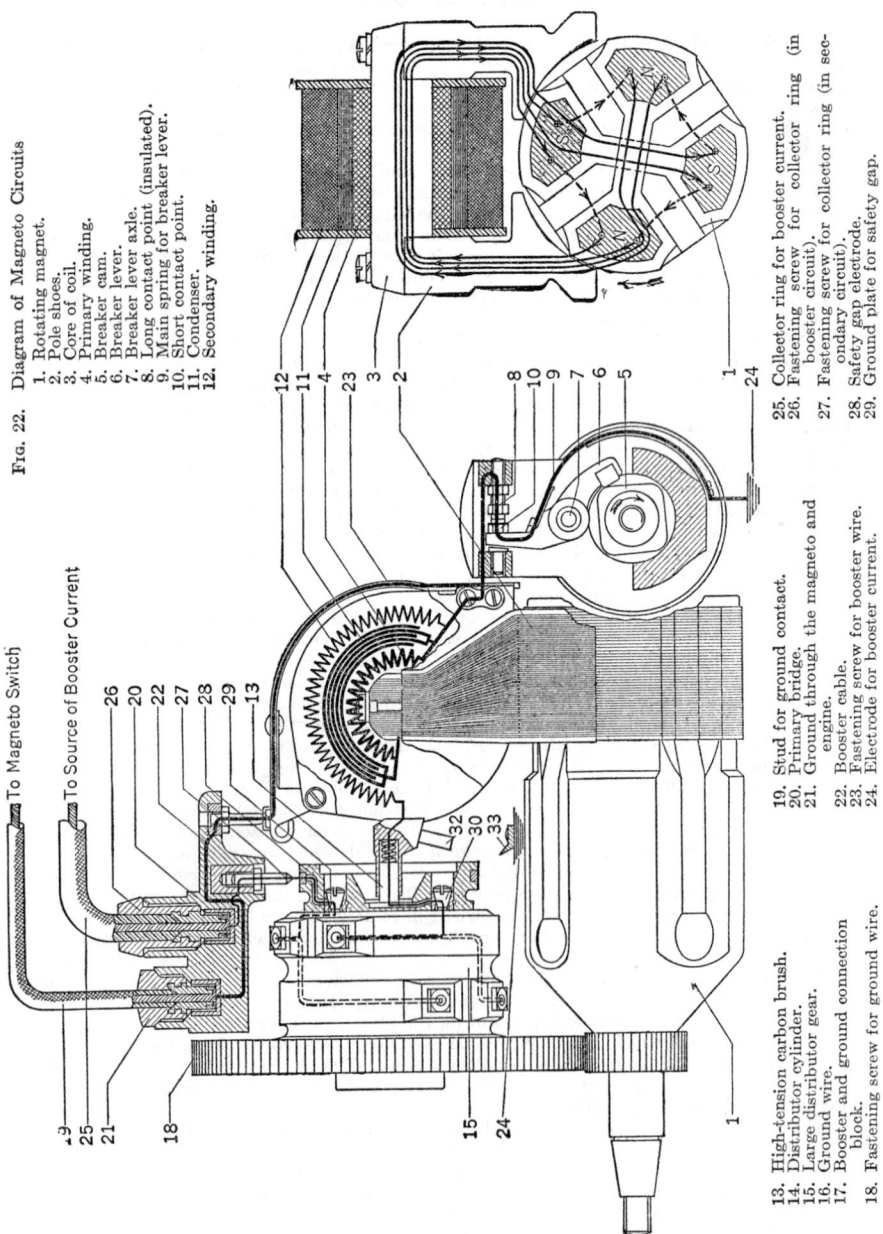

Fig. 22. Diagram of Magneto Circuits

1. Rotating magnet.
2. Pole shoes.
3. Core of coil.
4. Primary winding.
5. Breaker cam.
6. Breaker lever.
7. Breaker lever axle.
8. Long contact point (insulated).
9. Main spring for breaker lever.
10. Short contact point.
11. Condenser.
12. Secondary winding.
13. High-tension carbon brush.
14. Distributor cylinder.
15. Large distributor gear.
16. Ground wire.
17. Booster and ground connection block.
18. Fastening screw for ground wire.
19. Stud for ground contact.
20. Primary bridge.
21. Ground through the magneto and engine.
22. Booster cable.
23. Fastening screw for booster wire.
24. Electrode for booster current.
25. Collector ring for booster current.
26. Fastening screw for collector ring (in booster circuit).
27. Fastening screw for collector ring (in secondary circuit).
28. Safety gap electrode.
29. Ground plate for safety gap.

The magneto delivers sparking voltage to the plug approximately 30 deg before top dead center in most engines. A tolerance of ±4 deg of crankshaft rotation is allowed. In a typical engine this represents time values of 0.0004 sec and is the total of mechanical tolerances and electrical tolerances. The latter includes variation due to changes in the plug gap and the capacitance of the system.

The electrical tolerance includes the lag caused by inductance, capacitance, gap voltage, and energy required. It can amount to approximately 2 deg and can vary as much as 1 deg between adjacent cylinders. In the radial types of engines only one connecting rod is fastened to the crankshaft. The others are connected to the master rod. As a result the timing between cylinders is not even. The error in timing from this cause can be corrected by elaborate mechanical arrangements. By so designing the magnetic section of the magneto that the flux pattern is rectangular, variations in voltage at such odd timing intervals can be avoided.

With environmental temperatures ranging from −60 to 250 deg fahr, the primary design problems are those of lubrication and thermal expansion of different metals. A small air gap is desirable but difficult to maintain over such a range in temperature.

Flight at altitudes above 30,000 ft has been hampered by the ignition system. In the design of the magneto it has been necessary to increase the thickness of dielectrics, increase the spacing between points of high potential and ground, and round all corners. The last measure reduces voltage stress or gradient and delays the formation of corona. A safety gap is usually incorporated in the magneto to protect other portions of the system from high-voltage breakdown.

To reduce the effects of moisture, coils should be vacuum impregnated, all possible parts should be hermetically sealed, and insulating materials should be chosen for their ability to resist surface breakdown. Metal parts must be plated to avoid the corrosive effects of nitric acid. Without such plating some critical parts can be made inoperative in 12 hr of flight at high altitudes. Wells in the distributor block should be filled with a sealing compound to prevent the entrance of moisture around the cables.

Fundamental frequency of severe vibration is usually less than 200 cycles per second, and the magneto should be designed to avoid this frequency. During high-speed operation there is a tendency for breaker contacts to bounce. Moving parts should have very low inertia, and the retaining spring should exert a pressure of the order of 250 g.

Confining radio noise is most important at the magneto, as it is the source of the disturbing voltage signals. A metallic case that is fluidtight at all joints should enclose the magneto.

The magneto is mounted on the accessory section of the engine, several types of mounting pads being in use. To obtain proper timing the magneto drive is usually geared to the crankshaft. In some engine designs, the magneto or its high-voltage section is mounted on the nose section of the engine. The front location gives better cooling and shortens the high-voltage cables.

BOOSTER IGNITION. During starting the engine is rotated at speeds of 20 to 80 rpm. This is much lower than the cut-in speed of the magneto, and a supplementary source of ignition voltage is required. A popular solution is an induction coil whose high-voltage output is connected through the magneto distributor to the plugs. Later designs have substituted the secondary windings in the magneto for those in the coil. The external part then is a simple vibrator which delivers pulsating direct current from the electrical system to the magneto's primary system. Since one-half of the flux reversals from the externally supplied current will oppose the field of the permanent magnets in the magneto, care must be taken in the design of the magneto to avoid an undesirable reduction in residual magnetism.

SPARK PLUG. The spark plug (Fig. 23) uses the power and voltage from the magneto to produce the ignition spark. It is usually located in a recessed well in the cylinder head. Its metal shell is threaded into the cylinder head, and gas leakage is prevented by a gasket. The core is retained in the shell by a metal bushing, which is threaded into the shell. The core is the critical part of the plug. It is composed of the conductor or electrode and a surrounding cylinder of insulating material. Porcelain has been quite satisfactory as an insulating material for automotive plugs, but it cannot withstand the more severe conditions in a high output aircraft engine. Mica disks under compression are satisfactory for most operating conditions, but certain ceramics are more suitable for high altitudes.

In many designs shielding and cooling the plug are both accomplished by finned enclosures. Most failures of the high-voltage cable occur at the plug ferrule, which is the point where the cable meets its maximum temperature. It is also stressed mechanically when it is pulled into the tight-fitting elbow at the plug.

The gap setting of the plug has ranged from 0.008 to 0.035 in. Higher values are desirable, as fouling is less likely and a hotter spark results. Gaps of 0.015 in. were common in the 1930's, and 0.020 in. is now used. Gaps as high as 0.025 in. can be used at ordinary altitudes, but the voltages needed for this gap will cause premature failures in other parts of the system at high altitudes. Gaps of more than 0.030 in. are feasible with systems which use transformers at the plug.

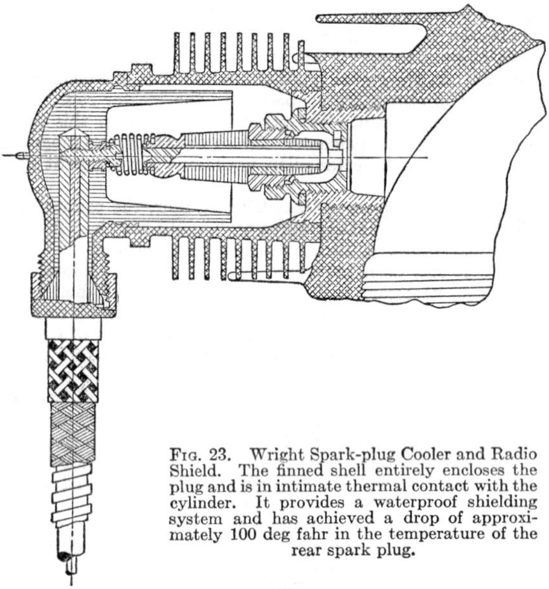

FIG. 23. Wright Spark-plug Cooler and Radio Shield. The finned shell entirely encloses the plug and is in intimate thermal contact with the cylinder. It provides a waterproof shielding system and has achieved a drop of approximately 100 deg fahr in the temperature of the rear spark plug.

CABLE. Transmission of ignition energy from the magneto to the plug is accomplished through ignition cable, which consists of a stranded conductor insulated with rubber or synthetic rubber. In some designs the rubber insulation is protected by lacquered braid whose purpose is to prevent damage from ozone and the resulting nitric acid.

The cable is subject to very severe operating and environmental conditions. It must withstand a maximum temperature of 350 deg fahr, and at the other extreme it must be flexible at temperatures of −40 deg. The lower value should be approximately −60 deg fahr, since at such temperatures plugs will be serviced but no material now available will withstand flexing at such a low temperature.

The insulation and its protective coating must withstand the corrosive effects of ozone and nitric acid, as well as the chemical action of oil, solvents, and aromatic gasolines.

Mechanically, the cable must withstand severe pulling and abrasion while it is being installed in the shielding conduit assembly. It will be severely compressed at plug ferrules. It and the harness should be so installed that no section is resonant at frequencies likely to be encountered on the engine. Rubber with lacquered-braid covering has not been entirely successful as insulating material. Glass braid with synthetic sealing compounds appears to be an improvement.

The cable will carry ignition voltages of 7.0 to 10.0 kv. Since the diameter of the cable must be kept small, an insulation of high dielectric strength is needed. At the same time a material and construction of low capacitance are desired. The energy required for charging the system is proportional to CE^2. A reduction in capacitance aids the design of the magneto.

Similar factors govern the choice of a conductor. It should be small in diameter to reduce capacitance. Conversely, reduction of diameter increases the voltage gradient. A third factor is the mechanical strength of the conductor.

Electrically, almost any material is suitable for a conductor. Resistance in the circuit is advantageous, as it reduces the crest current which follows breakdown across the plug gap, radio noise, and erosion of the plug points. Resistance to values of 100 ohms can be tolerated without affecting ignition efficiency. Most conductors use seven strands of 0.013-in. steel wire, giving an overall diameter of approximately 0.040 in. The interstices between strands are filled with a sealing compound to exclude moisture and to grip the rubber insulation more effectively.

For many years the overall diameter of the cable was 7 mm. Reduction of this diameter to 5 mm has proven advantageous for mechanical reasons. The cable receives less damage when it is pulled into shielding assemblies. The smaller diameter also reduces capacitance, since a larger portion of the shielding conduit is filled with air.

SHIELDING. Although the primary reason for shielding is the reduction of radio noise, it has other advantages, since it protects cable and aids cooling of the cable and plugs. Since shielding of cable increases its capacitance, it has the disadvantages of increasing the amount of current to be supplied by the magneto and the amount of eroding current which flows across the plug gap.

Basically, shielding is a fluidtight metallic enclosure around the magneto, cables, and plugs. It is bonded to the engine structure, usually at about 8-in. intervals. It reduces radiation of radio-noise-influence voltages by absorbing some of the energy and by neutralizing or localizing the remainder. The shielding with the cable becomes a concentric conductor; and, if the currents in the two parts are in opposite phase, their electromagnetic fields will counteract each other unless the wavelengths are short compared to the harness. For short wavelengths, added resistance in the conductor is required to make the current aperiodic in order to avoid radiation.

Shielding assemblies use as much rigid tubing as possible. The basic unit of an assembly for a radial engine is a ring of rigid tubing whose diameter approximates that of the crankcase. Flexible conduit sections extend from the magnetos to the ring and from the ring to each spark plug. Space within the conduit is made as great as possible because installation of the cables is easier and capacitance is decreased. The low-voltage control wiring which connects the primary side of the magneto to the switch in the pilot's cabin must also be carefully shielded and isolated from all other electrical circuits. Threaded couplings in the shielding assembly must be kept tight. Loosening a coupling will cause that section of the circuit from the coupling to the plug to radiate noise. Looseness at the magneto coupling will cause the whole assembly to radiate.

CONTROLS. The magneto type of ignition system requires only a switch for control. The switch is so connected that it grounds the ungrounded side of the contact points of the breaker. The breaker is thus short circuited, and there can be no pulsations of current in the magneto coil. Opening the switch allows the magneto to develop ignition voltage. To control the two magnetos on an engine, the switch has four positions: off, left, right, and both magnetos. Operation of the engine on each magneto gives a quick check of the condition of each half of the ignition system. Engine speed will drop about 5 per cent when using only one magneto. In multiengine aircraft the switches for each engine are combined into one assembly, and an emergency or master switch is provided to ground all breakers at once.

Advance and retard control is usually accomplished manually. It is needed only during the starting operation. The added control has been avoided in some installations by connecting the booster ignition through a lagging brush and thereby gaining the effect of a retarded spark during the starting operation only.

Power to operate a synchronization indicator has been taken from the magnetos. It is not easy to synchronize two or more engines exactly and to eliminate the annoying audible beat which accompanies two engines that are not quite synchronized. By connecting the primary circuit of a magneto on each of two engines to a sensitive indicator, a difference in speed is reflected by fluctuations of the needle. When the speeds are exactly alike, the pulses will cancel each other and the needle remain stationary.

Ignition requirements for the internal-combustion turbine type of engine are quite simple. An ignition spark is needed only during starting of the flames in the burners. Once started, no further ignition is needed. A simple battery ignition system is adequate.

32. BIBLIOGRAPHY

Chatfield, C. H., C. F. Taylor, and S. Ober, *The Airplane and Its Engine.* McGraw-Hill (1940)
Cobine, J. D., *Gaseous Conductors.* McGraw-Hill (1941).
Matson, R., *Aircraft Electrical Engineering.* McGraw-Hill (1943).
Morgan, H. K., *Aircraft Radio and Electrical Equipment.* Pitman (1941).
Sandretto, P. C., *Principles of Aeronautical Radio Engineering.* McGraw-Hill (1942).
Slepian, J., *Conduction of Electricity in Gases.* Westinghouse Elec. Co., Ed. Dept. Course No. 38, East Pittsburgh, Pa.
Veinott, C. G., *Fractional-horsepower Electric Motors.* McGraw-Hill (1939).
Webster, S. H., *Aircraft and Power Plant Accessory Equipment.* Musson Book Co., Toronto (1941).
Trans. AIEE, Vols. 60–64.
Trans. SAE, Vols. 49–53.
Publications, Bureau of Standards (1942–1946).
Instruments, Vols. 14–17.

ELECTRIC PROPULSION OF SHIPS

By W. N. Zippler

33. FUNDAMENTAL PROPULSION REQUIREMENTS

Practically all vessels are now propelled by "screw"-type propellers, the side wheel and stern wheel having generally passed out of use except for vessels operating in very shallow waters, e.g., in some parts of the Mississippi River. With existing propeller designs, the best efficiencies are obtained at relatively low numbers of revolutions per minute. Propellers seldom exceed a speed of 400 rpm and are generally operated between 80 and 200 rpm. The propeller may be driven by:

a. Direct connection to a reciprocating steam engine.

b. Reduction gears connected to steam turbine or turbines.

c. Direct or reduction gear connection to an electric motor which receives its power from a steam-turbine-driven electric generator.

d. Direct or reduction gear connection to a reversing internal-combustion engine.

e. Reduction gear, reverse gear, and clutch connection to a non-reversing internal-combustion engine.

f. Direct or reduction gear connection to an electric motor which receives its power from one or more electric generators, each direct connected to a non-reversing internal-combustion engine.

A ship may be equipped with one, two, three, or four propellers. Three-propeller designs are not often used in modern practice. The depth and speed of the vessel are, in general, the principal factors determining the number of propellers. As the designed speed is increased, the power required increases very rapidly, and in order not to obtain a poor propeller efficiency, the area of the propeller must increase with the power. The limit of area per propeller is determined by the diameter that can be suitably accommodated by the external dimensions of the vessel. It is advantageous to have the propeller submerged at all conditions of draft and yet not project below the keel of the vessel to interfere with "dry docking" or shallow-water operation; therefore it is necessary to provide more than one propeller. Maneuverability of a vessel is improved by the use of two propellers instead of one, because of the "turning" force that can be obtained by operating one propeller in the ahead direction and the other astern. Maneuverability may be further increased by the use of twin rudders with the twin propellers. No accurate data have been obtained indicating that there is a gain in this respect with four propellers instead of two.

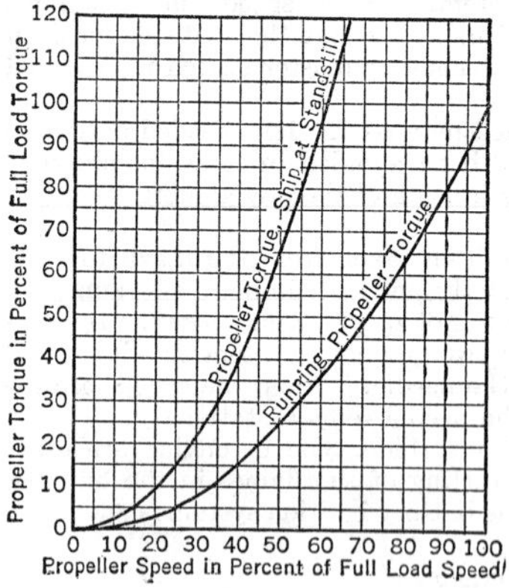

FIG. 1. Propeller Characteristics: Vessel Running Free and Vessel Stationary

PROPELLER CHARACTERISTICS. Figure 1 shows the relation between the speed and torque of a propeller when the ship is being driven at constant speeds in smooth water and when the ship is held stationary. It will be noted that the running propeller torque increases approximately as the square of the speed. The power to drive a ship, therefore, increases approximately as the cube of the speed. The exact values, however, vary with each propeller and hull design.

TORQUE REQUIRED TO STOP AND REVERSE. When the driving power is removed from a propeller with the ship traveling at a high speed, the propeller will continue to revolve in the same direction, driven by the water. In order to stop the ship in the shortest time or to maneuver quickly, it is necessary not only to stop the propeller against the action of the water but also to drive it in the reverse direction.

The relation between the propeller torque and speed during reversing has been determined experimentally by numerous model tests and ship trials. Figure 2 shows typical characteristic curves of propeller torque versus revolutions per minute, for both a normal medium-speed merchant vessel and a light vessel of high power and speed.

The relationship shown in Fig. 2 varies considerably for different vessels and is influenced by the propeller design, the rate of reversal, and the rate at which the vessel decelerates during the reversal period. Since the rate of reversal is dependent on the astern-torque versus revolutions-per-minute characteristics of the propelling machinery, the curves shown by Fig. 2 may also differ for two similar ships with similar propellers but with different propelling machinery.

During reversing, when ahead driving torque is removed, the propellers continue to rotate in the same direction, being driven by the water. In order to stop and reverse the propellers, astern torque must be available in excess of the maximum negative torque required by the propeller.

Comparison of curves *A* and *B* of Fig. 2 indicates the extent to which ship deceleration reduces the astern torque required. Curve *A* shows the torque which would be required if the vessel maintained its full-ahead speed while the propellers were reversed. An average medium-speed merchant vessel decelerates sufficiently during the reversal period so that the propeller torque is more nearly approximated by curve *B*.

For light, fast, high-powered vessels which decelerate rapidly and have propellers with wide blades, the torque–revolutions-per-minute relationship is more nearly represented by curve *C*.

When the propeller is turning in the ahead direction with negative torque, the energy which is supplied by the propeller must be absorbed in the propelling machinery. On Fig. 2 this energy is represented for curve *B* by the area 1–2–3–4.

The astern torque which should be provided to assure rapid reversal of the propellers rarely exceeds 100 per cent of full-ahead torque and is usually much lower. It is customary to provide 80 per cent for merchant ships.

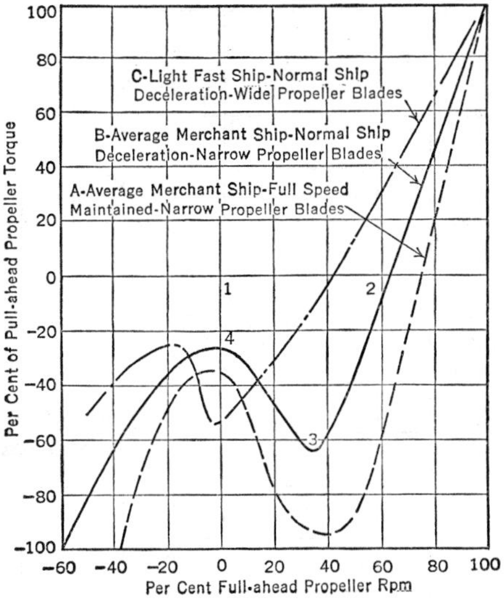

FIG. 2. Propeller Characteristics: Vessel Full Speed Ahead to Slow Speed Astern

VARIATION OF PROPELLER TORQUE WITH PITCHING OF SHIP. Elaborate tests have been made to determine the variation of the propeller torque when the ship is in a sea-way. An illustration of the variation which may be expected in moderately rough sea is shown in Fig. 3. Although the ship was pitching only 4 deg, the increase in torque varied from no load to 175 per cent of normal load.

These results were obtained on a cargo ship equipped with double reduction gearing. This turbine equipment did not have a governor, and the variation in torque may, in part, have been due to the variation in speed of the propelling machinery. Under worse conditions of sea, it may be expected that this variation in torque would be considerably increased.

It was observed that, when the propeller approached the surface of the water, the load was decreased, and as its submergence was increased, the load increased. The record was taken with the ship in ballast, and it was observed that the propeller did not break water. This variable condition of torque has been confirmed on electrically driven ships by observing the variation of load by electrical instruments.

The electric motor is an ideal piece of machinery for absorbing these shocks without deterioration. As there is no mechanical connection between the turbine and the propeller, the shock is absorbed in the air gap of the motor, it being necessary only to give the electrical machinery a sufficiently strong magnetic bond to hold the generator and motor together. This bond is increased by simply increasing the degree of excitation. In a-c drives, if the motors drop out of step with the generators, to bring them in step again it is necessary only to reduce the turbine speed and increase the excitation. Then the turbine speed can again be increased to normal. No harm is done if these motors are left out of step for short periods, as the generator is so proportioned that the increase in current cannot greatly exceed the normal.

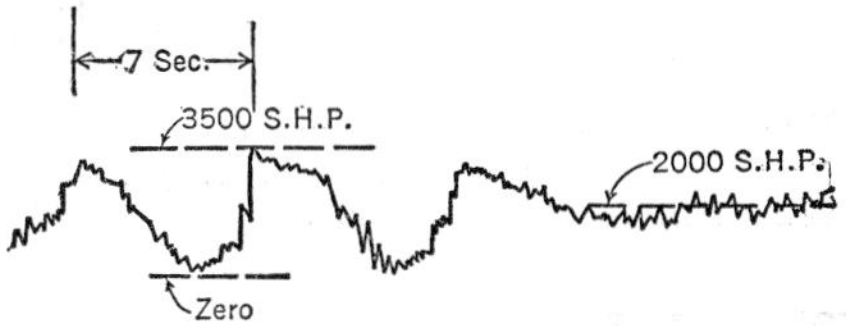

FIG. 3. Variation in Propeller S.H.P. with Vessel Pitching

TURNING. Another condition of maneuvering that affects the design of electric propelling machinery is "turning." Most direct-drive marine engines are not fitted with speed governors, the control of speed being entirely dependent upon the load and position of the throttle. A reciprocating engine or steam turbine, when operating with a fixed opening of the throttle, will develop approximately constant torque within a fairly wide range of speeds. Therefore, if the torque for a given number of revolutions varies as a result of rolling, pitching, or turning of the vessel, the speed of the propeller is inherently adjusted to maintain constant torque. Thus, without a governor, the power varies directly with the revolutions. In the electric drive the turbine is fitted with a governor which will hold the speed fairly constant, regardless of the load.

When turning a vessel which is driven by a turbine equipped with a constant-speed governor, the power in the screw on the outboard side of the turning circle drops, and that on the screw on the inboard side of the turning circle rises. After a short time the power on the outboard screw also rises. Figure 4 shows how, in a test of an actual electric-drive ship, the power first dropped on the outboard screw, reaching a minimum when the ship had turned through about 20 deg; it then began to rise, reaching its original value when the ship had turned through about 40 deg, and reaching a constant value when the ship had turned through about 140 deg. The power rose on the inboard screw from the beginning to the end of the turn. Thus, for this specific vessel, turning at 14 knots with the rudder at 25 deg, at the completion of the turn the increase of power on the outboard screw was 4.6 per cent, and on the inboard screw 53 per cent, making a total increase of power of 29 per cent.

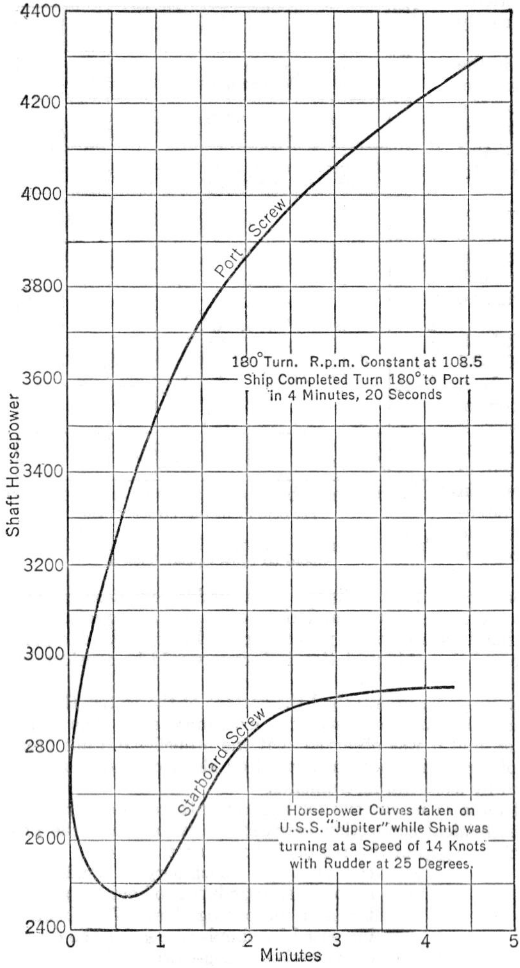

FIG. 4. Propeller Characteristics: Twin-screw Vessel Turning

Although these figures vary for vessels of different size and speed, the data give at least a rough indication of what may be expected in all prime movers equipped with constant-speed governors.

It is for this reason that means are provided on governed turbines to limit the steam flow, thus producing a decrease in speed when the power exceeds a predetermined value.

34. STEAM-TURBINE DRIVE

The relative light weight, small size, and high steam economy of high-speed steam turbines, compared with reciprocating engines, naturally suggest the use of such turbines for the propulsion of ships. Since the steam turbine has its best efficiency at a high speed, and the propeller its best efficiency at a low speed, some means of speed reduction must be provided to obtain the best overall efficiency. At present, two means of reduction are available, namely, mechanical reduction gears and "electric gearing." Mechanical gearing, as used on ships, generally consists of a single or double set of spur, herringbone, or spiral-toothed gears. Electric gearing consists of a high-speed generator directly connected to the turbine and electrically connected to a low-speed motor, which is direct connected to the propeller. Speed reductions as high as 68 to 1 have been obtained with mechanical gearing. Ships equipped with electric gearing are commonly referred to as being electrically propelled; and those equipped with mechanical reduction gearing as "geared turbine driven." There are many advantages and disadvantages for each type of reduction gearing.

The turbine has also permitted much higher steam pressures and steam temperatures than were ever considered for reciprocating engines. A recent merchant-marine installation has used steam conditions as high as 1425 psi gage and 850 deg fahr. Most modern merchant vessels now in service use steam conditions of approximately 450 psi gage and 750 deg fahr. The difficulty of determining the forces resulting from roll, pitch, vibration, etc., and the wide range and rapid change of power output required from each propelling plant have developed a state of conservatism that has slowed the acceptance of the higher pressures and temperatures used in land plants. A number of vessels built in Europe have used forced circulation boilers of the Lamont or Benson type with operating pressures of 1000 to 1200 psi gage.

The gain in fuel economy from the use of steam at high temperatures and pressures, together with high-speed turbines operating through double reduction gears, has resulted in the removal from service of practically all vessels equipped with direct-drive turbines. An outstanding example is the scrapping of the S.S. *Leviathan*, the largest vessel so far operated in the American merchant marine. The increased economy obtained in these modern propelling plants has resulted in more than doubling the driving power available at the propeller shaft per pound of fuel consumed, compared with that obtained from the direct-drive steam turbines of the S.S. *Leviathan*. Great reductions in weight and cost per horsepower have also resulted.

As is well known, superheat greatly improves the efficiency of steam turbines by reducing the friction and windage loss of turbine blade and disks. There is also a considerable gain due to increase in the available energy of the steam incident to increase in temperature. Steam consumption is reduced at least 1 per cent for every 12.5 deg fahr of superheat. With some installations there is a gain in boiler efficiency, caused by increased heating surface exposed to the flue gases. Superheated steam increases the life of the turbine blades and nozzles.

Because of the higher efficiency which results, superheated steam is now used for nearly all turbine-driven ships. Total temperatures higher than 850 deg fahr are still seldom used because special design and materials are required for the superheater, piping, and turbines.

TURBINES VERSUS RECIPROCATING STEAM ENGINES. Aside from questions of weight and cost, the higher steam economy of high-speed turbines is of itself ample in most cases to justify the adoption of the turbine for ship propulsion. The steam engine of the triple or quadruple expansion type, because of the limitations in size of the low-pressure cylinder, cannot be made with a steam-expansion ratio greater than about 16 to 1 or 20 to 1.

The reciprocating engine also cannot be used with highly superheated steam because of the difficulties in properly lubricating the cylinder walls at high temperatures and in obtaining a type of design and materials that can withstand the large stresses produced in the cylinder walls by the high pressures and temperatures.

The steam turbine has no such limitation and is capable of efficiently using steam expanded to any practicable extent. Where turbines are used, a vacuum of 28.5 in. is quite common, and 29.5 in. is being recorded on some vessels during winter months. The advantages due to high pressure, high superheat, and large expansions are indicated by the following tabulation of the available energy of a pound of steam when expanded from 600 psi boiler pressure, saturated (486 deg fahr), and at 850 deg fahr total temperature, to various degrees of vacuum:

600 lb pressure 486 deg fahr (saturated)	to atmospheric pressure = 206,000 ft-lb
" " " " " " "	to 24-in. vacuum = 272,000 ft-lb
" " " " " " "	to 26-in. vacuum = 288,000 ft-lb
" " " " " " "	to 28-in. vacuum = 311,000 ft-lb
" " " " " " "	to 29-in. vacuum = 334,000 ft-lb
" " " 850 deg fahr	to 29-in. vacuum = 428,000 ft-lb

The above figures for saturated steam give an indication of the saving that can be expected from the turbine which can efficiently use steam expanded to a low absolute pressure. The figures for steam at 850 deg fahr indicate that a considerably smaller quantity of superheated steam than saturated steam is required to pass through the system to produce the same horsepower output. (It should be noted that more fuel is required to produce a pound of superheated steam than of saturated steam.) This lesser quantity of steam results in less total wet weight of machinery, because of the smaller steam pipes, valves, condensate pumps, quantity of water in the system, etc. This weight saving is of material importance in a vessel, since every pound saved in machinery weight increases the revenue-producing load.

Because of these facts the reciprocating engine has passed into obsolescence where fuel consumption is a consideration.

REVERSING. Propelling machinery must, of course, be capable of backing a ship as well as driving it forward. With reciprocating steam engines, there is no difficulty in this respect. A steam turbine, however, cannot be reversed. Consequently, when a ship is equipped with turbines directly connected to the propeller shafts or connected thereto through mechanical gearing, each shaft must be equipped not only with a forward-driving turbine, but also with a turbine whose normal direction of rotation is opposite to that of the forward-driving turbine. Such turbines are called reversing turbines or astern turbines. When the ship is driven forward, the reversing turbine is driven backward, no steam being admitted to it. To reverse the direction of motion of the ship, the steam supply is cut off from the main turbine and admitted to the reversing turbine, the main turbine then being driven backward.

If it were essential to develop the same torque, power, and economy when going astern as ahead, the astern turbine unit would require approximately the same space as the ahead unit. In practice, however, a merchant vessel never goes astern for a long period of time, and the maximum speed desired is only a small percentage of that obtained for full speed ahead. The primary purpose of the backing power is to arrest the ahead motion. In view of these conditions, the astern turbine is physically much smaller than the ahead, is rated at much less horsepower, and has a very inferior efficiency. The quantity of steam required for developing the rated horsepower of the astern element may equal that required for full-speed, full-power-ahead operation.

With electric gearing, reversing turbines are, of course, unnecessary, since the direction of rotation of the motors connected to the propeller shafts can be reversed merely by operating suitable switches, and the torque astern is equal to the torque ahead. This is a decided advantage of electric propulsion over direct-connected or mechanically geared turbines.

Losses in Reversing Turbines. Experiment has shown that a turbine forced in a direction opposite to its normal direction of rotation has about ten times as much friction loss as when driven in its normal direction. This friction loss in a reversing turbine, which is always present when the ship is in motion, may amount to 1 per cent or more of the rating of the ahead turbine. This value varies with the rating and number of stages in the astern turbines.

TWO EFFICIENT SPEEDS. The propelling machinery of a war vessel should be capable of driving the ship efficiently not only at full speed but also at cruising speeds. This is also desirable in certain passenger ships. As the power required to propel a ship varies approximately as the cube of the speed, at cruising speeds the propelling machinery may be running at less than one-third of its maximum load. Since a steam turbine is most efficient at or near maximum load and at full speed, it follows that, when directly connected or mechanically geared turbines are employed, there is a sacrifice in efficiency at cruising speed, which is the speed at which a warship is driven by far the greater part of the time.

Cruising Turbines. To avoid this difficulty, it is possible to provide, in addition to the main turbines, an auxiliary turbine and reduction gear. This auxiliary propulsion turbine is designed to develop cruising power at a peripheral speed approximating the designed maximum of the main propulsion turbine. Clutches are sometimes provided for disconnecting the auxiliary turbines when the ship is driven at full speed by the main turbines. These auxiliary turbines are usually referred to as cruising turbines. In all modern naval and merchant vessels, where long periods of operation at reduced power are contemplated, cruising turbines are generally provided.

With a-c electric propulsion two efficient speeds are readily obtained without any such complication. All that is necessary is to change the ratio of the number of poles of the motors to the number of poles of the generators, which is readily accomplished by a simple switching operation. The induction motors on the U.S.S. *New Mexico*, which was the first electrically propelled battleship, having been put into commission in 1918, were arranged for pole changing.

Since the propeller torque required at half speed is only about one-quarter that required at full speed, the voltage applied to the motors at cruising speed can also be reduced, thus keeping the motors operating at relatively high efficiency at reduced speed.

GEAR LUBRICATION. In order to lubricate and carry away the heat generated in reduction gears, large volumes of lubricating oil must be continually circulated through the gear case. This requires storage tanks, piping, oil pumps, and oil coolers. Sea water is used for cooling the oil. A storage tank is usually located above the machinery so as to ensure an oil flow for several minutes upon failure of a pump, thus allowing sufficient time to stop the gears before damage.

Some vessels constructed during World War II are equipped with special-type pumps which are driven directly from the propeller shaft. This ensures lubrication whenever the gears are rotating and thus reduces the necessity for a large gravity-feed storage tank.

With turbine-electric drive a much smaller volume of lubricating oil is required because it is used only for lubricating the three or four turbine-generator bearings and the two motor bearings.

35. TURBINE-ELECTRIC DRIVE

The essential elements required for electric transmission between the turbines and the propellers of a ship are the generators, the motors, the cables connecting these two, and the switching and control devices.

ALTERNATING VERSUS DIRECT CURRENT. Where generators are driven by turbines, a-c apparatus is usually adopted for ship propulsion in preference to d-c apparatus. In a few cases, however, direct current has been used when small powers are involved, but this current is usually reserved for ships in which Diesel engines are used as prime movers.

Alternating current has many advantages over direct current, chiefly the following:

1. The combined transmission losses of the generator and motor, when a-c apparatus is used, are from 5 to 8 per cent less than with d-c apparatus.

2. High-speed prime movers and generators can be used with alternating current, whereas with direct current, when a high-speed steam turbine is used, it is necessary to have a reduction gear between the turbine and the d-c generator. High-speed apparatus is cheaper and lighter.

3. Alternating-current apparatus is more reliable than d-c and requires less maintenance, principally because of the lack of a commutator, thus eliminating the need for brushes and large, exposed current-carrying parts.

4. Higher voltages may be used for alternating current because of the absence of commutators and other uninsulated current-carrying parts.

For vessels requiring small powers, frequent and rapid reversals, and an accurate control of the speed from the pilothouse, d-c machinery is superior.

TYPES OF MOTORS SUITABLE FOR ELECTRIC DRIVE. Five types of a-c motors have been used for ship propulsion:

1. Induction motors equipped with slip rings and external resistors.

2. Induction motors without slip rings, but with two squirrel-cage rotor windings. A high-resistance rotor winding is placed in the slot near the periphery of the rotor and beneath it a low-resistance winding. During normal running, when the current alternates slowly in the rotor, the low-resistance winding carries the current. When, however, there is a high-frequency current flowing in the rotor, as during reversals of the propeller, the self-induction of the low-resistance winding forces current into the high-resistance winding, producing a high torque.

3. Induction motors without slip rings but with one squirrel-cage rotor winding, consisting of deep, narrow bars. The action of this motor during reversal is similar to that of the double squirrel-cage motor.

4. Induction motors equipped with slip rings but no external resistors. The rotor is wound with a low-resistance winding near its periphery, and beneath this a high-resistance squirrel-cage winding. The low-resistance winding is connected to the slip rings. To make the high-resistance winding carry the current during reversal of propeller, the low-resistance rotor winding is opened by means of contactor switches connected across the slip rings. If this motor is wound to give two different speeds, the slip rings can be made neutral for high-speed running and thus need not carry heavy current. All reversals are accomplished with low-speed connection.

5. Synchronous motors, the periphery of the poles being equipped with a low-resistance bar winding. With this type of motor the propeller is brought to rest during reversal by reversing the phases with the circuit "dead" and then applying a strong field on the motor, leaving the field off the generator. The propeller thus drives the synchronous motor as a generator, being loaded on the turbine generator. Sufficient torque is developed to bring the propeller approximately to rest. Field is now applied to the main generator and removed from the motor. If the turbine governor has previously been set for a slow speed, the motor will now reverse and quickly reach approximately synchronous speed, at which time the field is again applied. A small and simple air-cooled braking resistor can be switched in parallel with the generator during the stopping period. If

this is done, a higher torque can be developed with less excitation and with lower current flowing between the machines.

The following advantages are claimed for the synchronous motor over the induction motor:

1. There is a higher transmission efficiency due to operating at unity power factor.
2. Larger air gaps can be provided.
3. The generator and motor are lighter and cheaper.

All vessels constructed in recent years have been equipped with synchronous motors. Their decreased cost and weight and satisfactory performance have made it doubtful if any other available type will be used in the near future.

TYPES OF TURBINE GENERATORS SUITABLE FOR ELECTRIC DRIVE. The a-c propulsion turbine generator is of the conventional rotating field central-station type, except as follows:

a. It is designed for continuous operation at practically any speed from approximately 25 to 100 per cent of rating. In order to accomplish this requirement satisfactorily, it is necessary to design the turbine-generator rotating parts so that there will be no serious critical speeds within the operating range. In the central-station unit, because of operation at only one speed, it is necessary only to eliminate "criticals" in a narrow speed band.

b. The field windings must withstand the large current required to prevent collapse of the generator voltage when starting the propulsion motor with the turbine operating at approximately 25 per cent speed. As soon as the motor locks into synchronism with the generator, the field current is reduced to normal. Approximately double normal excitation is used during this starting period.

c. The governor must be designed to maintain approximately constant speed at any setting from about 25 to 100 per cent. Since a-c turbine generators are not usually operated in parallel, it is not essential to have the turbine speed governed within the limits required for central-station equipment.

d. It is also generally necessary, in a twin- or quadruple-propeller ship, to provide a steam-limit device on the turbine to prevent overloading the electric equipment to the "pull-out point" when the ship is turning at full power. The increased power requirements are discussed under Turning in Article 33. The steam limit simply limits the steam flow to the unit so that, when the overload occurs, the turbine generator and the propeller connected thereto will slow down, thus maintaining approximately constant torque.

The following are considered desirable turbine speeds (for the impulse type of turbine) for the corresponding generator output:

Horsepower	Rpm	Poles of Generator	Cycles per Second
6,000	5,600	2	93.3
7,000	3,600	2	60
10,000	3,000	2	50
16,000	2,400	2	40
29,000	1,800	4	60
65,000	1,200	4	40

CONTROL EQUIPMENT. The control equipment is usually placed in a wire-mesh enclosed cell with a steel or aluminum panel front. All the operating equipment and meters are located on the front of the panel. The switches, rheostats, current transformers, potential transformers, and other high-voltage parts are enclosed in the control cell. The entrance door to this cell is equipped with an interlock, which prevents application of excitation to the generators when the door is open.

All switches are of the air-break type, thus eliminating the fire hazard encountered with the oil-immersed type.

The connections are usually arranged so that the a-c generators cannot be operated in series or parallel, but either one, two, or four motors, depending on the number installed, can be operated from one generator. When two or four motors are operated from one generator, all motors must operate at the same speed in the same or opposite directions, because of the speed being determined by the setting of the turbine governor. The equipment in a four-propeller vessel is usually connected so that the two port motors are operated at the same speed and direction from a single control lever, and similarly for the starboard units. In all four-propeller vessels, it is customary to operate the port propellers as one unit and the starboard propellers as one unit, with alternate provisions for operating the port and starboard units either singly or from a single control lever.

When ready for operation, the turbine generator is started by opening the hand throttle. All electrical circuits are open, and the turbine-speed lever on the control board is at the minimum setting, which is usually about 25 per cent of rated speed. When the throttle is wide open, the governor takes charge and maintains approximately constant speed regardless of load.

After the "set-up" has been made, i.e., the selection of the motors and generators that are to be operated, the motor is started by operating the field control switch. The first step of this control closes the generator field circuit, applying approximately double normal current to the generator field windings. This condition is maintained until the motor or motors reach approximately constant speed; the field control lever is then moved to the next position, which applies current to the field of the synchronous motor. (This step is eliminated when induction motors are used.) As soon as the motor locks into synchronism, the load current decreases, and the field control lever is moved to the next position, which reduces the generator field excitation to normal.

From this point on, the speed is adjusted by operating the speed-control lever which adjusts the governor setting of the turbine, the motor speed following the turbine speed.

The speed-control lever and the field-control lever are interlocked, so that the turbine speed cannot be increased until the field-control lever is in the run position, and the field cannot be removed until the speed-control lever is in the minimum speed condition.

The motor-reversing switches are also interlocked, so that they cannot be operated until the turbine-speed lever is in the minimum-speed position and the field-control lever is in the "off" position.

The operation of the overload device opens the generator field circuit, and therefore under no condition of operation is it possible to open or close the main circuit switch without first removing excitation from the generator field.

VENTILATION OF ELECTRICAL APPARATUS. It is very important to ventilate the electrical apparatus in such a way that it will not collect moisture when shut down or running. To prevent this, the temperature of the apparatus must not be lower than that of the surrounding air. In order to prevent spray and water from being carried to the equipment, the ventilating air should be taken from the same room in which the apparatus is placed, except in Diesel engine rooms, where other conditions affect the situation. Suitable dampers can be provided to prevent the engine room from becoming uncomfortably cold during the winter by recirculating some of the heated air.

In most recently constructed electric-drive vessels, the closed system of ventilation is used, the air, after leaving the generators and motors, being passed around finned cooling tubes through which sea water is circulated. After passing the cooler, the air is returned to the generators or motors. Motor-driven fans are usually required for circulating the air in the motor-cooling system, because of the large size and low speed of the motors. Thus the same air is used continually. This method eliminates the large ducts leading from the upper decks of the ship and also ensures the use of clean dry air at all times.

In order to keep the temperature of the internal parts of the motors and generators above that of the surrounding air during shutdown, electric heaters are generally provided adjacent to the windings.

36. ADVANTAGES OF ELECTRIC PROPULSION

Some of the advantages of electric propulsion have already been noted. This subject has been very fully treated by W. L. R. Emmet (see Bibliography, Article 39).

RELATIVE WEIGHTS OF ELECTRIC DRIVE AND GEAR DRIVE. Under some conditions, when properly designed, the weight of the machinery required for electric drive is but little in excess of the weight required for single reduction gear drive. Generally, the full theoretical advantage of the electric drive cannot be realized, because the dimensions of most vessels at the stern are insufficient to accommodate the large motor diameter required by a large speed ratio without excessive slope of the propeller shaft.

EFFICIENCY OF ELECTRIC TRANSMISSION. The greatest advantage of the electric transmission, from an efficiency standpoint, is in the higher powers, because the efficiency of motors and generators increases with increased ratings at a greater rate than with mechanical gears.

In the 180,000-shaft-horsepower airplane carriers, the overall motor and generator efficiency at full power, including excitation, power input to vent fans, control losses, etc., is approximately 92.7 per cent. A reputable manufacturer of large-powered reduction gears guarantees an efficiency of approximately 97.5 per cent for a set of double reduction gears having approximately the same ratio as provided for the electric transmission. From test data taken on turbines operating in the reverse direction, we may expect a loss of approximately 1.2 per cent in the reversing turbine when steaming ahead. This leaves a net difference in efficiency of 3.6 per cent in favor of the geared drive at full power for this particular installation.

When operating at reduced power and speed, the electric drive shows a smaller reduction in efficiency than the gear drive, because of inherent characteristics of electric machinery, which has a nearly uniform efficiency within certain limits of voltage reduction. In central-station generating equipment, the efficiency at reduced load shows a greater reduction because the voltage and speed are maintained at a constant value. In addition to this advantage, the electric drive permits the operation of a lesser number of turbines at the reduced power by virtue of its ability to operate more than one propeller from the same turbine. In the equipment specifically mentioned above, the overall efficiency at approximately 30 per cent power is only 1/2 per cent less than at full power. Since in the turbine-electric drive the turbine speed is reduced in the same proportions as in a geared drive, the efficiency of the turbine is the same for both types of drive.

In the example cited above, the motors were of the induction type; if synchronous motors had been used, the difference in overall efficiency at full load would have been reduced to approximately 2.5 per cent in favor of the geared drive.

GAIN IN EFFICIENCY AT LOW SPEEDS BY OPERATING AT REDUCED VOLTAGE. Electric motors for shore purposes usually operate at approximately constant voltage regardless of the load carried. At light loads, therefore, they are relatively inefficient because of the iron losses remaining approximately constant. Motors used for propelling ships, however, usually obtain their power from a generator whose voltage can be varied to suit the load of the motor. At reduced speeds, therefore, in an electrically driven ship the voltage can be reduced; this reduction has the effect of materially reducing the losses in the iron laminations of both motor and generator. Thus, at reduced loads the percentage of power lost is nearly the same as at full load; whereas with gears the transmission loss remains quite high at reduced speeds. Also the loss of power in the gears increases when the gears have become worn, but the efficiency of electric machinery is not affected by use.

LOCATION OF PROPELLING MACHINERY. When vessels are equipped with turbine-electric propulsion, the driving motors can be placed as far aft as convenient, thus shortening the propeller shafting. The turbine-generator units can be placed near the boilers and at any desired level, reducing to a minimum the length of steam piping and other piping systems. The controlling mechanism can be placed at any convenient position.

EASE OF DISCONNECTION AND REPAIR OF PROPELLING MACHINERY. "In a geared equipment, each shaft has a system of turbines, gears, bearings, thrust-balancing devices, and lubricating systems all mechanically locked together. With high-speed machinery any kind of trouble with any of these parts will almost certainly necessitate the immediate stoppage of the whole system. To keep a high-speed turbine running out of balance or with bearing trouble is impossible, and the gearing part would present almost equal difficulty. In the event of mechanical trouble of such character, a ship would have to be stopped until the damage could be cleared. The work necessary to uncouple and disconnect any part of such very heavy apparatus would be a serious matter involving much time, including that required to stop the ship.

"If it was found impracticable to make this disconnection, and the damage was such that the shaft could not be allowed to revolve, it would be necessary to lock the shaft to hold the propeller. This locked propeller acts as a very serious resistance to a vessel traveling at even moderate speeds. In fact, such a dragging screw may add 20 per cent to the horsepower and would, in addition, materially reduce the maneuvering qualities of the vessel.

"In the electrically driven ship there is no mechanical connection of the shaft to anything but the rotors of motors. These are self-contained, iron-clad structures and cannot by any possibility be subject to mechanical interference. The shafts are subject to the same possibilities of bearing and thrust trouble as shafts in other ships, but the presence of the motors does not increase this danger and the speed being low it is remote in any case. With this equipment any motor, generator, or turbine, if in any kind of trouble, can be instantly disconnected without stopping the ship and with only a small loss from the highest speed capacity. Such a disconnection is made by simple switches."*

MUD IN CONDENSERS. In addition to the foregoing, there is a great advantage when maneuvering an electrically propelled vessel in shallow or muddy water in and around harbors, provided the ship is equipped with more than one turbine and condenser. Under the conditions only one turbine generator is needed to drive the ship, and if the condenser

---

* Quoted by permission from W. L. R. Emmet.

of this unit is plugged with mud, it can be immediately switched off and another turbine generator connected to the driving motors.

Where turbines are mechanically connected to their respective propellers, under similar conditions of operation, it is necessary to use all the turbines together with their condensers. If such a ship runs into mud, the plugging of even one of the condensers may seriously impair the maneuverability of the vessel. It has been reported that an electric-drive ship entering New York harbor had to shift main generators twice because of the plugging of her condensers with mud, and these shifts were made so quickly that they did not affect the operation of the ship at all.

BACKING POWER. "In geared turbine-driven ships, it is necessary to provide backing turbines which must run idle in the reverse direction when the ship is going ahead. These backing turbines involve complications which are very objectionable, and if these are reduced to a desirable minimum, the backing power will be greatly reduced, compared to that easily provided with electric drive. Experiment has shown that a turbine forced in an opposite direction involves about ten times as much friction loss as when driven in its normal direction. This loss, therefore, is very appreciable in the backing turbines of ships. There are also serious difficulties and dangers in high-speed apparatus incident to the abrupt and wide changes of temperature where steam is suddenly admitted to a cold reversing or ahead turbine. With electric drive, the turbine need never be stopped when the ship is under way." *

CRUISING ECONOMY. With the electric design, the number of motors and turbines used can, to a certain extent, be adapted to the demand for power, whereas with the other types all parts must be kept running. This gives a very important gain in economy at all speeds below the maximum. At 18 knots (in the 180,000-hp airplane carriers for the U. S. Navy), only one turbine is required to drive the ship, and electrical arrangements are made by which the turbine can be run at full speed instead of at half speed, as it would run if the ratio were fixed as by gearing. Thus, the steam efficiency of the turbine at 18 knots, a desirable cruising speed, is equal to the best attainable at any speed, and the overall efficiency from turbine input to power developed at propeller shaft is only slightly less than at full power. A cruising turbine will increase the efficiency over that developed by the main turbine at reduced speed, but in order to equal the electric drive, the efficiency of the cruising turbine must be sufficiently better than that of the main turbine to overcome the reduced efficiency of the reduction gear and the losses in the idling main turbine. Cruising economy gives increased cruising radius without renewal of fuel supply. This has always been considered a matter of the greatest importance in warships.

FLEXIBILITY OF INSTALLATION. In this respect, electric drive has an advantage over any other type of machinery in which the prime mover is mechanically connected to the propelling shaft. The main turbine generators may be placed in any part of the ship that is most desirable; they may be placed in compartments forward of each other, and they may be raised up enough to place the main condenser underneath them—in fact, there is practically no limit, other than headroom, to the position of the main turbine generators in the ship. This gives electric drive an enormous advantage over all other types of machinery and enables the naval constructor to give far more adequate protection to the ship and machinery against damage by torpedo or gunfire. Those parts of the machinery—the main motors—which it is necessary to connect mechanically to the shafts are comparatively small and take up only a little space, so that they can be placed in small isolated compartments which will not menace the ship in case of flooding. Also, the motors may be placed as far aft as the ship's structure will permit, which is generally much farther aft than for steam-driven turbines, and therefore the length of the main shafting may be reduced. This constitutes a big advantage, both because there is less liability to derangement of the shafting due to injury to the ship, and also because there is less danger to the ship since the shafting does not have to pierce a number of watertight bulkheads. In addition to advantages from the point of view of protection, there are also advantages from an engineering standpoint. The shorter lengths of shafting make it easier to keep the shafts in line; the grouping of boilers around the machinery permits short and direct steam pipes, with a consequent reduction in weight and complication and a smaller drop in steam pressure. The same statement may be made of practically all the other piping systems of the ship, such as feed lines, oil lines, and exhaust lines.†

SUPERHEAT AND REVERSING TURBINES. Whereas high superheat is beneficial to the operation and life of turbines, it has caused serious trouble in reversing turbines because of expansion strains set up by rapid changes in temperature. When steam expands without doing work, its temperature remains nearly constant, and if work is done upon

* Quoted by permission from W. L. R. Emmet.
† Partially quoted by permission from S. M. Robinson.

it by a reversed turning turbine, this temperature is considerably increased. With superheated steam these temperatures may be extreme and may be injurious to the main turbine as well as the reverse turbine, since it also must be used for reversal of direction and since the two often occupy the same casing.

Figure 5 illustrates the possible steam temperatures which may be reached when maneuvering a ship equipped with reversing turbines. The ahead turbine in this test was first run at full speed in the reverse direction in a high vacuum until a constant temperature was shown by the pyrometer. A little steam was then admitted and a constant temperature was again reached at 825 deg fahr, when more steam was admitted; the temperature then rose so quickly that the steam had to be shut off. In the General Electric Company's shops it has been discovered that the reversing wheels of marine turbines turn blue with heat when operated at normal speed in a vacuum of 20 in.

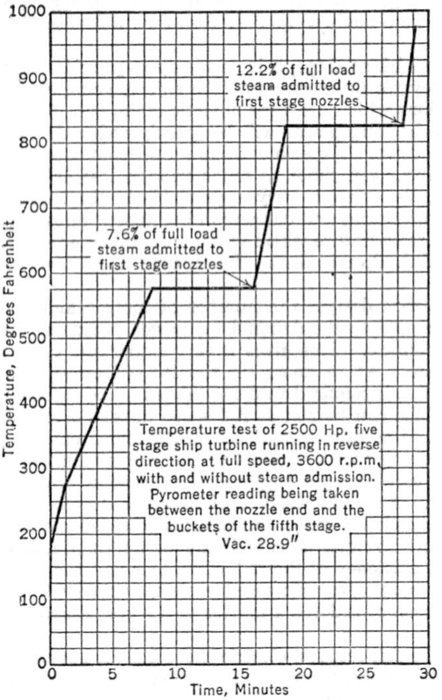

Fig. 5. Turbine Internal Temperatures When Rotated in Reverse Direction by External Force

When high superheat is used in combination with a reversing turbine, precautions must be adopted to preserve the turbine from damage. These complications are eliminated by the adoption of electric drive, which abolishes the astern turbine.

37. INTERNAL-COMBUSTION ENGINE DRIVE

GENERAL. Except for small pleasure boats and harbor craft, where the hazard of gasoline fuel does not outweigh the advantages of the internal-combustion gasoline engine, the heavy oil Diesel engine has in recent years progressed to a stage where its application as a ship-propulsion medium is established for all types of marine transportation, except the larger passenger and cargo ships. Diesel engines have been produced in a large range of horsepowers up to 6700 at 125 revolutions. Use of the Diesel engine is more widespread in merchant vessels of European registry than in those registered in the United States.

As in all other types of prime movers, the weight and space required per horsepower for Diesel engines decrease as the speed is increased. The use of reciprocating parts, however, limits the speeds below those obtainable with the steam turbine, but this limit is still in excess of best propeller performance. In order to obtain the advantage in weight, cost, and space saving of the higher-speed Diesels and the best efficiency of the propeller, a mechanical reduction gear or an electrical speed-reduction medium may be placed between the engine and the propeller.

The speed ratio between engine and propeller shaft for the larger-size engines generally does not exceed approximately 5 to 1. Although some small-horsepower Diesel-engine-driven vessels have been equipped with mechanical reduction gears, the usual installation is either slow-speed direct-drive engine or high-speed engines with electrical speed reduction.

The slow-speed direct-drive Diesel engine is connected directly to the propeller shaft. Therefore, when maneuvering the ship while entering or leaving port, the engine must be stopped, started, and reversed many times. The usual method of starting or reversing a Diesel is by operating it on compressed air until sufficient compression pressure is obtained in the cylinder to heat the injected fuel oil to the ignition point. During the docking of a large ship as many as thirty starts, including reversals of the propulsion engine, may be required. It is essential, therefore, that a large supply of compressed air be available, for without it the engine is incapable of being started and the motion of the vessel controlled. The necessity for this large amount of compressed air and the cost and weight of equipment necessary to furnish the air constitute one of the major disadvantages, in addition to the greater weight per horsepower, inherent in the direct-drive Diesel-engine installation. In a direct-connected engine it is very difficult to have the operating speed range of the engine free from criticals, so that it often becomes necessary to avoid operation at certain speeds.

Fuel consumption per shaft horsepower-hour for the Diesel engine is less than in the most modern high-pressure, high-temperature, steamship-propulsion plant, but the present cost of Diesel fuel oil is considerably higher than that of the grade of oil required by the steam plant. Because of this price difference, the fuel cost per shaft horsepower-hour of the Diesel engine is about equal to that of the steam plant or at least within debatable limits of equality with it. The price differential varies with location.

In contrast to the steam-turbine propelling plant, the Diesel engine consumes and contaminates its lubricating oil. As a result the Diesel plant requires more lubricating oil and purifying apparatus than is required by the steam-turbine installation.

Although fuel and lubricating-oil costs for any propulsion plant must be given serious consideration, along with the initial investment and operating costs, many other factors must be analyzed before final conclusions can be drawn for any particular ship.

Operating Cost. A review of the operating-cost data for several direct-drive Diesel-engine-driven cargo vessels indicates that the operating and maintenance costs for a Diesel ship of American registry is not materially different from that for a modern steam-turbine propulsion plant, when boilers and auxiliaries, ancillary to the installation, are considered. However, the maintenance cost of Diesel engines increases very rapidly when their cylinder sizes exceed a certain dimension because of the increased difficulty in handling the large parts in the limited confines of a ship's engine room. The overhaul of such engines usually necessitates the services of a contracting firm to provide additional men and equipment. This, of course, greatly raises the operating costs of the ship and frequently increases the ship's time in port.

The cost of large marine Diesel engines (4000 hp or over) in the United States is much greater than that of some other types of machinery of equal size because each engine is usually of special design and construction to suit a particular application, so that quantity production is not practical. Stationary or industrial engines, except in special cases, are not applicable to marine propulsion because of their greater weight and inability to operate continually at varying speeds.

DIESEL ELECTRIC DRIVE (ELECTRICAL SPEED REDUCTION). Diesel engines, when utilized in conjunction with electric generators and motors to propel a ship, are commonly known as Diesel-electric drive. With this type of drive several relatively low-horsepower high-speed Diesels, each directly connected to an electric generator, which in turn provide power to a slow-speed motor or geared electric motor connected to the ship's propeller shaft, accomplish the electric speed reduction previously mentioned. This type of installation overcomes several objections of the direct-drive Diesel plant; e.g., one outstanding advantage is that it eliminates the necessity for stopping and reversing the engines when maneuvering the ship.

MULTIPLE-ENGINE INSTALLATION. In either the a-c or d-c type of turbine-electric drive, it is customary to provide one turbine-driven generator of sufficient capacity to supply power to each propeller-shaft-driving motor. This is not true for the Diesel-electric type of drive. The relatively low-horsepower Diesel engine, like the automotive-type engines, lends itself to quantity production and consequent low cost per horsepower. These low-horsepower Diesel engines driving either a-c or d-c generators prove very advantageous to Diesel-electric drive, because several high-speed, light-weight engines, each connected to a generator, may be installed for providing power to a single propelling motor. This multiple-engine installation permits economical operation over a wide range in ship speeds. By proper selection of ship speeds the fuel consumption per horsepower can be maintained fairly constantly. For example, assume a single-propeller ship designed for a maximum of 10,000 shaft horsepower and powered with five approximately 2000-hp engine-driven generator sets. The fuel consumption per horsepower-hour for an output of 2000, 4000, 6000, or 8000 shaft horsepower, when operating with one, two, three, or four engines, would be fairly close to that required for the 10,000-shaft-horsepower output and much better than would be obtained with a single direct-drive engine operating at any one of the reduced powers. The present use of eight- and twelve-cylinder automotive engines for approximately the same power has proved conclusively that maintenance cost does not materially increase with increase in the number of cylinders. It is well known that the maintenance cost of an eight-cylinder automotive engine of modern design is no greater than that for the four-cylinder engine popular years ago. This is explained by the fact that the other economies of repetitive manufacture permit higher tooling costs and more highly developed parts.

Multiple-engine installation permits routine overhaul while at sea, because taking one engine and generator out of service has relatively little effect on the speed of the vessel. Repairs are facilitated, since replacement parts manufactured by quantity-production methods for high-speed engines are small, more accurate in dimensions, and of better materials than those for large-bore slow-speed engines. The quantity production of small engine parts also reduces the cost to a point where it is more economical to replace the part than to attempt to repair the damaged or worn component.

EXCITERS. In addition to the propulsion generators and motors, one or more exciters are usually installed. These exciters are either driven by the propulsion generator engines or are separate Diesel-engine-driven units. The exciters frequently operate at constant voltage and in addition to supplying excitation current to the propulsion generators and motors may be utilized to supply direct current to other auxiliaries of the ship, including lighting.

AIR SUPPLY. In order to utilize, in a Diesel-electric installation, the smallest and most efficient motors and generators, it is of vital importance that ample dry uncontaminated air be supplied to these units to maintain them within the maximum allowable temperatures and to prevent excessive hot spots in the insulation. Forced ventilation is usually necessary to accomplish this objective, and the source of air, the design of weather deck intakes, the discharge features, and the connecting air ducts require careful consideration. The air supply may be:

a. Taken from the engine room through the motors and generators and discharged to the open deck.

b. Taken through a plenium chamber on the weather deck, through air ducts to the motors and generators, and discharged directly to the engine room.

c. A combination of methods *a* and *b*.

d. The closed system of circulation.

Method *a* is objectionable because the air in a Diesel engine room is usually contaminated with considerable oil vapor which, when forced through the motors and generators, deposits in the ventilation passages within the windings and in time deteriorates the insulation. Method *b* is objectionable because of the required size of the plenium chamber and baffles and the difficulty of designing them to prevent the entrance of salt spray during bad weather. In addition, the high-temperature air discharging from the motors and generators increases the ambient temperature of the engine room and consequently causes discomfort to operating personnel. A combination of *a* and *b*, taking air from the weather deck through the motors and generators and discharging back to the weather deck, is satisfactory provided the design of the intake and exhaust features ensures against the entrance of salt spray. Method *d*, the closed system, is preferable where large volumes of air are required. Experience has shown that it is difficult to construct a topside intake and exhaust system that will be free of salt spray under all conditions of operation. With the closed system the air is continuously circulated through the motors and generators and attached air coolers. No air is taken from the outside. These coolers consist of finned tubes through which sea water is circulated. The air is forced over these finned tubes, either by a fan built into the armature or rotor of the propulsion generators and motors or by separate motor-driven fans, if the speed of the motor or generator being cooled is not high enough to provide efficient operation of a built-in fan. With this arrangement care must be taken, especially for d-c machinery, to assure that carbon dust in large quantities is not picked up by the cooling air and constantly recirculated, causing closure of ventilation passages in the windings. This condition may be overcome by (1), providing filters, but this method requires additional space and increases the weight of the installation, or (2), providing an enlarged cross-section at one point in the air passage, which will so reduce the air velocity that most of the carbon dust will drop to the bottom, where it can be removed readily.

SPEEDS. When using d-c machinery, the speed ratio between the engine and propeller is readily changed from that required to develop maximum power to zero, and the propeller motor can be reversed with very little difficulty. The procedure is so easily accomplished that, in most d-c Diesel-electric installations, the speed of the propeller is adjusted and controlled directly on the bridge instead of by the usual method of passing the order by means of a mechanical or electric telegraph from the bridge to the engine room, where the engineer makes the actual speed adjustment.

For d-c installations the shunt-wound, separately excited, d-c propulsion generator is directly connected to each Diesel engine, and a shunt motor directly connected to the propeller shaft. The propeller motor may be of the single or the double type. The double type consists of two electrically independent armatures on a single shaft with two bearings and two electrically independent field structures. If one motor, except shaft or bearings, is damaged for any reason, the vessel can still operate on the other motor at a reduced speed.

The first few installations of d-c Diesel-electric drive were designed for single-speed operation of the engine under governor control. Experience soon indicated that maintenance difficulties were encountered as a result of operating the engine at constant speed during the idling or reduced-power periods. Development of an adjustable, wide-range, constant-speed governor permitted satisfactory speed adjustment over a wide range, eliminated the previous difficulties, materially improved performance, and reduced maintenance costs. It is, therefore, now the accepted practice to operate the Diesel engine over a speed range from approximately one-third to rated.

This variable-speed method of operation permits the same system of control for Diesel-electric drive as for turbine-electric drive. The engines driving the generators are never reversed or started under load. This procedure reduces the starting air requirements, as the engine, once started, is not stopped until it is taken out of service or the vessel is secured at the dock.

SERIES AND PARALLEL OPERATION. For d-c Diesel-electric drive the generators may be operated either in series or in parallel. The most common method is to operate the generators in series, since by this method the division of load among the engines is not materially affected by a slight change in speed of one of the engines. Experience has shown that satisfactory parallel operation can be obtained by attention to engineering details and the installation of a properly designed, wide-range, adjustable-speed and torque type of governor with the correct speed droop. Satisfactory parallel operation has been obtained with an engine-speed range of 30 per cent to 100 per cent for any load condition that may be encountered in the normal operation of a vessel.

In the series method the armature circuits of the generators and motors are all connected in series, each generator being provided with a special switch, which has two positions. The "off" position opens the armature circuit of the associated generator and at the same time keeps the propulsion armature circuit complete for the remaining equipment. This will permit operation with one, two, three, or four generators in series, depending on the number available. In the parallel system a conventional-type generator switch is used for connecting and disconnecting each generator. All switching equipment is interlocked, so that it is impossible to open or close any of the main power circuits without first opening the generator field circuits, thus reducing the terminal voltage of the generator to residual and preventing heavy arcs by breaking large currents. The main power circuit switches are provided with arcing tips to facilitate renewal from any burning that may be encountered.

The shunt fields of the generators are operated in parallel, and each generator set-up switch has sufficient contacts to close the circuit to the generator field at the same time the generator armature is connected to the power circuit.

PROPELLER CONTROL. With the selected number of generating sets operating at idling speed, direction and speed of the propeller are controlled from the pilot house or engine room, as selected, by operating a lever or hand wheel for d-c machinery, which is connected to a potentiometer-type reversing generator field rheostat and an engine-governor speed-control device. The initial motion of the control lever or hand wheel applies field current to the selected generators for either the ahead or astern direction, depending on the direction of motion of the control lever or hand wheel. Further advancement of the lever or wheel increases the field until 75 per cent to 100 per cent of full field current is obtained. Further advancement of the control simultaneously increases the speed of the selected generators. In some designs full field is applied before the engine speed is increased. In other designs the field current is increased from some specified value, i.e., 75 per cent or better to 100 per cent, simultaneously with the increase in engine speed. The d-c machinery may be designed for continuous operation at any number of propeller revolutions per minute from zero to maximum.

Since a-c motors of the synchronous or induction type cannot be operated subsynchronously (described in detail under Turbine Electric Drive, Article 35), the lowest stable speed obtainable for continuous operation in an a-c plant is that corresponding to the idling speed of the engines. This is the primary reason for the use of direct current in tugs and similar vessels requiring long periods of maneuvering and slow-speed operation.

For the a-c system the generators must operate in parallel, and the propelling motor must operate at a fixed number of revolutions as determined by the speed of the engines.

The inability of the a-c plant to operate continuously at propeller revolutions per minute less than 25 to 30 per cent of maximum is not a serious handicap for an ocean-going vessel, when it is considered that at 30 per cent rpm a propeller will absorb only about 2.7 per cent of rated power and produce a ship's speed of about 30 to 35 per cent of maximum designed speed. (This statement is general, because the ship's speed, particularly at low powers, is very materially affected by the hull design, appendages, draft, and degree of underwater fouling.) Continuous operation of an ocean-going merchant vessel at such a slow speed would be extremely uneconomical. The slower vessel speeds required when maneuvering or docking may be obtained by regulating the "power" and "drifting" periods. Application and removal of power to the propeller in an electric-drive vessel with the prime movers rotating at idling speed under governor control are simple and rapid switching operations.

Unfortunately for the electrical system, the power required to drive a propeller does not vary directly as the revolutions but is generally rather close to the cube power. For example, for a typical installation:

Per Cent Power	Per Cent Revolutions per Minute
100	100
75	90.9
50	79.4
25	63.0
2.7	30
0.8	20

This cube characteristic of the propeller necessitates careful design of the electric machinery in multiengine installations in order to obtain a per horsepower-hour fuel consumption at partial power approximating that obtained at full power.

The speed-torque characteristic of Diesel engines, as presented by the manufacturers, indicates that full-load, full-speed torque is not available at idling speed. Some manufacturers specify a straight-line torque reduction from 100 per cent rpm to 30 per cent rpm, whereas others state that full-load torque is available from full speed to a lower speed, but that from this lower speed to idling speed the available torque gradually decreases.

This torque available and the torque required to drive the propeller at various revolutions per minute must be considered in the design and operation of the plant. No general design data applicable to all vessels can be given.

Since the generator is designed for the voltage required at full load, full speed, any increase in generator voltage to obtain a higher motor speed for the d-c series system when operating with reduced number of generators would mean an increase in the cost and weight of the generators. Therefore, in order to absorb full power at a voltage corresponding to the sum of the voltage ratings of the series operating generators, the motor field excitation is reduced for each generator removed from the circuit. This reduction increases the motor speed, thus adjusting the required propeller torque to equal the engine power available.

For the parallel d-c system the voltage across the propulsion motor for engine full speed would be the same, regardless of the number of generators operating, but the power available would be equal to the sum of the ratings of the generator units operating. With constant field on the shunt-type propulsion motor, the speed is approximately directly proportional to the applied voltage.

Full voltage would therefore produce full speed on the propeller and overload any number of engines less than the total. It is, therefore, necessary to decrease the maximum speed obtainable as the number of engines operating is reduced. The propeller revolutions may be reduced by any of the following methods:

a. Increasing the field excitation of the propulsion motor while maintaining full revolutions and full voltage on the generators.

b. Decreasing the generator excitation while maintaining full speed on engines and constant field on the propulsion motor.

c. Decreasing the engine speed while maintaining constant generator field and propulsion motor field.

d. Any combination of *a*, *b*, *c*.

Reduction of voltage by reducing the shunt field and maintaining engine speed would necessitate operating with a current exceeding the full-load rating if rated power is to be maintained on the engine. Either of these methods necessitates larger electric machinery. One method that has been used is to install a torque-limiting speed governor in the Diesel engine which automatically reduces the engine speed to a value where the available torque of the operating units matches the torque requirement of the propeller. The torque of the propeller, as previously stated, varies approximately as the square of the speed, and the available torque of the engine may remain constant or decrease with engine speed, as described in a previous paragraph. This decrease in speed and, in some instances, in engine torque as well makes it impossible to use the full horsepower rating of the remaining engines that are operating. Fortunately, the fuel consumption with constant torque and reduced engine speed is appreciably constant over a fairly wide range of speed. Therefore, even though full horsepower is not available, fuel consumption per horsepower-hour approaches that obtained at full load, full speed.

In the Diesel-electric a-c system, the speeds of the propeller and engine are directly related to each other by the ratio of the number of poles in the generator to the number of poles in the motor. Therefore, for an a-c installation the same method of controlling the output and speed of the propelling motor can be used as for a-c turbine-electric drive and d-c drive with constant excitation generator and constant excitation d-c shunt propulsion motor.

In a four-generator series d-c installation, the characteristics of the motor may not permit a sufficient reduction of the field to give the speed required for one-quarter power without the motor becoming unstable. This is overcome in a double-motor installation by using only one motor armature for the one generator condition or by building special motors which will be stable over a greater range of field excitation.

In order to keep the voltage to ground and between wires to a minimum in a double-motor series d-c installation, each half of the double motor may be connected between half the generators; i.e., in a four-generator installation with a double motor, the circuit is through two generators, one-half of the motor, the other two generators, then the other half of the motor.

The Diesel-electric drive introduces an additional power loss between the prime mover and the propeller, but this loss may be partially or even wholly made up in some cases by the better fuel consumption of the higher-speed engine and the increased efficiency of the lower-speed propeller. As the fuel consumption may be approximately the same per horsepower-hour at full, three-quarter, and one-half power for a four-engine installation, it requires only a very short period of operation at reduced power to effect a saving of fuel over that required for direct-drive Diesel equipment. In the direct-drive plant the fuel consumption increases quite rapidly per shaft horsepower-hour as the power and speed are reduced.

At the present time there are many very successful Diesel-electric d-c installations, the greatest number of applications being in tugs, small harbor craft, etc., where maneuverability and pilot-house control are of great importance. During World War II many Diesel-electric vessels were produced for the navy and operated by very inexperienced crews. The results were extremely gratifying.

MISCELLANEOUS APPLICATIONS OF DIESEL-ELECTRIC SYSTEM. The U. S. Army engineers have used Diesel-electric equipment in several of their **hopper dredges.** In this type of vessel the propeller and the dredge pump are each directly connected to a motor. When dredging, the propulsion generator is connected to the dredge-pump motor, and a smaller generating unit is connected to the propulsion motor. The speed of the vessel is very low when dredging; therefore the power required to operate the propelling motor is very small. After the hoppers are filled, the pump motor is shut down, and the propulsion generator connected to the propelling motor. This permits a reasonable speed of propulsion when proceeding to the dumping grounds with the minimum of idle equipment. This type of equipment makes it very easy to place the entire control of the dredging and maneuvering operations under one man on the bridge. Moreover, it is understood that the cost per cubic yard has been considerably reduced over that obtained with other types of equipment.

This double use of the propulsion generator has also been adopted in the **fire boat,** the propulsion generator being connected to operate the fire pumps when at a fire and the propulsion motor when on the way to the fire.

The U. S. Coast Guard has in operation several **ice-breaking vessels** of 10,000 shaft horsepower which make use of parallel operation of Diesel-electric machinery.

SUMMARY OF ADVANTAGES OF ELECTRIC REDUCTION GEAR. The advantages of the electric reduction gear for multiple-Diesel-engine installations are as follows:

a. Direct control of vessel on bridge.

b. Quicker reversal of propeller.

c. Greater torque during reversing.

d. Higher-speed Diesel engines.

e. Less air tanks and smaller compressors.

f. Small engines with small reciprocating parts.

g. Less chances of sea or port detention, since failure of one engine will not prevent operation of vessel and repairs may be made at sea without stopping vessel.

h. In some cases, fuel consumption at full load is same as for direct-drive Diesel.

i. Practically the same fuel consumption per shaft horsepower-hour at certain reduced speeds as at full speed.

j. Vessels requiring high port loads can utilize propulsion engines with consequent saving in auxiliary generator sets.

k. More flexible arrangement of units in ship, together with decreased alignment problems.

SUMMARY OF DISADVANTAGES OF ELECTRIC REDUCTION GEAR. The following disadvantages may be noted:

a. More equipment required than for direct-drive Diesel.

b. In the smaller horsepower ratings the initial cost and weight may be slightly more than for direct drive.

ELECTRIC DRIVE FOR AUXILIARIES. A source of considerable loss in the operation of many ships is the use of steam-driven auxiliaries. Both turbine and reciprocating steam-driven auxiliaries are relatively inefficient for low powers. It is for this reason that electric auxiliaries may be used advantageously to improve the operating efficiency of a ship's propulsion plant. One or two large turbine generators, with their individual condensers, will operate the many electric-driven auxiliaries required on a vessel more efficiently than individual steam-driven units can be operated. The initial cost of electric-driven auxiliaries is greater than that of steam, but, for auxiliaries that are operating continuously, the increased economy soon amortizes the difference in initial cost.

It is common practice, however, to make some of the largest auxiliaries steam driven in order to obtain exhaust steam for one or more of the stages of feed heating. If the amount of exhaust steam for these auxiliaries exceeds the feed-heating requirements, the excess may be discharged to a stage of the low-pressure propulsion turbine or to the propulsion condenser. If operation of any of these auxiliaries is essential in port, it is necessary to discharge the steam to the main condenser or install a separate condenser, together with the associated air ejector and circulating and condensate pumps. For best economy the exhaust steam should just equal that required for feed heating and all remaining auxiliaries electrically driven.

Some vessels have all auxiliaries electric driven and obtain all the steam for feed heating from appropriate extraction points in the low-pressure propulsion turbine. This works very well, since the amount of steam for feed heating varies directly as the steam requirements of the propulsion turbines.

Since the larger the turbine, the better the efficiencies, many attempts have been made to generate the electric power required for driving the various auxiliaries on the vessel at the same steam consumption per kilowatt of output as obtained for driving the ship. This has been accomplished in the following manner:

a. Direct-current generator driven from one of the speed-reduction gear shafts in a geared turbine drive.

b. Electric power taken directly from the propulsion generator in an a-c electric-drive vessel.

In case *a*, it is necessary to have a stand-by turbine generator which will automatically take over the electric load when the propulsion turbine speed is reduced below approximately 66 to 75 per cent of rated speed. This scheme necessitates an attached generator that will satisfactorily produce the required power at constant voltage over a speed range from 66 or 75 per cent to 100 per cent with a load range from zero to its rating.

In case *b*, special electrical equipment is installed that will provide over a specified speed range of the propulsion generator: (*a*) the correct voltage for the auxiliaries, (*b*) a constant voltage for the auxiliaries, (*c*) a constant voltage and frequency if it is essential to have the auxiliary operate at a constant speed, regardless of the speed of the propulsion unit.

For case *b*, as for case *a*, it is essential to install a stand-by generating set that will automatically take over the auxiliary electric load when the speed of the propulsion turbine is reduced below the predetermined speed. This special unit may consist of a steam turbine, an a-c motor, and a d-c generator mounted in tandem. The a-c motor receives its power from the propulsion generator over a speed range of 66 per cent to 100 per cent and drives the attached d-c generator, which is equipped with an automatic voltage regulator for maintaining constant voltage. When the speed of the propulsion generator drops below the designed minimum, steam is admitted to the turbine, the a-c motor is automatically disconnected from the propulsion generator, and the d-c generator is driven by the attached steam turbine.

One English organization has proposed operating the propulsion generator at constant voltage and frequency to obtain the electric power for operating the ship auxiliaries and to control the motion and speed of the vessel by the use of an adjustable pitch propeller. This suggestion has not been enthusiastically received because of the many moving parts which are required within the hub of the propeller and which cannot be serviced without drydocking the vessel.

SHIP-SERVICE ELECTRIC SYSTEM. Direct current was universally used for operating the ship's lighting and power system before World War II. It has been customary to provide 115-volt direct current for small merchant vessels, and a 230/115-volt, three-wire d-c system for large vessels. For ship lighting 115 volts is used, and 230 volts for all motors other than fractional horsepower.

In 1933 the U. S. Navy, primarily as a weight-saving measure, initiated a-c ship-service electric plants on naval vessels. A 440-volt three-phase 60-cycle power system was selected, with distribution to the three-phase, 115-volt lighting system by means of either groups of three single-phase or single three-phase transformers located at the appropriate load centers. In addition to reducing the weight and cost of the electric installation, the use of across-line starters and squirrel-cage motors has so materially reduced motor-maintenance cost that the equipment of all future large merchant vessels with an a-c ship-service electric plant is at present contemplated.

The d-c system will probably continue to be used for cargo vessels until an a-c motor or method of control can be developed that will equal or better the speed-torque characteristics of the present d-c motor type cargo winch.

38. TYPICAL ELECTRICALLY PROPELLED SHIPS

It will be of interest to note that from 1908 to the beginning of 1946 a total of almost 8,000,000 shaft horsepower of turbine-electric propulsion machinery manufactured in the United States was installed in more than 900 vessels of all types. Over 2,000,000 shaft horsepower of Diesel-electric propulsion machinery was installed between 1919 and 1946 in more than 500 vessels of all types. The average turbine-electric plant delivered about 8500 shaft horsepower, whereas the average output of the Diesel-electric plants was about 3600 shaft horsepower.

The following is a list of representative turbine-electric propelled vessels:

Name	Type	Number of Propellers	Total Shaft Horsepower, Full Speed	Type of Motors	Number of Ships in Class	Reference
U.S.S. *Tennessee*	Battleship	4	28,000	Induction	5	1
Rodman Wanamaker	Ferry	2	2,200	Induction	2	2
U.S.S. *Saratoga*	Aircraft carrier	4	180,000	Induction	2 *	3
California	Pass.-cargo	2	17,000	Synchronous	3	4
Chelan	Coast Guard	1	3,000	Synchronous	5	5
City of Saginaw	Car ferry	2	7,200	Induction	2	6
Oriente	Pass.-cargo	2	16,000	Synchronous	2 †	7
President Hoover	Cargo-pass.	2	26,500	Synchronous	2 ‡	8
Talamanca	Pass.-cargo	2	10,500	Synchronous	6	9
Corsair	Yacht	2	6,000	Synchronous	1	10
Louisiana	River boat	2	2,000	D-c special	1	13
Normandie	Pass.-cargo	4	160,000	Synchronous	1 §	11
Goethals	Hopper dredge	2	4,000	D-c special	1	13
U.S.S. *Suamico*	Tanker	1	6,600	Synchronous	488 ‖	13
U.S.S. *Buckley*	Destroyer escort	2	12,000	Synchronous	226 ‖	13
U.S.S. *Adm. W.S. Benson*	Troop transport	2	18,000	Synchronous	10 ‖	

NOTE. See table footnotes at end of next table.

The following is a list of representative Diesel-electric propelled vessels:

Name	Type	Number of Propellers	Total Shaft Horsepower, Full Speed	Type of Motors	Number of Ships in Class	Reference
Golden Gate	Ferry	¶	750	D-c shunt	2	12
Lake Tahoe	Ferry	2	1,400	D-c shunt	1	13
A. Mackenzie	Hopper dredge	2	1,600	Special	3	13
J. W. Van Dyke	Tanker	1	2,300	D-c shunt	1	14
N.Y.C. No. 34	Tug	¶	650	D-c shunt	1	15
ATF 96	Fleet tug	1	3,000	D-c shunt	22	13
Fresno	Ferry	¶	1,250	D-c shunt	6	16
Courageous	Cargo	1	4,000	D-c shunt	3	17
Firefighter	Fireboat	2	2,000	D-c shunt	1	13
Storis	Coast Guard cutter	1	1,800	D-c shunt	1	13
Sperry	Submarine tender	2	11,800	Synchronous	1	13
East Wind	Ice breaker	3	10,000	D-c shunt	8 ‖	13
U.S.S. *Evarts*	Destroyer escort	2	6,000	D-c shunt	168 ‖	13
U.S.S. *Grenadier*	Submarine	2	¶	D-c shunt	85 ‖	13

* The U.S.S. *Lexington* of this class was lost in the Battle of the Coral Sea, May 8, 1942.
† The *Morro Castle* of this class was burned in 1934.
‡ The *President Coolidge* of this class was lost in 1942.
§ The *Normandie* was severely damaged by fire and capsizing in February, 1942.
‖ Exclusive of losses and conversions during and after World War II.
¶ Data are not available or are restricted for security reasons.

39. BIBLIOGRAPHY

References to Specific Ships

1. Propelling Machinery of the U.S.S. *Tennessee*, *Marine Eng.* (Apr. 1921).
2. (No specific reference known.) Kennedy, A., Jr., and Frank V. Smith, Electric Drive Applied to Double-ended Ferryboats, *Marine Eng. & Shipping* (Oct., 1925).
3. Kranzfelder, Lt. Edgar, Propulsion Circuits of the U.S.S. *Lexington*, *J. Amer. Soc. Naval Eng.* (Nov. 1928).
4. Williams, Roger, Description and Trials of the S.S. *California*, *Trans. Soc. Naval Architects and Marine Eng.*, Vol. 36 (1928).
5. Newman, Comdr. Q. B., U.S. Coast Guard Cutters, *J. Amer. Soc. Naval Eng.* (Nov., 1928).
6. First Turbo-electric Car Ferry Launched at Manitowoc, *Marine Eng. & Shipping* (Sept. 1929).
7. *Morro Castle*, *Marine Eng. & Shipping* (Sept., 1930).
8. Presidents Hoover and Coolidge, *Marine Eng. & Shipping* (Aug., 1931).
9. *Talamanca*, *Marine Eng. & Shipping* (Jan., 1932).
10. *Corsair*, *Marine Eng. & Shipping* (Nov., 1930).
11. *Normandie*, *Shipbuilder & Marine Engine Builder* (June 1935).
12. Kennedy, A., Jr., and Frank V. Smith, Electric Drive Applied to Double-ended Ferryboats, *Marine Eng. & Shipping* (Oct., 1925).
13. No reference known.
14. *J. W. Van Dyke* Converted to Electric Drive, *Marine Eng. & Shipping* (July 1925).
15. *N.Y.C. No. 34*, *Marine Eng. & Shipping* (Feb. 1927).
16. *Fresno*—New Southern Pacific Electric Ferryboat, *Marine Eng. & Shipping* (June 1927).
17. Electric Motorship *Courageous*, *Marine Eng. & Shipping* (Jan. 1929).

General

Emmet, W. L. R., Electric Propulsion of Merchant Ships, *Gen. Elec. Rev.*, Vol. 23, p. 60 (Jan. 1920).
Kennedy, A., and R. A. Beekman, Some Phases of Ship Propulsion as Influenced by Modern Power-plant Engineering, *Gen. Elec. Rev.* Vol. 32, p. 626 (Nov. 1929).
Liston, J., Turbine-electric Drive for Modern Merchant Ships, *Gen. Elec. Rev.*, Vol. 23, p. 6 (Jan. 1920).
Robinson, S. M., Electric Ship Propulsion, *Gen. Elec. Co. Bull.* GEA-1526 (May 1932).
Smith, F. V., The Future Trend in Ship Electrification, *Gen. Elec. Rev.*, Vol. 32, p. 107 (Feb. 1929).
Smith, F. V., Diesel-electric Drive Modernized, *Marine Eng. & Shipping Rev.* (June 1944).
Smith, F. V., Propulsion Generators and Motors on Turbine-electric-Drive Ships, *Marine Eng. & Shipping Rev.* (Sept. 1944).
Smith, F. V., Control of Electric-drive Ships, *Marine Eng. & Shipping Rev.* (Nov. 1944).
Smith, F. V., Excitation Systems on Turbine-electric-drive Ships, *Marine Eng. & Shipping Rev.* (Dec. 1944).
Wooler, R. G., Maintenance of Geared-turbine and Turbo-electric-drive Units, *Marine Eng. & Shipping Rev.* (May, 1944).
Bowes Electric Drive for Ships, *Marine Eng. & Shipping Rev.* (Aug. 1944).

RAILWAY SIGNALING

By Robert B. Elsworth

40. GENERAL

The importance of railway operation, the financial values, and the weights and speeds involved make it essential that railway signaling systems be devised, constructed, and maintained to be thoroughly substantial and reliable. Basic fundamentals of railway signaling require that the closed circuit and gravity "drop-away" be used to the greatest extent possible, so that a broken wire or part will cause the signals to assume their most restrictive position and so that train movements will be stopped as quickly as is compatible with safety.

The more important features of railway signaling are treated herein, but details, which change from year to year, are properly covered by the Manual of the Signal Section of the Association of American Railroads, by the Principles and Practices published by that Section, by the circulars and handbooks issued by the manufacturers of electric signal apparatus, and by textbooks listed in the Bibliography, Section 17, Article 53.

SIGNAL INDICATIONS. Originally the day signal indications were given by means of a wooden arm placed at different angles with the supporting mast; night indications, by means of red or white lights. Red is still continued as a satisfactory night indication and universally indicates stop. White is not entirely satisfactory for the proceed night indication. It may occasionally be confused with other while lights on the roadway or adjacent thereto. A broken red lens may permit a white light when red is intended. With improvements in the manufacture of colored glass, yellow and green now give satisfactory indications and are largely used.

Improvements in the electric lamp bulb and in the design and location of reflector and lens permit the use of the daylight color light signals, position light signals, or a system combining both.

TRAIN WEIGHT AND ITS EFFECT. Because of great improvements in the efficiency and power of railway locomotives, the average weight of trains has been markedly increased. Passenger trains are operated with 18 or 20 steel cars. Freight trains are loaded in excess of 120 cars and 9000 tons.

The increased stopping distance necessary for these trains, when moving at relatively high speed, sometimes makes restricting signal indications three or four blocks in the rear of a stop signal desirable.

41. ASPECTS AND INDICATIONS

OPERATING SIGNALS. Signals used on railways may be audible, such as whistle or torpedo signals; they may be hand signals, such as are given by a colored flag, by a lighted lantern, or by the arm or hand; or they may be burning flares.

FIXED SIGNALS. Fixed signals are those of a fixed location which may indicate conditions affecting the movement of an engine, car, or train.

Mechanical Signals are fixed signals operated by a lever by means of a connecting pipe or wire.

Power Signals are fixed signals which may be operated by electric motor, compressed air, or gas.

SEMAPHORE SIGNALS. Semaphore signals may be either mechanically or power operated and give their various day aspects by different positions of an arm and their night aspects by different colors of light, generally green, yellow, or red.

LIGHT SIGNALS. Color Light Signals give their various aspects by light only, their aspects being identical with the night aspects of semaphore signals.

Position Light Signals give their various aspects by two or more white lights in various locations with respect to each other. The position of the lights has a similar aspect to the position of a signal arm, i.e., similar to the day indication of the semaphore signal.

Color Position Light Signals are a combination of both of the above systems and use both various positions and colors of lights in giving their different aspects.

SIGNAL ASPECTS. Different types of signals, although in use on different railways, follow a consistent system, and the different aspects may be readily read and understood by one familiar with any of the other types.

A **signal aspect** is the appearance of the signal to the engineman of an approaching train. A **signal indication** is the information conveyed by the aspect.

The signal aspect with its indication constitutes a rule governing the movement of an engine or train.

The Association of American Railroads has adopted and published in its Standard Code certain basic principles for guidance of each railway in the preparation of its signal rules. These principles include a note reading: "Aspects shown are typical. Each road should show the aspects and colors of light it uses."

Railways requiring signals have generally followed the fundamentals of this Code, using the type of signals, aspects, and indications particularly adapted to their local and operating conditions. For example, when an indication reading "Proceed, approaching next signal at medium speed" is required, the aspects under the different systems would be as shown by Fig. 1.

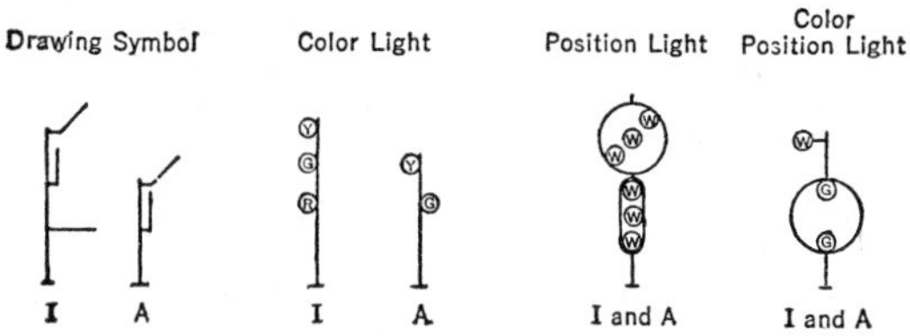

Fig. 1. Typical Railway Light Signal Aspects

The drawing symbol for all types of signals simulates the equivalent position of a semaphore signal arm. "A" refers to an automatic signal aspect; "I," to an equivalent interlocking signal aspect. The letters G, R, W, or Y within a circle indicate the color of signal light displayed: green, red, white. or yellow.

42. AUTOMATIC BLOCK SIGNALING

Definitions. A **block system** is a division of a section of railway into a series of consecutive blocks whereby trains, either moving in opposite directions or following each other in the same direction, may be segregated in separate sections of road. The blocks may be separated by block stations located at the beginning and end of each section.

A **Manual Block System** is one where information concerning the location of the opposing or preceding train is conveyed from station to station by telephone or telegraph. Instructions may be conveyed to the engineman of each train by written order, by fixed signal, or by other definite means.

A **Controlled Manual Block System** is a system whereby the telephone or telegraph communication is supplemented by electrical control. This may be effected by locks on the levers controlling the fixed signal or by a controlled magnetic clutch in the connection between the lever and the signal, which must be energized to permit the signal to clear.

ADVANTAGES OF BLOCK SIGNALING. Automatic block signaling is based on electrical control and electrical track circuits. This system has the advantages of permitting shorter blocks and thereby more flexible train operation. It eliminates the human element control, thereby furnishing greater reliability. The apparatus, methods of construction, and maintenance have been raised to such a high standard in the United States and Canada that a modern electric automatic block signal system, in 200,000 signal operations, permits but approximately 1 unnecessary stop indication to be given a train.

WAYSIDE AND CAB SIGNALS. An automatic block signal system may consist either of fixed signals located on the wayside adjacent to or above the tracks or signals located in the cab of the locomotive. Signals located on the locomotive are commonly known as cab signals and are treated in Article 46.

The preferable location of fixed automatic block signals is directly over or at the right (the engineman's side) of an approaching train.

LENGTH OF BLOCKS. The length of automatic signal blocks and the exact location of fixed signals depend upon a great many variables, including the most unfavorable stopping distance of a train on the basis of the weight and length of the train and the braking efficiency of the equipment.

Approach view, curves, grades, and protection for stations and switching points must also be considered in the location of wayside signals. Where traffic is light and relatively unimportant, automatic blocks may be based upon the most unfavorable stopping conditions of any train operated. Where traffic is heavy and important, shorter blocks provide greater track capacity and greater efficiency of operation. Under these conditions blocks shorter than the maximum stopping distance are used and additional signal indications applied, which require the train speed to be reduced gradually.

Approach signal distances include certain desirable tolerances in excess of the stopping distance of the train.

TRAIN BRAKING AND RETARDATION. Figure 2 shows typical curves to illustrate train braking distances, train retardation distances, and signal approach distances.

In curves *A* and *B*, Fig. 2, train speeds are plotted vertically and distances are plotted horizontally.

Curve *A* shows a typical train braking curve based on the actual distance required to stop a given train on level track at the different speeds based on a service application of the air brakes. These curves are relative in shape and can be readily drawn after five or six tests made on a given train at different speeds. After a set of curves has been established for different characters of trains, additional curves for different weights and speeds can be correctly interpolated.

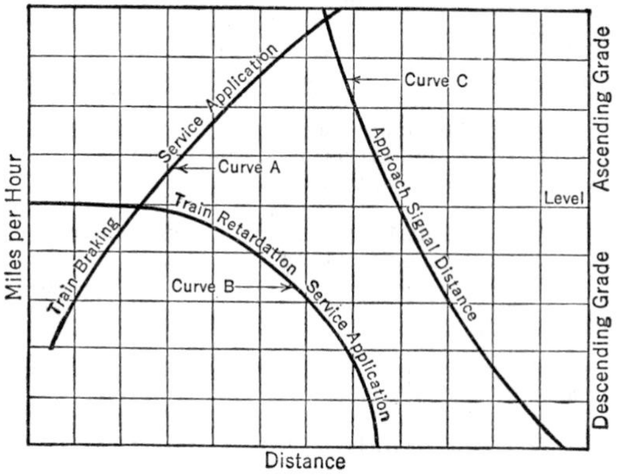

Fig. 2. Train Braking and Approach Signal Curves

Curve *B* illustrates typical train retardation curves, showing the relatively small retardation of the train during the early braking period and the relatively quick stopping of the train after the brakes have become fully effective. It will be noted that it takes approximately five-sixths of the stopping distance for a train to be reduced to one-half its speed, or if the original speed is 75 mph and medium speed is defined as 30 mph, which is a frequent standard, a few hundred feet will serve to bring the train to a stop after speed has been reduced to the medium point.

Curve *C* plots plus or minus grades vertically from the zero or level line and distances horizontally. This illustrates the corrections to be made for ascending or descending grades after the approach signal distance for level track has been established. The approach signal distance for level track should be something in excess of the actual stopping distance, depending upon the tolerance or margin of reliable working desired. In this particular case an approach signal distance of 7900 ft has been selected, and the corrections for the different grades can be readily found on the chart.

LOCATION OF SIGNALS. The degree of accuracy required for location of signals depends on economic conditions and on the importance, character, and variety of traffic. An extremely long block is undesirable because too much time would be lost if the automatic signal should give an unnecessary stop indication and the train be required to proceed through the block looking out for broken rail or other obstruction.

Figure 3 shows typical application of automatic signals. Where traffic is particularly important, some roads have added other aspects to the three indication arrangement.

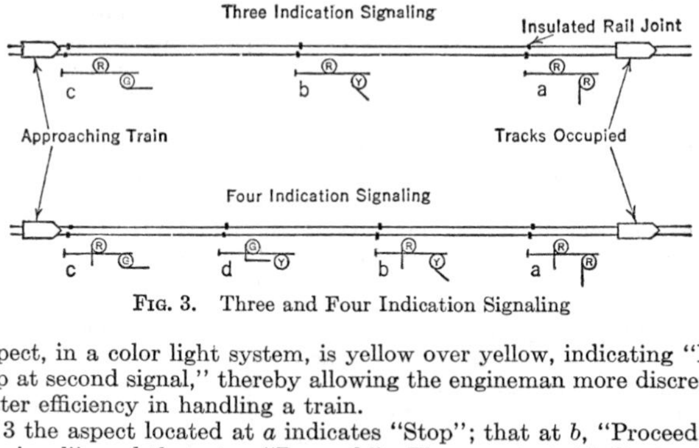

Fig. 3. Three and Four Indication Signaling

One such aspect, in a color light system, is yellow over yellow, indicating "Proceed, prepared to stop at second signal," thereby allowing the engineman more discretion and permitting greater efficiency in handling a train.

In Figure 3 the aspect located at *a* indicates "Stop"; that at *b*, "Proceed, prepared to stop at next signal"; and that at *c*, "Proceed." The aspect at *d* indicates that the next signal must be approached at not exceeding medium speed (generally 30 mph).

The rules of many roads permit a train, after having come to a stop, per aspect at *a*, to

"Proceed at a speed not exceeding that which will enable the train to stop short of train ahead, obstruction, or switch not properly lined, and look out for broken rail."

The letters within the circle (Fig. 3) refer to night colors of semaphore signals or to both day and night aspects of color light signals. With but the three block indications, one working unit only may be used. A separate non-changing lower unit is frequently used, as a marker light, to identify a signal more definitely and for protection in case one light is out. Where four or more indications are used, two operative units are desirable.

Control circuits for automatic signals are discussed in Article 50.

TYPES OF SIGNALS USED. The United States Interstate Commerce Commission requires that all railways operating under its supervision make extensive annual reports covering their methods of block signaling. This information is consolidated and issued annually by the Commission.

Table 1 is a summation of certain tables in that report and indicates the percentage and types of signals in use.

Table 1. Block Signaling on Railroads of the United States *

Automatic		Non-automatic		Total Automatic and Non-automatic		Total Passenger Lines Operated		Per Cent Block Signaled Miles of Passenger Track
Miles of Road	Miles of Track	Miles of Road	Miles of Track	Miles of Road	Miles of Track	Miles of Road	Miles of Track	
71,588.4	103,167.0	33,136.4	34,547.6	104,724.8	137,714.6	165,159.8	199,747.9	67

Kinds of Automatic Signals in Use

Semaphore		Light		Not Classified		Number of Automatic Block Sections
Miles of Road	Miles of Track	Miles of Road	Miles of Track	Miles of Road	Miles of Track	
28,967.8	39,929.2	41,464.5	61,036.3	1,156.1	2,201.5	85,552

* From the United States Interstate Commerce Commission Report for January 1, 1947.

43. ELECTRIC INTERLOCKING

DEFINITIONS AND TYPES. The standard code of the Association of American Railroads defines interlocking as, "An arrangement of switch, lock, and/or signal appliances so interconnected that their movements must precede each other in a predetermined order."

A later description appropriate to modern practice is, "An arrangement of signal appliances so interconnected as to insure the integrity of a route over which a train movement is to be authorized."

An assembly of such interlocked functions is known as an **interlocking plant.** Interlocking plants are generally used at important terminals, junctions, railway crossings, drawbridges, and other points where conflicting train movements are to be protected against, or where rapid operation of switches is of such importance as to warrant the installation, maintenance, and operation.

A **Mechanical Interlocking** is one arranged with the switches and signals operated by manual power. This is transmitted from a manually operated lever, located in the signal station, by means of sections of 1-in. pipe connected, together with couplings, plugs, and rivets, to the signal, switch, or derail units in or adjacent to the railway track.

In **Electromechanical Interlocking** part of the functions are mechanically operated and part are electrically operated. Generally the signals are of the electric type. Electric circuits are used for electric locking, approach indications, etc.

Power Interlocking is distinguished from mechanical interlocking in that compressed air or electricity is substituted for manual power for the operating of the various units.

Electropneumatic Interlocking employs both electricity and air as the operating agencies. This type of interlocking is used extensively at points where the units are reasonably close together and compressed air can be delivered economically at the units to be operated.

Compressed air is generally used at pressures of 50 to 100 psi.

Electrical energy for operation of the pneumatic control valves, electric locks, etc., is furnished at 12 to 16 volts, generally from a storage battery, either cycle charged or floated from a motor generator or rectifier.

The machine located in the signal station for the operation of electropneumatic interlocking may be of either the miniature lever or the push-button type.

ELECTRIC INTERLOCKING. **Electric Interlocking** is of three types: the miniature lever type, operated from a cabinet, in the signal station; the track diagram or panel type, operated from buttons or small levers mounted on a diagram showing the tracks, signals, switches, etc.; and the full automatic type.

Lever Type Interlocking. The lever-type interlocking machine is approximately one-half the length of a mechanical machine. The levers are generally of the slide or rotary types. The machine is equipped with circuit controllers, indication locks, lever locks for prevention of operating switches in advance or beneath a train, and auxiliary apparatus.

The mechanical locking of the electric lever machine is practically identical with that of the mechanical machine but of smaller size.

Figure 4 shows a typical electric interlocking layout for a main line cross-over and one turnout. The locking sheet for this arrangement is shown beneath the diagram at the left and the dog chart at the right. If a mechanical interlocking were provided at this point, additional levers and pipe connections would be required for a facing point plunger lock at each switch for the mechanical locking of the track switches in correct position before a movement over the route can be authorized by signal indication.

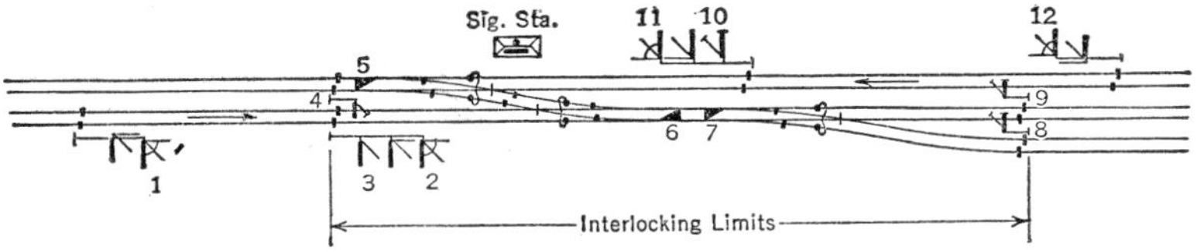

Locking Sheet

LEVER	LOCKS	LEVER	LOCKS
1		8	(7) 6 (6)
W	6 X 8 9	W	6 X 5
		W	(6) X (5)
2	3 5 6 7 9		
3	5 6 7 (7) 8 9	9	7 6 (6)
4	5 (5) 10 11	W	6 X 5
W	5 X 6 12	W	(6) X (5)
W	(5) X (6) 7 (7) 8 9		
		10	11 6 5
5		11	6 5
6		12	
7			

Dog Chart

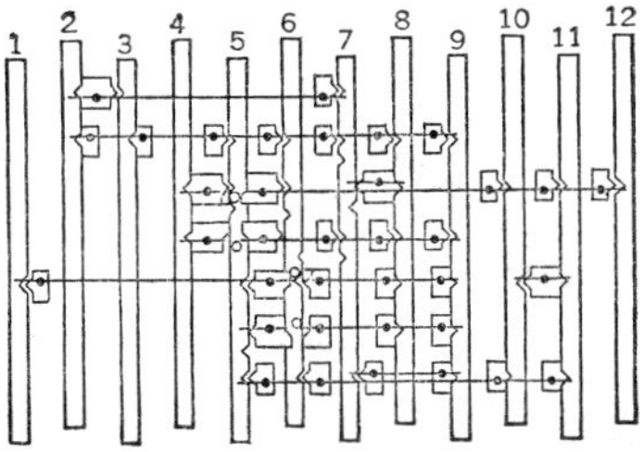

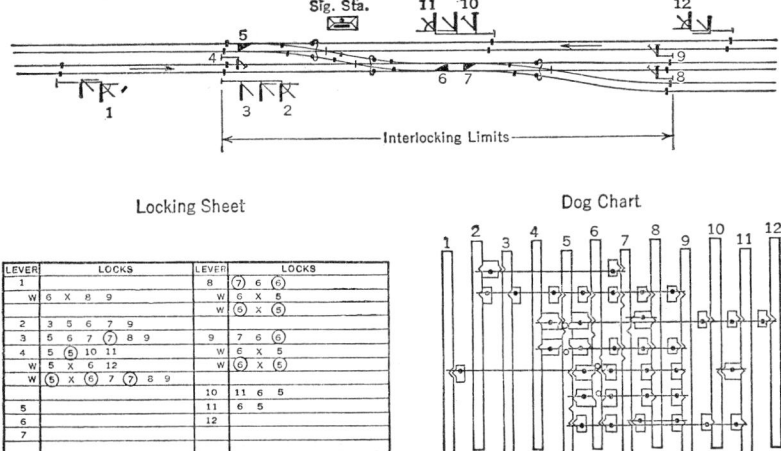

FIG. 4. Interlocking Layout with Mechanical Locking

The return indication current for lever-type electric interlocking may be provided by: (1) use of the same energy source as serves for control and operation—this is called "battery indication"; (2) use of polarized control in connection with a battery indication, or in connection with current of a character different from the operating current; (3) utilization of the momentum of the operating electric motor to generate the indication current after the motor has completed the movement of the apparatus—this is called dynamic indication; (4) a combination of a battery and dynamic indication.

At large terminal interlockings, where several hundred levers are required, the time in moving from lever to lever becomes a material item and often requires additional levermen. The mechanical locking becomes of such size that intermediate levers are required whose only function is to drive the locking bars, thereby increasing the labor and time required.

Panel-type Interlocking. Developments and improvements in the design of d-c relays to a point of unquestioned reliability have provided a means of simplifying electric interlocking and permitting use of the panel-type machine. The panel-type machine requires but a small control room with the operating buttons, levers, and indication lights mounted on a miniature track diagram. A panel-type machine for a terminal of moderate size can be operated by one man sitting in a chair in front of the cabinet. Circuit controls and relays are placed in a separate room and are a substitute for the mechanical locking, circuit controllers, electric locks, etc., of the lever-type machine.

The panel diagram may assume various shapes, the preferred one being of the wing type. This arrangement brings the ends of the diagram within the operator's reach and still permits a view past the end of the diagram.

One of the outstanding improvements in panel-type interlocking is the development of circuits and means for the automatic setting up of complete routes with the required signals upon the operator's manipulation of two knobs or buttons. These are located on the control panel in geographic correspondence to the train's entrance and exit points of the interlocking. This system eliminates the necessity of the operator having to identify and position the individual switches and signals involved in a route. Special operating features may be incorporated in the system, such as automatic selection of an alternate route in event the normal preferred route is not available, and the stick or non-stick operation of the signals. The advantages realized from this system of interlocking are simplification of the directing of trains, speeding up of operations, and minimization of the possibility of error in routing trains. The same maximum safety of operation is provided as with unit-operated interlocking. The system is equally adaptable to large or small layouts; however, the more complex the track arrangement and the more dense the traffic, the greater are the advantages derived from the simplicity and speed of operation which distinguishes this system.

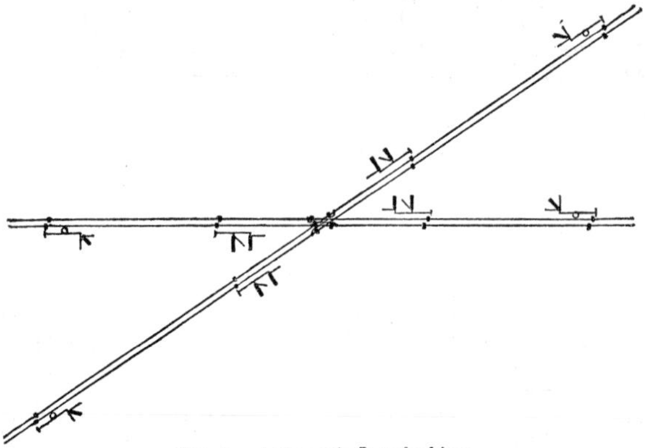

Fig. 5. Automatic Interlocking

Full Automatic Interlocking. Automatic interlocking, an example of which is shown in Fig. 5, is arranged to work entirely without manual control. Generally the first train approaching the point to be protected causes the signals on that track to assume the proceed position and holds the signals on a conflicting track in the stop position until after the first train has passed. This arrangement may be applied to locations where only signals are used. Sometimes derails and switches are operated from an automatic interlocking. Automatic interlocking is ideal for protecting railroad crossings, gauntlets, and junctions where traffic is such that the first train arriving may have preference of procedure.

Power for Interlocking. The power for operating the switches of an electric interlocking is generally at a normal voltage of 120 volts. Storage battery of different capacities from 80 to 400 amp-hr is provided. The size of battery depends upon the reliability of commercial power supply and the amount of emergency stand-by required. The signal-control circuits are operated from a separate battery of 12 to 16 volts. The storage battery may be cycle charged either from a gasoline or motor-generator set or from a rectifier. Many installations use but one set of battery floated from a motor generator or rectifier.

Where reliable power is available and it is not necessary to operate several switches at the same time, the interlocking may be operated directly from a rectifier with an independent primary battery as stand-by.

Wire Sizes. Tables 2 and 3 give minimum wire sizes for switches and motor- and solenoid-operated signals at different distances from the control point, as recommended by the General Railway Signal Company.

Auxiliary Equipment. Various relays, indicators, annunciators, emergency time-lever releases, and special circuits, in connection with or in addition to the essential interlocking apparatus, expedite the handling of traffic and meet special conditions.

Types of Locking. Approach Locking is electric locking effective while a train is approaching a signal which has given a proceed indication to that train, to prevent operation of levers or devices which would endanger that movement.

Route Locking is electric locking so arranged as to take effect when a train passes a signal to prevent operation which would endanger the train while it is within the limits of the route entered.

Table 2. Minimum Wire Size for Switches at Electric Interlocking (Individual Return System)

GRS Co. Switch Machines	Normal Voltage at tower	Size of Rail	Kind of Switch	Operating Current, Amperes	Operating Time, Seconds	Maximum Distance in Feet over Control Wire from Tower to Switch Machine with Various Sizes of Wire						
						No. 14	No. 12	No. 10	No. 9	No. 8	No. 6	No. 4
Model 2 Model 4A Model 5 Model 5A Model 5B	110 D-c	105 lb or less	SS or DRL	5	3	930	1,475	2,350	2,950	3,730	6,020	9,400
			DSS or MPF	7	3	595	950	1,510	1,910	2,400	3,870	6,040
		Over 105 lb	SS or DRL	6	3	795	1,260	2,000	2.530	3,170	5,130	8,000
			DSS or MPF	8.5	3	555	880	1,400	1,770	2,220	3,590	5,600
Model 4B	110 D-c	105 lb or less	SS or DRL	5.2	3	1050	1,650	2,750	3,350	4,250	6,700	10,700
			DSS or MPF	7.3	3	788	1,230	2,050	2,500	3,150	5,000	7,880
		Over 105 lb.	SS or DRL	6.25	3	1000	1,560	2,600	3,170	4,060	6,350	10,000
			DSS or MPF	8.75	3	655	1,020	1,700	2,050	2,600	4,150	6,550
Model 5 Model 5A Model 5B Low-speed	110 D-c	105 lb or less	SS or DRL	1.25	8–15	8710	13,850	22,100	27,800	35,100	55,800	88,500
			DSS or MPF	1.75	8–15	6215	9,880	15,710	19,840	25,050	39,800	63,150
		Over 105 lb	SS or DRL	1.5	8–15	7250	11,520	18,340	23,150	29,200	46,400	73,650
			DSS or MPF	2.1	8–15	5175	8,235	13,100	16,550	20,850	33,180	52,600
Model 5A Model 5B Medium-speed	110 D-c	105 lb or less	SS or DRL	3.3	8	1975	3,140	5,050	6,310	7,960	12,650	20,100
			DSS or MPF	4.6	8	1415	2,255	3,590	4,530	5,710	9,080	14,400
		Over 105 lb	SS or DRL	4.1	8	1590	2,530	4,030	5,080	6,410	10,200	16,170
			DSS or MPF	5.75	8	1130	1,800	2,870	3,620	4,570	7,270	11,500
Model 5 Model 5A	110 V, 60 Cy	Any size	SS or DRL DSS or MPF	8	3.5	595	940	1,500	1,900	2,380	3,840	6,000
Model 5B	110 V, 25 Cy	Any size	SS or DRL DSS or MPF	10	3.5	795	1,260	2,000	2,530	3,170	5,130	8,000
Model 5A Model 5B 111/1 Gear Ratio	150 D-c	105 lb or less	SS or DRL	3.8	4	1860	2,950	4,700	5,900	7,460	12,040	18,800
			DSS or MPF	5.4	4	1190	1,900	3,020	3,820	4,800	7,740	12,040
		Over 105 lb	SS or DRL	5	4	1590	2,520	4,000	5,060	6,340	10,260	16,000
			DSS or MPF	7	4	1110	1,760	2,800	3,540	4,440	7,180	11,200
Model 5A	220 D-c	105 lb or less	SS or DRL	2.5	3	3730	5,900	9,400	11,900	14,900	24,100	37,600
			DSS or MPF	3.5	3	2390	3,800	6,040	7,640	9,580	15,290	24,160
Model 5B		Over 105 lb	SS or DRL	3	3	3170	5,040	8,000	10,100	12,700	20,500	32,000
			DSS or MPF	4.2	3	2220	3,520	5,600	7,090	8,880	14,350	22,400

SS = single switch. DRL = derail. DSS = double slip switch. MPF = movable point frog. Operating time at maximum line resistance

Table 3. Motor and Solenoid Type Signals

GRS Co. Signal	Operating Current, Amperes	Holding Current, Amperes	Operating Time, Seconds	Maximum Distance in Feet over Control Wire from Tower to Signal with Various Sizes of Wire						
				No. 14	No. 12	No. 10	No. 9	No. 8	No. 6	No. 4
Model 2.........	3	0.14	4	1500	2350	3,900	4,760	6,000	9,500	15,000
Model 3.........	3	0.11	3	1500	2350	3,900	4,760	6,000	9,500	15,000
Model 2A, High..	0.82	0.25	8	6770	9000	15,000	18,000	23,000	36,600	58,000
Model 2A, Dwarf.	0.82	0.25	8	6770	9000	15,000	18,000	23,000	36,600	58,000
Solenoid Dwarf ..	4 to 5	0.17	1	1504	2410	4,000	4,815	6,050	9,650	16,500

Operating time at maximum line resistance.

Trailing Release Route Locking is arranged so that a switch may be operated for another movement as soon as a train has passed the switch without waiting for the completion of that movement. This is provided at points where traffic is sufficiently frequent and important to warrant the construction and maintenance costs.

44. CENTRALIZED TRAFFIC CONTROL

DESCRIPTION. Centralized traffic control is basically a series of small interlocking groups, including their switches and signals, with intervening automatic signals. This system may extend over a division or section of railway and is controlled electrically by one man at a panel-type operating machine at a suitable location. This arrangement dispenses with telephone and telegraph train orders and accomplishes a material saving in operating expense and a material increase in average operating speeds and efficiency. It also increases track capacity and may even reduce the number of running tracks required.

One of the earliest complete installations of this type is that on the Ohio Central Line of the New York Central Railroad, which has given satisfactory and efficient service since July 25, 1927. This installation extends for 40 miles between Stanley and Berwick, Ohio, and is being followed by other installations in many parts of the country.

A centralized traffic-control system, though basically an extended interlocking plant, differs from the conventional installation in several ways:

a. Lever locking is unnecessary, as the functions are intercontrolled electrically. Unit electrical interlocking protection is effective between adjacent controlled units or groups of units as well as within each unit.

b. Automatic block signals may be included to govern train movements between geographically separated switch groups. An automatic block system circuit control known as absolute permissive block is employed between groups of switches for single-track railways or for multiple tracks when traffic is required to operate in either direction on each track. Any standard automatic block signal system will suffice on multiple tracks with single-direction operation on each track.

c. Special circuits are employed to secure a high utilization of a few wires extending between the control machine and the wayside units of the railway for purposes of control and return indications.

TYPES. Centralized traffic-control systems are of two general types, coded and non-coded or unit wire. Each type has certain inherent characteristics and its own special scope of application. The **non-coded** or **unit wire system** requires one connecting wire from the machine to each controlled function and one common return wire for the system. A control group consists of a switch or cross-over with the associated protecting signals to govern train movements. The moving of a contactor on the control machine, if conditions are proper, will operate the corresponding switch or cross-over and permit the proper signal to authorize a train movement, either for the main track or for a siding as desired. The operator will receive, over the same line wire and common return, light indications assuring him that the switch or switches have moved to a position corresponding with the lever position, that the signal has cleared, and finally that the train has accepted the signal indication and is occupying the section of track in which the switch is located.

Coded systems may be of the one-circuit (two- or three-wire) or of the two-circuit arrangement (three- or four-wire). In either arrangement the respective circuits extend from the control office to the last group controlled.

In the one-circuit scheme the control line circuit is also used for the indication circuit.

In a two-line-circuit scheme the control line circuit is separate from the indication line circuit.

Either of these two schemes may provide **duplex operation,** which means that controls may be transmitted from the control office to any field location at the same time that indications are being transmitted from any field location to the control office. It has been most practical to provide duplex operation only in a two-line-circuit scheme.

Each control-lever position sets up an individual code of a fixed number of digits, which is impressed upon the control line circuit. The corresponding individual code at the wayside location is the only one that will respond to the impressed code.

Indications are transmitted in the same fundamental way over the indication line circuit.

Although more apparatus is required than with the unit-wire scheme, since two to four line wires only are required for almost any installation, it is apparent that a coded scheme is especially suited for extended layouts where control distances are comparatively long.

Of recent years the efficiency and expediency of centralized traffic control, particularly on single-track lines, have so proved themselves that upward to 6000 miles of road in the United States and Canada are equipped for this type of operation. Many individual installations comprise more than 100 miles of road. The capacity of such a system is largely determined by the number of wayside locations and functions controlled and the frequency of controls and indications that must be transmitted in order to meet operating requirements. This frequency is, of course, dependent on the density of traffic. In order to meet these increasing demands and realize fullest economies of operation, installation, and maintenance, carrier systems for use in conjunction with coded centralized traffic control have been developed. With a carrier system it is possible to divide an extensive centralized traffic-control territory into two or more sections, each operated independently from the same control office. Ordinarily, this would require a separate pair of line wires for each section. By means of the carrier, however, only one pair of wires is required for the entire territory. The section adjacent to the control office may be handled over the two wires by means of d-c codes. The second section may then be controlled over the same two wires by means of carrier currents superimposed on the line wires of the first section. A third section, or additional sections, can be handled by means of carrier currents of different frequencies superimposed on the line wires of the first and second sections. Control codes and/or indications to or from any or all sections may be transmitted simultaneously without interference with each other.

45. ELECTRIC CAR RETARDERS

THE HUMP. Where a large number of freight cars is to be regrouped, the throat track leading to the classification yard may be raised on what is known as a **hump.** This arrangement permits cars to be pushed slowly by a locomotive over the crest of the hump and detached singly or in groups as may be required for the various tracks. The cars are permitted to run by gravity from the crest of the hump to the various assigned tracks.

FUNCTION AND TYPES OF CAR RETARDERS. A car retarder performs the same function as the ordinary hand brakes operated by car riders in controlling the speed of cars on their journey from the hump to the various classification tracks. The speed must be kept under control so as to avoid undue shock when coupling with other cars. A retarder consists of braking apparatus placed adjacent to the tracks for the purpose of controlling the movement of the cars after they have been separated. This braking apparatus must be continually readjusted for the successive different weights and speeds of moving cars and may be operated by compressed air, hydraulic pressure, or electric motors.

ELECTRIC RETARDER. The electric type of retarder is operated by electric motors mounted in metal housings adjacent to the track and on special combination steel and wood ties.

The braking is effected by the friction of a steel shoe against the sides of the car wheels. The effectiveness of the braking power is increased with the height of the point where the brake shoe grasps the car wheel. The retarder brake shoes are placed as high with relation to the rail as is practicable and still present a level upper surface which will not contact car equipment.

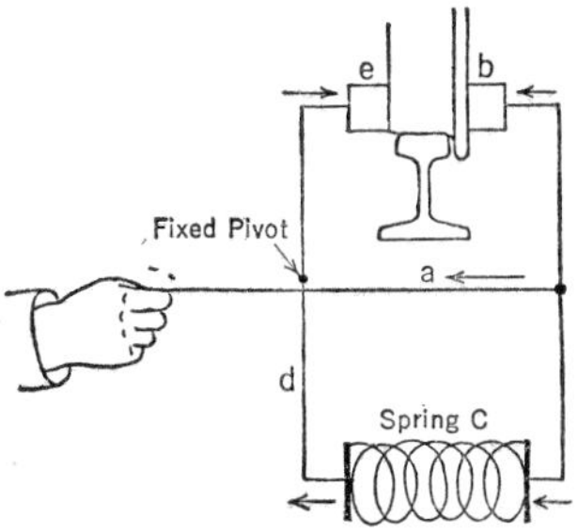

Fig. 6. Principle of Electric Car Retarder

The principle of the braking apparatus of the electric retarder is shown in Fig. 6. A pull on rod *a* forces shoe *b* against the wheel and compresses spring *c* against lever *d*, which forces shoe *e* against the other side of the wheel.

Figure 7 is a diagrammatic sketch of the electric retarder actuating mechanism. The left half of the diagram is in a horizontal plane, and the right half in a vertical plane. The drive gear is operated through a train of gears by a 5-hp motor of the type used on electric cranes.

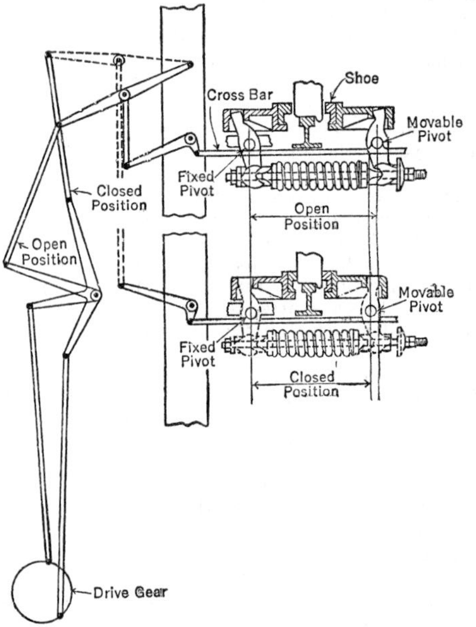

FIG. 7. Diagrammatic Sketch of Electric Car Retarder

A d-c motor is generally preferred to an a-c motor because of the simplicity in providing storage-battery reserve and because of the greater relative starting torque developed by a d-c motor of a given size.

The retarder-operating motors are operated on a nominal 240-volt circuit with a maximum demand of 120 amp. The average power consumption per car handled is approximately 0.07 kw-hr.

The switch points are thrown by a motor-operated switch mechanism at a nominal voltage of 120 and using between 7 and 8 amp. A series motor is used for switch operation. Complete movement of switch from one position to the other is made in 1.5 sec.

In addition to the braking feature of the retarder, a device known as a **skate** is generally placed on each track in advance of the last classification switch. This device, which is operated by motor for electric machines and by compressed air for the electropneumatic machines, will, when the lever is moved by the operator in the cabin, place an iron shoe on the rail upon which a car moving too fast will run. This shoe will slide along the track until the car is stopped. Skates are operated only occasionally and for emergency. After each use the skate must be brought back by hand and restored to the skate mechanism. The electric skate machine operates at 120 volts and between 2 and 3 amp.

It is desirable that electrical energy of sufficient capacity be available at all times when an electric car retarder is in operation; otherwise in case of a power failure heavily loaded cars or cars with explosives might run away and cause serious damage. Normal energy may be provided by an overcompounded motor-generator set operated continuously with a duplicate set for emergency, by a diverter pole motor-generator set with floating battery, by two sets of storage battery, cycle charged, or by a mercury-arc rectifier with separately charged storage battery for emergency stand-by.

Track circuits are short, about 50 ft in length, and arranged for rapid shunting. They eliminate the possibility of moving a switch between the wheels of a car. The beginning of the track section is placed sufficiently ahead of the switch so that necessary time is provided for the switch to complete its movement before the car wheels reach the switch points. Track circuit locking avoids derailments, damage, and delay. This subject is covered further under Track Circuits, Article 49.

ECONOMIC VALUE OF RETARDERS. From an economic study made on sixteen car-retarder installations by the Association of American Railroads it was found that the average saving per car handled varied anywhere from \$0.087 to \$0.55. The average estimated return on capital investment of the sixteen installations was 42.86 per cent.

Figure 8 shows a yard with twenty-five classification tracks in groups of two or three and with the retarders located fairly well down toward the separate tracks. The grades in this sketch are arranged for moving empty cars against the prevailing winds. This arrangement of grades and retarders provides a fairly fast operating yard.

By arranging the classification tracks in larger groups, possibly four, five, or six tracks together, fewer retarders would be required, with lower construction cost. With the larger groups more time is consumed in waiting for cars to clear each other, and slower car movement is inherent

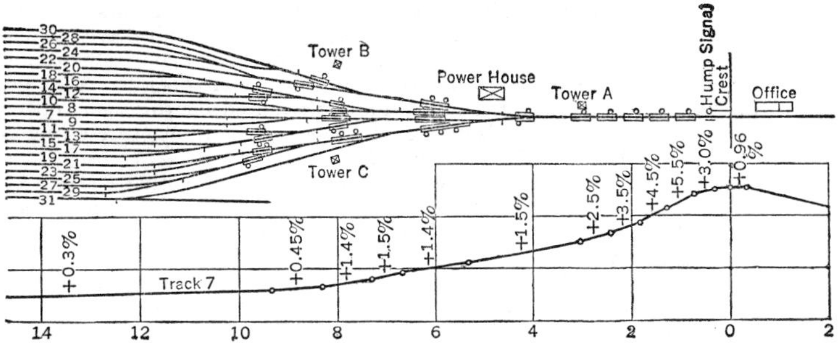

FIG. 8. Typical Layout of Electric Car Retarder

Each problem must be studied on the basis of present and expected speed and capacity requirements, and a correct balance must be reached between construction costs and operating costs.

46. CAB SIGNALS

DESCRIPTION. Cab signals are signals placed in the cab of the locomotive, where they may be readily observed by the engineman and fireman while operating the locomotive.

Cab signals may supplement wayside automatic signals or be a substitute for theml. When wayside automatic signals are not used, it is the general custom to provide wayside signals at interlockings to indicate the positive stop locations and for convenience in switching locomotives which may not be equipped with cab signals.

The cab signal system is a development of the automatic signal system, with the signal controls transmitted from the track to the locomotive.

TYPE OF CURRENT. Continuous indication in the cab is maintained by a 100-cycle track circuit current, fed into the running rails at the far or leaving end of each block and interrupted a definite number of times per minute, according to track and block conditions ahead. This current induces a voltage in apparatus on the engine and so causes other engine-carried apparatus to function to display the proper indication in the cab.

A frequency of 100 cycles is generally chosen in order to eliminate interference from the 25- and 60-cycle industrial, lighting, and propulsion systems.

CODE TRANSMITTER. The mechanism in the wayside system which interrupts the 100-cycle current and feeds it to the rails is called the code transmitter. It consists of a number of cam-operated contacts, the cams of which are driven through gears by a small induction motor. The number of projections on each cam determines the number of interruptions per minute which its set of contacts will produce. The circuit from the 100-cycle source to the track is made through one or another of these contacts, depending upon traffic conditions ahead. Therefore, when the code transmitter is operating, the 100-cycle track circuit is interrupted at a certain rate called the code frequency.

If coded rail current is required only for operation of the cab signal system, the code transmitter is normally at rest when the block is unoccupied. On the entrance of a train into a block, the track relay is de-energized, causing the code transmitter to become operative to code the rail current. As the train passes out of the block, the re-energization of the track relay shuts off power to the code transmitter, and it again comes to rest.

The track relay serves as an automatic switch to control the coded cab-signal circuits as well as the regular wayside signals if used. With the starting of the code transmitter the proper code for a given set of conditions is selected by circuits energized through the wayside control relays in advance, thereby impressing on the rails the code applicable to that set of conditions.

CAB SIGNAL INDICATION. The cab signal indication displayed depends upon the code frequency of the 100-cycle rail current. For example, the code frequency for a "clear" indication is 180 per minute; for "approach medium," 120 per minute; and for "approach," 75 per minute. If the block is occupied, the coded current is shunted by the train ahead and does not reach the following train, giving the "stop" or "restricting speed" indication in the cab. This indication is also given if the 100-cycle track-circuit current is flowing, but uncoded.

CODED RAIL CURRENT FOR WAYSIDE SIGNALS. Coded rail current may also be used instead of standard track circuits for the control of wayside signals in a manner similar to that for the control of cab signals. When coded track circuits are used, no signal control line wires are required, the wayside signal indication displayed depending upon the frequency of code in the track, which, in turn depends upon traffic conditions ahead. Since, with this type of control, the code transmitters are operated continuously, no changes in the wayside apparatus and circuits are required for cab-signal operation, and the necessity for approach energization of the code transmitter and resetting of the track circuit upon the exit of a train is eliminated.

LOCOMOTIVE EQUIPMENT. The equipment on the locomotive consists of the receiver, the filter and amplifier, the master relay, the acknowledging relays, the code selective relays and tuned circuits, the cab signal and warning whistle, the dynamotor and the acknowledging switch.

The **receiver** is the means by which the control is transmitted from the rails to the apparatus on the engine and is mounted over the rails just ahead of the forward wheels. The voltage induced in the receiver coils is first delivered to an electrical filter, which suppresses all frequencies except that of 100 cycles and renders the apparatus immune

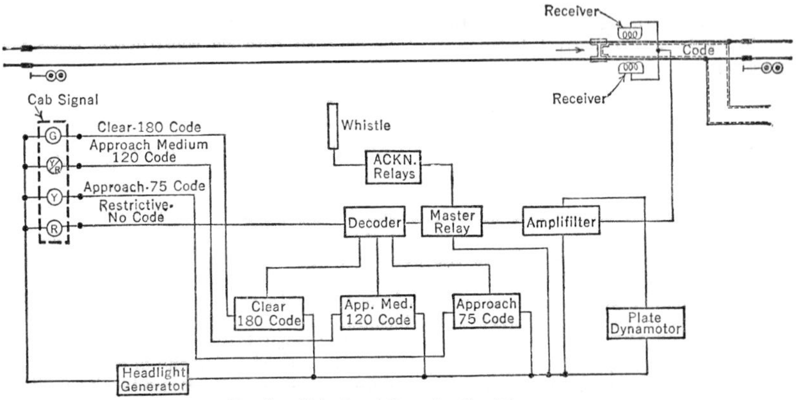

FIG. 9. Cab-signal Organization Diagram

to interference from other frequencies. The output of the filter is then delivered to the amplifier, which in turn delivers to the master relay a sufficiently increased amount of power for its operation.

The **master relay** is then periodically energized in opposite directions, and its armature moves back and forth at code frequency, alternately energizing with 32-volt direct current one half or the other of the primary winding of a decoding transformer. The energy delivered at the secondary of this transformer is supplied to the "clear," "approach medium," and "approach" control relays through two resonant and one untuned circuits. The "clear" control relay is energized through a circuit tuned to resonance at 3 cycles per second (180 code frequency); the "approach medium" control relay is energized through a circuit tuned to resonance at 2 cycles per second (120 code frequency); the "approach" control relay is energized at all three code frequencies, its circuit being untuned. When no 100-cycle current is flowing in the rails or when the 100-cycle current is not coded, none of the control relay circuits is energized.

The cab-signal lights are controlled over contacts on the control relays. The "clear" cab signal is lighted when the "clear" control relay is energized. The "approach medium" indication is received when the "approach medium" control relay is energized and the "clear" control relay is de-energized. The "approach" indication is received when the "approach" relay is energized and both the "clear" and "approach medium" relays are de-energized.

The **audible warning** magnet is energized over the same contacts as for the cab signal and in addition contacts of the acknowledging relays. When a change takes place in the cab signals to a more restrictive indication, the warning whistle sounds until acknowledged by reversal of the acknowledging switch, which is mounted in the cab of the locomotive within convenient reach of the engineman. Upon any subsequent change to a more restrictive indication the warning whistle will sound again and have to be acknowledged.

The dynamotor furnishes current at 350 volts for the plate circuits of the amplifier tubes, receiving power from the headlight generator at 32 volts.

Should more than four indications be required, higher code frequencies may be added with their associated tuned circuits.

Figure 9 gives a diagrammatic organization of the electrical apparatus required on a locomotive for the cab-signal system here described. In this diagram the amplifier and the filter are combined.

47. AUTOMATIC TRAIN STOP AND AUTOMATIC TRAIN CONTROL

Automatic train-stop and train-control devices are used to enforce or supplement wayside signals. They may, when used with cab signals, form a complete system without wayside signals.

The mechanical trip, such as is used in connection with metropolitan elevated or subway passenger trains, is a simple automatic stop which enforces obedience to a stop signal by raising a lever or arm from the roadway. This arm, when raised, will engage and operate a connection to the air-brake system and apply the brakes on the train. This system is not appropriate for a standard railway, which is subject to outside weather conditions and where the momentums and corresponding stopping distances are much greater.

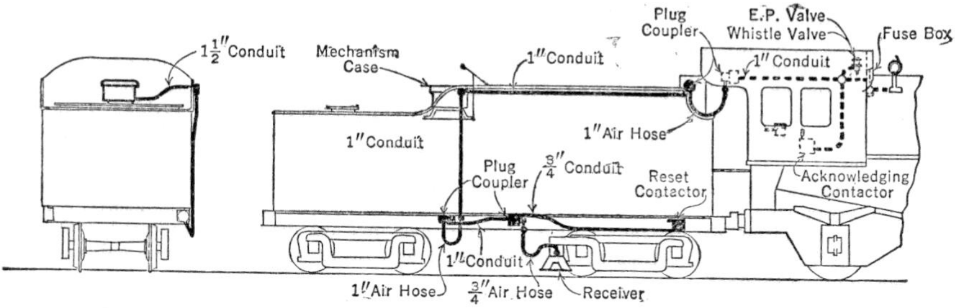

FIG. 10. Diagrammatic Location of Automatic Train-stop Apparatus and Connections

The **intermittent automatic train stop** is used primarily to enforce observance of wayside signals requiring a speed reduction or train stop. The engineman, if alert, after observing the wayside signal may operate an acknowledging lever to avoid a brake application and keep in his hands control of the train. If an engineman fails to observe a wayside signal giving a restricting indication and fails to operate the acknowledging device while passing such signal, the air brakes will be applied automatically and the train brought to a full stop. After a full stop has been made, the automatic control device may be released by the operation of a reset button, which may be located on the side of the locomotive tender, in the locomotive cab, or in other approved place.

Figure 10 is a diagrammatic sketch showing the location of the intermittent automatic-stop apparatus on a steam locomotive.

CONTROL DEVICE. The control device between the wayside and the locomotive is composed of two parts; one, carried by the locomotive and called a receiver, is securely fastened to the trucks of the locomotive and consists of an inverted U-shaped magnet with laminated cores, large pole pieces, and two coils. The other element is called an inductor. It consists of a U-shaped magnet with laminated cores and large pole pieces the same shape, size, and spacing as the pole pieces of the receiver. It is located on special ties with its pole faces 2 1/2 in. above the top of the running rail and its center line parallel with and usually 19 1/2 in. outside the gage line of rail. The receiver is adjusted so that, as the locomotive moves along the track, the pole faces of the receiver pass about 2 in. above and directly over the inductor pole faces.

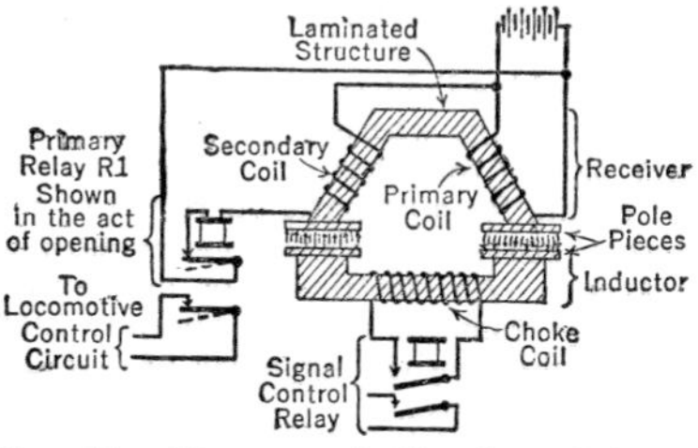

FIG. 11. Diagrammatic Sketch of Relative Location of Automatic Stop Inductor, Receiver, and Windings

Figure 11 is a diagrammatic sketch of the inductor and receiver locations and windings.

One of the two coils of the receiver is called the primary coil and, being constantly energized from a source of electrical energy, produces a strong magnetic field. The other coil, called the secondary coil, is connected to the same source of energy and in series with the coil and front contact of a relay, through which a current of about 13 ma flows normally.

The 32-volt locomotive headlight generator is generally used as a source of energy for the automatic-stop locomotive circuits.

The **wayside inductors** are provided with a choke coil, which is automatically controlled in such a way that, when a speed-restricting impulse is to be given from the wayside signal system, the coil is on open circuit, and when no impulse is to be given, the coil is closed on itself.

When the receiver carried by a locomotive approaches an inductor on open circuit, a surge of magnetic flux builds up in the secondary coil and produces a negative current in the relay of sufficient value to allow the relay to open. This in turn controls the automatic-stop circuits on the locomotive, which in their turn operate the automatic air-brake system.

CONTINUOUS TRAIN CONTROL. A continuous train-control system is operated inductively from alternating current impressed in the rails. The code impulses may be picked up inductively by the coils in the front of the locomotive in the manner previously described under Cab Signals, Article 46. This system differs from the intermittent system in that the locomotive apparatus is continuously controlled from the wayside, whereas the intermittent type of apparatus is controlled only at predetermined points or signal locations.

The continuous system may also, by means of contactors on a speed governor attached to the locomotive, be used to enforce maximum permissible speeds under the desired conditions. Both the intermittent and continuous systems are generally equipped with a whistle or bell which indicates to the engineman and other occupants of the cab when speed is to be reduced.

48. AUTOMATIC PROTECTION FOR HIGHWAYS AT RAILWAY CROSSINGS

A.A.R. RECOMMENDATIONS. Although each state has final authority to legislate concerning the kind and amount of protection to be provided in that state, the Association of American Railroads has established certain basic recommendations for the guidance of the various railroad and state commissions in the United States and Canada.

Because certain types of highway crossing signals had already been established by public authority in different sections of the country, it was necessary for the Association to include certain optional features.

The National Standards are designed to provide the greatest practicable reliability, a minimum of moving parts, and maximum protection from unfavorable weather conditions. The crossing bell previously used for horse-drawn vehicles has been generally discarded because it is neither sufficiently audible for automobile traffic nor so reliable as the flashing-light signal, which has no moving parts except the enclosed relay contacts.

The following recommendations are taken from the *Bulletin* issued under the date of July, 1935, by the Association of American Railroads:

Types. At crossings on heavily traveled highways where conditions justify, either of the following standard visible warning signals shall be installed:

a. Wigwag type.

b. Flashing-light type.

At crossings where wigwag or flashing-light signals are used, one shall be placed on each side of the track.

Circuits for automatic operation of wigwag or flashing-light signals shall be arranged so that crossing signals will operate until the rear of the train reaches or clears the crossing.

Aspect. An electrically or mechanically operated signal used for the protection of highway traffic at railroad highway grade crossings shall present toward the highway, when indicating the approach of a train, the appearance of a horizontally swinging red light and/or disk.

Mounting. The railroad standard highway crossing sign and the signal shall be mounted on the same post. Either a signal of the flashing-light type or one of the wigwag type may be used, but both shall not be placed on the same post.

Operating Time. Automatic signal devices used to indicate the approach of trains shall so indicate for not less than 20 sec before the arrival of the fastest train operated over the crossing.

NOTE: Local conditions may require a longer operating time; however, too long an operation by slow trains is undesirable.

Flashing-light Signals. The recommendations for flashing-light signals include:

Height. The lamps shall preferably be not less than 7 ft nor more than 9 ft above the surface of the highway.

Signal lights shall shine in both directions along the highway and shall be mounted horizontally 2 ft 6 in. centers. Lamps when arranged in pairs back to back shall open at the front and be designed so that the door will open to the side or downward.

Flashes. Lights shall flash alternately. The number of flashes of each light per minute shall be 30 minimum, 45 maximum.

Hoods and Backgrounds. Lamp units shall be properly hooded. Backgrounds, 20 in. in diameter, shall be painted black on both sides.

Range. When lamps are operated at normal voltage, the range, on tangent, shall be at least 300 ft on a clear day, with a bright sun at or near the zenith.

Spread. The beam spread shall be not less than 3 deg each side of the axial beam under normal conditions. This beam spread is interpreted to refer to the point at the angle mentioned where the intensity of the beam is 50 per cent of the axial beam under normal conditions.

Lenses or Roundels. Lenses or roundels shall be 5 3/8 in. minimum, 8 3/8 in. maximum.

Transmission Values (for red lenses and roundels). Transmission values based on A.A.R. standard scale shall be 150 to 220 where plain cover glass with reflector is used; 220 to 300 where signals are used without reflectors or where a ribbed Spreadlite lens is used in front of the reflector.

Short-range Indication. Signal shall display a satisfactory short-range indication.

Wigwag Signals. Among the recommendations for the wigwag type are:

Length of Stroke. Length of stroke is the length of chord which subtends the arc, determined by the center of the disk in its extreme positions, and shall be 2 ft 6 in.

Number of Cycles. Movement from one extreme to the other and back constitutes a cycle. The number of cycles per minute shall be 30 minimum, 45 maximum.

Direction of Lights. Signal lights shall shine in both directions along the highway.

Stop Sign. The recommendations of the Association include certain reflector unit signs and permit the use of an illuminated red stop sign which displays the word "stop" toward highway traffic while the lights are flashing.

Automatic Barriers. Other automatic devices, such as electrically operated gates and barriers, are in limited use. The automatic gate, most generally favored, blocks only the right half of the roadway to approaching vehicles, leaving the left half of the road unobstructed for vehicles which have started to cross to complete their movement over the tracks.

Highway barriers have been installed at a number of crossings. These are operated by electric motors and are automatically controlled. An obstruction in the highway is presented to a vehicle approaching the tracks. The device is so constructed that a vehicle may pass off the tracks over the barriers without inconvenience.

Highway crossing signal-control circuits generally follow the principles of standard railway signaling circuits.

49. TRACK CIRCUITS

Track circuits are a basic and important link in most railway signal systems. They make possible automatic signaling, cab signaling, approach track indications, switch and lever locking, etc.

Track circuits require correct and, in many cases, delicate adjustments to provide the correct amount of current to operate the relay properly and reliably under different ballast conditions and under different rail and car wheel shunting conditions.

DIRECT-CURRENT TRACK CIRCUITS. The track circuit consists primarily of a battery or other source of electrical energy, connections to the running rails of a railway, and connections from rails at the opposite end of the rail circuit to an instrument known as a relay, which controls contacts for other circuits. The contacts drop away by gravity if the relay is de-energized as the result of failure of power, an actual break in the running rails or wire connections, excessive leakage between the rails through the track ballast, or the track being occupied by a car or train, thereby providing a shunt connection between the rails completing the circuit without sufficient current passing through the relay to hold the contacts closed.

Track circuits, which in their turn control signal circuits, must operate continuously, for a momentary opening of the circuit due to power failure or other cause might change a signal indication from "proceed" to "stop" in the face of an approaching train. Such a change might result in too heavy application of train brakes and damage to the train.

A simple d-c track circuit is shown in Fig. 12. With this circuit two or three cells of battery are connected in multiple. Where primary battery is used, this is generally of the copper oxide-zinc-caustic soda type. The track relay magnet coils are generally wound to 2 or 4 ohms resistance. The adjustable resistance in the battery leads is varied to provide the correct amount of current at the relay for the different lengths of circuits and different ballast conditions. The resistance from rail to rail through the ground is known as ballast leakage resistance and varies materially with the nature and cleanliness

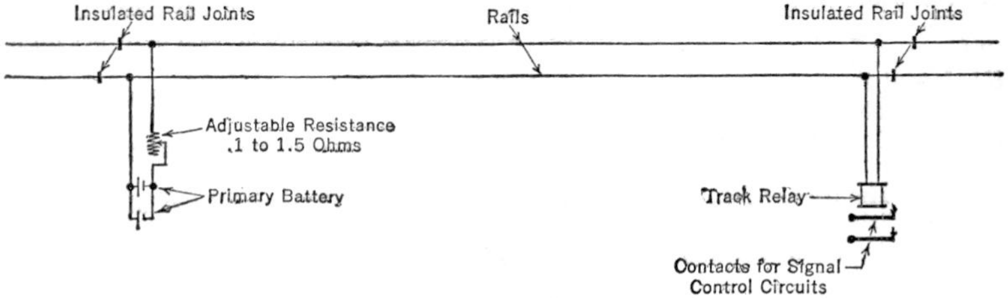

FIG. 12. Simple D-c Track Circuit

of the ballast, i.e., whether of clean broken stone, gravel, or cinders. It also varies with the number of highway crossings or station platforms in the circuit where dirt or planking may make contact with the rail and lower the resistance. The circuit also varies under wet or dry weather conditions and with the amount of brine drippings from refrigerator cars.

A track circuit with an average **ballast leakage** resistance of 4 ohms per 1000 ft will give good results. If the ballast resistance is less than 2 ohms per 1000 ft and the circuit is in excess of 3500 ft, special attention may be required. Generally, the shorter the circuit the more reliable will be the results. A single primary battery wet cell will operate a track circuit. Additional cells are generally connected in multiple to provide greater emergency supply and to extend the periods between battery renewals.

Figure 13 shows an equivalent track circuit with an a-c power supply, a rectifier, and a storage battery cell, floated between the rail connections to provide continuous current in case of power failures of short duration.

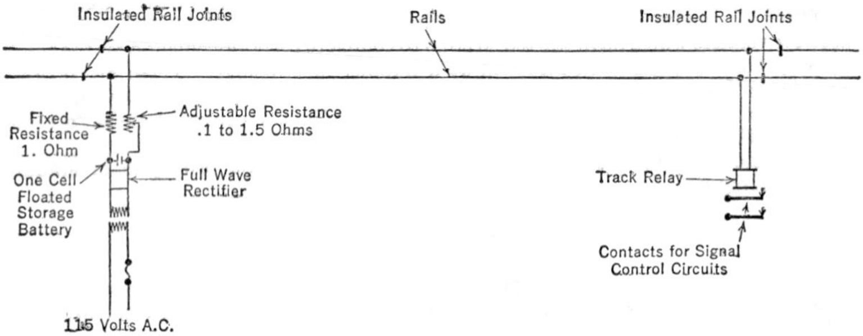

FIG. 13. Direct-current Track Circuit with A-c Current Supply

Figure 14 shows a track circuit especially designed to operate under poor rail shunting conditions where the rails may be rusty from little use, where cars with rusty wheels may be moved, or where light equipment makes a less positive contact. This circuit is particularly applicable to European railway conditions, where rail equipment is much lighter than in America and where standard track circuit shunting is questionable unless rail and wheels are bright from frequent use.

Figure 15 shows a **quick-acting track circuit** for car-retarder installations where, on account of the limited space available between the car retarder and the switch, the circuit must be restricted to about 50 ft in length. With this circuit it is not possible to start the control lever before the car passes on to the circuit and still permit the car wheels to pass on to a track switch when in mid-position.

Figure 16 shows track circuit for **automatic highway protection.** This figure shows circuits for flashing-light signals, but the basic circuit may be used for operation of automatic gates or other types of protection. (See Article 48.)

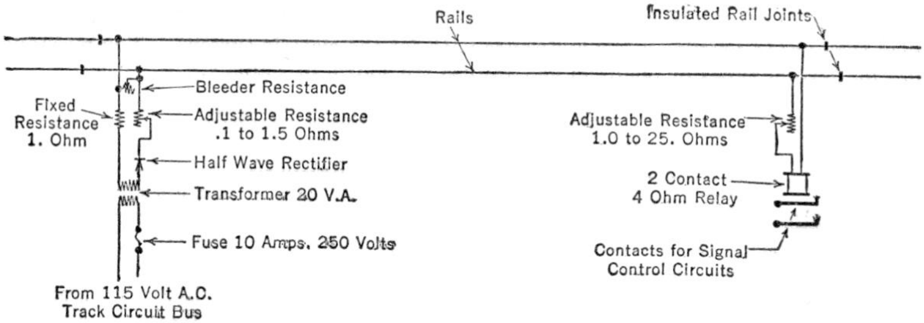

Fig. 14. Half-wave Track Circuit

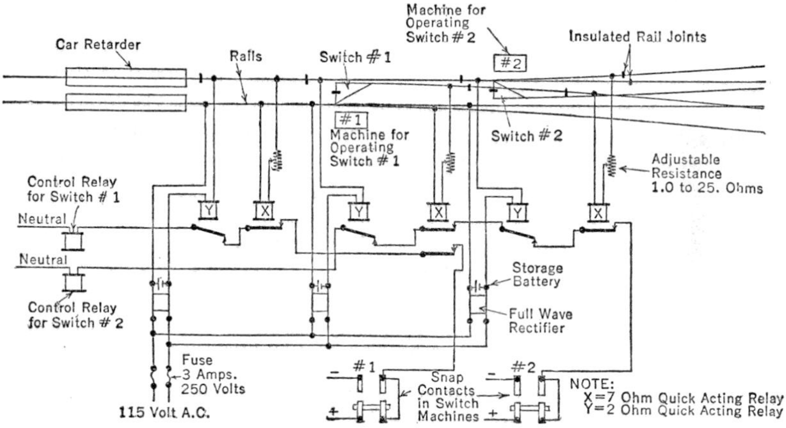

Fig. 15. Quick-acting Car-retarder Track Circuit

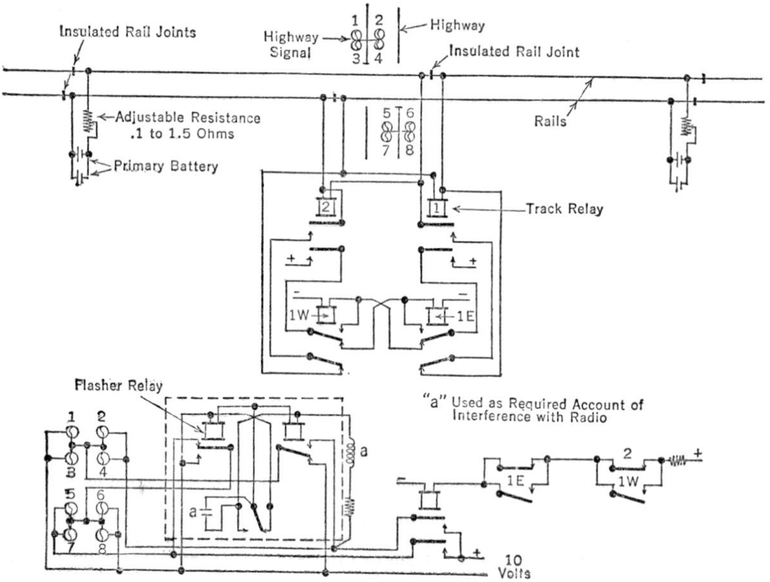

Fig. 16. Automatic Highway Protection Track Circuits

Rail Bonds. A reliable working track circuit requires that the adajcent rail ends be bonded together to provide a low-resistance path for the electric current. The resistance values for several types of bonds in general use are shown in Table 4.

Table 4. Resistance of Bonds to Signal Currents

Ohms per 1000 ft of Track (2 rails)

Bond Wires per Joint	30-ft Rails	33-ft Rails	39-ft Rails
Rail head welded bond 7 in. long	0.0067	0.0061	0.0051
Rail head plug bond 5 in. long	0.027	0.025	0.021
2 No. 6 copper 48 in. long	0.052	0.048	0.040
1 No. 6 copper and 1 No. 8 iron 48 in. long	0.089	0.082	0.069
2 No. 6 copper clad—40% 48 in. long	0.112	0.103	0.086
Stranded plug type copper core 52 in. long	0.147	0.134	0.112
2 No. 6 Copper clad—30% 48 in. long	0.150	0.138	0.115
2 No. 8 iron 48 in. long	0.315	0.291	0.242

Resistance of track circuit is determined by adding resistance of bonds per above table to resistance of rail, as follows: $\frac{2}{\text{weight of one rail per yd}}$ = resistance per 1000 ft of track. No allowance should be made for conductance due to rail splice bars as this cannot be depended upon.

ALTERNATING-CURRENT TRACK CIRCUITS. The d-c track circuit is economical to install and to maintain and will operate reliably under average rail conditions.

The reports of the United States Interstate Commerce Commission show approximately six times as many d-c as a-c track circuits in service in connection with railway automatic signals. Alternating-current track circuits require more energy, more expensive apparatus, and more careful adjustment and maintenance.

Many conditions necessitate a-c track circuits. This applies principally where the running rail is used for return of electric propulsion current to a powerhouse or where grounded current from adjacent electric railroads or from other sources may get on and use steam railway rails as a conductor. Frequently this foreign current can be reduced by the electric railway repairing its own bonding, thereby producing a lower resistance path of its own.

Where a-c electric propulsion is used and the rails provide the power return, a-c track circuits are required to be of a frequency which will not be affected by the frequency of the propulsion current.

In some special locations such as tunnels, because of dirt and continual moisture, it may be impracticable to keep a track circuit reasonably dry or clean. In these places a-c track circuits may give better results.

The A-c Track Relay, whether of the Vane or Rotor type, generally uses, for proper operation, electrical energy from a line connection in addition to the current passing through the rails. This is known as the local current, as distinguished from the current flowing through the rails, known as track current.

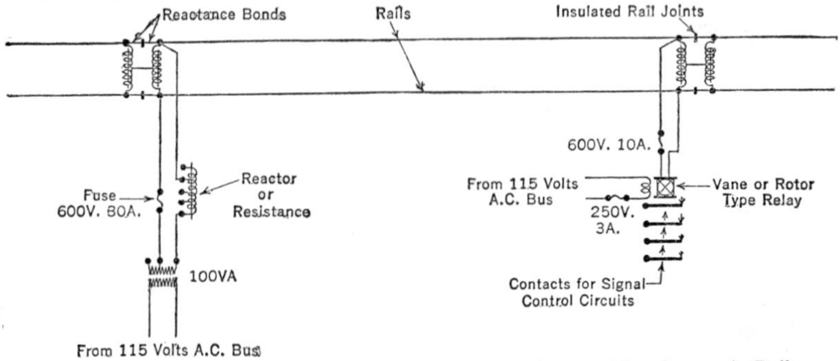

FIG. 17. Alternating-current Track Circuit with D-c Train-propulsion Current in Rails

Figure 17 shows a simple two-rail a-c **track circuit** with reactance bonds to permit the free flow of return d-c propulsion current around the insulated joints and between the parallel rails but with the a-c track-circuit current isolated to its particular section of track.

Where rails are not used for power return, the reactance bonds are omitted. For short circuits and where power return current in the rails is not large, a single-rail a-c circuit may be used. This confines the power return to one rail, omits the insulated joints in the power return rail, and omits the reactance bonds.

Alternating-current Circuit Calculations. Because of the complications of the a-c track circuit it is very difficult to arrive at the proper applied voltage and proper displacement angle between the two windings of the relay by cut-and-try adjustments in the field. Proper results can be obtained after the characteristics of the rail circuit and the relay have been determined by the vector method.

The track-circuit characteristics, ballast leakage resistance, and current lag behind voltage in rails may be determined by test of each track section, or by a typical circuit of a group of circuits having approximately similar characteristics.

Table 5 shows the electrical characteristics of several a-c track relays as manufactured by the General Railway Signal Company and by the Union Switch and Signal Company.

Table 5. Alternating-current Track Relay Characteristics

Two Element with Four Front and Four Back Contacts

General Railway Signal Co.

	Vane Type		Rotor Type		Rotor Type	
	Local Element	Track Element	Local Element	Track Element	Local Element	Track Element
Frequency	25	25	25	25	60	60
Normal volts	110	0.217	55	0.200	110	0.185
Amperes	0.300	0.350	0.230	0.550	0.250	0.300
Voltamperes	33.0	0.076	12.6	0.110	27.50	0.055
Watts	8.25	0.024	8.38	0.075	17.6	0.036
Per cent power factor	25	31	66.5	68.5	64.0	65.0
Phase angle lag	75.5°	72°	48° 20′	46° 45′	50°	49° 30′
Impedance, ohms	367	0.62	240	0.36	440	0.605
Resistance, ohms	44	0.10	107	0.18	107	0.18

Track phase current shown indicates minimum operating values at 90° or ideal phase displacement. To get working current at any other angle, figures in the table must be divided by the sine of that angle.

Union Switch and Signal Co.

	Vane Type		Vane Type		Vane Type	
	Local Element	Track Element	Local Element	Track Element	Local Element	Track Element
Frequency	25	25	60	60	100	100
Normal volts	110	0.36	110	0.30	110	0.41
Amperes	0.44	0.36	0.44	0.30	0.28	0.41
Voltamperes	48.4	0.13	48.4	0.09	30.8	0.168
Watts	24	0.092	23	0.082	14.5	0.151
Per cent power factor	49.5	70.7	47	91.0	47	90
For ideal phase relations track voltage should lag local voltage	....	73°	....	72°	....	72°

Above relays are with condenser tuning of track winding for long steam road track circuits.

Table 6 shows the characteristics of rail electrically bonded at the ends. The reference to "capacity" bonding indicates bonds with conductance equal to continuous rail. The column under Z gives the impedance in ohms. The column under A gives the phase angle of the current in the rail with respect to the voltage.

Vector Method. Figure 18 shows typical application by the vector method of determining the correct voltage, resistance, and reactance to be applied at the transformer end of the track circuit and the capacity to be applied in local line circuit at relay in order to assure the most reliable working of the circuit and relay.

The values for the relay used are for a 90-deg current phase displacement angle and must be corrected if the angle between the track and local element currents is less than 90 deg. This correction is made by dividing the ideal phase displacement current for the relay track element by the sine of the angle between the current in the track and local line elements of the relay. The result is the correct figure for the actual current through track element of relay.

The example shown on Fig. 18 is based on determined characteristics as follows: Energy 60 cycles; length of track section 4000 ft; rails 30 ft long, 110 lb per yd; relay, rotor type; working current of relay track element 0.375 amp, 0.24 volt, current lag 53° 7′; working

Table 6. Impedance of Bonded Rails to Signal Currents in Ohms per 1000 ft of Track

Weight of rail, Lb per Yd	Bonding	30-Ft. Rails						33-Ft Rails						39-Ft Rails					
		25 Cycle		60 Cycle		100 Cycle		25 Cycle		60 Cycle		100 Cycle		25 Cycle		60 Cycle		100 Cycle	
		Z	A	Z	A	Z	A	Z	A	Z	A	Z	A	Z	A	Z	A	Z	A
130	Capacity	0.10	70	0.19	74	0.30	77	0.10	70	0.19	74	0.30	77	0.10	70	0.19	74	0.30	77
	2-No. 6 copper	0.12	46	0.21	61	0.31	68	0.12	47	0.21	62	0.31	68	0.12	50	0.21	64	0.31	70
	2-No. 6 copper clad 40%	0.17	32	0.25	49	0.34	58	0.16	33	0.24	50	0.34	59	0.15	38	0.23	53	0.33	62
120	Capacity	0.10	69	0.20	74	0.30	77	0.10	69	0.20	74	0.30	77	0.10	69	0.20	74	0.30	77
	2-No. 6 copper	0.13	46	0.22	61	0.32	68	0.12	47	0.21	61	0.32	68	0.12	50	0.21	63	0.32	70
	2-No. 6 copper clad 40%	0.17	32	0.25	49	0.35	58	0.17	34	0.25	50	0.34	59	0.15	37	0.24	53	0.34	62
110	Capacity	0.10	68	0.20	73	0.31	76	0.10	68	0.20	73	0.31	76	0.10	68	0.20	73	0.31	76
	2-No. 6 copper	0.13	46	0.22	61	0.33	67	0.13	48	0.22	61	0.32	68	0.12	50	0.21	63	0.32	69
	2-No. 6 copper clad 40%	0.18	33	0.26	49	0.36	58	0.17	34	0.25	50	0.35	59	0.16	38	0.24	53	0.34	62
100	Capacity	0.11	68	0.21	73	0.32	76	0.11	68	0.21	73	0.32	76	0.11	68	0.21	73	0.32	76
	2-No. 6 copper	0.14	46	0.23	61	0.34	66	0.13	47	0.23	61	0.34	68	0.13	50	0.22	63	0.33	69
	2-No. 6 copper clad 40%	0.18	33	0.27	49	0.37	58	0.17	34	0.26	51	0.36	59	0.16	38	0.25	54	0.36	62
90	Capacity	0.11	67	0.22	73	0.33	75	0.11	67	0.22	73	0.33	75	0.11	67	0.22	73	0.33	75
	2-No. 6 copper	0.14	47	0.24	60	0.35	67	0.14	47	0.24	61	0.35	68	0.13	51	0.23	63	0.34	69
	2-No. 6 copper clad 40%	0.19	34	0.27	49	0.38	59	0.18	35	0.27	50	0.37	60	0.17	39	0.25	53	0.37	62

current relay local line element 0.25 amp, 110 volts, current lag 47° 23′.

Table 6 gives an impedance of 0.20 and an angle of 73 deg for this track section.

The ballast leakage resistance which must be determined for average unfavorable conditions of the track circuit is assumed, for purpose of this example, to be accumulated at equally distant points on the track circuit 1000 ft apart, as marked *A*, *B*, *C*, *D*, and *E*. The ballast leakage current at any point is computed by dividing the track voltage at that point by the ballast leakage resistance.

With the above data the desired track voltage at the relay is laid off as a voltage base line at a suitable scale to permit the closing of the diagram within the limits of the drawing. The relay control current, which has a phase angle of 53° 7′, is then laid off clockwise from this voltage base line at this angle as a current base line. These bases are shown in heavy lines in Fig. 18. The current used by the first section of ballast leakage is then laid off from the end of the relay control current line parallel to the relay voltage base line, this current being in phase with the same voltage applied at the relay. The resultant current is then the vector sum of these two currents. This value of current (scaled as 0.414 amp) is then used to determine the voltage drop in the rail to the next ballast leakage point, which is then added to the end of the relay voltage line, but at an angle from the last resultant current line equal to the phase angle of the rail, namely, 73 deg. The resultant voltage is used to determine the ballast leakage current at the second point, etc. The remaining ballast leakage currents and rail drops are similarly computed until the total current and the necessary track voltage at the energy-supply end of the circuit are determined, being for this vector 1.0124 amp and 0.665 volt. In order to get the most uniform and sensitive shunting with reliable working at all points on the circuit the limiting impedance is made up of resistance and reactance so that it will have the same value and phase angle as the relay. The voltage drop across this limiting device is then added to the rail voltage at an angle of 53 deg counterclockwise from the total current line

The vector sum of the rail voltage and volt drop across the limiting device gives the required transformer voltage.

The local winding voltage is shown 180 deg from the track transformer secondary voltage. Sufficient resistance and capacity should be placed in series with the local winding

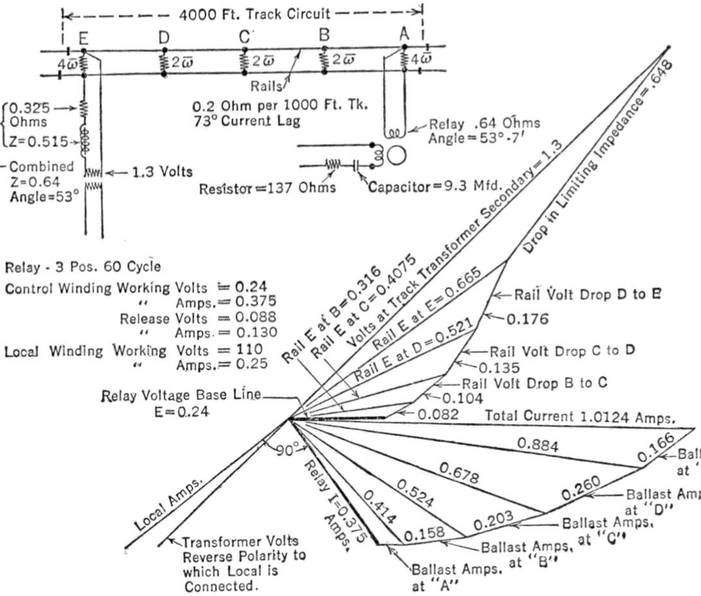

FIG. 18. Vector Method for Determining Values of A-c Track Circuit

so that there will be a 90-deg angle between the local current and the control current, at the minimum ballast leakage resistance where the greatest torque is required. The combination of resistance and capacity is required to get the 90-deg displacement and still have only the full line voltage applied to the local relay winding.

After computations have been made for a few typical circuits, curves can be plotted which will readily give the proper values for circuits with the same type of apparatus of other lengths.

CODED TRACK CIRCUITS. The coded track circuit constitutes the latest development in railway signaling as regards track circuits. Actually, it retains all the advantages of the steady-energy track circuit and has, in addition, other desirable attributes.

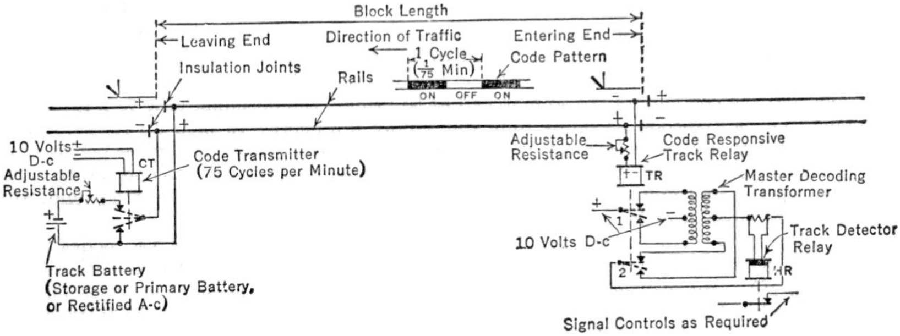

FIG. 19. Single-code Track Circuit

In the coded track circuit (Fig. 19), the current to the rails is uniformly interrupted by the code transmitter at a predetermined number of times (75) per minute so that the current in the rails pulses in alternate "on" and "off" periods. The track relay alternately picks up and releases in response to this coded current transmitted through the rails. The track relay is especially designed so that it will follow the coded pulses of current supplied to the rails by the coding apparatus at the highest code rate that may be used in the system. Too, it will respond only when energized with current of proper polarity, being in effect a polar relay. Direct-current energy is taken through the number 1 contact of the track relay to energize the primary winding of the master decoding transformer, one half of this winding being energized when the relay closes its front contact, and the other half when the relay closes its back contact. These pulses of current supplied to the primary winding cause an a-c output by the secondary winding. This a-c output is mechanically rectified by the track relay's number 2 contact, and, in the form of pulsating direct current, it energizes the track-detector relay HR. The signal circuits are controlled by the HR relay.

It is evident that, if the track relay is not alternately picking up and dropping, no a-c output will be delivered by the master transformer; hence relay HR will not be energized. Thus failure of pulsating current of sufficient strength to pick up the track relay, due to a train shunt, a broken rail, or other abnormalities, will cause a restricting indication to be displayed by the signal or signals controlled by relay HR. This is also true if the track relay should become steadily energized, as by foreign current.

A relay's "pick-up" value is appreciably more than its "drop-away" value. Therefore, it is evident that the detection of a train will be effected under higher resistance shunting conditions than would be possible with steady-energy track circuits, where it is necessary for the train shunt to be of sufficiently low resistance to rob the track relay of current to a point approaching or below its rated drop-away value. In other words, coded track circuits provide improved shunting sensitivity, improved sensitivity to broken rails, improved protection against foreign current and broken-down insulated joints. Other advantages are longer track circuits, often eliminating the necessity of cut sections in long blocks, and the elimination of line wires. The latter advantage is explained in the following discussion.

The circuit shown in Fig. 19 constitutes the basic circuit used for "detection" purposes only. It can provide control for only two signal indications, "stop" and "proceed." Other indications, such as "approach" and "approach medium," have to be provided by usual line circuits.

Multiple-indication Signaling with Coded Track Circuits. To secure multiple "proceed" indications without the use of line wires, codes of different rates, such as 75, 120, or 180 per minute, are applied to the rails at the battery end. The code rate in effect at a given time is governed by the indication being displayed by the next signal in advance.

Figure 20 shows such a typical arrangement. In this case the code allocations are as follows: 0 code, stop (red over red); 75-rate code, approach (yellow over red); 120-rate code, advance approach (yellow over yellow); and 180-rate code, clear (green over green).

At the battery end of the track circuit, one of the code transmitters is connected to the code transmitter repeater relay (CTPR) and drives it. The selection of transmitter so connected is made by the local circuits and depends on the indication displayed by signal 43, Fig. 20. The CTPR interrupts the current to the track at the rate corresponding to that of the transmitter by which it is being driven.

At the relay end of the track circuit, the 0 and the 75-rate code are detected and identified in the same manner as described for Fig. 19. Decoding units are provided for identifying the 120- and 180-rate codes. These decoding units are comprised of a transformer (with iron core), condensers, and a full-wave copper oxide rectifier. The primary of the transformer is connected in series with the condensers. The characteristics of the transformer and the condensers are so matched as to "tune" the decoding unit to a given frequency; i.e., a 120-code decoding unit will supply d-c energy to pick up its associated decoding relay (DR) only when it is being energized at the rate of 120 cycles per minute. The same applies to the 180-code decoding unit. By adding additional code transmitters at the battery end and a decoding unit for each additional code at the relay end, any practical number of codes may be added to the system.

Inverse Code. The codes described so far are known as direct codes, i.e., codes transmitted through the rails from the leaving end of the track circuit toward the entering end (see Fig. 19). By utilizing the "off" periods of the direct code, inverse codes may be transmitted in the opposite direction, i.e., from the entering end toward the leaving end. Inverse codes may be used for the control of approach lighting of signals and for the control of vital circuits, such as approach locking and rail/highway crossing protection.

A typical inverse-code circuit such as might be used in conjunction with the circuit in Fig. 20 is shown in Fig. 21. Relays are positioned as during the "off" period of the direct code and at the time of initial energization by the inverse code "on" period. At signal 41 location, a source of energy such as a battery is provided for the inverse code.

Two additional relays are provided, the TPB and TPA. When the code-responsive track relay (TR) drops during the "off" period of the direct code, energy is applied to the TPB and TPA. The TPA picks up immediately, disconnects the TR relay, and

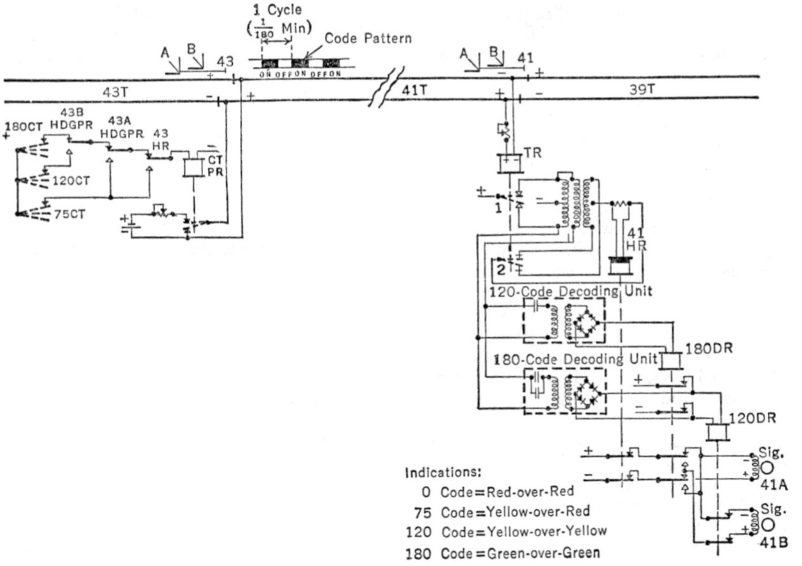

Fig. 20. Multiple-code Track Circuit

applies inverse-code current to the track. The TPB relay has slow pick-up characteristics; therefore the TPA is held energized until the TPB picks up, at which time the TPA drops and the inverse-code pulse is ended.

At signal 43, an approach relay (AR) is connected in multiple with the track leads when the CTPR relay is de-energized (direct-code "off" period). The inverse-code cur-

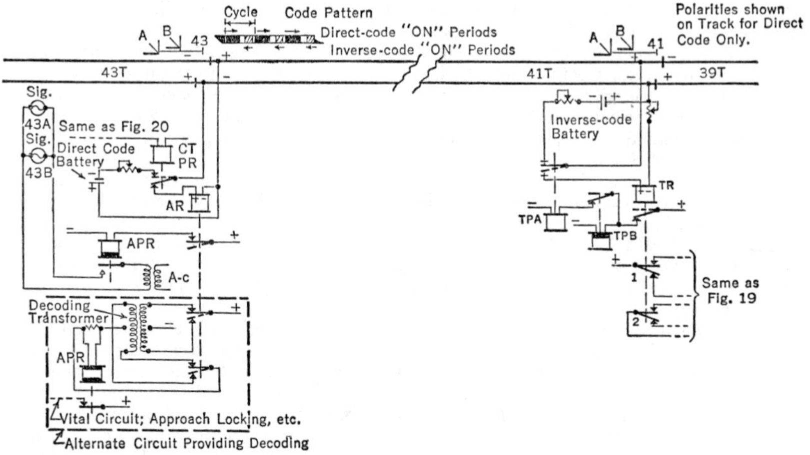

Fig. 21. Combined Multiple- and Inverse-code Track Circuit

rent, arriving at a time when the CTPR is making its back contact, energizes the AR relay, which in turn applies energy to the slow-release repeater relay APR. The APR relay holds up over the intervals between inverse-code pulses. The inverse-code circuit is operative at any of the direct-code code rates. A train shunt, broken rail, etc., has the same effect on the inverse-code circuit as on the direct-code circuit.

Where vital circuits are to be controlled, decoding (shown as an alternate) is desirable to safeguard against false energization of the AR, as from foreign current.

General. Experience has proven coded track-circuit control to be very flexible. Many variations in its execution have been developed, such as employing polarized codes for providing added controls or indications. Circuits providing a "follow-up" code, i.e., a code that follows a train into the block, are often used, such as for the purpose of cutting out the operation of crossing signals located midway in a block after the rear of the train has passed over the crossing.

Circuits have also been developed for use in single-track (double-direction running) territory. Such circuits, when used in centralized traffic-control territory, may be arranged to be normally de-energized.

Coded track circuits, employing a-c energy, may be used to operate cab signals. The various code rates will cause the desired indications to be displayed by the cab-signaling equipment. At the relay or entering end of the track circuit, the a-c code impulses are rectified to operate the code-responsive track relay.

50. CONTROL CIRCUITS

Control circuits for signals and for switch operation, indications and locking, and other railway signal circuits generally follow the closed-circuit principle.

A failure of energy or a break in a conductor or an instrument will, as far as may be arranged, cause the signals to assume a stop position and will prevent the moving of a switch, derail, or other device.

Circuits may be either d-c or a-c and either neutral or polarized. A complete metallic circuit is used, and later practice has been largely to avoid common return wires. Signal circuits are varied in design as may be applicable to a particular problem.

Figure 22 shows a typical arrangement of a two-wire polarized circuit for control of color light signals giving restricting indication for four blocks in the rear of a train.

This may be readily reduced to a three- or two-block arrangement by omitting the extra wiring and relays.

51. ELECTRICAL SIGNAL APPARATUS

Electrical signal apparatus is constructed, as far as practicable, so that a disarranged or broken part will cause the signals to give their most restrictive indication and so that train movements will be stopped or not started.

Included are such instruments as signal mechanisms, relays of various types, switch-operating machines, electric locks and releases, many types of circuit controllers and terminals, and a large number of other special items. They are generally manufactured by companies especially equipped for the design, building, and testing of this precision apparatus. Signal cables, wires, batteries, rectifiers, pipe, etc., are of a character especially adapted for signal work.

It is practicable to describe herein only a few of the most important items, such as color light signal mechanisms, relays, and signal rectifiers. Primary batteries for signaling are described in Section 7, Articles 10 to 15.

SEARCHLIGHT-TYPE SIGNALS. Searchlight-type signals have come into the favor of many railroads which use color light signals because of their powerful indication, extreme reliability, ease of maintenance, and universal application either as high or as dwarf signals.

This signal consists of an outer case equipped with lens, hood, aligning peep sights, and background. The mechanism slides into the outer case and is locked into place. Where plug-type couplers are used for the wire connections to the mechanism, signals of this sort can be taken out of service, the mechanism changed, and the signal restored to service within 2 min. This is of material assistance at congested railroad traffic points. The mechanism encloses the moving armature, on which are mounted three colored disks for a three-indication signal. An ellipsoidal silvered reflector gathers the light from a precision bayonet base lamp, the filament of which is located at one focal point of the ellipsoid and transmits the light through the color disk, which is located near the other focal point. The beam assumes the color of the disk through which it shines, thus assuring the correct indication at all times, whether from light within the signal or from external light, as from a locomotive headlight, reflected by the reflector. The vane-type

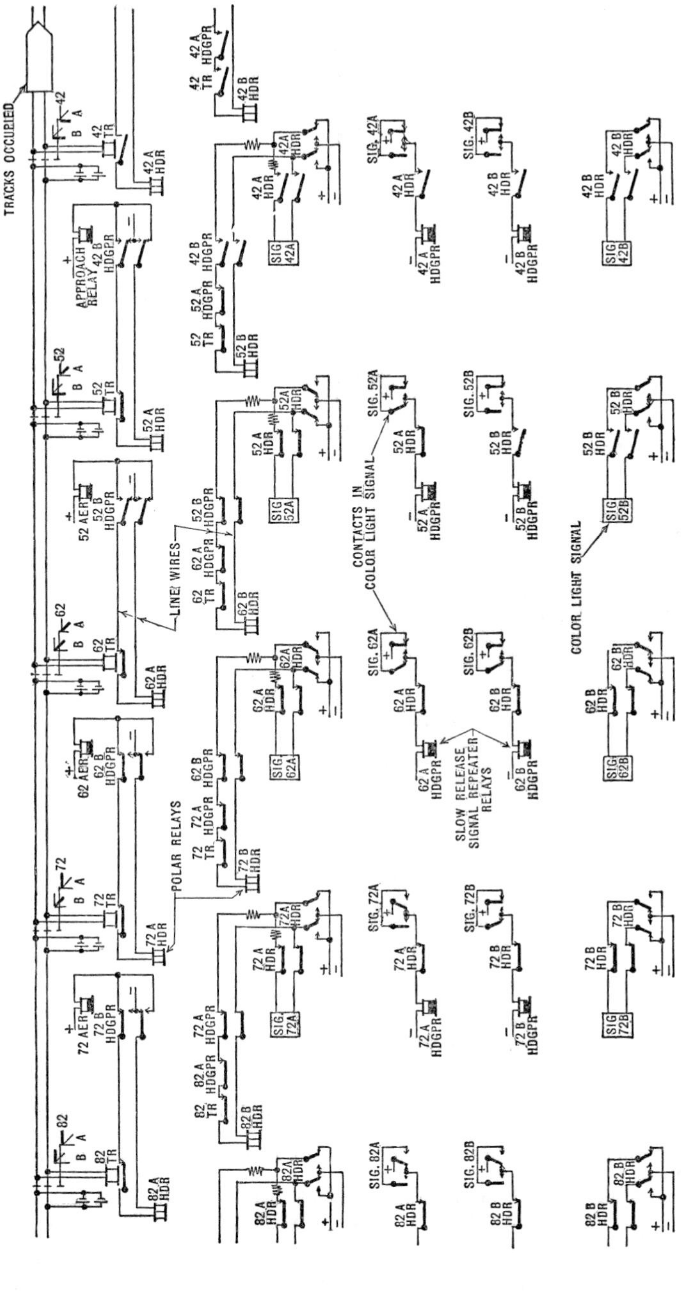

FIG. 22. Polarized Signal-control Circuits

armature actuates checking contacts which give a positive check at all times on the indications displayed. These contacts may be used to carry current for the control of other signals, relays, or indicators as required.

Figure 23 shows a cross-section through a searchlight signal. The dotted lines illustrate the light rays passing through the small colored disk. Figure 24 is a sketch of the spectacle which holds the 1-in. colored roundels and shows the shaft and counterweight.

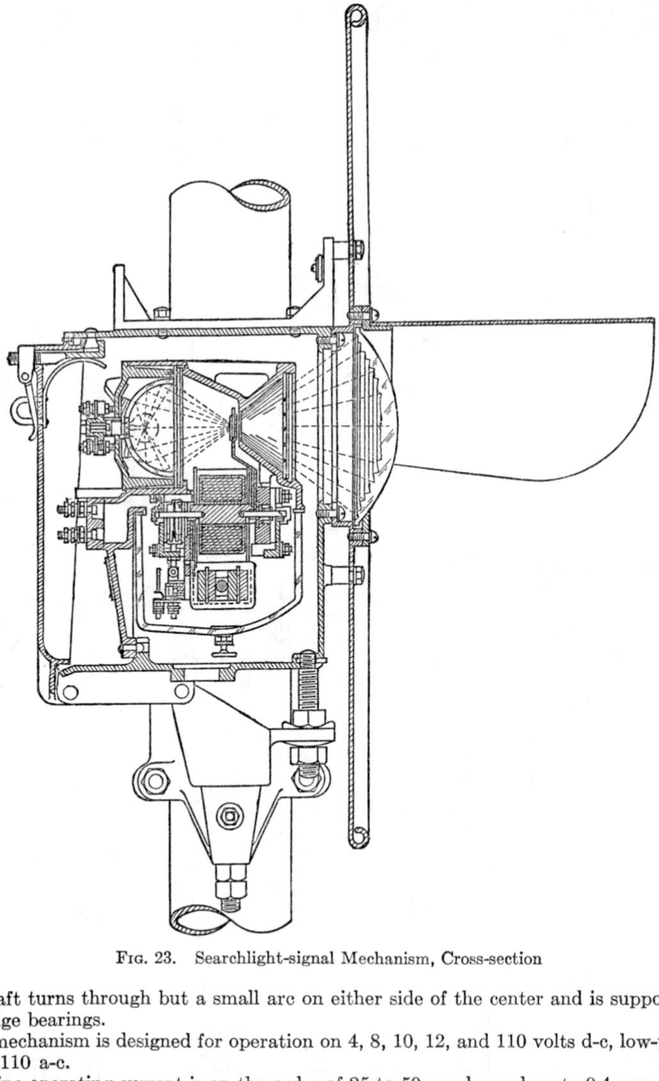

FIG. 23. Searchlight-signal Mechanism, Cross-section

This shaft turns through but a small arc on either side of the center and is supported on knife-edge bearings.

The mechanism is designed for operation on 4, 8, 10, 12, and 110 volts d-c, low-voltage, 55 and 110 a-c.

The line operating current is on the order of 25 to 50 ma d-c and up to 0.4 amp on low-voltage a-c. The signal lamps consume from 3 to 25 watts, depending on the signal application and the lens system used.

Beam candlepower is the measure of the light intensity in a beam of light by comparison with the standard international candle and can be accurately determined in a photometric laboratory for a given lamp and lens combination.

RANGE. As applied to light signals, range is usually understood to mean the distance on a tangent, in bright sunlight, at which the indications are clear and distinct to a person of average eyesight. Unless the range can be expressed in terms which will be equivalent to some known quantity, e.g., beam candlepower, and accepted generally, it is not a satisfactory means of drawing comparisons between different types of signals and various lamp and lens combinations.

An empirical formula, namely: Range = $\sqrt{2000 \times \text{B.C.P.}}$ (white light) has been found to work out satisfactorily as a means of expressing maximum beam candlepower in terms of range in feet. The formula is based on visibility in bright sunlight, using the green indication. Red gives about the same visibility, and yellow somewhat better. The formula does not apply to purple.

The range of the searchlight signal for any known beam candlepower may be quickly ascertained by referring to the curve on Fig. 25 which has been plotted from calculations using the above formula. Beam candlepower was accurately measured in a photometric laboratory, and averages which were the result of a number of tests were taken. In making the measurements, beams of white light were used, the color disk being omitted, since the formula takes care of the transmission factor through color. The formula was developed for temperate climates; allowance should be made for the tropics.

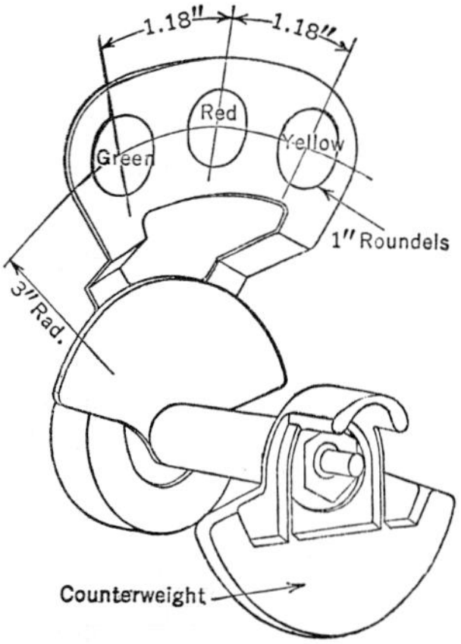

Fig. 24. Searchlight-signal Shaft and Spectacle

RELAYS. Relays are perhaps the most important and are the most generally used of signal instruments. Absolute dependence is placed on relay performance. They are incorporated in practically all signal circuits.

Direct-current relays, because of their greater simplicity, are generally used where practicable.

Essentially a d-c relay is a device equipped with an electromagnet consisting of an iron core, horseshoe shaped, around the two legs of which are coils. When electrical energy is applied to these coils, the magnetic force created picks up a bar of iron which is suitably pivoted. On this bar of iron, known as an armature, are contact fingers which close secondary electrical circuits. When the energy is cut off the coils, these secondary electrical circuits are opened, and others are closed as the armature drops away or releases. The contacts making when the armature picks up are called front contacts; the contacts making when the armature drops or releases are called back contacts. The armature drops by gravity; this is a necessary requirement so that, if anything happens to cut energy off the relay, it will release immediately.

Relays are equipped with a varying number of contacts to control a varying number of circuits, depending upon the applications. They are usually supplied in sizes of two, four, six, and eight contacts.

Those relays whose armatures pick up and drop away are called "neutral" relays. Relays which have a polar armature in addition to a neutral armature are called neutral-polar relays. When current of one polarity flows through the coils, the polar armature pivots in one direction; when current of the other polarity flows through the same coils, the polar armature pivots in the other direction.

There are relays also for special applications, such as flashing relays for the control of highway crossing signals, power-transfer relays to switch from one source of power to another in case of power failure, interlocking relays for the control of automatic interlockings at railroad crossings, and retained neutral polar relays for certain block signaling applications. They all operate on the basic principles outlined above.

There are relays designed for use in important signal circuits and others for use only in less important circuits. Important circuits include track, locking, traffic, and switch- and signal-control circuits. Other circuits include centralized traffic control and communicating circuits and circuits for control of indicating lights on control machines. In both classes relays are constructed with characteristics applicable to their specific uses. These may comprise only an operating characteristic, such as magnetic stick, slow release, slow pick-up, or quick acting, or they may be such as to identify the relay as being of a particular classification, such as neutral, retained neutral, retained neutral polar, biased neutral, polar, power-off (used for transferring a load from regular source of energy to a stand-by source), light-out (for detecting the breakage or burning out of a signal-lamp filament), two-rate charge control (for controlling the charging of a storage battery), overload (for the protection of motors, etc.), flashing (for the operation of flashing lights used in rail highway crossing signals and on control machines), interlocking (used in railway crossing and rail highway crossing protection), and numerous others. All operate on the basic principles previously described.

Other relays operating on different principles or in combination with the above principles may be thermal-time element, motor-operated time element, and some of the a-c types.

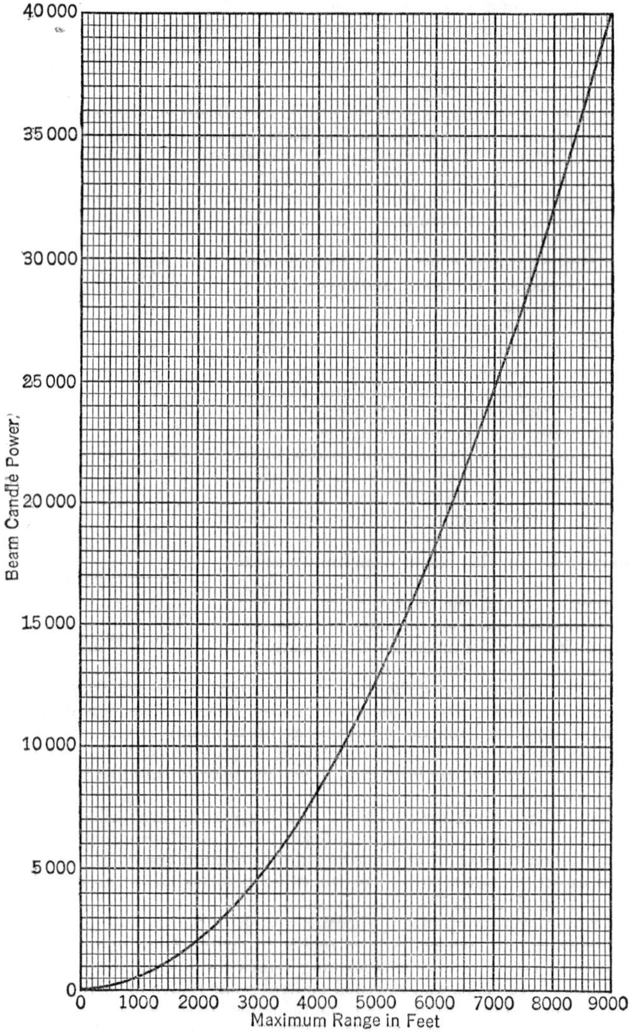

FIG. 25. Searchlight-signal Light Range

Earlier practice was to provide individual binding posts on the relay for wire connections. Of late years there has been considerable trend toward "plug-in" relays, so constructed that attachments are made by means of a male and female circuit connection. This arrangement has the advantages of permitting quick replacement and reduces to a minimum the possibility of errors in replacing circuits at time of change.

SIGNAL RECTIFIERS. Signal rectifiers are devices for changing alternating current into pulsating, unidirectional current.

During the past several years rectifiers of various types have been extensively used in railway signaling to charge storage batteries from an a-c line and to operate d-c signal relays and other devices directly from an a-c supply.

Dry-plate copper oxide rectifiers are extensively used in railway signaling since they possess the following advantages: (1) no moving parts; (2) reasonably high efficiency; (3) low maintenance cost; (4) ease of adjustment in connection with transformers, reactors, and resistors; (5) small housing space required. (See Section 11, Article 17.)

The basic principle of the **copper oxide rectifier** is that, when copper oxide has been formed on one face of a copper disk by special heat treatment, current flows more readily

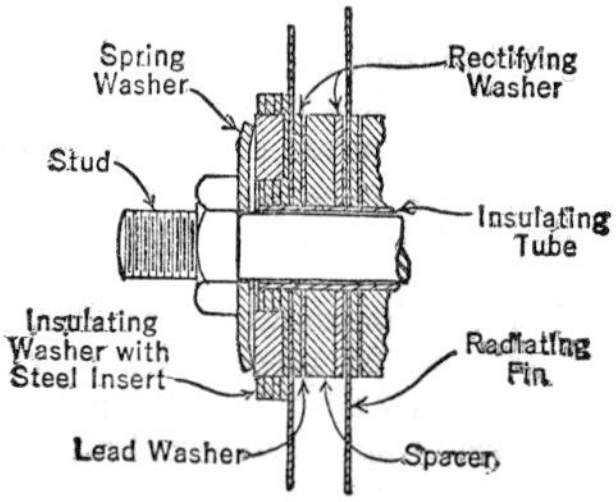

Fig. 26. Dry-plate Rectifier

from copper oxide to copper than in the opposite direction. Each disk, therefore, serves as an electrical valve in which rectification takes place at the junction of the copper and copper oxide. The action is entirely electronic, without any chemical action or decomposition of the rectifier elements.

Each rectifier unit comprises copper oxide disks or washers, lead washers or contacting surfaces, brass radiating washers or cooling fins, spacer and insulating washers all assembled on an insulating tube over a bolt as shown in Fig. 26. The number of copper oxide

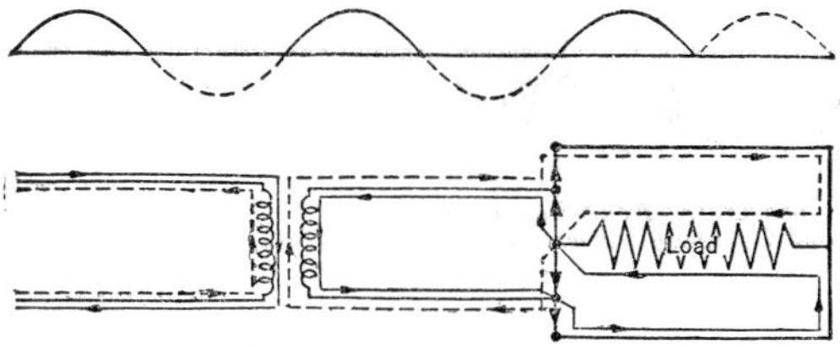

Fig. 27. Diagram of Current Flow in Dry-plate Rectifier

disks and their connection in series or multiple depend upon the output voltage and current for which a particular rectifier is designed. In order to secure full-wave rectification the disks in most of the rectifiers are connected to form a Wheatstone bridge circuit.

Figure 27 shows the manner in which the flow of current is directed by the valve action of four rectifying disks connected in a Wheatstone bridge circuit, so that current from both halves of the a-c wave flows through the load in the same direction.

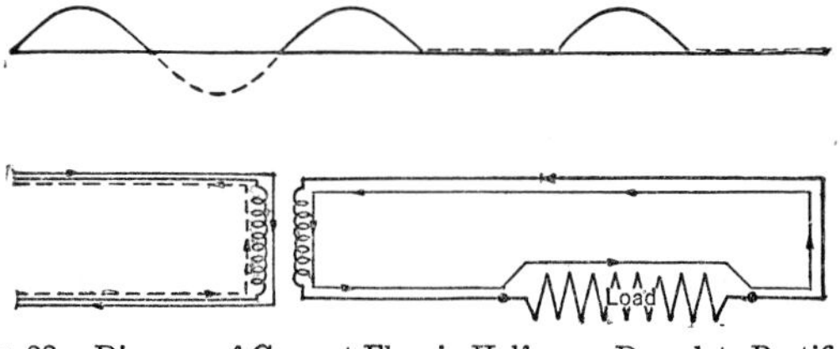

Fig. 28. Diagram of Current Flow in Half-wave Dry-plate Rectifier

Figure 28 shows the circuit arrangement of a half-wave rectifier which prevents the flow of current in one direction, so that current from only one-half of the a-c wave flows through the load.

52. RAILROAD RADIO

One of the more recent contributions to railroad operation has been the development of railroad radio. By means of it, two-way voice communication may be provided between fixed land stations and mobile stations (trains or locomotives), two or more mobile (or land) stations, or between the head and rear ends of a train. It may be used to assist in railroad operation and on harbor craft.

There are two general kinds of "railroad radio": space radiation and inductive carrier. The difference between the two is technical and is not apparent to the personnel using them. The inductive-carrier method requires the presence of wires paralleling the railroad to act as wave guides to couple a remote land station to the mobile stations as they stand or move along the railroad near the wires. The space-radiation method does not require the presence of wires.

On December 31, 1945, the Federal Communications Commission allocated a block of 60 frequencies extending from 158 to 162 megacycles to railroad space radio. These frequencies may be augmented with other specific frequencies to be shared with television channels located at points in the radio spectrum starting with the 44- to 50-megacycle channel and ending with the 210- to 216-megacycle channel. The electronic equipment for the space radio train communication system consists of an FM transmitter and receiver designed for operation on the above frequencies, operatable on 110/120 volts 60 cycles, and stations may be operated by persons having a second-class radio operator's license or by railroad employees who have qualified under FCC order No. 126.

The electronic equipment for the inductive-carrier train-communication system consists of a narrow-band FM transmitter and receiver for operation on 70 to 200 kilocycles and operatable on 110/120 volts 60 cycles. Inductive-carrier stations may be operated by unlicensed persons.

53. BIBLIOGRAPHY

Association of American Railroads, Signal Section Manual and Principles and Practices.
W. H. Elliott, *Blocks and Interlocking Signals* and *A.B.C. of Railroad Signals.*
General Railway Signal Company, *Electric Interlocking Handbook.*
E. E. King, *Railway Signaling.*
J. B. Latimer, *Railway Signaling in Theory and Practice.*
F. C. Lavarack, *Locking.*
Ohio Brass Company, *The Track Circuit Handbook.*
Union Switch and Signal Company, *Alternating Current Signaling.*

SECTION 18

ELECTROCHEMICAL AND ELECTROTHERMAL PROCESSES

* By W. A. Koehler.

ELECTROCHEMICAL PROCESSES

By W. A. Koehler

1. INTRODUCTION

In electrochemical processes, chemical changes take place at cell electrodes. These changes, called electrochemical reactions, always involve an interchange of electrons between the electrodes and ions; these ions may be in a solution or in a fused compound or in a gas.

Direct current is used for electrochemical processes. Sometimes this current is generated at the plant, but more commonly alternating current is purchased and converted to direct current by means of motor-generator sets or rectifiers. The installation of large automatic conveyor systems for plating in the automotive and hardware industries has called for generators with outputs up to 25,000 amp at 6 or 12 volts. The principal rectifiers used for electroplating are the copper oxide, copper sulfide, and selenium rectifiers, which are built in capacities up to 10,000 amp. At full load the efficiency of large generators is about 80 per cent, whereas that of rectifiers is about 70 per cent over long periods. With reduction in load generators decrease in efficiency more rapidly than rectifiers, so that at intermediate and widely variable loads rectifiers may prove more economical.

2. ELECTROPLATING

Electroplating has for its object the deposition of a layer of metal on another metal or non-metal (glass, porcelain, plastics) for the purpose of decorative effect, corrosion protection, building up worn parts, improving wear resistance, improving light reflectivity, or otherwise changing the physical or chemical properties of surfaces. All electroplating on a commercial scale is done in aqueous solutions; the object to be plated is made the cathode, or negative electrode, and the solution contains ions of the metal being plated. Usually the anode, or positive electrode, is made of the same metal as is being deposited on the cathode; it goes into solution electrolytically, thus maintaining the metal content of the plating solution. In chromium plating an anode of lead or iron is used, and the ions in solution must be replenished by the periodic addition of chromic acid to the plating bath.

CURRENT GENERATION. The current for electroplating is commonly supplied at 6 volts, with individual rheostats for each plating tank for proper current control. For barrel plating and chromium plating 12 volts are desirable. The tanks are always connected in parallel, so that any tank can be cut out without affecting the others. The current is usually supplied by motor-generator sets or rectifiers, such as the copper oxide rectifier.

Proper current density, commonly expressed in amperes per square foot, is important in order to obtain suitable deposits, so that in addition to a rheostat for each plating tank it is desirable to have also an ammeter and a voltmeter.

TANK CONSTRUCTION. Wooden tanks were formerly widely used in the plating industry and are still common for rinse tanks; if used for plating tanks, they are usually lined with asphalt or sheet lead. Steel tanks are used for cleaners and alkaline solutions; wood, steel, or concrete tanks lined with lead are used for solutions containing sulfuric or chromic acid. Rubber-lined tanks are proving very popular for acid and neutral baths, such as acid pickles, acid zinc, and copper solutions. Rubber linings are usually applied to steel tanks in the form of hard or soft rubber sheets cemented to the steel, or rubber is sprayed on to produce seamless coatings. Tanks with conducting surfaces, such as the lead-lined ones, may cause short circuits with loss of current and may deposit metal on the lining if it is cathode or liberate acid if the lining is anode.

Small articles, such as screws and small bolts, are plated in a barrel plater. In its present most common form the barrel consists of a horizontal non-conducting drum, usually hexagonal or octagonal in cross-section. The drum walls are perforated for admission of the solution and for passage of current. The articles to be plated are placed in the drum, which is immersed in the plating solution between two rows of anodes. The charge in the drum is electrically connected to the negative terminal; as plating proceeds, the drum revolves about its axis, thus tumbling the charge and causing uniform plating and at the same time some polishing. Larger items, such as faucets and stove legs, are hung on racks, which are hung on the cathode bars for still plating or on a rotating spider for agitation. For quantity production automatic plating machines are used; they clean, plate, and rinse the articles by means of a specially constructed conveyor.

REMOVAL OF OIL AND GREASE. If the article to be plated is not clean, the deposited metal will not adhere uniformly. All oil and grease must be removed. This can be done by several methods:

Alkaline Cleaners. Such cleaners usually contain two or more of the following ingredients: soda ash, sal soda, sodium hydroxide, trisodium phosphate, sodium silicate, and borax. A 10 per cent solution is of suitable strength. Proprietary preparations, usually made up of some of the above ingredients, are in common use. The solutions are used hot.

Organic Solvents. The article to be cleaned is washed with gasoline, benzene, carbon tetrachloride, or trichlorethylene. In the vapor-phase degreasing process, the article is hung in the vapors in a tank above boiling trichlorethylene. The vapors condense on the article, and as the liquid drips off it washes off all grease. This process is very popular because there is little loss of solvent and the cleaning action is thorough.

Electrolytic Degreasing. The article to be cleaned is made cathode in a tank containing an alkaline solution. A high current density, 30 to 35 amp per sq ft, is used, and there is brisk evolution of hydrogen on the article to be cleaned. The tank is usually made of iron and serves as the anode, and the solution is used hot, 50 deg cent (122 deg fahr) and up to boiling temperature. A cleaning operation of a few seconds is ordinarily sufficient. If articles containing zinc, tin, lead, or solder are cleaned, these metals may pass into solution and later be plated as a thin film on the article being cleaned. Such a film can be removed by reversing the current a few seconds so that the cleaned article becomes anode.

REMOVAL OF SCALE. In addition to removal of grease, all scale must be removed from the base metal before plating; the scale is usually a film of oxide.

Pickling Iron and Steel. Sulfuric and hydrochloric acids are well suited for pickling iron and steel, sulfuric acid being used more commonly. The action is a combination of solution of the scale and mechanical separation due to formation of bubbles of hydrogen between the scale and the metal. A 10 per cent by weight solution of sulfuric acid ($^1/_2$ pint or 1 lb commercial acid per gal) used hot is a suitable pickling solution for descaling. Sand particles imbedded in cast iron from a previous sand-blasting process can be removed by pickling in hydrofluoric acid.

Electrolytic Scale Removal. If iron or steel is covered with a thin layer of scale, the article may be made cathode in a 30 per cent solution of sulfuric acid heated to 60 deg cent (140 deg fahr). The scale is reduced to iron and there is little loss of metal, but the absorbed hydrogen makes the metal brittle. If the iron is made the anode in the same bath, there is no impairment of the physical properties of the iron, but there is loss of metal.

In the Bullard-Dunn descaling process [U. S. Patent 1,775,671 (Sept. 1930); C. G. Fink and T. H. Wilber, *Trans. Electrochem. Soc.*, Vol. 66, 251 (1934)], the scale is removed electrolytically by using the article to be plated as cathode; when it is clean, lead or tin is plated on the metal surface, automatically giving a protective surface. The lead or tin is supplied by using lead or tin anodes. If it is desired to remove the protective film before drying, the work is placed as anode in a reversing alkaline bath containing 90 grams per liter caustic soda and 30 grams per liter trisodium phosphate (12 and 4 oz per gal, respectively). The bath is used hot.

Sand Blasting. For cleaning castings a spray of sand or steel shot is blown against the casting with a blast of air, which removes the scale mechanically. It is important to protect the worker from the sand and fine dust that is formed. If too high an air pressure is used, some sand may be imbedded in the iron and will have to be removed before plating by pickling in hydrofluoric acid. Brass is treated with a "bright dip"; a satisfactory solution consists of approximately 58 per cent sulfuric acid, 7 per cent nitric acid, 0.2 per cent hydrochloric acid, the rest being water. Lampblack added to the dip aids in uniform attack of the acid.

Sodium Hydride Process. The descaling action is produced by NaH, dissolved in fused caustic soda, as a carrier, in a carbon steel tank. The sodium hydride concentration is maintained within the range of 1.5 to 2 per cent by weight. The bath is operated at 360 to 382 deg cent (680 to 720 deg fahr). When the article to be descaled is immersed in the hot bath, metallic oxides are reduced; when this process is complete, all action stops and there is no attack on the base metal. Mill scale, annealing and welding scale, drawing compounds, and small quantities of oil and grease are removed. No hydrogen embrittlement results from the process. All metals and alloys which do not react with fused caustic soda and whose physical characteristics are not adversely affected by the temperature of the bath are amenable to treatment. Most oxides forming scale are reduced to the metal; oxides of metals which form acid radicals are partially reduced, but the action is sufficient to produce entirely satisfactory results. No electric current is used for the process.

The sodium hydride is produced in the cell by reacting metallic sodium with hydrogen in open chambers or generators partially immersed in the bath. Sodium bricks weighing either 2 1/2 or 5 lb are fed at regular intervals to the generators, in which hydrogen or dissociated anhydrous ammonia is introduced at the bottom at the proper rate.

CHARACTER OF PLATED DEPOSITS. The character of the deposit depends upon the preparation and nature of the plating surface, the composition of the electrolyte, and the operating conditions. The brighter the surface is to begin with, the brighter will be the deposited metal for a given electrolyte and given operation conditions. A bright deposit is caused, in part at least, by a fine crystalline structure and crystal orientation. All electrodeposited metals are crystalline in structure; in some cases the deposited metal reproduces the structure of the base metal, even when the two metals belong to different crystallographic systems. Agitation of the bath and increased temperature favor the formation of large crystals. Small crystals are promoted by higher current densities, addition agents, e.g., glue, and low ionic metal concentration. Low ionic metal concentration is obtained by the use of two or more salts having a common ion (copper sulfate and sulfuric acid, nickel sulfate and ammonium sulfate) or by the use of compounds with a low dissociation constant, which usually calls for a cyanide of the metal.

CADMIUM. Cadmium plate provides an attractive corrosion protective coating for iron and steel and to some extent for copper and copper alloys. The resistance of cadmium to acids and corrosive salts is quite low but is good in alkaline solutions, which makes it suitable for plating laundry equipment, for hardware that comes in contact with floors, and for steel for reinforced concrete. It is also used to plate dissimilar metals that are to come in contact with each other, in order to prevent voltaic-couple corrosion. All electroplating of cadmium is done with alkaline solutions (acid solutions are used for cadmium electrowinning). Addition agents are usually used; these include glue, casein, goulac, turkon oil, and small amounts of nickel or cobalt.

Cadmium-plating Solution

	Grams per Liter	Ounces per Gallon
Cadmium oxide, CdO	30	4
Sodium cyanide, NaCN	90	12
Nickel sulfate, $NiSO_4 \cdot 7H_2O$	1	0.13
Goulac	12	1.6

Temperature: atmospheric
Current density: 1.6 to 5 amp per sq dm (15 to 45 amp per sq ft)

CHROMIUM. Chromium plating is used for two different purposes, necessitating different techniques. The most common form is a thin coating used as a bright decorative finish on metal articles. The other, called "hard" chromium plating, is used industrially as a heavy coating for purposes of heat resistance, wear resistance, low coefficient of friction, and corrosion resistance. Chromium plated on iron or steel for decorative purposes should always have an undercoat of copper and frequently has coats of copper and nickel, but the hard chromium deposits are usually plated directly on the basic metal.

Chromium metal is seldom used as anodes; insoluble lead anodes are used almost universally, antimonial lead being preferable to pure lead. The tanks are lead-lined steel; either they contain heating coils or the solution is piped out of the tank to a heater and returned. On account of the spray given off, the tanks should be provided with adequate exhaust facilities.

Chromium baths are very simple. The metal ion is supplied by chromic acid, CrO_3, and in addition there must be a catalyst acid radical, which is usually supplied by sulfuric acid. The ratio of chromic acid to the sulfate radical should be within the limits of 50 to 1 and 250 to 1, the ratio most commonly used being 100 to 1. Bath temperatures and current densities must be correlated within definite limits in order to obtain bright deposits. For example, for bath 1 below, a bright deposit can be obtained at 40 deg cent (100 deg fahr) at current densities from 5.7 to 15.5 amp per sq dm (50 to 144 amp per sq ft), whereas at 50 deg cent (122 deg fahr) the current densities fall between 10.3 and 31 amp per sq dm (96 to 298 amp per sq ft).

Chromium deposited at 45 to 55 deg cent (113 to 130 deg fahr) and 10 to 30 amp per sq dm is brittle and contains cracks that increase in size and number when heated, because of contraction of the chromium. Such a deposit may contain over 1 per cent of chromic oxide, Cr_2O_3. It has a hardness of about 900 Brinell. If the surface is etched, e.g., anodically, the cracks are enlarged and produce a surface which retains lubricating oil and thereby reduces wear. Such "porous" or "oil-absorbent" chromium has been used on cylinder walls and piston rings of aircraft and Diesel engines. This was a wartime development that promises extended peacetime application.

CHROMIUM-PLATING SOLUTIONS

	Grams per Liter	Ounces per Gallon
1. Chromic acid, CrO_3	250	33.5
Sulfate radical, (SO_4)	2.5	0.33
2. Chromic acid, CrO_3	400	53
Sulfate radical, (SO_4)	4	0.53

Temperature: 40 to 55° C (104 to 131° F)
Current density: 10 to 25 amp per sq dm (93 to 233 amp per sq ft)

If the chromium is deposited at about 85 deg cent (185 deg fahr) and at 40 to 120 amp per sq dm it is relatively soft, with a hardness of about 450 Brinell. The deposit does not crack appreciably on heating and contains only about 0.2 per cent chromic oxide.

COPPER. Practically all commercial copper plating is done in one or both of two types of baths: the acid copper sulfate solution and the copper cyanide or alkaline solution.

Copper Cyanide Solution. This solution is used for plating directly on iron or steel. The concentration of the copper ions is so low that there is no deposition by immersion; instead, a well-adhering copper deposit is formed electrolytically. The protective action of cyanide-deposited copper is poor, however, and the plating rate and efficiency are both low. The cyanide deposit is coated with copper from an acid copper bath or with nickel, chromium, or other metal.

COPPER ALKALINE SOLUTION

	Grams per Liter	Ounces per Liter
Cuprous cyanide, CuCN	22.5	3
Sodium cyanide, NaCN	30.0	4
Sodium carbonate, Na_2CO_3	10.0	1.5

Temperature: 30 to 40° C (86 to 104° F)
Current density: 0.3 to 0.5 amp per sq dm (2.8 to 4 amp per sq ft)

COPPER ROCHELLE CYANIDE SOLUTION

	Grams per Liter	Ounces per Gallon
Cuprous cyanide, CuCN	26	3.5
Sodium cyanide, NaON	35	4.6
Sodium carbonate, Na_2CO_3	30	4.0
Rochelle salt, $KNaC_4H_4O_6 \cdot 4H_2O$	45	6.0
Sodium hydroxide, NaOH, to give a *p*H of 12.6		

Temperature: 50 to 80° C (122 to 176° F)
Current density: 2 to 6 amp per sq dm (18.6 to 56 amp per sq ft) for bright deposits

For the rochelle cyanide solution the anode area should be at least double that of the cathode area, and iron anodes equivalent to about 5 per cent of the total area should be used.

Copper Acid Solution. The copper acid bath is one of the simplest the electroplater encounters. It is not used for plating directly on iron or steel, for it will not produce a firm and adhering deposit. Iron and steel must first be given a deposit from the copper cyanide solution. On account of the greater efficiency and higher plating rate possible with the acid bath, the work is usually transferred to this bath as soon as the iron has been covered with a thin layer of copper. The acid solution is also used for producing electrotypes.

COPPER ACID SOLUTION

	Grams per Liter	Ounces per Gallon
Copper sulfate, $CuSO_4 \cdot 5H_2O$	150 to 240	20 to 32
Sulfuric acid, H_2SO_4	45 to 100	6 to 13

For hard deposits temperatures below 25 deg cent (77 deg fahr) and current densities of 6 to 16 amp per sq dm (56 to 150 amp per sq ft) are used; for soft deposits the temperature is increased to 35 deg cent (95 deg fahr), current densities are reduced to 3 amp per sq dm (28 amp per sq ft), and the acid concentration is reduced to the lower limit given in the formula.

GOLD PLATING. This is used for decorative purposes and for resistance to corrosion. Jewelry, musical instruments, pen points, some table service ware, and special laboratory equipment, such as analytical weights and the insides of calorimetric bombs, are frequently gold plated.

Gold can be deposited satisfactorily on copper, brass, nickel, and silver. Steel is usually given an undercoat of one of these metals before gold plating.

Anodes of high-purity rolled gold or insoluble anodes of platinum, stainless steel, nichrome, or carbon are used. If gold anodes are used, the gold content of the solution tends to increase.

GOLD-PLATING SOLUTION

	Grams per Liter	
Metallic gold (as fulminate or cyanide)	2.1	5 dwt per gal
Potassium cyanide, KCN	15.0	2.0 oz per gal
Sodium phosphate, $Na_2HPO_4 \cdot 12H_2O$	4.0	0.5 oz per gal

Temperature: 60 to 82° C (140 to 180° F)
Current density: 0.1 to 1.0 amp per sq dm (1 to 10 amp per sq ft)

In the "salt-water process" for gold plating no external current is used. The plating cell is also a voltaic cell furnishing its own power. The gold solution is placed in a porous cup, which in turn is placed in a copper can containing salt water; around the porous cup but not touching it is placed a sheet of zinc bent in cylindrical form. This sheet extends above the tops of the porous cup and the copper can; the cathode bar, from which the article to be plated is suspended, rests on the top of the zinc sheet, completing the circuit of the cell.

GOLD SALT-WATER SOLUTION

	Grams per Liter	
Gold (as fulminate)	1.2	3.0 dwt per gal
Sodium ferrocyanide, $Na_4Fe(CN)_6 \cdot 10H_2O$	15	2.0 oz per gal
Sodium phosphate, $Na_2HPO_4 \cdot 12H_2O$	7.5	1.0 oz per gal
Sodium carbonate, Na_2CO_3	4.0	0.5 oz per gal
Sodium sulfite, Na_2SO_3	0.15	0.02 oz per gal

Salt-water gold plating is used where relatively thin coatings of gold are satisfactory. The cost of plating by this method is less than that of the regular gold-plating method, but the plate will not stand as hard wear.

In the dip gold-plating process the gold is plated by "deposition by immersion," also called "by replacement" and "contact plating." Good deposits are obtained only on copper and brass.

GOLD-DIP SOLUTION

	Grams per Liter	
Gold (as fulminate)	2.1	5 dwt per gal
Sodium carbonate, Na_2CO_3	45.0	6 oz per gal
Sodium cyanide, NaCN	30.0	4 oz per gal

Temperature: 80° C (176° F)

The solution should be boiled several hours in a stone crock or enameled kettle before use.

LEAD PLATING. Lead plating is used chiefly as a protective coating against corrosion. Two types of solutions are in use, the fluosilicate and the fluoborate. The former is the cheaper and is therefore used on large-scale operations, but the latter produces a finer-grained deposit and gives a satisfactory deposit on steel. Both solutions are difficult to prepare, especially on a small scale.

LEAD FLUOSILICATE SOLUTION (BETTS'S SOLUTION)

	Grams per Liter	Ounces per Gallon
Lead, as $PbSiF_6$	80	10.75
Fluosilicic acid, H_2SiF_6	150	20
Glue	0.2	0.025

Temperature: 35 to 40° C (95 to 105° F)
Current density: 0.54 to 8.6 amp per sq dm (5 to 80 amp per sq ft)

LEAD FLUOBORATE SOLUTION

	Grams per Liter	Ounces per Gallon
Basic lead carbonate, white lead $Pb(OH)_2 \cdot 2PbCO_3$	150	20
Hydrofluoric acid, HF (the 50% commercial acid)	240	32
Boric acid	105	14
Glue	0.2	0.025

Temperature: 25 to 40° C (77 to 105° F)
Current density: 1 to 2 amp per sq dm (9.3 to 19 amp per sq ft)

NICKEL PLATING. Nickel plating is one of the most important electroplating industries, for it serves as a final deposit itself and also as an undercoat for chromium and several other metals.

NICKEL SOLUTION (MODIFIED WATTS'S SOLUTION) *

	Grams per Liter	Ounces per Gallon
Nickel sulfate, $NiSO_4 \cdot 7H_2O$	240	32
Nickel chloride, $NiCl_2 \cdot 6H_2O$	45	6
Boric acid, H_3BO_4	30	4

Temperature: 50 to 60° C (122 to 140° F)
Current density: 2 to 5 amp per sq dm (19 to 47 amp per sq ft)

The *p*H of the solution should be 1.5 to 3.0 or 4.5 to 5.6.

SILVER PLATING. Silver is seldom plated directly on steel, although firm, adhering deposits are possible; steel is usually first given a deposit of copper, nickel, or tin. This is commonly followed by a strike solution,† which may consist of the following:

SILVER STRIKE FOR BRASS, NICKEL, SILVER, AND BRITANNIA

	Grams per Liter	
Silver cyanide, AgCN	6.5	0.8 troy oz per gal
Potassium cyanide, KCN	68	9 oz per gal

A silver strike solution is used at room temperature.

SILVER-PLATING SOLUTION, BRIGHT DEPOSITS

	Grams per Liter	
Silver chloride, AgCl	39	4.7 troy oz per gal
Potassium cyanide, KCN	70	9.3 oz per gal
Potassium carbonate	38	5.0 oz per gal

Temperature: 24 to 32° C (75 to 90° F)
Current density, 0.5 to 1.0 amp per sq dm (5 to 10 amp per sq ft)

Carbon bisulfide is added as a brightener.

TIN PLATING. Tin is electroplated from acid and from alkaline baths, the latter having been in use longer. All tin-plating solutions have good throwing power, that of the alkaline solution being greater. The alkaline bath has a higher resistance and will deposit considerably less tin per ampere-hour than the acid bath; the acid bath consequently is used more widely.

TIN ACID SOLUTION

	Grams per Liter	Ounces per Gallon
Stannous sulfate, $SnSO_4$	54	7
Sulfuric acid, H_2SO_4	100	13
Cresol sulfonic acid	100	13
β-Naphthol	1	0.13
Gelatin	2	0.26

Temperature: preferably 21 to 27° C (70 to 80° F)
Current density: 1.1 to 4.3 amp per sq dm (10 to 40 amp per sq ft)

---

* O. P. Watts, *Trans. Am. Electrochem. Soc.*, Vol. 29, 395 (1916).

† A preliminary plating solution low in metal ions, used where difficulty is experienced because of deposition by immersion if the article to be plated is placed directly in the regular plating solution.

TIN ALKALINE SOLUTION FOR STILL PLATING

	Grams per Liter	Ounces per Gallon
Sodium stannate, $Na_2SnO_3 \cdot 3H_2O$	90	12
Sodium hydroxide, NaOH	7.5	1
Sodium acetate, CH_3COONa	15	2
Hydrogen peroxide, H_2O_2 (100 vol)	0.5	1/16

Temperature: 60 to 80° C (140 to 176° F)
Current density: 1.08 to 2.70 amp per sq dm (10 to 25 amp per sq ft)

Continuous plating of tin on strip steel received much impetus during the early years of World War II and was responsible for much saving of tin. Because of the more careful control possible with the electrodeposit process, approximately 65 per cent as much tin will cover a piece of steel as is needed in the hot-dipping process. A diagrammatic view of a strip-tinning process is shown in Fig. 1.

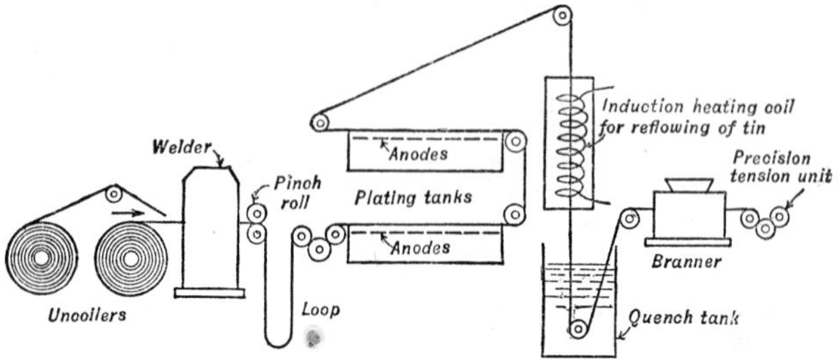

FIG. 1. Diagrammatic Representation of a Continuous Strip Tin-plating Line

ZINC PLATING.

ZINC ACID SULFATE SOLUTION

	Grams per Liter	Ounces per Gallon
Zinc sulfate, $ZnSO_4 \cdot 7H_2O$	360	48
Ammonium chloride, NH_4Cl	30	4
Sodium acetate, CH_3COONa	15	2
Glucose	120	16

Temperature: 20 to 30° C (68 to 86° F)
Current density: 1.5 to 3 amp per sq dm (14 to 28 amp per sq ft)

ZINC CYANIDE SOLUTION

	Grams per Liter	Ounces per Gallon
Zinc cyanide, $Zn(CN)_2$	60	8
Sodium cyanide, NaCN	52.5	7
Sodium hydroxide, NaOH	15	2
Sodium carbonate, Na_2CO_3	30	4
Sodium fluoride, NaF	7.5	1
Corn sugar	7.5	1
Gum arabic	1.1	0.15

Temperature: 40 to 50° C (104 to 122° F)
Current density: 4 amp per sq dm (37 amp per sq ft)

The acid bath has a higher plating rate and lower operating cost than the cyanide bath and is used widely for electrogalvanizing wire, strip steel, and wire screen. The cyanide bath has greater throwing power and produces brighter deposits.

BRASS PLATING. Brass plating is used for decorative effects and to obtain adhesion of rubber to steel. Many bath compositions have been published. A bath that is giving good results consists of the following:

BRASS SOLUTION *

	Grams per Liter	Ounces per Gallon
Copper cyanide, CuCN	26.2	3.5
Zinc cyanide, $Zn(CN)_2$	11.3	1.5
Total sodium cyanide, NaCN	45.0	6.0

Temperature: 27 to 35° C (80 to 95° F)
Current density, 1.0 amp per sq dm (0.9 amp per sq ft)

* L. C. Pan, *Monthly Rev. Amer. Electroplaters' Soc.* (July 1929).

The "total" cyanide in the formula refers to the amount of cyanide to be weighed out for the bath. Part of the sodium cyanide will combine to form $Na_2Cu(CN)_3$ and $Na_2Zn(CN)_4$; the remaining or "free" cyanide will amount to 7.5 grams per liter (1.0 oz per gal).

PLATING ON ALUMINUM. Aluminum can be plated with a variety of metals, but the surface requires special preparation, and it is important that the composition of the aluminum alloy be taken into consideration. (Practically all commercial aluminum is alloyed with small percentages of other metals.) The aluminum surface is first cleaned free from grease by means of organic solvents or alkali cleaners. The alkali cleaner must not be too strong, however, or it will mar the luster of the surface; commercial alkali cleaners used for other metals are too concentrated. A satisfactory cleaner contains sodium carbonate and trisodium phosphate in equal amounts, 15 grams per liter (2 oz per gal) of each. It is used at a temperature between 82 and 93 deg cent (180 and 200 deg fahr). The alkali cleaning is followed by an acid cleaning with a solution of 1 part 50 per cent hydrofluoric acid and 9 parts water. This treatment is followed by a cold-water rinse. The aluminum surface is then slightly roughened and usually plated with nickel before depositing the final coat. A roughening solution for 2S aluminum consists of the following:

ALUMINUM ROUGHENING SOLUTION

Nickel chloride, $NiCl_2 \cdot 6H_2O$	270 grams	36 oz
Hydrochloric acid, HCl, sp. gr. 1.18	200 cc	0.2 gal
Water	1000 cc	1.0 gal

Temperature: 32° C (90° F)

The duration of immersion is determined by experiment; the time required is in the neighborhood of 15 sec.

The undercoat of nickel can be applied by most commercial nickel baths. Chromium, copper, brass, and silver are plated commercially on top of the nickel. Chromium and zinc can also be plated directly on aluminum, but the chromium bath should be operated at double the usual current density.

3. ELECTROFORMING

Electroforming, called "galvanoplasty" in the older literature, has for its object the reproduction of objects by electrodeposition. It is used for making electrotypes for printing, the reproduction of medallions, the manufacture of phonograph records, and the production of tubes; the term is also commonly applied to the electroplating of non-metallic articles, such as plaster and leather articles and many other objects.

In electrotyping, the impression from the type is made in a wax mold or in lead. In the wax-molding process the wax consists of a combination of such materials as beeswax, ozokerite, rosin or rosin oil, and sometimes other oils to produce a mixture having suitable physical properties. This molding "compound" is melted and poured on a table into a layer $1/4$ to $3/8$ in. thick, and the type impression is made in this wax at a temperature of approximately 38 to 43 deg cent (100 to 110 deg fahr), the wax being first brushed with graphite to prevent the type from sticking to the wax. The wax surface is then made conducting by applying graphite in either the dry powdered form or in suspension in a liquid. In order to increase the conductivity still more, it is the practice in some plants to treat this surface with a copper sulfate solution and iron filings to produce a very thin layer of metallic copper on the surface.

In the lead-molding process the impression is made on a thin sheet of lead polished on one side. In order to facilitate removal of the lead from the type, it is lightly brushed with wax or a dilute solution of wax or grease in a solvent. Before plating, the lead mold is treated with a dilute solution of sodium or potassium dichromate or sodium sulfide. These solutions produce a thin deposit of lead chromate or lead sulfide, respectively, and prevent the electrodeposit from sticking to the lead.

The form, whether wax mold or lead mold, with its conducting surface is then immersed in an electroplating tank and connected to the cathode bar. The plating metal is commonly copper, but nickel and other metals are used successfully. After a thin layer of metal is deposited, the electrotype is removed, backed with a low-melting alloy, and dressed to exact type height. In some cases where the electrotype will receive a great deal of wear, it is the practice to plate a thin layer of chromium on the outer face of the newly formed copper or nickel electrotype.

In the plating of non-conducting articles, such as china, glass, plastics, plaster, leather, or wood, the surface, if porous, must first be made non-porous. This can be done by brushing or spraying the surface with shellac, varnish, or a lacquer, or by dipping the article in molten paraffin or beeswax. Excess wax is removed by melting or scraping. The surface must be made conducting. A wax surface is brushed with graphite or copper bronze powder. A varnished or lacquered surface is given a second coat; when tacky, graphite or copper bronze is brushed over it carefully. If copper bronze is used, it is desirable, but not necessary, to treat the surface with a dilute silver cyanide solution. The article is placed in a plating bath before the surface has an opportunity to dry.

The prepared article is suspended from wires in the plating bath; several wires attached to different parts of the object provide better current distribution. It is usually desirable to shift the wires during the plating operation to produce a more uniform deposit.

4. ELECTROREFINING OF METALS

In electrorefining, the metal to be refined is made anode in a suitable electrolyte and is plated out on a cathode which may be a thin sheet of the metal being refined or some other metal from which the deposit may later be removed. The more electropositive metals generally do not dissolve but remain adhering to the anode or finally drop off and form the anode mud or anode slime. The more electronegative metals dissolve readily and must be removed from the electrolyte continuously or intermittently.

COPPER REFINING. The refineries generally obtain the copper as blister-pigs. After the blister has been carefully drilled in order to obtain samples for analysis, it is

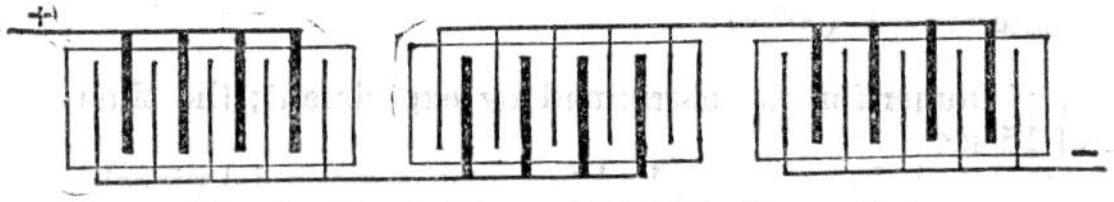

Fig. 2. Simple Form of Multiple Connection

melted and refined in an anode furnace. The molten charge is subjected to an air blast, after which the slag formed is skimmed from the surface. The blowing leaves the copper saturated with cuprous oxide, Cu_2O. The copper oxide is reduced by a poling process, in which long wooden poles are inserted in the melt. The refining process reduces the sulfur content to 0.005 per cent or less and adjusts the copper oxide content to about 2.5 per cent, the amount necessary to produce properly cast anodes. The anodes contain approximately 99.5 per cent copper.

Systems of Refining. In the multiple system of refining, the anodes and cathodes are placed in the tanks alternating with each other. Several systems of electrical contacts have been used. The simple form, shown in Fig. 2, was later modified by Walker to that shown in Fig. 3 and still later modified by Whitehead to the single-contact system shown in Fig. 4 and by Aubel to the connection shown in Fig. 5. Starting sheets of pure electrolytic copper serve as cathodes on which the purified copper is deposited. These starting sheets are made by electrodeposition of copper in thin sheets on slightly oiled or greased plates of rolled copper, from which the sheets can be removed by stripping.

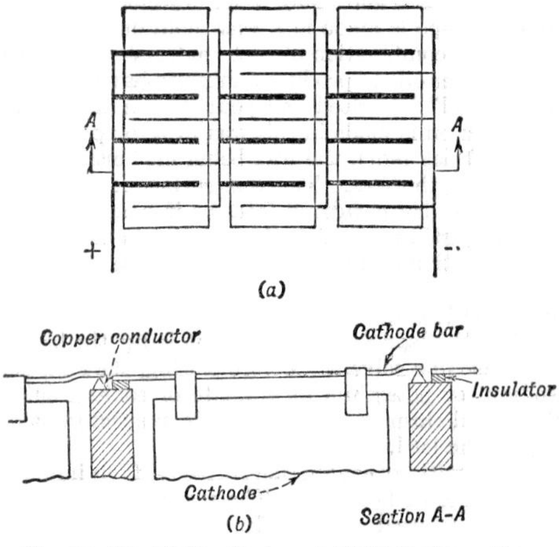

Fig. 3. The Walker System of Multiple Connection

In the series system of refining, only anodes are placed in the tanks, with no separate cathodes between them, as shown in Fig. 6. Only the two end electrodes are connected to the circuit, all the other electrodes serving as intermediate electrodes. As the current passes through the intermediate electrodes, the surface faces go into solution, and the pure copper is deposited on the backs of the next electrodes. The backs of the electrodes receiving the deposits are first coated with a conducting preparation, so that the copper deposited can be separated from the impure copper. When a certain amount of the impure copper has been dissolved, the electrodes are removed, and old impure copper is stripped from the pure and is melted and cast into new electrodes. The series system of copper refining is employed at the Laurel Hill, New York, plant of the Nichols Copper Company, where cast anodes are used,

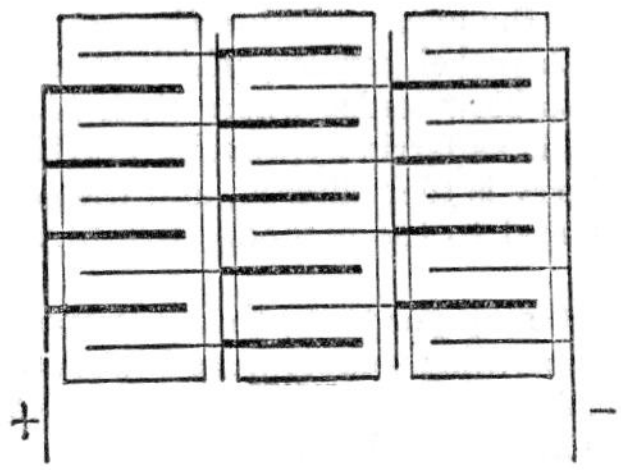

FIG. 4. The Whitehead Single-contact System

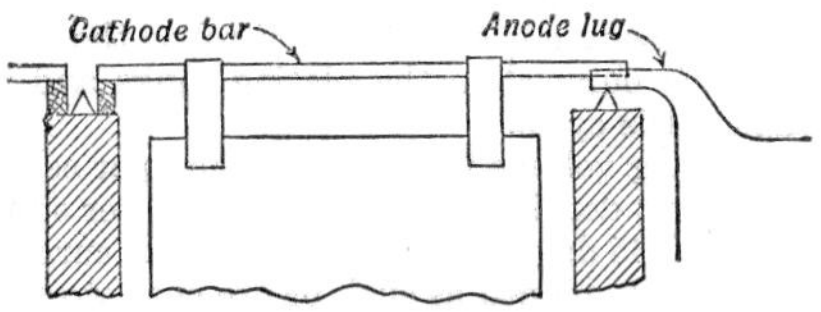

FIG. 5. The Aubel System of Connecting Electrodes between Tanks

and at one tank room of the Baltimore plant of the American Smelting and Refining Company, where rolled anodes are used.

Comparison of the Two Systems. The multiple system, it is maintained, will treat a relatively more impure copper than will the series system, and loss of precious metal is less. It adapts itself better to mechanical handling of the electrodes, and there is better circulation of the electrolyte.

With the series system, it is claimed, a smaller amount of copper is tied up in the plant, because of the smaller number of bus-bars required. Less scrap is produced, less floor space is needed, and less power is required per pound of copper produced. A series tank requires about 17 volts per tank and 76 amp; a multiple tank requires 0.18 to 0.40 volts per tank and 7000 to 10,000 amp.

FIG. 6. Series Connection

Tanks. Tanks are made of wood or of reinforced concrete. For the multiple system they are lined with about 1/8 in. of 6 per cent antimonial lead. For the series system of refining the lining must be non-conducting, and the tanks are lined with asphalt.

The Electrolyte. The electrolyte, as prepared, consists of approximately the following:

	Grams per Liter	Ounces per Gallon
Copper sulfate, $CuSO_4 \cdot 5H_2O$	137 to 169	18.3 to 22.5
Sulfuric acid, H_2SO_4	180 to 220	24 to 29.3

Free sulfuric acid is added daily to replace that removed by impurities and that which forms insoluble sulfates. About 0.02 to 0.03 gram per liter of chloride is added to precipitate the antimony, bismuth, and silver. Small amounts of glue or goulac are added as addition agents; they reduce the formation of nodules and produce dense firm deposits.

Operating Conditions. The electrolyte is circulated through the tanks to prevent stratification and to permit purification and heating before returning to the tanks. It is heated in a hot well so that it enters the tanks at about 60 deg cent (140 deg fahr); it leaves the tanks at about 50 deg cent (122 deg fahr) or an even lower temperature. Current densities range from 18 to 27 amp per sq ft of cathode area for the series system and 20 to 24 amp per sq ft for the multiple system. The cathodes contain approximately 99.98 per cent copper.

The slimes are a concentrated product of the insoluble impurities found in the anodes plus some copper. They have considerable value, in some cases exceeding $16,000 per ton of dry slimes, and care is taken to avoid loss. They are filtered, dried, and roasted at 300 deg cent (572 deg fahr) and leached with a 10 per cent sulfuric acid solution. The leached slimes are again filtered and then melted in a doré furnace, where the base metals are removed as a slag. Lead and antimony form a slag and are skimmed off. Selenium and tellurium are partly volatilized, but the greater part is removed in an alkaline flux. The main furnace produce is a bullion or doré, consisting of 97 to 99 per cent silver and 1 to 2 per cent gold, plus traces of other metals. The doré is cast into anodes and sent to the silver-refining plant.

NICKEL REFINING. The largest nickel refinery in the world is operated by the International Nickel Company at Port Colburn, Ontario. The tanks are made of concrete lined with mastic 1 1/2 in. thick. The anodes measure 27 by 36 in. below the lugs; there are 29 anodes and 30 cathodes to a tank. Cathode starting sheets are made of nickel prepared in special tanks, in which nickel is electrodeposited on aluminum starting sheets from which it is stripped. Each cathode in the refining tank is placed in a "box" consisting of spruce wood frames with canvas duck sides. As the anodes pass into solution, copper and iron also go into the electrolyte, so that all anolyte is removed from the tanks for purification and then returned to the cathode boxes. As the anolyte leaves the tanks, it is heated to 82 deg cent (180 deg fahr) and treated with nickel powder, which removes the copper by cementation. The clear electrolyte goes to blow tanks, where air is blown through the solution to oxidize the iron and precipitate it as ferric hydroxide, $Fe(OH)_3$. Nickel carbonate is added to neutralize the acid; the slurry is then filtered, and the clear solution is pumped to the plating tanks and is distributed to the cathode chambers to complete the cycle.

Operating voltage is about 2.4 volts per cell with 4,800 amp per tank; this corresponds to a current density of about 11.8 amp per sq ft. The operating temperature is about 57 deg cent (135 deg fahr), and the cathode current efficiency about 94 per cent. The nickel cathodes are shipped as such or are cut into squares ranging from 2 to 9 in. on a side and packed in wooden barrels, each holding from 1000 to 1500 lb.

The anode slimes contain nickel, copper, and precious metals. They are filtered in a wooden plate-and-frame filter press, calcined in a furnace, and then melted and cast into secondary or precious-metal anodes containing 24 per cent copper and 73 per cent nickel. The slimes from these secondary metals contain about 2 per cent of the platinum-group metals, and to prevent loss the anodes are wrapped in closely woven cotton duck. The secondary slimes are soldered in tin cans and shipped to the platinum refinery.

LEAD REFINING. The electrorefining of lead has been limited to some extent by the fact that pyrometallurgical methods of refining are cheap, simple, and capable of producing a product of 99.99 per cent lead plus bismuth.

The ore galena, PbS, is concentrated, roasted, and then smelted in a blast furnace with coke and a flux. The furnace products are lead bullion, containing most of the lead and some gold and silver; matte, made up of the sulfides of lead; copper, iron, and the precious metals; and slag.

The bullion, consisting of 98 per cent or more of lead and 1 to 2 per cent antimony, is cast into anodes; cathodes consist of lead starting sheets. The electrolyte is a lead fluosilicate solution containing 7 to 10 per cent lead, 3 to 5 per cent free fluosilicic acid, and 5 to 10 per cent combined fluosilicic acid. Glue or by-product sulfonates (goulac) are used as addition agents. The electrolyte is prepared at the plant by treating silica with hydrofluoric acid to form fluosilicic acid. This acid is then treated with granulated lead or commercial white lead, forming lead fluosilicate.

The refining tanks are commonly made of concrete lined with asphaltum or of wood lined with pitch.

The electrolyte is operated at 35 to 40 deg cent (95 to 104 deg fahr). The current density ranges from 15 to 18 amp per sq ft, requiring 0.34 to 0.6 volt per tank. Anodes have a life of about 10 days, during which time two crops of cathodes are obtained.

The cathodes are melted and subjected to a final pyrorefining for the removal of any remaining antimony, arsenic, and tin. Compressed air is blown through the molten lead, mixing the metal and oxidizing the tin, arsenic, and antimony; the oxides collect on the surface of the lead bath with some litharge.

SILVER REFINING. A large part of the silver produced is obtained from the slimes, which are a by-product from the electrorefining of baser metals, especially from the refining of copper, lead, nickel, and zinc. Other sources are the silver concentrates obtained when lead is desilvered by the Parkes process, silver-gold bullion of various composition, and scrap.

The doré metal obtained from copper-anode mud treatment, for instance, is cast into anodes. Doré contains about 95 per cent silver and 3 per cent gold and may contain also copper, lead, bismuth, cadmium, nickel, iron, tellurium, platinum, and palladium.

The cells are made of acid-proof stoneware or mastic-lined concrete.

The electrolyte is a neutral solution containing generally 60 grams silver per liter as silver nitrate and 30 to 40 grams copper per liter as cuprous nitrate. Part of the electrolyte is withdrawn each day and replaced by fresh electrolyte to keep down the impurities. The waste electrolyte is treated with copper, which precipitates the silver, and then with iron to precipitate the copper.

The slime, amounting to 1 to 2 per cent of the weight of the anodes, collects in bags surrounding the anodes or on canvas under the anodes if they are placed flat in a tray or basket. It contains all the gold, platinum, and palladium and is removed every few days. It is washed and treated with sulfuric acid to remove copper and silver, washed, dried, and cast into anodes for refining electrolytically by the Wohlwill process for gold refining.

Two types of cells are in use, the Moebius cell and the Thum cell. In the Moebius cell the anodes are hung vertically opposite silver or stainless steel cathodes. The deposits of silver are loosely adhering, and wooden scrapers move back and forth to remove the deposited crystals from the cathode. The silver falls to the bottom of the tank, where it is collected in a tray or on a burlap filter bottom. The crystals are washed and melted in graphite crucibles and cast into bars weighing 1,000 oz. The bars assay 999 fine.

The cells operate at 50 amp per sq ft, which requires about 450 amp per cell, at 2.7 volts.

In the Thum cell the anodes are placed flat on muslin cloth placed on duck on the bottom of a wooden or stoneware basket. A basket will hold five anodes about 8 by 12 in. each. The cathode consists of a slab of carbon or graphite covering the entire bottom of the cell. The muslin serves as a diaphragm to prevent gold slime from falling on the silver deposited on the cathode below. The current density is about the same as for the Moebius cell, but the voltage is greater (3 to 3.5 volts) because of the slime settling on the muslin below the anodes.

The Moebius cell requires less floor space and less energy per unit silver produced and consumes less nitric acid than the Thum cell. About 15 per cent anode scrap must be remelted, however, wheras all the anode is consumed in the Thum cell.

GOLD REFINING. Gold is refined by the Wohlwill process, first used in 1878. The anodes may be made from impure gold bullion or from the anode mud from the electrolytic refining of silver. The anodes generally contain 94 to 98 per cent gold and in addition silver, copper, lead, platinum, and palladium. Gold, copper, lead, platinum, and palladium dissolve anodically; silver, iridium, rhodium, and ruthenium remain in the slimes.

The anodes are suspended vertically from gold hooks; the cathodes consist of thin sheets of gold foil and are of the same shape as the anodes. A number of anodes and cathodes are connected in multiple, as in copper, nickel, and lead refining.

The electrolyte contains 7 to 8 per cent gold as gold chloride, $AuCl_3$, and 10 to 16 per cent hydrochloric acid. It is operated at about 70 deg cent (158 deg fahr) at a current density of 110 to 120 amp per sq ft, which requires about 1.3 to 1.5 volts per cell.

Part of the electrolyte is removed for purification. Gold is removed by sulfur dioxide or ferrous sulfate solution, and the platinum is precipitated by ammonium chloride as ammonium chlorplatinate. Copper is removed by scrap iron.

The cells are made of glazed porcelain or acid-proof stoneware.

The gold cathodes produced are 999.80 fine, or very nearly 24 karat.

BISMUTH REFINING. Practically all the bismuth produced in the United States is obtained from the anode mud or sponge at electrolytic lead and tin refineries. The washed and dried sponge is smelted in a furnace under reducing conditions, during which nearly all the arsenic is volatilized. After removal of the slag formed, the melt is heated under oxidizing conditions by having a stream of compressed air playing on the surface of the bath. Antimony and arsenic are volatilized as the oxides, after which bismuth and copper oxide form a slag on the surface. This latter scoria is given a furnace treatment with carbon, producing crude bismuth, which is cast into anodes and refined electrolytically. The anodes contain about 94 per cent bismuth, plus lead, gold, silver, and a little antimony and copper.

The electrolyzing cell is similar to the Thum cell used in silver refining. The electrolyte contains 5 to 6 grams per liter of bismuth chloride, $BiCl_3$, and about 1000 grams per liter of hydrochloric acid. The cell is operated at about 55 deg cent (130 deg fahr). The cathode consists of sheet lead resting on the bottom of the cell, from which the bismuth is removed periodically. The deposited bismuth must be free from lead and arsenic, for a considerable portion of the metal is used for medicinals.

SOLDER. Formerly all solder was made by melting together virgin lead and tin in the desired proportions; today the base of most solders comes from secondary materials, such as solder-bearing scrap, drosses, and residues.

Before the electrolytic refining, the impure solder is melted, given a partial purification, and then cast into anodes. A typical anode composition is: lead, 62 per cent; tin, 3.5 per cent; antimony, 2 per cent; copper, 0.05 per cent; arsenic, 0.25 per cent; iron, nil; gold, 0.05 oz per ton; and silver, 30 oz per ton. Cathode starting sheets are made by pouring molten solder from a trough down an inclined plane.

The electrolyte consists of a solution of lead fluosilicate and free hydrofluosilicic acid. The composition (grams per liter) is approximately as follows: lead, 30; silicon, 40; free H_2SiF_6, 65; total acid, 140. During operation the electrolyte is continually circulated through the electrolysis tanks and through a tank containing solder shot, which removes iron, nickel, and zinc, on to a heating tank, where it is heated to 40 deg cent (105 deg fahr), and then is returned to the original tank. The tank is made of concrete lined with pitch and mastic. The lead-tin ratio of the cathode is not definitely controlled but is allowed to wander with the input of the anodes; it is then blended with virgin tin or lead to meet the specifications demanded.

5. ELECTROMETALLURGY: ELECTROLYSIS OF AQUEOUS SOLUTIONS

In this process (also called electrowinning and hydrometallurgy) the ore is leached by a suitable solution, which later serves as an electrolyte for electrodeposition of the metal. During electrodeposition the original solution is regenerated and is used again for leaching.

COPPER. Electrometallurgy of copper is applied to low-grade ores, the copper content being generally between 1 and 2 per cent. For such ores the production costs are less than they are in pyrometallurgical processes, although the plants employing the latter processes use higher-grade ores.

The ores suitable for this process are generally found in arid regions and usually contain several of the following: malachite, $CuCO_3 \cdot Cu(OH)_2$; chalcopyrite, $CuFeS_2$; bornite, Cu_5FeS_4; chalcocite, Cu_2S; chalcanthite, $CuSO_4 \cdot 5H_2O$; brochantite, $CuSO_4 \cdot 3Cu(OH)_2$; and stacamite, $CuCl_2 \cdot 3Cu(OH)_2$.

The ore is crushed, ground, screened, and placed in large leaching tanks, each holding from 5000 to 10,000 tons. These tanks are commonly built of reinforced concrete lined with lead or mastic. The ore is leached with a sulfuric acid solution. There may be a dozen tanks, one being filled, some with ore being leached, others with ore being washed, and one being unloaded. A cycle of progressive leaching and washing is formed, the leach solution from one tank going on to the next until the copper load is such as to warrant sending the solution to the electrolytic plant or tank house. The electrolyte contains about 24 to 36 grams per liter of copper and 20 to 48 grams per liter of sulfuric acid. The concentration of the acid, the number of leaching tanks, and the cycle of leaching and washing varies from plant to plant.

The leach liquor, after leaving the tanks, goes to a purifying process and then to the tank house, where the copper is removed electrolytically. The tanks and arrangement of electrodes are similar to those found in copper-refining plants, except that the anodes are made of antimonial lead. Current densities range from 5.5 to 12 amp per sq ft, and current efficiencies vary from 65 to 90 per cent, going down with increase of ferrous sulfate in the electrolyte. The cathode copper is usually well above 99 per cent pure. During the electrolysis sulfuric acid is regenerated and is sent to the leaching tanks to complete the cycle.

ZINC. The electrolytic zinc process has a number of advantages over the retort method: it permits the treatment of low-grade zinc which cannot be treated by smelting; it produces a higher grade of metal; it allows a higher extraction of zinc; and it recovers a number of other metals in the ore which are ordinarily lost in the smelting operation.

The chief commercial ores of zinc are the sulfide ore, sphalerite, carrying zinc blend, ZnS; calamine, $(ZnOH)_2SiO_3$; and smithsonite, $ZnCO_3$. The average zinc content of the ores as mined is generally under 5 per cent.

The ore is ground and concentrated by flotation, which produces a concentrate ranging from 20 to 60 per cent zinc. The concentrate is roasted to convert the sulfide, ZnS, to the oxide and sulfate, ZnO and $ZnSO_4$; these compounds are soluble in the leaching acid. The roasting temperature ranges from 650 to 927 deg cent (1200 to 1700 deg fahr).

The roasted ore or calcine is leached with sulfuric acid, which, as in the electrometallurgy of copper, is the spent electrolyte from the tank house. Low-acid and high-acid processes are in use, the concentration of the acid being 100 to 130 grams per liter for the former and 280 grams per liter for the high-acid or Tainton process. Zinc, being relatively high in the electromotive series, must be deposited electrolytically from a relatively pure solution, and as a consequence the leaching process is accompanied and followed by a carefully regulated purification process. In this process iron, arsenic, antimony, silica, lead, copper, cadmium, nickel, and most of the cobalt are reduced to low concentrations. Manganese is not removed and deposits on the anodes in the electrolysis cells. The total zinc recovery seldom reaches 90 per cent.

The electrolysis cells are constructed of wood lined with lead, or of concrete lined with a mixture of sulfur and sand or with a coating of soft rubber. The anodes are made of pure lead or may contain 1 per cent silver. The zinc is deposited on aluminum starting sheets, from which it is stripped at the end of 12, 24, or 48 hr.

The electrolyte is introduced into the cells continuously and electrolyzed until the zinc concentration is down to a small amount. During the electrolysis sulfuric acid is regenerated, and the solution is sent on to the leaching process. Glue or some other addition agent is commonly used in the electrolysis tanks.

The cathodes are melted, during which hydrogen is evolved, for the cathodes contain many times their own volume of hydrogen. Regardless of the hydrogen, however, zinc oxide is formed, so that 3 to 4 per cent of the zinc has to be removed as dross.

Electrogalvanizing Wire from Leach Solution. The solution from the leaching process, instead of being sent to the tank house, can be sent to special plating cells for electrogalvanizing of iron wire. Coils of wire are butt-welded for continuous processing. The wire is first passed through molten lead for annealing, on top of which lies a covering bath of molten caustic. The wire is made cathodic in this caustic cell so that sodium is discharged on the wire, where it removes silicon, sulfur, and phosphorus and reduces the iron oxide to metal. The combined annealing and cleaning pan is 20 ft long. The plating cells are made of steel lined with lead and are about 55 ft long. Eight wires pass through the cell continuously. The current density is approximately 1000 amp per sq ft. The spent electrolyte, with its regenerated acid, is purified with zinc dust and returned to the leaching process.

CADMIUM. Cadmium is one of the few metals recovered solely as a by-product. Practically all cadmium recovered comes from zinc, copper, and lead ores. It is obtained in the bag-house condensation products from lead and copper refineries, in the blue powder in zinc distillation furnaces, and in the purification residues from electrolytic zinc plants. The purification residue, for example, is roasted or ground very fine and then leached with sulfuric acid in lead-lined tanks. This is followed by a somewhat involved chemical purification process.

The electrolysis cells, anodes, and cathodes are of a type similar to those used for zinc electrolysis. A low current density of 4 amp per sq ft at 2.6 volts reduces the tendency to treeing. The cells operate at 30 deg cent (86 deg fahr). Anodes are of lead; cathodes are starting sheets of aluminum.

The cathodes are melted under a thin layer of sodium hydroxide to prevent oxidation, and are then cast into bars. These bars are remelted, and the cadmium is cast into marketable shapes, a large part going into balls 2 in. in diameter for use in the plating industry.

MANGANESE. Manganese is now produced from high- and low-grade ores, and it is anticipated that much of the domestic ore once considered useless will become a reliable source of manganese. The ore is given a reducing roast to make the manganese soluble in sulfuric acid and is leached with spent electrolyte fortified with sulfuric acid, so that, after leaching, the solution contains 150 grams per liter of ammonium sulfate, 70 grams per liter of manganous sulfate, and a small amount of sulfuric acid. The leach solution is subjected to a careful process of purification, for it is relatively high in the electromotive series, so that most metallic impurities, being more electropositive, will deposit electrolytically with the manganese. Their presence, even in amounts that can be detected only spectrographically, may cause serious trouble in the cell room.

The purified solution is electrolyzed in wood cells, lined with a phenolic plastic. The anodes are made of lead containing 1 per cent silver. The cathodes are in canvas-walled compartments. Current density is 35 to 40 amp per sq ft at 5 volts. The metallic manganese is deposited on stainless steel starting sheets, from which it is stripped after washing and drying. The metal is 98.5 to 99.9 per cent pure.

6. ELECTROMETALLURGY: ELECTROLYSIS OF FUSED ELECTROLYTES

Some metals are difficult to reduce from their ores by reduction with carbon, and their ions cannot be deposited from aqueous solution. For such metals the electrolysis of the fused electrolyte is the more practical, usually the only practical, way of obtaining the metals. The more common metals thus obtained commercially are aluminum, beryllium, calcium, cerium, Misch metal, lithium, potassium, and sodium. The compounds more commonly employed for electrolysis are the halides, oxides, or hydrates.

ALUMINUM. Aluminum is obtained from bauxite, which is a mixture of the alpha-monohydrate, $Al_2O_3 \cdot H_2O$ and the trihydrate, gibbsite, $Al_2O_3 \cdot 3H_2O$, together with a number of impurities. The bauxite must be carefully purified, producing a white product, $Al_2O_3 \cdot 3H_2O$. This purification process is expensive and is a very important item in the cost of aluminum production. The trihydrate is calcined, resulting in a product containing 99.5 per cent or more of Al_2O_3.

Alumina has too high a melting point to be fused by itself commercially. Cryolite is more stable electrochemically, and the use of this material as a solvent was discovered and patented independently by Hall in the United States and Héroult in France. Cryolite melts at a little under 1000 deg cent (1832 deg fahr). At a little above its melting temperature it will dissolve 14 to 20 per cent alumina with a lowering in the melting point, so that the cells can be operated at 950 to 1000 deg cent (1742 to 1832 deg fahr).

The cell shown diagrammatically in Fig. 7 consists of a steel box lined with carbon

6 in. or more in thickness. This lining serves as a refractory and as a carrier of the electric current from the molten aluminum cathode to the collector plate attached to the negative bus-bar. When the cell is in operation, it contains a lower layer of purified molten aluminum; on top of this is the electrolyte of fused alumina and cryolite, with carbon anodes dipping into it. On top of the electrolyte is a layer of frozen crust. Fresh alumina is added periodically by breaking the crust, on top of which has been placed the proper amount of alumina.

During normal operation the voltage drop per cell, including the connectors, varies between 5 and 7 volts. Forty to 100 cells are connected in series, requiring a line emf of from 200 to 600 volts; each line of cells takes from 8000 to 30,000 amp.

The power required per pound of aluminum is 10 to 12 kwhr; the anode carbon consumption, 0.6 to 0.8 lb per pound of aluminum. The current efficiency varies between 75 and 90 per cent. The cathode product of aluminum is about 99.7 per cent pure.

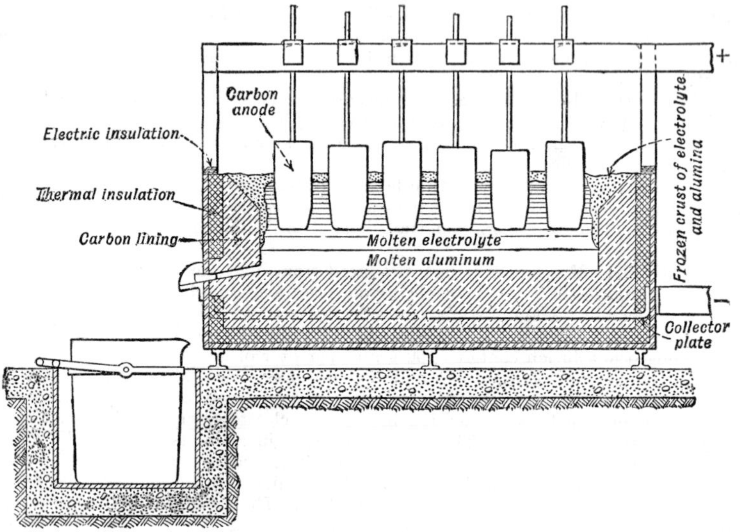

FIG. 7. An Aluminum Cell, in Section. (Courtesy, Aluminum Company of America.)

ALUMINUM REFINING. A process has been developed by Hoopes and his associates by which the aluminum metal from the reduction cells is refined electrolytically to a purity of 99.80 to 99.99 per cent. In some respects the cell is similar to the aluminum reduction cell, but the top carbon electrodes are negative and the carbon lining is positive. The cell contains three layers. A lower anodic layer consists of the aluminum to be refined, alloyed with copper and silicon to increase the density and lower the freezing point. On top of this rests the fused electrolyte, consisting of a mixture of cryolite, aluminum fluoride, barium fluoride, and alumina. Floating on this is the molten refined aluminum.

The cell operates at 5 to 7 volts and at about 20,000 amp.

BERYLLIUM. Beryllium can be obtained as a metal by electrolysis of the chloride or the oxyfluoride. The raw material from which these compounds are prepared chemically is the mineral beryl, which contains approximately 4 per cent beryllium. In the electrolysis the iron pot holding the fused electrolyte serves as cathode; graphite anodes are used.

Very little beryllium is produced as a pure metal, the greater part being used as a beryllium-copper alloy. This, however, is not produced electrolytically but is an electric furnace product.

CALCIUM. Calcium is produced by the electrolysis of fused calcium chloride, which must be dehydrated carefully before fusion. The cell container is made of graphite, although in most cases a crust of solidified electrolyte prevents direct contact with the fused charge. Graphite anodes are used. To prevent the calcium from burning, it is collected on a water-cooled cathode and removed from the cell as rapidly as it is formed.

The cells operate at a cathode current density of 80 to 100 amp per sq cm, requiring a total of 450 to 500 amp per cell. The heating effect of the current keeps the electrolyte in a molten condition at 780 to 800 deg cent (1436 to 1472 deg fahr). The cell requires 25 to 30 volts.

LITHIUM. Lithium is produced by the electrolysis of a mixture of lithium chloride and potassium chloride or sodium chloride. The alkali chlorides lower the operating temperature of the cell to about 450 deg cent (842 deg fahr) and also improve conditions in the cell, for without the presence of another alkali halide the amperage falls and electrolysis is brought to a standstill. The melting point of lithium is 186 deg cent (367 deg fahr).

Alloys of lithium can be prepared by adding the chlorides of certain metals as part of the fused electrolyte; a fused mixture of lithium, potassium, and calcium chlorides will produce a calcium-lithium alloy. Or, if fused lead is used for a cathode, a lithium-calcium-lead alloy is produced. It is also possible to alloy lithium with nickel, copper, and other heavy metals by using these alloying metals as anodes.

MAGNESIUM. Magnesium is prepared by electrolysis of fused magnesium chloride, with additions of sodium chloride and in some cases also of calcium and ammonium

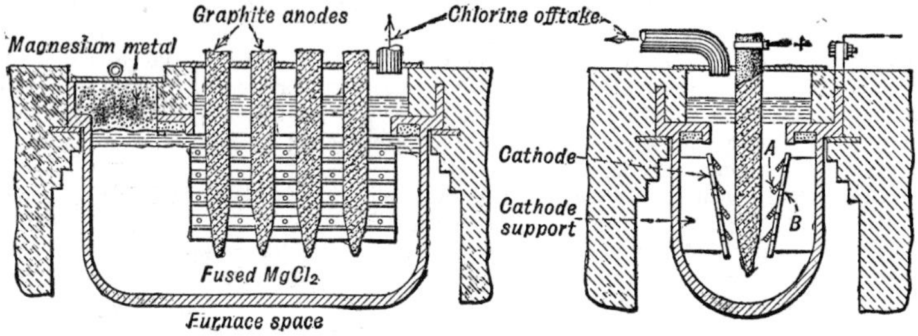

Fig. 8. Ward Magnesium Cell, Using Magnesium Chloride

chlorides. The magnesium chloride used for electrolysis is obtained in the United States from brine wells and from sea water. For methods used for separating the magnesium chloride from these raw materials see Koehler, *Applications of Electrochemistry*, 2nd ed., John Wiley (1944). The magnesium chloride must be carefully dehydrated.

The Ward cell is shown in Fig. 8. A central row of graphite anodes is flanked by two steel cathodes, one on each side. These cathodes have horizontal baffle plates A that deflect the magnesium through ports B into the inactive space back of the cathodes. This permits placing the anodes close together without danger of having the magnesium come in contact with the chlorine around the anodes. The magnesium produced is 99.9 per cent pure.

SODIUM. Most sodium is produced by electrolysis of a mixture of fused sodium chloride and sodium carbonate. The Downs cell is shown in Fig. 9. A central anode of

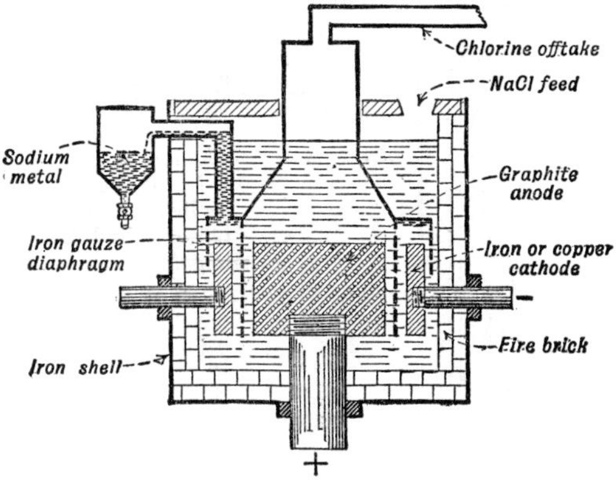

Fig. 9. Downs Sodium Cell, Using Sodium Chloride

carbon or graphite is surrounded by an iron gage diaphragm, which in turn is surrounded by an iron or copper cathode. The sodium rises from the cathodes and is collected in the collector ring, then goes up the riser to the reservoir. The chlorine rises from the central anode and is led off through the stoneware dome.

The cell operates at about 600 deg cent (1112 deg fahr). The current efficiency is somewhat under 80 per cent, and the energy consumption is about 15 kwhr per kilogram of sodium.

7. ELECTROLYSIS OF ALKALI HALIDES

When a solution of sodium chloride is electrolyzed, several different products may result, depending upon the design of the cell, the operating temperature, and the presence of accessory materials. It is thus possible to prepare from sodium chloride, by electrolysis, chlorine and caustic soda; sodium hypochlorite, $NaClO$; sodium chlorate, $NaClO_3$; or sodium perchlorate, $NaClO_4$.

SODIUM HYPOCHLORITE. Sodium hypochlorite is prepared by electrolyzing sodium chloride between carbon or platinum electrodes, the electrodes being spaced so close together that the anode product, chlorine, and the cathode product, sodium hydroxide, unite chemically to produce the hypochlorite: $Cl_2 + 2NaOH = NaClO + NaCl + H_2O$. Sodium hypochlorite is not prepared by this method on a large scale, but for small-scale production the method is convenient. For large-scale production gaseous chlorine is passed into a caustic soda solution. In the United States the Electro Chemical Company cell at one time found considerable popularity; in Europe the Kellner and Schuckert cells are commonly used.

A neutral, nearly saturated solution of sodium chloride is electrolyzed; the operating temperature of the cell is kept below 40 deg cent (105 deg fahr), and a relatively high current density is used. A liquor can be prepared containing 25 to 30 grams active chlorine per liter with an energy consumption of 6.2 kwhr per kilogram (2.2 lb) active chlorine, and a solution of 90 grams per liter active chlorine can be produced at 9.3 kwhr per kilogram.

CHLORINE AND CAUSTIC SODA. The preparation of chlorine and caustic soda is by far the most important process for the electrolysis of sodium chloride. In contrast to hypochlorite cells, the electrode products, chlorine and sodium hydroxide, are kept from interacting as much as possible. The various designs of cells in use aim to keep these products from interacting without introducing construction features that increase the total resistance unduly, and to obtain a high yield per kilowatthour and per unit of floor area occupied.

Three different types of cell construction are used for keeping the anode product (chlorine) and cathode product (caustic soda) from interacting. In diaphragm cells an asbestos-paper diaphragm separates anode and cathode products. In bell-jar cells the anode is surrounded by a bell jar; the cathode is outside the bell jar; and the chlorine gas liberated around the anode escapes out the top of the jar. In the mercury cell, mercury, as explained below, serves as an intermediate electrode between a compartment in which the sodium chloride is decomposed and a compartment in which caustic soda is generated.

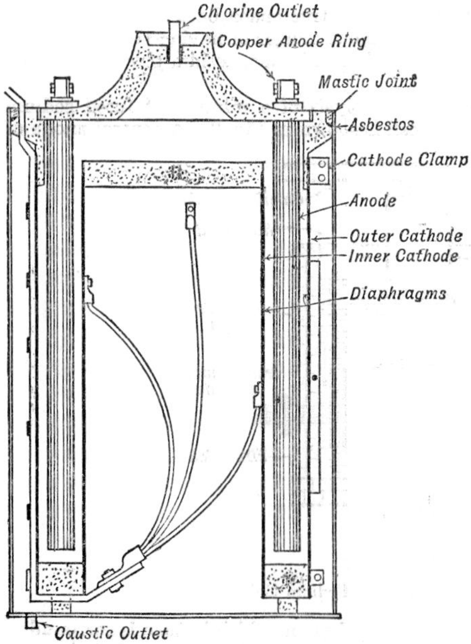

FIG. 10. Vorce Double-cathode Cell

In all these types of cells chlorine is liberated at the anode, and in diaphragm and bell-jar cells hydrogen ions from the water are discharged on the cathode; this leaves the sodium ion from the salt and the hydroxyl ion from the water to form caustic soda. In mercury cells, sodium ions are discharged on the mercury cathode. The sodium dissolves in the mercury and is liberated in another compartment as ions.

In diaphragm cells the diaphragms are always placed next to the cathode toward the anode side. The cathodes are made of steel, either perforated sheet steel or woven meshed wire; anodes are made of graphite.

Preparation of the Brine. The brine is obtained from salt wells or is prepared by dissolving rock salt. Calcium and magnesium salts, if present, must be removed, or their hydroxides formed in the cell will tend to clog the diaphragm. The saturated brine is treated with sodium carbonate; the calcium and magnesium carbonates formed are filtered off, and the solution is neutralized with hydrochloric acid. Sulfates, if present, are removed by precipitation with barium carbonate.

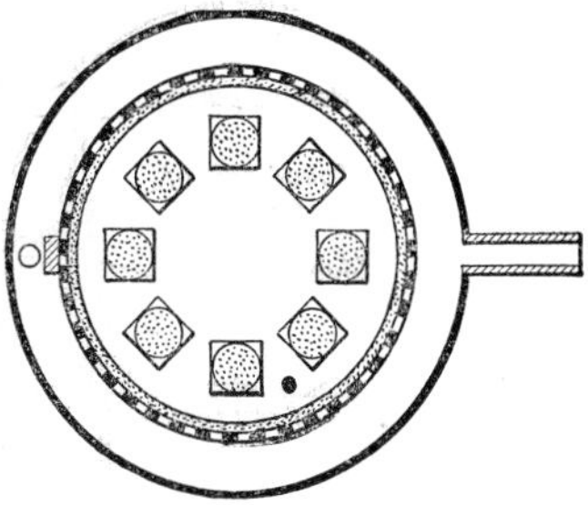

DIAPHRAGM CELLS. Diaphragm cells are extensively used in the United States. The cell feed in most cases consists of a saturated purified brine solution. The cells require from 1000 to 14,000 amp, depending upon the size and construction, with cathodic current densities usually under 75 amp per sq ft. The voltage per cell ranges from 3.25 to 3.7. Current efficiencies range from 92 to 97 per cent, energy efficiencies from 50 to 68 per cent. The caustic production varies from 0.78 to 0.93 lb per kwhr; the chlorine production, from 0.64 to 0.83 lb per kwhr.

The **Vorce cell** is a vertical cylindrical cell. The cathode is made of perforated sheet iron, with the asbestos diaphragm placed inside the cylinder. A circular row of anodes is placed inside the cylinder. In the double-cathode cell shown in Fig. 10, another perforated steel cathodic cylinder and diaphragm are placed inside the row of anodes.

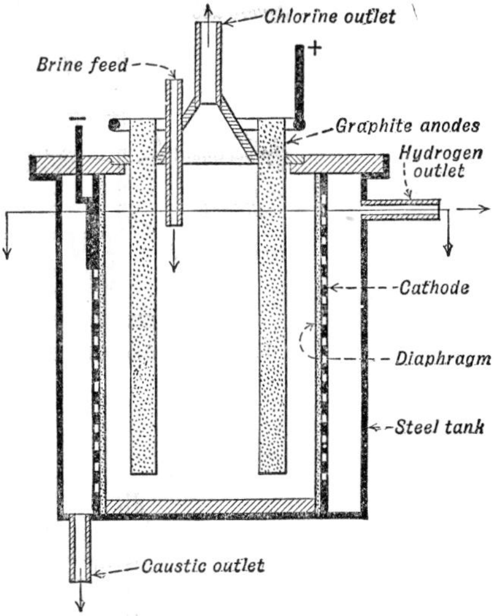

Fig. 11. Gibbs Chlorine-caustic Cell

The **Gibbs cell** (Fig. 11) and the **Wheeler cell** (Fig. 12) are forerunners of the Vorce cell. All three are of the vertical, cylindrical, diaphragm type. They differ in the nature of the brine feed and other structural details. In the Vorce cell the cover and its anode assembly are supported by the cylindrical cathode; in the Wheeler cell this weight is carried by the central column, made of impregnated cement. In all three cells as now constructed, the brine feed is through the cover at the top. In earlier cells considerable pains were taken to introduce the brine through the bottom and even into the central portion of the cell, in order to obtain even distribution. It was found, however, that the energetic "boiling" effect due to chlorine evolution had such an effective stirring action that no careful distribution was necessary.

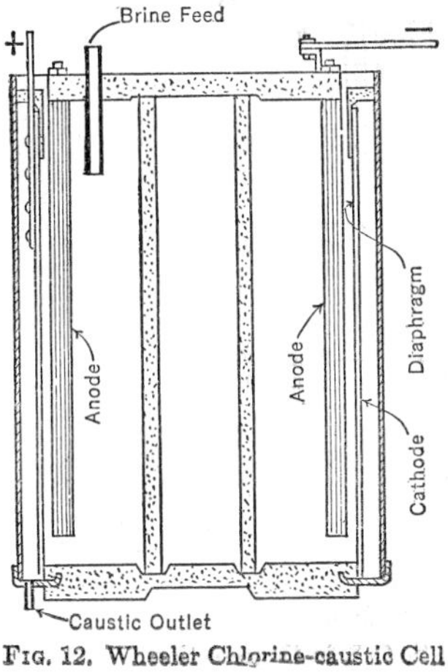

Fig. 12. Wheeler Chlorine-caustic Cell

The **Nelson cell** has a U-shaped cathode, as shown in Fig. 13, with a row of graphite anodes placed in the trough or channel formed. Steam introduced into the cathode compartment tends to wash down the caustic formed, removing it quickly from possible electrochemical action, but the effluent is diluted slightly.

The **Allen-Moore cell** also has a U-shaped cathode, but in the newer type KML cell two such rows of cathodes are placed close together in the same cell, as shown in Fig. 14, thus increasing the capacity per unit of floor space.

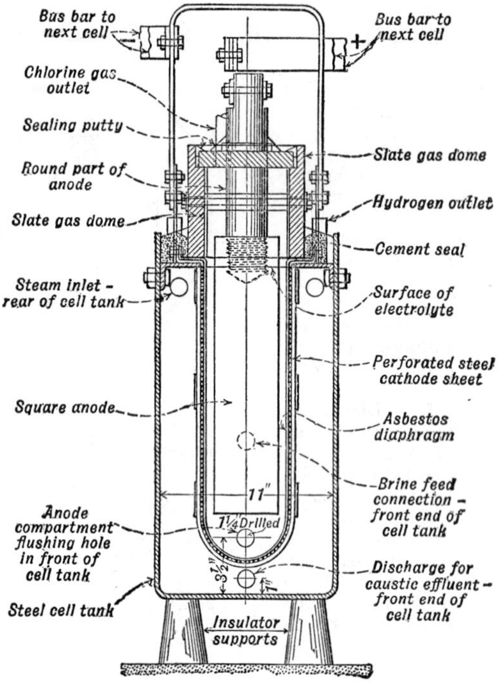

Fig. 13. Nelson Chlorine-caustic Cell, in Cross-section

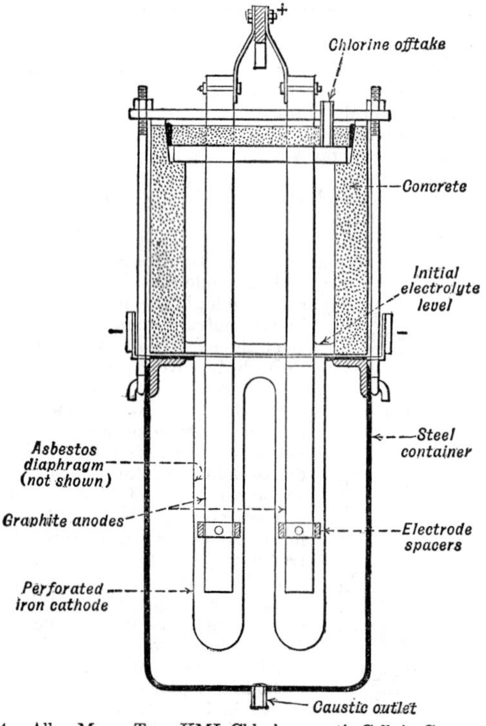

Fig. 14. Allen-Moore Type KML Chlorine-caustic Cell, in Cross-section

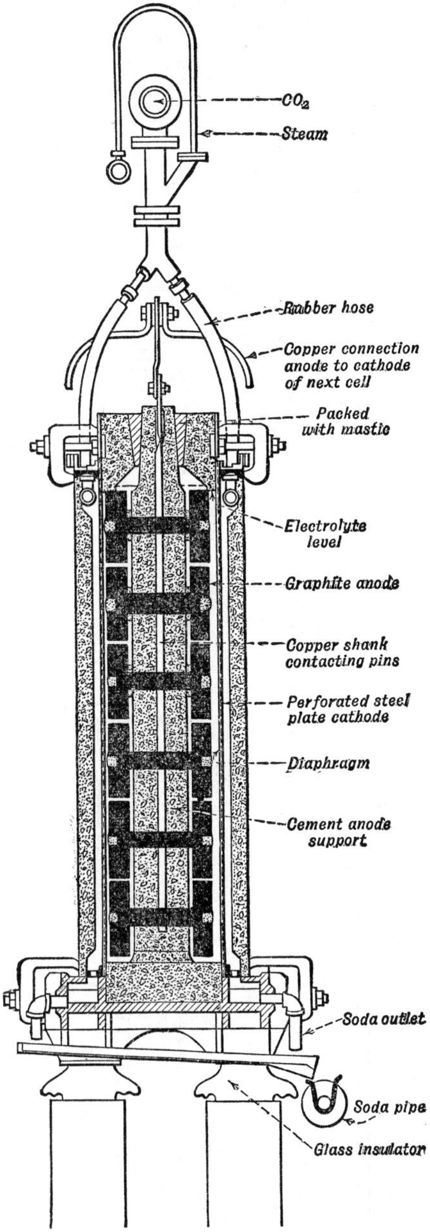

FIG. 15. Hargreaves-Bird Cell for Producing Chlorine and Sodium Carbonate. (Courtesy, West Virginia Pulp and Paper Company.)

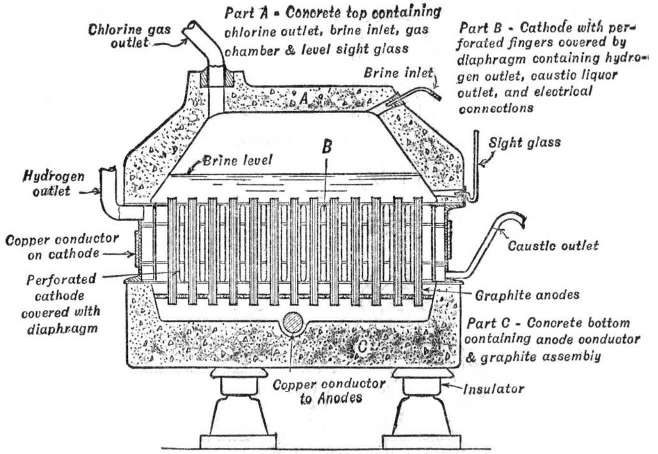

FIG. 16. Hooker Cell, in Section. (Courtesy, Hooker Electrochemical Company.)

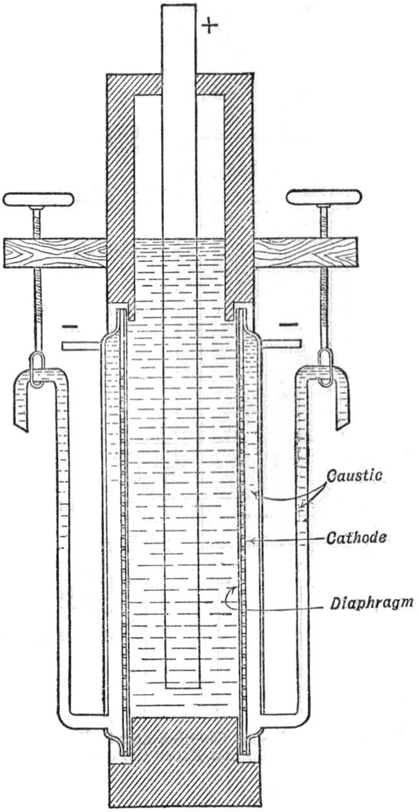

FIG. 17. Giordani-Pomilio Chlorine-caustic Cell

In the **Hargreaves-Bird cell** (Fig. 15) steam and carbon dioxide are introduced into the cathode compartment. The carbon dioxide unites with the caustic soda, forming sodium carbonate as a cell product. The cell consists of a cast-iron box lined on the bottom ends and roof with acid-proof bricks. Two parallel cathodes, each supporting a diaphragm, divide the cell longitudinally into three compartments. The middle or anode compartment contains the anodes of retort carbon. Cathodes are made of perforated steel or copper sheet or wire gauze.

The cells discussed above are of the "unsubmerged diaphragm" type; i.e., the cathode compartment is empty, permitting the caustic soda to run down the back sides of the cathodes and out openings in the bottoms of the cells.

The remaining diaphragm cells to be discussed are of the "submerged diaphragm" type; i.e., the cathode compartment is filled with caustic liquor, giving a back pressure to the brine on the other side of the diaphragm. In some cells the diaphragm is only partially submerged.

The **Hooker Type S cell** (Fig. 16) consists of three sections, bottom, middle, and top, placed one above the other. The concrete bottom and top sections serve for support and as containers for brine and chlorine. The active middle section consists of rows of flat anodes alternating with asbestos-covered wire-screen cathode chambers. This type of construction permits large electrode areas, 130 and 110 sq ft of active area for cathodes and anodes, respectively.

The **Giordani-Pomilio cell** (Fig. 17) was developed in Italy. The outlet of the gooseneck can be raised and lowered to obtain different heights of liquid level in the cathode compartments. This permits a variation in the static head, so that a steady flow can be maintained as the pores of the diaphragm become clogged.

LeSueur Cell. All the cells which have been discussed have vertical diaphragms and electrodes. Cells with horizontal diaphragms are also in use, the cathode being supported horizontally with the diaphragm resting on it. Suspended above this is the anode. The LeSueur cell (Fig. 18) is the only cell in America using horizontal or nearly horizontal

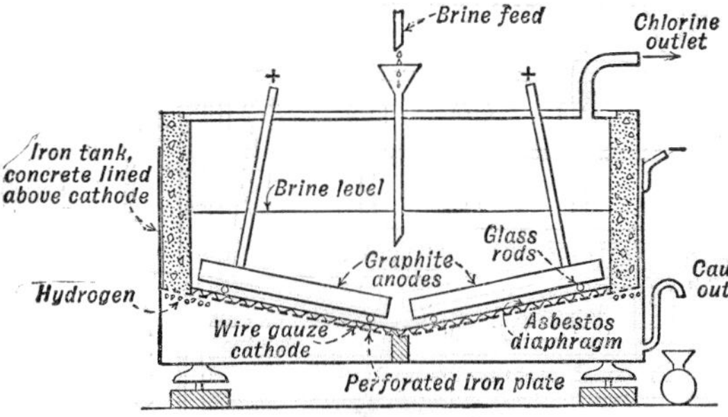

FIG. 18. LeSueur Chlorine-caustic Cell. (Courtesy, Brown Company.)

diaphragms. The cathode-diaphragm combination is sloped slightly to permit easy escape of hydrogen. This cell is also of interest as the first porous-diaphragm cell in commercial production of chlorine and caustic soda in the United States. Some of the original cells put into operation in 1893 are still in use, except for replacement parts.

MERCURY CELLS. A method for keeping anode and cathode products separated, other than using diaphragms or bell-jars, is to deposit sodium from the sodium chloride electrolyte in a mercury cathode. The sodium amalgam, containing approximately 0.2 per cent sodium, can then be brought into contact with water, which reacts with the sodium to form caustic soda.

Mercury cells have, up to the present, not found wide adoption in the United States, but in Germany all new installations since 1936 have been mercury cells. An outstanding advantage of mercury cells is that the caustic liquor may contain 24 to 70 per cent caustic soda, as compared to 9 to 16 per cent for diaphragm cells; this permits considerable saving in evaporation costs. In addition, no sodium chloride is mixed with the caustic, compared to approximately 15 per cent for diaphragm cells. Mercury losses range from 1 to 7 per cent per year.

The **Castner cell** (Fig. 19) consists of a slate box, 4 ft square and 6 in. deep. It is divided by two slate partitions into three compartments. The partitions are so placed that mercury forms a continuous layer in the bottom of all three compartments. The two end compartments contain graphite anodes and brine solutions. In these chambers the brine is decomposed electrolytically, and the liberated sodium dissolves in the mercury, which serves as cathodes in these compartments. In the middle compartment the mercury serves as anode, and suspended above the mercury is an iron-grid cathode. The compartment contains the caustic solution which is formed by the sodium leaving the mercury and entering the water. A gentle rocking of the cell helps some of the mercury to flow from outer to center compartments and back again, thus helping in the transfer of sodium from the outer to the center compartment and in the return of the partially denuded mercury.

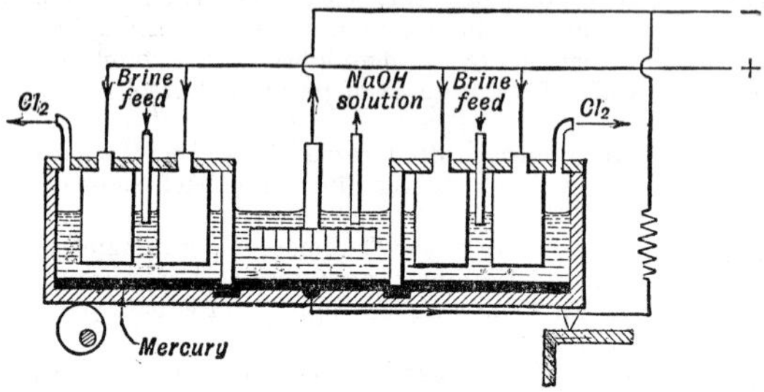

FIG. 19. Chlorine-caustic Cell, Intermediate Mercury Electrode

In the **Sorensen cell** (Fig. 20) the sodium is liberated on and absorbed by the mercury in the decomposing compartment and is then pumped to an oxidizing or denuder compartment, where the sodium enters water to form sodium hydroxide.

Investigations after World War II showed that some modification of this type of cell was very popular in Germany. Some of the largest cells are 25 in. wide, 8 in. deep, and 45 ft long, and operate at 22,000 amp.

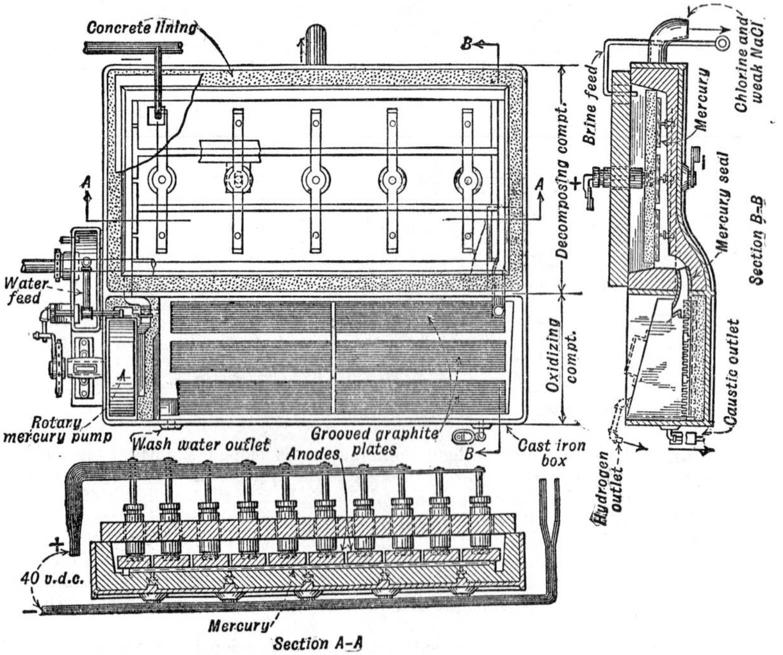

FIG. 20. Sorensen Chlorine-caustic Cell

A **German rotating-cathode mercury cell** is shown in Fig. 21 [*Chem. Met. Eng.*, Vol. 53, 113 ff. (1946)]. Five steel circular cathodes 6 ft in diameter are mounted vertically on a drum, on a shaft. The lower parts of the cathodes are immersed in mercury. The brine solution rests on the mercury. Six sets of graphite anodes are supported in the brine so that they are parallel to the upper parts of the cathode disks. The cathodes rotate at 7 rpm and become amalgamated as they pass through the mercury. The thin coating of mercury spreads over the entire surface of the cathodes, and when the current is on, the rotating cathodes pick up the sodium liberated by electrolysis of the brine. As the cathode moves out of the brine zone down into the mercury, the amalgam dissolves in the mercury and is replaced by mercury of lower sodium content from the bottom of the cell. The sodium amalgam is pumped to a decomposing tower, where the sodium combines with water to form caustic.

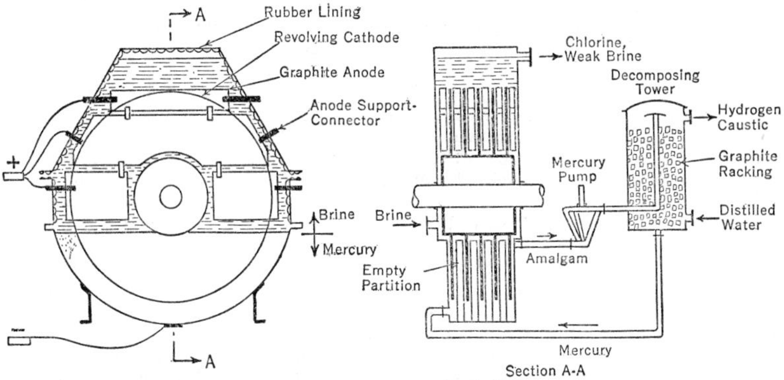

FIG. 21. Rotating-cathode Cell

Disposition of Cell Products. The chlorine is generally above 97 per cent pure, the remainder being carbon dioxide, hydrogen, carbon monoxide, and nitrogen. In some places all the chlorine is used at the plant for making bleach, particularly in paper mills. Much of the chlorine produced is sold in liquid form. For this purpose the chlorine is dried by passing it against a spray of sulfuric acid to remove moisture. By cooling and compressing it is liquefied and packaged in cylinders and drums or is shipped in tank cars or large tanks on barges.

The caustic liquor is concentrated by evaporation; the salt crystallizes and is separated. The liquor is further concentrated. Much caustic soda is shipped hot as liquid of 73 per cent in insulated tank cars. It is also shipped at atmospheric temperatures as a 50 to 55 per cent liquor or is evaporated to dryness and shipped solid in drums.

The hydrogen is usually wasted to the atmosphere, but in some plants is burned in an atmosphere of chlorine to form hydrochloric acid and in other plants is burned as fuel under the evaporators.

8. ELECTROLYTIC PRODUCTION OF HYDROGEN AND OXYGEN

Hydrogen and oxygen are produced electrolytically in over 200 electrolytic plants in the United States, about one-half of which are run by industrial users for making their own gases. Many users of oxygen or hydrogen or both find electrochemical preparation advantageous, for it offers an outlet for off-peak power.

Electrolytic hydrogen and oxygen are very pure; a purity of over 99.0 per cent is common, and 99.9 per cent is attained. The impurity in electrolytic hydrogen is mainly oxygen, and vice versa.

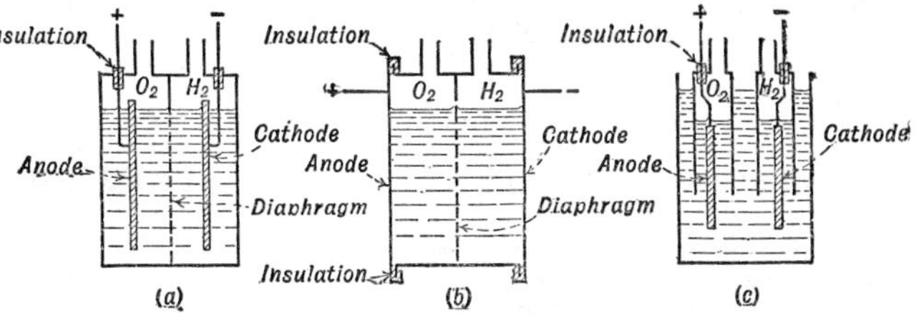

FIG. 22. Different Types of Cells for Producing Hydrogen and Oxygen Electrolytically

The cells are constructed of iron or steel, with asbestos diaphragms. The steel electrodes are commonly nickel plated, but cobalt plate is used by one manufacturer. Nickel and cobalt reduce anode corrosion and oxygen overvoltage on the anodes. Figure 22 shows diagrammatically three types of cell construction. In (*a*) and (*b*) asbestos diaphragms keep the oxygen and hydrogen separated; in (*c*) a bell-jar serves this purpose, although asbestos skirts usually are attached to the lower edges of the bells.

The electrolyte is commonly a 15 per cent sodium hydroxide solution or a 25 per cent potassium hydroxide solution. The production per kilowatthour is about 8 per cent greater for the potassium hydroxide solution, but sodium hydroxide is considerably cheaper and is used more commonly. Approximately 3 to 5 per cent alkali is removed with the gas per annum. The total voltage usually varies from 2 to 2.5. The current efficiency is almost 100 per cent.

THE ELECTROLABS CELL. This cell, also called the Levin cell, is of the unit type. The cathode is placed centrally with an anode on each side. The electrodes are cobalt plated. The cell is of the type shown in Fig. 22(*a*). Asbestos-cloth diaphragms attached to metal frames are placed between the anodes and cathodes. A 600-amp cell measures 34 by 47 by 8 1/2 in. and produces 4.8 and 8.9 cu ft of oxygen and hydrogen per hour, respectively.

THE KNOWLES CELL. This cell is of the bell-jar type, as shown in Fig. 22(*c*). A multiple cell tank consists of a sheet-steel tank which contains the anodes and cathodes suspended under steel gas-collecting bells. Electrodes are made of heavy steel plate 1/8 in. thick to prevent warping. The electrodes are hung from the collecting bells, electrically insulated from them, and each electrode is so placed that nearly the whole of it is lower than the bell itself. Asbestos diaphragms surround alternate electrodes and are suspended from the collecting sheet-steel bell above that electrode. Hydrogen and oxygen are taken off at the tops of alternate collecting bells. The gas bells are assembled in numbers according to the capacity desired.

A 3000-amp cell is approximately 4 ft long, 2 1/2 ft wide, and 2 1/2 ft deep. It operates at an emf of 2.125 to 2.25 volts and produces 24.2 cu ft of oxygen and 48.4 cu ft of hydrogen per hour.

SHRIVER FILTER-PRESS-TYPE ELECTROLYZER. This electrolyzer resembles the well-known recess-plate-type filter press. It consists of a series of cells formed by cast-iron bipolar electrode plates, which are recessed and separated from each other by rubber-bound asbestos diaphragms. These plates and diaphragms are clamped together by a steel screw activated by a capstan in a conventional filter-press frame. Each chamber forms a cell of the type shown in Fig. 22(*b*).

The cell with electrode plates 3 ft square takes 300 amp; the 60 cells require 120 volts. The generators have an output of 3.6 cu ft of oxygen and 7.2 cu ft of hydrogen per kilowatthour. They are built in sizes ranging from 100 to 12,000 cu ft of oxygen and double the amount of hydrogen per 24 hr.

PECHKRANZ ELECTROLYZER. This cell, widely used in Europe and to some extent in America, is also of the filter-press type. The electrodes are circular in shape, about 6 ft in diameter. The diaphragm is made of a sheet of pure nickel, perforated with fine holes less than 0.004 in. in diameter. There are 5000 perforations per square inch. The electrolyzer uses a potassium hydroxide solution and operates at 80 deg cent (176 deg fahr). The emf required is 2 to 2.5 volts per cell, and as many as 150 cells can be built up in one electrolyzer.

PRESSURE ELECTROLYZERS. If the confining spaces for the gases produced by an electrolyzer are limited, the pressure will gradually rise as the gases are generated; by proper construction gases can be generated at pressures of 200 atm, thus saving the cost of separate gas compressors. In addition, the voltage required is lower than for cells operated at atmospheric pressure. Difficulty is experienced in equalizing the pressures of the hydrogen and oxygen. Unequal pressures will cause passage of one of the gases through the diaphragm, thus contaminating the other; or the pressure may even break the diaphragm. The Noeggerath cell used in Germany produced 10 cu ft of hydrogen per kilowatthour.

9. ELECTROLYTIC OXIDATION AND REDUCTION

Strictly speaking, all anode processes are oxidation processes, and all cathode processes are reduction processes; in certain cases, however, advantage is taken of the nature of the reactions at the electrodes to produce new products by oxidation or reduction. In most oxidation reactions, oxygen or chlorine are liberated at the anode, and these nascent gases oxidize some material to produce a new compound; in reduction reactions nascent hydrogen is liberated at the cathode to reduce a material to form a new compound. The manufacture of only two compounds will be taken up here.

Only a few organic compounds are prepared commercially by electrolytic oxidation or reduction. A large number of compounds, however, have been prepared in the laboratory.

MANITOL AND SORBITOL FROM GLUCOSE. A commercial process for the electrolytic reduction of sugars to polyhydric alcohols has been developed by Creighton and is used especially for the manufacture of sorbitol and manitol. Glucose is a suitable raw material. It is reduced to its corresponding alcohol in alkaline aqueous solution. A cell contains amalgamated lead cathodes and lead anodes. Each anode is surrounded by an alundum diaphragm.

WHITE LEAD. The process consists essentially of electrolyzing a lead acetate solution in a cell with lead anodes and iron cathodes. The cell is built of reinforced concrete ("gunnite"), lined with an asphalt enamel. The anodes are encased in envelopes of linen duck, which act as diaphragms. The anolyte and catholyte each contain approximately 4 per cent sodium acetate. The anolyte in addition contains from 0.06 to 0.2 per cent sodium carbonate and approximately 0.05 per cent sodium bicarbonate. Lead dissolves at the anode electrolytically. The lead ions, meeting the carbonate ions from the catholyte and hydroxyl ions from the water, form basic lead carbonate or white lead.

10. OZONE

Ozone, O_3, can be produced from oxygen by heating the oxygen to a high temperature and then cooling it suddenly, by the action of ultraviolet, by the electrolysis of aqueous solutions, and by means of the silent or brush electric discharge. Only the third method is used commercially.

In producing ozone by electric discharge, air or oxygen is passed between two metal plates or concentric cylinders. In addition to the air gap left between the two plates or cylinders, a glass or mica plate or cylinder is usually also placed to reduce chances of direct arcing. Voltages used run up to 25,000 and frequencies up to 1000 cycles per second. The yield is almost proportional to the frequency. The air must be dried before passing into the ozonizer; this is accomplished by refrigeration or by passing it over calcium chloride or silica gel. The theoretical yield is 1200 grams ozone per kilowatthour. Commercial ozonizers show an efficiency of 5 per cent in air and 15 per cent in oxygen.

11. ATMOSPHERIC-NITROGEN FIXATION

The free nitrogen of the air can be combined into chemical compounds by several methods; two methods involving electrochemistry are (1) the arc process, and (2) the calcium cyanamide process. The second is the only commercial process in use to any extent.

ARC PROCESS. The arc process is not in use in America and is employed to only a limited extent in Norway, where electrical energy is relatively cheap. According to reports in the technical press, plants using the arc process are being built in Spain, but no details are available. When air is subjected to a high-tension arc discharge, nitric oxide is formed according to the following reaction:

$$N_2 + O_2 \rightleftharpoons 2NO - 43{,}100 \text{ cal}$$

The gases leaving the various types of arc furnaces contain 0.8 to 1.8 per cent nitrous oxide at a temperature of approximately 1000 deg cent (1832 deg fahr). The gases pass through heat interchangers; the cooled gases proceed to oxidation chambers, where they are oxidized to NO_2:

$$2NO + O_2 \rightleftharpoons 2NO_2 + 27{,}800 \text{ cal}$$

which in absorption and reaction towers is converted to 30 per cent nitric acid:

$$3NO_2 + H_2O \rightleftharpoons 2HNO_3 + NO$$

The various types of furnaces were constructed so as to produce many arcs per second; these were usually drawn out into long arcs or spread out into a sheet. The air, usually dried and preheated, was passed through the arcs.

Bradley-Lovejoy Process. This process was used in Niagara Falls, New York, early in the century. A hub with electrodes like spokes on a wheel rotated so that arcs were struck repeatedly between the moving electrodes and stationary ones. This produced sparks at the rate of 9250 per second. Direct current was used. A unit operated at 0.75 amp at 8000 volts.

Birkeland-Eyde Process. In this process, used in Norway, the furnace contains water-cooled copper electrodes about 1 cm apart, placed between poles of an electromagnet arranged so that a circular arc flame about 6 ft in diameter is formed. The air enters through perforations in the furnace, passes through the arc, and passes out through the circumference. A unit takes a current of approximately 200 amp at 5000 volts. The process is used at Notodden, Norway, where 25 furnaces are supplied with 100,000 kw in operation (Thompson, *Theoretical and Applied Electrochemistry*).

Schönherr-Hessberger Process. This process used a long filament arc in the furnace. The furnace unit consisted of four concentric vertical tubes over 20 ft long, so arranged that incoming air was preheated by the outgoing gases. A long steady arc was produced in the inner tube or reaction chamber.

Pauling Process. In the Pauling furnace the arc was formed between two electrodes so shaped and arranged that they made a V-shaped arc gap. When an arc was struck by special electrodes, an air blast from nozzles beneath blew it along the divergent electrodes. The arc had the appearance of a vertical sheet of flame. The furnace took 200 amp at 5000 volts.

CALCIUM CYANAMIDE. When gaseous nitrogen is passed over calcium carbide at elevated temperatures, calcium cyanamide is formed:

$$CaC_2 + N_2 \rightleftharpoons CaCN_2 + C + 97{,}800 \text{ cal}$$

Calcium carbide is made from lime and coke in arc furnaces. Nitrogen is obtained from the atmosphere by liquefaction and fractional distillation of air. Both the Lind and Claude processes are suitable. Gas liquefaction has been highly developed, and the cost of nitrogen production forms only about 3 per cent of the total energy consumption, including that used for carbide formation. In the fractional distillation pure nitrogen boils off, and with the use of rectification columns nitrogen gas of 99.99 per cent purity is obtained.

The carbide, ground to 100 mesh and finer and mixed with approximately 2 per cent calcium fluoride as catalyst, is placed in a vertical cylindrical furnace. A cylindrical vertical opening axially in the charge permits introduction of a carbon resistor. The furnace top is closed with a cover, and nitrogen gas from the liquid-air plant is admitted to the oven through the bottom. The nitriding action is exothermic and, once started, continues until completed. The crude furnace product contains 61 to 70 per cent cyanamide plus calcium oxide, free carbon, and a little unconverted calcium carbide.

Cyanamide is used as a fertilizer and can serve as a starting point in the manufacture of other chemicals, especially ammonia and sodium cyanide.

12. MISCELLANEOUS ELECTROCHEMICAL PROCESSES

ANODIZING OF METALS. A surface film of aluminum oxide can be formed on aluminum metal by making the surface an anode in a suitable electrolyte. Such films have been used in aluminum rectifiers and condensers. They also protect the surface against corrosion and abrasion and may have pronounced absorptive properties, so that they can be dyed almost any color. Chromic acid, oxalic acid, sulfuric acid, and boric acid plus ammonium borate are satisfactory electrolytes. A 15 per cent sulfuric acid solution operated at a temperature between 15 and 30 deg cent (60 and 86 deg fahr) is commonly used. A current density of 12 amp per sq ft (requiring approximately 15 volts) is used. Treatment time is from 10 to 15 min. Upon removal from the oxidizing bath the articles are washed in water, then dipped in dilute ammonia to neutralize any remaining acid, and again rinsed in water. The surface film is porous, but if porosity is not desired, the pores of the oxide coating can be sealed by treatment in hot or boiling water. Porous films are highly absorptive and permit coloring by dyes. The color penetrates the oxide coating throughout its depth and is not easily removed.

Other metals have also been anodized to produce resistant surfaces. For anodizing magnesium an acid mixture with a *p*H of 4.0 to 4.8, containing about 10 per cent sodium dichromate, $Na_2Cr_2O_7 \cdot 2H_2O$, and from 2 to 5 per cent monosodium phosphate, $NaH_2PO_4 \cdot H_2O$, is used. After cleaning or pickling, the magnesium surface is made anode at a current density of 5 to 10 amp per sq ft at a temperature of 50 deg cent (122 deg fahr) for a period of 30 to 60 min. Magnesium-anodized surfaces do not have corrosion and abrasion resistances comparable to those of aluminum-anodized surfaces.

ELECTROLYTIC POLISHING AND ETCHING. In electrolytic polishing and etching, the surface to be treated is made anode in a suitable electrolyte. Carbon or stainless steel is satisfactory for the cathode. The polishing of metals is apparently accomplished by anodic solution of high points of the surface. Corrosion products probably form a poorly conducting film which adheres to the depressions but exposes projections. The metal is leveled to a microscopic plane, which may have a degree of polish equal to or better than that obtained on a metallographic polishing wheel. The process was originally used for polishing metallographic specimens but later found other applications also. One electrolyte suitable for use with low-carbon steels contains 40 per cent sulfuric acid, 46 per cent phosphoric acid, 4 per cent dextrose, and 10 per cent water. Current densities range from 1.5 to 4.5 amp per sq in. (23 to 70 amp per sq dm) at temperatures of 28 to 40 deg cent (82 to 104 deg fahr).

ELECTROSTATIC SPRAYING. In the spraying of paints, lacquers, and porcelain enamels, a good share of the material does not reach its desired destination. In some cases salvage of wasted materials is costly; in others, impossible. The process consists of the charging of the spray particles in an electric field and the attraction of these particles to the object to be coated. The object to be sprayed is grounded and is surrounded by an electrical field of such nature as to impart a negative charge to the atomized particles entering the field. The charged particles are drawn to the object, which is charged positively. Power is supplied by a power pack which receives a 220-volt, 60-cycle single-phase current. The secondary voltage reaches a maximum of 100,000 volts, single-phase, half-wave, 60 cycles, with a current not exceeding 10 ma. The electrical field is produced by an electrode system composed of a series of fine copper wires, suspended parallel to the pieces being sprayed [James B. Willis, *J. Amer. Ceram. Soc.*, Vol. 28, 121 (1945)].

DEWATERING CLAYS. The clay-water suspension is placed in a cell with a revolving metal-drum anode and a wire-net cathode suspended beneath it. The voltage across the electrodes ranges from 80 to 180 volts. The clay particles, being negatively charged, move toward the drum and adhere to it. As the drum rotates, the clay is carried out of the suspension and is removed by a doctor-blade. The process has found favor in Germany and Czechoslovakia. One plant dewaters 78 tons of clay per day. The machines deliver the clay in continuous sheets $^1/_8$ in. thick and 5 ft wide, containing on an average 35 per cent of water.

ELECTRICAL LUBRICATION. In the manufacture of bricks and other heavy clay structural materials by the stiff mud process, the plastic clay is extruded through a die, and proper lubrication of the die with usual lubricants is difficult. By making the die negative electrically and by placing a positive electrode farther back in the clay mass, moisture will move to the negative die walls and in some cases provide excellent lubrication. Usually 200 volts direct current will produce the desired results [J. O. Everhart, *J. Amer. Ceram. Soc.*, Vol. 17, 272 (1934)].

WATER PURIFICATION. The cells for water purification consist of three compartments separated by two diaphragms. The two outer compartments contain the electrodes. Under the influence of direct current the positive ions pass through one diaphragm into the negative compartment, and the negative ions through the other diaphragm into the positive compartment, leaving the middle section relatively free of ions. Usually several cells are connected in series, and 220 volts direct current is necessary. With proper construction of the cells and careful operation it is possible to produce water with a conductivity lower than that of doubly distilled water.

WATER STERILIZATION. Water can be sterilized by suitably constructed electrolysis cells in which bacteria are killed by the nascent oxygen liberated at the anode. A cell for operation on 110 volts direct current consists of 21 plates. The cell is provided with cocks for flushing out sediment. The energy consumption is in the neighborhood of 0.2 kwhr per 1000 gal.

PRODUCTION OF DEUTERIUM (HEAVY HYDROGEN). When water is decomposed into its elements by electrolysis, heavy water remains in more concentrated form in the cell, while a greater portion of lighter hydrogen passes off as a gas. Fresh water is added to the cell to maintain the level until two or three times the volume of the cell has been electrolyzed; electrolysis is then continued to reduce the volume of the water in the cell. Sodium hydroxide or potassium hydroxide is added to the water to make it conducting. When the final concentration of the heavy water is obtained, it is neutralized with carbon dioxide and then distilled. If it is desired to separate the heavy hydrogen, the solution is electrolyzed, and the heavy hydrogen is collected.

As electrolysis proceeds in the cells, the hydrogen given off carries an appreciable amount of deuterium, so that for efficient operation this is burned, and the water thus obtained is electrolyzed in other cells.

The cells used are simple containers with two electrodes passing through the top. The electrodes are rods or strips of steel, either bare or nickel plated. Filter-press-type electrolyzers have also been used. Ordinary water contains about 175 deuterium atoms in 1,000,000 hydrogen atoms.

13. ELECTRIC-ARC FURNACES

In electric-arc furnaces heat is generated by electric arcs, and in most types of furnaces also by resisting heating in the charge as the current passes from one arc to another. This latter source of heating is small compared to that generated in the arc itself. Generally the furnace consists of a steel shell lined with a refractory material. Provisions are made for two or more carbon or graphite electrodes, a charging door or removable top, and pouring spout; as a rule, arrangements are provided for tilting the furnace.

The arc is what is called a low-tension arc, operating on voltages generally under 300. The temperature of the low-tension arc varies with conditions; with carbon electrodes it is about 3500 deg cent (6500 deg fahr). The low-tension arc gives the highest temperature obtainable, except that possible of attainment by atomic energy. In the low-tension arc the temperature is high enough to heat the charge of metals or metallic oxides to a point where it gives out a large quantity of electrons. Electrons are shot off from the electrodes during the negative half of the cycle and positive ions during the positive half, and their impact on the gaseous molecules increases their ionization enormously. This increases the conductance of the arc, thus producing a rise in temperature.

Electric-arc furnaces have found wide application in the steel industry for refining and making alloy steels and in the production of alloys. They are used for brass and bronze refining, copper refining, production of metals from ores, and cast-iron production. One of the earliest applications of electric-arc furnaces on a commercial scale was for the production of calcium carbide in 1892; arc furnaces were first used in the steel industry in 1904.

Arc furnaces, in common with other types of electric furnaces (resistance and induction furnaces), are amenable to automatic control; they make it easy to repeat a cycle; the energy is easy to meter; high temperatures are easily attainable; and there is no need for storage of fuel. This last item may also be a disadvantage, for in case of strikes or breakdown of generating or transmitting equipment there is no reserve fuel or energy supply to keep the furnaces going. Arc furnaces for steel refining have approximately the same initial cost as open-hearth furnaces of the same output. Electric energy at 1 cent a kilowatthour is more costly than combustible fuels. At 0.1 cent a kilowatthour it is cheaper than fuel oil at 5 cents a gallon or than natural gas at 40 cents per 1000 cu ft.

Electric furnaces require considerable accessory materials. A transformer is needed to step the voltage down from the line to the furnace voltage. A tap transformer with 6 to 12 taps is used for obtaining selective secondary voltages. A bus line and a flexible section are needed from transformer to furnace. Reactance and resistance help to maintain a steady arc and limit current fluctuations. For large furnaces the inherent reactance of the transformer, together with the resistance of the secondary line, usually supplies sufficient impedance for this purpose. For small transformers, below approximately 5000 kva, additional reactors are supplied in the primary line. Automatic regulation of the electrodes is common; for this purpose direct current is used, supplying power to small d-c motors. Switching mechanism, instruments, and meters are also needed. The electrode holders and door frames are water cooled, for which proper water supply and removal must be provided.

Electrodes are made either of amorphous carbon or of graphite. The selection is determined by several factors, including tradition. In favor of the graphite electrodes is the fact that, their conductance being four times that of amorphous carbon, the electrodes need be only half the diameter. The smaller electrodes permit a stronger roof construction, do not require as heavy a control mechanism, have but half the transportation costs, and are removed more easily if they break and fall into the batch. Graphite electrodes have no extra resistance losses at the joints and require no electrode joint compound. Because of their greater resistance to oxidation, the consumption is less than half that of amorphous carbon electrodes.

Among the disadvantages of graphite electrodes, as compared to amorphous carbon electrodes, is that they are more expensive per pound of electrode (but not necessarily per ton of steel), so that, if breakage occurs in transportation or handling, the loss is greater. Also, since they are more fragile, they break more easily. They have a higher heat conductivity than carbon electrodes.

There is a gradual tendency toward use of graphite electrodes in the steel industry. They are smaller and can be placed near the center, and they bore only one hole through the charge when melting cold scrap. Graphite electrodes are used exclusively for furnaces of 15-ton capacity and above. Carbon is seldom used for electrodes above 17 in. in diameter, but on account of its superior mechanical strength it is used for electrodes under 8 in. in diameter for steel furnaces. Ferroalloy furnaces are almost always operated with carbon electrodes, whereas non-ferrous furnaces are operated exclusively with graphite electrodes.

The Soderberg continuous electrode has been applied to arc furnaces, including the tilting type. An electrode shell is made of sheet iron of small gage. An electrode mixture of about the same composition as is used for ordinary carbon electrodes is tamped into the metal shell. As the electrode is slowly fed downward into the furnace, the green mixture becomes baked, so that the part of the electrode below the holder is a solid carbon electrode. When new electrode material is needed, a new section of steel casing is welded to the top of the old and is filled with the mixture.

CLASSIFICATION OF ARC FURNACES. In the **independent** or **indirect-arc** furnace the electrodes are placed in the furnace above the charge, and the arc forms between the electrodes. Heating from the arc to the charge is by radiation from the arc and from the hot furnace roof. This type of furnace is shown diagrammatically in Fig. 23.

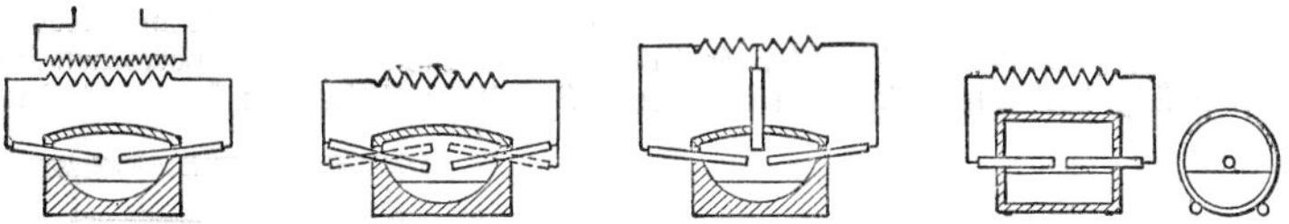

(a) Stassano Single Phase (b) Bassanese Single Phase (c) Rennerfelt 2 Phase (d) Detroit Rocking Single Phase

FIG. 23. Independent or Indirect-arc Furnaces

In **direct-arc** furnaces the arc is formed between the electrodes and the charge. Of these, the **series-arc** furnace has a non-conducting hearth (Fig. 24). By far the greatest

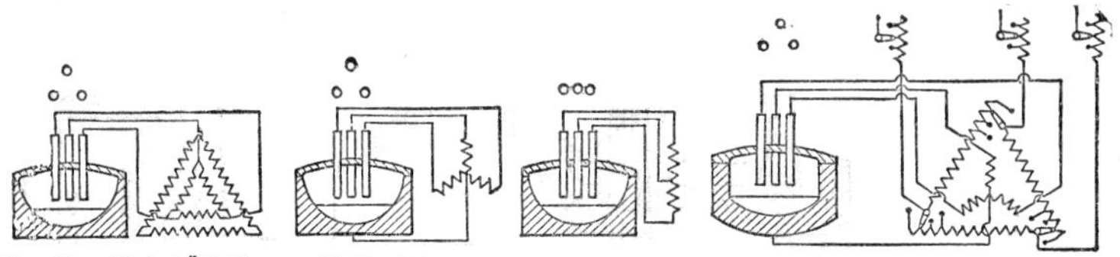

(a) Heroult and Swindell 3 Phase (b) Flat 3 Phase (c) Vom Baur 2 Phase (d) Lectromelt 3 Phase

FIG. 24. Direct-series-arc Furnaces

number of arc furnaces in use are of this type. There is also a **hearth-electrode** type (Fig. 25) and a **conducting-hearth** type (Fig. 26).

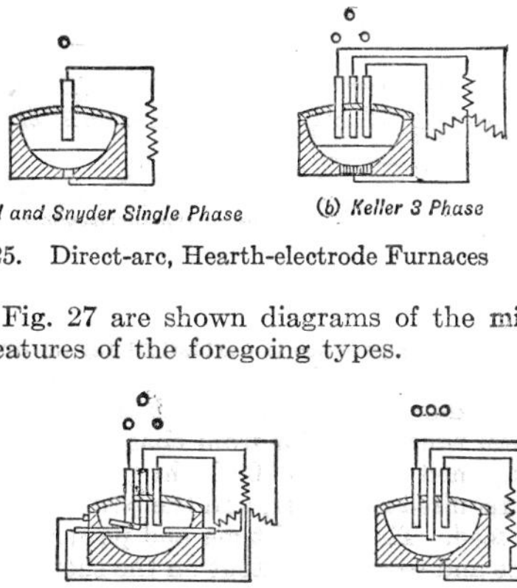

(a) Girod and Snyder Single Phase (b) Keller 3 Phase

FIG. 25. Direct-arc, Hearth-electrode Furnaces

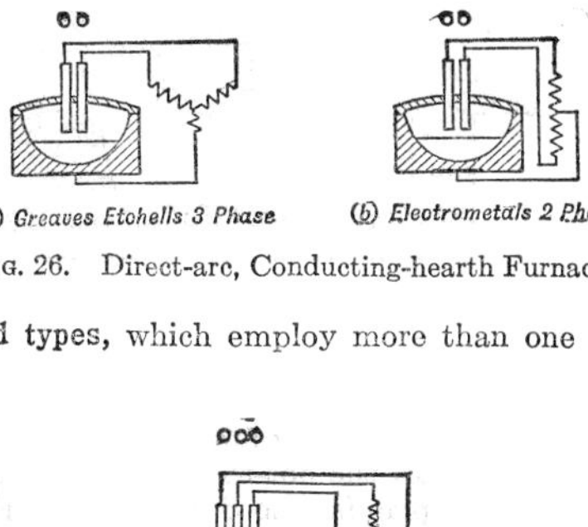

(a) Greaves Etchells 3 Phase (b) Electrometals 2 Phase

FIG. 26. Direct-arc, Conducting-hearth Furnaces

In Fig. 27 are shown diagrams of the **mixed types**, which employ more than one of the features of the foregoing types.

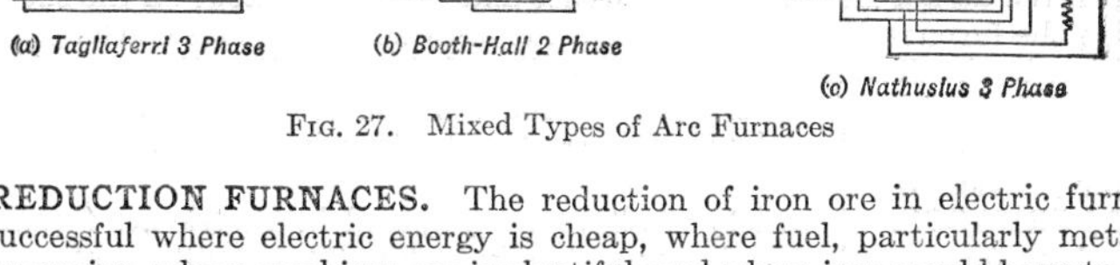

(a) Tagliaferri 3 Phase (b) Booth-Hall 2 Phase (c) Nathusius 3 Phase

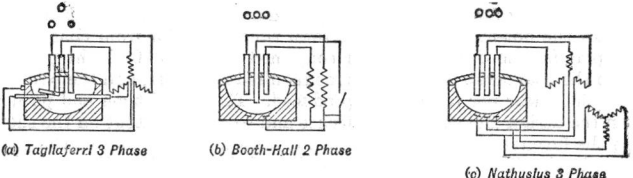

FIG. 27. Mixed Types of Arc Furnaces

ORE-REDUCTION FURNACES. The reduction of iron ore in electric furnaces has proved successful where electric energy is cheap, where fuel, particularly metallurgical coke, is expensive, where good iron ore is plentiful, and where iron would have to be transported great distances to supply the local demand. Electrical energy must substitute for two-thirds of the coke of the blast furnace, which makes the process too costly for general adoption. Electrothermic iron-smelting is practiced in Sweden, Norway, and Italy, and, before World War II, was carried on in Japan. In America electric reduction furnaces are being used in a few cases for reduction of non-ferrous ores.

ELECTRIC STEEL MAKING. The superiority of the electric-arc furnace in steel manufacture lies in the fact that it produces a purer and more uniform steel than fuel-fired furnaces. It is true that the crucible process has long produced a steel of excellent quality, but its operation is costly and the steel is made in small batches, so that large ingots must be formed by pouring several crucibles into one mold, with resultant non-uniformity. In the production of electric steel the electric furnace does not replace the Bessemer converter or the open-hearth furnace but supplements them.

Practically all steel refining is done in direct-series-arc furnaces, shown diagrammatically in Fig. 24. The two leading furnaces are the Heroult and the Moore Rapid Lectromelt. A schematic cross-sectional view of an installation of a Moore Rapid Lectromelt furnace is shown in Fig. 28. These furnaces are also built with a tilting top for rapid charging.

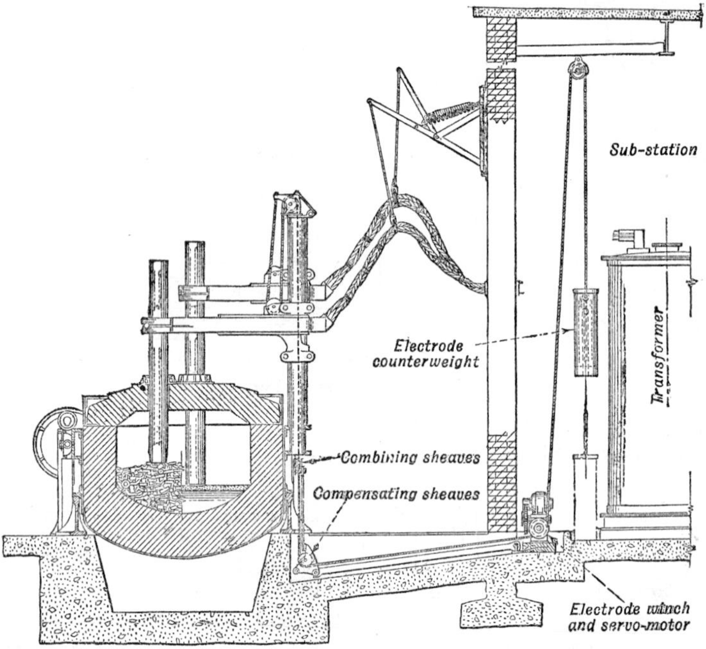

FIG. 28. Schematic Cross-sectional View of an Installation of a Moore Rapid Lectromelt Furnace

They are used for the hot-metal process, in which the process is carried as far as possible in the open-hearth furnace and the liquid steel then poured into the electric furnace for further refining. They are also used for the cold-scrap process for melting and refining scrap steel. In the hot-metal process the total time of a heat from charging the liquid metal to tapping is 4 to 5 hr; the power consumed is 100 to 200 kwhr per ton of steel produced. In the cold-scrap process the time required is greater, and the power consumption is 500 to 530 kwhr per ton of steel produced. The actual time and power required depend upon the design and size of the furnace and to a large extent on the shop practice of operating the furnaces. Furnaces are built in sizes up to 100 tons capacity.

In addition to the Heroult and the Moore Rapid Lectromelt well over a dozen other furnaces are or have been used, especially abroad. Some of these are shown diagramatically in Figs. 23 to 27. More detailed description is given in the references listed in the Bibliography, Article 15.

ALLOY FURNACES. A variety of alloys are prepared in electric-arc furnaces. The Detroit rocking furnace, shown diagrammatically in Fig. 23, is built up of a cylindrical steel shell lined with refractories. Two electrodes project horizontally into the furnace from the ends of the cylinder. The furnace is mounted on rollers and during operation is rocked backward and forward automatically by an electric motor. Rocking avoids superheating of the lining and ensures a very uniform product in any one heat. The Detroit furnace serves a variety of purposes. In addition to the large field of non-ferrous melting, it is used for producing high-alloy irons, gray iron, and stainless steel.

A variety of ferroalloys is produced in electric-arc furnaces. The furnaces are usually simple and are constructed at the plant where used. The furnace proper commonly consists of a rectangular box-like structure built of boiler plate lined with refractory brick, open at the top, of the type shown in Fig. 29. They usually have three vertical electrodes operating on three-phase current. Heating is by arcing and resistance.

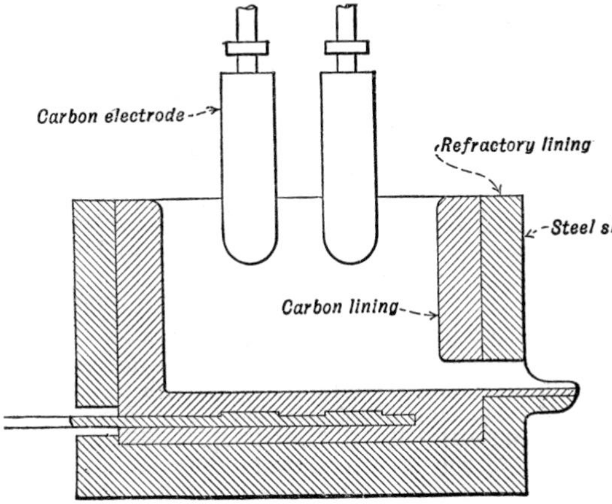

FIG. 29. General Arrangement of a Ferroalloy Furnace

Ferrosilicon. The furnace charge consists of steel or iron ore, silica as lumps of ganister or quartzite, and charcoal or coke. The slag boils up around the electrodes, where it is removed, being too viscous for tapping. The energy consumption is 2 to 6 kwhr per lb.

Ferrochrome. Ferrochrome is produced by reducing chromite ore with carbon in an arc furnace. If pure materials are available, no lime slag is necessary. Anthracite coal is a suitable reducing agent. The furnace product contains 60 to 75 per cent chromium, 1 to 3 per cent silicon, and up to 8 per cent carbon.

Ferrovanadium. The raw materials used are vanadium ore concentrates, smelted with iron and fluxes. Reducing agents may be coke, silicon, or ferrosilicon. High reduction temperatures are needed, and to avoid volatilization losses of the valuable metal the furnaces are closed at the top.

Ferromolybdenum. This alloy is produced from a molybdenum ore such as molybdenite, MoS_2, roasted and concentrated, iron or steel turnings or scrap, with carbon or silicon as a reducing agent and lime as a flux. An alloy containing 50 to 60 per cent molybdenum is usually produced, with a carbon content ranging from 0.25 to 2 per cent. Calcium molybdate, $CaMoO_4$, is coming into favor for production of ferromolybdenum, especially if a low carbon content of the alloy is desired.

Ferrotungsten. The raw materials are wolframite, $(Fe, Mn)WO_4$, or scheelite, $CaWO_4$, carbon or coal, iron ore or scrap iron, and a slag. The alloy produced commonly contains 78 to 84 per cent tungsten.

Ferromanganese. The raw materials are manganese ores, carbonaceous reducing agents, iron or iron ore, and in some cases a slag-forming agent. A standard grade of ferromanganese contains approximately 78 to 82 per cent manganese, 15 to 20 per cent iron, 6 to 8 per cent carbon, 1.00 per cent or less silicon, 0.35 per cent or less phosphorus, and 0.05 per cent or less sulfur. The furnaces are closed at the top, with provisions for collection and utilization of the gaseous products from the furnace.

14. ELECTRIC INDUCTION FURNACES

(See Articles 16–18)

In induction furnaces heat is generated in the charge itself, which serves as a secondary in a step-down transformer. As a result, high currents are generated in the charge, and melting or heating is rapid. Three types of induction furnaces are shown in Fig. 30.

The low-frequency, open-channel furnaces have not found much favor in the United States but are used commercially in Europe. The **Kjellin** and **Rochling-Rodenhauser** furnaces are examples of this type. The former uses a frequency of 5 cycles per second; the latter, 25. This requires special generating equipment. A difficulty encountered is the **pinch** effect, in which a contraction of the metal path caused by the current may completely break the circuit. It is necessary to retain a circuit of molten metal from one charge to another.

AJAX-WYATT FURNACE. In this furnace, shown diagramatically in Fig. 30(*b*), the secondary consists of a vertical loop of molten metal contained in a refractory channel. The bulk of the metal is above this secondary and exerts a static head on the liquid metal in the channel, thereby overcoming the pinch effect so troublesome in horizontal channel furnaces. The heating takes place in the V channel. The hot metal rises, and cooler metal from the main chamber takes its place. Stirring effect is also produced by the **motor effect** in the circuit. These furnaces have found wide application for brass melting. They are built in standard sizes, ranging from 200- to 3000-lb capacities, with power inputs of 30 to 240 kw. They operate on 25, 30, or 60 cycles alternating current on

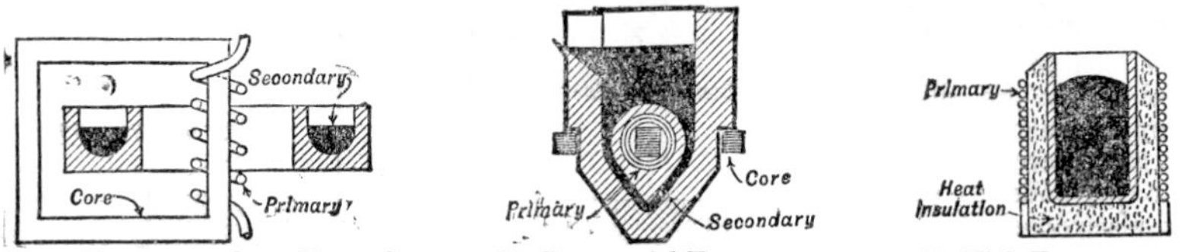

(*a*) Low Frequency, Open Channel (*b*) Commercial Frequency, Vertical Channel (*c*) High Frequency, Coreless

Fig. 30. Induction Furnaces

any commercial voltages, such as 220, 440, or 550 volts, single-phase, except that the 2000- to 3000-lb sizes use three-phase current. The furnaces are used for melting most non-ferrous alloys. Power consumption varies with the nature of the charge, ranging from 90 kwhr per ton of zinc to 305 kwhr per ton of copper.

AJAX-NORTHRUP HIGH-FREQUENCY FURNACE. This furnace, shown diagrammatically in Fig. 30(*c*), generates the heat directly in the metal in a crucible by producing eddy currents. It is necessary to have a supply of high-frequency current; 1000 cycles per second has become almost standard for commercial-size furnaces. This is supplied by a single-phase alternator, usually driven by a polyphase synchronous motor. The power factor is low, in some cases not above 0.1. The primary coil around the furnace is a flattened water-cooled copper tube. High-frequency induction furnaces have a wide variety of uses. They are used for production melting of steel, tool steel, stainless steel, iron, bronze, nickel silver, precious metals, and a variety of alloys. Non-conducting materials are heated in a special graphite crucible placed in the furnace, the heat being generated by the induced current in the crucible walls; and temperatures above 3000 deg cent (5432 deg fahr) can be obtained. Furnaces ranging in capacities up to 2200 lb are built. Power consumption ranges from 340 to 660 kwhr per ton, depending on the nature of the charge.

This type of furnace can also be used for heat-treating metals. The furnace induction coil is open at both ends so that objects can pass through it. Heat is generated in the outer surface and penetrates to the interior if sufficient time is allowed. The process lends itself especially to surface hardening; time cycles of only a few seconds are maintained by automatic regulation of power and split-second heating and quenching intervals.

15. BIBLIOGRAPHY

General References

Allmand, A. J., and H. J. T. Ellingham, *The Principles of Applied Electrochemistry*, 2nd ed. Edward Arnold, London (1924).
Engelhardt, Victor, *Handbuch der technischen Elektrochemie*. Akademische Verlagsgesellschaft M.B.H., Leipzig (1933).
Creighton, H. Jermain, *Principles of Electrochemistry*, 4th ed. John Wiley (1943).
Koehler, W. A., *Applications of Electrochemistry*, 2nd ed. John Wiley (1944).
Mantell, C. L., *Industrial Electrochemistry*, 2nd ed. McGraw-Hill (1940).
Thompson, Maurice de Kay, *Theoretical and Applied Electrochemistry*, Macmillan (1939).

Electroplating and Electroforming

Addicks, L., *Silver in Industry*, Reinhold (1940).
Blum, W., and G. B. Hogaboom, *Principles of Electroplating and Electroforming*, 2nd ed. McGraw-Hill (1930).
Freeman, B., and F. G. Hoppe, *Electroplating with Chromium, Copper and Nickel*, Prentice-Hall (1930).
Langbein, G., and W. T. Brannt, *Electro-Deposition of Metals*, 8th ed. H. C. Baird, New York (1924).
Richards, E. S., *Chromium Plating*, J. B. Lippincott, Philadelphia (1936).
Modern Electroplating, The Electrochemical Society, New York (1942).

Transactions of the Electrochemical Society (called *Transactions of the American Electrochemical Society* before November 1931), New York.

Monthly Review of the American Electroplaters' Society, Chicago. *Metal Cleaning and Finishing*, Pittsburgh, Pa.

Electrorefining

Nickel

Peek, R. L., *Eng. Mining J.*, Vol. 130, 482 (1930).
Yardlye, J. L. McK., *ibid.*, Vol. 114, 810 (1922).
Royal Ontario Nickel Commission Report, 1917.

Lead

Betts, A. G., *Lead Refining by Electrolysis*, John Wiley (1908).
Fingland, J. J., *Trans. Amer. Electrochem. Soc.*, Vol. 57, 179 (1930).
McIntyre, P. F., *Trans. Amer. Inst. Mining Met. Eng.*, Vol. 121, 271 (1936).

Silver

Addicks, L., *Silver in Industry.* Reinhold (1940).
Colcord, F. F., *Trans. Amer. Electrochem. Soc.*, Vol. 49, 351 (1926).
Easterbrooks, F. D., *Trans. Amer. Electrochem. Soc.*, Vol. 8, 125 (1905).
Griswold, G. G., *Trans. Amer. Electrochem. Soc.*, Vol. 35, 251 (1919).
Kern, E. F., *Met. Chem. Eng.*, Vol. 9, 443 (1911).
Annual Report of the Director of the (U.S.) Mint, p. 50 (1912).

Gold

Wohlwill, E., *Electrochem. Ind.*, Vol. 2, 221 (1904); and *Electrochem. Met. Ind.*, Vol. 6, 450 (1908).

Solder

Hermsdorf, R. P. E., and Max Heberlein, *Trans. Amer. Inst. Mining Met. Eng.*, Vol. 121, 289 (1936).

Electrometallurgy

Copper

Aldrich, H. W., and W. G. Scott, *Eng. Mining J.*, Vol. 128, 612 (1929).
Allen, A. W., *Eng. Mining J. Press*, Vol. 113, 1003, 1051; Vol 114, 184 (1922).
Baroch, C. T., *Trans. Amer. Electrochem. Soc.*, Vol. 57, 205 (1930).
Eichrodt, C. W., *Trans. Amer. Inst. Mining Met. Eng.*, p. 186 (1930).

Zinc

Archibald, M. W., and associates, *Trans. Can. Inst. Mining Met.*, Vol. 27, 306 (1924).
Huttl, J. B., *Eng. Mining J.*, Vol. 139, 42 (1938).
Stimmel, B. A., W. H. Hannay, and K. D. McBean, *Trans. Mining Met. Eng.*, Vol. 121, 540 (1936).
Tainton, V. C., and D. Bosqui, *Trans. Amer. Electrochem. Soc.*, Vol. 57, 241 (1930).
Woolf, W. G., and E. R. Crutcher, *Trans. Amer. Inst. Mining Met. Eng.*, Vol. 121, 527 (1936).

Electrogalvanizing

Tainton, U. C., *J. Amer. Zinc. Inst.*, Vol. 18, 42 (1937).
Winkler, L. W., *Wire and Wire Products*, Vol. 16, 687 (1941).

Manganese

Allen, G L., J. H. Jacobs, and J. W. Hunter, *Chem. Met. Eng.*, Vol. 53, 106 (1946).
Hammerquist, W L., *Steel*, Vol. 105, 42 (1939).

Aluminum

Edwards, J. D., F C. Frary, and Z. Jeffries, *The Aluminum Industry.* McGraw-Hill (1930).
Engelhardt, V., *Handb. tech. Elektrochem.*, Vol. 3, 303 (1934).
Frary, F. C., *Trans. Amer. Electrochem. Soc.*, Vol. 47, 275 (1925).
Frary, F. C., *Ind. Eng. Chem.*, Vol. 28, 146 (1936).

Calcium

Brace, P. H., *Trans. Amer. Electrochem. Soc.*, Vol. 37, 468 (1920); *Chem. Met. Eng.*, Vol. 25, 105 (1921).
Mantell, C. L., and Charles Hardy, *Calcium Metallurgy and Technology.* Reinhold, (1945).

Lithium

Osborg, H., *Trans. Electrochem. Soc.*, Vol. 66, 91 (1934); *Monograph on Lithium*, The Electrochemical Society, New York (1935).

Magnesium

Alico, John, *Introduction to Magnesium and Its Alloys*, Ziff-Davis Publishing Co., 350 Fifth Ave., New York (1945).
Gann, J. A., *Trans. Am. Inst. Chem. Eng.*, Vol. 24, 206 (1930); *Ind. Eng. Chem.*, Vol. 22, 694 (1930).
Harvey, W. G., *Trans. Amer. Electrochem. Soc.*, Vol. 47, 331 (1925).
Hunter, Ralph M., *Trans. Electrochem. Soc.*, Vol. 86, 21 (1945).
Killeffer, D. H., *A.C.S. News Ed.*, Vol. 19, 1189 (1941).
Kirkpatrick, S. D., *Chem. Met. Eng.*, Vol. 48, 76, 130 (1941).

General, Fused Electrolytes

Drossbach, Paul, *Elektrochemie geschmolzener Salze* Julius Springer, Berlin (1938) Edwards Brothers, Ann Arbor (1943).

Electrolytic Oxidation and Reduction

Bowman, R. G., *Trans. Amer. Inst. Mining Met. Eng.*, Vol. 73, 146 (1926).
Brockman, C. J., *Electro-organic Chemistry.* John Wiley (1926).
Creighton, H. J., *Trans. Electrochem. Soc.*, Vol. 75, 289 (1939).
Swan, Sherlock, Jr., *Trans. Amer. Electrochem. Soc.*, Vol. 69, 287 (1936); Vol. 77, 459 (1944).
Taylor, R. L., *Chem. Met. Eng.*, Vol. 44, 588 (1937).

Electric Furnaces

Moore, W. E., Twenty Years Advance in Electric Arc Furnaces for the Production of Iron and Steel, *Trans. Amer. Electrochem. Soc.*, Vol. 60, 65–85 (1931).
Paschkiss, V., *Industrial Electric Furnaces and Appliances.* Interscience Publishers (1945).
Rodenhauser, W., J. Schoenawa, and C. H. Vom Baur, *Electric Furnaces in the Iron and Steel Industry* John Wiley (1920).
Sisco, F. T., *The Manufacture of Electric Steel.* McGraw-Hill (1924).
Tess, T. J., The Modern Arc Furnaces, *Iron Steel Eng.*, 21 AF7 (1944).

INDUCTION AND DIELECTRIC HEATING

By J. P. Jordan

16. INDUCTION HEATING

The term induction heating covers the generation of heat in any conducting material by means of magnetically induced currents. Frequencies in use range from 60 cycles to 5,000,000 cycles, although the major applications use frequencies from 1000 cycles to 2,000,000 cycles. Since the depth of the zone in which heat is generated is a function of frequency, several different frequencies, together with means for generating them, are in general use.

GENERATING MEANS. Motor-generator Sets. For the generation of frequencies above 60 cycles and below 15,000 cycles the motor-generator set is almost universally used. Although generators for a great number of different frequencies have been made, the most popular frequencies have been 960 cycles, 1920 cycles, 3000 cycles, and 9600 cycles. In recent years very few 1920-cycle machines have been made, their jobs being taken over by either 960-cycle machines or 3000-cycle machines. Likewise, the manufacture of 9600-cycle machines in smaller sizes is largely being discontinued in favor of 9600-to-10,080-cycle machines of a new design.

The earlier type of generator was an inductor machine of the unipolar design operating at 3600 rpm. To reduce iron losses a stator was built up of quite thin laminations, that of the 9600-cycle machine being 0.007 in. thick. Since in the unipolar design the rotor carried no alternating flux, it was made of an iron forging with milled teeth. Air cooling was used universally. The smaller sizes were made in two-bearing models, whereas machines over 250 kw were generally four-bearing sets.

These machines had the advantage of being low in cost, relatively efficient in operation, and easily serviced because of their simple design. Because of the air blast required for cooling, however, they were extremely noisy and easily damaged by the build up of airborne dirt in the necessarily small air gaps. They were not easily water cooled because of the necessity of splitting the stator to permit removal of the annular field winding. They are still largely used, however, for 960-cycle generation and for high-power, higher-frequency machines where water cooling in a hydrogen atmosphere is the rule.

The new inductor generator now largely supplanting the above units in the smaller kilowatt ratings and two-bearing design for 9600-to-10,080-cycle and 3000-cycle operation is of the multipolar type with a chord wound field coil. Both the rotor and stator are made of thin laminated iron. Water cooling of this type of machine is greatly simplified, since a cored frame may be used. Totally enclosed units of this design eliminate the noise and service problems inherent in the air-cooled type.

Standard sizes now produced are as follows:

Frequency	Kilowatt Rating	
	Two-bearing	Four-bearing
9600 to 10,080 (multipolar)	7½, 15, 30, 50, 75, 100, 150	
9600 (unipolar)		700
3000 (multipolar)	100, 200	300, 500, 1250
960 (unipolar)	175, 250	

Electronic Oscillators. Power at frequencies above 15,000 cycles cannot be readily generated by means of rotating machines. Since experience during the past 10 years has shown the need for higher frequencies, the use of the electronic oscillator has greatly expanded.

The major element in any oscillator is the resonant circuit. Vacuum tubes or any of the various types of spark gaps could be used to excite the circuit, but an oscillator could not operate without some form of resonant circuit. Furthermore, both the inductor coil used for induction heating and the electrodes used for dielectric heating, as will be seen later, form part of this resonant circuit. Thus the following theory is of prime importance and should be understood thoroughly.

A resistor presents an impedance to the flow of any current which is equal to its resistance R. This impedance is not a function of frequency, remaining constant regardless of the type of current flowing, whether direct, alternating, or radio frequency. An inductor presents an impedance (inductive reactance), however, which varies directly with frequency: $X_L = 2\pi fL$, and a capacitor presents an impedance (capacitive reactance) which varies inversely with frequency: $X_c = 1/2\pi fC$.

If an inductance and a capacitance are connected in parallel, a curious situation exists in some one frequency at which X_L equals X_c. Since X_L varies directly and X_c varies inversely with frequency, there is one frequency at which equality will exist. This frequency is called the resonant frequency. To determine its value, X_L can be equated to X_c and solved for frequency to give

$$f_r = \frac{1}{2\pi\sqrt{Lc}} \tag{1}$$

If no resistance is present, further analysis would show that at the resonant frequency the current I in the line feeding the circuit shown in Fig. 1 would be zero:

$$I = \frac{E}{-X_L X_c/(jX_L - jX_c)} = E\frac{jX_L - jX_c}{-X_L X_c} = E\frac{0}{-X_c^2} = 0 \tag{2}$$

whereas the current I_L flowing in the inductor and the current I_c flowing in the capacitor would equal the applied voltage E, divided by their individual impedances X_L and X_c:

$$I_L = -I_c = \frac{E}{X_L} = -\frac{E}{X_c} \tag{3}$$

In eq. (2), the notation j indicates a phase angle of 90 deg between applied voltage and current, $+j$ indicating a lagging angle and $-j$ a leading angle. Thus it will be noted

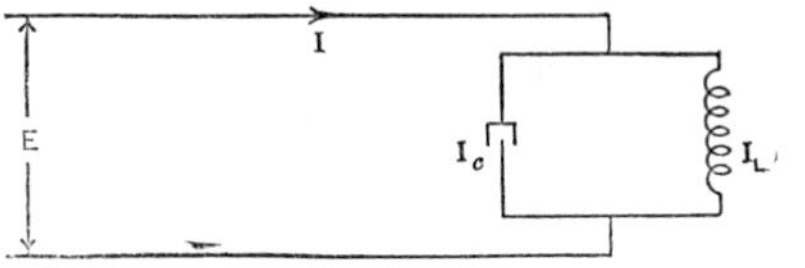

FIG. 1. Resonant Circuit

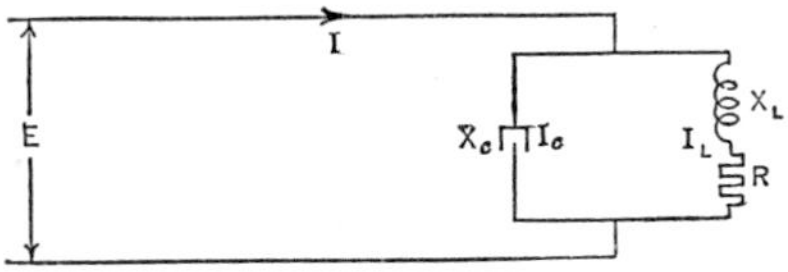

FIG. 2. Loaded Resonant Circuit

that the phenomenon of zero current input results from a cancellation of the currents through X_L and X_c, due to these currents being 180 deg out of phase at all times. When a resistance is inserted in either branch of the circuit as shown in Fig. 2, the current I no longer is zero but has a component, in phase with the applied voltage, sufficient to supply the energy loss in the resistance. A solution of this circuit at the resonant frequency gives

$$I = \frac{E}{L/Rc - j\sqrt{L/c}} \tag{4}$$

For small values of R, the input is still essentially resistive, and the voltage E times the current I (Fig. 2) will approximate the power dissipated in the resistance due to the current I_L. Note again, however, that **the current through the inductance L or the capacitance C is a function only of the applied voltage E and the impedance,** and does not flow through the external circuit. The product of this current I_L (or I_c) times the voltage E is termed the **circulating kva,** which can assume considerable magnitudes with very small values of input current I. This circuit is sometimes referred to in radio parlance as a tank, since it can be thought of as storing potential energy.

Assume now that a d-c supply is connected to this resonant circuit through a switch which can be operated very rapidly. If this switch were operated at the resonant frequency of the circuit, it is obvious from the equations given previously that high circulating currents would flow between the capacitor and the inductance, but the current through the switch would be only such as to supply the losses in the circuit.

Even though such a switch were available, this circuit has several disadvantages, including the fact that, if the circuit through the inductance is broken for any reason, the inductor is at full d-c potential.

By shunting the switch across the d-c supply and interposing a capacitor between the switch and the resonant circuit, however, similar results can be obtained.

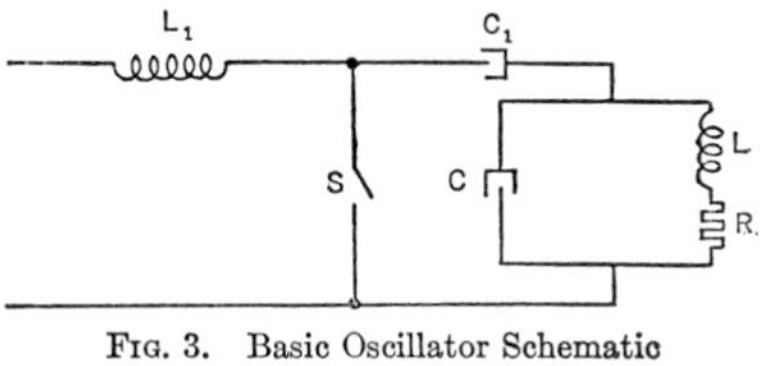

FIG. 3. Basic Oscillator Schematic

In Fig. 3, when switch S is open, the voltage builds up across C_1, through L and L_1. For the first instant of time, however, the inrush current causes nearly the full voltage to appear across the circuit L-C, and capacitor C becomes charged. If at this point switch S were closed, the applied voltage would drop, and capacitor C would discharge into inductance L in the resonant circuit. Meanwhile, the full d-c voltage would build up across L_1. When S is reopened, L would discharge into C, and the process would be repeated. Thus, if S were operated at the resonant frequency of the circuit L-C, the conditions previously noted would exist. That is, high circulating currents would flow through L and C, whereas the current through C_1 would be great enough only to supply the losses.

This process becomes clear if it is recalled that, at the resonant frequency, the input impedance of the resonant circuit is very high (zero current input with full voltage applied); and thus, if switch S were operated to supply short pulses of current at this frequency, the full voltage would appear across this high impedance. Capacitor C_1 acts to block any steady flow of direct current through inductance L but presents a very low impedance to the flow of the pulse of current feeding the rest of the circuit. Inductance L_1 is a coil or choke which serves to absorb the sharp current pulses and prevents them from affecting other elements in the circuit. This, then, is a satisfactory oscillator circuit—all that remains is to find an appropriate switch at speeds below 200,000 to 300,000 cycles per second. Figure 4 is an elementary diagram of a spark-gap unit as used commercially. For the higher frequencies the only satisfactory element is the vacuum tube.

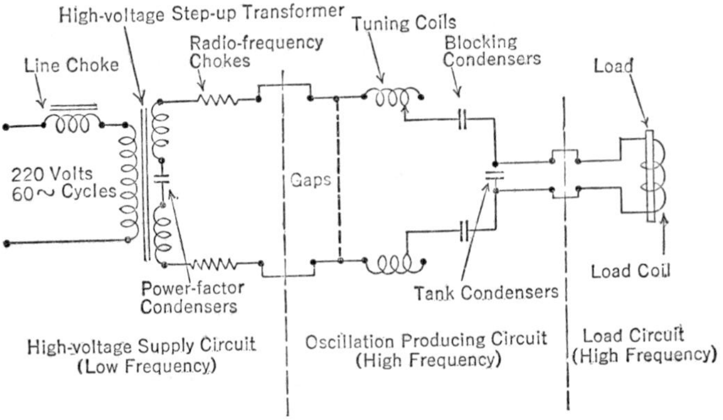

FIG. 4. Spark-gap Oscillator

The vacuum tube can be thought of as a contactor capable of operating at very high speeds. The tubes used for this application consist of three elements: an anode, a cathode, and a grid, all within an evacuated envelope. The cathode is generally a tungsten or thoriated tungsten element operating at very high temperatures to provide a source of electrons. When any other element of the tube is positive with respect to this cathode, these electrons are attracted to it and thus establish a current flow. (Note that the conventional concept of current flow from plus to minus polarities is opposite to the actual flow, or migration, of electrons.)

However, a wire-mesh screen, the grid, interposed between the anode and the cathode, can control this flow of electrons by either cancelling or aiding the positive voltage field set up by the anode. Thus, regardless of the plate voltage, if the grid is sufficiently negative, no current will flow, whereas, if the grid is positive or only slightly negative, the electrons will flow through the opening in the mesh to the anodes.

Before the vacuum tube will operate to maintain oscillation in the circuit, a proper grid voltage must be determined and a means found to create it. It has been determined that a combination of a negative d-c voltage applied between grid and cathode plus a radio-frequency voltage 180 deg out of phase with the anode-cathode voltage will give correct operation. The d-c voltage is usually obtained by permitting the grid to rectify a small portion of the applied radio-frequency voltage. A resistor connected between the grid and cathode as shown in Fig. 5 permits adjustment of this d-c voltage to the desired level.

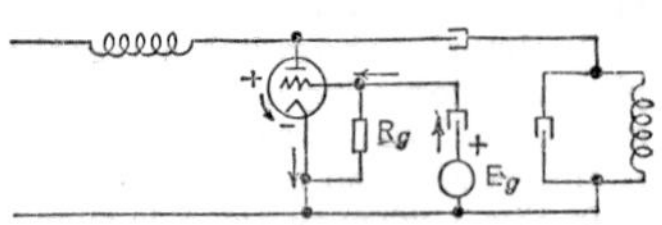

FIG. 5. Basic Vacuum-tube Oscillator Circuit

Several methods are used to obtain the r-f grid voltage from the resonant circuit, and this is the main point of difference in the oscillator circuits now being used by the various manufacturers of induction heating equipment.

In the coupled grid circuits (Fig. 6) the voltage is developed across a coil inductively coupled to a portion of the resonant circuit. In the **Colpitts circuit** (Fig. 7) this grid voltage is obtained by direct connection to a capacity divider forming part of the resonant circuit.

Figures 6 and 7 illustrate the major circuits in use. Section M represents schematically the mercury tube rectifier commonly used to convert the 60-cycle voltage output of the high-voltage transformer T to the high-voltage direct current utilized by the oscillator circuit S. The Colpitts circuit has the advantage of somewhat greater stability, since the

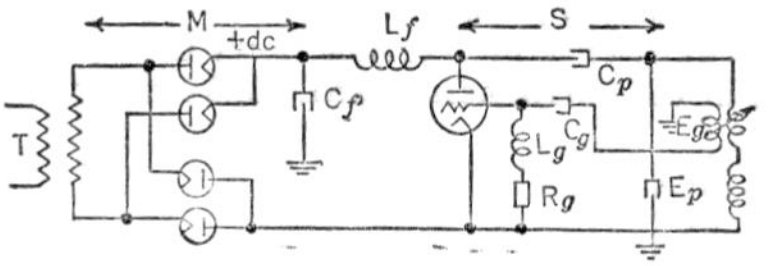

FIG. 6. Coupled-grid Oscillator Circuit

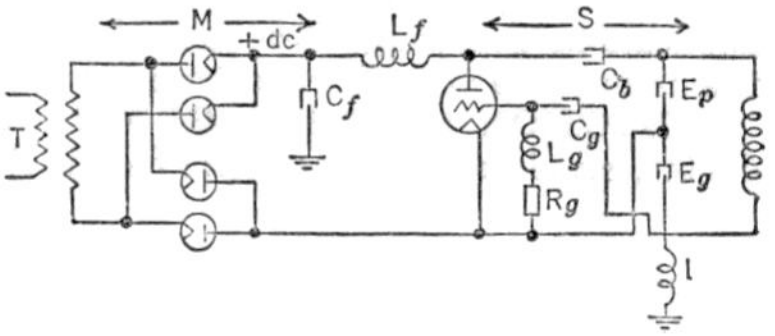

FIG. 7. Colpitts Oscillator Circuit

capacitor ratio E_p versus E_g is always fixed, thus providing a stiffer voltage source. The coupled grid circuit, however, affords a ready means of adjusting the amplitude of the grid voltage, which in some cases is advantageous.

In conclusion of this discussion of the oscillator circuit, it is important that several points be understood thoroughly because of their importance in the application of this circuit to induction and dielectric heating.

First, voltage and current in the resonant circuit have no relation to power output, being solely a function of voltage, frequency, and circuit constants. This current does not flow through the tubes.

Second, all direct current from the rectifier flows through the oscillator tubes. Since the direct current times the voltage equals the power output of the rectifier, this current is proportional to the power output of the oscillator circuit at any fixed voltage.

Third, the serviceable life of all oscillator tubes is determined by the life of the filaments, which in turn is governed by their operating temperature. Temperatures either too low or too high will seriously shorten the life of thoriated tungsten filaments. With pure tungsten filaments, the lower the temperature, the longer is the tube life. At reduced temperatures however, fewer electrons are emitted, thus limiting the permissible power output. It is always advisable to operate such tubes at the lowest filament voltage that does not affect their operating condition.

Fourth, in some types of equipment in use today the vacuum tube is replaced by a spark gap, either in air using tungsten electrodes or in mercury vapor. The basic concepts of the oscillator circuit apply equally well to such spark-gap equipments.

The sizes of electronic heaters as produced today have not yet reached standardization, although there is some effort to build units to fit the NEMA proposed standard ratings of 1/2, 1, 2, 5, 10, 20, and 50 kw. The rating of almost all vacuum-tube equipment is based on heat **output** expressed in kilowatts or Btu. All spark-gap equipment is rated on kilowatt **input,** the output varying from 25 to 75 percent of this value, depending on the type of load.

Equipment is obtainable today in the following approximate ratings:

Equipment	Frequency	Rating, Kilowatts
Induction heating		
Spark gap..........	40– 70 kc	6, 10, 20, 35, 75 (input)
Spark gap..........	100–400 kc	15, 30 (input)
Vacuum tube.......	400–550 kc	1, 2, 5, 10, 15, 20, 25, 50, 75, 100, 200 (output)
Vacuum tube.......	2 mc	2 1/2, 5, 20 (output)
Dielectric heating		
Vacuum tube.......	2– 40 mc	1/2, 1, 2, 5, 7 1/2, 15, 20, 25, 50, 100 (output)

Vacuum-tube oscillators, when loaded, usually operate at approximately 50 per cent efficiency from d-c input to radio-frequency output. However, since the rectifier losses and the filament power approximate 20 to 30 per cent of the output rating, the line demand at full load is usually 220 to 230 per cent of the output. Power factors are generally 95 per cent or better.

Although equipment is being manufactured to operate from a variety of power supplies, the majority of standard sets are being designed for 115 or 230 volts single phase for sizes rated below 2 kw, and for 230 or 460 volts, three phase 50/60 cycles for the larger units.

The filament voltage must be held within narrow limits for long life and satisfactory operation, the usual allowance being ±5 per cent for tungsten filaments and ±3 per cent for thoriated filaments. Also, if all other operating conditions are held constant, the power into a given load varies approximately as the square of the voltage. Experience has shown that for the majority of applications a maximum voltage drift during the day of 10 per cent, i.e., ±5 per cent of rated operating voltage, will not cause trouble. For applications requiring close control of temperatures in repetitive operations, however, the permissible voltage variation may be as low as 3 or 4 per cent.

INDUCTION HEATING THEORY. When an alternating current flows in any conductor, an alternating magnetic field is set up in the surrounding area. Likewise, when any conducting material is placed in an alternating magnetic field, a current flow is set up in that material. This current is such that the counter magnetic field generated by it will tend to cancel the existing field.

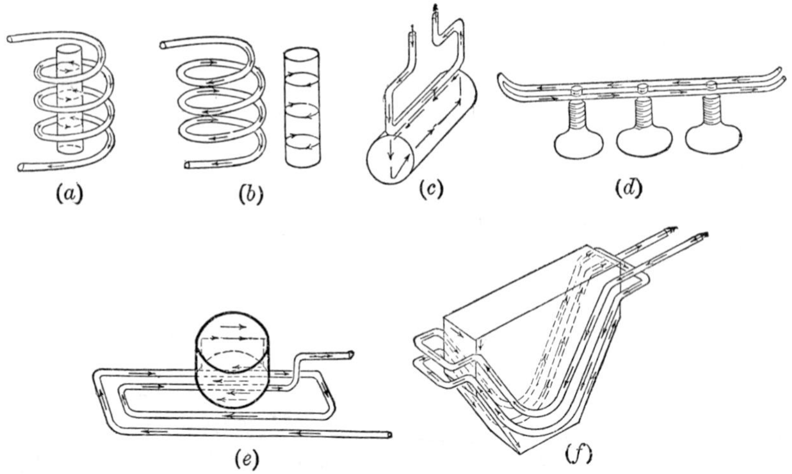

FIG. 8. Direction of Induced Current Flow in Heated Part

Since the external magnetic flux must penetrate the surface before reaching the interior of this conducting material, the greater part of the current flow will be near the surface. The intensity of the counter magnetic field set up by the current flow in the body is a function of the frequency; thus, as the frequency is increased, the current flowing on the surface becomes more effective in generating the total counter magnetic field required, and less current will flow in the layers below the surface. This is known as the skin effect, since it results in concentration of current flow on the surface or skin of the body. The currents themselves are called eddy currents, and the heat caused by the resistance of the materials to its flow is known as eddy-current loss. It will be noted that the intensity of the external field affects only the magnitude of the eddy-current flow and thus the rate of heating, whereas the frequency affects the depth to which these currents will penetrate.

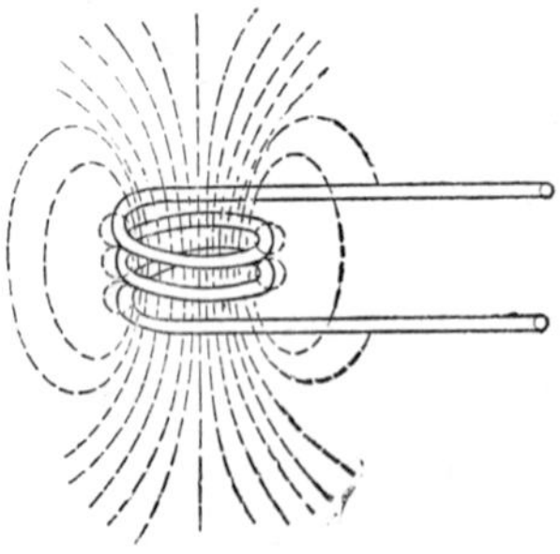

FIG. 9. Magnetic Flux Lines around Coil in Air

In induction heating the inductor coil or heater coil can be thought of as the primary of a transformer, with the charge being a single turn secondary due to the skin effect. Thus the load appears as a resistance in the heater coil. However, although this is true in all cases, in complicated applications it is often easier to think of the coil as setting up a magnetic field of a certain shape, which in turn causes current to flow in the charge. Note that these currents must flow in closed loops in the same plane as the coil currents: thus, if a coil is placed around a bar, as shown in Fig. 8, the current will flow in a closed loop around the surface of the bar in the same plane as in the coil, but if the same bar is placed close to the outside of the coil [Figs. 8(*b*) and (*c*)], the current will still flow in the bar as before, although with a considerably smaller magnitude because the magnetic flux density outside a coil is less than that within it. Figure 9 shows the magnetic flux lines existent in the space surrounding a multiturn coil.

The remaining sketches, Figs. 8(*d*), (*e*), and (*f*), serve to illustrate the application of the above concept to actual heating problems. In all cases the current flow is indicated in both the current and the coil.

In magnetic materials the hysteresis loss will create some heat, but this loss is generally small in comparison with the eddy-current loss.

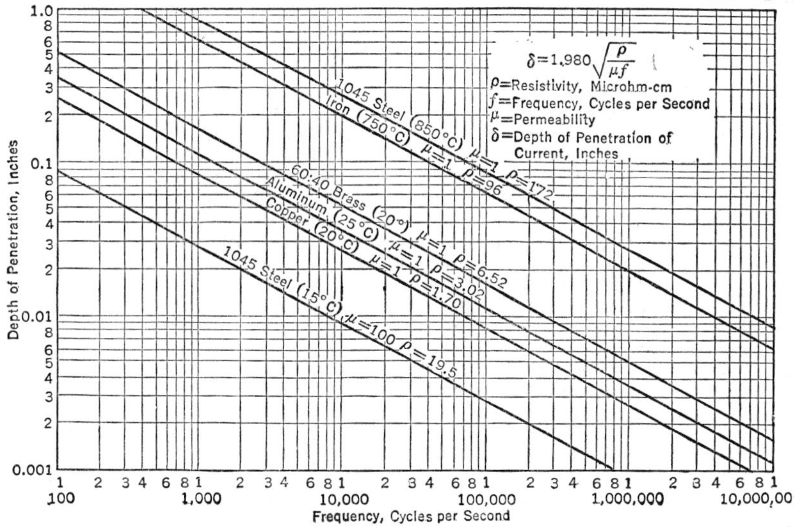

FIG. 10. Depth of Penetration *versus* Frequency of Induced Current

The rigorous equations for eddy-current losses are quite complex and of little general use. However, some approximate formulas will serve to show the relationship of the various parameters, although, since many of the factors are difficult to measure with any degree of accuracy, they cannot be applied to most practical problems. The amount of power dissipated as heat in the surface of a part, in terms of the magnetic flux density, the frequency, and the electrical characteristics in the metal being heated, is

$$\Delta P = \frac{H_t^2\sqrt{\rho\mu f}}{8\pi} \tag{5}$$

where ΔP = power dissipated as eddy currents.
H_t = tangential component of magnetic flux at surfaces of charge.
ρ = resistivity of charge.
μ = permeability of charge.
f = frequency.

Since the magnetic flux density H_t is proportional to the ampere-turns in the coil, the factor H_t^2 could be replaced with I^2N^2 (times a constant), where I equals the coil current and N is the effective number of turns in the coil.

The depth of penetration δ of the heat for any given frequency and material is

$$\delta = \frac{1}{2\pi}\sqrt{\frac{\rho}{\mu f}} \tag{6}$$

To be exact, this is the depth at which the eddy currents fall to a value equal to $1/e$ (37 per cent) times their magnitude at the surface.

Combining equations (5) and (6),

$$\Delta P = \frac{H_t^2\rho}{16\pi^2\delta} \tag{7}$$

Examining these equations, it can be seen that the rate of heat input is directly proportional to the square root of the frequency and the resistivity, and inversely proportional to the depth of penetration factor.

Also note that the permeability, which is a measure of the magnetic properties of the metal, has a value of 1 for all non-magnetic materials but can be of major importance when heating magnetic metals below the curie temperature (1400 deg fahr for low-carbon steel), above which magnetic properties effectively disappear.

Figure 10 shows the depth of penetration as a function of frequency for several metals.

COIL DESIGN. It must be borne in mind that the heat generated in a part is entirely due to the magnetic flux created by the inductor coils; only by changing the intensity of the flux pattern in the part can the rate of heat generation be affected. This magnetic flux, as stated above, is directly proportional to the ampere-turns in the coil, i.e., the coil current times the effective number of turns. Also, since the flux density is greatest at the conductors themselves, diminishing rapidly in the surrounding space, the closer the coil is to the part, the greater will be the heat generated.

Thus, if it is desirable to obtain high concentration of power in restricted areas, it is desirable to utilize single-turn coils carrying a very high current. Since the voltages at which the higher frequencies are generated are usually dictated by considerations other than inductor coil utilization, it is often necessary to utilize output transformers to provide the high current, low voltage necessary for single-turn coils. In the lower frequencies supplied by motor-generator sets up to 15,000 cycles, these transformers are of more or less conventional designs, utilizing water-cooled conductors with iron cores made of thin laminated steel. For the higher frequencies generated by electronic means, iron cores have not been used because of their high losses; consequently air-core transformers, utilizing single-turn secondaries for the most part, have been developed.

Multiturn coils either can be directly connected to the terminals of the motor-generator set, together with such capacity as may be needed to permit operation of the generator at unity power factor, or in electronic equipment, may be substituted as part of the inductance in the resonant circuit.

The following notes will aid in coil design, but it must be remembered that because of the complexity of the factors involved it is necessary in all but the simplest cases to determine the final design by actual trial and error.

1. The coil should roughly conform to the shape of the part if no sharp contours need be considered. Symmetry is important. The part should be centered in the coil as far as possible.
2. Sharp corners will heat first because of the concentration of flux and the lack of mass. Thus the coil should be furthest from the part at these points.
3. If dissimilar metals are being heated for brazing, etc., the magnetic flux must be concentrated on the slowest-heating metal; in general, magnetic steel heats more easily than any other material, with stainless steel, brass, copper, and silver following in the order of their resistances.
4. In brazing, the joint should be at the correct temperature before the brazing alloy melts, so that it will be drawn into the joints. Thus a concentration of heat on the brazing alloy should be avoided.
5. In hardening, double bank coils are sometimes necessary because of the current limitations of the equipment. Since the outside layer of turns is far less efficient than the inner layer, however, such coils should be used only if essential.
6. To obtain uniform heating on a symmetrical part, rotation of the part is advisable, so as to avoid the heating effects of the coil leads.
7. The type and size of coils that can be used on any equipment are determined by two factors: the power capability of the generator and, in electronic equipment, the maximum inductance allowable.

FIELD OF APPLICATION OF INDUCTION HEAT. The original cost of induction heating equipment is in general higher than the equivalent furnace or other type of heating equipment. This is particularly true of electronic equipment, where the cost often is several times that of any other equivalent form of heat. To offset this factor, however, efficiencies of operation are often significantly higher, that of motor-generator equipment averaging above 75 per cent and electronic equipment on the order of 50 per cent overall efficiency. Furthermore, it is often possible to heat only a localized area inductively, whereas with other means it might be necessary to heat a much greater mass.

Since the heat generated by induction heating can be accurately predetermined and closely controlled, it is often unnecessary to use highly skilled labor for operation of the unit, and spoilage due to the human element is materially reduced.

In many cases results which are unique can be obtained with induction heat. Heat can be confined to sharply controlled areas in a manner never before possible. In many cases this will permit a cheaper overall design of a product.

For high production parts induction heating permits complete mechanization of a heat-treatment line. Such equipment can be designed to fit directly in production lines, accepting parts from a prior operation and delivering them to the next without in any way slowing up the production.

Where very accurately controlled temperatures are required, as in aluminum brazing, or where uniform through heating of an entire part is necessary, as in normalizing or solution heat treatment, other types of heating equipment are usually preferable.

Because of the speed of heating obtainable, it is often possible to heat treat or braze parts after they have been finish-machined. This permits working the material in the soft state or before assembly.

CHOICE OF EQUIPMENT. The best equipment for any particular application may often be chosen on the basis of frequency, as determined by the operation desired or the size of the parts.

The great majority of all melting applications is carried out at frequencies of 960 or 3000 cycles. Commercial frequencies of 60 cycles are in wide use, employing iron-core induction furnaces, whereas the higher frequencies utilize coreless induction furnaces. Induction generators are used almost exclusively for this service. Mercury-tube inverters have recently been applied to such work and may develop into an important alternative means of generating such frequencies.

Preheating for forging is carried out at frequencies from 960 cycles to 10,000 cycles, depending on the size of the part. Billets with a diameter greater than approximately 2 in. are easily heated at 960 cycles. One-inch bars can be heated with 3000 cycles; smaller charges usually must be heated at frequencies of 9600 cycles or higher.

The best frequency for brazing, hardening, and annealing is determined primarily by the diameter of the part or the diameter of the sharpest contour about which it is desirable to localize heat. Thus a part with a minimum diameter of 2 in. or greater can be induction heated at a frequency of 3000 cycles. Parts with a minimum diameter of roughly $1/2$ in. can be treated satisfactorily with 9600 cycles; any parts smaller than this or having projections smaller than $1/2$ in. in diameter should preferably be treated with electronic equipment. Since the depth of heating varies inversely as the square root of frequency, relatively large changes in frequency in the range of 200,000 to 500,000 cycles produce very little modification in the heating effect. Thus any equipment operating in this range may be used.

TYPICAL APPLICATIONS. Selective hardening of the bearings on crankshafts was perhaps one of the earliest large applications of induction heating. With a frequency of 960 cycles for shafts larger than 2 in. in diameter and higher frequencies up to 10,800 cycles for shafts of smaller diameter only the bearing surfaces are hardened, thus permitting a greatly improved core metallurgy and much harder bearing surfaces than would be feasible if it were necessary to machine the shafts after hardening.

The hardening of gears after machining has in many cases resulted in a vastly superior product. Large gears and sprockets are successfully being hardened by use of 9600 cycles, either in a single heat or tooth by tooth. Smaller gears, such as are used in the automotive industry, can often be hardened by frequencies from 200,000 cycles to 500,000 cycles. In most such cases, however, true contour hardening is as yet unobtainable, because of the unequal heat conduction away from the root and tip of tooth. Power densities above 5 kw per sq in. of surface are usually used at frequencies below 15,000 cycles; surface hardening of most parts can be obtained at power densities of 1 kw per sq in. at frequencies above 200,000 cycles.

Elongated flat surfaces, such as rails, have been successfully surface hardened by a process known as scanning, in which the coil is moved along the work, closely followed by a quench. The higher frequencies obtained from electronic equipment are generally used for these applications. Power densities between 10 and 150 kw per sq in. are necessary, depending on the depth of hardening desired. Proper quenching in this application is very critical. It has been successfully accomplished by flowing a sheet of water behind the coil and preventing wetting of the area under the coil by use of a small air jet.

The continuous annealing of rods and tubes is being successfully carried out at both 10,800 cycles and 520,000 cycles, the higher frequency being used where it is necessary to heat small-diameter, non-magnetic tubing.

In recent years induction heat has been applied more widely where small lots are hardened, brazed, or soldered. Where various small parts are not treated, 10,000-cycle motor-generator sets are in common use. Where it is necessary to treat very small parts as well as larger ones, however, electronic equipment at higher frequencies is desirable. Likewise, where the quantity of any one item is very small and it is necessary to design a new inductor for each part, it is often more advisable to use higher-frequency electronic equipment where coil construction is easier because of the lower currents handled.

17. DIELECTRIC HEATING

THEORY. Dielectric heating is the name generally applied to the generation of heat in non-conducting materials by their losses when subjected to an alternating electric field. The term electrostatic heating is a misnomer, since it is impossible to generate heat with a static electric field.

Dielectric heating is a mass phenomenon peculiar to non-conducting materials and is thought to be due to molecular friction caused by the stresses set up in the molecules by the alternating electric field.

Dielectric heating is essentially a voltage phenomenon. The charge is placed between two plates, thus forming a capacitor, and a high-frequency voltage is applied. It is well known that losses occur even in the best capacitors in ordinary use because the dielectric material is not perfect. Thus, when relatively poor, i.e., high-power-factor dielectric materials are used, considerable power in the form of heat can be generated within this material, and the heat is distributed uniformly throughout the mass.

Another factor of importance to this study is the specific inductive capacity or dielectric constant. Different materials, when placed between identical electrodes, will result in different total capacitances. The dielectric constant is a measure of this effect and is the ratio of the capacitance between two electrodes, with the material as the dielectric, to the capacitance between the same electrodes in a vacuum.

Unlike induction heating, the heat loss in any insulating material due to dielectric losses can be calculated to a fair degree of accuracy once the power factor and dielectric constant of that material at the pertinent frequency are known. When an alternating voltage is applied to any two electrodes, an alternating potential gradient exists in the intervening space. If these electrodes are flat parallel plates (or any other symmetrical form), and if the insulating material of known dielectric constant fills the volume between them, as is generally the case, the total capacitance can be easily calculated. From the capacitance, voltage, and frequency, the impressed volt-amperes can be determined, which, multiplied by power factor, gives the heat generated in watts. Equations (8) and (9) present these factors in a convenient form and can be used for estimating purposes.

$$C = \frac{2248AK}{10^{10}d} \text{ (for parallel plates)} \tag{8}$$

where C = capacitance in microfarads.
A = area of one electrode in square inches.
d = distance between electrodes in inches.
k = dielectric constant.

$$W = \frac{2\pi fCE^2(PF)}{10^6} \tag{9}$$

where W = power loss as heat in watts.
f = frequency in cycles per second.
C = capacitance in microfarads.
E = applied voltage rms.
PF = power factor at frequency f.

Precautions in Applying Equations. In applying these equations the following precautions should be observed:

1. The power factor of most materials varies with frequency and temperature. Thus the power factor used in eqs. (8) and (9) must be measured at or near the operating frequency and at a known temperature.

2. The dielectric constant changes relatively slowly with frequency and temperature and can be taken from published tables with fair accuracy.

Equations (8) and (9) hold only for uniform electric fields. Since the field at the edges of the electrodes is always distorted, a good approximation is possible only when the minimum dimension of the electrode is large compared to the distance between plates. Also, to reduce non-uniform heating at the edges, it is advantageous to use plates somewhat larger than the material to be heated.

4. Because of corona and arcing effects, the maximum voltage applied to the electrodes should not be greater than approximately 14 to 15 kv (rms) with 2 or 3 kv (rms) per inch of separation as the maximum for smaller spacings.

5. An analysis of eqs. (8) and (9) will, for some materials, indicate the desirability of very high frequency. The tubes and equipment commercially available at this time, however, limit the maximum frequencies obtainable to approximately 200 megacycles for power outputs up to 100 watts, 40 megacycles for power outputs up to 20 kw, and 13 megacycles for the higher power ratings.

At the higher frequencies the maximum electrode dimensions should be limited to at least one-eighth of a wavelength to avoid standing waves, which would result in non-uniform heating. If longer electrodes are required, special precautions must be taken to assure uniform heating. A convenient formula for determining wavelengths in air is given in eq. (10).

$$\lambda = \frac{300}{f} \text{ (in air)} \tag{10}$$

where λ = wavelength in meters.
f = frequency in megacycles.

6. The charge to be heated must be of uniform analysis throughout and must contact each plate. An air gap between the electrodes and the charge results in a series capacitor effect and introduces serious errors in the foregoing equations. Since an air gap is often necessary to prevent arcing at the electrodes, the capacitance and voltage used in eq. (2) must be corrected in the ratio of the capacitance of the air gap to the capacitance of the volume occupied by the materials to be heated. In other words, the capacitance and voltage used must be those applied to the material itself, not necessarily the electrode capacitance or voltage.

The dielectric constant for most materials falls in the range of 2 to 6 but may vary from 1 for gases to 1000 for some ceramics. The power factors usually lie between 0.02 and 0.07 but may be as low as 0.00015 (mica, polystyrene, steatite) or as high as 0.15 (asbestos). Gases and absolutely pure water have power factors which are essentially zero and thus cannot be heated.

EQUIPMENT USED. Since the lowest frequency that can feasibly be used for dielectric heating falls above 1,000,000 cycles because of the tremendous voltage gradient that would be required at lower frequencies, only vacuum-tube oscillators can be used for these applications. It should be obvious that the oscillator previously described is an ideal power source for dielectric heating. The capacitor in the resonant circuit can be replaced with external electrodes and non-conducting materials heated by the high-frequency voltages obtainable. Or, if the total voltages and capacitances suitable for use in the oscillator do not agree with those required for a specific application, the load circuit may be separately resonated and then inductively coupled to the oscillator circuit. This is by far the most generally used means, since the electrical characteristics of most dielectric heating loads rarely agree with those required for efficient operation of most oscillators.

DIELECTRIC HEATING APPLICATIONS. The field of application of dielectric heating is somewhat restricted because of the high initial cost of the equipment and the higher cost of electric power when compared to other forms of heat generation. Dielectric heating equipment costs more than $1000 per kilowatt in sizes on the order of 1 kw, ranging downward to approximately $500 per kilowatt for 100-kw units. Overall operating efficiencies rarely exceed 50 per cent.

These factors preclude the use of dielectric heating to remove large volumes of water or to heat large masses, which can be done at a slower rate, utilizing cheaper forms of heat. However, dielectric heating has found wide use in the preheating of plastic charges before molding, the fabrication of certain types of high-cost plywoods, and the curing of certain rubber products, as well as in other industries where the ability to generate heat throughout a mass results in economies or improved products which make the higher costs unimportant.

18. BIBLIOGRAPHY

C. R. Burch and N. R. Davis, *An Introduction to the Theory of Eddy Current Heating.* Ernest Benn, London (1928).

Induction Heating—Motor Generators

Chestnut, F. T., Induction Heating: A History of Its Development, *Iron Age*, Vol. 155, pp. 46–53 (1945).
Osborn, H. B., Jr., *The Tocco Process.* The Ohio Crankshaft Co., Cleveland.
Strickland, H. A., Jr., Induction Heating for Forging, *Metals and Alloys*, Vol. 21, pp. 719–723 (1945).
Vaughan, J. T., and J. W. Williamson, Design of Induction Heating Coils for Cylindrical Magnetic Loads, *AIEE Tech. Paper*, 46–124 (May 1946); Design of Induction Heating Coils for Cylindrical Non-magnetic Loads, *Trans. AIEE*, Vol. 64, pp. 587–592, (1945).

Induction Heating—Oscillators

Babat, George, Some Peculiarities of Heating Steel by Induction, *Applied Physics*, Vol. 15, pp. 835–839 (1944).
Babat, George, and Michel Losinsky, High-frequency Heating for the Surface Hardening of Steel, *Revue generale de l'electricite*, Vol. 44, No. 16, pp. 495–510 (1938).
Baker, R. M., Induction Heating of Moving Strip, *Trans. AIEE*, Vol. 64, pp. 184–189 (1945).
Brown, G. H., Efficiency of Induction Heating Coils, *Electronics*, Vol. 17, pp. 124–129, 382–385 (1944).
Curtis, F. W., *Induction Heating*, John Wiley (1944); Coil Design for Successful Induction Heating, *Amer. Machinist*, Vol. 87, pp. 83–86 (1943).
Jordan, J. P., Better Brazed and Soldered Joints Made Possible by Induction Heating, *Product Eng.*, Vol. 14, p. 102 (1943); The Theory and Practice of Industrial Electronic Heating, *Gen. Elec. Rev.*, Vol. 46, pp. 675–683 (1943).
Prince, D. C., and F. B. Vogdes, Vacuum Tubes as Oscillation Generators. General Electric Co. (1929).
Setapen, H. M., Design, Preparation, and Use of Silver Brazed Joints, *Steel Processing*, p. 568 (Sept. 1944).

Dielectric Heating

Bierwirth, R. A., and Cyril N. Hoyler, Radio-frequency Heating Applied to Wood Gluing, *Proc. IRE*, Vol. 31, pp. 529–537 (1943).
Gillespie, H. C., Operating Experience with H. F. Heating, *Electronic Industries*, Vol. 3, pp. 80–83 (1944).
Jordan, J. P., Radio Interference and Electronic Heaters, *Radio News*, Vol. 5, p. 14.
Meharg, V. E., Heatronic Molding, *Modern Plastics*, Vol. 20, pp. 87–90 (1943).
Scott, G. W., Jr., The Role of Frequency in Industrial Dielectric Heating, *Trans. AIEE*, Vol. 64, pp. 558–562 (1945).
Stephen, J. L., and H. J. Holmquest, Drying Lumber with High-frequency Electric Fields, *Wood Products*, p. 10 (Oct. 1936); p. 15 (Nov. 1936).

ELECTRIC RESISTANCE HEATING UNITS AND FURNACES

By Willard Roth

19. MATERIALS

RESISTORS. Suitable resistor materials, preferably having high resistivity, low temperature coefficient of resistance, and the ability to retain physical and chemical stability at operating temperatures are essential for electric resistance heating. In commercial practice the materials which best answer these requirements are nickel-chromium, nickel-chromium-iron, and iron-chromium-aluminum alloys in the metals, and recrystallized silicon carbide in the non-metals. Data on these and other resistance materials are given in Table 1.

Table 1. Resistance Materials

Approximate Composition	Max. Working Temp., °F		Resistivity, Ohms per Sq Mil Ft	
	In Air	In Reducing Atmosphere	70° F	1500° F
80% Ni, 20% Cr	2000	2150	510	540
34% Ni, 20% Cr, bal. Fe	1700	2150	472	585
13% Cr, 3.5% Al, bal. Fe	1500	Not used	534	662
Molybdenum	Not used	3500	21	105
Platinum	3000	3000	60	203
Recrystallized silicon carbide	2800	2600	7.2×10^6	4.6×10^6

THERMAL AND ELECTRICAL INSULATION. Supporting materials for resistors or heating elements must be refractory, have high dielectric strength and mechanical strength at working temperatures, and have low coefficient of thermal expansion. Thermal insulating materials need to have low thermal conductivity, low thermal capacity, and chemical and physical stability at working temperatures.

Among the materials which meet these requirements to a degree are mica, asbestos products, woven glass, glass and rock wool, the refractory oxides, such as silica, alumina, magnesia, zirconia, and beryllium oxide, and the silicates of aluminum, magnesium, and zirconium. The common fire clays are largely combinations of alumina and silica, with impurities.

SMALL HEATING UNITS. These may be classified as contact or cartridge heaters where the heater is in direct contact with the material to be heated; strip heaters, where heat transfer is by a combination of radiation and natural or forced convection; air heaters.

where heat transfer is effected largely by forced convection; and immersion heaters for liquids. There are countless variations of these four classifications, and it is not within the scope of this discussion to cover all of them. The following examples are typical.

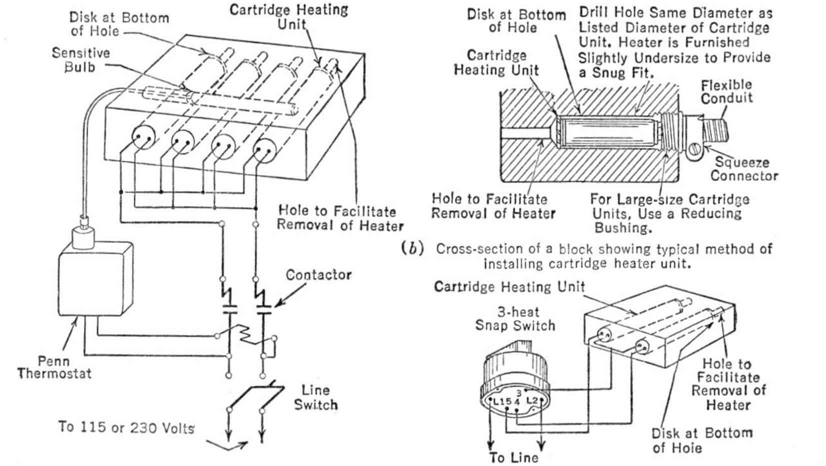

(*a*) Schematic arrangement of a section equipped with cartridge heaters and automatic temperature control.

(*b*) Cross-section of a block showing typical method of installing cartridge heater unit.

(*c*) Two cartridge heaters applied with connections for a 3-heater snap switch. (Diagram accompanies each switch because each is connected differently.)

Fig. 1. Cartridge Heaters

Cartridge Heaters. See Fig. 1. Usually cylindrical in shape, brass sheathed, fitting snugly into hole drilled into die or platten which is to be heated. Built for sheath temperatures up to 750 deg fahr. Typical sizes and ratings are as follows:

Diameter, Inches	Length, Inches	Watts	Volts
1/2	2 1/2 –8	75 to 150	115 or 230
5/8	2 3/8 –5 1/2	90 to 285	115 or 230
3/4	2 3/8 –6	200 to 350	115 or 230
1 5/16	3 5/8 –5 5/8	150 to 400	115 or 230
1 5/16	4 13/16–7 1/2	600 to 800	115 or 230

Strip Heaters. See Fig. 2. Resistors embedded in alumina, with steel or stainless steel sheath. For heating crane cabs, valve houses, pipe lines, process machinery, etc.

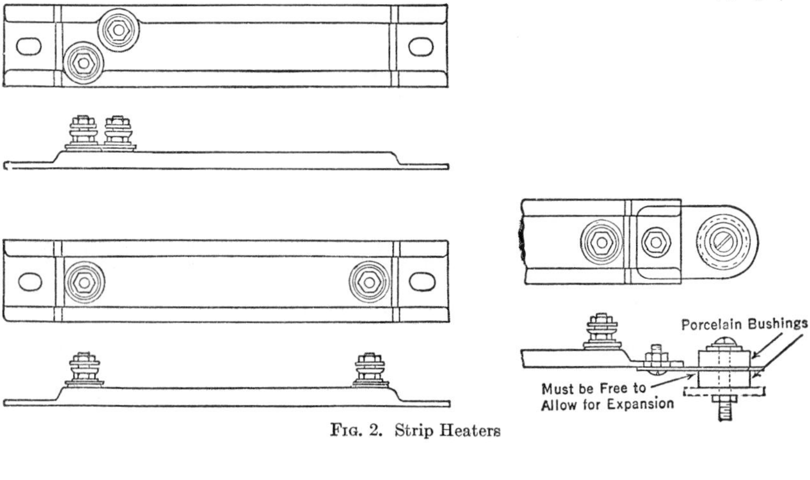

Fig. 2. Strip Heaters

Typical sizes and ratings are as follows:

Overall length, Inches	Width, Inches	Rating, Watts	
		750° Sheath	1200° Sheath
8	1 1/2	150	250
12	1 1/2	250	350
18	1 1/2	350	500
24	1 1/2	500	750
30 1/2	1 1/2	750	1000
36	1 1/2	1000	1500

Air Heaters. See Fig. 3. Used with fan to heat air. Resistor embedded in magnesium oxide, sheathed in finned tube. Useful for heating small enclosures, such as shop

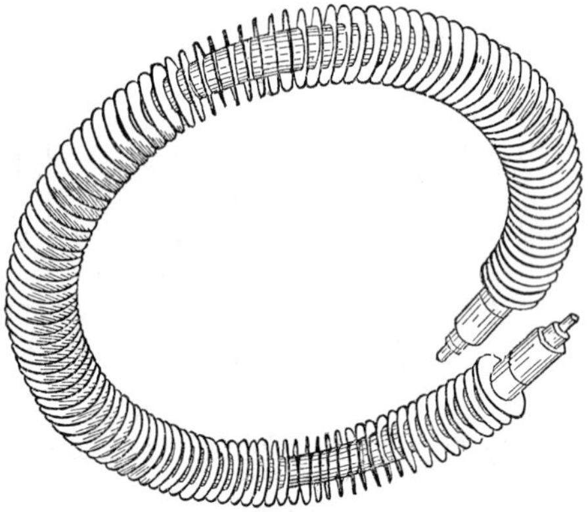

FIG. 3. Air Heaters

offices, guard houses, elevators, construction shacks. Typical performance data are as follows:

Rating, Watts	Volume Air Heated, Cubic Feet per Minute	Temperature Rise in Air, Degrees Fahrenheit	Air Velocity, Feet per Second
2000	120	70°	10
3000	212	50°	12
4000	212	65°	12

Units may used without heat on, for cooling ing in warm weather.

Immersion Heaters. See Fig. 4. Resistor embedded in compacted magnesium oxide and sheathed in copper, steel, or stainless steel tubing. Tubing is formed to suit application. For heating water, oils, and other liquids. Dissipation rates for water are approximately 35 watts per square inch of tube surface; for oil, 15 to 20 watts per square inch. Ratings for water heaters are from 0.60 to 10.0 kw; for oil, 1.0 to 8 kw.

20. INDUSTRIAL FURNACES

Industrial resistance furnaces have in recent years become precision metallurgical tools. Time-temperature cycles, both in heating and cooling, or quenching, may be automatically controlled. Atmospheres also are controlled, so that carbon content and surface finishes may remain unchanged in the heat-treating process. Decarburized skin, resulting from hot-rolling or forging processes, may be restored to the original carbon content. Surface oxides may be reduced by suitable atmospheres. This last is essential, for instance, in copper brazing or in sintering powdered metals.

Furnace types have been evolved for various methods of material handling. The box furnace, for instance, lends itself to toolroom heat treatments, where production is small in volume but varied. On the other hand, furnaces for mass production are likely to be

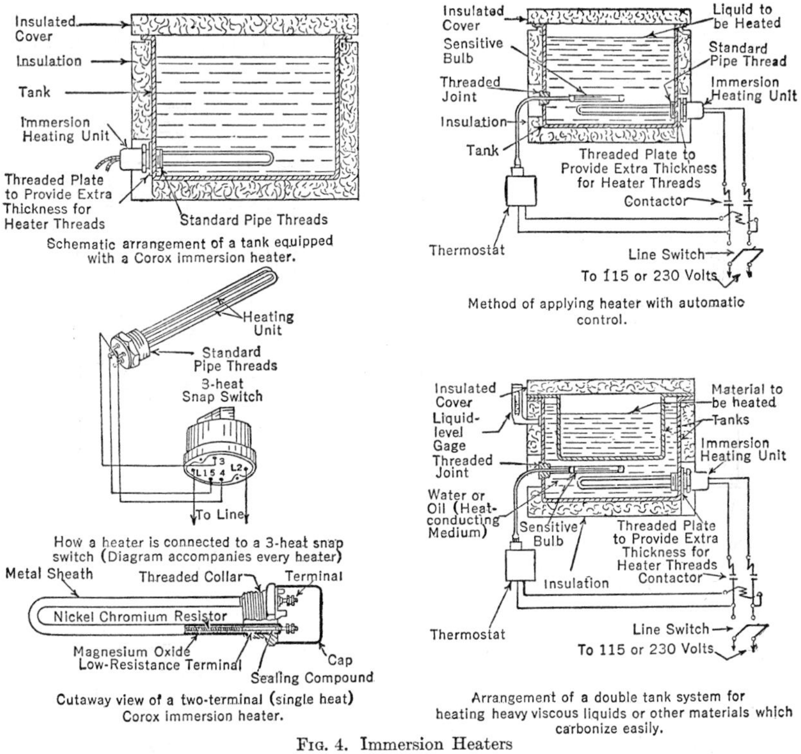

Schematic arrangement of a tank equipped with a Corox immersion heater.

Method of applying heater with automatic control.

How a heater is connected to a 3-heat snap switch (Diagram accompanies every heater)

Cutaway view of a two-terminal (single heat) Corox immersion heater.

Arrangement of a double tank system for heating heavy viscous liquids or other materials which carbonize easily.

FIG. 4. Immersion Heaters

continuous, with a minimum of manual handling. An example is the shaker hearth furnace, particularly adapted to the bright hardening of small parts, such as knife blades, springs, and bearing races. Another example is the hydraulic pusher hardening furnace, particularly adapted to bright hardening of parts too large to be tumbled into the quench, such as shafts, gears, and large bearing races.

Table 2 gives an abbreviated classification of furnace types and applications.

Table 2. Furnace Types

Type	Method of Handling	Type of Work	Suitable for Processes
Box	Batch	Small to medium	a, n, h, t, c
Box with cooling chamber	Batch	Small to medium	ba, bn, bh, s, fb
Pit	Batch	Long parts suspended, small parts in baskets	a, n, h, t, c, n, st
Car	Batch	Large and heavy	a, n, h
Bath	Batch	Small to medium	h, t, c, cn, st
Bell	Batch	Medium to large	ba, n
Rotary hearth	Continuous	Small to medium	h, t, n, a
Cast link-belt conveyor	Continuous	Small to medium	bh
Shaker hearth	Continuous	Small	bh
Mesh-belt conveyor	Continuous	Light-weight	ba, bn, fb, s
Roller hearth	Continuous	Medium to heavy	ba, bn, fb, m, s
Hydraulic pusher	Continuous	Heavy	ba, bn, fb, m, bh, st

a = Annealing.
ba = Bright annealing.
n = Normalizing.
Bn = Bright normalizing.
h = Hardening.
bh = Bright hardening.
t = Tempering.
c = Carburizing.
n = Nitriding.
cn = Cyaniding.
fb = Furnace brazing.
s = Sintering.
st = Solution treatment.
m = Malleablizing.

BOX-TYPE FURNACE. See Fig. 5. Temperature range 1200–2000 deg. fahr. Has electric heating elements on sides and under nickel-chromium alloy hearth plate. Temperature controlled by thermocouple in top. Principal uses are hardening, normalizing, and tempering, usually with controlled atmosphere. Performance data on typical sizes are as follows:

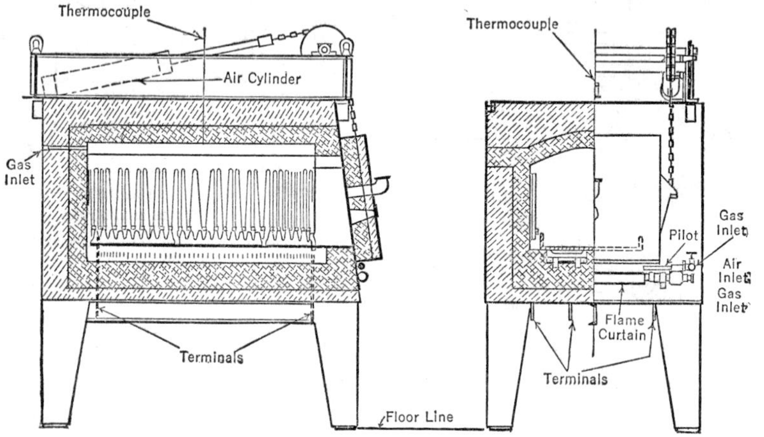

Fig. 5. Box-type Furnace

Working Dimensions, Inches			Rating, Kilowatts	Approx. Hardening Capacity, Lb per Hr at 1550° F	Standby Losses at 1550° F, Kilowatts
Width	Depth	Height			
8	18	5	7.5	25– 35	2.3
12	27	8	15.0	56– 79	3.9
15	30	10	20.0	78–110	4.5
18	36	13	27.0	112–157	5.0
24	54	18	45.0	224–314	7.3
36	72	24	75.0	448–628	10.7

Box furnaces with a temperature range of 1850–2500 deg, suitable for heat treatment of high-speed steels, utilize silicon carbide resistors and hearth plates.

BOX-TYPE FURNACE WITH COOLING CHAMBER. See Fig. 6. Temperature range 1200–2100 deg fahr. Has metallic heating elements on sides only, with silicon carbide hearth. Heating chamber is separated from cooling chamber by insulated door.

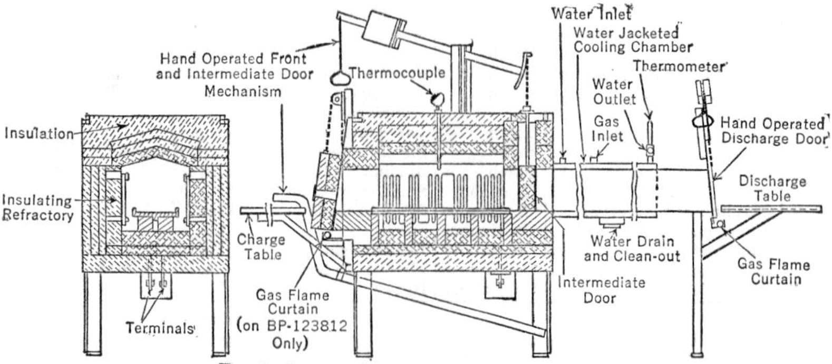

Fig. 6. Box-type Furnace with Cooling Chamber

Best suited to bright annealing, furnace brazing, and sintering, where production is small and varied, and parts are light enough to be handled on light-weight trays, which are pushed through manually. Heating chamber has room for one tray load, and cooling chamber has room for four. Typical sizes and data are as follows:

Tray Size, Inches		Loading Height, Inches	Rating, Kilowatts	Approx. Brazing Capacity, Including Trays, Lb per Hr at 2050° F	Standby Losses at 2050° F, Kilowatts
Width	Length				
6 1/2	20 1/2	6	18	35– 75	7.0
10 1/2	31 1/2	8	30	75–150	10.0
16 1/2	29 1/2	10	50	125–250	14.0

PIT-TYPE FURNACES. See Fig. 7. Temperature range 250–1400 deg fahr. Utilizes forced convection, with circulating fan in bottom and baffle to direct air over heating elements. Work is loaded into cylindrical container, which has grille bottom to permit air

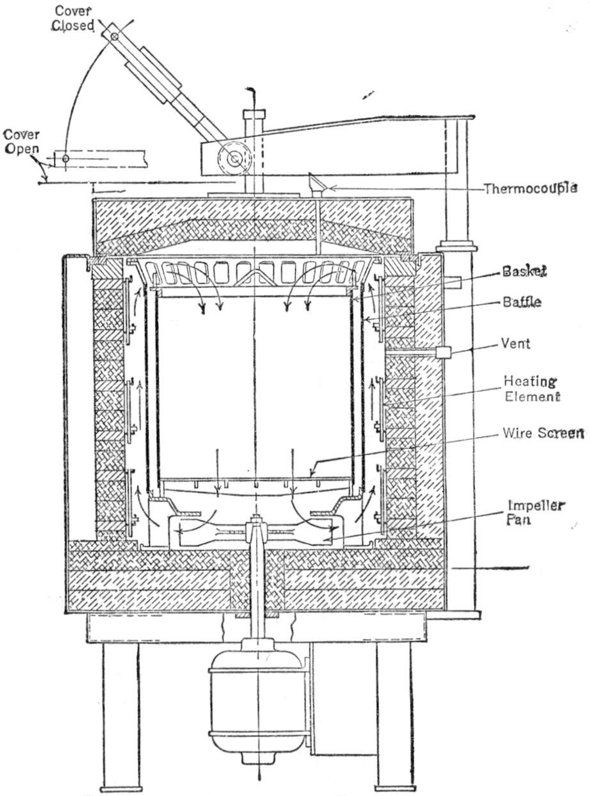

Fig. 7. Pit-type Furnace

passage. Particularly suited to drawing, stress relief, and the heat treatment of aluminum and magnesium alloys. Typical sizes and ratings are as follows:

Working Space, Inches		Rating, Kilowatts	Fan, Horsepower	Standby Losses at 1200° F, Kilowatts
Diameter	Depth			
15	24	25	2	3.8
20	24	45	5	4.5
20	36	45	5	5.0
25	24	60	7.5	5.0
25	36	60	7.5	5.5

Pit-type furnaces may be used also for heating long pieces, such as shafts, rolls, and gun barrels, at temperatures up to 2100 deg fahr. At the higher temperatures, heating is by radiation rather than forced convection. Hence the baffle and fan are eliminated.

Pit furnaces may also be used with protective atmospheres in scale-free drawing, dry-cyaniding, carburizing, and nitriding.

BELL-TYPE FURNACE. See Fig. 8. In effect, an inversion of the pit-type furnace. Utilizes both radiation and forced convection in heating. Work is loaded on base and covered by sheet-metal hood, which seals in the protective atmosphere. Furnace with electric heating elements mounted inside is set down over hood for heating, then removed to another base at the end of heating, leaving the charge to cool in the protective atmosphere under the hood. Thus the furnace remains hot, conserving its stored energy. Particularly suited to bright annealing of copper and steel strip and wire in coils or on spools and to nitriding. Usual temperatures are 1200 to 1500 deg fahr for steel annealing, 400 to 900 deg fahr for copper annealing, and 900 to 1000 deg fahr for nitriding.

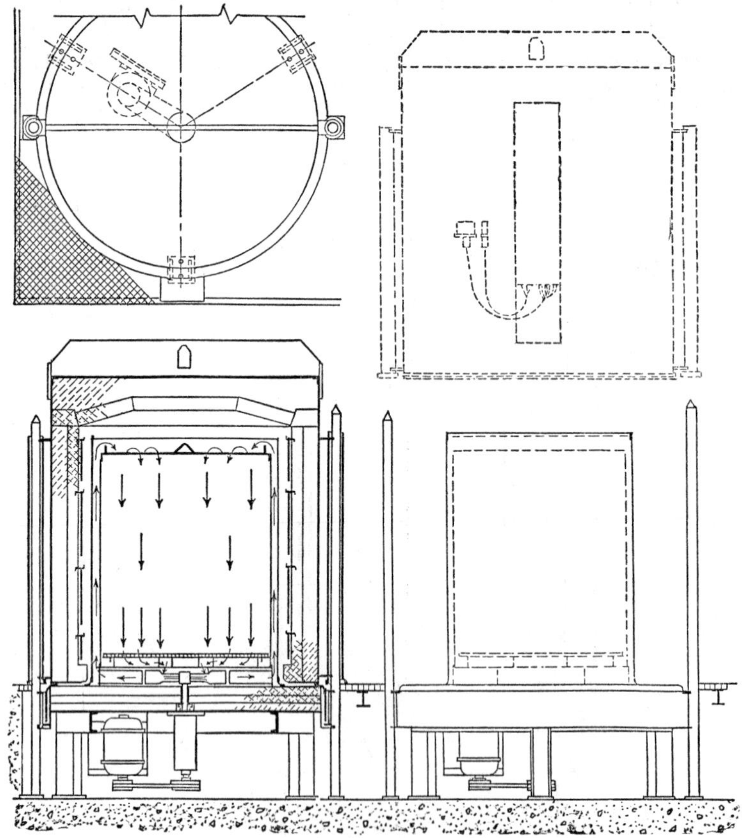

Fig. 8. Bell-type Furnace

Typical sizes and ratings for copper annealing are as follows:

Working Space, Inches		Rating, Kilowatts	Fan, Horsepower
Diameter	Height		
36	42	60.0	5.0
36	60	75	5.0
44	48	75	7.5
55	62	105	7.5

Energy consumption ranges from 150 to 200 kwhr per ton of steel, and 50 to 80 kwhr per ton of copper annealed.

ROTARY HEARTH FURNACE. See Fig. 9. Particularly adapted to the hardening and normalizing of steels when manual charging and discharging are desirable. The annular hearth rotates on a central shaft; the operator loads work into the charging door and it is heated as it progresses around, counterclockwise, to the discharge door. The two doors are adjacent, so that one operator usually loads and unloads. Heating is by radiation from heating elements on the inner walls. Liquid seals between hearth structure and furnace make the unit gastight, suitable for atmosphere control.

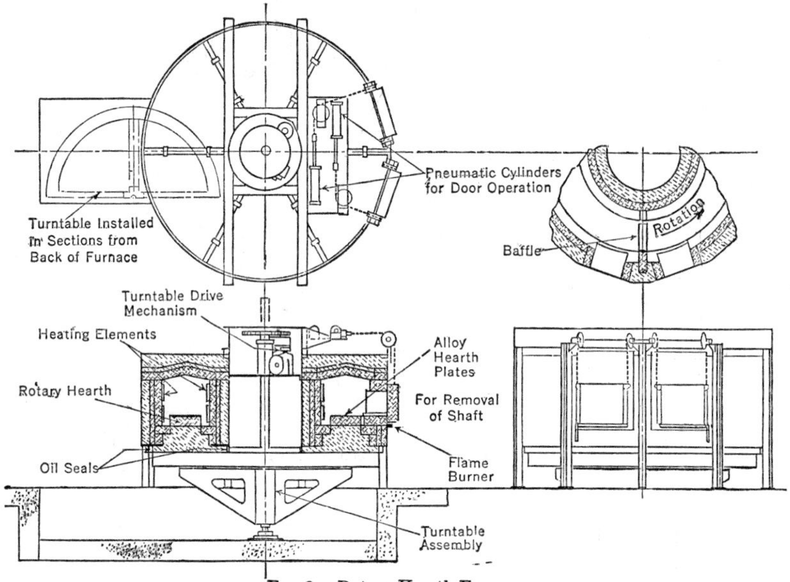

Fig. 9. Rotary Hearth Furnace

Typical size and performance data are as follows:

Working Space		Height, Inches	Rating, Kilowatts	Standby Losses, at 1500° F, Kilowatts	Approx. Production Capacity, Lb per Hr
Width, Inches	Mean Diameter, Feet and Inches				
12	5–0	12	70	20	200– 400
18	5–6	12	95	25	300– 600
18	8–0	14	175	33	500–1000
24	9–0	18	213	45	750–1500

CAST LINK-BELT CONVEYOR FURNACE. See Fig. 10. Adapted to continuous hardening of steels. Parts are loaded onto shaker charger, which deposits them on furnace belt. In turn the belt carries them through the furnace, dropping them through the quench chute onto the quench conveyor. Generally used for small parts, such as bolts, screws, small tools, bearing races. These furnaces are not suited to handling parts where nicking due to falling into the quench is objectionable, or where distortion results from tumbling into the quench.

Heating is by radiation from heating elements in the roof and under the cast link belt. Quench oil temperature is maintained by circulating the oil through a cooler outside the tank, with a thermostat in the cooling water supply. Speeds of the quench conveyor and furnace conveyor are adjusted through the variable speed drive.

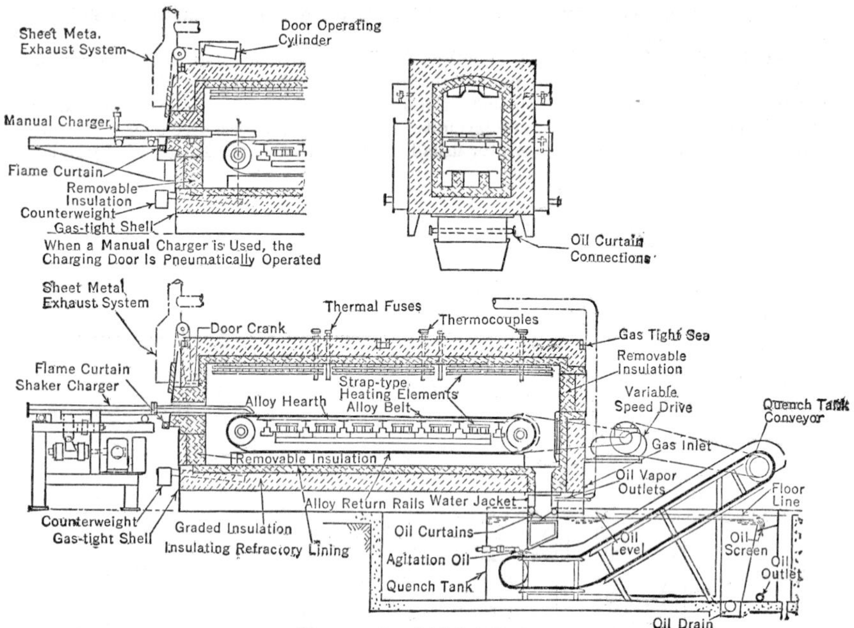

FIG. 10. Cast Link-belt Furnace

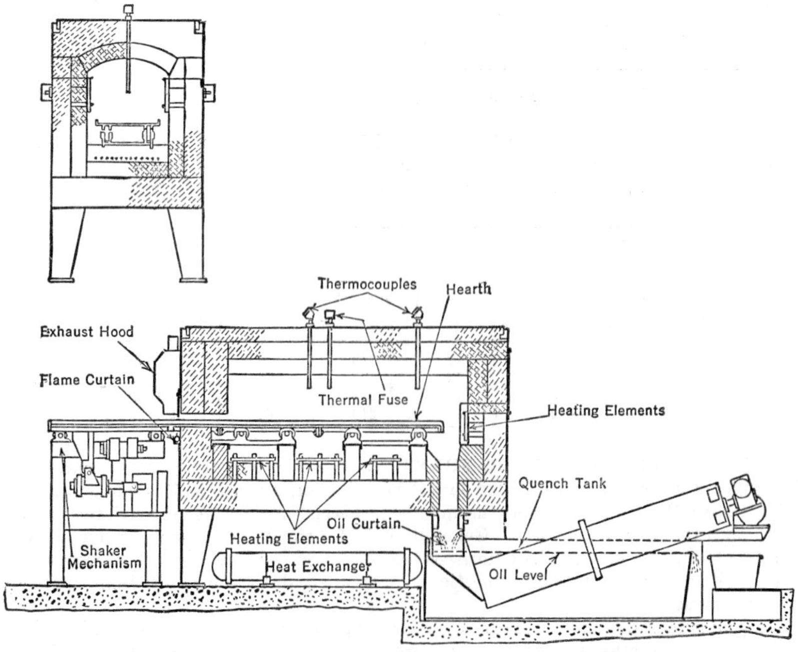

FIG. 11. Shaker Hearth Furnace

Typical sizes and performance data are as follows:

Belt Width, Inches	Heating Chamber Length (Effective), Feet and Inches	Rating Kilowatts	Standby Losses at 1550° F, Kilowatts	Approx. Productive Capacity Lb per Hr
16 7/8	8–6	60	19	250– 400
22 3/4	12–6	130	29	500– 800
34 3/8	12–6	176	38	750–1200

SHAKER HEARTH FURNACE. See Fig. 11. For continuous hardening of small, light-weight steel parts, such as knife blades, springs, small screws, which would catch in the joints of a link-belt conveyor furnace. Hearth is a smooth plate with upturned edges. It is given a reciprocating motion with abrupt deceleration on the forward stroke, so that work slides forward a fraction of an inch. Work falls off the end of the hearth into the quench and onto the quench conveyor, which is either the rotary drum type, as shown, or a close mesh-woven wire belt.

Typical sizes and performance data are as follows:

Hearth Width, Inches	Effective Length, Inches	Rating, Kilowatts	Standby Losses at 1550° F, Kilowatts	Approx. Productive Capacity, Lb per Hr
12	48	24	9.0	60–120
18	60	38	11.0	110–225
24	78	60	14.0	200–400

HYDRAULIC PUSHER FURNACE. See Fig. 12. For hardening steel parts, such as shafts and small ring gears, which must be quenched in a given position to avoid distortion or which must not be nicked by dropping into quench. Parts generally are larger than can be handled by the cast-link belt conveyor furnace and are loaded onto cast grid trays.

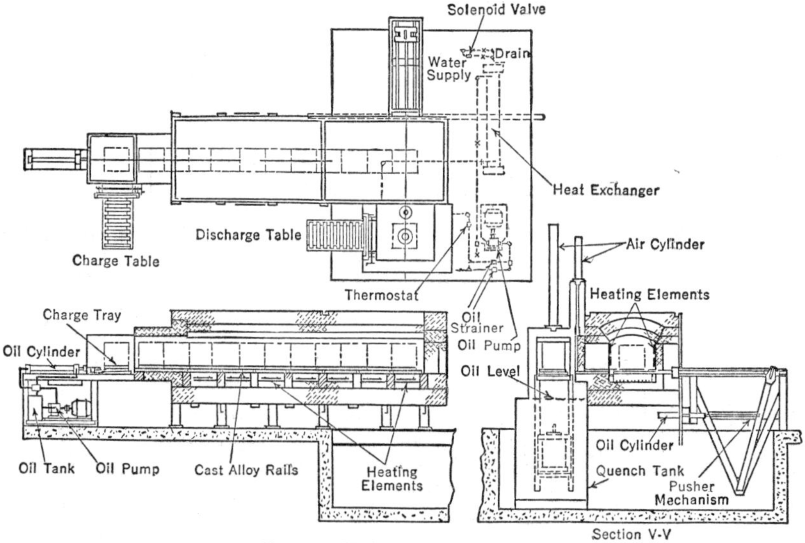

FIG. 12. Hydraulic Pusher Furnace

Loaded trays are pushed in a line through furnace, then pushed crosswise onto quenching-chamber platform. This platform and its load immediately are lowered into quenching liquid by an air cylinder which keeps them oscillating until the quench is completed. The whole sequence is automatically timed and executed; the operator merely charges and removes the loaded trays at proper signals.

Typical sizes and data are as follows:

Tray Size, Inches	Load Height, Inches	Number of Tray Positions	Rating, Kilowatts	Standby Loss at 1500° F, Kilowatts	Approx. Productive Capacity, Lb per Hr
10 × 10	16	9	55	14	100–200
15 × 15	16	9	96	17	240–480
20 × 20	20	10	140	23	460–920

Hydraulic pusher furnaces are used also for large-scale bright annealing of copper and steel products and in brazing heavy assemblies. Such furnaces have water-jacketed cooling chambers following the heating chamber, all in line.

MESH-BELT CONVEYOR FURNACES. See Fig. 13. Particularly suited to annealing, brazing, or sintering of light parts, which are carried through furnace and cooling

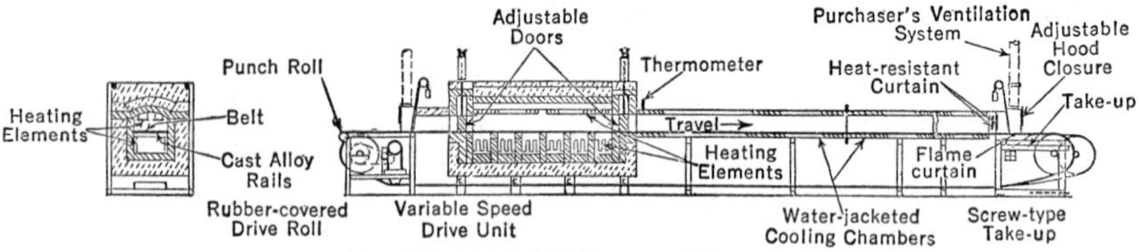

FIG. 13. Mesh-belt Conveyor Furnaces

chamber on an alloy wire-mesh belt. Heating is by radiation from heating elements in roof and on sides below belt. Variable speed drive and automatic takeup of belt stretch are provided.

Sizes and data on performance are as follows:

Belt Width, Inches	Heating-chamber Length, Feet and Inches	Rating, Kilowatts	Standby Losses at 2050° F, Kilowatts	Approx. Productive Capacity at 2050° F, Lb per Hr
8	4–6	30	13	50–100
12	8–0	60	22	100–200
20	8–0	100	24	200–400

ROLLER HEARTH FURNACES. See Fig. 14. For annealing, brazing, normalizing, and sintering of parts too heavy for the mesh-belt conveyor or for continuous annealing or normalizing of steel strip. Parts are carried through furnace on driven alloy rolls;

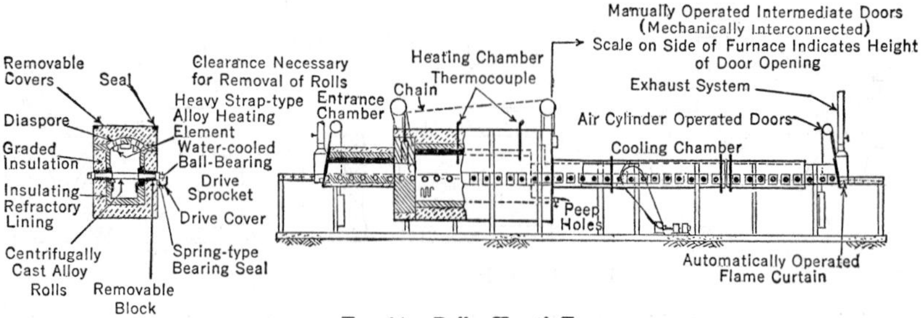

FIG. 14. Roller Hearth Furnaces

trays are used when necessary. End doors are normally closed, with automatic charge and discharge of trays. Variable speed drive is employed. Heating elements are in roof and on sides below rolls.

Typical sizes and performance data are as follows:

Load Width, Inches	Heating-chamber Length, Feet and Inches	Rating, Kilowatts	Standby Losses at 2050° F, Kilowatts	Approximate Productive Capacity at 2050° F, Lb per Hr
20	8–0	100	30	240– 480
26	9–8	185	35	460– 920
26	13–0	250	45	640–1280
26	17–2	325	55	920–1840

21. CONTROLLED-ATMOSPHERE GENERATORS

The majority of electric resistance furnaces now use controlled atmospheres for prevention of oxidation and, in the case of carbon steels, for prevention of decarburization. Moreover, such processes as gas carburizing, nitriding, and dry cyaniding require specific atmospheres peculiar to the process. Copper brazing and sintering depend on reduction of oxide films by the furnace atmosphere for the process.

Among the readily available gases, air, water vapor, and carbon dioxide are oxidizing and are decarburizing in their effect on steels. Water vapor and carbon dioxide, however, will not oxidize copper. Hydrogen, and carbon monoxide, on the other hand, are reducing to most metals.

The most commonly used atmospheres are produced through partial combustion of fuel gases and hence contain all the above component gases, both oxidizing and reducing, as well as some residual hydrocarbons, which are carburizing. Figure 15 illustrates what

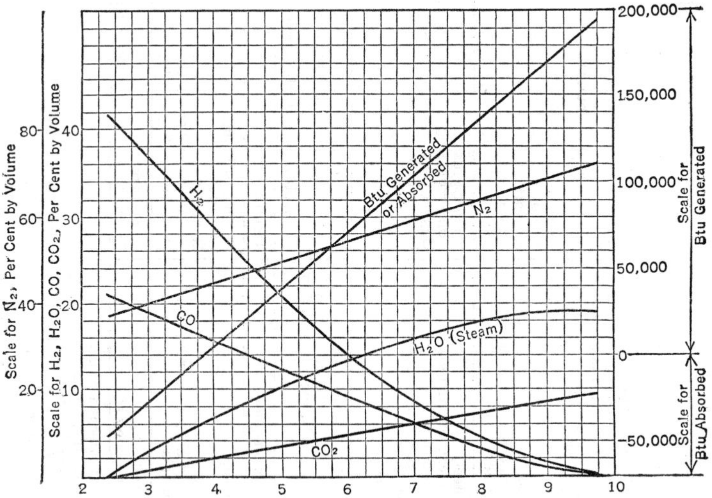

FIG. 15. Combustion of Methane

happens when methane, CH_4, the chief constituent of natural gas, is partly burned with various proportions of air. For instance, with 6 parts of air to 1 part of methane, the resultant gas contains approximately 4.5 per cent CO_2, 9.2 per cent CO, 14.0 per cent H_2, and the rest N_2 and water vapor, H_2O. If the water vapor is removed by a surface condenser to a dew point of 60 deg fahr or below, this gas is reducing to steels and to most other metals, chromium, zinc, and magnesium excepted. It is, however, decarburizing to the higher-carbon steels. If, however, the air to gas ratio is 2.70 to 1, then the resultant gas contains approximately 0.1 per cent CO_2, 20 per cent CO, and 40 per cent H_2, with moisture content equivalent to a dew point of approximately 15 deg fahr. This gas is carburizing to medium-carbon steels and approximately in equilibrium with the higher-carbon steels. It can, therefore, be used as a furnace atmosphere in hardening carbon

Table 3. Atmospheres Suitable for Heat Treatment of Different Metals

Heat Treatment	Metal Processed	Process	Temperature Range, °F	Time Cycle ("Long" if over 2 hr)	Required Surface	Recommended Atmosphere
Annealing	Low-carbon steels	Anneal	1200–1350	Long	Bright	A
	Medium- and high-carbon steels	Anneal (no decarburization)	1200–1450	Long	Bright	C or D *
	Alloy steels, medium and high carbon	Anneal (no decarburization)	1300–1600	Long	Bright or clean	C or D
	High-speed tool steels, including molybdenum, high speeds	Anneal (no decarburization)	1400–1600	Long	Bright or clean	C or D
	Stainless steels, chromium and nickel-chromium	Anneal	1800–2100	Long or short	Bright	F
	Copper	Anneal	400–1200	Long or short	Bright	B
	Various brasses	Anneal	800–1350	Long or short	Clean	B
	Copper-nickel alloys	Anneal	800–1400	Long or short	Bright	A
	Silicon-copper alloys	Anneal	1200–1400	Long or short	Bright	D
Brazing or soldering operations	Low-carbon steels	Copper brazing	2050	Short	Bright	A
	Medium-, high-carbon, and alloy steels	Copper brazing (no decarburization)	2050	Short	Bright	E
	Medium-, high-carbon and alloy steels	Copper brazing (no decarburization)	2050	Long	Bright	C
	High-carbon, high-chromium steels	Copper brazing (no decarburization)	2050	Short	Bright	F
	Stainless steels	Copper brazing	2050	Short	Bright	F
	Copper or brass	Phos-copper brazing or silver soldering	1200–1600	Short	Bright	B
Hardening	Medium- and high-carbon steels	Hardening (no decarburization)	1400–1800	Short	Bright or clean	E
	Alloy steels, medium and high carbon	Hardening (no decarburization)	1400–1800	Short	Bright or clean	E
	High-speed tool steels, including molybdenum	Hardening (no decarburization)	1800–2400	Short	Bright or clean	E
Tempering	All classes of ferrous metals	Tempering or drawing	400–1200	Short	Bright or clean	A or B *
Gas carburizing	Carburizing steels	Gas carburizing	1400–1800	Long	Clean	E (as a carrier)
Sintering of metals	Low-carbon ferrous metals	Reduction and sintering	1800–2050	Short	Bright or clean	E
	High-carbon and alloy ferrous metals	Reduction and sintering	1800–2050	Short	Bright or clean	E
	Nonferrous metals	Reduction and sintering	1400–1800	Short	Bright or clean	E
Normalizing	Low-carbon steels	Normalizing	1600–1850	Short	Bright or clean	A
	High-carbon and alloy steels	Normalizing (no decarburization)	1500–2000	Short	Bright or clean	C or D

A = Partly burned fuel gas at 6.0–1 air-gas ratio.
B = Partly burned fuel gas at 9.5–1 air-gas ratio.
C = Gas A with CO_2 and H_2O removed.
D = Gas B with CO_2 and H_2O removed.
E = Partly burned fuel gas at 2.75–1 air-gas ratio.
F = Dissociated ammonia.

Ratios are for 1000 Btu natural gas. For other fuel gases the proper ratios will be roughly proportional to the Btu content; i.e., 500-Btu fuel gases will require air-gas ratios equal to 40 to 50 per cent of the above values.

* Whether rich or lean depends upon individual applications.

steels or as a carrier gas in carburizing steels. It is, however, an active gas, and is not suitable for long-cycle annealing.

To obtain a comparatively neutral atmosphere, suitable for annealing without oxidation or decarburization, combustion is carried out at air-gas ratios of from 6.5–1 to 9.0–1; then the carbon dioxide is removed by absorption in the chemical monoethanolamine and the gas is dried to a dew point of −50 deg fahr. This gas, for the 9.0–1 ratio, will contain approximately 3 per cent of combustibles ($H_2 + CO$), and the balance is nitrogen. It is nearly inert and suitable for annealing all ranges of carbon steels without decarburization.

Another source of furnace atmospheres is dissociation of ammonia to produce 75 per cent H_2 and 25 per cent N_2. This gas is highly reducing and, when dry, is not decarburizing. It may be used to bright-anneal stainless steels and brasses, again provided it is kept dry. When mixed with small percentage of water vapor, it is strongly decarburizing and is therefore used in the annealing of some magnetic materials where decarburization is desirable.

Table 3 shows a classification of the various atmospheres applicable to specific processes.

22. TEMPERATURE CONTROL

The most common method of control is "on-off"; i.e., power to the heating elements is turned on or off in response to electrical contacts made in the control pyrometer, which is in turn responsive to the thermocouple in the furnace heating chamber. Thus the temperature varies a few degrees above and below the control setting as the power goes on and off. The amplitude of the variation depends on the sensitivity of the pyrometer, the thermocouple, and the furnace thermal characteristics. Generally it is within 5 or 10 deg fahr. This variation can be reduced when desirable by means of an interrupter, which turns power off a preset fraction of each minute, independent of the pyrometer control. This has the effect of decreasing input and reduces overshooting of temperature after the control has shut off.

Where very close control is desired, a saturable reactor is placed in series with the heating elements. The d-c excitation of the reactor is varied automatically by the control pyrometer. This has the effect of adjusting the voltage and hence the power input on the heating elements to just equal the demand on the furnace. By this means temperature can be held almost exactly at control point.

23. BIBLIOGRAPHY

Anderson, R. H., D. C. Dunn, and A. L. Roberts, Permeable Refractories for Furnace Construction, *Eng. Digest* (Amer. ed.), Vol. 2, p. 410 (1945).

Koebel, N. K., *Industrial Controlled Atmospheres.* Lindberg Engineering Company.

Peck, C. E., Application of Controlled Atmospheres to Heat Treating, *Metal Progress* (June 1942).

Slowter, E. E., and B. W. Gonser, Gases for Controlled Atmospheres, *Metals and Alloys* (June 1937 and Feb., Mar., July 1938).

Trinks, W., *Industrial Furnaces.* John Wiley (Vol. I, 1934; Vol. II, 1942).

ELECTRICAL PRECIPITATION OF SUSPENDED PARTICLES

By L. M. Roberts

The smoke, dust, fume, and liquids arising from metallurgical, cement, power, petroleum, chemical, and other industries present a problem both from the standpoint of civic nuisance and from that of economic conservation. The removal of such suspended matter from gas or air streams by means of electric discharges was commercially demonstrated by F. G. Cottrell in 1906. His process is called the Cottrell electrical precipitation process or, more simply, the Cottrell process. Engineering and technical developments have established this process as an important tool of many varied industries.

24. OPERATING PRINCIPLES

Electrical precipitation is based, fundamentally, on the physical fact that an electrically charged body experiences a force in the presence of an electrical field.

In the operation of an electrical precipitator, a high unidirectional difference of potential is maintained, and a corona current flow is set up between two spaced sets of electrodes between which the gas to be cleaned flows. One set of these electrodes, known as the discharge electrodes, is usually of small cross-sectional area to facilitate corona discharge, whereas the opposing or collecting electrodes have extended surfaces and are non-discharging. The high voltage impressed across the electrodes produces a unipolar corona discharge at the discharge electrode and a large supply of gas ions of the same polarity as the discharge electrode. The ions thus formed attach themselves to the suspended particles, and the resultant charged particles are attracted to the collecting electrode, which is of opposite polarity. The concentration of ions present between the electrodes normally is much greater than the concentration of suspended particles, so that the size, weight, or number of particles in most cases has little effect on the charging and precipitation of such particles.

In practice, it has been found that best results are secured by maintaining a negative polarity on the discharge electrodes. This condition is related to the spark-over voltage, which is altered when the distance between electrodes is reduced. In a point-plane discharge, for example, at sparking distances less than 2.3 mm this sparking voltage is less for the point negative than for the point positive, but for distances greater than 2.3 mm, the sparking voltage is greater for the point negative than for the point positive. In precipitators the distance between electrodes is greater than 2.3 mm, and the negative discharge electrode is preferred because of the higher operating voltage possible. Also, with negative polarity, the corona discharge points are more numerous and much more evenly spaced.

In the electrical precipitation process the following steps are necessary:

1. Charging the particles by means of gaseous ions or electrons.
2. Precipitating or transporting the charged particles through the gas to the collecting electrode by the force exerted on the charged particles by the electric field.
3. Discharging the charged particles.
4. Removing the precipitated material from the electrode to a suitable receptacle, such as a hopper.

25. EQUIPMENT

In order to fulfill the foregoing requirements a precipitation installation must comprise two groups of equipment: (1) the electrical equipment for generation of the proper electrical energy and (2) the precipitator in which the gases are treated.

ELECTRICAL EQUIPMENT. The electrical equipment consists primarily of a high-tension transformer, a rectifier for converting the high-voltage alternating current to unidirectional current, and suitable control equipment.

Rectifying Equipment. There are two types of this equipment in general use in America, namely: (1) the synchronous mechanical rectifier, which is the simplest mechanical device for converting alternating current to unidirectional current, and (2) the electronic tube rectifier.

The mechanical rectifier for full-wave rectification is in effect a synchronously driven switch comprising rotor and stator elements. Four metal shoes spaced 90 deg apart on a circle are mounted on fixed insulating arms of laminated plastic material and attached to the motor frame. Usually the top shoe is connected to the precipitator; the bottom shoe is grounded; and each side shoe is connected to one secondary terminal of the transformer. Rotating within these shoes with a clearance of $^1/_8$ to $^1/_4$ in. and mounted on the motor shaft is a rotor with four tips at 90-deg intervals mounted on plastic arms or a plastic disk. Two adjacent tips of the rotor are electrically connected. The rotor advances one-half turn per complete cycle of alternating voltage or, for 60-cycle current, runs at a rate of 1800 rpm.

Where half-wave rectification is used, a second rotor and stator mounted on the same motor shaft act as a synchronous commutating switch in the high-tension circuit.

The motor for driving the mechanical rectifier is of the shaded pole induction type, which starts as an induction motor but runs as a synchronous motor at full speed. The motor may be for 220 or 440 volts, two or three phase, 25 to 60 cycles.

A newer development in rectifiers, as applied to precipitation, is the electronic tube type. In this type a "hot cathode" tube is generally used. For full-wave rectification, two or four tubes per transformer are required. For half-wave rectification only one tube is required, but two or four tubes may be used to supply two commutated half-waves to two separate precipitator sections.

Transformer. The transformers may vary in capacity from 2 $^1/_2$ to 25 kva, depending on the requirements of the precipitator installation. Standard designs call for a single-phase 25- or 60-cycle, 200-, 400-, 500-volt primary, 35- to 75-kv secondary, with a number of primary taps to afford a range of secondary voltages. The coils are immersed in oil or non-inflammable liquids, such as Inerteen or Pyranol. The transformers are internally surge protected. Radio interference coils are provided to prevent interference with radio reception.

Switchboard. The switchboard or control panel carries the switches, meters, and control instruments for the rectifier and the low-tension circuit of the transformer.

THE PRECIPITATOR. The precipitator in which the gases are treated consists of a shell fitted with suitable inlet and outlet gas connections and equipped with a hopper or some other type of precipitate-collecting chamber. Means are provided for electrically insulating the discharge electrode support frame from the collecting electrodes.

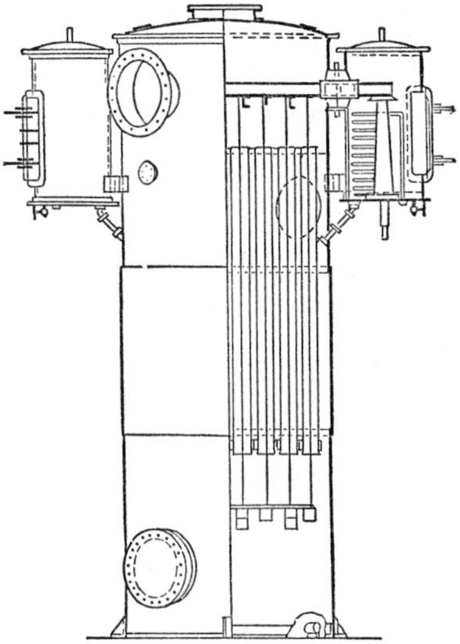

Fig. 1. Pipe-type Precipitator

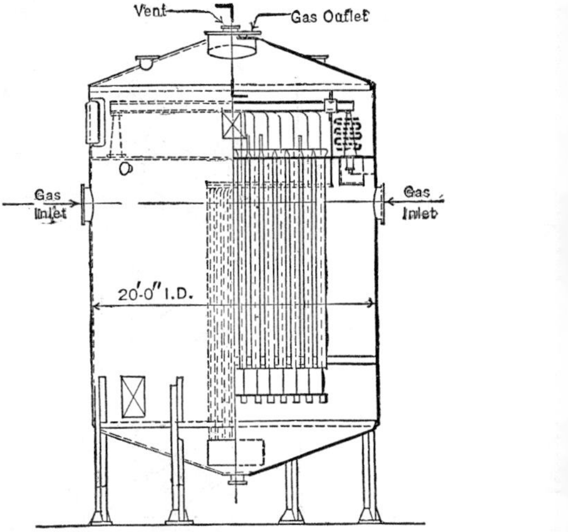

Fig. 2. Precipitator with Wetting Means to Remove Precipitate

Precipitators can be classified under two main headings, namely, the pipe type and the plate type. The pipe type is usually used for liquids or free-running precipitates and volatilized fumes, whereas the plate type is more generally used for dusts.

Pipe-type Precipitators. The pipe-type precipitator usually consists of a number of pipes nested in a header plate within a shell or between headers, with a discharge electrode suspended axially in the center of each pipe. A typical pipe-type precipitator is shown in Fig. 1. The collecting tubes may vary from 4 1/2 to 12 in. in diameter and from 4 to 15 ft long. The maximum size of a pipe precipitator is limited by mechanical and structural considerations, and for larger volumes of gas several units may be required. The materials used for construction are determined by the nature of the gas treated and the suspended matter removed. Non-corroding materials are used where necessary.

Liquid suspended matter precipitated from the gas in passing through the pipes is collected on the interior walls and settles in a lower header,

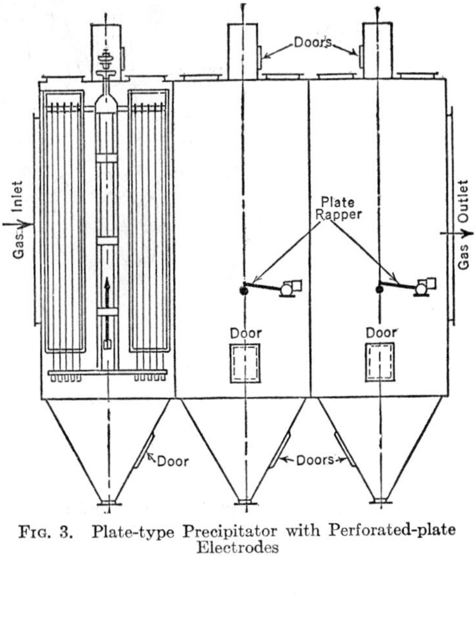

Fig. 3. Plate-type Precipitator with Perforated-plate Electrodes

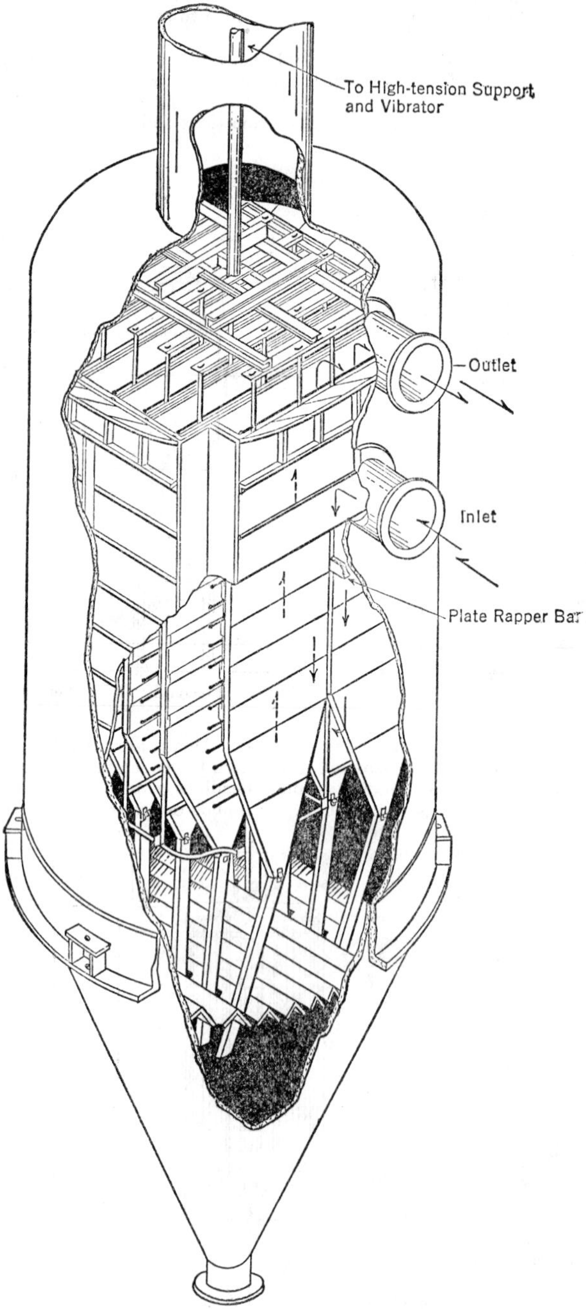

FIG. 4. Vertical-flow Duct-type Precipitator

from which it is removed through drains or hoppers. Where the precipitated material is not free flowing, rappers or shakers are provided to remove it to the hoppers. In one type the pipes are continually wetted by a water film which facilitates the removal of the precipitate. A diagram of this type of precipitator is shown in Fig. 2.

Usually, in commercial practice, the gas passes upward through the pipes; in special cases, however, the precipitators are operated "down-flow."

Plate-type Precipitators. The plate- or duct-type precipitator is made up primarily of a series of parallel collecting plates encased in a shell, thereby forming ducts or gas passages through which the gas passes. A series of discharge electrodes is suspended in the center of these ducts from an insulated framework. The discharge electrodes may be suspended either vertically or horizontally.

The plate-type precipitator may be further subdivided into groups, dependent on the type of collecting electrodes used. The selection of the type of electrodes depends on the nature of the problem. The types in general use are:

a. Solid steel, sometimes corrugated.

b. Concrete, also called graded-resistance, with conductors imbedded in the center.

c. Rod-curtain, in which curtains of small rods or pipes are hung close together to form the collecting electrode.

d. Perforated plate, in which the collecting electrodes are in the form of narrow boxes with perforated sides.

e. Pocket type, in which the precipitated material is trapped in pockets on the side of a hollow plate, the bottom of which terminates in a pipe for removal of the dust to the hopper.

The plate-type unit is usually equipped with collecting electrode rappers or vibrators to aid in the removal of the dust to the hoppers. Depending on the nature of the dust or fume collected, rappers for the discharge electrodes may be required, in which case an insulator is provided in the drive shaft so that the rapper may rest on the high-tension frame without short-circuiting the current. Both rappers may operate continuously or at stated intervals. With the latter arrangement an automatic timer is usually supplied.

A typical plate-type precipitator embodying perforated-plate collecting electrodes is shown in Fig. 3. In this type the gases pass through the ducts horizontally. In Fig. 4 a vertical-flow duct-type unit is shown, in which pocket-type electrodes are used.

26. TYPICAL INSTALLATION

A typical installation of the equipment used in the Cottrell process is shown schematically in Fig. 5. In such an installation the electrical equipment is housed in a substation,

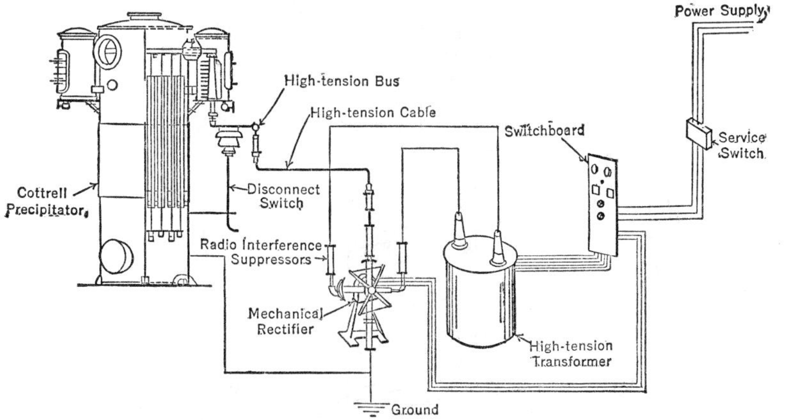

FIG. 5. Equipment Used in the Cottrell Process

which may be located at a convenient point, preferably near the precipitator. A high-tension feeder cable may be used between the rectifier set and the precipitator if it is desired for reasons of convenience to separate the two by a considerable distance.

Figure 6 shows an installation in which the precipitator and electrical equipment are housed in one shell. In this particular case electronic rectifying equipment is used. This type of equipment is particularly adaptable to small-volume installations.

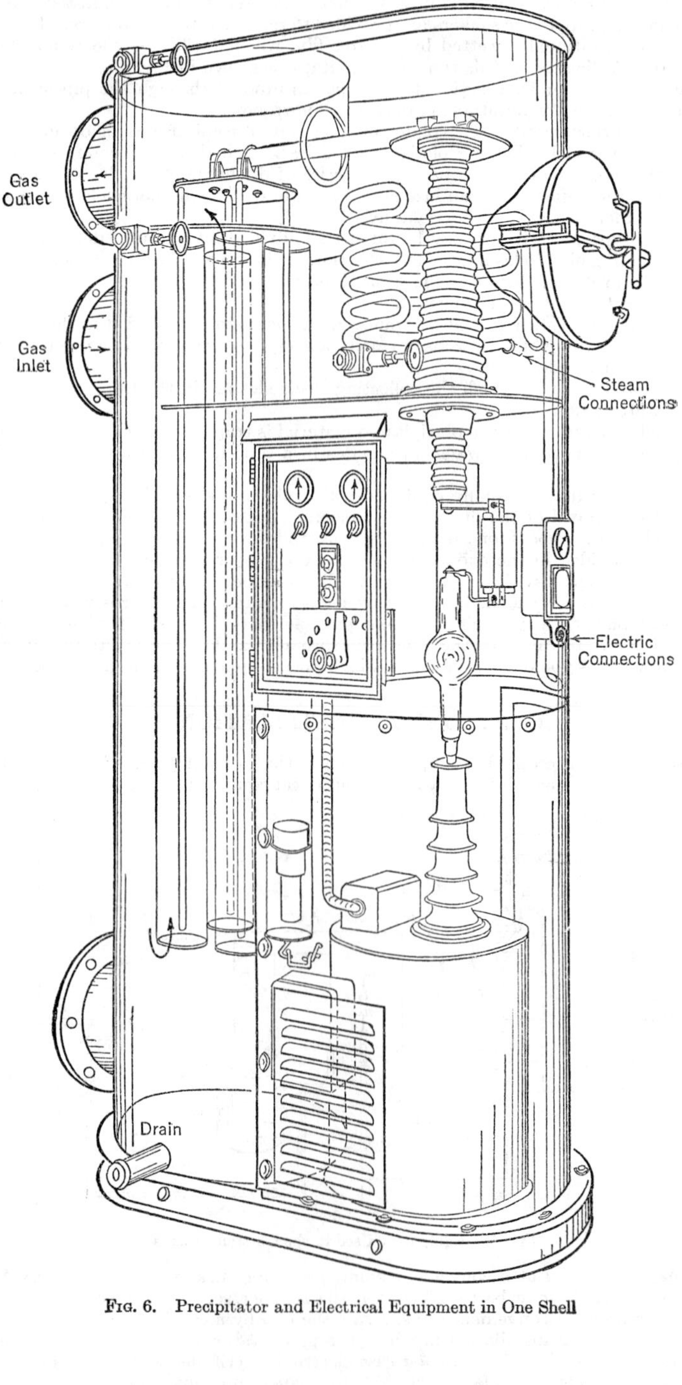

FIG. 6. Precipitator and Electrical Equipment in One Shell

27. OPERATING CHARACTERISTICS

One of the advantages of electrical precipitation is that it may be applied to practically any kind of gas and to a large variety of aerosols. It is operative in equipment that can be varied in both general and detailed design in order to fit the peculiar requirements of a specific application. The limits in which the process is applicable are as follows:

1. It is effective on both suspended solid and liquid particles.
2. There are no limits with regard to the size of the suspended solid or liquid particles which can be precipitated. Typical examples of the ranges of dispersoid sizes ordinarily encountered are shown in the following table.

DIAMETERS OF PARTICLES, MICRONS

Cement dust	5–100	NH_4Cl fume	0.1–1
Talc dust	10	ZnO fume	0.05
Silica dust	5	Condensed zinc dust	2
Sprayed zinc dust	15	Tobacco smoke	0.01–0.15
Coal dust	10	Pigments	0.02–5
Flour-mill dust	15	Sprayed dry milk	0.1–3
Alkali fume	1–5	Powdered coal ash	1–150

3. Percentage removals can be varied to suit economic requirements. They can be raised to a degree where an engineering test will hardly detect the presence of suspended particles in the treated gases.
4. No limitations are placed on the concentration or number of particles in the gases to be cleaned.
5. Operating temperature may, under some conditions, be as high as 1100 deg fahr. Particles have been removed from gases at 125 lb pressure, but this is not necessarily the limiting pressure.
6. The equipment can take various forms to fit space limitations and means of tying into other equipment. Materials of construction can be selected to meet requirements of temperature or pressure and corrosion resistance.
7. Method of removal of precipitated materials can be varied to suit particular requirements.

The precipitator sets up practically no resistance to flow of air or gas. Its introduction into a gas stream, therefore, raises no added problems in draft maintenance or suitable gas flow.

The power requirements for a precipitator depend upon the precipitator size, the composition and density of the gas, and the characteristics of the suspended matter. The power is roughly proportional to the precipitator size and therefore increases with the gas volume handled and with the required cleaning efficiency. The precipitator voltage is closely proportional to the gas density and may also depend markedly on the characteristic of the dispersoid, low-conductivity dispersoids causing a great reduction in the sparking voltage.

Representative values for the electric power requirements of precipitators vary from about 0.1 kva per 1000 cu ft per min of gas cleaned for large-size precipitators at about 90 to 95 per cent efficiency to about 0.7 kva per 1000 cu ft per min for small, very-high-efficiency precipitators.

28. RECOVERIES AND EFFICIENCIES

The velocity of the gas through the precipitator and the time of treatment or the time the particles remain in the electric field determines the efficiency of precipitation. Electrical precipitation equipment can be designed to give practically any degree of gas cleanliness required. In certain cases 90 per cent removal is all that is required. In others, where the demand is for practically complete removal or recovery of dispersoids, 99.9 per cent removals can be achieved.

Typical values for time of treatment and related efficiencies for various dispersoids are shown in Table 1.

Table 1. Approximate Gas Velocity and Time of Treatment for Several Typical Applications

Application	Suspended Matter	Type Precipitator	Gas Velocity, Feet per Second	Time of Treatment, Seconds	Efficiency, Per Cent
Zinc ore sintering........	Dust and fume	Rod Curtain	3.20	6.30	95
Zinc ore roasting.........	Dust and fume	Steel Plate	4.25	3.20	95
Catalytic clay............	Dust	Tulip Steel Plate	4.25	4.25	95
Pulverized coal fly ash....	Dust	Perf. Steel Plate	4.85	2.50	95
Acid mist................	H_2SO_4 mist	Pipe	4.75	2.53	95
Tar fog..................	Tar and oil	Pipe	9.10	.99	95
Metallurgical fume.......	Fume	Pipe	4.25	3.53	95
Iron blast furnace........	Dust and fume	Pipe, Water Flushed	11.50	1.30	95

The effect of time of treatment on the efficiency of removal for a particular dust is shown in Table 2.

Table 2. Typical (Approximate) Gas Velocities and Time of Treatment for Different Efficiencies of Hot Roaster Gas Cleaning

Efficiency, Per Cent	Type Precipitator	Duct Width, Inches	Duct Length, Feet	Gas Velocity, Feet per Second	Time of Treatment, Seconds
90	Plate	8	13.5	5.62	2.40
95	Plate	8	13.5	4.25	3.20
98	Plate	8	13.5	3.10	4.35

29. COMMERCIAL INSTALLATIONS

In general, commercial electrical precipitation installations may be classified into three general groups, depending on the purpose for which they are provided.

1. For the collection of valuable suspended materials from gases before they are wasted to the atmosphere.

2. For the elimination of a hazard or nuisance that may result from the presence of suspended matter in gases which are discharged to atmosphere.

3. For the removal of suspended matter before utilization of the gases for industrial purposes in which the presence of the dispersoid is objectionable.

The process has had wide application in a large number of industries. In some cases the material removed from the gas is a dry dust or fume; in others, a liquid, such as acid or tar.

Some of the more typical installations will be described.

DETARRING GAS. In the manufacture of coke oven or carburetted water gas, it is necessary to remove the tar and oil from the gas before it passes to the purifiers, saturators, light oil scrubbers, etc. In this application relatively high efficiencies are necessary to provide a clean gas for subsequent processing.

For this service the pipe-type precipitator, such as is shown in Fig. 1, has been commonly used in American practice.

CLEANING OF IRON BLAST FURNACE GAS. The utilization of the heating value of the gas produced in an iron blast furnace is an important economic phase of the steel industry. In order for the gas to be used efficiently, it must be cleaned of the suspended matter carried by it from the furnace. Precipitators have been found to be especially adaptable for this service.

For cleaning this gas, a pipe-type precipitator is generally used in which the surfaces of the collecting electrodes are continually wetted with a water film for removal of the precipitated suspended matter. (See Fig. 2.) The first precipitator for this service was installed in 1929, and since then more than 80 units have been installed in the United States and foreign countries.

SULFURIC ACID. Electric precipitators have been used in the manufacture of sulfuric acid for a considerable time. In this application the precipitator is used for the collection of sulfuric acid mist from exit gases coming from acid concentrators and for the purification of sulfur dioxide bearing gas, i.e., removal of acid mist and other suspended matter in contact sulfuric acid plants.

For this application pipe-type precipitators are generally used. Because of the corrosive conditions, the precipitators are built of lead, carbon, or other corrosion-resistant materials.

ROASTER GAS CLEANING. In the roasting of zinc or iron sulfide ore, the gases evolved containing sulfur dioxide represent considerable value. In order that these gases may be used efficiently, they must be cleaned. These gases are usually treated in a plate-type precipitator and at temperatures of from 700 to 900 deg fahr to prevent acid condensation in the equipment. The precipitators for this service may be brick lined or heat insulated and are normally equipped with collecting and discharge electrode rappers.

COLLECTION OF CARBON BLACK. In the manufacture of carbon black by the furnace process, the black produced is usually collected in a precipitator installation. The application of precipitation to this industry is unique in that a principle not generally used is embodied. The carbon particles are particularly fine as they leave the furnace and are therefore very difficult to collect. Under the influence of the electric field in a precipitator, however, they combine to form large agglomerates. The density of these agglomerates is very low, and they are carried along even by very low gas velocities. In order to collect these particles, low-velocity cyclones which effect the final separation are provided. The collection equipment, therefore, comprises a precipitator for agglomerating and cyclones for removal of the agglomerates from the gas stream.

The precipitators are usually of the horizontal-flow plate type, although down-flow plate-type and the pipe-type units are sometimes used. The gases are treated at about 450 deg fahr, and since they may carry sulfur compounds, special materials of construction are provided.

PAPER INDUSTRY. In the manufacture of paper by either the soda or the sulfate process, the wood treating liquors, comprising considerable organic matter, are concentrated and finally burned in a furnace. In the burning of this material, considerable sodium compounds are volatilized and also mechanically carried by the gas stream. These dispersoids can be returned to the cyclic system for reuse and therefore represent considerable economic value. Also, this material, being of very fine particle size and therefore having considerable obscuring power, causes a very undesirable stack tail which creates a nuisance around the vicinity of the mill. The problem is, therefore, one of abating a nuisance and also of recovering a chemical.

For this service vertical-flow plate precipitators are used. In view of the corrosive nature of the gases, the precipitator shell is built of ceramic materials.

PETROLEUM INDUSTRY. In the manufacture of gasoline by the fluid catalytic cracking process, the cracking is done by contacting the material to be cracked with a suspended powdered catalyst which promotes the reaction. In the reaction the catalyst becomes coated with carbon, which must be removed before the catalyst can be reused. This is done by mixing the catalyst with hot air in a regenerator, wherein the carbon coating is burned. The flue gas resulting from the combustion of the carbon which is vented to atmosphere carries a considerable amount of catalyst in suspension. Since the catalyst represents an appreciable economic value, it must be recovered from the gases.

For this problem horizontal-flow plate-type precipitators are used on the larger installation, and vertical-flow pocket-type precipitators in the smaller-volume plants. Inasmuch as the catalyst is relatively expensive, high-efficiency equipment is warranted, and consequently efficiencies of 99.6 per cent are the rule.

COLLECTION OF METALLURGICAL FUMES. Numerous plants in the non-ferrous metallurgical industry are using electrical precipitation for the collection of metallic compounds, such as tin, lead, zinc, gold, silver, and cadmium, from gases coming from reverberatory furnace, blast furnaces, kilns, sintering machines, and other smelting and refining equipment. These installations are operated primarily for the collection of the metallic values and therefore play an important role in the conservation of materials.

REMOVAL OF FLY ASH FROM PULVERIZED COAL FIRED BOILERS. Electric precipitators have been used since 1925 f r the removal of fly ash from gases coming from pulverized coal fired boilers. In this application precipitators are used to abate an annoying nuisance.

The horizontal-flow perforated plate or guarded resistance plate is generally used for this purpose. These precipitators are usually guaranteed for efficiencies between 90 and 97 per cent.

In January, 1946, precipitation equipment was in use or under construction in 158 different power stations. The number of precipitators per station ranged from one to eleven; the gas volumes handled from 25,000 to 1,500,000 cu ft per min.

CEMENT KILN DUST. Many of the portland cement plants collect dust and fume from gases coming from rotary kilns producing cement clinkers. The precipitators used for this service are of the horizontal-flow plate type encased in either a steel or a concrete shell.

30. THE PRECIPITRON

The Precipitron, shown diagrammatically in Fig. 7, is used largely for removing dust, including pollen, from ventilating systems. It operates on the same basic principle as the Cottrell system but applies it differently. The ionizing wire is made positive instead of negative as in the Cottrell precipitator. This entails a slight sacrifice of ionizing efficiency but gains a 10 to 1 reduction in ozone formation. The dust-ionizing and dust-collecting functions are separated. In the ionizing section, shown at the left in Fig. 7, the ionizing wires are held between tubes oppositely charged to the wires. A dust particle, as it passes between a wire and a tube, is charged positively and is swept into the collector section, which consists of parallel plates of opposite charge, 5000 volts to ground. Here the strong field drives the positively charged particle out of its line of flight to the negatively charged plates.

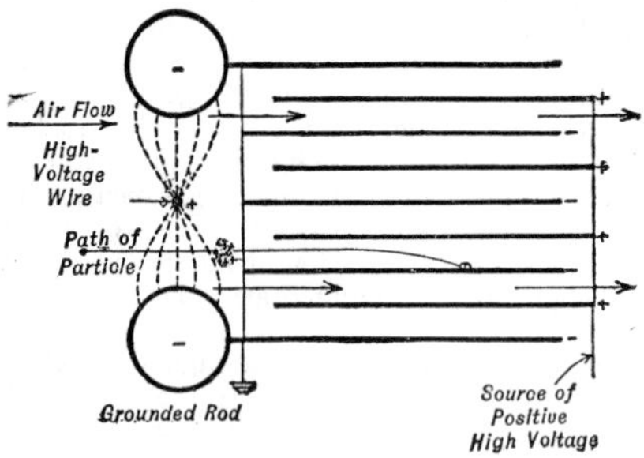

Fig. 7. Schematic Diagram of the Precipitron for Electrostatic Air Cleaning. (Courtesy, Westinghouse Electric and Manufacturing Company.)

Energy is supplied by a power pack, which consists of a high-voltage transformer, vacuum-tube rectifier, and capacitors to smooth out the pulsating d-c voltage.

The Precipitron is used for cleaning air for office buildings, stores, auditoriums, libraries, hotels, and restaurants, as well as in manufacturing establishments where clean air is necessary.

31. BIBLIOGRAPHY

Hohlfeld, M., Das Niederschlagen des Rauchs durch Elektricitäts, *Arch. gesamte Naturlehre* (K. W. Kastner), Vol. 2, p. 200 (1824).
Guitard, C. F., Condensation by Electricity, *Mechanic Mag.*, Vol. 53, p. 346 (1850).
Lodge, Sir Oliver, The Electric Disposition of Dust, Smoke, etc., *J. Soc. Chem. Ind.*, Vol. 5, p. 57 (1886).
Cottrell, F. G., The Electrical Precipitation of Suspended Particles, *J. Ind. Eng. Chem.*, Vol. 3, p. 542 (1911).
Schmidt, W. A., The Control of Dust in the Portland Cement Industry, *Proc. 8th Intern. Congr. Applied Chem.*, Vol. 5, p. 117 (1912).
White, Alfred H., The Electrical Separation of Tar from Coal Gas, *Amer. Gaslight J.*, Vol. 97, p. 210 (1912).
Howard, W. H., Fume Precipitation at Garfield, *Bull. Amer. Instit. Min. Eng.*, Vol. 49, p. 540 (1914).
Cottrell, F. G., Problems in Smoke, Fume, and Dust Abatement, *Smithsonian Inst. Report for 1913, Pub.* 1307, p. 653 (1914).
Strong, W. W., Theory of Electric Precipitation, *Proc. AIEE*, Vol. 34, p. 220 (1915).
Nesbit, A. F., Theoretical and Experimental Consideration of Electrical Precipitation, *Proc. AIEE*, Vol. 34, p. 507 (1915).
Bradley, Linn, Practical Application of Electrical Precipitation, *Proc. AIEE*, Vol. 34, p. 523 (1915).
Schmidt, W. A., Cottrell Processes of Electrical Precipitation, *Trans. Can. Min. Inst.*, Vol. 18, p. 110 (1915).
Aldrich, C. H., Treatment of Silver Furnace Fume by the Cottrell Process, *Trans. Amer. Electrochem. Soc.*, Vol. 28, p. 119 (1915).
Meston, A. F., Electrical Precipitation in the Chemical Industries, *Elec. J.*, p. 248 (June, 1917).
Carothers, J. N., Electric Furnace Smelting of Phosphate Rock and Use of Cottrell Precipitator in Collecting Volatilized Phosphoric Acid, *J. Ind. Eng. Chem.*, Vol. 10, p. 35 (1918).
Heimrod, A. A., and H. D. Egbert, The Cottrell Processes in the Sulphuric Acid Industry, *Chem. Met. Eng.*, Vol. 19, p. 309 (1918).
Landolt, P. E., and H. M. Pier, Air Cleaning by Cottrell Electrical Precipitation Processes, *Bull. Amer. Soc. Htg. Ventilating Eng.* (Jan. 1920).
Kee, W. J., and F. H. Viets, Cottrell Electrical Precipitator Installation at Plant of the American Acid Co., Medford, Mass., *Chem. Age* (Feb. 1921).
Anderson, E., Recent Progress in Electrical Precipitation. *Chem. Met. Eng.*, Vol. 25, p. 1555 (1922).
Landolt, P. E., Application of Electrical Precipitation to Chemical Engineering Processes, *Chem. Met. Eng.*, Vol. 29, No. 13 (1923).
Wood, R. U., Gas Cleaning with Cottrell Electrical Precipitation Process, *Glass Ind.* (June 1923).
Landolt, P. E., Application of Electrical Precipitation to Chemical Engineering, *Chem. Met. Eng.* (Sept. 24, 1923).
Schmidt, Walter A., Electrical Precipitation in Retrospect, *J. Ind. Eng. Chem.*, Vol. 16, No. 10, p. 1038 (1924).
Sultzer, N. W., Recovery of Gold at the U. S. Assay Office, New York, *Chem. Age* (Jan. 1925).
Sultzer, N. W., The Cleaning of Combustible Gases with the Cottrell Electrical Precipitation Processes, *Report of Chemical Committee Amer. Gas Assoc.* (1926).
Sultzer, N. W., and C. W. Hedberg, The Removal of Tar from Gases, *A.G.A. Monthly* (Dec. 1934).
Pier, H. M., and A. N. Crowder, Catching Pulverized Coal Ash at the Trenton Channel Plant by Cottrell Electrical Precipitation, *Power* (May 31, 1927).
Hedberg, C. W., Blast Furnace Gas Cleaning, *Iron and Steel Eng.* (July 1929).

Landolt, P. E., Cottrell Process for Sulphuric Acid Recovery, *Petroleum Mech. Eng.* (1930).
Hedberg, C. W., Design of Gas Cleaning Installations, *Ind. Eng. Chem.* (April 1937).
Hedberg, C. W., and L. M. Roberts, Developments in Electrical Precipitation in Steel Plants, *Iron Steel Eng.* (July 1940).
Roberts, L. M., Electrical Precipitation of Solids from Kraft Stack Gases, *Paper Trade J.* (Sept. 11, 1941).
Beaver, C. E., Cottrell Electrical Precipitation Equipment: Some Technical and Engineering Features, Recent Developments and Application in the Chemical Field, *Trans. AI Ch E*, Vol. 42, No. 2 (April 1946).

ELECTRIC WELDING

By Wendell F. Hess

32. THE NATURE OF A WELD AND THE GENERATION OF HEAT

Welding consists of the joining of two or more pieces of metal by the application of heat and sometimes of pressure. Welds accompanied by fusion are the most common, and these ordinarily do not require the application of pressure. When pressure is applied to a weld in which fusion takes place, the molten metal must be confined within a solid mass of the metal, as in electric resistance spot welding. A weld may take place without fusion if the metal is heated under pressure to such a temperature that a new system of grains is produced or grain growth of the existing system takes place. A weld made at subfusion temperatures is frequently referred to as a plastic weld, because of the fact that steel, when heated above a critical temperature, becomes plastic. Forge or other pressure welding, including electric resistance butt welding, is usually carried out at subfusion temperatures.

Although welding may be carried out by heating the entire parts to be welded to the proper temperature, as in a forge or furnace, most of the success of modern welding processes is related to their ability to concentrate the heat effectively on the surfaces which it is desired to join. Not only must a welding process be capable of heating very localized areas, but it must also be capable of supplying heat in sufficiently large amounts to bring the surfaces rapidly to the welding temperature. The more rapidly the heat can be supplied, the less will be dissipated to the surrounding metal, and the more efficient will be the welding process. The very extensive application of the electric welding processes is due to their pronounced ability to supply sufficient quantities of heat under controlled conditions to localized areas of metal. Of the other methods for the generation of heat, chemical or mechanical, only a limited number of the gas-oxygen flames are sufficiently intense to be useful as welding processes. Of these, only the oxyacetylene flame produces a sufficiently high temperature to weld satisfactorily the metals of higher melting temperature, such as the steels. The oxyhydrogen flame may be used for lower-melting-temperature metals, such as aluminum. Mechanical methods for the generation of heat, like most chemical methods, have not found practical application as welding processes, because the generation of heat was insufficient either in magnitude or in concentration. Friction may under certain circumstances produce sufficient heat to make a weld. A steel tube may be revolved at high speed against a friction plate until its temperature has been raised to the plastic range. In this condition the end of the tube may be spun around and closed. The quality of the closure may be under suspicicion, however, because of the possibility of trapped oxide in the joint. This brings out another point with regard to successful welding, namely, the ability of the preparation or process to provide properly cleaned surfaces as well as sufficient intensity or concentration of heat.

Two other methods for the generation of heat find use in welding processes. One of these, thermit welding, involves the liberation of the molecular heat of formation of aluminum oxide. The replacement of iron by aluminum in the oxide results in the production of sufficient heat to raise the temperature of the resultant iron to more than double its normal melting temperature. This excess heat may be used to bring the edges of parts to be joined to the fusion temperature, provided auxiliary heat has been originally given to the parts in a pre-heating operation. The other process, atomic-hydrogen welding, involves the heat of recombination of single hydrogen atoms to form diatomic gas molecules. In this process dissociation of the gas is brought about by passing it through an electric arc. Thus the process may be included in the electric welding methods. After performing its primary function in this process, the hydrogen gas burns in the surrounding air, effectively shielding the molten metal from oxidation and the introduction of nitrogen. The combustion also provides some preheat to the plates which are being welded and helps to retard cooling.

Of the electrical methods for the generation of heat, two are of great importance; the electric arc and electrical resistance. The electric arc is particularly suitable as a source of energy for welding because the heat may be effectively concentrated. The fact that the temperature of the electric arc is from two to four times the melting point of the most widely used and most extensively welded metal, steel, ensures the rapid delivery of large amounts of heat to the surfaces being welded. Arcs in which from 5 to 10 kw are developed in close proximity to a surface no larger in diameter than a lead pencil are commonly used for welding. In certain cases much higher concentrations of electric power are possible. In the electrical resistance method of welding, heat is generated in the weld zone because of the passage of the welding current through the resistance of that zone. The power expended is measured as the product of the square of the current multiplied by the resistance and is expressed in watts. The electrical resistances contributing to the development of heat are the body resistances of the material and the electrical resistances at the various contacting surfaces.

33. ELECTRIC-ARC WELDING

TYPES OF ARC WELDING. Two different kinds of arc welding are employed, carbon-arc welding and metal-arc welding.

Carbon-arc Welding. In carbon-arc welding, an arc (usually several tenths of an inch in length) is established between a carbon or graphite electrode and the two pieces of metal which it is desired to join. The carbon arc is manipulated with one hand, and, if necessary, a filler rod is melted into the joint. The manipulation in this case is similar to that in gas welding, in which a torch is held in one hand and a filler rod in the other. The carbon arc is not frequently employed for the welding of steel because of the tendency of the arc stream to wander over the surface of the steel and thus not to concentrate the heat at the point desired. A comparatively long carbon arc of relatively high voltage (about 40 volts), combined with a high magnitude of current, has been successfully employed for the very rapid arc welding of copper and some of its alloys.

The difficulty experienced with manipulation of the carbon electrode, because of its tendency to wander over the surface, has been avoided in an automatic application of carbon-arc welding by superimposing a magnetic field on the arc stream parallel to its intended direction from the carbon electrode to the metal plates being welded. Under the action of this longitudinal magnetic field any tendency for the conducting particles of the arc to wander away from the axis of the magnetic field is attended by a mechanical force tending to make the conducting particle turn back to a circumferential path with respect to the direction of the arc stream. The direction of travel of the conducting particles along the arc stream is thus analogous to the air currents in a tornado. For this reason the company developing this process has given it the name, Electronic Tornado. The magnetically controlled carbon arc has been used extensively only in an automatic process, since the mass of the equipment required to produce the magnetic field is such as to preclude the use of this process in manual welding. Special features are incorporated in the automatic equipment to provide gaseous shielding of the molten metal from the oxygen and nitrogen of the air, to provide additional filler metal if required, and to provide slag protection over the solidifying weld metal.

Metal-arc Welding. In metal-arc welding, the arc is established between a steel rod called a welding electrode and the steel parts to be joined. In the early application of this process the electrode was a bare mild steel rod. In order to protect the weld metal from the action of the oxygen and nitrogen of the air at high temperatures, the shortest practicable arc length was maintained between the end of the welding rod and the work. In spite of all the precautions exercised to maintain a short welding arc, welds made using bare electrodes in this process were not comparable in ductility to the plate material which was being welded. Great improvement in the ductility of weld metal and in the ease of manipulation of the metallic arc was achieved when the practice of using covered electrodes was developed. Other attempts were made to improve the ductility of weld metal by welding in a reducing atmosphere, such as was provided by hydrogen gas. The use of this gas increased the cost of the welding operation and rendered the welding arc objectionably unstable. More recently, an inert gas, argon or helium, has been found highly satisfactory for certain operations, which will be described later.

ARC CHARACTERISTICS. Contrary to the behavior of the resistance elements of an electric circuit, in which the voltage drop increases with the current, an electric arc possesses a drooping volt-ampere characteristic. In other words the voltage drop across the arc decreases as the arc current increases. This type of circuit element is inherently unstable if supplied from a constant-potential source of electrical energy. It becomes necessary in that case to use a resistance element in series with the arc of such magnitude that the voltage drop of the entire combined circuit tends to rise as the magnitude of current increases. Such a stabilizing resistance is accompanied by the necessity of dissipating a very large amount of energy in the resistor. Experience has shown that a minimum of approximately 60 volts is necessary for the satisfactory starting of a d-c arc. If the arc characteristics are such that the voltage drop at a satisfactory value of current is approximately 30 volts, then as much power must be dissipated in the series stabilizing resistance as is dissipated in the arc itself. If the lowest standard constant-potential system of 120 volts were used to supply the above-mentioned arc, three times as much power would have to be dissipated in the resistance as was usefully employed in the arc.

The arc power, in a d-c arc, is measured by the product of the current through the arc and the voltage drop across the arc. The voltage drop across the arc consists of fixed amounts at each terminal and a variable amount proportional to the length of the arc. Experiment has shown that the voltage drop at the positive terminal of a carbon or bare metal electrode arc is much higher than the voltage drop at the negative terminal. This accounts for the development of greater heat at the positive terminal, which is commonly experienced in the operation of carbon and bare metal electrode arcs. This condition explains the general use, with the carbon and bare electrode arc, of polarity which makes the work positive and the electrode negative. The plates being welded have so much larger mass than the electrode that they require the greater heat input for proper fusion. This polarity, work positive, is referred to as straight polarity. The magnitudes of power corresponding to the terminal drops of voltage are most effectively utilized, because they are delivered exactly where needed, to melt the electrode at one terminal of the arc and to melt the plate at the other terminal of the arc in preparation for receiving the deposited molten metal. The energy corresponding to the voltage, which varies with the arc length, is also effective to a considerable degree and permits some variation in the arc energy by manipulation of arc length. This variable voltage across the arc stream also serves as a measure of arc length and thus as a basis for the automatic control of arc welding. Although it might be considered desirable to increase the arc power by lengthening the arc stream, and although moderate lengthening of the arc does increase the melting rate, it is possible with most welding equipment to draw such a long arc as to decrease the fusion of the base metal because of spreading the arc energy over the surface.

Although the maximum heat is developed at the positive electrode in the case of a carbon or bare low-carbon-steel electrode, the reverse condition is sometimes true where the electrode is provided with a flux coating. With certain ionizing elements obtained with coated electrodes, satisfactory operation of the arc requires that the electrode be connected to the positive terminal of the supply, and in this case greater heat appears to be developed at the negative terminal. This is evident from the greater depth of fusion or penetration into the plate material secured with this polarity. Although some coated electrodes operate equally well with either polarity and are also suited for operation with a-c power supply, other coated electrodes operate satisfactorily with only one polarity and are very erratic and unstable in operation if the other polarity is used.

POWER SOURCE REQUIREMENTS. As mentioned above, convenient starting of an electric arc requires a minimum of 60 volts. Thus, if the arc is to be fed from a constant-potential system, 60 volts must be the minimum circuit voltage. If more than 60 volts is available, the arc will start even more readily, but the electrical efficiency of the system will decrease. Hence it is not economical to supply welding arcs from a constant-potential system having a voltage greater than approximately 60 volts. Since a constant-potential system requires the addition of a stabilizing resistance, with consequent loss of energy, there has been a strong tendency to develop special arc welding generators capable of providing the necessary voltage for starting but having a sufficiently drooping volt-ampere characteristic to permit operation of the electric arc in a stable manner without the use of a stabilizing resistance.

When alternating current instead of direct current is used for arc welding, the power source requirements are essentially the same as those for direct current. However, for satisfactory starting of an a-c arc, because of its inherently less stability, a minimum voltage of about 80 is required. If a-c arcs were to be supplied from a constant-potential system, a stabilizing reactor could be used rather than a resistance, and this would eliminate to a large degree the inefficiency of the system. Nevertheless it is more customary to provide a drooping characteristic a-c power source within the welding transformer, since a transformer is usually required, because of the fact that it is desirable to reduce the standard power voltages of 120 or 240 volts to approximately 80 volts before applying them to arc welding. This is desirable from the standpoint both of safety and of the size of the conductors required to carry the current.

TYPES OF CURRENT. As indicated above, both direct current and alternating current are used for metal-arc welding. The a-c arc is inherently less stable than the d-c arc. The arc becomes extinguished and re-established twice during each cycle of the a-c system. During the time when the alternating current is extinguished, the ionization within the arc stream tends to disappear rapidly, particularly in the bare electrode arc, in which conduction is electronic to a rather large degree. Thus, in all the earlier applications of welding involving bare electrodes, direct current was considered the only satisfactory source of power. Introduction of coated electrodes for arc welding, however, made it possible to add ionizing constituents to the arc stream. The conductivity of the arc stream with these heavier ionized particles tends to persist during the period of extinction of the arc. Heavier magnitudes of welding current also tend to increase the stability of the arc. Thus the larger sizes of covered electrodes tend to operate more satisfactorily on alternating current. Alternating current for arc welding tends to make the power supply more difficult when it is desired to use long leads from the established position of the welding machines to points perhaps several hundred feet away on the construction site. With alternating current, the reactance of such long leads becomes a serious problem, since too great a proportion of the voltage drop becomes absorbed in the reactance. Hence it becomes necessary to minimize the reactance by using a two-conductor cable for the greater part of the distance between the welding machine and the work. In d-c welding this problem does not arise, and a single-conductor cable, with current returned through the structure, is perfectly satisfactory. The problem of lead reactance does not arise in manufacturing operations within a shop, where the welding machines may be placed close to the work. In this type of work, when welding with direct current on heavy plates, the problem of arc disturbance by the magnetic field of the current returning in the plate becomes serious. With a-c arcs this phenomenon of magnetic disturbance is not troublesome.

In spite of the greater simplicity of the welding transformer as compared with the rotating machine usually required for the production of direct current for welding, the use of d-c welding equipment still persists to a great extent. This is due to its greater versatility in such matters as the welding of very thin materials using low magnitudes of arc current, the ease of securing deep penetration by means of certain electrodes, and the greater versatility in the use of long leads.

Since welding machines are actually engaged in supplying an arc for only 10 to 50 per cent of the working period, the standby losses of the welding machine are important. The a-c welding transformer tends to show considerably lower standby losses than the rotating motor generator that is usually required for d-c welding. Even during the actual welding operation, the electrical efficiency of the transformer is considerably higher than that of the motor generator. Furthermore, the maintenance of the rotating machinery required for direct current is likely to be higher than that on the static welding transformer. The advent of electrodes suitable for the production of high-quality welds, combined with the characteristics of ease of arc manipulation, strongly favor greater use of a-c arc welding.

34. WELDING MACHINES AND EQUIPMENT

The voltage at the terminals of the arc in the metal-arc process is usually from 15 to 35 volts, the lower values being used with bare electrodes and the higher with covered electrodes. In order to start or strike a d-c arc, at least 60 volts must be available for quick and easy starting. With a-c arcs a somewhat higher voltage, approaching 80 volts, is desirable. Since standard distribution systems are almost universally of the a-c type, motor-generator sets are employed to deliver direct current at lower than standard distribution voltages, which are just sufficient for convenient starting of the arc. Two different systems of power supply from motor-generator sets are commonly employed, one known as the constant-potential or multiple-operator set, and the other as the variable-voltage or single-operator set.

CONSTANT-POTENTIAL OR MULTIPLE-OPERATOR SYSTEM. In this system a motor-generator set is employed, which has a d-c generator of a capacity sufficient for supplying 10 or more welding arcs. The generator is usually designed to operate at from 50 to 75 volts and is flat-compounded so as to deliver approximately constant voltage regardless of load. Each welding operator must be provided with a current-regulating resistor in his individual welding circuit. It is obvious that with this arrangement the power losses in the resistor will be very greatly reduced below what they would be if the open-circuit voltage were 125 or 250. Even with this more favorable operating voltage, however, the power losses in the resistor will be a major contributing factor to the low electrical efficiency of this system. For example, with a 60-volt constant-potential generator supplying a 30-volt arc, half the energy delivered will be wasted in the resistor, and the maximum electrical efficiency of the arc circuit will be limited to 50 per cent. If the efficiency of the motor-generator set is also taken into consideration, the overall electrical efficiency will be less than 40 per cent even with continuous operation, which is never possible in practice, and will be much lower in actual operating practice, because of the no-load losses of the motor-generator set.

VARIABLE-VOLTAGE OR SINGLE-OPERATOR SYSTEM. In this system power for each welding circuit is supplied by an individual generator, which may be motor, belt, or engine driven. The generator is designed as a variable-voltage machine, such as by differential compounding, so that the open-circuit voltage may be from 50 to 100 volts, but, when the arc is struck and current is drawn from the generator, the output voltage drops to that required for the arc. With such a generator it is not necessary to provide a resistor in series with the arc, resulting in a saving of power which would be wasted in the constant-potential system. With the variable-voltage welding generator, welding current is adjusted by various means, such as changing the shunt field rheostat, changing the amount of diverter resistance which is in parallel with the series field, or shifting brushes, depending upon the design of the individual generator. A certain amount of reactance in the circuit is desirable in order to stabilize the arc, and either this may be embodied in the design of the generator itself, or else a separate reactor may be employed.

Each of several manufacturers has placed on the market types of welding apparatus comprising single-operator d-c generators with characteristics for which various advantages are claimed and realized.

MULTIPLE- VERSUS SINGLE-OPERATOR SYSTEMS. The multiple-operator system usually comprises one or more large generators connected by a heavy-current bus system to the welding area. The current-regulating resistors provided for each operator in the welding area are attached to the same welding current bus system. Such a system comprises a relatively permanent installation and is thus primarily suited to conditions where a large number of welding operators are working in a comparatively small area continuously, such as in a welding shop of a manufacturing company or a shipbuilding way. The multiple-operator system may make use of the diversity in time of use of the different welding arcs. Thus, a 1500-amp generator may be perfectly suitable for supplying, not five 300-amp arcs, but perhaps ten such arcs, in regular welding production. Even twice that number of arcs may be supplied in cases of less frequent usage, such as for tack welding in the assembling of structures. Maintenance costs for the fewer number of larger generators used in the multiple-operator system are likely to be very much lower than in the single-operator system, where each operator draws his welding current from a separate generator. The somewhat lower electrical efficiency of the multiple-operator system, resulting from the use of the load-regulating resistors required in each arc circuit, becomes a less important factor as the cost of electric power declines. Although it might be expected that the fewer number of larger units required in the multiple-operator system would result in an installation of lower first cost, the permanent-type installation, including the bus system, combined with the relatively low unit cost of the single-operator sets resulting from the mass production of these units, tends to offset this possible advantage of the multiple-operator system.

If the welding operations are scattered, or if portability is a factor, then the **single-operator** set should be chosen. These factors are of such general importance that the vast majority of arc-welding operations are carried on with single-operator sets. The single-operator system also provides ease of expansion of welding operations and flexibility in the transfer of equipment to meet increased needs in one department or at one location by shifting equipment from one place to another within a manufacturing plant.

35. WELDED JOINTS

Fusion welding may be applied to five basic types of joints, namely, butt, corner, edge, lap, and tee. These are illustrated in Fig. 1. Either fillet welds or groove welds are commonly employed to make these joints, although butt welds in the sheet material are sometimes made by means of bead welds, and lap joints may occasionally be made by use of plug welds. The different types of welds are illustrated in Fig. 2.

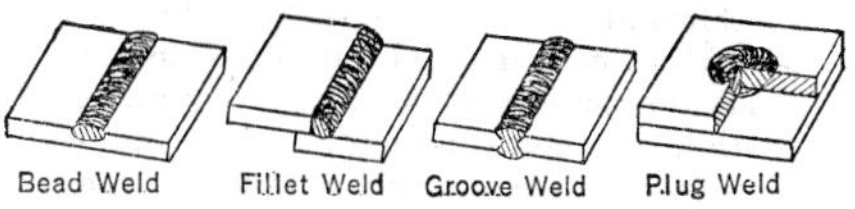

FIG. 1.

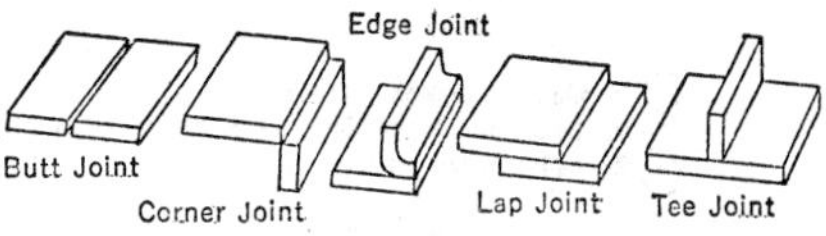

FIG. 2.

Many different types of preparation are employed in making groove welds. Typical examples of these are illustrated in Fig. 3. Square-edge preparation is suitable only for plate thicknesses less than $^1/_4$ in. In order to secure completely fused welds for plate thicknesses of $^1/_4$ in. and greater, the various types of preparation illustrated are used. The single types of preparation are generally employed for the moderate thicknesses where the saving of metal deposition by the double type of preparation is not important. The single preparations are also used when it is convenient to weld from one side only. The V-type preparations can be made by oxyacetylene flame cutting. The U- and J-type preparations are more economical of deposited metal, which is important in heavy plate welding, but generally require a machining operation for their preparation.

In order to facilitate the specification of various types of weld and plate preparation, as well as to indicate the size of weld on a particular design drawing, the American Welding Society has standardized a system of symbols to be used on drawings. These may be found in the American Welding Society Handbook.

Groove welds in plate are made in four different positions. The speed of welding, method of manipulation, and degree of skill required for their accomplishment vary with these positions, which are designated flat, horizontal, vertical, and overhead. They may be briefly described as follows:

1. Flat position. The plates are located in an approximately horizontal plane, and the weld metal is deposited from the upper side.
2. Horizontal position. The plates are located in an approximately vertical plane with the welding groove approximately horizontal.
3. Vertical position. The plates are placed in an approximately vertical plane with the welding groove approximately vertical.
4. Overhead position. The plates are located in an approximately horizontal plane, and the weld metal is deposited from the under side.

36. WELDING ELECTRODES

The metal-arc process of welding was developed and became widely adopted using electrodes of bare steel wire. It was recognized, however, that the ductility of arc-welded deposits made with these electrodes was much less than might be desired. It was demonstrated that very appreciable percentages of the carbon and manganese in the steel weld rod were lost as a result of exposure to atmospheric oxygen while traversing the electric arc and being deposited in the molten pool. It was also demonstrated that, during the arc-welding process, using bare electrodes, very large increases in the nitrogen content of the steel took place. Trapped metal oxides and nitrides introduced during welding were held responsible for the very serious lack of ductility of bare-electrode-deposited metal.

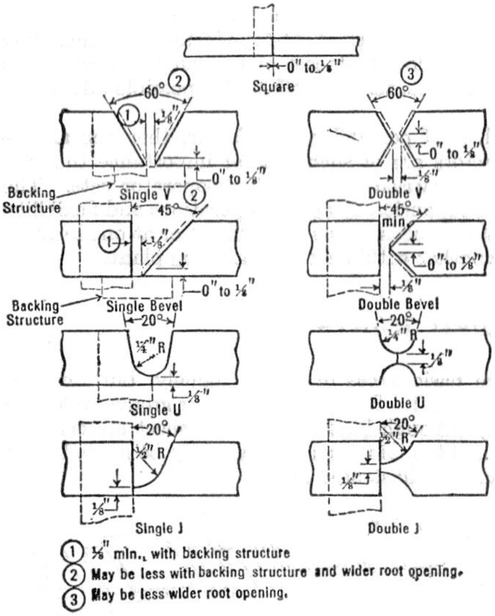

FIG. 3.

COVERED ELECTRODES. Covered electrodes have largely superseded bare electrodes for arc welding, because of the very great improvement in ductility of deposited metal which can be obtained with this type of electrode. Early in the attempt to improve the quality of electrodes, it was found possible to raise the stability of the electric arc, and thus the ease of manipulation and deposition of weld metal, by introduction of a very thin coating of certain inert materials, such as titanium oxide. A very thin coating of this material was shown also to increase the melting rate of the electrode. Titanium oxide thus became important very early as an arc stabilizer, and this characteristic makes it an important constituent in many present-day electrodes. It is not necessary to use pure titanium oxide in electrode coatings, since several natural mineral compounds, such as rutile, containing approximately 96 per cent titanium oxide, are available. Although the composition of electrode coatings has been held as a trade secret by most manufacturers, these coatings usually contain certain amounts of material for the following purposes:

1. A flux for the removal of metal oxide.
2. A binder, such as sodium silicate.
3. A ferroalloy, such as ferromanganese, for deoxidation purposes.
4. An arc stabilizer, e.g., titanium oxide.
5. Asbestos or cotton fiber or other cellulose to produce a lighter slag.

Types of Covered Electrodes. Electrodes for the welding of mild steel are of three principal types, varying in the position of welding for which they are suitable, as well as in other respects, such as the polarity with which they may be used, the ductility to be expected in the deposited metal, and the shape and type of deposit, as follows:

The **high-cellulose type,** containing 35 to 45 per cent cellulose in the form of digested wood pulp. Small amounts of sodium silicate, ferromanganese, titanium oxide, and asbestos are also included. The burning of the cellulose releases large volumes of gaseous products of combustion which provide an effective shielding of the arc stream from the effects of atmospheric oxygen and nitrogen. The slag resulting from the deposition of a welding electrode must be carefully controlled with respect to its melting point, fluidity and refractoriness upon cooling to permit easy removal. As a result of the effectiveness of its coating, this electrode produces deposits high in ductility. It operates satisfactorily only upon reversed polarity direct current, that is, with the work negative. It is characterized by deep fusion into the base metal or good penetration. The organic material in the coatings of these electrodes is responsible for splatter loss, or the tendency for small globules of metal to be expelled from the arc stream. This loss of metal is particularly noticeable when currents are above normal. For example, when welding with a certain brand of this type of electrode, normal welding with about a 25-volt arc and 175-amp current, using $^3/_{16}$-in. electrode at a speed of 78 sec per ft, resulted in a splatter loss of 19.7 per cent. Increasing the current to 200 amp raised the splatter loss to 25.1 per cent, and increasing the current further to 225 amp at a welding speed of 62 sec per ft resulted in a splatter loss of 34.0 per cent.

The organic compounds tend to promote porosity in weld metals by favoring gas absorption and particularly tend to introduce large volumes of hydrogen into the deposited metal. This has the especially objectionable effect of promoting cracking of the base metal underneath the weld deposit in medium-carbon and alloy steels. To avoid this difficulty, newer electrodes for the welding of these steels have been developed in which the coating is very similar to that used on stainless steel electrodes, known as the lime-type coating and containing appreciable quantities of calcium carbonate.

The **rutile-type electrode** has a coating which contains approximately 40 per cent of a high-grade rutile and usually sodium silicate, ferromanganese, and some feldspar, mica, and lime, but not over about 8 per cent cellulose. As might be expected from its composition, this electrode is characterized by extremely high arc stability. It is the easiest electrode for a beginner to operate. It is also characterized by rather shallow penetration into the base plate and is preferably used on straight polarity. Because of its lack of tendency to penetrate the base plate and generally more viscous weld metal, this electrode is suitable for filling accidentally large gaps between adjacent plates due to faulty plate preparation. The deposits possess reasonable ductility but are not so good in this respect as the cellulosic type of electrode. This electrode may be used in all positions and also with alternating current, since it is so highly stabilized. When deposited as fillet welds, this electrode produces noticeably convex deposits, in comparison with the flat deposit produced by the cellulosic type of electrode.

The **iron oxide type of electrode** is sometimes referred to as the **"hot-rod" type,** because of the noticeably higher fluidity of the deposited metal. As a result of this characteristic, the electrode is suitable principally for work in the flat position, although it may be used for the production of horizontal fillet welds. This electrode is not at all suitable for vertical or overhead welding. Its coating contains essentially 20 per cent iron oxide, although this exists in several forms which are properly balanced to give the desired properties. The coating also contains a variety of silicates, such as magnesium silicate, feldspar, and silica, as well as asbestos and not over 8 per cent cellulose. This electrode may be used on either reversed or straight polarity, with a slight difference in the operating characteristics resulting from the change in polarity. When deposited in groove welds or in fillet welds, it produces noticeably smooth concave deposits.

Classification of Electrodes. A system of classification of electrodes has been standardized by the American Welding Society. The electrodes are classified on the basis of usability, with respect to position of welding and the ultimate tensile strength of all-weld metal specimens in the stress-relieved condition. Thus the three types of electrodes mentioned above, namely, the high-cellulose, the rutile, and the iron oxide types, are

AWS Classes E6010, E6012, and E6020, respectively. Alternating-current electrodes with characteristics similar to those of E6010 and E6012 are designated E6011 and E6013. Improvements in E6013 type have resulted in making this electrode in small sizes suitable for the arc welding of thin sheet metal. The E6020 electrode, mentioned above as the iron oxide type, is suitable for use either on direct current or on alternating current. Many special electrodes have been developed for the various alloy steels, stainless steels, and also for the non-ferrous alloys.

Although the covered electrodes are somewhat more expensive than bare electrodes, their increased melting rate tends in many cases to compensate to a considerable degree for their increased cost. Thus a welding operator using the covered type may deposit sufficiently more pounds of electrode to make the total cost of deposited metal very little greater for the covered than for bare electrode welding. This condition has largely elimimated the use of bare electrodes, even for classes of work where the quality of the deposited metal produced by the bare electrode would be sufficient for the purpose. On the other hand, the much greater ductility of the deposited metal produced by covered electrodes is a compelling reason for their use, even though the cost of the deposited metal may be considerably higher.

37. OTHER METHODS OF ARC PROTECTION

GASEOUS PROTECTION. A number of methods have been developed to provide for making the weld within a gaseous medium which protects the weld from the oxygen and nitrogen in the surrounding air and which may also reduce the oxide on the part being welded. A typical example is the process which employs a continuous stream of hydrogen or other suitable gas around the arc. In the case of hydrogen, the arc drop is approximately 60 volts, as compared with 20 volts when the arc is operated in air. Consequently, for the same current some three times as much heat is developed at the arc. This method has been employed in a semiautomatic arc-welding process in which the electrode is fed from a coil by an electric motor drive controlled to maintain constant arc voltage. In hydrogen gas, however, the arc tends to operate in a somewhat wild and unstable manner, and the welds obtainable from the process have not shown sufficient superiority to justify its commercial acceptance.

ATOMIC-HYDROGEN WELDING. The atomic-hydrogen process successfully employs hydrogen-gas shielding. In this process the welding is not accomplished directly by means of the heat of the electric arc. The arc is established between the tips of two tungsten wires. The current is alternating and is usually supplied from a single-phase transformer connected to an ordinary distribution system. The open-circuit voltage is approximately 300 volts, and the arc voltage during welding is of the order of 100 volts. The stream of hydrogen gas is passed through annular orifices around each electrode. Energy supplied by the electric arc dissociates the hydrogen molecules into free atoms. When the hydrogen in the atomic form has travelled a short distance beyond the arc, it has cooled sufficiently to recombine into the more stable molecular form, and the energy thus released constitutes the heat actually employed for welding. If the arc is brought into close proximity to a steel surface, the recombination process tends to be concentrated upon the surface, where it is most effective in raising the temperature to that required for welding. The surface acts to bring about the sharp drop in hydrogen-gas temperature required to cause recombination. After giving up its heat of recombination from the atomic to the molecular state, the hydrogen gas envelops the surfaces being welded, protecting the metal from the action of the atmospheric oxygen and nitrogen. The hydrogen gas then burns in air at a short distance from the weld, providing some preheating and postheating effect. The hydrogen gas has a further effect of protecting the tungsten electrodes from the rapid oxidation which would take place if such an arc were operated in air. Since the plates being welded do not constitute a part of the path of the current used in the atomic-hydrogen arc, this process possesses the freedom and flexibility of manipulation provided by any gas torch process.

Applications of Atomic-hydrogen Welding. The atomic-hydrogen process was developed while the use of bare electrodes still predominated in ordinary electric arc welding. The remarkable ductility of mild steel welds made by the atomic-hydrogen process, when compared with welds made by bare electrode arc welding, seemed to provide a very great future for atomic-hydrogen welding in the field of mild steel. However, the subsequent development of covered electrodes, providing the desired ductility without the necessity for special equipment, and without the comparatively high cost of supplying the hydrogen gas, prevented the wide use of atomic-hydrogen welding for ordinary steels. Nevertheless the protection afforded the weld in the atomic-hydrogen process makes it suitable for a wide variety of ferrous and non-ferrous alloys. It is particularly versatile in these cases, since it is not necessary to use special flux-coated electrodes, such as would be required for metal-arc welding. Bare rods of the same composition as the plates being welded are all that are usually required. This process is particularly suitable for the welding of nickel and its high alloys but is not suitable for welding ordinary copper, since hydrogen attacks the oxygen dissolved in ordinary copper.

HELIARC WELDING. A somewhat more recent development in gaseous protected arcs is known as the heliarc method, in which the arc operates in a stream of helium or argon gas. In this process the arc is maintained between a tungsten electrode and the work to be welded. The current density employed is such that the tungsten electrode is consumed at a very low rate. This process has been found particularly suitable for welding the readily oxidizable low-melting magnesium alloys and has also been found useful for the fusion welding of aluminum alloys and for the thinner gages of stainless steel.

SUBMERGED MELT WELDING. In this process a bare metal electrode plows through a thick mass of powdered fluxing material which serves both as a shield for the arc and as a blanket to permit the metal already deposited to cool slowly and to protect the deposited metal from oxidation. This process is usually employed with a very much higher power level than is commonly employed with metal-arc welding, permitting much greater depth of penetration and deposition of metal than are found practicable in metal-arc welding, and permitting travel speeds considerably higher than are employed in ordinary metal-arc welding. The much greater energy input used with this process creates a relatively large pool of molten flux material which completely submerges the arc, so that it is not necessary for the operator to wear special eye protection against the intense light radiation from the arc. Since the arc is invisible to the operator, this process cannot readily be controlled manually but must be employed in a process in which the feeding of the electrode is controlled automatically to maintain constant voltage across the terminals of the arc. The speed of butt welding of plates of the order of $^3/_4$ in. in thickness is of the order of 30 times faster with this process than with manual arc welding using multiple pass deposits. This factor has been of great importance in the large-scale production of pressure vessels and in shipbuilding.

38. AUTOMATIC ARC WELDING

The process of metal-arc welding may be speeded up considerably by feeding the welding electrode to the arc from coils through motor-driven feed rolls. The process is made automatic by controlling the speed of the motor to maintain constant arc voltage and hence constant arc length. The conversion of bare electrode arc welding to an automatic process was comparatively simple in comparison with the problem of making the covered electrode process automatic. The difficulty arises from the necessity of feeding current through the electrode-coating material, which is non-conducting. Perhaps the commonest of the several expedients which have been tried is to provide a cutting tool to slit the coating and permit a series of narrow fingers to conduct current to the core wire. Another method is to prepare electrodes in comparatively long lengths, approximately 4 ft, with screwed couplings to which a flexible arc-welding cable may be clamped. It is then necessary for the operator to screw on additional lengths of electrode and also to move the cable connections from one coupling to the next. Still another method involves wrapping flux impregnated paper around a bare electrode close to the point of feeding into the arc stream. Just behind the wrapping mechanism are the sliding electric contacts which carry current to the electrode. Other methods have involved grooving or deforming the outer surface of the electrode to contain the fluxing material and still provide an external metal contact. These and other methods involving an impregnation of wire mesh in the coating have been found to be an unsatisfactory compromise between good electrical contact and sufficient space for the proper amount of welding flux.

ADVANTAGES AND DISADVANTAGES OF AUTOMATIC METAL-ARC WELDING. The automatic metal-arc welding process provides greater speed of welding through continuous uninterrupted operation, obtained by feeding electrode material from a long coil. The frequent interruptions in operation which are necessary in manual arc welding to permit changing electrodes represent an appreciable percentage of the total welding time. The interruption in arc time is effectively lengthened by the careful manipulation required in restarting the arc, to secure proper fusion at the point of restarting. The automatic process tends to be somewhat more economical, since it avoids the wastage resulting from discarding approximately 10 per cent of the electrodes used for manual arc welding, as stub ends remaining in the electrode holder.

The disadvantages of automatic arc welding are principally the greater first cost of the installation and the requirement for greater precision in the preparation of plates for welding. The automatic arc-welding equipment cannot compensate readily for even small differences in plate separation and width of welding groove.

Automatic welding equipment may be provided for the feeding of single electrodes. In regular production many of the advantages of automatic welding are lost with this method of operation, unless the welding operation involves just the amount of deposited metal which is provided by a single electrode. The single-electrode feeding mechanisms are also useful for the testing of electrodes and for other research applications where it is desired to eliminate the personal factors involved in manual manipulation.

Another disadvantage of automatic arc welding is the general requirement that the work be carried to the machine, rather than the machine to the work, and also that provision be made for depositing metal in a downward position. This latter consideration is not always a serious disadvantage, however, since even with manual welding the speed of deposition in the downward position is sufficiently greater, and the quality of the deposited metal is sufficiently improved, to justify the installation of proper equipment to permit most of the welding to be done in the downward position. The lack of portability of most automatic arc-welding equipment makes manual welding necessary for many applications, such as the erection of steel buildings and bridge construction. In the construction of trusses and girders in the shop, however, there is an opportunity for automatic welding. Although the initial installation of automatic machinery is relatively expensive, the saving in time and the superior nature of the welds generally justify the method in mass production. Automatic welding machines differ in many of their details. Essentially, the electrode material is fed automatically to the arc, and the arc is moved along the seam to be welded by a mechanical drive mechanism. The mechanism for feeding the electrode is embodied in a part designated the "welding head." Sometimes two or more welding heads follow one another in welding a seam, or two different seams of a single weldment are welded simultaneously by the operation of two welding heads. For example, in the building up of plate girders using automatic welding equipment, it is desirable from the viewpoint of distortion to weld both sides of the web to the flange at the same time.

The current density used with automatic arc welding may be much higher than that used with manual welding for those types of equipment which provide for the feeding of current to the electrode close to the arc. The current density for manual arc welding is limited by the heating of the electrode, which must be fed through its entire length for the length of time required to consume the electrode. This greater current density is one of the reasons for the greater speed of automatic arc welding. The possibility of operating continuously for long periods without the necessity for rest also contributes to the speed of automatic welding. Often one man can run continuously more than one automatic machine. In view of the lack of strain on the welding operator, older men or less rugged individuals may successfully operate automatic arc-welding machines. Furthermore, long periods of experience, required for the development of manipulatory skill for high-grade manual arc welding, are not required with automatic equipment.

39. AUXILIARY EQUIPMENT

ELECTRODE HOLDERS. A great deal of effort has been expended in the development of electrode holders to permit easy removal and replacement of electrodes, since this operation represents an appreciable part of the time and effort of the welding operator. For easy access into various corner locations it is customary to be able to clamp the electrodes not only at right angles to the handle but also at an acute angle with the axis of the handle. Some types of electrode holder have been especially constructed with insulation and guards to permit laying the electrode holder down on a grounded metal plate surface without short-circuiting when no electrode, or only a short stub end, remains in the holder.

CHIPPING TOOLS. Another operation which involves the expenditure of a considerable portion of the welding operator's time is the removal of scale or slag from a previously deposited layer of weld metal, in order to permit the deposition of a succeeding layer or to permit restarting to complete a previously deposited pass of weld metal. Light-weight pneumatic tools have been developed for quickly and conveniently loosening the slag from the weld metal surface. These light-weight tools reduce arm fatigue in the welding operator, thus avoiding any interference with the steadiness with which he is able to hold the electric arc.

POSITIONING EQUIPMENT. Since welding in a downward direction, such that the weld is in a horizontal plane, permits much higher deposition rates, as well as tends toward greater soundness and density of the weld deposit, and also because welds made in this position are neater in appearance and higher in mechanical properties, it becomes economical to provide special fixtures for holding and revolving the work so as to bring the welds into the above-mentioned position. Such equipment, which is built in a wide variety of sizes to handle from the smallest to comparatively large weldments, usually consists of a table provided with suitable slots for bolting the weldment to it, together with tilting mechanisms to incline the table at any of a wide variety of angles. The table is also arranged to be revolved at a speed which is adjustable to permit continuous welding at the top of a circumferential seam in a cylindrical vessel. Positioning equipments are frequently used with automatic welding machines, as well as for holding and revolving parts which are to be manually welded.

40. RESIDUAL STRESSES AND DISTORTION

The rate of heating in electric-arc welding is so rapid as to create steep thermal gradients in the vicinity of the weld. These result in thermal expansion gradients which are so large that they cause the metal in the vicinity of the weld to become plastically upset in compression. In the thermal contraction which follows, only a portion of the upsetting is relieved plastically, and the remainder of the weld shrinkage produces residual elastic stresses and distortion. The shrinkage of the weld and immediately adjacent metal results in the production of longitudinal and transverse tensile stresses (11).* The greater the freedom of the structure to permit weld-metal contractions, the lower will be the magnitude of the residual stresses. When butt welds are made in flat plates, the plates themselves provide sufficient restraint of weld contraction to result in residual stresses, approaching the yield strength of the weld metal, in a direction along the center line of the weld, except where the weld approaches the edges of the plates (12). Transverse stresses of high magnitude exist near the center of the weld but fall off rapidly with distance away from the weld, unless the edges of the plate being welded are restrained from transverse contraction by virtue of their attachment to other rigid parts of the structure. Under certain conditions the rigidity of the structure may be sufficient to cause the welds to crack upon cooling.

The careful avoidance of such severe conditions of restraint and the planning of a proper sequence of welding to ensure maximum possible freedom of contraction are an important part of the forethought required to produce a successful welded structure.

Proper planning of weld sequence to minimize weld stresses is only one of the problems arising from weld shrinkage. When welds are made in locations that are not symmetrical with respect to the remainder of the structure, they tend to pull on the remainder of the structure and cause objectionable distortion. In order to avoid distortion, it is frequently advantageous to attach the structure being welded to a much more rigid body while the weld is being deposited. This relieves more of the weld shrinkage by plastic flow and thus tends to minimize distortion. In some cases it is possible to predistort the weld structure in the opposite direction to that in which weld shrinkage would tend to deflect it. If the welded structure is left clamped in this position until the weld has cooled, the final structure will have the desired alignment.

The deposition of weld metal in small beads of low energy input tends to minimize the magnitude of transverse stresses built up when it is necessary to make welds under conditions of severe transverse restraint.

Preheating of the plates adjacent to a weld before the beginning of a welding operation will tend to reduce the longitudinal tension stresses if the amount of previous deformation has been properly estimated. The ability to estimate accurately the amount of prior deformation is a matter of judgment based on experience. If a number of identical or nearly identical structures are to be built, it is possible after one or more tries to fabricate welded structures to remarkably close dimensional tolerances. Not only is it possible to reduce distortion by preliminary mechanical deformation, but also it is possible to use thermal expansion properly applied locally to accomplish the same result. In order that these methods may have maximum effectiveness, it is necessary to maintain the mechanical or thermal deformation until the weld metal has cooled to the temperature of the surrounding plates.

When weld metal which has appreciable ductility is deposited, the presence of residual stresses in a welded structure does not lower its strength properties. This is due to the fact that, when external loads are applied, portions of the structure containing high residual stresses tend to relieve themselves by local plastic flow, thus distributing the load stresses over the remainder of the structure uniformly. There are certain types of service in which plastic flow tends to be prevented, however, and in these cases it may be desirable to re-

* Numbers in parentheses refer to items in the Bibliography, Section 45.

lieve the residual stresses. For example, plastic flow is restricted in low-temperature service, in very rigid structures tending to create a high degree of triaxiality of stress, and in structures subjected to very rapidly applied loads. Residual stresses may also be objectionable in a structure if subsequent machining operations are performed which tend to relieve a portion of the residual stresses, thus unbalancing the system of stresses within the structure and producing further objectionable distortion as a result of machining. Residual stresses may also tend to accelerate corrosion and thus be objectionable when corrosion is a serious problem in the life of the structure.

THERMAL STRESS RELIEVING. The residual stresses in a welded steel structure may be largely eliminated by heating the entire structure in a furnace to a temperature within the range from 1100 to 1250 deg fahr (15). The entire structure must be held in the furnace at this temperature for a period of time and subsequently cooled slowly in the furnace. It is customary to use a heating time of approximately 1 hr per in. of thickness. The range of temperatures mentioned above constitutes the highest practicable without transforming most steels into the plastic condition in which not only would the grain structure of the metal be altered, but also the welded structure would become so plastic as to be incapable of supporting itself and maintaining its shape. The ability of a structure to be relieved of residual stresses by the thermal treatment just described is due to two factors. The first is that the elastic limits of most steels drop very rapidly in this temperature range. Thus ordinary steel, having an elastic limit of approximately 34,000 psi at room temperature, suffers a drop in elastic limit to about 8000 psi at 1200 deg fahr. Thus, as soon as a structure reaches this temperature, all residual stresses in excess of 8000 psi are immediately relieved by plastic flow. A second factor, which still further reduces the magnitude of residual stresses to something less than 2000 psi, is the phenomenon of creep, i.e., continuous plastic flow of metal under the action of stress at elevated temperatures.

Residual stresses may also be relieved in a structure by subjecting the entire structure to stresses causing plastic deformation. After the stresses in the structure have been raised so that all parts tend to flow plastically, any residual stresses previously existing will have been relieved. This method of relieving residual stresses is not adaptable to most structures, however, although the residual stresses in the longitudinal seams of a pressure vessel may be considerably relieved in this manner by the application of sufficient internal hydraulic pressure.

PEENING. Peening is a term applied to the extensive hammering of the surface of deposited metal, e.g., with a pneumatic hammer, immediately after welding. Although this process has frequently been used for the prevention of weld-metal cracking in the deposition of numerous layers of weld metal in making a joint in heavy plate material, and also for the correction of distortion caused by welding, it has only recently been shown to be effective for relief of weld contraction stresses (14). It tends to produce compressive stresses in the surface of the last pass of weld metal to which it is applied and thus tends to prevent weld-metal cracking. By putting sufficient compressive stresses in the surface of a multipass welded joint, it is possible to balance the weld shrinkage tensile stresses with peened surface compressive stresses and thus correct distortion.

In order to relieve residual stresses, peening must be applied not only after the early passes of a multipass arc weld, but particularly after the last pass. It is even more effective if applied not only to the surface of the weld metal, but also to a strip about 2 in. wide on either side of the weld, after the weld has been completed and cooled. In many cases it is difficult to control the intensity of peening in a sufficiently reliable manner to secure consistent results. If peening is too light, it is ineffective; and if too heavy, the result is likely to be a surface full of fine cracks. Peening must of course be done with a blunt-nosed tool to minimize surface fracture. Peening may be more satisfactorily controlled when it is used for the correction of distortion, since it is possible to observe, after the deposition of each layer of weld metal, when sufficient peening has been applied.

41. QUALIFICATION OF WELDING OPERATORS

In order to guarantee the quality of welding in important structures a careful series of tests is usually required, both to ensure that the procedure to be used is capable of producing welds of adequate strength and dependability, and also to test each operator's ability to perform welding under the specified procedure which will be acceptable in quality for the service intended. Details of these procedure-qualification tests and operator-qualification tests may be found in the latest edition of the Welding Handbook of the American Welding Society. The testing methods used for operator qualification have been greatly simplified so as to test only the operator's ability to produce sound welds with good fusion to the base metal in various welding positions, and not the entire procedure and quality of welding materials. The operator qualification tests have also been designed to require a minimum of machining.

42. NON-DESTRUCTIVE TESTING

Radiographic inspection of welds is required for certain kinds of work, particularly in the important classes of pressure vessels. Radiographic inspection is well suited to determination of the soundness of deposited metal. Excessive porosity and lack of fusion determined in this manner may be removed by chipping, and the joints rewelded so as to ensure dependable operation of the structure. Magnetic inspection methods may also be used to indicate the completeness of fusion and the absence of cracks in welded joints. This method is considerably less expensive and more rapid than radiographic inspection and is equally well suited to the detection of cracks in welds. The non-destructive methods of testing just mentioned have exercised a great influence in developing the confidence of users in welded structures. The fact that the better types of pressure vessels are now constructed universally by welding is a definite indication of the successful methods of control which have been developed for the fusion arc-welding process.

43. ELECTRIC RESISTANCE WELDING

Electric resistance welding includes a group of welding processes wherein heat is obtained from the resistance to the flow of an electric current, and the weld is consummated by pressure. The generation of heat takes place not only within the body of the material because of its electrical resistivity, but also at the contacting surfaces of the parts to be welded. It is desirable to restrict the high temperatures required to the immediate vicinity of the surfaces to be welded. This localization of heat is accomplished by taking advantage of surface contact resistance, by restriction of cross-section of part at the weld, or by providing a sufficiently high power level to burn clear the series of short circuits produced by bringing the metals into light contact repeatedly. The heat is further confined to the immediate vicinity of the weld by employing the highest practicable power level for the minimum time required to produce a satisfactory weld. In addition to its advantage in localizing the heat, a high power level results in the minimum total expenditure of energy to make the weld, because of avoidance of a prolonged dissipation of heat into the adjacent metal. However, the use of high power levels for short periods of time introduces two important requirements, one in the matter of control, and the other in the matter of power supply.

High power level results in very rapid rises in temperature of the parts to be welded. Unless the factors responsible for the generation of heat are kept under precise control to deliver exactly the amount of energy required to make a proper weld, either no weld at all or a badly overheated weld may result. The generation of heat in an electrical resistance is a function of the square of the current, the resistance, and the time. In order to reproduce a resistance weld consistently, it becomes necessary to control the magnitude of the current, the time of flow of the current, and the resistance through which the current flows. Consistent delivery of the proper magnitude of current involves careful attention to the voltage regulation of the power system which supplies the welding current.

If numerous welding machines are operated from the same power supply, there is likelihood that the voltage drop caused by the operation of one machine may seriously affect the current delivered by another machine, if two or more machines should happen to operate simultaneously. The correction of this situation involves expensive reinforcement of the power supply by use of heavy cables, high voltages, and much larger transformers than would normally be required by the heat generated in these parts, in order to reduce the reactance and consequent voltage drop in the cables and transformers. Another solution to this problem involves elaborate interlocking equipment to prevent the operation of more than one machine at one time. Such an arrangement is theoretically possible in many cases, because of the very low duty cycle of many resistance welding operations.

The time of current flow may be controlled by either mechanical or electrical timing devices. The magnitude of the resistance through which the current flows is determined by the cross-sectional area of the current path, the type of material used, its surface condition, and the pressure with which the parts are held in contact. The resistance to the flow of current is inversely proportional to the force holding the parts in contact.

The power supply is important not only from the standpoint of properly controlling the magnitude of the welding current, but also from the standpoint of its possible interference with other forms of electrical service, particularly electric illumination. The frequent or periodic withdrawal of currents of large magnitude for short periods of time may result in objectionable flicker of lighting circuits connected to the same power system. Unless properly provided for, this type of interference may extend to other customers of the power company, therefore being doubly objectionable. If only one large resistance welding machine, or one particularly large machine in a group of relatively small ones, is to be provided for, it is possible to use a motor generator set to avoid interference with the rest of the electrical system. One reason for the relatively large magnitudes of resistance welding-machine currents is the comparatively low power factor of these machines. With a motor-generator set it is possible to reduce the current drawn from the system by use of a polyphase motor to distribute the load uniformly over the different phases of the power system. If a synchronous motor is used, operation near unity power factor will tend to reduce still further the magnitude of the current drawn from the system, and hence the voltage drops in the various parts of the system. The fly-wheel effect of the rotating motor generator will also tend to reduce the severity of the shock on the electrical system resulting from the sudden withdrawal of a high magnitude of welding current. The use of a motor-generator set is a very costly method of providing power supply for resistance welding and is therefore justifiable only under special circumstances. It is also not at all suitable when more than one large resistance welding machine is to be operated, because of the large voltage regulation of the comparatively small generator in the motor-generator set. Operation of more than one resistance welding machine from a motor generator would require a system of interlocking to avoid simultaneous operation.

The various considerations mentioned have forced the introduction of stored-energy resistance welding machines for conditions where it is desired to operate large batteries of these machines in relatively close proximity. A conspicuous example of this type of condition existed in the aircraft industry when it was organized for the mass production of military aircraft. The problem in this case was further complicated by the necessity of locating aircraft plants in regions which were not served by very heavy electric power utility systems. The stored-energy machines, as their name implies, are arranged to take power from an electric power system, and usually from all phases of a polyphase system, over a relatively long period of time and at a low rate, in order to be able to deliver the much larger magnitudes of current required by the welding process for short periods of time. The power-supply problem for the aircraft industry was also complicated by the machines of relatively large capacity required to weld the high-conductivity aluminum alloys. In the paragraphs which follow, the various resistance welding processes will be defined and described as to equipment, power supply, controls, importance and magnitude of the welding variables, special problems, and range of applicability.

SPOT WELDING Spot welding is a resistance welding process whereby welds are made between two or more overlapping sheets of metal by pressing them together between two electrodes arranged to conduct current to the outer surfaces of the overlapped sheets. The tips of one or both of the electrodes are restricted in area to approximately the diameter of spot weld desired. Sufficiently high electrode force must be applied to localize the current properly and consistently reproduce the resistance of its path for successive welds. If insufficient pressure is applied, the surface-contact resistance will be erratic in magnitude and hence the welds will be inconsistent in strength. Another difficulty which may be experienced if the pressure is of lower than proper magnitude is the generation of heat at too rapid a rate at the contact or faying surfaces. When heat is generated too rapidly at the faying surfaces, the metal may be brought to the melting point and expelled violently from between the sheets, before the adjacent metal within the body of the part to be welded is sufficiently heated to make a satisfactory weld. This condition of early flashing and expulsion of metal tends to make the final weld strength inconsistent. If more than sufficient pressure is applied in resistance welding, the usual result is excessive distortion of the parts being welded. Hence the pressure to be used in resistance welding is a compromise between that required to effect a satisfactory weld consistent in strength, and that which, because of its high magnitude, causes objectionable distortion.

The proper electrode force to be used with various materials is determined, as mentioned above, by the difficulty in securing intimacy of contact. For harder materials and heavier gages, greater electrode forces are required. In some of the high-strength materials the need for high electrode force introduces a serious problem in the matter of electrode deformation as a result of making a series of consecutive welds. This deformation, frequently called "mushrooming," if allowed to proceed very far, results in changing the current density and hence the ability of a weld to be properly fused by the particular value of welding current found satisfactory for the first welds made in the series. When electrodes become deformed in this manner, they must be removed from the welding machine and redressed to the proper tip contour, usually in a lathe. Much development has gone into the problem of making electrodes of the highest possible strength consistent with satisfactory electrical conductivity. The electrode material, usually a copper alloy, because of the necessity for high electrical conductivity, cannot be hardened appreciably by most alloy additions without seriously affecting its electrical conductivity. In order to prevent overheating of the electrodes and consequent acceleration of mushrooming, spot-welding electrodes are usually water cooled. The combination of high electrical conductivity and water cooling tends to keep the electrode surfaces sufficiently cool to avoid overheating of the surfaces of the metal sheets being welded. It is fortunate that the high electrical conductivity of the electrode material necessary to minimize the generation of heat within the electrode is also accompanied by high thermal conductivity, which is of great value in dissipating the heat unavoidably generated at the electrode contacting surfaces.

In building a resistance welding machine, it is necessary to mount the electrodes in such a way that work pieces of the desired size and shape may be inserted between the electrodes. This involves providing a certain throat depth, which is the distance in a welding machine from the center line of the electrodes to the nearest point of interference for flat work or sheets. For more complicated shapes, such as box-type structures, unobstructed work clearance must be provided at right angles to the throat depth between the arms or frame of the welding machine. In a spot-welding machine the product of throat depth and throat opening just described is a throat area which determines to a large degree the inductance, and hence reactance, of the secondary circuit of the welding machine. This determines the voltage required to produce the desired welding current in the secondary circuit of the welding machine. Since the welding current is usually obtained from the power-supply system through a welding transformer built into the machine, the higher the secondary voltage required, the higher will be the magnitude of primary current demand for a given welding current. This higher current demand will be reflected in higher cost of cable installation and greater size and cost of distribution transformers, together with power cost of losses in cables and transformers.

These considerations point to the very great importance of careful selection of welding equipment to avoid too large an area in the secondary circuit of the welding machine. It is therefore preferable to determine at the time of installation the maximum size of part which it is necessary to weld in the immediately foreseeable future, and not to overestimate for some remote future possibility. In addition to the penalty of excessive power losses and higher cost of power supply resulting from overestimation of the proper size of resistance welding machines, there is also a severe penalty in the first cost of the welding machine itself. This is due to the fact that the cost is largely determined by the force which must be applied by the welding electrodes and the throat depth at which this force must be applied, since this determines the size and strength of the machine required.

In some machines the entire lower arm, extending out for the full throat depth, is unsupported in order to permit placing work, such as a cylindrical tube, over the whole length of the arm. In this type of machine, called the "rocker-arm" type, it is also customary to have the entire upper arm pivoted from a point within the main frame of the welding machine behind the throat depth. This permits raising the entire upper arm to allow easy entrance of the work. In other machines the entire frame is rigid right out to the point of attachment of the welding electrodes. This is called a "press-type" welder. Retraction of the upper electrode is provided in order to admit entrance of the work.

The application of electrode force in most resistance welding machines is controlled either by pneumatic pressure or by hydraulic pressure. Operation of the machine to permit entrance and removal of the work is easily provided, either pneumatically or by means of a motor drive. It is customary to provide a number of welding "heats," or settings of welding current, either by taps on the welding transformer or by taps on a separate auto-transformer built into the welding machine. Large changes in the range of welding current are commonly provided by switching two or more primary windings in series and parallel combinations.

Series Spot Welding. When in the fabrication of extended structures, such as rail cars, it becomes difficult to reach both sides of the structure in order to make spot welds, it is possible to use two electrodes on the exterior surface of the structure and to back up the parts being welded by a heavy copper bar extending between the positions contacted by the two electrodes. In this way two spot welds may be made simultaneously by passing the current from one electrode through the sheet to the copper backing bar and then back through the sheets to the other electrode. These two electrodes are connected to the same welding transformer. Series spot welding permits a great reduction in the reactance of the secondary circuit of the welding machine, since the electrical connections from the welding transformer may be brought close together to the two electrodes, which are applied to one side of the parts being welded. This method is also applicable if it is desired to make only one spot weld, by substituting for the other restricted area electrode one of sufficiently large size so that no weld will be produced.

Control Equipment. Many spot-welding machines have been and still are controlled by electromagnetic contactors designed to close and open the welding circuits under control of mechanical, electrical, or electromagnetic relays, or electronic timing circuits. The development of metal tubes of the ionized gas conduction type, having sufficient current-conducting capacity to be used directly in the primary circuit of the welding machine, has led to the widespread adoption of these devices for the switching and control of welding circuits. The rapid switching of welding circuits by means of electromagnetic contactors presents a serious duty cycle for these devices, resulting from frequent arcing at the contacts due to opening large currents in heavily inductive circuits. These devices have a further objection from the standpoint of welding operation, in that it is difficult to synchronize the closure of the circuit properly with the voltage wave of the power supply. This results in a variety of unpredictable transient currents at the start of current flow for different welds, with consequent appreciable differences in the energy delivered to these welds and variations in the consistency of weld strength. With the gaseous conduction switching devices, called ignitrons, it is possible always to initiate the welding current at the same phase relation with respect to the voltage wave of the power supply, thus assuring consistent successive welding currents. A further great advantage of the ignitron tube for handling spot-welding currents is the possibility of almost infinite adjustment in the magnitude of welding current by phase control of the point of initiation of the welding discharge in each half-cycle of the passage of welding current. With magnetic contactors there is, moreover, some difficulty in obtaining satisfactory consistency in timing the duration of current flow. This problem also is much simpler and capable of more precise control when ignitron tubes are used for the control of welding current. The period of conduction of a welding current is usually timed by an electronic circuit involving the charge of a condenser with a regulated magnitude of current. This results in building up the charge on the condenser in a definite, controllable, and reproducible time. Positive timing of welding circuits within one cycle of the usual 60-cycle power system is therefore possible.

STORED-ENERGY WELDING MACHINES. Stored-energy machines for electric resistance spot welding were introduced into this country by the Sciaky brothers. These machines were of the electromagnetic type, in which energy is stored in a magnetic field as the current is built up in a coil to produce that field. A second coil, the secondary of the welding machine, is linked with the field set up in the primary coil. Interruption of the current in the primary winding of the magnetic inductor, or special transformer, tends to cause collapse of the magnetic field, which in turn induces the welding current in the secondary coil linked with the same magnetic circuit. The power supply for this type of welding machine consists of a polyphase rectifier capable of providing several hundred amperes of direct current at a sufficient voltage to cause rapid building up of the magnetic field in the core of the welding transformer.

The magnitude of the welding current induced in the secondary circuit is controlled by an adjustable current relay which may be set to interrupt the direct current building up in the primary winding when it reaches any predetermined value. The operation of this welding machine is simple in principle, but the problem of interrupting the current flowing in such a highly inductive circuit proved difficult. It was first solved by use of a series of electromagnetic contactors operating in rapid succession, the opening of each contactor acting to insert additional resistance in the primary circuit. By the time the last contactor completely opened the circuit, the current had been reduced by the insertion of resistance to a safe value for final interruption. Since maintenance was required to keep the series of contactors in proper operation, another method of interruption was developed whereby these contactors are eliminated. This consisted of a high-speed pneumatically operated switch, across which was provided a bank of condensers of enough capacity to absorb energy from the magnetic field for a sufficient period of time to allow the contacts of the high-speed switch to separate far enough to avoid arcing. Stored-energy welding machines were adopted rapidly by the aircraft industry as a solution to the almost impossible problem of providing adequate power supplies for a larger number of resistance welding machines in regions remote from centers of power development and distribution.

Stimulated by the rapid acceptance of the electromagnetic stored-energy machines, several other companies became engaged in the development of electrostatic stored-energy machines. In these machines a polyphase rectifier delivers rather smaller currents than are required by electromagnetic stored-energy welders, but at much higher voltages, up to 3000 volts, to a bank of condensers. The magnitude of the welding current of these machines is adjusted by controlling the charging voltage and the number of condensers charged. These machines are simple in principle, but rather complicated control circuits are required to take care of charging and discharging the condensers, together with such other features as providing for the establishment of definite charging voltages independent of line voltage fluctuation.

BATTERY SPOT WELDER. A still more recent development in stored-energy machines is the use of storage batteries to provide spot-welding currents running up into the tens of thousands of amperes. Polyphase rectifiers are provided to deliver several hundred amperes at from 4 to 8 volts, depending upon the number of storage cells connected in series. The ampere capacity of the charging units is determined by the magnitude of the welding current and the speed of operation desired. The control unit is arranged to turn off the charging unit or to turn it on at one or more different rates, depending upon the battery voltage. The high discharge rate and comparatively high charging rates employed in battery welding machines necessitated the introduction of water cooling in the storage cells and in their connections. Standard batteries are unnecessarily bulky for this type of service, in which large areas of plates of thinner and hence more compact construction could be employed to provide high ampere capacity, without the unnecessarily high ampere-hour capacity usually provided in standard batteries for other types of service.

No welding transformer is provided with this type of welding machine, the welding current coming directly from the storage batteries. This arrangement constituted one of the greatest problems in the development of this type of equipment, since it became necessary to close and open rapidly a circuit carrying several tens of thousands of amperes. This problem has been effectively solved by providing a rather large area of thin carbon plate pressed between copper plates by a sufficiently high pneumatically applied force. A system of magnetically operated valves provides for the rapid application and removal of force on the carbon plate. The operation of this device provides an essentially square wave of direct current for spot welding. With this wave form, spot welds may be made in aluminum alloys at lower values of current than the peak values required with condenser discharge spot-welding machines.

SPOT-WELDING ELECTRODES. As mentioned above, the electrode materials used for spot welding are a compromise between the desire for high strength to resist deformation or change of shape and the desire, on the other hand, for high electrical and thermal conductivity to permit operation of the electrodes at the lowest possible temperature. This operating condition tends to preserve the surface condition of the materials being welded and to give the electrodes longer life before redressing is necessary. With some materials low electrode temperature is desirable to prevent fouling and subsequent sticking of the electrodes to the surface of the work, as a result of the pick-up and alloying of the sheet material with the electrode surface. Pure copper electrodes have high electrical conductivity but insufficient mechanical strength to resist mushrooming. On the other hand, copper-tungsten-alloy electrodes or tungsten-alloy-tipped electrodes have sufficient hardness to resist deformation but have such high electrical resistivity as to generate too much heat at the electrode-to-sheet contact surfaces. This may result in burning or otherwise damaging the sheet surfaces and in rapid deterioration of the electrodes by pick-up of the sheet material.

The shape of the electrode surface in contact with the work is of great importance in the production of satisfactory spot welds. Electrodes are usually made of cylindrical material and are hollow to permit internal water cooling. Flat surfaces may be provided at the ends for contacting the sheets to be welded. However, in order to localize the current properly, it is necessary to reduce the diameter of the flat contacting surface to not more than about 10 per cent greater than the desired diameter of the spot welds. This is accomplished by machining the ends of the electrodes in the form of truncated cones, the sides of which make an angle of 60 or 45 deg with the axis of the electrodes.

Although flat-surfaced electrodes may properly be used for spot welding, a difficult problem of practical importance is the alignment of the two electrodes to give uniform contact pressure over the entire surface of electrode contact. Unless special care is exercised, the electrodes will tend to exert higher pressure on one edge than on the rest of the surface. In view of this situation, it has been found possible to secure much better results in practice by machining the contacting tip of the electrode in the form of a sphere of comparatively large radius. For the thinner gages of the softer materials (below 0.040 in.), 4-in.-radius domes have been found satisfactory. This radius has also been found suitable for considerably heavier gages (up to 0.080 in.) for the relatively soft structural aluminum alloys. For the higher-strength steels and for heavier gages, the radii of the spheres of contact should be increased to 6, 8, or even 10 in. Even with these large radii, the alignment problem noted in the use of the flat electrodes is largely eliminated. This results in distinctly more symmetrical welds, because of the more intimate contact provided at the center of the weld area.

The surface appearance of the sheet is also improved by the dome-shaped impressions, as compared with the irregular impressions produced by improperly aligned flat-surfaced electrodes. With the softer structural aluminum alloys, it is not necessary to restrict the dome-shaped surface to a diameter of not more than 10 per cent greater than the desired weld size. It is, on the other hand, possible to permit the dome-shaped surface to extend to the full diameter of the cylindrical electrode tip. If this type of unrestricted dome is used with high-strength steels or other non-ferrous alloys, very porous welds will result, because of the tendency of the electrodes to press into the sheet surface, when making a weld, only sufficiently to become supported on the cold ring of metal outside the weld area. If sound welds are to be produced, the electrode tips must be able to sink into the sheet surfaces sufficiently to compensate for weld shrinkage by plastic deformation at right angles to the sheet surface.

Table 1 shows typical values of welding variables for the spot welding of low-carbon steel. The electrode-face diameter shown in the table is that of the flat surface in contact with the sheet. Somewhat better welds may be made if a dome-shaped surface having a radius of from 4 to 10 in. is employed, the smaller radius being for the smaller face diameter, and the larger radius for the larger face diameter to be used with the heavier sheet thicknesses. It is not necessary to use a large number of radii, since 4-, 6-, 8-, or 10-in. radii will be all the changes required to cover the thickness ranges from the 0.014- to 0.125-in. sheet thicknesses. These electrode-face diameters have been selected to be slightly larger than the desirable weld size in this material. An approximation to the desired size may be obtained by taking the square root of the sheet thickness. It is customary, however, to use welds somewhat larger than the figure thus obtained. The electrode forces have been selected to produce approximately 20,000 psi over the flat face in contact with the sheet. Although these figures are satisfactory for low-carbon steels, the alloy steels and stainless steels require higher electrode forces. The values of weld time shown in the table may be used not only for low-carbon steel but also for low-alloy steels. These times were selected from wide experience with the spot welding of a large variety of steels and after experimental investigation to determine the shortest time in which a satisfactory spot weld could be produced in the corresponding thickness. In the usual run of materials, the attempt to use shorter times tends toward the production of flashing beneath the surfaces of the electrodes contacting the sheets. Longer weld times may of course be used but are not necessary for the production of sound welds in the usual variety of low-carbon steels. The welding currents shown in this table are only approximations, since some freedom in the actual magnitude of current must be provided to take care of variations in the surface condition of sheet material and in the operating characteristics of different welding machines. The values of current here shown will serve to illustrate the magnitudes required for welding different sheet thicknesses and will thus be of assistance in the selection of equipment for particular applications.

Table 1. Typical Values of Welding Variables for Low-carbon Steel

Sheet Thickness		Electrode-Face Diameter, Inches	Electrode Force, Pounds	Weld Time, Cycles (60/Sec)	Welding Current, Amperes (Approx.)
Inches	U.S.S. Gage				
0.014	29	1/8	250	2	8,000
0.021	25	5/32	350	3	10,000
0.031	22	3/16	500	4	13,000
0.044	19	1/4	700	6	16,000
0.062	16	5/16	1,000	10	19,000
0.078	14	3/8	1,250	15	21,000
0.094	13	7/16	1,500	20	22,000
0.109	12	1/2	1,750	30	23,000
0.125	11	9/16	2,000	45	24,000

SPECIAL PROBLEMS. Surface Preparation. For best-quality spot welds, the surface of the material must be clean and free from scale or oxide. When hot-rolled steel sheets are to be used, it is customary to require pickling to remove scale. After pickling, in order to preserve the clean surfaces of the steel, it is customary to oil the sheets. For most consistent and reliable spot welding, it is necessary to degrease the sheets before spot welding. If this is not done, the electrodes tend to become coated with a carbonaceous film in the center of the contacting surface, which gradually reduces the conductivity at this part of the electrode, forcing the current to enter the sheet only at the outer periphery of contact. When this condition is continued, toroidal and finally crescent-shaped welds are produced. In addition to the hot-rolled and pickled surface condition, steel may be welded in the cold-rolled finish or after grit blasting.

In the spot welding of aluminum and magnesium alloys, the material is usually obtained with a surface of such high resistance that satisfactory spot welding cannot be performed until the oxide has been removed and the surface resistance reduced to a low magnitude.

Magnesium sheet material for spot welding should not be given the customary chrome-pickled treatment but should be obtained in the plain or oil-finished condition. The reason is that the vigorous chemical stripping action which must be performed to remove the chrome-pickled surface leaves the surface in an undesirably pitted condition. Methods have been developed for the proper chemical surface preparation of magnesium-alloy sheet for spot welding (3). The surface may also be abrasively cleaned for spot welding, but this method is less satisfactory in production, because of the additional cost and complication, as compared with chemical surface treatment. Furthermore, sticking of the electrodes to the sheet in the spot welding of magnesium alloys occurs much more rapidly if mechanical abrasive surface treatment is used.

In the spot welding of aluminum alloys, either mechanical abrasive treatment, such as wire brushing, or a chemical treatment may be used. The mechanical treatments were generally used, however, only before the development of satisfactory chemical methods. Information concerning the development of satisfactory chemical methods of surface preparation may be found in the literature (8).

One of the best methods of determining whether surface preparation for spot welding has been properly carried out is to measure the surface-contact resistance. Although these methods were developed for determining the satisfactoriness of chemical methods of preparation of aluminum alloys for spot welding, they have proved equally applicable to magnesium alloys and to steel. It has been found that reliable surface-contact resistance measurements of a material may be made by placing two pieces of it between a pair of copper-alloy electrodes having 4-in.-radius spherical contacting surfaces. Arrangement is made to press these electrodes together with a force of 1000 lb. For satisfactory surface preparation resistance values under the above conditions will be measured in the range from 10 to 100 microhms. The simplest method of measuring resistances of this order of magnitude is a Kelvin double bridge. The high precision of laboratory instruments, however, is not required for this purpose. Simple adaptations giving results to two significant figures are all that are needed. The general subject of surface-contact resistance and its importance in spot welding has been the subject of a recent paper by R. A. Wyant (6). This paper also includes methods of operation and descriptions of apparatus for surface-contact-resistance measurements, together with a list of further references in this field.

Spot Welding of Scaly Steel. Although it is generally preferable to remove the black mill scale ordinarily found on hot-rolled steel before spot welding, this operation is sometimes considered costly for certain structural applications where only a small number of spot welds are required near the ends of hot-rolled sections, such as plates, angles, and channels. To meet this condition, it has been found possible to weld right through the black mill scale of hot-rolled steel by applying a special welding sequence involving relatively slow preheating of the material to a sufficient temperature to render the scale conducting. It is then possible to apply a heavy welding current and produce a satisfactorily fused spot weld. The mill scale on the faying surfaces is apparently dissolved in the weld, but this much oxide does not seem to affect the strength of the spot welds seriously. The scale remains on the electrode contacting surface after the weld has been completed. Several hundred welds may be made without the necessity of changing electrodes for redressing, but this number will be less if the mill scale is contaminated by the presence of rust. Rust spots tend to develop in scaly steel when it has been exposed for a period of time to the atmosphere. Fresh mill scale is therefore much less troublesome. Since a matter of seconds of flow of preheat current appears to be required to render the scale conducting, there is some question as to how far this method of welding can be carried into the thinner gages of steel, where the weld period is usually a comparatively small fraction of a second. For gages of material between $^3/_{16}$ and $^1/_2$ in., the weld period is of the order of seconds, and the employment of an additional few seconds for preheating is not such a serious production limitation. Dome-shaped electrode tips of restricted diameter have been found preferable for this type of welding. A special control circuit must be employed to provide not only a weld interval but also a preheat interval. Although at the date of this writing only a limited amount of work has been done to develop the possibilities of the welding of scaly steel, it appears that the practical importance of this problem may lead to further investigation and possible improvements in this process.

Spot Welding of Hardenable Steels. Spot welds made in the usual manner on steels containing 0.15 per cent or more carbon are found to be brittle, so that they tend to fail easily under suddenly applied loads or upon the application of loads which tend to pull the welds apart at right angles to the sheet surfaces. This brittleness is due to the fact that the cooling rates associated with spot welding are so severe as to produce martensitic structures of a maximum hardness determined by the carbon content of the steel. It is true that these structures, resulting from rapid quenching from temperatures up to and above the melting point of steel, to low temperatures of the order of only a few hundred degrees, are the most brittle structures ordinarily produced in metals; nevertheless, these same structures, when tempered by heating to a temperature below that which will cause recrystallization, become among the toughest steel structures which can be produced.

It is a metallurgical axiom that the maximum toughness can be produced by tempering a fully hardened steel to the maximum possible temperature level below that which will cause recrystallization and rehardening. This principle led to the development of the method of tempering spot welds within the welding machines by passing a second pulse of current, following that which produced the weld by a sufficient interval to allow the metal in the weld and adjacent zones to be quenched to martensite (9).

It was found that the duration of the tempering current could be as short as, or even shorter than, the time required to make a satisfactory weld in any given thickness range. If a much shorter time than that used for welding were employed for tempering, however, the operation would be somewhat more critical, because of the very steep thermal gradient set up by the short heating time. For this reason it was found desirable to use a tempering time either as long as, or in some cases twice as long as, the welding time. A longer time permitted more uniform tempering over the spot-welded surface, and this factor was of somewhat more than average importance when spot welding the thickness ranges from $^1/_{16}$ to $^1/_8$ in.

From the metallurgical principles governing the production of steel castings having maximum toughness, it might be supposed that a grain-refinement treatment should be given to spot welds before the tempering operation, in order to break up the dendritic structure resulting from the solidification of the weld metal. Such a treatment has been found to be definitely advantageous for materials $^1/_4$ in. thick and heavier and for carbon contents as low as 0.15 per cent. Experience has also shown that the simple tempering treatment, involving only one magnitude of postheat current, is sufficient to produce the best possible mechanical properties for sheets $^1/_8$ in. thick and less. The property of the spot welds which is most sensitive to the difference between those made with and without tempering is the normal tension strength, i.e., the strength measured at right angles to the surface of the welded sheets. In low-carbon steels the normal tension strength may be from 50 to 75 per cent of the shear strength. In untempered spot welds having carbon contents above 0.15 per cent, the normal tension strength may be only from 5 to 20 per cent of the shear strength. Tempering, applied within the welding machine, is able to raise the normal tension strength to 30 to 60 per cent of the shear strength. For steels containing more than 0.25 per cent carbon, even the shear strength is improved by tempering. For example, with about 0.45 per cent carbon, the shear strength may be more than doubled, and the normal tension strength increased more than six times. Tempering in this case is able to transform an entirely useless structure to one of maximum serviceability. By this process it is possible to produce excellent spot welds in steels of any hardenability, which means in this case any carbon content.

A recent important application involved the spot welding of steel containing 0.70 per cent carbon. Without tempering, the spot welds were extremely brittle to the extent that they were unable to withstand even the slightest shock. After tempering, these spot welds were able to resist severe mechanical deformation involved in a mechanical forming operation after welding. Automatic tempering within the welding machine opens to spot welding the entire field of fabrication of low-alloy structural steels and aircraft steels.

A considerable body of experimental data has been accumulated to permit the ready adaptation of tempered spot welding to manufacturing production. General relations have been established which permit the selection of welding conditions with a minimum of special developmental effort on the part of the user. For a summary of this work, see the paper by W. D. Doty and W. J. Childs (10).

Range of Applicability. Spot welding is applicable to all thicknesses of sheet and plate, from the thinnest up to approximately $^1/_2$ in. in thickness. High-speed production, however, is generally limited to thicknesses up to $^1/_8$ in. The upper limitation of $^1/_2$ in. is satisfactory for steel and the higher-resistivity alloys. With high-conductivity aluminum alloy, it is extremely difficult to obtain practical equipment for welding even two pieces $^1/_4$ in. in thickness, and most commercial equipment will not weld two pieces greater than 0.10 in. Practically all materials that are obtainable in sheet form can be spot welded, with the possible exception of copper, which has so high an electrical conductivity that it is difficult to generate sufficient heat to produce fusion.

PROJECTION WELDING. Projection welding is a resistance welding process similar to spot welding, in which the currents are localized to produce welds at predetermined points by the design of the parts to be welded. The localization of current is usually accomplished by projections, or intersections. Whereas in spot welding the current is localized by the shape of the electrode and the force with which it is pressed against the sheet, in projection welding the electrodes may be in contact with a relatively large area of surface, and the problem of deformation of electrodes becomes much less serious. If it is desired to make a cluster of welds or a circular weld of considerable size, such that very high pressure has to be exerted by the electrode, the electrode itself may be made of a harder copper alloy to resist deformation, without regard to its lower electrical conductivity. The lower electrical conductivity is not a serious disadvantage because of the much larger area of contact which can usually be secured with this type of welding. When two sheets of dissimilar thickness are joined by projection welding, it is best to punch the projection into the thicker sheet, since a projection in the thinner material is likely to collapse and become ineffective in localizing the current before the heavier sheet becomes sufficiently heated to produce a properly fused weld. The making of a cluster of welds is quite common in projection welding and is called multiple projection welding. Sufficient current may be passed through the relatively large electrode to complete all the cluster of welds simultaneously. This type of weld is particularly advantageous for the attachment of stamped parts, such as brackets, to the surfaces of flat sheets. It becomes a comparatively simple matter to punch the proper number of projections accurately located on the stamped part when it is being formed. This obviates the necessity of a separate stamping operation to produce the projections.

The equipment for projection welding is very similar to that for spot welding. Since it is customary to make several welds simultaneously whenever possible, however, the press-type machine is frequently used, since it is a type of construction permitting much higher electrode forces, and since it is usually provided with platens, to which specially shaped projection-welding dies may be attached conveniently. Such machines are usually also of higher electrical capacity, as well as higher mechanical capacity, than would customarily be used for spot welding the same gages of material. The controls used for projection welding are generally of the same type as those for spot welding, except that it is quite common to pass the welding current in more than one impulse, to permit initial softening of projections which come into first contact, followed by later discharges of welding current to complete the entire group of welds.

Range of Applicability. Projection welding is capable of being employed throughout practically the entire range of materials and thicknesses for which spot welding is applicable. There is a general tendency to use projection welds for the medium and heavier gages of material, since the localization of current is simpler in this case than in spot welding. As mentioned previously, whenever parts are to be formed by stamping, projection welding is a much easier and more natural process to apply. Projection welding is advantageous when there is a small edge distance in overlapping parts, since in this case it is difficult to produce good-quality spot welds. Projection welding also lends itself to odd- or other-than-circular-shaped welds, it being frequently desirable to arrange elongated projections or even circular projections when required. Although projections are usually formed by a stamping operation, there are other cases where machining of projections may be more practicable. This is particularly true in attaching threaded connections to pressure vessels, when the entire connection is to be machined and a circular projection is desired.

SEAM WELDING. Seam welding is a resistance welding process whereby overlapping spot welds are made progressively along a joint by means of circular electrodes. The circular seam-welding wheels roll along the overlapping edges to be welded, and the control circuit is arranged to pass current at sufficiently close intervals to produce the desired degree of overlapping of the spot welds. The primary purpose of a seam-welded joint is to produce liquid- or air-tight containers from comparatively thin sheet metal. When the wheel is driven at constant speed, very uniform rows of spot welds can be produced. The spot welds can also be made very much more rapidly than would be possible with an ordinary spot-welding machine.

A common welding cycle on 0.040-in. steel would make 10 ft of joint per minute, with 10 spots per inch, or 1200 welds per minute. For certain types of construction, it is desirable to make two or more rows of spot welds, or two or more seam welds, spaced 2 or more in. apart. In this case the reactance of the secondary circuit can be kept at a low value, and two seam welds can be made simultaneously by connecting the two wheels to the same transformer. The current passes from one wheel through the sheets to a heavy copper backing bar. The current then passes back through the sheets to the other seam-welding wheel and thence to the transformer, thus making two welds simultaneously with the same transformer.

For special types of construction requiring more than two seam welds, machines have been made combining four or five transformers to make eight or ten seam welds at one time. Although seam-welding wheels are usually 8 or 10 in. in diameter, they may be made smaller or larger to meet special geometric conditions set up by the design of the parts being welded. Seam-welding wheels are customarily water cooled, either externally or internally. Although it is most common to have the wheels roll continuously, under certain circumstances it has been found desirable to give an intermittent motion to the seam-welding wheel to allow it to remain still and hold pressure on the work while the weld is being made and for a very short interval thereafter. This type of drive is not so common in seam welding as it is when the same type of equipment is used to space spots uniformly along a joint, and it is desired to move the wheel between welds as rapidly as possible. If welds are made with a rapidly moving wheel, the tendency is for the pressure to be removed from the weld before the metal has solidified, resulting in porous welds. In some materials the tendency to make porous welds because of this condition requires unduly slow movement of the seam-welding wheel when continuous rotation is employed.

The power supply for seam welding is particularly important, since a momentary drop in voltage of as little as 5 or 10 per cent may result in insufficient fusion to produce properly overlapping spots, with resultant leakage in the welded product. Care must be taken in calculating voltage drops at the seam welder terminals due to other resistance welding loads on the same power system. In order to minimize the disturbances of adjacent resistance welding equipments, it is always desirable to balance these as uniformly as possible on the different phases of the polyphase power-supply system.

The controls for seam welding must be arranged to switch the circuit on for periods of one or more cycles of a 60-cycle system, separated by off periods of the same order of magnitude. Although seam-welding machines were originally controlled with magnetic contactors, this type of service represents extremely severe duty for such a switch. Hence it is now almost universal practice to use electronic switching devices of the ignitron type for this service. For very thin materials, of the order of 0.010 in. in thickness, it may be possible to provide sufficient energy to make a single weld on $1/2$ cycle of an a-c system. Under this condition it is not necessary to interrupt the alternating current, but merely to increase the speed of travel of the wheels to provide the proper spacing between adjacent welds. Under these conditions travel speeds of the order of 30 ft per min may be employed with the usual 60-cycle power system.

Range of Applicability of Seam Welding. Seam welding is generally applicable to most metals up to a thickness of about $1/8$ in. Both the electrical and mechanical requirements become severe when this process is attempted for greater thicknesses. The electrical requirements for the seam welding of aluminum alloys become serious with sheets greater than about $1/16$ in. in thickness, because of the higher electrical conductivity and hence greater current required for making welds in this material. Seam welding is also not generally applicable to the high-strength structural aluminum alloys if high joint efficiencies are desired, because of the effect of the welding heat in overaging these alloys, resulting in an objectionable loss of strength of the material. Seam-welding wheels may be mounted to produce welds either parallel to the length of the arm supporting the wheel or perpendicular to the direction of the seam-welding wheel arms. The former arrangement is suitable for welding longitudinal seams in cylindrical vessels, whereas the latter arrangement is used for most general operations, including the welding of dished or flanged heads into cylindrical pressure vessels. Although most seam-welding machines are designed to make welds in comparatively straight lines, special arrangements have been made to permit the wheels to swivel about the supporting axis, thus making it possible to seam-weld around corners. This swivel arrangement permits rapid changing from the longitudinal to the transverse welding direction.

ROLL SPOT WELDING. Roll spot welding is the making of separated spot welds with circular electrodes of exactly the same type as those used for seam welding. When it is desired to weld long seams with uniformly spaced spot welds, e.g., in the fabrication of aircraft wings, seam-welding wheels may be arranged to roll along these joints at constant speed or with a series of interrupted movements, the control circuits providing welding current at properly timed intervals to secure the desired spacing of spot welds. This type of equipment permits making many more welds than can be produced with ordinary spot-welding equipment, where it is necessary to raise the electrodes after each weld and move the parts to the proper place for making the succeeding one.

UPSET WELDING. Upset welding is a resistance welding process wherein the weld is made simultaneously over the entire area of abutting surfaces or progressively along the joint with the aid of rolls or clamps which force the abutting surfaces together. The pressure is applied before heating starts and is maintained throughout the heating period. This process was originally called butt welding because of the type of joint which is made by this process, as distinguished from the lap joint made by the most common resistance welding process, which is spot welding. After the abutting ends of wires, bars, or plates are pressed together by the mechanical clamping devices, actuated either manually or by pneumatic or hydraulic means, sufficient current is passed to heat the abutting edges to the plastic condition. When this temperature has been reached, the applied force is able to squeeze the parts together, resulting in an appreciable enlargement of the section in the vicinity of the welds. The upsetting which is responsible for making a proper weld when the steel is in the plastic condition is responsible for the present name of the process and for an intermediate term applied to it—upset butt welding. In order to obtain the best-quality welds by this process, it is necessary that the parts be carefully faced by a machine cutting operation, so as to make intimate and uniform contact over the entire abutting surfaces before the heating operation commences. This is necessary to prevent the formation of oxide on the abutting surfaces, which may later become entrapped within the weld. When no particular precautions are exercised to ensure intimacy in contact during welding, the quality of welds made by this process suffers in strength because of the entrapment of oxides. The enlargement of the section at the weld and the large amount of heating produced behind the weld in this process are considered objectionable for some metals.

Equipment for Upset Welding. The equipment for this process consists of a suitably rigid frame containing fixed and moving platens, the former usually insulated from the frame of the machine, so that the welding transformer may be connected with one lead to the fixed and one to the moving platen. The platens are provided with suitable clamping devices for holding the work to be welded while the parts are moved together during the upsetting operation. When the sections to be welded become greater than a few square inches, the power requirements for this process become extremely great. The process is therefore generally used for sections less than 1 sq in. A magnetic contactor is usually provided for switching the primary circuit of the welding transformer, and the duration of current flow may be controlled either manually or by mechanical or electric timing devices. In most of the earlier applications of this process the interruption of the welding current was controlled manually by the operator, who observed the temperature of the surfaces of the abutting parts. When it became apparent that a sufficient temperature for welding had been reached and upsetting was proceeding properly, the operator opened the primary circuit of the welding transformer, usually by removing his finger from a push-button switch. The upset welding process is now used only under limited circumstances where very ordinary quality of welded joints is satisfactory or for the welding of tubes, where it is desired not to leave any loose material within the tubes. The continuous form of upset welding is still the basis for the production of a very large amount of high-quality electric-welded steel tubes. In this process, however, the weld more nearly approaches in quality and character that made by the flash-welding process, which will be described next.

FLASH WELDING. Flash welding is a resistance welding process for making a butt joint, wherein the weld is made over the entire area of abutting surfaces, with the heating provided by a series of discontinuous short circuits rapidly burned clear by having adequate power immediately available, and the pressure is applied only after heating is substantially completed. In this process the parts are clamped to the fixed and moving platens and the power circuit is energized, after which the moving platen starts to bring the parts together very slowly. If the power supply is adequate, flashing starts at the first instant of contact, resulting in the short circuit being immediately burned clear. As other parts of the pieces make contacts, further short circuits follow in rapid succession. This operation results in the emission of a stream of sparks of oxidizing molten metal and is termed flashing. It is essential that the movement of the parts together be very slow at the start, when the surfaces are cold, in order that the available power supply will be able to keep the flashing operation continuous and avoid butting. Any short circuit which is not cleared within 1 or 2 cycles is referred to as butting, and this usually results in very poor welds. If a very high power level is used at the start of flashing to ensure avoidance of the butting condition, the available power level will be too high when the surfaces have become heated near the end of the flashing period and coarse flashing of large droplets of metal from the surfaces will result, leaving comparatively large craters exposed to the action of atmospheric oxygen. The surfaces of these craters will be oxidized, and some of these oxides will be trapped in the final weld, rendering it inadequate in strength. After the flashing operation has proceeded for a period of time, the moving platen is arranged to bring the parts together suddenly for the purpose of squeezing out molten metal and oxides on the surfaces and consummating the weld. The heating value of the current used during the flashing process is so much less than that which would be required for an ordinary butt weld that the parts are not heated very much behind the welding surfaces. It therefore becomes necessary in many cases to continue the current flow into the upset interval for a brief period to permit proper forging of the weld. When the short circuit resulting from upset occurs, the magnitude of current increases greatly, and if this current is continued for any appreciable period of time, metallurgical damage to the metal in the vicinity of the weld may result. For some materials it is therefore critically important to maintain the current after upset for a period only long enough to permit proper forging, and not long enough to produce metallurgical damage. For some materials, e.g., the structural aluminum alloys, the matter of cutting off the current at the proper interval after upset starts becomes so critical as to make difficult the production of consistent welds. In this case it has been found highly desirable to make provision for automatically reducing the magnitude of current at the moment of upset.

In order to provide the necessary fine flashing at the end of a flashing period, careful attention must be given to the use of the minimum possible available power level at the end of the flashing period. If only a single power level is available, it becomes necessary to make special provision for easy starting of the flashing operation. This may be done by chamfering the ends of the parts to produce a reduced section of metal during the starting interval, or the parts may intentionally be prepared so as to be misaligned at the start of the flashing operation. Another method of taking care of this situation is to arrange for two different power levels during the flashing operation, a high level to permit easy starting and a proper lower level to obtain proper flashing at the end of the flashing cycle. Still another method is employed to start the flashing operation without too high a power level: preheating the parts before starting the flashing operation. This preheating operation is usually employed for the welding of relatively heavy sections by the flash-welding process. It consists of bringing the parts into butting contact under light pressure with power supplied for short periods of time, until a considerable body of heat has been built up in the abutting surfaces. The flashing operation may then be started easily on the hot surfaces with the employment of the power level which would produce satisfactory flashing in the latter part of the flashing cycle. The preheating operation is usually controlled manually, after which the control of the flashing operation is turned over to the machine. Equipment has been developed, involving electronic timing and phase control, to remove the control of the preheating operation from the welding operator, and it is likely that the future tendency in flash welding will be in this direction.

44. CABLES

Arc-welding cables are usually insulated with a 60 per cent rubber compound and are provided with a cotton or paper separator over the conductor. Standard dimensions and carrying capacities are given in Table 2.

Table 2. Rubber-sheathed Arc-welding Cable

Nominal Size, AWG	Nominal Number of No. 34 Wires	Minimum Outside Diameter, Inches	Amperes	Volts Drop per 100 Ft
8	420	0.390	50	3.71
6	665	0.400	75	3.58
4	1064	0.495	100	3.18
3	1323	0.500	150	3.70
2	1666	0.560	200	3.92
1	2107	0.625	250	3.88
1/0	2646	0.675	300	3.72
2/0	3325	0.750	375	3.68
3/0	4256	0.815	450	3.51
4/0	5320	0.900	550	3.41
250,000	6384	0.950	600	3.12
300,000	7581	1.030	650	2.86

Copper temperature: 60° C; ambient temperature: 40° C.

Yield load factors of 32 per cent for No. 2 to 23 per cent for 3/0, the usual sizes.

Resistance welding cables for automatic machines are usually water cooled by circulating water either through a hollow core in the conductor or between the conductor and a loose sheath or hose.

For non-automatic resistance welding, arc-welding cables are commonly used. The current ratings of Table 1 would not apply, the current depending on the relative length of on and off periods.

45. BIBLIOGRAPHY

1. *Welding Handbook.* American Welding Society, 29 West 39th Street, New York.
2. Hess, W. F., L. L. Merrill, E. F. Nippes, Jr., and A. P. Bunk, The Measurement of Cooling Rates Associated with Arc Welding and Their Application to the Selection of Optimum Welding Conditions, *Welding J. Research Supplement,* Vol. 22, 377s–422s (Sept. 1943).
3. Hess, W. F., T. B. Cameron, and D. J. Ashcraft, The Surface Preparation of Magnesium Alloy Sheet for Spot Welding, *Welding J. Research Supplement* (Feb. 1947).
4. Hess, W. F., R. A. Wyant, B. L. Averbach, and F. J. Winsor, An Investigation of Electrode Pressure-cycles and Current Wave-forms for Spot Welding Alclad 24S-T, *Welding J. Research Supplement,* Vol. 25, 148s–162s (March 1946).
5. Hess, W. F., R. A. Wyant, B. L. Averbach, and F. J. Winsor, Some Observations of Spot Weld Consistency in Aluminum Alloys, *Welding J. Research Supplement,* Vol. 25, 201s–222s (April 1946).
6. Wyant, Robert A., Measurement and Effect of Contact Resistance in Spot Welding, *Elec. Eng.* (*Trans.*), Vol. 65, 26–34 (Jan. 1946).
7. Hess, W. F., R. A. Wyant, and F. J. Winsor, The Spot Welding of Ten Aluminum Alloys in the 0.040-inch Gage, Including XB75S-T (Bare & Alclad), R-301-T, Alclad 24S-T81, 24S-T (Bare & Alclad), 3S-1/2H, 14S-T, 52S-1/2H, and 61S-T, *Welding J. Research Supplement,* Vol. 25, 467s–484s (August 1946).
8. Hess, W. F., R. A. Wyant, and B. L. Averbach, Surface Treatment at Room Temperature of Aluminum Alloys for Spot Welding, *Welding J. Research Supplement,* Vol. 23, 417s-435s (Sept., 1944).
9. Hess, W. F., W. D. Doty, and W. J. Childs, The Spot Welding of NE-8715, NE-8630 and SAE-4340 Steels in the 0.125-inch Thickness, *Welding J. Research Supplement,* Vol. 24, 521s-530s (Oct. 1945).
10. Doty, W. D., and Wylie Childs, A Summary of the Spot Welding of High-tensile Carbon and Low Alloy Steels, *Welding J. Research Supplement,* Vol. 25, 624s–630s (Oct. 1946).
11. DeGarmo, E. Paul, J. L. Meriam, and Finn Jonassen, Residual Stresses in Intersecting Butt Welds, *Welding J. Research Supplement,* Vol. 25, 451s–463s (August 1946).
12. DeGarmo, E. Paul, J. L. Meriam, and Finn Jonassen, The Effect of Weld Length upon the Residual Stresses of Unrestrained Butt Welds, *Welding J. Research Supplement,* Vol. 25, 485s–487s (August 1946).
13. Jonassen, Finn, J. L. Meriam, and E. Paul DeGarmo, Reduction of Residual Welding Stresses by the Use of Austenitic Electrode, *Welding J. Research Supplement,* Vol. 25, 489s–491s (Sept. 1946).
14. Jonassen, Finn, E. Paul DeGarmo, and J. L. Meriam, The Effect of Peening upon Residual Welding Stresses, *Welding J. Research Supplement,* Vol. 25, 616s–623s (Oct. 1946).
15. McDowell and Cunick, The Effect of Time and Temperature on the Relief of Residual Stresses in Low-alloy Steels, *Welding J. Research Supplement,* Vol. 23, 481s–486s (Oct. 1944).

SECTION 19

RURAL ELECTRIFICATION DISTRIBUTION SYSTEMS

BY

CARL D. DIMITY

RURAL ELECTRIFICATION DISTRIBUTION SYSTEMS

By Carl D. Dimity

1. SYSTEMS

Distribution for rural electrification presents such different problems from urban distribution and is so often designed by organizations with less technical background that it warrants special treatment.

A three-phase four-wire primary distribution system operating at 12,500/7200 Y, with single-phase service lines at 7200 volts, phase wire to neutral, is considered a nominal standard for rural electrification.

Single-phase 7200/230/115-volt transformers for each consumer connect with a secondary system consisting of three bare wires, one a grounded neutral, running to the service pole and three weatherproof wires from this pole to the consumer's premises. Such a system is shown in Fig. 1.

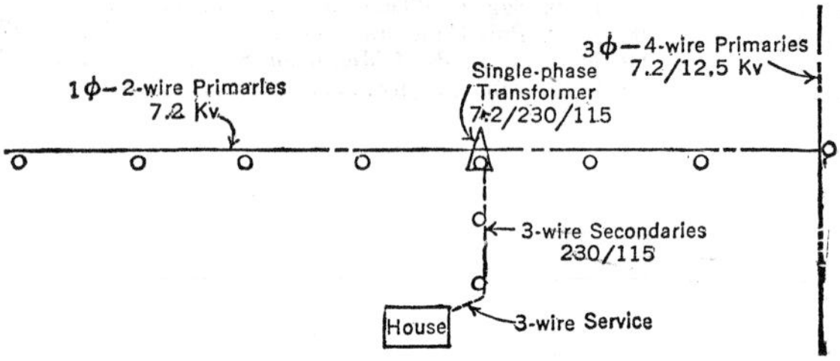

Fig. 1. Usual Distribution System for Rural Electrification

Where secondary circuits are needed on the same poles as the primary lines, the general practice is to use the multi-grounded primary neutral as the neutral of the secondary system, the common neutral not being supported on insulators.

The first step for the designing engineer is to determine from maps the approximate total distance of the line and available route, or routes, for the line to take in order to serve all potential loads in the area. Studies should then be made to determine the most economical combination of conductor sizes required to serve this load.

2. SYSTEM STUDIES

CALCULATION OF VOLTAGE REGULATION ON DISTRIBUTION SYSTEM. The calculation of voltage regulation should be made on the basis of the ultimate system. The actual number of potential (unsigned) members along the line that may eventually be expected to take service should be included in obtaining the peak demand used in determining voltage regulation.

In the voltage-regulation study provision should be made for including future extensions. The location of future lines and the number and distribution of future consumers should be based upon the best information available at the time the study is made.

The maximum demand curves (Fig. 2) include consumptions of 500 kwh per month per consumer and 5000 consumers. These curves were obtained from averages of a large number of systems in operation and are sufficiently accurate for preparing voltage-regulation studies. Wire factors give voltage regulation. The per cent voltage regulation, as calculated by this procedure, is the ratio of the voltage change from light load to peak load, expressed in per cent of the voltage at peak load.

As a basis for the preparation of a voltage-regulation study of a rural power-distribution system, the following information and material relative to that system should be on hand:

1. A circuit diagram or map showing the schematic layout of the system, phasing, etc. (Fig. 3).

2. The number of signed and potential (unsigned) consumers on each section of a balanced system. This includes the estimated potential consumers on future lines.

3. The number of signed and potential (unsigned) consumers on each section of an unbalanced system. This includes the estimated potential consumers on future lines.
4. The size of conductor, number of phases, and location of any substantially large concentrated load.
5. Actual maximum and minimum voltages on the supply side of the substation. Actual maximum and minimum voltages on load side of substation. If metering point, actual maximum and minimum voltages.

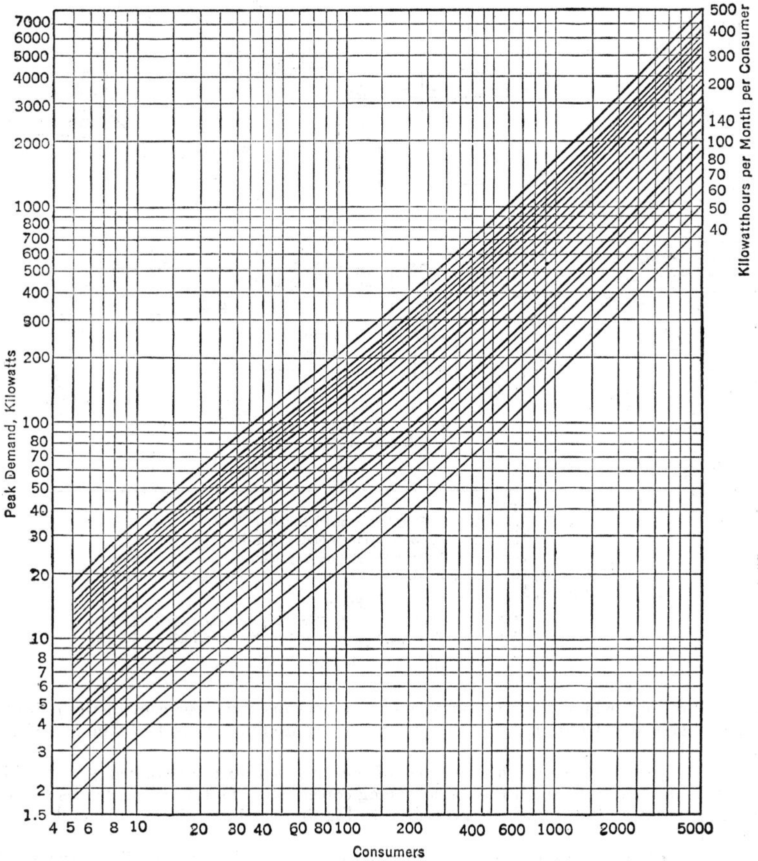

Fig. 2. Average Maximum Demand for Various Numbers of Consumers

On existing systems and projects for which plans and specifications are being prepared, it may be assumed that 75 per cent of the potential (unsigned) consumers will eventually take service. For areas which will ultimately be served but for which consumers have not been signed, it may be assumed that 85 per cent of these potential consumers will eventually take service.

The voltage regulation in each section is the same as that which would exist if one-half of the load in the section being considered and all of the load beyond this section were concentrated at the end of the section away from the source of supply. This assumption avoids the necessity of calculating a load center for each section and does not introduce appreciable error unless the load in the section is very unevenly distributed.

The ratio of light load to peak load has been taken as 0.20 to 0.25.

If we refer to Table 1 as an example, the procedure for design would be as follows:

Column 1. Starting at the farthest ends of the system from the substation, designate the section being considered by letters corresponding to the points previously marked on the map (Fig. 3) to indicate the ends of the sections. Thus A5 B5–A6 B6 designates the section of V phase line between points 5 and 6.

Column 2. Show the total number of signed consumers in the section.

Column 3. Show the number of potential (unsigned) consumers in the section.

Column 4. Insert the sum of the signed consumers and three-fourths of the unsigned (potential) consumers; i.e., Column 2 plus three-fourths of Column 3. This is in accordance with the assumption that service is to be provided to the signed and 75 per cent of the unsigned prospective consumers.

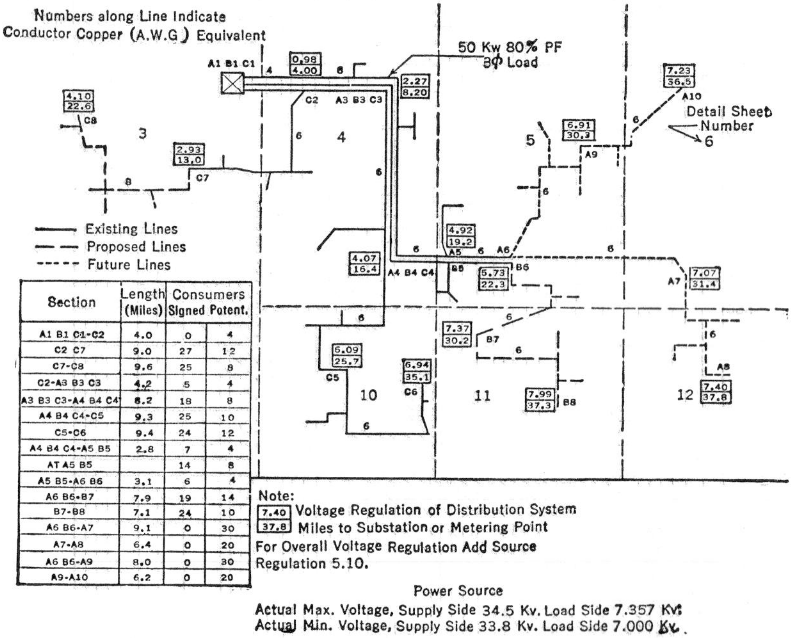

Section	Length (Miles)	Consumers	
		Signed	Potent.
A1 B1 C1-C2	4.0	0	4
C2 C7	9.0	27	12
C7-C8	9.6	25	8
C2-A3 B3 C3	4.2	5	4
A3 B3 C3-A4 B4 C4	8.2	18	8
A4 B4 C4-C5	9.3	25	10
C5-C6	9.4	24	12
A4 B4 C4-A5 B5	2.8	7	4
AT A5 B5		14	8
A5 B5-A6 B6	3.1	6	4
A6 B6-B7	7.9	19	14
B7-B8	7.1	24	10
A6 B6-A7	9.1	0	30
A7-A8	6.4	0	20
A6 B6-A9	8.0	0	30
A9-A10	6.2	0	20

FIG. 3. Typical Schematic Circuit Diagram

For future lines where there are no signed members insert 85 per cent of the potential consumers shown in Column 3, whether or not the future lines are to be taken off existing or proposed lines.

Column 5. Show the ultimate number of consumers who are supplied power which must flow all the way through the section being considered. Obtain these figures by adding the figures in Column 4 which, according to Fig. 3, pertain to sections beyond the one being considered.

Column 6. Show the equivalent number of consumers who are supplied through the section being considered. Obtain these figures by adding one-half of the ultimate consumers (Column 4) to the number beyond this section (Column 5).

Column 7. Show the average kilowatthour consumption per month on the project as a whole, i.e., including all consumers.

Column 8. Enter the peak kilowatt demand for the number of consumers shown in Column (6), or the peak kilowatt demand for a concentrated, large power load. For a distributed load, read the peak kilowatt demand directly from the maximum demand curve (Fig. 2) for the number of consumers shown in Column 6 and the kilowatthour consumption shown in Column 7.

Column 9. Show the total length, in miles, of the section being considered.

Column 10. Enter the kilowatt-miles, which is the product of the figures in Column 8 and those in Column 9.

Column 11. Note the conductor size used in the section and its copper equivalent. Select the size on a trial basis, usually assuming that the voltage regulation of the primary line should not exceed 8 per cent from the source of power to the consumer's transformer and 5 per cent from the primary line to the consumer's meter.

Table 1. Typical Design Sheet for Distribution System

VOLTAGE REGULATION SHEET

Project: Somestate 14 Jones

Sheet 1 of 1 Sheets

Submitted by: John Doe — Date: November 1, 19—

Checked by: Richard Roe — Date: November 3, 19— — Phase (s): A B C

Section	Number of Consumers					Kwh per Month	Kw Peak	Length of Section, Miles	Kw Miles	Conductor Size, Cu Equiv.	Phases	Kv	Wire Factor	Per Cent Voltage Regulation						
	Signed	Potential	Ultimate	Beyond This Section	Equivalent This Section									Distribution System: This Section: Concentrated Load	Distribution System: This Section: Distributed Load	Distribution System: This Section: Sum	Distribution System: Total	Source	Over-all	At Point
(1)	(2)	(3)	(4)	(5)	(6)	(7)	(8)	(9)	(10)	(11)	(12)		(13)	(14)	(15)	(16)	(17)	(18)	(19)	(20)
A9–A10	0	20	17.0	0	8.50	140	9.5	6.2	58.9	6	1	7.2	5.36		0.32		7.23	5.10	12.33	A10
A6 B6–A9	0	30	25.5	17.0	29.75	140	27.5	8.0	220.0	6	1	7.2	5.36		1.18		6.91	5.10	12.01	A9
A7–A8	0	20	17.0	0	8.50	140	9.5	6.4	60.8	6	1	7.2	5.36		0.33		7.40	5.10	12.50	A8
A6 B6–A7	0	30	25.5	17.0	29.75	140	27.5	9.1	250.3	6	1	7.2	5.36		1.34		7.07	5.10	12.17	A7
B7–B8	24	10	31.5	0	15.75	140	16.4	7.1	116.0	6	1	7.2	5.36		0.62		7.99	5.10	13.09	B8
A6 B6–B7	19	14	29.5	31.5	46.25	140	38.7	7.9	305.7	6	1	7.2	5.36		1.64		7.37	5.10	12.47	B7
A5 B5–A6 B6	6	4	9.0	146.0	150.50	140	98.0	3.1	303.8	6	V	7.2	2.68		0.81		5.73	5.10	10.83	A6 B6
At A5 B5	14	8	20.0																	
A4 B4 C4–A5 B5	7	4	10.0	175.0	180.00	140	113.0	2.8	316.4	6	V	7.2	2.68		0.85		4.92	5.10	10.02	A5 B5
C5–C6	24	12	33.0	0	16.50	140	16.8	9.4	157.9	6	1	7.2	5.36		0.85		6.94	5.10	12.04	C6
A4 B4 C4–C5	25	10	32.5	33.0	49.25	140	40.5	9.3	376.7	6	1	7.2	5.36		2.02		6.09	5.10	11.19	C5
A3 B3 C3–A4 B4 C4	18	8	24.0	250.5	262.50	140	156.0	8.2	1279	6	3	7.2	1.41		1.80		4.07	5.10	9.17	A4 B4 C4
At A3 B3 C3	Concentrated Load 3ϕ at 80% PF						50.0													
C2–A3 B3 C3	For Concentrated Load at A3 B3 C3						50.0	4.2	210.0	6	3	7.2	1.52	0.32						
C2–A3 B3 C3	5	4	8.0	274.5	278.50	140	164.0	4.2	638.8	6	3	7.2	1.41		0.97	1.29	2.27	5.10	7.37	A3 B3 C3
C7–C8	25	8	31.0	0	15.50	140	16.0	9.6	153.6	8	1	7.2	7.63		1.17		4.10	5.10	9.20	C8
C2–C7	27	12	36.0	31.0	49.00	140	40.4	9.0	363.6	6	1	7.2	5.36		1.95		2.93	5.10	8.03	C7
A1 B1 C1–C2	For Concentrated Load at A3 B3 C3						50.0	4.0	200.0	4	3	7.2	1.06	0.21						
A1 B1 C1–C2	0	4	3.0	349.5	351.00	140	200.0	4.0	800.0	4	3	7.2	0.96		0.77	0.98	0.98	5.10	6.08	C2
On A1 B1 C1				352.5													0	5.10	5.10	A1 B1 C1

Column 12. Indicate here the number of phases in the section of line under consideration and the line-to-ground kilovolts.

Column 13. Take these values from the table of wire factors (Table 2) for the conductor size, number of phases, and voltage given in Columns 11 and 12. Note that these factors apply only to multigrounded systems. For V phase lines the wire factor is one-half that of the single-phase wire factor (Table 2).

Table 2. Wire Factors (for Voltage Regulation) for Multigrounded Neutral Lines Only

Single-phase Balanced Systems, 4-ft Spacing

Voltage L-G	Domestic Consumers—90% PF (Lag)							Motor Loads—80% PF (Lag)						
	2,400	4,800	6,900	7,200	7,620	12,000	13,200	2,400	4,800	6,900	7,200	7,620	12,000	13,200
Conductor														
0 Cu equiv.	19.1	4.77	2.31	2.12	1.90	0.763	0.630	23.2	5.79	2.80	2.58	2.30	0.928	0.766
1 Cu equiv.	21.8	5.43	2.64	2.42	2.16	0.872	0.720	26.4	6.57	3.18	2.93	2.61	1.06	0.872
2 Cu equiv.	25.6	6.38	3.10	2.84	2.54	1.02	0.848	30.4	7.60	3.69	3.38	3.02	1.22	1.01
4 Cu equiv.	34.8	8.72	4.21	3.86	3.46	1.39	1.14	40.2	10.0	4.86	4.45	3.98	1.60	1.34
6 Cu equiv.	48.2	12.1	5.83	5.36	4.78	1.93	1.59	54.0	13.5	6.55	6.00	5.36	2.16	1.79
8 Cu equiv.	68.6	17.1	8.32	7.63	6.82	2.74	2.27	74.8	18.7	9.12	8.32	7.42	2.99	2.48
9 1/2 Cu equiv.	88.8	22.2	10.7	9.84	8.80	3.54	2.94	95.2	24.0	11.6	10.6	9.44	3.82	3.17
11 Cu equiv.	124.0	31.0	15.0	13.8	12.3	4.96	4.10	132.0	32.9	15.9	14.7	13.1	5.28	4.37
4 Steel *	153.6	38.5	18.6	17.0	15.2	6.16	5.09	164.0	40.8	19.8	18.2	16.3	6.55	5.42
6 Steel *	195.2	48.8	23.7	21.7	19.4	7.81	6.44	204.8	51.2	24.8	22.7	20.3	8.24	6.78

Three-phase Balanced Systems, 4.69-ft Spacing

Voltage	Domestic Consumers—90% PF (Lag)							Motor Loads—80% PF (Lag)						
L-G	2,400	4,800	6,900	7,200	7,620	12,000	13,200	2,400	4,800	6,900	7,200	7,620	12,000	13,200
L-L	4,160	8,300	11,950	12,450	13,150	20,800	22,850	4,160	8,300	11,950	12,450	13,150	20,800	22,850
Conductor														
4/0 Cu equiv.	2.74	0.690	0.334	0.306	0.274	0.110	0.091	3.52	0.880	0.426	0.392	0.350	0.141	0.117
3/0 Cu equiv.	3.30	0.832	0.400	0.368	0.328	0.133	0.110	4.14	1.03	0.502	0.458	0.410	0.166	0.137
2/0 Cu equiv.	3.78	0.944	0.458	0.420	0.375	0.151	0.125	4.70	1.18	0.568	0.522	0.466	0.188	0.156
0 Cu equiv.	4.42	1.10	0.534	0.490	0.438	0.177	0.146	5.30	1.33	0.642	0.590	0.526	0.212	0.175
1 Cu equiv.	5.15	1.29	0.623	0.572	0.510	0.206	0.170	6.10	1.53	0.738	0.678	0.606	0.244	0.202
2 Cu equiv.	6.03	1.51	0.730	0.670	0.598	0.242	0.200	7.00	1.75	0.848	0.778	0.695	0.284	0.230
4 Cu equiv.	8.64	2.16	1.05	0.960	0.856	0.346	0.286	9.36	2.40	1.16	1.06	0.952	0.382	0.316
6 Cu equiv.	12.7	3.18	1.54	1.41	1.26	0.507	0.420	13.7	3.42	1.66	1.52	1.36	0.547	0.451
8 Cu equiv.	19.2	4.78	2.32	2.13	1.90	0.766	0.633	20.2	5.06	2.45	2.25	2.01	0.808	0.668
9 1/2 Cu equiv.	25.6	6.45	3.10	2.84	2.54	1.02	0.848	26.8	6.70	3.24	2.98	2.66	1.07	0.880
11 Cu equiv.	37.0	9.20	4.47	4.10	3.66	1.48	1.22	38.2	9.52	4.62	4.24	3.79	1.52	1.26
4 Steel *	46.2	11.5	5.60	5.14	4.58	1.85	1.53	48.3	12.1	5.85	5.37	4.79	1.93	1.60
6 Steel *	59.7	14.8	7.19	6.59	5.88	2.38	1.97	61.4	15.4	7.42	6.82	6.11	2.46	2.03

* Three strand at 10 amp

NOTES. *Balanced Systems.* For sections of V phase use one-half the single-phase wire factor for balanced systems.

Unbalanced Systems. (1) For sections of single phase and each phase of V phase sections use the single-phase wire factor for balanced systems. (2) For each phase of sections of three phase use three times the three-phase wire factor for balanced systems.

Balanced and Unbalanced Systems. (1) For sections of single phase connected to delta circuits, or with ungrounded neutral, use two times the three-phase wire factor for balanced systems, with proper line-to-line voltage. (2) To obtain wire factors for voltages not shown, assume that the wire factor varies inversely as the square of the voltage. (3) The per cent regulation in a section is determined from the formula:

$$\text{Per cent regulation} = \frac{\text{Peak kilowatts} \times \text{Length in miles} \times \text{Wire factor}}{1000}$$

Column 14. Consider the per cent voltage regulation due to concentrated power loads separately for each section. Use this column for tabulating that portion of the per cent voltage regulation in the section due to a large power load beyond the section. Obtain the values by applying to the values in Columns 10 and 13 this equation:

$$\text{Per cent voltage regulation} = \frac{\text{Kilowatt-miles} \times \text{Wire factor}}{1000}$$

Column 15. Enter the per cent voltage regulation in the section due to distributed load. Obtain these values also by applying the above equation to the values in Columns 10 and 13.

Column 16. For sections in which the voltage regulation is partially due to concentrated (large power) loads and partially to distributed loads, show the total regulation in the section.

Column 17. Show the per cent voltage regulations at the far end of each section. Find these figures by starting with the section nearest to the source and summing up the voltage regulations in all the sections between the source and the section being considered, inclusive. The voltage regulations thus calculated, therefore, apply at the far ends of the sections considered.

Column 17 completes the voltage-regulation calculations for the distribution system. It is now necessary to take account of the voltage regulation of the source.

Voltage Regulation of Power Supply. The design of a rural electrical distribution system is based primarily on its own voltage regulation. However, the voltage regulation of a rural distribution system includes the regulation of all links in the power supply back to the source, and this regulation must be taken into consideration if the actual regulation of the distribution system is to be known.

The peak loads of the transmission line or tie line supplying the distributor's system may not occur at the same time as the peak load at the distributor's substation or metering point. Also the generation or transmission voltage may be varied to correct voltage regulation. For these reasons the maximum regulation of the supply will not necessarily be coincident with the maximum voltage regulation to be expected on the distributor's system. It is necessary to make this determination under the most unfavorable condition, which occurs when the sum of the maximum voltage regulations of the distribution system and the various parts of the power supply is a maximum.

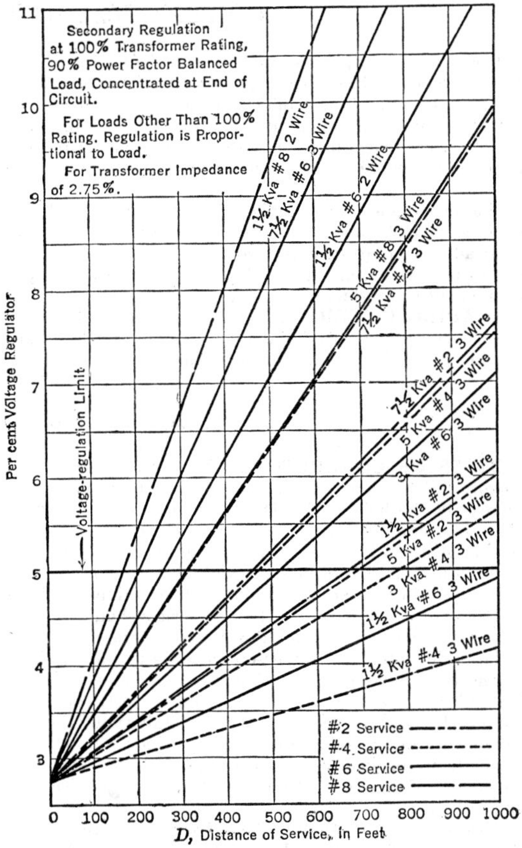

Fig. 4. Voltage Regulation of Secondaries

As peak loads and consequently voltage regulation will vary with the season, the time of year having the heaviest demand should be used in computing the overall voltage regulation.

If the distribution system is supplied from a substation, the voltage regulation foı various loads of the substation transformer may be obtained from the manufacturer. For convenience the regulation of the substation transformers is to be included with the regulation of the power supply.

The regulation on the high-voltage side of the substation or metering point should be obtained. This regulation is to take into account the effect of present and future loads on the distribution lines. In the event this information cannot be obtained, it should be determined by the design engineer.

Column 18. Enter the sum of the voltage regulation of the substation transformers and the regulation on the high-voltage side of the substation.

Column 19. Enter the sum of Columns 17 and 18.

Column 20. Show the point at which the calculated overall voltage regulation applies. The letters designate the ends farthest from the source of the respective sections of Column 1. Finally, consider the voltage drop in the secondary distribution, which may be derived directly from Fig. 4.

3. MECHANICAL DESIGN

Conductors and poles must have sufficient strength with a predetermined safety factor to withstand the loads due to the line itself and stresses imposed by ice and wind loads.

The poles must have sufficient height and be so located as to provide adequate ground clearance at either maximum loading or maximum temperature condition. The conductor ground clearance for railroad tracks and wire line crossings, as well as from buildings and other objects, must meet the requirements of the National Electric Safety Code (NESC).

The poles must have ample strength to withstand the stresses or load imposed on the component parts by the line itself, including the load set up by the tension in conductors at dead-end points, compression stresses due to guying tension, transfer loads due to angles in the line, and vertical stresses due to the weight of conductors and the vertical component of conductor tension.

LOADING DISTRICTS. Three general degrees of loading due to weather conditions are recognized and are designated as heavy, medium, and light loading. For details see Section 14, Article 24, particularly Table 2 and Fig. 1.

4. CONDUCTORS

TYPES OF CONDUCTORS. For rural distribution lines economical construction usually calls for long spans. In order to obtain proper ground clearances without unduly increasing the height of poles, conductors of high tensile strength are usually essential. This does not hold true in all cases, however, as some solid hard-drawn conductors on certain sizes and classes of poles will cost less per mile than high-strength conductors where a higher class of pole is required.

Bare conductors are generally used for primary line voltages and secondary voltages. Conductors of corrosion-resisting materials should be used.

Three-wire strand is the most rugged of the various conductor designs. The wires are of large diameter in comparison to wires in similar-size seven-wire stranded conductors. Large wires have greater resistance to wear and mechanical abrasion and have greater life under attacks of corrosion, erosion, or electrical burning. They are less susceptible to damage both in service and during installation. Three-wire conductors, because of the irregularity of the surface exposed to winds, show little or no tendency to vibrate even when strung on long spans with high stringing tensions. They are practically free from all danger of fracturing through fatigue and service or the destructive effect of chafing at insulator ties.

Three types of conductors have been considered standard for the design and construction of rural distribution lines. These are hard-drawn copper, copperweld copper, and aluminum cable steel reinforced (ACSR). There are, of course, other types, such as steel conductors, Amerductor, and solid and stranded copperweld conductors.

The conductivity of all types of conductors is usually based on their copper equivalent.

COPPER. Hard-drawn copper conductors have the advantage of maximum transmission capacity for a given power loss and voltage drop. The relatively small diameter of copper conductors, solid or stranded, as compared with their load-carrying capacity, permits a minimum of projected area to wind and ice loads. This gives a greater factor of safety for the poles and requires only a minimum of guying against transverse loading.

Copper is a homogeneous material and follows definite laws in regard to sagging characteristics and temperature coefficient, making sag calculations for any assumed or known loading conditions relatively simple. Because of its comparatively low ratio of strength to weight, copper necessarily requires greater sag for a given factor of safety or given span length when compared with copperweld or ASCR conductors. This, in addition, means that higher poles or shorter spans must be used to provide adequate ground clearance at maximum temperature conditions.

COPPERWELD COPPER. Three-wire stranded copperweld-copper conductors are used for distribution systems, particularly for long-span construction. Conductances and strengths are well suited for economical line design. Copperweld-copper conductors combine the high conductance of copper with the high strength of copperweld-steel strand.

Type A conductors are composed of one extra-high-strength copperweld wire and two hard-drawn copper wires. This type of conductor has been used exclusively on rural distribution lines, both utilities and REA. Other types include Type C, composed of one 40 per cent conductivity copperweld wire with two hard-drawn copper wires, and Type D, composed of two copperweld wires and one copper wire.

Copperweld-copper conductors are protected against corrosion by the copper surface on each wire. The wire has a thick welded-on exterior of pure copper, giving it a life equivalent to that of copper wires. The copper-to-copper contact throughout the conductor eliminates the possibility of galvanic action or electrolytic deterioration within the cable.

COPPERWELD. Three-wire stranded copperweld conductors are ideal for overhead lines, where conductor size is determined mainly by conductance and mechanical strength. These conductors provide low-cost construction through long spans and small sags, and adequate conductance for many applications, such as tap lines of limited length.

In comparison with Types A, C, and D copperweld-copper conductors, conductors with all copperweld wires have higher tensile strength and lower conductance, but from a structural and performance standpoint the two types are similar. The main use of smaller-size conductors is for tap lines of limited length, not requiring high conductance and carrying capacity. The relatively small overall diameter of these conductors presents a minimum of projected area to ice and wind loading.

ACSR. Aluminum cable steel reinforced is universally adaptable for use as a distribution-line conductor. As a result of its high strength, minimum sag is required, and shorter poles may therefore be used for a given span length. Armor rods are installed at the insulator supports in order to absorb stresses and to prevent fracturing of the aluminum strand through fatigue. The armor rod serves to prevent burning of the aluminum strands from flashovers. Armor rods are standard equipment for aluminum-cable steel-reinforced (ACSR) conductors.

ACSR has the disadvantage, however, of having a greater projected area exposed to wind loads, which may in some cases necessitate a pole of higher class to give the required factor of safety.

REFERENCE. Mechanical properties for all these conductors can be found in Section 14, Articles 56 and 57.

Electrical characteristics of all these conductors can be found in the tables of Section 14, Article 54.

5. POLES AND LINE HARDWARE

POLES. Poles should be of the kind and lengths specified for ASA class. (See Section 14, Article 28.) Most economically used throughout the United States are the western red cedar, northern white cedar, southern yellow pine, Douglas fir, lodgepole pine, and any acceptable equivalent of the same classification. Western red and northern white cedar poles should have an incised butt treatment with creosote oil. Southern yellow pine, Douglas fir, and lodgepole pines should have a full-length pressure treatment with creosote oil.

Table 3 gives physical characteristics and dimensions of southern yellow pine, western red cedar, northern white cedar, and lodgepole pine.

Pole loading calculations can, at best, yield only approximate results, since there will usually be a slight movement of the pole at the ground line. As a result, the calculated fiber stress may differ from the actual value. To determine the length of an unguyed span for a given height, kind, and class of pole, assuming the pole will be set in firm soil, it is necessary to calculate the bending moment of the pole at the ground line, which will usually be the point of failure. Such calculations may be performed by means of eqs. (62 to 68), Section 14, Article 28.

Table 3. Wood Poles (Dimensions and Resisting Moments)

Pole Length, Feet	ASA Class	Ground-line Distance from Butt, Feet	Minimum Top Circumference, Inches	Northern White Cedar			Lodgepole Pine			Southern Yellow Pine			Western Red Cedar		
				Minimum Circumference		Resisting Moment, Pound-feet	Minimum Circumference		Resisting Moment, Pound-feet	Minimum Circumference		Resisting Moment, Pound-feet	Minimum Circumference		Resisting Moment, Pound-feet
				6 Ft from Butt, Inches	At Ground Line, Inches		6 Ft from Butt, Inches	At Ground Line, Inches		6 Ft from Butt, Inches	At Ground Line, Inches		6 Ft from Butt, Inches	At Ground Line, Inches	
25	1	5.0	27	43.5	44.4	83,137	36.0	36.5	84,728	34.5	34.9	83,046	38.0	38.6	85,026
25	2	5.0	25	41.0	41.8	69,412	33.5	34.0	68,483	32.5	32.9	69,570	35.5	36.1	69,553
25	3	5.0	23	38.0	38.8	55,514	31.0	31.4	53,943	30.0	30.4	54,884	33.0	33.5	55,580
25	4	5.0	21	35.5	36.3	45,460	29.0	29.4	44,278	28.0	28.4	44,749	30.5	31.0	44,043
25	5	5.0	19	32.5	33.2	34,779	27.0	27.4	35,843	26.0	26.4	35,946	28.5	29.0	36,057
25	6	5.0	17	30.0	30.7	27,499	25.0	25.4	28,553	24.0	24.4	28,380	26.0	26.5	27,512
25	7	5.0	15	28.0	28.7	22,467	23.0	23.4	22,325	22.0	22.4	21,957	24.5	25.0	23,100
30	1	5.5	27	47.5	47.9	104,451	39.0	39.3	105,760	37.5	37.7	104,680	41.0	41.3	104,146
30	2	5.5	25	44.5	44.9	86,029	36.5	36.7	86,129	35.0	35.2	85,204	38.5	38.8	86,355
30	3	5.5	23	41.5	41.9	69,911	34.0	34.2	69,699	32.5	32.7	68,310	35.5	35.8	67,833
30	4	5.5	21	38.5	38.9	55,944	31.5	31.7	55,504	30.0	30.2	53,810	33.0	33.3	54,591
30	5	5.5	19	35.5	35.8	43,607	29.0	29.2	43,381	28.0	28.2	43,811	30.5	30.7	42,776
30	6	5.5	17	33.0	33.3	35,094	27.0	27.2	35,064	26.0	26.2	35,135	28.5	28.7	34,949
30	7	5.5	15	30.5	30.8	27,769	25.0	25.2	27,884	24.0	24.2	27,686	26.5	26.7	28,140
35	1	6.0	27	50.5	50.5	122,400	41.5	41.5	124,535	40.0	40.0	125,030	43.5	43.5	121,692
35	2	6.0	25	47.5	47.5	101,856	38.5	38.5	99,434	37.5	37.5	103,021	41.0	41.0	101,893
35	3	6.0	23	44.0	44.0	80,959	36.0	36.0	81,293	35.0	35.0	83,761	38.0	38.0	81,123
35	4	6.0	21	41.0	41.0	65,503	33.5	33.5	65,506	32.0	32.0	64,016	35.5	35.5	66,142
35	5	6.0	19	38.0	38.0	52,150	31.0	31.0	51,908	30.0	30.0	52,747	32.5	32.5	50,751
35	6	6.0	17	35.0	35.0	40,748	28.5	28.5	40,335	27.5	27.5	40,629	30.5	30.5	41,947
35	7	6.0	15	32.5	32.5	32,625	26.5	26.5	32,426	25.5	25.5	32,393	28.0	28.0	32,454

40	1	6.0	27	53.5	53.5	145,535	44.0	44.0	148,425	42.0	42.0	144,738	46.0	46.0	143,902
40	2	6.0	25	50.0	50.0	118,800	41.0	41.0	120,088	39.5	39.5	120,400	43.5	43.5	121,692
40	3	6.0	23	46.5	46.5	95,558	38.0	38.0	95,609	37.0	37.0	98,956	40.5	40.5	98,210
40	4	6.0	21	43.5	43.5	78,230	35.5	35.5	77,953	34.0	34.0	76,784	37.5	37.5	77,962
40	5	6.0	19	40.0	40.0	60,826	33.0	33.0	62,617	31.5	31.5	61,062	34.5	34.5	60,709
40	6	6.0	17	37.0	37.0	48,141	30.5	30.5	49,437	29.0	29.0	47,646	32.0	32.0	48,444
40	7	6.0	15	....	....	....	28.0	28.0	38,249	27.0	27.0	38,453	....	....	....
45	1	6.5	27	56.0	55.6	163,355	46.0	45.8	167,396	44.0	43.8	164,157	48.5	48.2	165,551
45	2	6.5	25	52.5	52.2	135,182	43.0	42.8	136,609	41.5	41.3	137,621	45.5	45.2	136,523
45	3	6.5	23	49.0	48.7	109,772	40.0	39.8	109,850	38.5	38.3	109,757	42.5	42.3	111,896
45	4	6.5	21	45.5	45.2	87,765	37.0	36.8	86,834	36.0	35.8	89,637	39.5	39.3	89,736
45	5	6.5	19	42.0	41.7	68,915	34.5	34.3	70,313	33.0	32.8	68,939	36.5	36.3	70,715
45	6	6.5	17	....	....	....	32.0	31.8	56,030	30.5	30.3	54,345	....	....	....
45	7	6.5	15	....	....	....	29.5	29.3	43,828	28.5	28.3	44,279	....	....	....
50	1	7.0	27	58.5	57.8	183,523	48.0	47.5	186,736	46.0	45.6	185,238	50.5	50.0	184,800
50	2	7.0	25	55.0	54.3	152,162	45.0	45.6	165,213	43.0	42.6	151,031	47.5	47.0	153,492
50	3	7.0	23	51.5	50.9	125,331	42.0	41.6	125,437	40.0	39.6	121,317	44.5	44.0	125,936
50	4	7.0	21	47.5	46.9	98,045	39.0	38.6	100,209	37.5	37.1	99,761	41.0	40.6	98,939
50	5	7.0	19	44.0	43.4	77,692	36.0	35.6	78,614	34.5	34.2	78,148	38.0	37.6	78,587
50	6	7.0	17	....	....	....	33.5	33.1	63,188	32.0	31.7	62,232	....	....	....
50	7	7.0	15	....	....	....	31.0	30.6	49,925	29.5	29.2	48,639	....	....	....
55	1	7.5	27	61.0	60.0	205,286	49.5	48.8	202,491	47.5	46.9	201,537	52.5	51.7	204,297
55	2	7.5	25	57.5	56.5	171,416	46.5	45.8	167,396	44.5	43.9	165,284	49.5	48.8	171,811
55	3	7.5	23	53.5	52.6	138,314	43.5	42.9	137,569	41.5	40.9	133,661	46.0	45.3	137,432
55	4	7.5	21	49.5	48.6	109,097	40.5	39.9	110,679	39.0	38.5	111,486	42.5	41.8	107,975
55	5	7.5	19	46.0	45.2	87,765	37.5	36.9	87,543	36.0	35.5	87,402	39.5	38.9	87,025
55	6	7.5	17	....	....	....	34.5	34.0	68,483	33.5	33.0	70,207	....	....	....

ASA ultimate fiber stresses, pounds per square inch:

Northern white cedar	3,600
Lodgepole pine	6,600
Southern yellow pine (creosoted)	7,400
Western red cedar	5,600

Resisting moment, $Mr = 0.000264\ FC^3$ pound-feet

where F = fiber stress, pounds per square inch.

C = pole circumference at ground line, inches.

CROSS-ARMS. All cross-arms should be untreated and unpainted and of lumber in accordance with EEI specifications or of other approved materials. (See Section 14, Article 30.)

INSULATORS. All insulators except service supports should be wet-processed porcelain, or equivalent, and should conform with AIEE standards for insulators. Neutral and secondary spools should be of standard 3-in.-overall-diameter wet-processed porcelain, or equiv alent. Service wire spools should be of standard 2 1/4-in.-overall-diameter dry- or wet-processed porcelain, or equivalent. (See Section 14, Article 31.)

HARDWARE. All hardware, machine bolts, carriage bolts, etc., should conform to EEI specifications. All bolts should be furnished with lock nuts and should be long enough without projecting more than 1 1/2 in. at the free end. All hardware, whether steel or malleable iron parts, should be hot-dipped galvanized, conforming to ASTM specifications. (See Section 14, Article 32.)

6. GUYS

Guys may be designed to support both tensile and compressive stresses, in which event they are usually termed braces. More frequently they support tensile stresses only, such a guy being usually a steel strand.

There are two situations in which it is necessary to have the formulas for computing the stresses in pole guys. One of these is a dead end, and the other is a change in direction or angle in the line.

In the schematic diagram (Fig. 5) of a dead-end anchor guy the horizontal distance from the center of the pole to the point where the guy enters the ground is defined as the

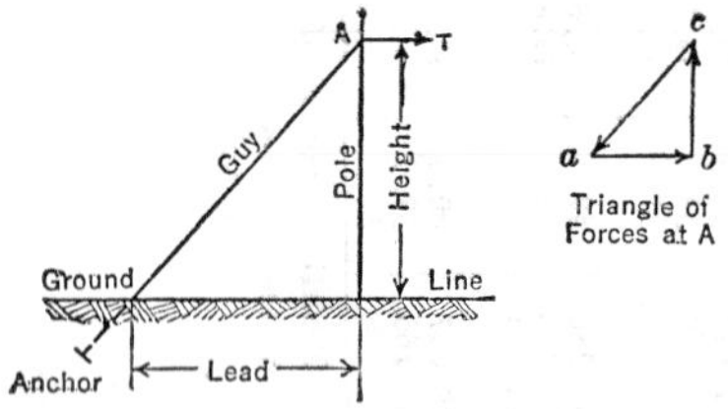

Fig. 5. Dead-end Anchor Guy

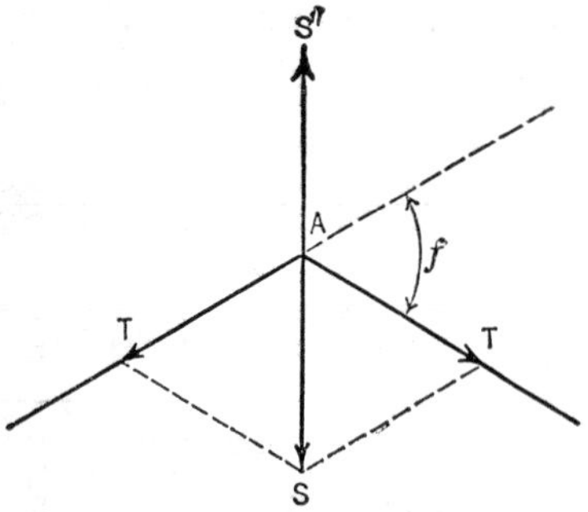

Fig. 6. Plan View of Angle Pole and Diagram of Forces

lead, and the vertical distance from the ground line of the pole up to the point A at the center of the resultant horizontal pull of the conductors is defined as the height. There are three forces at the point A, which are in equilibrium, as shown in the triangle of forces in the diagram. The force denoted by the line ab is the total tension T of the conductors, acting on the pole at the point A. The force denoted by the line bc is the compression in the pole, tending to thrust it down into the ground. The force denoted by the line ca is the tension in the guy strand. The formula for the tension in the guy, which is denoted by S, is expressed as follows:

$$S = T\sqrt{1 + (\text{Height}/\text{Lead})^2} \tag{1}$$

Figure 6 shows the plan view of an angle pole, indicated at the point A in the diagram. The tensions T in the conductors and the directions in which they act are denoted by AT, and the angle by which one of them diverges from the tangent, by f. The vector sum S of the two tensions T is shown as AS and acts in the same horizontal plane. This tension is given by the following formula:

$$S = 2T \sin \frac{f}{2} \tag{2}$$

This must be resisted by an equal but opposite force AS' exerted by the guy wire. As, however, the guy wire will probably be at a vertical angle, as in Fig. 5, the actual tension therein will be greater than S and may be derived from the triangle of forces, as in Fig. 5.

7. TRANSFORMERS

Most rural distribution transformers are single-phase and of the oil-cooled type. Two types of mounting arrangements have been standardized on rural distribution lines, the EEI-NEMA and the REA standard mounting, both of which provide flexibility in the transformer location on the pole. REA standard mountings are shown in Figs. 7 and 8.

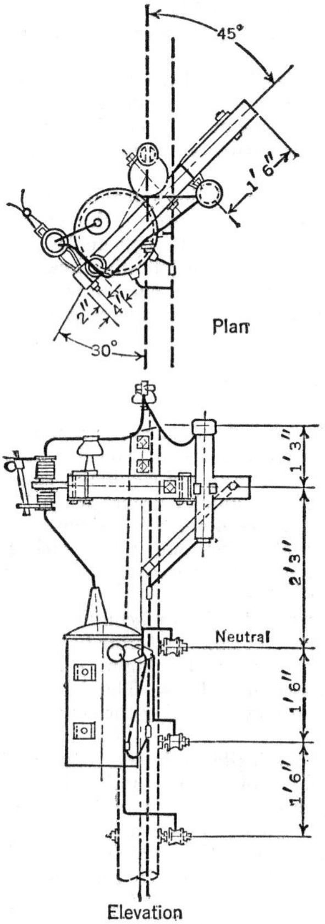

FIG. 7. REA Transformer Mounting

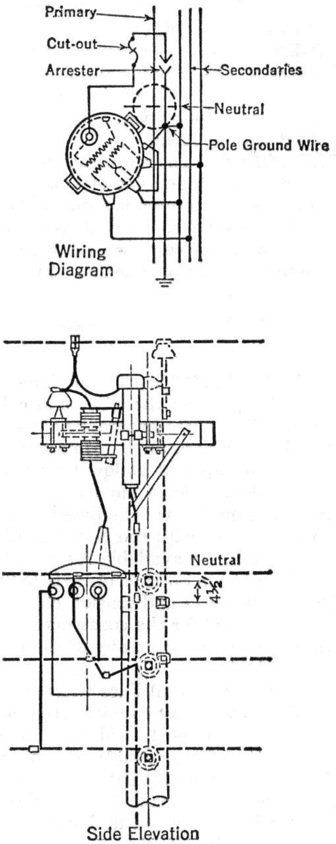

FIG. 8. REA Transformer Mounting

8. STAKING POLES

RIGHT-OF-WAY. It is desirable to have all right-of-way and easements for a given line secured before final designs, plans, and specifications are released for staking and construction.

STAKING. After the route of the line has been established, the size and type of conductor selected, and the loading conditions determined, the next step is to stake pole locations.

Control Points. In the layout of any line there will always be certain control points where poles are required. Such points are ends of the line, angles in the line, taps to branch lines, transformer locations, crossings, or sharp changes in the contour of the ground, such as the top of a peak or cliff. Some of these control points will be definitely fixed. In other cases there will be some leeway, and the engineer can use his ingenuity to obtain the most advantageous location within the limit permissible. It is advisable to locate such control points before staking intermediate pole locations. A transformer-pole location becomes an absolute control point when the shifting of a transformer to the next pole would make the secondary circuit too long. If a house is well back from the road, the transformer pole becomes an absolute control point, except where additional poles must be used to support the house service line.

Economic Pole Spacing. The spacing of poles should be such that, with the 18-ft clearance of wires above ground recommended by the National Safety Code, the cost of poles, cross-arms, insulators, and wires will be a minimum. The longer the span, the taller must be the poles, and in extreme cases the wire size or material may depend on the span. Normal spans of 300 ft are not uncommon on rural lines. The most economical length will vary with the locality.

DEAD-END SPANS. Dead-end spans are those which are dead-ended at both ends. Usually there will be very few dead-end spans in a line. In rough country it is sometimes best to cross a valley with one long span, or a river crossing or a deep ravine may make it necessary to use a long span. When abnormally long spans occur, it is usually advantageous to dead-end them.

ANGLE POLES. The fewer the angles, the more stable the line and the more economical it is from the standpoint of both labor cost and maintenance. Although a certain number of angle structures are unavoidable, the use of frequent small angles should be avoided by running each tangent as far as possible rather than following long sweeping curves. Where it is necessary for a line to follow a highway which makes a wide sweeping curve, the line may follow one side of the road around the curve or may cross the road.

9. SAGS AND TENSIONS

SAG CALCULATIONS. The designer must determine in advance the amount of sag and tension to be given the wires and cables of a particular distribution line at a given temperature. The three hazardous elements in the calculations of conductor sag are ice, wind, and low temperatures. Maximum stress in the conductor occurs when the wind and ice loads are applied at low temperatures. It is, therefore, the designer's function to determine the sag for a particular conductor at a stringing temperature such that the stress under the most severe assumed loading conditions will not exceed certain prescribed values in accordance with the National Electric Safety Code.

Many methods are available for calculating sags and tensions of conductors. These are given at length in Section 14, Article 25, and conductor manufacturers have also made available standard sag and tension charts for design (final) and stringing (initial) for all conductor sizes used for the construction of rural distribution lines. The design engineer can obtain such charts from any of the manufacturers now supplying the several types of rural line conductors.

Ruling Span. On rural lines there is a considerable flexibility in the entire system of attachment of conductors (poles, cross-arms, insulator pins, etc.), and quite small deflections of these supports will equalize relatively large differences of tension between spans. It is, therefore, possible to assume a uniform tension between dead-end supports and substitute a single calculation of sag for the set of calculations which would otherwise be required for all the spans between those supports. This is done by selecting a "ruling span," which is an assumed design span approximately representing the mechanical performance of a section of line.

If the actual spans are known, the ruling span is given by the following formula:

$$\text{Ruling span} = \sqrt{\frac{S_1^3 + S_2^3 + S_3^3 \cdots S_n^3}{S_1 + S_2 + S_3 \cdots S_n}} \tag{3}$$

where S_1, S_2, etc., are the several spans.

For practical purposes an exact determination of the ruling span is not necessary. The following empirical formula gives a fairly close estimate, and generally a standard ruling-span chart to the nearest 25-ft span length may be used:

$$\text{Ruling span} = \text{Average span} + 2/3\,(\text{Maximum span} - \text{Average span}) \tag{4}$$

Sags and tensions for the ruling-span length can be taken from initial or final dead-end sag charts such as are described below, and the sag for other spans can be calculated from the following relation:

$$\frac{\text{Ruling-span sag}}{\text{Ruling span}^2} = \text{Constant} \tag{5}$$

$$\text{Constant} \times \text{Span}^2 = \text{Sag}$$

TYPES OF SAG AND TENSION CHARTS. Four types of sag and tension charts or tables are used for the design and construction of overhead lines. Each has its particular usefulness. The following is a brief description of each type and its application. For details see Section 14, Article 25.

Final Charts give sag and tension data for a conductor under final operating conditions. These charts are frequently known as final loci charts. The curves represent final sag and tension values calculated for dead-end spans of different lengths. The charts are based on the assumption that the conductor has been subjected to the maximum tension used in the line design.

These charts are used:

1. As a basis for final ruling-span calculations.
2. For designing dead-end spans.
3. For stringing dead-end spans when conductors are prestressed, i.e., have been subjected during the stringing operation to the maximum tension assumed in the line design.

Initial Charts give sags and tensions for new conductor, i.e., conductor which has not been prestressed. They show somewhat less sag, to allow for the non-elastic stretch in the conductor. This allowance is such that, after the assumed maximum loading has been imposed, the resulting final sags will conform to the corresponding final charts.

These charts are used:

1. As a basis for initial ruling-span charts.
2. For stringing dead-end spans when conductors are not prestressed. This method is useful when the tension necessary to prestress the conductor is somewhat greater than can be safely handled with the regular construction equipment.

Final Ruling-span Tables or charts give final sags and tensions for spans throughout a line with the same normal tension in all spans. The sags and tensions for the ruling span are taken from a final chart, described above, and for other spans the sags are based on the same tension as that of the ruling span, in accordance with eq. (5). Being final charts, they show sags for conductors which have been prestressed. They are applicable for the ordinary type of line having spans of various lengths between dead-ends.

These tables or charts are used:

1. For locating structures and determining clearances.
2. For stringing the usual type of line, where conductors are prestressed before sagging to final values.

Initial Ruling-span Tables or charts give sags and tensions for the stringing of conductors which have not been prestressed. The sags and tensions for the ruling span may be taken directly from an initial chart, described above; for other spans, sags are based on the same tensions as that of the ruling span, in accordance with eq. (5). This type of chart is applicable mainly for large conductors, particularly when the prestressing tensions are greater than can be handled with available construction equipment.

These tables or charts are used for stringing conductors which have not been prestressed or subjected to tensions higher than the initial values.

TENSION LIMITATIONS. Sag and tension charts are based on limits for conductor tensions as specified in the line design. These limits depend on the type of line, grade of construction, code regulations which apply, and any special requirements for the particular line.

Code regulations commonly limit tensions under condition of maximum assumed ice and wind loading to 60 per cent of rated conductor strength, but the line designs are often based on tensions of 40 to 50 per cent of conductor strength for greater operating safety.

Limitations for the normal tension of conductors at 60 deg fahr are also generally included in line designs. Code regulations specify final normal tensions not to exceed 25 per cent of rated conductor strength, with corresponding initial tensions not to exceed 35 per cent of rated conductor strength, except for three-wire strands, where final normal tensions of only 30 per cent of rated conductor strength are permitted.

ELASTIC AND NON-ELASTIC STRETCH. Tensile stressing of a conductor is accompanied by corresponding elongation of the conductor. For the initial stressing the elongation consists of both elastic and non-elastic stretch. The non-elastic stretch is removed by the initial loading, and for later loadings not in excess of the initial loading the elongation is elastic stretch only.

The elastic stretch in a conductor is directly proportional to the stress applied to the conductor, the ratio of unit stress to unit elastic stretch being the modulus of elasticity for the material.

The non-elastic stretch or permanent set is an adjustment which takes place in a conductor, tending to remove inequalities which may be in the conductor due to the coiling and handling of the wire during its manufacture and installation. This non-elastic stretch accounts for the difference between the final and initial sags of a conductor. One of the

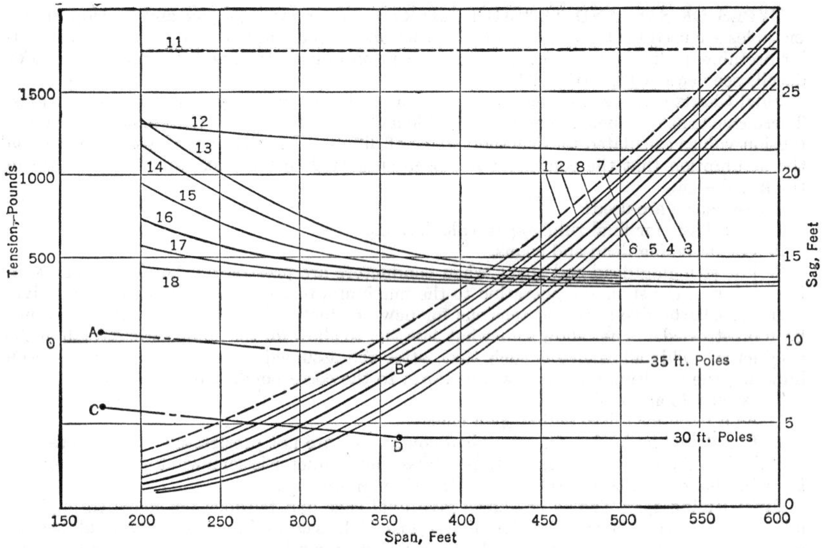

FIG. 9. Final Sag Chart

basic factors affecting conductor sag is the actual length of conductor in the span. This conductor length for the final sag condition differs from the length which determines the initial sag by the amount of non-elastic stretch occurring between the initial tension and the tension corresponding to the maximum loading.

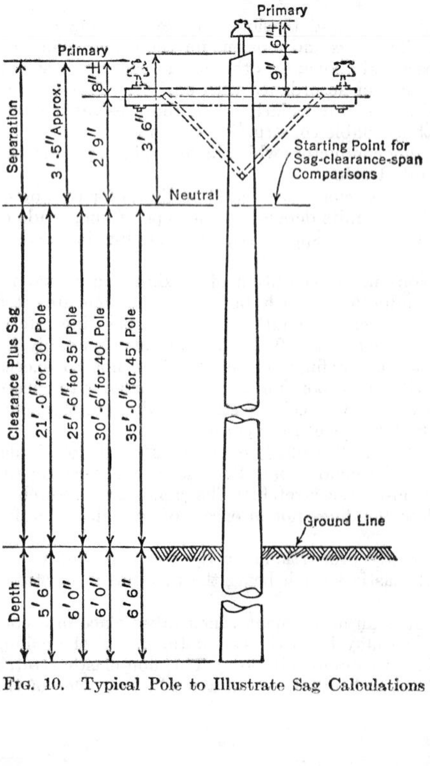

FIG. 10. Typical Pole to Illustrate Sag Calculations

MAXIMUM LEVEL-GROUND SPAN. Maximum level-ground spans can be determined graphically by plotting on the final sag chart (Fig. 9) for the conductor selected and extending a straight line commencing at the point representing the basic maximum span length (as specified by Rule 232-A-2, NESC) and maximum sag or permissible clearance (as specified by Rule 232-A, Table 1, NESC), and then extending to the right and downward with a slope equal to 0.1 ft less sag (permissible clearance) for each 10 ft of ground span length (as specified by Rule 232-B-1-a-1, NESC) subject to a minimum sag which equals the maximum permissible sag minus the maximum sag increase (as specified by Rule 232-B-1-a-3, NESC); from this latter point, the line continues horizontally at a constant sag to the right. The point at which the line intersects the sag curve representing the final unloaded sag at 60 deg fahr is the maximum level-ground span.

Consider, for example, a 35-ft pole which is set 6.0 ft into the ground, on which the conductor is supported 3.5 ft below the top, and assume that the conductor is a neutral, which requires a basic clearance of 15 ft in rural districts. The sum of these last three dimensions is 24.5 ft, and the difference between this figure and the height of the pole is 10.5 ft, which is the maximum permissible final sag of the conductor on 35-ft poles (see Fig. 10). The maximum basic span for heavy loading is 175 ft as specified in Rule 232-A-2 of the NESC. Plot at point *A* 10.50 ft of sag at a span length of 175 ft on the final sag chart (Fig. 9) of the conductor, which is assumed to be No. 2, 3-strand hard-drawn copper wire; then draw from this point a straight line sloping downward to the right at 0.1 ft less sag for each 10-ft increase in ground span length. Continue this line until it reaches a sag equal to the 10.50 feet minus 1.88 ft, or 8.62 ft (point *B*); then extend the line horizontally to the right at this constant sag. The figure 1.88 ft is the maximum sag increase (or maximum clearance increase) for this conductor (as specified by Rule 232-B-1-a-3, NESC). The point of intersection of this line with the final sag curve at 60 deg fahr, which is 368 ft, is the maximum level-ground span for this conductor on 35-ft poles in heavy loading districts.

Interpretation of Rule 232-B-1-a-3, NESC, for 35-ft Pole

Maximum permissible sag: 10.50 ft at 175 ft
Maximum sag increase: 1.88

8.62

	Span, Feet	Permissible Sag, Feet	
Point *A*	175	10.50	Sag decreases 0.1 ft for each 10-ft increase of span
	200	10.25	
	300	9.25	
	325	9.00	
Point *B*	363	8.62	

From here on, sag remains constant.

Consider, for example, a 30-ft pole which is set 5.5 ft into the ground and on which the conductor is supported 3.5 ft below the top, assuming that the conductor is a neutral which requires a basic clearance of 15 ft in rural districts. The sum of the last three dimensions equals 24.0 ft, and the difference between this figure and the height of the pole is 6.0 ft, which is the maximum permissible final sag of this conductor on 30-ft poles (Fig. 10). Following the same procedure as outlined above, plot at Point *C* 6.0 ft of sag at a span of 175 ft on the final sag chart (Fig. 9) of the conductor. Then draw from this point a straight line sloping downward until it reaches a sag equal to 6.0 minus 1.88 ft, or 4.12 ft. Then extend it horizontally to the right at a constant sag. The point of intersection of this line with the final sag curve at 60 deg fahr is 298 ft, which is the maximum level-ground span for this conductor on 30-ft poles in heavy loading districts.

Ground conditions will often determine the span lengths where there is a depression or a rise at the center of the span. Usually it will be necessary to determine the rise or depression at only one point, at either approximately the center or the quarter point of the span.

10. POLE CONFIGURATIONS

Typical pole top assemblies for rural-distribution construction of REA type are shown in Figs. 11 to 17. Various types of pole framing have been used in the construction of

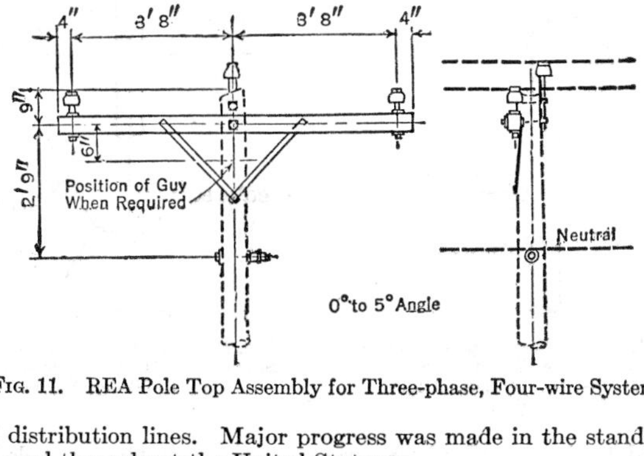

FIG. 11. REA Pole Top Assembly for Three-phase, Four-wire System

overhead rural distribution lines. Major progress was made in the standardization lines which are now used throughout the United States.

Figure 11 shows three-phase, four-wire pole top assembly for tangent and up to 5 deg angle construction. Figure 12 shows single-phase tangent construction for angles 0 to 5 deg.

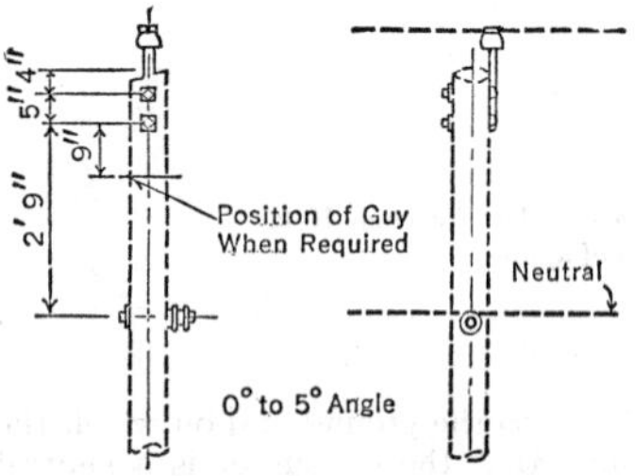

FIG. 12. REA Pole Top Assembly Single-phase Tangent Construction

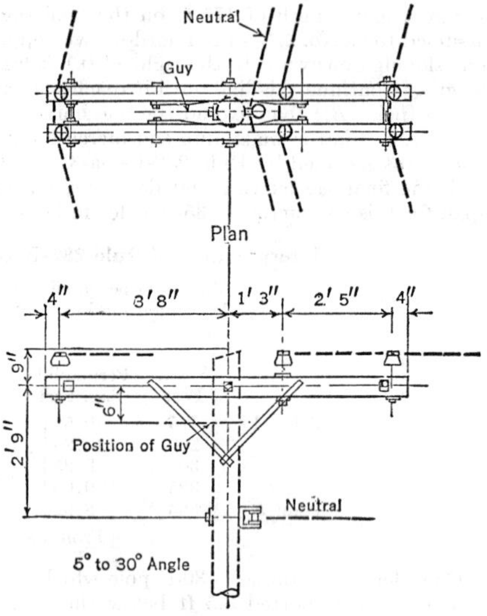

FIG. 13. REA Pole Top Assembly, Three-phase Four-wire System for 5- to 30-deg Angle Construction

Figures 13 and 14 show three-phase, four-wire, and single-phase, two-wire assemblies for 5 to 30 deg angle construction.

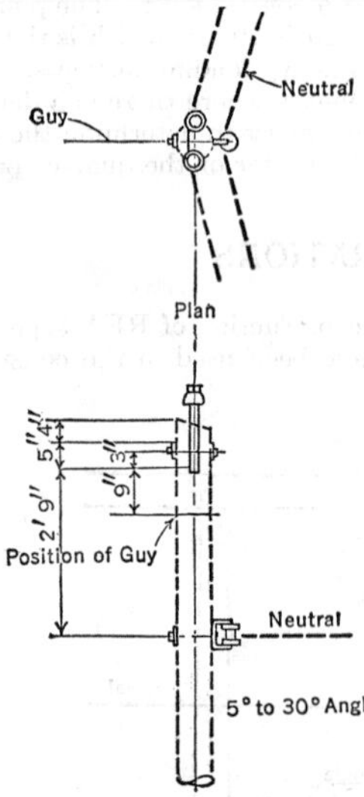

FIG. 14. REA Pole Top Assembly, Single-phase Two-wire System for 5- to 30-deg Angle Construction

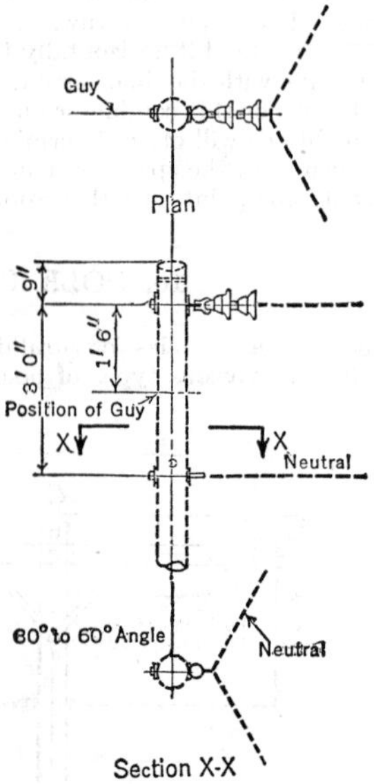

FIG. 15. REA Pole Top Assembly, Single-phase Two-wire System for 30- to 60-deg Angles

Figures 15 and 16 show single-phase, two-wire pole assembly for angles 30 to 60 deg and 60 to 90 deg.

Figure 17 is a schematic drawing showing three-phase, four-wire tangent cross-arm assembly to vertical construction for 30 to 60 deg angles.

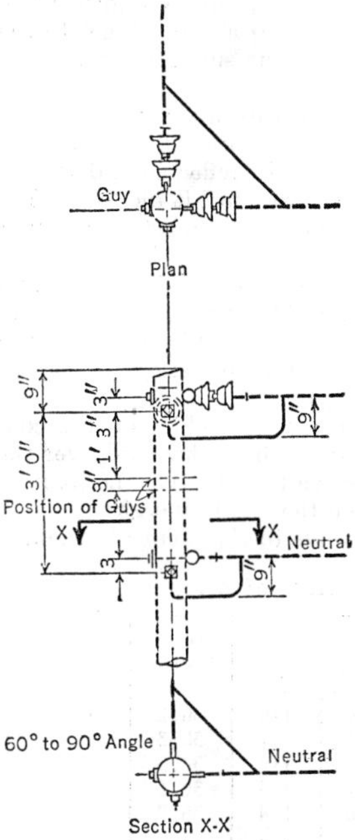

Fig. 16. REA Pole Top Assembly, Single-phase Two-wire Systems for 60- to 90-deg Angles

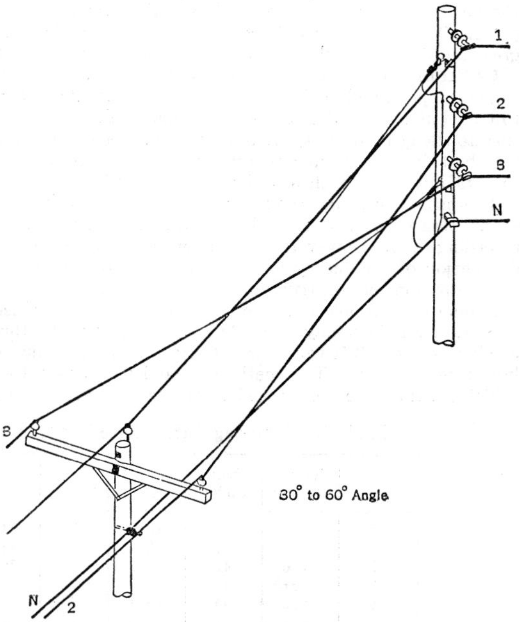

Fig. 17. REA Pole Top Assembly, Three-phase Four-wire System, Tangent Cross-arm, Vertical Construction for 30- to 60-deg Angles

11. METHODS OF MEASURING SAG

There are two methods of measuring sags in the erecting of rural line conductors. The simpler is known as the sighting method, and the other is termed the oscillation or stop-watch method.

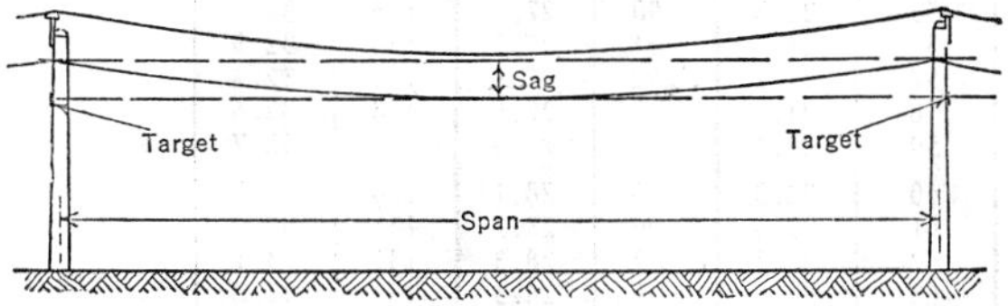

Span on Level Ground

Fig. 18

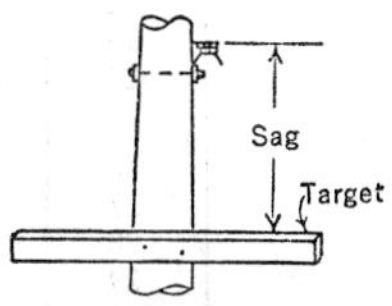

Fig. 19

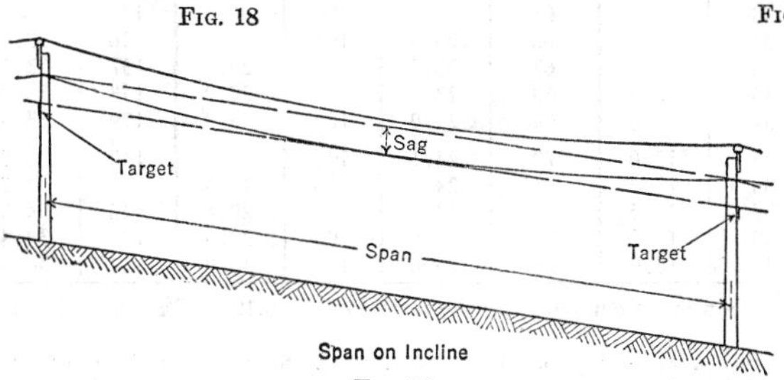

Span on Incline

Fig. 20

Figs. 18, 19, 20. Sighting Sags

SIGHT METHOD. Attach a target consisting of a lath or other wood strip to the poles at each end of the span, so that the distance between the upper edge of the target and the conductor is equal to the required sag. The conductor is properly sagged when the low point is in line with the top of the targets when sighting from one target to the other. The spans selected for checking sags should have adjacent supports at not too great a difference in elevation. (See Figs. 18, 19, and 20.)

OSCILLATION OR STOP-WATCH METHOD. Strike the wire a sharp blow close to one support, and at the same time start the watch. Striking the wire will cause a wave to travel along the wire to the far support. There this wave will be reflected and will return to the near support, where it will again be reflected, and so on until it dies out. The length of time, in seconds, required for the wave to return to the near support corresponds to a definite sag in inches, which can be calculated.

This principle holds true regardless of span length, size and type of conductor, or tension. Measurements with a large number of returns are desirable to minimize errors in recording time. However, the number of returns which can be discerned is limited by the dissipation of the energy and is dependent to some extent on the characteristics of the conductor and the construction features of the line.

A man on the pole may feel the return wave by placing a finger lightly on the conductor, or readings may be made from the ground by throwing a light dry cord over the conductor about 3 ft from the support. This cord may also be used to give the impulse initiating the wave. This method is particularly useful when the line is energized.

Table 4 is a time table based on the tenth return of the wave to the point of origin.

Table 4. Time-sag Table, Based on Tenth Return of Wave

Sag, Inches	Time, Seconds	Sag, Inches	Time, Seconds	Sag, Inches	Time, Seconds	Sag, Inches	Time, Seconds
5	6.4	40	18.2	75	24.9	110	30.2
6	7.0	41	18.4	76	25.1	111	30.3
7	7.6	42	18.7	77	25.3	112	30.5
8	8.1	43	18.9	78	25.4	113	30.6
9	8.6	44	19.1	79	25.6	114	30.7
10	9.1	45	19.3	80	25.7	115	30.9
11	9.5	46	19.5	81	25.9	116	31.0
12	10.0	47	19.7	82	26.1	117	31.1
13	10.4	48	19.9	83	26.2	118	31.3
14	10.8	49	20.1	84	26.4	119	31.4
15	11.1	50	20.4	85	26.5	120	31.5
16	11.5	51	20.6	86	26.7	121	31.7
17	11.9	52	20.8	87	26.8	122	31.8
18	12.2	53	21.0	88	27.0	123	31.9
19	12.5	54	21.1	89	27.2	124	32.0
20	12.9	55	21.3	90	27.3	125	32.2
21	13.2	56	21.5	91	27.5	126	32.3
22	13.5	57	21.7	92	27.6	127	32.4
23	13.8	58	21.9	93	27.8	128	32.6
24	14.1	59	22.1	94	27.9	129	32.7
25	14.4	60	22.3	95	28.1	130	32.8
26	14.7	61	22.5	96	28.2	131	32.9
27	15.0	62	22.7	97	28.3	132	33.1
28	15.2	63	22.8	98	28.5	133	33.2
29	15.5	64	23.0	99	28.6	134	33.3
30	15.8	65	23.2	100	28.8	135	33.4
31	16.0	66	23.4	101	28.9	136	33.6
32	16.3	67	23.6	102	29.1	137	33.7
33	16.5	68	23.7	103	29.2	138	33.8
34	16.8	69	23.9	104	29.3	139	33.9
35	17.0	70	24.1	105	29.5	140	34.0
36	17.3	71	24.3	106	29.6	141	34.2
37	17.5	72	24.4	107	29.8	142	34.3
38	17.7	73	24.6	108	29.9	143	34.4
39	18.0	74	24.8	109	30.0	144	34.5

Values in this table were calculated from the formula $D = 0.12075T^2$, where D = sag in inches and T = time in seconds.

Values for fifth return wave may be calculated from the formula $D = 0.483T^2$, or the above table may be used by dividing the time by 2.

Values for third return wave may be calculated from the formula $D = 1.3417T^2$.

When measuring or checking sags, it is desirable to choose a span of average length, i.e. one approximately equal to the ruling span in the section being sagged or checked.

In short sections of five spans or less it is usually sufficient to check the sag in one span near the midpoint, but on longer sections the sag should be checked in not less than two spans. One of these spans should be selected near the back, and the other near the pulling end of the section.

The required sag for the particular span length being checked and at the prevailing temperature should always be determined from applicable stringing tables.

12. BIBLIOGRAPHY

Publications of:
Copper Wire Engineering Association (Washington, D. C.).
Rural Electrification Administration, Dept. of Agriculture (Washington, D. C.).
American Society for Testing Materials (Philadelphia, Pa.).
American Standards Association (New York).
Edison Electrical Institute (New York).

INDEX

NOTE. The double numbers refer to both section and page numbers. For example, 1-38 indicates page 38 of Section 1.

ELECTRICAL ENGINEERS' HANDBOOK

ELECTRIC POWER

Fourth Edition

SECTION

ELECTRIC COMMUNICATION AND ELECTRONICS

Fourth Edition

SECTION